HEAT AND MASS TRANSFER

THE RICHARD D. IRWIN SERIES IN HEAT TRANSFER

Heat Transfer
by Anthony F. Mills, 1992

Basic Heat and Mass Transfer
by Anthony F. Mills, 1995

Heat and Mass Transfer
by Anthony F. Mills, 1995

HEAT AND MASS TRANSFER

ANTHONY F. MILLS
University of California at Los Angeles

IRWIN

Chicago • Bogota • Boston • Buenos Aires • Caracas
London • Madrid • Mexico City • Sydney • Toronto

Computer Graphics by Geograf, by permission of Geocomp Corporation, 66 Commonwealth Avenue, Concord, MA 01742

Sponsoring editor: Elizabeth Jones
Developmental editor: Kelley Butcher
Editorial coordinator: Christine Wright
Project editor: Gladys True
Production manager: Ann Cassady
Designer: John Rokusek/Laurie Entringer
Art coordinator: Heather Burbridge
Artist: David Wood
Cover designer: Joshua Paul Carr
Compositor: Publication Services, Inc.
Typeface: 10/12 Times Roman
Printer: R. R. Donnelley & Sons Company

Library of Congress Cataloging-in-Publication Data

Mills, Anthony F.
Heat and mass transfer / Anthony F. Mills.
p. cm. — (Irwin heat transfer series)
Includes bibliographical references and index.
ISBN 0-256-11443-9
1. Heat—Transmission. 2. Mass transfer. I. Title. II. Series.
TJ260.M518 1995
621.402′2—dc20 94–146

Printed in the United States of America
2 3 4 5 6 7 8 9 0 DO 1 0 9 8 7 6 5

To Brigid
For your patience and understanding.

PREFACE

Heat and Mass Transfer has been written for undergraduate and first-year graduate students in mechanical engineering programs. Apart from the usual lower-division mathematics and science courses, the preparation required of the student is introductory courses in fluid mechanics and thermodynamics, and preferably the usual junior-level engineering mathematics course. The ordering of the material and the pace at which it is presented have been carefully chosen so that the beginning student can proceed from the most elementary concepts to those that are more difficult. As a result, the book should prove to be quite versatile. It can be used as the text for an introductory course during the junior or senior year, although the coverage is sufficiently comprehensive for use in a more advanced course and as a reference work in laboratory and design courses. Throughout, the emphasis is on engineering calculations, and each topic is developed to a point that will provide the student with the tools needed to practice the art of design. In addition, the student is introduced to design methodology for heat and mass transfer equipment.

Heat and Mass Transfer is planned to complement *Heat Transfer,* which was published in 1992. The results of a market survey indicate that, whereas a majority of schools prefer a heat transfer text containing no coverage of mass transfer, a large minority of schools consider it essential to include some mass transfer in a heat transfer course sequence. *Heat and Mass Transfer* was developed by adding three chapters on mass transfer to *Heat Transfer*. As a result, *Heat and Mass Transfer* contains 11 chapters and three appendixes:

Chapter 1: Elementary Heat Transfer

Chapter 2: Steady One-Dimensional Heat Conduction

Chapter 3: Multidimensional and Unsteady Conduction

Chapter 4: Convection Fundamentals and Correlations

Chapter 5: Convection Analysis

Chapter 6: Thermal Radiation

Chapter 7: Condensation, Evaporation, and Boiling

Chapter 8: Heat Exchangers

Chapter 9: Elementary Mass Transfer

Chapter 10: High Mass Transfer Rate Theory

Chapter 11: Mass Exchangers

Appendix A: Property Data

Appendix B: Units, Conversion Factors, and Mathematics

Appendix C: Charts

The mathematics and physics of heat transfer and mass transfer are very similar. Yet most instructors agree that students find mass transfer to be a much more difficult subject than heat transfer. Students have less experience and poorer intuition regarding species concentrations and mass fluxes than they do regarding temperature and heat flux. Also, mass transfer analysis is applied to a large variety of physical and chemical systems, and it does take time to become familiar with the essential physics and chemistry. For these reasons, I believe that students should be introduced to mass transfer only after they have developed a proficiency in heat transfer. When the two subjects are taught simultaneously, the student is exposed to too many new concepts in too short a time interval, and may become needlessly confused. For example, once the student is fully familiar with the Prandtl and Nusselt numbers (having used them in homework exercises; for example) introduction of the Schmidt and Sherwood numbers is easily accomplished. Also, when the student is familiar with the pertinent mathematics, more attention can be paid to the physics and chemistry of the problem. For example, the student who is familiar with ε–N_{tu} relations for heat exchangers is better prepared to learn about the more complex mass exchangers and simultaneous heat and mass exchangers.

Chapter 1 is a brief but self-contained introduction to heat transfer. All problem solving can be performed without use of the property data in Appendix A. It serves to give the student an overview of the subject and to provide material needed in subsequent chapters. For example, the analysis of single-stream heat exchangers in Section 1.6 introduces the student to a most important engineering application of heat transfer, but also provides a result essential for heat convection in Chapter 4; most texts refer ahead to a subsequent heat exchanger chapter for this result, which is most unsatisfactory. But perhaps most significantly, the student can be given meaningful and interesting engineering problems at the earliest opportunity and thereby develop an enthusiastic interest in the subject. Chapters 2 and 3 present a relatively conventional treatment of heat conduction, except that an introduction to moving-boundary problems is included. The treatment of finite-difference numerical methods for conduction has been kept relatively brief and focused.

Most instructors agree that heat convection is a particularly difficult topic to teach in a first heat transfer course, and many hold strong opinions as to the most appropriate approach to follow. In keeping with the overall philosophy of the book, my objective in Chapter 4 is to develop the student's ability to calculate convective heat transfer coefficients. In a brief introduction, the physics of convection is explained and the heat transfer coefficient is defined. Dimensional analysis using the Buckingham pi theorem is used to quickly introduce the required dimensional groups and to allow a discussion of the important role played by laboratory experiments. The rather large number of correlation formulas that follow might be overwhelming if it were not for the associated computer program CONV. The instructor can discuss selected configurations in class as time allows; the student is able, with the aid of CONV, to reliably calculate heat transfer coefficients, and skin friction coefficients or pressure drop, for a much wider range of configurations. Analysis of convection is deferred to Chapter 5. High-speed flows are treated first in Section 5.2 since an understanding of the recovery temperature concept enhances the student's problem-solving capabilities. Each of the topics in Sections 5.3 through 5.8 is essentially self-contained, and the instructor can select as few or as many as required. My own opinion, however, is that convection analysis belongs more properly in a graduate-level course; it is more difficult, and of less practical use, than the analysis of conduction, radiation, heat exchangers, and so on for the beginning student.

The treatment of thermal radiation in Chapter 6 has some special features. Radiation properties are initially defined on a total basis, and the shape factor is introduced as a simple geometrical concept without discussion of the concept of radiation intensity. This approach gives the student the base required to solve most simple engineering radiation exchange problems. Only subsequently need the student tackle the more difficult directional and spectral aspects of radiation. I believe that this approach is justified because an understanding of the directional and spectral aspects of radiation does not lead to a significant increase in engineering problem-solving capability at this level. For gas radiation, the ubiquitous Hottel charts have been replaced by the more accurate models developed by Edwards; the accompanying computer program makes their use particularly simple.

The treatment of condensation and evaporation heat transfer in Chapter 7 is reasonably original, while the treatment of pool boiling is quite conventional. Forced-convection boiling and condensation is taken far enough for the student to be able to calculate both pressure drop and heat transfer. Heatpipes are dealt with in some detail, enabling the student to calculate the wicking limit and to analyze the performance of simple gas-controlled heatpipes. The thermal analysis of heat exchangers in Chapter 8 is also conventional. The student is then taken further into the design process than is customary. Additional topics include the calculation of exchanger pressure drop, thermal-hydraulic design, heat transfer surface selection for compact heat exchangers, and economic analysis leading to the calculation of the benefit-cost differential associated with heat recovery operations. A simple design program, HEX2, serves to introduce the student to computer-aided design of heat exchangers.

Chapter 9 considers diffusion in a stationary medium, and low mass transfer rate convection. As was the case for heat convection in Chapter 4, mass convection is introduced using dimensional analysis and the Buckingham pi theorem. The analogy between low mass transfer rate convection and heat transfer to an impermeable surface is thoroughly exploited. Simultaneous heat and mass transfer is considered, with an emphasis on problems involving evaporation of water, such as the wet- and dry-bulb psychrometer. Diffusion and chemical reaction in porous catalysts are analyzed in order to provide results needed for the study of automobile catalytic converters in Chapter 11. The chapter closes with methods for calculating transport properties, with special emphasis on gas mixtures. Chapter 10 is considerably more advanced than Chapter 9. Velocities and fluxes in a mixture or solution are rigorously defined and the general species conservation equation derived. Important problems such as diffusion with one component stationary and combustion of volatile hydrocarbon fuel droplets are carefully analyzed. The Couette-flow model is introduced to obtain blowing factors for high mass transfer rate convection. These factors are then improved to account for both flow geometry and variable properties to give a unique engineering problem-solving facility. Mass, momentum, and heat transfer in a constant laminar boundary layer on a flat plate are rigorously analyzed for a binary mixture, and complements the analysis of momentum and heat transfer in a pure fluid given in Chapter 5. The chapter closes with a generalized formulation of steady convective heat and mass transfer, based on the well-known contributions of D. B. Spalding.

Chapter 11 deals with mass exchangers, such as catalytic converters and gas scrubbers; particle removal equipment, such as filters and electrostatic precipitators; and simultaneous heat and mass exchangers, such as humidifiers and cooling towers. All the analyses are based on low mass transfer rate theory, and the high mass transfer rate theory of Chapter 10 is not a prerequisite to the study of Chapter 11. Humidifiers and cooling towers are usually considered within the scope of mechanical engineering, but mass exchangers have been traditionally the province of chemical engineers. Particle removal equipment has not been emphasized in either discipline. However, environmental concerns are now having profound effects on the design of mechanical engineering systems. Tuning and operating of the modern automobile engine is now dictated by requirements of the catalytic converter used to reduce exhaust emissions. A significant portion of the investment in a modern coal-fired central power plant is for equipment used to clean the stack gas, including a scrubber to remove sulfur oxides, and cyclones, baghouses, or electrostatic precipitators to remove particulates. Incinerators for destroying toxic wastes have even more stringent requirements on stack gas cleanup. A major concern of nuclear engineers has become the design of safety systems to prevent escape of radioactive gases and particulates into the atmosphere in the event of an accident involving damage to the reactor core. Major aerospace engineering projects, such as the space station, require complex life-support systems that involve a variety of mass exchangers. My goal in Chapter 11 is to prepare mechanical, aerospace, and nuclear engineers for effective participation in the design of the systems described above, complementing rather than competing with chemical engineers on the design team.

The extent to which engineering design should be introduced in a heat transfer course is a controversial subject. In the recent past, the practice at most universities in the United States has been to teach heat transfer as an engineering science course, and textbooks have generally reflected this philosophy. However, due partly to pressure from ABET and the engineering profession, there is now a move toward including more design in introductory heat transfer courses. Some educators are of the opinion that this should be done by including "open-ended" exercises for the student. In surveying examples of this approach, I have found that, whereas many of these exercises are indeed excellent and challenging problems for the student, they are seldom good vehicles for teaching the elements of design methodology. The fact that a problem does not have a unique answer does not, in itself, ensure that it will provide a satisfactory design experience. It is my opinion that the student can be best introduced to design methodology through an increased emphasis on equipment such as heat and mass exchangers. It is in the context of such equipment that students can be introduced to design topics such as synthesis, parametric studies, trade-offs, optimization, economics, and material or health constraints. The thermal-hydraulic design of a heat exchanger or cooling tower should surely be regarded as an essential topic in a heat transfer course. Thus, in *Heat and Mass Transfer* I present a more extensive coverage of heat and mass exchangers than is found in comparable textbooks. If this material is taught, I am confident that ABET guidelines will be met; but, more importantly, I believe that engineering undergraduates are better served by exposure to this material, even if it means studying somewhat less heat transfer science.

Based on my experience at UCLA, I suggest two possible strategies for teaching the mass transfer component of this text. In a one-semester introductory heat and mass transfer course, I would teach most of Chapter 9 and selected single-stream exchangers from Chapter 11; for example, the automobile catalytic converter and the adiabatic humidifier. Alternatively, in a two-quarter or two-semester sequence, I would teach Chapter 9, portions of Chapter 10, and most of Chapter 11.

When writing a heat transfer textbook, the author must take into account that the student's preparation in mathematics varies from school to school. It is imperative, nonetheless, that engineering courses follow up with the use of topics taught in mathematics courses, or else the student will surely never become proficient. The more advanced mathematics used in *Heat and Mass Transfer* is as follows.

1. First- and second-order linear ordinary differential equations
2. Sets of n linear algebraic equations
3. The use of separation of variables to solve partial differential equations, Fourier series, and special functions
4. Elementary numerical methods including integration, Newton-Raphson iteration, matrix inversion or Gauss-Siedel iteration, and integration of nonlinear ordinary differential equations
5. Finite-difference solution methods for partial differential equations

When using classical mathematics in the text, I have given complete details so as to reinforce what the student learned in previous mathematics courses. Of course, an instructor can choose to omit the details if appropriate. However, when using the numerical methods listed under item 4 above (which occurs almost exclusively in more advanced examples and exercises), I have assumed that the student is either able to write an appropriate computer program or has access to standard subroutines on a programmable calculator or computer. The number of programming languages in use, the wide variety of software available, and rapid changes in the field make such an approach essential. Computing resources vary from school to school. I was particularly impressed by the situation at the University of Auckland, where I taught from 1983 to 1985. In their sophomore year, engineering students had lectures on elementary numerical methods accompanied by a series of laboratory sessions, where they implemented these methods and, at the same time, became familiar with the school's computing resources. Thus, when teaching at the junior or senior level, the instructor could assume that the students were fully ready to use such methods as needed.

A unique feature of *Heat and Mass Transfer* is that it has a fully integrated package of computer software. With the aid of generous grants from the IBM Corporation, the School of Engineering and Applied Science at UCLA has been able to equip a number of computer classrooms and laboratories. Also, the Chancellor's Committee on Instructional Improvement made funds available for software development to enable these computers to be used effectively for undergraduate instruction. These funds assisted development of the software package, and this support is gratefully acknowledged. The software is intended to serve primarily as a tool for the student, both at college and after graduation as a practicing engineer. Most of the programs are designed to reduce the effort required to obtain reliable numerical results, and thereby increase the efficiency and effectiveness of the engineer. I have found the impact of the software on the educational process to be encouraging. It is now possible to assign more meaningful and interesting problems, because the students need not get bogged down in lengthy calculations. Parametric studies, which are the essence of engineering design, are relatively easily performed. Of course, computer programs are not a substitute for a proper understanding. My practice has been to require the student to perform various hand calculations, using the computer to give immediate feedback. For example, the student does not have to wait a week or two until homework is returned to find that a calculated convective heat transfer coefficient was incorrect because a property table was misread.

The *Heat Transfer* software package has been well received, and the additional programs that accompany *Heat and Mass Transfer* should prove even more useful. The calculations required to solve practical mass transfer problems are often very long and tedious, owing to the complexity of the analysis and the need to evaluate many thermodynamic and transport properties of mixtures. The software provided radically reduces the effort required to obtain reliable numerical results. For example, with the interactive programs provided, the student can easily perform parametric studies of the performance of scrubbers and cooling towers, and thereby develop a much deeper understanding of these systems than has been possible hitherto. Furthermore, at UCLA I have seen how programs such as SCRUB and CTOWER have certainly increased the students' interest in such equipment: previously the lengthy calculations required all but killed their enthusiasm for the subject. In providing this software to the student, it is not my intent to discourage students from writing their own computer programs. But, during a typical semester or quarter course, students do not have time to write more than two or three computer programs of any substance. Hopefully, the *Heat and Mass Transfer* software will demonstrate the value of computer programs for routine engineering calculations, and encourage the student and engineer to write such programs when circumstances suggest that they will be cost-effective.

Some of the material in *Heat and Mass Transfer,* mostly in the form of examples and exercises, has been taken from an earlier text which was co-authored by my ex-colleagues at UCLA, D. K. Edwards and V. E. Denny (*Transfer Processes*, 1st ed., Holt, Rinehart & Winston, 1973; 2nd ed., Hemisphere–McGraw-Hill, 1979). I have also used material on radiation heat transfer from a more recent text by D. K. Edwards (*Radiation Heat Transfer Notes,* Hemisphere, 1981). D. B. Spalding introduced me to the subject of mass transfer as a student at the Imperial College of Science and Technology, London: his influence is surely in evidence in this text. I gratefully acknowledge the contributions of these gentlemen, both to this book and to my professional career. The computer software for *Heat and Mass Transfer* was expertly written by Baek Youn and Hae-Jin Choi, with able assistance from Benjamin Tan. I also wish to acknowledge the contributions made by many others. The late D. N. Bennion provided a chemical engineering perspective to some of the material on mass exchangers. R. Greif of the University of California, Berkeley, and J. H. Lienhard V of the Massachusetts Institute of Technology provided detailed critiques of the mass transfer chapters. Reviewers commissioned by Richard D. Irwin, Inc., suggested numerous improvements, most of which I was able to incorporate in the final manuscript. My students have been most helpful—in particular, S. W. Hiebert, R. Tsai, B. Cowan, E. Myhre, B. H. Chang, D. C. Weatherly, A. Gopinath, J. I. Rodriguez, B. P. Dooher, M. A. Friedman, and C. Yuen. My special thanks to the secretarial staff at UCLA and the University of Auckland—in particular, Phyllis Gilbert, Joy Wallace, and Julie Austin, for their enthusiastic and expert typing of the manuscript.

A. F. Mills

ACKNOWLEDGMENTS

Richard D. Irwin, Inc., would like to thank the following reviewers for their contribution to the development of *Heat and Mass Transfer*.

Martin Crawford, University of Alabama—Birmingham

Prakash R. Damshala, University of Tennessee—Chattanooga

Tom Diller, Virginia Polytechnic Institute and State University

Glenn Gebert, Utah State University

Clarke E. Hermance, University of Vermont

John H. Lienhard V, Massachusetts Institute of Technology

Jennifer Linderman, University of Michigan—Ann Arbor

Robert J. Ribando, University of Virginia

Jamal Seyed-Yagoobi, Texas A&M University—College Station

The author worked hard to improve the book based on the comments of each reviewer. Please feel free to contact us with suggestions for improvement of future editions. We hope you enjoy using the book.

Elizabeth Jones
Sponsoring Editor

Kelley Butcher
Senior Developmental Editor

NOTES TO THE INSTRUCTOR AND STUDENT

These notes have been prepared to assist the instructor and student and should be read before the text is used. Topics covered include conventions for artwork and mathematics, the format for example problems, organization of the exercises, comments on the thermophysical property data in Appendix A, and a guide for use of the accompanying computer software.

ARTWORK

Conventions used in the figures are as follows.

———▶ (solid arrowhead)	Conduction or convection heat flow
∿∿∿▶ (wavy arrow)	Radiation heat flow
———▷ (open arrowhead)	Fluid flow
———➤	Species flow
———	Temperature or concentration profile

MATHEMATICAL SYMBOLS

Symbols that may need clarification are as follows.

$\simeq$ Nearly equal

$\sim$ Of the same order of magnitude

$|_x$ All quantities in the term to the left of the bar are evaluated at x

EXAMPLES

Use of a standard format for engineering problem solving is a good practice. The format used for the examples in *Heat and Mass Transfer,* which is but one possible approach, is as follows.

Problem statement

Solution

Given:

Required:

Assumptions: 1.
2. etc.

Sketch (when appropriate)

Analysis (diagrams when appropriate)

Properties evaluation

Calculations

Results (tables or graphs when appropriate)

Comments

1.

2. etc.

It is always assumed that the problem statement precedes the solution (as in the text) or that it is readily available (as in the *Solutions Manual*). Thus, the *Given* and *Required* statements are concise and focus on the essential features of the problem. Under *Assumptions*, the main assumptions required to solve the problem are listed; when appropriate, they are discussed further in the body of the solution. A sketch of the physical system is included when the geometry requires clarification; also, expected temperature and concentration profiles are given when appropriate. (Schematics that simply repeat the information in the problem statements are used sparingly. I know that many instructors always require a schematic. My view is that students need to develop an appreciation of when a figure or graph is necessary, because artwork is usually an expensive component of engineering reports. For example, I see little use for a schematic that shows a 10 m length of straight 2 cm–O.D. tube.) The analysis may consist simply of listing some formulas from the text, or it may require setting up a differential equation and its solution. Strictly speaking, a property should not be evaluated until its need is identified by the analysis. However, in routine calculations, such as evaluation of convective heat transfer coefficients, it is often convenient to list all the property values taken from an Appendix A table in one place. The calculations then follow with results listed, tabulated, or graphed as appropriate. Under *Comments*, the significance of the results can be discussed, the validity of assumptions further evaluated, or the broader implications of the problem noted.

In presenting calculations for the examples in *Heat and Mass Transfer*, I have rounded off results at each stage of the calculation. If additional figures are retained for the complete calculations, discrepancies in the last figure will be observed. Since many of the example calculations are quite lengthy, I believe my policy will facilitate checking a particular calculation step of concern. As is common practice, I have generally given results to more significant figures than is justified, so that these results can be conveniently used in further calculations. It is safe to say that no engineering heat transfer calculation will be accurate to within 1%, and that most experienced engineers will be pleased with results accurate to within 10% or 20%. Thus, preoccupation with a third or fourth significant figure is misplaced (unless required to prevent error magnification in operations such as subtraction).

EXERCISES

The diskette logo next to an exercise statement indicates that it can be solved using the *Heat and Mass Transfer* software, and that the sample solution provided to the instructor has been prepared accordingly. There are many additional exercises that can be solved using the software but that do not have the logo designation. These exercises are intended to give the student practice in hand calculations, and thus the sample solutions were also prepared manually.

The exercises have been ordered to correspond with the order in which the material is presented in the text, rather than in some increasing degree of difficulty. Since the range of difficulty of the exercises is considerable, the instructor is urged to give students guidance in selecting exercises for self-study. Answers to all exercises are listed in the *Solutions Manual* provided to instructors. Odd- and even-numbered exercises are listed separately; the instructor may choose to give either list to students to assist self-study.

PROPERTY DATA

A considerable quantity of property data has been assembled in Appendix A. Key sources are given as references or are listed in the bibliography. Since *Heat and Mass Transfer* is a textbook, my primary objective in preparing Appendix A was to provide the student with a wide range of data in an easily used form. Whenever possible, I have used the most accurate data that I could obtain, but accuracy was not always the primary concern. For example, the need to have consistent data over a wide range of temperature often dictated the choice of source. All the tables are in SI units, with temperature in kelvins. The computer program UNITS can be used for conversions to other systems of units. Appendix A should serve most needs of the student, as well as of the practicing engineer, for doing routine calculations. If a heat transfer research project requires accurate and reliable thermophysical property data, the prudent researcher should carefully check relevant primary data sources.

SOFTWARE

The *Heat and Mass Transfer* software has menus that describe the content of each program. The programs are also described at appropriate locations in the text. The input format and program use are demonstrated in example problems in the text. Use of the text index is recommended for locating the program descriptions and examples. There is a one-to-one correspondence between the text and the software. In principle, all numbers generated by the software can be calculated manually from formulas, graphs, and data given in the text. Small discrepancies may be seen when interpolation in graphs or property tables is required, since some of the data are stored in the software as polynomial curve fits.

The software facilitates self-study by the student. Practice hand calculations can be immediately checked using the software. When programs such as CONV, PHASE, BOIL, and SCRUB are used, properties evaluation and intermediate calculation steps can also be checked when the final results do not agree.

Since there is a large thermophysical property database stored in the software package, the programs can also be conveniently used to evaluate these properties for other purposes. For example, in CONV both the wall and fluid temperatures can be set equal to the desired temperature to obtain property values required for convection calculations. We can even go one step further when evaluating a convective heat transfer coefficient from a new correlation not contained in CONV: if a corresponding item is chosen, the values of relevant dimensionless groups can also be obtained from CONV, further simplifying the calculations.

CONTENTS

CHAPTER

APPENDIX

HEAT AND MASS TRANSFER

CHAPTER

1

ELEMENTARY HEAT TRANSFER

CONTENTS

1.1 INTRODUCTION

The process of heat transfer is familiar to us all. On a cold day we put on more clothing to reduce heat transfer from our warm body to cold surroundings. To make a cup of coffee we may plug in a kettle, inside which heat is transferred from an electrical resistance element to the water, heating the water until it boils. The engineering discipline of **heat transfer** is concerned with methods of calculating **rates** of heat transfer. These methods are used by engineers to design components and systems in which heat transfer occurs. Heat transfer considerations are important in almost all areas of technology. Traditionally, however, the discipline that has been most concerned with heat transfer is mechanical engineering because of the importance of heat transfer in energy conversion systems, from coal-fired power plants to solar water heaters.

Many *thermal design* problems require reducing heat transfer rates by providing suitable *insulation*. The insulation of buildings in extreme climates is a familiar example, but there are many others. The space shuttle has thermal tiles to insulate the vehicle from high-temperature air behind the bow shock wave during reentry into the atmosphere. Cryostats, which maintain the cryogenic temperatures required for the use of superconductors, must be effectively insulated to reduce the cooling load on the refrigeration system. Often, the only way to ensure protection from severe heating is to provide a fluid flow as a heat "sink." Nozzles of liquid-fueled rocket motors are cooled by pumping the cold fuel through passages in the nozzle wall before injection into the combustion chamber. A critical component in a fusion reactor is the "first wall" of the containment vessel, which must withstand intense heating from the hot plasma. Such walls may be cooled by a flow of helium gas or liquid lithium.

A common thermal design problem is the transfer of heat from one fluid to another. Devices for this purpose are called *heat exchangers*. A familiar example is the automobile radiator, in which heat is transferred from the hot engine coolant to cold air blowing through the radiator core. Heat exchangers of many different types are required for power production and by the process industries. A power plant, whether the fuel be fossil or nuclear, has a *boiler* in which water is evaporated to produce steam to drive the turbines, and a *condenser* in which the steam is condensed to provide a low back pressure on the turbines and for water recovery. The condenser patented by James Watt in 1769 more than doubled the efficiency of steam engines then being used and set the Industrial Revolution in motion. The common vapor cycle refrigeration or air-conditioning system has an *evaporator* where heat is absorbed at low temperature and a *condenser* where heat is rejected at a higher temperature. On a domestic refrigerator, the condenser is usually in the form of a tube coil with cooling *fins* to assist transfer of heat to the surroundings. An oil refinery has a great variety of heat transfer equipment, including rectification columns and thermal crackers. Many heat exchangers are used to transfer heat from one process stream to another, to reduce the total energy consumption by the refinery.

Often the design problem is one of *thermal control*, that is, maintaining the operating temperature of temperature-sensitive components within a specified range. Cooling of all kinds of electronic gear is an example of thermal control. The development of faster computers is now severely constrained by the difficulty of controlling the temperature of very small components, which dissipate large amounts of heat. Thermal control of temperature-sensitive components in a communications satellite orbiting the earth is a particularly difficult problem. Transistors and diodes must not overheat, batteries must not freeze, telescope optics must not lose alignment due to thermal expansion, and photographs must be processed at the proper temperature to ensure high resolution. Thermal control of space stations of the future will present even greater problems, since reliable life-support systems also will be necessary.

From the foregoing examples, it is clear that heat transfer involves a great variety of physical phenomena and engineering systems. The phenomena must first be understood and quantified before a methodology for the thermal design of an engineering system can be developed. Chapter 1 is an overview of the subject and introduces key topics at an elementary level. In Section 1.2, the distinction between the subjects of heat transfer and thermodynamics is explained. The first law of thermodynamics is reviewed, and closed- and open-system forms required for heat transfer analysis are developed. Section 1.3 introduces the three important modes of heat transfer: **heat conduction**, **thermal radiation**, and **heat convection**. Some formulas are developed that allow elementary heat transfer calculations to be made. In practical engineering problems, these modes of heat transfer usually occur simultaneously. Thus, in Section 1.4, the analysis of heat transfer by combined modes is introduced. The remainder of Chapter 1 deals with changes that occur in engineering systems as a result of heat transfer processes. In Section 1.5, the first law is applied to a very simple model closed system to determine the temperature response of the system with time. In Section 1.6, the first law for an open system is used to determine the change in temperature of a fluid flowing through a simple heat exchanger. Appropriate relations are developed that allow the design engineer to evaluate the exchanger performance. Finally, in Section 1.7, the International System of units (SI) is reviewed, and the units policy that is followed in the text is discussed.

1.2 HEAT TRANSFER AND ITS RELATION TO THERMODYNAMICS

When a hot object is placed in cold surroundings, it cools: the object loses internal energy, while the surroundings gain internal energy. We commonly describe this interaction as a *transfer of heat* from the object to the surrounding region. Since the caloric theory of heat has been long discredited, we do not imagine a "heat substance" flowing from the object to the surroundings. Rather, we understand that internal energy has been transferred by complex interactions on an atomic or subatomic scale. Nevertheless, it remains common practice to describe these interactions as transfer, transport, or flow, of heat. The engineering discipline of heat transfer is concerned with calculation of the rate at which heat flows within a medium, across an interface, or from one surface to another, as well as with the calculation of associated temperatures.

It is important to understand the essential difference between the engineering discipline of heat transfer and what is commonly called thermodynamics. Classical thermodynamics deals with systems in equilibrium. Its methodology may be used to calculate the energy required to change a system from one equilibrium state to another, but it cannot be used to calculate the rate at which the change may occur. For example, if a 1 kg ingot of iron is quenched from 1000°C to 100°C in an oil bath, thermodynamics tells us that the loss in internal energy of the ingot is mass (1 kg) × specific heat (~450 J/kg K) × temperature change (900 K), or approximately 405 kJ. But thermodynamics cannot tell us how long we will have to wait for the temperature to drop to 100°C. The time depends on the temperature of the oil bath, physical properties of the oil, motion of the oil, and other factors. An appropriate heat transfer analysis will consider all of these.

Analysis of heat transfer processes does require using some thermodynamics concepts. In particular, the **first law of thermodynamics** is used, generally in particularly simple forms since work effects can often be ignored. The first law is a statement of the *principle of conservation of energy*, which is a basic law of physics. This principle can be formulated in many ways by excluding forms of energy that are irrelevant to the problem under consideration, or by simply redefining what is meant by energy. In heat transfer, it is common practice to refer to the first law as the *energy conservation principle* or simply as an *energy* or *heat balance* when no work is done. However, as in thermodynamics, it is essential that the correct form of the first law be used. The student must be able to define an appropriate system, recognize whether the system is *open* or *closed*, and decide whether a steady state can be assumed. Some simple forms of the energy conservation principle, which find frequent use in this text, follow.

A closed system containing a fixed mass of a solid is shown in Fig. 1.1. The system has a volume V [m^3], and the solid has a density ρ [kg/m^3]. There is heat transfer into the system at a rate of $\dot{Q}$ [J/s or W], and heat may be generated within the solid, for example, by nuclear fission or by an electrical current, at a rate $\dot{Q}_v$ [W]. Solids may be taken to be incompressible, so no work is done by or on the system. The principle of conservation of energy requires that over a time interval Δt [s],

$$\begin{array}{c}\text{Change in internal energy} \\ \text{within the system}\end{array} = \begin{array}{c}\text{Heat transferred} \\ \text{into the system}\end{array} + \begin{array}{c}\text{Heat generated} \\ \text{within the system}\end{array}$$

$$\Delta U = \dot{Q}\Delta t + \dot{Q}_v\Delta t \tag{1.1}$$

Dividing by Δt and letting Δt go to zero gives

$$\frac{dU}{dt} = \dot{Q} + \dot{Q}_v$$

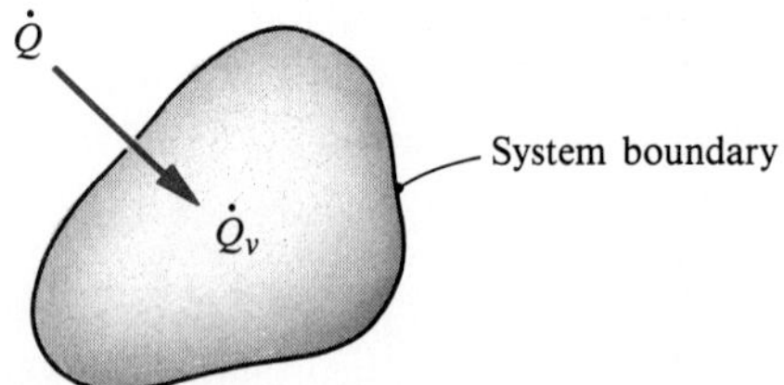

Figure 1.1 Application of the energy conservation principle to a closed system.

The system contains a fixed mass (ρV); thus, we can write $dU = \rho V du$, where u is the specific internal energy [J/kg]. Also, for an incompressible solid, $du = c_v \, dT$, where c_v is the constant-volume specific heat [J/kg K], and T [K] is temperature. Since the solid has been taken to be incompressible, the constant-volume and constant-pressure specific heats are equal, so we simply write $du = c \, dT$ to obtain

$$\rho V c \frac{dT}{dt} = \dot{Q} + \dot{Q}_v \tag{1.2}$$

Equation (1.2) is a special form of the first law of thermodynamics that will be used often in this text. It is written on a *rate* basis; that is, it gives the rate of change of temperature with time. For some purposes, however, it will prove convenient to return to Eq. (1.1) as a statement of the first law.

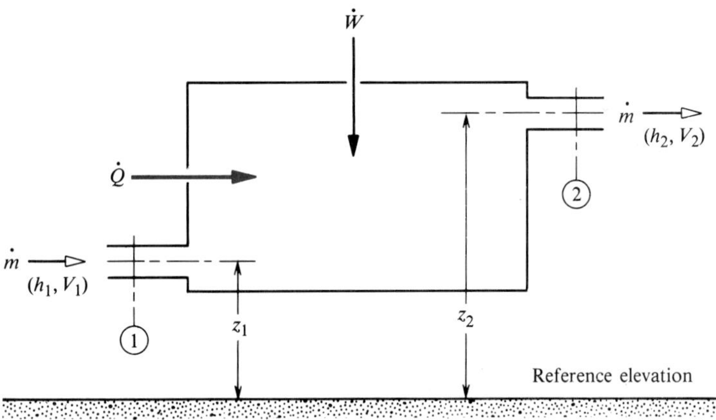

Figure 1.2 Application of the energy conservation principle to a steady-flow open system.

Figure 1.2 shows an *open* system, for which a particularly useful form of the first law is the **steady-flow energy equation**. It is used widely in the thermodynamic analysis of equipment such as turbines and compressors. The equation is

$$\dot{m}\Delta\left(h + \frac{V^2}{2} + gz\right) = \dot{Q} + \dot{W} \tag{1.3}$$

where $\dot{m}$ [kg/s] is the mass flow rate, h [J/kg] is the specific enthalpy, V [m/s] is velocity, g [m/s^2] is the gravitational acceleration, z is elevation [m], $\dot{Q}$ [W] is the rate of heat transfer, as before, and $\dot{W}$ [W] is the rate at which external (shaft) work is done on the system.[1] Notice that the sign convention here is that external work done *on* the system is positive; the opposite sign convention is also widely used. The symbol ΔX means $X_{\text{out}} - X_{\text{in}}$, or the change in X. Equation (1.3) applies to a pure

[1] Equation (1.3) has been written as if h, V, and z are uniform in the streams crossing the control volume boundary. Often such an assumption can be made; if not, an integration across each stream is required to give appropriate average values.

substance when conditions within the system, such as temperature and velocity, are unchanging over some appropriate time interval. Heat generation within the system has not been included. In many types of heat transfer equipment, no external work is done, and changes in kinetic and potential energy are negligible; Eq. (1.3) then reduces to

$$\dot{m}\Delta h = \dot{Q} \tag{1.4}$$

The specific enthalpy h is related to the specific internal energy u as

$$h = u + Pv \tag{1.5}$$

where P [N/m^2 or Pa] is pressure, and v is specific volume [m^3/kg]. Two limit forms of Δh are useful. If the fluid enters the system at state 1 and leaves at state 2:

1. For ideal gases with $Pv = RT$,

$$\Delta h = \int_{T_1}^{T_2} c_p \, dT \tag{1.6a}$$

where R [J/kg K] is the gas constant and c_p [J/kg K] is the constant-pressure specific heat.

2. For incompressible liquids with $\rho = 1/v =$ constant

$$\Delta h = \int_{T_1}^{T_2} c_v \, dT + \frac{P_2 - P_1}{\rho} \tag{1.6b}$$

where $c_v = c_p$. The second term in Eq. (1.6*b*) is usually negligible.

Equation (1.4) is the usual starting point for the heat transfer analysis of steady-state open systems.

The *second law of thermodynamics* tells us that if two objects at temperatures T_1 and T_2 are connected, and if $T_1 > T_2$, then heat will flow spontaneously and irreversibly from object 1 to object 2. Also, there is an entropy increase associated with this heat flow. As T_2 approaches T_1, the process approaches a reversible process, but simultaneously the rate of heat transfer approaches zero, so the process is of little practical interest. All heat transfer processes encountered in engineering are irreversible and generate entropy. With the increasing realization that energy supplies should be conserved, efficient use of available energy is becoming an important consideration in thermal design. Thus, the engineer should be aware of the irreversible processes occurring in the system under development and understand that the optimal design may be one that minimizes entropy generation due to heat transfer and fluid flow. Most often, however, energy conservation is simply a consideration in the overall economic evaluation of the design. Usually there is an important trade-off between energy costs associated with the operation of the system and the capital costs required to construct the equipment.

1.3 MODES OF HEAT TRANSFER

In thermodynamics, *heat* is defined as energy transfer due to temperature gradients or differences. Consistent with this viewpoint, thermodynamics recognizes only two modes of heat transfer: *conduction* and *radiation*. For example, heat transfer across a steel pipe wall is by conduction, whereas heat transfer from the sun to the earth or to a spacecraft is by thermal radiation. These modes of heat transfer occur on a molecular or subatomic scale. In air at normal pressure, conduction is by molecules that travel a very short distance ($\sim 0.65 \mu$m) before colliding with another molecule and exchanging energy. On the other hand, radiation is by photons, which travel almost unimpeded through the air from one surface to another. Thus, an important distinction between conduction and radiation is that the energy carriers for conduction have a short *mean free path*, whereas for radiation the carriers have a long mean free path. However, in air at the very low pressures characteristic of high-vacuum equipment, the mean free path of molecules can be much longer than the equipment dimensions, so the molecules travel unimpeded from one surface to another. Then heat transfer by molecules is governed by laws analogous to those for radiation.

A fluid, by virtue of its mass and velocity, can transport momentum. In addition, by virtue of its temperature, it can transport energy. Strictly speaking, *convection* is the transport of energy by motion of a medium (a moving solid can also convect energy in this sense). In the steady-flow energy equation, Eq. (1.3), convection of internal energy is contained in the term $\dot{m}\Delta h$, which is on the left-hand side of the equation, and heat transfer by conduction and radiation is on the right-hand side, as $\dot{Q}$. However, it is common engineering practice to use the term *convection* more broadly and describe heat transfer from a surface to a moving fluid also as convection, or *convective heat transfer*, even though conduction and radiation play a dominant role close to the surface, where the fluid is stationary. In this sense, convection is usually regarded as a distinct mode of heat transfer. Examples of convective heat transfer include heat transfer from the radiator of an automobile or to the skin of a hypersonic vehicle. Convection is often associated with a change of phase, for example, when water boils in a kettle or when steam condenses in a power plant condenser. Owing to the complexity of such processes, boiling and condensation are often regarded as distinct heat transfer processes.

The hot water home heating system shown in Fig. 1.3 illustrates the modes of heat transfer. Hot water from the furnace in the basement flows along pipes to radiators located in individual rooms. Transport of energy by the hot water from the basement is true convection as defined above; we do not call this a heat transfer process. Inside the radiators, there is convective heat transfer from the hot water to the radiator shell, conduction across the radiator shell, and both convective and radiative heat transfer from the hot outer surface of the radiator shell into the room. The convection is *natural* convection: the heated air adjacent to the radiator surface rises due to its buoyancy, and cooler air flows in to take its place. The radiators are heat exchangers. Although commonly used, the term *radiator* is misleading since

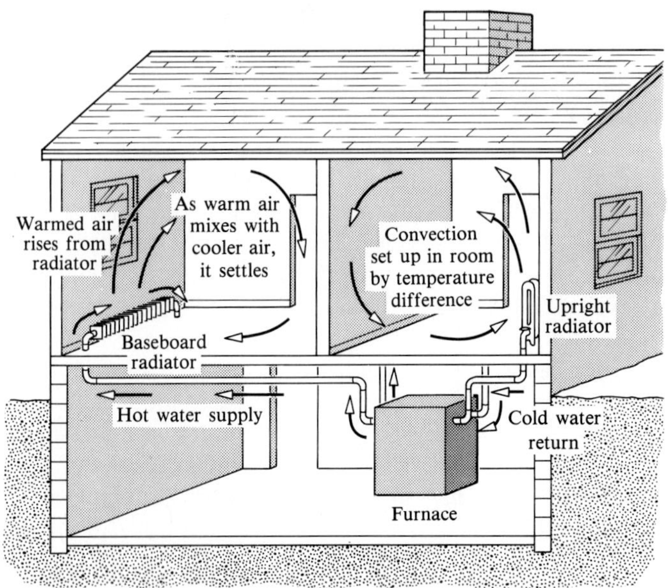

Figure 1.3 A hot-water home heating system illustrating the modes of heat transfer.

heat transfer from the shell surface can be predominantly by convection rather than by radiation (see Exercise 1–19). Heaters that transfer heat predominantly by radiation are, for example, electrical resistance wire units.

Each of the three important subject areas of heat transfer will now be introduced: conduction, in Section 1.3.1; radiation, in Section 1.3.2; and convection, in Section 1.3.3.

1.3.1 Heat Conduction

On a microscopic level, the physical mechanisms of conduction are complex, encompassing such varied phenomena as molecular collisions in gases, lattice vibrations in crystals, and flow of free electrons in metals. However, if at all possible, the engineer avoids considering processes at the microscopic level, preferring to use *phenomenological laws*, at a macroscopic level. The phenomenological law governing heat conduction was proposed by the French mathematical physicist J. B. Fourier in 1822. This law will be introduced here by considering the simple problem of one-dimensional heat flow across a plane wall—for example, a layer of insulation.[2] Figure 1.4 shows a plane wall of surface area A and thickness L, with its face at $x = 0$ maintained at temperature T_1 and the face at $x = L$ maintained at T_2. The heat flow $\dot{Q}$ through the wall is in the direction of decreasing temperature: if

[2] In thermodynamics, the term *insulated* is often used to refer to a *perfectly* insulated (zero-heat-flow or adiabatic) surface. In practice, insulation is used to *reduce* heat flow and seldom can be regarded as perfect.

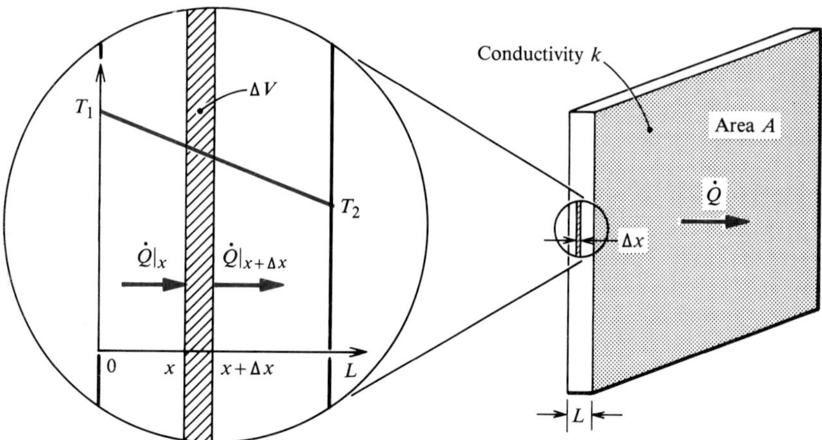

Figure 1.4 Steady one-dimensional conduction across a plane wall, showing the application of the energy conservation principle to an elemental volume Δx thick.

$T_1 > T_2$, $\dot{Q}$ is in the positive x direction.[3] The phenomenological law governing this heat flow is **Fourier's law of heat conduction**, which states that in a homogeneous substance, the local heat flux is proportional to the negative of the local temperature gradient:

$$\frac{\dot{Q}}{A} = q \qquad \text{and} \qquad q \propto -\frac{dT}{dx} \tag{1.7}$$

where q is the heat flux, or heat flow per unit area perpendicular to the flow direction [W/m^2], T is the local temperature [K or °C], and x is the coordinate in the flow direction [m]. When dT/dx is negative, the minus sign in Eq. (1.7) gives a positive q in the positive x direction. Introducing a constant of proportionality k,

$$q = -k\frac{dT}{dx} \tag{1.8}$$

where k is the **thermal conductivity** of the substance and, by inspection of the equation, must have units [W/m K]. Notice that temperature can be given in kelvins or degrees Celsius in Eq. (1.8): the temperature gradient does not depend on which of these units is used since one kelvin equals one degree Celsius (1 K = 1°C). Thus, the units of thermal conductivity could also be written [W/m °C], but this is not the recommended practice when using the SI system of units. The magnitude of the thermal conductivity k for a given substance very much depends on its microscopic structure and also tends to vary somewhat with temperature; Table 1.1 gives some selected values of k.

[3] Notice that this $\dot{Q}$ is the heat flow in the x direction, whereas in the first law, Eqs. (1.1)–(1.4), $\dot{Q}$ is the heat transfer into the whole system. In linking thermodynamics to heat transfer, some ambiguity in notation arises when common practice in both subjects is followed.

Table 1.1 Selected values of thermal conductivity at 300 K (~25°C).

Material	k W/m K
Copper	386
Aluminum	204
Brass (70% Cu, 30% Zn)	111
Mild steel	64
Stainless steel, 18–8	15
Mercury	8.4
Concrete	1.4
Pyrex glass	1.09
Water	0.611
Neoprene rubber	0.19
Engine oil, SAE 50	0.145
White pine, perpendicular to grain	0.10
Polyvinyl chloride (PVC)	0.092
Freon 12	0.071
Cork	0.043
Fiberglass (medium density)	0.038
Polystyrene	0.028
Air	0.027

Note: Appendix A contains more comprehensive data.

Figure 1.4 shows an elemental volume ΔV located between x and $x + \Delta x$; ΔV is a closed system, and the energy conservation principle in the form of Eq. (1.2) applies. If we consider a steady state, then temperatures are unchanging in time and $dT/dt = 0$; also, if there is no heat generated within the volume, $\dot{Q}_v = 0$. Then Eq. (1.2) states that the net heat flow into the system is zero. Since heat is flowing into ΔV across the face at x, and out of ΔV across the face at $x + \Delta x$,

$$\dot{Q}|_x = \dot{Q}|_{x+\Delta x}$$

or

$$\dot{Q} = \text{Constant}$$

But from Fourier's law, Eq. (1.8),

$$\dot{Q} = qA = -kA\frac{dT}{dx}$$

The variables are separable: rearranging and integrating across the wall,

$$\frac{\dot{Q}}{A}\int_0^L dx = -\int_{T_1}^{T_2} k\,dT$$

where $\dot{Q}$ and A have been taken outside the integral signs since both are constants. If the small variation of k with temperature is ignored for the present we obtain

$$\dot{Q} = \frac{kA}{L}(T_1 - T_2) = \frac{T_1 - T_2}{L/kA} \tag{1.9}$$

Comparison of Eq. (1.9) with Ohm's law, $I = E/R$, suggests that $\Delta T = T_1 - T_2$ can be viewed as a driving potential for flow of heat, analogous to voltage being the driving potential for current. Then $R \equiv L/kA$ can be viewed as a **thermal resistance** analogous to electrical resistance.

If we have a composite wall of two slabs of material, as shown in Fig. 1.5, the heat flow through each layer is the same:

$$\dot{Q} = \frac{T_1 - T_2}{L_A/k_A A} = \frac{T_2 - T_3}{L_B/k_B A}$$

Rearranging,

$$\dot{Q}\left(\frac{L_A}{k_A A}\right) = T_1 - T_2$$

$$\dot{Q}\left(\frac{L_B}{k_B A}\right) = T_2 - T_3$$

Adding eliminates the interface temperature T_2:

$$\dot{Q}\left(\frac{L_A}{k_A A} + \frac{L_B}{k_B A}\right) = T_1 - T_3$$

or

$$\dot{Q} = \frac{T_1 - T_3}{L_A/k_A A + L_B/k_B A} = \frac{\Delta T}{R_A + R_B} \tag{1.10a}$$

Using the electrical resistance analogy, we would view the problem as two resistances in series forming a **thermal circuit**, and immediately write

$$\dot{Q} = \frac{\Delta T}{R_A + R_B} \tag{1.10b}$$

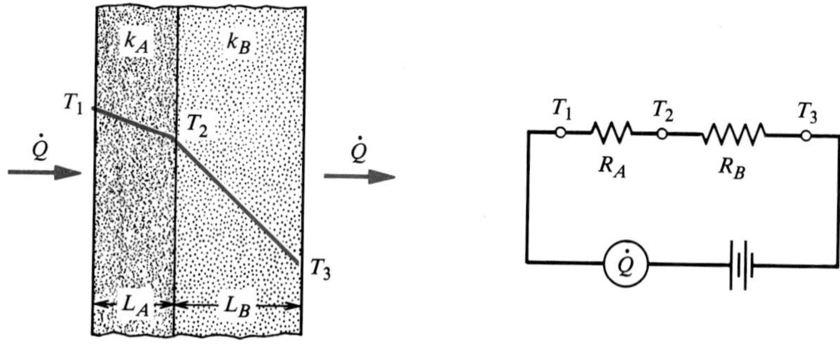

Figure 1.5 The temperature distribution for steady conduction across a composite plane wall and the corresponding thermal circuit.

EXAMPLE 1.1 Heat Transfer through Insulation

A refrigerated container is in the form of a cube with 2 m sides and has 5 mm–thick aluminum walls insulated with a 10 cm layer of cork. During steady operation, the temperatures on the inner and outer surfaces of the container are measured to be $-5°C$ and $20°C$, respectively. Determine the cooling load on the refrigerator.

Solution

Given: Aluminum container insulated with 10 cm–thick cork.

Required: Rate of heat gain.

Assumptions: 1. Steady state
2. One-dimensional heat conduction (ignore corner effects)

Equation (1.10) applies:

$$\dot{Q} = \frac{\Delta T}{R_A + R_B} \qquad \text{where } R = \frac{L}{kA}$$

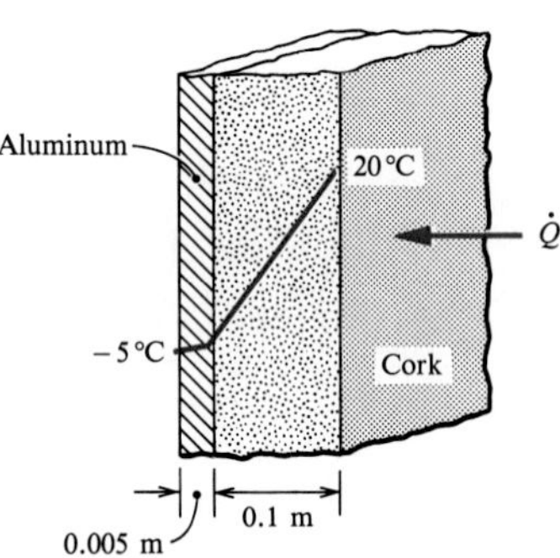

Let subscripts A and B denote the aluminum wall and cork insulation, respectively. Table 1.1 gives $k_A = 204$ W/m K, $k_B = 0.043$ W/m K. We suspect that the thermal resistance of the aluminum wall is negligible, but we will calculate it anyway. For one side of area $A = 4$ m^2, the thermal resistances are

$$R_A = \frac{L_A}{k_A A} = \frac{(0.005 \text{ m})}{(204 \text{ W/m K})(4 \text{ m}^2)} = 6.13 \times 10^{-6} \text{ K/W}$$

$$R_B = \frac{L_B}{k_B A} = \frac{(0.10 \text{ m})}{(0.043 \text{ W/m K})(4 \text{ m}^2)} = 0.581 \text{ K/W}$$

Since R_A is five orders of magnitude less than R_B, it can be ignored. The heat flow for a temperature difference of $T_1 - T_2 = 20 - (-5) = 25$ K, is

$$\dot{Q} = \frac{\Delta T}{R_B} = \frac{25 \text{ K}}{0.581 \text{ K/W}} = 43.0 \text{ W}$$

For six sides, the total cooling load on the refrigerator is $6.0 \times 43.0 = 258$ W.

Comments

1. In the future, when it is obvious that a resistance in a series network is negligible, it can be ignored from the outset (no effort should be expended to obtain data for its calculation).
2. The assumption of one-dimensional conduction is good because the 0.1 m insulation thickness is small compared to the 2 m–long sides of the cube.

3. Notice that the temperature difference $T_1 - T_2$ is expressed in kelvins, even though T_1 and T_2 were given in degrees Celsius.
4. We have assumed perfect thermal contact between the aluminum and cork; that is, there is no thermal resistance associated with the interface between the two materials (see Section 2.2.2).

1.3.2 Thermal Radiation

All matter and space contains electromagnetic radiation. A particle, or *quantum*, of electromagnetic energy is a photon, and heat transfer by radiation can be viewed either in terms of electromagnetic waves or in terms of photons. The flux of radiant energy incident on a surface is its **irradiation**, G [W/m^2]; the energy flux leaving a surface due to emission and reflection of electromagnetic radiation is its **radiosity**, J [W/m^2]. A **black surface** (or **blackbody**) is defined as a surface that absorbs all incident radiation, reflecting none. As a consequence, all of the radiation leaving a black surface is emitted by the surface and is given by the **Stefan-Boltzmann law** as

$$J = E_b = \sigma T^4 \tag{1.11}$$

where E_b is the **blackbody emissive power**, T is absolute temperature [K], and σ is the Stefan-Boltzmann constant ($\simeq 5.67 \times 10^{-8}$ W/m^2 K^4). Table 1.2 shows how $E_b = \sigma T^4$ increases rapidly with temperature.

Table 1.2 Blackbody emissive power σT^4 at various temperatures.

Surface Temperature K	Blackbody Emissive Power W/m^2
300 (room temperature)	459
1000 (cherry-red hot)	56,700
3000 (lamp filament)	4,590,000
5760 (sun temperature)	62,400,000

Figure 1.6 shows a convex black object of surface area A_1 in a black isothermal enclosure at temperature T_2. At equilibrium, the object is also at temperature T_2, and the radiation flux incident on the object must equal the radiation flux leaving:

$$G_1 A_1 = J_1 A_1 = \sigma T_2^4 A_1$$

Hence

$$G_1 = \sigma T_2^4 \tag{1.12}$$

and is uniform over the area. If the temperature of the object is now raised to T_1, its radiosity becomes σT_1^4 while its irradiation remains σT_2^4 (because the enclosure reflects no radiation). Then the net radiant heat flux through the surface, q_1, is the

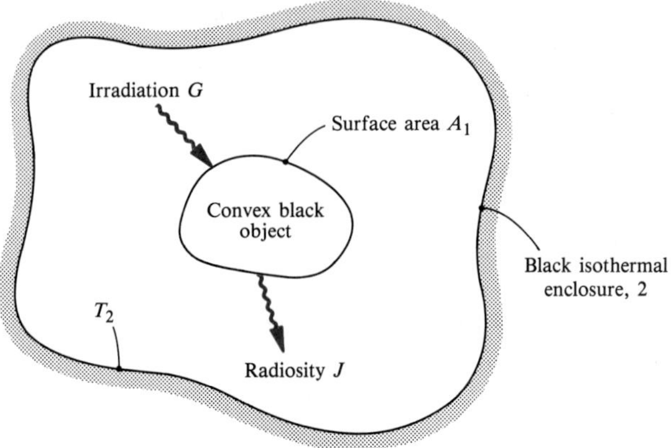

Figure 1.6 A convex black object (surface 1) in a black isothermal enclosure (surface 2).

radiosity minus the irradiation:

$$q_1 = J_1 - G_1 \tag{1.13}$$

or

$$q_1 = \sigma T_1^4 - \sigma T_2^4 \tag{1.14}$$

where the sign convention is such that a net flux away from the surface is positive. Equation (1.14) is also valid for two large black surfaces facing each other, as shown in Fig. 1.7.

The blackbody is an ideal surface. Real surfaces absorb less radiation than do black surfaces. The fraction of incident radiation absorbed is called the absorptance (or absorptivity), α. A widely used model of a real surface is the **gray surface**, which is defined as a surface for which α is a constant, irrespective of the nature of the incident radiation. The fraction of incident radiation reflected is the **reflectance** (or reflectivity), ρ. If the object is opaque, that is, not transparent to electromagnetic radiation, then

$$\rho = 1 - \alpha \tag{1.15}$$

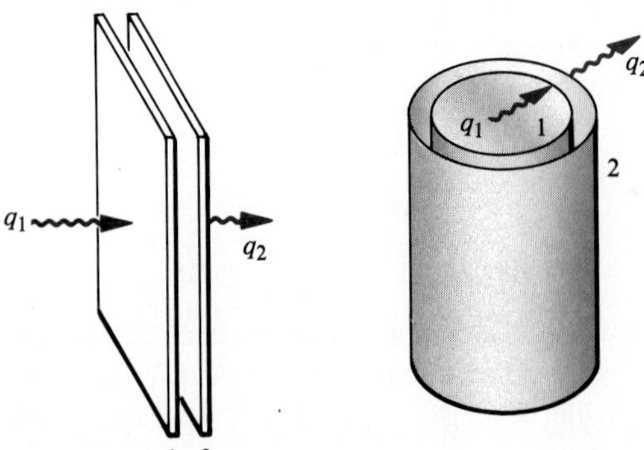

Figure 1.7 Examples of two large surfaces facing each other.

Table 1.3 Selected approximate values of emittance, ε (total hemispherical values at normal temperatures).

Surface	Emittance, ε
Aluminum alloy, unoxidized	0.035
Black anodized aluminum	0.80
Chromium plating	0.16
Stainless steel, type 312, lightly oxidized	0.30
Inconel X, oxidized	0.72
Black enamel paint	0.78
White acrylic paint	0.90
Asphalt	0.88
Concrete	0.90
Soil	0.94
Pyrex glass	0.80

Note: More comprehensive data are given in Appendix A. Emittance is very dependent on surface finish; thus, values obtained from various sources may differ significantly.

Real surfaces also emit less radiation than do black surfaces. The fraction of the blackbody emissive power σT^4 emitted is called the **emittance** (or emissivity), ε.[4] A gray surface also has a constant value of ε, independent of its temperature, and, as will be shown in Chapter 6, the emittance and absorptance of a gray surface are equal:

$$\varepsilon = \alpha \quad \text{(gray surface)} \tag{1.16}$$

Table 1.3 shows some typical values of ε at normal temperatures. Bright metal surfaces tend to have low values, whereas oxidized or painted surfaces tend to have high values. Values of α and ρ can also be obtained from Table 1.3 by using Eqs. (1.15) and (1.16).

If heat is transferred by radiation between two gray surfaces of finite size, as shown in Fig. 1.8, the rate of heat flow will depend on temperatures T_1 and T_2 and emittances ε_1 and ε_2, as well as the geometry. Clearly, some of the radiation leaving surface 1 will not be intercepted by surface 2, and vice versa. Determining the rate of heat flow is usually quite difficult. In general, we may write

$$\dot{Q}_{12} = A_1\mathscr{F}_{12}(\sigma T_1^4 - \sigma T_2^4) \tag{1.17}$$

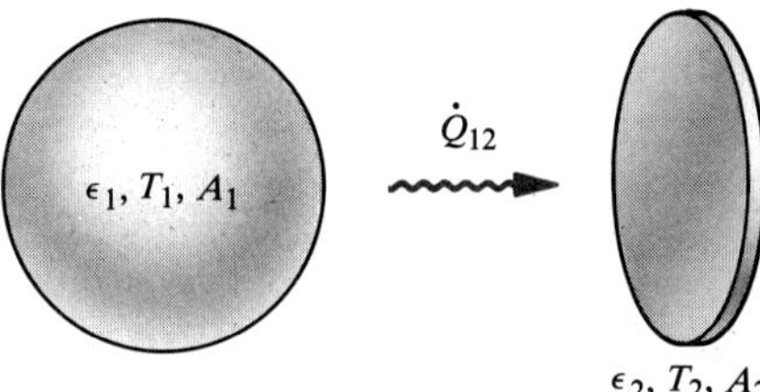

Figure 1.8 Radiation heat transfer between two finite gray surfaces.

[4] Both the endings *-ance* and *-ivity* are commonly used for radiation properties. In this text, *-ance* will be used for surface radiation properties. In Chapter 6, *-ivity* will be used for gas radiation properties.

where $\dot{Q}_{12}$ is the net radiant energy interchange (heat transfer) from surface 1 to surface 2, and $\mathscr{F}_{12}$ is a **transfer factor**, which depends on emittances and geometry. For the special case of surface 1 surrounded by surface 2, where either area A_1 is small compared to area A_2, or surface 2 is nearly black, $\mathscr{F}_{12} \simeq \varepsilon_1$ and Eq. (1.17) becomes

$$\dot{Q}_{12} = \varepsilon_1 A_1(\sigma T_1^4 - \sigma T_2^4) \qquad \textbf{(1.18)}$$

Equation (1.18) will be derived in Chapter 6. It is an important result and is often used for quick engineering estimates.

The T^4 dependence of radiant heat transfer complicates engineering calculations. When T_1 and T_2 are not too different, it is convenient to linearize Eq. (1.18) by factoring the term $(\sigma T_1^4 - \sigma T_2^4)$ to obtain

$$\begin{aligned}\dot{Q}_{12} &= \varepsilon_1 A_1 \sigma(T_1^2 + T_2^2)(T_1 + T_2)(T_1 - T_2) \\ &\simeq \varepsilon_1 A_1 \sigma(4T_m^3)(T_1 - T_2)\end{aligned}$$

for $T_1 \simeq T_2$, where T_m is the mean of T_1 and T_2. This result can be written more concisely as

$$Q_{12} \simeq A_1 h_r(T_1 - T_2) \qquad \textbf{(1.19)}$$

where $h_r = 4\varepsilon_1\sigma T_m^3$ is called the **radiation heat transfer coefficient** [W/m^2 K]. At 25°C (= 298 K),

$$h_r = (4)\varepsilon_1(5.67 \times 10^{-8}\ \mathrm{W/m^2\ K^4})(298\ \mathrm{K})^3$$

or

$$h_r \simeq 6\varepsilon_1 \mathrm{W/m^2\ K}$$

This result can be easily remembered: The radiation heat transfer coefficient at room temperature is about six times the surface emittance. For $T_1 = 320$ K and $T_2 = 300$ K, the error incurred in using the approximation of Eq. (1.19) is only 0.1%; for $T_1 = 400$ K and $T_2 = 300$ K, the error is 2%.

EXAMPLE 1.2 Heat Loss from a Transistor

An electronic package for an experiment in outer space contains a transistor capsule, which is approximately spherical in shape with a 2 cm diameter. It is contained in an evacuated case with nearly black walls at 30°C. The only significant path for heat loss from the capsule is radiation to the case walls. If the transistor dissipates 300 mW, what will the capsule temperature be if it is (i) bright aluminum and (ii) black anodized aluminum?

Solution

Given: 2 cm–diameter transistor capsule dissipating 300 mW.

Required: Capsule temperature for (i) bright aluminum and (ii) black anodized aluminum.

Assumptions: Model as a small gray body in large, nearly black surroundings.

Equation (1.18) is applicable with

$$\dot{Q}_{12} = 300\,\text{mW}$$

$$T_2 = 30°\text{C} = 303\,\text{K}$$

and T_1 is the unknown.

$$\dot{Q}_{12} = \varepsilon_1 A_1(\sigma T_1^4 - \sigma T_2^4)$$

$$0.3\text{ W} = (\varepsilon_1)(\pi)(0.02\text{ m})^2[\sigma T_1^4 - (5.67 \times 10^{-8}\text{ W/m}^2\,\text{K}^4)(303\text{ K})^4]$$

Solving,

$$\sigma T_1^4 = 478 + \frac{239}{\varepsilon_1}$$

(i) For bright aluminum ($\varepsilon = 0.035$ from Table 1.3),

$$\sigma T_1^4 = 478 + 6828 = 7306\text{ W/m}^2$$

$$T_1 = 599\text{ K }(326°\text{C})$$

(ii) For black anodized aluminum ($\varepsilon = 0.80$ from Table 1.3),

$$\sigma T_1^4 = 478 + 298 = 776\text{ W/m}^2$$

$$T_1 = 342\text{ K }(69°\text{C})$$

Comments

1. The anodized aluminum gives a satisfactory operating temperature, but a bright aluminum capsule could not be used since 326°C is far in excess of allowable operating temperatures for semiconductor devices.
2. Note the use of kelvins for temperature in this radiation heat transfer calculation.

1.3.3 Heat Convection

As already explained, *convection* or *convective heat transfer* is the term used to describe heat transfer from a surface to a moving fluid, as shown in Fig. 1.9. The surface may be the inside of a pipe, the skin of a hypersonic aircraft, or a water-air interface in a cooling tower. The flow may be *forced*, as in the case of a liquid pumped

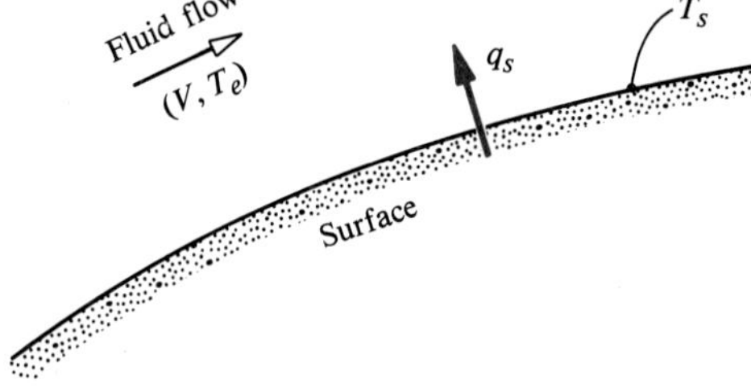

Figure 1.9 Schematic of convective heat transfer to a fluid at temperature T_e, flowing at velocity V past a surface at temperature T_s.

through the pipe or air on the flight vehicle propelled through the atmosphere. On the other hand, the flow could be *natural* (or *free*), driven by buoyancy forces arising from a density difference, as in the case of a natural-draft cooling tower. Either type of flow can be *internal*, such as the pipe flow, or *external*, such as flow over the vehicle. Also, both forced and natural flows can be either *laminar* or *turbulent*, with laminar flows being predominant at lower velocities, for smaller sizes, and for more viscous fluids. Flow in a pipe becomes turbulent when the dimensionless group called the **Reynolds number**, $\mathrm{Re}_D = VD/\nu$, exceeds about 2300, where V is the velocity [m/s], D is the pipe diameter [m], and ν is the kinematic viscosity of the fluid [m^2/s]. Heat transfer rates tend to be much higher in turbulent flows than in laminar flows, owing to the vigorous mixing of the fluid. Figure 1.10 shows some commonly encountered flows.

The rate of heat transfer by convection is usually a complicated function of surface geometry and temperature, the fluid temperature and velocity, and fluid thermophysical properties. In an external forced flow, the rate of heat transfer is approximately proportional to the difference between the surface temperature T_s and the temperature of the free stream fluid T_e. The constant of proportionality is called the **convective heat transfer coefficient** h_c:

$$q_s = h_c \Delta T \tag{1.20}$$

where $\Delta T = T_s - T_e$, q_s is the heat flux from the surface into the fluid [W/m^2], and h_c has units [W/m^2 K]. Equation (1.20) is often called *Newton's law of cooling* but is a definition of h_c rather than a true physical law. For natural convection, the situation is more complicated. If the flow is laminar, q_s varies as $\Delta T^{5/4}$; if the flow is turbulent, it varies as $\Delta T^{4/3}$. However, we still find it convenient to define a heat transfer coefficient by Eq. (1.20); then h_c varies as $\Delta T^{1/4}$ for laminar flows and as $\Delta T^{1/3}$ for turbulent ones.

An important practical problem is convective heat transfer to a fluid flowing in a tube, as may be found in heat exchangers for heating or cooling liquids, in condensers, and in various kinds of boilers. In using Eq. (1.20) for internal flows, $\Delta T = T_s - T_b$, where T_b is a properly averaged fluid temperature called the **bulk temperature** or mixed mean temperature and is defined in Chapter 4. Here it is sufficient to note that enthalpy in the steady-flow energy equation, Eq. (1.4), is also the bulk value, and T_b is the corresponding temperature. If the pipe has a uniform wall temperature T_s along its length, and the flow is laminar ($\mathrm{Re}_D \lesssim 2300$), then sufficiently far from the pipe entrance, the heat transfer coefficient is given by the exact relation

$$h_c = 3.66 \frac{k}{D} \tag{1.21}$$

where k is the fluid thermal conductivity and D is the pipe diameter. Notice that the heat transfer coefficient is directly proportional to thermal conductivity, inversely proportional to pipe diameter, and—perhaps surprisingly—independent of flow velocity. On the other hand, for fully turbulent flow ($\mathrm{Re}_D \gtrsim 10{,}000$), h_c is given

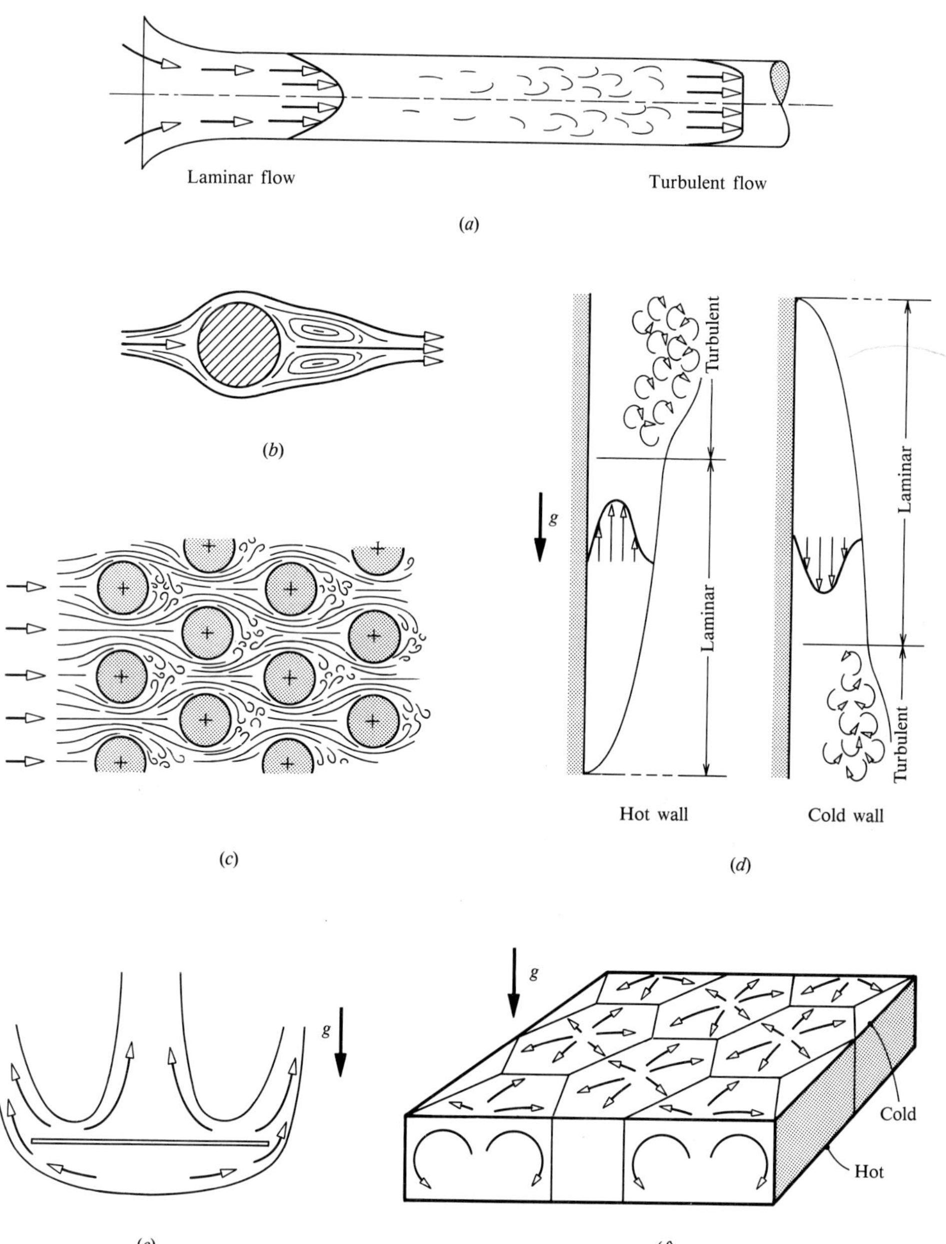

Figure 1.10 Some commonly encountered flows. (*a*) Forced flow in a pipe, $\mathrm{Re}_D \simeq 50{,}000$. The flow is initially laminar because of the "bell-mouth" entrance but becomes turbulent downstream. (*b*) Laminar forced flow over a cylinder, $\mathrm{Re}_D \simeq 25$. (*c*) Forced flow through a tube bank as found in a shell-and-tube heat exchanger. (*d*) Laminar and turbulent natural convection boundary layers on vertical walls. (*e*) Laminar natural convection about a heated horizontal plate. (*f*) Cellular natural convection in a horizontal enclosed fluid layer.

approximately by the following, rather complicated correlation of experimental data:

$$h_c = 0.023 \frac{V^{0.8} k^{0.6} (\rho c_p)^{0.4}}{D^{0.2} \nu^{0.4}} \tag{1.22}$$

In contrast to laminar flow, h_c is now strongly dependent on velocity, V, but only weakly dependent on diameter. In addition to thermal conductivity, other fluid properties involved are the kinematic viscosity, ν; density, ρ; and specific heat, c_p. In Chapter 4 we will see how Eq. (1.22) can be rearranged in a more compact form by introducing appropriate dimensionless groups. Equations (1.21) and (1.22) are only valid at some distance from the pipe entrance and indicate that the heat transfer coefficient is then independent of position along the pipe. Near the pipe entrance, heat transfer coefficients tend to be higher, due to the generation of large-scale vortices by upstream bends or sharp corners and the effect of suddenly heating the fluid.

Figure 1.11 shows a natural convection flow on a heated vertical surface, as well as a schematic of the associated variation of h_c along the surface. Transition from a laminar to a turbulent boundary layer is shown. In gases, the location of the transition is determined by a critical value of a dimensionless group called the **Grashof number**. The Grashof number is defined as $\mathrm{Gr}_x = (\beta \Delta T) g x^3/\nu^2$, where $\Delta T = T_s - T_e$, g is the gravitational acceleration [m/s^2], x is the distance from the bottom of the surface where the boundary layer starts, and β is the volumetric coefficient of expansion, which for an ideal gas is simply $1/T$, where T is absolute temperature [K]. On a vertical wall, transition occurs at $\mathrm{Gr}_x \simeq 10^9$. For air, at normal temperatures, experiments show that the heat transfer coefficient for natural convection on a vertical wall can be approximated by the following formulas:

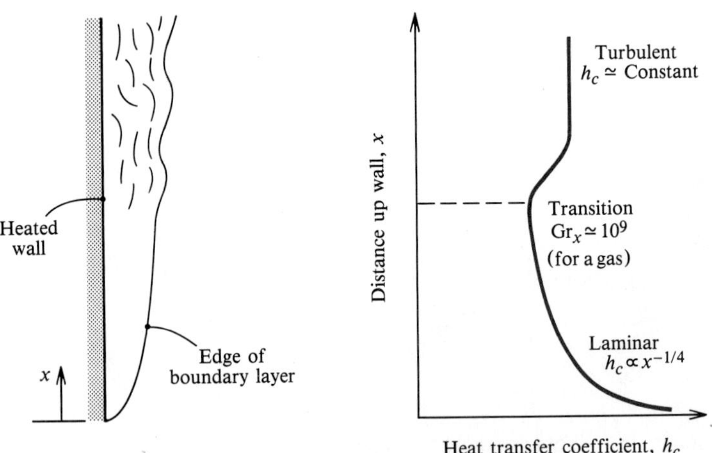

Figure 1.11 A natural-convection boundary layer on a vertical wall, showing the variation of local heat transfer coefficient. For gases, transition from a laminar to turbulent flow occurs at a Grashof number of approximately 10^9; hence, $x_{\mathrm{tr}} \simeq [10^9 \nu^2/\beta \Delta T]^{1/3}$.

$$\text{Laminar flow:} \quad h_c = 1.07(\Delta T/x)^{1/4} \text{ W/m}^2\text{ K} \qquad 10^4 < \text{Gr}_x < 10^9 \tag{1.23a}$$

$$\text{Turbulent flow:} \quad h_c = 1.3(\Delta T)^{1/3} \text{ W/m}^2\text{ K} \qquad 10^9 < \text{Gr}_x < 10^{12} \tag{1.23b}$$

Since these are dimensional equations, it is necessary to specify the units of h_c, ΔT, and x, which are [W/m^2 K], [K], and [m], respectively. Notice that h_c varies as $x^{-1/4}$ in the laminar region but is independent of x in the turbulent region.

Usually the engineer requires the total heat transfer from a surface and is not too interested in the actual variation of heat flux along the surface. For this purpose, it is convenient to define an average heat transfer coefficient $\overline{h}_c$ for an *isothermal* surface of area A by the relation

$$\dot{Q} = \overline{h}_c A(T_s - T_e) \tag{1.24}$$

so that the total heat transfer rate, $\dot{Q}$, can be obtained easily. The relation between $\overline{h}_c$ and h_c is obtained as follows: For flow over a surface of width W and length L, as shown in Fig. 1.12,

$$d\dot{Q} = h_c(T_s - T_e)W\,dx$$

$$\dot{Q} = \int_0^L h_c(T_s - T_e)W\,dx$$

or

$$\dot{Q} = \left(\frac{1}{A}\int_0^A h_c\,dA\right)A(T_s - T_e), \qquad \text{where } A = WL,\ dA = W\,dx \tag{1.25}$$

if $(T_s - T_e)$ is independent of x. Since T_e is usually constant, this condition requires an isothermal wall. Thus, comparing Eqs. (1.24) and (1.25),

$$\overline{h}_c = \frac{1}{A}\int_0^A h_c\,dA \tag{1.26}$$

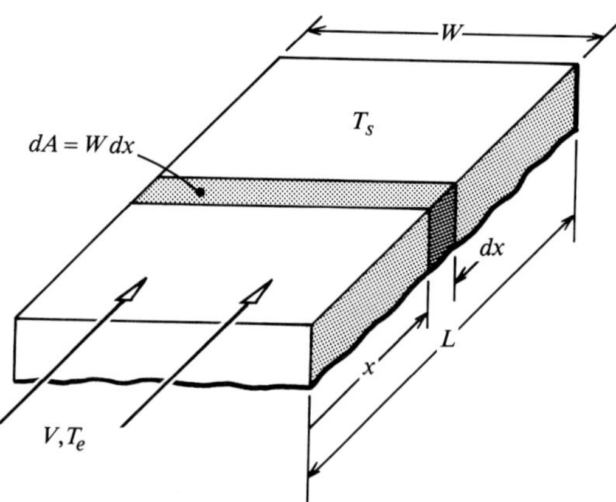

Figure 1.12 An isothermal surface used to define the average convective heat transfer coefficient $\overline{h}_c$.

Table 1.4 Orders of magnitude of average convective heat transfer coefficients.

Flow and Fluid	$\overline{h}_c$ W/m^2 K
Free convection, air	3–25
Free convection, water	15–100
Forced convection, air	10–200
Forced convection, water	50–10,000
Forced convection, liquid sodium	10,000–100,000
Condensing steam	5000–50,000
Boiling water	3000–100,000

The surface may not be isothermal; for example, the surface may be electrically heated to give a uniform flux q_s along the surface. In this case, defining an average heat transfer coefficient is more difficult and will be dealt with in Chapter 4. Table 1.4 gives some order-of-magnitude values of average heat transfer coefficients for various situations. In general, high heat transfer coefficients are associated with high fluid thermal conductivities, high flow velocities, and small surfaces. The high heat transfer coefficients shown for boiling water and condensing steam are due to another cause: as we will see in Chapter 7, a large enthalpy of phase change (latent heat) is a contributing factor.

The complexity of most situations involving convective heat transfer precludes exact analysis, and *correlations* of experimental data must be used in engineering practice. For a particular situation, a number of correlations from various sources might be available, for example, from research laboratories in different countries. Also, as time goes by, older correlations may be superseded by newer correlations based on more accurate or more extensive experimental data. Heat transfer coefficients calculated from various available correlations usually do not differ by more than about 20%, but in more complex situations, much larger discrepancies may be encountered. Such is the nature of engineering calculations of convective heat transfer, in contrast to the more exact nature of the analysis of heat conduction or of elementary mechanics, for example.

EXAMPLE 1.3 Heat Loss through Glass Doors

The living room of a ski chalet has a pair of glass doors 2.3 m high and 4.0 m wide. On a cold morning, the air in the room is at 10°C, and frost partially covers the inner surface of the glass. Estimate the convective heat loss to the doors. Would you expect to see the frost form initially near the top or the bottom of the doors? Take $\nu = 14 \times 10^{-6}$ m^2/s for the air.

Solution

Given: Glass doors, width $W = 4$ m, height $L = 2.3$ m.

Required: Estimate of convective heat loss to the doors.

Assumptions:
1. Inner surface isothermal at $T_s \simeq 0°\text{C}$
2. The laminar to turbulent flow transition occurs at $\text{Gr}_x \simeq 10^9$.

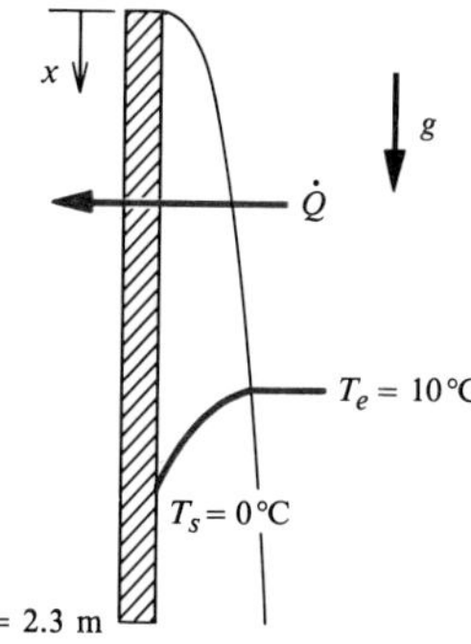

Equation (1.24) will be used to estimate the heat loss. The inner surface will be at approximately 0°C since it is only partially covered with frost. If it were warmer, frost couldn't form; and if it were much colder, frost would cover the glass completely. There is a natural convection flow down the door since $T_e = 10°\text{C}$ is greater than $T_s = 0°\text{C}$. Transition from a laminar boundary layer to a turbulent boundary layer occurs when the Grashof number is about 10^9. For transition at $x = x_{\text{tr}}$,

$$\text{Gr} = 10^9 = \frac{(\beta \Delta T) g x_{\text{tr}}^3}{\nu^2}; \qquad \beta = 1/T \text{ for an ideal gas}$$

$$x_{\text{tr}} = \left[\frac{10^9 \nu^2}{(\Delta T/T) g}\right]^{1/3} = \left[\frac{(10^9)(14 \times 10^{-6}\,\text{m}^2/\text{s})^2}{(10/278)(9.81\,\text{m/s}^2)}\right]^{1/3} = 0.82\ \text{m}$$

where the average of T_s and T_e has been used to evaluate β. The transition is seen to take place about one third of the way down the door.

We find the average heat transfer coefficient, $\bar{h}_c$, by substituting Eqs. (1.23*a*,*b*) in Eq. (1.26):

$$\bar{h}_c = \frac{1}{A}\int_0^A h_c\, dA; \qquad A = WL, \qquad dA = W\,dx$$

$$= \frac{1}{L}\int_0^L h_c\, dx$$

$$= \frac{1}{L}\left[\int_0^{x_{\text{tr}}} 1.07(\Delta T/x)^{1/4}\, dx + \int_{x_{\text{tr}}}^L 1.3(\Delta T)^{1/3}\, dx\right]$$

$$= (1/L)[(1.07)(4/3)\Delta T^{1/4} x_{\text{tr}}^{3/4} + (1.3)(\Delta T)^{1/3}(L - x_{\text{tr}})]$$

$$= (1/2.3)[(1.07)(4/3)(10)^{1/4}(0.82)^{3/4} + (1.3)(10)^{1/3}(2.3 - 0.82)]$$

$$= (1/2.3)[2.19 + 4.15]$$

$$= 2.75\ \text{W/m}^2\,\text{K}$$

Then, from Eq. (1.24), the total heat loss to the door is

$$\dot{Q} = \bar{h}_c A \Delta T = (2.75\ \text{W/m}^2\,\text{K})(2.3 \times 4.0\ \text{m}^2)(10\ \text{K}) = 253\ \text{W}$$

Comments

1. The local heat transfer coefficient is larger near the top of the door, so that the relatively warm room air will tend to cause the glass there to be at a higher temperature than further down the door. Thus, frost should initially form near the bottom of the door.
2. In addition, interior surfaces in the room will lose heat by radiation through the glass doors.

1.4 COMBINED MODES OF HEAT TRANSFER

Heat transfer problems encountered by the design engineer almost always involve more than one mode of heat transfer occurring simultaneously. For example, consider the nighttime heat loss through the roof of the house shown in Fig. 1.3. Heat is transferred to the ceiling by convection from the warm room air, and by radiation from the walls, furniture, and occupants. Heat transfer across the ceiling and its insulation is by conduction, across the attic crawlspace by convection and radiation, and across the roof tile by conduction. Finally, the heat is transferred by convection to the cold ambient air, and by radiation to the nighttime sky. To consider realistic engineering problems, it is necessary at the outset to develop the theory required to handle *combined modes* of heat transfer.

1.4.1 Thermal Circuits

The electrical circuit analogy for conduction through a composite wall was introduced in Section 1.3.1. We now extend this concept to include convection and radiation as well. Figure 1.13 shows a two-layer composite wall of cross-sectional area A with the layers A and B having thickness and conductivity L_A, k_A and L_B, k_B, respectively. Heat is transferred from a hot fluid at temperature T_i to the inside of the wall with a convective heat transfer coefficient $h_{c,i}$, and away from the outside of the wall to a cold fluid at temperature T_o with heat transfer coefficient $h_{c,o}$.

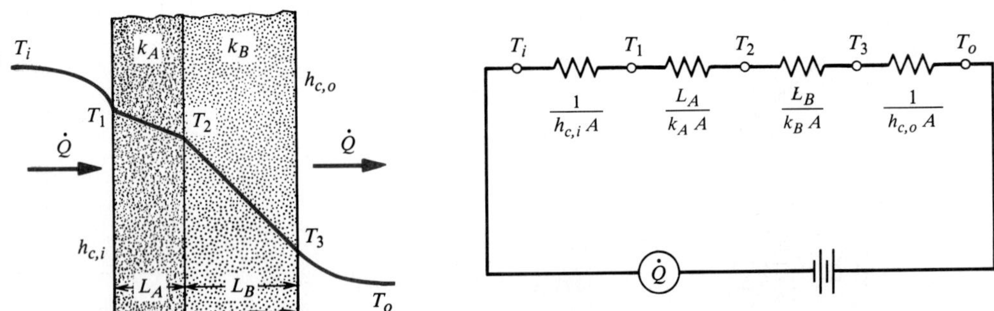

Figure 1.13 The temperature distribution for steady heat transfer across a composite plane wall, and the corresponding thermal circuit.

Newton's law of cooling, Eq. (1.20), can be rewritten as

$$\dot{Q} = \frac{\Delta T}{1/h_c A} \tag{1.27}$$

with $1/h_c A$ identified as a convective thermal resistance. At steady state, the heat flow through the wall is constant. Referring to Fig. 1.13 for the intermediate temperatures,

$$\dot{Q} = \frac{T_i - T_1}{1/h_{c,i}A} = \frac{T_1 - T_2}{L_A/k_A A} = \frac{T_2 - T_3}{L_B/k_B A} = \frac{T_3 - T_o}{1/h_{c,o}A} \tag{1.28}$$

Equation (1.28) is the basis of the thermal circuit shown in Fig. 1.13. The total resistance is the sum of four resistances in series. If we define the **overall heat transfer coefficient** U by the relation

$$\dot{Q} = UA(T_i - T_o) \tag{1.29}$$

then $1/UA$ is an overall resistance given by

$$\frac{1}{UA} = \frac{1}{h_{c,i}A} + \frac{L_A}{k_A A} + \frac{L_B}{k_B A} + \frac{1}{h_{c,o}A} \tag{1.30a}$$

or, since the cross-sectional area A is constant for a plane wall,

$$\frac{1}{U} = \frac{1}{h_{c,i}} + \frac{L_A}{k_A} + \frac{L_B}{k_B} + \frac{1}{h_{c,o}} \tag{1.30b}$$

Equation (1.29) is simple and convenient for use in engineering calculations. Typical values of U [W/m^2 K] vary over a wide range for different types of walls and convective flows.

Figure 1.14 shows a wall whose outer surface loses heat by both convection and radiation. For simplicity, assume that the fluid is at the same temperature as the surrounding surfaces, T_o. Using the approximate linearized Eq. (1.19),

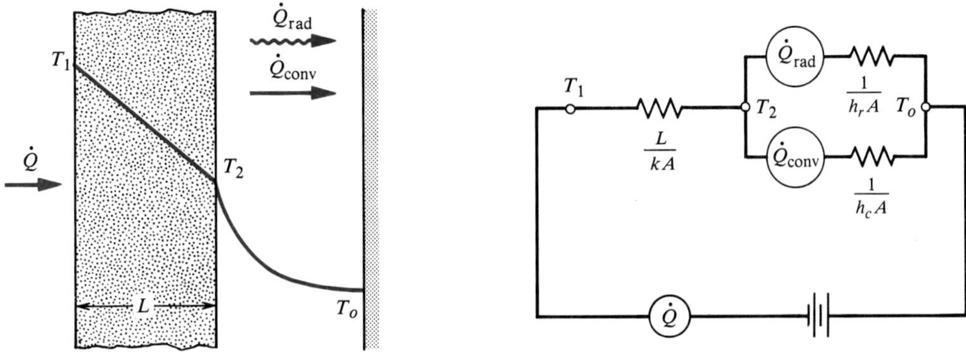

Figure 1.14 A wall that loses heat by both conduction and radiation; the thermal circuit shows resistances in parallel.

$$\dot{Q}_{\text{rad}} = \frac{\Delta T}{1/h_r A} \tag{1.31}$$

with $1/h_r A$ identified as a radiative thermal resistance. We now have two resistances in parallel, as shown in Fig. 1.14. The sum of the resistances is

$$\sum R = \frac{L}{kA} + \frac{1}{h_c A + h_r A}$$

or

$$\frac{1}{UA} = \frac{L}{kA} + \frac{1}{(h_c + h_r)A} \tag{1.32}$$

so that the convective and radiative heat transfer coefficients can simply be added. However, often the fluid and surrounding temperatures are not the same, or the simple linearized representation of radiative transfer [Eq. (1.19)] is invalid, so the thermal circuit is then more complex. When appropriate, we will write $h = h_c + h_r$ to account for combined convection and radiation.[5]

EXAMPLE 1.4 Heat Loss through a Composite Wall

The walls of a sparsely furnished single-room cabin in a forest consist of two layers of pine wood, each 2 cm thick, sandwiching 5 cm of fiberglass insulation. The cabin interior is maintained at 20°C when the ambient air temperature is 2°C. If the interior and exterior convective heat transfer coefficients are 3 and 6 W/m^2 K, respectively, and the exterior surface is finished with a white acrylic paint, estimate the heat flux through the wall.

Solution

Given: Pine wood cabin wall insulated with 5 cm of fiberglass.

Required: Estimate of heat loss through wall.

Assumptions:
1. Forest trees and shrubs are at the ambient air temperature, $T_e = 2$°C.
2. Radiation transfer inside cabin is negligible since inner surfaces of walls, roof, and floor are at approximately the same temperature.

From Eq. (1.29), the heat flux through the wall is

$$q = \frac{\dot{Q}}{A} = U(T_i - T_o)$$

From Eqs. (1.30) and (1.32), the overall heat transfer coefficient is given by

$$\frac{1}{U} = \frac{1}{h_{c,i}} + \frac{L_A}{k_A} + \frac{L_B}{k_B} + \frac{L_C}{k_C} + \frac{1}{(h_{c,o} + h_{r,o})}$$

---

[5] Notice that the notation used for this combined heat transfer coefficient, h, is the same as that used for enthalpy. The student must be careful not to confuse these two quantities. Other notation is also in common use, for example, α for the heat transfer coefficient and i for enthalpy.

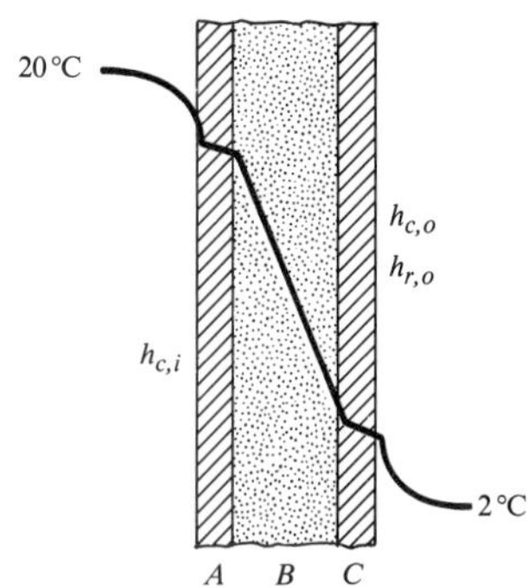

The thermal conductivities of pine wood, perpendicular to the grain, and of fiberglass are given in Table 1.1 as 0.10 and 0.038 W/m K, respectively. The exterior radiation heat transfer coefficient is given by Eq. (1.19) as

$$h_{r,o} = 4\varepsilon\sigma T_m^3$$

where $\varepsilon = 0.9$ for white acrylic paint, from Table 1.3, and $T_m \simeq 2°\text{C} = 275$ K (since we expect the exterior resistance to be small). Thus,

$$\begin{aligned} h_{r,o} &= 4(0.9)(5.67 \times 10^{-8}\ \text{W/m}^2\ \text{K}^4)(275\ \text{K})^3 \\ &= 4.2\ \text{W/m}^2\ \text{K} \end{aligned}$$

$$\begin{aligned} \frac{1}{U} &= \frac{1}{3} + \frac{0.02}{0.10} + \frac{0.05}{0.038} + \frac{0.02}{0.10} + \frac{1}{6 + 4.2} \\ &= 0.333 + 0.200 + 1.316 + 0.200 + 0.098 \\ &= 2.15\ (\text{W/m}^2\ \text{K})^{-1} \\ U &= 0.466\ \text{W/m}^2\ \text{K} \end{aligned}$$

Then the heat flux $q = U(T_i - T_o) = 0.466(20 - 2) = 8.38$ W/m^2.

The thermal circuit is shown below.

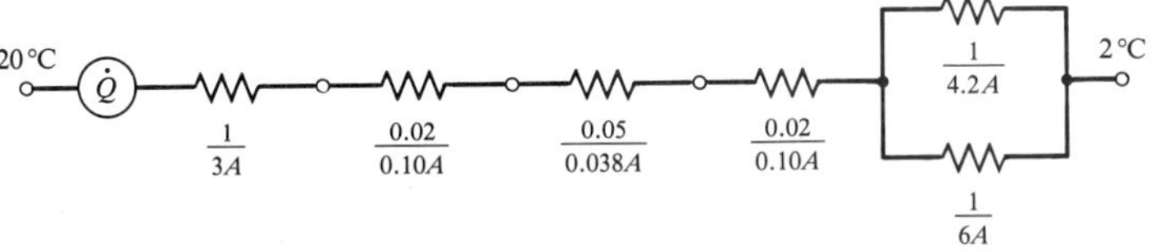

Comments

1. The outside resistance is seen to be $0.098/2.15 \simeq 5\%$ of the total resistance; hence, the outside wall of the cabin is only about 1 K above the ambient air, and our assumption of $T_m = 275$ K for the evaluation of $h_{r,o}$ is adequate.

2. The dominant resistance is that of the fiberglass insulation; therefore, an accurate calculation of q depends mainly on having accurate values for the fiberglass thickness and thermal conductivity. Poor data or poor assumptions for the other resistances have little impact on the result.

1.4.2 Surface Energy Balances

Section 1.4.1 assumed that the energy flow $\dot{Q}$ across the wall surfaces is continuous. In fact, we used a procedure commonly called a *surface energy balance*, which is used in various ways. Some examples follow. Figure 1.15 shows an opaque solid that is losing heat by convection and radiation to its surroundings. Two imaginary surfaces are located on each side of the real solid-fluid interface: an s-surface

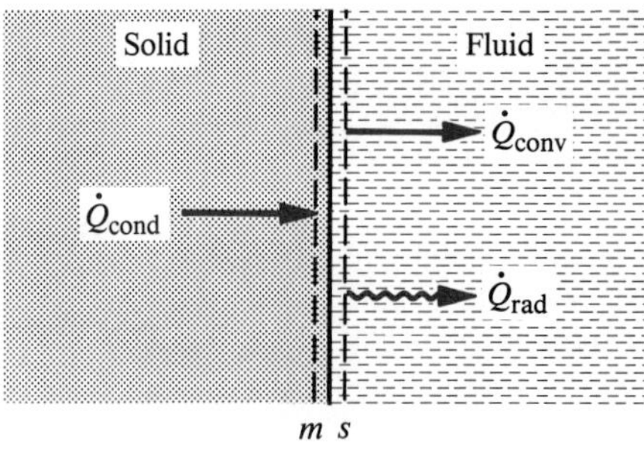

Figure 1.15 Schematic of a surface energy balance, showing the m- and s- surface in the solid and fluid, respectively.

in the fluid just adjacent to the interface, and an m-surface in the solid located such that all radiation is emitted or absorbed between the m-surface and the interface. Thus, energy is transferred across the m-surface by conduction only. (The choice of s and m to designate these surfaces follows an established practice. In particular, the use of the s prefix is consistent with the use of the subscript s to denote a surface temperature T_s in convection analysis.) The first law as applied to the closed system located between m- and s-surfaces requires that $\sum \dot{Q} = 0$; thus,

$$\dot{Q}_{cond} - \dot{Q}_{conv} - \dot{Q}_{rad} = 0 \tag{1.33}$$

or, for a unit area,

$$q_{cond} - q_{conv} - q_{rad} = 0 \tag{1.34}$$

where the sign convention for the fluxes is shown in Fig. 1.15. If the solid is isothermal, Eq. (1.33) reduces to

$$\dot{Q}_{conv} + \dot{Q}_{rad} = 0 \tag{1.35}$$

which is a simple energy balance on the solid. Notice that these surface energy balances remain valid for unsteady conditions, in which temperatures change with time, because the mass contained between the s- and m-surfaces is negligible and cannot store energy.

EXAMPLE 1.5 Air Temperature Measurement

A machine operator in a workshop complains that the air-heating system is not keeping the air at the required minimum temperature of 20°C. To support his claim, he shows that a mercury-in-glass thermometer suspended from a roof truss reads only 17°C. The roof and walls of the workshop are made of corrugated iron and are not insulated; when the thermometer is held against the wall, it reads only 5°C. If the average convective heat transfer coefficient for the suspended thermometer is estimated to be 10 W/m^2 K, what is the true air temperature?

Solution

Given: Thermometer reading a temperature of 17°C.

Required: True air temperature.

Assumptions: Thermometer can be modeled as a small gray body in large, nearly black surroundings at 5°C.

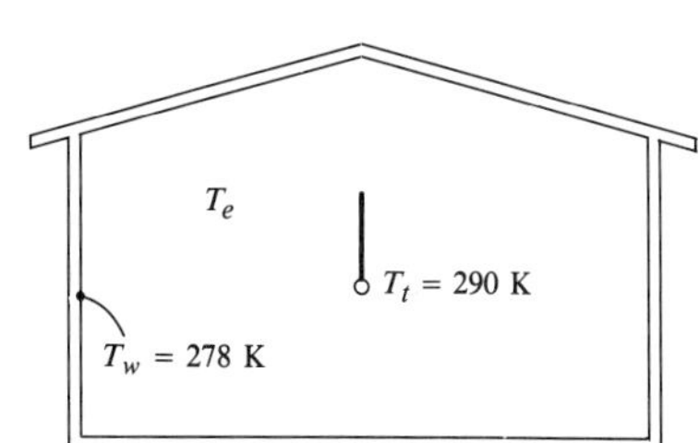

Let T_t be the thermometer reading, T_e the air temperature, and T_w the wall temperature. Equation (1.35) applies,

$$\dot{Q}_{\text{conv}} + \dot{Q}_{\text{rad}} = 0$$

since at steady state there is no conduction within the thermometer. Substituting from Eqs. (1.24) and (1.18),

$$\overline{h}_c A(T_t - T_e) + \varepsilon\sigma A(T_t^4 - T_w^4) = 0$$

From Table 1.3, $\varepsilon = 0.8$ for pyrex glass. Canceling A,

$$10(290 - T_e) + (0.8)(5.67)(2.90^4 - 2.78^4) = 0$$

Solving,

$$T_e = 295 \text{ K} \simeq 22°\text{C}$$

Comments

1. Since $T_e > 20°$C, the air-heating system appears to be working satisfactorily.
2. Our model assumes that the thermometer is completely surrounded by a surface at 5°C: actually, the thermometer also receives radiation from machines, workers, and other sources at temperatures higher than 5°C, so that our calculated value of $T_e = 22°$C is somewhat high.

1.5 TRANSIENT THERMAL RESPONSE

The heat transfer problems described in Examples 1.1 through 1.5 were *steady-state* problems; that is, temperatures were not changing in time. In Example 1.2, the transistor temperature was steady with the resistance (I^2R) heating balanced by the radiation heat loss. *Unsteady-state* or *transient* problems occur when temperatures change with time. Such problems are often encountered in engineering practice, and the engineer may be required to predict the temperature-time response of a system involved in a heat transfer process. If the system, or a component of the system, can be assumed to have a spatially uniform temperature, analysis involves a relatively simple application of the energy conservation principle, as will now be demonstrated.

1.5.1 The Lumped Thermal Capacity Model

If a system undergoing a transient thermal response to a heat transfer process has a nearly uniform temperature, we may ignore small differences of temperature within the system. Changes in internal energy of the system can then be specified in terms of changes of the assumed uniform (or average) temperature of the system. This approximation is called the **lumped thermal capacity** model.[6] The system might be

[6] The term *capacitance* is also used, in analogy to an equivalent electrical circuit.

a small solid component of high thermal conductivity that loses heat slowly to its surroundings via a large external thermal resistance. Since the thermal resistance to conduction in the solid is small compared to the external resistance, the assumption of a uniform temperature is justified. Alternatively, the system might be a well-stirred liquid in an insulated tank losing heat to its surroundings, in which case it is the mixing of the liquid by the stirrer that ensures a nearly uniform temperature. In either case, once we have assumed uniformity of temperature, we have no further need for details of the heat transfer within the system—that is, of the conduction in the solid component or the convection in the stirred liquid. Instead, the heat transfer process of concern is the interaction of the system with the surroundings, which might be by conduction, radiation, or convection.

Governing Equation and Initial Condition

For purposes of analysis, consider a metal forging removed from a furnace at temperature T_0 and suddenly immersed in an oil bath at temperature T_e, as shown in Fig. 1.16. The forging is a closed system, so the energy conservation principle in the form of Eq. (1.2) applies. Heat is transferred out of the system by convection. Using Eq. (1.24) the rate of heat transfer is $\overline{h}_c A(T - T_e)$, where $\overline{h}_c$ is the heat transfer coefficient averaged over the forging surface area A, and T is the forging temperature. There is no heat generated within the forging, so that $\dot{Q}_v = 0$. Substituting in Eq. (1.2):

$$\rho V c \frac{dT}{dt} = -\overline{h}_c A(T - T_e)$$

$$\frac{dT}{dt} = -\frac{\overline{h}_c A}{\rho V c}(T - T_e) \tag{1.36}$$

which is a first-order ordinary differential equation for the forging temperature, T, as a function of time, t. One initial condition is required:

$$t = 0: \quad T = T_0 \tag{1.37}$$

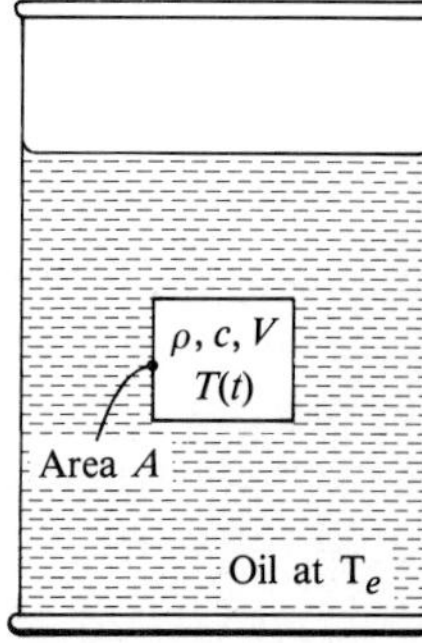

Figure 1.16 A forging immersed in an oil bath for quenching.

Solution for the Temperature Response

A simple analytical solution can be obtained provided we assume that the bath is large, so T_e is independent of time, and that $\overline{h}_c A/\rho V c$ is approximated by a constant value independent of temperature. The variables in Eq. (1.36) can then be separated:

$$\frac{dT}{T - T_e} = -\frac{\overline{h}_c A}{\rho V c}\, dt$$

Writing $dT = d(T - T_e)$, since T_e is constant, and integrating with $T = T_0$ at $t = 0$,

$$\int_{T_0}^{T} \frac{d(T - T_e)}{T - T_e} = -\frac{\overline{h}_c A}{\rho V c} \int_0^t dt$$

$$\ln \frac{T - T_e}{T_0 - T_e} = -\frac{\overline{h}_c A}{\rho V c} t$$

$$\frac{T - T_e}{T_0 - T_e} = e^{-(\overline{h}_c A/\rho V c)t} = e^{-t/t_c} \tag{1.38}$$

where $t_c = \rho V c/\overline{h}_c A$ [s] is called the **time constant** of the process. When $t = t_c$, the temperature difference $(T - T_e)$ has dropped to be 36.8% of the initial difference $(T_0 - T_e)$. Our result, Eq. (1.38), is a relation between two dimensionless parameters: a dimensionless temperature, $T^* = (T - T_e)/(T_0 - T_e)$, which varies from 1 to 0; and a dimensionless time, $t^* = t/t_c = \overline{h}_c A t/\rho V c$, which varies from 0 to ∞. Equation (1.38) can be written simply as

$$T^* = e^{-t^*} \tag{1.39}$$

and a graph of T^* versus t^* is a single curve, as illustrated in Fig. 1.17.

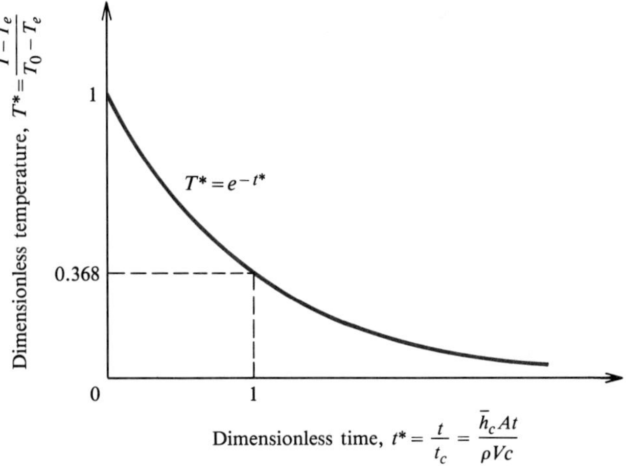

Figure 1.17 Lumped thermal capacity temperature response in terms of dimensionless variables T^* and t^*.

Methods introduced in Chapter 2 can be used to deduce directly from Eqs. (1.36) and (1.37) that T^* must be a function of t^* alone [i.e., $T^* = f(t^*)$] without solving the equation. Of course, the solution also gives us the form of the function. Thus, the various parameters, $\overline{h}_c$, c, ρ, and so on, only affect the temperature response in the combination t^*, and not independently. If both $\overline{h}_c$ and c are doubled, the temperature at time t is unchanged. This dimensionless parameter t^* is a dimensionless group in the same sense as the Reynolds number, but it does not have a commonly used name.

Validity of the Model

We would expect our assumption of negligible temperature gradients within the system to be valid when the internal resistance to heat transfer is small compared with the external resistance. If L is some appropriate characteristic length of a solid body, for example, V/A (which for a plate is half its thickness), then

$$\frac{\text{Internal conduction resistance}}{\text{External convection resistance}} \simeq \frac{L/k_s A}{1/\overline{h}_c A} = \frac{\overline{h}_c L}{k_s} \tag{1.40}$$

where k_s is the thermal conductivity of the solid material. The quantity $\overline{h}_c L/k_s$ [W/m^2 K][m]/[W/m K] is a dimensionless group called the **Biot number**, Bi. More exact analyses of transient thermal response of solids indicate that, for bodies resembling a plate, cylinder, or sphere, $\text{Bi} < 0.1$ ensures that the temperature at the center will not differ from that at the surface by more than 5%; thus, $\text{Bi} < 0.1$ is a suitable criterion for determining if the assumption that the body has a uniform temperature is justified. If the heat transfer is by radiation, the convective heat transfer coefficient in Eq. (1.40) can be replaced by the approximate radiation heat transfer coefficient h_r defined in Eq. (1.19).

In the case of the well-stirred liquid in an insulated tank, it will be necessary to evaluate the ratio

$$\frac{\text{Internal convection resistance}}{\text{External resistance}} \simeq \frac{1/h_{c,i} A}{1/UA} = \frac{U}{h_{c,i}} \tag{1.41}$$

where U is the overall heat transfer coefficient, for heat transfer from the inner surface of the tank, across the tank wall and insulation, and into the surroundings. If this ratio is small relative to unity, the assumption of a uniform temperature in the liquid is justified.

The approximation or model used in the preceding analysis is called a lumped thermal capacity approximation since the thermal capacity is associated with a single temperature. There is an electrical analogy to the lumped thermal capacity model, owing to the mathematical equivalence of Eq. (1.36) to the equation governing the voltage in the simple resistance-capacitance electrical circuit shown

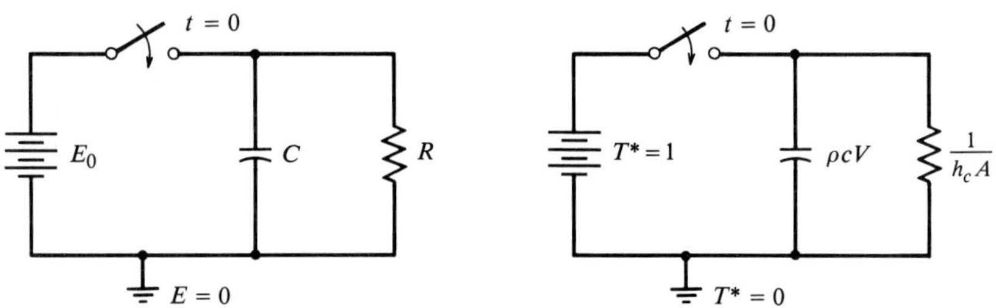

Figure 1.18 Equivalent electrical and thermal circuits for the lumped thermal capacity model of temperature response.

in Fig. 1.18,

$$\frac{dE}{dt} = -\frac{E}{RC} \tag{1.42}$$

with the initial condition $E = E_0$ at $t = 0$ if the capacitor is initially charged to a voltage E_0. The solution is identical in form to Eq. (1.38),

$$\frac{E}{E_0} = e^{-t/RC}$$

and the time constant is RC, the product of the resistance and capacitance [or $C/(1/R)$, the ratio of capacitance to conductance, to be exactly analogous to Eq. (1.38)].

EXAMPLE 1.6 Quenching of a Steel Plate

A steel plate 1 cm thick is taken from a furnace at 600°C and quenched in a bath of oil at 30°C. If the heat transfer coefficient is estimated to be 400 W/m^2 K, how long will it take for the plate to cool to 100°C? Take k, ρ, and c for the steel as 50 W/m K, 7800 kg/m^3, and 450 J/kg K, respectively.

Solution

Given: Steel plate quenched in an oil bath.

Required: Time to cool from 600°C to 100°C.

Assumptions: Lumped thermal capacity model valid.

First the Biot number will be checked to see if the lumped thermal capacity approximation is valid. For a plate of width W, height H, and thickness L,

$$\frac{V}{A} \simeq \frac{WHL}{2WH} = \frac{L}{2}$$

where the surface area of the edges has been neglected.

$$\text{Bi} = \frac{\overline{h}_c(L/2)}{k_s}$$

$$= \frac{(400\ \text{W/m}^2\ \text{K})(0.005\ \text{m})}{50\ \text{W/m K}}$$

$$= 0.04 < 0.1$$

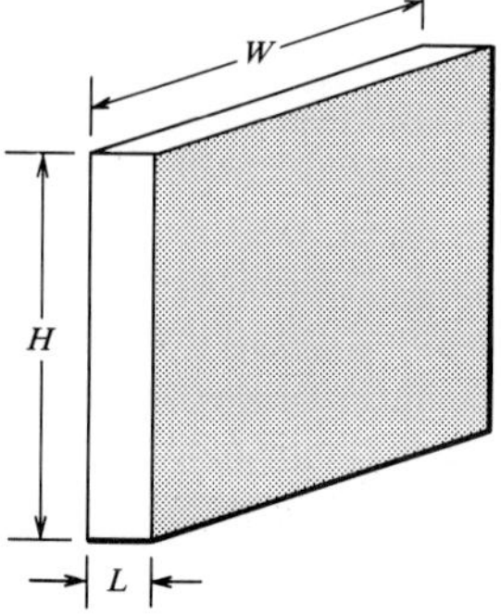

so the lumped thermal capacity model is applicable. The time constant t_c is

$$t_c = \frac{\rho V c}{\overline{h}_c A} = \frac{\rho(L/2)c}{\overline{h}_c} = \frac{(7800\ \text{kg/m}^3)(0.005\ \text{m})(450\ \text{J/kg K})}{(400\ \text{W/m}^2\ \text{K})} = 43.9\ \text{s}$$

Substituting $T_e = 30°\text{C}$, $T_0 = 600°\text{C}$, $T = 100°\text{C}$ in Eq. (1.38) gives

$$\frac{100 - 30}{600 - 30} = e^{-t/43.9}$$

Solving,

$$t = 92\ \text{s}$$

Comments

The use of a constant value of h_c may be inappropriate for heat transfer by natural convection or radiation (see Section 1.5.2).

1.5.2 Combined Convection and Radiation

The analysis of Section 1.5.1 assumes that the heat transfer coefficient was constant during the cooling period. This assumption is adequate for forced convection but is less appropriate for natural convection, and when thermal radiation is significant. Equation (1.23) shows that the natural convection heat transfer coefficient $\overline{h}_c$ is proportional to $\Delta T^{1/4}$ for laminar flow and to $\Delta T^{1/3}$ for turbulent flow. The temperature difference $\Delta T = T - T_e$ decreases as the body cools, as does $\overline{h}_c$. Radiation heat transfer is proportional to $(T^4 - T_e^4)$ and hence cannot be represented exactly by Newton's law of cooling. We now extend our lumped thermal capacity analysis to allow both for a variable convective heat transfer coefficient and for situations where both convection and radiation are important.

Governing Equation and Initial Condition

Figure 1.19 shows a body that loses heat by both convection and radiation. For a small gray body in large, nearly black surroundings also at temperature T_e, the radiation heat transfer is obtained from Eq. (1.18) as $\dot{Q} = \varepsilon A \sigma (T^4 - T_e^4)$. As in

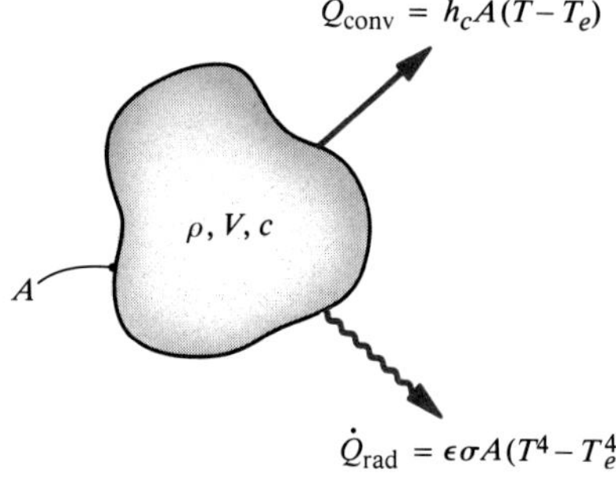

Figure 1.19 Schematic of a body losing heat by convection and radiation for a lumped thermal capacity model.

Section 1.5.1, the energy conservation principle Eq. (1.2) becomes

$$\rho V c \frac{dT}{dt} = -\bar{h}_c A(T - T_e) - \varepsilon A\sigma(T^4 - T_e^4)$$

or

$$\frac{dT}{dt} = -\frac{\bar{h}_c A}{\rho V c}(T - T_e) - \frac{\varepsilon A\sigma}{\rho V c}(T^4 - T_e^4) \tag{1.43}$$

The initial condition is again

$$t = 0: \quad T = T_0 \tag{1.44}$$

This first-order ordinary differential equation has no closed-form analytical solution even when the convective heat transfer coefficient $\bar{h}_c$ is constant as in forced convection. However, Eq. (1.43) can be solved easily using a numerical integration procedure. For this purpose, it can be rearranged as

$$\frac{dT}{dt} + \frac{hA}{\rho V c}(T - T_e) = 0 \tag{1.45}$$

$$h = \bar{h}_c + h_r = B(T - T_e)^n + \sigma\varepsilon(T^2 + T_e^2)(T + T_e) \tag{1.46}$$

where $(T^4 - T_e^4)$ has been factored, as was done in deriving Eq. (1.19). For forced convection, $n = 0$, $B = \bar{h}_c$; for laminar natural convection $n = 1/4$ and B is a constant [for example, for a plate of height L, Eq. (1.23a) gives $B = (4/3)(1.07)/L^{1/4}$]. Equation (1.46) defines a total heat transfer coefficient that accounts for both convection and radiation and changes continuously as the body cools. To put Eq. (1.45) in dimensionless form, we use the dimensionless variables introduced in Section 1.5.1:

$$T^* = \frac{T - T_e}{T_0 - T_e}, \qquad t^* = \frac{t}{t_c} \tag{1.47a,b}$$

The definition of the time constant t_c poses a problem since h is not a constant as before. We choose to define t_c in terms of the value of h at time $t = 0$, when the body temperature is T_0,

$$t_c = \frac{\rho V c}{h_0 A} = \frac{\rho V c}{[B(T_0 - T_e)^n + \sigma\varepsilon(T_0^2 + T_e^2)(T_0 + T_e)]A} \tag{1.48}$$

Equation (1.45) then becomes

$$\frac{dT^*}{dt^*} + \frac{h}{h_0}T^* = 0 \tag{1.49}$$

with the initial condition

$$t^* = 0: \quad T^* = 1 \tag{1.50}$$

Computer Program LUMP

Numerical integration is appropriate for this problem. The computer program LUMP has been prepared accordingly. LUMP solves Eq. (1.49), that is, it obtains the temperature response of a body that loses heat by convection and/or radiation, based on the lumped thermal capacity model. The required input constant B is defined in Eq. (1.46). Any consistent system of units can be used. The output can be obtained either as a graph or as numerical data.

EXAMPLE 1.7 Quenching of an Alloy Sphere

A materials processing experiment under microgravity conditions on the space shuttle requires quenching in a forced flow of an inert gas. A 1 cm–diameter metal alloy sphere is removed from a furnace at 800°C and is to be cooled to 500°C by a flow of nitrogen gas at 25°C. Determine the effect of the convective heat transfer coefficient on cooling time for $10 < \overline{h}_c < 100$ W/m² K. Properties of the alloy include: $\rho = 14,000$ kg/m³; $c = 140$ J/kg K; $\varepsilon = 0.1$. The surrounds can be taken as nearly black at 25°C.

Solution

Given: A metal alloy sphere to be quenched.

Required: Effect of convective heat transfer coefficient on cooling time.

Assumptions:
1. Lumped thermal capacity model valid.
2. Constant convective heat transfer coefficient.

The computer code LUMP can be used to solve this problem.
The required inputs are:

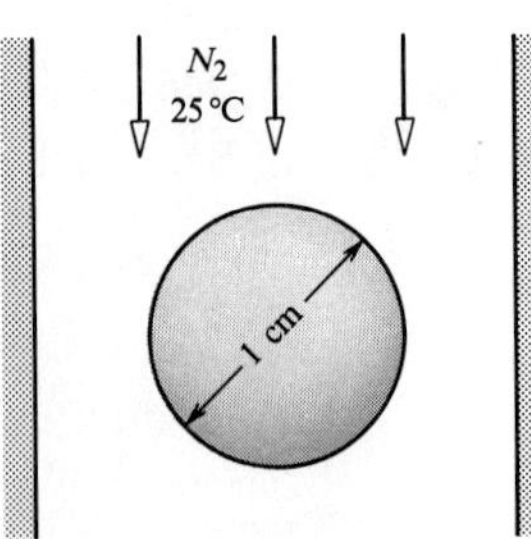

T_0 and $T_e = 1073, 298$
$B = \overline{h}_c = 10$ (repeat for 20, 30, 50, 100)
$n = 0$
$\sigma = 5.67 \times 10^{-8}$
$\varepsilon = 0.1$
Final value of t^*: try $t^* = 1$

The required dimensionless temperature is

$$T^* = \frac{T - T_e}{T_0 - T_e} = \frac{773 - 298}{1073 - 298} = 0.613$$

and the code is used to obtain the corresponding dimensionless time t^*. For a sphere $V/A = (\pi D^3/6)/(\pi D^2) = D/6$, so that the time constant is

$$t_c = \frac{\rho V c}{h_0 A} = \frac{\rho(D/6)c}{h_0} = \frac{(14,000)(0.01/6)(140)}{\bar{h}_c + (5.67 \times 10^{-8})(0.1)(1073^2 + 298^2)(1073 + 298)}$$

$$= \frac{3267}{\bar{h}_c + 9.64} \text{ s}$$

The actual time is $t = t^* t_c$. Results obtained using LUMP are tabulated below.

$\bar{h}_c$ W/m^2 K	t^*	t_c s	t s
10	0.59	166	98
20	0.55	110	61
30	0.53	82	43
50	0.52	55	29
100	0.51	30	15

Comments

1. Only two significant figures have been given since high accuracy is not warranted for the problem.
2. The heat transfer coefficient does not have a strong effect on t^*. Why?
3. For the lumped thermal capacity model to be valid, the Biot number should be less than 0.1. The worst case is with $\bar{h}_c = 100$ W/m^2 K at time $t = 0$, giving $h_0 = 109.6$ and $0.1 > (109.6)(0.01/6)/k_s$, that is, $k_s > 1.8$ W/m K, which certainly will be true for a metal alloy.

1.6 HEAT EXCHANGERS

In Section 1.5, we considered problems in which the temperature of a system changed with time as a result of heat transfer between the system and its surroundings. We now consider problems in which the temperature of a fluid changes as it flows through a passage as a result of heat transfer between the passage walls and the fluid. These problems are encountered in the analysis of heat exchanger performance. A *heat exchanger* is a device that facilitates transfer of heat from one fluid stream to another. Power production, refrigeration, heating and air conditioning, food processing, chemical processing, oil refining, and the operation of almost all vehicles depends on heat exchangers of various types. The analysis and design of heat exchangers is the subject of Chapter 8. The analysis of a very simple heat exchanger configuration is presented here to introduce some of the basic concepts underlying heat exchanger analysis and associated terminology. These concepts will prove useful in the development and application of heat transfer theory in chapters preceding Chapter 8—particularly in Chapters 4 and 5, which deal with convection.

1.6.1 Single- and Two-Stream Exchangers

One important classification of heat exchangers is into **single-stream exchangers** and **two-stream exchangers**. A single-stream exchanger is one in which the temperature of only one stream changes in the exchanger; examples include many types of evaporators and condensers found in power plants and refrigeration systems. A power plant condenser is shown in Fig. 1.20. A two-stream exchanger is one in which the temperatures of both streams change in the exchanger; examples include radiators and intercoolers for automobile engines, and oil coolers for aircraft engines. Figure 1.21 shows an oil cooler, which has a **counterflow** configuration; that is, the streams flow in opposite directions in the exchanger.

In the analysis of heat exchangers, a useful first step is to draw a sketch of the expected fluid temperature variations along the exchanger. Figure 1.22*a* is such a sketch for the power plant condenser. The hot stream is steam returning from the turbines, which condenses at a constant temperature T_H. This is the saturation temperature corresponding to the pressure maintained in the condenser shell. The cold stream is cold water from a river, ocean, or cooling tower, and its temperature

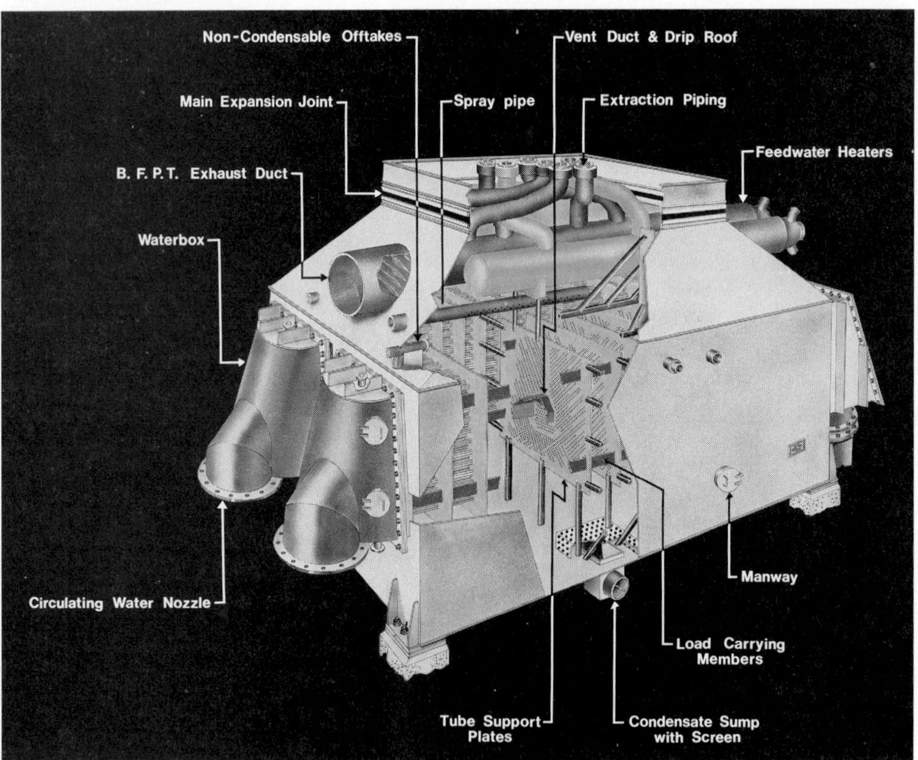

Figure 1.20 A power plant condenser. (Courtesy Senior Engineering Co. [formerly Southwestern Engineering], Los Angeles, California.)

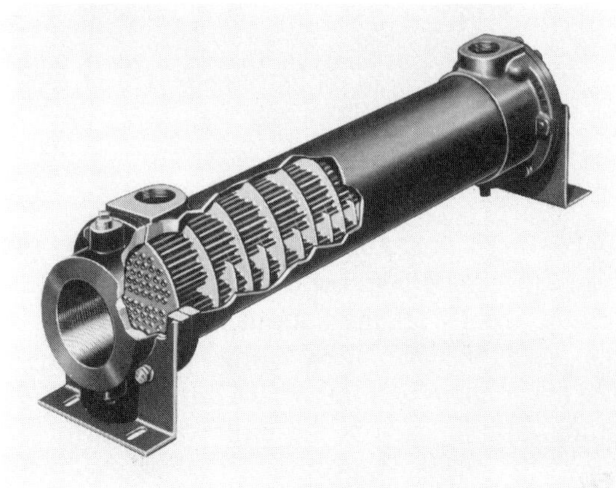

Figure 1.21 A small single-pass shell-and-tube two-stream heat exchanger, typically used for cooling oil or water. (Photograph courtesy of the Young Radiator Company, Racine, Wis.)

T_C increases as it flows through the exchanger. Figure 1.22*b* shows the sketch for the oil cooler. The hot stream is oil from the engine, and the cold stream is coolant water. Notice that in this counterflow configuration, the cold stream can leave the exchanger at a higher temperature than the hot stream!

A point that might confuse the beginning student is that there are actually two streams in many single-stream exchangers. The definition simply requires that the temperature of only one stream changes in a single-stream exchanger. It is this feature that makes the analysis of single-stream exchangers particularly simple, as will now be demonstrated. In Section 1.5, the system analysis was based on the

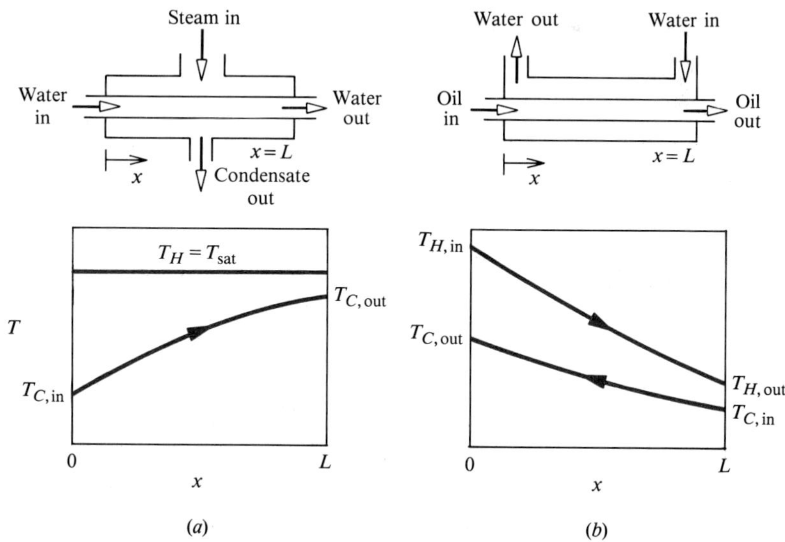

Figure 1.22 Temperature variations along heat exchangers. (*a*) A power plant condenser. (*b*) A counterflow oil cooler.

energy conservation principle in the form of the first law of thermodynamics applied to a *closed system*. In contrast, the system analysis that follows is based on the first law applied to an *open system*.

1.6.2 Analysis of a Condenser

Figure 1.23*a* shows a simple single-tube condenser. Pure saturated vapor enters the shell at the top and condenses on a single horizontal tube. The condensate forms a thin film on the outside of the tube, drops off the bottom, and leaves the shell through a drain. The vapor condenses at the saturation temperature corresponding to the pressure in the shell. Hence, the condensate film surface temperature is $T_{sat}(P)$. Figure 1.23*b* shows the temperature variation across the tube wall and the corresponding thermal circuit. The enthalpy of condensation is transferred by conduction across the thin condensate film, by conduction across the tube wall, and by convection into the coolant. As a result, the coolant temperature rises as it gains energy flowing along the tube. The vapor flow rate is denoted $\dot{m}_H$ [kg/s] and the coolant flow rate $\dot{m}_C$ (the *hot* and *cold* streams, respectively).

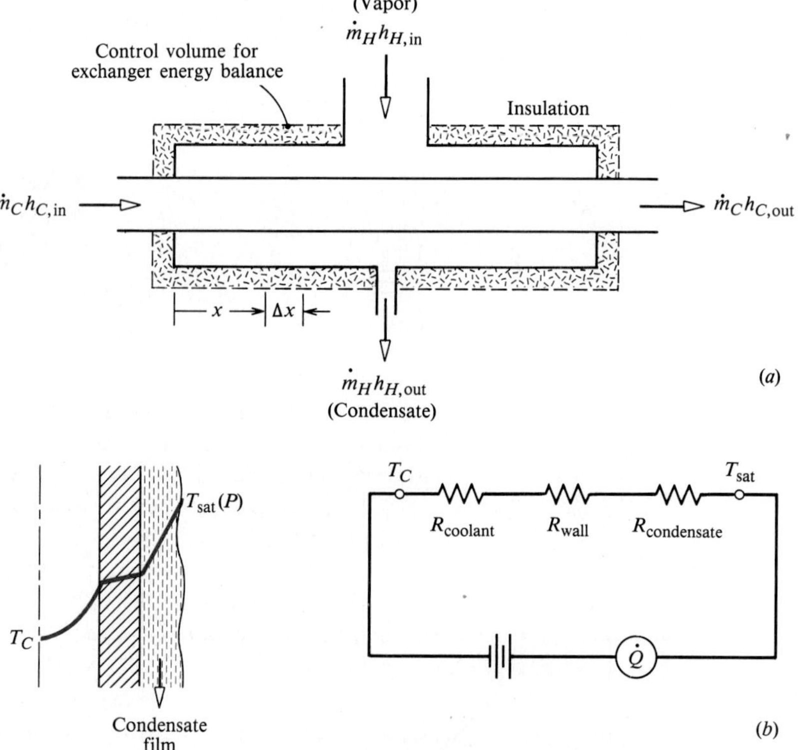

Figure 1.23 (*a*) Schematic of a single-tube condenser. (*b*) The temperature variation across the tube wall and the thermal circuit for heat transfer across the tube wall.

The Exchanger Energy Balance

An energy balance on the exchanger as a whole is formulated by writing down the steady-flow energy equation for a control volume enclosing the exchanger (the dashed line in Fig. 1.23*a*). If the exchanger is well insulated, there is no heat loss to the surroundings, and Eq. (1.4) requires that the enthalpy inflow equal the enthalpy outflow:

$$\dot{m}_H h_{H,\text{in}} + \dot{m}_C h_{C,\text{in}} = \dot{m}_H h_{H,\text{out}} + \dot{m}_C h_{C,\text{out}}$$

where h is specific enthalpy [J/kg] and subscripts "in" and "out" denote inlet and outlet values, respectively. Rearranging gives

$$\dot{m}_C(h_{C,\text{out}} - h_{C,\text{in}}) = \dot{m}_H(h_{H,\text{in}} - h_{H,\text{out}}) \qquad \textbf{(1.51)}$$

If we assume a constant specific heat for the coolant and that the condensate leaves at the saturation temperature, Eq. (1.51) becomes

$$\dot{m}_C c_{pC}(T_{C,\text{out}} - T_{C,\text{in}}) = \dot{m}_H h_{\text{fg}} \qquad \textbf{(1.52)}$$

where h_{fg} is the enthalpy of vaporization for the vapor. When the coolant flow rate $\dot{m}_C$ and inlet temperature $T_{C,\text{in}}$ are known, Eq. (1.52) relates the coolant outlet temperature $T_{C,\text{out}}$ to the amount of vapor condensed $\dot{m}_H$.

Governing Equation and Boundary Condition

To determine the variation of coolant temperature along the exchanger, we make an energy balance on a differential element of the exchanger Δx long and so derive a differential equation with x as the independent variable and T_C as the dependent variable. When the steady-flow energy equation, Eq. (1.4), is applied to the control volume of length Δx, shown in Fig. 1.24 as a dotted line, the contribution to $\dot{Q}$ due to x-direction conduction in the coolant is small and can be neglected. Thus, the

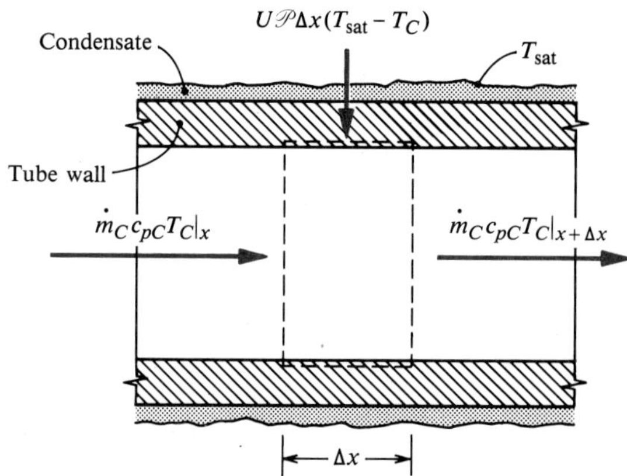

Figure 1.24 An elemental control volume Δx long for application of the steady-flow energy equation to a condenser coolant stream.

coolant flow rate times its enthalpy increase must equal the heat transfer across the tube wall:

$$\dot{m}_C c_{pC} (T_C|_{x+\Delta x} - T_C|_x) = U\mathscr{P}\Delta x (T_{\text{sat}} - T_C)$$

where U [W/m^2 K] is the overall heat transfer coefficient for heat transfer from the vapor to the coolant, and $\mathscr{P}$ [m] is the perimeter of the tube wall. Thus, $\mathscr{P}\Delta x$ is the surface area of the tube element with length Δx. For a circular tube $\mathscr{P} = \pi D$ where D is the pipe diameter. Dividing by Δx,

$$\dot{m}_C c_{pC} \left(\frac{T_C|_{x+\Delta x} - T_C|_x}{\Delta x} \right) = U\mathscr{P} (T_{\text{sat}} - T_C)$$

and letting $\Delta x \to 0$, gives

$$\dot{m}_C c_{pC} \frac{dT_C}{dx} = U\mathscr{P} (T_{\text{sat}} - T_C)$$

Rearranging,

$$\frac{dT_C}{dx} - \frac{U\mathscr{P}}{\dot{m}_C c_{pC}} (T_{\text{sat}} - T_C) = 0 \tag{1.53}$$

Equation (1.53) is a first-order ordinary differential equation for $T_C(x)$; it requires one boundary condition, which is

$$x = 0: \quad T_C = T_{C,\text{in}} \tag{1.54}$$

Temperature Variation

To integrate Eq. (1.53), let $\theta = T_{\text{sat}} - T_C$; then $dT_C/dx = -d\theta/dx$, and the equation becomes

$$\frac{d\theta}{dx} + \frac{U\mathscr{P}}{\dot{m}_C c_{pC}} \theta = 0$$

If U is assumed constant along the exchanger, the solution is

$$\theta = Ae^{-(U\mathscr{P}/\dot{m}_C c_{pC})x}$$

where A is the integration constant. Substituting for θ and using the boundary condition, Eq. (1.54) gives the integration constant:

$$T_{\text{sat}} - T_{C,\text{in}} = Ae^0 = A$$

Thus, the solution of Eq. (1.53) is

$$T_{\text{sat}} - T_C = (T_{\text{sat}} - T_{C,\text{in}})e^{-(U\mathscr{P}/\dot{m}_C c_{pC})x} \tag{1.55}$$

which is the desired relation $T_C(x)$, showing an exponential variation along the exchanger. Of particular interest is the coolant outlet temperature $T_{C,\text{out}}$ which is obtained by letting $x = L$, the length of the exchanger, in Eq. (1.55):

$$T_{\text{sat}} - T_{C,\text{out}} = (T_{\text{sat}} - T_{C,\text{in}})e^{-(U\mathscr{P}L/\dot{m}_C c_{pC})} \tag{1.56}$$

Exchanger Performance Parameters

The product of perimeter and length $\mathscr{P}L$ is the area of the heat transfer surface. The exponent in Eq. (1.56) is, of course, dimensionless,

$$\left[\frac{U\mathscr{P}L}{\dot{m}_C c_{pC}}\right] = \frac{[\text{W/m}^2\,\text{K}][\text{m}][\text{m}]}{[\text{kg/s}][\text{J/kg K}]} = \left[\frac{\text{W s}}{\text{J}}\right] = 1$$

since a watt is a joule per second. This dimensionless group is called the **number of transfer units**, with abbreviation NTU and symbol N_{tu}.[7] For a given $\dot{m}_C c_{pC}$, the larger U, $\mathscr{P}$, or L, the greater the NTU of the exchanger. Thus, the NTU can be viewed as a measure of the heat transfer "size" of the exchanger. Equation (1.56) can then be rearranged as

$$\frac{T_{sat} - T_{C,out}}{T_{sat} - T_{C,in}} = e^{-N_{tu}} \tag{1.57}$$

Thus, if T_{sat}, $T_{C,in}$, and the NTU of the exchanger are known, $T_{C,out}$ can be calculated. But we find it convenient to rearrange Eq. (1.57) by subtracting each side from unity to obtain

$$1 - \frac{T_{sat} - T_{C,out}}{T_{sat} - T_{C,in}} = 1 - e^{-N_{tu}}$$

or

$$\frac{T_{C,out} - T_{C,in}}{T_{sat} - T_{C,in}} = 1 - e^{-N_{tu}} \tag{1.58}$$

Now, even if the exchanger were infinitely long, the maximum outlet temperature of the coolant would be T_{sat} (see Fig. 1.22*a*). Thus, the left-hand side of Eq. (1.58) is the ratio of the actual temperature rise of the coolant $(T_{C,out} - T_{C,in})$ divided by the maximum possible rise for an infinitely long exchanger $(T_{sat} - T_{C,in})$ and can be viewed as the **effectiveness** of the exchanger, for which we use the symbol ε. Our result is therefore

$$\varepsilon = 1 - e^{-N_{tu}} \tag{1.59}$$

Equation (1.59) indicates that the larger the number of transfer units of the exchanger, the higher its effectiveness. Although a high effectiveness is desirable, as the length of an exchanger increases, so does the cost of materials for its construction and the pumping power required by the coolant flow. Thus, the goal of the design engineer is to maximize the effectiveness subject to the constraints of construction (capital) costs and power (operating) costs. In practice, values of ε between 0.6 and 0.9 are typical.

---

[7] NTU is also widely used as the symbol for number of transfer units.

EXAMPLE 1.8 Performance of a Steam Condenser

A steam condenser is 4 m long and contains 2000, 5/8 inch nominal-size, 18 gage brass tubes (1.59 cm O.D., 1.25 mm wall thickness). In a test 120 kg/s of coolant water at 300 K is supplied to the condenser, and when the steam pressure in the shell is 10,540 Pa, condensate is produced at a rate of 3.02 kg/s. Determine the effectiveness of the exchanger and the overall heat transfer coefficient. Take the specific heat of the water to be 4174 J/kg K.

Solution

Given: A shell-and-tube steam condenser.

Required: The effectiveness, ε, and overall heat transfer coefficient, U.

Assumptions: U is constant along the exchanger so that Eq. (1.59) applies.

The hot-stream temperature T_H is the saturation temperature corresponding to the given steam pressure of 10,540 Pa; from steam tables (see Table A.12*a* in Appendix A of this text) $T_{\text{sat}} = 320.0$ K. We first find the coolant water outlet temperature from the exchanger energy balance Eq. (1.52):

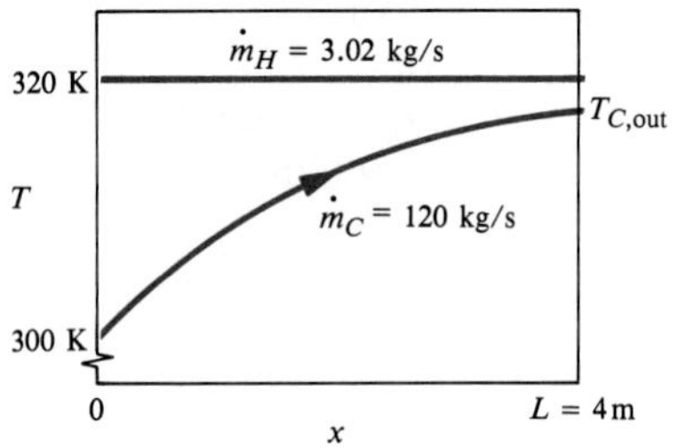

$$\dot{m}_C c_{pC}(T_{C,\text{out}} - T_{C,\text{in}}) = \dot{m}_H h_{fg}$$

From steam tables, the enthalpy of vaporization at $T_{\text{sat}} = 320$ K is $h_{fg} = 2.389 \times 10^6$ J/kg.

$$(120 \text{ kg/s})(4174 \text{ J/kg K})(T_{C,\text{out}} - 300 \text{ K}) = (3.02 \text{ kg/s})(2.389 \times 10^6 \text{ J/kg})$$

Solving gives $T_{C,\text{out}} = 314.4$ K.

The effectiveness, ε, is then obtained from Eq. (1.58) as

$$\varepsilon = \frac{T_{C,\text{out}} - T_{C,\text{in}}}{T_{\text{sat}} - T_{C,\text{in}}} = \frac{314.4 - 300}{320 - 300} = 0.720$$

and the number of transfer units, from Eq. (1.59), is

$$N_{tu} = \ln\frac{1}{1-\varepsilon} = \ln\frac{1}{1-0.720} = 1.27 = \frac{U\mathscr{P}L}{\dot{m}_C c_{pC}}$$

Solving for the $U\mathscr{P}L$ product,

$$U\mathscr{P}L = 1.27\dot{m}_C c_{pC} = (1.27)(120 \text{ kg/s})(4174 \text{ J/kg K}) = 6.36 \times 10^5 \text{ W/K}$$

If we choose to base the overall heat transfer coefficient on the outside of the tubes, then, for N tubes, the heat transfer area $\mathscr{P}L$ is

$$\mathscr{P}L = N\pi DL = (2000)(\pi)(1.59 \times 10^{-2} \text{ m})(4 \text{ m}) = 400 \text{ m}^2$$

Hence, $U = U\mathscr{P}L/\mathscr{P}L = 6.36 \times 10^5/400 = 1590 \text{ W/m}^2\text{ K}$

Comments

We could have performed these calculations by considering a single tube of the tube bundle, for which the coolant flow is (120/2000) kg/s and the heat transfer area is simply πDL. But common practice is always to consider the exchanger as a whole, as we have done here.

1.6.3 Other Single-Stream Exchangers

Simple evaporators and boilers are also single-stream exchangers, where the cold stream is an evaporating or boiling liquid and the hot stream supplies the enthalpy of vaporization. Such exchangers will be analyzed in Chapter 8. Heat transfer to a fluid stream may also be a concern in problems that do not involve heat exchangers. The exhaust gas stack cooled by a crosswind, shown in Fig. 1.25, can also be viewed as a single-stream heat exchanger, since only the exhaust gas temperature changes with location up the stack. Thus, the analysis of Section 1.6.2, properly interpreted, applies (see Exercise 1–52). Single-stream heat exchanger theory also will be used in Chapters 4 and 5 in the examination of convective heat transfer in internal flows.

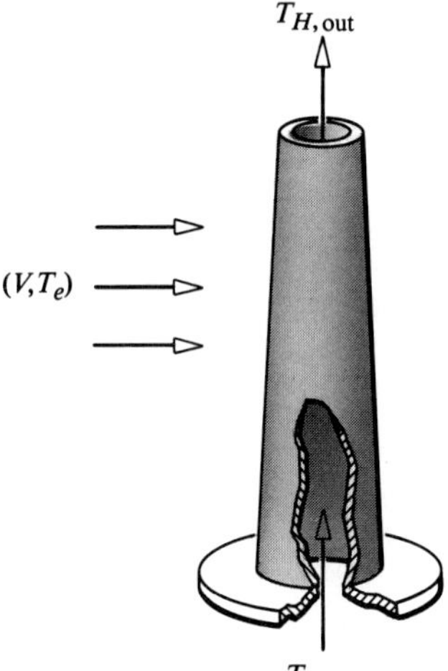

Figure 1.25 An exhaust gas stack cooled by a crosswind.

1.7 DIMENSIONS AND UNITS

Dimensions are physical properties that are measurable—for example, length, time, mass, and temperature. A system of *units* is used to give numerical values to dimensions. The system most widely used throughout the world in science and industry is the International System of units (SI), from the French name *Système International d'Unités*. This system was recommended at the General Conference on Weights and Measures of the International Academy of Sciences in 1960 and was adopted by the U.S. National Bureau of Standards in 1964. In the United States, the transition from the older English system of units to the SI system has been slow and is not complete. The SI system is used in science education, by engineering professional societies, and by many industries. However, engineers in some more mature industries still prefer to use English units, and, of course, commerce and trade in the United States remains dominated by the English system. We buy pounds of vegetables, quarts of milk, drive miles to work, and say that it is a hot day when the temperature exceeds 80°F. (Wine is now sold in 750 ml bottles, though, which is a modest step forward!)

In this text, we will use the SI system, with which the student has become familiar from physics courses. For convenience, this system is summarized in the tables of Appendix B. Base and supplementary units, such as length, time, and plane angle, are given in Table B.1*a*; and derived units, such as force and energy, are given in Table B.1*b*. Recognized non-SI units (e.g., hour, bar) that are acceptable for use with the SI system are listed in Table B.1*c*. Multiples of SI units (e.g., kilo, micro) are defined in Table B.1*d*. Accordingly, the property data given in the tables of Appendix A are in SI units. The student should review this material and is urged to be careful when writing down units. For example, notice that the unit of temperature is a kelvin (not Kelvin) and has the symbol K (not °K). Likewise, the unit of power is the watt (not Watt). The symbol for a kilogram is kg (not KG). An issue that often confuses the student is the correct use of Celsius temperature. Celsius temperature is defined as $(T - 273.15)$ where T is in kelvins. However, the unit "degree Celsius" is equal to the unit "kelvin" ($1°C = 1$ K).

Notwithstanding the wide acceptance of the SI system of units, there remains a need to communicate with those engineers (or lawyers!) who are still using English units. Also, component dimensions, or data for physical properties, may be available only in English or cgs units. For example, most pipes and tubes used in the United States conform to standard sizes originally specified in English units. A 1 inch nominal-size tube has an outside diameter of 1 in. For convenience, selected dimensions of U.S. commercial standard pipes and tubes are given in SI units in Appendix A as Tables A.14*a* and A.14*b*, respectively. The engineer must be able to convert dimensions from one system of units to another. Table B.2 in Appendix B gives the conversion factors required for most heat transfer applications. The program UNITS is based on Table B.2 and contains all the conversion factors in the table. With the input of a quantity in one system of units, the output is the same quantity in the alternative units listed in Table B.2. It is recommended that the student or engineer perform all problem solving using the SI system so as to efficiently use the Appendix A property data and the computer software. If a problem is stated in English units, the data should be converted to SI units using UNITS; if a customer requires results in units other than SI, UNITS will give the required values.

1.8 CLOSURE

Chapter 1 had two main objectives:

1. To introduce the three important modes of heat transfer, namely, conduction, radiation, and convection.

2. To demonstrate how the first law of thermodynamics is applied to engineering systems to obtain the consequences of a heat transfer process.

For each mode of heat transfer, some working equations were developed, which, though simple, allow heat transfer calculations to be made for a wide variety of problems. Equations (1.9), (1.18), and (1.20) are probably the most frequently used equations for thermal design. An electric circuit analogy was shown to be a useful aid for problem solving when more than one mode of heat transfer is involved. In applying the first law to engineering systems, both a closed system and an open system were considered. In the first case, the variation of temperature with time was determined for a solid of high conductivity or a well-stirred fluid. In the second case, the variation of fluid temperature with position along a heat exchanger tube was determined.

The student should be familiar with some of the Chapter 1 concepts from previous physics, thermodynamics, and fluid mechanics courses. A review of texts for such courses is appropriate at this time. Many new concepts were introduced, however, which will take a little time and effort to master. Fortunately, the mathematics in this chapter is simple, involving only algebra, calculus, and the simplest first-order differential equation, and should present no difficulties to the student. After successfully completing a selection of the following exercises, the student will be well equipped to tackle subsequent chapters.

A feature of this text is an emphasis on real engineering problems as examples and exercises. Thus, Chapter 1 has somewhat greater scope and detail than the introductory chapters found in most similar texts. With the additional material, more realistic problems can be treated in subsequent chapters. Conduction problems in Chapters 2 and 3 have more realistic convection and radiation boundary conditions. Convective heat transfer coefficients for flow over tube bundles in Chapter 4 are calculated in the appropriate context of a heat exchanger. Similarly, condensation heat transfer coefficients in Chapter 7 can be discussed in the context of condenser performance. Throughout the text are exercises that require application of the first law to engineering systems, for it is always the consequences of a heat transfer process that motivate the engineer's concern with the subject.

Two computer programs accompany Chapter 1. The program LUMP calculates temperature response using the lumped thermal capacity model of Section 1.5. When heat loss is by convection and radiation simultaneously, the problem does not have an analytical solution. However, a numerical solution is easily obtained; LUMP demonstrates the value of writing a computer program in such situations. It is most important that the engineer be aware of the potential of the PC as an engineering tool and take the initiative to use the PC when appropriate. The program UNITS is a simple units conversion tool that allows unit conversions to be made quickly and reliably.

EXERCISES

1–1. Solve the following ordinary differential equations:

(i) $\dfrac{dy}{dx} + \beta y = 0$

(ii) $\dfrac{dy}{dx} + \beta y + \alpha = 0$

(iii) $\dfrac{d^2y}{dx^2} - \lambda^2 y = 0$

(iv) $\dfrac{d^2y}{dx^2} + \lambda^2 y = 0$

(v) $\dfrac{d^2y}{dx^2} - \lambda^2 y + \alpha = 0$

where α, β, and λ are constants.

1–2. A low-pressure heat exchanger transfers heat between two helium streams, each with a flow rate of $\dot{m} = 5 \times 10^{-3}$ kg/s. In a performance test the cold stream enters at a pressure of 1000 Pa and a temperature of 50 K, and exits at 730 Pa and 350 K.

(i) If the flow cross-sectional area for the cold stream is 0.019 m^2, calculate the inlet and outlet velocities.

(ii) If the exchanger can be assumed to be perfectly insulated, determine the heat transfer in the exchanger. For helium, $c_p = 5200$ J/kg K.

1–3. A shell-and-tube condenser for an ocean thermal energy conversion and fresh water plant is tested with a water feed rate to the tubes of 4000 kg/s. The water inlet and outlet conditions are measured to be $P_1 = 129$ kPa, $T_1 = 280$ K; and $P_2 = 108$ kPa, $T_2 = 285$ K.

(i) Calculate the heat transferred to the water.

(ii) If saturated steam condenses in the shell at 1482 Pa, calculate the steam condensation rate.

For the feed water, take $\rho = 1000$ kg/m^3, $c_v = 4192$ J/kg K. (Steam tables are given as Table A.12*a* in Appendix A.)

1–4. A pyrex glass vessel has a 5 mm–thick wall and is protected with a 1 cm–thick layer of neoprene rubber. If the inner and outer surface temperatures are 40°C and 20°C, respectively, and the total surface area of the vessel is 400 cm^2, calculate the rate of heat loss from the vessel. Also calculate the temperature of the interface between the glass and the rubber, and carefully sketch the temperature profile through the composite wall.

1–5. In the United States, insulations are often specified in terms of their thermal resistance in $[\text{Btu/hr ft}^2\ °\text{F}]^{-1}$, called the "R" value.

(i) What is the R value of a 10 cm–thick layer of fiberglass insulation?
(ii) How thick a layer of cork is required to give an R value of 18?
(iii) What is the R value of a 2 cm–thick board of white pine?

1–6. A picnic icebox is 40 cm long, is 20 cm high and deep, and is insulated with 2 cm–thick polystyrene foam insulation. If the ambient air temperature is 30°C, estimate how much ice will melt in 8 hours. Use an enthalpy of melting for water of 335 kJ/kg.

1–7. A composite wall has a 6 cm layer of fiberglass insulation sandwiched between 2 cm–thick white pine boards. If the inner and outer surface temperatures are 20°C and 0°C, respectively, calculate the heat flow per unit area across the wall. Also calculate the wood-fiberglass interface temperatures, and accurately draw the temperature profile through the wall.

1–8. A freezer is 1 m wide and deep and 2 m high, and must operate at −10°C when the ambient air is at 30°C. What thickness of polystyrene is required if the load on the refrigeration unit should not exceed 200 W? Assume that the outer surface of the insulation is approximately at the ambient air temperature and that the base of the freezer is perfectly insulated.

1–9. A very effective insulation can be made from multiple layers of thin aluminized plastic film separated by rayon mesh and evacuated to a very low pressure ($\sim 10^{-5}$ torr). Such "superinsulation" can be used for insulating storage tanks holding cryogenic liquids. On a space station, a 1 m–O.D. spherical tank contains saturated nitrogen at 1 atm pressure. What thickness of a superinsulation having an effective thermal conductivity of 9×10^{-6} W/m K is required to have a boil-off rate of less than 2 mg/s when the ambient temperature is 250 K? The boiling point of nitrogen is 77.4 K, and its enthalpy of vaporization is 0.200×10^6 J/kg.

1–10. A blackbody radiates to a surrounding black enclosure. If the body is maintained at 100 K above the enclosure temperature, calculate the net radiative heat flux leaving the body when the enclosure is at 80 K, 300 K, 1000 K, and 5000 K.

1–11. An astronaut is at work in the service bay of a space shuttle and is surrounded by walls that are at −100°C. The outer surface of her space suit has an area of 3 m^2 and is aluminized with an emittance of 0.05. Calculate her rate of heat loss when the suit's outer temperature is 0°C. Express your answer in watts and kcal/hr.

1–12. An electronic device is contained in a cylinder 10 cm in diameter and 30 cm long. It operates inside an unpressurized module of an orbiting space station. The device dissipates 60 W, and its temperature must not exceed 80°C when the module walls are at −80°C. What value of emittance should be specified for the surface coating of the cylinder?

1–13. A high-vacuum chamber has its walls cooled to −190°C by liquid nitrogen. A sensor in the chamber has a surface area of 10 cm^2 and must be maintained at a temperature of 25°C. Plot a graph of the power required versus emittance of the sensor surface.

1–14. Consider a 3 m length of tube with a 1.26 cm inside diameter. Determine the convective heat transfer coefficient when

(i) water flows at 2 m/s.
(ii) oil (SAE 50) flows at 2 m/s.
(iii) air at atmospheric pressure flows at 20 m/s.

Thermophysical property data at 300 K are as follows:

	ρ kg/m^3	ν m^2/s	k W/m K	c_p J/kg K
Water	996	0.87×10^{-6}	0.611	4178
SAE 50 oil	883	570×10^{-6}	0.145	1900
Air at 1 atm	1.177	15.7×10^{-6}	0.0267	1005

1–15. Consider flow of water at 300 K in a long pipe of 1 cm inside diameter. Plot a graph of the heat transfer coefficient versus velocity over the range 0.01 to 100 m/s. Repeat for air at 1 atm and 300 K. Use the property values given in Exercise 1–14.

1–16. A 1 m–high vertical wall is maintained at 310 K, when the surrounding air is at 1 atm and 290 K. Plot the local heat transfer coefficient as a function of location up the wall. Take $\nu = 15.7 \times 10^{-6}$ m^2/s for air. Also calculate the convective heat loss per meter width of wall.

1–17. A 2 m–high vertical surface is maintained at 15°C when exposed to stagnant air at 1 atm and 25°C. Plot a graph showing the variation of the local heat transfer coefficient, and calculate the convective heat transfer for a 3 m width of wall. Take $\nu = 15.0 \times 10^{-6}$ m^2/s for air.

1–18. A thermistor is used to measure the temperature of an air stream leaving an air heater. It is located in a 30 cm square duct and records a temperature of 42.6°C when the walls of the duct are at 38.1°C. What is the true temperature of the air? The thermistor can be modeled as a 3 mm–diameter sphere of emittance 0.7. The convective heat transfer coefficient from the air stream to the thermistor is estimated to be 31 W/m^2 K.

1–19. A room heater is in the form of a thin vertical panel 1 m long and 0.7 m high, with air allowed to circulate freely on both sides. If its rating is 800 W, what will the average panel surface temperature be when the room air temperature is 20°C? The emittance of the surface is 0.85. Take $\nu_{air} = 17.5 \times 10^{-6}$ m²/s.

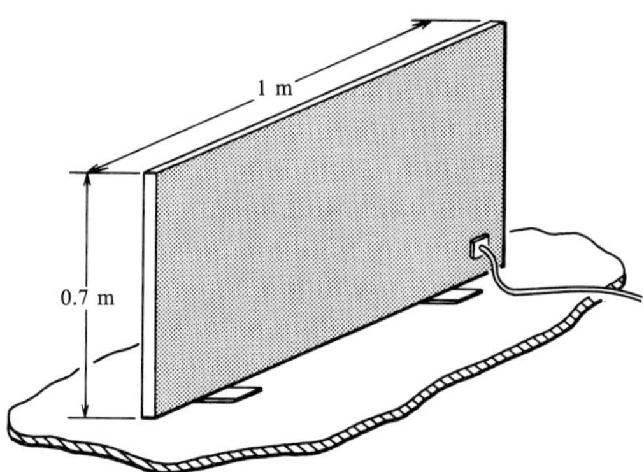

1–20. An electric water heater has a diameter of 1 m and a height of 2 m. It is insulated with 6 cm of medium-density fiberglass, and the outside heat transfer coefficient is estimated to be 8 W/m² K. If the water is maintained at 65°C and the ambient temperature is 20°C, determine

(i) the rate of heat loss.
(ii) the monthly cost attributed to heat loss if electricity costs 8 cents/kilowatt hour.

1–21. A 1 cm–diameter sphere is maintained at 60°C in an enclosure with walls at 35°C through which air at 40°C circulates. If the convective heat transfer coefficient is 11 W/m² K, estimate the rate of heat loss from the sphere when its emittance is

(i) 0.05.
(ii) 0.85.

1–22. Estimate the heating load for a building in a cold climate when the outside temperature is −10°C and the air inside is maintained at 20°C. The 350 m² of walls and ceiling are a composite of 1 cm–thick wallboard (k = 0.2 W/m K), 10 cm of vermiculite insulation (k = 0.06 W/m K), and 3 cm of wood (k = 0.15 W/m K). Take the inside and outside heat transfer coefficients as 7 and 35 W/m² K, respectively.

1–23. If a 2.5 × 10 m shaded wall in the building of Exercise 1–22 is replaced by a window, compare the heat loss through the wall if it is

(i) 0.3 cm–thick glass (k = 0.88 W/m K).
(ii) double-glazed with a 0.6 cm air gap between two 0.3 cm–thick glass panes.
(iii) the original wall.

1–24. Rework Exercise 1–19 for a panel 0.7 m high and 1.5 m long that is rated at 1 kW.

1–25. A mercury-in-glass thermometer used to measure the air temperature in an enclosure reads 15°C. The enclosure walls are all at 0°C. Estimate the true air temperature if the convective heat transfer coefficient for the thermometer bulb is estimated to be 12 W/m^2 K.

1–26. A tent is pitched on a mountain in an exposed location. The tent walls are opaque to thermal radiation. On a clear night the outside air temperature is −1°C, and the effective temperature of the sky as a black radiation sink is −60°C. The convective heat transfer coefficient between the tent and the ambient air can be taken to be 8 W/m^2 K. If the temperature of the outer surface of a sleeping bag on the tent floor is measured to be 10°C, estimate the heat loss from the bag in W/m^2,

(i) if the emittance of the tent material is 0.7.
(ii) if the outer surface of the tent is aluminized to give an emittance of 0.2.

For the sleeping bag, take an emittance of 0.8 and a convective heat transfer coefficient of 4 W/m^2 K. Assume that the ambient air circulates through the tent.

1–27. A schematic of a convective heat transfer coefficient meter is as shown. This meter is intended for use in situations where the air temperature T_e is known and the surrounding surfaces are also at temperature T_e. The two sensors that make up the meter are identical except that one has a surface coating of emittance $\varepsilon_1 = 0.9$, and the other has an emittance of $\varepsilon_2 = 0.1$. For equal power inputs to each sensor and $T_e = 300$ K, the measured surface temperatures are $T_{s1} = 321$ K and $T_{s2} = 336$ K. Determine $\overline{h}_c$.

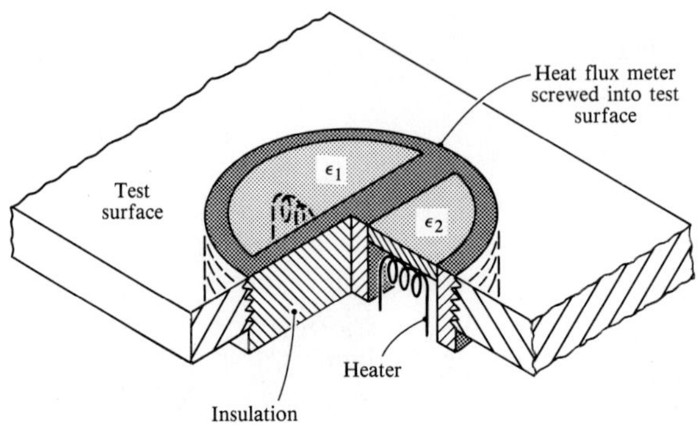

1–28. The horizontal roof of a building is surfaced with black tar paper of emittance 0.96. On a clear, still night the air temperature is 5°C, and the effective temperature of the sky as a black radiation sink is −60°C. The underside of the roof is well insulated.

(i) Estimate the roof surface temperature for a convective heat transfer coefficient of 5 W/m^2 K.

(ii) If the wind starts blowing, giving a convective heat transfer coefficient of 20 W/m^2 K, what is the new roof temperature?

(iii) Repeat the preceding calculations for aluminum roofing of emittance 0.15.

1–29. A chemical reactor has a 5 mm–thick mild steel wall and is lined inside with a 2 mm–thick layer of polyvinylchloride. The contents are at 80°C, and the ambient air is at 20°C. The inside thermal resistance is negligible ($h_{c,i}$ very large), and the outside heat transfer coefficient for combined convection and radiation is 7 W/m^2 K.

(i) Draw the thermal circuit.

(ii) Plot a graph of the temperature profile through the wall.

(iii) Calculate the rate of heat loss for a surface area of 10 m^2.

1–30. To prevent misting of the windscreen of an automobile, recirculated warm air at 37°C is blown over the inner surface. The windscreen glass (k = 1.0 W/m K) is 4 mm thick, and the ambient temperature is 5°C. The outside and inside heat transfer coefficients are 70 and 35 W/m^2 K, respectively.

(i) Determine the temperature of the inside surface of the glass.

(ii) If the air inside the automobile is at 20°C, 1 atm, and 80% relative humidity, will misting occur? (Refer to your thermodynamics text for the principles of psychrometry.)

1–31. The horizontal roof of a building is coated with tar of emittance 0.94. On a cloudy, still night the air temperature is 5°C, and the convective heat transfer coefficient between the air and the roof is estimated to be 4 W/m^2 K.

(i) If the effective temperature of the sky as a black radiation sink is −10°C, determine the roof temperature. Assume that the under surface of the roof is well insulated.

(ii) If a wind starts blowing, resulting in a convective heat transfer coefficient of 12 W/m^2 K, what is the new roof temperature?

(iii) Repeat the preceding calculations for aluminum roofing of emittance 0.15.

1–32. An alloy cylinder 3 cm in diameter and 2 m high is removed from an oven at 200°C and stood on its end to cool in ambient air at 20°C. Give a rough estimate of the time for the cylinder to cool to 100°C. For the alloy, take ρ = 8600 kg/m^3, c = 340 J/kg K, k = 110 W/m K, and ε = 0.74; for air, take $\nu = 23 \times 10^{-6}$ m^2/s.

1–33. A thermometer is used to check the temperature in a freezer that is set to operate at −5°C. If the thermometer initially reads 25°C, how long will it take for the reading to be within 1°C of the true temperature? Model the thermometer bulb as a 4 mm–diameter mercury sphere surrounded by a 2 mm–thick shell of glass. For mercury, take ρ = 13,530 kg/m^3, c = 140 J/kg K; and for glass ρ = 2640 kg/m^3, c = 800 J/kg K. Use a heat transfer coefficient of 15 W/m^2 K.

1–34. A thermocouple junction bead is modeled as a 1 mm–diameter lead sphere (ρ = 11,340 kg/m^3, c = 129 J/kg K) and is initially at a room temperature of 20°C. If the thermocouple is suddenly immersed in ice water to serve as a reference junction, what will be the error in indicated temperature corresponding to 1, 2, and 3 times the time constant of the thermocouple? If the heat transfer coefficient is calculated to be 2140 W/m^2 K, what are the corresponding times?

1–35. A hot-water cylinder contains 150 liters of water. It is insulated, and its outer surface has an area of 3.5 m^2. It is located in an area where the ambient air is 25°C, and the overall heat transfer coefficient between the water and the surrounds is 1.0 W/m^2 K, based on outer surface area. If there is a power failure, how long will it take the water to cool from 65°C to 40°C? Take the density of water as 980 kg/m^3 and its specific heat as 4180 J/kg K.

1–36. An aluminum plate 10 cm square and 1 cm thick is immersed in a chemical bath at 50°C for cleaning. On removal, the plate is shiny bright and is allowed to cool in a vertical position in still air at 20°C. Estimate how long the plate will take to cool to 30°C by

(i) assuming a constant heat transfer coefficient evaluated at the average ΔT of 20 K.
(ii) allowing exactly for the $\Delta T^{1/4}$ dependence of h_c given by Eq. (1.23a).

For air, take ν = 16.5 × 10^{-6} m^2/s, and for aluminum, take k = 204 W/m K, ρ = 2710 kg/m^3, c = 896 J/kg K.

1–37. A 2 cm–diameter copper sphere with a thermocouple at its center is suddenly immersed in liquid nitrogen contained in a Dewar flask. The temperature response is determined using a digital data acquisition system that records the temperature every 0.05 s. The maximum rate of temperature change dT/dt is found to occur when T = 92.5 K, with a value of 19.8 K/s.

(i) Using the lumped thermal capacity model, determine the corresponding heat transfer coefficient.
(ii) Check the Biot number to ensure that the model is valid.
(iii) The fact that the cooling rate is a maximum toward the end of the cooldown period is unusual; what must be the reason?

The saturation temperature of nitrogen at 1 atm pressure is 77.4 K. Take ρ = 8930 kg/m^3, c = 235 J/kg K, and k = 450 W/m K for copper at 92.5 K.

1–38. A 1 cm–diameter alloy sphere is to be heated in a furnace maintained at 1000°C. If the initial temperature of the sphere is 25°C, calculate the time required for the sphere to reach 800°C

(i) if the gas in the furnace is circulated to give a convective heat transfer coefficient of 100 W/m^2 K.
(ii) if there is no forced convection, and the free-convection heat transfer coefficient is given by $\overline{h}_c \sim 5\Delta T^{1/4}$ W/m^2 K for ΔT in kelvins.

Properties of the alloy include ρ = 4900 kg/m^3, c = 400 J/kg K, ε = 0.45.

1–39. A material sample, in the form of a 1 cm–diameter cylinder 10 cm long, is removed from a boiling water bath at 100°C and allowed to cool in air at 20°C. If the free-convection heat transfer coefficient can be approximated as $\bar{h}_c = 3.6\Delta T^{1/4}$ W/m^2 K for ΔT in kelvins, estimate the time required for the sample to cool to 25°C. For the sample properties take $\rho = 2260$ kg/m^3, $c = 830$ J/kg K, $\varepsilon = 0.77$.

1–40. Two small blackened spheres of identical size—one of aluminum, the other of an unknown alloy of high conductivity—are suspended by thin wires inside a large cavity in a block of melting ice. It is found that it takes 4.8 minutes for the temperature of the aluminum sphere to drop from 3°C to 1°C, and 9.6 minutes for the alloy sphere to undergo the same change. If the specific gravities of the aluminum and alloy are 2.7 and 5.4, respectively, and the specific heat of the aluminum is 900 J/kg K, what is the specific heat of the alloy?

1–41. A mercury-in-glass thermometer is to be used to measure the temperature of a high-velocity air stream. If the air temperature increases linearly with time, $T_e = \alpha t +$ constant, perform an analysis to determine the error in the thermometer reading due to its thermal "inertia." Evaluate the error if the inside diameter of the mercury reservoir is 3 mm, its length is 1 cm, and the glass wall thickness is 0.5 mm, when the heat transfer coefficient is 60 W/m^2 K and the air temperature increases at a rate of

(i) 1°C per minute.
(ii) 1°C per second.

Property values for mercury are $\rho = 13{,}530$ kg/m^3, $c = 140$ J/kg K; for glass $\rho = 2640$ kg/m^3, $c = 800$ J/kg K.

1–42. Under high-vacuum conditions in the space shuttle service bay, radiation is the only significant mode of heat transfer. Set $\bar{h}_c = 0$ in Eq. (1.43) and obtain an analytical solution for the lumped thermal capacity model thermal response. Also, identify a dimensionless group analogous to the Biot number that can be used to determine if the model is valid. (*Hint:* A table of standard integrals found in mathematics handbooks may be of assistance.)

1–43. A thermocouple is immersed in an air stream whose temperature varies sinusoidally about an average value with angular frequency ω. The thermocouple is small enough for the Biot number to be less than 0.1, but the convective heat transfer coefficient is high enough for radiation heat transfer to be negligible compared to convection.

(i) Set up the differential equation governing the temperature of the thermocouple.
(ii) Solve the differential equation to obtain the amplitude and phase lag of the thermocouple temperature response.
(iii) The thermocouple can be modeled as a 2 mm–diameter lead sphere ($\rho = 11{,}340$ kg/m^3, $c = 129$ J/kg K). If the air temperature varies as $T = 320 + 10\sin t$, for T in kelvins and t in seconds, calculate the amplitude and phase lag of the thermocouple for heat transfer coefficients of 30 and 100 W/m^2 K.

1–44. A system consists of a body in which heat is continuously generated at a rate $\dot{Q}_v$, while heat is lost from the body to its surroundings by convection. Using the lumped thermal capacity model, derive the differential equation governing the temperature response of the body. If the body is at temperature T_0 when time $t = 0$, solve the differential equation to obtain $T(t)$. Also determine the steady-state temperature.

1–45. Electronic components are often mounted with good heat conduction paths to a finned aluminum base plate, which is exposed to a stream of cooling air from a fan. The sum of the mass times specific heat products for a base plate and components is 5000 J/K, and the effective heat transfer coefficient times surface area product is 10 W/K. The initial temperature of the plate and the cooling air temperature are 295 K when 300 W of power are switched on. Find the plate temperature after 10 minutes.

1–46. A reactor vessel's contents are initially at 290 K when a reactant is added, leading to an exothermic chemical reaction that releases heat at a rate of 4×10^5 W/m^3. The volume and exterior surface area of the vessel are 0.008 m^3 and 0.24 m^2, respectively, and the overall heat transfer coefficient between the vessel contents and the ambient air at 300 K is 5 W/m^2 K. If the reactants are well stirred, estimate their temperature after

(i) 1 minute.
(ii) 10 minutes.

Take $\rho = 1200$ kg/m^3 and $c = 3000$ J/kg K for the reactants.

1–47. A carbon steel butane tank weighs 4.0 kg (empty) and has a surface area of 0.22 m^2. When full it contains 2 kg of liquified gas. Butane gas is drawn off to a burner at a rate of 0.05 kg/h through a pressure-reducing valve. If the ambient temperature is 55°C, estimate the steady temperature of the tank and the time taken for 80% of the temperature drop to occur. Take the sum of the convective and radiative heat transfer coefficients from the tank to the surroundings as 5 W/m^2 K. Property values for butane are $c = 2390$ J/kg K and $h_{fg} = 3.86 \times 10^5$ J/kg; for the steel $c = 434$ J/kg K.

1–48. A 2.5 m–diameter, 3.5 m–high milk storage tank is located in a dairy factory in Onehunga, New Zealand, where the ambient temperature is 30°C. The tank has walls of stainless steel 2 mm thick and is insulated with a 7.5 cm–thick layer of polyurethane foam. The tank is filled with milk at 4°C and is continuously stirred by an impeller driven by an electric motor that consumes 400 W of power. What will the milk temperature be after 24 hours? For the milk, take $\rho = 1034$ kg/m^3, $c = 3894$ J/kg K; for the insulation, $k = 0.026$ W/m K; and for the outside heat transfer coefficient, $h = 5$ W/m^2 K. The impeller motor efficiency can be taken as 0.75.

1–49. Steam is condensed on a bundle of 400 tubes through which there is a flow of cold water. The tubes have an outside diameter of 2 cm, and the overall heat transfer coefficient based on outside area is 600 W/m^2 K. The water flow rate is 0.2 kg/s per tube, and it enters at 290 K. If the outlet temperature is 350 K when the condenser is operating at atmospheric pressure, how much steam is condensed on each tube, and how long is the tube bundle? The specific heat of water at 320 K is 4174 J/kg K, and the enthalpy of vaporization for steam at atmospheric pressure is 2.257×10^6 J/kg.

1–50. A 3 m–long stage of a multistage flash vaporization desalination plant operates at 3×10^4 Pa. The condenser tube bundle comprises 1300, 3 cm–O.D. titanium tubes through which cooling water flows at 60 kg/s.

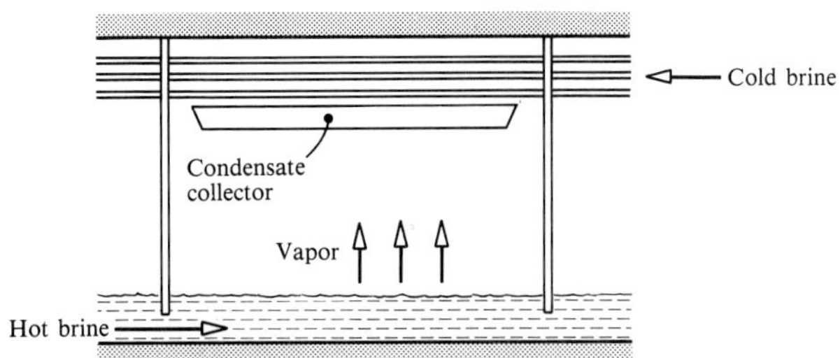

If the cooling water (sea water) enters the stage at 335 K, at what temperature does it exit? At what rate does the stage produce fresh water? Take the overall heat transfer coefficient based on the tube outside area as 1200 W/m^2 K and the specific heat of sea water as 4200 J/kg K.

1–51. A power station operating on a mercury-steam binary cycle has mercury condensing at 530 K on the outside of 6 cm–O.D. steel tubes. Water boils inside the tubes at 500 K. If the overall heat transfer coefficient based on tube outside area is 20,000 W/m^2 K, how many tubes 3 m long are required for a turbine steam consumption of 25 kg/s? For the water, take $h_{fg} = 1.827 \times 10^6$ J/kg.

1–52. A sheet iron exhaust gas stack is 50 cm in diameter and 10 m high. Exhaust gas at a rate of 1.0 kg/s enters the base of the stack at a temperature of 600 K. A strong wind blows across the stack, and the ambient air temperature is 300 K. If the overall heat transfer coefficient is 12 W/m^2 K, find the outlet temperature of the exhaust. Use $c_p = 1200$ J/kg K for the exhaust gas.

1–53. A geothermal power plant uses isobutane as the secondary working fluid. After expanding through the turbine, isobutane vapor condenses in a shell-and-tube condenser at 325 K. The condenser coolant is water at 305 K supplied from a cooling tower at a rate of 500 kg/s. The condenser shell contains 4000 tubes of 25 mm O.D. and 2 mm wall thickness; the overall heat transfer coefficient based on tube outside area is estimated to be 450 W/m^2 K. If it is desired to have a condenser effectiveness of 80%, determine

(i) the outlet water temperature.
(ii) the number of transfer units required.
(iii) the length of the tube bundle.

Also, if the thermal efficiency of the power cycle is 30%, determine the power output of the turbine. For water, take $c_p = 4174$ J/kg K, $\rho = 995$ kg/m^3.

1–54. In a test of a shell-and-tube steam condenser, 140 kg/s of coolant water at 300 K is supplied to the tubes. When the pressure in the shell is maintained at 0.010 MPa, the condensate flow rate is measured to be 3.0 kg/s. Determine

(i) the outlet water temperature.
(ii) the heat exchanger effectiveness.
(iii) the number of transfer units.
(iv) the $U\mathscr{P}L$ product.

For water, take $c_p = 4174$ J/kg K.

1–55. An air-cooled R-12 air-conditioning system condenser is required to condense 0.011 kg/s of saturated R-12 vapor at 320 K. The frontal area of the exchanger is 0.4 m^2, and air at 295 K enters the exchanger at 2 m/s.

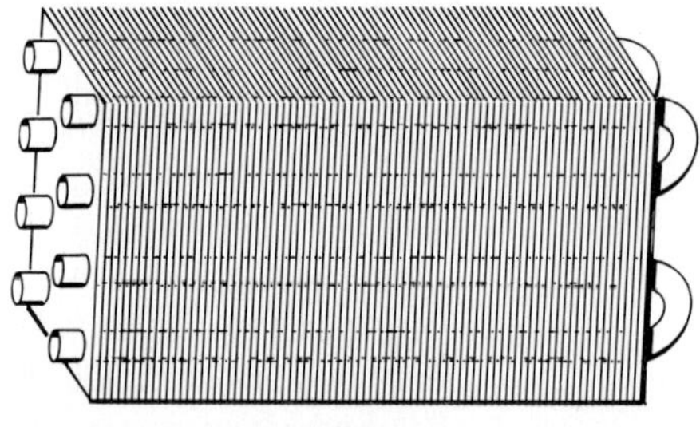

(i) If the overall heat transfer coefficient is estimated to be 40 W/m^2 K, determine the heat transfer area required.
(ii) Plot a graph of the required heat transfer area for an air inlet temperature in the range 290–305 K.
(iii) If the overall heat transfer coefficient increases approximately as air velocity to the 0.6 power, plot a graph of the required heat transfer area for an air velocity range of 2–10 m/s.

Take $h_{fg} = 0.1237 \times 10^6$ J/kg for R-12; $\rho = 1.177$ kg/m^3 and $c_p = 1005$ J/kg K for air.

1–56. Derive conversion factors for the following units conversions.

(i) Enthalpy of vaporization, Btu/lb to J/kg
(ii) Specific heat, Btu/lb °F to J/kg K
(iii) Density, lb/ft^3 to kg/m^3
(iv) Dynamic viscosity, lb/ft hr to kg/m s
(v) Kinematic viscosity, ft^2/hr to m^2/s
(vi) Thermal conductivity, Btu/hr ft °F to W/m K
(vii) Heat flux, Btu/hr ft^2 to W/m^2

1–57. In the United States, gas and liquid flow rates are commonly expressed in cubic feet per minute (CFM) and gallons per minute (GPM), respectively.

(i) For air at 1 atm and 300 K ($\rho = 1.177$ kg/m^3), prepare a table showing flow rates in m^3/s and kg/s corresponding to 1, 10, 100, 1000, and 10,000 CFM.
(ii) For water at 300 K ($\rho = 996$ kg/m^3), prepare a table showing flow rates in m^3/s and kg/s corresponding to 1, 10, 100, 1000, and 10,000 GPM.

1–58. In January 1989 the barometric pressure reached 31.84 inches of mercury at Northway, Alaska, a record for North America. On the other hand, a typical barometric pressure for Denver, Colorado, is 24.4 inches of mercury.

(i) What are these pressures in mbar and pascals?
(ii) At what temperature does water boil at these pressures?

1–59. Specify the following in the English system of units (Btu, hr, ft, °F or °R):

(i) The Stefan-Boltzmann constant
(ii) The radiation heat transfer coefficient at 25°C
(iii) The free-convection formulas given by Eqs. (1.23*a*) and (1.23*b*).

1–60. Convert the problem statement of Example 1.3 to the English system of units, work the problem in English units, and convert your answer back to SI units.

1–61. Convert the problem statement of Example 1.4 to the English system of units, work the problem in English units, and convert your answers back into SI units.

1–62. Convert the problem statement of Example 1.5 to the English system of units, work the problem in English units, and convert your answers back into SI units.

1–63. Convert the problem statement of Example 1.7 to the English system of units, work the problem in English units, and convert your answers back into SI units.

1–64. Convert the problem statement of Example 1.8 to the English system of units, work the problem in English units, and convert your answers back into SI units.

1–65. Check the dimensions of Eq. (1.22) in

(i) SI units.
(ii) English units.

1–66. Write a computer program to solve Eq. (1.45). That is, write your own version of LUMP. (LUMP is based on trapezoidal integration, with Newton's method used to solve the resulting nonlinear equation at each time step. Heun's explicit formula is used for the initial guess of the Newton iteration.)

CHAPTER

2

STEADY ONE-DIMENSIONAL HEAT CONDUCTION

CONTENTS

2.1 INTRODUCTION

In this chapter we analyze problems involving steady one-dimensional heat conduction. By **steady** we mean that temperatures are constant with time; as a result, the heat flow is also constant with time. By **one-dimensional** we mean that temperature is a function of a single "dimension" or spatial coordinate. One-dimensional conduction can occur in a number of geometrical shapes. In Section 1.3.1, one-dimensional conduction across a plane wall was examined, with temperature as a function of Cartesian coordinate x only; that is, $T = T(x)$. Conduction in cylinders or spheres is one-dimensional when temperature is a function of only the radial coordinate r and does not vary with polar angle and axial distance, in the case of the cylinder, or with polar or azimuthal angles, in the case of the sphere; that is, $T = T(r)$. Analysis of steady one-dimensional heat conduction problems involves the solution of very simple ordinary differential equations to give algebraic formulas for the temperature variation and heat flow. Thus, if at all possible, engineers like to approximate, or *model*, a practical heat conduction problem as steady and one-dimensional, even though temperatures might vary slowly with time or vary a little in a second coordinate direction.

A wide range of practical heat transfer problems involve steady one-dimensional heat conduction. Examples include most heat insulation problems, such as the refrigerated container of Example 1.1, the prediction of temperatures in a nuclear reactor fuel rod, and the design of cooling fins for electronic gear. Often, complex systems involving two- or three-dimensional conduction can be divided into subsystems in which the conduction is one-dimensional. Cooling of integrated circuit components can often be satisfactorily analyzed in this manner.

In Section 2.2, Fourier's law of heat conduction is briefly revisited. The physical mechanisms of heat conduction are discussed, and the applicability of Fourier's law at the interface between two solids is examined. Conduction across plane walls has already been treated in Section 1.3.1; thus, in Section 2.3, we restrict our attention to conduction across cylindrical and spherical shells and include the effect of **heat generation** within the solid. Section 2.4 deals with the class of problems known as **fin** problems, including familiar cooling fins and an interesting variety of mathematically similar problems.

There is a common methodology to the analyses in Chapter 2. Each analysis begins with the application of the first law, Eq. (1.2), to a closed-system volume element and introduction of Fourier's law, to obtain the governing differential equation. This equation is then integrated to give the temperature distribution, with the constants of integration found from appropriate boundary conditions. Finally, the heat flow is obtained using Fourier's law.

2.2 FOURIER'S LAW OF HEAT CONDUCTION

Fourier's law of heat conduction was introduced in Section 1.3.1. A general statement of this law is: The conduction heat flux in a specified direction equals the negative of the product of the medium thermal conductivity and the temperature derivative in that direction. In Chapter 2, we are concerned with one-dimensional conduction. In

Cartesian coordinates, with temperature varying in the x direction only,

$$q = -k\frac{dT}{dx} \tag{2.1}$$

Recall from Section 1.3.1 that the negative sign ensures that the heat flux q is positive in the positive x direction. In cylindrical or spherical coordinates, with temperature varying in the r direction only,

$$q = -k\frac{dT}{dr} \tag{2.2}$$

Equation (2.2) is the form of Fourier's law required for Section 2.3.

2.2.1 Thermal Conductivity

Table 1.1 gave a brief list of thermal conductivities to illustrate typical values for gases, liquids, and solids. Appendix A gives more complete tabulated data. The relevant tables are:

Table A.1	Solid metals
Table A.2	Solid dielectrics (nonmetals)
Table A.3	Insulators and building materials
Table A.4	Solids at cryogenic temperatures
Table A.7	Gases
Table A.8	Dielectric liquids
Table A.9	Liquid metals
Table A.13	Liquid solutions

Additional data may be found in the literature, for example, References [1] through [4]. In the case of commercial products, such as insulations, data can be obtained from the manufacturer.

The engineer needs conductivity data to solve heat conduction problems (as was seen in Chapter 1) and is usually not too concerned about the actual physical mechanism of heat conduction. However, conductivity data for a given substance are often sparse or nonexistent, and then a knowledge of the physics of heat conduction is useful to interpolate or extrapolate what data are available. Unfortunately, the physical mechanisms of conduction are many and complicated, and it is possible to develop simple theoretical models for gases and pure metals only. A brief account of some of the more important aspects of the conduction mechanisms follows.

Gases

The kinetic theory model gives a reliable basis for determining the thermal conductivity of a gas. Molecules are in a state of random motion. When collisions occur, there is an exchange of energy that results in heat being conducted down a temperature gradient from a hot region to a cold region. The simplest form of kinetic theory gives

$$k = \frac{1}{3} c \mathscr{N} \ell_t \left(m c_v + \frac{1}{2} \mathscr{k} \right) \quad [\text{W/m K}] \qquad \textbf{(2.3)}$$

where c is the average molecular speed [m/s], $\mathscr{N}$ is the number of molecules per unit volume [m^{-3}], ℓ_t is the transport mean free path [m], m is the mass of the molecule [kg], c_v is the constant-volume specific heat [J/kg K], and $\mathscr{k}$ is the Boltzmann constant, 1.38054×10^{-23} J/K. The product $c\,\mathscr{N}\ell_t$ is virtually independent of pressure in the vicinity of atmospheric pressure; thus, conductivity is also independent of pressure. Table A.7 gives data for a nominal pressure at 1 atm but can be used for pressures down to about 1 torr (1/760 of an atmosphere, or 133.3 Pa). Notice also that the average molecular speed is higher and the mean free path is longer for small molecules; thus, conductivities of gases such as hydrogen and helium are much greater than those of xenon and the refrigerant R-12.

Dielectric Liquids

In liquids (excluding liquid metals such as mercury at normal temperatures and sodium at high temperatures), the molecules are relatively closely packed, and heat conduction occurs primarily by longitudinal vibrations, similar to the propagation of sound. The structure of liquids is not well understood at present, and there are no good theoretical formulas for conductivity.

Pure Metals and Alloys

The primary mechanism of heat conduction is the movement of free electrons. A smaller contribution is due to the transfer of atomic motions by lattice vibrations, or waves; however, this contribution is unimportant except at cryogenic temperatures. In alloys, the movement of free electrons is restricted, and thermal conductivity decreases markedly as alloying elements are added. The lattice wave contribution then becomes more important but is difficult to predict owing to the variable effects of heat treatment and cold working.

Dielectric Solids

Heat conduction is almost entirely due to atomic motions being transferred by lattice waves; hence, thermal conductivity is very dependent on the crystalline structure of the material. Many building materials and insulators have anomalously low values of conductivity because of their porous nature. The conductivity is an effective value for the porous medium and is low due to air or a gas filling the interstices and pores, through which heat is transferred rather poorly by conduction and radiation. Expanded plastic insulations, such as polystyrene, have a large molecule refrigerant gas filling the pores to reduce the conductivity.

In general, thermal conductivity is temperature-dependent. Fortunately, the variation in typical engineering problems is small, and it suffices to use an appropriate average value. An exception is solids at cryogenic temperatures, as an examination of Table A.4 will show. Another exception is gases in the vicinity of the critical point. Special care must be taken in such situations.

2.2.2 Contact Resistance

In Section 1.3.1, heat conduction through a composite wall was analyzed. Figure 2.1*a* shows the interface between two layers of a composite wall, with the surface of each layer assumed to be perfectly smooth. Two mathematical surfaces, the u- and s-surfaces, are located on each side of and infinitely close to the real interface, as shown. The first law of thermodynamics applied to the closed system located between the u- and s-surfaces requires that

$$\dot{Q}|_u = \dot{Q}|_s \tag{2.4}$$

since no energy can be stored in the infinitesimal amount of material in the system. Considering a unit area and introducing Fourier's law gives

$$-k_A \left.\frac{dT}{dx}\right|_u = -k_B \left.\frac{dT}{dx}\right|_s \tag{2.5}$$

Also, since the distance between the u- and s-surfaces is negligible, thermodynamic equilibrium requires

$$T_u = T_s \tag{2.6}$$

as shown on the temperature profile (for $k_A > k_B$). For perfectly smooth surfaces, there is no thermal resistance at the interface.

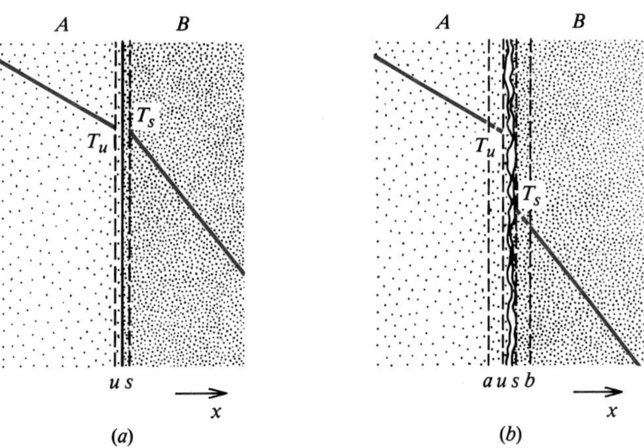

Figure 2.1 Interfaces between two layers of a composite wall. (*a*) Smooth surfaces. (*b*) Rough surfaces.

Figure 2.1*b* shows a more realistic situation, in which each surface has some degree of roughness. The solid materials are in contact at relatively few places, and the gaps may contain a fluid or, in some applications, a vacuum. The heat flow in the interface region is complicated: the conduction is three-dimensional as the heat tends to "squeeze" through the contact areas, and there are parallel paths of conduction and radiation through the gaps. The u- and s-surfaces are located just on either side of a somewhat arbitrarily defined interface location. In addition, a- and b-surfaces are located just far enough from the interface for the heat conduction to be one-dimensional. No temperature profile is shown between the a- and b-surfaces since no unique profile $T(x)$ exists there; instead, the temperature profiles are extrapolated from the bulk material to the interface as shown, thereby defining the temperatures and temperature gradients at the u- and s-surfaces. As was the case for the perfectly smooth surfaces, the first law requires

$$-k_A \left.\frac{dT}{dx}\right|_u = -k_B \left.\frac{dT}{dx}\right|_s \tag{2.7}$$

but now there is no continuity of temperature at the interface; that is, $T_u \neq T_s$. The thermal resistance to heat flow at the interface is called the **contact resistance** and is usually expressed in terms of an **interfacial conductance** h_i [W/m^2 K], defined in an analogous manner to Newton's law of cooling, namely,

$$\dot{Q} = h_i A(T_u - T_s) \tag{2.8}$$

or

$$-k_A \left.\frac{dT}{dx}\right|_u = h_i(T_u - T_s) = -k_B \left.\frac{dT}{dx}\right|_s \tag{2.9}$$

Figure 2.2 shows the contact resistance added to the thermal circuit of Fig. 1.5.

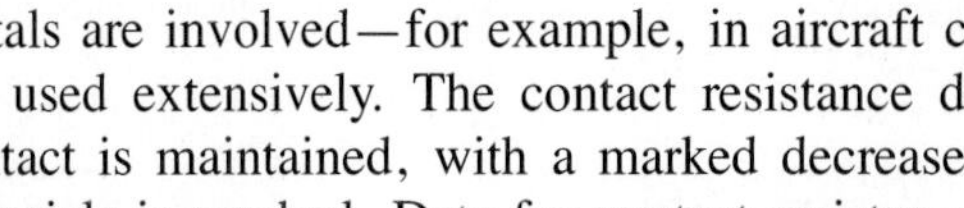

Figure 2.2 A contact resistance in a thermal circuit.

There is always a contact resistance to conduction across real solid-solid interfaces. The contact resistance can be the dominant thermal resistance when high-conductivity metals are involved—for example, in aircraft construction, where aluminum alloys are used extensively. The contact resistance depends on the pressure with which contact is maintained, with a marked decrease once the yield point of one of the materials is reached. Data for contact resistances are, unfortunately, sparse and unreliable. Table 2.1 does, however, show some representative values. Additional data can be found in the literature [5,6,7].

Table 2.1 Typical interfacial conductances (at moderate pressure and usual finishes, unless otherwise stated).

Interface	h_i W/m^2 K
Ceramic-ceramic	500–3000
Ceramic-metals	1500–8500
Graphite-metals	3000–6000
Stainless steel–stainless steel	1700–3700
Aluminum-aluminum	2200–12,000
Stainless steel–aluminum	3000–4500
Copper-copper	10,000–25,000
Rough aluminum–aluminum (vacuum conditions)	~150
Iron-aluminum	4000–40,000

2.3 CONDUCTION ACROSS CYLINDRICAL AND SPHERICAL SHELLS

Steady one-dimensional conduction in cylinders or spheres requires that temperature be a function of only the radial coordinate r. The analysis of steady heat flow across a plane wall in Section 1.3.1 was particularly simple because the flow area A did not change in the flow direction. In the case of a cylindrical or spherical shell, the area for heat flow changes in the direction of heat flow. For a cylindrical shell of length L, the area for heat flow is $A = 2\pi rL$; for a spherical shell, it is $A = 4\pi r^2$. In both cases, A increases with increasing r.

2.3.1 Conduction across a Cylindrical Shell

Figure 2.3 shows a cylindrical shell of length L, with inner radius r_1 and outer radius r_2. The inner surface is maintained at temperature T_1 and the outer surface is maintained at temperature T_2. An elemental control volume is located between radii r and $r + \Delta r$. If temperatures are unchanging in time and $\dot{Q}_v = 0$, the energy conservation principle, Eq. (1.2), requires that the heat flow across the face at r equal that at the face $r + \Delta r$

$$\dot{Q}|_r = \dot{Q}|_{r+\Delta r}$$

that is,

$$\dot{Q} = \text{Constant, independent of } r$$

Using Fourier's law in the form of Eq. (2.2),

$$\dot{Q} = Aq = 2\pi rL\left(-k\frac{dT}{dr}\right)$$

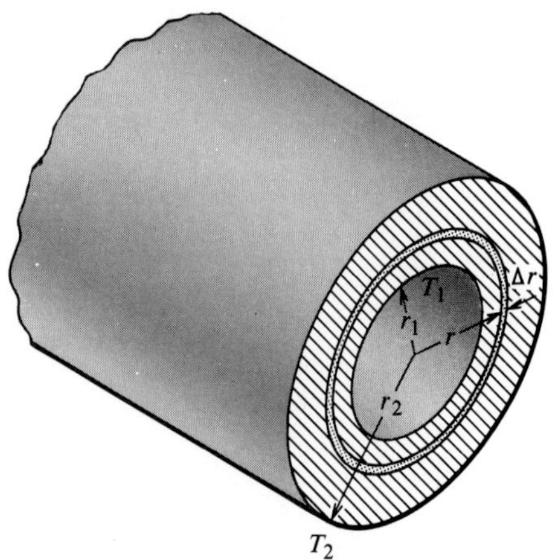

Figure 2.3 A cylindrical shell showing an elemental control volume for application of the energy conservation principle.

Dividing by $2\pi kL$ and assuming that the conductivity k is independent of temperature gives

$$\frac{\dot{Q}}{2\pi kL} = -r\frac{dT}{dr} = \text{Constant} = C_1 \tag{2.10}$$

which is a first-order ordinary differential equation for $T(r)$ and can be integrated easily:

$$\frac{dT}{dr} = -\frac{C_1}{r}$$

$$T = -C_1 \ln r + C_2 \tag{2.11}$$

Two boundary conditions are required to evaluate the two constants; these are

$$r = r_1: \quad T = T_1 \tag{2.12a}$$

$$r = r_2: \quad T = T_2 \tag{2.12b}$$

Substituting in Eq. (2.11) gives

$$T_1 = -C_1 \ln r_1 + C_2$$

$$T_2 = -C_1 \ln r_2 + C_2$$

which are two algebraic equations for the unknowns C_1 and C_2. Subtracting the second equation from the first:

$$T_1 - T_2 = -C_1 \ln r_1 + C_1 \ln r_2 = C_1 \ln(r_2/r_1)$$

or

$$C_1 = \frac{T_1 - T_2}{\ln(r_2/r_1)}$$

Using either of the two equations then gives

$$C_2 = T_1 + \frac{T_1 - T_2}{\ln(r_2/r_1)} \ln r_1$$

Substituting back in Eq. (2.11) and rearranging gives the temperature distribution as

$$\frac{T_1 - T}{T_1 - T_2} = \frac{\ln(r/r_1)}{\ln(r_2/r_1)} \tag{2.13}$$

which is a logarithmic variation, in contrast to the linear variation found for the plane wall in Section 1.3.1. The heat flow is found from Eq. (2.10) as $\dot{Q} = 2\pi kLC_1$, or

$$\dot{Q} = \frac{2\pi kL(T_1 - T_2)}{\ln(r_2/r_1)} \tag{2.14}$$

Equation (2.14) is again in the form of Ohm's law, and the thermal resistance of the cylindrical shell is

$$R = \frac{\ln(r_2/r_1)}{2\pi kL} \tag{2.15}$$

When $r_2 = r_1 + \delta$ and $\delta/r_1 \ll 1$, Eq. (2.15) reduces to the resistance of a slab, $\delta/2\pi r_1 kL = \delta/kA$.

It is now possible to treat composite cylindrical shells with convection and radiation from either side without any further analysis. Figure 2.4 shows the cross section of an insulated pipe of length L, through which flows superheated steam and

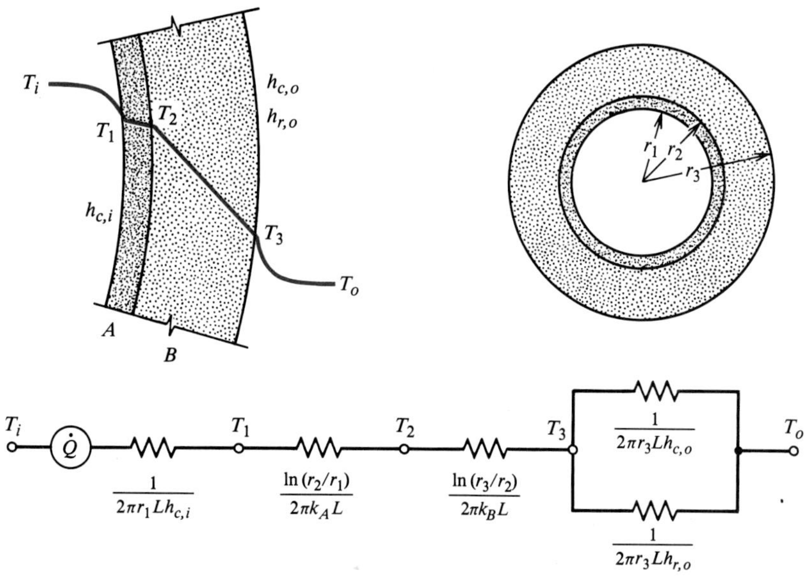

Figure 2.4 An insulated steam pipe showing the temperature distribution and thermal circuit.

which loses heat by convection and radiation to its surroundings. The thermal circuit is also shown, with Eq. (2.15) used for the conductive resistances of the pipe and insulation. Notice that in contrast to the plane wall case, the area for heat flow is different on each side of the composite wall: on the inside it is $2\pi r_1 L$, and on the outside it is $2\pi r_3 L$. Again we define an *overall heat transfer coefficient* by Eq. (1.29):

$$\dot{Q} = UA(T_i - T_o) = \frac{T_i - T_o}{1/UA} \tag{2.16}$$

Then, summing the resistances in the thermal network,

$$\frac{1}{UA} = \frac{1}{2\pi r_1 L h_{c,i}} + \frac{\ln(r_2/r_1)}{2\pi k_A L} + \frac{\ln(r_3/r_2)}{2\pi k_B L} + \frac{1}{2\pi r_3 L(h_{c,o} + h_{r,o})} \tag{2.17}$$

The area A need not be specified since all we need is the UA product. However, often a value of U will be quoted based on either the inside or outside area; then the appropriate area must be used in Eqs. (2.16) and (2.17).

EXAMPLE 2.1 Heat Loss from an Insulated Steam Pipe

A mild steel steam pipe has an outside diameter of 15 cm and a wall thickness of 0.7 cm. It is insulated with a 5.3 cm–thick layer of 85% magnesia insulation. Superheated steam at 500 K flows through the pipe, and the inside heat transfer coefficient is 35 W/m^2 K. Heat is lost by convection and radiation to surroundings at 300 K, and the sum of outside convection and radiation coefficients is estimated to be 8 W/m^2 K. Find the rate of heat loss for a 20 m length of pipe.

Solution

Given: Steam pipe with 85% magnesia insulation.

Required: Heat loss for 20 m length if h_o = 8 W/m^2 K.

Assumptions: Steady one-dimensional heat flow.

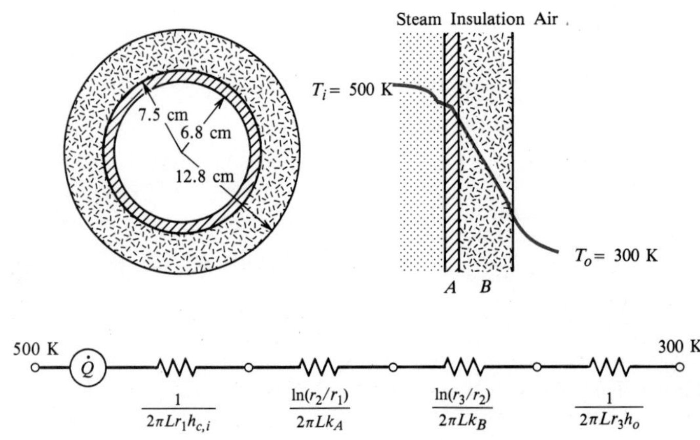

Equation (2.16) applies, with Eq. (2.17) used to obtain the UA product,

$$\dot{Q} = UA(T_i - T_o)$$

$$\frac{1}{UA} = \frac{1}{2\pi L}\left(\frac{1}{r_1 h_{c,i}} + \frac{\ln(r_2/r_1)}{k_A} + \frac{\ln(r_3/r_2)}{k_B} + \frac{1}{r_3 h_o}\right)$$

Tables A.1b and A.3 in Appendix A give the variation of conductivity with temperature for 1010 steel and magnesia, respectively. As a first step, we guess that the steel is close to the steam temperature (500 K), and since most of the temperature drop will be across the magnesia insulation, its average temperature will be about $(500 + 300)/2 = 400$ K. The corresponding conductivity values are $k_A = 54$ W/m K and $k_B = 0.073$ W/m K.

$$\frac{1}{UA} = \frac{1}{(2)(\pi)(20)}\left(\frac{1}{(0.068)(35)} + \frac{\ln(0.075/0.068)}{54} + \frac{\ln(0.128/0.075)}{0.073} + \frac{1}{(0.128)(8)}\right)$$

$$= \frac{1}{125.7}(0.42 + 0.002 + 7.32 + 0.98)$$

$$UA = 14.4 \text{ W/K}$$

$$\dot{Q} = UA\Delta T = (14.4)(500 - 300) = 2880 \text{ W}$$

Since the resistance of the steel wall is negligible, we do not need to check our guess for its conductivity. For the magnesia insulation, we estimate its average temperature by examining the relevant segment of the thermal circuit. For convenience, the thermal resistance of the insulation is split in half to estimate an average temperature $\overline{T}$:

$$\overline{T} - T_o = \dot{Q}\left[\left(\frac{1}{2}\right)\frac{\ln(r_3/r_2)}{2\pi L k_B} + \frac{1}{2\pi L r_3 h_o}\right]$$

$$\overline{T} - 300 = (2880)\left[\left(\frac{1}{2}\right)\frac{7.32}{125.7} + \frac{0.98}{125.7}\right]$$

$$= 106 \text{ K}$$

$$\overline{T} = 406 \text{ K}$$

A look at Table A.3 shows that our guess of 400 K introduced an error of less than 1%, so there is no need to calculate a new value of $\dot{Q}$ using an improved k value.

Comments

1. After one has gained some experience with this type of calculation, the problem can be simplified by ignoring the small resistance of the steel pipe.
2. In practice, the outside heat transfer coefficient varies somewhat around the circumference of the insulation, and the conduction is not truly one-dimensional. For an engineering calculation, we simply use an average value for h_o.

2.3.2 Critical Thickness of Insulation on a Cylinder

The insulation on the large steam pipe in Example 2.1 was installed because it reduced the heat loss. However, adding a layer of insulation to a cylinder does not necessarily reduce the heat loss. When the outer radius of the insulation r_o is small, there is the possibility that the added thermal resistance of the insulation is less than the reduction of the outside resistance $1/2\pi r_o L h_o$ due to the larger value of the area for convective and radiative heat transfer, $2\pi r_o L$. This phenomenon is often used for cooling electronic components that must dissipate I^2R heating. Figure 2.5 shows a resistor with an insulation sheath of inner radius r_i and outer radius r_o. Since the resistor usually has a relatively high thermal conductivity, we will assume it is isothermal at temperature T_i. The ambient air temperature is T_e, and the outside heat transfer coefficient is h_o. There are two resistances in series; denoting the total resistance as R the heat flow is

$$\dot{Q} = \frac{T_i - T_e}{R} = \frac{T_i - T_e}{\ln(r_o/r_i)/2\pi Lk + 1/2\pi L r_o h_o} \tag{2.18}$$

$\dot{Q}$ will have a maximum value when the total resistance R has a minimum value; differentiating R with respect to r_o:

$$\frac{dR}{dr_o} = \frac{1}{2\pi L}\left(\frac{1}{r_o k} - \frac{1}{r_o^2 h_o}\right)$$

which equals zero when the outer radius of the insulation equals the **critical radius,**

$$r_o = r_{\text{cr}} = \frac{k}{h_o} \tag{2.19}$$

To check whether r_{cr} gives a minimum resistance, we differentiate again and evaluate at $r_o = r_{\text{cr}}$:

$$\frac{d^2R}{dr_o^2} = \frac{1}{2\pi L}\left(-\frac{1}{r_o^2 k} + \frac{2}{r_o^3 h_o}\right)$$

and

$$\left.\frac{d^2R}{dr^2}\right|_{r_o = r_{\text{cr}}} = \frac{1}{2\pi L}\left(-\frac{h_o^2}{k^3} + \frac{2h_o^2}{k^3}\right) = \frac{h_o^2}{2\pi L k^3} > 0$$

as required.

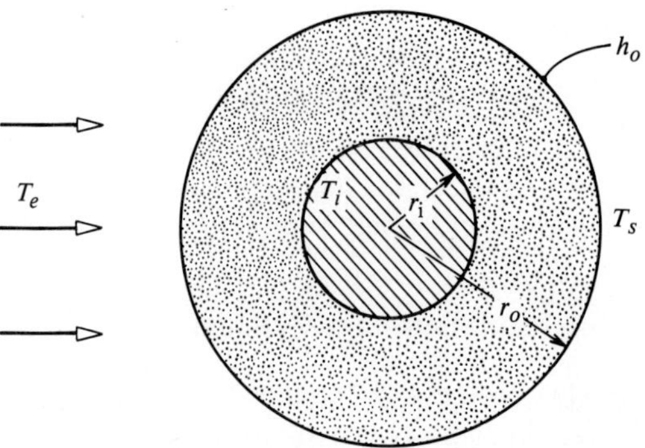

Figure 2.5 An electric resistor with an insulation sheath.

Equation (2.19) is an often-used formula for the critical radius, but it is only an approximate estimate since the heat transfer coefficient was assumed to be independent of r_o. In general, we can write $h_o = \alpha r_o^{-n}$ where, for example, $n = 1/2$ for laminar forced convection. Thus, the total resistance R is more correctly written as

$$R = \frac{\ln(r_o/r_i)}{2\pi Lk} + \frac{1}{2\pi L\alpha r_o^{1-n}}$$

hence,

$$\frac{dR}{dr_o} = \frac{1}{2\pi L}\left(\frac{1}{r_o k} + \frac{(n-1)r_o^{n-2}}{\alpha}\right)$$

which equals zero when

$$r_o = r_{\mathrm{cr}} = \left(\frac{\alpha}{(1-n)k}\right)^{1/(n-1)} \tag{2.20}$$

For $n = 1/2$, $r_{\mathrm{cr}} = (k/2\alpha)^2$. For natural convection, the situation is more complex: not only do we have $n = 1/4$, but Newton's law of cooling is invalid, with $h_o \propto \Delta T^{1/4}$. In the case of radiation, h_r can be taken to be independent of r_o, but $h_r \propto T_s^3$ [for small $(T_s - T_e)$]. Exercises 2–17 and 2–18 examine these situations. Fortunately, from a practical standpoint, a precise value of r_{cr} is not needed. Because $\dot{Q}$ is a maximum at r_{cr}, the heat loss is not sensitive to the precise value of r when r is in the vicinity of r_{cr}.

EXAMPLE 2.2 Cooling of an Electrical Resistor

A 0.5 W, 1.5 MΩ graphite resistor has a diameter of 1 mm and is 20 mm long; it has a thin glass sheath and is encapsulated in micanite (crushed mica bonded by a phenolic resin). The micanite serves both as additional electrical insulation and to increase the heat loss. It can be assumed that 50% of the I^2R heating is dissipated by combined convection and radiation from the outer surface of the micanite to surroundings at 300 K with $h_o = 16$ W/m^2 K; the remainder is conducted through copper leads to a circuit board. If the conductivity of micanite is 0.1 W/m K, what radius will give the maximum cooling effect, and what is the corresponding resistor temperature?

Solution

Given: Cylindrical graphite resistor encapsulated in micanite.

Required: Critical radius of micanite insulation, and the resistor temperature.

Assumptions:
1. The resistance of the glass sheath is negligible.
2. The outside heat transfer coefficient h_o is constant.
3. The resistor temperature is uniform.

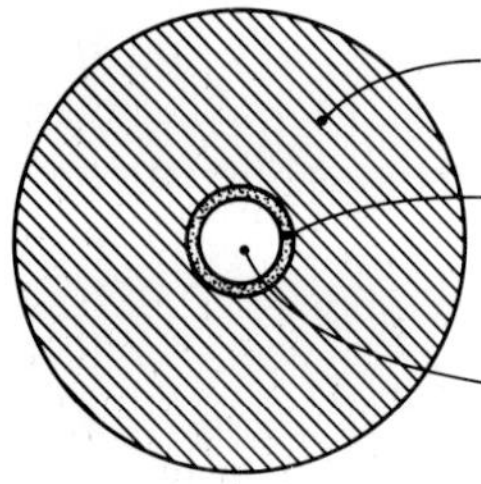

The critical radius is given by Eq. (2.19):

$$r_{cr} = \frac{k}{h_o} = \frac{0.1}{16} = 0.00625 \text{ m}$$

$$= 6.25 \text{ mm}$$

From Eq. (2.18),

$$T_i - T_e = \frac{\dot{Q}}{2\pi L}\left(\frac{\ln(r_o/r_i)}{k} + \frac{1}{r_o h_o}\right)$$

$$T_i - 300 = \frac{(0.5)(0.5)}{(2)(\pi)(0.020)}\left(\frac{\ln(6.25/0.5)}{0.1} + \frac{1}{(0.00625)(16)}\right)$$

$$= 70.1 \text{ K}$$

Hence, $T_i = 300 + 70.1 = 370.1$ K.

Comments

1. To obtain a more accurate result, additional data are required, particularly for the variation of h_o with radius.
2. Section 2.3.4 shows how to check the validity of assumption 3.
3. In general, h_o (and hence r_{cr}) vary around the circumference of the insulation. For engineering purposes, we ignore this complication and use an average value for h_o.

2.3.3 Conduction across a Spherical Shell

Figure 2.6 shows a spherical shell of inner radius r_1 and outer radius r_2. The inner surface is maintained at temperature T_1 and the outer surface at T_2. An elemental control volume is located between radii r and $r + \Delta r$. As was shown for the cylindrical shell in Section 2.3.1, energy conservation applied to the control volume requires that the heat flow $\dot{Q}$ be constant, independent of r if the temperatures are unchanging in time and $\dot{Q}_v = 0$. Using Fourier's law, Eq. (2.2),

$$\dot{Q} = Aq = 4\pi r^2\left(-k\frac{dT}{dr}\right)$$

Dividing by $4\pi k$ and assuming that the conductivity k is independent of temperature gives

$$\frac{\dot{Q}}{4\pi k} = -r^2\frac{dT}{dr} = \text{Constant} = C_1 \tag{2.21}$$

$$\frac{dT}{dr} = -\frac{C_1}{r^2}$$

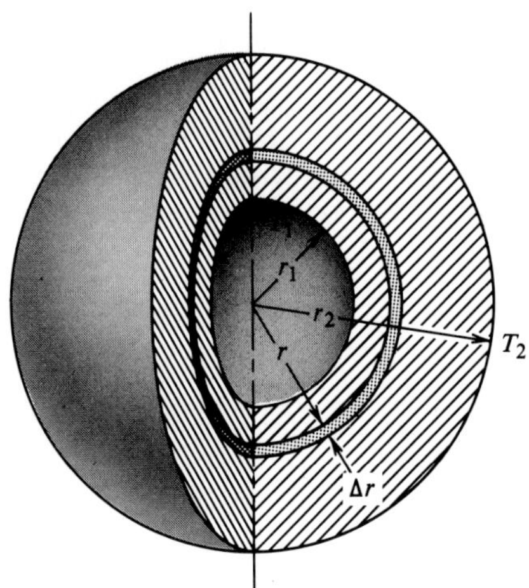

Figure 2.6 A spherical shell showing an elemental control volume for application of the energy conservation principle.

Integrating,

$$T = \frac{C_1}{r} + C_2 \tag{2.22}$$

The boundary conditions required to evaluate the two constants are

$$r = r_1: \quad T = T_1 \tag{2.23a}$$

$$r = r_2: \quad T = T_2 \tag{2.23b}$$

Substituting in Eq. (2.22) and solving for C_1 and C_2 gives

$$C_1 = \frac{T_1 - T_2}{1/r_1 - 1/r_2}; \qquad C_2 = T_1 - \frac{T_1 - T_2}{1 - r_1/r_2}$$

Then substituting back in Eq. (2.22) and rearranging gives the temperature distribution as

$$\frac{T_1 - T}{T_1 - T_2} = \frac{1/r_1 - 1/r}{1/r_1 - 1/r_2} \tag{2.24}$$

The heat flow is found from Eq. (2.21) as $\dot{Q} = 4\pi k C_1$, or

$$\dot{Q} = \frac{4\pi k(T_1 - T_2)}{1/r_1 - 1/r_2} \tag{2.25a}$$

Equation (2.25a) can be used to build up thermal circuits for composite spherical shells, as was done for composite cylindrical shells in Section 2.3.1. The thermal resistance of a spherical shell is

$$R = \frac{1/r_1 - 1/r_2}{4\pi k} \tag{2.25b}$$

EXAMPLE 2.3 Determination of Thermal Conductivity

To measure the effective thermal conductivity of an opaque honeycomb material for an aircraft wall, a spherical shell of inner radius 26 cm and outer radius 34 cm was constructed and a 100 W electric light bulb placed in the center. At steady state, the temperatures of the inner and outer surfaces were measured to be 339 and 311 K respectively. What is the effective conductivity of the material?

Solution

Given: Spherical shell containing a 100 W heat source.

Required: Thermal conductivity of shell material.

Assumptions: 1. Steady state.
2. Spherical symmetry, $T = T(r)$.

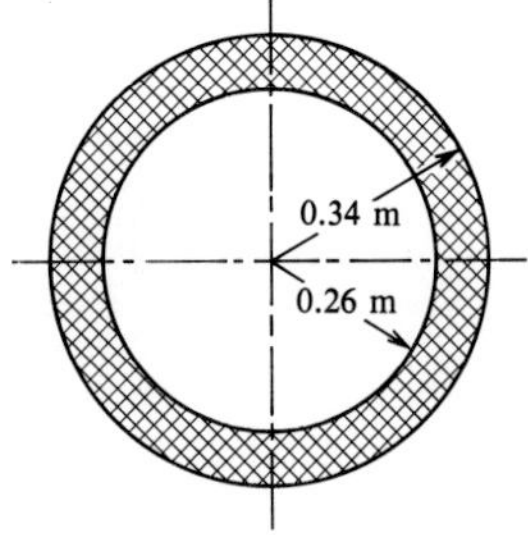

Eq. (2.25*a*) applies, with $\dot{Q}$, T_1, T_2 known and k the unknown:

$$100 = \frac{4\pi k(339 - 311)}{1/0.26 - 1/0.34}$$

Solving, $k = 0.257$ W/m K.

Comments

The large thermal resistance of the honeycomb results in a relatively large temperature difference across it, which is easy to measure accurately. The same method would not be practical for determining the conductivity of a metal shell.

2.3.4 Conduction with Internal Heat Generation

In some situations, the thermal behavior of a body is affected by internally generated or absorbed thermal energy. The most common example is I^2R heating associated with the flow of electrical current I in an electrical resistance R. Other examples include fission reactions in the fuel rods of a nuclear reactor, absorption of radiation in a microwave oven, and emission of radiation by a flame. We will use the symbol $\dot{Q}_v'''$ [W/m^3] for the heat generation rate per unit volume.[1] As an example, consider internal heat generation in a solid cylinder of outer radius r_1, as might occur in an electrical wire or a nuclear fuel rod. Figure 2.7 shows an elemental volume located between radii r and $r + \Delta r$. Applying the energy conservation principle, Eq. (1.2), requires

[1] The triple prime indicates "per unit volume" (per length dimension cubed). In Section 1.2, the symbol $\dot{Q}_v$ [W] was used for the heat generated within a system and is related to $\dot{Q}_v'''$ [W/m^3] as $\dot{Q}_v = \int_V \dot{Q}_v''' \, dV$.

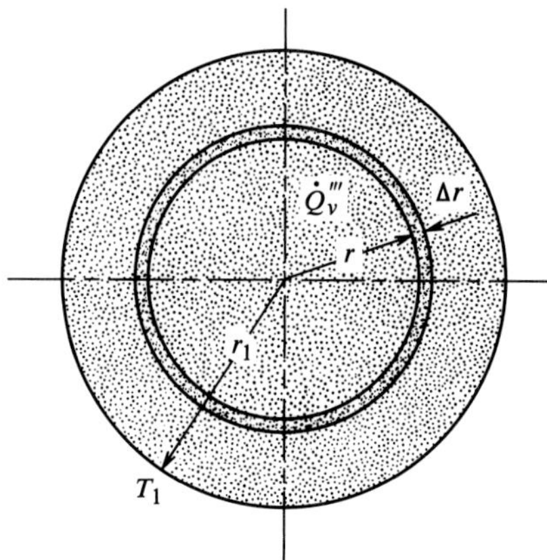

Figure 2.7 A solid cylindrical rod with internal heat generation.

that

$$\dot{Q}|_r - \dot{Q}|_{r+\Delta r} + \dot{Q}_v'''2\pi rL\Delta r = 0$$

if temperatures are steady. Dividing by Δr and rearranging gives

$$\frac{\dot{Q}|_{r+\Delta r} - \dot{Q}|_r}{\Delta r} = \dot{Q}_v'''2\pi rL$$

which for $\Delta r \to 0$ becomes

$$\frac{d\dot{Q}}{dr} = 2\pi rL\dot{Q}_v'''$$

Introducing Fourier's law, $\dot{Q} = Aq = 2\pi rL[-k\,(dT/dr)]$, and assuming k constant gives

$$\frac{d}{dr}\left(r\frac{dT}{dr}\right) = -\frac{\dot{Q}_v'''}{k}r \tag{2.26}$$

which is a second-order linear ordinary differential equation for $T(r)$.

Two boundary conditions are required; the first comes from symmetry:

$$r = 0: \quad \frac{dT}{dr} = 0 \tag{2.27a}$$

To obtain a result of some generality, we will take as the second boundary condition a specified temperature on the outer surface of the cylinder:

$$r = r_1: \quad T = T_1 \tag{2.27b}$$

In a typical engineering problem, T_1 might not be specified; however, we shall see that the result will be in a form suitable for problem solving. Integrating Eq. (2.26) once gives

$$r\frac{dT}{dr} = -\frac{1}{2}\frac{\dot{Q}_v'''}{k}r^2 + C_1$$

or

$$\frac{dT}{dr} = -\frac{1}{2}\frac{\dot{Q}_v'''}{k}r + \frac{C_1}{r}$$

Applying the first boundary condition, Eq. (2.27*a*),

$$0 = 0 + \frac{C_1}{0}, \qquad \text{or } C_1 = 0$$

Integrating again,

$$T = -\frac{1}{4}\frac{\dot{Q}_v'''}{k}r^2 + C_2$$

Applying the second boundary condition, Eq. (2.27*b*) allows C_2 to be evaluated:

$$T_1 = -\frac{1}{4}\frac{\dot{Q}_v'''}{k}r_1^2 + C_2$$

Substituting back gives the desired temperature distribution, $T(r)$:

$$T - T_1 = \frac{1}{4}\frac{\dot{Q}_v'''}{k}\left(r_1^2 - r^2\right) \tag{2.28}$$

The maximum temperature is at the centerline of the cylinder. Setting $r = 0$ in Eq. (2.28) gives

$$T_{\text{max}} - T_1 = \frac{1}{4}\frac{\dot{Q}_v''' r_1^2}{k} \tag{2.29}$$

The use of this result is illustrated in the following example.

EXAMPLE 2.4 Temperature Distribution in a Nuclear Reactor Fuel Rod

Uranium oxide fuel is contained inside 0.825 cm–I.D., 0.970 cm–O.D. Zircaloy-4 tubes. The tubes have a 1.75 cm pitch in a square array. The power averaged over the volume including the space between the fuel rods is 152.4 W/cm³. At a specific location along the bundle the coolant water is at 400 K and the convective heat transfer coefficient h_c is 1.0×10^4 W/m² K. If the interfacial conductance between the fuel and the tube, h_i, is 6000 W/m² K, determine the maximum temperature in the fuel rods.

Solution

Given: Nuclear reactor fuel rod.

Required: Maximum rod temperature at location where the coolant water is at $T_e = 400$ K.

Assumptions: Steady one-dimensional heat flow.

We cannot immediately use Eq. (2.29) to obtain T_{max} because the surface temperature of the fuel rod is unknown. We proceed as follows: first we calculate $\dot{Q}_v'''$ in the fuel itself,

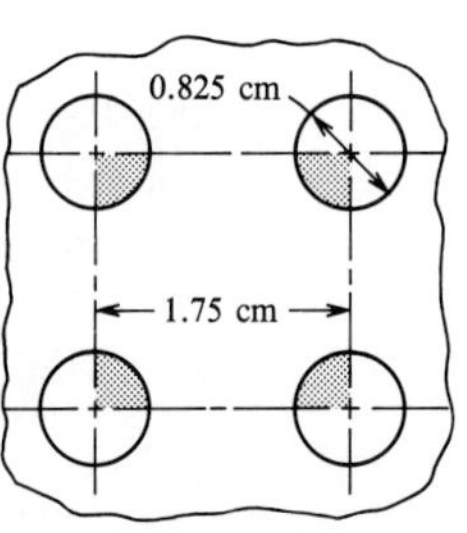

(*a*)

$$\dot{Q}_v''' = 152.4\frac{\text{Volume of array}}{\text{Volume of fuel}}$$

$$= 152.4\frac{(1.75)^2}{(\pi/4)(0.825)^2}$$

$$= 873 \text{ W/cm}^3 = 8.73 \times 10^8 \text{ W/m}^3$$

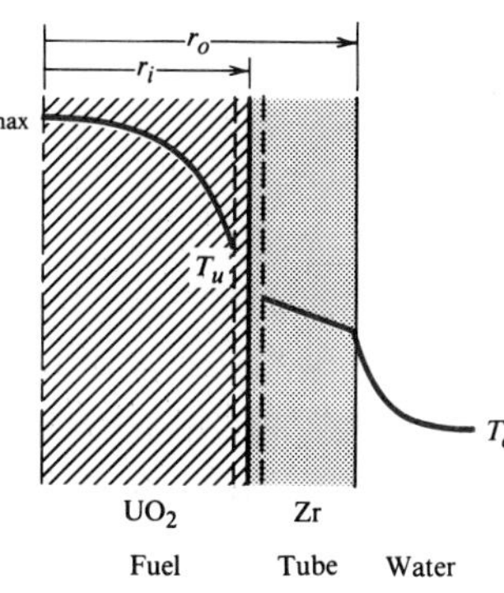

(*b*)

Next we find the temperature of the outer surface of the fuel rod. For unit length of rod, the heat flow across the outer surface of the Zircaloy tube is $\dot{Q} = \dot{Q}_v'''(\pi D^2/4)(1)$:

$$\dot{Q} = (8.73 \times 10^8)(\pi/4)(0.825 \times 10^{-2})^2(1)$$

$$= 46{,}700 \text{ W/m}$$

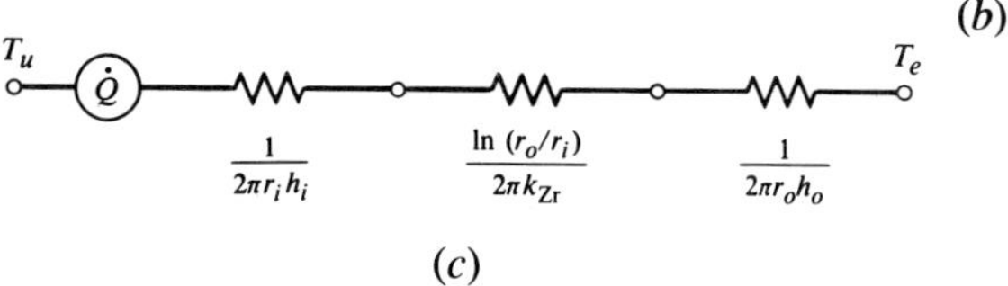

(*c*)

From the thermal circuit, as shown,

$$T_u = T_e + \dot{Q}\sum R$$

$$= 400 + 46{,}700\left[\frac{1}{(2\pi)(0.00413)(6000)} + \frac{\ln(0.485/0.413)}{2\pi k_{Zr}} + \frac{1}{(2\pi)(0.00485)(10^4)}\right]$$

$$= 400 + 46{,}700(0.00642 + 0.0256/k_{Zr} + 0.00328)$$

As a guess, we take the mean temperature of the tube to be 600 K; from Table A.1*b*, the conductivity of Zircaloy-4 is 17.2 W/m K, and

$$T_u = 400 + 46{,}700(0.00642 + 0.00149 + 0.00328) = 923 \text{ K}$$

Now Eq. (2.29) can be used to obtain T_{max}. If we guess a mean temperature of 1500 K for the uranium oxide, Table A.2 gives $k_{UO_2} = 2.6$ W/m K, and

$$T_{max} = T_u + \frac{1}{4}\frac{\dot{Q}_v'''r_i^2}{k_{UO_2}} = 923 + \frac{(8.73 \times 10^8)(0.00413)^2}{(4)(2.6)} = 2355 \text{ K}$$

To check if our guessed mean temperatures are appropriate, we first determine the mean temperature of the tube. From the thermal circuit,

$$\overline{T}_{tube} \simeq 400 + (923 - 400)\frac{(0.00328 + 0.00149/2)}{(0.00642 + 0.00149 + 0.00328)} = 588 \text{ K}$$

which is close enough to our guess of 600 K. The mean temperature of the fuel rod is

$$\overline{T}_{UO_2} \simeq \frac{923 + 2355}{2} = 1639 \text{ K}$$

and at this temperature, $k_{UO_2} = 2.5$ W/m^2 K. The new value of T_{max} is

$$T_{max} = 923 + \frac{(8.73 \times 10^8)(0.00413)^2}{(4)(2.5)} = 2412 \simeq 2400 \text{ K}$$

Comments

1. Since the k-value of UO_2 is given to only two significant figures, no further iteration is warranted.
2. Notice that the conductivity of the zirconium alloy Zircaloy-4 is lower than that of pure zirconium.
3. The largest thermal resistance in the circuit is at the fuel-cladding interface. The accuracy of the result depends primarily on our ability to obtain a reliable value of h_i. In fact, it could be argued that the second iteration for $T_{\max}$ was unwarranted due to uncertainty in the value of h_i.

2.4 FINS

Heat transfer from a system can be increased by extending the surface area through the addition of fins. Fins are used when the convective heat transfer coefficient h_c is low, as is often the case for gases such as air, particularly under natural-convection conditions. Common examples are the cooling fins on electronics components, on the cylinders of air-cooled motorcycles and lawnmowers, and on the condenser tubes of a home refrigerator. Figure 2.8 shows a variety of fin configurations. A careful examination of an automobile radiator will show how it is designed to provide a large exterior surface.

Fins are added to increase the h_cA product and hence decrease the convective thermal resistance $1/h_cA$. But the added area is not as efficient as the original surface area since there must be a temperature gradient along the fin to conduct the heat. Thus, for cooling, the average temperature difference $(T_s - T_e)$ is lower on a finned surface compared with the unfinned surface, and an appropriate thermal resistance for a fin is $1/h_cA\eta_f$, where A is the surface area of the fin and η_f is the *efficiency* of the fin $(0 < \eta_f < 1)$. For short fins of high thermal conductivity, η_f is large, but as the fin length increases, η_f decreases. Our objective here is to analyze heat flow in a fin to determine the temperature variation along the fin and, hence, to evaluate its efficiency η_f. Because fins are thin in one direction, it can be assumed that the temperature variation in this direction is negligible; this key assumption allows the conduction along the fin to be treated as if it were one-dimensional, which greatly simplifies the analysis.

2.4.1 The Pin Fin

Simple *pin fins*, such as those used to cool electronic components, will be analyzed to develop the essential concepts of fin theory. The first law is used to derive the governing differential equation, which, when solved subject to appropriate boundary conditions, gives the temperature distribution along the fin. The heat loss from the fin is then obtained and put in dimensionless form as the fin efficiency.

Figure 2.8 Some heat sinks incorporating fins for cooling of standard packages for integrated circuits. (Photograph courtesy of EG&G Wakefield Engineering, Wakefield, Mass.)

Governing Equation and Boundary Conditions

Consider the pin fin shown in Fig. 2.9. The cross-sectional area is $A_c = \pi R^2$ where R is the radius of the pin, and the perimeter $\mathscr{P} = 2\pi R$. Both A_c and R are uniform, that is, they do not vary along the fin in the x direction. The energy conservation principle, Eq. (1.2), is applied to an element of the fin located between x and $x + \Delta x$. Heat can enter and leave the element by conduction along the fin and can also be lost by convection from the surface of the element to the ambient fluid at temperature T_e. The surface area of the element is $\mathscr{P}\Delta x$; thus,

$$qA_c|_x - qA_c|_{x+\Delta x} - h_c\mathscr{P}\Delta x(T - T_e) = 0$$

Dividing by Δx and letting $\Delta x \to 0$ gives

$$-\frac{d}{dx}(qA_c) - h_c\mathscr{P}(T - T_e) = 0 \tag{2.30}$$

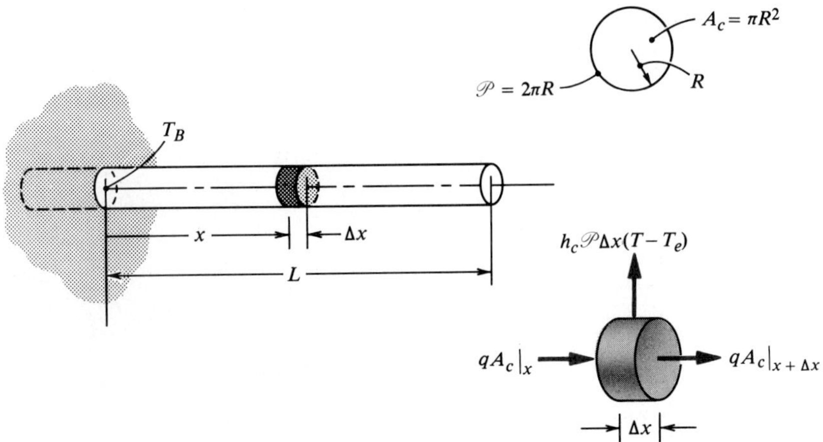

Figure 2.9 A pin fin showing the coordinate system, and an energy balance on a fin element.

For the pin fin, A_c is independent of x; using Fourier's law $q = -k\, dT/dx$ with k constant gives

$$kA_c \frac{d^2T}{dx^2} - h_c \mathscr{P}(T - T_e) = 0 \tag{2.31}$$

which is a second-order ordinary differential equation for $T = T(x)$. Notice that modeling of the conduction along the fin as one-dimensional has caused the convective heat loss from the sides of the fin to appear in the differential equation, in contrast to the problems dealt with in Section 2.3, where convection became involved as a boundary condition.

Next, boundary conditions for Eq. (2.31) must be specified. Since we wish to examine the performance of the fin itself, it is appropriate to take its base temperature as known; that is,

$$T|_{x=0} = T_B \tag{2.32}$$

At the other end, the fin loses heat by Newton's law of cooling:

$$-A_c k \left.\frac{dT}{dx}\right|_{x=L} = A_c h_c (T|_{x=L} - T_e) \tag{2.33a}$$

where the convective heat transfer coefficient here is, in general, different from the one for the sides of the fin because the geometry is different. However, because the area of the end, A_c, is small compared to the side area, $\mathscr{P}L$, the heat loss from the end is correspondingly small and usually can be ignored. Then Eq. (2.33a) becomes

$$\left.\frac{dT}{dx}\right|_{x=L} \simeq 0 \tag{2.33b}$$

and this boundary condition is simpler to use than Eq. (2.33a). An even simpler result can be obtained if the temperature distribution along the fin is assumed identical to that for an infinitely long fin, for which the appropriate boundary condition is

$$\lim_{x \to \infty} T = T_e \tag{2.33c}$$

Figure 2.10 illustrates these boundary conditions.

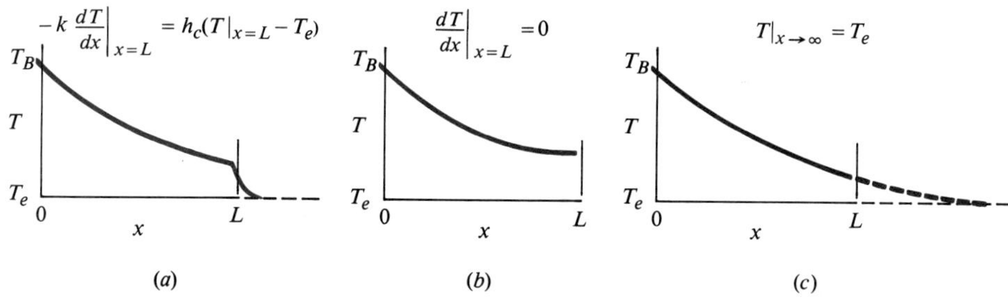

Figure 2.10 Three tip boundary conditions for the pin fin analysis. (*a*) Heat loss by convection. (*b*) Insulated tip. (*c*) Infinitely long fin.

Temperature Distribution

We will use Eq. (2.33*b*) for the second boundary condition as a compromise between accuracy and simplicity of the result. For mathematical convenience, let $\theta = T - T_e$ and $\beta^2 = h_c \mathscr{P}/kA_c$; then Eq. (2.31) becomes

$$\frac{d^2\theta}{dx^2} - \beta^2\theta = 0 \tag{2.34}$$

For β a constant, Eq. (2.34) has the solution

$$\theta = C_1 e^{\beta x} + C_2 e^{-\beta x}$$

or

$$\theta = B_1 \sinh \beta x + B_2 \cosh \beta x$$

The second form proves more convenient; thus, we have

$$T - T_e = B_1 \sinh \beta x + B_2 \cosh \beta x \tag{2.35}$$

Using the two boundary conditions, Eqs. (2.32) and (2.33*b*) give two algebraic equations for the unknown constants B_1 and B_2,

$$T_B - T_e = B_1 \sinh(0) + B_2 \cosh(0); \qquad B_2 = T_B - T_e$$

$$\left.\frac{dT}{dx}\right|_{x=L} = \beta B_1 \cosh \beta L + \beta B_2 \sinh \beta L = 0; \qquad B_1 = -B_2 \tanh \beta L$$

Substituting B_1 and B_2 in Eq. (2.35) and rearranging gives the temperature distribution as

$$\frac{T - T_e}{T_B - T_e} = \frac{\cosh \beta(L - x)}{\cosh \beta L}, \qquad \text{where } \beta = \left(\frac{h_c \mathscr{P}}{kA_c}\right)^{1/2} \tag{2.36}$$

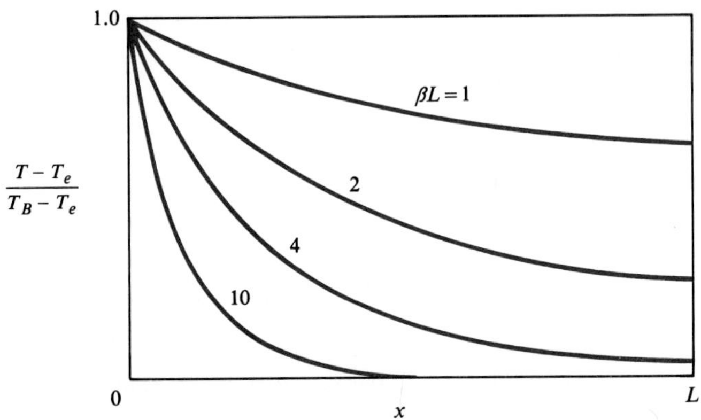

Figure 2.11 Fin temperature distributions calculated from Eq. (2.36).

Figure 2.11 shows a plot of Eq. (2.36). When β is small—for example, if the fin is made of aluminum and has a high thermal conductivity—the temperature T does not drop much below the base temperature T_B. For β large, T approaches the fluid temperature at the tip of the fin.[2]

Heat Loss

The heat dissipated from the fin can be found by integrating the heat loss over the side surface of the fin (there is no heat loss from the fin tip):

$$\dot{Q} = \int_0^L h_c \mathscr{P}(T - T_e)dx \tag{2.37}$$

with T obtained from Eq. (2.36). Substituting gives

$$\dot{Q} = \frac{h_c \mathscr{P}(T_B - T_e)}{\cosh \beta L} \int_0^L \cosh \beta(L - x)dx$$

To simplify the integration, let $\xi = \beta(L - x)$; then $dx = -d\xi/\beta$ and

[2] A_c in $\beta = (h_c \mathscr{P}/kA_c)^{1/2}$ is the cross-sectional area of the fin. The subscript c denotes "cross section" and not "convection" as in the heat transfer coefficient h_c. The area for convective heat loss is the surface area of the fin, $\mathscr{P}L$.

$$\begin{aligned}\dot{Q} &= \frac{(h_c\mathscr{P}/\beta)(T_B - T_e)}{\cosh \beta L}\left[-\int_{\beta L}^{0} \cosh \xi \, d\xi\right] \\ &= \frac{h_c\mathscr{P}}{\beta}(T_B - T_e)\left[-\frac{\sinh 0 - \sinh \beta L}{\cosh \beta L}\right] \\ &= \frac{h_c\mathscr{P}}{\beta}(T_B - T_e)\tanh \beta L \end{aligned} \qquad \textbf{(2.38)}$$

A less obvious alternative, but usually a more convenient way to find the heat dissipation, is to apply Fourier's law at the base of the fin:

$$\dot{Q} = -kA_c \left.\frac{dT}{dx}\right|_{x=0} \qquad \textbf{(2.39)}$$

Substituting from Eq. (2.36),

$$\begin{aligned}\dot{Q} &= -kA_c(T_B - T_e)\frac{[(d/dx)\cosh \beta(L - x)]_{x=0}}{\cosh \beta L} \\ &= -kA_c(T_B - T_e)\frac{[-\beta \sinh \beta(L - x)]_{x=0}}{\cosh \beta L} \\ &= kA_c\beta(T_B - T_e)\tanh \beta L \end{aligned} \qquad \textbf{(2.40)}$$

Since $\beta^2 = h_c\mathscr{P}/kA_c$, Eqs. (2.38) and (2.40) give the same result, which is to be expected since there is no heat loss from the end of the fin.

Fin Efficiency

Let us now put Eq. (2.38) in **dimensionless** form by dividing through by $h_c\mathscr{P}L(T_B - T_e)$:

$$\frac{\dot{Q}}{h_c\mathscr{P}L(T_B - T_e)} = \frac{1}{\beta L}\tanh \beta L \qquad \textbf{(2.41)}$$

The dimensions of the left-hand side of this equation are [W]/[W/m^2 K][m][m][K] = 1, as desired. The right-hand side must also be dimensionless since β has dimensions [m^{-1}] and the group βL has dimensions [m^{-1}][m] = 1. (Of course, βL must be dimensionless to be the argument of the tanh function.) Now $h_c\mathscr{P}L(T_B - T_e)$ is the rate at which heat would be dissipated if the entire fin surface were at the base temperature T_B; in reality, there is a decrease in temperature along the fin, and the actual heat loss is less. Thus, the left-hand side of Eq. (2.41) can be viewed as the ratio of the actual heat loss to the maximum possible and is termed the **fin efficiency**, η_f. The right-hand side is a function of the dimensionless parameter βL only; we will set $\beta L = \chi$ as a *fin parameter*, and then Eq. (2.41) can be written in the compact form

$$\eta_f = \frac{\tanh \chi}{\chi} \qquad \textbf{(2.42)}$$

When χ is small, η_f is near unity; when χ is larger than about 4, $\tanh\chi \simeq 1$ and $\eta_f \simeq 1/\chi$. Since $\chi = \beta L = (h_c \mathscr{P} L^2/kA_c)^{1/2}$, a small value of χ corresponds to relatively short, thick fins of high thermal conductivity, whereas large values of χ correspond to relatively long, thin fins of poor thermal conductivity. When χ is small, T does not fall much below T_B, and the fin is an efficient dissipator of heat. However, it is most important to understand that a thick fin with an efficiency of nearly 100% usually is not optimal from the viewpoint of heat transferred per unit weight or unit cost. The concept of fin efficiency refers only to the ability of the fin to transfer heat per unit area of exposed surface. Figure 2.12 shows a plot of Eq. (2.42). Use of dimensionless parameters has allowed the heat dissipation to be given by a single curve: different curves are not required for fins of various materials or lengths or for different values of the heat transfer coefficient. Likewise, storage of this information in a computer software package is efficient.

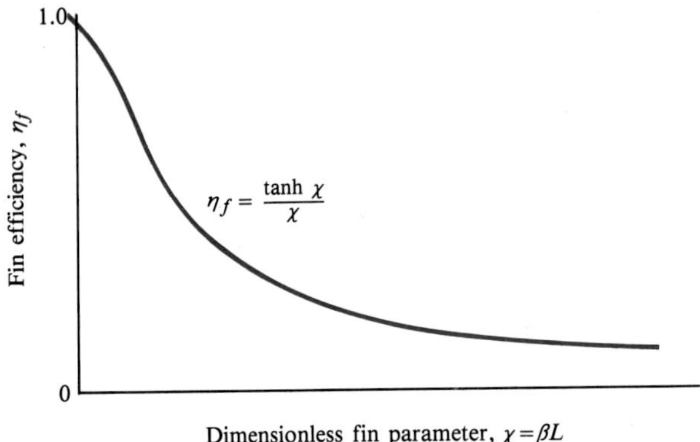

Figure 2.12 Efficiency of a pin fin as given by Eq. (2.42).

Straight Rectangular Fins

Although the pin fin shown in Fig. 2.9 was used for the purposes of this analysis, the results apply to any fin with a cross-sectional area A_c and perimeter $\mathscr{P}$ constant along the fin. The straight rectangular fin shown in Fig. 2.13 has a width W and thickness $2t$. The cross-sectional area is $A_c = 2tW$, and the perimeter is $\mathscr{P} = 2(W + 2t)$. For $W \gg t$, the ratio $\mathscr{P}/A_c$ is simply $1/t$, and $\beta = (h_c/kt)^{1/2}$.

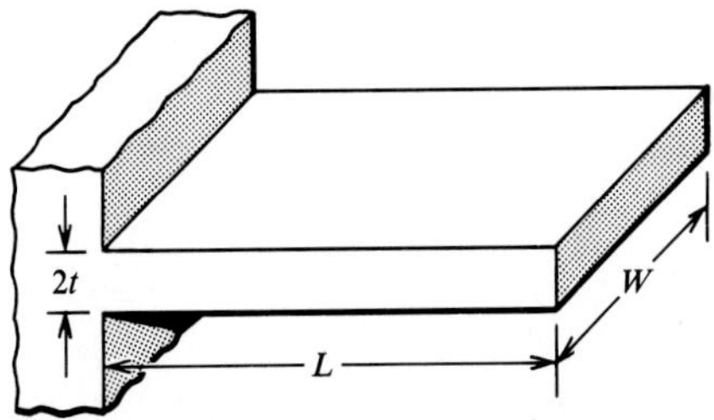

Figure 2.13 A straight rectangular fin.

Computer Program FIN1

The program FIN1 calculates the temperature distribution, fin efficiency, and base heat flow of straight rectangular fins. There are three options for the tip boundary condition: (1) infinitely long fin, (2) insulated, and (3) convective heat loss. The analysis for option 2 was given above; the analyses for options 1 and 3 are given as Exercises 2–30 and 2–31, respectively. For all three options, η_f is defined in terms of an isothermal fin heat loss of $\dot{Q} = h_c \mathscr{P} L(T_B - T_e)$. Use of FIN1 is illustrated in the example that follows.

EXAMPLE 2.5 Fins to Cool a Transistor

An array of eight aluminum alloy fins, each 3 mm wide, 0.4 mm thick, and 40 mm long, is used to cool a transistor. When the base is at 340 K and the ambient air is at 300 K, how much power do they dissipate if the combined convection and radiation heat transfer coefficient is estimated to be 8 W/m² K? The alloy has a conductivity of 175 W/m K.

Solution

Given: Aluminum fins to cool a transistor.

Required: Power dissipated by 8 fins.

Assumptions: 1. Heat transfer coefficient constant along fin.
2. Heat loss from fin tip negligible.

For one fin,

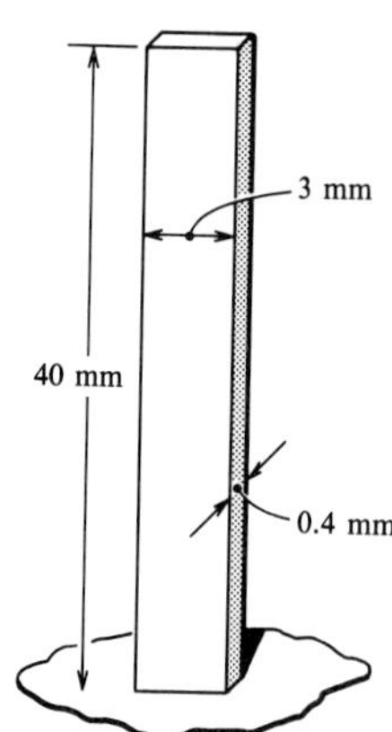

$$A_c = (0.003)(0.0004) = 1.2 \times 10^{-6} \text{ m}^2$$

$$\mathscr{P} = 2(0.003 + 0.0004) = 6.8 \times 10^{-3} \text{ m}$$

$$\beta^2 = \frac{h\mathscr{P}}{kA_c}$$

$$= \frac{(8.0 \text{ W/m}^2 \text{ K})(6.8 \times 10^{-3} \text{ m})}{(175 \text{ W/m K})(1.2 \times 10^{-6} \text{ m}^2)}$$

$$= 259 \text{ m}^{-2}$$

$$\beta = 16.1 \text{ m}^{-1}$$

$$\chi = \beta L = (16.1 \text{ m}^{-1})(0.040 \text{ m}) = 0.644$$

Substituting in Eq. (2.42),

$$\eta_f = \frac{1}{0.644} \tanh(0.644) = \frac{1}{0.644} \frac{e^{2(0.644)} - 1}{e^{2(0.644)} + 1} = 0.881$$

The side surface area of one fin is $\mathscr{P}L = (6.8 \times 10^{-3})(0.040) = 2.72 \times 10^{-4} \text{ m}^2$. If each fin were 100% efficient, it would dissipate

$$h(\mathscr{P}L)(T_B - T_e) = (8)(2.72 \times 10^{-4})(340 - 300) = 8.70 \times 10^{-2} \text{ W}$$

Since the fins are only 88.1% efficient,

$$\dot{Q} = (0.881)(8.70 \times 10^{-2}) = 7.67 \times 10^{-2} \text{ W}$$

For 8 fins, $\dot{Q}_{\text{total}} = (8)(7.67 \times 10^{-2}) = 0.613$ W.

Solution using FIN1

The required input is:

Boundary condition = 2
Half-thickness, length, and width = 0.0002, 0.040, 0.003
Thermal conductivity = 175
Heat transfer coefficient = 8
Base temperature and ambient temperature = 340, 300
x-range for plot = 0.0, 0.04

FIN1 gives the output:

$\eta_f = 0.881$
$\dot{Q} = 7.67 \times 10^{-2}$ (watts)

Comments

1. Any consistent system of units can be used with FIN1. Since SI units were used here, the heat flow is in watts.
2. Notice the use of $h = h_c + h_r$ to account for radiation.

2.4.2 Fin Resistance and Surface Efficiency

It is useful to have an expression for the **thermal resistance** of a pin fin for use in thermal circuits. Equation (2.38) can be rewritten as

$$\dot{Q} = \frac{T_B - T_e}{1/[(h_c\mathcal{P}/\beta)\tanh \beta L]} \tag{2.43}$$

Thus, the thermal resistance of a pin fin is

$$R_{\text{fin}} = \frac{1}{(h_c\mathcal{P}/\beta)\tanh \beta L} = \frac{1}{h_c\mathcal{P}L\eta_f} \tag{2.44}$$

Notice that this thermal resistance accounts for both conduction along the fin and convection into the fluid. There are two parallel paths for heat loss from a finned surface—one through the fins and one through the area between the fins, as shown in Fig. 2.14. The respective conductances are thus additive; however, quite often the heat loss through the area between the fins is negligible.

The *total surface efficiency* η_t of a surface with fins of fin efficiency η_f is obtained by adding the unfinned portion of the surface area at 100% efficiency to the surface area of the fins at efficiency η_f:

$$A\eta_t = (A - A_f) + \eta_f A_f \tag{2.45}$$

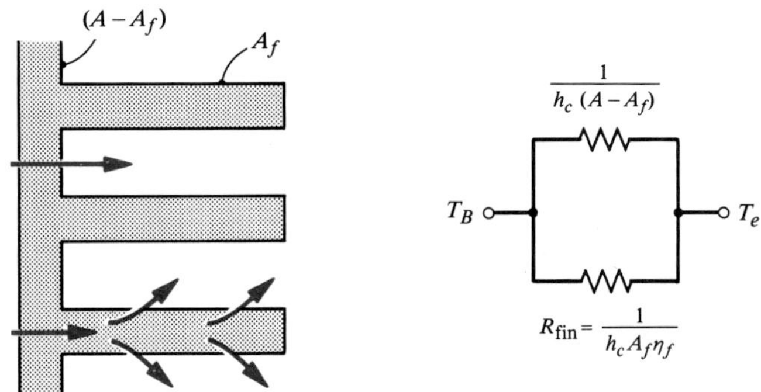

Figure 2.14 A finned surface showing the parallel paths for heat loss.

where A_f is the surface area of the fins and A is the total heat transfer surface area, including the fins and exposed tube or other surface. Solving for η_t,

$$\eta_t = 1 - \frac{A_f}{A}(1 - \eta_f) \tag{2.46}$$

The corresponding thermal resistance of the finned surface is then

$$R = \frac{1}{h_c A \eta_t} \tag{2.47}$$

Design calculations for the finned surfaces used in heat exchangers, such as automobile radiators, are conveniently made using Eq. (2.47).

2.4.3 Other Fin Type Analyses

The key feature of the fin analysis presented in Section 2.4.1 was that the thinness of the fin allowed us to ignore the temperature variation across the fin and, hence, to account for the convective loss from the surface directly in the differential equation for $T(x)$. The same assumption is valid for *extended surfaces* unrelated to cooling fins, and the results obtained in Section 2.4.1 are directly or indirectly applicable to these surfaces.

Sometimes it is quite obvious that the situation is similar to that for a cooling fin. For example, Fig. 2.15 shows a thermocouple installation used to measure the temperature of a hot air stream. The thermocouple junction is at a lower temperature than the air since the conduction heat flow along the thermocouple wires to the colder

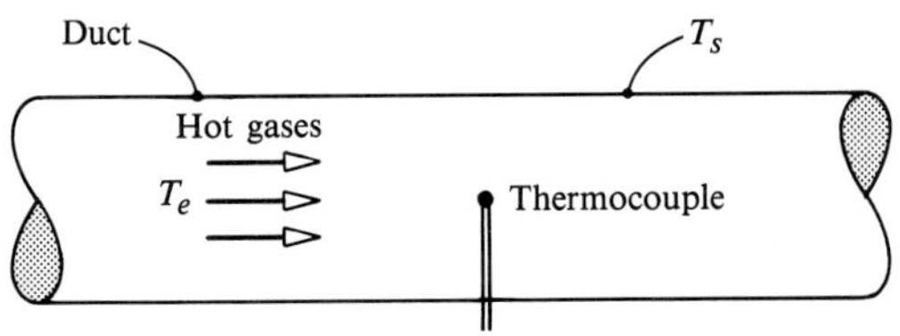

Figure 2.15 A thermocouple immersed in a fluid stream.

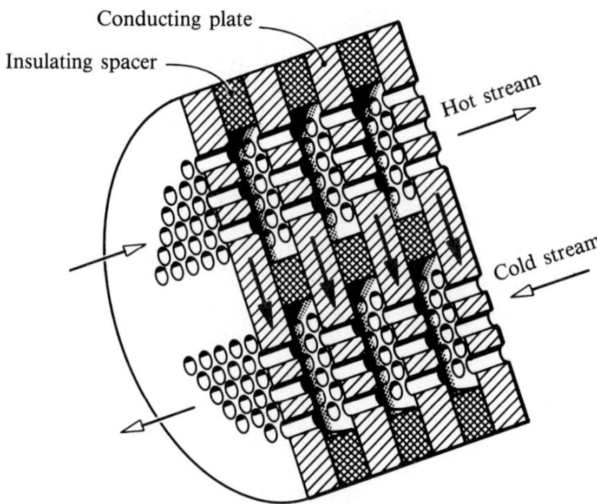

Figure 2.16 An element of a perforated-plate heat exchanger showing heat conduction along the plates.

wall must be balanced by convection from the air. The temperature variation along the thermocouple is identical to that for a pin fin, so Eq. (2.36), with appropriate choices for the kA_c product, can be used to determine the error expected in the thermocouple reading.

Sometimes it is not obvious that the situation resembles that for a cooling fin, yet the assumption of negligible temperature variation in the thin direction of a wire or plate gives a differential equation similar to Eq. (2.30). The perforated plates in the heat exchanger shown in Fig. 2.16 can be treated as fins since the temperature variation across the plates is small compared to the temperature variation along the plates between the hot and cold streams. The copper conductors on the circuit board shown in Fig. 2.17 can be treated as fins, as can the circuit board between the conductors. The examples that follow relate to Figs. 2.15 and 2.16, and Exercise 2–52 is based on Fig. 2.17.

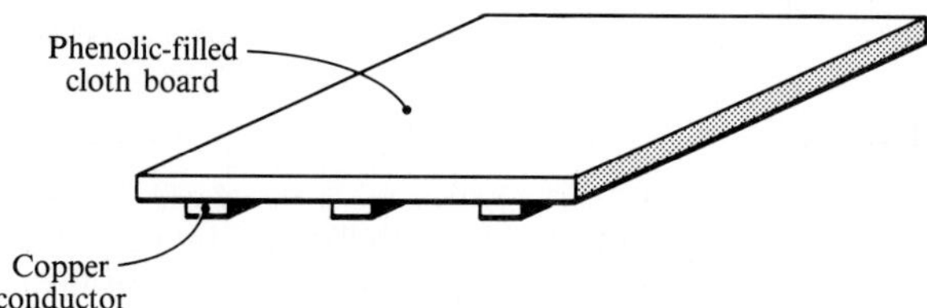

Figure 2.17 Copper conductors on a circuit board.

EXAMPLE 2.6 Error in Thermocouple Readings

Duplex thermocouple leads have both wires embedded in polyvinyl electrical insulation. One available size has wires of 0.25 mm diameter in insulation with an outside perimeter of 1.5 mm and is to be used in the situation depicted in Fig. 2.15. The air temperature is 350 K, and the wall temperature is 300 K. What length of immersion is required for the error in the thermocouple reading to be 0.1 K when the heat transfer coefficient on the perimeter is approximately 30 W/m^2 K? The wires are (i) copper and constantan (type T), (ii) iron and constantan (type J), and (iii) chromel and alumel (type K).

Solution

Given: Duplex thermocouple leads for types T, J, and K thermocouples.

Required: Length of immersion for a specified error.

Assumptions: Temperature variation across lead is small compared to the variation along the lead.

This is a fin-type problem since the temperature variation across the lead is small compared to the 50 K variation along the lead. The effective β^2 is

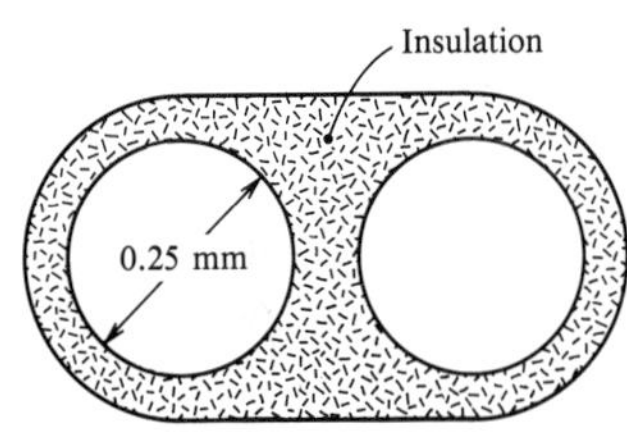

$$\beta^2 = \frac{h_c \mathscr{P}}{\sum kA_c}$$

where $h_c = 30$ W/m^2 K, $\mathscr{P} = 1.5 \times 10^{-3}$ m, and $\sum kA_c$ must be evaluated for the thermal resistances of the two wires and the insulation in *parallel*. For each wire, $A_c = (\pi/4)(0.25 \times 10^{-3})^2 = 4.91 \times 10^{-8}$ m^2, and for the insulation $A_c \simeq 10 \times 10^{-8}$ m^2. Thermal conductivity and kA values are given in the following table.

Wire Material	k W/m K	kA_c W m/K
Copper	385	19×10^{-6}
Constantan (55% Cu, 45% Ni)	23	1.1×10^{-6}
Iron	73	3.6×10^{-6}
Chromel-P (90% Ni, 10% Cr)	17	0.83×10^{-6}
Alumel (95% Ni, 2% Mn, 2% Al)	48	2.36×10^{-6}
Insulation	0.1	0.01×10^{-6}

The contribution of the insulation to $\sum kA_c$ is seen to be negligible; hence its precise shape or composition is unimportant. The temperature of the thermocouple junction (located at $x = L$) is given by Eq. (2.36) as

$$\frac{T_L - T_e}{T_B - T_e} = \frac{-0.1}{300 - 350} = \frac{1}{\cosh \beta L}$$

Solving,

$$\cosh \beta L = 500$$

$$\beta L = 6.91 \text{ from cosh tables, or use } \cosh x = (1/2)(e^x + e^{-x})$$

$$L = 6.91/\beta$$

Evaluating β for each thermocouple pair gives the following results:

Thermocouple Type	$\sum kA_c$ W m/K	β m^{-1}	L cm
T	20.1×10^{-6}	47.3	14.6
J	4.7×10^{-6}	97.8	7.1
K	3.2×10^{-6}	119	5.8

Comments

1. Type T thermocouples are to be avoided when conduction along the wires may cause a significant error.
2. There are other criteria for choosing thermocouple pairs, including operating temperature range, emf output, and corrosion resistance.
3. Type T thermocouples are widely used because the component wires are relatively free from inhomogeneities; hence, calibration charts are reliable.

EXAMPLE 2.7 A Perforated Plate Heat Exchanger

Perforated-plate heat exchangers are used in cryogenic refrigeration systems. The fluids flow through high-conductivity perforated plates separated by insulating spacers. Heat is transferred from the hot stream to the cold stream by conduction along the plates, as shown in Fig. 2.16. A counterflow helium to helium unit has 0.5 mm–thick rectangular aluminum plates with 0.9 mm–diameter holes in a square array of pitch 1.3 mm. The plate length exposed to each stream is 20 mm, and the plate width is 80 mm. The spacer is 4 mm wide and 0.86 mm thick. If the heat transfer coefficient is 400 W/m^2 K for both streams, calculate the overall heat transfer coefficient. Take $k = 200$ W/m K for the aluminum.

Solution

Given: A perforated-plate heat exchanger.

Required: Overall heat transfer coefficient U.

Assumptions:
1. The plates are thin enough for a fin-type analysis to be valid.
2. Heat flow along the plates is one-dimensional.

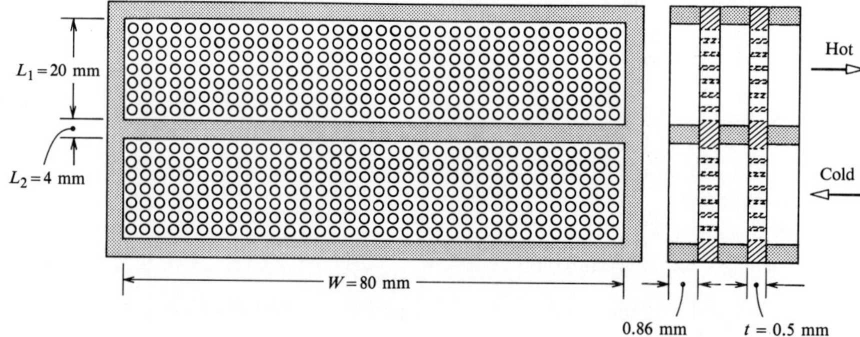

At first sight, the heat transfer process might appear complicated. However, after a little thought, it should become clear that heat is transferred from the hot to the cold stream through three resistances in series: a fin on the hot side, the aluminum plate between the spacers, and a fin on the cold side. Referring to the sketch, the product of overall heat transfer coefficient and area for a single plate is

$$\frac{1}{UA} = \frac{1}{h_c \mathscr{P} L_1 \eta_f} + \frac{1}{ktW/L_2} + \frac{1}{h_c \mathscr{P} L_1 \eta_f}$$

With the thermal resistances identified, the remaining complexity is associated with calculating the various relevant geometric parameters. These are the area A for convective heat transfer to the plates, and the perimeter $\mathscr{P}$ and cross-sectional area A_c, which appear in the fin parameter β. The calculation will be broken down into four steps.

1. *The convective heat transfer area A.* The area A should include both sides of the plate as well as the inside area of the perforations. The number of perforations per unit area is the reciprocal of the pitch p squared; thus, for either stream,

$$\begin{aligned} A &= \left[2WL_1 - \frac{2WL_1}{p^2}\left(\frac{\pi d^2}{4}\right)\right] + \frac{WL_1}{p^2}(\pi\, dt) \\ &= WL_1\left[2 - \frac{\pi}{2}\left(\frac{d}{p}\right)^2 + \frac{\pi\, dt}{p^2}\right] \\ &= (0.08)(0.02)\left[2 - \frac{\pi}{2}\left(\frac{9}{13}\right)^2 + \frac{(\pi)(9)(5)}{(13)^2}\right] \\ &= 0.00333\ \text{m}^2 \end{aligned}$$

2. *The fin perimeter $\mathscr{P}$ and cross-sectional area A_c.* The portions of the plate exposed to the gas flow are fins of efficiency η_f, and fin parameter $\beta^2 = h_c\mathscr{P}/kA_c$. Both $\mathscr{P}$ and A_c will be approximated as average values. The perimeter $\mathscr{P}$ is then just the surface area per unit length of fin:

$$\mathscr{P} = \frac{A}{L_1} = \frac{0.00333}{0.02} = 0.167\ \text{m}$$

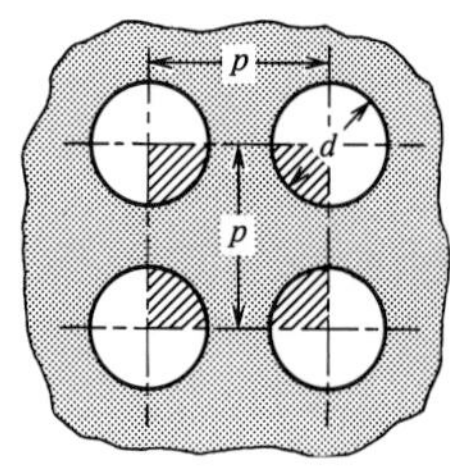

The cross-sectional area A_c for heat conduction along the fin will be taken as $A_c = Wt(1 - \varepsilon_v)$ where ε_v is the volume void fraction of the plate, which is equal to the cross-sectional area of the holes per unit area of plate:

$$\varepsilon_v = \frac{4(\pi d^2/16)}{p^2} = \frac{\pi}{4}\left(\frac{9}{13}\right)^2 = 0.376$$

$$A_c = (0.08)(0.0005)(1 - 0.376) = 2.50 \times 10^{-5}\ \text{m}^2$$

(Actually, heat conduction along the fin is two-dimensional since the heat flow is "squeezed" between the perforations. Exact analysis of the problem shows that the use of this estimate of the effective area A_c is conservative; the true value is a few percent higher.)

3. *The fin efficiency* η_f. The fin parameter β can now be calculated:

$$\beta^2 = \frac{h_c\mathscr{P}}{kA_c} = \frac{(400)(0.167)}{(200)(2.50 \times 10^{-5})} = 1.336 \times 10^4$$

$$\beta = 115.6; \qquad \chi = \beta L_1 = (115.6)(0.02) = 2.31$$

If we assume that the fin efficiency is given by Eq. (2.42):

$$\eta_f = \tanh \chi/\chi = 0.9805/2.31 = 0.424$$

4. *The overall heat transfer coefficient.* Finally, we calculate the thermal resistances of the fins and the spacer as

$$\frac{1}{h_c\mathscr{P}L_1\eta_f} = \frac{1}{(400)(0.167)(0.02)(0.424)} = 1.765 \text{ K/W}$$

$$\frac{1}{ktW/L_2} = \frac{1}{(200)(0.0005)(0.08)/(0.004)} = 0.50 \text{ K/W}$$

The overall heat transfer coefficient is then obtained from

$$\frac{1}{UA} = \frac{1}{h_c\mathscr{P}L_1\eta_f} + \frac{1}{ktW/L_2} + \frac{1}{h_c\mathscr{P}L_1\eta_f}$$

$$= 1.765 + 0.50 + 1.765 = 4.03 \text{ K/W}$$

$$U = \frac{1}{(4.03)(0.00333)} = 74.5 \text{ W/m}^2 \text{ K}$$

Comments

1. Most of the work in this solution involves geometry calculations; the heat transfer problem is a simple one of three thermal resistances in series.

2. Perhaps the problem has been oversimplified. In fact, Eq. (2.42) does not give the correct fin efficiency. As the gas flows through the exchanger, its temperature will not remain uniform. Low-velocity laminar flow is typical for these exchangers; hence, the gas does not mix between plates. The gas closer to center heats more rapidly since the temperature difference is larger; after a few plates, the temperature difference between the plate and gas, $T - T_e$, becomes constant along the plate. The plate can still be viewed as a fin since the heat flow along the fin is essentially one-dimensional, but Eq. (2.42) is no longer valid since the analysis assumed that T_e was constant along the fin. The new analysis is required as Exercise 2–69 and gives a 15% lower fin efficiency.

2.4.4 Fins of Varying Cross-Sectional Area

Many fin profiles encountered in practice have a cross-sectional area A_c that varies along the length of the fin, and the analysis for these is more difficult than for the simple pin fin of Section 2.4.1. In this section, the annular (or radial) fin is analyzed. The results for a variety of other fin profiles are also given.

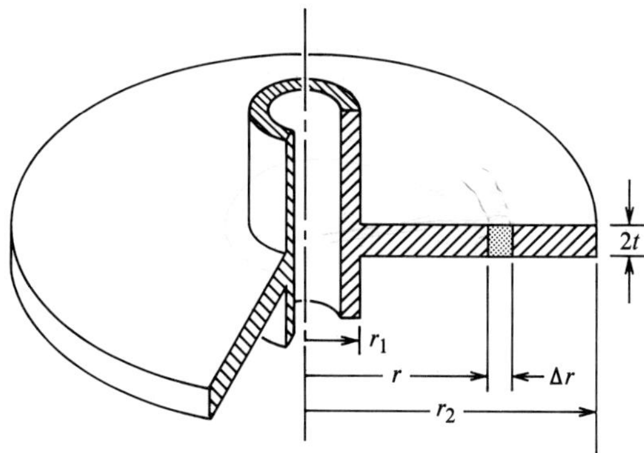

Figure 2.18 An annular fin of uniform thickness.

Governing Equation and Boundary Conditions

Figure 2.18 shows an annular fin of uniform thickness $2t$, as might be found on the outside of a tube. Such fins have extensive application in liquid-gas heat exchangers such as air-cooled evaporators for refrigeration systems. The energy conservation principle, Eq. (1.2), applied to a differential element between radii r and $r + \Delta r$ requires that

$$q(2\pi r)(2t)\big|_r - q(2\pi r)(2t)\big|_{r+\Delta r} - h_c(2)2\pi r\Delta r(T - T_e) = 0$$

Dividing through by $4\pi\Delta r$ and letting $\Delta r \to 0$,

$$-\frac{d}{dr}(rtq) - h_c r(T - T_e) = 0$$

Substituting Fourier's law, $q = -k(dT/dr)$, and dividing by (tk) gives

$$\frac{d}{dr}\left(r\frac{dT}{dr}\right) - \frac{h_c r}{tk}(T - T_e) = 0 \qquad \textbf{(2.48)}$$

Note that this equation also could have been obtained directly from Eq. (2.30) by substituting $A_c = 4\pi rt$ and $\mathcal{P} = 2\pi r$, since A_c and $\mathcal{P}$ had not yet been given constant values. Equation (2.48) can be rearranged as

$$r^2\frac{d^2T}{dr^2} + r\frac{dT}{dr} - \beta^2 r^2(T - T_e) = 0$$

where $\beta^2 = h_c/tk$; then introducing new variables $z = \beta r$ and $\theta = (T - T_e)/(T_B - T_e)$ gives

$$z^2\frac{d^2\theta}{dz^2} + z\frac{d\theta}{dz} - z^2\theta = 0 \qquad \textbf{(2.49)}$$

Suitable boundary conditions are as used for the pin fin in Section 2.4.1, that is, a specified base temperature and zero heat flow through the tip of the fin,

$$r = r_1: \quad T = T_B$$

$$r = r_2: \quad \frac{dT}{dr} = 0$$

or

$$z = z_1 = \beta r_1: \quad \theta = 1 \tag{2.50a}$$

$$z = z_2 = \beta r_2: \quad \frac{d\theta}{dz} = 0 \tag{2.50b}$$

Temperature Distribution

Equation (2.49) is a modified Bessel's equation of zero order and has the solution

$$\theta = C_1 I_0(z) + C_2 K_0(z) \tag{2.51}$$

where I_0 and K_0 are zero-order modified Bessel functions of the first and second kinds, respectively. Properties of Bessel functions are given in Appendix B. Applying the boundary conditions and using the differentiation formulas in Appendix B,

$$1 = C_1 I_0(z_1) + C_2 K_0(z_1)$$

$$0 = C_1 I_1(z_2) - C_2 K_1(z_2)$$

since $dI_0/dz = I_1$, $dK_0/dz = -K_1$, where I_1 and K_1 are first-order modified Bessel functions. Solving for C_1 and C_2,

$$C_1 = \frac{K_1(z_2)}{F(z_1, z_2)}, \qquad C_2 = \frac{I_1(z_2)}{F(z_1, z_2)}$$

where

$$F(z_1, z_2) \equiv I_0(z_1)K_1(z_2) + I_1(z_2)K_0(z_1)$$

Substitution in Eq. (2.51) gives the temperature distribution along the fin.

Heat Loss and Efficiency

Next we obtain the heat dissipated by the fin and its efficiency. The heat flow through the base of the fin is

$$\dot{Q} = -kA_c \left.\frac{dT}{dr}\right|_{r=r_1} = -k(2\pi r_1)(2t)(T_B - T_e)\beta \left.\frac{d\theta}{dz}\right|_{z=z_1}$$

since $dT = (T_B - T_e)d\theta$ and $dr = dz/\beta$. Differentiating Eq. (2.51) gives

$$\left.\frac{d\theta}{dz}\right|_{z=z_1} = C_1 I_1(z_1) - C_2 K_1(z_1)$$

and hence

$$\dot{Q} = k(4\pi r_1 t)(T_B - T_e)\beta[C_2 K_1(\beta r_1) - C_1 I_1(\beta r_1)]$$

The maximum possible heat loss is from an isothermal fin and is simply the product of the heat transfer coefficient, surface area, and temperature difference: $(h_c)(2)(\pi r_2^2 - \pi r_1^2)(T_B - T_e)$. The fin efficiency η_f is the ratio of the actual heat loss to that for an isothermal fin and can be rearranged as

$$\eta_f = \frac{(2r_1/\beta)}{(r_2^2 - r_1^2)} \frac{K_1(\beta r_1)I_1(\beta r_2) - I_1(\beta r_1)K_1(\beta r_2)}{K_0(\beta r_1)I_1(\beta r_2) + I_0(\beta r_1)K_1(\beta r_2)} \tag{2.52}$$

Equation (2.52) can be evaluated using the tables of Bessel functions given in Appendix B.

Other Fin Profiles: Computer Program FIN2

A variety of fin profiles are used in practice. Table 2.2 gives the efficiency of a selection of straight fins, annular fins, and spines. To facilitate the calculation of heat flow and fin mass, the surface area per unit width S' and profile area A_p are given for straight fins, and the surface area S and volume V are given for annular fins and spines. The efficiencies for items 1, 5, 6, and 7 were obtained using the boundary condition of zero heat flow through the tip. For thick rectangular fins, a simple approximate rule to account for heat loss from the tip is to add half the fin thickness to the fin length L for the straight fin and to the outer radius r_2 for the annular fin.

The computer program FIN2 calculates the efficiency, base heat flow, and mass for the 10 fin profiles listed in Table 2.2. For straight fins, the heat flow and mass are per unit width of fin. Its use is illustrated in Examples 2.8 and 2.9.

Cooling Fin Design

The proper design of cooling fins is an optimization problem: usually the objective is to minimize the amount of material in the fins in order to minimize either weight or cost. Exercises 2–39 and 2–71 show how optimal dimensions can be found for a given fin shape, and Exercises 2–63 and 2–65 illustrate that there is an optimal fin shape. The engineer is also free to choose the fraction of area covered by fin "footprints." This is a more difficult problem because as the fins are moved closer to each other, the value of the heat transfer coefficient h_c changes in a complicated way. There is always the question of whether fins should be provided at all. Exercise 2–62 shows that when the heat transfer coefficient is large, adding fins can actually reduce the heat loss. The conduction resistance in the fin can exceed the decrease in convective resistance due to the increased surface area. A useful rule is not to use fins unless $k/h_c t > 5$.

Table 2.2 Fins of various shapes: Efficiency, surface area per unit width (S'), and profile area (A_p) for straight fins; surface area (S) and volume (V) for annular fins and spines: $\beta = (h_c/kt)^{1/2}$.

Straight Fins		
1. Rectangular $y = t$	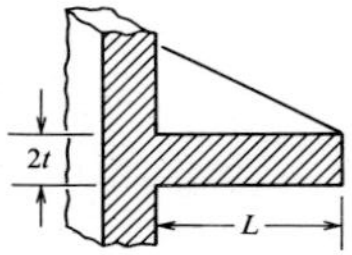	$\eta_f = \dfrac{1}{\beta L}\tanh\beta L \qquad S' = 2L, \qquad A_p = 2tL$
2. Parabolic $y = t(1-x/L)^{1/2}$	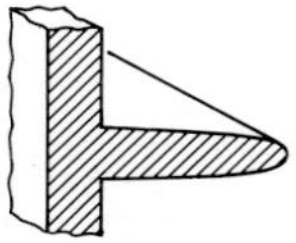	$\eta_f = \dfrac{1}{\beta L}\dfrac{I_{2/3}\left(\frac{4}{3}\beta L\right)}{I_{-1/3}\left(\frac{4}{3}\beta L\right)}$ $\quad S' = tB + (t^2/2L)\ln(2L/t + B)$, $\quad B = \sqrt{1 + 4L^2/t^2}, \qquad A_p = \frac{4}{3}tL$
3. Triangular $y = t(1 - x/L)$	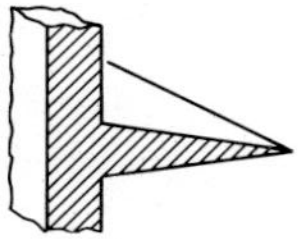	$\eta_f = \dfrac{1}{\beta L}\dfrac{I_1(2\beta L)}{I_0(2\beta L)}$ $\quad S' = 2\sqrt{t^2 + L^2}$, $\quad A_p = tL$
4. Parabolic $y = t(1-x/L)^2$	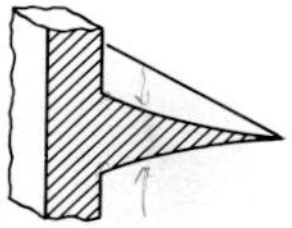	$\eta_f = \dfrac{2}{\sqrt{4(\beta L)^2 + 1} + 1}$ $\quad S' = LB + (L^2/2t)\ln(2t/L + B)$, $\quad B = \sqrt{1 + 4t^2/L^2}, \qquad A_p = \frac{2}{3}tL$
Annular Fins		
5. Rectangular $y = t$	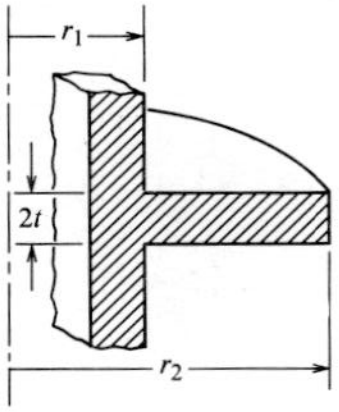	$\eta_f = \dfrac{(2r_1/\beta)}{(r_2^2 - r_1^2)}\dfrac{K_1(\beta r_1)I_1(\beta r_2) - I_1(\beta r_1)K_1(\beta r_2)}{K_0(\beta r_1)I_1(\beta r_2) + I_0(\beta r_1)K_1(\beta r_2)}$ $\quad S = 2\pi(r_2^2 - r_1^2), \qquad V = 2\pi(r_2^2 - r_1^2)t$

6. Hyperbolic $y = t(r_1/r)$	2t	$\eta_f = -\dfrac{2r_1/\beta}{(r_2 + r_1)}\,\dfrac{I_{2/3}\left(\frac{2}{3}\beta r_1\right)I_{-2/3}\left(\frac{2}{3}\beta r_2\sqrt{r_2/r_1}\right) - I_{2/3}\left(\frac{2}{3}\beta r_2\sqrt{r_2/r_1}\right)I_{-2/3}\left(\frac{2}{3}\beta r_1\right)}{I_{1/3}\left(\frac{2}{3}\beta r_1\right)I_{2/3}\left(\frac{2}{3}\beta r_2\sqrt{r_2/r_1}\right) - I_{-2/3}\left(\frac{2}{3}\beta r_2\sqrt{r_2/r_1}\right)I_{-1/3}\left(\frac{2}{3}\beta r_1\right)}$ $S = 2\pi r_1\left\{C - B + (t/2)\ln\dfrac{(C - t)(B + t)}{(C + t)(B - t)}\right\}$ $\quad B = \sqrt{r_1^2 + t^2}$ $V = 4\pi t r_1(r_2 - r_1)$ $\quad C = \sqrt{(r_2^2/r_1)^2 + t^2}$
Spines (Circular Cross Section)		
7. Pin $y = t$	2t L	$\eta_f = \dfrac{1}{\sqrt{2}\beta L}\tanh(\sqrt{2}\beta L), \quad S = 2\pi t L, \quad V = \pi t^2 L$
8. Parabolic $y = t(1-x/L)^{1/2}$		$\eta_f = \dfrac{2}{\left(\frac{4}{3}\sqrt{2}\beta L\right)}\,\dfrac{I_1\left(\frac{4}{3}\sqrt{2}\beta L\right)}{I_0\left(\frac{4}{3}\sqrt{2}\beta L\right)},$ $\quad S = (t^4\pi/6L^2)\left\{(4L^2/t^2 + 1)^{3/2} - 1\right\}$ $V = (\pi/2)t^2 L$
9. Triangular $y = t(1 - x/L)$		$\eta_f = \dfrac{4}{(2\sqrt{2}\beta L)}\,\dfrac{I_2(2\sqrt{2}\beta L)}{I_1(2\sqrt{2}\beta L)}, \quad S = \pi t\sqrt{L^2 + t^2}, \quad V = (\pi/3)t^2 L$
10. Parabolic $y = t(1 - x/L)^2$		$\eta_f = \dfrac{2}{\sqrt{8/9(\beta L)^2 + 1} + 1}, \quad V = (\pi/5)t^2 L$ $S = (\pi L^3/16t)\left\{AB - (L/4t)\ln[(4tB/L) + A]\right\}$ $A = 1 + (8t^2/L^2), \quad B = \sqrt{1 + (4t^2/L^2)}$

EXAMPLE 2.8 Cooling Fin for a Transistor

An aluminum annular fin is used to cool a transistor. The inner and outer radii are 5 mm and 20 mm, respectively, and the thickness is 0.2 mm. Calculate its efficiency and the heat dissipated when its base is at 380 K, the ambient air temperature is 300 K, and the estimated heat transfer coefficient is 8.2 W/m^2 K. Take the conductivity of aluminum as 205 W/m K.

Solution

Given: Aluminum annular fin.

Required: Efficiency and heat dissipated.

Assumptions: 1. Heat transfer coefficient constant over the fin surface.
2. Heat loss from tip negligible.

For an annular fin, $\beta^2 = h_c/kt$:

$$\beta^2 = \frac{(8.2)}{(205)(0.1 \times 10^{-3})} = 400 \text{ m}^{-2}$$

$$\beta = 20 \text{ m}^{-1}$$

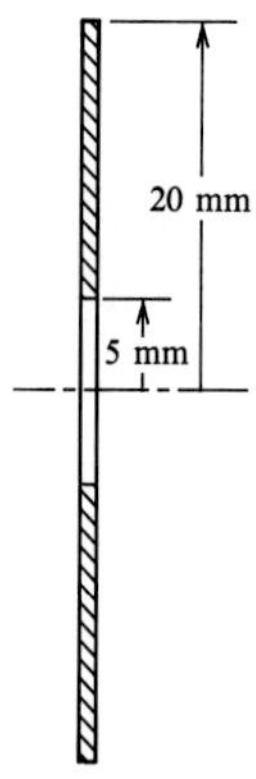

The fin effectiveness is given by Eq. (2.52):

$$\eta_f = \frac{(2r_1/\beta)}{(r_2^2 - r_1^2)} \frac{K_1(\beta r_1)I_1(\beta r_2) - I_1(\beta r_1)K_1(\beta r_2)}{K_0(\beta r_1)I_1(\beta r_2) + I_0(\beta r_1)K_1(\beta r_2)}$$

$$\beta r_1 = (20)(0.005) = 0.1; \qquad \beta r_2 = (20)(0.020) = 0.4$$

From Appendix B, Table B.3*b*, the required values of Bessel functions are:

βr	I_0	I_1	K_0	K_1
0.1	1.0025	0.0501	2.4271	9.8538
0.4		0.2040		2.1843

Substituting in Eq. (2.52),

$$\eta_f = \frac{(2)(0.005)/(20)}{(0.020^2 - 0.005^2)} \frac{(9.8538)(0.2040) - (0.0501)(2.1843)}{(2.4271)(0.2040) + (1.0025)(2.1843)} = 0.944$$

The heat dissipation is the efficiency times the dissipation for an isothermal fin:

$$\begin{aligned} \dot{Q} &= \eta_f(h_c)(2)(\pi)(r_2^2 - r_1^2)(T_B - T_e) \\ &= (0.944)(8.2)(2)(\pi)(0.020^2 - 0.005^2)(380 - 300) \\ &= 1.46 \text{ W} \end{aligned}$$

Solution using FIN2

The required input in SI units is:

Item number = 5
Thermal conductivity and density of the fin = 205, 2700
Heat transfer coefficient = 8.2
Base temperature and ambient temperature = 380, 300
t = 0.0001
r_1 and r_2 = 0.005, 0.020

FIN2 gives the following output:

Fin efficiency = 0.944
Base heat flow = 1.459 (watts)
Mass of fin = 6.362×10^{-4} (kilograms)

Comments

1. Notice that Table B.3*b* gives e^{-x} times $I_0(x)$ and $I_1(x)$ to simplify the tabulation.
2. The high efficiency suggests that the thickness of such fins is determined by rigidity rather than by heat transfer considerations.
3. Any consistent system of units can be used in FIN2. Since SI units were used here, the base heat flow is in watts, and the mass of the fin is in kilograms.

EXAMPLE 2.9 Heat Loss from a Parabolic Fin

A straight Duralumin fin has a parabolic profile $y = t(1 - x/L)^2$, with t = 3 mm and L = 20 mm. Determine the heat dissipation by the fin when its base temperature is 500 K and it is exposed to fluid at 300 K with a heat transfer coefficient of 2800 W/m² K. Also calculate the fin mass.

Solution

Given: Straight fin with a parabolic profile.

Required: Heat dissipation, mass.

Assumptions: Constant heat transfer coefficient over fin surface.

From Table A.1*b*, the conductivity of Duralumin at a guessed average fin temperature of (1/2)(500 + 300) = 400 K is 187 W/m K. Using Table 2.2, item 4,

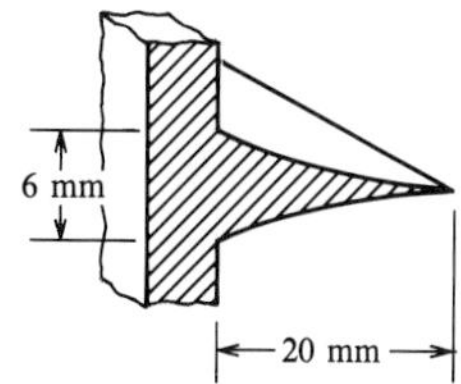

$$\beta = \left(\frac{h_c}{kt}\right)^{1/2} = \left[\frac{2800}{(187)(0.003)}\right]^{1/2} = 70.65\,\text{m}^{-1}$$

$$\beta L = (70.65)(0.02) = 1.413$$

$$\eta_f = \frac{2}{[4(\beta L)^2 + 1]^{1/2} + 1} = \frac{2}{[4(1.413)^2 + 1]^{1/2} + 1} = 0.500$$

$$B = \left[1 + 4\left(\frac{t}{L}\right)^2\right]^{1/2} = \left[1 + 4\left(\frac{0.003}{0.02}\right)^2\right]^{1/2} = 1.044$$

$$S' = LB + \left(\frac{L^2}{2t}\right) \ln\left(\frac{2t}{L} + B\right)$$

$$= (0.02)(1.044) + \frac{0.02^2}{2 \times 0.003} \ln\left(\frac{2 \times 0.003}{0.02} + 1.044\right)$$

$$= 0.0406 \text{ m}$$

For a unit width of fin,

$$\dot{Q} = h_c S'(T_B - T_e)\eta_f = (2800)(0.0406)(500 - 300)(0.500) = 11{,}370 \text{ W/m}$$

From Table A.1*a* the density of Duralumin is 2770 kg/m^3; thus,

$$\text{Fin mass} = A_p \rho = \frac{2}{3} t L \rho = \left(\frac{2}{3}\right)(0.003)(0.02)(2770) = 0.1108 \text{ kg/m}$$

Solution using FIN2

The required input in SI units is:

Item number = 4
Thermal conductivity and density of the fin = 187, 2770
Heat transfer coefficient = 2800
Base temperature and ambient temperature = 500, 300
$t = 0.003$
$L = 0.02$

FIN2 gives the following output:

Fin efficiency = 0.500
Base heat flow = 11,370 (watts/meter)
Mass of fin = 0.1108 (kilograms/meter)

Comments

Exercise 2–71 shows that this fin profile gives the maximum heat loss for a given weight of any profile.

2.4.5 The Similarity Principle and Dimensional Analysis

To conclude our analysis of fins we use the pin fin problem of Section 2.4.1 to illustrate the **similarity principle** and **dimensional analysis**. These are important concepts used in the analysis of more complex heat transfer problems. Equation (2.42) showed that the fin performance could be expressed as a relation between just two dimensionless parameters: the fin efficiency, $\eta_f = \dot{Q}/h_c\mathscr{P}L(T_B - T_e)$, and a fin parameter $\chi = (h_c\mathscr{P}L^2/kA_c)^{1/2}$. The similarity principle for this problem is simply the statement that η_f is a function of χ only. Thus, for example, if $\mathscr{P}$ and A_c are both doubled, η_f remains the same. We say that all pin fins with the same value of χ are *similar*, even though their sizes, materials, or heat transfer coefficients may be quite different.

The dimensionless groups relevant to a given problem are required for use of the similarity principle. We can deduce these dimensionless groups without actually solving the governing equations, as was done in Section 2.4.1. For this purpose, we use *dimensional analysis*, for which a number of methods are available. The pin fin problem will be used to demonstrate a method that requires a transformation of variables to make the governing equation and boundary conditions dimensionless. The first step is to choose dimensionless forms of the independent variable x and the dependent variable T. For x, an obvious choice is $\xi = x/L$, where L is the length of the fin; ξ then varies from zero to unity as x varies from zero to L. For T we will choose $\theta = (T - T_e)/(T_B - T_e)$; θ has a value of unity at the fin base and will approach zero at the tip of an infinitely long fin. Next, we transform the problem statement into the new variables. The rules of the transformation are

$$x = L\xi \qquad T = (T_B - T_e)\theta + T_e$$

$$dx = L\,d\xi \qquad dT = (T_B - T_e)\,d\theta$$

and Eq. (2.31) becomes

$$kA_c\frac{(T_B - T_e)}{L^2}\frac{d^2\theta}{d\xi^2} - h_c\mathscr{P}(T_B - T_e)\theta = 0$$

or

$$\frac{kA_c}{L^2}\frac{d^2\theta}{d\xi^2} - h_c\mathscr{P}\theta = 0$$

The boundary condition Eq. (2.32) becomes

$$\xi = 0: \quad \theta = 1 \qquad \textbf{(2.53a)}$$

and the boundary condition Eq. (2.33*b*) becomes

$$\xi = 1: \quad \frac{(T_B - T_e)}{L}\frac{d\theta}{d\xi} = 0$$

or

$$\frac{d\theta}{d\xi} = 0 \qquad \textbf{(2.53b)}$$

The differential equation is now put in dimensionless form. Dividing by kA_c/L^2,

$$\frac{d^2\theta}{d\xi^2} - \frac{h_c\mathscr{P}L^2}{kA_c}\theta = 0$$

or

$$\frac{d^2\theta}{d\xi^2} - \chi^2\theta = 0; \qquad \chi = \beta L, \qquad \beta^2 = \frac{h_c\mathscr{P}}{kA_c} \tag{2.54}$$

and we see that the fin parameter χ appears quite naturally in the dimensionless form of the governing equation. The boundary conditions are already dimensionless. Equation (2.54) is a differential equation for θ as a function of ξ and contains one dimensionless parameter, χ; the boundary conditions contain no further parameters. Thus, the solution must be of the form

$$\theta = \theta(\xi, \chi)$$

We now transform Eq. (2.37) for the rate of heat dissipation:

$$\dot{Q} = \int_0^L h_c\mathscr{P}(T - T_e)\,dx$$

$$= h_c\mathscr{P}L(T_B - T_e)\int_0^1 \theta\,d\xi$$

or

$$\frac{\dot{Q}}{h_c\mathscr{P}L(T_B - T_e)} = \eta_f = \int_0^1 \theta\,d\xi \tag{2.55}$$

For this simple case, the fin efficiency appears quite naturally as the dimensionless form of the heat dissipation rate. Although $\theta = \theta(\xi, \chi)$, the definite integral in Eq. (2.55) is not a function of ξ, so that

$$\eta_f = \eta_f(\chi) \tag{2.56}$$

which corresponds to the analytical solution, Eq. (2.42).

More complex heat transfer problems are often governed by differential equations that are difficult or impossible to solve analytically. Use of the above procedure allows the most concise form of the solution to be determined, which can be used as a basis for correlating experimental data or the results of numerical solutions. Also, when properly used, dimensional analysis facilitates the estimation of errors incurred in making simplifying assumptions. Such an approach is especially important for the analysis of convective heat transfer, as will be shown in Chapter 5. As a rather simple example of error estimation, consider the pin fin problem when the tip heat loss is not neglected. The appropriate second boundary condition is then Eq. (2.33a), which transforms into

$$\xi = 1: \quad -k\frac{(T_B - T_e)}{L}\frac{d\theta}{d\xi} = h_c(T_B - T_e)\theta$$

or

$$\frac{d\theta}{d\xi} = -\text{Bi}\theta \tag{2.57}$$

The Biot number $\mathrm{Bi} = h_c L/k$ was discussed in Section 1.5.1. This boundary condition introduces a second parameter into the problem; the temperature distribution must now be of the form

$$\theta = \theta(\xi, \chi, \mathrm{Bi}) \tag{2.58}$$

and the fin efficiency is a function of both χ and Bi. Using physical intuition, we would expect the tip heat loss to be significant only when the tip temperature is slightly less than that of the base, that is, when the fin parameter χ is small. Then the ratio of the heat loss from the tip to the heat loss from the sides is on the order of the area ratio $A_c/\mathscr{P}L$. At first sight, this area ratio may appear to be a new dimensionless parameter, but it is simply Bi/χ^2. The analytical solution for this case is given as Exercise 2–37. It confirms that the tip loss is significant only for small χ, in which case the fractional tip loss is approximately Bi/χ^2.

2.5 CLOSURE

In this chapter, the first law of thermodynamics and Fourier's law of heat conduction were used to solve a variety of steady one-dimensional heat conduction problems. Simple analytical results were obtained for conduction through cylindrical and spherical shells, and these can be used to build up thermal circuits for more complicated problems. The concept of a critical radius of insulation for a cylinder was introduced, which showed that insulation should not be added to a small cylinder (or sphere) for the purpose of reducing heat loss. The temperature distribution in a cylinder with internal heat generation was obtained, and the result was applied to calculation of the maximum temperature in a nuclear reactor fuel rod.

Much of the chapter dealt with the very important subject of extended surfaces, or fins. Cooling fins are widely used to reduce convective heat transfer resistance, particularly when gases are involved. The fin efficiency of a pin fin and annular fin were obtained by analysis. The result for an annular fin is obtained in terms of Bessel functions, which may be new to the student. Use of these functions is similar to use of the familiar trigonometric functions, and Appendix B conveniently specifies the required differentiation rules and provides tables. The efficiencies for eight additional fin profiles are given in Table 2.2 for engineering use. The key assumption in the analysis of cooling fins was that the temperature variation across the fin can be ignored. The same assumption is valid for a variety of other extended surfaces—for example, thermocouples and copper conductors. In some cases, the results of the cooling fin analyses can be used directly; in others, a new but similar analysis is required. The discussion of fins concluded by using the pin fin problem to demonstrate the use of dimensional analysis and the principle of similarity, both very important concepts that will be used throughout this text.

Two computer programs were introduced in Chapter 2. FIN1 is primarily an instructional aid and allows the student to explore the effect of fin parameters and boundary conditions on the temperature profile along a rectangular fin and on the fin efficiency. FIN2 is an engineering tool that gives the fin efficiency and fin mass for 10 different fin profiles, including straight fins, annular fins, and spines.

REFERENCES

1. Touloukian, Y. S., and Ho, C. Y., eds., *Thermophysical Properties of Matter. Vol. 1, Thermal Conductivity of Metallic Solids; Vol. 2, Thermal Conductivity of Nonmetallic Solids*, Plenum Press, New York (1972).

2. Vargaftik, N. B., *Tables of Thermophysical Properties of Liquids and Gases,* 2nd ed., Hemisphere Publishing Corp., Washington, D.C. (1975).

3. Desai, D. P., Chu, T. K., Bogaard, R. H., Ackerman, M. W., and Ho, C. Y., *CINDAS Special Report. Part I: Thermophysical Properties of Carbon Steels; Part II: Thermophysical Properties of Low Chromium Steels; Part III: Thermophysical Properties of Nickel Steels; Part IV: Thermophysical Properties of Stainless Steels,* Purdue University, West Lafayette, Ind., September (1976).

4. American Society of Heating, Refrigerating and Air Conditioning Engineers, *ASHRAE Handbook of Fundamentals*, ASHRAE, New York (1981).

5. Schneider, P. J., "Conduction," Chap. 4 in *Handbook of Heat Transfer Fundamentals,* 2nd ed., eds. Rohsenow, W. M., Hartnett, J. P., and Ganic, E. N., McGraw-Hill, New York (1985).

6. Veziroglu, T. N., "Correlation of thermal contact conductance experimental results," *Prog. Astronaut. Aeronaut.* 20, 879–907, Academic Press, New York (1967).

7. Wheeler, R. F., "Thermal conductance of fuel element materials," *USAEC Report HW–60343,* April (1959).

8. Jakob, M., *Heat Transfer,* vol. 1, John Wiley & Sons, New York (1949).

EXERCISES

2–1. The thermal conductivity of a solid may often be assumed to vary linearly with temperature, $k = k_0[1 + a(T - T_0)]$, where $k = k_0$ at a reference temperature T_0 and a is a constant coefficient. Consider a solid slab, $0 < x < L$, with the face at $x = 0$ maintained at temperature T_1. Determine the temperature profile across the slab, $T = T(x)$, in terms of T_1 and the heat flux q. Sketch profiles for zero, positive, and negative coefficients.

2–2. A typical polystyrene coffee cup has a wall thickness of 2 mm. We are all familiar with the insulation properties of such a cup from the time required for coffee to cool. To obtain the same insulating effect, how thick a layer of each of the following materials is required?

(i) polyvinylchloride
(ii) paper

(iii) oak wood
(iv) stainless steel
(v) brass
(vi) copper
(vii) diamond (type IIb)

2–3. A long, flat slab is made of n pairs of square bars of different thermal conductivities, k_A and k_B. Determine the effective thermal conductivity of the slab

(i) across the width.
(ii) through the thickness.

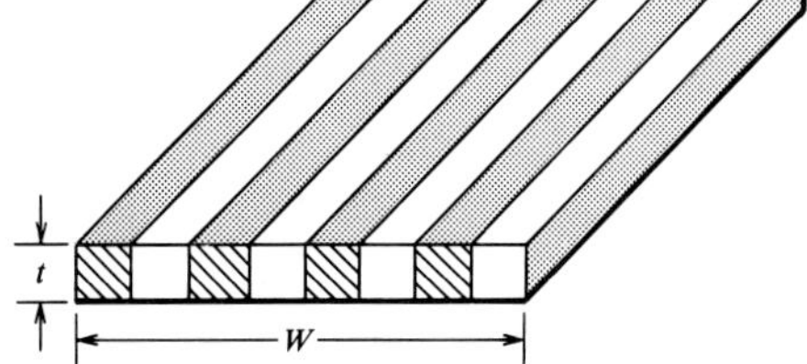

2–4. Use Eq. (2.3) to estimate the thermal conductivity of

(i) air at 300 K and 1 atm pressure.
(ii) air at 400 K and 1 atm pressure.
(iii) air at 300 K and 0.01 atm pressure.
(iv) helium at 300 K and 1 atm pressure.

Compare your results to the data given in Table A.7, and discuss the effects of temperature and pressure. The average molecular speed c and transport mean free path ℓ_t may be calculated from the following formulas:

$$c = \left(\frac{8\kappa T}{\pi m}\right)^{1/2}; \qquad \ell_t = \nu\left(\frac{9\pi m}{8\kappa T}\right)^{1/2}$$

2–5. An aluminum/aluminum interface may have an interfacial conductance in the range 150–12,000 W/m^2 K (the lower-limit value corresponds to vacuum conditions). Consider conduction from an aluminum channel to an aluminum plate, as shown. The channel is 3 mm thick, and the plate is 1 mm thick. Plot a graph of the ratio of interfacial thermal resistance to total thermal resistance (channel, interface, plate) for this range of interfacial conductances. Take k = 200 W/m K for the aluminum.

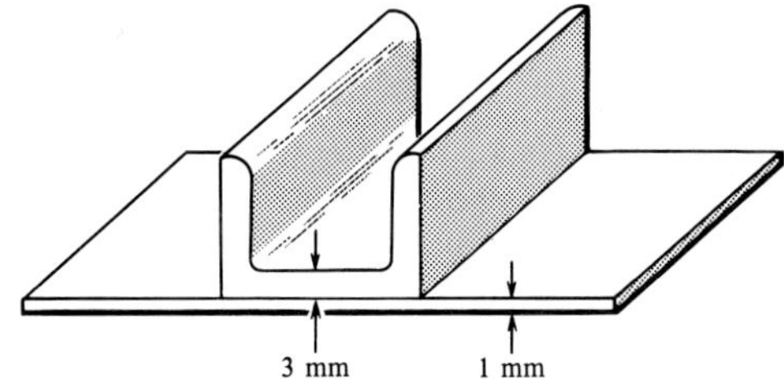

2–6. A horizontal fireclay brick partition, 20 cm thick, is covered by a 5 cm–thick layer of medium-density (ρ = 28 kg/m^3) fiberglass insulation. The under surface of the brick is maintained at 540 K, and the upper surface of the fiberglass is maintained at 300 K. Calculate the heat flow per unit area. Evaluate the k values of each layer at the average temperature of the layer.

2–7. A composite wall consists of a 1 mm–thick stainless steel plate, 2 cm of 4-ply laminated asbestos paper, and 2 cm of 8-ply laminated asbestos paper. The stainless steel surface is maintained at 380 K while the other side loses heat to ambient air at 300 K, with a combined convection and radiation heat transfer coefficient of 5 W/m^2 K. Estimate the heat flow per unit area. Evaluate the k values of the asbestos layers at suitable average temperatures.

2–8. A 2 m–long cylindrical chemical reactor has an inside diameter of 5 cm, has a 1 cm–thick stainless steel shell, and is insulated on the outside by a layer of medium-density ($\rho = 28$ kg/m^3) fiberglass 5 cm thick. The ambient air is at 25°C, and the surroundings can be assumed large and black. The convective heat transfer coefficient between the insulation and air is 6 W/m^2 K, and the emittance of the insulation is 0.8. At steady state the outer surface of the insulation is measured at 32°C. Draw the thermal circuit and determine the temperature of the inner surface of the stainless steel shell. Take $k = 16$ W/m K for the stainless steel.

2–9. A 4 cm–O.D., 2 mm–wall-thickness stainless steel tube is insulated with a 5 cm–thick layer of cork. Chilled milk flows through the tube. At a given location the milk temperature is 5°C when the ambient temperature is 25°C. If the inside and outside heat transfer coefficients are estimated to be 50 and 5 W/m^2 K, respectively, calculate the rate of heat gain per meter length of tube.

2–10. Saturated steam at 200°C flows through an AISI 1010 steel tube with an outer diameter of 10 cm and a 4 mm wall thickness. It is proposed to add a 5 cm–thick layer of 85% magnesia insulation. Compare the heat loss from the insulated tube to that from the bare tube when the ambient air temperature is 20°C. Take outside heat transfer coefficients of 6 and 5 W/m^2 K for the bare and insulated tubes, respectively.

2–11. A hollow cylinder, of inner and outer diameters 3 and 5 cm, respectively, has an inner surface temperature of 400 K. The outer surface temperature is 326 K when exposed to fluid at 300 K with an outside heat transfer coefficient of 27 W/m^2 K. What is the thermal conductivity of the cylinder?

2–12. Superheated steam at 500 K flows in Schedule 40 steel pipe of nominal size 6 in. Determine the effect of adding magnesia insulation to the pipe as a function of insulation thickness and outside heat transfer coefficient. Assume an inside heat transfer coefficient of 7000 W/m^2 K and surroundings at 300 K. Prepare a graph of heat loss per unit length as a function of insulation thickness, with the outside heat transfer coefficient ($10 < h_o < 100$ W/m^2 K) as a parameter.

2–13. A thermal conductivity cell consists of concentric thin-walled copper tubes with an electrical heater inside the inner tube and is used to measure the conductivity of granular materials. The inner and outer radii of the annular gap are 2 and 4 cm. In a particular test the electrical power to the heater was 10.6 W per meter length, and the inner and outer tube temperatures were measured to be 321.4 K and 312.7 K, respectively. Calculate the thermal conductivity of the sample.

2–14. Refrigerant-12 at $-35°C$ flows in a copper tube of 8 mm outer diameter and 1 mm wall thickness. It has been suggested that a foam insulation with an aluminized outer surface be used to reduce heat leakage to the refrigerant. If the inside and outside heat transfer coefficients are taken as 300 and 5 W/m^2 K, respectively, and the surrounding air is at 20°C, plot a graph of heat leakage per meter versus insulation thickness. Should the insulation be used? Take $k = 0.035$ W/m K for the insulation.

2–15. A 1 in Schedule 10 copper pipe carries 10 GPM of brine at $-5°C$. The ambient air is at 20°C and has a dewpoint of 10°C. How thick a layer of insulation of thermal conductivity 0.2 W/m K is required to prevent condensation on the outside of the insulation? Take the outside heat transfer coefficient as 11.0 W/m^2 K.

2–16. A 1.5 mm–diameter wire is to be insulated with a plastic material of thermal conductivity 0.37 W/m K. It is found by experiment that a given current will heat the bare wire to 40°C when the ambient air temperature is 20°C. Plot the wire temperature and the surface temperature of the insulation as a function of insulation thickness for the same current. The emittance of the insulation is 0.9, and that of the bare wire is 0.07. The convective heat transfer coefficient can be calculated from an approximate relation for a horizontal cylinder in air at normal temperature, $\overline{h}_c = 1.3(\Delta T/D)^{1/4}$ W/m^2 K, for ΔT in kelvins and D in meters.

2–17. A 6 mm–O.D. tube is to be insulated with an insulation of thermal conductivity 0.08 W/m K and a very low surface emittance. Heat loss is by natural convection, for which the heat transfer coefficient can be taken as $\overline{h}_c = 1.3(\Delta T/D)^{1/4}$ W/m^2 K for $\Delta T = T_s - T_e$ in kelvins and the diameter D in meters. Determine the critical radius of the insulation and the corresponding heat loss for a tube surface temperature of 350 K, and an ambient temperature of 300 K.

2–18. A 1 mm–diameter resistor has a sheath of thermal conductivity $k = 0.12$ W/m K and is located in an evacuated enclosure. Determine the radius of the sheath that maximizes the heat loss from the resistor when it is maintained at 450 K and the enclosure is at 300 K. The surface emittance of the sheath is 0.85.

2–19. A 1 mm–diameter resistor, for an electronic component on a space station, is to have a sheath of thermal conductivity 0.1 W/m K. It is cooled by forced convection with $\overline{h}_c \simeq 1.1D^{-1/2}$ W/m^2 K, for diameter D in meters, and by radiation with $q_{\text{rad}} = \sigma\varepsilon(T_s^4 - T_e^4)$. Determine the radius of the sheath that maximizes the heat loss when the resistor is at 400 K and the surroundings are at 300 K. Take the value of the surface emittance ε as

(i) 0.9.
(ii) 0.5.

2–20. An experimental boiling water reactor is spherical in shape and operates with a water temperature of 420 K. The shell is made from nickel alloy steel ($k = 21$ W/m K) and has an inside radius of 0.7 m with a wall thickness of 7 cm. The reactor is surrounded by a layer of concrete 20 cm thick. If the outside heat transfer coefficient is 8 W/m^2 K and the ambient air is at 300 K, what are the temperatures of the internal and external surfaces of the concrete? Also, if the reactor operates at a power level of 30 kW, what fraction of the power generated is lost by heat transfer through the shell? The resistance to heat flow from the water to the shell can be taken to be negligible.

2–21. (i) Derive an expression for the relation between heat loss and temperature difference across the inner and outer surfaces of a hollow sphere, the conductivity of which varies with temperature in manner given by $k = k_0[1 + a(T - T_0)]$, where T_0 is a reference temperature.
(ii) Find the corresponding result for a hollow cylinder.
(iii) Compare the expressions derived for parts (i) and (ii) for the special case of the outside radius becoming infinite. Explain the different values obtained.

2–22. A 5 cm–high stainless steel truncated cone has a base diameter of 10 cm and a top diameter of 5 cm. The sides are insulated, and the base and top temperatures are 100°C and 50°C, respectively. Assuming one-dimensional heat flow, estimate the heat flow. Take $k = 15$ W/m K for the stainless steel.

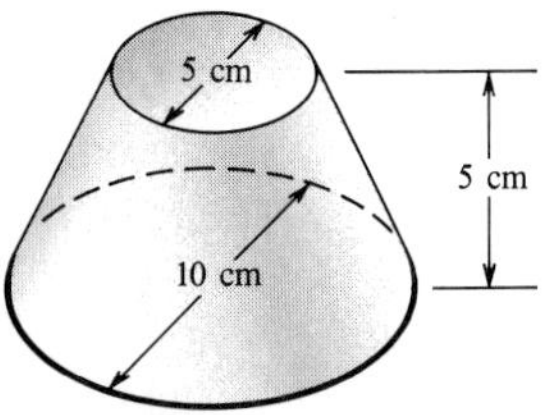

2–23. The concept of a critical radius for maximum heat loss developed for a cylinder in Section 2.3.2 also applies to a sphere. Derive expressions for critical radius for the following conditions:

(i) The outside heat transfer coefficient has a constant value h_o.
(ii) The outside heat transfer coefficient is proportional to $r_o^{-1/2}$.

2–24. (i) Heat is generated uniformly in a plate $2L$ thick at a rate $\dot{Q}_v'''$ W/m^3. If the surfaces are maintained at temperature T_s, determine the temperature distribution across the plate.
(ii) A 1 cm–thick stainless steel plate is heated by an electric current giving an I^2R internal heat generation of 1×10^6 W/m^3. The plate is cooled by an air stream at temperature $T_e = 300$ K, and the air velocity is adjusted to give a maximum plate temperature of 360 K. What is the average convective heat transfer coefficient?

2–25. An electrical current of 15 A flows in an 18 gage copper wire (1.02 mm diameter). If the wire has an electrical resistance of 0.0209 Ω/m, calculate

(i) the rate of heat generation per meter length of wire.
(ii) the rate of heat generation per unit volume of copper.
(iii) the heat flux across the wire surface at steady state.

2–26. An electrical cable has a 2 mm–diameter copper wire encased in a 4 mm–thick insulator of conductivity 0.2 W/m K. The cable is located in still air at 25°C, and the convective heat transfer coefficient can be approximated as $h_c =$

$1.3(\Delta T/D)^{1/4}$ W/m^2 K. If the temperature limit for the insulator is 150°C, determine the maximum I^2R losses that can be allowed in the wire.

(i) Ignore thermal radiation.
(ii) Account for thermal radiation if the emittance of the insulator is 0.8.

2–27. Determine the allowable current in a 10 gage (2.59 mm diameter) copper wire that is insulated with a 1 cm–O.D. layer of rubber. The outside heat transfer coefficient is 20 W/m^2 K, and the ambient air is at 310 K. The allowable maximum temperature of the rubber is 380 K. Take $k = 0.15$ W/m K for the rubber and an electrical resistance of 0.00328 Ω/m for the copper wire.

2–28. An explosive is to be stored in large slabs of thickness $2L$ clad on both sides with a protective sheath. The rate at which heat is generated within the explosive is temperature-dependent and can be approximated by the linear relation $\dot{Q}_v''' = a + b(T - T_e)$, where T_e is the prevailing ambient air temperature. If the overall heat transfer coefficient between the slab surface and the ambient air is U, show that the condition for an explosion is $L = (k/b)^{1/2}\tan^{-1}[U/(kb)^{1/2}]$. Determine the slab thickness if $k = 0.9$ W/m K, $U = 0.20$ W/m^2 K, $a = 60$ W/m^3, $b = 6.0$ W/m^3 K.

2–29. On the flight of Apollo 12, plutonium oxide ($Pu^{238}O_2^{16}$) was used to generate electrical power. Heat was generated uniformly through the loss of kinetic energy from alpha particles emitted by the Pu^{238}. Consider a sphere of plutonium oxide of 3 cm diameter covered with thermo-electric elements for converting heat to electricity. The physical properties of these elements (tellurides) and heat rejection considerations suggest that the surface of the sphere be at 200°C. On the other hand, the ceramic nature of the plutonium oxide allows a maximum temperature of 1750°C. With these constraints, determine

(i) the maximum allowable volumetric heating rate.
(ii) the electrical power generated, assuming a thermal efficiency of 4%.

Take $k_{PuO_2} = 4$ W/m K.

2–30. Show that the temperature distribution along an infinitely long pin fin is given by

$$\frac{T - T_e}{T_B - T_e} = e^{-\beta x}$$

Also find the heat dissipated by determining the base heat flow. Compare this result to Eq. (2.40) for an insulated tip and discuss.

2–31. Show that the temperature distribution along a short pin fin, for which heat loss from the tip cannot be neglected, is

$$\frac{T - T_e}{T_B - T_e} = \frac{\cosh \beta(L - x) + (h_c/\beta k)\sinh \beta(L - x)}{\cosh \beta L + (h_c/\beta k)\sinh \beta L}$$

where the heat transfer coefficient h_c is the same on the tip and sides.

2–32. A copper tube has a 2 cm inside diameter and a wall thickness of 1.5 mm. Over the tube is an aluminum sleeve of 1.5 mm thickness having 100 pin fins per centimeter length. The pin fins are 1.5 mm in diameter and are 4 cm long. The fluid inside the tube is at 100°C, and the inside heat transfer coefficient is 5000 W/m^2 K. The fluid outside the tube is at 250°C, and the heat transfer coefficient on the outer surface is 7 W/m^2 K. Calculate the heat transfer per meter length of tube. Take $k = 204$ W/m K for the aluminum.

2–33. A gas turbine rotor has 54 AISI 302 stainless steel blades of dimensions $L = 6$ cm, $A_c = 4 \times 10^{-4}$ m^2, and $\mathscr{P} = 0.1$ m. When the gas stream is at 900°C, the temperature at the root of the blades is measured to be 500°C. If the convective heat transfer coefficient is estimated to be 440 W/m^2 K, calculate the heat load on the rotor internal cooling system.

2–34. Aluminum alloy straight rectangular fins for cooling a semiconductor device are 1 cm long and 1 mm thick. Investigate the effect of choice of tip boundary condition on heat loss as a function of convective heat transfer coefficient. Use $k = 175$ W/m K for the alloy and a range of h_c values from 10 to 200 W/m^2 K.

2–35. Inconel-X-750 straight rectangular fins are to be used in an application where the fins are 2 mm thick and the convective heat transfer coefficient is 300 W/m^2 K. Investigate the effect of tip boundary condition on estimated heat loss for fin lengths L varying from 6 mm to 20 mm. Take $T_B = 800$ K, $T_e = 300$ K, $k = 18.8$ W/m K.

2–36. A straight rectangular fin has a constant incident radiation heat flux q_{rad} W/m^2 on one side from a high-temperature source and loses heat by convection from both sides. If $L = 10$ cm, half-thickness $t = 5$ mm, $k = 30$ W/m K, $q_{\text{rad}} = 30{,}000$ W/m^2, and $h_c = 100$ W/m^2 K, determine the base temperature to give a tip temperature of 400 K when the ambient fluid is at 300 K.

2–37. If the heat loss from the tip of a pin fin is not neglected, show that the fin efficiency is given by

$$\eta_f = \frac{\sinh \chi/\chi + \zeta \cosh \chi}{(1 + \zeta)(\cosh \chi + \zeta\chi \sinh \chi)}$$

where $\chi = \beta L = (h_c\mathscr{P}/kA_c)^{1/2}L$ and $\zeta = A_c/\mathscr{P}L$. By comparing the heat loss with that given by Eq. (2.40), show that tip loss is important only for small values of χ and is then of order ζ. Was this condition met in Example 2.5?

2–38. A heat sink assembly capable of mounting 36 power transistors may be idealized as a 15 cm cube containing four rows of 24 aluminum fins per row, each fin being 15 cm wide, 2.5 cm high, and 2 mm thick. A fan is an integral part of the assembly and blows air at a velocity that gives a heat transfer coefficient of 50 W/m^2 K. If the manufacturer's transistor temperature limit is 360 K, specify the allowable power dissipation per transistor. The mean air temperature is 310 K. If the rise in air temperature is limited to 10 K, specify the required capacity of the fan in m^3/min.

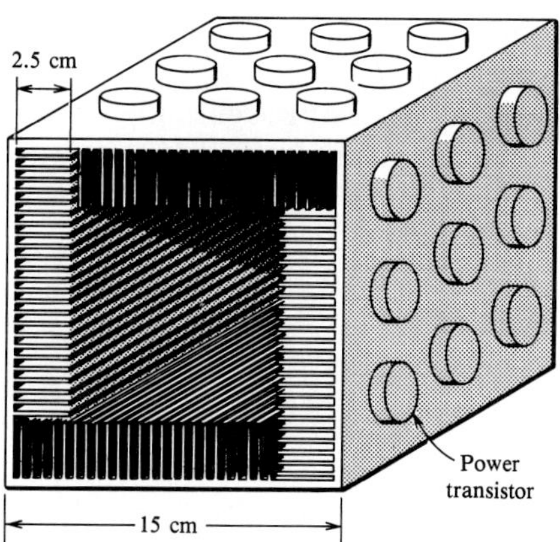

2–39. The dimensions of a straight rectangular fin can be optimized to give maximum heat transfer for a given mass. If the fin has thickness $2t$ and length L, the mass per unit width is ρA_p, where $A_p = 2tL$ is the profile area. For negligible tip heat loss, show that the heat flow $\dot{Q}$ is a maximum when

$$\tanh \xi = \frac{3\xi}{\cosh^2 \xi}$$

where $\xi = \beta A_p/2t$ and $\beta^2 = h_c/kt$. Hence show that the optimal dimensions are

$$\frac{L}{t} = 1.419(k/h_c t)^{1/2}$$

Heat transfer coefficients of 150 W/m^2 K are typical for air-cooled reciprocating aircraft engines. What is the optimal length of 1 mm–thick rectangular fins if made from

(i) mild steel?
(ii) aluminum?

2–40. Some cooling fins lose heat predominantly by radiation. Show that the solution for the temperature distribution along a long pin fin, which loses heat by radiation only, can be given as an integral that can be evaluated numerically. Also find the heat loss from a pin fin of 1 cm diameter, 10 cm long, when the base temperature is 1000 K and the surroundings are black at 300 K. The thermal conductivity of the fin material is 10 W/m K. Assume a constant transfer factor $\mathscr{F}$ along the fin, with a value of 0.8.

2–41. A long gas turbine blade receives heat from combustion gases by convection and radiation. If emission from the blade can be neglected ($T_s \ll T_e$), determine the temperature distribution along the blade. Assume

(i) the blade tip is insulated.
(ii) the heat transfer coefficient on the tip equals that on the blade sides.

The cross-sectional area of the blade may be taken to be constant.

2–42. A 60 cm–long, 3 cm–diameter AISI 1010 steel rod is welded to a furnace wall and passes through 20 cm of insulation before emerging into the surrounding air. The furnace wall is at 300°C, and the air temperature is 20°C. Estimate the temperature of the bar tip if the heat transfer coefficient between the rod and air is taken to be 13 W/m^2 K.

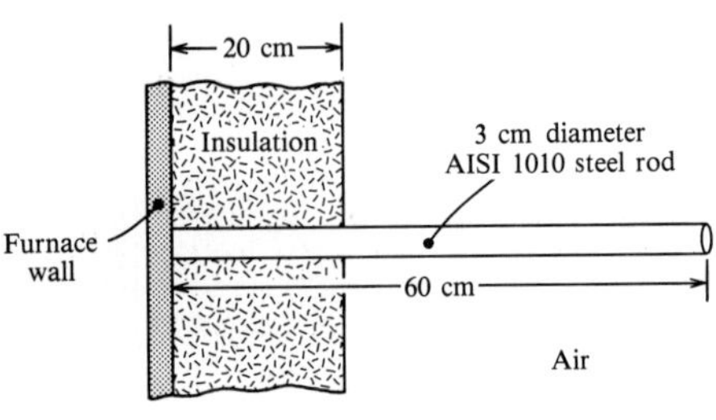

2–43. Jakob [8] suggests that Eq. (2.40) for the heat loss from a fin can be corrected to account for heat loss from the tip by adding to the length L the "tip size" $(A_c/\mathscr{P})$. Select a number of test cases and use FIN1 to check the validity of this rule.

2–44. A rectangular fin has a length of 2 cm, a width of 4 cm, and a thickness of 1 mm. It has a base temperature of 120°C and is exposed to air at 20°C with a convective heat transfer coefficient of 20 W/m^2 K. Determine the fin efficiency, heat loss and tip temperature for each of the three tip boundary conditions given by Eqs. (2.33*a*, *b*, *c*). Take the fin material as

(i) aluminum, $k = 220$ W/m K.
(ii) stainless steel, $k = 15$ W/m K.

2–45. A test technique for measuring the thermal conductivity of copper-nickel alloys is based on the measurement of the tip temperature of pin fins made from the alloys. The standard fin dimensions are a diameter of 5 mm and a length of 20 cm. The test fin and a reference brass fin ($k = 111$ W/m K) are mounted on a copper base plate in a wind tunnel. The test data include $T_B = 100$°C, $T_e = 20$°C, and tip temperatures of 64.2°C and 49.7°C for the brass and test alloy fins, respectively.

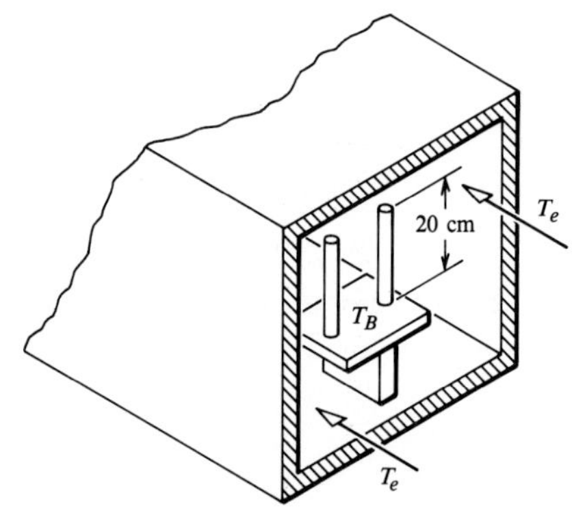

(i) Determine the conductivity of the test alloy.
(ii) If the conductivity should be known to ±1.0 W/m K, how accurately should the tip temperatures be measured?

2–46. Exercise 2–39 requires an analytical proof that the dimensions of a straight rectangular fin that result in a maximum heat transfer for a given weight are $(L/t) = 1.419(k/h_c t)^{1/2}$. For an aluminum alloy fin ($k = 175$ W/m K) and a convective heat transfer coefficient of 200 W/m^2 K, use FIN1 to check this result using a 2 mm–thick fin as a base case.

2–47. The term *fin effectiveness* is used in two ways in the heat transfer literature. It can be a synonym for fin efficiency, and it may be defined as the ratio of the fin heat transfer rate to the rate that would exist without the fin, denoted ε_f. An AISI 302 stainless steel ($k = 15$ W/m K) straight rectangular fin is 1 cm wide, is 2 mm thick, and is cooled by an air flow giving a convective heat transfer coefficient on the sides and tip of 25 W/m^2 K. Calculate $\dot{Q}$ as a function of fin length, and hence prepare a graph or table of ε_f versus L. Comment on the significance of this result to the design of such fins.

2–48. A mercury-in-glass thermometer is to be used to measure the temperature of a hot gas flowing in a duct. To protect the thermometer, a pocket is made from a 7 mm–diameter, 0.7 mm–wall-thickness stainless steel tube, with one end sealed and the other welded to the duct wall. The small gap between the thermometer and the pocket wall is filled with oil to ensure good thermal contact and the thermometer bulb is in contact with the sealed end. The gas stream is at 320°C and the duct wall is at 240°C. How long should the pocket be for the error in the thermometer reading to be less than 2°C? Take the thermal conductivity of the stainless steel as 15 W/m K, and the convective heat transfer coefficient on the outside of the pocket as 30 W/m^2 K.

2–49. A pool of liquid is heated by an immersed long, thin wire of length L and diameter d, through which an electrical current I is passed. The end of the wire at $x = 0$ is at temperature T_1 and at $x = L$, T_2. Assuming the convective heat transfer coefficient h_c to be constant along the wire, determine the x-location at which the maximum wire temperature occurs.

2–50. Consider a heat barrier consisting of a 2 mm–thick brass plate to which 3 mm copper tubing is soldered. The tubes are spaced 10 cm apart. Cooling water passed through the tubes keeps them at approximately 315 K. The underside of the brass wall is insulated with a 1.5 cm–thick asbestos layer, which in turn contacts a hot wall at 600 K.

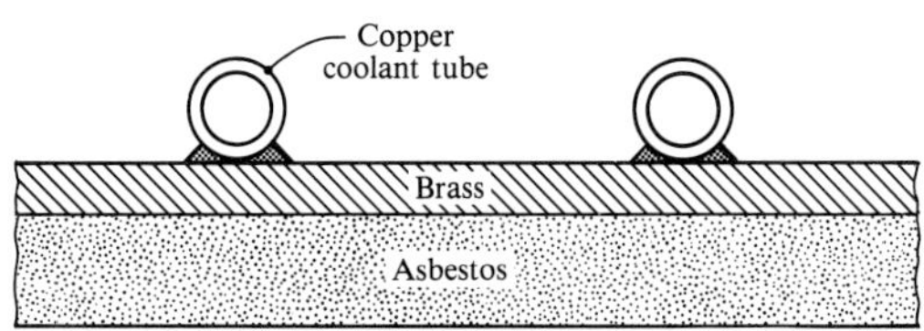

Assuming that the heat transfer from the cold side of the brass plate is negligible, estimate the temperature of the hottest spot on the brass wall. Take $k = 0.16$ W/m K for the asbestos and 111 W/m K for the brass.

2–51. A pressure transducer is connected to a high-temperature furnace by a copper tube "pigtail" of 3 mm outer diameter and 0.5 mm wall thickness. If the furnace operates at 1000 K and the transducer must not exceed 340 K, how long should the tube be? Take the ambient temperature as 300 K and assume a heat transfer coefficient of 30 W/m^2 K.

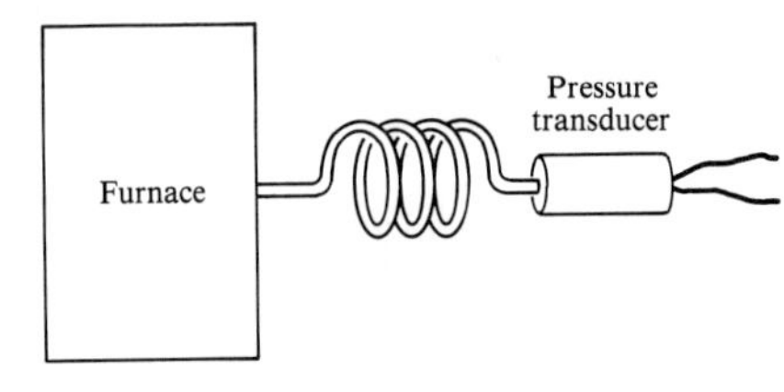

2–52. An encapsulated semiconductor chip is connected by a brass lead 5 mm long and with a 1.25 × 0.25 mm cross section, to the end of a copper conductor on a circuit board, as shown in Fig. 2.17. The board is 10 cm wide and 1.5 mm thick and has a conductivity of 0.2 W/m K. The conductors are 2 mm wide and 0.75 mm thick and are spaced at 12 mm intervals along the board. If 60% of the power generated by the chip is to be dissipated by the board, what is the allowable rating for the chip if its temperature is not to exceed 350 K? The average cooling air temperature is 310 K, and the convective heat transfer coefficient between the air and the board is estimated to be 5 W/m^2 K. Take k = 386 W/m K for the copper.

2–53. A copper-constantan (45% Ni) thermocouple is constructed from 24 gage (0.510 mm diameter) wire and protrudes into a steam chamber. The steam is at 320 K, and the chamber wall is at 300 K. The wires are bare and well separated.

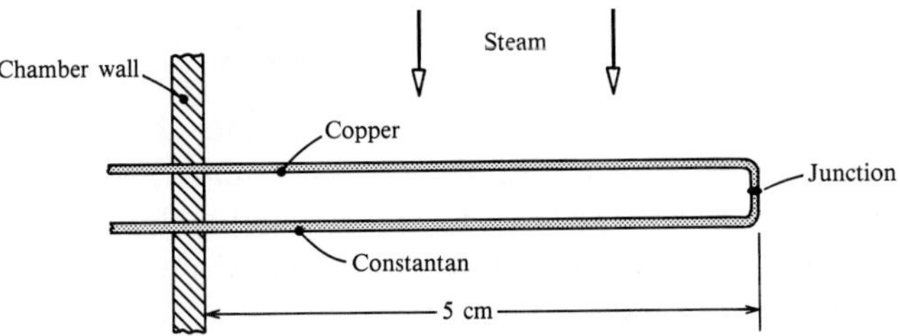

If the length of protrusion is 5 cm, calculate the error due to conduction along the wires for a heat transfer coefficient of 100 W/m^2 K.

2–54. An electrical current is passed through a horizontal copper rod 2 mm in diameter and 30 cm long, located in an air stream at 20°C. If the ends of the rod are also maintained at 20°C and the convective heat transfer coefficient is estimated to be 30 W/m^2 K, determine the maximum current that can be passed if the midpoint temperature is not to exceed 50°C.

(i) Ignore thermal radiation.
(ii) Include the effect of thermal radiation.

For the copper rod, take k = 386 W/m K, ε = 0.8, and electrical resistivity of 1.72×10^{-8} Ωm.

2–55. A stainless steel tube of 1 cm outer diameter and 1 mm wall thickness receives a uniform radiative heat flux of 100 W/cm^2 from a high-temperature plasma over 180° of its outer circumference; the remaining 180° is insulated. Water at 300 K flows through the tube, and the inside heat transfer coefficient is 1000 W/m^2 K.

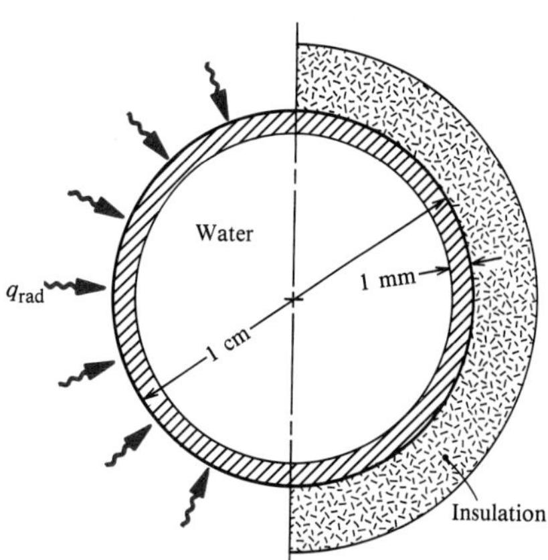

(i) Determine the wall temperature distribution around the tube.

(ii) If there is also a volumetric heat source of 50 W/cm^3 in the tube wall due to neutron absorption, find the new temperature distribution.

Take $k = 20.0$ W/m K for the stainless steel.

2–56. The absorber of a simple flat-plate solar collector with no coverplate consists of a 2 mm–thick aluminum plate with 6 mm–diameter aluminum water tubes spaced at a pitch of 10 cm, as shown.

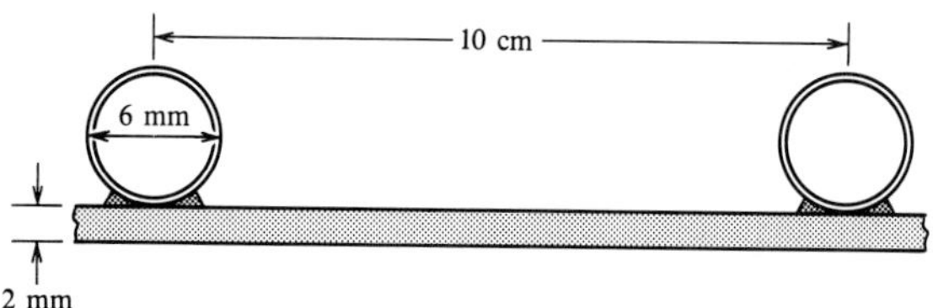

On a clear summer day near the ocean, the air temperature is 20°C, and a steady wind is blowing. The solar radiation absorbed by the plate is calculated to be 680 W/m^2, and the convective heat transfer coefficient is estimated to be 14 W/m^2 K. If coolant water enters the collector at 7×10^{-3} kg/meter width of collector, and the collector is 3 m long, estimate the outlet water temperature. For the aluminum take $k = 200$ W/m K and $\varepsilon = 0.20$. (*Hint:* Evaluate the heat lost by reradiation using Eq. (1.19) with a constant value of h_r corresponding to a guessed average plate temperature. Then the single-stream heat exchanger analysis of Section 1.6 applies with appropriate interpretation of effectiveness and number of transfer units.)

2–57. The tip of a soldering iron consists of a 4 mm–diameter copper rod, 5 cm long. If the tip must operate at 350°C when the ambient air temperature is 20°C, determine the base temperature and heat flow. The heat transfer coefficient from the rod to the air is estimated to be about 10 W/m^2 K. Take $k = 386$ W/m K for the copper.

2–58. A skin panel for an actively cooled hypersonic aircraft has square passages through which supercritical hydrogen flows before being used as fuel in a scramjet engine. Possible dimensions are shown for a panel made from Inconel-X-750 nickel alloy. If the heating load is 100 kW/m^2 and the inside convective heat transfer coefficient is 6000 W/m^2 K, estimate the maximum skin temperature at a location along the panel where the bulk coolant temperature is 100 K.

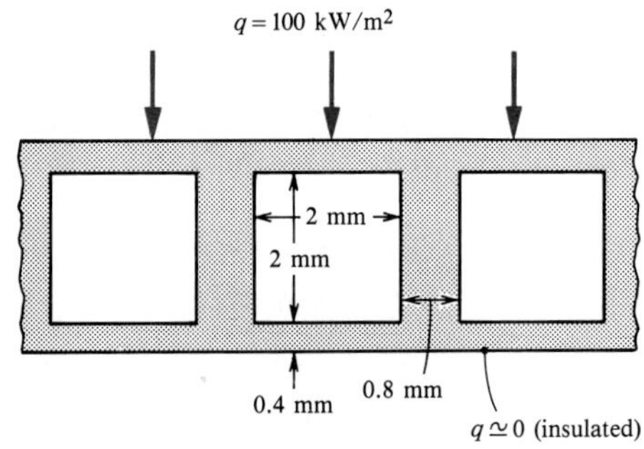

2–59. An exhaust stack thermocouple is inserted into a 20 cm–long well made of 5 mm–O.D., 0.7 mm–wall-thickness AISI 316 stainless steel tube and is held in good thermal contact with the sealed end by a spring-loaded plug. The convective heat transfer coefficient for the exhaust gases flowing across the tubing is estimated to be 65 W/m^2 K. If the thermocouple reading is 221°C when the stack walls are at 178°C, determine the gas temperature.

2–60. A 2 cm–O.D. stainless steel tube with a 1 mm wall thickness receives a radiative heat flux from a high-temperature gas distributed as $q = q_0 \cos\phi$ over 180° of its circumference, as shown. The remaining 180° is insulated. Water at 320 K flows inside the tube, and the inside heat transfer coefficient is 800 W/m^2 K. Determine the wall temperature distribution around the tube for $q_0 = 10^5$ W/m^2. Take $k = 18.0$ W/m K for the stainless steel.

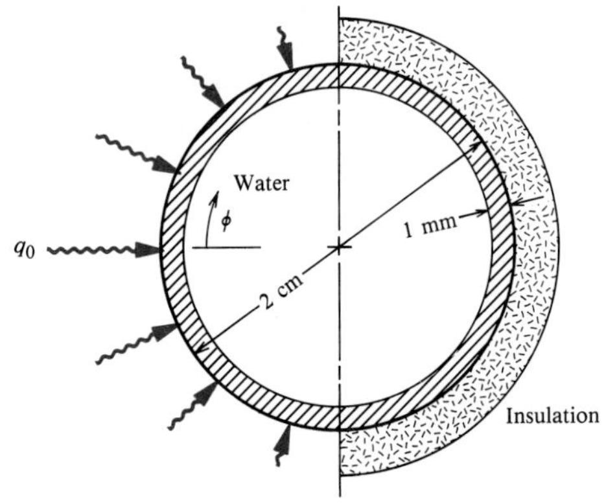

2–61. The convective heat transfer coefficient around a cylinder held perpendicular to a flow varies in a complicated manner. A test cylinder to investigate this behavior consists of a 0.001 in–thick, 12.7 mm–wide stainless steel heater ribbon (cut from shim stock) wound around a 2 cm–O.D., 2 mm–wall-thickness Teflon tube. A single thermocouple is located just underneath the ribbon and measures the local ribbon temperature $T_s(\theta)$.

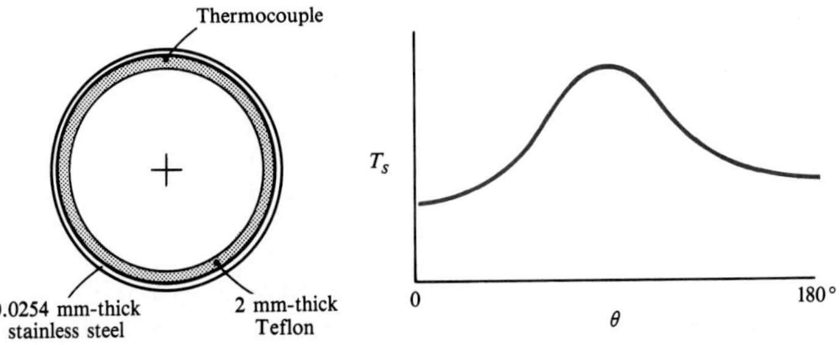

The cylinder is installed in a wind tunnel, and a second thermocouple is used to measure the ambient air temperature T_e. The power input to the heater is metered, from which the electrical heat generation per unit area $\dot{Q}/A$ can be calculated (A is the surface area of one side of the ribbon).

As a first approximation, the local heat transfer coefficient $h_c(\theta)$ can be obtained from

$$h_c(\theta) = \frac{\dot{Q}/A}{T_s(\theta) - T_e}$$

Hence, by rotating the cylinder with the power held constant, the variation of h_c can be obtained from the variation of T_s: where $T_s(\theta)$ is low, $h_c(\theta)$ is high, and vice versa. A typical variation of $T_s(\theta)$ is shown in the graph. A problem with this technique is that conduction around the circumference of the tube causes the local heat flux $q_s(\theta)$ to not exactly equal $\dot{Q}/A$.

(i) Derive a formula for $h_c(\theta)$ that approximately accounts for circumferential conduction.

(ii) The following table gives values of $T_s(\theta)$ in a sector where circumferential conduction effects are expected to be large. Use these values together with $\dot{Q}/A = 5900$ W/m^2 and $T_e = 25$°C to estimate the conduction effect at $\theta = 110°$.

Angle (degrees)	T_s (°C)
100	65.9
110	65.7
120	64.4

(iii) Comment on the design of the cylinder. Would a 3 mm–thick brass tube, directly heated by an electric current, be a suitable alternative?

Use $k = 15$ W/m K for the stainless steel and 0.38 W/m K for Teflon.

2–62. It is proposed to redesign a cast-iron channel to have fins in the streamwise direction of approximately triangular cross section, with height, base width, and pitch all of 2 cm. Determine the effect of adding the fins on the surface thermal resistance if the heat transfer coefficient on both the finned and unfinned surfaces is

(i) 1000 W/m^2 K.
(ii) 8000 W/m^2 K.

2–63. Referring to Example 2.9, show that straight rectangular and triangular fins with the same base thickness and mass per unit width as the parabolic fin, dissipate less heat.

2–64. A straight Duralumin fin has a parabolic profile $y = t(1 - x/L)^2$ with $t = 3$ mm and $L = 10$ mm. Determine the heat dissipation by the fin when the base temperature is 400 K and it is exposed to fluid at 300 K with a heat transfer coefficient of 40 W/m^2 K. Also calculate the fin mass.

2–65. (i) An aluminum pin fin has a diameter of 4 mm and a length of 20 mm. Calculate the heat dissipated when the base temperature is 600 K, the fluid temperature is 400 K, and the heat transfer coefficient is 100 W/m^2 K. Take k = 180 W/m K.

(ii) Compare the heat dissipation obtained above with that which would be obtained with parabolic and triangular spines (items 8, 9, 10 of Table 2.2) of the same base area and equal mass.

2–66. A perforated-plate heat exchanger has a cross section as shown, with D_1 = 5.00 cm, D_2 = 5.30 cm, and D_3 chosen to give equal flow areas for each stream. The plates are 1 mm–thick aluminum with 1.5 mm–diameter holes taking up 30% of the plate area. If the convective heat transfer coefficient is 300 W/m^2 K, determine the overall heat transfer coefficient. Take k = 190 W/m K for the aluminum.

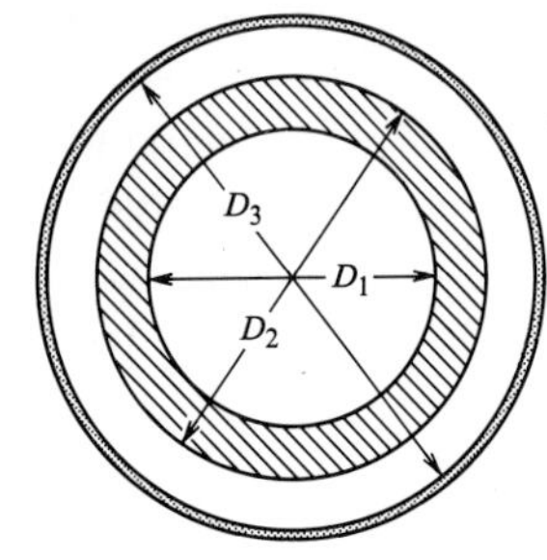

2–67. A steel heat exchanger tube of 2 cm outer diameter is fitted with a steel spiral annular fin of 5 cm outer diameter and 0.2 mm thickness, wound at a pitch of 3 mm. If the outside heat transfer coefficient is 20 W/m^2 K on the bare tube and 15 W/m^2 K on the finned tube, determine the reduction in the outside thermal resistance achieved by adding the fins. Take k_{steel} = 42 W/m K.

2–68. A thin metal disk is insulated on one side and exposed to a jet of hot air at temperature T_e on the other. The periphery at $r = R$ is maintained at a uniform temperature T_R. If the convective heat transfer coefficient h_c can be taken to be constant over the disk, obtain an expression for the temperature at the center of the disk.

2–69. Referring to comment 2 of Example 2.7, obtain the temperature distribution along a fin for which the temperature difference $[T(x) - T_e]$ is constant. Assume the tip is insulated. If the fin efficiency is defined as $\eta_f = \dot{Q}/h_c\mathcal{P}L(T_B - \overline{T}_e)$, show that $\eta_f = 1/[(\chi^2/3) + 1]$. Calculate the value of η_f appropriate to Example 2.7.

2–70. A transistor has a cylindrical cap of radius R and height L.

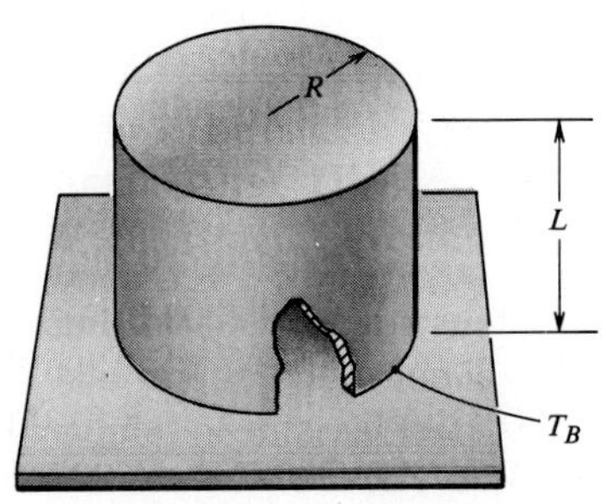

(i) Show that the heat dissipated is

$$\dot{Q} = 2\pi kRt(T_B - T_e)\beta\left[\frac{I_0(\beta R)\sinh\beta L + I_1(\beta R)\cosh\beta L}{I_0(\beta R)\cosh\beta L + I_1(\beta R)\sinh\beta L}\right]$$

where the metal thickness is t, $\beta = (h_c/kt)^{1/2}$, and the heat transfer coefficient on the sides and top is assumed to be the same.

(ii) The cap is fabricated from steel ($k = 50$ W/m K) of thickness 0.3 mm and has a height of 9 mm and a diameter of 8 mm. Air at 310 K blows over the cap to give a heat transfer coefficient of 25 W/m^2 K. What is the base temperature for 400 mW dissipation? Compare your answer to the manufacturer's allowable limit of 370 K.

2–71. The straight fin with a parabolic profile $y = t(1 - x/L)^2$ is of particular interest since it can give the maximum heat loss for a given weight of any profile.

(i) By substituting appropriate values of A_c and $\mathscr{P}$ in Eq. (2.30), show that the governing differential equation is

$$z^2\frac{d^2T}{dz^2} + 2z\frac{dT}{dz} - \beta^2L^2(T - T_e) = 0; \qquad \beta = (h_c/kt)^{1/2}$$

where $z = L - x$. This is an *Euler* equation.

(ii) Show that a solution of the differential equation that satisfies the boundary conditions $T = T_B$ at $x = 0$ and $T = T_e$ at $x = L$ is

$$\frac{T - T_e}{T_B - T_e} = \left(1 - \frac{x}{L}\right)^p; \qquad p = -\frac{1}{2} + \frac{1}{2}(1 + 4\beta^2L^2)^{1/2}$$

(iii) If $p = 1$, the profile is linear. Since q_x is then constant, this fin proves to be the fin of least material and hence gives the maximum heat dissipation for a given weight. Show that the efficiency of such a fin is 50% (see Example 2.9).

2–72. An annular fin of uniform thickness has an inner radius of 2 cm, an outer radius of 4 cm, and a thickness of 2 mm. The material is steel with $k = 60$ W/m K. It is cooled by air at 20°C, giving a convective heat transfer coefficient of 24 W/m^2 K. When the fin base is at 110°C, determine the rate at which heat is dissipated by the fin.

2–73. A wall has its surface maintained at 180°C and is in contact with a fluid at 80°C. Find the percent increase in the heat dissipation if triangular fins are added to the surface. The fins are 6 mm thick at the base, are 30 mm long, and are spaced at a pitch of 15 mm. Assume that the heat transfer coefficient is 20 W/m^2 K for both the plain and finned surfaces, and that the fin material thermal conductivity is 50 W/m K.

2–74. A thin metal disk is insulated on one side, and the other side is exposed to a high-temperature radiation source and convective cooling. Obtain an expression for the difference between the temperature at the center and at the outer edge. Ignore reradiation.

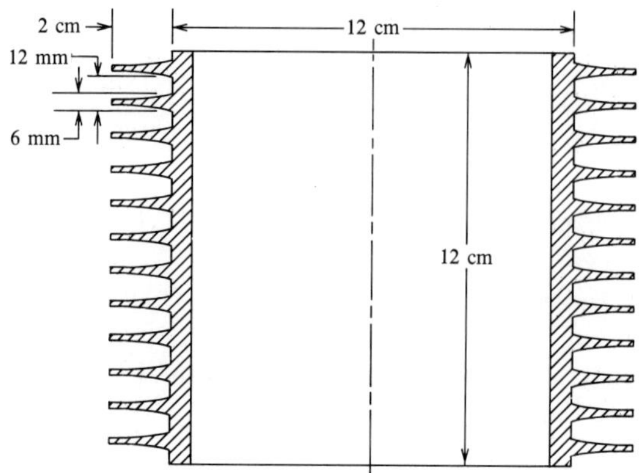

2–75. Cylinders of air-cooled internal combustion engines are provided with cooling fins owing to the large amount of heat that must be dissipated. A two-stroke motorcycle engine has a cast aluminum alloy 195 cylinder of height 12 cm and outside diameter 12 cm, with hyperbolic fins of base width 6 mm, pitch 12 mm, and length 20 mm. In a test simulating a road speed of 90 km/h, the fin base temperatures are measured to average 485 K for ambient air at 300 K. If the heat transfer coefficient is estimated to be 60 W/m^2 K, determine the heat loss from the cylinder. Under these conditions, how does the efficiency and mass of the hyperbolic fin compare with the rectangular fin of the same base and length?

2–76. So-called "compact" heat exchanger cores often consist of finned passages between parallel plates. A particularly simple configuration has square passages with the effective fin length equal to half the plate spacing L. In a particular application with $L = 5$ mm, a convective heat transfer coefficient of 160 W/m^2 K is expected. If 95% efficient fins are desired, how thick should they be if the core is constructed from

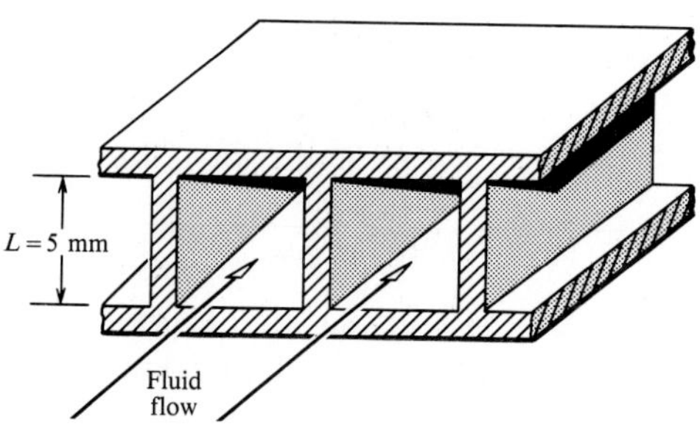

(i) an aluminum alloy with $k = 180$ W/m K?
(ii) mild steel with $k = 64$ W/m K?
(iii) a plastic with $k = 0.33$ W/m K?

Discuss the significance of your results to the design of such cores.

2–77. A steel heat exchanger tube of 2 cm outer diameter is fitted with a steel spiral annular fin of 4 cm outer diameter, thickness 0.4 mm, and wound at a pitch of 3 mm. If the outside heat transfer coefficient is 20 W/m^2 K on the original bare tube and 15 W/m^2 K on the finned tube, determine the reduction in the outside thermal resistance achieved by adding the fins. Take $k_{steel} = 42$ W/m K.

CHAPTER

3

MULTIDIMENSIONAL AND UNSTEADY CONDUCTION

CONTENTS

3.1 INTRODUCTION

The analyses of steady one-dimensional heat conduction in Chapter 2 were relatively simple but nevertheless gave results that are widely used by engineers for the design of thermal systems. In general, however, heat conduction can be **unsteady**; that is, temperatures change with time. An example is heat flow through the cylinder wall of an automobile engine. Also, heat conduction can be **multidimensional**; that is, temperatures vary significantly in more than one coordinate direction. An example is heat loss from a hot oil line buried underneath the ground. Heat conduction can also be simultaneously unsteady and multidimensional, for example, when a rectangular block forging is quenched.

In Chapter 2, each new analysis commenced with the application of the first law to an elemental volume to yield the governing differential equation. In Chapter 3, our approach will be different. We will derive a partial differential equation that governs the temperature distribution in a solid under very general conditions; we will then start each analysis by choosing the form of this equation appropriate to the problem under consideration. This **general heat conduction equation** is derived in Section 3.2, where boundary and initial conditions as well as solution methods are discussed. Solution methods are broadly divided into two groups: (1) classical mathematical methods, and (2) numerical methods. Classical mathematical methods are demonstrated in Section 3.3 for multidimensional steady conduction, and in Section 3.4 for unsteady conduction. In particular, the method of separation of variables is used, which leads to the need to construct Fourier series expansions. Section 3.5 introduces the analysis of a special class of conduction problems, in which there is a moving boundary—for example, solidification of ice from water. Numerical methods commonly used to solve the heat conduction equation include the finite-difference method, the finite-element method, and the boundary-element method. In Section 3.6, use of the finite-difference method is demonstrated for steady two-dimensional conduction, unsteady one-dimensional conduction, and a moving-boundary problem.

The classical mathematical methods used to solve the heat conduction equation might at first appear intimidating to the student. However, these methods rely on concepts normally studied in freshman- and sophomore-level mathematics courses, such as partial differentiation, integration, and second-order ordinary differential equations. Sufficient detail is given in the analyses for the student to proceed step by step without having to refer to a text on advanced engineering mathematics. Those students who have already had a junior- or senior-level engineering mathematics course should find the mathematics straightforward (and even perhaps old-fashioned!).

3.2 THE HEAT CONDUCTION EQUATION

In this section, the energy conservation principle and Fourier's law of heat conduction are used to derive various forms of the differential equation governing the temperature distribution in a stationary medium. The types of boundary and initial conditions encountered in practical problems are then discussed and classified. Finally, various methods available for solving the equation are introduced.

3.2.1 Fourier's Law as a Vector Equation

Chapter 2 used one-dimensional forms of Fourier's law of conduction. In general, the temperature in a body may vary in all three coordinate directions, which requires a more general form of Fourier's law. For simplicity, we will restrict our attention to *isotropic* media, for which the conductivity is the same in all directions. Most materials are isotropic. Exceptions include timber, which has different values of thermal conductivity k along and perpendicular to the grain, and pyrolytic graphite, for which the value of k can vary by an order of magnitude in different directions. For an isotropic medium, Fourier's law in terms of Cartesian coordinates is

$$q_x = -k\frac{\partial T}{\partial x}; \qquad q_y = -k\frac{\partial T}{\partial y}; \qquad q_z = -k\frac{\partial T}{\partial z} \tag{3.1}$$

where q_x is the component of the heat flux in the x direction, $\partial T/\partial x$ is the *partial derivative* of $T(x, y, z, t)$ with respect to x, and so on. As indicated in Fig. 3.1, Eq. (3.1) can be written more compactly in vector form as

$$\mathbf{q} = -k\nabla T \tag{3.2}$$

where $\mathbf{q}$ is the conduction heat flux vector, and ∇T is the gradient of the scalar temperature field. In Cartesian coordinates,

$$\mathbf{q} = \mathbf{i}q_x + \mathbf{j}q_y + \mathbf{k}q_z$$

$$\nabla T = \mathbf{i}\frac{\partial T}{\partial x} + \mathbf{j}\frac{\partial T}{\partial y} + \mathbf{k}\frac{\partial T}{\partial z}$$

where $\mathbf{i}$, $\mathbf{j}$, and $\mathbf{k}$ are the unit vectors in the x, y, and z directions, respectively.

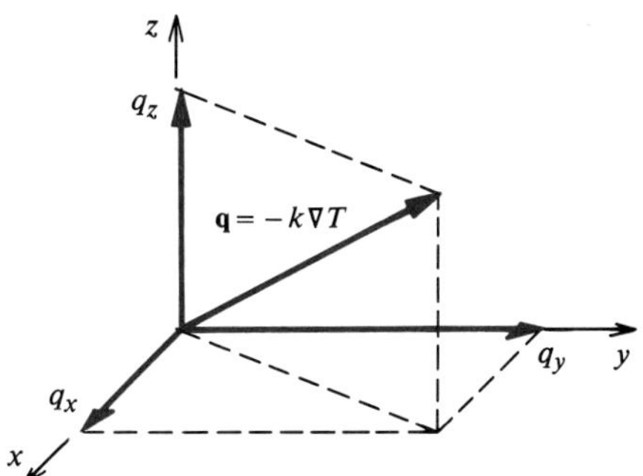

Figure 3.1 Vector representation of Fourier's law of heat conduction.

3.2.2 Derivation of the Heat Conduction Equation

The general heat conduction equation will first be derived in a Cartesian coordinate system. Subsequently, the result will be written in cylindrical and spherical coordinates.

Derivation in Cartesian Coordinates

Figure 3.2 depicts an elemental volume Δx by Δy by Δz located in a solid. The energy conservation principle, Eq. (1.2), applied to the elemental volume as a closed system gives

$$\rho(\Delta x \Delta y \Delta z)c\frac{\partial T}{\partial t} = \dot{Q} + \dot{Q}_v \quad \textbf{(3.3)}$$

where the time derivative is a partial derivative, since T is also a function of the spatial coordinates x, y, and z.

The term $\dot{Q}$ represents heat transfer across the volume boundaries by conduction. The rate of heat inflow across the face at x is

$$q_x|_x \Delta y \Delta z$$

and the outflow across the face at $x + \Delta x$ is

$$q_x|_{x+\Delta x} \Delta y \Delta z$$

The net inflow in the x direction is then

$$(q_x|_x - q_x|_{x+\Delta x})\Delta y \Delta z$$

The outflow heat flux can be expanded in a Taylor series as

$$q_x|_{x+\Delta x} = q_x|_x + \frac{\partial q_x}{\partial x}\Delta x + \text{Higher-order terms}$$

Substituting and dropping the higher-order terms gives the net inflow in the x direction as

$$-\frac{\partial q_x}{\partial x}\Delta x \Delta y \Delta z$$

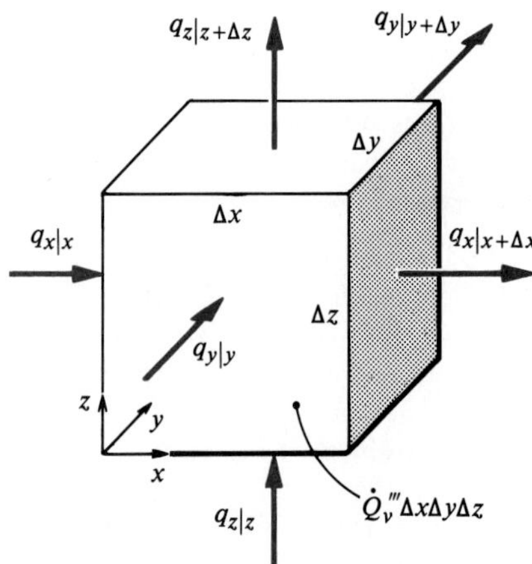

Figure 3.2 Three-dimensional Cartesian elemental volume in a solid for derivation of the heat conduction equation.

Similar terms arise from conduction in the y and z directions. Thus, the net heat transfer into the volume by conduction is

$$\left(-\frac{\partial q_x}{\partial x}-\frac{\partial q_y}{\partial y}-\frac{\partial q_z}{\partial z}\right)\Delta x\Delta y\Delta z$$

The rate of generation of thermal energy within the volume $\dot{Q}_v$ is simply

$$\dot{Q}_v'''\Delta x\Delta y\Delta z$$

where $\dot{Q}_v'''$ [W/m^3] is the rate of internal or volumetric heat generation introduced in Section 2.3.4. Substituting in Eq. (3.3) and dividing by $\Delta x\Delta y\Delta z$ gives

$$\rho c\frac{\partial T}{\partial t}=-\left(\frac{\partial q_x}{\partial x}+\frac{\partial q_y}{\partial y}+\frac{\partial q_z}{\partial z}\right)+\dot{Q}_v'''$$

Introducing Fourier's law, Eq. (3.1), for q_x, q_y, and q_z,

$$\rho c\frac{\partial T}{\partial t}=\frac{\partial}{\partial x}\left(k\frac{\partial T}{\partial x}\right)+\frac{\partial}{\partial y}\left(k\frac{\partial T}{\partial y}\right)+\frac{\partial}{\partial z}\left(k\frac{\partial T}{\partial z}\right)+\dot{Q}_v''' \tag{3.4}$$

Notice that the thermal conductivity k has been left inside the derivatives since, in general, k is a function of temperature. However, we usually simplify heat conduction analysis by taking k to be independent of temperature; k is then also independent of position, and Eq. (3.4) becomes

$$\rho c\frac{\partial T}{\partial t}=k\left(\frac{\partial^2 T}{\partial x^2}+\frac{\partial^2 T}{\partial y^2}+\frac{\partial^2 T}{\partial z^2}\right)+\dot{Q}_v''' \tag{3.5}$$

When there is no internal heat generation, $\dot{Q}_v'''=0$, and Eq. (3.5) reduces to

$$\frac{\partial T}{\partial t}=\alpha\left(\frac{\partial^2 T}{\partial x^2}+\frac{\partial^2 T}{\partial y^2}+\frac{\partial^2 T}{\partial z^2}\right) \tag{3.6}$$

where $\alpha\equiv k/\rho c$ [m^2/s] is a thermophysical property of the material called the **thermal diffusivity**. Table 3.1 gives selected values of the thermal diffusivity. Additional data are given in Appendix A, as are values for k, ρ, and c, from which α can be calculated. Equation (3.6) is called **Fourier's equation** (or the *heat* or *diffusion equation*) and governs the temperature distribution $T(x,y,z,t)$ in a solid. The relevance of the thermal diffusivity can be seen in Fourier's equation: when there is no internal heat generation, it is the only physical property that influences temperature changes in the solid. The thermal diffusivity is the ratio of thermal conductivity to a volumetric heat capacity: the larger α, the faster temperature changes will propagate through the solid.

For a timewise steady state, $\partial/\partial t=0$, and Eq. (3.5) reduces to *Poisson's equation:*

$$\frac{\partial^2 T}{\partial x^2}+\frac{\partial^2 T}{\partial y^2}+\frac{\partial^2 T}{\partial z^2}=-\frac{\dot{Q}_v'''}{k} \tag{3.7}$$

Table 3.1 Selected values of thermal diffusivity at 300 K (~25°C). (Other values may be calculated from the data given in Appendix A.)

Material	α $m^2/s \times 10^6$
Copper	112
Aluminum	84
Brass, 70% Cu, 30% Zn	34.2
Air at 1 atm pressure	22.5
Mild steel	18.8
Mercury	4.43
Stainless steel, 18-8	3.88
Fiberglass (medium density)	1.6
Concrete	0.75
Pyrex glass	0.51
Cork	0.16
Water	0.147
Engine oil, SAE 50	0.086
Neoprene rubber	0.079
White pine, perpendicular to grain	0.071
Refrigerant R-12	0.056
Polyvinylchloride (PVC)	0.051

Note: This table should be read as $\alpha \times 10^6$ m^2/s = Listed value; for example, for copper $\alpha = 112 \times 10^{-6}$ m^2/s.

Finally, for a steady state and no internal heat generation, $\partial/\partial t = 0, \dot{Q}_v''' = 0$, so

$$\frac{\partial^2 T}{\partial x^2} + \frac{\partial^2 T}{\partial y^2} + \frac{\partial^2 T}{\partial z^2} = 0 \tag{3.8}$$

which is **Laplace's equation**.

The Fourier, Poisson, and Laplace equations are *partial differential equations* and have been thoroughly studied by mathematicians [1,2,3]. They are important because each is the governing equation for many different physical phenomena in fields as diverse as heat conduction, mass diffusion, electrostatics, and fluid mechanics. Solutions of Laplace's equation are called *potential* or *harmonic* functions.

Other Coordinate Systems

The solution of partial differential equations is simpler when boundary conditions are specified on *coordinate surfaces*, for example, x = Constant in the Cartesian coordinate system. Thus, for conduction problems in cylindrical or spherical bodies, the Cartesian coordinate system is inappropriate. Such problems require heat conduction equations in terms of the cylindrical and spherical coordinate systems shown in Fig. 3.3. We can proceed in a number of ways. The most direct approach

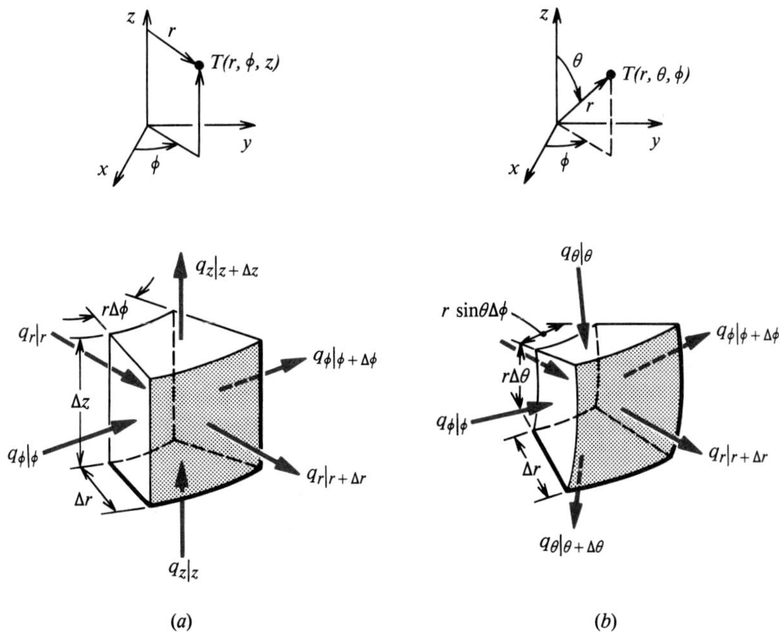

Figure 3.3 (*a*) Cylindrical coordinate system, r, ϕ, z, and an elemental volume. (*b*) Spherical coordinate system, r, θ, ϕ, and an elemental volume.

is to repeat our derivation using appropriate volume elements, as shown in Fig. 3.3. Alternatively, we can transform the equations already derived in Cartesian coordinates to cylindrical or spherical coordinates; Exercises 3–1, 3–2, and 3–4 illustrate these procedures. To present the results in compact form, we introduce the *del-squared* or *Laplacian operator*. In Cartesian coordinates,

$$\nabla^2 = \frac{\partial^2}{\partial x^2} + \frac{\partial^2}{\partial y^2} + \frac{\partial^2}{\partial z^2}$$

and Eq. (3.5) becomes

$$\rho c \frac{\partial T}{\partial t} = k\nabla^2 T + \dot{Q}_v''' \tag{3.9}$$

Writing the heat conduction equation in any coordinate system then simply requires the proper expression for ∇^2. For cylindrical coordinates r, ϕ, z,

$$\nabla^2 = \frac{1}{r}\frac{\partial}{\partial r}\left(r\frac{\partial}{\partial r}\right) + \frac{1}{r^2}\frac{\partial^2}{\partial \phi^2} + \frac{\partial^2}{\partial z^2} \tag{3.10}$$

For spherical coordinates r, θ, ϕ,

$$\nabla^2 = \frac{1}{r^2}\frac{\partial}{\partial r}\left(r^2\frac{\partial}{\partial r}\right) + \frac{1}{r^2 \sin\theta}\frac{\partial}{\partial \theta}\left(\sin\theta\frac{\partial}{\partial \theta}\right) + \frac{1}{r^2 \sin^2\theta}\frac{\partial^2}{\partial \phi^2} \quad \textbf{(3.11)}$$

Two additional useful relations are

$$\frac{1}{r}\frac{\partial}{\partial r}\left(r\frac{\partial}{\partial r}\right) = \frac{\partial^2}{\partial r^2} + \frac{1}{r}\frac{\partial}{\partial r} \quad \textbf{(3.12}\boldsymbol{a}\textbf{)}$$

$$\frac{1}{r^2}\frac{\partial}{\partial r}\left(r^2\frac{\partial}{\partial r}\right) = \frac{\partial^2}{\partial r^2} + \frac{2}{r}\frac{\partial}{\partial r} \quad \textbf{(3.12}\boldsymbol{b}\textbf{)}$$

As a final comment regarding the heat conduction equation, it is noted that Fourier's law, Eq. (3.2), is a vector equation; thus, the heat conduction equation is most efficiently derived using the methods of vector calculus. Such a derivation is required as Exercise 3–3.

3.2.3 Boundary and Initial Conditions

In solving heat conduction problems in Chapters 1 and 2, boundary and initial conditions were used to evaluate integration constants. We now classify the types of boundary and initial conditions required to solve heat conduction problems.

Boundary Conditions

In Chapters 1 and 2, it was shown how practical heat conduction problems involve adjacent regions that may be quite different. For example, in Example 2.4, a uranium oxide nuclear fuel rod is enclosed in a Zircaloy-4 sheath and cooled by flowing water. Heat is generated within the fuel and flows by conduction to the fuel-sheath interface, by conduction across the sheath to the sheath-water interface, and by convection into the water. To analyze such problems, it is necessary to specify thermal conditions at solid-solid and solid-liquid interfaces. In general, it is required that both the heat flux and the temperature be continuous across an interface (although when a contact resistance model is used, as described in Section 2.2.2, the effect is to have a discontinuity in temperature). Thus, the solutions of the heat conduction equation in each region are *coupled.*

When analyzing more difficult heat transfer problems, we often find it convenient to uncouple the regions and consider each region independently. The boundary condition is then simply one of specified temperature. Considering the coordinate surface $x = L$ and referring to Fig. 3.4a,

$$T|_{x=L} = T_s \quad \textbf{(3.13)}$$

which is called a **first-kind** or **Dirichlet** boundary condition.

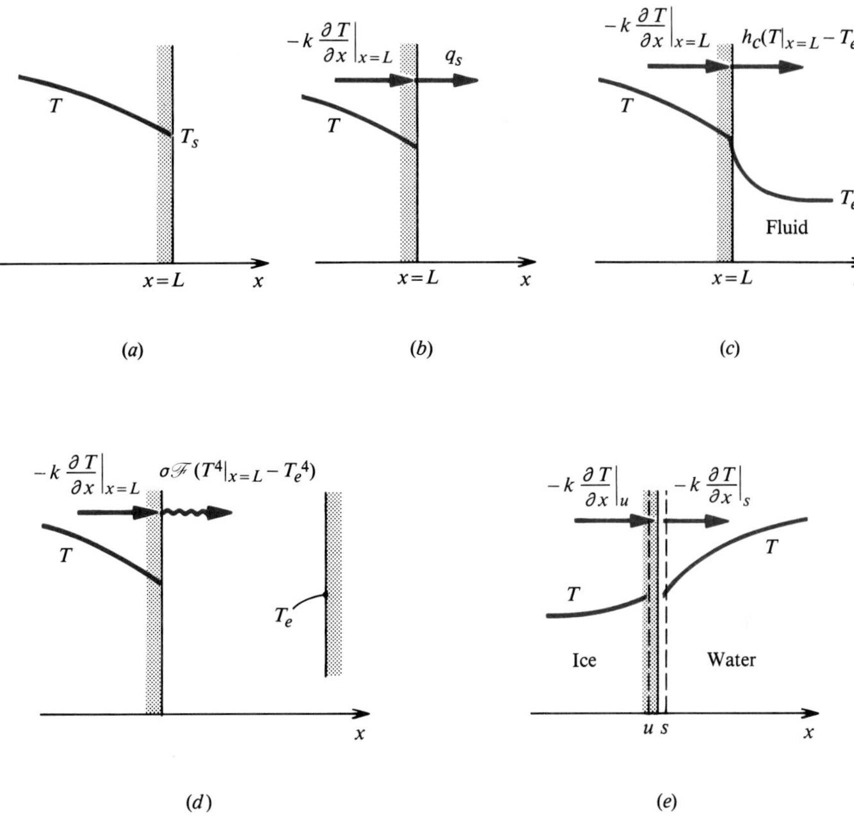

Figure 3.4 Types of boundary conditions. (*a*) First kind, or Dirichlet. (*b*) Second kind, or Neumann. (*c*) Third kind, or mixed. (*d*) Fourth kind, or radiation. (*e*) Phase change.

Sometimes it is convenient to apply a boundary condition at a surface where the heat flux is known. Referring to Fig. 3.4*b*,

$$-k\frac{\partial T}{\partial x}\bigg|_{x=L} = q_s \tag{3.14}$$

which is called a **second-kind** or **Neumann** boundary condition. Often the known heat flux is zero, such as at a plane of symmetry, or approximately zero, such as when the adjoining region is a good insulator.

Another commonly encountered situation is one in which the adjacent region is a fluid and we wish to describe heat transfer to the fluid using Newton's law of cooling. Referring to Fig. 3.4*c*,

$$-k\frac{\partial T}{\partial x}\bigg|_{x=L} = h_c\left(T|_{x=L} - T_e\right) \tag{3.15}$$

which is the **third-kind** or **mixed** boundary condition. Notice that Eq. (3.15) involves both the value of the dependent variable and that of its derivative at the boundary. Similarly, if there is heat loss by thermal radiation into adjacent surroundings (see Fig. 3.4*d*), using Eq. (1.17),

$$-k\frac{\partial T}{\partial x}\bigg|_{x=L} = \sigma\mathscr{F}\left(T|^4_{x=L} - T_e^4\right) \tag{3.16}$$

which is a **fourth-kind** or **radiation** boundary condition.[1] (Note, however, that some texts refer to the third-kind boundary condition as a radiation boundary condition, a practice with historical precedent but preferably avoided.)

There are other, more complex boundary conditions encountered in practice. The contact resistance described in Section 2.2.2 is one example. Another example is when there is a change of phase at the interface, such as ice melting or steam condensing. Figure 3.4*e* represents the surface of a block of ice melting in warm water. The *s*- and *u*-surfaces are on either side and infinitesimally close to the actual water-ice interface. Thus, the *s*-surface is in water, and the *u*-surface is in solid ice. Only the temperature is the same at the *s*- and *u*-surfaces; in general, the physical properties and temperature gradients are different. If the ice is melting at a rate per unit area $\dot{m}''$ [kg/m^2 s] and the interface is imagined to be fixed in space, then ice flows toward the interface, and water flows away at the melting rate. Application of the steady-flow energy equation to the control volume bounded by the *u*- and *s*-surfaces gives

$$\dot{m}''(h_s - h_u) = -k\frac{\partial T}{\partial x}\bigg|_u - \left(-k\frac{\partial T}{\partial x}\bigg|_s\right) \tag{3.17}$$

for $\dot{m}''$ positive. Equation (3.17) can be rearranged as

$$k\frac{\partial T}{\partial x}\bigg|_s = k\frac{\partial T}{\partial x}\bigg|_u + \dot{m}''h_{fs} \tag{3.18}$$

where $h_{fs} = h_s - h_u$ is the enthalpy of fusion of the ice. Equation (3.18) can be interpreted as stating that the heat conducted from the warm water to the interface must balance both the heat conducted away from the interface into the cold ice and the heat required to melt the ice.

Initial Conditions

Transient heat conduction problems usually require specification of an **initial condition,** which simply means that the temperature throughout the region must be known at some instant in time before its subsequent variation with time can be

[1] This simple form is strictly valid only when the surroundings are isothermal and have a uniform emittance. More general situations are treated in Chapter 6.

determined. An exception is when the temperature varies periodically, for example, conduction in a spacecraft orbiting the earth. Then a periodic condition must be imposed on the solution.

3.2.4 Solution Methods

During the 19th century, considerable progress was made in developing mathematical methods for solving the various forms of the heat conduction equation. The first major contribution was by J. Fourier. His book, published in 1822 [4], developed the use of the method of **separation of variables**, which leads to the need to express an arbitrary function in a **Fourier series expansion**. Subsequently, **transform methods**—particularly the use of the *Laplace transform*—as well as other methods of classical mathematics were widely used. The treatise on heat conduction by Carslaw and Jaeger [5] contains an extensive compilation of solutions obtained using classical mathematical methods. Use of these methods usually requires that (1) the bounding surfaces be of relatively simple shape, (2) the boundary conditions be of simple mathematical form, and (3) the thermophysical properties be constant. Notwithstanding these limitations, many analytical results have wide engineering utility. Analytical solutions are useful benchmarks for checking the accuracy of numerical methods of solution. Also, the exercise of obtaining an analytical solution gives valuable insight into the essential features of heat transfer by conduction.

More recently, numerical methods, including **finite-difference** and **finite-element** methods, have been developed that allow solutions to be easily obtained for problems involving unusual shapes, complicated boundary conditions, and variable thermophysical properties. An early example was the application of the numerical *relaxation* method to steady heat conduction by H. Emmons in 1943 [6]. However, with the advent of the modern high-speed computer in the 1960s, numerical methods have been greatly improved. The wide availability of the personal computer in the 1980s has led to the marketing of versatile computer programs. These can be used to solve a great variety of heat conduction problems without requiring the user to have detailed knowledge of the numerical methods involved.

Some other methods have been used to obtain solutions to heat conduction problems. Two graphical methods, the *flux plotting* method and the *Schmidt plot*, are described in some heat transfer texts. The first is used for steady two-dimensional conduction and involves the free-hand sketching of isotherms and lines of heat flow; the latter is used for transient conduction and is the graphical equivalent to a finite-difference numerical method. A number of *analog* methods have also been used. Since Laplace's equation also governs electrical potential fields, two-dimensional steady conduction problems have been solved by making voltage and current measurements in appropriate shapes cut out of graphite-coated paper of high electrical resistance [7]. Before digital computers were developed, the analog computer or "differential analyzer" was used to solve heat conduction problems [8].

Methods of mathematical analysis are demonstrated in Sections 3.3 through 3.5, and finite-difference methods are dealt with in Section 3.6.

3.3 MULTIDIMENSIONAL STEADY CONDUCTION

One-dimensional steady conduction was dealt with in Chapter 2. Although the simple analytical results obtained are very useful, they have obvious limitations. Often the heat flow is multidimensional, that is, in two or three directions. For example, consider the furnace shown in Fig. 3.5. If the insulation is thin compared to the furnace dimensions, the assumption of one-dimensional heat flow is adequate (see Example 1.1). High-temperature furnaces, however, require thick insulation to reduce heat loss; in these furnaces, the heat flow through the edges is two-dimensional, and through the corners it is three-dimensional. Multidimensional steady conduction with no internal heat generation is governed by Laplace's equation. The classical approach to solving Laplace's equation is the *separation of variables* method; Section 3.3.1 uses a simple two-dimensional problem to demonstrate this approach. Often we are concerned with conduction between two isothermal surfaces, all other surfaces present being adiabatic. A **conduction shape factor** can be defined for such configurations, and a compilation of useful shape factors is given in Section 3.3.3.

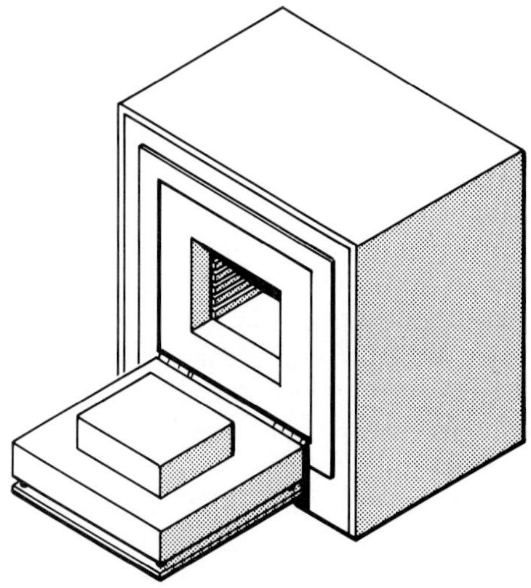

Figure 3.5 An insulated furnace.

3.3.1 Steady Conduction in a Rectangular Plate

The relatively simple problem of two-dimensional steady heat conduction in a rectangular plate will be used to demonstrate the method of separation of variables for solving Laplace's equation.

The Governing Equation and Boundary Conditions

Figure 3.6 depicts a thin rectangular plate with negligible heat loss from its surface. Temperature variations across the plate in the z direction are assumed to be zero ($\partial^2 T/\partial z^2 = 0$), and the thermal conductivity is assumed to be constant.

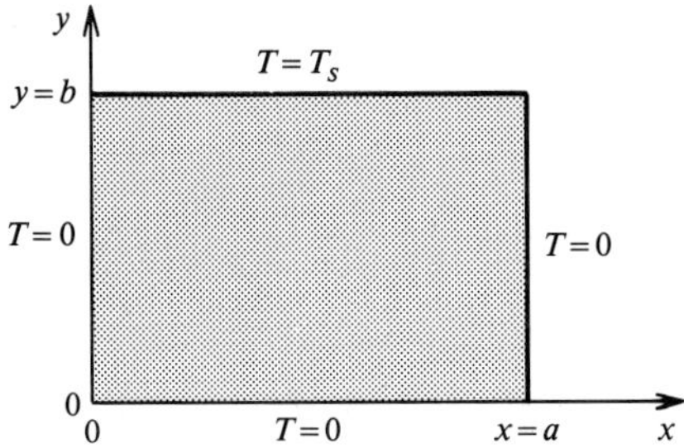

Figure 3.6 Boundary conditions for two-dimensional steady conduction in a rectangular plate.

The temperature distribution $T(x, y)$ is then governed by the two-dimensional form of Laplace's equation:

$$\frac{\partial^2 T}{\partial x^2} + \frac{\partial^2 T}{\partial y^2} = 0 \tag{3.19}$$

To demonstrate the method of solution, we will first consider a model problem and later demonstrate how our solution can also be used to construct solutions for practical situations. Thus, we choose the boundary conditions shown in Fig. 3.6, namely,

$$\begin{aligned} x = 0,\quad & 0 < y < b: \quad T = 0 \\ y = 0,\quad & 0 < x < a: \quad T = 0 \\ x = a,\quad & 0 < y < b: \quad T = 0 \end{aligned} \tag{3.20a}$$

$$y = b,\quad 0 < x < a: \quad T = T_s \tag{3.20b}$$

The zero values for the boundary conditions, Eqs. (3.20*a*), will simplify the problem without loss of generality. Such boundary conditions are *homogeneous*. In general, a differential equation or boundary condition is said to be homogeneous if a constant times a solution is also a solution, that is, if T_1 is a solution, CT_1 is also a solution. The boundary condition Eq. (3.20*b*) looks simple, but we shall see that it actually makes solution of the problem quite difficult.

Solution for the Temperature Distribution

To use the method of separation of variables, we first assume that the function $T(x, y)$ can be expressed as the product of a function of x only, $X(x)$, and a function of y only, $Y(y)$:

$$T(x, y) = X(x)Y(y) \tag{3.21}$$

Substitution in Eq. (3.19) gives

$$Y\frac{d^2X}{dx^2} + X\frac{d^2Y}{dy^2} = 0$$

where total derivatives have replaced the partial derivatives, since X is a function of independent variable x only, and Y is a function of independent variable y only. Rearranging,

$$-\frac{1}{X}\frac{d^2X}{dx^2} = \frac{1}{Y}\frac{d^2Y}{dy^2} \tag{3.22}$$

Since each side of Eq. (3.22) is a function of a single independent variable, the equality can hold only if both sides are equal to a constant, which can be positive, negative, or zero. Using hindsight, the constant will be chosen to be a positive number λ^2. The reason for this choice will be discussed later. Two ordinary differential equations are thus obtained,

$$\frac{d^2X}{dx^2} + \lambda^2 X = 0 \qquad \frac{d^2Y}{dy^2} - \lambda^2 Y = 0$$

which have the solutions

$$X = B\cos\lambda x + C\sin\lambda x \qquad Y = De^{-\lambda y} + Ee^{\lambda y}$$

Substituting in Eq. (3.21) gives a tentative solution for $T(x, y)$:

$$T(x, y) = (B\cos\lambda x + C\sin\lambda x)(De^{-\lambda y} + Ee^{\lambda y}) \tag{3.23}$$

Applying boundary conditions Eqs. (3.20*a*) and taking care to ensure that the x- and y- dependences remain,

$$x = 0: \quad B(De^{-\lambda y} + Ee^{\lambda y}) = 0; \qquad \text{thus}, B = 0$$

$$y = 0: \quad C\sin\lambda x(D + E) = 0; \qquad \text{thus}, E = -D$$

$$x = a: \quad CD\sin\lambda a(e^{-\lambda y} - e^{\lambda y}) = -2CD\sin\lambda a\sinh\lambda y = 0$$

This requires that $\sin\lambda a = 0$, which has the roots $\lambda_n = n\pi/a$, for $n = 0, 1, 2, 3, \ldots$. These values of λ are called the **eigenvalues** or *characteristic values* of the problem. There is a distinct solution for each eigenvalue, each with its own constant. Writing the constant $-2CD$ for the nth solution as A_n,

$$T_n(x, y) = A_n \sin\frac{n\pi x}{a}\sinh\frac{n\pi y}{a}; \qquad n = 0, 1, 2, 3, \ldots \tag{3.24}$$

Equation (3.19) is a *linear* differential equation, so its general solution is a sum of the series of solutions given by Eq. (3.24):

$$T(x, y) = \sum_{n=1}^{\infty} A_n \sin\frac{n\pi x}{a}\sinh\frac{n\pi y}{a} \tag{3.25}$$

where the solution for $n = 0$ has been deleted since $\sinh 0 = 0$. We now apply the last boundary condition, Eq. (3.20*b*), which requires that at $y = b$,

$$T|_{y=b} = T_s = \sum_{n=1}^{\infty} A_n \sin\frac{n\pi x}{a}\sinh\frac{n\pi b}{a} \tag{3.26}$$

that is, the constant T_s must be expressed in terms of an infinite series of sine functions, or a *Fourier series*.

Construction of a Fourier Series Expansion

In general, the temperature distribution along the boundary $y = b$ will be some arbitrary function $f(x)$, and the required expansion is then

$$f(x) = \sum_{n=1}^{\infty} C_n \sin \frac{n\pi x}{a}; \qquad C_n = A_n \sinh \frac{n\pi b}{a} \tag{3.27}$$

The constants C_n are determined as follows. Multiply Eq. (3.27) by $\sin n\pi x/a$, and integrate term by term from $x = 0$ to $x = a$:

$$\int_0^a f(x) \sin \frac{n\pi x}{a} dx = \int_0^a C_1 \sin \frac{\pi x}{a} \sin \frac{n\pi x}{a} dx + \cdots + \int_0^a C_n \sin^2\left(\frac{n\pi x}{a}\right) dx$$
$$+ \cdots + \int_0^a C_m \sin \frac{m\pi x}{a} \sin \frac{n\pi x}{a} dx + \cdots \tag{3.28}$$

Using standard integral tables, we find:

$$\int_0^a \sin \frac{n\pi x}{a} \sin \frac{m\pi x}{a} dx = \left[\frac{\sin(n\pi x/a - m\pi x/a)}{2(n\pi/a - m\pi/a)} - \frac{\sin(n\pi x/a + m\pi x/a)}{2(n\pi/a + m\pi/a)}\right]_0^a$$
$$= 0 \text{ for } n \neq m$$

$$\int_0^a \sin^2\left(\frac{n\pi x}{a}\right) dx = \frac{a}{2n\pi}\left[\frac{n\pi x}{a} - \frac{1}{2}\sin\frac{2n\pi x}{a}\right]_0^a = \frac{a}{2}$$

Thus, only the nth term on the right-hand side of Eq. (3.28) remains, and solving for C_n gives

$$C_n = \frac{2}{a}\int_0^a f(x) \sin \frac{n\pi x}{a} dx \tag{3.29}$$

The set of sine functions $\sin \pi x/a$, $\sin 2\pi x/a, \ldots, \sin n\pi x/a, \ldots$ is said to be **orthogonal** over the interval $0 \le x \le a$, because the integral of $\sin m\pi x/a \sin n\pi x/a$ is zero if $m \neq n$. As shown in texts on applied mathematics [2,3], if the function $f(x)$ is piecewise continuous, it can always be expressed in terms of a uniformly converging series of orthogonal functions. The cosine function is also orthogonal over an appropriate interval, as are many other functions, including Bessel functions and Legendre polynomials.

Temperature Distribution for $f(x) = T_s$

For the function $f(x)$ equal to a constant value T_s, the integral in Eq. (3.29) can be evaluated analytically:

$$C_n = \frac{2}{a}\left[-\frac{T_s a}{n\pi}\cos\frac{n\pi x}{a}\right]_0^a = T_s\frac{2}{n\pi}[1 - (-1)^n]$$

and from Eq. (3.27),

$$A_n = \frac{C_n}{\sinh(n\pi b/a)} = T_s \frac{2[1-(-1)^n)]}{n\pi \sinh(n\pi b/a)}$$

Substituting into Eq. (3.25) gives the desired temperature distribution for steady conduction in a rectangular plate:

$$T(x, y) = T_s \sum_{n=1}^{\infty} \frac{2[1-(-1)^n]}{n\pi \sinh(n\pi b/a)} \sin\frac{n\pi x}{a} \sinh\frac{n\pi y}{a} \tag{3.30}$$

Lines of constant temperature, or *isotherms*, are shown in Fig. 3.7. The temperature discontinuities at the top corners are physically unrealistic since an infinite heat flow is implied. In fact, the heat flow across the plate edge at $y = b$, evaluated from Eq. (3.30) using Fourier's law, is infinite. Thus, the isotherms in the vicinity of these corners correspond to a mathematical problem only; in real physical problems, there might be a very marked variation in temperature along the edges near these corners, but there cannot be an actual discontinuity.

Since the variables in the partial differential equation, Eq. (3.19), did separate, and because a solution could be found to satisfy the boundary conditions, the method of solution has been successful. If a negative constant is chosen for Eq. (3.22), the boundary conditions cannot be satisfied. Use of a negative constant just reverses the roles of the independent variables x and y, and the negative constant is appropriate if the temperature T_s is specified on the edge $x = a$.

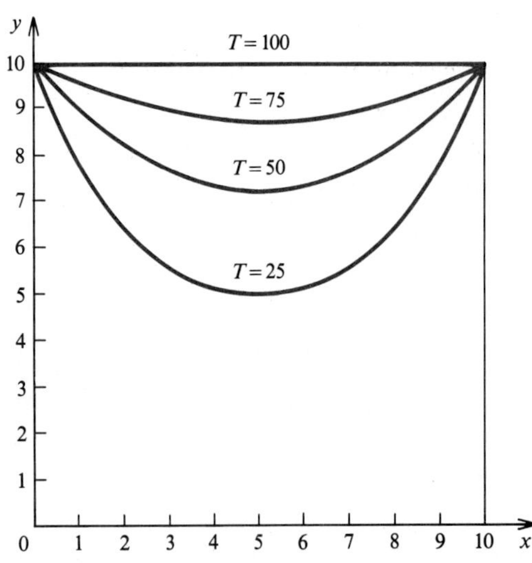

Figure 3.7 Isotherms for conduction in a rectangular plate obtained from Eq. (3.30); $a = b = 10$, $T_s = 100$.

Generalization using the Principle of Superposition

Laplace's equation is a *linear* differential equation. A useful consequence of this property is that the solution of a problem with complicated boundary conditions can be constructed by adding solutions for problems having simpler boundary conditions.

To illustrate the procedure, consider a problem where the plate temperature is specified along two edges. If the boundary conditions Eqs. (3.20) are replaced by those shown in Fig. 3.8,

$$x = 0, \quad 0 < y < b: \quad T = 0$$
$$y = 0, \quad 0 < x < a: \quad T = 0$$
$$x = a, \quad 0 < y < b: \quad T = f_1(y)$$
$$y = b, \quad 0 < x < a: \quad T = f_2(x)$$

the *superposition principle* can be used as follows. Let $T(x, y) = T_1(x, y) + T_2(x, y)$, where T_1 and T_2 satisfy

$$\frac{\partial^2 T_1}{\partial x^2} + \frac{\partial^2 T_1}{\partial y^2} = 0 \qquad \frac{\partial^2 T_2}{\partial x^2} + \frac{\partial^2 T_2}{\partial y^2} = 0$$

$$x = 0,\ y = 0,\ y = b: T_1 = 0 \qquad x = 0,\ y = 0,\ x = a: T_2 = 0$$

$$x = a: T_1 = f_1(y) \qquad y = b: T_2 = f_2(x)$$

Addition of the equations and boundary conditions shows that $T = T_1 + T_2$ satisfies the original problem. Clearly, this approach can be extended to a problem where the temperature is specified on three or all four sides.

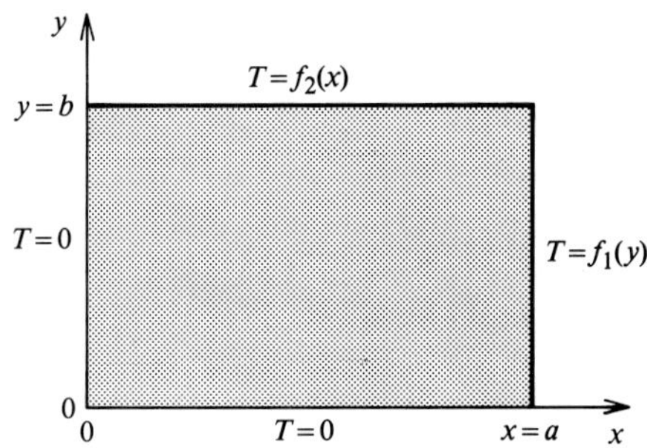

Figure 3.8 Rectangular plate with nonhomogeneous boundary conditions specified on two edges.

EXAMPLE 3.1 Heat Flow across a Neoprene Rubber Pad

A long neoprene rubber pad of width $a = 2$ cm and height $b = 4$ cm is a component of a spacecraft structure. Its sides and bottom are bonded to a metal channel at temperature $T_e = 20°\text{C}$, and the temperature distribution along the top can be approximated as a simple sine curve, $T = T_e + T_m \sin(\pi x/a)$, where $T_m = 80$ K. Determine the heat flow across the pad per meter length.

Solution

Given: Long rubber pad with a rectangular cross section.

Required: Heat flow across pad for given boundary conditions.

Assumptions: Two-dimensional, steady conduction.

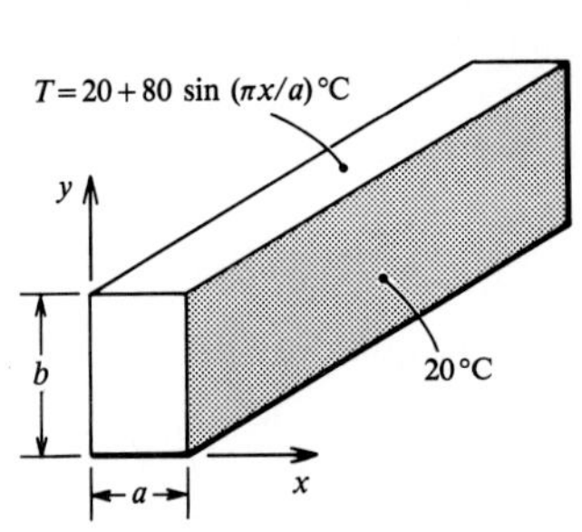

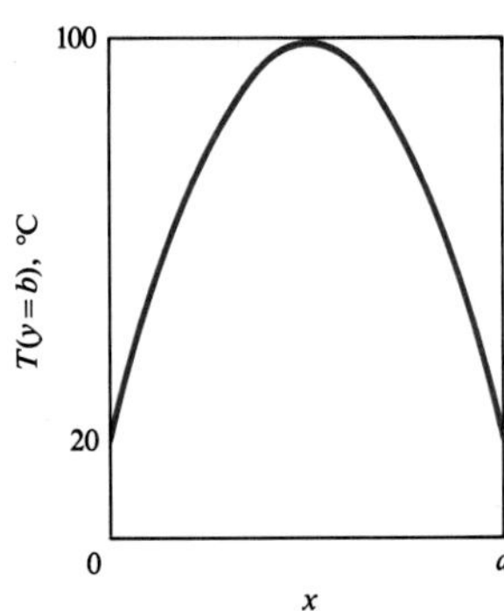

Conservation of energy requires that at steady state, the heat flow across the top surface equal the heat flow out across the sides and bottom; thus,

$$\dot{Q} = (1)\int_0^a q_y\big|_{y=b}\,dx = \int_0^a -k\frac{\partial T}{\partial y}\bigg|_{y=b} dx$$

The sides and bottom of the pad can be taken to have a temperature of 20°C. Thus, T is the solution of Laplace's equation that satisfies the boundary conditions

$$x = 0,\ x = a,\ y = 0:\quad T = T_e$$

$$y = b:\quad T = T_e + T_m \sin\frac{\pi x}{a}$$

Let $T_1 = T - T_e$; then T_1 satisfies the boundary conditions

$$x = 0,\ x = a,\ y = 0:\quad T_1 = 0$$

$$y = b:\quad T_1 = T_m \sin\frac{\pi x}{a}$$

Equation (3.25) satisfies the homogeneous boundary conditions. Applying the boundary condition at $y = b$ gives

$$T_1\big|_{y=b} = T_m \sin\frac{\pi x}{a} = \sum_{n=1}^{\infty} A_n \sin\frac{n\pi x}{a}\sinh\frac{n\pi b}{a}$$

which can be satisfied if

$$A_1 = \frac{T_m}{\sinh(\pi b/a)};\qquad A_2 = A_3 = \cdots = 0$$

That is, only the first term of the infinite series is required. Hence,

$$T - T_e = T_1 = T_m \sin\frac{\pi x}{a}\frac{\sinh(\pi y/a)}{\sinh(\pi b/a)}$$

$$\frac{\partial T}{\partial y}\bigg|_{y=b} = T_m\frac{(\pi/a)\sin(\pi x/a)}{\tanh(\pi b/a)}$$

$$\dot{Q} = \int_0^a -k\frac{\partial T}{\partial y}\bigg|_{y=b} dx = \frac{T_m k}{\tanh(\pi b/a)}(\cos\pi - \cos 0) = -\frac{2T_m k}{\tanh(\pi b/a)}\ \text{W/m}$$

For $T_m = 80$ K, $k = 0.19$ W/m K (Table A.2), and $b/a = 2.0$, the heat flow is

$$\dot{Q} = -\frac{(2)(80)(0.19)}{\tanh 2\pi} = -\frac{30.4}{1.0000} = -30.4\ \text{W/m}$$

Comments

Notice that in the limit of large b/a, $\tanh \pi b/a \to 1$ and $\dot{Q}$ is independent of b/a; even for a square pad, $b/a = 1$, $\tanh \pi = 0.996$. In the opposite limit of $b/a \to 0$, we can use the expansion $\tanh x = x - x^3/3 + \cdots$ to obtain

$$\dot{Q} = -\frac{2T_m ka}{\pi b} = -\frac{(2)(80)(0.19)(0.02)}{(\pi)(0.04)} = -4.83\ \text{W/m}$$

which is the result for one-dimensional heat flow across a thin wall. For a 1 m length of pad,

$$\dot{Q} = \int_0^a q\,dx = -\int_0^a \frac{k}{b}\left(T_e + T_m \sin\frac{\pi x}{a} - T_e\right)dx$$

$$= -\frac{kT_m}{b}\int_0^a \sin\frac{\pi x}{a}dx = -\frac{2T_m ka}{\pi b}$$

which agrees with the result obtained above.

3.3.2 Steady Conduction in a Rectangular Block

Consider a rectangular block with boundary conditions, which cause the temperature to vary in all three coordinate directions, x, y, and z. For an assumed constant thermal conductivity, the temperature distribution $T(x, y, z)$ is governed by the three-dimensional form of Laplace's equation:

$$\frac{\partial^2 T}{\partial x^2} + \frac{\partial^2 T}{\partial y^2} + \frac{\partial^2 T}{\partial z^2} = 0 \qquad \textbf{(3.31)}$$

Again the method of separation of variables can be used and $T(x, y, z)$ taken to have the form

$$T(x, y, z) = X(x)Y(y)Z(z)$$

The analysis proceeds as for the rectangular plate in Section 3.3.1, but the mathematics is very cumbersome, even for the simplest of boundary conditions. In practice, one cannot expect a simple behavior of the boundary conditions over a three-dimensional object, and the numerical methods of solution described in Section 3.6 are more appropriate than analytical methods.

Table 3.2 Shape factors for steady-state conduction for use in Eq. (3.32), $\dot{Q} = kS\,\Delta T$; $\Delta T = T_1 - T_2$. (See also the bibliography for Chapter 3.)

Configuration	Shape Factor
1. Plane wall 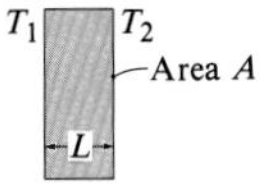 T_1 T_2 Area A L	$\dfrac{A}{L}$
2. Concentric cylinders 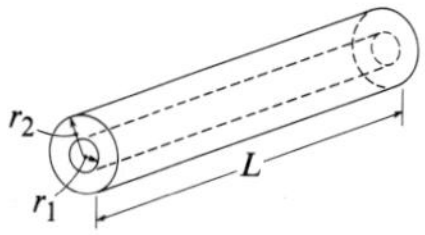 r_2 r_1 L $L \gg r_2$	$\dfrac{2\pi L}{\ln(r_2/r_1)}$ Note there is no steady-state solution for $r_2 \to \infty$, i.e., for a cylinder in an infinite medium.
3. Concentric spheres 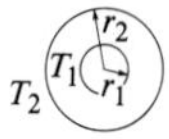 T_1 T_2 r_1 r_2	(*a*) $\dfrac{4\pi}{1/r_1 - 1/r_2}$ (*b*) $4\pi r_1$ for $r_2 \to \infty$
4. Eccentric cylinders 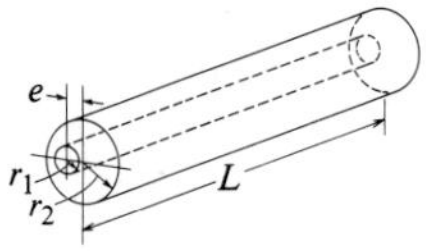 e r_1 r_2 L $L \gg r_2$	$\dfrac{2\pi L}{\cosh^{-1}\left(\dfrac{r_2^2 + r_1^2 - e^2}{2r_1 r_2}\right)}$
5. Concentric square cylinders 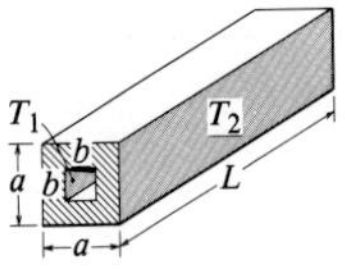 T_1 T_2 a b L $L \gg a$	$\dfrac{2\pi L}{0.93\ln(a/b) - 0.0502}$ for $\dfrac{a}{b} > 1.4$ $\dfrac{2\pi L}{0.785\ln(a/b)}$ for $\dfrac{a}{b} < 1.4$
6. Concentric circular and square cylinders 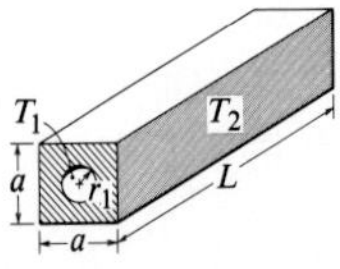 T_1 T_2 a r_1 L	$\dfrac{2\pi L}{\ln(0.54a/r)}$ $a > 2r$

Table 3.2 *(Concluded)*

Configuration	Shape Factor
7. Buried sphere Medium at infinity also at T_2	$\dfrac{4\pi r_1}{1 - r_1/2h}$ For $h \to \infty$, the result for item 3(b) is recovered
8. Buried cylinder Medium at infinity also at T_2 $\quad L \gg r_1$	$\dfrac{2\pi L}{\cosh^{-1}(h/r_1)}$ $\dfrac{2\pi L}{\ln(2h/r_1)}$ for $h > 3r_1$ For $h/r_1 \to \infty$, $S \to 0$ since steady flow is impossible
9. Buried rectangular beam Medium at infinity also at T_2 $\quad L \gg h, a, b$	$2.756L\left[\ln\left(1 + \dfrac{h}{a}\right)\right]^{-0.59}\left(\dfrac{h}{b}\right)^{-0.078}$
10. The edge of adjoining walls	$0.54W$ for $W > L/5$ (W is the inner edge)
11. The corner of three adjoining walls	$0.15L$ for $W > L/5$
12. Disk area on the adiabatic surface of a semi-infinite solid 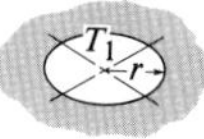 Medium at infinity at T_2	$4r$

3.3.3 Conduction Shape Factors

Many multidimensional conduction problems involve heat flow between two surfaces, each of uniform temperature, with any other surfaces present being adiabatic. The conduction *shape factor*, S, is defined such that the heat flow between the surfaces, $\dot{Q}$, is

$$\dot{Q} = kS\Delta T \tag{3.32}$$

where k is the thermal conductivity and ΔT is the difference in surface temperatures; S is seen to have the dimensions of length. The results we have already obtained for one-dimensional conduction can also be expressed in terms of the shape factor. For example, a plane slab of area A and thickness L has $S = A/L$ from Eq. (1.9). Table 3.2 lists shape factors for various configurations.

Some points to note when using Table 3.2 are:

1. There is no internal heat generation: $\dot{Q}_v''' = 0$.
2. The thermal conductivity, k, is constant.
3. The two surfaces should be isothermal. If these temperatures are not prescribed, but are intermediate temperatures in a series thermal circuit, the isothermal condition may not be satisfied. The surfaces will generally be isothermal when the component in question has the dominant thermal resistance. Example 3.3 illustrates this point.
4. Special care must be taken with the configurations involving an infinite medium. For example, in item 7, not only the plane surface but also the medium at infinity must be at temperature T_2.
5. Item 8 is often used incorrectly for calculating heat loss or gain from buried pipelines. It is essential that the deep soil be at the same temperature as the surface, which is a condition seldom met in reality. Also, the buried pipeline problem often involves transient conduction.
6. The shape factors given in items 10 and 11 were developed by the physicist I. Langmuir and coworkers in 1913 for calculating the heat loss from furnaces. Example 3.2 illustrates their use.

EXAMPLE 3.2 Heat Loss from a Laboratory Furnace

A small laboratory furnace is in the form of a cube and is insulated with a 10 cm layer of fiberglass insulation, with an inside edge 30 cm long. If the only significant resistance to heat flow across the furnace wall is this insulation, determine the power required for steady operation at a temperature of 600 K when the outer casing temperature is 350 K. The thermal conductivity of the fiberglass insulation at the mean temperature of 475 K is approximately 0.11 W/m K.

Solution

Given: Insulated laboratory furnace.

Required: Power required for operation at 600 K.

Assumptions: 1. Steady state.
2. The outer convective resistance is negligible.

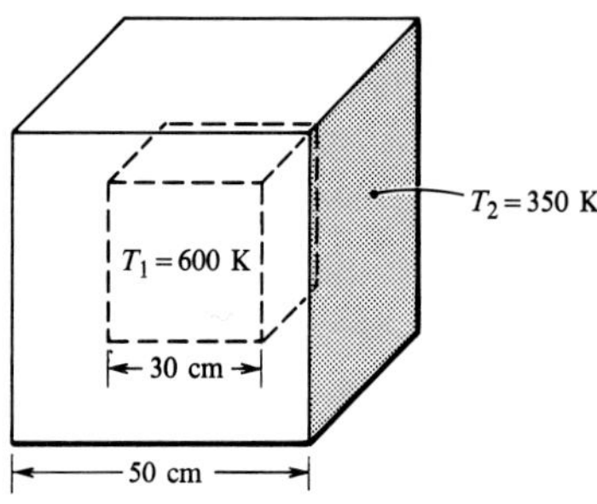

The shape factors given as items 1, 10, and 11 of Table 3.2 should be used, assuming independent parallel paths for heat flow through the 6 sides, 12 edges and 8 corners of the enclosure. Thus, if L is the insulation thickness and W the inside edge,

$$\begin{aligned}\dot{Q} &= k\Delta T\, S \\ &= k(T_1 - T_2)[6W^2/L + (12)(0.54)W + (8)(0.15)L] \\ &= (0.11)(600 - 350)[(6)(0.3)^2/(0.1) + (12)(0.54)(0.3) + (8)(0.15)(0.1)] \\ &= (0.11)(250)[5.40 + 1.94 + 0.12] \\ &= 205 \text{ W}\end{aligned}$$

Comments

If an effective area for heat flow A_{eff} is defined by the equation for one-dimensional heat flow,

$$\dot{Q} = \frac{kA_{\text{eff}}(T_1 - T_2)}{L}$$

then $A_{\text{eff}} = 6W^2 + (12)(0.54)WL + (8)(0.15)L^2 = 0.746 \text{ m}^2$. Notice that A_{eff} is significantly less than either the arithmetic average of the inner and outer areas, 1.02 m^2, or the value midway through the wall, 0.96 m^2.

EXAMPLE 3.3 Heat Loss from a Buried Oil Line

An oil pipeline has an outside diameter of 30 cm and is buried with its centerline 1 m below ground level in damp soil. The line is 5000 m long, and the oil flows at 2.5 kg/s. If the inlet temperature of the oil is 120°C and the ground level soil is at 23°C, estimate the oil outlet temperature and the heat loss (i) for an uninsulated pipe, and (ii) if the pipe is insulated with a 15 cm layer of insulation with conductivity $k = 0.03$ W/m K. Take the soil thermal conductivity as 1.5 W/m K and the oil specific heat as 2000 J/kg K.

Solution

Given: Buried oil pipeline.

Required: Oil outlet temperature and heat loss if (i) uninsulated, (ii) insulated.

Assumptions:
1. Steady state.
2. Deep ground temperature same as surface temperature.
3. Negligible resistance of the pipe wall and for convection from the oil.
4. Isothermal surface exposed to soil.

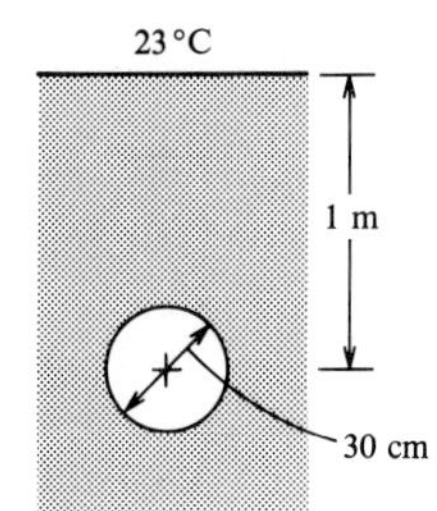

(i) Equation (3.32), $\dot{Q} = kS\Delta T$, can be used for the heat loss, with the shape factor obtained from Table 3.2, item 8:

$$\frac{h}{r_1} = \frac{1.0}{0.15} = 6.67 > 3$$

Hence,

$$S = \frac{2\pi L}{\ln(2h/r_1)} = \frac{(2\pi)(5000)}{\ln(2 \times 1/0.15)} = 12{,}130 \text{ m}$$

The temperature difference between the oil and the ground surface ΔT decreases continuously in the flow direction, so how do we use Eq. (3.32)? We recognize that this problem is similar to the single-stream heat exchanger analyzed in Section 1.6 because the ground temperature is constant along the pipeline. We could easily perform an analysis similar to that of Section 1.6.2; however, with a little thought, we can simply reinterpret the results of Section 1.6.2 for the problem at hand. Equation (1.59), for the effectiveness, is

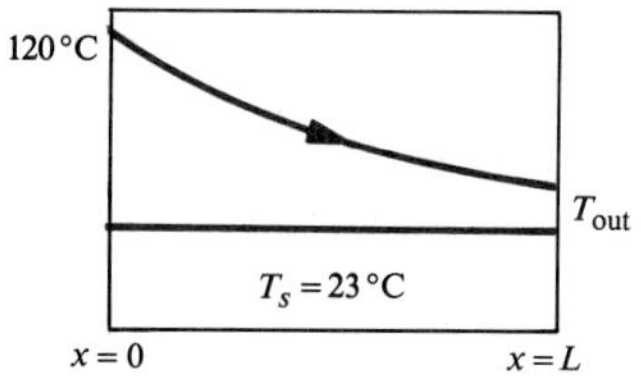

$$\varepsilon = 1 - e^{-N_{tu}}$$

where now

$$\varepsilon = \frac{\text{Actual temperature change}}{\text{Maximum possible temperature change}} = \frac{T_{in} - T_{out}}{T_{in} - T_s}$$

To specify the number of transfer units, $N_{tu} = U\mathscr{P}L/\dot{m}c_p$, recall that the heat transfer for an element of exchanger Δx long was

$$\Delta\dot{Q} = U\mathscr{P}\Delta x \Delta T$$

while for the pipeline it is

$$\Delta\dot{Q} = k\Delta S\Delta T; \qquad \Delta S = \frac{2\pi\Delta x}{\ln(2h/r_1)}$$

Thus, $U\mathscr{P}\Delta x$ can be replaced by $k\Delta S$, or $U\mathscr{P}L$ by kS:

$$N_{tu} = \frac{U\mathscr{P}L}{\dot{m}c_p} = \frac{kS}{\dot{m}c_p} = \frac{(1.5)(12{,}130)}{(2.5)(2000)} = 3.64$$

$$\varepsilon = 1 - e^{-3.64} = 0.974$$

$$T_{out} = T_{in} - \varepsilon(T_{in} - T_s) = 120 - 0.974(120 - 23) = 25.5°\text{C}$$

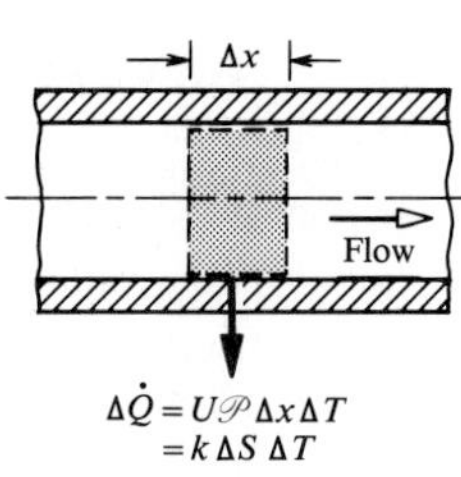

The heat loss is equal to the enthalpy given up by the oil,

$$\dot{Q} = \dot{m}c_p(T_{in} - T_{out}) = (2.5)(2000)(120 - 25.5) = 472 \text{ kW}$$

(ii) This problem cannot be solved exactly using the shape factor concept, since, when the insulation is added to the pipe, the outer surface of the insulation will not be isothermal. However, to get some idea of the effect of the insulation, we will assume an isothermal surface. Then for two resistances in series,

$$\frac{1}{U\mathscr{P}L} = \frac{\ln(r_2/r_1)}{2\pi k_{\text{ins}}L} + \frac{1}{k_{\text{soil}}S}$$

$$\frac{\ln(r_2/r_1)}{2\pi k_{\text{ins}}L} = \frac{\ln(0.30/0.15)}{(2\pi)(0.03)(5000)} = 7.35 \times 10^{-4}\ (\text{W/K})^{-1}$$

$$\frac{h}{r_2} = \frac{1.0}{0.3} = 3.33 > 3;\ \text{hence}$$

$$S = \frac{2\pi L}{\ln(2h/r_2)} = \frac{(2\pi)(5000)}{\ln(2 \times 1/0.3)} = 16{,}560\ \text{m}$$

$$\frac{1}{k_{\text{soil}}S} = \frac{1}{(1.5)(16{,}560)} = 0.403 \times 10^{-4}\ (\text{W/K})^{-1}$$

$$\frac{1}{U\mathscr{P}L} = (7.35 + 0.40)10^{-4}; \qquad U\mathscr{P}L = 1290\ \text{W/K}$$

$$N_{\text{tu}} = \frac{1290}{(2.5)(2000)} = 0.258; \qquad \varepsilon = 1 - e^{-0.258} = 0.227$$

$$T_{\text{out}} = 120 - 0.227(120 - 23) = 98.0^\circ\text{C}$$

$$\dot{Q} = (2.5)(2000)(120 - 98.0) = 110\ \text{kW}$$

Comments

The pipeline is located on a tropical island where the annual ground temperature variation is relatively small, so a steady-state analysis is reasonably valid. Problems concerning the freezing of buried water pipes often require a transient analysis because of ground temperature variations (see Exercise 3–32).

3.4 UNSTEADY CONDUCTION

In unsteady or *transient* conduction, temperature is a function of both time and spatial coordinates. In the absence of internal heat generation, the temperature response of a body is governed by Fourier's equation. Again the method of separation of variables is useful, and examples of its use are given in Sections 3.4.1 and 3.4.3. The method, however, fails under certain circumstances—for example, when the medium extends to infinity. Then possible methods include the use of *Laplace transforms* or a **similarity** transformation of the partial differential equation into an ordinary differential equation. The latter method is demonstrated in Section 3.4.2. Analytical results of unsteady conduction tend to be complicated and awkward to use. Thus, where possible, approximate solutions of adequate accuracy will be indicated. Often the results are conveniently presented in graphical form for rapid engineering calculations.

3.4.1 The Slab with Negligible Surface Resistance

Figure 3.9 shows a slab $2L$ thick. It is initially at a uniform temperature T_0, and at time $t = 0$ the surfaces at $x = +L$ and $x = -L$ are suddenly lowered to temperature T_s. Such a situation is encountered in practice when a poorly conducting slab is suddenly immersed in a liquid for which the convective heat transfer coefficient is very large, that is, under conditions where the convective resistance to heat transfer is negligible. Note that this situation is the opposite limit to that considered in Section 1.5, for which the lumped thermal capacity model was applicable. In that case, the Biot number $\text{Bi} = h_c L/k$ had to be small; for this case, the Biot number must be large. In addition to its practical utility, the solution of this problem allows a simple demonstration of some important features of transient conduction.

The Governing Equation and Conditions

With no internal heat generation and an assumed constant thermal conductivity, Fourier's equation, Eq. (3.6), applies. Since the temperature does not vary in the y and z directions,

$$\frac{\partial T}{\partial t} = \alpha \frac{\partial^2 T}{\partial x^2} \tag{3.33}$$

which is the governing differential equation for $T(x, t)$ and must be solved subject to the initial condition

$$t = 0: \quad T = T_0 \tag{3.34a}$$

The symmetry of the problem allows the differential equation to be solved in the region $0 \le x \le L$, for which the appropriate boundary conditions are

$$x = 0: \quad \frac{\partial T}{\partial x} = 0 \tag{3.34b}$$

$$x = L: \quad T = T_s \tag{3.34c}$$

Equation (3.34*b*) follows from the symmetry of the temperature profile about the plane $x = 0$, or equivalently, from the condition that there can be no heat flow across this plane.

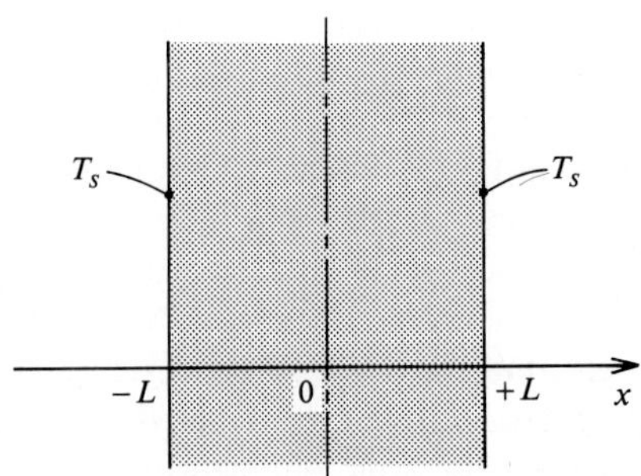

Figure 3.9 Coordinate system for the analysis of unsteady conduction in an infinite slab.

Dimensional Analysis

Before an attempt is made to solve Eq. (3.33), it is helpful to perform a dimensional analysis. We begin by constructing suitable dimensionless independent and dependent variables. Suitable choices for x and T are $\eta = x/L$ and $\theta = (T - T_s)/(T_0 - T_s)$, since η and θ will both vary between zero and unity. However, there is no obvious time scale for the problem that can be used to make t dimensionless, so we will let the differential equation itself indicate an appropriate choice. Transforming Eq. (3.33) with

$$x = L\eta; \qquad dx = L\,d\eta$$

$$T = (T_0 - T_s)\theta + T_s; \qquad dT = (T_0 - T_s)d\theta$$

gives

$$(T_0 - T_s)\frac{\partial\theta}{\partial t} = \frac{\alpha(T_0 - T_s)}{L^2}\frac{\partial^2\theta}{\partial\eta^2}$$

or

$$\frac{\partial\theta}{\partial(\alpha t/L^2)} = \frac{\partial^2\theta}{\partial\eta^2}$$

Thus, we choose $\zeta = t/(L^2/\alpha)$ as dimensionless time, and the result is

$$\frac{\partial\theta}{\partial\zeta} = \frac{\partial^2\theta}{\partial\eta^2} \tag{3.35}$$

The initial and boundary conditions transform into

$$\zeta = 0: \quad \theta = 1 \tag{3.36a}$$

$$\eta = 0: \quad \frac{\partial\theta}{\partial\eta} = 0 \tag{3.36b}$$

$$\eta = 1: \quad \theta = 0 \tag{3.36c}$$

There are no parameters in the transformed statement of the problem, so the solution is simply $\theta(\zeta, \eta)$. The dimensionless time variable ζ is also commonly called the **Fourier number**, $\text{Fo} = \alpha t/L^2$. Notice that the Fourier number can also be written as $\text{Fo} = t/t_c$; $t_c = L^2/\alpha$ is a characteristic time (time constant) for this conduction problem. The behavior of the solution will depend on the value of t relative to t_c, that is, on whether Fo is much smaller than unity, of order unity, or much larger than unity. The relation $\theta = \theta(\zeta, \eta)$ is a statement of the similarity principle for this problem, a concept introduced in Section 2.4.5. We now know that the solution for the dimensionless temperature θ will be a function of dimensionless time ζ and dimensionless position η only.

Solution for the Temperature Response

As in Section 3.3.1, for steady conduction in a rectangular plate, the method of separation of variables will be used to solve the partial differential equation. We assume that the function $\theta(\zeta, \eta)$ can be expressed as the product of a function of ζ

only, $Z(\zeta)$, and a function of η only, $H(\eta)$:

$$\theta(\zeta, \eta) = Z(\zeta)H(\eta) \tag{3.37}$$

Substitution in Eq. (3.35) yields

$$H\frac{dZ}{d\zeta} = Z\frac{d^2H}{d\eta^2}$$

where total derivatives have replaced partial derivatives, since Z is a function of independent variable ζ only, and H is a function of independent variable η only. Rearranging gives

$$\frac{1}{Z}\frac{dZ}{d\zeta} = \frac{1}{H}\frac{d^2H}{d\eta^2} \tag{3.38}$$

Since each side of Eq. (3.38) is a function of a single independent variable, the equality can hold only if both sides are equal to a constant. To satisfy the boundary conditions, this constant must be a negative number, which will be written $-\lambda^2$. Two ordinary differential equations are obtained:

$$\frac{dZ}{d\zeta} + \lambda^2 Z = 0 \qquad \frac{d^2H}{d\eta^2} + \lambda^2 H = 0$$

which have the solutions

$$Z = C_1 e^{-\lambda^2\zeta} \qquad H = C_2 \cos\lambda\eta + C_3 \sin\lambda\eta$$

Hence,

$$\theta(\zeta, \eta) = e^{-\lambda^2\zeta}(A\cos\lambda\eta + B\sin\lambda\eta)$$

where $A = C_1C_2$ and $B = C_1C_3$. Applying boundary condition Eq. (3.36*b*),

$$\left.\frac{\partial\theta}{\partial\eta}\right|_{\eta=0} = e^{-\lambda^2\zeta}(-A\lambda\sin\lambda\eta + B\lambda\cos\lambda\eta)_{\eta=0} = 0$$

which requires that $B = 0$. Next, applying boundary condition Eq. (3.36*c*) gives

$$\theta|_{\eta=1} = Ae^{-\lambda^2\zeta}\cos\lambda\eta|_{\eta=1} = 0$$

which requires that $\cos\lambda = 0$, or $\lambda_n = (n + 1/2)\pi$, $n = 0, 1, 2, 3, \ldots$. The λ_n are the eigenvalues for this problem, and the solution corresponding to the nth eigenvalue may be written as

$$\theta_n(\zeta, \eta) = A_n e^{-(n+1/2)^2\pi^2\zeta}\cos\left(n + \frac{1}{2}\right)\pi\eta \tag{3.39}$$

The general solution to Eq. (3.35) is the sum of the series of solutions given by Eq. (3.39):

$$\theta(\zeta, \eta) = \sum_{n=0}^{\infty} A_n e^{-(n+1/2)^2\pi^2\zeta}\cos\left(n + \frac{1}{2}\right)\pi\eta \tag{3.40}$$

The constants A_n are determined from the initial condition, Eq. (3.36*a*):

$$\theta|_{\zeta=0} = \sum_{n=0}^{\infty} A_n \cos\left(n + \frac{1}{2}\right)\pi\eta = 1$$

That is, a constant must be expressed in terms of an infinite series of cosine functions, which again is a Fourier series expansion. More generally, $\theta|_{\zeta=0} = f(\eta)$ is an arbitrary function of η, and the required expansion is then

$$f(\eta) = \sum_{n=0}^{\infty} A_n \cos\left(n + \frac{1}{2}\right)\pi\eta \tag{3.41}$$

Determination of the constants A_n follows the procedure demonstrated in Section 3.3.1. The result is

$$A_n = 2\int_0^1 f(\eta)\cos\left(n + \frac{1}{2}\right)\pi\eta \, d\eta \tag{3.42}$$

For $f(\eta) = 1$, $A_n = \dfrac{2}{(n+1/2)\pi}\left[\sin\left(n + \dfrac{1}{2}\right)\pi\eta\right]_0^1 = \dfrac{2(-1)^n}{(n+1/2)\pi}$

Substituting in Eq. (3.40) gives the solution for the temperature distribution,

$$\theta(\zeta, \eta) = \sum_{n=0}^{\infty} \frac{2(-1)^n}{(n+1/2)\pi} e^{-(n+1/2)^2\pi^2\zeta} \cos\left(n + \frac{1}{2}\right)\pi\eta \tag{3.43}$$

or, in terms of temperature $T(x, t)$, Fourier number $\mathrm{Fo} = \alpha t/L^2$, and x/L,

$$\frac{T - T_s}{T_0 - T_s} = \sum_{n=0}^{\infty} \frac{2(-1)^n}{(n+1/2)\pi} e^{-(n+1/2)^2\pi^2 \mathrm{Fo}} \cos\left(n + \frac{1}{2}\right)\pi\frac{x}{L} \tag{3.44}$$

which is plotted in Fig. 3.10.

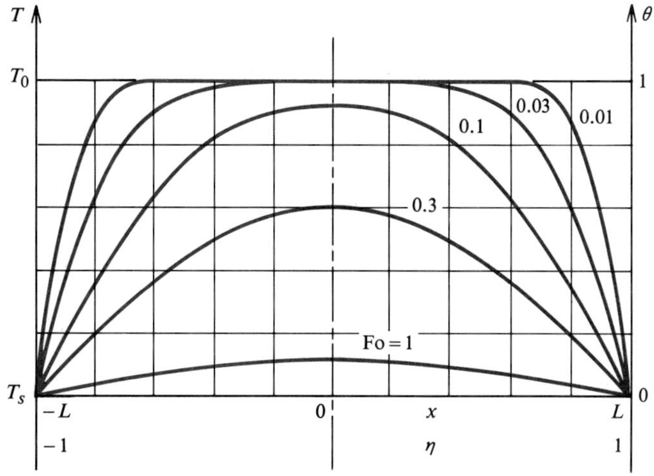

Figure 3.10 Temperature response in a slab with negligible surface resistance, calculated from Eq. (3.44).

The Surface Heat Flux

Of particular interest is the rate of heat transfer out of the slab. The surface heat flux at $x = L$ is obtained from Fourier's law and Eq. (3.44) as

$$q_s = -k\frac{\partial T}{\partial x}\bigg|_{x=L}$$

$$= -k(T_0 - T_s)\sum_{n=0}^{\infty}\frac{2(-1)^n}{(n+1/2)\pi}e^{-(n+1/2)^2\pi^2\mathrm{Fo}}\frac{\partial}{\partial x}\left[\cos\left(n+\frac{1}{2}\right)\pi\frac{x}{L}\right]_{x=L}$$

$$= -k(T_0 - T_s)\sum_{n=0}^{\infty}\frac{2(-1)^n}{(n+1/2)\pi}e^{-(n+1/2)^2\pi^2\mathrm{Fo}}\left[\frac{-(n+1/2)\pi(-1)^n}{L}\right]$$

$$= \frac{2k(T_0 - T_s)}{L}\sum_{n=0}^{\infty}e^{-(n+1/2)^2\pi^2\mathrm{Fo}} \tag{3.45}$$

or

$$q_s = \frac{2k(T_0 - T_s)}{L}\left[e^{-(\pi/2)^2\mathrm{Fo}} + e^{-(3\pi/2)^2\mathrm{Fo}} + e^{-(5\pi/2)^2\mathrm{Fo}} + e^{-(7\pi/2)^2\mathrm{Fo}} + \cdots\right]$$

Table 3.3 gives the first four terms in the series for values of the Fourier number $\mathrm{Fo} = \alpha t/L^2$ equal to 0.01, 0.1, 0.2, 0.3, and 1.0. It can be seen that the series converges rapidly unless the Fourier number is very small. For $\mathrm{Fo} \gtrsim 0.2$, only the first term in the series need be retained, and

$$q_s \simeq \frac{2k(T_0 - T_s)}{L}e^{-(\pi/2)^2\mathrm{Fo}} \tag{3.46}$$

with an error of less than 2%. Since $\mathrm{Fo} = \alpha t/L^2$, Eq. (3.46) is a solution valid for *long times*.

For small values of the Fourier number (that is, soon after the slab is immersed in the liquid), the series converges slowly, and sufficient terms must be retained for an accurate result. However, in the limit $\mathrm{Fo} \to 0$, there is the simple mathematical result:

Table 3.3 The first four terms of the series of Eq. (3.45) for the surface heat flux of a slab with negligible surface resistance.

Fo	$e^{-(\pi/2)^2\mathrm{Fo}}$	$e^{-(3\pi/2)^2\mathrm{Fo}}$	$e^{-(5\pi/2)^2\mathrm{Fo}}$	$e^{-(7\pi/2)^2\mathrm{Fo}}$
0.01	0.9756	0.8009	0.5396	0.2985
0.1	0.7813	0.1085	0.0021	$\sim 10^{-5}$
0.2	0.6105	0.0118	$\sim 10^{-6}$	
0.3	0.4770	0.0013		
1.0	0.0848	$\sim 10^{-10}$		

$$\sum_{n=0}^{\infty} e^{-(n+1/2)^2\pi^2 \text{Fo}} = \frac{1}{2\pi^{1/2}\,\text{Fo}^{1/2}}; \qquad \text{Fo} \to 0$$

Substituting in Eq. (3.45) gives

$$q_s \simeq \frac{k(T_0 - T_s)}{(\pi \alpha t)^{1/2}} \tag{3.47}$$

which can be used for Fo $\lesssim 0.05$. Equation (3.47) is thus a valid solution for *short times*. Figure 3.10 shows that for sufficiently short times, the temperature changes in a thin region near the surface only, while the temperature of the slab interior remains unaffected. The heat conduction process is confined to this thin region, and the thickness of the slab is of no consequence: notice that L does not appear in Eq. (3.47). It is this fact that motivates the analysis of the semi-infinite solid in Section 3.4.2.

Nonsymmetrical Boundary Conditions

In our analysis, we specified the same temperature on both sides of the slab. Symmetry allowed the problem to be solved in the half slab, $0 \le x \le L$; upon transformation, the boundary condition at $\eta = 1$ was $\theta = 0$, that is, homogeneous. If we now specify a temperature T_s at $x = L$ and $T_{s'}$ at $x = -L$, as shown in Fig. 3.11, how do we proceed? It is not possible to define a dimensionless temperature θ such that $\theta = 0$ on both boundaries, and without homogeneous boundary conditions, it is not possible to obtain an eigenvalue problem as before. To get around this hurdle, we reduce the problem to the **superposition** of a steady and a transient problem. Let $T(x,t) = T_1(x) + T_2(x,t)$ such that

$$\frac{d^2T_1}{dx^2} = 0 \qquad\qquad \frac{\partial T_2}{\partial t} = \alpha \frac{\partial^2 T_2}{\partial x^2}$$

$$x = L: \quad T_1 = T_s \qquad\qquad x = L, -L: \quad T_2 = 0$$

$$x = -L: \quad T_1 = T_{s'} \qquad\qquad t = 0: \quad T_2 = T_0 - T_1$$

$$T_1 = \frac{T_s - T_{s'}}{2}\frac{x}{L} + \frac{T_s + T_{s'}}{2} \qquad\qquad T_2 = \sum_{n=0}^{\infty} A_n e^{-(n+1/2)^2\pi^2 \text{Fo}} \cos\left(n + \frac{1}{2}\right)\pi \frac{x}{L}$$

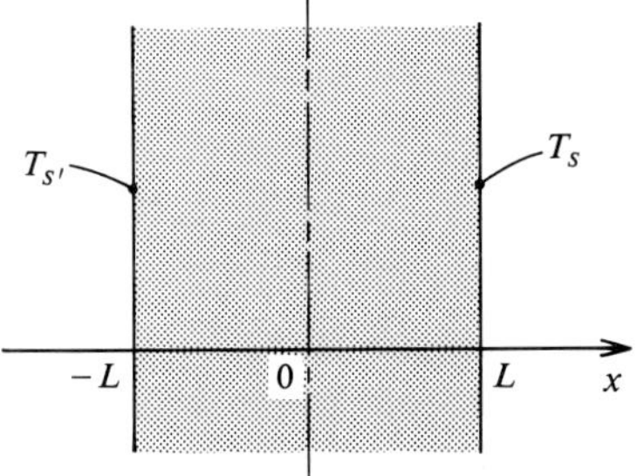

Figure 3.11 Schematic for the analysis of unsteady conduction in a slab with unequal surface temperatures.

and the constants A_n must be determined from

$$\sum_{n=0}^{\infty} A_n \cos\left(n + \frac{1}{2}\right)\pi\frac{x}{L} = T_0 - \frac{T_s - T_{s'}}{2}\frac{x}{L} - \frac{T_s + T_{s'}}{2} \tag{3.48}$$

Figure 3.12 illustrates this superposition technique.

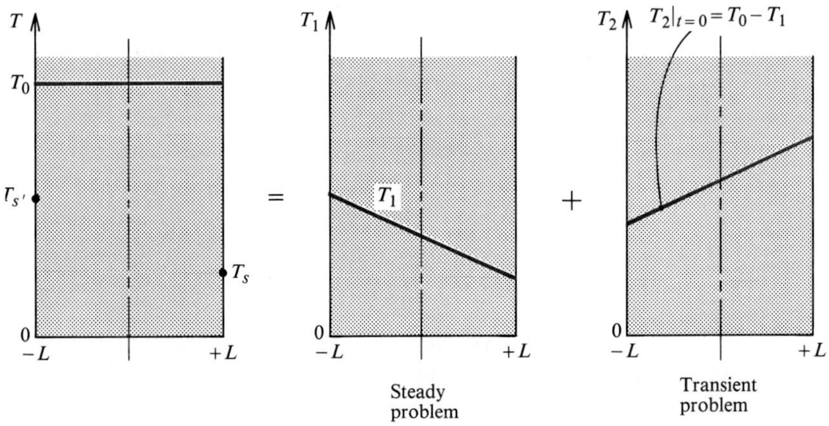

Figure 3.12 Schematic of superposition of solutions for a slab with unequal surface temperatures.

EXAMPLE 3.4 Transient Steam Condensation on a Concrete Wall

In a postulated nuclear reactor accident scenario, a concrete wall 20 cm thick at an initial temperature of 20°C is suddenly exposed on both sides to pure steam at atmospheric pressure. If the thermal resistance of the condensate flowing down the wall is negligible, estimate the rate of steam condensation on 160 m² wall area after (i) 10 s, (ii) 10 min, and (iii) 3 h.

Solution

Given: Concrete wall suddenly exposed to steam.

Required: Rate of steam condensation at various times.

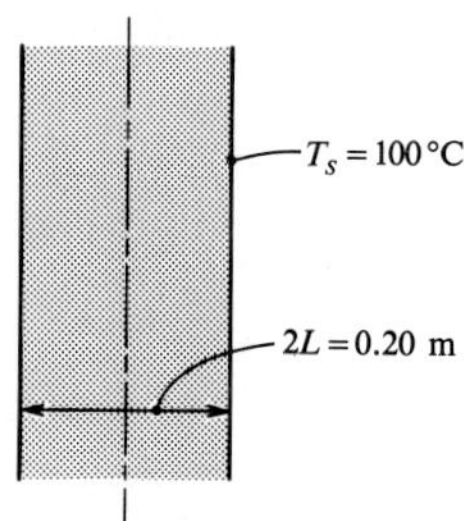

Assumptions: 1. Thermal resistance of condensate negligible.
2. Pure steam, T_{sat} (1 atm) = 100°C.

The first step is to calculate the time constant for the wall, $t_c = L^2/\alpha$. From Table A.3, we take $\alpha = 0.75 \times 10^{-6}$ m²/s for concrete. Then

$$t_c = \frac{L^2}{\alpha} = \frac{(0.1)^2}{0.75 \times 10^{-6}} = 1.33 \times 10^4 \text{ s (3.7 h)}$$

(i) $t = 10$ s:

$\text{Fo} = t/t_c = 10/1.33\times10^4 = 7.5\times10^{-4} < 0.05$. Thus, Eq. (3.47) applies. The condensation rate is $\dot{m} = q_s A/h_{fg}$ where h_{fg} is the enthalpy of vaporization of steam at 100°C. From steam tables (for example, Table A.12*a*), $h_{fg} = 2.257 \times 10^6$ J/kg. Also required is the conductivity of concrete, which from Table A.3 is 1.4 W/m K.

$$\dot{m} = \frac{q_s A}{h_{fg}} = \frac{k(T_s - T_0)A}{(\pi\alpha t)^{1/2}h_{fg}} = \frac{(1.4)(100-20)(160)}{[\pi(0.75\times10^{-6})(10)]^{1/2}(2.257\times10^6)} = 1.64 \text{ kg/s}$$

(ii) $t = 10$ min:

$\text{Fo} = t/t_c = (10)(60)/1.33 \times 10^4 = 0.0451 < 0.05$. Thus, Eq. (3.47) remains valid. Since q_s and, hence, $\dot{m}$ decrease like $t^{-1/2}$, the condensation rate is now

$$\dot{m} = 1.64\left(\frac{600}{10}\right)^{-1/2} = 0.212 \text{ kg/s}$$

(iii) $t = 3$ h:

$\text{Fo} = t/t_c = (3)(3600)/1.33 \times 10^4 = 0.812 > 0.2$. Equation (3.46) now applies, and the condensation rate is

$$\dot{m} = \frac{2k(T_s - T_0)A}{Lh_{fg}}e^{-(\pi/2)^2\text{Fo}} = \frac{(2)(1.4)(100-20)(160)}{(0.10)(2.257\times10^6)}e^{-(\pi/2)^2(0.812)} = 0.0214 \text{ kg/s}$$

Comments

These estimates may be regarded as upper limits: in practice the steam is likely to contain some noncondensables, such as air, and condense at a temperature lower than 100°C. The problem is then one involving simultaneous heat and mass transfer.

3.4.2 The Semi-Infinite Solid

In Section 3.4.1, we saw that for sufficiently short times, temperature changes did not penetrate far enough into the slab for the thickness of the slab to have any effect on the heat conduction process. Such situations are encountered in practice. One method of case-hardening tool steel involves rapid quenching from a high temperature for a short time. Only the metal close to the surface is rapidly cooled and hardened; the interior cools slowly after the quenching process and remains ductile. Thus, during the quenching process, the interior temperature remains unchanged, and the precise thickness and shape of the tool is irrelevant. The conduction process is confined to a thin region near the surface into which temperature changes have penetrated. Thus, it is useful to have formulas giving the temperature distribution and heat flow for various kinds of boundary conditions that are applicable to these *penetration* problems.

The Governing Equation and Conditions

An appropriate model for penetration problems is transient conduction in a semi-infinite solid, as shown in Fig. 3.13. If we assume constant thermal conductivity, no internal heat generation, and negligible temperature variations in the y and z directions, Eq. (3.6) again reduces to

$$\frac{\partial T}{\partial t} = \alpha \frac{\partial^2 T}{\partial x^2} \qquad \textbf{(3.49)}$$

If the solid is initially at a uniform temperature T_0, the appropriate initial condition is

$$t = 0: \quad T = T_0 \qquad \textbf{(3.50}\textbf{\textit{a}}\textbf{)}$$

The left face of the solid is suddenly raised to temperature T_s at time zero and held at that value. The two required boundary conditions are then

$$x = 0: \quad T = T_s \qquad \textbf{(3.50}\textbf{\textit{b}}\textbf{)}$$

$$x \to \infty: \quad T \to T_0 \qquad \textbf{(3.50}\textbf{\textit{c}}\textbf{)}$$

It is the second boundary condition, Eq. (3.50c), that makes this problem different from the slab problem of Section 3.4.1.

Solution for the Temperature Distribution

It might at first appear that the *separation of variables* solution method can be used once again. As in the slab analysis of Section 3.4.1, the variables are separable in the differential equation, Eq. (3.49). However, a necessary requirement for completing the solution is that the boundary conditions of the eigenvalue problem be specified on *coordinate surfaces*, and $x = \infty$ is not a coordinate surface of the Cartesian coordinate system. Instead, we proceed as follows.

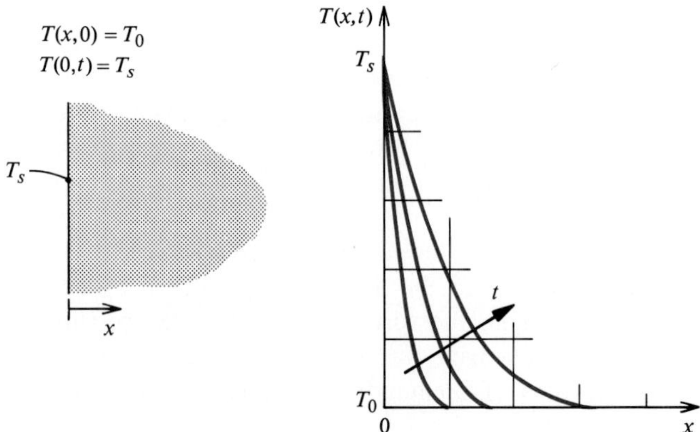

Figure 3.13 Coordinate system and expected temperature profiles for transient conduction in a semi-infinite solid with a step change in surface temperature.

We first normalize the dependent variable $T(x, t)$ for algebraic convenience. Defining $\theta = (T - T_0)/(T_s - T_0)$, the problem statement transforms into

$$\frac{\partial \theta}{\partial t} = \alpha \frac{\partial^2 \theta}{\partial x^2} \tag{3.51}$$

$$t = 0: \quad \theta = 0 \tag{3.52a}$$

$$x = 0: \quad \theta = 1 \tag{3.52b}$$

$$x \to \infty: \quad \theta \to 0 \tag{3.52c}$$

This mathematical problem can be reduced to the solution of an ordinary differential equation for θ. The independent variable in this equation is called a **similarity variable** and is an appropriate combination of x and t. One way to discover the combination is based on physical reasoning and dimensional considerations. Figure 3.13 shows the expected temperature profiles as a function of time as the heat penetrates into the solid. We can arbitrarily define some penetration depth δ, for example, the location where $\theta = 0.01$ or 0.001. Then δ is clearly a function of time t and thermal diffusivity α: the larger α, the deeper the penetration at a given time. Notice that the problem statement, Eqs. (3.51) and (3.52), contains no other quantities on which δ can depend. Time has units [s], and thermal diffusivity has units [m^2/s]. The only combination of these two variables that will give the required units of [m] for the penetration depth is $(\alpha t)^{1/2}$. For the temperature profile to be a function of a single variable, distances into the solid must be *scaled* to this penetration depth. Thus, we will choose

$$\eta = \frac{x}{(4\alpha t)^{1/2}} \tag{3.53}$$

where the factor of $4^{1/2}$ has been inserted for future algebraic convenience. Eq. (3.51) is transformed by careful use of the rules of partial differentiation. The required differential operators are

$$\frac{\partial \theta}{\partial t} = \frac{d\theta}{d\eta}\frac{\partial \eta}{\partial t}\bigg|_x = \frac{d\theta}{d\eta}\left(-\frac{x}{2t(4\alpha t)^{1/2}}\right)$$

$$\frac{\partial \theta}{\partial x} = \frac{d\theta}{d\eta}\frac{\partial \eta}{\partial x}\bigg|_t = \frac{d\theta}{d\eta}\left(\frac{1}{(4\alpha t)^{1/2}}\right)$$

$$\frac{\partial^2 \theta}{\partial x^2} = \frac{1}{(4\alpha t)^{1/2}}\frac{d^2\theta}{d\eta^2}\frac{\partial \eta}{\partial x}\bigg|_t = \frac{1}{(4\alpha t)}\frac{d^2\theta}{d\eta^2}$$

Notice that we have written the derivatives of θ with respect to η as total derivatives since we have assumed that θ is a function of η alone. Substituting in Eq. (3.51) gives

$$-\frac{x}{2t(4\alpha t)^{1/2}}\frac{d\theta}{d\eta} = \frac{\alpha}{(4\alpha t)}\frac{d^2\theta}{d\eta^2}$$

which simplifies to

$$-2\eta \frac{d\theta}{d\eta} = \frac{d^2\theta}{d\eta^2} \tag{3.54}$$

Equation (3.54) is a second-order nonlinear ordinary differential equation, which requires two boundary conditions. Transforming Eqs. (3.52) gives

$$\eta = 0: \quad \theta = 1, \qquad \text{since } \eta = 0 \text{ when } x = 0 \tag{3.55a}$$

$$\eta \to \infty: \quad \theta = 0, \qquad \text{since } \eta \to \infty \text{ when } x \to \infty, \text{ or } t \to 0 \tag{3.55b}$$

Since both the transformed equation and conditions do not depend on x or t, the *similarity transformation* has been successful. Let $d\theta/d\eta = p$; then Eq. (3.54) becomes the first-order equation

$$-2\eta p = \frac{dp}{d\eta}$$

or

$$\frac{dp}{p} = -2\eta\, d\eta = -d\eta^2$$

Integrating once,

$$p = \frac{d\theta}{d\eta} = C_1 e^{-\eta^2}$$

Integrating again,

$$\theta = C_1 \int_0^\eta e^{-u^2} du + C_2 \tag{3.56}$$

where u is a dummy variable for the integration. The integration constants are evaluated from the boundary conditions, Eq. (3.55). From Eq. (3.55a), $\theta = 1$ at $\eta = 0$; hence, $C_2 = 1$. From Eq. (3.55b), $\theta \to 0$ as $\eta \to \infty$,

$$0 = C_1 \int_0^\infty e^{-u^2} du + 1$$

The definite integral is given by standard integral tables as $\pi^{1/2}/2$; hence,

$$0 = C_1 \frac{\pi^{1/2}}{2} + 1 \qquad \text{or} \qquad C_1 = -\frac{2}{\pi^{1/2}}$$

Substituting back in Eq. (3.56) gives the temperature distribution as

$$\theta = \frac{T - T_0}{T_s - T_0} = 1 - \frac{2}{\pi^{1/2}} \int_0^\eta e^{-u^2} du \tag{3.57}$$

The function $(2/\pi^{1/2})\int_0^{\eta} e^{-u^2}\,du$ is called the **error function**, erf η. For our purposes, it is more convenient to use the **complementary error function**, erfc $\eta = 1 - \text{erf}\ \eta$, which is tabulated as Table B.4 in Appendix B. Then

$$\frac{T - T_0}{T_s - T_0} = \text{erfc}\,\frac{x}{(4\alpha t)^{1/2}} \tag{3.58}$$

Equation (3.58) is plotted as Fig. 3.14. Since the temperature profiles at any time fall on a single curve when plotted as $\theta(\eta)$, they are said to be *self-similar*, which is why η is called the *similarity variable* for the problem.

The Surface Heat Flux

The heat flux at the surface is found from Eq. (3.58) using Fourier's law and the chain rule of differentiation,

$$\begin{aligned} q_s &= -k\left.\frac{\partial T}{\partial x}\right|_{x=0} \\ &= -k(T_s - T_0)\frac{\partial}{\partial x}\left[1 - \frac{2}{\pi^{1/2}}\int_0^{\eta} e^{-u^2}\,du\right]_{x=0} \\ &= -k(T_s - T_0)\left[-\frac{2}{\pi^{1/2}}e^{-\eta^2}\frac{1}{(4\alpha t)^{1/2}}\right]_{\eta=0} \end{aligned}$$

$$q_s = \frac{k(T_s - T_0)}{(\pi\alpha t)^{1/2}} \tag{3.59}$$

which is identical to the short-time solution for the slab problem, Eq. (3.47).

Solutions of the semi-infinite solid problem for other kinds of boundary conditions are also useful. A selection follows.

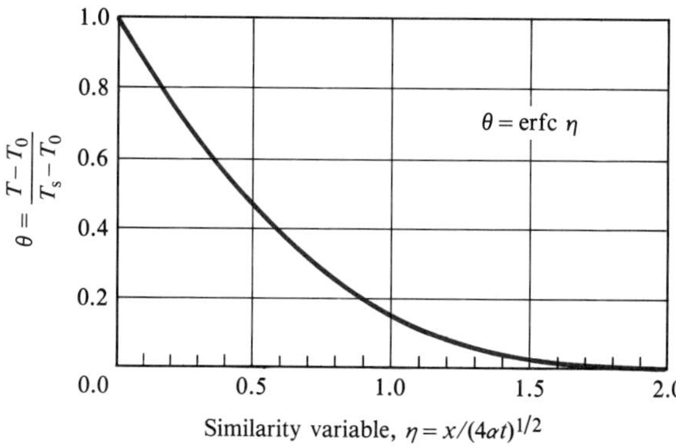

Figure 3.14 Dimensionless temperature response in a semi-infinite solid with a step change in surface temperature, Eq. (3.58).

Constant Surface Heat Flux

If at time $t = 0$ the surface is suddenly exposed to a constant heat flux q_s—for example, by radiation from a high-temperature source—the resulting temperature response is

$$T - T_0 = \frac{q_s}{k}\left[\left(\frac{4\alpha t}{\pi}\right)^{1/2} e^{-x^2/4\alpha t} - x \operatorname{erfc}\frac{x}{(4\alpha t)^{1/2}}\right] \tag{3.60}$$

This temperature response is shown in Fig. 3.15.

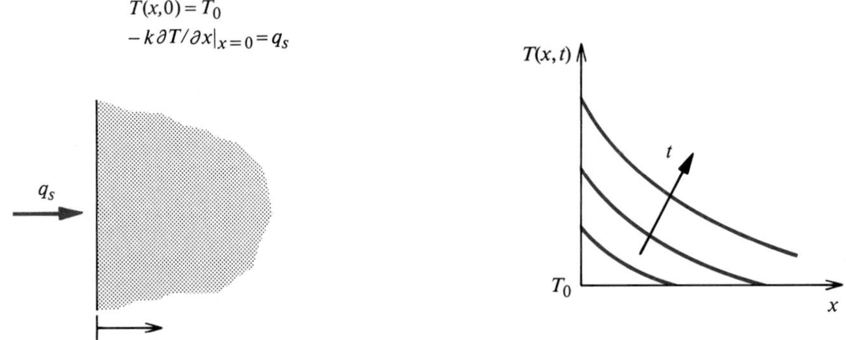

Figure 3.15 Temperature response in a semi-infinite solid exposed to a constant surface heat flux, Eq. (3.60).

Convective Heat Transfer to the Surface

If at time $t = 0$ the surface is suddenly exposed to a fluid at temperature T_e, with a convective heat transfer coefficient h_c, the resulting temperature response is

$$\frac{T - T_0}{T_e - T_0} = \operatorname{erfc}\frac{x}{(4\alpha t)^{1/2}} - e^{h_c x/k + (h_c/k)^2 \alpha t}\operatorname{erfc}\left(\frac{x}{(4\alpha t)^{1/2}} + \frac{h_c}{k}(\alpha t)^{1/2}\right) \tag{3.61}$$

as shown in Fig. 3.16.

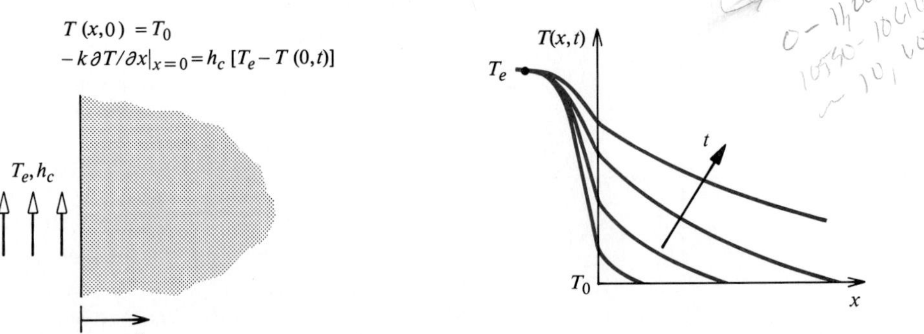

Figure 3.16 Temperature response in a semi-infinite solid suddenly exposed to a fluid, Eq. (3.61).

Surface Energy Pulse

If an amount of energy E per unit area is released instantaneously on the surface at $t = 0$ (e.g., if the surface is exposed to an energy pulse from a laser), and none of this energy is lost from the surface, the resulting temperature response is

$$T - T_0 = \frac{E}{\rho c(\pi \alpha t)^{1/2}} e^{-x^2/4\alpha t} \tag{3.62}$$

as shown in Fig. 3.17.

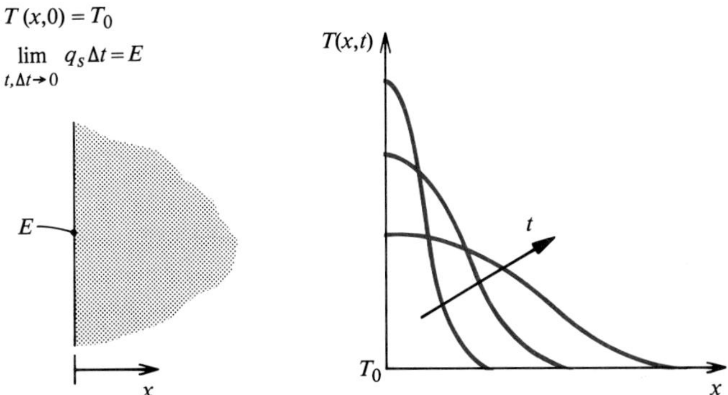

Figure 3.17 Temperature response in a semi-infinite solid after an instantaneous release of energy on the surface, Eq. (3.62).

Periodic Surface Temperature Variation

The surface temperature varies periodically as $(T_s - T_0) = (T_s^* - T_0)\sin \omega t$, as shown in Fig. 3.18. The resulting temperature response is

$$\frac{T - T_0}{T_s^* - T_0} = e^{-x(\omega/2\alpha)^{1/2}} \sin[\omega t - x(\omega/2\alpha)^{1/2}] \tag{3.63}$$

Notice how the amplitude of the temperature variation decays into the solid exponentially, while a phase lag $x(\omega/2\alpha)^{1/2}$ develops.

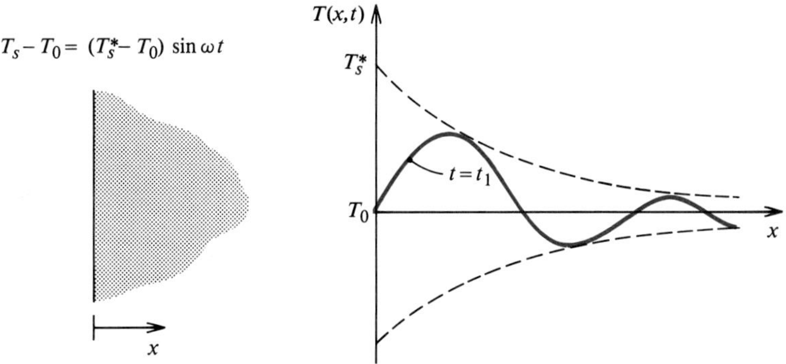

Figure 3.18 Periodic surface temperature variation for a semi-infinite solid: instantaneous temperature profile at $t = t_1$, Eq. (3.63).

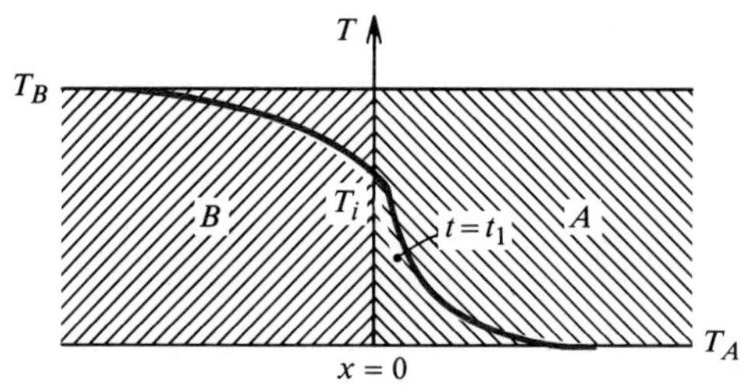

Figure 3.19 Contact of two semi-infinite solids: instantaneous temperature profile at $t = t_1$ for $k_B > k_A$.

Contact of Two Semi-Infinite Solids

Now consider two semi-infinite solids, A and B, of different materials with uniform temperatures T_A and T_B, brought together at time $t = 0$, as shown in Fig. 3.19. The solution of this problem shows that the interface temperature T_i is constant in time and is given by

$$\frac{T_A - T_i}{T_i - T_B} = \frac{k_B}{k_A}\left(\frac{\alpha_A}{\alpha_B}\right)^{1/2} = \left(\frac{(k\rho c)_B}{(k\rho c)_A}\right)^{1/2} \tag{3.64}$$

with corresponding erfc function temperature distributions in each solid. For example, if solid A is a carbon steel with $k = 48$ W/m K and $\alpha = 13.3 \times 10^{-6}$ m^2/s at 100°C, and solid B is neoprene rubber with $k = 0.19$ W/m K and $\alpha = 0.079 \times 10^{-6}$ m^2/s at 0°C, the interface temperature T_i is calculated to be 95.1°C. This is much closer to the initial temperature of the steel than to that of the rubber. Equation (3.64) shows why a high-conductivity material at room temperature feels colder to the touch than does a low-conductivity material at the same temperature.

The instantaneous heat flux at the surface $q_s(t)$ can be found from Eqs. (3.61) through (3.63) by applying Fourier's law. The derivation of Eq. (3.63) is given as Exercise 3–30. The other solutions are best obtained using Laplace transforms [5,9].

Computer Program COND1

COND1 calculates the thermal response of a semi-infinite solid initially at temperature T_0. There is a choice of five boundary conditions imposed at time $t = 0$:

1. The surface temperature is changed to T_s.
2. A heat flux q_s is imposed on the surface.
3. The surface is exposed to a fluid at temperature T_e, with a convective heat transfer coefficient h_c.
4. An amount of energy E is released instantaneously at the surface.
5. The surface temperature varies periodically as $T_s - T_0 = (T_s^* - T_0)\sin\omega t$.

Plotting options include $T(x)$, $T_s(t)$, or $q_s(t)$, which can be chosen appropriately. The analysis for boundary condition 1 was given in Section 3.4.2. The temperature responses for boundary conditions 2 through 5 are given as Eqs. (3.60) through (3.63), respectively.

EXAMPLE 3.5 Cooling of a Concrete Slab

A thick concrete slab initially at 400 K is sprayed with a large quantity of water at 300 K. How long will the location 5 cm below the surface take to cool to 320 K?

Solution

Given: Hot concrete slab sprayed with water at 300 K.

Required: Rate of cooling 5 cm below surface.

Assumptions: 1. The rate of spraying is sufficient to maintain the surface at 300 K.
2. The slab can be treated as a semi-infinite solid.

Equation (3.58) applies and can be written as

$$\theta = \operatorname{erfc} \eta \qquad \text{where } \theta = \frac{T - T_0}{T_s - T_0},\ \eta = \frac{x}{(4\alpha t)^{1/2}}$$

$$\theta = \frac{320 - 400}{300 - 400} = 0.8$$

Thus, $0.8 = \operatorname{erfc} \eta$; from Table B.4, $\eta = 0.179$.

$$t = \frac{x^2}{4\alpha\eta^2}$$

From Table A.3, α for concrete is 0.75×10^{-6} m^2/s. Thus, the required time is

$$t = \frac{(0.05)^2}{(4)(0.75 \times 10^{-6})(0.179)^2} = 2.60 \times 10^4 \text{ s} \simeq 7 \text{ h}$$

Comments

1. A temperature penetration depth δ_t may be defined as the location where the tangent to the temperature profile at $x = 0$ intercepts the line $T = 400$ K, as shown in the figure. The temperature gradient at $x = 0$ is found by differentiating Eq. (3.58):

$$-\left.\frac{\partial T}{\partial x}\right|_{x=0} = \frac{T_s - T_0}{(\pi\alpha t)^{1/2}} \equiv \frac{T_s - T_0}{\delta_t}$$

Hence, $\delta_t = 1.772(\alpha t)^{1/2} = 1.772(0.75 \times 10^{-6} \times 2.60 \times 10^4)^{1/2} = 0.247$ m.

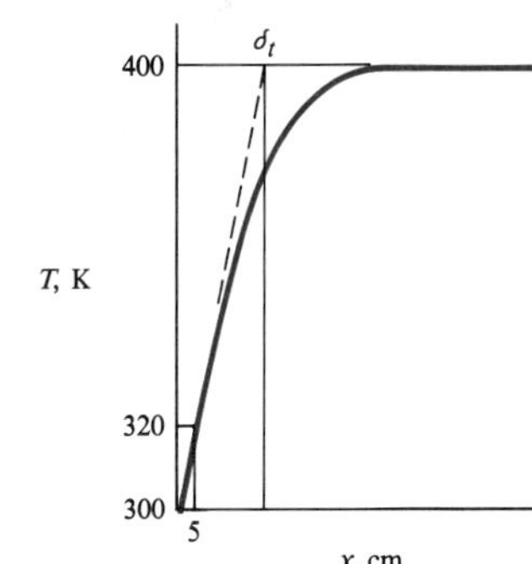

2. Check t using COND1.

EXAMPLE 3.6 Radiative Heating of a Firewall

A 15 cm–thick concrete firewall has 4 mm–thick steel sheathing. The wall is suddenly exposed to a radiant heat source that can be approximated as a blackbody at 1000 K. How long will it take for the surface to reach 500 K if the initial temperature of the wall is 300 K?

Solution

Given: Concrete firewall exposed to radiant heat source.

Required: Temperature response of surface.

Assumptions:
1. Wall can be modeled as a semi-infinite solid.
2. Negligible temperature drop across steel sheathing.
3. An absorptance $\alpha = 0.9$ for a heavily oxidized steel surface; negligible radiation emitted by the wall.

Equation (3.60) evaluated at $x = 0$ is

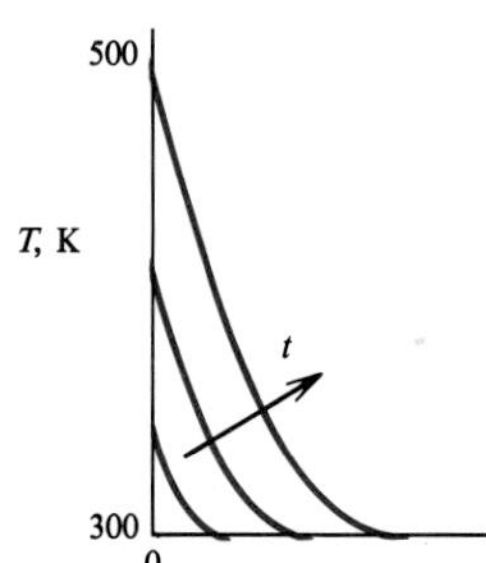

$$T_s - T_0 = \frac{q_s}{k}\left(\frac{4\alpha t}{\pi}\right)^{1/2}$$

or

$$t = \frac{\pi}{\alpha}\left[\frac{(T_s - T_0)k}{2q_s}\right]^2$$

From Table A.3, the required concrete properties are $k = 1.4$ W/m K and $\alpha = 0.75 \times 10^{-6}$ m²/s.

Since the surface temperature is low compared to the radiation source temperature, we can neglect radiation emitted by the surface, so

$$q_s \simeq \alpha\sigma T^4 = (0.9)(5.67 \times 10^{-8})(1000)^4 = 51.0 \times 10^3 \text{ W/m}^2$$

$$t = \frac{\pi}{0.75 \times 10^{-6}}\left[\frac{(500 - 300)(1.4)}{(2)(51.0 \times 10^3)}\right]^2 = 31.6 \text{ s}$$

Comments

1. At most, $\varepsilon\sigma T_s^4 = (0.9)(5.67 \times 10^{-8})(500^4) = 3.19 \times 10^3$ W/m², which is only 6% of q_s and is justifiably neglected in making this engineering estimate.

2. To check the temperature drop across the steel sheathing of thickness L, assume quasi-steady conduction; $\Delta T = q_s L/k$. For $L = 0.004$ m, $k = 59$ W/m K for AISI 1010 carbon steel; thus, $\Delta T = (51.0 \times 10^3)(0.004)/(59) = 3.5$ K, which is small.

3. To check whether the assumption of a semi-infinite solid is valid, we estimate a penetration depth δ_t:

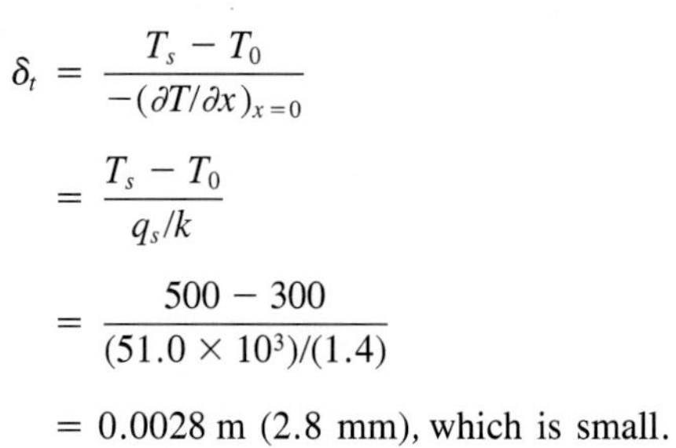

$$\delta_t = \frac{T_s - T_0}{-(\partial T/\partial x)_{x=0}} = \frac{T_s - T_0}{q_s/k} = \frac{500 - 300}{(51.0 \times 10^3)/(1.4)}$$

$= 0.0028$ m (2.8 mm), which is small.

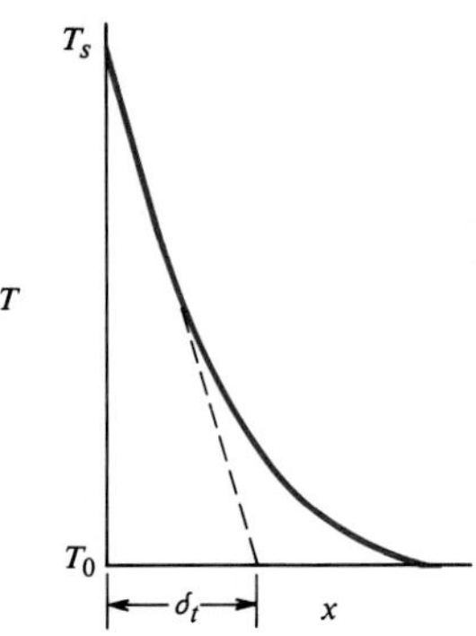

Solution using COND1

The required input in SI units is:

Boundary condition = 2
Plot option = 2 (T_s versus t)
t-range for plot = 0, 60
Thermal conductivity = 1.4
Thermal diffusivity = 0.75×10^{-6}
Initial temperature T_0 = 300
Surface heat flux q_s = 51.0×10^3

From the graph of T_s versus t, $t \simeq 32$ (seconds) when $T_s = 500$ (kelvins).

Comments

1. A more accurate answer can be obtained from COND1 by adjusting the t-range.
2. Any consistent units can be used in COND1. With SI units, temperature may be in kelvins or degrees Celsius.

EXAMPLE 3.7 Thermal Response of Soil

On a tropical island, a large refrigerated shed has been operating at 5°C for a number of years. It is then put out of service; a wood floor is removed, and ambient air at 27°C is allowed to circulate freely through the shed. How long will it take for the ground 1 m below the surface to reach 15°C? Assume a convective heat transfer coefficient of 3.0 W/m^2 K, and use thermal properties of a wet soil ($k = 2.6$ W/m K, $\alpha = 0.45 \times 10^{-6}$ m^2/s).

Solution

Given: Ground initially at 5°C, exposed to air at 27°C.

Required: Temperature response 1 m below surface.

Assumptions: 1. Semi-infinite solid model valid.
2. The initial temperature is uniform (i.e., 5°C for an appreciable distance below the surface).

Equation (3.61) applies:

$$\frac{T - T_0}{T_e - T_0} = \operatorname{erfc}\frac{x}{(4\alpha t)^{1/2}} - e^{h_c x/k + (h_c/k)^2 \alpha t} \operatorname{erfc}\left(\frac{x}{(4\alpha t)^{1/2}} + \frac{h_c}{k}(\alpha t)^{1/2}\right)$$

$$\frac{T - T_0}{T_e - T_0} = \frac{15 - 5}{27 - 5} = 0.4545; \qquad \frac{h_c}{k} = \frac{3.0}{2.6} = 1.154 \text{ m}^{-1}, \qquad x = 1 \text{ m}$$

An iterative solution is required, but how do we make a reasonable first guess for t? A lower limit is obtained if we use only the first term of Eq. (3.61), which corresponds to $h_c \to \infty$, $T_s = T_e$.

$$0.4545 = \operatorname{erfc} \eta$$

From Table B.4, $\eta = 0.53$.

$$0.53 = \frac{x}{(4\alpha t)^{1/2}} = \frac{1}{(4 \times 0.45 \times 10^{-6} t)^{1/2}}; \qquad t = 2.0 \times 10^6 \text{ s}$$

The actual time will be greater than 2.0×10^6 s; taking $t = 4 \times 10^6$ s as a first guess, the following table summarizes the results:

Time, t s $\times 10^{-6}$	$\dfrac{T - T_0}{T_e - T_0}$
4	0.368
5	0.415
6	0.452
6.05	0.4540
6.06	0.4545

Hence, $t = 6.06 \times 10^6$ s (~70 days)

Comments

1. If the solution is done by hand, care must be taken to evaluate the erfc function accurately. COND1 will perform the required calculations rapidly and reliably.
2. The long time required suggests that problems involving conduction into the ground are almost always *transient* problems.

EXAMPLE 3.8 Temperature Fluctuations in a Diesel Engine Cylinder Wall

A thermocouple is installed in the 5 mm–thick cylinder wall of a stationary diesel engine, 1 mm below the inner surface. In a particular test, the engine operates at 1000 rpm, and the thermocouple reading is found to have a mean value of 322°C and an amplitude of 0.79°C. If the temperature variation can be assumed to be approximately sinusoidal, estimate the amplitude and phase difference of the inner-surface temperature variation. Take $\alpha = 12.0 \times 10^{-6}$ m^2/s and $k = 40$ W/m K for the carbon steel wall.

Solution

Given: Thermocouple installed in diesel engine cylinder wall.

Required: Amplitude and phase difference of inner-surface temperature.

Assumptions: Model wall as a semi-infinite solid.

At first it would appear that the analysis of Section 3.4.2 does not apply to this problem. The cylinder wall is not very thick, and since the outer surface is cooled, there will be a temperature gradient through the wall. However, if the temperature wave is damped out in a very short distance from the surface, Eq. (3.63) can be used to estimate the amplitude decay and phase lag:

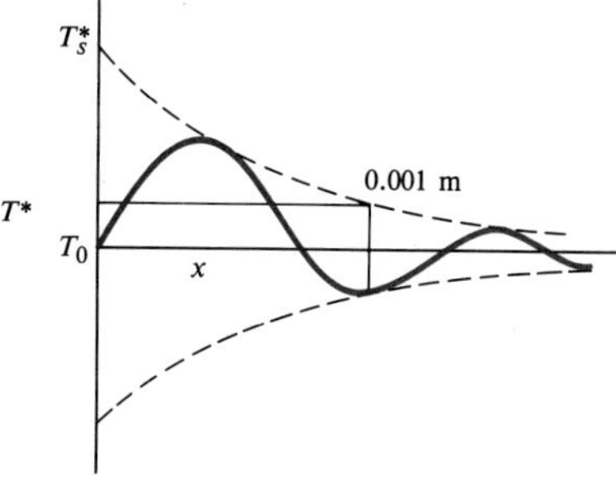

$$\frac{T - T_0}{T_s^* - T_0} = e^{-x(\omega/2\alpha)^{1/2}} \sin[\omega t - x(\omega/2\alpha)^{1/2}]$$

$$\omega = 2\pi\left(\frac{1000}{60}\right) = 104.7 \text{ rad/s}; \qquad (\omega/2\alpha)^{1/2} = \left(\frac{104.7}{(2)(12.0 \times 10^{-6})}\right)^{1/2} = 2089 \text{ m}^{-1}$$

If $(T^* - T_0)$ is the amplitude of the temperature variation,

$$\frac{T^* - T_0}{T_s^* - T_0} = e^{-x(\omega/2\alpha)^{1/2}} = e^{-(0.001)(2089)} = 0.124$$

Thus,

$$T_s^* - T_0 = \frac{T^* - T_0}{0.124} = \frac{0.79}{0.124} = 6.38°\text{C}$$

The phase lag is $x(\omega/2\alpha)^{1/2} = 2.09$ rad $= 120$ degrees.

Comments

Use COND1 to examine some spatial and temporal temperature profiles.

3.4.3 Convective Cooling of Slabs, Cylinders, and Spheres

We now consider the more general problem of transient conduction in three common shapes: the infinite slab, the infinite cylinder, and the sphere, with surface cooling (or heating) by convection. The slab problem will be analyzed first, and the results will be generalized to the cylinder and sphere.

Analysis for the Slab

In Section 3.4.1, we considered the temperature response $T(x, t)$ of a slab suddenly immersed in a fluid under conditions where the convective heat transfer resistance is negligible, that is, $\text{Bi} = h_c L/k$ is large. On the other hand, the lumped thermal capacity model of Section 1.5 applies when the conduction resistance in the slab is negligible, that is, the Biot number is small. We now consider the general case where the convection and conduction resistances are of comparable magnitudes, giving a Biot number of order unity. The slab and coordinate system are shown in

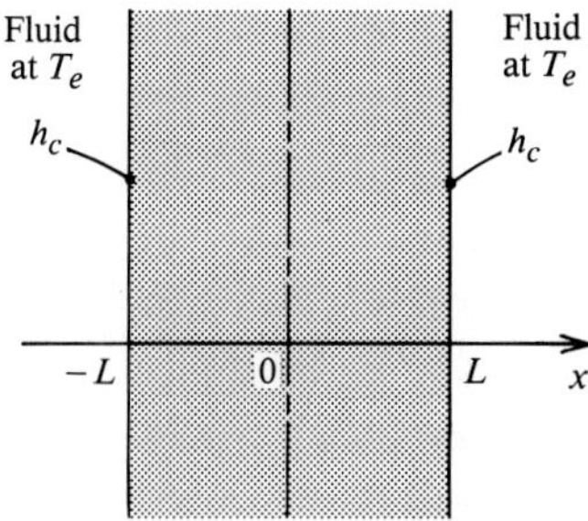

Figure 3.20 Schematic of a slab suddenly immersed in a fluid.

Fig. 3.20. We again define $\eta = x/L$ and $\zeta = \alpha t/L^2$, but now we define the dimensionless temperature θ in terms of the initial slab temperature T_0 and fluid temperature T_e as $\theta = (T - T_e)/(T_0 - T_e)$. The analysis proceeds as in Section 3.4.1 to obtain

$$\theta(\zeta, \eta) = e^{-\lambda^2\zeta}(A\cos\lambda\eta + B\sin\lambda\eta)$$

The boundary condition, Eq. (3.36*b*), is as before; namely, $\partial\theta/\partial\eta|_{\eta=0} = 0$, so $B = 0$ and

$$\theta(\zeta, \eta) = Ae^{-\lambda^2\zeta}\cos\lambda\eta \tag{3.65}$$

However the second boundary condition is now obtained from the requirement that the heat conduction at the surface of the solid equal the heat convection into the fluid:

$$-k\left.\frac{\partial T}{\partial x}\right|_{x=L} = h_c(T|_{x=L} - T_e)$$

which transforms into

$$-\left.\frac{k}{L}\frac{\partial\theta}{\partial\eta}\right|_{\eta=1} = h_c\theta|_{\eta=1}$$

or

$$-\left.\frac{\partial\theta}{\partial\eta}\right|_{\eta=1} = \mathrm{Bi}\,\theta|_{\eta=1} \tag{3.66}$$

We see once again how the Biot number $\mathrm{Bi} = h_cL/k$ occurs naturally when the convective boundary condition is put in dimensionless form. Substituting Eq. (3.65) into Eq. (3.66) gives

$$Ae^{-\lambda^2\zeta}\lambda\sin\lambda = \mathrm{Bi}\,Ae^{-\lambda^2\zeta}\cos\lambda$$

or

$$\cot\lambda = \frac{\lambda}{\mathrm{Bi}} \tag{3.67}$$

which is a transcendental equation with an infinite number of roots or eigenvalues.

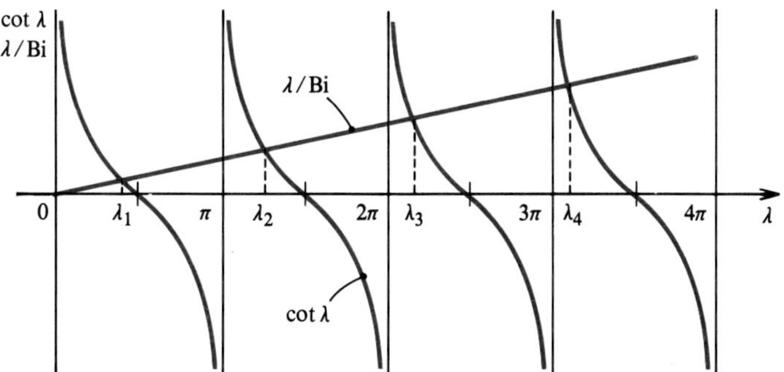

Figure 3.21 Graphical solution of the transcendental equation, Eq. (3.67): $\cot \lambda = \lambda/\text{Bi}$.

Figure 3.21 shows a plot of Eq. (3.67); for a given value of the Biot number, the eigenvalues λ_n $(n = 1, 2, 3, \ldots)$ can be calculated. To each of the eigenvalues corresponds a solution with eigenfunction $\cos \lambda_n \eta$ and arbitrary constant A_n. Their sum is the general solution:

$$\theta(\zeta, \eta) = \sum_{n=1}^{\infty} A_n e^{-\lambda_n^2 \zeta} \cos \lambda_n \eta \tag{3.68}$$

The constants A_n are evaluated from the initial condition, Eq. (3.36*a*):

$$\theta(0, \eta) = \sum_{n=1}^{\infty} A_n \cos \lambda_n \eta = 1$$

Thus, a Fourier series expansion in terms of the eigenfunctions $\cos \lambda_n \eta$ is required. The details are required as Exercise 3–46, and the result is

$$A_n = \frac{2 \sin \lambda_n}{\lambda_n + \sin \lambda_n \cos \lambda_n} \tag{3.69}$$

The solution in terms of temperature, Fourier number, and x/L is

$$\frac{T - T_e}{T_0 - T_e} = \sum_{n=1}^{\infty} e^{-\lambda_n^2 \text{Fo}} \frac{2 \sin \lambda_n}{\lambda_n + \sin \lambda_n \cos \lambda_n} \cos \lambda_n \frac{x}{L} \tag{3.70}$$

Figure 3.22 shows a plot of Eq. (3.70) for $\text{Bi} = 3.0$. Notice how the tangents to the temperature curves at the surface all intersect at a common point, the location of which is given by boundary condition, Eq. (3.66).

In addition to the temperature distribution, it is often useful to know the **fractional energy loss** Φ, which is the actual energy loss in time t divided by the total loss in cooling completely to the ambient temperature. An energy balance on unit area of the half slab gives

$$\Phi = \frac{\int_0^t q_s \, dt}{\rho c L (T_0 - T_e)} = \frac{\rho c L (T_0 - \overline{T})}{\rho c L (T_0 - T_e)} = 1 - \frac{\overline{T} - T_e}{T_0 - T_e}$$

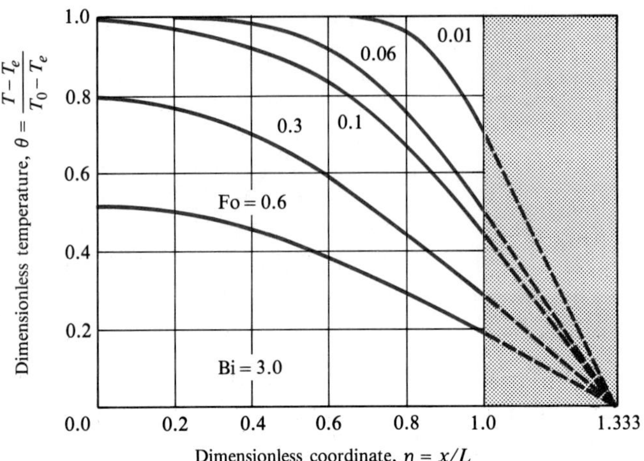

Figure 3.22 Temperature profiles for convective cooling of a slab calculated from Eq. (3.70); Bi = 3.0.

where $\overline{T}$ is the volume-averaged temperature. Evaluating $\overline{T}$ from Eq. (3.70) gives

$$\Phi = 1 - \sum_{n=1}^{\infty} e^{-\lambda_n^2 \mathrm{Fo}} \frac{2 \sin \lambda_n}{\lambda_n + \sin \lambda_n \cos \lambda_n} \frac{\sin \lambda_n}{\lambda_n} \tag{3.71}$$

Figure 3.23 shows a plot of Eq. (3.71) for various values of the Biot number.

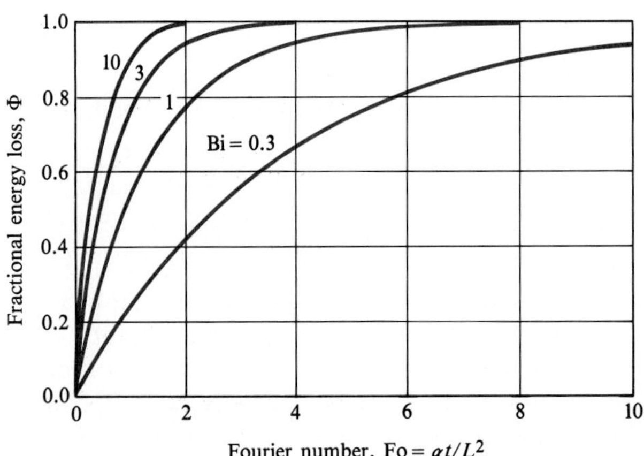

Figure 3.23 Fractional energy loss Φ as a function of Fourier number $\alpha\, t/L^2$ for convective cooling of a slab calculated from Eq. (3.71). Biot number Bi = 0.3, 1, 3, and 10.

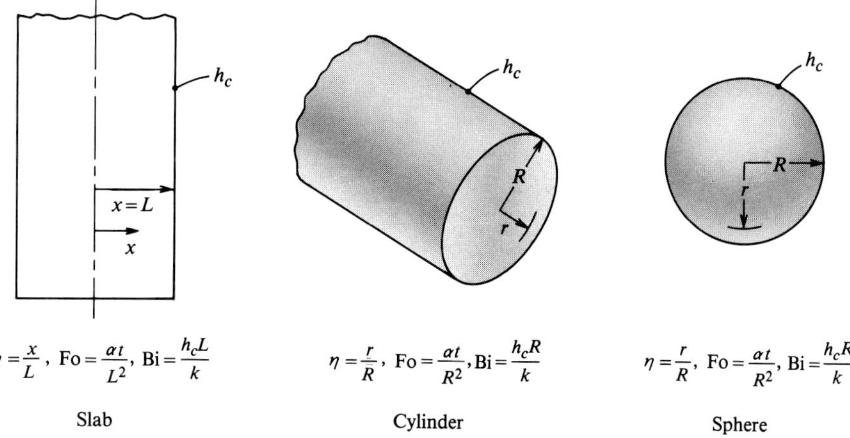

Figure 3.24 Schematic for the generalized solution of the temperature response of a convectively cooled slab, cylinder, or sphere.

Generalized Form of the Solution

It is convenient to have a general form of the solution applicable to a slab, cylinder, or sphere. Referring to Fig. 3.24,

$$\theta = \sum_{n=1}^{\infty} A_n e^{-\lambda_n^2 \mathrm{Fo}} f_n(\lambda_n \eta) \tag{3.72}$$

$$\Phi = 1 - \overline{\theta} = 1 - \sum_{n=1}^{\infty} A_n e^{-\lambda_n^2 \mathrm{Fo}} B_n \tag{3.73}$$

with the eigenvalues given by

Slab: $\mathrm{Bi}\cos\lambda - \lambda\sin\lambda = 0$ **(3.74*a*)**

Cylinder: $\lambda J_1(\lambda) - \mathrm{Bi}\, J_0(\lambda) = 0$ **(3.74*b*)**

Sphere: $\lambda\cos\lambda + (\mathrm{Bi} - 1)\sin\lambda = 0$ **(3.74*c*)**

For the slab, $\eta = x/L$, $\mathrm{Fo} = \alpha t/L^2$, and $\mathrm{Bi} = h_c L/k$, as before; for the cylinder and sphere, $\eta = r/R$, $\mathrm{Fo} = \alpha t/R^2$, and $\mathrm{Bi} = h_c R/k$. Table 3.4 gives A_n, f_n, and B_n. In Eq. (3.74*b*) J_0 and J_1 are Bessel functions of the first kind, of orders 0 and 1, respectively. These functions are defined and tabulated in Appendix B.

Notice that the characteristic lengths used to define the Biot number are the slab half-width, L, and cylinder or sphere radius, R. Recall that in the lumped thermal capacity analysis of Section 1.5, we defined the characteristic length as volume/area, V/A. For a slab, $V/A = L$, so the Biot numbers are the same. However, for the cylinder and sphere, $V/A = R/2$ and $R/3$, respectively, and the Biot number definitions are different.

Table 3.4 The constants A_n and B_n and the function f_n for the transient thermal response of slabs, cylinders, and spheres.

Geometry	$A_n(\lambda_n)$	$B_n(\lambda_n)$	$f_n(\lambda_n \eta)$
Slab	$2\dfrac{\sin\lambda_n}{\lambda_n + \sin\lambda_n \cos\lambda_n}$	$\dfrac{\sin\lambda_n}{\lambda_n}$	$\cos\left(\lambda_n \dfrac{x}{L}\right)$
Cylinder	$2\dfrac{J_1(\lambda_n)}{\lambda_n\,[J_0^2(\lambda_n) + J_1^2(\lambda_n)]}$	$2\dfrac{J_1(\lambda_n)}{\lambda_n}$	$J_0\left(\lambda_n \dfrac{r}{R}\right)$
Sphere	$2\dfrac{\sin\lambda_n - \lambda_n\cos\lambda_n}{\lambda_n - \sin\lambda_n\cos\lambda_n}$	$3\dfrac{\sin\lambda_n - \lambda_n\cos\lambda_n}{\lambda_n^3}$	$\dfrac{\sin[\lambda_n(r/R)]}{\lambda_n(r/R)}$

Computer Program COND2

The generalized form of the solution described above is implemented in COND2. The program computes the eigenvalues from Eqs. (3.74) using Newton's method. Up to 40 eigenvalues are calculated in order to meet a specified accuracy of 10^{-4} in the dimensionless temperature θ. For very short times, more than 40 eigenvalues are required to obtain the desired accuracy. Thus, for Fourier number Fo $< 10^{-3}$, COND1 should be used, since the semi-infinite solid model is quite appropriate for such short times. The output can be obtained as numerical data, a plot of the temperature profile, $\theta(\eta)$, or a plot of the fractional energy loss as a function of time, $\Phi(\text{Fo})$.

Approximate Solutions for Long Times

In Section 3.4.1, it was shown that the series solution converged rapidly for long times, and for Fo > 0.2, only the first term of the series need be retained for 2% accuracy. Usually we are most interested in the temperature at the center of the body ($x = 0$ or $r = 0$), where the response is slowest. Denoting the dimensionless center temperature as $\theta_c = (T_c - T_e)/(T_0 - T_e)$ and retaining only the first term of Eq. (3.72) gives

$$\theta_c = A_1 e^{-\lambda_1^2 \text{Fo}}, \qquad \text{Fo} > 0.2 \tag{3.75}$$

since $f_1(\lambda_1\eta) = 1$ for $\eta = 0$. When only the first term of the series is retained, the *shape* of the temperature distribution is unchanging with time. Thus, the temperature at any location is simply related to the center temperature as

$$\theta = \theta_c f_1(\lambda_1\eta), \qquad \text{Fo} > 0.2 \tag{3.76}$$

Similarly, retaining only one term in Eq. (3.73), the fractional energy loss is

$$\Phi = 1 - B_1\theta_c, \qquad \text{Fo} > 0.2 \tag{3.77}$$

Table 3.5 gives values of λ_1^2, A_1, and B_1 as a function of Biot number.

Table 3.5 Coefficients in the one-term approximation for convective cooling of slabs, cylinders, and spheres.

Slab

Bi	λ_1^2	A_1	B_1	Bi	λ_1^2	A_1	B_1
0.02	0.01989	1.0033	0.9967	2	1.160	1.180	0.8176
0.04	0.03948	1.0066	0.9934	4	1.600	1.229	0.7540
0.06	0.05881	1.0098	0.9902	6	1.821	1.248	0.7229
0.08	0.07790	1.0130	0.9871	8	1.954	1.257	0.7047
0.10	0.09678	1.016	0.9839	10	2.042	1.262	0.6928
0.2	0.1873	1.031	0.9691	20	2.238	1.270	0.6665
0.4	0.3519	1.058	0.9424	30	2.311	1.272	0.6570
0.6	0.4972	1.081	0.9192	40	2.321	1.272	0.6521
0.8	0.6257	1.102	0.8989	50	2.371	1.273	0.6490
1.0	0.7401	1.119	0.8811	100	2.419	1.273	0.6429
				∞	2.467	1.273	0.6366

Cylinder

Bi	λ_1^2	A_1	B_1	Bi	λ_1^2	A_1	B_1
0.02	0.03980	1.0051	0.9950	2	2.558	1.338	0.7125
0.04	0.07919	1.010	0.9896	4	3.641	1.470	0.6088
0.06	0.1182	1.015	0.9844	6	4.198	1.526	0.5589
0.08	0.1568	1.020	0.9804	8	4.531	1.553	0.5306
0.10	0.1951	1.025	0.9749	10	4.750	1.568	0.5125
0.2	0.3807	1.049	0.9526	20	5.235	1.593	0.4736
0.4	0.7552	1.094	0.9112	30	5.411	1.598	0.4598
0.6	1.037	1.135	0.8753	40	5.501	1.600	0.4527
0.8	1.320	1.173	0.8430	50	5.556	1.601	0.4485
1.0	1.577	1.208	0.8147	100	5.669	1.602	0.4401
				∞	5.784	1.602	0.4317

Sphere

Bi	λ_1^2	A_1	B_1	Bi	λ_1^2	A_1	B_1
0.02	0.05978	1.0060	0.9940	2	4.116	1.479	0.6445
0.04	0.1190	1.012	0.9881	4	6.030	1.720	0.5133
0.06	0.1778	1.018	0.9823	6	7.042	1.834	0.4516
0.08	0.2362	1.024	0.9766	8	7.647	1.892	0.4170
0.10	0.2941	1.030	0.9710	10	8.045	1.925	0.3952
0.2	0.5765	1.059	0.9435	20	8.914	1.978	0.3500
0.4	1.108	1.116	0.8935	30	9.225	1.990	0.3346
0.6	1.599	1.171	0.8490	40	9.383	1.994	0.3269
0.8	2.051	1.224	0.8094	50	9.479	1.996	0.3223
1.0	2.467	1.273	0.7740	100	9.673	1.999	0.3131
				∞	9.869	2.000	0.3040

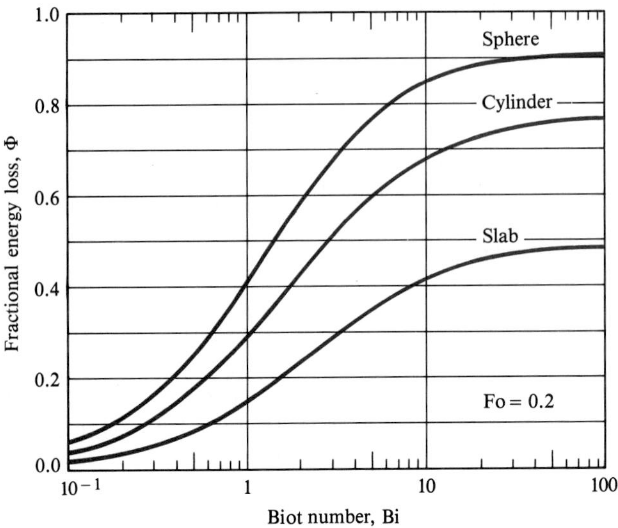

Figure 3.25 Fractional energy loss Φ at a Fourier number (dimensionless time) of Fo $= 0.2$ for the slab, cylinder, and sphere: effect of Biot number Bi. The one-term approximation is valid above each curve.

The single-term approximation is useful because it is valid during much of the cooling period. Figure 3.25 shows the fractional energy loss Φ at Fo $= 0.2$ as a function of Bi for the three shapes. The fraction of the cooling period over which the approximation is valid increases as the Biot number and surface-area/volume ratio (sphere-cylinder-slab) decrease. The single-term approximation is also used for problems in which the fluid temperature T_e changes slowly with time. It is then useful to define an *interior heat transfer coefficient* for conduction into the body for use in a thermal-circuit representation of the complete system. This concept is developed and used in Exercises 3–65 through 3–68.

Temperature Response Charts

Graphs of the various solutions for convective cooling, known as *temperature response charts*, were indispensable to engineers until handheld calculators and personal computers became standard tools. Charts for the series solution were given by Gurney and Lurie [10] as early as 1923, but better examples are now available. Perhaps the best widely available selection of charts is found in the *Handbook of Heat Transfer Fundamentals* [11]. In addition to charts for the convective cooling problem discussed here, Reference [11] gives charts for numerous other cooling and heating problems. For example, there are charts for radiation heating, conduction with internal heat generation, and composite solids. These charts were first published by Schneider in 1963 [12].

Heisler published charts for the one-term approximate solution in 1947 [13]. Heisler-type charts are found in many textbooks and are widely used. However, such charts have three major limitations:

1. The charts are invalid for Fo < 0.2.
2. The charts are often difficult to read accurately for Fo $\lesssim 1$.
3. The cooling process is nearly complete over a large region of the charts. For example, for a sphere with Bi $\gtrsim 1$, it is 90% complete for Fo $\gtrsim 1$.

Appendix C contains temperature response charts for convective cooling. Two types of charts for the slab, cylinder, and sphere are given. Figure C.1 gives the temperature response of the body center ($x = 0$ or $r = 0$), and Fig. C.2 gives the fractional energy loss Φ as a function of time. Both figures are based on the complete solution and are thus valid for all values of Fourier number. The curves for Bi $= 1000$ correspond to negligible convective resistance and, hence, to a prescribed surface temperature $T_s = T_e$.

Problem-Solving Strategy

Our solutions for convective cooling (or heating) of slabs, cylinders, and spheres can be organized in the form of a problem-solving strategy. The basic problem is one where the heat transfer coefficient is known, and the temperature or fractional heat loss must be calculated at a given time. Then we may proceed as follows:

1. Calculate the lumped thermal capacity model Biot number.[2] If Bi < 0.1, the lumped thermal capacity solution of Section 1.5 applies.
2. If Bi > 0.1, calculate the Fourier number. If Fo < 0.05, the semi-infinite solid solution, Eq. (3.61), applies. A handheld calculator or COND1 can be used.
3. If Bi > 0.1 and $0.05 <$ Fo < 0.2, the complete series solution, Eqs. (3.72) and (3.73), applies. COND2 or the temperature response charts in Appendix C should be used.[3]
4. If Bi > 0.1 and Fo > 0.2, the long-time approximate solution, Eqs. (3.75) through (3.77), applies. COND2 can be used if available. Otherwise, the temperature response charts or a handheld calculator will suffice.

Sometimes a problem may be posed in such a manner that the foregoing procedure cannot be followed exactly (as will be the case in Examples 3.9 and 3.10). Also, as in all engineering problem solving, the required accuracy of a particular calculation should be viewed in the context of the complete problem. It is of little value to obtain θ or Φ to even two-figure accuracy when the model is a poor simulation of

---

[2] Bi $= h_c(V/A)/k$; $V/A = L$ for an infinite slab, $R/2$ for an infinite cylinder, and $R/3$ for a sphere.

[3] For a slab, the required value of Bi was calculated in step 1. For a cylinder and sphere, the required values are two and three times larger, respectively.

the real engineering problem. The major source of error here is in the specification of the heat transfer coefficient. Not only is it difficult to specify a value of h_c with less than 10% error, but in many cooling problems, h_c is not a constant as assumed by the model. For example, when the cooling is by natural convection, Eq. (1.23) shows that h_c is proportional to $(T_s - T_e)$ to the 1/4 power for laminar flow, and to the 1/3 power for turbulent flow. Thus, h_c must be estimated at some average temperature difference.

EXAMPLE 3.9 Annealing of Steel Plate

When steel plates are thinned by rolling, periodic reheating is required. Plain carbon steel plate 8 cm thick, initially at 440°C, is to be reheated to a minimum temperature of 520°C in a furnace maintained at 600°C. If the sum of the convective and radiative heat transfer coefficients is estimated to be 200 W/m² K, how long will the reheating take? Take $k = 40$ W/m K and $\alpha = 8.0 \times 10^{-6}$ m²/s for the steel.

Solution

Given: Steel plate, thickness $2L = 8$ cm.

Required: Temperature response of the center of the plate.

Assumptions: The heat transfer coefficient is constant at 200 W/m² K.

We first calculate the time constant for the heating process:

$$t_c = \frac{L^2}{\alpha} = \frac{(0.04)^2}{8 \times 10^{-6}} = 200 \text{ s}$$

which tells us the order of magnitude of the time required, namely, a few minutes (not seconds and not hours). Next we calculate the Biot number:

$$\text{Bi} = \frac{hL}{k} = \frac{(200)(0.040)}{40} = 0.2$$

Since $\text{Bi} > 0.1$, the lumped thermal capacity model should not be used. Since we cannot calculate the Fourier number yet (it is the answer to the problem), the complete series solution will be used. The minimum temperature is at the center of the plate, and the desired value of θ is

$$\theta_c = \frac{T_c - T_e}{T_0 - T_e} = \frac{520\text{–}600}{440\text{–}600} = 0.50$$

Using the temperature response chart, Fig. C.1*a* in Appendix C,

$$\text{Fo} = 3.9; \qquad t = t_c\,\text{Fo} = (200)(3.9) = 780 \text{ s (13 min)}$$

Since $\text{Fo} > 0.2$, we could have used the one-term approximation, Eq. (3.75):

$$\theta_c = 0.5 = A_1 e^{-\lambda_1^2 \text{Fo}}$$

From Table 3.5 for $\text{Bi} = 0.2$, $\lambda_1^2 = 0.1873$ and $A_1 = 1.031$. Solving, $\text{Fo} = 3.86$.

Solution using COND2

The required inputs are:

Geometry = 1 (slab)
Bi = 0.2
Output option = 2 (θ vs. η plot)
Fo = (must guess and iterate)
η range = 0, 1

A few iterations will give Fo = 3.9 for $\theta_c = 0.50$.

EXAMPLE 3.10 A Pebble Bed Air Heater

A pebble bed for storing thermal energy in a solar heating system has pebbles that can be approximated as 6 cm–diameter spheres. The bed is initially at 350 K before cold air at 280 K is admitted to the bed. If the heat transfer coefficient is 80 W/m^2 K, how long will it take the pebbles at the inlet of the bed to lose 90% of their available energy? Take $k = 1.6$ W/m K and $\alpha = 0.7 \times 10^{-6}$ m^2/s for the pebbles.

Solution

Given: Hot pebbles suddenly exposed to a cold air stream.

Required: Time for pebbles at inlet to lose 90% of their available energy.

Assumptions: Pebbles are spherical.

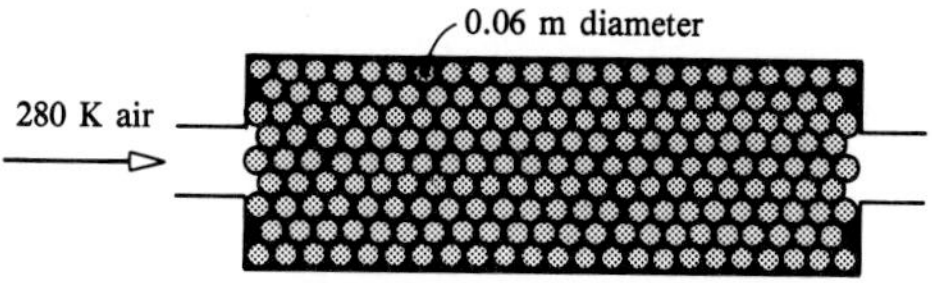

We first calculate the time constant for the cooling process:

$$t_c = \frac{R^2}{\alpha} = \frac{(0.03)^2}{0.7 \times 10^{-6}} = 1286 \text{ s (21 min)}$$

Next we calculate the lumped thermal capacity model Biot number:

$$\text{Bi} = \frac{h_c(R/3)}{k} = \frac{(80)(0.03/3)}{1.6} = 0.5$$

Since Bi > 0.1, the lumped thermal capacity method cannot be used. Although we suspect that Fo > 0.2 for $\Phi = 0.9$, we can conveniently use the complete series solution given as the temperature response chart, Fig. C.2*c* of Appendix C. For $\text{Bi} = h_cR/k = (3)(0.5) = 1.5$ and $\Phi = 0.9$, the chart gives $\text{Bi}^2\,\text{Fo} = 1.5$. Thus,

$$\text{Fo} = 1.5/\text{Bi}^2 = 1.5/(1.5)^2 = 0.67$$

$$t = t_c\,\text{Fo} = (1286)(0.67) = 860 \text{ s } (\sim 14 \text{ min})$$

Since Fo > 0.2, we could have used the one-term approximate solution as follows. From Eq. (3.77), $\Phi = 1 - B_1\theta_c$. For Bi = 1.5, interpolation in Table 3.5 gives $B_1 = 0.70$. Thus,

$$0.9 = 1 - (0.70)\theta_c \qquad \text{or} \qquad \theta_c = 0.143$$

By Eq. (3.75), $\theta_c = A_1 e^{-\lambda_1^2 \mathrm{Fo}}$. For Bi = 1.5, interpolation in Table 3.5 gives $\lambda_1^2 = 3.33$, $A_1 = 1.38$.

$$0.143 = (1.38)e^{-(3.33)\,\mathrm{Fo}} \qquad \text{or} \qquad \mathrm{Fo} = 0.68$$

which agrees well with the chart solution.

Solution using COND2

The required input is:

Geometry = 3 (sphere)
Bi = 1.5
Output option = 1 (Numerical data), or 3 (Φ vs. Fo plot)
Fo range = 0, 1

Fo = 0.67 for $\Phi = 0.9$.

Comments

The relatively short time of 14 min does not mean that warm air cannot be obtained for a long period. The bed is a regenerative heat exchanger (see Section 8.5): a temperature "wave" passes slowly through the bed, and the useful operating time is the time taken for this wave to break through the outlet end of the bed.

3.4.4 Product Solutions for Multidimensional Unsteady Conduction

Consider a long rectangular bar, with sides $2L_1$ and $2L_2$ wide, that is initially at temperature T_0 and suddenly immersed in a fluid at temperature T_e. The heat transfer coefficients on the sides are h_{c1} and h_{c2}. The task is to determine the temperature distribution $T(x, y, t)$. Again, a dimensionless temperature $\theta = (T - T_e)/(T_0 - T_e)$ is defined. The governing equation and appropriate initial and boundary conditions in a coordinate system such that $-L_1 \le x \le L_1$, $-L_2 \le y \le L_2$ are

$$\frac{\partial \theta}{\partial t} = \alpha\left(\frac{\partial^2\theta}{\partial x^2} + \frac{\partial^2\theta}{\partial y^2}\right) \tag{3.78}$$

$$t = 0: \quad \theta = 1$$

$$x = 0: \quad \frac{\partial\theta}{\partial x} = 0; \qquad y = 0: \quad \frac{\partial\theta}{\partial y} = 0$$

$$x = L_1: \quad -k\frac{\partial\theta}{\partial x} = h_{c1}\theta; \qquad y = L_2: \quad -k\frac{\partial\theta}{\partial y} = h_{c2}\theta$$

If the *separation of variables* method is to be used, one might assume a product solution of the form

$$\theta(t, x, y) = \mathscr{T}(t)X(x)Y(y)$$

where the functions $\mathscr{T}(t)$, $X(x)$, and $Y(y)$ are to be determined as before. However, it will now be shown that the solution can be expressed as the product of known solutions for the infinite slab. Consider two slabs of thickness $2L_1$ and $2L_2$, for which the dimensionless temperature governing equations and boundary conditions are

$$\theta_1 = \frac{T_1 - T_e}{T_0 - T_e} \qquad \theta_2 = \frac{T_2 - T_e}{T_0 - T_e}$$

$$\frac{\partial \theta_1}{\partial t} = \alpha \frac{\partial^2 \theta_1}{\partial x^2} \qquad \frac{\partial \theta_2}{\partial t} = \alpha \frac{\partial^2 \theta_2}{\partial y^2} \tag{3.79a,b}$$

$$t = 0: \quad \theta_1 = 1 \qquad t = 0: \quad \theta_2 = 1$$

$$x = 0: \quad \frac{\partial \theta_1}{\partial x} = 0 \qquad y = 0: \quad \frac{\partial \theta_2}{\partial y} = 0$$

$$x = L_1: \quad -k\frac{\partial \theta_1}{\partial x} = h_{c1}\theta_1 \qquad y = L_2: \quad -k\frac{\partial \theta_2}{\partial y} = h_{c2}\theta_2$$

The product of the solutions of these two problems satisfies the original problem. Let

$$\theta(t, x, y) = \theta_1(t, x)\theta_2(t, y)$$

Then

$$\frac{\partial^2 \theta}{\partial x^2} = \theta_2 \frac{\partial^2 \theta_1}{\partial x^2}; \qquad \frac{\partial^2 \theta}{\partial y^2} = \theta_1 \frac{\partial^2 \theta_2}{\partial y^2} \tag{3.80a,b}$$

$$\frac{\partial \theta}{\partial t} = \theta_1 \frac{\partial \theta_2}{\partial t} + \theta_2 \frac{\partial \theta_1}{\partial t} \tag{3.81}$$

Substituting Eqs. (3.79*a,b*) into Eq. (3.81),

$$\frac{\partial \theta}{\partial t} = \alpha \left(\theta_1 \frac{\partial^2 \theta_2}{\partial y^2} + \theta_2 \frac{\partial^2 \theta_1}{\partial x^2} \right)$$

Then substituting from Eqs. (3.80*a,b*) gives

$$\frac{\partial \theta}{\partial t} = \alpha \left(\frac{\partial^2 \theta}{\partial x^2} + \frac{\partial^2 \theta}{\partial y^2} \right)$$

which is the original differential equation, Eq. (3.78). Also, the initial and boundary conditions become

$$t = 0: \quad \theta(0, x, y) = \theta_1(0, x)\theta_2(0, y) = (1)(1) = 1$$

$$x = 0: \quad \frac{\partial \theta}{\partial x} = \theta_2 \frac{\partial \theta_1}{\partial x} = \theta_2 \times 0 = 0$$

$$y = 0: \quad \frac{\partial \theta}{\partial y} = \theta_1 \frac{\partial \theta_2}{\partial y} = \theta_1 \times 0 = 0$$

$$x = L_1: \quad -k\frac{\partial \theta}{\partial x} = \theta_2 \left(-k\frac{\partial \theta_1}{\partial x}\right) = \theta_2(h_{c1}\theta_1) = h_{c1}\theta$$

$$y = L_2: \quad -k\frac{\partial \theta}{\partial y} = \theta_1 \left(-k\frac{\partial \theta_2}{\partial y}\right) = \theta_1(h_{c2}\theta_2) = h_{c2}\theta$$

which are the original conditions. Figure 3.26 shows a schematic of the product solution.

Shapes amenable to product solutions of this type are shown in Table 3.6. Langston [14] has recently shown how the product rule can be applied to obtain the fractional energy loss. If the shape is formed by the intersection of two bodies, for example, the short cylinder of item 4 in Table 3.6, the fractional energy loss is

$$\Phi = \Phi_1 + \Phi_2(1 - \Phi_1) = \Phi_1 + \Phi_2 - \Phi_1\Phi_2 \tag{3.82a}$$

where subscripts 1 and 2 refer to the infinite slab and infinite cylinder, respectively. If the shape is formed by the intersection of three bodies, for example, the rectangular block of item 6 in Table 3.6, then

$$\Phi = \Phi_1 + \Phi_2(1 - \Phi_1) + \Phi_3(1 - \Phi_1)(1 - \Phi_2) \tag{3.82b}$$

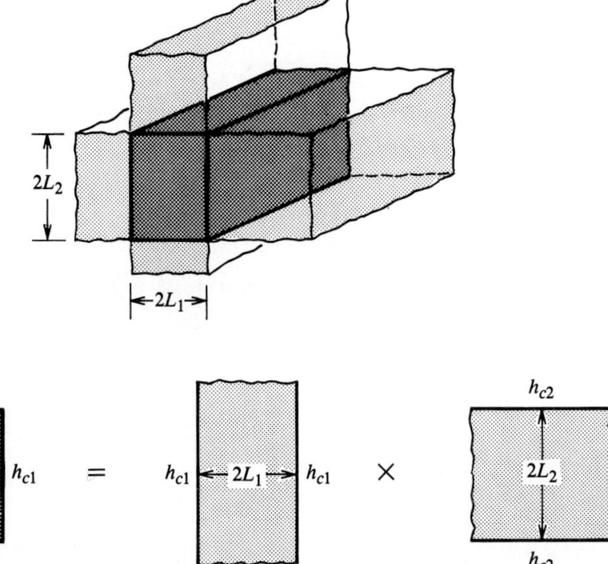

Figure 3.26 Schematic of the product solution procedure for the temperature response of a convectively cooled long rectangular bar.

Table 3.6 Shapes amenable to product solutions: *Caution:* The dimensionless temperatures S, P, and C are evaluated at the same value of actual time (not Fourier number).

Basic solutions

Semi-infinite solid $S(x,t)$	Infinite plane slab $P(x,t)$	Infinite cylinder $C(r,t)$
Eq. (3.61); $1 - \dfrac{(T - T_0)}{(T_e - T_0)}$	Eq. (3.70)	Eq. (3.72)

1. Infinite rectangular bar
$\theta = P(x,t)P(y,t)$

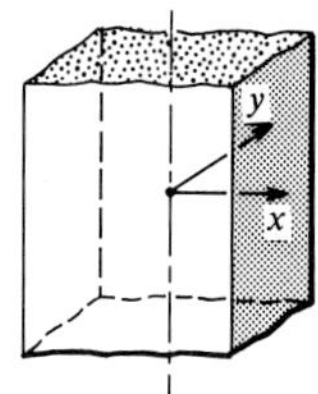

2. Semi-infinite plate
$\theta = S(x,t)P(y,t)$

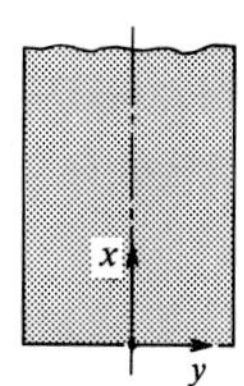

3. Semi-infinite cylinder
$\theta = S(x,t)C(r,t)$

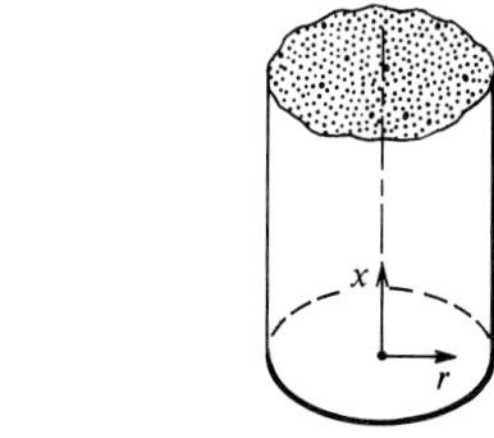

4. Finite cylinder
$\theta = P(x,t)C(r,t)$

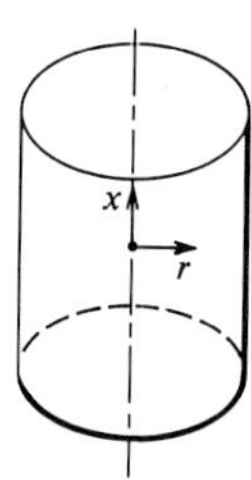

5. Semi-infinite rectangular bar
$\theta = S(z,t)P(x,t)P(y,t)$

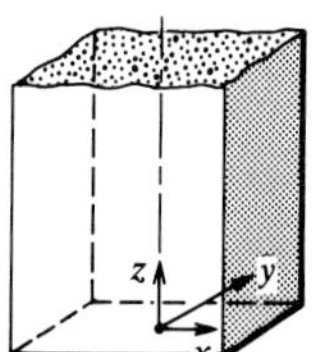

6. Rectangular block
$\theta = P(x,t)P(y,t)P(z,t)$

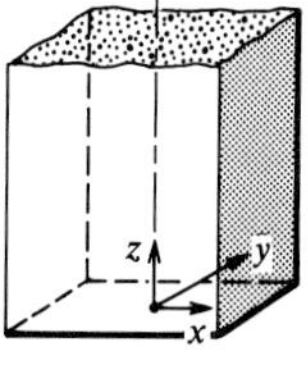

7. Two-dimensional corner
$\theta = S(x,t)S(y,t)$

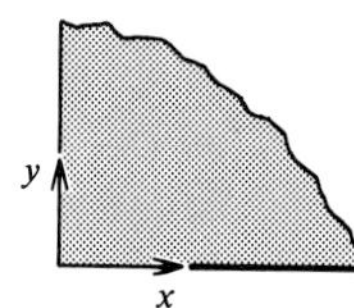

8. Three-dimensional corner
$\theta = S(x,t)S(y,t)S(z,t)$

9. Finite-width corner
$\theta = P(x,t)S(y,t)S(z,t)$

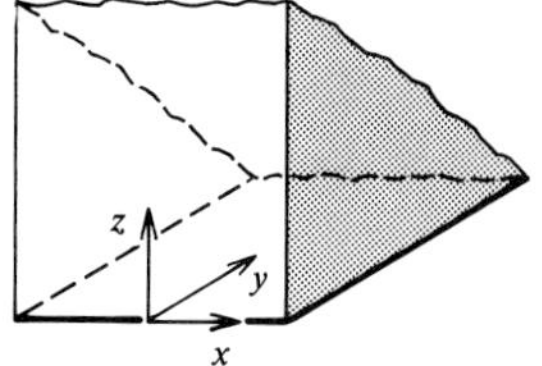

Notice that these product solutions are not applicable when

1. The initial temperature of the body is nonuniform.
2. The fluid temperature T_e is not the same on all sides of the body.
3. The surface boundary conditions are not of the third kind.

The requirement that the initial temperature of the body be uniform is, of course, the specified initial condition for the solutions given by Eqs. (3.72) and (3.73). However, the beginning student often overlooks this point in problem solving.

EXAMPLE 3.11 Sterilization of a Can of Vegetables

A can of vegetables 10 cm in diameter and 8 cm high is to be sterilized by immersion in saturated steam at 105°C. If the initial temperature is 40°C, what will be the minimum temperature in the can after 80 minutes? Also, calculate the total heat transfer to the can in this period.

Solution

Given: Can of vegetables to be sterilized in steam.

Required: (i) Minimum temperature in can after 80 minutes, and (ii) total heat transfer.

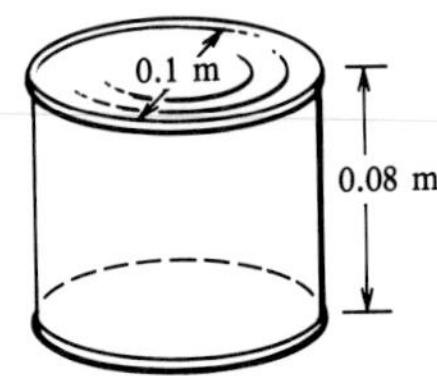

Assumptions:
1. The heat transfer coefficient for condensing steam is very large; hence, Bi $\rightarrow \infty$.
2. No circulation inside can.
3. Thermal diffusivity of contents approximates that of water.

(i) Since Bi $\rightarrow \infty$, the lumped thermal capacity model cannot be used. Item 4 of Table 3.6 applies, with the minimum temperature at the center of the can. The thermal diffusivity is evaluated at a guessed average temperature of 360 K; using the data in Table A.8 of Appendix A, $k = 0.676$ W/m K, $\rho = 967$ kg/m^3, $c = 4200$ J/kg K; hence, $\alpha = k/\rho c = (0.676)/(967)(4200) = 0.166 \times 10^{-6}$ m^2/s. The two Fourier numbers are:

Slab with 4 cm half-width: $$\text{Fo}_1 = \frac{\alpha t}{L^2} = \frac{(0.166 \times 10^{-6})(4800)}{(0.04)^2} = 0.498$$

Infinite cylinder of 5 cm radius: $$\text{Fo}_2 = \frac{\alpha t}{R^2} = \frac{(0.166 \times 10^{-6})(4800)}{(0.05)^2} = 0.319$$

Then $\theta_c = P(0,t)C(0,t)$, where $P(0,t)$ and $C(0,t)$ can be obtained from Fig. C.1*a* and *b* of Appendix C or by using COND2. Using the charts with Bi $= 1000$, $P(0,t) = 0.38$ and $C(0,t) = 0.27$.

$$\theta_c = \frac{T_c - T_e}{T_0 - T_e} = P(0,t)C(0,t)$$

$$\frac{T_c - 105}{40 - 105} = (0.38)(0.27); \qquad \text{solving, } T_c = 98.3^\circ\text{C}$$

(ii) The total heat transfer is related to the fractional energy gain as

$$Q = \Phi \rho c V(T_e - T_0)$$

where V is the volume of the can and $\Phi = \Phi_1 + \Phi_2 - \Phi_1\Phi_2$ from Eq. (3.82*a*). Using COND2 or Fig. C.2*a* and *b* with Bi $= 50$, $\Phi_1 = 0.75$, $\Phi_2 = 0.88$. Thus,

$$\Phi = 0.75 + 0.88 - (0.75)(0.88) = 0.97$$

$$Q = (0.97)(967)(4200)(\pi)(0.05)^2(0.08)(105 - 40) = 161 \text{ kJ}$$

Comments

Although the heat transfer coefficient for condensing steam is large (see Chapter 7), it is not infinite. Using the curves for the largest values of Bi available in the charts gives satisfactory estimates for this problem.

3.5 MOVING-BOUNDARY PROBLEMS

There are many engineering problems that involve heat conduction in a region bounded by a moving surface. Examples include solidification of a melt to form a crystal, growth of a vapor bubble in a superheated liquid, and ablation of a heat shield on a reentry vehicle. There is usually either a phase change or chemical reaction at the surface. In the case of a phase change, a boundary condition of the form of Eq. (3.18) is appropriate. For a chemical reaction, the boundary condition is considerably more complicated and may involve mass transfer as well as heat transfer considerations. In this brief introduction to the analysis of *moving-boundary problems*, exact solutions to two rather simple problems are demonstrated.

3.5.1 Solidification from a Melt

Consider a liquid, initially at its solidification temperature T_s, suddenly exposed to a surface at temperature $T_i < T_s$ located at $z = 0$, as shown in Fig. 3.27. A solidification front then moves upward, and the expected temperature profile when the front is at $z = s$ is shown. Let $\theta = (T - T_i)/(T_s - T_i)$; assuming constant properties, the temperature in the solid is governed by Fourier's equation,

$$\frac{\partial \theta}{\partial t} = \alpha \frac{\partial^2 \theta}{\partial z^2} \tag{3.83}$$

with initial and boundary conditions

$$t = 0: \quad \theta = 1 \tag{3.84a}$$

$$z = 0: \quad \theta = 0 \tag{3.84b}$$

$$z = s: \quad \theta = 1 \tag{3.84c}$$

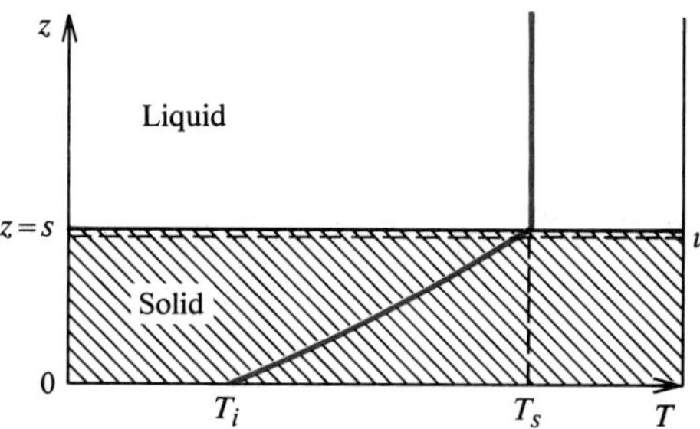

Figure 3.27 Solidification from a liquid: coordinate system and instantaneous temperature profile.

The unknown location of the interface is determined from the interface energy balance, Eq. (3.18). Since there is no temperature gradient in the liquid phase, Eq. (3.18) reduces to

$$0 = k \left.\frac{\partial T}{\partial z}\right|_u + \dot{m}'' h_{\text{fs}} \tag{3.85}$$

where $\dot{m}''$ [kg/m^2 s] is the solidification rate and is negative to be consistent with Eq. (3.18). With respect to the interface, the liquid flows across in the negative z direction. The speed at which the front moves is

$$V = \frac{ds}{dt} = -\frac{\dot{m}''}{\rho} \tag{3.86}$$

Substituting for T and $\dot{m}''$ in Eq. (3.85) gives

$$k(T_s - T_i) \left.\frac{\partial \theta}{\partial z}\right|_u - \rho h_{\text{fs}} V = 0 \tag{3.87}$$

Based on our experience with the semi-infinite solid problem of Section 3.4.2, we might suspect that the solution has the form

$$\theta = A \operatorname{erf} \frac{z}{(4\alpha t)^{1/2}} \tag{3.88}$$

We know Eq. (3.88) satisfies the differential equation; the task is to see if it can fit the initial and boundary conditions. Equations (3.84*a*) and (3.84*b*) are satisfied since $\operatorname{erf} \infty = 1$ and $\operatorname{erf} 0 = 0$; Eq. (3.84*c*) requires that

$$A \operatorname{erf} \frac{s}{(4\alpha t)^{1/2}} = 1$$

which is possible only if s is proportional to $t^{1/2}$. To simplify the algebra, we set $s = \lambda(4\alpha t)^{1/2}$, where λ is a constant yet to be determined. Then $A = 1/\operatorname{erf}\lambda$, so that Eq. (3.88) gives the temperature distribution $T(z, t)$ as

$$\frac{T - T_i}{T_s - T_i} = \theta = \frac{\operatorname{erf}[z/(4\alpha t)^{1/2}]}{\operatorname{erf}\lambda}$$

Also,

$$V = \frac{ds}{dt} = \lambda\left(\frac{\alpha}{t}\right)^{1/2} \tag{3.89}$$

the speed of the front.

The constant λ is now determined from the interface energy balance, Eq. (3.87). After some algebra (see Section 3.4.2 for differentiation of the error function), we obtain

$$\pi^{1/2}\lambda e^{\lambda^2} \operatorname{erf}\lambda = \frac{c(T_s - T_i)}{h_{\text{fs}}} \tag{3.90}$$

The right-hand side of this equation is a dimensionless group that characterizes heat transfer during phase change; it is usually called the **Jakob number**, Ja. Equation (3.90) is an implicit relation for $\lambda = \lambda(\text{Ja})$ but is easily solved by specifying a range of λ values and then calculating the corresponding values of Ja. Figure 3.28 was prepared in this way.

Often the Jakob number is quite small; for example, when ice forms from water with $T_s - T_i = 10$ K, the Jakob number is 0.058. For small Jakob numbers, e^{λ^2} erf λ can be approximated as $(2/\pi^{1/2})\lambda$; then

$$\lambda = \left(\frac{\text{Ja}}{2}\right)^{1/2}; \qquad V = \left(\frac{\text{Ja}\,\alpha}{2t}\right)^{1/2} \tag{3.91}$$

This result corresponds to a linear temperature variation across the solid, as can be seen by substituting $d\theta/dz = 1/s$ and $V = ds/dt$ in Eq. (3.87) to give

$$\frac{ds}{dt} = \frac{\alpha\,\text{Ja}}{s}$$

or

$$s\,ds = \alpha\,\text{Ja}\,dt$$

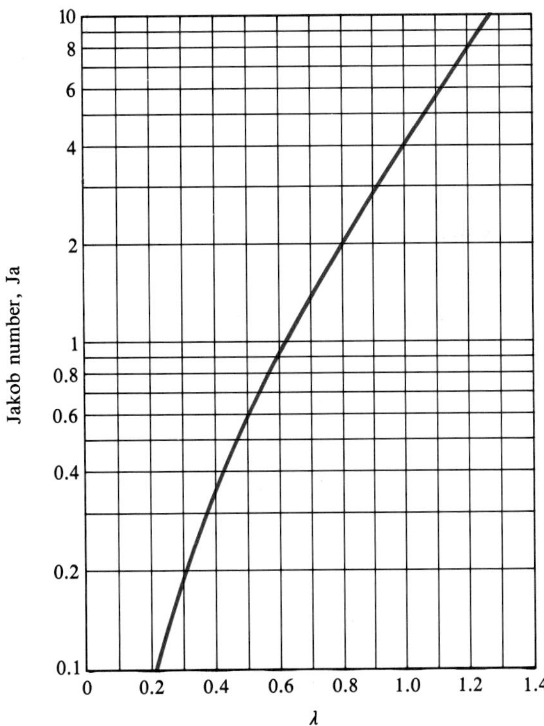

Figure 3.28 Solution of Eq. (3.90), namely, $\pi^{1/2}\lambda e^{\lambda^2}$ erf λ = Ja. The Jakob number Ja $= c(T_s - T_i)/h_{fs}$.

Integrating,

$$s^2 = 2\alpha \,\mathrm{Ja}\, t \tag{3.92}$$

Equation (3.92) is called a *parabolic* growth law by metallurgists. Substituting back in Eq. (3.86) gives

$$V = \left(\frac{\mathrm{Ja}\,\alpha}{2t}\right)^{1/2}$$

which is the same as Eq. (3.91). Notice that if, at the outset, we had simply assumed steady-state conduction, a linear temperature variation would have been obtained. Thus, the conduction can be said to be *quasi steady*, even though temperatures in the solid are changing with time. In terms of the physics of the problem, we are neglecting the heat loss required to cool the solid below the solidification temperature, T_s, in comparison with the enthalpy of fusion. The solid is said to be *subcooled* since its temperature is below the solid-liquid equilibrium value, T_s.

In closing, it is worthwhile to note that although the analytical solution Eq. (3.91) is for a very simple phase change problem, the result that the interface moves at a speed proportional to $(\alpha/t)^{1/2}$ is found to apply in other geometries and in more complex situations (for example, bubble growth in some boiling heat transfer processes).

EXAMPLE 3.12 An Ice-Making Process

In an ice-making process, water at 5°C flows over a plate cooled by refrigerant at −10°C. When the ice layer is 2 mm thick, it is scraped off the plate, and the process repeats itself. Estimate the time interval required between scrapings. Take the overall heat transfer coefficient from the refrigerant to the plate-ice interface as 900 W/m^2 K. The enthalpy of fusion of ice is 335 kJ/kg.

Solution

Given: An ice-making process.

Required: Time to form an ice layer 2 mm thick.

Assumptions:
1. One-dimensional solidification.
2. Quasi-steady conduction across the ice layer.
3. Sensible heat loss of the water cooling from 5°C to 0°C is small compared with the enthalpy of fusion.

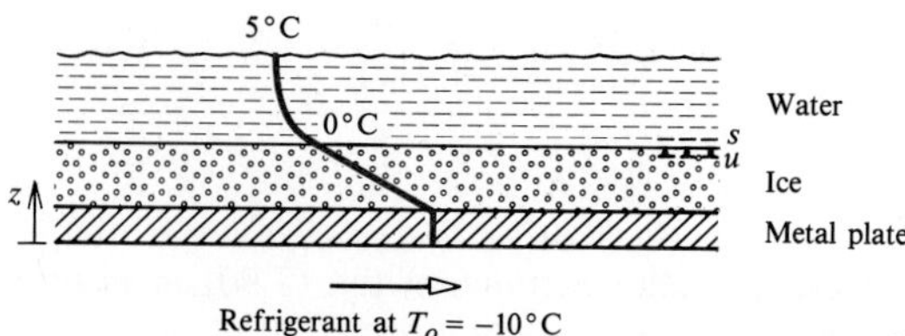

We have shown that the Jakob number for ice formation is small; hence, the conduction across the ice layer can be taken as quasi steady. We cannot use Eq. (3.92) for the thickness of the layer since T_i is not a constant in this problem. However, we can proceed in a similar manner. The surface energy at the ice-water interface is

$$k\left.\frac{\partial T}{\partial z}\right|_s = k\left.\frac{\partial T}{\partial z}\right|_u + \dot{m}''h_{\mathrm{fs}}$$

$$k\left.\frac{\partial T}{\partial z}\right|_s \simeq 0 \qquad \text{if the sensible heat loss of the water is neglected}$$

$$k\left.\frac{\partial T}{\partial z}\right|_u = -q = \frac{T_s - T_o}{s/k + 1/U} \qquad \text{for two resistances in series}$$

$$\dot{m}'' = -\rho\frac{ds}{dt}$$

Substituting above,

$$\rho h_{\mathrm{fs}}\frac{ds}{dt} = \frac{T_s - T_o}{s/k + 1/U}$$

This differential equation is easily solved. Separating variables and rearranging,

$$\frac{T_s - T_o}{\rho h_{\mathrm{fs}}}dt = \left(\frac{s}{k} + \frac{1}{U}\right)ds$$

Integrating with $s = 0$ at $t = 0$,

$$\frac{T_s - T_o}{\rho h_{\mathrm{fs}}}t = \frac{s^2}{2k} + \frac{s}{U}$$

From Table A.2 for ice at 0°C, $\rho = 910$ kg/m^3, $k = 2.22$ W/m K.

$$\frac{(0 - (-10))}{(910)(335 \times 10^3)}t = \frac{(0.002)^2}{(2)(2.22)} + \frac{0.002}{900}; \qquad t = 95 \text{ s}$$

Comments

1. A more complete solution for this type of problem is given by London and Seban [15].
2. The student should think about the design of an ice-making machine based on this process.

3.5.2 Steady-State Melting Ablation

Ablative heat shields are widely used to protect structures from high heat fluxes. Heat shields are made from a great variety of materials, including refractory metals such as tungsten, Teflon, graphite, and silica-phenolic composites. The heat transfer analysis of a simple melting ablator is straightforward if it can be assumed that the

Figure 3.29 A silica-phenolic nosecone after arc-jet testing, showing evidence of melting ablation. (Photograph courtesy of Mr. J. Courtney, TRW Systems, Redondo Beach.)

liquid melt is removed as fast as it is formed. On a reentry vehicle, the friction and pressure forces will cause the melt to flow backward over the vehicle, leaving only a thin film of negligible thermal resistance. The heat shield in Fig. 3.29 shows evidence of undergoing melting ablation. Alternatively, in some situations, gravity forces may be sufficient to ensure that the liquid drains quickly from the solid surface.

We consider the one-dimensional model of melting ablation shown in Fig. 3.30. It is assumed that the imposed heat flux q_s is a constant and is unaffected by the ablation process (most often q_s is affected by the ablation, and a coupled problem involving both the gas and solid phases must be considered). After time $t = 0$, there will be an initial transient as temperatures in the solid rise until a quasi-steady state is attained, with the surface temperature T_s at $x = 0$ equal to the melting temperature. The melting rate is again denoted $\dot{m}''$ [kg/m^2 s], and the surface recedes at a constant

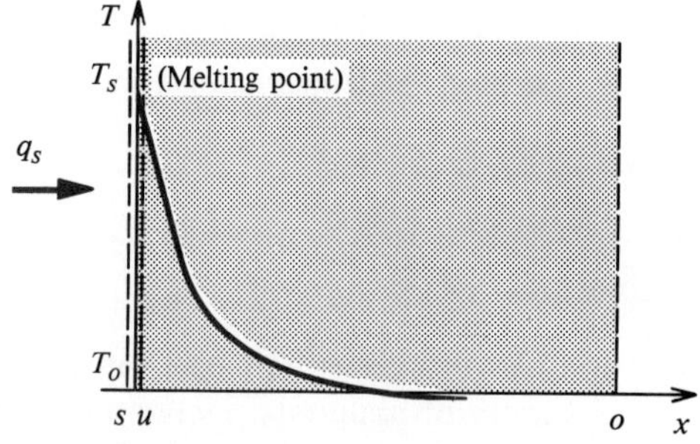

Figure 3.30 Schematic of one-dimensional melting ablation.

speed $\dot{m}''/\rho$, where ρ is the density of the solid. The relation between $\dot{m}''$ and q_s can be obtained without knowing the temperature profile in the solid, as will now be demonstrated.

Figure 3.31 shows a control volume located between s- and o-surfaces. The s-surface is in the gas phase just adjacent to the solid. The o-surface is a fixed distance from the interface and is located deep enough into the solid for the temperature gradient there to be negligible. A u-surface is located in the solid phase adjacent to the gas-solid interface. Molten material is imagined to leave through the sides of the control volume between the s-surface and the interface at the rate $\dot{m}''$ [kg/m^2 s]. Mass conservation requires that solid material should flow across the o-surface also at the rate $\dot{m}''$. The steady-flow energy equation, Eq. (1.4), applied to the s-o control volume of cross-sectional area A, requires that:

$$\text{Mass flow rate} \times \text{Change in enthalpy} = \text{Net rate of heat inflow}$$

$$\dot{m}''A(h_s - h_o) = q_s A - 0$$

where h_s is the enthalpy of molten material at the interface temperature T_s, and h_o is the enthalpy of solid material at temperature T_o. The enthalpy of phase change is h_{fs}; then $h_s - h_u = h_{fs}$, where h_u is the enthalpy of solid material at temperature T_s $(= T_u)$, and

$$q_s = \dot{m}''[h_{fs} + (h_u - h_o)]$$

or

$$\dot{m}'' = \frac{q_s}{h_{fs} + (h_u - h_o)} \tag{3.93}$$

If the specific heat of the solid is taken to be constant, the final result is

$$\dot{m}'' = \frac{q_s}{h_{fs} + c_p(T_s - T_o)} \tag{3.94}$$

The speed at which the solid surface recedes is simply $\dot{m}''/\rho$.

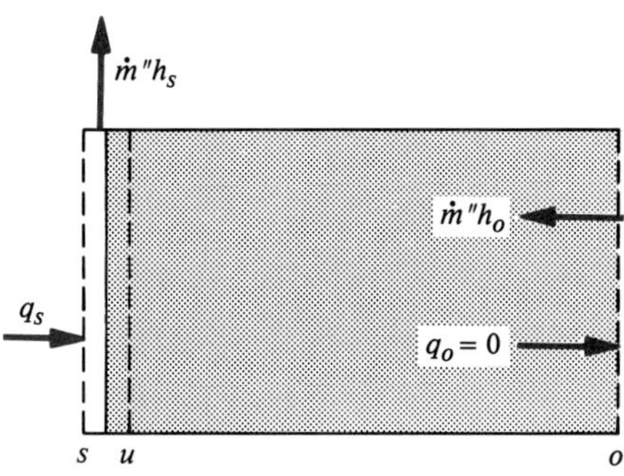

Figure 3.31 Control volume for the application of mass and energy conservation principles to steady-state melting ablation.

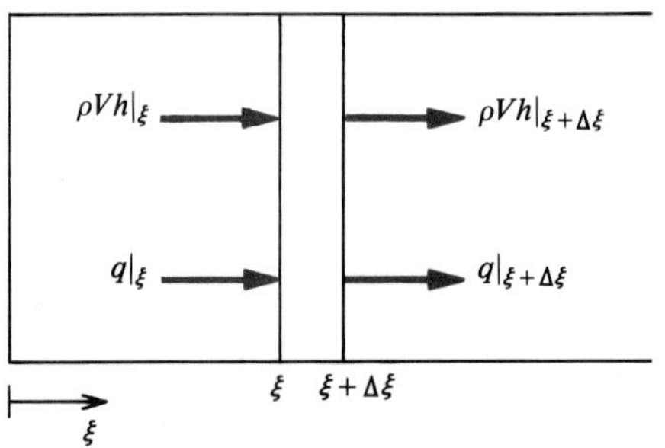

Figure 3.32 Elemental control volume in a coordinate system located on the moving boundary for one-dimensional ablation.

To obtain the temperature profile in the solid, it is necessary to solve an appropriate form of the heat conduction equation. It is convenient to use first principles to derive a special form of the equation for which the coordinate axes are located on the moving surface, as shown in Fig. 3.32. The coordinate ξ is taken to be positive measured into the solid, so that the velocity V at which the solid moves (from right to left) has a negative value. For steady ablation, the temperature at any location ξ is unchanging with time. Application of the steady-flow energy equation, Eq. (1.4), to the elemental control volume $A\Delta\xi$ located between ξ and $\xi + \Delta\xi$ requires that:

Mass flow rate $\times$ Change in enthalpy = Net rate of heat inflow by conduction

$$\rho VA\left[h\,\big|_{\xi+\Delta\xi} - h\big|_{\xi}\right] = qA\,\big|_{\xi} - qA\big|_{\xi+\Delta\xi}$$

Dividing by $A\Delta\xi$ and letting $\Delta\xi \to 0$ gives

$$\rho V\frac{dh}{d\xi} = -\frac{dq}{d\xi}$$

But

$$q = -k\frac{dT}{d\xi} \qquad \text{and} \qquad dh = c_p dT$$

Substituting and rearranging,

$$\frac{d}{d\xi}\left(k\frac{dT}{d\xi}\right) - \rho Vc_p\frac{dT}{d\xi} = 0 \tag{3.95}$$

Taking the thermal conductivity k to be constant and $c_p = c_v = c$ gives the equation governing the temperature distribution $T(\xi)$:

$$\frac{d^2T}{d\xi^2} - \frac{V}{\alpha}\frac{dT}{d\xi} = 0; \qquad \alpha = \frac{k}{\rho c} \tag{3.96}$$

Appropriate boundary conditions for this second-order ordinary differential equation are

$$\xi = 0: \quad T = T_s; \quad \xi \to \infty: \quad T \to T_o \tag{3.97a,b}$$

The solution of Eq. (3.96) is

$$T = C_1 e^{(V/\alpha)\xi} + C_2$$

and when the constants are evaluated from the boundary conditions, the result is

$$\frac{T - T_o}{T_s - T_o} = e^{(V/\alpha)\xi}; \qquad (V \text{ is negative}) \tag{3.98}$$

That is, the temperature profile shows a simple exponential decay. The heat flux into the solid at the u-surface can be obtained using Fourier's law:

$$q_u = -k\left.\frac{dT}{d\xi}\right|_{\xi=0} = -k(T_s - T_o)\left(\frac{V}{\alpha}\right) = -\rho c V(T_s - T_o) \tag{3.99}$$

An energy balance on the control volume located between the s- and u-surfaces will show that q_s differs from q_u by the enthalpy of phase change absorbed at the interface.

Notice that Eq. (3.96) could have been derived from the heat conduction equation for one-dimensional transient conduction, Eq. (3.33), by an appropriate change of variables (see Exercise 3–73).

EXAMPLE 3.13 Ablation of Stainless Steel

In an experiment to study laser heating, a stainless steel component is exposed to a powerful laser beam, and a jet of argon gas impinging on the surface sweeps away the molten metal. The component is initially at 300 K, and after a short transient the surface is measured to recede at a rate of 230 μm/s. Estimate the total heat flux to the surface and the penetration of the thermal response into the solid. The stainless steel properties include $\rho = 7800$ kg/m^3, $c = 600$ J/kg K, $\alpha = 4.0 \times 10^{-6}$ m^2/s, $h_{fs} = 2.7 \times 10^5$ J/kg, and a melting temperature of 1670 K.

Solution

Given: Rate of surface recession of ablating stainless steel.

Required: Surface heat flux and depth of penetration of thermal response.

Assumptions:
1. One-dimensional heat flow.
2. Steady-state ablation.
3. The melt layer is thin enough for there to be a negligible temperature drop across it.
4. No chemical reactions, since the argon is inert.

The surface heat flux can be obtained from Eq. (3.94). Solving for q_s,

$$\begin{aligned} q_s &= \dot{m}''[h_{fs} + c_p(T_s - T_o)] \\ &= \{(7800)(230 \times 10^{-6})[2.7 \times 10^5 + 600(1670 - 300)]\} \\ &= 1.96 \times 10^6 \text{ W/m}^2 \ (1.96 \text{ MW/m}^2) \end{aligned}$$

Referring to the figure, we we can take δ_t to be an estimate of the thermal penetration. Differentiating Eq. (3.98),

$$\left.\frac{dT}{d\xi}\right|_{\xi=0} = \frac{(T_s - T_o)V}{\alpha}$$

$$\delta_t = \frac{\alpha}{-V} = \frac{4.0 \times 10^{-6}}{230.0 \times 10^{-6}} = 0.017 \text{ m (1.7 cm)}$$

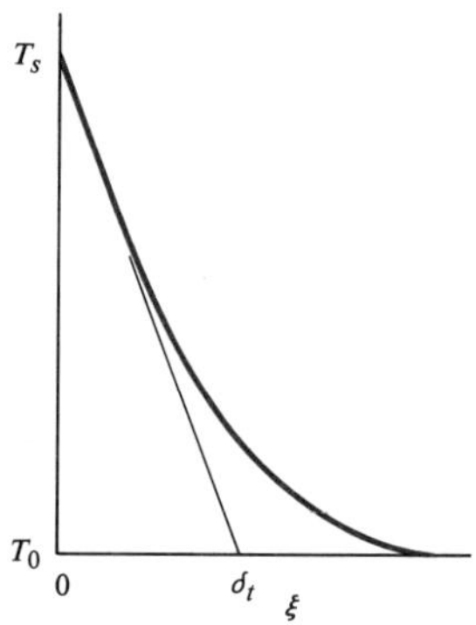

Comments

The depth of thermal penetration is seen to be small.

3.6 NUMERICAL SOLUTION METHODS

Although many simple steady-state and transient heat conduction problems can be solved analytically, solutions for more complex problems are best obtained numerically. Numerical solution methods are particularly useful when the shape of the solid is irregular, when thermal properties are temperature- or position-dependent, and when boundary conditions are nonlinear. Numerical methods commonly used include the *finite-difference method*, the *finite-element method*, and the *boundary-element method*. The finite-difference method was the first numerical method to be used extensively for heat conduction. It remains a popular method, not because it is superior to other methods for heat conduction, but because it is easier to implement and is also the most useful numerical solution method for heat convection problems.

The first step in a finite-difference solution procedure is to discretize the spatial and time coordinates to form a *mesh of nodes*. Next, finite-difference approximations are made to the derivatives appearing in the heat conduction equation to convert the *differential* equation to an algebraic *difference* equation. Alternatively, the difference equation can be constructed by applying the energy conservation principle directly to a volume element surrounding the node. In steady-state problems, a set of linear algebraic equations is obtained with as many unknowns as the number of nodes in the mesh. These equations can be solved by matrix inversion or by iteration. For transient conduction, temperatures at the current time step may be found directly using values at the preceding time step. In some formulations, iteration may be required, since values at the current time step are also involved. In the first applications of finite-difference methods to heat conduction, the calculations were done by hand, limiting consideration to a coarse mesh with relatively few nodes. Because the accuracy of a finite-difference approximation increases with number of nodes, these solutions were inadequate.

The availability of mainframe digital computers in the late 1950s completely changed the picture, and by the 1960s, engineers could use as fine a grid as was necessary to meet their requirements. The 1970s saw increased use of the programmable calculator to obtain finite-difference solutions to heat conduction problems, which meant that an engineer would write a computer program for the specific problem under consideration. In the 1980s and 1990s, powerful personal computers have become available. Furthermore, there is no longer the need, nor is it cost-effective, for engineers to write their own computer programs to implement a finite-difference solution method. There are many standard computer programs available for this purpose. Some examples are listed at the end of this chapter [16–19]. Thus, in Section 3.6, finite-difference methods are presented with the modest objective of giving the student an appreciation of the essential ideas involved, so that the available computer programs can be used intelligently. Any serious endeavor to develop engineering tools based on finite-difference solution procedures should be preceded by an appropriate course in numerical analysis.

Finite-element methods are widely used in structural mechanics and are also applicable to heat conduction problems. An object is divided into discrete spatial regions called *finite elements*. The most common two-dimensional element is the triangle, and the most common three-dimensional element is the tetrahedron. The finite-element method allows the heat conduction equation to be satisfied in an average sense over the finite element; thus, the elements can be much larger than the control volumes used in finite-difference methods. The use of triangles or tetrahedrons for elements allows the approximation of complex and irregularly shaped objects. Application of the method leads to a set of algebraic equations, which are solved by matrix inversion or iteration. Compared to finite-difference methods, the formulation of these equations is considerably more involved and requires more effort, as does writing a computer program to implement the procedure. However, once written, finite-element computer programs tend to be more versatile than their finite-difference counterparts. Choice of which method to use is perhaps dictated by the objective rather than by the intrinsic virtues of the method. For example, calculation of temperature variations in solids is often required for the purpose of determining thermal stresses. Since the finite-element method is preeminent for stress calculations, many standard computer codes use the finite-element method to calculate both temperatures and stresses in one package.

A newer development is the boundary-element method, in which the boundary of the object is divided into finite elements, and there is no need to consider the interior at all. Although this method is attractive for simpler heat conduction problems, its versatility and generality have yet to be established.

3.6.1 A Finite-Difference Method for Two-Dimensional Steady Conduction

Consider two-dimensional steady conduction with volumetric heat generation and constant thermal properties. Using Cartesian coordinates, Fig. 3.33 shows a finite control volume Δx by Δy of unit depth surrounding node (m, n) within the solid. The m and n indices denote x and y locations, respectively, in a uniform mesh of node points. For convenience, we will use compass directions to denote the faces of the element (N, S, E, and W). The energy conservation principle, Eq. (1.2), applied to the finite control volume reduces to

$$0 = \dot{Q} + \dot{Q}_v$$

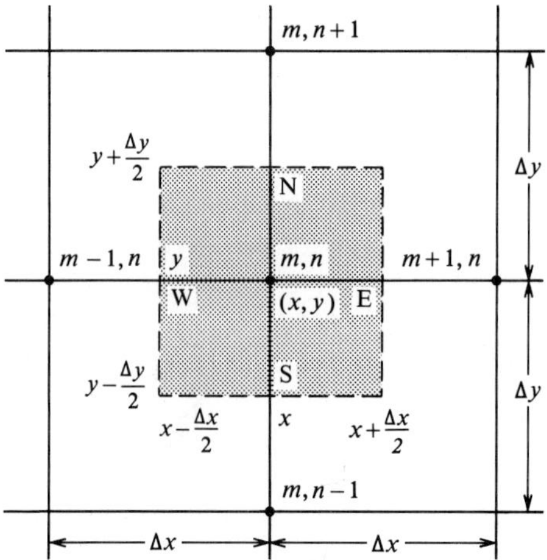

Figure 3.33 Finite control volume Δx by Δy by 1 surrounding node (m,n) at the location (x,y) used to derive the difference equation for two-dimensional steady conduction.

The heat transfer $\dot{Q}$ is by conduction across the four faces of the element; thus,

$$0 = \dot{Q}_x|_W + \dot{Q}_y|_S - \dot{Q}_x|_E - \dot{Q}_y|_N + \Delta\dot{Q}_v \tag{3.100}$$

since the sign convention in Eq. (1.2) requires heat transfer into the system to be positive. The heat conduction across the left-hand face of the element is

$$\dot{Q}_x|_W = -k\left.\frac{\partial T}{\partial x}\right|_W \Delta y \cdot 1$$

To approximate the derivative of $T(x, y)$ we will assume a linear temperature gradient between the nodes $(m - 1, n)$ and (m, n); then

$$\dot{Q}_x|_W = -k\frac{T_{m,n} - T_{m-1,n}}{\Delta x}\Delta y$$

which is seen to be positive for $T_{m-1,n} > T_{m,n}$. Similarly, for the right-hand, bottom, and top faces,

$$\dot{Q}_x|_E = -k\frac{T_{m+1,n} - T_{m,n}}{\Delta x}\Delta y$$

$$\dot{Q}_y|_S = -k\frac{T_{m,n} - T_{m,n-1}}{\Delta y}\Delta x$$

$$\dot{Q}_y|_N = -k\frac{T_{m,n+1} - T_{m,n}}{\Delta y}\Delta x$$

The internal heat generation is simply $\dot{Q}_v'''$ times the volume of the element:

$$\Delta\dot{Q}_v = \dot{Q}_v'''\Delta x\Delta y \cdot 1$$

Substituting in Eq. (3.100), multiplying by $\Delta x/k\Delta y$, and rearranging,

$$2(1+\beta)T_{m,n} = T_{m-1,n} + T_{m+1,n} + \beta(T_{m,n-1} + T_{m,n+1}) + \frac{\dot{Q}_v'''}{k}\Delta x^2 \quad \textbf{(3.101)}$$

where $\beta = (\Delta x/\Delta y)^2$ is a geometric mesh factor. For a square mesh $\Delta x = \Delta y$, $\beta = 1$, and no internal heat generation,

$$4T_{m,n} = T_{m,n+1} + T_{m+1,n} + T_{m,n-1} + T_{m-1,n} \quad \textbf{(3.102)}$$

which simply states that the temperature at each node is the arithmetic average of the temperatures at the four nearest neighboring nodes.

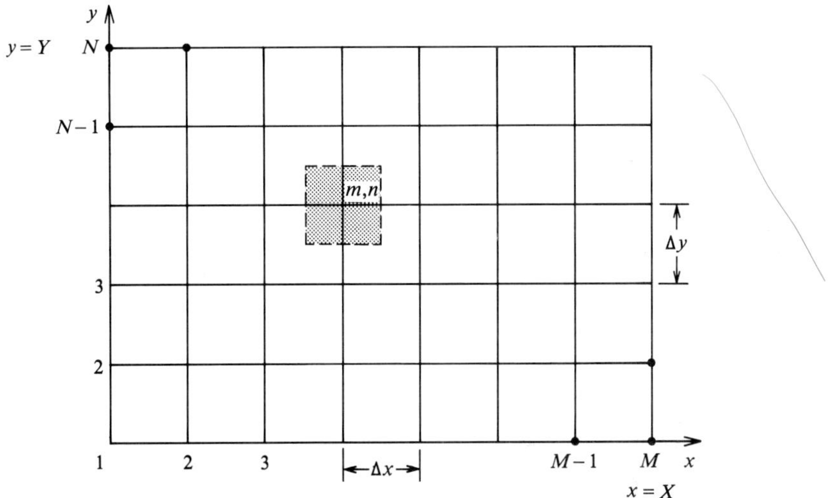

Figure 3.34 Nodal mesh for steady conduction in a rectangular plate.

Boundary Conditions

Figure 3.34 shows a complete mesh for a rectangular plate of dimensions X, Y. There are M node points in the x direction and N node points in the y direction. If the boundary condition is one of prescribed temperature, the temperatures at the boundary nodes are known. If the boundary condition is one of prescribed heat flux or convection, then the finite-difference form of the condition is obtained by making an energy balance on a finite control volume adjacent to the boundary. For example, consider the convection boundary condition shown in Fig. 3.35, and assume $\dot{Q}_v''' = 0$ for simplicity,

$$\begin{matrix}\text{Net heat conduction} \\ \text{into the volume}\end{matrix} + \begin{matrix}\text{Heat convection across} \\ \text{face at } x = 0\end{matrix} = 0$$

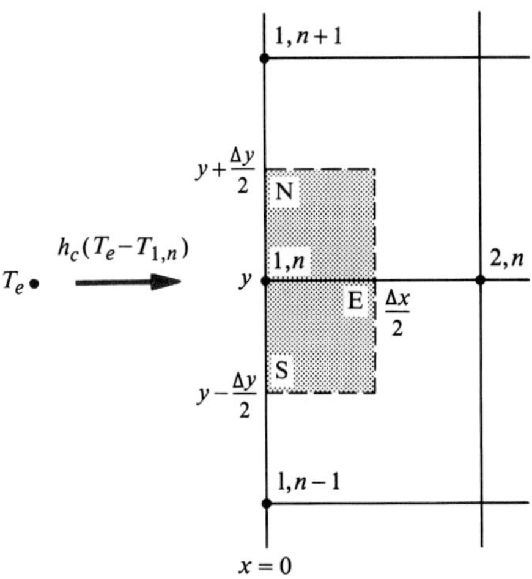

Figure 3.35 Finite control volume used to derive the difference equation for a convective boundary condition at $x = 0$: two-dimensional steady conduction.

As before,

$$\dot{Q}_x|_E = -k\frac{T_{2,n} - T_{1,n}}{\Delta x}\Delta y$$

$$\dot{Q}_y|_S = -k\frac{T_{1,n} - T_{1,n-1}}{\Delta y}\frac{\Delta x}{2}$$

$$\dot{Q}_y|_N = -k\frac{T_{1,n+1} - T_{1,n}}{\Delta y}\frac{\Delta x}{2}$$

and

$$\dot{Q}_x|_0 = h_c(T_e - T_{1,n})\Delta y$$

Substituting in the energy balance and taking $\Delta x = \Delta y$,

$$T_{2,n} + \frac{1}{2}(T_{1,n-1} + T_{1,n+1}) - 2T_{1,n} + \frac{h_c\Delta x}{k}(T_e - T_{1,n}) = 0$$

A mesh Biot number is defined as $\text{Bi} = h_c\Delta x/k$; then solving for $T_{1,n}$ gives

$$T_{1,n} = \frac{1}{2 + \text{Bi}}\left[T_{2,n} + \frac{1}{2}(T_{1,n-1} + T_{1,n+1}) + \text{Bi}\,T_e\right] \tag{3.103}$$

Table 3.7 gives similar results for a variety of boundary conditions, including interior and exterior corners. In this table, a simplified node numbering scheme is used for clarity.

Table 3.7 Finite-difference approximations for steady-state conduction, square mesh.

	Configuration	Mesh	Equation
1.	Interior node	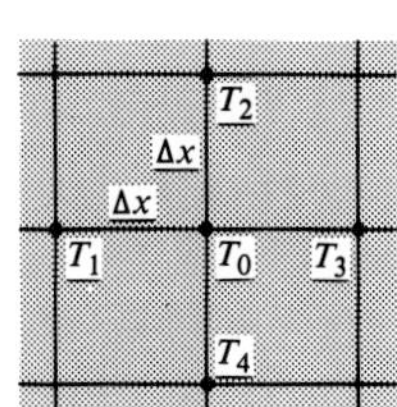	$T_0 = \frac{1}{4}[T_1 + T_2 + T_3 + T_4]$
2.	Plane surface, convection	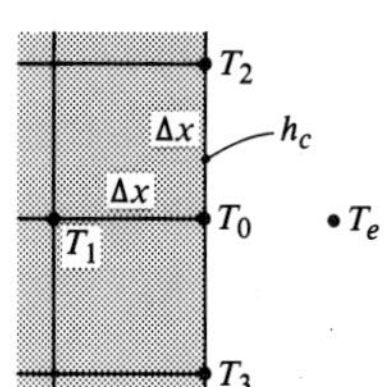	$T_0 = \frac{1}{2 + \mathrm{Bi}}\left[T_1 + \frac{1}{2}(T_2 + T_3) + \mathrm{Bi}T_e\right]$ $\mathrm{Bi} = \frac{h_c \Delta x}{k}$
3.	Plane wall, known heat flux	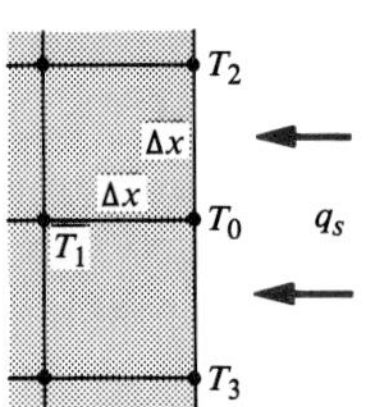	$T_0 = \frac{1}{2}T_1 + \frac{1}{4}(T_2 + T_3) + \frac{q_s \Delta x}{2k}$ (for an adiabatic surface or plane of symmetry set $q_s = 0$)
4.	Exterior corner, convection	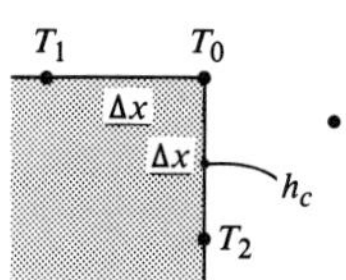	$T_0 = \frac{1}{1 + \mathrm{Bi}}\left[\frac{1}{2}(T_1 + T_2) + \mathrm{Bi}T_e\right]$ $\mathrm{Bi} = \frac{h_c \Delta x}{k}$
5.	Interior corner, convection	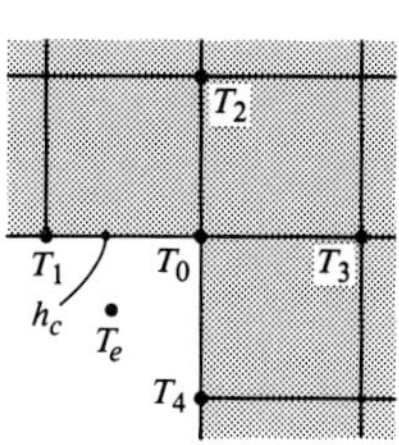	$T_0 = \frac{1}{3 + \mathrm{Bi}}\left[T_2 + T_3 + \frac{1}{2}(T_1 + T_4) + \mathrm{Bi}T_e\right]$ $\mathrm{Bi} = \frac{h_c \Delta x}{k}$
6.	Interior node near a curved non-isothermal surface	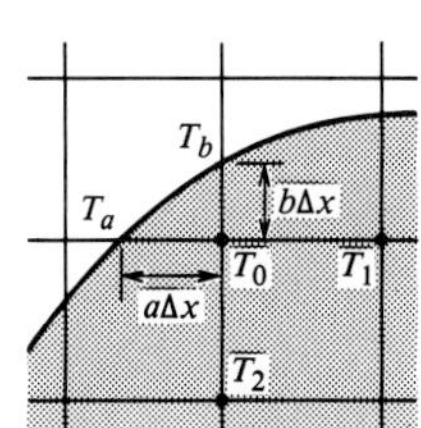	$T_0 = \frac{1}{(1/a + 1/b)}\left[\frac{T_1}{1 + a} + \frac{T_2}{1 + b} + \frac{T_a}{a(1 + a)} + \frac{T_b}{b(1 + b)}\right]$

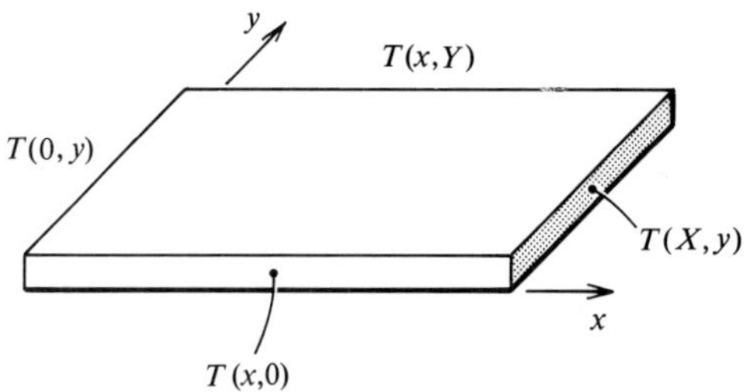

Figure 3.36 Schematic of rectangular plate with prescribed edge temperatures.

Solution Procedures

Consider the simplest case of a rectangular plate with prescribed boundary temperatures, as shown in Fig. 3.36. Referring to Fig. 3.34, the plate has dimensions X and Y, with M and N nodes in the x and y directions, respectively; then $\Delta x = X/(M-1)$, $\Delta y = Y/(N-1)$. Equation (3.102) is a set of $(M-2)\times(N-2)$ linear algebraic equations in the $(M-2)\times(N-2)$ unknown interior nodal temperatures $T_{m,n}$. If a coarse mesh is chosen so that there are relatively few nodes, *matrix inversion* or *Gaussian elimination* can be used to solve the equation set. In writing a computer program to effect the solution, all that is required is to call on a standard subroutine. However, unless very carefully implemented, these direct solution methods can be seriously affected by round-off error for finer meshes. Also, these methods tend not to take advantage of the sparseness of the matrix generated by the simple finite-difference approximations described here. More importantly, the accuracy of a finite-difference method will increase as the mesh size is reduced, provided round-off error in the numerical computations is not introduced. Thus, typically many nodes will be used, perhaps 100 or more; then, direct solution by matrix inversion or Gaussian elimination is uneconomical, and iterative methods are preferred. The simplest such method is *Gauss-Seidel* iteration, which proceeds as follows:

1. A reasonable initial guess $T^0_{m,n}$ is made for each of the unknown interior nodal temperatures. This is iteration zero.

2. New values $T^1_{m,n}$ are calculated by applying Eq. (3.102) to each node sequentially. Initial T^0 values or, if available, new T^1 values are substituted in the right-hand side of the equation.

3. The mesh is swept repeatedly until, at iteration k, the temperatures at each node are seen to change by less than a prescribed small amount,

$$\left| 1 - \frac{T^k_{m,n}}{T^{k-1}_{m,n}} \right| < \varepsilon \tag{3.104}$$

   The finite-difference solution is then said to have *converged* to the exact solution of the difference equation.

The system of linear algebraic equations, Eq. (3.102), is *diagonally dominant*; that is, when written in matrix form, the largest elements on each row of the coefficient matrix are on the main diagonal. The Gauss-Seidel procedure applied to such a system always converges uniformly to the proper solution and never becomes unstable; however, for a large number of equations, the convergence can be slow. Many methods have been devised to obtain faster convergence, for example, the *alternating-direction implicit* and *successive over-relaxation methods*. Such methods are described in the references listed in the bibliography at the end of the text.

Surface Heat Flux

Once the temperature field is obtained, it is sometimes necessary to determine the heat flux at a boundary surface. We make an energy balance on a finite control volume adjacent to the boundary, as shown in Fig. 3.37, and proceed as in the derivation of the boundary condition, Eq. (3.103).

$$\begin{matrix}\text{Net heat conduction} \\ \text{into the volume}\end{matrix} + \begin{matrix}\text{Heat flux across} \\ \text{face at } x = 0\end{matrix} = 0$$

As before,

$$\dot{Q}_x|_E = -k\frac{T_{2,n} - T_{1,n}}{\Delta x}\Delta y$$

$$\dot{Q}_y|_S = -k\frac{T_{1,n} - T_{1,n-1}}{\Delta y}\frac{\Delta x}{2}$$

$$\dot{Q}_y|_N = -k\frac{T_{1,n+1} - T_{1,n}}{\Delta y}\frac{\Delta x}{2}$$

and

$$\dot{Q}_x|_0 = q_s\Delta y$$

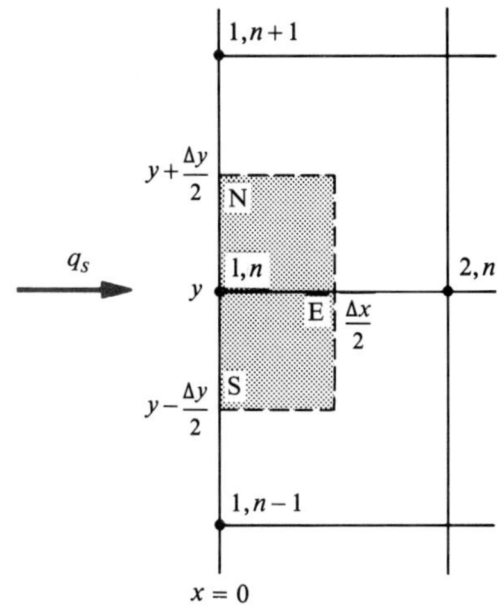

Figure 3.37 Finite control volume used to derive a difference equation for the surface heat flux: steady two-dimensional conduction.

Substituting in the energy balance, taking $\Delta x = \Delta y$, and solving for q_s gives

$$q_s = \frac{k}{\Delta x}\left[2T_{1,n} - T_{2,n} - \frac{1}{2}(T_{1,n-1} + T_{1,n+1})\right] \tag{3.105}$$

Equation (3.105) is simply item 3 of Table 3.7 rearranged to give q_s. Equation (3.105) is quite general; however, for a convective boundary condition, the surface heat can be just as easily calculated from Newton's law of cooling as $q_s = h_c(T_e - T_{1,n})$.

EXAMPLE 3.14 Steady Conduction in a Square Plate

An 8 × 8 cm square plate has one edge maintained at 100°C; the other three edges are maintained at 0°C. Use the finite-difference method to determine the temperature distribution in the plate. Compare the result with the exact solution given in Section 3.3.1.

Solution

Given: Square plate with edge temperatures prescribed.

Required: Steady-state temperature distribution using the finite-difference method.

Assumptions: Temperatures are constant across the thickness of the plate to give a two-dimensional problem.

The figure shows the mesh and the prescribed temperatures along the edges. There are nine interior nodes, but symmetry about the centerline results in only six unknown temperatures, which are labeled $T_1, T_2, \ldots, T_6$ as shown. From Eq. (3.102), these temperatures are given by

$$T_1 = \frac{1}{4}(T_3 + T_2 + 0 + 100)$$

$$T_2 = \frac{1}{4}(T_4 + T_1 + T_1 + 100)$$

$$T_3 = \frac{1}{4}(T_5 + T_4 + 0 + T_1)$$

$$T_4 = \frac{1}{4}(T_6 + T_3 + T_3 + T_2)$$

$$T_5 = \frac{1}{4}(0 + T_6 + 0 + T_3)$$

$$T_6 = \frac{1}{4}(0 + T_5 + T_5 + T_4)$$

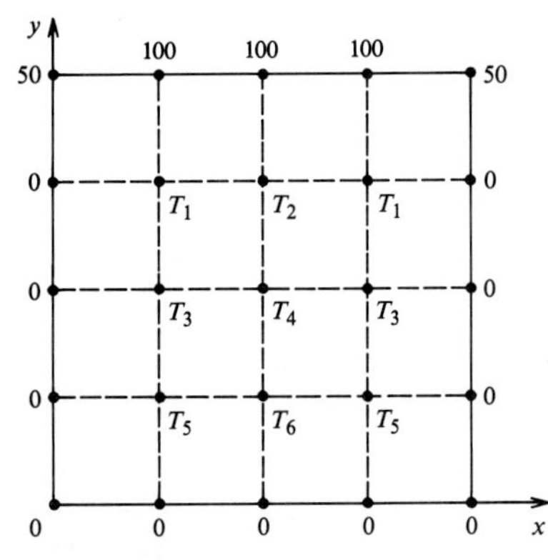

For an initial guess, a linear variation in y is assumed, and the preceding equations are evaluated in order, using latest available values. For example, when T_2 is evaluated, the new value of T_1 just obtained is used. The process is repeated until the convergence is satisfactory. The following table shows 16 iterations. Also shown is the exact solution obtained by evaluating Eq. (3.30).

Temperature	T_1	T_2	T_3	T_4	T_5	T_6
Initial guess, °C	75	75	50	50	25	25
Iteration level:						
$k = 1$	56.25	65.62	32.81	39.06	14.45	16.99
$k = 2$	49.61	59.57	25.78	32.03	10.69	13.35
$k = 4$	44.61	54.43	20.51	26.76	8.02	10.70
$k = 8$	42.97	52.79	18.86	25.11	7.20	9.88
$k = 16$	42.86	52.68	18.75	25.00	7.14	9.82
Exact solution	43.20	54.05	18.20	25.00	6.80	9.54
Percent error	0.80	2.54	3.01	0.00	5.08	2.93

Comments

1. Notice that the center temperature, T_4, is the average of the edge temperatures.
2. At the corners where the temperature is discontinuous, the average value of 50°C can be assigned.

3.6.2 Finite-Difference Methods for One-Dimensional Unsteady Conduction

Consider one-dimensional unsteady conduction with no internal heat generation and constant properties. Figure 3.38 shows a finite control volume $\Delta x \cdot 1 \cdot 1$ surrounding node m at location x in the solid. Figure 3.39 shows the mesh where the time coordinate is discretized in steps Δt and the index i denotes time. The energy conservation principle, Eq. (1.1), is applied to the finite control volume over a time interval Δt, from time step i to step $(i + 1)$:

$$\Delta U = \dot{Q}\Delta t \tag{3.106}$$

The increase in internal energy from time step i to step $(i + 1)$ is

$$\Delta U = \rho c(\Delta x \cdot 1 \cdot 1)(T_m^{i+1} - T_m^i)$$

The conduction over the time interval Δt is taken as

$$\dot{Q}_x|_W \Delta t = -k\frac{T_m^i - T_{m-1}^i}{\Delta x}(1 \cdot 1)\Delta t$$

$$\dot{Q}_x|_E \Delta t = -k\frac{T_{m+1}^i - T_m^i}{\Delta x}(1 \cdot 1)\Delta t$$

where the fluxes have been evaluated at time step i. Substituting in Eq. (3.106),

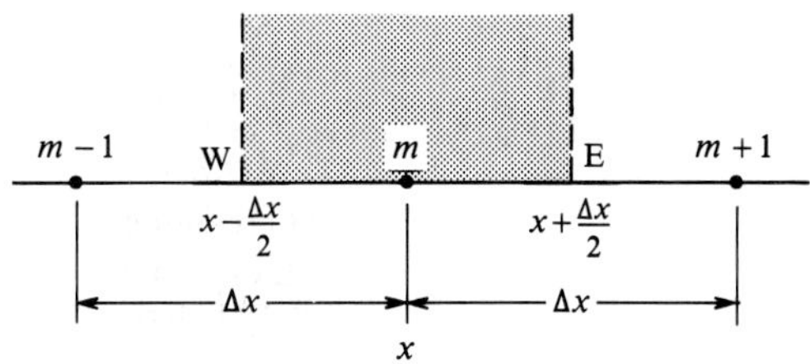

Figure 3.38 Finite control volume Δx by 1 by 1 surrounding node m at location x used to derive the difference equation for one-dimensional unsteady conduction.

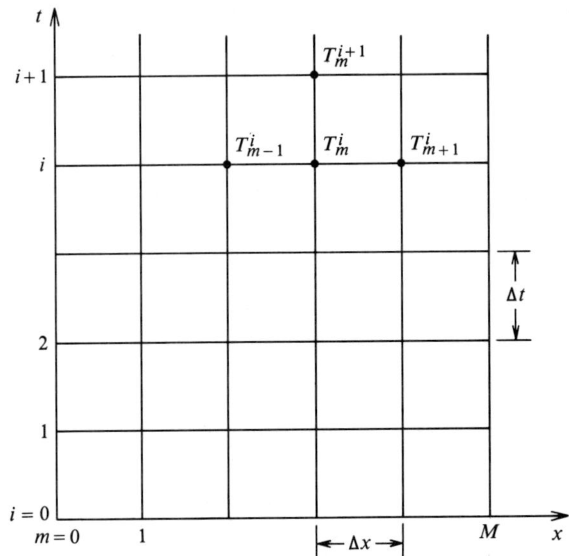

Figure 3.39 Nodal mesh for one-dimensional unsteady conduction.

dividing by Δt, and rearranging gives

$$T_m^{i+1} = \text{Fo}(T_{m-1}^i + T_{m+1}^i) + (1 - 2\,\text{Fo})T_m^i \tag{3.107}$$

where $\text{Fo} = \alpha\Delta t/\Delta x^2$ is the mesh Fourier number. Equation (3.107) is an explicit relation for T_m^{i+1}, the temperature of node m at the $(i+1)th$ time step, in terms of temperatures at the previous time step. In this *explicit* method, no iteration is required. Using given initial temperatures at time $t = 0$, T_m^0, the new temperatures, T_m^1, can be calculated. Temperatures at the boundary nodes 0 and M are fixed by the specified boundary conditions. The process is then repeated, marching forward in time.

Although simple, the explicit method has a major drawback: the allowable size of time step is limited by stability requirements. To avoid divergent oscillations in the solution, the coefficient for T_m^i in Eq. (3.107) must not be negative. That is,

$$\text{Fo} \le \frac{1}{2} \tag{3.108}$$

Since the spatial discretization step Δx is usually chosen to give a desired spatial resolution of the temperature profile, Eq. (3.108) sets a limit on the time step:

$$\Delta t \le \frac{\Delta x^2}{2\alpha} \tag{3.109}$$

If the boundary condition is other than that of prescribed temperature, the associated stability requirement is more stringent than Eq. (3.109). Consider the convective boundary condition illustrated in Fig. 3.40. An energy balance requires that, during the time step Δt,

$$\begin{matrix}\text{Increase in internal}\\ \text{energy within volume}\end{matrix} = \begin{matrix}\text{Net conduction}\\ \text{into volume}\end{matrix} + \begin{matrix}\text{Convection across}\\ \text{boundary}\end{matrix}$$

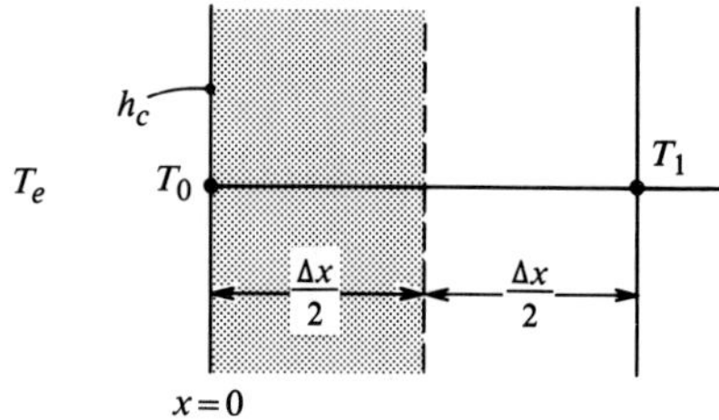

Figure 3.40 Finite control volume used to derive the difference equation for a convective boundary condition at $x = 0$: one-dimensional unsteady conduction.

or

$$\rho c \frac{\Delta x}{2}(1 \cdot 1)(T_0^{i+1} - T_0^i) = k(1 \cdot 1)\frac{T_1^i - T_0^i}{\Delta x}\Delta t + h_c(1 \cdot 1)(T_e^i - T_0^i)\Delta t$$

Rearranging and solving for the new nodal temperature, T_0^{i+1},

$$T_0^{i+1} = 2\,\text{Fo}(T_1^i + \text{Bi}T_e^i) + (1 - 2\,\text{Fo} - 2\,\text{Fo}\,\text{Bi})T_0^i \tag{3.110}$$

where again $\text{Bi} = h_c\Delta x/k$. For stability, the coefficient of T_0^i must not be negative:

$$1 - 2\,\text{Fo} - 2\,\text{Fo}\,\text{Bi} \geq 0$$

or

$$\text{Fo} \leq \frac{1}{2(1 + \text{Bi})} \tag{3.111}$$

Since the Biot number is positive, Eq. (3.111) is always a more stringent stability condition than Eq. (3.108).

The stability criterion limits Δt to no more than $(\Delta x)^2/2\alpha$, and less if the mesh Biot number is not small. Thus, if we wish to improve accuracy by halving the mesh size Δx, the time step Δt must be divided by four. For accurate solutions using a fine spatial mesh size, alternative methods are available. The *implicit* method evaluates the conduction fluxes at time step $(i + 1)$ rather than at step i:

$$\dot{Q}_x|_W\Delta t = -k\frac{T_m^{i+1} - T_{m-1}^{i+1}}{\Delta x}(1 \cdot 1)\Delta t$$

$$\dot{Q}_x|_E\Delta t = -k\frac{T_{m+1}^{i+1} - T_m^{i+1}}{\Delta x}(1 \cdot 1)\Delta t$$

The new form of Eq. (3.107) is

$$T_m^{i+1} = \frac{\text{Fo}(T_{m+1}^{i+1} + T_{m-1}^{i+1}) + T_m^i}{1 + 2\,\text{Fo}} \tag{3.112}$$

There are three unknown temperatures in each nodal equation. The set of algebraic equations is *tridiagonal*; that is, when written in matrix form, all the elements of the coefficient matrix are zero except for those that are on or to either side of the main diagonal. To advance the solution through each time step, Gauss-Seidel iteration works well.[4] At time step $(i + 1)$, the nodal equations are swept repeatedly

[4] Direct methods, such as successive substitution, may also be used when the equation set is tridiagonal.

Table 3.8 Finite-difference approximations for one-dimensional unsteady conduction.

Item	Configuration	Explicit Form and Stability Criterion	Implicit Form
1.	Δx, Δx, T_1, T_0, T_2 Interior node	$T_0^{i+1} = \text{Fo}(T_1^i + T_2^i) + (1 - 2\text{Fo})T_0^i$ $\text{Fo} = \dfrac{\alpha \Delta t}{\Delta x^2} \leq \dfrac{1}{2}$	$T_0^{i+1} = \dfrac{\text{Fo}(T_1^{i+1} + T_2^{i+1}) + T_0^i}{1 + 2\text{Fo}}$
2.	h_c, Δx, T_1, T_0, T_e Convection at surface	$T_0^{i+1} = 2\text{Fo}(T_1^i + \text{Bi}T_e^i) + (1 - 2\text{Fo} - 2\text{FoBi})T_0^i$ $\text{Fo} \leq \dfrac{1}{2(1 + \text{Bi})}$; $\text{Bi} = \dfrac{h_c \Delta x}{k}$	$T_0^{i+1} = \dfrac{2\text{Fo}(T_1^{i+1} + \text{Bi}T_e^{i+1}) + T_0^i}{1 + 2\text{Fo} + 2\text{FoBi}}$
3.	Δx, T_1, T_0, q_s Known surface heat flux	$T_0^{i+1} = 2\text{Fo}\left(T_1^i + \dfrac{q_s^i \Delta x}{k}\right) + (1 - 2\text{Fo})T_0^i$ $\text{Fo} \leq \dfrac{1}{2}$	$T_0^{i+1} = \dfrac{2\text{Fo}[T_1^{i+1} + (q_s^{i+1}\Delta x/k)] + T_0^i}{1 + 2\text{Fo}}$
4.	Adiabatic, Δx, T_1, T_0 Adiabatic surface, or plane of symmetry	$T_0^{i+1} = 2\text{Fo}T_1^i + (1 - 2\text{Fo})T_0^i$ $\text{Fo} \leq \dfrac{1}{2}$	$T_0^{i+1} = \dfrac{2\text{Fo}T_1^{i+1} + T_0}{1 + 2\text{Fo}}$

Table 3.9 Finite-difference approximations and stability criteria for two-dimensional unsteady conduction, square mesh. For an adiabatic surface or plane of symmetry, set $q_s = 0$ in item 3.

Configuration	Mesh	Equations	Stability criterion
1. Interior node	T_1, T_2, T_3, T_4, T_0, Δx, Δx	Explicit: $T_0^{i+1} = \text{Fo}(T_1^i + T_2^i + T_3^i + T_4^i) + (1 - 4\text{Fo})T_0^i$; Implicit: $T_0^{i+1} = \dfrac{\text{Fo}(T_1^{i+1} + T_2^{i+1} + T_3^{i+1} + T_4^{i+1}) + T_0^i}{1 + 4\text{Fo}}$	$\text{Fo} \le \dfrac{1}{4}$
2. Plane surface, convection	T_1, T_2, T_3, T_0, Δx, Δx, h_c, T_e	Explicit: $T_0^{i+1} = 2\text{Fo}\left[T_1^i + \frac{1}{2}(T_2^i + T_3^i) + \text{Bi}T_e^i\right] + (1 - 4\text{Fo} - 2\text{FoBi})T_0^i$; Implicit: $T_0^{i+1} = \dfrac{2\text{Fo}\,[T_1^{i+1} + (1/2)(T_2^{i+1} + T_3^{i+1}) + \text{Bi}T_e^{i+1}] + T_0^i}{1 + 2\text{Fo}(2 + \text{Bi})}$	$\text{Fo} \le \dfrac{1}{2(2 + \text{Bi})}$
3. Plane surface, known heat flux	T_1, T_2, T_3, T_0, Δx, Δx, q_s	Explicit: $T_0^{i+1} = 2\text{Fo}\left[T_1^i + \frac{1}{2}(T_2^i + T_3^i) + \frac{q_s^i \Delta x}{k}\right] + (1 - 4\text{Fo})T_0^i$; Implicit: $T_0^{i+1} = \dfrac{2\text{Fo}\,[T_1^{i+1} + (1/2)(T_2^{i+1} + T_3^{i+1}) + (q_s^{i+1}\Delta x/k)] + T_0^i}{1 + 4\text{Fo}}$	$\text{Fo} \le \dfrac{1}{4}$
4. Exterior corner, convection	T_1, T_0, T_2, Δx, Δx, h_c, T_e	Explicit: $T_0^{i+1} = 2\text{Fo}(T_1^i + T_2^i + 2\text{Bi}T_e^i) + (1 - 4\text{Fo} - 4\text{FoBi})T_0^i$; Implicit: $T_0^{i+1} = \dfrac{2\text{Fo}(T_1^{i+1} + T_2^{i+1} + 2\text{Bi}T_e^{i+1}) + T_0^i}{1 + 4\text{Fo}(1 + \text{Bi})}$	$\text{Fo} \le \dfrac{1}{4(1 + \text{Bi})}$
5. Interior corner, convection	T_1, T_2, T_3, T_4, T_0, h_c, T_e	Explicit: $T_0^{i+1} = \frac{4}{3}\text{Fo}\left[\frac{1}{2}(T_1^i + T_4^i) + T_2^i + T_3^i + \text{Bi}T_e^i\right] + \left(1 - 4\text{Fo} - \frac{4}{3}\text{FoBi}\right)T_0^i$; Implicit: $T_0^{i+1} = \dfrac{(4/3)\text{Fo}\,[(1/2)(T_1^{i+1} + T_4^{i+1}) + T_2^{i+1} + T_3^{i+1} + \text{Bi}T_e^{i+1}] + T_0^i}{1 + 4\text{Fo}[1 + (1/3)\text{Bi}]}$	$\text{Fo} \le \dfrac{3}{4(3 + \text{Bi})}$

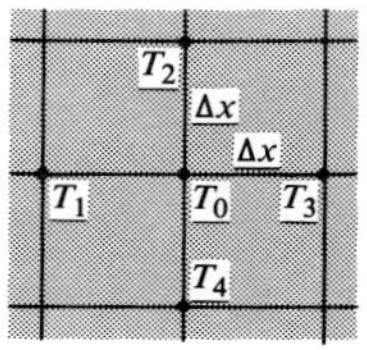

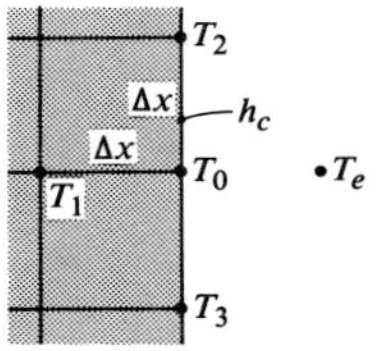

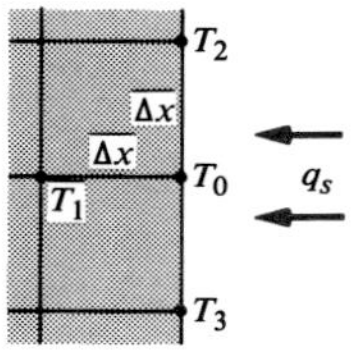

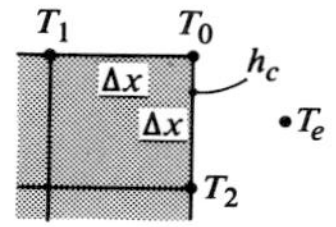

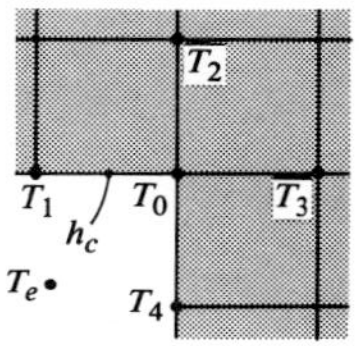

until convergence to sufficient accuracy is obtained. The implicit method is unconditionally stable, and the choice of the time step size Δt is dictated by accuracy rather than stability considerations. As mentioned in Section 3.6.1, there are iteration schemes that give faster convergence than Gauss-Seidel iteration, and these may be found in numerical methods texts. When the boundary condition is other than that of prescribed temperature, an energy balance must be used at the boundary control volumes, as was shown for the explicit method, but with spatial derivatives evaluated at time step $(i + 1)$. Table 3.8 lists both explicit and implicit forms for a variety of boundary conditions. Table 3.9 lists corresponding results for two-dimensional unsteady conduction.

In addition to the explicit and implicit methods, a third method often used is the *Crank-Nicolson* method. Whereas the explicit method evaluates conduction fluxes at the old time step i, and the implicit method uses the new time step $(i + 1)$, the Crank-Nicolson method uses an average of the values at time steps i and $(i + 1)$. The nodal equation is then more complicated (see Exercise 3–90). For a given mesh size, the Crank-Nicolson method gives more accurate results than either the explicit or implicit methods. Although oscillations can occur, they never become unstable.

EXAMPLE 3.15 Convective Heating of a Resin Slab

An 8 cm–thick slab of resin is to be cured under an array of air jets at 100°C, as shown in the accompanying sketch. If the initial temperature of the resin is 20°C, determine the temperature of the back face after one hour. Take the heat transfer coefficient as 40 W/m^2 K, and for the resin $\rho = 2600$ kg/m^3, $c = 800$ J/kg K, $k = 1.0$ W/m K.

Solution

Given: Slab, convectively heated on one face.

Required: Back face temperature after one hour.

Assumptions:
1. The back face is well insulated.
2. The heat transfer coefficient h_c is uniform over the surface, and ρ and c are constant.
3. Edge losses are negligible.

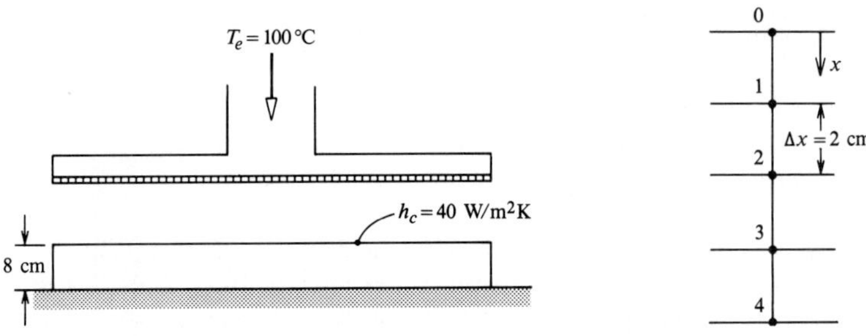

The explicit finite-difference method will be used. Let $\Delta x = 2$ cm; then the mesh size–based Biot number is

$$\text{Bi} = \frac{h_c \Delta x}{k} = \frac{(40)(0.02)}{1.0} = 0.8$$

The stability criterion, Eq. (3.111), is

$$\text{Fo} \leq \frac{1}{2(1 + \text{Bi})} = \frac{1}{2(1 + 0.8)} = 0.277$$

Choose Fo $= 0.25$, so that the time step is

$$\Delta t = \frac{\text{Fo}(\Delta x)^2}{\alpha} = \frac{(0.25)(0.02)^2}{1.0/(2600 \times 800)} = 208 \text{ s } (3.47 \text{ min})$$

Equation (3.107) for the interior nodes becomes

$$\begin{aligned} T_m^{i+1} &= [1 - (2)(0.25)]T_m^i + 0.25(T_{m-1}^i + T_{m+1}^i) \\ &= 0.5T_m^i + 0.25(T_{m-1}^i + T_{m+1}^i) \end{aligned}$$

and Eq. (3.110) for the surface node is

$$\begin{aligned} T_0^{i+1} &= (2)(0.25)(T_1^i + 0.8T_e) + [1 - 2(0.25) - 2(0.25)(0.8)]T_0^i \\ &= 0.5(T_1^i + 0.8T_e) + 0.1T_0^i \end{aligned}$$

To obtain an appropriate equation for node 4 at the adiabatic surface, we simply set Bi $= 0$ in Eq. (3.110) to obtain

$$T_4^{i+1} = (2)(0.25)T_3^i + [1 - (2)(0.25)]T_4^i = 0.5(T_3^i + T_4^i)$$

The initial condition is $T = 20$°C; thus, the temperatures at the first time step are

$$\begin{aligned} T_0^1 &= 0.5[20 + (0.8)(100)] + 0.1(20) &&= 52 \\ T_1^1 &= 0.5(20) + 0.25(20 + 20) &&= 20 \\ T_2^1 &= 0.5(20) + 0.25(20 + 20) &&= 20 \\ T_3^1 &= 0.5(20) + 0.25(20 + 20) &&= 20 \\ T_4^1 &= 0.5(20 + 20) &&= 20 \end{aligned}$$

and at the second time step,

$$\begin{aligned} T_0^2 &= 0.5[20 + (0.8)(100)] + 0.1(52) &&= 55.2 \\ T_1^2 &= 0.5(20) + 0.25(52 + 20) &&= 28.0 \\ T_2^2 &= 0.5(20) + 0.25(20 + 20) &&= 20 \\ T_3^2 &= 0.5(20) + 0.25(20 + 20) &&= 20 \\ T_4^2 &= 0.5(20 + 20) &&= 20 \end{aligned}$$

and so on. The results for 20 time steps obtained using a programmable hand calculator are given in the following table. Also shown is the surface temperature T_0, calculated using computer program COND2, which is essentially exact.

Time step	Time min	T_0 °C	T_1 °C	T_2 °C	T_3 °C	T_4 °C	T_0 (exact) °C
0	0.00	20.00	20.00	20.00	20.00	20.00	20.00
1	3.47	52.00	20.00	20.00	20.00	20.00	46.32
2	6.93	55.20	28.00	20.00	20.00	20.00	53.30
3	10.40	59.52	32.80	22.00	20.00	20.00	57.68
4	13.87	62.35	36.78	24.20	20.50	20.00	60.85
5	17.33	64.63	40.03	26.42	21.30	20.25	63.32
6	20.80	66.48	42.78	28.54	22.32	20.78	65.33
7	24.27	68.04	45.14	30.54	23.49	21.55	67.01
8	27.73	69.37	47.22	32.43	24.77	22.52	68.45
9	31.20	70.55	49.06	34.21	26.12	23.64	69.71
10	34.67	71.58	50.72	35.90	27.52	24.88	70.82
11	38.13	72.52	52.23	37.51	28.96	26.20	71.81
12	41.60	73.37	53.62	39.05	30.41	27.58	72.72
13	45.07	74.15	54.92	40.53	31.86	28.99	73.54
14	48.53	74.87	56.13	41.96	33.31	30.43	74.31
15	52.00	75.55	57.27	43.34	34.75	31.87	75.02
16	55.47	76.19	58.36	44.68	36.18	33.31	75.69
17	58.93	76.80	59.40	45.97	37.59	34.75	76.32
18	62.40	77.38	60.39	47.23	38.97	36.17	76.92
19	65.87	77.93	61.35	48.46	40.34	37.57	77.50
20	69.33	78.47	62.27	49.65	41.67	38.95	78.05

Comments

Notice that, for this crude calculation using $\Delta x = 2$ cm, the stability criterion is not a significant limitation on the time step. However, if Δx were reduced to, say, 0.5 cm to improve spatial resolution of the temperature profile, Δt becomes only 13 s. Even this time step poses no problem to present-day personal computers.

EXAMPLE 3.16 Quenching of a Slab with Nucleate Boiling

A 4 cm–thick slab of steel initially at 500 K is immersed in a water bath at 310 K and 1atm. Under these conditions, *nucleate boiling* (see Chapter 7) occurs on the slab surface; the heat transfer coefficient is very large and is strongly dependent on temperature difference. An appropriate empirical equation for h_c under these conditions is $h_c = 140(T - T_{\text{sat}})^2$ W/m^2 K, where T_{sat} is the saturation temperature (boiling point). Determine the temperature profile across the slab for a period of 30 s. For the steel, take $k = 54$ W/m K, $\alpha = 1.5 \times 10^{-5}$ m^2/s.

Solution

Given: Slab immersed in water; nucleate boiling on surface.

Required: Temperature profiles.

Assumption: Edge effects negligible to give a one-dimensional problem.

The implicit finite-difference method will be used for this problem, and the results will be obtained using a computer. Since the problem is symmetrical about the center plane of the slab, node $M = 21$ is located on the center plane as shown. Choosing a time step of 1 s, the mesh Fourier number is

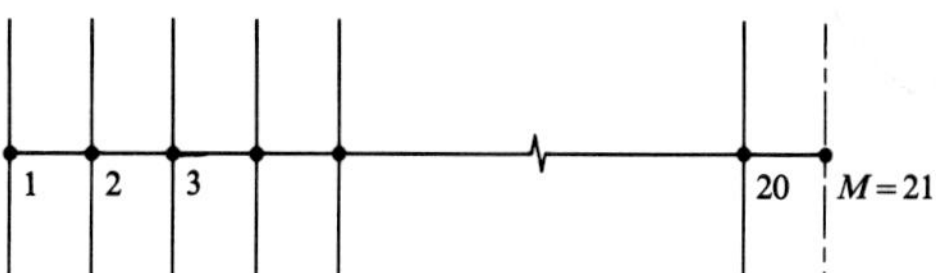

$$\text{Fo} = \frac{\alpha \Delta t}{\Delta x^2} = \frac{(1.5 \times 10^{-5})(1)}{(0.02/20)^2} = 15$$

Temperatures at the interior nodes $m = 2, \ldots, M - 1$ are given by Eq. (3.112):

$$T_m^{i+1} = \frac{1}{1 + (2)(15)}\left[15(T_{m+1}^{i+1} + T_{m-1}^{i+1}) + T_m^i\right]$$

$$= \frac{1}{31}\left[15(T_{m+1}^{i+1} + T_{m-1}^{i+1}) + T_m^i\right]$$

The temperature at node M is given in Table 3.8, item 4:

$$T_M^{i+1} = \frac{1}{1 + (2)(15)}\left[(2)(15)T_{M-1}^{i+1} + T_M^i\right]$$

$$T_{21}^{i+1} = \frac{1}{31}\left[30T_{20}^{i+1} + T_{21}^i\right]$$

The temperature at the surface node $m = 1$ is given in Table 3.8, item 2, with T_e replaced by T_{sat}:

$$T_1^{i+1} = \frac{2(15)(T_2^{i+1} + \text{Bi}\, T_{\text{sat}}) + T_1^i}{1 + 2(15) + (2)(15)\,\text{Bi}}$$

The Biot number is not a constant in this problem and, when the implicit formulation is used, must be evaluated at the current time step:

$$\text{Bi} = \frac{h_c \Delta x}{k} = \frac{140(T_1^{i+1} - T_{\text{sat}})^2(0.02/20)}{54} = 2.59 \times 10^{-3}(T_1^{i+1} - T_{\text{sat}})^2$$

Thus,

$$T_1^{i+1} = \frac{30(T_2^{i+1} + 2.59 \times 10^{-3}(T_1^{i+1} - T_{\text{sat}})^2 T_{\text{sat}}) + T_1^i}{31 + 7.78 \times 10^{-2}(T_1^{i+1} - T_{\text{sat}})^2}$$

Since T_1^{i+1} appears on both sides of this equation, it should be solved for by iteration. A flow diagram for a simple program based on Gauss-Seidel and Newton iteration follows. Sample results are given in the accompanying table.

Node	1	6	11	16	21
Location x, cm	0	0.5	1.0	1.5	2.0
Nodal temperatures, K:					
$t = 0$ s	500.0	500.0	500.0	500.0	500.0
1	394.9	471.0	492.0	497.6	498.8
2	390.8	451.2	481.2	492.8	495.6
5	387.6	425.8	455.1	472.4	478.0
10	385.4	409.4	429.3	442.4	447.0
20	382.5	393.3	402.3	408.2	410.2
30	380.5	385.6	389.8	392.6	393.5

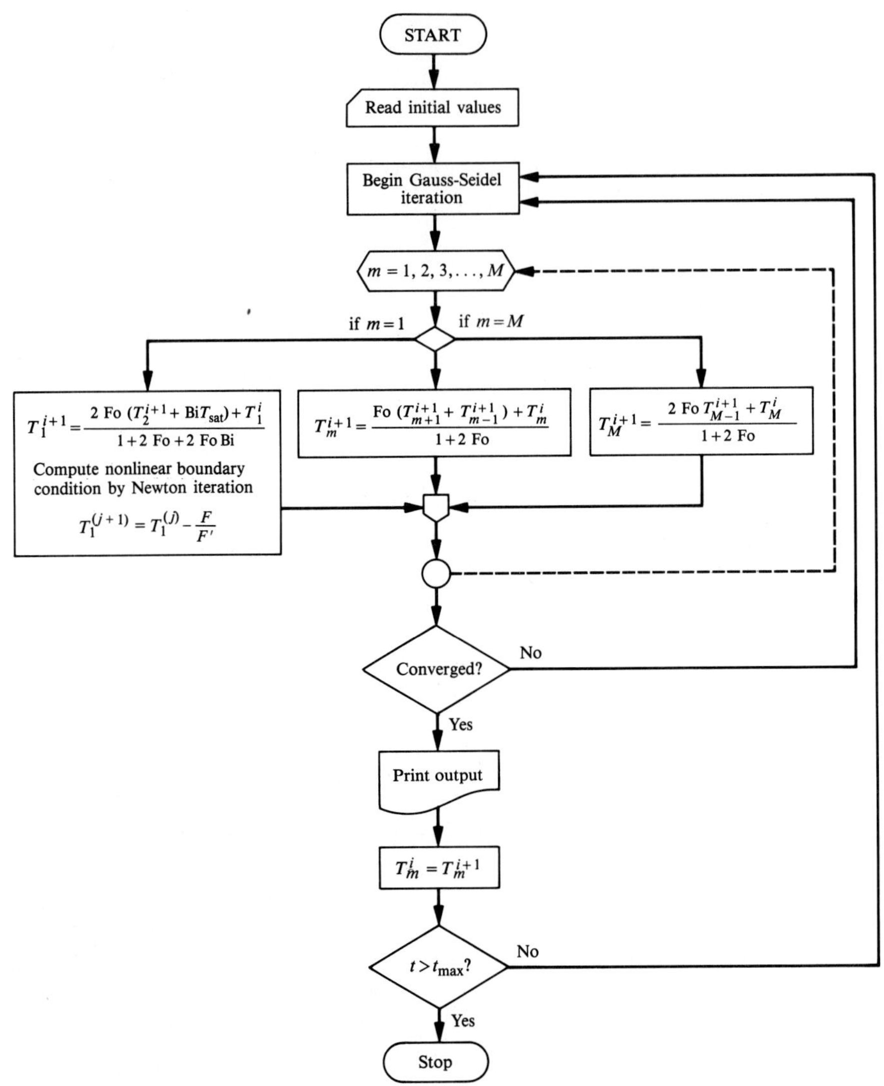

Comments

1. The effect of Δx and Δt on the accuracy of the solution should be explored.
2. The task of writing a computer program is given as Exercise 3–92.

3.6.3 Resistance-Capacitance (*RC*) Formulation

A useful alternative to the formulations for finite-difference numerical methods presented in the previous sections is the **resistance-capacitance** formulation. Consider multidimensional unsteady conduction with internal heat generation and variable thermal properties. Figure 3.41 shows the node of interest, m, surrounded by a volume

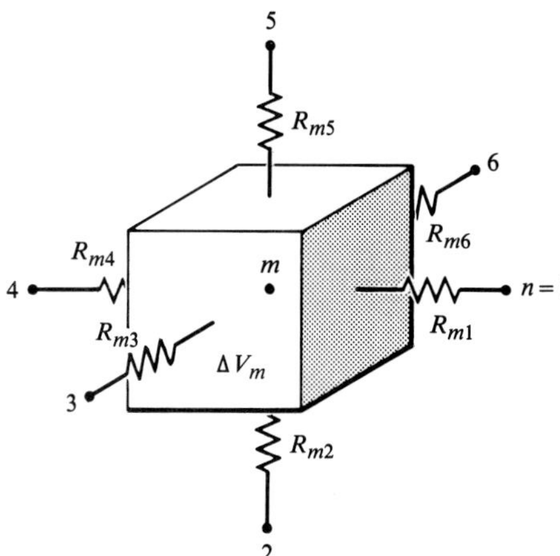

Figure 3.41 Finite control volume ΔV_m surrounding node m, with thermal resistances R_{mn}: resistance-capacitance formulation.

ΔV_m of thermal capacitance $C_m = \rho_m c_m \Delta V_m$. The surrounding nodes are denoted n, and conduction from the volume ΔV_n into volume ΔV_m is expressed in terms of a thermal resistance R_{mn}. An energy balance on volume ΔV_m over the time interval Δt gives

$$C_m \frac{T_m^{i+1} - T_m^i}{\Delta t} = \sum_n \frac{T_n^i - T_m^i}{R_{mn}} + \dot{Q}_v''' \Delta V_m \tag{3.113}$$

where the superscript refers to the time step, as before. Equation (3.113) is an explicit formulation. Although written for an interior node, Eq. (3.113) can also be applied to boundary nodes with the resistances R_{mn} evaluated accordingly. The concepts involved here are, of course, identical to those introduced for elementary thermal networks in Sections 1.4 and 2.3. Notice that when properties are temperature-dependent, the capacitance C_m, resistances R_{mn}, and source $\dot{Q}_v'''$ are evaluated at time step i to preserve the explicit formulation. If an implicit formulation is desired, the driving potentials $(T_n - T_m)$ must be evaluated at time step $(i + 1)$. However, to have a *fully implicit* formulation, C_m, R_{mn}, and $\dot{Q}_v'''$ must also be evaluated at time step $(i + 1)$, which can slow down the required iteration procedure. Thus, the explicit formulation is usually preferred when this resistance-capacitance representation is used.

Solving Eq. (3.113) for T_m^{i+1} gives

$$T_m^{i+1} = \left(1 - \frac{\Delta t}{C_m} \sum_n \frac{1}{R_{mn}}\right) T_m^i + \left(\dot{Q}_v''' \Delta V_m + \sum_n \frac{T_n^i}{R_{mn}}\right) \frac{\Delta t}{C_m} \tag{3.114}$$

Again, a satisfactory condition for stability is that the coefficient of T_m^i should not be negative:

$$1 - \frac{\Delta t}{C_m} \sum_n \frac{1}{R_{mn}} \geq 0 \tag{3.115}$$

The resistance-capacitance representation is usually applied to problems that have complicated boundary conditions, that require a variation in the size and shape of volumes ΔV_m, or for which thermal properties are temperature-dependent. Thus, the stability criterion must be evaluated for each node, and a time step chosen such that

$$\Delta t \leq \left(\frac{C_m}{\sum_n (1/R_{mn})} \right)_{\min} \tag{3.116}$$

That is, it is the most restrictive nodal equation that controls the allowable time step.

Table 3.10 gives the volumes ΔV_m and resistances R for interior nodes in the Cartesian, cylindrical, and spherical coordinate systems shown in Fig. 3.42. Table 3.11 shows examples for boundary nodes in Cartesian coordinates.

The resistance-capacitance representation is also widely applied to systems-level thermal analysis. Figure 3.43 shows how a complicated system is divided into elements. At this level, the representation is equivalent to the lumped thermal capacity approach of Section 1.5. The major advantage of the RC representation is the versatility of resulting computer programs. The equation-solving routines can be written

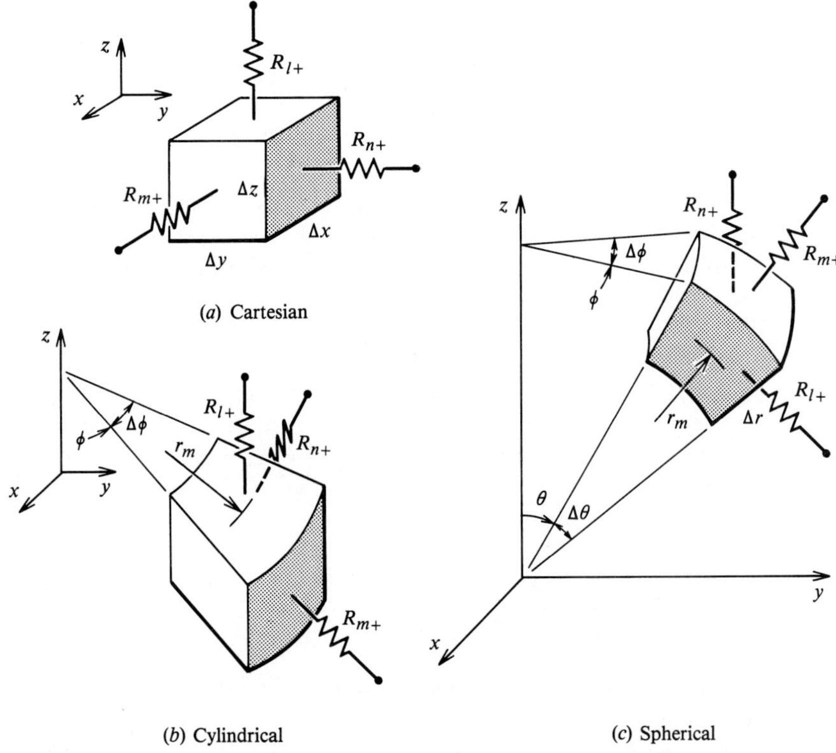

Figure 3.42 Finite control volumes ΔV_m and resistances R in Cartesian, cylindrical, and spherical coordinates for the resistance-capacitance formulation.

Table 3.10 Internal nodal resistances in Cartesian, cylindrical, and spherical coordinates.

	Coordinate System		
	Cartesian	Cylindrical	Spherical
Coordinates	x, y, z	r, ϕ, z	r, ϕ, θ
Indices	m, n, l	m, n, l	m, n, l
Volume element	$\Delta x \Delta y \Delta z$	$r_m \Delta r \Delta\phi \Delta z$	$r_m^2 \sin\theta \Delta r \Delta\phi \Delta\theta$
R_{m+}	$\dfrac{\Delta x}{\Delta y \Delta z k}$	$\dfrac{\Delta r}{(r_m + \Delta r/2)\Delta\phi \Delta z k}$	$\dfrac{\Delta r}{(r_m + \Delta r/2)^2 \sin\theta \Delta\phi \Delta\theta k}$
R_{m-}	$\dfrac{\Delta x}{\Delta y \Delta z k}$	$\dfrac{\Delta r}{(r_m - \Delta r/2)\Delta\phi \Delta z k}$	$\dfrac{\Delta r}{(r_m - \Delta r/2)^2 \sin\theta \Delta\phi \Delta\theta k}$
R_{n+}	$\dfrac{\Delta y}{\Delta x \Delta z k}$	$\dfrac{r_m \Delta\phi}{\Delta r \Delta z k}$	$\dfrac{\Delta\phi \sin\theta}{\Delta r \Delta\theta k}$
R_{n-}	$\dfrac{\Delta y}{\Delta x \Delta z k}$	$\dfrac{r_m \Delta\phi}{\Delta r \Delta z k}$	$\dfrac{\Delta\phi \sin\theta}{\Delta r \Delta\theta k}$
R_{l+}	$\dfrac{\Delta z}{\Delta x \Delta y k}$	$\dfrac{\Delta z}{r_m \Delta\phi \Delta r k}$	$\dfrac{\Delta\theta}{\sin(\theta + \Delta\theta/2)\Delta r \Delta\phi k}$
R_{l-}	$\dfrac{\Delta z}{\Delta x \Delta y k}$	$\dfrac{\Delta z}{r_m \Delta\phi \Delta r k}$	$\dfrac{\Delta\theta}{\sin(\theta - \Delta\theta/2)\Delta r \Delta\phi k}$

in terms of the very general capacitance C_m and resistances R_{mn}, with specialization of these quantities relegated to subroutines prepared for a particular problem. It is usual practice to also allow for the possibility of radiation heat transfer between the volume elements; Eq. (3.113) then becomes

$$C_m \frac{T_m^{i+1} - T_m^i}{\Delta t} = \sum_n \frac{T_{n_i} - T_m^i}{R_{mn}^{\text{cond}}} + \sum_n \frac{T_n^i - T_m^i}{R_{mn}^{\text{rad}}} + \dot{Q}_v''' \Delta V_m \tag{3.117}$$

where R_{mn}^{cond} and R_{mn}^{rad} are conduction and radiation resistances, respectively. Referring to Section 1.3.2, Eq. (1.17) becomes

$$\begin{aligned} \dot{Q}_{nm} &= A_n \mathscr{F}_{nm} \sigma (T_n^4 - T_m^4) \\ &= A_n \mathscr{F}_{nm} \sigma (T_n^2 + T_m^2)(T_n + T_m)(T_n - T_m) \end{aligned}$$

Hence,

$$R_{mn}^{\text{rad}} = \frac{1}{A_n \mathscr{F}_{nm} \sigma (T_n^2 + T_m^2)(T_n + T_m)} \tag{3.118}$$

In general, the transfer factors $\mathscr{F}_{nm}$ depend on the geometry and emittances of all the surfaces making up an enclosure and are difficult to calculate. However, computer programs are available for this purpose (or simple approximations are used).

Table 3.11 Control volumes and resistances for two-dimensional Cartesian coordinates, $\Delta x = \Delta y$.

Item	Configuration	Control Volume and Resistances
1.	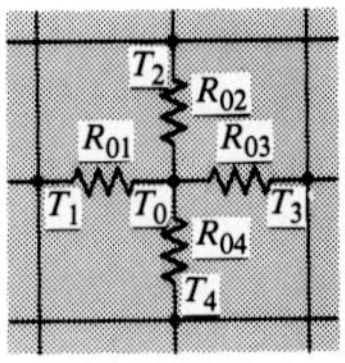 Interior node	$\Delta V = (\Delta x)^2$ $R_{01} = R_{02} = R_{03} = R_{04} = \frac{1}{k}$
2.	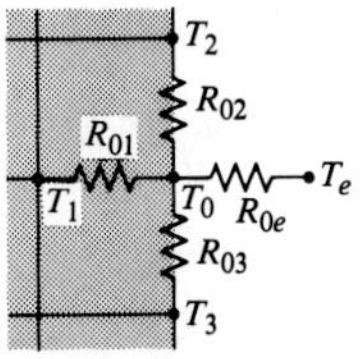 Plane surface, convection	$\Delta V = \frac{(\Delta x)^2}{2}$ $R_{01} = \frac{1}{k}; \quad R_{02} = R_{03} = \frac{2}{k}$ $R_{0e} = \frac{1}{h_c \Delta x}$
3.	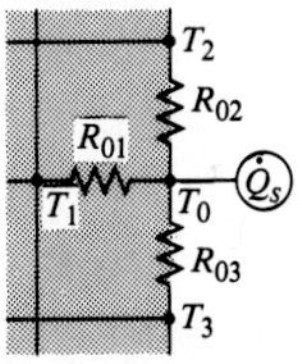 Plane surface, known heat flux	$\Delta V = \frac{(\Delta x)^2}{2}$ $R_{01} = \frac{1}{k}; \quad R_{02} = R_{03} = \frac{2}{k}$ $\dot{Q}_s = q_s \Delta x$
4.	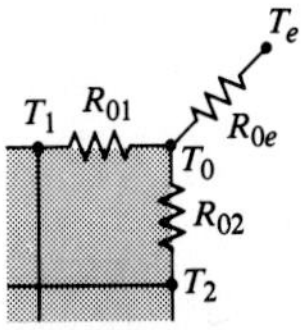 Exterior corner, convection	$\Delta V = \frac{(\Delta x)^2}{4}$ $R_{01} = R_{02} = \frac{2}{k}$ $R_{0e} = \frac{1}{h_c \Delta x}$
5.	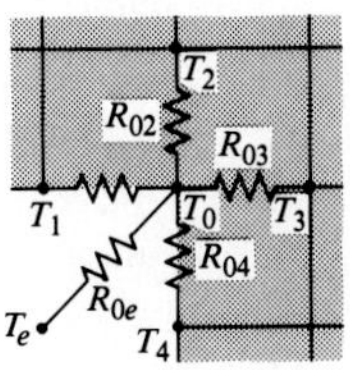 Interior corner, convection	$\Delta V = \frac{3(\Delta x)^2}{4}$ $R_{01} = R_{04} = \frac{2}{k}$ $R_{02} = R_{03} = \frac{1}{k}$ $R_{0e} = \frac{1}{h_c \Delta x}$

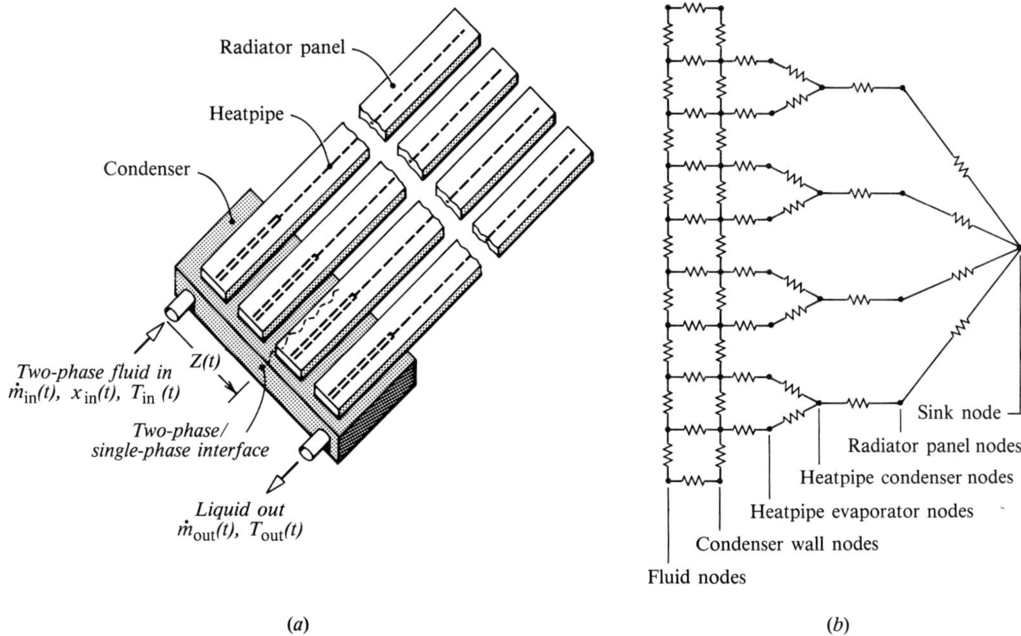

Figure 3.43 *RC* representation of a spacecraft heat rejection system. (*a*) The physical system. (*b*) The nodal network.

Computer programs based on the resistance-capacitance representation have found very wide industrial use, particularly in the aerospace industry. Popular examples include SINDA [16] and MITAS [17]. Although simple in concept, these programs have been refined over many years and are efficient, reliable, and convenient to use. Other software that can be used for conduction calculations include PHOENICS[18] and COMPACT[19].

EXAMPLE 3.17 Asymmetrical Heating of a Cylindrical Rod

A long, 9 cm–diameter ceramic rod, initially at 20°C, is exposed to a radiation heat flux on one side such that $q_s = 5000 \cos\phi$, $270° < \phi < 90°$, and is insulated on the other side, $q_s = 0$, $90° < \phi < 270°$. Determine the temperature response of the rod. For the ceramic, take $\rho = 3000$ kg/m^3, $k = 5$ W/m K, and $c = 800$ J/kg K.

Solution

Given: Ceramic rod heated on one side, insulated on the other.

Required: Temperature response, $T(r, \phi, t)$.

Assumptions:
1. Constant properties.
2. No axial variation of temperature.

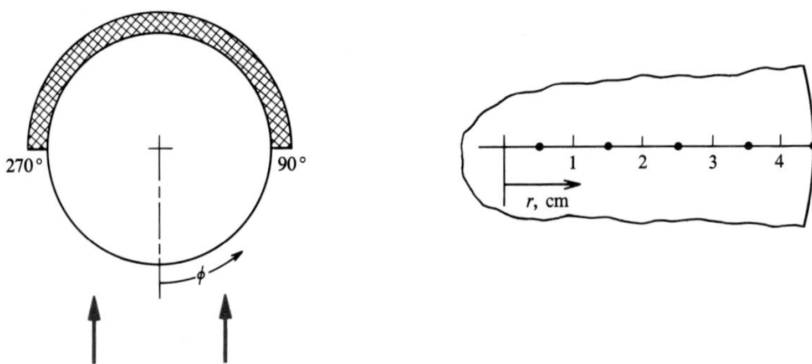

The explicit resistance-capacitance formulation described in Section 3.6.3 will be used. Equation (3.114) gives the temperatures at the interior nodes; for $\dot{Q}_v''' = 0$,

$$T_m^{i+1} = \left(1 - \frac{\Delta t}{C_m}\sum_n \frac{1}{R_{mn}}\right)T_m^i + \frac{\Delta t}{C_m}\sum_n \frac{T_n^i}{R_{mn}}$$

Choosing $\Delta r = 1$ cm, $\Delta\phi = 30° = 0.5236$ rad, and $\Delta z = 1$ m, Table 3.10 gives the volume element as

$$\Delta V_m = r_m \Delta r \Delta\phi \Delta z = r_m(0.01)(0.5236)(1) = 5.236 \times 10^{-3} r_m$$

and $C_m = \rho c \Delta V_m = (3000)(800)(5.236 \times 10^{-3})r_m = 1.257 \times 10^4 r_m$. Also from Table 3.10, the nodal resistances are

$$R_{m+} = \frac{\Delta r}{(r_m + \Delta r/2)\Delta\phi\Delta z k} = \frac{0.01}{(r_m + 0.005)(0.5236)(1)(5)} = \frac{3.820 \times 10^{-3}}{r_m + 0.005}$$

$$R_{m-} = \frac{\Delta r}{(r_m - \Delta r/2)\Delta\phi\Delta z k} = \frac{0.01}{(r_m - 0.005)(0.5236)(1)(5)} = \frac{3.820 \times 10^{-3}}{r_m - 0.005}$$

$$R_{n+} = \frac{r_m \Delta\phi}{\Delta r \Delta z k} = \frac{r_m(0.5236)}{(0.01)(1)(5)} = 10.47 r_m$$

$$R_{n-} = R_{n+} = 10.47 r_m$$

A surface node for $270° < \phi < 90°$ is shown in the accompanying figure. For this half-volume,

$$C_m = \frac{1}{2}(1.257 \times 10^4)(0.045)$$

$$= 282.8$$

$$R_{01} = \frac{\Delta r}{(r_0 - \Delta r/2)\Delta\phi\Delta z k} = \frac{0.01}{(0.045 - 0.005)(0.5236)(1)(5)} = 9.549 \times 10^{-2}$$

$$R_{n+} = \frac{r_0 \Delta\phi}{(\Delta r/2)(\Delta z)(k)} = \frac{(0.045)(0.5236)}{(0.005)(1)(5)} = 0.9425$$

$$R_{n-} = R_{n+} = 0.9425$$

$$\dot{Q}_s = q_s r_s \Delta\phi \Delta z = (5000\cos\phi)(0.045)(0.5236)(1) = 117.8\cos\phi$$

The same values of C_m, R_{01}, R_{n+}, and R_{n-} apply to the surface nodes for $90° < \phi < 270°$, but $\dot{Q}_s = 0$.

The stability criterion for the explicit method is given by Eq. (3.116):

$$\Delta t \leq \left(\frac{C_m}{\sum_n (1/R_{mn})} \right)_{\min}$$

The minimum value is obtained for $r_m = 0.005$ m, for which

$$\frac{C_m}{\sum_n (1/R_{mn})} = \frac{(1.257 \times 10^4)(0.005)}{\dfrac{(0.005 + 0.005)}{(3.820 \times 10^{-3})} + \dfrac{(0.005 - 0.005)}{(3.820 \times 10^{-3})} + \dfrac{1}{(10.47)(0.005)} + \dfrac{1}{(10.47)(0.005)}}$$

$$= 1.54 \text{ s}$$

Hence, choose $\Delta t = 1.0$ s.

Sample results are shown below.

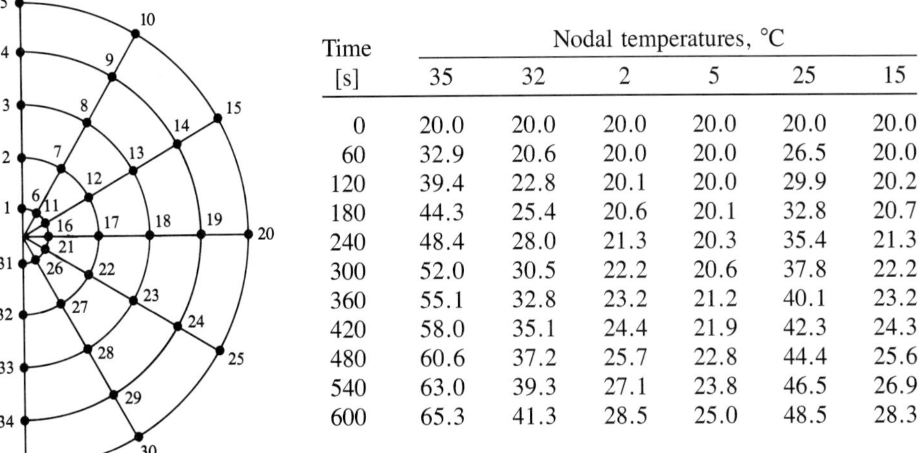

Time [s]	Nodal temperatures, °C 35	32	2	5	25	15
0	20.0	20.0	20.0	20.0	20.0	20.0
60	32.9	20.6	20.0	20.0	26.5	20.0
120	39.4	22.8	20.1	20.0	29.9	20.2
180	44.3	25.4	20.6	20.1	32.8	20.7
240	48.4	28.0	21.3	20.3	35.4	21.3
300	52.0	30.5	22.2	20.6	37.8	22.2
360	55.1	32.8	23.2	21.2	40.1	23.2
420	58.0	35.1	24.4	21.9	42.3	24.3
480	60.6	37.2	25.7	22.8	44.4	25.6
540	63.0	39.3	27.1	23.8	46.5	26.9
600	65.3	41.3	28.5	25.0	48.5	28.3

3.6.4 A Finite-Difference Method for Moving-Boundary Problems

Finite-difference methods have proven invaluable for analyzing moving-boundary problems. One important application has been the calculation of the thermal response of ablative heat shields on reentry vehicles. It is doubtful whether Project Apollo could have been successful without the use of such methods for the design of the heat shield on the command module. As an example, we will consider a simple one-dimensional transient ablation problem. The coordinate system is fixed to the surface so that the solid can be imagined to flow at a negative velocity V through the plane at $x = 0$, as shown in Fig. 3.44. A finite control volume $\Delta x \cdot 1 \cdot 1$ surrounds node m at location x. An energy balance on the control volume over time interval Δt requires that

$$\begin{matrix}\text{Increase in internal} \\ \text{energy within volume}\end{matrix} = \begin{matrix}\text{Net conduction} \\ \text{into volume}\end{matrix} + \begin{matrix}\text{Net inflow} \\ \text{of enthalpy}\end{matrix} \qquad \textbf{(3.119)}$$

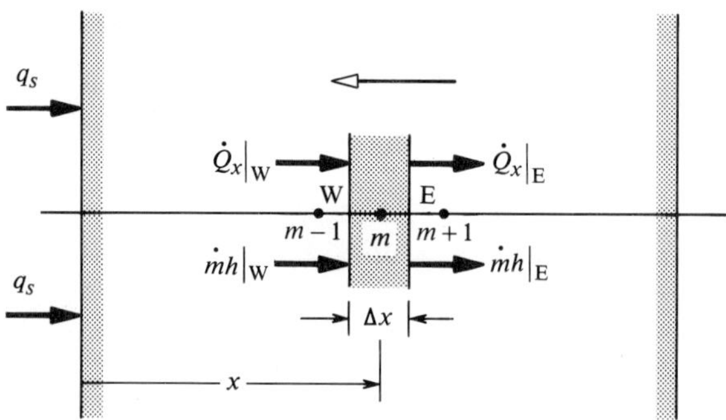

Figure 3.44 Finite control volume Δx by 1 by 1 surrounding node m at location x used to derive the difference equation for one-dimensional transient ablation.

Notice that Eq. (3.119) is a special form of the *unsteady-flow* energy equation for an open system. The increase in internal energy from time step i to step $(i + 1)$ is

$$\Delta U = \rho c(\Delta x \cdot 1 \cdot 1)(T_m^{i+1} - T_m^i)$$

To obtain an implicit formulation, the conduction heat flows are evaluated at time step $(i + 1)$:

$$\dot{Q}_x|_W \Delta t = -k \frac{T_m^{i+1} - T_{m-1}^{i+1}}{\Delta x}(1 \cdot 1)\Delta t$$

$$\dot{Q}_x|_E \Delta t = -k \frac{T_{m+1}^{i+1} - T_m^{i+1}}{\Delta x}(1 \cdot 1)\Delta t$$

The enthalpy flow poses a special problem. Perhaps the obvious approach is to write

$$\dot{m}h|_W \Delta t = \rho c V \frac{T_{m-1}^{i+1} + T_m^{i+1}}{2}(1 \cdot 1)\Delta t$$

$$\dot{m}h|_E \Delta t = \rho c V \frac{T_m^{i+1} + T_{m+1}^{i+1}}{2}(1 \cdot 1)\Delta t$$

where the enthalpy flows across the planes at $x - \Delta x/2$ and $x + \Delta x/2$ have been evaluated in terms of the average temperature of adjacent nodes. This approach leads to what is called a *central-difference* scheme. On physical grounds, however, the enthalpy flow across a plane can be influenced by the temperature on the upwind side only; the temperature on the downwind side should not have an influence. Thus, the *upwind-difference* scheme requires that the enthalpy flows be written as

$$\dot{m}h|_W \Delta t = \rho c V T_m^{i+1} (1 \cdot 1)\Delta t$$

$$\dot{m}h|_E \Delta t = \rho c V T_{m+1}^{i+1} (1 \cdot 1)\Delta t$$

for velocity V negative, as is the case for our problem. Accuracy and stability considerations dictate which scheme is most appropriate. In fact, it is a common practice to use a hybrid scheme in which central differencing is used when the velocity V is small, and upwind differencing is used when V is large. For simplicity, we will use the central-difference scheme only. Substituting in Eq. (3.119) and rearranging gives

$$T_m^{i+1} = \frac{\mathrm{Fo}(T_{m+1}^{i+1} + T_{m-1}^{i+1}) - (1/2)\,\mathrm{Fo}\,\mathrm{Pe}(T_{m+1}^{i+1} - T_{m-1}^{i+1}) + T_m^i}{1 + 2\,\mathrm{Fo}} \tag{3.120}$$

where Pe is the mesh *Peclet number*, which is defined as $\mathrm{Pe} = V\Delta x/\alpha$. It is, in fact, the mesh Peclet number that determines whether central or upwind differencing should be used: upwind differencing should be used when $|\mathrm{Pe}| > 2$.

Boundary Conditions

Formulation of the finite-difference forms of the boundary conditions is similar to the procedure used for the implicit method described in Section 3.6.2. For the front face, shown in Fig. 3.45*a*, an energy balance requires that

$$\rho c\left(\frac{\Delta x}{2}\cdot 1\cdot 1\right)(T_0^{i+1} - T_0^i) = \left(\dot{Q}_x|_0 - \dot{Q}_x|_E\right)\Delta t + (\dot{m}h|_0 - \dot{m}h|_E)\,\Delta t$$

$$\dot{Q}_x|_0 = q_s(1\cdot 1); \qquad \dot{Q}_x|_E = -k\frac{T_1^{i+1} - T_0^{i+1}}{\Delta x}(1\cdot 1)$$

$$\dot{m}h|_0 = \rho c V T_0^{i+1}(1\cdot 1); \qquad \dot{m}h|_E = \rho c V\frac{T_1^{i+1} + T_0^{i+1}}{2}(1\cdot 1)$$

Substituting and rearranging,

$$T_0^{i+1} = \frac{2\,\mathrm{Fo}\left(T_1^{i+1} + q_s\Delta x/k\right) - \mathrm{Fo}\,\mathrm{Pe}\,T_1^{i+1} + T_0^i}{1 + 2\,\mathrm{Fo} - \mathrm{Fo}\,\mathrm{Pe}} \tag{3.121}$$

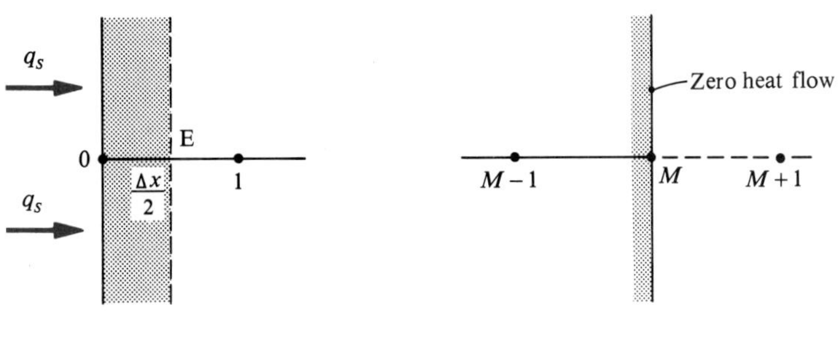

Figure 3.45 Nodal network and finite control volume for formulation of finite-difference forms of boundary conditions for one-dimensional transient ablation. (*a*) Front-face node. (*b*) Node M.

Referring to Fig. 3.45*b* we place node $m = M$ at $x = L$, for L chosen sufficiently large to ensure zero heat flow. An appropriate finite difference form of this boundary condition can be derived in a number of ways. For example, the temperature at the *mirror-image* node $(M + 1)$ can be set equal to the temperature at node $(M - 1)$ to give a zero temperature gradient at $x = L$,

$$T_{M+1}^{i+1} = T_{M-1}^{i+1}$$

Substituting in Eq. (3.120) for $m = M$,

$$T_M^{i+1} = \frac{2\,\text{Fo}\,T_{M-1}^{i+1} + T_M^i}{1 + 2\,\text{Fo}} \tag{3.122}$$

Notice that this result is also given as item 4 of Table 3.8. Example 3.18 illustrates the implementation of this finite-difference solution procedure.

EXAMPLE 3.18 Dust Erosion of a Plastic Heat Shield

A very thick plastic heat shield, initially at 0°C, is exposed simultaneously to a heat flux of 160 kW/m^2 from a high-temperature radiation source and to a dust blast that erodes the surface at a rate of 0.1 mm/s. Determine the temperature response of the shield. For the plastic, take property values of $\rho = 1200$ kg/m^3, $k = 0.3$ W/m K, and $\alpha = 0.015 \times 10^{-6}$ m^2/s.

Solution

Given: Transient conduction with a specified surface ablation rate.

Required: Temperature profiles $T(x, t)$

Assumptions: **1.** One-dimensional conduction.
2. Constant properties.
3. The heat flow does not penetrate deeper than 2 mm.

Choose $\Delta x = 0.1$ mm, $L = 2$ mm, and $\Delta t = 0.1$ s;
then the mesh size–based Fourier number is

$$\text{Fo} = \frac{\alpha \Delta t}{\Delta x^2} = \frac{(0.015 \times 10^{-6})(0.1)}{(1 \times 10^{-4})^2} = 0.15$$

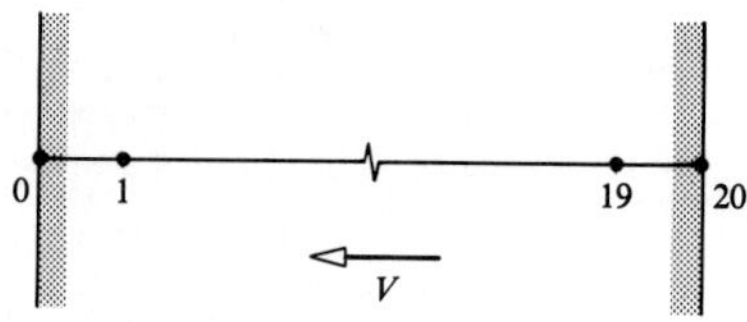

and the mesh size–based Peclet number is

$$\text{Pe} = \frac{V \Delta x}{\alpha} = \frac{(-10^{-4})(10^{-4})}{0.015 \times 10^{-6}} = -0.667$$

Equation (3.120) for the interior nodes $m = 1, 2, \ldots, 19$ is

$$T_m^{i+1} = \frac{0.15(T_{m+1}^{i+1} + T_{m-1}^{i+1}) - (1/2)(0.15)(-0.667)(T_{m+1}^{i+1} - T_{m-1}^{i+1}) + T_m^i}{1 + (2)(0.15)}$$

$$= 0.1539 T_{m+1}^{i+1} + 0.07690 T_{m-1}^{i+1} + 0.7692 T_m^i$$

Equation (3.121) for the front-face node, $m = 0$, is

$$T_0^{i+1} = \frac{(2)(0.15)[T_1^{i+1} + (160 \times 10^3)(10^{-4})/(0.3)] - (0.15)(-0.667)T_1^{i+1} + T_0^i}{1 + (2)(0.15) - (0.15)(-0.667)}$$

$$= 11.43 + 0.2857T_1^{i+1} + 0.7143T_0^i$$

Equation (3.122) for node $m = 20$, is

$$T_{20}^{i+1} = \frac{(2)(0.15)T_{19}^{i+1} + T_{20}^i}{1 + (2)(0.15)}$$

$$= 0.2308T_{19}^{i+1} + 0.7692T_{20}^i$$

As in Example 3.16, the Gauss-Seidel method can be used to solve the nodal equations. Sample temperature profiles are shown in the following table. Also given is the quasi-steady solution obtained from Eqs. (3.98) and (3.99). First, from Eq. (3.99),

$$q_u = -\rho c V(T_s - T_0)$$

$$T_s = T_0 - \frac{q_u}{\rho c V} = T_0 - \frac{\alpha q_u}{kV} = 0 - \frac{(0.015 \times 10^{-6})(160 \times 10^3)}{(0.3)(-10^{-4})} = 80°\text{C}$$

and from Eq. (3.98),

$$T = T_0 + (T_s - T_0)e^{(V/\alpha)\xi} = 0 + (80 - 0)e^{-10^{-4}\xi/0.015\times 10^{-6}} = 80e^{-6667\xi}\ °\text{C}$$

Time	Nodal temperatures, °C										
s	0	2	4	6	8	10	12	14	16	18	20
0	0.0	→									
1	49.9	3.7	0.2	0.0	→						
2	61.9	8.4	0.8	0.1	0.0	→					
3	68.0	11.8	1.6	0.2	0.0	→					
4	71.6	14.0	2.3	0.3	0.0	→					
5	74.0	15.6	2.9	0.5	0.1	0.0	→				
6	75.6	16.7	3.3	0.6	0.1	0.0	→				
8	77.5	18.1	4.0	0.8	0.1	0.0	→				
10	78.5	18.9	4.4	1.0	0.2	0.0	→				
15	79.6	19.7	4.8	1.1	0.3	0.1	0.0	→			
20	79.9	19.9	4.9	1.2	0.3	0.1	0.0	→			
25	80.0	20.0	5.0	1.2	0.3	0.1	0.0	→			
∞	80.0	21.1	5.6	1.5	0.4	0.1	0.0	→			

Comments

The mesh size Peclet number is 0.667, so use of the central-difference scheme for the convective term was appropriate.

3.7 CLOSURE

The temperature distribution in a solid is governed by the general heat conduction equation. This partial differential equation can be solved using classical mathematical methods or using numerical methods. In either case, considerable effort is required to obtain the solution for a particular problem.

The use of the classical separation-of-variables method was demonstrated for both steady multidimensional conduction and unsteady one-dimensional conduction. The method of superposition of solutions was used to build up the solution of a problem with complicated boundary conditions from solutions for simple boundary conditions. A useful product rule allows the temperature response of a number of shapes of finite dimensions to be obtained as a product of the responses of simpler shapes with infinite dimensions. For example, the response for a finite-length cylinder is obtained from the response of an infinite cylinder and an infinite slab. The conduction shape factor is convenient for calculating two-dimensional heat conduction between two isothermal surfaces. Solutions for conduction into a semi-infinite solid are always applicable for times short enough for the penetration of the thermal response to be small compared to the body dimensions. The computer program COND1 is a useful tool for such calculations. Determining the temperature response of convectively cooled (or heated) slabs, cylinders, and spheres is made simple by the computer program COND2 and by the availability of results in graphical form as temperature response charts.

When the shape of a solid is irregular, or when boundary conditions are complex, solutions to the heat conduction equation are best obtained numerically. Standard computer programs for this purpose are widely available, and such programs should be used for any serious thermal design activity. Thus, numerical methods were not presented in great detail. Only the finite-difference method was considered, and then only with the objective of conveying the essential ideas involved. Actually, the heat conduction equation is one of the easiest equations to solve numerically, and even the simplest methods yield satisfactory results. At this level, the ideas involved are almost intuitive. However, any serious effort to develop versatile and efficient computer programs to solve the heat conduction equation should be preceded by an appropriate course in numerical analysis, so that questions concerning stability, rate of convergence, and accuracy are properly handled.

Many heat conduction problems involve a moving boundary. Both analytical and numerical solution methods were demonstrated for two such problems. Numerical methods are particularly relevant because some important moving-boundary problems, such as the ablation of a heat shield on a reentry vehicle, can involve strongly varying thermal properties, complicated surface boundary conditions, and chemical reactions. We chose to use coordinate axes fixed to the moving surface for the ablation problem in Sections 3.5.2 and 3.6.4. This approach has the effect of giving a *convection* term in the governing differential equation, Eq. (3.96) or (3.119), even though a solid rather than a fluid is involved. Similar terms will arise in the analysis of convection in Chapter 5.

REFERENCES

1. Haberman, R., *Elementary Applied Partial Differential Equations*, 2nd ed., Prentice-Hall, Englewood Cliffs, N.J. (1987).

2. Boyce, W. E., and DiPrima, R. C., *Elementary Differential Equations and Boundary Value Problems*, 4th ed., John Wiley & Sons, New York (1986).

3. Kreyszig, E., *Advanced Engineering Mathematics*, 4th ed., John Wiley & Sons, New York (1979).

4. Fourier, J., *The Physical Theory of Heat*, Dover Publications, New York (1955). (Originally published in 1822.)

5. Carslaw, H. S., and Jaeger, J. C., *Conduction of Heat in Solids*, 2nd ed., Clarendon Press, Oxford (1959).

6. Emmons, H. W., "The numerical solution of heat conduction problems," *Trans. ASME*, 65, 607–615 (1943).

7. Kayan, C. F., "An electrical geometrical analogue for complex heat flow," *Trans. ASME*, 67, 713–718 (1945).

8. Karplus, W. J., and Soroka, W. W., *Analog Methods: Computation and Simulation*, 2nd ed., McGraw-Hill, New York (1959).

9. Myers, G. E., *Analytical Methods in Heat Conduction*, Genium Publishing Corporation, Schenectady, N.Y. (1987).

10. Gurney, H. P., and Lurie, J., "Charts for estimating temperature distributions in heating and cooling solid shapes," *Ind. Eng. Chem.*, 15, 1170–1172 (1923).

11. Rohsenow, W. M., Hartnett, J. P., and Ganic, E. N., eds., *Handbook of Heat Transfer Fundamentals*, 2nd ed., McGraw-Hill, New York (1985).

12. Schneider, P. J., *Temperature Response Charts*, John Wiley & Sons, New York (1963).

13. Heisler, M. P., "Temperature charts for induction and constant temperature heating," *Trans. ASME*, 69, 227–236 (1947).

14. Langston, L. S., "Heat transfer from multidimensional objects using one-dimensional solutions for heat loss," *Int. J. Heat Mass Transfer*, 25, 149–150 (1982).

15. London, A. L., and Seban, R. A., "Rate of ice formation," *Trans. ASME*, 65, 711–778 (1943).

16. SINDA (Systems Improved Numerical Differencing Analyzer), Lockheed Engineering and Management Services Company, Inc.

17. MITAS (Martin Marietta Interactive Thermal Analysis System), Martin Marietta Corporation.

18. PHOENICS (Parabolic, Hyperbolic or Elliptic Numerical Code Series), CHAM Ltd., 40 High Street, Wimbledon, London SW 19A4, England.

19. COMPACT, Innovative Research, Inc., 7846 Ithaca Lane, N., Maple Grove, Minn.

EXERCISES

3–1. Derive the general heat conduction equation in cylindrical coordinates by applying the first law to the volume element shown in Fig. 3.3*a*.

3–2. Derive the general heat conduction equation in spherical coordinates by applying the first law to the volume element shown in Fig. 3.3*b*.

3–3. Use the methods of vector calculus to derive the general heat conduction equation. (*Hint:* Apply the first law to a volume V with surface S, and use the Gauss divergence theorem to convert the surface integral of heat flow across S to a volume integral over V.)

3–4. The cylindrical and spherical coordinate systems are examples of *orthogonal curvilinear coordinates*. In general, we can denote these coordinates by u_1, u_2, u_3, which are defined by specifying the Cartesian coordinates x, y, z as

$$x = x(u_1, u_2, u_3)$$

$$y = y(u_1, u_2, u_3)$$

$$z = z(u_1, u_2, u_3)$$

A coordinate system is orthogonal when the three families of surfaces u_1 = Const, u_2 = Const, u_3 = Const are orthogonal to one another. The figure shows an elemental parallepiped whose faces coincide with planes u_1 or u_2 or u_3 = Const, with edge lengths $h_1 du_1$, $h_2 du_2$, $h_3 du_3$ where h_1, h_2, h_3 are called the *metric coefficients*. The length of a diagonal is given by

$$ds^2 = h_1^2 du_1^2 + h_2^2 du_2^2 + h_3^2 du_3^2$$

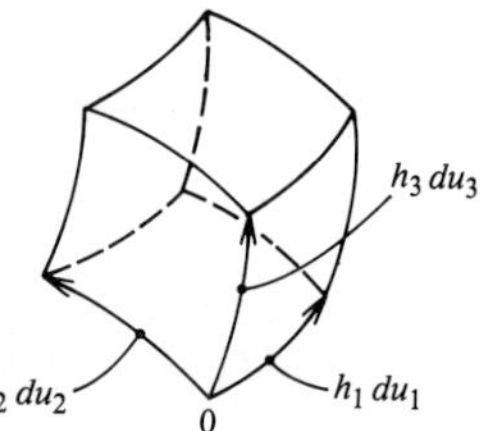

In terms of these coordinates, the components of the temperature gradient are

$$\frac{1}{h_1}\frac{\partial T}{\partial u_1}, \qquad \frac{1}{h_2}\frac{\partial T}{\partial u_2}, \qquad \frac{1}{h_3}\frac{\partial T}{\partial u_3}$$

The divergence of the heat flux is

$$\nabla \cdot \mathbf{q} = \frac{1}{h_1 h_2 h_3}\left[\frac{\partial}{\partial u_1}(h_2 h_3 q_1) + \frac{\partial}{\partial u_2}(h_3 h_1 q_2) + \frac{\partial}{\partial u_3}(h_1 h_2 q_3)\right]$$

and $\nabla^2 T$ is

$$\nabla^2 T = \frac{1}{h_1 h_2 h_3}\left[\frac{\partial}{\partial u_1}\left(\frac{h_2 h_3}{h_1}\frac{\partial T}{\partial u_1}\right) + \frac{\partial}{\partial u_2}\left(\frac{h_3 h_1}{h_2}\frac{\partial T}{\partial u_2}\right) + \frac{\partial}{\partial u_3}\left(\frac{h_1 h_2}{h_3}\frac{\partial T}{\partial u_3}\right)\right]$$

(i) Identify the metric coefficients for the cylindrical coordinate system, and hence write down $\nabla^2 T$ in cylindrical coordinates.
(ii) Repeat for the spherical coordinate system.

3–5. One face of a block of ice is observed to recede at a rate of 0.22 mm/min. If the ice has been melting for some time, calculate the temperature gradient in the water adjacent to the ice surface. The density and enthalpy of fusion of ice at 0°C are 910 kg/m^3 and 0.335×10^6 J/kg, respectively.

3–6. A thin rectangular plate, $0 \le x \le a, 0 \le y \le b$, with negligible heat loss from its sides, has a linear temperature variation along the edge at $y = b$ given by $T = 20 + 100(x/a)$°C. The other three edges are maintained at 20°C. Determine the temperature distribution $T(x, y)$.

3–7. A thin rectangular plate $0 \le x \le a, 0 \le y \le b$ has the following temperature distribution around its boundary:

$$x = 0,\ 0 < y < b : T = 300 \text{ K}$$

$$x = a,\ 0 < y < b : T = 300 + 100 \sin(\pi y/b) \text{ K}$$

$$y = 0,\ 0 < x < a : T = 300 \text{ K}$$

$$y = b,\ 0 < x < a : T = 300 + 200 \sin(\pi x/a) \text{ K}$$

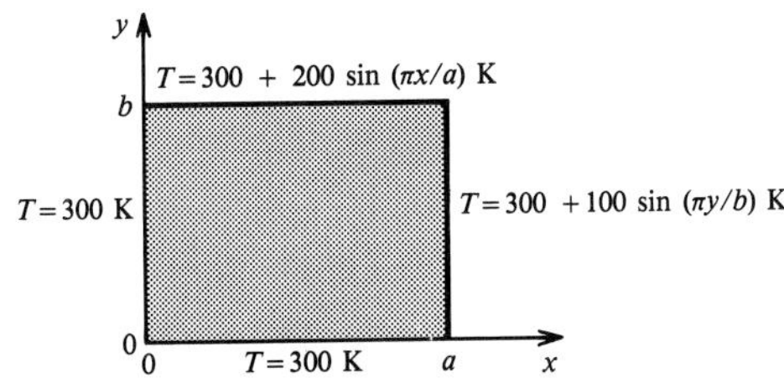

(i) Determine the steady-state temperature distribution if there is negligible heat loss from the sides.
(ii) If $a = b$, determine the center temperature.

3–8. A thin, 20 cm–square plate, with negligible heat loss from its sides, has three edges maintained at 20°C, and a fourth edge has a temperature distribution given by $T = 20(1 + \sin \pi x/L)$°C, where $L = 20$ cm, and x [cm] is measured from one corner.

(i) Determine the temperature at the midpoint of the plate.
(ii) If the plate is 3 mm–thick stainless steel with $k = 16$ W/m K, determine the rate of heat supply required to maintain a steady state.

3–9. A thin rectangular plate, $0 \le x \le a, 0 \le y \le b$, has the following boundary conditions:

$$x = 0,\ 0 < y < b : T = 300 \text{ K}$$

$$x = a,\ 0 < y < b : T = 300 \text{ K}$$

$$y = 0 :\ 0 < x < a : q_y = 0 \text{ (insulated)}$$

$$y = b :\ 0 < x < a : T = 300 + 300 \sin(\pi x/a)$$

(i) Determine the temperature distribution in the plate if it has negligible heat loss from its surface.
(ii) If $a = b$, find the center temperature.

3–10. A thin rectangular plate, $0 \le x \le a, 0 \le y \le b$, has the following temperature distribution around its boundary:

$$x = 0,\ 0 < y < b : T = 300 \text{ K}$$

$$x = a,\ 0 < y < b : T = 400 \text{ K}$$

$$y = 0,\ 0 < x < a : T = 320 \text{ K}$$

$$y = b,\ 0 < x < a : T = 380 \text{ K}$$

Determine the temperature distribution in the plate if it has negligible heat loss from its surface.

3–11. A thin rectangular plate, $0 \le x \le a, 0 \le y \le b$, with negligible heat loss from its sides, has the following boundary conditions:

$$x = 0,\ 0 < y < b : T = 300 \text{ K}$$

$$x = a,\ 0 < y < b : T = 400 \text{ K}$$

$$y = 0,\ 0 < x < a : q_y = 0 \text{ (insulated)}$$

$$y = b,\ 0 < x < a : T = 500 \text{ K}$$

(i) Determine the steady-state temperature distribution.
(ii) If $a = b$, determine the center temperature.

3–12. An 8 × 8 cm square plate with negligible heat loss from its sides has boundary conditions as indicated. Obtain an analytical solution for the temperature distribution, and evaluate the center temperature.

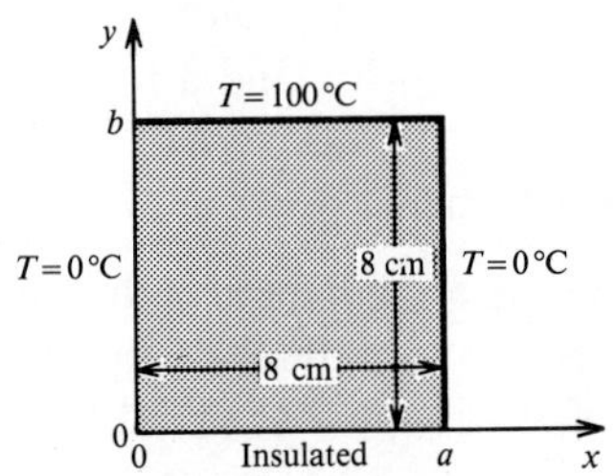

3–13. (i) Show that the heat dissipated by a straight rectangular fin of thickness $2t$, allowing for two-dimensional heat conduction, with boundary conditions $T = T_B$ at $x = 0$ and $\partial T/\partial x = 0$ at $x = L$, is

$$\dot{Q}_{2D} = 8kW(T_B - T_e) \sum_{n \text{ odd}} \frac{\tanh n\pi t/2L}{n\pi\left(\dfrac{n\pi}{2}\dfrac{t}{\text{Bi}\,L}\tanh\dfrac{n\pi t}{2L} + 1\right)}; \quad \text{Bi} = \frac{h_c t}{k}$$

(ii) Write a computer program to evaluate $\dot{Q}_{2D}/\dot{Q}_{1D}$, with $\dot{Q}_{1D}$ given by Eq. (2.40). Explore the error incurred by using the one-dimensional fin model as a function of Bi and t/L.

3–14. In a natural convection test rig, a glass partition 1 cm thick and 4 cm high separates two copper plates. The lower plate is maintained at 340 K, while the upper plate is maintained at 300 K. The air temperature on each side of the partition is 320 K, and the heat transfer coefficient is 6.0 W/m^2 K.

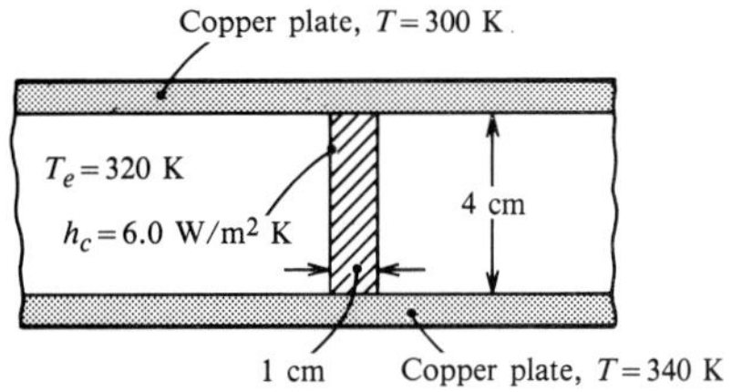

Determine the heat flow at the bottom and the top of the partition

(i) assuming one-dimensional conduction.
(ii) assuming two-dimensional conduction.

Take $k = 0.78$ W/m K for the glass. You will need the roots of Eq. (3.67), which are given in the following table for an appropriate range of Biot number.

Bi	λ_1	λ_2	λ_3	λ_4
0.01	0.100	3.145	6.285	9.426
0.02	0.141	3.148	6.286	9.427
0.05	0.222	3.157	6.291	9.430
0.1	0.311	3.173	6.299	9.435
0.2	0.433	3.204	6.315	9.446
0.5	0.653	3.292	6.362	9.477
1.0	0.861	3.426	6.437	9.529

3–15. Radioactive wastes dissipating 500 W are temporarily stored in a 2 m–diameter spherical container, the center of which is buried 5 m below the ground in a location where the soil is relatively dry. If the ground-level temperature does not exceed 30°C at any time, estimate the maximum temperature the container might attain. Take k for the soil as 0.6 W/m K.

3–16. Exhaust gas from a furnace flows at 4.50 kg/s up a concrete chimney 26 m high, 90 cm square on the inside, with a wall thickness of 30 cm. The gas inlet temperature is 500 K, and the ambient air temperature is 300 K. Determine the gas outlet condition for wind conditions that give an outside heat transfer coefficient of approximately 12 W/m^2 K. Take the inside heat transfer coefficient as 15 W/m^2 K, the specific heat of the gas mixture as 1100 J/kg K, and the concrete conductivity as 1.13 W/m K.

3–17. A small kiln for firing ceramic products has inside dimensions 1 m wide, 1 m high, and 1.5 m deep. The walls are 50 cm thick and made of zirconia brick. At steady operating conditions the inside temperature is 700°C, and the ambient air temperature is 25°C. If the outside heat transfer coefficient is approximately 5 W/m^2 K, estimate the heat loss from the kiln. Take $k = 2.4$ W/m K for zirconia brick.

3–18. The heat flow between two concentric cylinders of radii r_1 and r at temperatures T_1 and T, respectively, is, from Eq. (2.14),

$$\dot{Q} = \frac{2\pi k L(T_1 - T)}{\ln(r/r_1)} \quad \text{or} \quad T = T_1 - \frac{\dot{Q}}{2\pi k L} \ln \frac{r}{r_1}$$

which can be interpreted as the temperature due to a line source of strength $\dot{Q}$ at $r = 0$, expressed in terms of the temperature T_1 at a reference radius r_1. Derive the shape factor S given by item 8 of Table 3.2 by superimposing the temperature fields of a line source $\dot{Q}$ below the ground and a line sink $-\dot{Q}$ above the ground, in a mirror image position as shown. Define an excess temperature $T - T_2$, where T_2 is the ground surface temperature and show that the isotherms are concentric circles with origins at $x = 0$, $y = a(1+c)/(1-c)$, and radii $2c^{1/2}a/(1-c)$, where c is a constant parameter. Also show that the ground surface is an isotherm.

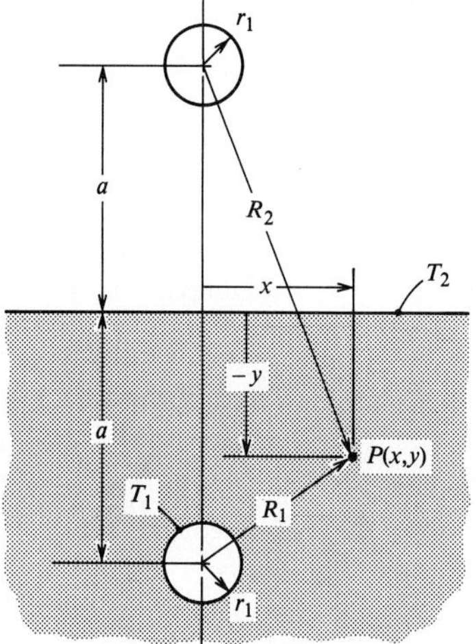

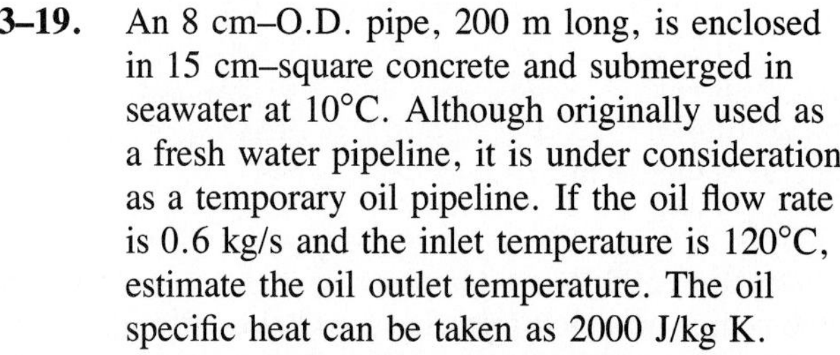

3–19. An 8 cm–O.D. pipe, 200 m long, is enclosed in 15 cm–square concrete and submerged in seawater at 10°C. Although originally used as a fresh water pipeline, it is under consideration as a temporary oil pipeline. If the oil flow rate is 0.6 kg/s and the inlet temperature is 120°C, estimate the oil outlet temperature. The oil specific heat can be taken as 2000 J/kg K.

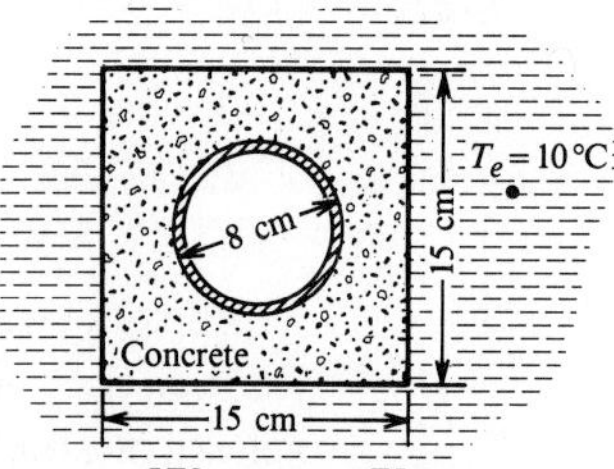

3–20. A 15 cm–O.D. pipe is buried with its centerline 1 m below the surface of the ground. An oil of specific gravity 0.8 and specific heat 1950 J/kg K flows in the pipe at 0.5 m^3/min. If the ground surface temperature is 25°C and the pipe wall temperature is 95°C, estimate the oil temperature drop, in kelvins per meter. Take $k = 1$ W/m K for the soil.

3–21. A buried insulated power cable has an outside diameter of 3 cm and is 1 m below the surface of the ground. What is the maximum allowable dissipation per unit length if the outer surface of the insulation must not exceed 350 K when the ground surface and the deep soil are at 300 K? Take $k = 1$ W/m K for the soil.

3–22. A 25 cm–diameter oil line is buried with its centerline 60 cm below the ground. Assuming a soil temperature of 10°C and a soil thermal conductivity of 0.8 W/m K, estimate the steady-state heat loss from the pipe (in W/m) when the oil is at 90°C.

(i) Ignore the inside thermal resistance.
(ii) Recalculate your result if the inside heat transfer coefficient is 300 W/m^2 K.

3–23. A 20 cm–O.D. steam pipe is buried 1.5 m below the ground surface in dry soil. Steam flows through the pipe at 1.2 kg/s. Assuming a ground surface temperature of 15°C and saturated steam at 1.1×10^5 Pa, estimate the steam condensation rate per 100 m of pipeline.

3–24. A small laboratory oven is cubical in shape with an inside edge 20 cm long. It is insulated with 6 cm of medium-density fiberglass. What power supply is required to maintain an interior temperature of 440 K when the ambient temperature is 20°C and the outside heat transfer coefficient is 7 W/m^2 K?

3–25. A 20 cm–O.D. pipe is buried 1 m below the surface of the ground. Hot water flows in the pipe at 500 gal/min. Assuming a pipe wall temperature of 70°C, estimate the length of pipe in which the water temperature decreases by 1°C when the ground surface is at 10°C. Take the soil thermal conductivity as 1 W/m K.

3–26. A slab $2L$ thick, initially at a uniform temperature T_0, has its surfaces suddenly lowered to temperature T_s. Simultaneously a volumetric heat source $\dot{Q}_v'''$ is activated within the slab. Show how you would analyze this problem to determine the temperature response. (*Hint:* Try a superposition of a steady and a transient problem.)

3–27. Show why the constant for Eq. (3.38) cannot be positive or zero.

3–28. A thick slab of pyrex glass, initially at 350 K, has the temperature on one surface suddenly dropped to 300 K. Calculate the time needed for the location 1 cm below the surface to decrease in temperature by 5 K.

3–29. On Thanksgiving Day in St. Louis a sudden storm reduces the ambient air to −15°C. If the ground was at a uniform temperature of 20°C before the storm, what is the surface temperature after 4 hours, and to what depth below the surface will the freezing temperature have penetrated? Take $k = 1.2$ W/m K, $\alpha = 0.40 \times 10^{-6}$ m^2/s, $h_c = 20$ W/m^2 K. Ignore any phase change effects.

3–30. Derive Eq. (3.63) by assuming a solution that has the functional form $T = C \exp(-px) \sin(\omega t - qx) + D$, and determine C, p, and q by substituting back into the governing equation.

3–31. To determine whether a new composite material can withstand thermal cycling without degrading, a thick slab is subjected to heating, which causes the surface temperature to vary in an approximate sinusoidal manner with an amplitude of 200°C and a period 50 s. If the thermal conductivity and diffusivity of the composite are 10 W/m K and 2.0×10^{-6} m²/s, respectively, determine the amplitude and phase lag of the temperature variation 1 cm below the surface.

3–32. Water pipes are to be buried in a geographical area that has a mean winter temperature of about 5°C but is subject to sudden drops in air temperature to about −10°C for a maximum duration of 48 hours. Estimate the minimum depth of the pipes needed to prevent freezing. Take $\alpha = 0.5 \times 10^{-6}$ m²/s for the soil.

3–33. A steel firewall is suddenly exposed to a radiant heat source that can be approximated as a blackbody at 1000 K. How long will it take for the surface to reach 500 K if the initial temperature of the wall is 300 K?

3–34. The *integral method* can be used to solve the problem of conduction in a semi-infinite solid subject to a step change in surface temperature. If it is assumed that temperature changes penetrate only to $x = \delta$, the first law applied to a control volume of unit cross-sectional area located between $x = 0$ and $x = \delta$ gives

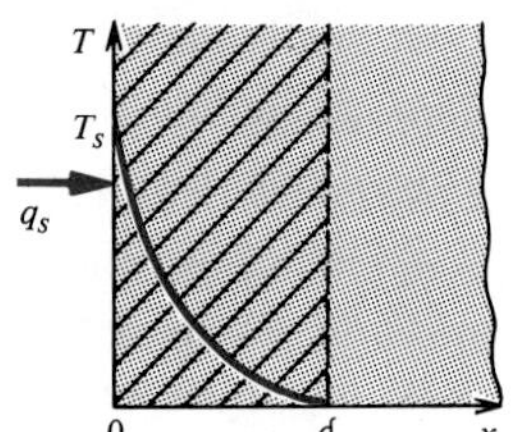

$$\dot{Q} = \frac{dU}{dt} \quad \text{or} \quad q_s - 0 = \frac{d}{dt}\int_0^\delta \rho u\,dx = \frac{d}{dt}\int_0^\delta \rho c(T - T_0)dx \qquad \textbf{(1)}$$

where the initial temperature T_0 is chosen as the datum state for internal energy.

(i) Assume that the temperature profile is parabolic, $T = A + Bx + Cx^2$ with constants obtained from boundary conditions, and hence show that

$$\frac{T - T_0}{T_s - T_0} = \left(1 - \frac{x}{\delta}\right)^2; \qquad q_s = \frac{2k}{\delta}(T_s - T_0) \qquad \textbf{(2a, b)}$$

(ii) By substituting Eqs. (2) in Eq. (1), show that the thermal penetration δ satisfies the differential equation

$$\frac{d\delta}{dt} = \frac{6\alpha}{\delta} \qquad \textbf{(3)}$$

(iii) Finally, show that the heat flow into the solid is given by

$$q_s = \frac{k(T_s - T_0)}{(3\alpha t)^{1/2}}$$

and compare this result with the exact solution, Eq. (3.59).

3–35. At a ski resort in the Sierra Nevada mountains, the winter ground-surface temperature typically varies between −10 and 12°C each day, while the in-depth soil temperature is +1°C. How far below the surface should a water pipe be buried to prevent freezing?

3–36. A thermocouple is installed in the cylinder wall of a two-stroke internal combustion engine 1.0 mm below the inner surface. In a particular test the engine operates at 2500 rpm, and the thermocouple reading is found to have a mean value of 290.0°C and an amplitude of 1.08°C. If the temperature variation is assumed to be sinusoidal, estimate the amplitude of the cylinder wall surface temperature variation and the phase difference. Take $\alpha = 12.0 \times 10^{-6}$ m²/s for the carbon steel wall.

3–37. A thin, flat electrical resistance strip heater is sandwiched between a firebrick wall and a thick AISI 1010 steel plate, both initially at 300 K. The heater is switched on and generates heat at a rate of 50,000 W/m². Plot the temperature of the steel surface adjacent to the heater as a function of time for 10 seconds.

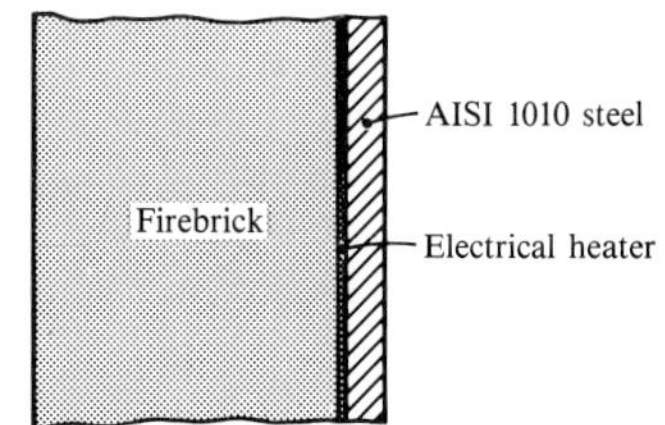

3–38. A long rod is well insulated along its sides. A heater varies the temperature of one end face sinusoidally from 100 to 200°C with a period of 90.9 s. Two thermocouples, 10 and 70 cm from the heated end, record temperatures. The phase difference between the peak temperatures is found to be 15.0 min. If the density of the rod material is 8300 kg/m³ and its specific heat 470 J/kg K, what is its thermal conductivity?

3–39. Estimate the depth below the ground at which the annual temperature variation will be 10% of that at the surface for

(i) dry soil, $k = 1.0$ W/m K.
(ii) wet soil, $k = 2.0$ W/m K.

Repeat for the diurnal temperature variation.

3–40. A turbine component in the preburner section of a turbopump for a rocket motor is made from Inconel X-750. A "thermal barrier" coating of 0.3 mm–thick YSZ (yittria stabilized zirconia) is plasma-sprayed on the surface. This coating has a high reflectance and can significantly reduce the radiation heat loads on cooled surfaces. Upon start-up, the surfaces are initially at 30°C and are suddenly exposed to hot gases at 800°C, with an estimated effective heat transfer coefficient of 180 W/m² K. Plot the surface and interface temperature as a function of time. Take $k = 5.0$ W/m K for YSZ.

3–41. Associated with the annual variation of the weather, the average ground surface temperature at a given location will have a maximum in summer and a minimum in winter. If you dig a hole exactly six months after the maximum occurs, how deep will you have to dig to reach the peak of the resulting temperature response, and what will be the percent decay of the amplitude? Take $\alpha = 1.2 \times 10^{-6}$ m²/s for the soil diffusivity.

3–42. The temperature of the outer 2 mm of skin on the forearm can be taken to be 32°C when the ambient air is at 25°C. If your arm suddenly contacts a slab of aluminum at 100°C, what is the skin surface temperature until blood flow responds to the change in conditions? Repeat for slabs of 18-8 stainless steel, pyrex glass, and Teflon. For skin tissue take $k = 0.37$ W/m K, $\alpha = 0.1 \times 10^{-6}$ m²/s.

3–43. During vulcanization a tire carcass is heated by exposure to saturated steam at 150°C on both sides. If the carcass is 2 cm thick and has an initial temperature of 20°C, how long will it take for the centerplane to reach 130°C? Take $\alpha = 0.06 \times 10^{-6}$ m²/s for the rubber.

3–44. An AISI 1010 carbon steel slab 5 cm thick is insulated on one side and is initially at 700 K. To cool the slab, it is exposed to jets of water at 300 K which give a heat transfer coefficient of 2000 W/m² K. What is the temperature of the back face after

(i) 5 minutes?
(ii) 1 hour?

Repeat for a chrome brick slab.

3–45. A sphere of 1 cm diameter at 320 K is suddenly immersed in an air stream at 280 K. If the average heat transfer coefficient is 100 W/m² K, determine the time required for the sphere to lose 90% of its energy content. Consider four materials:

(i) cork
(ii) Teflon
(iii) AISI 302 stainless steel
(iv) pure aluminum

3–46. Derive Eq. (3.69); that is, show that the constants A_n in the solution for the temperature response of a slab with finite surface resistance are given by

$$A_n = \frac{2 \sin \lambda_n}{\lambda_n + \sin \lambda_n \cos \lambda_n}$$

where λ_n are the roots of $\cot \lambda = \lambda / \text{Bi}$.

3–47. The nozzle of an experimental rocket motor is fabricated from 5 mm–thick alloy steel. The combustion gases are at 2100 K, and the effective heat transfer coefficient is 6000 W/m² K. If the nozzle is initially at 300 K and the maximum allowable operating temperature for the steel is specified as 1300 K, what is the allowable duration of firing? To obtain a conservative estimate, neglect the heat loss from the outer surface of the nozzle. Take $k = 43$ W/m K, $\alpha = 12.0 \times 10^{-6}$ m²/s for the steel.

3–48. If the pebble bed air heater of Example 3.10 were packed with 6 cm iron spheres, what would be the result?

3–49. A 2 cm–thick ceramic slab has a thin metal sheath for protection. The contact resistance between the ceramic and metal gives an interfacial conductance of 1600 W/m² K. The slab is initially at 300 K and is suddenly exposed to a hot air stream at 1500 K with a convective heat transfer coefficient of 800 W/m^2 K. How long will it take for the center of the slab to reach 1200 K? The ceramic properties are

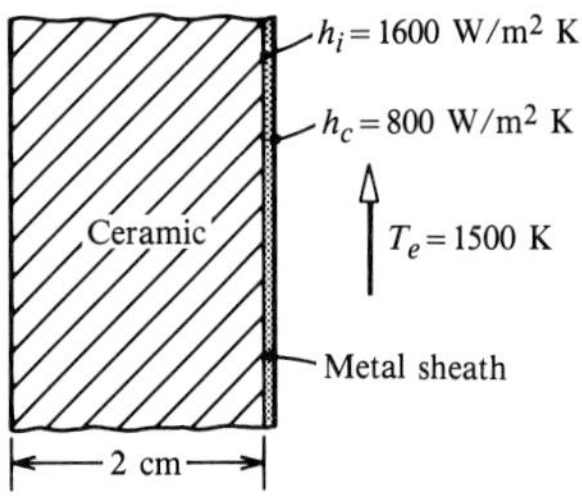

$$\rho = 2600\,\text{kg/m}^3$$
$$c = 1150\,\text{J/kg K}$$
$$k = 3.0\,\text{W/m K}$$

3–50. A long bar of AISI 316 stainless steel has an 8 cm × 8 cm square cross section. It is hot-rolled at 620°C and then cooled by cold air jets at 30°C, giving a heat transfer coefficient of 400 W/m^2 K. Determine the time required for the center temperature to decrease to 130°C.

3–51. Find the time required to heat the center of a 5 × 10 × 20 cm clay brick from 550°C to 1500°C in a convective oven at 1700°C. Take the heat transfer coefficient to be 100 W/m^2 K, and $k = 1.7$ W/m K, $\alpha = 0.35 \times 10^{-6}$ m^2/s for the brick.

3–52. A 2 cm–thick slab of cloth-filled thermoplastic resin, initially at 20°C, is placed in a press with walls maintained at 100°C by condensing steam. As soon as the center temperature reaches 85°C, the slab is removed and allowed to cool in 20°C air until the center temperature falls to 40°C. How long does each stage of the process take? Use $k = 0.6$ W/m K, $\rho = 1700$ kg/m^3, and $c = 1200$ J/kg K. The heat transfer coefficient for the cooling process can be taken as 7 W/m^2 K.

3–53. A can of beer at 300 K is placed in a refrigerator that maintains an air temperature of 277 K. The can is 8 cm in diameter and 12 cm high. The outside heat transfer coefficient is 5 W/m^2 K. After 6 hours the beer is removed from the refrigerator and poured into a glass. Estimate the beer temperature.

3–54. A physicist working for your company has proposed the following simple procedure for measuring the thermal conductivity of soil. An electrically heated sphere 6 cm in diameter is to be buried 30 cm below the surface of a very large box of the soil. The sphere will be maintained at 30°C while the laboratory air-conditioner maintains the surface of the soil at 20°C. The power input to the heater required to maintain a steady state will be measured and Table 3.2, item 7 used to calculate k. Evaluate the practicality of this procedure.

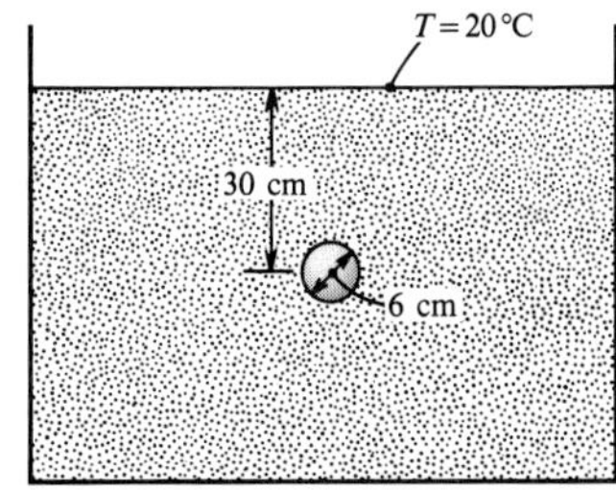

3–55. A 4 kg beef roast, roughly spherical in shape with a diameter of 20 cm, is removed from a 5°C refrigerator and placed in a 150°C oven. It is desired to raise the center temperature to 70°C. How long should the roast be cooked? Assume that the meat has the thermal properties of water and that the heat transfer coefficient is 12 W/m^2 K. If the roast lost 0.4 kg of its mass by evaporation, make a rough estimate of the enthalpy of vaporization absorbed and compare this to the sensible enthalpy change of the roast.

3–56. A blank for a telescope mirror is a 25 cm–diameter, 5 cm–thick disk of glass. The blank is at room temperature, 20°C, and is placed in an oven at 420°C for stress relieving. If the heat transfer coefficient is 12 W/m^2 K, how long will it be before the minimum temperature in the glass is 400°C? Take k = 1.09 W/m K, $\alpha = 0.51 \times 10^{-6}$ m^2/s for the glass.

3–57. A 15 cm–diameter, 30 cm–long 18-8 stainless steel billet at 20°C is placed in an oil bath at 300°C. The heat transfer coefficient may be taken to be 400 W/m^2 K. Determine the center temperature after 500 s have elapsed

(i) using the lumped thermal capacity model.
(ii) assuming the billet is long compared to its diameter.
(iii) accounting for the finite length of the cylinder.

3–58. Plywood is to be cured between plates maintained at 105°C by condensing steam. How long will it take to cure a sheet 1 cm thick if the minimum temperature must reach 95°C and its initial temperature is 25°C?

3–59. Rework Exercise 3–51 for h_c = 200 W/m^2 K.

3–60. An egg, which may be modeled as a 4 cm–diameter sphere with the thermal properties of water, is initially at 5°C and is immersed in boiling water. Determine the temperature at the center of the egg after

(i) 4 minutes.
(ii) 7 minutes.

Take the outside heat transfer coefficient as 1200 W/m^2 K.

3–61. A large rectangular safe has a 10 cm–thick asbestos insulation. In a fire the outside of the insulation is estimated to be 800°C. How long must the fire last to destroy papers that char at 150°C contained in the safe? At the outbreak of the fire the safe was at 20°C. Take $\alpha = 0.4 \times 10^{-6}$ m^2/s for the asbestos.

3–62. A cylindrical AISI 302 stainless steel pin 2 cm in diameter and 10 cm high initially at 20°C is suddenly exposed to saturated steam at 1 atm pressure. A thermocouple is located on the centerline of the cylinder 1 cm from the top of the pin. What temperature will the thermocouple record after

(i) 10 s?
(ii) 100 s?

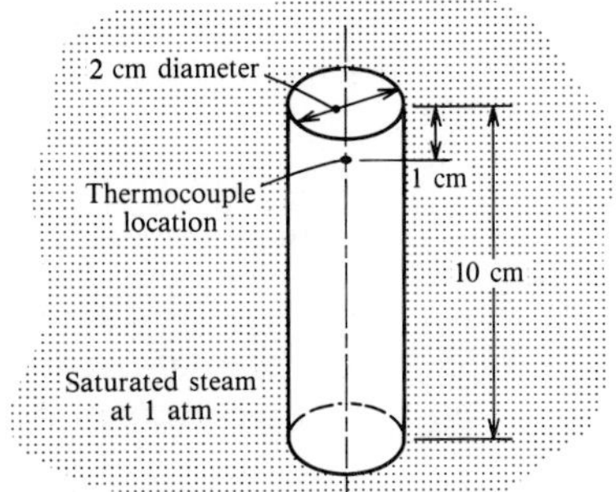

3–63. Repeat Exercise 3–62 for a 2 cm–square pin.

3–64. A 20 cm–long, 25 cm–O.D. fused silica glass cylinder initially at 30°C is placed in a furnace at 800°C. If the heat transfer coefficient is estimated to be 20 W/m^2 K, prepare a graph of the center temperature as a function of time.

3–65. An *interior heat transfer coefficient h* can be defined for heat conduction out of a convectively cooled slab of thickness $2L$ as

$$h = \frac{-k(\partial T/\partial x)|_{x=L}}{\overline{T} - T_s}$$

with a corresponding dimensionless **Nusselt number** $\text{Nu} = h(2L)/k$. Show that for $\text{Fo} > 0.2$, Nu has a constant value

$$\text{Nu} = \frac{2\lambda_1^2 \sin^2 \lambda_1}{\sin^2 \lambda_1 - \sin \lambda_1 \cos \lambda_1}$$

which for $\text{Bi} \to \infty$ is $\pi^2/2 = 4.934$.

3–66. Repeat Exercise 3–65 for a cylinder.

3–67. Repeat Exercise 3–65 for a sphere.

3–68. A regenerative heat exchanger (see Section 8.6) has a matrix in the form of parallel plates of plastic 6 mm thick, at a spacing of b mm. If the convective heat transfer coefficient for the fluid flow is approximately $4k/b$, where k is the gas conductivity, estimate the overall heat transfer coefficient for heat transfer from the fluid into the plates. Tabulate your results for $b = 1, 2, \ldots, 10$ mm. Take $k = 0.15$ W/m K for the plastic and 0.10 W/m K for the fluid.

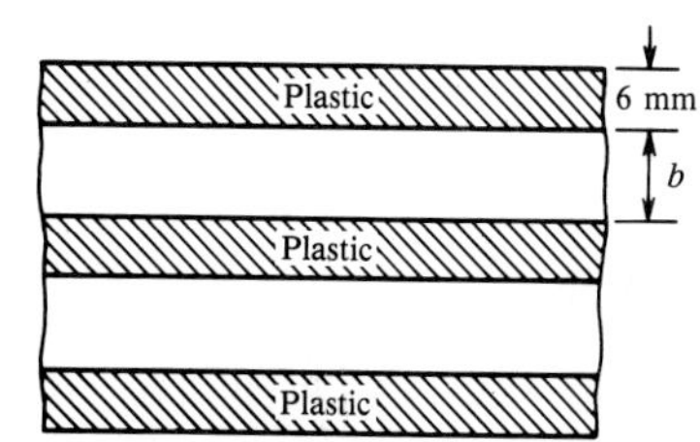

3–69. Transient heat conduction in citrus fruits is of concern to farmers who must develop strategies to prevent freezing during cold weather. A relatively simple procedure for determining the thermal diffusivity of a citrus fruit involves installing a thermocouple at the center of the fruit and determining the temperature response when the fruit is placed in a rapidly stirred water bath maintained at 0°C. The following table gives some typical data for a 6.8 cm–diameter grapefruit. Estimate the thermal diffusivity and compare your value to that of water at 10°C.

t, min	0	5	10	15	20	25	30	35	40	45
T_c,°C	20.0	20.0	19.6	18.0	15.8	13.0	10.8	8.4	7.2	5.8

3–70. (i) A slab of thickness L is initially at a uniform temperature T_0. The face at $x = 0$ is perfectly insulated. At time $t = 0$, a constant heat flux q_s is imposed on the face at $x = L$. Show that the temperature response is

$$\frac{T - T_0}{q_s L/k} = \frac{3x^2 - L^2}{6L^2} + \frac{\alpha t}{L^2} - \frac{2}{\pi^2} \sum_{n=1}^{\infty} \frac{(-1)^n}{n^2} e^{-(\alpha n^2 \pi^2/L^2)t} \cos \frac{n\pi x}{L}$$

(ii) If a negative heat flux equal to $-q_s$ is subsequently imposed at time $t = t_1$, show that the temperature response for $t > t_1$ is

$$\frac{T - T_0}{q_s L/k} = \frac{\alpha t_1}{L^2} - \frac{2}{\pi^2} \sum_{n=1}^{\infty} \frac{(-1)^n}{n^2} e^{-(\alpha n^2 \pi^2/L^2)t} \left[1 - e^{(\alpha n^2 \pi^2/L^2)t_1}\right] \cos \frac{n\pi x}{L}$$

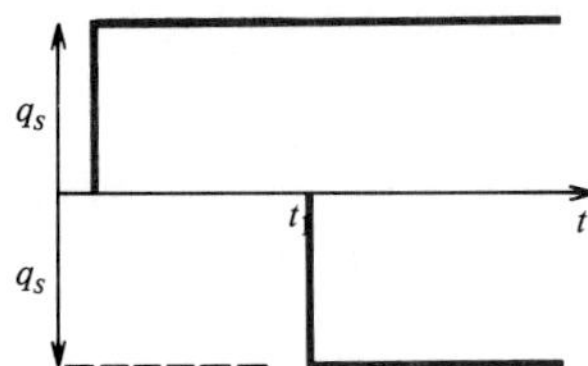

(*Hint:* Use the principle of superposition for a linear differential equation.)

3–71. (i) A slab of thickness L is initially at a uniform temperature T_0. The back face at $x = 0$ is perfectly insulated. At time $t = 0$, a laser deposits an amount of energy E per unit area on the face at $x = L$. Determine the temperature response. (*Hint:* Let $\theta(x, t)$ be the temperature response for a unit uniform surface heat flux obtained by setting $q_s = 1$ in the result of Exercise 3–70, part (i). Then, following part (ii) of Exercise 3–70, show that the required temperature response is

$$T - T_0 = E \frac{\partial \theta(x, t)}{\partial t}$$

by imposing a heat flux q_s for time Δt and then letting $\Delta t \to 0$ such that $q_s \Delta t = E$.)

(ii) Hence show that the time for the back-face temperature rise to equal half of its maximum rise is $t_{1/2} = 1.39L^2/\pi^2\alpha$.

3–72. The analysis of Exercise 3–71 forms the basis of a technique for measuring the thermal diffusivity of thin samples. In an experiment on a 1 mm–thick laminate, a CO_2 gas laser is used to deposit approximately 0.2×10^{-4} J/m^2 of energy over a period of 200 nanoseconds. The back-face temperature is measured using a type K thermocouple, and the μV-time trace is as shown. Estimate the thermal diffusivity of the sample.

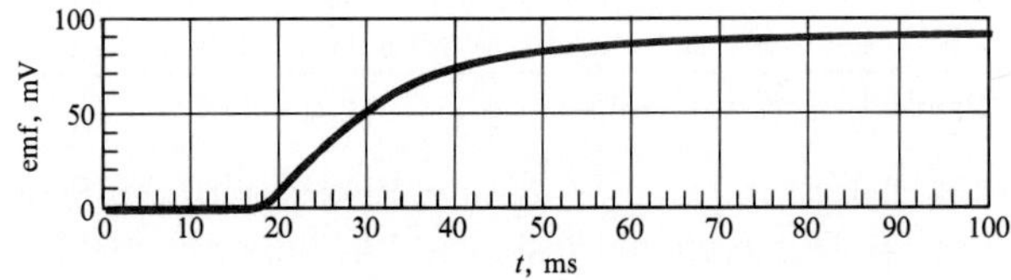

3–73. Derive Eq. (3.96) from the one-dimensional transient heat conduction equation, Eq. (3.33), by an appropriate change of independent variables.

3–74. A wire of perimeter $\mathscr{P}$ and cross-sectional area A_c emerges from a die at a temperature ΔT above ambient with a velocity V. Specify the steady-state temperature distribution along the wire if the exposed length of wire is long. Make assumptions that allow an analytical solution to be obtained, and discuss their validity.

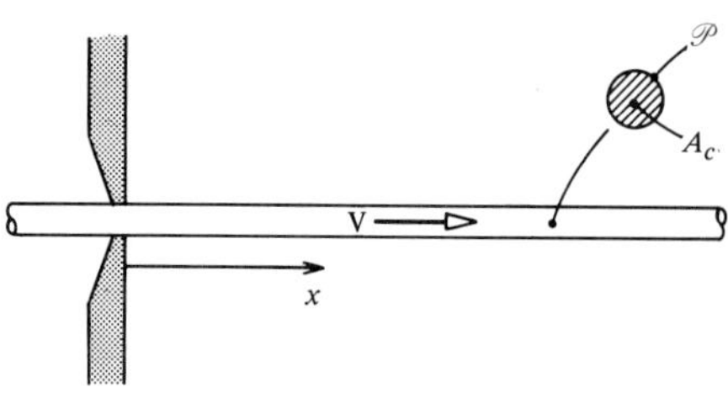

3–75. A common situation in metallurgical practice is solidification from a melt in a cylindrical mold. If the Jakob number, Ja, is small, we can assume quasi-steady conduction in the cylindrical shell of solidified melt, as well as in the mold wall. Show that the time required to solidify the complete melt is given by

$$\frac{\alpha t}{r_2^2} = \frac{1}{4\,\mathrm{Ja}}\left[1 + \frac{2}{\mathrm{Bi}}\right]$$

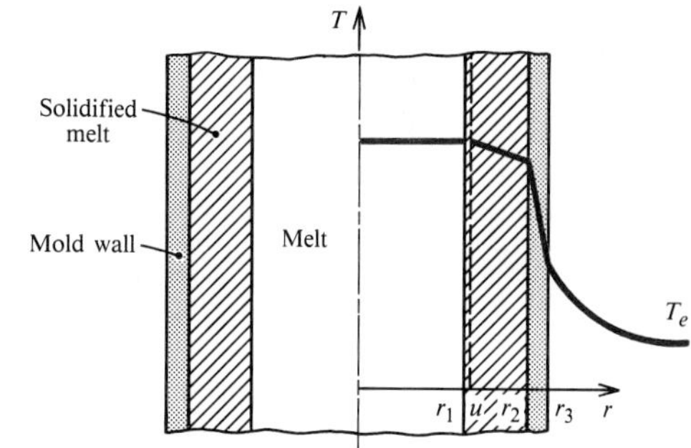

Here the Biot number is a pseudo value that accounts for both the outside resistance and the mold wall,

$$\mathrm{Bi} = \frac{U_o r_2}{k}$$

$$\frac{1}{2\pi r_2 L U_o} = \frac{\ln(r_3/r_2)}{2\pi k_w L} + \frac{1}{2\pi r_3 L h_o}$$

where k and k_w are the conductivities of the solidified melt and mold wall, respectively.

3–76. A steel camshaft blank 6 cm in diameter and 90 cm long is to be cast in a shell mold of resin-bonded sand 1.5 cm thick. The molten steel is poured at 1850 K, and the ambient temperature is 300 K. Estimate the time required for the melt to solidify. The fusion temperature of the steel is 1805 K, and the enthalpy of fusion is 2.72×10^5 J/kg. For the solidified steel, take $k = 38.0$ W/m K, $\alpha = 4.9 \times 10^{-6}$ m^2/s, and c $= 775$ J/kg K. For the shell mold, take $k = 0.8$ W/m K, an interfacial conductance $h_i = 1200$ W/m^2 K, and an outside convective plus radiative heat transfer coefficient of $h_o = 30$ W/m K. For the melt take $c_l = 880$ J/kg K.

3–77. Repeat the analysis of Exercise 3–75 for the melt in the form of a slab of width $2L$ and a sphere of radius r_2. Show that the general result is

$$\mathrm{Fo} = \frac{1}{2\,\mathrm{Ja}(n+1)}\left(1 + \frac{2}{\mathrm{Bi}}\right)$$

where $n = 0, 1, 2$ for the slab, cylinder, and sphere, respectively. The characteristic length in the Fourier and Biot numbers is L for the slab and r_2 for the cylinder and sphere.

3–78. A problem that is mathematically similar to the steady-state melting ablation problem analyzed in Section 3.5.2 is steady-state transpiration cooling. Transpiration cooling is a process in which a fluid is injected through a porous wall whose outer surface is subjected to a severe thermal environment. As the relatively cold fluid passes through the wall, its temperature increases as it absorbs thermal energy: if the flow rate is sufficient, the outer surface of the wall can be maintained at an acceptable temperature. Consider the situation where the outer surface of a porous wall is subjected to a steady heat load of q_2 W/m^2 from a high-temperature radiation source. Fluid is injected through the wall at a rate $\dot{m}''$ kg/m^2 s from a reservoir (plenum chamber) where the temperature is T_o. Cooling coils are brazed to the inlet face of the wall, and an additional cooling q_1 W/m^2 can be supplied by passing refrigerant through the coils.

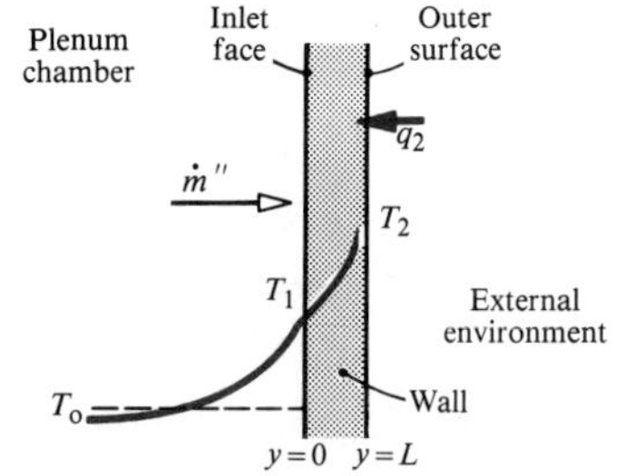

(i) Through application of the steady flow energy equation, show that the temperature distributions $T(y)$ in the reservoir and $T_m(y)$ in the wall are governed by

$$\frac{d^2T}{dy^2} - \frac{\dot{m}''c_p}{k}\frac{dT}{dy} = 0 \qquad -\infty < y < 0$$

$$\frac{d^2T_m}{dy^2} - \frac{\dot{m}''c_p}{k_{\text{eff}}}\frac{dT_m}{dy} = 0 \qquad 0 < y < L$$

where k is the fluid conductivity and k_m is an effective conductivity of the fluid-solid mixture in the wall. Assume that the fluid is in thermal equilibrium with the solid matrix at each location y in the wall.

(ii) Solve the governing equations subject to appropriate boundary conditions to show that the temperature distributions in the reservoir and wall are

$$T = T_o + (q_2/\dot{m}''c_p)e^{\dot{m}''c_p(y/k - L/k_m)} - (q_1/\dot{m}''c_p)e^{\dot{m}''c_p y/k}; \qquad y \le 0$$

$$T_m = T_o - (q_1/\dot{m}''c_p) + (q_2/\dot{m}''c_p)e^{\dot{m}''c_p(y/k_m - L/k_m)}; \qquad 0 \le y \le L$$

3–79. The feasibility of a thermal protection system for a fireman based on transpiration cooling is to be investigated. Such a system would be useful in fighting severe fires—for example, a fire that might occur in an oil refinery. Suppose that air at 20°C can be continuously delivered to a 4 mm–thick asbestos cloth suit through a flexible hose, which may itself be protected by transpiration. Assume that no other cooling is provided and that the inner surface of the suit should be maintained at 30°C. The heat load can be taken as blackbody radiation from flames at 700°C. Determine the required air velocity at the inner surface of the suit and discuss the feasibility of the design. Take $k = 0.2$ W/m K for the asbestos cloth.

3–80. A Teflon slab initially at 300 K is exposed simultaneously to a heat flux of 100 kW/m^2 from a high-temperature radiation source, and to a dust blast that erodes the surface at a rate of 80 μm/s. Determine the steady-state surface temperature of the slab and the penetration of the thermal response into the slab.

3–81. Write a computer program to solve Example 3.14.

3–82. Write a computer program to solve the problem of Example 3.14, but allow for a $N \times N$ mesh. Explore the effect of N on the accuracy and rate of convergence of the solution. Take $a/b = 2$.

3–83. Modify the computer program of Exercise 3–82 to solve Exercise 3–10 numerically.

3–84. An 8×8 cm square plate has one edge maintained at 100°C, whereas the other three edges are exposed to a fluid at 0°C with a heat transfer coefficient of 10 W/m^2 K. Modify the computer program of Exercise 3–82 to obtain the temperature distribution in the plate.

3–85. Write a computer program to calculate the shape factor given as item 5 of Table 3.2. Take $a/b = 2.0$. Check the formulas given in the table.

3–86. Derive the finite-difference approximation formulas for steady conduction given as items 3, 4, and 5 of Table 3.7.

3–87. For steady conduction, derive the finite-difference approximation for an interior node near a curved nonisothermal surface. (See item 6 of Table 3.7.)

3–88. Write a computer program to solve for the temperature distribution and fin efficiency of a straight rectangular fin, allowing for two-dimensional conduction. Compare your numerical results with the analytical formula given in Exercise 3–13 for $t = 0.5$ cm, $L = 3$ cm, $k = 1.0$ W/m^2 K, $h_c = 4$ W/m^2 K.

3–89. Modify the finite-difference approximation formula for a convective boundary condition, given as Eq. (3.103), to include the effect of internal heat generation $\dot{Q}_v'''$.

3–90. The Crank-Nicolson method has been widely used for finite-difference numerical solution of unsteady heat conduction problems. Whereas the explicit method evaluates conduction fluxes at the old time step, and the implicit method uses the new time step, this method uses an average of the old and the new. Derive the nodal equations corresponding to Eqs. (3.107) and (3.112).

3–91. Write a computer program to solve Example 3.15. Allow for an arbitrary mesh size Δx and time step Δt. Investigate the effect of Δx and Δt on the accuracy of the solution.

3–92. Write a computer program to implement the flow diagram shown in Example 3.16. Rework the example for

(i) $h_c = 280(T - T_{sat})^2$ W/m^2 K.
(ii) $h_c = 600(T - T_{sat})^{1.8}$ W/m^2 K.

3–93. For one-dimensional unsteady conduction, derive the implicit form of the finite-difference formula for the convective boundary condition given as item 2 of Table 3.8.

3–94. For one-dimensional unsteady conduction, derive both the explicit and implicit forms of the finite-difference formula for a boundary where the surface heat flux is known (item 3 of Table 3.8). In the case of the explicit form, also derive the stability criterion.

3–95. For two-dimensional unsteady conduction, derive both the explicit and implicit forms of the finite-difference approximation for a convective boundary condition at a plane surface. For the explicit case, also derive the stability criterion. (See Table 3.9, item 2.)

3–96. For two-dimensional unsteady conduction, derive both the explicit and implicit forms of the finite-difference approximation for a known heat flux boundary condition at a plane surface. For the explicit case, also derive the stability criterion. (See Table 3.9, item 3.)

3–97. For two-dimensional unsteady conduction, derive both the explicit and implicit forms of the finite-difference approximation for a convective boundary condition at an exterior corner. For the explicit case, also derive the stability criterion. (See Table 3.9, item 4.)

3–98. For two-dimensional unsteady conduction, derive both the explicit and implicit forms of the finite-difference approximation for a convective boundary condition at an interior corner. For the explicit case, also derive the stability criterion. (See Table 3.9, item 5.)

3–99. Write a computer program to solve Example 3.17.

3–100. For the moving boundary problem of Section 3.6.4, use upwind differencing to derive finite-difference approximation formulas for interior nodes and the front-face node.

3–101. Write a computer program to solve Example 3.18 using the Gauss-Siedel method. Allow the problem parameters, and Δx and Δt, to be input values. Check the performance of the program against the results given in the text.

3–102. Rework Example 3.18 for erosion rates of

(i) 0.2 mm/s.
(ii) 0.3 mm/s.
(iii) 0.4 mm/s.

3–103. A large slab of polyvinylchloride 1 cm thick is in contact with fluid on either side. On one side the heat transfer coefficient is 20 W/m^2 K, and on the other it is 40 W/m^2 K. Initially both fluids are at 20°C, and the system is in thermal equilibrium. Suddenly the fluid temperatures are raised to 100°C. Determine the time required for the center of the plate to reach 80°C.

3–104. A Pyrex glass tube, inner radius 2 cm and outer radius 3 cm, is part of a laboratory rig to study high-pressure combustion phenomena. It is initially at 20°C and suddenly admits a flow of gases at 600°C, while the outer surface remains exposed to air at 20°C. The inner and outer heat transfer coefficients are estimated to be 400 and 20 W/m^2 K, respectively.

(i) Set up an *RC* formulation of the problem using cylindrical coordinates. Use an explicit form and consider carefully the stability criteria imposed by the inner and surface boundary conditions.
(ii) Estimate numerically the time required for the tube temperatures to become steady.

3–105. A large slab of 5 cm–thick Pyrex glass is to be cooled slowly from an initial uniform temperature of 400°C by an array of air jets impinging on each side of the plate. The ducting of the air is such that on one side the air is at 20°C, whereas on the other side it is at 70°C. The convective heat transfer coefficient on both sides can be taken as 30 W/m^2 K. Determine the temperature at the center of the slab after 3 hours have elapsed.

CHAPTER

4

CONVECTION FUNDAMENTALS AND CORRELATIONS

CONTENTS

4.1 INTRODUCTION

In the preceding chapters, we have seen that engineering calculations of heat conduction are based on analysis. A model is formulated, and the resulting differential or algebraic equations are solved. The nature of engineering calculations for heat convection is different. The most frequent task an engineer performs is the estimation of heat transfer coefficients from correlations of experimental data, because the differential equations governing convection can be analytically solved only for the simplest of flows. The engineer must rely on experimental data for most of the great variety of flows encountered in practice.

Chapter 4 has been organized accordingly. In Section 4.2, the fundamentals of convection are discussed rather briefly, and the focus is on those concepts required by the engineer to use heat transfer coefficient correlations effectively. The subsequent sections present correlations for a wide range of commonly encountered flows. Analyses of convection are deferred to Chapter 5, because although these analyses may give insight into the nature of convection, their mastery is not essential to the use of heat transfer coefficient correlations.

Calculation of heat transfer coefficients using a handheld calculator can be tedious. The correlations are often complicated functions, and fluid properties must be evaluated by interpolating in tables. Also, the risk of making a careless error is significant. Thus, an ideal tool is a computer program in which the large database is efficiently stored and that executes the calculations rapidly and reliably. Section 4.8 describes the computer program CONV, which was developed for this purpose. CONV should prove indispensable and will probably be used more frequently than any other program in the software package. Before CONV is used for a particular problem, the relevant section of the text in Chapter 4 should be consulted to ensure that an appropriate correlation is used.

4.2 FUNDAMENTALS

This section presents the fundamentals of convection required to understand and use correlations of the convective heat transfer coefficient. First, the convective heat transfer coefficient is defined for three types of flow. In each case, the physics of the flow is described in detail to give the student an appreciation of the nature of convection. Next, dimensional analysis based on the Buckingham pi theorem is used to derive the dimensionless groups pertinent to convection, such as the Reynolds, Grashof, and Nusselt numbers. Experimental data for convective heat transfer coefficients are best correlated in terms of the pertinent dimensionless groups. Heat transfer to flow over a cylinder is used to demonstrate the approach. Finally, we address the awkward problem of how to evaluate temperature-dependent fluid properties in dimensionless groups. Experience has shown that unless this problem is tackled head-on at the outset, it tends to remain a source of confusion for the student.

4.2.1 The Convective Heat Transfer Coefficient

Fluid motion past a surface increases the rate of heat transfer between the surface and the fluid. We are all aware of how a brisk wind increases our discomfort on a cold day. A flowing fluid transports thermal energy by virtue of its motion, and it does so very effectively. Resulting convective heat transfer coefficients can be very large. To define the heat transfer coefficient, we consider three types of flow: (1) an external forced flow, (2) an internal natural-convection flow, and (3) an internal forced flow.

Forced Flow over a Cylinder

Figure 4.1 shows isotherms around a heated cylinder mounted transversely in a wind tunnel. As the air velocity increases, the isotherms on the windward side move closer together. At a sufficiently high velocity, the heated fluid is confined to a very thin **thermal boundary layer**, which has approximately the same thickness as the hydrodynamic boundary layer. The flow on the leeward side of the cylinder is more complex, and when flow separation occurs, large-scale vortices are shed from the cylinder. Fourier's law is not only valid in a stationary solid, but it also applies in a moving fluid. Thus, as the isotherms move closer together and the

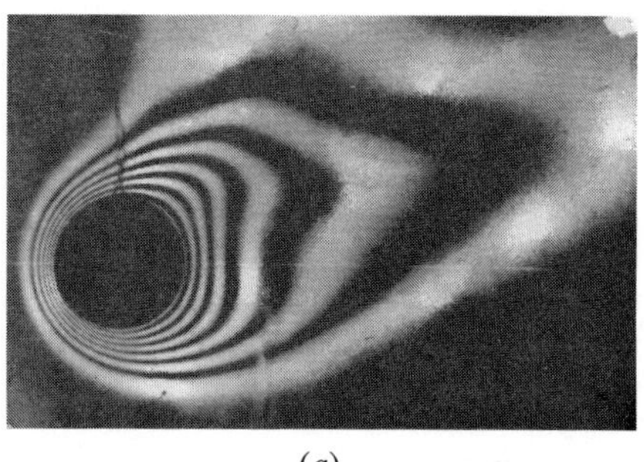

(*a*)

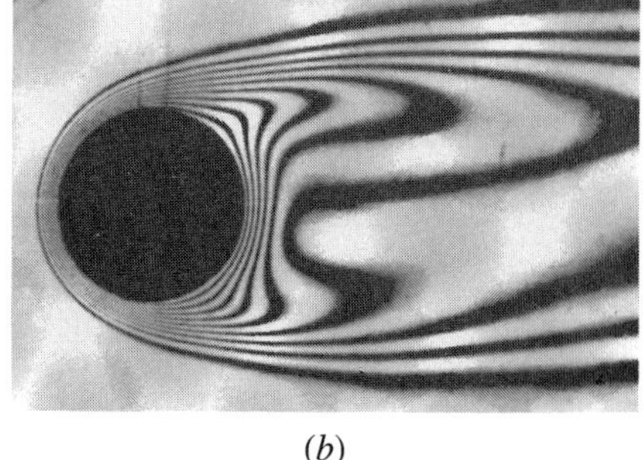

(*b*)

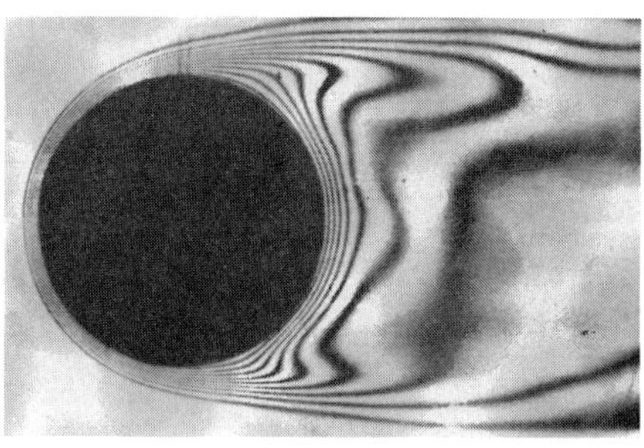

(*c*)

Figure 4.1 Flow over a heated cylinder: the effect of velocity on isotherms, (*a*) Re = 23, (*b*) Re = 120, (*c*) Re = 597. (Photograph by E. Soehngen, courtesy Professor J. P. Holman, Southern Methodist University, Dallas.)

temperature gradient normal to the cylinder surface increases, so does the resulting heat conduction. Indeed, the fluid in contact with the cylinder is stationary due to viscous action, and heat can be transferred from the cylinder surface into the fluid by conduction only. Thus, the local convective heat transfer rate is given by

$$q_s = -k \frac{\partial T}{\partial y}\bigg|_{y=0} \tag{4.1}$$

where y is the coordinate direction normal to the surface and k is the fluid thermal conductivity. On the other hand, Newton's law of cooling, Eq. (1.20), defined the convective heat transfer coefficient by the relation

$$q_s = h_c(T_s - T_e) \tag{4.2}$$

where T_s is the surface temperature and T_e is the fluid temperature in the free stream. Combining Eqs. (4.1) and (4.2),

$$h_c = \frac{-k(\partial T/\partial y)|_{y=0}}{T_s - T_e} \tag{4.3}$$

The effect of increasing the fluid velocity is to steepen the temperature gradient at the surface, thereby increasing the heat transfer coefficient.

Natural Convection between Two Horizontal Plates

Figure 4.2 shows an internal natural-convection flow. A layer of fluid of thickness L is confined between two isothermal plates and heated from below. For small values of the difference between the plate temperatures $(T_H - T_C)$, the fluid is stationary, and the heat transfer through the layer is by conduction only. However, if $(T_H - T_C)$ is increased to a critical value, the fluid becomes unstable, and a cellular flow pattern is set up. The circulation of fluid in a cell convects warm fluid upward and cold fluid downward, and, to accommodate the higher rate of heat transfer across the layer, the isotherms adjacent to each plate move closer together. If $(T_H - T_C)$ is further increased, there are transitions to increasingly more complex flows until cellular flow is replaced by a chaotic turbulent motion. The core of the fluid layer is then almost isothermal, with the major temperature variation confined to a very thin *viscous sublayer* adjacent to each plate, where viscosity serves to damp the turbulence motions. The engineer is usually concerned with heat transfer from one plate to the other, as, for example, in a covered flat-plate solar collector, rather than from one of the plates into the fluid. Thus, it is usual to define the average heat transfer coefficient for such configurations in terms of $(T_H - T_C)$, that is,

$$\overline{h}_c = \frac{\dot{Q}/A}{T_H - T_C} \tag{4.4}$$

rather than in terms of some fluid temperature. An average heat transfer coefficient

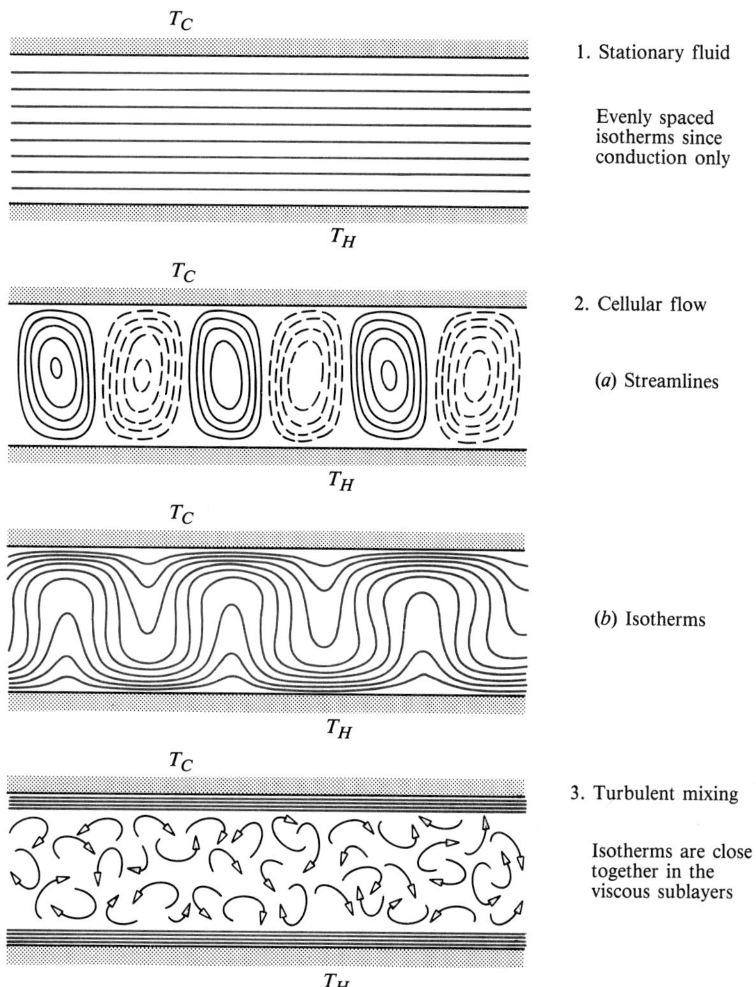

Figure 4.2 Natural-convection flow regimes for a layer of fluid between two horizontal isothermal plates. The temperature difference $(T_H - T_C)$ increases from regime (1) through regime (3). Cellular flow streamlines and isotherms courtesy Professor G. Mallinson, University of Auckland.

is appropriate since for cellular flow h_c is not constant over each plate. As before, the heat flow can be expressed in terms of conduction in the fluid adjacent to the plate surfaces at $y = 0$ and $y = L$:

$$\dot{Q} = \int_A q_s \, dA = \int_A -k \left.\frac{\partial T}{\partial y}\right|_{y=0} dA = \int_A -k \left.\frac{\partial T}{\partial y}\right|_{y=L} dA \tag{4.5}$$

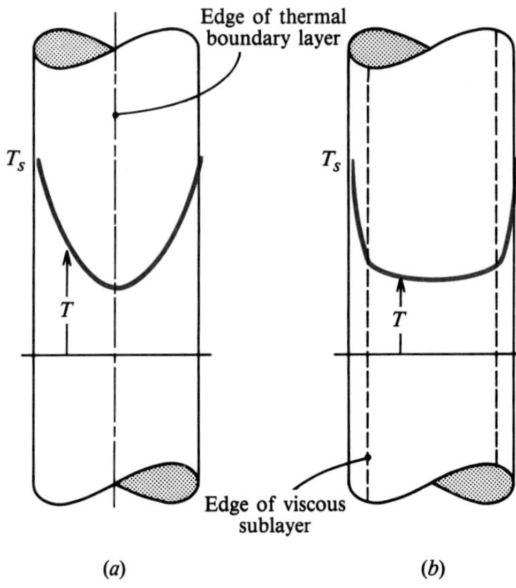

Figure 4.3 Temperature profiles for laminar and turbulent flow in a tube. (*a*) Laminar flow. (*b*) Turbulent flow.

Flow in a Tube

Figure 4.3*a* shows a temperature profile for laminar flow in a tube some distance from the entrance. In this case, the thermal boundary layer extends to the centerline of the tube, and its thickness is independent of velocity. As a result, the temperature gradient at the wall is also independent of velocity, as is the heat transfer coefficient (cf Eq. 1.21). Turbulent flow is shown in Fig. 4.3*b*. In this case, an increase in velocity produces a more vigorous turbulent mixing in the core of flow, and there is a resulting thinning of the viscous sublayer adjacent to the wall, with an increase in the heat transfer coefficient (cf Eq. 1.22). For an internal flow, such as flow in a tube or annulus, it is customary to define the heat transfer coefficient in terms of the *bulk* temperature, which is the temperature the fluid would attain at a given axial location if it were diverted into an adiabatic mixing chamber and thoroughly mixed, as shown in Fig. 4.4 for flow in a tube. The velocity profile $u(r)$ will be parabolic if the flow is laminar, or much flatter if the flow is turbulent. The mass flow rate $\dot{m}$ [kg/s] is obtained by integrating over the cross section,

$$\dot{m} = \int_0^R \rho u 2\pi r \, dr \tag{4.6}$$

and, by mass conservation, is the same entering and leaving the chamber. Application of the steady-flow energy equation, Eq. (1.4), to the chamber requires simply that the rate at which enthalpy, h, enters and leaves the chamber be equal:

$$\int_0^R \rho u h 2\pi r \, dr = \int_0^R \rho u h_b 2\pi r \, dr$$

where h_b is the bulk enthalpy. If we assume constant properties (ρ and c_p) and an

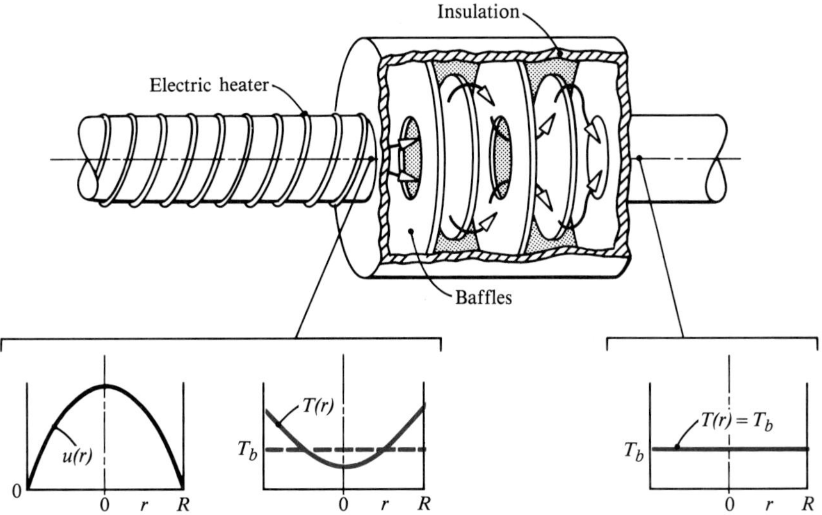

Figure 4.4 An adiabatic mixing chamber for measuring the bulk temperature.

enthalpy datum state of $h = 0$ at $T = 0$, we can write $h = c_pT$, and

$$\int_0^R \rho u c_p T 2\pi r\, dr = \int_0^R \rho u c_p T_b 2\pi r\, dr = c_p T_b \dot{m}$$

from Eq. (4.6), since T_b is not a function of r. Canceling c_p gives

$$T_b = \frac{\int_0^R \rho u T 2\pi r\, dr}{\dot{m}} \tag{4.7}$$

which allows the bulk temperature to be calculated if the velocity and temperature profiles across the flow are known. Now consider an element of tube Δx long, as shown in Fig. 4.5, and once again apply the steady-flow energy equation:

$$q_s 2\pi R \Delta x = \int_0^R \rho u c_p T 2\pi r\, dr\bigg|_{x+\Delta x} - \int_0^R \rho u c_p T 2\pi r\, dr\bigg|_x$$

where q_s is the wall heat flux. Using Eq. (4.7),

$$q_s 2\pi R \Delta x = \dot{m} c_p T_b|_{x+\Delta x} - \dot{m} c_p T_b|_x$$

Dividing by Δx and letting $\Delta x \rightarrow 0$,

$$q_s 2\pi R = \dot{m} c_p \frac{dT_b}{dx} \tag{4.8}$$

If we define the heat transfer coefficient in terms of bulk temperature, that is,

$$h_c = \frac{q_s}{T_s - T_b} \tag{4.9}$$

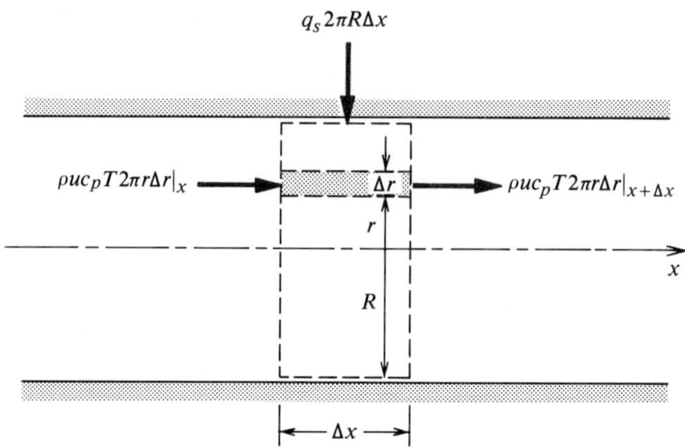

Figure 4.5 An energy balance on an element of tube Δx long.

then Eq. (4.8) becomes

$$h_c(T_s - T_b)2\pi R = \dot{m} c_p \frac{dT_b}{dx} \tag{4.10}$$

If the wall temperature is constant along the tube, there is a single dependent variable $T_b(x)$, and this equation can be integrated directly. It would be less convenient had h_c been defined in terms of, for example, the centerline temperature.

For a tube of length L and T_s constant, Eq. (4.10) can be integrated with $T_b = T_{b,\text{in}}$ at $x = 0$, and $T_b = T_{b,\text{out}}$ at $x = L$, to give

$$T_{b,\text{out}} = T_s - (T_s - T_{b,\text{in}})e^{-\overline{h}_c 2\pi RL/\dot{m} c_p} \tag{4.11}$$

where $\overline{h}_c$ is an average heat transfer coefficient defined by

$$\overline{h}_c = \frac{1}{L}\int_0^L h_c \, dx \tag{4.12}$$

T_s and $T_b(x)$ are plotted in Fig. 4.6. The definition of average heat transfer coef-

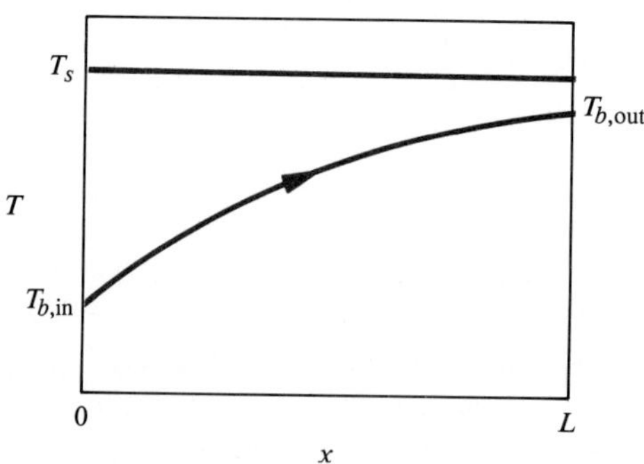

Figure 4.6 Heating of a fluid flowing through a tube with a constant wall temperature: the characteristic exponential temperature variation.

ficient is seen to be equivalent to Eq. (1.26) for an external flow on an isothermal wall. Equation (4.11) will prove useful for problem solving. It is essentially the same as the equation for the single-stream heat exchanger, Eq. (1.59). The isothermal tube can be viewed as a heat exchanger, with the number of transfer units $N_{tu} = \overline{h}_c 2\pi RL/\dot{m}c_p$.

Table 4.1 summarizes the three definitions of the convective heat transfer coefficient.

Table 4.1 Summary of defining equations for the convective heat transfer coefficient.

Flow Configuration	Heat Transfer Coefficient Defining Equation	Comments
External flow over a surface	$q_s = h_c(T_s - T_e)$ (4.2)	T_e is the free-stream temperature.
Natural flow in an enclosure	$\frac{\dot{Q}}{A} = \overline{h}_c(T_H - T_C)$ (4.4)	T_H and T_C are the hot and cold surface temperatures, respectively.
Internal flow in a duct	$q_s = h_c(T_s - T_b)$ (4.9)	T_b is the bulk temperature defined by Eq. (4.7).

4.2.2 Dimensional Analysis

Experimental data for convective heat transfer coefficients, as well as the results of analysis, can be conveniently and concisely organized as relationships between dimensionless groups of the pertinent variables. The Reynolds and Grashof numbers introduced in Chapter 1 are examples of such groups. The simplest method for selecting dimensionless groups appropriate to a given problem is dimensional analysis using the **Buckingham pi theorem** and the **method of indices**. It is assumed that the student has already had an introduction to fluid mechanics, in which the methodology of dimensional analysis is applied to viscous flows. The application of this methodology to convective heat transfer is somewhat more difficult, owing to the larger number of variables involved.

Before commencing the dimensional analysis of convective heat transfer, a short review of the dimensionless groups pertinent to simple viscous flows is appropriate. The dimensionless group characterizing a viscous flow is the **Reynolds number**, Re:

$$\text{Re} = \frac{\rho VL}{\mu} = \frac{VL}{\nu} \tag{4.13}$$

where L is a characteristic length of the configuration, and for external flows, V is usually the velocity of the undisturbed free stream. For internal flows, ρV [kg/m^{2}s] is taken as the *mass velocity* $G = \dot{m}/A_c$, where $\dot{m}$ is the mass flow rate and A_c is the cross-sectional area for flow. If the density can be assumed constant, then $G = \rho u_b$, where u_b is the *bulk velocity*. The SI units for dynamic viscosity μ are [kg/m s],[1] and the units of kinematic viscosity ν (= μ/ρ) are [m²/s]. Momentum transfer to a wall is made dimensionless as the **skin friction coefficient**, C_f:

$$C_f = \frac{\tau_s}{(1/2)\rho V^2} \tag{4.14}$$

where τ_s [N/m²] is the wall shear stress. In pipe flows, the pressure gradient is made dimensionless as the **friction factor**, f, for incompressible flow:

$$f = \frac{\Delta P/L}{(1/2)\rho V^2/D} \tag{4.15}$$

where ΔP [N/m²] is the pressure drop over a length L of a pipe with diameter D. This friction factor is called the *Darcy* friction factor. Some texts use the *Fanning* friction factor, which is one fourth the Darcy value. *We will use the Darcy friction factor exclusively.* If the flow in the pipe is hydrodynamically fully developed, that is, if the velocity profile is unchanging with axial position, then f and C_f are simply related as

$$f = 4C_f \tag{4.16}$$

which can be derived from the force balance shown in Fig. 4.7. The pressure drop across an orifice can be made dimensionless as the **Euler number**, Eu:[2]

$$\text{Eu} = \frac{\Delta P}{G^2/\rho} = \frac{\Delta P}{\rho V^2} \tag{4.17}$$

To introduce the dimensionless groups of convective heat transfer, we will consider a number of sample flows.

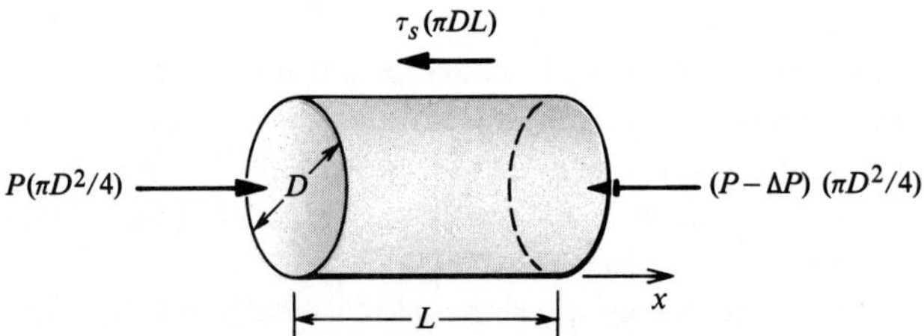

Figure 4.7 A force balance on a length of pipe, used to relate the friction factor to the skin friction coefficient.

[1] Strictly speaking, the SI units for dynamic viscosity are *pascal seconds* [Pa s]. However, the equivalent units [kg/m s] and [N s/m²] are more widely used in engineering practice. For dimensional analysis, the most convenient form is [kg/m s], and this form will be used throughout this text.

[2] In some texts the Euler number is defined as Eu = $\Delta P/(1/2)\rho V^2$. Thus, the definition of the Euler number should be checked when consulting relevant literature.

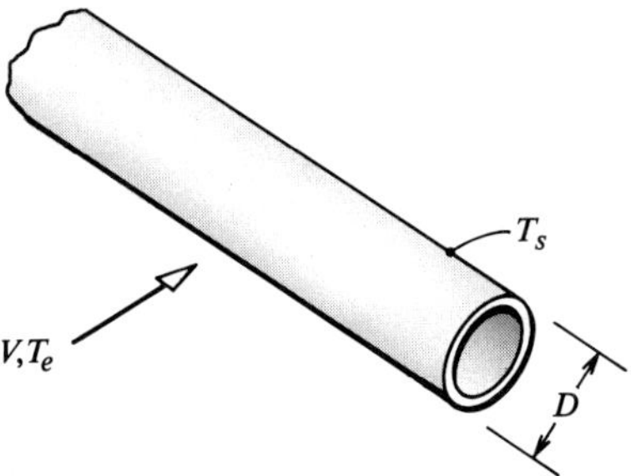

Figure 4.8 Schematic of forced flow across a cylinder.

Flow across a Cylinder

External forced flow across a long cylinder is shown in Fig. 4.8 and was discussed in detail in Section 4.2.1. The flow field is expected to depend on the upstream velocity V, diameter D, and the fluid density ρ and dynamic viscosity μ. The average wall heat flux is expected to depend on these variables, as well as on the temperature difference, $\Delta T = T_s - T_e$, and the fluid specific heat c_p and thermal conductivity k. Thus, the expected functional dependence of $\overline{q}_s$ is

$$\overline{q}_s = f(V, D, \rho, \mu, \Delta T, c_p, k)$$

The units of the variables may be written as follows:

$\overline{q}_s$	V	D	ρ	μ	ΔT	c_p	k
$\dfrac{\text{W}}{\text{m}^2}$	$\dfrac{\text{m}}{\text{s}}$	m	$\dfrac{\text{kg}}{\text{m}^3}$	$\dfrac{\text{kg}}{\text{m s}}$	K	$\dfrac{\text{W s}}{\text{kg K}}$	$\dfrac{\text{W}}{\text{m K}}$

Since we only use SI units in this text, we need not distinguish between the general concept of *dimensions* and a specific system of *units*. Notice that the units of c_p are written [W s/kg K] rather than [J/kg K] so as to utilize the set of *primary* dimensions or units kg, m, s, K, W. There are five primary dimensions and eight variables; thus, according to the Buckingham pi theorem, $(8 - 5) = 3$ independent dimensionless groups or products can be formed from the variables. The most familiar method of obtaining these groups is the *method of indices*, which is used as follows. We begin by writing any dimensionless group Π as a product of the variables, each raised to an unknown exponent,

$$\Pi = \overline{q}_s^a V^b D^c \rho^d \mu^e \Delta T^f c_p^g k^h$$

and substitute the appropriate dimensions:

$$\Pi = \left[\frac{\text{W}}{\text{m}^2}\right]^a \left[\frac{\text{m}}{\text{s}}\right]^b [\text{m}]^c \left[\frac{\text{kg}}{\text{m}^3}\right]^d \left[\frac{\text{kg}}{\text{m s}}\right]^e [\text{K}]^f \left[\frac{\text{W s}}{\text{kg K}}\right]^g \left[\frac{\text{W}}{\text{m K}}\right]^h$$

For Π to be dimensionless, the sum of exponents of each primary dimension must add to zero:

$$\begin{aligned}
\text{kg:} \quad & d + e - g = 0 \\
\text{m:} \quad & -2a + b + c - 3d - e - h = 0 \\
\text{s:} \quad & -b - e + g = 0 \\
\text{K:} \quad & f - g - h = 0 \\
\text{W:} \quad & a + g + h = 0
\end{aligned}$$

This constitutes a set of five linear algebraic equations in eight unknowns. A basic theorem of linear algebra states that the number of distinct solutions is equal to the number of unknowns minus the number of linearly independent equations. The number of linearly independent equations is given by the *rank* of the coefficient matrix, which for this problem can be shown to equal five. Thus, there are $8 - 5 = 3$ distinct solutions, as stated by the pi theorem.

To obtain these solutions, we are free to choose values for three exponents, and we let the equations give the values for the remaining five. Because the dimensionless groups initially obtained depend on our choice, we are faced with a dilemma. We might proceed as follows. Based on fluid mechanics experience, we know that the Reynolds number characterizes the flow field; hence, set $b = 1$, $a = g = 0$:

$$\begin{aligned}
d + e &= 0 \\
1 + c - 3d - e - h &= 0 \\
-1 - e &= 0 \\
f - h &= 0 \\
h &= 0
\end{aligned}$$

Hence, $h = 0$, $f = 0$, $e = -1$, $d = 1$, $c = 1$, which gives the Reynolds number

$$\Pi_1 = \frac{VD\rho}{\mu}$$

Next, since we would like to have q_s in the dependent variable group, we set $a = 1$ and we do not set $f = 0$ so as to obtain the heat transfer coefficient $q_s/\Delta T$ in this group. Thus, somewhat arbitrarily, we set $b = g = 0$:

$$\begin{aligned}
d + e &= 0 \\
-2 + c - 3d - e - h &= 0 \\
e &= 0 \\
f - h &= 0 \\
1 + h &= 0
\end{aligned}$$

Hence, $h = -1$, $f = -1$, $e = 0$, $d = 0$, $c = 1$, which gives

$$\Pi_2 = \frac{\overline{q}_s D}{\Delta T k}$$

which is the average **Nusselt number**, $\overline{\text{Nu}}$.

To obtain a third group, we recognize that fluid properties affect heat transport in the fluid and thus attempt to obtain a dimensionless group of properties by setting $g = 1$, $a = b = 0$:

$$\begin{aligned} d + e - 1 &= 0 \\ c - 3d - e - h &= 0 \\ -e + 1 &= 0 \\ f - 1 - h &= 0 \\ 1 + h &= 0 \end{aligned}$$

Hence, $h = -1$, $f = 0$, $e = 1$, $d = 0$, $c = 0$, which gives

$$\Pi_3 = \frac{c_p \mu}{k}$$

which is the **Prandtl number**, Pr. Thus, the result of the dimensionless analysis can be written as

$$\overline{\mathrm{Nu}} = f(\mathrm{Re}, \mathrm{Pr}) \tag{4.18}$$

The ratio $q_s/\Delta T$ is the convective heat transfer coefficient; thus, the Nusselt number may be viewed as a dimensionless heat transfer coefficient. In general,

$$\mathrm{Nu} = \frac{h_c L}{k} \tag{4.19}$$

where the characteristic length L may be the distance along a flat plate, the diameter of a cylinder in cross-flow, or pipe diameter for flow in a pipe. The Prandtl number can be written as

$$\mathrm{Pr} = \frac{c_p \mu}{k} = \frac{\nu}{k/\rho c_p} = \frac{\nu}{\alpha} \tag{4.20}$$

where it is seen that this fluid properties group is the ratio of the kinematic viscosity to thermal diffusivity. Indeed, the kinematic viscosity is more appropriately termed the *momentum diffusivity*, so that the Prandtl number is the ratio of momentum and thermal diffusivities. Table 4.2 gives values of the Prandtl number for various commonly encountered fluids. There are three broad groups: liquid metals, with $\mathrm{Pr} \ll 1$; gases, with $\mathrm{Pr} \sim 1$; and oils, with $\mathrm{Pr} \gg 1$. The value for water ranges from about 1 to 10, depending on temperature. It will be seen that the convective heat transfer characteristics of a fluid are very much dependent on its Prandtl number.

In the dimensional analysis, we could have obtained an alternative Π_2 as follows. Again, $a = 1$, $f \neq 0$, but now we choose $c = h = 0$:

$$\begin{aligned} d + e - g &= 0 \\ -2 + b - 3d - e &= 0 \\ -b - e + g &= 0 \\ f - g &= 0 \\ 1 + g &= 0 \end{aligned}$$

Table 4.2 Prandtl number values for various fluids ($\mathrm{Pr} = c_p \mu / k = \nu / \alpha$).

Fluid	Temperature K	Prandtl Number
Sodium	1000	0.0038
Mercury	500	0.012
Lithium	700	0.031
Argon	400	0.67
Air	300	0.69
Saturated steam	373	0.98
Liquid ammonia	270	1.49
Water	460	0.98
	360	2.00
	275	12.9
Liquid refrigerant-12	250	4.6
Therminol 60	350	31.3
Ethylene glycol	300	151
SAE 50 oil	400	154
	300	6600

Hence, $g = -1$, $f = -1$, $b = -1$, $d = -1$, $e = 0$, which gives

$$\Pi_2' = \frac{\overline{q}_s}{\Delta T \rho c_p V}$$

which is the average **Stanton number**, $\overline{\mathrm{St}}$. Again, writing $\overline{q}_s / \Delta T = \overline{h}_c$, we have

$$\overline{\mathrm{St}} = \frac{\overline{h}_c}{\rho c_p V} = \frac{\overline{h}_c}{c_p G} \tag{4.21}$$

The Stanton number is an alternative dimensionless heat transfer coefficient and is related to the Nusselt number as St = Nu/RePr. Of course, other alternatives can be obtained by combining the preceding groups. For example, the **Peclet number**,

$$\mathrm{Pe} = \mathrm{RePr} = \frac{VL}{\alpha} \tag{4.22}$$

is often used for creeping external flow (Re $\simeq$ 1) and laminar internal flows.

High-Speed Flow

In the foregoing dimensional analysis, five primary dimensions were used. We will now rework the analysis using only four primary dimensions: kg, s, m, and K. Then, since power equals force times velocity,

$$[\mathrm{W}] = [\mathrm{N}]\left[\frac{\mathrm{m}}{\mathrm{s}}\right] = \left[\frac{\mathrm{kg\ m}}{\mathrm{s}^2}\right]\left[\frac{\mathrm{m}}{\mathrm{s}}\right] = \left[\frac{\mathrm{kg\ m}^2}{\mathrm{s}^3}\right]$$

and the units of the variables that involved watts are now

$\overline{q}_s$	c_p	k
$\dfrac{\text{kg}}{\text{s}^3}$	$\dfrac{\text{m}^2}{\text{s}^2\,\text{K}}$	$\dfrac{\text{kg m}}{\text{s}^3\,\text{K}}$

$$\Pi = \left[\frac{\text{kg}}{\text{s}^3}\right]^a \left[\frac{\text{m}}{\text{s}}\right]^b [\text{m}]^c \left[\frac{\text{kg}}{\text{m}^3}\right]^d \left[\frac{\text{kg}}{\text{m s}}\right]^e [\text{K}]^f \left[\frac{\text{m}^2}{\text{s}^2\ \text{K}}\right]^g \left[\frac{\text{kg m}}{\text{s}^3\ \text{K}}\right]^h$$

$$\begin{aligned} \text{kg:} \quad & a + d + e + h = 0 \\ \text{s:} \quad & -3a - b - e - 2g - 3h = 0 \\ \text{m:} \quad & b + c - 3d - e + 2g + h = 0 \\ \text{K:} \quad & f - g - h = 0 \end{aligned}$$

According to the pi theorem, there are now $(8-4) = 4$ independent dimensionless groups. The Reynolds, Nusselt, and Prandtl numbers can be obtained as before, but now we must choose the fourth group. We could reason as follows. In eliminating the watt as a primary dimension by using the form power equals force times velocity, we have recognized that kinetic energy can be converted into thermal energy; thus, the fourth group might involve the kinetic energy of the flow. Choosing $b = 2$, $a = d = e = 0$,

$$\begin{aligned} h &= 0 \\ -2 - 2g - 3h &= 0 \\ 2 + c + 2g + h &= 0 \\ f - g - h &= 0 \end{aligned}$$

Hence, $h = 0$, $g = -1$, $c = 0$, $f = -1$, which gives

$$\Pi_4 = \frac{V^2}{c_p \Delta T}$$

Since V^2 in the numerator of Π_4 is associated with kinetic energy of the fluid, we choose to introduce a factor of 1/2 and define the **Eckert number**, Ec, as

$$\text{Ec} = \frac{(1/2)V^2}{c_p \Delta T} \qquad \textbf{(4.23)}$$

This result appears satisfactory and perhaps could be used for correlating heat transfer data for high-speed flows. But experience tells us that the conversion of kinetic energy into thermal energy in an incompressible flow is due to the action of viscous stresses, and yet viscosity does not appear in the Eckert number. Indeed, this phenomenon is called *viscous dissipation*. So we try again and choose $b = 2$, $a = d = g = 0$:

$$
\begin{aligned}
e + h &= 0 \\
-2 - e - 3h &= 0 \\
2 + c - e + h &= 0 \\
f - h &= 0
\end{aligned}
$$

Hence, $h = -1$, $e = 1$, $f = -1$, $c = 0$, which gives

$$\Pi_4' = \frac{V^2 \mu}{k \Delta T}$$

This product is the **Brinkman number**, Br, and is most appropriate for characterizing viscous dissipation. It is easily seen that

$$\text{Br} = 2 \frac{(1/2)V^2}{c_p \Delta T} \cdot \frac{c_p \mu}{k} = 2\text{EcPr} \tag{4.24}$$

That is, the Eckert-Prandtl number product is equivalent to the Brinkman number. Thus, for a high-speed flow,

$$
\begin{aligned}
\overline{\text{Nu}} &= f(\text{Re, Pr, Br}) \\
&= F(\text{Re, Pr, EcPr})
\end{aligned} \tag{4.25}
$$

Natural-Convection Flow on a Vertical Plate

A vertical plate of height L at temperature T_s is immersed in a fluid at temperature T_e, as shown in Fig. 4.9. Before postulating the functional dependence of the average heat flux, it is necessary to first examine the underlying physics. In natural convection, motion is caused by density variations. From Archimedes' principle, the buoyancy force per unit volume acting on the heated fluid adjacent to the wall is $(\rho - \rho_e)g$, where ρ is the local fluid density, ρ_e is the ambient fluid density, and g is the acceleration due to gravity. The buoyancy force per unit mass is then

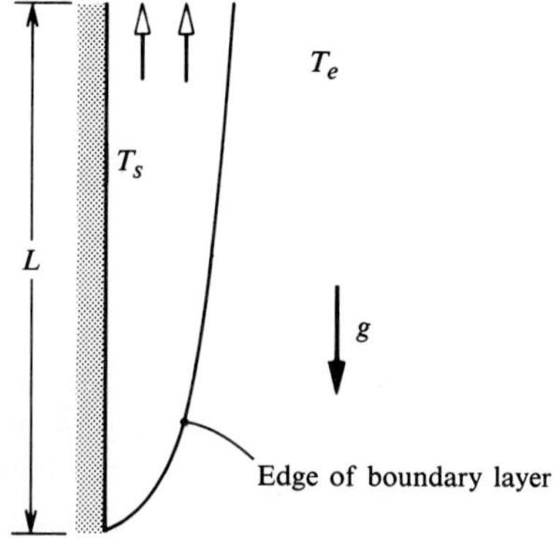

Figure 4.9 Schematic of a natural-convection boundary layer on a vertical plate.

$(\rho - \rho_e)g/\rho$. If β [K^{-1}] is the volumetric coefficient of thermal expansion, then $\beta(T - T_e) \simeq -(\rho - \rho_e)/\rho$. In terms of β, the buoyancy force per unit mass becomes $-(T - T_e)\beta g$, and it is this force that causes the natural-convection motion.

We now let $\Delta T = T_s - T_e$ and suggest the following functional dependence for the average heat flux, $\overline{q}_s$:

$$\overline{q}_s = f(\Delta T, \beta, g, \rho, \mu, k, c_p, L)$$

$$\Pi = \overline{q}_s^a \Delta T^b \beta^c g^d \rho^e \mu^f k^g c_p^h L^i$$

which, in terms of the primary dimensions kg, m, s, K, and W, is

$$\Pi = \left[\frac{\text{W}}{\text{m}^2}\right]^a [\text{K}]^b [\text{K}^{-1}]^c \left[\frac{\text{m}}{\text{s}^2}\right]^d \left[\frac{\text{kg}}{\text{m}^3}\right]^e \left[\frac{\text{kg}}{\text{m s}}\right]^f \left[\frac{\text{W}}{\text{m K}}\right]^g \left[\frac{\text{W s}}{\text{kg K}}\right]^h [\text{m}]^i$$

Equating the sum of exponents of each primary dimension gives

$$\begin{aligned}
&\text{kg:} & e + f - h &= 0 \\
&\text{m:} & -2a + d - 3e - f - g + i &= 0 \\
&\text{s:} & -2d - f + h &= 0 \\
&\text{K:} & b - c - g - h &= 0 \\
&\text{W:} & a + g + h &= 0
\end{aligned}$$

There are $(9 - 5) = 4$ independent dimensionless groups. At this point, we use experience gained in analyzing forced convection to immediately choose the Nusselt number $\text{Nu} = q_s L/k\Delta T$ as the dimensionless heat flux, and the Prandtl number $\text{Pr} = c_p\mu/k$ as a fluid properties group. Also, by inspection, $\Pi_3 = \beta\Delta T$ is obviously dimensionless and independent of Nu and Pr. Thus, there remains but one group to obtain, which we can ensure is independent from the other groups by choosing $d = 1$, $a = b = g = 0$:

$$\begin{aligned}
e + f - h &= 0 \\
1 - 3e - f + i &= 0 \\
-2 - f + h &= 0 \\
-c - h &= 0 \\
h &= 0
\end{aligned}$$

Hence, $c = 0$, $f = -2$, $e = 2$, $i = 3$, which gives

$$\Pi_4 = \frac{g\rho^2 L^3}{\mu^2} = \frac{gL^3}{\nu^2}$$

Thus,

$$\overline{\text{Nu}} = f\left(\text{Pr}, \beta\Delta T, \frac{gL^3}{\nu^2}\right) \quad \textbf{(4.26)}$$

The groups Π_3 and Π_4 do not have names because both theory and experiment show that natural convection depends on the product $\Pi_3\Pi_4$ rather than on each group independently.[3] This group is the familiar **Grashof number**, Gr:

$$\mathrm{Gr} = \frac{\beta \Delta T g L^3}{\nu^2} \tag{4.27}$$

Thus,

$$\overline{\mathrm{Nu}} = f(\mathrm{Gr}, \mathrm{Pr}) \tag{4.28}$$

In fact, there is a further simplification for either very high Prandtl-number fluids (e.g., oils) or very low Prandtl-number fluids (e.g., liquid metals). For $\mathrm{Pr} \gg 1$, the Nusselt number is found to depend on the product of Grashof and Prandtl numbers; this group is the **Rayleigh number**, Ra:

$$\mathrm{Ra} = \mathrm{GrPr} = \frac{\beta \Delta T g L^3}{\nu \alpha} \tag{4.29}$$

$$\overline{\mathrm{Nu}} = f(\mathrm{Ra}) \ ; \qquad \mathrm{Pr} \gg 1 \tag{4.30}$$

For $\mathrm{Pr} \ll 1$, the Nusselt number is found to depend on the product of Grashof number and Prandtl number squared; this group is the **Boussinesq number**, Bo:

$$\mathrm{Bo} = \mathrm{GrPr}^2 = \frac{\beta \Delta T g L^3}{\alpha^2} \tag{4.31}$$

$$\overline{\mathrm{Nu}} = f(\mathrm{Bo}) \ ; \qquad \mathrm{Pr} \ll 1 \tag{4.32}$$

Natural Convection in an Inclined Enclosure

Figure 4.10 shows an inclined enclosure of the kind found in flat-plate solar collectors. The flow regimes for a horizontal enclosure were discussed in Section 4.2.1. We let $\Delta T = T_H - T_C$ and suggest the following functional dependence for

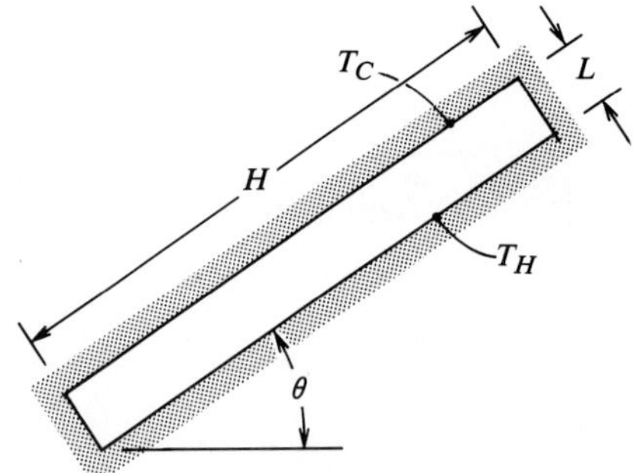

Figure 4.10 Schematic of an inclined enclosure.

[3] Since the buoyancy force per unit mass is $\beta g \Delta T$, one could argue that the product βg should be taken as a single variable; then the Grashof number is obtained directly.

the average heat flux $\overline{q}_s$:

$$\overline{q}_s = f(\Delta T, \beta, g, \rho, \mu, k, c_p, L, H, \theta)$$

There are now 11 variables in 5 primary dimensions, giving $11 - 5 = 6$ dimensionless groups. Since we have already performed a dimensional analysis of natural convection, we can immediately write down the four groups just obtained,

$$\frac{\overline{q}_s L}{\Delta T k}, \quad \frac{c_p \mu}{k}, \quad \beta \Delta T, \quad \frac{g L^3}{\nu^2}$$

leaving two groups to be obtained. The **aspect ratio**, H/L, is obviously dimensionless, as is angle θ; these are independent of the four groups listed. Thus, the desired result is

$$\overline{\text{Nu}} = f\left(\text{Pr}, \beta \Delta T, \frac{g L^3}{\nu^2}, \frac{H}{L}, \theta\right) \tag{4.33}$$

Notice that the characteristic length, in both the Nusselt number and the group gL^3/ν^2, was chosen to be L, the spacing of the plates forming the enclosure, and not H. This choice is obvious for the Nusselt number since heat is transferred across the enclosure. For the group gL^3/ν^2, the choice is also obvious for a horizontal enclosure ($\theta = 0$) because the buoyancy force acts in the vertical direction, so that L characterizes the flow length of the motion. For a vertical enclosure ($\theta = 90°$), the choice proves to be inappropriate for some flow regimes. *Whenever there is the possibility for ambiguity in the choice of appropriate length scale for dimensionless groups, the groups should be subscripted accordingly* (e.g., $\overline{\text{Nu}}_L$, Gr_L, Re_D, etc.). Figure 4.11 illustrates this practice for natural convection on the curved surface of a cylinder.

In general, dimensionless transfer coefficients can be *local* values at a particular location on a surface, or *average* values over a surface. In the preceding discussion of dimensional analysis, only average values were considered. In the remainder of

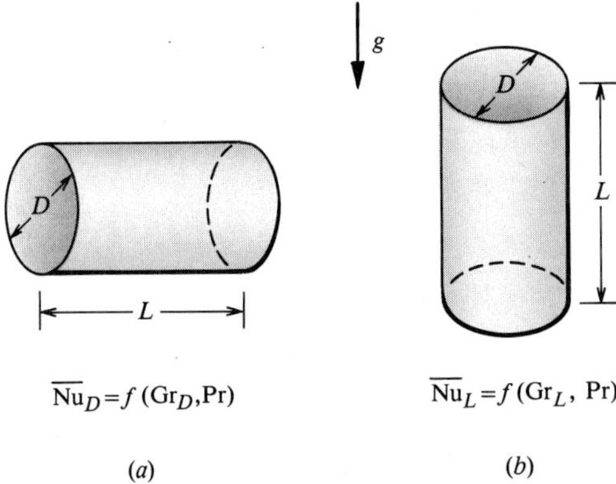

Figure 4.11 Characteristic lengths for natural convection on a cylinder.

the chapter, both local and average values will be of concern. The bar notation will be used to denote average coefficients, for example, $\overline{\text{Nu}}$ and $\overline{C}_f$; local values will simply be Nu, C_f, and so on.

The preceding dimensional analyses of forced and natural convection have served to introduce the dimensionless groups commonly used in the study of heat convection. Table 4.3 summarizes these groups. Although the application of this methodology of dimensional analysis to simple problems is quite straightforward, its successful application to heat convection problems requires some physical insight and experience. Dimensional analysis based on the Buckingham pi theorem will not yield useful

Table 4.3 Summary of the major dimensionless groups characterizing momentum transfer and convective heat transfer.

Group	Definition	Use
Skin friction coefficient	$C_f = \dfrac{\tau_s}{(1/2)\rho V^2}$	External flows
Friction factor	$f = \dfrac{\Delta P}{(L/D)(1/2)\rho V^2}$	Internal flows
Euler number	$\text{Eu} = \dfrac{\Delta P}{\rho V^2}$	Flow through orifices
Reynolds number	$\text{Re} = \dfrac{VL}{\nu}$	Forced flows
Nusselt number	$\text{Nu} = \dfrac{h_c L}{k}$	Forced and natural flows
Stanton number	$\text{St} = \dfrac{h_c}{\rho c_p V}$	Forced flows
Prandtl number	$\text{Pr} = \dfrac{c_p \mu}{k} = \dfrac{\nu}{\alpha}$	Forced and natural flows
Grashof number	$\text{Gr} = \dfrac{\beta \Delta T g L^3}{\nu^2}$	Natural flows with Pr $\sim$ 1
Peclet number	$\text{Pe} = \dfrac{VL}{\alpha}$	Laminar internal flows, creeping external flows
Rayleigh number	$\text{Ra} = \dfrac{\beta \Delta T g L^3}{\nu \alpha}$	Natural flows with Pr $\gg$ 1
Boussinesq number	$\text{Bo} = \dfrac{\beta \Delta T g L^3}{\alpha^2}$	Natural flows with Pr $\ll$ 1
Brinkman number	$\text{Br} = \dfrac{V^2 \mu}{k \Delta T}$	Flows with viscous dissipation

results unless it is accompanied by careful thought. Nevertheless, it has proven to be a valuable tool in many areas of engineering. Exercises 4–4 through 4–7 are further examples of the application of the pi theorem to heat convection problems.

4.2.3 Correlation of Experimental Data

Most engineering calculations of convective heat transfer use heat transfer coefficients obtained from experimental data. Dimensional analysis has proven to be an invaluable tool for the efficient planning of experiments and organization of the resulting data. Once again we will use flow across a heated cylinder to illustrate the procedure. It is relatively simple to perform an experiment to obtain the average heat transfer coefficient in an air flow, as will now be described.

A 3 cm–diameter copper tube containing an electrical heater is shown in Fig. 4.12. Thermocouples are located as shown to measure the surface temperature T_s and the air temperature T_e. Use of a copper tube with a high thermal conductivity ensures an isothermal surface. The experiments are conducted in a conveniently available wind tunnel. In addition to temperatures, required measurements are the power input to the heater $\dot{Q}$, and air speed V. If A is the heated area of the cylinder, the average heat transfer coefficient is simply

$$\overline{h}_c = \frac{\dot{Q}/A}{T_s - T_e}$$

The air speed V is varied over the tunnel operating range while the power is adjusted to hold T_s constant, at about 10 K above the air temperature. In this manner, secondary effects due to variation of fluid properties are eliminated.

Sample data for h_c versus V are shown in Fig. 4.13. As presented in the figure, the data are of limited utility. The graph applies only to heat transfer from a 3 cm–diameter cylinder to air. To extend the utility of the data, we use the result of our dimensional analysis, Eq. (4.18):

$$\overline{\text{Nu}} = f(\text{Re}, \text{Pr})$$

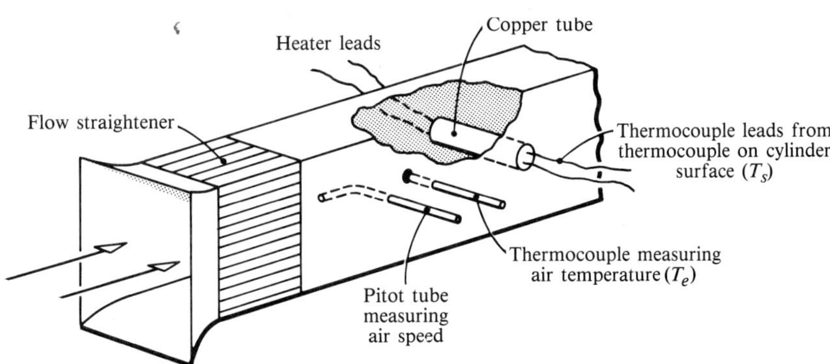

Figure 4.12 An experimental rig for investigating forced-convection heat transfer from a cylinder.

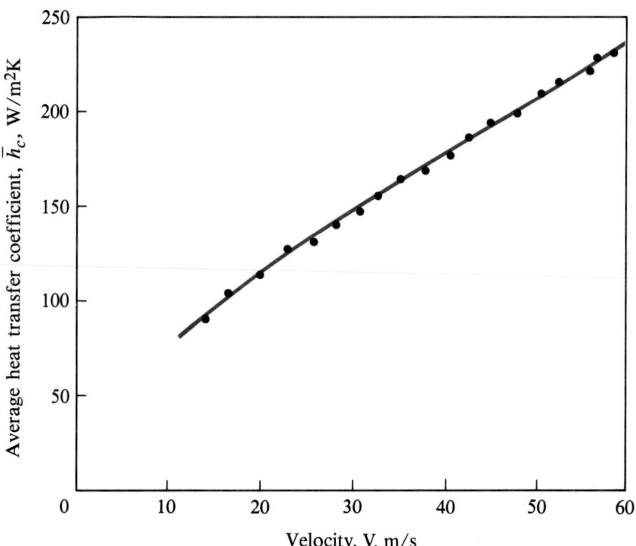

Figure 4.13 Average heat transfer coefficient versus velocity of air flow over a 3 cm–diameter cylinder.

Table A.7 in Appendix A shows that the Prandtl number of air is constant over a wide range of temperature, equal to 0.69. Thus, we present the data as

$$\overline{\text{Nu}} = f(\text{Re})\ ; \qquad \text{Pr} = 0.69 \tag{4.34}$$

or

$$\frac{\bar{h}_c D}{k} = f\left(\frac{VD}{\nu}\right)$$

with the fluid properties k and ν evaluated at the average of T_s and T_e. The result is shown in Fig. 4.14. Since a simple power law of the form $\text{Nu} = C_1\ \text{Re}^n$ would be a convenient correlation formula for the data, Fig. 4.14 is a log-log plot on which a power law will be a straight line with slope n and intercept C_1. A least squares linear regression analysis shows that $n = 0.63$ is a good fit. Figure 4.14 applies to any combination of cylinder diameter and air velocity, provided that the Reynolds number is in the indicated range of $1.5 \times 10^4 < \text{Re} < 10^5$. For example, it applies to a 12 cm–diameter cylinder provided the air speed does not exceed $(3/12)(60) = 15$ m/s to ensure that $\text{Re} < 10^5$. This important observation is a statement of the *similarity principle*, which is the basis of modeling. The engineer often exploits the similarity principle to perform experiments more conveniently. For example, a small model can be tested in a wind tunnel, provided the air speed is increased accordingly to give the desired Reynolds number range. Notice also that the Reynolds number range of the data in Fig. 4.14 can be extended by testing larger- and smaller-diameter cylinders. The similarity principle, which was introduced in Chapters 2 and 3 for heat conduction, is discussed further in Chapter 5.

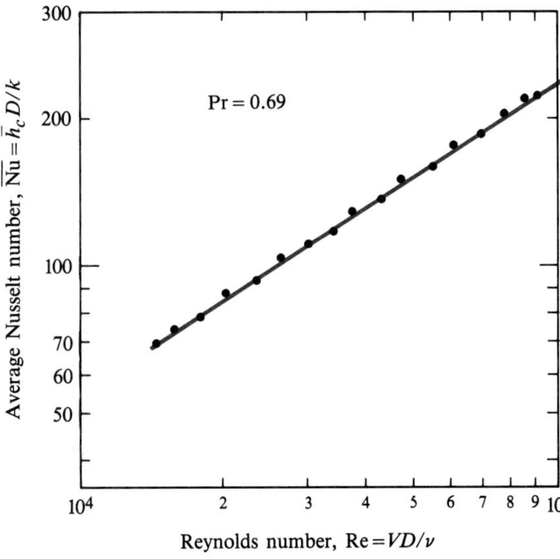

Figure 4.14 Average Nusselt number versus Reynolds number for flow over a cylinder: air at 300 K, 1 bar.

Figure 4.14 applies to any fluid whose Prandtl number is approximately 0.7. An examination of Table A.7 shows that this condition is met by many gases, including carbon dioxide, helium, and hydrogen. Notice that these gases can have values of k, μ, and c_p that are quite different from those for air, yet their Prandtl numbers are almost equal. To estimate heat transfer coefficients for fluids with Prandtl numbers not equal to 0.7, further experiments must be performed. Examination of Table 4.2 suggests experiments with a liquid metal such as mercury, cold water, and Therminol 60 to obtain a large Prandtl number range. However, experiments with mercury and other liquid metals are difficult to perform. Mercury is highly toxic, sodium ignites spontaneously in air, and there are materials compatibility problems. Liquid metals are not routinely used in experiments unless a specific application requiring a liquid metal is contemplated, for example, in a sodium-cooled nuclear reactor. Thus, we will restrict our attention to higher values of Prandtl number. Figure 4.15*a* shows the results of a series of experiments. It can be seen that for each fluid, the data are correlated well with straight lines of slope 0.63. Thus, we can now seek a power law correlation of the form

$$\overline{\text{Nu}} = C_2 \text{Re}^{0.63} \text{Pr}^m$$

Using the correlations for each fluid shown in Fig. 4.15*a*, $\log(\overline{\text{Nu}}/\text{Re}^{0.63})$ is plotted versus log Pr in Fig. 4.15*b*, where it is seen that $m = 0.36$, $C_2 = 0.19$. Thus, the recommended correlation is

$$\overline{\text{Nu}} = 0.19 \text{Re}^{0.63} \text{Pr}^{0.36} \qquad \textbf{(4.35)}$$

which is valid for $10^4 \leq \text{Re} \leq 10^5$, $0.69 < \text{Pr} < 31.3$. Modest extrapolation outside these Reynolds and Prandtl number ranges would be warranted. Of course, the preferred procedure to determine the constant C_2 and exponents n and m is a multivariable least squares fit of all the data points using a standard computer subroutine.

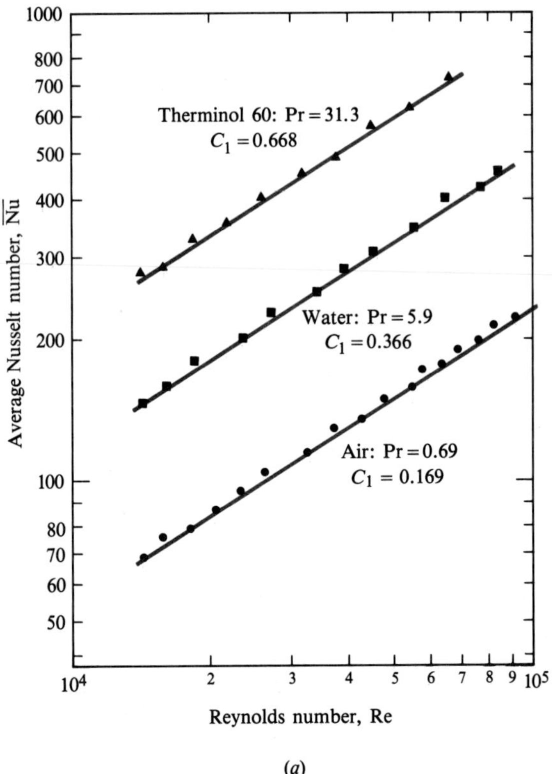

(a)

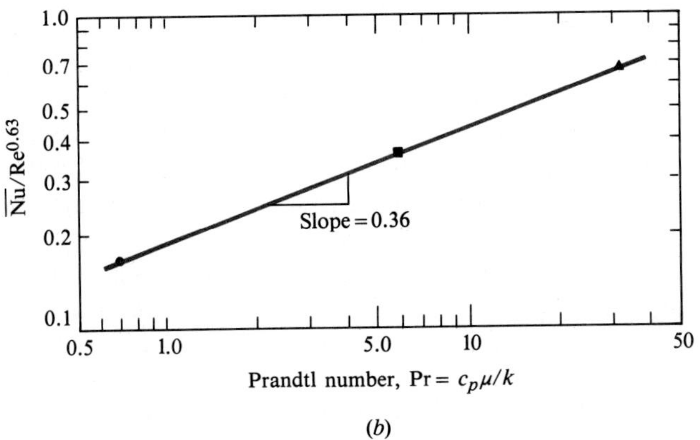

(b)

Figure 4.15 (*a*) Average Nusselt number versus Reynolds number for flow across a cylinder with three fluids: (1) air at 300 K, 1 bar; (2) water at 300 K; (3) Therminol 60 at 350 K. The constant C_1 is defined by the relation $\overline{\text{Nu}} = C_1\text{Re}^{0.63}$. (*b*) The ratio $\overline{\text{Nu}}/\text{Re}^{0.63}$ versus Prandtl number for flow over cylinder; $1.5 \times 10^4 < \text{Re} < 10^5$.

As mentioned earlier, most engineering calculations of convective heat transfer use heat transfer coefficients obtained from experimental data. For some simple flows (e.g., laminar flow in tubes), we have exact analytical solutions of the governing differential equations. In the process of obtaining these solutions, the appropriate dimensionless groups arise quite naturally; thus, the final results are also presented in terms of the commonly used groups. Examples of such analyses are given in Chapter 5.

The use of numerical methods to solve the governing differential equations is rapidly increasing. This activity is often called *computational fluid dynamics (CFD)*, but in relation to convective heat transfer, the subject has broader scope than this name implies. The computer codes that execute the numerical computations yield results that, in many ways, resemble the data obtained from physical experiments. The term *numerical experiments* is a good description of the activity. As for physical experiments, the data are organized in terms of dimensionless groups whenever possible. However, since the numerical results are obtained by solving model-governing differential equations, these equations can be used to determine the appropriate dimensionless groups, as was done for the pin fin problem in Section 2.4.5.

4.2.4 Evaluation of Fluid Properties

The fluid properties appearing in the dimensionless groups pertinent to convective heat transfer are density ρ [kg/m^3], dynamic viscosity μ [kg/m s] or kinematic viscosity ν [m^2/s], thermal conductivity k [W/m K], specific heat c_p [J/kg K], and the volume-expansion coefficient β [l/K]. Data for these properties for selected fluids are given in Tables A.7–A.10 of Appendix A and can be seen to be temperature-dependent to a greater or lesser extent. For liquids, ρ is essentially constant, whereas for gases, $\rho \propto T^{-1}$. For liquids such as water and oils, the viscosity varies markedly with temperature. For water, the viscosity more than doubles as the temperature changes from 330 to 290 K. For gases, the dynamic viscosity and conductivity increase with temperature, approximately as $\mu \propto T^{0.7}$, $k \propto T^{0.7}$. The specific heat does not vary much unless there is a change in molecular structure.

In elementary fluid mechanics, the problems the student faces, such as evaluation of the Reynolds number for calculating drag on an immersed object, usually involve an isothermal fluid; thus, the temperature at which the kinematic viscosity must be evaluated is obvious. But in convective heat transfer, there is always a temperature difference between the surface and the bulk or free-stream fluid, so the question arises as to what is the proper temperature to use when calculating fluid properties in dimensionless groups. There is no simple answer. Strictly speaking, the property variation itself is an additional problem parameter to be characterized by an additional dimensionless group. For example, the friction factor for flow in a pipe depends on the viscosity ratio μ_s/μ_b where μ_s and μ_b are the viscosity values at the wall and bulk temperatures, respectively. Such **variable-property effects** can be determined only by careful experiment or detailed analysis. For engineering calculations, the effects of variable properties are usually approximately accounted for as follows.

External Flows

For external flows, all properties are evaluated at a *reference temperature* T_r, where

$$T_r = T_s - \alpha(T_s - T_e) \tag{4.36a}$$

Unless otherwise stated, the value of α should be taken as 1/2, that is, the reference temperature is the arithmetic mean of the surface and free-stream temperatures. This value of the reference temperature is also called the **mean film temperature**, perhaps because the boundary layer is sometimes imagined to be a thin film of stagnant fluid adjacent to the surface.

Internal Flows

For internal flows, the reference temperature approach can also be used, with T_r given by

$$T_r = T_s - \alpha(T_s - T_b) \tag{4.36b}$$

and with α taken as 1/2 unless otherwise stated. For liquids with an essentially constant density, the reference temperature approach is straightforward to use, but for gases with a variable density, the evaluation of Reynolds and Stanton numbers is awkward. In an internal flow, the mass velocity ρV is always $\dot{m}/A_c$, and if ρ varies across the duct, the separation of ρ and V so as to evaluate ρ at the reference temperature is best avoided. Thus, for internal flows, the *property ratio* or *temperature ratio* approach is often used. For liquids, a viscosity or Prandtl number ratio is used since viscosity varies more than any other property. For gases, a temperature ratio is used since density, viscosity, and conductivity are all well-behaved functions of absolute temperature. We can write

$$\frac{f}{f_b} = \left(\frac{\mathrm{Pr}_s}{\mathrm{Pr}_b}\right)^m \quad \text{or} \quad \left(\frac{\mu_s}{\mu_b}\right)^m \quad \text{or} \quad \left(\frac{T_s}{T_b}\right)^m \tag{4.37}$$

$$\frac{\mathrm{Nu}}{\mathrm{Nu}_b} = \left(\frac{\mathrm{Pr}_s}{\mathrm{Pr}_b}\right)^n \quad \text{or} \quad \left(\frac{\mu_s}{\mu_b}\right)^n \quad \text{or} \quad \left(\frac{T_s}{T_b}\right)^n \tag{4.38}$$

where f_b and Nu_b are evaluated using bulk properties.

A further complication for internal flows is that both the wall temperature T_s and the bulk temperature T_b may vary along the duct as heat is added or removed. In calculating local values of friction factor or Nusselt number, the temperatures at the location are used to evaluate properties. In calculating average values for a duct of length L, it is usually adequate to use arithmetic averages of the inlet and outlet temperatures.

In the sections that follow, specific instructions on the evaluation of fluid properties will be given for each correlation or group of correlations.

4.3 FORCED CONVECTION

Forced flows may be *internal* or *external*. In an internal flow, such as in a heat exchanger tube, the flow is forced by a fan if the fluid is a gas, or a pump if it is a liquid. An external flow over a model in a wind tunnel is forced by the tunnel fan. Alternatively, the surface may move through a stationary fluid; an example is the flight of a hypersonic vehicle. In Section 4.3, we consider only simple forced-convection flows over smooth surfaces. More complicated forced-convection flows are dealt with in Section 4.5, and the effect of surface roughness will be discussed in Section 4.7. Unless otherwise noted, the formulas presented are correlations of experimental data.

4.3.1 Forced Flow in Tubes and Ducts

Fully Developed Flow in Round Tubes (or Pipes)

For laminar flow sufficiently far from the entrance of a tube or pipe, where the flow is *hydrodynamically fully developed* and has the characteristic parabolic velocity profile of *Poiseuille* flow, some simple analytical results are available. The friction factor has a constant value

$$f = \frac{64}{\mathrm{Re}_D}; \qquad \mathrm{Re}_D = \frac{GD}{\mu} \tag{4.39}$$

where D is the tube diameter and G is the mass velocity ($G = \dot{m}/A_c$). Note the use of the subscript D to indicate that the characteristic length in the Reynolds number is the tube diameter. If the wall temperature is uniform, for example, if steam is condensing on the outside of the tube wall, then sufficiently far downstream of where heating starts, the flow becomes *thermally fully developed*, the shape of the temperature profile is unchanging, and the Nusselt number has a constant value given by Eq. (1.21) rearranged into dimensionless form:

$$\mathrm{Nu}_D = 3.66 \tag{4.40}$$

If, on the other hand, the heat flux through the tube wall is uniform, for example, if the tube is wound with an electrical resistance wire at constant pitch, then

$$\mathrm{Nu}_D = \frac{48}{11} = 4.364 \tag{4.41}$$

Equations (4.39) and (4.41) are derived in Chapter 5.

Transition to turbulence takes place at $\mathrm{Re}_D \simeq 2300$, although the turbulence becomes fully established only for $\mathrm{Re}_D > 10{,}000$. For hydrodynamically fully developed flow, the friction factor can be obtained from a Moody chart (given later in this chapter as Fig. 4.49) or, for a smooth wall, from Petukhov's formula [1]:

$$f = (0.790 \ln \text{Re}_D - 1.64)^{-2}; \qquad 10^4 < \text{Re}_D < 5 \times 10^6 \tag{4.42}$$

Alternatively, there is a less accurate power law formula:

$$f = 0.184\text{Re}_D^{-0.2}; \qquad 4 \times 10^4 < \text{Re}_D < 10^6 \tag{4.43}$$

In contrast to laminar flow, the effect of wall boundary condition (e.g., whether or not the wall is at a uniform temperature or the heat flux is uniform along the tube) is unimportant for turbulent flow of all fluids except low-Prandtl-number liquid metals. For thermally fully developed flow in a smooth tube with $\text{Pr} > 0.5$, a simple power law formula is

$$\text{Nu}_D = 0.023\text{Re}_D^{0.8}\text{Pr}^{0.4}; \qquad \text{Re}_D > 10{,}000 \tag{4.44}$$

which is in fact Eq. (1.22) in dimensionless form.[4] If more accurate results are desired, Gnielinski's formula is recommended [2]:

$$\text{Nu}_D = \frac{(f/8)(\text{Re}_D - 1000)\text{Pr}}{1 + 12.7(f/8)^{1/2}(\text{Pr}^{2/3} - 1)}; \qquad 3000 < \text{Re}_D < 10^6 \tag{4.45}$$

where the friction factor must be calculated from Eq. (4.42). Equation (4.45) agrees with most available experimental data within 20%; an example is shown in Fig. 4.16. Below $\text{Re}_D \simeq 10{,}000$, the turbulence can be intermittent and the correlation is less reliable. For low-Prandtl-number liquid metals, Notter and Sleicher [4] recommend,

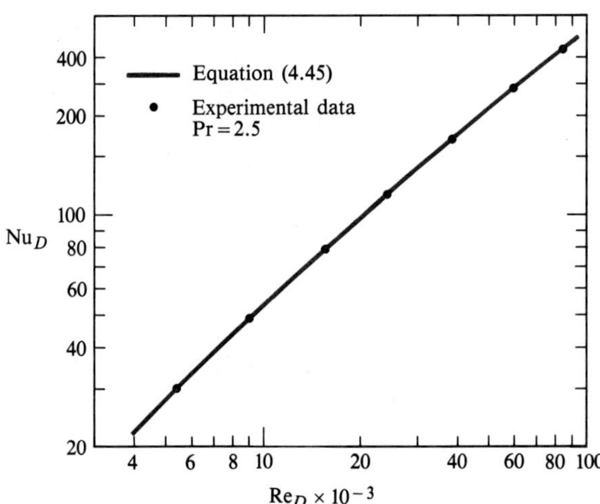

Figure 4.16 Turbulent flow in tubes: comparison of Eq. (4.45) with experimental data of Sparrow and Ohadi [3].

[4] There are many power law formulas similar to Eq. (4.44) but with different constants, Prandtl-number exponents, and schemes for accounting for variable properties. The *Dittus-Boelter*, *McAdams*, and *Colburn* formulas, all developed in the 1930s, are widely used but are often misquoted in the literature. Equation (4.44) should be viewed as the formula Colburn might have proposed if log-log slide rules (or handheld calculators!) had been in common use at that time.

for a uniform wall temperature,

$$\mathrm{Nu}_D = 4.8 + 0.0156\mathrm{Re}_D^{0.85}\mathrm{Pr}^{0.93}; \qquad \begin{matrix} 0.004 < \mathrm{Pr} < 0.01 \\ 10^4 < \mathrm{Re}_D < 10^6 \end{matrix} \tag{4.46}$$

and for a uniform wall heat flux,

$$\mathrm{Nu}_D = 6.3 + 0.0167\mathrm{Re}_D^{0.85}\mathrm{Pr}^{0.93}; \qquad \begin{matrix} 0.004 < \mathrm{Pr} < 0.01 \\ 10^4 < \mathrm{Re}_D < 10^6 \end{matrix} \tag{4.47}$$

Entrance Effects

Near the entrance of a tube, the friction and rate of heat transfer are generally higher than far downstream, where the velocity and temperature profiles are fully developed. A *hydrodynamic entrance length* L_{ef} can be defined as the distance required for the friction factor to decrease to within 5% of its fully developed value f_∞. If the flow is laminar, and if fluid enters the tube through a smooth, rounded entrance as shown in Fig. 4.17, the velocity profile is initially uniform, and analysis gives

$$\frac{L_{ef}\ (5\%)}{D} \simeq 0.05\mathrm{Re}_D \tag{4.48}$$

Similarly, a *thermal entrance length* L_{eh} can be defined as the distance required for the Nusselt number to decrease to within 5% of its fully developed value Nu_∞. If at $x = 0$ the flow is laminar and is already fully developed hydrodynamically

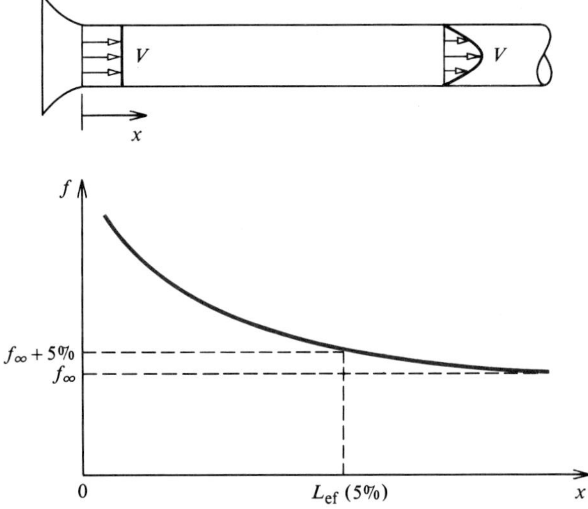

Figure 4.17 Laminar flow in a tube with a smooth, rounded entrance: definition of the hydrodynamic entrance length.

(i.e., the velocity profile is parabolic), and heating commences with a uniform wall temperature then analysis gives

$$\frac{L_{\text{eh}}\ (5\%)}{D} = 0.017\text{Re}_D\text{Pr} \tag{4.49}$$

The corresponding average Nusselt number for a tube of length L is [5]

$$\overline{\text{Nu}}_D = 3.66 + \frac{0.065(D/L)\text{Re}_D\text{Pr}}{1 + 0.04[(D/L)\text{Re}_D\text{Pr}]^{2/3}}; \qquad \text{Re}_D \lesssim 2300 \tag{4.50}$$

which is seen to have the asymptote $\text{Nu}_D = 3.66$ as $L/D \rightarrow \infty$. Equation (4.49) shows that thermal entrance lengths tend to be very short for low-Prandtl-number liquid metals but long for high-Prandtl-number oils. For example, Fig. 4.18 shows oil of $\text{Pr} = 200$ at $\text{Re}_D = 100$ being cooled in a heat exchanger, in which each tube is 100 diameters long. From Eq. (4.48), the hydrodynamic entrance length is 5 diameters, so an assumption of fully developed hydrodynamics throughout would be appropriate. However, the thermal entrance length from Eq. (4.49) is 340 diameters; therefore, the heat transfer is not fully developed. In fact, Eq. (4.50) gives $\overline{\text{Nu}}_D = 9.15$, which is 2.5 times larger than the fully developed value.

Turbulent flows with simply defined hydrodynamics at a tube entrance are seldom encountered in engineering practice. Most often there is a sharp 90° edge, a bend, or an elbow, as shown in Fig. 4.19. The corresponding hydrodynamic entrance lengths vary from about 10–15 diameters, when no large-scale eddies are present, to about 30–40 diameters, when there are large-scale eddies. Similarly, the thermal entrance length depends on entrance configuration as well as Prandtl number. At usual Reynolds numbers, the thermal entrance length can be less than 5 diameters for high-Prandtl-number oils and low-Prandtl-number liquid metals, provided there are no large-scale eddies. For a fluid with Prandtl number of order unity, including gases and water at higher temperatures, the thermal entrance length varies between 15 and 40 diameters. Figure 4.20 shows some typical variations of the local heat transfer coefficient in the entrance region for some practical entrance configurations, based

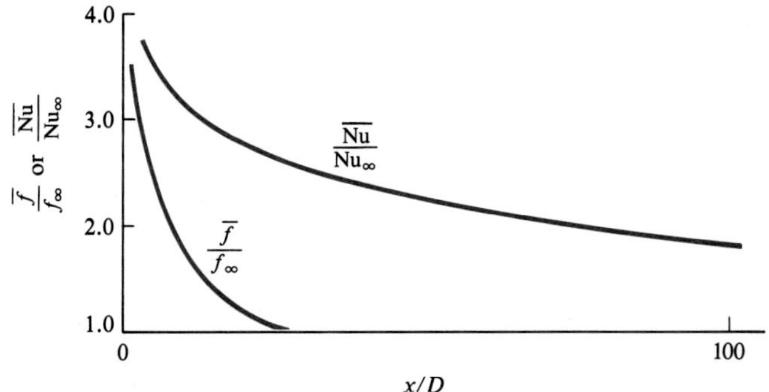

Figure 4.18 Laminar flow of oil in a heat exchanger tube: variation of friction factor and Nusselt number for $\text{Re}_D = 100$, $\text{Pr} = 200$, $L/D = 100$.

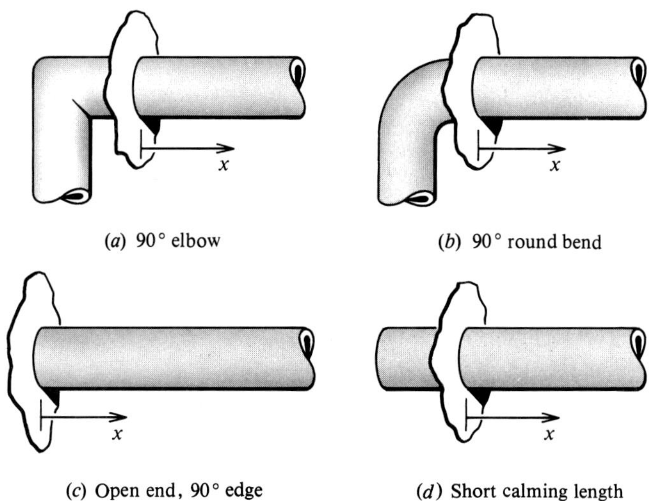

Figure 4.19 Various tube entrance configurations used in practice. (*a*) 90° elbow. (*b*) 90° round bend. (*c*) Open end, 90° edge. (*d*) Short calming length.

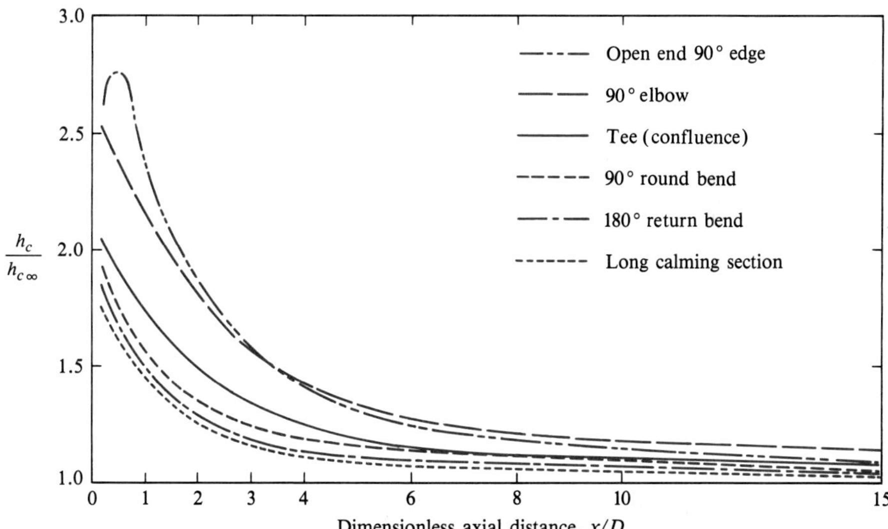

Figure 4.20 Entrance region heat transfer for turbulent flow of gases with various entrance configurations; $h_{c\infty}$ is the heat transfer coefficient far from the entrance, x is the distance from where heating commences, and D is the tube diameter [6].

Table 4.4 Effect of entrance configuration on average heat transfer for turbulent pipe flow: the ratio of average Nusselt number to the fully developed value, $\overline{\text{Nu}}/\text{Nu}_\infty$, for gas flow or Pr about unity [6].

	Pipe Length, in Diameters									
Entrance Configuration	2	4	6	8	10	20	40	80	160	320
Long calming section	1.49	1.34	1.26	1.21	1.17	1.10	1.06	1.03	1.01	1.01
Open end, 90° edge	2.36	1.95	1.73	1.60	1.54	1.32	1.18	1.09	1.05	1.02
90° elbow	2.15	1.86	1.68	1.57	1.49	1.32	1.18	1.09	1.05	1.02
Tee (confluence)	1.77	1.56	1.44	1.36	1.31	1.19	1.10	1.06	1.03	1.01
90° round bend	1.63	1.44	1.34	1.28	1.24	1.16	1.10	1.05	1.03	1.01
180° return bend	1.54	1.37	1.28	1.23	1.19	1.12	1.08	1.04	1.02	1.01

on experiments with air. Table 4.4 shows corresponding average Nusselt numbers for various tube lengths; the table applies to gases and other fluids with Prandtl numbers of about unity.

Use of average Nusselt numbers for internal flows requires some care. Consider first an isothermal wall. If the flow is laminar, Eq. (4.50) may be used to determine the average heat transfer coefficient in Eq. (4.12). If the flow is turbulent and $\text{Pr} \sim 1$, Eq. (4.45) together with Table 4.4 can be used. For turbulent flows and $\text{Pr} \gg 1$ or $\text{Pr} \ll 1$, entrance effects can be neglected unless the tube is very short: then Eq. (4.45) can be used directly. Examples 4.1 and 4.2 illustrate the procedure. For a nonisothermal wall, such as that found in the two-stream heat exchanger illustrated in Fig. 1.22*b*, the problem is more difficult and will be discussed in Chapter 8.

Flow in Ducts of Various Cross Sections

Table 4.5 gives friction factors and Nusselt numbers for fully developed laminar flow in ducts of various cross sections. For noncircular ducts, the length scale in the Reynolds and Nusselt numbers is the **hydraulic diameter**, $D_h = 4A_c/\mathscr{P}$ where A_c is the cross-sectional area for flow and $\mathscr{P}$ is the wetted perimeter. (For a round tube, $D_h = D$ since $A_c = \pi D^2/4$ and $\mathscr{P} = \pi D$.)

Entrance lengths and the variation of Nusselt number in the thermal entrance length are similar to those for a round tube. For example, the average Nusselt number for flow between isothermal parallel plates of length L is [5]

$$\overline{\text{Nu}}_{D_h} = 7.54 + \frac{0.03(D_h/L)\text{Re}_{D_h}\text{Pr}}{1 + 0.016[(D_h/L)\text{Re}_{D_h}\text{Pr}]^{2/3}}; \qquad \text{Re}_{D_h} \lesssim 2800 \tag{4.51}$$

where the hydraulic diameter is simply twice the spacing of the plates. Transition Reynolds numbers are a little different from those for a round tube, with a value of 2800 being more appropriate for flow between parallel plates.

For turbulent flow, the correlations for a round tube can be used, with the diameter D replaced by the hydraulic diameter D_h. The same correlations can be used for all turbulent duct flows because the viscous sublayer around the perimeter of the duct is

Table 4.5 Nusselt numbers and the product of friction factor times Reynolds number for fully developed laminar flow in ducts of various cross-sections.

Cross Section	Nu_{D_h} Constant Axial Wall Heat Flux	Nu_{D_h} Constant Axial Wall Temperature	fRe_{D_h}
Equilateral triangle	3.1	2.4	53
Circle	4.364	3.657	64
Square (1 × 1)	3.6	2.976	57
Rectangle (1 × 1.4)	3.8	3.1	59
1 × 2	4.1	3.4	62
1 × 3	4.8	4.0	69
1 × 4	5.3	4.4	73
1 × 8	6.5	5.6	82
∞	8.235	7.541	96
Heated / ∞ / Insulated	5.385	4.861	96

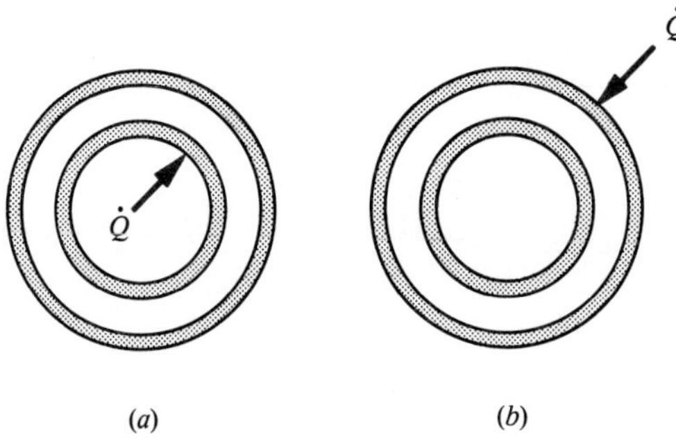

Figure 4.21 Schematic of annular ducts. (*a*) Heat transfer through the inner wall. (*b*) Heat transfer through the outer wall.

very thin, and the velocity and temperature are nearly uniform across the core fluid. Since the viscous sublayer is the major resistance to momentum and heat transfer, the precise shape of the core fluid is not critical. For the annular ducts shown in Fig. 4.21, greater accuracy can be obtained if the Nusselt number given by Eq. (4.45) is multiplied by the correction factors recommended by Petukov and Roizen [7]. For heat transfer through the inner wall with the outer wall insulated, the factor is

$$0.86\left(\frac{D_i}{D_o}\right)^{-0.16} \tag{4.52a}$$

where D_i and D_o are the inner and outer diameters, respectively. For the inner wall insulated and heat transfer through the outer wall, the factor is

$$1 - 0.14\left(\frac{D_i}{D_o}\right)^{0.6} \tag{4.52b}$$

When Eq. (4.52) is used, the appropriate heat transfer area is that of the heated wall only.

Reynolds Analogy

The physical processes of momentum and heat transfer in turbulent flow are very similar. The turbulent eddy that transports momentum from the core fluid to the fluid near the wall also transports heat. For fluids with Prandtl number values close to unity, the resistance of the viscous sublayer to transfer of momentum is almost the same as the resistance to heat transfer. Thus, we would expect a simple relation between friction and heat transfer. Using Eqs. (4.43) and (4.44),

$$\frac{f}{8} = \frac{C_f}{2} = \mathrm{StPr}^{0.6} \tag{4.53}$$

For Pr = 1,

$$\frac{C_f}{2} = \mathrm{St}$$

which is the famous *Reynolds analogy* between momentum and heat transfer, first proposed by O. Reynolds in 1874.

Variable-Property Effects

Recommended exponents m and n for property and temperature ratio corrections are given in Table 4.6. These data are for pipe flow but are also approximately valid for other duct shapes.

Table 4.6 Exponents for property and temperature ratio corrections for use in Eqs. (4.37) and (4.38): flow in tubes.

Type of Flow	Fluid	Wall Condition	m	n
Laminar	Liquids (μ_s/μ_b)	Heating	0.58	−0.11
		Cooling	0.50	−0.11
	Gases (T_s/T_b)	Heating and cooling	1	0
Turbulent	Liquids (μ_s/μ_b)	Heating	0.25	−0.11
		Cooling	0.25	−0.25
	Gases (T_s/T_b)	Heating	−0.2	−0.55
		Cooling	−0.1	0.0

EXAMPLE 4.1 Laminar Flow of Oil

SAE 50 oil flows at 0.007 kg/s through a 1 cm–I.D. tube, 1.5 m long. The wall temperature is 300 K, and the inlet oil bulk temperature is 377 K. Estimate the average heat transfer coefficient.

Solution

Given: SAE 50 oil flowing inside a tube.

Required: Average heat transfer coefficient.

Assumptions: The flow is hydrodynamically fully developed at the commencement of cooling.

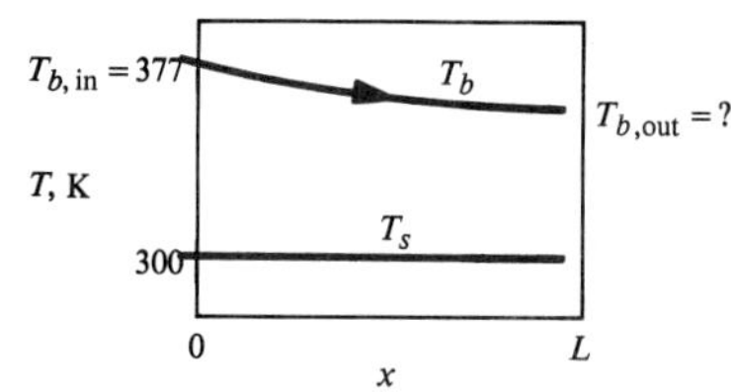

The average Nusselt number is to be evaluated using the bulk temperature for fluid properties, with a subsequent viscosity ratio correction for variable property effects from Table 4.6. The bulk temperature of the oil decreases along the tube, but an arithmetic average of the inlet and outlet values is usually adequate. There remains a difficulty, however. We do not know the oil outlet temperature since it depends on the heat transfer coefficient we are attempting to calculate. We must guess and check later. Therefore, expecting an oil temperature drop of 10° to 20°C, we take the average of the inlet and outlet bulk oil temperatures to be 370 K. From Table A.8, $k = 0.137$ W/m K, $\mu = 1.89 \times 10^{-2}$ kg/m s, $c_p = 2200$ J/kg K, and Pr $= 300$.

We first calculate the tube cross-sectional area and the Reynolds number:

$$A_c = \pi \frac{D^2}{4} = \pi \left(\frac{0.01^2}{4} \right) = 7.854 \times 10^{-5} \text{ m}^2$$

$$\text{Re}_D = \frac{(\dot{m}/A_c)D}{\mu} = \frac{(0.007/7.854 \times 10^{-5})(0.01)}{1.89 \times 10^{-2}} = 47.2$$

Since $\text{Re}_D < 2300$, the flow is laminar, and Eq. (4.50) applies:

$$\overline{\text{Nu}}_D = 3.66 + \frac{(0.065)(D/L)\text{Re}_D\text{Pr}}{1 + 0.04[(D/L)\text{Re}_D\text{Pr}]^{2/3}}$$

$$= 3.66 + \frac{(0.065)(0.01/1.5)(47.2)(300)}{1 + 0.04[(0.01/1.5)(47.2)(300)]^{2/3}} = 3.66 + 3.35 = 7.01$$

We now correct for variable properties. From Table 4.6, $n = -0.11$, and from Table A.8, $\mu_s = \mu(300 \text{ K}) = 50.3 \times 10^{-2}$ kg/m s:

$$\left(\frac{\mu_s}{\mu_b} \right)^{-0.11} = \left(\frac{50.3 \times 10^{-2}}{1.89 \times 10^{-2}} \right)^{-0.11} = 0.697$$

The corrected Nusselt number is $\overline{\text{Nu}}_D = (7.01)(0.697) = 4.89$. Thus,

$$\bar{h}_c = \left(\frac{k}{D} \right) \overline{\text{Nu}}_D = \left(\frac{0.137}{0.01} \right) 4.89 = 66.9 \text{ W/m}^2 \text{ K}$$

To check the guessed average bulk temperature, we must find the outlet oil temperature. Since the wall temperature T_s is constant, Eq. (4.11) applies.

$$T_{b,\text{out}} = T_s - (T_s - T_{b,\text{in}})e^{-\bar{h}_c 2\pi RL/\dot{m}c_p}$$

$$= 300 - (300 - 377)e^{-(66.9)(2\pi)(0.005)(1.5)/(0.007)(2200)}$$

$$= 300 - (300 - 377)e^{-0.205} = 363 \text{ K}$$

The guessed average temperature was thus appropriate, and no iteration is required.

Solution using CONV

The required input is:

Configuration number = 2 (tubes: laminar flow)
Fluid = 2 (SAE 50 engine oil)
$T_s = 300$
$T_b = 370$
P = Any value for a liquid
$D = 0.01$
$L = 1.5$
$\dot{m} = 0.007$

The output is:

SAE 50 oil properties at 370 K
Re = 47.2
f = 7.00
$\Delta P/L$ = 3303 Pa/m
Nu = 4.89
h_c = 67.0 W/m^2 K

Comments

1. Input to CONV must be in SI units with temperatures in kelvins.
2. Notice that for a liquid, any value for pressure can be input into CONV.
3. The tube is essentially a single-stream heat exchanger with the number of transfer units $N_{tu} = \bar{h}_c 2\pi RL/\dot{m}c_p = 0.205$. It is good practice to calculate N_{tu} and think in terms of its value. If $N_{tu} \ll 1$, $T_{b,\text{out}} \simeq T_{b,\text{in}}$; if $N_{tu} > 3$ or 4, $T_{b,\text{out}} \simeq T_s$.

EXAMPLE 4.2 Turbulent Flow of Air

Air flows at 0.11 kg/s through a 1 cm–wide, 0.5 m–high channel of a plate-type heat exchanger. The channel is 0.8 m long and its walls are at 600 K. If the pressure is 1 atm and the average of the inlet and outlet bulk air temperatures is estimated to be 400 K, determine the average heat transfer coefficient. The entrance has a 90° edge.

Solution

Given: Air flowing between parallel plates.

Required: Average heat transfer coefficient, $\bar{h}_c$.

Assumptions: The data for tube flow in Table 4.4 can be used to correct for entrance effects.

At 400 K, 1 atm, air properties from Table A.7 are: k = 0.0331 W/m K, Pr = 0.69, and $\mu = 22.5 \times 10^{-6}$ kg/m s. Thus,

$$\text{Hydraulic diameter } D_h = \frac{4A_c}{\mathscr{P}} = \frac{4 \times (0.01)(0.5)}{2 \times (0.5 + 0.01)}$$

$$= 0.0196 \text{ m}$$

$$\text{Re}_{D_h} = \frac{(\dot{m}/A_c)D_h}{\mu} = \frac{(0.11/0.005)(0.0196)}{22.5 \times 10^{-6}}$$

$$= 19{,}160 > 2800$$

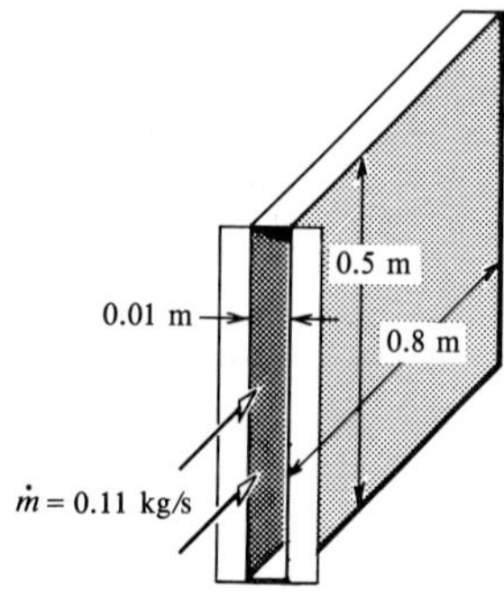

The flow is turbulent, so that Eqs. (4.42) and (4.45) apply:

$$f = (0.790 \ln \mathrm{Re}_{D_h} - 1.64)^{-2} = (0.790 \ln 19{,}160 - 1.64)^{-2} = 0.0264$$

$$\mathrm{Nu}_{D_h} = \frac{(f/8)(\mathrm{Re}_{D_h} - 1000)\mathrm{Pr}}{1 + 12.7(f/8)^{1/2}(\mathrm{Pr}^{2/3} - 1)} = \frac{(0.0264/8)(19{,}160 - 1000)(0.69)}{1 + 12.7(0.0264/8)^{1/2}(0.69^{2/3} - 1)} = 49.2$$

We now correct for variable-property effects. Table 4.6 gives $n = -0.55$:

$$\left(\frac{T_s}{T_b}\right)^{-0.55} = \left(\frac{600}{400}\right)^{-0.55} = 0.800$$

$$\mathrm{Nu}_{D_h} = (49.2)(0.800) = 39.4$$

To correct for entrance effects, Table 4.4 applies approximately. For item 2 and $L/D_h = 41$,

$$\overline{\mathrm{Nu}}/\mathrm{Nu}_\infty = 1.18$$

$$\overline{\mathrm{Nu}}_{D_h} = (39.4)(1.18) = 46.5$$

$$\bar{h}_c = (k/D_h)\overline{\mathrm{Nu}}_{D_h} = (0.0331/0.0196)(46.5) = 78.5 \text{ W/m}^2\text{ K}$$

Comments

1. Notice that the friction factor f is *not* corrected for variable property effects before substitution in Eq. (4.45) to calculate Nu_{D_h}.
2. Use CONV to check Re_{D_h} and Nu_{D_h}; note that the friction factor given by CONV has been corrected for variable property effects.

4.3.2 External Forced Flows

In this section, various external forced flows are considered and correlations for skin friction and heat transfer given. The heat transfer correlations all apply to an *isothermal surface*; other wall boundary conditions are discussed at the end of the section.

Flow along a Flat Plate

Figure 4.22 shows a schematic of flow along a flat plate. A laminar boundary layer forms from the leading edge, and transition to turbulent flow usually occurs at a value of $\mathrm{Re}_x = u_e x/\nu$ where u_e is the free-stream velocity, in the range 50,000–500,000, for x measured from the leading edge. Higher values are associated with careful wind-tunnel tests, and lower values are more characteristic of practical situations where such factors as surface roughness and vibration are present. If the Reynolds number were based on an appropriate thickness of the boundary layer, the transition value would be of the same order as that given in Section 4.3.1 for pipe flow. Both the local shear stress τ_{sx} and the local heat transfer coefficient h_{cx} vary along the plate as shown, and a constant asymptotic value is never attained; that is, conditions are similar to the entrance region of internal duct flows.

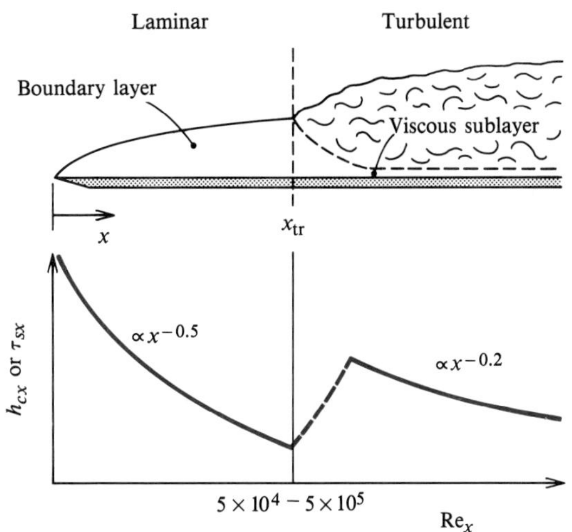

Figure 4.22 Forced flow along a flat plate, showing boundary layer growth and the resulting variation of the local shear stress and heat transfer coefficient.

For laminar flow, the local skin friction coefficient at a distance x from the leading edge is

$$C_{fx} = 0.664\mathrm{Re}_x^{-1/2} \tag{4.54}$$

where the subscript x on the Reynolds number again indicates the length scale on which it is based. The average shear stress for a plate of length L can be obtained from the average skin friction coefficient,

$$\overline{C}_f = \frac{1}{L}\int_0^L C_{fx}\,dx = 1.328\mathrm{Re}_L^{-1/2}; \qquad 10^3 < \mathrm{Re}_L \lesssim 5\times10^5 \tag{4.55}$$

where the Reynolds number is now based on plate length, L. The local Nusselt number, $\mathrm{Nu}_x = h_{cx}x/k$, is given by

$$\mathrm{Nu}_x = 0.332\mathrm{Re}_x^{1/2}\mathrm{Pr}^{1/3}; \qquad \mathrm{Pr} > 0.5 \tag{4.56}$$

To obtain a correlation for the average heat transfer coefficient, we cannot simply average the Nusselt number, since it contains x. Instead, we must evaluate

$$\overline{h}_c = \frac{1}{L}\int_0^L h_{cx}\,dx = \frac{1}{L}\int_0^L \left(\frac{k}{x}\right)(0.332)\left(\frac{u_e x}{\nu}\right)^{1/2}\mathrm{Pr}^{1/3}\,dx$$

to obtain

$$\overline{\mathrm{Nu}} = \frac{\overline{h}_c L}{k} = 0.664\mathrm{Re}_L^{1/2}\mathrm{Pr}^{1/3}; \qquad \mathrm{Pr} > 0.5 \tag{4.57}$$

For low-Prandtl-number liquid metals, an appropriate correlation is

$$\overline{\mathrm{Nu}} = 1.128\mathrm{Re}_L^{1/2}\mathrm{Pr}^{1/2}; \qquad \mathrm{Pr} \ll 1 \tag{4.58}$$

Equations (4.54) through (4.58) are all based on exact analysis and have been confirmed by experiment. The analysis is given in Chapter 5.

For turbulent flow, the local skin friction coefficient is given by simple power law expressions,

$$C_{fx} = 0.0592\mathrm{Re}_x^{-1/5}; \qquad 10^5 < \mathrm{Re}_x < 10^7 \tag{4.59a}$$

$$C_{fx} = 0.026\mathrm{Re}_x^{-1/7}; \qquad 10^6 < \mathrm{Re}_x < 10^9 \tag{4.59b}$$

or, if greater accuracy is required, White's formula may be used [8]:

$$C_{fx} = \frac{0.455}{(\ln 0.06\mathrm{Re}_x)^2}; \qquad 10^5 < \mathrm{Re}_x < 10^9 \tag{4.60}$$

In these expressions, x is the distance measured from the *virtual origin* of the turbulent boundary layer, shown in Fig. 4.23; for flow along a flat plate, this can be taken to be the leading edge with sufficient accuracy for most engineering purposes. To determine the total drag, an average skin friction coefficient is required. If transition is assumed to occur abruptly at x_{tr}, the average shear stress on a plate of length L is

$$\overline{\tau}_s = \frac{1}{L}\left[\int_0^{x_{\mathrm{tr}}} \tau_s(\text{laminar})\,dx + \int_{x_{\mathrm{tr}}}^{L} \tau_s(\text{turbulent})\,dx\right]$$

Dividing by $(1/2)\rho u_e^2$,

$$\overline{C}_f = \frac{1}{L}\left[\int_0^{x_{\mathrm{tr}}} C_{fx}(\text{laminar})\,dx + \int_{x_{\mathrm{tr}}}^{L} C_{fx}(\text{turbulent})\,dx\right]$$

and substituting from Eqs. (4.54) and (4.59a),

$$\overline{C}_f = \frac{1}{L}\left[\int_0^{x_{\mathrm{tr}}} 0.664\mathrm{Re}_x^{-1/2}\,dx + \int_{x_{\mathrm{tr}}}^{L} 0.0592\mathrm{Re}_x^{-1/5}\,dx\right]$$

It is convenient to integrate with respect to Re_x rather than x:

$$\mathrm{Re}_x = \frac{u_e x}{\nu}; \qquad d\mathrm{Re}_x = \frac{u_e}{\nu}\,dx \qquad \text{or} \qquad dx = \frac{\nu}{u_e}\,d\mathrm{Re}_x$$

$$\overline{C}_f = \frac{\nu}{u_e L}\left[\int_0^{\mathrm{Re}_{\mathrm{tr}}} 0.664\mathrm{Re}_x^{-1/2}\,d\mathrm{Re}_x + \int_{\mathrm{Re}_{\mathrm{tr}}}^{\mathrm{Re}_L} 0.0592\mathrm{Re}_x^{-1/5}\,d\mathrm{Re}_x\right]$$

$$= \frac{1}{\mathrm{Re}_L}\left[(2)(0.664)\mathrm{Re}_{\mathrm{tr}}^{1/2} + (5/4)(0.0592)\left(\mathrm{Re}_L^{4/5} - \mathrm{Re}_{\mathrm{tr}}^{4/5}\right)\right]$$

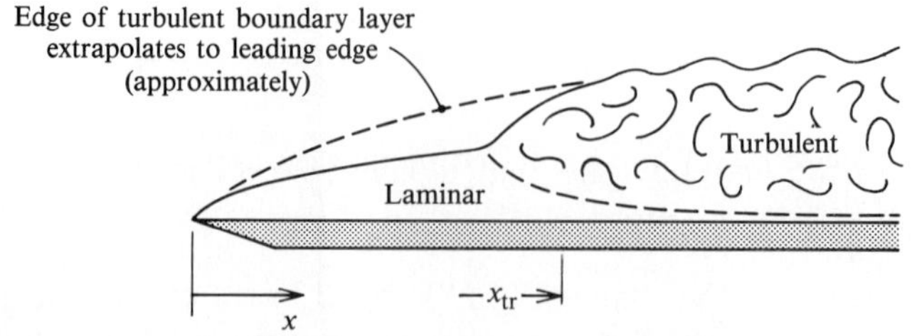

Figure 4.23 Virtual origin of a turbulent boundary layer on a flat plate.

which can be rearranged as

$$\overline{C}_f = 1.328\mathrm{Re}_{\mathrm{tr}}^{-1/2}\left(\frac{\mathrm{Re}_{\mathrm{tr}}}{\mathrm{Re}_L}\right) + 0.0740\mathrm{Re}_L^{-1/5}\left[1 - \left(\frac{\mathrm{Re}_{\mathrm{tr}}}{\mathrm{Re}_L}\right)^{4/5}\right] \tag{4.61}$$

Equation (4.61) is accurate for $\mathrm{Re}_L < 10^7$. For higher Reynolds numbers, the integration can be extended using Eq. (4.59*b*). Alternatively, although Eq. (4.60) cannot be integrated analytically, it can be integrated numerically and the result curve-fitted. When used in place of Eq. (4.59), the resulting expression for $\overline{C}_f$ is

$$\overline{C}_f = 1.328\mathrm{Re}_{\mathrm{tr}}^{-1/2}\left(\frac{\mathrm{Re}_{\mathrm{tr}}}{\mathrm{Re}_L}\right) + \frac{0.523}{\ln^2 0.06\mathrm{Re}_L} - \left(\frac{\mathrm{Re}_{\mathrm{tr}}}{\mathrm{Re}_L}\right)\frac{0.523}{\ln^2 0.06\mathrm{Re}_{\mathrm{tr}}} \tag{4.62}$$

and is accurate for $\mathrm{Re}_L < 10^9$. Equation (4.62) is recommended for general use. The average skin friction coefficient $\overline{C}_f$ is also the *drag coefficient*, since the total viscous drag force on a plate of width W and length L is $F = \overline{C}_f(1/2)\rho u_e^2 WL$.

For heat transfer across a turbulent boundary layer, the local Nusselt number is given by White [9] as

$$\mathrm{Nu}_x = \frac{(C_{fx}/2)\mathrm{Re}_x\mathrm{Pr}}{1 + 12.7(C_{fx}/2)^{1/2}(\mathrm{Pr}^{2/3} - 1)} \tag{4.63}$$

with C_{fx} given by Eq. (4.59*a*); this form is valid for $0.5 < \mathrm{Pr} < 2000$, $5 \times 10^5 < \mathrm{Re}_x < 10^7$. Alternatively, there is a simpler power law expression recommended by Whitaker [10],

$$\mathrm{Nu}_x = 0.029\mathrm{Re}_x^{0.8}\mathrm{Pr}^{0.43} \tag{4.64}$$

which is valid for $0.7 < \mathrm{Pr} < 400$, $5 \times 10^5 < \mathrm{Re}_x < 3 \times 10^7$. To determine the average Nusselt number, we first evaluate the average heat transfer coefficient:

$$\overline{h}_c = \frac{1}{L}\left[\int_0^{x_{\mathrm{tr}}} h_{cx}(\text{laminar})\,dx + \int_{x_{\mathrm{tr}}}^{L} h_{cx}(\text{turbulent})\,dx\right]$$

Substituting Eqs. (4.56) and (4.64),

$$\overline{h}_c = \frac{1}{L}\left[\int_0^{x_{\mathrm{tr}}} (k/x)0.332\mathrm{Re}_x^{1/2}\mathrm{Pr}^{2/3}\,dx + \int_{x_{\mathrm{tr}}}^{L} (k/x)0.029\mathrm{Re}_x^{0.8}\mathrm{Pr}^{0.43}\,dx\right]$$

from which $\overline{\mathrm{Nu}} = \overline{h}_c L/k$ is obtained as

$$\overline{\mathrm{Nu}} = 0.664\mathrm{Re}_{\mathrm{tr}}^{1/2}\mathrm{Pr}^{1/3} + 0.036\mathrm{Re}_L^{0.8}\mathrm{Pr}^{0.43}\left[1 - \left(\frac{\mathrm{Re}_{\mathrm{tr}}}{\mathrm{Re}_L}\right)^{0.8}\right] \tag{4.65}$$

For the flat plate, there is an analogy between momentum and heat transfer for both laminar and turbulent flow. For laminar flow using the relation $\mathrm{St}_x = \mathrm{Nu}_x/\mathrm{Re}_x \mathrm{Pr}$ and Eqs. (4.54) and (4.56),

$$\mathrm{St}_x = \left(\frac{C_{fx}}{2}\right)\mathrm{Pr}^{-2/3}; \qquad \mathrm{Pr} > 0.5 \tag{4.66}$$

For turbulent flow, using Eqs. (4.59*a*) and (4.64),

$$\mathrm{St}_x = \left(\frac{C_{fx}}{2}\right)\mathrm{Pr}^{-0.57}; \qquad 0.7 < \mathrm{Pr} < 400 \tag{4.67}$$

is a good approximation. Equation (4.63) itself can also be viewed as a form of the analogy.

Flow across a Cylinder

The flow pattern around a cylinder depends very much on the Reynolds number VD/ν, where V is the velocity of the undisturbed flow. Figure 4.24 gives the main flow regimes, showing in particular the shedding of vortices, separation of the boundary layer, and at higher Reynolds numbers, transition from a laminar to a

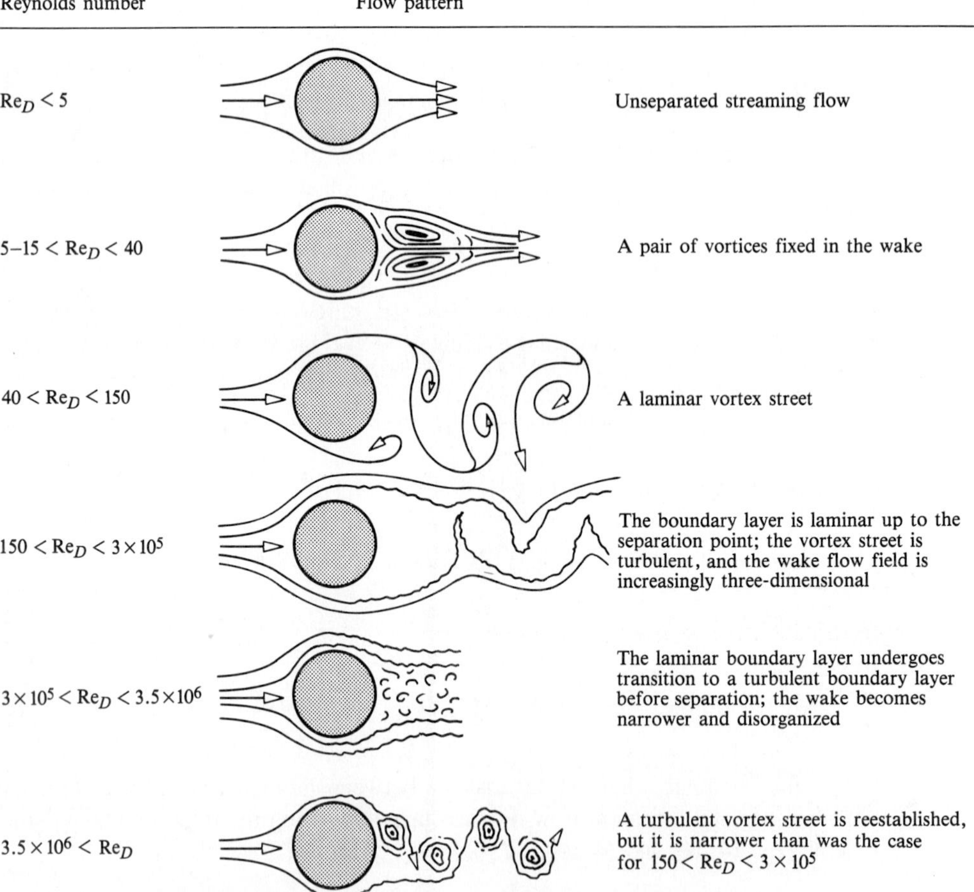

Figure 4.24 The main flow regimes for flow across a cylinder.

turbulent boundary layer before separation. The local skin friction coefficient varies in a complicated fashion around the cylinder. However, in practice, what is usually required is the total drag force F on the cylinder due to both friction, or *viscous drag*, and pressure imbalance, or *form drag*. The drag force can be obtained from the drag coefficient C_D:

$$C_D = \frac{F}{(1/2)\rho V^2 A_f} \tag{4.68}$$

where A_f is the area of the cylinder *normal* to the flow, that is, DL for a cylinder of diameter D and length L. Figure 4.25 is a graph of C_D versus Re_D. At low Reynolds numbers, viscous drag predominates, and a useful formula is

$$C_D = 1 + \frac{10}{\mathrm{Re}_D^{2/3}}; \qquad 1 < \mathrm{Re}_D < 10^4 \tag{4.69}$$

Above about $\mathrm{Re}_D = 10^3$, the flow separates at $\theta \simeq 80°$, and the form drag dominates to give $C_D \simeq 1.2$, nearly independent of Reynolds number. At about $\mathrm{Re}_D = 2 \times 10^5$, there is a transition from a laminar to a turbulent boundary layer, which has the effect of moving the separation point to the rear of the cylinder (up to $\theta = 130°$), thereby causing a significant decrease in the form drag. Figure 4.26 is a schematic showing laminar and turbulent boundary layer separation.

The corresponding variation of the local heat transfer coefficient around the cylinder is also very complicated, as might be expected. Figure 4.27 shows some experimental data. The Nusselt number is high on the front of the cylinder, where the

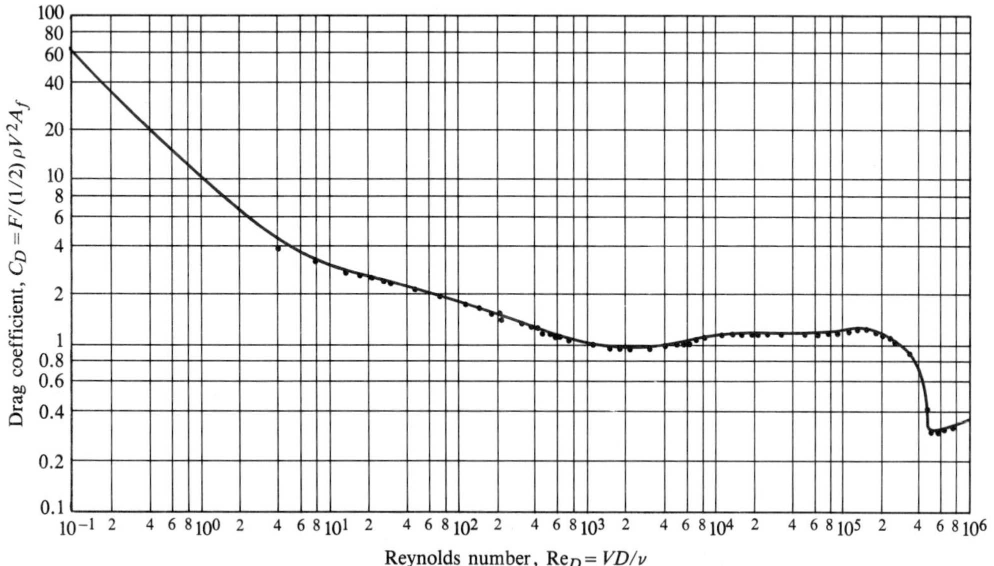

Figure 4.25 The drag coefficient $C_D = F/(1/2)\rho V^2 A_f$ for flow across a cylinder [11]. (Adapted with permission.)

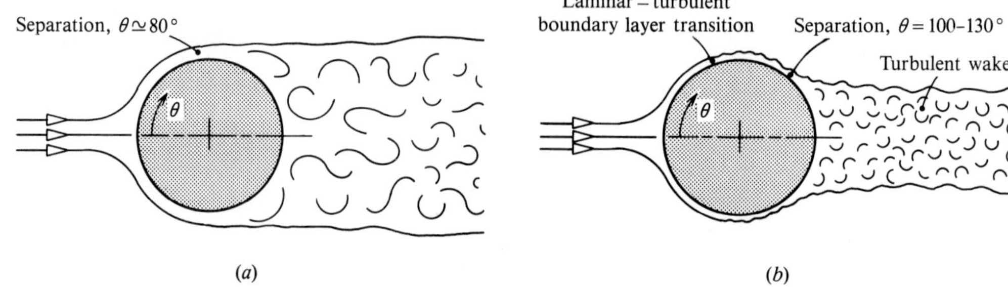

Figure 4.26 Schematic showing the difference between (*a*) laminar and (*b*) turbulent boundary layer separation for flow across a cylinder.

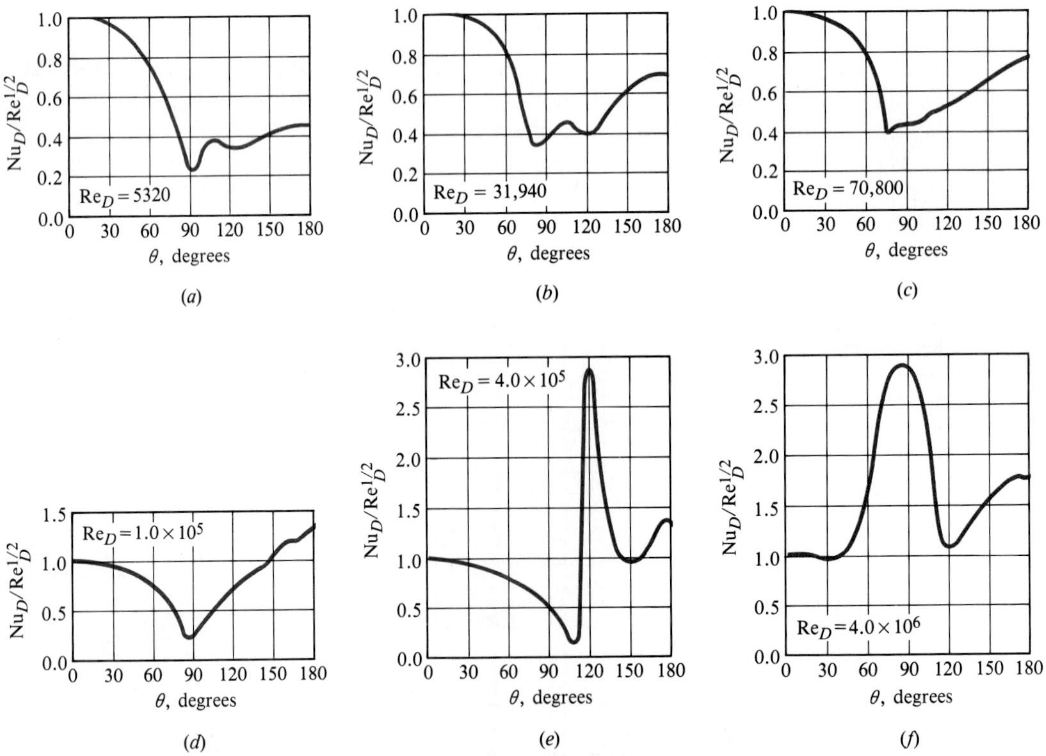

Figure 4.27 Variation of local heat transfer coefficient around a cylinder in a cross-flow of air [12,13,14]. Notice that Eq (4.70) gives $\mathrm{Nu}/\mathrm{Re}_D^{1/2} \simeq 1.00$ at the forward stagnation line.

boundary layer is thin, and decreases as the boundary layer grows in thickness around the cylinder. When the flow does not separate, the Nusselt number decreases steadily toward the rear of the cylinder. For $\mathrm{Re}_D \gtrsim 10^3$, the flow separates at about $\theta = 80°$, and there is a Nusselt number minimum in the separated region. On the rear of the cylinder, there can be a local maximum at the point of reattachment of the flow; for example, Fig. 4.27*b* shows a local maximum at 105°. For $\mathrm{Re}_D \gtrsim 2 \times 10^5$, there is a transition from laminar to turbulent flow in the boundary layer that delays separation. Figure 4.27*e* shows two Nusselt number minima, the first just before transition, at about $\theta = 110°$, and the second in the separated region, at about $\theta = 150°$. At $\mathrm{Re}_D = 4.0 \times 10^6$, Fig. 4.27*f* shows the corresponding Nusselt number minima at 30° and 120°, respectively. The local Nusselt number at the forward stagnation line can be obtained by analysis [15]. For $\mathrm{Pr} > 0.5$,

$$\mathrm{Nu}_D = 1.15\mathrm{Re}_D^{1/2}\mathrm{Pr}^{1/3} \tag{4.70}$$

The average Nusselt number for $\mathrm{Pr} > 0.5$ is given by a rather complicated correlation suggested by Churchill and Bernstein [16]:

$$\overline{\mathrm{Nu}_D} = 0.3 + \frac{0.62\mathrm{Re}_D^{1/2}\mathrm{Pr}^{1/3}}{[1 + (0.4/\mathrm{Pr})^{2/3}]^{1/4}}; \qquad \mathrm{Re}_D < 10^4 \tag{4.71a}$$

$$\overline{\mathrm{Nu}_D} = 0.3 + \frac{0.62\mathrm{Re}_D^{1/2}\mathrm{Pr}^{1/3}}{[1 + (0.4/\mathrm{Pr})^{2/3}]^{1/4}}\left[1 + \left(\frac{\mathrm{Re}_D}{282{,}000}\right)^{1/2}\right]; \tag{4.71b}$$

$$2 \times 10^4 < \mathrm{Re}_D < 4 \times 10^5$$

$$\overline{\mathrm{Nu}_D} = 0.3 + \frac{0.62\mathrm{Re}_D^{1/2}\mathrm{Pr}^{1/3}}{[1 + (0.4/\mathrm{Pr})^{2/3}]^{1/4}}\left[1 + \left(\frac{\mathrm{Re}_D}{282{,}000}\right)^{5/8}\right]^{4/5}; \tag{4.71c}$$

$$4 \times 10^5 < \mathrm{Re}_D < 5 \times 10^6$$

However, for very low Reynolds numbers, Nakai and Okazaki [17] recommend

$$\overline{\mathrm{Nu}_D} = \frac{1}{0.8237 - \ln(\mathrm{Re}_D\mathrm{Pr})^{1/2}}; \qquad \mathrm{Re}_D\mathrm{Pr} < 0.2 \tag{4.72}$$

Flow over a Sphere

The flow pattern around a sphere is somewhat similar to that for a cylinder except that it does not exhibit the same regular eddy shedding phenomena. Figure 4.28 shows a graph of C_D versus Re_D and is very similar to Fig. 4.25 for a cylinder. For very low Reynolds-number *creeping* flows, *Stokes' law* is valid [18]:

$$C_D = \frac{24}{\mathrm{Re}_D}; \qquad \mathrm{Re}_D < 0.5 \tag{4.73}$$

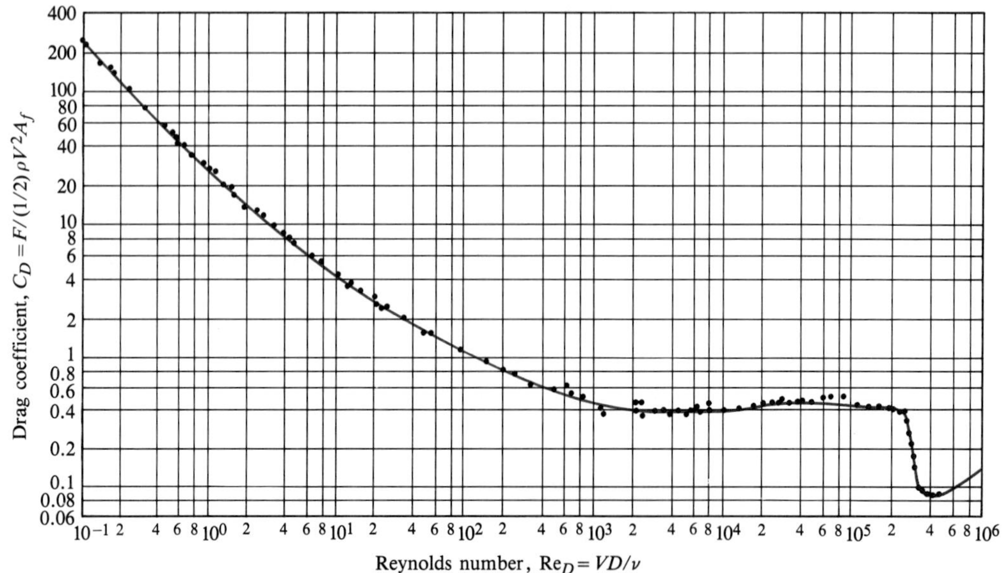

Figure 4.28 The drag coefficient $C_D = F/(1/2)\rho V^2 A_f$ for flow across a sphere [11]. (Adapted with permission.)

For somewhat higher Reynolds numbers, a useful correlation is

$$C_D \simeq \frac{24}{\mathrm{Re}_D}\left(1 + \frac{\mathrm{Re}_D^{2/3}}{6}\right); \qquad 2 < \mathrm{Re}_D < 500 \tag{4.74}$$

At still higher Reynolds numbers, *Newton's law* applies, and C_D is a constant, equal to approximately 0.44 in the range $500 < \mathrm{Re}_D < 2 \times 10^5$.

The local Nusselt number at the forward stagnation point can be obtained by analysis [15]. For $\mathrm{Pr} > 0.5$,

$$\mathrm{Nu}_D = 1.32\mathrm{Re}_D^{1/2}\mathrm{Pr}^{1/3} \tag{4.75}$$

Whitaker [10] recommends that the average Nusselt number for $0.7 < \mathrm{Pr} < 380$ be calculated from

$$\overline{\mathrm{Nu}}_D = 2 + (0.4\mathrm{Re}_D^{1/2} + 0.06\mathrm{Re}_D^{2/3})\mathrm{Pr}^{0.4}; \qquad 3.5 < \mathrm{Re}_D < 8 \times 10^4 \tag{4.76}$$

Notice that Eq. (4.76) has a lower limit of $\overline{\mathrm{Nu}}_D = 2$, which corresponds to conduction from a sphere into stationary infinite surrounds. Figure 4.29 illustrates this important point. From Section 2.3.3, the heat flow by conduction across a spherical shell is given by

$$\dot{Q} = \frac{4\pi k(T_1 - T_2)}{1/r_1 - 1/r_2} \tag{4.77}$$

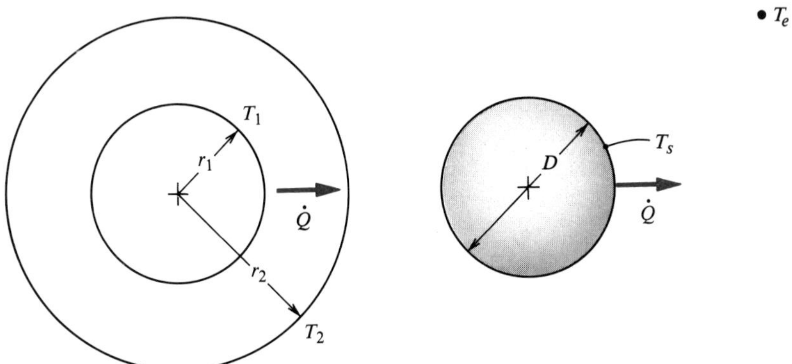

Figure 4.29 Schematic for conduction across a spherical shell, and from a sphere into stationary infinite surrounds.

which for $D = 2r_1$ and $r_2 \to \infty$ is

$$\dot{Q} = \frac{2k}{D}\pi D^2(T_1 - T_2) = h_c A(T_s - T_e)$$

Hence, the heat transfer coefficient is

$$h_c = \frac{2k}{D} \qquad \text{and} \qquad \text{Nu}_D = 2 \tag{4.78}$$

Variable Property Effects

All the correlations given in this section for external forced flows should be used with fluid properties evaluated at the mean film temperature. An exception is Eq. (4.76), for which better results are obtained if fluid properties are evaluated at the bulk temperature, with a viscosity ratio correction $(\mu_s/\mu_e)^n$, $n = -0.25$, applied to the convection contribution.

Other Wall Boundary Conditions

All the foregoing correlations for external flows are valid for an isothermal surface. Correlations are also available for a uniform wall heat flux, for which the local Nusselt numbers tend to be higher than their counterparts for an isothermal surface. For laminar flow along a uniformly heated flat plate, analysis gives the local Nusselt number [15,19]

$$\text{Nu}_x = 0.453\text{Re}_x^{1/2}\text{Pr}^{1/3}; \qquad \text{Pr} > 0.5 \tag{4.79}$$

which is 36% higher than the values obtained from Eq. (4.56) for an isothermal surface. For turbulent gas flow along a uniformly heated flat plate, Kays and Crawford [15] recommend for the local Nusselt number

$$\mathrm{Nu}_x = 0.030\mathrm{Re}_x^{0.8}\mathrm{Pr}^{0.4}; \qquad 0.5 < \mathrm{Pr} < 400, \quad 5\times10^5 < \mathrm{Re}_x < 5\times10^6 \tag{4.80}$$

which is only 4% higher than the isothermal plate result, Eq. (4.64).

Appropriately averaged Nusselt numbers are far less sensitive to the wall boundary condition. For a nonisothermal wall, Eq. (1.26), the definition of an average heat transfer coefficient, used to derive Eq. (4.57), namely,

$$\bar{h}_c = \frac{1}{A}\int_0^A h_c\,dA \qquad \left(= \frac{1}{L}\int_0^L h_{cx}\,dx \text{ for a rectangular surface}\right)$$

is of little practical utility. When the temperature difference $(T_s - T_e)$ is not constant, heat transfer from the plate cannot be calculated from the simple formula used for isothermal surfaces, namely,

$$\dot{Q} = \bar{h}_c A(T_s - T_e) \tag{4.81}$$

which was introduced as Eq. (1.24). For the special case of a uniformly heated wall, the average heat transfer coefficient is preferably defined in terms of $\overline{(T_s - T_e)}$ as

$$\bar{h}_c = \frac{q_s}{\overline{(T_s - T_e)}}; \qquad \overline{T_s - T_e} = \frac{1}{L}\int_0^L (T_s - T_e)\,dx \tag{4.82}$$

Such a definition is particularly useful when resistance thermometry is used to measure the average temperature of the wall. Since $\mathrm{Nu}_x = q_s x/k(T_s - T_e)$, Eq. (4.79) can be integrated to obtain

$$0.453k(V/\nu)^{1/2}\mathrm{Pr}^{1/3}\overline{(T_s - T_e)} = \frac{q_s}{L}\int_0^L x^{1/2}\,dx = \frac{2}{3}q_s L^{1/2}$$

$$\overline{\mathrm{Nu}}_L = \frac{q_s L}{k\overline{(T_s - T_e)}} = 0.680\mathrm{Re}_L^{1/2}\mathrm{Pr}^{1/3} \tag{4.83}$$

which is only 2.3% higher than the isothermal wall result, Eq. (4.57). For turbulent flow, the difference is even smaller. As a general rule, the average heat transfer coefficient for external flows over uniformly heated surfaces defined by Eq. (4.82) can be taken to be equal to that for an isothermal surface [19]. For surfaces that are neither isothermal nor uniformly heated, the advanced texts on convective heat transfer listed in the bibliography at the end of the text should be consulted.

EXAMPLE 4.3 Heat Loss from a Hut Roof

A wind blows at 8 m/s over the 5 m square flat roof of a hut that is part of an Antarctic research station. If the ambient air is at approximately 250 K, estimate the average heat transfer coefficient. Use a transition Reynolds number of 10^5.

Solution

Given: Wind blowing over roof of a hut.

Required: Average heat transfer coefficient.

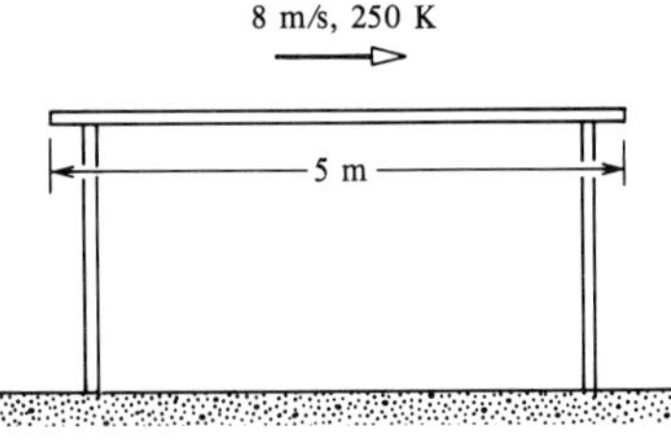

Assumptions:
1. Roof can be modeled as a flat plate.
2. $Re_{tr} = 10^5$.
3. Properties can be evaluated at 250 K since we do not expect a large temperature difference between the roof and air.

Equation (4.65) applies. From Table A.7, the required air properties at 250 K are $k = 0.0235$ W/m K, $\nu = 11.42 \times 10^{-6}$ m²/s, Pr = 0.69. The Reynolds number at the end of the roof is $Re_L = VL/\nu = (8)(5)/(11.42 \times 10^{-6}) = 3.50 \times 10^6$. Since $Re_{tr} = 0.1 \times 10^6$, only the first 3% of the boundary layer is laminar; the remainder is turbulent. Using Eq. (4.65),

$$\overline{Nu} = 0.664 Re_{tr}^{1/2} Pr^{1/3} + 0.036 Re_L^{0.8} Pr^{0.43} \left[1 - \left(\frac{Re_{tr}}{Re_L} \right)^{0.8} \right]$$

$$= 0.664(0.1 \times 10^6)^{1/2}(0.69)^{1/3} + 0.036(3.5 \times 10^6)^{0.8}(0.69)^{0.43} \left[1 - \left(\frac{0.1}{3.5} \right)^{0.8} \right]$$

$$= 186 + 4968 = 5150$$

$$\overline{h}_c = (k/L)\overline{Nu}_L = (0.0235/5)5150 = 24.2 \text{ W/m}^2 \text{ K}.$$

Comments

1. The precise location of transition in this problem is unimportant. The boundary layer could have been taken as turbulent from the leading edge with little effect on the result.
2. Use CONV to check $\overline{h}_c$.

EXAMPLE 4.4 Cooling of a Molten Droplet of Aluminum

During arc welding of aluminum, molten metal droplets are ejected. Some of these droplets are hot enough ($\gtrsim$2300 K) to ignite and form sparks. Most droplets are ejected at lower temperatures and simply cool down. If a particular molten droplet has a diameter of 0.5 mm, an initial temperature of 1700 K, and an initial velocity of 1 m/s, estimate its initial rate of cooling in air at 300 K. Liquid aluminum properties at 1700 K are estimated to be $\rho = 2100$ kg/m³, $c_p = 1100$ J/kg K, $\varepsilon = 0.20$.

Solution

Given: Molten aluminum droplet at 1700 K.

Required: Initial rate of cooling, dT/dt.

Assumptions: **1.** Model as a small gray object in large surroundings to calculate q_{rad}.
2. The lumped thermal capacity model is applicable.
3. Negligible formation of solid oxide on the droplet surface.

Equation (4.76) will be used to calculate the average Nusselt number with air properties evaluated at the mean film temperature of (1700 + 300)/2 = 1000 K. From Table A.7, k = 0.0672 W/m K, ν = 117.3 × 10⁻⁶ m²/s, Pr = 0.70. The Reynolds number is

$$\text{Re}_D = \frac{VD}{\nu} = \frac{(1)(0.0005)}{117.3 \times 10^{-6}} = 4.26$$

Substituting in Eq. (4.76),

$$\overline{\text{Nu}}_D = 2 + (0.4\text{Re}_D^{1/2} + 0.06\text{Re}_D^{2/3})\text{Pr}^{0.4}$$

$$= 2 + [0.4(4.26)^{1/2} + 0.06(4.26)^{2/3}](0.70)^{0.4} = 2.85$$

$$\bar{h}_c = \left(\frac{k}{D}\right)\overline{\text{Nu}}_D = \left(\frac{0.0672}{0.0005}\right)2.85 = 383 \text{ W/m}^2\text{ K}$$

$$q_{\text{conv}} = \bar{h}_c(T_s - T_e) = (383)(1700 - 300) = 5.37 \times 10^5 \text{ W/m}^2$$

The radiation heat loss can be estimated using Eq. (1.18) for a small gray object in large surroundings:

$$q_{\text{rad}} = \sigma\varepsilon(T_s^4 - T_e^4) = (5.67 \times 10^{-8})(0.2)(1700^4 - 300^4) = 0.946 \times 10^5 \text{ W/m}^2$$

Taking $k \simeq 200$ W/m K for molten aluminum, the Biot number based on convective heat transfer from Eq. (1.40) is

$$\text{Bi} = \frac{\bar{h}_c(D/6)}{k_s} = \frac{(383)(0.0005/6)}{(200)} \simeq 1.6 \times 10^{-4} \ll 0.1$$

Thus, the lumped thermal capacity model is certainly valid. From Section 1.5.2,

$$\rho c V\frac{dT}{dt} = -(q_{\text{conv}} + q_{\text{rad}})A$$

Substituting $V = \pi D^3/6$, $A = \pi D^2$ gives

$$\frac{dT}{dt} = -\frac{6(q_{\text{conv}} + q_{\text{rad}})}{\rho c D} = -\frac{6(5.37 + 0.95)(10^5)}{(2100)(1100)(0.0005)} = -3280 \text{ K/s}$$

Comments

1. Such small droplets cool off very fast.
2. Although the temperature is high, the radiation contribution is relatively small. The reason is the low value of ε and the large value of h_c (due to the small size).
3. Use CONV to check the value of h_c.
4. Is natural convection significant for such a hot droplet? Section 4.4.3 will deal with mixed natural and forced convection.

5. Recall that a viscosity ratio correction for variable properties, $(\mu_s/\mu_e)^n$, $n = -0.25$, applied to the convection contribution should give a better result than use of the mean film temperature (CONV uses the mean film temperature).

6. Some solid oxide will form on the droplet surface. However, the heat of oxidation liberated will be small unless the droplet temperature is close to the ignition value of 2300 K.

4.4 NATURAL CONVECTION

A heated fluid rises. Density differences and the earth's gravitational field act to produce a **buoyancy force**, which drives the flow. Such flows are called *natural*, *free*, or *buoyant convection*. Whenever a fluid is heated or cooled in a gravitational field, there is the possibility of natural convection. Density differences can also be caused by composition gradients. For example, moist air rises mainly due to the lower density of the water vapor present in the mixture. A related phenomenon is flow induced by a centripetal acceleration field, as in rotating machinery. We will restrict our attention to pure fluids in the earth's gravity field. For a pure fluid, density gradients can be related to temperature gradients through the volumetric coefficient of expansion, β. For an ideal gas, $\beta = 1/T$, where T is *absolute* temperature. Data for β of liquids are given in Table A.10.

Natural convection flows can be either *external* or *internal*. External flows include flow up a heated wall and the plume rising above a power plant stack. Internal flows are found between the cover plate and absorbing surface of a solar collector and inside hollow insulating walls. Velocities associated with natural convection are relatively small, not much more than 2 m/s. Thus, natural-convection heat transfer coefficients tend to be much smaller than those for forced convection. For gases, these coefficients are of the order of only 5 W/m^2 K, and the engineer must be careful to always check if simultaneous radiation heat transfer is significant to the thermal design. Since there is no obvious characteristic velocity of a natural convection flow, the Reynolds number of forced convection does not play a role. It is replaced by the Grashof or Rayleigh number.

4.4.1 External Natural Flows

In this section, various *external* natural convection flows are considered, and correlations are given for heat transfer from isothermal surfaces. Other wall boundary conditions are discussed at the end of the section.

Flow on a Vertical Wall

Figure 4.30 shows a natural-convection boundary layer on a vertical plate. A laminar boundary layer forms at the lower end, and transition to a turbulent boundary layer occurs at a critical value of the Rayleigh number $\mathrm{Ra}_x = \beta \Delta T g x^3/\nu\alpha \simeq 10^9$.

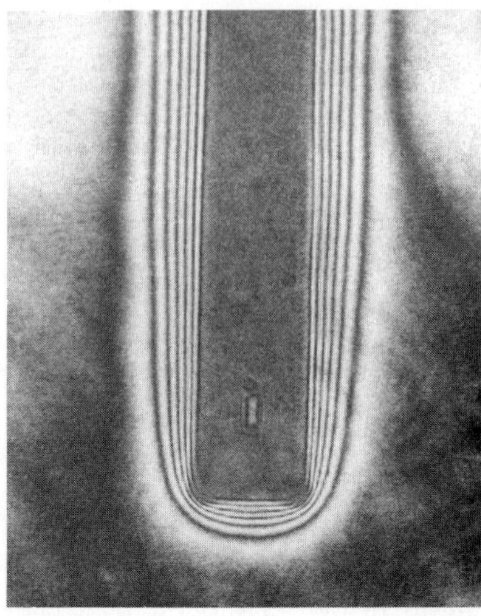

Figure 4.30 A thermal boundary layer on a heated vertical plate. (Interferogram courtesy of Professor J. Gryzagoridis, University of Cape Town.)

Since $\text{Ra}_x = \text{Gr}_x\text{Pr}$, where the Grashof number $\text{Gr}_x = \beta\Delta T g x^3/\nu^2$, for *gases* with $\text{Pr} \simeq 1$, transition can be said to occur at $\text{Gr}_x \simeq 10^9$ (see Section 1.3.3). Following Churchill and Usagi [21], we define a Prandtl number function Ψ as

$$\Psi = \left[1 + \left(\frac{0.492}{\text{Pr}}\right)^{9/16}\right]^{-16/9} \tag{4.84}$$

Churchill and Chu [22] correlated the average Nusselt number for laminar flow on a plate with a sharp leading edge and of height L as

$$\overline{\text{Nu}}_L = 0.68 + 0.670(\text{Ra}_L\Psi)^{1/4}; \qquad \text{Ra}_L \lesssim 10^9 \tag{4.85}$$

and for turbulent flow,

$$\overline{\text{Nu}}_L = 0.68 + 0.670(\text{Ra}_L\Psi)^{1/4}(1 + 1.6 \times 10^{-8}\text{Ra}_L\Psi)^{1/12}; \quad 10^9 \lesssim \text{Ra}_L < 10^{12} \tag{4.86}$$

Notice that for a large Rayleigh number, Eq. (4.86) shows that the heat transfer coefficient is independent of plate height L. Equations (4.85) and (4.86) do not match exactly at $\text{Ra}_L \sim 10^9$.

Flow on a Horizontal Cylinder

Figure 4.31 is an interferogram showing isotherms around an isothermal heated cylinder in air. A plume of heated air is seen to rise above the cylinder, and the boundary layer is rather thick compared to the forced-flow case, which is typical. The flow is laminar for $\text{Ra}_D = \text{Gr}_D\text{Pr} \lesssim 10^9$. Churchill and Chu [23] correlated the average Nusselt number as

$$\overline{\text{Nu}}_D = 0.36 + \frac{0.518\text{Ra}_D^{1/4}}{[1 + (0.559/\text{Pr})^{9/16}]^{4/9}}; \qquad 10^{-6} < \text{Ra}_D \lesssim 10^9 \tag{4.87}$$

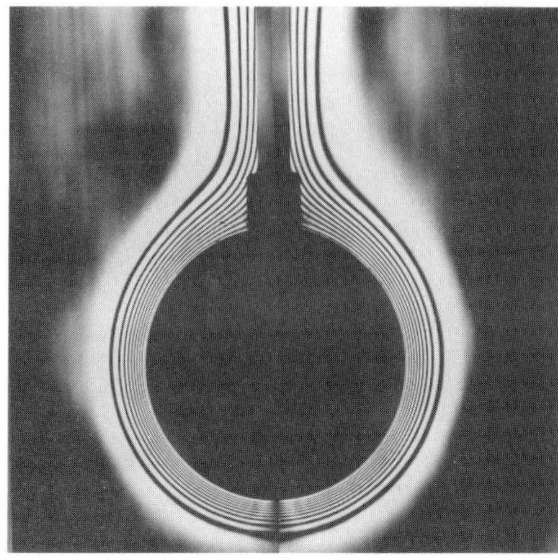

Figure 4.31 A thermal boundary layer on a heated horizontal cylinder. (Interferogram courtesy of Professor U. Grigŭll and W. Haŭf, Technische Universität, München.)

When $\mathrm{Ra}_D \gtrsim 10^9$, there is a transition from a laminar to a turbulent boundary layer, and the rate of increase of Nusselt number with Rayleigh number is greater. The recommended correlation then is

$$\overline{\mathrm{Nu}}_D = \left\{0.60 + 0.387\left[\frac{\mathrm{Ra}_D}{[1 + (0.559/\mathrm{Pr})^{9/16}]^{16/9}}\right]^{1/6}\right\}^2; \qquad \mathrm{Ra}_D \gtrsim 10^9 \quad \textbf{(4.88)}$$

Equations (4.87) and (4.88) do not match exactly at $\mathrm{Ra}_D \sim 10^9$.

Flow on a Sphere

For fluids with Prandtl number of order unity, which includes all gases, Yuge [24] gives the average Nusselt number as

$$\overline{\mathrm{Nu}}_D = 2 + 0.43\mathrm{Ra}_D^{1/4}; \qquad 1 < \mathrm{Ra}_D < 10^5 \quad \textbf{(4.89)}$$

A more general formula valid for $\mathrm{Pr} > 0.5$ is due to Churchill [25]:

$$\overline{\mathrm{Nu}}_D = 2 + \frac{0.589\mathrm{Ra}_D^{1/4}}{[1 + (0.469/\mathrm{Pr})^{9/16}]^{4/9}}; \qquad \mathrm{Ra}_D \lesssim 10^{11} \quad \textbf{(4.90)}$$

Objects of Arbitrary Shape

For a laminar natural convection boundary layer on an object of any shape, in fluids other than those for which $\mathrm{Pr} \ll 1$, Lienhard [26] suggests that the average Nusselt number is approximately

$$\overline{\mathrm{Nu}}_L = 0.52\mathrm{Ra}_L^{1/4} \quad \textbf{(4.91)}$$

where the characteristic length L is the length of the boundary layer, for example, $L = \pi D/2$ for a cylinder or sphere.

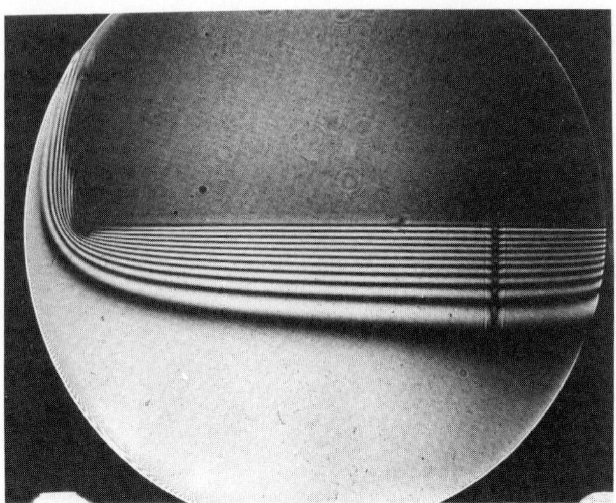

Figure 4.32 The thermal boundary layer below a 10 cm × 30 cm heated horizontal plate: $\mathrm{Ra}_L = 3.33 \times 10^6, \mathrm{Nu}_L = 15.0$. The pin marks the plate center. Air rises from some distance below the plate until, in the vicinity of the plate, it flows sideways and then around the corner, carrying the heat away in a vertical plume. There are 10 isotherms visible; these are 3°C apart adjacent to the plate where the air is hot (and index of refraction is lower), decreasing to 2° apart as the ambient air is approached. (Photograph courtesy of Professor D. K. Edwards, University of California, Irvine.)

Heated Horizontal Plate Facing Down, or Cooled Horizontal Plate Facing Up

Figure 4.32 shows an interferogram of the flow field underneath a heated horizontal plate. For a square plate of side length L, Kadambi and Drake [28] give the average Nusselt number as

$$\overline{\mathrm{Nu}}_L = 0.82\mathrm{Ra}_L^{1/5}; \qquad 10^5 < \mathrm{Ra}_L < 10^{10} \tag{4.92}$$

and for an infinitely long strip of width L, Fugii and Imura [29] recommend

$$\overline{\mathrm{Nu}}_L = 0.58\mathrm{Ra}_L^{1/5}; \qquad 10^6 < \mathrm{Ra}_L < 10^{11} \tag{4.93}$$

Alternatively, there is a more general correlation developed by Hatfield and Edwards [27] that applies to all aspect ratios and also allows for adiabatic extensions. If, as shown in Fig. 4.33, L is the length of the shorter side, W the length of the longer side, and L_a the length of adiabatic extensions on the shorter side, then

$$\overline{\mathrm{Nu}}_L = 6.5\left[1 + 0.38\frac{L}{W}\right]\left[(1+X)^{0.39} - X^{0.39}\right]\mathrm{Ra}_L^{0.13} \tag{4.94}$$

$$X = 13.5\mathrm{Ra}_L^{-0.16} + 2.2\left(\frac{L_a}{L}\right)^{0.7}$$

which is based on experimental data for $10^6 < \mathrm{Ra}_L < 10^{10}$, $0.7 < \mathrm{Pr} < 4800$, and $0 < L_a/L < 0.2$.

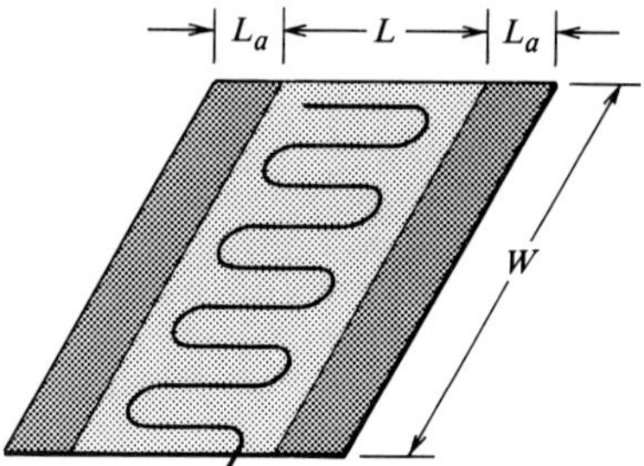

Figure 4.33 Schematic of a heated horizontal plate facing downward, showing adiabatic extensions on the shorter side.

Heated Horizontal Plate Facing Up, or Cooled Horizontal Plate Facing Down

In contrast to the previous case, the flow for the configuration of Fig. 4.34 is unstable. McAdams [30] recommends

$$\overline{\mathrm{Nu}}_L = 0.54\mathrm{Ra}_L^{1/4}; \qquad 10^5 < \mathrm{Ra}_L < 2 \times 10^7 \tag{4.95}$$

where L is the side length of a square plate, or the length of the shorter side of a rectangular plate. For $\mathrm{Ra}_L > 10^7$, turbulent thermals rise irregularly above the plate and result in an average Nusselt number independent of plate size or shape [30]:

$$\overline{\mathrm{Nu}}_L = 0.14\mathrm{Ra}_L^{1/3}; \qquad 2 \times 10^7 < \mathrm{Ra}_L < 3 \times 10^{10} \tag{4.96}$$

Since the characteristic length L cancels, this result can also be written as

$$\frac{h_c(\nu\alpha/g)^{1/3}}{k} = 0.14\left(\frac{\Delta\rho}{\rho}\right)^{1/3} \tag{4.97}$$

where the left-hand side can be viewed as a Nusselt number with a characteristic length $L = (\nu\alpha/g)^{1/3}$. These formulas are based on experimental data for air but may be used for any fluid with $\mathrm{Pr} > 0.5$.

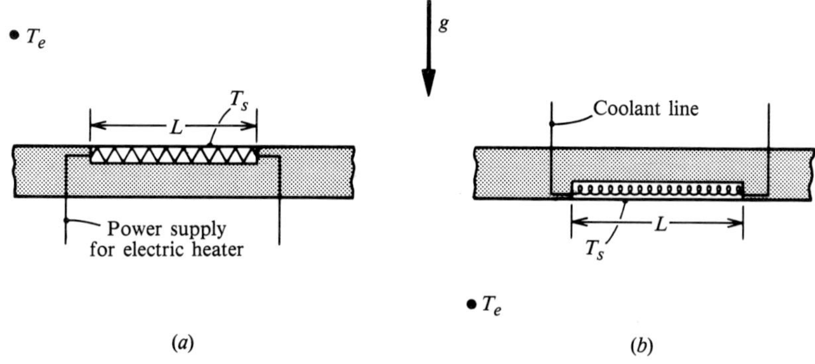

Figure 4.34 Schematic of (*a*) a heated plate facing upward, and (*b*) a cooled plate facing downward. The flow is unstable and the streamlines are unsteady.

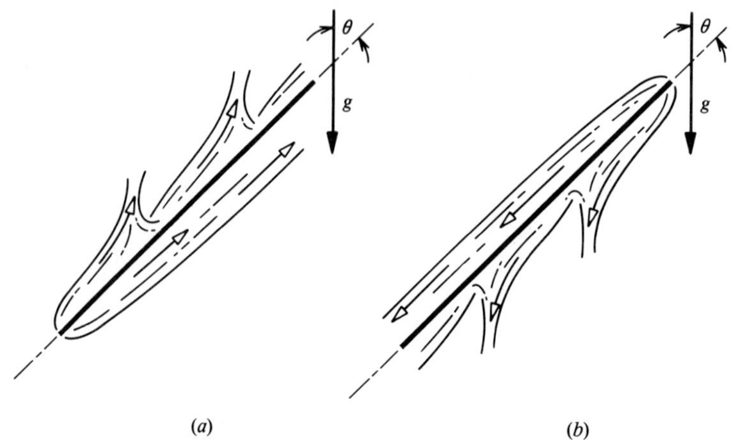

Figure 4.35 Flow patterns around (a) heated and (b) cooled inclined plates.

Inclined Plates

Figure 4.35 shows flow patterns around heated and cooled inclined plates. Provided θ is not too close to 90°, one would expect Eq. (4.85) for a vertical wall to apply to inclined walls, with g replaced by $g\cos\theta$. For the lower side of hot plates and the upper side of cold plates, the limit is $\theta \simeq 88°$ for $10^5 < \mathrm{Gr}_L < 10^{11}$; for the upper side of hot plates and the lower side of cold plates, the limit is $\theta \simeq 60°$.

Variable Property Effects

In the correlations of Section 4.4.1, all properties, including β, should be evaluated at the *mean film temperature*. Note that there are correlations for natural convection that require β to be evaluated at T_e, but this is not the case for the correlations presented here.

Other Wall Boundary Conditions

All the foregoing correlations are for an isothermal surface. Except for the laminar boundary layer on a vertical wall, there are few results available for nonisothermal surfaces. For a uniform wall heat flux, use of Eq. (4.82) and an isothermal wall correlation will give an engineering estimate of the average wall temperature.

EXAMPLE 4.5 Heat Loss from a Solar Power Plant Central Receiver

The central receiver of a solar power plant is in the form of a cylinder 7 m in diameter and 13 m high. A tower supports the receiver high above the ground, and solar radiation is reflected to the receiver by rows of reflectors at ground level. If the operating temperature of its surface is 700 K, calculate the convective heat loss into still ambient air at 300 K. If the total irradiation of the collector is 20 MW, express the loss as a percentage of the irradiation.

Solution

Given: Vertical cylinder at 700 K.

Required: Convective heat loss into still ambient air.

Assumptions: 1. Curvature effects are negligible.
2. The free-convection boundary layer commences at the bottom of the cylinder.

Depending on whether the flow is laminar or turbulent, either Eq. (4.85) or Eq. (4.86) should apply. Evaluate all properties at a mean film temperature of 500 K; $\beta = 1/T_r = 1/500$, $k = 0.0389$ W/m K, $\nu = 37.3 \times 10^{-6}$ m²/s, Pr $= 0.69$. The Rayleigh number is

$$\text{Ra}_L = \text{Gr}_L\text{Pr} = \frac{\beta \Delta T g L^3}{\nu^2}\text{Pr}$$

$$= \frac{(1/500)(400)(9.81)(13)^3(0.69)}{(37.3 \times 10^{-6})^2}$$

$$= 8.55 \times 10^{12}$$

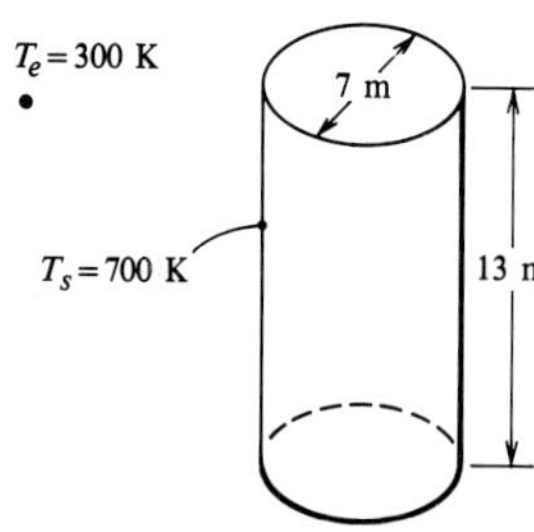

This Rayleigh number is somewhat larger than the value of 10^{12} quoted as the upper limit of validity for Eq. (4.86). But $\bar{h}_c$ is almost independent of L for large values of Ra_L, so the result will not be too much in error. Using Eq. (4.86),

$$\Psi = \left[1 + \left(\frac{0.492}{\text{Pr}}\right)^{9/16}\right]^{-16/9} = \left[1 + \left(\frac{0.492}{0.69}\right)^{9/16}\right]^{-16/9} = 0.343$$

$$\text{Ra}_L\Psi = (8.55 \times 10^{12})(0.343) = 2.93 \times 10^{12}$$

$$\overline{\text{Nu}_L} = 0.68 + 0.670(\text{Ra}_L\Psi)^{1/4}(1 + 1.6 \times 10^{-8}\text{Ra}_L\Psi)^{1/12}$$

$$= 0.68 + 0.670(2.93 \times 10^{12})^{1/4}[1 + (1.6 \times 10^{-8})(2.93 \times 10^{12})]^{1/12} = 2150$$

$$\bar{h}_c = (k/L)\overline{\text{Nu}_L} = (0.0389/13)(2150) = 6.43 \text{ W/m}^2\text{ K}$$

$$\dot{Q} = \bar{h}_c A(T_s - T_e) = (6.43)(\pi)(7)(13)(700 - 300) = 7.35 \times 10^5 \text{ W} = 0.735 \text{ MW}$$

$$\frac{\dot{Q}_{\text{loss}}}{\dot{Q}_{\text{gross}}} = \frac{0.735}{20} = 3.7\%$$

Solution using CONV

The required input in SI units is:

Configuration number = 9 (vertical wall: turbulent flow)
Fluid number = 21
$T_s = 700$
$T_e = 300$
$P = 1.013 \times 10^5$
$L = 13$

The output is:

Air properties at 500 K, 1.013×10^5 Pa
$\text{Gr} = 1.241 \times 10^{13}$
$\text{Ra} = 8.57 \times 10^{12}$
$\overline{\text{Nu}} = 2150$
$\bar{h}_c = 6.43\ \text{W/m}^2\ \text{K}$

Comments

1. The thermal plume above the receiver carries away 0.735 MW of thermal energy. Can you propose a method for recovering some of this waste heat?

2. For $\beta \Delta T$ not much less than unity, Barrow and Sitharamarao [31] recommend that the Nusselt number be increased by a factor of $(1 + 0.6\beta\Delta T)^{1/4}$. For this problem, the correction is 10%.

EXAMPLE 4.6 Heat Loss from a Steam Pipe

A 30 cm–O.D. horizontal steam pipe has an outer surface temperature of 500 K and is located in still air at 300 K. Calculate the average heat transfer coefficient and the convective heat loss per meter length of pipe.

Solution

Given: Natural convection from a horizontal steam pipe.

Required: $\bar{h}_c$ and heat loss per meter.

Evaluate air properties at the mean film temperature of $(500+300)/2 = 400$ K. From Table A.7, $k = 0.0331$ W/m K, $\nu = 25.5 \times 10^{-6}$ m²/s, $\text{Pr} = 0.69$, and $\beta = 1/400$ for an ideal gas. The Rayleigh number is

$$\text{Ra}_D = \text{Gr}_D\text{Pr} = \frac{\beta \Delta T g D^3}{\nu^2}\text{Pr} = \frac{(1/400)(200)(9.81)(0.3)^3(0.69)}{(25.5 \times 10^{-6})^2} = 1.41 \times 10^8$$

Equation (4.87) applies:

$$\overline{\text{Nu}}_D = 0.36 + \frac{0.518\text{Ra}_D^{1/4}}{[1 + (0.559/\text{Pr})^{9/16}]^{4/9}} = 0.36 + \frac{0.518(1.41 \times 10^8)^{1/4}}{[1 + (0.559/0.69)^{9/16}]^{4/9}} = 42.9$$

$$\bar{h}_c = \left(\frac{k}{D}\right)\overline{\text{Nu}}_D = \left(\frac{0.0331}{0.3}\right)42.9 = 4.73\ \text{W/m}^2\ \text{K}$$

$$\dot{Q} = \bar{h}_c A(T_s - T_e) = (4.73)(\pi)(0.3)(1)(500 - 300) = 892\ \text{W/m}$$

We will check this result using the general correlation for laminar natural flows, Eq. (4.91).

For $L = \pi D/2 = 0.47$ m,

$$\overline{\mathrm{Nu}}_L = 0.52(\mathrm{Gr}_L\mathrm{Pr})^{1/4} = (0.52)\left(\frac{(1/400)(200)(9.81)(0.47)^3(0.69)}{(25.5 \times 10^{-6})^2}\right)^{1/4} = 79.3$$

$$\overline{h}_c = \left(\frac{k}{L}\right)\mathrm{Nu}_L = \left(\frac{0.0331}{0.47}\right)79.3 = 5.58 \text{ W/m}^2\text{ K}$$

Comments

1. The more approximate Eq. (4.91) gives a value of $\overline{h}_c$ that is 18% higher than that from Eq. (4.87).
2. Use CONV to check $\overline{h}_c$.

4.4.2 Internal Natural Flows

Figure 4.36 shows a selection of enclosures in which natural convection is of engineering concern—for example, in flat-plate solar collectors, wall cavities, and window glazing. The horizontal layer heated from below was discussed in Section 4.2.1, where an appropriate definition of the convective heat transfer coefficient was shown to be

$$\overline{h}_c = \frac{\dot{Q}/A}{T_H - T_C}$$

The length parameter commonly used to define the Nusselt number is the plate spacing L. If the temperature difference $(T_H - T_C)$ is less than the critical value

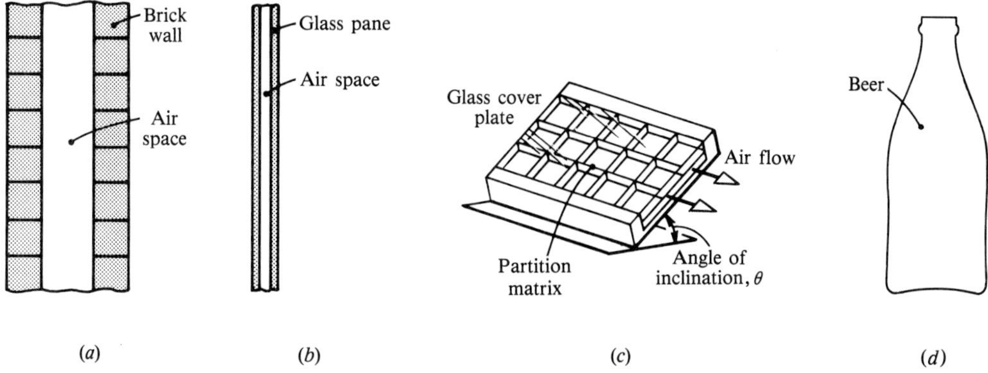

Figure 4.36 Enclosures. (*a*) A double wall with an air gap. (*b*) Double window glazing. (*c*) A flat-plate solar collector with a partition to suppress natural convection. (*d*) Sterilization of beer by condensing steam.

required for the fluid to become unstable, then heat transfer across the layer is by conduction only, and from Eq. (1.9),

$$\dot{Q} = \frac{kA}{L}(T_H - T_C)$$

or

$$\bar{h}_c = \frac{k}{L}, \qquad \overline{\mathrm{Nu}_L} = 1$$

Thus, correlations for the Nusselt number always have a lower limit of $\overline{\mathrm{Nu}_L} = 1$, corresponding to pure conduction. In Section 4.2.1, it was indicated that a horizontal layer heated from below becomes unstable at a critical value of $(T_H - T_C)$. In dimensionless form, the criterion for instability and the onset of cellular convection is a critical value of the Rayleigh number,

$$\mathrm{Ra}_L = \frac{g\beta(T_H - T_C)L^3}{\nu\alpha} = 1708$$

As the temperature difference $(T_H - T_C)$ and, hence, the Rayleigh number increases, there are transitions to increasingly more complex flow patterns until finally the flow in the core is turbulent. In the case of a vertical layer of fluid contained between parallel plates maintained at different temperatures, circulation occurs for any $\mathrm{Ra}_L > 0$; however, heat transfer is essentially by pure conduction for $\mathrm{Ra}_L < 10^3$. As the Rayleigh number is increased, the circulating flow develops and cells are formed. At $\mathrm{Ra}_L \simeq 10^4$ the flow changes to a boundary layer type with a boundary layer flowing upward on the hot wall and downward on the cold wall, while the fluid in the core region remains relatively stationary. At $\mathrm{Ra}_L \simeq 10^5$ vertical rows of horizontal vortices develop in the core; and at $\mathrm{Ra}_L \simeq 10^6$, the flow in the core finally becomes turbulent.

The marked changes in flow pattern with changes in Rayleigh number are characteristic of internal natural convection in all shapes of enclosures. Thus, it would be unreasonable to seek a single simple correlation formula valid over wide Rayleigh and Prandtl number ranges. Simple power law–type formulas are usually valid for small ranges of Ra; more general formulas are usually quite complex. Thus, only a few configurations will be considered here.

Heat transfer across thin air layers is of considerable engineering importance. Referring to Fig. 4.37, correlations recommended by Hollands and coworkers for aspect ratios of $H/L > 10$ are as follows.

1. $0 \le \theta < 60°$ [32]:

$$\overline{\mathrm{Nu}_L} = 1 + 1.44\left[1 - \frac{1708}{\mathrm{Ra}_L \cos\theta}\right]\left\{1 - \frac{1708(\sin 1.8\theta)^{1.6}}{\mathrm{Ra}_L \cos\theta}\right\} + \left[\left(\frac{\mathrm{Ra}_L \cos\theta}{5830}\right)^{1/3} - 1\right] \tag{4.98}$$

where if either of the terms in square brackets is negative, it must be set equal to zero. Equation (4.98) is valid for $0 < \mathrm{Ra}_L < 10^5$.

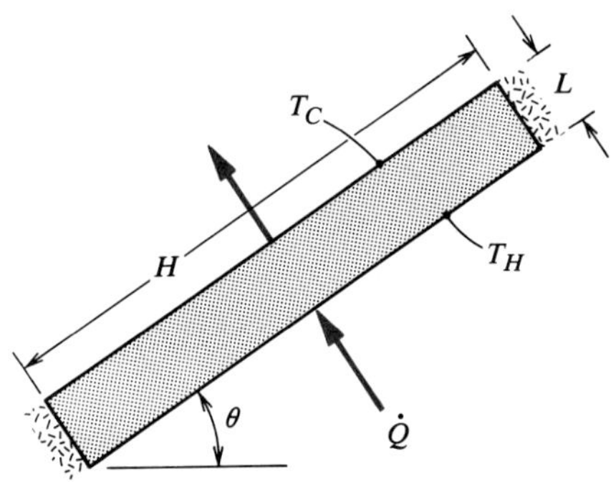

Figure 4.37 Schematic of a large-aspect-ratio inclined enclosure. The angle θ is measured from the horizontal.

2. $\theta = 60°$ [33]:

$$\overline{\mathrm{Nu}}_{L60°} = \max\{\mathrm{Nu}_1,\ \mathrm{Nu}_2\} \tag{4.99}$$

where

$$\mathrm{Nu}_1 = \left\{1 + \left[\frac{0.0936\mathrm{Ra}_L^{0.314}}{1 + \{0.5/[1 + (\mathrm{Ra}_L/3160)^{20.6}]^{0.1}\}}\right]^7\right\}^{1/7}$$

$$\mathrm{Nu}_2 = \left(0.104 + \frac{0.175}{H/L}\right)\mathrm{Ra}_L^{0.283}$$

and is valid for $0 < \mathrm{Ra}_L < 10^7$.

3. $60° < \theta < 90°$ [33]:

$$\overline{\mathrm{Nu}}_L = \left(\frac{90 - \theta}{30}\right)\overline{\mathrm{Nu}}_{L60°} + \left(\frac{\theta - 60}{30}\right)\overline{\mathrm{Nu}}_{L90°} \tag{4.100}$$

4. $\theta = 90°$ [33]:

$$\overline{\mathrm{Nu}}_{L90°} = \max\{\mathrm{Nu}_1,\ \mathrm{Nu}_2,\ \mathrm{Nu}_3\} \tag{4.101}$$

where

$$\mathrm{Nu}_1 = 0.0605\mathrm{Ra}_L^{1/3}$$

$$\mathrm{Nu}_2 = \left\{1 + \left[\frac{0.104\mathrm{Ra}_L^{0.293}}{1 + (6310/\mathrm{Ra}_L)^{1.36}}\right]^3\right\}^{1/3}$$

$$\mathrm{Nu}_3 = 0.242\left(\frac{\mathrm{Ra}_L}{H/L}\right)^{0.272}$$

and is valid for $10^3 < \mathrm{Ra}_L < 10^7$; for $\mathrm{Ra}_L \le 10^3$, $\overline{\mathrm{Nu}}_{L90°} \simeq 1$.

For liquids of moderate Prandtl number, such as water, Eq. (4.98) can also be used for Ra < 10^5. For higher Rayleigh numbers, the Globe and Dropkin correlation [34] may be used for horizontal layers:

$$\overline{\mathrm{Nu}}_L = 0.069\mathrm{Ra}_L^{1/3}\mathrm{Pr}^{0.074}; \qquad 3 \times 10^5 < \mathrm{Ra}_L < 7 \times 10^9 \qquad \textbf{(4.102)}$$

Also, for horizontal layers of air, that is, $\theta = 0°$, the range of validity of Eq. (4.98) extends to $\mathrm{Ra}_L = 10^8$.

Data are available in the literature for inclined layers of small aspect ratio but have not been correlated in a satisfactory manner.

In vertical cavities of small aspect ratio, with the horizontal surfaces insulated, as shown in Fig. 4.38, the following correlations due to Berkovsky and Polevikov [35] may be used for fluids of any Prandtl number.

1. $2 < H/L < 10$:

$$\overline{\mathrm{Nu}}_L = 0.22\left(\frac{\mathrm{Pr}}{0.2 + \mathrm{Pr}}\mathrm{Ra}_L\right)^{0.28}\left(\frac{H}{L}\right)^{-1/4}; \qquad \mathrm{Ra}_L < 10^{10} \qquad \textbf{(4.103}\textbf{\textit{a}}\textbf{)}$$

2. $1 < H/L < 2$:

$$\overline{\mathrm{Nu}}_L = 0.18\left(\frac{\mathrm{Pr}}{0.2 + \mathrm{Pr}}\mathrm{Ra}_L\right)^{0.29}; \qquad 10^3 < \frac{\mathrm{Pr}}{0.2 + \mathrm{Pr}}\mathrm{Ra}_L \qquad \textbf{(4.103}\textbf{\textit{b}}\textbf{)}$$

The flow and convective heat transfer in small-aspect-ratio cavities can depend on the temperature variation along the separating walls and, hence, on conduction in the walls and on radiation exchange within the cavity.

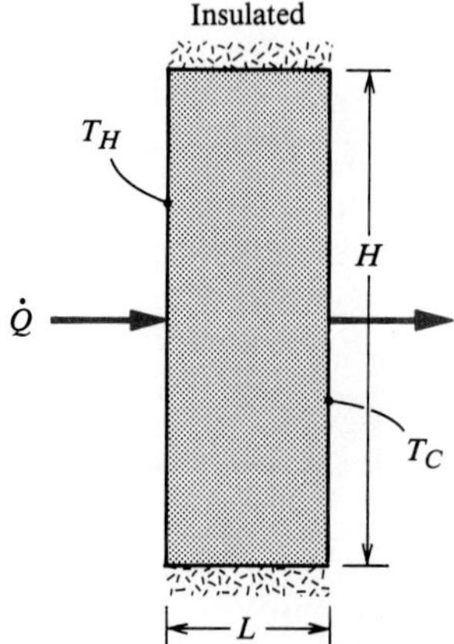

Figure 4.38 Schematic of a small-aspect-ratio vertical enclosure.

Concentric Cylinders and Spheres

Figure 4.39 shows isotherms for natural convection between concentric cylinders, with the inner cylinder heated and the outer cylinder cooled. The correlations recommended by Raithby and Hollands [36] for natural convection between concentric

Figure 4.39 Isotherms for natural convection between two concentric cylinders: $D_o/D_i = 3.9$, $L = 2.9$ cm, $\Delta T = 14.5$ K. (Interferogram courtesy of Professor U. Grigŭll and W. Haŭf, Technische Universität, München.)

cylinders and spheres are in the form of an effective thermal conductivity for use in the equations for conduction between concentric cylinders and spheres [Eqs. (2.14) and (2.25), respectively].

1. Concentric cylinders:

$$\frac{k_{\text{eff}}}{k} = 0.386\left(\frac{\text{Pr}}{0.861 + \text{Pr}}\right)^{1/4} \text{Ra}_{\text{cyl}}^{1/4}; \qquad 10^2 < \text{Ra}_{\text{cyl}} < 10^7 \tag{4.104}$$

where

$$\text{Ra}_{\text{cyl}} = \frac{[\ln(D_o/D_i)]^4}{L^3\left(D_i^{-3/5} + D_o^{-3/5}\right)^5}\text{Ra}_L$$

and $L = (D_o - D_i)/2$ is the gap width.

2. Concentric spheres:

$$\frac{k_{\text{eff}}}{k} = 0.74\left(\frac{\text{Pr}}{0.861 + \text{Pr}}\right)^{1/4} \text{Ra}_{\text{sph}}^{1/4}; \qquad 10^2 < \text{Ra}_{\text{sph}} < 10^4 \tag{4.105}$$

where

$$\text{Ra}_{\text{sph}} = \frac{L}{(D_o D_i)^4}\frac{\text{Ra}_L}{\left(D_i^{-7/5} + D_o^{-7/5}\right)^5}$$

Equations (4.104) and (4.105) are valid only when $k_{\text{eff}}/k \geq 1$.

Variable Property Effects

All properties should be evaluated at a reference temperature $T_r = (1/2)(T_H + T_C)$.

EXAMPLE 4.7 Heat Loss through a Double Wall

A vertical double wall 3 m high has an air gap 10 cm thick. Estimate the convective heat transfer coefficient when the wall faces are at 305 K and 295 K, and the pressure is 1 atm. Compare the convective and radiative losses across the gap.

Solution

Given: Vertical air gap, 10 cm thick.

Required: Convective and radiative heat transfer across gap.

Assumptions: The facing surfaces are black.

Evaluate air properties at a mean fluid temperature of 300 K. From Table A.7, $k = 0.0267$ W/m K, $\rho = 1.177$ kg/m^3, $\nu = 15.7 \times 10^{-6}$ m^2/s, Pr $= 0.69$, and $\beta = 1/300$. The Rayleigh number and aspect ratio are

$$\mathrm{Ra}_L = \mathrm{Gr}_L\mathrm{Pr} = \frac{(\beta\Delta T)gL^3\mathrm{Pr}}{\nu^2}$$

$$= \frac{(10/300)(9.81)(0.1)^3(0.69)}{(15.7 \times 10^{-6})^2}$$

$$= 9.15 \times 10^5$$

$$\frac{H}{L} = \frac{3}{0.1} = 30$$

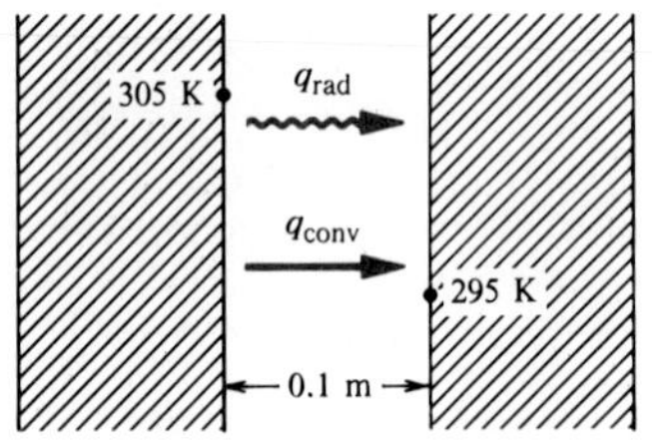

Equation (4.101) applies:

$$\mathrm{Nu}_1 = 0.0605(9.15 \times 10^5)^{1/3} = 5.87$$

$$\mathrm{Nu}_2 = \left\{1 + \left[\frac{0.104(9.15 \times 10^5)^{0.293}}{1 + (6310/9.15 \times 10^5)^{1.36}}\right]^3\right\}^{1/3} = 5.81$$

$$\mathrm{Nu}_3 = 0.242\left(\frac{9.15 \times 10^5}{30}\right)^{0.272} = 4.01$$

$$\overline{\mathrm{Nu}}_L = \max\{5.87, 5.81, 4.01\} = 5.87$$

$$\bar{h}_c = \left(\frac{k}{L}\right)\overline{\mathrm{Nu}}_L = \left(\frac{0.0267}{0.1}\right)5.87 = 1.57\ \mathrm{W/m^2\ K}$$

$$q_{\mathrm{conv}} = \bar{h}_c\Delta T = (1.57)(305 - 295) = 15.7\ \mathrm{W/m^2}$$

For black surfaces, the radiative transfer is given by Eq. (1.13):

$$q_{\mathrm{rad}} = \sigma T_1^4 - \sigma T_2^4 = (5.67 \times 10^{-8})(305^4 - 295^4) = 61.3\ \mathrm{W/m^2}$$

$$q_{\mathrm{total}} = q_{\mathrm{conv}} + q_{\mathrm{rad}} = 15.7 + 61.3 = 77.0\ \mathrm{W/m^2}$$

Comments

1. Use CONV to check $\bar{h}_c$.

2. Two possible methods of reducing the heat loss are as follows:
   (*a*) The gap could be filled with fiberglass insulation, with conductivity $k = 0.048$ W/m K, $q_{\text{total}} = (k/L)\Delta T = (0.048/0.1)(10) = 4.8$ W/m^2.
   (*b*) Since the radiation loss dominates, the walls could be lined with aluminum foil of $\varepsilon = 0.04$. In Chapter 6, the radiation transfer factor is shown to be $\mathscr{F}_{12} = \varepsilon/(2 - \varepsilon) = 0.0204$. Using Eq. (1.17), $q_{\text{rad}} = \mathscr{F}_{12}\sigma(T_1^4 - T_2^4) = 1.25$ W/m^2, $q_{\text{total}} = 15.7 + 1.25 = 17.0$ W/m^2.
   Economic considerations will have an impact on the final choice.

EXAMPLE 4.8 Convection Loss in a Flat-Plate Solar Collector

A flat-plate solar collector 2 m long and 1 m wide is inclined 60° to the horizontal. The cover plate is separated from the absorber plate by an air gap 2 cm thick. If the average temperatures of the cover and absorber plates are 305 K and 335 K, respectively, estimate the convective heat loss. Take the pressure as 1 atm.

Solution

Given: Horizontal air gap, 2 cm thick.

Required: Convective heat transfer.

Assumptions: Surfaces are isothermal.

Evaluate properties at a mean film temperature of 320 K. From Table A.7, $k = 0.0281$ W/m K, $\nu = 17.44 \times 10^{-6}$ m^2/s, Pr = 0.69, and $\beta = 1/T_r = 1/320$. The Rayleigh number and aspect ratio are

$$\text{Ra}_L = \text{Gr}_L\text{Pr} = \frac{(\beta\Delta T)gL^3\text{Pr}}{\nu^2}$$

$$= \frac{(30/320)(9.81)(0.02)^3(0.69)}{(17.44 \times 10^{-6})^2}$$

$$= 16{,}700$$

$$\frac{H}{L} = \frac{2}{0.02} = 100$$

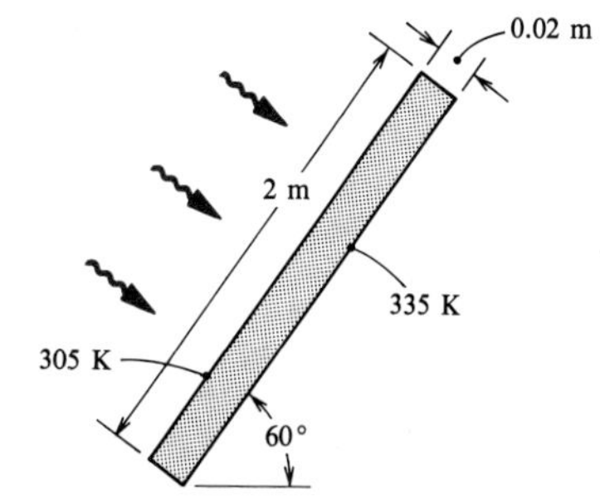

Equation (4.99) applies:

$$\text{Nu}_1 = \left\{1 + \left[\frac{(0.0936)(16{,}700)^{0.314}}{1 + \{0.5/[1 + (16{,}700/3160)^{20.6}]^{0.1}\}}\right]^7\right\}^{1/7} = 1.95$$

$$\mathrm{Nu}_2 = \left(0.104 + \frac{0.175}{100}\right)(16{,}700)^{0.283} = 1.66$$

$$\overline{\mathrm{Nu}}_{L60^\circ} = \max\{1.95, 1.66\} = 1.95$$

$$\bar{h}_c = \left(\frac{k}{L}\right)\overline{\mathrm{Nu}} = \left(\frac{0.0281}{0.02}\right)1.95 = 2.74 \text{ W/m}^2\text{ K}$$

$$\dot{Q} = h_c A \Delta T = (2.74)(2 \times 1)(30) = 164 \text{ W}$$

Comments

1. Use CONV to check $\bar{h}_c$.
2. It may be desirable to place a honeycomb structure between the plates to suppress natural convection.

4.4.3 Mixed Forced and Natural Flows

In treating forced flows in Section 4.3, it was implicitly assumed that buoyancy effects were negligible. But heat transfer requires a temperature gradient, and if this gradient causes the fluid to be unstable, there is the possibility of the buoyancy force contributing to the flow. Figure 4.40 shows some possible situations, where it is shown that the buoyancy force may *assist* or *oppose* the forced flow, or perhaps act perpendicularly to the forced flow. Such **mixed flows** are obviously very complex and cannot be dealt with in detail here. In this section, criteria for deciding whether forced or natural convection dominates will be discussed, and correlations and data for some very simple mixed flows will be presented.

External Flows

We have seen that the dimensionless group that characterizes a forced flow is the Reynolds number, whereas for a natural flow it is the Grashof or Rayleigh number. To gain some insight into the significance of the Grashof number, consider natural convection on a vertical isothermal wall, as shown in Fig. 4.41. If we follow a fluid element along the depicted streamline and ignore the viscous forces, then, as a rough approximation, the gain in kinetic energy of the fluid element must equal the work done by the buoyancy force:

$$\frac{1}{2}\rho V^2 \sim g\Delta\rho L$$

or,

$$V \sim \left(\frac{\Delta\rho}{\rho} gL\right)^{1/2} \tag{4.106}$$

where the factor of 1/2 can be ignored in making this *estimate* of the characteristic

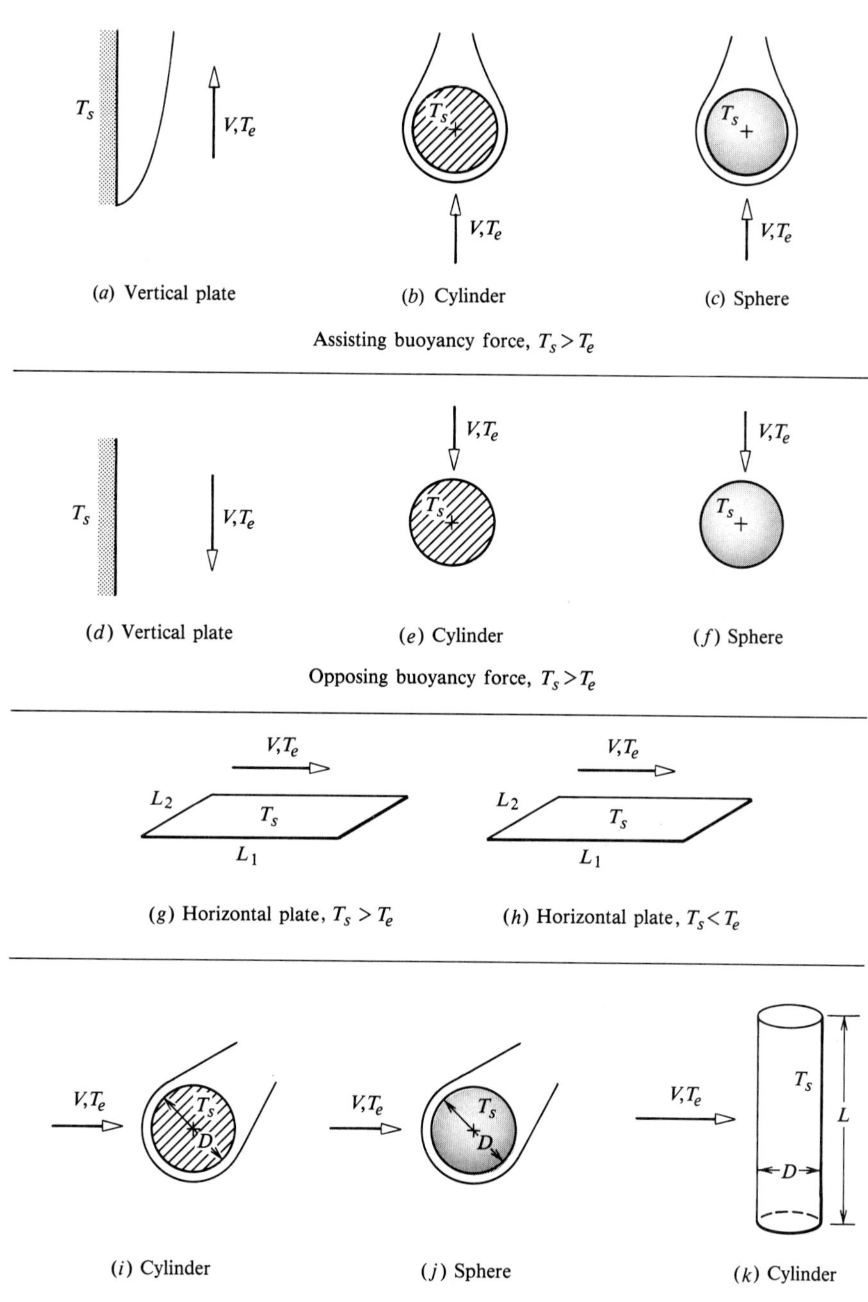

Figure 4.40 Schematic of possible mixed forced- and natural-convection flows.

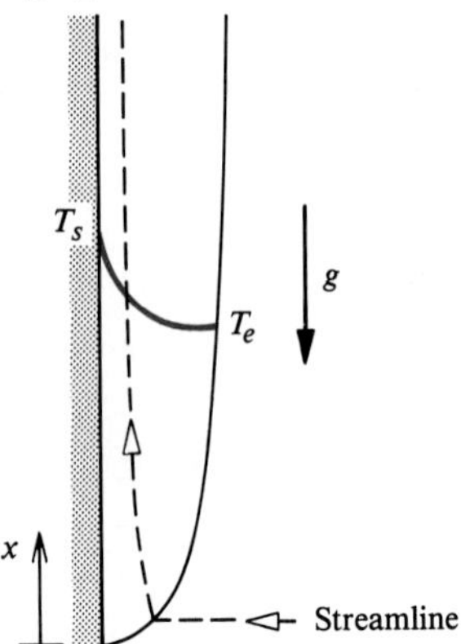

Figure 4.41 A streamline in a laminar natural convection boundary layer on a vertical isothermal wall.

natural convection velocity V. A Reynolds number based on this velocity and plate length L is thus

$$\mathrm{Re} = \frac{[(\Delta\rho/\rho)gL]^{1/2}L}{\nu} \tag{4.107}$$

and

$$\mathrm{Re}^2 = \frac{(\Delta\rho/\rho)gL^3}{\nu^2} = \mathrm{Gr} \tag{4.108}$$

So the Grashof number is simply the square of a Reynolds number based on an estimate of the free-convection velocity, V. The numerical value of V given by Eq. (4.106) will be somewhat larger than the maximum velocity in the natural-convection boundary layer but will be of the same order of magnitude. Since we have ignored viscous forces in estimating V, we would expect Eq. (4.106) to be less accurate for very viscous fluids, that is, if the Prandtl number is large. In the limit $\mathrm{Pr} \to \infty$, the fluid motion is determined by a balance of buoyancy and viscous forces, and the characteristic velocity must be estimated accordingly.

To ascertain whether forced or natural convection dominates, or whether there is mixed convection, we must compare the velocities associated with each: for forced convection, the appropriate velocity is either the free-stream velocity for external flows, or the bulk velocity for pipe and duct flows. For fluids of Prandtl number of order unity or less, the velocity given by Eq. (4.106) is appropriate; equivalently, the Grashof number for natural convection can be compared to the square of the Reynolds number. For high-Prandtl-number fluids, the Grashof number divided by the cube root of the Prandtl number should be compared to the square of the Reynolds number. The classification in Table 4.7 has been prepared accordingly. Care must be taken to use appropriate length parameters. For example, in Fig. 4.40*b*, the proper parameter is the cylinder diameter D, and in Fig. 4.40*k*, it is D for forced convection and the tube length L for natural convection. In the case of the horizontal rectangular plate in Fig. 4.40*g*, the length parameter is L_1 for forced convection and L_2 for natural convection, if $L_2 < L_1$ (although if the flow is turbulent, L_2 is of no consequence).

Table 4.7 Criteria for mixed convection in external flows.

	$\mathrm{Pr} \lesssim 1$	$\mathrm{Pr} \to \infty$
Forced convection	$(\mathrm{Gr}/\mathrm{Re}^2) \ll 1$	$(\mathrm{Gr}/\mathrm{Pr}^{1/3}\mathrm{Re}^2) \ll 1$
Mixed convection	$(\mathrm{Gr}/\mathrm{Re}^2) \simeq 1$	$(\mathrm{Gr}/\mathrm{Pr}^{1/3}\mathrm{Re}^2) \simeq 1$
Natural convection	$(\mathrm{Gr}/\mathrm{Re}^2) \gg 1$	$(\mathrm{Gr}/\mathrm{Pr}^{1/3}\mathrm{Re}^2) \gg 1$

Internal Flows

Mixed convection in duct flows is more complicated than in external flows. In vertical ducts, an aiding buoyancy force increases heat transfer in laminar flows but usually decreases heat transfer in turbulent flows. Similarly, an opposing buoyancy force decreases heat transfer in laminar flow but usually increases heat transfer in turbulent flow. Heating a fluid flowing in a horizontal pipe produces a secondary motion in which fluid circulates upward on the vertical walls and downward in the central region, with a resulting increase in heat transfer. Cooling reverses the direction of circulation and similarly increases the heat transfer.

Figure 4.42*a* and *b* indicates the mixed convection regime, defined as a flow for which the Nusselt number deviates by more than 10% from the pure forced- or natural-convection value. For horizontal flow, in Fig. 4.42*b*, the boundary between mixed and natural convection is omitted since it has yet to be satisfactorily established.

In situations where it is not obvious that forced or natural convection dominates, the criteria given here, namely, Table 4.7 for external flows and Fig. 4.42 for internal flows, must be used to ascertain whether special correlations for mixed convection are necessary.

Correlations

Some appropriate correlations for mixed convection in external flows recommended by Churchill [38] are as follows.

1. Laminar or turbulent boundary layer flows with an assisting buoyancy force on vertical plates, cylinders, or spheres (Fig. 4.40*a*, *b*, *c*):

$$(\overline{\mathrm{Nu}} - \overline{\mathrm{Nu}}_0)^3 = (\overline{\mathrm{Nu}}_f - \overline{\mathrm{Nu}}_0)^3 + (\overline{\mathrm{Nu}}_n - \overline{\mathrm{Nu}}_0)^3 \qquad \textbf{(4.109)}$$

where the subscripts f and n refer to forced and natural convection, respectively, and $\overline{\mathrm{Nu}}_0 = 0, 0.3$, and 2 for plates, cylinders, and spheres, respectively. Equation (4.109) is well established for laminar flows but is expected to be less reliable for turbulent flows. Also, it should be noted that a buoyancy force can suppress transition of a laminar forced-convection boundary layer.

2. Boundary layer flows with an opposing buoyancy force (Fig. 4.40*d*, *e*, *f*):

$$(\overline{\mathrm{Nu}} - \overline{\mathrm{Nu}}_0)^3 = \left|(\overline{\mathrm{Nu}}_f - \overline{\mathrm{Nu}}_0)^3 - (\overline{\mathrm{Nu}}_n - \overline{\mathrm{Nu}}_0)^3\right| \qquad \textbf{(4.110)}$$

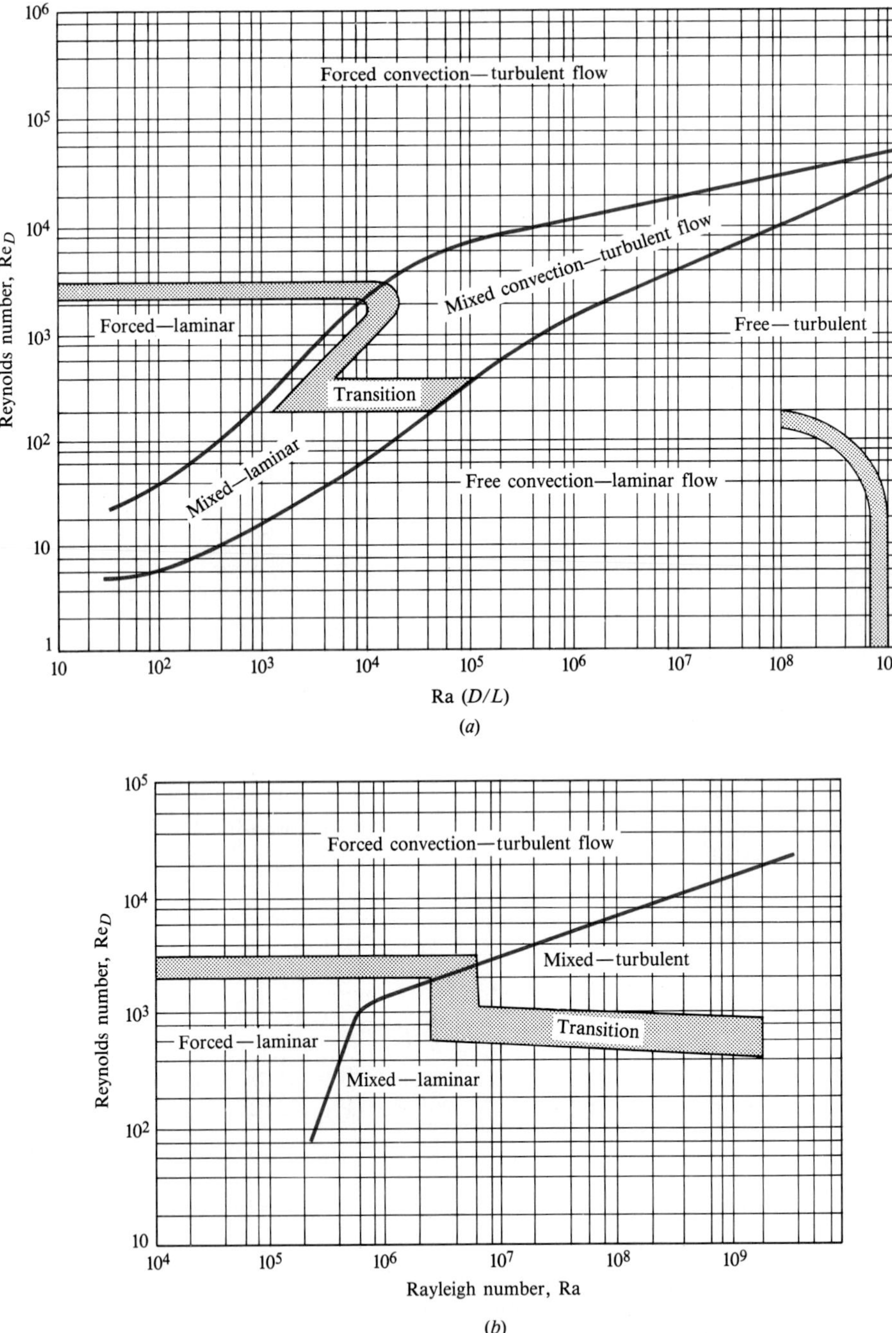

Figure 4.42 (*a*) Mixed forced- and natural-convection flow regimes in vertical tubes; $10^{-2} < \mathrm{Pr}D/L < 1$. (*b*) Mixed forced- and natural-convection flow regimes in horizontal tubes; $10^{-2} < \mathrm{Pr}D/L < 1$ [37]. (Adapted with permission.)

3. Forced boundary layer flow along a horizontal plate with a transverse buoyancy force (Fig. 4.40*g*, *h*):

$$\overline{\text{Nu}}^{7/2} = \overline{\text{Nu}}_f^{7/2} \pm \overline{\text{Nu}}_n^{7/2}; \qquad L_1 < L_2 \tag{4.111}$$

where the positive sign is applicable when the fluid on the upper side is heated or the lower side is cooled, and vice versa for the negative sign. Since separation of the boundary layer occurs before $\overline{\text{Nu}}_n$ exceeds $\overline{\text{Nu}}_f$, the limit of applicability of Eq. (4.111) is $\overline{\text{Nu}} > 0$.

4. Cross-flow over an immersed cylinder or sphere with a transverse buoyancy force (Fig. 4.40*i*, *j*):

$$(\overline{\text{Nu}} - \overline{\text{Nu}}_0)^4 = (\overline{\text{Nu}}_f - \overline{\text{Nu}}_0)^4 + (\overline{\text{Nu}}_n - \overline{\text{Nu}}_0)^4 \tag{4.112}$$

These combining rules for mixed convection, Eqs. (4.109) through (4.112), are clearly rather crude; however, they are adequate for most engineering purposes. Correlations for mixed convection in internal flows are more complicated; some useful results may be found in handbooks listed in the bibliography for Chapter 4.

EXAMPLE 4.9 Cooling of an Electronics Package

The sides of an electronics package are rectangular aluminum plates 70 cm wide and 40 cm high and act as a heat sink for the contents. The plate temperature must not exceed 340 K. If air at 300 K is blown upward over the plates at 1 m/s, estimate the power each plate dissipates by convective heat transfer.

Solution

Given: Mixed forced and natural convection on a vertical plate.

Required: Convective heat transfer rate.

Assumptions: The boundary layer starts at the bottom of the plate.

Evaluate air properties at a mean film temperature of 320 K: $\beta = 1/T_r = 1/320$, $k = 0.0281$ W/m K, $\nu = 17.44 \times 10^{-6}$ m^2/s, and Pr $= 0.69$. We must first calculate average Nusselt numbers assuming there is only forced or only natural convection.

$$\text{Re}_L = \frac{(1)(0.4)}{17.44 \times 10^{-6}} = 2.29 \times 10^4 \quad \text{(laminar flow)}$$

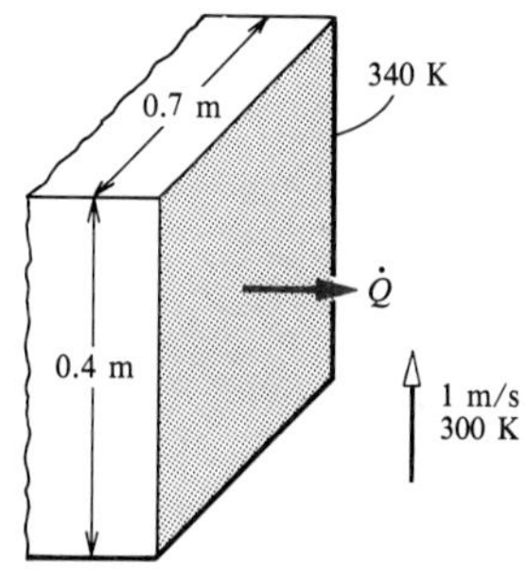

Using Eq. (4.57),

$$\overline{\text{Nu}}_f = 0.664\text{Re}_L^{1/2}\text{Pr}^{1/3} = 0.664(2.29 \times 10^4)^{1/2}(0.69)^{1/3}$$

$$= 88.8$$

$$\text{Gr}_L = \frac{\beta \Delta T g L^3}{\nu^2} = \frac{(40/320)(9.81)(0.4)^3}{(17.44 \times 10^{-6})^2}$$

$$= 2.58 \times 10^8$$

$$\text{Ra}_L = \text{Gr}_L\text{Pr} = (2.58 \times 10^8)(0.69) = 1.780 \times 10^8 < 10^9 \quad \text{(laminar)}$$

Using Eqs. (4.84) and (4.85),

$$\Psi = \left[1 + \left(\frac{0.492}{0.69}\right)^{9/16}\right]^{-16/9} = 0.3426$$

$$\overline{\mathrm{Nu}}_n = 0.68 + 0.670(1.780 \times 10^8 \times 0.3426)^{1/4} = 59.9$$

For mixed convection, we use Eq. (4.109) with $\mathrm{Nu}_0 = 0$:

$$\overline{\mathrm{Nu}} = (\overline{\mathrm{Nu}}_f^3 + \overline{\mathrm{Nu}}_n^3)^{1/3} = (88.8^3 + 59.9^3)^{1/3} = 97.1$$

$$\bar{h}_c = \left(\frac{k}{L}\right)\overline{\mathrm{Nu}} = \left(\frac{0.0281}{0.4}\right)97.1 = 6.82 \text{ W/m}^2 \text{ K}$$

$$\dot{Q} = \bar{h}_c A \Delta T = (6.82)(0.4 \times 0.7)(340 - 300) = 76.4 \text{ W}$$

Comments

1. Since the heat transfer coefficient is relatively low (~ 7 W/m^2 K), radiative heat transfer will play an important role, particularly if the aluminum surface is black-anodized to give a high emittance.
2. Use of vertical fins would markedly improve the convective heat transfer.
3. Use CONV to check $\overline{\mathrm{Nu}}_f$ and $\overline{\mathrm{Nu}}_n$.

EXAMPLE 4.10 Heat Loss from a Workshop Roof

A wind blows at 2 m/s over the 10 m square horizontal roof of a workshop on a sunny day. The ambient air temperature is 295 K, and the roof temperature is measured to be approximately 315 K. Estimate the convective heat loss from the roof. Take $\mathrm{Re}_{\mathrm{tr}} = 10^5$.

Solution

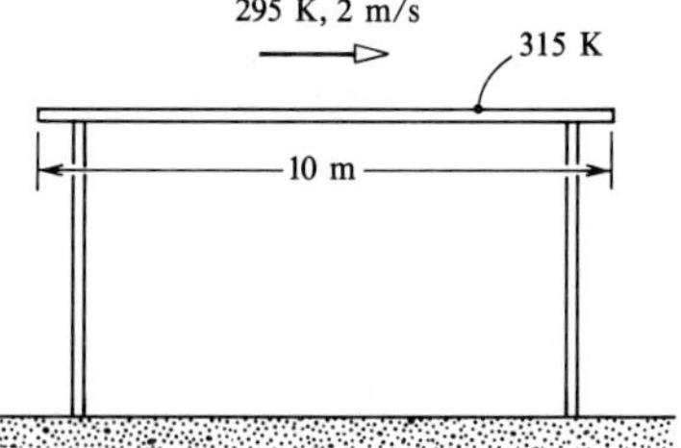

Given: Mixed forced and natural convection on a horizontal surface.

Required: Convective heat transfer rate.

Assumptions: The wind direction is normal to one edge of the roof.

Evaluate fluid properties at a mean film temperature of 305 K: $\beta = 1/T_r = 1/305$, $k = 0.0270$ W/m K, $c_p = 1005$ J/kg K, $\nu = 16.1 \times 10^{-6}$ m^2/s, and Pr $= 0.69$. The Reynolds number is

$$\mathrm{Re}_L = \frac{VL}{\nu} = \frac{(2)(10)}{16.1 \times 10^{-6}} = 1.242 \times 10^6$$

For forced convection, $\overline{\text{Nu}}$ is given by Eq. (4.65):

$$\overline{\text{Nu}}_f = 0.664\text{Re}_{\text{tr}}^{1/2}\text{Pr}^{1/3} + 0.036\text{Re}_L^{0.8}\text{Pr}^{0.43}\left[1 - \left(\frac{\text{Re}_{\text{tr}}}{\text{Re}_L}\right)^{0.8}\right]$$

$$= 0.664(10^5)^{1/2}(0.69)^{1/3} + 0.036(1.242 \times 10^6)^{0.8}(0.69)^{0.43}\left[1 - \left(\frac{0.1}{1.242}\right)^{0.8}\right]$$

$$= 186 + 1996 = 2182$$

The Grashof number is

$$\text{Gr}_L = \frac{\beta \Delta T g L^3}{\nu^2} = \frac{(20/305)(9.81)(10)^3}{(16.1 \times 10^{-6})^2} = 2.48 \times 10^{12}$$

Thus, Eq. (4.96) applies for natural convection:

$$\overline{\text{Nu}}_n = 0.14\text{Ra}_L^{1/3} = 0.14(\text{Gr}_L\text{Pr})^{1/3} = 0.14[(2.48 \times 10^{12})(0.69)]^{1/3} = 1674$$

For mixed convection, Eq. (4.111) applies:

$$\overline{\text{Nu}}^{7/2} = \overline{\text{Nu}}_f^{7/2} + \overline{\text{Nu}}_n^{7/2} = (2182)^{7/2} + (1674)^{7/2} = 4.85 \times 10^{11} + 1.92 \times 10^{11}$$

$$= 6.77 \times 10^{11}$$

$$\overline{\text{Nu}} = 2400$$

$$\bar{h}_c = \left(\frac{k}{L}\right)\overline{\text{Nu}} = \left(\frac{0.0270}{10}\right)2400 = 6.48 \text{ W/m}^2\text{ K}$$

$$\dot{Q} = \bar{h}_c A \Delta T = (6.48)(10 \times 10)(315 - 295) = 13.0 \text{ kW}$$

Comments

1. Use CONV to check $\overline{\text{Nu}}_f$ and $\overline{\text{Nu}}_n$.
2. Equation (4.96) has been used outside its stated range of applicability ($\text{Ra}_L < 3 \times 10^{10}$).

4.5 TUBE BANKS AND PACKED BEDS

Flow through tube banks (bundles) or packed beds in heat exchangers are forced flows and have characteristics of both internal and external flows. The flow is confined in a shell or duct to give an overall internal flow behavior, but the flow over a particular tube or packing piece has similarities to an external flow.

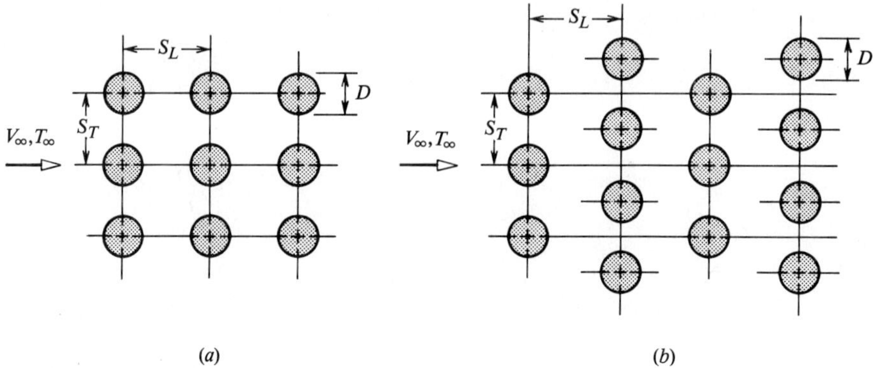

Figure 4.43 Tube bank configurations. (*a*) Aligned. (*b*) Staggered.

4.5.1 Flow through Tube Banks

Cross-flow over tube banks occurs in shell-and-tube heat exchangers, which are widely used as evaporators and condensers in power and refrigeration systems and for a great variety of applications in the process industries. Often the tubes are finned, but here attention will be restricted to smooth tubes. Both aligned and staggered tube banks are used, as shown in Fig. 4.43. The flow inside a tube bank is complex, involving boundary layer separation on each tube and interactions of the resulting wakes with adjacent wakes and downstream tubes. Figure 4.43 shows the geometric parameters of a tube bank, which include the tube diameter D and the transverse and longitudinal pitches, S_T and S_L, respectively. The number of rows of tubes transverse to the flow is N.

To calculate heat transfer, a procedure similar to that of Gnielinski et al. [39] is recommended. The heat transfer coefficient on the first row of tubes is higher than for a single tube in cross-flow, as the fluid must accelerate to pass through the spaces between adjacent tubes. However, if the Reynolds number is based on an average velocity in the space between two adjacent tubes, that is, defined by the relation

$$\frac{\overline{V}}{V_0} = \frac{S_T}{S_T - (\pi/4)D} \tag{4.113}$$

where V_0 is the velocity of the fluid in the empty cross section of the shell or duct, then Eq. (4.71) for a single tube in cross-flow applies.

The heat transfer coefficient increases from the first to about the fifth row of a tube bank. The average Nusselt number for a tube bank with 10 or more rows may be calculated from

$$\overline{\mathrm{Nu}}_D^{10+} = \Phi \overline{\mathrm{Nu}}_D^{1} \tag{4.114}$$

where $\overline{\mathrm{Nu}}_D^{1}$ is the Nusselt number for the first row, and Φ is an *arrangement factor*. We define the dimensionless transverse pitch as $P_T = S_T/D$, the dimensionless longitudinal pitch as $P_L = S_L/D$, and a factor ψ as

$$\psi = 1 - \frac{\pi}{4P_T} \qquad \text{if } P_L \geq 1 \tag{4.115a}$$

$$\psi = 1 - \frac{\pi}{4P_T P_L} \qquad \text{if } P_L < 1 \tag{4.115b}$$

Then the arrangement factors are

$$\Phi_{\text{aligned}} = 1 + \frac{0.7}{\psi^{1.5}} \frac{S_L/S_T - 0.3}{(S_L/S_T + 0.7)^2} \tag{4.116}$$

$$\Phi_{\text{staggered}} = 1 + \frac{2}{3P_L} \tag{4.117}$$

For tube banks of fewer than 10 rows, a simple interpolation formula applies:

$$\overline{\text{Nu}_D} = \frac{1 + (N - 1)\Phi}{N} \overline{\text{Nu}_D^1} \tag{4.118}$$

Properties are to be evaluated at the average mean film temperature for gases; for liquids, properties are first evaluated at the average bulk temperature, $(T_{\text{in}} + T_{\text{out}})/2$, and then a Prandtl number correction factor is used with $n = -0.25$ for heating and $n = -0.11$ for cooling.

Zukauskus [40] recommends that the pressure drop in a tube bank be calculated as

$$\Delta P = N\chi\left(\frac{\rho V_{\max}^2}{2}\right) f \tag{4.119}$$

where the friction factor f and correction factor χ are given in Fig. 4.44*a* and *b*. Notice that $\chi = 1$ for a square aligned configuration and for an equilateral-triangle staggered configuration. The velocity $V_{\max}$ is the maximum velocity in the tube bank. For an aligned bank, the maximum velocity occurs between adjacent tubes of a transverse row; hence,

$$\frac{V_{\max}}{V_0} = \frac{S_T}{S_T - D} \tag{4.120a}$$

For a staggered bank, $V_{\max}$ may occur between adjacent tubes of a transverse row or of a diagonal row. Then

$$\frac{V_{\max}}{V_0} = \max\left\{\frac{S_T}{S_T - D}, \frac{S_T/2}{[S_L^2 + (S_T/2)^2]^{1/2} - D}\right\} \tag{4.120b}$$

Properties for the pressure drop calculation are to be evaluated at the average bulk temperature.

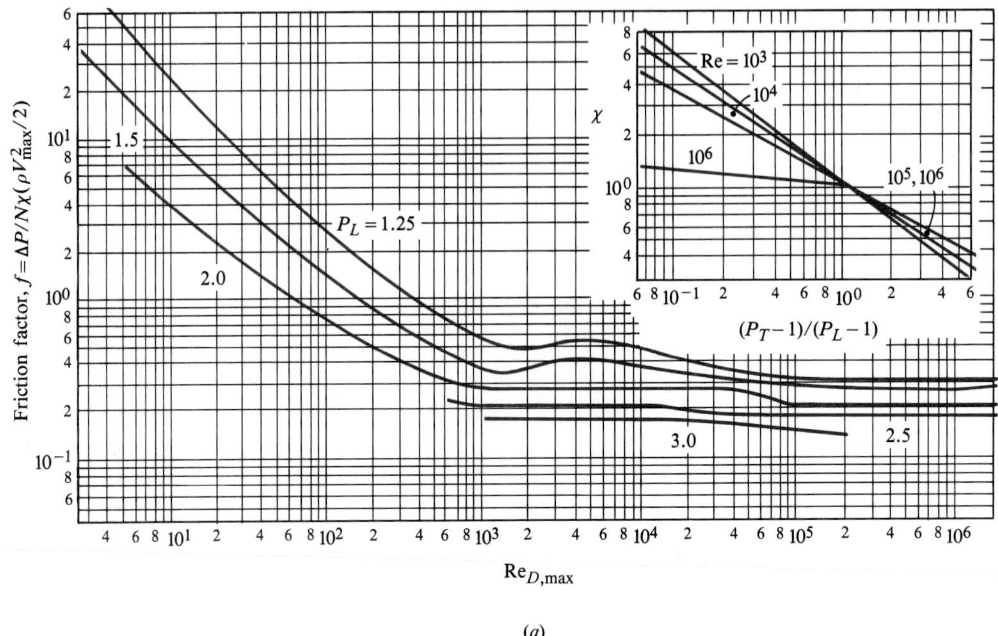

(*a*)

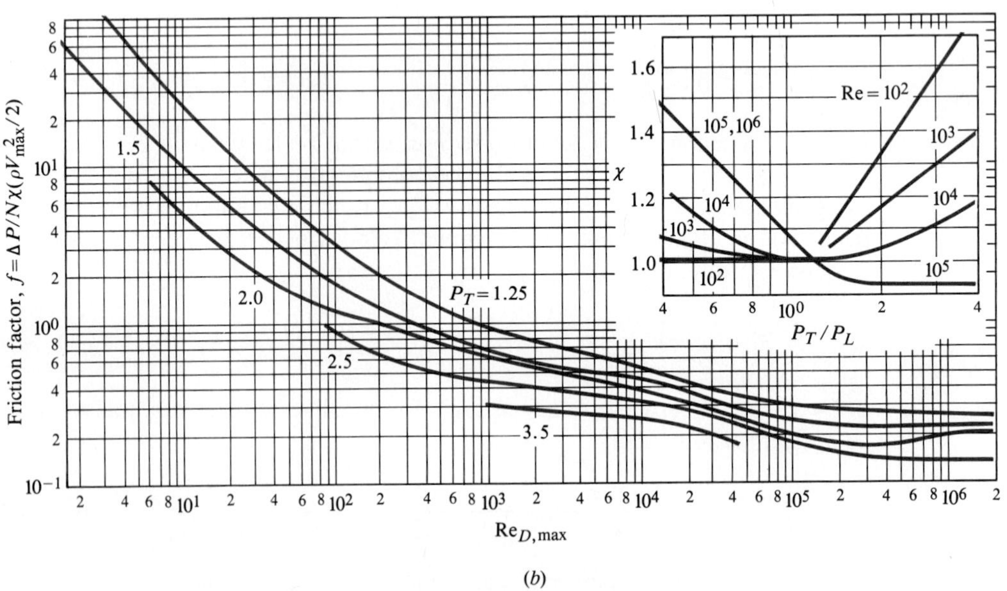

(*b*)

Figure 4.44 Friction factor f and correction factor χ in Eq. (4.119). (*a*) For an aligned tube bank. (*b*) For a staggered tube bank [40]. (Adapted with permission.)

EXAMPLE 4.11 A Tube Bank Air Heater

Air at 290 K and 1 atm flows at 2.0 kg/s down a duct of cross section 1 m × 0.4 m. The duct contains a tube bank of 1 m–long, 15 mm–O.D. tubes in a staggered arrangement, with both transverse and longitudinal pitches of 30 mm. If steam at atmospheric pressure is condensed in the tubes, how many rows of tubes are required to heat the air to 324 K? Also estimate the pressure drop.

Solution

Given: Air flowing through a bank of tubes, inside which steam condenses at 1 atm pressure.

Required: Number of tubes needed to heat air to 324 K, and the pressure drop.

Assumptions: Negligible condensing side and wall thermal resistances; hence, the outside wall temperature can be taken to be P_{sat} (1 atm) = 373 K.

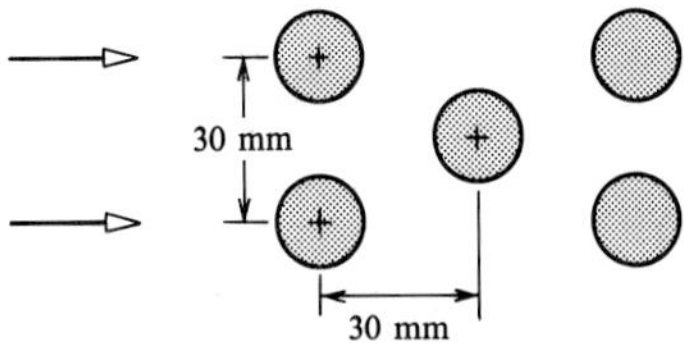

To calculate heat transfer, the air properties are evaluated at the mean film temperature based on the tube wall temperature, and the average of the inlet and outlet bulk air temperatures:

$$T_r = \frac{1}{2}\left(373 + \frac{290 + 324}{2}\right) = 340 \text{ K}$$

From Table A.7, $k = 0.0294$ W/m K, $\nu = 19.32 \times 10^{-6}$ m²/s, $\rho = 1.042$ kg/m³, Pr = 0.69, and $c_p = 1007$ J/kg K.

The empty duct velocity is $V_0 = \dot{m}/\rho A_c = (2.0)/(1.042)(0.4) = 4.80$ m/s. Using Eq. (4.113) for the first row of tubes,

$$\overline{V} = V_0 \frac{S_T}{S_T - (\pi/4)D} = (4.80)\frac{30}{30 - (\pi/4)15} = 7.90 \text{ m/s}$$

$$\text{Re}_D = \frac{\overline{V}D}{\nu} = \frac{(7.90)(0.015)}{19.32 \times 10^{-6}} = 6130$$

Using Eq. (4.71*a*),

$$\overline{\text{Nu}}_D^1 = 0.3 + \frac{0.62\text{Re}_D^{1/2}\text{Pr}^{1/3}}{[1 + (0.4/\text{Pr})^{2/3}]^{1/4}} = 0.3 + \frac{0.62(6130)^{1/2}(0.69)^{1/3}}{[1 + (0.4/0.69)^{2/3}]^{1/4}} = 37.9$$

We will assume that more than 10 tube rows are required and check later. Using Eqs. (4.117) and (4.114),

$$P_L = \frac{S_L}{D} = \frac{30}{15} = 2$$

$$\Phi_{\text{staggered}} = 1 + \frac{2}{3P_L} = 1 + \frac{2}{(3)(2)} = 1.333$$

$$\overline{\text{Nu}}_D^{10+} = \Phi\overline{\text{Nu}}_D^1 = (1.333)(37.9) = 50.5$$

$$\overline{h}_c = \frac{\overline{\text{Nu}}_D k}{D} = \frac{(50.5)(0.0294)}{0.015} = 99.0 \text{ W/m}^2\text{ K}$$

Since the tube wall temperature is constant through the bank, this air heater is a single-stream heat exchanger, and Eq. (1.59) applies. The required effectiveness is

$$\varepsilon = \frac{T_{\text{out}} - T_{\text{in}}}{T_s - T_{\text{in}}} = \frac{324 - 290}{373 - 290} = 0.410$$

Hence, the required number of transfer units is

$$N_{\text{tu}} = \ln\frac{1}{1-\varepsilon} = \ln\frac{1}{1-0.410} = 0.528 = \frac{\overline{h}_c A}{\dot{m} c_p}$$

The required transfer area is then

$$A = \frac{0.528\dot{m}c_p}{\overline{h}_c} = \frac{0.528(2.0)(1007)}{99.0} = 10.74 \text{ m}^2$$

The number of tubes per row is the duct width divided by the transverse pitch:

$$\frac{W}{S_T} = \frac{0.4}{0.030} = 13.3 \simeq 13$$

Then

$$\begin{aligned} A &= \text{(Number of rows)(Number of tubes per row)(Area of one tube)} \\ &= N(13)(\pi)(0.015)(1) \\ &= 0.613N \end{aligned}$$

Hence, $N = 10.74/0.613 = 17.5 \simeq 18$ rows (> 10).

The pressure drop is obtained from Eq. (4.119) and Fig. 4.44*b*. From Eq. (4.120*b*),

$$\begin{aligned} \frac{V_{\max}}{V_0} &= \max\left\{\frac{S_T}{S_T - D}, \frac{S_T/2}{[S_L^2 + (S_T/2)^2]^{1/2} - D}\right\} \\ &= \max\left\{\frac{30}{30-15}, \frac{30/2}{[30^2 + (30/2)^2]^{1/2} - 15}\right\} \\ &= \max\{2, 0.809\} = 2 \end{aligned}$$

To calculate pressure drop, properties are evaluated at the average bulk temperature:

$$T_r = \frac{1}{2}(290 + 324) = 307 \text{ K}$$

From Table A.7, $\rho = 1.151$ kg/m^3, $\nu = 16.27 \times 10^{-6}$ m^2/s.

$$V_0 = \frac{\dot{m}}{\rho A} = \frac{2.0}{(1.151)(0.4)} = 4.34 \text{ m/s}$$

$$V_{\max} = (2)(4.34) = 8.68 \text{ m/s}$$

$$\text{Re}_{D,\max} = \frac{(8.68)(0.015)}{16.27 \times 10^{-6}} = 8000$$

$$P_L = P_T = 30/15 = 2.0, \qquad P_T/P_L = 1.0$$

Using Fig. 4.44*b*, $f = 0.38$, $\chi = 1.0$.

$$\Delta P = N\chi\left(\frac{\rho V_{\max}^2}{2}\right)f = (18)(1.0)\left[\frac{(1.151)(8.68)^2}{2}\right](0.38) = 297 \text{ Pa}$$

Comments

1. Use CONV to check $\overline{h}_c$ and ΔP.
2. In general, there may be an uncomfortably large discrepancy in ΔP given by CONV, because CONV contains an approximate curve fit to the data in Fig. 4.44. However, the original data are not too reliable, so greater precision is perhaps not warranted.
3. Read the discussion following Eq. (4.11) regarding the use of the single-stream heat exchanger equation.
4. The pressure drop is less than 1% of the total pressure; thus, its effect on density can be ignored. When the pressure drop is a significant fraction of the total pressure, an iterative calculation procedure is required.
5. Notice the different property evaluation schemes for heat transfer and pressure drop. However, the difference is of little practical significance owing to the large expected error in f values obtained from Fig. 4.44.

EXAMPLE 4.12 A Tube Bank Water Heater

A tube bank is 30 rows deep and has 15 mm–O.D. tubes in a staggered arrangement, with a transverse pitch of 24 mm and a longitudinal pitch of 15 mm. Steam at 2×10^5 Pa condenses inside the tubes. In a performance test, 304.0 K water enters the tube bank at 2 m/s and leaves at 316.0 K. Estimate the average heat transfer coefficient of the tube bank, and the pressure drop.

Solution

Given: Water flowing through a bank of tubes, inside which steam condenses at 2×10^5 Pa.

Required: Average heat transfer coefficient $\overline{h}_c$ and pressure drop ΔP.

Assumptions: The tube wall temperature is approximately constant through the bundle.

For a liquid flowing through a tube bank, properties are evaluated at the average of the inlet and outlet bulk temperatures, with a Prandtl number ratio correction applied later.

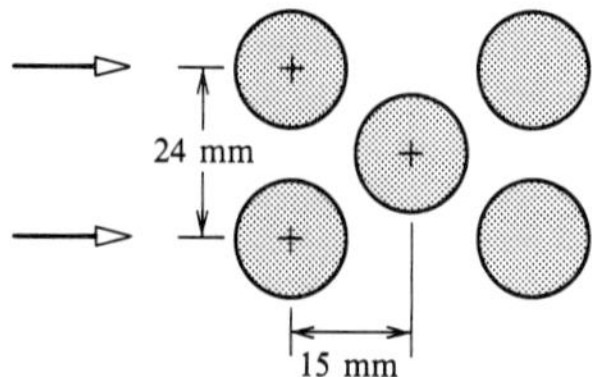

$$T_r = \frac{304 + 316}{2} = 310 \text{ K}$$

From Table A.8, water properties are $k = 0.628$ W/m K, $\rho = 993$ kg/m^3, $c_p = 4174$ J/kg K, $\nu = 0.70 \times 10^{-6}$ m^2/s, and Pr $= 4.6$. Using Eq. (4.113) for the first row of tubes,

$$\overline{V} = V_0 \frac{S_T}{S_T - (\pi/4)D} = (2)\frac{24}{24 - (\pi/4)(15)} = 3.93 \text{ m/s}$$

$$\text{Re}_D = \frac{\overline{V}D}{\nu} = \frac{(3.93)(0.015)}{0.70 \times 10^{-6}} = 84{,}200$$

Using Eq. (4.71*b*),

$$\overline{\text{Nu}}_D^{1} = 0.3 + \frac{0.62\text{Re}_D^{1/2}\text{Pr}^{1/3}}{[1 + (0.4/\text{Pr})^{2/3}]^{1/4}}\left[1 + \left(\frac{\text{Re}_D}{282{,}000}\right)^{1/2}\right]$$

$$= 0.3 + \frac{0.62(84{,}200)^{1/2}(4.6)^{1/3}}{[1 + (0.4/4.6)^{2/3}]^{1/4}}\left[1 + \left(\frac{84{,}200}{282{,}000}\right)^{1/2}\right] = 443$$

$$P_L = \frac{S_L}{D} = \frac{15}{15} = 1.00$$

$$\Phi_{\text{staggered}} = 1 + \frac{2}{3P_L} = 1 + \frac{2}{(3)(1.0)} = 1.667$$

$$\overline{\text{Nu}}_D^{10+} = \Phi\overline{\text{Nu}}_D^{1} = (1.667)(443) = 738$$

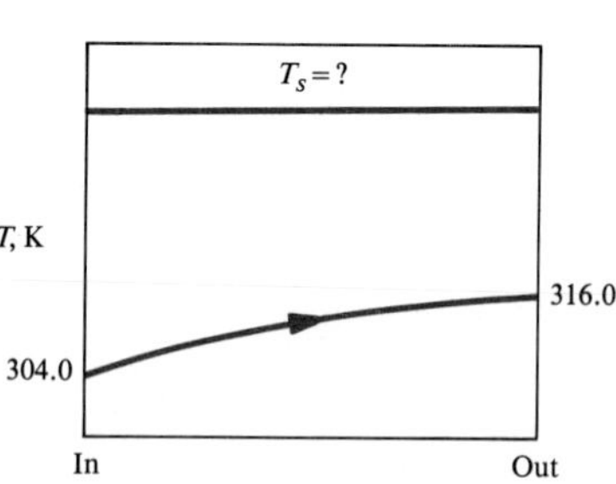

To obtain the Prandtl number ratio correction, $(\text{Pr}_s/\text{Pr}_b)^{-1/4}$, we need an estimate of the tube wall temperature. For this purpose, we assume a constant tube wall temperature through the bank so that the equations for a single-stream exchanger can be used. The water-side number of transfer units is $N_{\text{tu}} = \overline{h}_c A/\dot{m}c_p$. The uncorrected value of $\overline{\text{Nu}}_D$ is used to estimate $\overline{h}_c$:

$$\overline{h}_c = \left(\frac{k}{D}\right)\overline{\text{Nu}}_D = \left(\frac{0.628}{0.015}\right)738 = 3.09 \times 10^4 \text{ W/m}^2 \text{ K}$$

A = (Number of rows)(Number of tubes per row)(Area of one tube)

$= (30)(1/0.024)(\pi)(0.015)(1)$ for 1 m^2 of cross section

$= 58.9 \text{ m}^2$

$\dot{m} = \rho V A_c = (995)(2)(1) = 1990$ kg/s (using ρ at 304 K)

$$N_{\text{tu}} = \frac{(3.09 \times 10^4)(58.9)}{(1990)(4174)} = 0.219$$

Using Eq. (1.59),

$$\varepsilon = 1 - e^{-N_{\text{tu}}} = 1 - e^{-0.219} = 0.197 = \frac{T_{\text{out}} - T_{\text{in}}}{T_s - T_{\text{in}}}$$

$$T_s = T_{\text{in}} + (1/\varepsilon)(T_{\text{out}} - T_{\text{in}}) = 304 + (1/0.197)(316 - 304) = 364.9 \text{ K}$$

At 364.9 K, Table A.8 gives Pr = 1.93; hence,

$$(\text{Pr}_s/\text{Pr}_b)^{-1/4} = (1.93/4.6)^{-1/4} = 1.24$$

Thus, the corrected Nusselt number and heat transfer coefficients are

$$\overline{\mathrm{Nu}}_D = (1.24)(738) = 917$$

$$\bar{h}_c = (0.628/0.015)(917) = 3.84 \times 10^4 \text{ W/m}^2\text{ K}$$

Since the correction is quite large, a second iteration is perhaps justified.

$$N_{tu} = 0.273; \qquad \varepsilon = 0.239; \qquad T_s = 354; \qquad \mathrm{Pr}_s = 2.24$$

$$(\mathrm{Pr}_s/\mathrm{Pr}_b)^{-1/4} = 1.20; \qquad \overline{\mathrm{Nu}}_D = 886; \qquad \bar{h}_c = 3.71 \times 10^4 \text{ W/m}^2\text{ K}$$

The pressure drop is obtained from Eqs. (4.119) and (4.120*b*) and Fig. 4.44*b*:

$$\frac{S_T}{S_T - D} = \frac{24}{24 - 15} = 2.67$$

$$\frac{S_T/2}{[S_L^2 + (S_T/2)^2]^{1/2} - D} = \frac{24/2}{[15^2 - (24/2)^2]^{1/2} - 15} = 2.85$$

$$V_{max} = 2.85V_0 = (2.85)(2) = 5.70 \text{ m/s}$$

$$\mathrm{Re}_{D,max} = \frac{(5.70)(0.015)}{0.70 \times 10^{-6}} = 1.22 \times 10^5$$

$$P_L = 15/15 = 1; \qquad P_T = 24/15 = 1.6; \qquad P_T/P_L = 1.6$$

From Fig. 4.44*b*, $f = 0.21$, $\chi = 0.95$; thus,

$$\Delta P = N\chi\left(\frac{\rho V_{max}^2}{2}\right)f = (30)(0.95)\left[\frac{(993)(5.70)^2}{2}\right](0.21) = 0.97 \times 10^5 \text{ Pa}$$

Comments

1. In contrast to the previous example, the tube wall temperature could not be assumed equal to the saturation temperature of the steam: in this situation, the water-side resistance is comparable to the steam-side resistance, and the latter cannot be ignored.
2. Use CONV to check $\bar{h}_c$ and ΔP.
3. For a liquid, the properties used to calculate V_{max} and Re_D for heat transfer and pressure drop are the same (for a gas, they are different; see Example 4.11).

4.5.2 Flow through Packed Beds

Packed beds of solid particles, such as the pebble bed shown in Fig. 4.45, are often used as heat exchangers or for storage of thermal energy. The pebbles are heated by passing a hot fluid through the bed, and the energy stored in the pebbles is subsequently extracted by passing a cold fluid through the bed. The objective may be to simply transfer heat from one fluid stream to another: then the bed is termed a *regenerative* heat exchanger, and the performance of such an exchanger will be analyzed in Chapter 8. Alternatively, the objective may be to store thermal energy

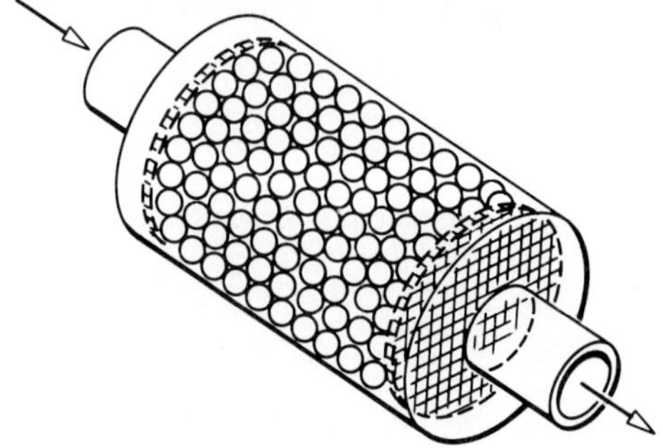

Figure 4.45 A packed pebble bed.

for a specific period of time. For example, in a solar home-heating system, the bed is heated during the day, when solar energy is available, and extracted during the night to heat the home. Packed beds are also very widely used as mass exchangers in process engineering practice, where a great variety of particle shapes are encountered.

The **void fraction** ε_v of a packed bed is defined as

$$\varepsilon_v = \frac{\text{Bed volume} - \text{Packing volume}}{\text{Bed volume}} \tag{4.121}$$

Values of ε_v between 0.3 and 0.5 are typical. The surface area for transfer depends on the particle size and shape and the void fraction. The **specific surface area** a [m^{-1}] of a packed bed is the wetted or transfer area per unit volume of bed:

$$a = \frac{\text{Total surface area of particles}}{\text{Bed volume}} \tag{4.122}$$

If a bed consists of particles of volume V_p and surface area A_p, then

$$a = \frac{A_p}{V_p}(1 - \varepsilon_v) \tag{4.123}$$

For example, for a spherical particle of diameter d_p, $A_p/V_p = \pi d_p^2/(1/6)\pi d_p^3 = 6/d_p$, and $a = 6(1 - \varepsilon_v)/d_p$. In general, the specific surface area can be calculated if the geometry of the particles and the void fraction in the bed are known. The hydraulic diameter of the bed D_h is defined in an analogous manner as for duct flow:

$$D_h = \frac{\text{Volume of bed available for flow}}{\text{Wetted surface in bed}} = \frac{\text{Void volume/unit volume}}{\text{Wetted surface/unit volume}} = \frac{\varepsilon_v}{a} \tag{4.124}$$

Using Eq. (4.123),

$$D_h = \left(\frac{\varepsilon_v}{1 - \varepsilon_v}\right)\frac{V_p}{A_p} \tag{4.125}$$

We next define a characteristic length and velocity for flow through packing. For the characteristic length $\mathcal{L}$, we choose $6D_h$; then, if we also define the effective particle diameter as

$$d_p = 6\frac{V_p}{A_p} \tag{4.126}$$

which yields the actual diameter for a spherical particle, the characteristic length becomes

$$\mathcal{L} = d_p\left(\frac{\varepsilon_v}{1-\varepsilon_v}\right) \tag{4.127}$$

For the characteristic velocity, we choose the average velocity of the fluid flowing in the void space, $\mathcal{V}$. The **superficial velocity** in the bed V is defined as the velocity in the bed if no packing were present:

$$V = \frac{\dot{m}}{\rho A_c} \tag{4.128}$$

where $\dot{m}$ [kg/s] is the mass flow, ρ is the fluid density, and A_c is the cross-sectional area of the bed. The superficial velocity can also be viewed as the velocity immediately upstream or downstream of the packing. Since the average cross-sectional area available for flow is simply $\varepsilon_v A_c$, the characteristic velocity is

$$\mathcal{V} = \frac{\dot{m}}{\rho\varepsilon_v A_c} \tag{4.129}$$

We would expect correlations of pressure drop and heat transfer based on $\mathcal{L}$ and $\mathcal{V}$ to be valid for a variety of particle shapes.

The pressure drop across a packed bed can be obtained from the *Ergun* equation [41]:

$$\frac{dP}{dx} = \frac{150\mu\mathcal{V}}{\mathcal{L}^2} + \frac{1.75\rho\mathcal{V}^2}{\mathcal{L}}; \qquad 1 < \text{Re} < 10^4 \tag{4.130}$$

The first term of Eq. (4.130) accounts for the viscous drag, and the second term accounts for form drag. The constants are based on experimental data for many shapes of particles, but the equation is most accurate for spherical particles. For flow of gases in a packed bed, an appropriate heat transfer correlation is [10]

$$\text{Nu} = (0.5\text{Re}^{1/2} + 0.2\text{Re}^{2/3})\text{Pr}^{1/3}; \qquad 20 < \text{Re} < 10^4 \tag{4.131}$$

Equation (4.131) can also be used for liquids of moderate Prandtl number with fair accuracy. The constants in Eq. (4.131) are based on experimental data for spheres and short cylinders, as well as for commercial packings used in mass-transfer operations, such as Raschig rings and Berl saddles. Variable-property effects can be accounted for using a viscosity ratio with $n = -0.14$. Entrance effects are negligible for packed beds; thus, Eq. (4.131) gives both the local and average Nusselt number.

Finally, we note that often the transfer perimeter $\mathscr{P}$ [m] is often used to characterize the surface area for transfer in a packed bed:

$$\mathscr{P} = \frac{\text{Total surface area of particles}}{\text{Bed length}}$$

It follows from Eq. (4.122) that

$$\mathscr{P} = aA_c$$

where A_c is the cross-sectional area of the bed.

Perforated-Plate Packings

A novel form of packing consists of perforated plates separated by gaskets of low thermal conductivity, as shown in Fig. 4.46. Heat exchangers using such packing are popular for cryogenic refrigeration systems, where the characteristic low gas flow rates require axial conduction effects to be reduced (see Section 8.5.4). Geometrical parameters that may affect pressure drop and heat transfer include the open-area ratio ε_p, plate spacing p, plate thickness t, and the hole diameter d. The Reynolds number is defined in terms of flow through the holes: $\text{Re} = Gd/\mu$, where G is the mass velocity through the holes. Recommended correlations for pressure drop and heat transfer are based on data for air of Shevyakova and Orlov [42]. The pressure drop across a single plate is correlated in terms of the *Euler number*, $\text{Eu} = \Delta P \rho/G^2$, and is given by

$$\text{Eu} = 8.17\text{Re}^{-0.55}(1.707 - \varepsilon_p)^2; \qquad 20 < \text{Re} < 150 \tag{4.132a}$$

$$\text{Eu} = 0.5(1.707 - \varepsilon_p)^2; \qquad 150 < \text{Re} < 3000 \tag{4.132b}$$

The heat transfer is correlated in terms of a Stanton number, $\text{St} = h_c/Gc_p$. The heat

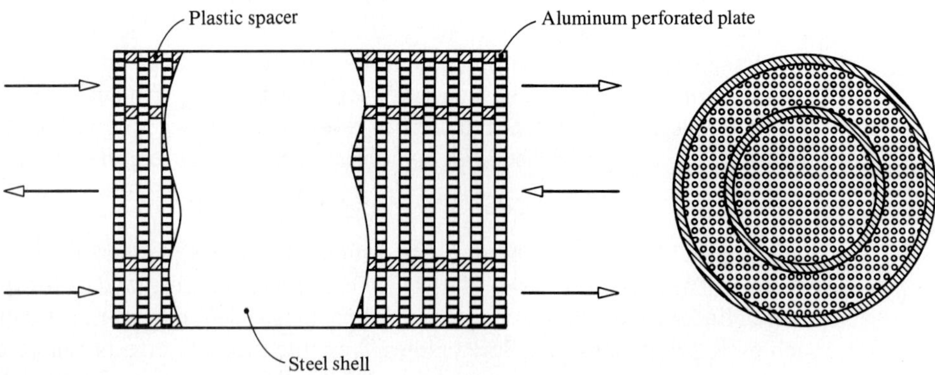

Figure 4.46 A perforated-plate heat exchanger.

transfer coefficient is based on the specific surface area a, which includes both the plate faces and hole interiors:

$$\mathrm{St} = C\mathrm{Re}^n\mathrm{Pr}^{-2/3}; \qquad 300 < \mathrm{Re} < 3000; \qquad \mathrm{Pr} > 0.5 \qquad \textbf{(4.133)}$$

where

$$C = 3.6 \times 10^{-4}[(1 - \varepsilon_p)\varepsilon_p - 0.2]^{-2.07}$$

$$n = -4.36 \times 10^{-2}\varepsilon_p^{-2.34}$$

These correlations are valid for plates 0.5 mm thick, with $0.3 < \varepsilon_p < 0.6, 0.4 < p < 1.6$ mm, and $0.625 < d < 1.65$ mm. The relative location of holes in adjacent plates is arbitrary.

EXAMPLE 4.13 A Pebble Bed Thermal Store

A pebble bed is used to store thermal energy in a solar energy utilization system. The bed is 2 m long, 1 m in diameter, and the pebbles can be approximated as 2 cm–diameter spheres packed with a void fraction of 0.44. During the storage phase of the cycle, 0.8 kg/s of air at 360 K and 1 atm comes from a solar collector. If the bed is initially at a uniform temperature of 300 K, determine the initial rate of heat storage in the bed.

Solution

Given: Hot air flowing through a packed bed of pebbles.

Required: Initial rate of heat storage.

Assumptions: Bed initially at a uniform temperature of 300 K.

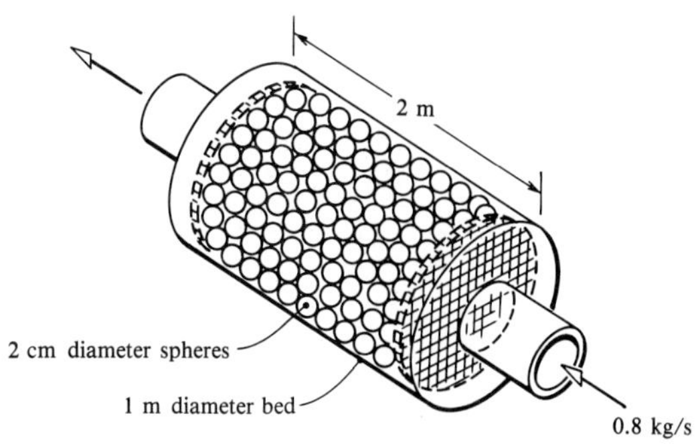

Guess an outlet air temperature of ~300 K, and evaluate air properties at an average bulk air temperature of (1/2)(360 + 300) = 330 K. From Table A.7, $k = 0.0287$ W/m K, $\rho = 1.073$ kg/m^3, $\mu = 19.71 \times 10^{-6}$ kg/m s, $c_p = 1006$ J/kg K, and Pr = 0.69. Equation (4.131)

applies. The characteristic velocity and length are

$$\mathscr{V} = \frac{\dot{m}}{\rho A_c \varepsilon_v} = \frac{0.8}{(1.073)(\pi/4)(1^2)(0.44)} = 2.16 \text{ m/s}$$

$$\mathscr{L} = \frac{d_p \varepsilon_v}{1-\varepsilon_v} = \frac{(0.02)(0.44)}{1-0.44} = 0.0157 \text{ m}$$

The Reynolds number is

$$\text{Re} = \frac{\mathscr{V}\mathscr{L}\rho}{\mu} = \frac{(2.16)(0.0157)(1.073)}{19.71 \times 10^{-6}} = 1848$$

Using Eq.(4.131),

$$\begin{aligned} \text{Nu} &= (0.5\text{Re}^{1/2} + 0.2\text{Re}^{2/3})\text{Pr}^{1/3} \\ &= [0.5(1848)^{1/2} + 0.2(1848)^{2/3}](0.69)^{1/3} = 45.6 \end{aligned}$$

Correcting for variable-property effects,

$$\begin{aligned} \mu_b &= 19.71 \times 10^{-6} \text{ kg/m s} \\ \mu_s &= 18.43 \times 10^{-6} \text{ kg/m s} \\ (\mu_s/\mu_b)^{-0.14} &= (18.43/19.71)^{-0.14} = 1.009 \\ \text{Nu} &= (1.009)(45.6) = 46.0 \\ h_c &= (k/\mathscr{L})\text{Nu} = (0.0287/0.0157)(46.0) = 84.1 \text{ W/m}^2\text{K} \end{aligned}$$

At time $t = 0$, the pebbles are at a uniform temperature; thus, Eq. (1.59) for a single-stream heat exchanger applies, with $N_{\text{tu}} = h_c \mathscr{P} L/\dot{m}c_p$.

For spherical particles, the specific surface area is

$$a = \frac{6(1-\varepsilon_v)}{d_p} = \frac{6(1-0.44)}{0.02} = 168 \text{ m}^{-1}$$

and the transfer perimeter $\mathscr{P}$ is

$$\mathscr{P} = aA_c = (168)(\pi/4)(1)^2 = 131.9 \text{ m}$$

$$N_{\text{tu}} = \frac{(84.1)(131.9)(2)}{(0.8)(1006)} = 27.6$$

which is very large. Our guess of $T_{\text{out}} \simeq 300$ K is certainly correct.

The rate of heat storage is equal to the rate at which the air gives up energy:

$$\dot{Q} = \dot{m}c_p(T_{\text{in}} - T_{\text{out}}) = (0.8)(1006)(360 - 300) = 48.3 \text{ kW}$$

Comments

1. As the bed heats up, the temperature along the bed will no longer be uniform, and calculation of the rate of heat storage requires an analysis of the transient response of the bed, as will be shown in Section 8.6.

2. Use CONV to check h_c.

EXAMPLE 4.14 A Perforated-Plate Heat Exchanger

A perforated-plate two-stream heat exchanger for a helium refrigeration system has 0.5 mm–thick plates spaced 1.0 mm apart. The holes are 1.0 mm in diameter and are arranged in square arrays with an open-area ratio of 0.4. The flow rate and cross-sectional area for each stream are 0.02 kg/s and 0.004 m^2, respectively. Determine the pressure drop per plate and the heat transfer coefficient. Evaluate the helium properties at 200 K, 1 atm.

Solution

Given: Helium flowing through a perforated-plate heat exchanger.

Required: Pressure drop per plate and heat transfer coefficient.

Assumptions: Fluid properties may be evaluated at 200 K.

Using Table A.7 for helium at 200 K, $k = 0.116$ W/m K, $\rho = 0.244$ kg/m^3, $c_p = 5200$ J/kg K, $\mu = 15.6 \times 10^{-6}$ kg/m s, and Pr $= 0.70$.

The Reynolds number is calculated as follows:

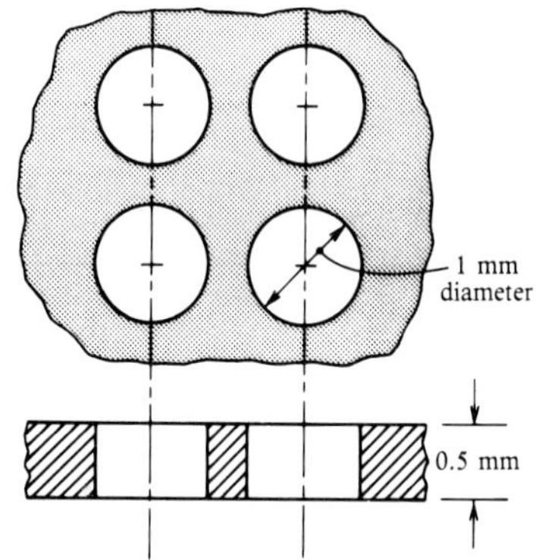

$$\text{Hole area} = \varepsilon_p A_c$$

$$= (0.4)(0.004) = 0.0016 \text{ m}^2$$

$$\text{Mass velocity } G = 0.02/0.0016$$

$$= 12.5 \text{ kg/m}^2 \text{ s}$$

$$\text{Re} = \frac{Gd}{\mu} = \frac{(12.5)(0.001)}{15.6 \times 10^{-6}} = 801$$

Equation (4.132*b*) gives the Euler number:

$$\text{Eu} = 0.5(1.707 - \varepsilon_p)^2 = 0.5(1.707 - 0.4)^2 = 0.854$$

$$\Delta P = \frac{\text{Eu}G^2}{\rho} = \frac{(0.854)(12.5)^2}{0.244} = 547 \text{ Pa}$$

The Stanton number is obtained from Eq. (4.133).

$$C = 3.6 \times 10^{-4}[(1 - \varepsilon_p)\varepsilon_p - 0.2]^{-2.07} = 3.6 \times 10^{-4}[(1 - 0.4)0.4 - 0.2]^{-2.07}$$

$$= 0.282$$

$$n = -4.36 \times 10^{-2}\varepsilon_p^{-2.34} = -(4.36 \times 10^{-2})(0.4)^{-2.34} = -0.372$$

$$\text{St} = C\text{Re}^n\text{Pr}^{-2/3} = 0.282(801)^{-0.372}(0.7)^{-2/3} = 2.97 \times 10^{-2}$$

$$h_c = \text{St}Gc_p = (2.97 \times 10^{-2})(12.5)(5200) = 1930 \text{ W/m}^2 \text{ K}$$

Comments

1. Usually in exchanger design we are interested in the $h_c\mathscr{P}$ product in order to calculate the number of transfer units, N_{tu}. The perimeter $\mathscr{P}$ is calculated as follows:

$$\text{Plate face area} = (\text{Flow area})(2 \text{ faces})(1 - \text{Open-area ratio})$$
$$= (0.004)(2)(1 - 0.4)$$
$$= 0.0048 \text{ m}^2$$

$$\text{Hole interior area} = (\text{Number of holes})(\pi\, d\, t)$$
$$= \left[\frac{A_c \varepsilon_p}{(\pi/4)d^2}\right] \pi\, d\, t = \frac{4A_c \varepsilon_p t}{d}$$
$$= \frac{(4)(0.004)(0.4)(0.0005)}{0.001} = 0.0032 \text{ m}^2$$

$$\text{Area/plate} = 0.0048 + 0.0032 = 0.008 \text{ m}^2$$

The perimeter is the heat transfer area per unit length; since the plate spacing is 1 mm, the pitch is 1.5 mm and $\mathscr{P} = 0.008/0.0015 = 5.33$ m.

2. Alternatively, the specific surface area a is often quoted for packed bed exchangers: $a = \mathscr{P}/A_c = (5.33)/(0.004) = 1333 \text{ m}^{-1}$.

3. Use CONV to check ΔP and h_c.

4.6 ROTATING SURFACES

The design of cooling systems for rotating machinery, such as turbines and electric motors, as well as the design of rotating heat exchangers, high-speed gas bearings, and a variety of other equipment, requires data for convective heat transfer in rotating systems. Many complex configurations are encountered, and there is an extensive literature on the subject. In this section, correlations are presented only for some very simple configurations.

4.6.1 Rotating Disks, Spheres, and Cylinders

Figure 4.47 shows a disk, sphere, and cylinder rotating in an infinite quiescent fluid. For the disk, there is a transition from laminar to turbulent flow at a Reynolds number $\text{Re}_r = \Omega r_{\text{tr}}^2/\nu \simeq 2.4 \times 10^5$, where r_{tr} is the radius at which transition occurs and Ω [s^{-1}] is the angular velocity. In the laminar region, the local Nusselt number at radius r is given by Edwards et al. [5] as

$$\text{Nu}_r = \frac{0.585\text{Re}_r^{1/2}}{0.6/\text{Pr} + 0.95/\text{Pr}^{1/3}} \tag{4.134}$$

and is valid for all values of Prandtl number. Notice that the resulting heat transfer coefficient is independent of r. For the turbulent region of the disk (if present), the local Nusselt number based on data of Cobb and Saunders [43] is

$$\text{Nu}_r = 0.021\text{Re}_r^{0.8}\text{Pr}^{1/3}; \qquad \text{Re}_r \gtrsim 2.4 \times 10^5; \qquad \text{Pr} > 0.5 \tag{4.135}$$

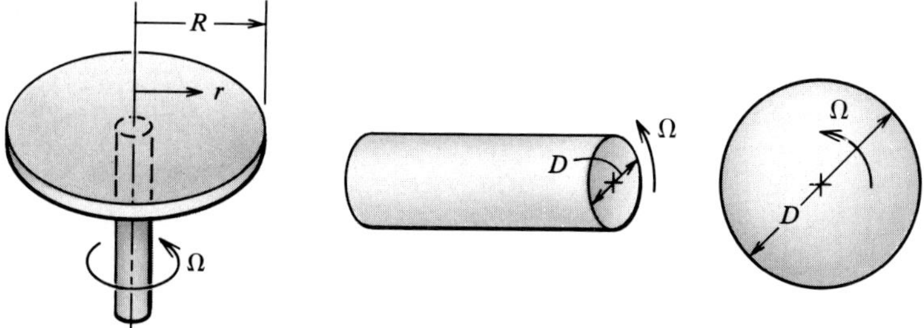

Figure 4.47 Schematic of a rotating disk, cylinder, and sphere.

The correlation given for air has been extrapolated to other fluids by including an ad hoc Prandtl number correction. The average Nusselt number can be found by integration in a similar way as for the flat plate in Section 4.3.2.

For spheres and cylinders, the behavior of the local Nusselt number is complicated; therefore, only correlations for the average value $\overline{\mathrm{Nu}}_D$ will be given. If the Reynolds number is defined as $\mathrm{Re}_D = \Omega D^2/\nu$, then for a sphere in fluids of $\mathrm{Pr} > 0.7$, Kreith et al. [44] recommend

$$\overline{\mathrm{Nu}}_D = 0.43\mathrm{Re}_D^{0.5}\mathrm{Pr}^{0.4}; \qquad 10^2 < \mathrm{Re}_D < 5 \times 10^5 \tag{4.136a}$$

$$\overline{\mathrm{Nu}}_D = 0.066\mathrm{Re}_D^{0.67}\mathrm{Pr}^{0.4}; \qquad 5 \times 10^5 < \mathrm{Re}_D < 7 \times 10^6 \tag{4.136b}$$

For a cylinder,

$$\overline{\mathrm{Nu}}_D = 0.133\mathrm{Re}_D^{2/3}\mathrm{Pr}^{1/3}; \quad \mathrm{Re}_D < 4.3 \times 10^5, \qquad 0.7 < \mathrm{Pr} < 670 \tag{4.137}$$

Mixed natural- and forced-convection effects may be expected to become significant for $\mathrm{Re}_D < 4.7(\mathrm{Gr}_D^3/\mathrm{Pr})^{0.137}$.

EXAMPLE 4.15 Heat Loss from a Centrifuge

The coverplate of a thermostatically controlled centrifuge is a horizontal disk 40 cm in diameter. The centrifuge rotates at 18,000 revolutions per minute. What is the convective heat loss from the coverplate when it is at 305 K and the ambient air is at 295 K?

Solution

Given: Heated horizontal disk rotating at 18,000 rpm.

Required: Convective heat transfer to ambient air.

Assumptions: Still ambient air.

Evaluate fluid properties at a mean film temperature of 300 K, $k = 0.0267$ W/m K, $\nu = 15.66 \times 10^{-6}$ m²/s, Pr = 0.69. The angular velocity is

$$\Omega = (2\pi)(18{,}000/60) = 1885 \text{ s}^{-1}$$

If the transition Reynolds number is taken as 2.4×10^5, the radius at which transition occurs is

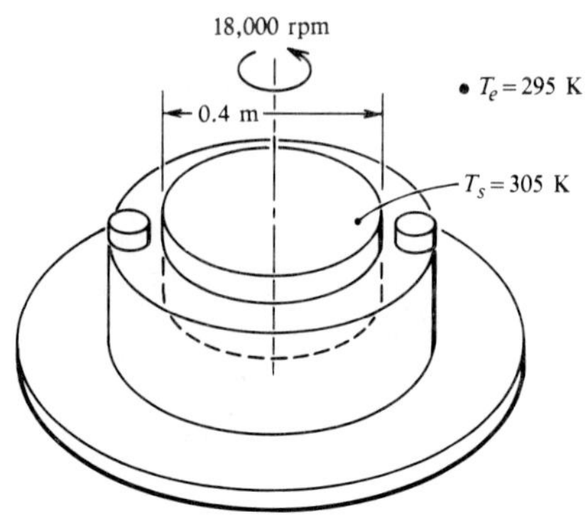

$$r_{tr} = \left(\frac{2.4 \times 10^5 \nu}{\Omega}\right)^{1/2}$$

$$= \left(\frac{(2.4 \times 10^5)(15.66 \times 10^{-6})}{1885}\right)^{1/2}$$

$$= 0.0447 \text{ m}$$

In the laminar region, the heat transfer coefficient is obtained from Eq. (4.134) and is independent of r:

$$\frac{h_c r}{k} = \frac{0.585(\Omega r^2/\nu)^{1/2}}{0.6/\text{Pr} + 0.95/\text{Pr}^{1/3}}$$

$$h_c = \frac{(0.585)k(\Omega/\nu)^{1/2}}{0.6/\text{Pr} + 0.95/\text{Pr}^{1/3}} = \frac{(0.585)(0.0267)(1885/15.66 \times 10^{-6})^{1/2}}{0.6/0.69 + 0.95/(0.69)^{1/3}} = 88.1 \text{ W/m}^2 \text{ K}$$

$$\dot{Q}_{\text{laminar}} = h_c A \Delta T = (88.1)(\pi)(0.0447)^2(10) = 5.53 \text{ W}$$

In the turbulent region, Eq. (4.135) applies:

$$\frac{h_c r}{k} = 0.021\left(\frac{\Omega r^2}{\nu}\right)^{0.8} \text{Pr}^{1/3}$$

$$h_c = 0.021k(\Omega/\nu)^{0.8}\text{Pr}^{1/3}r^{0.6}$$

$$\dot{Q}_{\text{turbulent}} = \int_{r_{tr}}^{R} h_c \Delta T 2\pi r \, dr$$

$$= 0.021k(\Omega/\nu)^{0.8}\text{Pr}^{1/3}\Delta T(2\pi)\int_{r_{tr}}^{R} r^{1.6} \, dr$$

$$= 0.021k(\Omega/\nu)^{0.8}\text{Pr}^{1/3}\Delta T(2\pi/2.6)(R^{2.6} - r_{tr}^{2.6})$$

$$= (0.021)(0.0267)(1885/15.66 \times 10^{-6})^{0.8}(0.69)^{1/3}(10)(2\pi/2.6)(0.2^{2.6} - 0.0447^{2.6})$$

$$= 520 \text{ W}$$

$$\dot{Q}_{\text{total}} = \dot{Q}_{\text{laminar}} + \dot{Q}_{\text{turbulent}} = 5 + 520 = 525 \text{ W}$$

Comments

1. An upper limit on the radiative loss is for a black surface,

$$\dot{Q}_{\text{rad}} = h_r A \Delta T \simeq (6)(\pi/4)(0.4)^2(10) = 7.5 \text{ W}$$

   which is 1.5% of the convective loss.

2. Check using CONV and $\dot{Q} = \overline{h}_c A \Delta T$.

EXAMPLE 4.16 Heat Loss from a Shaft

A 3 cm–diameter shaft rotates at 15,000 rpm in air at 1 atm pressure and 300 K. If the shaft temperature is estimated to be 420 K, calculate the heat loss per unit length.

Solution

Given: A hot cylinder rotating in air.

Required: Heat loss per unit length.

Assumptions: 1. No end effects.
2. The shaft temperature is uniform.

Evaluate fluid properties at the mean film temperature of 360 K; $k = 0.0306$ W/m K, $\nu = 21.30 \times 10^{-6}$ m²/s, Pr $= 0.69$. The angular velocity is

$$\Omega = (15{,}000/60)(2\pi) = 1571 \text{ s}^{-1}$$

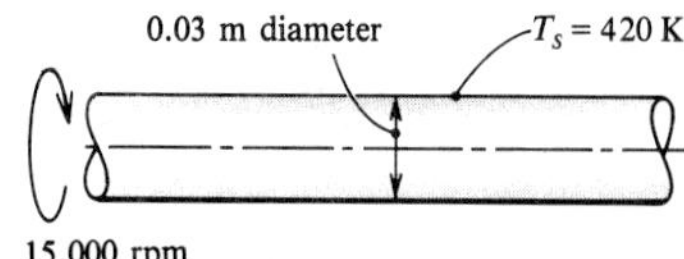

Hence, the Reynolds number is

$$\mathrm{Re}_D = \frac{\Omega D^2}{\nu} = \frac{(1571)(0.03)^2}{21.3 \times 10^{-6}} = 6.64 \times 10^4$$

Equation (4.137) applies:

$$\overline{\mathrm{Nu}}_D = 0.133\mathrm{Re}_D^{2/3}\mathrm{Pr}^{1/3} = (0.133)(6.64 \times 10^4)^{2/3}(0.69)^{1/3} = 193$$

$$\overline{h}_c = (k/D)\overline{\mathrm{Nu}}_D = (0.0306/0.03)(193) = 197 \text{ W/m}^2 \text{ K}$$

$$\dot{Q} = \overline{h}_c A\Delta T = (197)(\pi)(0.03)(1)(420 - 300) = 2.23 \text{ kW}$$

Comments

1. Check to see if free convection is negligible:

$$\mathrm{Gr}_D = \frac{\beta\Delta T g D^3}{\nu^2} = \frac{(120/360)(9.81)(0.03)^3}{(21.3 \times 10^{-6})^2} = 1.946 \times 10^5$$

$$4.7\left(\frac{\mathrm{Gr}_D^3}{\mathrm{Pr}}\right)^{0.137} = 4.7\frac{(1.946 \times 10^5)^3}{(0.69)^{0.137}} = 738 < \mathrm{Re}_D$$

Free convection can be ignored.

2. Use CONV to check h_c.

4.7 ROUGH SURFACES

All the correlations presented so far are for smooth surfaces. In practice, surfaces are often rough, perhaps due to poor surface finish, corrosion, or deposits. Sometimes surfaces are roughened on purpose to increase the heat transfer in a turbulent flow. For example, transverse ribs may be provided on the outside of the fuel element cladding in a gas-cooled nuclear reactor. Figure 4.48 shows two applications of transverse ribs. Such surfaces are called **enhanced surfaces**.

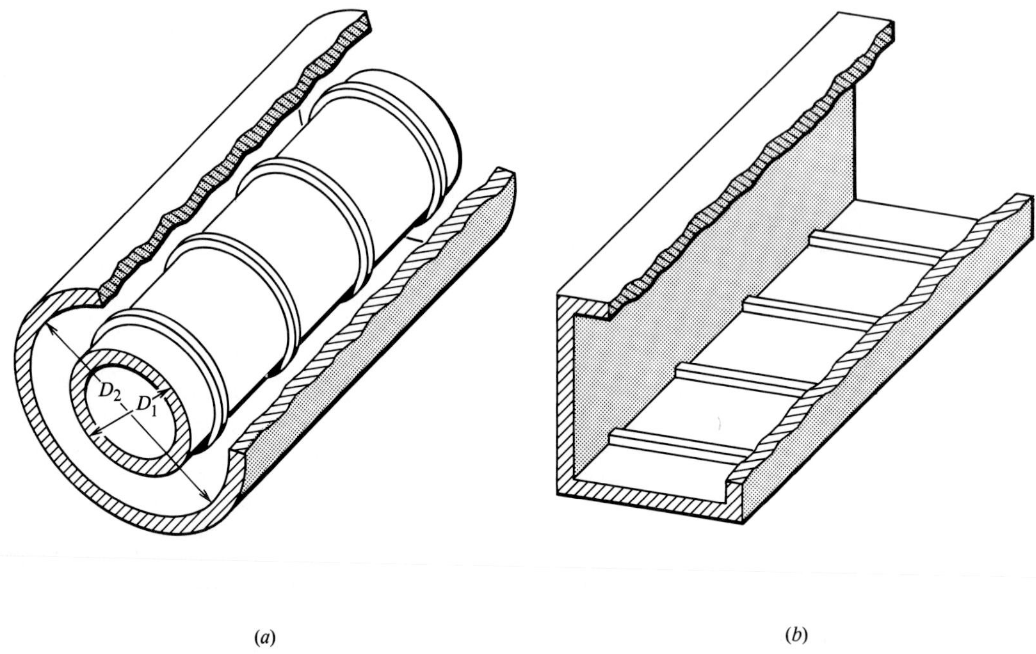

Figure 4.48 Use of transverse ribs to augment convective heat transfer. (*a*) Flow in an annulus with the inner surface heated (e.g., in a gas cooled nuclear reactor core). (*b*) Flow in a square channel with heating on one side only (e.g., in a cooling panel of an actively cooled scram-jet engine inlet).

4.7.1 Effect of Surface Roughness

Wall roughness has no effect on skin friction or heat transfer in laminar flows, provided that the height of the roughness elements is small compared to an appropriate length scale—for example, the pipe diameter or boundary layer thickness. However, in turbulent flows, the roughness elements need only protrude outside the thin viscous sublayer to have a significant effect on skin friction and heat transfer. For example, if water at 300 K flows through a 3 cm–diameter pipe at 5 m/s, the thickness of the viscous sublayer is only about 20 μm. The effect of roughness must be accounted for if the mean roughness height significantly exceeds this value. The effect of roughness is to increase skin friction and heat transfer, *but the effect on skin friction is always greater*, except for high-Pr fluids at low Re. Reynolds-type analogies between friction and heat transfer [e.g., Eq. (4.53)] are invalid for rough walls. The increase in skin friction is due to form drag on the roughness elements: vortex shedding causes a higher pressure on the front than on the rear of the elements. But there is no heat transfer analog to form drag, and on a rough surface heat must still be transferred by molecular conduction through small viscous sublayers on and between the roughness elements.

When surfaces are intentionally roughened to increase heat transfer rates, the design needs to be carefully optimized, since the pumping power required is also increased.

Skin Friction Correlations

Figure 4.49 shows an adaptation of the widely used **Moody chart**, which gives the friction factor for flow in rough pipes. The characteristic velocity V is the *bulk* velocity u_b; for a constant density flow, $u_b = \dot{m}/\rho A_c = G/\rho$. In the *fully rough* regime, the friction factor is independent of Reynolds number $\text{Re}_D = u_b D/\nu$, that is, independent of viscosity, since the viscous drag is negligible compared to form drag. Between the *hydrodynamically smooth* regime and the fully rough regime is the *transitionally rough* regime. The pioneering experimental work was done by J. Nikuradse in 1933 [46], who used a surface uniformly coated with closely packed sand grains of mean diameter k_s. As a consequence, other roughness patterns are often characterized in terms of an **equivalent sand grain roughness** k_s, which gives the same friction factor in the fully rough regime. Table 4.8 gives values of equivalent sand grain roughness for selected rough surfaces, and these values can be used to calculate the parameter k_s/D of the Moody chart. Whereas use of the equivalent sand grain roughness ensures that the correct friction factor is obtained in the fully rough regime, the values of f in the transitionally rough regime given by the Moody chart are applicable only to commercial surfaces, for which there is a wide range of protuberance sizes. Regular roughness patterns, such as Nikuradse's sand grain roughness, exhibit a different behavior in the transitionally rough regime.

Zigrang and Sylvester [49] recommend an explicit formula that can be used in place of the Moody chart:[5]

$$f = \left\{-2.0\log\left[\frac{(k_s/R)}{7.4} - \frac{5.02}{\text{Re}_D}\log\left(\frac{k_s/R}{7.4} + \frac{13}{\text{Re}_D}\right)\right]\right\}^{-2} \qquad \textbf{(4.138)}$$

which has as its asymptote in the fully rough regime Nikuradse's famous formula [46],

$$f = [1.74 + 2.0\log(R/k_s)]^{-2} \qquad \textbf{(4.139)}$$

We can define a dimensionless sand grain size by the relations

$$k_s^+ = \frac{u_b k_s}{\nu}\left(\frac{f}{8}\right)^{1/2} \quad \text{for a pipe} \qquad \textbf{(4.140a)}$$

$$= \frac{u_e k_s}{\nu}\left(\frac{C_{fx}}{2}\right)^{1/2} \quad \text{for a plate} \qquad \textbf{(4.140b)}$$

The criteria for the flow regimes are then as follows:

$0 < k_s^+ \leq 5$: hydrodynamically smooth

$5 < k_s^+ < 60$: transitionally rough

$60 < k_s^+$: fully rough

The rationale underlying Eqs. (4.140) will become evident following study of the analysis of turbulent flows.

[5] Note that the logarithms in Eqs. (4.138) and (4.139) are to base 10.

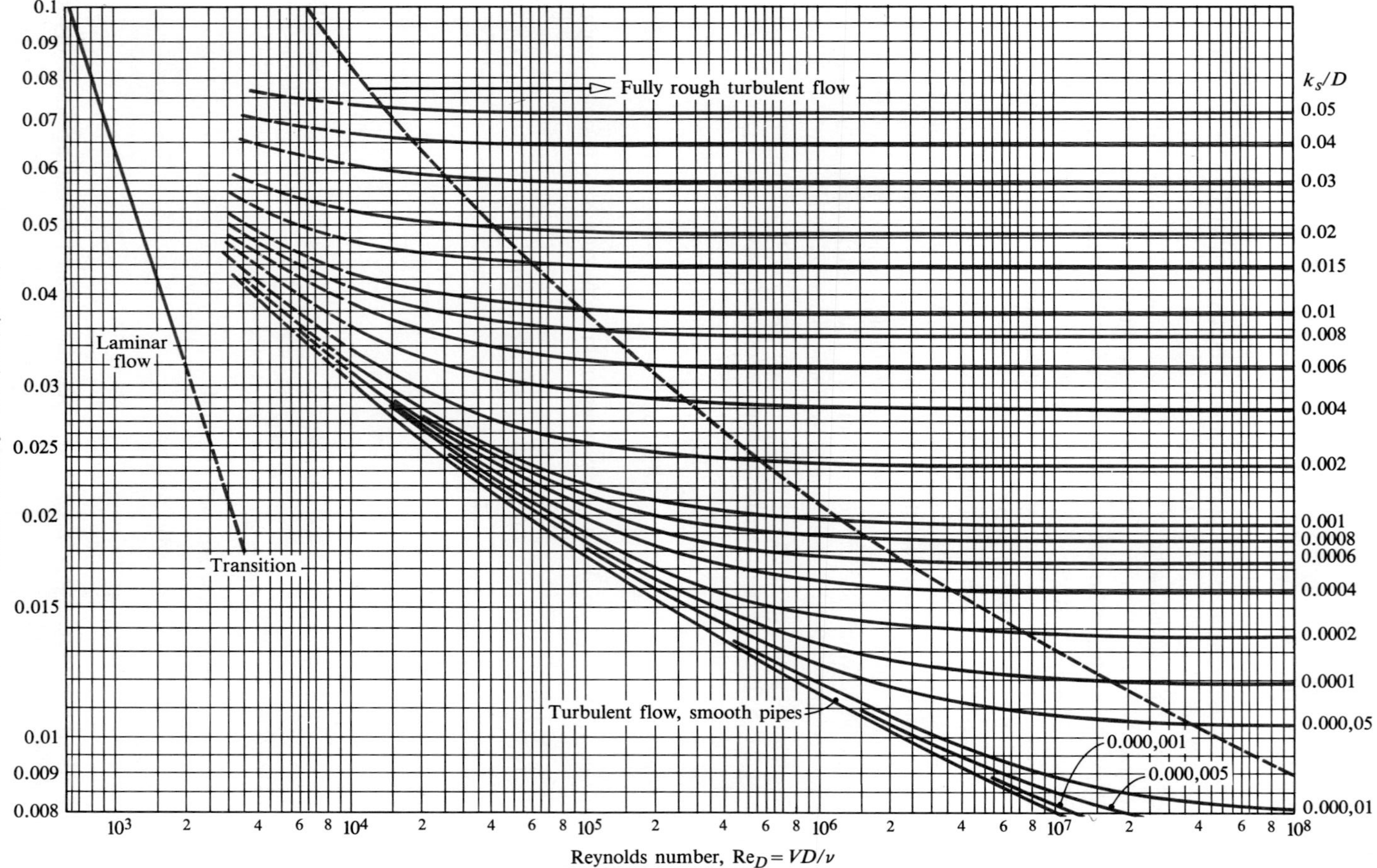

Figure 4.49 Friction factor for commercial pipes (Moody chart). The parameter k_s is the equivalent sand grain roughness: a commercial pipe with equivalent sand grain roughness k_s will have the same value of f in the fully rough regime as a pipe roughened with closely packed sand grains of average diameter k_s [45]. (Adapted with permission.)

Table 4.8 Equivalent sand grain roughness k_s for various surfaces [47,48].

Roughness Pattern	D mm	p mm	h mm	k_s mm
1. Spheres—staggered array	4.1	10	4.1	0.492
	4.1	20	4.1	1.68
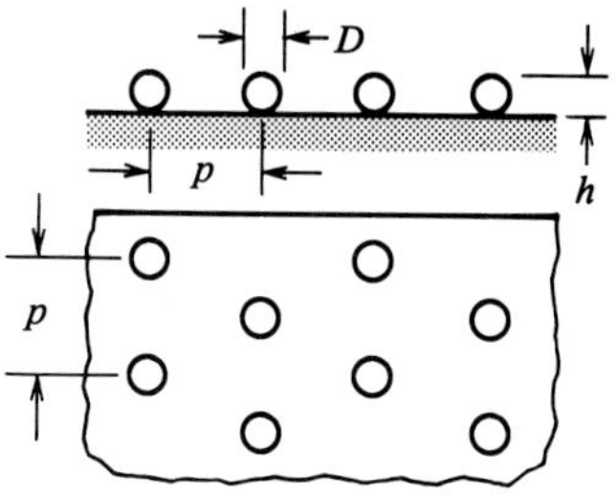	4.1	10	4.1	9.96
	4.1	6	4.1	10.6
	4.1	Densest	4.1	1.55
	2.1	10	2.1	0.903
	2.1	5	2.1	5.19
2. Spherical segments—staggered array	8.0	40	2.6	0.0468
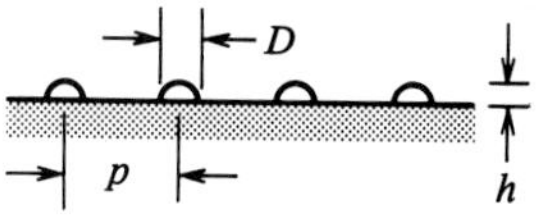	8.0	30	2.6	0.0884
	8.0	20	2.6	0.723
	8.0	Densest	2.6	2.48
3. Cones—staggered array	8.0	40	3.75	0.173
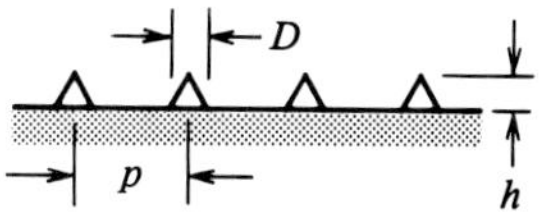	8.0	30	3.75	0.458
	8.0	20	3.75	1.77
4. Welded steel				
New				0.04–0.10
Uniformly rusted				0.13–0.41
Cleaned after long use				0.10–0.20
5. Commercial tubes				
Glass				0.00031
Drawn tubing				0.0015
Steel or wrought iron				0.05
Asphalted cast iron				0.12
Galvanized steel				0.1–0.3
Cast iron				0.26
Concrete				0.3–3
Riveted steel				0.9
6. Transverse square ribs 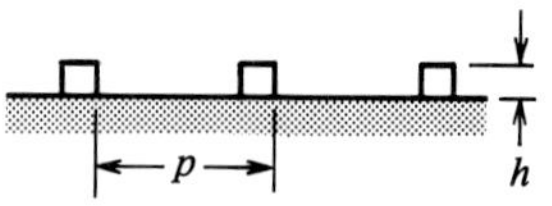	$2 < p/h < 6.3$: $k_s = h\exp[3.4 - 3.7(p/h)^{-0.73}]$; $6.3 < p/h < 20$: $k_s = h\exp[3.4 - 0.42(p/h)^{0.46}]$			

For flow along a rough plate, Mills and Hang [50] recommend for the fully rough regime

$$C_{fx} = \left(3.476 + 0.707 \ln \frac{x}{k_s}\right)^{-2.46}, \qquad 150 < \frac{x}{k_s} < 1.5 \times 10^7 \qquad \textbf{(4.141)}$$

and

$$\overline{C}_f = \left(2.635 + 0.618 \ln \frac{L}{k_s}\right)^{-2.57}, \qquad 150 < \frac{L}{k_s} < 1.5 \times 10^7 \qquad \textbf{(4.142)}$$

when boundary layer transition occurs at the leading edge. Reliable correlations are unavailable for the transitionally rough regime.

Heat Transfer Correlations

The effect of roughness on heat transfer is very much dependent on roughness pattern. A characteristic height h is defined for the roughness patterns (e.g., the height of a rib for transverse square ribs, or the sand grain size for sand grain roughness). In the fully rough regime, heat transfer correlations adapted from the work of Dipprey and Sabersky [51,52] are

$$\text{St} = \frac{f/8}{0.9 + (f/8)^{1/2}[g(h^+, \text{Pr}) - 7.65]} \qquad \text{for a pipe} \qquad \textbf{(4.143a)}$$

$$\text{St}_x = \frac{C_{fx}/2}{0.9 + (C_{fx}/2)^{1/2}[g(h^+, \text{Pr}) - 7.65]} \qquad \text{for a flat plate} \qquad \textbf{(4.143b)}$$

where the function $g(h^+, \text{Pr})$ is given in Table 4.9 for a variety of roughness patterns. The dimensionless characteristic height h^+ is defined similarly to the definition of

Table 4.9 The function $g(h^+,\text{Pr})$ for use in Eq. (4.143): heat transfer to rough walls [52,53].

Roughness Pattern, h	$g(h^+, \text{Pr})$	Prandtl Number Range
1. Sand grain indentation: $h = k_s$, the sand grain diameter	$4.8(h^+)^{0.2}\text{Pr}^{0.44}$	$1 < \text{Pr} < 6$
2. Transverse square ribs—height h, pitch p: $10 < p/h < 40, 0.02 < h/R$, or $\delta < 0.08$ where R is the pipe radius and δ the boundary layer thickness $h = h$ the rib height	$4.3(h^+)^{0.28}\text{Pr}^{0.57}$	$0.7 < \text{Pr} < 40$
3. General (hence, not too accurate): $h =$ Mean height of protrusions	$0.55(h^+)^{1/2}(\text{Pr}^{2/3} - 1) + 9.5$	$\text{Pr} > 0.5$

k_s^+ by Eqs. (4.140); alternatively, since k_s^+ is usually calculated first, one can simply use $h^+ = (k_s^+/k_s)h$. Simple Stanton number correlations are not available for the transitionally rough regime.

Variable-property effects for rough walls are not well established. The research literature contains a few correlations for specific applications.

EXAMPLE 4.17 Flow of Helium in a Roughened Tube

Helium at 5×10^5 Pa pressure flows in a 4 cm–I.D. tube at 60 m/s. Calculate the pressure gradient and heat transfer coefficient for fully developed conditions when (i) the tube wall is smooth, and (ii) the tube is artificially roughened with transverse ribs of height 0.8 mm and pitch 8 mm. Evaluate fluid properties at 500 K.

Solution

Given: Helium flow in a tube.

Required: Pressure gradient and heat transfer coefficient for (i) smooth wall and (ii) transverse rib roughness.

Assumptions:
1. Helium behaves as an ideal gas.
2. Fully developed conditions (no entrance effect).

For an ideal gas, k, c_p, and μ are independent of pressure, and the helium properties at 1 atm pressure in Table A.7 can be used to obtain properties at 5×10^5 Pa: $k = 0.205$ W/m K, $\rho = 0.481$ kg/m^3, $\nu = \mu/\rho = 58.7 \times 10^{-6}$ m^2/s, and Pr $= 0.72$. The Reynolds number is $\text{Re}_D = u_b D/\nu = (60)(0.04)/58.7 \times 10^{-6} = 4.09 \times 10^4$.

(i) For a smooth wall, Eq. (4.42) gives the friction factor:

$$f = (0.790 \ln \text{Re}_D - 1.64)^{-2} = (0.790 \ln 4.09 \times 10^4 - 1.64)^{-2} = 0.0219$$

$$\frac{dP}{dx} = \left(\frac{f}{D}\right)\frac{1}{2}\rho u_b^2 = \left(\frac{0.0219}{0.04}\right)(0.5)(0.481)(60)^2 = 474 \text{ N/m}^2 \text{ per meter}$$

Equation (4.45) gives the Nusselt number:

$$\text{Nu}_D = \frac{(f/8)(\text{Re}_D - 1000)\text{Pr}}{1 + 12.7(f/8)^{1/2}(\text{Pr}^{2/3} - 1)} = \frac{(0.0219/8)(40{,}900 - 1000)(0.72)}{1 + 12.7(0.0219/8)^{1/2}(0.72^{2/3} - 1)} = 90.5$$

$$h_c = (k/D)\text{Nu}_D = (0.205/0.04)(90.5) = 464 \text{ W/m}^2 \text{ K}$$

(ii) For the transverse rib roughness $h = 0.8$ mm, $p = 8$ mm, $p/h = 10$.

The equivalent sand grain size is obtained from Table 4.8, item 6:

$$k_s = h \exp[3.4 - 0.42(p/h)^{0.46}] = 0.8 \times 10^{-3} \exp[3.4 - 0.42(10)^{0.46}]$$
$$= 7.14 \times 10^{-3} \text{ m}$$

We will assume fully rough conditions and check later. Equation (4.139) gives the friction factor as

$$f = [1.74 + 2.0 \log(R/k_s)]^{-2} = [1.74 + 2.0 \log(0.02/0.00714)]^{-2} = 0.144$$

The dimensionless equivalent sand grain size is obtained from Eq. (4.140*a*):

$$k_s^+ = \left(\frac{u_b k_s}{\nu}\right)\left(\frac{f}{8}\right)^{1/2} = \frac{(60)(7.14 \times 10^{-3})}{58.7 \times 10^{-6}}\left(\frac{0.144}{8}\right)^{1/2} = 979 > 60$$

and the conditions are fully rough as assumed.

$$\frac{dP}{dx} = (f/D)(1/2)\rho u_b^2 = (0.144/0.04)(0.5)(0.481)(60)^2 = 3120 \text{ N/m}^2 \text{ per meter}$$

Equation (4.143*a*) gives the Stanton number, with $g(h^+,\text{Pr})$ from Table 4.9, item 2:

$$h^+ = (h/k_s)k_s^+ = (0.8/7.14)979 = 109.7$$

$$g(h^+, \text{Pr}) = 4.3(h^+)^{0.28}\text{Pr}^{0.57} = 4.3(109.7)^{0.28}(0.72)^{0.57} = 13.29$$

$$\text{St} = \frac{f/8}{0.9 + (f/8)^{1/2}[g(h^+, \text{Pr}) - 7.65]} = \frac{0.144/8}{0.9 + (0.144/8)^{1/2}[13.29 - 7.65]} = 0.0109$$

$$\text{Nu}_D = \text{St}\text{Re}_D\text{Pr} = (0.0109)(4.09 \times 10^4)(0.72) = 320$$

$$h_c = (k/D)\text{Nu}_D = (0.205/0.04)(320) = 1640 \text{ W/m}^2 \text{ K}$$

Comments

1. The effect of the ribs is to increase the pressure drop (3120/474) = 6.6 times and the heat transfer (1640/464) = 3.5 times.
2. Check using CONV.

EXAMPLE 4.18 Air Flow over a Sandblasted Flat Plate

Air at 1 atm pressure flows at 20 m/s over 2 m–long flat plate that has been sandblasted to give a surface of equivalent sand grain roughness size of 1 mm. Compare the drag force and average heat transfer coefficient with that for a smooth plate. Evaluate fluid properties at 300 K, and take Re_{tr} = 100,000 and 50,000 for the smooth and rough plates, respectively.

Solution

Given: Air flowing along a roughened flat plate.

Required: Comparison of drag force and $\bar{h}_c$ with smooth-plate values.

Assumptions: 1. Re_{tr} = 100,000 and 50,000 for the smooth and rough plates, respectively.
2. For the heat transfer calculation, the roughness pattern can be approximated as similar to a sand grain indentation pattern.

Air properties at 300 K and 1 atm are: $k = 0.0267$ W/m K, $\rho = 1.177$ kg/m^3, $\nu = 15.7 \times 10^{-6}$ m^2/s, $c_p = 1005$ J/kg K, and Pr = 0.69. The Reynolds number at the end of the plate is

$$\text{Re}_L = \frac{VL}{\nu} = \frac{(20)(2)}{15.7 \times 10^{-6}} = 2.55 \times 10^6$$

(i) *Smooth plate:*

Equation (4.62) gives the average skin friction coefficient:

$$\overline{C_f} = 1.328\text{Re}_{\text{tr}}^{-1/2}\left(\frac{\text{Re}_{\text{tr}}}{\text{Re}_L}\right) + \frac{0.523}{\ln^2 0.06Re_L} - \left(\frac{\text{Re}_{\text{tr}}}{\text{Re}_L}\right)\frac{0.523}{\ln^2 0.06\text{Re}_{\text{tr}}}$$

$$= 1.328(10^5)^{-1/2}\left(\frac{0.1}{2.55}\right) + \frac{0.523}{\ln^2 0.06(2.55\times 10^6)} - \left(\frac{0.1}{2.55}\right)\frac{0.523}{\ln^2 0.06(10^5)}$$

$$= 0.00356$$

The drag force is

$$F = \overline{C_f}\left(\frac{1}{2}\rho\ V^2\right)WL = (0.00356)(0.5)(1.177)(20)^2(1)(2) = 1.68 \text{ N per meter width}$$

Equation (4.65) gives the average Nusselt number:

$$\overline{\text{Nu}} = 0.664\text{Re}_{\text{tr}}^{1/2}\text{Pr}^{1/3} + 0.036\text{Re}_L^{0.8}\text{Pr}^{0.43}\left[1 - \left(\frac{\text{Re}_{\text{tr}}}{\text{Re}_L}\right)^{0.8}\right]$$

$$= (0.664)(10^5)^{1/2}(0.69)^{1/3} + 0.036(2.55\times 10^6)^{0.8}(0.69)^{0.43}\left[1 - \left(\frac{0.1}{2.55}\right)^{0.8}\right]$$

$$= 3970$$

$$\overline{h}_c = (k/L)\overline{\text{Nu}} = (0.0267/2)(3970) = 53.0 \text{ W/m}^2\text{ K}$$

(ii) *Rough plate:*

Check the dimensionless equivalent sand grain roughness at the end of the plate. At $x = L =$ 2 m, Eq. (4.141) gives

$$C_{fx} = (3.476 + 0.707\ln L/k_s)^{-2.46} = (3.476 + 0.707\ln 2/10^{-3})^{-2.46} = 0.00468$$

From Eq. (4.140*b*),

$$k_s^+ = \left(\frac{u_e k_s}{\nu}\right)\left(\frac{C_{fx}}{2}\right)^{1/2} = \frac{(20)(1\times 10^{-3})}{15.7\times 10^{-6}}\left(\frac{0.00468}{2}\right)^{1/2} = 61.6 > 60$$

that is, fully rough. Since C_{fx} decreases with x, the complete turbulent region will be fully rough if the end of the plate is fully rough. We can assume transition at the leading edge without incurring a significant error, because

$$\frac{\text{Re}_{\text{tr}}}{\text{Re}_L} = \frac{0.05\times 10^6}{2.55\times 10^6} = 0.0196 \simeq 2\%$$

Equation (4.142) gives the average skin friction coefficient:

$$\overline{C_f} = (2.635 + 0.618\ln L/k_s)^{-2.57} = (2.635 + 0.618\ln 2/10^{-3})^{-2.57} = 0.00597$$

and the drag force is

$$F = \overline{C_f}\left(\frac{1}{2}\rho V^2\right)WL = 0.00597(0.5)(1.177)(20)^2(1)(2) = 2.81 \text{ N per meter width}$$

Equation (4.143*b*) gives the local Stanton number:

$$\mathrm{St}_x = \frac{C_{fx}/2}{0.9 + (C_{fx}/2)^{1/2}[g(h^+, \mathrm{Pr}) - 7.65]}$$

Approximating the roughness pattern as item 1 in Table 4.9,

$$g(h^+, \mathrm{Pr}) = 4.8(k_s^+)^{0.2}\mathrm{Pr}^{0.44}, \qquad \text{where } k_s^+ = \left(\frac{u_e k_s}{\nu}\right)\left(\frac{C_{fx}}{2}\right)^{1/2}$$

$$= (4.8)\left[\frac{(20)(0.001)}{15.7 \times 10^{-6}}\right]^{0.2}(0.69)^{0.44}\left(\frac{C_{fx}}{2}\right)^{0.1}$$

$$= 17.0\left(\frac{C_{fx}}{2}\right)^{0.1}$$

$$\mathrm{St}_x = \frac{C_{fx}/2}{0.9 + 17.0(C_{fx}/2)^{0.6} - 7.65(C_{fx}/2)^{1/2}}$$

The average Stanton number is obtained by integration:

$$\overline{\mathrm{St}} = \frac{1}{L}\int_0^L \mathrm{St}_x\,dx = \frac{1}{2}\int_0^2 \frac{C_{fx}/2\,dx}{0.9 + 17.0(C_{fx}/2)^{0.6} - 7.65(C_{fx}/2)^{1/2}}$$

Equation (4.141) gives $C_{fx}/2 = (0.5)[3.476 + 0.707\ln(x/0.001)]^{-2.46}$: substituting and integrating numerically for $L = 2$ m gives $\overline{\mathrm{St}} = 0.00304$. The average heat transfer coefficient is

$$\bar{h}_c = \rho c_p u_e \overline{\mathrm{St}} = (1.177)(1005)(20)(0.00304) = 71.9\ \mathrm{W/m^2\,K}$$

Solution using CONV

The required input for the rough wall case is:

Configuration number = 25 (fully rough flat plate)
Fluid number = 21 (air)
$T_s = 300$
$T_e = 300$
$P = 1.013 \times 10^5$
$L = 2$
$k_s = 0.001$
Roughness pattern = 1 (sand grain indentation)
$u_e = 20$

The output is:

Air properties at 300 K, 1.013×10^5 Pa
$\mathrm{Re}_L = 2.56 \times 10^6$
$k_s^+ = 61.8$
$C_{fL} = 4.68 \times 10^{-3}$
$\overline{C}_f = 5.97 \times 10^{-3}$

$$\tau_{sL} = 1.103 \text{ N/m}^2$$
$$\overline{\tau}_s = 1.41 \text{ N/m}^2$$
$$\text{St}_L = 2.39 \times 10^{-3}$$
$$\overline{\text{St}} = 2.97 \times 10^{-3}$$
$$\text{Nu}_L = 4210$$
$$\overline{\text{Nu}} = 5230$$
$$h_{cL} = 56.3 \text{ W/m}^2\text{ K}$$
$$\overline{h}_c = 69.8 \text{ W/m}^2\text{ K}$$

Comments

1. The effect of the roughness is to increase the drag force $(2.81/1.68) = 1.67$ times and the heat transfer $(71.9/53.0) = 1.36$ times.
2. Since numerical integration is required, the use of a computer is indicated. CONV performs this integration.
3. Note that in CONV we put $T_e = T_s = 300$ K, the given temperature, for evaluation of fluid properties.

4.8 THE COMPUTER PROGRAM CONV

CONV calculates the heat transfer coefficient (and the pressure gradient or wall shear stress, where appropriate) for the 25 flow configurations listed in Table 4.10. CONV has a menu of 5 gases and 10 dielectric liquids, with thermophysical properties based on the data in Tables A.7, A.8, A.10, and A.13*a* of Appendix A. Liquid metals are not included because of the need to use special correlations for most configurations. Properties are evaluated according to the rules described in Section 4.2.4. Configurations 1 through 24 require straightforward evaluation of algebraic correlation formulas. Configuration 25 requires numerical integration of the local Stanton number to obtain the average value, as described in Example 4.18.

4.9 CLOSURE

The objective of this chapter was to develop expertise in the calculation of heat transfer coefficients and, where appropriate, in the calculation of pressure drop or skin friction as well. Flows of concern generally can be classified in three ways: (1) forced versus natural flow, (2) internal versus external flows, and (3) laminar versus turbulent flows. Relevant parameters, rules for evaluation of fluid properties, and even the definition of the heat transfer coefficient may differ from case to case. Dimensional analysis based on the Buckingham pi theorem was used to obtain the dimensionless groups pertinent to a given flow. The student should now be thoroughly familiar with dimensionless groups such as the Nusselt, Stanton, Prandtl, Reynolds, Grashof, and Rayleigh numbers, as well as those that were used less frequently.

The pi theorem and method of indices have served their purpose; in Chapter 5, more powerful methods of dimensional analysis will be demonstrated. A wide range of flows were discussed and appropriate correlations presented. The computer code CONV should have proven to be a most useful tool for executing the required calculations.

The correlations presented in Chapter 4 are but a small fraction of those that can be found in the heat transfer literature. The flow configurations were chosen to include those most commonly encountered in thermal design, and also to illustrate the wide range of possibilities. For many of the cases considered, the literature contains many alternative correlations; those presented here often reflect a compromise between simplicity and accuracy. The question of accuracy is a particularly vexing one. Very few statements concerning the accuracy of a particular correlation have been made in Chapter 4, because it is almost impossible to do so unambiguously. This point requires further elaboration.

Sometimes a correlation is based on data obtained from a single test rig. The engineer will usually report an accuracy reflecting the scatter in the data, such as a standard deviation or equivalent measure. But this accuracy reflects only the random error in the experiment. *Systematic* error due to various causes, such as faulty instrumentation or measurement techniques, is not reflected in this reported accuracy. More importantly, the test rig will usually match the primary variables of the model flow only; there are invariably secondary or "nuisance" variables present as well. For example, consider the wind-tunnel test of heat transfer from a cylinder described in Section 4.2.3. Unless the tunnel working section is sufficiently high, there will be blockage effects due to the flow being constrained as it accelerates past the cylinder. Also, unless the tunnel is sufficiently wide, there will be significant effects of vortices generated by the interaction of the boundary layer on the side walls with the cylinder. Time (and money) is seldom invested to carefully explore the effects of such secondary variables. The point is that the effects of secondary variables might be quite different in the particular engineering problem under consideration, and these effects are often not accounted for in the engineer's statement of accuracy.

In contrast, some correlations are based on data obtained by many different engineers using a great variety of test rigs. The stated possible error for such correlations is usually relatively large because of the effects of secondary variables, and often it is too conservative. First, there is the question of whether the data have been thoroughly screened to eliminate data of inferior quality. Second, there may be a particular data set obtained under conditions that match the problem of concern very closely and therefore should be more appropriate than the general correlation.

The bottom line is that the correlations in Chapter 4 should be adequate for most thermal design purposes. If accuracy is of particular concern, for example, in a research project, then the original literature should be studied carefully to obtain a meaningful accuracy estimate.

Table 4.10 The correlations contained in CONV. Evaluate all properties at the mean film temperature unless otherwise specified. All the heat transfer correlations are for isothermal walls. However, (1) item 1 can also be used for a uniform wall heat flux, and (2) values of $\overline{\text{Nu}}$ for external flows can also be used for a uniform wall heat flux, provided Eq. (4.82) is used to define the average heat transfer coefficient.

<table>
<tr><th>Item No</th><th>Configuration</th><th>Correlations</th><th>Comments</th></tr>
<tr><td>1</td><td>Turbulent flow in smooth ducts with fully developed hydrodynamics and heat transfer</td><td>$f = (0.790 \ln \text{Re}_{D_h} - 1.64)^{-2}; \quad 10^4 < \text{Re}_{D_h} < 5 \times 10^6 \quad (4.42)$

$\text{Nu}_{D_h} = \dfrac{(f/8)(\text{Re}_{D_h} - 1000)\text{Pr}}{1 + 12.7(f/8)^{1/2}(\text{Pr}^{2/3} - 1)}; \quad 3000 < \text{Re}_{D_h} < 10^6 \quad (4.45)$
$0.5 < \text{Pr}$</td><td rowspan="3">$D_h = \dfrac{4A_c}{\mathscr{P}} (= D \text{ for a circular tube})$

Exponents for property and temperature ratio corrections for duct flows (subscripts s and b refer to wall and bulk values, respectively):
<table><tr><th>Type of Flow</th><th>Fluid</th><th>Wall Condition</th><th>f m</th><th>Nu n</th></tr><tr><td>Laminar</td><td>Liquids (μ_s/μ_b)</td><td>Heating</td><td>0.58</td><td>−0.11</td></tr><tr><td></td><td></td><td>Cooling</td><td>0.50</td><td>−0.11</td></tr><tr><td></td><td>Gases (T_s/T_b)</td><td>Heating</td><td>1</td><td>0</td></tr><tr><td></td><td></td><td>Cooling</td><td>1</td><td>0</td></tr><tr><td>Turbulent</td><td>Liquids (μ_s/μ_b)</td><td>Heating</td><td>0.25</td><td>−0.25</td></tr><tr><td></td><td></td><td>Cooling</td><td>0.25</td><td>−0.11</td></tr><tr><td></td><td>Gases (T_s/T_b)</td><td>Heating</td><td>−0.2</td><td>−0.55</td></tr><tr><td></td><td></td><td>Cooling</td><td>−0.1</td><td>0.0</td></tr></table>
$D_h = \dfrac{4A_c}{\mathscr{P}} = 2 \times$ Plate spacing</td></tr>
<tr><td>2</td><td>Laminar flow in a pipe with fully developed hydrodynamics</td><td>$f = \dfrac{64}{\text{Re}_D}; \quad \text{Re}_D < 2300 \quad (4.39)$

$\overline{\text{Nu}}_D = 3.66 + \dfrac{0.065(D/L)\text{Re}_D\text{Pr}}{1 + 0.04\,[(D/L)\text{Re}_D\text{Pr}]^{2/3}}; \quad \text{Re}_D < 2300 \quad (4.50)$</td></tr>
<tr><td>3</td><td>Laminar flow between parallel plates with fully developed hydrodynamics</td><td>$f = \dfrac{96}{\text{Re}_{D_h}}; \quad \text{Re}_{D_h} < 2800$ (Table 4.5)

$\overline{\text{Nu}}_{D_h} = 7.54 + \dfrac{0.03(D_h/L)\text{Re}_{D_h}\text{Pr}}{1 + 0.016\left[(D_h/L)\text{Re}_{D_h}\text{Pr}\right]^{2/3}}; \quad \text{Re}_{D_h} < 2800 \quad (4.51)$</td></tr>
<tr><td>4</td><td>Laminar boundary layer on a flat plate</td><td>$\overline{C}_f = 1.328\text{Re}_L^{-1/2}; \quad 10^3 < \text{Re}_L \lesssim 5 \times 10^5 \quad (4.55)$
$\overline{\text{Nu}} = 0.664\text{Re}_L^{1/2}\text{Pr}^{1/3}; \quad 10^3 < \text{Re}_L \lesssim 5 \times 10^5, \quad \text{Pr} > 0.5 \quad (4.57)$</td><td></td></tr>
<tr><td>5</td><td>Turbulent boundary layer on a smooth flat plate</td><td>$\overline{C}_f = 1.328\text{Re}_{\text{tr}}^{-1/2}\left(\dfrac{\text{Re}_{\text{tr}}}{\text{Re}_L}\right) + \dfrac{0.523}{\ln^2 0.06\text{Re}_L} - \left(\dfrac{\text{Re}_{\text{tr}}}{\text{Re}_L}\right)\dfrac{0.523}{\ln^2 0.06\text{Re}_{\text{tr}}} \quad (4.62)$
$\text{Re}_{\text{tr}} < \text{Re}_L < 10^9$

$\overline{\text{Nu}} = 0.664\text{Re}_{\text{tr}}^{1/2}\text{Pr}^{1/3} + 0.036\text{Re}_L^{0.8}\text{Pr}^{0.43}\left[1 - (\text{Re}_{\text{tr}}/\text{Re}_L)^{0.8}\right] \quad (4.65)$
$\text{Re}_{\text{tr}} < \text{Re}_L < 3 \times 10^7$
$0.7 < \text{Pr} < 400$</td><td>$\text{Re}_{\text{tr}} = 50{,}000 - 500{,}000$. Lower values are characteristic of practical situations where disturbing factors such as roughness and vibration are present</td></tr>
</table>

(Continued)

Table 4.10 *(Continued)*

6	Flow across a cylinder	$C_D = 1 + \frac{10}{\mathrm{Re}_D^{2/3}}; \quad 1 < \mathrm{Re}_D < 10^4$ (4.69)	C_D = Drag force/$(1/2)\rho V^2 A_f$
		$\mathrm{Nu}_D = 1.15\mathrm{Re}_D^{1/2}\mathrm{Pr}^{1/3}; \quad \mathrm{Pr} > 0.5$ (4.70)	Stagnation line local Nusselt number
		$\overline{\mathrm{Nu}}_D = 0.3 + \frac{0.62\mathrm{Re}_D^{1/2}\mathrm{Pr}^{1/3}}{[1+(0.4/\mathrm{Pr})^{2/3}]^{1/4}}; \quad \mathrm{Re}_D < 10^4, \quad \mathrm{Pr} > 0.5$ (4.71*a*)	
		$= 0.3 + \frac{0.62\mathrm{Re}_D^{1/2}\mathrm{Pr}^{1/3}}{[1+(0.4/\mathrm{Pr})^{2/3}]^{1/4}}\left[1 + \left(\frac{\mathrm{Re}_D}{282{,}000}\right)^{1/2}\right]$ (4.71*b*) $2 \times 10^4 < \mathrm{Re}_D < 4 \times 10^5$	
		$= 0.3 + \frac{0.62\mathrm{Re}_D^{1/2}\mathrm{Pr}^{1/3}}{[1+(0.4/\mathrm{Pr})^{2/3}]^{1/4}}\left[1 + \left(\frac{\mathrm{Re}_D}{282{,}000}\right)^{5/8}\right]^{4/5}$ (4.71*c*) $4 \times 10^5 < \mathrm{Re}_D < 5 \times 10^6$	
		$\overline{\mathrm{Nu}}_D = \frac{1}{0.8237 - \ln(\mathrm{Re}_D\mathrm{Pr})^{1/2}}; \quad \mathrm{Re}_D\mathrm{Pr} < 0.2$ (4.72)	
7	Flow across a sphere	$C_D = \frac{24}{\mathrm{Re}_D}; \quad \mathrm{Re}_D < 0.5$ (4.73)	
		$C_D \simeq \frac{24}{\mathrm{Re}_D}\left(1 + \frac{\mathrm{Re}_D^{2/3}}{6}\right); \quad 2 < \mathrm{Re}_D < 500$ (4.74)	
		$C_D \simeq 0.44; \quad 500 < \mathrm{Re}_D < 2 \times 10^5$ (Fig. 4.28)	
		$\mathrm{Nu}_D = 1.32\mathrm{Re}_D^{1/2}\mathrm{Pr}^{1/3}; \quad \mathrm{Pr} > 0.5$ (4.75)	Stagnation point local Nusselt number
		$\overline{\mathrm{Nu}}_D = 2 + (0.4\mathrm{Re}_D^{1/2} + 0.06\mathrm{Re}_D^{2/3})\mathrm{Pr}^{0.4}$ (4.76) $3.5 < \mathrm{Re}_D < 8 \times 10^4, \quad 0.7 < \mathrm{Pr} < 380$	Use mean film temperature; or better results can be obtained using a viscosity ratio correction with $n = -1/4$ applied to the convection contribution

8	Laminar natural-convection boundary layer on a vertical wall	$\overline{\mathrm{Nu}}_L = 0.68 + 0.670(\mathrm{Ra}_L\Psi)^{1/4}; \quad \mathrm{Ra}_L < 10^9 \quad (4.85)$ $\Psi = \left[1 + \left(\frac{0.492}{\mathrm{Pr}}\right)^{9/16}\right]^{-16/9}$	
9	Turbulent natural-convection boundary layer on a vertical wall	$\overline{\mathrm{Nu}}_L = 0.68 + 0.670(\mathrm{Ra}_L\Psi)^{1/4}(1 + 1.6 \times 10^{-8}\mathrm{Ra}_L\Psi)^{1/12} \quad (4.86)$ $10^9 < \mathrm{Ra}_L < 10^{12}$	Ψ defined in item 8
10	Natural convection on a horizontal cylinder	$\overline{\mathrm{Nu}}_D = 0.36 + \frac{0.518\mathrm{Ra}_D^{1/4}}{[1 + (0.559/\mathrm{Pr})^{9/16}]^{4/9}}; \quad 10^{-4} < \mathrm{Ra}_D \lesssim 10^9 \quad (4.87)$ $\overline{\mathrm{Nu}}_D = \left\{0.60 + 0.387\left[\frac{\mathrm{Ra}_D}{[1 + (0.559/\mathrm{Pr})^{9/16}]^{16/9}}\right]^{1/6}\right\}^2; \quad \mathrm{Ra}_D \gtrsim 10^9 \quad (4.88)$	
11	Natural convection on a sphere	$\overline{\mathrm{Nu}}_D = 2 + \frac{0.589\mathrm{Ra}_D^{1/4}}{[1 + (0.469/\mathrm{Pr})^{9/16}]^{4/9}}; \quad \mathrm{Ra}_D \lesssim 10^{11}; \quad \mathrm{Pr} > 0.5 \quad (4.90)$	
12	Natural convection on a heated horizontal plate facing down, or a cooled plate facing up	$\overline{\mathrm{Nu}}_L = 6.5\left[1 + 0.38\frac{L}{W}\right]\left[(1 + X)^{0.39} - X^{0.39}\right]\mathrm{Ra}_L^{0.13} \quad (4.94)$ $X = 13.5\mathrm{Ra}_L^{-0.16} + 2.2\left(\frac{L_a}{L}\right)^{0.7}$ $10^6 < \mathrm{Ra}_L < 10^{10}; \quad 0.7 < \mathrm{Pr} < 4800; \quad 0 < L_a/L < 0.2$	W is the length of the longer side, L is the length of the shorter side, L_a is the length of adiabatic extensions
13	Natural convection on a heated horizontal plate facing up, or a cooled plate facing down	$\overline{\mathrm{Nu}}_L = 0.54\mathrm{Ra}_L^{1/4}; \quad 10^5 < \mathrm{Ra}_L < 2 \times 10^7 \quad (4.95)$ $\overline{\mathrm{Nu}}_L = 0.14\mathrm{Ra}_L^{1/3}; \quad 2 \times 10^7 < \mathrm{Ra}_L < 3 \times 10^{10} \quad (4.96)$	L is the length of the shorter side

(Continued)

Table 4.10 *(Continued)*

14	Natural convection across thin enclosures ($H/L > 10$)	$0 \le \theta < 60°$: $\overline{Nu}_L = 1 + 1.44\left[1 - \frac{1708}{Ra_L \cos\theta}\right]\left\{1 - \frac{1708(\sin 1.8\theta)^{1.6}}{Ra_L \cos\theta}\right\} + \left[\left(\frac{Ra_L \cos\theta}{5830}\right)^{1/3} - 1\right]; \quad 0 < Ra_L < 10^5$ (4.98)	All properties to be evaluated at $T_r = (1/2)(T_H + T_C)$ If a term in square brackets is negative, set equal to zero
		$\theta = 60°$: $\overline{Nu}_{L60°} = \max\{Nu_1, Nu_2\}; \quad 0 < Ra_L < 10^7$ (4.99) $Nu_1 = \left\{1 + \left[\frac{0.0936 Ra_L^{0.314}}{1 + \frac{0.5}{[1 + (Ra_L/3160)^{20.6}]^{0.1}}}\right]^7\right\}^{1/7}$ $Nu_2 = \left(0.104 + \frac{0.175}{H/L}\right) Ra_L^{0.283}$	Strictly speaking, these correlations are for gases only; however, Eq. (4.98) can also be used for liquids of moderate Prandtl number.
		$60° < \theta < 90°$: $\overline{Nu}_L = \left(\frac{90 - \theta}{30}\right)\overline{Nu}_{L60°} + \left(\frac{\theta - 60}{30}\right)\overline{Nu}_{L90°}$ (4.100)	
		$\theta = 90°$: $\overline{Nu}_{L90°} = \max\{Nu_1, Nu_2, Nu_3\}; \quad 10^3 < Ra_L < 10^7$ (4.101) $Nu_1 = 0.0605 Ra_L^{1/3}$ $Nu_2 = \left\{1 + \left[\frac{0.104 Ra_L^{0.293}}{1 + (6310/Ra_L)^{1.36}}\right]^3\right\}^{1/3}$ $Nu_3 = 0.242\left(\frac{Ra_L}{H/L}\right)^{0.272}$	For $Ra_L \le 10^3$, $\overline{Nu}_{L90°} \simeq 1$

15	Natural convection across vertical cavities with insulated horizontal surfaces, $1 < H/L < 10$	$2 < H/L < 10$: $\overline{\mathrm{Nu}}_L = 0.22\left(\dfrac{\mathrm{Pr}}{0.2+\mathrm{Pr}}\mathrm{Ra}_L\right)^{0.28}\left(\dfrac{H}{L}\right)^{-1/4}; \quad \mathrm{Ra}_L < 10^{10}$ (4.103*a*) $1 < H/L < 2$: $\overline{\mathrm{Nu}}_L = 0.18\left(\dfrac{\mathrm{Pr}}{0.2+\mathrm{Pr}}\mathrm{Ra}_L\right)^{0.29}; \quad 10^3 < \dfrac{\mathrm{Pr}}{0.2+\mathrm{Pr}}\mathrm{Ra}_L$ (4.103*b*)	
16	Natural convection between concentric cylinders	$\dfrac{k_{\mathrm{eff}}}{k} = 0.386\left(\dfrac{\mathrm{Pr}}{0.861+\mathrm{Pr}}\right)^{1/4}\mathrm{Ra}_{\mathrm{cyl}}^{1/4}; \quad 10^2 < \mathrm{Ra}_{\mathrm{cyl}} < 10^7$ (4.104) $\mathrm{Ra}_{\mathrm{cyl}} = \dfrac{[\ln(D_o/D_i)]^4}{L^3\left(D_i^{-3/5}+D_o^{-3/5}\right)^5}\mathrm{Ra}_L; \quad L = (D_o - D_i)/2$	k_{eff} is the effective thermal conductivity for use in Eq. (2.14) Valid only for $k_{\mathrm{eff}}/k > 1$
17	Natural convection between concentric spheres	$\dfrac{k_{\mathrm{eff}}}{k} = 0.74\left(\dfrac{\mathrm{Pr}}{0.861+\mathrm{Pr}}\right)^{1/4}\mathrm{Ra}_{\mathrm{sph}}^{1/4}; \quad 10^2 < \mathrm{Ra}_{\mathrm{sph}} < 10^4$ (4.105) $\mathrm{Ra}_{\mathrm{sph}} = \dfrac{L}{(D_oD_i)^4}\dfrac{\mathrm{Ra}_L}{\left(D_i^{-7/5}+D_o^{-7/5}\right)^5}; \quad L = (D_o - D_i)/2$	k_{eff} is the effective thermal conductivity for use in Eq. (2.25*a*) Valid only for $k_{\mathrm{eff}}/k > 1$
18	Flow through tube banks	$\overline{\mathrm{Nu}}_D^{10+} = \Phi\overline{\mathrm{Nu}}_D^1; \quad N \geq 10$ (4.114) $\overline{\mathrm{Nu}}_D = \dfrac{1+(N-1)\Phi}{N}\overline{\mathrm{Nu}}_D^1; \quad N < 10$ (4.118) $\Delta P = N\chi\left(\dfrac{\rho V_{\max}^2}{2}\right)f$ (4.119)	$\overline{\mathrm{Nu}}_D^1 = \overline{\mathrm{Nu}}_D$ from item 6 with Re_D based on the average velocity in the space between two adjacent tubes (Eq. 4.113) Φ from Eqs. (4.116) and (4.117). $V_{\max}$ is the maximum velocity in the tube bank; Re_D for f based on $V_{\max}$; f and χ from Fig. 4.44*a*,*b*

(Continued)

Table 4.10 *(Concluded)*

19	Flow through packed beds	$\frac{dP}{dx} = \frac{150\mu\mathscr{V}}{\mathscr{L}^2} + \frac{1.75\rho\mathscr{V}^2}{\mathscr{L}}; \quad 1 < \mathrm{Re} < 10^4$ $\mathrm{Nu} = (0.5\mathrm{Re}^{1/2} + 0.2\mathrm{Re}^{2/3})\mathrm{Pr}^{1/3}; \quad 20 < \mathrm{Re} < 10^4$ $0.5 < \mathrm{Pr} < 20$	(4.130) (4.131)	$d_p = 6\frac{V_p}{A_p}; \quad \mathscr{L} = d_p\left(\frac{\varepsilon_v}{1-\varepsilon_v}\right)$ $\mathscr{V} = \frac{\dot{m}}{\rho\varepsilon_v A_c}$ Viscosity ratio with $n = -0.14$ Specific surface area $a = \frac{A_p}{V_p}(1-\varepsilon_v)$
20	Flow through perforated plates	$\mathrm{Eu} = 8.17\mathrm{Re}^{-0.55}(1.707 - \varepsilon_p)^2; \quad 20 < \mathrm{Re} < 150$ $= 0.5(1.707 - \varepsilon_p)^2; \quad 150 < \mathrm{Re} < 3000$ $\mathrm{St} = C\mathrm{Re}^n\mathrm{Pr}^{-2/3}; \quad 300 < \mathrm{Re} < 3000; \quad \mathrm{Pr} > 0.5$ $C = 3.6 \times 10^{-4}[(1-\varepsilon_p)\varepsilon_p - 0.2]^{-2.07}$ $n = -4.36 \times 10^{-2}\varepsilon_p^{-2.34}; \quad 0.3 < \varepsilon_p < 0.6$	(4.132*a*) (4.132*b*) (4.133)	$\mathrm{Re} = Gd/\mu$, where G is the mass velocity through the holes of diameter d $\mathrm{Eu} = \Delta P\rho/G^2; \quad \Delta P$ for a single plate ε_p = Plate open-area ratio Eq. (4.133) valid for gases and liquids of moderate Prandtl number
21	Rotating disk in a quiescent fluid	$\mathrm{Nu}_r = \frac{0.585\mathrm{Re}_r^{1/2}}{0.6/\mathrm{Pr} + 0.95/\mathrm{Pr}^{1/3}}; \quad \mathrm{Re}_r < 2.4 \times 10^5$ $\mathrm{Nu}_r = 0.021\mathrm{Re}_r^{0.8}\mathrm{Pr}^{1/3}; \quad \mathrm{Re}_r > 2.4 \times 10^5; \quad \mathrm{Pr} > 0.5$	(4.134) (4.135)	
22	Rotating sphere in a quiescent fluid	$\overline{\mathrm{Nu}}_D = 0.43\mathrm{Re}_D^{0.5}\mathrm{Pr}^{0.4}; \quad 10^2 < \mathrm{Re}_D < 5 \times 10^5$ $0.7 < \mathrm{Pr}$ $\overline{\mathrm{Nu}}_D = 0.066\mathrm{Re}_D^{0.67}\mathrm{Pr}^{0.4}; \quad 5 \times 10^5 < \mathrm{Re}_D < 7 \times 10^6$ $0.7 < \mathrm{Pr}$	(4.136*a*) (4.136*b*)	
23	Horizontal rotating cylinder in a quiescent fluid	$\overline{\mathrm{Nu}}_D = 0.133\mathrm{Re}_D^{2/3}\mathrm{Pr}^{1/3}; \quad \mathrm{Re}_D < 4.3 \times 10^5$ $0.7 < \mathrm{Pr} < 670$	(4.137)	Lower limit on Re_D due to natural convection effects is $4.7(\mathrm{Gr}_D^3/\mathrm{Pr})^{0.137}$

24	Turbulent flow in a rough pipe	$f = \left\{-2.0\log\left[\frac{(k_s/R)}{7.4} - \frac{5.02}{\mathrm{Re}_D}\log\left(\frac{k_s/R}{7.4} + \frac{13}{\mathrm{Re}_D}\right)\right]\right\}^{-2}$ (4.138) $\mathrm{St} = \frac{f/8}{0.9 + (f/8)^{1/2}[g(h^+, \mathrm{Pr}) - 7.65]}; \quad k_s^+ > 60$ (4.143a)	Options for $g(h^+, \mathrm{Pr})$: 1. Sand grain indentation: $4.8(h^+)^{0.2}\mathrm{Pr}^{0.44}; \quad 1 < \mathrm{Pr} < 6$ 2. Rectangular transverse ribs, $10 < p/h < 40$: $4.3(h^+)^{0.28}\mathrm{Pr}^{0.57}; \quad 0.7 < \mathrm{Pr} < 40$ 3. General: $0.55(h^+)^{1/2}(\mathrm{Pr}^{2/3} - 1) + 9.5; \ \mathrm{Pr} > 0.5$
25	Turbulent boundary layer on a full rough flat plate	$C_{fx} = \left(3.476 + 0.707\ln\frac{x}{k_s}\right)^{-2.46}$ (4.141) $150 < \frac{x}{k_s} < 1.5 \times 10^7, \quad k_s^+ > 60$ $\overline{C}_f = \left(2.635 + 0.618\ln\frac{L}{k_s}\right)^{-2.57}$ (4.142) $150 < \frac{L}{k_s} < 1.5 \times 10^7, k_s^+ > 60$ $\mathrm{St}_x = \frac{C_{fx}/2}{0.9 + (C_{fx}/2)^{1/2}[g(h^+, \mathrm{Pr}) - 7.65]}; \quad k_s^+ > 60$ (4.143b)	

REFERENCES

1. Petukhov, B. S., "Heat transfer and friction in turbulent pipe flow with variable physical properties," in *Advances in Heat Transfer*, vol. 6., eds. J. P. Hartnett and T. F. Irvine, Academic Press, New York (1970).

2. Gnielinski, V., "New equations for heat and mass transfer in turbulent pipe and channel flow," *Int. Chemical Engineering*, 16, 359–368 (1976).

3. Sparrow, E. M., and Ohadi, M. M., "Numerical and experimental studies of turbulent heat transfer in a tube," *Numerical Heat Transfer*, 11, 461–476 (1977).

4. Notter, R. H., and Sleicher, C. A., "A solution to the turbulent Graetz problem—III. Fully developed and entry region heat transfer rates," *Chem. Eng. Science*, 27, 2073–2093 (1972).

5. Edwards, D. K., Denny, V. E., and Mills, A. F., *Transfer Processes*, 2nd ed., Hemisphere, Washington, D.C. (1979).

6. Mills, A. F., "Experimental investigation of turbulent heat transfer in the entrance region of a circular conduit," *J. Mech. Eng. Sci.*, 4, 63–77 (1962).

7. Petukhov, B. S., and Roizen, L. I., "Generalized relationships for heat transfer in a turbulent flow of a gas in tubes of annular section," *High Temp.* (USSR), 2, 65–68 (1964).

8. White, F. M., *Viscous Fluid Flow*, McGraw-Hill, New York (1974).

9. White, F. M., *Heat Transfer*, Addison-Wesley, Reading, Mass. (1984).

10. Whitaker, S., "Forced convection heat transfer correlations for flow in pipes, past flat plates, single cylinders, single spheres, and for flow in packed beds and tube bundles," *AIChE Journal*, 18, 361–371 (1972).

11. Schlichting, H., *Boundary Layer Theory*, 7th ed., McGraw-Hill, New York (1979).

12. Van Meel, D. A., "A method for the determination of local convective heat transfer from a cylinder placed normal to an air stream," *Int. J. Heat Mass Transfer*, 5, 715–722 (1962).

13. Giedt, W. H., "Investigation of variation of point unit heat transfer coefficient around a cylinder normal to an air stream," *Trans. ASME*, 71, 375–381 (1949).

14. Achenbach, E., "Total and local heat transfer from a smooth circular cylinder in crossflow at high Reynolds number," *Int. J. Heat Mass Transfer*, 18, 1387–1396 (1975).

15. Kays, W. M., and Crawford, M. E., *Convective Heat and Mass Transfer*, 2nd ed., McGraw-Hill, New York (1980).

16. Churchill, S. W., and Bernstein, M., "A correlating equation for forced convection from gases and liquids to a circular cylinder in crossflow," *J. Heat Transfer*, 99, 300–306 (1977).

17. Nakai, S., and Okazaki, T., "Heat transfer from a horizontal circular wire at small Reynolds and Grashof numbers—I. Pure convection," *Int. J. Heat Mass Transfer*, 18, 387–396 (1975).

18. Bird, R. B., Stewart, W. E., and Lightfoot, E. N., *Transport Phenomena*, 2nd ed., John Wiley & Sons, New York (1962). [Eq. (2.6–14) rewritten in terms of drag coefficient.]

19. Reynolds, W. C., Kays, W. M., and Kline, S. J., "Heat transfer in the turbulent incompressible boundary layer, III. Arbitrary wall temperature and heat flux," NASA Memo 12-3-58 W (1958).

20. Mills, A. F., "Average Nusselt numbers for external flows," *J. Heat Transfer*, 101, 734–735 (1979).

21. Churchill, S. W., and Usagi, R., "A general expression for the correlation of rates of transfer and other phenomena," *AIChE Journal,* 18, 1121–1128 (1972).

22. Churchill, S. W., and Chu, H. H. S., "Correlating equations for laminar and turbulent free convection from a vertical plate," *Int. J. Heat Mass Transfer*, 18, 1323–1329 (1975).

23. Churchill, S. W., and Chu, H. H. S., "Correlating equations for laminar and turbulent free convection from a horizontal cylinder," *Int. J. Heat Mass Transfer*, 18, 1049–1053 (1975).

24. Yuge, T., "Experiments on heat transfer from spheres including combined natural and forced convection," *J. Heat Transfer*, 82, 214–220 (1960).

25. Churchill, S. W., "Free convection around immersed bodies," *Hemisphere Heat Exchanger Design Handbook*, ed. G. F. Hewitt, §2.5.7, Hemisphere, Washington, D.C. (1990).

26. Lienhard, J. H., "On the commonality of equations for natural convection from immersed bodies," *Int. J. Heat Mass Transfer*, 16, 2121–2123 (1973).

27. Hatfield, D. W., and Edwards, D. K., "Edge and aspect ratio effects on natural convection from the horizontal heated plate facing downwards," *Int. J. Heat Mass Transfer*, 24, 1019–1024 (1981). (Constants C_2, C_3, and C_4 have been corrected.)

28. Kadambi, V., and Drake, R. M., "Free convection heat transfer from horizontal surfaces for prescribed variations in surface temperature and mass flow through the surface," *Technical Report, Mech. Eng. HT-1*, Princeton University (1960).

29. Fugii, T., and Imura, H., "Natural convection heat transfer from a plate with arbitrary inclination," *Int. J. Heat Mass Transfer*, 15, 755–767 (1972).

30. McAdams, W. H., "Heat Transmission," 3rd ed., McGraw-Hill, New York (1954).

31. Barrow, H., and Sitharamaroa, T. L., "The effect of variable β on free convection," *Brit. Chem. Eng.*, 16, 704–705 (1971).

32. Hollands, K. G. T., Unny, T. E., Raithby, G. D., and Konicek, L., "Free convective heat transfer across inclined air layers," *J. Heat Transfer*, 98, 189–193 (1976).

33. ElSherbiny, S. M., Raithby, G. D., and Hollands, K. G. T., "Heat transfer by natural convection across vertical and inclined air layers," *J. Heat Transfer*, 104, 96–102 (1982).

34. Globe, S., and Dropkin, D., "Natural convection heat transfer in liquids confined between two horizontal plates," *J. Heat Transfer*, 81, 24–28 (1959).

35. Berkovsky, B. M., and Polevikov, V. K., "Numerical study of problems on high-intensive free convection," in *Heat Transfer and Turbulent Buoyant Convection*, eds. D. B. Spalding and N. Afgan, Hemisphere, Washington, D.C. (1977), pp. 443–455.

36. Raithby, G. D., and Hollands, K. G. T., "A general method of obtaining approximate solutions to laminar and turbulent free convection problems," in *Advances in Heat Transfer*, vol. 11, eds. J. P. Hartnett and T. F. Irvine, Jr., Academic Press, New York (1975).

37. Metais, B., and Eckert, E. R. G., "Forced, mixed, and free convection regimes," *J. Heat Transfer*, 86, 295–296 (1964).

38. Churchill, S. W., "Combined free and forced convection around immersed bodies," *Hemisphere Heat Exchanger Design Handbook*, ed. G. F. Hewitt, §2.5.9., Hemisphere, Washington, D.C. (1990).

39. This text; developed from Gnielinski, V., Zukauskas, A., and Skrinska, A., "Banks of plain and finned tubes," *Hemisphere Heat Exchanger Design Handbook*, ed. G. F. Hewitt, §2.5.3., Hemisphere, Washington, D.C. (1990).

40. Zukauskas, A., and Ulinskas, R., "Efficiency parameters for heat transfer in tube banks," *Heat Transfer Engineering*, 6, No. 2, 19–25 (1985).

41. Ergun, S., "Fluid flow through packed columns," *Chem. Eng. Prog.*, 48, 89–94 (1952).

42. Shevyakova, S. A., and Orlov, V. K., "Study of hydraulic resistance and heat transfer in perforated-plate heat exchangers," *Inzhenerno-Fizicheskii Zhurnal*, 45, 32–36 (1983).

43. Cobb, E. C., and Saunders, O. A., "Heat transfer from a rotating disc," *Proc. Roy. Soc. London*, ser. A, 236, 343–351 (1956).

44. Kreith, F., Roberts, L. G., Sullivan, J. A., and Sinha, S. N., "Convection heat transfer and flow phenomena of rotating spheres," *Int. J. Heat Mass Transfer*, 6, 881–895 (1963).

45. Moody, L. F., "Friction factors for pipe flow," *Trans. ASME*, 66, 671–684 (1944).

46. Nikuradse, J., "Laws of flow in rough pipes," NACA TM 1292 (1950). (English translation of VDI-Forschungsheft 361, 1933.)

47. Coleman, H. W., Hodge, B. K., and Taylor, R. P., "A re-evaluation of Schlichting's surface roughness experiment," *J. Fluids Engineering*, 106, 60–65 (1984).

48. Dalle Donne, M., and Meyer, L., "Turbulent convective heat transfer from rough surfaces with two-dimensional rectangular ribs," *Int. J. Heat Mass Transfer*, 20, 583–620 (1977).

49. Zigrang, D. J., and Sylvester, N. D., "Explicit approximations to the solution of Colebrook's friction factor equation," *AIChE Journal*, 28, 514–515 (1982).

50. Mills, A. F., and Xu Hang, "On the skin friction coefficient for a fully rough flat plate," *J. Fluids Engineering*, 105, 364–365 (1983).

51. Dipprey, D. F., and Sabersky, R. H., "Heat and momentum transfer in smooth and rough tubes at various Prandtl numbers," *Int. J. Heat Mass Transfer*, 6, 329–353 (1963).

52. Wassel, A. T., and Mills, A. F., "Calculation of variable property turbulent friction and heat transfer in rough pipes," *J. Heat Transfer*, 101, 469–474 (1979).

53. Yaglom, A. M., and Kader, B. A., "Heat and mass transfer between a rough wall and turbulent fluid flow at high Reynolds and Peclet numbers," *J. Fluid Mechanics*, 62, part 3, 601–623 (1974).

54. Buchberg, H., Catton, I., and Edwards, D. K., "Natural convection in enclosed spaces—A review of application to solar energy collection," *J. Heat Transfer*, 98, 182–188 (1976).

EXERCISES

4–1. A laminar flow in a 2 cm–I.D. tube has the following velocity and temperature profiles:

$$u = 0.1[1 - (r/0.01)^2] \text{ m/s}$$

$$T = 400 - 3 \times 10^6 \, (1.875 \times 10^{-5} - 0.25r^2 + 624r^4) \text{ K}$$

for r in meters. Determine the bulk temperature.

4–2. The following table gives velocity and temperature profiles for a turbulent flow of air between parallel plates 4 cm apart. The coordinate y is measured from the wall, and the profiles are symmetrical about the centerplane. Determine the bulk velocity and the bulk temperature.

y, cm	u, m/s	T, K	y, cm	u, m/s	T, K
0.0	0.0	373.0	1.2	21.4	302.6
0.2	16.3	322.4	1.4	21.8	301.0
0.4	18.2	315.2	1.6	22.1	299.8
0.6	19.4	310.7	1.8	22.2	299.1
0.8	20.2	307.3	2.0	22.3	298.8
1.0	20.9	304.7			

4–3. For fully developed flow in a duct, show that the friction factor is four times the skin friction coefficient, $f = 4C_f$.

4–4. A heated horizontal cylinder rotates rapidly about its axis in still air. Use dimensional analysis to obtain dimensionless groups pertinent to heat transfer.

4–5. The central receiver of a solar power plant is in the form of a vertical cylinder, of height H and diameter D. Experimental data have been obtained for the effect of a crosswind on the convective heat loss from the receiver. Suggest appropriate dimensionless groups for correlating the data.

4–6. Consider the packing of a perforated-plate heat exchanger for a cryogenic refrigeration system. Geometric parameters include the hole diameter d, hole pitch p (or, alternatively, the open-area ratio ε_p), the plate thickness t, and plate spacing s. Expected parameter values for a hydrogen flow include a mass velocity based on a hole cross-sectional area of 6×10^{-3} kg/m^2 s, a pressure of 100 Pa, temperatures of 10–350 K, and surface-to-gas temperature differences of 10 K. Hole diameters are in the range 1–3 mm. Suggest appropriate dimensionless groups for correlating experimental data for pressure drop and heat transfer.

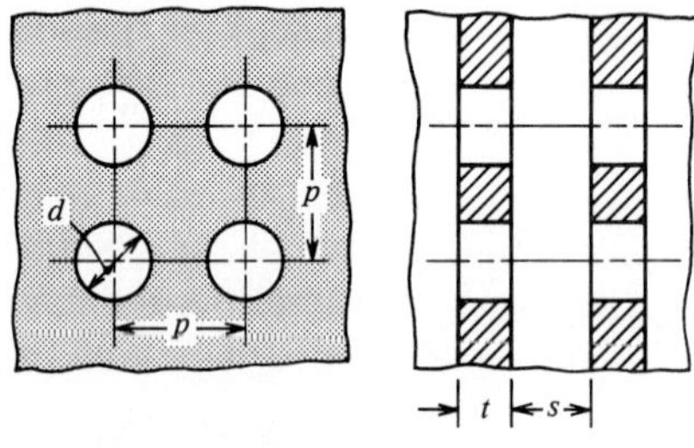

4–7. Consider a constant property, low-speed flow at velocity u_e along a heated flat plate. Use dimensional analysis to find the dimensionless groups governing the growth of the hydrodynamic boundary layer thickness δ along the plate, that is, as a function of coordinate x measured from the leading edge. Obtain the corresponding result for the thermal boundary layer thickness Δ.

4–8. Convective heat transfer is sometimes modeled as conduction across a thin stagnant layer of fluid adjacent to the surface. If this thickness is denoted δ_f, then Eq. (1.9) for conduction across a plane slab gives $q_s = (k/\delta_f)\Delta T$. Comparing with Eq. (4.2) or (4.9), $h_c = k/\delta_f$ or $\delta_f = k/h_c$. Calculate the equivalent stagnant film thickness for

(i) water at 300 K flowing at 5 m/s in a 2 cm–I.D. tube.

(ii) air in free convection on a vertical plate heated to 310 K in ambient air at 290 K and 1 atm, at a position 20 cm above the bottom of the plate.

Use the heat transfer coefficient formulas given in Section 1.3.3.

4–9. An experiment to study heat transfer from an elliptical cylinder was performed in a wind tunnel. Heat loss from a cylinder of major and minor axes 8 and 6 cm, respectively, to an air flow normal to the minor axis of the cylinder was measured. Heat loss from the cylinder consisted of both radiation and convection. The measured total heat loss, and a calculation of the radiation contribution for each test condition, are listed here. (The radiation calculation was based on an emittance of 0.8 for the cylinder surface.) Use the data to estimate the convective heat loss from a geometrically similar cylinder of 12 cm major axis at 30°C to a 4.5 m/s air flow at 23°C.

Test	V m/s	T_s K	T_e K	q_{rad} W/m^2	q_{total} W/m^2
1	1.37	327.5	304.3	133	585
2	2.83	309.4	301.5	42	261
3	3.54	304.7	291.6	65	553
4	6.10	306.1	301.4	23	261
5	8.63	302.8	297.9	24	403
6	8.50	311.5	305.9	32	410
7	8.81	309.2	304.4	26	402
8	8.32	299.5	292.7	35	552
9	11.5	308.2	304.5	18	466
10	11.4	307.0	301.8	26	548
11	14.8	306.3	302.4	19	546
12	17.5	302.2	298.8	16	577

4–10. A common problem in the design of electronic equipment is the need to control the temperatures of a number of heat-producing devices arranged in a symmetrical matrix. One effective method is to mount the devices on studs and to pump a liquid coolant over the studs. In a particular design there are 252 devices mounted on cylindrical studs, with generated heat fluxes as high as 5 W/cm^2 measured at the stud surface. Refrigerant-113 is a suitable coolant, having appropriate dielectric properties. To obtain design data, a model unit was made by attaching 1 cm–diameter, 4 cm–high copper studs, in a 2 cm–pitch square array, to a thin, electrically heated plate. Results of experiments are as follows. If the bulk temperature of the R-113 was 300 K, suggest a correlation of the data suitable for engineering use. Also comment on the adequacy of the experiment.

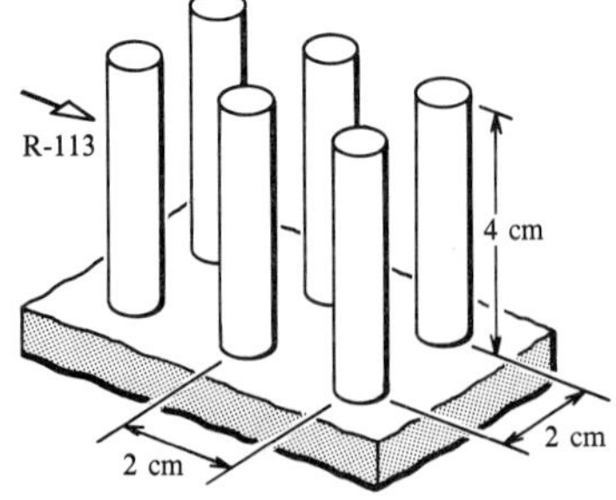

Velocity V, m/s	1.0	1.6	2.4	3.1	3.6
Heat transfer coefficient h_c, W/m^2 K	1810	2430	3150	3730	4130

4–11. Consider a 10 m length of 2 cm–I.D. tube. What is the average convective heat transfer coefficient and pressure gradient inside the tube when the tube wall is at 320 K and water enters at 300 K, 1 atm pressure, and flows at a velocity of 3 m/s?

4–12. Consider a 10 m length of 2 cm–I.D. tube. What is the average heat transfer coefficient and pressure gradient inside the tube when the tube wall is at 320 K, and SAE 50 oil enters at 300 K, 1 atm pressure, and flows at a velocity of 3 m/s?

4–13. Consider a 10 m length of 2 cm–I.D. tube. What is the average heat transfer coefficient and pressure gradient inside the tube when the tube wall is at 320 K, and mercury enters at 300 K, 1 atm pressure, and flows at a velocity of 3 m/s?

4–14. Consider a 10 m length of 2 cm–I.D. tube. What is the average heat transfer coefficient and pressure gradient inside the tube when the tube wall is at 320 K, and air enters at 300 K, 1 atm pressure, and flows at a velocity of 3 m/s?

4–15. Consider a 10 m length of 2 cm–I.D. tube. What is the average heat transfer coefficient and pressure gradient inside the tube when the tube wall is at 320 K, and helium enters at 300 K, 1 atm pressure, and flows at a velocity of 3 m/s?

4–16. Therminol 60 is under consideration as a high-temperature coolant. Prepare tables of heat transfer coefficient and pressure gradient as a function of average temperature for flow at 3 m/s in a 5 cm–I.D. pipe. Ignore entrance effects. Take $T_s = T_b$.

4–17. Liquid lithium flows at 5 m/s in a 1 cm–I.D. tube. Calculate the Nusselt number and heat transfer coefficient

(i) if the wall temperature is uniform.
(ii) if the wall heat flux is uniform.

Evaluate properties at 800 K.

4–18. Air at 1×10^5 Pa and 300 K flows at 2 m/s through a 1 cm–I.D. pipe, 1 m long. An electric resistance heater surrounds the last 20 cm of the pipe and supplies a constant heat flux to raise the bulk temperature of the air to 330 K. What power input is required?

4–19. SAE 50 oil flows in a 60 m–long, 25 mm–I.D. tube at 0.5 kg/s. The oil enters the tube at 300 K, and the walls of the tube are maintained at 370 K. Determine the exit bulk temperature of the oil.

4–20. Calculate the Nusselt number and heat transfer coefficient for liquid sodium flowing at 3 m/s in a 5 cm–I.D. pipe if

(i) the wall temperature is uniform.
(ii) the wall heat flux is uniform.

Evaluate properties at 1000 K.

4–21. Aqueous solutions of ethylene glycol are used as "antifreeze" working fluids in many heat exchangers. For a 10 m–long, 3 cm–I.D. tube, prepare graphs of pressure drop and heat transfer coefficient for bulk velocities in the range 0.2–4.0 m/s. Use the properties in Table A.13*a* for 20, 30, 40, 50, and 60% glycol by mass. Assume $T_s = T_b = 270$ K.

4–22. Crude oil of specific gravity 0.862 flows at 0.2 kg/s through a 1.2 cm–I.D. horizontal tube. At a temperature of 370 K the pressure drop in a length of 3 m is 31.6 kPa. Calculate the dynamic and kinematic viscosities of the oil. Also, if the conductivity and specific heat of the oil are approximately the same as that for SAE 50 oil, estimate the convective heat transfer coefficient.

4–23. SAE 50 oil flows at 3 m/s between 0.25 m–long parallel plates spaced 3 mm apart. If the plates are at 400 K and the average of the inlet and outlet bulk temperatures is 360 K, determine the average heat transfer coefficient and pressure drop.

4–24. SAE 50 oil flows at 4 m/s through a 1 cm–square cross-section duct, 3 m long. If the oil enters at 290 K and the walls are at 330 K, determine the pressure drop, the average heat transfer coefficient, and the outlet bulk temperature.

4–25. Air flows at a rate of 0.05 kg/s through a 3 cm–I.D., 4 cm–O.D. annular duct, 9 m long. The air enters at 245 K, 10 atm pressure, and is heated by saturated steam condensing in the inner tube of the annulus at 1.1×10^5 Pa. Determine the average heat transfer coefficient, the air outlet temperature, and the pressure drop. The outer surface can be taken as insulated.

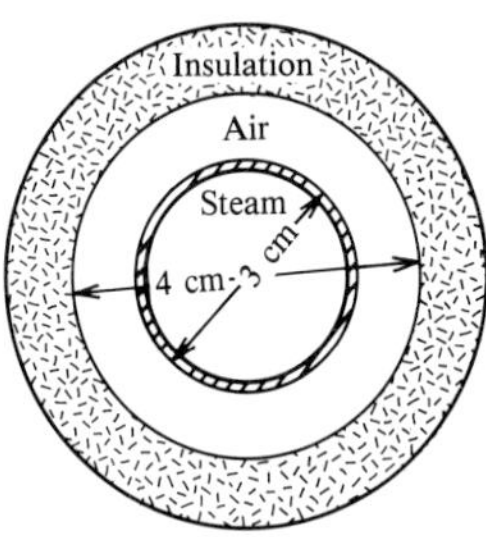

4–26. Air flows at 0.05 kg/s through a 3 cm–I.D., 4 cm–O.D. annular duct, 9 m long. The air enters at 240 K, 1 atm pressure, and is heated by saturated steam condensing in the inner tube of the annulus at 1.1×10^5 Pa. Determine the average heat transfer coefficient, the air outlet temperature, and the pressure drop. The outer surface can be taken to be insulated.

4–27. In the design of a shell-and-tube heat exchanger, the tube diameter must be chosen to satisfy a number of constraints. In particular, there is a trade-off between heat transfer and pressure drop considerations, since both increase as the tube diameter decreases. It is required to design an exchanger to heat 0.01 kg/s of helium from 100 to 300 K, and we wish to explore the effect of tube inside diameter in the range 3–30 mm for

(i) laminar flow.
(ii) turbulent flow.

In each case the flow cross-sectional area will be fixed, 0.007 m^2 for laminar flow and 3×10^{-4} m^2 for turbulent flow. The nominal pressure is 10^5 Pa.

Prepare graphs of the heat transfer coefficient, the $h_c\mathscr{P}$ product, and pressure gradient as a function of tube size for each case. Assume fully developed conditions, and evaluate all properties at 200 K.

4–28. Air at 400 K and 1250 kPa flows at 40 m/s along a 10 cm–long flat plate maintained at 300 K. Determine the drag force and heat transfer per unit width.

4–29. The roof of a building is flat and is 20 meters wide and long. When the wind speed over the roof is 10 m/s, determine

(i) the convective heat gain on a clear night when the roof temperature is 280 K and the air temperature is 290 K.
(ii) the convective heat loss on a hot, sunny day when the roof temperature is 320 K and the air temperature is 300 K.

4–30. An oil pan under an automobile is 30 cm wide, 70 cm long, and 15 cm deep. Estimate the total heat loss from the five exposed sides when the oil is at 120°C and the automobile is traveling at 80 km/h through ambient air at 20°C.

4–31. Insulating wet suits worn by scuba divers may be made of 3 mm–thick foam neoprene ($k = 0.05$ W/m K), which traps next to the skin a layer of water, say 1 mm thick. Estimate the rate of heat loss from a 1.8 m–tall diver swimming at 8 km/h in 286 K water if his skin temperature does not fall below 297 K.

4–32. A small submarine is to be as silent as possible. A proposed design requires the use of a thermoelectric generator rejecting heat directly to the hull, thus eliminating cooling water pumping machinery. An area 10 m wide and 10 m long at the bow is available for heat rejection. If the submarine cruises at 4 m/s and must reject 15 MW to 15°C water, estimate the average temperature of the heat rejection surface.

4–33. Consider flow along a flat plate. For free-stream velocities in the range 0.1–100 m/s determine the average heat transfer coefficient and drag force per unit width on a 1 m length of plate for

(i) air at 1 atm.
(ii) air at 0.01 atm.

Use a transition Reynolds number of 10^5, and evaluate properties at 295 K.

4–34. Air at -10°C blows at 10 m/s over the exterior of a 30 cm–O.D. pipeline spanning a river gorge. The line is heated to 40°C to reduce the viscosity of the liquid inside. What would be the heat loss per 100 m of pipe if the line was uninsulated?

4–35. What is the rise velocity of 1 mm–diameter oxygen bubbles 3 m below the surface of a water pool at 300 K?

4–36. A 6 mm–diameter hailstone falls at its terminal velocity through air at 287 K and 1 atm. Calculate the average heat transfer coefficient.

4–37. A 1.905 cm–diameter cylinder spans a small wind tunnel and is exposed to air at 97 kPa and 290 K flowing at 20 m/s. The cylinder contains an electrical heater over a length of 12 cm. Determine the stagnation line temperature and average cylinder temperature when the power input to the heater is 60 W. Also determine the drag force on the 12 cm length of cylinder. The cylinder wall is very thin and has a low conductivity.

4–38. A hot-film sensor to measure ocean speeds consists of a thin (1000 Å) platinum film deposited on a 3 cm–long quartz cylinder with outside diameter of 9 mm. A known current is passed through the platinum film, and the resistance of the film is measured using an AC resistance bridge. From the known temperature dependence of resistivity for platinum, an average temperature of the sensor surface is determined. With the sensor perpendicular to the ocean flow, a power input of 1.101 W to the sensor is measured when the sensor is at 21.35°C and the ambient sea water temperature is 20.54°C.

(i) Estimate the flow speed.
(ii) If the individual temperature measurements are accurate only to ±0.01°C, what is the resulting expected error in flow speed?
(iii) If bio-fouling results in deposits with thermal conductivity of approximately 1 W/m K, what thickness of deposit would give a 5% error in flow speed?

4–39. Water at 300 K flows at 2 m/s over a 3 cm–diameter sphere maintained at 340 K.

(i) Determine the drag force on the sphere.
(ii) Determine the ratio of the stagnation point heat transfer coefficient to the average value.
(iii) Determine the heat loss from the sphere.

4–40. A horizontal aluminum rod, 1 cm in diameter and 10 cm long, has both ends maintained at 363 K. Air at 293 K and 1 bar is blown over the rod at 2.5 m/s. Calculate

(i) the temperature at the midplane of the rod.
(ii) the heat dissipated by the rod.

4–41. What is the cost per day, per meter length of pipe, of energy lost from an uninsulated 4 cm–O.D. hot water pipe with a surface temperature of 45°C exposed to air at 10°C blowing at 2 m/s normal to the pipe axis? Electric power in Los Angeles costs about 8 cents/kW h.

4–42. An alloy sphere 1 cm in diameter is suspended by a fine wire at the center of a laboratory furnace, the inside wall of which is maintained at 700 K. Helium at 350 K and 1.1 bar pressure is blown through the furnace at 5 m/s. The interior of the furnace may be assumed black, and the emittance of the alloy is 0.40. Calculate the steady-state temperature of the sphere.

4–43. An ice sensor is located on an airplane wing flying at 50 m/s through air at 61 kPa and $-26°C$. A heater is required to heat the sensor surface to 0°C to ascertain from the temperature-time response whether ice is present (if ice is present, the $T - t$ curve has a plateau corresponding to the supply of enthalpy of melting). If the sensor is in the form of a 2 cm–diameter disk, what is the minimum power required? The front of the wing can be modeled as a 30 cm–diameter cylinder with the sensor located on the stagnation line.

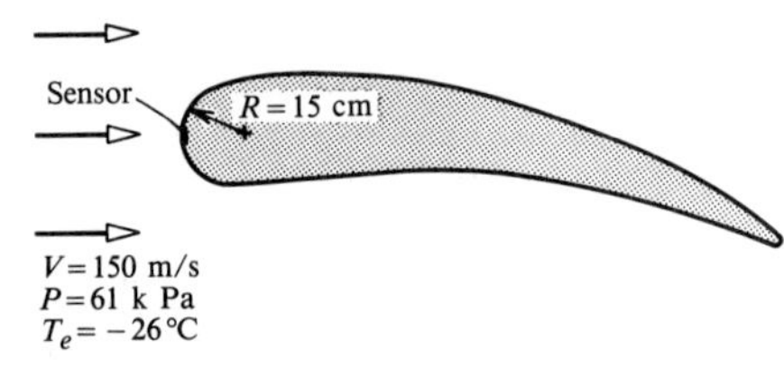

4–44. A 0.12 mm–diameter platinum wire 3 mm long is used for a constant-temperature (200°C) hot-wire anemometer to measure 20°C air velocities in the range 0.3 to 7 m/s. Tabulate the wire current versus air velocity. A value of 10×10^{-8} Ω m for the electrical resistivity of platinum should be used.

4–45. Consider a length of horizontal pipe of outside diameter 3 cm. What is the average convective heat transfer coefficient on the outside of the pipe when the pipe is at 320 K and

(i) it is immersed in still air at 300 K, 2 atm pressure.
(ii) it is immersed in still water at 300 K.

4–46. A 100 liter covered tank full of water at 290 K is to be heated to 340 K by means of steam condensing inside a 1 cm–O.D. copper steam coil having 10 turns of 0.5 m diameter. If the steam-side thermal resistance is negligible and the tank well insulated, estimate the time required for heating and the total amount of steam condensed. Assume a steam pressure of 1 atm.

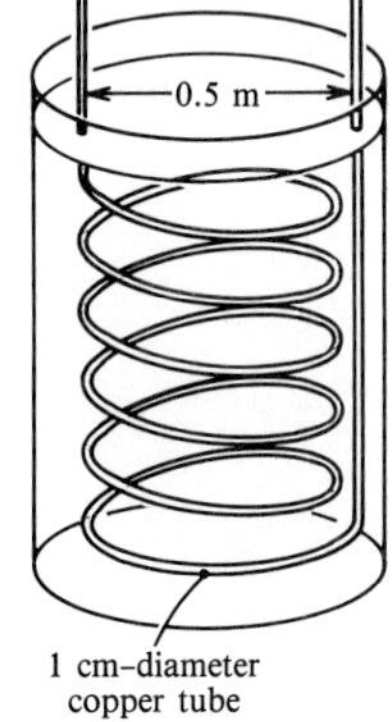

4–47. An uninsulated aluminum overhead electrical transmission line has a diameter of 2 cm and a resistance of 8.33×10^{-5} Ω/m. If the current flowing is 600 A, determine the surface temperature of the wire on a still day with a low cloud cover when the air temperature is 290 K. The emittance of the wire can be taken as 0.7.

4–48. Consider a sphere of 1 mm diameter in air at 300 K and 1 atm pressure.

(i) At what air velocity will it have a heat transfer coefficient 50% greater than the conduction limit value ($\mathrm{Nu}_D = 2.0$)?
(ii) If the air is stationary so that only natural convection occurs, at what temperature will it have a heat transfer coefficient 50% greater than the conduction limit value?

Repeat for water at 300 K.

4–49. A 60 W light bulb can be modeled as a 5 cm–diameter sphere. If the surface temperature is measured to be 135°C when the ambient air is at 20°C, estimate the fraction of the lamp power that is lost by

(i) natural convection.
(ii) radiation.

Take $\varepsilon = 0.8$ for the glass.

4–50. W. H. McAdams [30] recommends a simple correlation for natural convection from horizontal cylinders:

$$\overline{\mathrm{Nu}_D} = 0.53\,\mathrm{Ra}_D^{1/4}$$

for $\mathrm{Pr} > 0.5$, $10^3 < \mathrm{Gr}_D < 10^9$. Compare the values of Nusselt number given by this formula with those given by Eq. (4.87) for Pr = 0.7, 10, and 100.

4–51. A vertical plate 1 m high is immersed in a stagnant fluid. The plate is maintained at 350 K, and the fluid temperature ranges from 280 to 340 K. Determine the average heat transfer coefficient if the fluid is

(i) air at 1 atm.
(ii) air at 0.01 atm.
(iii) water.
(iv) SAE 50 oil.

4–52. Compare the convective heat loss from the top and bottom of a 10 cm–square plate immersed in water at 300 K when the plate temperature is

(i) 310 K.
(ii) 350 K.

4–53. Compare the convective heat loss from the top and the bottom of a heated 20 cm–square plate located in air at 1 atm pressure and 300 K when the plate temperature is

(i) 320 K.
(ii) 400 K.

4–54. A heated horizontal plate is 10 cm wide and 20 cm long and has 2 cm adiabatic extensions. What is the convective heat loss from the underside of the plate when it is at 320 K and is exposed to air at 300 K?

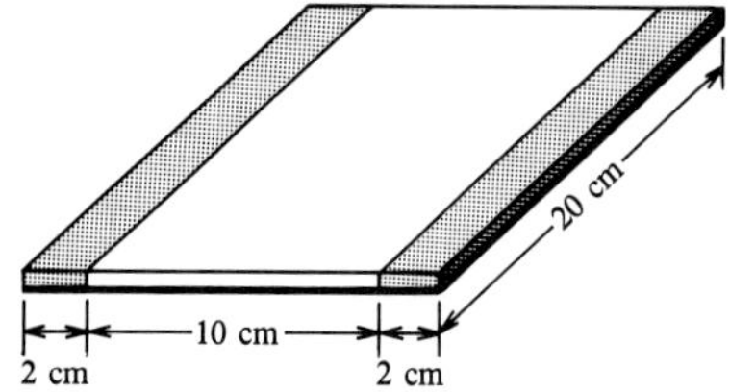

4–55. A steam supply pipe carries saturated steam at 1.2 MPa. The pipe has a 15 cm outer diameter and is lagged with 6 cm of 85% magnesia insulation. Determine the heat loss per meter length of pipe and the surface temperature of the insulation on a windless day when the air temperature is 20°C.

4–56. A 20 cm–square plate is inclined at 45° to the vertical in air at 1 atm and 300 K. If the plate temperature is 320 K, determine the convective heat loss from each side.

4–57. The sphere-cone shown in the sketch is maintained at 340 K in water at 300 K. Estimate the rate of heat loss if the top is well insulated.

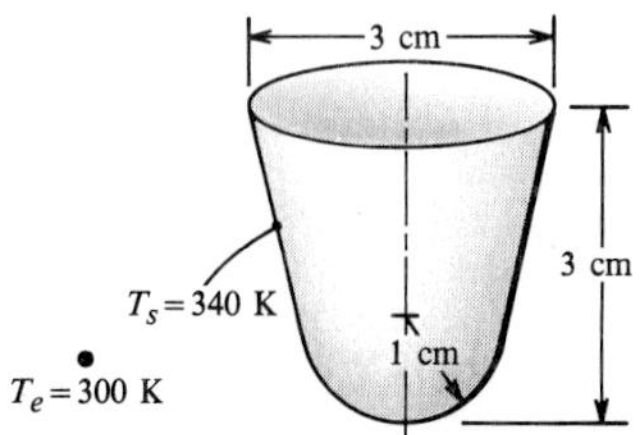

4–58. A 1 m–high double-glazed window has an air gap of 1 cm. In a test the facing glass surfaces were measured to be at 14.2 and −10.6°C. Calculate the convective heat transfer across the gap.

4–59. A flat-plate solar collector has an absorber plate at 350 K and a cover plate at 310 K. If the air gap is 5 cm, determine the convective heat flux across the gap when the collector is inclined at angles from the horizontal of

(i) 0°.
(ii) 30°.
(iii) 60°.

4–60. An aluminum roof 30 m long and 10 m wide is inclined at an angle of 30°. A 2 cm–thick plasterboard ($k = 0.06$ W/m K) ceiling is located below the roof, giving a 5 cm air gap. On a clear winter night condensation on the outside of the roof maintains its temperature at 5°C, while heating within the room maintains the lower surface of the ceiling at 15°C. Neglecting the effects of structural members between the ceiling and the roof, estimate the heat loss through the ceiling.

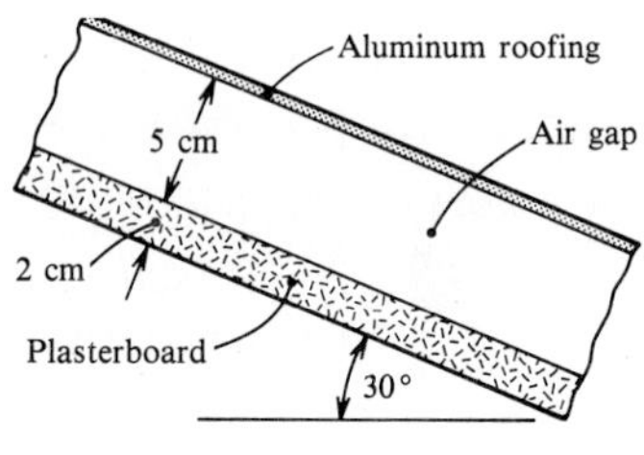

4–61. A long, 1 m–high, 30 cm–thick air cavity has insulated top and bottom surfaces. Calculate the convective heat transfer per meter when the side surfaces are at 450 and 350 K.

4–62. A horizontal cylindrical annulus has an outer diameter of 18 cm and an inner diameter of 10 cm. If the outer surface is maintained at 320 K and the inner surface at 280 K, determine the effective conductivity of enclosed air at 1 atm and the convective heat transfer per unit length.

4–63. Two concentric spheres of outer and inner diameters 18 cm and 10 cm, respectively, are separated by air at 1 atm pressure. If the outer surface and inner surface are maintained at 320 K and 280 K, respectively, determine the effective conductivity of the air and the convective heat transfer.

4–64. Buchberg et al. [54] proposed the following three-regime correlation for heat transfer across high-aspect-ratio inclined layers:

$$\overline{\mathrm{Nu}_L} = 1 + 1.446\left[1 - \frac{1708}{\mathrm{Ra}_L \cos\theta}\right] \qquad 1708 < \mathrm{Ra}_L \cos\theta < 5900$$

$$= 0.229(\mathrm{Ra}_L \cos\theta)^{0.252} \qquad 5900 < \mathrm{Ra}_L \cos\theta < 9.23 \times 10^4$$

$$= 0.157(\mathrm{Ra}_L \cos\theta)^{0.285} \qquad 9.23 \times 10^4 < \mathrm{Ra}_L \cos\theta < 10^6$$

where the term in square brackets is set equal to zero if negative. Compare the Nusselt numbers given by this correlation at $\theta = 30°$ and $50°$ for air at normal temperatures, with the values given by the correlations of Hollands et al., Eqs. (4.98)–(4.101).

4–65. Space heating in buildings accounts for about 20% of the total energy consumed in the United States. Energy conservation during cold weather requires reducing heat loss through large glass areas. Compare the heat loss through 3 mm–thick window glass for the two configurations shown in the sketch. Assume 600 m^2 of glass, an indoor temperature of 21°C, and an outdoor temperature of 3°C. The window drape has a thickness of 1 mm and a conductivity of 0.07 W/m K. To include thermal radiation, assume that the window glass and drape are opaque and that all surfaces have an emittance of 1.0.

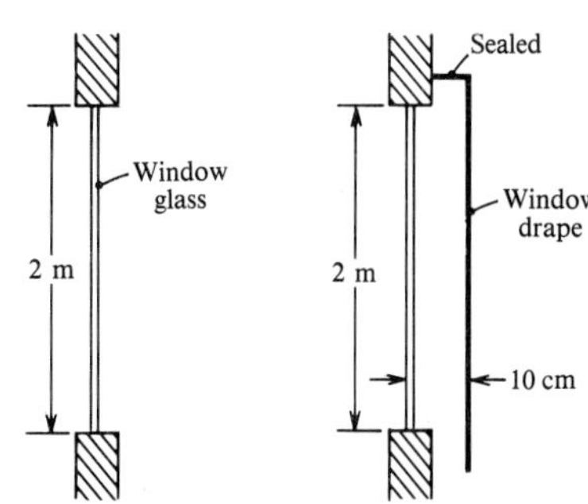

4–66. The sketch shows a test cell used to study natural convection in a circular cross-section cavity, 15.2 cm in diameter. The cold-plate assembly can be moved to vary the thickness of the liquid layer. The complete test cell can be rotated to incline the liquid layer at any angle θ. Referring to the thermal circuit, the rubber layers serve as heat flux meters. These meters are calibrated by rotating the

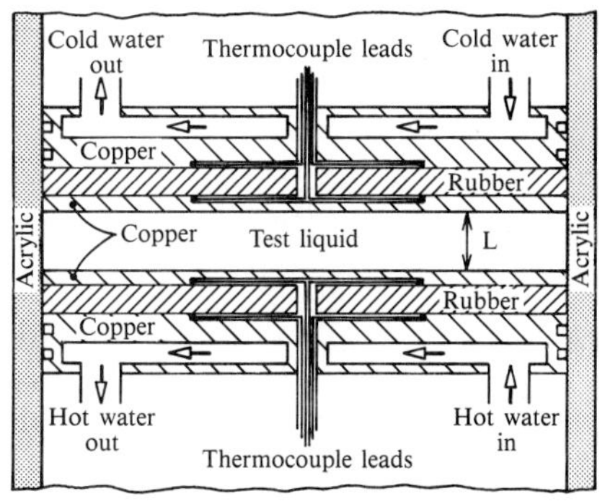

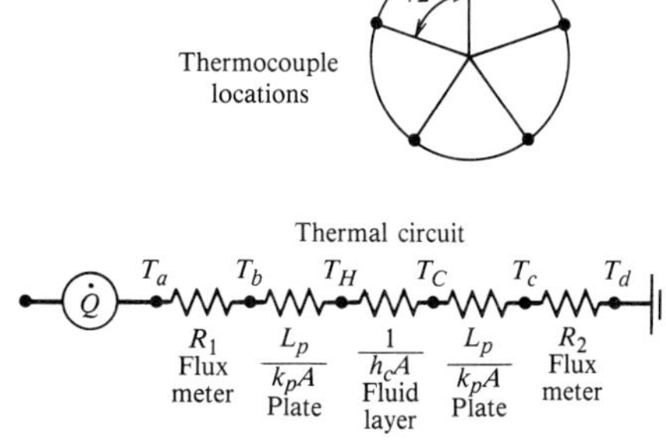

cell to $\theta = 180°$, where the hot plate is above the cold plate, the liquid layer is stable, and hence heat transfer through the liquid is by conduction only. Its thermal resistance is then simply $1/h_c A = L/kA$, and is known. A series of tests measuring the temperatures T_a, T_b, T_c, and T_d allows the unknown meter resistances R_1 and R_2 to be determined as 0.346 and 0.327 K/W, respectively. Notice that the thermal resistances of the copper plates are negligible, so that $T_b = T_H$, $T_c = T_C$. It is advantageous to operate the system with the average fluid layer temperature approximately equal to the ambient temperature in order to minimize errors due to heat exchange with the surroundings. The plate spacing is varied to obtain a range of flow conditions. Typical results for water are shown in the table. Prepare graphs of Nusselt number versus Rayleigh number, and make comparisons with the correlations given in Section 4.4.2. (*Hint:* Use CONV to assist in the data processing.)

Angle, θ	L mm	T_a °C	T_b °C	T_c °C	T_d °C
0°	4	49.9	31.4	22.6	3.3
	7	52.6	33.5	23.5	3.7
	15	51.1	31.9	21.4	1.7
	30	53.3	33.8	22.3	2.3
90°	4	51.1	33.0	20.4	3.4
	15	50.6	32.3	19.9	2.3
	30	51.6	32.9	20.4	2.3
120°	4	51.6	33.3	20.3	3.4
	7	51.6	33.2	20.6	3.4
	15	50.6	32.3	19.2	2.2

4–67. A 3 cm–diameter horizontal cylinder is located transverse to the flow in a wind tunnel. The cylinder is maintained at 400 K in an air stream at 300 K and 1 atm. Determine the convective heat loss per meter length for air speeds in the range 0.05–0.30 m/s.

4–68. Water at 320 K flows at 0.15 m/s along a 10 cm–long, 1 m–wide flat plate. Determine the average heat transfer coefficient for plate temperatures in the range 280–360 K.

4–69. Consider a vertical fluid flow upward in a tube of diameter 2 cm and height 20 cm. The wall is maintained at 310 K, and the inlet temperature is 300 K. Below what velocity must mixed-convection effects be considered for the following fluids?

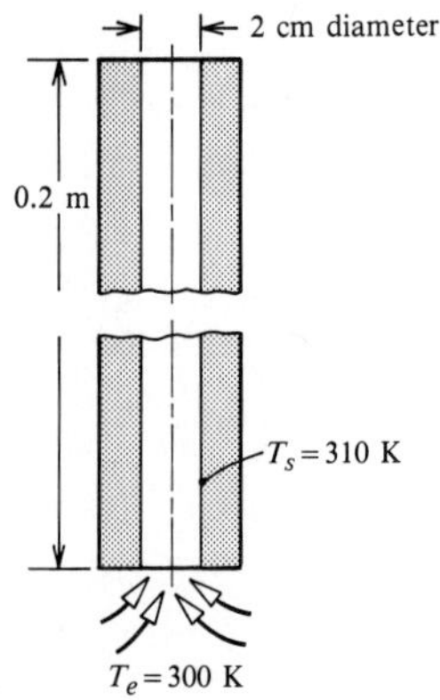

(i) Water

(ii) SAE 50 oil

(iii) Therminol 60

(iv) Air

(v) Mercury

4–70. Water flows through a tube bank 10 rows deep, with 20 mm–O.D. tubes in an aligned arrangement of transverse pitch 30 mm and longitudinal pitch 25 mm. In a performance test the superficial velocity is 0.8 m/s, the water inlet and outlet temperatures are 324 and 336 K, respectively, and the tube outer wall temperature is approximately 453 K. Estimate the average heat transfer coefficient for the tube bank and the pressure drop across the bank.

4–71. An air heater consists of a staggered tube bank in which waste hot water flows inside the tubes, with air flow through the bank perpendicular to the tubes. There are 30 rows of 15 mm–O.D. tubes, with transverse and longitudinal pitches of 28 and 32 mm, respectively. The air is at 1 atm and flows at 5.36 kg/s in a duct of 1.0 m square cross section. Preliminary design calculations for this heat exchanger suggest average tube surface and bulk air temperatures of approximately 350 K and 310 K, respectively. Estimate the average heat transfer coefficient and pressure drop across the bank.

4–72. Air at 280 K and 1 atm pressure flows at 3.0 kg/s along a duct of cross section 1 m $\times$ 0.6 m. The duct contains a tube bank of 1 m–long, 10 mm–O.D. tubes in a staggered arrangement, with longitudinal pitch 24 mm and transverse pitch 20 mm. Steam at 1.29×10^5 Pa condenses inside the tubes. How many rows of tubes are required to heat the air to 360 K, and what is the pressure drop? Make reasonable assumptions.

4–73. A tube bank is 30 rows deep and has 20 mm–O.D. tubes in a staggered arrangement. Steam condenses in the tubes at 1.6×10^5 Pa, and air enters the bank at 4 m/s. Investigate the effect of transverse and longitudinal pitch on the average heat transfer coefficient and pressure drop. Take 25 mm $\leq S_T, S_L \leq$ 50 mm, and a bulk air temperature of 330 K. Prepare graphs with S_T as abscissa and S_L as a parameter.

4–74. Water enters a 20-row-deep tube bank at 1.5 m/s. The tubes have a 25 mm outside diameter and are in an aligned arrangement. Investigate the effect of transverse and longitudinal pitch on the average heat transfer coefficient for $30 \leq S_T, S_L \leq 60$ mm. Take a tube wall temperature of 350 K and a bulk water temperature of 300 K. Prepare graphs with S_T as abscissa and S_L as a parameter.

4–75. Helium at 250 K and 1 atm flows at 0.3 kg/s along a duct of cross section 1 m $\times$ 0.5 m. The duct contains a tube bank of 1 m–long, 15 mm–O.D. tubes in a staggered arrangement, with both transverse and longitudinal pitches of 30 mm. If steam is condensed in the tubes at 1.1×10^5 Pa, how many rows of tubes are required to heat the gas to 350 K? Also estimate the pressure drop.

4–76. A tube bank is 25 rows deep and has 16 mm–O.D., 1 mm–wall thickness brass tubes in a staggered arrangement, with a transverse pitch of 26 mm and a longitudinal pitch of 17 mm. Steam at 1.7×10^5 Pa condenses inside the tubes, and SAE 50 oil at 300 K enters the bank with a velocity of 1.23 m/s. If the steam-side heat transfer coefficient is estimated to be 8000 W/m^2 K, calculate the outlet oil temperature and the pressure drop.

4–77. Surplus 9 mm steel ball bearings are packed with a void fraction of 0.38 in a regenerative air heater. The bed cross-sectional area is 0.1 m^2 and the air flow is 0.05 kg/s. At a location along the bed where the balls are at 1500 K and the air is at 1000 K and 1 atm, determine the pressure gradient and heat transfer coefficient. Also calculate the specific area and transfer perimeter of the bed.

4–78. A perforated-plate heat exchanger for a hydrogen refrigeration system has flow cross-sectional areas of 0.002 m^2 for each 0.005 kg/s stream. The holes are 0.9 mm in diameter in square arrays with an open-area ratio of 0.35. The plates are 0.5 mm thick and are spaced 0.8 mm apart. At a location where the pressure is 0.8 atm and temperature is 200 K, determine the pressure drop per plate and heat transfer coefficient.

4–79. A regenerative air heater is to have a cross-sectional area of 0.2 m^2 and an air flow of 0.1 kg/s. Surplus steel ball bearings are to be used for the packing. A special feature of the design is that the bed must be relatively thin and a relatively large pressure drop can be accommodated. Thus, the use of small-diameter balls is indicated. Investigate the effect of ball diameter ($0.5 < d_p < 10$ mm) and volume void fraction ($0.3 < \varepsilon_v < 0.5$) on the heat transfer coefficient, the heat transfer coefficient times specific area product, and pressure gradient. Evaluate properties at 1000 K and 110 kPa.

4–80. A regenerative heater has a cross-sectional area of 0.1 m^2 and is packed with pebbles, which can be approximated as 1 cm–diameter spheres. The void fraction is 0.41. Helium flows at 0.11 kg/s through the bed. At a location where the pebbles are at 1000 K and the helium is at 600 K and 1 atm, determine the heat transfer coefficient and the pressure gradient. Also calculate the specific area and transfer perimeter of the bed.

4–81. A two-stream helium-to-helium heat exchanger is an essential component of a cryogenic refrigeration system used to cool superconducting magnets on a magnetically levitated train. A perforated-plate packing is under consideration in the design process, and the effect of hole diameter d and open-area ratio ε_p are to be investigated. The plates are 0.5 mm thick and are spaced 1.0 mm apart. The holes are arranged in square arrays. The helium flow rate and cross-sectional area for each stream are 0.018 kg/s and 0.005 m^2, respectively. Prepare graphs of the heat transfer coefficient, the $h_c\mathscr{P}$ product, and pressure drop per plate, with hole diameter as abscissa and open-area ratio as a parameter. Evaluate the helium properties at 170 K, 1 atm.

4–82. A pebble bed is used to preheat 5 kg/s of air for a blast furnace. The bed is 2 m in diameter, is 0.5 m long, and is packed with pebbles that can be approximated as 2.2 cm–diameter spheres with a void fraction of 0.43. Calculate the heat transfer coefficient and pressure gradient at a location where the pebbles are at 1400 K and the air is at 1200 K and 1 atm.

4–83. An experimental high-temperature gas nuclear reactor is 8 cm in diameter and 24 cm long. It contains 8 mm–diameter spherical uranium oxide pellets, each with a 0.8 mm–thick graphite sheath, packed with a volume void fraction of

0.41. Helium enters the bed at 380 K, 1 atm with a velocity of 20 m/s. Calculate the maximum pellet surface temperature if the fission heat release is 3.6×10^7 W/m^3 within the uranium oxide fuel.

4–84. A rotary chemical reactor is in the form of a cylinder 5 m long and 1 m diameter, which rotates around a horizontal axis at 10 revolutions per minute. If the outer surface is measured to be at 380 K when the ambient air temperature is 300 K, estimate the convective heat loss from the cylindrical surface.

4–85. Estimate the heat loss from one side of a 1 m–diameter disk at 330 K rotating in still air at 300 K and 1 atm, if the rate of rotation is

(i) 100 rpm.
(ii) 10,000 rpm.

4–86. In a material processing experiment on a space station, a 1 cm–diameter alloy sphere spins at 3000 rpm in a nitrogen-filled enclosure at 800 K and 1 atm. The sphere is maintained at 1000 K by focusing a beam of infrared radiation on the sphere. At what rate must radiant energy be absorbed by the sphere at steady state? The emittance of the sphere is 0.15.

4–87. To estimate the effect of surface roughness on the skin drag of a high-speed underwater vehicle, model the vehicle as a flat plate 2.5 m long traveling at 70 km/h. What is the percentage increase in skin drag over the smooth-wall value if the equivalent sand grain roughness of the surface is 0.25 mm? Assume a sea temperature of 300 K.

4–88. An 8 cm–I.D., 2000 m–long pipeline with an equivalent sand grain roughness of 0.1 mm contains hot water at 350 K flowing from an elevation of 20 m to one of 60 m above sea level. The pump is 80% efficient and consumes 45 kW of power. What is the water flow rate and the pressure difference across the pump?

4–89. Air at 1 atm flows in a 4 cm–I.D. tube at 40 m/s. Investigate the effect of transverse square ribs on the heat transfer coefficient and pressure drop. Let the rib height vary from 0.2 mm to 2 mm for pitch/height ratios of 10, 15, and 20. Prepare graphs with rib height as abscissa and pitch/height as a parameter. Evaluate the air properties at 300 K.

4–90. Helium at 4×10^5 Pa pressure flows in a 3 cm–I.D. tube at 50 m/s. The tube is artificially roughened with transverse square ribs 0.6 mm high. Investigate the effect of pitch/height ratio on pressure gradient for $2 < p/h < 20$. Evaluate the helium properties at 400 K.

4–91. Water flows at 3 m/s in a badly fouled heat exchanger tube of 5 cm inside diameter. The deposits on the tube surface do not have a simple pattern, but the protrusions do have a mean height of 2.5 mm. Pressure drop measurements indicate an equivalent sand grain roughness of 3.2 mm. Estimate the pressure gradient and heat transfer coefficient at a location where the tube wall is at 360 K and the bulk water temperature is 400 K.

4–92. Air at 1 atm pressure flows at 40 m/s over a 1 m–long flat plate that has a sand grain indentation roughness pattern. The air is at 300 K, and the plate is maintained at 400 K. Investigate the effect of equivalent sand grain roughness size on the drag force and average heat transfer coefficient for 1 mm $< k_s <$ 5 mm.

4–93. Water flows under pressure at 6 m/s in a 4 cm–I.D. tube that is artificially roughened with transverse ribs of height 0.4 mm and pitch 4 mm. At a location where the wall and bulk fluid temperatures are 700 K and 300 K, respectively, calculate the equivalent sand grain roughness k_s, and k_s^+, f, τ_s, Nu, h_c, and q_s. Evaluate all properties at the mean film temperature.

CHAPTER

5

CONVECTION ANALYSIS

CONTENTS

5.1 INTRODUCTION

The analysis of convection is a challenging task. The general equations governing convection are formidable, far more so than the general heat conduction equation. Exact analytical solutions can be obtained for the simplest flows only—for example, *Couette flow* and fully developed laminar flow in a tube—and such flows exhibit few of the essential features of convection. Approximate analytical methods, such as the *integral method* for boundary layers, work well for simple flows but become awkward and inaccurate when extended to more complicated flows. Series expansion methods have met with limited success. The governing equations for convection are a set of partial differential equations describing conservation of mass, momentum, and energy. The major mathematical difficulty is that the momentum equation is *nonlinear*, and exact solution of such equations generally requires numerical methods. In some cases (e.g., *self-similar* boundary layers), the partial differential equations can be reduced to ordinary differential equations, which can be solved by rather simple numerical techniques, such as Runge-Kutta integration, or by iteration. In general, however, the partial differential equations must be solved using finite-difference or finite-element methods.

For laminar flows, the basic physics is simple, involving only conservation principles, Newton's law of viscosity, and Fourier's law of heat conduction. However, the flows of engineering concern are more often turbulent. Understanding the nature of turbulent flows remains one of the unsolved problems of physics. Only recently, with the aid of supercomputers, has there been significant progress at a basic level. Thus, out of necessity, engineers have relied on simple empirical models to describe turbulent transport of momentum and energy. Development of these models has relied heavily on experimental data, in particular, measured velocity and temperature profiles. An appreciation of the pertinent experimental data is essential. What engineers call turbulent flow theory is a partial theory only; but in science and engineering, it is possibly unique in its careful blending of theory and empiricism.

The difficulties just described dictate the organization of Chapter 5. In Sections 5.2 and 5.3, two simple analytical solutions are presented. In both cases, the governing equations are derived from first principles. The analysis of high-speed Couette flow in Section 5.2 leads to the *recovery factor* concept, which allows generalization of the flat plate heat transfer correlations in Chapter 4 to include high-speed flow. In Section 5.4, the equations governing a forced-convection laminar boundary layer on a flat plate are derived, again from first principles. An exact analytical solution is obtained for the limit of zero Prandtl number, which is useful for liquid metals. The *integral method* is used to obtain approximate analytical solutions for both forced- and natural-convection boundary layers. The forced-convection laminar boundary layer is also solved exactly as a self-similar flow, for which the mathematics involved are more demanding.

Sections 5.2 through 5.4 can be regarded as an introduction to convection analysis. The remaining sections contain more advanced material. Turbulent flow is dealt with in Section 5.5. Empirical turbulent transport models are developed and are then used for the analysis of a turbulent boundary layer on a flat plate and of turbulent flow in a tube. Section 5.6 introduces the concepts of similarity and modeling for convection, using the flows analyzed in Sections 5.2 through 5.4 as examples. The general conservation equations governing mass, momentum, and energy are presented in Section 5.7. Cartesian coordinates are used, but the results are also written in vector form for conciseness and generality. Finally, in Section 5.8, the methods of *scale analysis* are demonstrated. The boundary layer equations are derived from the general conservation equations, and the natural-convection boundary layer equations are scaled to obtain the functional dependence of the Nusselt number.

5.2 HIGH-SPEED FLOWS

The work required to sustain the motion of a real fluid is converted irreversibly into heat by the action of viscosity, a process we call *viscous dissipation.* If the fluid is very viscous, significant heat can be produced at relatively low speeds, for example, in the extrusion of a plastic or in an oil-lubricated journal bearing. However, for a less viscous fluid such as air, significant heat is produced only at very high speeds. For flight through the atmosphere, *aerodynamic heating* becomes significant above a Mach number of about 3. Thus, an aluminum skin is adequate for a supersonic airliner such as the Concorde, whereas nickel or titanium alloys are used for hypersonic vehicles. In a compressible fluid such as air, heating by compression is also important in the hypersonic flow regime. The intense heating experienced by a blunt-nosed space vehicle upon reentry into the atmosphere is due primarily to the compression of the oncoming air by the bow shock wave.

5.2.1 A Couette Flow Model

Considerable insight into the phenomenon of aerodynamic heating can be obtained through analysis of a simple **Couette flow.** Figure 5.1 shows a fluid confined between two infinite parallel plates spaced L apart. The lower plate is stationary, and the upper plate moves with a uniform velocity u_e. There is no pressure gradient in the x direction. The upper plate may be imagined to be a conveyor belt, or it could be the outer of two concentric cylinders whose diameters are large compared to the gap between them. If the lower plate is maintained at a temperature T_s different from

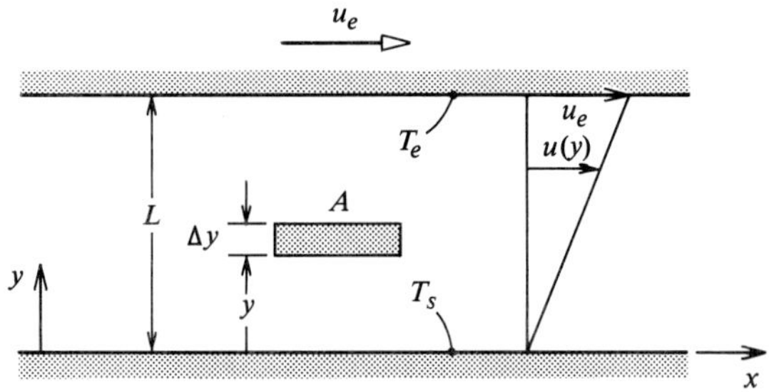

Figure 5.1 Schematic of simple Couette flow.

that of the upper plate, T_e, there will be heat conduction across the fluid layer. Also, if the upper plate velocity u_e is sufficiently large, there will be a significant heat generation in the fluid as a consequence of viscous dissipation. For simplicity, we will assume laminar flow and constant fluid properties: in this problem, the relevant properties are the viscosity μ and conductivity k.

The procedure we follow is common to all analyses of forced convection in this chapter. The first step is to solve the fluid mechanics problem. The momentum conservation principle is used to derive the differential equation governing the velocity field. Solution of this equation subject to the given boundary conditions shows that the velocity profile $u(y)$ has a simple linear form. The next step is to solve the heat transfer problem. The energy conservation principle is used to derive the differential equation governing the temperature field. Solution of this equation subject to the given boundary conditions gives the temperature profile $T(y)$. Finally, Fourier's law is used to obtain the heat flux at the plate surface.

The Fluid Mechanics Problem

At steady state, the only forces acting in the fluid are shear forces due to the action of viscosity. According to Newton's law of viscosity, the shear stress on a plane at location y in a laminar flow is

$$\tau = -\mu \frac{du}{dy} \tag{5.1}$$

where μ is the *dynamic viscosity* with units [kg/m s], and our sign convention is such that the stress is applied on the fluid of greater y from below.[1] Since du/dy is positive, Eq. (5.1) states that τ is in the negative x direction, which is appropriate since the fluid below exerts a drag on the fluid of greater y. Of course, there is an equal and opposite shear stress exerted by the fluid above on the fluid of lesser y. The forces on an element of fluid $A\Delta y$ located between planes y and $y + \Delta y$ are shown in Fig. 5.2, in terms of both the shear stress τ and the velocity gradient du/dy.

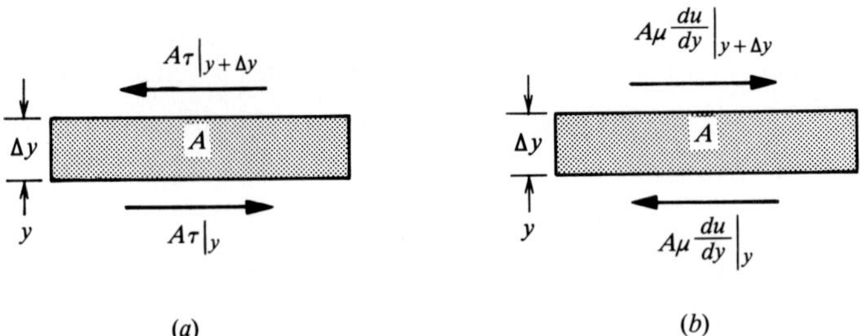

Figure 5.2 Force balance on a fluid element in a Couette flow. (*a*) In terms of the shear stress τ. (*b*) In terms of the velocity gradient du/dy.

[1] The opposite sign convention is used in many texts.

The momentum conservation principle reduces to a simple force balance:

$$A\mu \left.\frac{du}{dy}\right|_{y+\Delta y} - A\mu \left.\frac{du}{dy}\right|_{y} = \text{Rate of change of momentum} = 0$$

Hence,

$$\mu \left.\frac{du}{dy}\right|_{y+\Delta y} = \mu \left.\frac{du}{dy}\right|_{y} = \text{Constant} = C_1$$

or

$$\frac{du}{dy} = \frac{C_1}{\mu}$$

When the viscosity is assumed constant, integration yields

$$u = \frac{C_1}{\mu} y + C_2 \tag{5.2}$$

The constants are evaluated using the real fluid no-slip condition on each plate:

$$y = 0: \quad u = 0 \tag{5.3a}$$

$$y = L: \quad u = u_e \tag{5.3b}$$

to obtain

$$u = \frac{u_e}{L} y \tag{5.4}$$

The Couette flow velocity profile $u(y)$ is seen to have a simple linear form.

The Heat Transfer Problem

To obtain the differential equation governing the temperature profile in the fluid, we apply the first law of thermodynamics to an elemental control volume located between the planes y and $y + \Delta y$, as shown in Fig. 5.3. The other dimensions of the control volume will be taken as Δx and Δz, although, since there are no changes in temperature or velocity in the x and z directions, these dimensions could be taken as finite. The control volume is an open system, so that the steady-flow energy equation, Eq. (1.3), applies. Fluid flows in the left face of the volume at a

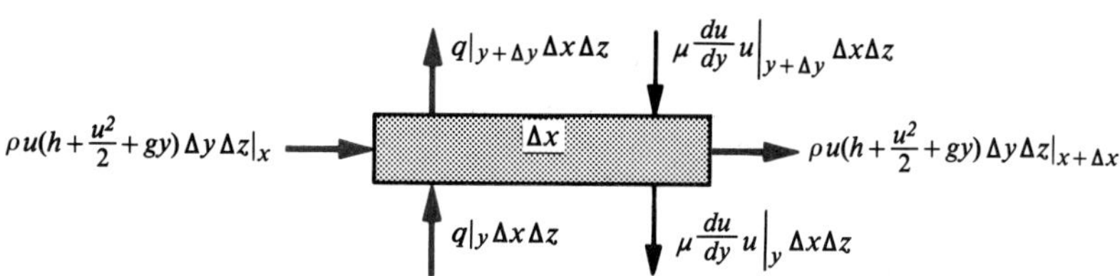

Figure 5.3 Application of the first law of thermodynamics to an elemental control volume in a Couette flow.

rate $\dot{m} = \rho u \Delta y \Delta z$ and out the right face at the same rate. Equation (1.3) becomes

$$\rho u \Delta y \Delta z \left[\left(h + \frac{u^2}{2} + gy \right)_{x+\Delta x} - \left(h + \frac{u^2}{2} + gy \right)_x \right] = 0 = \dot{Q} + \dot{W}$$

since T (and hence enthalpy h) and velocity u are functions of y only. The contributions to $\dot{Q}$ are the rates at which heat is transferred by conduction.

Into the system at y: $\dot{Q}|_y = q|_y \Delta x \Delta z$

Out of the system at $y + \Delta y$: $\dot{Q}|_{y+\Delta y} = q|_{y+\Delta y} \Delta x \Delta z$

The contributions to $\dot{W}$ are the rates at which work is done due to shear forces.

By the system at y: $\dot{W}|_y = \mu \frac{du}{dy} u \Big|_y \Delta x \Delta z$

On the system at $y + \Delta y$: $\dot{W}|_{y+\Delta y} = \mu \frac{du}{dy} u \Big|_{y+\Delta y} \Delta x \Delta z$

Substituting into $\dot{Q} + \dot{W} = 0$ and canceling the area $\Delta x \Delta z$ gives

$$(q|_y - q|_{y+\Delta y}) + \left(\mu \frac{du}{dy} u \Big|_{y+\Delta y} - \mu \frac{du}{dy} u \Big|_y \right) = 0$$

(recall that our steady-flow energy equation sign convention is that heat transfer *into* the system, and work done *on* the system, are positive).

Dividing by Δy and taking limits as $\Delta y \to 0$, we obtain

$$-\frac{dq}{dy} + \mu \left(\frac{du}{dy} \right)^2 = 0$$

since du/dy is a constant. Substituting Fourier's law $q = -k\,dT/dy$ then gives

$$k \frac{d^2 T}{dy^2} + \mu \left(\frac{du}{dy} \right)^2 = 0 \tag{5.5}$$

if k is assumed constant. Notice that Eq. (5.5) is just the steady one-dimensional heat conduction equation with a heat source $\mu(du/dy)^2$, the viscous dissipation. The viscous dissipation is always positive, reflecting the fact that mechanical work is converted irreversibly into heat. There is no reverse process whereby heat is converted into work by the action of viscosity.

Equation (5.4) gives $du/dy = u_e/L$; substituting in Eq. (5.5),

$$k \frac{d^2 T}{dy^2} + \frac{\mu u_e^2}{L^2} = 0 \tag{5.6}$$

Equation (5.6) is easily integrated:

$$\frac{d^2 T}{dy^2} = -\frac{\mu u_e^2}{k L^2}$$

$$T = -\frac{\mu u_e^2}{2k} \left(\frac{y}{L} \right)^2 + C_1 y + C_2$$

The integration constants are evaluated from the boundary conditions:

$$y = 0: \quad T = T_s \tag{5.7a}$$

$$y = L: \quad T = T_e \tag{5.7b}$$

to obtain the temperature profile as

$$\frac{T_s - T}{T_s - T_e} = \frac{y}{L}\left[1 - \frac{\mu u_e^2}{2k(T_s - T_e)}\left(1 - \frac{y}{L}\right)\right] \tag{5.8}$$

The heat flux at the lower plate can now be found using Fourier's law:

$$q_s = -k\left.\frac{dT}{dy}\right|_0 = \frac{k(T_s - T_e)}{L} - \frac{\mu u_e^2}{2L} \tag{5.9}$$

The first term is simply the conduction heat flux across a slab of thickness L; the second term gives the contribution of viscous dissipation to the heat flux.

Temperature Profiles

To examine the temperature profiles given by Eq. (5.8), consider the situation where the stationary plate temperature, T_s, is higher than that of the moving plate, T_e. Figure 5.4 shows how the temperature profiles depend on the dimensionless parameter $\mu u_e^2/k(T_s - T_e)$, which is the *Brinkman number*, introduced in Section 4.2.2. The effect of increasing the plate speed is to increase the Brinkman number and the magnitude of the viscous dissipation heat source, as evidenced by the change in shape of the temperature profile. For Br $\rightarrow 0$, the viscous heating is negligible, and the temperature profile is linear, as for simple conduction across a slab. The heat transfer is away from the hot plate. For Br $= 2$, the heat transfer from the hot plate is zero, and for Br > 2, heat is transferred from the fluid to the hot plate.

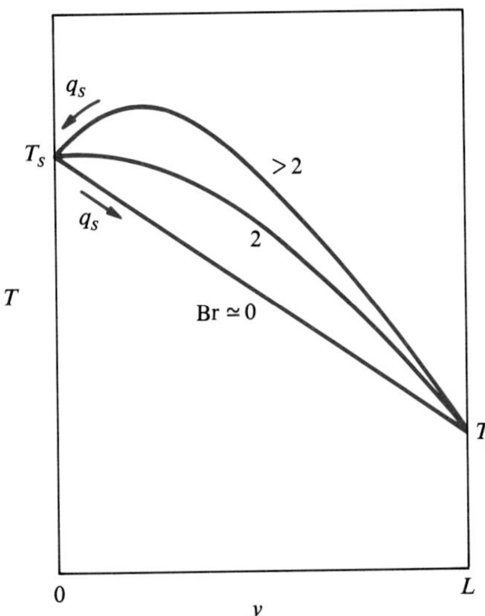

Figure 5.4 Temperature profiles for a Couette flow with $T_s > T_e$, showing the effect of Brinkman number.

5.2.2 The Recovery Factor Concept

Equation (5.9) for the wall heat flux can be rearranged as

$$q_s = \frac{k}{L}\left[T_s - \left(T_e + \mathrm{Pr}\frac{u_e^2}{2c_p}\right)\right] \tag{5.10}$$

With a little imagination, we can see how this result can be applied to a real boundary layer. Equation (5.10) shows that the driving potential for heat transfer when viscous heating is present is not $(T_s - T_e)$ but $T_s - [T_e + \mathrm{Pr}(u_e^2/2c_p)]$. If we define the *adiabatic wall temperature for a Couette flow* as $T_{aw} = T_e + \mathrm{Pr}(u_e^2/2c_p)$, then the effect of viscous dissipation on heat transfer can be accounted for by simply replacing T_e with T_{aw}. The adiabatic wall temperature is also the temperature the wall attains if $q_s = 0$, that is, is adiabatic. Thus, for real boundary layer flows with significant viscous heating, we simply replace Newton's law of cooling $q_s = h_c(T_s - T_e)$ with

$$q_s = h_c(T_s - T_{aw}) \tag{5.11}$$

and define a **recovery factor** r such that

$$T_{aw} = T_e + r\frac{u_e^2}{2c_p} \tag{5.12}$$

For the Couette flow, $r = \mathrm{Pr}$. More advanced analysis shows that the recovery factor for a laminar boundary layer on a flat plate is $r \simeq \mathrm{Pr}^{1/2}$, while experiment shows that for turbulent boundary layers $r \simeq \mathrm{Pr}^{1/3}$. If Eq. (5.12) is rewritten as

$$c_pT_{aw} = c_pT_e + r\frac{u_e^2}{2}$$

the recovery factor can be viewed as the fraction of kinetic energy of the free stream fluid, which is *recovered* as thermal energy in the fluid adjacent to an adiabatic wall. If the Prandtl number, and hence r, are less than unity, this concept seems reasonable. However, for most liquids, the Prandtl number, and hence r, are greater than unity, so the concept is too simplistic. Most high-speed flows of interest involve gases, particularly air, which explains the appeal of the recovery concept.

For engineering calculations of heat transfer in high-speed boundary layers, the formulas given in Section 4.3.2 for external flows are applicable if the free-stream temperature is replaced by the adiabatic wall temperature T_{aw}, with the appropriate recovery factor, namely, $\mathrm{Pr}^{1/2}$ for laminar flows and $\mathrm{Pr}^{1/3}$ for turbulent flows. Evaluation of fluid properties poses a special problem. For gases, fluid properties can be evaluated at the *Eckert reference temperature* [1]

$$T_r = T_e + 0.5(T_s - T_e) + 0.22(T_{aw} - T_e) \tag{5.13}$$

which is valid for both laminar and turbulent flows. Note also that most oils have viscosities that vary strongly with temperature; thus, more careful analysis allowing for a temperature-dependent viscosity may be needed for oils when temperature differences are large.

The student is cautioned that the recovery factors given here are, strictly speaking, accurate only for gas flows over flat, impermeable surfaces [2].

EXAMPLE 5.1 Temperature of a Helicopter Rotor

The main rotor of a helicopter has a radius of 5 m, has a blade width of 20 cm, and rotates at 380 rpm. Estimate the maximum temperature the rotor can attain if the forward speed is 90 m/s through air at 300 K.

Solution

Given: Helicopter flying at 90 m/s, rotor rotating at 380 rpm.

Required: An estimate of the maximum rotor temperature.

Assumptions: 1. Model the rotor as a flat plate.
2. The radiation heat loss and conduction within the rotor are negligible.

The rotor temperature varies with position and time in a complicated manner due to aerodynamic heating. The maximum temperature will occur at the rotor tip, where the surface energy balance is given by Eq. (1.34) as

$$q_{\text{cond}} - q_{\text{conv}} - q_{\text{rad}} = 0$$

An upper bound on the rotor surface temperature can be estimated by neglecting the radiation and conduction losses; then

$$q_{\text{conv}} = h_c(T_s - T_{\text{aw}}) = 0 \qquad \text{or} \qquad T_s = T_{\text{aw}}$$

From Eq. (5.12),

$$T_{\text{aw}} = T_e + r\frac{u_e^2}{2c_p}$$

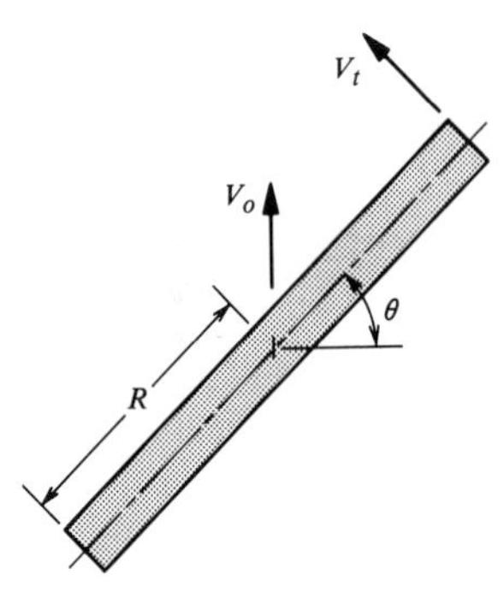

The tangential velocity component at the tip of the rotor is

$$V_t = R\omega = (5)(2\pi)(380/60) = 199.0 \text{ m/s}$$

The maximum speed is attained by the tip when $\theta = 0$ and is

$$u_e = V_0 + V_t = 90 + 199 = 289 \text{ m/s}$$

To evaluate the recovery factor r, we must check to see if the boundary layer is laminar or turbulent:

$$\text{Re} \simeq \frac{u_e L}{\nu} = \frac{(289)(0.20)}{(15.66 \times 10^{-6})} = 3.69 \times 10^6$$

That is, the boundary layer is turbulent over almost the whole blade in the tip region. Thus,

$$r \simeq \text{Pr}^{1/3} = (0.69)^{1/3} = 0.884$$

$$T_s = T_{\text{aw}} = 300 + \frac{(0.884)(289)^2}{(2)(1005)} = 300 + 36.7 = 336.7 \text{ K}$$

Comments

1. To check if radiation losses are negligible, we estimate the convective heat transfer coeffcient. Assuming that a turbulent boundary layer starts at the leading edge, Eq. (4.65) gives

$$\overline{\mathrm{Nu}} = 0.036\mathrm{Re}_L^{0.8}\mathrm{Pr}^{0.43} = 0.036(3.69 \times 10^6)^{0.8}(0.69)^{0.43} = 5500$$

$$\overline{h}_c = (k/L)\overline{\mathrm{Nu}} = (0.0267/0.20)(5500) = 734 \text{ W/m}^2\text{ K}$$

where properties for air at 300 K have been used. The surface energy balance for convection-radiation equilibrium is

$$q_{\text{conv}} + q_{\text{rad}} = 0$$

Since $(T_s - T_e)$ is small, q_{rad} can be linearized as $4\sigma\varepsilon T_m{}^3(T_s - T_e) \simeq 7(T_s - T_e)$ for $T_m = 315$ K, $\varepsilon = 1$.

$$\overline{h}_c(T_s - T_{\text{aw}}) + 7(T_s - T_e) = 0$$

$$734(T_s - 336.7) + 7(T_s - 300) = 0$$

$$T_s = 336.4 \text{ K}$$

which is called the *convection-radiation equilibrium temperature*. The effect of radiation is negligible in this situation.

2. The conduction losses are more difficult to estimate since the problem is both unsteady and two-dimensional.

5.3 LAMINAR FLOW IN A TUBE

Figure 5.5 shows a steady fluid flow in a tube sufficiently far downstream for entrance effects to be negligible. The flow is laminar, which usually requires a Reynolds number $u_b D/\nu < 2300$. Constant-property fully developed laminar flow in a tube is a simple flow, for which we can obtain the heat transfer coefficient by exact analysis. For this forced-convection problem, we follow the general procedure introduced in Section 5.2. The fluid mechanics problem is first solved to obtain the characteristic parabolic velocity profile $u(r)$. The student will no doubt already have encountered this flow in a physics or fluid mechanics course, where it was called *Poiseuille flow*.

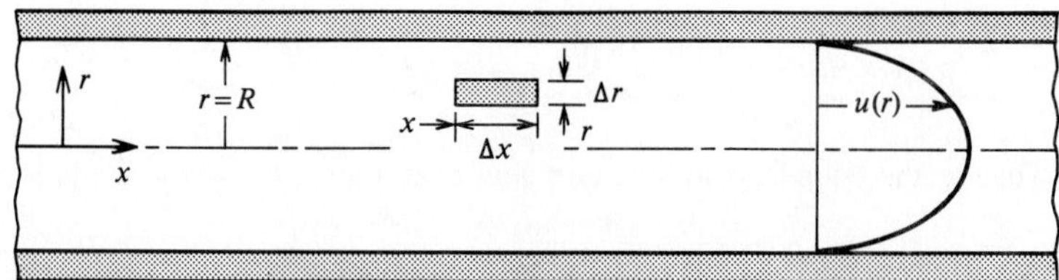

Figure 5.5 Schematic of fully developed laminar flow in a tube, showing the coordinate system and characteristic parabolic velocity profile.

The energy conservation principle is then used to derive the differential equation governing the temperature field. For uniform heating of the tube wall, this equation is solved for the temperature field $T(x,r)$ and the heat transfer coefficient and Nusselt number so obtained.

5.3.1 Momentum Transfer in Hydrodynamically Fully Developed Flow

At a long distance from the tube entrance, the flow "forgets" the precise nature of its initial velocity profile, and, irrespective of the entrance configuration (elbow, flare, etc.), there will be a characteristic axial velocity profile $u(r)$ unchanging along the tube (if the fluid properties remain constant). Correspondingly, there will be no velocity component in the radial direction. Also, the axial pressure gradient required to sustain the flow against the viscous forces will be constant along the tube. The flow is then said to be **hydrodynamically fully developed.**

The Governing Differential Equation

Since velocities are unchanging with time and unchanging along the tube, the momentum conservation principle reduces to a simple force balance. Accordingly, a force balance is made on an elemental fluid element, Δx long, located between radii r and $r + \Delta r$, as shown in Figs. 5.5 and 5.6*a*. The pressure and viscous forces acting

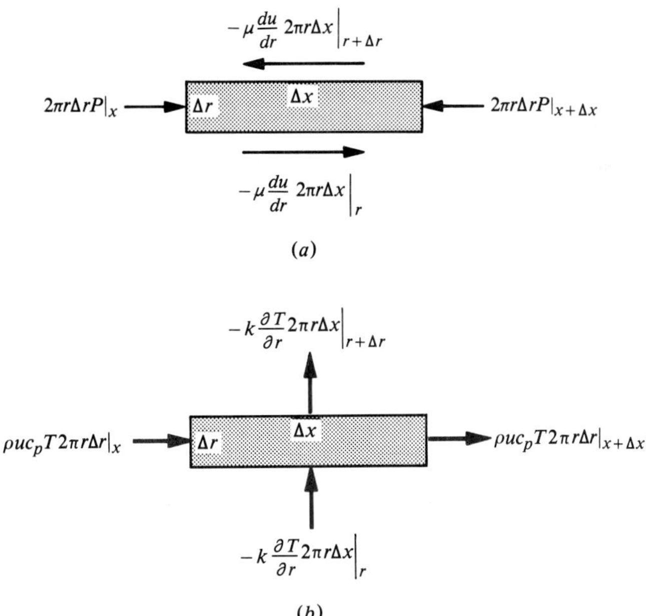

Figure 5.6 (*a*) Force balance on a fluid element for fully developed laminar flow in a tube. (*b*) The first law of thermodynamics applied to an elemental control volume for fully developed heat transfer in a tube.

on the element are in balance, and their sum equals zero:

$$[2\pi r \Delta r P]_x + \left[2\pi r \Delta x\left(-\mu\frac{du}{dr}\right)\right]_r - [2\pi r \Delta r P]_{x+\Delta x} - \left[2\pi r \Delta x\left(-\mu\frac{du}{dr}\right)\right]_{r+\Delta r} = 0$$

Dividing by $2\pi \Delta r \Delta x$ and rearranging,

$$\frac{r\mu(du/dr)|_{r+\Delta r} - r\mu(du/dr)|_r}{\Delta r} = \frac{rP|_{x+\Delta x} - rP|_x}{\Delta x}$$

Letting $\Delta r, \Delta x \to 0$ then gives

$$\frac{d}{dr}\left(r\mu\frac{du}{dr}\right) = r\frac{dP}{dx} \qquad \textbf{(5.14)}$$

which is the desired differential equation. Since dP/dx is a constant for hydrodynamically fully developed flow, it is a second-order ordinary differential equation for $u(r)$, and the two boundary conditions required are

$$r = 0: \quad \frac{du}{dr} = 0, \qquad \text{by symmetry} \qquad \textbf{(5.15a)}$$

$$r = R: \quad u = 0, \qquad \text{no slip at the wall} \qquad \textbf{(5.15b)}$$

Solution for the Velocity Profile

Integrating Eq. (5.14) once gives

$$r\mu\frac{du}{dr} = \frac{1}{2}r^2\frac{dP}{dx} + C_1$$

or

$$\frac{du}{dr} = \frac{r}{2\mu}\frac{dP}{dx} + \frac{C_1}{r\mu}$$

Using the first boundary condition, Eq. (5.15*a*), $C_1 = 0$. Integrating again,

$$u = \frac{r^2}{4\mu}\frac{dP}{dx} + C_2$$

The second boundary condition, Eq. (5.15*b*), gives C_2:

$$0 = \frac{R^2}{4\mu}\frac{dP}{dx} + C_2$$

and substituting back gives the velocity profile $u(r)$:

$$u = \frac{R^2}{4\mu}\left(-\frac{dP}{dx}\right)\left[1 - \left(\frac{r}{R}\right)^2\right] \qquad \textbf{(5.16)}$$

The velocity profile is seen to be parabolic in shape. The maximum velocity is on the centerline, where $r = 0$, and is

$$u_{\max} = \frac{R^2}{4\mu}\left(-\frac{dP}{dx}\right) \tag{5.17}$$

The bulk velocity u_b was introduced in Section 4.2.2 and is simply the area-weighted average velocity,

$$u_b = \frac{1}{\pi R^2}\int_0^R u2\pi r\,dr$$

Substituting from Eq. (5.16) for u and using Eq. (5.17) gives

$$u_b = \frac{1}{2}u_{\max} = \frac{R^2}{8\mu}\left(-\frac{dP}{dx}\right) \tag{5.18}$$

The velocity profile $u(r)$ can then be conveniently expressed in terms of the bulk velocity as

$$u = 2u_b\left[1 - \left(\frac{r}{R}\right)^2\right] \tag{5.19}$$

In practice, the mass flow rate $\dot{m}$ [kg/s] is usually known; then the bulk velocity is obtained from the relation $\dot{m} = \rho u_b A_c$.

The Friction Factor and Skin Friction Coefficient

The Darcy friction factor for flow in a tube was defined by Eq. (4.15); writing $\Delta P/L$ as $-dP/dx$ and $D = 2R$ gives

$$f = \frac{(-dP/dx)(2R)}{(1/2)\rho u_b^2}$$

Substituting for dP/dx from Eq. (5.18),

$$f = \frac{64\mu}{(2R)\rho u_b} = \frac{64}{\mathrm{Re}_D} \tag{5.20}$$

where the Reynolds number $\mathrm{Re}_D = (2R)\rho u_b/\mu$; this result is the same as Eq. (4.39). The shear stress exerted by the fluid on the wall can be obtained from the velocity profile Eq. (5.19) by using Newton's law of viscosity:

$$\tau_s = -\mu\left.\frac{du}{dr}\right|_{r=R} = -\mu(2u_b)\left(-\frac{2r}{R^2}\right)_{r=R} = \frac{4\mu u_b}{R} \tag{5.21}$$

The skin friction coefficient C_f is then

$$C_f = \frac{\tau_s}{(1/2)\rho u_b^2} = \frac{4\mu u_b/R}{(1/2)\rho u_b^2} = \frac{16\mu}{(2R)\rho u_b} = \frac{16}{\mathrm{Re}_D} = \frac{f}{4} \tag{5.22}$$

Recall that the relation $f = 4C_f$ can also be obtained by a simple force balance (Exercise 4–3).

5.3.2 Fully Developed Heat Transfer for a Uniform Wall Heat Flux

In Section 5.3.1, we saw that at distances sufficiently far from a tube entrance, the velocity profile does not change along the tube and has a characteristic parabolic form; the wall shear stress does not change either. If the tube wall is heated such that the wall heat flux is constant with axial position (e.g., by winding an electrical heating wire around a tube at a constant pitch), then the bulk fluid temperature increases steadily in the flow direction at a rate dependent on the power input to the heater. At a distance sufficiently far from where heating commences, the heat transfer coefficient for a constant-property fluid becomes constant, and the *shape* of the temperature profile also does not change along the tube: the flow is then said to be **thermally fully developed.**

The Governing Differential Equation

To derive the differential equation governing the temperature profile, we apply the first law of thermodynamics to an elemental control volume Δx long located between radii r and $r + \Delta r$, as shown in Figure 5.6*b*. We will assume constant fluid properties, low-speed flow so that viscous dissipation can be ignored, and negligible change in potential energy. Since the specific heat is assumed constant, we will write $h = c_p T$ for convenience, which implies an enthalpy datum state of $h = 0$ at $T = 0$. The steady-flow energy equation, Eq. (1.4), applies and requires that the net outflow of enthalpy equal the heat conducted into the volume. The required quantities are as follows:

Rate of enthalpy inflow at x: $\rho u c_p T 2\pi r \Delta r|_x$

Rate of enthalpy outflow at $x + \Delta x$: $\rho u c_p T 2\pi r \Delta r|_{x+\Delta x}$

Rate of heat conduction in at r: $-k\dfrac{\partial T}{\partial r}2\pi r \Delta x\Big|_r$

Rate of heat conduction out at $r + \Delta r$: $-k\dfrac{\partial T}{\partial r}2\pi r \Delta x\Big|_{r+\Delta r}$

Heat conduction in the x direction is neglected, as it is expected to be small compared to the enthalpy flow. We will return to this point after completing the analysis. Equation (1.4), $\dot{m}\Delta h = \dot{Q}$, becomes

$$\rho u c_p T 2\pi r \Delta r|_{x+\Delta x} - \rho u c_p T 2\pi r \Delta r|_x = -k\frac{\partial T}{\partial r}2\pi r \Delta x\Big|_r + k\frac{\partial T}{\partial r}2\pi r \Delta x\Big|_{r+\Delta r}$$

Rearranging and dividing by $2\pi \Delta x \Delta r$ gives

$$\frac{\rho u c_p r(T|_{x+\Delta x} - T|_x)}{\Delta x} = \frac{k\,[r(\partial T/\partial r)|_{r+\Delta r} - r(\partial T/\partial r)|_r]}{\Delta r}$$

since $u = u(r)$ and is not a function of x for hydrodynamically fully developed flow; also, the fluid properties ρ, c_p, and k have been assumed constant. Let $\Delta x, \Delta r \to 0$;

then

$$\rho u c_p r \frac{\partial T}{\partial x} = k \frac{\partial}{\partial r}\left(r \frac{\partial T}{\partial r}\right) \tag{5.23}$$

or

$$u \frac{\partial T}{\partial x} = \frac{\alpha}{r} \frac{\partial}{\partial r}\left(r \frac{\partial T}{\partial r}\right) \tag{5.24}$$

where $\alpha = k/\rho c_p$ is the thermal diffusivity of the fluid. Substituting from Eq. (5.19) for the velocity $u(r)$ gives

$$2u_b\left[1 - \left(\frac{r}{R}\right)^2\right] \frac{\partial T}{\partial x} = \frac{\alpha}{r} \frac{\partial}{\partial r}\left(r \frac{\partial T}{\partial r}\right) \tag{5.25}$$

which is a partial differential equation since T is a function of both x and r. Fortunately, we can show that for fully developed heat transfer and uniform heating, $\partial T/\partial x$ is a constant, so that Eq. (5.25) reduces to an ordinary differential equation.

The Axial Temperature Gradient

To determine $\partial T/\partial x$, we apply the steady-flow energy equation to a slice of tube Δx long, as shown in Fig. 5.7.

$$\int_0^R \rho u c_p T 2\pi r \, dr\big|_{x+\Delta x} - \int_0^R \rho u c_p T 2\pi r \, dr\big|_x = q_s 2\pi R \Delta x \tag{5.26}$$

But from the definition of bulk temperature, Eq. (4.7),

$$\int_0^R \rho u c_p T 2\pi r \, dr = \dot{m} c_p T_b = \rho u_b \pi R^2 c_p T_b \tag{5.27}$$

hence,

$$\rho u_b c_p \pi R^2 T_b\big|_{x+\Delta x} - \rho u_b c_p \pi R^2 T_b\big|_x = q_s 2\pi R \Delta x$$

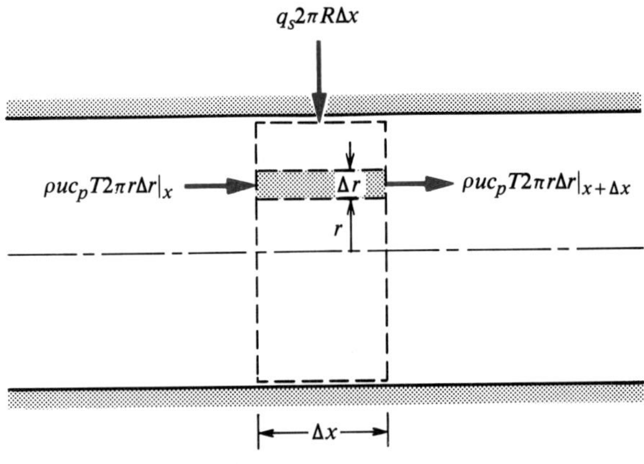

Figure 5.7 The steady-flow energy equation applied to flow in a length of tube Δx long.

Dividing by $\rho u_b c_p \pi R^2 \Delta x$ and letting $\Delta x \to 0$,

$$\frac{dT_b}{dx} = \frac{2q_s}{\rho u_b c_p R} = \text{Constant} \tag{5.28}$$

since q_s is a constant for a uniform wall heat flux. Figure 5.8 shows the expected temperature profiles for fully developed heat transfer. Far enough from the entrance, the *shape* of the profile does not change along the tube. Thus, radial temperature differences are constant; that is, $T_s - T_b$, $T_s - T(r)$, and $T(r) - T_b$ are independent of x. In particular,

$$T - T_b = \mathscr{R}(r) \tag{5.29}$$

a function of r only. Hence,

$$\frac{\partial T}{\partial x} - \frac{dT_b}{dx} = 0$$

or, using Eq. (5.28),

$$\frac{\partial T}{\partial x} = \frac{dT_b}{dx} = \frac{2q_s}{\rho u_b c_p R} \tag{5.30}$$

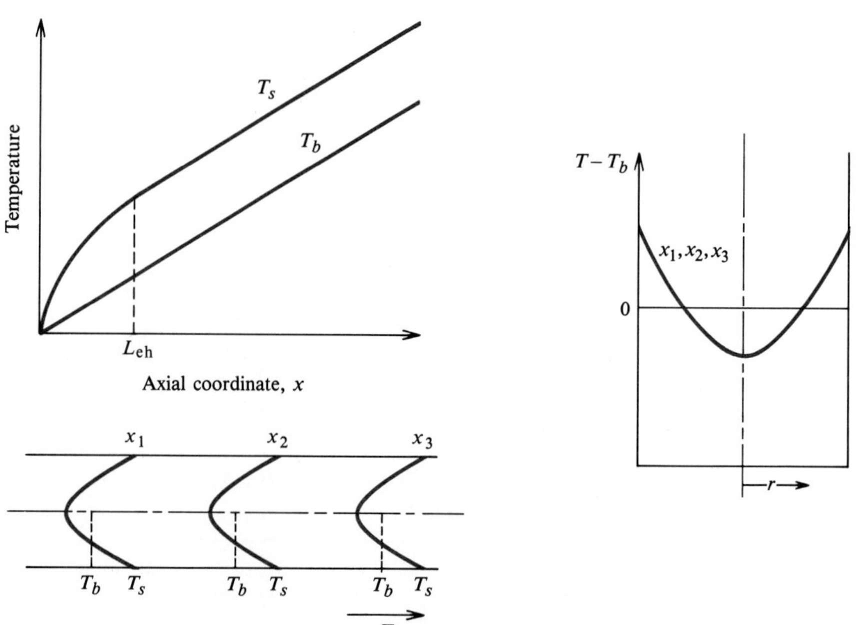

Figure 5.8 Temperature profiles for fully developed heat transfer in a tube with a uniform wall heat flux.

a constant. Also, differentiating Eq. (5.29) with respect to r,

$$\frac{\partial T}{\partial r} - 0 = \frac{d\mathscr{R}(r)}{dr}$$

that is, $\partial T/\partial r$ is not a function of x. By definition, the heat transfer coefficient is

$$h_c = \frac{q_s}{T_s - T_b} = \frac{-k(\partial T/\partial r)|_R}{T_s - T_b} \tag{5.31}$$

and since both the numerator and denominator are constants, so is h_c. Thus, we see that it is the unchanging shape of the temperature profile that gives the characteristic constant heat transfer coefficient for fully developed conditions.

Solution for the Radial Temperature Profile

From Eq. (5.29), we can write $T = T_b + \mathscr{R}(r)$ and substitute in Eq. (5.25) to obtain

$$2u_b\left[1 - \left(\frac{r}{R}\right)^2\right]\frac{dT_b}{dx} = \frac{\alpha}{r}\frac{d}{dr}\left(r\frac{d\mathscr{R}}{dr}\right)$$

where the partial derivatives have become total derivatives because $T_b = T_b(x)$ and $\mathscr{R} = \mathscr{R}(r)$. Then, using Eq. (5.28) and rearranging,

$$\left[1 - \left(\frac{r}{R}\right)^2\right]\frac{4q_s}{kR} = \frac{1}{r}\frac{d}{dr}\left(r\frac{d\mathscr{R}}{dr}\right) \tag{5.32}$$

which is an ordinary differential equation for $\mathscr{R}(r)$. Appropriate boundary conditions are

$$r = 0: \quad \frac{dT}{dr} = \frac{d\mathscr{R}}{dr} = 0 \quad \text{by symmetry} \tag{5.33a}$$

$$r = R: \quad T = T_s \quad \text{or} \quad \mathscr{R} = T_s - T_b \tag{5.33b}$$

Equation (5.33*b*) is used as the boundary condition at the wall since the specified constant heat flux condition has been used already in deriving the differential equation; the fact that T_s is unknown will present no difficulty. Integrating once,

$$r\frac{d\mathscr{R}}{dr} = \frac{4q_s}{kR}\left[\frac{r^2}{2} - \frac{r^4}{4R^2}\right] + C_1$$

$$\frac{d\mathscr{R}}{dr} = \frac{4q_s}{kR}\left[\frac{r}{2} - \frac{r^3}{4R^2}\right] + \frac{C_1}{r}$$

and the first boundary condition, Eq. (5.33*a*), requires that $C_1 = 0$. Integrating again,

$$\mathscr{R} = \frac{4q_s}{kR}\left[\frac{r^2}{4} - \frac{r^4}{16R^2}\right] + C_2$$

Evaluating C_2 from the second boundary condition, Eq. (5.33*b*), and noting that $T = T_b + \mathscr{R}$ from Eq. (5.29) gives

$$T = T_s - \frac{4q_s}{kR}\left[\frac{3R^2}{16} - \frac{r^2}{4} + \frac{r^4}{16R^2}\right] \quad \textbf{(5.34)}$$

a quartic form for the temperature profile $T(r)$. The bulk temperature can now be determined using its definition, Eq. (4.7), in the form

$$T_b = \frac{\int_0^R uT2\pi r\,dr}{\pi R^2 u_b}$$

Substituting from Eqs. (5.19) and (5.34) and simplifying gives

$$T_b = \frac{4}{R^2}\int_0^R \left[1 - \left(\frac{r}{R}\right)^2\right]\left[T_s - \frac{4q_s}{kR}\left(\frac{3R^2}{16} - \frac{r^2}{4} + \frac{r^4}{16R^2}\right)\right] r\,dr$$

$$= T_s - \frac{11}{24}\frac{q_s R}{k}$$

or

$$T_s - T_b = \frac{11}{24}\frac{q_s R}{k} \quad \textbf{(5.35)}$$

Equation (5.35) shows that the difference between the wall and bulk temperatures is directly proportional to the wall heat flux and tube radius and inversely proportional to the fluid conductivity.

The Heat Transfer Coefficient and Nusselt Number

Since the temperature profile is in terms of the wall heat flux q_s, we do not need to use Fourier's law to find q_s and hence the heat transfer coefficient. The heat transfer coefficient is simply $h_c = q_s/(T_s - T_b)$; thus, using Eq. (5.35),

$$h_c = \frac{24k}{11R} = \frac{48k}{11D}, \qquad \text{where } D = 2R$$

and the Nusselt number $\text{Nu}_D = h_c D/k$ is

$$\text{Nu}_D = \frac{48}{11} = 4.364 \quad \textbf{(5.36)}$$

which is the result given in Section 4.3.1 as Eq. (4.41). This result is perhaps surprising. The Nusselt number is not a function of the Reynolds and Prandtl numbers, as might have been expected from the dimensional analysis of Section 4.2.2. However, like the skin friction coefficient, Eq. (5.22), the Stanton number, St = Nu/RePr, is inversely proportional to Reynolds number. The physical reason for why the heat transfer coefficient does not depend on velocity for this flow should become clear after boundary layer flows are analyzed in Section 5.4.

As a final comment, notice that Eq. (5.30) states that $\partial T/\partial x$ is a constant. Hence, the contribution of x-direction conduction to $\dot{Q}$ in the steady-flow energy equation is not just small, as was expected, but is in fact zero for this situation of fully developed heat transfer with a uniform wall heat flux. The rate of heat conduction into the elemental control volume at x exactly equals the rate of heat conduction out at $x + \Delta x$.

EXAMPLE 5.2 An Oil Heater

A special-purpose oil flows at 1.81×10^{-2} kg/s inside a 1 cm–diameter tube that is heated electrically at a rate of 76 W/m. At a particular location where the flow and heat transfer are fully developed, the wall temperature is 370 K. Determine (i) the oil bulk temperature, (ii) the centerline temperature, (iii) the axial gradient in bulk temperature, and (iv) the heat transfer coefficient. For properties, take $k = 0.139$ W/m K, $\rho = 854$ kg/m^3, $c_p = 2120$ J/kg K, and $\nu = 41 \times 10^{-6}$ m^2/s.

Solution

Given: Oil flowing inside a uniformly heated tube.

Required: (i) T_b, (ii) T_c, (iii) dT_b/dx, (iv) h_c.

Assumptions: Fully developed flow and heat transfer.

First check the Reynolds number:

$$u_b = \frac{\dot{m}}{\rho \pi R^2} = \frac{(1.81 \times 10^{-2})}{(854)(\pi)(0.005)^2} = 0.270 \text{ m/s}$$

$$\text{Re} = \frac{u_b D}{\nu} = \frac{(0.270)(0.01)}{(41 \times 10^{-6})} = 66$$

that is, laminar.

(i) From Eq. (5.35),

$$T_b = T_s - \frac{11}{24}\frac{q_s R}{k}$$

To obtain the wall heat flux q_s, the heat input per unit length must be divided by the tube perimeter πD:

$$q_s = \frac{76}{(\pi)(0.01)} = 2419 \text{ W/m}^2$$

Hence,

$$T_b = 370 - \frac{11}{24}\frac{(2419)(0.005)}{(0.139)} = 330.1 \text{ K}$$

(ii) Substituting $r = 0$ in Eq. (5.34),

$$T_c = T_s - \frac{4q_s}{kR}\frac{3R^2}{16} = T_s - \frac{3}{4}\frac{q_s R}{k}$$

$$T_c = 370 - \frac{3}{4}\frac{(2419)(0.005)}{(0.139)} = 304.7 \text{ K}$$

(iii) From Eq. (5.28),

$$\frac{dT_b}{dx} = \frac{2q_s}{\rho u_b c_p R} = \frac{(2)(2419)}{(854)(0.270)(2120)(0.005)} = 1.98 \text{ K/m}$$

(iv) From Eq. (5.36),

$$\text{Nu}_D = \frac{h_c D}{k} = 4.364$$

$$h_c = \frac{k\text{Nu}_D}{D} = \frac{(0.139)(4.364)}{(0.01)} = 60.7 \text{ W/m}^2\text{ K}$$

5.4 LAMINAR BOUNDARY LAYERS

The analysis of laminar-flow boundary layers has played a central role in the development of convection theory. The results of such analysis have practical utility, but, perhaps more importantly, the laminar boundary layer is the simplest flow that exhibits the essential features of convection. There is a price to be paid: the analysis that follows is significantly more difficult than the preceding analyses of Couette and tube flow. Exact solution of the governing differential equations requires numerical rather than analytical methods. Fortunately, the approximate *integral method* can be used to obtain accurate results in many situations. Also, formulation and use of the integral method gives added insight into the physics of the problem. The boundary layer for forced flow along a flat plate will be analyzed first and given considerable attention. Subsequently, the boundary layer for natural convection on a vertical wall will be analyzed.

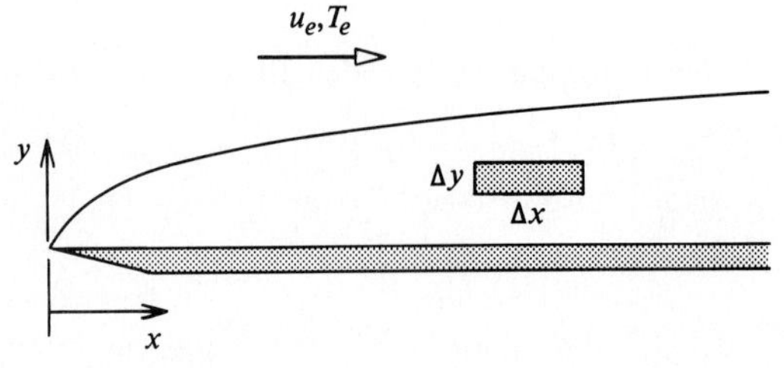

Figure 5.9 Schematic of a laminar boundary layer on a flat plate, showing the coordinate system.

5.4.1 The Governing Equations for Forced Flow along a Flat Plate

Figure 5.9 shows a schematic of the laminar boundary layer for forced flow along a flat plate, with a constant free-stream velocity u_e and constant free-stream temperature T_e. To perform an analysis of momentum and heat transfer, we need to first derive differential equations governing conservation of mass, momentum, and energy. For simplicity, we will assume steady flow, constant fluid properties, and negligible viscous dissipation. Since the density is constant, this is an *incompressible* flow. Our elemental control volume is Δx by Δy by unity.

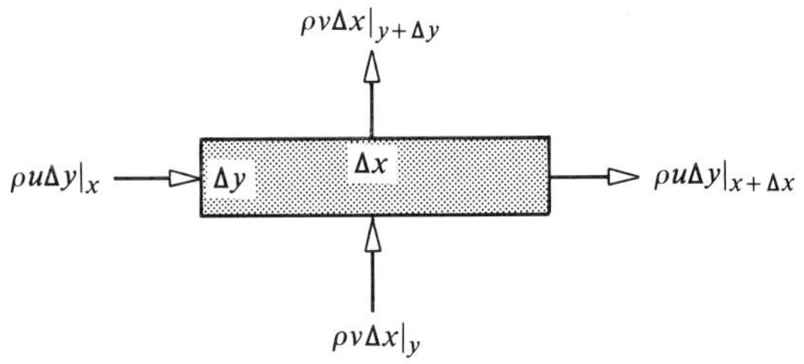

Figure 5.10 Conservation of mass applied to an elemental control volume in a laminar boundary layer on a flat plate.

Referring to Figs. 5.9 and 5.10, conservation of mass requires that the mass inflow minus the mass outflow equal zero:

$$\rho u \Delta y|_x + \rho v \Delta x|_y - \rho u \Delta y|_{x+\Delta x} - \rho v \Delta x|_{y+\Delta y} = 0$$

Dividing by $\rho \Delta x \Delta y$ and rearranging gives

$$\frac{u|_{x+\Delta x} - u|_x}{\Delta x} + \frac{v|_{y+\Delta y} - v|_y}{\Delta y} = 0$$

and letting $\Delta x, \Delta y \to 0$,

$$\frac{\partial u}{\partial x} + \frac{\partial v}{\partial y} = 0 \tag{5.37}$$

Equation (5.37) is the **mass conservation** or **continuity equation**.

Referring to Fig. 5.11, Newton's second law applied to the control volume requires that the net outflow of x-direction momentum equal the sum of the viscous forces

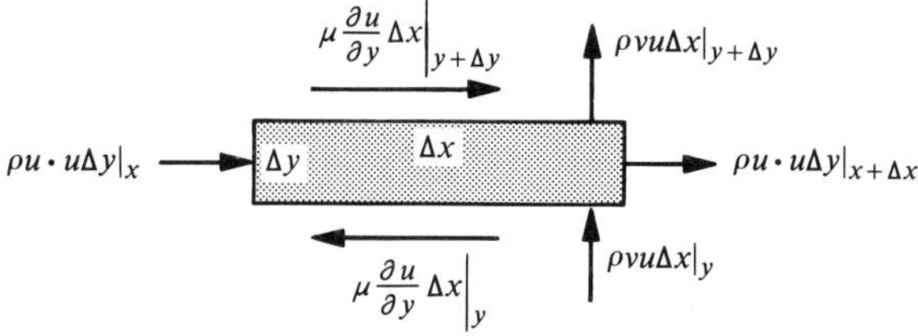

Figure 5.11 Newton's second law applied to an elemental control volume in a laminar boundary layer on a flat plate.

acting in the x-direction:[2]

$$\rho u^2\Delta y|_{x+\Delta x} + \rho vu\Delta x|_{y+\Delta y} - \rho u^2\Delta y|_x - \rho vu\Delta x|_y = \mu\frac{\partial u}{\partial y}\Delta x|_{y+\Delta y} - \mu\frac{\partial u}{\partial y}\Delta x|_y$$

Dividing by $\rho\Delta x\Delta y$ and rearranging gives

$$\frac{u^2|_{x+\Delta x} - u^2|_x}{\Delta x} + \frac{vu|_{y+\Delta y} - vu|_y}{\Delta y} = \nu\frac{(\partial u/\partial y)|_{y+\Delta y} - (\partial u/\partial y)|_y}{\Delta y}$$

Letting Δx and $\Delta y \to 0$,

$$\frac{\partial}{\partial x}(u^2) + \frac{\partial}{\partial y}(vu) = \nu\frac{\partial^2 u}{\partial y^2} \tag{5.38}$$

or

$$2u\frac{\partial u}{\partial x} + u\frac{\partial v}{\partial y} + v\frac{\partial u}{\partial y} = \nu\frac{\partial^2 u}{\partial y^2}$$

Multiplying the continuity equation by u and subtracting it from Eq. (5.38) gives

$$u\frac{\partial u}{\partial x} + v\frac{\partial u}{\partial y} = \nu\frac{\partial^2 u}{\partial y^2} \tag{5.39}$$

which is the **momentum conservation equation**. Notice that there are no pressure forces acting on the fluid since there is no x-direction pressure gradient in the flow outside the boundary layer for flow along a flat plate, and pressure changes across the thin boundary layer are negligible. Also, the viscous shear stress $-\mu\,\partial u/\partial x$ has been ignored since it is negligible in a thin boundary layer. Scaling arguments to justify these assumptions are given in Section 5.8.

Referring to Fig. 5.12, the steady-flow energy equation, Eq. (1.4), requires that the net outflow of enthalpy equal the heat conducted into the volume. Writing $h = c_pT$ again,

$$\rho uc_pT\Delta y|_{x+\Delta x} - \rho uc_pT\Delta y|_x + \rho vc_pT\Delta x|_{y+\Delta y} - \rho vc_pT\Delta x|_y$$
$$= -k\frac{\partial T}{\partial y}\Delta x|_y + k\frac{\partial T}{\partial y}\Delta x|_{y+\Delta y}$$

Dividing by $\rho c_p\Delta x\Delta y$ and rearranging,

$$\frac{uT|_{x+\Delta x} - uT|_x}{\Delta x} + \frac{vT|_{y+\Delta y} - vT|_y}{\Delta y} = \alpha\left[\frac{(\partial T/\partial y)|_{y+\Delta y} - (\partial T/\partial y)|_y}{\Delta y}\right]$$

Letting $\Delta x, \Delta y \to 0$,

$$\frac{\partial}{\partial x}(uT) + \frac{\partial}{\partial y}(vT) = \alpha\frac{\partial^2 T}{\partial y^2} \tag{5.40}$$

[2] Alternatively, we could require that the net *inflow* of x-direction momentum equal the sum of the viscous forces acting in the x-direction to *oppose* the flow. Since the flow in the boundary layer is decelerating, the latter statement is more in line with our physical intuition.

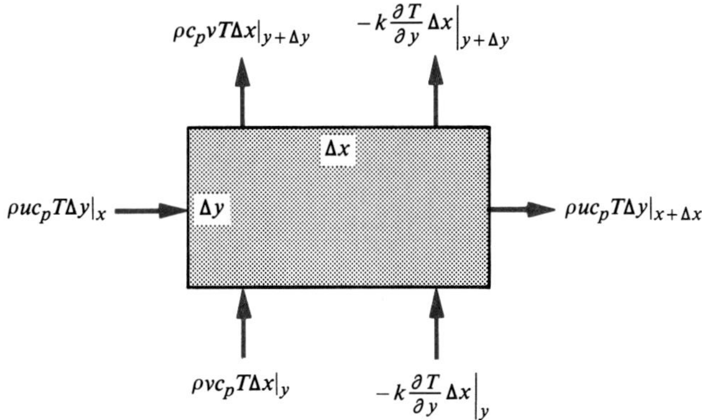

Figure 5.12 The first law of thermodynamics applied to an elemental control volume in a laminar boundary layer on a flat plate.

or

$$T\frac{\partial u}{\partial x} + u\frac{\partial T}{\partial x} + T\frac{\partial v}{\partial y} + v\frac{\partial T}{\partial y} = \alpha\frac{\partial^2 T}{\partial y^2} \tag{5.41}$$

Multiplying the continuity equation by T and subtracting from Eq. (5.41) gives the **energy conservation equation**

$$u\frac{\partial T}{\partial x} + v\frac{\partial T}{\partial y} = \alpha\frac{\partial^2 T}{\partial y^2} \tag{5.42}$$

Notice that conduction in the x direction, $-k\,\partial T/\partial x$, has been ignored since, like the shear stress $-\mu\,\partial u/\partial x$, it is negligible in a thin boundary layer. Scaling arguments to justify this assumption are given in Section 5.8. In addition, since a low-speed incompressible flow has been assumed, there are no terms that account for viscous dissipation and compression heating.

Equations (5.37), (5.39), and (5.42), subject to appropriate boundary conditions, are to be solved to obtain the velocity and temperature profiles and the associated rates of momentum and heat transfer.

5.4.2 The Plug Flow Model

We first consider a simple model in which the velocity, u, is assumed constant through the boundary layer at its free-stream value u_e, that is, as if the flow were inviscid. This model is often called a *plug* or *slug* flow model. Then,

$$\frac{\partial u}{\partial x} = \frac{du_e}{dx} = 0$$

and from Eq. (5.37),

$$\frac{\partial v}{\partial y} = 0, \qquad v = \text{Constant} = 0 \quad \text{since } v\big|_{y=0} = 0$$

Thus, the energy equation Eq. (5.42) becomes

$$u_e \frac{\partial T}{\partial x} = \alpha \frac{\partial^2 T}{\partial y^2} \tag{5.43}$$

If the plate temperature is uniform, appropriate boundary conditions are

$$y = 0: \quad T = T_s \qquad \text{and} \qquad x = 0, y \to \infty: \quad T = T_e \tag{5.44}$$

where, like u_e, T_e is not a function of x. We now define a new independent variable $\zeta = x/u_e$ (which has the dimensions of time), and the mathematical problem becomes

$$\frac{\partial T}{\partial \zeta} = \alpha \frac{\partial^2 T}{\partial y^2} \tag{5.45}$$

$$y = 0: \quad T = T_s \qquad \text{and} \qquad \zeta = 0, y \to \infty: \quad T = T_e \tag{5.46}$$

This problem is identical to the problem of heat conduction in a semi-infinite solid, analyzed in Section 3.4.2. Replacing x by y, t by $\zeta = x/u_e$, and T_0 by T_e in Eq. (3.58) gives

$$\frac{T - T_e}{T_s - T_e} = \operatorname{erfc} \eta; \qquad \eta = \frac{y}{(4\alpha x/u_e)^{1/2}} \tag{5.47}$$

and from Eq. (3.59),

$$q_s = \frac{k(T_s - T_e)}{(\pi \alpha x/u_e)^{1/2}} \tag{5.48}$$

The local heat transfer coefficient is then

$$h_{cx} = \frac{q_s}{T_s - T_e} = \frac{k}{\pi^{1/2}} \left(\frac{u_e}{\alpha x} \right)^{1/2} \tag{5.49}$$

and the local Nusselt number is

$$\mathrm{Nu}_x = \frac{h_{cx} x}{k} = \frac{1}{\pi^{1/2}} \left(\frac{u_e x}{\alpha} \right)^{1/2} = \frac{1}{\pi^{1/2}} \left(\frac{u_e x}{\nu} \right)^{1/2} \left(\frac{\nu}{\alpha} \right)^{1/2}$$

or

$$\mathrm{Nu}_x = 0.564 \mathrm{Re}_x^{1/2} \mathrm{Pr}^{1/2} (= 0.564 \mathrm{Pe}_x^{1/2}) \tag{5.50}$$

which is the basis for the correlation recommended for heat transfer to liquid metals in Section 4.3.2. Typical velocity and temperature profiles in a liquid-metal boundary layer are shown in Fig. 5.13, where it can be seen that the velocity is nearly constant throughout the region where the temperature changes from its wall to free-stream value. Or, to put it another way, the *hydrodynamic boundary layer* is appreciably thinner than the *thermal boundary layer*. It is the characteristic high thermal conductivity of liquid metals that causes this behavior, and this is evidenced in the characteristic low values of Prandtl number (0.001–0.05). In fact, Eq. (5.50) can be shown to be an exact solution in the limit of $\mathrm{Pr} \to 0$.

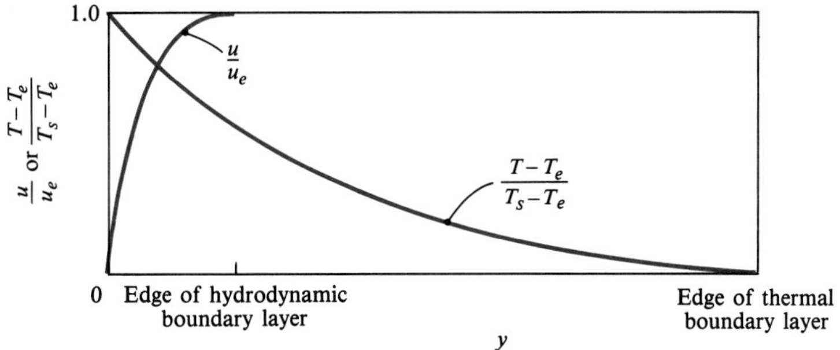

Figure 5.13 Velocity and temperature profiles for a liquid-metal laminar boundary layer.

5.4.3 Integral Solution Method

For fluids with Prandtl number of order unity or greater, such as gases, water, and oils, the simple plug flow model is inadequate. The hydrodynamic boundary layer is of the same or greater thickness than the thermal boundary layer. Before considering exact solutions that require numerical methods, we will consider the approximate **integral method**, which yields accurate results for the flat plate problem.

The Fluid Mechanics Problem

The *integral form* of the momentum conservation equation can be derived from first principles or by integrating the differential equation across the boundary layer; the former approach will be used here. Figure 5.14 shows an elemental control volume of unit depth located between x and $x + \Delta x$ and extending to a location $y = Y$, where Y is greater than the boundary layer thickness δ. Equating the net momentum outflow to the viscous force exerted by the wall on the fluid,

$$\int_0^Y \rho u^2 dy|_{x+\Delta x} + \rho v u|_Y \Delta x - \int_0^Y \rho u^2 dy|_x = -\mu \left.\frac{\partial u}{\partial y}\right|_0 \Delta x$$

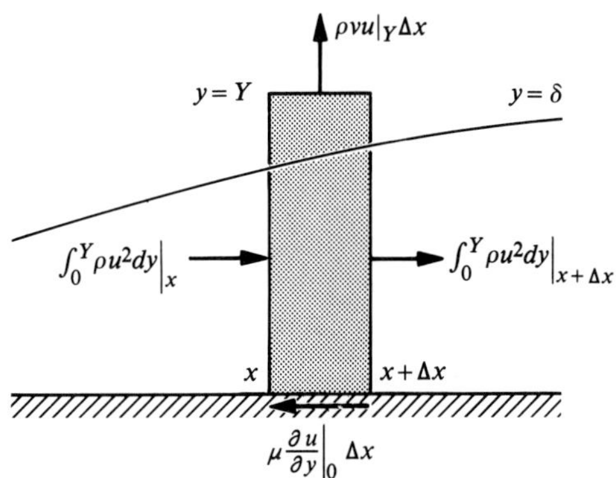

Figure 5.14 Elemental control volume for the integral method application of Newton's second law to a boundary layer on a flat plate.

Dividing by $\rho\Delta x$ and letting $\Delta x \to 0$ gives

$$\frac{d}{dx}\int_0^Y u^2 dy + vu|_Y = -\nu \left.\frac{\partial u}{\partial y}\right|_0$$

Now $u|_Y = u_e$, and from the continuity equation Eq. (5.37),

$$\frac{\partial v}{\partial y} = -\frac{\partial u}{\partial x}, \quad v(Y) = v|_0 - \int_0^Y \frac{\partial u}{\partial x} dy = -\int_0^Y \frac{\partial u}{\partial x} dy$$

Substituting back,

$$\frac{d}{dx}\int_0^Y u^2 dy - \int_0^Y u_e \frac{\partial u}{\partial x} dy = -\nu \left.\frac{\partial u}{\partial y}\right|_0$$

Interchanging the order of integration and differentiation in the second term, which allows the partial derivative to be replaced by a total derivative, and rearranging gives

$$\frac{d}{dx} u_e^2 \int_0^Y \left(1 - \frac{u}{u_e}\right)\left(\frac{u}{u_e}\right) dy = \nu \left.\frac{\partial u}{\partial y}\right|_0$$

But the integrand is zero for $y > \delta$, so we can now let $Y \to \infty$:

$$\frac{d}{dx} u_e^2 \int_0^\infty \left(1 - \frac{u}{u_e}\right)\left(\frac{u}{u_e}\right) dy = \nu \left.\frac{\partial u}{\partial y}\right|_0 \quad \textbf{(5.51)}$$

which is the integral form of the momentum conservation equation.

It is convenient to define some characteristic thickness parameters of the boundary layer. Referring to Fig. 5.15, the deficit in mass flow due to retardation in the boundary layer is set equal to u_e times the *displacement thickness* δ_1:

$$\delta_1 u_e = \int_0^\infty (u_e - u) dy$$

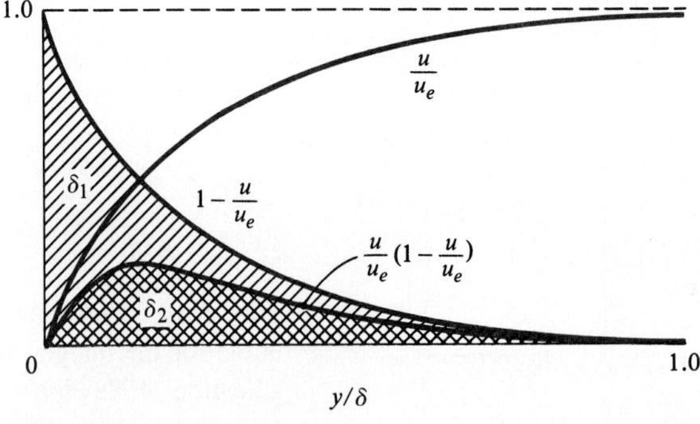

Figure 5.15 Displacement and momentum thickness integrals for a boundary layer.

or

$$\delta_1 = \int_0^\infty \left(1 - \frac{u}{u_e}\right) dy \tag{5.52}$$

That is, δ_1 is the distance the flow outside the boundary layer is displaced due to the presence of the wall. Similarly, the deficit in momentum is set equal to u_e^2 times the *momentum thickness* δ_2:

$$\delta_2 u_e^2 = \int_0^\infty u(u_e - u)dy$$

or

$$\delta_2 = \int_0^\infty \left(1 - \frac{u}{u_e}\right)\left(\frac{u}{u_e}\right) dy \tag{5.53}$$

Equation (5.51) can then be rewritten as

$$\frac{d\delta_2}{dx} = \frac{\mu(\partial u/\partial y)|_0}{\rho u_e^2}$$

or, introducing the skin-friction coefficient $C_{fx} = \tau_s/(1/2)\rho u_e^2$, where $\tau_s = \mu\, \partial u/\partial y|_0$ is the shear stress exerted on the wall by the fluid from above,

$$\frac{d\delta_2}{dx} = \frac{C_{fx}}{2} \tag{5.54}$$

This simple equation states that the rate of increase of the momentum deficit in the boundary layer is equal to one-half the skin friction coefficient. Equation (5.54) is exact, but it cannot be used without knowledge of the velocity profile $u(x,y)$. An approximate solution to the equation can be obtained if the shape of the velocity profile is assumed. This approach, called the *profile method*, was pioneered by T. von Kármán in 1921 [3]. If a simple algebraic form is used for the profile, Eq. (5.54) can be solved analytically. For example, consider a cubic polynomial,

$$\frac{u}{u_e} = a + b\frac{y}{\delta} + c\left(\frac{y}{\delta}\right)^2 + d\left(\frac{y}{\delta}\right)^3 \tag{5.55}$$

where $\delta(x)$ is the boundary layer thickness, defined simply by the condition $u \simeq u_e$ to sufficient accuracy outside the boundary layer. The constants are determined from four appropriate boundary conditions. Two of these are obvious:

$$y = 0: \quad u = 0; \qquad y = \delta: \quad u = u_e$$

A third comes from the requirement that the velocity profile be smooth at the boundary layer edge, that is,

$$y = \delta: \quad \frac{\partial u}{\partial y} = 0$$

The fourth comes from the differential form of the momentum equation, Eq. (5.39). At the wall, both u and v are zero; hence, $\partial u^2/\partial y^2$ is zero also, that is,

$$y = 0: \quad \frac{\partial^2 u}{\partial y^2} = 0$$

Evaluating the constants gives the velocity profile as

$$\frac{u}{u_e} = \frac{3}{2}\frac{y}{\delta} - \frac{1}{2}\left(\frac{y}{\delta}\right)^3 \tag{5.56}$$

which is shown in Fig. 5.16. The momentum thickness can now be evaluated in terms of δ,

$$\delta_2 = \int_0^\delta \left[1 - \frac{3}{2}\frac{y}{\delta} + \frac{1}{2}\left(\frac{y}{\delta}\right)^3\right]\left[\frac{3}{2}\frac{y}{\delta} - \frac{1}{2}\left(\frac{y}{\delta}\right)^3\right] dy = \frac{39}{280}\delta$$

and the skin friction coefficient as

$$\rho u_e^2 \frac{C_{fx}}{2} = \mu \left.\frac{\partial u}{\partial y}\right|_0 = \mu u_e \frac{\partial}{\partial y}\left[\frac{3}{2}\frac{y}{\delta} - \frac{1}{2}\left(\frac{y}{\delta}\right)^3\right]\Bigg|_{y=0} = \frac{3}{2}\frac{\mu u_e}{\delta}$$

Substituting in Eq. (5.54) and rearranging gives an ordinary differential equation for δ:

$$\delta \frac{d\delta}{dx} = \frac{140}{13}\frac{\nu}{u_e} \tag{5.57}$$

which upon integration with $\delta = 0$ at $x = 0$ gives

$$\delta = 4.64(\nu x/u_e)^{1/2} \tag{5.58}$$

and, in particular, the skin friction coefficient becomes

$$C_{fx} = \frac{0.646}{\text{Re}_x^{1/2}} \tag{5.59}$$

Equation (5.59) is accurate within 3% of the exact numerical solution of the differential momentum conservation equation, which was given as Eq. (4.54). However,

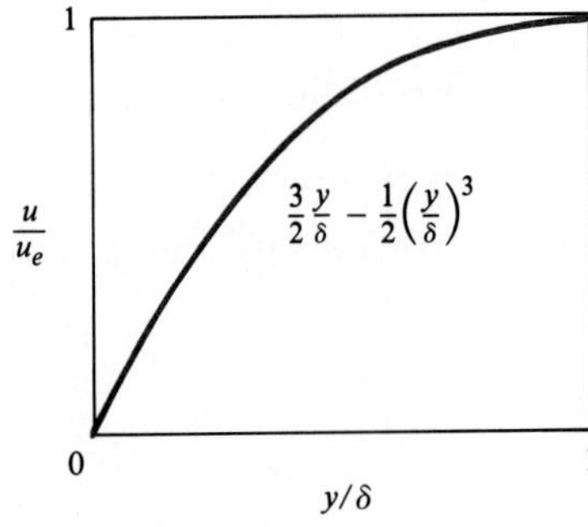

Figure 5.16 Cubic velocity profile for integral method analysis.

this accuracy is fortuitous. The higher the order of the polynomial used to represent the velocity profile, the greater the accuracy we might expect. Thus, we did not attempt to use a linear profile, and a quadratic is also not adequate. But if a quartic profile is used, the accuracy is also less than for a cubic! (See Exercise 5–20.)

The Heat Transfer Problem

We now derive the integral form of the energy conservation equation. Figure 5.17 shows an elemental control volume Δx long extending to $y = Y$, where Y is greater than the thermal boundary layer thickness Δ. In general, $\Delta \neq \delta$. The steady-flow energy equation, Eq. (1.4), requires that the net enthalpy outflow from the volume equal the heat transfer from the wall:

$$\int_0^Y \rho u c_p T\, dy|_{x+\Delta x} + \rho v c_p T|_Y \Delta x - \int_0^Y \rho u c_p T\, dy|_x = -k \left.\frac{\partial T}{\partial y}\right|_0 \Delta x$$

Again, we have neglected x-direction conduction, and y-direction conduction is zero outside the thermal boundary layer. Dividing by $\rho c_p \Delta x$ and letting $\Delta x \to 0$,

$$\frac{d}{dx}\int_0^Y uT\, dy + vT|_Y = -\alpha \left.\frac{\partial T}{\partial y}\right|_0$$

Now $T|_Y = T_e$, and from the continuity equation, $v(Y) = -\int_0^Y (\partial u/\partial x)dy$; thus,

$$\frac{d}{dx}\int_0^Y uT\, dy - \int_0^Y T_e \frac{\partial u}{\partial x} dy = -\alpha \left.\frac{\partial T}{\partial y}\right|_0$$

Interchanging the order of integration and differentiation in the second term and rearranging gives

$$\frac{d}{dx}\int_0^Y u(T - T_e)dy = -\alpha \left.\frac{\partial T}{\partial y}\right|_0$$

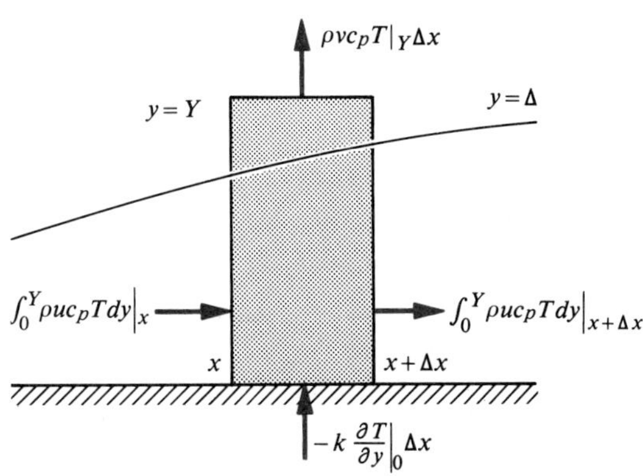

Figure 5.17 Elemental control volume for the integral method application of the first law of thermodynamics to a boundary layer on a flat plate.

The integrand is zero for $y > \Delta$, so we can let $y \to \infty$:

$$\frac{d}{dx}\int_0^\infty u(T - T_e)dy = -\alpha \left.\frac{\partial T}{\partial y}\right|_0 \tag{5.60}$$

which is the integral form of the energy conservation equation. It can be rearranged as

$$\frac{d}{dx}\int_0^\infty \left(\frac{u}{u_e}\right)\left(\frac{T - T_e}{T_s - T_e}\right)dy = \frac{-k(\partial T/\partial y)|_0}{\rho u_e c_p(T_s - T_e)} \tag{5.61}$$

The *energy thickness* Δ_2 is defined as

$$\Delta_2 = \int_0^\infty \left(\frac{u}{u_e}\right)\left(\frac{T - T_e}{T_s - T_e}\right)dy \tag{5.62}$$

and the local Stanton number is

$$\mathrm{St}_x = \frac{h_{cx}}{\rho c_p u_e} = \frac{q_s}{\rho c_p u_e(T_s - T_e)}$$

Thus, Eq. (5.61) can be compactly written as

$$\frac{d\Delta_2}{dx} = \mathrm{St}_x \tag{5.63}$$

Use of the profile method to solve Eq. (5.63) requires consideration of whether the thermal boundary layer thickness Δ is greater than or less than the hydrodynamic boundary layer thickness δ. For high-conductivity fluids, such as liquid metals, we would expect $\Delta \gg \delta$, whereas for highly viscous fluids, such as oils, we would expect $\Delta \ll \delta$. Examination of the differential conservation equations shows that for $\nu = \alpha$, Eq. (5.39) is identical to Eq. (5.42), with T replacing u. Thus, for similar boundary conditions, the velocity and temperature profiles are identical, and $\Delta = \delta$, which is a statement of the Reynolds analogy for a laminar boundary layer. For $\nu = \alpha$ the Prandtl number is unity, and thus $\Delta \simeq \delta$ for gases. As will be shown in Section 5.8, careful scaling of the differential conservation equations shows that $\Delta/\delta \simeq \mathrm{Pr}^{-1/3}$ for $\mathrm{Pr} \gg 1$. Also, the thermal boundary layer will be much thinner than the hydrodynamic boundary layer if the heating commences a long distance from the leading edge of the plate.

To show the power of analytical methods based on the integral conservation equations, we will now solve the *unheated starting length* problem, shown in Fig. 5.18. For $0 < x < \xi$, the plate is insulated and hence is at temperature T_e; heating commences at $x = \xi$ such that for $x > \xi$, the plate is isothermal at temperature T_s. We restrict our attention to fluids of Prandtl number unity or greater, so that always $\Delta < \delta$. As for the velocity profile, a cubic polynomial is chosen for the temperature profile and the constants chosen to satisfy analogous boundary conditions:

$$y = 0: \quad T = T_s, \quad \frac{\partial^2 T}{\partial y^2} = 0$$

$$y = \Delta: \quad T = T_e, \quad \frac{\partial T}{\partial y} = 0$$

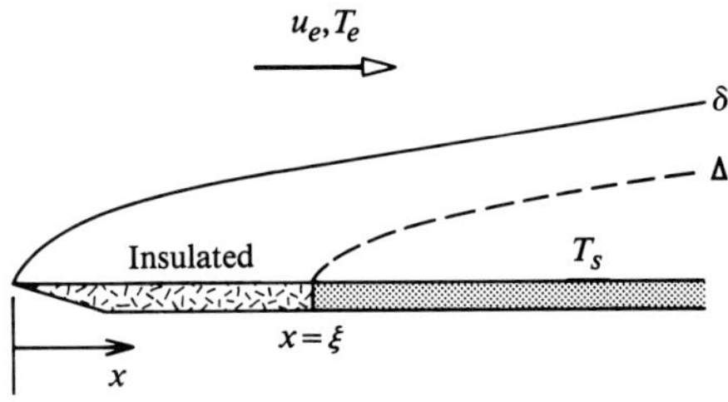

Figure 5.18 Coordinate system for the unheated starting length problem.

to give

$$\frac{T - T_e}{T_s - T_e} = 1 - \frac{3}{2}\frac{y}{\Delta} + \frac{1}{2}\left(\frac{y}{\Delta}\right)^3 \tag{5.64}$$

Then

$$\Delta_2 = \int_0^\infty \left[\frac{3}{2}\frac{y}{\delta} - \frac{1}{2}\left(\frac{y}{\delta}\right)^3\right]\left[1 - \frac{3}{2}\frac{y}{\Delta} + \frac{1}{2}\left(\frac{y}{\Delta}\right)^3\right] dy = 3\delta\left(\frac{r^2}{20} - \frac{r^4}{280}\right)$$

where $r \equiv \Delta/\delta$. Differentiating with respect to x,

$$\frac{d\Delta_2}{dx} = 3\delta\left(\frac{r}{10} - \frac{r^3}{70}\right)\frac{dr}{dx} + 3\left(\frac{r^2}{20} - \frac{r^4}{280}\right)\frac{d\delta}{dx} \tag{5.65}$$

Also

$$\text{St}_x = \frac{-k(\partial T/\partial y)|_{y=0}}{\rho u_e c_p(T_s - T_e)} = \frac{3\alpha}{2u_e\Delta} \tag{5.66}$$

Substituting into Eq. (5.63),

$$3\delta\left(\frac{r}{10} - \frac{r^3}{70}\right)\frac{dr}{dx} + 3\left(\frac{r^2}{20} - \frac{r^4}{280}\right)\frac{d\delta}{dx} = \frac{3\alpha}{2u_e\Delta}$$

To obtain an analytical solution, it is necessary to neglect the two highest-order terms in r. For $\Delta \ll \delta$, $r \ll 1$, and certainly such a simplification is justified. Also, when $\Delta \simeq \delta$ and $r \simeq 1$, either because the unheated starting length was short or because the location in question is far along the plate, $dr/dx \simeq 0$, and the error introduced cannot be large. Deleting these terms and rearranging,

$$2r^2\delta^2\frac{dr}{dx} + r^3\delta\frac{d\delta}{dx} = \frac{10\alpha}{u_e}$$

Substituting for δ^2 and $d\delta/dx$ from Eq. (5.57) and rearranging,

$$r^3 + 4r^2x\frac{dr}{dx} = \frac{13}{14\text{Pr}}$$

which is an ordinary differential equation for $r(x)$. Let $\zeta = r^3$; then

$$r^2\frac{dr}{dx} = \frac{1}{3}\frac{d}{dx}(r^3) = \frac{1}{3}\frac{d\zeta}{dx}$$

and the differential equation becomes

$$\zeta + \frac{4}{3}x\frac{d\zeta}{dx} = \frac{13}{14\text{Pr}}$$

which is easily seen to have the solution

$$\zeta = r^3 = C_1 x^{-3/4} + \frac{13}{14\text{Pr}}$$

The constant C_1 is evaluated from the initial condition $\Delta = 0$ at $x = \xi$, that is, $r = 0$ at $\xi = 0$, to obtain

$$r = 0.976\text{Pr}^{-1/3}[1 - (\xi/x)^{3/4}]^{1/3} \tag{5.67}$$

Finally, using Eqs. (5.58), (5.66), and (5.67), and the relation $\text{Nu}_x = \text{St}_x\text{Re}_x\text{Pr}$,

$$\text{Nu}_x = \frac{0.331\text{Re}_x^{1/2}\text{Pr}^{1/3}}{[1 - (\xi/x)^{3/4}]^{1/3}} \tag{5.68}$$

Notice that for $\xi = 0$, Eq. (5.68) is almost identical to Eq. (4.56), which is based on an exact numerical solution of the differential conservation equations for an isothermal plate. This agreement is reassuring, but it is also fortuitous, since our integral method is an approximate one. The various assumptions give compensating errors. Figure 5.19 shows typical variations in h_{cx} calculated from Eq. (5.68).

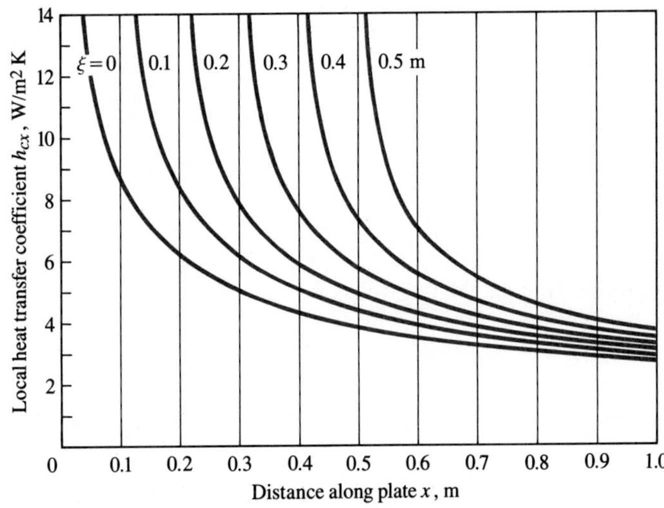

Figure 5.19 Effect of an unheated starting length on the local heat transfer coefficient for a laminar boundary layer on a flat plate. Air at 1 atm, 300 K; $u_e = 2$ m/s.

EXAMPLE 5.3 A Plate with an Unheated Starting Length

Air at 280 K and 1 atm pressure flows at 10 m/s along a flat plate. The plate is insulated for the first 10 cm from its leading edge and thereafter is maintained at 320 K. At a distance of 20 cm from the leading edge, determine δ, δ_2, C_{fx}, τ_s, r, Δ, Δ_2, Nu_x, St_x, h_{cx}, and q_s. Take $\mathrm{Re}_{tr} = 200{,}000$.

Solution

Given: Air flowing over a flat plate with an unheated starting length.

Required: At $x = 20$ cm: δ, δ_2, C_{fx}, τ_s, r, Δ, Δ_2, Nu_x, St_x, h_{cx}, q_s.

Assumptions: A transition Reynolds number of 200,000.

Evaluate all properties at a mean film temperature of $(280 + 320)/2 = 300$ K. Then, from Table A.7, $k = 0.0267$ W/m K, $\rho = 1.177$ kg/m^3, $c_p = 1005$ J/kg K, $\nu = 15.66 \times 10^{-6}$ m^2/s, Pr $= 0.69$. Check the Reynolds number at $x = 0.2$ m,

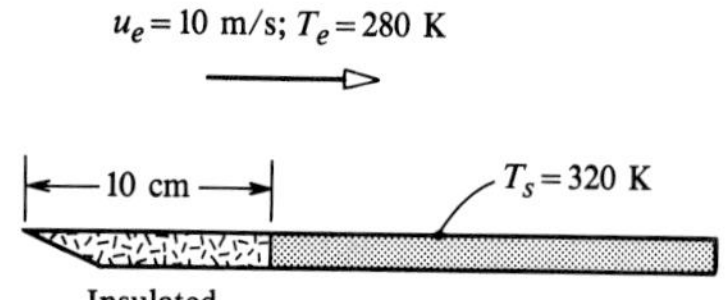

$$\mathrm{Re}_x = \frac{(10)(0.2)}{(15.66 \times 10^{-6})} = 127,700 < 200,000$$

The boundary layer is laminar. Equation (5.58) gives the boundary layer thickness, δ:

$$\delta = 4.64\left(\frac{\nu x}{u_e}\right)^{1/2} = 4.64\left(\frac{15.66 \times 10^{-6} \times 0.2}{10}\right)^{1/2}$$

$$= 2.60 \times 10^{-3} \text{ m (2.60 mm)}$$

The boundary layer momentum thickness is then

$$\delta_2 = \frac{39\delta}{280} = \frac{(39)(2.60 \times 10^{-3})}{280} = 3.62 \times 10^{-4} \text{ m}$$

Equation (5.59) gives the local skin friction coefficient as

$$C_{fx} = \frac{0.646}{\mathrm{Re}_x^{1/2}} = \frac{0.646}{(127{,}700)^{1/2}} = 0.00181$$

Hence, the local shear stress is

$$\tau_s = \frac{1}{2}\rho u_e^2 C_{fx} = \left(\frac{1}{2}\right)(1.177)(10)^2(0.00181) = 0.106 \text{ N/m}^2$$

The ratio of the thermal-to-hydrodynamic boundary layer thickness $r = \Delta/\delta$ is obtained from Eq. (5.67) as

$$r = 0.975\mathrm{Pr}^{-1/3}[1 - (\xi/x)^{3/4}]^{1/3} = (0.975)(0.69)^{-1/3}[1 - (0.1/0.2)^{3/4}]^{1/3}$$

$$= 0.817$$

Hence, $\Delta = r\delta = (0.817)(2.60 \times 10^{-3}) = 2.12 \times 10^{-3}$ m (2.12 mm).

The energy thickness Δ_2 is

$$\Delta_2 = 3\delta\left(\frac{r^2}{20} - \frac{r^4}{280}\right) = (3)(2.60 \times 10^{-3})\left(\frac{0.816^2}{20} - \frac{0.816^4}{280}\right)$$

$$= 2.47 \times 10^{-4} \text{ m}$$

Equation (5.68) gives the local Nusselt number:

$$\text{Nu}_x = \frac{0.331\text{Re}_x^{1/2}\text{Pr}^{1/3}}{[1 - (\xi/x)^{3/4}]^{1/3}} = \frac{(0.331)(127{,}700)^{1/2}(0.69)^{1/3}}{[1 - (0.1/0.2)^{3/4}]^{1/3}} = 141$$

Hence, the local Stanton number is

$$\text{St}_x = \frac{\text{Nu}_x}{\text{Re}_x\text{Pr}} = \frac{141}{(127{,}700)(0.69)} = 0.00160$$

Finally, the local heat transfer coefficient and heat flux are

$$h_{cx} = (k/x)\text{Nu}_x = (0.0267/0.2)(141) = 18.8 \text{ W/m}^2\text{ K}$$

$$q_s = h_{cx}(T_s - T_e) = 18.8(320 - 280) = 752 \text{ W/m}^2$$

Comments

If the plate had been isothermal at 320 K, Nu_x at $x = 20$ cm would be given by Eq. (5.68) for $\xi = 0$, with the value 105. The effect of the unheated starting length and thinner thermal boundary layer is to increase the Nusselt number by 35%.

5.4.4 Self-Similar Solutions

We now obtain an exact numerical solution for the laminar boundary layer on a flat plate with a constant free-stream velocity u_e.

The Fluid Mechanics Problem

The governing equations are the mass and momentum conservation equations, Eqs. (5.37) and (5.39), respectively:

$$\frac{\partial u}{\partial x} + \frac{\partial v}{\partial y} = 0 \qquad \textbf{(5.37)}$$

$$u\frac{\partial u}{\partial x} + v\frac{\partial u}{\partial y} = \nu\frac{\partial^2 u}{\partial y^2} \qquad \textbf{(5.39)}$$

Appropriate boundary conditions are:

$$y = 0: \quad u = 0, v = 0 \qquad \textbf{(5.69a)}$$

$$y \to \infty, x = 0: \quad u = u_e \qquad \textbf{(5.69b)}$$

Equation (5.69*a*) states that there is no slip at the wall, and that the wall is impermeable; Eq. (5.69*b*) states that the free-stream velocity is u_e.

Equations (5.37) and (5.39) are partial differential equations with independent variables x and y. At first sight, one might think that direct numerical solution using finite-difference methods is necessary to obtain the solution. However, as was the case for the problem of transient conduction in a semi-infinite solid, analyzed in Section 3.4.2, a similarity variable can be found, which reduces the problem to the solution of a single ordinary differential equation. If the normalized velocity profiles u/u_e are to be similar, distances in the y direction must scale with the boundary layer thickness δ, which can be arbitrarily defined as the location where $u/u_e = 0.99$. Then the velocity profiles plotted as u/u_e versus y/δ will fall on a single curve irrespective of x location. The dependence of δ on x, and hence the required form of the similarity variable, can be deduced in a number of ways. We have done so already in the integral analysis of Section 5.4.3: Eq. (5.58) states that δ is proportional to $x^{1/2}$. Thus, the similarity variable should have the form $y/x^{1/2}$. The transformation is effected as follows. It is convenient to first introduce a stream function ψ such that

$$u = \frac{\partial \psi}{\partial y}, \qquad v = -\frac{\partial \psi}{\partial x} \tag{5.70}$$

which when substituted in Eq. (5.37) gives

$$\frac{\partial^2 \psi}{\partial y \partial x} - \frac{\partial^2 \psi}{\partial x \partial y} = 0 \tag{5.71}$$

Thus, mass conservation is automatically satisfied, and we no longer need be concerned with Eq. (5.37). Next, we make a transformation of independent variables $x, y \rightarrow \xi, \eta$. The required forms of the differential operators are obtained from the rules of partial differentiation:

$$\frac{\partial}{\partial x} = \left.\frac{\partial}{\partial \eta}\right|_\xi \left.\frac{\partial \eta}{\partial x}\right|_y + \left.\frac{\partial}{\partial \xi}\right|_\eta \left.\frac{\partial \xi}{\partial x}\right|_y \tag{5.72a}$$

$$\frac{\partial}{\partial y} = \left.\frac{\partial}{\partial \eta}\right|_\xi \left.\frac{\partial \eta}{\partial y}\right|_x + \left.\frac{\partial}{\partial \xi}\right|_\eta \left.\frac{\partial \xi}{\partial y}\right|_x \tag{5.72b}$$

Our choice for the similarity variable η is $y(u_e/2\nu x)^{1/2}$, where the constant $(u_e/2\nu)^{1/2}$ has been introduced for future algebraic convenience. Since we expect the transformed equation to be a function of η only, we simply put $\xi = x$ at this time. Equations (5.72) then become

$$\frac{\partial}{\partial x} = \left.\frac{\partial}{\partial \eta}\right|_\xi \left[-\frac{y}{2x}\left(\frac{u_e}{2\nu x}\right)^{1/2}\right] + \left.\frac{\partial}{\partial \xi}\right|_\eta (1) = \left(-\frac{\eta}{2x}\right)\left.\frac{\partial}{\partial \eta}\right|_x + \left.\frac{\partial}{\partial x}\right|_\eta \tag{5.73a}$$

$$\frac{\partial}{\partial y} = \left.\frac{\partial}{\partial \eta}\right|_\xi \left(\frac{u_e}{2\nu x}\right)^{1/2} + \left.\frac{\partial}{\partial \xi}\right|_\eta (0) = \left(\frac{u_e}{2\nu x}\right)^{1/2} \left.\frac{\partial}{\partial \eta}\right|_x \tag{5.73b}$$

Next we suggest that the stream function can be written as a product of functions of x and η only:

$$\psi(x, y) = g(x) f(\eta) \tag{5.74}$$

and choose $g(x)$ such that the dimensionless velocity u/u_e is a function of η alone. Since

$$u = \frac{\partial \psi}{\partial y} = g(x)\left(\frac{u_e}{2\nu x}\right)^{1/2} \frac{df}{d\eta}$$

choosing $g(x) = (2u_e \nu x)^{1/2}$ will give

$$\frac{u}{u_e} = \frac{df}{d\eta} \equiv f' \tag{5.75}$$

as desired, where the prime notation for a derivative with respect to η has been introduced for compactness. Also,

$$\psi = g(x)f(\eta) = (2u_e \nu x)^{1/2} f \tag{5.76}$$

The velocity component v is then

$$\begin{aligned} v = -\frac{\partial \psi}{\partial x} &= (2u_e \nu x)^{1/2} \frac{\eta}{2x} f' - (u_e \nu/2x)^{1/2} f \\ &= (u_e \nu/2x)^{1/2}(\eta f' - f) \end{aligned} \tag{5.77}$$

and the required derivatives of velocity component u are

$$\frac{\partial u}{\partial x} = u_e\left(-\frac{\eta}{2x}\right) f'' \tag{5.78}$$

$$\frac{\partial u}{\partial y} = u_e\left(\frac{u_e}{2\nu x}\right)^{1/2} f'' \tag{5.79}$$

$$\frac{\partial^2 u}{\partial y^2} = u_e\left(\frac{u_e}{2\nu x}\right) f''' \tag{5.80}$$

Substituting in Eq. (5.39) and simplifying yields

$$f''' + ff'' = 0 \tag{5.81}$$

which is the expected ordinary differential equation for f as a function of η alone. It is called the Blasius equation after H. Blasius, who first solved it in 1908 [4]. The boundary conditions transform as follows:

$$u = 0 \text{ at } y = 0 \text{ becomes } f'(0) = 0 \tag{5.82a}$$

$$u \to u_e \text{ as } y \to \infty \text{ becomes } f' \to 1 \text{ as } \eta \to \infty \tag{5.82b}$$

$$v = 0 \text{ at } y = 0 \text{ becomes } f(0) = 0 \tag{5.82c}$$

The last condition is obtained from Eq. (5.77), since $f'(0) = 0$ from Eq. (5.82*a*).

The transformed momentum equation is a third-order nonlinear ordinary differential equation, which must be solved numerically. A number of methods are quite suitable for this purpose. A particularly simple method involves formal integration followed by iteration [5]. Equation (5.81) is rearranged as

$$\frac{f'''}{f''} = -f$$

Integrating,

$$\ln f'' = -\int_0^\eta f\,d\eta + C_1$$

$$f'' = C_1 e^{-\int_0^\eta f\,d\eta}$$

Integrating again,

$$f' = \int_0^\eta C_1 e^{-\int_0^\eta f\,d\eta}\,d\eta + C_2$$

Applying the boundary conditions of Eqs. (5.82*a*) and (5.82*b*) gives $C_2 = 0$ and

$$C_1 = \frac{1}{\int_0^\infty e^{-\int_0^\eta f\,d\eta}\,d\eta} \tag{5.83}$$

Integrating once more,

$$f = C_1 \int_0^\eta \int_0^\eta e^{-\int_0^\eta f\,d\eta}\,d\eta\,d\eta + C_3 \tag{5.84}$$

and from the boundary condition of Eq. (5.82*c*), $C_3 = 0$.

The above procedure may look complicated, but only integration is involved. Dummy variables have not been used to identify the repeated integrations because, in programming the solution procedure, they are not required. Use of the trapezoidal rule is quite adequate for the numerical integrations. Since we cannot integrate to infinity numerically, a maximum value of $\eta = \eta_{max} < \infty$ must be chosen. A possible calculation procedure is as follows.

1. Choose η_{max}, the number of integration steps N, and convergence criterion ε.
2. Calculate $\Delta\eta = \eta_{max}/N$.
3. Set $\eta_1 = 0$, and calculate $\eta_{i+1} = \eta_i + \Delta\eta$ for $i = 1, 2, \ldots, N$.
4. As an initial guess set $f_i = \eta_i$.
5. Calculate $g_i = \int_0^{\eta_i} f\,d\eta$; $h_i = e^{-g_i}$; $p_i = \int_0^{\eta_i} h\,d\eta$; $q_i = \int_0^{\eta_i} p\,d\eta$.
6. Calculate $C_1 = 1/p_N$.
7. Calculate $f_i'' = C_1 h_i$, $f_i' = C_1 p_i$, $f_i = C_1 q_i$.
8. Check convergence: $|\,1 - f_i^{new}/f_i^{old}\,| < \varepsilon$?, $i = 1, 2, \ldots, N$. If not, go back to step 5.

Numerical experiment will show that $\eta_{max} = 6$ is quite adequate, and 100 integration steps will give four-figure accuracy in the solution. The initial guess of $f = \eta$ is arbitrary: the procedure will converge with any initial guess for f. Table 5.1 gives the results, and Fig. 5.20 shows a comparison of the u velocity profile with experiment.

Table 5.1 Solution of the Blasius equation. The skin friction coefficient is obtained as $C_{fx} = 2^{1/2}f''(0)\mathrm{Re}_x^{-1/2} = 0.664\mathrm{Re}_x^{-1/2}$.

$\eta = y\sqrt{u_e/2\nu x}$	f	$f' = u/u_e$	f''
0.0	0.0	0.0	0.46960
0.1	0.00235	0.04696	0.46956
0.2	0.00939	0.09391	0.46931
0.3	0.02113	0.14081	0.46861
0.4	0.03755	0.18761	0.46725
0.5	0.05864	0.23423	0.46503
0.6	0.08439	0.28058	0.47173
0.7	0.11474	0.32653	0.45718
0.8	0.14967	0.37196	0.45119
0.9	0.18911	0.41672	0.44363
1.0	0.23299	0.46063	0.43438
1.1	0.28121	0.50354	0.42337
1.2	0.33366	0.54525	0.41057
1.3	0.39021	0.58559	0.39598
1.4	0.45072	0.62439	0.37969
1.5	0.51503	0.66147	0.36180
1.6	0.58296	0.69670	0.34249
1.7	0.65430	0.72993	0.32195
1.8	0.72887	0.76106	0.30045
1.9	0.80644	0.79000	0.27825
2.0	0.88680	0.81669	0.25567
2.2	1.05495	0.86330	0.21058
2.4	1.23153	0.90107	0.16756
2.6	1.41482	0.93060	0.12861
2.8	1.60328	0.95288	0.09511
3.0	1.79557	0.96905	0.06771
3.2	1.99058	0.98037	0.04637
3.4	2.18747	0.98797	0.03054
3.6	2.38559	0.99289	0.01933
3.8	2.58450	0.99594	0.01176
4.0	2.78388	0.99777	0.00687
4.2	2.98355	0.99882	0.00386
4.4	3.18338	0.99940	0.00208
4.6	3.38329	0.99970	0.00108
4.8	3.58325	0.99986	0.00054
5.0	3.78323	0.99994	0.00026
5.2	3.98322	0.999971	0.000119
5.4	4.18322	0.999988	0.000052
5.6	4.38322	0.999995	0.000022
5.8	4.58322	0.999998	0.000009
6.0	4.78322	0.999999	0.000003

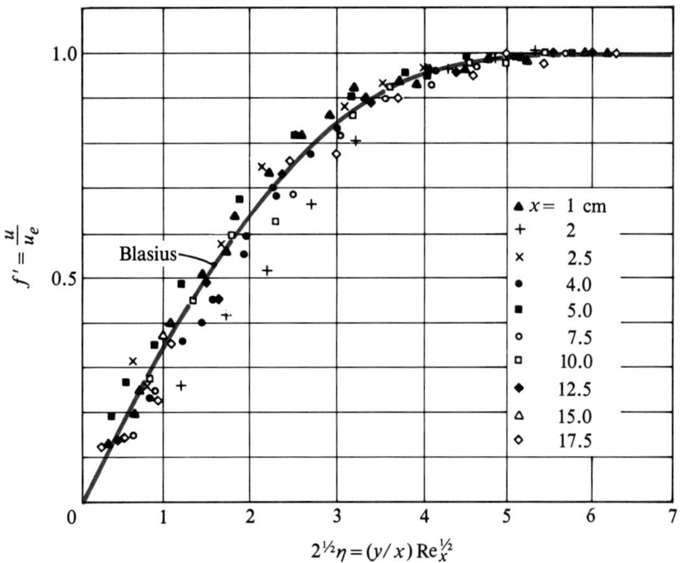

Figure 5.20 Comparison of theory with experiment for the velocity profile in a laminar boundary layer on a flat plate; air, u_e = 8 m/s. (Data of M. Hansen, NACA TM 585, 1930.)

The shear stress exerted on the wall by the fluid τ_s is

$$\begin{aligned}\tau_s &= \mu \left.\frac{\partial u}{\partial y}\right|_{y=0} \\ &= \mu \left.\frac{\partial u}{\partial \eta}\right|_0 \frac{\partial \eta}{\partial y} \\ &= \mu u_e f''(0)(u_e/2\nu x)^{1/2}\end{aligned} \tag{5.85}$$

Table 5.1 shows that $f''(0) = 0.4696$.[3] The local skin friction coefficient is given by

$$C_{fx} = \frac{\tau_s}{(1/2)\rho u_e^2} = \frac{2^{1/2} f''(0)}{\mathrm{Re}_x^{1/2}} = 0.664\mathrm{Re}_x^{-1/2} \tag{5.86}$$

which was given in Chapter 4 as Eq. (4.54). The velocity profile $u/u_e = f'$ given in Table 5.1 can be used to estimate the boundary layer thickness δ. If the edge of

[3] In some texts the constant $2^{1/2}$ is not included in the definitions of η and f. The resulting form of the Blasius equation is $2f''' + ff'' = 0$, which has the solution $f''(0) = 0.4696/2^{1/2} = 0.3321$.

the boundary layer is taken to be where $u \simeq 0.99u_e$, then $\eta \simeq 3.5 = \delta(u_e/2\nu x)^{1/2}$. Hence,

$$\frac{\delta}{x} \simeq \frac{5}{\mathrm{Re}_x^{1/2}} \tag{5.87}$$

Notice that for Prandtl's concept of a "thin" boundary layer to be valid, that is, $\delta/x \ll 1$, the Reynolds number must be greater than about 100.

The student might be somewhat uncomfortable with the way in which the required form of the similarity variable for this *self-similar* solution was "lifted" from the results of the integral analysis. A more popular method for identifying the similarity variable is based on the scaling arguments given in Section 5.8. Simple dimensional analysis, as was used in Section 4.2.2, is not adequate, although *vectorial dimensional analysis* can be used (see Exercise 5–50). But, in essence, the form of the similarity variable is simply a mathematical property of the partial differential equation system under consideration, and, in general, the identification of similarity variables should be viewed as a problem in mathematics.

The Heat Transfer Problem

Once the fluid mechanics problem of obtaining the velocity components, u and v, is solved, solution of the heat transfer problem is relatively straightforward. The energy conservation equation is Eq. (5.42):

$$u\frac{\partial T}{\partial x} + v\frac{\partial T}{\partial y} = \alpha\frac{\partial^2 T}{\partial y^2} \tag{5.42}$$

which we will solve subject to the boundary conditions

$$y = 0: \quad T = T_s, \text{ a constant} \tag{5.88a}$$

$$y \to \infty, x = 0: \quad T = T_e, \text{ a constant} \tag{5.88b}$$

In most situations, the free-stream temperature T_e is a constant, but often in practice the wall temperature T_s is not constant. However, we need to impose the condition of an isothermal wall in order to obtain a self-similar solution. For convenience, we introduce a normalized temperature $\theta = (T - T_e)/(T_s - T_e)$, so that the problem statement becomes

$$u\frac{\partial \theta}{\partial x} + v\frac{\partial \theta}{\partial y} = \alpha\frac{\partial^2 \theta}{\partial y^2} \tag{5.89}$$

$$y = 0: \quad \theta = 1 \tag{5.90a}$$

$$y \to \infty, x = 0: \quad \theta = 0 \tag{5.90b}$$

Equation (5.89) can be transformed in exactly the same way as the momentum equation, using the same expressions for the differential operators and velocity components. The result is

$$\theta'' + \mathrm{Pr} f \theta' = 0; \qquad \mathrm{Pr} = \nu/\alpha \tag{5.91}$$

$$\theta(0) = 1 \tag{5.92a}$$

$$\theta(\infty) = 0 \tag{5.92b}$$

Equation (5.91) is a second-order linear ordinary differential equation. Since f is a known function (tabulated in Table 5.1), Eq. (5.91) can be integrated as

$$\frac{\theta''}{\theta'} = -\mathrm{Pr} f$$

$$\theta = C_1 \int_0^\eta e^{-\mathrm{Pr}\int_0^\eta f\,d\eta}\,d\eta + C_2 \tag{5.93}$$

From Eqs. (5.92), $\theta(0) = 1$, $C_2 = 1$, and using $\theta(\infty) = 0$ gives

$$C_1 = \frac{-1}{\int_0^\infty e^{-\mathrm{Pr}\int_0^\eta f\,d\eta}\,d\eta} \tag{5.94}$$

Using the Blasius equation, Eq. (5.81), Eq. (5.93) can be simplified as follows:

$$f''' + ff'' = 0$$

$$f = -\frac{f'''}{f''}$$

$$\int_0^\eta f\,d\eta = -\ln f'' + C$$

Substituting in Eq. (5.93) and using Eq. (5.94),

$$\theta = 1 - \frac{\int_0^\eta (f'')^{\mathrm{Pr}}\,d\eta}{\int_0^\infty (f'')^{\mathrm{Pr}}\,d\eta} \tag{5.95}$$

Using the data for $f''(\eta)$ given in Table 5.1, the temperature profile $\theta(\eta)$ is easily calculated for a given value of the Prandtl number. Results are shown in Fig. 5.21. The thickness of the thermal boundary layer is seen to increase with decreasing Prandtl number: for liquid metals, it is very large.

The wall heat flux is given by Fourier's law:

$$q_s = -k\left.\frac{\partial T}{\partial y}\right|_0$$

$$= -k\left.\frac{\partial T}{\partial \eta}\right|_0 \frac{\partial \eta}{\partial y}$$

$$= -k(T_s - T_e)\theta'(0)\left(\frac{u_e}{2\nu x}\right)^{1/2}$$

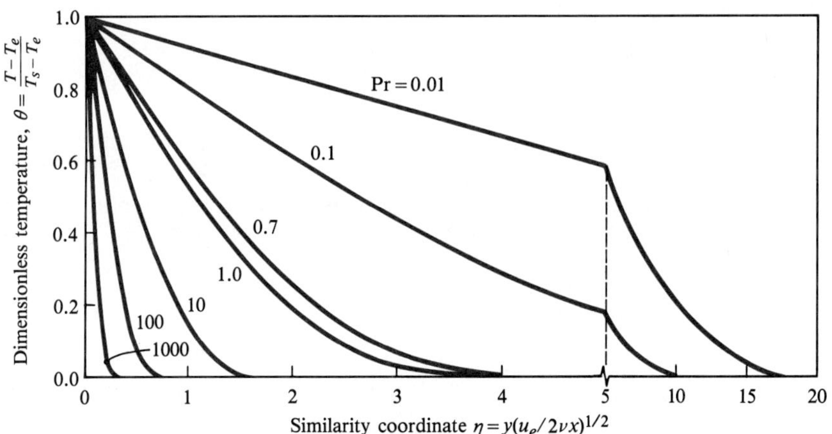

Figure 5.21 Effect of Prandtl number on temperature profiles for a laminar boundary layer on a flat plate.

The local Nusselt number is

$$\mathrm{Nu}_x = \frac{h_c x}{k} = \frac{q_s x}{(T_s - T_e)k} = -\frac{\theta'(0)\mathrm{Re}_x^{1/2}}{2^{1/2}} \tag{5.96}$$

The slope of the normalized temperature profile at the wall, $\theta'(0)$, is simply the constant C_1 given by Eq. (5.94) and is obtained as numerical data, in the form of a function of Prandtl number, as given in Table 5.2. For $\mathrm{Pr} > 0.5$, $\theta'(0)$ can be accurately approximated as $-0.4696\mathrm{Pr}^{1/3}$; thus,

$$\mathrm{Nu}_x = 0.332\mathrm{Re}_x^{1/2}\mathrm{Pr}^{1/3}; \qquad \mathrm{Pr} > 0.5 \tag{5.97}$$

which was Eq. (4.56) in Chapter 4. Notice that for Prandtl number equal to unity, Eqs. (5.81) and (5.91) are of similar form, with f' and θ analogous; that is, the normalized velocity and temperature profiles are identical. This result is simply an alternative statement of Reynolds analogy for a laminar boundary layer on a flat plate.

Table 5.2 The dimensionless wall temperature gradient $\theta'(0)$ for the self-similar laminar boundary layer on a flat plate.

Pr	$-\theta'(0)$	Pr	$-\theta'(0)$
0.003	0.04154	1.0	0.4696
0.01	0.07296	3.0	0.6860
0.03	0.1194	10	1.0297
0.1	0.1980	100	2.2229
0.72	0.4179	1000	4.7961

Significance of Self-Similar Solutions

Although exact, this self-similar solution is less useful than the integral method solution obtained in Section 5.4.3. For the self-similar solution, the wall must be isothermal from the leading edge; the integral method solution allows for an unheated starting length. Furthermore, the integral method result can be used as the basis of a simple method for calculating heat transfer for any arbitrary variation of wall temperature. Notwithstanding the limitations of self-similar solutions, they have played a very important role in convection theory.

Since the 1950s, self-similar solutions have been obtained for a great variety of boundary layer problems. In some important practical situations, the real problem is indeed self-similar—for example, at the stagnation line on a cylinder or the stagnation point of an axisymmetric body. The heat transfer correlations given as Eqs. (4.70) and (4.75) are based on exact self-similar solutions and are of great utility. Integral methods, being approximate, have their limitations for more complex boundary layer flows—for example, over curved surfaces and when the wall is not impermeable. Modern practice, since the 1970s, is to use direct numerical solution of the governing partial differential equations employing finite-difference methods. Standard computer codes are available for this purpose; however, use of such codes is often not economical. Thus, even today, it might be cost-effective to use self-similar solutions from the heat transfer literature even though the conditions for self-similarity are not exactly satisfied. Often a boundary layer can be assumed to be *locally* self-similar (that is, at a particular location) for this purpose. Also, new problems that have self-similar solutions continue to be encountered and useful results obtained.

5.4.5 Natural Convection on an Isothermal Vertical Wall

Figure 5.22 shows a schematic of a natural-convection boundary layer on a vertical, heated wall. If the Rayleigh number is less than about 10^9, the flow in the boundary layer will be laminar. For this natural-convection problem, the analysis strategy we used for forced-convection problems needs to be changed. First, we cannot assume

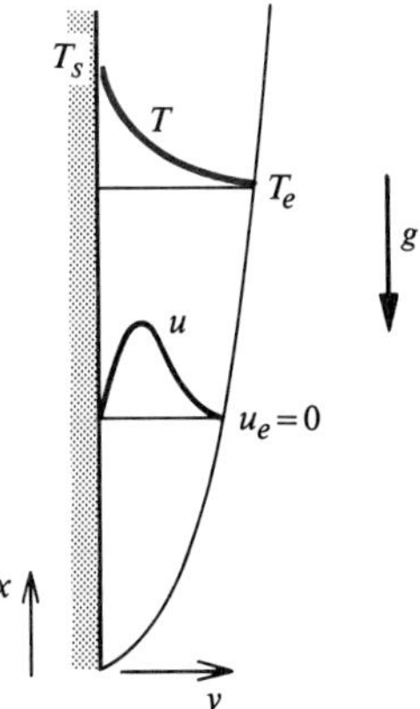

Figure 5.22 Schematic of a laminar natural-convection boundary layer on a vertical wall showing the coordinate system and expected velocity and temperature profiles.

all fluid properties to be constant. The flow is driven by a buoyancy force, which arises due to density gradients adjacent to the heated wall. Second, we cannot solve the fluid mechanics problem without reference to the heat transfer problem, since it is the temperature gradients adjacent to the wall that cause the density gradients. The differential equations governing the velocity and temperature profiles are coupled and must be solved simultaneously. The integral method of analysis gives a particularly good result for this problem.

Integral Conservation Equations

In deriving the integral form of the momentum conservation equation, the buoyancy force acting on the fluid must be included. According to Archimedes' principle, the buoyancy force per unit volume acting on an element of warmer fluid in the boundary layer is $g(\rho_e - \rho)$ directed vertically upward, where g is the gravitational acceleration, ρ_e is the fluid density outside the boundary layer, which is constant, and ρ is the density of the fluid element that varies across the boundary layer. In Fig. 5.23, Newton's second law of motion is applied to the elemental control volume of unit depth and Δx long extending to $y = Y$, where Y is greater than the boundary layer thickness. The net momentum outflow from the volume is equal to the buoyancy force minus the viscous drag force exerted by the wall:

$$\int_0^Y \rho u^2 dy|_{x+\Delta x} - \int_0^Y \rho u^2 dy|_x = \int_0^Y g(\rho_e - \rho) dy \Delta x - \mu \left.\frac{\partial u}{\partial y}\right|_0 \Delta x$$

Notice that there is no momentum flow across the boundary at $y = Y$ since $u(Y) = 0$. Dividing by Δx and letting $\Delta x \to 0$,

$$\frac{d}{dx}\int_0^Y \rho u^2 dy = \int_0^Y g(\rho_e - \rho) dy - \mu \left.\frac{\partial u}{\partial y}\right|_0$$

We now introduce the **Boussinesq approximation** by taking the density to be constant except in the buoyancy term; dividing by ρ then gives

$$\frac{d}{dx}\int_0^Y u^2 dy = \int_0^Y g\left(\frac{\rho_e - \rho}{\rho}\right) dy - \nu \left.\frac{\partial u}{\partial y}\right|_0 \tag{5.98}$$

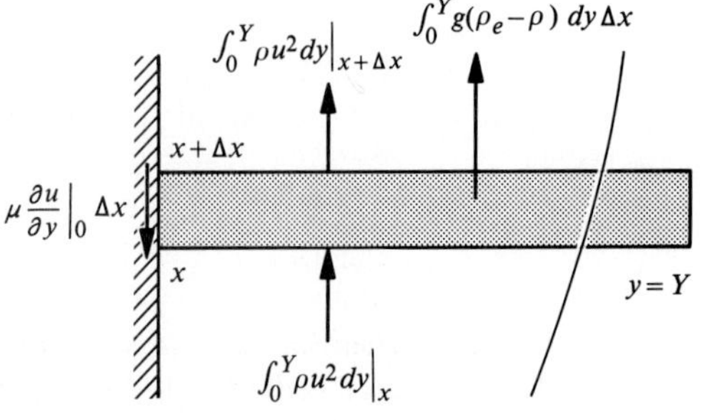

Figure 5.23 Elemental control volume for the integral method application of Newton's second law to a natural-convection boundary layer.

Since $\rho = \rho(P, T)$ we can write a Taylor expansion as $\rho_e - \rho = (\partial \rho / \partial T)_P (T_e - T) + (\partial \rho / \partial P)_T (P_e - P) +$ higher-order terms. The pressure variation across the boundary layer is negligible; thus, we can take $P = P_e$ to obtain

$$\frac{\rho_e - \rho}{\rho} = -\frac{1}{\rho}\left(\frac{\partial \rho}{\partial T}\right)_P (T - T_e) = \beta(T - T_e) \tag{5.99}$$

where β is the volumetric coefficient of thermal expansion. For an ideal gas, $\beta = 1/T$; selected data for liquids are given in Appendix A, Table A.10. Substituting Eq. (5.99) in Eq. (5.98) and letting $Y \to \infty$, since there is no contribution to the integrals for $y > Y$,

$$\frac{d}{dx}\int_0^\infty u^2 dy = \int_0^\infty g\beta(T - T_e)dy - \nu \left.\frac{\partial u}{\partial y}\right|_0 \tag{5.100}$$

The integral form of the energy conservation equation is identical to that for forced flow, Eq. (5.60):

$$\frac{d}{dx}\int_0^\infty u(T - T_e)dy = -\alpha \left.\frac{\partial T}{\partial y}\right|_0 \tag{5.101}$$

Simultaneous Solution of the Equations

The two ordinary differential equations, Eqs. (5.100) and (5.101), are *coupled* since the variable T appears in both; hence, they must be solved simultaneously. Boundary conditions for the velocity and temperature profiles are

$y = 0$: $\quad u = 0, T = T_s$, a constant for an isothermal wall

$y \to \infty$: $\quad u = 0, T = T_e$

We assume that the hydrodynamic and thermal boundary layers have the same thickness, $\Delta = \delta$, and assume the following forms for the velocity and temperature profiles:

$$\frac{u}{U} = \frac{y}{\delta}\left(1 - \frac{y}{\delta}\right)^2, \qquad \frac{T - T_e}{T_s - T_e} = \left(1 - \frac{y}{\delta}\right)^2 \tag{5.102a,b}$$

where U is a scaling velocity, which is a function of x only and is yet to be determined. Figure 5.24 shows these profiles; the maximum velocity is $0.148U$ at $y = \delta/3$. In the forced-convection problem, the known free-stream velocity was used to scale the velocity profile; here the scaling velocity U, as well as the boundary layer thickness δ, are unknowns. Our profiles give $\partial u/\partial y = 0$ and $\partial T/\partial y = 0$ at $y = \delta$ (i.e., they are smooth at the edge of the boundary layer). However, the velocity profile does not have the correct limiting value of $\partial^2 u/\partial y^2$ at the wall; also, the assumption of $\Delta = \delta$ is not valid for high-Prandtl-number fluids. But our method is an approximate one, and use of these profiles gives surprisingly good results. Substituting the profiles in Eqs. (5.100) and (5.101) and performing the indicated

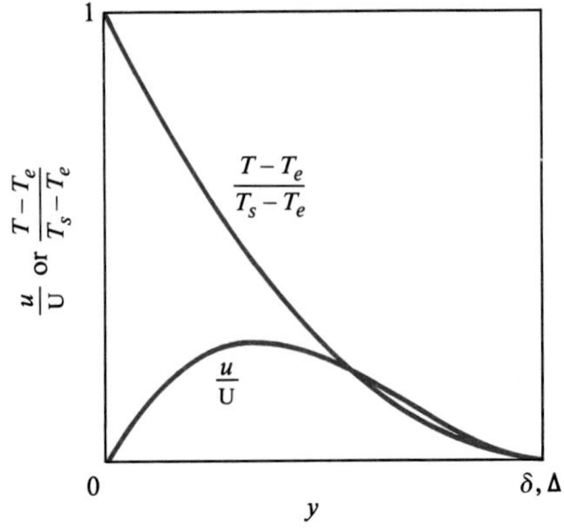

Figure 5.24 Velocity and temperature profiles for integral method analysis of laminar natural convection.

integrations and differentiations gives

$$\frac{d}{dx}\left(\frac{U^2\delta}{105}\right) = -\frac{\nu U}{\delta} + \frac{g\beta(T_s - T_e)\delta}{3} \tag{5.103}$$

$$\frac{d}{dx}\left(\frac{U\delta}{30}\right) = \frac{2\alpha}{\delta} \tag{5.104}$$

which are a pair of ordinary differential equations for the unknowns U and δ. We will assume power law variations for δ and U:

$$\delta = Dx^m; \qquad U = Xx^n$$

which give $\delta = 0$ and $u = 0$ at $x = 0$. Substituting in Eq. (5.103) gives

$$\frac{d}{dx}\left(\frac{X^2Dx^{2n+m}}{105}\right) = -\frac{\nu X x^{n-m}}{D} + \frac{g\beta(T_s - T_e)Dx^m}{3}$$

The x dependence cancels if $2n + m - 1 = n - m = m$, which requires $m = 1/4$, $n = 1/2$. Equations (5.103) and (5.104) then reduce to two algebraic equations for X and D:

$$\frac{5}{4}\frac{X^2D}{105} = -\frac{\nu X}{D} + \frac{g\beta(T_s - T_e)D}{3} \tag{5.105}$$

$$\frac{3}{4}\frac{XD}{30} = \frac{2\alpha}{D} \tag{5.106}$$

Solving,

$$X = \frac{80\alpha}{D^2} \tag{5.107}$$

$$D = 3.94\left[\frac{(20/21)\alpha^2 + \nu\alpha}{g\beta(T_s - T_e)}\right]^{1/4} \tag{5.108}$$

The wall heat flux is obtained from the temperature profile, Eq. (5.102*b*), as

$$q_s = -k\left.\frac{\partial T}{\partial y}\right|_0 = \frac{2k}{\delta}(T_s - T_e)$$

Thus,

$$\frac{h_{cx}}{k} = \frac{2}{\delta} = \frac{2}{Dx^{1/4}}$$

and

$$\mathrm{Nu}_x = \frac{h_{cx}x}{k} = \frac{2x^{3/4}}{D}$$

Substituting for D and rearranging gives

$$\mathrm{Nu}_x = 0.508\left[\frac{\mathrm{Pr}}{0.952 + \mathrm{Pr}}\right]^{1/4}\mathrm{Ra}_x^{1/4} \tag{5.109}$$

where $\mathrm{Ra}_x = \beta(T_s - T_e)gx^3/\nu\alpha$ is the Rayleigh number introduced in Section 4.2.2. This result agrees very well with exact numerical solutions to the differential conservation equations and was widely used before exact solutions became available.

EXAMPLE 5.4 A Natural-Convection Water Boundary Layer

A vertical plate at 320 K is immersed in water at 300 K. At a location 10 cm from the bottom of the plate, determine δ, U, Nu_x, h_{cx}, and q_s. Also plot the velocity and temperature profiles.

Solution

Given: Natural-convection boundary layer on a vertical plate.

Required: At $x = 10$ cm: δ, U, Nu_x, h_{cx}, q_s, $u(y)$, $T(y)$.

Assumptions: Laminar flow.

Properties will be evaluated at a mean film temperature of 310 K; from Table A.8, $k = 0.628$ W/m K, $\rho = 993$ kg/m^3, $\nu = 0.70 \times 10^{-6}$ m^2/s, Pr $= 4.6$. Also, $\alpha = \nu/\mathrm{Pr} = 1.52 \times 10^{-7}$ m^2/s. From Table A.10*b*, $\beta = 3.62 \times 10^{-4}$ K^{-1}. The Rayleigh number is checked to see if the flow is laminar:

$$\mathrm{Ra}_x = \frac{\beta(T_s - T_e)gx^3}{\nu\alpha} = \frac{(3.62 \times 10^{-4})(320 - 300)(9.81)(0.1)^3}{(0.70 \times 10^{-6})(1.52 \times 10^{-7})}$$

$$= 6.68 \times 10^8 < 10^9 \quad \text{(laminar)}$$

The boundary layer thickness is $\delta = Dx^{1/4}$, where D is given by Eq. (5.108):

$$D = 3.94\left[\frac{(20/21)\alpha^2 + \nu\alpha}{g\beta(T_s - T_e)}\right]^{1/4}$$

$$= 3.94\left[\frac{(20/21)(1.52 \times 10^{-7})^2 + (0.70 \times 10^{-6})(1.52 \times 10^{-7})}{(9.81)(3.62 \times 10^{-4})(320 - 300)}\right]^{1/4}$$

$$= 4.57 \times 10^{-3} \text{ m}^{3/4}$$

$$\delta = (4.57 \times 10^{-3})(0.1)^{1/4} = 2.57 \times 10^{-3} \text{ m (2.57 mm)}$$

The scaling velocity is $U = Xx^{1/2}$, where X is given by Eq. (5.107):

$$X = \frac{80\alpha}{D^2} = \frac{(80)(1.52 \times 10^{-7})}{(4.57 \times 10^{-3})^2} = 0.582 \text{ m}^{1/2}\text{/s}$$

$$U = (0.582)(0.1)^{1/2} = 0.184 \text{ m/s}$$

Equation (5.109) gives the local Nusselt number:

$$\text{Nu}_x = 0.508\left[\frac{\text{Pr}}{0.952 + \text{Pr}}\right]^{1/4}\text{Ra}_x^{1/4}$$

$$= 0.508\left[\frac{4.6}{0.952 + 4.6}\right]^{1/4}(6.68 \times 10^8)^{1/4} = 77.9$$

Hence, the local heat transfer coefficient and heat flux are

$$h_{cx} = (k/x)\text{Nu}_x = (0.628/0.1)(77.9) = 489 \text{ W/m}^2\text{ K}$$

$$q_s = h_c(T_s - T_e) = 489(320 - 300) = 9790 \text{ W/m}^2$$

The velocity and temperature profiles are shown in the accompanying diagram.

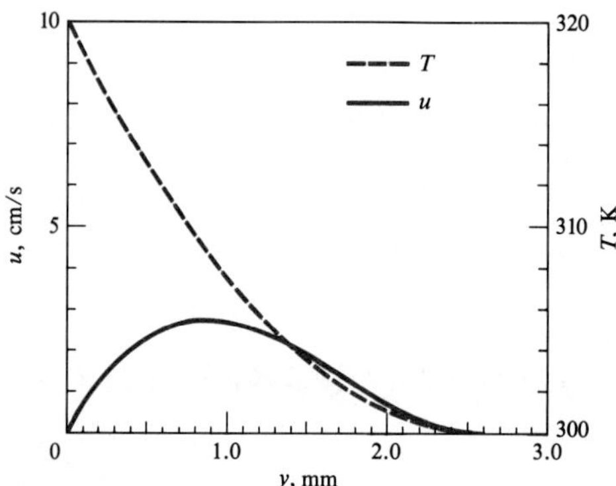

Comments

Notice that CONV does not give a value of the local heat transfer coefficient h_{cx}, since it uses Eq. (4.85) for $\overline{\text{Nu}}_L$ and is based on experimental data.

5.5 TURBULENT FLOWS

Reynolds, in 1883, introduced the concept of a critical Reynolds number, $\mathrm{Re} = u_b D/\nu$, for transition from laminar to turbulent flow in a pipe. A nominal value of 2300 is usually quoted, but in the absence of disturbances, the value can be much higher. The Reynolds number range from 2300 to about 10,000 is rather troublesome since the turbulence tends to have an intermittent character. Experimental data for skin friction and heat transfer in this regime are not easily reproduced; thus, this regime is usually avoided in engineering design, if at all possible. In contrast, in external flows, transition from laminar to turbulent flow often occurs on the surface of interest, and both the location of transition and the region immediately downstream of transition may play important roles in determining overall drag and heat transfer. Wind tunnel tests show that transition on a smooth, flat plate occurs at a critical Reynolds number $\mathrm{Re} = u_e x/\nu$ in the range of 300,000 to 500,000. In practice, factors such as free-stream turbulence and surface roughness tend to lower the critical Reynolds number, and values in the range 50,000 to 100,000 are often used for engineering calculations.

Turbulent flow is characterized by a complex eddying motion, which causes fluctuations of the velocity components, pressure, temperature, and in compressible flows, density as well. Consider a trace of the streamwise velocity component u taken with a hot-wire anemometer probe, as shown in Fig. 5.25. The instantaneous velocity u at time t_0 can be written as the sum of a mean component $\overline{u}$ and a fluctuating component u':

$$u = \overline{u} + u'; \qquad \overline{u}(t_0) = \frac{1}{\tau}\int_{t_0-\tau/2}^{t_0+\tau/2} u\, dt \tag{5.110}$$

where the interval τ is longer than the period of any significant fluctuation but is much shorter than any mean flow time scale. The average value of the fluctuating component is zero, since

$$\overline{u} = \overline{\overline{u} + u'} = \overline{\overline{u}} + \overline{u'} = \overline{u} + \overline{u'}; \qquad \overline{u'} = 0 \tag{5.111}$$

A mean flow may be steady, that is, at a given location $\overline{u}$ is not a function of time, but the instantaneous velocity $u = \overline{u} + u'$ is unsteady in a turbulent flow. The

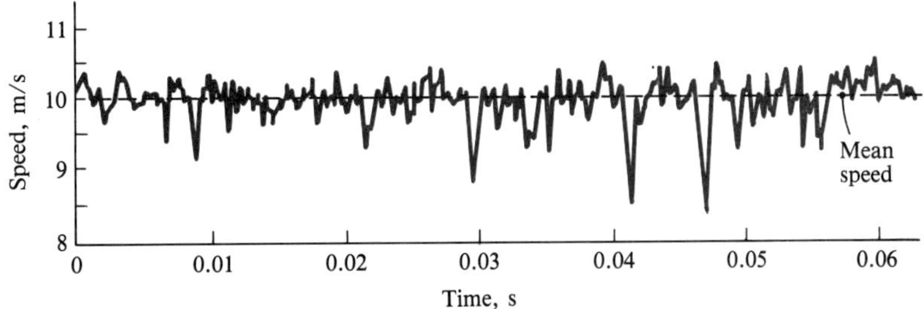

Figure 5.25 Hot-wire anemometer measurement of a turbulent velocity component.

mean component in Fig. 5.25 is steady. The fluctuating component in Fig. 5.25 can be counted to have a frequency of about 10^3 Hz, and its amplitude is about 5% to 10% of the mean value. Correlations of fluctuating components prove to be of considerable importance: consider, for example, the velocity components u and v, with mean values $\overline{u}$ and $\overline{v}$, respectively. The average of the product uv is

$$\overline{uv} = \overline{(\overline{u} + u')(\overline{v} + v')} = \overline{u}\,\overline{v} + \overline{\overline{u}v'} + \overline{u'\overline{v}} + \overline{u'v'}$$

But the average of a constant times a fluctuation whose average is zero, is also zero. That is, $\overline{\overline{u}v'} = 0$, $\overline{u'\overline{v}} = 0$. Hence,

$$\overline{uv} = \overline{u}\,\overline{v} + \overline{u'v'} \tag{5.112}$$

If u' and v' are uncorrelated, then $\overline{u'v'}$ is zero. However, if u' always tends to have the same or opposite sign as v', then $\overline{u'v'}$ is clearly nonzero.

It is quite impractical to attempt to predict a trace such as that shown in Fig. 5.25 by solving the equations governing mass and momentum conservation in a time-dependent flow. Thus, engineers have used experimental data to develop simple models of turbulent transport as a basis for the semi-empirical analysis of the behavior of mean quantities such as mean velocity and mean temperature, as well as the wall shear stress and heat transfer. The turbulent eddies and associated mixing motion are very effective in transporting momentum and heat if velocity or temperature gradients exist. The process is somewhat analogous to transport in a gas as described by the kinetic theory of gases, with eddies replacing molecules as momentum and thermal energy carriers, and a **mixing length** replacing the mean free path. One of the most popular and most successful approaches to modeling turbulent transport, the *Prandtl mixing length theory,* is based on such an analogy.

A large amount of experimental data is available for turbulent flows: the skin friction and heat transfer correlations given in Section 4.3 are examples of such data. Detailed measurements also have been made of mean velocity and mean temperature profiles, of the turbulent fluctuations, and of correlations of fluctuating components, such as $\overline{u'v'}$ and $\overline{T'v'}$. Such data form the essential basis of our knowledge about turbulent flows and have proven indispensable for the successful development and refinement of models for turbulent transport. Even today we cannot use theory alone to predict skin friction and heat transfer for the simplest of flows, for example, the flat plate boundary layer or fully developed pipe flow.

5.5.1 The Prandtl Mixing Length and the Eddy Diffusivity Model

Momentum Transport

In a turbulent flow, we can define an *effective viscosity* as the sum of the molecular viscosity μ and a turbulent viscosity μ_t. The turbulent viscosity accounts for momentum transport by eddies. A model of turbulent transport is required to determine μ_t, and a key step in the development of such models came in 1925 with Prandtl's mixing length hypothesis [6]. Its origin can be seen in the kinetic theory of gases, which gives the molecular viscosity as

$$\mu = \left(\frac{1}{3}\right)(\rho)(\text{Mean free path})(\text{Mean molecular speed}) \tag{5.113}$$

By analogy, the first part of the mixing length hypothesis is

$$\mu_t = \rho \ell v_t \tag{5.114}$$

where ℓ is the *mixing length*, and v_t is a characteristic turbulent speed. However, whereas the mean free path and mean molecular speed are a function of thermodynamic quantities such as temperature and pressure, ℓ and v_t can be expected to vary throughout the flow and have values dependent on the flow velocity and geometry.

Prandtl reduced the number of unknowns in Eq. (5.114) to one by introducing a second part of the hypothesis to estimate v_t. Fig. 5.26 shows the mean velocity profile adjacent to a wall. An eddy moving upward has a positive v' and comes from near the wall, where $\overline{u}$ is less. Such an eddy will be surrounded by fluid of higher $\overline{u}$; thus, the difference between the eddy's x component of velocity u and $\overline{u}$ is a negative value of u'.[4] Since the eddy has moved a distance ℓ,

$$u' \simeq -\ell \frac{\partial \overline{u}}{\partial y} \qquad \text{for } v' > 0 \tag{5.115}$$

and if the eddies are moving randomly,

$$|u'| \simeq |v'| \simeq |w'| \simeq \ell \left| \frac{\partial \overline{u}}{\partial y} \right| \tag{5.116}$$

Thus, if the absolute magnitude of the fluctuating components is taken as the characteristic turbulent speed v_t,

$$v_t = \ell \left| \frac{\partial \overline{u}}{\partial y} \right| \tag{5.117}$$

The absolute magnitude of the mean velocity gradient is used in order to generalize our result to coordinate systems in which the mean velocity gradient may be negative.

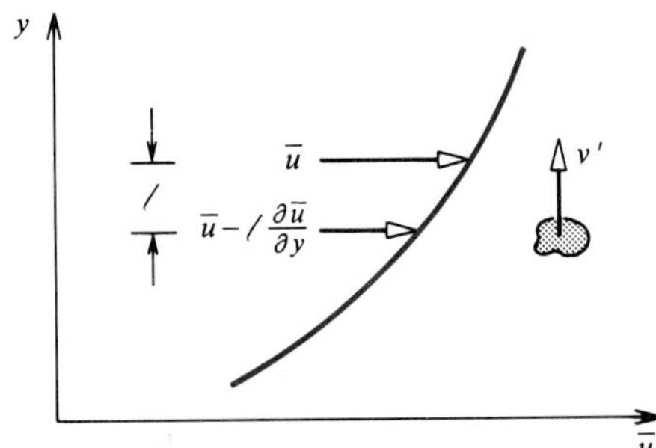

Figure 5.26 Prandtl's mixing length concept for a simple shear flow $\overline{u}(y)$.

[4] Notice that, in this shear flow, the correlation $\overline{u'v'}$ will be negative.

Substituting in Eq. (5.114),

$$\mu_t = \rho \ell^2 \left| \frac{\partial \overline{u}}{\partial y} \right| \tag{5.118}$$

The remaining task is to specify ℓ. If the mixing length model is to be useful, a fairly simple specification of ℓ must work for a variety of flow geometries. Prandtl's original suggestion for flow near any wall was a simple linear relation,

$$\ell = \kappa y \tag{5.119}$$

where $\kappa < 1$ since eddies cannot move through a solid wall; κ is often called the *von Kármán constant,* and it will be shown that experiments suggest a value $\kappa \simeq 0.4$. Specification of ℓ for external boundary layers and pipe flows will be discussed in Sections 5.5.2 and 5.5.3, respectively.

Equation (5.118) gives the turbulent viscosity: in analogy to the molecular kinematic viscosity ν, we can divide μ_t by ρ to obtain the **eddy viscosity** or *eddy diffusivity of momentum,* ε_M:

$$\varepsilon_M = \ell^2 \left| \frac{\partial \overline{u}}{\partial y} \right| \tag{5.120}$$

Also, the total shear stress in the x direction on a plane of constant y is

$$\tau = -(\mu + \mu_t)\frac{\partial \overline{u}}{\partial y} = -\rho(\nu + \varepsilon_M)\frac{\partial \overline{u}}{\partial y} \tag{5.121}$$

where our sign convention is as discussed in Section 5.2.1.

Heat Transport

In a turbulent flow, the transport of heat is similar to the transfer of momentum. An eddy moving toward the wall transports momentum by virtue of its velocity surplus u'; likewise, if a hot fluid flows over a cold wall, the same eddy transports thermal energy by virtue of its temperature surplus T'. Hence, by analogy to Eq. (5.121), we write the total heat flux normal to the wall as the sum of molecular and turbulent contributions:

$$q_y = -(k + k_t)\frac{\partial \overline{T}}{\partial y} = -\rho c_p(\alpha + \varepsilon_H)\frac{\partial \overline{T}}{\partial y} \tag{5.122}$$

where k_t is the turbulent thermal conductivity and ε_H is the eddy diffusivity of heat. Also, we define a **turbulent Prandtl number,** Pr_t, in an analogy to the molecular Prandtl number, Pr:

$$\mathrm{Pr} = \frac{\nu}{\alpha}; \qquad \mathrm{Pr}_t = \frac{\varepsilon_M}{\varepsilon_H} \tag{5.123}$$

Whereas ℓ (and hence ε_M) is found to vary throughout the flow, and in particular increases rapidly with distance away from a wall, Pr_t can often be taken as a constant of order unity. This result might have been expected from the physics involved, since eddies transport both momentum and heat.

5.5.2 Forced Flow along a Flat Plate

The analysis of even the simplest turbulent flow problem is made difficult by the need to use experimental data to develop an appropriate turbulent transport model and to indicate features that allow simplifications to be made. The general strategy for forced convection is similar to that used for constant-property laminar flows in Sections 5.2 through 5.4 and involves solving the fluid mechanics problem prior to solving the heat transfer problem.

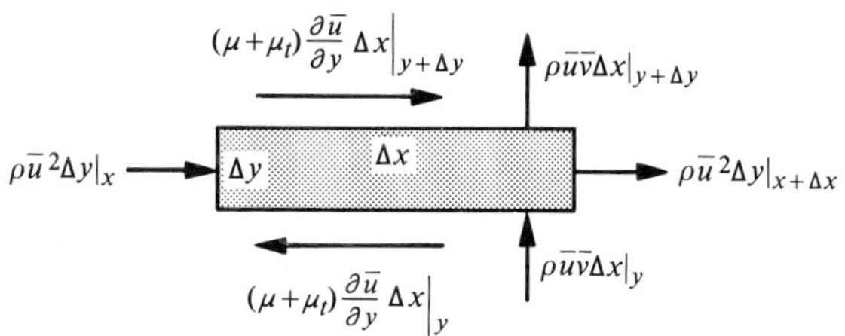

Figure 5.27 Elemental control volume for application of Newton's second law in a turbulent boundary layer on a flat plate.

The Momentum Conservation Equation

Fig. 5.27 shows a fluid element Δx by Δy by unity in a turbulent boundary layer on a flat plate. Paralleling the derivation for the laminar case in Section 5.4.1, application of Newton's second law gives

$$\rho\bar{u}^2\Delta y|_{x+\Delta x} + \rho\bar{u}\,\bar{v}\Delta x|_{y+\Delta y} - \rho\bar{u}^2\Delta y|_x - \rho\bar{u}\,\bar{v}\Delta x|_y = (\mu + \mu_t)\frac{\partial \bar{u}}{\partial y}\Delta x|_{y+\Delta y} - (\mu + \mu_t)\frac{\partial \bar{u}}{\partial y}\Delta x|_y$$

where $\bar{u}$ and $\bar{v}$ are mean velocity components. The effect of turbulence has been accounted for through use of the turbulent viscosity μ_t. *Use of the bar notation for a mean quantity is now discontinued since all the dependent variables are time averages*. Proceeding as in Section 5.4.1 and using Eq. (5.121) gives the differential momentum conservation equation as

$$u\frac{\partial u}{\partial x} + v\frac{\partial u}{\partial y} = \frac{\partial}{\partial y}\left[(\nu + \varepsilon_M)\frac{\partial u}{\partial y}\right] \quad \textbf{(5.124)}$$

where the essential difference from Eq. (5.39) for the laminar case is that $(\nu + \varepsilon_M)$ is left inside the $\partial/\partial y$ differential because ε_M is a function of distance from the wall. The next step is to specify a suitable expression for the eddy viscosity ε_M.

Development of the Eddy Viscosity Model

The eddy viscosity is given by Eq. (5.120) as $\varepsilon_M = \ell^2|\partial u/\partial y|$, and if we use Prandtl's simple idea of $\ell = \kappa y$,

$$\varepsilon_M = \kappa^2 y^2\left|\frac{\partial u}{\partial y}\right| \quad \textbf{(5.125)}$$

At this point, we cannot proceed without use of experimental data: we need to ascertain whether or not Eq. (5.125) is indeed valid and what value of κ is appropriate. Soon after researchers commenced to make detailed measurements of mean velocity profiles, they discovered that, near the wall, velocity profiles measured at different values of Reynolds number $\text{Re}_x = xu_e/\nu$ collapse onto a single curve when plotted in terms of appropriate dimensionless variables. These variables are called the *inner variables* of the turbulent boundary layer and are formulated by recognizing that the square root of shear stress divided by density has the units of velocity: hence, a scaling velocity $v^* = (\tau_s/\rho)^{1/2} = u_e(C_f/2)^{1/2}$ called the **friction velocity** can be defined. The dimensionless variables are then

$$u^+ = \frac{u}{v^*}; \qquad y^+ = \frac{yv^*}{\nu} \tag{5.126}$$

The essential idea here is that the velocity profile depends on the shear stress at the wall, $\tau_s(x)$, the distance from the wall, and the fluid properties ν and ρ. To examine the velocity profile near the wall, it is convenient to use a semilog plot, u^+ versus $\ln y^+$; Fig. 5.28 shows a selection of typical data. The function $u^+(y^+)$ in this **inner region** near the wall is often called the *universal velocity profile*. Adjacent to the wall, the data appear to follow a linear relationship $u^+ = y^+$; at some distance further away, a straight line region is in clear evidence, corresponding to a logarithmic relationship $u^+ = A \ln y^+ + C$. To see how these features of the velocity profiles relate to Prandtl's mixing length ideas, it is necessary to also examine the variation of shear stress across the boundary layer. Figure 5.29 shows a typical shear stress distribution obtained from an experiment. It can be seen that in the inner region, the

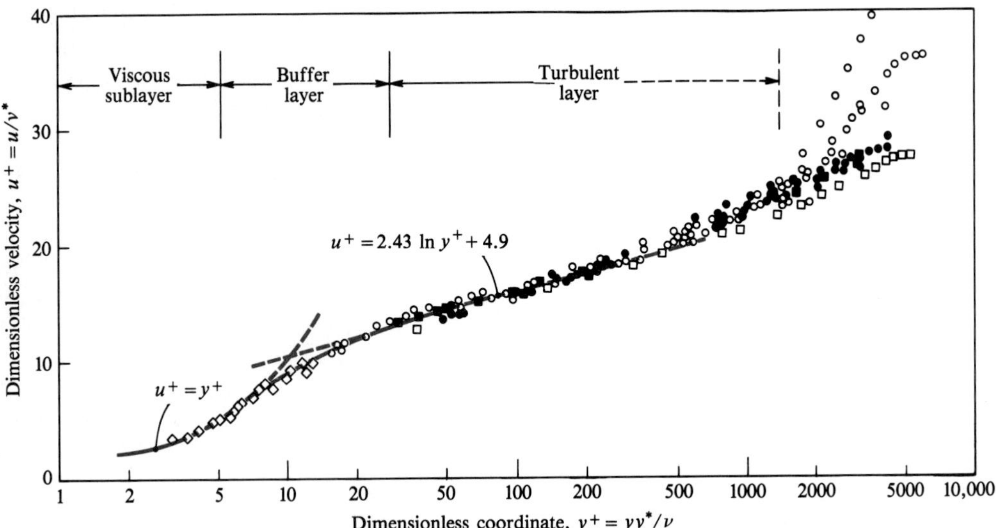

Figure 5.28 Turbulent boundary layer velocity profiles plotted in inner variable coordinates. The data include profiles for flow along a flat plate, flows with a pressure gradient, and flow in a tube.

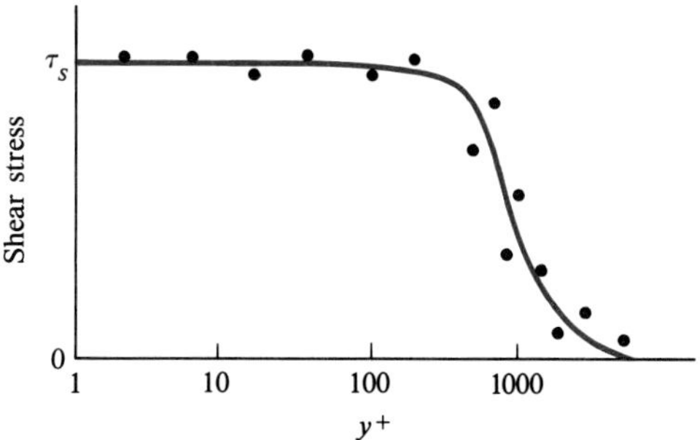

Figure 5.29 Shear stress profile across a turbulent boundary layer on a flat plate.

shear stress is approximately constant at the wall value τ_s. To explore the significance of this feature, we substitute Eq. (5.121) in Eq. (5.124) to obtain

$$u\frac{\partial u}{\partial x} + v\frac{\partial u}{\partial y} = -\frac{1}{\rho}\frac{\partial \tau}{\partial y} \tag{5.127}$$

For τ to be constant, the left-hand side of this equation must be zero; that is, changes of momentum of the fluid in the inner region are negligible. Thus, in the inner region the differential momentum conservation equation simplifies to

$$\tau = -\rho(\nu + \varepsilon_M)\frac{\partial u}{\partial y} = \text{Constant} = -\tau_s$$

where τ_s is the shear stress exerted by the fluid on the wall and is positive. Rearranging,

$$\left(1 + \frac{\varepsilon_M}{\nu}\right)\frac{\partial(u/(\tau_s/\rho)^{1/2})}{\partial(y(\tau_s/\rho)^{1/2}/\nu)} = 1$$

or

$$\left(1 + \frac{\varepsilon_M}{\nu}\right)\frac{du^+}{dy^+} = 1 \tag{5.128}$$

The partial derivative has been replaced by a total derivative since Fig. 5.28 shows that u^+ is only a function of y^+ in the constant-shear-stress inner region. Immediately adjacent to the wall we might expect that the turbulence is sufficiently weak to be damped out by the action of molecular viscosity, and, indeed, experiments show that velocity fluctuations do become very small very close to the wall. Then $\varepsilon_M \ll \nu$, and Eq. (5.128) becomes

$$\frac{du^+}{dy^+} = 1$$

Integrating with $u^+ = 0$ at $y^+ = 0$ gives

$$u^+ = y^+ \tag{5.129}$$

and Fig. 5.28 shows Eq. (5.129) to be valid up to about $y^+ = 5$; this region is called the **viscous sublayer**. Somewhat further from the wall we would expect turbulent transport of momentum to dominate any molecular contribution; in this **turbulent layer** $\varepsilon_M \gg \nu$, and Eq. (5.128) becomes

$$1 = \frac{\varepsilon_M}{\nu}\frac{du^+}{dy^+}$$

But from Eq. (5.125),

$$\varepsilon_M = \kappa^2 y^2 \left|\frac{\partial u}{\partial y}\right| = \nu\kappa^2 y^{+2}\left|\frac{du^+}{dy^+}\right|$$

Hence,

$$1 = \left(\kappa y^+ \frac{du^+}{dy^+}\right)^2$$

or

$$du^+ = \frac{dy^+}{\kappa y^+} \tag{5.130}$$

Integrating,

$$u^+ = \frac{1}{\kappa}\ln y^+ + C \tag{5.131}$$

The straight line seen in Fig. 5.28 has a slope of 2.43, and thus $\kappa = 0.41$. The most appropriate value of the von Kármán constant κ has been in some dispute for many years, and values from 0.40 to 0.44 have been widely used. Equation (5.131) is often called the **law of the wall**.

As the wall is approached from the turbulent layer, the effect of molecular viscosity is to reduce ℓ more rapidly than indicated by Prandtl's linear relation. In 1951, E. R. van Driest suggested a simple exponential *damping factor* for ℓ [7]:

$$\ell = \kappa y\left[1 - \exp\left(-\frac{y^+}{y_t^+}\right)\right] \tag{5.132}$$

where a value of $y_t^+ = 26$ gives velocity profiles that agree well with experimental measurements for flow over a flat plate with no blowing or suction. Equation (5.132) gives a smooth transition into the viscous sublayer where $\ell \to 0$. The region between the viscous sublayer and the turbulent region is often called the **buffer layer**. The precise specification of ℓ in this region proves to be important for the analysis of heat transfer, and the van Driest formula is but one of many that have been proposed.

We next discuss the variation of mixing length in the **wake region** outside the inner region, where the shear is not constant. Figure 5.28 shows that the velocity profiles far from the wall do not fall on a single curve when plotted as u^+ versus $\ln y^+$. However, if the velocity profiles are plotted as u^+ versus y/δ, where δ is the boundary layer thickness, then the profiles do fall on a single curve, except very near the wall. It is common practice to represent this result as a plot of $u^+ - u_e^+$ versus y/δ, as shown in Fig. 5.30. The equation of the curve is called the *velocity defect law*, and we term $u^+ - u_e^+$ and y/δ the **outer variables**. The outer variables

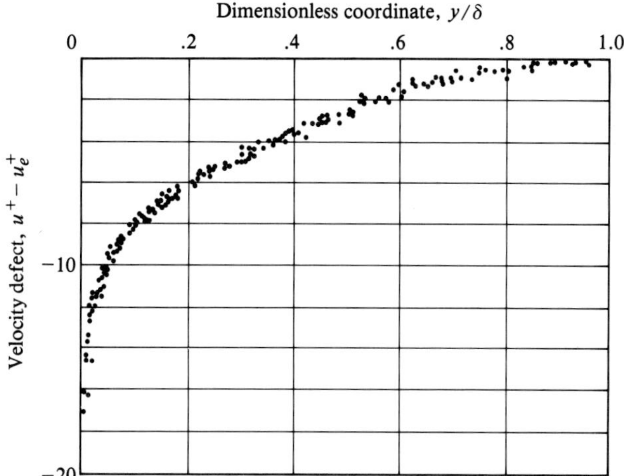

Figure 5.30 Flat plate turbulent boundary layer velocity profiles plotted in outer variable coordinates. The data include profiles for both smooth and rough surfaces.

correlate the velocity profiles in both the wake region and the turbulent layer, but not in the buffer layer or viscous sublayer. *Both the inner and outer variables correlate the profiles in the turbulent layer; hence, this layer is also called the overlap region.* The velocity profile shown in Fig. 5.30 does not have a simple shape. However, the deviation of the actual velocity profile from the law of the wall, when normalized by the maximum deviation at $y = \delta$, is a function of y/δ only and can be well represented by the $\sin^2$ function. This result is called the **law of the wake** introduced by D. Coles in 1956 [8]. An appropriate expression for the velocity profile in the outer region is then

$$u^+ = A \ln y^+ + C + 2A\Pi \sin^2\left(\frac{\pi}{2}\frac{y}{\delta}\right) \tag{5.133}$$

where $\Pi \simeq 0.5$, $A = 1/0.41$, and $C = 5.0$ (Coles recommended $\Pi = 0.55$). Notice that u^+ depends on both y^+ and y in Eq. (5.133). Since the shear is not constant in the wake region, it is not possible by simple analysis to deduce the mixing length variation $\ell(y/\delta)$ that corresponds to the law of the wake, as was done for the law of the wall. However, very simple variations of ℓ prove quite adequate. Based on experiment, a simple rule for ℓ in the outer region is [9]

$$\frac{y}{\delta} \leq \frac{\lambda}{\kappa}: \quad \ell = \kappa y \text{ or } \frac{\ell}{\delta} = \kappa\frac{y}{\delta} \tag{5.134a}$$

$$\frac{y}{\delta} > \frac{\lambda}{\kappa}: \quad \frac{\ell}{\delta} = \lambda, \text{ a constant} \tag{5.134b}$$

with $\kappa = 0.41$ and $\lambda = 0.09$. The boundary layer thickness δ used here is defined as the location where $u = 0.99u_e$. Notice that this result states that the length is

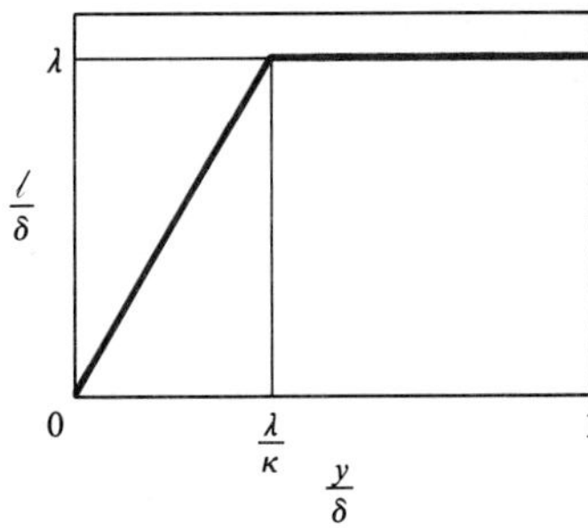

Figure 5.31 Variation of mixing length across the outer region of a turbulent boundary layer.

proportional to the boundary layer thickness in the wake region. A similar result holds true for turbulent wakes behind bluff objects and for turbulent jets. Figure 5.31 shows a schematic of the mixing length variation.

Solution of the Momentum Equation

With ℓ, and equivalently ε_M through Eq. (5.120), specified across the boundary layer, it is possible to solve the differential momentum conservation equation, Eq. (5.124), using numerical methods to obtain velocity profiles, and the wall shear stress as a function of position along the plate. The finite-difference numerical method has been used widely for this purpose. The numerical results for the wall shear stress can be used to develop a skin friction correlation. On the other hand, the skin friction on a flat plate can be quite accurately obtained in a simple manner by solving the integral momentum equation, as will now be demonstrated.

The integral momentum equation in the form of Eq. (5.54) is valid for both laminar and turbulent boundary layers on an impermeable flat plate. To introduce dimensionless variables, we define $\mathrm{Re}_x = u_e x/\nu$ and $\mathrm{Re}_{\delta_2} = u_e \delta_2/\nu$, and Eq. (5.54) becomes

$$\frac{C_{fx}}{2} = \frac{d\mathrm{Re}_{\delta_2}}{d\mathrm{Re}_x} \tag{5.135}$$

or, integrating,

$$\mathrm{Re}_x = 2\int_0^{Re_{\delta_2}} \frac{d\mathrm{Re}_{\delta_2}}{C_{fx}} \qquad \text{for } \delta_2 = 0 \text{ at } x = 0 \tag{5.136}$$

The integral can be evaluated provided we supply a relation $C_{fx} = C_f(\mathrm{Re}_{\delta_2})$. The contribution to the momentum thickness δ_2 from the buffer layer and viscous sublayer is negligible, since we have already seen that the constant shear in the inner region implies negligible changes in fluid momentum. Thus, δ_2 can be calculated accurately using the outer region velocity profile, Eq. (5.133), and its definition, Eq. (5.53). Some lengthy algebra gives (see Exercise 5–37)

$$\frac{\delta_2}{\delta} = \frac{3.66}{u_e^+} - \frac{23.6}{u_e^{+2}} \tag{5.137}$$

and u_e^+ is obtained from Eq. (5.133) by substituting $y = \delta$:

$$u_e^+ = \frac{1}{0.41} \ln \delta^+ + 7.44 \tag{5.138}$$

where $\delta^+ = \delta v^*/\nu$. Eliminating δ between Eqs. (5.137) and (5.138) gives

$$\mathrm{Re}_{\delta_2} = u_e^+ \left(\frac{3.66}{u_e^+} - \frac{23.6}{u_e^{+2}} \right) e^{0.41(u_e^+ - 7.44)} \tag{5.139}$$

which is an implicit function for $C_f(\mathrm{Re}_{\delta_2})$ since $u_e^+ = (2/C_f)^{1/2}$. Values of C_f, Re_{δ_2}, and Re_δ are shown in Table 5.3. Following the example of F. White [10], the data are approximated by a simple power law to obtain an analytical result:

$$C_f \simeq 0.013\mathrm{Re}_{\delta_2}^{-1/6} \tag{5.140}$$

Substituting Eq. (5.140) in Eq. (5.136) then gives

$$\mathrm{Re}_x = 2 \int_0^{\mathrm{Re}_{\delta_2}} \frac{d\mathrm{Re}_{\delta_2}}{0.013\mathrm{Re}_{\delta_2}^{-1/6}}$$

Integrating and rearranging,

$$\mathrm{Re}_{\delta_2} = 0.0152\mathrm{Re}_x^{6/7} \tag{5.141}$$

$$C_{fx} = 0.026\mathrm{Re}_x^{-1/7} \tag{5.142}$$

which agrees with Eq. (4.59*b*). With the fluid mechanics problem complete, we now turn to the heat transfer problem.

Table 5.3 *Law of the wake* computations for the turbulent boundary layer on a flat plate.

u_e^+	C_f	Re_{δ_2}	Re_δ
20	0.00500	427	3,450
25	0.00320	3,690	33,500
30	0.00222	29,900	312,000
35	0.00163	241,000	2,830,000
40	0.00125	1,930,000	25,100,000

The Energy Conservation Equation

The steady-flow energy equation, Eq. (1.4), applied to the elemental control volume shown in Fig. 5.32 requires that the net outflow of enthalpy equal the heat conducted into the volume. Writing $\overline{h} = c_p\overline{T}$,

$$\rho\overline{u}c_p\overline{T}\Delta y|_{x+\Delta x} - \rho\overline{u}c_p\overline{T}\Delta y|_x + \rho\overline{v}c_p\overline{T}\Delta x|_{y+\Delta y} - \rho\overline{v}c_p\overline{T}\Delta x|_y = -(k + k_t)\frac{\partial\overline{T}}{\partial y}\Delta x|_y + (k + k_t)\frac{\partial\overline{T}}{\partial y}\Delta x|_{y+\Delta y}$$

where, as for laminar flow, constant-property low-speed flow is assumed, and the x-direction molecular conduction and turbulent transport are neglected. The bar

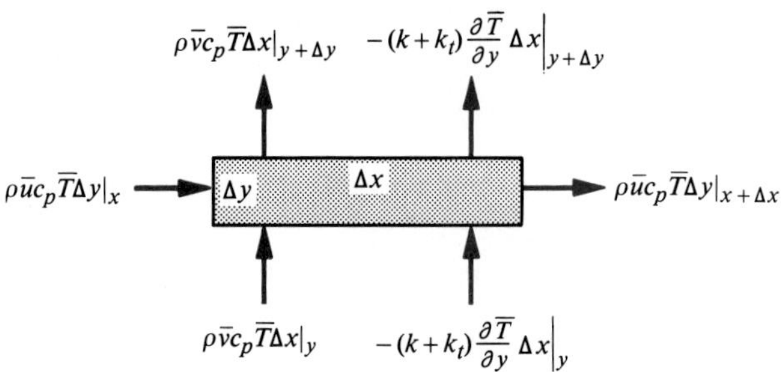

Figure 5.32 Elemental control volume for application of the first law of thermodynamics in a turbulent boundary layer.

notation for mean quantities can be discarded since turbulent transport is accounted for through use of the turbulent conductivity k_t. Proceeding as for the laminar case in Section 5.4.1 and using Eq. (5.122) gives the differential energy conservation equation as

$$u\frac{\partial T}{\partial x} + v\frac{\partial T}{\partial y} = \frac{\partial}{\partial y}\left[(\alpha + \varepsilon_H)\frac{\partial T}{\partial y}\right] \tag{5.143}$$

which is similar in form to Eq. (5.124) for momentum conservation. Since $\Pr = \nu/\alpha$ and $\Pr_t = \varepsilon_M/\varepsilon_H$, Eq. (5.143) can be rearranged as

$$u\frac{\partial T}{\partial x} + v\frac{\partial T}{\partial y} = \frac{\partial}{\partial y}\left[\left(\frac{\nu}{\Pr} + \frac{\varepsilon_M}{\Pr_t}\right)\frac{\partial T}{\partial y}\right] \tag{5.144}$$

The Turbulent Prandtl Number

Specification of the mixing length ℓ and hence the eddy viscosity ε_M has already been discussed. We still need to specify the turbulent Prandtl number $\Pr_t$. A turbulent eddy moving transverse to the mean flow can be expected to lose heat at a rate different from the rate at which it loses momentum. In particular, a liquid-metal eddy with its high molecular conductivity will lose heat faster than an air or water eddy; hence, turbulent transport of heat will be less effective for liquid metals. In fact, a limit of $\Pr_t \to \infty$ as $\Pr \to 0$ might be expected. Experimental data for air show a value of $\Pr_t$ somewhat less than unity in the turbulent layer but increasing to values of about 2 in the buffer layer. If we desire to accurately determine temperature profiles, it is necessary to carefully specify the variation of $\Pr_t$ across the boundary layer. However, for the purpose of determining heat transfer, the use of $\Pr_t =$ Constant $= 0.9$ or 1.0 is quite satisfactory for all fluids except liquid metals.

Solution of the Energy Equation

As was the case for the differential momentum equation, Eq. (5.124), it is possible to solve Eq. (5.144) using finite-difference or similar numerical methods. The momentum equation must be solved first, or simultaneously, to obtain the velocity components u and v, as well as the velocity gradient $\partial u/\partial y$, which is required to calculate ε_M. However, it is possible to obtain an analytical solution of reasonable accuracy if some simplifying assumptions are made, as will now be demonstrated. Experiment shows that if the plate is isothermal, the heat flux, like the shear stress, is constant in the inner region. Furthermore, the ratio q/τ is approximately constant across the whole boundary layer.

$$\frac{q}{\tau} = \frac{-\rho c_p(\alpha + \varepsilon_H)(\partial T/\partial y)}{-\rho(\nu + \varepsilon_M)(\partial u/\partial y)} \simeq \frac{q_s}{-\tau_s} \tag{5.145}$$

where τ_s is the shear stress exerted on the wall. Integrating with $u = 0$ and $T = T_s$ at the wall,

$$T_s - T = \frac{q_s}{c_p\tau_s}\int_0^u \frac{\nu + \varepsilon_M}{\nu/\text{Pr} + \varepsilon_M/\text{Pr}_t}du$$

Introducing dimensionless variables gives

$$T^+ = \int_0^{u^+} \frac{\varepsilon^+ du^+}{1/\text{Pr} + (1/\text{Pr}_t)(\varepsilon^+ - 1)} \tag{5.146}$$

where

$$T^+ = (T_s - T)\rho c_p v^*/q_s \tag{5.147}$$

$$\varepsilon^+ = 1 + (\varepsilon_M/\nu) \tag{5.148}$$

are dimensionless temperature and total viscosity, respectively. For a chosen ε^+ profile and Pr_t, the dimensionless temperature T^+ can be calculated; a typical result is shown in Fig. 5.33. As for the velocity profile, there is a logarithmic region in this **temperature law of the wall** for $y^+ \gtrsim 30$, but in contrast to the velocity law of the

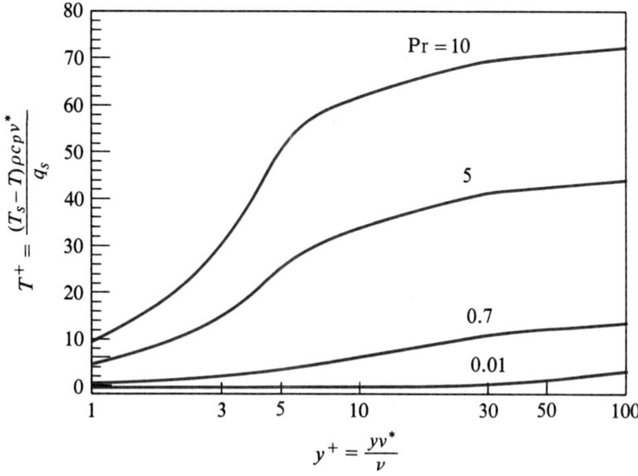

Figure 5.33 Temperature law of the wall $T^+(y^+, \text{Pr})$ for a turbulent boundary layer on a flat plate.

wall, the molecular Prandtl number is an additional parameter. If Eq. (5.146) is integrated to the edge of the boundary layer, we obtain

$$T_e^+ = \int_0^{u_e^+} \frac{\varepsilon^+ du^+}{1/\mathrm{Pr} + (1/\mathrm{Pr}_t)(\varepsilon^+ - 1)} \tag{5.149}$$

and since $u_e^+ = (2/C_{fx})^{1/2}$, the Stanton number is

$$\mathrm{St}_x = \frac{q_s}{(T_s - T_e)\rho c_p u_e} = \left(\frac{C_{fx}}{2}\right)^{1/2} \frac{1}{T_e^+} \tag{5.150}$$

It is simple to numerically integrate Eq. (5.149) using our recommended mixing length specification to obtain ε^+. But if we restrict our attention to fluids of Prandtl number in the range 0.5 to 30, a rather simple specification of ε^+ suggested by T. von Kármán [11] gives a good estimate of heat transfer. The specification of ε^+ and the corresponding velocity profile obtained using Eq. (5.128) are

$$0 < y^+ \le 5: \quad \varepsilon^+ = 1 \quad u^+ = y^+ \tag{5.151a}$$

$$5 \le y^+ < 30: \quad \varepsilon^+ = 0.2y^+ \quad u^+ = 5.0 \ln y^+ - 3.05 \tag{5.151b}$$

$$30 \le y^+: \quad \varepsilon^+ = 0.4y^+ \quad u^+ = 2.5 \ln y^+ + 5.5 \tag{5.151c}$$

This ε^+ profile is shown in Fig. 5.34, where it is seen to be discontinuous at $y^+ = 30$; the corresponding mixing length specification is shown in Fig. 5.35. Substituting Eq. (5.151) in (5.149) gives

$$T_e^+ = \int_0^5 \frac{dy^+}{1/\mathrm{Pr}} + \int_5^{30} \frac{dy^+}{1/\mathrm{Pr} + (1/\mathrm{Pr}_t)(0.2y^+ - 1)} + \int_{14}^{u_e^+} \frac{\varepsilon^+ du^+}{1/\mathrm{Pr} + (1/\mathrm{Pr}_t)(\varepsilon^+ - 1)}$$

$$= 5\mathrm{Pr} + 5\ln(5\mathrm{Pr} + 1) + \mathrm{Pr}_t[(2/C_{fx})^{1/2} - 14]$$

where Eq. (5.128) has been used to replace $\varepsilon^+ du^+$ by dy^+ for $y^+ < 30$, and at $y^+ = 30$, $u^+ = 2.5 \ln 30 + 5.5 = 14$. In the buffer layer, Pr_t can be taken to be

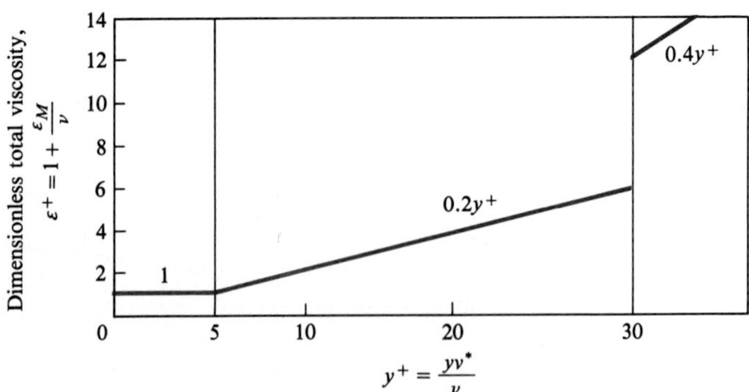

Figure 5.34 Dimensionless total viscosity profile according to von Kármán, as given by Eq. (5.151).

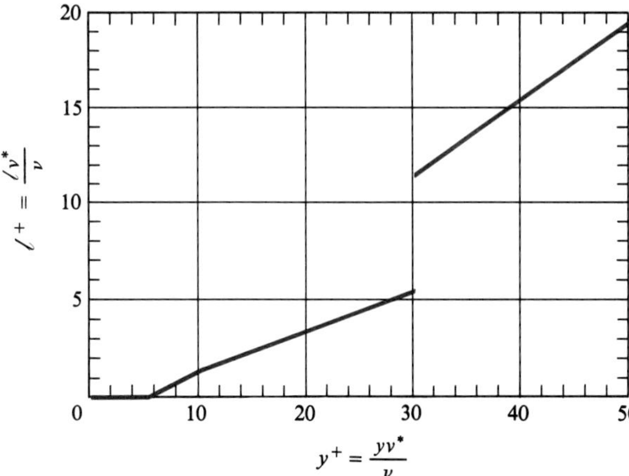

Figure 5.35 Mixing length distribution across a turbulent boundary layer corresponding to Eq. (5.151).

1.0, and in the turbulent layer, $(1/\text{Pr} - 1/\text{Pr}_t)$ can be neglected compared to $\varepsilon^+/\text{Pr}_t$. Then, using Eq. (5.150),

$$\text{St}_x = \frac{(C_{fx}/2)^{1/2}}{5\text{Pr} + 5\ln(5\text{Pr} + 1) + \text{Pr}_t[(2/C_{fx})^{1/2} - 14]} \tag{5.152a}$$

The denominator can be viewed as the sum of three thermal resistances in series, for the viscous sublayer, the buffer layer, and the remainder of the boundary layer. The relative magnitudes of these resistances depend on the molecular Prandtl number and Reynolds number. Equation (5.152*a*) used in conjunction with a skin friction relation such as Eq. (5.142) gives quite a good agreement with experiment for Prandtl numbers in the range $0.5 < \text{Pr} < 30$.

It is useful to slightly rearrange Eq. (5.152*a*) as

$$\text{St}_x = \frac{C_{fx}/2}{\text{Pr}_t + (C_{fx}/2)^{1/2}[5\text{Pr} + 5\ln(5\text{Pr} + 1) - 14\text{Pr}_t]} \tag{5.152b}$$

In this form, the result is more easily compared to Eq. (4.63) and to the Reynolds analogy, Eq. (4.67).

EXAMPLE 5.5 A Turbulent Air Boundary Layer on a Flat Plate

Air at 280 K and 1 atm pressure flows at 50 m/s along a flat plate maintained at 320 K. At a distance of 0.5 m from the leading edge, determine C_{fx}, τ_s, v^*, u_e^+, δ_2, St_x, Nu_x, h_{cx}, and q_s. Also plot the temperature profile.

Solution

Given: A turbulent boundary layer on an isothermal flat plate.

Required: C_{fx}, τ_s, v^*, u_e^+, δ_2, St_x, Nu_x, h_{cx}, q_s, and $T(y)$ at $x = 0.5$ m.

Assumptions: 1. $\text{Pr}_t = 1.0$ in buffer layer, 0.9 in turbulent core.
2. A boundary layer trip causes turbulent flow from the leading edge at $x = 0$.

Evaluate all properties at a mean film temperature of 300 K. From Table A.7, $k = 0.0267$ W/m K, $\rho = 1.177$ kg/m³, $c_p = 1005$ J/kg K, $\nu = 15.66 \times 10^{-6}$ m²/s, $\text{Pr} = 0.69$. The Reynolds number at $x = 0.5$ m is

$$\text{Re}_x = \frac{u_e x}{\nu} = \frac{(50)(0.5)}{15.66 \times 10^{-6}} = 1.596 \times 10^6$$

The boundary layer is turbulent. Equation (5.142) gives the local skin friction coefficient as

$$C_{fx} = 0.026\text{Re}_x^{-1/7} = (0.026)(1.596 \times 10^6)^{-1/7} = 3.38 \times 10^{-3}$$

Hence, the shear stress is

$$\tau_s = C_{fx}\left(\frac{1}{2}\rho u_e^2\right) = (3.38 \times 10^{-3})(0.5)(1.177)(50)^2 = 4.97 \text{ N/m}^2$$

The friction velocity v^* can now be calculated:

$$v^* = (\tau_s/\rho)^{1/2} = (4.97/1.177)^{1/2} = 2.06 \text{ m/s}$$

and the dimensionless free stream velocity u_e^+ is then

$$u_e^+ = u_e/v^* = 50/2.06 = 24.3$$

The momentum thickness δ_2 is easily obtained from $\text{Re}_{\delta_2} = u_e\delta_2/\nu$; using Eq. (5.141),

$$\text{Re}_{\delta_2} = 0.0152\text{Re}_x^{6/7} = (0.0152)(1.596 \times 10^6)^{6/7} = 3.15 \times 10^3$$

$$\delta_2 = \frac{\nu \text{Re}_{\delta_2}}{u_e} = \frac{(15.66 \times 10^{-6})(3.15 \times 10^3)}{50} = 9.87 \times 10^{-4} \text{ m } (\sim 1 \text{ mm})$$

Equation (5.152*a*) gives the local Stanton number as

$$\text{St}_x = \frac{(C_{fx}/2)^{1/2}}{5\text{Pr} + 5\ln(5\text{Pr}+1) + \text{Pr}_t[(2/C_{fx})^{1/2} - 14]}; \qquad \text{Pr} = 0.69; \qquad \text{Pr}_t = 0.9$$

$$(C_{fx}/2)^{1/2} = (3.38 \times 10^{-3}/2)^{1/2} = 4.11 \times 10^{-2}; \qquad (2/C_{fx})^{1/2} = 24.3 \; (= u_e^+)$$

$$\text{St}_x = \frac{4.11 \times 10^{-2}}{3.45 + 7.46 + 9.27} = 2.04 \times 10^{-3}$$

Hence, the local Nusselt number, heat transfer coefficient, and heat flux are

$$\text{Nu}_x = \text{St}_x\text{Re}_x\text{Pr} = (2.04 \times 10^{-3})(1.596 \times 10^6)(0.69) = 2240$$

$$h_{cx} = \text{Nu}_x(k/x) = (2240)(0.0267/0.5) = 119.8 \text{ W/m}^2\text{ K}$$

$$q_s = h_{cx}(T_s - T_e) = (119.8)(320 - 280) = 4790 \text{ W/m}^2$$

To obtain the temperature profile, Eq. (5.151) is substituted into Eq. (5.146), with $\text{Pr}_t = 1.0$ in the buffer layer and $\text{Pr}_t = 0.9$ in the turbulent layer.

$$0 < y^+ \leq 5: \qquad T^+ = \int_0^{y^+} \frac{dy^+}{1/\text{Pr}} = \text{Pr}\, y^+$$

$$5 < y^+ < 30: \qquad T^+ = 5\text{Pr} + \int_5^{y^+} \frac{dy^+}{(1/\text{Pr}) + (0.2y^+ - 1)} = 5\text{Pr} + 5\ln\left[\text{Pr}\left(\frac{y^+}{5} - 1\right) + 1\right]$$

$$30 \leq y^+: \qquad T^+ = 5\text{Pr} + 5\ln(5\text{Pr} + 1) + \int_{30}^{y^+} \frac{dy^+}{(1/0.9)(0.4y^+)}$$

$$= 5\text{Pr} + 5\ln(5\text{Pr} + 1) + 2.25\ln(y^+/30)$$

where $(1/\text{Pr} - 1/\text{Pr}_t)$ has been neglected compared to $\varepsilon^+/\text{Pr}_t$ in the turbulent layer. We need y and T in terms of the dimensionless y^+ and T^+ in order to plot $T(y)$:

$$y = \left(\frac{\nu}{v^*}\right)y^+ = \left(\frac{15.66 \times 10^{-6}}{2.06}\right)y^+ = 7.60 \times 10^{-6} y^+$$

$$T_s - T = \left(\frac{q_s}{\rho c_p v^*}\right)T^+ = \left[\frac{4790}{(1.177)(1005)(2.06)}\right]T^+ = 1.966T^+$$

Finally, for an approximate estimate of the boundary layer thickness, we evaluate the velocity profile Eq. (5.151c) at $u = u_e$:

$$u_e^+ = 24.3 = 2.5\ln\delta^+ + 5.5$$

Solving, $\delta^+ = 1845$.

$$\delta = (\nu/v^*)\delta^+ = (7.60 \times 10^{-6})(1845) = 1.4 \times 10^{-2} \text{ m } (1.4 \text{ cm})$$

A graph of T versus $\ln y$ is appropriate and follows.

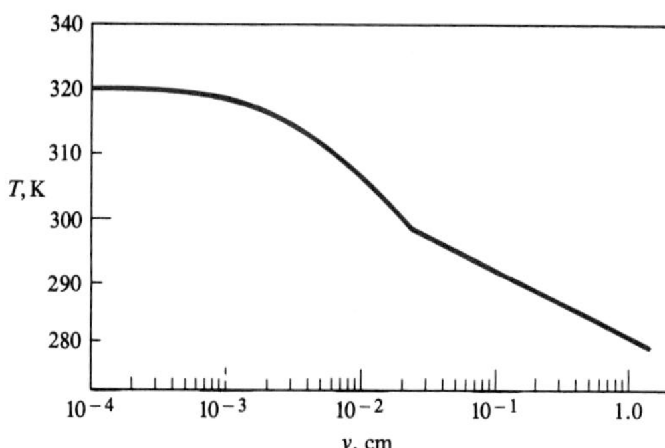

Comments

1. Notice that the friction velocity v^* is on the order of 1 m/s, and the dimensionless free-stream velocity u_e^+ typically has a value between 20 and 30.

2. The temperature profile is seen to have a discontinuity in slope at the edge of the turbulent core. This feature is physically unrealistic but is of no great concern.

5.5.3 Flow in a Tube

As was the case for flow along a flat plate, the fluid mechanics problem is solved prior to solving the heat transfer problem. Under usual circumstances in tube flow, the turbulence is established at Reynolds numbers greater than about 10,000. We do not consider lower Reynolds numbers, where the turbulence may be intermittent.

The Momentum Conservation Equation

For fully developed turbulent flow in a tube with constant properties, the differential momentum conservation equation is

$$\frac{d}{dr}\left[r(\mu + \mu_t)\frac{du}{dr}\right] = r\frac{dP}{dx} \tag{5.153}$$

which differs from the laminar flow equation, Eq. (5.14), only in the replacement of μ by $\mu + \mu_t$. Dividing by ρr gives

$$\frac{1}{r}\frac{d}{dr}\left[r(\nu + \varepsilon_M)\frac{du}{dr}\right] = \frac{1}{\rho}\frac{dP}{dx} \tag{5.154}$$

Specification of ε_M allows integration of the equation to obtain the velocity profile and the friction factor as a function of Reynolds number.

The Eddy Viscosity

Turbulent flow in a tube has an inner region adjacent to the wall in which the turbulence structure is essentially identical to that for flow along a flat plate. The same inner region velocity profile $u^+(y^+)$, mixing length variation $\ell(y)$, and eddy viscosity variation $\varepsilon_M(y)$ can be used. Deviations from the logarithmic velocity profile become discernible for $y/R > 0.15$; however, the deviations are small. For most purposes, the use of the logarithmic profile in this region, called the *turbulent core*, is adequate, even though a nonzero velocity gradient at the center of the pipe is implied. Alternatively, the mixing length or eddy viscosity in the turbulent core can be specified and the velocity profile calculated. Using his own experimental data, J. Nikuradse [12] developed a useful mixing length expression:

$$\frac{\ell}{R} = 0.14 - 0.08\left(1 - \frac{y}{R}\right)^2 - 0.06\left(1 - \frac{y}{R}\right)^4 \tag{5.155}$$

which has the limits of $\ell = 0.4y$ for $y/R \to 0$, and $\ell/R = 0.14$ for $y/R \to 1$. Equation (5.155) is plotted in Fig. 5.36. A very simple eddy viscosity expression for the core was proposed by J. O. Hinze [13]:

$$\frac{\varepsilon_M}{\nu} = 0.035(f/8)^{1/2}\mathrm{Re}_D \tag{5.156}$$

where f is the Darcy friction factor.

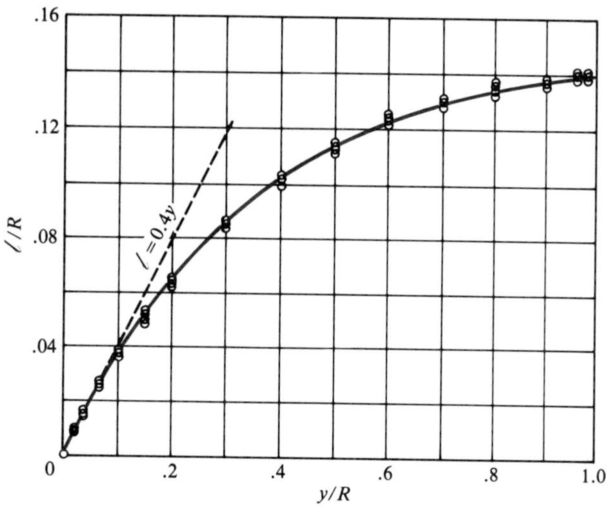

Figure 5.36 Nikuradse's mixing length data and distribution, Eq. (5.155).

Solution of the Momentum Equation

We proceed by regarding the pressure gradient as known and constant; then, using the definitions of f, C_f, and v^*,

$$f = \left(-\frac{dP}{dx}\right)D/(1/2)\rho u_b^2; \quad C_f = \frac{\tau_s}{(1/2)\rho u_b^2}; \quad v^* = \left(\frac{\tau_s}{\rho}\right)^{1/2}$$

the pressure gradient and friction velocity are related as

$$-\frac{dP}{dx} = \frac{2\tau_s}{R} = \frac{2v^{*2}\rho}{R}$$

Substituting in the momentum equation, Eq. (5.154),

$$\frac{1}{r}\frac{d}{dr}\left[r(\nu + \varepsilon_M)\frac{du}{dr}\right] = -\frac{2v^{*2}}{R} \tag{5.157}$$

where the friction velocity v^* is now known, and integrating with boundary conditions $r = 0$: $du/dr = 0$ and $r = R$: $u = 0$ gives the velocity profile:

$$u = -\frac{v^{*2}}{R}\int_R^r \frac{r\,dr}{(\nu + \varepsilon_M)}$$

Introducing the inner variables u^+ and y^+ defined by Eq. (5.126), where $y = R - r$ is measured from the tube wall, and the dimensionless total viscosity $\varepsilon^+ = 1 + \varepsilon_M/\nu$, gives the dimensionless velocity profile:

$$u^+ = \frac{1}{R^+}\int_0^{y^+} \frac{(R^+ - y^+)dy^+}{\varepsilon^+} \tag{5.158}$$

where $R^+ = v^*R/\nu$. The bulk velocity u_b is

$$u_b = \frac{1}{\pi R^2}\int_0^R u2\pi r\,dr \qquad \text{or} \qquad u_b^+ = \frac{2}{R^{+2}}\int_0^{R^+} u^+(R^+ - y^+)dy^+ \tag{5.159}$$

and, since $\tau_s = \rho v^{*2}$,

$$\frac{C_f}{2} = \frac{f}{8} = \frac{v^{*2}}{u_b^2} = \frac{1}{u_b^{+2}} \tag{5.160}$$

The Reynolds number is

$$\text{Re}_D = \frac{u_b(2R)}{\nu} = 2u_b^+R^+ \tag{5.161}$$

For a specified ε^+ profile, we can now choose a range of R^+ values and numerically integrate Eqs. (5.158) and (5.159) to obtain u_b^+. The corresponding values of f and Re_D can then be calculated from Eqs. (5.160) and (5.161), respectively, to yield a graph of f versus Re_D. However, our ε^+ profile is based on measured velocity profiles for tube flow, so for this simple flow we can use a selected velocity profile directly in Eq. (5.159). Let us assume that the logarithmic profile is valid from the wall to the centerline; then, using $1/\kappa = 2.5, C = 5.5$,

$$\begin{aligned} u_b^+ &= \frac{2}{R^{+2}}\int_0^{R^+}(2.5\ln y^+ + 5.5)(R^+ - y^+)dy^+ \\ &= 2.5\ln R^+ + 1.75 \end{aligned}$$

But

$$\frac{f}{8} = \frac{1}{u_b^{+2}} \qquad \text{and} \qquad R^+ = \frac{1}{2}\text{Re}_D(f/8)^{1/2}$$

Thus, using base 10 logarithms,

$$\frac{1}{f^{1/2}} = 2.035\log(\text{Re}_D f^{1/2}) - 0.913 \tag{5.162}$$

In 1935, L. Prandtl [14] adjusted these constants to give a better fit to experimental data to obtain his well-known formula,

$$\frac{1}{f^{1/2}} = 2.0\log(\text{Re}_D f^{1/2}) - 0.8 \tag{5.163}$$

Notice that Eq. (5.163) is an implicit relation for f, which makes it difficult to evaluate f when given Re_D. Petuhkov's formula, Eq. (4.42), is an explicit relation and is easier to use.

The Energy Conservation Equation

For fully developed heat transfer with constant fluid properties and negligible viscous dissipation, the differential energy conservation equation is

$$\rho u c_p r \frac{\partial T}{\partial x} = \frac{\partial}{\partial r}\left[r(k + k_t)\frac{\partial T}{\partial r}\right] \tag{5.164}$$

Equation (5.164) differs from the laminar flow equation, Eq. (5.23), only in that the molecular conductivity k has been replaced by the total conductivity $(k + k_t)$, which is a function of r and hence is kept inside the derivative $\partial/\partial r$. Rearranging,

$$u\frac{\partial T}{\partial x} = \frac{1}{r}\frac{\partial}{\partial r}\left[r(\alpha + \varepsilon_H)\frac{\partial T}{\partial r}\right] \tag{5.165}$$

As for the flat plate boundary layer, a constant value of order unity is adequate for the turbulent Prandtl number, and with $\varepsilon_H = \varepsilon_M/\mathrm{Pr}_t$, and ε_M already specified, the problem can be solved.

Solution of the Energy Equation

Equation (5.165) can be solved in the same way as for the laminar flow problem in Section 5.3.2, except now the integrals must be evaluated numerically. However, as was the case for the flat plate in Section 5.5.2, an approximate result can be obtained quite easily using the three-layer eddy viscosity specification. We assume a constant heat flux across the viscous sublayer and buffer layer. With $y = R - r$ and $y \ll R$, where R is the tube radius, Eq. (5.164) near the wall can be written as

$$\rho c_p u\frac{\partial T}{\partial x} \simeq -\frac{\partial q}{\partial y}, \qquad \text{where} \quad q = -(k + k_t)\frac{\partial T}{\partial y}$$

so that a constant heat flux implies that streamwise convection in these layers is negligible. Then

$$q = -(k + k_t)\frac{\partial T}{\partial y} = -\rho c_p(\alpha + \varepsilon_H)\frac{\partial T}{\partial y} = \text{Constant} = q_s$$

Integrating with $T = T_s$ at $y = 0$ gives

$$\rho c_p(T_s - T) = q_s \int_0^y \frac{dy}{\alpha + \varepsilon_H} = q_s \int_0^y \frac{dy}{\nu/\mathrm{Pr} + \varepsilon_M/\mathrm{Pr}_t}$$

Introducing the dimensionless variables y^+, u^+, ε^+, and $T^+ = (T_s - T)\rho c_p v^*/q_s$:

$$T^+ = \int_0^{y^+} \frac{dy^+}{1/\mathrm{Pr} + (1/\mathrm{Pr}_t)(\varepsilon^+ - 1)} \tag{5.166}$$

Substituting values of ε^+ from Eq. (5.151) and $\text{Pr}_t = 1.0$ in the buffer layer, the temperature profile is obtained as

$$0 < y^+ \leq 5: \quad T^+ = \text{Pr}\, y^+ \tag{5.167a}$$

$$5 < y^+ \leq 30: \quad T^+ = 5\text{Pr} + 5\ln\left[\text{Pr}\left(\frac{y^+}{5} - 1\right) + 1\right] \tag{5.167b}$$

In the turbulent core, we assume a linear heat flux distribution $q = q_s(1 - y/R)$, which for a uniform wall heat flux is equivalent to assuming that the velocity profile is flat for $y^+ > 30$, and this is not unreasonable considering the characteristic shape of the turbulent velocity profile. Then

$$q_s\left(1 - \frac{y}{R}\right) = -\rho c_p(\alpha + \varepsilon_H)\frac{\partial T}{\partial y}$$

Introducing dimensionless variables and integrating from $y^+ = 30$ gives

$$T^+ - T^+|_{y^+=30} = \int_{30}^{y^+} \frac{(1 - y^+/R^+)\,dy^+}{1/\text{Pr} + (1/\text{Pr}_t)(\varepsilon^+ - 1)}$$

Then, with specification of $\varepsilon^+ = 0.4y^+$ from Eq. (5.151c), and assuming $1/\text{Pr} - 1/\text{Pr}_t \ll \varepsilon^+/\text{Pr}_t$, which essentially neglects molecular conduction in the turbulent core,

$$T^+ - T^+|_{y^+=30} = \frac{\text{Pr}_t}{0.4}\left[\ln\left(\frac{y^+}{30}\right) - \frac{(y^+ - 30)}{R^+}\right] \tag{5.167c}$$

Using Eq. (5.167b) and taking $\text{Pr}_t = 0.9$ in the turbulent core gives the centerline temperature for $R^+ \gg 30$ as

$$T_c^+ \simeq 5\text{Pr} + 5\ln(5\text{Pr} + 1) + 2.25\left[\ln\left\{\frac{\text{Re}_D}{60}\left(\frac{f}{8}\right)^{1/2}\right\} - 1\right] \tag{5.168}$$

The Stanton number is

$$\text{St} = \frac{q_s}{\rho c_p u_b(T_s - T_b)} = \frac{1}{u_b^+ T_b^+} = \left(\frac{f}{8}\right)^{1/2}\frac{1}{T_b^+} \tag{5.169}$$

and thus,

$$\text{St} = \left(\frac{f}{8}\right)^{1/2}\frac{1}{T_c^+}\left(\frac{T_b^+}{T_c^+}\right)^{-1} \quad \text{and} \quad \frac{T_b^+}{T_c^+} = \frac{T_s - T_b}{T_s - T_c}$$

Substituting from Eq. (5.168) and rearranging,

$$\text{St} = \frac{f/8}{\left(\dfrac{T_s - T_b}{T_s - T_c}\right)\left(\dfrac{f}{8}\right)^{1/2}\left\{5\text{Pr} + 5\ln(5\text{Pr} + 1) + 2.25\left[\ln\left\{\dfrac{\text{Re}_D}{60}\left(\dfrac{f}{8}\right)^{1/2}\right\} - 1\right]\right\}} \tag{5.170}$$

This result is more complicated than the corresponding result for a flat plate, Eq. (5.152), because of the need to define the heat transfer coefficient in terms of the bulk temperature rather than the centerline temperature. However, the turbulent temperature profile is rather flat, particularly at higher Prandtl numbers, so that the factor $(T_s - T_b)/(T_s - T_c)$ is close to unity. It can be evaluated by substituting the temperature profile, Eqs. (5.167*a,b,c*), into the definition of bulk temperature:

$$T_b = \frac{1}{u_b \pi R^2} \int_0^R uT 2\pi (R-y)dy$$

or

$$T_b^+ = \frac{2}{u_b^+ R^{+2}} \int_0^{R^+} u^+ T^+ (R^+ - y^+) dy^+ \tag{5.171}$$

The ratio $(T_s - T_b)/(T_s - T_c)$ is relatively insensitive to the choice of ε^+ profile: typical results are given in Table 5.4. Equation (5.170) agrees well with experiment for $\mathrm{Re}_D > 10{,}000$ and $0.5 < \mathrm{Pr} < 30$.

Table 5.4 The ratio $(T_s - T_b)/(T_s - T_c)$ for fully developed turbulent flow in a pipe.

	Re_D			
Pr	10^4	10^5	10^6	10^7
10^{-2}	0.589	0.639	0.738	0.813
10^{-1}	0.692	0.761	0.823	0.864
1	0.865	0.877	0.897	0.912
10	0.958	0.962	0.963	0.966
10^2	0.992	0.993	0.993	0.994
10^3	1.000	1.000	1.000	1.000

The heat transfer process can be viewed as involving three resistances in series, represented by the additive terms in the denominator of Eq. (5.170). The relative magnitudes of these resistances depend on the molecular Prandtl number and Reynolds number. Since $(f/8)^{1/2} \propto \mathrm{Re}^{-0.1}$ from Eq. (4.43), the resistance of the turbulent core increases with Reynolds number. Also, for $\mathrm{Pr} \gg 1$, the resistances of the viscous and buffer layers dominate, whereas for $\mathrm{Pr} \ll 1$, the resistances of these layers play a minor role. Temperature profiles shown in Fig. 5.37 reflect this behavior: for high Pr, the profiles are very steep near the wall, whereas for low Pr, the temperature variation extends far into the turbulent core. It is clear that accurate heat transfer calculations for high-Pr fluids require precise specification of ε_H close to the wall, and for low-Pr fluids, precise specification of ε_H in the core is required. Notice also that molecular conduction in the turbulent core cannot be neglected for very low-Pr liquid metals.

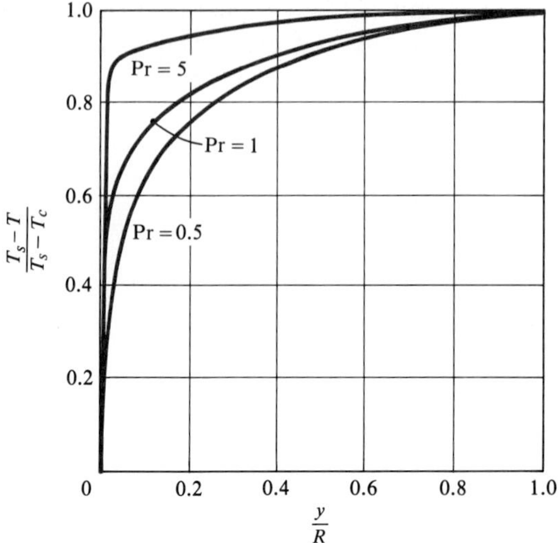

Figure 5.37 Temperature profiles for turbulent flow in a tube: effect of Prandtl number.

EXAMPLE 5.6 Turbulent Flow of Water in a Tube

Water flows at 4 m/s in a 5 cm–I.D. tube. At a particular location, the wall temperature is 340 K and the bulk water temperature is 300 K. (i) Determine f, τ_s, dP/dx, v^*, u_b^+. (ii) Compare St values given by Eq. (4.45) and Eq. (5.170). (iii) Plot ε^+ profiles according to Eq. (5.151), and compare the centerline value with the value given by Eq. (5.156).

Solution

Given: Water flowing in a pipe.

Required: (i) f, τ_s, dP/dx, v^*, u_b^+; (ii) St; (iii) ε^+ profile.

Assumptions:
1. $\Pr_t = 1.0$ in buffer layer, 0.9 in turbulent core.
2. Fully developed heat transfer (no entrance effects).

Evaluate all fluid properties at a mean film temperature of 320 K. From Table A.8, $k = 0.641$ W/m K, $\rho = 989$ kg/m^3, $c_p = 4174$ J/kg K, $\nu = 0.59 \times 10^{-6}$ m^2/s, Pr = 3.8.

(i) The Reynolds number is

$$\text{Re}_D = \frac{u_b D}{\nu} = \frac{(4)(0.05)}{0.59 \times 10^{-6}} = 3.39 \times 10^5$$

that is, turbulent flow. The friction factor is obtained from Eq. (4.42) as

$$f = [0.790 \ln(\text{Re}_D - 1.64)^{-2} = [0.790 \ln(3.39 \times 10^5) - 1.64]^{-2} = 1.411 \times 10^{-2}$$

Hence, the wall shear stress is

$$\tau_s = (C_f/2)(\rho u_b^2) = (f/8)(\rho u_b^2) = (1.411 \times 10^{-2}/8)(989)(4)^2 = 27.9 \text{ N/m}^2$$

A force balance on the fluid for an element of tube Δx long relates shear stress to pressure gradient:

$$\tau_s \pi D \Delta x = -\Delta P(\pi D^2/4) \qquad \text{or} \qquad \frac{dP}{dx} = -\frac{4\tau_s}{D}$$

$$\frac{dP}{dx} = -\frac{(4)(27.9)}{0.05} = -2230 \text{ Pa/m}$$

The friction velocity v^* and dimensionless bulk velocity u_b^+ are

$$v^* = (\tau_s/\rho)^{1/2} = (27.9/989)^{1/2} = 0.168 \text{ m/s}$$

$$u_b^+ = u_b/v^* = 4/0.168 = 23.8$$

(ii) Equation (4.45) relates the Nusselt number to the friction factor:

$$f/8 = 1.411 \times 10^{-2}/8 = 1.764 \times 10^{-3}$$

$$\text{Nu}_D = \frac{(f/8)(\text{Re}_D - 1000)\text{Pr}}{1 + 12.7(f/8)^{1/2}(\text{Pr}^{2/3} - 1)} = \frac{(1.764 \times 10^{-3})(3.39 \times 10^5 - 1000)(3.8)}{1 + 12.7(1.764 \times 10^{-3})^{1/2}(3.8^{2/3} - 1)} = 1283$$

$$\text{St} = \frac{\text{Nu}}{\text{RePr}} = \frac{1283}{(3.39 \times 10^5)(3.8)} = 9.96 \times 10^{-4}$$

Equation (5.170) gives the Stanton number as

$$\text{St} = \frac{f/8}{\left(\dfrac{T_s - T_b}{T_s - T_c}\right)\left(\dfrac{f}{8}\right)^{1/2}\left\{5\text{Pr} + 5\ln(5\text{Pr} + 1) + 2.25\left[\ln\left\{\dfrac{\text{Re}_D}{60}\left(\dfrac{f}{8}\right)^{1/2}\right\} - 1\right]\right\}}$$

Interpolating in Table 5.4,

$$\text{St} = \frac{1.764 \times 10^{-3}}{(0.94)(4.20 \times 10^{-2})(19.0 + 14.98 + 10.06)} = 1.01 \times 10^{-3}$$

(iii) The dimensionless tube radius is

$$R^+ = \frac{v^* R}{\nu} = \frac{(0.168)(0.025)}{0.59 \times 10^{-6}}$$

$$= 7120$$

Equation (5.151) gives the three-layer ε^+ specification:

$$0 < y^+ \le 5: \qquad \varepsilon^+ = 1$$

$$5 \le y^+ < 30: \qquad \varepsilon^+ = 0.2y^+$$

$$30 \le y^+ < 7120: \qquad \varepsilon^+ = 0.4y^+$$

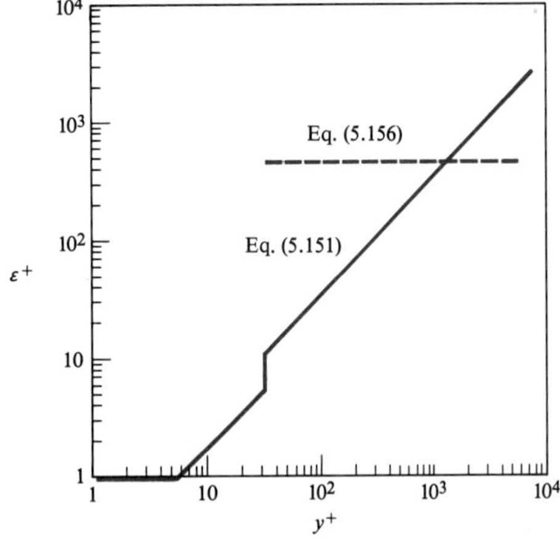

At $y^+ = R^+ = 7120$, $\varepsilon^+ = 2850$. On the other hand, Eq. (5.156) gives a constant value in the turbulent core of

$$\varepsilon^+ = 1 + 0.035(f/8)^{1/2}\text{Re}_D = 1 + (0.035)(1.764 \times 10^{-3})^{1/2}(3.39 \times 10^5) = 499$$

Comments

1. The two values of St differ by only 1.4%.
2. Equation (5.151) gives values of ε^+ near the pipe center that are too large, but the effect on shear and heat transfer is negligible.

5.5.4 More Advanced Turbulence Models

The algebraic mixing length and eddy viscosity expressions used in Sections 5.5.2 and 5.5.3 are adequate for the simple flows that were considered. However, for more complicated flows, such as those having recirculating regions, a more sophisticated approach is required. In recent years, considerable effort has been expended to develop more generally applicable turbulence models. Instead of using a simple algebraic specification of the mixing length, these models generally require the solution of additional partial differential equations, of similar form to the momentum and energy equation, in order to obtain the eddy viscosity. For example, one popular method solves equations governing the kinetic energy of the turbulence and its dissipation rate. Although considerable progress has been made by using such models, much remains to be done. Further development of such models will remain an important research area for many more years. The student should consult advanced texts and the journal literature for their current status.

5.6 SIMILARITY AND MODELING

The concept of *similarity* is very important in heat transfer theory. This concept was introduced for conduction along a fin in Section 2.4.5, and for transient conduction in a slab in Section 3.4.1. It is the principle underlying the use of dimensionless groups for correlation of experimental data, as illustrated for flow over a cylinder in Section 4.2.3. Also, it is the principle underlying the use of models to simulate full-scale equipment. For the principle of similarity to be used, the dimensionless groups pertinent to the problem must be obtained. These are best obtained by making the governing equations and boundary conditions dimensionless. Such a procedure was used in the conduction analyses mentioned above, and it will now be demonstrated for convection.

5.6.1 Dimensionless Equations and Boundary Conditions

Three examples will be considered, namely, the laminar boundary layer on a flat plate, fully developed tube flow, and Couette flow. In each case the governing equations and boundary conditions will be made dimensionless in order to obtain the relevant dimensionless groups.

The Constant-Property Laminar Boundary Layer on a Flat Plate

From Section 5.4, the governing equations and boundary conditions are

Mass: $$\frac{\partial u}{\partial x} + \frac{\partial v}{\partial y} = 0 \tag{5.37}$$

x momentum: $$u\frac{\partial u}{\partial x} + v\frac{\partial u}{\partial y} = \nu\frac{\partial^2 u}{\partial y^2} \tag{5.39}$$

Energy: $$u\frac{\partial T}{\partial x} + v\frac{\partial T}{\partial y} = \alpha\frac{\partial^2 T}{\partial y^2} \tag{5.42}$$

$$y = 0: \quad u = 0, v = 0, T = T_s \tag{5.172a}$$

$$x = 0, y \to \infty: \quad u = u_e, T = T_e \tag{5.172b}$$

We choose the length of the plate L as the characteristic length, the free-stream velocity u_e as the characteristic velocity, and $(T_s - T_e)$ as the characteristic temperature difference for this problem. To render the equations dimensionless, we simply scale all lengths with L, all velocities with u_e, and all temperature differences with $(T_s - T_e)$. We define the dimensionless variables:

$$x^* = \frac{x}{L}; \qquad y^* = \frac{y}{L}$$

$$u^* = \frac{u}{u_e}; \qquad v^* = \frac{v}{u_e}$$

$$T^* = \frac{T - T_e}{T_s - T_e}$$

The rules for the transformation of variables are

$$\frac{\partial}{\partial x} = \frac{\partial}{\partial x^*}\frac{dx^*}{dx} = \frac{1}{L}\frac{\partial}{\partial x^*}; \qquad \frac{\partial}{\partial y} = \frac{1}{L}\frac{\partial}{\partial y^*}$$

$$u = u_e u^*, \qquad v = u_e v^*, \qquad T - T_e = (T_s - T_e)T^*$$

The mass conservation equation becomes

$$\frac{u_e}{L}\frac{\partial u^*}{\partial x^*} + \frac{u_e}{L}\frac{\partial v^*}{\partial y^*} = 0$$

or

$$\frac{\partial u^*}{\partial x^*} + \frac{\partial v^*}{\partial y^*} = 0 \tag{5.173}$$

The x-momentum equation becomes

$$\frac{u_e^2}{L}u^*\frac{\partial u^*}{\partial x^*} + \frac{u_e^2}{L}v^*\frac{\partial u^*}{\partial y^*} = \nu\frac{u_e}{L^2}\frac{\partial^2 u^*}{\partial y^{*2}}$$

or

$$u^*\frac{\partial u^*}{\partial x^*} + v^*\frac{\partial u^*}{\partial y^*} = \frac{\nu}{u_e L}\frac{\partial^2 u^*}{\partial y^{*2}} \quad \textbf{(5.174)}$$

Since T_e is a constant, it can be subtracted from T throughout the energy equation, which after transformation becomes

$$\frac{u_e(T_s - T_e)}{L}u^*\frac{\partial T^*}{\partial x^*} + \frac{u_e(T_s - T_e)}{L}v^*\frac{\partial T^*}{\partial y^*} = \frac{\alpha(T_s - T_e)}{L^2}\frac{\partial^2 T^*}{\partial y^{*2}}$$

or

$$u^*\frac{\partial T^*}{\partial x^*} + v^*\frac{\partial T^*}{\partial y^*} = \frac{\alpha}{u_e L}\frac{\partial^2 T^*}{\partial y^{*2}} \quad \textbf{(5.175)}$$

The boundary conditions Eqs. (5.172) become

$$y^* = 0: \quad u^* = 0, v^* = 0, T^* = 1 \quad \textbf{(5.176}a\textbf{)}$$

$$y^* \to \infty, x^* = 0: \quad u^* = 1, T^* = 0 \quad \textbf{(5.176}b\textbf{)}$$

In this dimensionless statement of the problem there are only two parameters, both dimensionless groups. In Eq. (5.174) there is the Reynolds number $\text{Re} = u_e L/\nu$, and in Eq. (5.175) there is the Peclet number $u_e L/\alpha$. However, we usually choose to replace the Peclet number by the product of Reynolds and Prandtl numbers, $\text{Pe} = \text{RePr}$, $\text{Pr} = \nu/\alpha$, since the Prandtl number is simply a fluid property.

Our primary purpose in solving these conservation equations is to obtain the skin friction and heat transfer. The shear stress exerted by the fluid on the wall is

$$\tau_s = \mu \left.\frac{\partial u}{\partial y}\right|_0$$

$$= \mu\frac{u_e}{L}\left.\frac{\partial u^*}{\partial y^*}\right|_{y^*=0}$$

The dimensionless skin friction coefficient C_{fx} is defined as

$$\frac{C_{fx}}{2} = \tau_s^* = \frac{\tau_s}{\rho u_e^2}$$

Thus,

$$\frac{C_{fx}}{2} = \frac{\nu}{u_e L}\left.\frac{\partial u^*}{\partial y^*}\right|_{y^*=0} = \frac{1}{\text{Re}}\left.\frac{\partial u^*}{\partial y^*}\right|_{y^*=0} \quad \textbf{(5.177)}$$

The heat transfer across the wall is

$$q_s = -k \left.\frac{\partial T}{\partial y}\right|_0 = \frac{k(T_s - T_e)}{L}\left(-\left.\frac{\partial T^*}{\partial y^*}\right|_{y^*=0}\right)$$

The dimensionless Stanton number for heat transfer is defined as

$$\mathrm{St}_x = q_s^* = \frac{q_s}{(\rho c_p u_e)(T_s - T_e)} = -\frac{1}{\mathrm{RePr}}\left.\frac{\partial T^*}{\partial y^*}\right|_{y^*=0} \quad \textbf{(5.178)}$$

We have shown that the dimensionless mass and momentum conservation equations together contain only one parameter, the Reynolds number. The dimensionless boundary conditions and skin friction coefficient introduce no further parameters. It follows that the dimensionless velocity and hence dimensionless skin friction are functions only of the dimensionless independent variables and Reynolds number. In particular, the average skin friction coefficient

$$\frac{\overline{C_f}}{2} = \overline{\tau}_s^* = \int_0^1 \tau_s^* dx^* \quad \textbf{(5.179)}$$

is given by

$$\frac{\overline{C_f}}{2} = \frac{\overline{C_f}}{2}(\mathrm{Re}) \quad \textbf{(5.180)}$$

The dimensionless energy equation contains two dimensionless parameters, the Reynolds and Prandtl numbers, appearing as the product RePr = Pe, the Peclet number. However, the Reynolds number itself is an independent parameter through the coupling of the energy equation to the momentum equation via the velocity field (u^*, v^*). The average Stanton number is

$$\overline{\mathrm{St}} = \overline{q}_s^* = \int_0^1 q_s^* dx^*$$

Thus,

$$\overline{\mathrm{St}} = \overline{\mathrm{St}}(\mathrm{Re}, \mathrm{Pr}) \qquad \text{or} \qquad \overline{\mathrm{St}} = \overline{\mathrm{St}}(\mathrm{Pe}, \mathrm{Pr}) \quad \textbf{(5.181)}$$

We see that the dimensionless groups characterizing a convection problem can be obtained from the governing equations without actually solving the equations. Thus, even in performing an experimental investigation, it is good practice to attempt to write down the equations governing the problem and so obtain candidate dimensionless parameters. With this approach, it is less likely that an essential variable will be omitted than when using dimensionless analysis based on the Buckingham pi theorem. Furthermore, it will be shown in Section 5.8 that by careful choice of characteristic lengths, velocities, and other properties, it is possible to deduce when various dimensionless parameters have negligible influence and can be discarded.

Laminar Flow in a Tube

We next consider steady laminar flow in a tube with fully developed hydrodynamics, constant properties, and a uniform wall temperature. The energy equation is Eq. (5.25):

$$2u_b\left[1 - \left(\frac{r}{R}\right)^2\right]\frac{\partial T}{\partial x} = \frac{\alpha}{r}\frac{\partial}{\partial r}\left(r\frac{\partial T}{\partial r}\right) \tag{5.25}$$

We must now recognize that there are two characteristic lengths for this problem: the length of tube, L, and the radius, R (or diameter, D). We might expect that, in general, heat transfer in a short, fat tube will be different from that in a long, thin tube. Appropriate dimensionless variables are then

$$x^* = \frac{x}{L}, \qquad r^* = \frac{r}{R}, \qquad T^* = \frac{T - T_b}{T_s - T_b}$$

Equation (5.25) transforms into

$$\frac{2u_b(T_s - T_b)}{L}\left[1 - r^{*2}\right]\frac{\partial T^*}{\partial x^*} = \frac{\alpha(T_s - T_b)}{R^2}\frac{1}{r^*}\frac{\partial}{\partial r^*}\left(r^*\frac{\partial T^*}{\partial r^*}\right)$$

or

$$\frac{u_bD^2}{\alpha L}\left[1 - r^{*2}\right]\frac{\partial T^*}{\partial x^*} = \frac{2}{r^*}\frac{\partial}{\partial r^*}\left(r^*\frac{\partial T^*}{\partial r^*}\right) \tag{5.182}$$

where $(u_bD^2/\alpha L)$ is dimensionless. The boundary conditions introduce no further dimensionless parameters, so that the solution of Eq. (5.182) must give

$$\overline{\text{Nu}}_D = \frac{\overline{q}_s(2R)}{k(T_s - T_b)} = -2\int_0^1 \left.\frac{\partial T^*}{\partial r^*}\right|_{r^*=R^*} dx^* = \overline{\text{Nu}}_D\left(\frac{u_bD^2}{\alpha L}\right) \tag{5.183}$$

The dimensionless group $u_bD^2/\alpha L$ is the **Graetz number**, which at first appears new. But the correlation for this problem when the wall temperature is uniform is Eq. (4.50),

$$\overline{\text{Nu}}_D = 3.66 + \frac{0.065(D/L)\text{Re}_D\text{Pr}}{1 + 0.04[(D/L)\text{Re}_D\text{Pr}]^{2/3}}$$

and

$$(D/L)\text{Re}_D\text{Pr} = \frac{D}{L}\frac{u_bD}{\nu}\frac{\nu}{\alpha} = \frac{u_bD^2}{\alpha L}$$

Thus, the group has in fact been encountered before, but as a product of more familiar groups, the Reynolds and Prandtl numbers and a geometric parameter, D/L.

Recall that for fully developed heat transfer, the Nusselt number is simply a constant; for example, for a uniform wall heat flux, Eq. (5.36) gives $\mathrm{Nu}_D = 48/11$. The energy equation for this problem was Eq. (5.32),

$$\left[1 - \left(\frac{r}{R}\right)^2\right]\frac{4q_s}{kR} = \frac{1}{r}\frac{d}{dr}\left(r\frac{d\mathcal{R}}{dr}\right)$$

where $\mathcal{R} = T - T_b$. Transforming this equation with $r^* = r/R$ and $T^* = \mathcal{R}/(T_s - T_b)$ gives

$$[1 - r^{*2}]\frac{4q_s R}{(T_s - T_b)k} = \frac{1}{r^*}\frac{d}{dr^*}\left(r^*\frac{dT^*}{dr^*}\right)$$

or

$$2[1 - r^{*2}]\mathrm{Nu}_D = \frac{1}{r^*}\frac{d}{dr^*}\left(r^*\frac{dT^*}{dr^*}\right) \qquad \textbf{(5.184)}$$

The transformed boundary conditions introduce no further parameters, so that Eq. (5.184) is simply an equation for the Nusselt number, which has a unique solution, $\mathrm{Nu}_D = \text{Constant}$.

Couette Flow

As a last example, we consider the high-speed Couette flow analyzed in Section 5.2.1, for which the energy equation was Eq. (5.5):

$$k\frac{d^2T}{dy^2} + \mu\left(\frac{du}{dy}\right)^2 = 0$$

An appropriate characteristic length is the distance between plates, L, and a characteristic velocity is that of the moving plate, u_e. The dimensionless variables are then

$$y^* = \frac{y}{L}; \qquad u^* = \frac{u}{u_e}; \qquad T^* = \frac{T - T_e}{T_s - T_e}$$

and Eq. (5.5) transforms into

$$\frac{k(T_s - T_e)}{L^2}\frac{d^2T^*}{dy^{*2}} + \frac{\mu u_e^2}{L^2}\left(\frac{du^*}{dy^*}\right)^2 = 0$$

or

$$\frac{d^2T^*}{dy^{*2}} = -\frac{\mu u_e^2}{k(T_s - T_e)}\left(\frac{du^*}{dy^*}\right)^2 = -\mathrm{Br}\left(\frac{du^*}{dy^*}\right)^2 \qquad \textbf{(5.185)}$$

where $\mathrm{Br} = \mu u_e^2/k(T_s - T_e)$ is the Brinkman number. The boundary conditions

Eqs. (5.8) transform into

$$y^* = 0: \quad T^* = 1; \qquad y^* = 1: \quad T^* = 0 \tag{5.186a,b}$$

Thus, the dimensionless temperature profile $T^*(y^*)$ depends only on the Brinkman number. In particular, the adiabatic wall temperature, which is the special case when the temperature gradient at the lower plate is zero, occurs at a unique value of the Brinkman number; the analysis of Section 5.2.1 gave Br $= 2$.

5.6.2 Modeling

Engineers perform a variety of experimental work. An experiment may have the simple objective of measuring thermophysical properties such as thermal conductivity or emittance. Another type of experiment is the evaluation of prototype equipment, for example, the testing of an automobile engine coupled to a dynamometer. But the most interesting and challenging experiments are those in which the engineer chooses to test a model of a system under consideration, because it is either impossible or impractical to test the system itself. Perhaps the most well-known application of modeling is the use of wind tunnels for aerodynamic testing. The advance of the art of aeronautics would have been severely handicapped had wind tunnels not been widely used for model testing.

The concept of using a model to obtain experimental data was introduced in Section 4.2.3. In the context of similarity theory, the basic principle underlying modeling can be stated as follows. If the same dimensionless differential equations subject to the same dimensionless boundary conditions govern both the modeling experiment and the full-scale prototype, the dimensionless solutions are identical. Then, if the dimensionless parameters involved are matched, the dimensionless wall transfer rates (e.g., the skin friction coefficient and Nusselt number) will be identical. For example, if the skin friction on a one-fifth scale model automobile is to be measured in a wind tunnel, the air speed must be five times the design value in order to match the Reynolds number, Re $= VL/\nu$. Since the Mach number will be five times larger than the design value, care must be taken to ensure that it is small enough to imply incompressible flow.

In planning an experiment, the engineer must carefully examine the dimensionless parameters involved to avoid acquiring unnecessary data. For example, if heat transfer for laminar flow in a duct of unusual cross section is to be determined experimentally, then Eq. (5.183), with diameter D replaced by hydraulic diameter D_h, suggests that the **Graetz number** $u_b D_h^2/\alpha L$ is the only dimensionless parameter involved. Thus, the experiment should be planned so as to vary the Graetz number over the same range as encountered in the prototype system. This can be done by varying any of the variables, u_b, D_h, α, L, whichever is most convenient. If the prototype application involves a number of different fluids with different thermal diffusivities, it may be possible to test just one fluid in the experiment and obtain the desired range by varying u_b and/or L. The end result is simply a graph of $\overline{\text{Nu}}$ versus $u_b D_h^2/\alpha L$, as shown in Fig. 5.38.

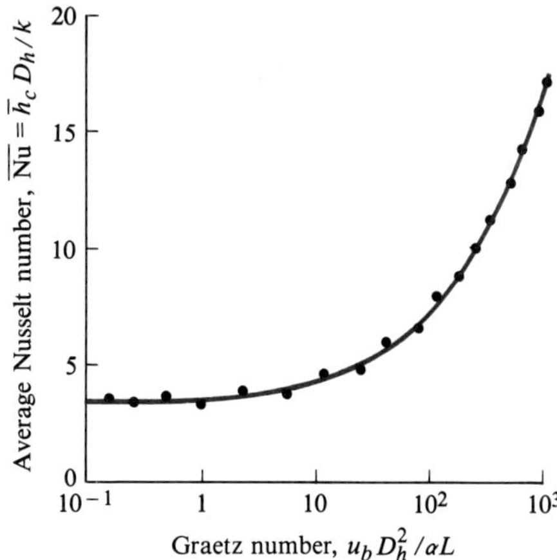

Figure 5.38 Correlation of convective heat transfer data for laminar flow in a duct using the Graetz number.

Often, exact modeling of a particular system is not feasible. The student might consider the problem of obtaining a heat transfer data base for the design of the solar power plant central receiver of Example 4.5, or for the heat shield of a vehicle entering the atmosphere after an interplanetary mission.

5.7 THE GENERAL CONSERVATION EQUATIONS

Each time a new convection problem is to be studied, the experienced engineer need not start from scratch. The principles that apply to any problem are conservation of mass, momentum, and energy. It is advantageous to formulate general governing equations based on these principles; for a specific problem, the engineer then need only delete terms in the equations that are zero or negligible.

5.7.1 Conservation of Mass

For simplicity, we consider a Cartesian control volume element $\Delta x \Delta y \Delta z$ fixed in space, as shown in Fig. 5.39. The rate at which mass is stored in the volume equals the net rate of inflow of mass across the control volume boundary. Mass can be stored within the volume by a change in density; the rate of storage is

$$\frac{\partial \rho}{\partial t} \Delta x \Delta y \Delta z$$

Mass can cross the control volume boundaries by convection. The gross rate of inflow is

$$\rho u|_x \Delta y \Delta z + \rho v|_y \Delta x \Delta z + \rho w|_z \Delta x \Delta y$$

and the gross rate of outflow is

$$\rho u|_{x+\Delta x} \Delta y \Delta z + \rho v|_{y+\Delta y} \Delta x \Delta z + \rho w|_{z+\Delta z} \Delta x \Delta y$$

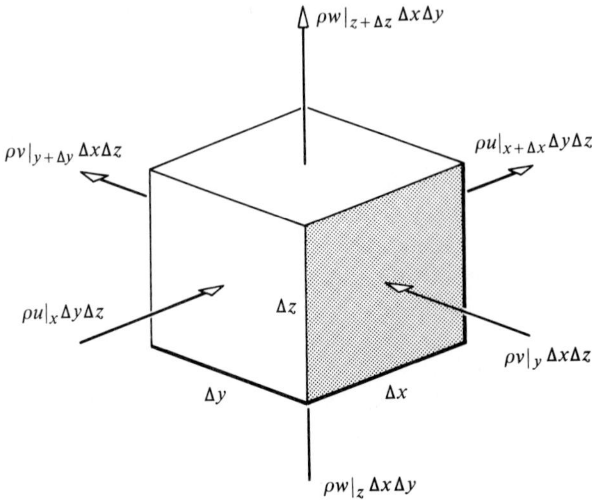

Figure 5.39 Conservation of mass applied to a Cartesian elemental control volume.

The net rate of inflow is found by subtracting the outflow from the inflow; expanding a quantity evaluated at $x + \Delta x$, $y + \Delta y$, or $z + \Delta z$ in a Taylor series; and canceling terms that subtract out. For example, the x-direction terms become

$$\begin{aligned}\rho u|_x\Delta y\Delta z - \rho u|_{x+\Delta x}\Delta y\Delta z &= \rho u|_x\Delta y\Delta z - \left[\rho u|_x + \frac{\partial}{\partial x}(\rho u)\Delta x + \frac{1}{2}\frac{\partial^2}{\partial x^2}(\rho u)\Delta x^2 + \cdots\right]\Delta y\Delta z \\ &= -\frac{\partial}{\partial x}(\rho u)\Delta x\Delta y\Delta z - \frac{1}{2}\frac{\partial^2}{\partial x^2}(\rho u)\Delta x^2\Delta y\Delta z + \cdots\end{aligned}$$

with similar results in the other directions. Now substitute in our statement of the conservation principle, divide throughout by the volume $\Delta x\Delta y\Delta z$, and let Δx, Δy, and Δz go to zero. Only first-order terms remain as the convective contribution; thus,

$$\frac{\partial\rho}{\partial t} = -\frac{\partial}{\partial x}(\rho u) - \frac{\partial}{\partial y}(\rho v) - \frac{\partial}{\partial z}(\rho w)$$

or

$$\frac{\partial\rho}{\partial t} + \frac{\partial}{\partial x}(\rho u) + \frac{\partial}{\partial y}(\rho v) + \frac{\partial}{\partial z}(\rho w) = 0 \tag{5.187}$$

Using the notation of vector calculus, Eq. (5.187) can be rewritten compactly as

$$\frac{\partial\rho}{\partial t} + \nabla\cdot(\rho\mathbf{v}) = 0 \tag{5.188}$$

where $\mathbf{v} = \mathbf{i}u + \mathbf{j}v + \mathbf{k}w$ and $\nabla = \mathbf{i}(\partial/\partial x) + \mathbf{j}(\partial/\partial y) + \mathbf{k}(\partial/\partial z)$ are the velocity vector and del operator for unit vectors **i, j, k** in the x, y, and z directions, respectively. For coordinate systems other than the Cartesian system, the divergence $\nabla\cdot$ can be expressed accordingly.

Equation (5.187) can be rearranged by performing the indicated differentiation to obtain

$$\frac{\partial \rho}{\partial t} + u\frac{\partial \rho}{\partial x} + v\frac{\partial \rho}{\partial y} + w\frac{\partial \rho}{\partial z} = -\rho\left(\frac{\partial u}{\partial x} + \frac{\partial v}{\partial y} + \frac{\partial w}{\partial z}\right) \quad \textbf{(5.189)}$$

or

$$\frac{D\rho}{Dt} = -\rho\nabla \cdot \mathbf{v} \quad \textbf{(5.190)}$$

The operator D/Dt is called the *substantial derivative* and is the time derivative for an observer moving with the fluid. Consider flow inside the cylinder of an automobile engine; in a time Δt two things can happen. First, the compression of the gas by the piston causes the density at a given location to change by the amount

$$\frac{\partial \rho}{\partial t}\Delta t$$

Second, the swirling gas can flow to a colder region: in time Δt, it moves a distance $\Delta x = u\Delta t$, $\Delta y = v\Delta t$, $\Delta z = w\Delta t$, and the density change due to this flow is

$$\frac{\partial \rho}{\partial x}\Delta x + \frac{\partial \rho}{\partial y}\Delta y + \frac{\partial \rho}{\partial t}\Delta z = \frac{\partial \rho}{\partial x}u\Delta t + \frac{\partial \rho}{\partial y}v\Delta t + \frac{\partial \rho}{\partial z}w\Delta t$$

Thus, an observer moving with the gas observes a total density change

$$D\rho = \frac{\partial \rho}{\partial t}\Delta t + \frac{\partial \rho}{\partial x}u\Delta t + \frac{\partial \rho}{\partial y}v\Delta t + \frac{\partial \rho}{\partial z}w\Delta t$$

Divide by Δt and let Δt approach zero:

$$\frac{D\rho}{Dt} = \frac{\partial \rho}{\partial t} + u\frac{\partial \rho}{\partial x} + v\frac{\partial \rho}{\partial y} + w\frac{\partial \rho}{\partial z} \quad \textbf{(5.191)}$$

or

$$\frac{D\rho}{Dt} = \frac{\partial \rho}{\partial t} + \mathbf{v} \cdot \nabla\rho \quad \textbf{(5.192)}$$

For the special case of a fluid of constant density, Eq. (5.190) becomes

$$\nabla \cdot \mathbf{v} = 0 \quad \textbf{(5.193)}$$

Although no fluid is truly incompressible, Eq. (5.193) can often be used in engineering analysis with negligible error.

5.7.2 Conservation of Momentum

Choice of Control Volume

An *Eulerian* control volume is one that is fixed in space. Such a control volume was used for the preceding derivation of the mass conservation and for the derivation of the momentum equation for a boundary layer in Section 5.4.1. A *Lagrangian* control volume contains a fixed mass of fluid and moves with the fluid. We choose to use the Lagrangian approach here since the principle of conservation of momentum can be stated in its simplest form, that is, the time rate of change of momentum equals the sum of forces acting. Since the control volume contains a fixed mass and moves with the fluid, the time rate of change of momentum is the mass times acceleration. The acceleration is given by the substantial derivative introduced in Section 5.7.1; hence the rate of change of momentum is

$$\rho \Delta x \Delta y \Delta z \frac{D\mathbf{v}}{Dt}$$

where for the x component

$$\frac{Du}{Dt} = \frac{\partial u}{\partial t} + u\frac{\partial u}{\partial x} + v\frac{\partial u}{\partial y} + w\frac{\partial u}{\partial z}$$

Surface Forces

Figure 5.40 shows all the x-direction surface forces acting on a cube $\Delta x \Delta y \Delta z$, including forces due to both pressure and viscous stresses. To understand the sign and subscript notation for the viscous stresses, consider, for example, the shear stress τ_{yx}. This stress is applied to the fluid of greater y from below. The first subscript, y, denotes that the stress acts on a plane of constant y; the second subscript denotes that the stress is in the x direction. Similarly, the normal stress τ_{xx} is applied to the fluid of greater x from below: the first subscript denotes that the stress acts on a plane of constant x, and the second subscript denotes that the stress is in the x direction. Notice that τ_{xx} does not include the hydrostatic pressure P. The net x-direction surface force is

$$(\tau_{xx} + P)_x \Delta y \Delta z + (\tau_{yx})_y \Delta x \Delta z + (\tau_{zx})_z \Delta x \Delta y$$
$$-[(\tau_{xx} + P)_{x+\Delta x} \Delta y \Delta z + (\tau_{yx})_{y+\Delta y} \Delta x \Delta z + (\tau_{zx})_{z+\Delta z} \Delta x \Delta y]$$

Expanding the incremental quantities in Taylor series, canceling terms, dividing by $\Delta x \Delta y \Delta z$, and letting Δx, Δy, and Δz go to zero gives

$$\left(-\frac{\partial P}{\partial x} - \frac{\partial \tau_{xx}}{\partial x} - \frac{\partial \tau_{yx}}{\partial y} - \frac{\partial \tau_{zx}}{\partial z} \right) \Delta x \Delta y \Delta z$$

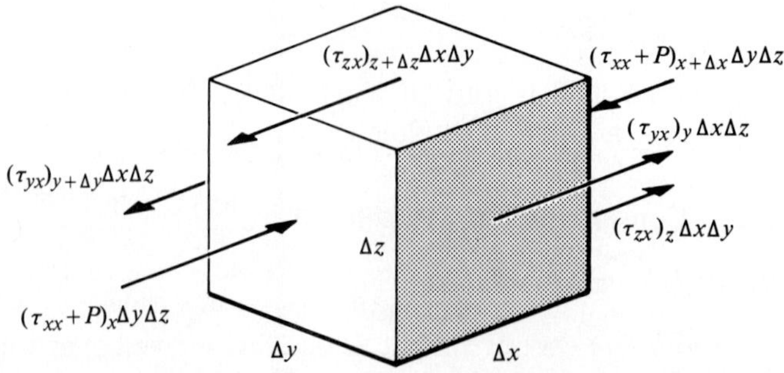

Figure 5.40 The x-direction forces acting on a Cartesian element $\Delta x \Delta y \Delta z$.

Volume Forces

Volume forces may also act on a fluid, for example, gravitational and electromagnetic forces; only the former will be considered here. If the gravitational force per unit mass is denoted by the vector **g**, then the resulting force acting on the fluid within the control volume is $\mathbf{g}\rho\Delta x\Delta y\Delta z$, and its x component is $g_x\rho\Delta x\Delta y\Delta z$.

The Momentum Conservation Equation

We now equate the time rate of change of momentum to the sum of the forces acting. For the x direction,

$$\rho\Delta x\Delta y\Delta z\frac{Du}{Dt} = \left(-\frac{\partial P}{\partial x} - \frac{\partial \tau_{xx}}{\partial x} - \frac{\partial \tau_{yx}}{\partial y} - \frac{\partial \tau_{zx}}{\partial z} + \rho g_x\right)\Delta x\Delta y\Delta z$$

or

$$\rho\frac{Du}{Dt} = -\frac{\partial P}{\partial x} - \frac{\partial \tau_{xx}}{\partial x} - \frac{\partial \tau_{yx}}{\partial y} - \frac{\partial \tau_{zx}}{\partial z} + \rho g_x \quad \textbf{(5.194}\textbf{\textit{a}}\textbf{)}$$

Similar equations can be obtained for the y and z directions:

$$\rho\frac{Dv}{Dt} = -\frac{\partial P}{\partial y} - \frac{\partial \tau_{xy}}{\partial x} - \frac{\partial \tau_{yy}}{\partial y} - \frac{\partial \tau_{zy}}{\partial z} + \rho g_y \quad \textbf{(5.194}\textbf{\textit{b}}\textbf{)}$$

$$\rho\frac{Dw}{Dt} = -\frac{\partial P}{\partial z} - \frac{\partial \tau_{xz}}{\partial x} - \frac{\partial \tau_{yz}}{\partial y} - \frac{\partial \tau_{zz}}{\partial z} + \rho g_z \quad \textbf{(5.194}\textbf{\textit{c}}\textbf{)}$$

The three equations can be summarized in matrix form:

$$\rho\frac{D}{Dt}\begin{bmatrix} u \\ v \\ w \end{bmatrix} = -\begin{bmatrix} P+\tau_{xx} & \tau_{yx} & \tau_{zx} \\ \tau_{xy} & P+\tau_{yy} & \tau_{zy} \\ \tau_{xz} & \tau_{yz} & P+\tau_{zz} \end{bmatrix}\begin{bmatrix} \partial/\partial x \\ \partial/\partial y \\ \partial/\partial z \end{bmatrix} + \rho\begin{bmatrix} g_x \\ g_y \\ g_z \end{bmatrix} \quad \textbf{(5.195)}$$

The convention for matrix multiplication of the operator is that row i of the desired one-column vector results from operating on the term in row i and column j of the matrix by the differential operator in row j of the operator vector, and summing the terms for $j = 1, 2, 3$. The matrix is called the *stress tensor*.

The Stokes Hypothesis

To use Eq. (5.195), it is necessary to express the viscous stresses in terms of the velocity gradients. Recall that a Newtonian fluid is one in which the shear stress is linearly proportional to the velocity gradient in a simple Couette-type shear flow. Stokes generalized the concept to a flow in which the velocity varies in more than one coordinate direction by assuming that the shear stress is linearly related to the rate of angular deformation. Most common fluids, such as water and air, are Newtonian, but many important fluids are not, for example, blood and polymer solutions.

Figure 5.41 shows how angular deformation occurs in simple shear and in the more general case. In time Δt, the upper left corner of what originally was a square of fluid will move a distance $(\partial u/\partial y)\Delta y\Delta t$ farther than the lower left corner. This

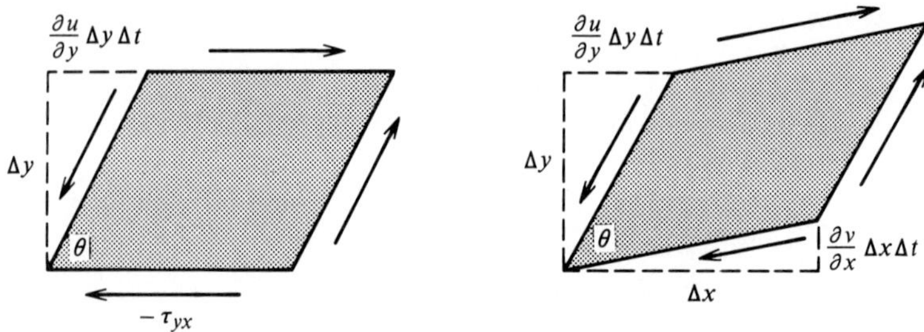

Figure 5.41 Rate of angular deformation in one and two dimensions.

distance divided by Δy gives the change in angle θ. Similarly, the movement of the lower right corner causes an angular change of $(\partial v/\partial x)\Delta t$. The combined rate of change of angle θ is then

$$-\frac{\Delta\theta}{\Delta t} = \frac{\partial u}{\partial y} + \frac{\partial v}{\partial x}$$

This angular rate of change was brought about by (or gives rise to) the negative shear stresses shown in Fig. 5.41. By the Stokes hypothesis,

$$\tau_{yx} = -\mu\left(\frac{\partial u}{\partial y} + \frac{\partial v}{\partial x}\right) = \tau_{xy} \qquad \textbf{(5.196a)}$$

Similarly,

$$\tau_{zx} = -\mu\left(\frac{\partial u}{\partial z} + \frac{\partial w}{\partial x}\right) = \tau_{xz} \qquad \textbf{(5.196b)}$$

$$\tau_{zy} = -\mu\left(\frac{\partial v}{\partial z} + \frac{\partial w}{\partial y}\right) = \tau_{yz} \qquad \textbf{(5.196c)}$$

It can be shown, by some rather involved analysis involving coordinate transformations, that the compatible relations for the normal viscous stresses are

$$\tau_{xx} = -\mu\left(2\frac{\partial u}{\partial x} - \frac{2}{3}\nabla\cdot\mathbf{v}\right) \qquad \textbf{(5.197a)}$$

$$\tau_{yy} = -\mu\left(2\frac{\partial v}{\partial y} - \frac{2}{3}\nabla\cdot\mathbf{v}\right) \qquad \textbf{(5.197b)}$$

$$\tau_{zz} = -\mu\left(2\frac{\partial w}{\partial z} - \frac{2}{3}\nabla\cdot\mathbf{v}\right) \qquad \textbf{(5.197c)}$$

In some fluids, such as liquids with small gas bubbles, a *second coefficient of viscosity* is needed. It adds a term proportional to $\nabla\cdot\mathbf{v}$ to each normal stress. But in most situations, we can simply approximate the normal viscous stresses as

$$\tau_{xx} = -2\mu\frac{\partial u}{\partial x} \qquad \textbf{(5.198a)}$$

$$\tau_{yy} = -2\mu \frac{\partial v}{\partial y} \tag{5.198b}$$

$$\tau_{zz} = -2\mu \frac{\partial w}{\partial z} \tag{5.198c}$$

If Eqs. (5.196) and (5.198) are substituted in Eq. (5.195) and a constant density and viscosity assumed, the resulting three components of the momentum conservation equation are

$$\rho\left(\frac{\partial u}{\partial t} + u\frac{\partial u}{\partial x} + v\frac{\partial u}{\partial y} + w\frac{\partial u}{\partial z}\right) = -\frac{\partial P}{\partial x} + \mu\left(\frac{\partial^2 u}{\partial x^2} + \frac{\partial^2 u}{\partial y^2} + \frac{\partial^2 u}{\partial z^2}\right) + \rho g_x \tag{5.199a}$$

$$\rho\left(\frac{\partial v}{\partial t} + u\frac{\partial v}{\partial x} + v\frac{\partial v}{\partial y} + w\frac{\partial v}{\partial z}\right) = -\frac{\partial P}{\partial y} + \mu\left(\frac{\partial^2 v}{\partial x^2} + \frac{\partial^2 v}{\partial y^2} + \frac{\partial^2 v}{\partial z^2}\right) + \rho g_y \tag{5.199b}$$

$$\rho\left(\frac{\partial w}{\partial t} + u\frac{\partial w}{\partial x} + v\frac{\partial w}{\partial y} + w\frac{\partial w}{\partial z}\right) = -\frac{\partial P}{\partial z} + \mu\left(\frac{\partial^2 w}{\partial x^2} + \frac{\partial^2 w}{\partial y^2} + \frac{\partial^2 w}{\partial z^2}\right) + \rho g_z \tag{5.199c}$$

Equations (5.199) can be written in the compact form

$$\rho\frac{D\mathbf{v}}{Dt} = -\nabla P + \mu\nabla^2\mathbf{v} + \rho\mathbf{g} \tag{5.200}$$

These equations are often called the **Navier-Stokes equations.**

5.7.3 Conservation of Energy

Control Volume

We return to the Eulerian viewpoint and consider a Cartesian elemental control volume $\Delta x \Delta y \Delta z$ located in a pure fluid. The energy conservation principle requires that the rate at which energy is stored within the volume equal the net rate of inflow of energy by convection, plus the rate of heat transfer across the boundary, plus the rate at which work is done on the fluid within the volume, plus the rate at which energy is produced within the volume.

Energy Storage

Energy can be stored within the volume as internal energy and as kinetic energy of the mass as a whole. The rate of energy storage per unit time is thus

$$\frac{\partial}{\partial t}\left(\rho u + \frac{1}{2}\rho v^2\right)\Delta x \Delta y \Delta z$$

where u is the specific internal energy,[5] and $v^2 = \mathbf{v}\cdot\mathbf{v} = u^2 + v^2 + w^2$.

[5] Care must be taken not to confuse the specific internal energy u with the velocity component u in this derivation.

Energy Inflow

Energy can flow into the control volume by convection. As mass flows across the volume boundaries, it brings with it energy $[u + (1/2)v^2]$ per unit mass. For example, the convection across the y face per unit time is

$$\left[\rho v\left(u + \frac{1}{2}v^2\right)\right]_y \Delta x \Delta z$$

and the convection out across the $y + \Delta y$ face is

$$\left[\rho v\left(u + \frac{1}{2}v^2\right)\right]_{y+\Delta y} \Delta x \Delta z$$

The net rate of inflow for the two faces is

$$-\frac{\partial}{\partial y}\left[\rho v\left(u + \frac{1}{2}v^2\right)\right] \Delta x \Delta y \Delta z$$

Similar terms arise for the other two pairs of faces. The sum of all these terms can be written

$$-\nabla \cdot \left[\rho \mathbf{v}\left(u + \frac{1}{2}v^2\right)\right] \Delta x \Delta y \Delta z$$

Heat Transfer

Heat is transferred into the control volume by conduction. The rate of heat inflow across the x face by conduction is

$$\left(-k\frac{\partial T}{\partial x}\right)_x \Delta y \Delta z$$

and the rate of outflow across the $x + \Delta x$ face is

$$\left(-k\frac{\partial T}{\partial x}\right)_{x+\Delta x} \Delta y \Delta z$$

The net rate of inflow for the two faces is

$$\frac{\partial}{\partial x}\left(k\frac{\partial T}{\partial x}\right) \Delta x \Delta y \Delta z$$

Again, the other two pairs of faces give similar terms. The sum for all faces can be written

$$\nabla \cdot k \nabla T \Delta x \Delta y \Delta z$$

Work

The rate at which work is done by the volume force $\mathbf{g}$ is simply the vector dot product of force and velocity:

$$\rho \mathbf{g} \cdot \mathbf{v} \Delta x \Delta y \Delta z = (\rho g_x u + \rho g_y v + \rho g_z w) \Delta x \Delta y \Delta z$$

Work can also be done by the surface forces acting on a volume element. The vector force acting on the x face is the first column of the stress tensor times the area of the face. The rate at which work is done *on* the element at this face is the vector dot product of velocity and force. Denote the first, second, and third columns of the stress tensor as $\mathbf{F}_x$, $\mathbf{F}_y$, and $\mathbf{F}_z$, respectively. The subscripts denote the direction of the normal to the face on which the force acts. Then, for example,

$$\mathbf{F}_x = \mathbf{i}(P + \tau_{xx}) + \mathbf{j}(\tau_{xy}) + \mathbf{k}(\tau_{xz})$$

and the rate at which work is done on the element by this force is

$$\mathbf{v} \cdot \mathbf{F}_x \Delta y \Delta z = [(P + \tau_{xx})u + \tau_{xy}v + \tau_{xz}w]\Delta y \Delta z$$

At the $x + \Delta x$ face, work is done *by* the element, on the fluid of greater x, at the rate

$$(\mathbf{v} \cdot \mathbf{F}_x)_{x+\Delta x} \Delta y \Delta z$$

The net rate at which work is done *on* the element for these two faces is

$$-\frac{\partial}{\partial x}(\mathbf{v} \cdot \mathbf{F}_x)\Delta x \Delta y \Delta z$$

The net work done on the element for all six faces is then

$$-\left[\frac{\partial}{\partial x}(\mathbf{v} \cdot \mathbf{F}_x) + \frac{\partial}{\partial y}(\mathbf{v} \cdot \mathbf{F}_y) + \frac{\partial}{\partial z}(\mathbf{v} \cdot \mathbf{F}_z)\right] \Delta x \Delta y \Delta z$$

Heat Generation

Internal heat generation may arise due to resistive (I^2R) heating, neutron and fission fragment slowing, or photon absorption and emission. These sources may be represented by an equivalent volumetric source term $\dot{Q}_v'''$ [W/m^3], and the production rate within the elemental volume is

$$\dot{Q}_v''' \Delta x \Delta y \Delta z$$

Total Energy Equation

Substituting all the preceding terms in our statement of conservation of energy and dividing by the volume $\Delta x \Delta y \Delta z$ gives the **total energy equation,**

$$\frac{\partial}{\partial t}\rho\left(u + \frac{1}{2}v^2\right) = -\nabla \cdot \left[\rho \mathbf{v}\left(u + \frac{1}{2}v^2\right)\right] + \nabla \cdot k\nabla T + \rho \mathbf{g} \cdot \mathbf{v} - \frac{\partial}{\partial x}(\mathbf{v} \cdot \mathbf{F}_x) - \frac{\partial}{\partial y}(\mathbf{v} \cdot \mathbf{F}_y) - \frac{\partial}{\partial z}(\mathbf{v} \cdot \mathbf{F}_z) + \dot{Q}_v''' \qquad \textbf{(5.201)}$$

With the aid of the mass conservation equation, Eq. (5.188), Eq. (5.201) can be written in terms of the substantial derivative as

$$\rho\frac{D}{Dt}\left(u + \frac{1}{2}v^2\right) = \nabla \cdot k\nabla T + \rho \mathbf{g} \cdot \mathbf{v} - \frac{\partial}{\partial x}(\mathbf{v} \cdot \mathbf{F}_x) - \frac{\partial}{\partial y}(\mathbf{v} \cdot \mathbf{F}_y) - \frac{\partial}{\partial z}(\mathbf{v} \cdot \mathbf{F}_z) + \dot{Q}_v''' \qquad \textbf{(5.202)}$$

Thermal Energy Equations

Since Eq. (5.202) accounts for conservation of both thermal and mechanical energy, it is not the most convenient starting point for the solution of most heat transfer problems. We prefer to have a conservation equation that isolates thermal phenomena, which can be derived by subtracting the conservation equation for mechanical energy from the total energy equation, as follows.

The mechanical energy equation is obtained by taking the dot product of velocity $\mathbf{v}$ and the momentum equation to obtain the rate at which mechanical work is done. In Cartesian coordinates, we add u times the x-momentum equation, v times the y-momentum equation, and w times the z-momentum equation. The result is

$$\rho\frac{D}{Dt}\left(\frac{1}{2}v^2\right) = -\mathbf{v}\cdot\frac{\partial \mathbf{F}_x}{\partial x} - \mathbf{v}\cdot\frac{\partial \mathbf{F}_y}{\partial y} - \mathbf{v}\cdot\frac{\partial \mathbf{F}_z}{\partial z} + \rho\mathbf{v}\cdot\mathbf{g} \tag{5.203}$$

Using the product rule of differentiation in Eq. (5.202) and subtracting Eq. (5.203) gives

$$\rho\frac{Du}{Dt} = \nabla\cdot k\nabla T - \frac{\partial \mathbf{v}}{\partial x}\cdot\mathbf{F}_x - \frac{\partial \mathbf{v}}{\partial y}\cdot\mathbf{F}_y - \frac{\partial \mathbf{v}}{\partial z}\cdot\mathbf{F}_z + \dot{Q}_v''' \tag{5.204}$$

Substituting for $\mathbf{F}_x$, $\mathbf{F}_y$, and $\mathbf{F}_z$ using Eqs. (5.196) and (5.198) and rearranging gives

$$\rho\frac{Du}{Dt} = -P\nabla\cdot\mathbf{v} + \nabla\cdot k\nabla T + \mu\Phi + \dot{Q}_v''' \tag{5.205}$$

where the quantity $\mu\Phi$ represents the portion of the mechanical work that is irreversibly converted into heat by the viscous stresses. In terms of the velocity components,

$$\Phi = 2\left[\left(\frac{\partial u}{\partial x}\right)^2 + \left(\frac{\partial v}{\partial y}\right)^2 + \left(\frac{\partial w}{\partial z}\right)^2\right] + \left(\frac{\partial u}{\partial y} + \frac{\partial v}{\partial x}\right)^2 + \left(\frac{\partial u}{\partial z} + \frac{\partial w}{\partial x}\right)^2 + \left(\frac{\partial v}{\partial z} + \frac{\partial w}{\partial y}\right)^2 \tag{5.206}$$

The quantity $\mu\Phi$ is the viscous dissipation; Φ is often simply called the **dissipation function.** The term $-P\nabla\cdot\mathbf{v}$ represents the portion of the mechanical work that is reversibly converted into heat by compression.

We often prefer to work in terms of enthalpy h rather than internal energy u. Substituting the thermodynamic relation $h = u + P/\rho$ into Eq. (5.205) and rearranging,

$$\rho\frac{Dh}{Dt} = \frac{DP}{Dt} + \nabla\cdot k\nabla T + \mu\Phi + \dot{Q}_v''' \tag{5.207}$$

For the special case of an ideal gas, $dh = c_p\,dT$, and Eq. (5.207) becomes

$$\rho c_p\frac{DT}{Dt} = \frac{DP}{Dt} + \nabla\cdot k\nabla T + \mu\Phi + \dot{Q}_v''' \tag{5.208}$$

For an incompressible liquid with specific heat $c = c_p = c_v$, we go back to Eq. (5.205): since $\nabla\cdot\mathbf{v} = 0$,

$$\rho c\frac{DT}{Dt} = \nabla\cdot k\nabla T + \mu\Phi + \dot{Q}_v''' \tag{5.209}$$

Equations (5.205) through (5.209) are all forms of the **thermal energy equation** for a Newtonian fluid. One of these equations is the usual starting point of a convection analysis.

5.7.4 Use of the Conservation Equations

Solving the general conservation equations, even by direct numerical means, is difficult and usually impractical. Fortunately, however, many problems of engineering interest are adequately described by simplified forms of these equations, and these forms often can be solved more easily. Sections 5.2 through 5.4 contained examples of such problems involving laminar flows, and there are many more. Earlier in this chapter we set up the governing equations from first principles; now we are in a position to obtain the governing equations by simply deleting superfluous terms in the general equations. The following discussion applies directly to laminar flows. In the case of turbulent flows, the remarks apply only to the time-averaged equations. For example, on average, a turbulent flow might be steady and two-dimensional, but there are instantaneous fluctuations of all three velocity components.

The manner in which a particular problem is posed may immediately imply that some terms are identically zero. For example, if a timewise steady-state solution is sought, the time derivatives are set equal to zero. Some resulting classes of simplified flows are:

1. Constant density
2. Constant transport properties
3. Timewise steady flow (or quasi-steady flow)
4. Two-dimensional flow
5. One-dimensional flow
6. Fully developed flows

Terms may also be negligibly small. Often, intuition or experimental evidence will suffice to make a decision. But a more rigorous approach is to use order-of-magnitude estimates based on scaling arguments, as will be shown in Section 5.8. Some classes of flow that result are:

1. Creeping flows
2. Forced flows
3. Natural flows
4. Boundary layer flows
5. Low-speed flows

A complete mathematical statement of a convection problem requires specification of boundary conditions. Each boundary condition is based on a physical statement or principle. The student can examine the analyses of Sections 5.2 through 5.5 for examples.

5.8 SCALE ANALYSIS

For the discussion of the concept of similarity in Section 5.6, the equations governing a problem were made dimensionless. Variables were made dimensionless by dividing by convenient characteristic values; for example, the laminar boundary layer equations were made dimensionless by dividing all lengths by the plate length, L, and all velocities by the free-stream velocity, u_e. As a result, the orders of magnitude of the resulting dimensionless variables may differ. The variable $x^* = x/L$ varies from 0 to 1 along the plate, but $y^* = y/L$ varies from 0 to $\delta/L \ll 1$ across the boundary layer. To perform a **scale analysis,** we forsake the simplicity of making all lengths dimensionless with L, all velocities dimensionless with u_e, and so on, instead choosing characteristic values that ensure that all dimensionless variables are of order of magnitude unity or less. For example, an obvious choice for the coordinate y in a boundary layer is the boundary layer thickness, δ. A scale analysis can serve a number of purposes. It can indicate which terms in the governing equations are negligible; it can also tell something about the functional form of the solution. Two examples follow.

5.8.1 Forced-Convection Laminar Boundary Layers

One of the most important applications of scale analysis is the derivation of the boundary layer equations from the general conservation equations. We proceed as follows.

The Momentum Equation

We assume that the flow is steady ($\partial/\partial t = 0$), laminar, and two-dimensional ($\partial/\partial z, w = 0$); that the fluid properties ρ and μ are constant; and that the plate is horizontal, $g_x = 0, g_y = -g$. The mass and momentum conservation equations, Eqs. (5.187) and (5.199), then reduce to

$$\frac{\partial u}{\partial x} + \frac{\partial v}{\partial y} = 0 \quad \textbf{(5.210)}$$

$$\rho u\frac{\partial u}{\partial x} + \rho v\frac{\partial u}{\partial y} = -\frac{\partial P}{\partial x} + \mu\left[\frac{\partial^2 u}{\partial x^2} + \frac{\partial^2 u}{\partial y^2}\right] \quad \textbf{(5.211)}$$

$$\rho u\frac{\partial v}{\partial x} + \rho v\frac{\partial v}{\partial y} = -\frac{\partial P}{\partial y} + \mu\left[\frac{\partial^2 v}{\partial x^2} + \frac{\partial^2 v}{\partial y^2}\right] - \rho g \quad \textbf{(5.212)}$$

A plate of length L with fluid flowing over it is shown in Fig. 5.42. The plate is located in a wind tunnel and, by appropriately contouring the tunnel ceiling, the pressure variation along the plate can be controlled. We expect a layer of fluid,

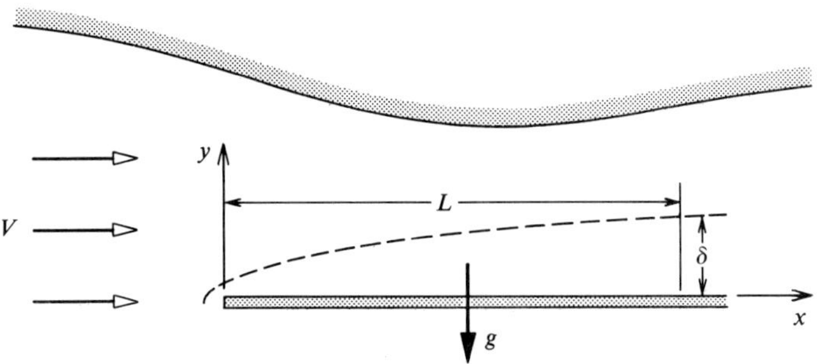

Figure 5.42 A flat plate in a contoured wind tunnel.

moving more slowly than the oncoming stream, to build up adjacent to the plate due to the action of viscous shear stresses. At $x = L$, the end of the plate, this layer of fluid is characterized by a thickness δ. We wish to investigate situations where the condition $\delta \ll L$ is met, which corresponds to Prandtl's concept of a thin *boundary layer* in which the fluid velocity changes from the free-stream value to zero.

Our first step is to scale the variables in the conservation equations in such a manner that the resulting dimensionless variables are of order of magnitude unity or less. The choices for lengths x, y, and the velocity component u are obvious:

$$x^* = \frac{x}{L}; \qquad y^* = \frac{y}{\delta}; \qquad u^* = \frac{u}{V}$$

An appropriate scale for the velocity component v is not obvious, so we assign a scaling velocity v_S such that $v^* = v/v_S$ and let the equations give the appropriate order of magnitude for v_S.

Pressure variations arise due to changes in the hydrostatic pressure, which would exist if the plate were not present, and changes in velocity due to the contouring of the wind tunnel. The hydrostatic pressure variation would exist even if there were no flow. Setting u and v identically equal to zero in Eq. (5.212) gives

$$0 = \frac{\partial P_0}{\partial y} + \rho g$$

Integrating gives the hydrostatic pressure variation:

$$P_0 = P_{0,0} - \rho g y \tag{5.213}$$

where $P_{0,0}$ is the hydrostatic pressure at $y = 0$. We now assign a scaling pressure P_S and define a dimensionless pressure $P^* = (P - P_0)/P_S$. Again, the equations will indicate an appropriate order of magnitude for P_S. There are three unknowns at this point: δ, v_S, and P_S.

To transform Eqs. (5.210) through (5.212), we substitute

$$u = Vu^*; \qquad v = v_S v^*; \qquad P = P_0 + P_S P^*$$

and the differential operators become

$$\frac{\partial}{\partial x} = \frac{\partial}{\partial x^*}\frac{dx^*}{dx} = \frac{1}{L}\frac{\partial}{\partial x^*}$$

$$\frac{\partial}{\partial y} = \frac{\partial}{\partial y^*}\frac{dy^*}{dy} = \frac{1}{\delta}\frac{\partial}{\partial y^*}$$

The mass conservation equation transforms into

$$\frac{V}{L}\frac{\partial u^*}{\partial x^*} + \frac{v_S}{\delta}\frac{\partial v^*}{\partial y^*} = 0 \tag{5.214}$$

In Fig. 5.42, visualize measuring the velocity u along a horizontal traverse very close to the plate surface, starting at the leading edge and progressing along to $x = L$. The velocity u will change from the value V to about zero in a length of about L; thus, $\partial u^*/\partial x^*$ has order of magnitude unity and is negative. If we now choose

$$v_S = \frac{\delta}{L}V \tag{5.215}$$

then Eq. (5.214) gives $\partial v^*/\partial y^*$ an order of magnitude unity, and it is positive. Since y^* varies from 0 to 1 across the boundary layer, it also follows that v^* has order of magnitude unity or less. Next, the x-momentum equation is transformed, using Eq. (5.215) for v_S:

$$\frac{\rho V^2}{L}u^*\frac{\partial u^*}{\partial x^*} + \frac{\rho V^2}{L}v^*\frac{\partial u^*}{\partial y^*} = -\frac{P_S}{L}\frac{\partial P^*}{\partial x^*} + \mu\left[\frac{V}{L^2}\frac{\partial^2 u^*}{\partial x^{*2}} + \frac{V}{\delta^2}\frac{\partial^2 u^*}{\partial y^{*2}}\right]$$

or

$$u^*\frac{\partial u^*}{\partial x^*} + v^*\frac{\partial u^*}{\partial y^*} = -\frac{P_S}{\rho V^2}\frac{\partial P^*}{\partial x^*} + \frac{\mu}{VL\rho}\left[\frac{\partial^2 u^*}{\partial x^{*2}} + \frac{L^2}{\delta^2}\frac{\partial^2 u^*}{\partial y^{*2}}\right] \tag{5.216}$$

Similarly, the y-momentum equation becomes

$$u^*\frac{\partial v^*}{\partial x^*} + v^*\frac{\partial v^*}{\partial y^*} = -\frac{P_S}{\rho V^2}\frac{L^2}{\delta^2}\frac{\partial P^*}{\partial y^*} + \frac{\mu}{VL\rho}\left[\frac{\partial^2 v^*}{\partial x^{*2}} + \frac{L^2}{\delta^2}\frac{\partial^2 v^*}{\partial y^{*2}}\right] \tag{5.217}$$

Recall that at the outset we said we were going to investigate situations where the condition $\delta/L \ll 1$ is met; accordingly, we will now assume $L/\delta \gg 1$ and examine the consequences. Inspecting Eq. (5.216), we note that $\partial^2 u^*/\partial y^{*2}$ has order of magnitude unity, since $\partial u^*/\partial y^*$ is on the order of magnitude unity near the plate and is near 0 at $y = \delta$ ($y^* = 1$). Similarly, $\partial^2 u^*/\partial x^{*2}$ has order of magnitude unity and thus can be neglected compared to $(L^2/\delta^2) \cdot (\partial^2 u^*/\partial y^{*2})$. The viscous term in Eq. (5.216) is then on the order of magnitude $(\mu/VL\rho)(L^2/\delta^2)$, and the acceleration terms on the left side of Eq. (5.216) have order of magnitude unity. Considering first the situation where $\partial P/\partial x$ is zero, that is, the tunnel ceiling is parallel to the plate, we conclude that $(\mu/VL\rho)(L^2/\delta^2)$ must be of order unity for the viscous term to balance the acceleration terms. Thus, we have succeeded in estimating δ:

$$\frac{\delta}{L} = \frac{1}{\mathrm{Re}^{1/2}}; \qquad \mathrm{Re} = \frac{VL\rho}{\mu} \tag{5.218}$$

To see if this conclusion remains valid when there is a pressure gradient $\partial P/\partial x$, we see what form Eq. (5.216) takes outside the boundary layer by dropping the viscous terms:

$$u^*\frac{\partial u^*}{\partial x^*} + v^*\frac{\partial u^*}{\partial y^*} = -\frac{P_S}{\rho V^2}\frac{\partial P^*}{\partial x^*} \tag{5.219}$$

If we have chosen the scaling velocity V sensibly, that is, to have order of magnitude comparable to the highest velocity in the flow field, the terms on the left-hand side of Eq. (5.219) are of order no greater than unity. Then $(P_S/\rho V^2)(\partial P^*/\partial x^*)$ can be of order no greater than unity, and the appropriate choice for P_S is $P_S = \rho V^2$. Thus, when there is a pressure gradient driving the flow, a scaling velocity V of order comparable to the highest velocity in the flow field will ensure that our scaling arguments are valid.

In the y-momentum equation, Eq. (5.217), $\partial^2 v^*/\partial x^{*2} \ll (L^2/\delta^2)(\partial^2 v^*/\partial y^{*2})$ and can be neglected; the remaining terms are rearranged to obtain

$$\frac{P_S}{\rho V^2}\frac{\partial P^*}{\partial y^*} = \frac{1}{\mathrm{Re}}\left\{\frac{\partial^2 v^*}{\partial y^{*2}} - \left[u^*\frac{\partial v^*}{\partial x^*} + v^*\frac{\partial v^*}{\partial y^*}\right]\right\} \tag{5.220}$$

The pressure variations across the boundary layer are of order 1/Re, that is, very small, so that the pressure that exists in the free stream outside the boundary layer $(y > \delta)$ impresses itself through the boundary layer. Thus, the partial derivative $\partial P^*/\partial x^*$ in Eq. (5.216) can be replaced by the total derivative dP^*/dx^*, where P is the free-stream pressure. Outside the boundary layer, Eq. (5.219) holds, but as long as the wall lies approximately in the x direction, the term $v^*(\partial u^*/\partial y^*)$ can be dropped. Denoting the free-stream velocity as u_e and returning to dimensional form,

$$u_e\frac{du_e}{dx} = -\frac{1}{\rho}\frac{dP}{dx}$$

or

$$u_e\,du_e + \frac{dP}{\rho} = 0 \tag{5.221}$$

which is the Euler equation. The x-momentum equation for a laminar boundary layer can therefore be written as

$$u^*\frac{\partial u^*}{\partial x^*} + v^*\frac{\partial u^*}{\partial y^*} = u_e^*\frac{du_e^*}{dx^*} + \frac{\partial^2 u^*}{\partial y^{*2}}$$

or, in dimensional form,

$$u\frac{\partial u}{\partial x} + v\frac{\partial u}{\partial y} = u_e\frac{du_e}{dx} + \nu\frac{\partial^2 u}{\partial y^2} \tag{5.222}$$

We have assumed that δ/L is small. Unless the density is very low, as for rocket-propelled flight in the upper atmosphere, or unless the length L is quite small, as for a small particle, or unless the viscosity is very large, as for some oils, the Reynolds number is indeed large, and δ/L is thus small. For example, consider air at normal conditions flowing at 3 m/s along a flat plate only 30 cm long. Then $\text{Re} = VL/\nu = (3)(0.3)/(15.7 \times 10^{-6}) = 57{,}300$.

The Energy Equation

To develop the thermal energy equation for a laminar boundary layer, we consider the plate to be heated such that a surface temperature of T_s is maintained when the free-stream temperature is T_e. Analogous to the hydrodynamic boundary layer thickness δ, we define a thermal boundary layer thickness Δ, which, in general, is not equal to δ. For example, with a liquid metal of high thermal conductivity, we would expect temperature gradients to exist quite far into the free stream, that is, $\Delta \gg \delta$. In addition to the assumptions made already, we assume low-speed flow and ignore viscous dissipation and compressibility effects. For constant thermal conductivity Eq. (5.208) then reduces to

$$\rho u c_p \frac{\partial T}{\partial x} + \rho v c_p \frac{\partial T}{\partial y} = k\left[\frac{\partial^2 T}{\partial x^2} + \frac{\partial^2 T}{\partial y^2}\right] \tag{5.223}$$

Temperature variations across the thermal boundary layer are scaled with $(T_s - T_e)$, $T^* = (T - T_e)/(T_s - T_e)$, and y is scaled with Δ, $y_T^* = y/\Delta$. Upon transforming Eq. (5.223), we obtain

$$\frac{\rho c_p (T_s - T_e)V}{L}\left(u^* \frac{\partial T^*}{\partial x^*} + \frac{\delta}{\Delta} v^* \frac{\partial T^*}{\partial y_T^*}\right) = \frac{k(T_s - T_e)}{\Delta^2}\left[\frac{\partial^2 T^*}{\partial y_T^{*2}} + \frac{\Delta^2}{L^2}\frac{\partial^2 T^*}{\partial x^{*2}}\right]$$

Dividing through by $\rho c_p (T_s - T_e)V/L$, and dropping the last term in square brackets for $\Delta^2 \ll L^2$,

$$u^* \frac{\partial T^*}{\partial x^*} + \frac{\delta}{\Delta} v^* \frac{\partial T^*}{\partial y_T^*} = \frac{1}{\text{RePr}}\frac{L^2}{\Delta^2}\frac{\partial^2 T^*}{\partial y_T^{*2}} \tag{5.224}$$

Thus, scaling arguments have allowed us to neglect streamwise heat conduction compared to heat conduction across the boundary layer. We can also deduce the behavior of Δ/δ as a function of Prandtl number. Using Eq. (5.218), Eq. (5.224) becomes

$$u^* \frac{\partial T^*}{\partial x^*} + \frac{\delta}{\Delta} v^* \frac{\partial T^*}{\partial y_T^*} = \frac{1}{\text{Pr}}\frac{\delta^2}{\Delta^2}\frac{\partial^2 T^*}{\partial y_T^{*2}} \tag{5.225}$$

Case 1: Pr ≪ 1. Since the dimensionless terms are of order of magnitude unity or less, comparing the right-hand side of Eq. (5.225) with the first term of the left-hand side shows $(1/\text{Pr})(\delta^2/\Delta^2)$ should be of order of magnitude unity; thus,

$$\frac{\Delta}{\delta} = 0\left(\frac{1}{\text{Pr}^{1/2}}\right); \qquad \text{Pr} \ll 1 \tag{5.226}$$

Note that then δ/Δ is very small, and the second term on the left-hand side of Eq. (5.225) is negligible. Deletion of this term gives the plug flow model used in Section 5.4.2.

Case 2: Pr ≫ 1. Now δ/Δ must be large for the conduction term on the right-hand side of Eq. (5.225) to be of the same order as the left-hand side of the equation. A somewhat lengthy argument is required to establish the Pr dependence. Since δ/Δ is very large, the entire thermal boundary layer is deep within the velocity boundary layer, very close to the wall. The x-momentum equation tells us that in this region, $\partial u^*/\partial y^*$ is nearly constant, so that u^* is of order $y^* \leq \Delta/\delta$. From the mass conservation equation, v^* very close to the wall is of order $y^{*2} \leq \Delta^2/\delta^2$. Thus, the entire left-hand side of Eq. (5.225) is of order Δ/δ, and it follows that

$$\frac{\Delta}{\delta} = 0\left(\frac{1}{\text{Pr}^{1/3}}\right); \qquad \text{Pr} \gg 1 \tag{5.227}$$

We see that it is the Prandtl number that determines the relative thicknesses of the momentum and thermal boundary layers. When Pr is large, the kinematic viscosity $\nu = \mu/\rho$ is much larger than the thermal diffusivity $\alpha = k/\rho c_p$, and the effect of the wall retarding the flow penetrates more deeply into the flow than the thermal effect of the heated wall. The reverse is true for small Pr.

Wall Transfer Rates

From our order-of-magnitude estimates of δ and Δ, we can also estimate the order of magnitude of the wall shear stress and heat flux. The shear stress τ_{yx} is

$$\begin{aligned}\tau_{yx} &= -\mu\left(\frac{\partial u}{\partial y} + \frac{\partial v}{\partial x}\right)\\ &= -\frac{\mu V}{\delta}\left(\frac{\partial u^*}{\partial y^*} + \frac{\partial v^*}{\partial x^*}\right)\end{aligned}$$

Since $\partial v^*/\partial x^* \ll \partial u^*/\partial y^*$ in a boundary layer, the shear stress exerted by the fluid on the wall becomes

$$\tau_s = \frac{\mu V}{\delta}\frac{\partial u^*}{\partial y^*}\bigg|_{y^*=0}$$

or

$$\tau_s \sim \frac{\mu V}{\delta} \tag{5.228}$$

Similarly,

$$q_s \sim \frac{k(T_s - T_e)}{\Delta} \tag{5.229}$$

Introducing the definitions of skin friction coefficient and Nusselt number gives

$$\frac{C_f}{2} = \frac{\tau_s}{\rho V^2} \sim \frac{1}{\mathrm{Re}^{1/2}} \tag{5.230}$$

$$\mathrm{Nu} = \frac{q_s L}{k(T_s - T_e)} \sim \mathrm{Re}^{1/2}\mathrm{Pr}^{1/2}; \qquad \mathrm{Pr} \ll 1 \tag{5.231}$$

$$\sim \mathrm{Re}^{1/2}\mathrm{Pr}^{1/3}; \qquad \mathrm{Pr} \gg 1 \tag{5.232}$$

The arguments presented in this section can be applied to the conservation equations in their general form. The resulting laminar boundary layer equations are summarized in Table 5.5.

Table 5.5 The conservation equations, in boundary layer form, for steady two-dimensional planar or axisymmetric laminar boundary layer flows.

Mass:	$\frac{\partial}{\partial x}(\rho u r^{\varepsilon}) + \frac{\partial}{\partial y}(\rho v r^{\varepsilon}) = 0$
	$\varepsilon = 0$ for planar flows, $\varepsilon = 1$ for axisymmetric flows with radius r measured from the axis of symmetry.
Momentum:	$\rho u \frac{\partial u}{\partial x} + \rho v \frac{\partial u}{\partial y} = -\frac{dP}{dx} + \frac{\partial}{\partial y}(\mu \frac{\partial u}{\partial y}) + \rho g_x$
Total energy:	$\rho u \frac{\partial H}{\partial x} + \rho v \frac{\partial H}{\partial y} = \frac{\partial}{\partial y}\left(k \frac{\partial T}{\partial y}\right) + \frac{\partial}{\partial y}\left(\mu u \frac{\partial u}{\partial y}\right) + \rho u g_x$
	$H = h + (1/2)u^2$
Thermal energy:	$\rho u \frac{\partial h}{\partial x} + \rho v \frac{\partial h}{\partial y} = u \frac{dP}{dx} + \frac{\partial}{\partial y}\left(k \frac{\partial T}{\partial y}\right) + \mu \left(\frac{\partial u}{\partial y}\right)^2$
	or
	$\rho u c \frac{\partial T}{\partial x} + \rho v c \frac{\partial T}{\partial y} = k \frac{\partial^2 T}{\partial y^2} + \mu \left(\frac{\partial u}{\partial y}\right)^2$ (constant properties)

5.8.2 Natural-Convection Laminar Boundary Layer on a Vertical Wall

As a second example of scale analysis, we examine the familiar natural-convection laminar boundary layer on a vertical wall, as shown in Fig. 5.43. Our objective is to estimate the thickness of the thermal boundary layer and hence gain insight into the behavior of the Nusselt number. As was the case in Section 5.8.1, we could begin with the complete conservation equations and for a thin boundary layer deduce that (1) pressure gradients across the boundary layer are small and hence fluid momentum changes in the y direction can be neglected, (2) the shear stress $-\mu \partial u / \partial x$ in the x-momentum equation can be neglected, and (3) the conduction flux $-k \partial T / \partial x$ in the energy equation can be neglected. But since we have already performed such an exercise, we will instead commence with boundary layer equations.

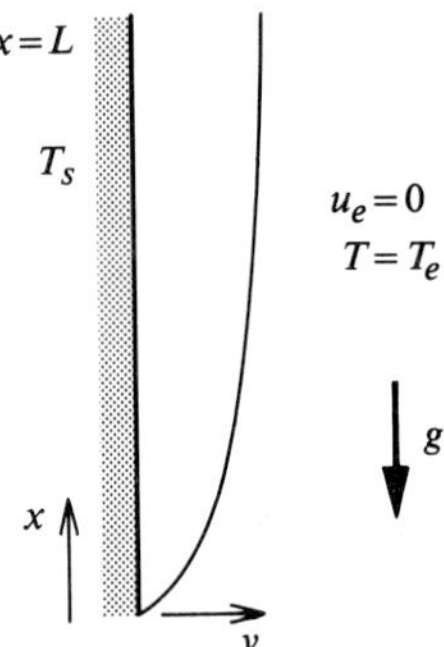

Figure 5.43 Schematic of a natural-convection boundary layer on a vertical wall, showing the coordinate system.

The Governing Equations

The general conservation equations, Eqs. (5.189), (5.199), and (5.208), reduce to

Mass:
$$\frac{\partial}{\partial x}(\rho u) + \frac{\partial}{\partial y}(\rho v) = 0 \quad \textbf{(5.233)}$$

x momentum:
$$\rho u \frac{\partial u}{\partial x} + \rho v \frac{\partial u}{\partial y} = -\frac{dP}{dx} + \frac{\partial}{\partial y}\left(\mu \frac{\partial u}{\partial y}\right) - \rho g \quad \textbf{(5.234)}$$

Thermal energy:
$$\rho u c_p \frac{\partial T}{\partial x} + \rho v c_p \frac{\partial T}{\partial y} = \frac{\partial}{\partial y}\left(k \frac{\partial T}{\partial y}\right) \quad \textbf{(5.235)}$$

where we have assumed that viscous dissipation and compressibility effects are negligible in the energy equation. Notice that the pressure gradient in the momentum equation is a total derivative, and that the subscript x on the gravitational acceleration g has been dropped since it is now unnecessary.

Far from the wall, velocity gradients are zero, and Eq. (5.234) reduces to

$$0 = -\frac{dP}{dx} - \rho_e g \qquad \text{or} \qquad -\frac{dP}{dx} = \rho_e g$$

Substituting back in Eq. (5.234) gives

$$\rho u \frac{\partial u}{\partial x} + \rho v \frac{\partial u}{\partial y} = \frac{\partial}{\partial y}\left(\mu \frac{\partial u}{\partial y}\right) + g(\rho_e - \rho) \quad \textbf{(5.236)}$$

Implicit in this step is the Archimedes principle, which was used directly in Section 5.4.5 to obtain the buoyancy force as $g(\rho_e - \rho)$. We now assume constant properties except for the density difference in the buoyancy force. Since $\rho = \rho(P, T)$ we can write

$$\rho_e - \rho = \left(\frac{\partial \rho}{\partial T}\right)_P (T_e - T) + \left(\frac{\partial \rho}{\partial P}\right)_T (P_e - P) + \text{ Higher-order terms}$$

and in the boundary layer we can assume $P \simeq P_e$ to obtain

$$\frac{\rho_e - \rho}{\rho} = -\frac{1}{\rho}\left(\frac{\partial \rho}{\partial T}\right)_P (T - T_e) = \beta(T - T_e) \quad \textbf{(5.237)}$$

The preceding simplifications are often called the *Boussinesq approximation,* and the resulting governing equations are

$$\frac{\partial u}{\partial x} + \frac{\partial v}{\partial y} = 0 \tag{5.238}$$

$$u\frac{\partial u}{\partial x} + v\frac{\partial u}{\partial y} = \nu\frac{\partial^2 u}{\partial y^2} + g\beta(T - T_e) \tag{5.239}$$

$$u\frac{\partial T}{\partial x} + v\frac{\partial T}{\partial y} = \alpha\frac{\partial^2 T}{\partial y^2} \tag{5.240}$$

Scale Analysis

We proceed as in Section 5.8.1 and define dimensionless variables:

$$x^* = \frac{x}{L}, \quad y_T^* = \frac{y}{\Delta}, \quad u^* = \frac{u}{u_S}, \quad v^* = \frac{v}{(\Delta/L)u_S}, \quad T^* = \frac{T - T_e}{T_s - T_e}$$

where the scaling of v follows from the continuity equation exactly as in Section 5.8.1. In contrast to the forced-convection boundary layer, there is no prescribed velocity that can be used for scaling purposes; hence, the appropriate magnitude of u_S must be determined from the equations. Notice that the thermal boundary layer thickness Δ is used to scale y in all the equations, rather than using the hydrodynamic boundary layer thickness δ in the mass and momentum conservation equations. Expected velocity and temperature profiles are shown in Fig. 5.44. For $\Pr \ll 1$, δ and Δ are, in fact, of similar magnitude since the buoyancy force produces a flow throughout the thermal boundary layer, although the maximum velocity does occur close to the wall. For $\Pr \gg 1$, Δ is much less than δ since viscous forces exert a drag on the fluid outside the thermal boundary layer. However, it is the buoyancy force that is driving the flow, and the maximum velocity, of order u_S, must occur at a value of $y < \Delta$. Thus, it is necessary to scale y by Δ in the mass and momentum

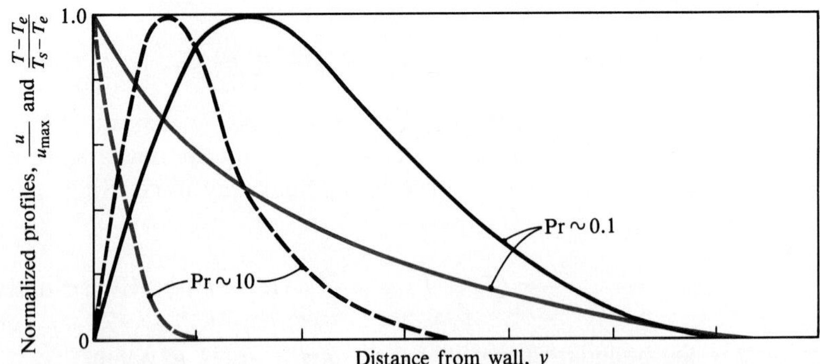

Figure 5.44 Velocity and temperature profiles for a laminar natural-convection boundary layer on a vertical wall: effect of Prandtl number.

conservation equations in order to have $\partial u^*/\partial y^*$ and $\partial v^*/\partial y^*$ of order of magnitude unity. The scaled momentum and energy equations are

$$\frac{u_S^2}{L}\left(u^*\frac{\partial u^*}{\partial x^*} + v^*\frac{\partial u^*}{\partial y_T^*}\right) = \frac{\nu u_S}{\Delta^2}\frac{\partial^2 u^*}{\partial y_T^{*2}} + g\beta(T_s - T_e)T^* \tag{5.241}$$

$$\frac{u_S(T_s - T_e)}{L}\left(u^*\frac{\partial T^*}{\partial x^*} + v^*\frac{\partial T^*}{\partial y_T^*}\right) = \frac{\alpha(T_s - T_e)}{\Delta^2}\frac{\partial^2 T^*}{\partial y_T^{*2}} \tag{5.242}$$

Consider first the high-Prandtl-number limit. For such fluids, $\Delta \ll \delta$: heat transfer is controlled by the inner part of the boundary layer, where viscous forces balance buoyancy forces and inertia (acceleration) forces are negligible. Then, from Eq. (5.241),

$$\frac{\nu u_S}{\Delta^2} \sim g\beta(T_s - T_e) \qquad \text{or} \qquad \frac{u_S}{\Delta^2} \sim \frac{g\beta(T_s - T_e)}{\nu} \tag{5.243}$$

Also, from Eq. (5.242),

$$\frac{u_S(T_s - T_e)}{L} \sim \frac{\alpha(T_s - T_e)}{\Delta^2} \qquad \text{or} \qquad u_S\Delta^2 \sim \alpha L \tag{5.244}$$

Solving Eqs. (5.243) and (5.244) gives

$$\frac{\Delta}{L} \sim \left(\frac{\alpha\nu}{g\beta(T_s - T_e)L^3}\right)^{1/4}; \qquad u_S \sim \left(\frac{g\beta(T_s - T_e)L}{\nu/\alpha}\right)^{1/2} \tag{5.245}$$

The dimensionless group $g\beta(T_s - T_e)L^3/\alpha\nu$ is the Rayleigh number, Ra; thus,

$$\frac{\Delta}{L} \sim \frac{1}{\text{Ra}^{1/4}} \tag{5.246}$$

Consider next the low-Prandtl-number limit. For such fluids the high thermal diffusivity suggests that the heat transfer is controlled by the outer part of the boundary layer, where inertia forces balance the buoyancy forces and viscous forces are negligible. From Eq. (5.241),

$$\frac{u_S^2}{L} \sim g\beta(T_s - T_e) \qquad \text{or} \qquad u_S \sim [g\beta(T_s - T_e)L]^{1/2} \tag{5.247}$$

Substituting in Eq. (5.244) gives

$$\frac{\Delta}{L} \sim \left(\frac{\alpha^2}{g\beta(T_s - T_e)L^3}\right)^{1/4} \qquad \text{or} \qquad \frac{\Delta}{L} \sim \frac{1}{\text{Bo}^{1/4}} \tag{5.248}$$

where Bo is the Boussinesq number. To develop these two limit cases, the inertia and viscous forces were neglected in turn. Now consider the case where these forces are of comparable magnitude. Then, from Eq. (5.241),

$$\frac{u_S^2}{L} \sim \frac{\nu u_S}{\Delta^2} \qquad \text{or} \qquad u_S\Delta^2 \sim \nu L \tag{5.249}$$

Comparison with Eq. (5.242) shows that $\nu/\alpha = \text{Pr} \sim 1$ for this situation. Furthermore, since it is the buoyancy force that is always driving the flow, the inertia and viscous forces must also be comparable in magnitude to the buoyancy force. Again using Eq. (5.241),

$$\frac{u_S^2}{L} \sim \frac{\nu u_S}{\Delta^2} \sim g\beta(T_s - T_e) \tag{5.250}$$

Solving for Δ and u_S gives

$$\frac{\Delta}{L} \sim \left(\frac{\nu^2}{g\beta(T_s - T_e)L^3}\right)^{1/4}; \qquad u_S \sim (g\beta\Delta T L)^{1/2} \tag{5.251}$$

The dimensionless group $g\beta(T_s - T_e)L^3/\nu^2$ is the familiar Grashof number; thus,

$$\frac{\Delta}{L} \sim \frac{1}{\text{Gr}^{1/4}} \tag{5.252}$$

Scale analysis has shown how the magnitude of the Prandtl number determines whether the Rayleigh, Boussinesq, or Grashof number characterizes natural convection. Since $\text{Ra} = \text{GrPr}$ and $\text{Bo} = \text{GrPr}^2$, for $\text{Pr} \sim 1$ all three of these dimensional groups are equivalent. Also, scale analysis has shown that a correct interpretation of these groups is a parameter relating the thermal boundary layer thickness to distance up the wall. The Reynolds number plays a similar role for the hydrodynamic boundary layer thickness of a forced-convection boundary layer. Since these groups are obtained by comparing forces acting on the fluid, one often sees their interpretation as *ratios* of these forces—for example, the Rayleigh number as the ratio of the buoyancy forces to viscous forces. Such an idea is erroneous because the length scale for Ra is L, rather than an appropriate thickness of the boundary layer.

The Nusselt Number

Using our estimate of the thermal boundary layer thickness Δ, we can examine the behavior of the Nusselt number. The wall heat flux is

$$q_s = -k\left.\frac{\partial T}{\partial y}\right|_{y=0} = -\frac{k(T_s - T_e)}{\Delta}\left.\frac{\partial T^*}{\partial y_T^*}\right|_{y_T^*=0}$$

Hence,

$$q_s \sim \frac{k(T_s - T_e)}{\Delta}$$

$$\text{Nu} \sim \frac{L}{\Delta} \tag{5.253}$$

Using Eqs. (5.246), (5.252), and (5.248) for Δ/L gives

$$\mathrm{Pr} \gg 1: \quad \mathrm{Nu} \sim \mathrm{Ra}^{1/4} \ (= \mathrm{Gr}^{1/4}\mathrm{Pr}^{1/4}) \tag{5.254a}$$

$$\mathrm{Pr} \simeq 1: \quad \mathrm{Nu} \sim \mathrm{Gr}^{1/4} \tag{5.254b}$$

$$\mathrm{Pr} \ll 1: \quad \mathrm{Nu} \sim \mathrm{Bo}^{1/4} \ (= \mathrm{Gr}^{1/4}\mathrm{Pr}^{1/2}) \tag{5.254c}$$

Such relations are useful in correlating experimental data. The similarity principle (or, equivalently, dimensional analysis using the Buckingham pi theorem) simply gives the result $\mathrm{Nu} = f(\mathrm{Gr}, \mathrm{Pr})$. Scale analysis has indicated the form of the function.

5.9 CLOSURE

The early part of this chapter introduced the student to the analysis of convection by considering the simplest flows only. The analyses should have provided insight into some of the essential features of convection, such as the effect of viscous dissipation in a high-speed flow, the nature of fully developed flow in a duct, and the important role played by the growth and thickness of boundary layers. The analytical results show why some of the correlations in Chapter 4 took their particular forms.

The latter part of the chapter may be viewed as preparation for advancing further in the subject of convection. Turbulence models, the concepts of similarity and modeling, the general conservation equations, and the methods of scale analysis should all be familiar to the student before he or she proceeds to consult the vast literature on convection or uses packaged computer programs for solving convection problems. Otherwise, the student will surely be overwhelmed by the literature and will be unable to obtain reliable results from computer programs.

The evolution of computer use for convection analysis makes an interesting story. In the late 1950s, the then-new digital computer was used to solve the ordinary differential equations of self-similar flows. During the 1960s, efficient numerical methods were developed to solve the partial differential equations of boundary layer and duct flows; by the end of the decade such solutions were routine. The 1970s saw attention directed toward solving the general conservation equations governing flows with vortices, shock waves, and other complicated features. Many of these flows remain very challenging today, requiring sophisticated numerical methods and large, fast computers. The most recent development has been the increased availability of packaged computational fluid dynamics (CFD) software, for example PHOENICS [15], FLUENT [16], FLOW3D [17], and COMPACT [18]. Versions are available even for the personal computer. Such programs vary in degree of generality, but all can be used to solve a wide variety of flows, even by users having little knowledge of (and sometimes no access to) the numerical methods incorporated in the program. Sophisticated interactive programming allows new geometric configurations and boundary conditions to be easily implemented, and standard procedures are prescribed to ensure a properly converged solution. Unquestionably, the availability of such programs implies that portions of the convection literature are now obsolete.

On the other hand, the difficulties and subtleties of convection make it impossible to use such programs blindly. When new problems are attempted, the first solutions obtained are often far from accurate. A good understanding of the fundamentals is required, as well as a familiarity with benchmark experimental data and previous analytical and numerical solutions of established validity.

REFERENCES

1. Eckert, E. R. G., "Engineering relations for heat transfer and friction in high-velocity laminar and turbulent boundary layer flow over surfaces with constant pressure and temperature," *Trans. ASME,* 78, 1273–1284 (1956).
2. Wortman, A., Mills, A. F., and Soo Hoo, G., "The effect of mass transfer on recovery factors in laminar boundary layer flows," *Int. J. Heat Mass Transfer,* 15, 443–456 (1972).
3. von Kármán, T., "Über laminare und turbulente Reibung," *ZAMM,* 1, 233–252 (1921); see also NACA TM 1092 (1946).
4. Blasius, H., "Grenzschichten in Flüssigkeiten mit kleiner Reibung," *Z. Math. u. Phys.,* 56, 1 (1908). (English translation in NACA TM 1256.)
5. Piercy, N. A. V., and Preston, J. H., "A simple solution of the flat plate problem of skin friction and heat transfer," *Phil. Mag.,* 21, 995–1005 (1936).
6. Prandtl, L., "Bericht über Untersuchungen zur ausgebildeten Turbulenz," *ZAMM,* 5, 136 (1925).
7. van Driest, E. R., "On turbulent flow near a wall," *J. Aero. Sci.,* 23, 1007–1011 (1951).
8. Coles, D., "The law of the wake in the turbulent boundary layer," *J. Fluid Mechanics,* 1, part 2, 191–226 (1956).
9. Escudier, M. P., "The distribution of the mixing length in turbulent flows near walls," Imperial College, London, *Mechanical Engineering Department Report TWF/TN/1* (1965).
10. White, F. M., *Viscous Fluid Flow,* McGraw-Hill, New York (1974).
11. von Kármán, T., "The analogy between fluid friction and heat transfer," *Trans. ASME,* 61, 705–710 (1939).
12. Nikuradse, J., "Laws of flow in rough pipes," NACA TM 1292 (1950). (English translation of VDI-Forschungsheft 361, 1933.)
13. Hinze, J. O., *Turbulence,* 2nd ed., McGraw-Hill, New York, (1975).
14. Prandtl, L., "The mechanics of viscous fluids," in W. F. Durand, *Aerodynamic Theory,* vol. 3 (1935).

15. PHOENICS (Parabolic, Hyperbolic, or Elliptic Numerical Integration Code Series); available from CHAM Ltd., 40 High Street, Wimbledon, London, SW19A4, England.

16. FLUENT; available from Creare, Etna Road, P.O. Box 71, Hanover, NH 03755.

17. FLOW3D; available from Computer Fluid Dynamic Services, Harwell Laboratory, UKAEA, Oxfordshire, OX11 ORA, England.

18. COMPACT; available from Innovative Research Inc., 7846 Ithaca Lane N., Maple Grove, MN 55369-8549.

19. Noto, K., and Matsumoto, R., "Turbulent heat transfer by natural convection along an isothermal vertical flat surface," *J. Heat Transfer,* 97, 621–624 (1975).

20. Lykoudis, P. S., "Non-dimensional numbers as ratios of characteristic times," *Int. J. Heat Mass Transfer,* 33, 1560–1570 (1990).

EXERCISES

5–1. Two 1 m–long concentric cylinders form an annular gap 1 mm wide. The outer cylinder is stationary. The inner cylinder has a radius of 10 cm and rotates at 1000 rpm. Determine the power dissipated by viscous dissipation in the gap if the fluid is

(i) air,
(ii) water,
(iii) SAE 50 oil,

all at 300 K and 1 atm pressure.

5–2. Prepare a table of recovery factors for the following fluids:

(i) Air at 300 K, 1 atm
(ii) Saturated steam at 380 K
(iii) Water at 300 K
(iv) Water at 456 K
(v) SAE 50 oil at 350 K
(vi) Liquid mercury at 600 K.

In each case, calculate the values for a Couette flow, a laminar boundary layer on a flat plate, and a turbulent boundary layer.

5–3. A fighter aircraft's mission necessitates a high-speed penetration behind enemy lines and beneath radar detection altitude. Prior to making this run, the aircraft cruises at high altitude, and most of the structure—in particular the wings—reaches a uniform temperature of 273 K. By modeling a wing as a 16 m × 2.5 m flat plate, estimate:

(i) the rate of heat input to the wing during the initial stages of a high-speed run, at 1500 km/h.
(ii) the fraction of the heat input that can be attributed to viscous dissipation.

Take the ambient air to be at 303 K and 1 bar.

5–4. At an altitude of 30,000 m the atmospheric pressure is 1197 Pa, and the temperature is 227 K. For a turbulent boundary layer on a flat plate, prepare a graph of adiabatic wall temperature versus Mach number for $0 < M < 5$. Also calculate the convection-radiation equilibrium temperature at $x = 0.5$ m for a surface emittance of 0.9. Assume transition takes place at the leading edge.

5–5. Consider low-speed, hydrodynamically fully developed flow in a parallel-plate duct. For $x < 0$, both plates are insulated, and the fluid is at temperature T_1; for $x > 0$, one plate is maintained at $T = T_2 > T_1$, and the other remains insulated. Draw sketches of

(i) $T_b(x)$
(ii) $T(y)$ at various values of x

to illustrate the development of the temperature field.

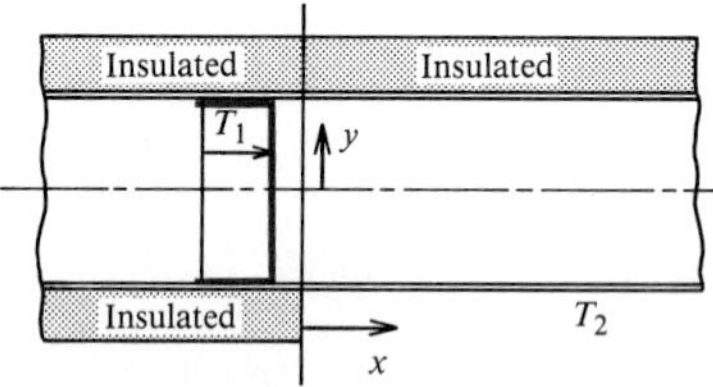

5–6. Consider low-speed, constant-property, fully developed laminar flow between two parallel plates at $y = \pm b$. The plates are electrically heated to give a uniform wall heat flux. Show that

(i) the velocity profile is $u = (3/2)u_b[1 - (y/b)^2]$.
(ii) the friction factor is $f = 96/\mathrm{Re}_{D_h}$.
(iii) the Nusselt number is $\mathrm{Nu} = 140/17 = 8.24$.

5–7. As in Exercise 5–6, consider constant-property, fully developed laminar flow between parallel plates, but now one plate is insulated, the other is isothermal, and the velocity is high enough for viscous dissipation to be significant. Determine the temperature profile.

5–8. Consider low-speed, constant-property, fully developed laminar flow between parallel plates, with one plate insulated and the other uniformly heated. Show that the Nusselt number is $\mathrm{Nu} = 140/26 = 5.385$.

5–9. Consider laminar flow of high-viscosity oil in a circular tube with a uniform wall temperature. If viscous heating is significant, determine the temperature profile a long distance from the inlet. Assume constant properties.

5–10. A gas flows between parallel porous plates. The same gas is blown through one wall and exhausted through the other, such that the normal velocity component v_s is equal at both walls. At a location far from the entrance, determine

(i) the velocity profile
(ii) the temperature profile

if both walls are isothermal at different temperatures T_1 and T_2. Assume constant-property, low-speed flow.

5–11. Lithium enters a parallel-plate duct of spacing $2b$, with a uniform velocity U and a uniform temperature T_0. The walls are maintained at a uniform temperature T_s. If laminar plug flow is assumed, determine the temperature distribution downstream of the entrance. Also, if $\mathrm{Re}_{D_h} = 1500$ and $\mathrm{Pr} = 0.03$, determine the Nusselt number variation along the duct. (*Hint:* There is an analogous conduction problem in Chapter 3.)

5–12. Show that for fully developed laminar flow in an annular duct the velocity profile is given by

$$u = \frac{R_o^2}{4\mu}\left(-\frac{dP}{dx}\right)\left[1 - \left(\frac{r}{R_o}\right)^2 + \frac{1-\gamma^2}{\ln(1/\gamma)}\ln\frac{r}{R_o}\right]$$

where $\gamma = R_i/R_o$ and R_i, R_o are the inner and outer radii, respectively. Hence show that the bulk velocity is

$$u_b = \frac{R_o^2}{8\mu}\left(-\frac{dP}{dx}\right)\left[1 + \gamma^2 - \frac{1-\gamma^2}{\ln(1/\gamma)}\right]$$

and the location where the maximum velocity occurs is given by

$$\frac{r}{R_o} = \left(\frac{1-\gamma^2}{2\ln(1/\gamma)}\right)^{1/2}$$

5–13. Calculate the pressure drop per meter length when air at 300 K, 1 atm pressure flows at 7×10^{-4} kg/s through an annulus of inner and outer radii 6 and 10 mm, respectively.

5–14. Derive the differential momentum conservation equation for a laminar boundary layer on a flat plate using a Lagrangian approach; that is, apply Newton's second law of motion to an element of fluid of fixed mass moving along a streamline.

5–15. Derive the integral momentum equation for a laminar boundary layer on a flat plate, Eq. (5.51), by integrating the differential momentum equation, Eq. (5.39), across the boundary layer.

5–16. Determine the skin friction coefficient for a laminar boundary layer on a flat plate by assuming a quartic polynomial for the velocity profile when using the von Kármán profile method to solve the integral momentum equation.

5–17. Air at 300 K and 1 atm pressure flows along a flat plate at 2 m/s. For $x < 5$ cm, $T_s = 300$ K, whereas for 5 cm $< x <$ 20 cm, $T_s = 330$ K. Calculate the heat loss from the plate and compare your result with the heat loss if the plate were isothermal at 330 K. Assume a laminar boundary layer.

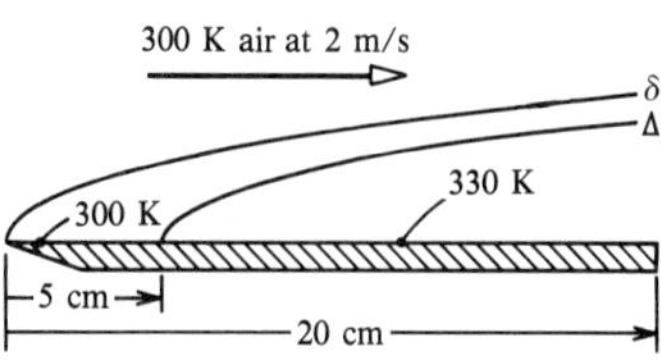

5–18. Equation (5.50) gives the local Nusselt number for plug flow along an isothermal flat plate as

$$\mathrm{Nu}_x = 0.564\,\mathrm{Pe}_x^{1/2}$$

(i) Using Eq. (3.60), show that the corresponding result for a uniformly heated plate is

$$\mathrm{Nu}_x = 0.886\,\mathrm{Pe}_x^{1/2}$$

(ii) Using Eqs. (1.26) and (4.82) where appropriate, obtain corresponding expressions for $\overline{\mathrm{Nu}}_L$ and compare the results.

5–19. Water at 290 K flows at 1 m/s along a flat plate. For $x < 10$ cm, $T_s = 290$ K, whereas for $10 < x < 30$ cm, $T_s = 310$ K. Calculate the local heat flux at $x = 0.3$ m and compare your result with the heat flux if the plate were isothermal at 310 K.

5–20. Determine the skin friction for a laminar boundary layer on a flat plate by assuming a quadratic velocity profile $u/u_e = 2(y/\delta) - (y/\delta)^2$ when using the von Kármán profile method to solve the integral momentum equation. Also determine the Nusselt number for an isothermal surface using a quadratic temperature profile $(T - T_e)/(T_s - T_e) = 1 - 2(y/\Delta) + (y/\Delta)^2$.

5–21. Repeat Exercise 5–20 with sine function velocity and temperature profiles, namely, $u/u_e = \sin(\pi y/2\delta)$; $(T - T_e)/(T_s - T_e) = 1 - \sin(\pi y/2\Delta)$.

5–22. Write a computer program to solve the Blasius equation, Eq. (5.81), by formal integration and iteration. Use trapezoidal rule integration, $\eta_{\max} = 6$, and 100 integration steps.

5–23. Extend the computer program of Exercise 5–22 to include a calculation of $\theta'(0)$. Hence, prepare a graph of $\mathrm{Nu}_x\, 2^{1/2}/\mathrm{Re}_x^{1/2}$ versus Pr. Take special care with your choice of $\eta_{\max}$ and the number of integration steps for $\mathrm{Pr} \ll 1$ and $\mathrm{Pr} \gg 1$.

5–24. Air at 300 K and 1 atm flows along a flat plate at 3 m/s. At a location 0.2 m from the leading edge, plot the u and v velocity profiles using the exact solution to the Blasius equation. Also determine the boundary layer thickness, if it is defined as the location where $u = 0.99u_e$.

5–25. Use the computer program written for Exercise 5–23 to obtain the temperature profile in Exercise 5–24 when the plate temperature is 400 K.

5–26. The Runge-Kutta method can be used to solve the Blasius equation. The third-order nonlinear equation is written as three equivalent first-order equations,

$$f' = g$$
$$g' = h$$
$$h' = -fh$$

which can be solved by integrating from $\eta = 0$ if $f(0)$, $g(0)$, and $h(0)$ are known. Since $h(0)$ is not known, a guess must be made. After integrating to a sufficiently large value of η, say $\eta_{\max} = 6$, $g = f'$ will approach a constant value, not equal to the required value of 1. If $g > 1$, a lower value of $h(0) = f''(0)$ must be guessed, and vice versa. Write the required computer program using a standard subroutine for Runge-Kutta or equivalent integration. Devise a suitable scheme for automating the successive guesses.

5–27. (i) Show that the energy conservation equation for a high-speed constant property laminar boundary layer flow along a flat plate is

$$u\,\frac{\partial T}{\partial x} + v\,\frac{\partial T}{\partial y} = k\,\frac{\partial^2 T}{\partial y^2} + \frac{\mu}{\rho c}\left(\frac{\partial u}{\partial y}\right)^2$$

by including the effect of viscous dissipation in the derivation of Eq. (5.42).

(ii) Show how this equation can be transformed into self-similar form to obtain $\phi'' + f\,\mathrm{Pr}\phi + 2\,\mathrm{Pr}(f'')^2 = 0$, where $\phi = (T - T_e)/(u_e^2/2c)$.

(iii) Obtain a particular solution of this equation for an adiabatic wall, that is, satisfying $\phi'(0) = 0$ and $\phi(\infty) = 0$, by formally integrating using an integrating factor $\exp(\int f\,\mathrm{Pr}\,d\eta)$.

(iv) Using the profiles of f and f'' given in Table 5.1 (or obtained in Exercises 5–22 or 5–26), write a computer program to obtain $\phi'(0)$, which is the recovery factor r. Hence show that $r \simeq \mathrm{Pr}^{1/2}$ for $\mathrm{Pr} \sim 1$.

5–28. A plate emerges from a slot at a speed u_s m/s into a stagnant ambient fluid as shown in the figure.

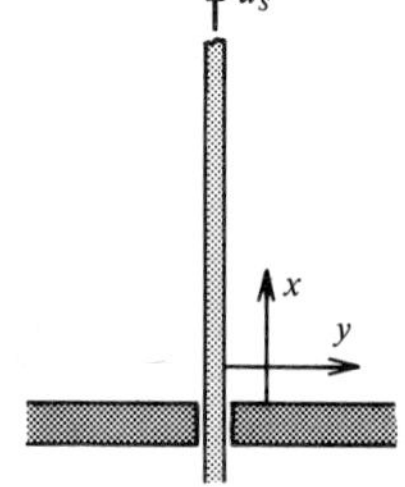

(i) Draw a sketch of the streamlines and velocity profiles adjacent to the plate in the vicinity of the slot.

(ii) Assuming laminar flow, write down appropriate forms of the mass, momentum, and energy equations for $\mathrm{Re} \gg 1$.

(iii) Show that a self-similar solution can be obtained for the velocity field, and discuss its similarities/differences to the Blasius solution.

(iv) If the plate is hotter than the fluid, investigate the possibility of a self-similar solution for the temperature field that might have engineering utility.

5–29. Develop an iterative solution procedure for the fluid mechanics problem of Exercise 5–28, and write a computer program to execute the procedure. Hence, show that the skin friction coefficient is given by $C_{fx} = 0.887\,\mathrm{Re}_x^{-1/2}$.

5–30. Repeat the integral method analysis of Section 5.4.5, but allow the wall temperature to vary as $(T_s - T_e) = Cx^\gamma$. Show that the new expressions for X and D are

$$X = \frac{60\alpha}{D^2}\frac{4}{3+5\gamma}; \qquad D^4 = \frac{720}{Cg\beta}\left[\frac{12\alpha^2}{21}\frac{5+3\gamma}{(3+5\gamma)^2} + \frac{\alpha\nu}{3+5\gamma}\right]$$

Hence, show that if the wall heat flux is uniform, the wall temperature increases proportional to $x^{1/5}$.

5–31. A vertical plate at 320 K is located in air at 300 K and 1 atm. At a location 10 cm from the bottom of the plate determine δ, U, Nu_x, h_{cx} and q_s.

5–32. Consider natural convection from the underside of a heated horizontal cylinder. Derive an integral form of the boundary layer momentum equation valid for laminar or turbulent flow.

5–33. The integral method solution for laminar natural convection on a vertical wall given in Section 5.4.5 shows that the local heat transfer coefficient is proportional to $x^{-1/4}$, where x is measured from the leading edge. Use this result to derive a formula for the local Nusselt number consistent with the correlation for the average Nusselt number, Eq. (4.85). Compare your result with Eq. (5.109).

5–34. The equations governing an unsteady boundary layer flow along a flat plate are

$$\frac{\partial \rho}{\partial t} + \frac{\partial}{\partial x}(\rho u) + \frac{\partial}{\partial y}(\rho v) = 0 \qquad \textbf{(1)}$$

$$\rho\frac{\partial u}{\partial t} + \rho u\frac{\partial u}{\partial x} + \rho v\frac{\partial u}{\partial y} = \frac{\partial}{\partial y}(-\tau_{xy}) \qquad \textbf{(2)}$$

$$\rho\frac{\partial h}{\partial t} + \rho u\frac{\partial h}{\partial x} + \rho v\frac{\partial h}{\partial y} = \frac{\partial}{\partial y}(-q_y) \qquad \textbf{(3)}$$

In a turbulent flow we would expect Newton's and Fourier's laws to apply instantaneously, that is, $\tau_{xy} = -\mu\, \partial u/\partial y$; $q_y = -k\, \partial T/\partial y$. The momentum and energy equations, Eqs. (2) and (3), can be written as

$$\rho\frac{\partial \phi}{\partial t} + \rho u\frac{\partial \phi}{\partial x} + \rho v\frac{\partial \phi}{\partial y} = \frac{\partial}{\partial y}F \qquad \textbf{(4)}$$

Multiplying the continuity equation, Eq. (1), by ϕ and adding to Eq. (4) gives

$$\frac{\partial}{\partial t}(\rho\phi) + \frac{\partial}{\partial x}(\rho u\phi) + \frac{\partial}{\partial y}(\rho v\phi) = \frac{\partial}{\partial y}F \qquad \textbf{(5)}$$

Equation (5) is a suitable starting point for deriving time-averaged conservation equations for a turbulent boundary layer on a flat plate. All terms should be written as sums of mean and fluctuating components [see Eq. (5.110)]. Perform the necessary algebra for a constant-density, steady, mean flow. Identify the eddy diffusivities in the time-averaged equations.

5–35. Show that the inner-region eddy viscosity profile consistent with van Driest's mixing length expression, Eq. (5.132), is

$$\varepsilon^+ = 1 + \frac{\varepsilon}{\nu} = \frac{1}{2} + \frac{1}{2}\left\{1.0 + 4\kappa^2 y^{+2}[1.0 - \exp(-y^+/y_t^+)]^2\right\}^{1/2}$$

where for a turbulent boundary layer on a flat plate the shear stress can be taken to be constant across the inner region.

5–36. Consider turbulent flow of air along a porous wall. Air is also blown through the wall giving a normal velocity v_s at the wall. Show that Prandtl's mixing length hypothesis leads to a law of the wall of the form

$$\frac{2}{v_s^+}(1 + v_s^+ u^+)^{1/2} = \frac{1}{\kappa} \ln \frac{y}{y_0}$$

where y_0 is the value of y where $u^+ = -(v_s^+)^{-1}$.

5–37. Derive Eq. (5.137), which gives the momentum thickness of a turbulent boundary layer according to Coles' law of the wake.

5–38. Air at 300 K and 1 atm flows along a flat plate at a velocity of 120 m/s. At a distance 3.0 m from the leading edge, plot the velocity profile using the law of the wall and the law of the wake with $\kappa = 0.41$ and $C = 5.0$.

5–39. Noto and Matsumoto [19] proposed an eddy viscosity profile of the form

$$\frac{\varepsilon_M}{\nu} = 0.4y^+[1 - \exp(-0.0017y^{+2})]$$

Using $\mathrm{Pr}_t = 0.9$, show that the Stanton number for a turbulent boundary layer on a flat plate in the limit $\mathrm{Pr} \to \infty$ is

$$\mathrm{St}_x = 0.0753(C_{fx}/2)^{1/2}\,\mathrm{Pr}^{-2/3}$$

(For high Prandtl numbers the thermal boundary layer is very thin; thus, we can assume a constant heat flux across the thermal boundary layer and retain only the first term in a Taylor series expansion of ε^+.)

5–40. Calculate ε_M and y at $y^+ = 26$ for 330 K water flowing in a tube of 2.5 cm inside diameter at a bulk velocity of 2 m/s.

5–41. Consider flow at 10 m/s in a 2 cm–I.D. tube of the following fluids:

(i) Water at 300 K
(ii) Air at 300 K and 1 atm
(iii) Mercury at 400 K.

Calculate y and ε_H/α at $y^+ = 26$ using the van Driest eddy viscosity profile given in Exercise 5–35, $\kappa = 0.41$, and ε_H/α at $y^+ = R^+$ using Hinze's eddy viscosity expression for the core, Eq. (5.156). Assume a turbulent Prandtl number of 0.9.

5–42. Using van Driest's eddy viscosity profile given in Exercise 5.35, $\kappa = 0.41$ and $\mathrm{Pr}_t = 1.0$, show that the Stanton number for turbulent flow in a pipe in the limit as Prandtl number goes to infinity is

$$\mathrm{St} = 0.113(C_f/2)^{1/2}\,\mathrm{Pr}^{-3/4}$$

(For large Prandtl numbers the temperature profile near the wall is very steep. Thus, we can assume a constant heat flux near the wall, $T_b = T_c$, and retain only the first term in a Taylor series expansion of ε^+.)

5–43. Assuming the logarithmic velocity profile is valid across a parallel-plate channel, obtain the relation between friction factor and Reynolds number. Plot a graph of $f(\mathrm{Re}_{D_h})$ and compare it with a similar graph based on Eq. (5.162) with Re_D replaced by Re_{D_h}. Take $\kappa = 0.4$, $C = 5.5$.

5–44. Consider fully developed turbulent flow between parallel plates at a Reynolds number of $\mathrm{Re} = u_b D_h/\nu = 3 \times 10^6$. If one wall is insulated and the other is uniformly heated, determine the temperature profile. Use the von Kármán total viscosity profile.

5–45. Repeat Exercise 5–44 when the insulated wall is rough with an equivalent sand grain roughness of $k_s/D_h = 0.01$. Use the von Kármán total viscosity profile. (*Hint:* Assume that the zero shear plane divides the flow into two regions that do not influence each other, and assume that the bulk velocity u_b is the same in each region. Then use the Moody chart to determine the width of each region.)

5–46. Introduce appropriate dimensionless variables into the thermal energy equation, Eq. (5.209) to show that viscous dissipation for flow in a rectangular cross-section duct is negligible for $\mathrm{Br} \ll 1$, where Br is the Brinkman number.

5–47. Starting with the total energy equation, Eq. (5.202), derive the thermal energy equation, Eq. (5.205).

5–48. Starting with the thermal energy equation in the form of Eq. (5.205), derive the alternative forms, Eqs. (5.207), (5.208), and (5.209).

5–49. Table 5.5 gives the mass conservation equation for an axisymmetric boundary layer as

$$\frac{\partial}{\partial x}(\rho u r) + \frac{\partial}{\partial y}(\rho v r) = 0$$

where coordinate x is measured along the surface and r is the radius from the axis of symmetry. Derive this equation from first principles.

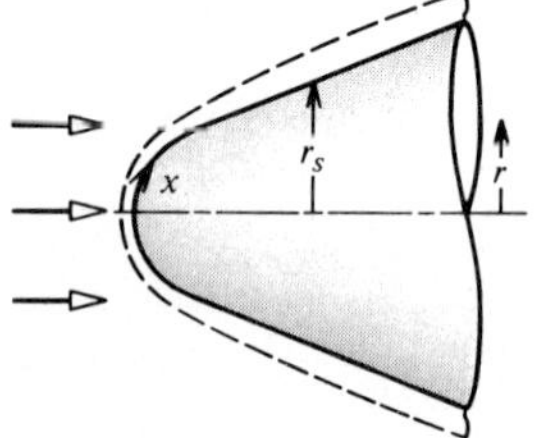

5–50. *Vectorial* dimensional analysis allows lengths measured in different coordinate directions to be independent variables. Properties such as kinematic viscosity then also have a vectorial character. Use this type of dimensional analysis to deduce the form of the similarity variable for the laminar boundary layer on a flat plate.

5–51. The variable property laminar boundary layer equations may be written as

$$\frac{\partial}{\partial x}(\rho u) + \frac{\partial}{\partial y}(\rho v) = 0$$

$$\rho u \frac{\partial u}{\partial x} + \rho v \frac{\partial u}{\partial y} = \frac{\partial}{\partial y}\left(\mu \frac{\partial u}{\partial y}\right) - \frac{dP}{dx}$$

$$\rho u \frac{\partial h}{\partial x} + \rho v \frac{\partial h}{\partial y} = u \frac{dP}{dx} + \frac{\partial}{\partial y}\left(k \frac{\partial T}{\partial y}\right) + \mu \left(\frac{\partial u}{\partial y}\right)^2$$

Show that total enthalpy $H = h + u^2/2$ is governed by

$$\rho u \frac{\partial H}{\partial x} + \rho v \frac{\partial H}{\partial y} = \frac{\partial}{\partial y}\left[\frac{\mu}{\Pr}\frac{\partial H}{\partial y} + \mu\left(1 - \frac{1}{\Pr}\right)\frac{\partial}{\partial y}\left(\frac{u^2}{2}\right)\right]$$

What can be said about the total enthalpy profile for a fluid of unity Prandtl number?

5–52. Use scale analysis of the equations governing the natural-convection laminar boundary layer on a vertical wall to estimate:

(i) the ratio of the hydrodynamic and thermal boundary layer thicknesses for a high-Prandtl-number fluid.
(ii) the ratio of the inner viscous layer and thermal boundary layer thicknesses for a low-Prandtl-number fluid.

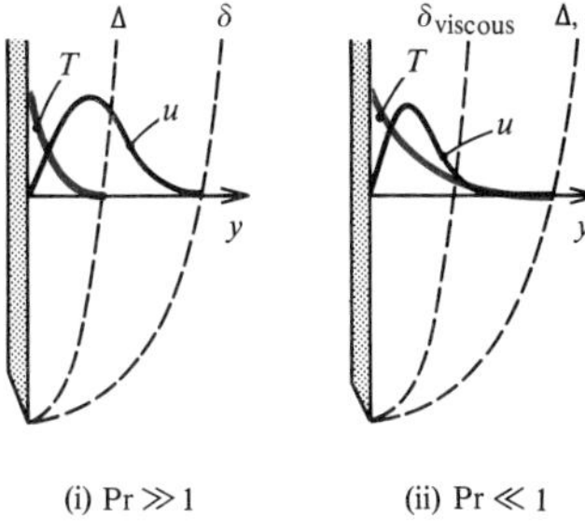

5–53. Using Table 5.5, the thermal energy equation for a high-speed laminar boundary layer on a flat plate can be written as

$$\rho u c_p \frac{\partial T}{\partial x} + \rho v c_p \frac{\partial T}{\partial y} = \frac{\partial}{\partial y}\left(k\frac{\partial T}{\partial y}\right) + \mu\left(\frac{\partial u}{\partial y}\right)^2$$

Use scale analysis to ascertain under what conditions viscous dissipation is negligible for flow of a gas.

5–54. A planar laminar jet is formed by fluid exiting a narrow slit $2b_0$ wide into a large reservoir containing the same fluid at a uniform pressure, and temperature T_e. The bulk velocity of the fluid in the slot is u_0 and is at a uniform temperature T_0.

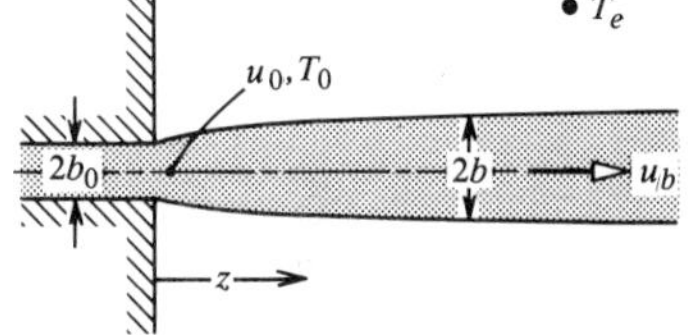

(i) Write down appropriate boundary layer forms of the mass, momentum, and energy conservation equations, ignoring gravity.
(ii) Show that the axial momentum and axial thermal convection integrated across the jet is independent of axial coordinate z.
(iii) By appropriate scaling, determine the dependence of the jet half-width b and bulk velocity u_b on z.
(iv) Similarly, determine the ratio b_T/b, where b_T is the width of region where there are significant temperature gradients, and the dependence of the jet bulk temperature on z.

5–55. It is often useful to interpret dimensionless groups as ratios of characteristic times. Such times include the following:

Convective time: $t_c \sim L/V$

Free-fall time: $t_f \sim [L/g(\Delta\rho/\rho)]^{1/2}$

Momentum diffusion time: $t_M \sim L^2/\nu$
Heat diffusion time: $t_H \sim L^2/\alpha$

Express the Reynolds, Peclet, Prandtl, Grashof, Rayleigh, and Boussinesq numbers in terms of these characteristic times. (See also reference [18].)

5–56. *Acoustic levitation* may be used to locate and transport samples for containerless materials processing in space. The acoustic field also induces "streaming" flow around the sample; streamlines are shown for a cylindrical sample. The processing requires heating a sample in a furnace and subsequent quenching outside the furnace. In order to predict heating and quenching rates, convective heat transfer coefficient data for the acoustic streaming flow are required. An experiment was performed in a 6.3 × 6.3 × 50 cm chamber with a 1018 Hz, 155 decibel acoustic field. A spherical thermistor (Omega model 33020) of diameter 2.286 mm was used to model a spherical sample. By measuring both the voltage and current across the thermistor, its resistance and power dissipation could be calculated, and from the calibration of the thermistor (temperature versus resistance) its temperature was deduced. The air in the chamber was maintained at 22 ± 1°C. The accompanying table lists a data set. Process the data by performing the following calculations.

Run	I mA	R Ω	T °C
1	3.5	2220	32.0
2	3.6	2160	32.8
3	12.0	650	64.0
4	4.2	1840	36.5
5	4.1	1910	35.5
6	5.6	1390	43.8

(i) Calculate the convective heat loss $\dot{Q}_{\text{conv}}$ by subtracting the radiation heat loss $\dot{Q}_{\text{rad}}$ from the total heat loss. Take $\varepsilon = 1$ for the thermistor.
(ii) Calculate $\overline{h}_c$ and $\overline{\text{Nu}}$, and obtain an average value of $\overline{\text{Nu}}$ for the data.
(iii) The acoustic flow is a laminar boundary layer; from Section 5.6 we expect a correlation of the form

$$\text{Nu}_D = C\,\text{Re}_D^{1/2} \qquad (\text{Pr} = 0.7,\ \text{fixed})$$

The sound field can be characterized by the amplitude A and frequency f of the waves. Suggest an appropriate definition of Reynolds number.

(iv) Obtain a value for the constant C.

(v) A small thermistor was chosen for this experiment to minimize the effect of natural convection (which would not occur in space). Is the assumption of negligible natural convection justified?

The amplitude of a sound wave A is related to frequency f [Hz] and intensity β[dB] as $A = 10^{(\beta/20)-6}/\pi f(2\rho a)^{1/2}$, where ρ and a are the density and speed of sound of the medium, respectively.

CHAPTER

6

THERMAL RADIATION

CONTENTS

6.1 INTRODUCTION

Radiation can be viewed either in terms of electromagnetic waves or in terms of transport of photons. The medium involved can be either *nonparticipating* or *participating*. Nonparticipating media include outer space and atmospheric air over short distances, in which photons can travel almost unimpeded from one surface to another. Radiation heat exchange between such surfaces depends only on surface temperatures, surface radiation properties, and the geometry of the configuration. Participating media include combustion gases containing H_2O and CO_2, as well as gases containing *aerosols*, such as dust, soot, and small liquid droplets. Radiation exchange between surfaces separated by a participating medium depends also on the radiation properties of the medium. In the case of a participating gas species, the emissivity and absorptivity depend strongly on temperature.

The first half of Chapter 6, Sections 6.2 through 6.4, is concerned with the engineering calculation of radiation exchange between surfaces separated by a nonparticipating medium. Section 6.2 reviews the physics of radiation and defines the *gray surface* model. Section 6.3 is the key section of the chapter. It deals with methods for calculating radiation heat exchange between surfaces, first for black surfaces and then for diffuse gray surfaces. Section 6.4 extends the diffuse gray surface model to problems involving solar and atmospheric radiation by allowing for a different value of absorptance to short-wavelength solar radiation.

The second half of Chapter 6, Sections 6.5 through 6.7, deals with more advanced topics. Section 6.5 examines *directional* characteristics of surface radiation. The *shape factor* concept, used in Section 6.3, is developed rigorously, and some simple problems involving *specularly* reflecting (mirror-like) surfaces are analyzed. *Spectral* characteristics of surface radiation are further examined in Section 6.6, with the limited objective of demonstrating how to calculate *total hemispherical* radiation properties from experimental data for spectral values. Finally, in Section 6.7, the subject of radiation transfer through gases is introduced. Data for total radiation properties are given, and analytical methods are developed for some simple radiation exchange problems in enclosures containing combustion gases.

Three computer programs accompany Chapter 6. RAD1 calculates shape factors for some simple configurations. RAD2 solves the problem of radiation heat exchange in an enclosure of diffuse gray surfaces. RAD3 calculates the radiation properties of combustion gases. The examples used to illustrate the theory in this chapter are diverse, including a solar collector, a spacecraft, and the combustion chamber of a hydrogen-fueled scramjet.

6.2 THE PHYSICS OF RADIATION

Electromagnetic radiation is characterized by its wavelength. *Thermal radiation* is associated with thermal agitation of molecules, that is, with atomic or molecular transitions, and has wavelengths in the range of about 0.1 to 100μm. In Section 6.2.2, the ideal *black surface* (blackbody) is defined. Also in this section, *Planck's law*, which describes the spectrum of radiant energy emitted by a black surface, is reviewed. It is subsequently related to the *Stefan-Boltzmann law*, which gives the total radiant energy emitted by a black surface. Section 6.2.3 deals with real opaque surfaces, and the important concept of a *gray surface* is introduced. In Section 6.3, the gray surface is used as a model of real surfaces for the engineering analysis of radiant heat exchange between surfaces.

6.2.1 The Electromagnetic Spectrum

All matter continuously emits electromagnetic radiation, which travels through a vacuum at the speed of light, $c_0 = 3 \times 10^8$ m/s. Radiation exhibits both a wave nature (e.g., interference effects) and a particle nature (e.g., the photoelectric effect). The wavelength of the radiation λ is related to its frequency ν_f and speed of propagation c as

$$c = \nu_f \lambda \tag{6.1}$$

The units used for λ are micrometers, or microns ($1\mu = 10^{-6}$ m); meters; or angstroms (1 Å$= 10^{-10}$ m). The unit used for ν_f is the hertz (1 Hz $= 1$ s^{-1}). The wavenumber ν of the radiation is related to its wavelength and frequency as

$$\nu = \frac{1}{\lambda} = \frac{\nu_f}{c} \tag{6.2}$$

The units for ν are usually cm^{-1}; hence ν [cm^{-1}] $= 10^4/\lambda$ [μm]. Figure 6.1 shows the *electromagnetic spectrum*. Thermal effects are associated with radiation in the band of wavelengths from about 0.1 to 100 μm, and visible radiation is in the very narrow band from about 0.4 to 0.7 μm.

According to quantum mechanics, radiation interacts with matter in discrete quanta called *photons*, with each photon having an energy E given by

$$E = \hbar \nu_f \tag{6.3}$$

where $\hbar = 6.626 \times 10^{-34}$ J s is *Planck's constant*. Each photon also has a momentum p given by

$$p = \frac{\hbar \nu_f}{c} \tag{6.4}$$

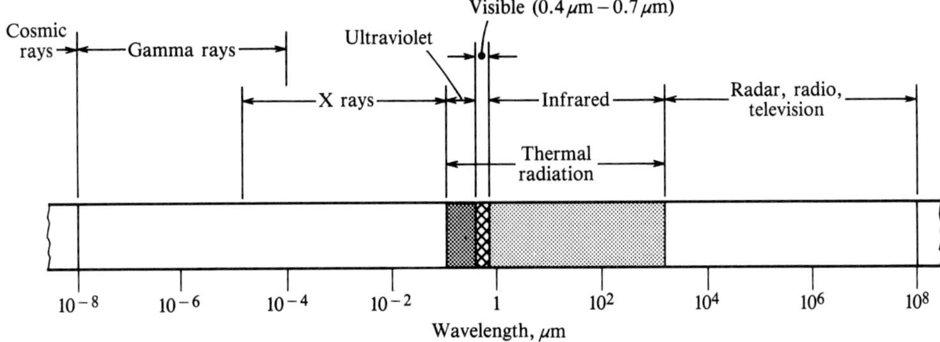

Figure 6.1 The electromagnetic spectrum.

In most solids and liquids, the radiation emitted by one molecule is strongly absorbed by surrounding molecules. Thus, the radiation emitted or absorbed by these liquids or solids involves only a layer of molecules close to the surface; for metals, this layer is a few molecules thick, whereas for nonmetals, it is about a few micrometers thick. For these materials, radiation emission and absorption can be regarded as *surface* phenomena. On the other hand, in gas mixtures containing species such as water vapor and carbon dioxide, or in a semitransparent solid, the absorption is weak, and radiation leaving the body can originate from anywhere in the body. Emission and absorption of radiation are then *volumetric* phenomena. Radiation transfer between surfaces through a nonparticipating medium is relatively simple to analyze and is the primary concern of Chapter 6.

6.2.2 The Black Surface

The black surface (or blackbody), introduced in Section 1.3.2, is an ideal surface. It is defined as a surface that absorbs all radiation falling upon it, irrespective of wavelength or angle of incidence: no radiation is reflected. An obvious consequence of this definition is that all the radiation leaving the surface is emitted by the surface. A subtler consequence is that no surface can emit more radiation than a black surface at a given temperature or wavelength, as will be derived in Section 6.2.3. A black surface also has no preferred direction for emission of radiation; that is, the emission is **diffuse**.

Many surfaces encountered in engineering can be approximated as being black; furthermore, the black surface is a useful standard against which to compare real surfaces, as was seen when the gray surface was defined in Section 1.3.2. A black surface can be closely approximated by a hole into a cavity, as shown in Fig. 6.2. Radiation entering the hole is trapped, since each reflection absorbs part of the remaining energy. A black surface appears black because it absorbs all visible radiation falling upon it and reflects none. However, there are many surfaces that absorb nearly all incident thermal radiation but do not appear black. These surfaces reflect sufficient radiation in the visible range for the eye to detect; examples are ceramics such as aluminum or magnesium oxides. Conversely, surfaces that appear black to the eye may reflect radiation outside the visible spectrum.

A black surface emits a spectrum of radiant energy. We define the **monochromatic emissive power** for a black surface, $E_{b\lambda}$, as the radiant energy emitted per unit area of surface per unit wavelength; it commonly has the units W/m$^2\,\mu$m. The spectral

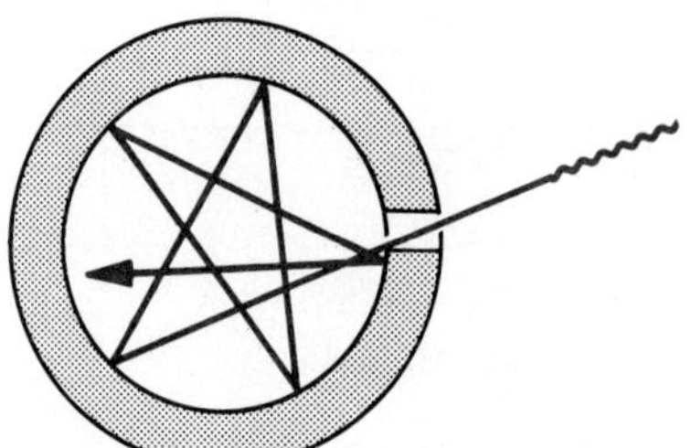

Figure 6.2 A cavity: the hole approximates a blackbody.

distribution of $E_{b\lambda}$, that is, $E_{b\lambda}$ as a function of wavelength λ, is given by **Planck's law**,

$$E_{b\lambda} = \frac{C_1 \lambda^{-5}}{e^{C_2/\lambda T} - 1} \tag{6.5}$$

where for λ in μm, $C_1 = 3.742 \times 10^8$ W μm^4/m^2, and $C_2 = 1.4389 \times 10^4$ μm K.[1] Figure 6.3 shows a plot of Planck's law for various temperatures: as the temperature increases, the rate of energy emission also increases, and the peak of the distribution shifts to shorter wavelengths. *Wien's displacement law* is an equation relating the peak wavelength to temperature:

$$\lambda_{\max} T = 2897.6 \ \mu\text{m K} \tag{6.6}$$

which explains the change in color of a surface from red to white as it is heated. [The derivation of Eq. (6.6) is required as Exercise 6–2.] Notice also that the visible radiation band coincides with the peak of the $E_{b\lambda}$ distribution corresponding to the sun's temperature. The human eye has evolved to be most sensitive to solar radiation reflected from surrounding objects.

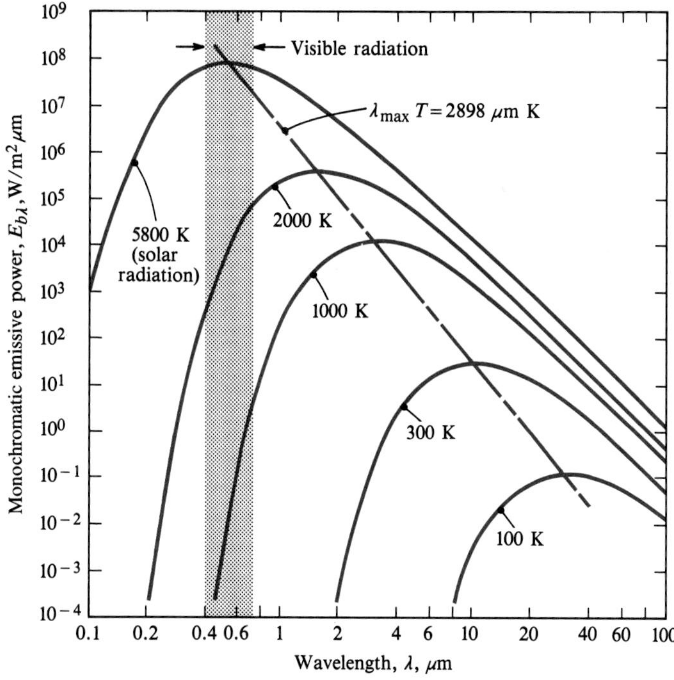

Figure 6.3 Monochromatic emissive power of a black surface at various temperatures according to Planck's law, Eq. (6.5).

[1] In SI-consistent units, $E_{b\lambda}$ has units [W/m^3], $C_1 = 3.742 \times 10^{-16}$ W m^2, and $C_2 = 1.4389 \times 10^{-2}$ m K.

The radiant energy emitted by a surface at all wavelengths is its **total emissive power**, denoted E_b for a black surface. It has a fourth-power temperature dependence given by the **Stefan-Boltzmann law**,

$$E_b = \sigma T^4 \tag{6.7}$$

where σ is 5.6697×10^{-8} W/m^2 K^4. Stefan proposed this law in 1879 based on experimental data, but it can also be derived from Planck's law by integrating over all wavelengths:

$$E_b = \int_0^\infty E_{b\lambda}\, d\lambda \tag{6.8}$$

This integration is required as Exercise 6–3.

6.2.3 Real Surfaces

Most surfaces of engineering importance are opaque (not transparent), although there are some important exceptions, such as glass and plastic films. If radiation falls on a real opaque surface, some will be absorbed, and the remainder will be reflected. The fraction of the incident radiation that is absorbed or reflected depends on the material and surface condition, the wavelength of the incident radiation, and the angle of incidence. There is also a small dependence on the surface temperature. However, for simple engineering calculations, it is often adequate to assume that appropriate average values can be used. If the absorptance (absorptivity) α is defined as the fraction of all incident radiation absorbed, and the reflectance (reflectivity) ρ is defined as the fraction of all incident radiation reflected, then for an opaque surface,

$$\alpha + \rho = 1 \tag{6.9}$$

Now consider an isothermal evacuated enclosure at temperature T, as shown in Fig. 6.4. If a black object of surface area A_b is placed in the enclosure and allowed to come to equilibrium, its temperature will also be T, and it will absorb as much radiation as it emits. Since, according to the Stefan-Boltzmann law, it emits $\sigma T^4 A_b$, it also absorbs $\sigma T^4 A_b$. By definition, a black surface absorbs all radiation falling upon it, so $\sigma T^4 A_b$ is also the radiation incident on the object. Since the shape, size, and location of the object is arbitrary, the irradiation G on any surface in the enclosure is

$$G = \sigma T^4 \tag{6.10}$$

If an object of surface area A and absorptance α is also placed in the enclosure and allowed to come to equilibrium, then it emits as much radiation as it absorbs, and an energy balance requires that

$$\text{Absorption} = AG\alpha = AE = \text{Emission} \tag{6.11}$$

where E is the total emissive power of the surface. We now define the total emittance (emissivity) ε of a real surface as the ratio of its emissive power to that of a black

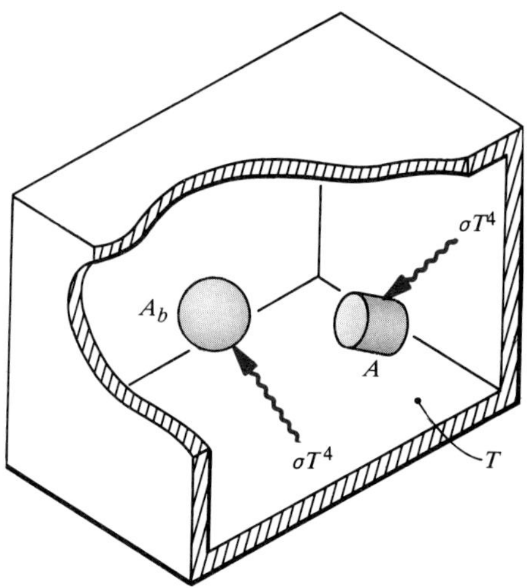

Figure 6.4 An isothermal enclosure at temperature T containing a black object of surface area A_b and a gray object of surface area A.

surface at the same temperature, $\varepsilon = E/E_b$; then $E = \varepsilon E_b$, and substituting Eq. (6.10) in Eq. (6.11) gives

$$A\sigma T^4\alpha = A\varepsilon E_b = A\varepsilon\sigma T^4$$

or

$$\alpha = \varepsilon \quad \textbf{(6.12)}$$

As was mentioned earlier, the absorptance of a real surface depends on the wavelength of the incident radiation: hence, the absorptance depends on the temperature of the radiation source. Likewise, the emittance of a real surface depends on the wavelength of the emitted radiation, so it also is a function of temperature. Equation (6.12) states only that the absorptance equals the emittance when the surface is at the same temperature as the enclosure. However, α and ε vary rather weakly with temperature, so it is convenient to define a **gray surface** as one for which α and ε are constant and equal over the range of source and surface temperatures of concern. Much useful engineering analysis can be performed by modeling real surfaces as gray surfaces.[2] An important exception consists of problems involving solar radiation, which will be discussed in Section 6.4.

In summary, we will model real opaque surfaces as gray surfaces, the radiation properties of which can be characterized by a single average property value [since given α, ε, or ρ, the other two can be obtained from Eqs. (6.9) and (6.12)]. An appropriate table of property data is one that gives the *total hemispherical emittance*, since these values also have been properly averaged over all directions (e.g., Table 1.3 and Tables A.5*a*, *b* in Appendix A).[3]

---

[2] The gray surface will be defined more rigorously in Section 6.6.

[3] The significance of the terms *total* and *hemispherical* will be explained in Section 6.6.

6.3 RADIATION EXCHANGE BETWEEN SURFACES

An analysis of radiation heat exchange between black surfaces is simple because a black surface reflects none of the energy falling on it. In simple configurations, such as two large plates facing each other, all radiation leaving one surface is intercepted by the other. The shape (view) factor is then unity. In more complex configurations, only a portion of the radiation leaving one surface is intercepted by a second surface, and the shape factor is less than unity. Calculation of shape factors from first principles is a difficult problem involving solid geometry and surface integration. Thus, in Section 6.3.2, shape factors are simply compiled, and rules are given for their use. Derivation of shape factors is deferred to Section 6.5. Section 6.3.3 introduces an electrical network analogy to radiant energy exchange, which is shown to be a useful conceptual aid.

The analysis of radiation exchange between real surfaces is based on the **diffuse gray surface** model. When only two surfaces form an enclosure, algebraic formulas can be obtained for the radiation exchange, as shown in Section 6.3.4. If there are many surfaces involved, the problem becomes one of solving a system of simultaneous linear algebraic equations, for which standard methods and computer subroutines are available. The computer program RAD2, introduced in Section 6.3.5 for this purpose, is based on the Gauss-Jordan elimination method.

6.3.1 Radiation Exchange Between Black Surfaces

Consider a convex black object (1) in a black isothermal enclosure (2) as shown in Fig. 6.5. When both are in equilibrium at temperature T_2, the radiation incident on and absorbed by surface 1 is $A_1\sigma T_2^4$ in order to balance the emission of $A_1\sigma T_2^4$. If the temperature of the object is raised to T_1, then the emission becomes $A_1\sigma T_1^4$ while the absorption remains $A_1\sigma T_2^4$. Thus, the rate at which energy is lost by radiation from the object is $A_1(\sigma T_1^4 - \sigma T_2^4)$. Since this energy must be gained by the enclosure, we can say that the net rate of heat transfer from the object to the enclosure,

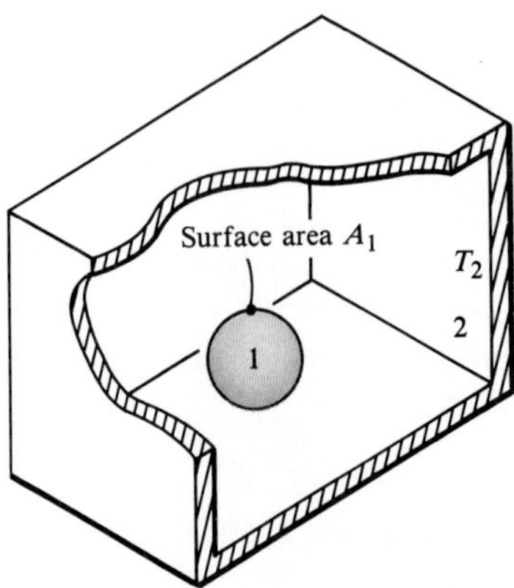

Figure 6.5 A convex black object, 1, in a black enclosure, 2.

or the net radiant energy exchange, is

$$\dot{Q}_{12} = A_1(\sigma T_1^4 - \sigma T_2^4) \tag{6.13}$$

Object 1 could be the inner sphere of two spheres or the inner cylinder of two long coaxial cylinders. If surfaces 1 and 2 are large, parallel walls facing each other, with areas $A_1 = A_2 = A$, then the same argument is valid: Eq. (6.13) still applies, but is more appropriately written as

$$\frac{\dot{Q}_{12}}{A} = \sigma T_1^4 - \sigma T_2^4 \tag{6.14}$$

The preceding situations were particularly simple to analyze because all the radiation leaving surface 1 reached surface 2. Now consider radiation exchange between two finite black surfaces A_1 and A_2, as shown in Fig. 6.6. By inspection, only part of the radiation leaving surface 1 is intercepted by surface 2 and vice versa. We define the **shape factor** (or **view factor**) F_{12} as the fraction of energy leaving A_1 that is intercepted by A_2; likewise, F_{21} is the fraction of energy leaving A_2 that is intercepted by A_1. The shape factor is a geometrical concept and depends only on the size, shape, and orientation of the surfaces. Radiation leaves surface 1 at the rate of $E_{b1}A_1$ [W]; the portion that is intercepted by surface 2 is then $E_{b1}A_1F_{12}$. Likewise, the radiation leaving surface 2 that is intercepted by surface 1 is $E_{b2}A_2F_{21}$. Since both surfaces are black, all incident radiation is absorbed, and the net radiant energy exchange is

$$\dot{Q}_{12} = E_{b1}A_1F_{12} - E_{b2}A_2F_{21} \tag{6.15}$$

If both surfaces are at the same temperature, the second law of thermodynamics requires that there be no net energy exchange; that is, if $E_{b1} = E_{b2}$, $\dot{Q}_{12} = 0$ and

$$A_1F_{12} = A_2F_{21} \tag{6.16}$$

which is the **reciprocal rule** for shape factors. Note that since the shape factors

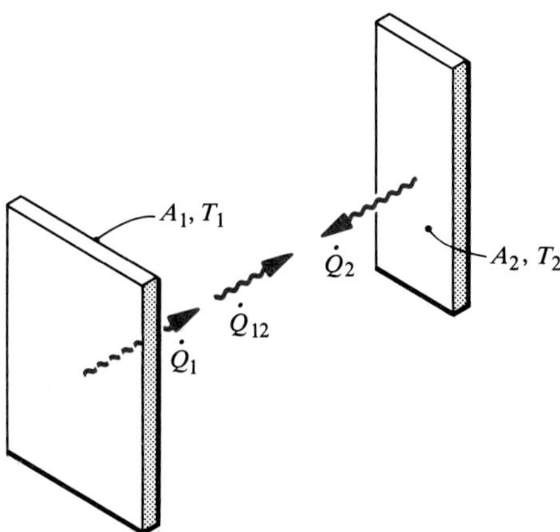

Figure 6.6 Radiation energy exchange between two finite surfaces.

depend only on geometry, this relationship is true even when the surfaces are at different temperatures. Substituting in Eq. (6.15),

$$\dot{Q}_{12} = A_1F_{12}(E_{b1} - E_{b2}) = A_1F_{12}(\sigma T_1^4 - \sigma T_2^4) \tag{6.17}$$

The reciprocal rule can also be derived using a purely geometrical argument, as will be shown in Section 6.5.2.

EXAMPLE 6.1 Heat Gain by an Ice Rink

A circular ice rink is 20 m in diameter and is to be temporarily enclosed in a hemispherical dome of the same diameter. The ice is maintained at 270 K, and on a particular day the inner surface of the dome is measured to be 290 K. Estimate the radiant heat transfer from the dome to the rink if both surfaces can be taken as black.

Solution

Given: Circular ice rink enclosed in a hemispherical dome.

Required: Radiant heat transfer from dome to ice.

Assumptions: Both ice and dome are black.

Denote the rink as surface 1 and the dome as surface 2. The area of the rink is $A_1 = (\pi/4)(20)^2 = 314.2$ m^2. All the radiation leaving the rink is intercepted by the dome; thus, Eq. (6.13) applies:

$$\begin{aligned}\dot{Q}_{12} &= A_1(\sigma T_1^4 - \sigma T_2^4)\\ &= (314.2)[(5.67)(2.70)^4 - (5.67)(2.90)^4]\\ &= (314.2)[301.3 - 401.0]\\ &= -31.3 \times 10^3 \text{ W}\end{aligned}$$

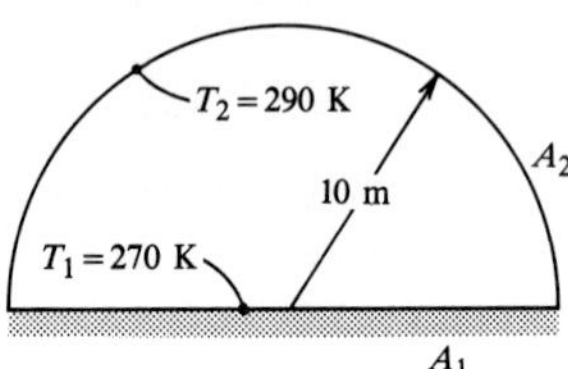

But $\dot{Q}_{12} = -\dot{Q}_{21}$; hence $\dot{Q}_{21} = 31.3 \times 10^3$ W (31.3 kW).

Comments

Notice that we could have started with $\dot{Q}_{21} = A_2F_{21}(\sigma T_2^4 - \sigma T_1^4)$ and used the reciprocal rule to write $A_2F_{21} = A_1F_{12} = A_1$, obtaining the same result.

6.3.2 Shape Factors and Shape Factor Algebra

Determining shape factors generally requires the evaluation of a double surface integral, which is not easy. However, shape factors have been determined for a great variety of configurations and are available in the form of formulas and graphs. Table 6.1 gives formulas for shape factors of some simple 2-D and 3-D configurations; some of these can be written down by inspection. Item 1 consists of a long cylinder

Table 6.1 Shape factors for two- and three-dimensional configurations.

Configuration	Shape Factors
Two-dimensional	
1. Concentric cylinders	$F_{12} = 1; \quad F_{21} = \frac{A_1}{A_2}$ $F_{22} = 1 - F_{21} = 1 - \frac{A_1}{A_2}$
2. Long duct with equilateral triangular cross-section 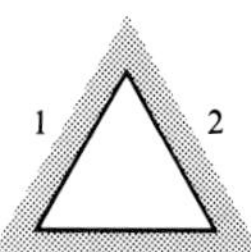	$F_{12} = F_{13} = \frac{1}{2}$
3. Long parallel plates of equal width	$F_{12} = F_{21} = \left[1 + \left(\frac{c}{a}\right)^2\right]^{1/2} - \left(\frac{c}{a}\right)$
4. Long adjacent plates	$F_{12} = \frac{1}{2}\left\{1 + \left(\frac{c}{a}\right) - \left[1 + \left(\frac{c}{a}\right)^2\right]^{1/2}\right\}$
5. Long symmetrical wedge	$F_{12} = F_{21} = 1 - \sin\frac{\alpha}{2}$
6. Long cylinder parallel to a large plane area	$F_{12} = \frac{1}{2}$

(Continued)

Table 6.1 ***(Continued)***

Configuration	Shape Factors
7. Long cylinder parallel to a plate 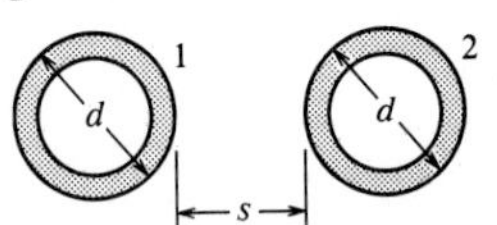	$F_{12} = \frac{r}{b-a}\left(\tan^{-1}\frac{b}{c} - \tan^{-1}\frac{a}{c}\right)$
8. Long adjacent parallel cylinders of equal diameters 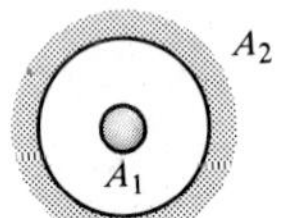	Let $X = 1 + \frac{s}{d}$, then $F_{12} = F_{21} = \frac{1}{\pi}\left[(X^2 - 1)^{1/2} + \sin^{-1}\frac{1}{X} - X\right]$
Three-dimensional	
9. Concentric spheres 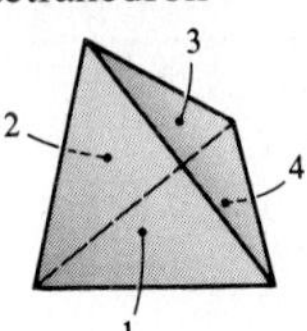	$F_{12} = 1;\quad F_{21} = \frac{A_1}{A_2}$ $F_{22} = 1 - F_{21} = 1 - \frac{A_1}{A_2}$
10. Regular tetrahedron	$F_{12} = F_{13} = F_{14} = \frac{1}{3}$
11. Sphere near a large plane area 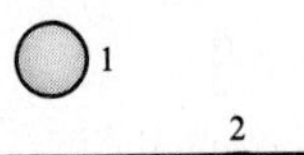	$F_{12} = \frac{1}{2}$
12. Small area perpendicular to the axis of a surface of revolution 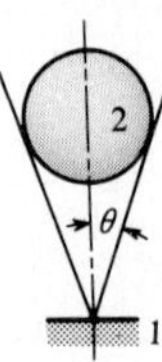	$F_{12} = \sin^2\theta$

Table 6.1 *(Concluded)*

Configuration	Shape Factors
13. Area on the inside of a sphere 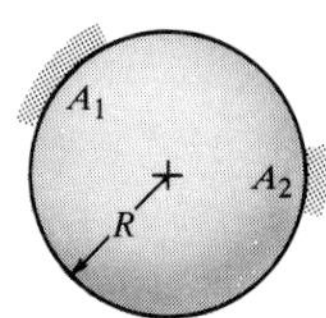	$F_{12} = \dfrac{A_2}{4\pi R^2}$
14. Coaxial parallel disks 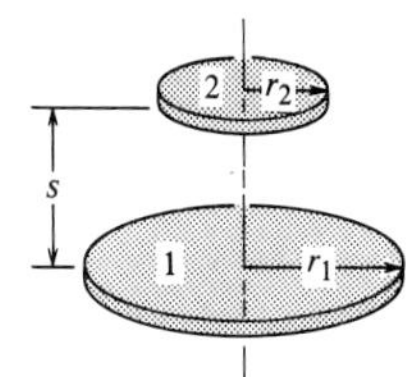	Let $R_1 = \frac{r_1}{s}$, $R_2 = \frac{r_2}{s}$; $X = 1 + \dfrac{1 + R_2^2}{R_1^2}$; then $F_{12} = \frac{1}{2}\left\{X - \left[X^2 - 4\left(\frac{R_2}{R_1}\right)^2\right]^{1/2}\right\}$
15. Opposite rectangles 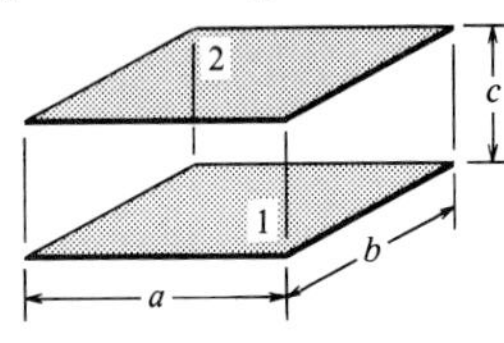	Let $X = \frac{b}{c}$, $Y = \frac{a}{c}$; then $F_{12} = \dfrac{2}{\pi XY}\left\{\ln\left[\dfrac{(1+X^2)(1+Y^2)}{1+X^2+Y^2}\right]^{1/2} - X\tan^{-1}X - Y\tan^{-1}Y + X(1+Y^2)^{1/2}\tan^{-1}\dfrac{X}{(1+Y^2)^{1/2}} + Y(1+X^2)^{1/2}\tan^{-1}\dfrac{Y}{(1+X^2)^{1/2}}\right\}$
16. Adjacent rectangles 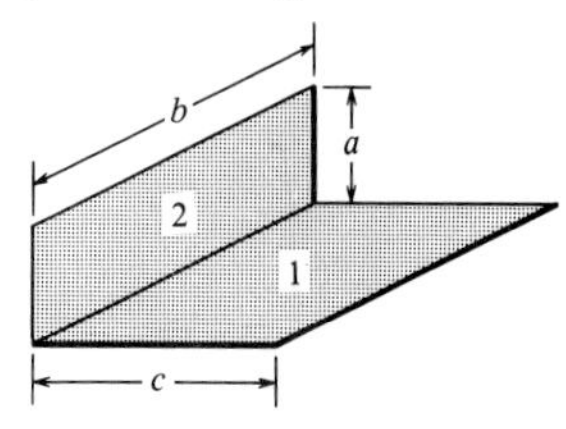	Let $L = \frac{c}{b}$, $N = \frac{a}{b}$; then $F_{12} = \dfrac{1}{\pi L}\left\{L\tan^{-1}\dfrac{1}{L} + N\tan^{-1}\dfrac{1}{N} - (N^2+L^2)^{1/2}\tan^{-1}(N^2+L^2)^{-1/2} + \dfrac{1}{4}\ln\left[\left(\dfrac{(1+L^2)(1+N^2)}{1+L^2+N^2}\right)\left(\dfrac{L^2(1+L^2+N^2)}{(1+L^2)(L^2+N^2)}\right)^{L^2}\left(\dfrac{N^2(1+L^2+N^2)}{(1+N^2)(L^2+N^2)}\right)^{N^2}\right]\right\}$

of surface area A_1 surrounded by a cylinder of area A_2. Since all the radiation leaving the inner cylinder is incident on the outer cylinder, $F_{12} = 1$. Use of the reciprocal rule then gives $F_{21} = A_1/A_2$. The radiation leaving the outer cylinder that is not intercepted by the inner cylinder returns to itself; thus, $F_{22} = 1 - A_1/A_2$, that is, surface 2 "sees" itself because it is concave. Identical relations hold for the spheres shown as item 9. Item 2 is a long duct with an equilateral triangular cross section; by symmetry, $F_{12} = F_{13} = 1/2$. Similarly, for the regular tetrahedron shown as item 10, the shape factor from one side to another is 1/3. If a small area, 1, is adjacent and perpendicular to a large area, 2, $F_{12} \simeq 1/2$, and this is the limit form for the shape factors given for items 4 and 16. Item 6 shows a long cylinder (1) parallel to a large plane area (2), for which $F_{12} \simeq 1/2$, and the same result holds for the sphere shown as item 11. If a small area, 1, is perpendicular to the axis of any surface of revolution, 2, then $F_{12} = \sin^2 \theta$, where θ is the angle between the axis and the limiting ray tangent to or at the edge of the surface. Item 12 shows an example. Finally, the shape factor from an area A_1 on the inside of a sphere of radius R to another area A_2 also on the inside of the sphere is $F_{12} = A_2/4\pi R^2$, as shown for item 13.

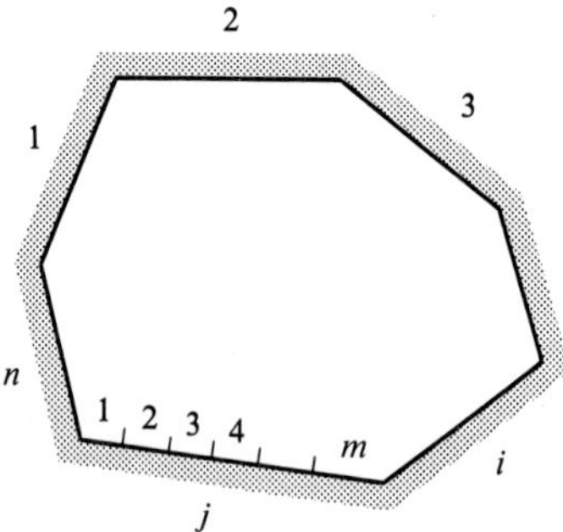

Figure 6.7 An enclosure of n surfaces, with surface j subdivided into m surfaces.

The utility of these shape factor formulas and graphs can be significantly extended by the use of shape factor algebra, including the reciprocal rule, Eq. (6.16), and the **summation rule**, which was just introduced to obtain F_{22} for concentric cylinders. More generally, if surface i is one surface in an enclosure of n surfaces, as shown in Fig. 6.7, then all energy leaving surface i must be intercepted by some surface of the enclosure (including i itself if it is concave); thus,

$$\sum_{j=1}^{n} F_{ij} = 1 \tag{6.18}$$

And if surface j is subdivided into m subareas, as is also shown in Fig. 6.7,

$$F_{ij} = \sum_{k=1}^{m} F_{ik} \tag{6.19}$$

Figure 6.8 shows a simple application of this relation.

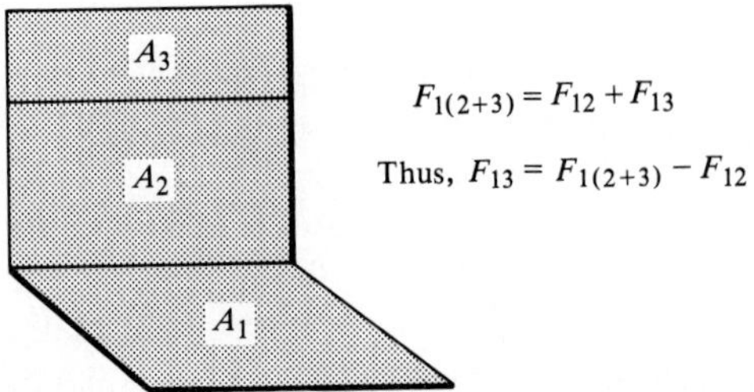

Figure 6.8 Use of Eq. (6.19) to determine shape factors for rectangles at 90°.

If Eq. (6.19) is multiplied by area A_i and the reciprocal rule is applied to each term, two more useful relations are obtained:

$$A_j F_{ji} = \sum_{k=1}^{m} A_k F_{ki} \tag{6.20}$$

$$F_{ji} = \frac{\sum_{k=1}^{m} A_k F_{ki}}{\sum_{k=1}^{m} A_k} \tag{6.21}$$

As another application, consider two opposing rectangles subdivided as shown in Fig. 6.9*a*. We introduce the notation $\mathcal{G}_{ij} = A_i F_{ij}$ for convenience. By symmetry, $\mathcal{G}_{14} = \mathcal{G}_{32}$ or, using the reciprocal rule,

$$\mathcal{G}_{14} = \mathcal{G}_{23} \tag{6.22}$$

The shape factor F_{14} can now be determined by considering the radiation leaving surfaces 1 and 2 that is intercepted by surfaces 3 and 4. Using Eq. (6.19),

$$\begin{aligned} \mathcal{G}_{(1+2)(3+4)} &= \mathcal{G}_{1(3+4)} + \mathcal{G}_{2(3+4)} \\ &= \mathcal{G}_{13} + \mathcal{G}_{14} + \mathcal{G}_{23} + \mathcal{G}_{24} \\ &= \mathcal{G}_{13} + 2\mathcal{G}_{14} + \mathcal{G}_{24} \end{aligned}$$

Hence,

$$\mathcal{G}_{14} = \frac{1}{2}[\mathcal{G}_{(1+2)(3+4)} - \mathcal{G}_{13} - \mathcal{G}_{24}] \tag{6.24}$$

which can be evaluated directly using item 15 of Table 6.1. Equations (6.22) and (6.24) also hold for the adjacent rectangles shown in Fig. 6.9*b*, but when the sides are not of equal size, the proof is more difficult.

$$A_1F_{14} = \frac{1}{2}[A_{(1+2)}F_{(1+2)(3+4)} - A_1F_{13} - A_2F_{24}]$$

$$\mathcal{G}_{14} = \frac{1}{2}[\mathcal{G}_{(1+2)(3+4)} - \mathcal{G}_{13} - \mathcal{G}_{24}]$$

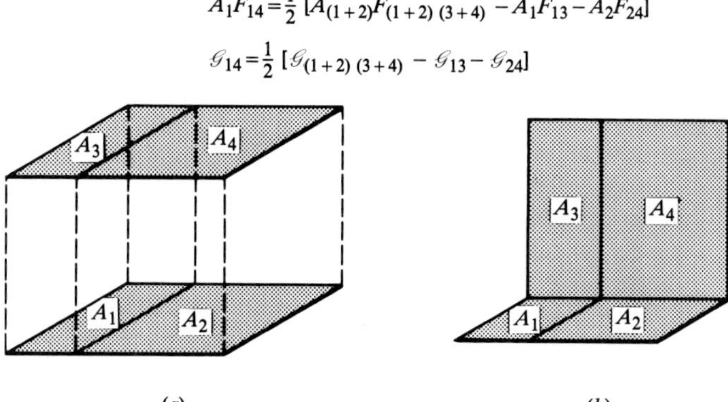

Figure 6.9 Applications of shape factor algebra to opposing and adjacent rectangles.

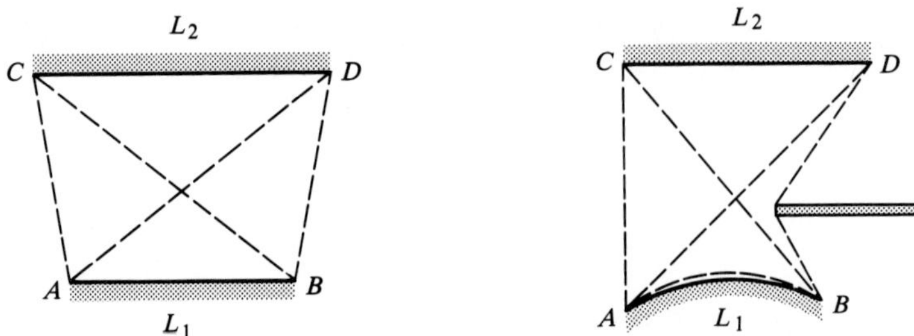

Figure 6.10 The "string rule" for shape factors of two-dimensional configurations.

Hottel's "string rule" [1] is a useful method for determining shape factors to two-dimensional configurations. Referring to Fig. 6.10, F_{12} is given by

$$F_{12} = \frac{1}{2L_1}[AD + BC - AC - BD] \tag{6.25}$$

where the diagonal distances AD, BC and the side distances AC, BD are to be evaluated as the lengths of strings stretched tightly between the respective vertices. The derivation of the string rule is given as Exercise 6–9.

Computer Program RAD1

RAD1 calculates shape factors using the formulas listed in Table 6.1.

Shape Factor Graphs

Figure C.3 in Appendix C gives shape factors in graphical form for items 14, 15, and 16 of Table 6.1, namely, coaxial parallel disks, opposed rectangles, and adjacent rectangles.

EXAMPLE 6.2 Shape Factor Determination

A 20 cm–diameter disk is coaxial and 20 cm distant from a 40 cm–I.D., 20 cm–long cylindrical tube. Determine the shape factor from the disk to the tube.

Solution

Given: Disk A_1 and tube A_2.

Required: Shape factor F_{12}.

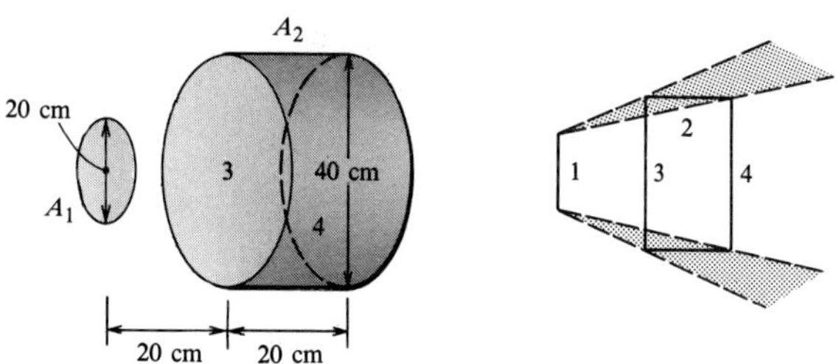

Locate hypothetical disk surfaces 3 and 4 as shown; then

$$F_{12} = F_{13} - F_{14}$$

since the radiation passing through surface 3, but not through surface 4, is intercepted by surface 2. Item 14 of Table 6.1 gives the required shape factors.

$$F_{13}: \quad R_1 = \frac{10}{20} = 0.5; \quad R_3 = \frac{20}{20} = 1; \quad X = 1 + \frac{1 + (1)^2}{(0.5)^2} = 9$$

$$F_{13} = \frac{1}{2}\left\{X - \left[X^2 - 4\left(\frac{R_3}{R_1}\right)^2\right]^{1/2}\right\} = \frac{1}{2}\left\{9 - \left[81 - 4\left(\frac{1}{0.5}\right)^2\right]^{1/2}\right\} = 0.469$$

$$F_{14}: \quad R_1 = \frac{10}{40} = 0.25; \quad R_4 = \frac{20}{40} = 0.5; \quad X = 1 + \frac{1 + (0.5)^2}{(0.25)^2} = 21$$

$$F_{14} = \frac{1}{2}\left\{21 - \left[441 - 4\left(\frac{0.5}{0.25}\right)^2\right]^{1/2}\right\} = 0.192$$

Thus, $F_{12} = F_{13} - F_{14} = 0.469 - 0.192 = 0.277$.

Comments

RAD1 could be used to obtain F_{13} and F_{14}.

6.3.3 Electrical Network Analogy for Black Surfaces

Equation (6.17) gives the net radiation energy exchange between two finite black surfaces. If the blackbody emissive power $E_b = \sigma T^4$ is viewed as a potential (rather than the temperature, T), then the linear form of Eq. (6.17) suggests an electrical analogy; for surfaces i and j,

$$\dot{Q}_{ij} = \frac{E_{bi} - E_{bj}}{1/A_i F_{ij}} \tag{6.26}$$

Thus, E_b is equivalent to electrical potential, $\dot{Q}$ is equivalent to current, and $1/A_i F_{ij}$

is equivalent to electrical resistance: $R_{ij} = 1/A_iF_{ij}$ will be termed a *space radiation resistance* between *nodes i* and *j*. As an illustration of a simple series circuit, consider two large, black parallel walls, 1 and 2, separated by a thin black plate, 3, as shown in Fig. 6.11. At steady state, there can be no energy stored in the plate, so $\dot{Q}_1 = \dot{Q}_{13} = \dot{Q}_{32} = -\dot{Q}_2$. From the equivalent circuit,

$$\dot{Q}_1 = \frac{E_{b1} - E_{b2}}{1/A_1F_{13} + 1/A_3F_{32}} \qquad \text{with } A_1 = A_2 = A_3 = A;\ F_{13} = F_{32} = 1$$

Thus,

$$\dot{Q}_1 = \frac{1}{2}A(\sigma T_1^4 - \sigma T_2^4) \tag{6.27}$$

That is, the effect of the plate is to halve the radiation energy exchange, and the plate is thus called a **radiation shield**. For m black shields,

$$\frac{\dot{Q}_1}{A} = \frac{1}{m+1}(\sigma T_1^4 - \sigma T_2^4) \tag{6.28}$$

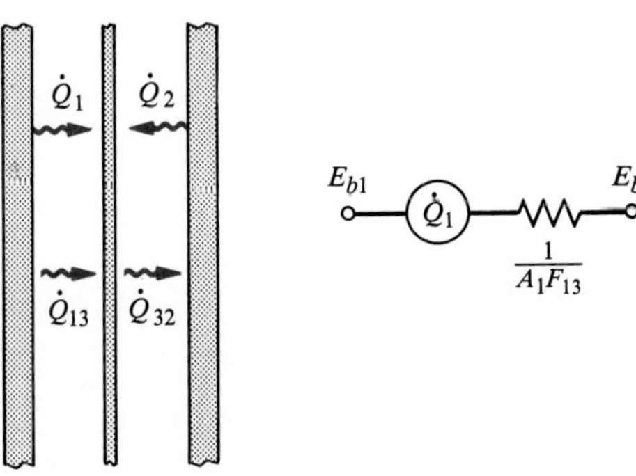

Figure 6.11 Radiation exchange between two parallel black surfaces separated by a black shield.

As another illustration, consider an enclosure of n black surfaces. The net radiant energy leaving surface i, $\dot{Q}_i$, can be obtained from Eq. (6.17) by summing over all surfaces forming the enclosure:

$$\dot{Q}_i = \sum_{j=1}^{n} A_iF_{ij}(E_{bi} - E_{bj}) = \sum_{j=1}^{n} \frac{E_{bi} - E_{bj}}{1/A_iF_{ij}}, \qquad i = 1, 2, \ldots, n \tag{6.29}$$

$\dot{Q}_i$ can also be termed the *radiation heat transfer* from surface i. When many surfaces are involved, the equivalent circuit becomes a complicated network of resistances. Figure 6.12 shows a cylindrical enclosure of isothermal surfaces consisting of a base (1), side walls (2), and a roof (3). Notice that in the network, there is no need to include a resistance $1/A_2F_{22}$: although $F_{22} \neq 0$, since the concave cylinder sees itself, there is no current flowing in the resistance because the cylinder has a uniform temperature and experiences no heat exchange with itself. Application of Kirchhoff's

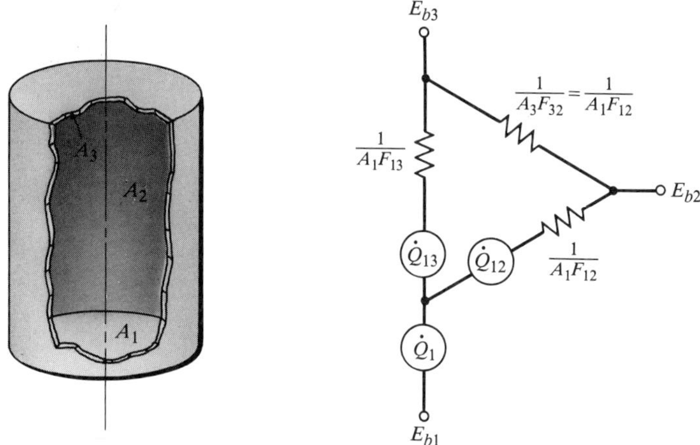

Figure 6.12 Equivalent electrical network for an enclosure of three black surfaces.

current law to node 1, for example, gives

$$\dot{Q}_1 = \dot{Q}_{12} + \dot{Q}_{13}$$

which is equivalent to Eq. (6.29).

EXAMPLE 6.3 Heat Loss from a Melt

A graphite block has a cylindrical cavity 10 cm in diameter and serves as a crucible for laboratory experiments. It is heated from below, and the side walls are well insulated. The cavity is filled with melt at 600 K to 5 cm below the opening. What will be the rate of heat loss from the melt by radiation if the surrounds are at 300 K and all surfaces are approximated as being black?

Solution

Given: Cylindrical crucible containing a melt at 600 K.

Required: Heat loss by radiation.

Assumptions: 1. Side walls are adiabatic.
2. All surfaces are black, $\varepsilon = \alpha = 1$.

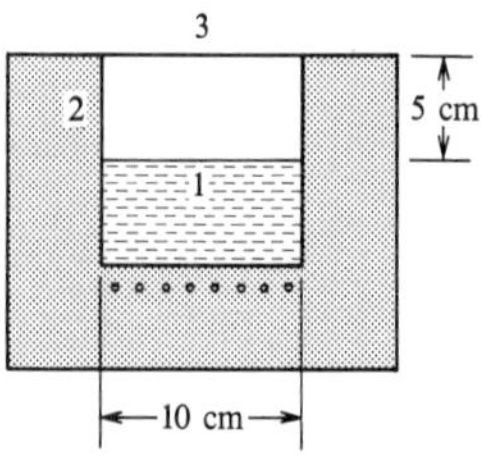

Let surface 1 be the top of the melt, surface 2 be the exposed side walls of the cavity, and surface 3 be the surrounds: surface 3 can be relocated at the opening of the cavity as shown without altering the radiation exchange problem. Equation (6.29) written for A_1 gives the heat flow through surface 1:

$$\dot{Q}_1 = A_1F_{12}(E_{b1} - E_{b2}) + A_1F_{13}(E_{b1} - E_{b3})$$

$$A_1 = (\pi/4)D^2 = (\pi/4)(0.1)^2 = 0.00785 \text{ m}^2$$

Using Table 6.1 or RAD1, item 14, $F_{13} = 0.38$, and the summation rule,

$$F_{12} = 1 - F_{13} = 0.62$$

The blackbody emissive powers are

$$E_{b1} = \sigma T_1^4 = (5.67)(6.00)^4 = 7348 \text{ W/m}^2$$

$$E_{b3} = \sigma T_3^4 = (5.67)(3.00)^4 = 459 \text{ W/m}^2$$

Since T_2 is unknown, E_{b2} cannot be calculated directly. Instead, we make use of the fact that $\dot{Q}_2 = 0$ since the sidewalls are adiabatic; Eq. (6.29) written for A_2 is

$$\dot{Q}_2 = A_2F_{21}(E_{b2} - E_{b1}) + A_2F_{23}(E_{b2} - E_{b3}) = 0$$

Now $F_{21} = F_{23}$ by symmetry; solving for E_{b2},

$$E_{b2} = \frac{1}{2}(E_{b1} + E_{b3}) = \frac{1}{2}(7348 + 459) = 3903 \text{ W/m}^2 \text{ (and } T_2 = \left(\frac{E_{b2}}{\sigma}\right)^{1/4} = 512.2 \text{ K)}$$

Thus,

$$\begin{aligned} \dot{Q}_1 &= (0.00785)[0.62(7348 - 3903) + 0.38(7348 - 459)] \\ &= (0.00785)[2136 + 2618] \\ &= 37.3 \text{ W} \end{aligned}$$

Comments

Take special note of how the surrounds are replaced by the hypothetical surface 3 as a conceptual aid.

6.3.4 Radiation Exchange between Two Diffuse Gray Surfaces

Radiation exchange between black surfaces is rather easy to analyze because all the radiation falling on a surface is absorbed. When nonblack surfaces are involved, the radiation can be reflected back and forth many times. These multiple reflections must be properly accounted for, which complicates the analysis. We will limit our attention to isothermal opaque gray surfaces, so that the radiative properties of each surface can be characterized by a single value of emittance ε. We will also assume that the surfaces are diffuse emitters and that the reflection is also diffuse. (*Specular reflection*, for which angles of incidence and reflection are equal, will be treated in Section 6.5.)

The Ray Tracing Method

To see how multiple reflections affect radiative heat transfer, consider two large, parallel gray surfaces facing each other, as shown in Fig. 6.13. We first trace how rays of radiant energy leaving surface 1 are absorbed and reflected. For $A_1 = A_2 = A$,

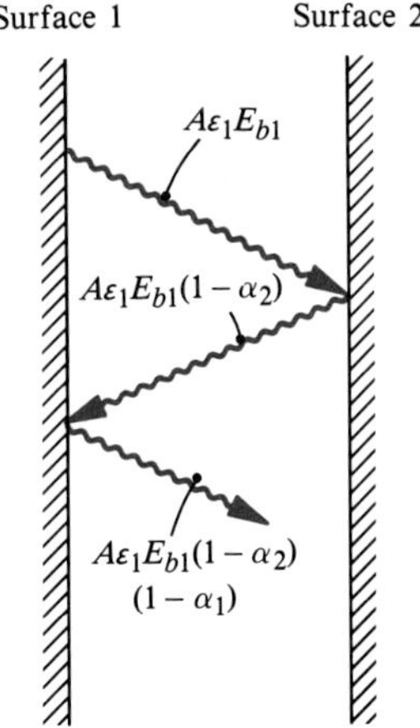

Figure 6.13 Radiation exchange between two parallel gray surfaces: ray tracing.

Surface 1 emits $A\varepsilon_1 E_{b1}$
Surface 2 absorbs $A\varepsilon_1 E_{b1}\alpha_2$
Surface 2 reflects $A\varepsilon_1 E_{b1}(1-\alpha_2)$
Surface 1 absorbs $A\varepsilon_1 E_{b1}(1-\alpha_2)\alpha_1$
Surface 1 reflects $A\varepsilon_1 E_{b1}(1-\alpha_2)(1-\alpha_1)$
Surface 2 absorbs $A\varepsilon_1 E_{b1}(1-\alpha_2)(1-\alpha_1)\alpha_2$
Surface 2 reflects $A\varepsilon_1 E_{b1}(1-\alpha_2)(1-\alpha_1)(1-\alpha_2)$
Surface 1 absorbs $A\varepsilon_1 E_{b1}(1-\alpha_2)(1-\alpha_1)(1-\alpha_2)\alpha_1$

and so on. Let $\gamma = (1-\alpha_2)(1-\alpha_1)$; then radiation emitted by surface 1 and absorbed by itself is

$$A\varepsilon_1 E_{b1}(1+\gamma+\gamma^2+\cdots)(1-\alpha_2)\alpha_1 = \frac{A\varepsilon_1 E_{b1}(1-\alpha_2)\alpha_1}{1-\gamma}$$

Similarly, the radiation emitted by surface 2 that is absorbed by surface 1 is found to be

$$A\varepsilon_2 E_{b2}(1+\gamma+\gamma^2+\cdots)\alpha_1 = \frac{A\varepsilon_2 E_{b2}\alpha_1}{1-\gamma}$$

Thus, the net radiant energy leaving surface 1 is

$$\dot{Q}_1 = A\varepsilon_1 E_{b1} - \frac{A\varepsilon_1 E_{b1}(1-\alpha_2)\alpha_1}{1-\gamma} - \frac{A\varepsilon_2 E_{b2}\alpha_1}{1-\gamma}$$

Rearranging with $\varepsilon = \alpha$ gives the net radiant energy exchange as

$$\dot{Q}_{12} = \dot{Q}_1 = \frac{A(E_{b1}-E_{b2})}{1/\varepsilon_1 + 1/\varepsilon_2 - 1} \tag{6.30}$$

The Energy Balance Method

Although the preceding method of analysis shows how multiple reflections affect radiative heat transfer, it is not a practical method for more complicated configurations. A more useful method is based on conservation of energy rather than on ray tracing. Our starting point is the definitions of irradiation and radiosity introduced in Section 1.3.2 for gray surfaces. Figure 6.14 shows a gray surface with an irradiation G [W/m^2], that is, G accounts for all the radiation incident on the surface. The radiosity of the surface, J [W/m^2], is all the radiation that leaves the surface, whether emitted or reflected; that is, $J = \varepsilon E_b + \rho G$. The heat flux through the surface can be written in two ways. First, referring to the imaginary s-surface located just above the real surface,

$$q = J - G \tag{6.31}$$

which uses the sign convention that q is positive away from the surface. Second, referring to the imaginary u-surface located just below the real surface and indefinitely close to it, so that absorption and emission take place *below* the u surface,

$$q = \varepsilon E_b - \alpha G \tag{6.32}$$

That is, the radiative heat transfer from a gray surface can be expressed either as radiosity minus irradiation, or as emission minus absorption. The radiative heat transfer can also be expressed in terms of radiosity and blackbody emissive power, as follows. Starting with the definition of J,

$$J = \varepsilon E_b + \rho G$$

substitute for G from Eq. (6.31):

$$J = \varepsilon E_b + \rho(J - q)$$

Solving for q,

$$\rho q = \varepsilon E_b - (1 - \rho)J$$

and substituting $1 - \rho = \alpha = \varepsilon$ gives

$$q = \frac{\varepsilon}{1 - \varepsilon}(E_b - J) \tag{6.33}$$

Also, $\dot{Q} = qA$, so

$$\dot{Q} = \frac{\varepsilon A}{1 - \varepsilon}(E_b - J) \tag{6.34}$$

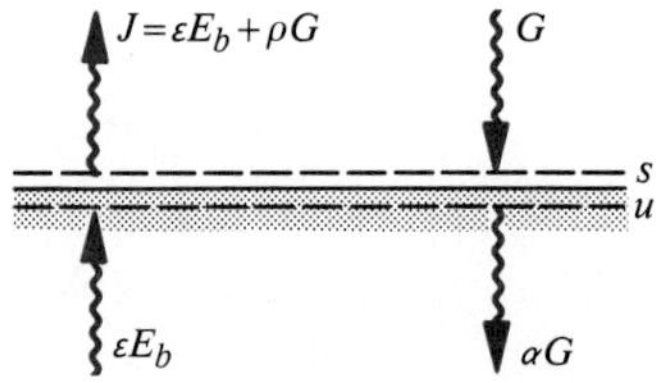

Figure 6.14 Radiant heat fluxes for a gray surface. The imaginary s- and u-surfaces are adjacent to the real interface, with all absorption and emission taking place below the u-surface.

Equation (6.34) gives the net radiant energy leaving a gray surface in terms of its blackbody emissive power, $E_b = \sigma T^4$, and its radiosity, J.

Now consider an enclosure formed by two long concentric cylinders or concentric spheres, as shown in Fig. 6.15. The inner surface has area A_1 and is at uniform temperature T_1; the outer surface has area A_2 and is at a uniform temperature T_2. The limit case is two large parallel surfaces facing each other. By symmetry, the irradiation and radiosity of the surfaces are also uniform. The fraction of energy leaving surface 1 that is intercepted by surface 2 is $J_1 A_1 F_{12}$, where the shape factor F_{12} is the same as that introduced for black surfaces in Section 6.3.1, because we have assumed that our gray surfaces are *diffuse* emitters and reflectors. Likewise, the radiant energy leaving surface 2 that is intercepted by surface 1 is $J_2 A_2 F_{21}$; thus, the net radiant energy exchange is

$$\dot{Q}_{12} = J_1 A_1 F_{12} - J_2 A_2 F_{21}$$

and using the reciprocal rule,

$$\dot{Q}_{12} = A_1 F_{12}(J_1 - J_2) \tag{6.35}$$

Also, from Eq. (6.34), the net radiant energy leaving each surface is

$$\dot{Q}_1 = \frac{\varepsilon_1 A_1}{1 - \varepsilon_1}(E_{b1} - J_1); \qquad \dot{Q}_2 = \frac{\varepsilon_2 A_2}{1 - \varepsilon_2}(E_{b2} - J_2) \tag{6.36a,b}$$

and conservation of energy requires

$$\dot{Q}_1 = \dot{Q}_{12} = -\dot{Q}_2 \tag{6.37}$$

These equations can be solved for the radiant energy exchange as follows:

$$\dot{Q}_1\left(\frac{1 - \varepsilon_1}{\varepsilon_1 A_1}\right) = E_{b1} - J_1$$

$$\dot{Q}_{12}\left(\frac{1}{A_1 F_{12}}\right) = J_1 - J_2$$

$$-\dot{Q}_2\left(\frac{1 - \varepsilon_2}{\varepsilon_2 A_2}\right) = J_2 - E_{b2}$$

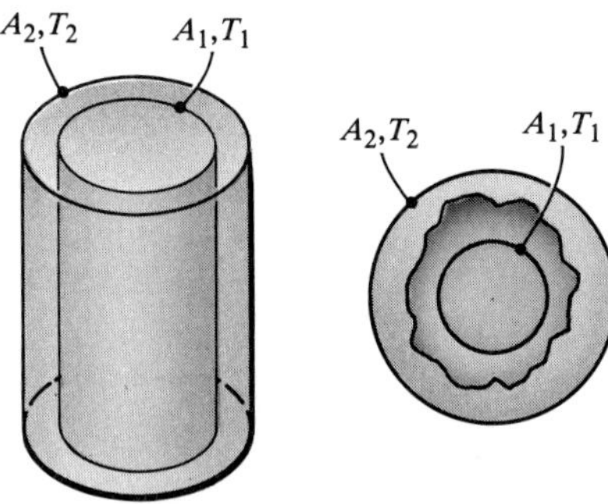

Figure 6.15 An enclosure formed by two long concentric cylinders or two concentric spheres.

Adding to eliminate the radiosities, and rearranging using Eq. (6.37) gives

$$\dot{Q}_{12} = \frac{E_{b1} - E_{b2}}{\dfrac{1-\varepsilon_1}{\varepsilon_1 A_1} + \dfrac{1}{A_1 F_{12}} + \dfrac{1-\varepsilon_2}{\varepsilon_2 A_2}} \tag{6.38}$$

If the temperatures of both surfaces are specified, then $E_{b1} = \sigma T_1^4$, $E_{b2} = \sigma T_1^4$, and $\dot{Q}_{12}$ can be calculated.

Special Forms of Eq. (6.38)

Equation (6.38) is a very useful formula. It can be further simplified since, for the configurations considered, $F_{12} = 1$; thus,

$$\dot{Q}_{12} = \frac{\varepsilon_1 A_1}{1 + \dfrac{\varepsilon_1 A_1}{\varepsilon_2 A_2}(1-\varepsilon_2)}(E_{b1} - E_{b2}) \tag{6.39}$$

When $(\varepsilon_1 A_1/\varepsilon_2 A_2)(1-\varepsilon_2)$ is small compared to unity, perhaps because $\varepsilon_1 A_1/\varepsilon_2 A_2$ is small or $(1-\varepsilon_2)$ is small, Eq. (6.39) becomes

$$\dot{Q}_{12} = \varepsilon_1 A_1(\sigma T_1^4 - \sigma T_2^4) \tag{6.40}$$

which was introduced as Eq. (1.18). Equation (6.40) is of great practical importance since it can be applied to the common situation of a small object in large, nearly black surrounds. Notice also that for two large parallel walls facing each other, $F_{12} = 1$ and $A_1 = A_2 = A$, and Eq. (6.38) becomes equivalent to Eq. (6.30),

$$\frac{\dot{Q}_{12}}{A} = \frac{\sigma T_1^4 - \sigma T_2^4}{1/\varepsilon_1 + 1/\varepsilon_2 - 1} \tag{6.41}$$

Electrical Network Analogy

An electrical network analogy is also useful for gray surfaces. Equation (6.35) can be rewritten as

$$\dot{Q}_{12} = \frac{J_1 - J_2}{1/A_1 F_{12}} \tag{6.42}$$

and the *space resistance* $R_{12} = 1/A_1 F_{12}$ is the same as for black surfaces, but the potential difference across it is the difference in radiosities rather than the difference in blackbody emissive powers. Equation (6.34) can be rewritten as

$$\dot{Q} = \frac{E_b - J}{(1-\varepsilon)/\varepsilon A} \tag{6.43}$$

which defines a *surface resistance* $R = (1-\varepsilon)/\varepsilon A$, with a corresponding potential difference of blackbody emissive power minus radiosity. Considering again the enclosure of two gray surfaces, Eq. (6.37) interpreted in terms of a circuit simply states that the current flowing through the space and the two surface resistances is the same; that is, the resistances are in series, and the circuit is as shown in Fig. 6.16. Equation (6.38) is then immediately obtained by inspection.

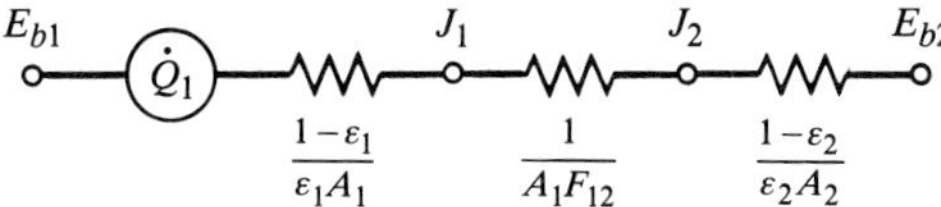

Figure 6.16 Equivalent electrical circuit for radiation exchange in an enclosure of two gray surfaces.

We now return to the radiation shield problem shown in Fig. 6.11. For gray surfaces, the equivalent circuit is shown in Fig. 6.17, and the heat flow is

$$\frac{\dot{Q}_{12}}{A} = \frac{\sigma T_1^4 - \sigma T_2^4}{\dfrac{1-\varepsilon_1}{\varepsilon_1} + 1 + \dfrac{2(1-\varepsilon_3)}{\varepsilon_3} + 1 + \dfrac{1-\varepsilon_2}{\varepsilon_2}} \tag{6.44}$$

A shield with a low emittance is desirable to maximize the surface resistances $(1-\varepsilon_3)/\varepsilon_3$.

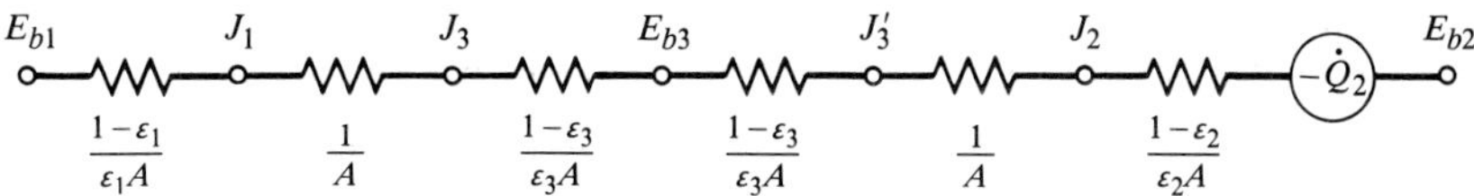

Figure 6.17 Equivalent electrical circuit for a gray radiation shield between parallel gray surfaces.

The major results obtained in this section are summarized as items 1–3 in Table 6.2 in the form of the *radiation transfer factor* $\mathscr{F}_{12}$ defined by Eq. (1.17), namely,

$$\dot{Q}_{12} = A_1 \mathscr{F}_{12}(\sigma T_1^4 - \sigma T_2^4) \tag{6.45}$$

Table 6.2 Transfer factors $\mathscr{F}_{12}$. (Item 4 is derived in Section 6.3.5.)

Configuration	Transfer Factor $\mathscr{F}_{12}$
1. Two large parallel walls facing each other	$\dfrac{1}{1/\varepsilon_1 + 1/\varepsilon_2 - 1}$
2. Concentric cylinders or spheres: A_1 surrounded by A_2	$\dfrac{\varepsilon_1}{1 + (\varepsilon_1 A_1/\varepsilon_2 A_2)(1-\varepsilon_2)}$
3. Small object A_1 in large, nearly black surrounds	ε_1
4. Surfaces A_1 and A_2 with the enclosure completed by refractory surface A_3	$\dfrac{1}{\dfrac{1-\varepsilon_1}{\varepsilon_1} + \dfrac{(1-\varepsilon_2)A_1}{\varepsilon_2 A_2} + \dfrac{1}{F_{12} + 1/(1/F_{13} + A_1/A_3 F_{32})}}$

The transfer factors for concentric cylinders and spheres are reasonably accurate even when the surfaces are not concentric, provided that the lack of symmetry does not give a markedly nonuniform radiosity distribution on either surface.

EXAMPLE 6.4 Boil-off from a Cryogenic Dewar Flask

Liquid oxygen is stored in a thin-walled spherical container, 96 cm in diameter, which in turn is enclosed in a concentric container 100 cm in diameter. The surfaces facing each other are plated and have an emittance of 0.05, and the space in between is evacuated. The inner surface is at 95 K, and the outer surface is at 280 K. (i) What is the oxygen boil-off rate? (ii) If a thin radiation shield also of emittance 0.05 is placed midway between the containers, what is the new boil-off rate?

Solution

Given: Spherical Dewar flask containing liquid oxygen.

Required: Effect of radiation shield on boil-off rate, $\dot{m}$ [kg/s].

Assumptions: 1. A perfect vacuum, and hence no conduction across the space between the shells.
2. Negligible conduction through filler neck.
3. Gray diffuse surfaces.

(i) We first calculate the heat leakage into the container. The equivalent circuit is as shown. The surface areas and blackbody emissive powers are as follows:

Inner sphere:

$$A_1 = \pi(0.96)^2 = 2.895 \text{ m}^2$$

$$E_{b1} = (5.67)(0.95)^4 = 4.6 \text{ W/m}^2$$

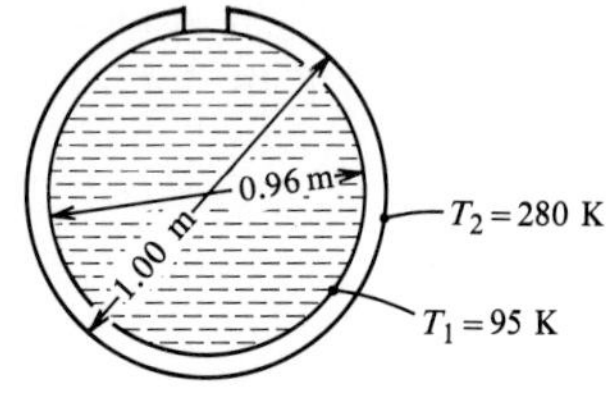

Outer sphere:

$$A_2 = \pi(1.0)^2 = 3.142 \text{ m}^2$$

$$E_{b2} = (5.67)(2.8)^4 = 348.5 \text{ W/m}^2$$

$$E_{b1} \circ\!\!-\!(\dot{Q}_1)\!-\!\underset{R_1}{\text{WW}}\!-\!\overset{J_1}{\circ}\!-\!\underset{R_{12}}{\text{WW}}\!-\!\overset{J_2}{\circ}\!-\!\underset{R_2}{\text{WW}}\!-\!\circ\, E_{b2}$$

The resistances are

$$R_1 = \frac{1-\varepsilon_1}{\varepsilon_1 A_1} = \frac{1-0.05}{(0.05)(2.895)} = 6.563$$

$$R_{12} = \frac{1}{A_1 F_{12}} = \frac{1}{(2.895)(1)} = 0.345$$

$$R_2 = \frac{1-\varepsilon_2}{\varepsilon_2 A_2} = \frac{1-0.05}{(0.05)(3.142)} = 6.047$$

$$\sum R = 6.563 + 0.345 + 6.047 = 13.0$$

The heat transfer between the surfaces is then

$$\dot{Q}_{12} = \frac{E_{b1} - E_{b2}}{\sum R} = \frac{4.6 - 348.5}{13.0} = -26.5 \text{ W}$$

The boil-off rate $\dot{m}$ is $-\dot{Q}_{12}/h_{fg}$. From Table A.8, the enthalpy of vaporization of oxygen is 0.213×10^6 J/kg; thus,

$$\dot{m} = -\frac{\dot{Q}_{12}}{h_{fg}} = \frac{26.5}{0.213 \times 10^6} = 1.24 \times 10^{-4} \text{ kg/s}$$

that is, about 0.1 gram per second.

(ii) The new equivalent circuit is as shown. The shield area is $A_3 = (\pi)(0.98)^2 = 3.017 \text{ m}^2$.

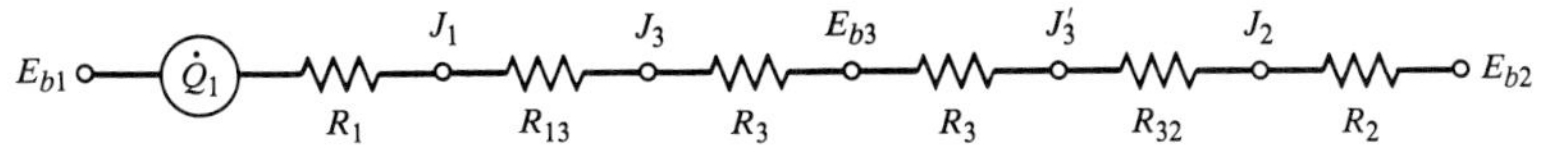

E_{b1} — $\dot{Q}_1$ — R_1 — J_1 — R_{13} — J_3 — R_3 — E_{b3} — R_3 — J_3' — R_{32} — J_2 — R_2 — E_{b2}

The new resistances are

$$R_3 = \frac{1 - \varepsilon_3}{\varepsilon_3 A_3} = \frac{1 - 0.05}{(0.05)(3.017)} = 6.298$$

$$R_{13} = \frac{1}{A_1 F_{13}} = \frac{1}{(2.895)(1)} = 0.345 \ (= R_{12})$$

$$R_{32} = \frac{1}{A_3 F_{32}} = \frac{1}{(3.017)(1)} = 0.331$$

Thus,

$$\sum R = 6.563 + 0.345 + (2)(6.298) + 0.331 + 6.047 = 25.9$$

$$\dot{Q}_{12} = \frac{4.6 - 348.5}{25.9} = -13.3 \text{ W}; \qquad \dot{m} = 6.24 \times 10^{-5} \text{ kg/s}$$

Comments

The effect of the radiation shield is to halve the boil-off rate.

6.3.5 Radiation Exchange between Many Diffuse Gray Surfaces

It is possible to obtain algebraic formulas for radiative heat transfer between diffuse gray surfaces only when the configuration is particularly simple, such as those considered in Section 6.3.4. The general problem of determining the radiation exchange in an enclosure formed by n gray diffuse surfaces requires the solution of n linear algebraic equations. Cramer's rule, matrix inversion, or successive substitution methods can be used. For this purpose, it is useful to formulate a standard calculation procedure, subject to the following restrictions:

1. Each surface is opaque and gray.
2. The emission from each surface is diffuse.
3. The reflection from each surface is diffuse.
4. The radiosity of each surface is uniform, which requires a uniformly irradiated and isothermal surface.

The first three of these restrictions have been stated already; the fourth is implicit in the analyses of Section 6.3.4 due to symmetry. Figure 6.18 shows two surfaces of uniform temperature forming a wedge. Since the shape factor $F_{\Delta A_1 A_2}$ is not uniform and increases towards the vertex of the wedge, it follows that the irradiation on surface 1 increases toward the vertex, as does the radiosity $J_1 = \varepsilon\sigma T_1^4 + (1-\varepsilon)G_1$. However, restriction 4 is not a serious one: any surface can be subdivided to obtain surfaces of sufficiently uniform radiosity, depending on the required accuracy of the result and whether or not the work required to calculate the shape factors and organize the calculations is justified. The calculation method is independent of the number of surfaces.

Consider an enclosure of n surfaces. The radiation incident on the ith surface is

$$\begin{aligned} A_i G_i &= J_1 A_1 F_{1i} + J_2 A_2 F_{2i} + J_3 A_3 F_{3i} + \cdots \\ &= J_1 A_i F_{i1} + J_2 A_i F_{i2} + J_3 A_i F_{i3} + \cdots \qquad \text{(using the reciprocal rule)} \\ &= A_i \sum_{k=1}^{n} J_k F_{ik} \end{aligned}$$

or

$$G_i = \sum_{k=1}^{n} J_k F_{ik}$$

the irradiation of surface i. The radiosity of surface i is $J_i = \varepsilon_i E_{bi} + (1-\varepsilon_i)G_i$; substituting for G_i gives

$$J_i = \varepsilon_i E_{bi} + (1-\varepsilon_i)\sum_{k=1}^{n} J_k F_{ik}; \qquad i = 1, 2, \ldots, n \tag{6.46}$$

If the temperatures of all the surfaces are specified, then $E_{bi} = \sigma T_i^4$, and Eq. (6.46)

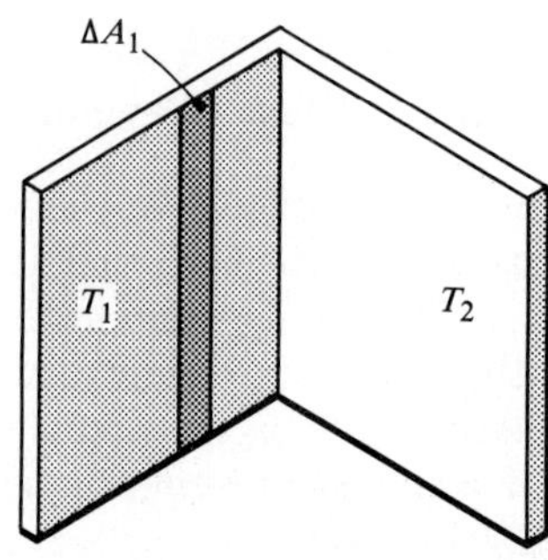

Figure 6.18 A wedge formed from two isothermal gray surfaces.

is a system of n linear equations in the n unknowns J_i. Upon solving and obtaining the J_i, the radiant heat transfer from each surface is obtained from Eq. (6.34) as

$$\dot{Q}_i = \frac{\varepsilon_i A_i}{1 - \varepsilon_i}(E_{bi} - J_i) \tag{6.47}$$

If the temperatures of some surfaces and the heat transfer from other surfaces are specified, Eq. (6.47) must be used to eliminate the unknown values of E_{bi} in Eq. (6.46).

Equivalent Electrical Network

It is often useful to draw the equivalent network as a conceptual aid and, in the case of simple problems, to assist in their solution. For example, consider an enclosure of three surfaces with one surface being adiabatic, that is, well insulated. Note that adiabatic surfaces are often called **refractory** surfaces in thermal radiation problems, since a historically important topic has been the design of furnaces in which a surface lined with refractory brick can usually be taken to be adiabatic. Figure 6.19 shows a simple configuration and the equivalent network. Recognizing that the space resistances R_{13} and R_{32} are in series and that R_{12} and $(R_{13} + R_{32})$ are in parallel, the net exchange between surfaces 1 and 2 is

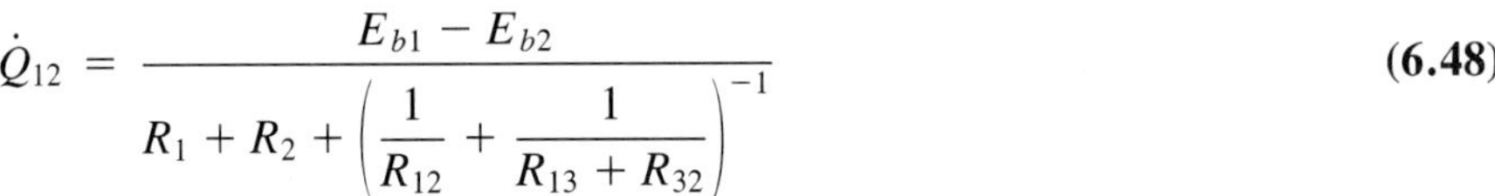

$$\dot{Q}_{12} = \frac{E_{b1} - E_{b2}}{R_1 + R_2 + \left(\dfrac{1}{R_{12}} + \dfrac{1}{R_{13} + R_{32}}\right)^{-1}} \tag{6.48}$$

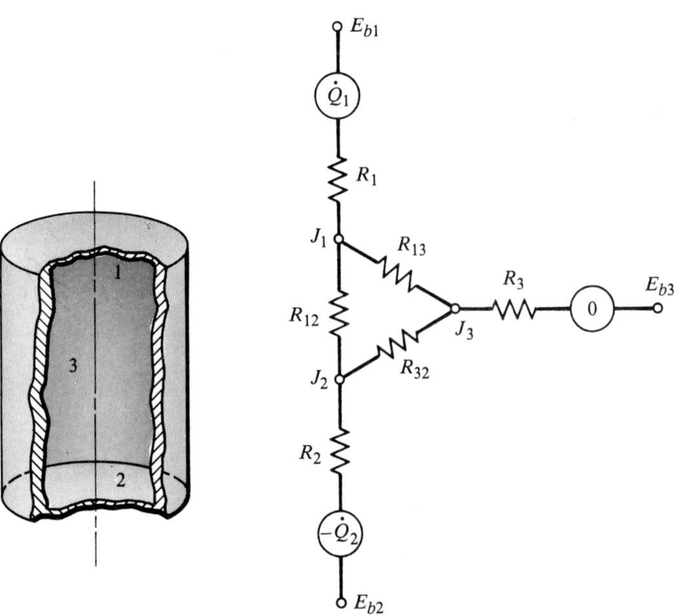

Figure 6.19 Equivalent electrical network for an enclosure formed by three gray surfaces; surface 3 is a refractory (adiabatic) surface.

When surfaces 1 and 2 are the only surfaces in an enclosure through which there is heat flow (other surfaces, if present, are adiabatic), it is convenient to determine the net exchange between surfaces 1 and 2, which we designate $\dot{Q}_{12}$, as was done in Section 6.3.4. When there are more than two nonadiabatic surfaces, it is more useful to determine the heat flow through surface i, as given by Eq. (6.47). Usually it is of little value to calculate $\dot{Q}_{ij} = (J_i - J_j)/(1/A_i F_{ij})$, which is the heat flow in the network between nodes J_i and J_j.

The Computer Program RAD2

RAD2 solves the problem of radiation exchange between many gray surfaces. For each surface, either the temperature or heat flux can be specified. The resulting system of n linear algebraic equations derived from Eqs. (6.46) and (6.47) are solved using Gauss-Jordan elimination. This method might give a large truncation error for some ill-conditioned matrices but will always give a solution if the matrix is nonsingular. The required matrix of shape factors should be calculated first using RAD1. RAD2 will check to ensure that input shape factors satisfy the reciprocal and summation rules.

EXAMPLE 6.5 Radiant Transfer in a Furnace

A long furnace used for stress relieving and annealing is 3 m × 3 m in cross section, with side walls and roof at 1700 K and 1400 K, respectively. What is the radiant heat transfer to the work floor when it is at 600 K? All the surfaces can be taken to be gray and diffuse and have an emittance of 0.5.

Solution

Given: Long furnace with a square cross section.

Required: Radiant heat transfer to work floor.

Assumptions: 1. Diffuse gray surfaces.
2. 2-D geometry so that Table 6.1, item 3 can be used for shape factors.

Equation (6.46) is used for each surface in turn, with the side walls treated as a single surface for convenience:

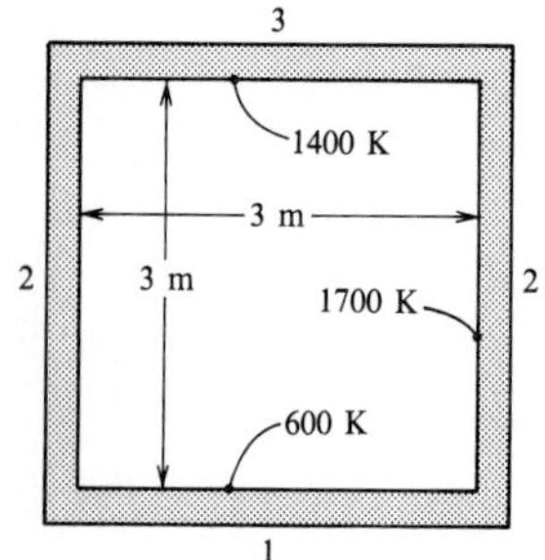

$$J_1 = \varepsilon_1 E_{b1} + (1 - \varepsilon_1)[J_1 F_{11} + J_2 F_{12} + J_3 F_{13}]$$

$$J_2 = \varepsilon_2 E_{b2} + (1 - \varepsilon_2)[J_1 F_{21} + J_2 F_{22} + J_3 F_{23}]$$

$$J_3 = \varepsilon_3 E_{b3} + (1 - \varepsilon_3)[J_1 F_{31} + J_2 F_{32} + J_3 F_{33}]$$

The emittances, blackbody emissive powers, and shape factors are as follows.

$$\varepsilon_1 = \varepsilon_2 = \varepsilon_3 = 0.5$$

$$E_{b1} = \sigma T_1^4 = (5.67 \times 10^{-8})(600)^4 = 7.348 \text{ kW/m}^2$$

$$E_{b2} = \sigma T_2^4 = (5.67 \times 10^{-8})(1700)^4 = 473.6 \text{ kW/m}^2$$

$$E_{b3} = \sigma T_3^4 = (5.67 \times 10^{-8})(1400)^4 = 217.8 \text{ kW/m}^2$$

From Table 6.1, item 3,

$$F_{13} = (1 + 1^2)^{1/2} - 1 = 0.414 = F_{31} = F_{22}$$

Using the summation rule

$$F_{12} = 1 - F_{11} - F_{13} = 1 - 0 - 0.414 = 0.586 = F_{32}$$

Using the reciprocal rule

$$F_{21} = (A_1/A_2)F_{12} = 0.5F_{12} = (0.5)(0.586) = 0.293 = F_{23}$$

Substituting in the radiosity equations gives

$$J_1 = (0.5)(7.348) + (1 - 0.5)[0 + 0.586J_2 + 0.414J_3]$$

$$J_2 = (0.5)(473.6) + (1 - 0.5)[0.293J_1 + 0.414J_2 + 0.293J_3]$$

$$J_3 = (0.5)(217.8) + (1 - 0.5)[0.414J_1 + 0.586J_2 + 0]$$

Rearranging,

$$J_1 - 0.293J_2 - 0.207J_3 = 3.674$$

$$0.147J_1 - 0.793J_2 + 0.147J_3 = -236.$$

$$0.207J_1 + 0.293J_2 - J_3 = -108.9$$

Solving,

$$J_1 = 166.4 \text{ kW/m}^2$$

$$J_2 = 376.2 \text{ kW/m}^2$$

$$J_3 = 253.6 \text{ kW/m}^2$$

From Eq. (6.47) written for surface 1

$$\frac{\dot{Q}_1}{A_1} = \frac{\varepsilon_1}{1 - \varepsilon_1}(E_{b1} - J_1)$$

$$= \frac{0.5}{1 - 0.5}(7.348 - 166.4)$$

$$= -159.0 \text{ kW/m}^2$$

which is the radiant heat flux across the work floor.

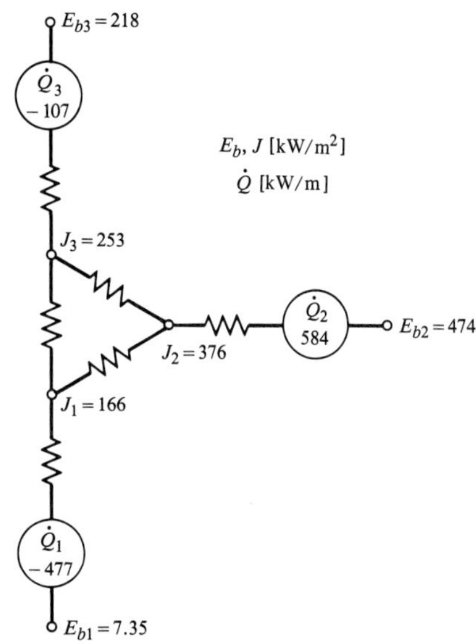

Solution using RAD2

The required input is

Surface	ε	A m^2/m	T K	q W/m^2
1	0.5	3	600	—
2	0.5	6	1700	—
3	0.5	3	1400	—

where the areas are per unit length of furnace.

$F_{13} = 0.414$ can be obtained from RAD1, and the remaining shape factors calculated as before. The shape factor matrix is

$$F_{11} = 0 \qquad F_{12} = 0.586 \qquad F_{13} = 0.414$$

$$F_{21} = 0.293 \qquad F_{22} = 0.414 \qquad F_{23} = 0.293$$

$$F_{31} = 0.414 \qquad F_{32} = 0.586 \qquad F_{33} = 0$$

RAD2 gives the following output:

Surface	Temperature K	Heat Flux W/m^2	Radiosity W/m^2
1	600	-1.5903×10^5	1.6638×10^5
2	1700	97,392	3.7617×10^5
3	1400	$-35{,}750$	2.5357×10^5

$$\sum \dot{Q}_i \simeq 10^{-10}\ \text{W}$$

Comments

1. The heat loss through the roof is excessive: better insulation is indicated.
2. The equivalent network for the furnace is shown in the accompanying figure; the heat flows are per unit length of furnace.

EXAMPLE 6.6 A Radiant Heater Panel

A radiant heater panel used in a high-vacuum process consists of a row of cylindrical electrical heating elements 1 cm in diameter, 150 cm long, spaced at a 3 cm pitch, and backed by a well insulated wall to act as a reflector and reradiator. The panel has dimensions 30 × 150 cm and is located 30 cm above a workpiece, which is also 30 × 150 cm. The heater elements are rated at 5 kW each. Estimate the operating temperature of the elements when the workpiece and surroundings are at 300 K. Take the emittances of the elements and back wall to be 0.9 and 0.8, respectively; the workpiece and surroundings can be assumed black.

Solution

Given: Radiant heating panel.

Required: Operating temperature of the 5 kW heater elements.

Assumptions:
1. Diffuse gray surfaces.
2. The radiosities of the heating elements and insulated back wall are uniform.
3. Negligible end losses, to give an upper-bound (conservative) estimate of the operating temperature.

Since their emittance is high and pitch not too small, the radiosity of the heating elements is relatively uniform around their periphery and thus can be represented by a single node, 1. Likewise, since the back (refractory) wall also has a high emittance, if it is not too close to the elements and is a good conductor, it too will have a relatively uniform radiosity and can be represented by a single node, 3. The workpiece and surroundings are both at the same temperature and black, so they can be represented by a single node, 2. Thus, we have the problem shown in Fig. 6.19, for which Eq. (6.48) applies:

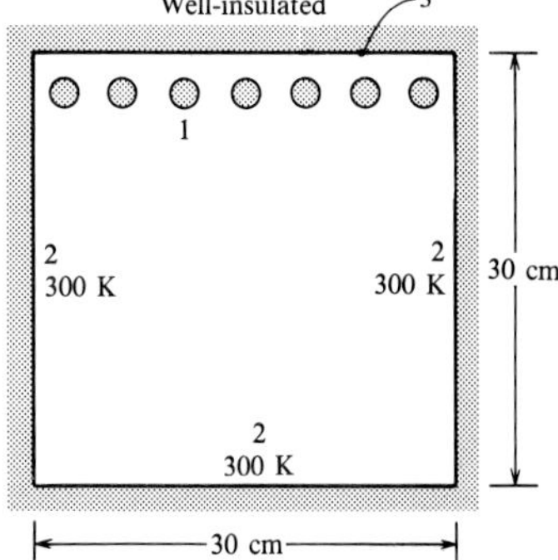

$$\dot{Q}_{12} = \frac{E_{b1} - E_{b2}}{R_1 + R_2 + \left(\dfrac{1}{R_{12}} + \dfrac{1}{R_{13} + R_{32}}\right)^{-1}}$$

The panel is relatively long, so the shape factor F_{11} between one cylinder and two adjacent cylinders is twice the shape factor given by item 8 of Table 6.1:

$$F_{11} = \frac{2}{\pi}\left[(X^2 - 1)^{1/2} + \sin^{-1}\frac{1}{X} - X\right]; \qquad X = 1 + \frac{s}{d} = \frac{d+s}{d} = \frac{3}{1} = 3$$

$$= \frac{2}{\pi}\left[(3^2 - 1)^{1/2} + \sin^{-1}\frac{1}{3} - 3\right] = 0.107$$

By symmetry, and from the summation rule,

$$F_{12} \simeq F_{13} = \frac{1}{2}(1 - F_{11}) = 0.446$$

Using the reciprocal rule,

$$F_{31} = \left(\frac{A_1}{A_3}\right)F_{13} = \left(\frac{\pi d}{d+s}\right)F_{13} = \left(\frac{\pi}{3}\right)F_{13} = 0.468$$

and from the summation rule,

$$F_{32} = 1 - F_{31} - F_{33} = 1 - 0.468 - 0 = 0.532$$

Equation (6.48) can be solved for the unknown heater element temperature T_1:

$$\sigma T_1^4 = \sigma T_2^4 + \dot{Q}_{12}\left[R_1 + R_2 + \left(\frac{1}{R_{12}} + \frac{1}{R_{13} + R_{32}}\right)^{-1}\right]$$

Number of elements = 30/3 = 10; $\dot{Q}_{12}$ = (10)(5000) = 50,000 W

$$A_1 = (10)(\pi)(0.01)(1.5) = 0.471 \text{ m}^2$$

$$A_2 = (3)(0.3)(1.5) = 1.35 \text{ m}^2$$

$$A_3 = (0.3)(1.5) = 0.45 \text{ m}^2$$

$$R_1 = \frac{1-\varepsilon_1}{\varepsilon_1 A_1} = \frac{1-0.9}{(0.9)(0.471)} = 0.236; \qquad R_2 = \frac{1-\varepsilon_2}{\varepsilon_2 A_2} = \frac{1-1}{(1)(1.35)} = 0$$

$$R_{12} = \frac{1}{A_1 F_{12}} = \frac{1}{(0.471)(0.446)} = 4.76 = R_{13}$$

$$R_{32} = \frac{1}{A_3 F_{32}} = \frac{1}{(0.45)(0.532)} = 4.18$$

$$\sigma T_1^4 = (5.67 \times 10^{-8})(300)^4 + (50{,}000)\left[0.236 + \left(\frac{1}{4.76} + \frac{1}{4.76 + 4.18}\right)^{-1}\right]$$

$$T_1 = 1311 \text{ K}$$

Solution using RAD2

The required input is

Surface	ε	A m^2	T K	q W/m^2
1	0.9	0.471	—	1.062×10^5
2	1.0	1.35	300	—
3	0.8	0.45	—	0

where $q_1 = \dot{Q}_{12}/A_1 = 50{,}000/0.471 = 1.062 \times 10^5$ W/m^2, and $q_3 = 0$ since surface 3 is a refractory surface. The additional shape factors required are

$$F_{21} = \frac{A_1}{A_2}F_{12} = \frac{0.471}{1.35}(0.446) = 0.156; \qquad F_{23} = \frac{A_3}{A_2}F_{32} = \frac{0.45}{1.35}(0.532) = 0.177$$

$$F_{22} = 1 - F_{21} - F_{23} = 1 - 0.156 - 0.177 = 0.667$$

The shape factor matrix is then

$F_{11} = 0.107$ $\quad F_{12} = 0.446$ $\quad F_{13} = 0.446$

$F_{21} = 0.156$ $\quad F_{22} = 0.667$ $\quad F_{23} = 0.177$

$F_{31} = 0.468$ $\quad F_{32} = 0.532$ $\quad F_{33} = 0$

RAD2 gives the following output:

Surface	Temperature K	Heat Flux W/m^2	Radiosity W/m^2
1	1310.9	1.062×10^5	1.557×10^5
2	300.0	−37,068	459
3	1065.5	0	73,093

$$\sum \dot{Q}_i = -21 \simeq 0 \text{ W}$$

Comments

A number of assumptions were made to simplify this problem: how confident are you that the result is satisfactory?

6.3.6 Radiation Transfer through Passages

An interesting engineering problem is the calculation of radiation heat transfer through narrow passages. Applications include regenerative heat exchangers and cracks in a layer of insulation. In general, such problems involve all three modes of heat transfer: conduction, convection, and radiation. However, in some cases, only radiation is important. An example is a crack in a layer of high-temperature insulation for which both axial conduction and convection in the gas-filled crack and conduction in the poorly conducting insulation are negligible: then the walls of the crack can be assumed to be perfectly insulated, that is, they are refractory surfaces. Such problems can be formulated as enclosures of gray diffuse surfaces by subdividing the walls into a number of nodes and using the methods of Section 6.3.5. In practice, more sophisticated methods are favored. The discrete sums of Eq. (6.46) for the wall nodes can be converted into integrals to yield an integral equation, which then can be solved approximately or numerically. But modern practice is to use the *Monte Carlo* method, because of the ease with which it can be generalized to apply to nongray or specular surfaces. As applied to radiation exchange problems, this method is simply a computerized statistical sampling approach to ray tracing.

Results for various passage shapes can be presented in terms of a transfer factor $\mathscr{F}_{12}^b$, where surfaces 1 and 2 are the ends of the passage, each assumed black, such that $\dot{Q}_{12} = A_1\mathscr{F}_{12}^b(\sigma T_1^4 - \sigma T_2^4)$. Figure 6.20 gives $\mathscr{F}_{12}^b$ for diffuse, refractory-walled passages as a function of L/D_h, where L is the passage length and D_h is its hydraulic diameter ($D_h = 4A_c/\mathscr{P}$, where A_c is the cross-sectional area and $\mathscr{P}$ is the perimeter: for a cylinder, D_h is simply the diameter; and for a slot, D_h is twice the wall spacing). When the end surfaces, 1 and 2, are not black, the transfer factor is given by

$$\mathscr{F}_{12} = \frac{1}{\dfrac{1-\varepsilon_1}{\varepsilon_1} + \dfrac{1}{\mathscr{F}_{12}^b} + \dfrac{1-\varepsilon_2}{\varepsilon_2}} \tag{6.49}$$

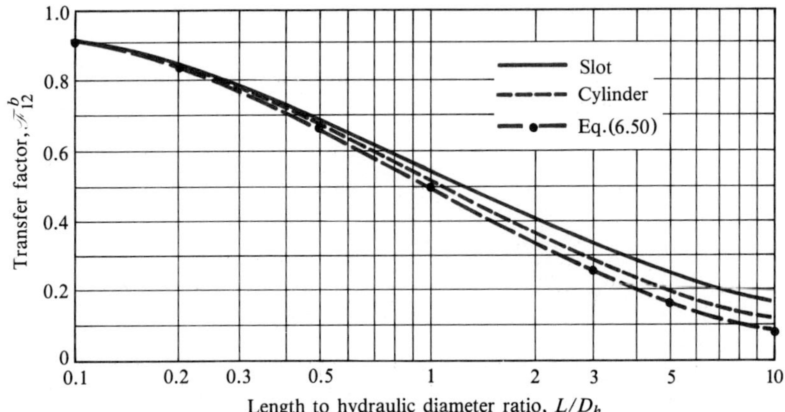

Figure 6.20 The transfer factor $\mathscr{F}_{12}^b$ for diffuse refractory-walled passages of length L and hydraulic diameter D_h [2].

If the passage wall is treated as a single-node refractory surface of area A_3, that is, of uniform radiosity, then Eq. (6.48) applies and gives

$$A_1\mathscr{F}_{12}^b = \frac{1}{R_{12}} + \frac{1}{R_{13} + R_{32}}$$

$$= A_1F_{12} + \frac{1}{1/A_1F_{13} + 1/A_3F_{32}}$$

Following Edwards [2], we set $A_1 = A_2 = A_c$ and use symmetry and shape factor algebra to give

$$\mathscr{F}_{12}^b = F_{12} + \frac{1}{2}F_{13}$$

$$= (1 - F_{13}) + \frac{1}{2}F_{13}$$

$$= 1 - \frac{1}{2}F_{13}$$

$$= 1 - \frac{1}{2}\frac{A_3}{A_c}F_{31}$$

Equation (6.48) will be accurate for a short passage, that is, $L/D_h \ll 1$; then F_{33} is small so that $F_{31} = F_{32} \simeq 1/2$, and

$$\mathscr{F}_{12}^b \simeq 1 - \frac{1}{4}\frac{A_3}{A_c} = 1 - \frac{1}{4}\frac{\mathscr{P}L}{A_c} = 1 - \frac{L}{D_h}$$

But for small L/D_h, $1 - L/D_h = 1/(1 + L/D_h)$, where the latter expression is preferred, since then $\mathscr{F}_{12}^b$ will also have the correct asymptote of zero as $L/D_h \to \infty$. Thus,

$$\mathscr{F}_{12}^b \simeq \frac{1}{1 + L/D_h} \tag{6.50}$$

a result that is independent of passage shape. Equation (6.50) is also plotted in Fig. 6.20, where it is seen to be quite accurate even for values of L/D_h larger than unity.

EXAMPLE 6.7 Heat Loss through a Crack

Estimate the heat leak through a 2 mm crack in a 4 cm–thick layer of insulation sandwiched between steel plates at 800 K and 400 K. Take the emittance of the plates as 0.6.

Solution

Given: 2 mm–wide crack in a layer of insulation.

Required: Heat transfer through crack by radiation.

Assumptions: 1. The crack walls are refractory surfaces and reflect diffusely.
2. The steel plates are diffuse gray surfaces.

Equation (6.49) applies:

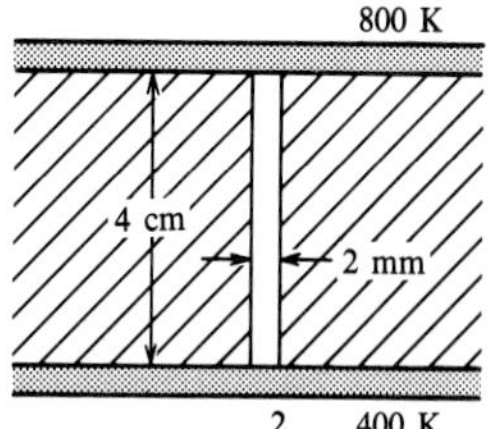

$$\mathscr{F}_{12} = \frac{1}{(1-\varepsilon_1)/\varepsilon_1 + 1/\mathscr{F}_{12}^b + (1-\varepsilon_2)/\varepsilon_2}$$

To obtain the transfer factor $\mathscr{F}_{12}^b$ from Fig. 6.20, we evaluate L/D_h, where L is the crack length and D_h its hydraulic diameter:

$$D_h = 2 \times \text{Wall spacing} = (2)(0.002) = 0.004 \text{ m}$$

$$L/D_h = 0.04/0.004 = 10$$

From Fig. 6.20, $\mathscr{F}_{12}^b \simeq 0.16$. Hence,

$$\mathscr{F}_{12} = \frac{1}{(1-0.6)/0.6 + 1/0.16 + (1-0.6)/0.6} = 0.132$$

The heat leak for a meter of crack is

$$\begin{aligned}\dot{Q}_{12} &= A_1\mathscr{F}_{12}\sigma(T_1^4 - T_2^4)\\ &= (1)(0.002)(0.132)(5.67 \times 10^{-8})(800^4 - 400^4)\\ &= 5.75 \text{ W/m}\end{aligned}$$

Comments

For a square meter of insulation of $k \sim 0.1$ W/m K, the design heat flow is $\dot{Q} = (k/L)\Delta T = (0.1/0.04)(400) = 1000$ W. If there are 20 meters of cracks, the heat leak is $(20)(5.75)/(1000) = 11.5\%$ of the design heat flow.

6.4 SOLAR RADIATION

Methods for calculating absorption of solar radiation are required for the design of solar collectors, temperature control systems for spacecraft, and air-conditioning systems for transit cars and buildings. In the case of a spacecraft, the solar radiation flux at the edge of the atmosphere is required and is easily determined. However, for a solar collector, the incident solar radiation depends on many variables, including the time of the day, the season, the latitude, and the weather conditions. The practice then is to use standard data found in appropriate design handbooks. In Section 6.4, the diffuse gray surface model used in Section 6.3 is extended to problems involving solar radiation. A different value of absorptance to short-wavelength solar radiation is allowed, while retaining a constant value for all long-wavelength radiation characteristic of surfaces at lower temperatures. This simple model gives results that are adequate for most engineering purposes.

6.4.1 Solar Irradiation

The *solar constant* G_0 is the average flux of solar energy incident on the outer fringes of the earth's atmosphere when the earth is at its mean distance from the sun of 1.495×10^{11} m (1 astronomical unit) and has a commonly used value of 1.353 kW/m^2. Owing to the elliptical nature of the earth's orbit, the actual flux varies from 1.31 kW/m^2 in June to 1.40 kW/m^2 in January. The uncertainty in these values is about 1 or 2%, and they are continually revised as new data from spacecraft become available. The total emissive power of the sun equals that of a blackbody at 5762 K, but its spectral distribution differs somewhat from blackbody behavior.[4] As the solar radiation passes through the earth's atmosphere, some energy is scattered by gas molecules and aerosols, and some is absorbed by gas molecules, particularly by CO_2 and H_2O. The solar radiation incident on the earth's surface thus consists of both a direct component and a diffuse component of scattered radiation. The sum of these is the solar irradiation, G_s, and its value for a horizontal surface is termed the *insolation*. Figure 6.21 shows a schematic of the absorption and scattering process in the atmosphere, and Fig. 6.22 shows the resulting effect on the spectral distribution of radiant energy. The *solar altitude* is the angle of the sun above the horizon. At lower solar altitudes, the path length of the radiation through the atmosphere is larger, and there is more attenuation. Thus, insolation depends on the hour of the day, the time of the year, and latitude. Calculation of solar irradiation on surfaces oriented at an arbitrary angle to the sun's rays—for example, the walls of a building—is a complex geometrical problem. Careful design of equipment such as solar collectors also requires that the solar irradiation be separated into its direct and diffuse components, which is a complicated task.

[4] The actual spectral distribution based on measurements made by NASA is readily available; see, for example, Garg [3].

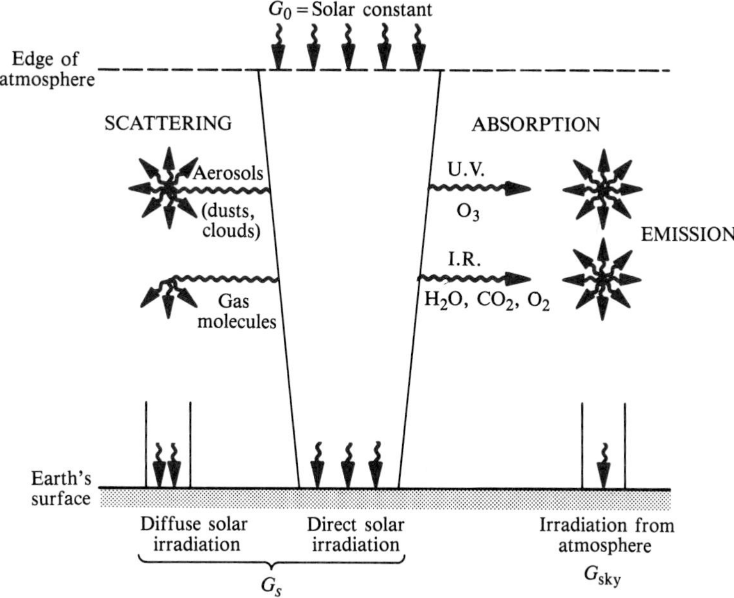

Figure 6.21 Schematic of clear sky absorption and scattering of solar radiation, and the components of irradiation incident on the earth's surface.

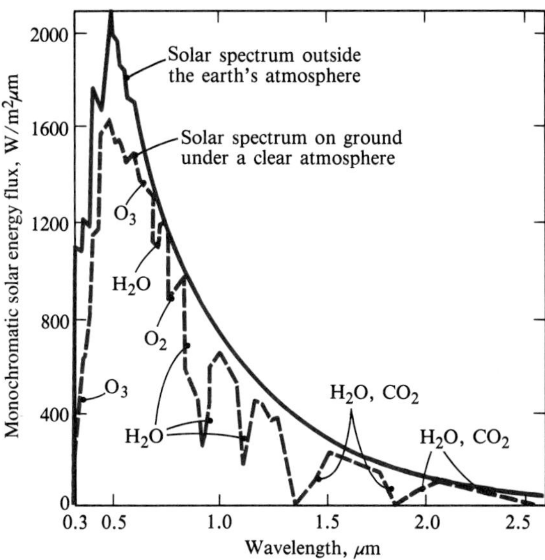

Figure 6.22 Solar spectra outside the earth's atmosphere and on the ground.

EXAMPLE 6.8 Effective Temperature of the Sun

If the sun has a diameter of 1.39×10^6 km and is assumed to radiate like a blackbody, what is its effective temperature?

Solution

Given: The sun.

Required: Estimate of temperature.

Assumptions: 1. The sun radiates like a blackbody.
2. The solar constant $G_0 = 1353$ W/m^2.

If the sun radiates like a blackbody at temperature T_s, the rate at which radiant energy leaves the sun is

$$\begin{aligned}\dot{Q} &= A_{\text{sun}}\sigma T_s^4 \\ &= (\pi)(1.39 \times 10^9)^2(5.67 \times 10^{-8})T_s^4 \\ &= 3.44 \times 10^{11} T_s^4\end{aligned}$$

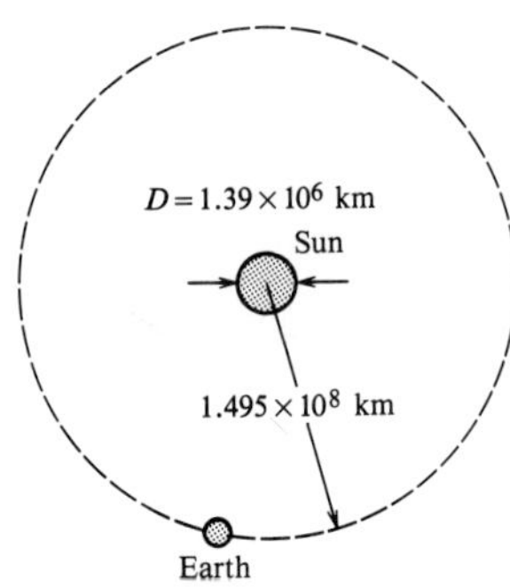

The radiant energy flux at the earth's mean distance from the sun is $\dot{Q}$ divided by the area of a sphere with radius equal to one astronomical unit:

$$G_0 = 1353 = \frac{3.44 \times 10^{11} T_s^4}{(4)(\pi)(1.495 \times 10^{11})^2}$$

Solving, $T_s = 5765$ K.

Comments

There are other criteria that could be used to define an effective temperature of the sun, so a number of values are in common use.

6.4.2 Atmospheric Radiation

Molecules in the atmosphere, chiefly CO_2 and H_2O, not only absorb solar radiation but also absorb radiation from the earth's surface. The radiation emitted by surfaces on earth is mainly from sources at 250 to 320 K, which produce radiation of wavelengths from about 4 to 40 μm with the peak at about 10μm. Atmospheric emission and absorption is mainly at wavelengths of 5 to 8μm and above 13μm, and this radiation is not distributed like blackbody radiation. Nevertheless, for engineering calculations, it is possible to approximate the emission from the atmosphere or sky as a fraction of blackbody radiation corresponding to the temperature of the air near the ground, T_e. The sky emittance ε_{sky} is defined such that the sky radiosity is

$$J_{\text{sky}} = \varepsilon_{\text{sky}}\sigma T_e^4 \tag{6.51}$$

Brunt [4] recommends an approximate correlation for ε_{sky} that accounts for atmospheric CO_2, H_2O, and dust. For a clear sky,

$$\varepsilon_{sky} \simeq 0.55 + 1.8(P_{H_2O}/P)^{1/2} \leq 1 \qquad \textbf{(6.52)}$$

where P_{H_2O} is the partial pressure of water vapor in the atmosphere, and P is the total atmospheric pressure. Alternatively, based on data obtained at a number of locations in the United States, Berdahl and Fromberg [5] recommend the following correlations for a clear sky:

$$\text{Nighttime:} \quad \varepsilon_{sky} \simeq 0.741 + 0.0062 T_{DP}\ (^\circ\text{C}) \qquad \textbf{(6.53a)}$$

$$\text{Daytime:} \quad \varepsilon_{sky} \simeq 0.727 + 0.0060 T_{DP}\ (^\circ\text{C}) \qquad \textbf{(6.53b)}$$

where T_{DP} is the dewpoint temperature. Cloud cover has a strong effect on atmospheric radiation, but the effect decreases in importance with cloud elevation because higher clouds are usually colder than low clouds.

An alternative approach that is occasionally used is to define an effective sky temperature, T_{sky}, assuming that it emits radiation like a blackbody. The effective sky temperature is always lower than the air temperature. When there is a low cloud cover we may assume $T_{sky} \simeq T_e$.

EXAMPLE 6.9 Calculation of Sky Emittance and Effective Sky Temperature

For a clear night sky with ambient air at 25°C, 10^5 Pa, and relative humidity of 50%, what is the sky emittance and effective sky temperature?

Solution

Given: Atmospheric conditions at night.

Required: Sky emittance, ε_{sky}, and effective sky temperature, T_{sky}.

Assumptions: No clouds.

Estimates can be made using either Eq. (6.52) or Eq. (6.53*a*). At 25°C = 298.15 K, steam tables (see Table A.12*a*) give P_{sat} = 3168 Pa; thus,

$$P_{H_2O} = (RH)(P_{sat}) = (0.5)(3168) = 1584 \text{ Pa}$$

Going back to the steam tables, P_{sat} = 1584 Pa corresponds to a temperature of 13.9°C, which is the dewpoint. From Eq. (6.52),

$$\varepsilon_{sky} = 0.55 + 1.8(1584/100{,}000)^{1/2} = 0.777$$

and from Eq. (6.53*a*),

$$\varepsilon_{sky} = 0.741 + (0.0062)(13.9) = 0.827$$

The effective sky temperature is obtained by equating blackbody emissive power to the actual sky radiosity:

$$\sigma T_{sky}^4 = J_{sky} = \varepsilon_{sky} \sigma T_e^4$$

or

$$T_{sky} = [\varepsilon_{sky} T_e^4]^{1/4}$$

Using the value of ε_{sky} obtained from Eq. (6.52),

$$T_{sky} = [(0.777)(298.15)^4]^{1/4} = 279.9 \text{ K}$$

Comments

The different results given by Eqs. (6.52) and (6.53*a*) indicate that these equations are approximate in nature.

6.4.3 Solar Absorptance and Transmittance

The absorptance of real surfaces depends on the wavelength of the incident radiation. In calculations of radiant exchange for gray surfaces, the values of absorptance used are suitable averages over the longer wavelengths characteristic of terrestrial applications, that is, in the range 300–2000 K, perhaps. For many surfaces, the average absorptance to short-wavelength solar radiation is very different from its value for longer wavelengths. For example, white epoxy paint has an absorptance to 270 K radiation of 0.85, whereas its absorptance to solar radiation is only 0.25. Thus, for engineering analysis involving solar radiation, it is necessary to modify the gray surface model by allowing for a different value of absorptance to solar radiation. Table A.5*a* in Appendix A gives values of the solar absorptance α_s for this purpose. A high value of α_s coupled with a low value of emittance at terrestrial temperatures is desirable for a surface that is required to collect solar heat. The reverse is true for a surface that is required to reject solar heat. Table 6.3 lists some surface finishes suitable for solar heat collection and rejection.

Table 6.3 Some surface finishes suitable for solar heat collection or rejection: total hemispherical emittance ε at ~300 K, and solar radiation absorptance α_s.

Surface Finish	ε	α_s
Black chrome electrodeposited on copper, aluminum, stainless steel, or mild steel	0.1–0.2	0.94–0.97
Black nickel electrodeposited on mild steel	0.15	0.90
Copper oxide, chemical conversion of copper sheet surface	0.10–0.15	0.87
Blue stainless steel	0.20	0.89
Nickel, Tabor solar absorber, electro-oxidized on copper, 110–30	0.05	0.85
4022 aluminum–201 nickel(TI)	0.21	0.97
Electroblack	0.05	0.86
PbS, vacuum deposit	0.20	0.98
Aluminum, vacuum-deposited on Mylar	0.03	0.10
Aluminum, hard-anodized	0.80	0.03
White epoxy paint	0.85	0.25
White potassium zirconium silicate spacecraft coating	0.87	0.13
Teflon	0.85	0.12

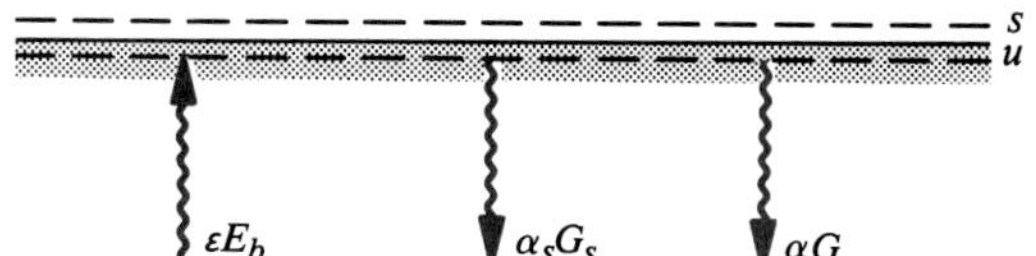

Figure 6.23 Radiant heat fluxes through a surface with solar radiation: absorption and emission takes place below the u-surface.

The heat flux through a surface receiving solar radiation is best formulated by referring to fluxes through a u-surface, as shown in Fig. 6.23. A surface energy balance requires that

$$q = \varepsilon E_b - \alpha_s G_s - \alpha G \tag{6.54}$$

where G_s is the solar irradiation, and the irradiation G includes radiation received from the sky (with radiosity J_{sky}) and from surrounding surfaces:

$$G_i = \sum_{k=1}^{n} J_k F_{ik}$$

The transmittance of real semitransparent materials is also dependent on wavelength; hence, the transmittance to short-wavelength visible radiation can be quite different from that for long-wavelength infrared radiation. The most important practical example is glass, which has a very high transmittance at wavelengths below 2 μm but is nearly opaque at wavelengths greater than 3.4 μm. Thus, solar radiation can readily enter a greenhouse and be absorbed by the plants, while the radiation emitted by the plants is reflected and absorbed by the glass. Table 6.4 gives the long-wavelength ($\lambda > 2.8\mu$m) and short-wavelength ($\lambda < 2.8\mu$m) transmission for some glazing materials.[5]

Table 6.4 Long-wavelength ($\lambda > 2.8\mu$m) and short-wavelength ($\lambda < 2.8\mu$m) transmission for selected glazing materials [3].

Cover Material	Long-Wavelength Transmission (%)	Solar Radiation Transmission (%) Versus Angle of Incidence (Degrees)					
		0	14	30	45	60	67
Glass, double-strength, 3 mm	3	89	89	88	85	82	74
Flat fiberglass, 25 mil	12	87	84	82	80	73	57
Corrugated fiberglass, 40 mil	8	83	83	82	80	62	43
Polycarbonate, 1.58 mm	6	86	85	84	83	80	72
Polyester, weatherable, 5 mil	32	88	88	88	86	81	72
Polyethylene, 4 mil, UV-resistant	80	92	91	88	86	82	67

[5] It is beyond the scope of this text to develop the theory of radiation transmission through a medium such as glass, for which the refractive index $n = c/c_0$ is not unity. The student can consult the advanced texts listed in the bibliography at the end of the book.

EXAMPLE 6.10 Temperature of an Airplane Roof

An airplane is left parked in the sun on a calm day. Calculate the roof temperature if the solar irradiation is 870 W/m^2, the air temperature is 24°C, the sky emittance is 0.77, and the top surface is (i) weathered aluminum alloy 75S-T6, and (ii) white epoxy paint. Take the convective heat transfer coefficient as $h_c = 1.24\Delta T^{1/3}$ W/m^2 K.

Solution

Given: Airplane parked outdoors.

Required: Roof temperature if it is (i) bare aluminum, and (ii) covered with white epoxy paint.

Assumptions:
1. The underside of the roof is well insulated.
2. The sky emittance is $\varepsilon_{\text{sky}} = 0.77$.
3. The roof–to–sky shape factor can be taken as unity.

Let surface 1 be the plane roof. The surface energy balance for a unit area is given by Eq. (1.34) as

$$q_{\text{cond}} - q_{\text{conv}} - q_{\text{rad}} = 0$$

where $q_{\text{cond}} = 0$ since the underside of the roof can be taken to be well insulated. The net radiation q_{rad} is obtained using Eq. (6.54):

$$q_{\text{rad}} = \varepsilon E_b - \alpha_s G_s - \alpha G$$

where $E_b = \sigma T_1^4$, $G_s = 870$ W/m^2, and $G \simeq J_{\text{sky}} = \varepsilon_{\text{sky}}\sigma T_e^4$, since $F_{1\text{sky}}$ is approximately unity; so

$$q_{\text{rad}} = \varepsilon\sigma T_1^4 - \alpha_s(870) - \alpha(0.77)\sigma(297)^4$$

The convective heat loss is

$$q_{\text{conv}} = h_c(T_1 - T_e) = 1.24(T_1 - T_e)^{1/3}(T_1 - T_e) = 1.24(T_1 - 297)^{4/3}$$

Substituting in the surface energy balance,

$$1.24(T_1 - 297)^{4/3} + \varepsilon\sigma T_1^4 - \alpha_s(870) - \alpha(0.77)\sigma(297)^4 = 0$$

Setting $\varepsilon = \alpha$ and rearranging gives

$$1.24(T_1 - 297)^{4/3} + \varepsilon(5.67 \times 10^{-8}T_1^4 - 340) - 870\alpha_s = 0$$

Radiation properties are obtained from Table A.5*a*. Substituting these values and solving numerically using Newton's method, or drawing a graph, gives the desired result:

	ε	α_s	T_1
(*a*) Weathered aluminum	0.20	0.54	364 K (91°C)
(*b*) White paint	0.85	0.25	312 K (39°C)

Comments

Use of white paint will reduce the load on the cabin air-conditioning system.

EXAMPLE 6.11 Temperature Control of a Spacecraft

A cubical spacecraft is one astronomical unit distant from the sun, and one face is exactly perpendicular to the sun's rays. It is coated with stripes of aluminum paint over a fraction F_1 of its area and with vacuum-deposited aluminum over a fraction $F_2 = 1 - F_1$. What value of F_1 will give an equilibrium temperature for the spacecraft of 300 K?

Solution

Given: Spacecraft coated with strips of aluminum paint over vacuum-deposited aluminum.

Required: Fraction of coated area to give a desired equilibrium temperature.

Assumptions:
1. Only one side of the spacecraft sees the sun.
2. Each side is isothermal.
3. The solar constant is 1353 W/m^2.

Table A.5*a* gives the required values of the emittance, ε, and solar absorptance, α_s:

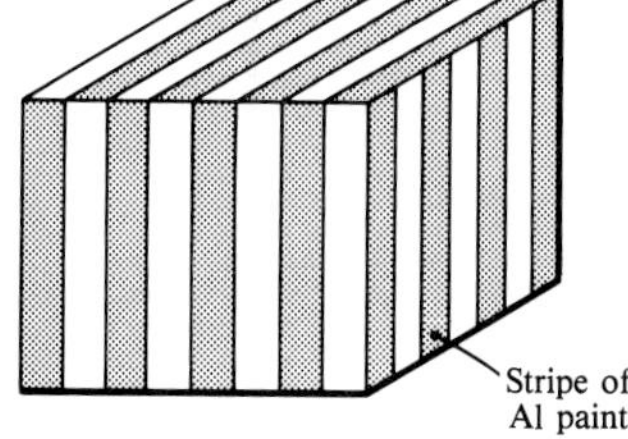

Al paint: $\varepsilon_1 = 0.20; \alpha_{s1} = 0.27$

Vacuum-deposited Al: $\varepsilon_2 = 0.03; \alpha_{s2} = 0.10$

If only one side of the spacecraft sees the sun, then an energy balance on the spacecraft of side length L gives

Absorption of solar radiation = Emission of long-wavelength radiation

$$(F_1\alpha_{s1} + F_2\alpha_{s2})G_0L^2 = (F_1\varepsilon_1 + F_2\varepsilon_2)\sigma T^4(6L^2)$$

$$\frac{F_1\alpha_{s1} + F_2\alpha_{s2}}{F_1\varepsilon_1 + F_2\varepsilon_2} = \frac{6\sigma T^4}{G_0} = \frac{(6)(5.67 \times 10^{-8})(300)^4}{1353} = 2.04$$

$F_2 = 1 - F_1$; hence,

$$\frac{F_1\alpha_{s1} + F_2\alpha_{s2}}{F_1\varepsilon_1 + F_2\varepsilon_2} = \frac{F_1(\alpha_{s1} - \alpha_{s2}) + \alpha_{s2}}{F_1(\varepsilon_1 - \varepsilon_2) + \varepsilon_2} = \frac{0.17F_1 + 0.10}{0.17F_1 + 0.03} = 2.04$$

Solving, $F_1 = 0.219$, $F_2 = 0.781$.

EXAMPLE 6.12 A Flat-Plate Solar Collector

A flat-plate solar collector used to heat water is 1 m wide and 3 m long and has no coverplate. The absorbing surface is selective, with an emittance of 0.14 and a solar absorptance of 0.93. Water enters the collector at 25°C and a flow rate of 0.015 kg/s. Estimate the outlet water temperature and the collector efficiency at noon when the solar irradiation is 800 W/m^2 and the air temperature is 20°C. Use a sky emittance of 0.6 and an overall heat transfer coefficient for heat transfer from the absorbing surface to the bulk water as $U = 25.0$ W/m^2 K. The relation $h_c = 1.1(T_s - T_e)^{1/3}$ W/m^2 K can be used to estimate the convective heat transfer from the absorbing surface to the ambient air. For the water, take $c_p = 4180$ J/kg K.

Solution

Given: Flat-plate solar collector with no coverplate.

Required: Outlet water temperature and collector efficiency.

Assumptions:
1. Steady state.
2. A well-insulated collector base.
3. Shape factor from collector to sky of unity.

The collector is a single-stream heat exchanger. Paralleling the development in Section 1.6, the steady-flow energy equation applied to an element of the absorber plate W wide and Δx long gives

$$\dot{m}c_p\Delta T_C = U(T_s - T_C)W\Delta x$$

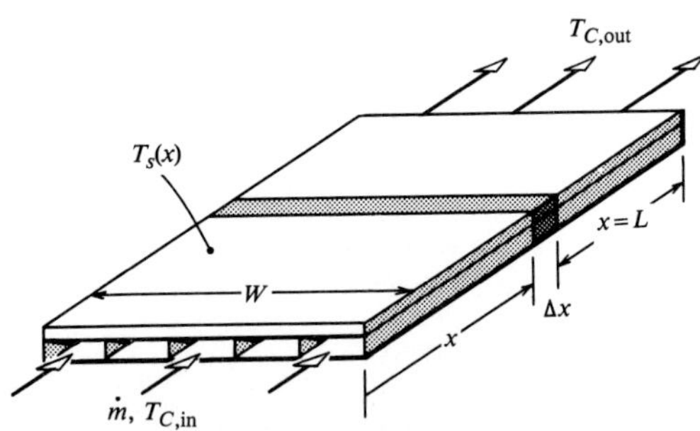

where $T_s(x)$ is the temperature of the absorbing surface and T_C is the water temperature (the subscript C is used since the water is the *coolant* stream for this heat exchanger). Dividing by Δx and letting $\Delta x \to 0$,

$$\dot{m}c_p\frac{dT_C}{dx} = U(T_s - T_C)W \tag{1}$$

Equation (1) is a first-order ordinary differential equation and requires one boundary condition, namely,

$$T_C = T_{C,\text{in}} \qquad \text{at } x = 0 \tag{2}$$

The absorber surface temperature T_s is determined from a surface energy balance on the absorber. The energy fluxes are shown in the sketch, taken to be positive as indicated.

$$\begin{matrix}\text{Net radiant energy}\\ \text{flux across } u\text{-surface}\end{matrix} - \begin{matrix}\text{Convective heat}\\ \text{loss to ambient air}\end{matrix} = \begin{matrix}\text{Heat transfer}\\ \text{to water}\end{matrix}$$

$$q_{\text{rad}} - q_{\text{conv}} = U(T_s - T_C) \tag{3}$$

From Eq. (6.54),

$$q_{\text{rad}} = \alpha_s G_s + \alpha G - \varepsilon E_b$$
$$= \alpha_s G_s + \alpha\varepsilon_{\text{sky}}\sigma T_e^4 - \varepsilon\sigma T_s^4$$

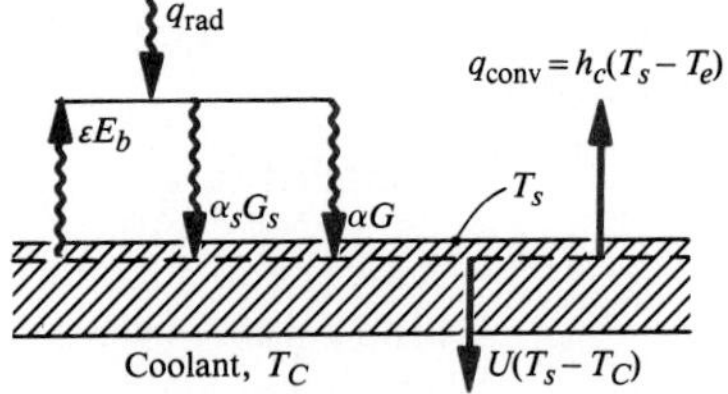

if the shape factor from the collector to the sky is assumed to be 1. Also,

$$q_{\text{conv}} = h_c(T_s - T_e)$$
$$= 1.1(T_s - T_e)^{1/3}(T_s - T_e)$$
$$= 1.1(T_s - T_e)^{4/3}$$

Substituting in Eq. (3) with $\alpha = \varepsilon$,

$$\alpha_s G_s + \varepsilon\varepsilon_{\text{sky}}\sigma T_e^4 - \varepsilon\sigma T_s^4 - 1.1(T_s - T_e)^{4/3} = U(T_s - T_C) \tag{4}$$

The absorber surface temperature should be regarded as the unknown in Eq. (4). Numerical methods are required to solve the problem. For example, the Runge-Kutta method can be used to solve Eq. (1), with T_s evaluated at each Δx step from Eq. (4) using Newton's method. Substituting numerical values, we have

$$\text{Eq. (1)}: \qquad (0.015)(4180)\frac{dT_C}{dx} = (25.0)(T_s - T_C)(1)$$

$$\frac{dT_C}{dx} = 0.399(T_s - T_C)$$

$$\text{Eq. (2)}: \qquad x = 0: \quad T_C = 298 \text{ K}$$

$$\text{Eq. (4)}: \qquad (0.93)(800) + (0.14)(0.6)(5.67 \times 10^{-8})(293)^4 - (0.14)(5.67 \times 10^{-8})T_s^4$$

$$-1.1(T_s - 293)^{4/3} = 25.0(T_s - T_C)$$

$$779 - 7.94 \times 10^{-9}T_s^4 - 1.1(T_s - 293)^{4/3} = 25.0(T_s - T_C)$$

The graph shows the resulting variation of T_C and T_s along the collector. In particular, $T_{C,\text{out}} = 323.5$ K. The collector efficiency is defined as the heat transfer to the water divided by the radiation incident upon the collector:

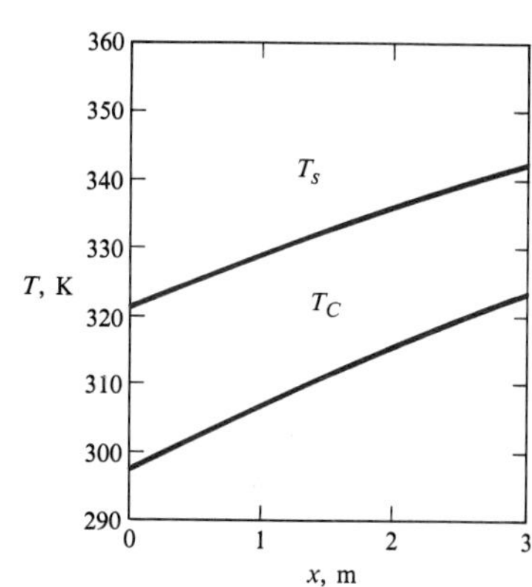

$$\eta = \frac{\dot{m}c_p(T_{C,\text{out}} - T_{C,\text{in}})}{(G_s + \varepsilon_{\text{sky}}\sigma T_e^4)WL}$$

$$= \frac{(0.015)(4180)(323.5 - 298.0)}{[800 + (0.6)(5.67 \times 10^{-8})(293)^4](1)(3)}$$

$$= 50.8\%$$

Comments

A good introduction to the thermal design of solar collectors is given by Kreith and Bohn [6].

6.5 DIRECTIONAL CHARACTERISTICS OF SURFACE RADIATION

In Section 6.3, consideration was restricted to gray surfaces that were assumed to emit and reflect diffusely, that is, in no preferred direction. The shape factor F_{12} was defined as the fraction of radiant energy leaving surface 1 that is intercepted by surface 2. Formulas for F_{12} were given, but how such formulas are obtained was not shown. We will now proceed to examine the directional characteristics of surface radiation more carefully. In Section 6.5.1, the concept of *radiation intensity* is introduced, which allows *Lambert's law* for diffuse surfaces to be stated. In Section 6.5.2, a general formula for the shape factor is derived, and examples of its use are subsequently given. In Section 6.5.3, the directional properties of real surfaces are briefly discussed; in particular, the radiation exchange problem is discussed for surfaces that reflect **specularly**, that is, with angle of reflection equal to angle of incidence, as for a mirror.

6.5.1 Radiation Intensity and Lambert's Law

Recall that a differential plane angle $d\alpha$ is the arc length ds on a circle divided by the radius of the circle, that is, $d\alpha = ds/R$, as shown in Fig. 6.24*a*. Similarly, a differential solid angle $d\omega$ is the area elcment dA_s on a sphere divided by the sphere radius squared:

$$d\omega = \frac{dA_s}{R^2} \tag{6.55}$$

as shown in Fig. 6.24*b*. A solid angle is dimensionless, but it is said to be measured in *steradians* (sr), just as a plane angle is said to be measured in radians. Notice that the area of a surface element on a sphere of unit radius has the same numerical value as the corresponding solid angle. A good grasp of the solid angle concept is essential to the understanding of radiation interchange between surfaces.

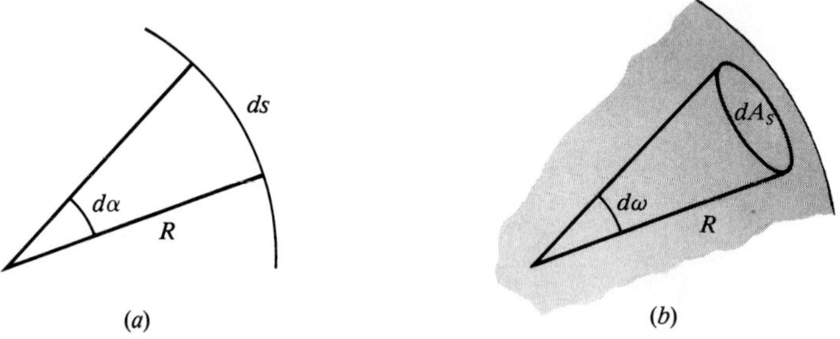

Figure 6.24 Plane and solid angles. (*a*) Differential plane angle. (*b*) Differential solid angle.

A circular surface-area element dA is shown in Fig. 6.25*a*. Radiant heat leaves dA in all directions: the **radiation intensity** I^+ is a quantity introduced to describe how this heat flow varies with respect to direction, that is, how it is distributed with respect to zenith angle θ and azimuthal angle ϕ in a spherical coordinate system. Consider how this heat flow might be measured. If a sensor of area dA_s is placed on the surface of a hemisphere surrounding the radiating area dA as shown, then the heat flow to the sensor will be proportional to the solid angle $d\omega$ it occupies. For a spherical surface, $dA_s = R\,d\theta\,R\sin\theta\,d\phi$, and thus

$$d\omega = dA_s/R^2 = \sin\theta\,d\theta\,d\phi \tag{6.56}$$

It is perhaps less obvious that the heat flow to the sensor will also depend on angle θ. Figure 6.25*b* shows how the radiating area "seen" by the sensor decreases as θ increases from 0° to 90° as a simple cosine variation $dA\cos\theta$. When the sensor is at

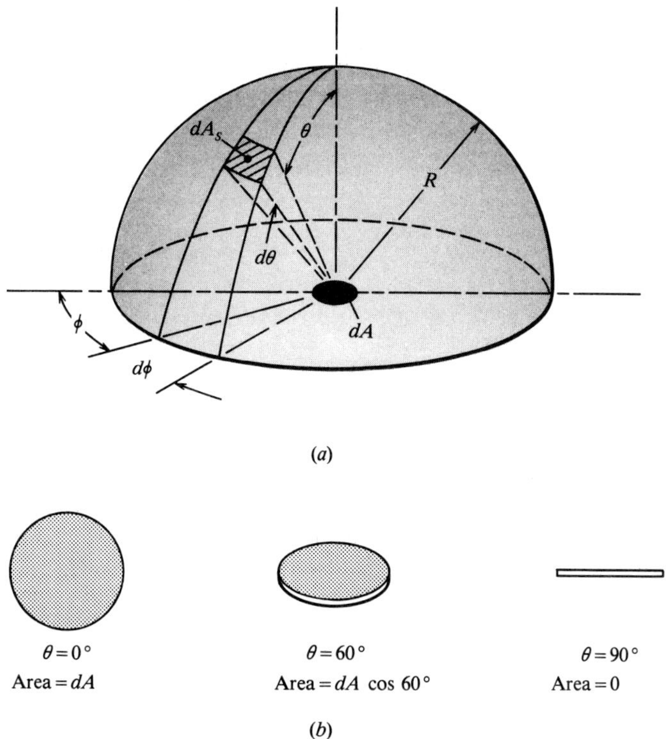

$\theta = 0°$
Area $= dA$

$\theta = 60°$
Area $= dA \cos 60°$

$\theta = 90°$
Area $= 0$

(b)

Figure 6.25 (*a*) Elemental areas used to define intensity. (*b*) The radiating area seen by a sensor at angle θ.

0°, the maximum radiation is received, whereas at 90°, no radiation can be received. The intensity of radiation *leaving* the surface in a given direction $I^+(\theta, \phi)$ is defined as the heat flow from the surface per unit solid angle, per unit area projected normal to the direction. Thus, if our sensor records a heat flow $d\dot{Q}(\theta, \phi)$,

$$I^+(\theta, \phi) = \frac{d\dot{Q}}{dA \cos\theta \, d\omega} \text{ W/m}^2 \text{ sr} \tag{6.57}$$

If $I^+(\theta, \phi)$ is independent of direction, the radiation is said to obey **Lambert's law**; this is the condition for diffuse radiation. Also, Lambert's law implies that radiation in an isothermal black enclosure is *isotropic*. In illumination engineering, the intensity defined by Eq. (6.57) is analogous to the *brightness* of the surface. Imagine a planar light source on the ceiling of a room: as the viewer's eye moves from 0° to 90°, the brightness of the light does not vary if Lambert's law is obeyed, although the amount of light received decreases proportional to $\cos\theta$ until, at 90°, the light cannot be seen. (Note that in some older texts, intensity is defined in terms of actual, rather than projected, area. Lambert's law then states that the intensity is the value at $\theta = 0$ multiplied by $\cos\theta$.)

From Eq. (6.57), the heat flux $q^+(\theta, \phi)$ leaving the surface is

$$q^+(\theta, \phi) = \frac{d\dot{Q}}{dA} = I^+ \cos\theta \, d\omega$$

Substituting for $d\omega$ from Eq.(6.56) and integrating over the hemisphere gives the radiosity J:

$$J = \int_0^{2\pi} \int_0^{\pi/2} I^+(\theta, \phi) \cos\theta \sin\theta \, d\theta \, d\phi \tag{6.58}$$

For a diffuse surface, I^+ is a constant; noting that $\int_0^{2\pi} \int_0^{\pi/2} \cos\theta \sin\theta \, d\theta \, d\phi = \pi$ gives

$$J = \pi I^+ \tag{6.59}$$

A black surface radiates diffusely with a radiosity $J = E_b$; thus

$$E_b = \pi I_b^+ \tag{6.60}$$

That is, the intensity of radiation from a black surface I_b^+ is simply $\sigma T^4/\pi$.

The intensity of radiation *incident* on a surface from a given direction is defined in a similar manner: $I^-(\theta, \phi)$ is the heat flow to the surface per unit solid angle, per unit area projected normal to the direction of incidence. Then the irradiation G is given by the integral

$$G = \int_0^{2\pi} \int_0^{\pi/2} I^-(\theta, \phi) \cos\theta \sin\theta \, d\theta \, d\phi \tag{6.61}$$

and for diffuse incident radiation,

$$G = \pi I^- \tag{6.62}$$

Intensity also can be defined on a spectral basis, as will be shown in Section 6.6.2.

EXAMPLE 6.13 Calculation of Irradiation from Intensity

Two surfaces, each of area 1 cm^2, are located 10 cm apart, as shown in the sketch. Surface 1 is black and is maintained at 1000 K. Calculate the rate at which surface 2 is irradiated by surface 1.

Solution

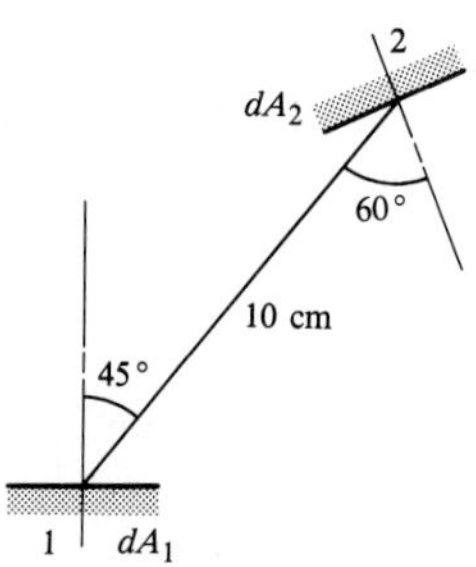

Given: Two surfaces, 1 and 2.

Required: $d\dot{Q}_{12}$, the irradiation of surface 2 by surface 1.

Assumptions: Since both surfaces are small compared to the distance between them, they can be approximated by elemental areas dA_1 and dA_2.

Surface 1 is black and is thus a diffuse emitter: the intensity of radiation leaving A_1 is given by Eq. (6.60) as

$$I_1^+ = \frac{E_{b1}}{\pi} = \frac{\sigma T_1^4}{\pi} = \frac{(5.67 \times 10^{-8})(1000)^4}{\pi} = 1.805 \times 10^4 \text{ W/m}^2\text{ sr}$$

The heat flow in the solid angle $d\omega_{12}$ subtended by surface dA_2 at dA_1 is obtained from Eq. (6.57) as

$$d\dot{Q}_{12} = I_1^+ \, dA_1 \cos\theta_1 \, d\omega_{12}$$

But from Eq. (6.55),

$$d\omega_{12} = \frac{dA_s}{R^2} = \frac{dA_2 \cos\theta_2}{R^2}$$

since dA_s is the projection of dA_2 perpendicular to the direction of the incident radiation from dA_1. Thus,

$$d\dot{Q}_{12} = \frac{I_1^+ \, dA_1 \, dA_2 \cos\theta_1 \cos\theta_2}{R^2} = \frac{(1.805 \times 10^4)(10^{-4})(10^{-4})(0.707)(0.500)}{(0.1)^2}$$

$$= 6.38 \times 10^{-3} \text{ W } (6.38 \text{ mW})$$

6.5.2 Shape Factor Determination

The shape factor F_{12} introduced in Section 6.3.2 was defined as the fraction of radiation leaving a black surface A_1 that is intercepted by surface A_2. Since a black surface emits diffusely, the shape factor is a geometrical concept that depends only on the size, shape, and orientation of the surfaces. In Section 6.3.4, the same shape factor was used for interchange between diffuse gray surfaces. In general, the shape factor F_{12} applies to any surface for which the intensity I^+ obeys Lambert's law. To derive a formula for the shape factor, we first consider the radiation leaving an elemental area dA_1 that is intercepted by finite area A_2, as shown in Fig. 6.26. If the element dA_2 subtends a solid angle $d\omega$ at element dA_1, Eq. (6.57) gives the radiation intercepted by dA_2 as

$$d\dot{Q} = I_1^+ \, dA_1 \cos\theta_1 \, d\omega$$

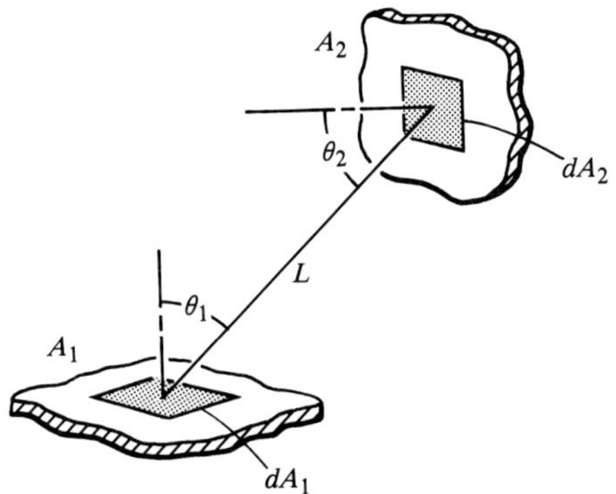

Figure 6.26 Geometry for derivation of the shape factor formula.

Substituting $I_1^+ = J_1/\pi$ from Eq. (6.59), and integrating over the solid angle subtended by area A_2 at dA_1, gives the radiation leaving dA_1 that is intercepted by surface A_2:

$$d\dot{Q}_{12} = \frac{J_1\, dA_1}{\pi} \int_{\omega_{12}} \cos\theta_1\, d\omega$$

The radiation leaving dA_1 in all directions is

$$d\dot{Q}_1 = \frac{J_1\, dA_1}{\pi} \int_0^{2\pi} \cos\theta_1\, d\omega = J_1\, dA_1$$

The shape factor F_{12} is the fraction of the radiation leaving dA_1 that is intercepted by area A_2, $F_{12} = d\dot{Q}_{12}/d\dot{Q}_1$, or

$$F_{12} = \frac{1}{\pi} \int_{\omega_{12}} \cos\theta_1\, d\omega \tag{6.63}$$

Equation (6.63) is convenient for application to simple configurations, as illustrated in Example 6.14. Introduction of $d\omega = dA_2 \cos\theta_2/L^2$ into Eq. (6.63) gives an alternative form:

$$F_{12} = \frac{1}{\pi} \int_{A_2} \frac{\cos\theta_1 \cos\theta_2}{L^2} dA_2 \tag{6.64}$$

To obtain the shape factor for a finite area A_1, F_{12} given by Eq. (6.64) is averaged over area A_1:

$$F_{12} = \frac{1}{A_1} \int_{A_1} \int_{A_2} \frac{\cos\theta_1 \cos\theta_2}{\pi L^2} dA_2\, dA_1 \tag{6.65}$$

From the symmetry of the problem, we can also write

$$F_{21} = \frac{1}{A_2} \int_{A_2} \int_{A_1} \frac{\cos\theta_2 \cos\theta_1}{\pi L^2} dA_1\, dA_2 \tag{6.66}$$

It follows from Eqs. (6.65) and (6.66) that

$$\int_{A_1} \int_{A_2} \frac{\cos\theta_1 \cos\theta_2}{\pi L^2} dA_2 dA_1 = A_1 F_{12} = A_2 F_{21} \tag{6.67}$$

which is both a formula for the shape factor and a statement of the reciprocal rule for shape factors. The summation rule follows directly from Eq. (6.63). Notice that in deriving Eq. (6.63), the heat flows $d\dot{Q}_{12}$ and $d\dot{Q}_1$ are indicated as differential elements of heat flow because they are leaving the elemental area dA_1. However, perhaps we could have written $d^2\dot{Q}$ since this heat flow is also a flow in an elemental solid angle $d\omega$; such is the practice in some texts.

Unless the geometry is very simple, analytical evaluation of the double integral in Eq. (6.67) is a challenging mathematical problem. Computer codes for the evaluation of Eq. (6.67) are available, for example, TRASYS [7].

EXAMPLE 6.14 Shape Factor between an Elemental Area and a Disk

Determine the shape factor F_{12} for the surfaces shown in the following sketch. Surface 2 is a disk of diameter D, and surface 1 is small ($A_1 \ll A_2$), parallel to surface 2, and a distance H away.

Solution

Given: Elemental area 1 parallel to disk 2.

Required: Shape factor F_{12}.

To use Eq. (6.63), we write $d\omega = \sin\theta_1\, d\theta_1\, d\phi$, to give

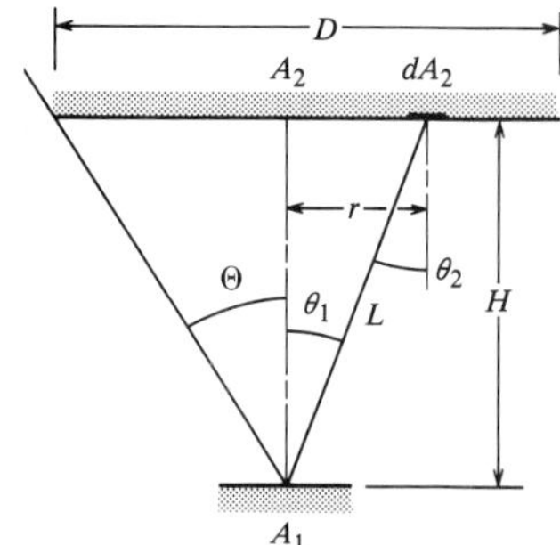

$$F_{12} = \frac{1}{\pi}\int_{\omega_{12}} \cos\theta_1\, d\omega$$

$$= \frac{1}{\pi}\int_0^{2\pi}\int_0^{\Theta} \cos\theta_1 \sin\theta_1\, d\theta_1\, d\phi$$

$$= 2\int_0^{\Theta} \cos\theta_1 \sin\theta_1\, d\theta_1$$

$$= \sin^2\Theta$$

$$\sin\Theta = \frac{D/2}{[(D/2)^2 + H^2]^{1/2}}; \qquad \text{thus} \qquad F_{12} = \frac{D^2}{D^2 + 4H^2}$$

Alternatively, to use Eq. (6.64), we note that $\theta_1 = \theta_2 = \theta$, $\cos\theta = H/L$, $L^2 = r^2 + H^2$, $dA_2 = 2\pi r\, dr$.

$$F_{12} = \frac{1}{\pi}\int_{A_2} \frac{\cos\theta_1 \cos\theta_2}{L^2}\, dA_2 = 2H^2\int_0^{D/2} \frac{r\, dr}{(r^2 + H^2)^2} = \frac{D^2}{D^2 + 4H^2}$$

EXAMPLE 6.15 Shape Factor between Areas on a Sphere

Determine the shape factor between two areas A_1 and A_2 on the inside of a sphere.

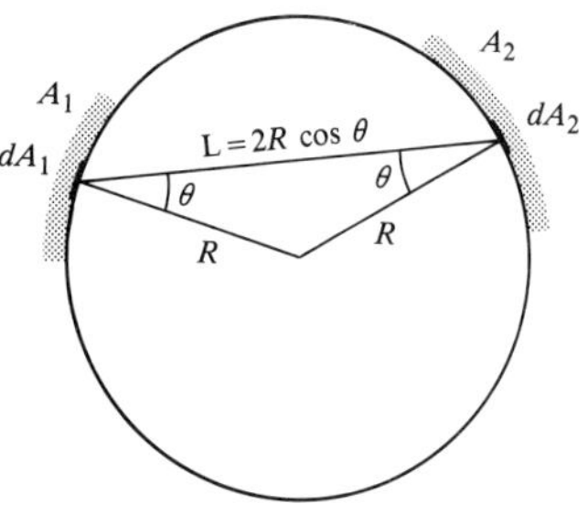

Solution

Given: Two areas on the inside of a sphere, A_1 and A_2.

Required: Shape factor F_{12}.

The figure shows that $\theta_1 = \theta_2 = \theta$, and $L = 2R\cos\theta$. Substituting into Eq. (6.65),

$$F_{12} = \frac{1}{A_1}\int_{A_1}\int_{A_2} \frac{\cos^2\theta}{4\pi R^2 \cos^2\theta}\, dA_2\, dA_1 = \frac{1}{A_1}\int_{A_1}\int_{A_2} \frac{dA_2\, dA_1}{4\pi R^2}$$

$$= \frac{A_2}{4\pi R^2} \left(= \frac{A_2}{\text{Total sphere area}} \right)$$

Comments

Notice that the shape factor is independent of the relative location of the areas A_1 and A_2: any location on the inner surface of a sphere "sees" all other locations in the same way.

6.5.3 Directional Properties of Real Surfaces

A diffuse gray surface obeys Lambert's law; that is, the intensity of radiation leaving the surface is independent of direction, which requires that the emittance and reflectance are independent of angle. The emittance of a real surface usually has negligible variation with azimuthal angle ϕ, but it does vary somewhat with zenith angle θ. Experimental data for the directional variation of emittance are sparse; Fig. 6.27

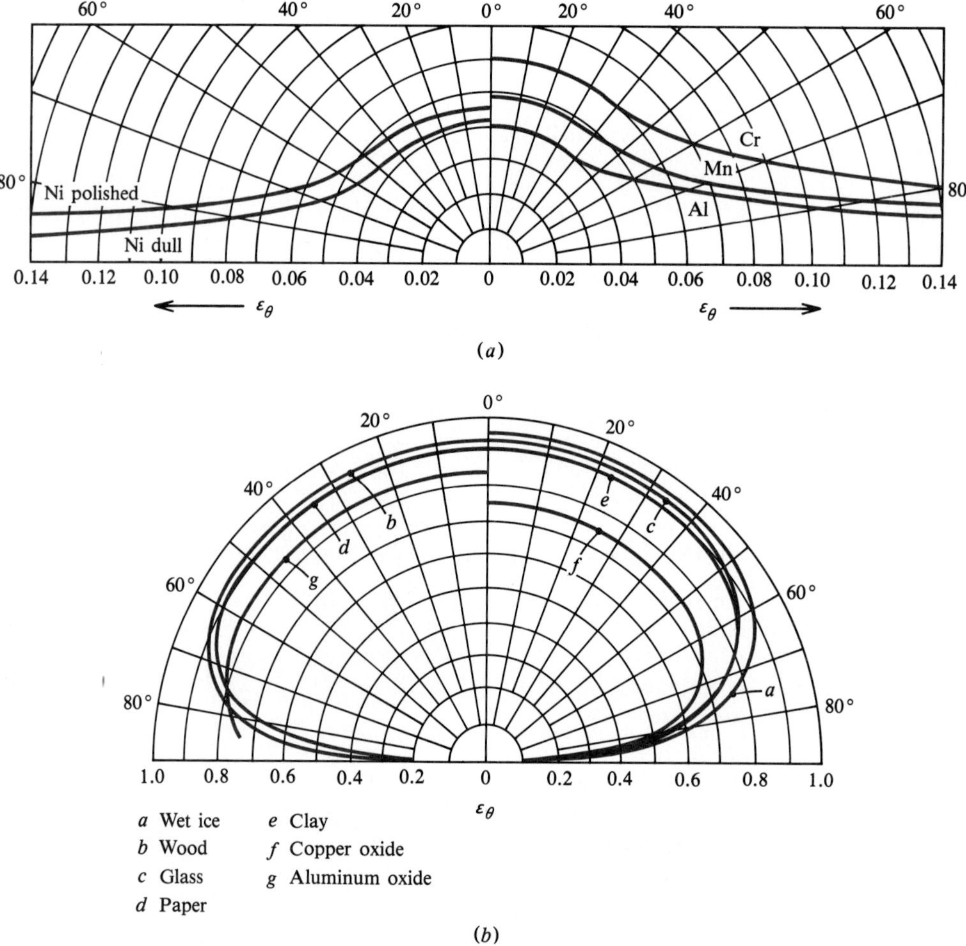

Figure 6.27 Variation of directional emittance with zenith angle θ. (*a*) Metals. (*b*) Nonmetals [8].

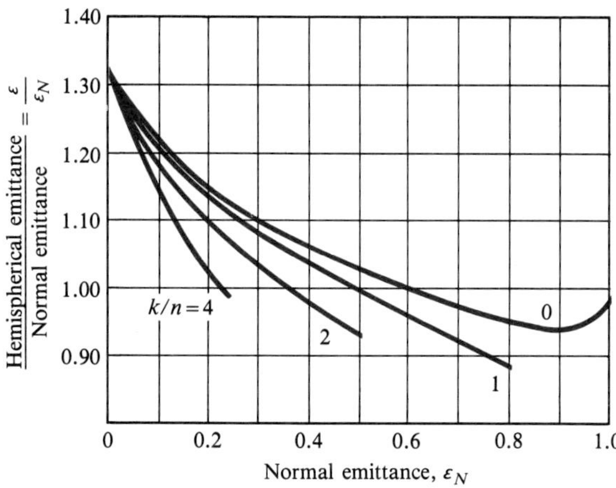

Figure 6.28 The ratio of hemispherical to normal emittance as a function of *k/n* (= absorptive index/refractive index) [2].

shows data for some selected surfaces. Electromagnetic theory gives the results shown in Fig. 6.28, where ε is the *hemispherical emittance*, that is, the emittance averaged over a hemisphere of solid angle 2π steradians above the surface, and ε_N is the emittance in the direction normal to the surface. The parameter *k/n* is the ratio of the absorptive index to the refractive index of the medium: for metals, a value of 1 can be used; 0 is more appropriate for nonmetals. A simple rule based on Fig. 6.28 is $\varepsilon/\varepsilon_N \simeq 1.2$ for low-emittance polished metals, and $\varepsilon/\varepsilon_N \simeq 0.96$ for high-emittance nonmetallic surfaces. Notice that it is the hemispherical emittance that is used in the diffuse gray surface model.

Specular Reflectors

Of some interest are surfaces for which reflection is not perfectly diffuse. Highly polished metal surfaces reflect thermal radiation in the same manner as a mirror reflects visible light, that is, with the angle of reflection equal to the angle of incidence. Such a surface is termed a *specular* reflector. Mixed specular and diffuse reflection can occur. For example, the surface of a gloss enamel paint tends to reflect specularly, while the underlying pigment particles reflect diffusely. The calculation of radiation exchange when some surfaces are specular reflectors is not as simple as for diffuse reflectors, unless the geometry is very simple. In the case of infinite parallel walls, all radiation reflected by one wall reaches the other directly, irrespective of whether the reflection is diffuse or specular: Eq. (6.41) and Table 6.2, item 1 are thus valid for specular surfaces as well. The same holds true for infinite concentric cylinders and concentric spheres when the inner surface is a specular reflector: Eq. (6.39) and Table 6.2, item 2 are valid for this situation. However, when the outer surface is specular and the inner surface is specular or diffuse, a new result is obtained, which can be derived as follows.

The inner surface is denoted A_1 and the outer surface A_2. We will first use ray tracing to obtain the result. Using Eq. (6.32) for a gray surface,

$$\dot{Q}_{12} = \varepsilon_1 A_1 E_{b1} - \varepsilon_1 A_1 G_1 \tag{6.68}$$

All radiation leaving surface 1 is intercepted by surface 2. In addition, all radiation *specularly* reflected by surface 2 returns to surface 1 (the angle of incidence equals the angle of reflection). Thus, the radiation incident on surface 1 due to radiation emitted by surface 1 is

$$\varepsilon_1 E_{b1} A_1(\rho_2 + \rho_1\rho_2^2 + \cdots) = \varepsilon_1 E_{b1} A_1 \frac{\rho_2}{1-\rho_1\rho_2} \tag{6.69}$$

The radiation incident on surface 1 due to radiation emitted by surface 2 is

$$\varepsilon_2 E_{b2} A_2 F_{21}(1 + \rho_1\rho_2 + \cdots) = \varepsilon_2 E_{b2} A_1 \frac{1}{1-\rho_1\rho_2} \tag{6.70}$$

where the reciprocal rule $A_2F_{21} = A_1F_{12} = A_1$ has been used for the diffuse emission from surface 2. Substituting for A_1G_1 in Eq. (6.68) gives

$$\dot{Q}_{12} = \varepsilon_1 A_1 E_{b1} - \varepsilon_1\left(\frac{\varepsilon_1 A_1 E_{b1}\rho_2}{1-\rho_1\rho_2} + \frac{\varepsilon_2 E_{b2} A_1}{1-\rho_1\rho_2}\right)$$

Setting $\rho = 1 - \alpha = 1 - \varepsilon$ and rearranging,

$$\dot{Q}_{12} = \frac{A_1(E_{b1} - E_{b2})}{1/\varepsilon_1 + 1/\varepsilon_2 - 1} \tag{6.71}$$

which is identical to Eq. (6.41), the result for infinite parallel walls. Alternatively, this result can be obtained by applying the energy conservation principle to obtain G_1. The radiation emitted (diffusely) by surface 2 and intercepted by surface 1 is $\varepsilon_2 E_{b2} A_2 F_{21}$ (= $\varepsilon_2 E_{b2} A_1$ from the reciprocal rule for diffuse radiation shape factors). The radiation emitted and reflected by surface 1 that returns to surface 1 is $A_1(\varepsilon_1 E_{b1} + \rho_1 G_1)\rho_2$, since *all* radiation leaving surface 1 and reflected by surface 2 returns to surface 1. Hence, the radiation incident on surface 1, A_1G_1, is

$$A_1G_1 = \varepsilon_2 E_{b2} A_1 + A_1(\varepsilon_1 E_{b1} + \rho_1 G_1)\rho_2$$

$$G_1 = \frac{\varepsilon_2 E_{b2} + \varepsilon_1\rho_2 E_{b1}}{1-\rho_1\rho_2}$$

Substituting for G_1 in Eq. (6.68),

$$\dot{Q}_{12} = \varepsilon_1 A_1\left(E_{b1} - \frac{\varepsilon_2 E_{b2} + \varepsilon_1\rho_2 E_{b1}}{1-\rho_1\rho_2}\right) = \frac{A_1(E_{b1} - E_{b2})}{1/\varepsilon_1 + 1/\varepsilon_2 - 1}$$

as before.

If an enclosure contains only a few specular plane surfaces, the mirror image concept can be used to modify the formulation developed in Section 6.3.5 for diffuse gray surface enclosures. The important idea is that a ray coming from a diffuse surface i and reflected by a specular surface m to a diffuse surface k, is equivalent to an uninterrupted straight-line ray from the mirror image of i to k, as shown in Fig. 6.29. With more than one specular surface, multiple reflections must be accounted for, and the method is feasible only if these can be easily added. However, in advanced engineering practice, the designer might wish to consider many complicating factors in addition to specular components of reflectance. For example, it might be desirable

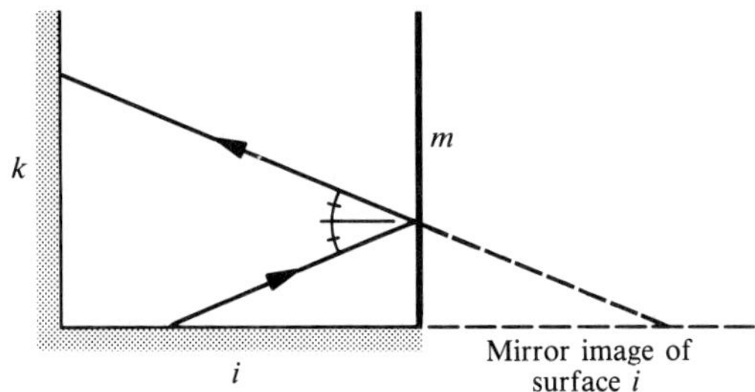

Figure 6.29 Use of the mirror image concept for a single specularly reflecting surface.

to consider directional and spectral variations of emittance. The *Monte Carlo* ray tracing method is then the most popular numerical method used.

Specular Reflecting Passages

A situation in which the effect of specular reflection is particularly important is the transmission of radiation along a passage with specular side walls. For a square or circular cross section, the problem is straightforward; but for a slot, **polarization** of the reflected radiation plays an important role. Radiation is polarized in the sense of having two wave components at right angles to each other and to the propagation direction; for black radiation, the two components are equal. Effects of polarization on radiant heat transfer are dealt with in more advanced texts. Figure 6.30 shows

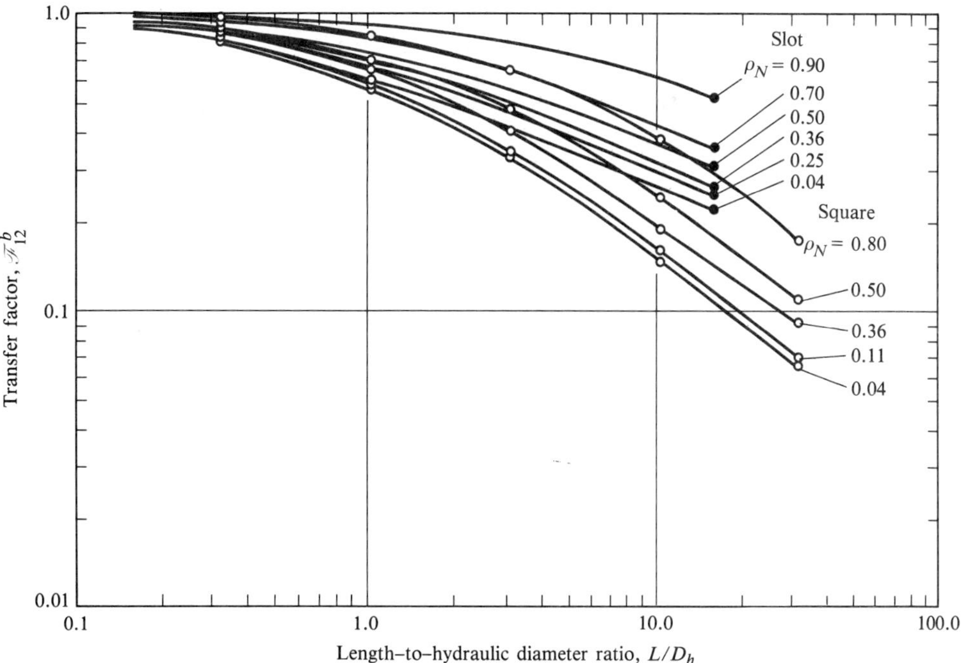

Figure 6.30 Transfer factor $\mathscr{F}_{12}^b$ for specular refractory walled passages of length L and hydraulic diameter D_h [2].

the transfer factor $\mathscr{F}_{12}^b$ for the slot and square passage with refractory side walls. A comparison with Fig. 6.20 for diffuse walls leads to the following observations:

1. The diffuse wall transfer factor does not depend on side wall reflectance (because there is no difference between diffusely reflected and diffusely reemitted radiation).
2. If the specular reflectance is unity, the transfer factor is 1.0, no matter how long the passage.
3. With specular reflection, the transfer factor is more sensitive to passage cross section.

EXAMPLE 6.16 Heat Gain by a Liquid Nitrogen Line

Liquid nitrogen flows through a 3 cm–O.D. tube, which is contained in an evacuated 5 cm–I.D. cylindrical shell. The nitrogen is at its normal boiling point of 77.4 K, and the shell inner surface is at 260 K. Estimate the heat gain per meter length if the emittances of the surfaces are 0.04 and (i) the surfaces are diffuse reflectors, (ii) the surfaces are specular reflectors.

Solution

Given: Dewar-type insulated liquid nitrogen line.

Required: Heat gain per meter length for diffuse and specular coatings.

Assumptions: 1. Emittance independent of temperature.
2. Completely diffuse or completely specular surfaces.

(i) For diffuse surfaces, item 2 of Table 6.2 gives the transfer factor as

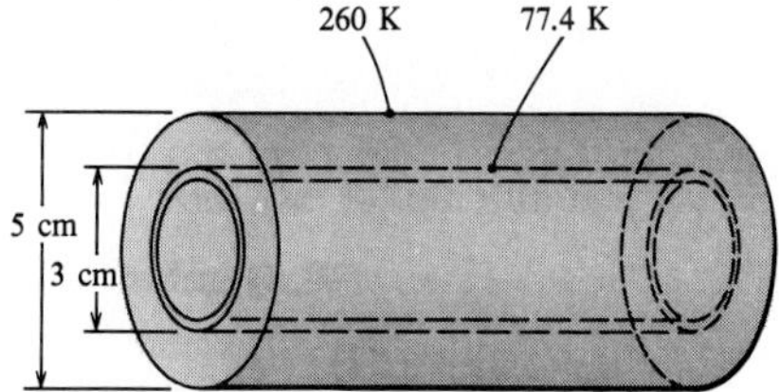

$$\mathscr{F}_{12} = \frac{\varepsilon_1}{1 + (\varepsilon_1 A_1/\varepsilon_2 A_2)(1 - \varepsilon_2)}$$

where the tube is surface 1 and the shell is surface 2. Using Eq. (6.45),

$$\dot{Q}_{12} = A_1\mathscr{F}_{12}(\sigma T_1^4 - \sigma T_2^4) = \frac{A_1\varepsilon_1\sigma(T_1^4 - T_2^4)}{1 + (\varepsilon_1 A_1/\varepsilon_2 A_2)(1 - \varepsilon_2)}$$

$$= \frac{(\pi)(0.03)(1)(0.04)(5.67 \times 10^{-8})(77.4^4 - 260^4)}{1 + (0.04/0.04)(3/5)(1 - 0.04)}$$

$$= -0.615 \text{ W/m}$$

(ii) For specular surfaces, Eq. (6.71) gives

$$\dot{Q}_{12} = \frac{A_1(E_{b1} - E_{b2})}{1/\varepsilon_1 + 1/\varepsilon_2 - 1} = \frac{(\pi)(0.03)(1)(5.67 \times 10^{-8})(77.4^4 - 260^4)}{1/0.04 + 1/0.04 - 1}$$

$$= -0.494 \text{ W/m}$$

Comments

The heat gain for specular surfaces is 20% less than for diffuse surfaces.

EXAMPLE 6.17 Radiation Transmission through a Spacecraft Shutter

A louvered shutter for temperature control of a spacecraft is 60 cm wide and 31 cm high. When fully open, the passages are 3 cm high and 6 cm long. The louvers are 1 mm thick. The spacecraft interior can be taken to be a black enclosure at 300 K, and the exterior is outer space at ~0 K. Determine the heat loss through the shutter (i) if the louvers are diffuse reflectors, and (ii) if the louvers have a coating that reflects specularly with a normal reflectance of 0.9.

Solution

Given: Louvered shutter for spacecraft temperature control.

Required: Effect of wall reflection properties on heat loss.

Assumption: The louvers have a low thermal conductivity, so they can be taken to be adiabatic.

The heat loss through the shutter is given by

$$\dot{Q}_{12} = A_1 \mathscr{F}_{12}^b \sigma (T_1^4 - T_2^4)$$

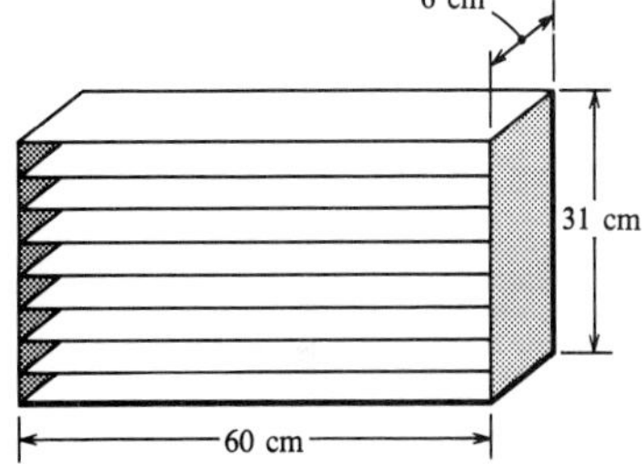

where the transfer factors $\mathscr{F}_{12}^b$ are obtained from Fig. 6.20 for diffuse reflection and from Fig. 6.30 for specular reflection. The hydraulic diameter of the slot-shaped passages between the louvers is

$$D_h = (2)(3) = 6 \text{ cm}$$

$$L/D_h = 6/6 = 1.0$$

The number of passages is $31/(3+0.1) = 10$, so that the area for radiant energy transmission is $A_1 = (10)(0.03)(0.6) = 0.18 \text{ m}^2$.

(i) Diffuse reflection: From Fig. 6.20, $\mathscr{F}_{12}^b \simeq 0.54$.

$$\dot{Q}_{12} = (0.18)(0.54)(5.67 \times 10^{-8})(300^4 - 0^4) = 44.6 \text{ W}$$

(ii) Specular reflection: From Fig. 6.30, $\mathscr{F}_{12}^b \simeq 0.91$, for $\rho_N = 0.90$.

$$\dot{Q}_{12} = (0.18)(0.91)(5.67 \times 10^{-8})(300^4 - 0^4) = 75.2 \text{ W}$$

Comments

1. The effect of the specularly reflecting coating is to increase the radiation transmission by 69%.
2. How the louver transmits as a function of angular rotation also depends significantly on whether the reflection is specular or diffuse.

6.6 SPECTRAL CHARACTERISTICS OF SURFACE RADIATION

In Section 6.3, the assumption of a gray surface allowed the calculation of radiant energy exchange without the need to consider the spectral characteristics of emitted or reflected radiation. Only the *total* energy emitted or reflected at all wavelengths was involved; thus, properties such as emittance were implicitly assumed to be appropriate average values. In order to solve simple problems involving solar radiation in Section 6.4, we crudely approximated the spectral distribution of radiant energy by using values of the absorptance that were different for solar and terrestrial radiation, otherwise treating the surface as if it were gray. We now proceed to examine the spectral characteristics of surface radiation more carefully. In Section 6.6.1, Planck's law is restated and the fractional functions defined. In Section 6.6.2, spectral-directional properties are defined and a fundamental statement of *Kirchhoff's law* given. Hemispherical and total hemispherical properties are obtained by appropriate averaging.

6.6.1 Planck's Law and Fractional Functions

Planck's law, Eq. (6.5), gives the spectral distribution of the monochromatic emissive power for a black surface, that is, $E_{b\lambda}$ as a function of λ:

$$E_{b\lambda} = \frac{C_1\lambda^{-5}}{e^{C_2/\lambda T} - 1} \tag{6.5}$$

where for λ in μm, $C_1 = 3.742 \times 10^8$ W μm^4/m^2, and $C_2 = 1.4389 \times 10^4$ μm K. The fraction of the total blackbody emissive power for wavelengths between 0 and λ is called the **external fractional function** f_e:

$$f_e(\lambda, T) = \frac{\int_0^\lambda E_{b\lambda}(\lambda, T) d\lambda}{\sigma T^4} \tag{6.72}$$

Substituting from Eq. (6.5) and then changing the integration variable from λ to $\zeta = C_2/\lambda T$,

$$f_e(\lambda T) = \frac{C_1}{\sigma C_2^4} \int_\zeta^\infty \frac{\zeta^3\, d\zeta}{e^\zeta - 1} \tag{6.73}$$

Table 6.5 shows this function. To obtain the fraction of power between two wavelengths, we simply subtract the fraction below the lower value from the fraction below the higher value:

$$\Delta f = f(\lambda T)_1 - f(\lambda T)_2$$

The use of the external fraction is demonstrated in Example 6.18. At this point, it is also appropriate to introduce the **internal fractional function** f_i, which is defined as

$$f_i(\lambda, T) = \frac{\int_0^\lambda [\partial E_{b\lambda}(\lambda, T)/\partial T]\, d\lambda}{4\sigma T^3} \tag{6.74}$$

The function $f_i(\lambda T)$ is also shown in Table 6.5. The importance of the internal function will be demonstrated in Section 6.6.2.

Table 6.5 Values of λT at specified values of the external and internal fractions of blackbody radiation [2].

f	Δf	$\sum \Delta f$	$(\lambda T)_e$ μm K $\times 10^{-4}$	$(\lambda T)_i$ μm K $\times 10^{-4}$	f	Δf	$\sum \Delta f$	$(\lambda T)_e$ μm K $\times 10^{-4}$	$(\lambda T)_i$ μm K $\times 10^{-4}$
0.0025	0.005	0.005	0.1230	0.1073	0.51	0.02	0.52	0.416	0.326
0.0075	0.005	0.010	0.1395	0.1209	0.53	0.02	0.54	0.428	0.334
0.0125	0.005	0.015	0.1495	0.1291	0.55	0.02	0.56	0.440	0.342
0.0175	0.005	0.02	0.1573	0.1352	0.57	0.02	0.58	0.453	0.351
0.025	0.01	0.03	0.1662	0.1423	0.59	0.02	0.60	0.467	0.361
0.035	0.01	0.04	0.1762	0.1501	0.61	0.02	0.62	0.482	0.371
0.045	0.01	0.05	0.1848	0.1565	0.63	0.02	0.64	0.497	0.382
0.055	0.01	0.06	0.1922	0.1625	0.65	0.02	0.66	0.513	0.393
0.07	0.02	0.08	0.202	0.1701	0.67	0.02	0.68	0.531	0.405
0.09	0.02	0.10	0.214	0.1791	0.69	0.02	0.70	0.550	0.417
0.11	0.02	0.12	0.225	0.1874	0.71	0.02	0.72	0.570	0.430
0.13	0.02	0.14	0.235	0.1949	0.73	0.02	0.74	0.593	0.446
0.15	0.02	0.16	0.245	0.202	0.75	0.02	0.76	0.616	0.462
0.17	0.02	0.18	0.254	0.209	0.77	0.02	0.78	0.642	0.479
0.19	0.02	0.20	0.263	0.216	0.79	0.02	0.80	0.671	0.498
0.21	0.02	0.22	0.272	0.222	0.81	0.02	0.82	0.704	0.521
0.23	0.02	0.24	0.281	0.229	0.83	0.02	0.84	0.741	0.546
0.25	0.02	0.26	0.290	0.235	0.85	0.02	0.86	0.783	0.574
0.27	0.02	0.28	0.299	0.242	0.87	0.02	0.88	0.825	0.609
0.29	0.02	0.30	0.308	0.248	0.89	0.02	0.90	0.899	0.653
0.31	0.02	0.32	0.317	0.255	0.91	0.02	0.92	0.982	0.690
0.33	0.02	0.34	0.326	0.261	0.93	0.02	0.94	1.090	0.774
0.35	0.02	0.36	0.335	0.268	0.945	0.01	0.95	1.200	0.846
0.37	0.02	0.38	0.344	0.275	0.955	0.01	0.96	1.298	0.909
0.39	0.02	0.40	0.353	0.281	0.965	0.01	0.97	1.433	0.997
0.41	0.02	0.42	0.363	0.288	0.975	0.01	0.98	1.63	1.123
0.43	0.02	0.44	0.373	0.295	0.9825	0.005	0.985	1.87	1.28
0.45	0.02	0.46	0.384	0.303	0.9875	0.005	0.990	2.22	1.43
0.47	0.02	0.48	0.394	0.310	0.9925	0.005	0.995	2.56	1.70
0.49	0.02	0.50	0.405	0.318	0.9975	0.005	1.000	7.34	2.49

EXAMPLE 6.18 Radiation Wavelengths of Practical Significance

Determine the wavelength range encompassing the 10–90% fraction of radiant energy emitted by black surfaces at the following temperatures: (i) a room temperature of 300 K, (ii) a cherry red–hot temperature of 1000 K, (iii) the temperature of a tungsten lamp filament, 3000 K, and (iv) the sun's temperature, 5700 K.

Solution

Given: Black surfaces at various temperatures.

Required: Wavelength range encompassing the 10–90% fraction of the emitted radiation.

From Table 6.5, the values of $(\lambda T)_e$ corresponding to $f = 0.1$ and $f = 0.9$ are

$$f = 0.1: \quad (\lambda T)_e = 0.220 \times 10^4 \ \mu\text{m K}$$

$$f = 0.9: \quad (\lambda T)_e = 0.940 \times 10^4 \ \mu\text{m K}$$

and $\Delta f_e = f_e(\lambda T)_2 - f_e(\lambda T)_1 = 0.9 - 0.1 = 0.8$ (80%).

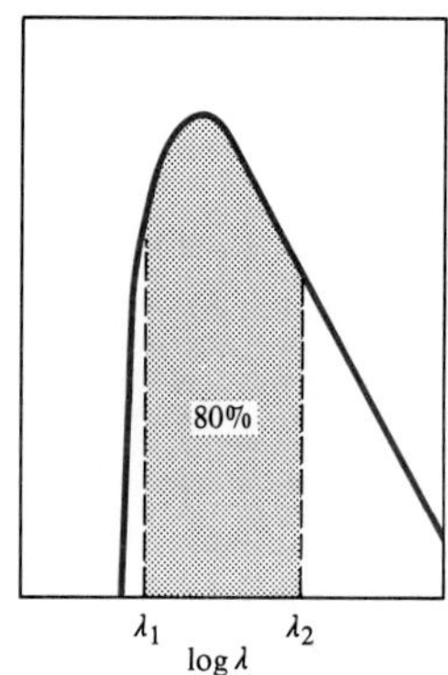

(i) $T = 300$ K:

$$\lambda_1 = (\lambda T)_1/T = 0.220 \times 10^4/300 = 7.33 \ \mu\text{m}$$

$$\lambda_2 = (\lambda T)_2/T = 0.940 \times 10^4/300 = 31.3 \ \mu\text{m}$$

(ii), (iii), (iv). Similarly, for $T = 1000$, 3000, and 5700 K, and the results are summarized below.

Source Temperature K	Wavelength Range μm
300	7.33–31.3
1000	2.20–9.40
3000	0.733–3.13
5700	0.386–1.65

Comments

1. As the temperature increases, the wavelength range of practical concern for radiant energy transfer shifts to shorter wavelengths and becomes narrower.
2. At room temperature, wavelengths of concern are on the order of 10 μm; at the sun's temperature, wavelengths are on the order of 1 μm.

6.6.2 Spectral Properties

Surface radiation properties, such as the emittance, depend on the nature of the surface. The chemical composition, physical structure (as perhaps modified by heat treatment or cold work), and roughness all influence the emittance. Nuclear irradiation may also be used to modify the surface. Once the surface is defined, its thermodynamic state is essentially fixed by its temperature, T_s; thus, the emittance may depend on T_s as well. In addition, the emittance depends on the direction and wavelength of the emitted radiation. The absorptance, reflectance, and transmittance may also depend on the state of polarization of the incident radiation, but this aspect will not be considered here.

For a given surface, we define **spectral-directional** radiation properties at temperature T_s. For example, the spectral-directional emittance of a surface at temperature T_s is the ratio of the intensity of radiation emitted at wavelength λ in the direction (θ, ϕ) to the intensity of radiation of the same wavelength emitted by a black surface at the same temperature:

$$\varepsilon(\theta, \phi, \lambda, T_s) = \frac{I_\lambda^+(\theta, \phi, \lambda, T_s)}{I_{b\lambda}^+(\lambda, T_s)} \tag{6.75}$$

There are similar definitions for the spectral-directional absorptance, reflectance, and transmittance. The *principle of detailed balancing* of statistical thermodynamics requires that

$$\varepsilon(\theta, \phi, \lambda, T_s) = \alpha(\theta, \phi, \lambda, T_s) \tag{6.76}$$

which is usually taken as the most fundamental statement of **Kirchhoff's law**. The net heat flux through a surface is

$$q = \frac{\dot{Q}}{A} = \int_0^\infty \int_0^{2\pi} \int_0^{\pi/2} \left\{ \varepsilon(\theta, \phi, \lambda, T_s) I_{b\lambda}^+(\lambda, T_s) - \alpha(\theta, \phi, \lambda, T_s) I_\lambda^-(\theta, \phi, \lambda) \right\} \cos\theta \sin\theta \, d\theta \, d\phi \, d\lambda \tag{6.77}$$

If the Monte Carlo method is to be used to calculate radiation exchange between surfaces, it is relatively straightforward to allow for directional and spectral variations in surface properties. However, often the engineer must use the simpler diffuse gray surface approximation, either because time or funds are limited or because the necessary data for the spectral-directional properties are unavailable for the surfaces of interest. Then suitable averaged surface properties are required.

Total Hemispherical Emittance

A **hemispherical property** is a value directionally averaged over the 2π steradian hemisphere of solid angle above the surface. A **total property** is a value spectrally averaged over all wavelengths from zero to infinity. Thus, the *total hemispherical emittance* is

$$\varepsilon(T_s) = \frac{1}{\sigma T_s^4} \int_0^\infty \int_0^{2\pi} \int_0^{\pi/2} \varepsilon(\theta, \phi, \lambda, T_s) I_{b\lambda}(\lambda, T_s) \cos\theta \sin\theta \, d\theta \, d\phi \, d\lambda \tag{6.78}$$

It is convenient to separate the averaging into two steps. In Section 6.5.3, we saw that, except for mirror-like surfaces, directional variations of the properties are not large, and data for such variations are sparse. More data are available for the hemispherical properties. Thus, as a first step, we directionally average by integrating with respect to θ and ϕ to give the *spectral* (or *monochromatic*) *hemispherical emittance* as

$$\varepsilon_\lambda(\lambda, T_s) = \frac{1}{\pi}\int_0^{2\pi}\int_0^{\pi/2} \varepsilon(\theta, \phi, \lambda, T_s)\cos\theta \sin\theta \, d\theta \, d\phi \tag{6.79}$$

and subsequently average with respect to λ to give

$$\varepsilon(T_s) = \frac{1}{\sigma T_s^4}\int_0^{\infty} \varepsilon_\lambda(\lambda, T_s)\pi I_{b\lambda}(\lambda, T_s) d\lambda \tag{6.80}$$

But from Eq. (6.72), the differential external fractional function is

$$df_e(\lambda T_s) = \frac{\pi I_{b\lambda}(\lambda, T_s) d\lambda}{\sigma T_s^4}$$

Thus,

$$\varepsilon(T_s) = \int_0^{1} \varepsilon_\lambda(\lambda, T_s) df_e(\lambda T_s) \tag{6.81}$$

Notation

There is no standard notation for the various surface properties, namely, spectral-directional, hemispherical, spectral, and total properties. The notation used here is consistent with many texts but has been simplified to avoid triple subscripts. An unsubscripted symbol, (e.g., ε) is a total hemispherical property and may be written $\varepsilon(T_s)$ when it is necessary to indicate its dependence on surface temperature. But unsubscripted ε is also used to denote the spectral-directional emittance: in this case, it is always written $\varepsilon(\theta, \phi, \lambda, T_s)$ to indicate its dependence on these parameters. Symbols subscripted λ [e.g., ε_λ or $\varepsilon_\lambda(\lambda, T_s)$] are spectral values that have been hemispherically averaged. Note also that subscripts θ and N are used for directional total properties in Section 6.5.3.

Total Hemispherical Absorptance

Averaging the spectral-directional absorptance (or reflectance) presents a more difficult problem, since the incident radiation is not unique but depends on the nature of the enclosure. If the enclosure surrounding the surface is taken to be black and isothermal at temperature T_e, then

$$\alpha(T_s, T_e) = \frac{1}{\sigma T_e^4}\int_0^{\infty}\int_0^{2\pi}\int_0^{\pi/2} \alpha(\theta, \phi, \lambda, T_s) I_{b\lambda}(\lambda, T_e)\cos\theta \sin\theta \, d\theta \, d\phi \, d\lambda \tag{6.82}$$

Notice that the important temperature here is T_e, which characterizes the spectrum of the incident radiation; temperature T_s characterizes the surface thermodynamic state and usually has a less significant effect on $\alpha(T_s, T_e)$. Again, the averaging is conveniently done in two steps:

$$\alpha_\lambda(\lambda, T_s) = \frac{1}{\pi}\int_0^{2\pi}\int_0^{\pi/2} \alpha(\theta, \phi, \lambda, T_s)\cos\theta\sin\theta\, d\theta\, d\phi \tag{6.83}$$

$$\alpha(T_s, T_e) = \frac{1}{\sigma T_e^4}\int_0^\infty \alpha_\lambda(\lambda, T_s)\pi I_{b\lambda}(\lambda, T_e) d\lambda = \int_0^1 \alpha_\lambda(\lambda, T_s) d f_e(\lambda T_e) \tag{6.84}$$

Kirchhoff's law requires $\alpha_\lambda(\lambda, T_s) = \varepsilon_\lambda(\lambda, T_s)$, so that data for only one of these properties are required.[6]

The heat flux through a surface at temperature T_s surrounded by a black isothermal enclosure at temperature T_e is

$$\begin{aligned} q &= \int_0^\infty\int_0^{2\pi}\int_0^{\pi/2} \{\varepsilon(\theta, \phi, \lambda, T_s) I_{b\lambda}^+(\lambda, T_s) \\ &\qquad - \alpha(\theta, \phi, \lambda, T_s) I_{b\lambda}^-(\lambda, T_e)\} \cos\theta\sin\theta\, d\theta\, d\phi\, d\lambda \\ &= \varepsilon(T_s)\sigma T_s^4 - \alpha(T_s, T_e)\sigma T_e^4 \end{aligned} \tag{6.85}$$

and we see that the emittance ε used for diffuse gray surfaces in Section 6.3 is correctly interpreted as the total hemispherical value.

Internal Emittance

As T_e approaches T_s, $\alpha(T_s, T_e)$ does not approach $\varepsilon(T_s)$. Edwards [2] shows how to obtain the correct result. When T_e approaches T_s, Eq. (6.85) may be written

$$\begin{aligned} q &= \int_0^\infty \alpha_\lambda(\lambda, T_s)\{\pi I_{b\lambda}(\lambda, T_s) - \pi I_{b\lambda}(\lambda, T_e)\}\, d\lambda \\ &= \int_0^\infty \alpha_\lambda(\lambda, T_s)\frac{\partial E_{b\lambda}(T_s)}{\partial T_s}(T_s - T_e) d\lambda \end{aligned} \tag{6.86}$$

The *internal total hemispherical emittance* is defined as

$$\varepsilon^i(T_s) = \frac{1}{4\sigma T_s^3}\int_0^\infty \alpha_\lambda(\lambda, T_s)\frac{\partial E_{b\lambda}(T_s)}{\partial T_s} d\lambda \tag{6.87}$$

Using Eq. (6.74) gives $\varepsilon^i(T_s)$ in terms of the internal fractional function:

$$\varepsilon^i(T_s) = \int_0^1 \alpha_\lambda(\lambda, T_s) d f_i(\lambda T_s) \tag{6.88}$$

[6] Often the irradiation is from a source that cannot be reasonably approximated by a black or gray surface, for example, a mercury lamp. Exercises 6–71 and 6–72 show how to calculate total absorptance in such situations.

Substituting Eq. (6.87) in Eq. (6.86) gives

$$q = \varepsilon^i(T_s)4\sigma T_s^3(T_s - T_e) = \alpha^i(T_s)4\sigma T_s^3(T_s - T_e) \tag{6.89}$$

and we see that the internal total hemispherical emittance is the appropriate limit for an isothermal enclosure. The situation where T_s is nearly equal to T_e occurs often inside enclosures such as rooms and space vehicles: hence the use of the adjective *internal*.

The Diffuse Gray Surface Approximation

We are now in a position to define a *gray* surface rigorously: it is a surface for which the spectral radiation properties are independent of wavelength. In the case of a *diffuse gray* surface, $\varepsilon_\lambda(\lambda, T_s)$ and $\alpha_\lambda(\lambda, T_s)$ are independent of wavelength. It follows that when these properties are spectrally averaged, the resulting properties are independent of temperature, that is, $\varepsilon(T_s)$ and $\alpha(T_s, T_e)$ are independent of temperature. Notice that, in general, the radiation incident on a surface is not directionally and spectrally distributed as blackbody radiation, and thus $\alpha(T_s, T_e)$ as given by Eq. (6.84) is not necessarily an appropriate value of the total hemispherical absorptance. However, in using the diffuse gray surface approximation we take $\alpha = \varepsilon$. The engineer must then simply choose a suitable common value for the conditions under consideration.

Spectral Property Data

Measuring surface radiation properties clearly requires considerable effort owing to the great variety of surfaces encountered in engineering practice. The choice of which property to measure depends on both the desired application and convenience. Data available in the heat transfer literature might be spectral or total values, directional

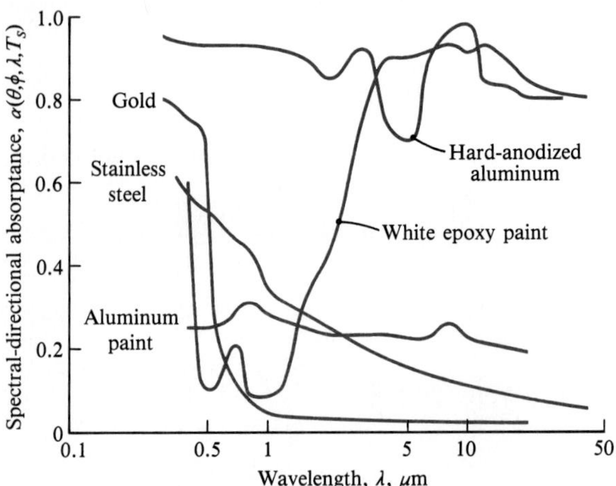

Figure 6.31 Spectral absorptance $\alpha_{\lambda,N}$ as a function of wavelength λ for selected surfaces. $T_s \simeq 300$ K, and $\theta = 25°$.

or hemispherical, and the total values may be internal or external averages. Usually, either the emittance or reflectance is measured, and the other properties are obtained by subtraction or from Kirchhoff's law. Table A.6*a* gives normal-incidence spectral and total absorptances for metals, from which approximate hemispherical values can be obtained using Fig. 6.28, with a value of $k/n = 1$ being appropriate for metals. Table A.6*b* and Fig. 6.31 give spectral absorptances of some selected surfaces. The data in Tables A.6*a* and A.6*b* are for a surface at room temperature, $T_s \simeq 300$ K. However, for many engineering purposes, spectral values of α_λ (or ε_λ) can be taken to be independent of temperature over quite a wide temperature range. Hence, if ε_λ increases with decreasing wavelength, ε will increase with temperature owing to the relatively greater emission at smaller wavelengths with increasing temperature.

EXAMPLE 6.19 Calculation of the Total Absorptance of White Epoxy Paint

Calculate the total absorptance of a surface at 300 K coated with white epoxy paint and exposed to a black radiation source at 2000 K. Use the data in Table A.6*b*.

Solution

Given: White epoxy paint surface at 300 K.

Required: Total absorptance to black radiation at 2000 K.

Assumptions: Incident radiation at 25° from the normal to the surface, so as to permit use of Table A.6*b*.

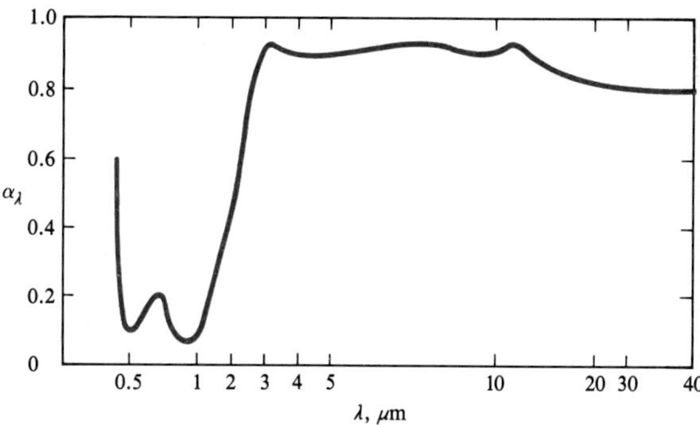

The data for α_λ of white epoxy paint from Table A.6*b* are plotted in the accompanying graph. We will assume that, for a nonconductor such as epoxy paint, the hemispherically averaged absorptance is approximately equal to the value at 25°. Then Eq. (6.84) gives

$$\alpha(T_s, T_e) = \frac{1}{\sigma T_e^4}\int_0^\infty \alpha_\lambda(\lambda, T_s)\pi I_{b\lambda}(\lambda, T_e)d\lambda = \int_0^1 \alpha_\lambda(\lambda, T_s)df_e(\lambda T_e)$$

Using Table 6.5 and the graph of α_λ, the following table is constructed to effect integration using the trapezoidal rule.

f	Δf	$(\lambda T)_e$ μm K	λ μm	$\alpha_\lambda(\lambda, 300\ \text{K})$
0.0025	0.005	0.1230×10^4	0.615	0.16
0.0075	0.005	0.1395×10^4	0.698	0.21
0.0125	0.005	0.1495×10^4	0.748	0.15
.				
.				
.				
0.9925	0.005	2.56×10^4	12.8	0.92
0.9975	0.005	7.34×10^4	36.7	0.80

$$\alpha(300\ \text{K}, 2000\ \text{K}) = \sum \alpha_\lambda(\lambda, 300\ \text{K})\Delta f_e = 0.53$$

Comments

Since epoxy paint is a nonconductor, the hemispherical total absorptance will be at most 4% lower than this value.

EXAMPLE 6.20 Radiant Heat Transfer inside a Spacecraft

An evacuated space inside a spacecraft is cylindrical in shape with a diameter of 20 cm and a height of 30 cm. The cylindrical walls and top surface can be taken to be black and are maintained at 280 K. The base is hard-anodized aluminum. Estimate the heat loss through the base when it is 300 K.

Solution

Given: Cylindrical enclosure, anodized aluminum base.

Required: Heat flow through base.

Assumptions: Walls and top surface are perfectly black.

Equation (6.89) gives the heat flux through the base surface as

$$q \simeq \varepsilon^i(T_s)4\sigma T_s^3(T_s - T_e)$$

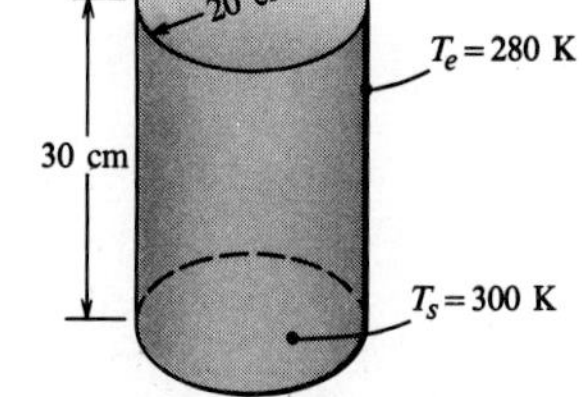

We will first assume that the data in Table A.6*b* for absorptance to radiation incident at 25° can be used to approximate the hemispherically averaged value, and we will make a correction later. From Eq. (6.88),

$$\varepsilon^i(T_s) = \int_0^1 \alpha_\lambda(\lambda, T_s)df_i(\lambda T_s)$$

Using Tables 6.5 and A.6*b*, the following table is constructed to effect integration using the trapezoidal rule:

f	Δf	$(\lambda T)_i$ μm K	λ μm	$\alpha_\lambda(\lambda, 300\text{ K})$
0.0025	0.005	0.1073×10^4	3.58	0.82
0.0075	0.005	0.1209×10^4	4.03	0.74
0.0125	0.005	0.1291×10^4	4.30	0.73
.				
.				
.				
0.9925	0.005	1.70×10^4	56.7	0.80
0.9975	0.005	2.49×10^4	83	0.80

$$\varepsilon^i(300\text{ K}) = \sum \alpha_\lambda(\lambda, 300\text{ K})\Delta f_i$$

$$= 0.87$$

For a high-emittance surface, the hemispherical average will be about 0.03 lower than this value; hence, we have

$$\varepsilon^i(300\text{ K}) = 0.87 - 0.03 = 0.84$$

$$\dot{Q} = qA = \varepsilon^i(T_s)4\sigma T_s^3(T_s - T_e)(\pi D^2/4)$$

$$= (0.84)(4)(5.67 \times 10^{-8})(300)^3(300 - 280)(\pi/4)(0.2)^2$$

$$= 3.2\text{ W}$$

Comments

Note that Table A.5*a* gives $\varepsilon = 0.80$ for hard-anodized aluminum.

6.7 RADIATION TRANSFER THROUGH GASES

In the preceding sections of Chapter 6, we considered radiation exchange between surfaces separated by perfectly transparent or *nonparticipating* media. Such media include a vacuum, as well as air at normal temperatures and pressures. Symmetrical gas molecules, such as N_2 and O_2, emit and absorb negligible radiation unless temperatures are high enough for electronic excitation or ionization to occur—for example, behind the bow shock wave of a reentry vehicle. Gas species with nonsymmetrical molecules can emit and absorb radiation; the most important of such species are H_2O, CO_2, CO, SO_2, and NH_3. (However, the small amounts of H_2O in atmospheric air means that its presence is usually ignored unless the path length is long, e.g., on the order of tens of meters.) Gas radiation is particularly significant to the design of furnaces and combustion chambers with carbon, hydrogen, or hydrocarbon fuels. A gas can also participate in the radiation exchange process by virtue of containing an *aerosol*. Such aerosols include liquid droplets, dust particles, and, of particular importance, soot particles in combustion products. Not only can radiation be absorbed and emitted by aerosols, but radiation can also be *scattered*. The participation of aerosols in radiation exchange will not be treated here.

Section 6.7.1 derives the *equation of transfer* for radiation propagating through a participating medium. Section 6.7.2 presents data and a methodology for evaluating total gas properties for CO_2, H_2O, and CO_2–H_2O mixtures. In Section 6.7.3, the effective and mean beam length concepts are formulated, and a simple prescription is given for evaluating the mean beam length. Section 6.7.4 analyzes radiation exchange between an isothermal nongray gas and a black enclosure, and Section 6.7.5 analyzes radiation exchange between an isothermal gray gas and a gray enclosure. Finally, in Section 6.7.6, an approximate result is given for radiation exchange between a nongray gas and a single-surface gray enclosure.

6.7.1 The Equation of Transfer

Consider a beam of radiation leaving a surface dA_s and propagating in direction x through a nonscattering medium, as shown in Fig. 6.32. The intensity of the beam is attenuated due to absorption by the medium and is augmented as a result of emission by the medium. A spectral energy balance on an elemental volume Δx thick requires that

$$I_\lambda \Delta\lambda|_{x+\Delta x} - I_\lambda \Delta\lambda|_x = \left[\begin{array}{c}\text{Emission per}\\ \text{unit length}\end{array} - \begin{array}{c}\text{Absorption per}\\ \text{unit length}\end{array}\right]\Delta\lambda\Delta x \tag{6.90}$$

The absorption per unit length is $\kappa_\lambda I_\lambda$, where κ_λ is the *spectral absorption coefficient* of the medium and has units m^{-1}. If the medium is in local thermodynamic equilibrium, the emission per unit length must be $\kappa_\lambda I_{b\lambda}$, so that if there is complete thermodynamic equilibrium, with $I_\lambda = I_{b\lambda}$, absorption will equal emission. Substituting in Eq. (6.90),

$$I_\lambda \Delta\lambda|_{x+\Delta x} - I_\lambda \Delta\lambda|_x = (\kappa_\lambda I_{b\lambda} - \kappa_\lambda I_\lambda)\Delta\lambda\Delta x$$

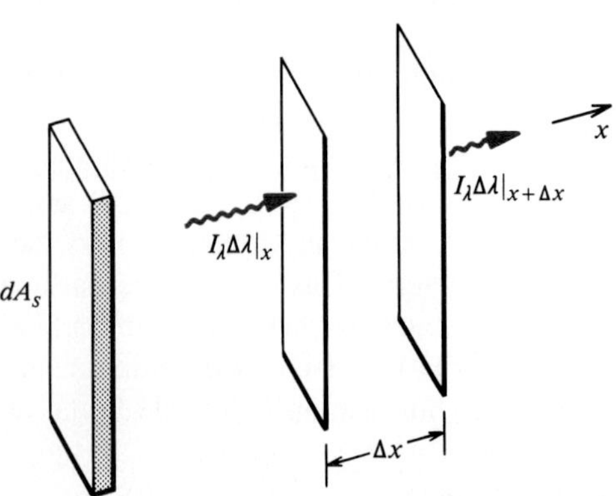

Figure 6.32 Attenuation of a beam of radiation in a nonscattering medium.

Dividing by $\Delta\lambda\Delta x$ and letting $\Delta x \to 0$ gives

$$\frac{dI_\lambda}{dx} = -\kappa_\lambda I_\lambda + \kappa_\lambda I_{b\lambda} \tag{6.91}$$

which is the *equation of transfer* for a nonscattering medium on a spectral basis.

For an isothermal medium, $I_{b\lambda}$ and κ_λ are constant so that Eq. (6.91) can be integrated to give

$$I_\lambda(x) = I_{\lambda,s}e^{-\kappa_\lambda x} + I_{b\lambda}\left(1 - e^{-\kappa_\lambda x}\right) \tag{6.92}$$

where $I_{\lambda,s}$ is the spectral intensity of the radiation leaving the surface at $x = 0$. In particular, at $x = L$,

$$I_\lambda(L) = I_{\lambda,s}e^{-\kappa_\lambda L} + I_{b\lambda}\left(1 - e^{-\kappa_\lambda L}\right) \tag{6.93}$$

The first term on the right-hand side is the radiation intensity at $x = L$ due to transmission of radiation leaving the surface at $x = 0$; the second term is the radiation intensity at $x = L$ due to emission along the path. Thus, we write

$$I_\lambda(L) = \tau_{g\lambda} I_{\lambda,s} + \varepsilon_{g\lambda} I_{b\lambda} \tag{6.94}$$

where $\tau_{g\lambda} = e^{-\kappa_\lambda L}$ and $\varepsilon_{g\lambda} = 1 - e^{-\kappa_\lambda L}$ are the *spectral* (or monochromatic) *transmissivity* and *emissivity*, respectively, of gas along the path $(0, L)$. Since there is no reflection, the *spectral absorptivity* $\alpha_{g\lambda} = 1 - \tau_{g\lambda}$, and Kirchhoff's law $\varepsilon_{g\lambda} = \alpha_{g\lambda}$ is obeyed, as required by the formulation of the equation of transfer.

In general, the absorption coefficient κ_λ is strongly dependent on wavelength, as will be discussed for gaseous mixtures in Section 6.7.2. Nevertheless, it is often possible to perform satisfactory engineering calculations using total properties averaged over all wavelengths. Equation (6.93) can be written on a total basis as

$$I(L) = I_s e^{-\kappa L} + I_b\left(1 - e^{-\kappa L}\right) \tag{6.95}$$

or

$$I(L) = \tau_g I_s + \varepsilon_g I_b \tag{6.96}$$

For continuum radiation, where κ_λ varies smoothly with wavelength, as is the case for a cloud containing a distribution of particle sizes, the calculation of a total property is straightforward. The total emissivity is simply

$$\varepsilon_g = \frac{\int_0^\infty I_\lambda \, d\lambda}{\int_0^\infty I_{b\lambda}\, d\lambda} = \int_0^1 \left[1 - e^{-\kappa_\lambda L}\right] df_e(\lambda T) \tag{6.97}$$

where f_e is the external fractional function. For the radiation emitted by a molecular gas, the averaging process to obtain ε_g is more complicated.

6.7.2 Gas Radiation Properties

Gaseous radiation is particularly relevant to heat transfer in furnaces and combustion chambers burning carbon, hydrogen, or hydrocarbon fuels. Absorption and emission by carbon dioxide and water vapor is of particular concern due to their relatively high concentrations in the products of combustion, and due to their strong radiation

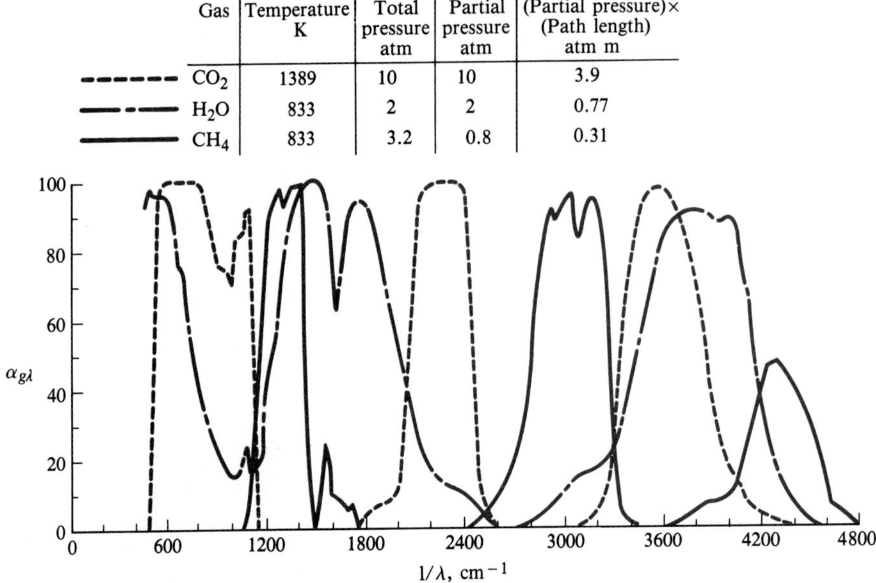

Figure 6.33 Spectral absorptivity $\alpha_{g\lambda}$ as a function of wavenumber $1/\lambda$, for water vapor, carbon dioxide, and methane [2].

absorption. Figure 6.33 shows some spectral absorptivity data for CO_2, H_2O, and CH_4. The absorption (or emission) does not take place continuously over the entire spectrum but rather occurs in a number of moderately wide *bands* of relatively strong absorption. The radiation properties are obviously nongray. Accurate radiation exchange calculations generally require that the nongray behavior be properly accounted for, albeit approximately, using *band models*. However, often satisfactory engineering calculations can be made on a total basis, and for this purpose total gas properties are required. The absorption bands of gases are actually arrays of lines at discrete wavelengths. The averaging process to obtain total gas properties is complicated, and only the results will be presented here. The *Hottel charts*, first presented by H. Hottel in 1927, have been widely used but are of limited accuracy [9,10,11]. More recently, Edwards and Matavosian [12,13] have presented new charts and calculation procedures that give more accurate results, and these are presented here.

Total Emissivity Charts

The total emissivity of a gas, ε_g, is a function of its temperature T_g due to nongrayness and density variation. Also, since the absorption coefficient κ is defined on a length basis, emissivity depends on partial pressure of the emitting gaseous species, P_a, times the thickness of the gas layer, L. The total emissivity can also depend on total pressure, P, due to a phenomenon called *collision line broadening*. Figures 6.34*a* and 6.35*a* gives ε_g for CO_2 and H_2O in N_2 at an **equivalent broadening**

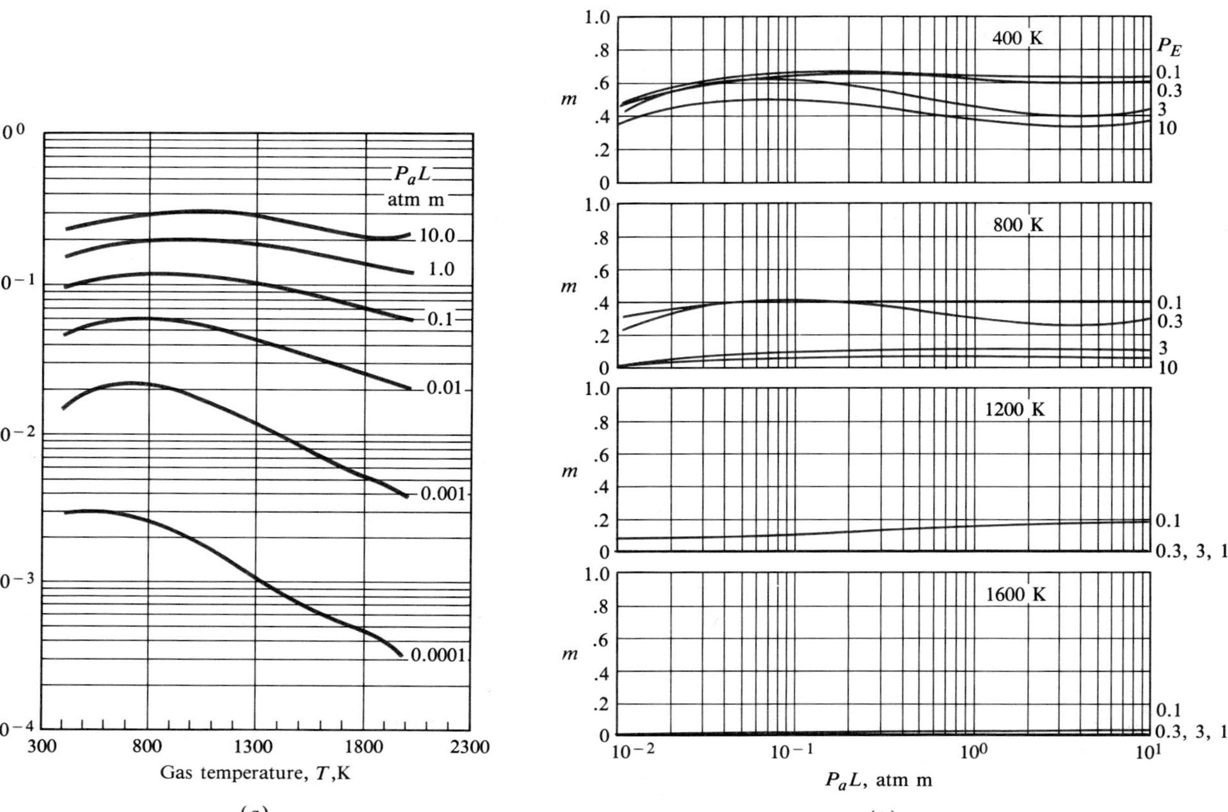

Figure 6.34 Gas radiation properties for CO_2 in N_2. *(a)* Gas emissivity ε_g at an equivalent broadening pressure ratio $P_E = 1$. *(b)* Total pressure scaling exponent m.

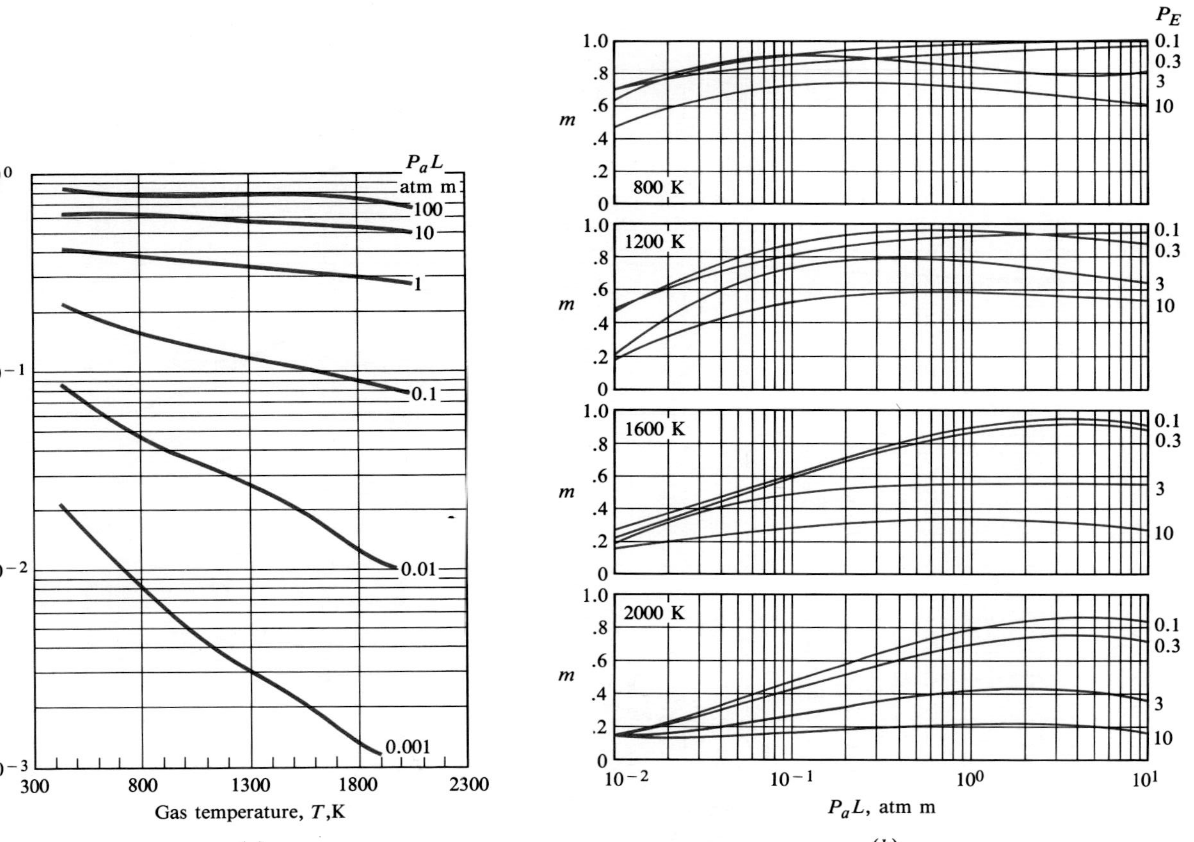

Figure 6.35 Gas radiation properties for H_2O in N_2. (*a*) Gas emissivity ε_g at an equivalent broadening pressure ratio $P_E = 1$. (*b*) Total pressure scaling exponent m.

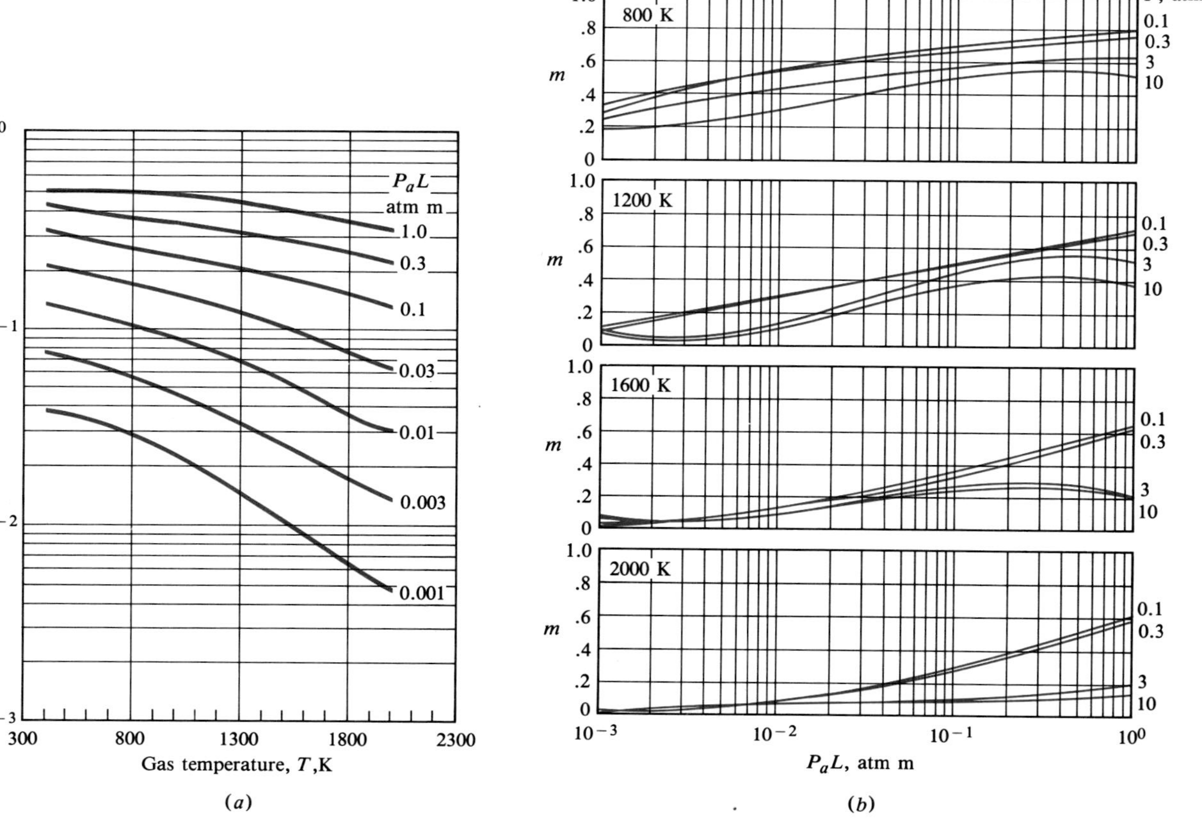

Figure 6.36 Gas radiation properties for a mixture of 10% CO_2, 10% H_2O, and 80% N_2 by volume. P_a is the partial pressure of H_2O only. (*a*) Gas emissivity at 1 atm total pressure. (*b*) Total pressure scaling exponent m.

pressure ratio P_E of unity; P_E is defined as

$$P_E = \left\{ \frac{P}{P_0} \left[1 + (b-1)\frac{P_a}{P} \right] \right\}^n \tag{6.98}$$

where P_0 is one atmosphere; b and n for CO_2 and H_2O are, respectively,

$$CO_2: \quad b = 1.3; \quad n = 0.8$$

$$H_2O: \quad b = 8.6(T_0/T_g)^{1/2} + 0.5; \quad n = 1$$

where $T_0 = 100$ K. Figure 6.36*a* gives ε_g for a mixture of 10% CO_2, 10% H_2O and 80% N_2, by volume, at a total pressure of one atmosphere. For this *equimolar* CO_2–H_2O mixture, P_a is defined as the partial pressure of H_2O only. Changing the relative amount of N_2 or the presence of other nonradiating species, such as O_2, has little effect on ε_g; also, the precise relative amount of CO_2 is of secondary importance, since emission by H_2O dominates. Hence, Fig. 6.36*a* can be used for the combustion products of most hydrocarbon fuels.

The effect of total pressure can be accounted for by using a simple scaling rule. For CO_2 and H_2O,

$$\varepsilon_g(T_g, P_aL, P_E) = \varepsilon_g(T_g, P_aL', P_E = 1) \tag{6.99a}$$

$$P_aL' = P_aLP_E^m \tag{6.99b}$$

For the CO_2–H_2O mixture, the rule is of the same form, with P_E replaced by P [atm], the total pressure. The exponent m is given in Figs. 6.34*b*, 6.35*b*, and 6.36*b*.

Total Absorptivity

The total absorptivity of a gas at temperature T_g to radiation from a black source at temperature T_s can be estimated using a second scaling law. For CO_2 and H_2O,

$$\alpha_g(T_g, T_s, P_aL, P_E) = (T_g/T_s)^{1/2}\varepsilon_g(T_s, P_aL'_\alpha, P'_E) \tag{6.100a}$$

$$P_aL'_\alpha = P_aL(T_s/T_g)^r \tag{6.100b}$$

$$P'_E = P_E(T_g/T_s)^s \tag{6.100c}$$

Again, P_E is replaced by total pressure P for the CO_2–H_2O mixture. The exponents r and s are given in Table 6.6. The two-step scaling required to obtain the absorptivity is illustrated in Examples 6.21 and 6.22.

Table 6.6 Path and pressure exponents for gas total absorptivity scaling.

Gas	Path Exponent r	Pressure Exponent s
CO_2	1.0	2.4
H_2O	1.5	1.6
CO_2/H_2O	1.5	1.6

The Computer Program RAD3

RAD3 calculates total gas properties using the data and methodology recommended by Edwards and Matavosian. The original data were curve-fitted, and these curve fits were also used to prepare Figs. 6.34 through 6.36. Thus, hand calculations using these figures should agree with the output of RAD3. The output of RAD3 is ε_{g1}, ε_{g2}, α_{g1}, and α_{g2}, where subscripts 1 and 2 refer to emission or absorption for path lengths L and $2L$, respectively. Example 6.26 describes the use of these properties for the calculation of radiation exchange between an isothermal nongray gas and a single gray surface enclosure. Note that RAD3 sets $P_E = 10$ if $P_E > 10$, in order to calculate the total pressure scaling exponent m: extrapolation of the data for m is not recommended (see Example 6.26).

EXAMPLE 6.21 Total Properties of Hydrogen Combustion Products

Exhaust gases from a combustor burning hydrogen are at 1200 K and 2 atm pressure, and they contain 10% (by volume) H_2O, the remainder being N_2. Calculate the total emissivity of the gas and its total absorptivity to black radiation from a surface at 800 K. Use a path length of 0.5 m.

Solution

Given: Mixture of H_2O and N_2.

Required: Total radiation properties.

Assumptions: The scaling laws of Section 6.7.2 are valid.

We first calculate the equivalent broadening pressure ratio from Eq. (6.98):

$$P_E = \left\{\frac{P}{P_0}\left[1 + (b-1)\frac{P_a}{P}\right]\right\}^n$$

$$P_0 = 1 \text{ atm}$$

$$b = 8.6(T_0/T_g)^{1/2} + 0.5 = 8.6(100/1200)^{1/2} + 0.5 = 2.98$$

$$n = 1$$

$$P_E = \left\{\frac{2}{1}\left[1 + (2.98-1)\frac{(0.1)(2.0)}{2.0}\right]\right\}^1 = 2.40$$

Next we calculate P_aL' from Eq. (6.99*b*):

$$P_aL' = P_aLP_E^m$$

$P_aL = (0.1)(2.0)(0.5) = 0.1$ atm m; from Fig. 6.35*b*, the exponent m is 0.74; thus

$$P_aL' = (0.1)(2.40)^{0.74} = 0.191 \text{ atm m}$$

Then, from Eq. (6.99*a*) and Fig. 6.35*a*,

$$\varepsilon_g = \varepsilon_g(T_g, P_aL') = \varepsilon_g(1200, 0.191) = 0.16$$

To determine the total absorptivity, we first calculate P'_E from Eq.(6.100c);

$$P'_E = P_E(T_g/T_s)^s, \qquad s = 1.6 \text{ from Table 6.6}$$

$$P'_E = 2.40(1200/800)^{1.6} = 4.59$$

We then calculate $P_aL'_\alpha$ from Eq. (6.100b):

$$P_aL'_\alpha = P_aL(T_s/T_g)^r, \qquad r = 1.5 \text{ from Table 6.6}$$

$$P_aL'_\alpha = 0.1(800/1200)^{1.5} = 0.0544 \text{ atm m}$$

Equation (6.100a) for α_g, requires that we next calculate

$$\varepsilon_g(T_s, P_aL'_\alpha, P'_E) = \varepsilon_g(T_s, P_aL''_\alpha, P'_E = 1)$$

$$P_aL''_\alpha = P_aL'_\alpha(P'_E)^{m'}$$

For $T_s = 800$ K, $P_aL'_\alpha = 0.0544$ atm m and $P'_E = 4.59$, Fig. 6.35b gives $m' = 0.81$.

$$P_aL''_\alpha = (0.0544)(4.59)^{0.81} = 0.187 \text{ atm m}$$

Figure 6.35a gives $\varepsilon_g(800{,}0.187) = 0.20$. Equation (6.100$a$) gives

$$\alpha_g = (T_g/T_s)^{1/2}(0.20) = (1200/800)^{1/2}(0.20) = 0.24$$

Comments

1. This result would remain valid if the exhaust gases contained oxygen as a result of using excess air in the combustion process.
2. Interpolation on the graphs can be rather difficult, and use of RAD3 is thus recommended: RAD3 gives $\varepsilon_g = 0.164$, $\alpha_g = 0.237$.

EXAMPLE 6.22 Total Properties of a Hydrocarbon Fuel Combustion Products

Exhaust gas from a combustor burning a hydrocarbon fuel is at 1600 K and 3 atm pressure. The composition can be approximated as 10% CO_2, 10% H_2O, and 80% N_2, by volume. Calculate the total emissivity for a path length of 0.34 m and the total absorptivity for radiation from a black wall at 800 K.

Solution

Given: Combustion products of a hydrocarbon fuel.

Required: Total gas emissivity and absorptivity.

Assumptions: Scaling laws of Section 6.7.2 are valid.

We first calculate P_aL' from Eq. (6.99b), where for the CO_2–H_2O mixture, P_E is replaced by total pressure, P.

$$P_aL' = P_aLP^m$$

$P_a = (0.1)(3) = 0.3$ atm; $P_aL = (0.3)(0.34) = 0.102$ atm m. From Fig. 6.36b, $m = 0.26$.

$$P_aL' = (0.102)(3)^{0.26} = 0.136 \text{ atm m}$$

Then, from Eq. (6.99a) and Fig. 6.36a,

$$\varepsilon_g = \varepsilon_g(T_g, P_aL') = \varepsilon_g(1600, 0.136) = 0.19$$

To determine the total absorptivity, we first calculate P' from Eq. (6.100c), with total pressure P replacing P_E:

$$P' = P(T_g/T_s)^s, \qquad s = 1.6 \text{ from Table 6.6}$$

$$P' = 3(1600/800)^{1.6} = 9.1 \text{ atm}$$

We then calculate $P_aL'_\alpha$ from Eq.(6.100b):

$$P_aL'_\alpha = P_aL(T_s/T_g)^r, \qquad r = 1.5 \text{ from Table 6.6}$$

$$P_aL'_\alpha = (0.102)(800/1600)^{1.5} = 0.036 \text{ atm m}$$

Equation (6.100a) for α_g requires that we next calculate

$$\varepsilon_g(T_s, P_aL'_\alpha, P') = \varepsilon_g(T_s, P_aL''_\alpha, P' = 1)$$

$$P_aL''_\alpha = P_aL'_\alpha(P')^{m'}$$

For $T_s = 800$ K, $P_aL'_\alpha = 0.036$ atm m, $P' = 9.1$ atm, Fig. 6.36b gives $m' = 0.41$.

$$P_aL''_\alpha = (0.036)(9.1)^{0.41} = 0.0890 \text{ atm m}$$

Figure 6.36a gives $\varepsilon_g(800, 0.0890) = 0.24$. Then Eq. (6.100$a$) gives

$$\alpha_g = (T_g/T_s)^{1/2}(0.240) = (1600/800)^{1/2}(0.24) = 0.34$$

Comments

1. This result would be unaffected if the relative amount of CO_2 were somewhat different. An equimolar mixture of CO_2 and H_2O in the products requires that the overall chemical formula of the hydrocarbon fuel be C_nH_{2n}, which is not strictly true for most hydrocarbon fuels.
2. RAD3 gives $\varepsilon_g = 0.194$, $\alpha_g = 0.353$.

6.7.3 Effective Beam Lengths for an Isothermal Gas

We will develop the concept of an effective beam length for radiation in a participating medium on a total basis; however, identical results can be obtained on a band or spectral basis. Consider two surface elements dA_1 and dA_2 of an enclosure containing an isothermal gas at temperature T_g, as shown in Fig. 6.37. The contribution to irradiation on surface dA_1 due to radiation received in the cone of solid angle $d\omega$ subtended by dA_2 is

$$dG_1 = I_1^- \cos\theta_1 \, d\omega \qquad \textbf{(6.101)}$$

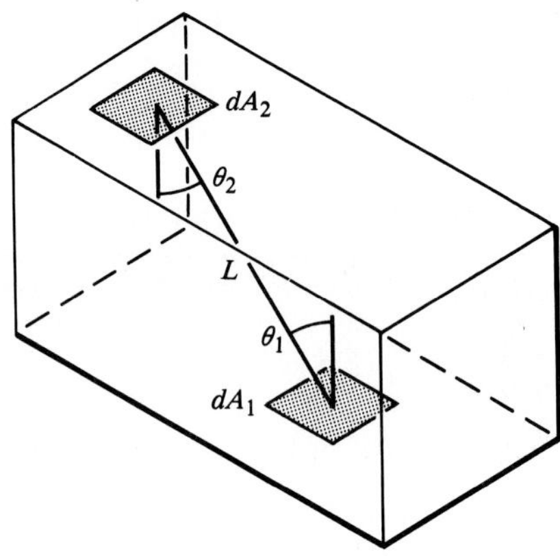

Figure 6.37 Elemental surfaces for radiation in an enclosure containing an isothermal gas.

Equation (6.95) gives

$$I_1^- = I_2^+ e^{-\kappa L} + I_{bg}\left(1 - e^{-\kappa L}\right)$$

Thus,

$$dG_1 = \left[I_2^+ e^{-\kappa L} + I_{bg}\left(1 - e^{-\kappa L}\right)\right] \cos\theta_1 \, d\omega \tag{6.102}$$

For a finite surface A_2, Eq. (6.102) is integrated over the solid angle containing surface A_2:

$$dG_1 = \int_{\omega_2} \left[I_2^+ e^{-\kappa L} + I_{bg}\left(1 - e^{-\kappa L}\right)\right] \cos\theta_1 \, d\omega \tag{6.103}$$

Introducing $d\omega = dA_2 \cos\theta_2 / L^2$, $I_2^+ = J_2/\pi$, $I_{bg} = E_{bg}/\pi$ gives

$$dG_1 = \int_{A_2} \left[J_2 e^{-\kappa L} + E_{bg}\left(1 - e^{-\kappa L}\right)\right] \frac{\cos\theta_1 \cos\theta_2}{\pi L^2} dA_2 \tag{6.104}$$

For a finite area A_1, Eq. (6.104) is averaged over area A_1:

$$G_{12} = \frac{1}{A_1} \int_{A_1} \int_{A_2} \left[J_2 e^{-\kappa L} + E_{bg}\left(1 - e^{-\kappa L}\right)\right] \frac{\cos\theta_1 \cos\theta_2}{\pi L^2} dA_2 \, dA_1 \tag{6.105}$$

where G_{12} is the contribution to irradiation of surface 1 by radiation along beams from surface 2. The beam length L varies over the surface: we define a *constant effective beam length* $\mathcal{L}_{12}$ such that

$$G_{12} = \left[J_2 e^{-\kappa \mathcal{L}_{12}} + E_{bg}\left(1 - e^{-\kappa \mathcal{L}_{12}}\right)\right] \frac{1}{A_1} \int_{A_1} \int_{A_2} \frac{\cos\theta_1 \cos\theta_2}{\pi L^2} dA_2 \, dA_1$$

Using Eq. (6.65), which defines the shape factor F_{12},

$$G_{12} = \left[J_2 e^{-\kappa \mathcal{L}_{12}} + E_{bg}\left(1 - e^{-\kappa \mathcal{L}_{12}}\right)\right] F_{12} \tag{6.106}$$

and comparing Eqs. (6.105) and (6.106) shows that the effective beam length $\mathcal{L}_{12}$

is given by

$$e^{-\kappa\mathscr{L}_{12}}F_{12} = \frac{1}{A_1}\int_{A_1}\int_{A_2} e^{-\kappa L}\frac{\cos\theta_1\cos\theta_2}{\pi L^2}\,dA_2\,dA_1 \tag{6.107}$$

Sometimes is it possible to model a furnace or combustion chamber as a single-surface enclosure, that is, with a uniform wall temperature and emittance: then, with $A_1 = A_s$, $F_{12} = 1$, Eq. (6.106) becomes

$$G_s = \left[J_s e^{-\kappa\mathscr{L}_m} + E_{bg}\left(1 - e^{-\kappa\mathscr{L}_m}\right)\right] \tag{6.108}$$

where $\mathscr{L}_m$ is called the **mean beam length** of the enclosure and is given by

$$e^{-\kappa\mathscr{L}_m} = \frac{1}{A_s}\int_{A_s}\int_{A_s'} e^{-\kappa L}\frac{\cos\theta\cos\theta'}{\pi L^2}\,dA_s'\,dA_s \tag{6.109}$$

The mean beam length is simply the effective beam length for a complete enclosure.

The introduction of an effective or mean beam length will be useful only if simple rules for their evaluation can be formulated. In general, $\mathscr{L}$ depends on the *optical depth* of the gas, which is the absorption coefficient times a characteristic path length. In the limit $\kappa \to 0$, $\mathscr{L}$ is independent of κ, as can be seen by making the approximation $e^{-x} \simeq 1 - x$ in either Eq. (6.107) or Eq. (6.109), to give *geometric* effective or mean beam lengths $\mathscr{L}^0$,

$$\mathscr{L}_{12}^0 = \lim_{\kappa\to 0}\mathscr{L}_{12} = \frac{1}{A_1F_{12}}\int_{A_1}\int_{A_2}\frac{\cos\theta_1\ \cos\theta_2}{\pi L}\,dA_2\,dA_1 \tag{6.110a}$$

$$\mathscr{L}_m^0 = \lim_{\kappa\to 0}\mathscr{L}_m = \frac{1}{A_s}\int_{A_s}\int_{A_s'}\frac{\cos\theta\ \cos\theta'}{\pi L}\,dA_s'\,dA_s \tag{6.110b}$$

Equation (6.110*b*) can be integrated to give the very simple result,

$$\mathscr{L}_m^0 = \frac{4V_g}{A_s} \tag{6.111}$$

where V_g is the volume of the gas in the enclosure. For a long duct of cross-sectional area A_c and perimeter $\mathscr{P}$, $\mathscr{L}_m^0 = 4A_c/\mathscr{P} = D_h$, the hydraulic diameter. The geometric mean beam length proves to be a good approximation for the actual mean beam length. For a sphere, the error is at worst 5.2% (high); for an infinite cylinder or slab, use of $\mathscr{L}_m = 0.9\mathscr{L}_m^0$ for $\kappa\mathscr{L}_m^0 > 0.1$ gives an error of less than 7%. Table 6.7 presents formulas for the geometric effective beam length $\mathscr{L}_{12}^0$ for opposite and adjacent rectangles. As for the mean beam length, $\mathscr{L}_{12}$ is not much less than $\mathscr{L}_{12}^0$. The rules of shape factor algebra illustrated in Fig. 6.9 also apply to geometric effective beam lengths, as can be seen by rewriting Eq. (6.110*a*) as

$$\mathscr{L}_{ij}^0 A_i F_{ij} = \int_{A_i}\int_{A_j}\frac{\cos\theta_i\cos\theta_j}{\pi L}\,dA_j\,dA_i = \mathscr{L}_{ji}^0 A_j F_{ji}$$

Thus, to use Figure 6.9, set $\mathscr{G}_{ij}$ equal to $\mathscr{L}_{ij}^0 A_i F_{ij}$. The rules $\mathscr{G}_{ij} = \mathscr{G}_{ji}$ and $\mathscr{G}_{i(j+k)} = \mathscr{G}_{ij} + \mathscr{G}_{ik}$ follow.

Table 6.7 Geometric effective beam lengths $\mathscr{L}_{12}^0$ for opposite and adjacent rectangles [2,14]. (See Table 6.1 for F_{12}.)

1. Opposite rectangles

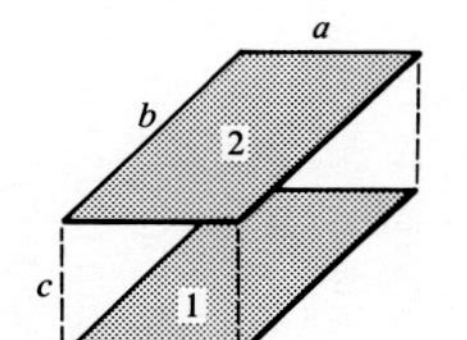

$X = a/c; Y = b/c$

$$\mathscr{L}_{12}^0 F_{12} = \frac{4c}{\pi XY}\left\{XY \tan^{-1}\frac{XY}{(1+X^2+Y^2)^{1/2}} + X\ln\frac{X+(1+X^2+Y^2)^{1/2}}{[X+(1+X^2)^{1/2}](1+Y^2)^{1/2}}\right.$$

$$\left. + Y\ln\frac{Y+(1+X^2+Y^2)^{1/2}}{[Y+(1+Y^2)^{1/2}](1+X^2)^{1/2}} + (1+X^2)^{1/2} + (1+Y^2)^{1/2} - (1+X^2+Y^2)^{1/2} - 1\right\}$$

2. Adjacent rectangles

$X = a/c; Y = b/c$

$$\mathscr{L}_{12}^0 F_{12} = \frac{c}{3\pi X}\left\{3X^2\ln\frac{[1+(1+X^2)^{1/2}](X^2+Y^2)^{1/2}}{X[1+(1+X^2+Y^2)^{1/2}]} + 3Y^2\ln\frac{[1+(1+Y^2)^{1/2}](X^2+Y^2)^{1/2}}{Y[1+(1+X^2+Y^2)^{1/2}]}\right.$$

$$+3X^2[(1+X^2+Y^2)^{1/2} - (X^2+Y^2)^{1/2} - (1+X^2)^{1/2}]$$

$$+3Y^2[(1+X^2+Y^2)^{1/2} - (X^2+Y^2)^{1/2} - (1+Y^2)^{1/2}]$$

$$\left. +(1+X^2)^{3/2} + (1+Y^2)^{3/2} + (X^2+Y^2)^{3/2} - (1+X^2+Y^2)^{3/2} + 2X^3 + 2Y^3 - 1\right\}$$

EXAMPLE 6.23 Mean Beam Length in a Tube Bank

A tube bank has 4 cm–O.D. tubes with centers forming equilateral triangles of side length 7 cm. Determine the mean beam length for combustion products of coke at 1200 K and 6 atm total pressure.

Solution

Given: Tube bank with tubes in a triangular pattern.

Required: Mean beam length for coke combustion products.

Assumptions: Stoichiometric combustion.

The first step is to calculate the geometric mean beam length from Eq. (6.111):

$$\mathscr{L}_m^0 = \frac{4V_g}{A_s}$$

$$V_g = \left[\frac{1}{2}(7)(7^2 - 3.5^2)^{1/2} - \frac{1}{2}(\pi/4)(4)^2\right](1)$$

$$= 21.22 - 6.28 = 14.94 \text{ cm}^3$$

$$A_s = (3)(1/6)(\pi)(4)(1) = 6.28 \text{ cm}^2$$

$$\mathscr{L}_m^0 = \frac{4(14.94)}{6.28} = 9.51 \text{ cm}$$

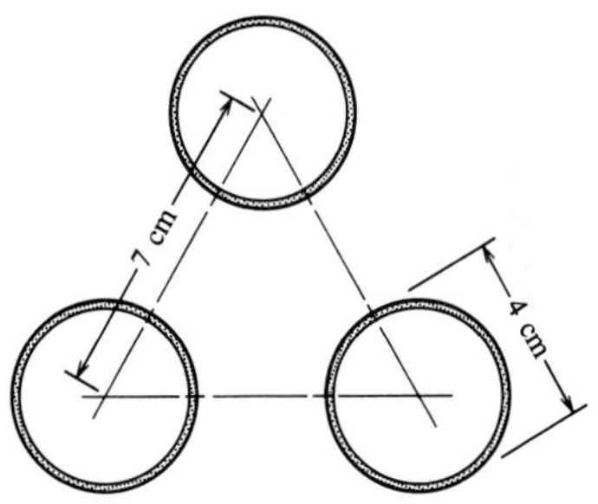

Next we must see if the product of absorption coefficient times path length is greater than or less than 0.1.

$$\varepsilon_g = 1 - e^{-\kappa L}, \qquad \text{or} \qquad \kappa L = \ln\left(\frac{1}{1-\varepsilon_g}\right)$$

where $L = \mathscr{L}_m^0$ for this purpose. The combustion reaction is

$$C + O_2 + (3.76N_2) \rightarrow CO_2 + (3.76N_2)$$

Assuming complete combustion, the partial pressure of CO_2 is

$$P_{CO_2} = (1/4.76)(6) = 1.26 \text{ atm} = P_a$$

RAD3 can be used to calculate ε_g. The required input is:

2: CO_2 in N_2

$P = 6$

$P_a = 1.62$

$T_g = 1200$

$T_s =$ any value

$\mathscr{L}_m^0 = 0.0951$

The output is:

$$\varepsilon_g = 0.121$$

$$\kappa L = \ln\left(\frac{1}{1-\varepsilon_g}\right) = 0.13 > 0.1$$

Hence, $\mathscr{L}_m \simeq 0.9\mathscr{L}_m^0 = (0.9)(0.0951) = 0.086$ m.

6.7.4 Radiation Exchange between an Isothermal Gas and a Black Enclosure

In a furnace or combustion chamber, turbulence ensures good mixing, and temperature variations are confined to thin boundary layers adjacent to the walls. Thus, as a first approximation, it is reasonable to assume that the gas as a whole is isothermal. Also, due to oxidation and soot deposits, the walls can be approximated as black surfaces when hydrocarbon fuels are used. Consider the enclosure of n black surfaces containing an isothermal gas at temperature T_g shown in Fig. 6.38. The radiant heat flux across the ith surface is

$$q_i = E_{bi} - \sum_{k=1}^{n} G_{ik} \tag{6.112}$$

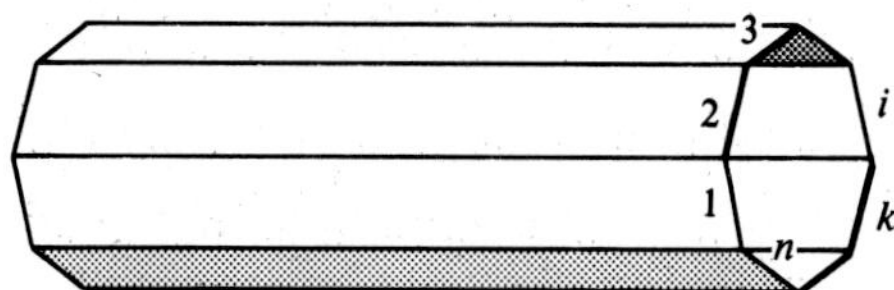

Figure 6.38 An enclosure of n black surfaces containing an isothermal gas.

Equation (6.106) can be written for the (i,k) black-surface pair as

$$G_{ik} = F_{ik}(\tau_g^{ik}E_{bk} + \varepsilon_g^{ik}E_{bg})$$

where $\tau_g^{ik} = e^{-\kappa\mathscr{L}_{ik}}$, $\varepsilon_g^{ik} = 1 - e^{-\kappa\mathscr{L}_{ik}}$. Substituting in Eq. (6.112),

$$q_i = E_{bi} - \sum_{k=1}^{n} F_{ik}(\tau_g^{ik}E_{bk} + \varepsilon_g^{ik}E_{bg}) \tag{6.113}$$

which are n linear equations in the n unknown heat fluxes or wall temperatures. The total gas emissivity ε_g^{ik} is for gas at temperature T_g over an effective beam length of $\mathscr{L}_{ik}$; the total transmissivity τ_g^{ik} is for radiation from a black source at temperature T_k transmitted by a gas at temperature T_g over an effective beam length of $\mathscr{L}_{ik}$.

For an enclosure consisting of a single surface at temperature T_s, Eq. (6.113) reduces to

$$q = E_{bs} - \tau_g E_{bs} - \varepsilon_g E_{bg}$$

or

$$q = \alpha_g \sigma T_s^4 - \varepsilon_g \sigma T_g^4 \tag{6.114}$$

where ε_g is the total gas emissivity at temperature T_g over the mean beam length of the enclosure, and α_g is the total gas absorptivity for radiation from a black source at temperature T_s absorbed over the mean beam length by a gas at temperature T_g.

EXAMPLE 6.24 A Kerosene Combustor

Exhaust gas from a kerosene-fueled combustor is at 1600 K and 3 atm pressure, with a composition of 10% CO_2, 10% H_2O, and 80% N_2 by volume. If the mean beam length is 0.34 m, calculate the radiant heat flux to walls at 800 K. Assume that the walls are black.

Solution

Given: A combustor burning a hydrocarbon fuel.

Required: Radiant heat flux to wall.

Assumptions: The walls are black.

The total emissivity and absorptivity for this combustor were calculated in Example 6.22; RAD3 gave

$$\varepsilon_g = 0.194$$

$$\alpha_g = 0.353$$

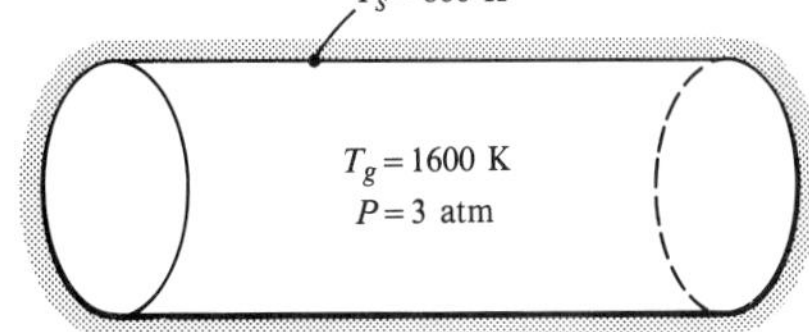

Equation (6.114) gives the radiant heat flux:

$$\begin{aligned} q &= \alpha_g \sigma T_s^4 - \varepsilon_g \sigma T_g^4 \\ &= (0.353)(5.67 \times 10^{-8})(800)^4 - (0.194)(5.67 \times 10^{-8})(1600)^4 \\ &= (8.2 - 72.1)10^3 \\ &= -63.9 \text{ kW/m}^2 \end{aligned}$$

Comments

Since the walls are relatively cold, radiation emitted by the walls and absorbed by the gas is of minor importance; hence, an accurate value of α_g is not required.

6.7.5 Radiation Exchange between an Isothermal Gray Gas and a Gray Enclosure

The calculation method of Section 6.7.4 applies to a nongray gas but is restricted to an enclosure of black surfaces to allow evaluation of the transmissivity τ_g^{ik}. When surface j is black, the spectral distribution of the emitted radiation is fixed by its temperature, and the transmissivity can be calculated using the rules given in Section 6.7.2. When surface j is gray, its radiosity has components of reflected radiation from other surfaces at different temperatures, and from the gas. Thus, the spectral distribution is complicated and unknown. In order to calculate the radiation exchange in an enclosure of gray surfaces containing an isothermal gas on a total basis, it is necessary to assume that the gas is gray as well, that is, $\varepsilon_g^{ik} = \alpha_g^{ik} = 1 - \tau_g^{ik}$. The heat flux across the ith surface is now

$$q_i = J_i - \sum_{k=1}^{n} G_{ik} \tag{6.115}$$

where

$$J_i = \varepsilon_i E_{bi} + (1 - \varepsilon_i) G_i \tag{6.116}$$

and

$$G_{ik} = F_{ik}(\tau_g^{ik} J_k + \varepsilon_g^{ik} E_{bg}) \tag{6.117}$$

Rearranging these equations into the form used for a gray enclosure in Section 6.3.5 gives

$$J_i = \varepsilon_i E_{bi} + (1 - \varepsilon_i) \sum_{k=1}^{n} F_{ik}[\tau_g^{ik} J_k + \varepsilon_g^{ik} E_{bg}] \tag{6.118}$$

$$\dot{Q}_i = \frac{\varepsilon_i A_i}{1 - \varepsilon_i}(E_{bi} - J_i) \tag{6.119}$$

Equations (6.118) and (6.119) can be solved for the unknown $\dot{Q}_i$ or T_i, if T_g and either $\dot{Q}_i$ or T_i are specified for each surface.

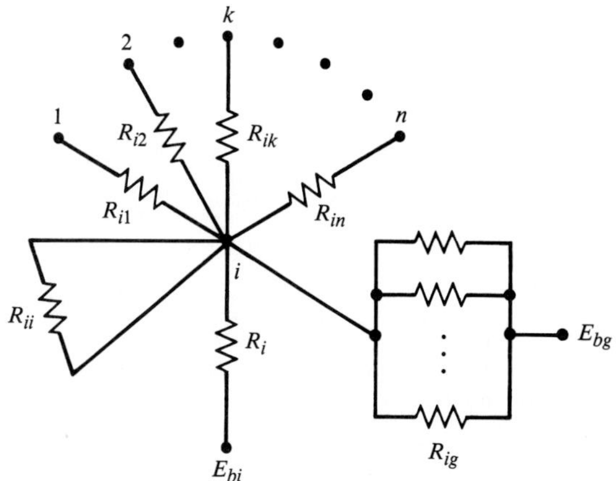

Figure 6.39 An isothermal gray gas in a gray enclosure: network connections for node i.

An equivalent electrical network can be constructed for the gray enclosure containing a gray gas. The surface resistances from Eq. (6.119) are as before,

$$R_i = \frac{1 - \varepsilon_i}{\varepsilon_i A_i} \tag{6.120}$$

Equations (6.115) and (6.117) can be arranged as

$$\dot{Q}_i = \sum_{k=1}^{n} A_i F_{ik} \tau_g^{ik} (J_i - J_k) + \sum_{k=1}^{n} A_i F_{ik} \varepsilon_g^{ik} (J_i - E_{bg}) \tag{6.121}$$

since for a gray gas $1 - \tau_g^{ik} = \alpha_g^{ik} = \varepsilon_g^{ik}$. The space resistances are thus

$$R_{ik} = \frac{1}{A_i F_{ik} \tau_g^{ik}} \tag{6.122}$$

and there is a parallel set of radiosity node to gas resistances:

$$R_{ig} = \frac{1}{\sum_{k=1}^{n} A_i F_{ik} \varepsilon_g^{ik}} \tag{6.123}$$

It is essential that R_{ii} resistances be included when F_{ii} is not zero, that is, when a surface sees itself. Figure 6.39 shows the network connections for node i, while Fig. 6.40 shows a complete network for a two-surface enclosure.

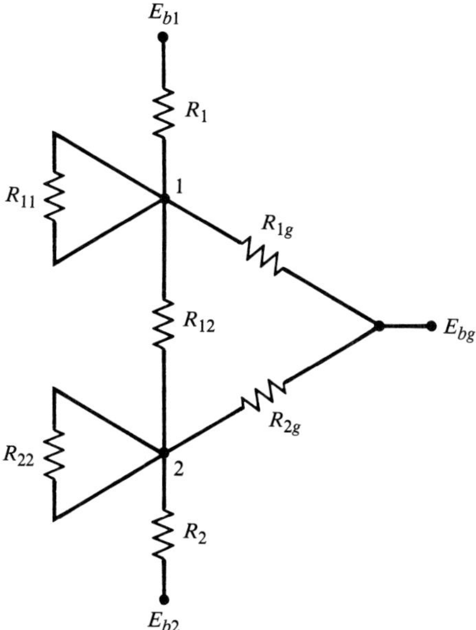

Figure 6.40 An isothermal gray gas in a two-gray-surface enclosure: the equivalent network.

EXAMPLE 6.25 Radiation Exchange in a Two-Surface Enclosure Containing a Gray Gas

An absorbing gas is contained between two large parallel plates, 1 and 2, with $T_1 = 1200$ K, $\varepsilon_1 = 0.8$; $T_2 = 800$ K, $\varepsilon_2 = 0.7$. The gas is assumed to be gray with $\varepsilon_g = 0.4$. Determine the effect of the gas on the radiation heat transfer between the two plates.

Solution

Given: Radiating gas contained between parallel plates.

Required: Effect of gas on radiation heat transfer.

Assumptions: 1. Gray gas, $\varepsilon_g^{12} = 1 - \tau_g^{12}$
2. Large plates, $F_{12} = 1$
3. Convection has a minor effect on the bulk gas temperature.

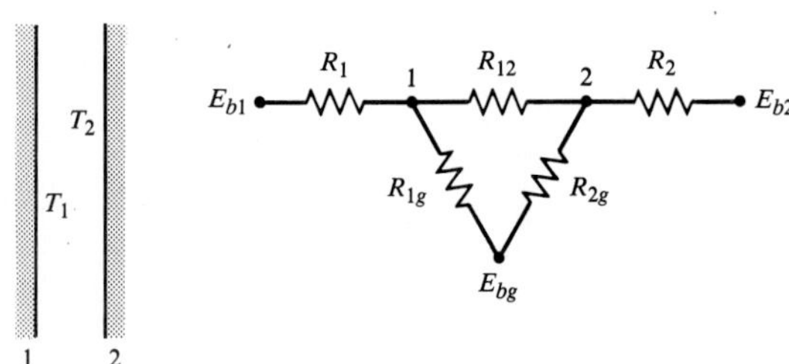

Referring to the equivalent network, the various resistances on a unit-area basis with $F_{12} = 1$ are

$$R_1 = \frac{1-\varepsilon_1}{\varepsilon_1} = \frac{1-0.8}{0.8} = 0.25; \qquad R_2 = \frac{1-0.7}{0.7} = 0.429$$

$$R_{12} = \frac{1}{1-\varepsilon_g^{12}} = \frac{1}{1-0.4} = 1.667$$

$$R_{1g} = \frac{1}{\varepsilon_g^{12}} = \frac{1}{0.4} = 2.5; \qquad R_{2g} = \frac{1}{\varepsilon_g^{12}} = \frac{1}{0.4} = 2.5$$

Also,

$$E_{b1} = \sigma T_1^4 = (5.67 \times 10^{-8})(1200)^4 = 117.6 \text{ kW/m}^2$$

$$E_{b2} = \sigma T_2^4 = (5.67 \times 10^{-8})(800)^4 = 23.2 \text{ kW/m}^2$$

Combining the space and gas resistances gives

$$R = \frac{1}{1/1.667 + 1/(2.5 + 2.5)} = 1.25$$

The heat transfer per unit area of plate is then

$$\frac{\dot{Q}_{12}}{A} = \frac{E_{b1} - E_{b2}}{\sum R} = \frac{117.6 - 23.2}{0.25 + 1.25 + 0.429} = 48.9 \text{ kW/m}^2$$

If there is no gas present, the heat transfer is given by Eq. (6.30):

$$\frac{\dot{Q}_{12}}{A} = \frac{E_{b1} - E_{b2}}{1/\varepsilon_1 + 1/\varepsilon_2 - 1} = \frac{117.6 - 23.2}{1/0.8 + 1/0.7 - 1} = 56.2 \text{ kW/m}^2$$

(or set $\varepsilon_g^{12} = 0$ above to obtain the same result).

Comments

1. The effect of the gas on the heat flux is not large, even though its emissivity is relatively large.
2. The gas acts somewhat like a radiation shield.

6.7.6 Radiation Exchange between an Isothermal Nongray Gas and a Single-Gray-Surface Enclosure

Equation (6.114) for radiation exchange between a *nongray* gas and a single-surface *black* enclosure is

$$\frac{\dot{Q}}{A} = \alpha_g \sigma T_s^4 - \varepsilon_g \sigma T_g^4 \tag{6.124}$$

where α_g is the gas absorptivity for radiation from a black source at temperature T_s absorbed by gas at temperature T_g. For radiation exchange between a single-surface *gray* enclosure and a *gray* gas, Eqs. (6.118) and (6.119) become, with $F_{11} = 1$,

$$J_1 = \varepsilon_1 E_{b1} + (1 - \varepsilon_1)(\tau_g^{11} J_1 + \varepsilon_g^{11} E_{bg}) \tag{6.125}$$

$$\frac{\dot{Q}_1}{A_1} = \frac{\varepsilon_1}{1 - \varepsilon_1}(E_{b1} - J_1) \tag{6.126}$$

Solving for J_1 from Eq. (6.125) and substituting in Eq. (6.126) gives

$$\frac{\dot{Q}_1}{A_1} = \frac{\varepsilon_1 \varepsilon_g^{11}}{1 - (1 - \varepsilon_1)\tau_g^{11}}(E_{b1} - E_{bg}) \tag{6.127}$$

Simplifying the notation to be consistent with Eq. (6.124) gives

$$\frac{\dot{Q}}{A} = \frac{\varepsilon_s \varepsilon_g \sigma T_s^4}{1 - (1 - \varepsilon_s)\tau_g} - \frac{\varepsilon_s \varepsilon_g \sigma T_g^4}{1 - (1 - \varepsilon_s)\tau_g} \tag{6.128}$$

where for a gray gas $1 - \tau_g = \varepsilon_g$. The dilemma to be resolved is that Eq. (6.124) applies only to black walls, that is, it does not allow for reflection of radiation by the walls, whereas Eq. (6.128) assumes a gray gas. Equation (6.128) gives unsatisfactory results when the wall temperature is close to, or greater than, the gas temperature. Fortunately, it is possible to obtain an exact solution for the problem of exchange between a nongray gas and a single-gray-surface enclosure [12]. A convenient and adequate approximation to this exact solution is

$$\frac{\dot{Q}}{A} = \frac{\varepsilon_s \alpha_{g1} \sigma T_s^4}{1 - (1 - \varepsilon_s)[(\alpha_{g2} - \alpha_{g1})/\alpha_{g1}]} - \frac{\varepsilon_s \varepsilon_{g1} \sigma T_g^4}{1 - (1 - \varepsilon_s)[(\varepsilon_{g2} - \varepsilon_{g1})/\varepsilon_{g1}]} \tag{6.129}$$

where α_{g1} is the gas absorptivity for the mean beam length of the enclosure, and α_{g2} is for two mean beam lengths (that is, including the effect of one reflection). The emissivities ε_{g1} and ε_{g2} are defined similarly. The absorptivities are evaluated for absorption of radiation from a black surface at temperature T_s. The exact solution accounts for an infinite number of reflections; the approximation of Eq. (6.129) retains the effect of the first reflection only. The result proves satisfactory for most engineering applications.

EXAMPLE 6.26 A Scramjet Combustor for a Hypersonic Aircraft

A Mach 10 hypersonic drone for testing a hydrogen-fueled scramjet engine has a combustor with a 0.40 m square cross section. The walls have an emittance of 0.2 and are cooled to a temperature of 800 K. At a location where the pressure is 8 atm and the combustion gases are at 2000 K, estimate the radiative heat transfer to the walls. Take the excess air ratio as 2.

Solution

Given: Hydrogen-fueled scramjet combustor.

Required: Radiative heat transfer to walls.

Assumptions:
1. The combustor is long enough to ignore end effects.
2. The one-reflection approximation, Eq. (6.129), is adequate.
3. The presence of O_2 in the combustion products has a negligible effect on gas radiation properties calculated from Fig. 6.35 for a H_2O–N_2 mixture.

The first task is to determine the mean beam length: For a long combustor, we assume a 2-D geometry.

$$\mathcal{L}_m^0 = \frac{4V_g}{A_s} = \frac{(4)(0.40)^2}{(4)(0.40)} = 0.40 \text{ m}$$

$$\mathcal{L}_m \simeq 0.9\mathcal{L}_m^0 = (0.9)(0.4) = 0.36 \text{ m}$$

Next the partial pressure of the water vapor must be calculated. The combustion reaction is

$$1 \text{ kmol } O_2 + 2 \text{ kmol } H_2 \rightarrow 2 \text{ kmol } H_2O$$

and, with 100% excess air, there are 1 kmol O_2 and 2×3.76 kmol N_2 in the products. Hence, for a total pressure of 8 atm, the partial pressure of H_2O is

$$P_{H_2O} = 8\left(\frac{2}{2 + 1 + (2)(3.76)}\right) = 1.52 \text{ atm} = P_a$$

$$P_aL = (1.52)(0.36) = 0.547 \text{ atm m}$$

The equivalent broadening pressure ratio is calculated from Eq. (6.98):

$$P_E = \left\{\frac{P}{P_0}\left[1 + (b-1)\frac{P_a}{P}\right]\right\}^n$$

$$P_0 = 1 \text{ atm}$$

$$b = 8.6(T_0/T_g)^{1/2} + 0.5 = 8.6(100/2000)^{1/2} + 0.5 = 2.42$$

$$n = 1$$

$$P_E = \left\{\frac{8}{1}\left[1 + (2.42 - 1)\frac{1.52}{8}\right]\right\}^1 = 10.16$$

Now P_aL' can be calculated from Eq. (6.99*b*):

$$P_aL' = P_aLP_E^m, \qquad m = 0.19 \text{ from Fig. 6.35}b$$

$$P_aL' = (0.547)(10.16)^{0.19} = 0.850 \text{ atm m}$$

Then, from Eq. (6.99a) with Fig. 6.35a,

$$\varepsilon_{g1} = \varepsilon_{g1}(T_g, P_aL') = \varepsilon_{g1}(2000, 0.850) = 0.24$$

Also, $2P_aL = (2)(0.547) = 1.09$ atm m.

$$2P_aL' = (2P_aL)P_E^m, \qquad m = 0.20 \text{ from Fig. 6.35}b$$

$$2P_aL' = (1.09)(10.16)^{0.20} = 1.73 \text{ atm m}$$

$$\varepsilon_{g2} = \varepsilon_{g2}(T_g, 2P_aL') = \varepsilon_{g2}(2000, 1.73) = 0.31$$

To determine the total absorptivities, we first calculate P_E' from Eq. (6.100c):

$$P_E' = P_E(T_g/T_s)^s; \qquad s = 1.6 \text{ from Table 6.6}$$

$$P_E' = 10.16(2000/800)^{1.6} = 44.0$$

We then calculate P_aL_α' from Eq. (6.100b):

$$P_aL_\alpha' = P_aL(T_s/T_g)^r; \qquad r = 1.5 \text{ from Table 6.6}$$

$$P_aL_\alpha' = 0.547(800/2000)^{1.5} = 0.138 \text{ atm m}$$

Equation (6.100a) for the absorptivity requires that we next find

$$\varepsilon_g(T_s, P_aL_\alpha', P_E') = \varepsilon_g(T_s, P_aL_\alpha'', P_E = 1)$$

$$P_aL_\alpha'' = P_aL_\alpha'(P_E')^{m'}$$

$P_E' = 44.0$ is out of the range of Fig. 6.35b. The highest value is $P_E = 10$; clearly, extrapolation is not straightforward. We will use the value of m for $P_E = 10$ and discuss the consequences. For $T_s = 800$ K, $P_aL_\alpha' = 0.138$ atm m, Fig. 6.35b gives $m' = 0.7$; thus $P_aL_\alpha'' = (0.138)(44.0)^{0.7} = 1.95$ atm m. Figure 6.35a gives $\varepsilon_{g1}(800{,}1.95) = 0.40$, and Eq. (6.100$a$) gives

$$\alpha_{g1} = (T_g/T_s)^{1/2}\varepsilon_{g1} = (2000/800)^{1/2}(0.40) = 0.63$$

Also, $2P_aL_\alpha' = (2)(0.138) = 0.276$ atm m, and Fig. 6.35b gives $m' = 0.7$; thus, $2P_aL_\alpha'' = 3.90$ atm m. Figure 6.35a gives ε_{g2} $(800{,}3.90) = 0.50$, and Eq. (6.100a) gives

$$\alpha_{g2} = (T_g/T_s)^{1/2}\varepsilon_{g2} = (2000/800)^{1/2}(0.50) = 0.79$$

With the required total properties evaluated, the wall heat flux can now be obtained from Eq. (6.129):

$$\frac{\dot{Q}}{A} = \frac{\varepsilon_s\alpha_{g1}\sigma T_s^4}{1-(1-\varepsilon_s)[(\alpha_{g2}-\alpha_{g1})/\alpha_{g1}]} - \frac{\varepsilon_s\varepsilon_{g1}\sigma T_g^4}{1-(1-\varepsilon_s)[(\varepsilon_{g2}-\varepsilon_{g1})/\varepsilon_{g1}]}$$

$$= \frac{(0.2)(0.63)(5.67\times10^{-8})(800)^4}{1-(1-0.2)[(0.79-0.63)/0.63]} - \frac{(0.2)(0.24)(5.67\times10^{-8})(2000)^4}{1-(1-0.2)[(0.31-0.24)/0.24]}$$

$$= (3.67 - 56.8)10^3 = -53.1 \text{ kW/m}^2$$

Comments

1. The convective heat transfer for this scramjet combustor has been estimated to be 300 kW/m^2; thus, the radiation contribution is about 15%. The importance of having a low wall emittance (absorptance) to reduce the radiation flux is clearly evident.

2. $\kappa L = \ln [1/(l - \varepsilon_g)] \simeq \ln[1/(1 - 0.24)] = 0.27 > 0.1$; hence, the assumption $\mathcal{L}_m \simeq 0.9\mathcal{L}_m^0$ is justified.

3. The low wall temperature (800 K) and small emittance (0.2) cause the absorbed radiation to play a minor role. Hence, the error introduced by not extrapolating Fig. 6.36*b* past $P_E = 10$ is small.

4. Check ε_{g1}, ε_{g2}, α_{g1}, and α_{g2} using RAD3. Note that RAD3 also sets $P_E = 10$ if $P_E \geq 10$ in order to calculate the exponent m.

6.8 CLOSURE

The physics of radiation is quite different from that of conduction and convection. Many new concepts were introduced in Chapter 6. Analysis of radiation transfer involved mainly algebra, rather than the solution of differential equations.

Sections 6.2 through 6.4 were concerned with the engineering calculation of radiation energy exchange between surfaces separated by a nonparticipating medium. Real engineering surfaces were modeled as diffuse gray surfaces. An exception was solar radiation, where allowance was made for a different value of absorptance to short-wavelength solar radiation. Engineering problem solving was aided by two computer programs. RAD1 calculates shape factors for often-used configurations. RAD2 solves for the radiation energy exchange in enclosures of diffuse gray surfaces; in practice, the number of surfaces used is limited by the work required to assemble the matrix of shape factors.

Sections 6.5 through 6.7 dealt with more advanced topics. Section 6.5 examined directional characteristics of surface radiation. The concept of radiation intensity, which was earlier avoided, was defined and used to rigorously derive shape factors. Effects of specular reflection were examined briefly, and, in particular, radiation transmission through cracks with either diffuse or specular reflecting walls was compared. Spectral characteristics of surface radiation were examined in Section 6.6, with the limited objective of demonstrating how total radiation properties may be obtained from spectral values. Total hemispherical radiation properties are required when the diffuse gray surface model is used, but for a particular surface, or surface coating, only spectral data may be available to the engineer. Chapter 6 closes with an introduction to radiation transfer through gases in Section 6.7. Both H_2O and CO_2 absorb and emit radiation strongly and are found in combustion products. A procedure and data for calculating total radiation properties for these species were given. Since the required calculations are complicated and lengthy, the computer program RAD3, which performs this task, proved to be a useful tool. Analytical methods were developed for some simple radiation exchange problems in enclosures containing combustion gases. In particular, the approximate formula given for a single-gray-surface enclosure containing a nongray gas was shown to be useful for making preliminary estimates of radiation exchange in furnaces and combustion chambers.

In keeping with the introductory nature of this text, the material in Chapter 6 has been selected to allow a wide range of engineering problems to be analyzed, without facing the difficult problem of nongray behavior head-on. In principle, it is not difficult to properly account for nongray behavior of both surfaces and participating gases. However, computer programs are required to make the calculations practical, and often spectral data are unavailable for the surfaces involved. Thus, the gray surface model should always be used to make an initial assessment of a problem. If such an assessment shows that an accurate estimate of radiation exchange is essential to develop a successful design, then advanced texts should be consulted. Another problem that was avoided is the participation of aerosols in radiation transfer. Of particular importance are the effects of soot in combustion chambers burning hydrocarbon fuels. This topic has received considerable attention in recent years, and much useful information is available in current journal papers.

REFERENCES

1. Hottel, H. C., and Sarofim, A. F., *Radiative Transfer,* McGraw-Hill, New York (1967).
2. Edwards, D. K., *Radiation Heat Transfer Notes,* Hemisphere, Washington, D.C. (1981).
3. Garg, H. P., *Treatise on Solar Energy, Vol. 1: Fundamentals of Solar Energy,* John Wiley & Sons, New York (1982).
4. Brunt, D., "Radiation in the atmosphere," *Quart. J. R. Meteorol. Soc.,* 66 (Suppl.), 34–40 (1940).
5. Berdahl, P., and Fromberg, R., "The thermal radiance of clear skies," *Solar Energy,* 29, 299–314 (1982).
6. Kreith, F., and Bohn, M. S., *Principles of Heat Transfer,* Chapter 11, Harper & Row, New York (1986).
7. TRASYS User's Manual, Version P22. Manual prepared for NASA Johnson Space Center by Lockheed Eng. Man. Services, April 1988.
8. Schmidt, E., and Eckert, E.R.G., "Über die Richtungsverteilung der Wärmestrahlung von Oberflachen," *Forsch. Gebiete Ingenieurw.,* 6, 175–183 (1935).
9. Hottel, H. C., "Heat transmission by radiation from non-luminous gases," *Trans. AIChE,* 19, 173–205 (1927).
10. Hottel, H. C., "Radiant Heat Transmission," Chapter 3 of *Heat Transmission,* by W. H. McAdams, 3rd ed., McGraw-Hill, New York (1954).

11. Hottel, H. C., and Egbert, R. B., "Radiant heat transmission from water vapor," *Trans. AIChE,* 38, 531–568 (1942).

12. Edwards, D. K., and Matavosian, R., "Scaling rules for total absorptivity and emissivity of gases," *J. Heat Transfer,* 106, 684–689 (1984).

13. Edwards, D. K., and Matavosian, R., "Emissivity data for gases," Section 5.5.5, *Hemisphere Handbook of Heat Exchanger Design,* ed. G. F. Hewitt, Hemisphere, New York (1990).

14. Oppenheim, A. K., and Bevans, J.T., "Geometric factors for radiant heat transfer through an absorbing medium in Cartesian coordinates," *J. Heat Transfer,* 82, 360–368 (1960).

EXERCISES

6–1. Calculate the blackbody emissive power E_b and the peak wavelength λ_{max} for surfaces at 300 K, 1000 K, 2000 K, and 5000 K.

6–2. Derive Wien's displacement law, Eq. (6.6), from Planck's law, Eq. (6.5).

6–3. Show that the Stefan-Boltzmann law, Eq. (6.7), can be obtained from Planck's law, Eq. (6.5), by integrating over all wavelengths. [*Hint:* Let $x = 1/\lambda T$, and hence $d\lambda = -(1/x^2 T)dx$ to effect the integration.]

6–4. Plot graphs of monochromatic emissive power versus wavelength for emission from black surfaces at

(i) 300 K.
(ii) 5800 K.

Approximately what fraction of the energy is in the visible range in each case?

6–5. A gray opaque surface at 500 K has an emittance of 0.3 and is exposed to a high-temperature heat source such that the irradiation on the surface is 30,000 W/m^2. Calculate the following heat fluxes:

(i) Absorbed flux
(ii) Reflected flux
(iii) Emitted flux
(iv) The radiosity

6–6. Useful approximations to Planck's law are the Wien and Rayleigh-Jeans spectral distributions, which are valid for very small and very large values of λT, respectively. These distributions are

$$E_{b\lambda} \simeq \frac{C_1}{\lambda^5} e^{-C_2/\lambda T} \quad \text{(Wien's law)}$$

$$E_{b\lambda} \simeq \frac{C_1}{C_2} \frac{T}{\lambda^4} \quad \text{(Rayleigh-Jeans law)}$$

Derive these distributions from Planck's law, Eq. (6.5), by in turn letting $C_2/\lambda T \gg 1$ and $C_2/\lambda T \ll 1$. Show all three laws on a graph of $\lambda^5 E_{b\lambda}$ versus λT.

6–7. Referring to the sketch, calculate the shape factors F_{14}, F_{41}, F_{23}, and F_{32}.

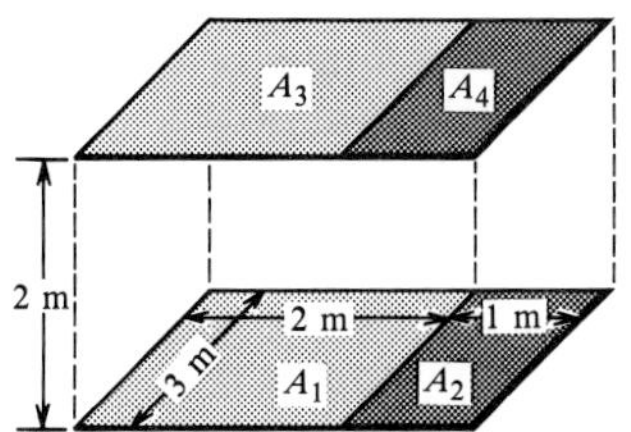

6–8. Referring to the sketch, calculate the shape factors F_{14}, F_{41}, F_{23}, and F_{32}.

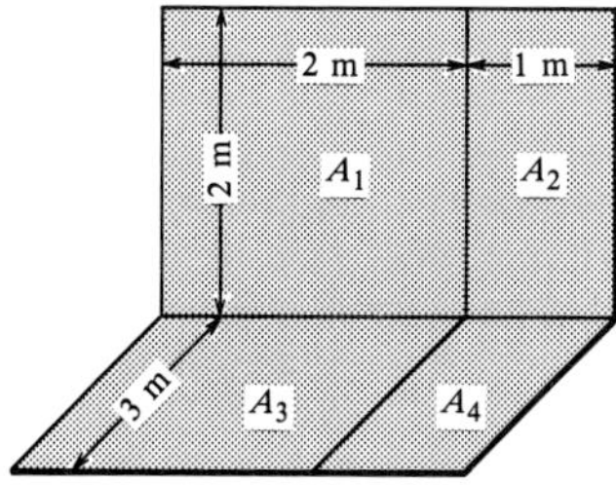

6–9. Prove the string rule for two-dimensional configurations, Eq. (6.25).

6–10. Derive the shape factors given as items 3, 4, and 5 of Table 6.1 using the string rule.

6–11. The string rule is convenient to use for 2-D configurations, for example, long furnaces. But how long must such a furnace be for the 2-D assumption to be accurate? Consider a furnace of width W, height H, and length L. For $W = H$ plot the percent error in the string rule shape factor between the roof and floor, as a function of W/L.

6–12. A cavity is in the form of a vertical cylinder 8 cm in diameter and 16 cm in length and is open at the top to black surroundings at 300 K. The bottom end is heated electrically and is maintained at 1900 K. If the side walls are at a uniform temperature of 1500 K, calculate the power input to the heater and the heat loss to the surroundings through the open end. Take the inner surfaces to be black.

6–13. Two 20 cm–diameter parallel disks are 10 cm apart. The facing sides of the two disks are maintained at 500 K and 400 K, while the back faces are well insulated. The disks are located in an enclosure with walls at 300 K. If all the surfaces can be taken to be black, determine

(i) the net radiation heat exchange between the disks.
(ii) the rate of radiative heat loss from each disk.

6–14. A circular cross-section heating duct of 50 cm diameter runs horizontally through the basement garage of a large apartment complex. The ambient air and walls of the garage are at 17°C. At a location where the surface temperature of the duct is 40°C, determine the radiation contribution to the total heat loss per meter length if

(i) the duct has an emittance of 0.7.
(ii) the duct is painted with aluminum paint with $\varepsilon = 0.2$.

6–15. Liquid oxygen flows through a tube of outside diameter 3 cm, the outer surface of which has an emittance of 0.03 and a temperature of 85 K. This tube is enclosed by a larger concentric tube of inside diameter 5 cm, the inner surface of which has an emittance of 0.05 and a temperature of 290 K. The space between the tubes is evacuated.

(i) Determine the heat gain by the oxygen per unit length of inner tube.
(ii) How much is the heat gain reduced if a thin-wall radiation shield with an emittance of 0.03 on each side is placed midway between the tubes?

6–16. Hot combustion gases flow in a duct whose walls are at 600 K. A thermocouple located in the center of the duct records a temperature of 873 K. If the emittance of the thermocouple is 0.8 and the heat transfer coefficient between the gas and the thermocouple is 110 W/m^2 K, determine the true gas temperature. How much gain in accuracy is obtained if the thermocouple is surrounded by a radiation shield in the form of a cylinder 1 cm in diameter and 4 cm long, made from a thin-wall stainless steel tube with its axis in the flow direction? Take the emittance of the shield as 0.6 and the heat transfer coefficient between the gas and the shield as 40 W/m^2 K.

6–17. A thermocouple is to be utilized to measure the temperature of exhaust gases from an automobile engine, and the effect of cylindrical radiation shields, installed as shown in the sketch, is to be investigated. The inner and outer shield diameters are 8 mm and 12 mm, respectively. The emittance of the thermocouple bead and the shields can be taken to be 0.4. If the exhaust gases are at 1200 K and the exhaust pipe walls are at 600 K, determine the temperature registered by the thermocouple for

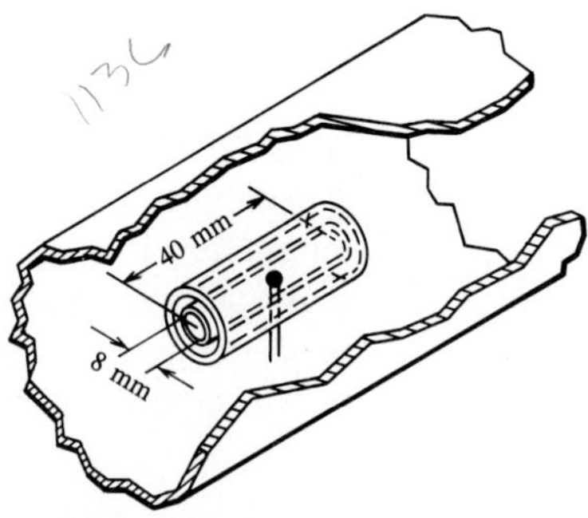

(i) no shield.
(ii) one shield.
(iii) two shields.

Take the convective heat transfer coefficients as 300 and 100 W/m^2 K on the thermocouple and shields, respectively. Assume the pipe walls are black, and make reasonable engineering simplifications.

6–18. A 1 mm–diameter spherical thermocouple bead is to be used to measure the temperature of a low-density air flow. It is located in a 4 cm–diameter tube and is surrounded by a 4 cm–long, 1 cm–diameter thin-walled cylindrical radiation shield lined up with the flow. The emittance of the thermocouple bead is 0.2, of the shield 0.05, and of the tube walls 0.8. The air is at a pressure of 10 torr and has velocity of 100 m/s. When the tube walls are at 300 K and the air temperature is 325 K, estimate the expected error in the thermocouple reading.

6–19. A heating panel is to be placed on the ceiling of a hospital room so that the patient will not be uncomfortable in 290 K air. The panel is painted with an off-white matte paint of emittance 0.88. As a design criterion it is required that the patient's face—modeled as a dry, black adiabatic surface—attains a 305 K equilibrium temperature. The panel temperature is 355 K and is located 2 m above the patient. The convective heat transfer coefficient for the face can be taken as 4.0 W/m^2 K, and it can be assumed that the room walls are black at the air temperature. If the panel is to be a circular disk, determine its diameter.

6–20. A cylindrical cavity, with a diameter of 10 cm and 20 cm deep, is located in an enclosure with black walls at 300 K. Calculate the radiative heat loss from the cavity for the following conditions:

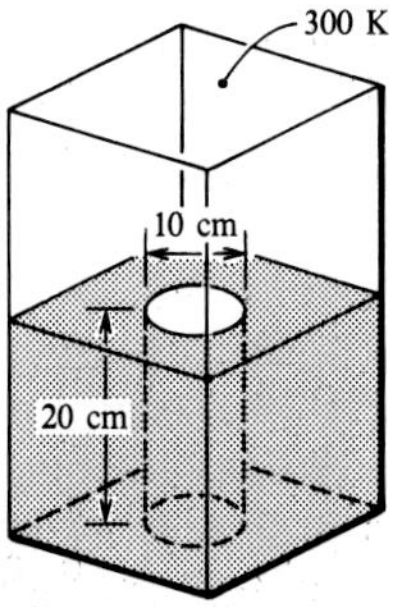

(i) All interior surfaces are black and are maintained at 1000 K.
(ii) All interior surfaces are diffuse-gray with emittance 0.5 and are maintained at 1000 K.
(iii) All interior surfaces are diffuse-gray with emittance 0.5; the base is maintained at 1000 K while the side walls are perfectly insulated.
(iv) As in case (iii), with the side walls having an emittance of 0.05.

300 K
10 cm
20 cm

6–21. A long furnace has a 3 m–square cross section. The roof is maintained at 2000 K by hot combustion gases, while the work floor is at 800 K. The side walls are well insulated with refractory brick. Calculate the radiant heat flux into the floor if the roof and side walls have an emittance of 0.7 and the floor has an emittance of 0.4.

6–22. A room is 3 m square and 2.5 m high. The walls can be taken as adiabatic and isothermal. The ceiling is at 35°C and has an emittance of 0.8, while the floor is at 20°C and has an emittance of 0.9. Denote the ceiling as surface 1, the floor 2, and the walls 3.

(i) Set up the radiosity equations. Determine and evaluate all the shape factors, and tabulate as a 3×3 array. Solve these equations to determine the heat flow into the floor, $\dot{Q}_2$.
(ii) Draw the radiation network. Use the network to obtain an expression for $\dot{Q}_2$, and solve for $\dot{Q}_2$ again.

6–23. Determine the heat transfer between the two surfaces shown in the sketch if $T_1 = 3000$ K, $\varepsilon_1 = 0.3$; $T_2 = 1500$ K, $\varepsilon_2 = 0.4$. Surface 1 is 1 m wide and 20 m long, and $A_2 = 2A_1$. Take the surroundings as black at 0 K.

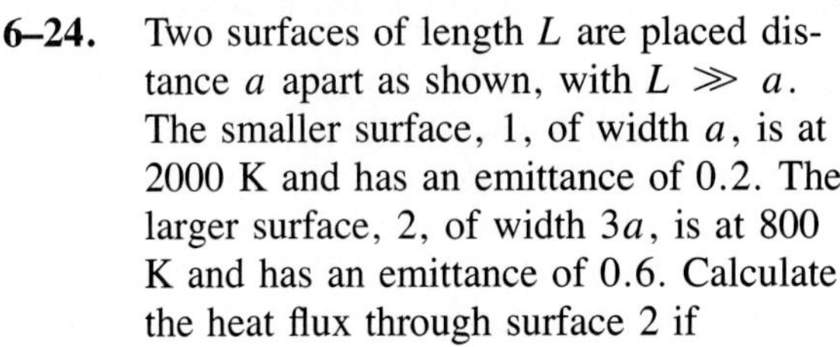

6–24. Two surfaces of length L are placed distance a apart as shown, with $L \gg a$. The smaller surface, 1, of width a, is at 2000 K and has an emittance of 0.2. The larger surface, 2, of width $3a$, is at 800 K and has an emittance of 0.6. Calculate the heat flux through surface 2 if

(i) the surroundings are black at 300 K.
(ii) the surfaces are joined by refractory surfaces.

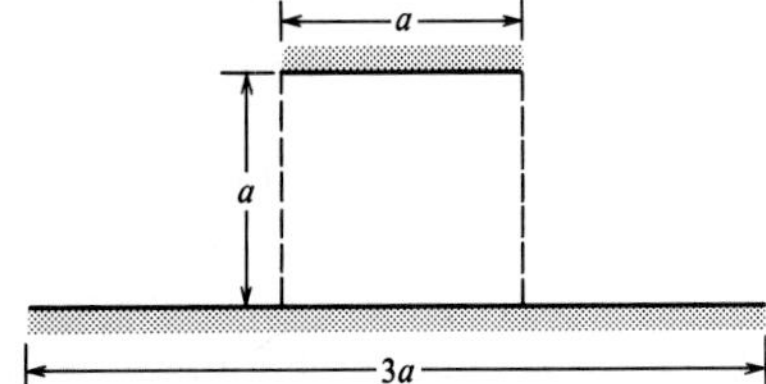

6–25. A long furnace used for an enameling process has a 3 m × 3 m cross section with the roof maintained at 1900 K, and the side walls are well insulated. What is the radiant heat transfer to the floor when it is at 600 K? All surfaces may be taken to be gray and diffuse, with emittances of 0.5.

(i) Assume the side walls are isothermal.
(ii) Divide the side walls into two surfaces and allow these surfaces to be isothermal at different temperatures.

6–26. A long furnace is 40 cm wide and 25 cm high. Its roof is maintained at 1100°C, and its side walls and floor are lined with refractory brick. Stainless steel strip 36 cm wide and 0.5 mm thick passes through the furnace at a speed of 0.3 m/s.

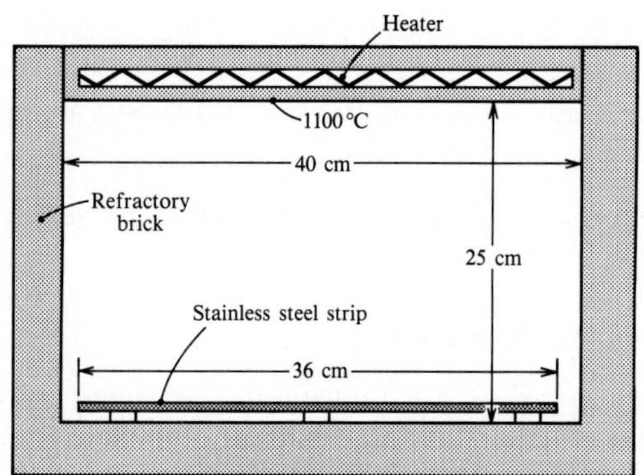

Estimate the length of furnace required to heat the strip from 100°C to 400°C. Take emittances for the roof and stainless steel as 0.8 and 0.25, respectively.

6–27. A long 90° wedge has 1 m sides, with surface 1 at 2000 K and surface 2 adiabatic. The surroundings can be taken as black at 0 K. If $\varepsilon_1 = 0.5$ and $\varepsilon_2 = 0.2$, determine the heat loss to the surroundings

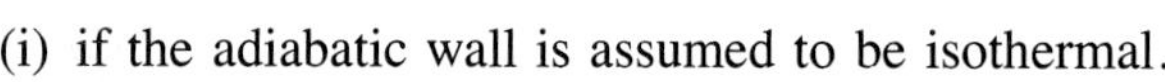

(i) if the adiabatic wall is assumed to be isothermal.
(ii) if the adiabatic wall is, in turn, subdivided into two, three, and four equal isothermal areas.

Plot your results.

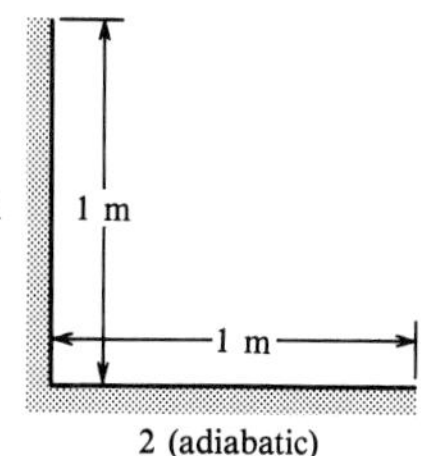

6–28. Two 3 cm–diameter thin-walled tubes run parallel, with their centers 5 cm apart. Through one tube flows hot combustion gases at 800 K, while through the other flows water at 300 K. If both tubes have an emittance of 0.8, calculate the net radiant energy absorbed by the water per meter length of tube. The surroundings are also at 300 K. Assume that both inside convective heat transfer coefficients are large.

6–29. A small furnace for heat-treating metal alloy samples in space has an inside diameter of 6 cm and a length of 12 cm, as shown. A 9 cm length of the wall is to be maintained at 800°C by an electrical heater. The remaining sidewall and endwall can be taken to be well insulated. The opening is exposed to black surroundings at 25°C. Owing to power supply limitations from a fuel cell source, a low operating power is desirable. Investigate the effect of heated wall emittance on the power required to balance radiative losses in steady operation.

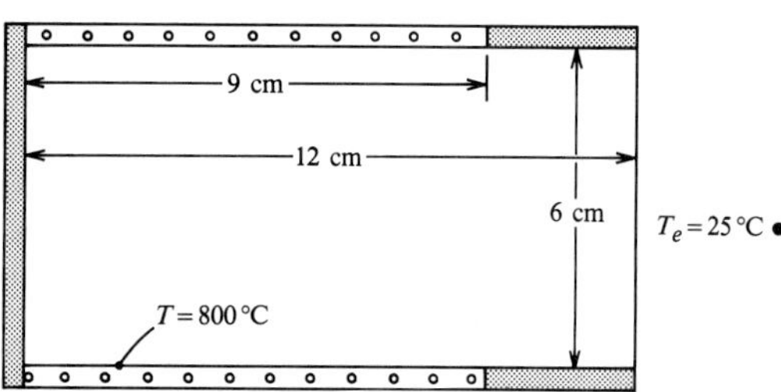

6–30. Two large parallel plates are 4 cm apart and are at temperatures 2000 K and 1000 K. If both have an emittance of 0.5, what is the heat transfer between the plates? If a thin sheet also of emittance 0.5 is placed between the plates, what is the new heat transfer rate? Why is the exact location of the sheet unimportant?

6–31. Two 15 cm–diameter parallel disks, 10 cm apart, have facing sides at 500 K and 400 K, and insulated backs. For gray disk surfaces with $\varepsilon = 0.5$ and black surroundings at 0 K, determine the net radiation exchange between the disks and the rate of heat loss from each disk.

6–32. The base of the conical cavity shown in the sketch is maintained at a temperature of 400 K, while the side walls are perfectly insulated. For gray diffuse surfaces, of emittances shown, determine the radiant heat transfer to the base for a top-surface temperature ranging from 800 K to 2000 K. Tabulate your results.

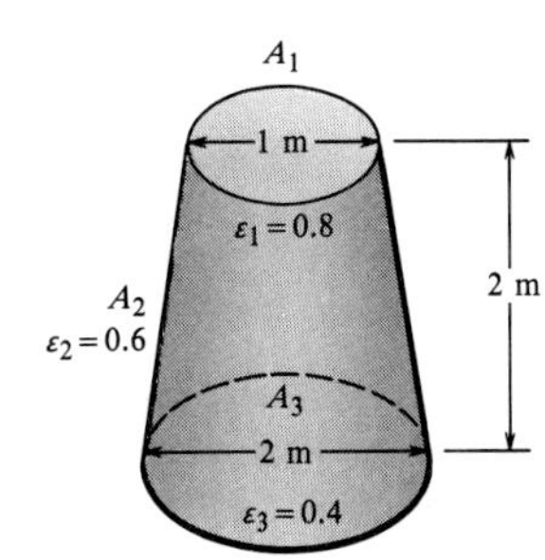

6–33. A cylindrical furnace has an inside diameter of 8 cm and is 16 cm high. The side walls are maintained at 1000 K, with power supplied by an electrical heater. The base and roof can be taken to be perfectly insulated. Determine the heat loss through the window in the roof when it is exposed to black surroundings at 300 K, for window diameters in the range 1 to 4 cm. The inside walls can be taken to be diffuse and gray with $\varepsilon = 0.6$; the transmittance of the window can be taken as unity.

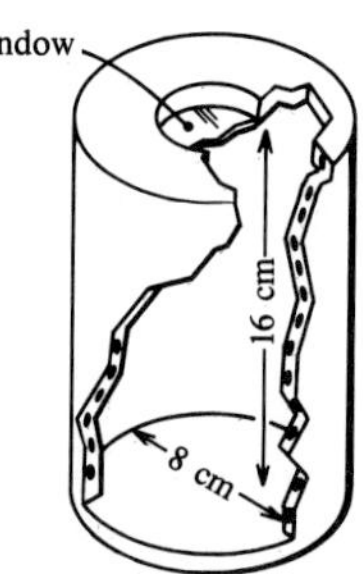

6–34. A square room of sides 8 m has a 3 m–diameter radiant heating panel centrally located on the ceiling, 3 m above the floor. The panel can be taken to be black and is maintained at 60°C. Calculate the irradiation on the floor due to the heater, at the center of the room, and in one corner.

6–35. A kiln has an average inside surface temperature of 2330 K and has a small 15×15 cm square opening in its 20 cm–thick walls. If the sides of the opening can be assumed to be adiabatic, determine the rate of radiant energy loss through the opening.

6–36. A sliding door on a furnace does not close entirely but leaves a gap 1.6 cm wide and 1 m high. The door is 20 cm thick and the emittance of the refractory ceramic is 0.8. What is the heat loss through the gap when the furnace is at 1800 K and the surroundings are at 300 K?

6–37. Thc planet Mars has a diameter of 6772 km, and it orbits the sun at a distance of 227.9×10^6 km. If the sun is assumed to radiate like a blackbody at 5760 K, and Mars has an albedo of 0.15 (reflects 15% of incident radiation back to space), estimate the average temperature of the Martian surface. Ignore the effects of the thin Martian atmosphere.

6–38. Estimate the sky emittance and effective sky temperature for a clear daytime sky when the ambient air is at 1.01 bar and 16.8°C and has a 60% relative humidity.

6–39. Estimate the sky emittance and effective sky temperature for a clear night sky when the ambient air is at 1 bar and 13°C and has a 30% relative humidity.

6–40. On a hot sunny day the corrugated iron roof of a work shed is measured to be 50°C when the ambient air is at 30°C and 80% relative humidity. Calculate the heat flux into the shed if h_c for the roof is estimated to be 20 W/m^2 K. Take $\varepsilon = 0.6$, $\alpha_s = 0.8$ for the iron, a shape factor from the roof to the sky of unity, and a solar irradiation of 900 W/m^2.

6–41. Will dew form on the top of a transit car parked on a still, clear night when the air temperature is 10°C and the relative humidity is 90%? Take the surface as

(i) weathered aluminum alloy 75S-T6.
(ii) white epoxy paint.

The convective heat transfer coefficient is approximately $h_c = 1.24\Delta T^{1/3}$ W/m^2 K, for ΔT in kelvins.

6–42. Estimate the equilibrium temperature of a spherical satellite 1 m in diameter, exposed to the sun, and in eclipse, at an altitude of 1000 km if it has an internal power dissipation of 300 W. Take the surface of the earth to be black at 19°C. The emittance of the satellite skin is 0.15, and its solar absorptance is 0.10.

6–43. Calculate the equilibrium temperature of a small, flat plate with its top face exposed to an unobstructed view of the sky while air at 298 K, 1 atm and 20% relative humidity flows along both sides. The bottom face sees black surroundings at 298 K. The solar irradiation is 800 W/m^2, and the average convective heat transfer coefficient is 20 W/m^2 K. Obtain solutions for three different kinds of surfaces:

(i) Representative of a very white paint, $\alpha_s = 0.2$, $\varepsilon = 0.9$
(ii) A metallic paint (aluminum), $\alpha_s = 0.3$, $\varepsilon = 0.3$
(iii) A black paint, $\alpha_s = 0.9$, $\varepsilon = 0.9$

6–44. A spacecraft is at 1 AU (astronomical unit) from the sun and can be assumed spherical and isothermal. Internal power dissipation is negligible, and it is desired to maintain an equilibrium temperature of 300 K. It is planned to use a checkered surface with a fraction F_1 of vacuum-deposited aluminum and the remainder $(1 - F_1)$ coated with white epoxy paint. Determine F_1.

6–45. A spherical spacecraft is 1 AU from the sun and can be assumed isothermal with negligible internal power dissipation. If the spacecraft is coated with Dow-Corning XP-310 aluminized silicone resin paint, determine its equilibrium temperature.

6–46. Calculate the irradiation of the area A_1 on a space vehicle as shown in the sketch. Surface A_2 is a solar cell array; the solar cells are at 50°C and can be taken to be black. (Area A_1 does not see the sun.)

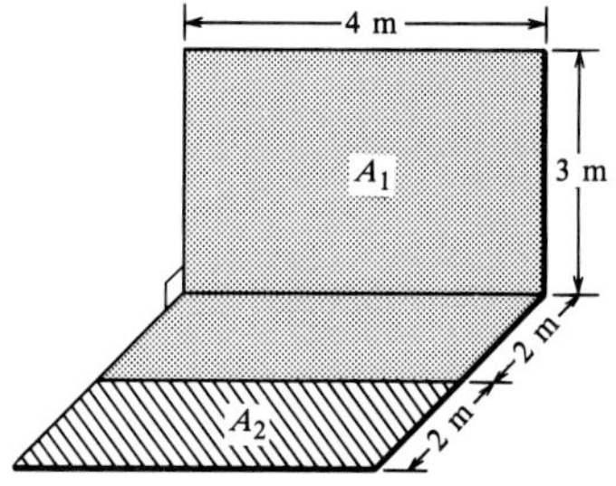

6–47. A flat-panel solar cell array on a three-axis-stabilized spacecraft in earth orbit has one side directly facing the sun and covered with solar cells; the other side is exposed to outer space at 0 K. The average solar absorptance on the front side is 0.8. The emittances of the front and back sides are 0.8 and 0.7, respectively. The solar cell operating efficiency is 15%, and the solar cell packing factor (the ratio of active solar cell area to total area) is 0.95. Determine the steady-state temperature of the panel.

6–48. A multilayer insulation (MLI) to be used on a spacecraft consists of 25 μm Kapton external foil sheets ($\alpha_s = 0.35$, $\varepsilon = 0.60$), aluminum-coated on the inside ($\varepsilon = 0.03$), sandwiching three aluminized Mylar sheets ($\varepsilon = 0.03$). The layers are separated by Dacron mesh to prevent contact of the sheets. If a MLI shield is faced to the sun at 1 AU distance and the inner surface is maintained at 300 K, determine the heat flux through the MLI and the outer surface temperature.

6–49. Write a computer program to solve the flat-plate solar collector problem of Example 6.12. Allow α_s, ε, $T_{C,\mathrm{in}}$, $\dot{m}_C$, G_s, T_e, $\varepsilon_{\mathrm{sky}}$, and U to be input parameters. Use the computer program to explore the effects of the various parameters on the collector efficiency.

6–50. Freezing of oranges in an orchard during winter can be catastrophic to the farmer. Estimate the rate at which an 8 cm–diameter orange hanging on a tree cools as a function of its temperature when the air has a temperature of 2°C and a relative humidity of 75%. Assume a convective heat transfer coefficient of 3 W/m^2 K, an emittance of orange peel of $\varepsilon_1 = 0.9$, and an emittance of surrounding leaves and grass of $\varepsilon_2 = 1.0$. Take the shape factor for the orange to surrounding leaves and grass as $F_{12} = 0.75$, and to the sky as $F_{13} = 0.25$. Estimate the temperature of the leaves and grass by requiring that they be adiabatic with $F_{23} = 0.5$. Will the convective heating when the orange is at 0°C exceed the radiative cooling? If not, will the use of fans to increase h_c from 3 to 12 W/m^2 K prevent freezing? For the orange, take $\rho = 900$ kg/m^3 and $c_p = 3600$ J/kg K. (*Hint:* First find T_2 from an energy balance on unit area $A_2 = 1$. Note that since $A_1F_{12} = A_2F_{21}$, F_{21} is negligible. Then find $\dot{Q}_1$ and dT_1/dt.)

6–51. Two surfaces, each of area 1 m^2, are located 8 m apart, as shown in the sketch. Surface 1 is black and is maintained at 800 K. Calculate the irradiation on surface 2 due to surface 1.

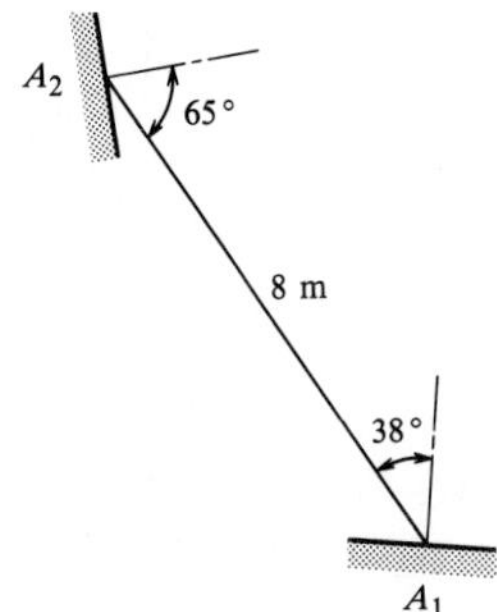

6–52. A small spherical thermocouple is located in the center of an open-ended circular tube of length H and radius R. Determine the shape factor for radiation transfer between the thermocouple and the inside wall of the tube.

6–53. Consider an elemental area dA_1 and a rectangle A_2 in a plane parallel to the plane of dA_1. The normal through dA_1 passes through a corner of the rectangle, as shown in the sketch. Show that the shape factor F_{12} is

$$F_{12} = \frac{1}{2\pi}\left[\frac{X}{(X^2+Z^2)^{1/2}}\sin^{-1}\frac{Y}{(X^2+Y^2+Z^2)^{1/2}} + \frac{Y}{(Y^2+Z^2)^{1/2}}\sin^{-1}\frac{X}{(X^2+Y^2+Z^2)^{1/2}}\right]$$

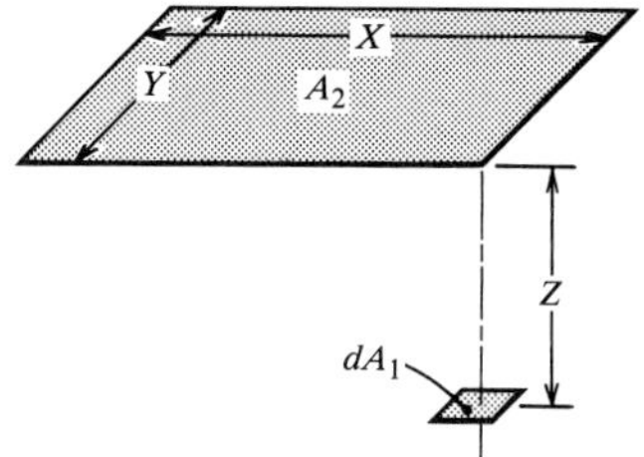

6–54. Liquid oxygen flows through a 2 cm–O.D. tube that is contained in an evacuated 4 cm–I.D. shell. The oxygen is at its normal boiling point of 90 K, and shell inner surface is at 250 K. Estimate the heat gain per meter length if the inner surface has an emittance of 0.03, the outer surface has an emittance of 0.5, and

(i) the surfaces are both diffuse.
(ii) the surfaces are both specular.
(iii) the inner surface is diffuse, and the outer surface is specular.

6–55. Liquid oxygen is contained in a thin-walled spherical container 48 cm in diameter, which in turn is enclosed in a concentric container 50 cm in diameter. The space in between is evacuated. The inner surface is at 95 K, and the outer surface is at 250 K. Determine the oxygen boil-off rate if

(i) both surfaces facing each other are gray with an emittance of 0.05 and reflect diffusely.
(ii) both surfaces reflect specularly.
(iii) the inner surface reflects diffusely, and the outer surface reflects specularly.

6–56. A louvered shutter on a spacecraft is 1 m wide and ~50 cm high. When fully open, the passages are 4 cm wide and 7 cm long. The louvers are 1 mm thick. The spacecraft interior can be taken to be a black enclosure at 320 K, and the exterior is outer space at ~0 K. Determine the heat loss through the shutter

(i) if the louvers are diffuse reflectors.
(ii) if the louvers have a coating that reflects specularly with a normal reflectance of 0.89.

6–57. Determine the wavelength range encompassing the 20–80% fraction of radiant energy emitted by black surfaces at the following temperatures:

(i) The normal boiling point of liquid oxygen, 90 K.
(ii) A room temperature of 300 K.
(iii) The normal boiling point of sodium, 1156 K.
(iv) The temperature of a tungsten lamp filament, 3000 K.
(v) A sun temperature of 5700 K.

6–58. Use the data for stainless steel in Table A.6*b* to determine the variation of total absorptance of a surface at room temperature to black radiation from a source at temperatures $300 < T_e < 1600$ K.

6–59. Use the data in Table A.6*a* and Fig. 6.28 to estimate the total hemispherical absorptance at $T_s = 300$ K to black radiation from a source at $T_e = 2000$ K for

(i) aluminum foil.
(ii) Inconel X rolled plate.
(iii) lapped 303 stainless steel.

6–60. Estimate the total hemispherical absorptance of a chromium-plated surface at 300 K when exposed to black radiation source at

(i) 300 K.
(ii) 1500 K.

Use the data in Table A.6*b*. Compare your results with values calculated from the formula in Table A.6*a*.

6–61. Estimate the total hemispherical absorptance for a gold-plated surface at 300 K exposed to radiation at 2000 K. Use the data in Table A6.*b*. Compare the result with the value calculated from the formula in Table A.6*a*.

6–62. Calculate the spectral absorptance to normally incident radiation for copper using the formula given in Table A.6*a*. Compare the result to the tabulated values for radiation incident at 25° given in Table A.6*b*.

6–63. Firebrick is often made of alumina and operates typically at 1600 K. Estimate the total emittance of firebrick at this temperature using the data for flame-sprayed alumina in Table A6.*b*.

6–64. Aluminum ingots are melted in a furnace for the purpose of alloying and recasting. The aluminum charge is covered with a layer of oxide (dross), which increases its absorptance and inhibits further oxidation. In practice the thickness of the dross must be carefully controlled so that it does not have too large a thermal resistance. Calculate the radiant heat transfer to 50 m^2 of dross surface at 1630 K from the furnace wall at 1600 K. Both the dross and the walls can be assumed to have the spectral characteristics at flame-sprayed alumina, as given in Table A.6*b*. Neglect furnace gas radiation.

6–65. Using the data in Table A.6*b*, calculate the absorptance of white epoxy paint for extraterrestrial solar radiation. Compare your result to value given in Table A.5*a*.

6–66. The effect of the earth's atmosphere on the solar spectrum can be roughly approximated by assuming (*a*) the sun radiates as a blackbody at 5760 K, (*b*) all radiation with wavelengths shorter than 0.4 μm and longer than 1.8 μm are absorbed, and (*c*) on a clear day 25% of the remaining radiation is absorbed.

(i) Estimate the solar irradiation on the earth's surface.
(ii) Prepare a table of λT versus fractional function for terrestrial solar radiation based on these assumptions.
(iii) Estimate the absorptance of white epoxy paint to terrestrial solar radiation.

6–67. Estimate the total hemispherical absorptance of a chromium-plated surface at 400 K to black radiation from a black source at the sun temperature of 5700 K.

6–68. An isothermal furnace with a small aperture approximating a blackbody is a useful device for calibrating radiation thermometers and heat flux meters. A thermostat controlling the power input to the furnace is to be selected to maintain the furnace at a nominal temperature of 1800 K, such that the variation in spectral intensity at 0.7 μm is less than 1%. What is the allowable variation in the furnace temperature?

6–69. A stainless steel surface at room temperature is exposed to radiation from a black source at 2000 K. Estimate its total hemispherical absorptance using the following information.

(i) Equation (6.84), Table A.6*b*, and Fig. 6.28
(ii) The formula in Table A.6*a*, and Fig. 6.28

6–70. Estimate the total hemispherical emittance of a gold-plated surface at 600 K. Use the data in Table A.6*b* and assume that the spectral emittance is independent of surface temperature in this temperature range.

6–71. In a metals processing experiment on the space shuttle, it is required to melt a spherical 0.5 cm–diameter gold sample acoustically levitated in a furnace maintained at 800°C. The additional heating required is supplied by a focused beam of radiation from a xenon arc lamp. The following table gives the spectral distribution of emissive power for the lamp. Available data show that the absorptance of gold near its melting point is not much different from room temperature values in the wavelength range of significance.

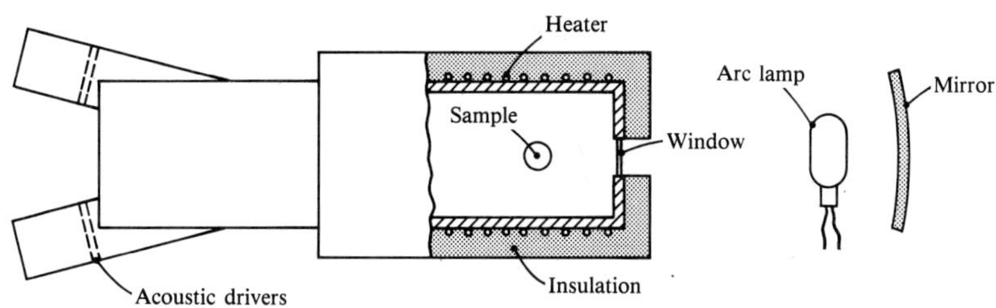

(i) Estimate an appropriate value of total hemispherical absorptance to be used in melting rate calculations.

(ii) In a proposed design the arc delivers 25 W to the sample, and the convective heat transfer coefficient on the sphere associated with a flow induced by the acoustical field is estimated to be 140 W/m^2 K. Can the sample be melted? If so, how long will it take to melt the sample once it reaches the melting temperature? The furnace walls can be taken to be black. The enthalpy of melting of gold is 6.44×10^4 J/kg.

$\lambda(\mu m)$	$\Delta I_\lambda/I_\lambda(\%)$	$\lambda(\mu m)$	$\Delta I_\lambda/I_\lambda(\%)$
0–0.25	2.49	0.60–0.70	12.12
0.25–0.35	3.50	0.70–0.80	9.36
0.35–0.40	4.58	0.80–1.00	24.86
0.40–0.45	5.25	1.00–1.50	13.23
0.45–0.50	5.63	1.50–2.00	4.24
0.50–0.60	10.94	2.00–2.50	1.04

6–72. Repeat part (i) of Exercise 6–71 for a mercury-xenon lamp with a spectral distribution of emissive power as given in the following table.

$\lambda(\mu m)$	$\Delta I_\lambda/I_\lambda(\%)$	$\lambda(\mu m)$	$\Delta I_\lambda/I_\lambda(\%)$
0–0.26	0	0.60–0.70	3.47
0.26–0.30	11.99	0.70–0.80	2.04
0.30–0.40	23.96	0.80–1.00	6.13
0.40–0.45	9.19	1.00–1.50	16.35
0.45–0.50	1.31	1.50–2.00	10.67
0.50–0.60	14.88		

6–73. A gas mixture at 1500 K and 1 atm contains 20% H_2O and 80% N_2 by volume. Calculate the total emissivity for path lengths in the range 5 cm to 1 m. Plot your result.

6–74. Exhaust gas from a furnace burning a hydrocarbon fuel is at 2 atm pressure. The composition can be approximated as 10% CO_2, 10% H_2O, and 80% N_2 by volume. Calculate the total emissivity as a function of temperature in the range 1100 to 1800 K. Also calculate the absorptivity to radiation from a black wall at 800 K. Take a path length of 0.5 m.

6–75. A spherical enclosure of 1 m diameter contains 90% N_2 and 10% H_2O by volume, at 3 atm and 1200 K. The walls are black.

(i) Calculate the gas emissivity.
(ii) Calculate the gas absorptivity for wall temperatures in the range 400–2000 K.

Plot the results.

6–76. Exhaust gas from a methane combustor is at 1400 K and 3 atm pressure for stoichiometric combustion with air. Estimate the radiant heat transfer to the walls of a 0.5 m–square duct maintained at 700 K. Assume that the walls are black. Repeat the calculation for an excess air ratio of 2.

6–77. A gray isothermal gas is contained in a long enclosure of square cross section, with 1 m sides. The walls are gray with emittance 0.60; one wall is at 350 K, and the others are refractory. The gas temperature is 3000 K, and its total absorption coefficient $\kappa = 0.3$. Calculate the rate of radiant heat transfer through the cold surface.

6–78. A cubical enclosure with 1 m sides contains 80% N_2, 10% H_2O, and 10% CO_2 by volume at 2.5 atm total pressure. The walls are black and are maintained at 600 K. If turbulence in the gas maintains an approximately uniform temperature, estimate the rate of temperature drop of the gas when it is at

(i) 1600 K.
(ii) 1000 K.

6–79. A kerosene combustor can be approximated as a long cylinder of 20 cm diameter. The exhaust gas contains 10% CO_2, 10% H_2O, and 80% N_2 by volume and is at 1400 K and 2 atm pressure. Calculate the radiant heat transfer to black walls at 700 K. Compare your result to an estimate of the convective heat transfer if the bulk velocity of the gas is 50 m/s.

6–80. Flue gas at 1100 K and 1 atm pressure contains 8% H_2O, 8% CO_2, and 84% N_2 by volume. The gas flows at 3 m/s along a 1 m–square flue with brick-lined walls. If the brick surface is at 1000 K, estimate the radiative and convective heat transfer to walls per meter length of flue. Take the brick surface to be black.

6–81. A cubical vessel has sides 2 m long with internal surfaces of emittance 0.8. The vessel contains a gas at 100 K that can be assumed gray with a total absorption coefficient of 0.26. The cylindrical walls are lined with refractory brick and can be assumed adiabatic. The upper surface is maintained at 1200 K while the base is at 800 K. Calculate the radiant heat transfer to the base, and compare your answer to the result if no gas were present.

6–82. Steam at 4 atm pressure flows between black parallel plates spaced 5 cm apart. At a location where the bulk steam temperature is 1200 K and the plates are at 800 K, calculate the radiative and convective heat transfer to the plates.

6–83. Two large, black parallel plates are 40 cm apart and are maintained at temperatures of 1200 and 800 K. Between the plates is a gas mixture at 1000 K and 1 atm pressure, containing 15% H_2O, 10% CO_2, and 75% N_2 by volume. Calculate the radiant heat transfer to the cold plate, and compare your answer to the result if no gas were present.

6–84. A gray isothermal gas at 1600 K is contained in a 3 m–diameter spherical shell maintained at 700 K. The total absorption coefficient of the gas is $0.3\ m^{-1}$. Determine the radiant heat transfer to the shell if the inner surface

(i) is black.
(ii) has an emittance of 0.5.

6–85. A direct-fired furnace is cylindrical in shape, with a diameter of 5 m and a height of 3 m. The side walls and roof are refractory, and the gas is at 1600 K, with an effective gray absorption coefficient of 0.5 m^{-1}. The load covers the furnace floor and has an emittance of 0.5. Determine the radiant heat transfer to the load when it is at 1000 K.

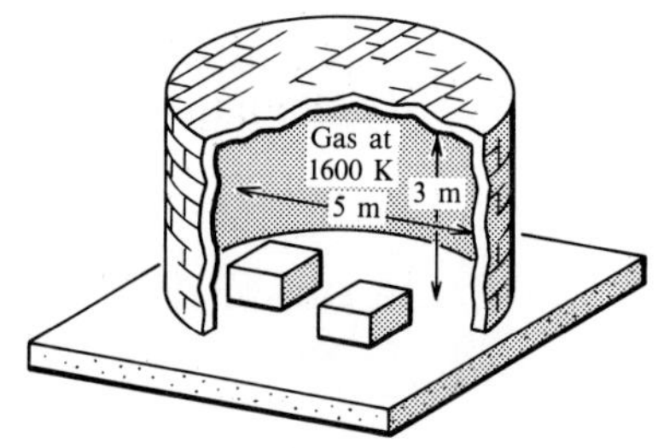

6–86. A direct-fired furnace is in the form of a cube with 3 m sides, with refractory side walls and roof. The furnace gases are at 1700 K and have an effective gray absorption coefficient of 0.3 m^{-1}. If the load covers the furnace floor and has an emittance of 0.7, determine the radiant heat transfer to the load when it is at 1100 K.

6–87. A hydrogen-fueled combustor can be approximated as a long cylinder of diameter 0.2 m. The combustion gases are at 2000 K and 3 atm pressure for stoichiometric combustion with air. If the walls are to be cooled to 900 K, determine the radiant heat transfer to the walls as a function of their surface emittance in the range $0.2 < \varepsilon < 1.0$. Plot your result.

6–88. A tube bank has 5 cm–diameter O.D. tubes in a square array at 10 cm pitch. Combustion products containing 11% H_2O, 11% CO_2, and 78% N_2, by volume, at 2 atm pressure flow across the tube bank. At a location in the bank where the gas temperature is 1200 K and the tube wall temperature is 500 K, calculate the radiant heat transfer to the tubes. Take $\varepsilon = 0.9$ for the tube surface. Also estimate the convective heat transfer if the gas velocity upstream of the bank is 8 m/s.

6–89. A hydrogen-fueled combustor has a 0.3 m–diameter circular cross section. The combustion products contain 14% H_2O and 86% N_2 by volume and are at 1500 K and 5 atm pressure. The combustor walls are maintained at 800 K. Calculate the radiative heat flux to the walls for wall emittances in the range 0.1 to 0.7.

6–90. A hydrogen-fueled combustor has a 0.5 m–square cross section. The walls have an emittance of 0.4 and are cooled by hydrogen from the fuel tank flowing in channels underneath the skin. The combustion products contain 15% H_2O and 85% N_2 by volume, and, at the location under consideration, are at 1600 K and 6 atm pressure. The total heat flux into the coolant cannot exceed 240 kW/m^2. What is the lowest allowable wall temperature if the convective heat transfer coefficient is estimated to be 200 W/m^2 K?

CHAPTER

7

CONDENSATION, EVAPORATION, AND BOILING

CONTENTS

7.1 INTRODUCTION

Condensation of vapors and the evaporation and boiling of liquids are commonplace in power and process engineering. In a fossil- or nuclear-fueled power plant, water is boiled in a boiler, and steam is condensed in a condenser. In an oil refinery, oil is evaporated in a distillation column to give a variety of vapor products, which are eventually condensed into liquid fuels such as gasoline and kerosene. In a multistage flash desalination plant, water vapor is produced by evaporation from brine and is subsequently condensed as pure water. Condensation processes require that the enthalpy of phase change be removed by a coolant, and evaporation and boiling processes require that this enthalpy be supplied from an energy source. Since the enthalpy of phase change is relatively large, particularly for water ($\sim 2.5 \times 10^6$ J/kg), the associated heat transfer rates are also usually large. In most industrial processes, the vapor and liquid phases both flow through the heat exchanger. Thus, the heat transfer to the phase interface is essentially a convective process, but it is often complicated by an irregular interface, for example, bubbles or drops.

In Section 7.2, *film condensation* is analyzed. Vapor condenses on a cooled surface, and the liquid drains off the surface under the action of gravity. An analysis of laminar film condensation on a vertical wall is presented first, followed by analyses of *wavy laminar* and *turbulent* film condensation on a vertical wall and laminar film condensation on a horizontal tube. The effects of vapor velocity and vapor superheat are analyzed using a Couette flow model of the vapor boundary layer. In Section 7.3 the reverse process is treated, in which a film of liquid flows down a heated wall while evaporation takes place from the film surface. Boiling on heater surfaces immersed in a pool of saturated liquid is dealt with in Section 7.4. After the regimes of *pool boiling* and the *boiling curve* are introduced, boiling inception is discussed, and correlations are presented for *nucleate boiling*, the *peak heat flux*, and *film boiling*.

In Section 7.5, two-phase flow regimes for forced flow in horizontal and vertical tubes are introduced, and correlations are presented for forced-convection boiling and condensation. Phase change at low pressures is analyzed in Section 7.6, and the significance of the *interfacial heat transfer resistance* is examined. Finally, in Section 7.7, a brief introduction to *heatpipes* is provided, and the performance of fixed-conductance and of variable-conductance heatpipes is analyzed. Two computer programs accompany Chapter 7, namely, PHASE for film condensation and evaporation, and BOIL for pool boiling. Both programs have a menu of six fluids: water, ammonia, nitrogen, mercury, R-12, and R-113.

7.2 FILM CONDENSATION

When a heat exchanger wall exposed to a vapor is cooled below the saturation temperature of the vapor, condensation will occur on the surface. If the vapor is pure, the saturation temperature corresponds to the total pressure; however, if there is a vapor–noncondensable gas mixture, the saturation temperature corresponds to the partial pressure of the vapor. Two distinct modes of condensation are possible. If the condensate wets a vertical wall, a complete film of liquid will cover the wall, and the film thickness will grow as it flows down the wall under the action of gravity: this mode is called **film condensation**. Since the enthalpy of condensation given up at the surface of the film must be transferred across the film to the wall, the film presents a thermal resistance to heat transfer. Such condensate films are very thin, and the corresponding heat transfer coefficients are relatively large. For example, when steam at a saturation temperature of 305 K condenses on a 2 cm–O.D. tube with a wall temperature of 300 K, the average film thickness is $\sim 50\,\mu$m, and the average heat transfer coefficient is 11,700 W/m^2 K. If the condensate flow rate is small because either the temperature difference is small or the wall is short, the surface of the film will be smooth and the flow laminar. At higher flow rates, waves will form on the surface to give what is called *wavy laminar* flow, and at yet higher flow rates, the flow becomes *turbulent*.

If the condensate does not wet the wall, because either it is dirty or it has been treated with a nonwetting agent, droplets of condensate nucleate at small pits and other imperfections on the surface, and they grow rapidly by direct vapor condensation upon them and by coalescence. When the droplets become sufficiently large, they flow down the surface under the action of gravity and expose bare metal in their tracks, where further droplet nucleation is initiated. This mode is called **dropwise condensation** and is shown in Fig. 7.1. Most of the heat transfer is through drops of less than 100 μm diameter. The thermal resistance of such drops is small; hence, heat transfer coefficients for dropwise condensation can be very high: values of up to 300,000 W/m^2 K have been measured. These high heat transfer coefficients are attractive to the design engineer; thus, considerable effort has been spent developing nonwetting heat exchanger surfaces, but no entirely satisfactory method has been found. If the surface is treated with a nonwetting agent such as stearic acid to promote dropwise condensation, the effect lasts but a few days until the promoter is washed off or oxidized. Continuous adding of the promoter to the vapor is expensive and contaminates the condensate. Bonding a polymer such as Teflon to the surface to promote dropwise condensation is expensive and adds an additional thermal resistance.

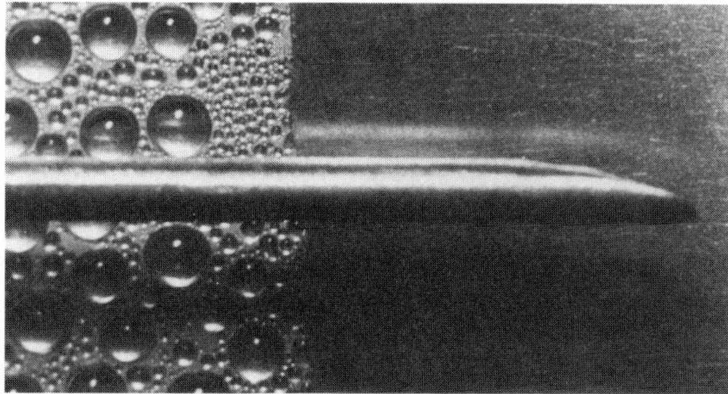

Figure 7.1 Steam condensing on a copper surface. The right-side surface is clean, giving filmwise condensation. The left side has a thin coating of cupric oleate, which causes dropwise condensation. The cylindrical temperature probe has a diameter of 1.7 mm. (Photograph courtesy of Professor J. W. Westwater, University of Illinois, Urbana.)

Gold plating is also expensive and often does not have a lasting effect. When an industrial or laboratory condenser is put into service for the first time, the initial model of condensation is *dropwise* unless particular care is taken to clean the system. But in a matter of hours or days, a mixed mode of dropwise and film condensation will develop, and finally the condensation will be entirely filmwise. Thus, current design practice is to always assume film condensation for a conservative estimate of heat transfer surface area.

7.2.1 Laminar Film Condensation on a Vertical Wall

Figure 7.2 depicts a vertical wall exposed to a saturated vapor at pressure P and saturation temperature $T_{sat} = T_{sat}(P)$. The wall could be flat or could be the outside surface of a vertical tube. If the surface is maintained at a temperature $T_w < T_{sat}$, vapor will continuously condense on the wall and, if the liquid phase wets the surface well, will flow down the wall in a thin film. Provided the condensation rate is not too large, there will be no discernible waves on the film surface, and the flow in the film will be laminar. The first step in the analysis of this *laminar film condensation* process is to examine the dynamics of the flow of a thin liquid film.

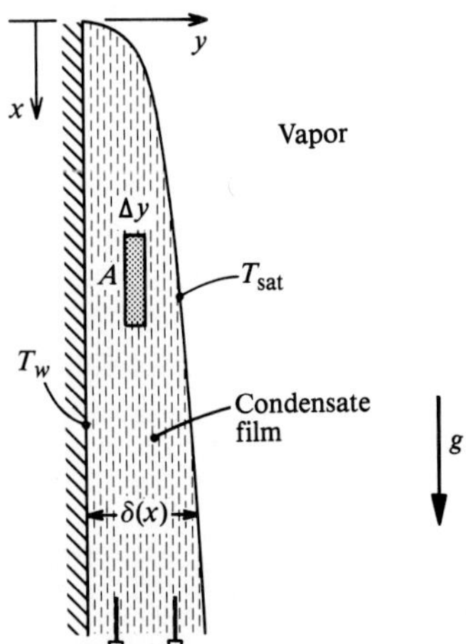

Figure 7.2 Coordinate system for film condensation on a vertical wall.

The Fluid Dynamics Problem

Since the velocity of a thin liquid film is relatively low, the forces required to accelerate the film are usually negligible, and the motion of the film is determined by a simple balance of buoyancy and viscous forces. Figure 7.3*a* depicts an element of fluid of area A, located between planes y and $y + \Delta y$, and the forces acting upon it. The buoyancy force is given by Archimedes' principle as $(\rho_l - \rho_v)gA\Delta y$ where ρ_l is the liquid density and ρ_v is the vapor density. The viscous forces are given by Newton's law of viscosity, with $\mu_l(\partial u/\partial y)|_y A$ acting upward and $\mu_l(\partial u/\partial y)|_{y+\Delta y} A$

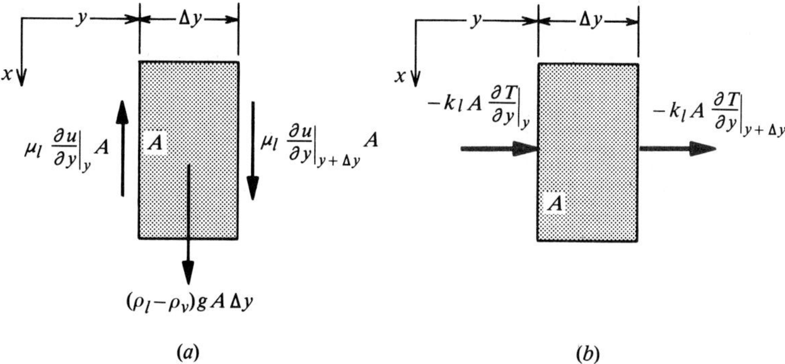

Figure 7.3 (*a*) Force balance on a fluid element $A\Delta y$ in a laminar falling film. (*b*) Energy balance on an elemental control volume $A\Delta y$ in a laminar falling film.

acting downward. Summing these forces gives

$$\mu_l \frac{\partial u}{\partial y}\bigg|_{y+\Delta y} A - \mu_l \frac{\partial u}{\partial y}\bigg|_y A + (\rho_l - \rho_v) g A \Delta y = 0$$

Dividing by $A\Delta y$ and assuming that the liquid viscosity μ is constant gives

$$\mu_l \frac{(\partial u/\partial y)\big|_{y+\Delta y} - (\partial u/\partial y)\big|_y}{\Delta y} + (\rho_l - \rho_v) g = 0$$

Then letting $\Delta y \to 0$,

$$\mu_l \frac{\partial^2 u}{\partial y^2} + (\rho_l - \rho_v) g = 0 \tag{7.1}$$

which is a differential equation for the velocity profile across the film, $u(y)$, at any particular value of x. Two boundary conditions are required. At the wall, there is the no-slip condition for a real fluid,

$$y = 0: \quad u = 0 \tag{7.2a}$$

and at the surface of the film, we assume that there is negligible drag of the vapor. If the film thickness is $\delta(x)$, the required boundary condition is

$$y = \delta: \quad \frac{\partial u}{\partial y} = 0 \tag{7.2b}$$

where $\delta(x)$ is a function still to be determined. The condition of negligible drag is applicable to many situations for which the vapor velocities are not too high, as will be discussed further in Section 7.2.4. Integrating Eq. (7.1) gives

$$\frac{\partial u}{\partial y} = -\frac{(\rho_l - \rho_v) g}{\mu_l} y + C_1$$

Applying the boundary condition Eq. (7.2*b*),

$$0 = -\frac{(\rho_l - \rho_v)g}{\mu_l}\delta + C_1$$

Solving for C_1 and substituting back,

$$\frac{\partial u}{\partial y} = \frac{(\rho_l - \rho_v)g}{\mu_l}(\delta - y)$$

Integrating again,

$$u = \frac{(\rho_l - \rho_v)g}{\mu_l}\left(\delta y - \frac{y^2}{2}\right) + C_2$$

Applying the boundary condition Eq. (7.2*a*) gives $C_2 = 0$. Rearranging,

$$u = \frac{(\rho_l - \rho_v)g\delta^2}{\mu_l}\left[\frac{y}{\delta} - \frac{1}{2}\left(\frac{y}{\delta}\right)^2\right] \tag{7.3}$$

Equation (7.3) gives the velocity profile $u(y)$, which is seen to be parabolic. The velocity is a maximum at the film surface and is obtained by setting $y = \delta$:

$$u_\delta = \frac{(\rho_l - \rho_v)g\delta^2}{2\mu_l} = \frac{g\delta^2}{2\nu_l}\frac{(\rho_l - \rho_v)}{\rho_l} \tag{7.4}$$

The *mass flow rate per unit width* of film is denoted Γ and has units kg/m s. It is obtained by integration of Eq.(7.3):

$$\Gamma = \int_0^\delta \rho_l u \, dy = \frac{g\delta^3(\rho_l - \rho_v)}{3\nu_l} \tag{7.5}$$

The Heat Transfer Problem

With the fluid dynamics of the film understood, we now turn to the heat transfer problem. Two special features of the problem are: (1) the film velocity is relatively low, and (2) temperature gradients in the *x* direction are negligible since both the wall and film surface are isothermal. Thus, in applying the steady-flow energy equation, Eq. (1.4), to the elemental control volume shown in Fig. 7.3*b*, the enthalpy convection term $\dot{m}\Delta h$ can be neglected, as can be the *x*-direction conduction contribution to $\dot{Q}$. Then the heat conducted out at y equals the heat conducted in at $y + \Delta y$:

$$-k_l A \left.\frac{\partial T}{\partial y}\right|_y = -k_l A \left.\frac{\partial T}{\partial y}\right|_{y+\Delta y}$$

Dividing through by $k_l A \Delta y$, for k_l assumed constant, and rearranging,

$$\frac{(\partial T/\partial y)|_{y+\Delta y} - (\partial T/\partial y)|_y}{\Delta y} = 0$$

Then letting $\Delta y \to 0$,

$$\frac{\partial^2 T}{\partial y^2} = 0 \tag{7.6}$$

Two boundary conditions are required. At the surface of the film, continuity of temperature requires that $T = T_{\text{sat}}$, where T_{sat} is the saturation temperature corresponding to the pressure P of the vapor:

$$y = \delta: \quad T = T_{\text{sat}} \tag{7.7a}$$

For the second boundary condition, we will assume that the wall is isothermal at temperature T_w:

$$y = 0: \quad T = T_w \tag{7.7b}$$

Integrating Eq. (7.6) twice,

$$T = C_1 y + C_2$$

and fitting the boundary conditions,

$$T - T_w = \frac{y}{\delta}(T_{\text{sat}} - T_w) \tag{7.8}$$

which is a linear temperature profile since this problem is identical to that of conduction across a plane slab, which was treated in Section 1.3.1. The heat flux into the wall is simply the heat flux across the film, which from Eq (1.9) is

$$k_l \left.\frac{\partial T}{\partial y}\right|_w = \frac{\dot{Q}}{A} = \frac{k_l(T_{\text{sat}} - T_w)}{\delta} \qquad \text{(ignoring the} - \text{sign)}$$

The local heat transfer coefficient h is defined as this heat flux divided by the temperature difference across the film $(T_{\text{sat}} - T_w)$:

$$h = \frac{k_l \left.(\partial T/\partial y)\right|_w}{T_{\text{sat}} - T_w} = \frac{k_l}{\delta} \tag{7.9}$$

Determination of the Film Thickness

To find how δ and hence h vary down the wall, we need to make mass and energy balances for an element of film Δx long, as shown in Fig. 7.4a. If the condensation rate is denoted $\dot{m}''$ [kg/m^2 s] and taken to be a positive quantity, then mass conservation for unit width of film requires

$$\dot{m}''\Delta x + \left.\int_0^{\delta(x)} \rho_l u \, dy\right|_x = \left.\int_0^{\delta(x)} \rho_l u \, dy\right|_{x+\Delta x}$$

Dividing by Δx and letting $\Delta x \to 0$,

$$\dot{m}'' = \frac{d}{dx}\int_0^{\delta(x)} \rho_l u \, dy = \frac{d\Gamma}{dx} \tag{7.10}$$

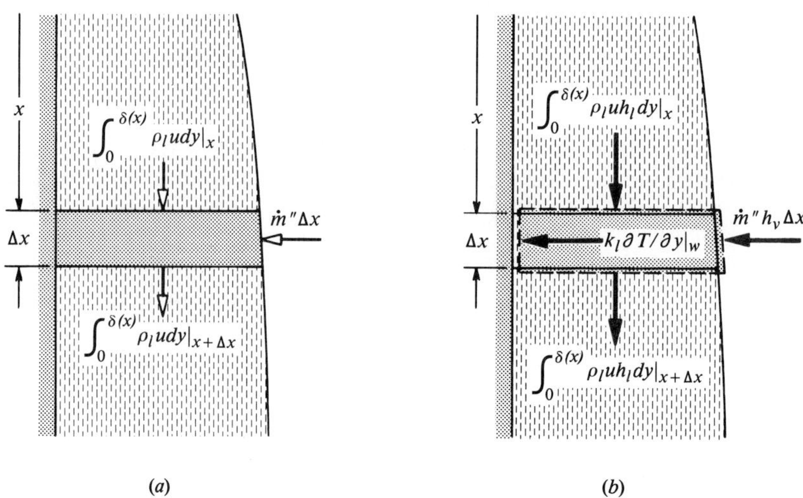

Figure 7.4 Mass and energy balances on an element of a falling film Δx long.

The steady-flow energy equation applied to the control volume shown in Fig. 7.4*b* requires that the enthalpy flowing into the volume equal the enthalpy flowing out plus the heat conducted to the wall:

$$\dot{m}'' h_v \Delta x + \int_0^{\delta(x)} \rho_l u h_l \, dy\Big|_x = \int_0^{\delta(x)} \rho_l u h_l \, dy\Big|_{x+\Delta x} + k_l \left.\frac{\partial T}{\partial y}\right|_w \Delta x$$

where h_v is the enthalpy of the vapor and h_l is the liquid enthalpy. Rearranging, dividing by Δx, and letting $\Delta x \to 0$,

$$k_l \left.\frac{\partial T}{\partial y}\right|_w = \dot{m}'' h_v - \frac{d}{dx}\int_0^{\delta(x)} \rho_l u h_l \, dy \tag{7.11}$$

Substituting Eq. (7.10) in Eq. (7.11) gives

$$k_l \left.\frac{\partial T}{\partial y}\right|_w = \frac{d}{dx}\int_0^{\delta(x)} \rho_l u (h_v - h_l) dy$$

which, for a constant liquid specific heat c_{pl}, becomes

$$k_l \left.\frac{\partial T}{\partial y}\right|_w = \frac{d}{dx}\int_0^{\delta(x)} \rho_l u [h_{fg} + c_{pl}(T_{sat} - T)] dy \tag{7.12}$$

The *subcooling* term $c_{pl}(T_{sat} - T)$ accounts for the sensible heat given up by the condensate as it cools below the saturation temperature T_{sat}. This term is usually quite small: for example, when steam condenses at 1 atm pressure and there is a 10 K temperature drop across the film, the subcooling term is at most 2% of the enthalpy of phase change h_{fg}. Thus, for simplicity, the subcooling term will be ignored for the present, and Eq. (7.12) can be rewritten using Eq. (7.5) as

$$k_l \left.\frac{\partial T}{\partial y}\right|_w = h_{\text{fg}} \frac{d}{dx} \int_0^{\delta(x)} \rho_l u \, dy = h_{\text{fg}} \frac{d\Gamma}{dx} \tag{7.13}$$

Now, from Eq. (7.9),

$$k_l \left.\frac{\partial T}{\partial y}\right|_w = \frac{k_l (T_{\text{sat}} - T_w)}{\delta}$$

and from Eq. (7.5),

$$\frac{d\Gamma}{dx} = \frac{g(\rho_l - \rho_v)}{\nu_l} \delta^2 \frac{d\delta}{dx}$$

Substituting in Eq. (7.13),

$$\frac{k_l (T_{\text{sat}} - T_w)}{\delta} = \frac{h_{\text{fg}} g(\rho_l - \rho_v)}{\nu_l} \delta^2 \frac{d\delta}{dx}$$

or

$$dx = \frac{h_{\text{fg}} g(\rho_l - \rho_v)}{k_l (T_{\text{sat}} - T_w) \nu_l} \delta^3 \, d\delta$$

If condensation begins at the top of the wall, $\delta = 0$ at $x = 0$, and integration is straightforward.

$$\int_0^x dx = \frac{h_{\text{fg}} g(\rho_l - \rho_v)}{k_l (T_{\text{sat}} - T_w) \nu_l} \int_0^\delta \delta^3 \, d\delta$$

$$x = \frac{h_{\text{fg}} g(\rho_l - \rho_v) \delta^4}{4 k_l (T_{\text{sat}} - T_w) \nu_l}$$

$$\delta = \left[\frac{4 x k_l (T_{\text{sat}} - T_w) \nu_l}{h_{\text{fg}} g(\rho_l - \rho_v)} \right]^{1/4} \tag{7.14}$$

The film thickness is seen to grow like $x^{1/4}$. Substituting for δ in Eq. (7.9) gives the heat transfer coefficient h,

$$h = \left[\frac{h_{\text{fg}} g(\rho_l - \rho_v) k_l^3}{4x (T_{\text{sat}} - T_w) \nu_l} \right]^{1/4} \tag{7.15}$$

which is seen to be proportional to $x^{-1/4}$ and $(T_{\text{sat}} - T_w)^{-1/4}$. The average heat transfer coefficient for a wall of height L is obtained by integration:

$$\overline{h} = \frac{1}{L} \int_0^L h \, dx = \frac{4}{3} h_{x=L}$$

$$\overline{h} = 0.943 \left[\frac{h_{\text{fg}} g(\rho_l - \rho_v) k_l^3}{L (T_{\text{sat}} - T_w) \nu_l} \right]^{1/4} \tag{7.16}$$

Equation (7.16) was first derived by W. Nusselt in 1916 [1]. The assumptions of negligible effects of fluid acceleration and heat convection are often called the *Nusselt assumptions* and have been used for many different falling film problems.

Effect of Liquid Subcooling

For liquids with relatively low enthalpies of phase change, such as many refrigerants, we should make a correction for the subcooling term, which was deleted in obtaining Eq. (7.13) from Eq. (7.12). If the subcooling term is retained in the analysis, the result is identical to Eq. (7.16) with h_{fg} replaced by a *modified latent heat* (enthalpy of phase change) h'_{fg} given by:

$$h'_{fg} = h_{fg} + \frac{3}{8} c_{pl}(T_{sat} - T_w) \tag{7.17a}$$

The derivation of Eq. (7.17*a*) is given as Exercise 7–4. More exact analysis, in which Nusselt assumptions are not invoked, gives [2]

$$h'_{fg} = h_{fg} + \left(0.683 - \frac{0.228}{\mathrm{Pr}_l}\right) c_{pl}(T_{sat} - T_w) \tag{7.17b}$$

Effect of Variable Fluid Properties

A reference temperature scheme to account for variable-property effects has been established by numerically solving the exact governing equations allowing for variable fluid properties [3]. The scheme requires that $h'_{fg} = h_{fg} + 0.35c_{pl}(T_{sat} - T_w)$ be used to account for liquid subcooling, with h_{fg} evaluated at the saturation temperature; all other liquid phase properties are evaluated at $T_r = T_w + \alpha(T_{sat} - T_w)$, where α is given in Table 7.1 for various fluids. Use of this scheme also accounts for the minor errors introduced through use of the Nusselt assumptions. For liquids not listed in Table 7.1, Eq. (7.17*b*) should be used for h'_{fg}, with h_{fg} evaluated at T_{sat} and all other properties evaluated at the mean film temperature.

Table 7.1 Values of α in the reference temperature $T_r = T_w + \alpha(T_{sat} - T_w)$ for laminar film condensation on a vertical wall [3].

Fluid	T_{sat} K	ΔT K	α
Carbon tetrachloride	290–320	5–20	0.07
Ethyl alcohol	300–350	5–20	0.12
n-propyl alcohol	330	10–20	0.15
n-butyl alcohol	290–330	20	0.25
t-butyl alcohol	330	10–20	0.29
Ethylene glycol	330	20	0.29
Glycerol	330	20	0.32
Water	280–380	1–30	0.33
Ammonia	270–320	20	0.61
Propane	350–370	20	1.00

EXAMPLE 7.1 Laminar Film Condensation of Steam

Saturated steam condenses on the outside of a 5 cm–diameter vertical tube, 50 cm high. If the saturation temperature of the steam is 302 K, and cooling water maintains the wall temperature at 299 K, calculate: (i) the average heat transfer coefficient, (ii) the total condensation rate, and (iii) the film thickness at the bottom of the tube.

Solution

Given: Film condensation of saturated steam on a vertical surface.

Required: (i) Average heat transfer coefficient $\overline{h}$, (ii) total condensation rate $\dot{m}$, and (iii) film thickness δ.

Assumptions:
1. Effect of tube curvature negligible.
2. Effect of liquid subcooling negligible.
3. Laminar flow.

(i) The average heat transfer coefficient is given by Eq. (7.16) with h_{fg} replaced by h'_{fg}:

$$\overline{h} = 0.943\left[\frac{h'_{fg}g(\rho_l-\rho_v)k_l^3}{L(T_{sat}-T_w)\nu_l}\right]^{1/4}$$

However, for water, we let $h'_{fg} \simeq h_{fg}$ and evaluate h_{fg} at the saturation temperature of 302 K; from Table A.12*a*, $h_{fg} = 2.432 \times 10^6$ J/kg, and $\rho_v = 0.03$ kg/m³. Table 7.1 gives $\alpha = 0.33$ for water, so the liquid phase properties are evaluated at $T_r = 299 + 0.33(302 - 299) = 300$ K: from Table A.8, $k_l = 0.611$ W/m K, $\rho_l = 996$ kg/m³, $\nu_l = 0.87 \times 10^{-6}$ m²/s.

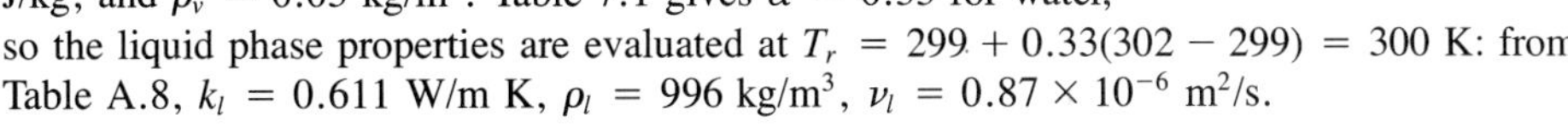

$$\overline{h} = 0.943\left[\frac{h_{fg}g(\rho_l-\rho_v)k_l^3}{L(T_{sat}-T_w)\nu_l}\right]^{1/4} = 0.943\left[\frac{(2.432\times10^6)(9.81)(996-0.03)(0.611)^3}{(0.5)(3)(0.87\times10^{-6})}\right]^{1/4}$$

$$= 7570 \text{ W/m}^2\text{ K}$$

(ii) The total condensation rate $\dot{m}$ is

$$\dot{m} = \frac{\dot{Q}}{h_{fg}} = \frac{\overline{h}\Delta TA}{h_{fg}} = \frac{(7570)(3)(\pi)(0.05)(0.5)}{(2.432\times10^6)} = 7.33\times10^{-4}\text{ kg/s}$$

(iii) The film thickness is obtained from Eq. (7.5): for $\rho_v \ll \rho_l$,

$$\delta = \left(\frac{3\nu_l\Gamma}{\rho_l g}\right)^{1/3}$$

The mass flow rate per unit width of film Γ is

$$\Gamma = \frac{\dot{m}}{\pi D} = \frac{(7.33\times10^{-4})}{(\pi)(0.05)} = 4.67\times10^{-3}\text{ kg/m s}$$

Hence,

$$\delta = \left[\frac{3(0.87\times10^{-6})(4.67\times10^{-3})}{(996)(9.81)}\right]^{1/3} = 1.08\times10^{-4}\text{ m (0.108 mm)}$$

Comments

1. After reading Section 7.2.2, check the film flow to see if it is laminar as assumed.
2. In practice, the wall temperature will seldom be constant along the tube (see Exercise 7–6).

7.2.2 Wavy Laminar and Turbulent Film Condensation on a Vertical Wall

The Reynolds number of a falling film is defined in terms of the mean velocity u_m and hydraulic diameter D_h of the film:

$$u_m = \frac{\Gamma}{\rho_l \times 1 \times \delta} = \frac{\Gamma}{\rho_l \delta}; \qquad D_h = \frac{4A_c}{\mathcal{P}} = \frac{4 \times 1 \times \delta}{1}$$

for a unit width of film and recognizing that the only wetted surface is the wall. Thus,

$$\text{Re} = \frac{\rho_l u_m D_h}{\mu_l} = \frac{\rho_l (\Gamma/\rho_l \delta) 4\delta}{\mu_l} = \frac{4\Gamma}{\mu_l} \tag{7.18}$$

The film Reynolds number is usually used to characterize the film flow.[1] Three regimes of film flow may be distinguished: *laminar*, *wavy laminar*, and *turbulent*. At low Reynolds numbers, the flow is laminar, and the surface of the film appears to be smooth. As the Reynolds number increases, ripples appear on the surface of the film, which at still higher Reynolds numbers develops into a complex three-dimensional wave pattern, as shown in Fig. 7.5. The waves cause some mixing of

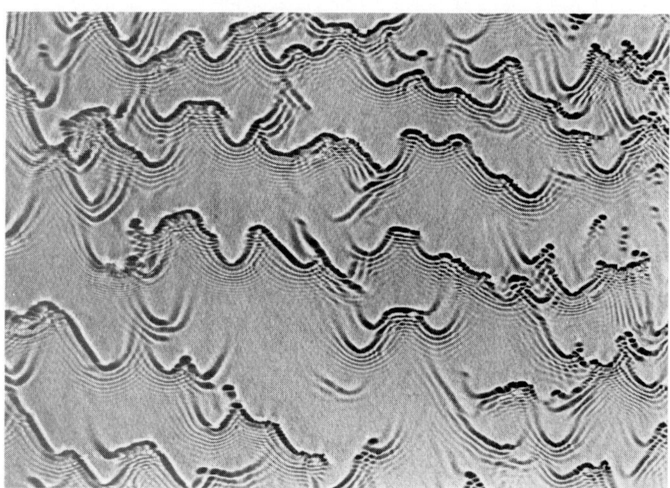

Figure 7.5 Surface waves on a laminar falling film, Reynolds number $\text{Re} = 4\Gamma/\mu_l = 280$. (Photograph courtesy of Mr. B. Dooher, University of California, Los Angeles.)

[1] Note that in USSR literature, the factor of 4 is omitted in the film Reynolds number definition.

the liquid, but the base flow remains laminar until, at a sufficiently high flow rate, shear-induced instabilities result in a transition to turbulent flow throughout the film. For water at ~300 K, the onset of wavy laminar flow is at Re $\simeq$ 30; transition to turbulent flow in the outer region of the film occurs at Re $\simeq$ 1000, and transition is complete in the inner region at Re $\simeq$ 1800. As was demonstrated in Section 7.2.1, analysis of flow in the laminar regime is straightforward, and excellent agreement between theory and experiment is obtained. On the other hand, rigorous analysis of flow in the wavy laminar and turbulent regimes is very difficult, owing to the complexity of the wave motion on the surface and the nature of turbulent flow. There are large, long-wavelength waves caused by gravity, and superimposed small, short-wavelength waves caused by surface tension, and both play a role in heat transport near the liquid surface. Thus, for engineering purposes, it is necessary to use experimental data for the wavy laminar and turbulent regimes. If the problem is formulated correctly, the use of such data is straightforward.

The first step is to reorganize the analysis of laminar film condensation given in Section 7.2.1. For algebraic simplicity, we assume $\rho_v \ll \rho_l$, which is valid except near the critical point, so that Eqs. (7.5) and (7.18) can be rearranged to give the film thickness δ in terms of the film Reynolds number as

$$\frac{\delta}{(\nu_l^2/g)^{1/3}} = \left(\frac{3}{4}\mathrm{Re}\right)^{1/3} \tag{7.19}$$

where the grouping $(\nu_l^2/g)^{1/3}$ is seen to have the dimensions of length. The same grouping is used as a length scale to define a Nusselt number for condensation,

$$\mathrm{Nu} = \frac{h(\nu_l^2/g)^{1/3}}{k_l} \tag{7.20}$$

since other possible length scales are less suitable. The film thickness δ is appropriate on physical grounds but is an unknown and varies down the wall. Use of the distance down the wall x is inappropriate since the local heat transfer coefficient depends only on the local film thickness and not on the prior history of the film. Then, substituting for δ from Eq. (7.9) and using Eq. (7.19) gives

$$\mathrm{Nu} = \left(\frac{3}{4}\mathrm{Re}\right)^{-1/3}; \qquad \mathrm{Re} < 30, \text{ laminar} \tag{7.21}$$

Equation (7.21) is valid for laminar films irrespective of wall temperature variation, whether or not the film has a zero or finite initial thickness, or, indeed, whether condensation or evaporation is taking place. It applies *locally* at a particular value of x and states that the local Nusselt number depends only on the local film Reynolds number or, equivalently, on the local value of Γ or δ.

The next step is to obtain relationships equivalent to Eq. (7.21) from experiment for wavy laminar and turbulent flows. Correlations of experimental data for water given by Chun and Seban [4] are

$$\mathrm{Nu} = 0.822\mathrm{Re}^{-0.22}; \qquad 30 < \mathrm{Re} < \mathrm{Re}_{tr}, \text{ wavy laminar} \tag{7.22}$$

$$\mathrm{Nu} = 3.8 \times 10^{-3}\mathrm{Re}^{0.4}\mathrm{Pr}_l^{0.65}; \qquad \mathrm{Re}_{tr} < \mathrm{Re}, \text{ turbulent} \tag{7.23}$$

where, for water, transition to a turbulent film takes place at

$$\mathrm{Re}_{\mathrm{tr}} = 5800\mathrm{Pr}_l^{-1.06} \tag{7.24}$$

Since liquid Prandtl numbers are strongly temperature-dependent, so is $\mathrm{Re}_{\mathrm{tr}}$. Equations (7.21), (7.22), and (7.23) are shown plotted in Fig. 7.6.

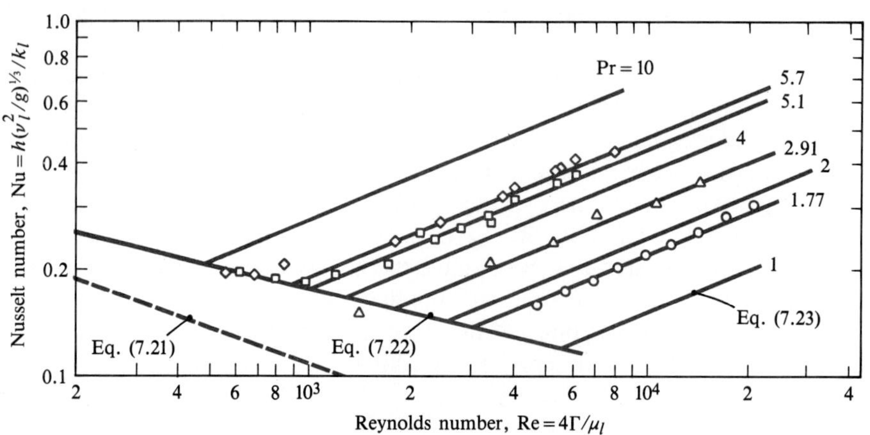

Figure 7.6 Local Nusselt number $\mathrm{Nu} = h(\nu_l^2/g)^{1/3}/k_l$ as a function of Reynolds number $\mathrm{Re} = 4\Gamma/\mu_l$, for the wavy laminar and turbulent regimes on a vertical surface: Eqs. (7.21), (7.22), and (7.23) and the experimental data for water of Chun and Seban [4].

To obtain a relation between average Nusselt number and film Reynolds number, we now rewrite the energy balance Eq. (7.13) using Eqs. (7.9), (7.18), and (7.20) as

$$\frac{dx}{(\nu_l^2/g)^{1/3}} = \frac{\mu_l h_{\mathrm{fg}}}{4k_l(T_{\mathrm{sat}} - T_w)\mathrm{Nu}} d\mathrm{Re}$$

or, introducing the Prandtl number $\mathrm{Pr}_l = c_{pl}\mu_l/k_l$ and the Jakob number $\mathrm{Ja}_l = c_{pl}(T_{\mathrm{sat}} - T_w)/h_{\mathrm{fg}}$,

$$\frac{dx}{(\nu_l^2/g)^{1/3}} = \frac{\mathrm{Pr}_l}{4\mathrm{Ja}_l}\frac{d\mathrm{Re}}{\mathrm{Nu}} \tag{7.25}$$

For an isothermal wall and condensation commencing at $x = 0$, Γ and $\mathrm{Re} = 0$, and Eq. (7.25) can be integrated down a wall of length L as

$$\frac{L}{(\nu_l^2/g)^{1/3}} = \frac{\mathrm{Pr}_l}{4\mathrm{Ja}_l}\int_0^{\mathrm{Re}_L}\frac{d\mathrm{Re}}{\mathrm{Nu}} \tag{7.26}$$

Finally, an overall energy balance on unit width of film from $x = 0$ to $x = L$ requires that the heat transfer to the wall equal the enthalpy of phase change given up by the condensate:

$$\bar{h}L(T_{\mathrm{sat}} - T_w) = \Gamma_L h_{\mathrm{fg}} = \frac{\mu_l h_{\mathrm{fg}}}{4}\mathrm{Re}_L$$

then

$$\overline{\mathrm{Nu}} = \frac{\overline{h}(\nu_l^2/g)^{1/3}}{k_l} = \frac{\mu_l h_{fg}(\nu_l^2/g)^{1/3}}{4k_l L(T_{sat} - T_w)}\mathrm{Re}_L$$

or

$$\overline{\mathrm{Nu}} = \frac{\mathrm{Pr}_l}{4\mathrm{Ja}_l}\frac{(\nu_l^2/g)^{1/3}}{L}\mathrm{Re}_L \tag{7.27}$$

Re_L can be obtained by integrating Eq. (7.26) using appropriate Nu(Re) formulas and, when substituted in Eq. (7.27), gives the desired result.

Laminar Film Condensation

Consider first the situation where $\mathrm{Re}_L < 30$, so that the film is always laminar. Then substitution of Eq. (7.21) into Eq. (7.26) gives

$$\frac{L}{(\nu_l^2/g)^{1/3}} = \frac{\mathrm{Pr}_l}{4\mathrm{Ja}_l}\left(\frac{3}{4}\right)^{1/3}\int_0^{\mathrm{Re}_L}\mathrm{Re}^{1/3}d\mathrm{Re}$$
$$= \frac{\mathrm{Pr}_l}{4\mathrm{Ja}_l}\left(\frac{3}{4}\right)^{4/3}\mathrm{Re}_L^{4/3}$$

or

$$\mathrm{Re}_L = \frac{4}{3}\left[\frac{4\mathrm{Ja}_l}{\mathrm{Pr}_l}\frac{L}{(\nu_l^2/g)^{1/3}}\right]^{3/4}$$

Substituting in Eq. (7.27) gives the average Nusselt number as

$$\overline{\mathrm{Nu}} = \frac{4}{3}\left[\frac{\mathrm{Pr}_l}{4\mathrm{Ja}_l}\frac{(\nu_l^2/g)^{1/3}}{L}\right]^{1/4} \tag{7.28}$$

Of course, Eq. (7.28) is simply Eq. (7.16) rearranged, with $\rho_v \ll \rho_l$.

Wavy Laminar Film Condensation

Next, consider the situation where $30 < \mathrm{Re}_L < \mathrm{Re}_{tr}$, with Re_{tr} for water given by Eq. (7.24). The integration in Eq. (7.26) is now performed for $0 < \mathrm{Re} < 30$ using Eq. (7.21) for Nu, and for $30 < \mathrm{Re} < \mathrm{Re}_L$ using Eq. (7.22):

$$\frac{L}{(\nu_l^2/g)^{1/3}} = \frac{\mathrm{Pr}_l}{4\mathrm{Ja}_l}\left[\left(\frac{3}{4}\right)^{1/3}\int_0^{30}\mathrm{Re}^{1/3}d\mathrm{Re} + \frac{1}{0.822}\int_{30}^{\mathrm{Re}_L}\mathrm{Re}^{0.22}d\mathrm{Re}\right]$$
$$= \frac{\mathrm{Pr}_l}{4\mathrm{Ja}_l}\left[\left(\frac{3}{4}\right)^{4/3}(30)^{4/3} + \frac{1}{(0.822)(1.22)}\left(\mathrm{Re}_L^{1.22} - 30^{1.22}\right)\right]$$
$$= 1.00\frac{\mathrm{Pr}_l}{4\mathrm{Ja}_l}\mathrm{Re}_L^{1.22} \tag{7.29}$$

The value of the coefficient 1.00 is fortuitous; however, the two constant terms cancel out because both Eqs. (7.21) and (7.22) give the same value of Nu at Re = 30. Solving,

$$\mathrm{Re}_L = \left[\frac{4\mathrm{Ja}_l}{\mathrm{Pr}_l}\frac{L}{(\nu_l^2/g)^{1/3}}\right]^{0.82}$$

Substituting in Eq. (7.27) gives the average Nusselt number as

$$\overline{\mathrm{Nu}} = \left[\frac{\mathrm{Pr}_l}{4\mathrm{Ja}_l}\frac{(\nu_l^2/g)^{1/3}}{L}\right]^{0.18} \tag{7.30}$$

Turbulent Film Condensation

Finally, consider the situation where $\mathrm{Re}_L > \mathrm{Re}_{\mathrm{tr}}$, that is, transition to a turbulent film takes place on the wall. First, from Eq. (7.29), the location of transition x_{tr} is given by

$$\frac{x_{\mathrm{tr}}}{(\nu_l^2/g)^{1/3}} = 1.00\frac{\mathrm{Pr}_l}{4\mathrm{Ja}_l}\mathrm{Re}_{\mathrm{tr}}^{1.22} \tag{7.31}$$

Equation (7.26) is now integrated using Eq. (7.23) for $\mathrm{Re}_{\mathrm{tr}} < \mathrm{Re} < \mathrm{Re}_L$:

$$\frac{L}{(\nu_l^2/g)^{1/3}} = \frac{\mathrm{Pr}_l}{4\mathrm{Ja}_l}\left\{\mathrm{Re}_{\mathrm{tr}}^{1.22} + \frac{1}{(3.8\times 10^{-3})\mathrm{Pr}_l{}^{0.65}}\int_{\mathrm{Re}_{\mathrm{tr}}}^{\mathrm{Re}_L}\mathrm{Re}^{-0.4}\,d\mathrm{Re}\right\}$$

$$= \frac{x_{\mathrm{tr}}}{(\nu_l^2/g)^{1/3}} + \frac{\mathrm{Pr}_l^{0.35}}{4(3.8\times 10^{-3})(0.6)\mathrm{Ja}_l}\left[\mathrm{Re}_L^{0.6} - \mathrm{Re}_{\mathrm{tr}}^{0.6}\right]$$

Solving for Re_L and substituting in Eq. (7.27),

$$\overline{\mathrm{Nu}} = \frac{\mathrm{Pr}_l}{4\mathrm{Ja}_l}\frac{(\nu_l^2/g)^{1/3}}{L}\left\{\frac{9.12\times 10^{-3}\mathrm{Ja}_l(L - x_{\mathrm{tr}})}{(\nu_l^2/g)^{1/3}\mathrm{Pr}_l^{0.35}} + \mathrm{Re}_{\mathrm{tr}}^{0.6}\right\}^{10/6} \tag{7.32}$$

which is the desired result.

Effect of Variable Properties

A reference temperature scheme to account for variable-property effects has not been established for wavy laminar and turbulent film condensation. Thus, the schemes developed for laminar film condensation described in Section 7.2.1 are tentatively recommended for use here.

Computer Program PHASE

Item 1 in the computer program PHASE calculates the average heat transfer coefficient and condensate flow rate per unit width for film condensation on an isothermal vertical surface according to the analysis of Section 7.2.2. Local heat transfer coefficients and transition Reynolds numbers for wavy laminar and turbulent flow can be calculated from the formulas given in Section 7.2.2, which are, strictly speaking, valid only for water. User-supplied constants and exponents for these formulas are also allowed. Variable-property and liquid subcooling effects are calculated following the recommendation given in Section 7.2.1. (The effect of vapor superheat is accounted for according to the recommendation in Section 7.2.4.) SI units, with temperature in kelvins, must be used.

There is a menu of six fluids: (1) water, (2) ammonia, (3) nitrogen, (4) mercury, (5) R-12, and (6) R-113.

EXAMPLE 7.2 Condensation of Steam on a Long Vertical Tube

In a multiple-effect evaporator, saturated steam at 27,150 Pa condenses on the outside of a 5 cm–diameter, 8 m–long vertical tube. If the tube outer wall can be taken to be at 320 K, calculate the average heat transfer coefficient and the total condensation rate.

Solution

Given: Film condensation of steam on a vertical surface.

Required: Average heat transfer coefficient $\overline{h}$ and total condensation rate $\dot{m}$.

Assumptions:
1. Effect of wall curvature negligible.
2. Effect of liquid subcooling negligible.
3. Isothermal surface.

We first check to see if transition to turbulent flow takes place on the wall. For this purpose, and for later use, we evaluate water properties at the reference temperature, $T_r = T_w + \alpha(T_{\text{sat}} - T_w)$, where, from Table 7.1, $\alpha = 0.33$ for water. From Table A.12*a*, $T_{\text{sat}} = 340.0$ K; thus,

$$T_r = 320 + 0.33(340 - 320) = 327 \text{ K}$$

From Table A.8, $k_l = 0.649$ W/m K, $\rho_l = 986$ kg/m^3, $c_{pl} = 4177$ J/kg K, $\nu_l = 0.53 \times 10^{-6}$ m^2/s, $\text{Pr}_l = 3.36$. Also, at the saturation temperature of 340 K, Table A.12*a* gives $h_{\text{fg}} = 2.341 \times 10^6$ J/kg. Then from Eq. (7.24)

$$\text{Re}_{\text{tr}} = 5800\text{Pr}_l^{-1.06} = 5800(3.36)^{-1.06} = 1605$$

From Eq. (7.31), $x_{\text{tr}} = (\nu_l^2/g)^{1/3}(\text{Pr}_l/4\text{Ja}_l)\text{Re}_{\text{tr}}^{1.22}$, where

$$(\nu_l^2/g)^{1/3} = [(0.53 \times 10^{-6})^2/9.81]^{1/3} = 3.06 \times 10^{-5} \text{ m}$$

$$\text{Ja}_l = c_{pl}(T_{\text{sat}} - T_w)/h_{\text{fg}} = (4177)(20)/(2.341 \times 10^6) = 0.0357$$

Hence, $x_{\text{tr}} = (3.06 \times 10^{-5})[3.36/(4)(0.0357)](1605)^{1.22} = 5.86$ m.

Transition is seen to take place about three-quarters of the way down the wall. Equation (7.32) gives the average Nusselt number:

$$\overline{\mathrm{Nu}} = \frac{\mathrm{Pr}_l}{4\mathrm{Ja}_l}\frac{(\nu_l^2/g)^{1/3}}{L}\left\{\frac{9.12\times 10^{-3}\mathrm{Ja}_l(L-x_{\mathrm{tr}})}{(\nu^2/g)^{1/3}\mathrm{Pr}_l^{0.35}} + \mathrm{Re}_{\mathrm{tr}}^{0.6}\right\}^{10/6}$$

$$= \frac{(3.36)(3.06\times 10^{-5})}{(4)(0.0357)(8)}\left\{\frac{(9.12\times 10^{-3})(0.0357)(8-5.86)}{(3.06\times 10^{-5})(3.36)^{0.35}} + 1605^{0.6}\right\}^{10/6}$$

$$= 0.1897$$

$$\bar{h} = \frac{\overline{\mathrm{Nu}}\, k_l}{(\nu_l^2/g)^{1/3}} = \frac{(0.1897)(0.649)}{3.06\times 10^{-5}} = 4020\ \mathrm{W/m^2\,K}$$

$$\dot{m} = \frac{\dot{Q}}{h_{fg}} = \frac{\bar{h}\Delta TA}{h_{fg}} = \frac{(4020)(20)(\pi)(0.05)(8)}{2.341\times 10^6} = 4.32\times 10^{-2}\ \mathrm{kg/s}$$

Solution using PHASE

The required input in SI units is:

Item 1
Y (correlations for water)
Fluid = 1 (H_2O)
Pressure = 27,150
Vapor temperature = 0 (gives saturation value)
T_w = 320
L = 8

PHASE gives the output:

Relevant property data
Re_L = 2080
Γ_L = 0.272 kg/m s
$\bar{h}$ = 4034 W/m² K

Comments

1. Using PHASE to obtain the condensation rate gives $\dot{m} = \Gamma\pi D = (0.272)(\pi)(0.05) = 4.27\times 10^{-2}$ kg/m s.

2. Notice that Nu values are $\sim 10^{-1}$, which is much smaller than the values for convection calculated in Chapter 4. Why?

7.2.3 Laminar Film Condensation on Horizontal Tubes

Shell-and-tube condensers, in which condensation takes place on a bank of horizontal tubes, are widely used in power plants and the process industries. Figure 7.7 depicts laminar film condensation on a single horizontal tube and a force balance on a liquid element. The buoyancy force per unit volume of liquid is now $(\rho_l - \rho_v)g \sin\phi$, with

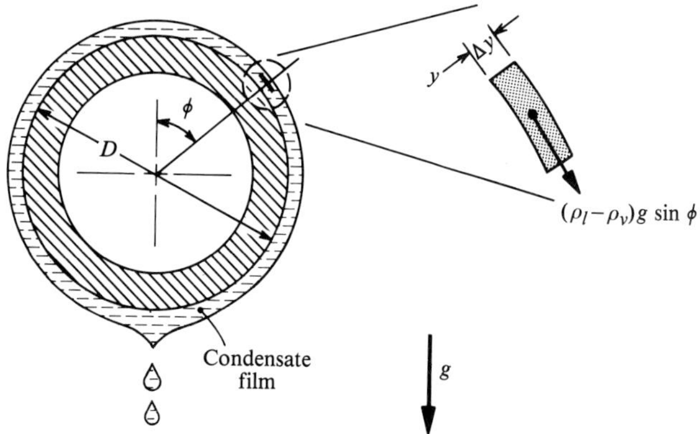

Figure 7.7 Coordinate system and gravitational force for film condensation on a horizontal condenser tube.

angle ϕ measured from the top of the tube. Thus, Eq. (7.1) becomes

$$\mu_l \frac{\partial^2 u}{\partial y^2} + (\rho_l - \rho_v)g \sin \phi = 0 \tag{7.33}$$

whereas Eq. (7.6) remains unchanged:

$$\frac{\partial^2 T}{\partial y^2} = 0 \tag{7.34}$$

Integrating Eq. (7.33) as before gives the mass flow rate per unit width as

$$\Gamma = \frac{(\rho_l - \rho_v)g \sin \phi \, \delta^3}{3\nu_l} \tag{7.35}$$

and the film Reynolds number $4\Gamma/\mu_l$ as

$$\text{Re} = \frac{4}{3} \frac{(\rho_l - \rho_v)g \sin \phi \, \delta^3}{\rho_l \nu_l^2} \tag{7.36}$$

Integrating Eq. (7.34) as before gives the heat transfer coefficient

$$h = \frac{q}{T_{\text{sat}} - T_w} = \frac{k_l \, (\partial T/\partial y)\big|_w}{T_{\text{sat}} - T_w} = \frac{k_l}{\delta} \tag{7.37}$$

where q has been taken to be positive for condensation. Hence, the energy balance Eq. (7.13) may be written as

$$q = k_l \left.\frac{\partial T}{\partial y}\right|_w = \frac{k_l(T_{\text{sat}} - T_w)}{\delta} = h_{\text{fg}} \frac{d\Gamma}{dx} \tag{7.38}$$

To proceed, the wall boundary condition must be specified: here we will obtain the solution for an isothermal wall T_w; the solution for a constant heat flux is given as Exercise 7–19. Equation (7.36) is solved for δ,

$$\delta = \left(\frac{3}{4}\frac{\rho_l \nu_l^2}{(\rho_l - \rho_v)g}\frac{\text{Re}}{\sin\phi}\right)^{1/3}$$

and, together with the relations $\Gamma = \mu_l \text{Re}/4$ and $x = (D/2)\phi$, is substituted in Eq. (7.38), to obtain

$$\frac{2k_l(T_{\text{sat}} - T_w)D}{\mu_l h_{\text{fg}}}\left(\frac{4}{3}\frac{(\rho_l - \rho_v)g}{\rho_l \nu_l^2}\right)^{1/3}\sin^{1/3}\phi \, d\phi = \text{Re}^{1/3} d\text{Re}$$

Integrating with Re $= 0$ at $\phi = 0$, and Re $= \text{Re}_\pi$ at $\phi = \pi$ gives

$$\text{Re}_\pi = \left[\left(\frac{4}{3}\right)\frac{2k_l(T_{\text{sat}} - T_w)D}{\mu_l h_{\text{fg}}}\left(\frac{4}{3}\frac{(\rho_l - \rho_v)g}{\rho_l \nu_l^2}\right)^{1/3}\int_0^\pi \sin^{1/3}\phi \, d\phi\right]^{3/4} \tag{7.39}$$

Notice that we choose to use Re rather than δ as the dependent variable. The Reynolds number Re is zero at $\phi = 0$ because Γ, the flow rate, is zero at the top of the tube by symmetry, whereas the film thickness δ is finite and unknown at $\phi = 0$.

An overall energy balance on half the tube gives the average heat transfer coefficient:

$$\overline{h}(\pi D/2)(T_{\text{sat}} - T_w) = \Gamma_\pi h_{\text{fg}} = \mu_l h_{\text{fg}} \text{Re}_\pi/4 \tag{7.40}$$

Substituting for Re_π from Eq. (7.39) and rearranging gives

$$\overline{h} = \left(\frac{4}{3\pi}\right)\left(\frac{1}{2}\right)^{1/4}\left(\int_0^\pi \sin^{1/3}\phi \, d\phi\right)^{3/4}\left[\frac{(\rho_l - \rho_v)g h_{\text{fg}} k_l^3}{\nu_l D(T_{\text{sat}} - T_w)}\right]^{1/4}$$

From mathematical tables, we can find that

$$\int_0^{\pi/2} \sin^n x \, dx = \frac{\pi^{1/2}}{2}\frac{\Gamma(n/2 + 1/2)}{\Gamma(1 + n/2)}, \qquad \text{for } n > -1$$

(Γ is the gamma function)

$$= 1.2946 \qquad \text{for } n = 1/3$$

Hence, $\int_0^\pi \sin^{1/3}\phi \, d\phi = 2\int_0^{\pi/2} \sin^{1/3}\phi \, d\phi = 2.5892$, and finally

$$\overline{h} = 0.728\left[\frac{(\rho_l - \rho_v)g h_{\text{fg}} k_l^3}{\nu_l D(T_{\text{sat}} - T_w)}\right]^{1/4} \tag{7.41}$$

Most commercial condensers condense steam on a large bundle of horizontal tubes, and condensate drips from one tube to the next, as shown in Fig. 7.8. If the analysis leading up to Eq. (7.41) is repeated for N tubes arranged vertically above one another, such that the condensate drips from one tube to the next, the result is

$$\overline{h} = 0.728\left[\frac{(\rho_l - \rho_v)g h_{\text{fg}} k_l^3}{N \nu_l D(T_{\text{sat}} - T_w)}\right]^{1/4} \tag{7.42}$$

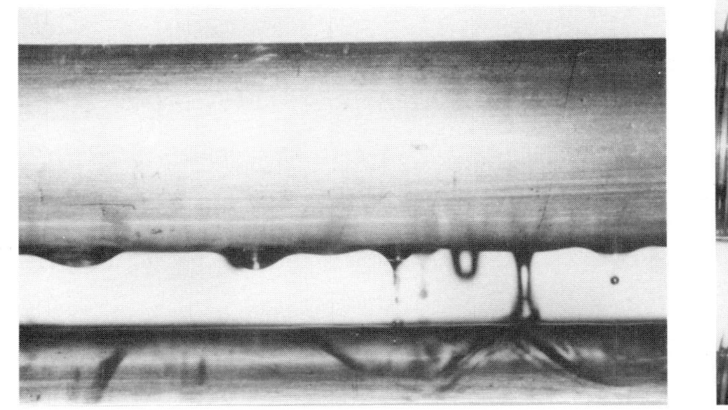

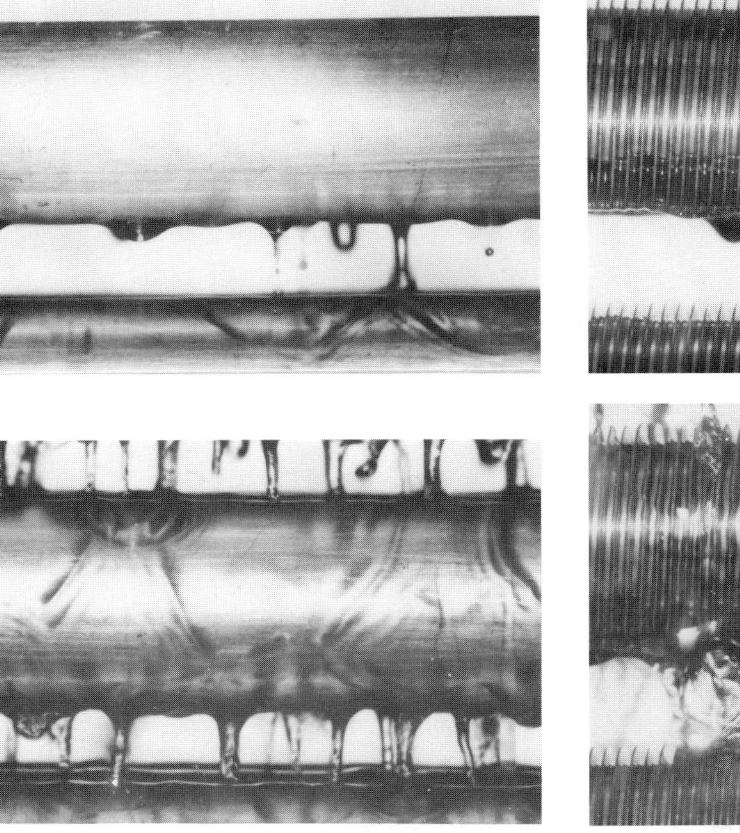

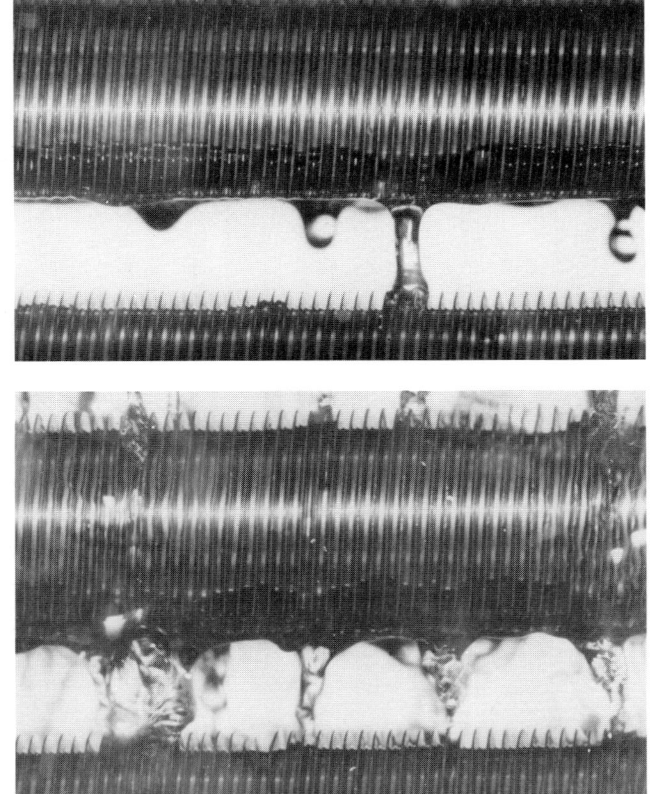

(a) (b)

Figure 7.8 Condensation of downward-flowing R-113 vapor on in-line bundles of horizontal tubes. (*a*) Smooth tubes, 15.9 mm diameter. $P = 0.12$ MPa, $\Delta T = 10$ K. 1st row, $\mathrm{Re}_D = 19$; 13th row, $\mathrm{Re}_D = 150$. (*b*) Low-fin tubes (see Exercise 7–29) 15.6 mm diameter, $P = 0.11$ MPa, $\Delta T = 3.5$ K. 1st row, $\mathrm{Re}_D = 50$; 13th row, $\mathrm{Re}_D = 550$. (Photographs courtesy of Professor H. Honda, Kyushu University, Kasuga.)

Effect of Variable Fluid Properties

The reference property scheme described for laminar film condensation on a vertical wall is also applicable to condensation on a horizontal tube.

Computer Program PHASE

Item 2 of PHASE calculates the average heat transfer coefficient and condensate flow rate per unit length, for laminar film condensation on isothermal horizontal tubes, based on the analysis of Section 7.2.3. Variable-property and liquid subcooling effects are calculated following the recommendations given in Section 7.2.1; vapor superheat effects are accounted for following the recommendations in Section 7.2.4.

EXAMPLE 7.3 Condensation of Refrigerant-12 on a Single Horizontal Tube

Saturated R-12 vapor at 320 K condenses on the outside of a 2 cm–O.D., 1 mm–wall-thickness horizontal brass tube, through which flows coolant water. At an axial location where the bulk water temperature is 295 K and the inside heat transfer coefficient is 4500 W/m^2 K, determine the average heat flux and condensation rate per unit length.

Solution

Given: Condensation of R-12 on a horizontal tube.

Required: Heat flux and condensation rate per unit length.

Assumptions: 1. The outer tube surface is isothermal.
2. Laminar film condensation.

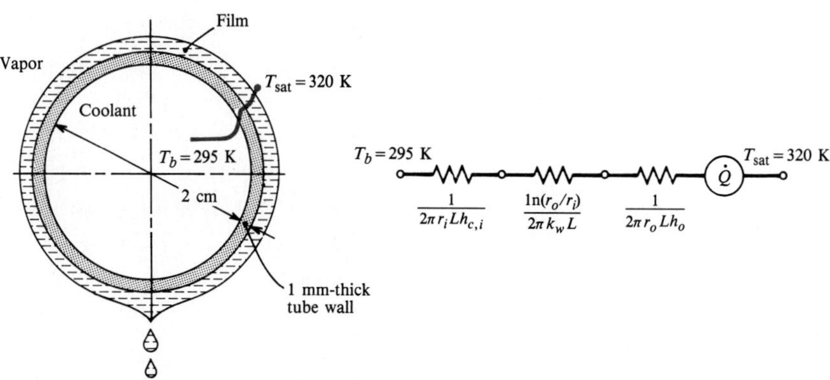

Referring to Section 2.3.1, the thermal circuit is as shown. Using Eq. (2.17), the overall heat transfer coefficient–area product is

$$\frac{1}{UA} = \frac{1}{2\pi r_i L h_{c,i}} + \frac{\ln(r_o/r_i)}{2\pi k_{\text{brass}} L} + \frac{1}{2\pi r_o L h_o}$$

If U is based on the outside area of the tube, $A = 2\pi r_o L$, and

$$\frac{1}{U} = \frac{1}{(r_i/r_o)h_{c,i}} + \frac{r_o \ln(r_o/r_i)}{k_{\text{brass}}} + \frac{1}{h_o}$$

$$= \frac{1}{(0.009/0.01)(4500)} + \frac{(0.01)\ln(0.01/0.009)}{111} + \frac{1}{h_o}$$

$$= 246.9 \times 10^{-6} + 9.49 \times 10^{-6} + \frac{1}{h_o}$$

where $k_{\text{brass}} = 111$ W/m K was obtained from Table A.1*a*. Hence,

$$\dot{Q}/A = U(T_{\text{sat}} - T_b) = \frac{320 - 295}{2.56 \times 10^{-4} + 1/h_o}$$

To evaluate the condensation heat transfer coefficient h_o, we first guess a wall temperature T_w of 300 K and evaluate liquid R-12 properties at the mean film temperature of 310 K. From Table A.8, $k_l = 0.0695$ W/m K, $\rho_l = 1263$ kg/m^3, $\nu_l = 0.193 \times 10^{-6}$ m^2/s, $c_{pl} = 995$ J/kg K, $\text{Pr}_l = 3.5$. From Table A.12*e*, at 320 K, $h_{\text{fg}} = 1.237 \times 10^5$ J/kg, and $\rho_v = 1/0.01531 = 65.3$ kg/m^3. Equation (7.17*b*) gives h'_{fg}:

$$h'_{\text{fg}} = h_{\text{fg}} + \left(0.683 - \frac{0.228}{\text{Pr}_l}\right) c_{pl}(T_{\text{sat}} - T_w)$$

$$= 1.237 \times 10^5 + \left(0.683 - \frac{0.228}{3.5}\right)(995)(320 - 300) = 1.360 \times 10^5 \text{ J/kg}$$

Substituting in Eq. (7.41), with h'_{fg} replacing h_{fg} to account for liquid subcooling,

$$h_o = 0.728\left(\frac{(\rho_l - \rho_v)g h'_{\text{fg}} k_l^3}{\nu_l D(T_{\text{sat}} - T_w)}\right)^{1/4}$$

$$= 0.728\left(\frac{(1263 - 65)(9.81)(1.360 \times 10^5)(0.0695)^3}{(0.193 \times 10^{-6})(0.02)(20)}\right)^{1/4}$$

$$= 1182 \text{ W/m}^2\text{K}$$

$$\frac{\dot{Q}}{A} = \frac{25}{(2.56 + 8.46)10^{-4}} = 2.269 \times 10^4 \text{ W/m}^2 \qquad \text{(based on tube outside area)}$$

We now calculate T_w and compare it to our guessed value of 300 K.

$$T_{\text{sat}} - T_w = \frac{\dot{Q}/A}{h_o} = \frac{2.269 \times 10^4}{1182} = 19.2 \text{ K}; \qquad T_w = 320 - 19.2 = 300.8 \text{ K}$$

For a second iteration, use of new values of h'_{fg} and properties is not justified; thus, $h_o = (19.2/20)^{-1/4}(1182) = 1194$ W/m^2 K; $\dot{Q}/A = 2.286 \times 10^4$ W/m^2 K. The condensation rate per unit length of tube is

$$\dot{m} = \frac{(\dot{Q}/A)\pi D}{h'_{\text{fg}}} = \frac{(2.286 \times 10^4)(\pi)(0.02)}{1.360 \times 10^5} = 1.056 \times 10^{-2} \text{ kg/s per meter}$$

Comments

1. Use PHASE to check h_o.
2. The film Reynolds number at the bottom of the tube is $(1/2)(4)(1.056 \times 10^{-2})/(1263)(0.193 \times 10^{-6}) = 87$. There is the possibility of wavy laminar flow, but precise criteria for ripple formation on R-12 films have yet to be developed.
3. Note the use of h'_{fg} to obtain $\dot{m}$ from $\dot{Q}/A$ [see Eq. (7.12)].

7.2.4 Effects of Vapor Velocity and Vapor Superheat

To obtain some insight into the effects of vapor velocity and vapor superheat, we return to the problem of laminar film condensation on a vertical wall, analyzed in Section 7.2.1. Figure 7.9 shows expected velocity and temperature profiles if the vapor flows down past the surface at a velocity U_e and has temperature T_e.

Effect of Vapor Velocity

To account for the effect of vapor drag on the liquid film, the boundary condition Eq. (7.2*b*) must be replaced by

$$y = \delta: \quad \mu_l \frac{\partial u}{\partial y} = \tau_s \tag{7.43}$$

where τ_s is obtained from an appropriate skin friction coefficient:

$$\tau_s = C_f \frac{1}{2} \rho_v U_e^2$$

It might appear that an appropriate expression for C_f can be obtained from Chapter 4, for example, Eq. (4.54) if the vapor boundary layer were laminar, or Eq. (4.60)

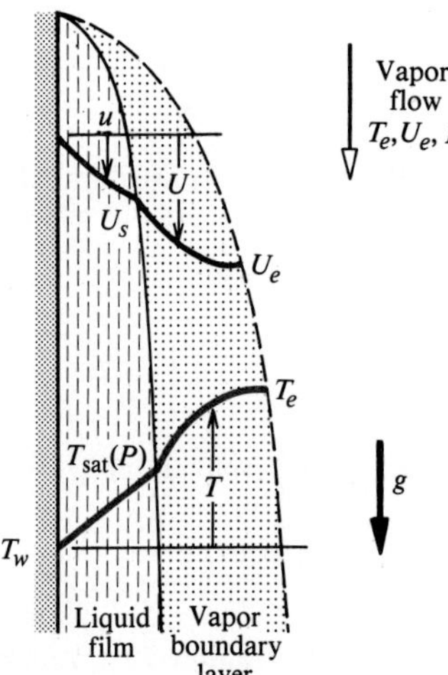

Figure 7.9 Velocity and temperature profiles for film condensation on a vertical wall from a flowing superheated vapor.

if it were turbulent. However, these relations are valid only for an impermeable wall, whereas the surface of the liquid film has a velocity component normal to it due to condensation. In fluid mechanics texts, it would be said that there is *suction* at the surface: for typical condensation problems, the suction velocity is relatively large and causes the boundary layer thickness to become almost constant quite close to the leading edge, and the vapor velocity U becomes essentially a function of y only. Figure 7.10 shows a model laminar Couette flow that will be used to determine shear stress for a boundary layer under strong suction. The free-stream velocity is U_e, and the surface velocity is U_s, which will be assumed to be independent of x. In fact, $U_s = u_\delta$ increases with x in the actual film condensation problem, but since we are concerned with situations where $u_\delta \ll U_e$, the error introduced by this assumption is small. An elemental control volume Δy thick, of area A located inside the vapor boundary layer, is shown in Fig. 7.10. Mass conservation requires that

$$\rho_v V\big|_y A = \rho_v V\big|_{y+\Delta y} A$$
$$\rho_v V = \text{Constant} = \rho_v V_s = -\dot{m}'' \tag{7.44}$$

where the condensation rate $\dot{m}''$ [kg/m^2 s] is again taken to be positive. Newton's second law requires that the sum of the forces acting on the volume equal the rate of change of momentum of the fluid flowing through the volume. In the x direction,

$$\mu_v \frac{dU}{dy}\bigg|_{y+\Delta y} A - \mu_v \frac{dU}{dy}\bigg|_y A = \rho_v VU\big|_{y+\Delta y} A - \rho_v VU\big|_y A$$

Substituting $\rho_v V = -\dot{m}''$ from Eq. (7.44), dividing by $A\Delta y$, and letting $\Delta y \to 0$,

$$\mu_v \frac{d^2U}{dy^2} = -\dot{m}'' \frac{dU}{dy} \tag{7.45}$$

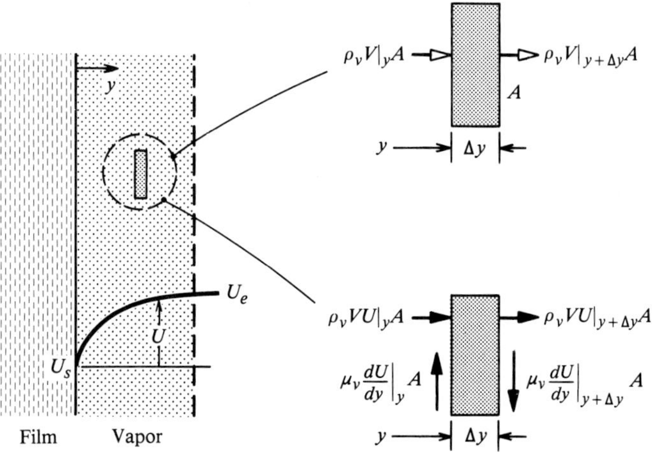

Figure 7.10 Model problem and elemental volumes for analysis of the effect of vapor velocity on condensation heat transfer.

which is the differential equation governing the vapor velocity $U(y)$ for constant vapor viscosity.[2] Appropriate boundary conditions are

$$y = 0: \quad U = U_s \tag{7.46a}$$

$$y \to \infty: \quad U \to U_e \tag{7.46b}$$

Rearranging Eq. (7.45) and integrating,

$$\frac{d^2U/dy^2}{dU/dy} = -\frac{\dot{m}''}{\mu_v}$$

$$\ln\frac{dU}{dy} = -\frac{\dot{m}''}{\mu_v}y + C_0$$

$$\frac{dU}{dy} = C_1 e^{-(\dot{m}''/\mu_v)y}$$

$$U = -C_1\frac{\mu_v}{\dot{m}''}e^{-(\dot{m}''/\mu_v)y} + C_2$$

Evaluating the integration constants from the boundary conditions gives the velocity profile as

$$\frac{U - U_s}{U_e - U_s} = 1 - e^{-(\dot{m}''/\mu_v)y} \tag{7.47}$$

and the shear stress at the film surface is then found to be

$$\tau_s = \mu_v \left.\frac{dU}{dy}\right|_{y=0} = \dot{m}''(U_e - U_s) \tag{7.48}$$

Thus, in this strong suction limit, *the shear stress exerted by the vapor on the liquid film is simply the momentum given up by the condensing vapor* as it decelerates from the free-stream velocity U_e to the film surface velocity U_s. It can be *many times greater* than the shear stress in the absence of suction.

To see when vapor drag might be important, we should compare τ_s with the shear stress at the wall when there is no vapor drag. Then a force balance on the film gives simply $\tau_w = (\rho_l - \rho_v)g\delta$; substituting for δ from Eq. (7.14) and rearranging gives

$$\frac{\tau_s}{\tau_w} = \left[\frac{\text{Ja}_l}{4\text{Pr}_l}\frac{\rho_l}{(\rho_l - \rho_v)}\frac{(U_e - U_s)^2}{gx}\right]^{1/2} \tag{7.49}$$

As expected, vapor drag becomes more important as U_e increases and as the quotient Ja_l/Pr_l increases (since $\text{Ja}_l/\text{Pr}_l = k_l(T_{\text{sat}} - T_w)/h_{fg}\mu_l$ scales the rate of condensation). Vapor drag becomes less important as x increases down the plate, and there is no direct effect of vapor density ρ_v, for $\rho_v \ll \rho_l$.

---

[2] Equation (7.45) can also be obtained from Eq. (5.39) by setting $u = U(y)$, $\rho v = -\dot{m}''$.

Substituting Eq. (7.48) in Eq. (7.43) gives the boundary condition required to account for vapor drag in the analysis of Section 7.2.1 when the condensation rate is sufficiently high; since the velocity profile is continuous, $U_s = u_\delta$, and

$$\mu_l \left.\frac{\partial u}{\partial y}\right|_{y=\delta} = \dot{m}''(U_e - u_\delta) \tag{7.50}$$

The algebra is now more complicated (see Exercise 7–25). For $\rho_v \ll \rho_l$ and $u_\delta \ll U_e$, the local heat transfer coefficient becomes [5]

$$h = \left[\frac{k_l^2 U_e}{8\nu_l x}\left\{1 + \left(1 + \frac{16\text{Pr}_l}{\text{Ja}_l}\frac{gx}{U_e^2}\right)^{1/2}\right\}\right]^{1/2} \tag{7.51}$$

Effect of Vapor Superheat

We turn now to the effect of vapor superheat. To account for the effect of vapor superheat in the analysis of Section 7.2.1, the energy balance on control volume must also include a term for heat transfer from the vapor to the surface of the film; Eq. (7.11) then becomes

$$k_l \left.\frac{dT}{dy}\right|_w = h_c(T_e - T_{\text{sat}}) + \dot{m}''h_v|_s - \frac{d}{dx}\int_0^{\delta(x)} \rho_l u h_l \, dy \tag{7.52}$$

where h_c is the convective heat transfer coefficient for the vapor flowing over the film surface. As was the case for the skin friction coefficient, the Nusselt number correlations in Chapter 4, Eqs. (4.56) and (4.64), cannot be used since the boundary layer is under strong suction. Again we will analyze a model laminar flow problem, as shown in Fig. 7.11, to obtain an estimate for h_c. Convection and conduction in the x direction are neglected, and the steady-flow energy equation applied to an

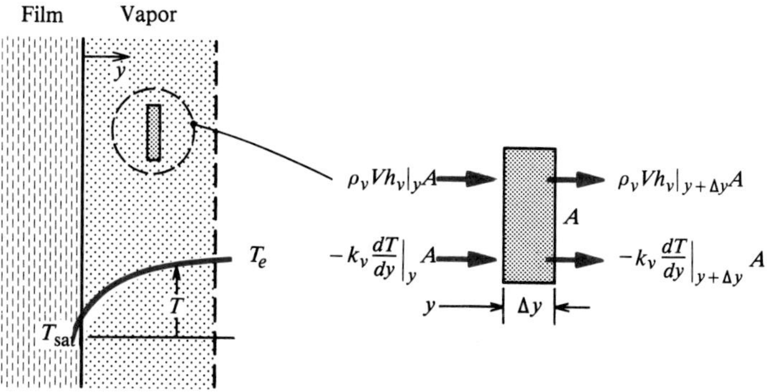

Figure 7.11 Model problem and elemental volume for analysis of the effect of vapor superheat on condensation heat transfer.

elemental control volume Δy thick of cross-sectional area A requires that

$$\rho_v VA\,[h_v|_{y+\Delta y} - h_v|_y] = -k_v A \left.\frac{dT}{dy}\right|_y + k_v A \left.\frac{dT}{dy}\right|_{y+\Delta y}$$

Dividing by $A\Delta y$, rearranging, and letting $\Delta y \to 0$ gives

$$k_v \frac{d^2T}{dy^2} = \rho_v V \frac{dh_v}{dy}$$

From Eq. (7.44), $\rho_v V = -\dot{m}''$, and noting that

$$\frac{dh_v}{dy} = \frac{dh_v}{dT}\frac{dT}{dy} = c_{pv}\frac{dT}{dy}$$

we obtain

$$\frac{d^2T/dy^2}{dT/dy} = -\frac{\dot{m}'' c_{pv}}{k_v} \tag{7.53}$$

which must be solved subject to the boundary conditions

$$y = 0: \quad T = T_{\text{sat}} \tag{7.54a}$$

$$y \to \infty: \quad T \to T_e \tag{7.54b}$$

The mathematical problem is seen to be identical to the vapor drag problem just solved, so the solution may be written down immediately after inspection of Eqs. (7.47) and (7.48). The temperature profile is

$$\frac{T - T_{\text{sat}}}{T_e - T_{\text{sat}}} = 1 - e^{(-\dot{m}'' c_{pv}/k_v)y} \tag{7.55}$$

and the heat transfer coefficient is

$$h_c = \frac{k_v(dT/dy)|_{y=0}}{T_e - T_{\text{sat}}} = \dot{m}'' c_{pv} \tag{7.56}$$

That is, in the limit of strong suction, *the heat transfer from the vapor to the film surface is simply the enthalpy given up by the condensing vapor* as it cools from the free-stream temperature T_e to the surface temperature T_{sat}. Substituting in Eq. (7.52) and rearranging using Eq. (7.10) gives

$$k_l \left.\frac{\partial T}{\partial y}\right|_w = \frac{d}{dx}\int_0^{\delta(x)} \rho_l u\big[\{h_{\text{fg}} + c_{pv}(T_e - T_{\text{sat}})\} + c_{pl}(T_{\text{sat}} - T)\big]\,dy \tag{7.57}$$

which replaces Eq. (7.12). Thus, it is seen that the effect of vapor superheat can be accounted for by simply adding $c_{pv}(T_e - T_{\text{sat}})$ to the enthalpy of phase change h_{fg}.

Effect of Variable Liquid Properties

The reference temperature scheme described in Section 7.2.1 was established for vapor velocities up to 60 m/s. Also, vapor superheat would not be expected to have any significant effect on the reference property scheme for the liquid film.

Effect of Turbulence in the Vapor Boundary Layer

Vapor boundary layers associated with film condensation are often laminar because condenser pressures are relatively low, and the strong suction has a marked effect in delaying transition from a laminar to a turbulent boundary layer. However, Eqs. (7.48) and (7.56) can be shown to be also valid for a turbulent boundary layer, so that the results obtained in this section are applicable whether the vapor boundary layer is laminar or turbulent.

EXAMPLE 7.4 Effect of Vapor Drag on Condensation of Refrigerant-12

Saturated R-12 vapor at 320 K flows at 14 m/s down along a vertical tube of 5 cm O.D. Coolant flowing inside the tube maintains the outside wall at 300 K. Estimate the effect of vapor drag on the local heat transfer coefficient at a location 10 cm from the top of the tube.

Solution

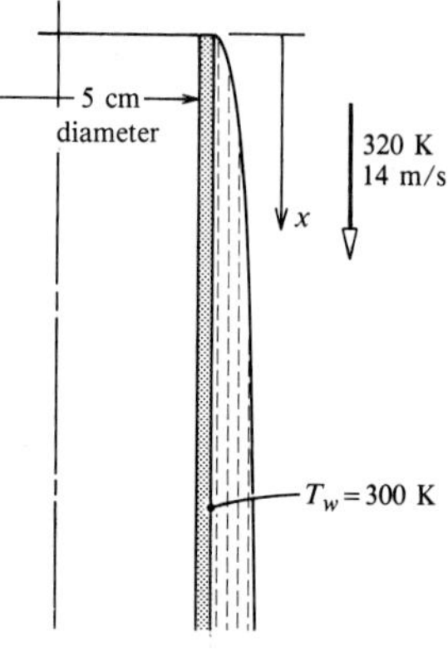

Given: Forced-flow condensation of R-12.

Required: Effect of vapor drag on h at $x = 10$ cm.

Assumptions: 1. Negligible effect of ripples.
2. Negligible effect of wall curvature.

The film flow will prove to be in the wavy laminar regime, but we assume an absence of ripples in order to obtain an approximate estimate of the increase in h due to vapor drag. Since Table 7.1 does not have an α value for R-12, liquid properties are evaluated at the mean film temperature of 310 K. From Table A.8: $k_l = 0.0695$ W/m K, $\rho_l = 1263$ kg/m^3, $\nu_l = 0.193 \times 10^{-6}$ m^2/s, $c_{pl} = 995$ J/kg K, $\text{Pr}_l = 3.5$. From Table A.12*e*, $h_{\text{fg}} = 1.237 \times 10^5$ J/kg. Equation (7.51) gives the local heat transfer coefficient as

$$h = \left[\frac{k_l^2 U_e}{8\nu_l x}\left\{1 + \left(1 + \frac{16\text{Pr}_l}{\text{Ja}_l}\frac{gx}{U_e^2}\right)^{1/2}\right\}\right]^{1/2}$$

We first evaluate (gx/U_e^2) and Pr_l/Ja_l:

$$\frac{gx}{U_e^2} = \frac{(9.81)(0.1)}{(14.0)^2} = 5.01 \times 10^{-3}$$

From Eq. (7.17*b*),

$$h'_{\text{fg}} = h_{\text{fg}} + \left(0.683 - \frac{0.228}{\text{Pr}_l}\right)c_{pl}(T_{\text{sat}} - T_w)$$

$$h'_{\text{fg}} = 1.237 \times 10^5 + \left(0.683 - \frac{0.228}{3.5}\right)(995)(320 - 300) = 1.360 \times 10^5 \text{ J/kg}$$

$$\frac{\text{Pr}_l}{\text{Ja}_l} = \frac{\rho_l \nu_l h'_{\text{fg}}}{k_l(T_{\text{sat}} - T_w)} = \frac{(1263)(0.193 \times 10^{-6})(1.360 \times 10^5)}{(0.0695)(20)} = 23.8$$

$$h = \left[\frac{(0.0695)^2(14)}{(8)(0.193 \times 10^{-6})(0.1)}\{1 + [1 + (16)(23.8)(5.01 \times 10^{-3})]^{1/2}\}\right]^{1/2}$$

$$= 1088 \text{ W/m}^2\text{ K}$$

In the absence of vapor drag, h is given by Eq. (7.15); for $\rho_v \ll \rho_l$ and h'_{fg} replacing h_{fg},

$$h = \left[\frac{h'_{fg} g \rho_l k_l^3}{4x(T_{sat} - T_w)\nu_l}\right]^{1/4} = \left[\frac{(1.360 \times 10^5)(9.81)(1263)(0.0695)^3}{(4)(0.1)(20)(0.193 \times 10^{-6})}\right]^{1/4} = 778 \text{ W/m}^2\text{ K}$$

Comments

The effect of vapor drag is to increase the heat transfer coefficient by 40%.

EXAMPLE 7.5 Effect of Vapor Superheat on Ammonia Condensation

Ammonia at 100°C and 1034 kPa is to be condensed on a 2 cm–O.D. horizontal tube maintained at 3°C. Estimate the condensation rate, and compare it to the condensation rate of saturated vapor at the same pressure.

Solution

Given: Ammonia condensing on a horizontal tube.

Required: Effcct of vapor superheat on condensation rate.

Assumptions: The strong-suction limit for h_c, Eq. (7.56), is valid for condensation on a horizontal tube.

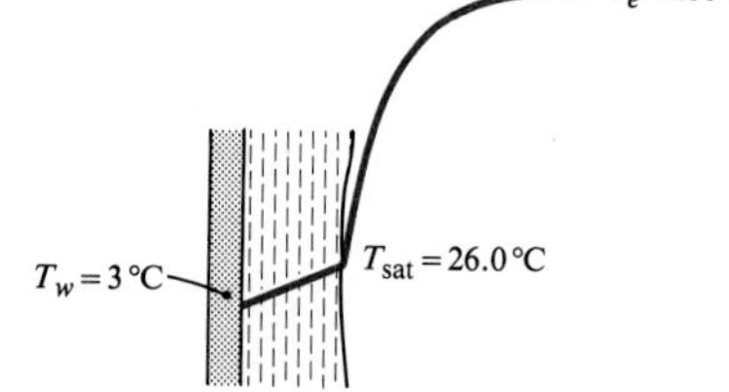

Equation (7.41) gives the heat transfer coefficient, provided we account for vapor superheat and liquid subcooling by replacing h_{fg} by h'_{fg}:

$$\overline{h} = 0.728\left[\frac{(\rho_l - \rho_v)g h'_{fg} k_l^3}{\nu_l D(T_{sat} - T_w)}\right]^{1/4}$$

where $h'_{fg} = h_{fg} + c_{pv}(T_e - T_{sat}) + 0.35 c_{pl}(T_{sat} - T_w)$.

From Table A.12*b* for $P_{sat} = 1034$ kPa, $T_{sat} = 299.2$ K (26.0°C), $h_{fg} = 1.163 \times 10^6$ J/kg. From Table 7.1, $\alpha = 0.61$ for ammonia; hence, $T_r = 276.1 + 0.61(299.2 - 276.1) \simeq 290$ K. From Table A.8, properties of liquid ammonia at 290 K are: $k_l = 0.522$ W/m K, $\rho_l = 616$ kg/m^3, $c_{pl} = 4800$ J/kg K, $\nu_l = 0.234 \times 10^{-6}$ m^2/s. From Table A.7, for ammonia vapor at a mean temperature of 340 K, $c_{pv} = 2230$ J/kg K. Since $\rho_v \ll \rho_l$, it need not be calculated accurately: for $P = 1034$ kPa and $T = 350$ K, the ideal gas law gives $\rho_v = (1034 \times 10^3)(17)/(8314)(340) = 6.2$ kg/m^3.

$$h'_{fg} = 1.163 \times 10^6 + 2230(100 - 26) + (0.35)(4800)(26 - 3)$$

$$= 1.163 \times 10^6 + 0.165 \times 10^6 + 0.039 \times 10^6 = 1.367 \times 10^6 \text{ J/kg}$$

$$\overline{h} = 0.728\left(\frac{(616 - 6)(9.81)(1.367 \times 10^6)(0.522)^3}{(0.234 \times 10^{-6})(0.02)(26 - 3)}\right)^{1/4} = 7420 \text{ W/m}^2\text{ K}$$

$$\dot{Q} = (\pi D)\bar{h}(T_{\text{sat}} - T_w) = (\pi)(0.02)(7420)(26 - 3) = 10{,}720 \text{ W per meter length}$$

$$\dot{m} = \frac{\dot{Q}}{h'_{\text{fg}}} = \frac{10{,}720}{1.367 \times 10^6} = 7.84 \times 10^{-3} \text{ kg/s per meter length}$$

Repeating the above calculations for saturated vapor gives

$$h'_{\text{fg}} = h_{\text{fg}} + 0.35 c_{pl}(T_{\text{sat}} - T_w) = (1.163 + 0.039)10^6 = 1.202 \times 10^6 \text{ J/kg}$$

$$\bar{h} = 7190 \text{ W/m}^2 \text{ K}; \quad \dot{Q} = 10{,}390 \text{ W per meter}; \quad \dot{m} = 8.64 \times 10^{-3} \text{ kg/s per meter}$$

Comments

1. The effect of vapor superheat is to increase the heat transfer coefficient $\bar{h}$ and the heat transfer $\dot{Q}$, but the condensation rate $\dot{m}$ is decreased. Why?
2. Use PHASE to check $\bar{h}$ and $\dot{m}$.

7.3 FILM EVAPORATION

Film or dropwise condensation takes place on a surface when it is cooled to a temperature below the saturation temperature, T_{sat}, corresponding to the pressure of the vapor. On the other hand, if a film of liquid is on a surface that is heated to a temperature above T_{sat}, evaporation from the surface of the film will occur. In industrial evaporators, evaporation may take place from a film falling down the insides of long vertical tubes or channels, as is shown in Fig. 7.12.

7.3.1 Falling Film Evaporation on a Vertical Wall

Heat transfer across an evaporating falling film is essentially identical to that for falling film condensation, except that it is in the opposite direction. There are, however, differences related to the circumstances in which these processes occur in industrial equipment. A major difference is that the film always starts with a finite thickness, and the film Reynolds number decreases down the wall as evaporation takes place; however, in many applications, the evaporation rate is small compared to the film flow rate, and the change in the Reynolds number might be small. Figure 7.13 shows the temperature profile across an evaporating turbulent falling film. Since $T_w > T_{\text{sat}}$, the liquid closer to the wall is superheated, and if the superheat is sufficient, *boiling* will occur on the wall (that is, bubbles will nucleate and grow on the wall), or bubbles entering with the feed liquid may *cavitate* (that is, grow explosively). In both cases, the film will be disrupted, and the heat transfer process will become much more complex. In Section 7.4.2 we will see how relatively small values of $(T_w - T_{\text{sat}})$ cause boiling. Often, a liquid evaporates into a vapor-gas mixture rather than into pure vapor; for example, water evaporates into an air stream in a great variety of industrial equipment. Then, because the total pressure is usually much higher than $P_{\text{sat}}(T_w)$, there is little possibility of boiling or cavitation. Since the evaporating vapor must diffuse away from the surface through the air, such

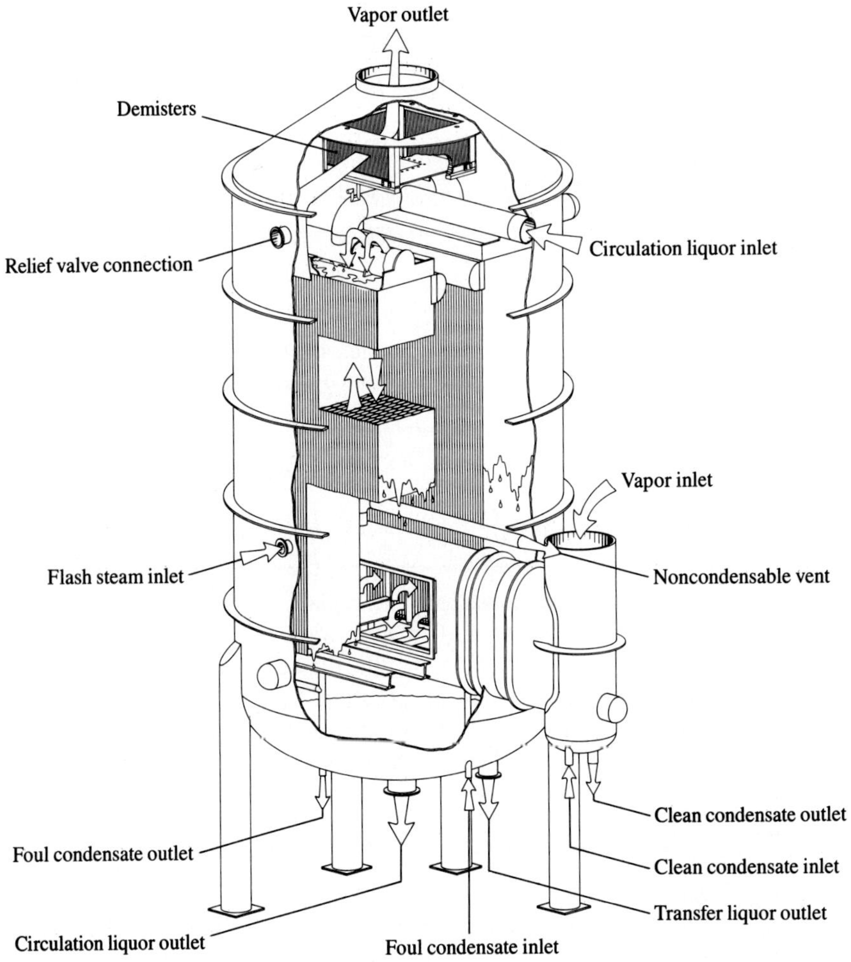

Vapor outlet
Demisters
Circulation liquor inlet
Relief valve connection
Vapor inlet
Flash steam inlet
Noncondensable vent
Clean condensate outlet
Foul condensate outlet
Clean condensate inlet
Transfer liquor outlet
Circulation liquor outlet
Foul condensate inlet

Figure 7.12 A falling-film evaporator used to concentrate black liquor in a pulp mill. (Courtesy Dr. J. W. Rauscher, Ahlstrom Recovery Inc., Roswell, Georgia.)

Liquid film
Vapor
T_w
T
T_{sat}

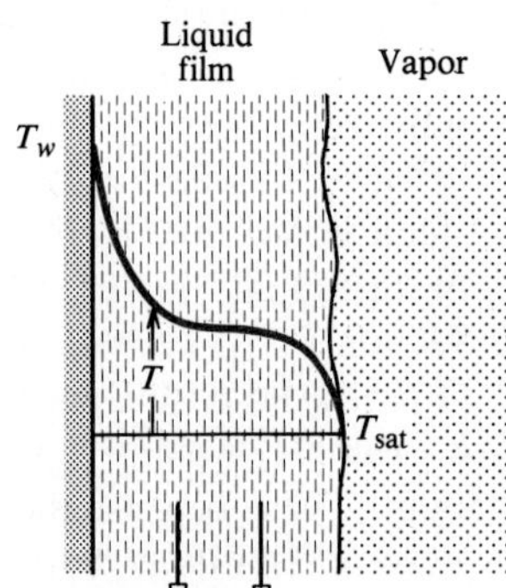

Figure 7.13 The temperature profile across an evaporating turbulent falling film.

equipment also involves a *mass transfer* process. In contrast to film condensation, where the vapor boundary layer is under strong suction, for film evaporation there is a *blowing* effect on the vapor boundary layer since the vapor velocity normal to the film is directed away from the surface. Whereas suction increases vapor drag and heat transfer from the vapor, blowing reduces vapor drag and heat transfer to the vapor. If the evaporation rate is high enough, both can be negligible.

Since evaporating films are most often initially wavy laminar or turbulent, our analysis follows Section 7.2.2. If vapor drag is negligible, then Eqs. (7.21) through (7.24) again apply to water, namely,

$$\mathrm{Nu} = \left(\frac{3}{4}\,\mathrm{Re}\right)^{-1/3} \qquad 0 < \mathrm{Re} < 30 \qquad \text{(laminar)}$$

$$\mathrm{Nu} = 0.822\mathrm{Re}^{-0.22} \qquad 30 < \mathrm{Re} < \mathrm{Re}_{\mathrm{tr}} \qquad \text{(wavy laminar)}$$

$$\mathrm{Nu} = 3.8 \times 10^{-3}\mathrm{Re}^{0.4}\mathrm{Pr}_l^{0.65} \qquad \mathrm{Re}_{\mathrm{tr}} < \mathrm{Re} \qquad \text{(turbulent)}$$

where $\mathrm{Re}_{\mathrm{tr}} = 5800\mathrm{Pr}_l^{-1.06}$. In fact, the turbulent correlation was developed using data for evaporating falling films. The energy balance Eq. (7.26) becomes

$$\frac{L}{(\nu_l^2/g)^{1/3}} = -\frac{\mathrm{Pr}_l}{4\mathrm{Ja}_l}\int_{\mathrm{Re}_0}^{\mathrm{Re}_L} \frac{d\mathrm{Re}}{\mathrm{Nu}} \tag{7.58}$$

where Re_0 is the initial film Reynolds number at $x = 0$, and the expression for the average Nusselt number becomes

$$\overline{\mathrm{Nu}} = \frac{\mathrm{Pr}_l}{4\mathrm{Ja}_l}\frac{(\nu_l^2/g)^{1/3}}{L}(\mathrm{Re}_0 - \mathrm{Re}_L) \tag{7.59}$$

Effect of Variable Fluid Properties

The reference property schemes for laminar film condensation in Section 7.2.1 are tentatively recommended for falling film evaporation.

Entrance Effects

Equations (7.22) and (7.23) are based on data measured some distance downstream from the liquid distributor, where entrance effects were considered to be negligible. Thus, the foregoing analysis should not be applied to a very short evaporator. For turbulent films, such entrance effects are negligible for distances greater than about 20 times the film thickness. For wavy laminar films, entrance effects persist somewhat further. In fact, there is a question as to whether a truly fully developed condition is ever attained, since the long-wavelength gravity waves on the surface of the film may require a very large distance to become fully established. However, this entrance effect is of secondary importance since it leads to relatively small changes in the Nusselt number.

Computer Program PHASE

Item 3 of PHASE calculates the average heat transfer coefficient and evaporation rate for a falling film according to the analysis of Section 7.3.1. Local heat transfer coefficients and transition Reynolds numbers for wavy laminar and turbulent flow can be calculated from the formulas given in Section 7.3.1, which are, strictly speaking, valid only for water. User-supplied constants and exponents for these formulas are also allowed. Variable-property and liquid superheat effects are calculated according to the recommendations given in Section 7.2.1.

EXAMPLE 7.6 Evaporation from a Falling Water Film

In an experimental rig, water is fed at 0.01 kg/s to the top of a vertical 5 cm–O.D. tube, 5 m high. The outside of the tube wall is maintained at 311 K by steam condensing inside the tube, and the saturation temperature corresponding to the system pressure is 308 K. Calculate the vapor production rate.

Solution

Given: A water film flow outside a heated tube.

Required: Vapor production rate.

Assumptions: The reference temperature scheme for laminar films can be used to evaluate properties.

We first obtain the necessary property values. At 308 K, Table A.12*a* gives $h_{fg} = 2.418 \times 10^6$ J/kg. From Table 7.1, $\alpha = 0.33$; hence, $T_r = 311 - 0.33(311 - 308) = 310$ K. From Table A.8: $k_l = 0.628$ W/m K, $\rho_l = 993$ kg/m^3, $c_{pl} = 4174$ J/kg K, $\nu_l = 0.70 \times 10^{-6}$ m^2/s, $\mathrm{Pr}_l = 4.6$. The initial film Reynolds number is

$$\mathrm{Re}_0 = \frac{4\Gamma_0}{\mu_l} = \frac{4\dot{m}_0}{\pi D \rho_l \nu_l} = \frac{(4)(0.01)}{(\pi)(0.05)(993)(0.70 \times 10^{-6})} = 366$$

and $\mathrm{Re}_{tr} = 5800(4.6)^{-1.06} = 1151 > 366$. We will assume that the film remains in the wavy laminar regime as it flows down the wall and check later. Then Eq. (7.58) becomes

$$\frac{L}{(\nu_l^2/g)^{1/3}} = -\frac{\mathrm{Pr}_l}{4\mathrm{Ja}_l}\int_{366}^{\mathrm{Re}_L} \frac{d\mathrm{Re}}{0.822\mathrm{Re}^{-0.22}} = -\frac{\mathrm{Pr}_l}{4\mathrm{Ja}_l(0.822)(1.22)}\left[\mathrm{Re}_L^{1.22} - 366^{1.22}\right]$$

Hence,

$$\mathrm{Re}_L^{1.22} = 1341 - \frac{4.01\mathrm{Ja}_l L}{(\nu_l^2/g)^{1/3}\mathrm{Pr}_l}$$

$$\mathrm{Ja}_l = \frac{c_{pl}(T_w - T_{sat})}{h_{fg}} = \frac{(4174)(3)}{(2.418 \times 10^6)} = 5.18 \times 10^{-3}$$

$$\left(\frac{\nu_l^2}{g}\right)^{1/3} = \left[\frac{(0.70 \times 10^{-6})^2}{9.81}\right]^{1/3} = 3.68 \times 10^{-5} \text{ m}$$

Substituting,

$$\mathrm{Re}_L^{1.22} = 1341 - \frac{(4.01)(5.18 \times 10^{-3})(5)}{(3.68 \times 10^{-5})(4.6)} = 727$$

and $\mathrm{Re}_L = 222$. The film remains in the wavy laminar regime.

The water flow rate at the bottom of the tube is

$$\dot{m}_L = \frac{\pi D \rho_l \nu_l \mathrm{Re}_L}{4} = \frac{(\pi)(0.05)(993)(0.70 \times 10^{-6})(222)}{4} = 6.05 \times 10^{-3}$$

The vapor production rate is obtained from a mass balance:

$$\dot{m}_v = \dot{m}_0 - \dot{m}_L = 0.01 - 0.00605 = 3.95 \times 10^{-3} \text{ kg/s}$$

Comments

Use PHASE to check $\dot{m}_v$ and Re_L.

7.4 POOL BOILING

In industrial boilers, boiling may take place in a stationary pool of liquid, or the liquid may boil as it flows through a tube. We will examine **pool boiling** in this section and defer consideration of *forced-convection boiling* to Section 7.5. In Section 7.4.1, the regimes of pool boiling are described, and the *boiling curve* is introduced. In succeeding sections, the various regimes are described in detail, and correlations are given for the boiling heat flux or boiling heat transfer coefficient.

7.4.1 Regimes of Pool Boiling

We have all watched a pot of water come to a boil on the kitchen stove. When heating begins, the bulk liquid and the pot wall are at a temperature lower than the saturation temperature, T_{sat}, corresponding to the pressure P, and vapor cannot coexist with the liquid phase. The liquid and wall are then *subcooled*. Initially, no vapor bubbles are seen; however, gradients in the index of refraction caused by temperature gradients near the heated wall reveal fluid motion associated with natural-convection heat transfer. Some time later, stagnant bubbles attached to the wall become visible: these are not bubbles of pure water vapor but are mostly air that has come out of solution. The solubility of air in water decreases with increasing temperature; thus, the water becomes supersaturated with air, and mass transfer occurs to minute air spaces trapped in scratches, pits, and crevices in the wall. These air bubbles act as nuclei for growth of vapor bubbles.

True boiling, which is the formation of pure vapor from superheated liquid, begins when the wall temperature T_w exceeds T_{sat} by a few degrees. The pot "sings" as vapor bubbles grow rapidly from nuclei on the superheated wall. The bubbles stop growing and collapse rapidly as they project through the thin natural-convection thermal boundary layer into the bulk subcooled liquid. At first, these bubbles are invisible to the eye, but as the bulk liquid temperature rises with continued heating, the growing and collapsing bubbles become large enough to be visible. This boiling is called *subcooled nucleate boiling*. With continued heating, the bulk liquid itself becomes mildly superheated: the bubbles grow from the wall nuclei until they are large enough for their buoyancy to tear them from the wall, and they rise to the pool surface. The heat transfer process is then one of transporting enthalpy of phase change by the vapor inside the bubbles; also, the bubbles sweep hot liquid away from the wall, causing a rapid increase in the heat transfer coefficient. **Saturated nucleate boiling** with isolated bubbles results, as shown in Fig. 7.14*a*.

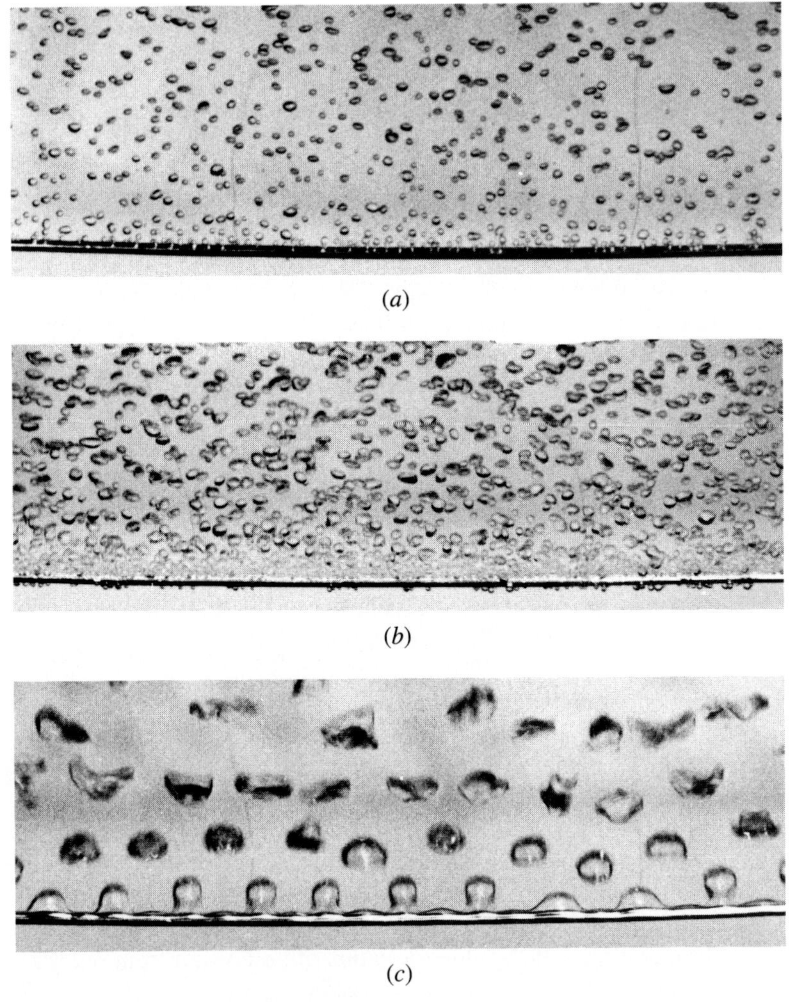

(*a*)

(*b*)

(*c*)

Figure 7.14 Pool boiling of saturated R-113 at atmospheric pressure on a horizontal platinum wire of 1 mm diameter. (*a*) Isolated bubbles, $q = 1 \times 10^4$ W/m^2, $\Delta T = 7$ K. (*b*) Slugs and columns, $q = 4.5 \times 10^4$ W/m^2, $\Delta T = 10$ K. (*c*) Film boiling, $q = 3.3 \times 10^4$ W/m^2, $\Delta T = 150$ K. (Photographs courtesy of Professor S. Nishio, University of Tokyo.)

Further heating produces a rapid increase in the number of active nucleation sites, and so many bubbles form that they merge to produce columns of vapor, which feed large slugs of vapor overhead, as shown in Fig. 7.14*b*. Finally, the vapor streams upward so fast that the liquid downflow to the surface is unable to sustain a higher evaporation rate. This condition is called the departure from nucleate boiling, or *boiling crisis*, and gives rise to a **peak heat flux** q_{max}. It is also often called the *burnout point*, which will be explained later. If the wall temperature is increased beyond the value corresponding to q_{max}, there is a transition regime where it becomes increasingly difficult for liquid to reach the wall, until a vapor film completely blankets the wall, preventing liquid from contacting the wall at all, as shown in Figure 7.14*c*. The corresponding heat fluxes are much lower than q_{max} since heat is transferred rather poorly across the low-conductivity vapor film, and there is a local minimum value q_{min}, characterizing this **film boiling**. Yet a further increase in wall temperature causes a modest increase in q, and if the temperature level is high enough, radiation heat transfer across the vapor film can become significant.

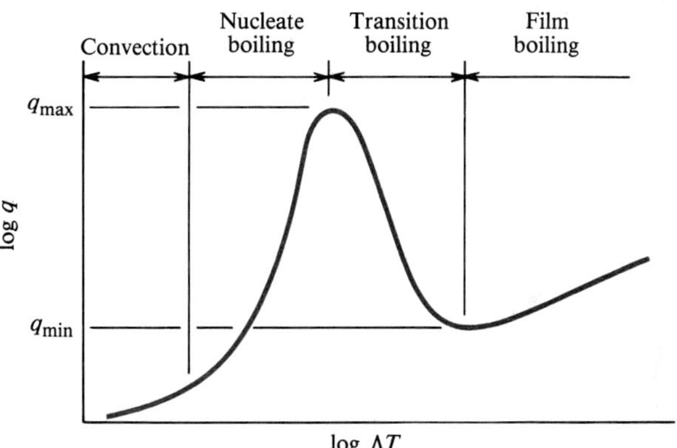

Figure 7.15 Heat flux, q, versus temperature difference, $\Delta T = T_w - T_{sat}$, for pool boiling: the *boiling curve*.

Figure 7.15 shows a typical *boiling curve*, in which the wall heat flux q is plotted versus $\Delta T = (T_w - T_{sat})$. We choose to plot q rather than the heat transfer coefficient because q_{max} and q_{min} can be relatively easily specified whereas the corresponding heat transfer coefficients cannot. The various regimes of boiling discussed above are indicated in Fig. 7.15. In practice, it is more often the case that the heat flux through the surface is controlled rather than the wall temperature; for example, when an electrical heater is used. If we imagine increasing the power input until q_{max} is reached, Fig. 7.16 shows that the situation becomes unstable, and a further small increase in q requires the wall temperature to jump from temperature T_A to a very high value in the film boiling regime, T_B, in order to effect the heat transfer. If T_B approaches or exceeds the melting point of the wall, the result can be catastrophic, and *burnout* is said to occur. Conversely, if during film boiling the heat flux is

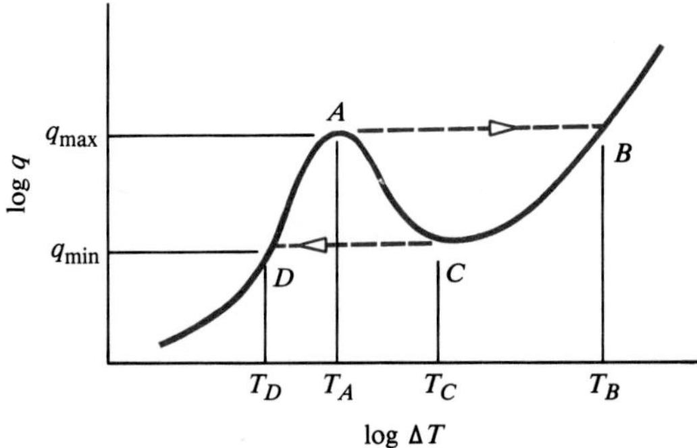

Figure 7.16 Unstable heating paths for heat-input-controlled pool boiling.

reduced, T_w decreases until q_{min} is reached. Figure 7.16 shows that the situation is again unstable, and a further decrease in q causes the wall temperature to jump from T_C to a very low value in the nucleate boiling regime, T_D.

7.4.2 Boiling Inception

Typical machined, drawn, or cast surfaces have a variety of pits, grooves, and crevices. When a liquid contacts the surface, surface tension forces prevent the liquid from entering the smaller cavities in which air or other gases are trapped. These cavities are the sites at which bubble nucleation occurs. To obtain an estimate of the superheat required to nucleate bubbles, we consider a force balance on a stationary spherical bubble, as shown in Fig. 7.17. The ambient pressure is denoted P_∞, P_v is the partial pressure of vapor inside the bubble, and P_g is the sum of the partial pressures of all the permanent gases inside the bubble. For static equilibrium, the surface tension force balances the net pressure force:

$$2\pi R_c\sigma = (P_v + P_g - P_\infty)\pi R_c^2$$

$$R_c = \frac{2\sigma}{P_v + P_g - P_\infty} \tag{7.60}$$

where σ is the surface tension and R_c is a **critical radius** of the bubble. It is reasonable to assume that the vapor in the bubble is in thermodynamic equilibrium

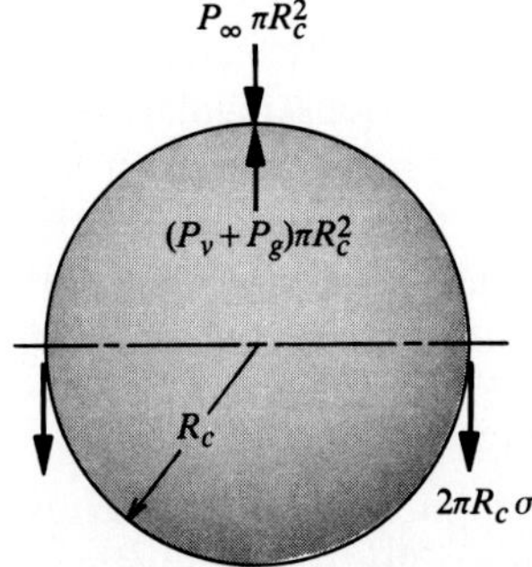

Figure 7.17 Force balance on a spherical bubble.

with the liquid; thus, the vapor partial pressure P_v can be taken as $P_{\text{sat}}(T_s)$, where T_s is the temperature of the liquid surrounding the bubble. The partial pressure of the gas, P_g, is more difficult to estimate: the gas is not likely to be in equilibrium with dissolved gas in the liquid phase, since mass diffusion rates in liquids are very low. This equilibrium bubble is unstable to small perturbations. A small decrease in size causes an increase in $2\sigma/R$. Since P_∞ may be regarded as fixed, there is a resulting increase in P_v to give a supersaturated vapor that starts to condense, allowing a further decrease in bubble size and eventual collapse of the bubble. Conversely, a small increase in bubble size causes a decrease in $2\sigma/R$ and a decrease in P_v to give superheated vapor. Liquid then starts to evaporate, and the bubble continues to grow.

Figure 7.18*a* shows a bubble that has emerged from a surface cavity. For simplicity, we assume a contact angle of 90°, so that the bubble radius is equal to the cavity radius. If the bubble is able to continue to grow and detach from the surface, we say that the bubble has *nucleated* and that the cavity is a *nucleation site*. Experience shows that for typical boiler wall surface finishes, nucleation sites with radii in the range 2.5–7.5 μm are present when the wall is first wetted, but prolonged boiling in deaerated water will usually deactivate the larger sites. Application of Eq. (7.60) to bubble nucleation by a cavity is made difficult by the fact that the liquid surrounding the bubble may not be at uniform temperature. Figure 7.18*b* shows the temperature profile expected for laminar natural convection. Also, the profile might be disturbed by the bubble as it emerges from the cavity. For a very simple model, we can assume that the liquid surrounding the bubble is at the wall temperature T_w, that is, $T_s = T_w$. If the gas pressure is ignored in Eq. (7.60), it can be rearranged as

$$P_{\text{sat}}(T_s) = P_\infty + \frac{2\sigma}{R_c} \tag{7.61}$$

Equation (7.61) can be used to estimate the wall temperature $T_w = T_s$ at which bubbles will nucleate from a cavity of radius R_c. The corresponding wall superheat is $T_w - T_{\text{sat}}(P_\infty)$. If gas is present in the bubbles, Eq. (7.60) shows that nucleation will occur at lower wall superheat.

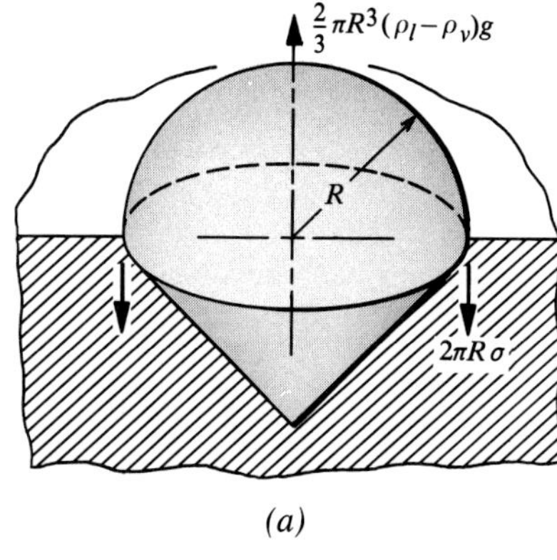

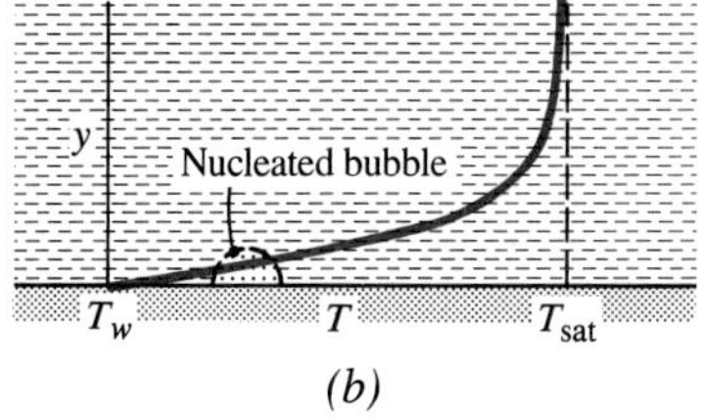

Figure 7.18 A hemispherical bubble attached to a horizontal surface. (*a*) Force balance. (*b*) Temperature profile in surrounding fluid.

The Computer Program BOIL

Item 1 in the program BOIL calculates the superheat required for boiling inception by a surface cavity of radius R_c, using Eq. (7.61). The fluid menu is identical to that contained in the computer program PHASE. SI units must be used, with temperature in kelvins.

EXAMPLE 7.7 Boiling Inception Superheat for Various Liquids

Estimate the superheat required for inception of boiling for a heater just below the surface of a pool of saturated liquid if the liquid is (i) water at 1 atm pressure, (ii) water at 300 K, and (iii) nitrogen at 1 atm pressure. Take the largest available nucleation sites to have a radius of 7.5 μm.

Solution

Given: Pool boiling.

Required: Boiling inception ΔT for (i) water at 1 atm, (ii) water at 300 K, and (iii) nitrogen at 1 atm.

Assumptions: 1. The largest available nucleation sites have a radius of 7.5 μm.
2. There is no dissolved gas.

Equation (7.61) gives the value of $P_{\text{sat}}(T_s)$ required to nucleate a bubble; the corresponding value of T_s is then found from vapor tables.

(i) Water at 1 atm:

$$P_{\text{sat}}(T_s) = P_\infty + \frac{2\sigma}{R_c}$$

$P_\infty = 1$ atm $= 1.0133 \times 10^5$ Pa; $T_{\text{sat}}(P_\infty) = 373.15$ K, from Table A.12*a*, and from Table A.11, σ ($T = 373.15$ K) is 58.9×10^{-3} N/m.

$$P_{\text{sat}}(T_s) = 1.0133 \times 10^5 + \frac{(2)(58.9 \times 10^{-3})}{(7.5 \times 10^{-6})} = 1.0133 \times 10^5 + 0.1571 \times 10^5$$

$$= 1.170 \times 10^5 \text{Pa}$$

Table A.12*a* gives the corresponding value of T_s as 377.2 K. Thus, the required superheat is

$$\Delta T = T_s - T_{\text{sat}}(P_\infty) = 377.2 - 373.15 = 4.1 \text{ K}$$

(ii) Water at 300 K:

$T_{\text{sat}}(P_\infty) = 300$ K, $P_\infty = 3533$ Pa from Table A.12*a*; $\sigma = 71.7 \times 10^{-3}$ N/m from Table A.11.

$$P_{\text{sat}}(T_s) = 3533 + \frac{2(71.7 \times 10^{-3})}{(7.5 \times 10^{-6})} = 3533 + 19{,}120 = 22{,}650 \text{ Pa}$$

$$T_s = 336 \text{ K}, \qquad \Delta T = 336 - 300 = 36 \text{ K}$$

(iii) Nitrogen at 1 atm:

$P_\infty = 1.0133 \times 10^5$ Pa, $T_{sat} = 77.4$ K from Table A.12*c*, $\sigma = 8.85 \times 10^{-3}$ N/m. The superheat is small in this case, and an accurate result cannot be obtained if Table A.12*c* is used to obtain T_s from $P_{sat}(T_s)$. Instead, we use the Clausius-Clapeyron relationship, which relates ΔP to ΔT along the saturation curve, and the ideal gas law:

$$\frac{\Delta P}{P} = \frac{h_{fg}M}{\mathscr{R}}\frac{\Delta T}{T^2}$$

$$\Delta T = \frac{T^2}{P}\frac{(\mathscr{R}/M)}{h_{fg}}\left(\frac{2\sigma}{R_c}\right)$$

At the normal boiling point of nitrogen, 77.4 K, $h_{fg} = 2.0 \times 10^5$ J/kg from Table A.12*c*. Hence,

$$\Delta T = \frac{(77.4)^2}{1.0133 \times 10^5}\frac{(8314/28)}{2.0 \times 10^5}\frac{(2)(8.85 \times 10^{-3})}{7.5 \times 10^{-6}} = 0.21 \text{ K}$$

Comments

1. In case (ii), a second iteration evaluating σ at $T_s = 336$ K is indicated.

2. Use BOIL to check these results.

7.4.3 Nucleate Boiling

As a matter of convenience, we define the heat transfer coefficient for boiling in terms of $T_w - T_{sat}(P_\infty)$, rather than $T_w - T_\infty$:

$$q = h(T_w - T_{sat}) \tag{7.62}$$

An appropriate characteristic length to form a Nusselt number for nucleate boiling is the size of a bubble as it breaks away from a wetted wall. Referring to Fig. 7.18*a*, the surface tension and buoyancy force are equated to give

$$2\pi R_b\sigma = \frac{2}{3}\pi R_b^3(\rho_l - \rho_v)g$$

where R_b is the bubble radius and subscripts l and v refer to the liquid vapor phase, respectively. Hence,

$$R_b = \left[\frac{3\sigma}{(\rho_l - \rho_v)g}\right]^{1/2}$$

and the characteristic length L_c can be taken simply as

$$L_c = \left[\frac{\sigma}{(\rho_l - \rho_v)g}\right]^{1/2}; \qquad \text{Nu} = \frac{hL_c}{k_l} \tag{7.63}$$

The rate of bubble growth depends on convective heat transfer through the liquid to the liquid-vapor interface required to supply the enthalpy of vaporization. Thus, we would expect the Prandtl number of the liquid Pr_l to be a relevant dimensionless group. Also, we saw in Section 7.2.2 how the Jakob number is relevant to phase change problems. For nucleate boiling, it is appropriate to define the Jakob number as

$$\mathrm{Ja}_l = \frac{c_{pl}(T_w - T_{\mathrm{sat}})}{h_{\mathrm{fg}}} \tag{7.64}$$

since the superheated liquid has $c_{pl}(T_w - T_{\mathrm{sat}})$ sensible heat, which can be given up to supply enthalpy of vaporization. Experimental data for nucleate boiling on a horizontal plate facing upward in a pool of liquid was correlated by W. M. Rohsenow [6] as

$$\mathrm{Nu} = \frac{\mathrm{Ja}^2}{C_{\mathrm{nb}}^3 \mathrm{Pr}_l^m} \tag{7.65}$$

where values of the constant C_{nb} and exponent m are given in Table 7.2 for a variety of combinations of liquid and heater wall materials. All properties are to be evaluated at T_{sat}. Notice that a factor-of-2 difference in C_{nb} results in a factor-of-8 difference in Nu. Equation (7.65) implies that the heat flux q is proportional to ΔT^3; this strong dependence on temperature difference is due to the rapid increase in active nucleation sites with increase in superheat. Equation (7.65) is not very accurate, however, and errors of 100% in q and 25% in ΔT are typical. Fortunately, the design engineer is more interested in not exceeding q_{max}, so as to avoid burnout, and usually does not require a precise value of the heat transfer coefficient in the nucleate boiling regime.

The computer program BOIL, item 2, calculates the heat flux for nucleate boiling using Rohsenow's correlation, Eq. (7.65). The menu of surfaces is based on Table 7.2.

Table 7.2 The constant C_{nb} and exponent m for the nucleate boiling correlation, Eq. (7.65): $\mathrm{Nu} = \mathrm{Ja}^2/C_{\mathrm{nb}}^3\mathrm{Pr}_l^m$.

Liquid	Surface	C_{nb}	m
Water	Copper, scored	0.0068	2.0
Water	Copper, polished	0.013	2.0
Water	Stainless steel, chemically etched	0.013	2.0
Water	Stainless steel, mechanically polished	0.013	2.0
Water	Stainless steel, ground and polished	0.008	2.0
Water	Brass	0.006	2.0
Water	Nickel	0.006	2.0
Water	Platinum	0.013	2.0
Carbon tetrachloride	Copper	0.013	4.1
Benzene	Chromium	0.010	4.1
n-pentane	Chromium	0.015	4.1
Ethanol	Chromium	0.0027	4.1
Isopropyl alcohol	Copper	0.0023	4.1
n-butyl alcohol	Copper	0.0030	4.1
35% K_2CO_3	Copper	0.0027	4.1

EXAMPLE 7.8 Nucleate Boiling of Water on Polished Copper

Determine the heat flux when water boils at 1 atm pressure on a polished copper surface at 390 K.

Solution

Given: Water boiling on a polished copper surface.

Required: Heat flux q.

Assumptions: Nucleate boiling regime.

Equation (7.65) applies: $\mathrm{Nu} = \mathrm{Ja}^2/C_{\mathrm{nb}}^3 \,\mathrm{Pr}_l^m$.

Evaluate all properties at $T_{\mathrm{sat}} = 373.15$ K: $\sigma = 58.9 \times 10^{-3}$ N/m, $h_{\mathrm{fg}} = 2.257 \times 10^6$ J/kg, $k_l = 0.681$ W/m K, $\rho_l = 958$ kg/m^3, $c_{pl} = 4212$ J/kg K, $\mathrm{Pr}_l = 1.76$.

$$\mathrm{Ja} = \frac{c_{pl}(T_w - T_{\mathrm{sat}})}{h_{\mathrm{fg}}} = \frac{(4212)(390 - 373.15)}{(2.257 \times 10^6)} = 3.145 \times 10^{-2}$$

From Table 7.2, $C_{\mathrm{nb}} = 0.013$, $m = 2.0$. Substitute in Eq. (7.65):

$$\mathrm{Nu} = \frac{(3.145 \times 10^{-2})^2}{(0.013)^3(1.76)^2} = 145$$

$$L_c = \left[\frac{\sigma}{(\rho_l - \rho_v)g}\right]^{1/2} = \left[\frac{58.9 \times 10^{-3}}{(958)(9.81)}\right]^{1/2} = 2.50 \times 10^{-3} \text{ m}$$

$$h = \left(\frac{k_l}{L_c}\right)\mathrm{Nu} = \left(\frac{0.681}{2.50 \times 10^{-3}}\right)(145) = 3.95 \times 10^4 \text{ W/m}^2\,\text{K}$$

$$q = h(T_w - T_{\mathrm{sat}}) = (3.95 \times 10^4)(390 - 373.15) = 6.66 \times 10^5 \text{ W/m}^2$$

Comments

1. Use BOIL to check q.
2. In general, ΔT for boiling inception and $q_{\max}$ should be calculated to bracket the nucleate boiling regime. For water at 1 atm, Example 7.7 shows that ΔT for inception is 4.1 K; Example 7.9 will show that $q < q_{\max}$.

7.4.4 The Peak Heat Flux

The peak heat flux is primarily determined by hydrodynamic considerations related to the maximum rate at which vapor can leave the wall. (An inability of the liquid to wet the wall may also play a role.) If we define a maximum vapor velocity $V_{\max}$, then the peak heat flux $q_{\max}$ is given by

$$q_{\max} \sim \rho_v V_{\max} h_{\mathrm{fg}} \tag{7.66}$$

$$\frac{q_{\max}}{\rho_v V_{\max} h_{\mathrm{fg}}} = C_{\max} \tag{7.67}$$

where C_{max} may depend on such factors as geometry. One way to estimate V_{max} is to equate the kinetic energy of the vapor to work done by buoyancy forces over the characteristic length L_c, defined by Eq. (7.63):

$$\frac{1}{2}\rho_v V_{max}^2 = g(\rho_l - \rho_v)L_c$$

Substituting for L_c from Eq. (7.63) gives

$$V_{max} \sim \left(\frac{\sigma(\rho_l - \rho_v)g}{\rho_v^2}\right)^{1/4} \tag{7.68}$$

An alternative viewpoint that is often used is to consider the stability of a column of vapor, as shown in Fig. 7.19. If the column is subject to disturbances of wavelength L_c, the column is unstable when its velocity is

$$V_H = \left(\frac{2\pi\sigma}{\rho_v L_c}\right)^{1/2} \tag{7.69}$$

If this velocity is used as an estimate of V_{max}, and we write $V_{max} \sim (\sigma/\rho_v L_c)^{1/2}$, Eq. (7.68) is once again obtained. This phenomenon is known as *Helmholtz instability*, and the derivation of Eq. (7.69) can be found in advanced fluid mechanics texts. It is Helmholtz instability that also causes a flag to flap in the wind. Substituting Eq. (7.68) in Eq. (7.67) gives

$$q_{max} = C_{max} h_{fg}[\sigma\rho_v^2(\rho_l - \rho_v)g]^{1/4} \tag{7.70}$$

Equation (7.70) was first derived by S. Kutateladze in the USSR [7] and N. Zuber in the United States [8]. For large, flat heaters, Lienhard and Dhir [9] report experimental data showing that C_{max} is approximately 0.15, and lower values are more appropriate when the size of the heater is less than about $2L_c$. We define a dimensionless parameter L^* as the ratio of a characteristic length of heater L to L_c:

$$L^* = \frac{L}{L_c} = \frac{L}{[\sigma/(\rho_l - \rho_v)g]^{1/2}} \tag{7.71}$$

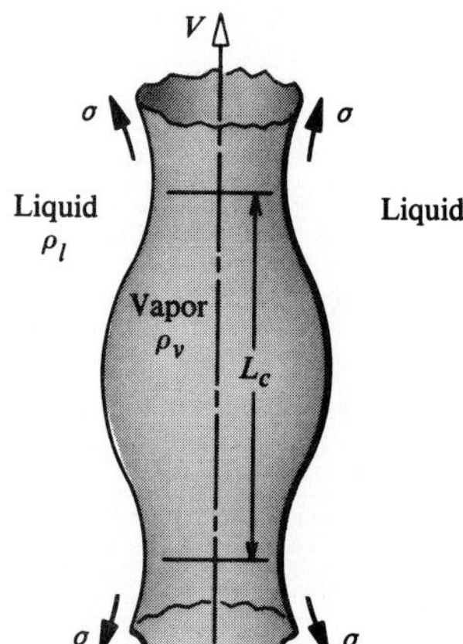

Figure 7.19 Schematic of Helmholtz instability of a vapor column.

Table 7.3 Values of C_{max} for Eq.(7.70) giving the peak heat flux [9]; $L^* = L/L_c = L/[\sigma/(\rho_l - \rho_v)g]^{1/2}$.

Geometry	C_{max}	Characteristic Heater Dimension	L^* Range
1. Infinite flat heater	0.15	Width or diameter	$L^* > 27$
2. Small flat heater	$0.15\dfrac{12\pi L_c^2}{\text{Area}}$	Width or diameter	$9 < L^* < 20$
3. Large horizontal cylinder	0.12	Cylinder radius	$L^* > 1.2$
4. Small horizontal cylinder	$0.12L^{*-1/4}$	Cylinder radius	$0.15 < L^* < 1.2$
5. Large sphere	0.11	Sphere radius	$4.26 < L^*$
6. Small sphere	$0.227L^{*-1/2}$	Sphere radius	$0.15 < L^* < 4.26$
7. Any large finite body	~ 0.12	—	

Table 7.3 gives values of C_{max} in terms of L^* for various heater shapes. All properties, including the vapor density, ρ_v, in Eq. (7.70) should be evaluated at T_{sat}. Notice that Eq. (7.70) does not allow the heat transfer coefficient corresponding to q_{max} to be calculated; hence, the superheat $(T_w - T_{sat})$ at q_{max} cannot be determined either. The computer program BOIL, item 3, calculates the peak heat flux for pool boiling using Eq. (7.70). A menu of seven configurations based on Table 7.3 is provided.

EXAMPLE 7.9 Peak Heat Flux for Pool Boiling of Water

Estimate the peak heat flux for boiling of water on a large, flat heater at saturation temperatures of (i) 100°C, (ii) 300 K.

Solution

Given: Pool boiling of water on a large, flat heater.

Required: Peak heat flux, q_{max}.

(i) At $T_{sat} = 100°\text{C} = 373.15$ K, $h_{fg} = 2.257\times10^6$ J/kg, $\rho_v = 0.598$ kg/m^3, $\sigma = 58.9\times10^{-3}$ N/m, $\rho_l = 958$ kg/m^3. Substituting in Eq. (7.70) with $C_{max} = 0.15$ from Table 7.3,

$$q_{max} = C_{max}h_{fg}[\sigma\rho_v^2(\rho_l - \rho_v)g]^{1/4}$$

$$= (0.15)(2.257\times10^6)[(58.9\times10^{-3})(0.598)^2(958 - 0.598)(9.81)]^{1/4}$$

$$= 1.27\times10^6 \text{ W/m}^2$$

(ii) At $T_{sat} = 300$ K, $h_{fg} = 2.437\times10^6$ J/kg, $\rho_v = 0.0255$ kg/m^3, $\sigma = 71.7\times10^{-3}$ N/m, $\rho_l = 996$ kg/m^3.

$$q_{max} = (0.15)(2.437\times10^6)[(71.7\times10^{-3})(0.0255)^2(996)(9.81)]^{1/4}$$

$$= 3.00\times10^5 \text{ W/m}^2$$

Comments

1. The decrease in q_{max} at lower values of T_{sat} is due to the marked decrease in vapor density, which limits the mass flow in the vapor columns.

2. Use BOIL to check q_{max}.

7.4.5. Film Boiling

The phenomenon of film boiling on immersed cylinders, spheres, and plates is very similar in nature to film condensation. This fact was first exploited by L. Bromley in 1950 [10]. Figure 7.20 shows film boiling on a sphere. Instead of a liquid film flowing down over the sphere, as in film condensation, a vapor film flows upward over the sphere. There are some differences, however: (1) Since usually $\rho_v \ll \rho_l$, vapor films tend to be much thicker than liquid films; (2) although we were able to ignore vapor drag on a liquid film in the analysis of film condensation, liquid drag on the vapor film is not negligible; and (3) on larger objects, the vapor-liquid interface can become Helmholtz-unstable—in particular, on a vertical wall the interface becomes unstable quite close to the leading edge. Notwithstanding, correlations of the Nusselt number for laminar film condensation can be used for laminar film boiling on small objects, with liquid properties replaced by vapor properties and slight adjustments made in multiplying constants. Therefore, for laminar film boiling, we write

$$\overline{h} = C_{fb}\left[\frac{(\rho_l \quad \rho_v) g h_{fg}' k_v^3}{\nu_v L (T_w - T_{sat})}\right]^{1/4} \tag{7.72}$$

The characteristic length L is related to the length of vapor film: for a horizontal cylinder or sphere, L can be taken as the diameter D. The constant C_{fb} can be taken as 0.62 for a horizontal cylinder [10], 0.67 for a sphere [11], and approximately 0.71 for a plane vertical surface. These constants are somewhat lower than their

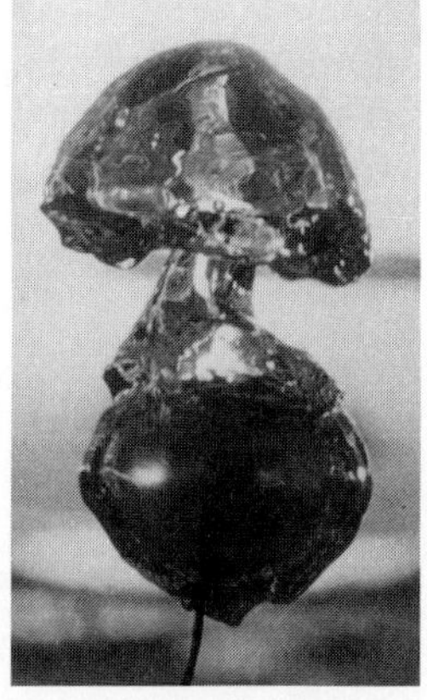

Figure 7.20 Film boiling of n-pentane on a 9.5 mm–diameter sphere. (Photograph courtesy of Professor T. K. Frederking, University of California, Los Angeles.)

counterparts for film condensation [cf. Eq. (7.41)] due to the influence of liquid drag on the vapor flow. The modified latent heat is taken as $h'_{fg} = h_{fg} + 0.35c_{pv}(T_w - T_{sat})$ to account for vapor superheating.

Equation (7.72) is valid for a smooth vapor-liquid interface. At high vapor flow rates, waves form on the liquid-vapor interface, and the vapor flow can become turbulent, similar to the behavior for film condensation discussed in Section 7.2.2. Such conditions are usually encountered unless L is small and, in particular, for film boiling of cryogenic liquids at atmospheric pressure. Based on experimental data for liquid nitrogen, Frederking and Clark [13] recommend

$$\overline{h} = 0.15\left[\frac{(\rho_l - \rho_v)g h'_{fg} k_v^2}{\nu_v(T_w - T_{sat})}\right]^{1/3}; \qquad \frac{L^3(\rho_l - \rho_v)g h'_{fg}}{k_v \nu_v(T_w - T_{sat})} > 5 \times 10^7 \tag{7.73}$$

where $h'_{fg} = h_{fg} + 0.50c_{pv}(T_w - T_{sat})$. Notice that Eq. (7.73) gives a heat transfer coefficient that is independent of the size of the surface and applies to spheres, cylinders, and vertical surfaces.

For film boiling on a horizontal plate, the length of vapor film is related to the spacing between detaching bubbles, which is determined by the instability of the liquid-vapor interface. This mechanism is commonly called *Taylor instability*. As shown in Fig. 7.21, the interface is unstable to a wavelength λ_T on the order of the characteristic length given by Eq. (7.63), $L_c = [\sigma/(\rho_l - \rho_v)g]^{1/2}$. For one-dimensional waves, $\lambda_T = 2\pi\sqrt{3}L_c$; for two-dimensional waves, $\lambda_T = 2\pi\sqrt{6}L_c$. We will simply take $L = L_c$ and absorb the constant in the final correlation for $\overline{h}$; then for a large horizontal plate, $C_{fb} = 0.425$ in Eq. (7.72).

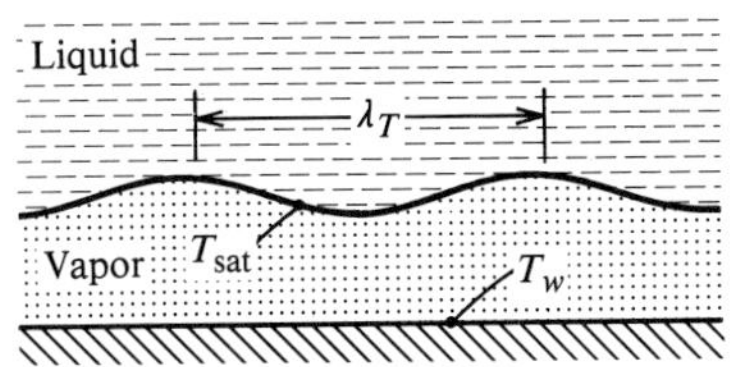

Figure 7.21 Schematic of film boiling on a horizontal plate showing an interface instability of wavelength λ_T.

Effect of Variable Properties

In Eqs. (7.72) and (7.73), h_{fg}, ρ_l, and σ are to be evaluated at T_{sat}; all other properties should be evaluated at the mean film temperature.

Minimum Heat Flux

If a surface operates in the film-boiling regime and q or ΔT is reduced, at a certain point the film breaks down, and the heat flux has a local minimum value q_{min}. If it is assumed that film breakdown occurs when the vapor generation rate becomes too low to sustain Taylor instability wave action on the interface, analysis yields [14]

$$q_{min} = C_{min}\rho_v h_{fg}\left[\frac{\sigma g(\rho_l - \rho_v)}{(\rho_l + \rho_v)^2}\right]^{1/4} \tag{7.74}$$

where $C_{\min} = 0.09$ for large horizontal surfaces [12], and

$$C_{\min} = 0.0464\left[\frac{18}{L^{*2}(2L^{*2}+1)}\right]^{1/4}$$

for horizontal wires [15], where $L^* = R/L_c$, and R is the wire radius. In Eq. (7.74), the vapor density ρ_v is evaluated at T_{sat}.

Effect of Radiation

Above about 600 K, radiation heat transfer across the vapor film becomes significant to film boiling. It is inaccurate to simply add the radiation transfer to the convective transfer, since the vapor film is thickened by the additional vapor produced. If radiation absorption in the vapor film is negligible, and if the absorptance of the liquid is unity, then from Eq. (1.19), the radiation transfer can be written in terms of a radiation heat transfer coefficient as

$$q_{\text{rad}} = h_r(T_w - T_{\text{sat}}); \qquad h_r = 4\sigma\varepsilon\left(\frac{T_w + T_{\text{sat}}}{2}\right)^3 \tag{7.75}$$

where ε is the wall emittance. If the boiling heat transfer coefficient is denoted h_b, then the total heat transfer coefficient h is simply

$$h = h_b + h_r \tag{7.76}$$

The similarity of film boiling to film condensation can be exploited to obtain a simple result for the effect of radiation on h_b. To the level of approximation desired, the result is independent of geometry, so for simplicity, laminar film boiling on a vertical wall will be considered. For film boiling, Eqs. (7.5) and (7.9) become

$$\Gamma = \frac{g\delta^3(\rho_l - \rho_v)}{3\nu_v} \tag{7.77}$$

$$h_b = \frac{k_v(\partial T/\partial y)\big|_w}{T_w - T_{\text{sat}}} = \frac{k_v}{\delta} \tag{7.78}$$

The energy balance Eq. (7.13), modified to include radiation becomes

$$k_v\left.\frac{\partial T}{\partial y}\right|_w + q_{\text{rad}} = h_{\text{fg}}\frac{d\Gamma}{dx} \tag{7.79}$$

Substituting Eqs. (7.75), (7.77), and (7.78) in Eq. (7.79) and rearranging,

$$\frac{k_v\nu_v(T_w - T_{\text{sat}})}{h_{\text{fg}}g(\rho_l - \rho_v)}dx = \frac{\delta^3 d\delta}{1 + h_r/(k_v/\delta)}$$

Integrating with $\delta = 0$ at $x = 0$ gives

$$\frac{4k_v\nu_v(T_w - T_{\text{sat}})x}{h_{\text{fg}}g(\rho_l - \rho_v)} = \delta^4\left[1 - \frac{4}{5}\frac{h_r}{k_v/\delta} + \frac{4}{6}\left(\frac{h_r}{k_v/\delta}\right)^2 - \frac{4}{7}\left(\frac{h_r}{k_v/\delta}\right)^3 + \cdots\right]$$

If the film thickness for no radiation is denoted δ_0, and the corresponding heat transfer coefficient is h_{b0} $(= k_v/\delta_0)$, then

$$\frac{1}{h_{b0}^4} = \frac{1}{h_b^4}\left[1 - \frac{4}{5}\frac{h_r}{h_b} + \frac{4}{6}\left(\frac{h_r}{h_b}\right)^2 - \frac{4}{7}\left(\frac{h_r}{h_b}\right)^3 + \cdots\right] \tag{7.80}$$

which is an implicit equation for h_b that can be solved by iteration. As a first guess, the approximate result $h_b = h_{b0} - (1/5)h_r$ can be used. Equation (7.80) was first given by E. Sparrow [16].

The computer program BOIL, item 4, calculates the average heat transfer coefficient for film boiling according to the recommendations made in this section. The effect of radiation on film boiling is not included.

EXAMPLE 7.10 Film Boiling of a Cryogenic Liquid

A pure copper sphere of 1.5 cm diameter is suddenly immersed in a Dewar flask of saturated liquid nitrogen at 1 atm pressure. Since the temperature difference is large, boiling commences in the film-boiling regime. When $\Delta T = T_w - T_{sat} = 185$ K, determine (i) the average heat transfer coefficient and (ii) the rate of change of the sphere temperature.

Solution

Given: Film boiling of liquid N_2 on a copper sphere.

Required: (i) Average heat transfer coefficient when $\Delta T = 185$ K, and (ii) the corresponding rate of change of the sphere temperature.

Assumptions: The lumped thermal capacity model is adequate for part (ii).

• $T_{sat} = 77.4$ K

1.5 cm

(i) Equation (7.73) gives the average heat transfer coefficient for film boiling in liquid nitrogen:

$$\bar{h} = 0.15\left[\frac{(\rho_l - \rho_v)gh'_{fg}k_v^2}{\nu_v(T_w - T_{sat})}\right]^{1/3}$$

At 1 atm pressure, Table A.12c gives $T_{sat} = 77.4$ K, the normal boiling point, and $h_{fg} = 0.20 \times 10^6$ J/kg. From Table A.8, $\rho_l = 809$ kg/m^3. Vapor properties are evaluated at the mean film temperature, $T_r = 77.4 + (185/2) = 170$ K. From Table A.7, $k_v = 0.0173$ W/m K, $c_{pv} = 1048$ J/kg K, $\rho_v = 2.0$ kg/m^3, $\nu_v = 5.7 \times 10^{-6}$ m^2/s.

$$h'_{fg} = h_{fg} + 0.5c_{pv}(T_w - T_{sat}) = 0.20 \times 10^6 + 0.5(1048)(185) = 0.297 \times 10^6 \text{ J/kg}$$

$$\bar{h} = 0.15\left[\frac{(809 - 2.0)(9.81)(0.297 \times 10^6)(0.0173)^2}{(5.7 \times 10^{-6})(185)}\right]^{1/3} = 131 \text{ W/m}^2\text{ K}$$

(ii) The Biot number for the sphere is

$$\text{Bi} = \frac{\bar{h}(R/3)}{k_{Cu}} = \frac{(131)(0.0075/3)}{(400)} = 0.8 \times 10^{-3} \ll 0.1$$

Thus, a lumped thermal capacity model of the sphere temperature response is adequate. Using Eq. (1.36),

$$\frac{dT}{dt} = -\frac{\bar{h}A}{c\rho V}(T - T_{sat})$$

At $T = 77.4 + 185 = 262.4$ K, pure copper properties from Table A.1 are $c = 374$ J/kg K, $\rho = 8933$ kg/m^3. Also, $A/V = \pi D^2/(\pi D^3/6) = 6/D$.

$$\frac{dT}{dt} = -\frac{(131)(6/0.015)(185)}{(374)(8933)} = -2.90 \text{ K/s}$$

Solution using BOIL

The required input is:

Item 4 (film boiling)
Geometry 4 (sphere)
Fluid = 3 (N_2)
$P = 1.013 \times 10^5$
$T_w = 262.4$
$L = 0.015$

BOIL gives the output:

Relevant property values
$\bar{h} = 131$ W/m^2 K
$q = 24{,}200$ W/m^2

Comments

1. Since Eq. (7.73) is based on experimental data for boiling N_2, the calculated heat transfer coefficient should be reliable.
2. BOIL does not use L_c for this case.

EXAMPLE 7.11 Film Boiling of Water on a Horizontal Plate

What heat flux can be rejected from a large horizontal plate by film boiling of water at 1 atm pressure, if the plate surface temperature is 827 K? The plate emittance is 0.8.

Solution

Given: Water boiling on a horizontal plate.

Required: Heat transfer rate q.

Assumptions:
1. Radiation absorption in the vapor film is negligible (check using the procedure given in Section 6.7).
2. The effect of radiation can be approximately estimated using the vertical wall result, Eq. (7.80).

We will first neglect radiation heat transfer; then Eq. (7.72) gives the average heat transfer coefficient as

$$\overline{h} = C_{\text{fb}}\left[\frac{(\rho_l - \rho_v)gh'_{\text{fg}}k_v^3}{\nu_v L(T_w - T_{\text{sat}})}\right]^{1/4}$$

The reference temperature for vapor properties is the mean film temperature, $T_r = (1/2)(373 + 827) = 600$ K: $k_v = 0.046$ W/m K, $c_{pv} = 2003$ J/kg K, $\nu_v = 58.5 \times 10^{-6}$ m²/s, $\rho_v = 0.366$ kg/m³. Also, at $T_{\text{sat}} = 373.15$ K, $h_{\text{fg}} = 2.257 \times 10^6$ J/kg, $\sigma = 58.9 \times 10^{-3}$ N/m, $\rho_l = 958$ kg/m³. The characteristic length to be used in Eq. (7.72) is L_c:

$$L = L_c = \left[\frac{\sigma}{(\rho_l - \rho_v)g}\right]^{1/2} = \left[\frac{(58.9 \times 10^{-3})}{(958)(9.81)}\right]^{1/2} = 2.50 \times 10^{-3} \text{ m (2.5 mm)}$$

The correction to the enthalpy of vaporization to account for vapor superheat is

$$h'_{\text{fg}} = h_{\text{fg}} + 0.35c_{pv}(T_w - T_{\text{sat}}) = 2.257 \times 10^6 + (0.35)(2003)(454) = 2.575 \times 10^6 \text{ J/kg}$$

Substituting in Eq. (7.72), with $C_{\text{fb}} = 0.425$ for a large horizontal surface,

$$\overline{h} = 0.425\left[\frac{(958)(9.81)(2.575 \times 10^6)(0.046)^3}{(58.5 \times 10^{-6})(2.5 \times 10^{-3})(827 - 373)}\right]^{1/4} = 184 \text{ W/m}^2\text{ K}$$

$$\overline{q} = \overline{h}(T_w - T_{\text{sat}}) = (184)(827 - 373) = 8.35 \times 10^4 \text{ W/m}^2$$

We should verify that this value of $\overline{q}$ is greater than $q_{\min}$, which is given by Eq. (7.74) with $C_{\min} = 0.09$. In Eq. (7.74), ρ_v is evaluated at T_{sat}: at 373.15 K, $\rho_v = 0.598$ kg/m³.

$$q_{\min} = C_{\min}\rho_v h_{\text{fg}}\left[\frac{\sigma g(\rho_l - \rho_v)}{(\rho_l + \rho_v)^2}\right]^{1/4}$$

$$= (0.09)(0.598)(2.257 \times 10^6)\left[\frac{(58.9 \times 10^{-3})(9.81)(958)}{(958)^2}\right]^{1/4} = 1.90 \times 10^4 \text{ W/m}^2$$

Thus, the film-boiling heat flux calculated above is about four times the minimum heat flux.

To include the effect of radiation transfer, we first calculate the radiation heat transfer coefficient h_r:

$$h_r = 4\sigma\varepsilon\left(\frac{T_w + T_{\text{sat}}}{2}\right)^3 = (4)(5.67 \times 10^{-8})(0.8)\left(\frac{827 + 373}{2}\right)^3 = 39.2 \text{ W/m}^2\text{ K}$$

Retaining two terms in the series of Eq. (7.80) and rearranging,

$$1 - \left(\frac{h_b}{h_{b0}}\right) = 1 - \left[1 - \frac{4}{5}\frac{h_r}{h_b} + \frac{4}{6}\left(\frac{h_r}{h_b}\right)^2\right]^{1/4}; h_{b0} = \overline{h} = 184 \text{ W/m}^2\text{ K}$$

Iteration is required to solve this equation. As a first guess, use $h_b = h_{b0} - (1/5)h_r = 184 - (0.2)(39.2) = 176.2$, and construct the following table.

h_b	l.h.s.	r.h.s.
176.2	0.0424	0.0384
177.0	0.0380	0.0383
176.8	0.0391	0.0383

Interpolating in the table,

$$h_b = 176.9 \text{ W/m}^2\text{ K}; \quad h = h_b + h_r = 176.9 + 39.2 = 216.1 \text{ W/m}^2\text{ K}$$
$$\overline{q} = h(T_w - T_{\text{sat}}) = (216.1)(827 - 373) = 9.81 \times 10^4 \text{ W/m}^2$$

Comments

1. The effect of radiation is to increase q by only 17%; hence, our approximate method for calculating the effect of radiation is adequate.
2. These film-boiling heat fluxes are only about 7% of the peak value calculated in Example 7.9.

7.5 FORCED-CONVECTION BOILING AND CONDENSATION

Forced-flow boiling and condensation in horizontal or vertical tubes and channels occurs in such diverse equipment as water-tube boilers, refrigerator evaporators and condensers, water-cooled nuclear reactors, and a great variety of process and chemical equipment. In a water-tube boiler, addition of heat changes the water flow to a two-phase flow of water and vapor, and finally to a pure vapor flow; in a tube condenser, the reverse process occurs. In each case, the two-phase flow is very complex, and many distinct flow patterns can be identified, such as *bubbly flow*, *slug flow*, and *annular flow*. Just predicting the flow pattern for specified system parameters is a difficult task. Reliable prediction of pressure drop and heat transfer is even more difficult, owing to the complexity of the flow patterns.

This section presents an introduction to the subject of internal-flow forced-convection boiling and condensation. The nature and scope of the subject is such that any serious engineering analysis and design should be preceded by a study of specialized texts on the subject, such as those listed in the bibliography at the end of this text. In Section 7.5.1, two-phase flow patterns are described, and flow maps are presented. These maps enable identification of flow regimes. A method of calculating pressure drop for two-phase flows is given in Section 7.5.2, based on the simple homogeneous flow model. In Section 7.5.3, the phenomena associated with forced convection boiling are described, and a calculation method is given that enables the heat transfer coefficient to be determined for a prescribed wall heat flux. The method is valid for both vertical and horizontal tubes or channels. Section 7.5.4 presents a calculation method for forced-convection condensation in horizontal tubes. Owing to the complexity of two-phase flows, these calculation methods are based on experimental data in the form of rather complicated correlations of the many parameters involved. Nevertheless, the student will find that the calculation methods are relatively simple to implement.

7.5.1 Two-Phase Flow Patterns

Figure 7.22 shows flow patterns characteristic of vertical cocurrent flow of liquid and a gas or vapor. The actual flow pattern encountered depends primarily on the flow velocity and the relative amounts of liquid and gas. A brief description of each regime follows.

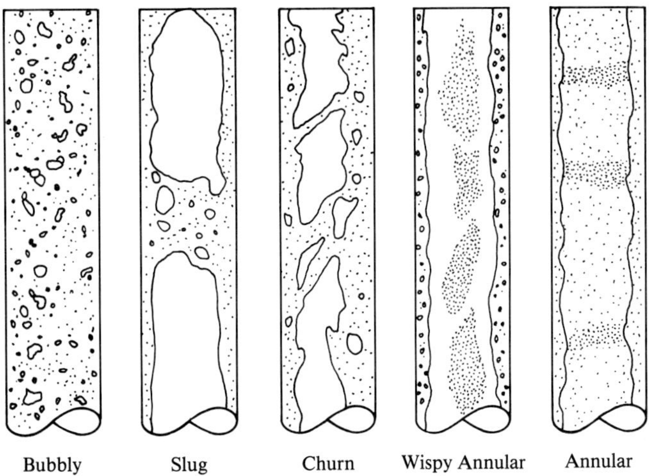

Figure 7.22 Flow patterns for vertical cocurrent flow of a liquid and a gas or vapor.

1. *Bubbly flow*. The gas or vapor phase is in the form of isolated bubbles in the liquid phase. The bubbles may be small and spherical, or large enough to develop a spherical cap shape.

2. *Slug flow*. Bubbles are approximately of the diameter of the tube, and a film of liquid runs down the tube wall. In between the bubbles are slugs of liquid, which may contain some small bubbles.

3. *Churn flow*. A highly irregular and often unsteady flow pattern consisting of large bubbles in the core, which continuously coalesce and break up. The liquid flow tends to be near the wall.

4. *Wispy annular flow*. A relatively thick liquid film that may contain small bubbles flows along the wall. The gas core contains a significant number of liquid droplets that form irregular clouds.

5. *Annular flow*. The phases are almost completely separated into a gas core and liquid film on the wall. However, there is some entrainment of droplets from the crests of waves that form on the film surface.

Our understanding of these various flow regimes, and of the conditions required for a transition from one regime to another, is far from complete. Nevertheless, flow "maps" are available that allow the flow regime to be identified. Before we present a sample of such a map, some definitions used in two-phase flow must be introduced.

The vapor mass quality x is defined as

$$x = \frac{\dot{m}_v}{\dot{m}_v + \dot{m}_l} \tag{7.81}$$

where $\dot{m}_v$ and $\dot{m}_l$ [kg/s] are the flow rates of vapor and liquid phases, respectively.

Mass velocities are defined in terms of the tube cross-sectional area A_c:

$$G = G_v + G_l; \qquad G_v = \frac{\dot{m}_v}{A_c}; \qquad G_l = \frac{\dot{m}_l}{A_c} \tag{7.82}$$

It follows that the mass flow rates in terms of mass velocities and quality are

$$\dot{m}_v = GA_c x; \qquad \dot{m}_l = GA_c(1 - x) \tag{7.83}$$

The *superficial velocity* j [m/s] is the volumetric flow divided by the tube cross-sectional area:

$$j = j_v + j_l; \qquad j_v = \frac{\dot{m}_v}{\rho_v A_c}; \qquad j_l = \frac{\dot{m}_l}{\rho_l A_c} \tag{7.84}$$

Figure 7.23 shows a map of flow regimes for upward cocurrent two-phase flow, based on observations on air-water and high-pressure steam-water flow in tubes of 1 to 3 cm diameter (extrapolation to other fluids or tube sizes is not advised). The axes are the superficial momentum fluxes $\rho_l j_l^2$ and $\rho_v j_v^2$ [kg/s^2 m]. These fluxes can be expressed in terms of mass velocity and mass quality as

$$\rho_l j_l^2 = \frac{G^2(1 - x)^2}{\rho_l}; \qquad \rho_v j_v^2 = \frac{G^2 x^2}{\rho_v} \tag{7.85}$$

We can see that annular flows require high vapor momentum fluxes, with wispy annular flow occurring at high liquid momentum fluxes. Bubbly flow occurs at high liquid momentum fluxes and low gas momentum fluxes, as might be expected. Slug and churn flows occur at low liquid momentum fluxes.

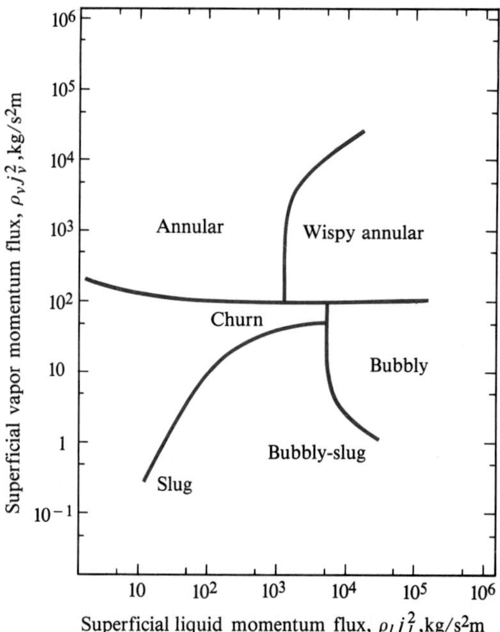

Figure 7.23 Flow regime map for upward cocurrent two-phase flow [17].

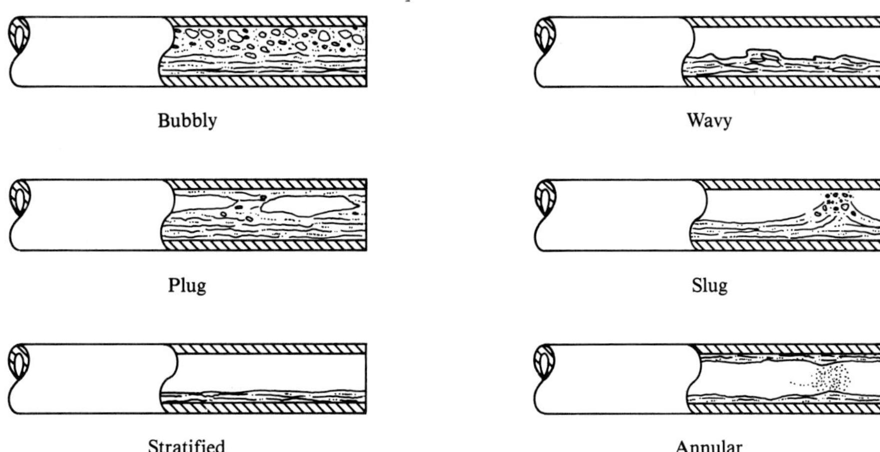

Figure 7.24 Flow patterns for horizontal cocurrent flow of a liquid and a gas or vapor.

For horizontal (or inclined) flows, Fig. 7.24 shows the flow patterns encountered. The influence of a gravity force perpendicular to the flow causes differences from the vertical case, especially at low flow rates. In bubbly and slug flows, the bubbles tend to flow in the upper half of the pipe. At low liquid and gas flow rates, there is a possibility of stratified flow, with the liquid flowing along the bottom of the tube with a relatively smooth surface. As the vapor velocity increases, waves form on the liquid surface and can become large enough to form liquid slugs, which wet the upper periphery of the pipe wall. A still higher vapor velocity results in annular flow, or the liquid can be dispersed into droplets entrained in the vapor flow. Figure 7.25 shows

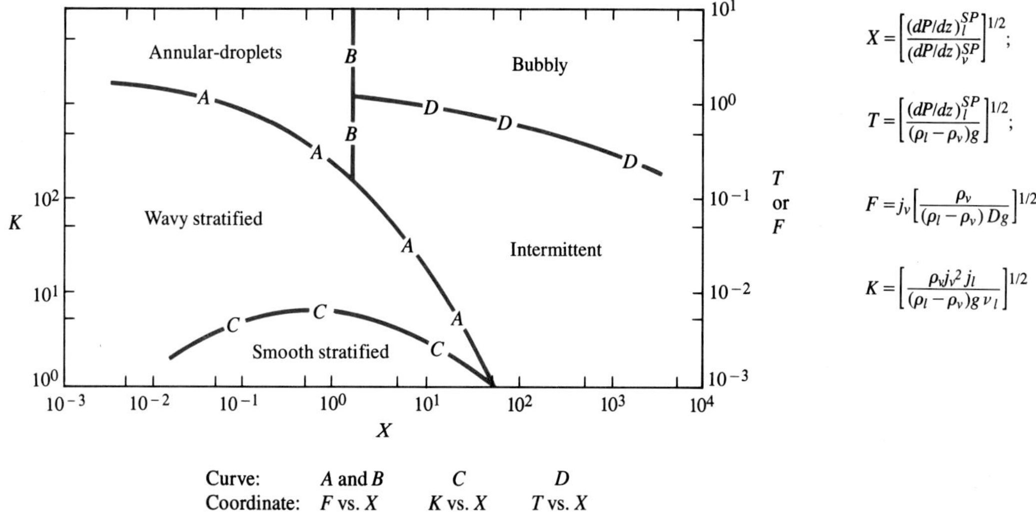

Figure 7.25 Two-phase flow map for horizontal flow [18].

a two-phase flow map for horizontal flow proposed by Taitel and Dukler [18]. The map gives the conditions for transitions between five flow regimes: smooth stratified, wavy stratified, intermittent (slug or churn), annular with droplets, and bubbly. The transition curves A, B, C, and D have different ordinates, as specified in the figure. Use of this flow map is demonstrated in Example 7.13.

Flow regime maps such as Figs. 7.23 and 7.25 are based on experimental data for adiabatic flows. With heat transfer the regimes may shift significantly; thus, these maps may prove unreliable when applied to evaporating or condensing flows.

EXAMPLE 7.12 A Vertical-Tube Water Evaporator

A vertical tube evaporator containing 100 5 cm–O.D., 2 mm–wall-thickness tubes generates 83 kg/s of steam. At a given location up the evaporator, the pressure is 1.17 MPa, and the mass quality is 10.0%. For cocurrent upward flow, identify the flow regime.

Solution

Given: Vertical cocurrent flow of water and steam.

Required: Flow regime.

Assumptions: The flow map, Fig. 7.23, applies.

Evaluate properties at saturation conditions. Using Tables A.12*a* and A.8, $T_{sat} = 460$ K, $\rho_v = 5.984$ kg/m^3, $\rho_l = 879$ kg/m^3.

$$G = \frac{\dot{m}}{A_c} = \frac{83}{(100)(\pi/4)(0.046)^2} = 499 \text{ kg/m}^2\text{ s}$$

Using Eq. (7.85),

$$\rho_l j_l^2 = \frac{G^2(1-x)^2}{\rho_l} = \frac{(499)^2(1-0.1)^2}{879} = 229$$

$$\rho_v j_v^2 = \frac{G^2 x^2}{\rho_v} = \frac{(499)^2(0.1)^2}{5.984} = 416$$

The flow map, Fig. 7.23, indicates the annular flow regime.

EXAMPLE 7.13 A Horizontal-Tube Refrigerant-12 Condenser

R-12 condenses in a horizontal copper tube of 1 cm–O.D. and 1 mm–wall-thickness. The mass flow rate is 0.0125 kg/s, and at a location where the pressure is 1.01 MPa, the mass quality is 60%. Identify the flow regime.

Solution

Given: Forced-convection condensation of R-12 in a horizontal tube.

Required: The flow regime.

Assumptions: The flow map, Fig. 7.25, applies.

$$A_c = (\pi/4)(D_i)^2 = (\pi/4)(0.008)^2 = 5.027 \times 10^{-5}\ \text{m}^2$$

$$G = \frac{\dot{m}}{A_c} = \frac{0.0125}{5.027 \times 10^{-5}} = 249\ \text{kg/m}^2\ \text{s}$$

$$G_v = xG = (0.6)(249) = 149.4\ \text{kg/m}^2\ \text{s}$$

$$G_l = (1 - x)G = (0.4)(249) = 99.6\ \text{kg/m}^2\ \text{s}$$

Evaluate properties at T_{sat}: from Table A.12*e*, T_{sat} (1.01 MPa) $= 315.0$ K, $\rho_v = 1/v_v = 1/0.01734 = 57.6\ \text{kg/m}^3$. From Table A.7, $\mu_v = 13.2 \times 10^{-6}$ kg/m s. From Table A.8, $\nu_l = 0.191 \times 10^{-6}\ \text{m}^2/\text{s}$; $\rho_l = 1242\ \text{kg/m}^3$.

To calculate the single-phase pressure gradients, we first evaluate the single-phase Reynolds numbers.

$$\text{Re}_v = \frac{G_v D}{\mu_v} = \frac{(149.4)(0.008)}{13.2 \times 10^{-6}} = 90,500$$

$$\text{Re}_l = \frac{G_l D}{\rho_l \nu_l} = \frac{(99.6)(0.008)}{(1242)(0.191 \times 10^{-6})} = 3360$$

The flow is turbulent; Eq. (4.42) gives the friction factor, and the pressure gradients are obtained from Eq. (4.15):

$$\left(\frac{dP}{dz}\right)_v^{\text{SP}} = f_v \frac{G_v^2}{2\rho_v D} = 0.0184 \frac{(149.4)^2}{(2)(57.6)(0.008)} = 446\ \text{Pa/m}$$

$$\left(\frac{dP}{dz}\right)_l^{\text{SP}} = f_l \frac{G_l^2}{2\rho_l D} = 0.0439 \frac{(99.6)^2}{(2)(1242)(0.008)} = 21.9\ \text{Pa/m}$$

The superficial velocities are

$$j_v = \frac{G_v}{\rho_v} = \frac{149.4}{57.6} = 2.59\ \text{m/s}$$

$$j_l = \frac{G_l}{\rho_l} = \frac{99.6}{1242} = 0.0802\ \text{m/s}$$

The flow map parameters are then

$$X = \left[\frac{(dP/dz)_l^{\text{SP}}}{(dP/dz)_v^{\text{SP}}}\right]^{1/2} = \left(\frac{21.9}{446}\right)^{1/2} = 0.22$$

$$T = \left[\frac{(dP/dz)_l^{\text{SP}}}{(\rho_l - \rho_v)g}\right]^{1/2} = \left[\frac{21.9}{(1242 - 58)(9.81)}\right]^{1/2} = 0.0434$$

$$F = j_v\left[\frac{\rho_v}{(\rho_l - \rho_v)Dg}\right]^{1/2} = 2.59\left[\frac{57.6}{(1242 - 58)(0.008)(9.81)}\right]^{1/2} = 2.04$$

$$K = \left[\frac{\rho_v j_v^2 j_l}{(\rho_l - \rho_v)g\nu_l}\right]^{1/2} = \left[\frac{(57.6)(2.59)^2(0.0802)}{(1242 - 58)(9.81)(0.191 \times 10^{-6})}\right]^{1/2} = 118$$

Since K lies above curve C and F lies above curve A, we conclude that the annular flow regime is present.

Comments

Taitel and Dukler [18] also show how to obtain flow regime maps for pipelines inclined to the horizontal.

7.5.2 Pressure Drop

The uncertainty and complexity of the flow patterns encountered with two-phase flow in tubes make the prediction of pressure drop very difficult. Much research has been done on this problem, with only partial success. Accurate prediction schemes tend to be highly empirical and hence limited in range of applicability. We will consider here only a simple method of calculating pressure drop, based on a *homogeneous model* of two-phase flow that ignores details of the flow patterns. More accurate calculation methods may be obtained from the bibliography at the end of the text.

The pressure gradient in a straight tube can be written as

$$\frac{dP}{dz} = \left(\frac{dP}{dz}\right)_F + \left(\frac{dP}{dz}\right)_G + \left(\frac{dP}{dz}\right)_M \tag{7.86}$$

where subscripts F, G, and M refer to pressure gradients due to *wall friction*, *gravity*, and *momentum changes*, respectively. The homogeneous model of two-phase flow assumes that the velocity of each phase is the same. Thus, in terms of the mass quality x, the density of the flow is

$$\rho = x\rho_v + (1 - x)\rho_l \tag{7.87}$$

The pressure gradient due to wall friction is calculated from

$$\left(\frac{dP}{dz}\right)_F = -\frac{f}{D}\frac{G^2}{2\rho} \tag{7.88}$$

The simplest method for evaluating the two-phase flow friction factor f is to assume that pure liquid is flowing at the mixture mass velocity, that is, at a Reynolds number $\text{Re} = GD/\mu_l$. This friction factor will generally be too low at qualities below about 70%, and too high at qualities above 70%. Of course, the limiting value for pure vapor, $x = 1$, is quite incorrect. To overcome this difficulty, a simple reference viscosity should be used,

$$\frac{1}{\mu_r} = \frac{x}{\mu_v} + \frac{1 - x}{\mu_l} \tag{7.89}$$

which mass-weights the inverse of viscosity and has the limits

$$x = 0: \quad \mu_r = \mu_l$$

$$x = 1: \quad \mu_r = \mu_v$$

The pressure gradient due to gravity is calculated from

$$\left(\frac{dP}{dz}\right)_G = -\rho g \sin\theta \tag{7.90}$$

where θ is the angle measured from the horizontal. Momentum changes can be large in evaporators and condensers. In an evaporator, velocity and momentum increase as the liquid vaporizes, requiring a pressure drop; in a condenser, there is a pressure recovery due to the decreasing velocity and momentum as the vapor condenses. In terms of momentum change,

$$\left(\frac{dP}{dz}\right)_M = -\frac{d}{dz}\left(\frac{G^2}{\rho}\right) \tag{7.91}$$

but since G is constant along the tube,

$$\left(\frac{dP}{dz}\right)_M = -G^2\frac{d}{dz}\left(\frac{1}{\rho}\right) = \left(\frac{G}{\rho}\right)^2\frac{d\rho}{dz} \tag{7.92}$$

In the case of an isothermal gas-liquid flow, such as an air-water flow, the mixture density ρ is nearly constant, and the preceding equations can be written in terms of a pressure drop over a length of tube L. However, for a liquid-vapor flow undergoing phase change, the large variation of ρ along the tube requires numerical integration in association with an energy balance to determine the quality x and, hence, ρ.

EXAMPLE 7.14 Pressure Gradient in a Vertical-Tube Water Evaporator

Calculate the pressure gradient in Example 7.12 if the heating rate per tube is 1.5×10^5 W per meter length.

Solution

Given: Vertical cocurrent flow of water and steam.

Required: Pressure gradient.

Assumptions:
1. Homogeneous flow model.
2. All properties may be evaluated at $T_{sat} = 460$ K.

From Example 7.12,

$$\text{I.D.} = 0.046 \text{ m}$$

$\dot{m} = 0.83$ kg/s $\qquad G = 499$ kg/m^2 s $\qquad x = 0.1$

$T_{sat} = 460$ K

$\rho_l = 879$ kg/m^3 $\qquad \rho_v = 5.984$ kg/m^3

From Tables A.8 and A.7,

$$\mu_l = 1.49 \times 10^{-4} \text{ kg/m s}, \ \mu_v = 16.0 \times 10^{-6} \text{ kg/m s}.$$

Equation (7.89) gives the reference viscosity:

$$\frac{1}{\mu_r} = \frac{x}{\mu_v} + \frac{1-x}{\mu_l} = \frac{0.1}{16.0 \times 10^{-6}} + \frac{0.9}{1.49 \times 10^{-4}}; \qquad \mu_r = 8.13 \times 10^{-5} \text{ kg/m s}$$

$$\text{Re} = \frac{GD}{\mu_r} = \frac{(499)(0.046)}{8.13 \times 10^{-5}} = 2.82 \times 10^5$$

The flow is turbulent. Using Eq. (4.42), the friction factor is

$$f = (0.790 \ln \text{Re}_D - 1.64)^{-2} = (0.790 \ln 2.82 \times 10^5 - 1.64)^{-2} = 1.467 \times 10^{-2}$$

The flow density is given by Eq. (7.87):

$$\rho = x\rho_v + (1-x)\rho_l = (0.1)(5.984) + (0.9)(879) = 792 \text{ kg/m}^3$$

The pressure gradient due to wall friction is given by Eq. (7.88):

$$\left(\frac{dP}{dz}\right)_F = -\frac{f}{D}\frac{G^2}{2\rho} = -\frac{1.467 \times 10^{-2}}{0.046}\frac{(499)^2}{(2)(792)} = -49.9 \text{ Pa/m}$$

The pressure gradient due to gravity is given by Eq. (7.90):

$$\left(\frac{dP}{dz}\right)_G = -\rho g \sin\theta = -(792)(9.81)(1) = -7770 \text{ Pa/m}$$

The pressure gradient due to momentum change is given by Eq. (7.92) as

$$\left(\frac{dP}{dz}\right)_M = \left(\frac{G}{\rho}\right)^2 \frac{d\rho}{dz} \simeq -\left(\frac{G}{\rho}\right)^2 \rho_l \frac{dx}{dz}$$

for $\rho \simeq (1-x)\rho_l$. An energy balance gives the gradient in mass quality: from Table A.12a, $h_{fg} = 1.990 \times 10^{-6}$ J/kg,

$$\frac{dx}{dz} = \frac{\dot{Q}/L}{\dot{m}h_{fg}} = \frac{1.5 \times 10^5}{(0.83)(1.990 \times 10^6)} = 0.0908 \text{ m}^{-1}$$

$$\left(\frac{dP}{dz}\right)_M = -\left(\frac{499}{792}\right)^2 (879)(0.0908) = -31.7 \text{ Pa/m}$$

Summing,

$$\frac{dP}{dz} = \left(\frac{dP}{dz}\right)_F + \left(\frac{dP}{dz}\right)_G + \left(\frac{dP}{dz}\right)_M = -49.9 - 7770 - 31.7 = -7850 \text{ Pa/m}$$

Comments

1. The wall friction is relatively small due to the large flow density and low velocity.
2. The pressure gradient due to acceleration is small due to the low velocity.
3. The pressure gradient due to gravity dominates under these conditions.

EXAMPLE 7.15 Pressure Gradient in a Horizontal-Tube Refrigerant-12 Condenser

Calculate the pressure gradient in Example 7.13 if the cooling rate is 500 W per meter length.

Solution

Given: Forced-convection condensation of R-12 in a horizontal tube.

Required: Pressure gradient.

Assumptions: 1. Homogeneous flow model.
2. All properties may be evaluated at T_{sat}.

From Example 7.13,

$$\begin{array}{lll} \text{I.D.} = 0.008 \text{ m} & A_c = 5.027 \times 10^{-5} \text{ m}^2 & \\ \dot{m} = 0.0125 \text{ kg/s} & G = 249 \text{ kg/m}^2 \text{ s} & x = 0.6 \\ P = 1.01 \text{ MPa} & T_{sat} = 315 \text{ K} & \\ \rho_l = 1242 \text{ kg/m}^3 & \rho_v = 57.6 \text{ kg/m}^3 & \\ \mu_l = 2.38 \times 10^{-4} \text{ kg/m s} & \mu_v = 13.2 \times 10^{-6} \text{ kg/m s} & \end{array}$$

$$\rho = x\rho_v + (1-x)\rho_l = (0.6)(57.6) + (0.4)(1242) = 531 \text{ kg/m}^3$$

$$\frac{1}{\mu_r} = \frac{x}{\mu_v} + \frac{1-x}{\mu_l} = \frac{0.6}{13.2 \times 10^{-6}} + \frac{0.4}{2.38 \times 10^{-4}}; \qquad \mu_r = 2.12 \times 10^{-5} \text{ kg/m s}$$

$$\text{Re} = \frac{GD}{\mu_r} = \frac{(249)(0.008)}{2.12 \times 10^{-5}} = 9.40 \times 10^4$$

The flow is turbulent. Equation (4.42) gives the friction factor:

$$f = (0.790 \ln \text{Re}_D - 1.64)^{-2} = (0.790 \ln 9.40 \times 10^4 - 1.64)^{-2} = 1.823 \times 10^{-2}$$

The pressure gradient due to wall friction is given by Eq. (7.88):

$$\left(\frac{dP}{dz}\right)_F = -\frac{f}{D}\frac{G^2}{2\rho} = -\frac{1.823 \times 10^{-2}}{0.008}\frac{(249)^2}{(2)(531)} = -133 \text{ Pa/m}$$

From Table A.12*e*, $h_{fg} = 0.1273 \times 10^6$ J/kg. An energy balance gives

$$\frac{dx}{dz} = -\frac{\dot{Q}/L}{\dot{m}h_{fg}} = -\frac{500}{(0.0125)(0.1273 \times 10^6)} = -0.314 \text{ m}^{-1}$$

$$\frac{d\rho}{dz} = (\rho_v - \rho_l)\frac{dx}{dz} = (57.6 - 1242)(-0.314) = 372 \text{ kg/m}^4$$

Substituting in Eq. (7.92) gives the pressure recovery:

$$\left(\frac{dP}{dz}\right)_M = \left(\frac{G}{\rho}\right)^2 \frac{d\rho}{dz} = \left(\frac{249}{531}\right)^2 (372) = 81.8 \text{ Pa/m}$$

Summing,

$$\frac{dP}{dz} = \left(\frac{dP}{dz}\right)_F + \left(\frac{dP}{dz}\right)_M = -133 + 82 = -51 \text{ Pa/m}$$

Comments

The deceleration pressure recovery is very significant.

7.5.3 Internal Forced-Convection Boiling

Figure 7.26 shows a schematic of flow patterns in a vertical evaporator tube. In the initial single-phase region, the liquid is heated to the saturation temperature. Further up the tube, the superheat exceeds that required for boiling inception, and nucleate boiling produces a bubbly flow region. With further vapor production, the bubbles coalesce to form a slug flow region, which, higher up the tube, gives way to annular flow. Depletion of the liquid film by evaporation and entrainment finally causes the film to dry out. The remaining droplets evaporate slowly until a single-phase vapor flow results. A similar scenario is encountered in a horizontal evaporator tube. Two modes of heat transfer dominate the overall process. The first is nucleate boiling from active sites on the heated wall, which is characterized by a strong dependence of the heat transfer coefficient on heat flux, as was the case for pool boiling. The second mode is forced-convection evaporation from a liquid film into the vapor core: for this mode, the heat transfer coefficient does not depend on heat flux but mainly depends on mass flow rate and vapor quality. In both modes, the tube thermal conductivity, k_w, plays a role.

Heat transfer coefficient correlation schemes for internal-flow forced-convection boiling tend to be complicated, and numerous alternatives may be found in the literature. The scheme recommended below was proposed by V. Klimenko [19] and has

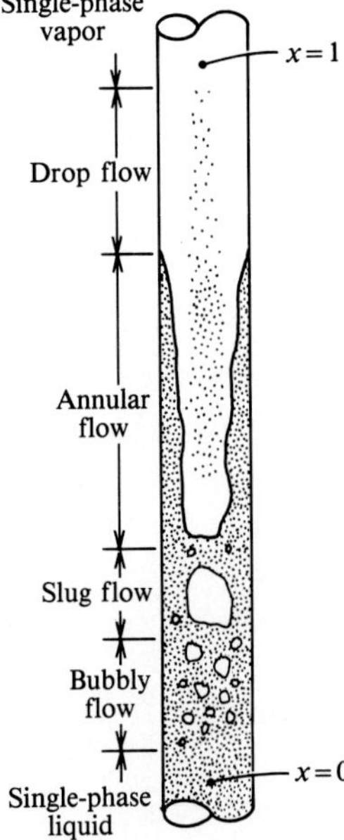

Figure 7.26 Flow patterns in a vertical evaporator tube, with cocurrent upward flow of liquid and vapor.

been selected as a compromise between simplicity and accuracy. It is valid only when the tube walls are wet ("dryout" does not occur). The first step is to determine whether the dominant mode is *nucleate boiling* or *film evaporation* by evaluating the parameter Φ:

$$\Phi = \frac{Gh_{fg}}{q}\left[1 + x\left(\frac{\rho_l}{\rho_v} - 1\right)\right]\left(\frac{\rho_v}{\rho_l}\right)^{1/3} \quad \textbf{(7.93)}$$

For $\Phi < 1.6 \times 10^4$, nucleate boiling dominates; for $\Phi > 1.6 \times 10^4$, film evaporation dominates.

The next step is to obtain the two-phase heat transfer coefficient h_{TP} from either of the following two correlations.

$\Phi < 1.6 \times 10^4$:

$$\text{Nu} = 7.4 \times 10^{-3} q^{*0.6} P^{*0.5} \text{Pr}_l^{-1/3}\left(\frac{k_w}{k_l}\right)^{0.15} \quad \textbf{(7.94)}$$

where

$$\text{Nu} = \frac{h_{TP}L_c}{k_l}; \qquad L_c = \left[\frac{\sigma}{g(\rho_l - \rho_v)}\right]^{1/2}$$

$$q^* = \frac{qL_c}{h_{fg}\rho_v\alpha_l}$$

$$P^* = \frac{P}{[\sigma g(\rho_l - \rho_v)]^{1/2}} = \frac{PL_c}{\sigma}$$

$\Phi > 1.6 \times 10^4$:

$$\text{Nu} = 0.087\text{Re}^{0.6}\text{Pr}_l^{1/6}\left(\frac{\rho_v}{\rho_l}\right)^{0.2}\left(\frac{k_w}{k_l}\right)^{0.09} \quad \textbf{(7.95)}$$

where

$$\text{Re} = \frac{VL_c}{\nu_l}; \qquad V = \frac{G}{\rho_l}\left[1 + x\left(\frac{\rho_l}{\rho_v} - 1\right)\right]$$

Properties are to be evaluated at T_{sat}.

The final step is to calculate a single-phase convective heat transfer coefficient h_{FC} based on an "all liquid" Reynolds number, $\text{Re} = GD/\mu_l$. Often h_{FC} is much smaller than h_{TP} and can be ignored; if it is not, a simple combining formula should be used:

$$h = (h_{TP}^3 + h_{FC}^3)^{1/3} \quad \textbf{(7.96)}$$

The preceding correlations represent experimental data with a mean absolute deviation of 12.9% for nine different fluids (water, refrigerants, and cryogens) over the following main parameter ranges: pressure 0.61–30.4 bar; mass velocity 50–2690 kg/m^2 s; vapor quality 0.017–1.00; tube diameter 1.63–41.3 mm. Notice that the vaporization rate depends on the tube wall thermal conductivity, which may be expected for nucleate boiling (surface finish may also have an effect). Figure 7.27 shows the wall conductivity effect clearly, despite the scatter in the data. Figure 7.28 shows the transition from the nucleate boiling to film evaporation regimes.[3]

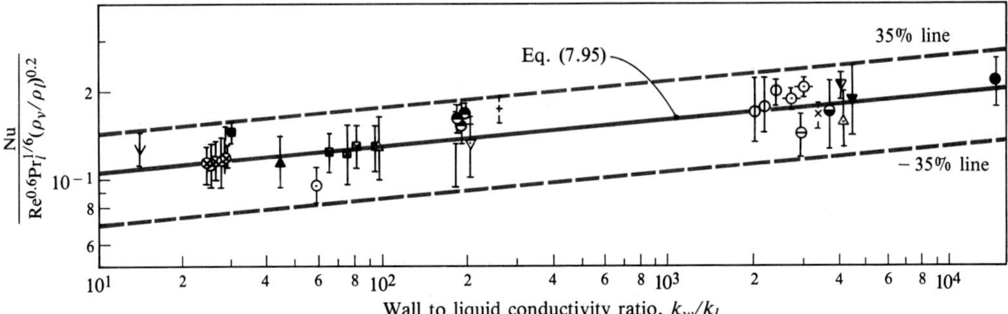

Figure 7.27 Effect of wall conductivity on Nusselt number for film evaporation in forced-convection boiling: experimental data compiled by V. Klimenko [19].

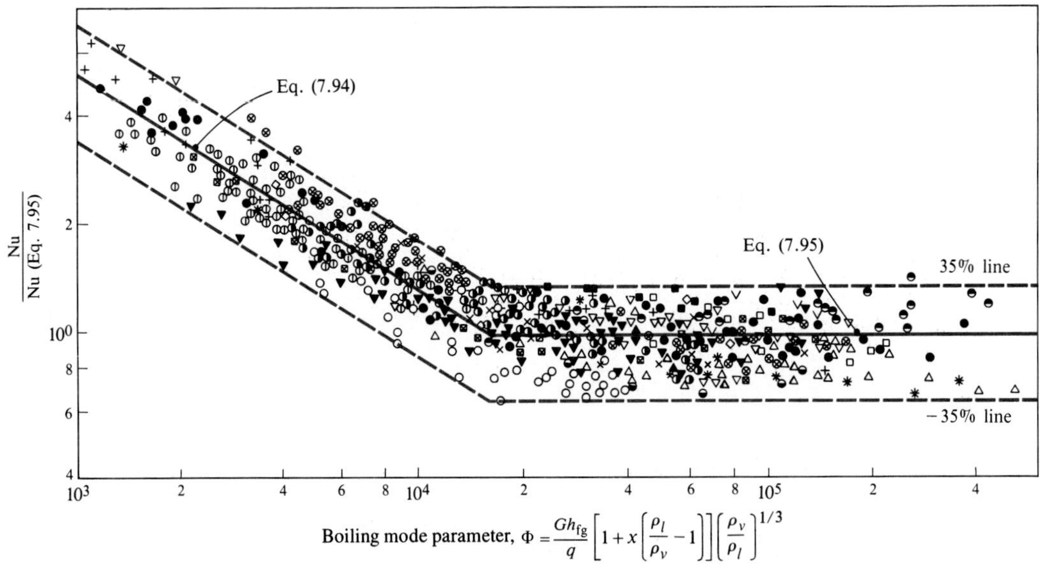

Figure 7.28 The transition from nucleate boiling to film evaporation in forced-convection boiling: experimental data compiled by V. Klimenko [19].

[3] Klimenko has recently published an improved form of his correlation that has an even wider range of applicability. See Klimenko, V. V., "A generalized correlation for two-phase forced flow heat transfer," *Int. J. Heat Mass Transfer*, 33, 2073–2088 (1990).

EXAMPLE 7.16 Heat Transfer in a Vertical-Tube Water Evaporator

Calculate the heat transfer coefficient in Example 7.12 if the tube material is AISI 304 stainless steel.

Solution

Given: Vertical cocurrent flow of water and steam.

Required: Heat transfer coefficient

Assumptions: Equations (7.93) through (7.96) apply.

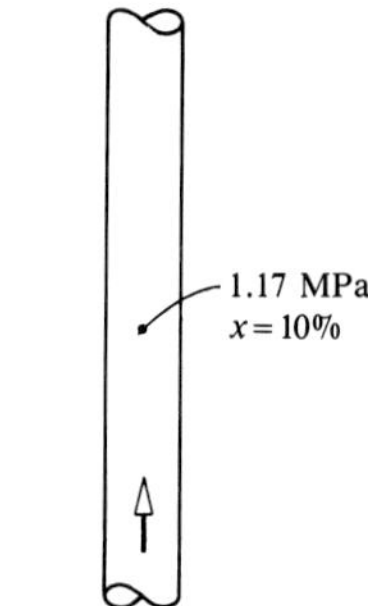

From Examples 7.12 and 7.14,

$$\begin{aligned} &\text{I.D.} = 0.046 \text{ m} \\ &\dot{m} = 0.83 \text{ kg/s} && G = 499 \text{ kg/m}^2\text{ s} && x = 0.1 \\ &P = 1.17 \text{ MPa} && T_{\text{sat}} = 460 \text{ K} \\ &\rho_l = 879 \text{ kg/m}^3 && \rho_v = 5.984 \text{ kg/m}^3 \\ &h_{\text{fg}} = 1.990 \times 10^6 \text{ J/kg K} \end{aligned}$$

Also, from Tables A.8 and A.11, $k_l = 0.670$ W/m K, $c_{pl} = 4400$ J/kg K, $\text{Pr}_l = 0.98$, $\sigma = 40.7 \times 10^{-3}$ N/m. Hence, $\alpha_l = 1.73 \times 10^{-7}$ m²/s.

The wall heat flux on the inside of the tube is

$$q = (\dot{Q}/L)/\pi D = \frac{1.5 \times 10^5}{\pi(0.046)} = 1.038 \times 10^6 \text{ W/m}^2$$

Equation (7.93) gives the parameter Φ:

$$\Phi = \frac{G h_{\text{fg}}}{q}\left[1 + x\left(\frac{\rho_l}{\rho_v} - 1\right)\right]\left(\frac{\rho_v}{\rho_l}\right)^{1/3}$$

$$= \frac{(499)(1.990 \times 10^6)}{1.038 \times 10^6}\left[1 + (0.1)\left(\frac{879}{5.984} - 1\right)\right]\left(\frac{5.984}{879}\right)^{1/3} = 2826$$

Since $\Phi < 1.6 \times 10^4$, nucleate boiling dominates, and Eq. (7.94) gives the Nusselt number

$$\text{Nu} = 7.4 \times 10^{-3}\, q^{*0.6} P^{*0.5} \text{Pr}_l^{-1/3}\left(\frac{k_w}{k_l}\right)^{0.15}$$

The characteristic length in Nu and q^* is L_c.

$$L_c = \left[\frac{\sigma}{g(\rho_l - \rho_v)}\right]^{1/2} = \left[\frac{40.7 \times 10^{-3}}{9.81(879 - 6)}\right]^{1/2} = 2.18 \times 10^{-3} \text{ m}$$

$$q^* = \frac{q L_c}{h_{\text{fg}} \rho_v \alpha_l} = \frac{(1.038 \times 10^6)(2.18 \times 10^{-3})}{(1.990 \times 10^6)(5.984)(1.73 \times 10^{-7})} = 1098$$

$$P^* = \frac{P}{[\sigma g(\rho_l - \rho_v)]^{1/2}} = \frac{1.17 \times 10^6}{[(40.7 \times 10^{-3})(9.81)(879 - 6)]^{1/2}} = 6.27 \times 10^4$$

$$\text{Pr}_l = 0.98$$

From Table A.1*b*, $k_w = 17.5$ W/m K.

$$\frac{k_w}{k_l} = \frac{17.5}{0.67} = 26.1$$

$$\text{Nu} = 7.4 \times 10^{-3}(1098)^{0.6}(6.27 \times 10^4)^{0.5}(0.98)^{-1/3}(26.1)^{0.15} = 203$$

$$h_{\text{TP}} = \left(\frac{k_l}{L_c}\right)\text{Nu} = \left(\frac{0.67}{2.18 \times 10^{-3}}\right)203 = 6.24 \times 10^4 \text{ W/m}^2\text{K}$$

If the forced-convection heat transfer coefficient h_{FC} is not negligible compared to h_{TP}, we need to use Eq. (7.96):

$$h = \left(h_{\text{TP}}^3 + h_{\text{FC}}^3\right)^{1/3}$$

To calculate h_{FC}, we need an estimate of the wall temperature; assuming h_{FC} is small,

$$T_w - T_b \simeq \frac{q}{h_{\text{TP}}} = \frac{1.038 \times 10^6}{6.24 \times 10^4} = 16.6 \text{ K}; \qquad T_w = 460 + 16.6 \simeq 477 \text{ K}$$

Using CONV, item 1, fluid 1, $T_w = 477$ K, $T_b = 460$ K, $P = 1.17\times10^6$ Pa, $D = 0.046$ m, $\dot{m} = 0.83$ kg/s, gives $h_{\text{FC}} = 4610$ W/m^2 K. Hence,

$$h = (62{,}400^3 + 4610^3)^{1/3} = 6.24 \times 10^4 \text{ W/m}^2 \text{ K}$$

Comments

The forced-convection heat transfer coefficient, h_{FC}, is negligible compared to h_{TP}.

7.5.4 Internal Forced-Convection Condensation

Most refrigerator condensers condense the refrigerant vapor inside long vertical, horizontal, or coiled tubes. Generally the flow is annular along most of the tube length, and condenser performance can be satisfactorily predicted over most of the practical range of operation by correlations based on an annular flow model. The following correlation scheme is recommended [20]. First, Reynolds numbers are calculated for each phase:

$$\text{Re}_l = \frac{G(1-x)D}{\mu_l}; \qquad \text{Re}_v = \frac{GxD}{\mu_v} \tag{7.97a,b}$$

and then the Nusselt numbers are obtained as follows:

$\text{Re}_v > 35{,}000$:

$$\text{Nu} = \frac{hD}{k_l} = 0.15\frac{\text{Pr}_l\text{Re}_l^{0.9}}{F}\left(\frac{1}{\chi} + \frac{2.85}{\chi^{0.476}}\right) \tag{7.98}$$

where χ is the *Martinelli* parameter,

$$\chi = \left(\frac{\mu_l}{\mu_v}\right)^{0.1}\left(\frac{1-x}{x}\right)^{0.9}\left(\frac{\rho_v}{\rho_l}\right)^{0.5} \tag{7.99}$$

and F is a function representing the thermal resistance of the annular film, which is in turbulent flow,

$$
\begin{aligned}
F &= 5\text{Pr}_l + 5\ln(1 + 5\text{Pr}_l) + 2.5\ln(0.0031\text{Re}_l^{0.812}) && 1125 < \text{Re}_l \\
&= 5\text{Pr}_l + 5\ln[1 + \text{Pr}_l(0.0964\text{Re}_l^{0.585} - 1)] && 50 < \text{Re}_l < 1125 \\
&= 0.707\text{Pr}_l\text{Re}_l^{0.5} && \text{Re}_l < 50
\end{aligned}
\tag{7.100}
$$

$\text{Re}_v < 35{,}000$: At lower vapor Reynolds numbers, annular flow cannot be sustained, and the flow is stratified if in a horizontal tube. The vapor quality is usually below about 20%.

$$
\text{Nu} = 0.728K\left[\frac{g\rho_l(\rho_l - \rho_v)D^3h'_{\text{fg}}}{\mu_l k_l(T_{\text{sat}} - T_w)}\right]^{1/4} \tag{7.101}
$$

where

$$
K = \left[1 - \left(\frac{1-x}{x}\right)\left(\frac{\rho_v}{\rho_l}\right)^{2/3}\right]^{-3/4}
$$

Equation (7.101) is based on a Nusselt-type analysis of film condensation on the inside wall of a horizontal tube.

Figure 7.29 shows a comparison of Eq. (7.98) with experimental data for R-22. In general, the correlation agrees with experiment to ±10%.

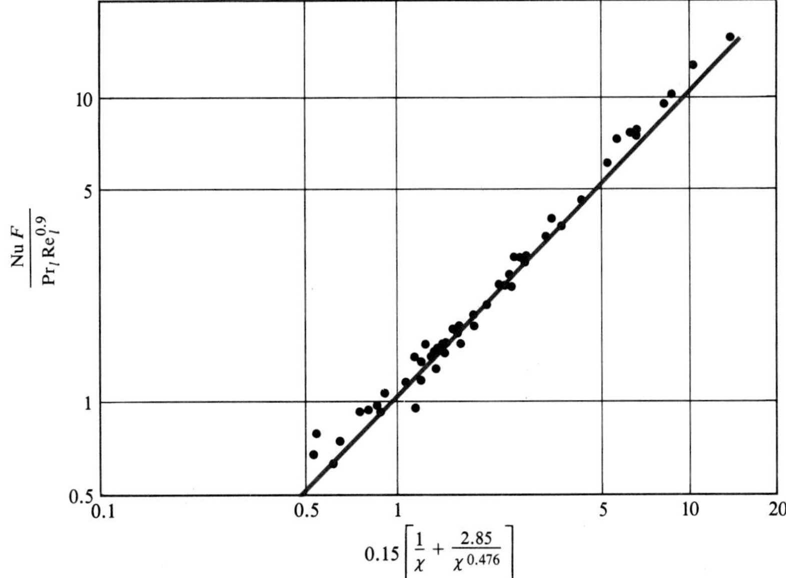

Figure 7.29 Forced-convection condensation in a tube: comparison of Eq. (7.98) with experiment [20].

EXAMPLE 7.17 Heat Transfer in a Horizontal-Tube Refrigerant-12 Condenser

Calculate the heat transfer coefficient in Example 7.13.

Solution

Given: Forced-convection condensation of R-12 in a horizontal tube.

Required: Heat transfer coefficient.

Assumptions: Equations (7.98) through (7.101) apply.

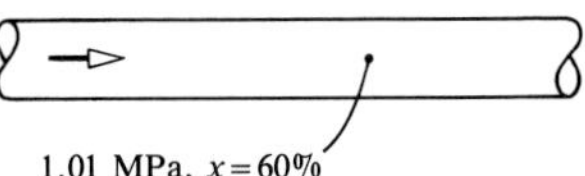

From Examples 7.13 and 7.15,

$$
\begin{array}{ll}
\text{I.D.} = 0.008 \text{ m} & A_c = 5.026 \times 10^{-5} \text{ m}^2 \\
\dot{m} = 0.0125 \text{ kg/s} & G = 249 \text{ kg/m}^2\text{ s} \qquad x = 0.6 \\
P = 1.01 \text{ MPa} & T_{sat} = 315 \text{ K} \\
\rho_l = 1242 \text{ kg/m}^3 & \rho_v = 57.6 \text{ kg/m}^3 \\
\mu_l = 2.38 \times 10^{-4} \text{ kg/m s} & \mu_v = 13.2 \times 10^{-6} \text{ kg/m s} \\
\text{Re}_l = 3360 & \text{Re}_v = 90{,}500 \\
\dot{Q}/L = 500 \text{ W/m} &
\end{array}
$$

Also, from Table A.8, $k_l = 0.0685$ W/m K, $\text{Pr}_l = 3.5$ at 315 K.
Since $\text{Re}_v > 35{,}000$, Eq. (7.98) gives the Nusselt number:

$$\text{Nu} = \frac{hD}{k_l} = 0.15\frac{\text{Pr}_l\text{Re}_l^{0.9}}{F}\left(\frac{1}{\chi} + \frac{2.85}{\chi^{0.476}}\right)$$

The Martinelli parameter χ is given by Eq. (7.99):

$$\chi = \left(\frac{\mu_l}{\mu_v}\right)^{0.1}\left(\frac{1-x}{x}\right)^{0.9}\left(\frac{\rho_v}{\rho_l}\right)^{0.5} = \left(\frac{2.38 \times 10^{-4}}{13.2 \times 10^{-6}}\right)^{0.1}\left(\frac{0.4}{0.6}\right)^{0.9}\left(\frac{57.6}{1242}\right)^{0.5} = 0.200$$

The F function for the thermal resistance of the annular film for $\text{Re}_l > 1125$ is given by Eq. (7.100):

$$
\begin{aligned}
F &= 5\text{Pr}_l + 5\ln(1 + 5\text{Pr}_l) + 2.5\ln(0.0031\text{Re}_l^{0.812}) \\
&= 5(3.5) + 5\ln(1 + 5 \times 3.5) + 2.5\ln[0.0031(3360)^{0.812}] \\
&= 34.1
\end{aligned}
$$

$$\text{Nu} = \frac{(0.15)(3.5)(3360)^{0.9}}{34.1}\left(\frac{1}{0.200} + \frac{2.85}{(0.200)^{0.476}}\right) = 256$$

$$h = \left(\frac{k_l}{D}\right)\text{Nu} = \left(\frac{0.0685}{0.008}\right)256 = 2190 \text{ W/m}^2\text{ K}$$

Comments

The wall heat flux is $q = (\dot{Q}/L)/\pi D = 500/\pi(0.008) = 1.99 \times 10^4$ W/m². Hence, $\Delta T = T_b - T_w = q/h = 1.99 \times 10^4/2190 = 9.1$ K, which is small.

7.6 PHASE CHANGE AT LOW PRESSURES

Analyses of phase change presented in Sections 7.2 through 7.5 assumed the existence of thermodynamic equilibrium at the liquid-vapor interface, allowing the liquid surface temperature to be related to the pressure of the vapor adjacent to the interface as

$$T_s = T_{\text{sat}}(P_v) \tag{7.102}$$

At very low pressures, the assumption of thermodynamic equilibrium is invalid, and Eq. (7.102) must be replaced.

7.6.1 Kinetic Theory of Phase Change

Early studies of nonequilibrium phase change were concerned with evaporation of substances into a vacuum. In 1882 H. Hertz proposed a hypothesis that can be stated as follows:

The maximum rate of evaporation of a substance cannot exceed the rate at which vapor molecules are incident on the interface at equilibrium.

In reference to Fig. 7.30, the kinetic theory of gases shows that the rate at which molecules are incident on a plane, $\mathscr{J}^-$, is simply

$$\mathscr{J}^- = \frac{\mathscr{N}c}{4} \text{ [molecules/m}^2\text{ s]} \tag{7.103}$$

where $\mathscr{N}$ [molecules/m^3] is the number density of the molecules, and c [m/s] is the average molecular speed. A physics or thermodynamics text will give the following relations:

$$\mathscr{N} = \frac{P\mathscr{A}}{\mathscr{R}T} \quad \text{(ideal gas equation of state)} \tag{7.104}$$

$$c = \left(\frac{8RT}{\pi}\right)^{1/2}; \qquad R = \frac{\mathscr{R}}{M} \tag{7.105}$$

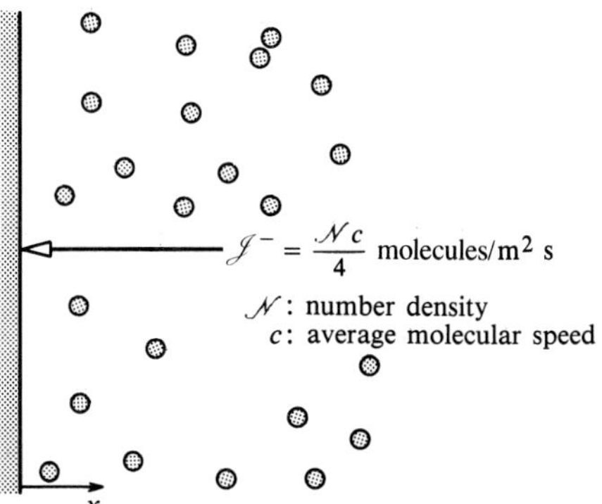

Figure 7.30 Molecular flux $\mathscr{J}^-$ incident on a plane from a stationary vapor.

where $\mathscr{R}$ [J/kmol K] is the universal gas constant, $\mathscr{A}$ [molecules/kmol] is Avogadro's number, and M [kg/kmol] is the molecular weight. The mass flux n [kg/m^2 s] is related to the molecule flux as

$$n = \frac{\mathscr{J} M}{\mathscr{A}} \tag{7.106}$$

so that the incident mass flux of the molecules is

$$\begin{aligned} n^- &= \frac{\mathscr{N} c}{4} \frac{M}{\mathscr{A}} \\ &= \frac{P \mathscr{A}}{\mathscr{R} T} \frac{(8RT/\pi)^{1/2}}{4} \frac{M}{\mathscr{A}} \end{aligned}$$

$$n^- = \frac{P}{(2\pi RT)^{1/2}} \text{ [kg/m}^2\text{ s]} \tag{7.107}$$

Consequently, the Hertz hypothesis states that the maximum rate of evaporation, $\mathscr{J}^+$, is

$$\mathscr{J}^+ = \frac{\mathscr{N}_s c_s}{4} \tag{7.108}$$

or on a mass basis,

$$n^+ = \frac{P_s}{(2\pi RT_s)^{1/2}} \tag{7.109}$$

where subscript s refers to vapor in equilibrium with a condensed phase surface at temperature T_s. In 1915, J. Knudsen showed that carefully purified mercury indeed evaporated into a vacuum at the maximum rate [21]. He introduced the concept of an evaporation coefficient σ_e to describe substances that evaporate at less than the maximum rate, and a condensation coefficient σ_c defined as the fraction of incident vapor molecules that actually condense; a fraction $(1 - \sigma_c)$ are specularly reflected. At equilibrium, the fluxes must balance: $\sigma_c \mathscr{J}^- = \sigma_e \mathscr{J}^+$, or

$$\sigma_c \frac{\mathscr{N}_s c_s}{4} = \sigma_e \frac{\mathscr{N}_s c_s}{4}$$

Hence, Knudsen concluded that $\sigma_c = \sigma_e = \sigma$, and σ is usually referred to as the *condensation coefficient*.

Now consider net condensation, and let the subscript v denote vapor molecules incident on the interface. On average, these molecules originate from a plane on the order of one mean free path from the interface, as shown in Fig. 7.31; at this location, the number density and average molecular speed are $\mathscr{N}_v$ and c_v, respectively. The net condensation rate is then

$$\begin{aligned} \mathscr{J} &= \text{Incident flux} - \text{Reflected flux} - \text{Emitted flux} \\ &= \mathscr{J}^- - (1 - \sigma_c)\mathscr{J}^- - \sigma_e \mathscr{J}^+ \end{aligned}$$

$$\mathscr{J} = \sigma \mathscr{J}^- - \sigma \mathscr{J}^+, \qquad \text{for } \sigma_c = \sigma_e = \sigma \tag{7.110}$$

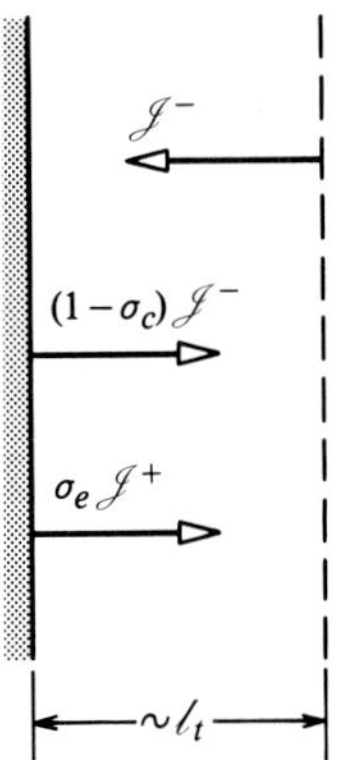

Figure 7.31 Molecular fluxes for net condensation.

According to the Hertz hypothesis, $\mathscr{J}^+ = \mathscr{N}_s c_s/4$. The incident flux is not given by Eq. (7.107) because the vapor is no longer in equilibrium. There is a velocity component normal to the surface due to condensation. The kinetic theory of gases shows that the incident flux can now be written as [22]

$$\mathscr{J}^- = \frac{\mathscr{N}_v c_v}{4}[e^{-\zeta^2} + \pi^{1/2}\zeta(1 + \text{erf } \zeta)] \tag{7.111}$$

where $\zeta = u/(2RT_v)^{1/2}$ and u is the velocity normal to the surface. For the small values of ζ typically encountered, Eq. (7.111) simplifies to

$$\mathscr{J}^- = \frac{\mathscr{N}_v c_v}{4}(1 + \pi^{1/2}\zeta); \qquad \zeta < 0.001 \tag{7.112}$$

Substituting in Eq. (7.110) gives the net condensation rate as

$$\mathscr{J} = \sigma\frac{\mathscr{N}_v c_v}{4}(1 + \pi^{1/2}\zeta) - \sigma\frac{\mathscr{N}_s c_s}{4}$$

$$= \sigma\left(\frac{\mathscr{N}_v c_v}{4} + \frac{\mathscr{J}}{2}\right) - \sigma\frac{\mathscr{N}_s c_s}{4}, \qquad \text{since } \mathscr{J} = \mathscr{N}_v u$$

or

$$\mathscr{J} = \frac{2\sigma}{2-\sigma}\left(\frac{\mathscr{N}_v c_v}{4} - \frac{\mathscr{N}_s c_s}{4}\right) \text{ [molecules/m}^2\text{ s]} \tag{7.113}$$

On a mass basis,

$$\dot{m}'' = \frac{2\sigma}{2-\sigma}\left(\frac{P_v}{(2\pi RT_v)^{1/2}} - \frac{P_s}{(2\pi RT_s)^{1/2}}\right) \text{ [kg/m}^2\text{ s]} \tag{7.114}$$

Equation (7.114) has been widely used for the interpretation of experimental results for phase change at low pressures. A major concern has been the value of the condensation coefficient σ for water and liquid metals. Early experiments showed that σ for water was perhaps as low as 0.04 [24], and this result had a considerable influence on later work. However, careful error analysis has invalidated experiments giving low values of σ for water, and it is now generally accepted that $\sigma \simeq 1.0$ for water vapor molecules, provided the water surface is not contaminated [24,25,26]. During normal boiling and condensation processes, the surface can be assumed clean. Early experimental data for the condensation of liquid-metal vapors also gave low values of σ, but it is now generally accepted that $\sigma = 1$ for metal vapors on a clean liquid surface, a result that should have been expected from Knudsen's 1915 evaporation experiment [27,28]. For $\sigma = 1$, Eq. (7.114) becomes

$$\dot{m}'' = 2\left[\frac{P_v}{(2\pi R T_v)^{1/2}} - \frac{P_s}{(2\pi R T_s)^{1/2}}\right] \tag{7.115}$$

which is the desired result. Equation (7.115) shows that for net condensation to occur, the pressure of the vapor adjacent to the liquid surface, P_v, must be greater than the saturation vapor pressure P_s corresponding to the liquid surface temperature T_s. Conversely, for evaporation, P_v is less than P_s.

EXAMPLE 7.18 Maximum Rate of Evaporation

Calculate the maximum rate of evaporation for water at 300 K and 400 K, and mercury at 400 K.

Solution

Given: (i) Water at 300 K, (ii) water at 400 K, (iii) mercury at 400 K.

Required: Maximum rate of evaporation into a vacuum, n^+.

Assumptions: 1. Ideal gas behavior.
2. Equation (7.109) gives the mass flux into a vacuum.

(**i**) Water at 300 K: Table A.12*a* gives P_{sat} (300 K) = 3533 Pa. Using Eq. (7.109):

$$n^+ = \frac{P_s}{(2\pi R T_s)^{1/2}} = \frac{3533 \text{ N/m}^2}{\{2\pi[(8314 \text{ J/kmol K})/(18 \text{ kg/kmol})](300 \text{ K})\}^{1/2}}$$
$$= 3.79 \text{ kg/m}^2 \text{ s}$$

(**ii**) Water at 400 K: Table A.12*a* gives P_{sat} (400 K) = 2.456×10^5 Pa.

$$n^+ = \frac{P_s}{(2\pi R T_s)^{1/2}} = \frac{2.456 \times 10^5}{[2\pi(8314/18)(400)]^{1/2}} = 228 \text{ kg/m}^2 \text{ s}$$

(**iii**) Mercury at 400 K: Table A.12*d* gives P_{sat} (400 K) = 140 Pa.

$$n^+ = \frac{P_s}{(2\pi R T_s)^{1/2}} = \frac{140}{[2\pi(8314/200.6)(400)]^{1/2}} = 0.434 \text{ kg/m}^2 \text{ s}$$

Alternatively, we can first obtain the molecule flux using Eq. (7.108):

$$\mathscr{J}^+ = \frac{\mathscr{N}_s c_s}{4}$$

where Eqs. (7.104) and (7.105) give $\mathscr{N}_s$ and c_s. For case (i), water at 300 K,

$$\mathscr{N}_s = \frac{P_s A}{\mathscr{R} T_s} = \frac{(3533 \text{ N/m}^3)(6.0225 \times 10^{26} \text{ molecules/kmol})}{(8314 \text{ J/kmol K})(300 \text{ K})}$$
$$= 8.531 \times 10^{23} \text{ molecules/m}^3$$

$$c_s = (8RT_s/\pi)^{1/2} = [(8/\pi)(8314/18)(300)]^{1/2} = 594 \text{ m/s}$$

$$\mathscr{J}^+ = (1/4)(8.531 \times 10^{23})(594) = 1.267 \times 10^{26} \text{ molecules/m}^2 \text{ s}$$

Equation (7.106) then gives the mass flux:

$$n^+ = \frac{\mathscr{J}^+ M}{A} = \frac{(1.267 \times 10^{26} \text{ molecules/m}^2 \text{ s})(18 \text{ kg/kmol})}{(6.0225 \times 10^{26} \text{ molecules/kmol})} = 3.79 \text{ kg/m}^2 \text{ s}$$

which agrees with our previous result.

Comments

The saturation vapor pressure is the most important factor in determining the maximum rate of evaporation. When the vapor pressure is low, so is the maximum evaporation rate.

7.6.2 Interfacial Heat Transfer Resistance

We have seen that for net condensation of saturated vapor, P_s must be less than P_v, that is, the temperature of the condensate surface must be less than the saturation temperature of the vapor. This temperature difference can be calculated from an **interfacial heat transfer coefficient**, which is defined as

$$h_i = \frac{\dot{m}'' h_{fg}}{T_{sat}(P_v) - T_s} \tag{7.116}$$

It is important to note that this definition is stated in terms of the vapor saturation temperature $T_{sat}(P_v)$, and not the temperature of the vapor molecules incident on the surface, T_v. The latter differs from the condensate surface temperature T_s by what is essentially the temperature "jump" defined for the slip flow regime of rarefied gas flow. This difference is negligible in most situations of concern. Also, since there is usually a negligible pressure gradient normal to the surface, the vapor undergoes supersaturation at constant pressure before condensing. The Clausius-Clapeyron relation relates vapor pressure and temperature along the saturation line:

$$\frac{dP}{dT} = \frac{\rho_v h_{fg}}{T} \tag{7.117}$$

if the pressure is low. Since the departure from equilibrium is usually small,

$$\frac{P_v - P_s}{T_{\text{sat}}(P_v) - T_s} \simeq \frac{\rho_v h_{\text{fg}}}{T} \tag{7.118}$$

Combining with Eqs. (7.115) and (7.116) and using $T_v \simeq T_s$ gives

$$h_i = \left(\frac{2}{\pi R T_s}\right)^{1/2} \frac{\rho_v h_{\text{fg}}^2}{T_s} \quad [\text{W/m}^2\ \text{K}] \tag{7.119}$$

which is the desired result. The corresponding interfacial thermal resistance is $1/h_i A_i$. Whether or not this resistance is significant depends on the situation, but the essential feature is that $1/h_i A_i$ is inversely proportional to pressure through the vapor density ρ_v.

EXAMPLE 7.19 Calculation of the Interfacial Heat Transfer Coefficient

Determine the interfacial heat transfer coefficient for saturated steam at 750 Pa.

Solution

Given: Saturated steam, $P = 750$ Pa.

Required: Interfacial heat transfer coefficient, h_i.

Assumptions: 1. Condensation coefficient $\sigma = 1.0$.
2. The linearization of Eq. (7.111) to give Eq. (7.112) is valid.

Equation (7.119) gives h_i:

$$h_i = \left(\frac{2}{\pi R T_s}\right)^{1/2} \frac{\rho_v h_{\text{fg}}^2}{T_s}$$

From Table A.12*a*, T_{sat} (750 Pa) $= 276.0$ K; $h_{\text{fg}} = 2.494 \times 10^6$ J/kg, $\rho_v = 0.00589$ kg/m^3. Also, $R = \mathscr{R}/M = (8314$ J/kmol K$)/(18$ kg/kmol$) = 461.9$ J/kg K.

$$h_i = \left[\frac{2}{\pi(461.9\ \text{J/kg K})(276\ \text{K})}\right]^{1/2} \frac{(0.00589\ \text{kg/m}^3)(2.494 \times 10^6\ \text{J/kg})^2}{(276\ \text{K})}$$

$$= 2.97 \times 10^5\ \text{W/m}^2\ \text{K}$$

Comments

The student should carefully check the units in Eq. (7.119), noting that 1 J $=$ 1 kg m^2/s^2.

7.6.3 Nusselt Analysis Including Interfacial Resistance

Figure 7.32 depicts laminar film condensation of a saturated vapor on an isothermal vertical wall, as well as the thermal circuit. There are two thermal resistances in series to be considered: the interfacial resistance, where the temperature drops from T_{sat} to T_s, and the film resistance, where the temperature drops from T_s to T_w. The analysis of laminar film condensation in Section 7.2.1 can be extended to include the interfacial resistance as follows. Equation (7.13) becomes

$$k_l \left.\frac{\partial T}{\partial y}\right|_w = h_{fg}\frac{d\Gamma}{dx} = h_i(T_{sat} - T_s)$$

and now

$$k_l \left.\frac{\partial T}{\partial y}\right|_w = \frac{k_l(T_s - T_w)}{\delta}$$

Also, $\rho_v \ll \rho_l$, so that

$$\frac{d\Gamma}{dx} = \frac{g\rho_l}{\nu_l}\delta^2\frac{d\delta}{dx}$$

Hence,

$$\frac{k_l(T_s - T_w)}{\delta} = \frac{h_{fg}g\rho_l}{\nu_l}\delta^2\frac{d\delta}{dx} = h_i(T_{sat} - T_s) \tag{7.120}$$

For convenience we define dimensionless variables:

$$\theta = \frac{T_{sat} - T}{T_{sat} - T_w}; \qquad x^* = \frac{x}{(\nu_l^2/g)^{1/3}}; \qquad \delta^* = \frac{\delta}{(\nu_l^2/g)^{1/3}}$$

Then

$$\frac{1 - \theta_s}{\delta^*} = \frac{\Pr_l}{\mathrm{Ja}_l}\delta^{*2}\frac{d\delta^*}{dx^*} = \mathrm{Nu}_i\theta_s \tag{7.121}$$

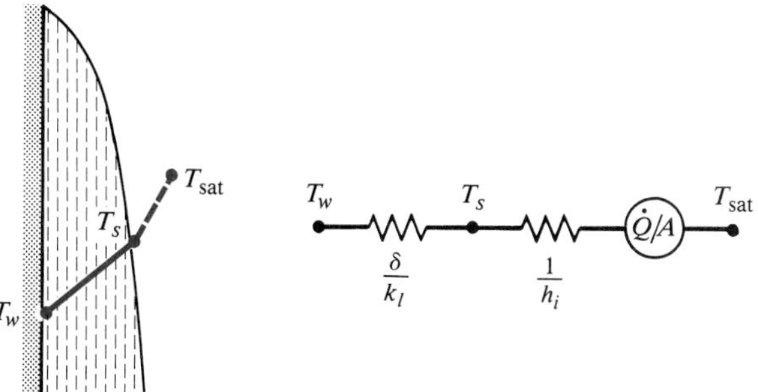

Figure 7.32 Schematic and thermal circuit for laminar film condensation on a vertical surface, including the interfacial resistance.

where

$$\frac{\text{Pr}_l}{\text{Ja}_l} = \frac{\mu_l h_{fg}}{k_l(T_{sat} - T_w)}; \qquad \text{Nu}_i = \frac{h_i(\nu_l^2/g)^{1/3}}{k_l}$$

Eliminating θ_s gives

$$1 = \frac{d\delta^*}{dx}\frac{\text{Pr}_l}{\text{Ja}_l}\left(\delta^{*3} + \frac{\delta^{*2}}{\text{Nu}_i}\right)$$

Assuming Nu_i constant and integrating with $\delta^* = 0$ at $x^* = 0$,

$$\frac{\text{Ja}_l}{\text{Pr}_l}x^* = \frac{\delta^{*4}}{4} + \frac{\delta^{*3}}{3\text{Nu}_i} \tag{7.122}$$

Equation (7.122) is an implicit relation for $\delta^*(x^*)$ that can be solved by trial and error or by using Newton's method. For a wall of height L, $\delta^*(L^*)$ can be obtained, and the average Nusselt number obtained from an energy balance on the film is

$$\overline{\text{Nu}} = \frac{\overline{q}(\nu_l^2/g)^{1/3}}{k_l(T_{sat} - T_w)} = \frac{\text{Pr}_l}{\text{Ja}_l}\left(\frac{\delta^{*3}}{3L^*}\right) \tag{7.123}$$

EXAMPLE 7.20 Laminar Film Condensation of Steam

In an experiment designed to investigate the condensation coefficient of water, saturated steam at 750 Pa is condensed on a vertical plate 10 cm high and maintained at 274.0 K. Determine the expected average heat flux if (i) $\sigma = 1.0$, (ii) $\sigma = 0.04$.

Solution

Given: Film condensation on a vertical plate, $P = 750$ Pa, $T_w = 274.0$ K.

Required: Average heat flux $\overline{q}$ for $\sigma = 1.0, 0.04$.

Assumptions: Laminar flow, $\text{Re}_f < 30$.

The dimensionless film thickness $\delta^*(L^*)$ must be determined from Eq. (7.122):

$$\frac{\text{Ja}_l}{\text{Pr}_l}L^* = \frac{\delta^{*4}}{4} + \frac{\delta^{*3}}{3\text{Nu}_i}$$

From Table A.12a, T_{sat} (750 Pa) $= 276.0$ K, $h_{fg} = 2.494 \times 10^6$ J/kg. Since the temperature difference across the film is small, it will be sufficiently accurate to evaluate water properties at $T_r = 275$ K; $k_l = 0.556$ W/m K, $\mu_l = 17.0 \times 10^{-4}$ kg/m s, $\nu_l = 1.70 \times 10^{-6}$ m²/s, $\text{Pr}_l = 12.9$.

$$\frac{\text{Pr}_l}{\text{Ja}_l} = \frac{\mu_l h_{fg}}{k_l(T_{sat} - T_w)} = \frac{(17.0 \times 10^{-4})(2.494 \times 10^6)}{(0.556)(276.0 - 274.0)} = 3813$$

$$\left(\frac{\nu_l^2}{g}\right)^{1/3} = \left[\frac{(1.70 \times 10^{-6})^2}{9.81}\right]^{1/3} = 6.66 \times 10^{-5} \text{ m}$$

$$L^* = \frac{L}{(\nu_l^2/g)^{1/3}} = \frac{0.10}{6.66 \times 10^{-5}} = 1501$$

(i) $\sigma = 1.0$:

From Example 7.19, $h_i = 2.97 \times 10^5$ W/m^2 K for $\sigma = 1.0$.

$$\text{Nu}_i = \frac{h_i(\nu_l^2/g)^{1/3}}{k_l} = \frac{(2.97 \times 10^5)(6.66 \times 10^{-5})}{0.556} = 35.5$$

Substituting in Eq. (7.122),

$$\frac{1501}{3813} = \frac{\delta^{*4}}{4} + \frac{\delta^{*3}}{106.5}$$

or

$$\delta^{*4} + 0.0376\delta^{*3} = 1.575$$

Solving by trial and error or Newton's method, $\delta^* = 1.111$.
Equation (7.123) gives the average Nusselt number:

$$\overline{\text{Nu}} = \frac{\text{Pr}_l}{\text{Ja}_l}\left(\frac{\delta^{*3}}{3L^*}\right) = 3813\left(\frac{1.111^3}{(3)(1501)}\right) = 1.161$$

$$\overline{q} = \frac{k_l \Delta T}{(\nu_l^2/g)^{1/3}}\overline{\text{Nu}} = \frac{(0.556)(2)}{6.66 \times 10^{-5}}(1.161) = 19{,}380 \text{ W/m}^2$$

(ii) $\sigma = 0.04$:

Combining Eqs. (7.114), (7.116), and (7.118) gives

$$h_i = \frac{\sigma}{2 - \sigma}h_i(\sigma = 1.0) = \frac{0.04}{2 - 0.04}(2.97 \times 10^5) = 6.05 \times 10^3 \text{ W/m}^2 \text{ K}$$

$$\text{Nu}_i = (6.05 \times 10^3)(6.66 \times 10^{-5})/(0.556) = 0.725$$

The equation for δ^* becomes

$$\delta^{*4} + 1.839\delta^{*3} = 1.575$$

$$\delta^* = 0.838; \quad \overline{\text{Nu}} = 0.498; \quad \overline{q} = 8320 \text{ W/m}^2$$

Notice also that if we ignore the interfacial resistance, $\text{Nu}_i = \infty$, and

$$\delta^{*4} = 1.575; \quad \delta^* = 1.120; \quad \overline{\text{Nu}} = 1.190; \quad \overline{q} = 19{,}870 \text{ W/m}^2$$

Comments

1. For the two values of σ, the difference in heat flux is (19,380−8320)/19,380 = 57%. Since the heat flux can be measured to at least 10% accuracy, a low value of σ such

as 0.04 would be readily observed in the experiment. Similar experiments have indeed shown that σ for water must be at least 0.35. However, the technique cannot yield accurate values of σ for σ approaching unity, so there is no reason to believe that σ is not unity.

2. If the interfacial resistance is ignored, the estimated heat flux is (19,870−19,380)/19,380 = 2.5% higher than the result for $\sigma = 1.0$. This discrepancy is very small.

7.6.4 Engineering Significance of the Interfacial Resistance

Whether or not the interfacial heat transfer resistance must be accounted for in a condenser or evaporator design depends primarily on two factors:

1. The pressure level
2. The relative magnitudes of other series resistances in the thermal circuit

Accepting that the condensation coefficient $\sigma = 1$, Example 7.20 shows that the interfacial resistance has negligible engineering significance for laminar film condensation of steam at the lowest pressure imaginable. Although the pressure is low and, hence, h_i is small, the magnitude of the thermal resistance of the liquid film is much larger than the interfacial resistance. For condensation of liquid-metal vapors of low pressures, however, the situation is quite different. Consider saturated mercury vapor at 430 K condensing on a 3 cm–O.D. horizontal tube with a temperature drop across the condensate film of 1 K. Equation (7.41) gives the heat transfer coefficient for the film as 2.58×10^5 W/m^2 K, whereas Eq. (7.119) gives $h_i = 3.44 \times 10^4$ W/m^{2}K. In this case, the interfacial resistance is about 8 times larger than the film resistance. Of course, in this situation, the tube wall and coolant thermal resistance must be considered as well. Figure 7.33 shows the thermal network. For water boiling inside the tube, a value of $h \approx 2 \times 10^4$ W/m^2 K is typical, and the tube wall could be 1 mm–thick stainless steel with thermal conductivity 20 W/m K, giving $k_w/\delta = 2 \times 10^4$ W/m^2 K. Examining the thermal resistances shown in Fig. 7.33, one can see that the boiling, tube wall, and interfacial resistances are all of comparable magnitude.

The interfacial resistance may be significant to *direct-contact* phase change processes. In a direct-contact condenser, vapor may condense directly on a falling liquid film or jet, or on a spray of droplets, and the heat transfer coefficient for heat transport into the liquid phase can be very high. Figure 7.34 shows a conceptual design of a turbulent falling film condenser suitable for a Claude-cycle ocean thermal energy conversion (OTEC) plant. Water from the deep ocean is typically below 280 K, and a saturation temperature in the condenser of 285 K is possible. Equation (7.119)

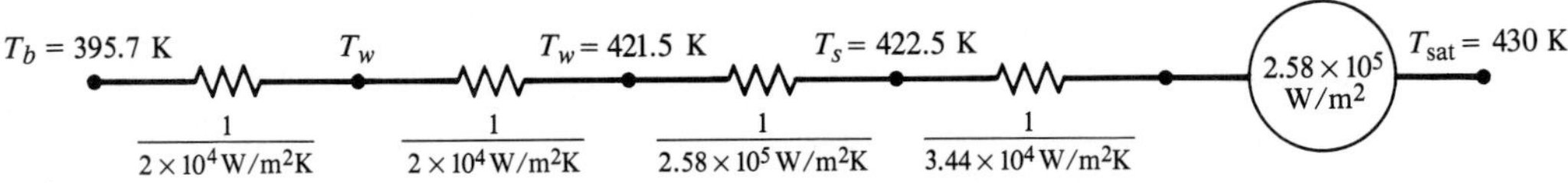

Figure 7.33 Thermal circuit for saturated mercury vapor condensing on a 3 cm–O.D. horizontal tube; $T_{sat} = 430$ K, $T_w = 421.5$ K.

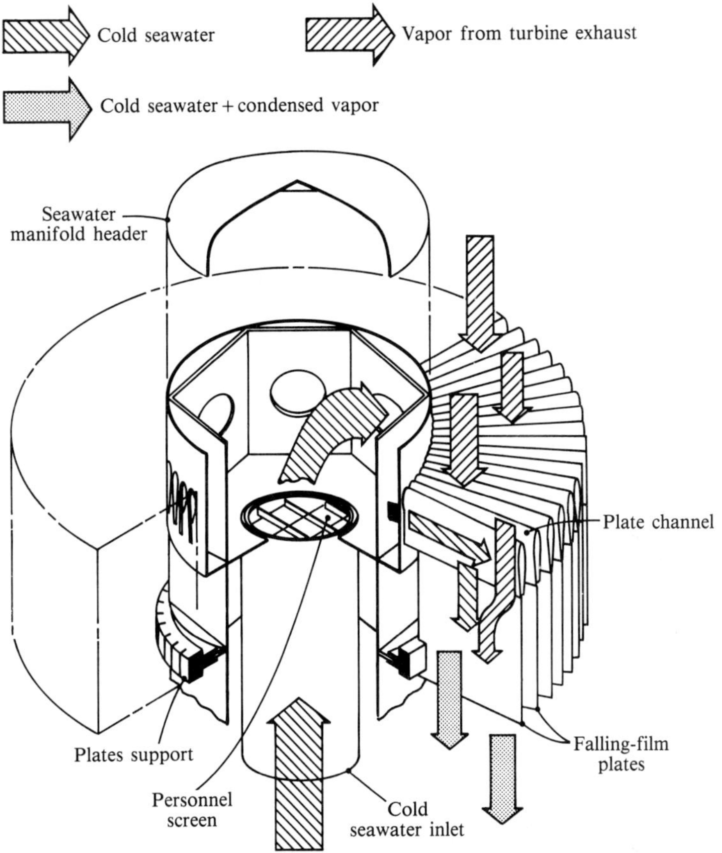

Figure 7.34 Conceptual design of a turbulent falling film condenser for a Claude-cycle ocean thermal energy conversion plant. (Courtesy Dr. A. T. Wassel, SAIC, Los Angeles.)

gives $h_i = 4.98 \times 10^5$ W/m^2 K, and the heat transfer coefficient for heat transport into the liquid film is estimated to be as high as 4×10^4 W/m^2 K. Thus, the interfacial resistance is approximately 10% of the total resistance.

Effect of Noncondensable Gas

In many condensers, the presence of a noncondensable gas in the vapor is unavoidable. For example, steam condensers of conventional power plants always contain some air due to inleakage and have exhauster systems to continually remove the air. In the Claude-cycle condenser, air can come out of solution from the coolant sea water. A small amount of gas has a marked detrimental effect on condenser performance. The gas is swept by condensing vapor to the condensate surface, where it accumulates. The partial pressure of the steam at the interface can be substantially lower than the total pressure in the condenser, and the condensate surface temperature will then be lower than the saturation temperature corresponding to the condenser pressure. The temperature drop available for transport of heat to the coolant is thereby reduced. This *noncondensable gas problem* is critical to the design and operation of many condensers, and it is a problem involving *mass transfer* since diffusion of the gas through the vapor plays a central role. When the thermal resistance due to noncondensable gas is dominant, the interfacial heat transfer resistance may play a negligible role, even though the pressure is low.

7.7 HEATPIPES

A **heatpipe** is an evaporator-condenser system in which the liquid is returned to the evaporator by capillary action. In its simplest form, it is a hollow tube with a few layers of wire screen along the wall to serve as a wick, as shown in Fig. 7.35. The screen is filled with a wetting liquid such as sodium or lithium for high-temperature applications, or with water, ammonia, or methanol for moderate-temperature applications. If one end of the heatpipe is heated and the other end is cooled, the liquid evaporates at the hot end and condenses at the cold end. As the liquid is depleted in the evaporator section, cavities form in the surface due to the liquid clinging to the wires of the screen. In the condenser section, meanwhile, the screen becomes flooded. The surface tension acting on the concave liquid-vapor interface in the evaporator causes the pressure to be higher in the vapor than in the liquid. This pressure is transmitted by the vapor to the flooded condenser section, where the vapor and liquid pressures are nearly equal, so that liquid is driven from the condenser to the evaporator through the wick. In a gravity field, the evaporator may be placed below the condenser to assist the liquid flow. If a wick is unnecessary, the device is more properly called a *Perkins tube*. Heatpipes are particularly attractive for space vehicle applications, where gravity fields are very weak, and an inherently reliable

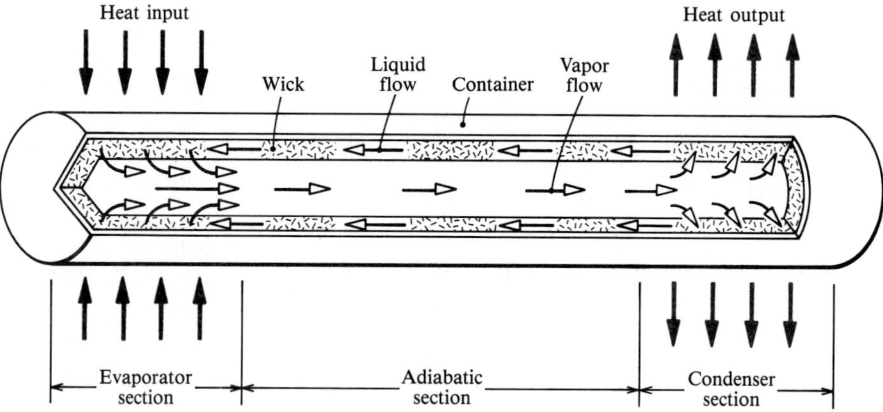

Figure 7.35 A simple heatpipe.

passive device is preferred. Of course, a pump could be used to return the liquid to the evaporator, but this would be an undesirable feature for many applications—in particular, space vehicle thermal control, where vibration and pump reliability would be major concerns.

The working fluids chosen for heatpipes have high enthalpies of vaporization. Thus, a small vapor flow along the tube can transport a large amount of thermal energy. The total temperature difference necessary is the sum of the differences required to conduct the heat through the evaporator wall and wick and through the condenser wall and wick, and the amount necessary to provide the vapor pressure difference that drives the vapor from the evaporator to the condenser. This temperature difference is quite small compared to the amount that would be necessary to transfer an equal amount of heat along the length of the pipe by conduction, even if the pipe were solid copper. In addition to a high enthalpy of vaporization, other desirable characteristics of working fluids include a high surface tension and a low viscosity, to improve capillary pumping in the wick, and a high liquid thermal conductivity, to reduce the temperature drops in the evaporator and condenser.

The wick must provide (1) surface pores to generate capillary pumping pressure, (2) internal flow passages for return of the liquid to the evaporator, and (3) a suitable heat flow path from the pipe inner wall to the liquid-vapor interface. Wick structures include wire screen, sintered metal, metal foam, metal felt, woven wire mesh, and axial grooves. Figure 7.36 shows some examples, and Table 7.4 lists the properties of some commonly used wicking materials. Wrapped screen wicks are simple to install and are widely used; however, the temperature drop across the wick tends to be large. Sintered powder metal wicks or axial grooves are preferred if the temperature drop across the wick is an important design constraint. A compromise is the screen-covered groove wick. The fine mesh screen gives a high capillary pumping pressure while the grooves give a low liquid flow resistance and a low resistance to heat flow across the wick. Graded-porosity wicks have a finer mesh in the evaporator to increase capillary pressure and a coarser mesh in the condenser to facilitate liquid flow.

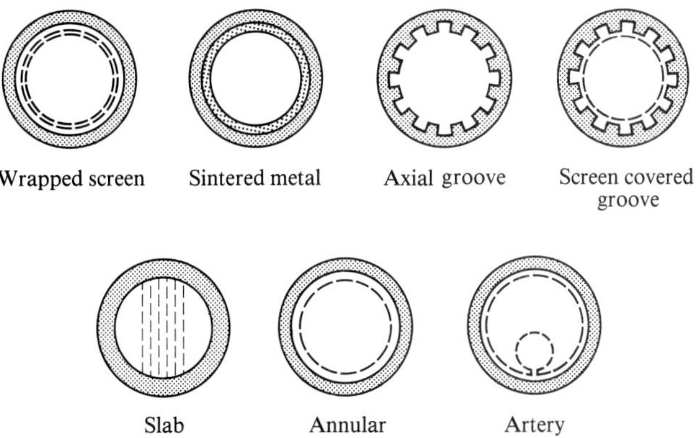

Figure 7.36 Examples of wick structures for heatpipes.

Table 7.4 Data for commonly used wick structures. (Courtesy of Mr. D. Antoniuk, TRW Systems Inc., Redondo Beach, California.)

Wick Type	Effective Pore Radius mm	Permeability $m^2 \times 10^{10}$	Porosity %
30 mesh screen	0.43	25	
100 mesh screen	0.12	1.8	58.5
200 mesh screen	0.063	0.55	67.6
Nickel felt	0.17	5.2	89.1
Nickel foam 210-5	0.23	36	94.4
33 μm–diameter stainless steel fibers	>0.034	2.0	80.8
75 μm–diameter stainless steel fibers	>0.041	11.6	82.8
Sintered fibers/powders	0.01–0.1	0.1–10	
Axial grooves	0.25–1.5	35–1250	
Open annulus	0.25–1.5	50–2000	
Open artery	0.50–1.5	300–3000	

Heatpipes that contain a single vapor are *fixed-conductance heatpipes*, since their thermal resistance is relatively insensitive to heat load. Also widely used are gas-loaded heatpipes, in which a small amount of noncondensable gas is added to the vapor to give a *variable-conductance heatpipe*. Figure 7.37 shows a schematic of a gas-loaded heatpipe with a wicked gas reservoir at the end of the condenser. When the heatpipe is off, that is, when there is no heat load, the evaporator is cold and the vapor pressure very low. The gas then expands to fill the heatpipe. As a heat load is applied, the evaporator temperature and vapor pressure increase, and the gas is pushed out of the evaporator toward the condenser. When the heatpipe is fully on, gas occupies only the very end of the condenser and the reservoir. Thus, the length of condenser in operation varies with heat load, and the heatpipe temperature can be maintained within a narrow temperature range over a wide variation of heat load and sink conditions. This feature is particularly attractive for some spacecraft applications, where it is important to keep electronic components within a prescribed temperature range. If a fixed-conductance heatpipe is used, as the heat load is reduced, the condenser temperature must drop closer to the sink temperature, and the evaporator temperature must fall accordingly. It is also easier to prevent gas-loaded heatpipes

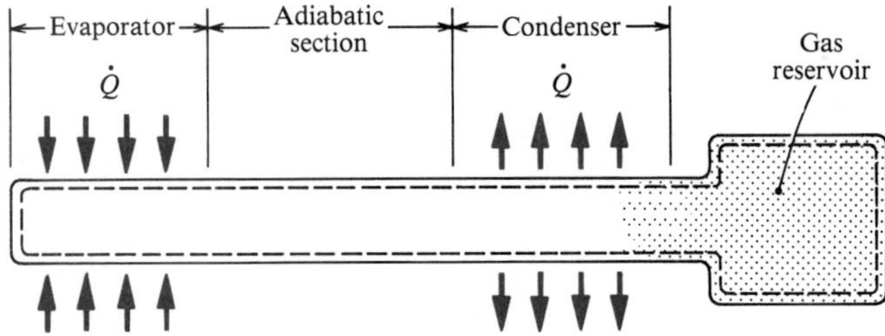

Figure 7.37 A gas-loaded heatpipe with a wicked gas reservoir.

on satellites from freezing during extended "cold soak" periods. By supplying only a small amount of residual heat to the evaporator, the evaporator temperature can be maintained above the freezing point of the liquid.

7.7.1 Capillary Pumping

Figure 7.38 shows the pressures acting inside an annular wicked heatpipe. During steady operation, the balance is

$$\begin{matrix}\text{Capillary} \\ \text{head}\end{matrix} - \begin{matrix}\text{Gravitational} \\ \text{head}\end{matrix} = \begin{matrix}\text{Liquid-phase} \\ \text{pressure drop}\end{matrix} + \begin{matrix}\text{Vapor-phase} \\ \text{pressure drop}\end{matrix} \qquad \textbf{(7.124)}$$

$$\Delta P_C - \Delta P_G = \Delta P_l + \Delta P_v$$

To simplify the analysis, the balance is made for an effective heatpipe length L_{eff}, from the midpoint of the evaporator to the midpoint of the condenser. The gravitational head ΔP_G is zero for a horizontal heatpipe and is negative if the condenser is located above the evaporator (i.e., has a *favorable tilt*). In general, the pressure drops ΔP_l and ΔP_v increase with heat load due to the increase in flow rate; hence, the required capillary head ΔP_C also increases. However, there is a maximum capillary pressure $(\Delta P_C)_{\max}$ that can be developed by a given liquid-wick combination. For a heatpipe to operate continuously, the required capillary pressure must not exceed this maximum value at any point along the pipe. If it does, the wick can dry out, a condition known as evaporator *burnout*. When burnout occurs, the load must be reduced to allow the wick to reprime.

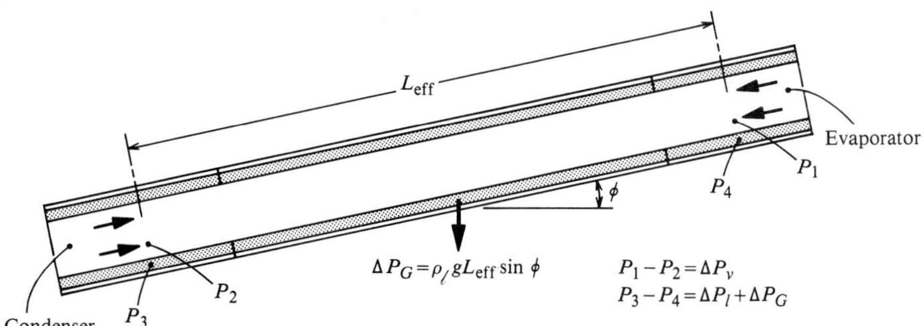

Figure 7.38 Pressures acting inside an annular wicked heatpipe with an adverse tilt.

Capillary Head

An elementary experiment in physics is the demonstration of a wetting liquid rising in a capillary tube, as shown in Fig. 7.39. A force balance gives the well-known result,

$$(\rho_l - \rho_v)gh = \frac{2\sigma \cos \theta}{r_p} \qquad \textbf{(7.125)}$$

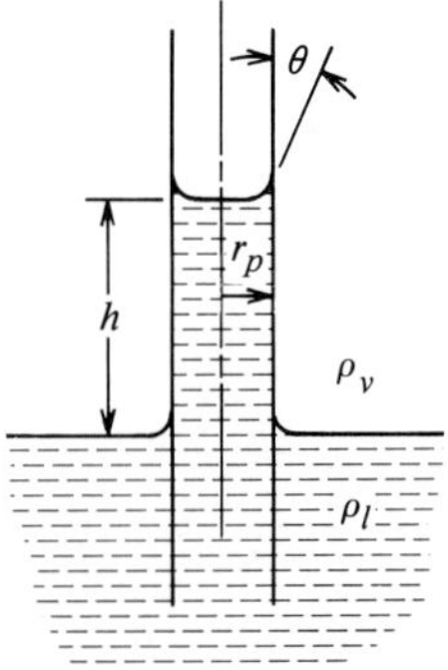

Figure 7.39 Rise of a wetting liquid in a capillary tube of radius r_p. Angle θ is the contact angle.

where θ is the *contact angle*. For wetting liquids, $0 < \theta < 90°$, and for nonwetting liquids, $\theta > 90°$. In heatpipes, it is essential to choose a liquid-wick pair that wets well. Figure 7.40 shows liquid menisci in the wick at the evaporator and condenser. Using the subscripts e to denote evaporator and c to denote condenser, the resultant capillary head available to pump the liquid along the wick is[4]

$$\Delta P_C = 2\sigma\left(\frac{\cos\theta_e}{r_p} - \frac{\cos\theta_c}{r_p}\right)$$

ΔP_C has a maximum value when $\cos\theta_e = 1$ and $\cos\theta_c = 0$:

$$(\Delta P_C)_{\max} = \frac{2\sigma}{r_p} \tag{7.126}$$

Table 7.4 gives the pore radius r_p for various homogeneous wick materials.

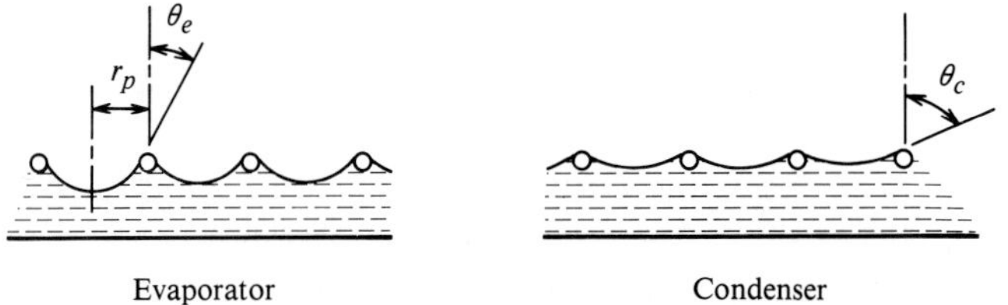

Figure 7.40 Liquid menisci in a heatpipe wick, showing a smaller radius of curvature in the evaporator.

Gravitational Head

The gravitational head is the hydrostatic pressure differential between the evaporator and condenser and may be positive, zero, or negative depending on their relative elevations. Referring to Fig. 7.38, the gravitational head for $\rho_v \ll \rho_l$ is

$$\Delta P_G = \rho_l g L_{\text{eff}} \sin\phi \tag{7.127}$$

[4] These contact angles θ_e and θ_c depend on the rates of evaporation and condensation (rather than on the intrinsic wetting behavior).

Liquid-Phase Pressure Drop

The pressure drop for liquid flowing through a homogeneous wick can be calculated from Darcy's law for flow through porous media:

$$\Delta P_l = \frac{\mu_l L_{\text{eff}} \dot{m}}{\kappa \rho_l A_w} \tag{7.128}$$

where $\dot{m}$ [kg/s] is the liquid flow rate, A_w is the cross-sectional area of the wick, and κ [m^2] is the *permeability* of the wick material. Values of κ are given in Table 7.4. The pressure drop for liquid flowing along open grooves can be roughly estimated from Eq. (4.39) for laminar flow in a pipe using the hydraulic diameter:

$$\Delta P_l = \frac{64}{\text{Re}_{D_{hl}}} \frac{L_{\text{eff}}}{D_{hl}} \left(\frac{1}{2} \rho_l V_l^2 \right); \qquad D_{hl} = 4 \frac{\text{Liquid flow area}}{\text{Wetted perimeter}} \tag{7.129}$$

At high vapor velocities, vapor drag may impede liquid flow in open grooves. A remedy is to cover the grooves with a screen to form a *composite* wick.

Vapor-Phase Pressure Drop

The vapor-phase pressure drop is generally much smaller than that for the liquid phase. Often it may be neglected, or a very approximate calculation may suffice. Provided the Mach number is less than about 0.3, incompressible flow may be assumed. The flow is usually laminar, and Eq. (4.39) for fully developed flow can be used:

$$\Delta P_v = \frac{64}{\text{Re}_{D_{hv}}} \frac{L_{\text{eff}}}{D_{hv}} \left(\frac{1}{2} \rho_v V_v^2 \right); \qquad D_{hv} = 4 \frac{\text{Vapor flow area}}{\text{Wetted perimeter}} \tag{7.130}$$

Note that the flow is not fully developed at each end of the heatpipe, and there is a normal velocity component associated with phase change that decreases the wall shear stress in the evaporator and increases it in the condenser ("blowing" in the evaporator and "suction" in the condenser are discussed in Sections 7.2 and 7.3). In addition, there is the pressure drop required to accelerate the vapor in the evaporator, and a pressure recovery due to deceleration in the condenser. However, these effects are self-compensating and are thus generally ignored in heatpipe design.

Wicking Limitation

Equation (7.124) written for the maximum capillary pressure available, $(\Delta P_C)_{\text{max}}$, gives the *wicking limitation* of the heatpipe. For a homogeneous wick,

$$\frac{2\sigma}{r_p} - \rho_l g L_{\text{eff}} \sin\phi = \frac{\mu_l L_{\text{eff}} \dot{m}_{\text{max}}}{\kappa \rho_l A_w} + \frac{64}{\text{Re}_{D_{hv}}} \frac{L_{\text{eff}}}{D_{hv}} \left(\frac{1}{2} \rho_v V_v^2 \right) \tag{7.131}$$

and

$$\dot{Q}_{\text{max}} = \dot{m}_{\text{max}} h_{\text{fg}} \tag{7.132}$$

If the vapor-phase pressure drop is neglected, Eqs. (7.131) and (7.132) can be rearranged as

$$\dot{Q}_{\max} = \left(\frac{\rho_l \sigma h_{fg}}{\mu_l}\right)\left(\frac{A_w \kappa}{L_{\text{eff}}}\right)\left(\frac{2}{r_p} - \frac{\rho_l g L_{\text{eff}} \sin\phi}{\sigma}\right) \quad \textbf{(7.133)}$$

The fluid properties combination in the first set of parentheses is the **figure of merit** $\mathcal{M}$ of the fluid, usually expressed in kW/cm^2:

$$\mathcal{M} = \frac{\rho_l \sigma h_{fg}}{\mu_l} \ [\text{kW/cm}^2] \quad \textbf{(7.134)}$$

$\mathcal{M}$ is a function of temperature and is plotted in Fig. 7.41 for a selection of heat-pipe fluids. The clear superiority of water in the intermediate temperature range is due to its high latent heat and surface tension, but there are also other criteria for fluid selection, such as materials compatibility. Liquid metals are indicated for high-temperature heatpipes, with lithium being the first choice above 1400 K.

Heatpipes are often characterized by a **heat transfer factor**, defined as the product of the maximum heat flow and effective pipe length for a horizontal pipe:

$$(\dot{Q}_{\max} L_{\text{eff}})_{\phi=0} = \left(\frac{\rho_l \sigma h_{fg}}{\mu_l}\right)(A_w \kappa)\left(\frac{2}{r_p}\right) = \frac{2\mathcal{M} A_w \kappa}{r_p} \ [\text{W m}] \quad \textbf{(7.135)}$$

In practice, the engineer attempts to improve this factor by choosing a wick material with large permeability κ and small pore radius r_p.

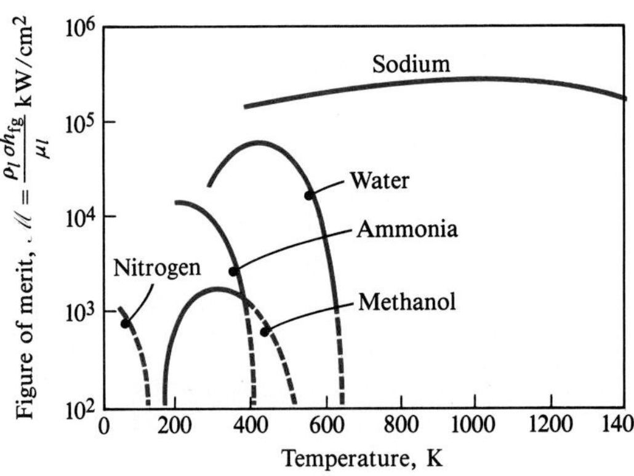

Figure 7.41 Figure of merit $\mathcal{M}$ for a number of heatpipe fluids [29].

EXAMPLE 7.21 An Ammonia Heatpipe

An ammonia heatpipe is constructed from an 0.5 inch–O.D. stainless steel tube and has an effective length of 1.40 m. The aluminum fibrous slab wick has a cross-sectional area of 4.7×10^{-5} m^2. To investigate the wick performance, an experiment was carried out in which the burnout heat load was determined as a function of heatpipe inclination. In these experiments, the adiabatic section was maintained at 22°C ± 2°C.

Heatpipe angle, degrees	0	0.3	0.5	0.7	0.9
$\dot{Q}_{max}$, W	94	74	55	38	23

Estimate the effective pore radius and Darcy permeability of the aluminum wick.

Solution

Given: Burnout heat load as a function of inclination for an ammonia heatpipe.

Required: Estimate of wick pore radius, r_p, and permeability, κ.

Assumptions: Negligible vapor flow pressure drop, so that Eq. (7.133) applies.

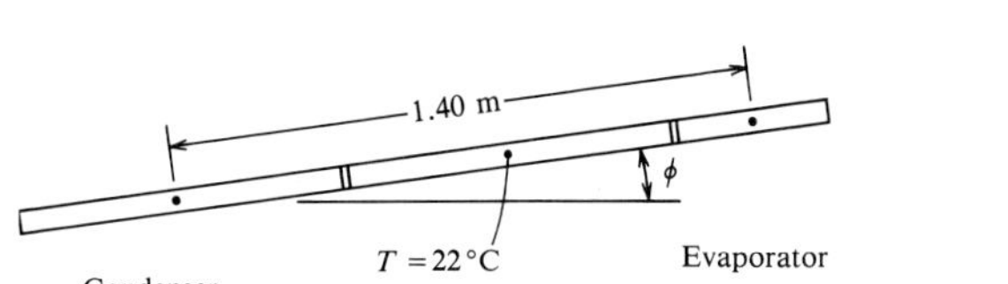

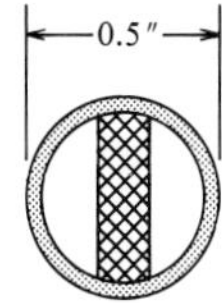

Equation (7.133) gives the burnout heat load:

$$\dot{Q}_{max} = \left(\frac{\rho_l \sigma h_{fg}}{\mu_l}\right)\left(\frac{A_w \kappa}{L_{eff}}\right)\left(\frac{2}{r_p} - \frac{\rho_l g L_{eff} \sin\phi}{\sigma}\right)$$

All ammonia properties will be evaluated at 22°C = 295 K. From Tables A.8, A.11, and A.12*b*, $\rho_l = 609$ kg/m^3, $\mu_l = 1.38 \times 10^{-4}$ kg/m s, $\sigma = 21 \times 10^{-3}$ N/m, $h_{fg} = 1.179 \times 10^6$ J/kg. Substituting above,

$$\dot{Q}_{max} = \left[\frac{(609)(21 \times 10^{-3})(1.179 \times 10^6)}{1.38 \times 10^{-4}}\right]\left[\frac{(4.70 \times 10^{-5})\kappa}{1.40}\right]\left[\frac{2}{r_p} - \frac{(609)(9.81)(1.40)\ \sin\phi}{21 \times 10^{-3}}\right]$$

$$= 3.67 \times 10^6\ \kappa\left(\frac{2}{r_p} - 3.98 \times 10^5\ \sin\phi\right)$$

$\dot{Q}_{max}$ is seen to be a linear function of $\sin\phi$, which suggests a plot of $\dot{Q}_{max}$ versus ϕ as shown ($\phi = \sin\phi$ for ϕ small). The intercept for $\dot{Q}_{max} = 0$ gives r_p, and κ can be determined from the slope.

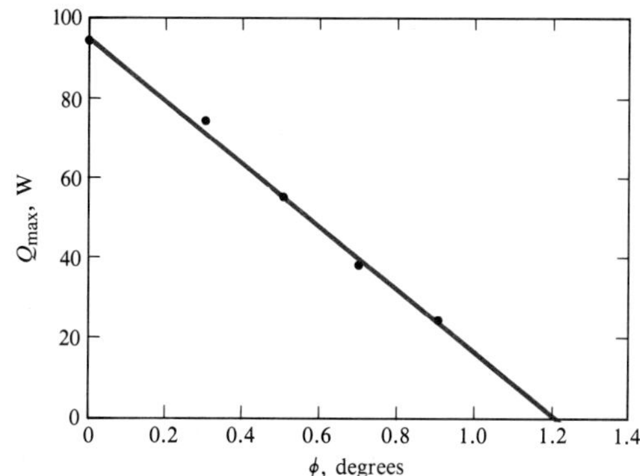

$$\dot{Q}_{max} = 0 \text{ at } \phi = 1.20 \text{ degrees} = 0.0209 \text{ radians}$$

$$r_p = 2/(3.98 \times 10^5) \sin\phi = 2/(3.98 \times 10^5)(0.0209) = 2.40 \times 10^{-4} \text{ m } (0.24 \text{ mm})$$

$$\frac{\dot{Q}_{max}}{\phi} = -\frac{95}{1.20} = -79.2 \text{ W/degree} = -4540 \text{ W/rad}$$

(from the intercept on the graph)

$$4540 = (3.67 \times 10^6)(3.98 \times 10^5)\kappa, \qquad \kappa = 31 \times 10^{-10} \text{ m}^2$$

Comments

The heat transfer factor $(\dot{Q}_{max}L_{eff})_{\phi=0}$ for this heatpipe is (95)(1.40) = 133 W m.

7.7.2 Sonic, Entrainment, and Boiling Limitations

Apart from the fundamental wicking or capillary limitation on heat transport by a heatpipe, there are other factors that may, under some circumstances, limit heat transport. The most important of these factors are choking of the vapor flow, entrainment of liquid by the vapor flow, and boiling of the liquid in the evaporator wick. Choking, or the *sonic limitation*, is influenced only by the vapor core size. The entrainment limitation is increased by using wicks with a smaller pore size at the vapor-liquid interface. The boiling limitation can be increased by using wicks with a high effective thermal conductivity. Figure 7.42 shows how these limitations affect the performance of a typical heatpipe.

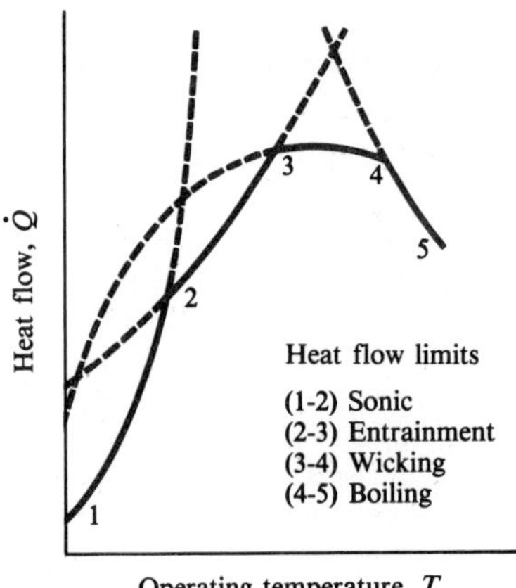

Figure 7.42 Schematic of limits on heatpipe performance.

Sonic Limitation

In a converging nozzle, a compressible gas accelerates to the sonic velocity due to area change; in a heatpipe, the vapor can accelerate to the sonic velocity by mass addition in the evaporator. If the vapor flow in a heatpipe is choked, a further decrease in the condenser temperature has no effect on the evaporator temperature; thus,

Table 7.5 Sonic limitation for metal vapor heat pipes [30].

Evaporator exit temperature, °C	Sonic heat flux limit, kW/cm^2			
	Cesium	Potassium	Sodium	Lithium
400	1.0	0.5	—	—
500	4.6	2.9	0.6	—
600	14.9	12.1	3.5	—
700	37.3	36.6	13.2	—
800			38.9	1.0
900			94.2	3.9
1000				12.0
1100				31.1
1200				71.0
1300				143.8

a *sonic limitation* to the axial heat transport per unit cross-sectional area of the core exists. The sonic limit is particularly relevant to the performance of liquid-metal heatpipes at the low end of their operating temperature range because of the low vapor density. Table 7.5 gives the sonic limitation for a selection of liquid metals. The sonic limitation may be encountered during start-up when the evaporator is cold, even though it may not be a factor at the design operating point.

Entrainment Limitation

If the momentum flux of the vapor flow is sufficiently high, the stress exerted by the vapor on the liquid surface can be sufficient to generate waves and entrain liquid droplets from the wave crests. Entrainment causes a substantial increase in the fluid circulation rate, and if the capillary pumping head is insufficient, a sudden dryout of the wick in the evaporator can occur. When there is significant entrainment, the droplets impinging at the condenser end may make an audible sound. The ratio of the vapor momentum flux to the surface tension forces restraining the liquid is proportional to the *Weber number*,

$$\mathrm{We} = \frac{\rho_v V_v^2 L}{\sigma} \tag{7.136}$$

where the characteristic length L is taken to be the hydraulic diameter of the wick surface pores. This diameter is equal to wire spacing for screen wicks, to twice the groove width for groove wicks, and to 0.82 times the sphere radius for packed spheres. A value of We $\sim$ 1 is taken to indicate the entrainment limit.

Boiling Limitation

The formation of bubbles in the evaporator wick is undesirable because hot spots can be formed that obstruct the liquid flow. Whereas the wicking, sonic, and entrainment limitations are limitations on the axial heat flux, the boiling limitation is a limitation on the radial heat flux in the evaporator. From a practical viewpoint, the engineer increases the boiling limitation by choosing a wick of high effective thermal conductivity and by providing an adequate heat transfer area in the evaporator. Bubble nucleation can be predicted by the theory given for pool boiling in Section 7.4.2. However, once again, there are problems related to reliable estimation of nucleation site size, the role played by dissolved gases, and the effect of temperature gradients in the liquid phase. The boiling limitation proves to be unimportant for liquid-metal heatpipes but can be a major problem for water heatpipes, since the water does not easily fill nucleation sites.

7.7.3 Gas-Loaded Heatpipes

Gas-loaded variable-conductance heatpipes were briefly described earlier. Figure 7.43 is a schematic of a heatpipe that has a gas reservoir of volume V_r connected to the end of the condenser. Notice that the wick extends into the reservoir. The important operating characteristics of such a heatpipe can be obtained from a simple model based on the following assumptions.

1. There is a flat front between the vapor and the gas that divides the condenser of length L_c into an active length L_a and an inactive length $(L_c - L_a)$.
2. The total pressure in the condenser P_c is constant and equals the reservoir pressure.
3. Axial conduction along the wall and wick is negligible, so that there is a step change in temperature of the heatpipe and its contents at the vapor-gas front.
4. Since it is assumed that there is pure vapor in the active-length, the active-length temperature T_a is the saturation temperature corresponding to pressure P_c: $T_a = T_{\text{sat}}(P_c)$.
5. Under steady operating conditions, the inactive condenser and reservoir are in thermal equilibrium with the heat rejection sink at temperature T_o.

The active length of the condenser is obtained from an inventory of the gas. The vapor pressure in both the inactive condenser and the reservoir is $P_{\text{sat}}(T_o)$. Using Dalton's law of partial pressures and the ideal gas law, the mass of gas is

$$w_g = \frac{P_c - P_{\text{sat}}(T_o)}{R_g T_o}[A_v(L_c - L_a) + V_r] \tag{7.137}$$

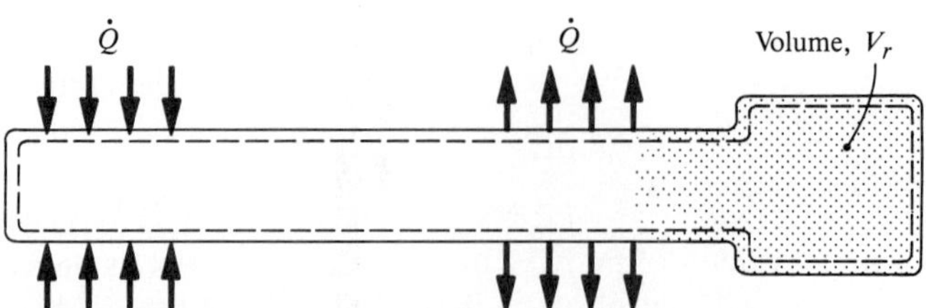

Figure 7.43 A gas-loaded heatpipe with a wicked reservoir.

where A_v is the cross-sectional area of the vapor core. Solving for L_a gives

$$L_a = L_c + \frac{V_r}{A_v} - \frac{w_g R_g T_o}{[P_c - P_{\text{sat}}(T_o)]A_v} \tag{7.138}$$

The heat rejected by the condenser can be written as

$$\dot{Q} = (U\mathscr{P})_c L_a (T_a - T_o) \tag{7.139}$$

where $U\mathscr{P}$ is the overall heat transfer coefficient–perimeter product for radial heat flow out of the condenser, and accounts for the thermal resistances of liquid-filled wick, wall, and convection or radiation to the sink at temperature T_o. Substituting Eq. (7.138) in Eq. (7.139) gives

$$\dot{Q} = (U\mathscr{P})_c (T_a - T_o)\left[L_c + \frac{V_r}{A_v} - \frac{w_g R_g T_o}{[P_c - P_{\text{sat}}(T_o)]A_v}\right] \tag{7.140}$$

The effect of the gas can be easily seen if we consider a situation where $V_r = 0$ (that is, no gas reservoir), as shown in Fig. 7.44. Recall that an objective of gas loading is to maintain a nearly constant evaporator temperature when the heat load drops below the design value. When there is no gas in the heatpipe and T_o is held constant, Eq. (7.140) shows that $(T_a - T_o)$ is nearly directly proportional to the heat load $\dot{Q}$, so that T_a must decrease linearly with $\dot{Q}$. But if there is an appropriate amount of gas present, a decrease in T_a decreases P_c, since $P_c = P_{\text{sat}}(T_a)$, and the last term in the square brackets of Eq. (7.140) increases as the gas expands. Thus, the effect of the gas is to limit the decrease in T_a required to match the decrease in heat load. The mass of gas added to the heatpipe depends on the size of reservoir, the sink temperature, and the desired design value for T_a. When the heatpipe is fully open, all the gas is in the reservoir, and the total pressure there is $P_c \simeq P_{\text{sat}}(T_a)$. Using the ideal gas law,

$$w_g = \frac{[P_c - P_{\text{sat}}(T_o)]V_r}{R_g T_o} \tag{7.141}$$

which allows the required mass of gas to be calculated for a chosen reservoir volume V_r. To see how the reservoir volume affects the performance of the heatpipe, it is convenient to divide the actual heat load $\dot{Q}$ by the maximum value within its control range, $\dot{Q}_{\text{max}}$.

$$\dot{Q}_{\text{max}} = (U\mathscr{P})_c (T_{a,\text{max}} - T_o) L_c \tag{7.142}$$

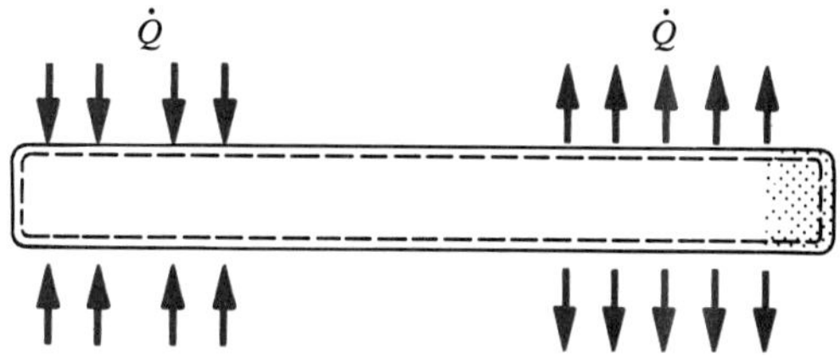

Figure 7.44 A gas-loaded heatpipe with no reservoir.

Dividing Eq. (7.140) by Eq. (7.142),

$$\frac{\dot{Q}}{\dot{Q}_{\max}} = \frac{T_a - T_o}{T_{a,\max} - T_o}\left[1 + \frac{V_r}{L_c A_v} - \frac{w_g R_g T_o}{[P_c - P_{\text{sat}}(T_o)]L_c A_v}\right] \tag{7.143}$$

With $V_c = L_c A_v$ denoting the vapor core volume, and using Eq. (7.141), Eq. (7.143) can be rearranged as

$$\frac{T_a - T_o}{T_{a,\max} - T_o} = \frac{\dot{Q}}{\dot{Q}_{\max}}\left[\frac{V_c/V_r}{1 + \dfrac{V_c}{V_r} - \dfrac{P_{\text{sat}}(T_{a,\max}) - P_{\text{sat}}(T_o)}{P_{\text{sat}}(T_a) - P_{\text{sat}}(T_o)}}\right] \tag{7.144}$$

The reservoir-to-core volume ratio, V_r/V_c, determines the sensitivity of the pipe. As V_r/V_c increases, V_c/V_r decreases, and the decrease in T_a for a given reduction in heat load $\dot{Q}$ decreases. This is best illustrated with a numerical example (see Example 7.22).

The model used in the preceding analysis is useful because, in most heatpipes, there is indeed a relatively sharp transition between the active and inactive regions of the condenser. Figure 7.45 shows axial variations of vapor pressure and temperature along a gas-loaded heatpipe. Axial conduction in the wall and wick tends to smear the temperature step, and there is some interpenetration of gas and vapor by diffusion. More exact analysis that accounts for mass diffusion is left to a mass transfer text.

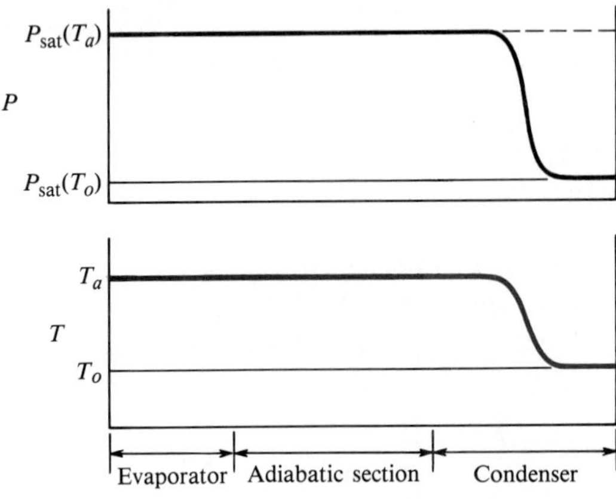

Figure 7.45 Axial variations of temperature and pressure along a gas-loaded heatpipe.

EXAMPLE 7.22 A Gas-Loaded Heatpipe for a Communications Satellite

A nitrogen-loaded methanol heatpipe is constructed from a 0.5 inch–O.D. stainless steel tube and is L-shaped, as shown in the sketch. A fibrous stainless steel slab wick extends into the gas reservoir. The cross-sectional area of the vapor space in the pipe is 50.3 mm^2, and the reservoir volume is 5.28×10^4 mm^3. The condenser length is 35 cm. When the sink temperature is 264 K, the heatpipe is fully open for an adiabatic section temperature of 283 K and a heat load of 50 W. Determine the mass of gas in the heatpipe, and prepare a graph of heat load versus adiabatic section temperature. In addition, vary the reservoir volume to demonstrate its effect on performance. At a reference temperature $T_r = 5°C$, methanol has a vapor pressure of 5330 Pa and enthalpy of vaporization of 1.18×10^6 J/kg.

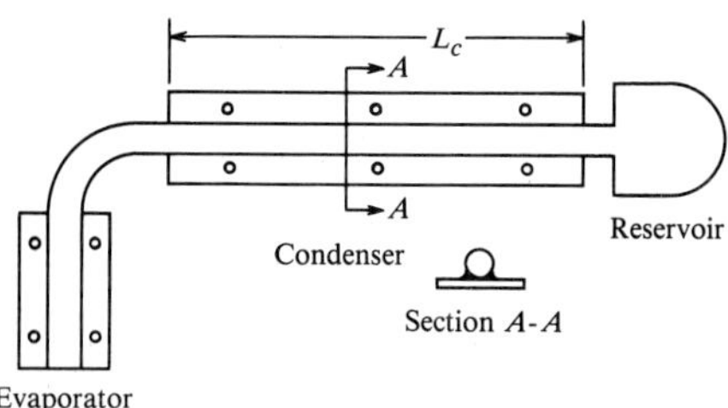

Solution

Given: Nitrogen-loaded methanol heatpipe, with a wicked gas reservoir.

Required: 1. Mass of gas in pipe, w_g.
2. $\dot{Q}$ versus T_a for three different reservoir volumes.

Assumptions: 1. A flat front between the vapor and gas.
2. Negligible pressure drop between adiabatic section and condenser.

When the heatpipe is fully open, the gas is contained in the reservoir, which is at $T_o = 264$ K. Using the ideal gas law,

$$w_g = \frac{[P_c - P_{\rm sat}(T_o)]V_r}{R_g T_o}$$

where the total pressure has been taken to be equal to the pressure in the active condenser, $P_c = P_{\rm sat}(T_a)$. The Clausius-Clapeyron relation will be used to obtain $P_{\rm sat}(T)$ from $P_{\rm sat}(T_r)$:

$$\frac{P}{P_r} \simeq \exp\left[-\frac{h_{\rm fg}}{R_v}\left(\frac{1}{T} - \frac{1}{T_r}\right)\right]$$

$$P_r = 5330 \text{ Pa}, \qquad T_r = 278.15 \text{ K}$$

$$h_{\rm fg} = 1.18 \times 10^6 \text{ J/kg}, \qquad R_v = 8314/32.04 = 259.5$$

$$P = 5330 \exp\left[-4548\left(\frac{1}{T} - 3.595 \times 10^{-3}\right)\right]$$

$$T = 283 \text{ K}, \qquad P = 7048 \text{ Pa}$$

$$T = 264 \text{ K}, \qquad P = 2217 \text{ Pa}$$

$$w_g = \frac{(7048 - 2217)(5.28 \times 10^{-5})}{(8314/28)(264)} = 3.25 \times 10^{-6} \text{ kg}$$

which is the required mass of gas. Rearranging Eq. (7.143) gives the part load performance as:

$$\frac{T_a - T_o}{T_{a,\max} - T_o} = \frac{\dot{Q}}{\dot{Q}_{\max}}\left[\frac{V_c/V_r}{1 + V_c/V_r - (w_g/V_r)R_g T_o/[P_c - P_{\rm sat}(T_o)]}\right]$$

$$V_c = L_c A_v = (0.35)(50.3 \times 10^{-6}) = 1.76 \times 10^{-5} \text{ m}^3$$

$$\frac{V_r}{V_c} = \frac{5.28 \times 10^{-5}}{1.76 \times 10^{-5}} = 3.00; \qquad \frac{V_c}{V_r} = 0.333$$

$$T_{a,\max} = 283\text{K}, \qquad T_o = 264\text{ K}$$

$$P_{\text{sat}}(T_o) = 2217\text{ Pa}, \qquad \frac{w_g R_g T_o}{V_r} = P_c(T_{a,\max}) - P_{\text{sat}}(T_o) = 7048 - 2217 = 4831\text{ Pa}$$

$$\frac{T_a - 264}{19} = \frac{\dot{Q}}{50}\left[\frac{0.333}{1 + 0.333 - 4831/(P_c - 2217)}\right]$$

For specified values of T_a, $P_c = P_{\text{sat}}(T_a)$ and $\dot{Q}$ can be calculated. The results are given in the table and graph that follow. Also given are results for $V_r/V_c = 2.0$ and 4.0.

T_a	P_c	$\dot{Q}[W]$		
K	Pa	$V_r/V_c = 2.0$	3.0	4.0
283	7048	50	50	50
282	6658	39.0	34.9	30.7
281	6286	28.0	19.6	11.2
280	5933	16.8	4.2	—
279	5598	5.6	—	—

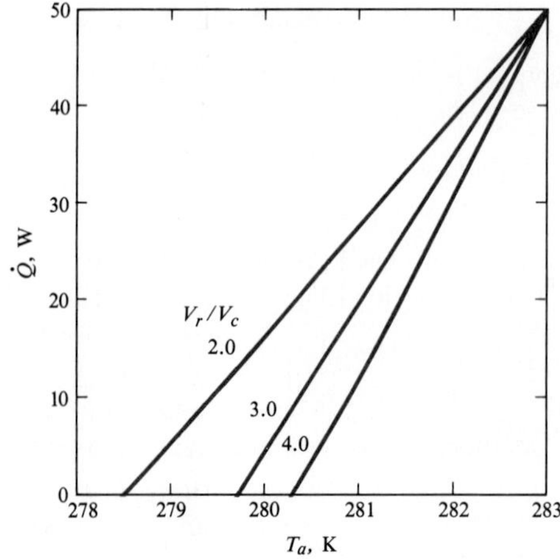

Comments

Notice that as V_r/V_c increases, the adiabatic section temperature becomes less sensitive to heat load.

7.8 CLOSURE

Heat transfer during phase change is essentially a convection process that is often made complicated by an irregular phase interface. Since enthalpies of phase change are large, the resulting heat transfer coefficients also tend to be large. Laminar film condensation and evaporation were satisfactorily analyzed using simplifying assumptions introduced by Nusselt in 1916. For situations where the film is wavy laminar or turbulent, we used empirical correlations for the local heat transfer coefficient and determined the change in film flow rate by satisfying mass and energy balances. The computer program PHASE allowed the rather complicated formulas to be evaluated efficiently and reliably.

We recognized two modes of boiling, namely, pool boiling and forced-convection boiling. The "boiling curve," that is, a graph of q versus $(T_w - T_{sat})$, defined the various regimes of pool boiling. Nucleate boiling is strongly influenced by surface condition but is relatively insensitive to surface geometry. On the other hand, the peak heat flux and film boiling depend primarily on hydrodynamic factors and hence are influenced by surface geometry. We considered only saturated pool boiling; the effect of liquid subcooling is left to advanced texts. Boiling heat transfer formulas for engineering use are based on crude physical models and dimensional analysis, with adjustments made by curve-fitting experimental data. The computer program BOIL allowed these formulas to be evaluated efficiently and reliably.

Forced-convection boiling and condensation in tubes involves a two-phase flow of liquid and vapor. There are a number of flow regimes possible, including bubbly flow, slug flow, churn flow, and annular flow. Flow maps were presented, which allow the flow regime to be identified for cocurrent vertical upward flow and for horizontal flow. Since the local pressure is required for calculating boiling or condensation heat transfer, we must be able to calculate pressure drop for a two-phase flow. The simple homogeneous flow model, which ignores the details of the flow regimes, was used for this purpose. Methods were given for calculating heat transfer coefficients for both boiling and condensation.

Phase change at low pressures was discussed briefly. We saw that the equations describing molecule fluxes at a phase change interface are similar to those used in Chapter 6 to describe photon fluxes at a surface. The effects of departure from thermodynamic equilibrium at the interface were seen to be significant only at very low pressures, such as those encountered for condensation of liquid-metal vapors.

The chapter concluded with a discussion of heatpipes. Both fixed-conductance pure-vapor heatpipes and variable conductance gas-loaded heatpipes were analyzed. The key equation governing the performance of fixed-conductance heatpipes was derived by balancing the capillary pressure generated by the wick against the gravitational head and the liquid- and vapor-phase pressure drops. Heatpipe working fluids were characterized by a *figure of merit* $\mathcal{M} = \rho_l \sigma h_{fg}/\mu_l$, and heatpipe performance was characterized by a *heat transfer factor* $(\dot{Q}_{max} L_{eff})_{\phi=0}$. The characteristics of gas-loaded heatpipes were derived using a simple model that assumes a flat front between the vapor and the gas in the condenser. A worked example illustrated the use of this model for a heatpipe used to control the temperature of a communications satellite.

REFERENCES

1. Nusselt, W., "Die Oberflachenkondensation des Wasserdampfes," *Z. Ver. D-Ing.*, 60, 541–546 (1916).

2. Sadasivan, P., and Lienhard, J. H., "Sensible heat correction in laminar film boiling and condensation," *J. Heat Transfer*, 109, 545–547 (1987).

3. Denny, V. E., and Mills, A. F., "Nonsimiliar solutions for laminar film condensation on a vertical surface," *Int. J. Heat and Mass Transfer*, 12, 965–979 (1969).

4. Chun, K. R., and Seban, R. A., "Heat transfer to evaporating liquid films," *J. Heat Transfer*, 93, 391–396 (1971).

5. Shekriladze, I. G., and Gomelauri, V. I., "Theoretical study of laminar film condensation of flowing vapour," *Int. J. Heat Mass Transfer*, 9, 581–591 (1966).

6. Rohsenow, W. M., "A method of correlating heat transfer data for surface boiling of liquids," *Trans. ASME*, 74, 969–976 (1952).

7. Kutateladze, S. S., "On the transition to film boiling under natural convection," *Kotloturbostroenie*, 3, 10 (1948).

8. Zuber, N., "Hydrodynamic aspects of nucleate boiling," Ph.D. dissertation, Department of Engineering, University of California, Los Angeles (1959). [Also AEC Report AECU-4439 (1959).]

9. Lienhard, J. H., and Dhir, V. K., "Hydrodynamic prediction of peak pool-boiling heat fluxes from finite bodies," *J. Heat Transfer*, 95, 152–158 (1973).

10. Bromley, L. A., "Heat transfer in stable film boiling," *Chem. Eng. Prog.*, 46, 221–227 (1950).

11. Frederking, T. H. K., and Daniels, D. J., "The relation between bubble diameter and frequency of removal from a sphere during film boiling," *J. Heat Transfer*, 88, 87–93 (1966).

12. Berenson, P., "Film-boiling heat transfer from a horizontal surface," *J. Heat Transfer*, 83, 351–358 (1961).

13. Frederking, T. H. K., and Clark, J. A., "Natural convection film boiling on a sphere," *Advances in Cryogenic Engineering*, 8, 501–506 (1962).

14. Zuber, N., "On the stability of boiling heat transfer," *Trans. ASME*, 80, 711–720 (1958).

15. Lienhard, J. H., and Wong, P. T. Y., "The dominant unstable wavelength and minimum heat flux during film boiling on a horizontal cylinder," *J. Heat Transfer*, 86, 220–226 (1964).

16. Sparrow, E. M., "The effect of radiation on film-boiling heat transfer," *Int. J. Heat Mass Transfer*, 7, 229–238 (1964).

17. Hewitt, G. F., and Roberts, D. N., "Studies of two-phase flow patterns by simultaneous X-ray and flash photography," AERE-M1259, HMSO (1969).

18. Taitel, Y., and Dukler, A. E., "A model for predicting flow regime transitions in horizontal and near horizontal gas-liquid flow" *AIChE Journal,* 22, 47–55 (1976).

19. Klimenko, V. V., "A generalized correlation for two-phase forced flow heat transfer," *Int. J. Heat Mass Transfer*, 31, 541–552 (1988).

20. Traviss, D. P., Rohsenow, W. M., and Baron, A. B., "Forced-convection condensation inside tubes: A heat transfer equation for condenser design," *ASHRAE Transactions*, 79, part 1, 157–165 (1973).

21. Knudsen, M., "Maximum rate of vaporization of mercury," *Ann. Phys.*, 47, 697 (1915).

22. Schrage, R. W., *A Theoretical Study of Interphase Mass Transfer*, Columbia University Press, New York (1953).

23. Alty, T. A., and Mackay, C. A., "The accommodation coefficient and the evaporation coefficient of water," *Proc. R. Soc.*, A149, 104–116 (1935).

24. Nabavian, K., and Bromley, L. A., "Condensation coefficient of water," *Chem. Engr. Sci.*, 18, 651–660 (1963).

25. Mills, A. F., and Seban, R. A., "The condensation coefficient of water," *Int. J. Heat Mass Transfer*, 10, 1815–1827 (1967).

26. Maa, J. R., "Evaporation coefficient of liquids," *I and EC Fund.*, 6, 504–518 (1967).

27. Narusawa, U., and Springer, G. S., "Measurement of the condensation coefficient of mercury by a molecular beam method," *J. Heat Transfer*, 97, 83–87 (1975).

28. Rohsenow, W. M., "Film condensation of liquid metals," *Trans. CSME*, 1, 5–12 (1972).

29. Dunn, P. D., and Reay, D. A., *Heat Pipes*, 3rd ed., Pergamon Press, Oxford (1982).

30. Chisholm, D., *The Heat Pipe*, Mills and Boon Limited, London, p. 34 (1971).

31. Labuntsov, D. A., "Heat transfer in film condensation of pure steam on vertical and horizontal surfaces and tubes," *Teploenergetika*, 72–89 (July 1957).

32. Shmerler, J. A., and Mudawwar, I., "Local heat transfer coefficient in wavy free-falling turbulent liquid films undergoing uniform sensible heating," *Int. J. Heat Mass Transfer,* 31, 67–77 (1988).

33. Mills, A. F., Hubbard, G. L., James, R. K., and Tan, C., "Experimental study of film condensation on horizontal grooved tubes," *Desalination,* 16, 121–133 (1975).

34. Kim, S., and Mills, A. F., "Condensation on coherent turbulent liquid jets," *J. Heat Transfer,* 111, 1068–1074 (1989).

35. Hsu, Y. Y., "On the size range of active nucleation cavities on a heating surface," *J. Heat Transfer*, 84, 207–216 (1962).

EXERCISES

7–1. Saturated ammonia vapor at 310 K condenses on a vertical surface maintained at 305 K. If the surface is 2 cm high, determine the average heat transfer coefficient and the rate of condensation per unit width.

7–2. Consider laminar film condensation from a saturated vapor at its normal boiling point on a vertical surface maintained at a temperature 2 K below the boiling point. If the onset of wavy laminar flow is taken to be Re = 30, determine the location down the surface at which ripples might be expected to be seen for the following fluids: water, ammonia, nitrogen, mercury, R-12, and R-113.

7–3. Saturated R-113 vapor at 0.110 MPa condenses on the outside of a 2 cm–O.D., 10 cm–high vertical tube maintained at 320 K. Determine the average heat transfer coefficient and rate of condensation.

7–4. Show that if the subcooling term is retained in Eq. (7.12), Eq. (7.16) remains valid if h_{fg} is replaced by a modified latent heat $h'_{\text{fg}} = h_{\text{fg}} + (3/8)c_{pl}(T_{\text{sat}} - T_w)$.

7–5. In practice, film condensation usually takes place on one side of a wall, and the enthalpy of condensation is taken up by a coolant flowing on the other side of the wall. Thus, there are three resistances in series in the thermal network: the condensate film resistance, the wall resistance, and the coolant convective resistance. If the sum of the latter two is much larger than the condensate film resistance, the heat flux along the wall will be nearly constant, rather than the wall temperature being constant, as assumed by Nusselt's analysis. Such a situation occurs in, for example, an air-cooled condenser. For laminar film condensation on a vertical wall with a constant heat flux q_w, show

(i) the film thickness increases proportional to $x^{1/3}$.
(ii) $\overline{\text{Nu}} = (4/3)^{4/3}\,\text{Re}_L^{-1/3}$, where $\overline{\text{Nu}} = q_w(\nu_l^2/g)^{1/3}/(\overline{T_{\text{sat}} - T_{wx}})k_l$, and show that this result is identical to that for the isothermal wall case.

7–6. Laminar film condensation occurs on a vertical wall with a temperature variation that can be approximated by a power law, $T_w - T_{\text{sat}} = ax^n$. Show that the average heat transfer coefficient for a wall of height L is

$$\overline{h} = \frac{4}{(3-n)}\left(\frac{n+1}{4}\right)^{1/4}\left[\frac{h_{\text{fg}}g(\rho_l - \rho_v)k_l^3}{L[T_{\text{sat}} - T_w(L)]\nu_l}\right]^{1/4}$$

where $T_w(L)$ is the wall temperature at $x = L$.

7–7. Consider laminar film condensation on the inside of a small-diameter vertical tube for which the film thickness cannot be assumed to be small compared to the tube radius. Extend the Nusselt-type analysis of Section 7.2.1 to this situation.

7–8. Saturated R-12 vapor at 1.005 MPa condenses on the outside of a 4 cm–diameter, 3 m–high vertical tube. Determine the average heat transfer coefficient and total condensation rate for wall temperatures in the range 295–315 K.

7–9. Saturated steam at 1 atm condenses on the outside of a 3 cm–diameter, 2 m–high vertical tube. Determine the average heat transfer coefficient and total condensation rate for wall temperatures in the range 340–370 K.

7–10. D.A. Labuntsov [31] recommends the following correlation of the local Nusselt number for turbulent falling films:

$$\mathrm{Nu} = 0.023\,\mathrm{Re}^{0.25}\,\mathrm{Pr}_l^{0.5}$$

Using a transition Reynolds number of 1600, show that the average Nusselt number for condensation commencing at the top of a vertical surface of length L, with $\mathrm{Re}_L > 1600$, is

$$\overline{\mathrm{Nu}} = \frac{\mathrm{Re}_L}{8090 + 58\mathrm{Pr}_l^{-0.5}(\mathrm{Re}_L^{0.75} - 253)}$$

Recalculate $\overline{\mathrm{Nu}}$ for Example 7.2 using this formula

7–11. New experimental data for heat transfer across a turbulent falling film has been correlated by Shmerler and Mudawwar [32] as

$$\mathrm{Nu} = 0.0106\,\mathrm{Re}^{0.3}\,\mathrm{Pr}_l^{0.63} \qquad 2.55 < \mathrm{Pr}_l < 6.87$$

Using this correlation as an alternative to Eq. (7.23), rework Example 7.2.

7–12. Saturated steam at 330 K condenses on the outside of a horizontal, 15 mm–O.D., 1 mm–wall-thickness copper tube, through which flows coolant water. At an axial location where the bulk water temperature is 310 K and the inside heat transfer coefficient is 7000 W/m^2 K, determine the rate of condensation per unit length of tube.

7–13. Saturated ammonia vapor at 310 K condenses on the outside of a horizontal, 2 cm–O.D., 1 mm–wall-thickness brass tube through which flows coolant water. At an axial location where the bulk coolant temperature is 290 K and the inside heat transfer coefficient is 6000 W/m^2 K, determine the heat transfer and condensation rates per unit length of tube.

7–14. Saturated mercury vapor at 0.010 MPa pressure condenses on a horizontal 2 cm–O.D., 1 mm–wall-thickness AISI 316 stainless steel tube through which flows pressurized water. At a location where the water bulk temperature is 490 K and the inside heat transfer coefficient is 12,000 W/m^2 K, determine the condensation heat transfer coefficient, the overall heat transfer coefficient, and the rate of condensation per unit length of tube.

7–15. Saturated R-12 vapor at 0.9 MPa pressure condenses on a horizontal 1.5 cm–O.D., 1 mm–wall-thickness brass tube through which coolant water flows at 2 m/s. At a location where the bulk water temperature is 290 K, determine the condensation heat transfer coefficient and the rate of condensation per unit length of tube.

7–16. To assist in the design of a steam condenser, parametric data for the condensation heat transfer coefficient is required. Saturated vapor at 320 K is to be condensed on a bank of 15 mm–O.D. horizontal tubes in a square array. Prepare a graph of the average heat transfer coefficient versus $(T_{sat} - T_w)$, with the number of tubes in a column, N, as a parameter. Let $(T_{sat} - T_w)$ vary from 1 to 5 K, and N vary from 1 to 5.

7–17. Consider condensation of saturated R-113 vapor on a horizontal tube. For $(T_{sat} - T_w) = 5$ K, determine the average heat transfer coefficient for tube diameters in the range 1 to 5 cm, and saturation temperatures from 320.71 to 420 K. Prepare a graph and discuss the results.

7–18. Hydrofluorocarbons (HFCs) and hydrochlorofluorocarbons (HCFCs) are likely replacements for chlorofluorocarbons (CFCs) as refrigerants and heat transfer fluids. HFCs do not contain chlorine and therefore have zero ozone depletion potential. In HCFCs the addition of the hydrogen causes the dissipation of nearly all the chlorine in the lower atmosphere before it can reach the ozone layer. The resulting ozone depletion potential varies from 2 to 10% of that for CFCs.

HCFC-123 (CF_3CHCl_2) has a boiling point of 27.9°C and is a suitable refrigerant for some types of chillers. Calculate the average heat transfer coefficient for saturated HCFC-123 vapor at 65°C condensing on the outside of a 2 cm–diameter horizontal tube with a wall temperature of 55°C. Compare your result to the value of $\overline{h}$ obtained for saturated R-113 vapor condensing at the same temperature. Estimated properties of HCFC-123 include P_{sat} (65°C) $= 0.222$ MPa, h_{fg} (65°C) $= 0.333$ MJ/kg, and at 60°C, $\rho_l = 1350$ kg/m^3, $k_l = 0.0672$ W/m K, $\mu_l = 4.49 \times 10^{-4}$ kg/m s.

7–19. Show that the average heat transfer coefficient for laminar film condensation on a horizontal tube with a constant heat flux q around its periphery is

$$\overline{h} = \frac{q}{\overline{T_w - T_{sat}}} = 0.615\left[\frac{(\rho_l - \rho_v)gh_{fg}k_l^3}{\nu_l D q}\right]^{1/3}$$

(*Hint:* The integral $\int_0^\pi (\phi/\sin\phi)^{1/3}d\phi$ must be evaluated carefully due to the singularity at $\phi = \pi$.)

7–20. For laminar film condensation on a horizontal tube with an isothermal wall, derive the differential equation governing the film thickness as a function of angle ϕ measured from the top of the tube. Obtain the limiting form of this equation as $\phi \to 0$, and hence solve for the film thickness at $\phi = 0$.

7–21. Show that the heat transfer coefficient for laminar film condensation on a horizontal tube can be expressed in terms of the Nusselt and film Reynolds numbers defined in Section 7.2.3 and Exercise 7–5 as

$$\overline{\mathrm{Nu}} = 1.209\,\mathrm{Re}_\pi^{-1/3} \qquad \text{uniform wall temperature}$$

$$\overline{\mathrm{Nu}} = 1.135\,\mathrm{Re}_\pi^{-1/3} \qquad \text{uniform wall heat flux}$$

7–22. Laminar film condensation occurs on a vertical, isothermal cone of height H and angle α, as shown in the sketch. Show that the mass flow $\dot{m}(= \Gamma\mathscr{P})$ is given by

$$\dot{m} = \frac{2\pi x \sin\alpha(\rho_l - \rho_v)g\cos\alpha\,\delta^3}{3\nu_l}$$

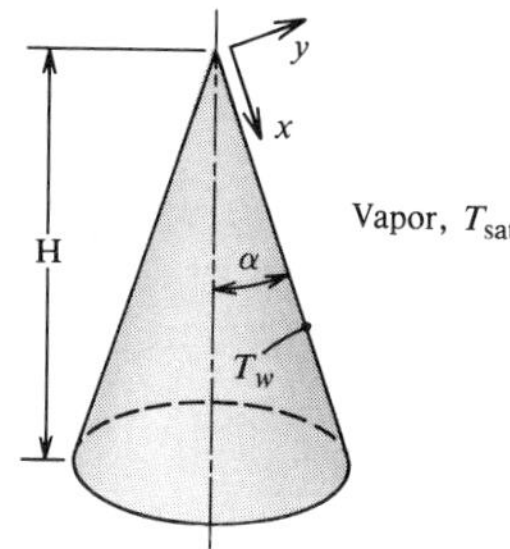

and that the energy balance for the film is

$$\frac{k_l}{\delta}(T_{sat} - T_w) = \frac{h_{fg}}{2\pi x \sin\alpha}\frac{d\dot{m}}{dx}$$

Hence show that the average heat transfer coefficient is given by

$$\overline{h} = 1.165\left[\frac{(\rho_l - \rho_v)g\cos^2\alpha\, h_{fg}k_l^3}{\nu_l H(T_{sat} - T_w)}\right]^{1/4}$$

7–23. Proceeding as in Exercise 7–22, show that the average heat transfer coefficient for laminar film condensation on an isothermal sphere is

$$\overline{h} = 0.828\left[\frac{(\rho_l - \rho_v)g\,h_{fg}k_l^3}{\nu_l D(T_{sat} - T_w)}\right]^{1/4}$$

(*Caution:* Incorrect values of the coefficient have appeared in the literature; the value of 0.828 given here is correct.)

7–24. A 15 mm–diameter copper sphere initially at 60°C is suddenly exposed to a flow of saturated steam at 1 atm pressure. Prepare a graph of the temperature-time response of the center of the sphere. Make reasonable simplifying assumptions.

7–25. Derive Eq. (7.51) for the effect of vapor drag on laminar film condensation on a vertical wall.

7–26. Saturated R-12 vapor at 1.50 MPa flows at 20 m/s down the outside of a 1 cm–O.D. vertical tube. Coolant inside the tube maintains the outside surface of the tube at 330 K. Estimate the effect of vapor drag on the local heat transfer coefficient at a location 5 cm from the top of the tube.

7–27. Steam at 10^4 Pa condenses on a 3 cm–O.D. horizontal tube maintained at 315 K. Determine the effect of vapor superheat on the average heat transfer coefficient and condensation rate, for superheats in the range 0–200 K. Graph your results.

7–28. Refrigerant-12 at 0.500 MPa condenses on a 2 cm–O.D. horizontal tube maintained at 280 K. Determine the effect of vapor superheat on the average heat transfer coefficient and condensation rate for superheats in the range 0–120 K.

7–29. In experiments to investigate the effect of surface enhancement on film condensation, saturated steam at 6000 Pa was condensed on a 19.1 mm–diameter horizontal grooved brass tube, with 14.2 grooves/cm (36 tpi Standard American Screw Thread). The construction of the tube is shown in the drawing.

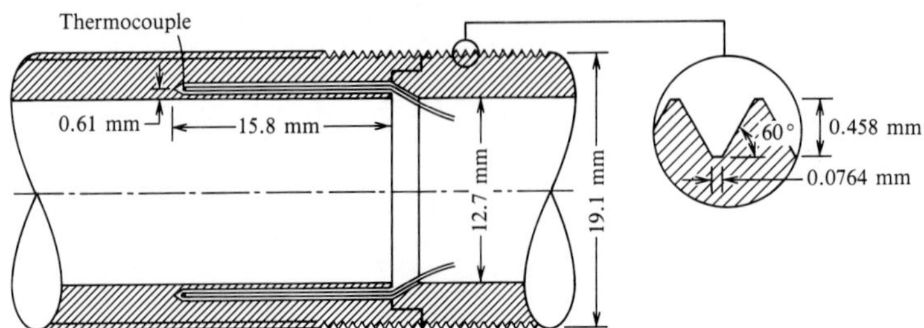

The average heat transfer coefficient was correlated as $\overline{h} = 89,700\Delta T^{-0.59}$ W/m^2 K, for a range of $\Delta T = T_{\text{sat}} - T_w$ from 1.1 K to 9 K. Prepare a graph comparing this result to the heat transfer coefficient for a smooth tube. What was the largest increase in $\overline{h}$ obtained? Details of the experiments may be found in reference [33].

7–30. Condenser tubes are often fluted to enhance the heat transfer. The flutes are contoured so that the radius of curvature $R(s)$ varies continuously with distance s over a length $0 < s < S$. The effect of surface tension and the varying curvature induce a pressure gradient that causes the condensate to flow from the ridges into the troughs in such a way as to form a uniform thin film of thickness δ. Perform a Nusselt-type analysis to show that the equations governing the flow are

$$\frac{d\Gamma}{ds} = \frac{h\Delta T}{h_{\text{fg}}}; \quad h = \frac{k}{\delta}$$

$$\frac{dP}{ds} = -\frac{3\mu\Gamma(s)}{\rho\delta^3}; \quad P(s) = \frac{\sigma}{R(s)}$$

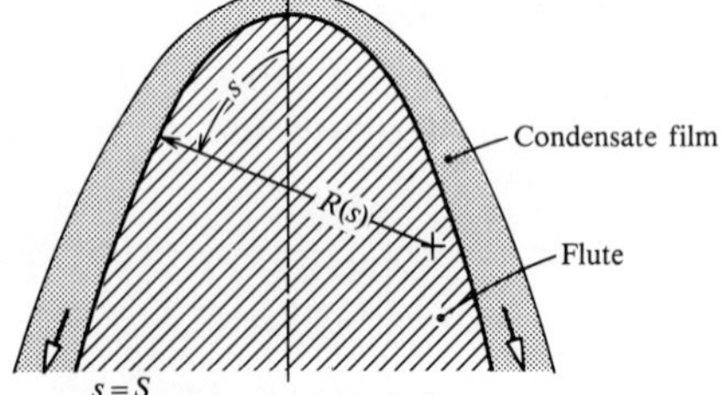

Tests are made using water to measure h versus ΔT. Use dimensional arguments to propose how these data should be correlated to allow application to ammonia condensation.

7–31. *Direct-contact* condensation involves condensation directly onto the cold liquid phase. James Watt's original steam condenser was of the direct-contact type, with steam condensing on a cold water spray. In an experiment, steam condenses on a vertical water jet. The nozzle diameter is 4 mm, and the jet collector is located 4 cm below the nozzle exit. In a particular test the water is supplied at 0.05 kg/s and 300 K, and the chamber pressure is maintained at 4200 Pa.

(i) Assuming that the jet flow can be modeled as laminar plug flow, estimate the rate of condensation of the steam on the jet. (*Hint:* There is an analogous heat conduction problem in Chapter 3.)

(ii) The jet is more likely to be turbulent. Kim and Mills [34] give the following correlation for the average Stanton number:

$$\overline{\text{St}} = 3.2\,\text{Re}_D^{-0.20}\,\text{Pr}^{-0.7}(L/D)^{-0.57}(\sigma D/\rho\nu^2)^{-0.19}$$

Using this correlation, make a new estimate of the condensation rate. (*Hint:* The turbulent jet can be modeled as a single-stream heat exchanger.)

7–32. An experimental vertical-tube evaporator consists of a single 4 cm–O.D. tube. The wall of the tube is maintained at 322 K, and the saturation temperature corresponding to the system pressure is 316 K. If water is fed at 0.028 kg/s to the top of the tube, over what length of tube is the film turbulent?

7–33. In a vertical-tube evaporator, liquid ammonia is distributed on the inside surfaces of 5 cm–I.D. tubes, at a rate of 0.1 kg/s per tube. Steam condenses on the outside of the tubes and maintains the tube wall temperature at 290 K. If the ammonia-side pressure is controlled to give a saturation temperature of 280 K, how long can the tubes be without having dryout of the film?

7–34. An array of 6 mm–square silicon chips on a ceramic baseplate is immersed in a pool of saturated R-113 at atmospheric pressure to provide cooling. If operation at 60% of q_{max} provides a sufficient safety factor, what is the allowable power level per chip, and what will be the chip operating temperature? Take $C_{nb} = 0.004$ and $m = 4.1$ in Rohsenow's nucleate boiling correlation.

7–35. Prepare as complete as possible a "boiling curve" in the form of a graph of $\log q$ versus $\log(T_w - T_{sat})$ for saturated water boiling on a large, mechanically polished stainless steel surface at 10 atm pressure.

7–36. A 1000 W stainless steel sheathed electric heater is to be designed for a water kettle. How large should the heat transfer surface be if the heater is required to operate at 50% of the burnout heat flux? What will the surface temperature of the heater be at that heat flux? Model as a large horizontal surface.

7–37. Y. Y. Hsu [35] proposed a criterion for an active nucleation site that requires the temperature of the tip of a nucleating bubble to at least equal the saturation temperature corresponding to the pressure in the bubble (see Fig. 7.18*b*). Use this criterion to determine the size of a cavity that will nucleate first for pool boiling on a 10 cm–diameter horizontal surface. Consider the following situations:

(i) water at 1 atm pressure.
(ii) water at a saturation temperature of 300 K.
(iii) R-113 at 1 atm pressure.

Assume a contact angle of 90° and a linear temperature profile near the wall with a slope equal to the wall value obtained from an appropriate natural-convection heat transfer correlation. Also use the Clausius-Clapeyron relation to relate ΔT to ΔP (see Example 7.7(iii)). Discuss the relevance of this criterion to engineering systems.

7–38. Water boils at 5 atm pressure on a 0.8 mm–diameter platinum rod heater. Determine

(i) the superheat required for boiling inception on 7.5 μm–radius nucleation sites.
(ii) the peak heat flux.
(iii) the rod temperature when the heat transfer is 50% of q_{max}.

7–39. A boiler is to be designed for a low-pressure condensation test rig and is required to supply 0.5×10^{-3} kg/s of saturated steam at 5000 Pa. How large must the surface area of the heater be if the heat flux is not to exceed 30% of the burnout heat flux? What will the surface temperature of the heater be at this heat flux if it is nickel-plated? Take $C_{nb} = 0.006$, and assume a large, flat surface.

7–40. A copper sphere is quenched in liquid nitrogen in an open Dewar flask. Calculate the average heat transfer coefficient and heat flux when $T_w - T_{sat} = 185$ K for spheres 1.5 and 2.5 cm in diameter.

7–41. Prepare a graph of q_{max} versus T_{sat} for water boiling on the outside of a 3 cm–diameter horizontal tube. Let $280 < T_{sat} < 550$ K. Explain the behavior of q_{max}.

7–42. Prepare a graph of q_{max} versus T_{sat} for R-12 boiling on the outside of a 3 cm–diameter horizontal tube. Let $250 < T_{sat} < 350$ K. Explain the behavior of q_{max}.

7–43. In a test rig, film boiling is established on the outside of a 1 cm–diameter horizontal tube immersed in water. If the tube wall temperature is 1000 K and the system pressure is 0.5 MPa, determine the heat transfer per unit length of tube. Take an emittance of 0.5 for the tube surface.

7–44. Prepare a graph of q_{min} versus T_{sat} for water boiling on

(i) a large horizontal surface.
(ii) a 1 mm–diameter horizontal wire.

Let $280 < T_{sat} < 500$ K. Explain any significant trends in the results.

7–45. Prepare graphs of q_{max} and q_{min} versus T_{sat} for mercury boiling on a large horizontal surface. Let $400 < T_{sat} < 800$ K. Explain any significant trends in the results.

7–46. Hydrofluorocarbons (HFCs) are being considered as replacements for chlorofluorocarbons (CFCs) as refrigerants (and as blowing agents, propellants, and cleaning agents). HFC134a (BP $= -26.5$°C) is a likely candidate to replace R-12 (BP $= -30.2$°C). Compare the boiling peak heat flux values for these two fluids at 1 atm pressure. Property values for saturated HFC134a at -26.5°C include $\rho_l \simeq 1380$ kg/m^3, $\rho_v = 5.05$ kg/m^3, $h_{fg} = 0.220 \times 10^6$ J/kg, $\sigma = 16.6 \times 10^{-3}$ N/m. (Property data courtesy of E. I. duPont de Nemours and Company, Wilmington, DE.)

7–47. Predict the probable flow patterns in a 3 cm–I.D. vertical steam boiler tube for system pressures of 30, 70, and 150 bar; mass qualities of 1, 10, and 50%; and mass velocities of 700 and 2500 kg/m^2 s.

7–48. Repeat Exercise 7–47 for a horizontal tube.

7–49. Estimate the pressure gradient due to wall friction for a steam-water flow of 10% mass quality in a 12 cm–I.D. tube at 18 bar, if the mass flow of steam is 11.1 kg/s.

7–50. Refrigerant-12 condenses inside a horizontal copper tube of 12 mm outside diameter and 1 mm wall thickness. The mass flow rate is 0.017 kg/s, and at a specific location the pressure is 1.6 MPa, the mass quality is 50%, and the cooling rate is 700 W/m length.

(i) What is the flow regime?
(ii) Estimate the pressure gradient.
(iii) Estimate the heat transfer coefficient.

7–51. A horizontal-tube evaporator contains 200 4 cm–O.D., 2 mm–wall-thickness AISI 316 stainless steel tubes and generates 100 kg/s steam. At a specific location along the evaporator where the heating rate per tube is 1.5×10^5 W/m length, the pressure is 1.1 MPa and the mass quality is 12%. Determine the flow regime, pressure gradient, and heat transfer coefficient.

7–52. Saturated water flows at 0.1 kg/s through a 2.5 cm–I.D. stainless steel vertical tube. At a given location the bulk water is at 550 K, the mass quality is 25%, and the wall heat flux is 135 kW/m^2. Estimate the tube wall temperature.

7–53. Refrigerant-113 condenses inside a horizontal copper tube of 10 mm outside diameter and 1 mm wall thickness. The mass flow rate is 0.008 kg/s, and at a particular location the pressure is 0.3 MPa and the mass quality is 60%. If the cooling rate is 400 W/m length, determine the flow regime, pressure gradient, and heat transfer coefficient.

7–54. A vertical-tube evaporator contains 50 4 cm–O.D., 2 mm–wall-thickness AISI 304 stainless steel tubes and generates 60 kg/s of steam. At a specific location up the evaporator the pressure is 0.932 MPa, the mass quality is 16%, and the heating rate per tube is 0.2×10^5 W/m length. Determine the flow regime, pressure drop, and heat transfer coefficient.

7–55. Calculate the maximum rate of evaporation for

(i) water at 280 K.
(ii) mercury at 500 K.
(iii) R-12 at 185 K.

7–56. Determine the interfacial heat transfer coefficient for saturated steam at

(i) 0°C.
(ii) 25°C.
(iii) 100°C.

Take $\sigma = 1.0$.

7–57. Saturated steam at 1000 Pa condenses on a vertical plate 12 cm high maintained at 275 K. Determine the condensation rate. Take $\sigma = 1.0$.

7–58. Saturated mercury vapor at 480 K condenses on a 2 cm–O.D., 1 mm–wall-thickness AISI 316 stainless steel horizontal condenser tube. Water at a bulk temperature of 460 K boils inside the tube, giving an inside heat transfer coefficient of 1.80×10^4 W/m^2 K.

(i) Draw the thermal circuit.
(ii) Estimate the heat transfer per unit length of tube.
(iii) Determine the fractional contribution of the interfacial resistance to the total thermal resistance.

7–59. When $\zeta = u/(2RT)^{1/2}$ is not very small, Eq. (7.114) is not accurate. R. W. Schrage [22] recommends the following form for higher condensation rates:

$$\dot{m}'' = \frac{\sigma}{1 - 0.523\sigma}\left(\frac{P_v}{(2\pi RT_v)^{1/2}} - \frac{P_s}{(2\pi RT_s)^{1/2}}\right) \qquad 0.001 < |\zeta| < 0.1$$

Evaluate the accuracy of Eq. (7.114) and this equation for $0 < \zeta < 0.1$ using the exact expression for $\mathcal{J}^-$ given by Eq. (7.111). Consider saturated mercury at 450 K with $\sigma = 1$ for this purpose.

7–60. Saturated mercury vapor at 450 K condenses on a vertical surface 1 cm high maintained at 449 K. Determine the condensation rate. Take $\sigma = 1.0$.

7–61. An ammonia heatpipe is constructed from a 2 cm–I.D. stainless steel tube and has an effective length of 1.0 m. The nickel foam (210-5) wick has a cross-sectional area of 6.0×10^{-5} m^2. Estimate the burnout heat load for an adiabatic section maintained at 300 K and an adverse tilt of

(i) 0°.
(ii) 0.5°.

Also calculate the heat transfer factor for the heatpipe.

7–62. A water heatpipe is constructed from a 3 cm–I.D. stainless steel tube and has an effective length of 0.70 m. The wick is made from 30 mesh screen and has a cross-sectional area of 10.0×10^{-5} m^2. Estimate the burnout heat load for an adiabatic section maintained at 500 K and an adverse tilt of

(i) 0°.
(ii) 2°.

7–63. Prepare a plot of the figure of merit $\mathcal{M}$ for heatpipe application over an appropriate temperature range for

(i) mercury.
(ii) R-12.
(iii) R-113.

7–64. A nitrogen heatpipe is constructed from a 2 cm–I.D. stainless steel tube and has an effective length of 0.50 m. The wick is made from nickel felt and has a cross-sectional area of 7×10^{-5} m^2. For an adiabatic section temperature in the vicinity of 85 K, prepare a plot of burnout heat load as a function of temperature. The heatpipe operates in a horizontal position.

7–65. A nitrogen-loaded methanol heatpipe is constructed from a 2 cm–I.D. stainless steel tube. A fibrous stainless wick is in the form of a 20 mm × 5 mm cross-section slab and extends into the gas reservoir. The condenser length is 40 cm, and the reservoir volume is 7×10^4 mm^3. The heatpipe is designed to be fully open for an adiabatic section temperature of 280 K, when the sink temperature is 260 K.

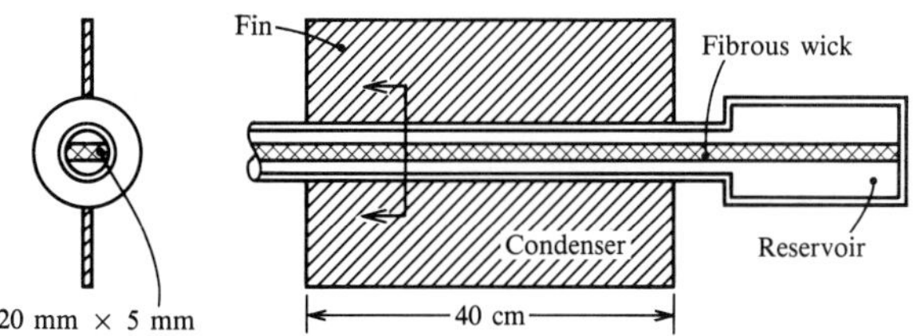

Determine the mass of gas in the heatpipe, and plot a graph of load versus adiabatic section temperature over an appropriate range of operation. At 5°C methanol has a vapor pressure of 5330 Pa and an enthalpy of vaporization of 1.18×10^6 J/kg.

7–66. A water heatpipe is 10 cm long and is made from a 2 cm–I.D. stainless steel tube. The wick consists of four layers of 30 mesh screen covered with one layer of 200 mesh screen, as shown in the sketch. Determine the maximum heat transport capability when the heat pipe is inclined at 10°, with the evaporator above the condenser, and the adiabatic section is at 100°C. The wire diameters of 30 mesh and 200 mesh screen are 0.86 mm and 0.126 mm, respectively.

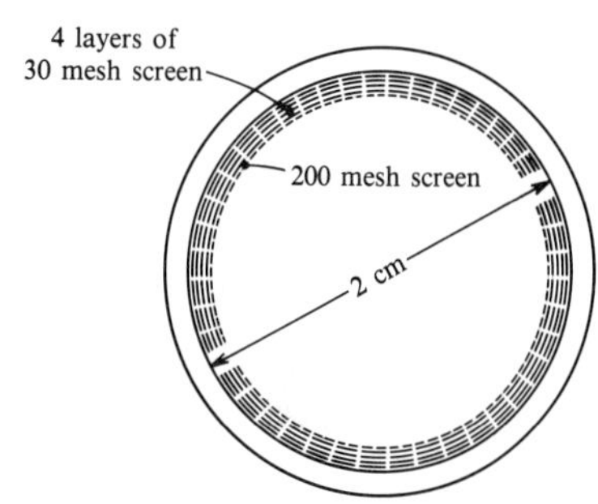

7–67. The condenser section of an ammonia heatpipe has three external radial fins per cm length. The fins are 5 cm in diameter, 0.3 mm thick, and made of aluminum. The pipe itself is aluminum with a 20 mm outside diameter, a 16 mm inside diameter, and a 0.5 mm–thick wick on the inner wall. The ammonia-filled wick has an effective thermal conductivity of 0.8 W/m K. If the outside convective heat transfer coefficient is 25 W/m^2 K, estimate the overall heat transfer coefficient from the condensing vapor to ambient cooling air. Make appropriate simplifying assumptions.

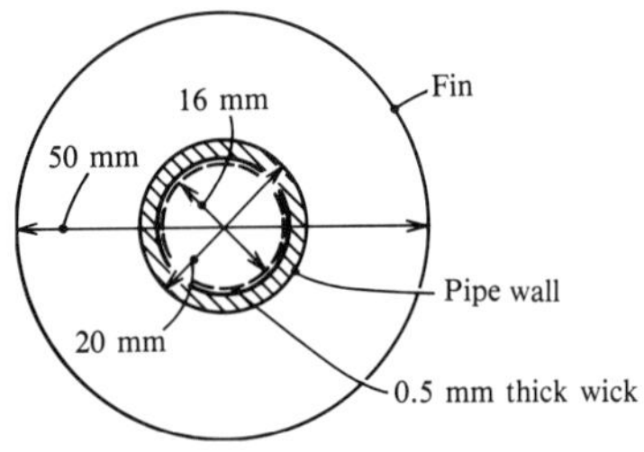

7–68. The schematic shows a method for cooling integrated circuits using a dielectric fluid, in this case R-113. The liquid R-113 boils on the surfaces of the ICs, and the vapor condenses on the condenser coil above the pool. An off-the-shelf unit was designed to operate at 1 atm pressure, and the condenser coil is a 2 m length of 10 mm–O.D., 8 mm–I.D. copper tube wound in a horizontal plane. A coolant water supply is available at a flow rate of 0.05 kg/s and at a temperature of 3°C. The ICs have a surface area of 24 mm^2 each and a maximum allowable operating temperature of 70°C.

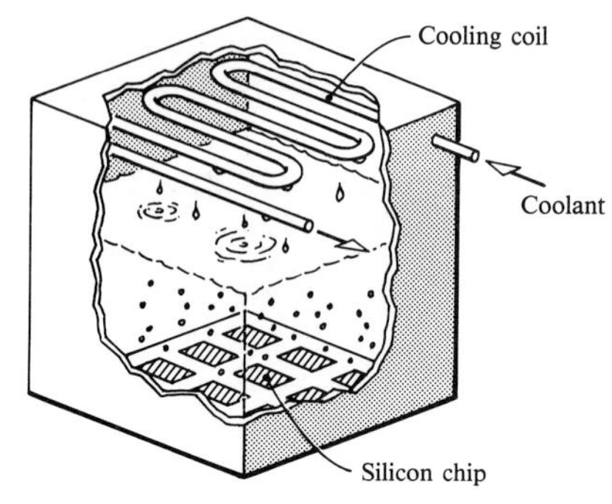

(i) Determine the rate at which heat can be absorbed by the coolant.
(ii) Determine the allowable power dissipation per circuit.
(iii) Hence specify the number of circuits that can be mounted in the unit.

Take $C_{nb} = 0.004$ and $m = 4.1$ in Rohsenow's nucleate boiling correlation.

7–69. An experimental rig for the study of nucleate boiling of R-113 on horizontal surfaces is shown in the sketch. The heater surface is the top of a 9 cm–diameter cylinder. The condenser coil is to be designed to allow steady operation at q_{max} for a system pressure of 0.105 MPa. Coolant water is available at 3°C, at a flow rate of 0.30 kg/s. If the coil is to be made from 1 cm–O.D., 1 mm–wall-thickness copper tubing wound in a horizontal plane, how long a tube will be required?

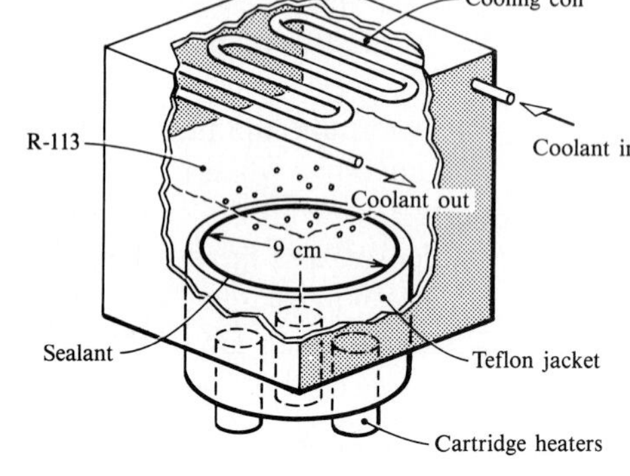

7–70. Thermosyphons (Perkins tubes) are to be used to transfer heat from a hot water stream flowing below a cold water stream. The tubes are 3 cm–O.D., 2 mm– wall-thickness stainless steel and are 2 m long. The lower and upper 50 cm lengths are immersed in the hot and cold streams, respectively, and the intervening 1 m length is insulated. At the location of concern the hot water stream is at 360 K, and the cold water stream is at 350 K. Both streams flow at 1 m/s. The working fluid in

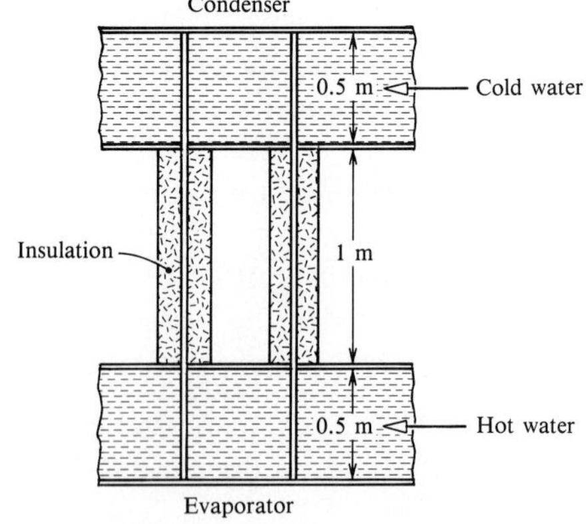

the tubes is water. At steady state, vapor condenses as a film in the condenser section of the tubes and evaporates as a film in the evaporator sections. The ends of the tubes can be taken to be adiabatic. Determine the rate of heat transfer per tube. Start-up of this device could be a problem: discuss.

7–71. A 19 mm–I.D., 25 mm–O.D. stainless steel Perkins tube is used to transfer heat from a hot stream of water at 360 K to colder water at 325 K. The hot stream flows at 0.3 m/s over the bottom 30 cm of the vertical tube, and the colder water exists under conditions of natural convection over the top 1 m of the 1.3 m–long tube.

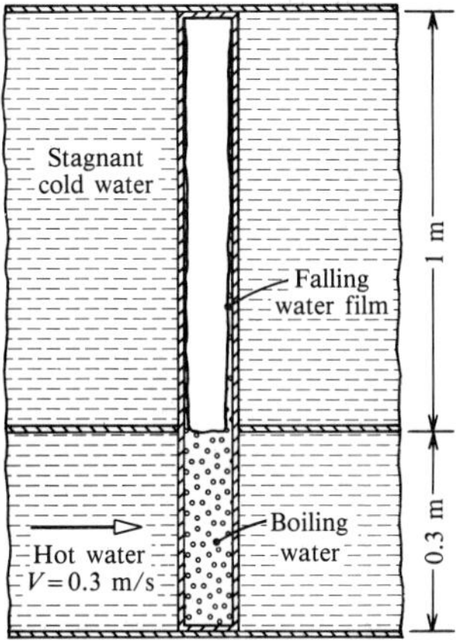

Within the tube is pure water and its vapor. The vapor condenses on the inside of the 1.0 m–long condenser section and falls as a film to maintain the level of water in the 0.3 m–long boiling section. What heat flow will the tube provide?

CHAPTER

8

HEAT EXCHANGERS

CONTENTS

8.1 INTRODUCTION

Heat exchangers were introduced in Section 1.6. Single- and two-stream exchangers were defined and explained, and a particular example of a single-stream exchanger—the steam condenser—was analyzed. Throughout the text we have referred to heat exchangers, since much of the heat transfer theory in Chapters 2 through 7 applies to heat exchangers. We are now in a position to apply this theory to the analysis and design of heat exchangers.

A great variety of heat exchangers are used. In Section 8.2, heat exchangers are classified with respect to flow configuration and heat transfer surface. In Section 8.3, the energy conservation principle is applied to an exchanger as a system to obtain *exchanger energy balances*. Also, the overall heat transfer coefficient concept is reviewed and extended. Section 8.4 presents an analysis of an evaporator, which parallels the analysis of a condenser in Section 1.6. The key section in this chapter is Section 8.5, in which two-stream steady-flow exchangers are analyzed, using both the *logarithmic mean temperature difference* and the *effectiveness-number of transfer units* formulations. **Regenerators** are unsteady two-stream exchangers in which the streams flow alternately through a matrix of substantial heat storage capacity. An example of regenerator analysis is given in Section 8.6.

Sections 8.3 through 8.6 deal with the **thermal analysis** of heat exchangers, that is, methods for calculating the heat transfer in the exchanger and the outlet fluid temperatures. Section 8.7 introduces the student to the broader subject of *heat exchanger design*. Two aspects of exchanger design are emphasized. First, the role played by pressure drop of the flow streams is examined, which leads to the concept of a **thermal-hydraulic design**: a heat exchanger is sized to give a required heat transfer performance, subject to a pressure drop constraint on one or both streams. Particular attention is given to **compact heat exchangers**, which have a large heat transfer surface area per unit volume. Second, some economic aspects of exchanger utilization are examined. In particular, operating costs are weighed against capital costs for waste heat recovery application.

Two computer programs accompany Chapter 8. HEX1 calculates thermal performance for nine flow configurations of steady two-stream exchangers. HEX2 introduces the student to computer-aided heat exchanger design. HEX2 performs the thermal-hydraulic design of a particular class of exchangers and also gives an economic evaluation of the resulting design.

8.2 TYPES OF HEAT EXCHANGERS

A great variety of heat exchangers are used in engineering practice. The geometry of the flow configuration, the type of heat transfer surface, and the materials of construction all vary according to the design requirements.

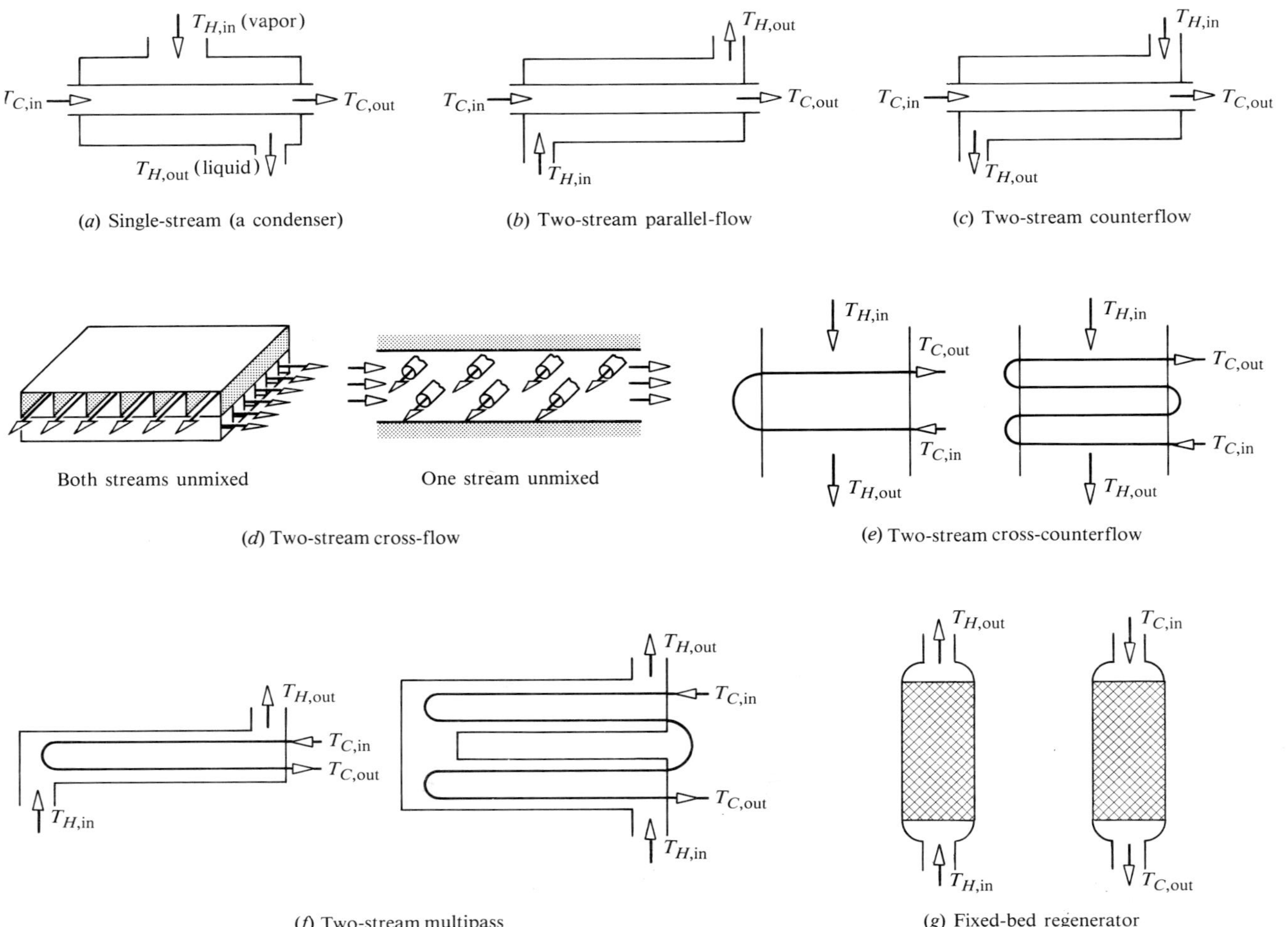

Figure 8.1 Schematics of common heat exchanger geometric flow configurations.

8.2.1 Geometric Flow Configurations

There are a number of possible geometric flow configurations for exchangers, the most important of which are the following.

Single-stream. The single-stream configuration was introduced in Section 1.6 and was defined as an exchanger in which the temperature of only one fluid changes; the direction in which the fluid flows is then immaterial. Examples include simple condensers and boilers. A simple condenser is illustrated in Fig. 8.1*a*.

Parallel-flow two-stream. The two fluids flow parallel to each other in the same direction. In its simplest form, this type of exchanger consists of two coaxial tubes, as shown in Fig. 8.1*b*. In practice, a large number of tubes are located in a shell to form what is known as a *shell-and-tube* exchanger, as shown in Fig. 8.2. The shell-and-tube type is used most often for liquids and for high pressures. The *plate* type, shown in Fig. 8.3, consists of multiple plates separated by gaskets and is more suitable for gases at low pressures. This configuration is often alternatively termed a *cocurrent* exchanger.

Counterflow two-stream. The fluids flow parallel to each other in opposite directions. A simple coaxial tube type is shown in Fig. 8.1*c*, but, as with the cocurrent case, shell-and-tube or plate exchangers are most commonly used. We shall see that, for a given number of transfer units, the effectiveness of a counterflow exchanger is higher than that of a parallel-flow exchanger. Hence, counterflow exchangers are preferred in practice. Examples include feed water preheaters for boilers, and oil coolers for aircraft. This configuration is often alternatively termed a *countercurrent* exchanger.

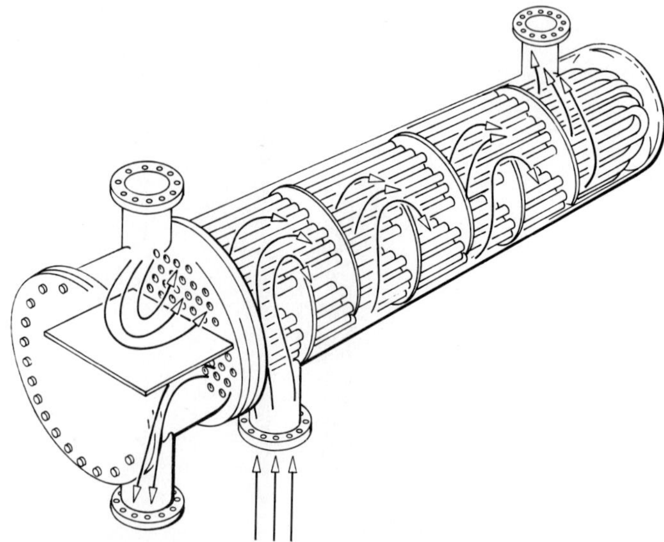

Figure 8.2 A two-tube-pass, one-shell-pass, shell-and-tube heat exchanger. The first tube pass gives parallel flow, and the second gives counterflow.

Figure 8.3 A plate-type heat exchanger during assembly. (Photograph courtesy of the Paul Mueller Company, Springfield, Missouri.)

Cross-flow two-stream. The two streams flow at right angles to each other, as shown in Fig. 8.1*d*. The hot stream may flow inside tubes arranged in a *bank* or *bundle*, and the cold stream may flow through the bank in a direction generally at right angles to the tubes. Either one or both of the streams may be *unmixed* as shown. This configuration is intermediate in effectiveness between parallel-flow and counterflow exchangers, but it is often simpler to construct owing to the relative simplicity of the inlet and outlet flow ducts. A common example is the automobile radiator shown in Fig. 8.4.

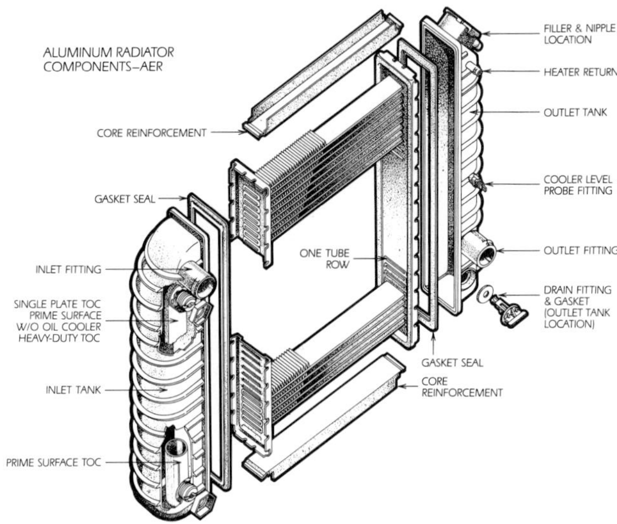

Figure 8.4 An automobile radiator. (Courtesy Harrison Radiator Division, General Motors, Lockport, New York.)

Cross-counterflow two-stream. In practice, exchanger flow configurations often approximate the idealizations shown in Fig. 8.1*e*; *two-* and *four-pass* types are shown, although more passes can be used. (In a two-pass exchanger, the tubes pass through the shell twice.) As the number of passes increases, the effectiveness approaches that of an ideal countercurrent exchanger.

Multipass two-stream. When the tubes of a shell-and-tube exchanger double back one or more times inside the shell, as shown in Fig. 8.1*f*, some passes are parallel flow while others are counterflow. The two-pass exchanger of this type is popular since only one end of the exchanger requires perforation to lead the tubes in and out, as shown schematically in Fig. 8.2.

Regenerators. The configurations discussed so far all involve steady flows and temperatures: such exchangers are traditionally called *recuperators*. On the other hand, in a *regenerator*, the two streams flow alternately through a *matrix* of substantial heat storage capacity. Heat transferred from the hot fluid is stored in the matrix, making its temperature rise, and is subsequently given up to the cold fluid when it, in turn, flows through the matrix, causing its temperature to fall. Regenerators can also have parallel-, counter-, or cross-flow configurations. The matrix may be in the form of a bed, as shown schematically in Fig. 8.1*g*, or in the form of a rotating wheel, as shown in Fig. 8.5. The bed

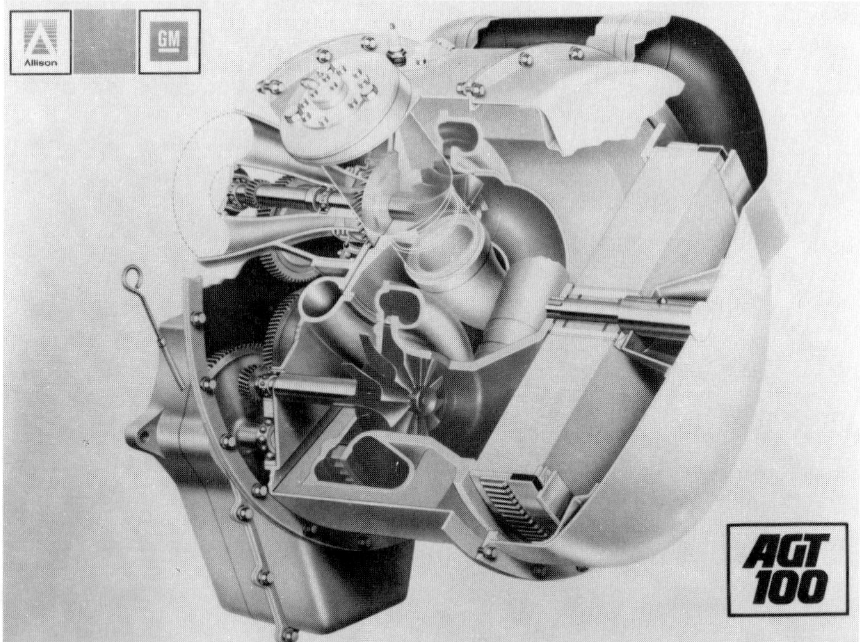

Figure 8.5 A rotary regenerator used to preheat combustion air for a gas turbine: the AGT 100 automotive engine. (Photograph courtesy of Allison Gas Turbine, Division of General Motors. Development sponsored by the U.S. Department of Energy and NASA–Lewis Research Center.)

type has been used extensively in high-temperature applications, such as for the preheating of combustion air by combustion products using a matrix of refractory bricks.

The configurations just described are often only approximated in practice but are useful idealizations, or *models*, that can be analyzed to obtain an understanding of the essential features of heat exchanger thermal behavior. However, often a configuration is encountered for which a simple idealization is not feasible; in such cases, purely empirical data must be used for its thermal design. Figure 8.6 shows a commonly encountered example consisting of a steam coil used to heat a tank of oil. If there is no stirrer in the tank, natural convection will tend to give a cross-counterflow of oil up the steam coil, but the magnitude of this flow is difficult to estimate.

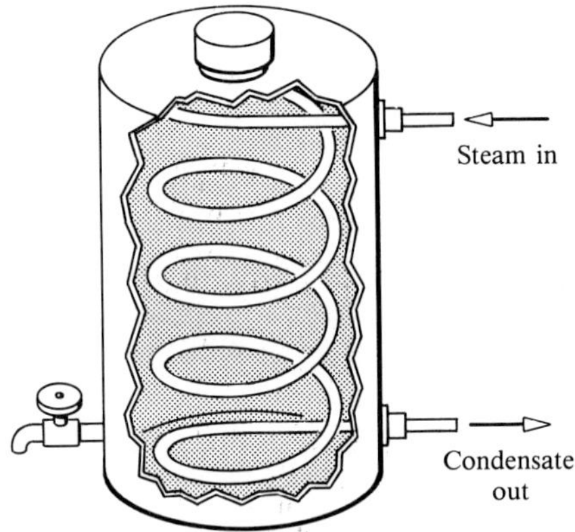

Figure 8.6 An immersed coil heater.

8.2.2 Fluid Temperature Behavior

Associated with the geometric flow patterns are characteristic temperature variations within the exchanger, as is shown for the simpler cases in Fig. 8.7. The subscript H is used to denote the hot stream, and C is used to denote the cold stream. In Fig. 8.7*a*, which depicts a simple condenser, T_H is constant while T_C increases along the exchanger. In the parallel-flow two-stream exchanger shown in Fig. 8.7*b*, the temperature difference for heat transfer $(T_H - T_C)$ decreases along the exchanger in the flow direction. Also, it is clear that the outlet temperature of the cold stream cannot exceed that of the hot stream, that is, $T_{C,\text{out}} < T_{H,\text{out}}$. In the counterflow two-stream exchanger, shown in Fig. 8.7*c*, the temperature difference $(T_H - T_C)$ may decrease or increase along the exchanger, or, as a special case, it may be constant. Notice that the outlet temperature of the cold stream can exceed the outlet temperature of the hot stream. The temperature patterns in cross-flow exchangers are more complex since the temperature varies in two directions, as shown for the cold stream in Fig. 8.7*d*. In a regenerator, the outlet temperatures vary with time, as shown in Fig. 8.7*e*. Often there is a combination of these temperature variation patterns in a single exchanger, as is the case for the parallel-flow steam generator shown in Fig. 8.7*f*.

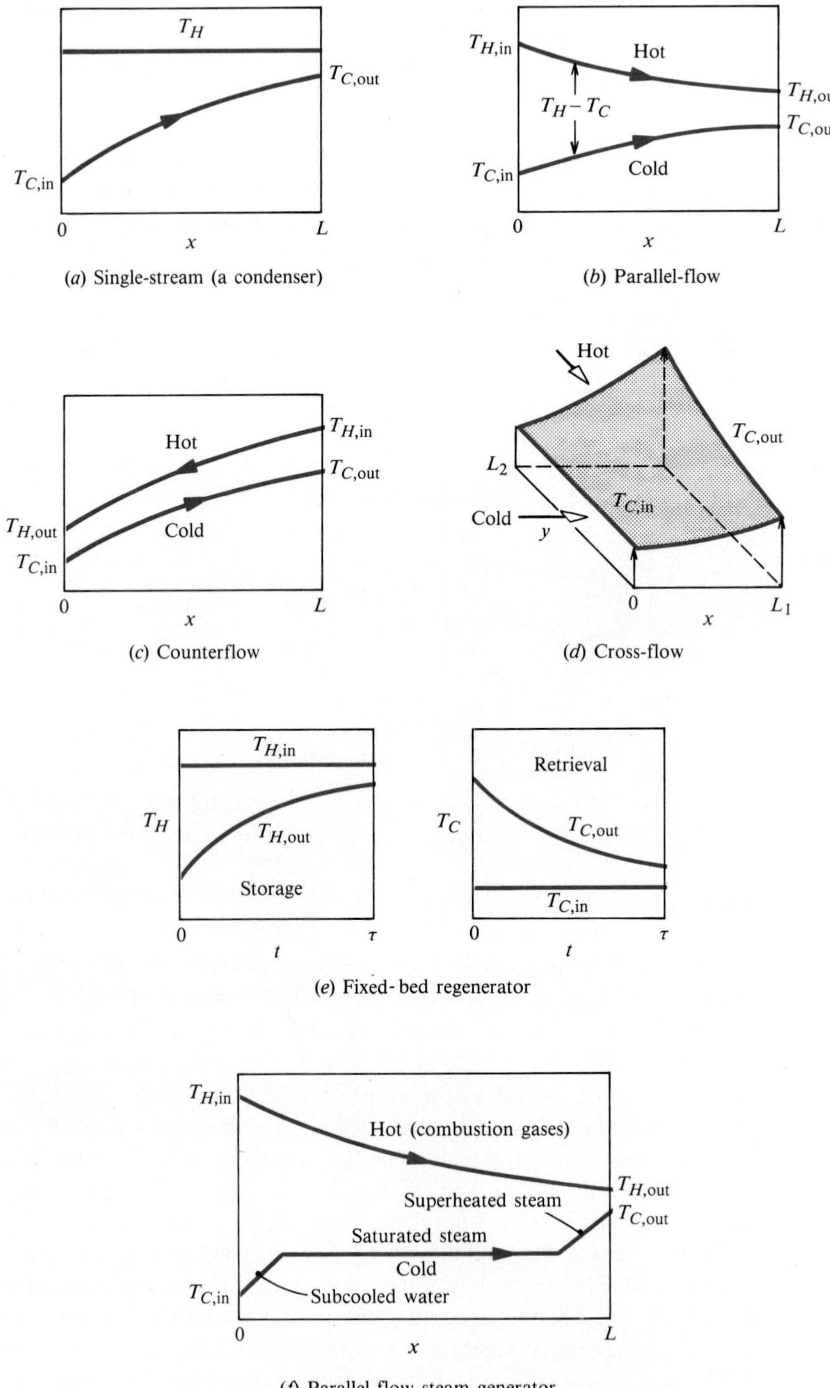

Figure 8.7 Characteristic fluid temperature variations for various exchanger configurations.

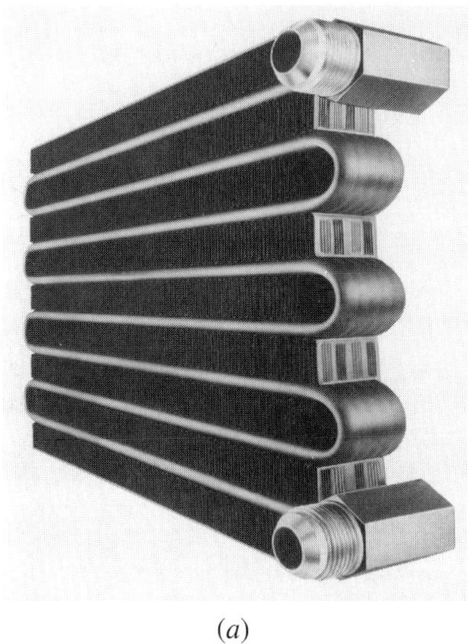

(*a*)

(*b*)

Figure 8.8 A cross-flow oil cooler with louvered fins on the air side and straight and triangular fins on the oil side. (Photograph courtesy of the Modine Manufacturing Company, Standard Industrial Products, Racine, Wisconsin.)

8.2.3 Heat Transfer Surfaces

The examples of heat exchangers shown in Figs. 8.2 through 8.6 show that the heat transfer surface can take many forms. The most common form is a plain tube, which is usually circular and straight but can be bent or coiled. When the heat transfer resistance on one side of the tube is much larger than that on the other, as is the case for a gas-to-liquid heat exchanger, the gas-side surface may be finned to increase the effective heat transfer area. Usually the fins are on the outside of the tube, but sometimes it is the internal surface that is provided with fins; an example is shown in Fig. 8.8. The manner in which fins are provided in plate- and compact-type heat exchangers can be quite complicated. Design engineers have shown considerable ingenuity in providing fins that increase the heat transfer area without incurring excessive pressure drop penalties, and that are inexpensive to fabricate. The automobile radiator in Fig. 8.4 is a good example.

8.2.4 Direct-Contact Exchangers

To complete this discussion of types of heat exchangers, it is appropriate to briefly mention an additional type of exchanger, in which there is *direct* contact between the two fluids. Direct-contact exchangers are more commonly used for mass transfer and simultaneous heat and mass transfer. A simple form of direct-contact heat exchanger is a water heater in which steam is bubbled through the water: the steam condenses and adds to the water inventory. Two-stream direct-contact heat exchangers may involve transfer between two immiscible liquids such as oil and water, as shown in Fig. 8.9. The expense of a heat transfer surface and surface-fouling problems are avoided, but the separation of the two liquids at each end of the exchanger is often difficult.

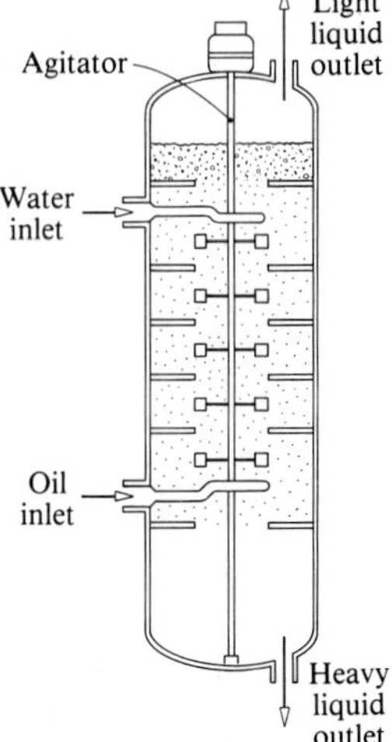

Figure 8.9 An oil-to-water direct-contact heat exchanger.

8.3 ENERGY BALANCES AND THE OVERALL HEAT TRANSFER COEFFICIENT

Before commencing the analysis of heat exchanger performance, we need to develop some supporting concepts. First, we will apply the steady-flow energy equation to the complete exchanger, in order to relate outlet stream temperatures to inlet stream temperatures. Next, the overall heat transfer coefficient concept introduced in Chapters 1 and 2 will be extended to include the effects of deposits fouling a heat exchanger tube, and to include the effect of augmenting the heat transfer surface with fins.

8.3.1 Exchanger Energy Balances

Figure 8.10 shows a coaxial-tube parallel-flow heat exchanger. The steady-flow energy equation is applied to a control volume enclosing the complete exchanger, as shown in the figure. It reduces to an enthalpy balance since no external work is done, there is no heat transfer into the system if the exchanger is well insulated, and changes in kinetic and potential energy are usually negligible. Denoting the hot and cold streams with subscripts H and C, respectively,

$$(\dot{m}_H h_H + \dot{m}_C h_C)_{x=0} = (\dot{m}_H h_H + \dot{m}_C h_C)_{x=L}$$

or

$$\dot{m}_H (h_{H,0} - h_{H,L}) = \dot{m}_C (h_{C,L} - h_{C,0}) = \dot{Q} \tag{8.1}$$

where $\dot{Q}$ is the heat transferred from the hot stream to the cold stream. If the specific heats can be assumed constant, then

$$(\dot{m}c_p)_H (T_{H,0} - T_{H,L}) = (\dot{m}c_p)_C (T_{C,L} - T_{C,0}) = \dot{Q} \tag{8.2}$$

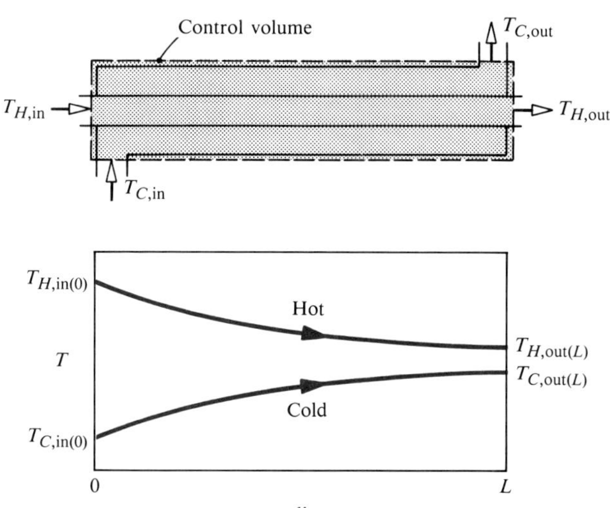

Figure 8.10 Control volume for an energy balance on a parallel-flow exchanger and the expected fluid temperature variations along the exchanger.

If, instead, the inlet and outlet states of each stream are denoted "in" and "out," respectively, then Eqs. (8.1) and (8.2) become

$$\dot{m}_H(h_{H,\text{in}} - h_{H,\text{out}}) = \dot{m}_C(h_{C,\text{out}} - h_{C,\text{in}}) = \dot{Q} \tag{8.3}$$

$$(\dot{m}c_p)_H(T_{H,\text{in}} - T_{H,\text{out}}) = (\dot{m}c_p)_C(T_{C,\text{out}} - T_{C,\text{in}}) = \dot{Q} \tag{8.4}$$

For a counterflow exchanger, it is convenient to adopt the convention that the flow rates $\dot{m}_H$ and $\dot{m}_C$ are positive, irrespective of direction. Then the **exchanger energy balance** in the form of Eqs. (8.3) and (8.4) applies to the counterflow configuration as well. Note also that, for multitube heat exchangers, it is usual to treat the exchanger as a whole rather than treating each tube separately; hence, $\dot{m}$ [kg/s] is the total flow rate through the exchanger of a given stream.

For a simple condenser that condenses saturated vapor, Eq. (8.3) becomes

$$\dot{m}_H h_{\text{fg}\,H} = (\dot{m}c_p)_C(T_{C,\text{out}} - T_{C,\text{in}}) \tag{8.5}$$

as was shown in Section 1.6; and for a simple evaporator generating saturated vapor,

$$(\dot{m}c_p)_H(T_{H,\text{in}} - T_{H,\text{out}}) = \dot{m}_C h_{\text{fg}\,C} \tag{8.6}$$

These exchanger energy balances, Eqs. (8.1)–(8.6), are also often called *overall* energy balances or, in the chemical engineering literature, *macroscopic* energy balances. Their use is straightforward: for example, in using Eq. (8.4), the flow rates, inlet temperatures, and one outlet temperature may be specified, and the unknown outlet temperature can then be found. Often they are used to find the maximum possible heat transfer in a given situation and to determine whether a specified goal is attainable. An example of this use follows.

EXAMPLE 8.1 Cooling Water Supply for a Steam Turbine Condenser

To obtain the required power output, a steam turbine must consume 11.5 kg/s of steam and exhaust to a back pressure of 4714 Pa. Will 300 kg/s of cooling water from a river at 290 K be sufficient to condense the steam?

Solution

Given: Cooling water supply.

Required: Evaluate adequacy for a steam turbine condenser.

From Table A.12a, at 4714 Pa, the saturation temperature T_H is 305.0 K and the enthalpy of vaporization h_{fgH} is 2.425×10^6 J/kg. The maximum possible outlet temperature of the cooling water is then 305.0 K and can be achieved in an infinitely long exchanger. Thus, the maximum energy that can be absorbed by the cooling water is

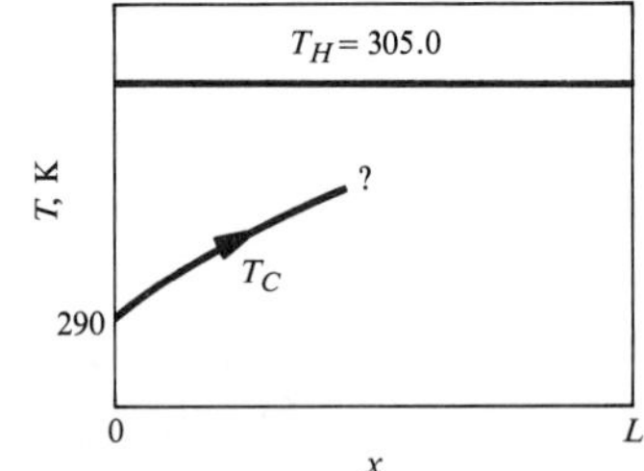

$$\begin{aligned}\dot{Q} &= (\dot{m}c_p)_C[T_{C,\text{out}}(\text{max}) - T_{C,\text{in}}] \\ &= (300)(4180)(305 - 290) \\ &= 1.88 \times 10^7 \text{ W}\end{aligned}$$

On the other hand, the enthalpy of vaporization to be given up by the condensing steam is

$$\dot{Q} = \dot{m}_H h_{fg\,H} = (11.5)(2.425 \times 10^6) = 2.79 \times 10^7 \text{ W}$$

which is greater than the amount that can be absorbed by the cooling water.

Comments

A larger flow of cooling water is required.

8.3.2 Overall Heat Transfer Coefficients

The overall heat transfer coefficient was introduced in Sections 1.4.1 and 2.3.1 for heat transfer across plane walls and cylindrical tube walls, respectively. Most of the heat exchangers discussed in Section 8.2 involve transfer of heat from one fluid to another across a plate or tube wall, with tube walls predominating. Figure 8.11 shows the cross section of a typical heat exchanger tube and the corresponding temperature profile and thermal circuit. For a clean tube, the product of overall heat transfer coefficient times perimeter is obtained from Eq. (2.17) by setting $A = \mathcal{P}L$ and canceling L throughout:

$$\frac{1}{U\mathcal{P}} = \frac{1}{h_{c,i}2\pi r_i} + \frac{\ln(r_o/r_i)}{2\pi k} + \frac{1}{h_{c,o}2\pi r_o} \quad \textbf{(8.7)}$$

where subscripts i and o denote the tube inside and outside walls, respectively. For an element of exchanger Δx long, the heat transfer rate is

$$\Delta\dot{Q} = U\mathcal{P}\,\Delta x(T_H - T_C) \quad \textbf{(8.8)}$$

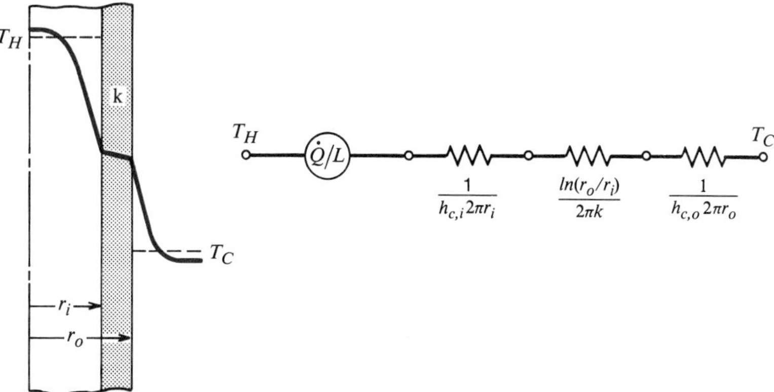

Figure 8.11 A local temperature profile and thermal circuit for heat flow through an exchanger tube. It is understood that the convective heat transfer coefficients may be average values: the bar notation is omitted consistent with our practice in Chapters 1–3.

The perimeter of the $U\mathscr{P}$ product in Eq. (8.7) can be chosen to be either $2\pi r_i$ or $2\pi r_o$. The corresponding value of U changes accordingly, since Eq. (8.7) specifies the $U\mathscr{P}$ product and not U itself. Often $h_{c,i}$ and $h_{c,o}$ are not accurately known, or the tube wall is relatively thin, so that a precise specification of $\mathscr{P}$ is unnecessary.

After a heat exchanger has been in service for some time, deposits may form on the heat transfer surfaces. Examples include soot deposits in furnace tubes, calcium or magnesium sulfate deposits in fresh water heaters, and rust. The heat transfer surfaces are then said to be *fouled:* Fig. 8.12 shows an extreme example. If the average thickness δ_f and thermal conductivity k_f of the deposit were known, one could estimate the additional thermal resistance per unit area as $R_f = \delta_f/k_f$. Generally,

Figure 8.12 An economizer tube fouled in a high-sulfur no. 6 fuel-oil exhaust. The deposits have a relatively high sulfur content, and the corroded fins are typical of sulfuric acid deposition. [Photograph courtesy of Dr. W. Marner, Jet Propulsion Laboratory, California Institute of Technology, Pasadena. See also W. J. Marner, "Gas-side fouling," *Mechanical Engineering,* 108, 3, 70–77 (March 1986).]

however, the extent of fouling is known only through a measured reduction in heat transfer performance of an exchanger, which is then attributed to a reduced value of the overall heat transfer coefficient. If U is the overall heat transfer coefficient for an unfouled heat exchanger, we can write

$$\frac{1}{U_f\mathscr{P}} = \frac{1}{U\mathscr{P}} + \frac{R_{fH}}{\mathscr{P}_H} + \frac{R_{fC}}{\mathscr{P}_C} \tag{8.9}$$

where U_f is the overall heat transfer coefficient for the fouled exchanger, and R_{fH} and R_{fC} are the hot stream and cold stream **fouling resistance**, respectively. Table 8.1 gives some representative values for fouling resistance per unit area. Clearly, the time-dependent nature of the fouling problem is such that it is very difficult to reliably estimate U values if fouling resistances are dominant.

Often one side of a tube wall is finned—for example, the air side of a water-air heat exchanger. The effect of the fins is conveniently included in the overall heat transfer coefficient through use of the fin efficiency. Thus, the overall heat transfer

Table 8.1 Recommended values of fouling resistances for heat exchanger design.

Fluid	Fouling Resistance, R_f $[\text{W/m}^2\ \text{K}]^{-1}$
Fuel oil	0.005
Transformer oil	0.001
Vegetable oils	0.003
Light gas oil	0.002
Heavy gas oil	0.003
Asphalt	0.005
Gasoline	0.001
Kerosene	0.001
Caustic solutions	0.002
Refrigerant liquids	0.001
Hydraulic fluid	0.001
Molten salts	0.0005
Engine exhaust gas	0.01
Steam (non–oil-bearing)	0.0005
Steam (oil-bearing)	0.001
Refrigerant vapors (oil-bearing)	0.002
Compressed air	0.002
Acid gas	0.001
Solvent vapors	0.001
Seawater	0.0005–0.001
Brackish water	0.001–0.003
Cooling tower water (treated)	0.001–0.002
Cooling tower water (untreated)	0.002–0.005
River water	0.001–0.004
Distilled or closed-cycle condensate water	0.0005
Treated boiler feedwater	0.0005–0.001

Table 8.2 Approximate overall heat transfer coefficients.

Heat Exchanger Duty	U, W/m^2 K
Gas to gas	10–30
Water to gas (e.g., gas cooler, gas boiler)	10–50
Condensing vapor-air (e.g., steam radiator, air heater)	5–50
Steam to heavy fuel oil	50–180
Water to water	800–2500
Water to other liquids	200–1000
Water to lubricating oil	100–350
Light organics to light organics	200–450
Heavy organics to heavy organics	50–200
Air-cooled condensers	50–200
Water-cooled steam condensers	1000–4000
Water-cooled ammonia condensers	800–1400
Water-cooled organic vapor condensers	300–1000
Steam boilers	10–40+ radiation
Refrigerator evaporators	300–1000
Steam-water evaporators	1500–6000
Steam-jacketed agitated vessels	150–1000
Heating coil in vessel, water to water	
Unstirred	50–250
Stirred	500–2000

coefficient for a clean tube is given by

$$\frac{1}{U\mathscr{P}} = \frac{1}{h_{c,i}2\pi r_i} + \frac{\ln(r_o/r_i)}{2\pi k} + \frac{1}{h_{c,o}(A_f/L)\eta_f + h_{c,o}A_p/L} \qquad \textbf{(8.10)}$$

where (A_f/L) is the surface area of fins per unit length of tube, η_f is the fin efficiency, and A_p/L is the *prime* (unfinned) surface area per unit length of tube. Some typical overall heat transfer coefficient values are listed in Table 8.2.

EXAMPLE 8.2 Overall Heat Transfer Coefficient for a Condenser

A brass condenser tube has a 30 mm outer diameter and 2 mm wall thickness. Sea water enters the tube at 290 K, and saturated low-pressure steam condenses on the outside of the tube. The inside and outside heat transfer coefficients are estimated to be 4000 and 8000 W/m^2 K, respectively, and a fouling resistance of 10^{-4} (W/m^2 K)$^{-1}$ on the water side is expected. Estimate the overall heat transfer coefficient based on inside area.

Solution

Given: Brass condenser tube with water-side fouling.

Required: Overall heat transfer coefficient U based on inside area.

Assumptions: The tube wall temperature is $\sim$300 K.

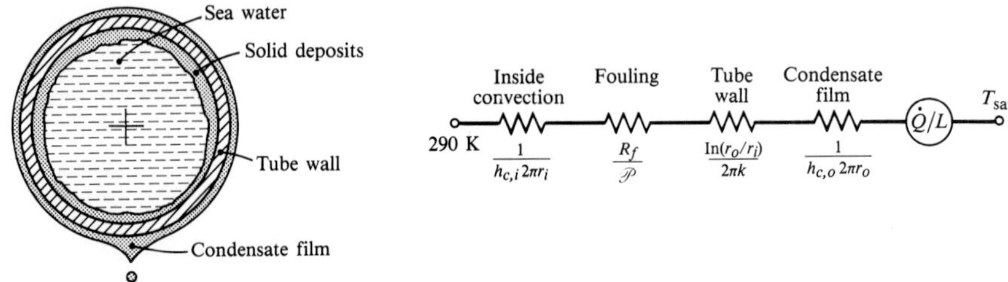

From Table A.1*a*, take k for brass as 111 W/m K. Equation (8.7) gives the $U\mathscr{P}$ product for a clean tube:

$$\frac{1}{U\mathscr{P}} = \frac{1}{h_{c,i}2\pi r_i} + \frac{\ln(r_o/r_i)}{2\pi k} + \frac{1}{h_{c,o}2\pi r_o}$$

$$= \frac{1}{(4000)(2\pi)(0.013)} + \frac{\ln(0.015/0.013)}{2\pi(111)} + \frac{1}{(8000)(2\pi)(0.015)}$$

$$= 10^{-3}(3.06 + 0.21 + 1.33) = 4.60 \times 10^{-3} \text{ (W/m K)}^{-1}$$

The inside perimeter is

$$\mathscr{P}_i = 2\pi r_i = (2\pi)(0.0130) = 0.0817 \text{ m}$$

Hence,

$$1/U = (0.0817)(4.60 \times 10^{-3}) = 3.76 \times 10^{-4} \text{ (W/m}^2\text{ K)}^{-1}; \qquad U = 2660 \text{ W/m}^2\text{ K}$$

Then, from Eq. (8.9) for the fouled tube,

$$\frac{1}{U_f} = \frac{1}{U} + R_{fC} = 10^{-4}(3.76 + 1); \qquad U_f = 2100 \text{ W/m}^2\text{ K}$$

Comments

1. The fouling reduces the overall heat transfer coefficient by 21%.
2. Due to fouling, the use of $2\pi r_i$ in the inside convective resistance may be inappropriate.

EXAMPLE 8.3 Overall Heat Transfer Coefficient of a Finned Tube

An aluminum tube has a 3 cm outside diameter and a wall thickness of 2 mm. It has 100 pin fins per cm length, of 1.5 mm diameter and 4 cm long. The inside and outside heat transfer coefficients are 5000 and 7 W/m^2 K, respectively. Calculate the overall heat transfer coefficient–perimeter product. Take $k = 204$ W/m K for the aluminum.

Solution

Given: Aluminum tube with pin fins.

Required: The $U\mathscr{P}$ product.

Assumptions: The outside heat transfer coefficient is constant over the fins and bare surface.

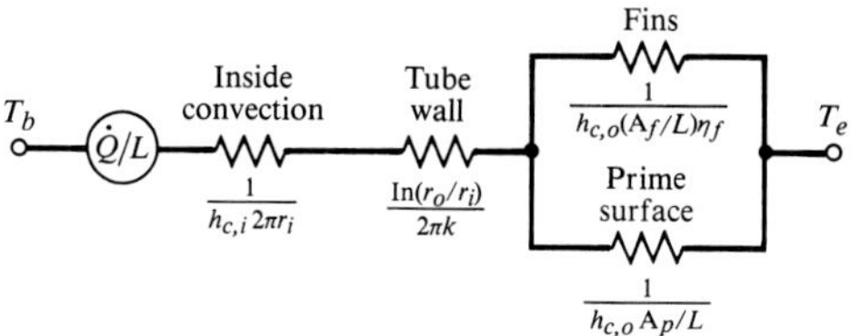

Equation (8.10) gives the $U\mathscr{P}$ product:

$$\frac{1}{U\mathscr{P}} = \frac{1}{h_{c,i}2\pi r_i} + \frac{\ln(r_o/r_i)}{2\pi k} + \frac{1}{h_{c,o}(A_f/L)\eta_f + h_{c,o}A_p/L}$$

If the pin fins have diameter d and length l,

$$(A_f/L) = (10{,}000)(\pi dl) = (10{,}000)(\pi)(0.0015)(0.04) = 1.885\,\text{m}$$

$$A_p/L = (2\pi r_o)(1) - (10{,}000)(\pi d^2/4) = (2)(\pi)(0.015) - (10{,}000)(\pi/4)(0.0015)^2$$

$$= 0.07658\text{ m}$$

$$\beta = \left(\frac{h_c\mathscr{P}}{kA_c}\right)^{1/2}; \qquad \text{for one fin,} \quad \frac{\mathscr{P}}{A_c} = \frac{\pi d}{\pi d^2/4} = \frac{4}{d} = \frac{4}{(0.0015)} = 2667\text{ m}^{-1}$$

$$\beta = \left(\frac{(7)(2667)}{204}\right)^{1/2} = 9.57\text{ m}^{-1}; \qquad \beta L = (9.57)(0.04) = 0.383$$

$$\eta_f = (1/\beta L)\tanh\beta L = (1/0.383)(0.366) = 0.956$$

$$\frac{1}{U\mathscr{P}} = \frac{1}{(5000)(2\pi)(0.013)} + \frac{\ln(0.015/0.013)}{(2\pi)(204)} + \frac{1}{(7)(1.885)(0.956) + (7)(0.07658)}$$

$$= 10^{-3}(2.45 + 0.11 + 76.04)$$

$$U\mathscr{P} = 12.7\text{ W/m K.}$$

Comments

1. Note that the symbol $\mathscr{P}$ has been used for both the tube perimeter and the fin perimeter.
2. Notwithstanding the presence of the pin fins, the inside and wall resistances are negligible.

8.4 SINGLE-STREAM STEADY-FLOW HEAT EXCHANGERS

In Section 1.6.1, single-stream exchangers were defined as exchangers along which the temperature of only one stream varies. In Section 1.6.2, the steam condenser was considered as an example of a single-stream exchanger, and an analysis was made to obtain its performance in terms of effectiveness and number of transfer units. We now present a similar analysis for an evaporator.

8.4.1 Analysis of an Evaporator

Figure 8.13*a* shows a simple single-tube evaporator, or *boiler*. Saturated liquid enters the shell from the bottom, and vapor leaves from the top. Hot combustion gases flow through the tubes. The enthalpy of vaporization is transferred by convection from the hot gases to the inner tube wall, by conduction across the tube wall, and by a complex boiling process into the liquid. Figure 8.13*a* also shows the temperature variation of the combustion gases along the tube, and Fig. 8.13*b* shows the temperature variation across the tube wall. In Chapter 7, we defined boiling heat transfer coefficients with respect to the saturation temperature of the liquid, T_{sat}; T_{sat} is obtained from steam tables as the saturation temperature corresponding to the pressure maintained in the evaporator shell. The vapor flow rate is denoted $\dot{m}_C$ and the combustion gas flow rate $\dot{m}_H$ (the cold and hot streams, respectively). The exchanger energy balance was derived in Section 8.3.1 as Eq. (8.6):

$$(\dot{m}c_p)_H(T_{H,\text{in}} - T_{H,\text{out}}) = \dot{m}_C h_{\text{fg}\,C}$$

When the gas flow rate $\dot{m}_H$ and inlet temperature $T_{H,\text{in}}$ are known, Eq. (8.6) relates the outlet gas temperature, $T_{H,\text{out}}$, to the amount of vapor produced, $\dot{m}_C$.

To obtain the variation of the gas temperature along the exchanger, we make an energy balance on a differential element of the exchanger Δx long. When the steady-flow energy equation, Eq. (1.4), is applied to the control volume Δx long (shown

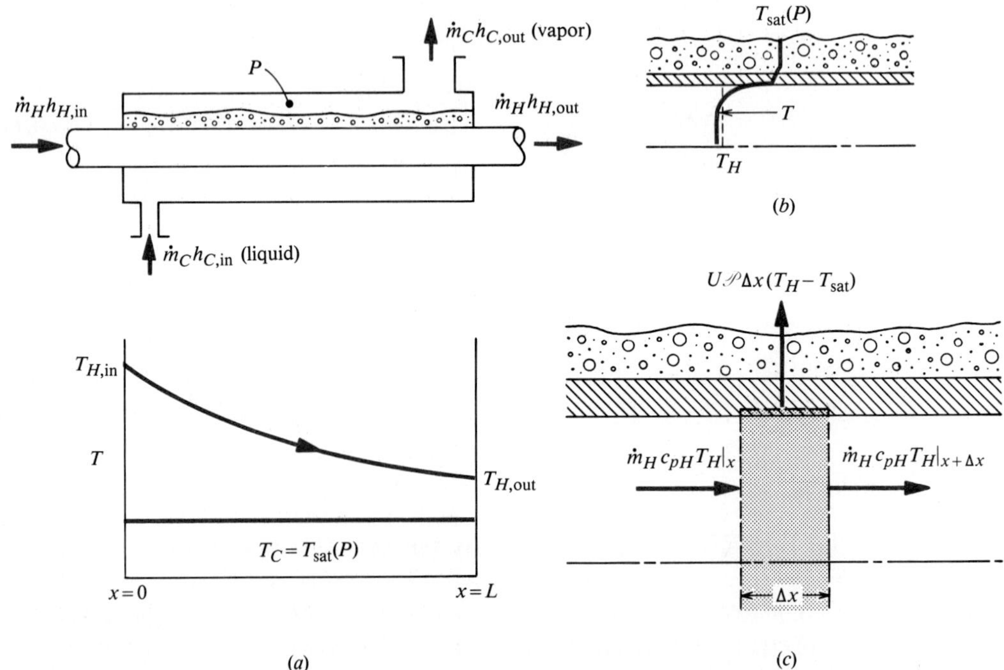

Figure 8.13 (*a*) A single-tube evaporator. (*b*) Temperature variation across an evaporator tube wall. (*c*) Elemental control volume for application of the steady-flow energy equation.

in Fig. 8.13*c* as a dotted line), the contribution due to *x*-direction conduction in the gas is small and can be neglected. The heat transfer across the tube wall must equal the gas flow rate times its enthalpy decrease: for the gas specific heat assumed to be constant,

$$U\mathscr{P}\Delta x(T_H - T_{\text{sat}}) = \dot{m}_H c_{pH}(T_H|_x - T_H|_{x+\Delta x}) \tag{8.11}$$

where U is the overall heat transfer coefficient for heat transfer from the gas to the steam, and $\mathscr{P} = \pi D$ is the perimeter of the tube wall (U and $\mathscr{P}$ must be based on the same diameter, either the I.D. or the O.D.). Dividing Eq. (8.11) by Δx, letting $\Delta x \to 0$, and rearranging,

$$\frac{dT_H}{dx} + \frac{U\mathscr{P}}{\dot{m}_H c_{pH}}(T_H - T_{\text{sat}}) = 0 \tag{8.12}$$

Equation (8.12) requires one boundary condition, which is

$$x = 0: \quad T_H = T_{H,\text{in}} \tag{8.13}$$

Integrating Eq. (8.12) and using Eq. (8.13) to evaluate the constant of integration gives

$$T_H - T_{\text{sat}} = (T_{H,\text{in}} - T_{\text{sat}})e^{-(U\mathscr{P}/\dot{m}_H c_{pH})x} \tag{8.14}$$

which shows an exponential decrease for $T_H(x)$. The outlet gas temperature is obtained by letting $x = L$ in Eq. (8.14):

$$T_{H,\text{out}} - T_{\text{sat}} = (T_{H,\text{in}} - T_{\text{sat}})e^{-U\mathscr{P}L/\dot{m}_H c_{pH}} \tag{8.15}$$

Equation (8.15) can be rearranged as

$$\frac{T_{H,\text{in}} - T_{H,\text{out}}}{T_{H,\text{in}} - T_{\text{sat}}} = 1 - e^{-U\mathscr{P}L/\dot{m}_H c_{pH}} \tag{8.16}$$

or

$$\varepsilon = 1 - e^{-N_{\text{tu}}} \tag{8.17}$$

where ε is the exchanger *effectiveness*, and N_{tu} is the *number of transfer units*. The effectiveness is the actual temperature decrease of the gas ($T_{H,\text{in}} - T_{H,\text{out}}$) divided by the maximum possible decrease that could be obtained in an infinitely long exchanger ($T_{H,\text{in}} - T_{\text{sat}}$). From the exchanger energy balance, the heat transfer is $\dot{Q} = \dot{m}_H c_{pH}(T_{H,\text{in}} - T_{H,\text{out}})$. Thus, the effectiveness can be viewed as the actual heat transfer divided by the maximum heat transfer obtainable in an infinitely long exchanger. The dimensionless group N_{tu} can be viewed as a measure of the heat transfer capability of the exchanger. Equation (8.17) indicates that the larger the number of transfer units, the higher the exchanger effectiveness. However, there are design tradeoffs to be considered, and in practice, values of ε between 0.6 and 0.9 are typical. Notice that Eq. (8.17) is identical to Eq. (1.59), which was derived for a condenser.

EXAMPLE 8.4 An Open-Cycle Ocean Thermal Energy Conversion Pilot Plant

In a pilot open-cycle ocean thermal energy conversion plant, 1 kg/s of warm sea water at 300 K enters an evaporator maintained at 2619 Pa. The water is injected through an array of nozzles to give an estimated transfer area and liquid-side heat transfer coefficient of 0.80 m^2 and 17,000 W/m^2 K, respectively. At what rate is vapor produced?

Solution

Given: Evaporator to flash-evaporate sea water at 300 K.

Required: Rate of vapor production.

Assumptions: The interfacial heat transfer resistance is negligible (see Section 7.6).

Equation (8.17) applies:

$$\varepsilon = 1 - e^{-N_{\text{tu}}}; \qquad N_{\text{tu}} = h_c A / \dot{m}_H c_{pH}$$

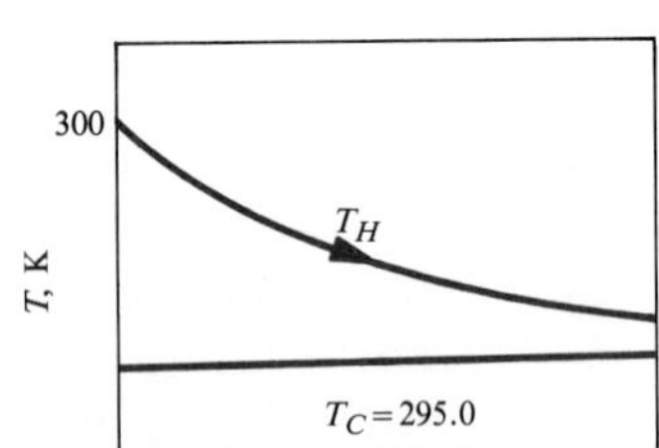

where h_c is the heat transfer coefficient for transfer of heat from the bulk water to the water-vapor interface, and $A = \mathscr{P}L$. Therefore,

$$N_{\text{tu}} = \frac{h_c A}{\dot{m}_H c_{pH}} = \frac{(17{,}000)(0.80)}{(1)(4178)} = 3.26$$

where c_{pH} was approximated by the value for pure water at 300 K using Table A.8.

$$\varepsilon = 1 - e^{-N_{\text{tu}}} = 1 - e^{-3.26} = 0.961$$

From the definition of effectiveness,

$$T_{H,\text{out}} = T_{H,\text{in}} - \varepsilon(T_{H,\text{in}} - T_{\text{sat}})$$

The saturation temperature corresponding to the evaporator pressure of 2619 Pa is 295.0 K from Table A.12*a*. Hence,

$$T_{H,\text{out}} = 300 - 0.961(300 - 295) = 295.2 \text{ K}$$

The rate of vapor production is obtained from the exchanger energy balance, Eq. (8.6):

$$(\dot{m}c_p)_H(T_{H,\text{in}} - T_{H,\text{out}}) = \dot{m}_C h_{\text{fg}C}$$

From Table A.12*a*, the enthalpy of vaporization at $T_{\text{sat}} = 295$ K is 2.449×10^6 J/kg. Thus,

$$(1)(4178)(300 - 295.2) = \dot{m}_C(2.449 \times 10^6)$$

$$\dot{m}_C = 8.20 \times 10^{-3} \text{ kg/s}$$

Comments

1. Examination of Table A.13*b* shows that the specific heat of sea water will be somewhat less than the pure water value.
2. Estimation of transfer area and heat transfer coefficient is not easy. In practice, we may often simply assume that a direct-contact evaporator of this type has an effectiveness in excess of 95%.

8.5 TWO-STREAM STEADY-FLOW HEAT EXCHANGERS

The most common type of heat exchanger is the two-stream steady-flow exchanger, with parallel, counter- or cross-flow of the two streams. Two methods will be used to analyze these exchangers: the logarithmic mean temperature method, in Section 8.5.1, and the effectiveness–number of transfer units method, in Section 8.5.2. Balanced-flow exchangers, for which the $\dot{m}c_p$ products for the streams are equal, are often required for air-conditioning systems and are examined in Section 8.5.3. If the mass flow rates through an exchanger are small, heat conduction along the tube walls can significantly affect the exchanger performance. In Section 8.5.4, dimensional analysis is used to derive an axial conduction parameter, and an approximate formula is given for the effect of axial conduction on a balanced-flow exchanger. Perforated-plate heat exchangers, which have inherently low axial conduction, are also discussed.

8.5.1 The Logarithmic Mean Temperature Difference

Figure 8.14*a* shows how the hot and cold stream temperatures, T_H and T_C, respectively, vary along heat exchangers for both parallel flow and counterflow. The temperature difference for heat transfer from the hot to the cold fluid, $T_H - T_C$, is seen to vary along the exchanger. In the case of parallel flow, for example, $T_H - T_C$ decreases continuously along the exchanger from the inlet to the outlet end. The total heat transfer in the exchanger will be written as

$$\dot{Q} = U\mathscr{P}L\Delta T_{\text{lm}} \tag{8.18}$$

where $U\mathscr{P}$ is the overall heat transfer coefficient $\times$ transfer perimeter product for the exchanger, L is the exchanger length, and ΔT_{lm} is an appropriate mean temperature difference between the hot and cold streams. In general, ΔT_{lm} must be determined by analysis. For parallel-flow and counterflow exchangers, the analysis is straightforward, provided some simplifying assumptions are made.

To determine ΔT_{lm}, we first consider a parallel-flow exchanger and make an energy balance on an exchanger element Δx long, as shown in Fig. 8.14*b*. Application of the steady-flow energy equation, Eq. (1.4), to a control volume containing either stream requires that

$$\Delta\dot{Q} = \dot{m}\Delta h = \dot{m}c_p\Delta T; \quad \Delta T = T|_{x+\Delta x} - T|_x \tag{8.19}$$

since no external work is done and changes of kinetic energy and potential energy are negligible. Thus, for the cold stream,

$$U\mathscr{P}\Delta x(T_H - T_C) = (\dot{m}c_p)_C\Delta T_C \tag{8.20a}$$

and for the hot stream

$$-U\mathscr{P}\Delta x(T_H - T_C) = (\dot{m}c_p)_H\Delta T_H \tag{8.20b}$$

For convenience, we let $C = \dot{m}c_p$ [J/K s], the **flow thermal capacity** of the stream; Eq. (8.19) shows that C is the amount of heat a stream gains or loses for a temperature

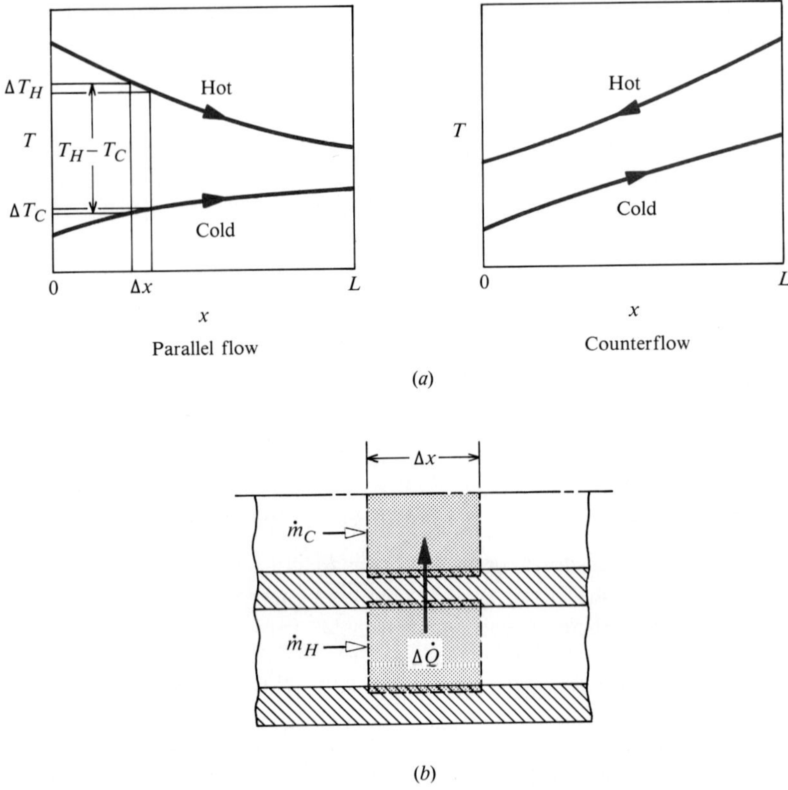

Figure 8.14 (*a*) Fluid temperature variations along parallel-flow and counterflow exchangers: notation for LMTD approach. (*b*) Elemental control volumes for analysis of a coaxial-tube parallel-flow exchanger.

change of 1 K. Dividing Eqs. (8.20) by Δx, rearranging, and letting $\Delta x \to 0$ gives

$$C_C \frac{dT_C}{dx} = U\mathscr{P}(T_H - T_C) \tag{8.21a}$$

$$C_H \frac{dT_H}{dx} = -U\mathscr{P}(T_H - T_C) \tag{8.21b}$$

Subtracting Eq. (8.21*a*) from Eq. (8.21*b*) and further rearranging,

$$\frac{d(T_H - T_C)}{T_H - T_C} = -U\mathscr{P}\left(\frac{1}{C_H} + \frac{1}{C_C}\right) dx$$

Integrating from $x = 0$ to $x = L$,

$$\ln \frac{T_{H,L} - T_{C,L}}{T_{H,0} - T_{C,0}} = -\int_0^L U\mathscr{P}\left(\frac{1}{C_H} + \frac{1}{C_C}\right) dx \tag{8.22}$$

If $U\mathscr{P}$ and the fluid specific heats are now assumed constant, then

$$\ln\frac{T_{H,L} - T_{C,L}}{T_{H,0} - T_{C,0}} = -U\mathscr{P}L\left(\frac{1}{C_H} + \frac{1}{C_C}\right) \quad \textbf{(8.23)}$$

But the exchanger energy balance, Eq. (8.2), is

$$(\dot{m}c_p)_H(T_{H,0} - T_{H,L}) = \dot{Q} = (\dot{m}c_p)_C(T_{C,L} - T_{C,0})$$

Hence,

$$C_H = \frac{\dot{Q}}{T_{H,0} - T_{H,L}}, \qquad C_C = \frac{\dot{Q}}{T_{C,L} - T_{C,0}}$$

Substituting in Eq. (8.23) and rearranging gives

$$\dot{Q} = U\mathscr{P}L\frac{(T_{H,L} - T_{C,L}) - (T_{H,0} - T_{C,0})}{\ln[(T_{H,L} - T_{C,L})/(T_{H,0} - T_{C,0})]} \quad \textbf{(8.24)}$$

Comparing Eqs. (8.24) and (8.18) gives the required formula for ΔT_{lm}:

$$\Delta T_{\text{lm}} = \frac{(T_H - T_C)_L - (T_H - T_C)_0}{\ln[(T_H - T_C)_L/(T_H - T_C)_0]} \quad \textbf{(8.25)}$$

ΔT_{lm} is called the **log mean temperature difference**, which is abbreviated as LMTD. If this analysis is repeated for a counterflow exchanger, exactly the same result is obtained.[1]

Often it is not appropriate to assume that the overall heat transfer coefficient is constant along the exchanger, perhaps due to entrance effects or due to fluid property variations. If concern is only with the effect of the entrance region on U, then U in Eq. (8.18) can be replaced by an average value, $\overline{U}$:

$$\dot{Q} = \overline{U}\mathscr{P}L\Delta T_{\text{lm}}; \qquad \overline{U} = \frac{1}{L}\int_0^L U\,dx \quad \textbf{(8.26)}$$

If fluid property variations are also important, then numerical integration of Eq. (8.22) is required, because U, C_H, and C_C all vary along the exchanger.

F Factor Charts

The concept of an appropriate mean temperature difference between the two streams has an appealing simplicity, and, as a result, the LMTD is widely used in engineering practice. For two-stream configurations other than the ideal coaxial type considered here, a correction factor F is applied to the LMTD for a *counterflow configuration* with the same inlet and outlet temperatures:

$$\dot{Q} = UAF\Delta T_{\text{lm}} \quad \textbf{(8.27)}$$

[1] In the limit $(T_H - T_C)_L \to (T_H - T_C)_0$ for counterflow exchangers, the common value equals the LMTD. This limit corresponds to *balanced* flow, $C_H = C_C$ (see Section 8.5.3).

Compilations of F factor charts may be found in References [1,2,3]; samples are given in Appendix C. These F factor charts are used in engineering practice, so the student should be familiar with them. However, there is a trend away from their use. Improved performance charts for two-stream exchangers have been presented recently by Turton et al. [4]. Also, the performance of two-stream exchangers can be expressed in terms of effectiveness and number of transfer units, as will be shown in Section 8.5.2. The ε–N_{tu} formulation is rapidly gaining popularity for various reasons, including its suitability for computer-aided design.

EXAMPLE 8.5 Counterflow Benzene Cooler

A coaxial-tube counterflow heat exchanger is to cool 0.03 kg/s of benzene from 360 K to 310 K with a counterflow of 0.02 kg/s of water at 290 K. If the inner tube outside diameter is 2 cm and the overall heat transfer coefficient based on outside area is 650 W/m^2 K, determine the required length of the exchanger. Take the specific heats of benzene and water as 1880 and 4175 J/kg K, respectively.

Solution

Given: Coaxial-tube counterflow exchanger.

Required: Exchanger length for specified performance.

Assumptions: U is constant along the exchanger.

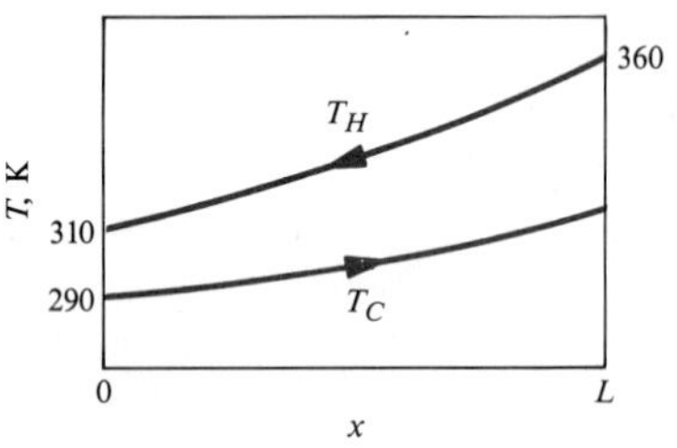

An exchanger energy balance is used to determine the required heat transfer and water outlet temperature. For a counterflow exchanger as shown,

$$(\dot{m}c_p)_H(T_{H,L} - T_{H,0}) = (\dot{m}c_p)_C(T_{C,L} - T_{C,0}) = \dot{Q}$$

$$(0.03)(1880)(360 - 310) = (0.02)(4175)(T_{C,L} - 290) = \dot{Q}$$

Solving, $\dot{Q} = 2820$ W, $T_{C,L} = 323.8$ K. Equation (8.25) gives the log mean temperature difference:

$$\Delta T_{\mathrm{lm}} = \frac{(T_H - T_C)_L - (T_H - T_C)_0}{\ln[(T_H - T_C)_L/(T_H - T_C)_0]} = \frac{36.2 - 20}{\ln(36.2/20)} = 27.3 \text{ K}$$

The tube outside perimeter is $\mathscr{P} = \pi D = (\pi)(0.02) = 0.0628$ m. Solving Eq. (8.18) for the exchanger length L gives

$$L = \frac{\dot{Q}}{U\mathscr{P}\Delta T_{\mathrm{lm}}} = \frac{2820}{(650)(0.0628)(27.3)} = 2.53 \text{ m}$$

Comments

A quick calculation of the arithmetic mean temperature difference provides a useful check of the LMTD, since it will not be very different. In this example, the arithmetic mean temperature difference is $(1/2)(36.2 + 20) = 28.1$ K.

EXAMPLE 8.6 Counterflow Oil Cooler

After a long time in service, a counterflow oil cooler is checked to ascertain if its performance has deteriorated due to fouling. In the test, SAE 50 oil flowing at 2.0 kg/s is cooled from 420 K to 380 K by a water supply of 1.0 kg/s at 300 K. If the heat transfer surface is 3.33 m^2 and the design value of the overall heat transfer coefficient is 930 W/m^2 K, how much has it been reduced by fouling?

Solution

Given: Performance data for a counterflow exchanger.

Required: Effect of fouling on overall heat transfer coefficient.

Assumptions: U is constant along the exchanger.

We first use an exchanger energy balance to find the heat transferred and the water outlet temperature. Evaluating c_{pH} at 400 K using Table A.8,

$$\begin{aligned}\dot{Q} &= (\dot{m}c_p)_H(T_{H,L} - T_{H,0})\\ &= (2.0)(2330)(420 - 380)\\ &= 186{,}400 \text{ W}\end{aligned}$$

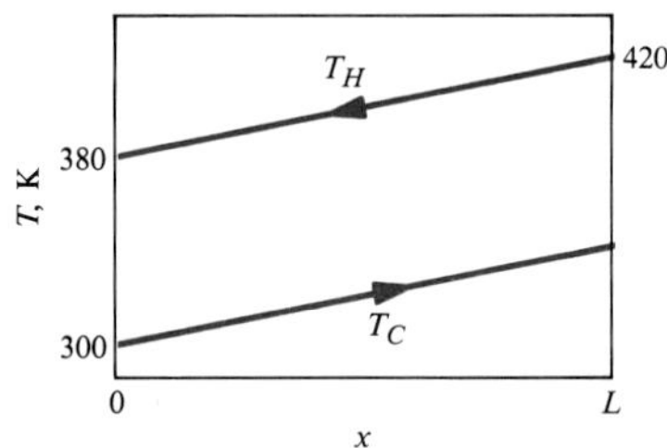

To find $T_{C,L}$, we guess an average water temperature of 320 K to evaluate c_{pC}

$$\begin{aligned}186{,}400 &= (\dot{m}c_p)_C(T_{C,L} - T_{C,0})\\ &= (1)(4174)(T_{C,L} - 300)\end{aligned}$$

Hence,

$$T_{C,L} = 344.7 \text{ K}$$

and our guessed average temperature is close enough. From Eq. (8.25), the LMTD is

$$\begin{aligned}\Delta T_{\text{lm}} &= \frac{(420 - 344.7) - (380 - 300)}{\ln[(420 - 344.7)/(380 - 300)]}\\ &= \frac{75.3 - 80}{\ln(75.3/80)} = 77.6 \text{ K}\end{aligned}$$

Then, from Eq. (8.18) and recognizing that $A = \mathscr{P}L$,

$$\dot{Q} = UA\Delta T_{\text{lm}} \qquad \text{or} \qquad U = \frac{\dot{Q}}{A\Delta T_{\text{lm}}} = \frac{186{,}400}{(3.33)(77.6)} = 721 \text{ W/m}^2\text{ K}$$

Comments

1. The reduction in U due to fouling is $(930 - 721)/(930) = 22.5\%$.
2. Can you think of any other reasons why the performance might have deteriorated?

8.5.2 Effectiveness and Number of Transfer Units

The LMTD formula for heat exchanger performance is useful only when inlet and outlet temperatures are known, either because they have been measured in a test or because they have been specified in a design. If it is desired to calculate inlet or outlet temperatures for a given exchanger and flow rates, use of the LMTD method requires an iterative solution procedure or specially constructed charts. Iteration can be avoided if exchanger performance is expressed in terms of effectiveness and number of transfer units, as was done for the single-stream exchanger.

Temperature variations along parallel-flow and counterflow exchangers are shown in Fig. 8.15. Inlet and outlet temperature are designated by subscripts "in" and "out," respectively, rather than by the locations $x = 0$ and $x = L$, as was done in the LMTD analysis. The exchanger effectiveness is now defined as the ratio of the actual heat transferred to the maximum possible amount of heat that could be transferred in an infinitely long *counterflow* exchanger. This definition is more general than that used for single-stream exchangers. From the exchanger energy balance, Eq. (8.4), the heat transferred is

$$\dot{Q} = C_H(T_{H,\text{in}} - T_{H,\text{out}}) = C_C(T_{C,\text{out}} - T_{C,\text{in}}) \qquad \textbf{(8.28)}$$

In an infinitely long counterflow exchanger with $C_C < C_H$, Fig. 8.15*b* shows that $T_{C,\text{out}} \rightarrow T_{H,\text{in}}$; then $\dot{Q}_{\text{max}} = C_C(T_{H,\text{in}} - T_{C,\text{in}})$. On the other hand, in an infinitely long parallel-flow exchanger with $C_C < C_H$, Fig. 8.15*a* shows $T_{C,\text{out}} \rightarrow T_{H,\text{out}} < T_{H,\text{in}}$, and the heat transfer $\dot{Q}$ will be less. Similarly, if $C_H < C_C$, the heat transfer will again be greater in an infinitely long counterflow exchanger, $\dot{Q}_{\text{max}} = C_H(T_{H,\text{in}} - T_{C,\text{in}})$. If we write

$$C_{\text{min}} = \min(C_H, C_C) \qquad \textbf{(8.29)}$$

then the maximum heat transfer in an exchanger of any configuration is

$$\dot{Q}_{\text{max}} = C_{\text{min}}(T_{H,\text{in}} - T_{C,\text{in}}) \qquad \textbf{(8.30)}$$

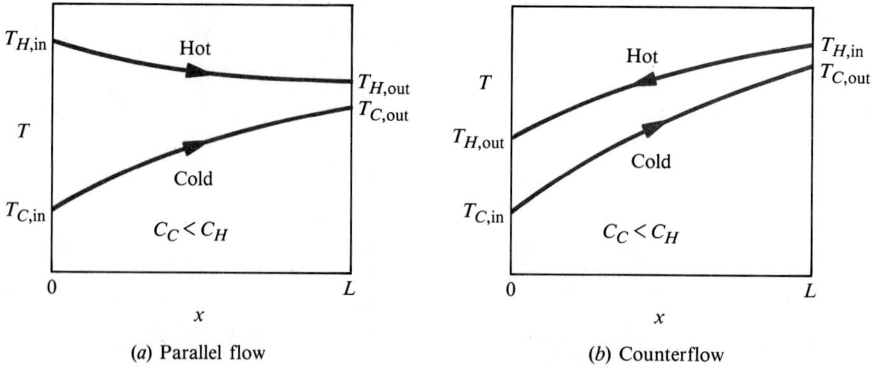

Figure 8.15 Fluid temperature variations along parallel-flow and counterflow exchangers: notation for the ε–N_{tu} approach.

Note that it is the stream with the smaller flow thermal capacity that limits the amount of heat that can be transferred. The actual heat transfer is expressed in terms of the enthalpy change of either stream, as given by Eq. (8.28); thus, the effectiveness is

$$\varepsilon = \frac{\dot{Q}}{\dot{Q}_{\max}} = \frac{C_H(T_{H,\text{in}} - T_{H,\text{out}})}{C_{\min}(T_{H,\text{in}} - T_{C,\text{in}})} = \frac{C_C(T_{C,\text{out}} - T_{C,\text{in}})}{C_{\min}(T_{H,\text{in}} - T_{C,\text{in}})} \tag{8.31}$$

Which of these two expressions is more convenient depends on the problem specification.

We first determine the effectiveness of a parallel-flow exchanger. Changing subscripts in Eq. (8.23) from "0, L" to "in, out" gives

$$\ln\frac{T_{H,\text{out}} - T_{C,\text{out}}}{T_{H,\text{in}} - T_{C,\text{in}}} = -U\mathcal{P}L\left(\frac{1}{C_H} + \frac{1}{C_C}\right) = \frac{-U\mathcal{P}L}{C_C}\left(1 + \frac{C_C}{C_H}\right) \tag{8.32}$$

The number of transfer units, N_{tu}, is defined as

$$N_{\text{tu}} = \frac{U\mathcal{P}L}{C_{\min}} \tag{8.33}$$

and the **capacity ratio** R_C as

$$R_C = \frac{C_{\min}}{C_{\max}} \quad (\leq 1) \tag{8.34}$$

Thus, if $C_C = C_{\min}$, Eq. (8.32) becomes

$$\frac{T_{H,\text{out}} - T_{C,\text{out}}}{T_{H,\text{in}} - T_{C,\text{in}}} = e^{-N_{\text{tu}}(1+R_C)} \tag{8.35}$$

But from Eq. (8.28),

$$T_{H,\text{out}} = T_{H,\text{in}} - R_C(T_{C,\text{out}} - T_{C,\text{in}})$$

and substituting in Eq. (8.35) and rearranging gives

$$\frac{(T_{H,\text{in}} - T_{C,\text{in}}) - R_C(T_{C,\text{out}} - T_{C,\text{in}}) - (T_{C,\text{out}} - T_{C,\text{in}})}{T_{H,\text{in}} - T_{C,\text{in}}} = e^{-N_{\text{tu}}(1+R_C)} \tag{8.36}$$

Also for $C_C = C_{\min}$, Eq. (8.31) gives

$$\varepsilon = \frac{T_{C,\text{out}} - T_{C,\text{in}}}{T_{H,\text{in}} - T_{C,\text{in}}} \tag{8.37}$$

so that Eq. (8.36) can be rewritten as

$$1 - (R_C + 1)\varepsilon = e^{-N_{\text{tu}}(1+R_C)} \tag{8.38}$$

Solving for ε gives

$$\varepsilon = \frac{1 - e^{-N_{\text{tu}}(1+R_C)}}{1 + R_C} \tag{8.39}$$

If the hot stream is chosen to have the minimum capacity, that is, $C_H = C_{\min}$, exactly the same result is obtained. For a given heat exchanger and flow rates, N_{tu} and R_C are known, so that the effectiveness ε can be calculated from Eq. (8.39), and the unknown outlet temperature can be determined. If, on the other hand, we are required to design a heat exchanger to have a specified outlet temperature, that is, a specified effectiveness, then the unknown is N_{tu}. Solving Eq. (8.39) for N_{tu} gives

$$N_{tu} = \frac{1}{1 + R_C} \ln \frac{1}{1 - (1 + R_C)\varepsilon} \tag{8.40}$$

Using Eq. (8.40) is equivalent to using the LMTD approach, Eqs. (8.18) and (8.25).

A similar analysis can be performed for a counterflow exchanger to give

$$\varepsilon = \frac{1 - e^{-N_{tu}(1-R_C)}}{1 - R_C e^{-N_{tu}(1-R_C)}} \tag{8.41}$$

and

$$N_{tu} = \frac{1}{1 - R_C} \ln \frac{1 - \varepsilon R_C}{1 - \varepsilon} \tag{8.42}$$

For a given N_{tu} and R_C, a counterflow exchanger is always more effective than a parallel-flow one, because a larger value of $(T_H - T_C)$ is sustained along a counterflow exchanger.

Figure 8.16 shows exchangers for which $C_C \ll C_H$; as a result, the temperature of the hot stream does not vary significantly along the exchanger. For $R_C \to 0$, both Eqs. (8.39) and (8.41) reduce to

$$\varepsilon = 1 - e^{-N_{tu}} \tag{8.43}$$

which is the same as the effectiveness of a single-stream exchanger. For $R_C = 0$, the temperature of one stream does not change through the exchanger, which satisfies the definition of a single-stream exchanger.

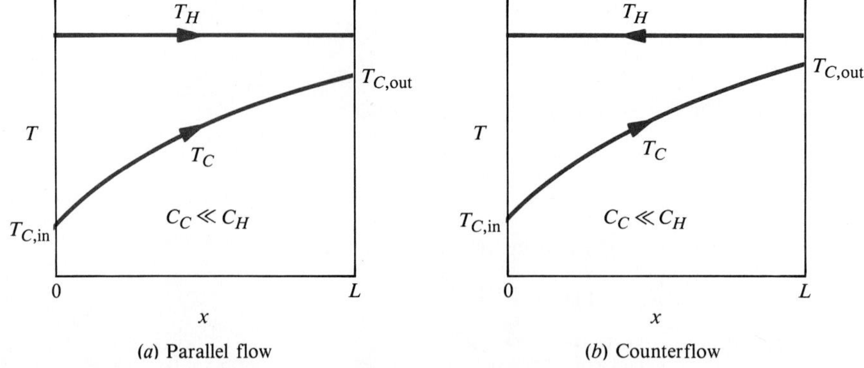

Figure 8.16 Fluid temperature variations along parallel-flow and counterflow exchangers for $C_C \ll C_H$.

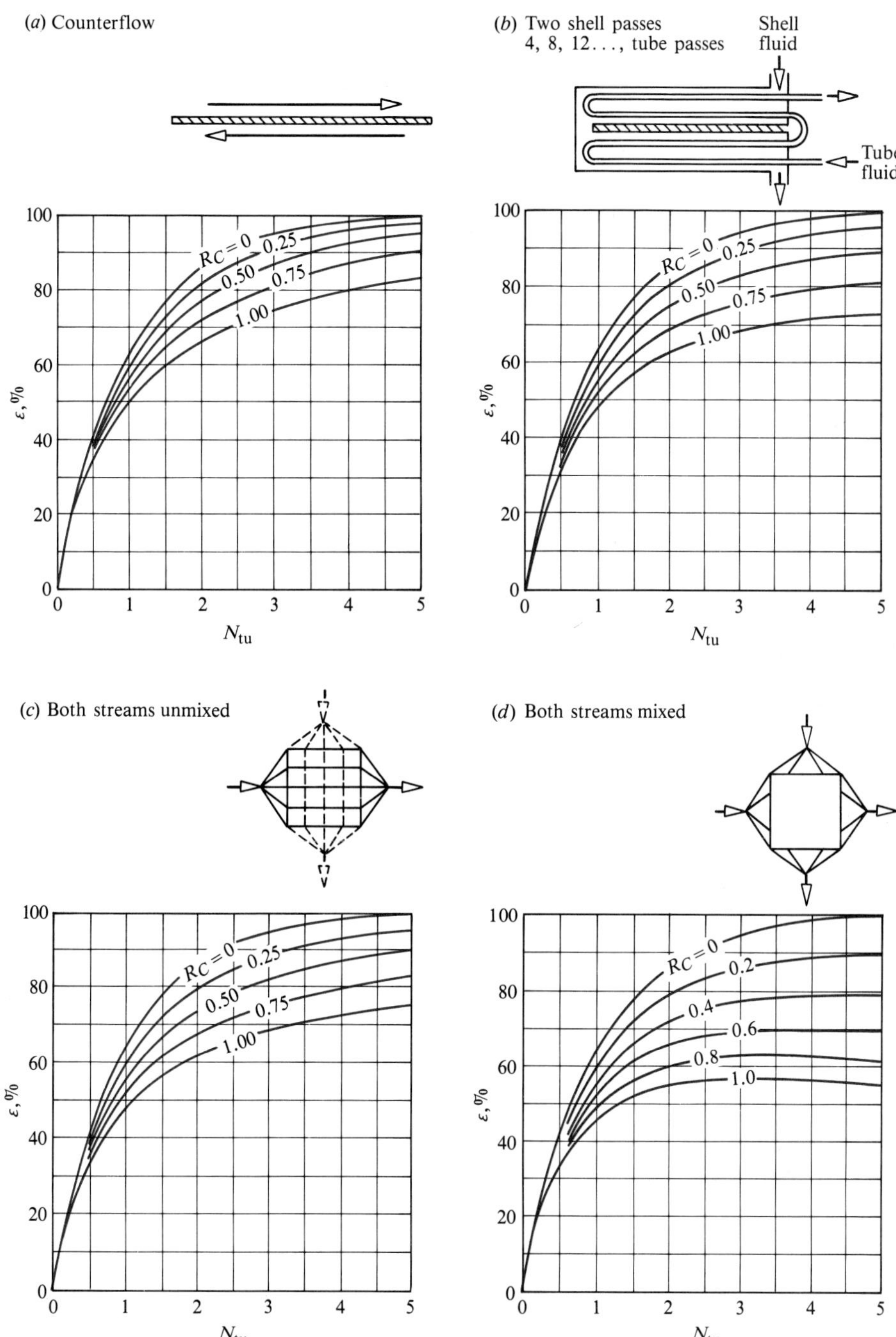

Figure 8.17 Effectiveness versus number of transfer units for four exchanger configurations. (*a*) Counterflow. (*b*) Multipass. (*c*) Cross-flow, both streams unmixed. (*d*) Cross-flow, both streams mixed.

Table 8.3*a* Effectiveness formulas for selected exchanger configurations.

Configuration	Effectiveness
1. Single-stream, and all exchangers when $R_C = 0$	$\varepsilon = 1 - \exp(-N_{tu})$
2. Parallel-flow	$\varepsilon = \dfrac{1 - \exp[-N_{tu}(1 + R_C)]}{1 + R_C}$
3. Counterflow	$\varepsilon = \dfrac{1 - \exp[-N_{tu}(1 - R_C)]}{1 - R_C \exp[-N_{tu}(1 - R_C)]}$
Single-pass cross-flow:	
4. Both fluids unmixed	$\varepsilon = 1 - \exp\left\{\dfrac{N_{tu}^{0.22}}{R_C}\left[\exp(-R_C N_{tu}^{0.78}) - 1\right]\right\}$
5. Both fluids mixed	$\varepsilon = \left[\dfrac{1}{1 - \exp(-N_{tu})} + \dfrac{R_C}{1 - \exp(-R_C N_{tu})} - \dfrac{1}{N_{tu}}\right]^{-1}$
6. C_{max} mixed, C_{min} unmixed	$\varepsilon = \dfrac{1}{R_C}\left\{1 - \exp\left[R_C\left(e^{-N_{tu}} - 1\right)\right]\right\}$
7. C_{max} unmixed, C_{min} mixed	$\varepsilon = 1 - \exp\left\{-\dfrac{1}{R_C}\left[1 - e^{-R_C N_{tu}}\right]\right\}$
Shell-and-tube:	
8. One shell pass; 2, 4, 6 tube passes[a]	$\varepsilon = \varepsilon_1 = 2\left\{1 + R_C + (1 + R_C^2)^{1/2}\dfrac{1 + \exp\left[-N_{tu}(1 + R_C^2)^{1/2}\right]}{1 - \exp\left[-N_{tu}(1 + R_C^2)^{1/2}\right]}\right\}^{-1}$
9. n shell passes; $2n, 4n, \ldots$ tube passes[a]	$\varepsilon = \left[\left(\dfrac{1 - \varepsilon_1 R_C}{1 - \varepsilon_1}\right)^n - 1\right]\left[\left(\dfrac{1 - \varepsilon_1 R_C}{1 - \varepsilon_1}\right)^n - R_C\right]^{-1}$

[a] In calculating ε_1, the N_{tu} *per shell pass* is used (i.e., N_{tu}/n).

Figure 8.17 gives ε–N_{tu} graphs for four two-stream exchanger configurations; Table 8.3*a* and *b* gives ε–N_{tu} and N_{tu}–ε relations for a larger variety of configurations. Extensive compilations of ε–N_{tu} graphs can be found in References [2,5]. It should be clearly understood that the effectiveness and LMTD formulations for two-stream exchangers are mathematically equivalent, and either is sufficient for the solution of a problem. Current practice tends to favor the effectiveness approach because both effectiveness and number of transfer units have a unique physical significance for a given exchanger and given flow thermal capacities. On the other hand, the log mean temperature difference depends also on the inlet temperatures of the streams and thus will vary according to the particular application. The design engineer using the effectiveness approach soon develops a "feel" for the number of transfer units and effectiveness to be desired or expected in a given situation.

Table 8.3b Formulas for number of transfer units for selected exchanger configurations.

Configuration	Number of Transfer Units
1. Single-stream, and all exchangers when $R_C = 0$	$N_{tu} = \ln \dfrac{1}{1-\varepsilon}$
2. Parallel-flow	$N_{tu} = \dfrac{1}{1+R_C} \ln \dfrac{1}{1-(1+R_C)\varepsilon}$
3. Counterflow	$N_{tu} = \dfrac{1}{1-R_C} \ln \dfrac{1-\varepsilon R_C}{1-\varepsilon}$
Single-pass cross-flow:	
4. Both fluids unmixed	
5. Both fluids mixed	
6. C_{max} mixed, C_{min} unmixed	$N_{tu} = -\ln[1 + (1/R_C)\ln(1-\varepsilon R_C)]$
7. C_{max} unmixed, C_{min} mixed	$N_{tu} = -(1/R_C)\ln[R_C \ln(1-\varepsilon) + 1]$
Shell-and-tube:	$N_{tu} = -(1+R_C^2)^{-1/2} \ln\left[\dfrac{E-1}{E+1}\right]$
8. One shell pass; 2, 4, 6 tube passes[a]	$E = \dfrac{2/\varepsilon - (1+R_C)}{(1+R_C^2)^{1/2}}$
9. n shell passes; $2n, 4n, \ldots$ tube passes[a]	$E = \dfrac{[2(F-R_C)/(F-1)] - (1+R_C)}{(1+R_C^2)^{1/2}}; \quad F = \left(\dfrac{\varepsilon R_C - 1}{\varepsilon - 1}\right)^{1/n}$

[a] Substituting E gives the N_{tu} *per shell pass* (i.e., the N_{tu} for the exchanger is $n \times$ this value).

The Computer Program HEX1

HEX1 calculates the thermal performance of heat exchangers for the nine flow configurations listed in Table 8.3. There are two options:

1. *The rating problem.* Known are the mass flow rates $\dot{m}$, overall heat transfer coefficient U, and heat transfer area A. Required is the heat exchanger effectiveness ε, which is calculated from the formulas in Table 8.3*a*. Also, for input inlet temperatures, HEX1 calculates outlet temperatures.

2. *The design or sizing problem.* Known are the mass flow rates $\dot{m}$, inlet temperatures, and one outlet temperature. Required is the number of transfer units N_{tu}, which is calculated from the formulas in Table 8.3*b*. In the case of items 4 and 5, for which there are no explicit formulas for $N_{tu}(\varepsilon, R_C)$, the $\varepsilon(N_{tu}, R_C)$ formulas from items 4 and 5 of Table 8.3*a* are solved for the lowest N_{tu} value using Newton's method. HEX1 also calculates the unknown outlet temperature.

EXAMPLE 8.7 Cooling of a Distillation Column Product Stream

A 4 kg/s product stream from a distillation column is to be cooled by a 3 kg/s water stream in a counterflow exchanger. The hot and cold stream inlet temperatures are 400 and 300 K, respectively, and the heat transfer area of the exchanger is 30 m^2. If the overall heat transfer coefficient is estimated to be 820 W/m^2 K, determine the product stream outlet temperature. The specific heat of the product stream can be taken to be 2500 J/kg K.

Solution

Given: Counterflow heat exchanger to cool distillation product stream.

Required: Performance, in particular $T_{H,\text{out}}$.

Assumptions: U is constant along the exchanger.

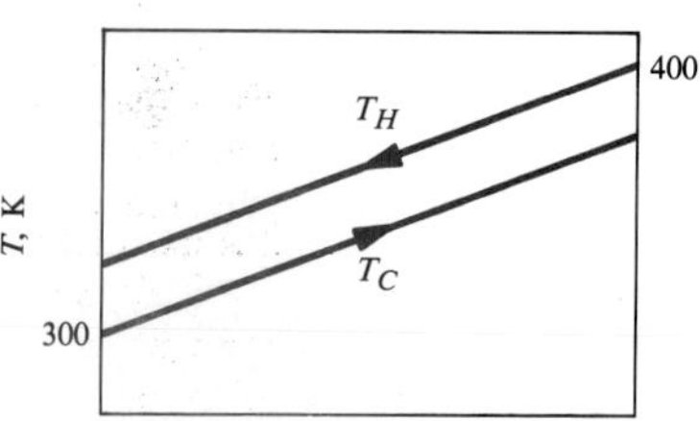

This is a *rating* problem since the $\dot{m}$ values, U, and A are known. We first calculate the effectiveness from Eq. (8.41):

$$\varepsilon = \frac{1 - e^{-N_{\text{tu}}(1-R_C)}}{1 - R_C e^{-N_{\text{tu}}(1-R_C)}}; \qquad R_C = \frac{C_{\min}}{C_{\max}}; \qquad N_{\text{tu}} = \frac{U\mathscr{P}L}{C_{\min}}$$

$$C_H = (\dot{m}c_p)_H = (4)(2500) = 10{,}000 \text{ W/K}$$

$$C_C = (\dot{m}c_p)_C = (3)(4180) = 12{,}540 \text{ W/K for } c_p \text{ evaluated at } \sim 330 \text{ K in Table A.8}$$

$$C_{\min} = \min(10{,}000,\ 12{,}540) = 10{,}000 \text{ W/K}$$

$$R_C = 10{,}000/12{,}540 = 0.797$$

$$N_{\text{tu}} = \frac{U\mathscr{P}L}{C_{\min}} = \frac{UA}{C_{\min}} = \frac{(820)(30)}{10{,}000} = 2.46$$

$$\varepsilon = \frac{1 - e^{-2.46(1-0.797)}}{1 - 0.797e^{-2.46(1-0.797)}} = 0.761$$

Equation (8.31) gives the effectiveness in terms of temperatures as

$$\varepsilon = \frac{C_H(T_{H,\text{in}} - T_{H,\text{out}})}{C_{\min}(T_{H,\text{in}} - T_{C,\text{in}})}$$

$$0.761 = \frac{10{,}000(400 - T_{H,\text{out}})}{10{,}000(400 - 300)}$$

Hence, $T_{H,\text{out}} = 323.9$ K.

The exchanger energy balance is

$$C_H(T_{H,\text{in}} - T_{H,\text{out}}) = C_C(T_{C,\text{out}} - T_{C,\text{in}})$$

$$10{,}000(400 - 323.9) = 12{,}540(T_{C,\text{out}} - 300)$$

Hence, $T_{C,\text{out}} = 360.7$ K, and our guessed average temperature of 330 K is satisfactory.

Solution using HEX1

The required input in SI units is:

c (counterflow)
$\dot{m}_C = 3; c_{pC} = 4180$
$\dot{m}_H = 4; c_{pH} = 2500$
$T_{C,\text{in}} = 300, T_{H,\text{in}} = 400$
U and A known
$U = 820; A = 30$

The output is:

$N_{\text{tu}} = 2.46$
$\varepsilon = 0.761$
$T_{C,\text{out}} = 360.7$
$T_{H,\text{out}} = 323.9$

Comments

Use of the ε–N_{tu} method is seen to be quite straightforward. All the student needs to do is to recognize from the given data whether it is a rating or design problem.

EXAMPLE 8.8 Cross-Flow Plate Exchanger Using the ε–N_{tu} Approach

A cross-flow plate heat exchanger is to be designed for waste heat recovery from the exhaust streams of a metallurgical process. A flow of 5 kg/s of exhaust gases enters the exchanger at 240°C and must be cooled to 120°C by 5 kg/s of air supplied at 20°C. If the overall heat transfer coefficient is estimated to be 40 W/m^2 K, determine the required heat transfer area if both streams are unmixed. The specific heat of the exhaust gases can be taken as 1200 J/kg K.

Solution

Given: Cross-flow plate heat exchanger for waste heat recovery.

Required: Heat transfer area A for specified performance.

Assumptions: U is constant over the plates.

This is a *design* or *sizing* problem since the $\dot{m}$ values and three temperatures are known. Equation (8.31) defines the effectiveness as

$$\varepsilon = \frac{C_H(T_{H,\text{in}} - T_{H,\text{out}})}{C_{\min}(T_{H,\text{in}} - T_{C,\text{in}})}$$

For an estimated 400 K average temperature of the air stream, $c_{pC} = 1010$ J/kg K from Table A.7, and the flow thermal capacities are

$$C_C = (\dot{m}c_p)_C = (5)(1010) = 5050 \text{ W/K}$$

$$C_H = (\dot{m}c_p)_H = (5)(1200) = 6000 \text{ W/K}$$

$$C_{\min} = 5050 \text{ W/K}; \qquad R_C = \frac{C_{\min}}{C_{\max}} = \frac{5050}{6000} = 0.842$$

Hence,

$$\varepsilon = \frac{6000(240 - 120)}{5050(240 - 20)} = 0.648$$

From Fig. 8.17*c*, the required number of transfer units is 2.0.

$$N_{tu} = \frac{UA}{C_{min}}; \qquad A = \frac{C_{min}N_{tu}}{U} = \frac{(5050)(2.0)}{40} = 253 \text{ m}^2$$

Solution using HEX1

The required inputs in SI units with temperatures in °C are:

d (crossflow, both streams unmixed)
$\dot{m}_C = 5; c_{pC} = 1010$
$\dot{m}_H = 5; c_{pH} = 1200$
$T_{C,\text{in}} = 20; T_{H,\text{in}} = 240$
T_{out} known
$T_{H,\text{out}} = 120$

The output is:

$\varepsilon = 0.648$
$N_{tu} = 1.956$
$T_{C,\text{out}} = 162.6$
$UA = 9877$; hence, $A = 9877/40 = 247 \text{ m}^2$

Comments

Notice that HEX1 accepts temperatures in either kelvins or degrees Celsius.

8.5.3 Balanced-Flow Exchangers

Often exchangers have hot and cold streams of approximately equal flow thermal capacity, that is, $C_H = C_C$, or $R_C = 1$. Such exchangers are said to have *balanced* flow. Figure 8.18 shows a heatpipe air preheater used to recover heat from the exhaust of a gas turbine. The exhaust gas stream does have a larger $\dot{m}$ and c_p than the incoming air stream, but $R_C = 1$ is not a bad assumption. In many air-conditioning systems, warm, fresh ambient air is first cooled by colder, stale air leaving the system, and the exchanger can be exactly balanced. If $R_C = 1$ is substituted in Eq. (8.39) for the effectiveness of a parallel-flow exchanger, the result is

$$\varepsilon = \frac{1}{2}\left(1 - e^{-2N_{tu}}\right) \tag{8.44}$$

If, however, $R_C = 1$ is substituted in Eq. (8.41) for a counterflow exchanger, the result is indeterminate; application of L'Hôpital's rule then gives

$$\varepsilon = \frac{N_{tu}}{1 + N_{tu}} \tag{8.45}$$

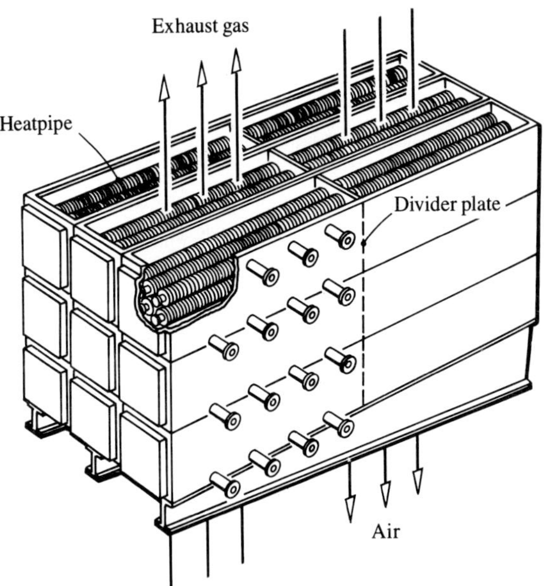

Figure 8.18 A heatpipe gas turbine recuperator.

The counterflow case is more interesting and of more practical importance than the parallel-flow case. Application of the steady-flow energy equation to a differential element of a balanced counterflow exchanger gives the differential equations

$$C_C \frac{dT_C}{dx} = U\mathscr{P}(T_H - T_C) \tag{8.46a}$$

$$C_H \frac{dT_H}{dx} = U\mathscr{P}(T_H - T_C) \tag{8.46b}$$

which are similar to Eqs. (8.21*a*,*b*) for the parallel-flow exchanger except for a sign change. Subtracting Eq. (8.46*a*) from (8.46*b*) with $C_C = C_H = C$ gives

$$\frac{d}{dx}(T_H - T_C) = 0 \tag{8.47}$$

Integrating,

$$T_H - T_C = \text{Constant} = T_{H,\text{in}} - T_{C,\text{out}} = T_{H,\text{out}} - T_{C,\text{in}}$$

as shown in Fig. 8.19. Substituting back in Eq. (8.46*a*),

$$\frac{dT_C}{dx} = \frac{U\mathscr{P}}{C}(T_{H,\text{in}} - T_{C,\text{out}}) \tag{8.48}$$

which states that T_C varies linearly with x if $U\mathscr{P}/C$ is constant along the exchanger. Similarly, T_H varies linearly with x.

Since $(T_H - T_C)$ is constant along a balanced-counterflow exchanger, the LMTD is equal to this constant difference. Notice that Eq. (8.25) is then indeterminate, and the logarithmic mean is simply equal to the constant value of $(T_H - T_C)$.

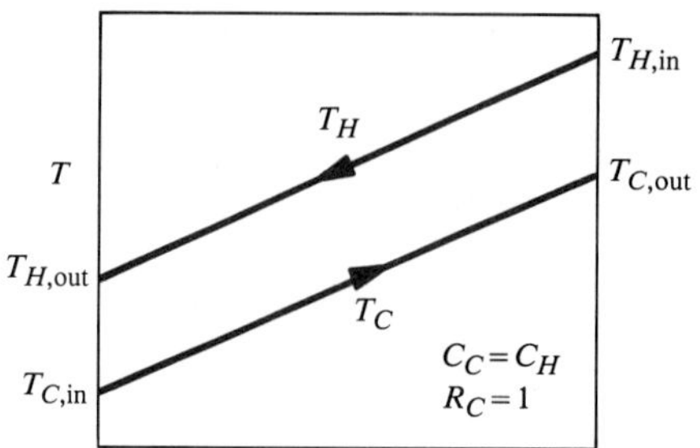

Figure 8.19 Fluid temperature variations along a balanced counterflow exchanger.

EXAMPLE 8.9 Recuperator for an Air-Conditioning System

In a solar-assisted air-conditioning system, 0.5 kg/s of ambient air at 270 K is to be preheated by the same amount of air leaving the system at 295 K. If a counterflow exchanger has an area of 30 m^2, and the overall heat transfer coefficient is estimated to be 25 W/m^2 K, determine the outlet temperature of the preheated air.

Solution

Given: Counterflow exchanger for balanced air flows.

Required: Performance, in particular $T_{C,\text{out}}$.

Assumptions: 1. Balanced flow, $C_{pC} = C_{pH}$.
2. U is constant along the exchanger.

This is a *rating* problem since the $\dot{m}$ values, U, and A are known. If we take $c_p = 1000$ J/kg K for air, the number of transfer units is

$$N_{\text{tu}} = \frac{UA}{\dot{m}c_p} = \frac{(25)(30)}{(0.5)(1000)} = 1.50$$

For balanced counterflow, Eq. (8.45) gives the effectiveness as

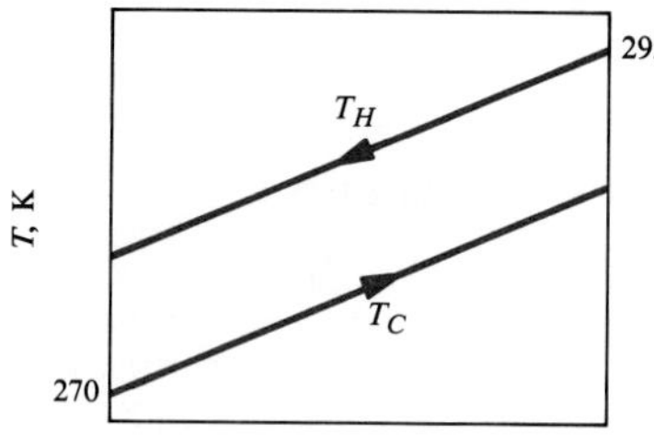

$$\varepsilon = \frac{N_{\text{tu}}}{1 + N_{\text{tu}}} = \frac{1.5}{1 + 1.5} = 0.6$$

$$T_{C,\text{out}} = T_{C,\text{in}} + \varepsilon(T_{H,\text{in}} - T_{C,\text{in}})$$

$$= 270 + 0.6(295 - 270)$$

$$= 285 \text{ K}$$

Comments

Use HEX1 to check N_{tu}, ε, and $T_{C,\text{out}}$.

EXAMPLE 8.10 Recuperator for a Gas Turbine

A flow of 0.1 kg/s of exhaust gases at 700 K from a gas turbine is used to preheat the incoming air, which is at the ambient temperature of 300 K. It is desired to cool the exhaust to 400 K, and it is estimated that an overall heat transfer coefficient of 30 W/m^2 K can be achieved in an appropriate exchanger. Determine the area required for a counterflow exchanger.

Solution

Given: Counterflow exchanger for a gas turbine.

Required: Area for specified performance.

Assumptions:
1. U is constant along the exchanger.
2. Balanced flow.
3. Specific heat of exhaust gases the same as for air, ~1000 J/kg K.

This is a *sizing* problem since the $\dot{m}$ values and three temperatures are known. Equation (8.31) gives the effectiveness as

$$\varepsilon = \frac{T_{H,\text{in}} - T_{H,\text{out}}}{T_{H,\text{in}} - T_{C,\text{in}}} = \frac{700 - 400}{700 - 300} = 0.750$$

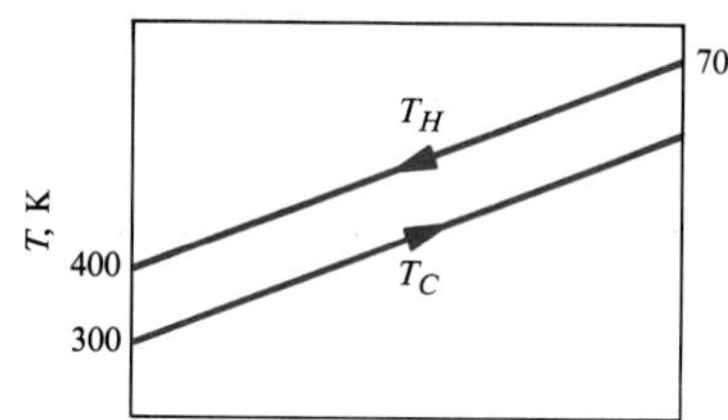

Equation (8.45) can be solved to give the number of transfer units:

$$\varepsilon = \frac{N_{\text{tu}}}{1 + N_{\text{tu}}}; \qquad N_{\text{tu}} = \frac{\varepsilon}{1 - \varepsilon} = \frac{0.750}{1 - 0.750} = 3.0$$

$$N_{\text{tu}} = \frac{UA}{C_{\min}} \quad \text{or} \quad A = \frac{CN_{\text{tu}}}{U} \simeq \frac{(0.1)(1000)(3.0)}{(30)} = 10 \text{ m}^2$$

Comments

1. Use HEX1 to check ε, N_{tu}, and the UA product.
2. This result is a rough estimate: the exchanger is not exactly balanced because both $\dot{m}$ and c_p are higher for the exhaust gases.
3. It would be impossible to cool the gases to 400 K in a parallel-flow exchanger: Eq. (8.44) shows that the maximum effectiveness of a balanced parallel-flow exchanger is 0.5.

8.5.4 Effect of Axial Conduction

The theory of two-stream heat exchangers developed in Sections 8.5.1 and 8.5.2 is the basis of most heat exchanger design procedures used in engineering practice. However, it is important to remember that the theory is based on a simple model that might be invalid under certain circumstances. Two key assumptions of the model are that heat conduction in the flow direction is negligible and that the overall heat transfer coefficient is constant along the exchangers. We examine the first of these assumptions here.

The parameter that determines whether or not axial conduction in the exchanger wall is negligible can be found by *scale analysis* of the governing differential equation. We now repeat the analysis of Section 8.5.1 but include axial conduction in the exchanger wall. For convenience, consider:

1. A balanced parallel-flow exchanger, $(\dot{m}c_p)_H = (\dot{m}c_p)_C = \dot{m}c_p$
2. Equal heat transfer coefficients, $h_{cH} = h_{cC} = h_c$
3. A thin-wall coaxial-tube heat exchanger, $T_w = T_w(x)$
4. Negligible resistance to heat flow across the tube wall

Then energy balances on the elemental control volumes shown in Fig. 8.20 are as follows.

I. Hot stream: $$\dot{m}c_pT_H|_x = h_c\mathscr{P}\Delta x(T_H - T_w) + \dot{m}c_pT_H|_{x+\Delta x}$$

II. Wall: $$-k_wA_w\frac{dT_w}{dx}\bigg|_x + h_c\mathscr{P}\Delta x(T_H - T_w) = -k_wA_w\frac{dT_w}{dx}\bigg|_{x+\Delta x} + h_c\mathscr{P}\Delta x(T_w - T_C)$$

III. Cold stream: $$\dot{m}c_pT_C|_x + h_c\mathscr{P}\Delta x(T_w - T_C) = \dot{m}c_pT_C|_{x+\Delta x}$$

Dividing by Δx and letting $\Delta x \to 0$,

$$\dot{m}c_p\frac{dT_H}{dx} = -h_c\mathscr{P}(T_H - T_w) \tag{8.49}$$

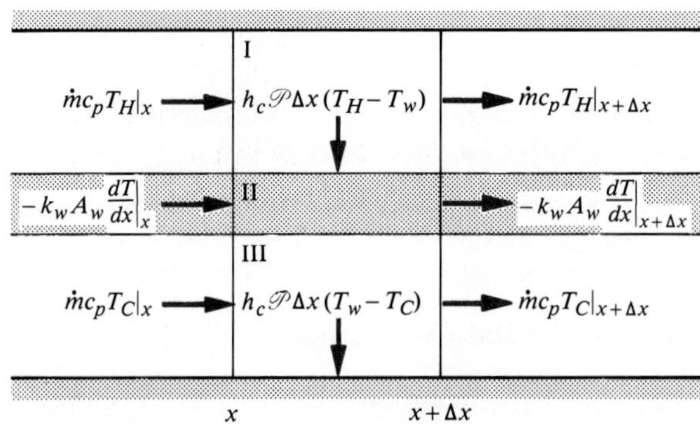

Figure 8.20 Elemental control volumes for the analysis of a parallel-flow exchanger, including axial conduction in the tube wall.

$$k_w A_w \frac{d^2 T_w}{dx^2} = -h_c \mathscr{P}(T_H - T_w) + h_c \mathscr{P}(T_w - T_C) \tag{8.50}$$

$$\dot{m} c_p \frac{dT_C}{dx} = h_c \mathscr{P}(T_w - T_C) \tag{8.51}$$

Substituting Eqs. (8.49) and (8.51) in Eq. (8.50) gives

$$k_w A_w \frac{d^2 T_w}{dx^2} = \dot{m} c_p \left(\frac{dT_H}{dx} + \frac{dT_C}{dx} \right) \tag{8.52}$$

We now introduce dimensionless variables T^* and x^*, defined as

$$T^* = \frac{T - T_{C,\text{in}}}{T_{H,\text{in}} - T_{C,\text{in}}}, \qquad x^* = \frac{x}{L}$$

$$\frac{k_w A_w (T_{H,\text{in}} - T_{C,\text{in}})}{L^2} \frac{d^2 T_w^*}{dx^{*2}} = \frac{\dot{m} c_p (T_{H,\text{in}} - T_{C,\text{in}})}{L} \left(\frac{dT_H^*}{dx^*} + \frac{dT_C^*}{dx^*} \right)$$

or

$$\frac{k_w A_w}{L} \frac{d^2 T_w^*}{dx^{*2}} = \dot{m} c_p \left(\frac{dT_H^*}{dx^*} + \frac{dT_C^*}{dx^*} \right) \tag{8.53}$$

The variables x^* and T^* vary from zero to unity; thus, the derivatives dT_H^*/dx^*, dT_C^*/dx, and d^2T_w/dx^{*2} have an order of magnitude of unity. It follows that the left-hand side of Eq. (8.53) will be negligible if

$$\frac{k_w A_w}{L} \ll \dot{m} c_p$$

or, introducing an axial conduction parameter λ,

$$\lambda = \frac{k_w A_w}{\dot{m} c_p L} \ll 1 \tag{8.54}$$

which is the criterion for axial conduction to be negligible.

A common situation where axial conduction cannot be neglected is when the flow rate $\dot{m}$ is very small. This is the case, for example, in heat exchangers for cryogenic refrigerators used to cool infrared radiation sensors. The effect of axial conduction is always to *reduce* the effectiveness of the exchanger. Exact solutions for effectiveness that account for axial conduction are available [6]. For a balanced-counterflow exchanger and $N_{\text{tu}} > 3$, an approximate formula based on these solutions is

$$\varepsilon = 1 - \frac{1}{1 + N_{\text{tu}}/(1 + \lambda N_{\text{tu}})} \tag{8.55}$$

which is valid even if the heat transfer coefficients for the two streams are markedly different [6].

Perforated-Plate Exchangers

Perforated-plate heat exchangers are often used when the effect of axial conduction must be reduced. Other attractive features of these exchangers is the large heat transfer surface per unit volume and weight, the uniform flow distribution at any cross section, and the high resistance to vibration and shock. Figure 8.21*a* shows

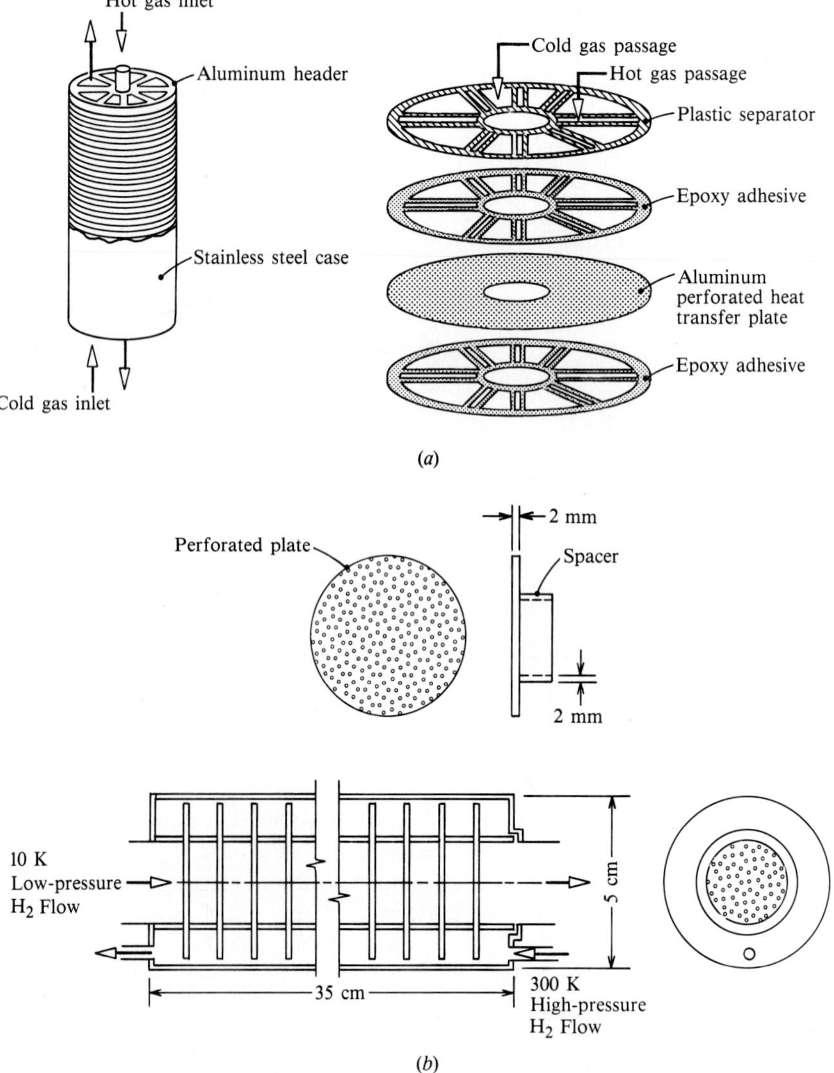

Figure 8.21 Perforated-plate heat exchangers. (*a*) As used in a small, on-board helium refrigerator of a superconducting magnetically levitated high-speed train in Japan. (*b*) As designed for a hydrogen refrigerator to cool infrared sensors for the Deep Space Network of NASA.

such an exchanger used in a small, on-board helium refrigeration system of a superconducting, magnetically levitated high-speed train in Japan. Figure 8.21*b* shows an exchanger for a hydrogen refrigerator designed to cool infrared sensors for the Deep Space Network of the U.S. National Aeronautical and Space Administration. The fin-type analysis required to calculate the overall heat transfer coefficient was given in Example 2.7, and sample correlations of pressure drop and heat transfer for perforated plates were presented in Section 4.5.2.

For these balanced-flow exchangers, the axial conduction parameter is defined as

$$\lambda = \frac{kA_c}{N\Delta L \dot{m} c_p} \tag{8.56}$$

where kA_c is conductivity × cross-sectional area for the spacers, N is the number of plates, and ΔL is the spacer thickness. The usual $\varepsilon - N_{\text{tu}}$ relations for two-stream exchangers do not apply to perforated-plate exchangers. Each plate can be viewed as a separate exchanger with temperature profiles as shown in Fig. 8.22. The high thermal conductivity of the plate prevents its temperature from varying in the flow direction. If axial conduction in the spacers and fluid is negligible, analysis of a balanced-counterflow exchanger gives [7]

$$\varepsilon = \frac{N(1-\beta)^2}{N(1-\beta)^2 + (1-\beta^2)}; \qquad \beta = \exp(-2N_{\text{tu}}/N) \tag{8.57}$$

The number of transfer units is $N_{\text{tu}} = NUA/\dot{m}c_p$, where A is the plate area exposed to one stream, and U is the corresponding overall heat transfer coefficient based on this area. For high-effectiveness exchangers, the effect of axial conduction can be obtained by subtracting a correction value $\Delta\varepsilon_c$ [7]:

$$\Delta\varepsilon_c = \frac{N}{1+N} - \frac{N(1+\lambda)}{1+N(1+2\lambda)} \tag{8.58}$$

For example, if $N = 20$ and $\lambda = 0.02$, $\Delta\varepsilon_c = 0.0166$. This correction may appear to be a small reduction, but when high effectiveness is required, it is significant.

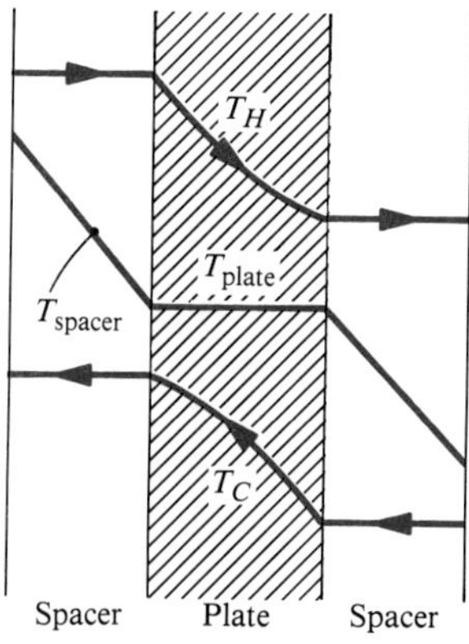

Figure 8.22 Axial temperature variations in a counterflow perforated-plate heat exchanger.

High-effectiveness exchangers are best described in forms of their *ineffectiveness* $(1 - \varepsilon)$. If $(1 - \varepsilon) = 0.05$, the increase due to conduction as calculated above is 33%. Figure 8.23 shows plots of Eqs. (8.57) and (8.58) for a 50-plate exchanger.

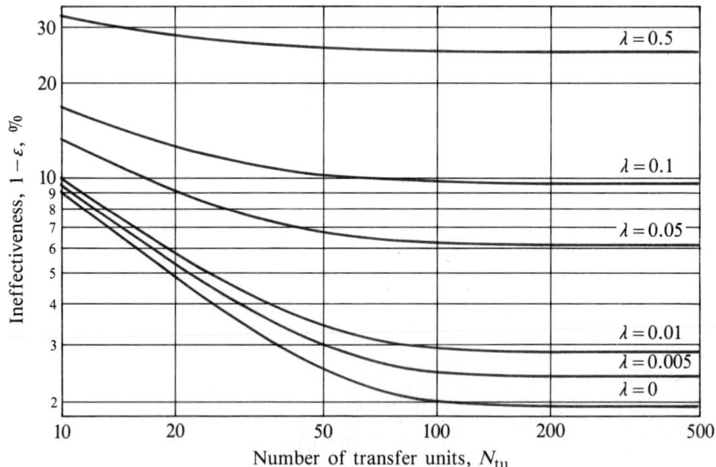

Figure 8.23 Ineffectiveness $(1 - \varepsilon)$ of a perforated-plate heat exchanger for $N = 50$ plates; λ is the axial conduction parameter, $kA_c/N\Delta L\dot{m}c_p$. (Courtesy Dr. J. Rodriguez, Jet Propulsion Laboratory, California Institute of Technology, Pasadena.)

EXAMPLE 8.11 Twin-Tube Recuperator for a Hydrogen Refrigerator

High-effectiveness balanced-counterflow heat exchangers are required for some types of cryogenic refrigeration systems. For laboratory development work, such an exchanger is conveniently made by soldering two copper tubes together and coiling the resulting twin tube. An available unit uses 10 mm–O.D., 1 mm–wall-thickness copper tubes and is 5 m long. It is proposed for use in a hydrogen adsorption refrigerator with a balanced flow of 1.0×10^{-5} kg/s of hydrogen. The cold stream enters at 11 K, and the hot stream enters at 310 K. Calculate the ineffectiveness of the exchanger for this application and the outlet temperature of the cold stream.

Solution

Given: A twin-tube exchanger for balanced hydrogen flows.

Required: Ineffectiveness, $(1 - \varepsilon)$, and $T_{C,\text{out}}$.

Assumptions:
1. Fully developed flows.
2. Fin effectiveness of the copper tube wall is unity.

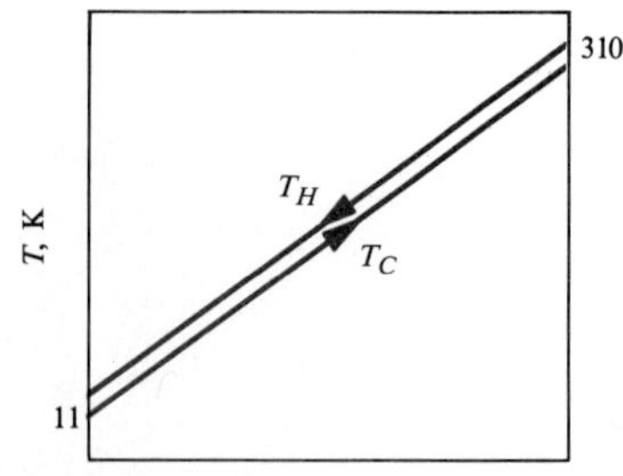

Properties for both streams will be evaluated at an estimated average temperature of 160 K; from Table A.7, $c_p = 16{,}120$ J/kg K, $k = 0.131$ W/m K, $\mu = 5.84 \times 10^{-6}$ kg/m s. The Reynolds number is

$$\mathrm{Re}_D = \frac{VD\rho}{\mu} = \frac{(\dot{m}/A_c)D}{\mu} = \frac{4\dot{m}}{\pi\mu D} = \frac{4(1 \times 10^{-5})}{\pi(5.84 \times 10^{-6})(0.008)} = 273$$

The flows are laminar; the heat transfer coefficient can be obtained from Table 4.5. Since $\Delta T = T_H - T_C$ is constant along a balanced-counterflow exchanger, the wall heat flux can be taken to be constant, so that Nu = 4.364.

$$h_c = \mathrm{Nu}(k/D) = 4.364(0.131/0.008) = 71.5 \text{ W/m}^2 \text{ K}$$

To estimate the overall heat transfer coefficient, we will assume a wall fin effectiveness of unity for the copper tubes and ignore the thermal resistance of the soldered joint (see Exercise 8–6). Then

$$\frac{1}{U\mathscr{P}} = \frac{1}{h_{c,H}\pi D} + \frac{1}{h_{c,C}\pi D} = \frac{2}{h_c \pi D} = \frac{2}{(71.5)\pi(0.008)}$$

$$U\mathscr{P} = 0.898 \text{ W/m K}$$

$$N_{\mathrm{tu}} = \frac{U\mathscr{P}L}{\dot{m}c_p} = \frac{(0.898)(5.0)}{(1.0 \times 10^{-5})(16{,}120)} = 27.9$$

Equation (8.54) gives the axial conduction parameter λ,

$$\lambda = \frac{k_w A_w}{\dot{m}c_p L} = \frac{(380)(\pi/4)(0.01^2 - 0.008^2)(2)}{(1.0 \times 10^{-5})(16{,}120)(5)} = 2.66 \times 10^{-2}$$

where a conservative value of 380 W/m K has been used for the conductivity of the copper tube wall. Equation (8.55) gives the effectiveness, including the effect of axial conduction:

$$\varepsilon = 1 - \frac{1}{1 + N_{\mathrm{tu}}/(1 + \lambda N_{\mathrm{tu}})} = 1 - \frac{1}{1 + 27.9/[1 + (0.0266)(27.9)]} = 0.9412$$

$$1 - \varepsilon = 1 - 0.9412 = 0.0588 \ (5.88\%)$$

The outlet temperature of the cold stream is obtained from

$$\varepsilon = 0.9412 = \frac{T_{C,\mathrm{out}} - T_{C,\mathrm{in}}}{T_{H,\mathrm{in}} - T_{C,\mathrm{in}}} = \frac{T_{C,\mathrm{out}} - 11}{310 - 11}$$

$$T_{C,\mathrm{out}} = 11 + 0.9412(310 - 11) = 292.4 \text{ K}$$

Comments

1. If axial conduction is ignored the effectiveness is given by Eq. (8.45),

$$\varepsilon = \frac{N_{\mathrm{tu}}}{1 + N_{\mathrm{tu}}} = \frac{27.9}{1 + 27.9} = 0.9654; \qquad 1 - \varepsilon = 0.0346$$

that is, the effect of axial conduction is to increase the ineffectiveness from 3.46% to 5.88%.

2. The large effect of axial conduction indicates that a copper heat transfer surface is not appropriate for this low hydrogen flow rate: the parameter λ could be reduced considerably by using stainless steel.
3. The estimated average temperature of 160 K for evaluation of properties is adequate.

EXAMPLE 8.12 Perforated-Plate Recuperator for a Hydrogen Refrigerator

A perforated-plate heat exchanger for a solid hydrogen refrigeration system has 100 plates with a flow cross-sectional area for each stream of 0.001 m^2. The plates are 2 mm–thick aluminum, and the spacers are 3 mm thick and have a footprint area of 0.0004 m^2 and a conductivity of 0.35 W/m K. Determine the ineffectiveness of the exchanger for balanced counterflow of hydrogen at 6×10^{-6} kg/s. Take the overall heat transfer coefficient as 60 W/m^2 K, based on a transfer area of 0.00216 m^2 per plate and a hydrogen specific heat of 13,600 J/kg K.

Solution

Given: Counterflow perforated-plate exchanger for a balanced flow of hydrogen.

Required: Effect of axial conduction on ineffectiveness $(1 - \varepsilon)$.

Assumptions: U is constant along the exchanger.

First, assume that axial conduction is negligible. The number of transfer units is

$$N_{\text{tu}} = \frac{NUA}{\dot{m}c_p} = \frac{(100)(60)(0.00216)}{(6 \times 10^{-6})(13{,}600)} = 158.8$$

Substituting in Eq. (8.57),

$$\beta = \exp(-2N_{\text{tu}}/N) = \exp[-2(158.8)/100] = 0.0417$$

$$\varepsilon = \frac{N(1-\beta)^2}{N(1-\beta)^2 + (1-\beta^2)} = \frac{100(1-0.0417)^2}{100(1-0.0417)^2 + (1-0.0417^2)} = 0.9892$$

$$1 - \varepsilon = 0.0108\ (1.08\%)$$

The axial conduction parameter is $\lambda = kA_c/N\Delta L\dot{m}c_p$. Since the conductivity of the spacer is low, we should also allow for axial conduction in the two streams of hydrogen. Taking $k \simeq 0.1$ W/m K for the hydrogen gives

$$\lambda = \frac{(0.35)(0.0004) + 2(0.1)(0.001)}{(100)(0.003)(6 \times 10^{-6})(13{,}600)} = 0.0139$$

The correction for axial conduction is given by Eq. (8.58):

$$\Delta\varepsilon_c = \frac{N}{1+N} - \frac{N(1+\lambda)}{1+N(1+2\lambda)} = \frac{100}{1+100} - \frac{100(1+0.0139)}{1+100[1+(2)(0.0139)]} = 0.0131$$

Hence, the true ineffectiveness is

$$1 - \varepsilon = 0.0108 + 0.0131 = 0.0239\ (2.39\%)$$

Comments

1. Axial conduction more than doubles the ineffectiveness of the exchanger, even though the configuration has inherently low axial conduction.
2. Our ad hoc allowance for axial conduction in the hydrogen may be inaccurate; more detailed analysis is required in order to properly assess the possible error.

8.6 REGENERATORS

The operation of regenerators was described in Section 8.2. These exchangers may be in the form of a fixed bed through which the hot and cold streams flow alternatively, or in the form of a rotating wheel with segmented manifolds for the hot and cold streams, as shown in Fig. 8.5. Fixed beds were introduced in 1816, when checkerboard arrays of bricks were used to preheat air for combustion in steel blast furnaces. The rotary regenerator was developed in 1919 and is used for waste heat recovery from gas turbines and air-conditioning systems.

The governing equations for regenerators are partial differential equations, since temperatures in the exchanger are a function of position and time. In general, these equations must be solved using numerical rather than analytical methods. Our approach here will be to derive the governing equations and to make the equations dimensionless so as to identify the pertinent dimensionless groups. Results will then be presented in the form of a chart in terms of these groups.

8.6.1 Balanced Counterflow Regenerators

Regenerators are most often used for balanced flow, that is, when the flow thermal capacities are equal, $C_H = C_C = C$. Figure 8.24 shows a schematic of a parallel-plate regenerator matrix. The half-thickness of the wall is δ, and we will assume that the matrix temperature, T_M, is constant across the wall. The matrix is exposed to cold

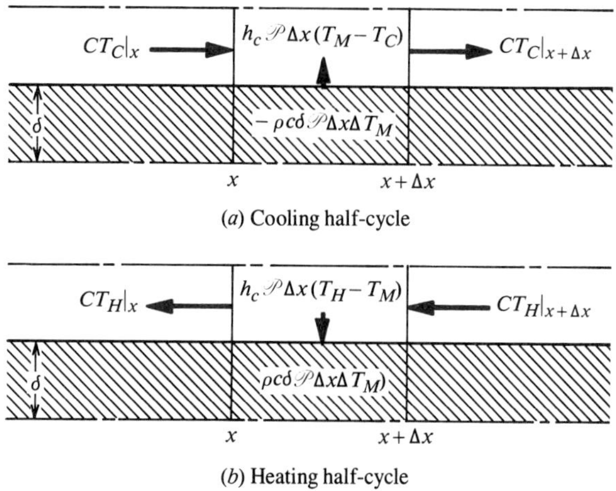

Figure 8.24 Elemental control volumes for the analysis of a regenerator.

flow from time $t = 0$ to τ, and to hot flow from $t = \tau$ to 2τ. Energy balances on the hot and cold streams for an element Δx long give

$$CT_C|_x + h_c\mathscr{P}\Delta x(T_M - T_C) = CT_C|_{x+\Delta x}$$

$$CT_H|_x + h_c\mathscr{P}\Delta x(T_H - T_M) = CT_H|_{x+\Delta x}$$

for counterflow. The heat transfer coefficients for both streams have been assumed equal. Also, energy storage in the fluid streams has been neglected, which in practice requires the fluid to be a gas. Dividing by Δx and letting $\Delta x \to 0$ gives

$$C\frac{\partial T_C}{\partial x} = h_c\mathscr{P}(T_M - T_C) \tag{8.59a}$$

$$C\frac{\partial T_H}{\partial x} = h_c\mathscr{P}(T_H - T_M) \tag{8.59b}$$

As the cold fluid flows through the exchanger, the matrix cools down. The heat capacity of a length of matrix Δx long is $\rho c\delta\mathscr{P}\Delta x$; thus, an energy balance over a time interval Δt, in which the matrix temperature change is ΔT_M, gives

$$-\rho c\delta\mathscr{P}\Delta x\Delta T_M = h_c\mathscr{P}\Delta x(T_M - T_C)\Delta t$$

for negligible axial conduction. Dividing by $\Delta x\Delta t$ and letting $\Delta t \to 0$ gives

$$-\rho c\delta\mathscr{P}\frac{\partial T_M}{\partial t} = h_c\mathscr{P}(T_M - T_C) \tag{8.60a}$$

Similarly, when the hot fluid heats up the matrix,

$$\rho c\delta\mathscr{P}\frac{\partial T_M}{\partial t} = h_c\mathscr{P}(T_H - T_M) \tag{8.60b}$$

Rewriting Eqs. (8.59a) and (8.60a) in terms of the dimensionless independent variables $x^* = x/L$ and $t^* = t/\tau$ gives, for $0 < t < \tau$,

$$C\frac{\partial T_C}{\partial x^*} = h_c\mathscr{P}L(T_M - T_C) = -C_R\frac{\partial T_M}{\partial t^*} \tag{8.61}$$

where C_R is the thermal capacity of the regenerator matrix divided by the exposure time. For a heat exchange matrix of mass W_M,

$$C_R = \frac{\rho c\delta\mathscr{P}L}{\tau} = \frac{cW_M}{\tau}$$

and is independent of matrix geometry. Dividing Eq. (8.61) by C then gives

$$\frac{\partial T_C}{\partial x^*} = N_{tu}(T_M - T_C) = -R_R\frac{\partial T_M}{\partial t^*} \tag{8.62}$$

where $N_{tu} = h_c\mathscr{P}L/C$ and $R_R = C_R/C$.

Similarly, for $\tau < t < 2\tau$,

$$C\frac{\partial T_H}{\partial x^*} = h_c\mathscr{P}L(T_H - T_M) = C_R\frac{\partial T_M}{\partial t^*} \tag{8.63}$$

$$\frac{\partial T_H}{\partial x^*} = N_{\rm tu}(T_H - T_M) = R_R\frac{\partial T_M}{\partial t^*} \tag{8.64}$$

Thus, there are two dimensionless parameters in the governing equations, $N_{\rm tu}$ and R_R. The required boundary conditions are

$$x = 0: \quad 0 < t < \tau; \; T_C = T_{C,\rm in} \tag{8.65a}$$

$$x = L: \quad \tau < t < 2\tau; \; T_H = T_{H,\rm in} \tag{8.65b}$$

There is no initial condition: instead, we require that the solution be periodic. The partial differential equations Eqs. (8.62) and (8.64) can be solved numerically. The solution in the form of heat transfer effectiveness $\varepsilon = (\overline{T}_{C,\rm out} - T_{C,\rm in})/(T_{H,\rm in} - T_{C,\rm in})$, where $\overline{T}_{C,\rm out}$ is an average over the half-cycle, is shown in Fig. 8.25.

If the exchanger is in the form of a wheel rotating at angular velocity ω rad/s, and if the matrix is exposed to hot flow from $\theta = 0$ to Θ and to cold flow for $\theta = \pi$ to $\pi + \Theta$, then the foregoing results apply, with t replaced by θ and τ replaced by Θ/ω; also, $C_R = cW_M/(\Theta/\omega)$.

Analytical solutions can be obtained for two limit cases, as follows.

Case 1: $R_R \gg N_{\rm tu}$. When the characteristic time τ is small, which in practice is obtained with a wheel-type exchanger rotating at a sufficiently high speed, Eqs. (8.62) and (8.64) show that $\partial T_M/\partial t^*$ goes to zero, that is, the matrix temperature is

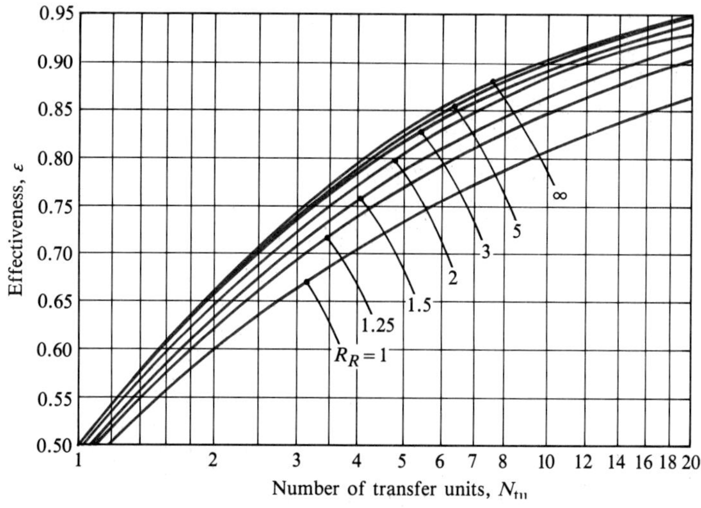

Figure 8.25 Effectiveness of a balanced-counterflow regenerator.

independent of time (or angle θ) and depends only on axial distance. Adding Eqs. (8.61) and (8.63) then gives

$$h_c \mathscr{P} L(T_H - T_C) = C\left(\frac{dT_H}{dx^*} + \frac{dT_C}{dx^*}\right)$$

But for balanced flow, $dT_H/dx^* = dT_C/dx^*$; thus,

$$h_c \mathscr{P} L(T_H - T_C) = 2C\frac{dT_H}{dx^*} = 2C\frac{dT_C}{dx^*} \tag{8.66}$$

which is equivalent to Eqs. (8.46) when $U = h_c/2$. Thus, the solution is the same as for a balanced-counterflow exchanger. From Eq. (8.45),

$$\varepsilon = \frac{N_{tu}}{1 + N_{tu}}; \qquad N_{tu} = \frac{h_c \mathscr{P} L}{2C} \tag{8.67}$$

When the hot and cold side heat transfer coefficients are not equal it can be shown that the relation $\varepsilon = N_{tu}/(1 + N_{tu})$ remains valid if

$$N_{tu} = \frac{U \mathscr{P} L}{C}; \qquad \frac{1}{U} = \frac{1}{h_{cH}} + \frac{1}{h_{cC}} \tag{8.68}$$

Note that the transfer area $\mathscr{P} L$ is for a single stream and is therefore $\Theta/2\pi$ of the total area in a wheel unit.

Case 2: $N_{tu} \gg R_R$. When h_c is large, we can set $T_H = T_M$, and from Eq. (8.64),

$$\frac{\partial T_H}{\partial x^*} = R_R \frac{\partial T_H}{\partial t^*} = \frac{\partial T_H}{\partial (t^*/R_R)} \tag{8.69}$$

which states that T_H is not a function of x^* or t^* independently but is a function only of the combination $x^* + (t^*/R_R)$. Similarly, T_C is a function of $x^* - (t^*/R_R)$ only. The result is that a temperature front moves like a shock wave through the regenerator. As hot fluid flows through the matrix, a hot front advances linearly with increasing time; this hot front retreats as the cold fluid flows back through the other way. In a real regenerator, the temperature change is not discontinuous due to the effect of a finite value of h_c and to axial conduction in the matrix.

The designer needs to know whether the hot front has sufficient time to pass all the way through the matrix. The requirement is $x^* = t^*/R_R$ for $x^* = 1$ and $t^* = 1$; that is,

$$R_R = \frac{cW_M}{\dot{m} c_p \tau} = 1 \tag{8.70}$$

Equation (8.70) gives the combination of length and half-period required for the front to just pass through the matrix. If R_R is much greater than unity, the matrix may be unnecessarily long, or the rate of rotation may be too fast.

EXAMPLE 8.13 A Packed-Bed Air Heater

Exercise 4–77 describes a regenerative air heater packed with 9 mm steel ball bearings. Relevant data include a cross-sectional area of 0.1 m^2, an air flow rate of $\dot{m} = 0.05$ kg/s, a volume void fraction $\varepsilon_v = 0.38$, a transfer perimeter $\mathscr{P} = 41.3$ m, and a heat transfer coefficient $h_c = 134$ W/m^2 K. Assuming balanced flow and equal heat transfer coefficients for both hot and cold flows, determine effectiveness as a function of half-period τ for an exchanger 5.0 cm long. Take the specific heat of the steel as 460 J/kg K.

Solution

Given: Packed-bed regenerator for heating air.

Required: Effectiveness ε as a function of half-period τ.

Assumptions: 1. Balanced flow.
2. Heat transfer coefficients equal for hot and cold streams.

The flow thermal capacity is $C = \dot{m}c_p = (0.05)(1130) = 56.5$ W/K, where c_p has been evaluated at 1000 K (the temperature of the air stream in Exercise 4–77). The matrix mass is $W_M = A(1 - \varepsilon_v)L\rho = (0.1)(1 - 0.38)(0.05)(7840) = 24.3$ kg.

$$N_{\text{tu}} = \frac{h_c \mathscr{P} L}{C} = \frac{(134)(41.3)(0.05)}{56.5} = 4.90$$

$$R_R = \frac{cW_M}{\tau C} = \frac{(460)(24.3)}{56.5\tau} = \frac{198}{\tau}$$

Using Fig. 8.25, the following table is constructed.

R_R	ε	τ, s
1	0.73	198
1.25	0.77	158
1.5	0.78	132
2	0.80	99
3	0.81	66
5	0.82	40
∞	0.83	0

Comments

The effectiveness increases with R_R until, as $R_R \to \infty$, it approaches the value for a two-stream balanced-counterflow exchanger, $\varepsilon = N_{\text{tu}}/(1 + N_{\text{tu}}) = 4.9/5.9 = 0.83$.

EXAMPLE 8.14 A Cold-Weather Mask

A face shield regenerator for use in the Antarctic is to be designed to reduce thermal stress in the nose. Assume that 1.0×10^{-3} kg of air is inhaled each 2 s. The ambient air is at −40°C, and the exhaled breath is at 33°C. Neglecting any effects due to condensation or evaporation of moisture, size a suitable matrix.

Solution

Given: A regenerator for the human nose.

Required: Thermal design, in particular, type and size of matrix.

Assumptions: 1. Balanced flow.
2. Negligible effects of condensation and evaporation of moisture.

The need to specify a half-period of 1 s is a fixed constraint on the design. In addition, the exchanger must be relatively small to be practical. We will choose a plastic sheet packing. Since our analysis assumes that the temperature across the matrix is uniform, the Biot number for the sheet should be less than 0.1 for the analysis to be valid. A matrix formed from a 200 μm–thick sheet wrapped in a spiral at pitch of 1 mm should be easy to construct, and it should give a small Bi value. We will explore whether these dimensions lead to a satisfactory design. Assuming laminar flow, since the gap width Δ is small, Table 4.5 gives a Nusselt number for a constant wall heat flux of 8.24 based on hydraulic diameter, which is 2Δ. Using air properties at an approximate average temperature of 0°C,

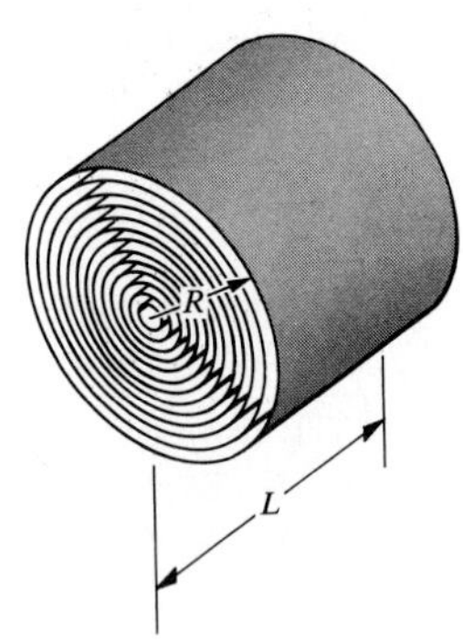

$$h_c = (k/D_h)\text{Nu} = (0.025/2\Delta)(8.24) = 0.103/\Delta$$

$$N_{tu} = \frac{h_c \mathscr{P} L}{\dot{m} c_p} = \frac{0.103}{(1.0 \times 10^{-3})(1009)} \frac{\mathscr{P} L}{\Delta} = 0.102 \frac{\mathscr{P} L}{\Delta}$$

From Table A.2, properties of polyethylene are $\rho = 960$ kg/m^3, $c = 2090$ J/kg K, and $k = 0.33$ W/m K. Thus, for a wall half-thickness of $\delta = 100 \times 10^{-6}$ m,

$$R_R = \frac{\rho c \delta}{\dot{m} c_p \tau} \mathscr{P} L = \frac{(960)(2090)(100 \times 10^{-6})}{(1.0 \times 10^{-3})(1009)(1)} \mathscr{P} L = 199 \mathscr{P} L$$

If the exchanger has a radius of R, then

$$\begin{aligned} \mathscr{P} &= \text{(Number of turns)(Average length of turn)(2 sides)} \\ &= (R/0.001)(\pi R)(2) \\ &= 6283 R^2 \end{aligned}$$

$$\Delta = 1 \times 10^{-3} - 0.2 \times 10^{-3} = 0.8 \times 10^{-3} \text{ m}$$

Hence,

$$N_{tu} = 8.01 \times 10^5 R^2 L$$

$$R_R = 1.25 \times 10^6 R^2 L$$

Before proceeding, we should check the Biot number of the packing:

$$\mathrm{Bi} = \frac{h_c\delta}{k_{\text{plastic}}} = \frac{(0.103)(100 \times 10^{-6})}{(0.8 \times 10^{-3})(0.33)} = 0.039 < 0.1$$

which is satisfactory.

Using Fig. 8.25, the following table can be constructed.

				L, cm		
R_R	R^2L, cm^3	N_{tu}	ε	R = 1 cm	2 cm	3 cm
1	0.8	0.64	<0.5	0.8		
2	1.6	1.28	0.55	1.6	0.4	
3	2.4	1.92	0.65	2.4	0.6	0.27
5	4.0	3.20	0.76	4.0	1.0	0.44
10	8.0	6.41	0.86	8.0	2.0	0.89

An effectiveness of 0.76 will give an average outlet cold air temperature of

$$\overline{T}_{C,\text{out}} = T_{C,\text{in}} + \varepsilon(T_{H,\text{in}} - T_{C,\text{in}}) = -40 + 0.76[33 - (-40)] = 15.5°\text{C}$$

which should be relatively comfortable. Hence, choosing $R_R = 5$, the matrix could be in the form of a cylindrical cartridge 2 cm in diameter and 4 cm long.

Comments

1. A final design may have somewhat different values of plastic sheet thickness, gap width, radius, and length, but the above result shows that a satisfactory solution can be obtained. Provision should be made for quick replacement of the cartridge should it plug with ice.
2. The Eskimo in the Arctic merely breathes through the fur of his parka!
3. This example is a D. K. Edwards original.

8.7 ELEMENTS OF HEAT EXCHANGER DESIGN

The examples of heat exchanger calculations given in Section 8.5 show that heat transfer considerations alone do not determine the dimensions of a heat exchanger. For example, in Example 8.8, the required heat transfer area was calculated, but further constraints must be specified in order to fix the overall dimensions and number of plates. The usual heat exchanger design problem requires the engineer to specify a unit that will have a given heat transfer performance—that is, effectiveness—subject to a number of constraints, which might include: (1) a low capital cost; (2) a low operating cost; (3) limitations on size, shape, or weight; and (4) ease of maintenance. The most important operating cost may be the power required to pump the fluids. For a liquid, this power is usually rather low and does not have a major impact on the design. For gases, the power required per unit mass of working fluid is great, so it is often a critical design constraint. The pumping power is simply volume flow rate times pressure drop, divided by the blower efficiency. Thus, the need to have a low operating cost translates into a pressure drop constraint on the design.

In general, the engineer has the freedom to choose the exchanger configuration (counterflow, cross-flow, multiple-pass, etc.), the type of heat transfer surface (coaxial tube, plate-fin, tube bank, etc.), and characteristic dimensions of the surface (tube diameter, spacing in a tube bank, etc.). A low pressure drop requires a large cross-sectional area for flow, although appropriate selection of configuration and heat transfer surface also plays a role. A possible design strategy is as follows:

1. Specify the required heat transfer effectiveness.
2. Specify the allowable pressure drop for one or both streams.
3. Choose a configuration.
4. Choose a type of heat transfer surface.
5. Choose the dimensions of the surface.
6. Calculate the resulting dimensions of the unit.
7. Evaluate the design with respect to constraints such as capital cost, size, weight, and maintenance.

The dimensions obtained for step 6 may not be unique. Also, for some configurations, such as the coaxial-tube exchanger, it will not be possible to choose the surface dimensions since these will be fixed by the heat transfer and pressure drop requirements.

The complete exchanger design problem is one of optimization, for which advanced mathematical and computational methods are available. These methods should be used in any serious endeavor. But before such sophisticated tools are used, the engineer must have a clear understanding of principles underlying the design process. Of particular importance are the impact of a pressure drop constraint and economic considerations, and these topics are the primary focus of this section. The calculation of pressure drop in exchangers is discussed in Section 8.7.1, and the impact of a pressure drop constraint on sizing a heat exchanger is discussed in Section 8.7.2. Section 8.7.3 deals with the criteria for the selection of heat transfer surfaces for **compact heat exchangers**, which are defined somewhat arbitrarily as exchangers with a heat transfer surface area per unit volume greater than 700 m^2/m^3. The exchanger itself may not be small. Compact heat exchangers are used for gas-to-gas and gas-to-liquid transfer, because the large surface area gives a high h_cA product on the gas side, even though the heat transfer coefficient h_c is typically low for gases. In Section 8.7.4, a brief introduction to the economic analysis of exchanger utilization is given. The chapter concludes with an example of computer-aided heat exchanger design, in Section 8.7.5.

8.7.1 Exchanger Pressure Drop

For single-phase flow inside tubes, across tube banks, and through packed beds, formulas and data for pressure drop are given in Chapter 4. Compact heat exchangers, however, need special consideration.

Flow inside Tubes

For flow inside the tubes of a shell-and-tube or a coaxial-tube exchanger, pressure drop calculations may be made using the formulas and data given in Section 4.3. Care must be taken to ascertain whether the flow is laminar or turbulent. In addition, the low-Reynolds-number turbulent flow regime ($\mathrm{Re} < 10{,}000$) is best avoided, if at all possible, owing to the uncertainty of the friction factor (and heat transfer coefficient) in this regime. Additional data may be found in the handbooks listed in the bibliography at the end of this text.

Flow across Tube Banks

Pressure drop across tube banks can be calculated from the formulas and graphs of Section 4.5. For a simple cross-flow exchanger, these data suffice. However, a complete calculation for the shell side of a shell-and-tube exchanger requires consideration of additional pressure drops across baffle windows and in the inlet and exit zones of the shell. Empirical formulas are used for this purpose. Examples may be found in handbooks, but often such data are company-proprietary, pertaining only to the manufacturer's own shell and baffle configurations.

Flow through Packed Beds

Packed beds are used in regenerators and also in some recuperators, such as the perforated-plate exchanger described in Example 4.14. Section 4.5 gives some appropriate formulas for pressure drop. The chemical engineering literature is a good source of additional data, owing to the extensive use of packed beds for mass exchangers.

Compact Heat Exchangers

Pressure drop is usually an important constraint on compact heat exchanger design, and for gas flows, these exchangers tend to have a large frontal area and short flow length. Figure 8.26 shows a schematic of a compact heat exchanger core. Following Kays and London [5], the total pressure drop between sections 1 and 2, ΔP, is written as the contraction pressure drop ΔP_{in} plus the core pressure drop ΔP_{core} minus the expansion pressure recovery ΔP_{out}:

$$\Delta P = \Delta P_{\mathrm{in}} + \Delta P_{\mathrm{core}} - \Delta P_{\mathrm{out}} \tag{8.71}$$

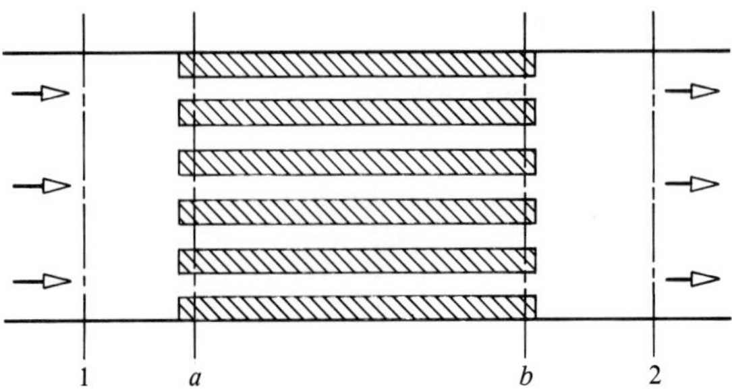

Figure 8.26 Schematic of a compact heat exchanger core.

The *inlet pressure drop* can be written as the sum of the pressure drop due to flow area change for an inviscid fluid, plus the irreversible pressure loss due to viscous effects. Assuming a constant density, since the pressure change is usually small compared to the total pressure,

$$\Delta P_{\text{in}} = \frac{1}{2}\rho_a V_a^2\left[1 - \left(\frac{A_c}{A_{\text{fr}}}\right)^2\right] + \frac{1}{2}\rho_a V_a^2 K_c \tag{8.72}$$

where A_c is the cross-sectional area for flow in the core, A_{fr} is the frontal area, and K_c is the *contraction coefficient*. Similarly, the outlet pressure rise is the sum of the pressure rise due to flow area change for an inviscid fluid, minus the pressure loss due to viscous effects. If A_c is constant and A_{fr} is the same at the inlet and outlet,

$$\Delta P_{\text{out}} = \frac{1}{2}\rho_b V_b^2\left[1 - \left(\frac{A_c}{A_{\text{fr}}}\right)^2\right] - \frac{1}{2}\rho_b V_b^2 K_e \tag{8.73}$$

where K_e is the *expansion coefficient*. There are two factors that contribute to the *core pressure drop*. First, there is the form and viscous drag of the heat transfer surface, and second, there is the pressure drop required to accelerate the fluid:

$$\Delta P_{\text{core}} = f\frac{1}{2}\rho_m V_m^2 (L/D_h) + (\rho_b V_b^2 - \rho_a V_a^2) \tag{8.74}$$

where ρ_m and V_m are suitable average values through the core, and L/D_h can be written as $A/4A_c$, where $A = \mathscr{P}L$ is the transfer surface area.

Introducing the mass velocity in the core, G [kg/m^2 s],

$$G = \rho_a V_a = \rho_m V_m = \rho_b V_b \tag{8.75}$$

and for a core–to–frontal area ratio $\sigma = A_c/A_{\text{fr}} < 1$,

$$V_a = \frac{V_1}{\sigma}; \qquad V_b = \frac{V_2}{\sigma} \tag{8.76}$$

Then, assuming $\rho_1 = \rho_a$ and $\rho_2 = \rho_b$,

$$G = \rho_1 \frac{V_1}{\sigma} = \rho_2 \frac{V_2}{\sigma} \tag{8.77}$$

Substituting Eqs. (8.72) through (8.74) in Eq. (8.71) and rearranging gives the *exchanger pressure drop equation*:

$$\frac{\Delta P}{P_1} = \frac{G^2}{2\rho_1 P_1}\left[(1 - \sigma^2 + K_c) + \frac{f}{4}\frac{\rho_1}{\rho_m}\frac{A}{A_c} + 2\left(\frac{\rho_1}{\rho_2} - 1\right) - \frac{\rho_1}{\rho_2}(1 - \sigma^2 - K_e)\right] \tag{8.78}$$

The contraction and expansion coefficients, K_c and K_e, are functions of geometry and, to a lesser extent, functions of the Reynolds number in the core. Figure 8.27 shows some sample data. The contraction coefficient K_c also accounts for momentum changes associated with the developing velocity profile in the hydrodynamic entrance length: thus, the friction factor f in Eq. (8.78) is evaluated for fully developed hydrodynamic conditions.

For flow across tube banks, it is a common practice to account for inlet and outlet viscous pressure losses in the friction factor f. Then, setting $K_c = K_e = 0$ in Eq. (8.78) and rearranging gives

$$\frac{\Delta P}{P_1} = \frac{G^2}{2\rho_1 P_1}\left[\frac{f}{4}\frac{A}{A_c}\frac{\rho_1}{\rho_m} + (1 + \sigma^2)\left(\frac{\rho_1}{\rho_2} - 1\right)\right] \tag{8.79}$$

For matrix-type surfaces, Eq. (8.78) also applies, but with σ replaced by the void fraction ε_v. Notice that in these equations, the frictional pressure drop $\Delta P/P_1$ is expressed in terms of a multiplier of the ratio of dynamic head to static head:

$$\frac{G^2}{2\rho_1 P_1} = \frac{V_a^2/2}{P_1/\rho_1} = \frac{\text{Dynamic head}}{\text{Static head}}$$

The appropriate value of the mean density ρ_m to be used in Eq. (8.78) or (8.79) depends on the variation of temperature through the exchanger. When the capacity ratio $R_C \simeq 1$, a simple average of ρ_1 and ρ_2 can be used for all configurations except parallel flow.

In some applications, such as an automobile radiator mounted at the front of the vehicle, the total pressure drop is as given by Eq. (8.78). However, in most applications, the working fluids are ducted to the exchanger and must be distributed through a *manifold* or *header* to the core. Not only must the header provide a uniform distribution of fluid to the core, but it must also provide a transition from the duct cross-sectional area to the exchanger frontal area with a minimum pressure loss. When there is a large mismatch of duct and exchanger cross-sectional areas, the pressure losses in the headers may exceed those of the core. In such situations, an optimal design of the exchanger requires that header pressure drop be included as a constraint.

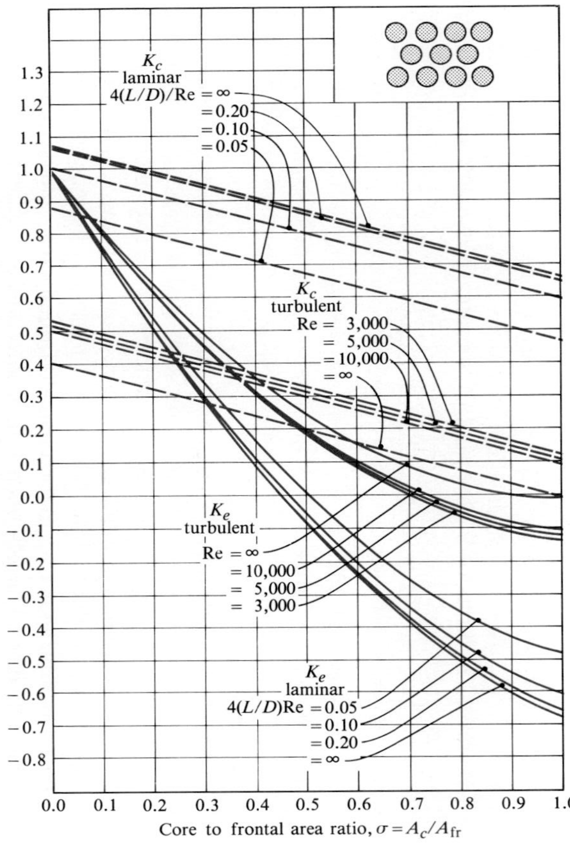

K_c
laminar
$4(L/D)/\text{Re} = \infty$
$= 0.20$
$= 0.10$
$= 0.05$
K_c
turbulent
Re = 3,000
= 5,000
= 10,000
= ∞
K_e
turbulent
Re = ∞
= 10,000
= 5,000
= 3,000
K_e
laminar
$4(L/D)\text{Re} = 0.05$
$= 0.10$
$= 0.20$
$= \infty$
0.0 0.1 0.2 0.3 0.4 0.5 0.6 0.7 0.8 0.9 1.0
1.3 1.2 1.1 1.0 0.9 0.8 0.7 0.6 0.5 0.4 0.3 0.2 0.1 0.0 −0.1 −0.2 −0.3 −0.4 −0.5 −0.6 −0.7 −0.8

Contraction and expansion coefficients, K_c and K_e
Core to frontal area ratio, $\sigma = A_c/A_{fr}$

(a)

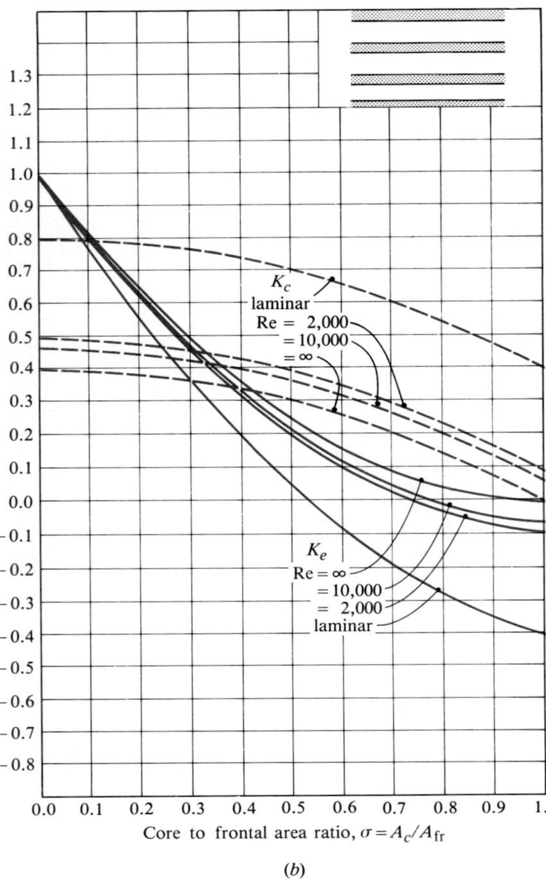

K_c
laminar
Re = 2,000
= 10,000
= ∞
K_e
Re = ∞
= 10,000
= 2,000
laminar
0.0 0.1 0.2 0.3 0.4 0.5 0.6 0.7 0.8 0.9 1.0
1.3 1.2 1.1 1.0 0.9 0.8 0.7 0.6 0.5 0.4 0.3 0.2 0.1 0.0 −0.1 −0.2 −0.3 −0.4 −0.5 −0.6 −0.7 −0.8

Contraction and expansion coefficients, K_c and K_e
Core to frontal area ratio, $\sigma = A_c/A_{fr}$

(b)

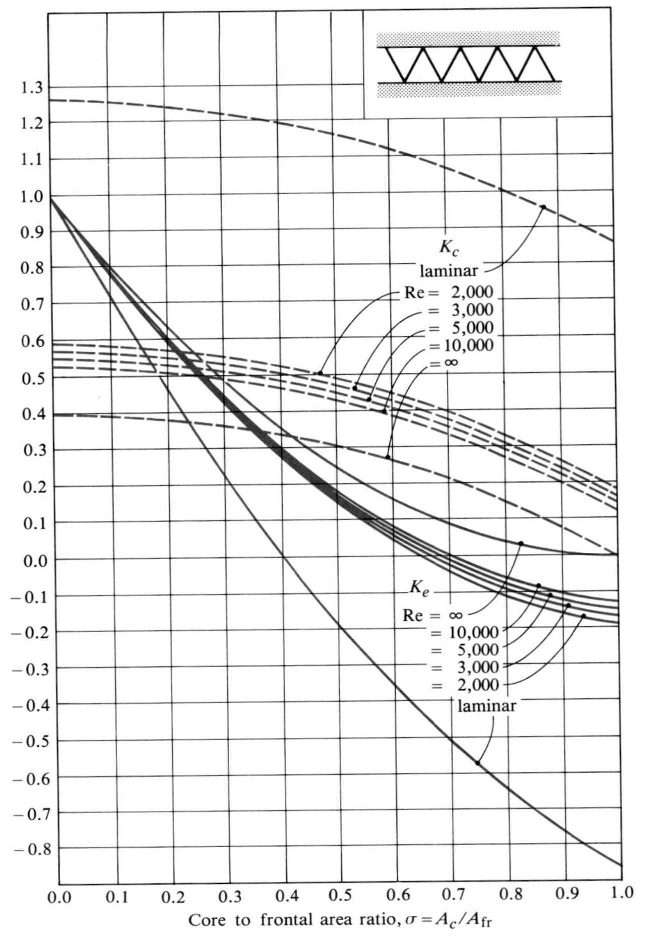

(c)

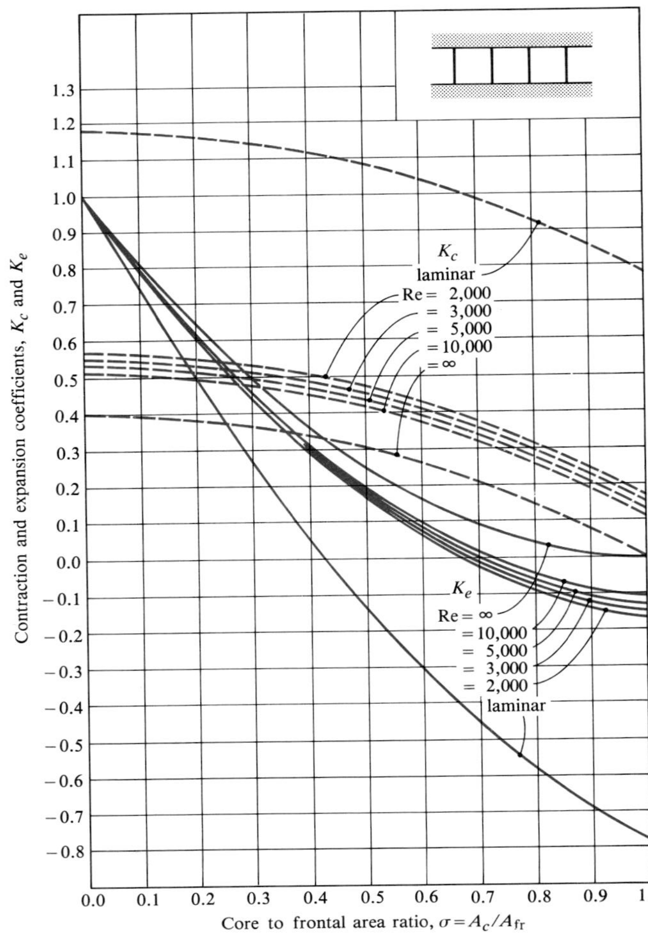

(d)

Figure 8.27 Contraction and expansion pressure loss coefficients for various exchanger cores (90° edges). (*a*) Circular tubes. (*b*) Parallel plates. (*c*) Triangular passages. (*d*) Square passages [5]. (Adapted with permission.)

EXAMPLE 8.15 Pressure Drop in a Multiple-Tube Exchanger

A multiple-circular-tube exchanger is constructed from 100 1 m–long, 6 mm–O.D., 0.5 mm–wall-thickness tubes in a square array, with a pitch of 8 mm. The frontal area of the exchanger is 0.0064 m^2. The tube walls are maintained at 350 K by condensing steam, while 5.66×10^{-3} kg/s of helium enters the tubes at 50 K and 15 kPa pressure. Determine the pressure drop of the helium flow.

Solution

Given: Multiple-circular-tube exchanger to heat helium by condensing steam.

Required: Pressure drop of helium flow.

Equation (8.78) gives the pressure drop. The cross-sectional area for helium flow is

$$A_c = (100)(\pi/4)(0.006 - 0.001)^2 = 0.00196 \text{ m}^2$$

The inlet helium density is

$$\rho_1 = \frac{PM}{\mathcal{R}T} = \frac{(15{,}000)(4)}{(8314)(50)} = 0.1443 \text{ kg/m}^3$$

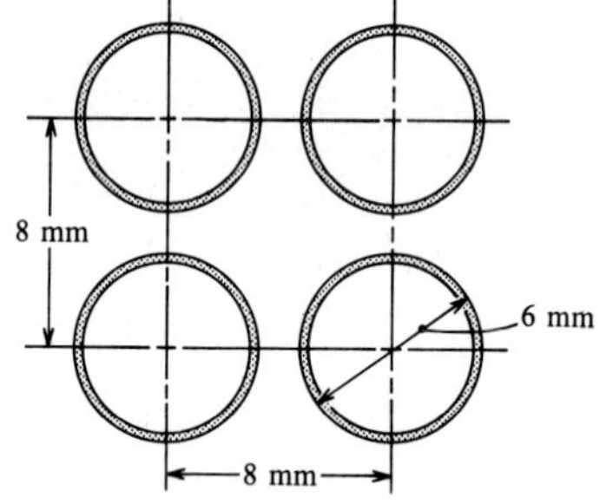

To calculate the outlet helium density, we must estimate the outlet temperature. The number of transfer units and effectiveness of this single-stream exchanger are thus required. The Reynolds number is

$$\text{Re} = \frac{\rho VD}{\mu} = \frac{\dot{m}D}{\mu A_c}$$

We will assume that the average bulk temperature of the helium is 200 K; then, from Table A.7, $\mu = 15.6 \times 10^{-6}$ kg/m s, $k = 0.116$ W/m K, $c_p = 5200$ J/kg K.

$$\text{Re} = \frac{(5.66 \times 10^{-3})(0.005)}{(15.6 \times 10^{-6})(0.00196)} = 926$$

The flow is laminar and the wall temperature is uniform; thus, from Table 4.5, Nu = 3.657.

$$h_c = \frac{k \,\text{Nu}}{D} = \frac{(0.116)(3.657)}{(0.005)} = 84.8 \text{ W/m}^2 \text{ K}$$

The heat transfer area is $A = (100)(\pi)(0.005)(1) = 1.57 \text{ m}^2$.

$$N_{\text{tu}} = \frac{h_c A}{\dot{m}c_p} = \frac{(84.8)(1.57)}{(5.66 \times 10^{-3})(5200)} = 4.52$$

Hence, the effectiveness is almost unity, and $T_{\text{out}} \simeq 350$ K, $T_{\text{avg}} \simeq (350 + 50)/2 = 200$ K, as assumed.

To use Eq. (8.78), we proceed as follows.

$$G = \frac{\dot{m}}{A_c} = \frac{5.66 \times 10^{-3}}{0.00196} = 2.89 \text{ kg/m}^2 \text{ s}$$

$$\frac{G^2}{2P_1\rho_1} = \frac{(2.89)^2}{2(15{,}000)(0.1443)} = 1.926 \times 10^{-3}$$

$$f = \frac{64}{\mathrm{Re}_D}\left(\frac{T_s}{T_b}\right)^{1.0} = \left(\frac{64}{926}\right)\left(\frac{350}{200}\right) = 0.121 \qquad \text{(using Tables 4.5 and 4.6)}$$

$$\frac{A}{A_c} = \frac{1.57}{0.00196} = 801$$

An iterative procedure is now required. After a few trials, $P_2 = 13.4$ kPa is seen to satisfy Eq. (8.78):

$$\rho_2 = \frac{P_2 M}{\mathcal{R} T_2} = \frac{(13{,}400)(4)}{(8314)(350)} = 0.0184 \text{ kg/m}^3$$

$$\rho_m = \frac{1}{2}(\rho_1 + \rho_2) = \frac{1}{2}(0.1443 + 0.0184) = 0.0814 \text{ kg/m}^3$$

$$\frac{f}{4}\frac{\rho_1}{\rho_m}\frac{A}{A_c} = \frac{0.121}{4}\left(\frac{0.1443}{0.0814}\right)(801) = 42.9$$

$$\sigma = \frac{A_c}{A_{\mathrm{fr}}} = \frac{0.00196}{0.0064} = 0.306$$

$$2\left(\frac{\rho_1}{\rho_2} - 1\right) = 2\left(\frac{0.1443}{0.0184} - 1\right) = 13.7$$

$$\frac{4(L/D)}{\mathrm{Re}} = \frac{4(1/0.005)}{926} = 0.864$$

From Fig. 8.27*a*, $K_c = 0.95$, $K_e = 0.27$.

$$(1 - \sigma^2 + K_c) = (1 - 0.306^2 + 0.95) = 1.9$$

$$\frac{\rho_1}{\rho_2}(1 - \sigma^2 - K_e) = \frac{0.1443}{0.0184}(1 - 0.306^2 - 0.27) = 5.0$$

Substituting in Eq. (8.78),

$$\frac{\Delta P}{P_1} = 1.926 \times 10^{-3}(1.9 + 42.9 + 13.7 - 5.0) = 0.103$$

$$\Delta P = (15.0)(0.103) = 1.55 \text{ kPa}$$

$$P_2 = P_1 - \Delta P = 15.0 - 1.55 \simeq 13.4 \text{ kPa}$$

Comments

1. The acceleration pressure drop is about one-third of the core viscous pressure drop in this situation.
2. Notice that the friction factor f varies appreciably along the exchanger, so that its evaluation at T_{avg} may be inadequate. A numerical integration of a differential momentum balance would give a more reliable result.

8.7.2 Thermal-Hydraulic Exchanger Design

Thermal-hydraulic design refers to the task of sizing a heat exchanger to give a required heat transfer effectiveness subject to a pressure drop constraint on one or both streams. The nature of the task depends on the geometric flow configuration of the exchanger, and it can also depend on whether the flow is laminar or turbulent; this is best illustrated by specific examples. In the case of the twin-tube exchangers to be covered in Examples 8.16 and 8.17, we will find that it is not possible to make an independent choice of the characteristic dimension of the heat transfer surface. On the other hand, for the plate-fin exchanger in Example 8.18, we will be free to choose this dimension, giving a family of exchangers that meet the heat transfer and pressure drop requirements. Criteria for choice of the characteristic dimension of the heat transfer surface, and for the choice of a particular type of surface, are discussed in Section 8.7.3.

EXAMPLE 8.16 A Laminar-Flow Twin-Tube Exchanger

A counterflow heat exchanger for a hydrogen cryogenic refrigeration system is to be fabricated by welding two steel tubes together. The system has a balanced flow of 6.0×10^{-6} kg/s. The cold stream enters at 11 K and must be heated to 300 K. The hot stream enters at 310 K. The average pressure on the cold side is 667 Pa; on the hot side, it is 10,130 Pa. The exchanger is to be reversible; that is, cold and hot streams are passed alternately on each side. If the allowable pressure drop on the cold side is 80 Pa, determine suitable dimensions for the exchanger.

Solution

Given: Balanced-counterflow twin-tube exchanger to heat 6.0×10^{-6} kg/s of hydrogen from 11 K to 300 K, with an allowable pressure drop of 80 Pa.

Required: Tube inside diameter and exchanger length.

Assumptions:
1. Negligible axial conduction (see Section 8.5.4).
2. Fully developed laminar flow.

For this balanced-flow exchanger, the required effectiveness is

$$\varepsilon = \frac{T_{C,\text{out}} - T_{C,\text{in}}}{T_{H,\text{in}} - T_{C,\text{in}}} = \frac{300 - 11}{310 - 11} = 0.9666$$

and from Eq. (8.45), the required number of transfer units is

$$N_{\text{tu}} = \frac{\varepsilon}{1-\varepsilon} = \frac{0.9666}{1-0.9666} = 28.9$$

By definition,

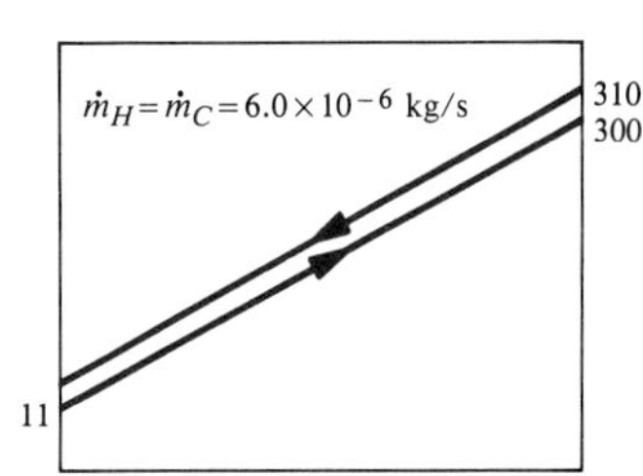

$$N_{tu} = \frac{U\mathscr{P}L}{C_{min}}; \qquad C_{min} = C_C = C_H = \dot{m}c_p$$

Solving for L,

$$L = \frac{\dot{m}c_pN_{tu}}{U\mathscr{P}}$$

Properties for both streams will be evaluated at an approximate average temperature of 160 K; from Table A.7, $c_p = 16{,}120$ J/kg K, $k = 0.131$ W/m K, $\mu = 5.84 \times 10^{-6}$ kg/m s.

$$L = \frac{(6.0 \times 10^{-6})(16{,}120)(28.9)}{U\mathscr{P}} = \frac{2.80}{U\mathscr{P}}$$

In order to proceed, we need to estimate the wall fin effectiveness: a value of 70% might be appropriate. Then, neglecting the resistance of the tubes where they are joined,

$$\frac{1}{U\mathscr{P}} = \frac{1}{0.7h_{c,H}\pi D} + \frac{1}{0.7h_{c,C}\pi D}$$

$$U\mathscr{P} = 0.35h_c\pi D$$

For laminar flow, the heat transfer coefficient can be obtained from Table 4.5. Since $\Delta T = T_H - T_C$ is almost constant along the exchanger, the wall heat flux can be taken to be constant, giving Nu $= 4.364$.

$$h_c = \text{Nu}(k/D) = 4.364(k/D)$$

$$U\mathscr{P} = (0.35)(4.364)(0.131)(\pi) = 0.629 \text{ W/m K}$$

$$L = \frac{2.80}{U\mathscr{P}} = \frac{2.80}{0.629} = 4.45 \text{ m}$$

The pressure drop is calculated using the friction factor:

$$\Delta P = f\left(\frac{L}{D}\right)\frac{1}{2}\rho V^2$$

From Table 4.5,

$$f = \frac{64}{\text{Re}_D} = \frac{64\mu}{\rho VD} \qquad \text{and} \qquad V = \frac{\dot{m}}{\rho A_c} = \frac{4\dot{m}}{\rho\pi D^2}$$

Hence if the acceleration pressure drop is neglected,

$$\Delta P = \frac{128L\dot{m}\mu}{\pi D^4\rho}; \qquad \text{solving,} \qquad D = \left(\frac{128L\dot{m}\mu}{\pi\Delta P\rho}\right)^{1/4}$$

On the cold side, the average pressure is 667 Pa,

$$\rho = \frac{PM}{\mathscr{R}T} = \frac{(667)(2.016)}{(8314)(160)} = 1.01 \times 10^{-3}\ \text{kg/m}^3$$

$$D = \left[\frac{128(4.45)(6.0 \times 10^{-6})(5.84 \times 10^{-6})}{\pi(80)(1.01 \times 10^{-3})}\right]^{1/4} = 16.7 \times 10^{-3}\ \text{m (16.7 mm)}$$

Thus, our preliminary estimate of the exchanger dimensions is a length of 4.45 m and a tube inner diameter of 16.7 mm. However, a number of the assumptions made in the calculation need to be validated.

1. *Is the flow laminar as assumed?* The Reynolds number is

$$\text{Re}_D = \frac{VD\rho}{\mu} = \frac{(\dot{m}/A_c)D}{\mu} = \frac{4\dot{m}}{\pi\mu D} = \frac{4(6.0 \times 10^{-6})}{\pi(5.84 \times 10^{-6})(0.017)} = 77 \quad \text{(laminar)}$$

2. *Are entrance effects negligible?* The L/D ratio is

$$L/D = 4.45/0.0167 = 266$$

Equation (4.49) gives the thermal entrance length as

$$\frac{L_{\text{eh}}}{D} \simeq 0.017\text{Re}_D\text{Pr} = (0.017)(77)(0.69) \simeq 1$$

so that the assumption of fully developed laminar flow is excellent.

3. *What is the tube wall fin effectiveness?* A 1 mm wall thickness will be used to check the assumed value of the tube wall fin effectiveness. The effective length is $\pi D/2 \simeq (\pi)(16.7 \times 10^{-3})/2 = 0.0262$ m. The heat transfer coefficient is $h_c = 4.364(k/D) = 4.364(0.131/0.0167) = 34.2$ W/m K, and $k = 17$ W/m K will be used for the steel. Then $\beta = (h_c\mathscr{P}/kA_c)^{1/2} = [(34.2)/(17)(0.001)]^{1/2} = 44.9\ \text{m}^{-1}$, $\beta L = (44.9)(0.0262) = 1.18$, $\eta_f = (1/\beta L)\tanh\beta L = (1/1.18)\tanh 1.18 = 0.70$, as assumed. (Fortunately! Had we not found $\eta_f = 0.7$, a second iteration would be required.)

4. *Is the acceleration pressure drop negligible?* The pressure drop required to accelerate the cold stream is equal to the change in momentum flux between inlet and outlet:

$$\Delta P(\text{Acceleration}) = (\rho V^2)_{\text{out}} - (\rho V^2)_{\text{in}} = G^2\left(\frac{1}{\rho_{\text{out}}} - \frac{1}{\rho_{\text{in}}}\right)$$

$$G = \frac{\dot{m}}{A_c} = \frac{6.0 \times 10^{-6}}{(\pi/4)(16.7 \times 10^{-3})^2} = 0.0274\ \text{kg/m}^2\ \text{s}$$

If the acceleration pressure drop is small, $P_{\text{in}} \simeq 667 + 40 = 707$ Pa, $P_{\text{out}} \simeq 667 - 40 = 627$ Pa.

$$\rho_{\text{in}} = \frac{(707)(2.016)}{(8314)(11)} = 0.0156\ \text{kg/m}^3; \qquad \rho_{\text{out}} = \frac{(627)(2.016)}{(8314)(300)} = 0.000507\ \text{kg/m}^3$$

$$\Delta P(\text{Acceleration}) = (0.0274)^2\left(\frac{1}{0.000507} - \frac{1}{0.0156}\right) = 1.4\ \text{Pa}$$

which is indeed small compared to the viscous pressure drop of 80 Pa.

Comments

1. Notice that the diameter D cancels in the $U\mathscr{P}$ product, so that the required number of transfer units fixes the exchanger length, independent of tube diameter. The pressure drop constraint then determines the allowable tube diameter: a tube inside diameter no smaller than 16.7 mm can be used.
2. The student is left to explore how to package and insulate the 4.45 m–long double tube.
3. Use FIN1 or FIN2 to check the tube wall fin effectiveness.
4. For a design problem, it is usually necessary to make a number of initial assumptions (sometimes bold!) to be able to calculate a tentative design. Subsequently, the validity of the initial assumptions can be checked and changes made, if necessary. Such an approach was demonstrated in this example. The student should learn not to agonize over the initial assumptions: only when numerical values for a tentative design are established can the validity of the assumptions be established.

EXAMPLE 8.17 A Turbulent-Flow Twin-Tube Exchanger

A counterflow twin-tube heat exchanger is to be used for balanced flow of air at 2×10^{-3} kg/s. The cold stream enters at 280 K and must be heated to 330 K; the hot stream enters at 340 K. If the average pressure in each stream is 1 atm and the allowable pressure drop for the cold stream is 9000 Pa, determine suitable dimensions. Copper tubes should be used to give a high tube wall fin effectiveness.

Solution

Given: Counterflow balanced twin-tube exchanger to heat 2×10^{-3} kg/s air from 280 K to 330 K, with an allowable pressure drop of 9000 Pa.

Required: Tube inside diameter and exchanger length.

Assumptions:
1. Fully developed turbulent flow.
2. A tube wall fin effectiveness of 100%.
3. Average pressure of 1 atm for both streams.

Proceeding as in Example 8.16,

$$\varepsilon = \frac{T_{C,\text{out}} - T_{C,\text{in}}}{T_{H,\text{in}} - T_{C,\text{in}}} = \frac{330 - 280}{340 - 280} = 0.833$$

$$N_{\text{tu}} = \frac{\varepsilon}{1 - \varepsilon} = \frac{0.833}{1 - 0.833} = 5.00$$

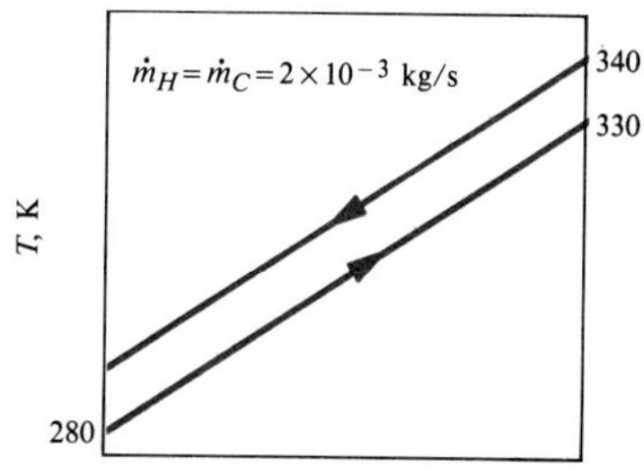

Evaluate properties for both air streams at 310 K: $c_p = 1005$ J/kg K, $k = 0.0274$ W/m K, $\rho = 1.141$ kg/m^3, $\mu = 18.87 \times 10^{-6}$ kg/m s, Pr $= 0.69$.

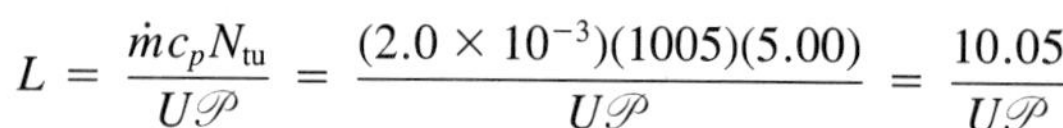

$$L = \frac{\dot{m}c_p N_{\text{tu}}}{U\mathscr{P}} = \frac{(2.0 \times 10^{-3})(1005)(5.00)}{U\mathscr{P}} = \frac{10.05}{U\mathscr{P}}$$

$U\mathscr{P} \simeq (1/2)h_c \pi D$ for a tube wall fin effectiveness of unity.

The high flow rate indicates that the flow will be turbulent. For simplicity, use Eq. (4.44):

$$h_c = \mathrm{Nu}_D(k/D) = 0.023\mathrm{Re}_D^{0.8}\mathrm{Pr}^{0.4}(k/D)$$

$$= 0.023\left(\frac{4\dot{m}}{\pi\mu D}\right)^{0.8}\mathrm{Pr}^{0.4}\left(\frac{k}{D}\right)$$

$$= 0.023\left(\frac{(4)(2\times10^{-3})}{\pi(18.87\times10^{-6})}\right)^{0.8}(0.69)^{0.4}(0.0274)D^{-1.8}$$

$$= 2.75\times10^{-2}D^{-1.8}$$

$$U\mathscr{P} = \frac{1}{2}(2.75\times10^{-2}D^{-1.8})(\pi D) = 4.32\times10^{-2}D^{-0.8}$$

$$L = \frac{10.05}{4.32\times10^{-2}D^{-0.8}} = 233D^{0.8} \qquad \textbf{(1)}$$

The pressure drop will give a second equation relating L and D:

$$\Delta P = f\left(\frac{L}{D}\right)\frac{1}{2}\rho V^2$$

Equation (4.43) gives $f = 0.184\mathrm{Re}_D^{-0.2}$ for turbulent flow.

$$\Delta P = 0.184\left(\frac{4\dot{m}}{\pi\mu D}\right)^{-0.2}\left(\frac{L}{D}\right)\left(\frac{1}{2}\right)\rho\left(\frac{4\dot{m}}{\pi D^2\rho}\right)^2$$

$$9000 = 0.184\left[\frac{4(2.0\times10^{-3})}{\pi(18.87\times10^{-6})}\right]^{-0.2}\left(\frac{1}{2}\right)(1.141)\left[\frac{4(2.0\times10^{-3})}{\pi(1.141)}\right]^2\frac{L}{D^{4.8}}$$

$$L = 4.59\times10^{10}D^{4.8} \qquad \textbf{(2)}$$

Solving Eqs. (1) and (2) gives the required dimensions of the exchanger:

$$233D^{0.8} = 4.59\times10^{10}D^{4.8}; \qquad D = 8.44\times10^{-3}\ \mathrm{m} \simeq 8.5\ \mathrm{mm}$$

$$L = 5.11\ \mathrm{m}$$

We can now calculate the Reynolds number to check if the flow is turbulent, as assumed.

$$\mathrm{Re}_D = \frac{4\dot{m}}{\pi\mu D} = \frac{4(2.0\times10^{-3})}{\pi(18.87\times10^{-6})(8.44\times10^{-3})} = 16{,}000 \quad \text{(turbulent)}$$

Comments

1. Equation (4.43) for f is not very accurate at this low Reynolds number; the problem could be reworked using Eq. (4.42).
2. The length-to-diameter ratio is $L/D = 5.11/0.00844 = 605$, which is much greater than the value of 30 required for fully developed heat transfer in a turbulent air flow.
3. In contrast to the laminar-flow exchanger of the previous example, heat transfer considerations alone do not fix the length of the exchanger. The heat transfer and pressure drop requirements together determine unique values of L and D.

EXAMPLE 8.18 A Cross-Flow Plate-Fin Exchanger

A cross-flow plate-fin heat exchanger with square cross-section passages is to be used for balanced flow of air, providing an effectiveness of 70%. When the flow rate is 0.330 kg/s, the allowable pressure drop is 330 Pa. Determine suitable dimensions for the exchanger. Evaluate fluid properties at 320 K and 1 atm, and neglect the thermal resistance of the plates and fins.

Solution

Given: Cross-flow plate-fin exchanger with square passages for balanced flow of air.

Required: Cross section and length to give $\varepsilon = 0.7$ at $\Delta P = 330$ Pa.

Assumptions:
1. Evaluate properties at 320 K and 1 atm.
2. Negligible thermal resistance of the plates and fins.
3. Negligible acceleration, entrance and exit pressure drops.

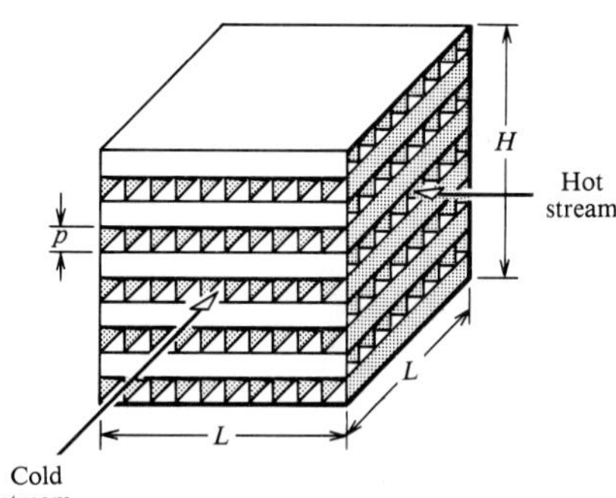

From Table A.7, air properties at 320 K and 1 atm are: $k = 0.0281$ W/m K, $\rho = 1.106$ kg/m^3, $c_p = 1006$ J/kg K, $\mu = 19.29 \times 10^{-6}$ kg/m s, Pr $= 0.69$.

This configuration corresponds to item 4 in Table 8.3 (cross-flow, both streams unmixed); solving by trial and error for $\varepsilon = 0.7$ gives $N_{\text{tu}} = 3.345$ (or use HEX1).

$$L = \frac{\dot{m}c_p N_{\text{tu}}}{U\mathscr{P}} = \frac{(0.33)(1006)(3.345)}{U\mathscr{P}} = \frac{1110}{U\mathscr{P}}$$

Let the pitch of the plates be p, and assume that the wall thickness $t \ll p$. Also, we will assume that the fin effectiveness is unity. Then the perimeter for one passage is $4p$, and the number of passages for each stream is $(1/2)(H/p)(L/p)$. Thus, $\mathscr{P} = 2HL/p$

$$L = \frac{1110}{(2HL/p)U}; \qquad L^2H = \frac{555p}{U} \tag{1}$$

The pressure drop is $\Delta P = f(L/D_h)(1/2)\rho V^2$; $D_h = p$, the side of a square passage; $V = \dot{m}/\rho A_c = 2\dot{m}/\rho LH$.

$$\Delta P = f\left(\frac{L}{p}\right)\frac{1}{2}\frac{4\dot{m}^2}{\rho L^2 H^2}; \qquad LH^2 = \frac{2f\dot{m}^2}{\rho p \Delta P} \tag{2}$$

Since the perimeter is the same for both streams,

$$\frac{1}{U} \simeq \frac{1}{h_{c,H}} + \frac{1}{h_{c,C}}; \qquad U = \frac{1}{2}h_c$$

Assume laminar flow, fully developed thermal conditions, and a uniform wall heat flux; then

Table 4.5 gives

$$h_c = \text{Nu}(k/D_h) = 3.6(k/p)$$

$$U = \left(\frac{1}{2}\right)(3.6)(k/p) = 1.8(k/p)$$

Substituting in Eq. (1) gives

$$L^2H = \frac{555p}{1.8(k/p)} = \frac{555p^2}{(1.8)(0.0281)} = 1.097 \times 10^4 p^2 \,[\text{m}^3] \qquad \textbf{(3)}$$

Table 4.5 also gives $f = 57/\text{Re}_{D_h}$.

$$\text{Re}_{D_h} = \frac{Vp\rho}{\mu} = \frac{(2\dot{m}/\rho LH)p\rho}{\mu} = \frac{2\dot{m}p}{LH\mu}$$

$$f = \frac{57LH\mu}{2\dot{m}p}$$

Substituting in Eq. (2) gives

$$H = \frac{57\mu\dot{m}}{\rho p^2 \Delta P} = \frac{(57)(19.3 \times 10^{-6})(0.33)}{(1.106)(330)p^2} = \frac{0.995 \times 10^{-6}}{p^2} \,[\text{m}] \qquad \textbf{(4)}$$

Choosing various values of pitch p and substituting into Eqs. (3) and (4) gives the following results. Values of Re_{D_h} are shown to confirm that the flow is laminar.

p mm	H m	L m	Re_{D_h}
1.0	0.995	0.105	327
1.2	0.691	0.152	391
1.4	0.508	0.206	456
1.6	0.389	0.268	524
1.8	0.307	0.340	589
2.0	0.249	0.420	655

Comments

1. Notice that Eqs. (3) and (4) are two equations in three unknowns: p, H, and L. Thus, we were free to vary p and then determine resulting values of H and L. All the exchangers in the table satisfy the heat transfer and pressure drop specification. In fact, all have the same frontal area, $LH = 0.1045 \text{ m}^2$, as can be deduced by eliminating p^2 between Eqs. (3) and (4). Often a nearly cubical shape will be preferred to simplify the manifold construction, which indicates a pitch somewhat less than 1.8 mm.

2. An essential difference between the twin-tube exchangers of the previous examples and the plate-fin exchanger is clear. In the case of the twin-tube exchangers, the characteristic dimensions of the flow cross-sectional area and the heat transfer surface were the same, that is, the tube diameter D. The result was that heat transfer and pressure drop requirements completely determined the exchanger. In contrast, the characteristic lengths for the plate-fin exchanger were L and H for the flow area and p for the heat transfer surface, which are independent and gave an extra degree of freedom with which to meet the heat transfer and pressure drop requirements.

8.7.3 Surface Selection for Compact Heat Exchangers

A large range of heat transfer surfaces is used in compact heat exchangers, including smooth tubes, externally finned tubes, internally enhanced tubes, and plate-fin surfaces. Examples are shown in Fig. 8.28. Useful insight into a strategy for selecting a surface appropriate to a given application can be obtained by writing down an approximate equation for the core velocity. Following Kays and London [5], we neglect the acceleration and inlet and outlet pressure drops, and we assume a fin effectiveness of unity. Then, from Eq. (8.78),

$$\frac{V_a^2/2}{P_1/\rho_1} \simeq \left(\frac{\Delta P/P_1}{N_{\text{tu}}^1}\right)\frac{\rho_m}{\rho_1}\frac{4\,\text{St}}{f} \tag{8.80}$$

where N_{tu}^1 is the number of transfer units for the one side of the exchanger under consideration. For a heat transfer area $A = \mathscr{P}L$ and flow cross-sectional area A_c, it is defined as

$$N_{\text{tu}}^1 = \frac{h_c A}{\dot{m} c_p} = \text{St}\frac{A}{A_c} \tag{8.81}$$

since the Stanton number is $\text{St} = h_c/c_p\rho V = h_c/c_p(\dot{m}/A_c)$. The number of transfer units of the exchanger is related to the one-side N_{tu} as follows:

$$N_{\text{tu}} = \frac{UA}{C_{\text{min}}}$$

$$\frac{1}{UA} = \frac{1}{(h_cA)_H} + \frac{1}{(h_cA)_C} \qquad \text{for a negligible wall and fin resistance}$$

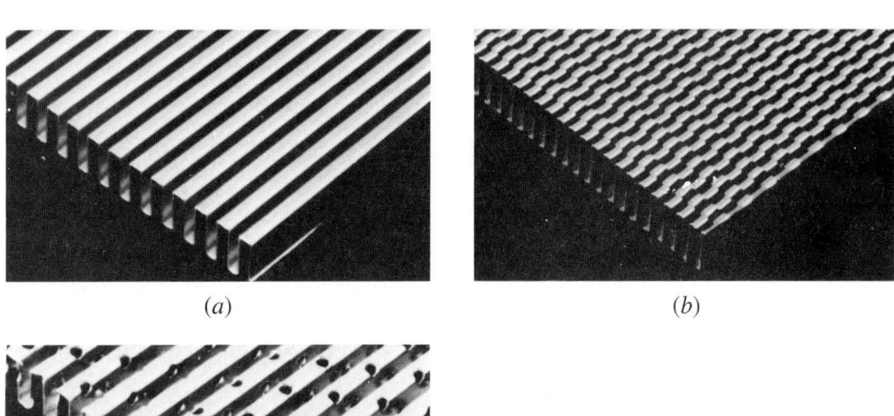

(*a*) (*b*)

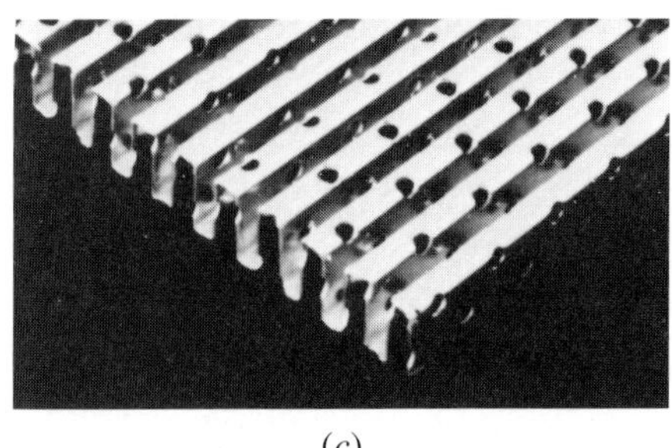

(*c*)

Figure 8.28 Some plate-fin surfaces used in compact heat exchangers. (*a*) Plain. (*b*) Serrated. (*c*) Perforated. (Photographs courtesy ALTEC International, Inc., LaCrosse.)

Substituting Eq. (8.81) and rearranging,

$$\frac{1}{N_{\text{tu}}} = \frac{1}{(C_H/C_{\min})N_{\text{tu}}^{1,H}} + \frac{1}{(C_C/C_{\min})N_{\text{tu}}^{1,C}} \tag{8.82}$$

For a given effectiveness ε and capacity flow rates, N_{tu} can be calculated and then N_{tu}^1 obtained from Eq. (8.82) by estimating or specifying a desirable ratio of hot side–to–cold side resistance.

The utility of Eq. (8.80) is based on the fact that, for a given fluid, St/*f* is a weak function of Reynolds number and, also, does not vary more than about fivefold for a wide range of surface types. Figure 8.29 shows representative data that allow St/*f* to be readily estimated for a preliminary calculation of the frontal area. As a first approximation, the Reynolds number dependence can be ignored and the result used to estimate Re for a second approximation. Since the velocity is proportional to the square root of St/*f*, only a $2\frac{1}{2}$-fold variation of velocity can be achieved by surface type selection. Also, if a small frontal area is a design requirement, Fig. 8.29 allows an appropriate surface type to be selected: the higher the value of St/*f*, the smaller the flow area A_c required. Since $A_{\text{fr}} = A_c/\sigma$, the frontal area will also be smaller, provided σ is not anomalously large. The ratio St/*f* can be viewed as an approximate indicator of the merit of the heat transfer surface. However, the fin effectiveness was assumed to be unity in this analysis, so that for values of η_f significantly lower than unity, the merit of a heat transfer surface must be evaluated more carefully (see, for example, Soland et al. [8]). Table 8.4*c* in Section 8.7.5 gives examples of correlations for both *f* and St for use in computer calculations.

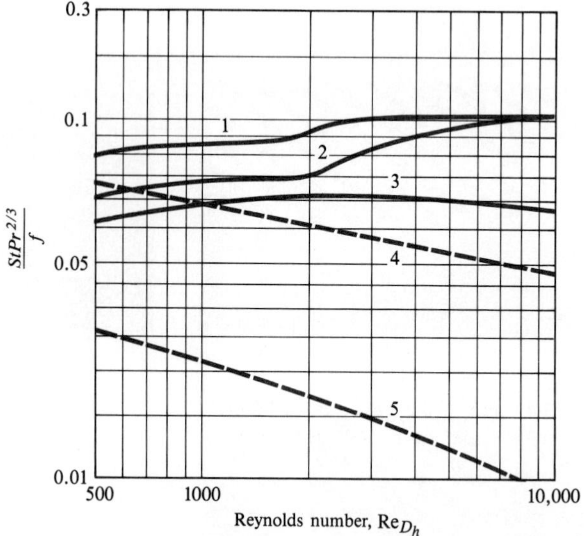

Figure 8.29 The ratio $\text{StPr}^{2/3}/f$ for five heat transfer surfaces. (1) Perforated plate-fin (13.95P). (2) Plain plate-fin (11.94T). (3) Louvered plate-fin (3/8–11.1). (4) Tube bank (S150.125). (5) Packed particle bed. [Numbers in parentheses are designations in *Compact Heat Exchangers*, W. M. Kays and A. L. London, McGraw-Hill, New York (any edition)].

EXAMPLE 8.19 Design of a Cross-Flow Plate-Fin Exchanger

Rework Example 8.18 using the equations developed in Section 8.7.3.

Solution

Given: Cross-flow plate-fin exchanger with square passages for balanced flow of air.

Required: Dimensions to give $\varepsilon = 0.7$ for $\Delta P = 330$ Pa.

Assumptions:
1. Properties evaluated at 320 K and 1 atm.
2. Negligible thermal resistance of plates and fins.
3. Negligible acceleration, entrance and exit pressure drops.

From Example 8.18, $\rho_m = \rho_1 = 1.106\ \mathrm{kg/m^3}$, Pr = 0.69, $P_1 = 1.013 \times 10^5$ Pa, and $N_{tu} = 3.345$. For balanced flow with $C_{min} = C_H = C_C$, Eq. (8.82) becomes

$$\frac{1}{N_{tu}} = \frac{1}{N_{tu}^{1,H}} + \frac{1}{N_{tu}^{1,C}} = \frac{2}{N_{tu}^{1}}$$

Hence, $N_{tu}^1 = 2N_{tu} = (2)(3.345) = 6.69$. Using Table 4.5,

$$\frac{\mathrm{St}}{f} = \left(\frac{\mathrm{Nu}}{\mathrm{Re}_{D_h}\mathrm{Pr}}\right)\left(\frac{1}{f}\right) = \left(\frac{3.6}{\mathrm{Re}_{D_h}\mathrm{Pr}}\right)\left(\frac{\mathrm{Re}_{D_h}}{57}\right) = \frac{0.063}{\mathrm{Pr}}$$

$$\frac{4\,\mathrm{St}}{f} = \frac{(4)(0.063)}{0.69} = 0.365$$

Rearranging Eq. (8.80),

$$V_a^2 = \frac{2\Delta P}{\rho_1 N_{tu}^1}\frac{4\,\mathrm{St}}{f} = \frac{(2)(330)(0.365)}{(1.106)(6.69)} = 32.6; \qquad V_a = 5.71\ \mathrm{m/s}$$

Using $\sigma = 0.5$ for thin fins and plates, the frontal velocity and area are

$$V_1 = (A_c/A_{fr})V_a = \sigma V_a = (0.5)(5.71) = 2.85\ \mathrm{m/s}$$

$$A_{fr} = \frac{\dot{m}}{\rho_1 V_1} = \frac{0.330}{(1.106)(2.85)} = 0.1045\ \mathrm{m^2}$$

At this point, we can see that specification of the *type* of heat transfer surface fixes the frontal area of the exchanger. To finalize the design, the pitch of the plates must be chosen and Eq. (3) of Example 8.18 can be used to obtain L and H.

$$A_{fr} = LH = 0.1045\ \mathrm{m^2}$$

$$L^2H = 1.097 \times 10^4 p^2\ \mathrm{m^3}$$

Hence, $L = 1.050 \times 10^5 p^2$. For example, for $p = 1.8$ mm $= 1.8 \times 10^{-3}$ m, $L = 0.340$ m, and $H = 0.307$ m, as before.

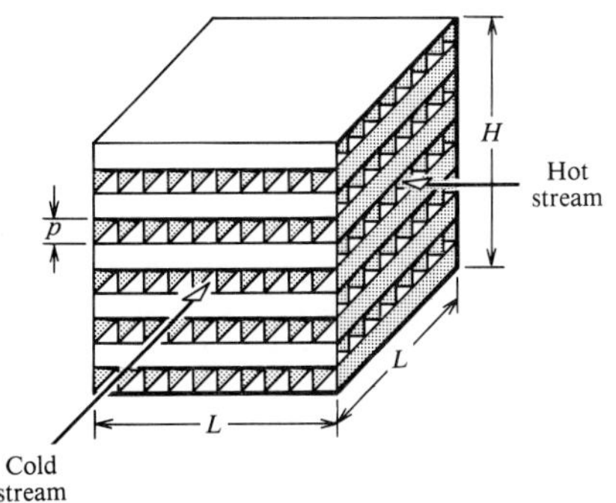

Comments

1. In this calculation procedure, the effect of heat transfer surface *type* on exchanger frontal area can be easily seen. The ratio St/f depends on the type of surface, and A_{fr} changes accordingly.
2. The volume of the exchanger is L^2H and hence increases proportionally to pitch squared. But, as mentioned in Example 8.18, a nearly cubical shape is often desired to simplify header construction.
3. The plate and fin areas are L^2H/p, which is proportional to pitch; hence, the weight of the exchanger core increases proportionally to pitch (for fixed plate and fin thickness).

8.7.4 Economic Analysis

The designer is often able to quantify both the costs and the benefits of a heat exchanger. Then optimization of the design requires maximization of the amount by which benefits exceed costs. Costs may be divided into three categories: *initial*, *annual*, and *operating* costs. The initial cost is the purchase price of the equipment. Annual costs include taxes, insurance, and maintenance. The major operating cost is usually the pump or fan power cost.

The initial cost, C_i, can be converted to an equivalent annual cost by assuming that the capital is borrowed from a bank at an annual interest rate i and must be repaid, or *amortized*, with n yearly payments.[2] The annual charge, or *annuity*, to repay a loan of \$1 over n years is

$$r = \frac{i}{1 - [1/(1+i)]^n} \tag{8.83}$$

If the exchanger is expected to have a salvage price C_s at the end of its working lifetime, the *present value* of the salvage price should be subtracted from the capital cost. The present value of \$1 received m years in the future is $[1/(1+i)]^m$; thus, the annual cost of the exchanger attributed to the initial cost is

$$\mathscr{C}_i = r\left[C_i - \left(\frac{1}{1+i}\right)^m C_s\right] \tag{8.84}$$

For example, the annual cost of a heat exchanger with initial cost \$10,000 and a salvage value after 15 years of \$1000, with a loan at 9% amortized over 15 years, is

$$\begin{aligned}\mathscr{C}_i &= \frac{0.09}{1 - (1/1.09)^{15}}\left[10{,}000 - \left(\frac{1}{1.09}\right)^{15} 1000\right]\\ &= 0.124(10{,}000 - 275)\\ &= \$1206\end{aligned}$$

Notice that the salvage value has little impact on the annual cost.

[2] In this economic analysis, the following notation scheme is used: C is a cost [\$], c is a unit cost, e.g., per unit heat transfer area [\$/m^2], $\mathscr{C}$ is an annual cost [\$/yr], and $\mathscr{c}$ is a unit annual cost, e.g., per unit heat transfer area [\$/m^2 yr].

For preliminary design purposes, it is often sufficient to assume that the initial cost is linearly proportional to the heat transfer surface area $\mathscr{P}L$. Referring to Fig. 8.30,

$$C_i = C_{if} + c_i \mathscr{P}L \tag{8.85}$$

where C_{if} is a fixed cost accounting for installation and other setup expenses, and c_i is the cost per unit transfer area. Substituting Eq. (8.85) in Eq. (8.84) gives

$$\mathscr{C}_i = r\left[C_{if} + c_i \mathscr{P}L - \left(\frac{1}{1+i}\right)^n C_s\right] \tag{8.86}$$

Since taxes and insurance are usually related to purchase price, these annual costs can also be taken to be linearly proportional to the transfer area:

$$\mathscr{C}_t = \mathscr{C}_{tf} + c_t \mathscr{P}L \tag{8.87}$$

Adding Eqs. (8.86) and (8.87) gives the total annual cost, $\mathscr{C}_a$, as

$$\mathscr{C}_a = \left[rC_{if} - r\left(\frac{1}{1+i}\right)^n C_s + \mathscr{C}_{tf}\right] + (rc_i + c_t)\mathscr{P}L \tag{8.88}$$

The operating cost associated with the pump or fan power for each stream is

$$\mathscr{C}_p = \frac{c_p \tau}{\eta_p} \cdot \frac{\dot{m}\Delta P}{\rho} \tag{8.89}$$

where c_p [\$/W h] is the unit pumping power cost, τ is the operating time per year [h/yr], and η_p is the pump or fan efficiency (including the efficiencies of the motor and drive box). The pressure drop can be expressed as a linear function of exchanger length:

$$\Delta P = \Delta P_0 + f\left(\frac{L}{D_h}\right)\frac{1}{2}\rho V^2$$

where ΔP_0 accounts for inlet and outlet losses and acceleration of a gas. Substituting in Eq. (8.89) gives the annual power cost as

$$\mathscr{C}_p = \frac{c_p \tau}{\eta_p} \cdot \frac{\dot{m}\Delta P_0}{\rho} + \frac{c_p \tau}{\eta_p} f\left(\frac{L}{D_h}\right)\frac{\dot{m}V^2}{2}$$

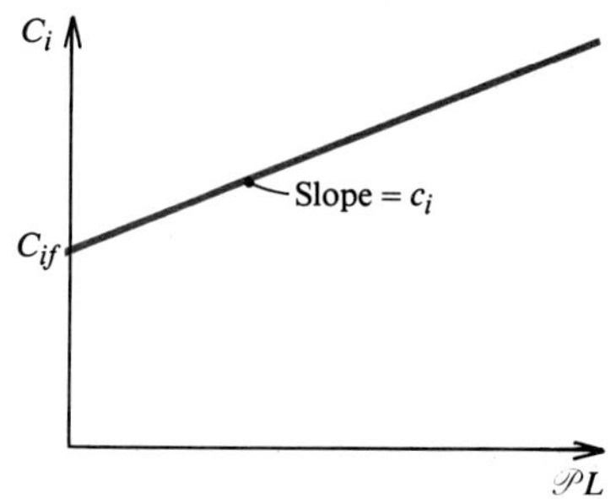

Figure 8.30 Initial cost C_i of a heat exchanger as a function of heat transfer surface area $\mathscr{P}L$.

or

$$\mathscr{C}_p = \mathscr{C}_{pf} + c_{pL}L \tag{8.90}$$

where c_{pL} is the annual cost of pumping power per unit length of exchanger.

In common process heating or heat recovery applications, the annual value of the benefit is directly proportional to the value of the heat transferred,

$$\mathscr{V} = v\dot{Q}\tau \tag{8.91}$$

where v [\$/W h] is the value of the heat per watt hour. On the other hand, if the exchanger is used in a power cycle for generation of mechanical or electrical power, the available energy transfer (the integral of heat transfer times Carnot efficiency) is a more appropriate measure of value. In such applications, a careful analysis of the system using the second law of thermodynamics is advisable.

Balanced-Counterflow Exchangers

Optimization of an exchanger design is generally an involved task requiring the use of appropriate computational tools. However, the important issues involved can be illustrated by a simple example problem. Consider a balanced-counterflow exchanger for which initial and operating costs are linear with length. We will specify a heat transfer surface and the cross-sectional area for each stream. The effectiveness and heat transfer increase with exchanger length, but so do the initial and operating costs; the problem is to determine the exchanger length that maximizes the benefit-cost differential.

Substituting for $\dot{Q}$ in Eq. (8.91) using the definition of effectiveness gives

$$\mathscr{V} = v\dot{m}c_p(T_{H,\text{in}} - T_{C,\text{in}})\varepsilon\tau \tag{8.92}$$

For algebraic convenience, we will write the total annual costs ($\mathscr{C}_a + \mathscr{C}_p$) as

$$\mathscr{C} = \mathscr{C}_f + cL \tag{8.93}$$

where $\mathscr{C}_f$ is the total of the fixed costs, and

$$c = (rc_i + c_t + c_{pL}/\mathscr{P})\mathscr{P} \tag{8.94}$$

is the total cost per unit length of exchanger. Figure 8.31 shows $\mathscr{V}$ and $\mathscr{C}$ plotted

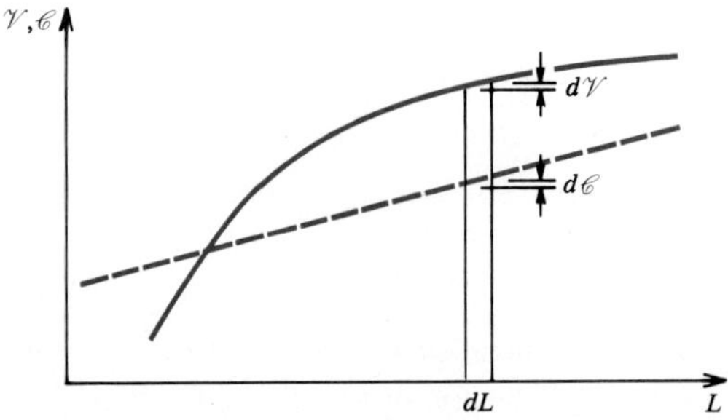

Figure 8.31 Annual benefit $\mathscr{V}$ and total annual costs $\mathscr{C}$ as a function of exchanger length.

versus L. The designer should maximize $(\mathscr{V} - \mathscr{C})$, that is, increase the exchanger length until the incremental annual cost $d\mathscr{C}$ equals the incremental annual value of the benefit $d\mathscr{V}$,

$$d\mathscr{V} = \upsilon \dot{m} c_p (T_{H,\text{in}} - T_{C,\text{in}})\tau \, d\varepsilon \tag{8.95}$$

But

$$d\varepsilon = \frac{d\varepsilon}{dN_{\text{tu}}} \frac{dN_{\text{tu}}}{dL} dL = \frac{d\varepsilon}{dN_{\text{tu}}} \frac{U\mathscr{P}}{\dot{m} c_p} dL$$

Hence,

$$d\mathscr{V} = \upsilon \tau (T_{H,\text{in}} - T_{C,\text{in}}) U\mathscr{P} \frac{d\varepsilon}{dN_{\text{tu}}} dL \tag{8.96}$$

Also

$$d\mathscr{C} = c \, dL = (r c_i + c_t + c_{pL}/\mathscr{P})\mathscr{P} \, dL \tag{8.97}$$

Equating Eqs. (8.96) and (8.97) gives

$$\frac{d\varepsilon}{dN_{\text{tu}}} = \frac{(r c_i + c_t + c_{pL}/\mathscr{P})}{\upsilon \tau (T_{H,\text{in}} - T_{C,\text{in}}) U} = \mathscr{E} \tag{8.98}$$

where $\mathscr{E}$ is a dimensionless *thermoeconomic* parameter. Examination of Fig. 8.31 shows that an exchanger should not be used at all if the length is such that $\mathscr{V}$ is less than $\mathscr{C}$. Equation (8.45) gives the effectiveness of a balanced-counterflow exchanger as

$$\varepsilon = \frac{N_{\text{tu}}}{1 + N_{\text{tu}}}$$

Hence

$$\frac{d\varepsilon}{dN_{\text{tu}}} = \frac{1}{(1 + N_{\text{tu}})^2} \tag{8.99}$$

Equating Eqs. (8.98) and (8.99) and solving for N_{tu} gives

$$N_{\text{tu,opt}} = \frac{1 - \mathscr{E}^{1/2}}{\mathscr{E}^{1/2}} \tag{8.100}$$

The optimal length is then obtained from the definition $N_{\text{tu}} = U\mathscr{P}L/\dot{m} c_p$.

Large integrated systems for heat and power supply with heat recovery exchangers are more difficult to optimize; see, for example, Linnhoff et al. [9,10].

EXAMPLE 8.20 Energy Conservation in a Brewery

In a bottle washing operation, 75°C wash water is dumped in a drain, and 80°C clean water is required. Currently, 20°C water is heated in an electric water heater, and the water requirement is 10,000 kg/h, 8 h/day, 250 days/yr. The cost of electricity is 8 cents/kW h.

In order to conserve energy, it has been proposed that a counterflow heat exchanger be installed to preheat the water feed to the electric water heater. The projected cost of the unit is \$30,000 plus \$1,800 per square meter of exchanger surface. The interest rate to amortize the investment over 15 years is 9% per annum. Taxes and insurance are expected to have a fixed cost of \$1100 per annum plus \$70/yr per square meter of exchanger surface. The overall heat transfer coefficient can be taken as 1200 W/m^2 K. What is the optimal heat transfer area of the exchanger, and what are the corresponding net annual savings?

Solution

Given: Counterflow exchanger for preheating feed water.

Required: Optimal heat transfer area and net annual savings.

Assumptions:
1. Balanced flow.
2. The exchanger has a negligible salvage value.
3. The additional pumping power requirements are negligible.
4. Exchanger maintenance costs are negligible.

First we calculate the thermoeconomic parameter, $\mathscr{E}$. From Eq. (8.98),

$$\mathscr{E} = \frac{rc_i + c_t + c_{pL}/\mathscr{P}}{\upsilon\tau(T_{H,\text{in}} - T_{C,\text{in}})U}$$

$$c_i = 1800 \text{ \$/m}^2 \qquad \upsilon = 8 \times 10^{-5} \text{ \$/W h}$$

$$c_t = 70 \text{ \$/m}^2\text{ yr} \qquad \tau = (250)(8) = 2000 \text{ h/yr}$$

$$c_{pL} = 0 \qquad U = 1200 \text{ W/m}^2\text{ K}$$

$$T_{H,\text{in}} - T_{C,\text{in}} = 75 - 20 = 55 \text{ K}$$

$$r = \frac{i}{1 - [1/(1+i)]^n} = \frac{0.09}{1 - (1/1.09)^{15}} = 0.124 \text{ yr}^{-1}$$

$$\mathscr{E} = \frac{(0.124)(1800) + 70}{(8 \times 10^{-5})(2000)(55)(1200)} = 0.0278$$

Substituting in Eq. (8.100),

$$N_{\text{tu,opt}} = \frac{1 - \mathscr{E}^{1/2}}{\mathscr{E}^{1/2}} = \frac{1 - (0.0278)^{1/2}}{0.0278^{1/2}} = 5.00$$

The heat exchanger area $\mathscr{P}L$ is found from the definition of N_{tu}:

$$\mathscr{P}L = \frac{\dot{m}c_pN_{\text{tu}}}{U} = \frac{(10{,}000/3600)(4180)(5.00)}{1200} = 48.4 \text{ m}^2$$

Also, from Eq. (8.45), the effectiveness for a balanced-flow exchanger is

$$\varepsilon = \frac{N_{\text{tu}}}{1 + N_{\text{tu}}} = \frac{5.00}{1 + 5.00} = 0.833$$

The annual value of the energy saved is given by Eq. (8.92):

$$\begin{aligned}\mathscr{V} &= v\dot{m}c_p(T_{H,\text{in}} - T_{C,\text{in}})\varepsilon\tau \\ &= (8 \times 10^{-5})(10{,}000/3600)(4180)(55)(0.833)(2000) \\ &= 85{,}100 \text{ \$/yr}\end{aligned}$$

If the salvage value of the exchanger is ignored, the total annual cost is given by Eq. (8.88) as

$$\begin{aligned}\mathscr{C}_a &= [rC_{if} + \mathscr{C}_{tf}] + (rc_i + c_t)\mathscr{P}L \\ &= [(0.124)(30{,}000) + 1100] + [(0.124)(1800) + 70](48.4) \\ &= 19{,}010 \text{ \$/yr}\end{aligned}$$

The extra pumping power requirements for the waste stream will add negligibly to the operating cost and have been ignored. Thus, the net annual savings are

$$\mathscr{V} - \mathscr{C}_a = 85{,}100 - 19{,}010 = 66{,}100 \text{ \$/yr}$$

Comments

1. The annual savings suggest that the exchanger is well worth installing.
2. Although the ratio of average annual savings to annual cost is $85{,}100/19{,}010 \simeq 4.5$, the last dollar of the \$19,010 annual cost gave exactly \$1 of benefit.

8.7.5 Computer-Aided Heat Exchanger Design: HEX2

An essential element of design is investigation of the effects of changes in the design's key parameters. Optimization is a special case of this process, in which a particular function of the parameters is maximized or minimized—for example, the benefit-cost differential was maximized in the economic design analysis of Section 8.7.4. The computer is an ideal tool for this purpose, and there is an increasing availability of computer software to aid the design engineer. To demonstrate the utility and potential of the computer for design calculations, a simple example is presented here to complete our introduction to heat exchanger design.

We will restrict our attention to gas-to-gas heat exchange in a single-pass cross-flow configuration with a plate-fin surface, similar to the exchanger analyzed in Examples 8.18 and 8.19. The flows will be balanced and unmixed. A typical application of such an exchanger is for heat recovery in an air-conditioning system. With these restrictions, it is possible to focus on the effects of such parameters as the type of plate-fin surface, the dimensions of the surface, and the wall and fin material. Table 8.4 gives dimensions, and correlations for f and St, for six plate-fin heat transfer surfaces. The design problem is to size the exchanger for a required heat transfer performance, subject to a pressure drop constraint. Subsequently, through economic analysis, the cost-benefit implications of the pressure drop constraint can be explored. The computer program HEX2 accomplishes these objectives.

Table 8.4a Six plate-fin heat transfer surfaces (all dimensions in mm).

1. Plain fin 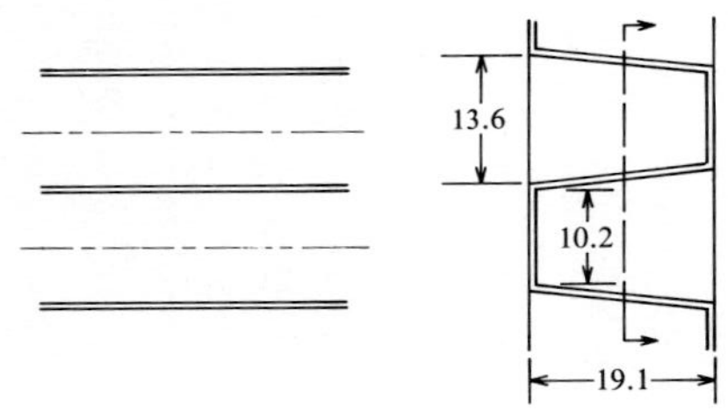	2. Plain fin 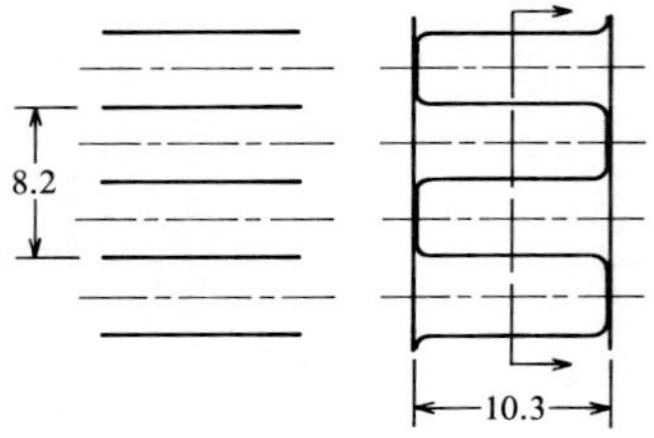	3. Plain fin 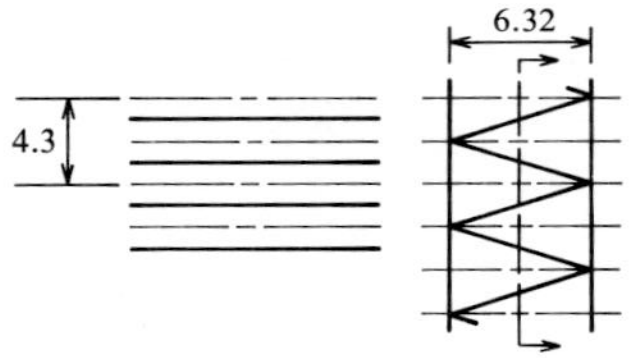
13.6; 10.2; 19.1	8.2; 10.3	4.3; 6.32
4. Plain fin 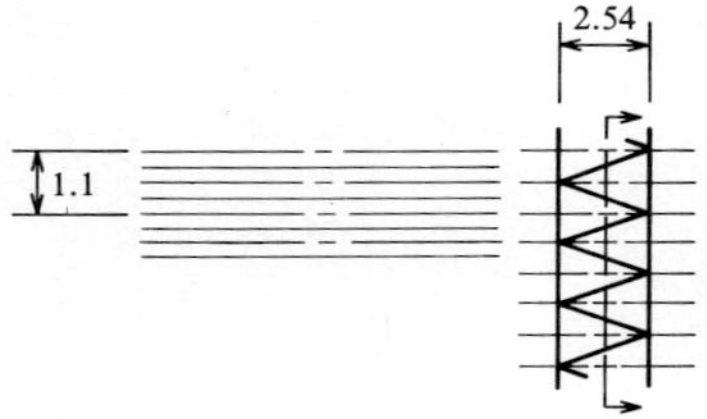	**5. Louvered fin** 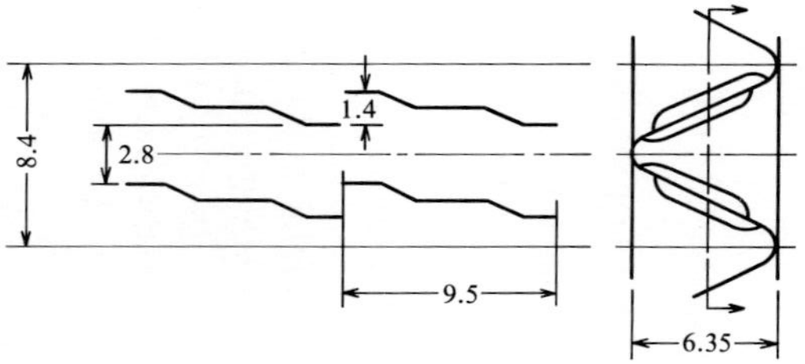	**6. Louvered fin** 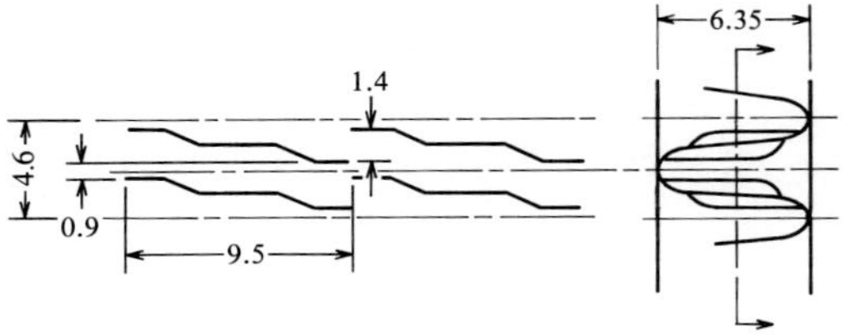
2.54; 1.1	8.4; 2.8; 1.4; 9.5; 6.35	6.35; 1.4; 4.6; 0.9; 9.5

Table 8.4*b* Geometrical data for six plate-fin heat transfer surfaces.

Item	Surface Type	Plate Spacing, b mm	Fins per Meter m^{-1}	Fin Height, L_f mm	Fin Thickness, t_f mm	$\frac{\text{Fin Area}}{\text{Total Heat Transfer Area}}$, A_f/A	Hydraulic Diameter, D_h mm	$\frac{\text{Heat Transfer Area}}{\text{Volume between Plates}}$, β' m^{-1}
1	Plain fin (2.0)	19.1	78.7	9.14	0.813	0.606	14.5	250
2	Plain fin (6.2)	10.3	244	5.02	0.254	0.728	5.54	665
3	Plain fin (11.94 T)	6.32	470	3.00	0.152	0.769	2.87	1290
4	Plain fin (46.45 T)	2.54	1829	1.24	0.0508	0.837	0.805	4371
5	Louvered fin (3/8–6.06)	6.35	239	2.39	0.152	0.640	4.45	840
6	Louvered fin (3/8–11.1)	6.35	437	2.96	0.152	0.756	3.08	1204

Notes: Numbers in parentheses are designations in *Compact Heat Exchangers* (W. M. Kays and A. L. London, McGraw-Hill, New York, any edition). The core-to-frontal-area ratio σ ($= A_c/A_{fr}$) can be calculated from $\sigma = b\beta' D_h/[8(b + t_w)]$, where t_w is the plate thickness.

Table 8.4c Friction factor and Stanton number data for six heat transfer surfaces (correlations of data given by Kays and London [5]).

Surface	Reynolds Number Range	Friction Factor, $f = a_0 + a_1\mathrm{Re}^{-1} + a_2\mathrm{Re}^{-0.5} + a_3\mathrm{Re}^{-0.2}$				Stanton Number, $\mathrm{StPr}^{2/3} = b_0 + b_1\mathrm{Re}^{-1} + b_2\mathrm{Re}^{-0.5} + b_3\mathrm{Re}^{-0.2}$			
		a_0	a_1	a_2	a_3	b_0	b_1	b_2	b_3
1	$4000 < \mathrm{Re} < 60{,}000$	-1.483×10^{-2}	172.60	−5.2079	0.4974	1.589×10^{-3}	−9.2526	0.3478	−0.005146
2	$800 < \mathrm{Re} < 12{,}000$	-7.782×10^{-2}	155.28	−9.7484	1.1947	-1.234×10^{-2}	17.426	−1.3971	0.1748
3	$300 < \mathrm{Re} < 10{,}000$	5.187×10^{-2}	64.348	0.1556	−0.1886	1.453×10^{-2}	0.3589	0.5123	−0.1037
4	$500 < \mathrm{Re} < 2000$	4.523×10^{-1}	14.611	13.601	−3.278	2.590×10^{-2}	−2.5448	1.0758	−0.2092
5	$500 < \mathrm{Re} < 10{,}000$	-2.732×10^{-4}	141.37	−7.392	1.216	-1.770×10^{-2}	11.491	−1.1057	0.2086
6	$500 < \mathrm{Re} < 10{,}000$	-2.340×10^{-1}	245.09	−18.574	3.090	-1.066×10^{-2}	7.8027	−0.5942	0.1343

Note: The Reynolds number is based on passage hydraulic diameter.

A schematic of the exchanger is shown in Fig. 8.32. HEX2 first asks for the following inputs:

Mass flow rate: $\dot{m}_H = \dot{m}_C = \dot{m}$ [kg/s]

Hot-stream inlet temperature: $T_{H,\text{in}}$ [K]

Cold-stream inlet temperature: $T_{C,\text{in}}$ [K]

Which outlet temperature is known, $T_{H,\text{out}}$ or $T_{C,\text{out}}$, and its value [K]

HEX2 then calculates the effectiveness ε from Eq. (8.31) and the required N_{tu} by solving the $\varepsilon - N_{\text{tu}}$ relation given by item 4 of Table 8.3*a* using Newton iteration. Also calculated is a reference temperature for evaluation of c_p, μ, and Pr, which is taken as the average of the average hot and cold stream temperatures,

$$T_r = \frac{1}{2}\left\{\frac{1}{2}(T_{H,\text{in}} + T_{H,\text{out}}) + \frac{1}{2}(T_{C,\text{in}} + T_{C,\text{out}})\right\} \tag{8.101}$$

HEX2 then asks for the following gas properties:

Molecular weight: M [kg/kmol]

Specific heat: c_p [J/kg K]

Viscosity: μ [kg/m s]

Prandtl number: Pr

Next the heat transfer surface is selected. Referring to Table 8.4*b* for typical data, HEX2 asks for the following inputs:

Plate spacing: b [mm]

Passage hydraulic diameter: D_h [mm]

Plate thickness: t_w [mm]

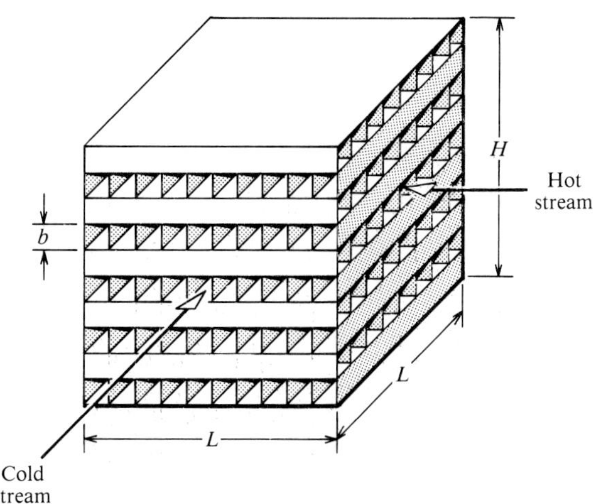

Figure 8.32 Schematic of a single-pass, cross-flow exchanger with a plate-fin surface.

Fin thickness: t_f [mm]
Fin height: L_f [mm]
Fin area/total area: A_f/A
Heat transfer area/volume between plates $= \beta'$ [m^{-1}]
Scale factor (e.g., a value of 2 will double all surface dimensions)
Plate and fin thermal conductivity: k [W/m K]

The performance of the heat transfer surface is supplied in the form of the correlations given in Table 8.4*c* (notice that these are independent of the scale factor):

Friction factor constants: a_0, a_1, a_2, a_3
$\mathrm{StPr}^{2/3}$ constants: b_0, b_1, b_2, b_3

Finally, the pressure drop data are supplied:

Choice of stream for pressure drop constraint: hot or cold
Inlet pressure: P_1 [Pa]
Allowable pressure drop: ΔP [Pa]
Choice of geometry (*c*) or (*d*) of Fig. 8.27 for the contraction and expansion coefficients, K_c and K_e

An iterative calculation procedure is required. To obtain a reasonable first guess, HEX2 uses the approximate Eq. (8.80) rearranged as

$$G^2 \simeq \frac{8\rho_m}{\mathrm{Pr}^{2/3}}\left(\frac{\Delta P}{N_{\mathrm{tu}}^1}\right)\left(\frac{\mathrm{StPr}^{2/3}}{f}\right) \tag{8.102}$$

The one-side N_{tu}^1 is estimated assuming a negligible wall and fin resistance; from Eq. (8.82) for balanced flow,

$$N_{\mathrm{tu}}^1 = 2N_{\mathrm{tu}} \tag{8.103}$$

Based on Fig. 8.29, $(\mathrm{StPr}^{2/3}/f)$ is set equal to 0.1. All other quantities on the right-hand side of Eq. (8.102) are known from the input data, and, hence, G can be calculated. This value is the initial guess for G. The calculation steps are now as follows:

1. Reynolds number:

$$\mathrm{Re} = \frac{GD_h}{\mu}$$

2. Core and frontal area:

$$A_c = \frac{\dot{m}}{G}; \qquad A_{\mathrm{fr}} = \frac{A_c}{\sigma}, \qquad \text{where } \sigma = \frac{b\beta' D_h}{8(b+t_w)}$$

3. Friction factor: f

4. Stanton number: $\mathrm{St} = (\mathrm{St}\mathrm{Pr}^{2/3})/\mathrm{Pr}^{2/3}$

5. Heat transfer coefficient: $h_c = Gc_p\mathrm{St}$

6. Fin effectiveness, Eq. (2.42): $\eta_f = (1/\beta L_f)\tanh\beta L_f$

7. Overall heat transfer coefficient:

$$\frac{1}{U} = \frac{2}{h_c[\eta_f(A_f/A) + (1 - A_f/A)]} + \frac{1}{k/t_w}$$

8. Required heat transfer area:

$$A = \frac{\dot{m}c_p N_{\mathrm{tu}}}{U}$$

9. Flow length:

$$L = \frac{A}{\mathscr{P}} = \frac{AD_h}{4A_c}$$

10. Contraction and expansion coefficients: K_c and K_e

11. Pressure drop: ΔP, calculated from Eq. (8.78).

This value of ΔP is then compared with the required value, and a new value of G is estimated using $\Delta P \propto G^2$. The next iteration commences with step 1 to obtain a new Reynolds number, *et seq*. The iteration process is repeated until the calculated ΔP agrees with the required value, to within a specified error tolerance.

HEX2 now asks whether an economic analysis is required. If the answer is yes, the following additional input is required:

Interest rate: i [%]
Loan period: n [yr]
Fixed cost for exchanger: C_{if} [$]
Salvage value of exchanger: C_s [$]
Working lifetime: m [yr]
Cost of heat transfer surface per unit area: c_i [$/m^2]
Fixed cost for taxes and insurance: $\mathscr{C}_{tf}$ [$/yr]
Cost of taxes, maintenance, and insurance per unit area: c_t [$/m^2 yr]
Value of heat: v [$/W h]
Operating time per year: τ [h/yr]
Unit power cost: c_p [$/W h]
Fan efficiency: η_f

HEX2 then calculates the benefit-cost differential $(\mathscr{V} - \mathscr{C})$ [$/yr].

EXAMPLE 8.21 A Balanced–Cross-Flow Plate-Fin Recuperator for an Air-Heating System

A recuperator is required for an air-heating system that processes 0.5 kg/s air. The hot air enters at 295 K, and the cold air enters at 270 K and is to be heated to 287 K. The inlet pressure for the hot air is 986 mbar. Determine the dimensions of a cross-flow exchanger, with the louvered-fin surface given as item 6 in Table 8.4*a*, that will give a pressure drop of 330 Pa on the hot side. The core is to be fabricated from AISI 1010 steel with a 0.300 mm plate thickness.

Solution

Given: Balanced–cross-flow exchanger with a louvered-fin heat transfer surface.

Required: Dimensions for a specified heat transfer performance and pressure drop.

Assumptions:
1. The density will be taken to be constant and equal to its average value ρ_m in the pressure drop equation; hence, the acceleration pressure drop is assumed to be negligible.
2. The exchanger will be sized for the cold stream.

The purpose of this example is to illustrate the calculation procedure of HEX2. Following the steps given in Section 8.7.5, we have

$$\begin{aligned}
\dot{m} &= 0.5 \text{ kg/s} \\
T_{H,\text{in}} &= 295 \text{ K} \\
T_{C,\text{in}} &= 270 \text{ K} \\
T_{C,\text{out}} &= 287 \text{ K}
\end{aligned}$$

Using HEX1, $\varepsilon = 0.680$, $N_{\text{tu}} = 2.92$, $T_{H,\text{out}} = 278$ K.

$$T_r = \frac{1}{2}\left\{\frac{1}{2}(295 + 278) + \frac{1}{2}(270 + 287)\right\} = 282.5\,\text{K}$$

Gas properties:

$$\begin{aligned}
M &= 29 \text{ kg/kmol} \\
c_p &= 1008 \text{ J/kg K} \\
\mu &= 17.65 \times 10^{-6} \text{ kg/m s} \\
\text{Pr} &= 0.69
\end{aligned}$$

For surface 6 of Table 8.4*b*,

$$\begin{aligned}
b &= 6.35 \text{ mm} = 6.35 \times 10^{-3} \text{ m} \\
D_h &= 3.08 \text{ mm} = 3.08 \times 10^{-3} \text{ m} \\
t_w &= 0.300 \text{ mm} = 0.300 \times 10^{-3} \text{ m} \\
t_f &= 0.152 \text{ mm} = 0.152 \times 10^{-3} \text{ m} \\
L_f &= 2.96 \text{ mm} = 2.96 \times 10^{-3} \text{ m} \\
A_f/A &= 0.756 \\
\beta' &= 1204 \\
\text{Scale factor} &= 1
\end{aligned}$$

k_w = 64 W/m K (from Table A.1*a*)
a_0, a_1, etc.: see Table 8.4*c*, item 6
b_0, b_1, etc.: see Table 8.4*c*, item 6

Pressure drop data:

Hot
$P_1 = 0.986 \times 10^5$ Pa
$\Delta P = 330$ Pa
Coefficients from Fig. 8.27*c*

The average density of the hot stream is

$$\rho_m = \frac{P_m M}{\mathscr{R} T_m} = \frac{(98{,}600 - 165)(29)}{(8314)(0.5)(295 + 278)} = 1.198 \text{ kg/m}^3$$

$$N_{\text{tu}}^1 = 2N_{\text{tu}} = (2)(2.92) = 5.84$$

$$\text{Pr}^{2/3} = (0.69)^{2/3} = 0.78$$

Substituting in Eq. (8.102) and setting $\text{StPr}^{2/3}/f = 0.1$,

$$G^2 \simeq \frac{8\rho_m}{\text{Pr}^{2/3}}\left(\frac{\Delta P}{N_{\text{tu}}^1}\right)\left(\frac{\text{StPr}^{2/3}}{f}\right) = \frac{(8)(1.198)}{0.78}\left(\frac{330}{5.84}\right)(0.1) = 69.5$$

$$G \simeq 8.34 \text{ kg/m}^2\text{ s}$$

The iterative calculation steps are:

1. $\text{Re} = \dfrac{GD_h}{\mu} = \dfrac{(8.34)(3.08 \times 10^{-3})}{17.65 \times 10^{-6}} = 1455$

2. $A_c = \dfrac{\dot{m}}{G} = \dfrac{0.5}{8.34} = 0.0600 \text{ m}^2$

$$\sigma = \frac{b\beta' D_h}{8(b + t_w)} = \frac{(6.35 \times 10^{-3})(1204)(3.08 \times 10^{-3})}{8(6.35 \times 10^{-3} + 0.3 \times 10^{-3})} = 0.443$$

$$A_{\text{fr}} = \frac{A_c}{\sigma} = \frac{0.0600}{0.443} = 0.135 \text{ m}^2$$

3. $f = -0.2340 + 245.09(1455)^{-1} - 18.574(1455)^{-0.5} + 3.090(1455)^{-0.2} = 0.168$

4. $\text{StPr}^{2/3} = -0.01066 + 7.803(1455)^{-1} - 0.5942(1455)^{-0.5} + 0.1343(1455)^{-0.2}$

$$= 0.0104$$

$$\text{St} = (0.0104/0.78) = 0.0133$$

5. $h_c = Gc_p\text{St} = (8.34)(1008)(0.0133) = 111.8 \text{ W/m}^2\text{ K}$

6. $\beta^2 = \dfrac{2h_c}{k_w t_f} = \dfrac{(2)(111.8)}{(64)(0.152 \times 10^{-3})} = 2.299 \times 10^4; \qquad \beta = 152$

$$\beta L_f = (152)(2.96 \times 10^{-3}) = 0.449; \qquad \eta_f = \tanh(0.449)/(0.449) = 0.938$$

7. $$\frac{1}{U} = \frac{2}{h_c[\eta_f(A_f/A) + (1 - A_f/A)]} + \frac{1}{k_w/t}$$

$$= \frac{2}{(111.8)[(0.938)(0.756) + (1 - 0.756)]} + \frac{1}{64/0.3 \times 10^{-3}}$$

$$U = 53.3 \text{ W/m}^2\text{ K}$$

8. $$A = \frac{\dot{m}c_p N_{\text{tu}}}{U} = \frac{(0.5)(1008)(2.92)}{(53.3)} = 27.6 \text{ m}^2$$

9. $$L = \frac{AD_h}{4A_c} = \frac{(27.6)(3.08 \times 10^{-3})}{(4)(0.0600)} = 0.354 \text{ m}$$

10. $K_c = 1.18, \qquad K_e = -0.06$

11. $$\frac{\Delta P}{P_1} \simeq \frac{G^2}{2\rho_m P_1}\left[(1 - \sigma^2 + K_c) + \frac{f}{4}\frac{A}{A_c} - (1 - \sigma^2 - K_e)\right]$$

$$= \frac{(8.34)^2}{2(1.198)(98{,}600)}\left\{[1 - (0.443)^2 + 1.18] + \frac{0.168}{4}\frac{27.6}{0.0600} - [1 - (0.443)^2 - (-0.06)]\right\}$$

$$= 2.94 \times 10^{-4}(1.98 + 19.32 - 0.86)$$

$$= 6.01 \times 10^{-3}$$

$$\Delta P = (98{,}600)(6.01 \times 10^{-3}) = 592 \text{ Pa}$$

This pressure drop is higher than the required value of 330 Pa. For a second iteration, we take

$$G = \left(\frac{330}{592}\right)^{1/2}(8.34) = 6.23 \text{ kg/m}^2\text{ s}$$

and obtain the following results:

1. $\text{Re} = 1086$
2. $A_c = 0.0803 \text{ m}^2, \qquad A_{\text{fr}} = 0.181 \text{ m}^2$
3. $f = 0.193$
4. $\text{StPr}^{2/3} = 0.0117, \qquad \text{St} = 0.0150$
5. $h_c = 94.1 \text{ W/m}^2\text{ K}$
6. $\beta = 139, \qquad \eta_f = 0.947$
7. $U = 45.2 \text{ W/m}^2\text{ K}$
8. $A = 32.0 \text{ m}^2$
9. $L = 0.307 \text{ m}$

10. K_c, K_e: no change

11. $\Delta P/P_1 = 3.34 \times 10^{-3}$; $\Delta P = 329$ Pa

The error in this estimate of ΔP is negligible. The height of the exchanger is

$$H = \frac{A_{\text{fr}}}{L} \simeq \frac{0.181}{0.307} = 0.590 \text{ m}$$

Comments

1. HEX2 calculates ρ_1, ρ_2, and ρ_m for use in the complete pressure drop equation, Eq. (8.78): the acceleration pressure drop is not neglected.

2. All input values are saved by HEX2 and become default values for subsequent use of the program.

EXAMPLE 8.22 Economic Analysis of a Recuperator for an Air-Heating System

An economic analysis is required for the recuperator of Example 8.21. The interest rate for a loan over 20 years is 8%, the fixed cost of the exchanger is \$300, and the cost of heat transfer surface per unit area is \$10/m^2. The fixed cost for insurance and taxes is \$20/yr, and the cost per unit area is 0.57\$/m^2 yr. The cost of electricity is 0.08 \$/kW h, and the heating system operates 16 hours per day for 190 days of the year. If the recuperator were not used, an electrical heater would have to make up the required heat: (i) Determine the benefit-cost differential for the exchanger sized in Example 8.21. (ii) Use HEX2 to map the benefit-cost differential as a function of cold-stream outlet temperature and hot stream pressure drop.

Solution

Given: Recuperator for an air-heating system.

Required: Economic analysis for a louvered-fin-surface cross-flow exchanger.

Assumptions: 1. A fan efficiency (including motor, etc.) of 65%.
2. Negligible salvage value.

(i) From Eq. (8.83), the annuity is

$$r = \frac{i}{1 - [1/(i+1)]^n} = \frac{0.08}{1 - (1/1.08)^{20}} = 0.102 \text{ yr}^{-1}$$

and, from Eq. (8.88), the total annual cost is

$$\begin{aligned}\mathscr{C}_a &= (rC_{if} + \mathscr{C}_{tf}) + (rc_i + c_t)A \\ &= [(0.102)(300) + 20] + [(0.102)(10) + 0.57](32.0) \\ &= 30.6 + 20 + 50.9 \\ &= 101.5 \text{ \$/yr}\end{aligned}$$

The operating cost is obtained from Eq. (8.89):

$$\mathscr{C}_p = \frac{c_p \tau}{\eta_p} \times \frac{\dot{m}\Delta P}{\rho} \simeq \frac{(0.08 \times 10^{-3})(16)(190)}{0.65} \times \frac{(2)(0.5)(330)}{1.198} = 103.1 \text{ \$/yr}$$

The annual value of the heat recovery is given by Eq.(8.91):

$$\begin{aligned}\mathscr{V} &= v\dot{Q}\tau \\ &= v\dot{m}c_p(T_{C,\text{out}} - T_{C,\text{in}})\tau \\ &= (0.08 \times 10^{-3})(0.5)(1008)(287 - 270)(16)(190) \\ &= 2084 \text{ \$/yr}\end{aligned}$$

The benefit-cost differential is then

$$\mathscr{V} - \mathscr{C} = 2084 - 103 - 102 = 1879 \text{ \$/yr}$$

(ii) Using HEX2, the following table of $(\mathscr{V} - \mathscr{C})$ can be prepared.

	$T_{C,\text{out}}$ [K]					
	286		287		288	
ΔP Pa	L m	$\mathscr{V} - \mathscr{C}$ \$/yr	L m	$\mathscr{V} - \mathscr{C}$ \$/yr	L m	$\mathscr{V} - \mathscr{C}$ \$/yr
270	0.247	1791	0.299	1901	0.372	2003
330	0.258	1775	0.313	1885	0.391	1988
390	0.267	1757	0.324	1869	0.407	1973

Comments

The table indicates that an increase in $T_{C,\text{out}}$ and a decrease in ΔP will further improve the benefit-cost differential.

8.8 CLOSURE

A great variety of heat exchangers used in engineering practice were classified according to geometric flow configuration. For purposes of analysis, we chose to differentiate single-stream and two-stream steady-flow exchangers, although the single-stream exchanger can be viewed as a limit case of the two-stream exchanger when the capacity ratio $R_C \to 0$. The thermal performance of steady-flow exchangers can be obtained using either the logarithmic mean temperature difference (LMTD) approach, or the effectiveness–number of transfer units (ε–N_{tu}) approach. The emphasis was on the latter procedure in this text. When flow rates are small, axial conduction effects may be significant and should be accounted for. The perforated-plate exchanger has inherently low axial conduction and can be used when a high effectiveness is required. The heat transfer in a regenerator is unsteady, and its performance is governed by a partial differential equation. The solution for the special case of balanced flow was conveniently presented as a chart of ε versus N_{tu}.

The design of heat exchangers requires that a specified heat transfer performance be achieved, subject to various constraints. The most important of these is often a pressure drop constraint, which, through the pumping power required, is related to operating costs. The capital cost of the exchanger depends on size and weight, and an economic analysis is used to trade off the competing objectives of low operating costs and low capital costs. Other design issues include fluid and exchanger material compatibility and ease of maintenance, but these issues are left to be dealt with by heat exchanger design handbooks.

The computer program HEX2 serves to introduce the student to computer-aided heat exchanger design. HEX2 sizes balanced cross-flow plate-fin heat exchangers to give a specified heat transfer performance, subject to a pressure drop constraint. In addition, HEX2 calculates the benefit-cost differential for the resulting design when used for heat recovery. Since HEX2 applies to a very limited class of exchangers and applications, it is not intended to be a general design tool. Rather, it was included here to give the student ideas about how to use computers to aid the design process.

REFERENCES

1. Bowman, R. A., Mueller, A. C., and Nagle, W. M., "Mean temperature difference in design," *Trans. ASME,* 62, 283–294 (1940).
2. Shah, R. K., and Mueller, A. C., "Heat exchanger basic thermal design methods," in Rohsenow, W. M., Harnett, J. P., and Ganić, E. N., eds., *Handbook of Heat Transfer Applications,* 2nd ed., Chap. 4, Part 1, McGraw-Hill, New York (1985).
3. Taborek, J., "Charts for mean temperature difference in industrial heat exchanger configurations," in Hewitt, G. F., coord. ed., *Hemisphere Handbook of Heat Exchanger Design,* Sec. 1.5, Hemisphere, New York (1990).
4. Turton, R., Ferguson, D., and Levenspiel, O., "Charts for the performance and design of heat exchangers," *Chemical Engineering,* 93, 81–88 (August 18, 1986).
5. Kays, W. M., and London, A. L., *Compact Heat Exchangers,* 3rd ed., McGraw-Hill, New York (1984).
6. Kroeger, P. G., "Performance deterioration in high effectiveness heat exchangers due to axial heat conduction effects," *Advances in Cryogenic Engineering,* 12, 363–372 (1966).
7. Sarangi, S., and Barclay, J. A., "An analysis of compact heat exchanger performance," *ASME,* Cryogenic Processes and Equipment—1984, New Orleans, December 1984, pp. 37–44.
8. Soland, J. G., Mack, W. M., Jr., and Rohsenow, W. M., "Peformance ranking of plate-fin heat exchanger surfaces," *J. Heat Transfer,* 100, 514–519 (1978).
9. Linnhoff, B., and Turner, J. A., "Heat-recovery networks: new insights yield big savings," *Chemical Engineering,* 88, 56–70 (November 2, 1981).

10. Linnhoff, B., Townsend, D. W., Boland, D., Hewitt, G. F., Thomas, B. E. A., Guy, R. H., and Marsland, R. H., *User Guide on Process Integration for the Efficient Use of Energy,* IChemE, Rugby (1982).

11. Edwards, D. K., and Matavosian, R., "Thermoeconomically optimum counterflow heat exchanger effectiveness," *J. Heat Transfer,* 104, 191–193 (1982).

EXERCISES

8–1. An oil supply of 1 kg/s is to be heated from 15°C to 70°C. A 0.3 kg/s supply of water at 95°C is available. Is this supply sufficient?

8–2. A boiler is required to generate 3.0 kg/s of steam at a pressure of 0.13 MPa. Waste heat is available in the form of 40 kg/s of exhaust gas at 300°C. Is this energy supply sufficient? Take $c_p = 1180$ J/kg K for the exhaust gas.

8–3. A 1.8 kg/s supply of saturated steam at 0.11 MPa is available for heating water in a bottle washing operation at a soft drink bottling plant. What is the maximum flow rate of water that can be heated from 20°C to 70°C?

8–4. A shell-and-tube heat exchanger is used to heat fuel oil with hot engine exhaust. The mild steel tubes are 1 ½ in nominal diameter and 14 gage. Design values of the inside and outside heat transfer coefficients are 40 and 100 W/m^2 K, respectively.

(i) Calculate the design value of the overall heat transfer coefficient based on tube outside area.

(ii) Using the fouling resistances given in Table 8.1, estimate the reduction in U expected during service due to fouling.

8–5. An air-cooled oil cooler has 1 cm–O.D., 1 mm–wall-thickness steel tubes, with 2 cm–O.D., 0.4 mm–thickness spiral fins at a pitch of 2 mm. If the inside and outside heat transfer coefficients are 200 and 25 W/m^2 K, respectively, determine the overall heat transfer coefficient-perimeter product. Take $k = 55$ W/m K for the steel.

8–6. A twin-tube heat exchanger for a hydrogen cryogenic refrigeration system is made by welding two 15 mm–O.D., 1 mm–wall-thickness tubes as shown, and forming them into a coil that is inserted in an evacuated container. If there is balanced flow of hydrogen gas at 2.0× 10^{-6} kg/s, and the tube material is stainless steel, estimate the overall heat transfer coefficient. Take $k = 15.0$ W/m K for the tube wall, and evaluate the hydrogen properties at 150 K and 2 torr.

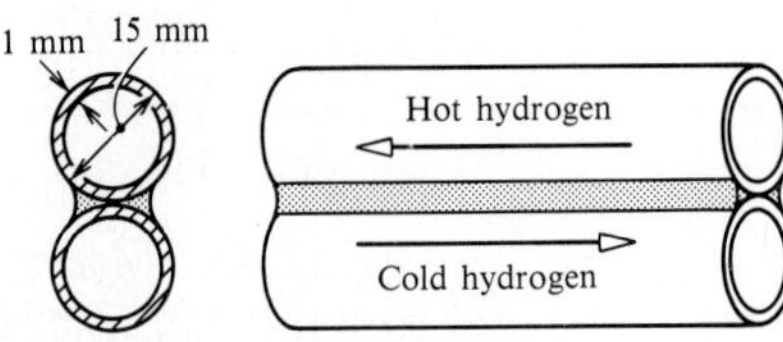

8–7. A coaxial-tube heat exchanger has an inner 316 stainless steel tube of 3 cm outside diameter and 2 mm wall thickness. Helium at 0.22 MPa flows at

80 m/s through the inner tube. Water flows through the outer tube at 0.97 kg/s, which has an inside diameter of 5 cm. A change in service conditions requires the overall heat transfer coefficient to be increased. It is proposed to artificially roughen the inner surface of the inner tube by fitting an insert that gives parallel ribs 1 mm high at a 10 mm pitch. Determine the percentage increase in the overall heat transfer coefficient that will be achieved. Take the bulk temperatures of the helium and water as 600 and 310 K, respectively.

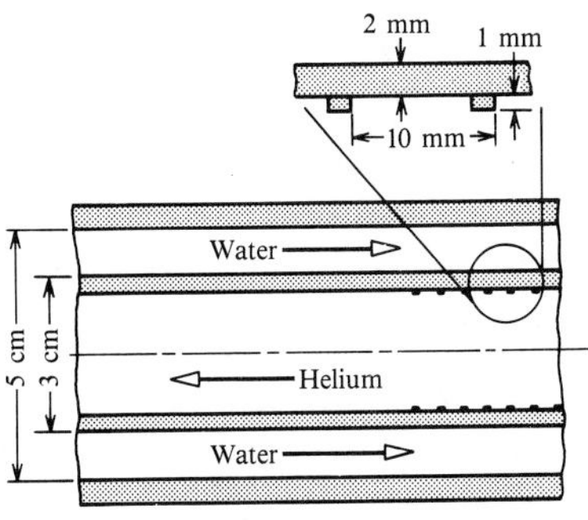

8–8. A 9 m–long stage of a multistage flash vaporization desalination plant operates at 27,150 Pa. The condenser tube bundle comprises 150 2 cm–O.D. titanium tubes with a 1 mm wall thickness, through which cooling sea water flows at 120 kg/s. If the cooling water enters the stage at 320 K, at what temperature does it exit? Take the outside heat transfer coefficient for film condensation to be 10,000 W/m^2 K. At what rate does the stage produce fresh water? Take k = 18 W/m K for the titanium.

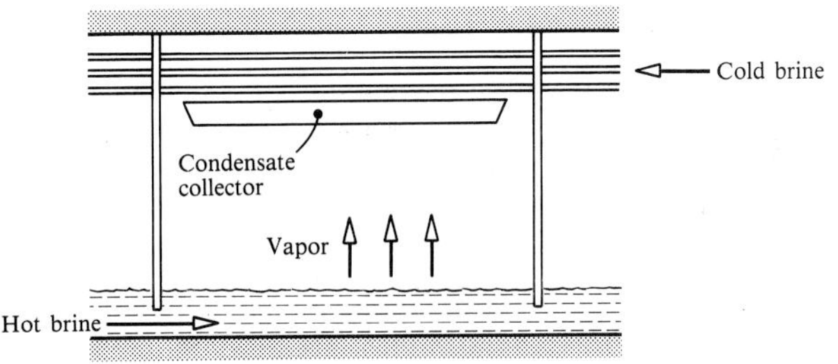

8–9. Hot brine from a geothermal well passes through 120 3 cm–O.D., 1 mm–wall-thickness, 16 m–long titanium alloy (k = 14 W/m K) tubes. Refrigerant-113 boils on the exterior of the tubes at 150°C, with an outside heat transfer coefficient of 20,000 W/m^2 K. The hot brine enters the tubes at 210°C, and the bulk velocity in each tube is 2 m/s. Calculate the following quantities.

(i) Re_D, the Reynolds number inside a tube
(ii) $h_{c,i}$, the inside heat transfer coefficient
(iii) U, the overall heat transfer coefficient
(iv) N_{tu}, the number of transfer units of the boiler
(v) ε, the exchanger effectiveness
(vi) $T_{H,\mathrm{out}}$, the hot stream outlet temperature
(vii) $\dot{Q}$, the heat transferred in the boiler
(viii) $\dot{m}_C$, the R-113 vapor production rate

For the brine, take k = 0.6 W/m K, ρ = 900 kg/m^3, c_p = 4000 J/kg K, and $\nu = 0.20 \times 10^{-6}$ m^2/s; for R-113, take $h_{fg} = 0.2 \times 10^6$ J/kg.

8–10. Performance tests have been performed on a brine/water tubular heat exchanger for geothermal energy utilization. The tubes are 2 m long and have a 10.26 mm inside diameter. Brine at 350 K is supplied at 0.134 kg/s per tube. Silica is found to deposit at a rate of 6.6×10^{-7} g/cm^2 min. If the initial overall heat transfer coefficient is 5000 W/m^2 K, and the effect of scale roughness is to increase the inside heat transfer coefficient by 60%, determine the overall heat transfer coefficient after 1000 hours of operation. Take the density and thermal conductivity of silica scale as 2200 kg/m^3 and 0.6 W/m K, respectively.

8–11. A single-pass shell-and-tube condenser is required to condense 1 kg/s of steam at 6224 Pa, by using cooling water at 295 K. If the allowable water temperature rise is 12 K, determine the length of a condenser containing 300 1 in, 18 gage brass tubes. Take the steam-side heat transfer coefficient as 6500 W/m^2 K, and allow for 0.2 mm–thick scale of conductivity 1.5 W/m K on the inside wall.

8–12. A one-shell-pass, two-tube-pass exchanger is to be used to condense 100 kg/s of steam at 4714 Pa using water at 290 K. If the expected overall heat transfer coefficient is 4000 W/m^2 K and the water temperature rise must not exceed 10 K, determine the heat transfer area required.

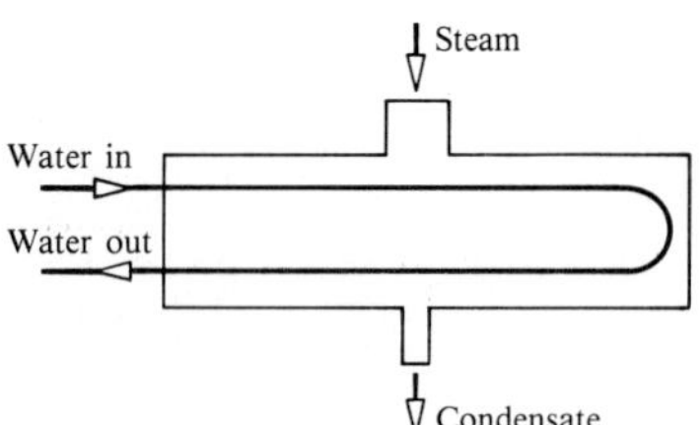

8–13. Saturated steam at 0.115 MPa is supplied to an air heater and condenses on 1 in, 18 gage horizontal brass tubes 0.7 m long. The tubes are staggered on 4 cm centers to give four rows, each with 15 tubes. Air enters the tubes at 1 atm, 285 K with a velocity of 2.0 m/s.

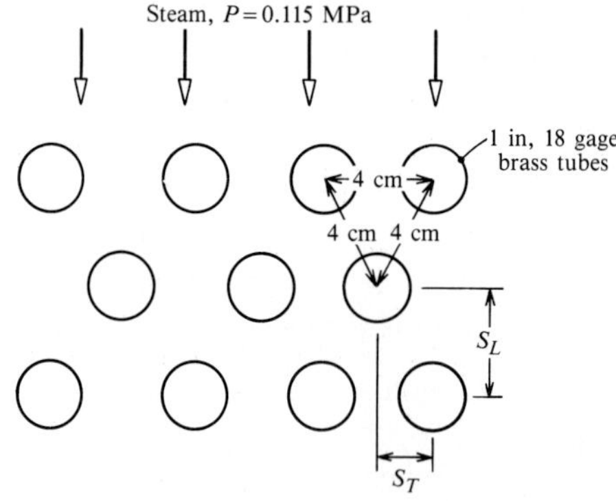

Determine the outlet air temperature, steam consumption rate, and air-side pressure drop.

8–14. A horizontal hot air duct in an attic has a 25 cm inside diameter and is 15 m long. The air velocity is 8 m/s. If the air enters at 35°C and the attic air temperature is 15 °C, what is the heat loss from the duct if it is uninsulated?

8–15. A power plant stack is 100 m high and has an inside diameter of 6 m. The wall is made of concrete and varies in thickness linearly from 50 cm at the bottom to 25 cm at the top. Exhaust gas enters the base of the stack at a rate of 100 kg/s and at a temperature of 600 K. Air at 300 K and 1 bar blows at 5 m/s across the stack. Determine the exit temperature of the stack gas.

8–16. Brine is to be evaporated in a multistage falling-film evaporator, and is fed to the top of a bundle of vertical 5 cm–O.D., 2 mm–wall-thickness stainless steel tubes 3 m high, at 0.010 kg/s per tube. Hot water at 340 K is fed to the bottom of the bundle and flows inside the tubes at 0.050 kg/s tube. If the saturation temperature in the stage is 330 K, what is the vapor production per tube?

8–17. Water flows at 1.2×10^{-3} kg/s through each of a number of parallel tubes bonded to the back face of a flat-plate solar collector. An approximate analysis of the collector indicates that heat will be collected at a rate of $1.3(90 - T)$ W/m per tube, where T is the local bulk temperature of the water, in degrees Celsius. It is desired to build a unit that can heat water from 15°C to 60°C.

(i) What is the required heat exchanger effectiveness?
(ii) How many transfer units are required?
(iii) How long should the unit be?

8–18. Vegetable oil is to be maintained at 50°C in a 3 m–diameter, 11 m–high tank. A water supply at 80°C is available, and a single helix coil of 5 cm–O.D. stainless steel tube has been proposed as the heat transfer surface. The tank is insulated with a 7 cm–thick medium-density fiberglass blanket, and the ambient temperature can go as low as 0°C. If the overall heat transfer coefficient from the water to the oil is estimated to be 160 W/m² K, determine the water flow rate and length of tube required for an 80% effective heat exchanger. Make reasonable assumptions for this preliminary design calculation.

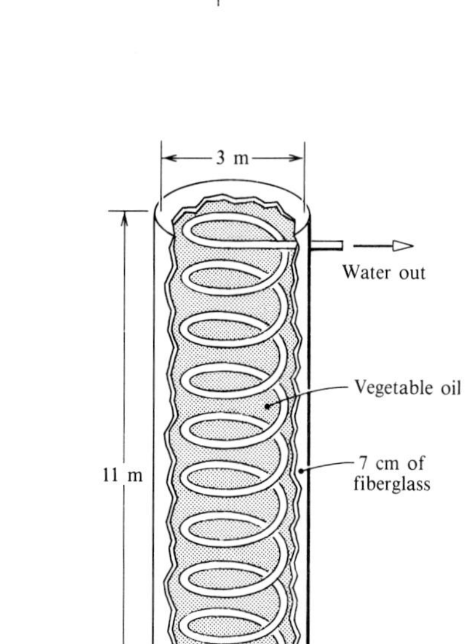

8–19. Exhaust gases exit a combustion test rig at 600°C and flow at 0.1 kg/s along a 5 m length of 4 cm–O.D., 2 mm–wall-thickness steel tube suspended from the roof of the laboratory. An estimate of the tube wall temperature is required for an ambient temperature of 22°C. Assume that the inside and wall resistances are negligible and that the heat loss from the outside is by natural convection and radiation ($\varepsilon = 0.9$ for the tube surface). (*Hint:* First show that the governing equation for this single steam exchanger can be transformed into the equation solved by LUMP. Then use LUMP to obtain the desired result.)

8–20. A coaxial tube exchanger is operated under the following conditions.

	$\dot{m}$ kg/s	c_p J/kg K	T_{in} K	T_{out} K
Cold fluid	0.125	4200	313	368
Hot fluid	0.125	2100	483	

(i) Determine T_{out} for the hot fluid.
(ii) Determine the exchanger effectiveness.
(iii) What is the ratio of required area for parallel versus counterflow operation?

8–21. A single-pass counterflow exchanger is required to cool 7000 kg/h of oil from 365 K to 330 K. Cooling water is available at 4000 kg/h and 290 K. If the overall heat transfer coefficient is 300 W/m^2 K, determine the surface area required. For the oil take $c_p = 2100$ J/kg K.

8–22. An economizer for a power plant is required to heat 8 kg/s of water from 340 K to 480 K. Flue gas is available at 25 kg/s and 800 K. Determine

(i) the outlet temperature of the flue gas.
(ii) the heat transfer area required for a counterflow exchanger if the overall heat transfer coefficient is 50 W/m^2 K.

Approximate the flue gas properties using those of air.

8–23. An aircraft oil cooler is to be designed to reduce the oil temperature from 390 K to 365 K. The oil (SAE 50) flow rate is 1.5 kg/s. If the overall heat transfer coefficient is 140 W/m^2 K and the entering air temperature is 310 K, find the necessary transfer area for

(i) counterflow.
(ii) parallel flow.

Assume balanced flow, $C_H = C_C$.

8–24. A two-shell-pass, four-tube-pass exchanger is available to cool 5 kg/s of liquid ammonia at 70°C, against 8 kg/s of water at 15°C. If the heat transfer area is 40 m^2 and the expected overall heat transfer coefficient is 2000 W/m^2 K when the ammonia is on the shell side, determine the outlet temperature of the ammonia.

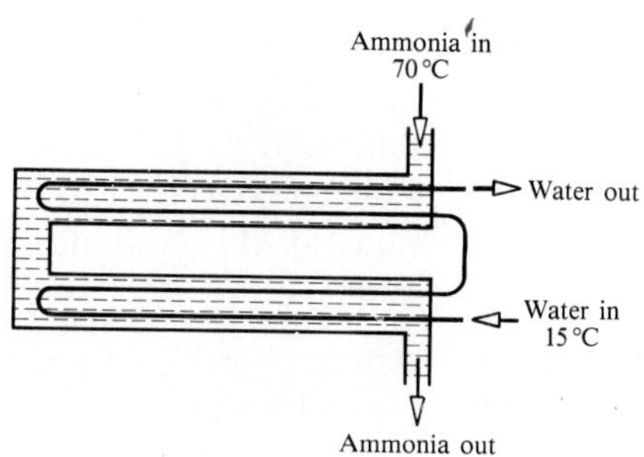

8–25. Flue gas in a power plant is used to preheat air in a cross-flow heat exchanger. The flue gas is available at 340°C and must not exit at less than 200°C. The plant requires 10 kg/s of air and the combustion reaction has a stoichiometric ratio of 15:1. If the air enters the exchanger at 40°C, what is its maximum exit temperature? If the overall heat transfer coefficient is 150 W/m^2 K, calculate the heat transfer surface area required if

(i) the air is unmixed and the flue gas is mixed.
(ii) both streams are unmixed.

Approximate the properties of flue gas with those for air.

8–26. Interstage cooling is used to reduce the compression work required in a multistage compressor of a helium-cooled nuclear reactor. In one such cooler, 2.5 kg/s of helium at 2.5 bar is cooled from 400 K to 310 K by 5.0 kg/s of cooling water at 300 K. It is proposed to have a single pass of helium through 1/2 in, 18 gage copper tubes at a velocity of 30 m/s, with the water in counterflow on the shell side. If the shell-side heat transfer coefficient is 8000 W/m^2 K, determine the number of tubes required and the length of the exchanger. Also calculate the helium pressure drop.

8–27. Show that the effectiveness of a cross-flow exchanger with the $C_{\max}$ stream mixed and the $C_{\min}$ stream unmixed is given by

$$\varepsilon = \frac{1}{R_C}\left\{1 - \exp\left[R_C\left(e^{-N_{\mathrm{tu}}} - 1\right)\right]\right\}$$

8–28. A diesel truck engine is turbocharged with 500 kg/h of air. The air is heated to 420 K by the compressor in the turbocharger. It is desired to cool the air to 392 K before it enters the engine. Water at 380 K is taken from the truck radiator at a rate of 1000 kg/h to provide cooling. An intercooler configured as a counterflow exchanger is used. The air-side heat transfer coefficient is 30 W/m^2 K, the water-side is 9000 W/m^2 K, and the air side is finned (η_f = 1) with a ratio of air to water area of 5 to 1. Determine the required effectiveness, the overall heat transfer coefficient based on water-side area, and the required water-side heat transfer area.

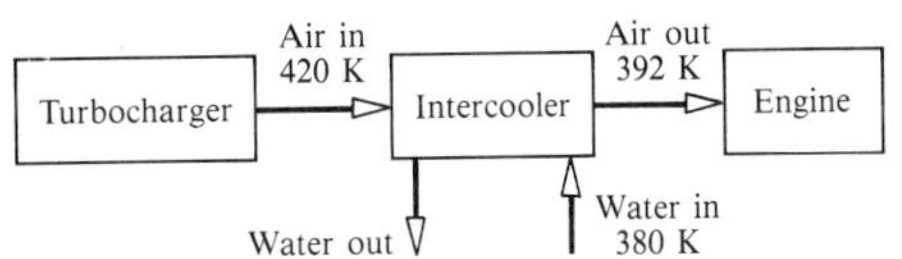

8–29. A heat exchanger is to be designed to cool 15 kg/s of a solution from 70°C to 40°C using 13 kg/s of water available at a temperature of 10°C. The specific heat of the solution is 3550 J/kg K. If the overall heat transfer coefficient may be taken to be 2000 W/m^2 K, calculate the required heat transfer areas for the following flow configurations:

(i) Parallel flow
(ii) Counterflow
(iii) A shell-and-tube exchanger with the hot solution in the shell for one shell pass and four tube passes
(iv) Same as (iii), but two shell passes

8–30. A 2.5 m–long coaxial-tube heat exchanger consists of a 6.38 mm–O.D., 4.52 mm–I.D. brass tube surrounded by a 9.63 mm–I.D. glass tube. The heat exchanger is tested in a laboratory with hot water inside the inner tube and cold water in the annulus. In a counterflow test the following data were recorded:

$$\dot{m}_C = 0.121 \text{ kg/s}; \; T_{C,\text{in}} = 23.5°\text{C}; \; T_{C,\text{out}} = 30.0°\text{C}$$

$$\dot{m}_H = 0.0562 \text{ kg/s}; \; T_{H,\text{in}} = 44.8°\text{C}; \; T_{H,\text{out}} = 31.3°\text{C}$$

Compare the effectiveness obtained in the test with the expected value.

8–31. A heat exchanger is to be designed for preheating secondary air supplied to an afterburner that reduces emissions from a gasoline engine. The exhaust gas flow rate varies from 20 to 150 kg/h, and the secondary air should be supplied at 10% of the exhaust mass flow rate. The exhaust gas leaves the afterburner at 850 K, the ambient air temperature is 290 K, and the secondary air should be heated to 780 K. A first design proposal is to construct a coaxial-tube counterflow exchanger by utilizing the exhaust pipe as the inner tube. If the exhaust pipe is 2 in schedule 10 steel pipe, and the proposed outer shell is 2½ in schedule 10 pipe, determine the required exchanger length if the outer shell is well insulated. Comment on the feasibility of the design.

8–32. In cardiac surgery it is often necessary to achieve a certain level of hypothermia just prior to the operation, with restoration to normal body temperature immediately thereafter. For this purpose a heat exchanger may be installed in the arterial inlet tube leading from a heart-lung machine. Important constraints on the design include the following:

(i) Blood flow must be upward to prevent trapping of gas bubbles.
(ii) There should be minimal agitation of the blood to prevent blood trauma.
(iii) Materials in contact with the blood must be inert, smooth, and easily sterilized.
(iv) The coolant should be tap water.

A stainless steel shell-and-tube exchanger installed vertically is proposed. Water is on the shell side, and there are sufficient baffles to assume overall counterflow. The shell contains 24 ¼ in 20 gage tubes in a square array at a pitch of ¾ in. A blood flow of 2000 cc/min is to be cooled from 37.2°C to 27.5°C, with a cooling water supply of 6000 cc/min at 15°C. The baffles give a water velocity transverse to the tubes of 0.15 m/s. What length of exchanger is required? Take blood properties as follows: $\rho = 993$ kg/m^3, $c_p = 3850$ J/kg K, $k = 0.52$ W/m K, $\mu = 3.7 \times 10^{-3}$ kg/m s.

8–33. Use L'Hopital's rule to derive Eq. (8.45) from Eq. (8.41).

8–34. Derive the $\varepsilon - N_{\text{tu}}$ relation for a counterflow exchanger, Eq. (8.41).

8–35. A cross-flow plate heat exchanger is to be designed for waste heat recovery from the exhaust stream in a copper smelting plant. A flow of 10 kg/s of exhaust gas enters the exchanger at 260°C, while 10 kg/s of air enters at 20°C. If the air is to be heated at 180°C and the overall heat transfer coefficient is estimated to be 50 W/m^2 K, determine the required heat transfer area for the following flow configurations:

(i) Both fluids unmixed
(ii) Both fluids mixed
(iii) $C_{\max}$ mixed, $C_{\min}$ unmixed
(iv) $C_{\max}$ unmixed, $C_{\min}$ mixed

The specific heat of the exhaust gas can be taken as 1220 J/kg K.

8–36. A counterflow heat exchanger for waste heat recovery consists of banks of finned heat-pipes, as shown in the sketch. The fins can be taken to have a fin effectiveness of unity, but the heatpipes have a thermal resistance that can be defined by $\dot{Q} = (T_e - T_c)/R_{\text{hp}}$. Derive the heat exchanger effectiveness for n discrete banks and balanced flow.

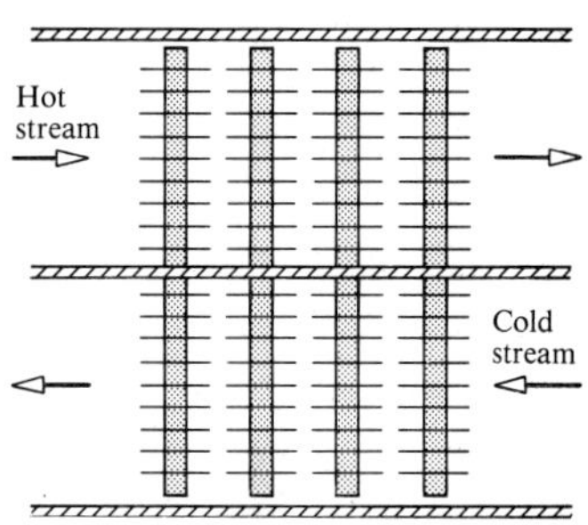

8–37. A counterflow heat exchanger is used to heat 6.0 kg/s of water from 35°C to 90°C by 14 kg/s of oil (c_p = 2100 J/kg K) supplied at 150°C. The design value of the overall heat transfer coefficient is 120 W/m^2 K. A second unit is to be built at another plant location; however, it has been proposed that the single exchanger be replaced by two smaller counterflow exchangers of equal heat transfer area. The two exchangers are to be connected in series on the water side and in parallel on the oil side. The oil flow is split equally between the two exchangers, and it may be assumed that the overall heat transfer coefficient for the smaller exchangers is also 120 W/m^2 K. Compare the transfer areas of the two arrangements.

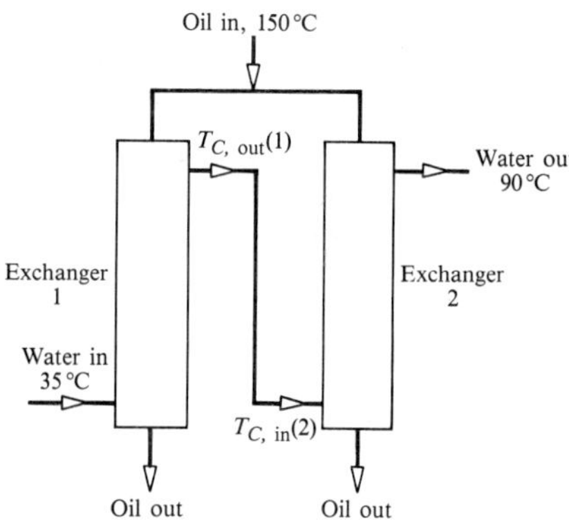

8–38. An oil is cooled in a coaxial-tube parallel-flow heat exchanger. A water flow of 0.1 kg/s enters the inner tube at 15°C and is heated to 40°C. The oil flows in the annulus and is cooled from 130°C to 55°C. It is proposed to cool the oil to a lower outlet temperature by increasing the length of the exchanger. Determine the following:

(i) The minimum temperature to which the oil may be cooled
(ii) The outlet oil temperature as a function of the fractional increase in exchanger length
(iii) The outlet temperature of each stream if the existing unit were switched to counterflow operation

(iv) The minimum temperature to which the oil may be cooled in counterflow operation
(v) The ratio of required length for counterflow to that for parallel flow as a function of outlet oil temperature

Neglect heat loss to the surroundings, and make any other reasonable assumptions.

8–39. Exhaust gases at 500°C from a 4 liter, 6-cylinder automotive engine, operating at 3000 rpm and 80% volumetric efficiency, are to be cooled in a coaxial-tube counterflow heat exchanger by 0.24 kg/s of cooling water supplied at 15°C. The exchanger is made from 15 m of 3 cm–O.D., 1 mm–wall-thickness 316 stainless steel tube inside a 4 cm–I.D. tube, with the gas flowing in the inside tube. Estimate the outlet temperature of the exhaust gas. Neglect heat losses to the surroundings.

8–40. A 6 kg/s flow of sulfuric acid (c_p = 1480 J/kg K) is to be cooled in a two-stage counterflow heat exchanger. The hot acid at 175°C is fed to a tank where it is stirred in contact with cooling coils; the continuous discharge from this tank at 89°C flows into a second stirred tank and leaves the second tank at 46°C. Cooling water at 20°C enters the cooling coil of the cold tank and leaves the cooling coil of the hot tank at 80°C. If the overall heat transfer coefficients are 52 and 38 W/m^2 K in the hot and cold tanks, respectively, determine the heat transfer surface areas required. Neglect heat losses to the surroundings, and assume that the acid temperature in each tank equals its outlet temperature.

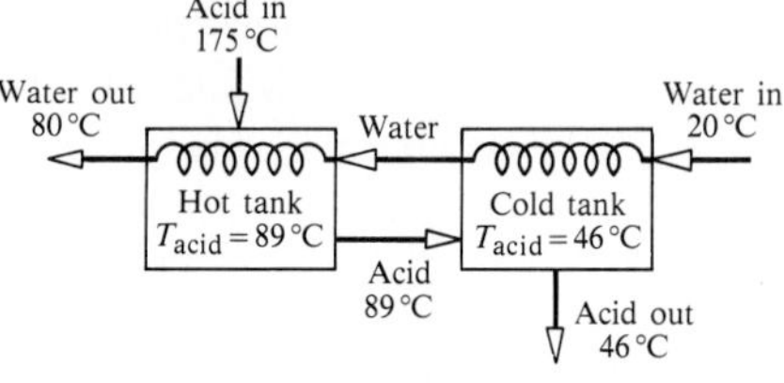

8–41. A one-shell-pass, two-tube-pass heat exchanger is used to cool a vegetable oil using river water. The design specifications are: U = 340 W/m^2 K, $T_{C,\text{in}}$ = 20°C, $T_{C,\text{out}}$ = 70°C, $T_{H,\text{in}}$ = 150°C, and $T_{H,\text{out}}$ = 60°C. After operating for one year, the performance is found to have deteriorated, and the oil is cooled only to 80°C. Fouling of the oil side is suspected; determine the apparent fouling resistance. Take c_p = 1800 J/kg K for the oil.

8–42. Air for a furnace is preheated from 20°C to 120°C in a preheater by flue gases available at 320°C. The furnace burns 8300 kg/h of coke (assumed to be 100% carbon) with 10% excess air. Using an overall heat transfer coefficient of 20 W/m^2 K for comparison purposes, determine the heat transfer area required for the following configurations:

(i) Counterflow
(ii) Single-pass cross-flow, both streams mixed
(iii) Single-pass cross-flow, air mixed, gas unmixed

8–43. A counterflow heat exchanger is to be designed to recover waste heat from a geothermal brine. A 7.2 kg/s brine flow at 340 K is to be cooled to 308 K by a 9 kg/s water supply at 300 K. The unfouled overall heat transfer coefficient is estimated to be 1500 W/m^2 K, and a brine-side fouling resistance of 0.002 (W/m^2 K)$^{-1}$ should be allowed for. What is the required heat transfer area? The brine specific heat has been measured to be 3480 J/kg K.

8–44. Oil is heated in a 4 m–long coaxial-tube heat exchanger, with oil in laminar flow in the inside tube and water in turbulent flow in the outer jacket. The inner tube has a 2 cm outside diameter and 1.5 mm wall thickness, and the outer jacket has a 3 cm inside diameter. The flow rates of oil and water are 1.1 and 2 kg/s, respectively. The oil enters at 10°C, and the water enters at 90°C. Estimate the oil outlet temperature. Approximate the oil properties by those for SAE 50 oil, and comment on the effects of oil property variation along the exchanger.

8–45. A heat exchanger is used to preheat hydrogen fuel for a hydrogen combustor. The combustor burns 8.7 kg/s of fuel, and the exhaust gas flow rate is 55.0 kg/s. The exchanger is located in a 1 m–square duct and is a cross-flow unit with two passes of the hydrogen through 1 cm–O.D., 1 mm–wall-thickness Inconel-X tubes arranged in a square array with a pitch of 2 cm. The exhaust gases enter the exchanger at 0.55 MPa and 1400 K, while the hydrogen enters at 7.6 MPa and 20 K. If the required outlet temperature of the hydrogen is 380 K, determine the number of tubes per pass. Approximate the exhaust gas properties using those for air. (*Hint:* Use the LMTD approach and Fig. C.4*d* of Appendix C.)

8–46. Rework Example 8.11 for AISI 304 stainless steel tubes.

8–47. A counterflow perforated-plate heat exchanger for a helium refrigeration system has 70 plates with flow cross-sectional areas of 0.002 and 0.004 m^2 for the high- and low-pressure streams, respectively. The plates are 0.4 mm thick, and the spacers are 0.8 mm thick with a footprint area of 0.001 m^2. The helium

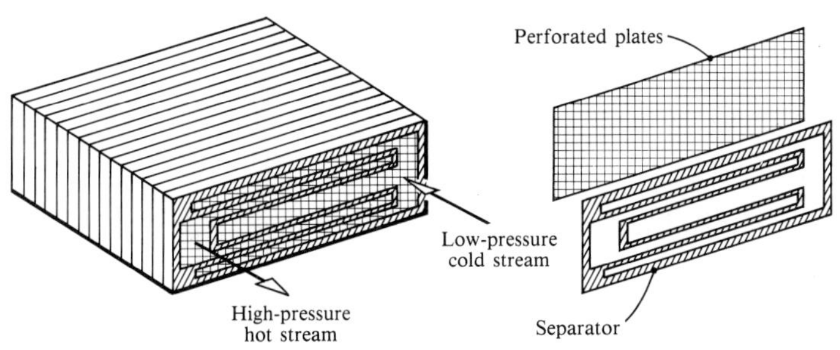

flow is balanced at 0.002 kg/s, giving an overall heat transfer coefficient of 460 W/m^2 K, based on a transfer area of 0.0083 m^2. Determine the effect of spacer thermal conductivity on exchanger ineffectiveness for k values in the range 0.1–2.0 W/m K.

8–48. A counterflow perforated-plate heat exchanger has balanced flows of nitrogen at 8×10^{-5} kg/s. The cold stream enters at 80 K, and the hot stream enters at 300 K, both at 1 atm. There are 30 aluminum plates each 2 mm thick, and the spacers ($k = 0.5$ W/m K) are 3 mm thick and have a footprint area of 5×10^{-4} m^2. The flow cross-sectional area for each stream is 0.0016 m^2, and the overall heat transfer coefficient is 40 W/m^2 K, based on a transfer area of 0.00312 m^2. Determine the exchanger ineffectiveness and the outlet temperatures.

8–49. A pebble bed regenerator is 1 m in diameter and is packed with pebbles that can be approximated as 2 cm–diameter spheres packed with a void fraction of 0.46. During the storage phase of the cycle, 0.9 kg/s of hot air at 360 K is supplied from a waste heat recovery system, and during the extraction phase an equal flow of ambient air at 280 K enters the bed. If the half-cycle time is 10 min, how long should the bed be to deliver the heated air at an average temperature of 350 K? Take $\rho = 2400$ kg/m^3 and $c_p = 790$ J/kg K for the pebbles.

8–50. A 2 m–diameter rotary generator is packed with stacks of AISI 1010 steel wire mesh to give a packing depth of 5 cm. The wire is 2 mm in diameter and is packed to give a void fraction of 0.5. The regenerator is used to preheat 0.3 kg/s of cold air at −10°C, using an equal flow of hot air at 20°C. Plot a graph of average air outlet temperature versus rate of rotation over an appropriate range.

8–51. A parallel-plate laminar flow packing is an attractive option for rotary regenerators, where it can be easily constructed by winding spaced sheets into a coil. Such packings have a relatively high ratio of Stanton number to friction factor. A proposed design for a 1.6 m–diameter wheel has 0.5 mm–thick polyethylene sheet spaced at 1 mm. The unit is to be used for recovering energy in an air-heating system and must operate with a balanced flow of 6.0 kg/s. Plot a graph of the regenerator effectiveness versus rate of rotation for packing lengths of 5, 10, and 20 cm. Evaluate all properties at 300 K.

8–52. A low-pressure helium exchanger has a flow area of 0.016 m^2. A test is conducted in which 2×10^{-3} kg/s of helium is heated from 20 K to 380 K, and the inlet and outlet pressures are measured to be 820 and 110 Pa, respectively. Determine the viscous contribution to the pressure drop.

8–53. A 1 m–cube compact heat exchanger core has the plain fin heat transfer surface given as item 3 of Table 8.4. On the hot side, 0.9 kg/s of air enters at 360 K and 1.09×10^5 Pa, and it exits at 320 K. Calculate the pressure drop.

8–54. Rework Example 8.19 for parallel plates without fins. Assume that transverse mixing is negligible.

8–55. A flow of 50 m^3/min of air at 30°C is to be cooled to 20°C against a balanced stream of air at 15°C.

(i) How many transfer units are required for a cross-flow exchanger with both streams unmixed?

(ii) Design an appropriate exchanger using a parallel-plate geometry with a gap width of 2 mm and a plate thickness of 0.2 mm. If the pressure drop for each stream should not exceed 40 Pa, specify the dimensions of the exchanger.

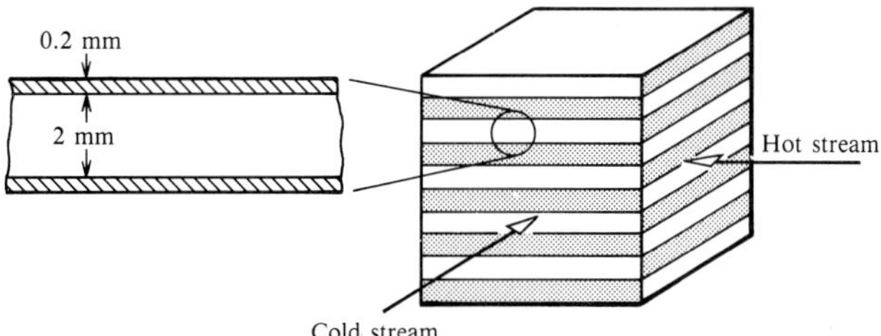

8–56. To cool a house in a hot but moderately dry climate, a flow of 40 m^3/min of ambient air at 40°C is to be cooled against a balanced stream of 15°C air from an evaporative cooler.

(i) How many transfer units are required for a counterflow exchanger to cool the air to 20°C?

(ii) Design an appropriate exchanger using a parallel-plate geometry with a plate pitch of 2.5 mm. Specify the frontal area required for each stream and the exchanger length if the pressure drop for each stream must not exceed 30 Pa. Assume that the plates are very thin.

8–57. A counterflow twin-tube heat exchanger is to be used for balanced flows of air at 3.0×10^{-3} kg/s. The hot stream enters at 360 K and must be cooled to 320 K; the cold stream enters at 290 K. If the average pressure in each stream is 1 atm and the allowable pressure drop for the hot stream is 10^4 Pa, determine suitable dimensions. Copper tubes may be assumed, giving a tube wall fin effectiveness of approximately unity.

8–58. To cool a small warehouse in Tucson, Arizona, a flow of 100 m^3/min of ambient air at 35°C is to be cooled to 26°C against a balanced stream of 18°C air from an evaporative cooler. Specify the dimensions of an appropriate cross-flow heat exchanger. Use a simple parallel-plate matrix with a plate thickness of 0.2 mm and a gap width of 2.5 mm. The allowable pressure drop for each stream is 50 Pa.

8–59. Repeat Exercise 8–58 using the plain fin surface given as item 4 of Table 8.4. Use a scale factor of 3 and AISI 1010 steel, and increase the allowable pressure drop to 1000 Pa.

8–60. A single-pass cross-flow heat exchanger is to be used as a recuperator to recover energy from the flue gases of a combustor by preheating the combustion air. Inconel-X-750 tubes of outer and inner diameters 60 and 54 mm, respectively, are arranged in a staggered array, with both transverse and longitudinal pitches of 120 mm. The exhaust gas flows through the tube bank, and the combustion air is inside the tubes. The combustor burns 1.2 kg/s of a fuel with stoichiometric ratio of 15.2, and 10% excess air is supplied. The air enters the tubes at 310 K, and the flue gas enters the tube bank at 1100 K. The inside of the tubes can be assumed clean, but a fouling resistance of 0.01 $(\mathrm{W/m^2\,K})^{-1}$ should be allowed for on the outside of the tubes. If the air is to be heated to 500 K and the pressure drop of the flue gas stream should not exceed 2000 Pa, suggest suitable dimensions for the tube bank. Approximate the flue gas properties by those for air. Use CONV to obtain $\overline{h}_c$ and ΔP.

8–61. In a batch process, 10,000 kg of 95°C wash water is dumped into a drain every hour in a plant that operates two eight-hour shifts per day, 310 days per year. At present, mains water at 18°C is heated by steam to 110°C to provide clean water to wash the next batch. The cost of the steam is estimated to be 0.7 cent/MJ of vaporization enthalpy. To reduce energy costs at the plant, it is proposed to pump the dirty wash water to an insulated holding tank for use to preheat the incoming water in a balanced-flow counterflow heat exchanger. The initial cost of the tank, pumps, and plumbing is estimated to be \$200,000, and the heat exchanger can be acquired at \$400/m^2 of heat transfer surface. Taxes and insurance are 15% of the exchanger cost, and capital can be loaned at 10% interest for 15 years. The exchanger can be designed to have an overall heat transfer coefficient of 1600 W/m^2 K. Should the proposal be adopted? If so, specify the optimal heat transfer surface area and the savings per year.

8–62. Rework the analysis in Section 8.7.4 for an unbalanced counterflow exchanger to obtain the relation between the optimal N_{tu} and the thermoeconomic parameter $\mathscr{E}$.

8–63. An oil supply of 1.25 kg/s must be heated from 15°C to 80°C during 7200 hours per year of operation. A 0.75 kg/s supply of water at 150°C is available for preheating, if worthwhile. The energy gained by preheating will save 1 cent/MJ, but the heat exchanger will cost \$250/m^2 yr to own and operate. If the exchanger design under consideration has an overall heat transfer coefficient of 170 W/m^2 K, should the preheater be installed? If so, how large should it be, what would it cost, and what would be the savings per year? Take c_p = 500 J/kg K for the oil.

8–64. In optimizing the design of a heater, it is sometimes appropriate to prescribe $\dot{m}_C$, $T_{C,\mathrm{in}}$, and $T_{C,\mathrm{out}}$ for the cold stream and $T_{H,\mathrm{in}}$ for the hot stream. The problem is to determine the optimal value of $\dot{m}_H$. We can write $T_{H,\mathrm{in}} = T_0 + \dot{m}_{\mathrm{fuel}}\Delta h_c/C_H$ if the hot stream is heated by burning a fuel, where Δh_c is the heat of combustion; then the annual cost of the fuel consumed is $\mathscr{C}_f = \dot{m}_{\mathrm{fuel}}\tau\Delta h_c v$. The cost of the exchanger remains as specified by Eq. (8.88), and U is assumed to be independent of $\dot{m}_H$. As the exchanger is lengthened, the cost goes up, but since the effectiveness increases, the value of the fuel used is reduced. Determine the relation between effectiveness ε and thermoeconomic parameter $\mathscr{E}$ that maximizes the benefit-cost differential. Consider two cases:

(i) $C_H = C_{max}$.
(ii) $C_H = C_{min}$.

Make appropriate simplifying assumptions. (See also reference [11].)

8–65. In a laundry, 67°C dirty wash water is dumped into the drain, and 70°C clean water is required. Presently 15°C water is heated in an electric hot water heater, and the electricity costs 9 cents/kW h. The water is required at a rate of 5000 kg/h, 12 hours per day, 312 days/yr. To conserve energy it is proposed to install a counterflow heat exchanger to preheat the feed to the electric water heater. The installation will cost \$20,000 plus \$900 per square meter of heat exchanger surface. The interest rate to amortize the investment over 12 years is 10% per annum. Taxes and insurance are expected to have a fixed cost of \$500 per annum plus \$50/yr per square meter of heat exchanger surface. If the overall heat transfer coefficient is estimated to be 1000 W/m^2 K, determine the optimal heat transfer area of the exchanger and the corresponding net annual savings.

8–66. Exhaust gases from a heat-treating furnace are presently discharged to the stack at the furnace temperature of 1100°C. Fuel with a lower heating value of 46.5 MJ/kg is used to fire the furnace and is presently consumed at a rate of 0.13 kg/s. The fuel costs 30 cents per kilogram. It is proposed to reduce fuel costs by preheating the 2 kg/s of combustion air above its present value of 15°C. A large rotary regenerator is planned, with a rate of rotation high enough to permit the assumption $R_R \gg N_{tu}$. The matrix has triangular passages with $f = 53.3/\mathrm{Re}_{D_h}$ and $\mathrm{Nu} = 3.1$ for laminar flow. The frontal areas for both streams are equal and are made large enough so that the larger Reynolds number is 200 in the 0.5 cm–hydraulic diameter passages. For both streams the properties can be taken as those for air at 1 atm and 800 K.

When the combustion air is preheated, the fuel consumption can be reduced. As a first approximation, neglect the change in C_H as the fuel flow is reduced, and assume balanced flow. The cost of the exchanger and all auxiliary equipment is projected to be \$20,000 plus \$100 per square meter of heat transfer surface. The cost of fan power is 10 cents/kW h, and the fan-motor unit efficiency can be taken to be 50%. The rate to amortize the investment over 15 years is 9%, and the salvage value of the exchanger after 20 years of use is \$1000. Taxes and insurance are expected to have a cost of \$1500 per annum plus \$40/yr per square meter of heat transfer surface. If the furnace operates 5000 hours per year, find the optimal length L of the exchanger, the annual fuel savings, and the net annual savings. Discuss the effect of the choice $\mathrm{Re} = 200$; how would the result change for $\mathrm{Re} = 100$ or $\mathrm{Re} = 400$? Also discuss the advantages of reducing the air flow to maintain a constant air-fuel ratio.

8–67. Referring to Examples 8.21 and 8.22, is there a combination of pressure drop and outlet temperature that maximizes the benefit-cost differential? If there is, comment on the feasibility of the design.

8–68. Referring to Examples 8.21 and 8.22, explore the effects of the characteristic dimension of the louvered fin surface by varying the scale factor from 0.5 to 2.0.

8–69. Referring to Examples 8.21 and 8.22, explore the effect of core material by considering the following materials:

(i) Aluminum alloy, $k = 180$ W/m K
(ii) Brass, $k = 111$ W/m K
(iii) AISI 302 stainless steel, $k = 15$ W/m K

Take $c_i = \$10/\text{m}^2$ for the aluminum, and $\$15/\text{m}^2$ for the brass and stainless steel.

8–70. Referring to Examples 8.21 and 8.22, explore the effect of type of heat transfer surface by reworking the problem for the louvered fin surface given as item 5 of Table 8.4. Notice that this surface has the same plate spacing as surface 6 but fewer fins and a larger hydraulic diameter.

8–71. Referring to Examples 8.21 and 8.22, explore the effect of type of heat transfer surface by reworking the problem for the plain fin surface given as item 3 of Table 8.4. Notice that this surface has approximately the same plate spacing as surface 6.

8–72. Referring to Example 8.22, explore the effect of annual interest rate of capital borrowed from the bank. Let i vary from 5 to 15% for a loan amortized over 20 years.

8–73. Referring to Example 8.22, explore the effect of the cost of electricity by varying c_p from 0.05 to 0.15 \$/kW h.

CHAPTER

9

ELEMENTARY MASS TRANSFER

CONTENTS

9.1 INTRODUCTION

Mass transfer may occur in a gas mixture, a liquid solution, or a solid solution. There are several physical mechanisms that can transport a chemical species through a phase and transfer it across phase boundaries. In this chapter we will consider only the two most important, which are **ordinary diffusion** and **convection.** Mass diffusion is analogous to heat conduction and occurs whenever there is a gradient in the concentration of a species. Mass convection is essentially identical to heat convection: a fluid flow that transports heat may also transport a chemical species. The similarity of the mechanisms of heat transfer and mass transfer results in the mathematics often being identical, a fact that can be exploited to advantage. For example, the solution to a problem of diffusion in a solid can often be written down immediately by referring to the analogous heat conduction problem. But, for the beginning student in particular, there are some significant differences between the subjects of heat and mass transfer. One difference is the much greater variety of physical and chemical processes that require mass transfer analysis, as illustrated by the following list.

1. Evaporation of water into air in a cooling tower
2. Drying of wood, paper, and textiles
3. Leakage of helium from the laser of a copying machine
4. Diffusion of carbon into iron during case-hardening of a gear wheel
5. Catalytic oxidation of carbon monoxide and unburnt hydrocarbons in an automobile catalytic converter
6. Measurement of humidity using wet and dry thermocouples
7. Combustion of pulverized coal in a power plant furnace
8. Combustion of kerosene droplets in a gas turbine combustion chamber
9. Combustion of iron in oxy-acetylene steel cutting
10. Absorption of sulfur dioxide into an alkaline solution in a power plant flue-gas scrubber
11. Doping of semiconductors for transistors
12. Discharge of a lead-acid battery
13. Power generation by a fuel cell
14. Aeration of sewage for biological treatment
15. Evaporation and condensation in gas-controlled heatpipes
16. Ablation of a heat shield on a reentry vehicle
17. Erosion of a graphite or rhenium nozzle in a solid-propellant rocket

18. Scrubbing of radioactive nuclides in a nuclear reactor pressure-suppression pool following an accident involving core degradation

19. Oxidation of Zircaloy fuel rod cans by steam following a loss-of-coolant accident in a boiling-water nuclear reactor

20. Separation of oxygen and nitrogen in the separation column of a gaseous oxygen plant

21. Blood oxygenation in a rotating-disk oxygenator during open-heart surgery

22. Desalination of brackish water by reverse osmosis

A second difference between mass and heat transfer is the extent to which essential features of a given process may depend on the particular chemical system involved, and on temperature and pressure. As an example of how a particular chemical system affects a given process, consider scrubbing of a chemical species from an air stream by water in a packed tower, as shown in Fig. 9.1. Inside the tower is a particle bed over which water is sprayed, while the air flows upward through the bed. If the species to be scrubbed is only slightly soluble, such as hydrogen sulfide, the process is controlled by diffusion into the water, and the precise details of the air flow are unimportant. However, if the species is ammonia, which is very soluble, the process is controlled by diffusion out of the air stream, and the precise details of the water flow are unimportant. As a second example, consider the combustion of a particle in air at a high temperature, as illustrated in Fig. 9.2. If it is a carbon particle,

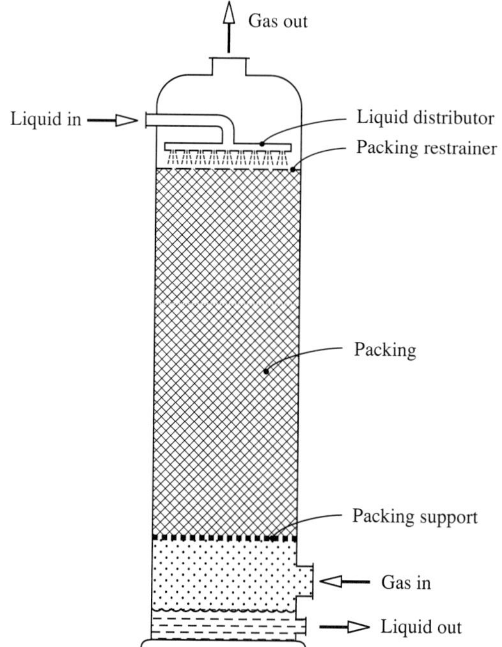

Figure 9.1 A packed-tower gas scrubber.

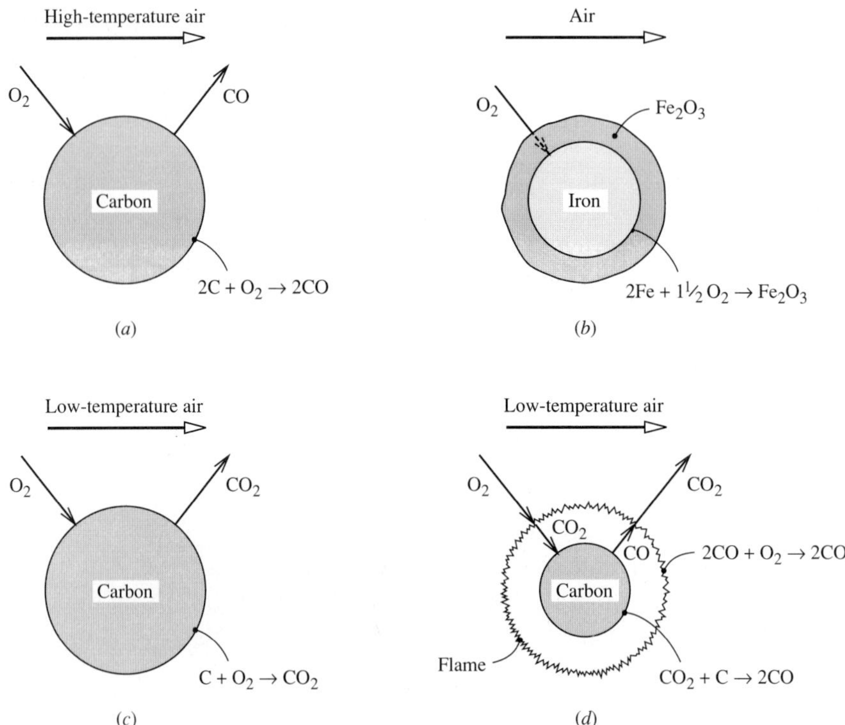

Figure 9.2 Some examples of particle oxidation. (*a*) High-temperature oxidation of carbon in air to form carbon monoxide. (*b*) Oxidation of iron in air to form solid ferrous oxide. (*c*) Oxidation of carbon at low temperatures to form carbon dioxide. (*d*) Oxidation of carbon at intermediate temperatures to form carbon dioxide.

the reaction occurs on the carbon surface to form gaseous carbon monoxide, which diffuses away into the air stream. However, if it is an iron particle, the reaction can produce solid iron oxide, which remains on the particle surface to impede the oxidation process. As an example of how temperature affects the process, consider the same carbon particle at a much lower temperature: the reaction product is then carbon dioxide, and not carbon monoxide. And at intermediate temperatures two reactions occur: on the surface carbon dioxide is reduced to carbon monoxide, while some distance away from the surface carbon monoxide reacts with oxygen to form carbon dioxide in a flame. Sometimes the essential features of the process will be well known and supported by experimental observation, but often physical and chemical thermodynamic data, and chemical kinetics data, must be employed to assist in the understanding of the process.

The foregoing discussion indicates that mass transfer might be both a challenging and interesting subject. Assuming the students (and instructor!) take up the challenge, the remainder of Chapter 9 contains the following material. In Section 9.2 concentration is defined, and special attention is paid to phase interfaces where the concentration of a chemical species is almost always discontinuous. Then Fick's law of diffusion, which is the mass transfer analogy to Fourier's law, is introduced. In Section 9.3 diffusion in stationary media is analyzed, exploiting where possible the analogy between diffusion and conduction. In Section 9.4 mass convection is introduced. Attention is restricted to so-called **low mass transfer rate** theory, for which there is an exact analogy between heat and mass transfer. In Section 9.5 processes involving simultaneous consideration of heat and mass transfer are analyzed. Examples considered include the wet- and dry-bulb psychrometer and combustion of carbon particles. In Section 9.6 mass transfer in porous catalysts is analyzed, and examples are given for automobile catalytic converters. Methods for calculating transport properties of gas mixtures, liquid solutions, and small particles are given in Section 9.7.

9.2 CONCENTRATIONS AND FICK'S LAW OF DIFFUSION

Concentration is to mass transfer what temperature is to heat transfer. Concentration is a measure of composition. A number of such measures are in common use, and they will be defined carefully in Section 9.2.1. In general, temperature is continuous across an interface, whereas, in general, concentration is discontinuous across an interface. Clearly, the concentrations of air on the two sides of a water-air interface are very different. Concentration differences across interfaces are given by thermodynamic data and relations, and a number of different physical situations are considered in Section 9.2.2. Fick's law of diffusion, which is analogous to Fourier's law of heat conduction, is introduced in Section 9.2.3. Other diffusion phenomena are briefly discussed in Section 9.2.4.

9.2.1 Definitions of Concentration

In a gas mixture, or liquid or solid solution, the local **concentration** of a mass species can be expressed in a number of ways. Figure 9.3 shows an elemental volume ΔV surrounding the location of concern. The problem is to specify the composition of the material within ΔV. One method would be to determine, somehow, the number of molecules of each species present and divide by ΔV to obtain the number of molecules per unit volume; hence the **number density** is defined.

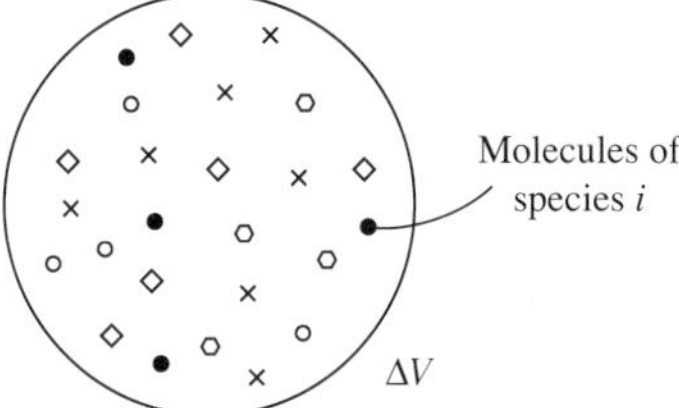

Figure 9.3 Elemental control volume used to define concentrations.

$$\text{Number density of species } i = \text{Number of molecules of } i \text{ per unit volume} \equiv \mathcal{N}_i \text{ molecules/m}^3 \quad \textbf{(9.1)}$$

Alternatively, if the total number of molecules of all species per unit volume is denoted as $\mathcal{N}$, then we define the **number fraction** of species i as

$$n_i \equiv \frac{\mathcal{N}_i}{\mathcal{N}}; \qquad \mathcal{N} = \sum \mathcal{N}_i \quad \textbf{(9.2)}$$

where the summation is over all species present, $i = 1, 2, \ldots, n$. Equations (9.1) and (9.2) describe *microscopic* concepts and are used, for example, when the kinetic theory of gases is used to describe transfer processes.

Whenever possible, it is more convenient to treat matter as a continuum. Then the smallest volume considered is sufficiently large for macroscopic properties such as pressure and temperature to have their usual meanings. For this purpose we also require *macroscopic* definitions of concentration. First, on a mass basis,

$$\textbf{Mass concentration} \text{ of species } i \equiv \text{Partial density of species } i \equiv \rho_i \text{ kg/m}^3 \quad \textbf{(9.3)}$$

The total mass concentration is the total mass per unit volume, that is, the density $\rho = \sum \rho_i$. The **mass fraction** of species i is defined as

$$m_i = \frac{\rho_i}{\rho} \quad \textbf{(9.4)}$$

Second, on a molar basis,[1]

$$\textbf{Molar concentration} \text{ of species } i \equiv \text{Number of moles of } i \text{ per unit volume} \equiv c_i \text{ kmol/m}^3 \quad \textbf{(9.5)}$$

If M_i [kg/kmol] is the molecular weight of species i, then

$$c_i = \frac{\rho_i}{M_i} \quad \textbf{(9.6)}$$

The total molar concentration is the molar density $c = \sum c_i$. The **mole fraction** of species i is defined as

$$x_i \equiv \frac{c_i}{c} \quad \textbf{(9.7)}$$

A number of important relations follow directly from these definitions. The mean molecular weight of the mixture or solution is denoted M and may be expressed as

$$M = \frac{\rho}{c} = \sum x_i M_i \quad \textbf{(9.8}a\textbf{)}$$

[1] The SI unit for the amount of a substance is the mole (the gram mole of the cgs system). However, to be consistent with using the kilogram as the unit for mass, it is more convenient to use the kilogram mole (symbol kmol) rather than the mole (symbol mol). We will use kmol exclusively.

or

$$\frac{1}{M} = \sum \frac{m_i}{M_i} \tag{9.8b}$$

There are summation rules:

$$\sum m_i = 1 \tag{9.9a}$$

$$\sum x_i = 1 \tag{9.9b}$$

It is often necessary to have the mass fraction of species i expressed explicitly in terms of mole fractions and molecular weights; this relation is

$$m_i = \frac{x_i M_i}{\sum x_j M_j} = x_i \frac{M_i}{M} \tag{9.10a}$$

and the corresponding relation for the mole fraction is

$$x_i = \frac{m_i/M_i}{\sum m_j/M_j} = m_i \frac{M}{M_i} \tag{9.10b}$$

Equations (9.10a) and (9.10b) are to be derived as Exercise 9–1.

Although all the above definitions are not strictly necessary for the engineering analysis of mass transfer, it will be seen later that the use of an appropriate definition of concentration often can simplify a particular analysis. Indeed, these measures of concentration are not the only ones in common use. For example, partial pressure is often used in dealing with ideal gas mixtures. Dalton's law of partial pressures states that

$$P = \sum P_i, \quad \text{where } P_i = \rho_i R_i T \tag{9.11}$$

Dividing partial pressure by total pressure and substituting $R_i = \mathscr{R}/M_i$ gives

$$\frac{P_i}{P} = \frac{\rho_i}{M_i} \frac{\mathscr{R}T}{P} = c_i \frac{\mathscr{R}T}{P} = x_i \frac{c\mathscr{R}T}{P} = x_i \tag{9.12}$$

Thus, for an ideal gas mixture, the mole fraction and partial pressure are equivalent measures of concentration (as also is the number fraction).

In combustion problems, we often need to know the mass fraction of oxygen in atmospheric air. A commonly used specification of the composition of dry air is given on a volume basis as 78.1% N_2, 20.9% O_2, and 0.9% A. (The next largest component is CO_2, at 0.03%.) Since equal volumes of gases contain the same number of moles, specifying composition on a volume basis is equivalent to specifying mole fractions. Thus, the mass fraction of O_2 is

$$m_{O_2} = \frac{x_{O_2} M_{O_2}}{x_{N_2} M_{N_2} + x_{O_2} M_{O_2} + x_A M_A}$$

$$= \frac{(0.209)(32)}{(0.781)(28) + (0.209)(32) + (0.009)(46)} = 0.231$$

Similarly, $m_{N_2} = 0.755$, $m_A = 0.014$. The value of $m_{O_2} = 0.231$ for dry air should be remembered: it will be used often.

9.2.2 Concentrations at Interfaces

Although temperature is continuous across a phase interface, concentrations are usually discontinuous. In order to clearly define concentrations at interfaces, we introduce imaginary surfaces, denoted u and s, on both sides of the real interface, each indefinitely close to the interface, as shown on Fig. 9.4*a* for water evaporating into an air stream. Thus, the liquid-phase quantities at the interface are subscripted u, and gas-phase quantities are subscripted s. If we ignore the small amount of air dissolved in the water, $x_{H_2O,u} = 1$. Notice that the subscript preceding the comma denotes the chemical species, and the subscript following the comma denotes location. To determine $x_{H_2O,s}$ we make use of the fact that, except in extreme circumstances, the water vapor and air mixture at the s-surface must be in thermodynamic equilibrium with the water at the u-surface. Equilibrium data for this system are found in

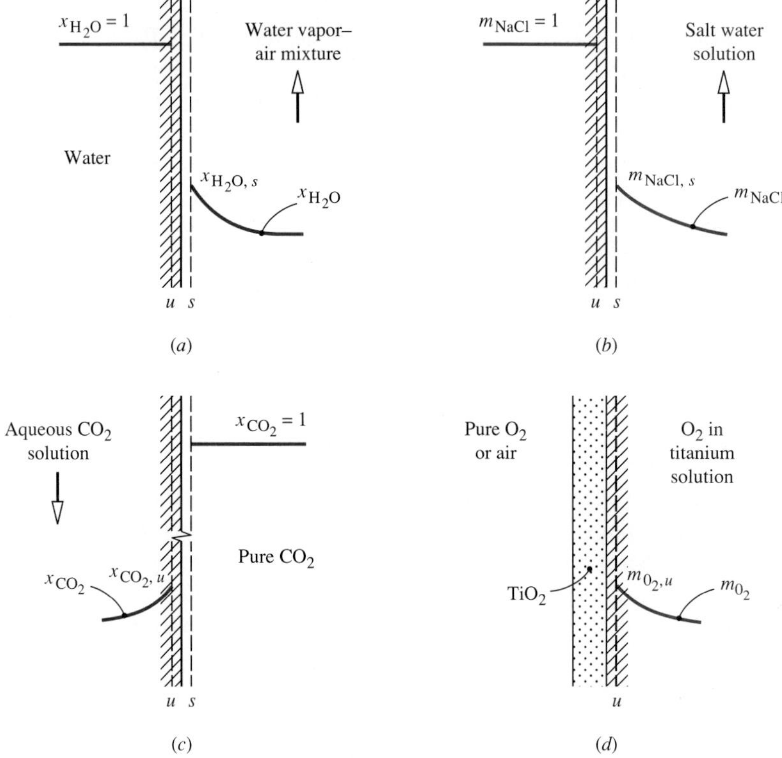

Figure 9.4 Concentrations at interfaces (*a*) water-air, (*b*) salt-water, (*c*) water–carbon dioxide, (*d*) the titanium-oxygen system.

conventional steam tables: the saturation vapor pressure of steam at the water temperature, $T_s(T_s = T_u)$ is the required partial pressure $P_{H_2O,s}$. With the total pressure P known, $x_{H_2O,s}$ is calculated as $P_{H_2O,s}/P$. If $m_{H_2O,s}$ is required, Eq. (9.10*a*) is used. For example, at $T_s = 310$ K, the saturation vapor pressure is 0.06224×10^5 Pa (Table A.12*a*); if the total pressure is 1 bar ($= 10^5$ Pa),

$$x_{H_2O,s} = \frac{0.06224 \times 10^5}{10^5} = 0.06224$$

$$m_{H_2O,s} = \frac{(0.06224)(18)}{(0.06224)(18) + (1 - 0.06224)(29)} = 0.0396$$

where air has been taken to have a molecular weight of 29.

As another example, consider a slab of salt dissolving in water, as shown in Fig. 9.4*b*. If the salt is pure, $m_{NaCl,u} = 1$. Again, thermodynamic equilibrium must exist between the salt in water solution at the s-surface and solid salt at the interface temperature. Equilibrium data of this kind is often referred to simply as *solubility* data, found in chemistry handbooks (usually in cgs units) and in Appendix A, Table A.24. For example, at 30°C the solubility of salt in water is 36.3 g/100 g water. Thus at 30°C = 303.15 K,

$$m_{NaCl,s} = \frac{36.3}{100 + 36.3} = 0.266$$

Many processes, particularly those encountered in chemical engineering practice, involve absorption of a gas into a liquid. Many gases are only sparingly soluble, and for such dilute solutions solubility data are conveniently represented by **Henry's law,** which states that the mole fraction of the gas at the s-surface is proportional to its mole fraction in solution at the u-surface, the constant of proportionality being the **Henry number,** He. For species i,

$$x_{i,s} = \mathrm{He}_i x_{i,u} \qquad \textbf{(9.13)}$$

The Henry number is inversely proportional to total pressure and is also a function of temperature. The product of Henry number and total pressure is the **Henry constant,** C_{He}, and for a given species is a function of temperature only:

$$\mathrm{He}_i P = C_{\mathrm{He}_i}(T) \qquad \textbf{(9.14)}$$

Selected data for the Henry constant are given in Table A.21. Figure 9.4*c* shows absorption of carbon dioxide from a stream of pure CO_2 at 3 bar pressure into water at 300 K. From Table A.21, $C_{He} = 1710$ bar; thus

$$\mathrm{He}_{CO_2} = \frac{1710}{3} = 570; \qquad x_{CO2,u} = \frac{1}{570} = 0.00175$$

where the small partial pressure of water vapor at the s-surface has been ignored. Advantage is taken of the increase in solubility with pressure in the carbonation of soft drinks. For highly soluble gases, such as sulfur dioxide and ammonia in water, a simple linear solubility relation does not hold, and solubility data are usually available in tables relating gas-phase partial pressure to liquid-phase mole fraction, in the form $P_{i,s} = P_{i,s}(x_{i,u}, T)$. Some such data are given in Table A.22.

Dissolution of gases into metals is characterized by varied and rather complex interface conditions. Provided temperatures are sufficiently high, hydrogen dissolution is reversible (similar to CO_2 absorption into water); hence, for example, titanium-hydrogen solutions can exist only in contact with a gaseous hydrogen atmosphere. As a result of hydrogen going into solution in atomic form, there is a characteristic square root relation,

$$m_{H_2,u} \propto P_{H_2,s}^{1/2}$$

The constant of proportionality is strongly dependent on temperature, as well as on the particular titanium alloy: for Ti-6Al-4V alloy it is twice that for pure titanium. Table A.23 gives selected data. In contrast to hydrogen, oxygen dissolution in titanium is irreversible and is complicated by the simultaneous formation of a rutile (TiO_2) scale on the surface, as shown in Fig. 9.4*d*. Provided some oxygen is present in the gas phase, the titanium-oxygen *phase diagram* (found in a metallurgy handbook) shows that $m_{O_2,u}$ in alpha-titanium is 0.143, a value essentially independent of temperature and O_2 partial pressure. Dissolution of oxygen in zirconium alloys has similar characteristics to those discussed above for titanium.

All the preceding examples of interface concentrations are situations where thermodynamic equilibrium can be assumed to exist at the interface. Sometimes thermodynamic equilibrium does not exist at an interface: a very common example is when a chemical reaction occurs at the interface, and temperatures are not high enough for equilibrium to be attained. Then the concentrations of the reactants and products at the s-surface are dependent both on the rate at which the reaction proceeds—that is, the *chemical kinetics*—as well as on mass transfer considerations. An example of such a situation is discussed in Section 9.3.3.

9.2.3 Fick's Law of Diffusion

It is convenient to introduce **Fick's law** as a phenomenological relation, and later examine the physical basis of the law in greater detail. In 1855 Fick proposed what has become known as his **first law,** a linear relation between the rate of diffusion of a chemical species and the local concentration gradient of that species. The concept of a linear relation between a flux and a corresponding driving force, or potential, had already been introduced by Newton in his law of viscosity, by Fourier in his law of heat conduction, and by Ohm in his law of electrical conduction. We will first restrict our attention to a stationary medium—for example, a solid or stagnant gas consisting of only two species, 1 and 2—and suggest the following form of Fick's law:

$$\mathbf{j}_1 = -\rho \mathscr{D}_{12} \nabla m_1 \tag{9.15}$$

where $\mathbf{j}_1$ [kg/m^2 s] is the diffusive mass flux of species 1, ρ is the local mixture or solution density [kg/m^3], m_1 is the mass fraction of species 1, and $\mathscr{D}_{12}$ [m^2/s] is the constant of proportionality, called the **binary diffusion coefficient,** or sometimes the **mass diffusivity.** (Notice that no comma separates the subscripts 1 and 2 on $\mathscr{D}_{12}$ since both 1 and 2 refer to chemical species. We use a comma only when the subsequent subscript refers to location.)

The corresponding law for species 2 is

$$\mathbf{j}_2 = -\rho\mathscr{D}_{21}\nabla m_2 \tag{9.16}$$

where it will be shown in Section 10.2.4 that $\mathscr{D}_{21} = \mathscr{D}_{12}$. As is the case with Fourier's law of heat conduction, Eq. (3.2), Fick's law is a vector equation, and the vector $\mathbf{j}_1$ is in the same direction as the negative gradient of concentration, $-\nabla m_1$. For a one-dimensional situation with a gradient of concentration of the z-direction only, Eq. (9.15) simplifies to

$$j_1 = -\rho\mathscr{D}_{12}\frac{dm_1}{dz} \tag{9.17}$$

Ordinary diffusion of a chemical species down its concentration gradient results from random motion of molecules of the species. Figure 9.5 illustrates this point for one-dimensional diffusion in a model binary system. Looking at the plane located at $z = Z$, random motion of the molecules results in more molecules of species 1 crossing the plane from left to right than from right to left. Hence there is a net flux of species 1 from left to right, that is, in the direction of decreasing concentration. A similar argument holds for species 2.

A valid question at this point is the suitability of Eq. (9.15) as a statement of Fick's law. After all, we have introduced six measures of the composition of a mixture (number density, number fraction, partial density, mass fraction, molar concentration, and mole fraction), so why should mass fraction be the appropriate driving potential for diffusion? And why should diffusion be expressed as a mass flux; would not a molar flux be more suitable? We can, simply by introducing the pertinent relations, derive alternative forms of Eq. (9.15). After we have carefully examined what we mean by a stationary medium, we may, for example, derive

$$\mathbf{J}_1 = -c\mathscr{D}_{12}\nabla x_1 \tag{9.18}$$

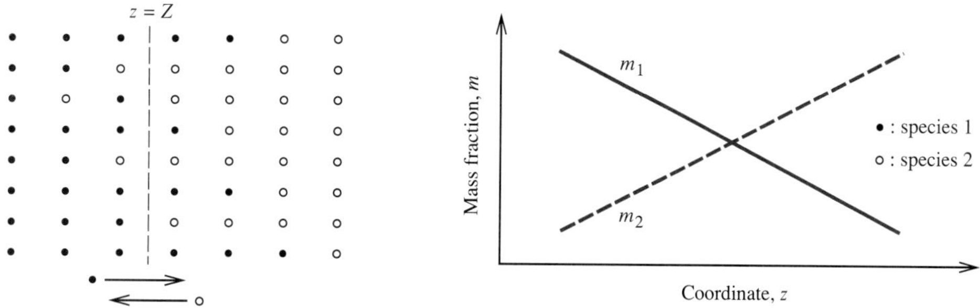

Figure 9.5 Schematic of one-dimensional diffusion in a model binary system.

where $\mathbf{J}_1$ [kmol/m^2 s] is the diffusive molar flux of species 1. On the other hand, if we were to simply write down the equation

$$\mathbf{j}_1 = -\mathscr{D}_{12}\nabla\rho_1 \tag{9.19}$$

we would find that, although the diffusion coefficient defined by Eq. (9.19) has the same dimensions as the one appearing in Eq. (9.15), it will in general have a different value. The problem is that $\rho\nabla m_1 \neq \nabla\rho_1$, unless ρ is constant. It is usually possible to assume a constant density for solids and dilute liquid solutions, and then Eq. (9.19) is indeed often used. On the other hand, it is seldom possible to assume a constant density for gas mixtures and concentrated liquid solutions, so either Eq. (9.15) or Eq. (9.18) should be used. It cannot be shown in a simple manner that Eq. (9.15) is indeed the most appropriate mathematical statement of Fick's law. There is no single physical mechanism of diffusion; in particular, there are radical differences between the mechanisms in solids, liquids, and gases. The problem must be examined in the light of physical theory, the results of irreversible thermodynamics, and experimental data. Here it will be stated, without demonstration, that the kinetic theory of gases shows that Eq. (9.15) is appropriate for binary gas mixtures at normal pressures; furthermore, experimental data show that it is appropriate for dilute liquid and solid solutions. Since the majority of engineering problems fall within these categories, it is convenient to accept Eq. (9.15) as a fundamental statement of Fick's law of diffusion.

Diffusion coefficients of gases at normal pressures are almost composition-independent, vary inversely with pressure, and increase with temperature. Liquid and solid diffusion coefficients are markedly concentration-dependent and generally increase with temperature. Table 9.1 shows some typical values, and Appendix A contains further data. Methods for calculating diffusion coefficients are given in Section 9.7. Notice that Table A.17*b*, for selected gas species in dilute mixture in air, and Table A.18, for diffusion in dilute aqueous solutions, give the **Schmidt number,** rather than the diffusion coefficient directly. The Schmidt number, Sc, is defined as $\text{Sc} = \mu/\rho\mathscr{D}$ or $\nu/\mathscr{D}$, and plays a role in convective mass transfer analogous to the role played by the Prandtl number in convective heat transfer. Since the mixture or solution is dilute, μ, ρ, and ν can be evaluated for pure air when using Table A.17*b*, and for pure water when using Table A.18. Notice also that gas-phase diffusion coefficients tend to be a few orders of magnitude larger than liquid- or solid-phase diffusion coefficients.

9.2.4 Other Diffusion Phenomena

Molecules in a gas mixture, and in a liquid or solid solution, can diffuse by mechanisms other than ordinary diffusion governed by Fick's law. *Thermal diffusion* is diffusion due to a temperature gradient and is often called the *Soret effect*. Thermal diffusion is usually negligible compared with ordinary diffusion, unless the temperature gradient is very large. However, there are some important processes that depend on thermal diffusion, the most well known being the large-scale separation of uranium isotopes. *Pressure diffusion* is diffusion due to a pressure gradient and is

Table 9.1 Selected data for binary diffusion coefficients.

System	Temperature K	Pressure atm	$\mathscr{D}_{12}$ m^2/s
Gas Mixtures			
H_2-air	300	1	7.77×10^{-5}
H_2-air	300	0.01	7.77×10^{-3}
CO_2-air	300	1	1.55×10^{-5}
CO_2-air	1000	1	1.24×10^{-4}
H_2O-air	373	1	3.71×10^{-5}
H_2O-air	300	0.01	2.57×10^{-3}
Dilute Liquid Solutions			
NH_3-water	300	—	2.18×10^{-9}
SO_2-water	300	—	1.67×10^{-9}
SO_2-water	400	—	8.09×10^{-9}
Sucrose-water	300	—	5.20×10^{-10}
CO_2-water	300	—	1.95×10^{-9}
CO_2-methanol	300	—	8.37×10^{-9}
CO_2-ethanol	300	—	3.88×10^{-9}
CO_2-propanol	300	—	2.73×10^{-9}
Solid Solutions			
He-Pyrex glass	300	—	8.70×10^{-13}
He-Pyrex glass	700	—	4.42×10^{-10}
H_2-steel	300	—	3.27×10^{-13}
H_2-steel	700	—	2.21×10^{-9}
O_2–alpha-phase Ti-6A1-4V alloy	1500	—	5.75×10^{-11}

also usually negligible unless the pressure gradient is very large. Pressure diffusion is the principle underlying the operation of a centrifuge. Centrifuges are used to separate liquid solutions and are increasingly being used to separate gaseous isotopes as well. *Forced diffusion* results from an external force field acting on a molecule. Gravitational force fields do not cause separation since the force per unit mass of molecule is constant. Forced diffusion occurs when an electrical field is imposed on an electrolyte (for example, in charging an automobile battery), on a semiconductor, or on an ionized gas (for example, in a neon tube or metal-ion laser). Depending on the strength of the electric field, rates of forced diffusion can be very large.

Some interesting diffusion phenomena occur in porous solids. When a gas mixture is in a porous solid, such as a catalyst pellet or silica-gel particle, the pores can be smaller than the mean free path of the molecules. Then the molecules collide with the wall more often than with other molecules. In the limit of negligible molecule-molecule collisions we have *Knudsen diffusion,* also called *free molecule flow* in the fluid mechanics literature. If the pore size approaches the size of a molecule, then Knudsen diffusion becomes negligible, and *surface diffusion,* in which adsorbed molecules move along the pore walls, becomes the dominant diffusion mechanism. Knudsen diffusion will be considered in Section 9.6.

Very small particles of 10^{-3}–10^{-1} μm size—for example, smoke, soot, and mist—behave much like large molecules. Ordinary diffusion of such particles is called *Brownian motion* and is described in most elementary physics texts. Diffusion of particles due to a temperature gradient is called *thermophoresis* and plays an important role for larger particles, typically in the size range 10^{-1}–1 μm. Diffusion of particles in a gas mixture due to concentration gradients of molecular species is called *diffusiophoresis*. *Forced diffusion* of a charged particle in an electrical field is similar to that for an ionized molecular species. Thermal and electrostatic precipitators are used to remove particles from power plant and incinerator stack gases, and depend on thermophoresis and forced diffusion, respectively, for their operation. Diffusion phenomena are unimportant for particles of size greater than about 1 μm in air at 1 atm; the motion of such particles is governed by the laws of Newtonian mechanics. Brownian motion is dealt with in Section 9.7, and in Chapter 11 electrostatic precipitators and particle filters are analyzed as examples of mass exchangers.

9.3 MASS DIFFUSION

Mass diffusion is analogous to heat conduction. The simplest mass diffusion problem is steady diffusion through a plane wall, which is analyzed from first principles in Section 9.3.1. Often the solution of a mass diffusion problem can be obtained directly from the solution of the analogous heat conduction problem. The solution for transient diffusion in a semi-infinite solid is obtained this way in Section 9.3.2. As mentioned in the introduction to this chapter, mass transfer problems usually differ from heat transfer problems in their variety and complexity of boundary conditions. Heterogeneous catalysis, analyzed in Section 9.3.3, is an example of a more complicated boundary condition.

9.3.1 Steady Diffusion through a Plane Wall

Figure 9.6 shows a plane solid wall of surface area A and thickness L. Imaginary u- and v-surfaces are located within the solid adjacent to the interfaces at $z=0$ and $z=L$, respectively, and the corresponding mass fractions of species 1 are $m_{1,u}$ and $m_{1,v}$, which are steady in time. We wish to determine the rate of transfer of species 1 through the wall. For example, the wall may be that of a hydrogen tank, with hydrogen leaking through the wall into the atmosphere. Note that our approach is similar to that used in Chapter 1 in analyzing elementary heat conduction: we initially take the mass fractions $m_{1,u}$ and $m_{1,v}$ as known, and later concern ourselves with how these are related to conditions on either side of the wall. An elemental control volume $\Delta V=A\Delta z$ is located between z and $z+\Delta z$; if the composition is unchanging in time, and if there are no chemical reactions producing or consuming species 1, then the **principle of conservation of mass species** requires that the flow of species 1, denoted $\dot{m}_1$ [kg/s], across the face at z must equal that across the face

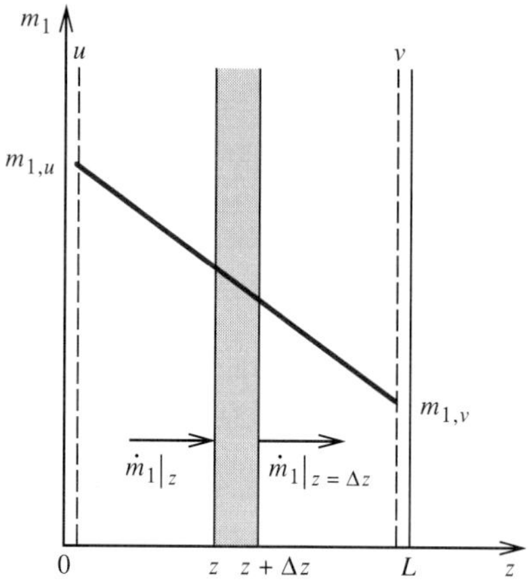

Figure 9.6 Schematic of steady diffusion through a plane wall.

at $z + \Delta z$,

$$\dot{m}_1|_z = \dot{m}_1|_{z+\Delta z}$$

or

$$\dot{m}_1 = j_1 A = \text{Constant} \tag{9.20}$$

We shall see that the mass flow of species 1, $\dot{m}_1$ [kg/s], is the mass transfer analog to the heat flow $\dot{Q}$ [J/s] of heat transfer. Substituting for j_1 from Fick's law, Eq. (9.17),

$$\frac{\dot{m}_1}{A} = -\rho\mathscr{D}_{12}\frac{dm_1}{dz} = \text{Constant}$$

Then integrating across the wall,

$$\frac{\dot{m}_1}{A}\int_0^L dz = -\int_{m_{1,u}}^{m_{1,v}} \rho\mathscr{D}_{12}\, dm_1$$

where $\dot{m}_1$ and A are outside the integral sign since both are constants. If any variations of ρ and $\mathscr{D}_{12}$ are ignored for the present,

$$\dot{m}_1 = \frac{\rho\mathscr{D}_{12}A}{L}(m_{1,u} - m_{1,v}) = \frac{m_{1,u} - m_{1,v}}{L/\rho\mathscr{D}_{12}A} \tag{9.21a}$$

Comparison of Eq. (9.21*a*) with Ohm's law, $I = E/R$, suggests that $\Delta m_1 = m_{1,u} - m_{1,v}$ can be viewed as a driving potential for flow of mass species 1, analogous to voltage difference being a driving potential for current (and temperature difference being the driving potential for heat flow); then $R = L/\rho\mathscr{D}_{12}A$ can be viewed as a *diffusion resistance* analogous to an electrical or conduction resistance. The mass flow through a slab can therefore be represented by the circuit shown in Fig. 9.7. Examination of the analysis in Section 1.3.1 leading to Eq. (1.9) shows the close analogy between

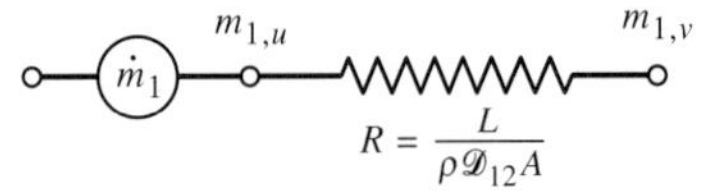

$$\dot{m}_1 \quad m_{1,u} \quad R = \frac{L}{\rho \mathscr{D}_{12} A} \quad m_{1,v}$$

Figure 9.7 Mass flow circuit resistance for diffusion across a plane wall.

heat conduction and mass diffusion; however, the fact that concentrations are usually discontinuous at phase interfaces, whereas temperature is not, means that circuits for species flow through diffusion resistances in series cannot be constructed in the same way as for heat or current flow (however, see Exercise 9–14).

If the preceding analysis is repeated on a molar basis, the result is

$$\dot{M}_1 = \frac{c\mathscr{D}_{12}A}{L}(x_{1,u} - x_{1,v}) \tag{9.21b}$$

where $\dot{M}_1$ [kmol/s] is the molar flow of species 1. Analysis of steady diffusion through a plane wall yields a simple result, as did the analogous problem of steady heat conduction through a plane wall. However, application of Eqs. (9.21) to mass transfer problems is made complicated by the need to use solubility data to obtain the concentrations at the u- and v-surfaces. Furthermore, as was seen in Section 9.2.2, solubility data obtained from different sources are expressed in a variety of forms.

Permeability

Consider gas-solid systems for which Henry's law applies. It is common practice to define the *solubility* $\mathscr{S}$ as the volume of solute gas (at STP of 0°C and 1 atm) dissolved in unit volume of solid, when the gas is at a partial pressure of 1 atm. Since 1 kmol of gas at STP occupies 22.414 m^3, it follows that the molar concentration of gas dissolved in a solid is related to the partial pressure of the adjacent gas as $c_{1,u} = \mathscr{S}(P_{1,s}\text{ atm})/22.414$ [kmol/m^3]. Also, the volume flow rate $\dot{V}_1$ is related to the molar flow rate $\dot{M}_1$ as $\dot{V}_1 = 22.414\dot{M}_1$. Then substituting in Eq. (9.21b) with c assumed constant gives the volume rate of gas flow across the wall as

$$\dot{V}_1 = \frac{\mathscr{P}_{12}A}{L}(P_{1,s} - P_{1,s'})\text{ m}^3\text{ (STP)/s} \tag{9.21c}$$

where $P_{1,s}$ and $P_{1,s'}$ are the partial pressures of species 1 in atmospheres on each side of the wall, and $\mathscr{P}_{12} = \mathscr{D}_{12}\mathscr{S}$ is the *permeability* of the gas in the solid. The units of permeability are [m^3 (STP)/m^2 s (atm/m)]. Physicists often prefer to tabulate data for permeability, rather than for solubility and diffusion coefficient separately, and use Eq. (9.21c) to calculate steady transfer across shells and membranes. Table A.23 contains selected solubility and permeability data for gases in solids. Notice that other units and definitions of solubility are used for gas-solid systems. Often, the solubility $\mathscr{S}$ is given in units of kmol/m^3 bar. Also, a dimensionless solubility coefficient is defined as $\mathscr{S}' = c_{1,u}/c_{1,s}$.

Two examples demonstrating the application of Eqs. (9.21) follow. The first example is worked on a mass basis and the second on a molar basis, the choice being determined by the form of the solubility data. Both examples concern leakage of a gas through the walls of a container. Such problems are encountered when gases with small molecules, for example, H_2 and He, are stored at high temperatures, owing to the relatively high diffusion coefficients (for solids). It is of interest to note that, although steady heat conduction through a plane wall is encountered in a great range of engineering problems, there are relatively few mechanical engineering mass transfer problems involving steady ordinary diffusion across a plane wall. In chemical engineering and the life sciences there are many examples of mass transfer across thin membranes (see Exercises 9–13 and 9–14). However, often the mass transfer mechanism is not simple ordinary diffusion governed by Fick's law, for example, reverse osmosis for desalination of saline water, or blood oxygenation in the human lung.

EXAMPLE 9.1 Loss of Hydrogen from a Storage Tank

A spherical steel tank of 1 liter capacity and 2 mm wall thickness is used to store hydrogen at 400°C. The initial pressure is 9 bar, and there is a vacuum outside the tank. Calculate the time required for the pressure to drop to 5 bar. Data for the hydrogen-steel (1-2) system include the following items.

Diffusion coefficient: $\mathscr{D}_{12} \simeq 1.65 \times 10^{-6} e^{-4630/T}\,\mathrm{m^2/s}$

Solubility relation: $m_{1,u} \simeq 2.09 \times 10^{-4} e^{-3950/T} P_{1,s}^{1/2}$ for T in kelvins and P in bars.

Solution

Given: High-temperature hydrogen gas stored in a 1 liter steel tank.

Required: Time required for pressure to drop from 9 bar to 5 bar.

Assumptions:
1. Quasi-steady diffusion through the steel.
2. Wall curvature effects are negligible.
3. The ideal gas law applies.

To obtain a differential equation governing the rate of change of pressure with time, we equate the rate of change of mass of H_2 stored in the tank to the rate at which H_2 diffuses across the tank wall. For an ideal gas the mass of gas in the tank, w, is

$$w = \frac{PV}{(\mathscr{R}/M)T}$$

Differentiating with respect to time and recognizing that V and T are constant gives

$$\frac{dw}{dt} = \frac{V}{(\mathscr{R}/M)T}\frac{dP}{dt}$$

or

$$\frac{dP}{dt} = \frac{\mathscr{R}T}{MV}\frac{dw}{dt} \quad \textbf{(1)}$$

Since the wall thickness is small compared with the tank diameter, Eq. (9.21*a*) for diffusion across a plane wall can be used.

$$\frac{dw}{dt} = -\dot{m}_1 = -\frac{\rho\mathscr{D}_{12}A}{L}(m_{1,u} - m_{1,v}) \qquad \textbf{(2)}$$

where $m_{1,u}$ and $m_{1,v}$ are H_2 mass fractions in the steel at the inner and outer surfaces of the tank, $A = 4\pi R^2$ is the surface area, and L is the wall thickness. Combining Eqs. (1) and (2),

$$\frac{dP}{dt} = -\frac{\mathscr{R}T}{MV}\frac{\rho\mathscr{D}_{12}A}{L}(m_{1,u} - m_{1,v})$$

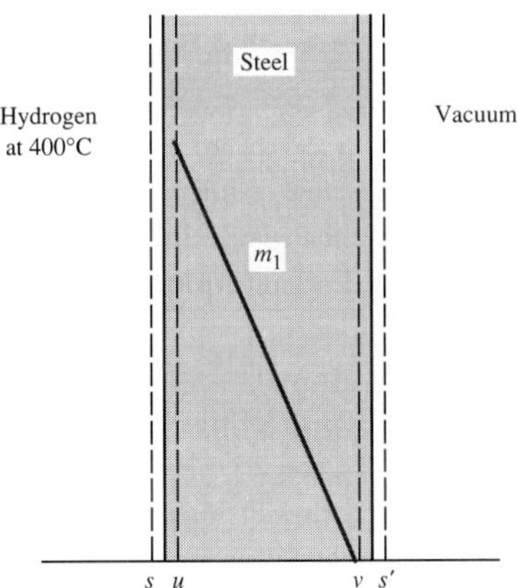

Since there is pure hydrogen in the tank, $P_{1,s} = P$, and for a constant temperature the solubility relation can be used to write $m_{1,u} = KP^{1/2}$; also, since there is a vacuum outside the tank, $P_{1,s'} = 0$, so that $m_{1,v} = 0$. Thus we can write

$$\frac{dP}{dt} = -BP^{1/2}; \qquad B = \frac{\mathscr{R}T}{MV}\frac{\rho\mathscr{D}_{12}A}{L}K$$

Integrating with $P = P_0$ at $t = 0$ gives

$$P = \left[P_0^{1/2} - \frac{Bt}{2}\right]^2 \qquad \text{or} \qquad t = \frac{2}{B}\left(P_0^{1/2} - P^{1/2}\right)$$

Next we evaluate the constant B. The tank volume is $(4/3)\pi R^3 = 10^{-3}$ m^3; hence $R =$ 6.20 cm and $A/V = 3/R$.

$$K = 2.09 \times 10^{-4}e^{-3950/673} = 5.90 \times 10^{-7}\ \text{bar}^{-1/2} = 1.87 \times 10^{-9}\ \text{Pa}^{-1/2}$$

$$\mathscr{D}_{12} = 1.65 \times 10^{-6}e^{-4630/673} = 1.70 \times 10^{-9}\ \text{m}^2/\text{s}$$

From Table A.1, the density of steel is $\rho \simeq 7800$ kg/m^3; thus

$$B = \frac{(8314)(673)(7800)(1.70 \times 10^{-9})(3)(1.87 \times 10^{-9})}{(2)(0.0620)(0.002)} = 1.68 \times 10^{-3}\ \text{Pa}^{1/2}/\text{s}$$

The time taken for the pressure to drop to 5 bar $= 5 \times 10^5$ Pa is, then,

$$t = \frac{2[(9 \times 10^5)^{1/2} - (5 \times 10^5)^{1/2}]}{(1.68 \times 10^{-3})} = 2.88 \times 10^5\ \text{s} \simeq 80\ \text{h}$$

Comments

1. Since atmospheric air contains very little hydrogen (0.5 parts per million by volume), the above result will be unchanged if the tank is located in a ventilated enclosure.
2. Strictly speaking, ρ in Eq. (2) is the density of the solution. However, since the amount of H_2 dissolved in the steel is very small, we simply use the steel density.

3. In assuming quasi-steady diffusion we are neglecting the change in the amount of H_2 stored in solution in the tank wall. Check this assumption by comparing the amount stored to the leakage rate.

4. Note the analogy to the lumped thermal capacity heat transfer analysis of Section 1.5. In Section 1.5 we used the energy conservation principle to derive the governing differential equation; here we used the mass conservation principle.

5. Since Henry's law does not apply to dissolution of H_2 in steel, the permeability method for calculating diffusion across a wall, Eq. (9.21*c*), is inappropriate.

6. Check the stresses in the tank wall at 9 bar: is the tank safe?

EXAMPLE 9.2 Loss of Helium from a He-Cd Laser

A helium-cadmium (blue) laser used in a copying machine printer contains He at a nominal pressure of 460 Pa. The glass outer shell has a volume of 150 cm^3, an area of 550 cm^2, and a wall thickness of 1.52 mm. At normal operating conditions the average temperature of the gas in the tube is 225°C and the shell temperature is 115°C. Estimate (i) the helium leak rate and (ii) the time required to lose 1% of the helium inventory. For the helium–shell glass (1-2) system the solubility coefficient and diffusion coefficient are

$$\mathscr{S}' = \frac{c_{1,u}}{c_{1,s}} \simeq 0.007 + 3.0 \times 10^{-5}(T\,°\text{C}); \qquad \mathscr{D}_{12} \simeq 1.40 \times 10^{-8} e^{-3280/(T\,K)}\ \text{m}^2/\text{s}$$

Solution

Given: Glass shell containing helium.

Required: (i) Helium leak rate. (ii) Time to lose 1% of helium inventory.

Assumptions: 1. Molar density c and diffusion coefficient $\mathscr{D}_{12}$ constant across glass shell.
2. Curvature effects are negligible.

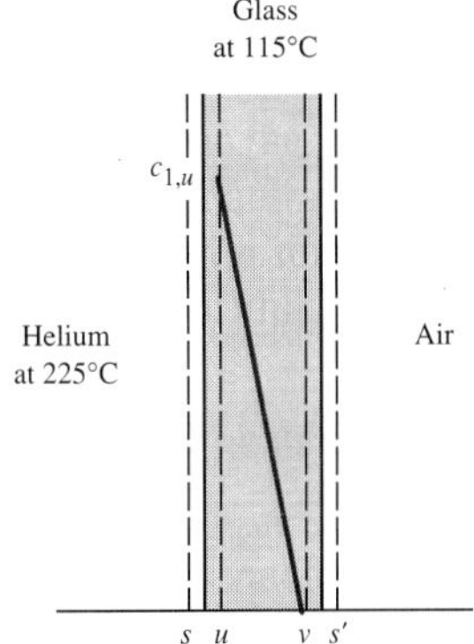

(i) Equation (9.21*b*) applies to quasi-steady diffusion of helium across the glass shell. It was derived assuming c and $\mathscr{D}_{12}$ constant. Since the helium concentration in the glass is very small, the molar density c can certainly be assumed constant. There is some variation of $\mathscr{D}_{12}$ due to the temperature gradient through the wall, but this variation will be ignored. The solubility data is given in terms of molar concentrations; thus it is convenient to rewrite Eq. (9.21*b*) as

$$\dot{M}_1 = \frac{\mathscr{D}_{12}A}{L}(c_{1,u} - c_{1,v}) \qquad \textbf{(1)}$$

The molar concentrations $c_{1,u}$ and $c_{1,v}$ are calculated first. At the s-surface there is pure helium at 115°C = 388 K, and 460 Pa. Using the ideal gas law,

$$c_{1,s} = \frac{P_{1,s}}{\mathscr{R}T_s} = \frac{(460 \text{ N/m}^2)}{(8314 \text{ J/kmol K})(388 \text{ K})} = 1.426 \times 10^{-4} \text{ kmol/m}^3$$

Substituting in the given solubility relation,

$$\frac{c_{1,u}}{c_{1,s}} = 0.007 + (3.0 \times 10^{-5})(115) = 1.05 \times 10^{-2}$$

Thus $c_{1,u} = (1.426 \times 10^{-4})(1.05 \times 10^{-2}) = 1.50 \times 10^{-6}$ kmol/m^3.

Since the outer shell is cooled by an air flow, the helium concentration at the s'-surface is kept at a very low value. Thus

$$c_{1,s'} \simeq 0 \qquad \text{and} \qquad c_{1,v} \simeq 0$$

Next the diffusion coefficient for helium in the glass at 115°C = 388 K is calculated.

$$\mathscr{D}_{12} = 1.40 \times 10^{-8} e^{-3280/388} = 2.98 \times 10^{-12} \text{ m}^2\text{/s}$$

Substituting in Eq. (1), the molar flow rate of helium through the shell is

$$\dot{M}_1 = \frac{(2.98 \times 10^{-12} \text{ m}^2\text{/s})(550 \times 10^{-4} \text{ m}^2)(1.50 \times 10^{-6} \text{ kmol/m}^3)}{(1.52 \times 10^{-3} \text{ m})}$$

$$= 1.62 \times 10^{-16} \text{ kmol/s}$$

(ii) The molar inventory in the laser is W = cV. The bulk helium is at approximately 225°C = 498 K and 460 Pa; hence the molar concentration is

$$c \simeq \frac{P}{\mathscr{R}T} = \frac{460}{(8314)(498)} = 1.111 \times 10^{-4} \text{ kmol/m}^3$$

$$\text{W} = (1.111 \times 10^{-4})(1.5 \times 10^{-4} \text{m}^3) = 1.67 \times 10^{-8} \text{ kmol}$$

Thus the time for 1% of the inventory to be lost is approximately

$$t \simeq \frac{(0.01)\text{W}}{\dot{M}_1} = \frac{(0.01)(1.67 \times 10^{-8})}{(1.62 \times 10^{-16})} = 1.03 \times 10^6 \text{ s } (\simeq 286 \text{ h})$$

Comments

1. In practice, such lasers are fitted with a high-pressure helium reservoir separated from the laser by a glass diaphragm, as shown. When the pressure falls below a set value, a control system switches on a heater, which raises the diaphragm temperature to allow helium to leak from the reservoir

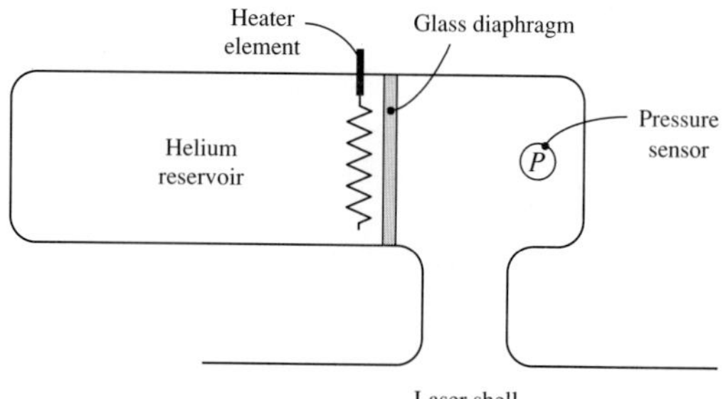

to the laser. When the pressure in the laser rises above a second set value, the heater is switched off.

2. The composition of glass used in practice varies considerably. Typically, B_2O_3, Al_2O_3, TiO_2, Na_2O, and so on are added to the SiO_2 in varying proportions. For example, Pyrex glass contains 81% SiO_2, 13% B_2O_3, 4% Na_2O, and 2% Al_2O_3. Helium solubility and diffusion coefficients depend on the glass composition: in particular, as the amount of B_2O_3 increases, $\mathscr{D}$ decreases markedly (about tenfold for an increase of 13% to 30%). Thus special care must be taken to ensure that appropriate data are used when calculating helium diffusion through glass. Conversely, a particular glass can be chosen to give a higher or lower diffusion rate.

3. See Exercise 9–12 for use of Eq. (9.21*c*) and permeability to solve this problem.

9.3.2 Transient Diffusion

The Mass Diffusion Equation

Transient diffusion in a stationary medium is governed by an equation that is analogous to the heat conduction equation derived in Section 3.2. For simplicity we will consider one-dimensional diffusion in Cartesian coordinates with no chemical reactions. Referring to Fig. 9.8, the principle of conservation of mass species applied to the elemental control volume ΔV located between z and $z + \Delta z$ requires that in a time interval Δt,

$$\text{Storage of species 1} = \text{Net inflow of species 1}$$

$$\rho_1 \Delta V|_{t+\Delta t} - \rho_1 \Delta V|_t = (\dot{m}_1|_z - \dot{m}_1|_{z+\Delta z})\Delta t$$

Substituting $\Delta V = A\Delta z$ and $\dot{m}_1 = j_1 A$, dividing through by $A\Delta z \Delta t$, and letting $\Delta z \Delta t \to 0$,

$$\frac{\partial \rho_1}{\partial t} = -\frac{\partial j_1}{\partial z}$$

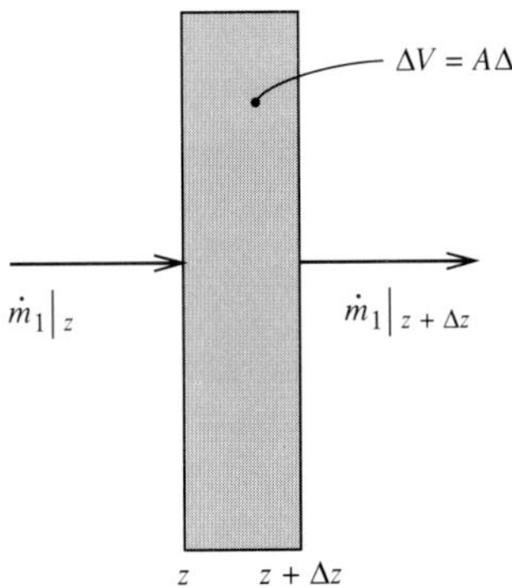

Figure 9.8 Elemental control volume used to derive the mass diffusion equation.

Introducing Fick's law of diffusion, Eq. (9.17), gives

$$\frac{\partial \rho_1}{\partial t} = \frac{\partial}{\partial z}\left(\rho \mathscr{D}_{12} \frac{\partial m_1}{\partial z}\right) \tag{9.22}$$

where the partial derivatives signify that ρ_1 and m_1 are functions of the two independent variables t and z. In the case of dilute solutions in solids, it is usually possible to assume that the density is constant; also, if the solid is isothermal, the diffusion coefficient $\mathscr{D}_{12}$ can also be taken to be a constant: then substituting $\rho_1 = \rho m_1$ and canceling ρ,

$$\frac{\partial m_1}{\partial t} = \mathscr{D}_{12} \frac{\partial^2 m_1}{\partial z^2} \tag{9.23}$$

which in some texts is called *Fick's second law of diffusion*. However, Eq. (9.23) is simply a statement of the principle of conservation of mass species rather than a new phenomenological law, and thus current practice is to drop this name and to simply call the phenomenological law of diffusion, Eq. (9.15), Fick's law. Equation (9.23) is of similar form to the one-dimensional transient heat conduction equation

$$\frac{\partial T}{\partial t} = \alpha \frac{\partial^2 T}{\partial z^2} \tag{9.24}$$

which is valid for constant thermal properties and no internal heat generation. We see also that both the mass diffusivity $\mathscr{D}_{12}$ and thermal diffusivity α have the same units [m^2/s]. It follows that, when the various assumptions made hold true, the solution of a diffusion problem may be obtained directly from the corresponding heat conduction problem. In the mathematics literature both the heat conduction equation and Fick's second law of diffusion are usually referred to as the **diffusion equation,**

$$\frac{\partial \phi}{\partial t} = \alpha \frac{\partial^2 \phi}{\partial z^2} \tag{9.25}$$

where ϕ is any scalar potential function.

Boundary conditions for mass diffusion are more complicated than those for heat conduction. In general concentrations are discontinuous at phase interfaces. Dirichlet-type boundary conditions for Eq. (9.23) must be specified at u-surfaces, not at s-surfaces. In the event that the s-surface concentration is known, solubility data are required to relate s- and u-surface concentrations (see Exercises 9–17 and 9–18). Third-kind boundary conditions involve mass convection at a surface described by the mass transfer analog to Newton's law of cooling, but are also made more complicated by a discontinuous concentration at the surface. Mass convection is introduced in Section 9.4, and the formulation of third-kind boundary conditions is explained in Section 9.4.5. Notice that there is no mass transfer analog to the fourth-kind (radiation) heat transfer boundary condition specified by Eq. (3.16). On the other hand, diffusion into a solid is often accompanied by a chemical reaction at the surface to give a more complicated boundary condition (see, for example, Exercises 9–20 and 9–61).

Transient Diffusion in a Semi-Infinite Solid

In Section 3.4.2 we encountered heat conduction problems where the conduction process is confined to a thin region near the surface into which temperature changes have penetrated. The typically low diffusion coefficients characterizing diffusion in solids ($\sim 10^{-9}$–10^{-12} m^2/s) results in analogous diffusion problems being quite common in mass transfer practice. Recall that the *penetration depth* for heat conduction was of the order $(\alpha t)^{1/2}$; analogously, it is of the order $(\mathscr{D}_{12}t)^{1/2}$ for mass diffusion. A well-known process is case-hardening of mild steel by packing the steel component in a carbonaceous material in a furnace (to increase the diffusion coefficient). The gem industry uses diffusion treatment to color clear stones. Figure 9.9*a* shows the cross section of a clear sapphire that was packed in titanium and iron oxide powders and heated for 600 hours near 2000°C: penetration of titanium and iron atoms is about 0.4 mm, which is sufficient to give the stone a brilliant blue color.

Figure 9.9*b* depicts a semi-infinite solid with species 1 in dilute solution in species 2. For example, the process might be case-hardening of steel with species 1 as carbon and species 2 as iron. The initial concentration of species 1 is $m_{1,0}$. At time $t = 0$ the concentration of species 1 at the u-surface is suddenly raised to $m_{1,u}$. The problem is to describe the concentration profiles of species 1 as a function of time and position, and to determine the rate of dissolution of species 1 as a function of time. Equation (9.23) is to be solved,

$$\frac{\partial m_1}{\partial t} = \mathscr{D}_{12}\frac{\partial^2 m_1}{\partial z^2} \tag{9.23}$$

with initial condition

$$m_1 = m_{1,0} \quad \text{at } t = 0$$

and boundary conditions

$$m_1 = m_{1,u} \quad \text{at } z = 0$$

$$m_1 \to m_{1,0} \quad \text{as } z \to \infty$$

The analogous heat conduction problem was solved in Section 3.4.2, and hence the solution can be written down immediately. From Eq. (3.58),

$$\frac{m_1 - m_{1,0}}{m_{1,u} - m_{1,0}} = \text{erfc}\frac{z}{(4\mathscr{D}_{12}t)^{1/2}} \tag{9.26}$$

Usually solid-phase diffusion coefficients are very low, and thus Eq. (9.26) finds wide utility in situations such as case-hardening, where the carbon penetrates only a short distance into the solid. A useful definition of the penetration depth δ_c of species 1 is the location where a tangent to the concentration profile at $z = 0$ intercepts the line $m_1 = m_{1,0}$, as shown in Fig. 9.9*c*. The concentration gradient at $z = 0$ is found by differentiation of Eq. (9.26),

$$-\left.\frac{\partial m_1}{\partial z}\right|_{z=0} = \frac{m_{1,u} - m_{1,0}}{(\pi\mathscr{D}_{12}t)^{1/2}} \equiv \frac{m_{1,u} - m_{1,0}}{\delta_c}$$

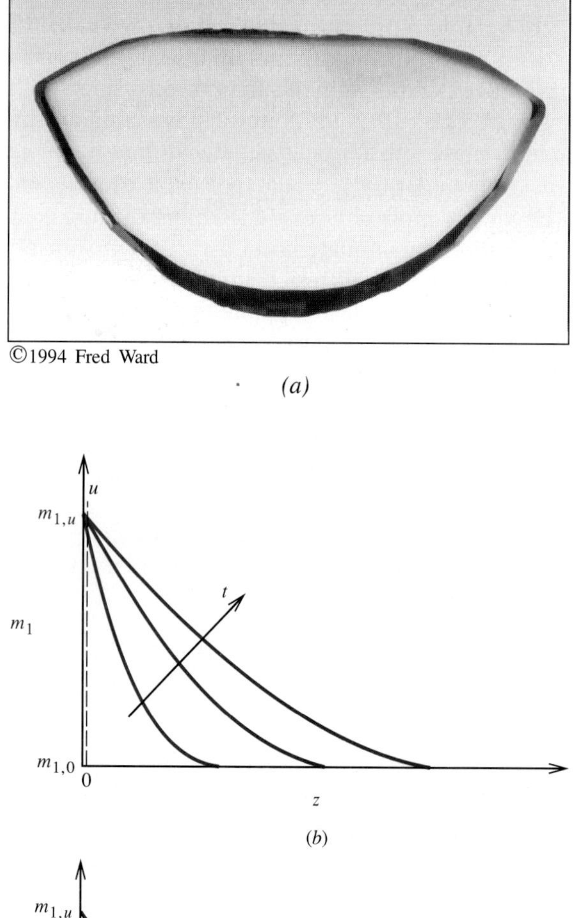

$m_{1,u}$

m_1

$m_{1,0}$ 0 $z = \delta_c$

z

(c)

Figure 9.9 Transient diffusion in a semi-infinite solid. (*a*) A clear sapphire heated for 600 h near 2050°C packed in titanium and iron oxide powders. (*b*) Expected concentration profiles. (*c*) Definition of penetration depth δ_c. (Photograph courtesy Gem Book Publishers, Bethesda, Maryland.)

and hence

$$\delta_c = 1.772(\mathscr{D}_{12}t)^{1/2} \sim (\mathscr{D}_{12}t)^{1/2} \tag{9.27}$$

That is, the penetration depth is proportional to the square root of both the diffusion coefficient and time. Note that other definitions of penetration depth are used, for example, the depth at which $(m_1 - m_{1,0})/(m_{1,u} - m_{1,0}) = 0.02$. However, the dependence on diffusion coefficient and time remains the same. The rate of dissolution per unit surface area is the diffusion flux $j_{1,u}$:

$$j_{1,u} = -\rho\mathscr{D}_{12}\left.\frac{\partial m_1}{\partial z}\right|_{z=0} = \frac{\rho\mathscr{D}_{12}(m_{1,u} - m_{1,0})}{(\pi\mathscr{D}_{12}t)^{1/2}} = \rho\left(\frac{\mathscr{D}_{12}}{\pi t}\right)^{1/2}(m_{1,u} - m_{1,0}) \tag{9.28}$$

and the rate of dissolution for a surface area A is $\dot{m}_1 = j_{1,u}A$. When the penetration depth is very small compared with the size of the solid, the shape of the solid has no effect on the diffusion process. Notice that $j_{1,u} \propto \mathscr{D}_{12}^{1/2}$: thus, the permeability approach used in Eq. (9.21*c*) for steady diffusion cannot be used for transient diffusion.

Transient Diffusion in Slabs, Cylinders, and Spheres

Transient heat conduction in slabs, cylinders, and spheres received considerable attention in Chapter 3, owing to the practical utility of the results obtained. As was the case for the semi-infinite solid, the solutions for analogous mass diffusion problems can be obtained from solutions presented in Chapter 3, and no further mathematical analysis is required. An example involving a liquid droplet is given as Exercise 9–18. However, the practical utility of these solutions for mass diffusion is rather limited. For most problems of engineering relevance involving mass transport in solids, the transport mechanism is not simple ordinary diffusion described by Fick's law with a constant diffusion coefficient. If Fick's law does apply, the diffusion coefficient usually varies significantly due to its dependence either on concentration or on temperature. But, more often, Fick's law does not apply, and there are additional complications. In particular, Fick's law is seldom applicable to the movement of moisture in porous solids or to diffusion of oxygen through oxide layers.

Perhaps the most important engineering problems involving transient mass transport in solids are the great variety of drying processes encountered in practice. Drying of coal, timber, paper, textiles, food, and pharmaceutical products are large-scale engineering operations that consume enormous amounts of energy. Drying of brush is a critical precursor to ignition in a brush fire and affects the propagation of such fires. Dehumidification of gases using desiccants, such as silica gel, depends on moisture transport in a porous solid. The student has surely had practical experience of drying processes, whether it was the baking of a cake or waiting for clothes to dry after falling into a river!

Figure 9.10 shows a typical *drying-rate curve* of a porous solid for constant drying conditions. When there is a large amount of moisture present, water moves

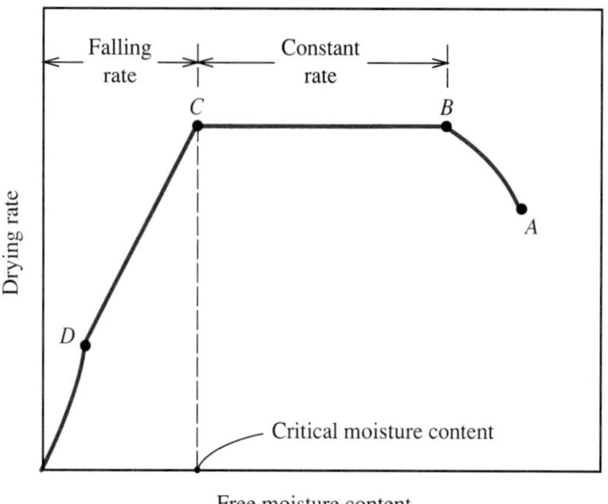

Figure 9.10 A characteristic drying-rate curve for a porous solid with constant drying conditions. (The free moisture content is the moisture in excess of the equilibrium content corresponding to the air condition.)

through the solid easily by capillary action, and the drying rate is controlled by evaporation from the wet surface. In turn, the evaporation is controlled by convective mass transfer of water vapor from the surface into an air flow, as well as by heat transfer to the surface to supply the enthalpy of vaporization. Such problems involve simultaneous convective heat and mass transfer, and are dealt with in Section 9.5. In the terminology of drying technology, this phase of a drying process is called the *constant-rate period,* shown as *B–C* in Fig. 9.10. Actually, when the solid is placed in a drying oven, there is an initial short transient *A–B* in which the drying rate increases as the solid temperature increases: at point *B* the solid has reached an equilibrium temperature, and thereafter the drying rate is constant.

Past point *C*, the moisture content of the solid is below a critical value: dry regions can form on the surface, and the location of evaporation recedes into the solid. This phase of drying is called the *falling-rate period,* since the drying rate steadily decreases. In this period, the drying rate tends to be controlled by moisture transport through the porous structure. The mechanisms of this transport are very complex: they are not well understood and are the subject of current research. Diffusion mechanisms such as ordinary, Knudsen, and surface diffusion may be important, depending on pore size. Liquid water may move by capillary action. There may be bulk motion of both water and gas due to pressure gradients induced by temperature gradients in the solid, and simultaneous heat conduction and mass transport must be considered. In some cases, these mechanisms act in opposite directions. Recent progress for wood is reported by Plumb et al. [1] and Perre et al. [2], and for silica gel by Pesaran and Mills [3]. Often the falling-rate period is divided into two portions by point *D*, at which the surface becomes completely dry. Depending on the nature of the porous solid, there may or may not be a discontinuity in slope at *D*, and, also, one of the two portions may not even exist.

Many texts present falling-rate drying analyses using a Fick's-type law with a constant effective diffusion coefficient to describe moisture transport: Exercise 9–19 is an example of such an approach. However, it is now generally accepted that such analyses usually lead to inconsistent results, and that the model is too simplified to be of much value. The bibliograpy for this chapter and Chapter 11 contain a variety of source material for the analysis and design of drying processes; most aspects are too specialized or advanced to be included in this introductory text.

EXAMPLE 9.3 Case Hardening of Steel

A mild steel rod with an initial carbon concentration 0.2% by weight is to be case-hardened. It is packed in a carbonaceous material and maintained at a high temperature. Thermodynamic data for the chemical system indicates an equilibrium concentration of carbon in iron at the phase interface of 1.5%. Calculate the time required for the location 1 mm below the surface to have a carbon mass concentration of 0.8%. The diffusion coefficient of carbon in steel at the process temperature is approximately 5.6×10^{-10} m^2/s.

Solution

Given: Mild steel exposed to carbon at high temperature.

Required: Time required to have a carbon mass fraction of 0.8% 1 mm below surface.

Assumptions: Since the penetration is small, the rod can be modeled as a semi-infinite solid.

Equation (9.26) for the instantaneous concentration profiles may be used with $m_{1,0} = 0.002$, $m_{1,u} = 0.015$, and a value of $m_1 = 0.008$ required at $z = 0.001$ m:

$$\frac{0.008 - 0.002}{0.015 - 0.002} = 0.462 = \operatorname{erfc}\left(\frac{z}{(4\mathscr{D}_{12}t)^{1/2}}\right)$$

Using Table B.4,

$$\frac{z}{(4\mathscr{D}_{12}t)^{1/2}} = 0.52$$

Solving,

$$t = \frac{z^2}{4\mathscr{D}_{12}(0.52)^2} = \frac{(0.001)^2}{(4)(5.6 \times 10^{-10})(0.52)^2} = 1650 \text{ s } (0.46 \text{ h})$$

Comments

1. The diffusion coefficient of carbon in iron increases exponentially with temperature. Hence, the process is carried out at high temperature to reduce the time required.
2. Notice that the u-surface concentration of carbon was specified in order to simplify the problem. Determination of $m_{1,u}$ requires solubility data, and possibly consideration of diffusion in the carbonaceous material.
3. The diffusion coefficient of carbon in steel depends both on the steel microstructure and the carbon concentration.

9.3.3 Heterogeneous Catalysis

The phenomenon of *heterogeneous catalysis* is often encountered in engineering problems. It is the process whereby a chemical reaction is promoted by contact of the reactants with a suitable surface, termed the *catalyst*. (The adjective *heterogeneous* means the reaction takes place on the surface, and is in contradistinction to the adjective *homogeneous,* which means that the reaction takes place within the medium.) For example, carbon monoxide and unburnt hydrocarbons can be oxidized in oxygen-rich exhaust gases of an automobile by passing the gases through a catalytic reactor in which the gases contact platinum metal. Also, the relatively cool metal of the nose of a hypersonic flight vehicle can act as a catalyst for the recombination of oxygen and nitrogen, the atomic species having resulted from dissociation in the high-temperature region behind the bow shock. The essential features of such processes can be demonstrated by analyzing a simple model experiment involving one-dimensional diffusion, as shown in Fig. 9.11. In Section 9.4, we shall see how the results obtained here can also be applied to the convection situations characteristic of typical applications.

The rate at which the catalytic reaction proceeds can often be expressed in a simple form as

$$\begin{matrix}\text{Rate of consumption}\\ \text{of species 1}\end{matrix} = \begin{matrix}\text{Rate}\\ \text{constant}\end{matrix} \times \begin{pmatrix}\text{Concentration of species 1}\\ \text{adjacent to the surface}\end{pmatrix}^n$$

The rate at which species 1 is consumed is equal to the rate at which it diffuses to the catalyst surface. Referring to Fig. 9.11, we can write on a mass basis

$$-j_{1,s} = k''(\rho_{1,s})^n \qquad \textbf{(9.29)}$$

where the rate constant k'' (the double prime signifies a surface reaction per unit area) depends on properties of the catalyst surface and on temperature, often as an *Arrhenius relation* $k'' = A\exp(-E_a/\mathscr{R}T)$, where A is the pre-exponential constant and E_a is the *activation energy.* The exponent n is the *order* of the reaction. Data for k'' and n are found in the chemical kinetics literature (though engineers often have to measure such data when developing a new process). The first-order reaction is particularly simple, inasmuch as the rate of reaction is directly proportional to the rate at which molecules of reactant collide with the surface. Appropriate units for k'' depend on the order of the reaction: for a first-order reaction, examination of Eq. (9.29) shows that

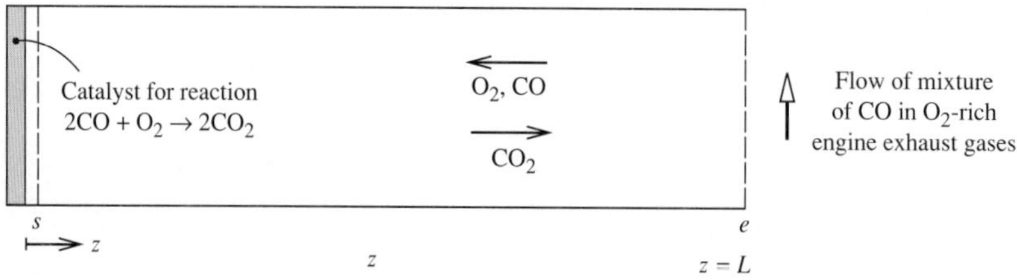

Figure 9.11 A model experiment for investigating heterogeneous catalysis with one-dimensional diffusion.

k'' has units [m/s]. The reaction

$$2CO + O_2 \rightarrow 2CO_2 \qquad \textbf{(9.30a)}$$

which is promoted in oxygen-rich automobile exhaust gases by platinum or cupric oxide in an *oxidation converter,* can be modeled as a first-order reaction. On the other hand, the reaction

$$2NO + 2CO \rightarrow N_2 + 2CO_2 \qquad \textbf{(9.30b)}$$

which is promoted in oxygen-free exhaust gases by platinum-rhodium alloy in a *three-way converter,* has a more complex behavior: it cannot be adequately modeled as a single reaction of first order. Since analysis of this problem is relatively simple for a single first-order reaction, we will focus our attention on CO removal in an oxidation converter.

Figure 9.11 shows a tube with length L, and over its mouth flows a mixture of oxygen-rich gases including CO. The surface at the base of the tube is a catalyst for the reaction given by Eq. (9.30a). We label CO as species 1. Now Fick's law, as we have defined it, is strictly valid only for binary mixtures—hence the use of a binary diffusion coefficient $\mathscr{D}_{12}$. However, when species 1 is in small concentration, we may safely define an **effective binary diffusion coefficient** for species 1 diffusing through the mixture, which we denote $\mathscr{D}_{1m}$. This diffusion coefficient is an appropriate average value for the various species in the mixture [see also Eq. (9.119)]. The flow across the mouth of the tube can be assumed to maintain the concentration of CO there equal to $m_{1,e}$, its value in the bulk flow, but is not rapid enough to induce spurious convection currents inside the tube. Thus transport of CO down the tube is by diffusion only.

The Governing Equation and Boundary Conditions

For a timewise steady state and no reactions in the gas phase (no homogeneous reactions), Eq. (9.20) applies,

$$\dot{m}_1 = j_1 A = \text{Constant}$$

or

$$\frac{dj_1}{dz} = 0$$

Substituting from Fick's law, Eq. (9.17), with $\mathscr{D}_{1m}$ replacing $\mathscr{D}_{12}$,

$$\frac{d}{dz}\left(-\rho\mathscr{D}_{1m}\frac{dm_1}{dz}\right) = 0 \qquad \textbf{(9.31)}$$

We now assume that the system is approximately isothermal so that $\mathscr{D}_{1m}$ is approximately constant, as is the mixture density ρ, since the range of species molecular weights is not large. Thus Eq. (9.31) becomes

$$\frac{d^2m_1}{dz^2} = 0 \qquad \textbf{(9.32)}$$

This differential equation is to be solved subject to the boundary conditions

$$z = 0:\ j_1 = -\rho\mathscr{D}_{1m}\left.\frac{dm_1}{dz}\right|_{z=0} = -k''\rho m_1|_{z=0} \quad \textbf{(9.33a)}$$

$$z = L:\ m_1 = m_{1,e} \quad \textbf{(9.33b)}$$

The boundary condition at $z = 0$ states that the rate at which CO diffuses to the surface must equal the rate at which it is consumed in the chemical reaction. It is a *third-kind* boundary condition.

Solution for the Concentration Profile and Reaction Rate

Integrating Eq. (9.32) yields a linear concentration profile,

$$m_1 = C_1 z + C_2$$

Application of the boundary conditions gives

$$-\rho\mathscr{D}_{1m}C_1 = -k''\rho C_2$$

$$m_{1,e} = C_1 L + C_2$$

Solving for C_1 and C_2 and substituting back gives the concentration profile as

$$\frac{m_1}{m_{1,e}} = \frac{1 + k''z/\mathscr{D}_{1m}}{1 + k''L/\mathscr{D}_{1m}} \quad \textbf{(9.34)}$$

The rate of reaction may be calculated most easily from Eq. (9.33*a*):

$$j_1 = -k''\rho m_1|_{z=0} = -k''\rho m_{1,s} = \frac{-k''\rho m_{1,e}}{1 + k''L/\mathscr{D}_{1m}}$$

Dividing by $\rho k''$ and dropping the minus sign from here on, we obtain the *removal rate* for species 1, $\dot{m}_1 = -j_1 A$, where A is the tube cross-sectional area:

$$\dot{m}_1 = \frac{m_{1,e}}{1/\rho k''A + L/\rho\mathscr{D}_{1m}A} \quad \textbf{(9.35)}$$

A circuit representation is given in Fig. 9.12*a* and shows the kinetics and diffusion resistances in series. Notice that the driving potential is $m_{1,e} = (m_{1,e} - 0)$. The zero potential appears because we assumed that the backward reaction ($2CO_2 \rightarrow 2CO + O_2$) is negligibly slow; that is, we wrote the reaction rate as $\rho k''m_{1,s} - 0 = \rho k''m_{1,s}$.

Two limiting cases can be distinguished, one in which the total resistance is dominated by the kinetics resistance, and the other in which the total resistance is dominated by the diffusion resistance.

1. *Rate-controlled limit*. If the reaction rate is slow, that is, $k'' \ll \mathscr{D}_{1m}/L$, the reaction is said to be *rate-controlled*. From Eq. (9.35),

   $$\dot{m}_1 = k''\rho m_{1,e}A$$

   and $m_{1,s} \rightarrow m_{1,e}$ in this limit.

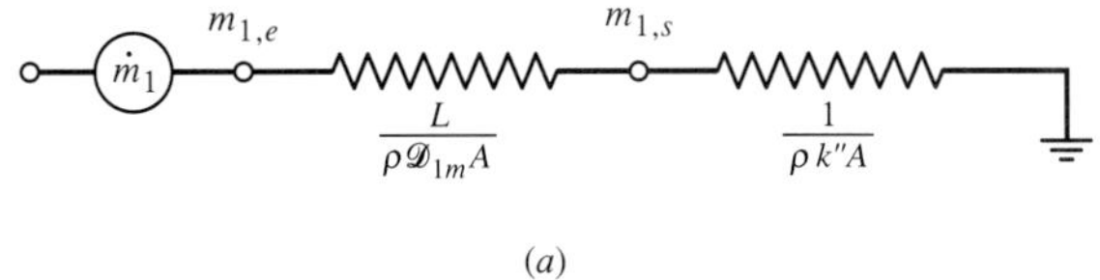

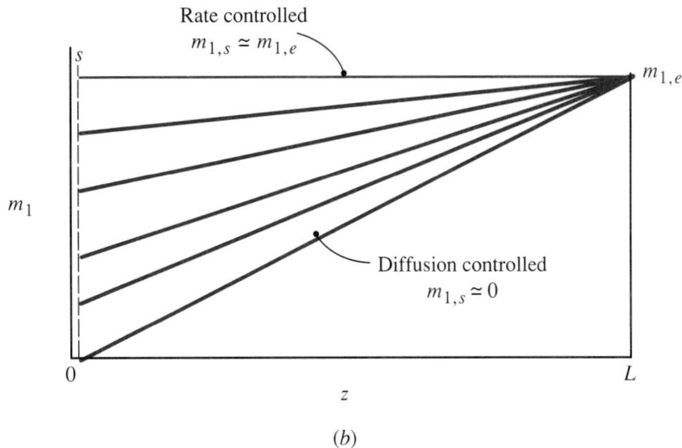

Figure 9.12 Heterogeneous catalysis with one-dimensional diffusion. (*a*) Mass flow circuit. (*b*) Concentration profiles.

2. *Diffusion-controlled limit.* If the reaction rate is fast, that is, $k'' \gg \mathscr{D}_{1m}/L$, the reaction is said to be *diffusion-controlled*. From Eq. (9.35),

$$\dot{m}_1 = \frac{\rho \mathscr{D}_{1m} A}{L} m_{1,e}$$

and it is seen that the rate at which the reaction proceeds is governed by the diffusion of species 1 to the catalyst. Also, $m_{1,s} \rightarrow 0$ in this limit. Concentration profiles are shown in Fig. 9.12*b*.

If the reaction is rate-controlled, it can be sped up by increasing k'', perhaps by increasing the temperature: decreasing the diffusion resistance by decreasing L will have no effect. Conversely, if the reaction is diffusion-controlled, it can be sped up by decreasing L; increasing k'' will have no effect. The concept of a reaction being rate-controlled or diffusion-controlled is relevant to all types of chemical reactions. For example, the metal worker of ancient times used bellows to blow air over his bed of burning charcoal, because the reaction was close to being diffusion-controlled, and the reaction rate and temperature could be increased by reducing the diffusion resistance for oxygen transport to the carbon surface. On the other hand, reactions in automobile catalytic converters are close to being rate-controlled, and it is important to operate these converters at high temperatures. A recent development is for automobiles to have two converters. One is located very close to the exhaust manifold, where it can heat up rapidly after the motor is started. A second converter is in a more convenient location, underneath the passenger compartment. This design is very effective in reducing pollutant emissions in the first minutes of operation of the automobile.

EXAMPLE 9.4 Catalytic Conversion of Carbon Monoxide

An experimental rig is set up to verify the theory described in Section 9.3.3. In a particular test, the tube is of length 10 cm and cross-sectional area 0.785 cm^2, the density of the gas mixture is 0.442 kg/m^3, and the diffusion coefficient of CO in the mixture is 97.8×10^{-6} m^2/s. When the mass fraction of CO in the gas mixture flowing over the tube is 0.001, the rate of removal of CO is measured to be 0.01855 ng/s. Determine the rate constant of the reaction on the particular catalyst used.

Solution

Given: Removal rate of CO due to catalytic reaction.

Required: Rate constant for the reaction.

Assumptions: 1. The flow of gas over the tube does not induce convection inside the tube.
2. The reaction is first-order.

Equation (9.35) gives the rate of removal of CO, $\dot{m}_1$. It can be solved for k'' to give

$$\frac{1}{k''} = \frac{\rho A m_{1,e}}{\dot{m}_1} - \frac{L}{\mathscr{D}_{1m}}$$

$$= \frac{(0.442)(0.785 \times 10^{-4})(0.001)}{(1.855 \times 10^{-11})} - \frac{0.1}{97.8 \times 10^{-6}}$$

$$= (1.870 - 1.022)10^3 = 0.848 \times 10^3$$

Hence $k'' = 1.179 \times 10^{-3}$ m/s.

Notice that this should be a fairly reliable determination of k'' since the kinetics and diffusion resistances are nearly of the same magnitude, namely,

$$\frac{1}{\rho k'' A} = \frac{1}{(0.442)(1.179 \times 10^{-3})(0.785 \times 10^{-4})} = 2.44 \times 10^7 \text{ (kg/s)}^{-1}$$

$$\frac{L}{\rho \mathscr{D}_{1m} A} = \frac{0.1}{(0.442)(97.8 \times 10^{-6})(0.785 \times 10^{-4})} = 2.95 \times 10^7 \text{ (kg/s)}^{-1}$$

Comments

1. If the diffusion resistance is much larger than the kinetics resistance, it is impossible to determine k'' accurately; in fact, if the objective is to determine rate constants accurately, then the diffusion resistance is usually made so small as to be negligible.
2. Suggest a possible technique for measuring the rate of removal of CO in this experiment.
3. A good textbook on chemical kinetics will give descriptions of methods commonly used to measure chemical kinetics data.

9.4 MASS CONVECTION

The terms **mass convection** or **convective mass transfer** are generally used to describe the process of mass transfer between a surface and a moving fluid, as shown in Fig. 9.13. The surface may be that of a falling water film in an air humidifier, of a coke particle in a gasifier, or of a silica-phenolic heat shield protecting a reentry vehicle. As was the case for heat convection, the flow can be *forced* or *natural, internal* or *external,* and *laminar* or *turbulent*. In addition, the concept of whether the mass transfer rate is *low* or *high* plays an important role: when mass transfer rates are low, there is a simple analogy between heat transfer and mass transfer that can be efficiently exploited in the solution of engineering problems.

9.4.1 The Mass Transfer Conductance

Analogous to convective heat transfer, the rate of mass transfer by convection is usually a complicated function of surface geometry and s-surface composition, the fluid composition and velocity, and fluid physical properties. For simplicity, we will restrict our attention to fluids that are either binary mixtures or solutions, or situations in which, although more than two species are present, diffusion can be adequately described using effective binary diffusion coefficients, as was discussed in Section 9.3.3. Referring to Fig. 9.13, we define the **mass transfer conductance** of species 1, $\mathcal{g}_{m1}$, by the relation

$$j_{1,s} = \mathcal{g}_{m1}\Delta m_1; \qquad \Delta m_1 = m_{1,s} - m_{1,e} \tag{9.36}$$

and the units of $\mathcal{g}_{m1}$ are seen to be the same as for mass flux [kg/m^2 s]. Equation (9.36) is of a similar form to Newton's law of cooling, Eq. (1.20), which defined the heat transfer coefficient h_c. Why we do not use a similar name and notation (e.g., mass transfer coefficient and h_m) will become clear later. On a molar basis, we define the **mole transfer conductance** of species 1, $\mathcal{G}_{m1}$, by a corresponding relation,

$$J_{1,s} = \mathcal{G}_{m1}\Delta x_1; \qquad \Delta x_1 = x_{1,s} - x_{1,e} \tag{9.37}$$

However, it usually proves advantageous to analyze convective transfer on a mass basis, rather than on a molar basis, so that Eq. (9.37) will be used infrequently.

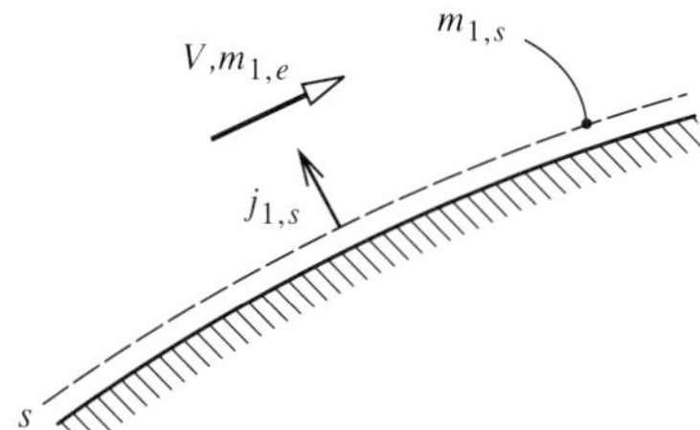

Figure 9.13 Notation for convective mass transfer into an external flow.

9.4.2 Low Mass Transfer Rate Theory

To introduce the concept of a low mass transfer rate, we consider, as an example, the evaporation of water into air, as shown in Fig. 9.14. The water-air interface might be the surface of a water reservoir, or the surface of a falling water film in a cooling tower or humidifier. In such situations the mass fraction of water vapor in the air is relatively small; the highest value is at the s-surface, but even if the water temperature is as high as 50°C, the corresponding value of $m_{H_2O,s}$ at 1 atm total pressure is only 0.077. From Eq. (9.36) the driving potential for diffusion of water vapor away from the interface is $\Delta m_1 = m_{1,s} - m_{1,e}$ and is small compared to unity, even if the free-stream air is very dry such that $m_{1,e} \simeq 0$. We then say that the mass transfer rate is *low* and the rate of evaporation of the water can be approximated as $j_{1,s}$; for a surface area A,

$$\dot{m}_1 \simeq j_{1,s}A \tag{9.38}$$

In contrast, if the water temperature approaches its boiling point, $m_{1,s}$ is no longer small, and of course, in the limit of $T_s = T_{BP}$, $m_{1,s} = 1$. The resulting driving potential for diffusion, Δm_1, is then large, and we say that the mass transfer rate is *high*. Then the evaporation rate cannot be calculated from Eq. (9.38), as will be explained in Section 10.2. For water evaporation into air, the error incurred in using low mass transfer rate theory is approximately $(1/2)\Delta m_1$, and a suitable criterion for application of the theory to engineering problems is $\Delta m_1 < 0.1$ or 0.2. A more general criterion is developed in Section 10.4.

A large range of engineering problems can be adequately analyzed assuming low mass transfer rates. These problems include cooling towers and humidifiers, as mentioned above, gas absorbers for sparingly soluble gases, and catalysis. In the case of catalysis, the *net* mass transfer rate is actually zero. Reactants diffuse toward the catalyst surface and the products diffuse away, but the catalyst only promotes the

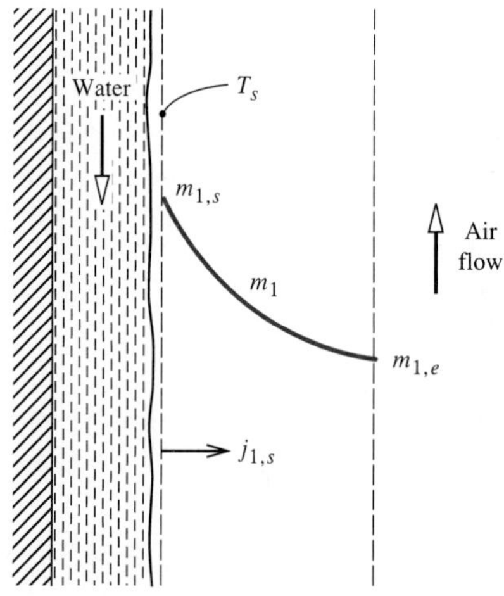

Figure 9.14 Schematic of evaporation from a falling water film into an air stream.

reaction and is not consumed. On the other hand, problems that are characterized by high mass transfer rates include condensation of steam containing a small amount of noncondensable gas, as occurs in most power plant condensers; combustion of volatile liquid hydrocarbon fuel droplets in diesel engines and oil-fired power plants; and ablation of phenolic-based heat shields on reentry vehicles. In this chapter we will restrict our attention to low mass transfer rate problems.

9.4.3 Dimensional Analysis

As an example of dimensional analysis of a mass convection problem, we consider flow through a particle bed, as shown in Fig. 9.15. Packed particle beds for heat transfer were discussed in Section 4.5.2. Such beds are used widely for mass transfer operations; for example, some types of automobile catalytic converters are in the form of packed beds of catalyst pellets. The flow field in the bed is expected to depend on the upstream superficial velocity, V, the bed volume void fraction, ε_v, and the effective particle diameter, d_p (defined in Section 4.5.2), as well as fluid density, ρ, and dynamic viscosity, μ. The locally averaged mass transfer conductance $\mathcal{g}_{m1}$ is expected to depend on the aforementioned variables and, in addition, the diffusion coefficient $\mathcal{D}_{12}$ for a binary mixture. The effect of bed diameter D will be negligible for $D \gg d_p$, and is ignored. Thus, the functional dependence of $\mathcal{g}_{m1}$ is

$$\mathcal{g}_{m1} = f(V, d_p, \rho, \mu, \mathcal{D}_{12}, \varepsilon_v)$$

From Appendix B, the SI units of these variables are

$\mathcal{g}_{m1}$	V	d_p	ρ	μ	$\mathcal{D}_{12}$	ε_v
$\frac{\text{kg}}{\text{m}^2\,\text{s}}$	$\frac{\text{m}}{\text{s}}$	m	$\frac{\text{kg}}{\text{m}^3}$	$\frac{\text{kg}}{\text{m s}}$	$\frac{\text{m}^2}{\text{s}}$	—

There are three primary dimensions or units—kg, m, and s—and seven variables. Using the Buckingham pi theorem, we expect $7 - 3 = 4$ independent dimensionless groups. We write

$$\Pi = \mathcal{g}_{m1}^a V^b d_p^c \rho^d \mu^e \mathcal{D}_{12}^f \varepsilon_v^g$$

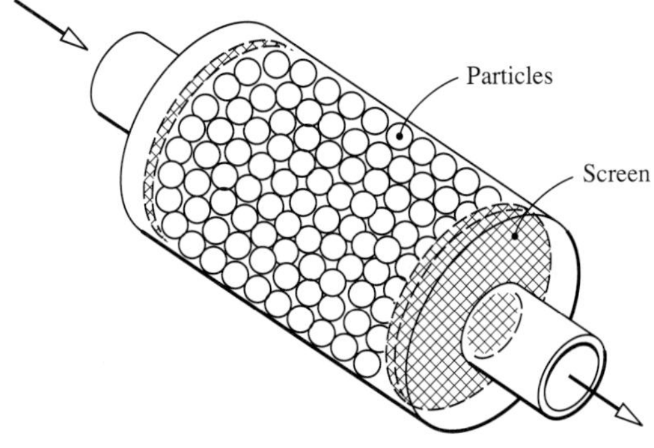

Figure 9.15 Flow through a packed particle bed.

and the sums of exponents of each primary dimension are

kg: $a + d + e = 0$

m: $-2a + b + c - 3d - e + 2f = 0$

s: $-a - b - e - f = 0$

The **Reynolds number** $\rho V d_p/\mu = \Pi_1$ can be found as for forced-convection heat transfer. We are free to choose values for four exponents; to obtain g_{m1} in a dimensionless group we set $a = 1,\ b = 0,\ e = 0,\ g = 0$:

$$
\begin{aligned}
1 + d &= 0 \qquad \text{hence } d = -1,\ f = -1,\ c = 1 \\
-2 + c - 3d + 2f &= 0 \\
-1 - f &= 0 \qquad \Pi_2 = \frac{g_{m1} d_p}{\rho \mathscr{D}_{12}}
\end{aligned}
$$

which is the **Sherwood number,** Sh (also called the mass transfer Nusselt number). The third group is found by seeking a fluid properties group that affects mass transfer in the fluid. Setting $e = 1,\ a = 0,\ b = 0,\ g = 0$,

$$
\begin{aligned}
d + 1 &= 0 \qquad \text{hence } d = -1,\ f = -1,\ c = 0 \\
c - 3d - 1 + 2f &= 0 \\
-1 - f &= 0 \qquad \Pi_3 = \frac{\mu}{\rho \mathscr{D}_{12}}
\end{aligned}
$$

which is the **Schmidt number,** Sc. The Schmidt number can also be written as $\nu/\mathscr{D}_{12}$, where $\nu = \mu/\rho$ is the kinematic viscosity. The last independent group is simply the dimensionless volume void fraction, ε_v. The result of the dimensional analysis can be written as

$$\text{Sh} = f(\text{Re}, \text{Sc}, \varepsilon_v) \tag{9.39}$$

The Sherwood number can be viewed as a dimensionless mass transfer conductance. In general,

$$\text{Sh} = \frac{g_{m1} L}{\rho \mathscr{D}_{12}} \tag{9.40}$$

where L is an appropriate characteristic length. As was the case for heat transfer, it is sometimes convenient to use a Stanton number. The **mass transfer Stanton number** is related to the Sherwood number as

$$\text{St}_m = \frac{\text{Sh}}{\text{Re}\,\text{Sc}}$$

Hence

$$\text{St}_m = \frac{g_{m1}}{\rho V} \tag{9.41}$$

Molar equivalents of Eqs. (9.40) and (9.41) are

$$\text{Sh} = \frac{\mathscr{G}_{m1} L}{c \mathscr{D}_{12}} \tag{9.42}$$

$$\text{St}_m = \frac{\mathcal{G}_{m1}}{cV} \tag{9.43}$$

Dimensional analysis of natural-convection mass transfer leads to the result

$$\text{Sh} = f(\text{Gr}, \text{Sc}) \tag{9.44}$$

where the Grashof number is expressed in terms of density difference across the boundary layer,

$$\text{Gr} = \frac{(\Delta\rho/\rho)gL^3}{\nu^2}$$

The term $(\Delta\rho/\rho)$ cannot be replaced by $\beta\Delta T$, as was done for heat transfer, since the density difference $\Delta\rho$ is due to a concentration gradient.[2]

Table 9.2 gives selected data for the Schmidt number for binary gas mixtures and dilute liquid solutions. Additional data may be found in Appendix A. Notice

Table 9.2 Selected Schmidt number data (the first species of each pair is in dilute concentration).

System	Temperature K	Schmidt Number
Dilute Gas Mixtures		
H_2-air	300	0.20
H_2O-air	300	0.61
H_2O-air	400	0.55
O_2-air	300	0.83
O_2-air	1000	0.77
CO_2-air	300	1.00
Heptane-air	300	2.1
Air-H_2O	300	0.48
Naphthalene-air	300	2.35
Dilute Liquid Solutions		
He-water	300	120
H_2O-water	300	340
SO_2-water	300	520
Isopropanol-water	300	730
Sucrose-water	300	1670
CO_2-water	287	813
CO_2-water	298	458
CO_2-water	313	236
CO_2-methanol	300	83
CO_2-ethanol	300	360
CO_2-propanol	300	890
CO_2–aqueous ethylene glycol solution, 0.20 mole fraction glycol	300	2700

[2] If there is simultaneous natural-convection heat and mass transfer—for example, cold water evaporating into warm air—there is a density difference $\Delta\rho$ due to both concentration and temperature gradients. Dimensional analysis then yields Sh = f(Gr, Sc, Pr). See Example 9.8.

that the Schmidt number is about unity for gas mixtures, whereas for liquids it is large. Strictly speaking, the Schmidt number should be subscripted the same as the diffusion coefficient, that is, Sc_{12} or Sc_{1m}, as the case might be. Also, the Sherwood and Stanton numbers should be subscripted the same as the mass transfer conductance, Sh_1, for example. Such practices will be followed only when it is necessary to avoid ambiguity.

9.4.4 The Analogy between Convective Heat and Mass Transfer

A close analogy exists between convective heat and convective mass transfer owing to the fact that conduction and diffusion in a fluid are governed by physical laws of identical mathematical form, that is, Fourier's and Fick's laws, respectively. As a result, in many circumstances the Sherwood or mass transfer Stanton number can be obtained in a simple manner from the Nusselt or heat transfer Stanton number for the same flow conditions. Indeed, in most gas mixtures Sh and St_m are nearly equal to their heat transfer counterparts. For dilute mixtures and solutions and low mass transfer rates, the rule for exploiting the analogy is simple: *The Sherwood or Stanton number is obtained by replacing the Prandtl number by the Schmidt number in the appropriate heat transfer correlation*. For example, in the case of fully developed turbulent flow in a smooth pipe, Eq. (4.44) is

$$\mathrm{Nu}_D = 0.023\,\mathrm{Re}_D^{0.8}\,\mathrm{Pr}^{0.4}; \qquad \mathrm{Pr} > 0.5$$

which for mass transfer becomes

$$\mathrm{Sh}_D = 0.023\,\mathrm{Re}_D^{0.8}\,\mathrm{Sc}^{0.4}; \qquad \mathrm{Sc} > 0.5 \tag{9.45}$$

Also, for natural convection from a heated horizontal surface facing upward, Eqs. (4.95) and (4.96) give

$$\overline{\mathrm{Nu}} = 0.54(\mathrm{Gr}_L\,\mathrm{Pr})^{1/4}; \qquad 10^5 < \mathrm{Gr}_L\,\mathrm{Pr} < 2\times 10^7 \quad \text{(laminar)}$$

$$\overline{\mathrm{Nu}} = 0.14(\mathrm{Gr}_L\,\mathrm{Pr})^{1/3}; \qquad 2\times 10^7 < \mathrm{Gr}_L\,\mathrm{Pr} < 3\times 10^{10} \quad \text{(turbulent)}$$

which for isothermal mass transfer with $\rho_s < \rho_e$ become

$$\overline{\mathrm{Sh}} = 0.54(\mathrm{Gr}_L\,\mathrm{Sc})^{1/4}; \qquad 10^5 < \mathrm{Gr}_L\,\mathrm{Sc} < 2\times 10^7 \quad \text{(laminar)} \tag{9.46a}$$

$$\overline{\mathrm{Sh}} = 0.14(\mathrm{Gr}_L\,\mathrm{Sc})^{1/3}; \qquad 2\times 10^7 < \mathrm{Gr}_L\,\mathrm{Sc} < 3\times 10^{10} \quad \text{(turbulent)} \tag{9.46b}$$

With evaporation, the condition $\rho_s < \rho_e$ will be met when the evaporating species has a smaller molecular weight than the ambient species, for example, water evaporating into air. Mass transfer correlations can be written down in a similar manner for almost all the heat transfer correlations given in Chapter 4.[3] There are some exceptions: for

[3] In chemical engineering practice, the analogy between convective heat and mass transfer is widely used in a form recommended by Chilton and Colburn in 1934, namely, $\mathrm{St}_m/\mathrm{St} = (\mathrm{Sc}/\mathrm{Pr})^{-2/3}$. The Chilton-Colburn form is of adequate accuracy for most external forced flows but is inappropriate for fully developed laminar duct flows.

example, there are no fluids with Schmidt numbers much less than unity, and thus there are no mass transfer correlations corresponding to those given for heat transfer to liquid metals with $\Pr \ll 1$. In most cases it is important for the wall boundary conditions to be of analogous form, for example, laminar flow in ducts. A uniform wall temperature corresponds to a uniform concentration $m_{1,s}$ along the s-surface, whereas a uniform heat flux corresponds to a uniform diffusive flux $j_{1,s}$.

A situation often encountered in mass transfer equipment that has no analogy in Chapter 4 is gas absorption in a falling film column. If the film falls down the inside of a vertical tube, or down parallel plates, and if there is an upward counterflow of gas, the following correlation of experimental data is recommended for the gas-side Sherwood number [4]:

$$\mathrm{Sh}_{D_h} = 0.00814\,\mathrm{Re}_{D_h}^{0.83}\,\mathrm{Sc}^{0.44}(4\Gamma/\mu_l)^{0.15} \qquad \textbf{(9.47)}$$

where Γ [kg/m s] is the film flow rate per unit width (see Section 7.2.2). Equation (9.47) is valid for $2000 < \mathrm{Re}_{D_h} < 20,000$, and $4\Gamma/\mu_l < 1200$. Equation (9.47) gives higher values of the Sherwood number than those given by Eq. (9.45) for turbulent flow in a smooth tube. The effect of waves on the film surface is to increase the convective transfer, much in the same way as does roughness of a solid surface.

Further insight into this analogy between convective heat and mass transfer can be seen by writing out Eqs. (4.44) and (9.45) as, respectively,

$$\frac{(h_c/c_p)D}{k/c_p} = 0.023\,\mathrm{Re}_D^{0.8}\left(\frac{\mu}{k/c_p}\right)^{0.4} \qquad \textbf{(9.48}\textbf{\textit{a}}\textbf{)}$$

$$\frac{g_m D}{\rho\mathscr{D}_{12}} = 0.023\,\mathrm{Re}_D^{0.8}\left(\frac{\mu}{\rho\mathscr{D}_{12}}\right)^{0.4} \qquad \textbf{(9.48}\textbf{\textit{b}}\textbf{)}$$

When cast in this form, the correlations show that the property combinations k/c_p and $\rho\mathscr{D}_{12}$ play analogous roles; these are **exchange coefficients** for heat and mass, respectively, both having units [kg/m s], which are the same as those for dynamic viscosity μ. Also, it is seen that the ratio of heat transfer coefficient to specific heat plays an analogous role to the mass transfer conductance, and has the same units [kg/m^2 s]. Thus it is appropriate to refer to the ratio h_c/c_p as the **heat transfer conductance,** g_h, and for this reason we chose not to refer to g_m as the mass transfer *coefficient*.

The Mass Transfer Biot Number

The heat transfer Biot number Bi was first introduced in Section 1.5.1. Equation (1.40) defined the Biot number as the ratio of the internal conduction resistance to the external convection resistance. When the Biot number is small (<0.1), the lumped thermal capacity model can be used to calculate the transient temperature response of a solid. Subsequently, the Biot number was introduced in Section 3.4.3: in the analysis of convective cooling of slabs, cylinders, and spheres, the Biot number appears naturally when the convection boundary condition is put in dimensionless form, for example, Eq. (3.66). For a slab, Bi $= h_c L/k$, where L is the slab half width.

Analogously, the mass transfer Biot number Bi_m is defined as the ratio of internal diffusion resistance to external convection resistance. However, the mathematical form of Bi_m is not the same as Bi because, whereas temperatures are continuous across phase interfaces, concentrations are generally discontinuous. For example, consider a spray of small water droplets injected into an air flow containing a small amount of a gas, species 1. If the droplets are too small for any liquid circulation to occur, the gas absorption process can be modeled as transient diffusion in a sphere with a convective boundary condition. We will denote water as species 2, and work on a molar basis. Following chemical engineering practice, gas- and liquid-phase mole fractions are denoted y and x, respectively: the advantage of making this distinction will become apparent in the analysis that follows. The initial concentration of species 1 dissolved in the droplet is $x_{1,0}$, and the bulk concentration of species 1 in the air is $y_{1,e}$. Referring to Fig. 9.16, the interface boundary condition is

$$-c\mathscr{D}_{12} \left.\frac{\partial x_1}{\partial r}\right|_{r=R} = \mathscr{G}_m(y_{1,s} - y_{1,e}) \tag{9.49}$$

To obtain a dimensionless form of Eq. (9.49) that is analogous to the corresponding heat transfer boundary condition, we introduce $\eta = r/R$ as a dimensionless radial coordinate; also, we must choose a dimensionless concentration that varies from an initial value of 1 when $x_1 = x_{1,0}$, to 0 when the droplet is in equilibrium with the gas phase. At equilibrium the concentration of the dissolved gas in the droplet will be uniform and, from Henry's law, Eq. (9.13), will equal $y_{1,e}/\mathrm{He}$. Thus, we define the dimensionless concentration as $\phi = (x_1 - y_{1,e}/\mathrm{He})/(x_{1,0} - y_{1,e}/\mathrm{He})$. Substituting in Eq. (9.49) gives

$$-\left.\frac{\partial\phi}{\partial\eta}\right|_{\eta=1} = \frac{\mathscr{G}_m R}{c\mathscr{D}_{12}} \frac{y_{1,s} - y_{1,e}}{x_{1,0} - y_{1,e}/\mathrm{He}}$$

$$= \frac{\mathscr{G}_m \mathrm{He}\, R}{c\mathscr{D}_{12}} \frac{x_{1,u} - y_{1,e}/\mathrm{He}}{x_{1,0} - y_{1,e}/\mathrm{He}}$$

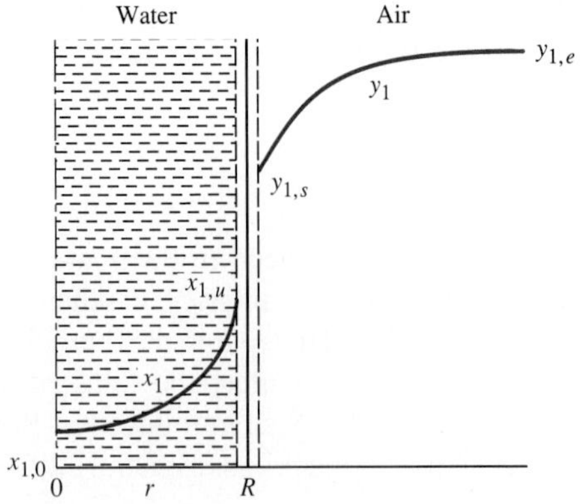

Figure 9.16 Gas absorption from an air mixture into a water droplet: concentration profiles for species 1.

and thus

$$-\left.\frac{\partial \phi}{\partial \eta}\right|_{\eta=1} = \mathrm{Bi}_m \,\phi\,|_{\eta=1}; \qquad \mathrm{Bi}_m = \frac{\mathcal{G}_m \mathrm{He}\, R}{c\mathscr{D}_{12}} \tag{9.50}$$

Table A.21 shows that Henry constants for sparingly soluble gases such as O_2 and CO_2 are very large: thus Bi_m for such gases also tend to be large. The convective resistance is then negligible, and the absorption process is controlled by diffusion in the liquid droplet. Clearly, a lumped capacity model is inappropriate for such situations. For a highly soluble gas such as ammonia, an effective Henry number calculated from the solubility data in Table A.22 is of order unity, and the convective resistance may not be negligible.

We have used gas absorption into a liquid to develop the concept of a mass transfer Biot number because the solubility data are conveniently represented by Henry's law. In the case of gas absorption into a solid, solubility data in the form of the solubility $\mathscr{S}$[m^3 solute gas (STP)/m^3 solid atm] are given in Table A.23. If molar concentration is used as the measure of concentration in the solid, it can be easily shown that the appropriate definition of the Biot number is $\mathrm{Bi} = \mathcal{G}_m(22.414/\mathscr{S}P\ [\mathrm{atm}])\mathscr{R}/\mathscr{D}_{12}$. Data for equilibrium between porous solids and a gas phase—for example, between wet wood and moist air, or between silica gel and moist air—are often given in graphical form or as analytical expressions. An effective Biot number at the moisture content of concern can be obtained from such data.

The Computer Program MCONV

MCONV calculates mass transfer rates according to the low mass transfer rate theory presented in Section 9.4.2, for the 12 flow configurations listed in Table 9.3. The transferred species is assumed to be in dilute concentration in air, so that the properties of the mixture can be approximated by pure air values at the mean film

Table 9.3 Menu of flow configurations in MCONV.

Item Number	Flow Configuration	Table 4.10 Designation
1	Turbulent flow in smooth ducts	1
2	Laminar flow in a pipe	2
3	Laminar flow between parallel plates	3
4	Laminar boundary layer on a flat plate	4
5	Turbulent boundary layer on a smooth flat plate	5
6	Flow across a cylinder	6
7	Flow across a sphere	7
8	Laminar natural convection on a vertical wall	8
9	Turbulent natural convection on a vertical wall	9
10	Natural convection on a heated horizontal plate facing down	12
11	Natural convection on a heated horizontal plate facing up	13
12	Counterflow of gas and a falling water film	Eq. (9.47)

temperature. Calculations can be made on either a mass or a molar basis. Two options are available.

1. *Transfer of an inert chemical species i.* A menu of 16 chemical species is available, with Schmidt numbers taken from Table A.17*b*.

2. *Evaporation of water or condensation of water vapor.* In this option the s-surface composition is calculated as the equilibrium value corresponding to the input surface temperature. In addition to the mass transfer rate, the convective and evaporative heat transfer rates are also calculated (see Section 9.5.1).

The use of option 1 is illustrated in Example 9.6, and the use of option 2 is illustrated in Example 9.8 of Section 9.5. Note that MCONV gives a warning if Δm_1 or Δx_1 is greater than 0.1, signifying that low mass transfer rate theory may be inadequate.

EXAMPLE 9.5 Sublimation of a Naphthalene Model

A useful method for determining convective heat transfer coefficients on surfaces of complicated shape is to cast models of naphthalene and expose the models to an air flow for a specified duration: the local mass loss by sublimation is obtained by weighing, or from the surface recession. In a particular experiment, the recession in one hour at the location of concern was 112 μm. The free-stream air velocity was 1 m/s, the pressure 1 bar, and both the model and the air temperatures were 300 K. Determine the convective heat transfer coefficient under the same air flow conditions if it is known that the flow is laminar. Properties of naphthalene include:

Solid density, $\rho = 1075 \text{ kg/m}^3$

Molecular weight, $M = 128.2$ kg/kmol

Schmidt number in air at 300 K, Sc = 2.35

Vapor pressure, $P = 3.631 \times 10^{13} \exp(-8586/T \text{ [K]})$ Pa

Solution

Given: Rate of recession of a naphthalene surface in a laminar air flow.

Required: Convective heat transfer coefficient for a similar air flow.

Assumptions: 1. Low mass transfer rate theory.
2. Laminar flow.

The rate of sublimation is equal to the rate at which naphthalene vapor can diffuse away from the surface. Denoting naphthalene as species 1, the recession rate $\dot{s}$ is

$$\dot{s} = \frac{j_{1,s}}{\rho_{\text{solid}}}; \qquad j_{1,s} = \mathcal{g}_{m1}(m_{1,s} - m_{1,e}),$$

where $m_{1,e} = 0$

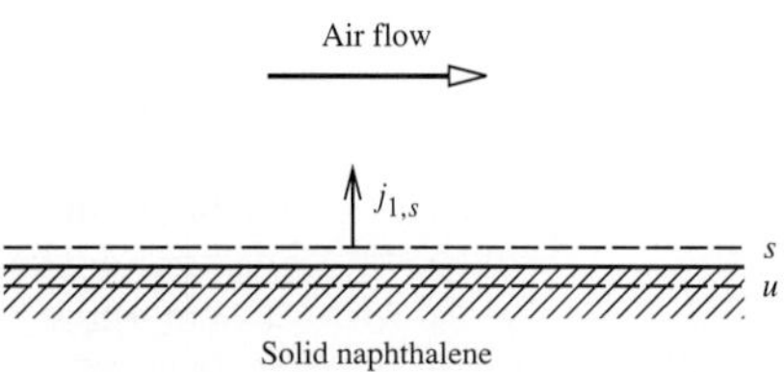

Solving,

$$\mathscr{g}_{m1} = \frac{\rho_{\text{solid}}\dot{s}}{m_{1,s}}$$

The mass fraction $m_{1,s}$ is obtained from the partial pressure as

$$m_{1,s} \simeq \left(\frac{P_{1,s}}{P}\right)\left(\frac{M_1}{M_{\text{air}}}\right)$$

since species 1 is in very small concentration.

$$P_{1,s} = 3.631 \times 10^{13} \exp(-8586/300) = 13.51 \text{ Pa}$$

Hence

$$m_{1,s} = \frac{13.51}{10^5}\frac{128.2}{29} = 5.97 \times 10^{-4}$$

and

$$\mathscr{g}_{m1} = \frac{(1075 \text{ kg/m}^3)(112 \times 10^{-6} \text{ m/h})(1/3600 \text{ h/s})}{(5.97 \times 10^{-4})} = 0.0560 \text{ kg/m}^2 \text{ s}$$

The mass transfer Stanton number is then

$$\text{St}_m = \frac{\mathscr{g}_{m1}}{\rho V} = \frac{0.0560 \text{ kg/m}^2 \text{ s}}{(1.163 \text{ kg/m}^3)(1 \text{ m/s})} = 0.0482$$

where ρ has been evaluated for pure air at 300 K and 1 bar.

We do not know the precise form of an appropriate heat or mass transfer correlation for this configuration; however, Chapter 4 shows that for forced-convection laminar boundary layer flows of air, $\text{St} \propto \text{Pr}^{-2/3}$. Thus, by analogy, $\text{St}_m \propto \text{Sc}^{-2/3}$, and hence

$$\text{St} = \text{St}_m(\text{Pr/Sc})^{-2/3} = (0.0482)(0.69/2.35)^{-2/3} = 0.109$$

$$h_c = \rho c_p V \,\text{St} = (1.163 \text{ kg/m}^3)(1005 \text{ J/kg K})(1 \text{ m/s})(0.109) = 127 \text{ W/m}^2 \text{ K}$$

where c_p has been evaluated for pure air at 300 K.

Comments

1. As a short cut we could simply use $(h_c/c_p)/\mathscr{g}_{m1} = (\text{Pr}/\text{Sc})^{-2/3}$.
2. Some examples of the use of this method to determine convective transfer coefficients may be found in references [5] and [6].
3. In older literature concerning sublimation of naphthalene, a value of Sc = 2.5 is often seen; also, there are slight variations in the vapor pressure data.

EXAMPLE 9.6 Scrubbing of Sulfur Dioxide from a Process Effluent

In a scrubber to remove SO_2 from a process effluent stream, the gas mixture flows countercurrent to a falling film of an aqueous alkaline solution in vertical tubes. The gas velocity is

1.8 m/s, the tube diameter is 5 cm, the liquid feed rate is 110 kg/h, and the column height is 4 m. The solubility of SO_2 in the alkaline solution is so high that the concentration of SO_2 in the gas phase adjacent to the liquid surface can be taken to be approximately zero, and the mass transfer process is "gas side–controlled." At a particular location in the column the pressure is 1 atm, the temperature is 300 K, and the bulk mole fraction of SO_2 in the gas stream is 0.03. Calculate the rate of removal of SO_2 per unit area of liquid-gas interface.

Solution

Given: Gas containing SO_2 flowing countercurrent to a falling film of aqueous alkaline solution.

Required: Rate of removal of SO_2 per unit area.

Assumptions: 1. Gas side–controlled absorption with $x_{SO_2,s} \simeq 0$.
2. Approximate liquid properties by those for pure water.
3. Approximate gas properties by those for pure air.

Since $x_{SO_2,s} \simeq 0$, the rate of mass transfer is simply the rate at which SO_2 can be transported from the bulk gas stream to the s-surface; transport in the liquid phase need not be considered. For this gas side–controlled absorption, Eq. (9.37), $J_{1,s} = \mathscr{G}_{m1}\Delta x_1$, gives the rate of SO_2 absorption per unit area of interface. In order to calculate the mole transfer conductance $\mathscr{G}_{m1}$, we first check the liquid and gas flow Reynolds numbers. For water at 300 K, $\mu = 8.67 \times 10^{-4}$ kg/m s. The water flow rate per unit perimeter is

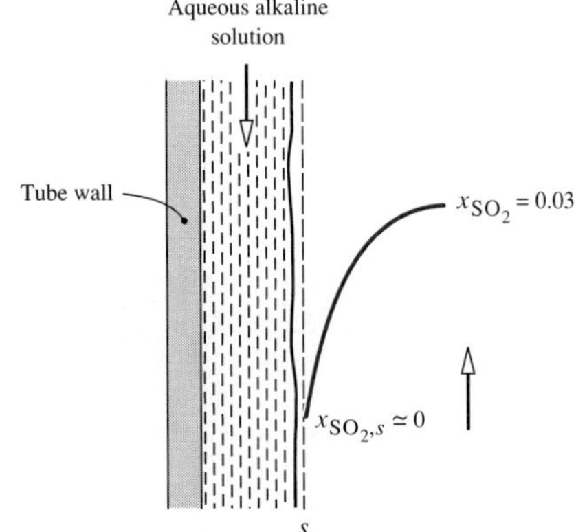

$$\Gamma = \dot{m}/\pi D = (110)(1/3600)/(\pi)(0.05)$$
$$= 0.1945 \text{ kg/m s}$$

and the film Reynolds number is

$$4\Gamma/\mu = (4)(0.1945)/(8.67 \times 10^{-4}) = 897$$

The film is in the wavy laminar regime (see Section 7.2.2).

For air at 300 K and 1 atm, $\rho = 1.177$ kg/m^3, $\nu = 15.66 \times 10^{-6}$ m^2/s, $c = \rho/M = 1.177/29 = 0.0406$ kmol/m^3. The gas-phase Reynolds number based on tube diameter is

$$\text{Re}_D = (1.8)(0.05)/15.66 \times 10^{-6} = 5747$$

The flow is just turbulent. The effect of the waves on the film surface is to increase the gas-side convective transfer. From Eq. (9.47),

$$\text{Sh}_D = 0.00814\,\text{Re}_D^{0.83}\,\text{Sc}^{0.44}(4\Gamma/\mu_l)^{0.15}$$

The diffusion coefficient of SO_2 in the mixture, $\mathscr{D}_{1m}$, is approximated by its value in air; from Table A.17a, $\mathscr{D}_{1m} = 12.6 \times 10^{-6}$ m^2/s, and the corresponding Schmidt number is

$$\text{Sc}_{1m} = \frac{\nu}{\mathscr{D}_{1m}} = \frac{15.66 \times 10^{-6}}{12.6 \times 10^{-6}} = 1.243$$

Thus

$$\text{Sh}_D = (0.00814)(5747)^{0.83}(1.243)^{0.44}(897)^{0.15} = 32.8$$

$$\mathscr{G}_{m1} = \frac{c\mathscr{D}_{1m}\text{Sh}_D}{D} = \frac{(0.0406 \text{ kmol/m}^3)(12.6 \times 10^{-6} \text{ m}^2\text{/s})(32.8)}{(0.05 \text{ m})}$$

$$= 3.36 \times 10^{-4} \text{ kmol/m}^2 \text{ s}$$

Then, from Eq. (9.37), with $\Delta x_1 = x_{1,s} - x_{1,b}$ for an internal flow,

$$J_{1,s} = \mathscr{G}_{m1}(x_{1,s} - x_{1,b}) = (3.36 \times 10^{-4})(0 - 0.03) = -1.01 \times 10^{-5} \text{ kmol/m}^2 \text{ s}$$

where the negative sign indicates that the flux $J_{1,s}$ is toward the liquid surface, that is, SO_2 is being absorbed.

Solution using MCONV

The required input is:

Configuration number = 12

Option = 1 (transfer of chemical species i)

Chemical species number = 7 (SO_2)

1. Interface temperature, T_s = 300
2. Bulk temperature, T_b = 300
3. Pressure, $P = 1.013 \times 10^5$
4. Hydraulic diameter, D_h = 0.05

   Water flow rate per unit perimeter, Γ = 0.1945

   Bulk velocity of air, V = 1.8
5. Option 2 (mole fractions)

$x_{1,s} = 0$

$x_{1,b} = 0.03$

The output gives:

$$J_{1,s} = -1.01 \times 10^{-5} \text{ kmol/m}^2 \text{ s}$$

Comments

1. If Eq. (9.45) is used for the Sherwood number, the result is

   $$\text{Sh} = 0.023\,\text{Re}_D^{0.8}\,\text{Sc}^{0.4} = 0.023(5747)^{0.8}(1.243)^{0.4} = 25.5$$

   which is 22% less than the value given by Eq. (9.47).
2. Check the complete output of MCONV.
3. Notice the similarity between gas side–controlled absorption and diffusion-controlled catalysis (Section 9.3.3).

9.4.5 The Equivalent Stagnant Film Model

Figure 9.17 shows flow past a surface (for example, salt, species 1, dissolving in water, species 2). The concentration of salt at the s-surface is $m_{1,s}$, the equilibrium

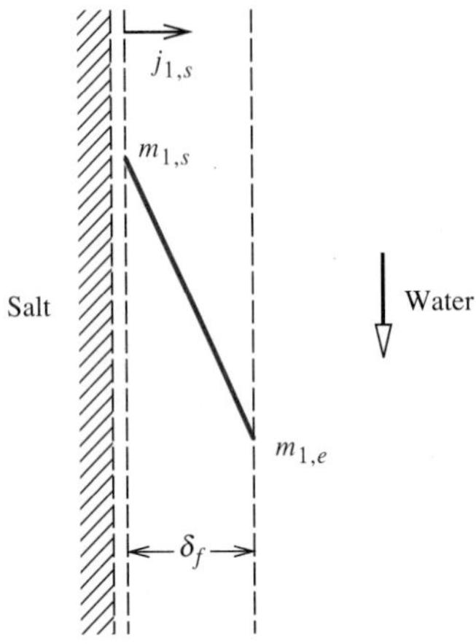

Figure 9.17 Equivalent stagnant-film model for convective mass transfer: salt dissolving in water.

value obtained from solubility data; at the e-surface in the external flow, the salt concentration is $m_{1,e}$. From Eq. (9.36), the rate at which salt is transferred into the water from a surface area A is

$$\dot{m}_1 = j_{1,s}A = g_{m1}A(m_{1,s} - m_{1,e})$$

If we now imagine the transfer process to be that of diffusion across an equivalent stagnant film of thickness δ_f, Eq. (9.21a) gives

$$\dot{m}_1 = \frac{\rho\mathscr{D}_{12}A}{\delta_f}(m_{1,s} - m_{1,e})$$

Comparing these two expressions for $\dot{m}_1$ gives

$$g_{m1} = \frac{\rho\mathscr{D}_{12}}{\delta_f} \tag{9.51}$$

and the corresponding diffusion resistance is $\delta_f/\rho\mathscr{D}_{12}A$.

Although the concept of an equivalent stagnant film is oversimplistic (there is, of course, some velocity profile in the fluid), it proves a useful model for the analysis of more complicated mass transfer problems. On a molar basis, the corresponding definition of an equivalent stagnant-film thickness is by the relation

$$\mathscr{G}_{m1} = \frac{c\mathscr{D}_{12}}{\delta_f} \tag{9.52}$$

As an example of the utility of the equivalent stagnant film concept, consider again the problem of heterogeneous catalysis analyzed in Section 9.3.3, except that now the reactants are flowing through a honeycomb matrix coated with the catalyst. If the actual process of convection and diffusion in the flow through the passages of the matrix is modeled as diffusion through a stagnant gas film of thickness

δ_f, then the tube length L in the analysis of Section 9.3.3 can simply be replaced by δ_f to give

$$\dot{m}_1 = \frac{m_{1,e}}{1/\rho k''A + \delta_f/\rho\mathscr{D}_{1m}A}$$

Since δ_f must be calculated from the mass transfer conductance using Eq. (9.51), this result is more conveniently written in terms of $\mathscr{g}_{m1}$ as

$$\dot{m}_1 = \frac{m_{1,e}}{1/\rho k''A + 1/\mathscr{g}_{m1}A} \quad \textbf{(9.53)}$$

Notice how the equivalent stagnant-film thickness δ_f does not appear in the final result: the concept was used simply to relate the result of Section 9.3.3 to a convective situation. Equation (9.53) also can be derived directly as follows. Since the rate at which species CO is consumed equals the rate at which it is transported to the surface,

$$\dot{m}_1 = k''\rho m_{1,s}A = -\mathscr{g}_{m1}(m_{1,s} - m_{1,e})A$$

Solving for $m_{1,s}$ and substituting back to obtain $\dot{m}_1$ gives Eq. (9.53).

EXAMPLE 9.7 Removal of Carbon Monoxide from an Automobile Exhaust

The packing for an automobile catalytic converter is in the form of a matrix with porous walls impregnated with catalyst. The passages are of square cross section with 1 mm sides, and the *effective* rate constant of the porous catalyst at the operating temperature of 800 K is 0.070 m/s. If the pressure in the reactor is 1.15×10^5 Pa and the mean molecular weight of the exhaust gases is 28, calculate the CO reduction rate at a location where the CO concentration is 0.187% by weight. Also calculate the equivalent stagnant-film thickness.

Solution

Given: Catalytic packing for CO removal from exhaust gases.

Required: CO reduction rate and equivalent stagnant-film thickness δ_f.

Assumptions:
1. Fully developed laminar flow in the matrix.
2. Approximate the CO diffusion coefficient by its value in air.

Equation (9.53) gives the rate at which CO is reduced:

$$\frac{\dot{m}_1}{A} = \frac{m_{1,e}}{1/\rho k'' + 1/\mathscr{g}_{m1}}$$

$$\mathscr{g}_{m1} = \frac{\rho\mathscr{D}_{1m}}{D_h}\text{Sh}$$

Sh $= 2.98$ from Table 4.5 for fully developed laminar flow in a square duct.

$$D_h = 4A_c/\mathscr{P} = 4(1^2)/4(1) = 1 \text{ mm}$$

$$\rho = \frac{P}{(\mathscr{R}/M)T} = \frac{1.15 \times 10^5}{(8314/28)(800)} = 0.484 \text{ kg/m}^3$$

From Table A.17a, approximating $\mathscr{D}_{1m}$ as $\mathscr{D}_{\text{CO}_\text{air}}$,

$$\mathscr{D}_{1m} = (106 \times 10^{-6})(1.013 \times 10^5/1.15 \times 10^5) = 93.4 \times 10^{-6} \text{ m}^2/\text{s}$$

$$g_{m1} = (0.484)(93.4 \times 10^{-6})(2.98)/(1 \times 10^{-3}) = 0.135 \text{ kg/m}^2\text{ s}$$

$$m_{1,e} = 0.00187$$

Substituting in Eq. (9.53),

$$\frac{\dot{m}_1}{A} = \frac{0.00187}{1/(0.484)(0.070) + 1/0.135} = 5.06 \times 10^{-5} \text{ kg/m}^2\text{ s}$$

Also, from Eq. (9.51),

$$\delta_f = \frac{\rho\mathscr{D}_{1m}}{g_{m1}}, \qquad \text{where } g_{m1} = \frac{\text{Sh}\,\rho\mathscr{D}_{1m}}{D_h}$$

Hence

$$\delta_f = \frac{D_h}{\text{Sh}} = \frac{1 \text{ mm}}{2.98} = 0.336 \text{ mm}$$

Comments

1. The effective rate constant takes into account the added surface area of the pores in a porous catalyst (see Section 9.6).
2. Physical chemists often simply guess a value for the equivalent stagnant-film thickness in a mass transfer process. This is a bad practice: it is relatively simple to obtain a reliable value from an appropriate correlation for the Sherwood number.

9.5 SIMULTANEOUS HEAT AND MASS TRANSFER

In many situations heat and mass transfer occur simultaneously at a phase interface. Whenever water evaporates into an air stream, the enthalpy of vaporization must be supplied by heat transfer from the bulk water, the air stream, or both. Thus, analysis of the performance of equipment such as cooling towers and humidifiers involves simultaneous consideration of heat and mass transfer processes. Combustion of a fuel oil droplet also requires simultaneous consideration of heat and mass transfer, since the rate of combustion depends both on the rate of diffusion of the fuel vapor and oxygen to the flame front, and on the rate of heat transfer from the flame to the droplet surface, which supplies the required enthalpy of vaporization. The rate at which an ablative heat shield loses mass depends on the rate of heat transfer to the shield. The combustion of carbon in the form of coke or coal also involves simultaneous heat and mass transfer: heat transfer considerations determine the temperature of the carbon and hence the mass transfer regime, that is, whether or not the reaction is rate- or diffusion-controlled, and the details of the kinetics mechanism and combustion products. In some simple situations engineering approximations can be made in order to *uncouple* the heat and mass transfer processes so that each can be analyzed separately: the problem of determining the heat loss from a well-stirred chemical dip tank is considered in Example 9.8. On the other hand, in most situations the heat and mass transfer processes are *coupled* in the sense that the equations governing the respective processes must be simultaneously solved. The wet- and dry-bulb psychrometer considered in Section 9.5.2 is such a situation, as are the combustion of a volatile hydrocarbon fuel droplet and the ablation of a heat shield.

Some simple examples of simultaneous heat and mass transfer involving low mass transfer rates are given in this section. Examples involving high mass transfer rates will be given in Chapter 10. The performance of exchangers that transfer both heat and mass is considered in Section 11.5. The use of the computer program MCONV, which was described in Section 9.4.4, is illustrated in Example 9.8.

9.5.1 Surface Energy Balances

Surface energy balances for heat transfer were introduced in Section 1.4. We now extend this concept to situations where there is simultaneous mass transfer. An important problem is evaporation of a liquid, for example, evaporation of water into air, as shown in Fig. 9.18. With H_2O denoted as species 1, the steady-flow energy equation, Eq. (1.4), applied to the control volume located between the u- and s-surfaces (Fig. 9.18*a*) requires that

$$\dot{m}\Delta h = \dot{Q}$$

$$\dot{m}(h_{1,s} - h_{1,u}) = A(q_{\text{cond}} - q_{\text{conv}} - q_{\text{rad}}) \tag{9.54}$$

where it has been recognized that only species 1 crosses the u- and s-surfaces. Also, the water has been assumed perfectly opaque so that all radiation is emitted or absorbed between the u-surface and the interface.[4]

If we restrict our attention to conditions for which low mass transfer rate theory is valid, we can write $\dot{m}/A \simeq j_{1,s} = g_{m1}(m_{1,s} - m_{1,e})$. Also, we can then calculate the convective heat transfer as if there were no mass transfer, and write $q_{\text{conv}} = h_c(T_s - T_e)$. Substituting in Eq. (9.54) with $q_{\text{cond}} = -k\partial T/\partial y\,|_u$, $h_{1,s} - h_{1,u} = h_{\text{fg}}$, and rearranging, gives

$$-k\left.\frac{\partial T}{\partial y}\right|_u = h_c(T_s - T_e) + g_{m1}(m_{1,s} - m_{1,e})h_{\text{fg}} + q_{\text{rad}} \tag{9.55}$$

[4] Strictly speaking, radiation is emitted or absorbed below the u-surface, since the u-surface is indefinitely close to the real interface. However, unless we are concerned about the detailed effect of emission and absorption on the temperature profile in the water very close to the interface, it is convenient to assume that the water is perfectly opaque so that heat is transferred across the u-surface by conduction only.

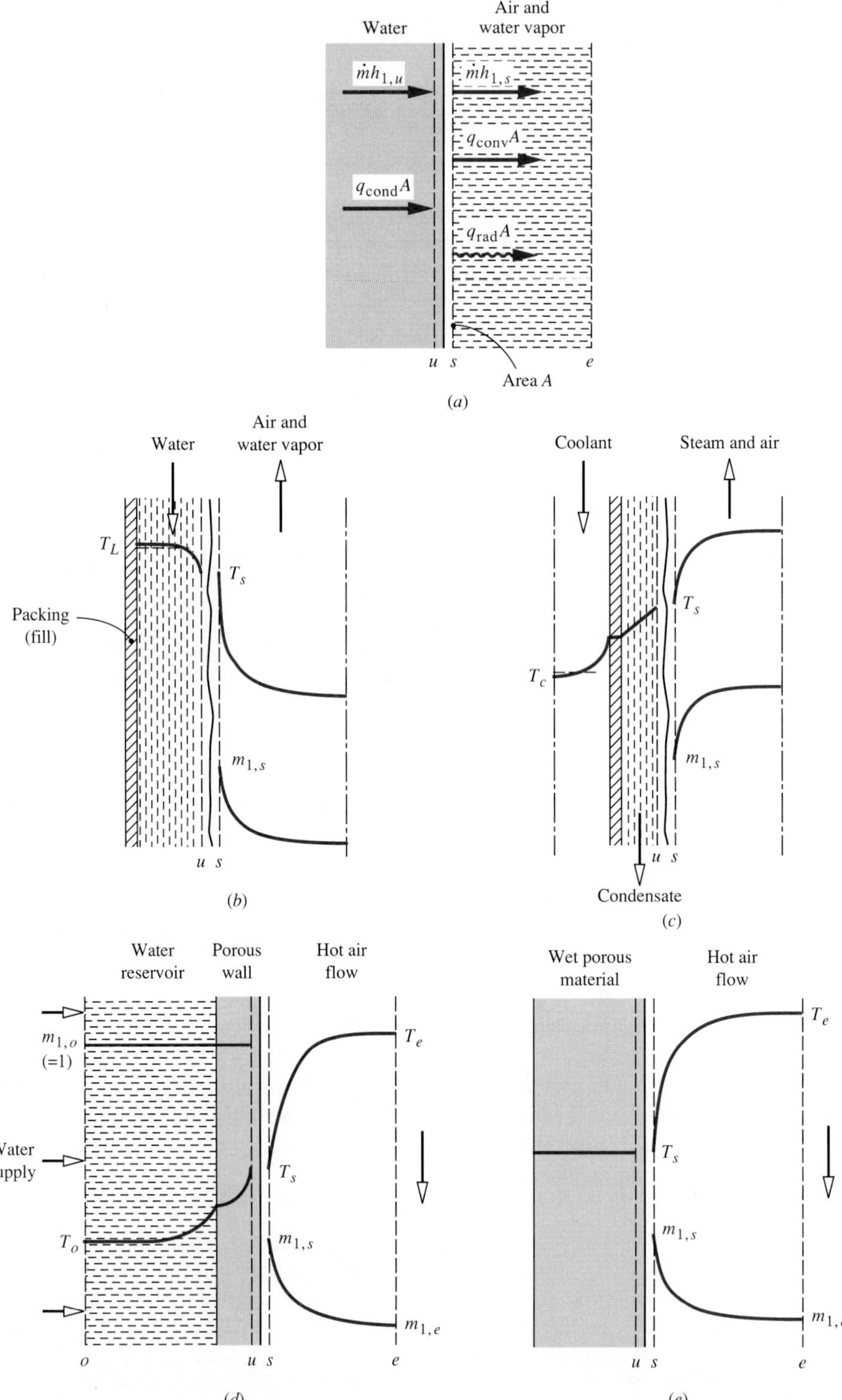

Figure 9.18 Evaporation of water into air. (*a*) The surface energy balance. (*b*) A falling water film in a cooling tower. (*c*) Film condensation from a steam-air mixture. (*d*) Sweat cooling. (*e*) Constant-rate drying of a porous material.

It is common practice to refer to the convective heat flux $h_c(T_s - T_e)$ as the *sensible* heat flux, whereas the term $\mathscr{g}_{m1}(m_{1,s} - m_{1,e})h_{fg}$ is called the *evaporative* or *latent* heat flux. Each of the terms in Eq. (9.55) can be positive or negative, depending on the particular situation. Also, the evaluation of the conduction heat flux at the u-surface, $-k\,\partial T/\partial y|_u$, depends on the particular situation. Four examples are shown in Fig. 9.18. For a water film flowing down a packing in a cooling tower (Fig. 9.18*b*), this heat flux can be expressed in terms of convective heat transfer from the bulk water at temperature T_L to the surface of the film, $-k\,\partial T/\partial y|_u = h_{cL}(T_L - T_s)$. If the liquid-side heat transfer coefficient h_{cL} is large enough, we can simply set $T_s \simeq T_L$, which eliminates the need to estimate h_{cL}. The evaporation process is then *gas side–controlled* (cf. Example 9.6). Figure 9.18*c* shows film condensation from a steam-air mixture on the outside of a vertical tube. In this case we can write $k\,\partial T/\partial y|_u = U(T_s - T_c)$, where T_c is the coolant bulk temperature. The overall heat transfer coefficient U includes the resistances of the condensate film, the tube wall, and the coolant. Sweat cooling is shown in Fig. 9.18*d*, with water from a reservoir (or *plenum chamber*) injected through a porous wall at a rate just sufficient to keep the wall surface wet. In this case, the conduction across the u-surface can be related to the reservoir conditions by application of the steady-flow energy equation to a control volume located between the o- and u-surfaces. Finally, Fig. 9.18*e* shows drying of a wet porous material (e.g., a textile or wood). During the constant-rate period of the process, evaporation takes place from the surface with negligible heat conduction into the solid; then $-k\,\partial T/\partial y|_u \simeq 0$. The term *adiabatic vaporization* is used to describe evaporation when $q_{\text{cond}} = 0$: constant-rate drying is one example, and the wet-bulb psychrometer analyzed in Section 9.5.2 is another.

Evaluation of Properties

Evaluation of composition-dependent properties, in particular the mixture specific heat and Prandtl number, poses a problem. In general, low mass transfer rates imply small composition variations across a boundary layer, and properties can be evaluated for a mixture of the free-stream composition at the mean film temperature. In fact, when dealing with evaporation of water into air, we will simply use the properties of dry air at the mean film temperature to give results of adequate engineering accuracy. If there are large composition variations across the boundary layer, as can occur in some catalysis problems, properties should be evaluated at the mean film composition and temperature. Methods for evaluating properties of mixtures and solutions are given in Section 9.7.

EXAMPLE 9.8 Heat Loss from a Chemical Dip Bath

A bath 1 m square containing a dilute aqueous solution is maintained at 330 K in still air at 1 atm, 300 K, and 15% relative humidity. Estimate the convective (sensible) and evaporative (latent) heat losses from the surface, if the bath is well stirred so that its surface temperature is also approximately 330 K. Assume that the vapor pressure can be approximated as that of water.

Solution

Given: Dip bath at 330 K evaporating into air at 300 K.

Required: Convective and evaporative heat losses.

Assumptions:
1. The vapor pressure of the aqueous solution can be approximated by data for pure water.
2. The ambient air is still.
3. Negligible difference between the bath surface and bulk temperatures.
4. Moist air transport properties can be approximated by those for dry air.

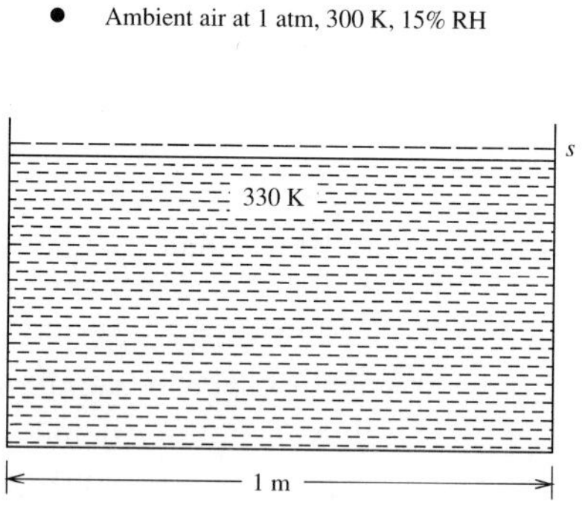

This is a natural-convection problem. To calculate the Grashof number $\mathrm{Gr} = (\Delta\rho/\rho)gL^3/\nu^2$, ρ_s and ρ_e are required. Denoting H_2O as species 1 and air as species 2, Table A.12*a* gives

$$P_{1,s} = P_{\mathrm{sat}}(T_s) = P_{\mathrm{sat}}(330\ \mathrm{K}) = 17{,}190\ \mathrm{Pa}$$

$$P_{1,e} = (\mathrm{RH})P_{\mathrm{sat}}(T_e) = (0.15)(3533) = 530\ \mathrm{Pa}$$

Assuming an ideal gas mixture,

$$\rho = \rho_1 + \rho_2 = \frac{P_1 M_1}{\mathscr{R}T} + \frac{P_2 M_2}{\mathscr{R}T}$$

$$\rho_s = \frac{(17{,}190)(18)}{(8314)(330)} + \frac{(101{,}330 - 17{,}190)(29)}{(8314)(330)} = 0.1128 + 0.8894 = 1.0022\ \mathrm{kg/m^3}$$

$$\rho_e = \frac{(530)(18)}{(8314)(300)} + \frac{(101{,}330 - 530)(29)}{(8314)(300)} = 0.003825 + 1.1720 = 1.1758\ \mathrm{kg/m^3}$$

$$\Delta\rho = \rho_e - \rho_s = 1.1758 - 1.0022 = 0.1736\ \mathrm{kg/m^3}$$

Also,

$$m_{1,s} = \frac{\rho_{1,s}}{\rho_s} = \frac{0.1128}{1.0022} = 0.1126; \qquad m_{1,e} = \frac{\rho_{1,e}}{\rho_e} = \frac{0.003825}{1.1758} = 0.003253$$

The Grashof number can be calculated with ν approximated by the value for pure air at a mean film temperature of $(1/2)(300 + 330) = 315$ K, and ρ as the mean of ρ_s and ρ_e, $\rho = (1/2)(1.176 + 1.002) = 1.089\ \mathrm{kg/m^3}$.

$$\mathrm{Gr}_L = \frac{(\Delta\rho/\rho)gL^3}{\nu^2} = \frac{(0.1736/1.089)(9.81)(1)^3}{(16.99 \times 10^{-6})^2} = 5.42 \times 10^9 > 10^9 \quad \text{(turbulent flow)}$$

Using Eqs. (4.96) and (9.46*b*) with the Prandtl number approximated as a pure air value of 0.69, and $\mathrm{Sc}_{12} = 0.61$ from Table 9.2,

$$\overline{\mathrm{Nu}}_L = 0.14(\mathrm{Gr}_L\,\mathrm{Pr})^{1/3} = 0.14[(5.42 \times 10^9)(0.69)]^{1/3} = 217$$

$$\overline{\mathrm{Sh}}_L = 0.14(\mathrm{Gr}_L\,\mathrm{Sc})^{1/3} = 0.14[(5.42 \times 10^9)(0.61)]^{1/3} = 209$$

$$\overline{h}_c = \frac{k}{L}\overline{\mathrm{Nu}}_L = \frac{(0.0277)(217)}{1} = 6.01 \ \mathrm{W/m^2\ K}$$

$$\overline{g}_{m1} = \frac{\rho \mathscr{D}_{12}}{L}\overline{\mathrm{Sh}}_L = \frac{\rho \nu}{\mathrm{Sc}_{12} L}\overline{\mathrm{Sh}}_L = \frac{(1.089)(16.99 \times 10^{-6})(209)}{(0.61)(1)} = 6.34 \times 10^{-3} \ \mathrm{kg/m^2\ s}$$

where k has been evaluated for pure air at the mean film temperature, and ρ again has been evaluated as the mean of ρ_s and ρ_e. Thus, the heat losses are

$$\dot{Q}_{\mathrm{conv}} = \overline{h}_c A(T_s - T_e) = (6.01)(1)(330 - 300) = 180 \ \mathrm{W}$$

$$\dot{Q}_{\mathrm{evap}} = \overline{g}_{m1} A(m_{1,s} - m_{1,e}) h_{fg}(T_s)$$

$$= (6.34 \times 10^{-3})(1)(0.1126 - 0.0033)(2.365 \times 10^{6}) = 1640 \ \mathrm{W}$$

Solution using MCONV

The required input is:

Configuration number = 11

Option = 2 (evaporation of water into air)

1. Interface temperature, T_s = 330
2. Ambient temperature, T_e = 300
3. Pressure, $P = 1.0133 \times 10^5$
4. Length, L = 1
5. Relative humidity, RH = 15

The output gives:

$$q_{\mathrm{evap}} = 1660 \ [\mathrm{W/m^2}]$$

$$q_{\mathrm{conv}} = 180 \ [\mathrm{W/m^2}]$$

Comments

1. The latent heat loss is an order of magnitude larger than the sensible heat loss because the air is very dry.
2. This result should be viewed as an upper bound: in reality, the surface temperature will be a little lower than the bulk temperature, in order to have heat transfer from the bulk liquid to the interface. Of course, the interface temperature may not be uniform either (see Exercise 9–44).
3. For natural-convection simultaneous heat and mass transfer, the simple analogy between heat and mass transfer given in Section 9.4.4 is not quite true. More exact analysis of this problem shows that the true Nusselt number is somewhat higher than 217 and the true Sherwood number is somewhat lower than 209. However, our result is quite adequate for engineering purposes (see footnote 2).
4. MCONV gives a warning that $\Delta m_1 = 0.109 > 0.10$. Low mass transfer rate theory is inaccurate, but more exact analysis shows that the error is approximately $(1/2)\Delta m_1$ expressed as a percentage, that is, approximately 5%.

5. It is of interest to calculate the radiation heat flux. For a water emittance of 0.9 from Table A.5*a*, and black surroundings at 300 K,

$$q_{\text{rad}} = (0.9)(5.67 \times 10^{-8})(330^4 - 300^4) = 192 \text{ W/m}^2$$

Thus, the radiation heat loss is a little larger than the sensible heat loss (though still much smaller than the latent heat loss).

9.5.2 The Wet- and Dry-Bulb Psychrometer

The wet- and dry-bulb psychrometer is widely used to measure the moisture content of air. In its simplest form, the air is made to flow over a pair of thermometers, one of which has its bulb covered by a wick whose other end is immersed in a small water reservoir. Evaporation of water from the wick causes the wet bulb to cool, and its steady-state temperature is a function of the air temperature measured by the dry bulb and the air humidity. Commonly, a *psychrometric chart* is used to deduce the air humidity from the two temperature readings. The *sling* psychrometer has thermometers mounted in a frame with a handle that allows the unit to be swung around the user's head. Alternatively, a fan is used to supply the air flow. Nowadays the common availability of digital readout thermocouples with built-in cold junctions has led to the increased use of thermocouple-based psychrometers. Since relatively small thermocouple junctions can be used, their response to changes in temperature and hence humidity is fast, and they are more appropriate than thermometer-based psychrometers in situations where the air state changes rapidly with time.

The principle of the wet- and dry-bulb psychrometer can be demonstrated using the mass convection theory developed in Section 9.4. Figure 9.19 shows a wet-bulb thermometer exposed to an air flow of temperature T_e (as measured by the dry bulb) and mass fraction of water vapor $m_{1,e}$. The surface of the moist wick is at temperature T_s, and the mass fraction of water vapor adjacent to the wick is $m_{1,s}$, the saturation value corresponding to temperature T_s. When the wet-bulb temperature is steady, there is negligible heat conduction into the thermometer, and from Eq. (9.55) the surface energy balance on the wet bulb requires that the heat transfer from the

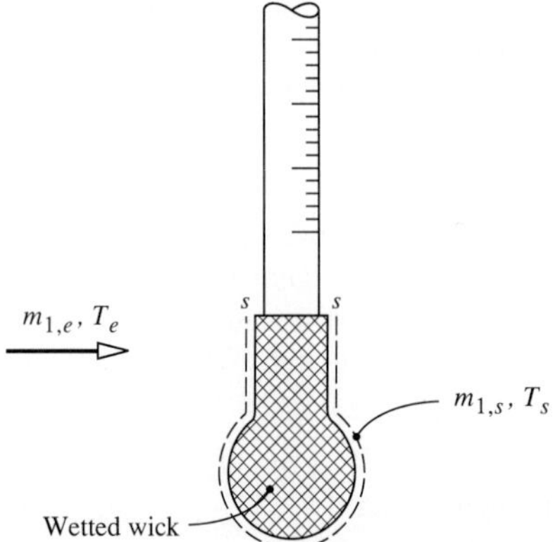

Figure 9.19 Wet bulb of a wet- and dry-bulb psychrometer.

air equal the evaporation rate times the enthalpy of vaporization:

$$\overline{h}_c(T_e - T_s)A = \overline{\mathcal{g}}_{m1}(m_{1,s} - m_{1,e})h_{fg}A$$

Radiation heat transfer from the surroundings has been ignored since, if the air velocity is sufficiently high, radiation can be neglected compared with convection. Also, the very small amount of heat given up by the water as it cools from the reservoir temperature to T_s is neglected in taking $-k\,\partial T/\partial y|_u = 0$. Rearranging gives

$$m_{1,e} = m_{1,s} - \frac{\overline{h}_c}{\overline{\mathcal{g}}_{m1} h_{fg}}(T_e - T_s) \tag{9.56}$$

For laminar forced flow over a cylinder, sphere, or similar shape, Stanton numbers for heat and mass transfer in gases can be approximated by Chilton-Colburn analogy power law relations,

$$\overline{\text{St}} = C\,\text{Re}^{-1/2}\,\text{Pr}^{-2/3} = \frac{\overline{h}_c}{\rho c_p V}$$

$$\overline{\text{St}}_m = C\,\text{Re}^{-1/2}\,\text{Sc}^{-2/3} = \frac{\overline{\mathcal{g}}_{m1}}{\rho V}$$

Thus

$$\frac{\overline{h}_c/c_p}{\overline{\mathcal{g}}_{m1}} = \left(\frac{\text{Pr}}{\text{Sc}}\right)^{-2/3}$$

and substituting into Eq. (9.56) gives

$$m_{1,e} = m_{1,s} - \frac{c_p}{h_{fg}}\left(\frac{\text{Pr}}{\text{Sc}_{12}}\right)^{-2/3}(T_e - T_s) \tag{9.57}$$

Strictly speaking, the specific heat and Prandtl number should be evaluated at an appropriate reference temperature and composition. However, if the water vapor is in small concentrations, it is sufficient to use pure air values at the mean film temperature. Also, it is sufficiently accurate to take $\text{Pr}/\text{Sc}_{12} \simeq (0.69/0.61) = 1.13$ and $(\text{Pr}/\text{Sc}_{12})^{-2/3} = 1/1.08$. The ratio of Prandtl to Schmidt number is a new dimensionless group that is relevant to simultaneous convective heat and mass transfer: it is called the **Lewis number**.[5] Equation (9.57) can thus be written as

$$m_{1,e} = m_{1,s} - \frac{c_{p,\text{air}}}{1.08 h_{fg}}(T_e - T_s) \tag{9.58}$$

Temperatures T_s and T_e are, respectively, the known measured wet- and dry-bulb temperatures, and $m_{1,e}$ is the unknown air moisture content. For T_s known, $m_{1,s}$ can

[5] Other definitions of the Lewis number are found in the literature, for example, the ratio of Schmidt to Prandtl number and the ratio of the heat to mass transfer conductance.

be obtained from steam tables in the usual way: such data can be represented by the relation

$$m_{1,s} = m_{1,s}(T_s, P) \tag{9.59}$$

where P is the total pressure. Equations (9.58) and (9.59) can be viewed as two equations in the unknowns $m_{1,e}$ and $m_{1,s}$.

Equation (9.58) is accurate only for humidity measurements in water vapor–air mixtures at low temperatures, that is, the usual conditions encountered in air-conditioning practice. The effect of ignoring radiation heat transfer in the analysis is explored in Exercise 9–41. The wet-bulb temperature recorded by a psychrometer is termed the *psychrometric* wet-bulb temperature and is in general not the same as the adiabatic saturation temperature or *thermodynamic* wet-bulb temperature.

The thermodynamic wet-bulb temperature is the temperature at which water, by evaporating into air in a constant pressure-mixing process, can bring the air to saturation adiabatically at the same temperature. The essential point is that the required enthalpy of vaporization exactly equals the enthalpy given up by the air as it cools down from its initial temperature to the wet-bulb temperature. The psychrometric and thermodynamic wet-bulb temperatures can be taken to be equal when the Lewis number is unity.[6] The Lewis number for water vapor–air mixtures is somewhat greater than unity, so that the two temperatures are never exactly equal; however, the difference is usually small enough to be ignored. The psychrometric charts that are widely used by air-conditioning engineers are based on the thermodynamic wet-bulb temperature.

If the wet- and dry-bulb temperatures are known, the calculation of moisture content using Eqs. (9.58) and (9.59) is straightforward, as will be seen in Example 9.9. If the unknown is the wet-bulb temperature, however, an iterative calculation is required. Such calculations must be made accurately to obtain a satisfactory result and are tedious. Air-conditioning engineers use psychrometric charts to avoid the problem. Alternatively, the computer program PSYCHRO can be used for this purpose. The principles of psychrometry and psychrometric charts may be found in various textbooks and handbooks, for example, [7,8,9,10].

The Computer Program PSYCHRO

PSYCHRO calculates psychrometric and thermodynamic properties of water vapor and air mixtures. Values of three properties are required to specify the thermodynamic state of the mixture. In PSYCHRO these are total pressure and two of the following: dry-bulb temperature, wet-bulb temperature, dewpoint temperature, relative humidity, vapor mass fraction, and humidity ratio. Temperatures may be input in either kelvins or degrees Celsius. Upon input of three property values, PSYCHRO calculates and displays the remaining properties, together with the mixture enthalpy, density, and specific volume, as well as the vapor mole fraction (see Table 9.4).

---

[6] Proof of this statement is left to more advanced texts.

Table 9.4 Psychrometric and thermodynamic properties of water vapor–air mixtures in PSYCHRO.

Symbol	Property	Units	Definition
P	Total pressure	Pa	$P = P_1 + P_2$, where P_1 and P_2 are the partial pressures of water vapor and dry air, respectively
T_{dry}	Dry-bulb temperature	K or °C	Mixture temperature
T_{wet}	Wet-bulb temperature	K or °C	Thermodynamic wet-bulb temperature
T_{dew}	Dewpoint temperature	K or °C	Saturation temperature corresponding to the partial pressure of water vapor in the mixture
RH	Relative humidity	%	$P_1/P_{sat}(T_{dry})$
m_1	Mass fraction of water vapor		$m_1 = \dfrac{\rho_1}{\rho} = \dfrac{\rho_1}{\rho_1 + \rho_2}$
ω	Humidity ratio		$\omega = \dfrac{\rho_1}{\rho_2} = \dfrac{m_1}{1 - m_1}$; also called the specific humidity ϕ
h	Mixture enthalpy	J/kg	$h = m_1 h_1 + m_2 h_2$ Datum states: $h_1 = 0$ for liquid water at 0°C; $h_2 = 0$ for air at 0°C
ρ	Mixture density	kg/m³	$\rho = \rho_1 + \rho_2$; $\rho_1 = \dfrac{P_1 M_1}{\mathscr{R} T_{dry}}, \rho_2 = \dfrac{P_1 M_2}{\mathscr{R} T_{dry}}$
v	Mixture specific volume	m³/kg	$v = \dfrac{1}{\rho}$
x_1	Mole fraction of water vapor		$x_1 = \dfrac{c_1}{c} = \dfrac{m_1/M_1}{m_1/M_1 + m_2/M_2}$

Two options for making subsequent changes are available:

1. Fixed total pressure and dry-bulb temperature
2. Fixed total pressure and water vapor concentration (mass fraction and humidity ratio)

PSYCHRO is based on data given in the ASHRAE *Handbook of Fundamentals* [10].

EXAMPLE 9.9 Humidity of an Air Flow

Air at 1010 mbar flows over a wet- and dry-bulb psychrometer. The dry bulb measures 31.8°C, and the wet bulb measures 26.8°C. Determine the relative humidity and humidity ratio.

Solution

Given: Wet- and dry-bulb temperatures of an air stream.

Required: Relative humidity and humidity ratio.

Assumptions: The air velocity is sufficiently high for radiation effects to be negligible.

Equation (9.58) gives the mass fraction of the water vapor in the air stream:

$$m_{1,e} = m_{1,s} - \frac{c_{p,\text{air}}}{1.08h_{\text{fg}}}(T_e - T_s)$$

$$T_e = 305.0 \text{ K}; \qquad T_s = 300.0 \text{ K}$$

From Table A.12*a*, $P_{1,s} = P_{\text{sat}}(300 \text{ K}) = 3533$ Pa.

$$x_{1,s} = P_{1,s}/P = 3533/101{,}000 = 0.0350$$

$$m_{1,s} = \frac{18x_{1,s}}{18x_{1,s} + 29x_{2,s}} = \frac{x_{1,s}}{x_{1,s} + (29/18)(1 - x_{1,s})}$$

$$= \frac{0.0350}{0.0350 + 1.61(1 - 0.0350)} = 0.0220$$

Also from Table A.12*a*, $h_{\text{fg}}(T_s) = 2.437 \times 10^6$ J/kg, and from Table A.7, $c_{p,\text{air}} = 1005$ J/kg K. Notice that the specific heat of air is almost constant at normal temperatures, and it is unnecessary to evaluate it at a precise reference temperature. Substituting in Eq. (9.58),

$$m_{1,e} = 0.0220 - \frac{1005}{(1.08)(2.437 \times 10^6)}(305 - 300) = 0.0220 - 0.0019 = 0.0201$$

The mole fraction of water vapor in the air is then

$$x_{1,e} = \frac{m_{1,e}/18}{m_{1,e}/18 + m_{2,e}/29} = \frac{m_{1,e}}{m_{1,e} + (18/29)(1 - m_{1,e})}$$

$$= \frac{0.0201}{0.0201 + 0.621(1 - 0.0201)} = 0.0320$$

and the partial pressure is

$$P_{1,e} = x_{1,e}P = (0.0320)(101{,}000) = 3230 \text{ Pa}$$

By definition, the relative humidity is $\text{RH} = P_{1,e}/P_{\text{sat}}(T_e)$; from Table A.12*a*, $P_{\text{sat}}(305.0 \text{ K}) = 4714$ Pa. Thus,

$$\text{RH} = P_{1,e}/P_{\text{sat}}(T_e) = 3230/4714 = 68.5\%$$

The humidity ratio ω is the mass of water vapor per mass of dry air:

$$\omega = \frac{\text{kg water vapor}}{\text{kg dry air}} = \frac{m_{1,e}}{1 - m_{1,e}} = \frac{0.0201}{1 - 0.0201} = 0.0205$$

Solution using PSYCHRO

The required input is:

Pressure, $P = 1.01 \times 10^5$

Dry-bulb temperature, $T_{\text{dry}} = 31.8°\text{C}$

Wet-bulb temperature, $T_{\text{wet}} = 26.8°\text{C}$

The output includes:

Relative humidity, RH $= 68.5\%$

Humidity ratio, $\omega = 0.0204$

Comments

1. Use MCONV to check $m_{1,e}$ and the energy balance at the s-surface of the wet bulb.

2. Compare these results to values given in a psychrometric chart.

EXAMPLE 9.10 Evaporation of a Water Droplet

A 50 μm–diameter water droplet initially at 315 K is injected into an air stream at 315 K, 1.050×10^5 Pa, and 50.5% RH. Estimate the droplet lifetime.

Solution

Given: A small water droplet.

Required: Lifetime after injection into an air stream.

Assumptions:
1. After a short initial transient the droplet temperature attains a constant value.
2. Radiation heat transfer is negligible.
3. The droplet is entrained into the air stream.

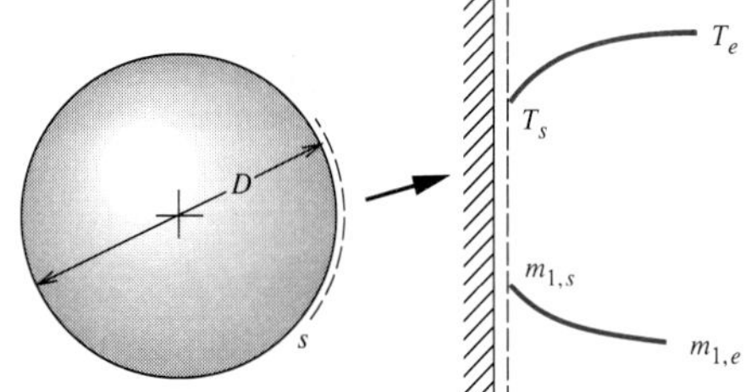

The above assumptions may appear to be rather bold! We will assume that they are valid and check later. Using Eq. (9.36), the evaporation rate of the droplet is

$$j_{1,s}A = g_{m1}(m_{1,s} - m_{1,e})A; \quad A = \pi D^2$$

If the droplets are entrained in the air flow, from Eq. (4.76) for Re $\to 0$, or from Eq. (4.78), the Sherwood number can be approximated by its lower-limit value of 2.

$$\text{Sh} = \frac{g_{m1}D}{\rho\mathcal{D}_{12}} \simeq 2; \qquad \text{hence } g_{m1} = \frac{2\rho\mathcal{D}_{12}}{D} \quad \text{and} \quad j_{1,s}A = \frac{2\rho\mathcal{D}_{12}A}{D}(m_{1,s} - m_{1,e})$$

A mass balance on a droplet requires that the rate of mass loss equal the evaporation rate:

$$\frac{d}{dt}\left(\frac{1}{6}\pi D^3\rho_l\right) = -j_{1,s}A$$

where ρ_l is the water density. Substituting for $j_{1,s}$ with $A = \pi D^2$, and differentiating,

$$\frac{dD}{dt} = -\frac{4\rho\mathcal{D}_{12}(m_{1,s} - m_{1,e})}{D\rho_l}$$

If D_0 is the initial diameter and τ is the droplet lifetime, then integrating with T_s constant gives

$$\int_{D_0}^{0} D\,dD = -\frac{4\rho\mathscr{D}_{12}(m_{1,s} - m_{1,e})}{\rho_l}\int_0^\tau dt$$

$$\frac{D_0^2}{2} = \frac{4\rho\mathscr{D}_{12}(m_{1,s} - m_{1,e})\tau}{\rho_l}$$

$$\tau = \frac{\rho_l D_0^2}{8\rho\mathscr{D}_{12}(m_{1,s} - m_{1,e})}$$

First we evaluate $m_{1,e}$; at 315 K, Table A.12a gives

$$P_{\text{sat}} = 0.08135 \times 10^5 \text{ Pa}$$

$$P_{1,e} = (\text{RH})P_{1,\text{sat}}(T_e) = (0.505)(0.08135 \times 10^5) = 0.0411 \times 10^5 \text{ Pa}$$

$$x_{1,e} = P_{1,e}/P = 0.0411/1.050 = 0.0391$$

$$m_{1,e} = \frac{0.0391}{0.0391 + (29/18)(1 - 0.0391)} = 0.0246$$

In order to evaluate $m_{1,s}$ we need to know the droplet surface temperature. After injection, the droplet will cool until the latent heat required for evaporation just balances the convective heat transfer from the air. For a precise result we should not use Eq. (9.58) to obtain T_s since the psychrometric wet-bulb temperature assumes a convective situation, whereas for an entrained droplet we have a purely diffusive situation with Nu = Sh = 2 and $(h_c/c_p)/g_{m1} = (\text{Pr}/\text{Sc})^{-1} \sim 1.13^{-1}$. Thus, Eq. (9.58) is replaced by

$$m_{1,s} = m_{1,e} + \frac{c_{p,\text{air}}}{1.13h_{\text{fg}}}(T_e - T_s) = 0.0246 + \frac{1006}{1.13h_{\text{fg}}(T_s)}(315 - T_s)$$

$$m_{1,s} = m_{1,s}(T_s, P)$$

Using Table A.12a and solving by iteration gives $T_s = 305.0$ K, $m_{1,s} = 0.0283$.

The properties required are as follows. At 305 K the density of water is $\rho_l = 995$ kg/m^3. Gas-phase properties are evaluated at the mean film temperature, $T_r = (1/2)(T_s + T_e) = (1/2)(305 + 315) = 310$ K. Using $\rho\mathscr{D}_{12} = \mu/\text{Sc}_{12}$, approximating μ as the pure air value of 18.87×10^{-6} kg/m s, and taking Sc = 0.61 for dilute water vapor–air mixtures gives $\rho\mathscr{D}_{12} = (18.87 \times 10^{-6})/0.61 = 3.09 \times 10^{-5}$ kg/m s.

$$\tau = \frac{\rho_l D_0^2}{8\rho\mathscr{D}_{12}(m_{1,s} - m_{1,e})} = \frac{(995)(50 \times 10^{-6})^2}{(8)(3.09 \times 10^{-5})(0.0283 - 0.0246)} = 2.72 \text{ s}$$

The various assumptions should now be checked.

1. *Duration of initial temperature transient*. The time required for the droplet to cool from its injection temperature of 315 K to 305 K can be estimated by calculating dT/dt from an energy balance at the average droplet temperature of 310 K.

$$\begin{matrix}\text{Rate of decrease in} \\ \text{droplet internal energy}\end{matrix} = \begin{matrix}\text{Rate at which latent} \\ \text{heat is supplied}\end{matrix} + \begin{matrix}\text{Sensible heat} \\ \text{transfer rate}\end{matrix}$$

$$-\rho_l(\pi D^3/6)c_v\frac{dT}{dt} = \pi D^2(q_{\text{evap}} + q_{\text{conv}})$$

The latent heat transfer is

$$q_{\text{evap}} = j_{1,s}h_{\text{fg}} = 2(\rho\mathscr{D}_{12}/D)(m_{1,s} - m_{1,e})h_{\text{fg}}$$

At 310 K, $P_{1,s} = 0.06224 \times 10^5$ Pa, $x_{1,s} = 0.0593$, $m_{1,s} = 0.0377$; $h_{\text{fg}} = 2.414 \times 10^6$ J/kg. Approximating $\rho\mathscr{D}_{12}$ as 3.09×10^{-5} kg/m s, as previously calculated,

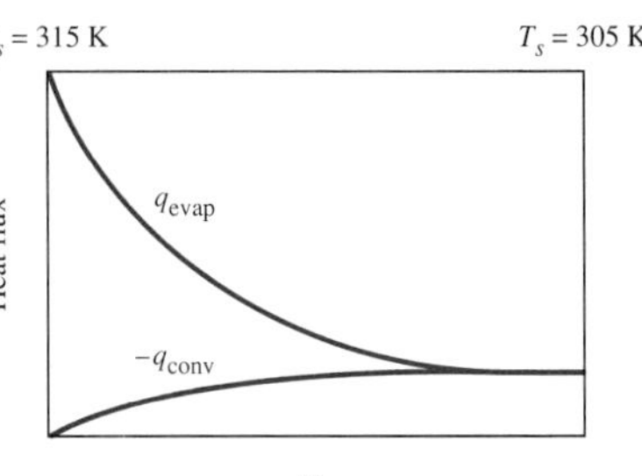

$$q_{\text{evap}} = 2[3.09 \times 10^{-5}/(50 \times 10^{-6})](0.0377 - 0.0246)(2.414 \times 10^6)$$
$$= 3.91 \times 10^4 \text{ W/m}^2$$

The sensible heat transfer is

$$q_{\text{conv}} = h_c(T_s - T_e)$$

$h_c = (k/D)\,\text{Nu}$, Nu $= 2$, $k = 0.0274$ W/m K for air at 310 K. Hence,

$$q_{\text{conv}} = 2[0.0274/(50 \times 10^{-6})](310 - 315) = -5.48 \times 10^3 \text{ W/m}^2$$

$$\frac{dT}{dt} = -(6/\rho_l c_v D)(q_{\text{evap}} + q_{\text{conv}})$$
$$= -[6/(995)(4174)(50 \times 10^{-6})](10^4)(3.91 - 0.548) = -971 \text{ K/s}$$

The duration of the 10 K temperature transient is of the order $10/971 \simeq 10^{-2}$ s, which is negligible compared to the lifetime of 2.7 s.

2. *Radiation heat transfer.* We will estimate the radiation heat transfer when the droplet is at 310 K. Assuming black surfaces for an upper bound,

$$q_{\text{rad}} = \sigma\varepsilon(T_s^4 - T_e^4) = (5.67 \times 10^{-8})(310^4 - 315^4) = -34.6 \text{ W/m}^2$$

which is two orders of magnitude less than the convection heat transfer, and is negligible.

3. *Droplet acceleration.* To estimate the time required for the droplet to be entrained, we apply Newton's second law of motion to the droplet:

$$\rho_l(\pi D^3/6)\frac{dV}{dt} = -C_D\left(\frac{1}{2}\rho V^2\right)(\pi D^2/4)$$

where V is the droplet velocity relative to the air stream. The drag coefficient C_D can be estimated from Stokes' law, Eq. (4.73), that is, $C_D = 24/\text{Re}_D$. Substituting and rearranging,

$$\frac{dV}{dt} = -\frac{18\mu}{\rho_l D^2}V$$

Integrating with $V = V_0$ at $t = 0$, and neglecting the change in D, gives

$$V = V_0 e^{-t/t_c}; \qquad t_c = \rho_l D^2/18\mu$$

where t_c is a time constant for the acceleration process. Using the same property values as before,

$$t_c = (995)(50 \times 10^{-6})^2/(18)(18.87 \times 10^{-6}) = 7.32 \times 10^{-3} \text{ s}$$

which is two orders of magnitude less than the evaporation time.

Comments

1. The fact that radiation has a negligible effect is due to the small size of the droplet. The convective heat transfer is inversely proportional to droplet size ($h_c = 2k/D$), while the radiation is independent of droplet size.
2. The small time constants for droplet cool-down and acceleration are also due to the small droplet size.
3. Use PSYCHRO to calculate the thermodynamic wet-bulb temperature and compare it to the value of T_s used in this example. Would it be an adequate approximation?
4. Use MCONV to check q_{conv} and q_{evap}.

9.5.3 Heterogeneous Combustion

Combustion is an important engineering problem in which the heat and mass transfer processes are coupled, and the equations governing each process often must be solved simultaneously. In diffusion-controlled combustion, the chemical kinetics are so fast that mass transfer considerations control the rate of combustion, while heat transfer considerations control the rate at which the heat of combustion can be removed, and hence the temperature of combustion. Since the transport properties, such as the diffusion coefficient, are temperature-dependent, the equations governing heat and mass transfer must be solved simultaneously. An example for a solid fuel will be analyzed below. More complex problems are the interesting phenomena of ignition and extinction of combustion, which are results of conflicting requirements of the mass transfer, chemical kinetics, and heat transfer involved. Such problems are dealt with in combustion texts.

We will consider the combustion of carbon particles in an air stream in the temperature range 1000–1600 K at 1 atm, for which the reaction is

$$C + O_2 \rightarrow CO_2$$

The combustion is diffusion-controlled; that is, the kinetics are so fast that the gas mixture at the s-surface is in chemical equilibrium with solid carbon. Equilibrium data for the reaction indicate that the resulting concentration of oxygen at the s-surface is essentially zero. Figure 9.20 shows a schematic of the situation. We will analyze the problem on a molar basis. The rate at which oxygen, species 1, diffuses across the s-surface is, from Eq. (9.37),

$$J_{1,s} = \mathcal{G}_{m1}(x_{1,s} - x_{1,e}) = -\mathcal{G}_{m1}x_{1,e} \qquad \textbf{(9.60)}$$

If the particles are spherical and small enough to be entrained in the air stream, the Sherwood number can be taken to be its lower-limit value of 2.

$$\text{Sh} = \frac{\mathcal{G}_{m1}D}{c\mathcal{D}_{1m}} = 2$$

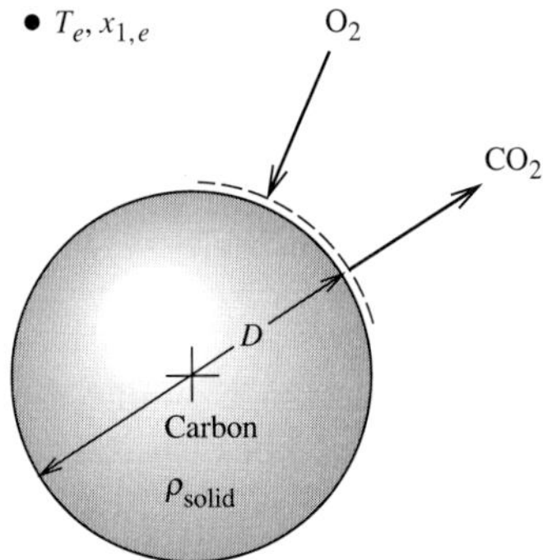

Figure 9.20 Schematic of a carbon particle oxidizing in an air stream.

Hence

$$\mathcal{G}_{m1} = \frac{2c\mathcal{D}_{1m}}{D}$$

and

$$J_{1,s} = -\frac{2c\mathcal{D}_{1m}x_{1,e}}{D} \tag{9.61}$$

Since one mole of carbon is consumed for each mole of oxygen crossing the s-surface, Eq. (9.61) also gives the rate at which carbon is consumed in kmol/m^2 s. A mass balance on the carbon particle requires that

$$\frac{d}{dt}\left(\frac{1}{6}\pi D^3 \rho_{\text{solid}}\right) = J_{1,s}M_C(\pi D^2) \tag{9.62}$$

where ρ_{solid} is the density of solid carbon and M_C is the molecular weight of element carbon. Differentiating and substituting from Eq. (9.61) then gives the rate of decrease of particle diameter as

$$\frac{dD}{dt} = -\frac{4c\mathcal{D}_{1m}M_C x_{1,e}}{\rho_{\text{solid}}D} \tag{9.63}$$

Since both c and $\mathcal{D}_{1m}$ depend on temperature, our next task is to obtain the particle temperature. If conduction into the particle is negligible, the surface energy balance requires that the rate at which heat is lost by convection and radiation equal the rate at which heat is liberated by the combustion reaction:

$$\dot{Q}_{\text{conv}} + \dot{Q}_{\text{rad}} = \dot{Q}_{\text{comb}}$$

or

$$h_c A(T_s - T_e) + \varepsilon\sigma A(T_s^4 - T_e^4) = (-J_{1,s})\Delta H_c A \tag{9.64}$$

where ΔH_c is the heat released per kmol of oxygen (or carbon) consumed. The expression for $\dot{Q}_{\text{rad}}$ is Eq. (1.18) for a small gray body in large surroundings, and assumes that the surroundings are at the same temperature as the gas. For an entrained spherical particle, the Nusselt number is also 2; thus $h_c = 2k/D$. Substituting and using Eq. (9.61) gives

$$\frac{2k}{D}(T_s - T_e) + \sigma\varepsilon(T_s^4 - T_e^4) = \frac{2c\mathscr{D}_{1m}x_{1,e}}{D}\Delta H_c \tag{9.65}$$

For a given particle diameter D, Eq. (9.65) can be solved for T_s by iteration. Alternatively, it makes more sense to specify a range of T_s values and solve directly for the corresponding values of D, as will be done in Example 9.11. Equation (9.65) shows that both q_{conv} and q_{comb} are inversely proportional to particle size, while q_{rad} is independent of particle size. Thus, as the particle burns and its size decreases, the radiation loss becomes relatively smaller and the combustion temperature increases. This feature is common to many types of particle combustion: the color brightness is seen to increase toward the end of the particle life.

The $c\mathscr{D}_{1m}$ product depends on temperature and hence particle size, so that Eq. (9.63) should be integrated numerically to obtain the particle lifetime. An approximate estimate can be obtained if a constant value of $c\mathscr{D}_{1m}$ is used. Rearranging Eq. (9.63) and integrating,

$$\int_{D_0}^{0} \frac{D\,dD}{c\mathscr{D}_{1m}} = -\frac{4M_C x_{1,e}}{\rho_{\text{solid}}} \int_0^{\tau} dt \tag{9.66}$$

where D_0 is the particle initial diameter and τ is its lifetime. Taking $c\mathscr{D}_{1m} = (c\mathscr{D}_{1m})_r$ at an appropriate reference temperature yields

$$\frac{1}{(c\mathscr{D}_{1m})_r}\frac{D_0^2}{2} = \frac{4x_{1,e}M_C\tau}{\rho_{\text{solid}}}$$

or

$$\tau = \frac{\rho_{\text{solid}}D_0^2}{8(c\mathscr{D}_{1m})_r M_C x_{1,e}} \tag{9.67}$$

Since $(1/c\mathscr{D}_{1m})_r$ is weighted with D in the integral, it should be evaluated at the temperature corresponding to a diameter somewhat larger than the mean value, perhaps $(0.5)^{1/2}D_0 \simeq 0.70D_0$.

Two comments close this section:

1. We chose to perform this analysis on a molar basis because the net rate of transfer of moles across the s-surface was zero: the molar fluxes of CO_2 and O_2 balance due to the stoichiometry of the reaction. Thus, low *mole* transfer rate theory applies and, in fact, is exact (as is the case of low *mass* transfer rate theory applied to catalysis). Had we analyzed the problem on a mass basis, low mass transfer rate theory would have been of adequate accuracy for most purposes, but not exact like the molar result.

2. In considering a small entrained particle so that the Sherwood and Nusselt numbers could be taken equal to 2, we have, in fact, analyzed a diffusion

rather than a convection problem. For larger particles there would be a relative velocity between the particle and air stream, and appropriate correlations should be used for Sh and Nu [e.g., Eq. (4.76)]. Then a numerical integration may be required to obtain the particle lifetime.

EXAMPLE 9.11 Low-Temperature Oxidation of a Carbon Particle

Hot carbon particles of initial diameter 300 μm are entrained into a gas stream containing 10% oxygen by volume at 500 K and 1 atm. Determine the particle temperature as a function of particle size, and hence estimate the time for a particle to be consumed. Take $\varepsilon = 0.9$ for the carbon and $\Delta H_c = 3.94 \times 10^8$ J/kmol carbon consumed.

Solution

Given: Carbon particles oxidizing in a gas stream containing 10% oxygen by volume.

Required: Particle temperature and lifetime.

Assumptions: 1. Diffusion-controlled oxidation, $C + O_2 \rightarrow CO_2$.
2. Gas properties can be approximated with air values.

We will assume that the particle temperature is in the range 1000–1600 K so that the analysis of Section 9.5.3 is valid. To avoid iteration, the particle energy balance, Eq. (9.65), is solved for particle diameter D:

$$D = \frac{(c\mathscr{D}_{1m}/k)x_{1,e}\Delta H_c - (T_s - T_e)}{(\sigma\varepsilon/2k)(T_s^4 - T_e^4)}$$

$T_e = 500$ K, $x_{1,e} = 0.10$, $\varepsilon = 0.9$, $\sigma = 5.67 \times 10^{-8}$ W/m² K⁴, and $\Delta H_c = 3.94 \times 10^8$ J/kmol. Substituting,

$$D = \frac{3.94 \times 10^7(c\mathscr{D}_{1m}/k) - (T_s - 500)}{(2.552/k)[(T_s \times 10^{-2})^4 - 625]}$$

Using $c = P/\mathscr{R}T = (1.0133 \times 10^5/8314T)$, and approximating $\mathscr{D}_{1m}$ as the value for O_2 in air from Table A.17*a* and k as the value for air from Table A.7, allows D to be obtained as a function of T_s, as given below. Properties are evaluated at the mean film temperature.

T_s, K	1000	1100	1200	1300	1400	1500	1600
D, μm	1420	824	470	252	128	45	—

The particle lifetime is given by Eq. (9.67) as

$$\tau = \frac{\rho_{\text{solid}} D_0^2}{8(c\mathscr{D}_{1m})_r M_C x_{1,e}}$$

where $(c\mathscr{D}_{1m})_r$ will be evaluated (somewhat arbitrarily) at a temperature corresponding to $0.7D_0 = 210$ μm. From the table, the required values are $T_s \simeq 1320$ K and $T_r = (500 + 1320)/2 = 910$ K: $c = 0.0134$ kmol/m³, $\mathscr{D}_{1m} = 129 \times 10^{-6}$ m²/s. Then, with $\rho_{\text{solid}} = 1810$

kg/m^3 from Table A.1,

$$\tau = \frac{(1810)(300 \times 10^{-6})^2}{8(0.0134)(129 \times 10^{-6})(12)(0.10)} = 9.82 \text{ s}$$

Comments

1. The particle temperature is in the range 1000–1600 K, so that the reaction regime is diffusion-controlled, as assumed.
2. By performing additional calculations, investigate the sensitivity of particle temperature to parameters such as T_e and $x_{1,e}$.
3. Notice that the particle temperature is independent of pressure.

9.6 MASS TRANSFER IN POROUS CATALYSTS

Porous solids are widely used in mass transfer operations to allow a large surface area per unit volume for transfer. An example is the engineering application of catalysis, such as hydrodesulfurization in petroleum refining or removal of CO, NO, and unburnt hydrocarbons from automobile exhausts. In common use are fixed-bed reactors where the catalyst is in the form of porous pellets of size ranging from 1 to 15 mm in diameter. For CO oxidation, a pellet of copper oxide on alumina may be used; the alumina provides a suitable support for the catalyst, as it is relatively inert and structurally stable at high temperatures. In one manufacturing procedure the pellet is prepared by first impregnating alumina powder with copper nitrate solution. When the powder is heated to 700 K, the nitrate decomposes, evolving gases and leaving cupric oxide. Finally, the pellet is formed by molding at high pressure. The resulting pellet has a density in the range 600–1200 kg/m^3, and a surface area ranging from 100 to 400 m^2/g. A pore diameter of 1 μm is typical.

9.6.1 Diffusion Mechanisms

Three mechanisms of diffusion can occur in porous solids: *ordinary diffusion, Knudsen diffusion,* and *surface diffusion*. Ordinary diffusion of gaseous species, as described by Fick's law, dominates when the pores are large and the gas relatively dense. When the pores are small or the gas density low, the molecules collide with pore walls more frequently than with each other. Then diffusion of molecules along the pore is described by the equations for free molecule flow and is called Knudsen diffusion. At intermediate pressures and pore sizes both types of collisions play an important role. Molecules are also *adsorbed* on the walls of the pores: these molecules are mobile and will diffuse along the walls in the direction of decreasing *surface concentration*. Surface diffusion is the dominant mechanism of transport for the smallest pores, for which ordinary diffusion and Knudsen diffusion rates are very small. Mass transport can be further complicated by the presence of gradients of total pressure and temperature in the porous solid.

Notwithstanding the complexity just described, relative success has been achieved in the analysis of mass transport within catalyst pellets by assuming that transport is governed by a linear relation,

$$J_1 \simeq -c\mathscr{D}_{1,\text{eff}}\nabla x_1 \tag{9.68}$$

where the subscript "eff" denotes an effective diffusivity that accounts for the presence of the solid material. The phenomena of ordinary diffusion and Knudsen diffusion are accounted for in an approximate manner, by simply assuming additive resistances,

$$\frac{1}{\mathscr{D}_{1,\text{eff}}} = \frac{1}{\mathscr{D}_{12,\text{eff}}} + \frac{1}{\mathscr{D}_{K1,\text{eff}}} \tag{9.69}$$

Surface diffusion can be ignored for catalyst pellets since the pores are relatively large.[7] The first of the two effective diffusivities is taken to be

$$\mathscr{D}_{12,\text{eff}} = \frac{\varepsilon_v}{\tau}\mathscr{D}_{12} \tag{9.70}$$

where $\mathscr{D}_{12}$ is the binary diffusion coefficient of Fick's law; ε_v is the *porosity,* or volume void fraction, of the pellet and accounts for the reduction in cross-sectional area for diffusion posed by the solid material; and τ is the *tortuosity factor* and accounts for the increased diffusion length due to the tortuous paths of real pores, and for the effects of constrictions and dead-end pores. The second is

$$\mathscr{D}_{K1,\text{eff}} = \frac{\varepsilon_v}{\tau}\mathscr{D}_{K1} \tag{9.71}$$

where $\mathscr{D}_{K1}$ is the Knudsen diffusion coefficient for species 1, which is given by free molecule flow theory [11] as

$$\mathscr{D}_{K1} = \frac{2}{3}r_e\overline{v}_1 \tag{9.72}$$

where r_e is the effective pore radius and $\overline{v}_1$ is the average molecular speed of species 1,[8]

$$\overline{v}_1 = \left(\frac{8\mathscr{R}T}{\pi M_1}\right)^{1/2} \tag{9.73}$$

Substituting in the value for the gas constant gives a dimensional equation,

$$\mathscr{D}_{K1} = 97r_e\left(\frac{T}{M_1}\right)^{1/2} \ \text{m}^2/\text{s} \tag{9.74}$$

for r_e in meters and T in kelvins.

[7] Surface diffusion plays an important role when water vapor is adsorbed by regular-density silica gels in air-drying operations.

[8] Both c and $\overline{v}$ are commonly used symbols for the average (mean) molecular speed. In Chapter 9 we will use $\overline{v}$ to avoid confusion since c is used for molar concentration.

At atmospheric pressure, ordinary diffusion is dominant for pore radii larger than about 2 μm and Knudsen diffusion is dominant for pores smaller than about 0.2 μm. Notice that Eq. (9.69) ensures that $\mathscr{D}_{1,\text{eff}} \rightarrow \mathscr{D}_{12,\text{eff}}$ in large pores for which $\mathscr{D}_{K1,\text{eff}}$ is large, and $\mathscr{D}_{1,\text{eff}} \rightarrow \mathscr{D}_{K1,\text{eff}}$ in small pores for which $\mathscr{D}_{K1,\text{eff}}$ is small.

9.6.2 Effectiveness of a Catalyst Pellet

When a chemical reaction takes place within a porous pellet, a concentration gradient is set up, and surfaces on pores deep within the pellet are exposed to lower reactant concentrations than surfaces near the pore openings. Since the rate at which a chemical reaction proceeds is dependent on reactant concentration, the average reaction rate throughout a catalyst pellet will be less than if all the available catalyst surface were exposed to reactant at the concentration prevailing at the exterior of the pellet. Figure 9.21 shows a spherical catalyst pellet of a fixed-bed reactor, where it is immersed in a gas flow containing species 1 in dilute concentration. The catalyst promotes a reaction that consumes species 1; for example, the catalyst might be cupric oxide or platinum, which promotes the reaction

$$2CO + O_2 \rightarrow 2CO_2$$

in oxygen-rich automobile exhaust gases; CO can be species 1. The pellet has radius R and catalytic surface area per unit volume of pellet a_p [m^{-1}]. The reaction will be taken to be of first order; thus, the reaction consumes $k''cx_1$ moles of species 1 per unit area–unit time. The gas flow maintains the concentration of species 1 at the exterior of the pellet at the value $x_{1,s}$.

The Governing Equation and Boundary Conditions

In order to set up the differential equation governing the concentration distribution of species 1, we apply the principle of conservation of species to an elemental control

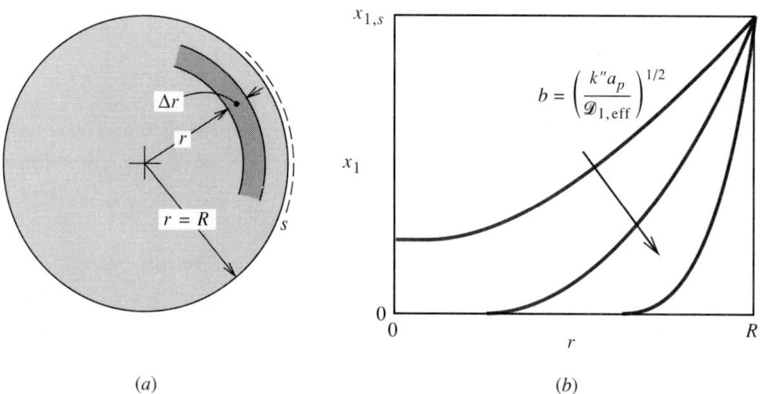

Figure 9.21 A spherical catalyst pellet. (*a*) Elemental control volume for application of the species conservation principle. (*b*) Concentration profiles.

volume located between spherical surfaces at radii r and $r + \Delta r$. At steady state

Net inflow of species 1 + Rate of production of species 1 = 0

$$[(4\pi r^2 J_1)_r - (4\pi r^2 J_1)_{r+\Delta r}] - k''cx_1 a_p 4\pi r^2 \Delta r = 0$$

Dividing by $4\pi\Delta r$ and letting $\Delta r \to 0$ gives

$$-\frac{d}{dr}(r^2 J_1) - r^2(k''cx_1 a_p) = 0$$

Substituting for J_1 from Eq. (9.68) and rearranging yields

$$\frac{1}{r^2}\frac{d}{dr}\left(r^2 c\mathscr{D}_{1,\text{eff}}\frac{dx_1}{dr}\right) - k''cx_1 a_p = 0$$

If the pellet is nearly isothermal and is at nearly constant total pressure, $c\mathscr{D}_{1,\text{eff}}$ may be assumed constant; then

$$\frac{1}{r^2}\frac{d}{dr}\left(r^2\frac{dx_1}{dr}\right) - \frac{k''a_p}{\mathscr{D}_{1,\text{eff}}}x_1 = 0 \tag{9.75}$$

This differential equation governs the concentration distribution of species 1; it must be solved subject to the boundary conditions

$r = 0$: x_1 bounded (not infinite)

$r = R$: $x_1 = x_{1,s}$

Solution for the Concentration Distribution and Reaction Rate

In order to simplify notation, we define $b = (k''a_p/\mathscr{D}_{1,\text{eff}})^{1/2}$ and then introduce a new dependent variable $\theta = rx_1$. Equation (9.75) becomes

$$\frac{d^2\theta}{dr^2} - b^2\theta = 0 \tag{9.76}$$

which has the general solution

$$\theta = C_1 \sinh br + C_2 \cosh br$$

or, since $\theta = rx_1$,

$$x_1 = \frac{C_1}{r}\sinh br + \frac{C_2}{r}\cosh br \tag{9.77}$$

Since $\sinh 0 = 0$ and $\cosh 0 = 1$, from the first boundary condition $C_2 = 0$. The constant C_1 is evaluated from the second boundary condition, and the resulting concentration distribution is

$$\frac{x_1}{x_{1,s}} = \frac{R}{r}\frac{\sinh br}{\sinh bR}; \qquad b = \left(\frac{k''a_p}{\mathscr{D}_{1,\text{eff}}}\right)^{1/2} \tag{9.78}$$

The rate at which species 1 is *consumed* within the catalyst pellet is given by $-4\pi r^2 J_1 \big|_{r=R}$, where

$$4\pi r^2 J_1\big|_{r=R} = -4\pi R^2 c\mathscr{D}_{1,\text{eff}} \left.\frac{dx_1}{dr}\right|_{r=R}$$

$$= 4\pi R c\mathscr{D}_{1,\text{eff}} x_{1,s}\left(1 - \frac{bR}{\tanh bR}\right) \tag{9.79}$$

The Pellet Effectiveness

If the diffusion coefficient were infinite, the complete internal surface would be exposed to reactant at concentration $x_{1,s}$; the rate of consumption would be

$$(4/3)\pi R^3 a_p k'' c x_{1,s}$$

The effectiveness $\eta_p (0 < \eta_p < 1)$ of the catalyst pellet is the ratio of the actual consumption rate divided by that for a pellet with an infinite diffusion coefficient:

$$\eta_p = \frac{-4\pi R c\mathscr{D}_{1,\text{eff}} x_{1,s}\left(1 - \dfrac{bR}{\tanh bR}\right)}{(4/3)\pi R^3 a_p k'' c x_{1,s}}$$

$$= \frac{3}{bR}\left(\frac{1}{\tanh bR} - \frac{1}{bR}\right) \tag{9.80}$$

Notice that this effectiveness is used in the same sense as the efficiency of a cooling fin, which is the ratio of actual heat transfer to that for a fin of infinite thermal conductivity. Following chemical engineering practice, we recast this result in final form by introducing the Thiele modulus Λ,

$$\Lambda = \frac{V_p}{S_p}\left(\frac{k'' a_p}{\mathscr{D}_{1,\text{eff}}}\right)^{1/2} = \frac{V_p}{S_p} b \tag{9.81}$$

where V_p is the pellet volume and S_p is the area of the pellet exterior. For a sphere, $V_p/S_p = (4/3)\pi R^3/4\pi R^2 = R/3$. Substituting in Eq. (9.80),

$$\eta_p = \frac{1}{\Lambda}\left(\frac{1}{\tanh 3\Lambda} - \frac{1}{3\Lambda}\right) \tag{9.82}$$

We know from the definition of effectiveness that when the Thiele modulus is small compared with unity, η_p approaches unity. When the Thiele modulus is large compared with unity, Eq. (9.82) reduces to

$$\eta_p \simeq \frac{1}{\Lambda} \tag{9.83}$$

For nonspherical pellets Eq. (9.82) may be applied approximately by using the appropriate value for V_p/S_p in the Thiele modulus. For example, a cylindrical pellet

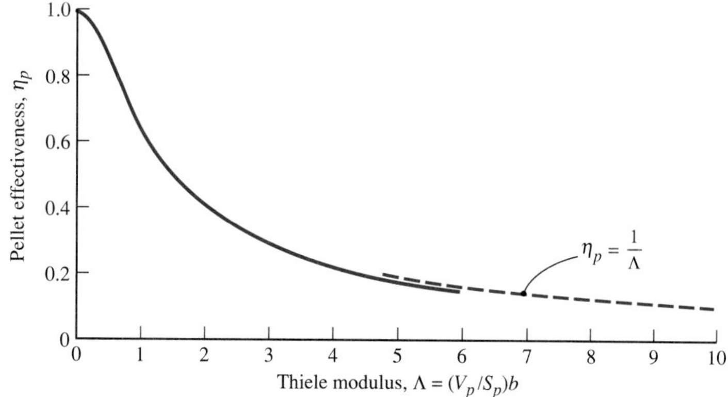

Figure 9.22 Pellet effectiveness as a function of the Thiele modulus, according to Eq. (9.82).

of radius R and length L will have

$$\frac{V_p}{S_p} = \frac{\pi R^2 L}{2\pi RL + 2\pi R^2} = \frac{RL}{2(L+R)}$$

Figure 9.22 shows a plot of pellet effectiveness versus Thiele modulus, as given by Eq. (9.82).

EXAMPLE 9.12 Effectiveness of a CuO-on-Alumina Pellet

The catalyst bed of an oxidation converter for an automobile is packed with 0.5 cm–diameter spherical pellets of CuO on alumina. The pellets have a volume void fraction $\varepsilon_v = 0.80$, tortuosity factor $\tau = 4.0$, average pore radius $r_e = 1\ \mu$m, and catalytic surface per unit volume $a_p = 7.12 \times 10^5$ cm^2/cm^3. Calculate the effectiveness of the pellet in promoting carbon monoxide oxidation at 800 K and 1 atm. The rate constant of the reaction is obtained from the chemical kinetics literature as $k'' = Ae^{-E_a/\mathcal{R}T}$, with $A = 50.6$ m/s and $E_a = 18.6$ kcal/mol.

Solution

Given: A 0.5 cm–diameter spherical catalyst pellet.

Required: Effectiveness for promoting CO oxidation at 800 K and 1 atm.

Assumptions:
1. The pellet is isothermal.
2. The reaction is of first order.
3. Equation (9.68) adequately describes diffusion inside the pellet.

Equation (9.82) gives the pellet effectiveness as

$$\eta_p = \frac{1}{\Lambda}\left[\frac{1}{\tanh 3\Lambda} - \frac{1}{3\Lambda}\right]$$

where $\Lambda = (V_p/S_p)(k''a_p/\mathscr{D}_{1,\text{eff}})^{1/2}$ is the Thiele modulus. We first calculate the effective diffusion coefficient $\mathscr{D}_{1,\text{eff}}$. From Eq. (9.70),

$$\mathscr{D}_{12,\text{eff}} = \frac{\varepsilon_v}{\tau}\mathscr{D}_{12} = \frac{0.8}{4.0}(1.06 \times 10^{-4}) = 2.12 \times 10^{-5} \text{ m}^2/\text{s}$$

where $\mathscr{D}_{12}$ is taken from Table A.17*a* as the value for a CO-air mixture at 800 K, 1 atm. From Eqs. (9.71) and (9.72),

$$\mathscr{D}_{K1,\text{eff}} = \frac{2}{3}\frac{\varepsilon_v}{\tau}r_e\bar{v}_1$$

$$r_e = 1\ \mu\text{m} = 10^{-6} \text{ m}$$

$$\bar{v}_1 = \left(\frac{8\mathscr{R}T}{\pi M_1}\right)^{1/2}$$

$$= \left\{\frac{(8)(8.314 \times 10^3 \text{ J/kmol K})(800 \text{ K})(1 \text{ N m/J})(1 \text{ kg m s}^{-2}/\text{N})}{(\pi)(28 \text{ kg/kmol})}\right\}^{1/2} = 778 \text{ m/s}$$

$$\mathscr{D}_{K1,\text{eff}} = \frac{2}{3}(0.8/4.0)(10^{-6} \text{ m})(778 \text{ m/s}) = 1.04 \times 10^{-4} \text{ m}^2/\text{s}$$

Then, from Eq. (9.69), the effective diffusion coefficient is

$$\frac{1}{\mathscr{D}_{1,\text{eff}}} = \frac{1}{\mathscr{D}_{12,\text{eff}}} + \frac{1}{\mathscr{D}_{K1,\text{eff}}} = \frac{1}{2.12 \times 10^{-5}} + \frac{1}{1.04 \times 10^{-4}}$$

$$\mathscr{D}_{1,\text{eff}} = 1.76 \times 10^{-5} \text{ m}^2/\text{s}$$

Next the rate constant is calculated:

$$k'' = Ae^{-E_a/\mathscr{R}T} \text{ m/s}$$

$$= 50.6\exp\left(-\frac{(18.6 \times 10^3 \text{ cal/mol})}{(1.987 \text{ cal/mol K})(800 \text{ K})}\right) = 50.6e^{-11.70} = 4.19 \times 10^{-4} \text{ m/s}$$

The product of the rate constant and catalytic surface area per unit volume is

$$k''a_p = (4.19 \times 10^{-4} \text{ m/s})(7.12 \times 10^5 \text{ cm}^2/\text{cm}^3)(10^2 \text{ cm/m}) = 2.98 \times 10^4 \text{ s}^{-1}$$

The Thiele modulus Λ may now be calculated:

$$\Lambda = \frac{R}{3}\left(\frac{k''a_p}{\mathscr{D}_{1,\text{eff}}}\right)^{1/2} = \frac{(0.25 \times 10^{-2} \text{ m})}{3}\left(\frac{2.98 \times 10^4 \text{ s}^{-1}}{1.76 \times 10^{-5} \text{ m}^2/\text{s}}\right)^{1/2} = 34.3$$

Since Λ is large, we may use Eq. (9.83) to obtain the effectiveness:

$$\eta_p \simeq \frac{1}{\Lambda} = \frac{1}{34.3} = 2.92 \times 10^{-2} \qquad \text{or } 2.92\%$$

Comments

1. Pellets are usually designed to have a low effectiveness, because in practice catalysts become "poisoned," reducing k''. For a new pellet, the reaction zone will be a thin

spherical shell just below the surface. As time goes on, the zone will move toward the center of the pellet as the outer surface area becomes poisoned.

2. An alternative catalyst support used in automobile catalytic converters is a ceramic matrix. The catalyst is impregnated into a thin porous alumina layer (*washcoat*) that is applied to the passage walls. A typical matrix has passages of hydraulic diameter about 1 mm, and the washcoat may be about 20 μm thick. Diffusion into the porous layer can then be analyzed in Cartesian coordinates (see Exercise 9–54).

9.6.3 Mass Transfer in a Pellet Bed

In Section 9.6.2 we were concerned only with diffusion inside the catalyst pellet. We now consider coupling between the transport of reactant to the exterior surface of the pellet on the one hand and transport within the pellet itself on the other, as occurs in a catalytic converter. We shall not be concerned with the overall performance of the converter; such matters belong more properly in Chapter 11, which deals with mass exchangers. Each pellet may be viewed as being surrounded by a gas flow containing a free-stream reactant concentration $x_{1,e}$. Mass transfer to the exterior surface of the pellet is described by Eq. (9.37), $J_{1,s} = \mathcal{G}_{m1}(x_{1,s} - x_{1,e})$, where $\mathcal{G}_{m1}$ is the mole transfer conductance and $x_{1,s}$ is the concentration of reactant at the s-surface shown in Fig. 9.23. As before, the pellet volume is V_p and the external surface is S_p; the catalytic area per unit volume is a_p. For a first-order reaction characterized by rate constant k'', the effectiveness at the specified temperature and pressure is η_p. We have two diffusive resistances in series, as shown in Fig. 9.24:

1. An outside convective resistance $R_o = 1/\mathcal{G}_{m1}S_p$ (based on the pellet exterior surface area S_p).

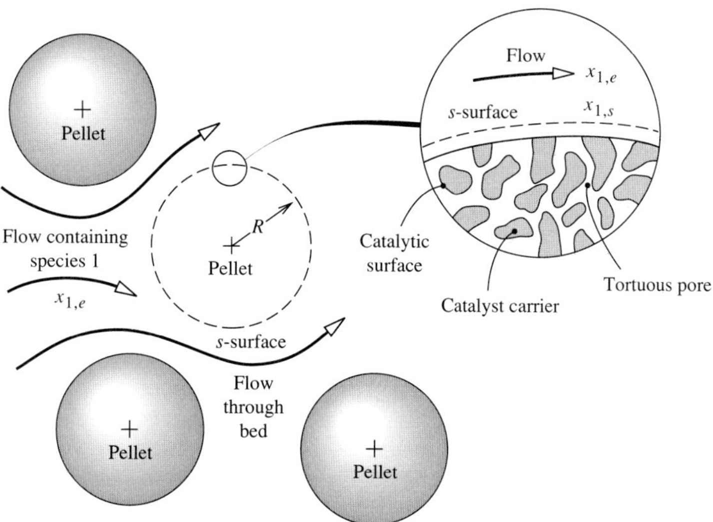

Figure 9.23 A porous catalyst pellet in a packed bed for the removal of species 1.

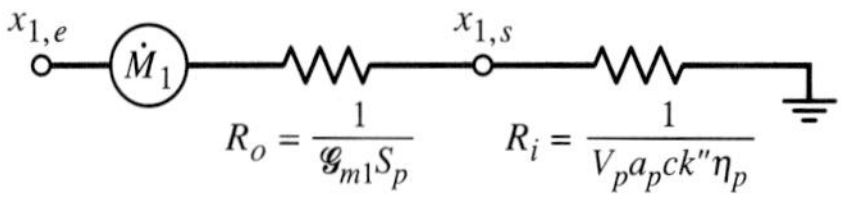

$$x_{1,e} \quad \dot{M}_1 \quad R_o = \frac{1}{\mathscr{G}_{m1}S_p} \quad x_{1,s} \quad R_i = \frac{1}{V_p a_p c k'' \eta_p}$$

Figure 9.24 Mass flow circuit for porous catalyst pellets in a fixed-bed catalytic reactor.

2. An inside resistance that accounts for both diffusion through the pores and the chemical reaction. The potential difference across the inside resistance is $x_{1,s}(= x_{1,s} - 0)$, and the molar current is $\dot{M}_1 = -J_{1,s}S_p = V_p a_p k'' c x_{1,s}\eta_p$. Since $R_i = x_{1,s}/\dot{M}_1$, we obtain $R_i = 1/V_p a_p c k'' \eta_p$.

The rate of reaction is

$$\dot{M}_1 = \frac{(x_{1,e} - 0)}{R_o + R_i} \tag{9.84}$$

Substituting for the resistances R_o and R_i and rearranging gives

$$\dot{M}_1 = \frac{x_{1,e}}{\dfrac{1}{V_p a_p \eta_p k'' c} + \dfrac{1}{\mathscr{G}_{m1}S_p}} \tag{9.85}$$

The equivalence of Eq. (9.85) and Eq. (9.35) is made clear if $\mathscr{G}_{m1}$ is replaced by $c\mathscr{D}_{12}/\delta$, where δ is the thickness of an equivalent stagnant film. Then

$$\dot{M}_1 = \frac{x_{1,e}}{\dfrac{1}{(V_p a_p/S_p)\eta_p k'' c S_p} + \dfrac{\delta}{c\mathscr{D}_{12}S_p}} \tag{9.86}$$

The quantity $(V_p a_p/S_p)$ is the area of catalyst per unit area of pellet exterior—that is, of s-surface—and η_p is the fraction of this area that is effective in promoting the reaction.

Equation (9.85) gives the rate of reaction for a single pellet. The rate of reaction per unit exterior surface area of the pellet is $\dot{M}_1/S_p$. When the pellets are packed in a bed, the surface area of the pellets can be obtained from either the specific surface area of the bed a or the transfer perimeter $\mathscr{P}$, both of which are defined in Section 4.5.2. Notice that in deriving Eq. (9.85), we have assumed $\mathscr{G}_{m1}$ and $x_{1,s}$ to be uniform around the pellet, in order to use the result for η_p obtained in Section 9.6.2. In a packed bed, both $\mathscr{G}_{m1}$ and $x_{1,s}$ will vary somewhat around the pellet; however, our result is adequate for engineering purposes.

EXAMPLE 9.13 A Pellet Bed for an Automobile Catalytic Converter

A catalytic converter attached to the exhaust system of an automobile is packed with 0.5 cm–diameter spherical CuO-on-alumina catalyst pellets. At a specific location in the bed the mole transfer conductance $\mathscr{G}_{m1}$ is 0.030 kmol/m^2 s, the pellet temperature is 800 K, and the free-stream concentration of CO is 5.0% by mass. Calculate the CO oxidation rate per unit volume if the bed void fraction is 30%.

Solution

Given: A packed-pellet-bed catalytic reactor to remove carbon monoxide from automobile exhaust.

Required: CO oxidation rate per unit volume of bed.

Assumptions: 1. Kinetics data for CO oxidation of CuO is as given in Example 9.12.
2. $\mathscr{G}_{m1}$ is uniform around the pellet.

The CO oxidation rate per unit volume is equal to the oxidation rate per unit pellet exterior area times the total surface area of pellets per unit volume. From Eq. (9.85) we have the oxidation rate of CO per unit pellet exterior area as

$$\frac{\dot{M}_1}{S_p} = \frac{x_{CO,e}}{\dfrac{1}{(V_p a_p/S_p)\eta_p k''c} + \dfrac{1}{\mathscr{G}_{m1}}}$$

For a spherical pellet,

$$\frac{V_p}{S_p} = \frac{R}{3} = \frac{0.25 \times 10^{-2}}{3} = 0.833 \times 10^{-3} \text{ m}$$

From Example 9.12,

$$\eta_p = 0.0292 \qquad \text{and} \qquad k''a_p = 2.98 \times 10^4 \text{ s}^{-1}$$

At 800 K and 1 atm,

$$c = \frac{(1.013 \times 10^5)}{(8.314 \times 10^3)(800)} = 1.523 \times 10^{-2} \text{ kmol/m}^3$$

Since N_2 predominates, we assume a mean molecular weight of 28 for the exhaust gases, $M_{CO} = 28$; thus, $x_{CO,e} = m_{CO,e} = 0.05$. Substituting above,

$$\begin{aligned}(V_p/S_p)\eta_p k''a_p c &= (0.833 \times 10^{-3} \text{ m})(0.0292)(2.98 \times 10^4 \text{ s}^{-1})(1.523 \times 10^{-2} \text{ kmol/m}^3) \\ &= 0.0110 \text{ kmol/m}^2 \text{ s}\end{aligned}$$

$$\frac{\dot{M}_1}{S_p} = \frac{0.05}{\dfrac{1}{0.0110} + \dfrac{1}{0.030}} = 4.02 \times 10^{-4} \text{ kmol/m}^2 \text{ s}$$

To find the total rate at which CO is oxidized per unit volume of the reactor, we multiply by the area-to-volume ratio A_s/V_{bed}. For N_p pellets in the bed,

$$\begin{aligned}\frac{A_s}{V_{\text{bed}}} &= \frac{N_p S_p}{V_{\text{bed}}} = \left(\frac{N_p V_p}{V_{\text{bed}}}\right)\left(\frac{S_p}{V_p}\right) = (1 - \varepsilon_{v,\text{bed}})\left(\frac{3}{R}\right) = (0.70)/(0.833 \times 10^{-3}) \\ &= 840 \text{ m}^2/\text{m}^3\end{aligned}$$

The CO consumed per unit volume is then

$$\frac{\dot{M}_1}{S_p}\frac{A_s}{V_{\text{bed}}} = (4.02 \times 10^{-4})(840) = 0.338 \text{ kmol/m}^3 \text{ s}$$

Comments

1. Correlations for convective transport and pressure drop in packed particle beds are given in Section 4.5.2.
2. The performance of the bed as a mass exchanger is analyzed in Section 11.3.1.

9.7 TRANSPORT PROPERTIES

The tables of transport property data given in Appendix A have proven useful for a wide range of engineering problem solving. However, in many situations the data are inadequate. For example, we often need binary diffusion coefficients for species pairs not listed in Tables A.17 and A.18 (which are restricted to air mixtures and aqueous solutions, respectively). We also often need the viscosity or thermal conductivity of a gas mixture of arbitrary composition. In this section we develop formulas that will be useful for such purposes. Very small particles (called aerosols when suspended in a gas) behave very much like large molecules. It is also useful to be able to estimate diffusion coefficients of such particles so as to exploit analogies between particle deposition and mass or heat transfer.

In Section 9.7.1 gas mixtures are considered. First, a kinetic theory model involving crude rigid spheres is used to give simple formulas for transport properties of pure species, and to give an understanding of the transport processes. Subsequently, the rigorous Chapman-Enskog kinetic theory and Lennard-Jones potential model are explained. Formulas are given for the viscosity and thermal conductivity of pure species, and for the binary diffusion coefficient. Then Wilke's mixture rules are given, which allow the viscosity and thermal conductivity of gaseous mixtures to be calculated. In Section 9.7.2, models for diffusion in dilute liquid solutions are briefly discussed, and selected empirical formulas for the binary diffusion coefficients are given. In Section 9.7.3, the phenonenon of Brownian motion of small particles is described, and formulas for the calculation of Brownian diffusion coefficients are given.

9.7.1 Gas Mixtures

Rigid-Sphere Kinetic Theory Model

We first consider a simple kinetic theory of gases in which the gas is uniform and the molecules are rigid spheres of diameter d. If molecule 1 is viewed as stationary, a collision will occur when molecule 2 moves into a sphere of radius d around molecule 1, as shown in Fig. 9.25. The area of this circle is termed the *collision cross section* $\sigma_c = \pi d^2$. The relative velocity of the molecules, $\Delta \mathbf{v} = \mathbf{v}_1 - \mathbf{v}_2$, is of the same order of magnitude as the average molecular speed, which may be obtained from a physics of thermodynamics text [11,12,13,14,15] as

$$\bar{v} = \left(\frac{8\ell T}{\pi m}\right)^{1/2} \tag{9.87}$$

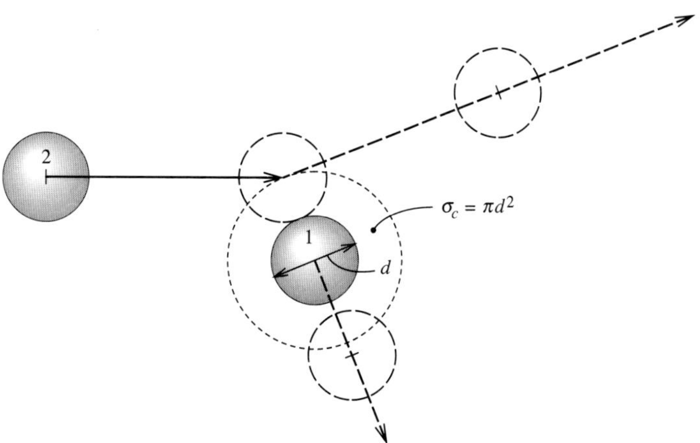

Figure 9.25 A molecule collision showing the collision cross section $\sigma_c = \pi d^2$.

where k is the Boltzmann constant and m is the mass of a molecule. It is quite simple to show that the average magnitude of the relative velocity between two colliding molecules is

$$\overline{\Delta v} = \sqrt{2}\bar{v} \tag{9.88}$$

During a time t, molecule 2 moves a distance $\overline{\Delta v} t$ relative to molecule 1. The number of collisions during this time is equal to the volume swept out, $\overline{\Delta v} t \pi d^2$, times the average number of molecules per unit volume, which is the number density $\mathcal{N}$. The average time between collisions is then

$$t_c = \frac{1}{\overline{\Delta v} \pi d^2 \mathcal{N}} = \frac{1}{\sqrt{2}\bar{v} \pi d^2 \mathcal{N}} \tag{9.89}$$

and during this time a molecule travels an average distance of

$$\ell = \bar{v} t_c = \frac{1}{\sqrt{2}\pi d^2 \mathcal{N}} = \frac{1}{\sqrt{2}\sigma_c \mathcal{N}} \tag{9.90}$$

which is the *collision mean free path*. Since $\mathcal{N} = P/kT$, ℓ is inversely proportional to pressure and to the collision cross section.

Detailed consideration of the flux of molecules per unit solid angle, and changes in the flux due to collisions, shows that the effective cross section affecting transport rates is

$$\sigma_t = (1 - \overline{\cos\theta})\sigma_c \tag{9.91}$$

where $\overline{\cos\theta}$ is the average angle that a molecule is deflected. If there were no deflection, $\theta = 0$, and we would not count the collision. If a collision resulted in perfect backscattering with $\theta = \pi$, then such a collision would have twice the effect as one that merely turned the molecule through $\pi/2$. For molecules of the same mass, $\overline{\cos\theta} = 2/3$, and if a light molecule of mass m_1 is scattered from a heavy molecule of mass m_2, $\overline{\cos\theta} = (2/3)(m_1/m_2)$ [16]. The *transport mean free path* is defined in terms of σ_t rather than σ_c,

$$\ell_t = \frac{1}{\sqrt{2}\sigma_t \mathcal{N}} \tag{9.92}$$

and is somewhat longer than the collision mean free path.

Self-Diffusion Coefficient

If the gas is a stationary isothermal mixture of two isotopes, say of uranium fluoride, the molecules have essentially the same mass m and diameter d. We wish to calculate the net transport of isotope 1 down its concentration gradient as shown in Fig. 9.26. The kinetic theory of gases gives the one-way flux of molecules across a plane in a uniform gas as (see also Section 7.6.1)

$$\mathcal{J} = \frac{1}{4}\bar{v}\mathcal{N} \tag{7.103}$$

The molecules of isotope 1 crossing the plane in the positive z direction come, on average, from a distance $b\ell_t$ below the plane, where b is of order magnitude unity; thus

$$\mathcal{J}_1^+ = \frac{1}{4}\bar{v}\mathcal{N}_1\big|_{z-b\ell_t} \tag{9.93a}$$

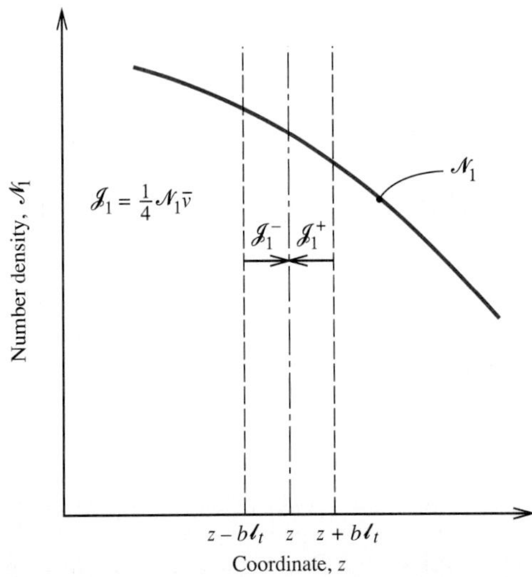

Figure 9.26 Diffusion of a gaseous isotope down a concentration gradient.

The molecules of species 1 crossing the plane in the negative z direction come from an average distance $b\ell_t$ above the plane,

$$\mathcal{J}_1^- = \frac{1}{4}\bar{v}\mathcal{N}_1\big|_{z+b\ell_t} \qquad \textbf{(9.93b)}$$

The net flux of molecules of species 1 crossing the plane is

$$\mathcal{J}_1 = \mathcal{J}_1^+ - \mathcal{J}_1^- = \frac{1}{4}\bar{v}\left[\mathcal{N}_1\big|_{z-b\ell_t} - \mathcal{N}_1\big|_{z+b\ell_t}\right] \qquad \textbf{(9.94)}$$

If the concentration gradient is small, a Taylor series expansion gives

$$\mathcal{N}_1\big|_{z-b\ell_t} = \mathcal{N}_1\big|_z + \frac{d\mathcal{N}_1}{dz}(-b\ell_t) + \cdots$$

$$\mathcal{N}_1\big|_{z+b\ell_t} = \mathcal{N}_1\big|_z + \frac{d\mathcal{N}_1}{dz}(+b\ell_t) + \cdots$$

Substituting in Eq. (9.94),

$$\mathcal{J}_1 = -\frac{b}{2}\bar{v}\ell_t\frac{d\mathcal{N}_1}{dz} \qquad \textbf{(9.95)}$$

Dividing by Avogadro's number $\mathcal{A}$ gives the molar flux of species 1 as

$$J_1 = -\frac{b}{2}\bar{v}\ell_t c\frac{dx_1}{dz} \qquad \textbf{(9.96)}$$

where the molar concentration c has been taken to be constant for the isothermal isotope mixture. Comparing Eq. (9.96) with Fick's law, Eq. (9.18), gives the self-diffusion coefficient

$$\mathcal{D}_{11} = \frac{b}{2}\bar{v}\ell_t \qquad \textbf{(9.97)}$$

Since molecules crossing the plane come from all angles, it is clear that $b < 1$; more careful analysis shows that an appropriate value of b for isotopes is 2/3 [15]. Then

$$\mathcal{D}_{11} = \frac{1}{3}\bar{v}\ell_t = \frac{1}{3}\left(\frac{8kT}{\pi m}\right)^{1/2}\left(\frac{1}{\sqrt{2}\sigma_t\mathcal{N}}\right) \qquad \textbf{(9.98)}$$

Substituting $k/m = \mathcal{R}/M$, $\mathcal{N} = c\mathcal{A} = P\mathcal{A}/\mathcal{R}T$, $\sigma_t = (1-\overline{\cos\theta})\pi d^2$, $\overline{\cos\theta} = 2/3$, gives

$$\mathcal{D}_{11} = 2\left(\frac{\mathcal{R}}{\pi}\right)^{3/2}\frac{T^{3/2}}{\mathcal{A}M^{1/2}d^2P} \qquad \textbf{(9.99)}$$

Equation (9.99) shows that the self-diffusion coefficient is proportional to temperature to the three-halves power, is inversely proportional to pressure, and decreases as the molecular weight and molecular diameter increase. However, the effective molecular diameter as it affects the collision process actually decreases with increasing temperature before becoming constant at high temperatures: thus the $T^{3/2}$ dependence is valid only at high temperatures.

Viscosity

If the gas is in simple shear flow, as shown in Fig. 9.27, molecules coming from below the plane at z have, on average, the velocity u at location $z - b\ell_t$. The momentum flux upward is

$$\mathscr{J}^+ m\left(u|_z - \frac{du}{dz}b\ell_t + \cdots\right)$$

The momentum flux downward is similarly

$$\mathscr{J}^- m\left(u|_z + \frac{du}{dz}b\ell_t + \cdots\right)$$

There is no net flow in the z direction, so $\mathscr{J}^+ = \mathscr{J}^- = (1/4)\overline{v}\mathscr{N}$. Then, consistent with the sign convention introduced in Section 5.2.1, the shear stress at the plane z is the difference between the upward and downward momentum fluxes:

$$\tau_{xz} = \frac{1}{4}\overline{v}\mathscr{N}m\left[-2b\ell_t\frac{du}{dz}\right] = -\frac{b}{2}\overline{v}\rho\ell_t\frac{du}{dz} \qquad \textbf{(9.100)}$$

Comparison with Newton's law of viscosity, $\tau = -\mu(du/dz)$, gives the viscosity as

$$\mu = \frac{b}{2}\overline{v}\rho\ell_t \qquad \textbf{(9.101)}$$

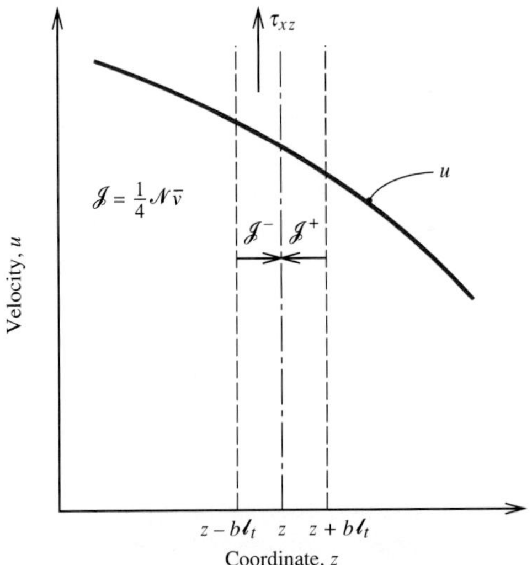

Figure 9.27 Transport of momentum in a simple shear flow of a gas.

Comparison with Eq. (9.97) shows that $\mu = \rho\mathscr{D}_{11}$, that is, the Schmidt number $\mathrm{Sc}_{11} = \mu/\rho\mathscr{D}_{11} = 1$. It follows that viscosity has the same pressure and temperature dependence as the $\rho\mathscr{D}$ product. If we again take $b = 2/3$, since the gas is uniform,

$$\mu = \frac{1}{3}\bar{v}\rho\ell_t \tag{9.102}$$

Equation (9.102) is the basis of the most convenient method for estimating the transport mean path; solving for ℓ_t,

$$\ell_t = 3\nu/\bar{v} = \nu\left(\frac{9\pi M}{8\mathscr{R}T}\right)^{1/2} \tag{9.103}$$

Thermal Conductivity

To complete this simple kinetic theory, we consider a uniform gas of monatomic molecules with an imposed temperature gradient, as shown in Fig. 9.28. Each molecule has, on an average, a kinetic energy of $(3/2)\mathscr{k}T$; however, the more energetic molecules cross the plane more frequently so that the energy transport per molecule proves to be $2\mathscr{k}T$ [15]. The net energy transport across the plane is

$$q = \frac{1}{4}\bar{v}\mathscr{N}\left[(2\mathscr{k}T)_{z-b\ell_t} - (2\mathscr{k}T)_{z+b\ell_t}\right] = -\frac{b}{2}\bar{v}\mathscr{N}\ell_t(2\mathscr{k})\frac{dT}{dz} \tag{9.104}$$

Comparing Eq. (9.104) with Fourier's law of heat conduction, $q = -k(dT/dz)$ with $b = 2/3$ gives the thermal conductivity as

$$k = \frac{1}{3}\bar{v}\rho\ell_t(2\mathscr{R}/M) \tag{9.105}$$

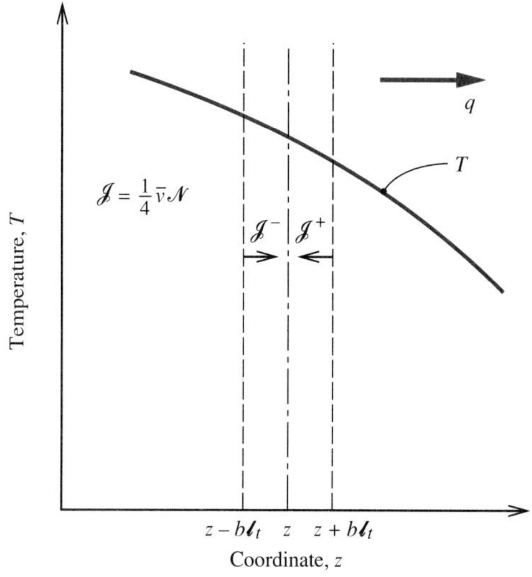

Figure 9.28 Transport of energy in a gas by monatomic molecules.

where the relations $\mathcal{N}m = \rho$ and $\mathcal{R}/M = k/m$ have been used. The ratio of conductivity to viscosity is then $2\mathcal{R}/M$. However, this rigid-sphere model result is not very accurate: more rigorous kinetic theory of gases gives this ratio as $5/2\mathcal{R}M$, which is higher because the faster, more energetic molecules have a longer-than-average mean free path. In the case of a polyatomic gas, transport of energy associated with the internal degrees of freedom must also be considered.

The Chapman-Enskog Kinetic Theory

The simple kinetic theory presented above provides a model, on a molecular scale, for physical reasoning about transport mechanisms in gases. The Chapman-Enskog kinetic theory of gases [17,18] is based both on a more realistic physical model and a rigorous mathematical formulation and solution. Some features of the theory that give an indication of the range of validity of the results are as follows. The density of the mixture must be low enough for three body collisions to occur with negligible frequency. Except at very low temperatures, pressures less than 100 atm are low enough to ensure that this condition is met. The model assumes monatomic molecules, but little error is introduced by applying the results to polyatomic gases. The viscosity and mass diffusivity are not appreciably affected by internal degrees of freedom. The thermal conductivity depends on the energy of both the translational and internal degrees of freedom: the so-called *Eucken correction* is introduced to account for the additional contribution of rotational and vibrational modes. Finally, the theory neglects higher than first-order spatial derivatives of temperature, pressure, concentration, and so on. Thus, the results are inapplicable when gradients change abruptly, for example, within a shock wave.

The forces acting between a pair of molecules during a collision are characterized by a potential energy of interaction ϕ: the functional form of ϕ is illustrated in Fig. 9.29. An empirical representation of the potential energy function that has proven

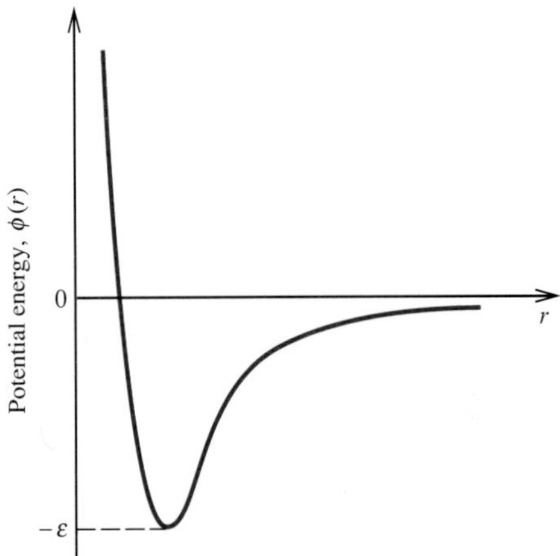

Figure 9.29 Potential energy for interaction as given by the Lennard-Jones 6–12 model.

fairly successful is the Lennard-Jones 6–12 potential model,

$$\phi(r) = 4\varepsilon\left[\left(\frac{\sigma}{r}\right)^{12} - \left(\frac{\sigma}{r}\right)^{6}\right] \tag{9.106}$$

where σ, the *collision diameter,* is the value of r for which $\phi(r) = 0$, and ε is the maximum energy of attraction between a pair of molecules. The model exhibits weak attraction, due to London dispersion forces at large separations (like r^{-6}), and strong repulsion due to electron cloud overlapping, at small separations (nearly like r^{-12}). Table A.26 lists values of σ and ε for a number of chemical species. In cases where the Lennard-Jones model parameters are not given in Table A.26, they may be estimated using the following rules:

$$\sigma \simeq 8.33V_c^{1/3} \simeq 11.8V_b^{1/3} \tag{9.107}$$

$$\varepsilon/k \simeq 0.75T_c \simeq 1.21T_b \tag{9.108}$$

where subscripts c and b refer to the critical state and boiling point, respectively, and V has units [m^3/kmol]. Table A.27 lists boiling point data for selected species.

The Lennard-Jones model describes a spherically symmetrical force field and hence is intended for use with nonpolar, nearly symmetrical molecules (for example, O_2, He, CO). Indeed, the Chapman-Enskog theory is, strictly speaking, valid only for molecules with spherically symmetrical force fields. Molecules with appreciable dipole moments (for example, H_2O, NH_3) or that are highly elongated (for example, C_3H_6, n-C_6H_{14}) interact with potentials that are angle-dependent. For polar molecules the Stockmayer potential model, which adds an angle-dependent factor to the Lennard-Jones expression, has been successfully used. However, for many practical purposes, the Lennard-Jones potential has been found adequate even for polar and elongated molecules. The usual practice is to *determine* the parameters σ and ε by matching theoretical viscosity predictions with experimental data. In this way experimental viscosity data are extrapolated outside the original temperature range; also, the same values of σ and ε usually (but not always) prove to be the best available for the estimation of thermal conductivity and mass diffusivity.

When the Lennard-Jones potential is used in the Chapman-Enskog kinetic theory of gases, the viscosity of a pure monatomic gas is given by

$$\mu = 2.67 \times 10^{-6}\frac{\sqrt{MT}}{\sigma^2\Omega_\mu} \text{ kg/m s} \tag{9.109}$$

where T is in kelvins and σ is in angstroms (1 angstrom = 1 Å = 10^{-10} m). The quantity Ω_μ is the *collision integral* and is tabulated in Table A.28; it is a weak function of temperature, becoming very nearly constant at high temperatures.

The Chapman-Enskog theory shows that, for a monatomic gas, the relation between thermal conductivity and viscosity is [17]

$$k = \frac{5}{2}c_v\mu, \qquad \left(c_v = \frac{3}{2}\frac{\mathcal{R}}{M}\right)$$

Thus

$$k_{\text{translational}} = 8.32 \times 10^{-2} \frac{\sqrt{T/M}}{\sigma^2 \Omega_k} \text{ W/m K} \tag{9.110}$$

The collision integral for thermal conductivity is identical to that for viscosity, $\Omega_k = \Omega_\mu$. Equation (9.110) is valid for monatomic gases; for polyatomic gases we add the *modified Eucken correction* to account for the contribution of the internal degrees of freedom:

$$k = k_{\text{translational}} + 1.32\left(c_p - \frac{5}{2}\frac{\mathcal{R}}{M}\right)\mu \tag{9.111}$$

We see that data for specific heats are required in order to calculate the thermal conductivity of polyatomic gases.

The mass diffusivity, in the form of the binary diffusion coefficient, is

$$\mathcal{D}_{12} = 1.86 \times 10^{-7} \frac{\sqrt{T^3\left(\dfrac{1}{M_1} + \dfrac{1}{M_2}\right)}}{\sigma_{12}^2 \Omega_{\mathcal{D}} P} \text{ m}^2\text{/s} \tag{9.112}$$

where P is in atmospheres. The intermolecular potential field for a pair of unlike molecules, species 1 and 2, is approximated as

$$\phi_{12}(r) = 4\varepsilon_{12}\left[\left(\frac{\sigma_{12}}{r}\right)^{12} - \left(\frac{\sigma_{12}}{r}\right)^{6}\right] \tag{9.113}$$

The collision integral for mass diffusivity $\Omega_{\mathcal{D}}$ differs from Ω_μ; values of $\Omega_{\mathcal{D}}$ are tabulated in Table A.28. The Lennard-Jones parameters σ_{12} and ε_{12} must be obtained from the empirical relations

$$\sigma_{12} = \frac{1}{2}(\sigma_1 + \sigma_2) \tag{9.114}$$

$$\varepsilon_{12} = \sqrt{\varepsilon_1 \varepsilon_2} \tag{9.115}$$

Mixture Rules

Next follows a prescription of how the viscosity and thermal conductivity of a gas mixture can be estimated from the values for the pure species. C. R. Wilke [19] simplified the rigorous kinetic theory prediction by introducing elements of the rigid-sphere model into the final result. The formulas are simple and have proven to be quite adequate.

$$\mu_{\text{mix}} = \sum_{i=1}^{n} \frac{x_i \mu_i}{\sum_{j=1}^{n} x_j \Phi_{ij}} \tag{9.116}$$

$$k_{\text{mix}} = \sum_{i=1}^{n} \frac{x_i k_i}{\sum_{j=1}^{n} x_j \Phi_{ij}} \tag{9.117}$$

where

$$\Phi_{ij} = \frac{\left[1 + (\mu_i/\mu_j)^{1/2} (M_j/M_i)^{1/4}\right]^2}{\sqrt{8}[1 + (M_i/M_j)]^{1/2}} \tag{9.118}$$

The important feature of these formulas is that the weighting is essentially with mole (number) fraction, as we would expect from simple kinetic theory.

Effective Binary Diffusion Coefficient

In a binary system, ordinary diffusion takes place whenever there is a concentration gradient, and is in the direction of decreasing concentration. It is tempting to intuitively generalize this result to a multicomponent system. However, as shown in advanced texts, in a multicomponent system it is possible for a species to move by ordinary diffusion up its concentration gradient. In a gas mixture, a light species such as H_2 or He can be pushed up its concentration gradient by collisions with heavier molecules in the mixture diffusing in that direction. Unfortunately, the exact theory of multicomponent diffusion is algebraically complicated and difficult to use. Thus, whenever possible, we try to solve engineering problems using the effective binary diffusion approach.

Fick's law, as we have defined it, is strictly valid only for binary mixtures—hence the use of a binary diffusion coefficient $\mathscr{D}_{12}$. If species 1 is in small concentration, we may safely define an *effective binary diffusion coefficient* for species 1 diffusing through the mixture, which we have denoted $\mathscr{D}_{1m}$ (see, for example, Section 9.3.3). This diffusion coefficient is an appropriate average value for interaction with the various species in the mixture or solution. For gas mixtures, a simple kinetic theory argument suggests

$$\mathscr{D}_{1m} \simeq \frac{(1 - x_1)}{\sum_{i=2}^{n} (x_i/\mathscr{D}_{1i})}, \qquad x_1 \ll 1 \tag{9.119}$$

This formula has a number of features:

1. For a true binary mixture of species 1 and 2, it gives the correct result of $\mathscr{D}_{1m} = \mathscr{D}_{12}$.
2. Similarly, in the limit of all the diffusion coefficients being equal, it recovers $\mathscr{D}_{1m} = \mathscr{D}_{12} = \mathscr{D}_{13} = \cdots$.
3. The addition of a small amount of a light gas (say species 3) to the mixture, small enough to leave m_1 essentially unaffected, does not have a significant effect on $\rho\mathscr{D}_{1m}$, even though $\mathscr{D}_{13}$ is much larger than the other $\mathscr{D}_{1i}$ values.

The Computer Programs GASMIX and AIRSTE

The computer program GASMIX calculates thermodynamic and transport properties of ideal gas mixtures. There is a menu of 36 species. The required inputs are temperature, pressure, and the mass fractions of the component species. Molecular weights and density are calculated using Table A.20. Specific heat is calculated from curve fits of the data listed in Table A.29. Transport properties are calculated using the Lennard-Jones potential model parameters listed in Tables A.26 and A.28.

The calculation of properties of mixtures containing water vapor poses a special problem, owing to the peculiar nature of the H_2O molecule. GASMIX is adequate for calculating properties of many mixtures containing water vapor, such as the combustion products of hydrocarbon fuels. However, for the widely encountered water vapor–air system, more accurate results are usually desirable. The computer program AIRSTE has been prepared for this purpose and is based on empirical data.

EXAMPLE 9.14 Thermal Conductivity of Methane

Calculate the thermal conductivity of methane (CH_4) at a temperature of 350 K and 1 atm pressure.

Solution

Given: Methane gas at 350 K, 1 atm.

Required: Thermal conductivity.

Assumptions: The Lennard-Jones potential model is adequate.

As a first step, we calculate the thermal conductivity of methane as if it were a monatomic molecule, and its viscosity. From Table A.20, $M = 16.0$. From Table A.26, $\sigma = 3.758$ Å and $\varepsilon/k = 149$ K; hence $kT/\varepsilon = 350/149 = 2.35$. From Table A.28, $\Omega_\mu = \Omega_k = 1.114$. Using Eqs. (9.110) and (9.109),

$$k_{\text{translational}} = \frac{8.32 \times 10^{-2}\sqrt{350/16.0}}{(3.758)^2(1.114)} = 2.47 \times 10^{-2} \text{ W/m K}$$

$$\mu = \frac{2.67 \times 10^{-6}\sqrt{(16.0)(350)}}{(3.758)^2(1.114)} = 1.27 \times 10^{-5} \text{ kg/m s}$$

The actual thermal conductivity of CH_4 is then obtained from Eq. (9.111) and includes the contributions of both the translational and internal degrees of freedom:

$$k = k_{\text{translational}} + 1.32\left(c_p - \frac{5}{2}\frac{\mathscr{R}}{M}\right)\mu$$

The specific heat c_p can be obtained from thermodynamic tables; alternatively, since many polyatomic molecules do not have their vibrational degrees of freedom excited at normal temperatures, a simple formula for c_p is

$$c_p = (5 + N_r)\left(\frac{1}{2}\frac{\mathscr{R}}{M}\right)$$

where N_r is the number of rotational degrees of freedom (zero for a monatomic gas, two for a linear molecule, and three for a nonlinear one). Thus $N_r = 3$ for CH_4 and

$$c_p - \frac{5}{2}\frac{\mathscr{R}}{M} = \frac{3}{2}\frac{\mathscr{R}}{M} = \frac{(3)(8314)}{(2)(16.0)} = 779.4 \text{ J/kg K}$$

Hence $k = 2.47 \times 10^{-2} + 1.32(779.4)(1.27 \times 10^{-5}) = 3.78 \times 10^{-2}$ W/m K.

Comment

Compare this result with the value given by GASMIX.

EXAMPLE 9.15 Diffusivity of Sulfur Dioxide in Air

Calculate the binary diffusion coefficient of sulfur dioxide (SO_2) in air at a temperature of 500 K and 1.2×10^5 Pa pressure.

Solution

Given: A SO_2-air mixture at 500 K and 1.2×10^5 Pa.

Required: Binary diffusion coefficient.

Assumptions: The Lennard-Jones potential model is satisfactory.

Equation (9.112) gives the binary diffusion coefficient as

$$\mathscr{D}_{12} = 1.86 \times 10^{-7} \frac{\sqrt{T^3(1/M_1 + 1/M_2)}}{\sigma_{12}^2 \Omega_{\mathscr{D}} P} \text{ m}^2\text{/s}$$

Using Tables A.20 and A.26,

Species	M	σ [Å]	$\frac{\varepsilon}{k}$ [K]
SO_2	64.1	4.112	335
Air	29.0	3.711	79

$$\sigma_{12} = \frac{1}{2}(\sigma_1 + \sigma_2) = \frac{1}{2}(4.112 + 3.711) = 3.912$$

$$\frac{\varepsilon_{12}}{k} = \sqrt{\left(\frac{\varepsilon_1}{k}\right)\left(\frac{\varepsilon_2}{k}\right)} = \sqrt{(335)(79)} = 163$$

$$\frac{kT}{\varepsilon} = \frac{500}{163} = 3.07$$

From Table A.28, $\Omega_{\mathscr{D}} = 0.943$. Thus,

$$\mathscr{D}_{12} = 1.86 \times 10^{-7} \frac{\sqrt{500^3 (1/64.1 + 1/29.0)}}{(3.912)^2(0.943)(1.2 \times 10^5/1.0133 \times 10^5)} = 27.2 \times 10^{-6} \text{ m}^2/\text{s}$$

Comment

GASMIX gives $\mathscr{D}_{12} = 27.25 \times 10^{-6}$ m²/s.

EXAMPLE 9.16 Viscosity of a Gaseous Fuel

A gaseous fuel has a composition of 70% (by volume) propane (C_3H_8) and 30% isobutane (iso-C_4H_{10}). Calculate the viscosity of the mixture at 300 K and 2 atm pressure.

Solution

Given: Mixture of propane and isobutane at 300 K and 2 atm.

Required: Viscosity.

Assumptions: 1. The Lennard-Jones potential model can be used for the pure species.
2. Wilke's mixture rule is adequate.

First we calculate the viscosity of the pure species C_3H_8 and iso-C_4H_{10} at 300 K from Eq. (9.109):

$$\mu = 2.67 \times 10^{-6} \frac{\sqrt{MT}}{\sigma^2 \Omega_\mu} \text{ kg/m s}$$

Using Tables A.20, A.26, and A.28:

Species	M	σ [Å]	$\frac{\varepsilon}{k}$ [K]	$\frac{kT}{\varepsilon}$	Ω_μ
C_3H_8	44	5.118	237	1.266	1.416
Iso-C_4H_{10}	58	5.278	330	0.909	1.667

$$\mu_1 = \frac{2.67 \times 10^{-6} \sqrt{(44)(300)}}{(5.118)^2(1.416)} = 8.27 \times 10^{-6} \text{ kg/m s}$$

$$\mu_2 = \frac{2.67 \times 10^{-6} \sqrt{(58)(300)}}{(5.278)^2(1.667)} = 7.58 \times 10^{-6} \text{ kg/m s}$$

The mixture rule is given by Eq. (9.116):

$$\mu_{\text{mix}} = \sum_{i=1}^{n} \frac{x_i \mu_i}{\sum_{j=1}^{n} x_j \Phi_{ij}} = \frac{x_1 \mu_1}{x_1 \Phi_{11} + x_2 \Phi_{12}} + \frac{x_2 \mu_2}{x_1 \Phi_{21} + x_2 \Phi_{22}}$$

where

$$\Phi_{ij} = \frac{\left[1 + \left(\frac{\mu_i}{\mu_j}\right)^{1/2}\left(\frac{M_j}{M_i}\right)^{1/4}\right]^2}{\sqrt{8}[1 + (M_i/M_j)]^{1/2}}; \qquad (\Phi_{11} = \Phi_{22} = 1)$$

$$\Phi_{12} = \frac{\left[1 + \left(\frac{8.27 \times 10^{-6}}{7.58 \times 10^{-6}}\right)^{1/2}\left(\frac{58}{44}\right)^{1/4}\right]^2}{\sqrt{8}[1 + (44/58)]^{1/2}} = 1.20$$

$$\Phi_{21} = \frac{\left[1 + \left(\frac{7.58 \times 10^{-6}}{8.27 \times 10^{-6}}\right)^{1/2}\left(\frac{44}{58}\right)^{1/4}\right]^2}{\sqrt{8}[1 + (58/44)]^{1/2}} = 0.83$$

Then with the mole fractions $x_1 = 0.7$ and $x_2 = 0.3$ the mixture viscosity is

$$\mu_{mix} = \frac{(0.7)(8.27 \times 10^{-6})}{(0.7)(1) + (0.3)(1.20)} + \frac{(0.3)(7.58 \times 10^{-6})}{(0.7)(0.83) + (0.3)(1)} = 8.04 \times 10^{-6} \text{ kg/m s}$$

Comments

1. Notice that viscosity is independent of pressure at normal pressures.
2. More accurate values of the molecular weights are 44.1 and 58.1 for C_3H_8 and iso-C_4H_{10}, respectively. However, the mixture rule is not of high accuracy, so the values used in the calculations are quite adequate.

9.7.2 Dilute Liquid Solutions

Table A.18 gives Schmidt numbers for various species in dilute solution in water, from which diffusion coefficients can be calculated in a simple manner. We now present formulas for diffusion coefficients in liquids of more general applicability. Unfortunately, there is no theory of diffusion in liquids of rigor comparable to the Chapman-Enskog kinetic theory of gases. An approximate theory that has been reasonably successful is based on a hydrodynamic interpretation of the *Nernst-Einstein equation,* which was originally used to describe Brownian motion of particles in liquids. The extrapolation to diffusion of molecules is surprisingly good. In its most fundamental form, the Nernst-Einstein equation is

$$\mathscr{D}_{12} = \frac{kT}{f_{12}} \tag{9.120}$$

where f_{12} [N s/m] is the *friction coefficient* of solute 1 in solvent 2, and is defined as the force required to give a molecule of species 1 unit velocity. If f_{12} is evaluated assuming creeping flow (Re $\ll$ 1) and no slip, Stokes' law, Eq. (4.73), applies and

$$\mathscr{D}_{12} = \frac{kT}{6\pi\mu_2 R_1} \tag{9.121}$$

where μ_2 is the viscosity of the pure solvent and R_1 is the radius of the diffusing molecule. Equation (9.121) shows that $\mathscr{D}_{12}$ should be inversely proportional to solvent viscosity and molecule size, and directly proportional to absolute temperature (in addition to the temperature dependence of μ_2). Equation (9.121) can be refined by allowing for slip between the molecule and solvent, but the result must still be empirically modified to give good agreement with measured values. Equation (9.121) is usually called the *Stokes-Einstein relation*. Wilke and Chang [20] proposed a useful formula, valid for dilute solutions of nondissociating solute 1 in solvent 2,

$$\mathscr{D}_{12} = 1.17 \times 10^{-16} \frac{(\phi_2 M_2)^{1/2} T}{\mu \tilde{V}_{1b}^{0.6}} \text{ m}^2/\text{s} \tag{9.122}$$

where $\tilde{V}_{1b}$ is the molar volume of the solute [m^3/kmol] as liquid at its normal boiling point, μ is the viscosity of the solution [kg/m s], and ϕ_2 is an "association parameter" for the solvent. Recommended values of ϕ_2 are 2.6 for water, 1.9 for methanol, 1.5 for ethanol, and 1.0 for heptane, ether, benzene, and other unassociated solvents. Equation (9.122) is usually accurate to $\pm 10\%$. For dilute aqueous solutions, the Othmer-Thakar [21] correlation is accurate to about 11% and is particularly simple:

$$\mathscr{D}_{12} = \frac{1.11 \times 10^{-13}}{\mu_2^{1.1} \tilde{V}_{1b}^{0.6}} \text{ m}^2/\text{s} \tag{9.123}$$

Table A.27 lists values of molar volumes at the normal boiling point for various liquids. Alternatively, Benson's formula can be used,

$$\tilde{V}_c / \tilde{V}_b = \rho_b / \rho_c = 0.422 P_c \text{ [atm]} + 1.981 \tag{9.124}$$

where the subscript c refers to the critical state. Tables of critical-state properties are readily available or they can be estimated. Another popular method for determining $\tilde{V}_b$ is to use the additive method of Le Bas; see, for example, Reid et al. [22].

The preceding recommendations are for *dilute* solutions only. The theory of diffusion in concentrated liquid solutions is beyond the scope of this text: it is difficult because (1) multicomponent diffusion must be considered, and (2) concentrated solutions are generally nonideal solutions, and the appropriate driving potential for diffusion is not concentration but the *chemical potential*. Thus, the subject of diffusion coefficients in concentrated solutions will not be discussed either. As a final comment on liquid diffusion coefficients, we note that all liquid diffusion coefficients are of the order 10^{-9}m^2/s at room temperature, irrespective of the nature of the solute or solvent.

There are no simple mixing rules similar to those for gases that allow the viscosity and thermal conductivity of a liquid solution to be obtained. Some complex methods of limited accuracy are available [22], but generally the engineer must rely on measured values. Table A.13 gives some examples.

EXAMPLE 9.17 Diffusivity of Oxygen and Chlorine in Water

Calculate the binary diffusion coefficients for oxygen and chlorine in dilute solution in water at 300 K using (i) the Wilke-Chang formula, (ii) the Othmer-Thakar formula, and (iii) Table A.18.

Solution

Given: O_2 and Cl_2 in dilute solution in water at 300 K.

Required: Binary diffusion coefficients using (i) the Wilke-Chang formula, (ii) the Othmer-Thakar formula, and (iii) Table A.18.

(i) Let species 2 be water. From Eq. (9.122), the Wilke-Chang formula is

$$\mathscr{D}_{12} = 1.17 \times 10^{-16} \frac{(\phi_2 M_2)^{1/2} T}{\mu \tilde{V}_{1b}^{0.6}} \ \mathrm{m^2/s}$$

Since the solution is dilute, μ can be taken as the value of viscosity for pure water; from Table A.8, $\mu = 8.67 \times 10^{-4}$ kg/m s. From Table A.27, $\tilde{V}_{1b} = 25.6 \times 10^{-3}$ and 48.4×10^{-3} m^3/kmol for oxygen and chlorine, respectively; also, for water $\phi_2 = 2.6$, $M_2 = 18$.

$$\text{Oxygen:} \quad \mathscr{D}_{12} = 1.17 \times 10^{-16} \frac{(2.6 \times 18)^{1/2}(300)}{(8.67 \times 10^{-4})(25.6 \times 10^{-3})^{0.6}} = 2.50 \times 10^{-9} \ \mathrm{m^2/s}$$

$$\text{Chlorine:} \quad \mathscr{D}_{12} = 1.17 \times 10^{-16} \frac{(2.6 \times 18)^{1/2}(300)}{(8.67 \times 10^{-4})(48.4 \times 10^{-3})^{0.6}} = 1.70 \times 10^{-9} \ \mathrm{m^2/s}$$

(ii) From Eq. (9.123), the Othmer-Thakar equation is

$$\mathscr{D}_{12} = \frac{1.11 \times 10^{-13}}{\mu_2^{1.1} \tilde{V}_{1b}^{0.6}}$$

$$\text{Oxygen:} \quad \mathscr{D}_{12} = \frac{1.11 \times 10^{-13}}{(8.67 \times 10^{-4})^{1.1}(25.6 \times 10^{-3})^{0.6}} = 2.34 \times 10^{-9} \ \mathrm{m^2/s}$$

$$\text{Chlorine:} \quad \mathscr{D}_{12} = \frac{1.11 \times 10^{-13}}{(8.67 \times 10^{-4})^{1.1}(48.4 \times 10^{-3})^{0.6}} = 1.59 \times 10^{-9} \ \mathrm{m^2/s}$$

(iii) Table A.18 gives Schmidt numbers for dilute solution in water at 300 K, and from Table A.8, $\nu = 0.87 \times 10^{-6}$ m^2/s for water at 300 K.

$$\text{Oxygen:} \quad \mathscr{D}_{12} = \frac{\nu}{\mathrm{Sc}_{12}} = \frac{0.87 \times 10^{-6}}{400} = 2.18 \times 10^{-9} \ \mathrm{m^2/s}$$

$$\text{Chlorine:} \quad \mathscr{D}_{12} = \frac{\nu}{\mathrm{Sc}_{12}} = \frac{0.87 \times 10^{-6}}{670} = 1.30 \times 10^{-9} \ \mathrm{m^2/s}$$

These results are summarized in the table below.

Source	$\mathscr{D}_{12}$, $m^2/s \times 10^9$	
	O_2-water	Cl_2-water
(i) Wilke-Chang	2.50	1.70
(ii) Othmer-Thakar	2.34	1.59
(iii) Table A.18	2.18	1.30

Comment

The discrepancies shown in the table reflect the expected uncertainty in liquid diffusion coefficient data.

9.7.3 Brownian Diffusion of Particles

Small particles play an important role in many areas of technology. When small particles are suspended in a gas, they are usually called *aerosols*. Two types of aerosols are usually identified. *Dispersion* aerosols are formed by the fracture or atomization of solids or liquids. *Condensation* aerosols result from the condensation of a vapor, or from reactions between gaseous species to form a nonvolatile product. Liquid aerosols are in the form of spherical droplets, but a wide variety of shapes are encountered for solid aerosols. Liquid droplet aerosols are usually called *mists,* solid dispersion aerosols are *dusts,* and solid condensation aerosols are *smokes*. Suspensions of small particles in liquids are also of concern; for example, scale formation in heat exchangers used in geothermal wells can result from deposition of silica particles. *Soot* particles play an important role in furnaces burning hydrocarbon fuels because of their large effect on radiation absorption and emission.

The size of spherical droplets can be characterized by the diameter d_p, but for irregular solid particles the size is more difficult to characterize: often the measure of particle size relates to the particular instrument used to determine the particle size distribution. Aerosol particles vary in size from molecular clusters of 10^{-3} μm size (diameter) to droplets and solid particles as large as 100 μm. Since the mean free path in air at 1 atm and 25°C is $\ell = 0.065$ μm, the particle *Knudsen number,* $\mathrm{Kn} = \ell/d_p$, varies from about 10^2 to 10^{-3}. As a result, the gas-particle interactions vary from being in the free molecule regime, through the transition regime, into the continuum regime. Inertial effects are very dependent on particle size: for a particle of unit specific gravity, inertial effects are unimportant for $d_p \leq 0.2$ μm and dominant for $d_p \geq 1.0$ μm. The larger particles settle out rapidly due to gravity, or larger particles in a flowing fluid cannot follow streamlines and thus impact on obstacles. For smaller particles, transfer from a flowing fluid to a surface is by diffusive mechanisms, whereas for larger particles it is primarily by sedimentation and impaction. Since the motion of large aerosol particles is governed by Newton's laws of motion, rather than by diffusion or convection, we will restrict our attention to smaller particles so as to be able to exploit the analogy to chemical species transfer and heat transfer.

The transport of small aerosol particles occurs in a similar manner to that of a molecular species. Small aerosols are convected by the mean flow, and in a turbulent flow they are transported by turbulent eddies in much the same way as a molecular species (of course, due to inertial effects, larger aerosols will not respond to the smallest turbulent fluctuations and may not follow streamlines in large turbulent eddies). Ordinary diffusion of aerosols is usually called *Brownian motion,* and, as might be expected, the Brownian diffusion coefficient decreases with increasing particle mass. Diffusion of aerosols due to a temperature gradient is called *thermophoresis* and often plays an important role. Concentration gradients of molecular species cause transport of aerosols by *diffusiophoresis*. *Forced diffusion* of a charged particle in an electrical field is similar to that for an ionized molecular species. Indeed, aerosol particles often carry electrostatic charges that affect both their transport and deposition. In this chapter attention is restricted to ordinary diffusion and convection. However, in Chapter 11 forced diffusion will be considered in connection with electrostatic precipitators for the removal of particulate matter from power plant and incinerator stack gases.

Brownian motion is governed by a Fick's-type law,

$$\mathcal{J} = -\mathcal{D}_p \nabla \mathcal{N} \quad \textbf{(9.125)}$$

where $\mathcal{J}$ [particles/m^2 s] is the particle flux, $\mathcal{N}$ [particles/m^3] is the particle number density, and $\mathcal{D}_p$ [m^2/s] is the Brownian diffusion coefficient. The Nernst-Einstein relation, introduced in Section 9.7.2 for diffusion of a solute in a liquid solution, relates the Brownian diffusion coefficient to the friction coefficient f as [23]

$$\mathcal{D}_p = \frac{kT}{f} \quad \textbf{(9.126)}$$

For particles much larger than the mean free path of the gas, the friction coefficient is obtained from the Stokes' drag law, Eq. (4.73), as

$$f = 3\pi\mu d_p; \qquad d_p \gg \ell \quad \textbf{(9.127)}$$

to give the Stokes-Einstein relation. However, Brownian diffusion is usually important for very small particles for which $d_p \simeq \ell$ or $d_p \ll \ell$. Cunningham introduced a *slip correction factor C* into Stokes' law to account for the slip that occurs between the rarefied gas and particle surface, to give

$$f = \frac{3\pi\mu d_p}{C} \quad \textbf{(9.128)}$$

Rarefied gas effects are characterized by the Knudsen number, which is the ratio of the mean free path to a characteristic length. For a particle of diameter d_p,

$$\mathrm{Kn} = \frac{\ell}{d_p} \quad \textbf{(9.129)}$$

An interpolation formula for the slip correction factor that has the proper form in the limits of free-molecule and Stokes flow is

$$C = 1 + \mathrm{Kn}\left(C_1 + C_2 e^{-C_3/\mathrm{Kn}}\right) \tag{9.130}$$

Based on experiment, Davies [24] recommends values of the constants for general use. If the mean free path for this purpose is defined[9] as

$$\ell = \frac{\nu}{0.499\bar{v}} \simeq \frac{2\nu}{\bar{v}} = \nu\left(\frac{\pi M}{2\mathscr{R}T}\right)^{1/2} \tag{9.131}$$

the values are $C_1 = 2.514$, $C_2 = 0.8$, and $C_3 = 0.55$. The precise values of these constants does depend on the nature of the molecule-particle collisions, and hence to some extent on physical properties of the particle material. Table A.30 gives values of C, $\mathscr{D}_p$, and $\mathrm{Sc}_p = \nu/\mathscr{D}_p$ for spherical particles in air at 1 atm and 300 K. For nonspherical particles, it is common practice to introduce a *dynamic shape factor* χ into Eq. (9.128),

$$f = \frac{3\pi\mu d_{pe}\chi}{C}; \qquad d_{pe} = (6m/\pi\rho_p)^{1/3} \tag{9.132}$$

where d_{pe} is the mass equivalent diameter. Table 9.5 gives some data for χ.

Brownian diffusion (or *motion*) also occurs in liquids, and indeed was discovered by Robert Brown in the early nineteenth century when he observed pollen grains in water [25]. It is usually in this context that the phenomenon is introduced in elementary physics texts. In a liquid, random molecular motion is on a length scale of the order of molecular size, and is thus much smaller than the particle size. Hence, for most purposes, Stokes' drag law can be used to obtain the friction coefficient f, as given by Eq. (9.127). *Macromolecules* (molecules with molecular weights of the order of 10,000 or greater) are important in many biological systems: for example, diffusion of proteins in aqueous solutions plays an important role in the life processes of animals and plants. Whether or not these macromolecules are modeled as solutes or as particles, the Stokes-Einstein relation applies. A useful approximation to this relation for macromolecules in dilute aqueous solutions is the Polson equation [26]

$$\mathscr{D}_{12} = \frac{9.40 \times 10^{-15}T}{\mu_2 M_1^{1/3}} \ \mathrm{m^2/s} \tag{9.133}$$

where T is in kelvins, μ_2 is the viscosity of water [kg/m s], and M_1 is the molecular weight of the macromolecule. Analysis of diffusion of smaller molecules in solutions containing macromolecules is made difficult by the presence of macromolecules. The problem is similar to that of multicomponent diffusion in gas mixtures and is beyond the scope of this text.

---

[9] In the use of the kinetic theory literature, a source of confusion is the different definitions of mean free path encountered. Care must be taken since the relevant definition is often not prominently displayed.

Table 9.5 Dynamic shape factor χ defined by Eq. (9.132).

Aerosol	$d_{pe} = (6m/\pi\rho_p)^{1/3}$ μm	χ
Sphere		1.0
Tetrahedron		1.17
Oblate spheroid		
Aspect ratio = 2		1.05
Aspect ratio = 5		1.24
Aspect ratio = 10		1.50
Prolate spheroid		
Aspect ratio = 2		1.05
Aspect ratio = 5		1.27
Aspect ratio = 10		1.60
Coal	0.56–4.27	1.88 ± 0.16
Quartz	0.65–1.85	1.84 ± 0.22
	> 4	1.23
Copper oxide	0.32	6.0
	0.70	8.5
Platinum oxide	0.038	1.1
	0.071	3.6
Uranium dioxide	0.19	2.6
	0.33	3.3
	0.55	1.2
U_3O_8	0.3–1	4–9
$(Pu_{0.14}U_{0.86})O_{2.22}$ aggregates (LFMBR fuel)		2–3
Sodium oxide aerosols formed in humid air		1.3
Iron oxide (Fe_2O_3) cluster aggregates	~ 1	4–9

EXAMPLE 9.18 Brownian Diffusion Coefficient of an Uranium Oxide Particle

Calculate the Brownian diffusion coefficient of a UO_2 aerosol particle of mass equivalent diameter $10^{-2}\mu$m in steam at 400 K and 1 atm. The dynamic shape factor χ can be taken as 4. Also determine the particle Schmidt number.

Solution

Given: UO_2 particle in steam.

Required: Brownian diffusion coefficient and particle Schmidt number.

Assumptions: The nonsphericity effect can be accounted for by using a dynamic shape factor $\chi = 4$.

For steam at 400 K and 1 atm, Table A.7 gives $\mu = 1.40 \times 10^{-5}$ kg/m s, $\nu = 25.2 \times 10^{-6}$ m²/s. From Eq. (9.131), the mean free path is

$$\ell \simeq \nu\left(\frac{\pi M}{2\mathcal{R}T}\right)^{1/2} = 25.2 \times 10^{-6}\left(\frac{18\pi}{2(8314)(400)}\right)^{1/2} = 7.35 \times 10^{-8} \text{ m} = 0.0735\ \mu\text{m}$$

The particle Knudsen number is

$$\text{Kn} = \frac{\ell}{d_{pe}} = \frac{0.0735}{0.01} = 7.35$$

Since this is the transitional flow regime, the friction coefficient is obtained from Eqs. (9.132) and (9.130) as

$$f = \frac{3\pi\mu d_{pe}\chi}{C}; \qquad C = 1 + \text{Kn}\left(2.514 + 0.8e^{-0.55/\text{Kn}}\right)$$

$$C = 1 + 7.35\left(2.514 + 0.8e^{-0.55/7.35}\right) = 24.9$$

$$f = \frac{(3\pi)(1.40 \times 10^{-5})(10^{-8})(4)}{24.9} = 2.12 \times 10^{-13} \text{ kg/s}$$

Equation (9.126) gives the Brownian diffusion coefficient,

$$\mathcal{D}_p = \frac{\ell T}{f}$$

From Table A.16, the Boltzmann constant $\ell = 1.38054 \times 10^{-23}$ J/K and 1 J/K = Nm/K = 1 kg m²/s² K; thus

$$\mathcal{D}_p = \frac{(1.38054 \times 10^{-23} \text{ kg m}^2/\text{s}^2\text{ K})(400 \text{ K})}{2.12 \times 10^{-13} \text{ kg/s}} = 2.61 \times 10^{-8} \text{ m}^2/\text{s}$$

The particle Schmidt number is

$$\text{Sc}_p = \frac{\nu}{\mathcal{D}_p} = \frac{25.2 \times 10^{-6}}{2.61 \times 10^{-8}} = 966$$

Comments

1. This Schmidt number is of a comparable magnitude to that of glycerol dissolved in water (see Table A.18).
2. Transport of UO_2 aerosols is of concern to nuclear engineers in analyzing a hypothetical loss-of-coolant accident (LOCA). The aerosols are formed when there is a core meltdown and are subsequently transported in steam (or steam-hydrogen-air mixtures) through the containment vessel.

9.8 CLOSURE

This chapter provided an introduction to the subject of mass transfer, with special emphasis on topics of concern to mechanical engineers. Simple analyses of mass transfer by diffusion were presented, with the governing equation derived by application of the principle of conservation of chemical species to an appropriate elemental control volume. Where possible, the analogy between mass diffusion and heat conduction was exploited. Consideration of mass convection was initially limited to situations for which low mass transfer rate theory is of acceptable accuracy. The analogy between heat convection and mass convection allowed convective heat transfer correlations from Chapter 4 to be adapted for convective mass transfer problems in Chapter 9. Particular attention was given to a variety of problems in which there was simultaneous heat and mass transfer. Mass transfer in porous catalysts was analyzed, with applications to catalysts used for automotive emissions control. To complete the chapter, methods for calculating transport properties of gas mixtures, liquid solutions, and small particles were presented.

Four computer programs were introduced in Chapter 9. The program MCONV is a useful tool for solving convection problems according to the theory presented in Sections 9.4 and 9.5. The program PSYCHRO calculates psychrometric and thermodynamic properties of water vapor and air mixtures. GASMIX calculates transport properties of gas mixtures and includes a menu of 36 chemical species. AIRSTE calculates transport properties of water vapor and air mixtures. Mass transfer calculations tend to require more effort than heat transfer calculations. Mass or mole fractions must be calculated, transport and thermodynamic properties can involve the use of mixture rules, and vapor pressure tables are frequently used. Thus, the use of computer programs to perform such calculations is particularly appropriate and can save the engineer considerable time and expense.

Chapter 10 is concerned with simultaneous diffusion and convection, which leads to high mass transfer rate theory for convection. However, in Chapter 11 all the analyses of mass transfer equipment are based on low mass transfer rate theory. Thus, if desired, the student can proceed directly to Chapter 11 and defer Chapter 10.

REFERENCES

1. Plumb, O. A., Spolek, G. A., and Olmstead, B. A., "Heat and mass transfer in wood during drying," *Int. J. Heat Mass Transfer,* 28, 1669–1678 (1985).

2. Perre, P., Moser, M., and Martin, M., "Advances in transport phenomena during convective drying with superheated steam and moist air," *Int. J. Heat Mass Transfer,* 36, 2725–2746 (1993).

3. Pesaran, A. A., and Mills, A. F., "Moisture transport in silica gel packed beds: I. Theoretical study," *Int. J. Heat Mass Transfer,* 30, 1037–1050 (1987).

4. Kafesjian, R., Plank, C. A., and Gerhard, E. R., "Liquid flow and gas phase mass transfer in wetted wall towers," *AIChE Journal,* 7, 463–466 (1961).

5. Saboya, F. E. M., and Sparrow, E. M., "Local and average transfer coefficients for one-row plate fin and tube heat exchanger configurations," *J. Heat Transfer,* 96, 265–272 (1974).

6. Goldstein, R. J., and Karni, J., "The effect of a wall boundary layer on local mass transfer from a cylinder in crossflow," *J. Heat Transfer,* 106, 260—267 (1984).

7. Cengel, Y. A., and Boles, M. A., *Thermodynamics: An Engineering Approach,* Chapter 13, McGraw-Hill, New York (1989).

8. Threlkeld, J. L., *Thermal Environmental Engineering,* 2nd ed., Prentice-Hall, Englewood Cliffs, N.J. (1970).

9. *Handbook of Air Conditioning, Heating and Ventilating,* eds. E. Stamper and R. L. Koral, 3rd ed., Industrial Press, New York (1979).

10. *ASHRAE Handbook: 1989 Fundamentals,* American Society of Heating, Refrigerating and Air Conditioning Engineers, Atlanta (1989).

11. Kennard, E. H., *Kinetic Theory of Gases,* McGraw-Hill, New York (1938).

12. Tien, C. L., and Lienhard, J., *Statistical Thermodynamics,* Holt, Rinehart & Winston, New York (1971).

13. Knuth, E. L., *Introduction to Statistical Thermodynamics,* McGraw-Hill, New York (1966).

14. Guggenheim, E. A., *Elements of the Kinetic Theory of Gases,* Pergamon Press, Oxford (1960).

15. Jeans, J., *An Introduction to the Kinetic Theory of Gases,* Cambridge University Press, New York (1940).

16. Edwards, D. K., Denny, V. E., and Mills, A. F., *Transfer Processes,* 2nd ed., Hemisphere/McGraw-Hill, Washington, D.C. (1979).

17. Hirschfelder, J. O., Curtiss, C. P., and Bird, R. B., *Molecular Theory of Gases and Liquids,* Wiley, New York (1954).

18. Chapman, S., and Cowling, T. G., *The Mathematical Theory of Nonuniform Gases,* 2nd ed., Cambridge University Press, New York (1964).

19. Wilke, C. R., "A viscosity equation for gas mixtures," *J. Chem. Phys.,* 18, 517–519 (1950).

20. Wilke, C. R., and Chang, P., "Correlation of diffusion coefficients in dilute solutions," *AIChE Journal,* 1, 264–270 (1955).

21. Othmer, D. F., and Thakar, M. S., "Correlating diffusion coefficients in liquids," *Ind. Eng. Chem.,* 45, 589–593 (1953).

22. Reid, R. C., Prausnitz, J. M., and Sherwood, T. K., *The Properties of Gases and Liquids,* 3rd ed., McGraw-Hill, New York (1977).

23. Einstein, A., *Investigations of the Theory of Brownian Movement,* Dover, New York (1956).

24. Davies, C. N., "Definitive equations for the fluid resistance of spheres," *Proc. Phys. Soc. Part A,* 57, 259–269 (1945).

25. Lavenda, B. H., "Brownian motion," *Scientific American,* 252, no. 2, 70–85 (1985).

26. Polson, A., "Some aspects of diffusion in solution and definition of colloidal particles," *J. Phys. Colloid Chem.,* 54, 649–652 (1950).

27. Won, Y. S., and Mills, A. F., "Correlation of the effects of viscosity and surface tension on gas absorption rates into freely falling turbulent liquid films," *Int. J. Heat Mass Transfer,* 25, 223–229 (1982).

28. Bartlett, E. P., Kendall, R. M., and Rindal, R. A., "An analysis of the coupled chemically reacting boundary layer and charring ablator, Part IV: A unified approximation for mixture transport properties for multicomponent boundary layer applications," NASA CR-1063 (1968).

EXERCISES

9–1. Derive Eqs. (9.10*a*) and (9.10*b*), namely,

$$m_i = \frac{x_i M_i}{\sum x_j M_j}; \qquad x_i = \frac{m_i/M_i}{\sum m_j/M_j}$$

9–2. (i) A mixture of noble gases contains equal mole fractions of helium, argon, and xenon. What is the composition in terms of mass fractions?
(ii) If the mixture contains equal mass fractions of He, Ar, and Xe, what are the corresponding mole fractions?

9–3. Methane is burned with 20% excess air. At 1250 K the equilibrium composition of the product is:

Species i:	CO_2	H_2O	O_2	N_2	NO
x_i:	0.0803	0.160	0.0325	0.727	0.000118

Determine the mean molecular weight M and gas constant R of the mixture, and the mass fraction of the pollutant nitric oxide in parts per million.

9–4. A closed cylindrical vessel containing stagnant air stands with its axis vertical. Each end is maintained at a uniform temperature with the base colder than the upper end; the cylindrical wall is insulated. Neglecting the variation in hydrostatic pressure, ascertain whether there are vertical gradients of

(i) partial density of oxygen.
(ii) partial pressure of oxygen.
(iii) mass fraction of oxygen.

9–5. Combustion products from a hydrogen burner contain 34.0% H_2O, 64.6% N_2, and 1.4% H_2 by volume at 600 K and 2 atm. Determine

(i) the mole fractions, partial pressures, and mass fractions of the components.
(ii) the mean molecular weight and gas constant of the mixture.
(iii) the density and molar concentration of the mixture.
(iv) the partial densities and molar concentrations of the components.

9–6. A 1 liter vessel contains 80% He and 20% N_2 by volume at 100 kPa and 300 K.

(i) Find the mass fraction and partial density of the helium.
(ii) If the cylinder is heated to 500 K, find the new values of P, m_{He}, and ρ_{He}.
(iii) If 2×10^{-5} kmol of helium are now added to the vessel while it is maintained at 500 K, find the final values of P, m_{He}, and ρ_{He}.

9–7. Water at 310 K is sprayed as a very fine mist into a vessel containing gas samples at 100 kPa. Determine the mass and mole fractions of dissolved gas in water collected at the bottom of the vessel if the gas is

(i) oxygen.
(ii) carbon dioxide.
(iii) hydrogen sulfide.
(iv) ammonia.

9–8. Plot a graph of the mass fraction of water vapor in saturated water vapor–air mixtures at 1 atm pressure for $273.15 \text{ K} < T < 373.15 \text{ K}$.

9–9. A 1 liter soda bottle at 290 K is at a pressure of 1.2 bar. If it was charged with CO_2 at 290 K and 5 bar pressure, what volume of gas was added?

9–10. A recent determination of the diffusivity of carbon in BCC iron gave values of 1.25×10^{-8} cm^2/s at 500°C, and 2.17×10^{-5} cm^2/s at 1000°C. If $\mathscr{D}$ can be represented by an expression of the form $\mathscr{D} = \mathscr{D}_0 \exp(-E_a/\mathscr{R}T)$, determine E_a in kJ/kmol, and $\mathscr{D}_0$. Also determine the diffusivity at 1000 K.

9–11. Hydrogen, at 500°C and 5 bar, flows at 10 m/s in a 2 cm–O.D., 0.5 mm–wall thickness steel tube located in an evacuated enclosure. Estimate the rate of hydrogen loss for a 10 m length of tube. Solubility and diffusion coefficient data for the hydrogen-steel system are given in Example 9.1.

9–12. From the data given in Example 9.2, tabulate the permeability of helium in the laser shell glass for $300 \text{ K} < T < 500 \text{ K}$. Using these data, recalculate the rate of helium loss through the shell. Why is the permeability approach inappropriate for hydrogen diffusion through steel as considered in Example 9.1?

9–13. A pharmaceutical product is to be protected from exposure to oxygen by vacuum wrapping with a 0.2 mm–thick polyethylene film of 400 cm^2 surface area. Oxygen that penetrates the packing is rapidly consumed by reaction with the product. If storage is at 30°C, calculate the rate at which oxygen is available for reaction with the product. The permeability of O_2 in polyethylene at 30°C can be taken as 4.17×10^{-12} m^3 (STP)/m^2 s (atm/m).

9–14. A 4 cm–diameter composite membrane, consisting of a 0.2 mm–thick film of vulcanized rubber and a 1 mm–thick layer of polyethylene, separates pure hydrogen at 1.085×10^5 Pa and 25°C from atmospheric air. Calculate the rate at which hydrogen leaks through the membrane. Permeabilities of H_2 in the rubber and polyethylene, respectively, are 3.42×10^{-11} and 6.53×10^{-12} m^3 (STP)/m^2 s (atm/m).

9–15. A worm gear cut from 0.1% C steel is to be carburized (case-hardened) at 930°C until the carbon content is raised to 0.45% C at a depth of 0.5 mm. The carburizing gas holds the u-surface concentration at 1% C by weight.

(i) Calculate the time required.
(ii) What time is required at the same temperature to double this depth of penetration?
(iii) What temperature is required to obtain 0.45% C at 1 mm, in the same time as it was attained at 0.5 mm for a temperature of 930°C?

Take the diffusion coefficient of carbon in the steel as

$$\mathscr{D} = 2.67 \times 10^{-5} e^{-17{,}400/T} \text{ m}^2\text{/s}$$

where T is in kelvins.

9–16. The Loschmidt apparatus for measuring the diffusion coefficient in a binary gas mixture consists of a long tube, $0 < z < 2L$, with a diaphragm at $z = L$. The pure gases are loaded into each half of the tube, and at time $t = 0$ the diaphragm is removed. The composition of the mixture at $z = 0$ is monitored as a function of time and compared with the theoretically expected value. Derive an expression for x_1 at $z = 0$, valid for $\mathscr{D}_{12}t/L^2 > 0.2$, suitable for this purpose.

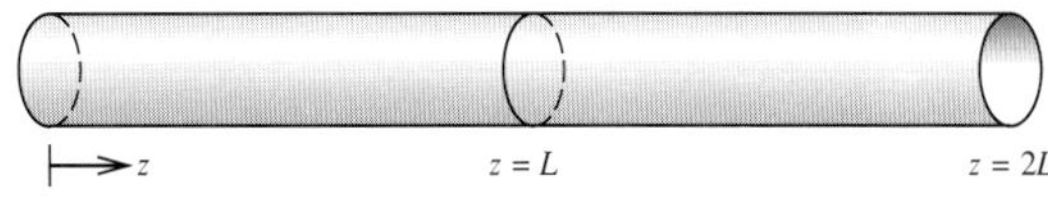

9–17. A 1 m–square, 1 m–deep pool of stagnant deaerated water at 290 K is suddenly exposed to pure oxygen at 2 atm pressure. Calculate the amount of O_2 absorbed after 12 hours has elapsed.

9–18. Water droplets, of 0.3 mm diameter, fall through a tower containing H_2S gas at 300 K and 10 atm.

(i) Use Henry's law to determine the u-surface H_2S mole fraction.
(ii) Determine the absorption rate at short times.
(iii) Determine the fractional approach to equilibrium as a function of time.

Assume no circulation inside the droplet, and refer to analogous heat conduction problems in Chapter 3.

9–19. Upon inspection at the warehouse, a consignment of fir planks was found to have been insufficiently dried. Tests showed a residual moisture content of 13.3% by weight. The planks were returned to the mill for further drying. If the drying oven is able to maintain a u-surface concentration of moisture of 6.6%, determine the time required for the center of 5 cm–thick planks to dry to 8.0% moisture content. Take the effective diffusion coefficient for moisture in fir under these conditions as 0.86×10^{-9} m²/s.

9–20. Experimental oxidation data for a titanium alloy, used to sheath hypersonic vehicle nose cones, indicates that the total weight gain can be described by a parabolic law of the form $w^2 = Ct$, where $C = 480\exp(-E_a/\mathcal{R}T)$ g²/cm⁴ s, and $E_a = 61{,}800$ cal/mol. Examination of oxidized samples shows that about 20% of the oxygen goes to form a rutile (TiO_2) surface scale. Assuming that the diffusion coefficient for oxygen in alpha titanium can be expressed as $\mathscr{D}_{12} = \mathscr{D}_0 \exp(-E_a/\mathcal{R}T)$, with E_a again 61,800 cal/mol, estimate $\mathscr{D}_0$. Show also that the parabolic oxidation law implies diffusion across the rutile scale characterized by a linear concentration gradient. Use an alloy density of 4400 kg/m³.

9–21. Naphthalene is used as a moth repellent in the form of spherical mothballs. If a 2 cm–diameter mothball is hung in a closet where the average temperature is 21°C, obtain a rough estimate of the lifetime of the mothball. Take $P = 1$ atm, and use data for naphthalene from Example 9.5.

9–22. At a particular location on a surface, the heat transfer coefficient to a laminar forced flow of 300 K water is 1000 W/m² K. Estimate the equivalent stagnant film thicknesses for

(i) heat transfer.
(ii) mass transfer for dissolution of sucrose.

9–23. A 0.3 mm–diameter carbon particle at 1650 K oxidizes in air at 1350 K and 1 atm pressure. At these temperatures, both the kinetic and diffusive resistances to mass transfer may be important, and the heterogeneous reaction is $C + O_2 \rightarrow CO_2$. The rate of combustion of carbon per unit surface area is $\Phi P_{O_2,s}^{1/2}$ where $\Phi = 1170\exp(-30{,}200/T_s)$ kmol/m² s Pa$^{-1/2}$ for T_s in kelvins and P in pascals (the s-surface is located in the gas phase adjacent to the solid carbon surface).

(i) Model the diffusion process using the equivalent stagnant film concept on a *molar* basis.
(ii) Obtain an expression for the molar rate of consumption of carbon in terms of the film thickness δ_f, the free-stream oxygen concentration $x_{O_2,e}$, and Φ.
(iii) If the particle is entrained in the air flow so that there is a negligible relative velocity, estimate δ_f.
(iv) Determine the initial rate of decrease of particle diameter.
(v) Determine the particle lifetime.

9–24. A common procedure for surface micromachining involves chemical etching of sacrificial layers. In one procedure, phosphosilicate glass is etched by hydrofluoric acid, with an overall reaction of

$$6HF + SiO_2 \rightarrow H_2SiF_6 + 2H_2O$$

When the etching depth is small, the reaction rate is rate-controlled; but for deep etching, as required to machine channels, diffusion of acid through the solution also plays a role. Assuming quasi-steady diffusion and a one-dimensional model, obtain an expression for the etching depth as a function of time. The reaction is approximately first-order in HF molar concentration with rate constant k''. Hence determine the time required to etch a channel 100 μm deep if $k''=2\times10^{-6}$ m/s, $\mathscr{D}_{1m}=2\times10^{-9}$ m^2/s, and the bulk concentration of HF in an aqueous solution is 7.0 kmol/m^3.

9–25. A proposed correlation for heat transfer from a jet to a disk normal to the flow has the form

$$\mathrm{Nu}_L = C\,\mathrm{Re}_L^{1/2}\,\mathrm{Pr}^{1/3}$$

where L is the distance from the jet exit to the disk. An experiment to determine the constant C measured the sublimation rate of a naphthalene disk exposed to an air jet. Pertinent data include pressure = 1 atm, temperature = 310 K, jet velocity = 10 m/s, L = 3 mm, for which the measured recession rate was 430 μm/h. Determine C.

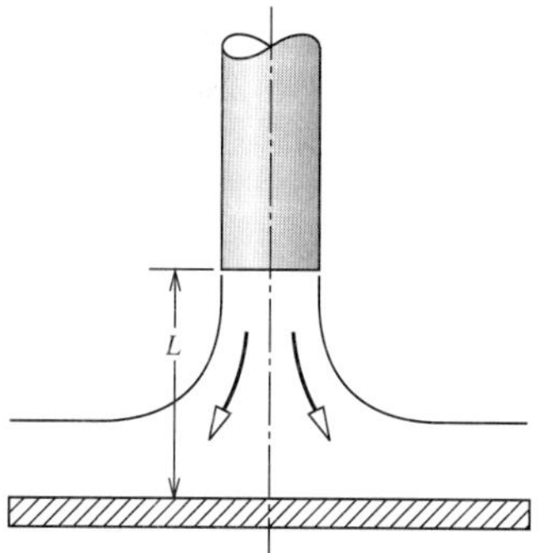

9–26. A new shape of particle for a packed-particle-bed catalytic reactor is under consideration. By means of a heat transfer experiment, it is found that, for superheated steam at 500 K and 1 atm, the heat transfer coefficient is 110 W/m^2 K at the design Reynolds number. If a stream of nitrogen containing 2% hydrogen by volume at 500 K flows through a bed at the same Reynolds number, estimate the maximum possible rate of formation of ammonia. The catalyst is iron promoted with ferric oxide. Property values at 500 K include Pr = 0.94, k = 0.0365 W/m K for the steam; and for the N_2—H_2—NH_3 mixture, Pr = 0.70, Sc_{H_2m} = 0.21, and $\mu = 25.1\times10^{-6}$ kg/m s.

9–27. A new catalyst for the oxidation of ammonia is being investigated. The catalyst is in the form of a 0.5 mm–diameter wire. The reactant gas is at 500 K and 1 atm pressure, and contains 12% ammonia by mass. If the flow across the wire is at a mass velocity of 0.3 kg/m^2 s, estimate the maximum rate at which ammonia can be oxidized per meter length of wire. Take the reactant gases to have properties similar to air.

9–28. A materials processing experiment on the space shuttle requires design data for convective heat transfer coefficients in an acoustically driven forced flow around a 1 cm–diameter sphere. A laboratory simulation is planned in Los Angeles, and to ensure acceptably small natural convection effects, $\mathrm{Re}^2/\mathrm{Gr}$ should be greater than 10 (there is negligible natural convection in the near zero-g environment on the shuttle). If the characteristic velocity of the forced flow is V, what is the minimum value of V that should be used in the simulation if

(i) a heat transfer experiment is performed with a 1 cm–diameter sphere at a temperature 4 K above the ambient air at 298 K, 1 atm?
(ii) a mass transfer experiment is performed with a 1 cm–diameter naphthalene sphere in air at 300 K, 1 atm?

9–29. In a field experiment to measure deposition rates of sulfur dioxide gas, the test surface is a 10 cm–square paper plate impregnated with potassium carbonate. The reaction between SO_2 and KCO_3 is rapid enough to ensure that the concentration of SO_2 at the s-surface is zero. Expected concentrations of SO_2 in the atmosphere at the test site are expected to be of the order of 5 parts per million by volume, and the average wind speed over the plate is 0.3 m/s. The ambient air is at 295 K and 1 atm, and the plate temperature is also 295 K. If at least 1 mg of SO_2 must be collected for a reliable result, what is the minimum time of exposure required?

9–30. In absorption towers, gases are often absorbed into a thin film of liquid flowing over a packing. A simple model of this process is a laminar film falling down a vertical wall. From Section 7.2.1, the surface velocity of the film is $u_\delta \simeq g\delta^2/2\nu_l$, where δ is the film thickness, and the characteristic small diffusion coefficient in liquids implies that the gas penetration is confined to a thin concentration boundary layer where the liquid velocity can be assumed constant at the surface value.

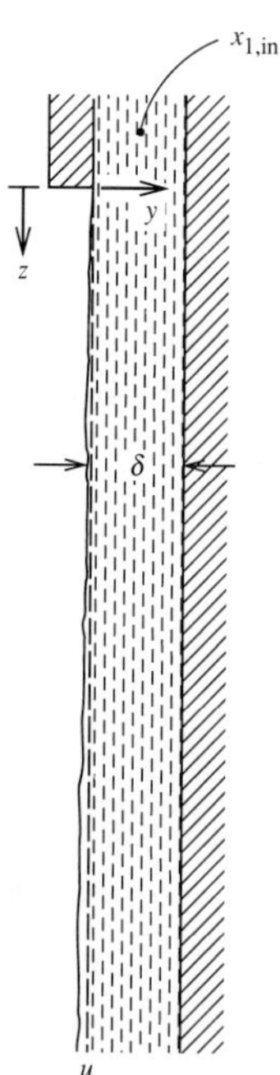

(i) Using a volume element fixed in space, show that the equation governing the steady-state concentration of absorbed gas in the concentration boundary layer is

$$u_\delta \frac{\partial x_1}{\partial z} = \mathscr{D}_{12} \frac{\partial^2 x_1}{\partial y^2}$$

(ii) By transforming to a coordinate system fixed on the film surface, show that the concentration distribution is

$$\frac{x_1 - x_{1,\text{in}}}{x_{1,u} - x_{1,\text{in}}} = \operatorname{erfc} \frac{y}{(4\mathscr{D}_{12} z/u_\delta)^{1/2}}$$

where $x_{1,\text{in}}$ is the mole fraction of the gas in the liquid at $z = 0$.

(iii) Show that the local Sherwood number is given by

$$\text{Sh}_z = \frac{1}{\pi^{1/2}} \text{Pe}_{mz}^{-1/2}$$

where Pe_{mz} is the local mass transfer Peclet number, $\text{Pe}_{mz} = u_\delta z/\mathscr{D}_{12}$.

9–31. Water at 300 K and initially pure, falls as a 0.36 mm–thick film on the inside wall of a vertical tube of 2 cm I.D. and 1 m length. If pure CO_2 at 1 atm pressure flows through the tube, calculate the rate at which CO_2 is absorbed. Also estimate the thickness of the concentration boundary layer at the bottom of the tube to check the assumption of a small penetration distance.

9–32. In an experiment, water at 300 K and initially pure, falls as a film on the inside wall of a vertical tube of 2 cm I.D. and 1 m length. The water flow rate is 0.04767 kg/s. Pure CO_2 at 1 atm flows through the tube. At this water flow rate, the film is in the turbulent regime, for which the following correlation can be used to obtain the mole transfer conductance for gas absorption into the film [27]:

$$\mathrm{St}_m = 6.79 \times 10^{-9}\,\mathrm{Re}^n\,\mathrm{Sc}^{-\alpha}\,\mathrm{Ka}^{-1/2}; \qquad \mathrm{Re} > 1100;$$

$$n = 3.49\,\mathrm{Ka}^{0.068}; \qquad \alpha = 0.36 + 2.43(\sigma N/m)$$

where $\mathrm{Ka} = \nu^4\rho^3 g/\sigma^3$ is the *Kapitza number,* the Stanton number is defined as $\mathrm{St}_m = \mathcal{G}_m/c(\nu g)^{1/3}$, and the Reynolds number is the film Reynolds number $\mathrm{Re} = 4\Gamma/\mu$ (defined in Chapter 7). Notice that, in addition to the usual Reynolds and Schmidt number dependence, the Stanton number also depends on surface tension (St_m increases with increasing σ).

(i) Calculate $\mathcal{G}_m$ for the conditions of the experiment.
(ii) Since the film is long, $\mathcal{G}_m$ can be assumed constant. Model the system as a single-stream exchanger, and hence obtain the outlet water bulk mole fraction of CO_2.
(iii) Determine the rate at which CO_2 is absorbed in the tube.

9–33. (i) Reconsider Exercise 9–18 for the situation where the tower contains air with a bulk H_2S mole fraction of 0.03. Estimate the mass transfer Biot number, and hence ascertain if the gas absorption process remains liquid side–controlled.
(ii) Reconsider Exercise 9–17 for water exposed to air at 2 atm pressure. Will the gas absorption process remain liquid side–controlled? (*Hint:* Estimate a reasonable gas-side conductance and a characteristic length for diffusion into the pool.)

9–34. In an experiment, three moist air samples at 101 kPa and 25.0°C have respective wet-bulb temperatures of 15.0, 20.0, and 23.0°C. Prepare a table giving the water vapor mass fraction and mole fraction, humidity ratio, and relative humidity of the three samples.

9–35. In a cooler-condenser performance test, the following data were obtained for the outlet steam-air mixture: $P = 22{,}300$ Pa, $T_{\mathrm{DB}} = 299.7$ K, $T_{\mathrm{WB}} = 301.0$ K. Determine the mass fraction of water vapor and the relative humidity.

9–36. Air at 290 K, 40% relative humidity blows at 5 m/s over a 10 m–square swimming pool near Albuquerque, New Mexico. If the ambient pressure is 870 mbar and the water surface temperature is measured to be 300 K, estimate the heat loss due to

(i) convective heat transfer.
(ii) evaporation.

9–37. A thick horizontal steel plate 1 m square has a temperature of 340 K. Water is spilled on the plate and forms a pool 0.2 mm deep. If the ambient air is still and is at a temperature of 310 K and a pressure of 1 atm, and it has a relative humidity of 30%, estimate the time required for the pool to evaporate.

9–38. A 1 m–square water bath is maintained at 320 K in surrounding still air at 290 K, 1 atm pressure, and 20% relative humidity. Estimate the convective, evaporative, and radiative heat losses.

9–39. A wet- and dry-bulb psychrometer is used to measure the composition of a mixture of methanol vapor and nitrogen. A sample at 100 kPa pressure gives readings of $T_{DB} = 9.2°C$ and $T_{WB} = 5.0°C$. Determine the mass fraction of methanol. The chemical formula of methanol is CH_3OH, and its vapor pressure is tabulated below. At 5.0°C the enthalpy of vaporization of methanol is 1.18×10^6 J/kg.

P, mm Hg:	1	10	40	100	400	760
P, Pa:	133	1330	5330	13,330	53,300	101,330
T, °C:	−44.0	−16.2	5.0	21.2	49.9	64.7

9–40. A wet- and dry-bulb psychrometer is used to measure the composition of a mixture of ammonia vapor and nitrogen. A sample at 15 atm pressure gives readings of $T_{DB} = 250.5$ K and $T_{WB} = 240.0$ K. Determine the mass fraction of ammonia.

9–41. Air at 990 mbar flows over a wet- and dry-bulb psychrometer: the dry thermocouple measures 310.1 K, and the wet thermocouple measures 305.3 K. Determine the humidity ratio and relative humidity,

(i) ignoring thermal radiation.
(ii) correcting for thermal radiation.

The instrument is located in a duct with walls maintained at 290 K, and the air velocity is 1 m/s. The thermocouples can be approximated as cylinders of diameters 1.0 and 3.0 mm for the dry and wet thermocouples, respectively. Use an emittance of 0.3 for the dry thermocouple and 0.9 for the wet wick.

9–42. A wet- and dry-bulb psychrometer is used to measure the concentration of ethanol vapor in an air stream at 10 atm total pressure. The dry bulb reads 350 K, and the wet bulb reads 340 K. What is the ethanol mass fraction? Use the data in Table A.17*b*. The spherical bulb diameters are 2 mm, and the gas velocity is 4 m/s. The surroundings are at 320 K.

9–43. Determine the rate of sublimation of a south-facing snowbank on Mammoth Mountain in the Sierra Nevada, when the ambient air is at −2°C and 30% RH.

Take the solar irradiation as 1000 W/m^2 and use a sky emittance of 0.75. The convective heat transfer coefficient for an equivalent surface has been estimated to be 6.0 W/m^2 K. Assume that heat transfer into the snowbank is negligible. Mammoth Mountain is at 2000 m altitude. Vapor pressure data for ice follows:

T, °C:	0	−2	−4	−6	−8	−10	−12	−14	−16	−18	−20
P_{sat}, Pa:	610	517	437	369	310	260	218	182	151	125	104

The enthalpy of sublimation is $h_{sg} = 2.827 \times 10^6$ J/kg at 0°C, and can be taken to be a constant over the temperature range of concern.

9–44. A 50 m × 20 m swimming pool in Los Angeles is maintained at 28°C during the winter. Typical nighttime conditions are a clear sky with still ambient air at 10°C, 1000 mbar, and 40% relative humidity. Calculate the convective, evaporative, and radiative heat losses when

(i) the pool is well stirred by swimmers.
(ii) the pool is unoccupied and the estimated water-side heat transfer coefficient is 100 W/m^2 K.

9–45. One method of retarding evaporation from water reservoirs in hot arid regions is to add a small amount of a large-molecular-weight alcohol, such as cetyl alcohol, to the water. The alcohol spreads to form a surface monolayer, which presents a significant resistance to evaporating water molecules. In a test, the water surface is at 23°C, and the ambient air is still at 1 atm, 20°C, and 20% RH. The measured rate of evaporation from a 1 m–square pool is 1.21×10^{-5} kg/m^2 s. What is the reduction in evaporation rate due to the cetyl-alcohol monolayer?

9–46. A strip of wet textile passes through a dryer. Hot air at 1 atm, 360 K and 1% RH is blown on the top surface of the textile through a perforated-plate distributor. When the mass velocity of mixture is 0.8 kg/m^2 s, design charts for the distributor indicate a heat transfer Stanton number of 0.08. Calculate the rate at which water is removed from the textile in the constant-drying-rate period of the drying process (see Fig. 9.10). The back surface of the textile in contact with a conveyer belt can be assumed adiabatic.

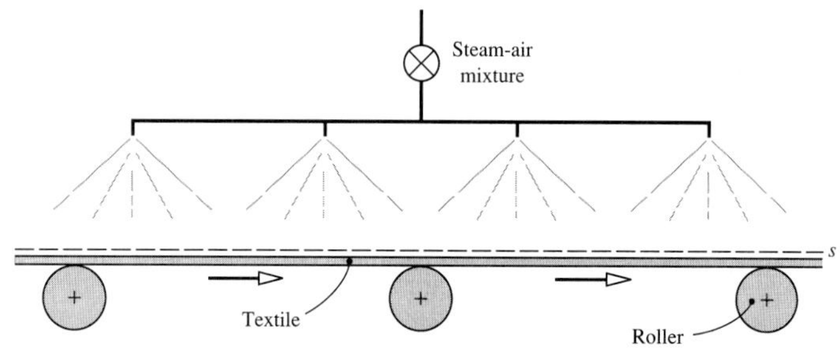

9–47. In a paper-drying unit the wet paper is on the outside of a rotating drum, the inside of which is maintained at 330 K by a hot water supply. Air at 305 K, 1 atm, and 40% RH is blown on the paper at a mass velocity of 0.3 kg/m^2 s. Design data for the configuration indicate a mass transfer Stanton number of 0.060. If the overall heat transfer coefficient from the hot water to the paper surface is estimated to be 50 W/m^2 K, calculate the paper drying rate in the constant-drying-rate period of the drying process (see Fig. 9.10).

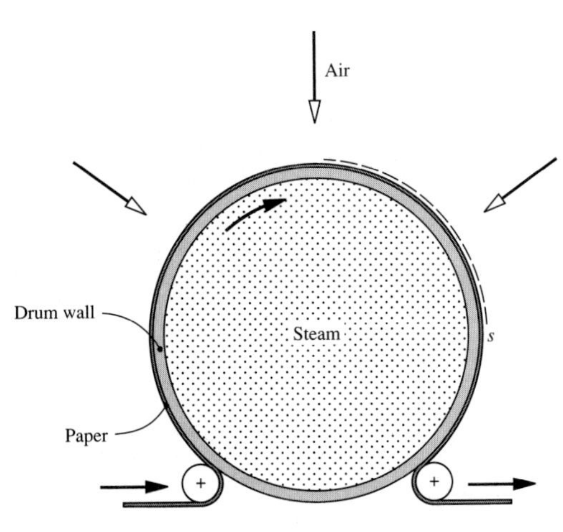

9–48. Air at 1 atm pressure, 360 K, and 1% RH flows at 10 m/s over a porous flat plate. Water flows through the plate at a rate just sufficient to keep the surface wet, and is supplied from a reservoir maintained at 280 K. Calculate the surface temperature, and the average evaporation rate for a 10 cm–long plate for

(i) a laminar boundary layer.
(ii) a turbulent boundary layer starting at the leading edge.

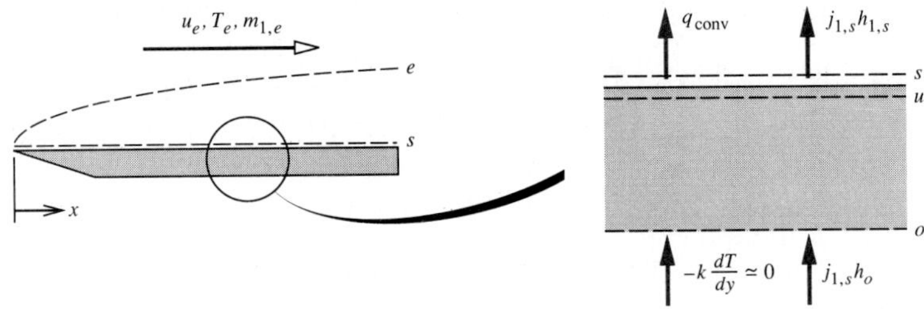

(*Hint:* First perform an energy balance on a control volume between the s-surface and an o-surface located far into the reservoir, where temperature gradients are negligible, to show that $h_c(T_e - T_s) = j_{1,s}[h_{fg} + c_{pl}(T_s - T_o)]$. Then show that the plate temperature is independent of location.) Neglect radiation heat transfer.

9–49. An operating hot tub loses heat continuously by a number of mechanisms, the most important being evaporative heat losses from the surface of the water, and into the air bubbling through the water. For steady operation, the heat loss must be balanced by adding heat to the circulating water. The design operating temperature for a small tub used indoors is 110°F (316.7 K) with the ambient air at 69°F (293.9 K), 1 atm, and 80% RH. The water circulation pump is rated at 1 kW, and its motor has an efficiency of 85%. Ambient air is blown into the pool at a rate of 7.3×10^{-3} kg/s. If the surface area of the water is 0.93 m^2 and the sides and bottom of the tub are well insulated, recommend a rating for the water heater.

9–50. The heat loss from a surface can be increased if it is kept wet. Consider air at 300 K, 1 atm, and 70% RH, flowing over a 3 cm–diameter sphere with a velocity of 2 m/s. Calculate the increase in heat transfer (wet versus dry) for sphere surface temperatures of 300, 305, 310, 320, and 330 K.

9–51. A 1 m–square wet towel is hung on a washline to dry on a day when there is a low overcast and no wind. The ambient air is at 21°C, 1 atm, and 50% RH. If the towel contains 0.5 kg of water, do you think the towel will dry in a reasonable time? Take $\varepsilon = 0.9$ for the wet towel.

9–52. Develop an appropriate model for raindrops falling from a cumulus cloud at the beginning of a summer thunderstorm. Write down equations governing both the motion and size of a droplet, and describe a scheme for solving the problem numerically. Discuss any assumptions made and list the data you might require.

9–53. A platinum wire is suspended in a steady stream of air. It is heated electrically, and its temperature is deduced from its electrical resistance and measured current and voltage drop. At a standard power input, air stream velocity, temperature, and pressure, the wire temperature is found to be 700 K.

(i) This apparatus is used for measuring small concentrations of combustible gas that may be present after emptying oil tanker tanks. In a particular test, in which the power input and air temperature, pressure, and velocity have the standard values, the wire temperature is found to be 805 K. Estimate the mass fraction of combustible gas in the air if the platinum is a powerful catalyst for the combustion reaction. Neglect thermal radiation effects.

(ii) Make a new estimate allowing for radiative effects using the following data: wire diameter = 0.2 mm, wire emittance = 0.91, velocity of air perpendicular to the wire = 3 m/s, air temperature = 310 K, and pressure = 1 atm.

(*Note:* This exercise is a D. B. Spalding original.)

9–54. Porous catalysts are sometimes manufactured in the form of a thin plate or a thin-walled matrix. Show that the effectiveness of such a plate for a first-order reaction, when only one side of the plate is exposed to the reactant flow, is $\eta_p = (1/\Lambda)\tanh\Lambda$, where Λ is the Thiele modulus.

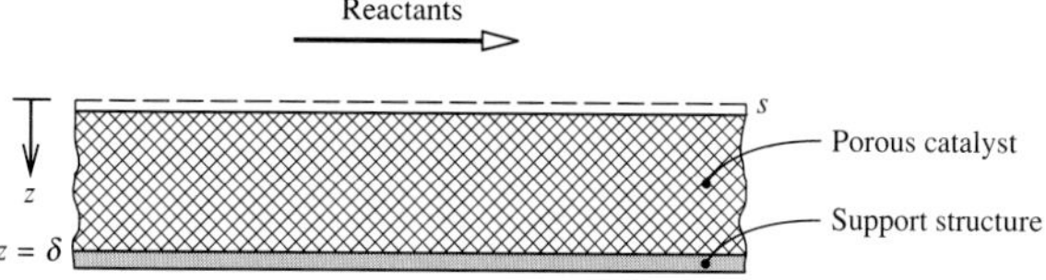

9–55. In designing a catalytic reactor, a plain wall was found to be inadequate, and thus it is proposed to bond a porous catalyst layer to the wall. If the layer thickness is 0.05 cm, the rate constant for a first-order reaction is $k''=4 \times 10^{-5}$ m/s, and $\mathscr{D}_{1,\text{eff}} = 1.4 \times 10^{-5}$ m^2/s, determine the area-to-volume ratio a_p m^2/m^3 required to achieve an effective rate constant based on s-surface area, of 0.07 m/s.

9–56. The catalyst bed of an automobile exhaust reactor is packed with CuO-on-alumina cylindrical pellets of 0.3 cm diameter and 0.6 cm long. The pellets have a volume void fraction of 0.60, tortuosity factor 3.5, average pore radius 0.7 μm, and catalytic surface area per unit volume of 6.81×10^5 cm^2/cm^3. Calculate the effectiveness of the pellet in promoting CO oxidation at 1 atm pressure and

(i) 700 K (idle).
(ii) 1000 K (full power).

The activation energy of the reaction is 18.6 kcal/mol, and the preexponential factor in the Arrhenius relation is 50.6 m/s.

9–57. Ammonia can be catalytically decomposed on a tungsten surface. When the ammonia is at a sufficiently high partial pressure, the catalyst surface is largely covered by adsorbed ammonia, and the reaction is consequently of zero order; that is, the reaction rate is independent of ammonia concentration and is given by $Ae^{-E_a/\mathscr{R}T}$ where A has units (moles reactant/unit area–unit time). Determine the concentration distribution in a spherical catalyst pellet promoting this reaction, assuming an isothermal pellet. Hence show if

$$\frac{V_p}{S_p}\left[\frac{k''a_p}{\mathscr{D}_{1,\text{eff}}c_{1,s}}\right]^{1/2}$$

is less than $\sqrt{2/3}$, the pellet effectiveness is equal to 1.0. (Warning: Of course, the reaction does not occur past the radius at which all the reactant has been consumed.)

9–58. A porous catalyst is used to burn fumes from a paint spray booth to environmentally safe CO_2 and H_2O. The reaction is first-order in mass fraction of fumes, with activation energy 18 kcal/mol and preexponential factor 10 m/s. The catalyst has specific surface area $a_p = 5 \times 10^5$ cm^2/cm^3 and average pore radius 0.5 μm, and is in the form of a matrix with 2 mm–thick walls and square cross-section passages with sides of 3 mm. The cross-sectional area of the reactor is large enough to ensure laminar flow through the passages. The catalyst operates at 500 K and 1 atm. At a location where the bulk mass fraction of fumes is 0.005, estimate the rate at which the fumes are oxidized per unit volume of reactor. For the fumes, take a molecular

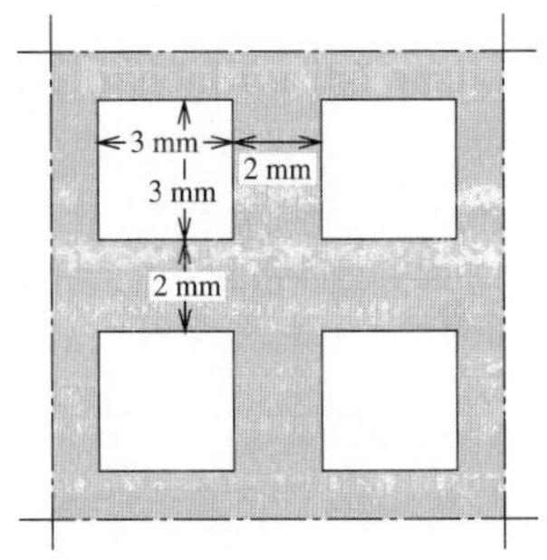

weight of 110 and an ordinary diffusion coefficient of 20×10^{-6} m^2/s. For the porous catalyst, take $\varepsilon_v = 0.7$, $\tau = 4.0$.

9–59. An automobile catalytic converter has cylindrical porous catalyst pellets of diameter 2.5 mm and length 5 mm. The pellets have a volume void fraction of 0.65, a tortuosity factor of 4.0, average pore radius 0.6 μm, and catalytic surface per unit volume 7.71×10^5 cm^2/cm^3. The catalyst promotes CO oxidation in a first-order reaction, with an activation energy of 19.1 kcal/mol and a pre-exponential factor in the Arrhenius relation equal to 43.3 m/s. Calculate the effectiveness of the pellet when the converter operates at 900 K and 1 atm.

9–60. To increase the rate of absorption of gas A into water, a reactant B is added to the water such that the homogeneous reaction A + B $\rightarrow$ C occurs. Analyze quasi-steady absorption into a pool of water of depth L for

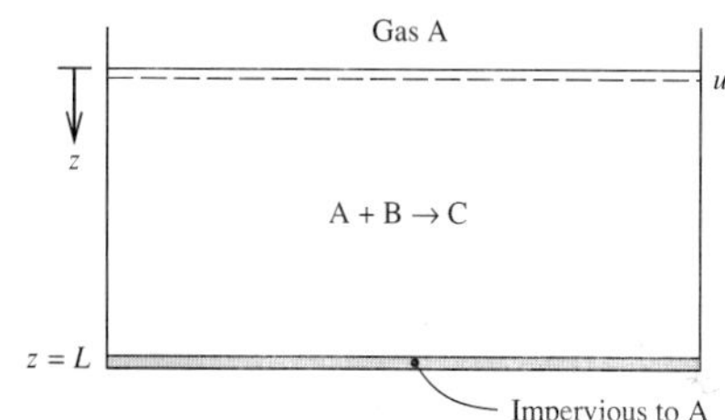

(i) a first-order reaction.
(ii) a zero-order reaction.

The bottom of the container is impervious to species A.

9–61. An important processing step in the production of semiconductors involves the formation of SiO_2 layers on silicon. These layers provide electrical insulation and dielectric properties necessary for semiconductor devices. Above 700°C, the layers form rapidly and follow exactly the topography of the underlying Si substrate. Develop a model for this *thermal oxidation* process with the following features:

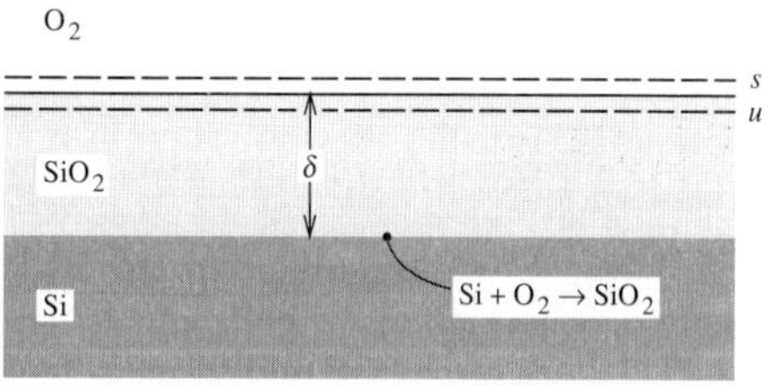

The solubility of O_2 in SiO_2 is given by the linear relation $c_{O_2,u} = KP_{O_2,s}$.

Quasi-steady diffusion of O_2 through the SiO_2 layer.

The heterogeneous reaction Si $+O_2 \rightarrow SiO_2$ occurs at the Si-SiO_2 interface and is first-order in molar concentration of O_2, with rate constant k''.

Hence show that the thickness-time relation is of the form

$$\frac{\delta}{k_L} + \frac{\delta^2}{k_P} = t$$

where k_L and k_P are so-called linear and parabolic rate constants for the overall oxidation process. Also solve for $\delta(t)$.

9–62. (i) Referring to Exercise 9–30, consider the situation where the solute gas is consumed in a first-order homogeneous reaction with rate constant k'''.

Show that the absorption rate is given by

$$J_{1,u} = cx_{1,u}\left(\mathscr{D}_{12}k'''\right)^{1/2}\left[\operatorname{erf}\left(k'''z/u_\delta\right) + \frac{e^{-k'''z/u_\delta}}{\left(\pi k'''z/u_\delta\right)^{1/2}}\right]$$

and that for a very fast reaction, the mole transfer conductance is

$$\mathscr{G}_{m1} = c\left(\mathscr{D}_{12}k'''\right)^{1/2}$$

(ii) A cross-flow falling-film reactor is to be designed for a chemical laser. Chlorine gas flows between walls wetted with basic hydrogen peroxide to produce singlet-delta oxygen according to the reaction

$$2KOH + H_2O_2 + Cl_2 \rightarrow 2H_2O + 2KCl + O_2\left({}^1\Delta\right)$$

The reaction is first-order, with a rate constant $k''' \simeq 10^8\ s^{-1}$. If the exchanger height is 1 m and the film surface velocity is 0.03 m/s, estimate the liquid-side mole transfer conductance and penetration distance. Take $\mathscr{D}_{12} \simeq 4 \times 10^{-10}\ m^2/s$ and $c = 36.0\ kmol/m^3$.

(*Hint:* Attempt this exercise after you have studied Laplace transforms in a mathematics or controls course.)

9–63. A sample of air at 300 K and 1 atm has an approximate composition of 23.2% O_2 and 76.8% N_2 by mass. Estimate the viscosity and thermal conductivity of the mixture using the Chapman-Enskog kinetic theory formulas of Section 9.7.1.

9–64. Estimate the diffusion coefficient of water vapor in air at 1 atm in the temperature range 0°C −100°C using

(i) the formula given in Table A.17*a*.
(ii) the Chapman-Enskog kinetic theory formula given in Section 9.7.1.

9–65. Estimate the diffusion coefficient in air at 25°C and 1 atm for

(i) helium.
(ii) methane.
(iii) carbon tetrachloride.

9–66. A useful approximate method for treating multicomponent diffusion is based on a bifurcation approximation for the binary diffusion coefficients. The approach is to correlate the binary diffusion coefficients in a particular system as

$$\mathscr{D}_{ij} = \frac{\overline{\mathscr{D}}}{F_iF_j}$$

where $\overline{\mathscr{D}}$ is a reference diffusion coefficient that contains the pressure and temperature dependence, and the F_i are *diffusion factors* that depend on species i properties only. If $\overline{\mathscr{D}}$ is taken to be the self-diffusion coefficient of O_2, then a simple correlation

$$F_i = \left(\frac{M_i}{26}\right)^{0.481}$$

has been shown to work well for mixtures of species formed from the elements C, H, O, and N, that is, for products of combustion of hydrocarbon fuels, and graphite or carbon-phenolic heat shields [28]. Check the validity of this correlation for a mixture of air, H_2, and H_2O at 400 K and 1.5 atm.

9–67. Estimate the diffusion coefficients for dilute aqueous solutions of methanol and n-propyl alcohol at 300 K using the Wilke-Chang and Othmer-Thakar formulas. Compare your results with values calculated from the Schmidt numbers given in Table A.18.

9–68. Estimate the diffusion coefficients for dilute aqueous solutions of ammonia in the temperature range 280 K to 320 K using the Wilke-Chang and Othmer-Thakar formulas. Compare your results with values calculated from the Schmidt number and temperature correction factor in Table A.18.

9–69. Aluminum sparks caused by arc welding or clashing electrical power cables are small droplets ($\sim$1 mm diameter) of molten aluminum surrounded by a flame in which aluminum vapor and atmospheric oxygen react to form aluminum oxide Al_2O_3 as a smoke of very small aerosol particles. Estimate the Brownian diffusion coefficient for Al_2O_3 particles in air at 3300 K, 1 atm. Assume a dynamic shape factor $\chi = 2.0$, and obtain results for mass equivalent diameters $d_{pe} = 10^{-3}$, 10^{-2}, and 10^{-1} μm.

9–70. For particle deposition from a flow onto a surface by Brownian diffusion, we can define a particle transfer conductance $\mathcal{G}_p$ that is analogous to the mole transfer conductance $\mathcal{G}_m$. Then the diffusive flux of particles across the s-surface is

$$\mathcal{J}_p = \mathcal{G}_p(\mathcal{N}_s - \mathcal{N}_e) \text{ particles/m}^2\text{s}$$

Since number density $\mathcal{N}$ has units [particles/m^3], $\mathcal{G}_p$ has units [m/s], that is, the same as a velocity. Usually we can assume that all particles reaching the s-surface are captured; thus, $\mathcal{N}_s = 0$ and the particle deposition rate is $-\mathcal{J}_p = \mathcal{G}_p\mathcal{N}_e$. The Sherwood number is $\text{Sh}_p = \mathcal{G}_pL/\mathcal{D}_p$. Air at 300 K and 1 atm flows over a 10 cm–long flat plate. Prepare a graph of the particle transfer conductance $\overline{\mathcal{G}}_p$ as a function of particle size, for air velocities of 0.3, 1, and 3 m/s. Use the data in Table A.30.

9–71. In a field experiment to measure "dry" deposition rates of atmospheric pollutants in the form of aerosols, the test surface is a 10 cm–diameter Teflon disk in the center of a Frisbee-shaped airfoil. A control surface is located nearby and is coated with naphthalene. Both surfaces are maintained at the same temperature. After exposure for 12 hours at a

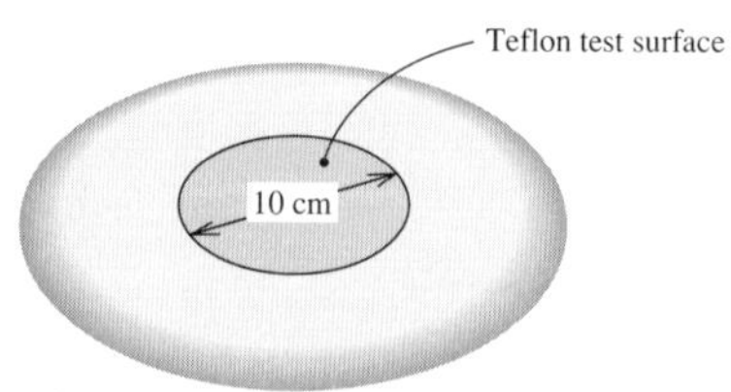

temperature of 300 K, the Teflon disk is washed in a small quantity of pure water to extract the pollutants, and the water is analyzed. Also, the weight loss of the naphthalene disk is determined. In a particular test, a deposition of 3560 ng of aerosol X is found, and the mass loss of naphthalene is 1.46 g. If aerosol X has a density of 2100 kg/m^3 and an average size of 0.05 μm, estimate the number density of aerosol X in the air at the test site.

9–72. Air at 1 atm pressure and 400 K flows at 20 m/s over a 10 cm–long flat plate. The air contains 10^{15} particles/m^3 of spherical aerosol particles of diameter 3×10^{-3} μm. Estimate the rate of deposition at the trailing edge for the following conditions:

(i) A laminar boundary layer.
(ii) A turbulent boundary layer starting at the leading edge.
(iii) A turbulent boundary layer starting at the leading edge of a plate roughened with 1 mm–diameter close-packed sand grains.

9–73. Experimental data give the diffusion coefficient of soybean protein in dilute aqueous solution at 20°C as 2.91×10^{-11} m^2/s. If the molecular weight of this protein is 361,800, use the Polson equation to estimate the diffusion coefficient, and compare your result to the experimental value.

9–74. In an experiment to investigate silica scaling in a geothermal energy power system, brine at 350 K flows at 0.366 kg/s in a 15.2 mm–I.D. pipe, 76 m long. At the inlet, the concentration of amorphous silica particles is 10^{19} particles/m^3, and their average diameter is 6 nm. Estimate

(i) the rate of scaling near the pipe inlet.
(ii) the outlet concentration of silica particles.

The silica density can be taken as 2200 kg/m^3.

CHAPTER

10

HIGH MASS TRANSFER RATE THEORY

CONTENTS

10.1 INTRODUCTION

In Chapter 9 we considered some simple mass diffusion and mass convection problems. In Section 9.3, transport of a chemical species was by diffusion only; that is, we restricted our attention to a stationary medium. Actually, we did not even bother to define precisely what we mean by *stationary*. Since we have seen that in a mixture the various species can move relative to each other, a suitable definition of *stationary* is not immediately obvious. What we had in mind was an imprecise idea based on physical experience. For example, diffusion of a trace species in a solid would qualify as diffusive transport. On the other hand, the dispersion of smoke issuing from a power plant stack is clearly dominated by convective motions due to wind currents and thermal buoyancy forces; molecular diffusion plays but a secondary role. Similarly, transport of a chemical species in the porous catalysts considered in Section 9.6 was by diffusion only. Convective mass transport was considered in Sections 9.4 and 9.5; however, we used a simple physics-based analogy to convective heat transfer. We did not formulate the governing equations in a rigorous manner. We considered only *low mass transfer rate theory,* which assumes that mass transport across the s-surface is by diffusion only. At higher mass transfer rates, there is also significant mass transport across the s-surface by convection, because there is a velocity component normal to the surface associated with net mass transfer. Unfortunately, the concepts involved in the analysis of diffusion in a moving medium, and in **high mass transfer rate** convection, are not simple. It was for this reason that the limit cases of diffusion in a stationary medium and low mass transfer rate convection were treated first, as a stepping-stone to the complete picture, with such details as the precise definition of a stationary medium ignored.

In this chapter we go back to square one and develop a rigorous and more complete mass transfer theory. In Section 10.2 velocities and fluxes in a mixture are carefully defined, a general species conservation equation is derived, and Fick's law is more precisely stated. In Section 10.3 some one-dimensional problems involving high mass transfer rate diffusion are analyzed. In Section 10.4 high mass transfer rate convection is introduced with the analysis of a simple Couette-flow model of real boundary layers. The results form the basis of an engineering problem-solving procedure that uses **blowing factors** to account for the effects of high mass transfer rates on mass transfer, skin friction, and heat transfer. Heat transfer in forced laminar boundary layers was analyzed in Section 5.4; in Section 10.5 this analysis is extended to include mass transfer, and exact self-similar solutions are obtained. The chapter closes with a general problem-solving procedure that can be used to solve a wide variety of convective heat and mass transfer problems.

10.2 VELOCITIES, FLUXES, AND THE SPECIES CONSERVATION EQUATION

When diffusion occurs in a moving medium, a chemical species is transported both by diffusion and by motion of the medium as a whole (i.e., convection). Thus, the flux of the species relative to stationary coordinate axes consists of two components, one due to diffusion and one due to convection. We now examine this *simultaneous diffusion and convection*. In Sections 10.2.1 and 10.2.2 velocities and fluxes are defined, and these allow the general species conservation equation to be derived in Section 10.2.3. We define Fick's law precisely in Section 10.2.4 after examining what we mean by a stationary medium.

10.2.1 Definitions of Velocities

In a multicomponent system the various species may move at different velocities. Let $\mathbf{v}_i$ denote the absolute velocity of species i, that is, the velocity relative to stationary coordinate axes. In this sense, the velocity is not that of an individual molecule of species i. Rather, it is the local average of the species, that is, the sum of the velocities of all molecules of species i within an elemental volume divided by the number of such molecules. Then the local **mass-average velocity, v,** is defined as

$$\mathbf{v} = \frac{\sum \rho_i \mathbf{v}_i}{\sum \rho_i} = \frac{\sum \rho_i \mathbf{v}_i}{\rho} = \sum m_i \mathbf{v}_i \tag{10.1}$$

The quantity $\rho\mathbf{v}$ is the local mass flux, that is, the rate at which mass passes through a unit area placed normal to the velocity vector $\mathbf{v}$. From Eq. (10.1), $\rho\mathbf{v} = \sum \rho_i \mathbf{v}_i$; that is, the local mass flux is the sum of the local species mass fluxes. The velocity $\mathbf{v}$ is the velocity that would be measured by a Pitot tube and corresponds to the velocity $\mathbf{v}$ used in considering pure fluids. Of particular importance is that this velocity $\mathbf{v}$ is the velocity field described by the Navier-Stokes equations and hence by Newton's second law of motion.

The local **molar-average velocity**, $\mathbf{v}^*$, is defined in an analogous manner:

$$\mathbf{v}^* = \frac{\sum c_i \mathbf{v}_i}{\sum c_i} = \frac{\sum c_i \mathbf{v}_i}{c} = \sum x_i \mathbf{v}_i \tag{10.2}$$

The quantity $c\mathbf{v}^*$ is the local molar flux, that is, the rate at which moles pass through a unit area placed normal to the velocity $\mathbf{v}^*$.

The velocity of a particular species relative to the mass- or molar-average velocity is termed a **diffusion velocity**, because a diffusion mechanism is required for a species to move relative to the average velocity. We define two such velocities:

$$\hat{\mathbf{v}}_i = \mathbf{v}_i - \mathbf{v} \equiv \text{Diffusion velocity of species } i \text{ relative to } \mathbf{v} \tag{10.3a}$$

$$\hat{\mathbf{v}}_i^* = \mathbf{v}_i - \mathbf{v}^* \equiv \text{Diffusion velocity of species } i \text{ relative to } \mathbf{v}^* \tag{10.3b}$$

Figure 10.1 illustrates these velocities.

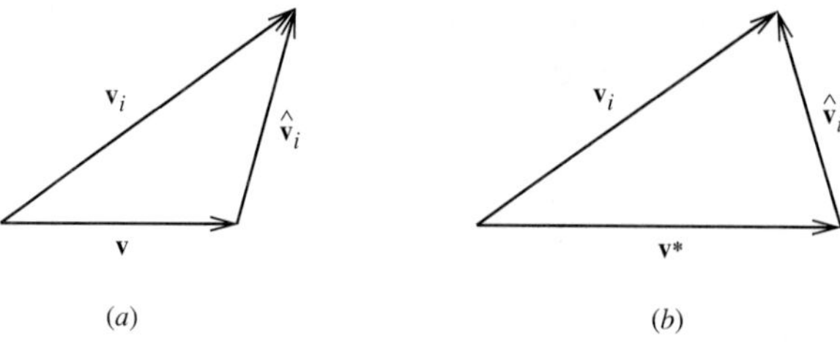

Figure 10.1 Velocity vectors: (*a*) mass basis, (*b*) molar basis.

EXAMPLE 10.1 Molar and Mass Average Velocities

A gas mixture contains 50% He and 50% O_2 by weight. The absolute velocities of each species are $\mathbf{v}_{\text{He}} = -9\mathbf{i}$ m/s, $\mathbf{v}_{O_2} = +9\mathbf{i}$ m/s, where $\mathbf{i}$ denotes the unit vector in the x direction. Determine the mass and molar average velocities.

Solution

Given: Mixture of He and O_2.

Required: Mass- and molar-average velocities, $\mathbf{v}$ and $\mathbf{v}^*$.

From Eq. (10.1) we find the mass-average velocity $\mathbf{v}$ as

$$\mathbf{v} = \sum_{i=1}^{2} m_i \mathbf{v}_i = (0.5)(-9\mathbf{i}) + (0.5)(+9\mathbf{i}) = 0$$

$-9\mathbf{i}$ m/s $+9\mathbf{i}$ m/s

The mole fractions are found from Eq. (9.10*b*):

$$x_{\text{He}} = \frac{m_{\text{He}}/M_{\text{He}}}{m_{\text{He}}/M_{\text{He}} + m_{O_2}/M_{O_2}} = \frac{0.5/4}{(0.5/4) + (0.5/32)} = 8/9$$

$$x_{O_2} = 1 - x_{\text{He}} = 1/9$$

Then, from Eq. (10.2), the mole-average velocity is

$$\mathbf{v}^* = \sum_{i=1}^{2} x_i \mathbf{v}_i = (8/9)(-9\mathbf{i}) + (1/9)(+9\mathbf{i}) = -7\mathbf{i} \text{ m/s}$$

Comment

It can be seen that although the mass-average velocity is zero, the molar-average velocity is large. Is the mixture stationary? We will soon be able to answer this question.

10.2.2 Definitions of Fluxes

The mass (or molar) flux of species i is a vector quantity giving the mass (or moles) of species i that pass per unit time through a unit area perpendicular to the vector.

Such fluxes may be defined relative to stationary coordinate axes or to either of the two local average velocities. We define the absolute mass and molar fluxes of species i, that is, relative to stationary coordinate axes, as

$$\text{Mass flux:} \quad \mathbf{n}_i = \rho_i \mathbf{v}_i \tag{10.4a}$$

$$\text{Molar flux:} \quad \mathbf{N}_i = c_i \mathbf{v}_i \tag{10.4b}$$

The mass diffusion flux relative to the mass-average velocity $\mathbf{v}$ of species i is

$$\mathbf{j}_i = \rho_i \hat{\mathbf{v}}_i = \rho_i(\mathbf{v}_i - \mathbf{v}) \tag{10.5a}$$

The molar diffusion flux relative to the molar-average velocity $\mathbf{v}^*$ of species i is

$$\mathbf{J}_i^* = c_i \hat{\mathbf{v}}_i^* = c_i(\mathbf{v}_i - \mathbf{v}^*) \tag{10.5b}$$

From a mathematical viewpoint, any one of these flux definitions is adequate for all diffusion situations; however, in a given situation there is usually one definition that leads to minimum algebraic complexity. An important example is when the convective transport present requires a solution of the conservation-of-momentum equation; the solution yields the mass-average velocity field, and it is then most convenient to use the mass flux relative to the mass-average velocity, that is, $\mathbf{j}_i$. Conditions of constant pressure and temperature, often encountered by chemical engineers, have often led to the choice of the absolute molar flux $\mathbf{N}_i$ because of simplifications that result from the molar density c being constant.

The definitions of fluxes lead directly to a number of important relations that can be easily derived. The absolute mass flux of the mixture (mass velocity) is

$$\mathbf{n} = \sum \mathbf{n}_i = \rho \mathbf{v} \tag{10.6a}$$

and on a molar basis,

$$\mathbf{N} = \sum \mathbf{N}_i = c\mathbf{v}^* \tag{10.6b}$$

Also,

$$\sum \mathbf{j}_i = \sum \mathbf{J}_i^* = 0 \tag{10.7}$$

$$\mathbf{N}_i = \frac{\mathbf{n}_i}{\mathbf{M}_i} \tag{10.8}$$

$$\mathbf{n}_i = \rho_i \mathbf{v} + \mathbf{j}_i = m_i \sum \mathbf{n}_i + \mathbf{j}_i \tag{10.9a}$$

$$\mathbf{N}_i = c_i \mathbf{v}^* + \mathbf{J}_i^* = x_i \sum \mathbf{N}_i + \mathbf{J}_i^* \tag{10.9b}$$

Notice that Eq. (10.9a) states that the absolute mass flux can be expressed as a sum of the convective flux $\rho_i \mathbf{v}$ and the diffusive flux $\mathbf{j}_i$. Equation (10.9b) can be interpreted similarly. Figure 10.2 illustrates the flux definitions.

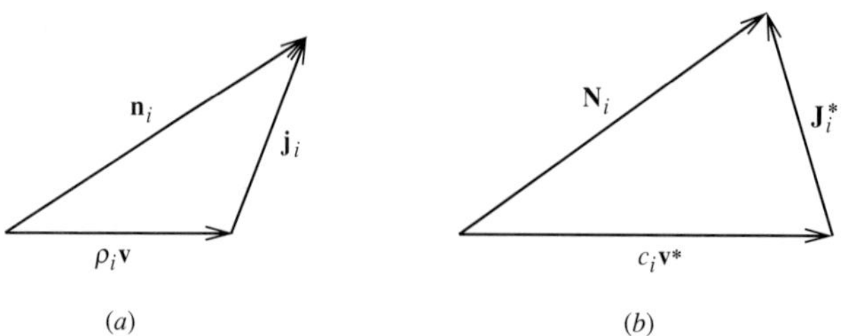

Figure 10.2 Flux vectors: (*a*) mass basis, (*b*) molar basis.

EXAMPLE 10.2 Evaporation of Water

A water surface is at 350 K and evaporates into air at 1 atm at a rate of 0.01 kg/m^2 s. Determine the convective and diffusive components of the water vapor flux across the s-surface.

Solution

Given: Water at 350 K evaporating at 0.01 kg/m^2 s into air at 1 atm.

Required: Convective and diffusive components of the water vapor flux across the s-surface.

Assumptions: Air is negligibly soluble in water.

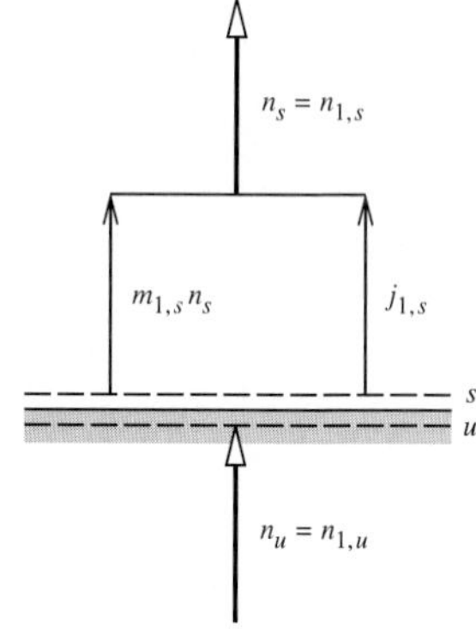

Let water be species 1 and air species 2. By mass conservation the absolute mass fluxes across the u- and s-surfaces are equal,

$$n_u = n_s = 0.01 \text{ kg/m}^2 \text{ s}$$

Using Eq. (10.9*a*), the absolute flux of species 1 across the s-surface can be written as a sum of convective and diffusive components,

$$n_{1,s} = m_{1,s}n_s + j_{1,s}$$

From PSYCHRO, for $T_s = 350$ K, $P = 1.013 \times 10^5$ Pa, and RH $= 100\%$, $m_{1,s} = 0.303$.
Hence, the convective component is

$$m_{1,s}n_s = (0.303)(0.01) = 0.00303 \text{ kg/m}^2 \text{ s}$$

Also, $n_{2,s} = 0$ since air is negligibly soluble in water. Hence,

$$n_s = n_{1,s} + n_{2,s} = n_{1,s} = 0.01 \text{ kg/m}^2 \text{ s}$$

$$j_{1,s} = n_s(1 - m_{1,s}) = (0.01)(1 - 0.303) = 0.00697 \text{ kg/m}^2 \text{ s}$$

Comment

The ratio of diffusive to convective components is equal to $(1 - m_{1,s})/m_{1,s}$. As $T \to T_{\text{BP}}$, $m_{1,s} \to 1$ and the diffusive component goes to zero.

10.2.3 The General Species Conservation Equation

We take an Eulerian viewpoint and consider a Cartesian coordinate volume element Δx by Δy by Δz, fixed in space, as shown in Fig. 10.3. Conservation of a chemical species i requires that the time rate of storage of species i within the volume equals the net rate of inflow of species i across the boundary plus the production rate of species i within the volume due to chemical reactions.

The time rate of storage of species i within the control volume is simply

$$\frac{\partial \rho_i}{\partial t}\Delta x \Delta y \Delta z$$

The gross rate of inflow of species i is

$$n_{ix}|_x \Delta y \Delta z + n_{iy}|_y \Delta x \Delta z + n_{iz}|_z \Delta x \Delta y$$

Similarly, the gross rate of outflow is

$$n_{ix}|_{x+\Delta x} \Delta y \Delta z + n_{iy}|_{y+\Delta y} \Delta x \Delta z + n_{iz}|_{z+\Delta z} \Delta x \Delta y$$

The net rate of inflow is found by subtracting the outflow from the inflow, expanding quantities evaluated at $x + \Delta x$, $y + \Delta y$, and $z + \Delta z$ in Taylor series, and retaining only first-order terms to obtain

$$-\left(\frac{\partial n_{ix}}{\partial x} + \frac{\partial n_{iy}}{\partial y} + \frac{\partial n_{iz}}{\partial z}\right)\Delta x \Delta y \Delta z$$

Species i may be produced at the rate $\dot{r}_i'''$ mass per unit volume per unit time [kg/m^3 s] due to homogeneous chemical reactions (reactions that take place *within*

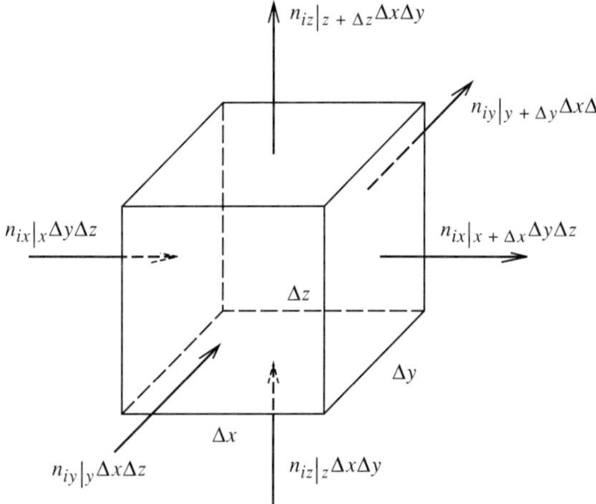

Figure 10.3 Cartesian elemental control volume for derivation of the species conservation equation.

the medium). The rate of production of i within the volume is then

$$\dot{r}_i'''\Delta x \Delta y \Delta z$$

Substituting in our conservation statement, dividing throughout by the volume $\Delta x \Delta y \Delta z$, and letting Δx, Δy, Δz go to zero,

$$\frac{\partial \rho_i}{\partial t} = -\left(\frac{\partial n_{ix}}{\partial x} + \frac{\partial n_{iy}}{\partial y} + \frac{\partial n_{iz}}{\partial z}\right) + \dot{r}_i''' \tag{10.10}$$

which is one form of the species conservation equation. Summation of Eq. (10.10) over all species i gives

$$\frac{\partial}{\partial t}\sum \rho_i = -\left(\frac{\partial}{\partial x}\sum n_{ix} + \frac{\partial}{\partial y}\sum n_{iy} + \frac{\partial}{\partial z}\sum n_{iz}\right) + \sum \dot{r}_i'''$$

Since $\sum \rho_i = \rho$, $\sum n_i = n$, and $\sum \dot{r}_i''' = 0$ (mass is conserved in chemical reactions), there results

$$\frac{\partial \rho}{\partial t} + \left(\frac{\partial n_x}{\partial x} + \frac{\partial n_y}{\partial y} + \frac{\partial n_z}{\partial z}\right) = 0 \tag{10.11}$$

For velocity components u, v, w in the x, y, and z coordinate directions, we have $n_x = \rho u$, $n_y = \rho v$, and $n_z = \rho w$; hence,

$$\frac{\partial \rho}{\partial t} + \left(\frac{\partial \rho u}{\partial x} + \frac{\partial \rho v}{\partial y} + \frac{\partial \rho w}{\partial z}\right) = 0 \tag{10.12a}$$

or, in vector form,

$$\frac{\partial \rho}{\partial t} + \nabla \cdot \rho \mathbf{v} = 0; \qquad \mathbf{v} = \mathbf{i}u + \mathbf{j}v + \mathbf{k}w \tag{10.12b}$$

where $\nabla \cdot$ is the divergence operator. Equation (10.12) is the mass conservation (or continuity) equation. Equation (10.10) can also be written in compact form using the divergence operator,

$$\frac{\partial \rho_i}{\partial t} + \nabla \cdot \mathbf{n}_i = \dot{r}_i''' \tag{10.13}$$

In addition, if we wish to write Eq. (10.13) in coordinate systems other than the Cartesian one, we need only to express $\nabla \cdot$ appropriately. For example, in spherical coordinates with spherical symmetry, we obtain

$$\frac{\partial \rho_i}{\partial t} + \frac{1}{r^2}\frac{\partial}{\partial r}\left(r^2 n_{ir}\right) = \dot{r}_i''' \tag{10.14}$$

The species equation can also be derived on a molar basis as

$$\frac{\partial c_i}{\partial t} + \nabla \cdot \mathbf{N}_i = \dot{R}_i''' \tag{10.15}$$

where $\dot{R}_i'''$ is the molar rate of production of species i [kmol/m^3 s]. Summing Eq. (10.15) over all species gives

$$\frac{\partial c}{\partial t} + \nabla \cdot (c\mathbf{v}^*) = \sum_i \dot{R}_i''' \tag{10.16}$$

where, in general, $\sum \dot{R}_i''' \neq 0$ since moles need not be conserved in chemical reactions.

The final step to obtain a differential equation governing the concentration distribution is to write the absolute flux as the sum of convective and diffusive fluxes, and introduce appropriate physical laws for the diffusive flux $\mathbf{j}_i$ (or $\mathbf{J}_i^*$). Also, if there are rate-controlled homogeneous chemical reactions, chemical kinetics expressions are required for the production term $\dot{r}_i'''$ (or $\dot{R}_i'''$).

10.2.4 A More Precise Statement of Fick's Law

We have noted how a mass species may be transported by convection and diffusion. Convection is of its nature a *bulk* motion and thus transports the mixture as a whole. A given species can be transported relative to this bulk motion if diffusion occurs. Thus, it is clear that a precise definition of Fick's first law must describe diffusion relative to an average velocity of the mixture. Our definitions of velocities and fluxes, Eqs. (10.1) through (10.5), were of a mathematical nature and depend on no physical laws. However, they were chosen so that the introduction of physics via Fick's first law is straightforward. We now propose that the law should be written

$$\mathbf{j}_1 = -\rho\mathscr{D}_{12}\nabla m_1 \tag{10.17}$$

that is, exactly as Eq. (9.15), but now $\mathbf{j}_1$ has been precisely defined as the mass flux of species 1 *relative to the mass-average velocity.* The corresponding law written for species 2 is

$$\mathbf{j}_2 = -\rho\mathscr{D}_{21}\nabla m_2$$

Since $\nabla m_1 = -\nabla m_2$ in a binary system, and from Eq. (10.7), $\mathbf{j}_1 + \mathbf{j}_2 = 0$, it then follows immediately that

$$\mathscr{D}_{12} = \mathscr{D}_{21} \tag{10.18}$$

The molar form of Fick's first law equivalent to Eq. (10.17) is

$$\mathbf{J}_1^* = -c\mathscr{D}_{12}\nabla x_1 \tag{10.19}$$

Algebraic manipulation will show that Eqs. (10.19) and (10.17) are mathematically equivalent. However, we see that the diffusive molar flux $\mathbf{J}_1^*$ is the molar flux *relative to the molar-average velocity.*

It is now possible to give a correct interpretation of what we mean by a stationary medium. If we are working in mass units, we require that the mass-average velocity be zero; if we are working in molar units, we require that the molar-average velocity be zero. Then, in a stationary medium, mass transport is by diffusion only. Since a

Pitot tube or anemometer measures the mass-average velocity, the interpretation of stationary as zero mass-average velocity is more in accord with our physical intuition. A zero molar-average velocity corresponds to a zero molecule average velocity, and is a concept more appropriate to interpretation at a molecular level.

Finally, the substitution of Fick's law into Eqs. (10.9*a*) and (10.9*b*), written for a binary system, yields two important relations:

$$\mathbf{n}_1 = \rho_1\mathbf{v} - \rho\mathscr{D}_{12}\nabla m_1 = m_1(\mathbf{n}_1 + \mathbf{n}_2) - \rho\mathscr{D}_{12}\nabla m_1 \qquad \mathbf{(10.20}\boldsymbol{a}\mathbf{)}$$

$$\mathbf{N}_1 = c_1\mathbf{v}^* - \rho\mathscr{D}_{12}\nabla x_1 = x_1(\mathbf{N}_1 + \mathbf{N}_2) - c\mathscr{D}_{12}\nabla x_1 \qquad \mathbf{(10.20}\boldsymbol{b}\mathbf{)}$$

We see that the absolute flux of a species can always be conveniently expressed as the sum of two components, one due to convection and the other due to diffusion.

10.3 HIGH MASS TRANSFER RATE DIFFUSION

Net mass transfer across a surface results in a velocity component normal to the surface, and an associated convective flux in the direction of mass transfer. If the mass transfer rate is high, this convection cannot be ignored, as was done in Chapter 9 in using low mass transfer rate theory. In the problems analyzed in this section, the only convection is that induced by the mass transfer process itself, and it is always in the direction of the mass transfer. The induced convective flow is called a *Stefan flow*. In this section we are not concerned with situations where there is also a forced- or natural-convection flow along the transfer surface: such situations are dealt with in Sections 10.4 and 10.5. The analyses that follow consider a variety of evaporation and combustion problems.

10.3.1 Diffusion with One Component Stationary

Diffusion in a binary mixture where one component is stationary occurs in many situations involving evaporation, condensation, or transpiration. To develop the concepts involved, a particular situation will be considered, that of a heatpipe into which has leaked a small amount of gas. Fluid flow and heat transfer in heatpipes were analyzed in Chapter 7. We now consider a feature of heatpipe operation that requires a mass transfer analysis.

One-Dimensional Model of a Heatpipe

We wish to determine the effect of a small amount of noncondensable gas added to the vapor in a fixed-conductance heatpipe. Corrosion may have generated gas within the heatpipe, or a construction defect may have allowed gas to leak into the heatpipe. A simple one-dimensional analysis gives insight into this phenomenon. We will consider the case of a heatpipe with the evaporator and condenser located at the *ends* of the heatpipe pipe only. (We shall see later that this is a poor design.) Figure 10.4 depicts an idealized model of such a heatpipe.

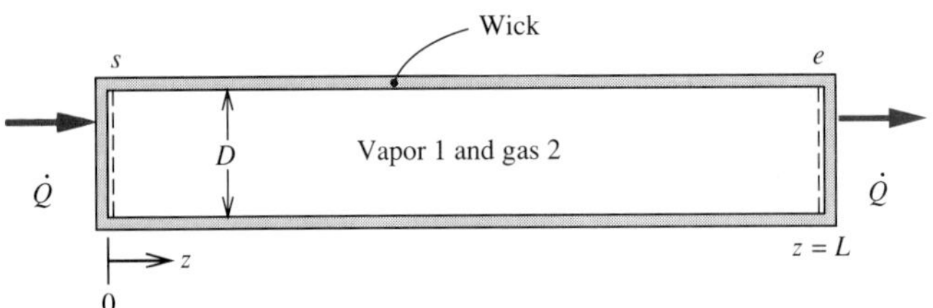

Figure 10.4 A heatpipe with the evaporator and condenser located at its ends.

The vapor is designated as species 1 and the gas as species 2. Heat is applied at the surface $z = 0$, and heat is removed at the surface $z = L$. The sides of the heatpipe are assumed insulated. The temperatures at $z = 0$ and $z = L$ are T_s and T_e, respectively. The corresponding mole fractions of vapor adjacent to the liquid surfaces are $x_{1,s}$ and $x_{1,e}$. These values are the equilibrium values obtained from the saturation vapor pressures corresponding to the liquid surface temperatures T_s and T_e, respectively. The reason for choosing subscripts s and e (rather than, say, E and C for evaporator and condenser) will become apparent later. The solubility of gas in the liquid is assumed zero, and thus species 1 is the only substance transferred across the interfaces. At steady state there is net transfer of vapor along the heatpipe, whereas conservation of species requires that the gas, species 2, remains stationary. The variations of pressure and absolute temperature along the heatpipe are small; thus, if an ideal gas mixture is assumed, the molar concentration c is virtually independent of position z. Advantage of this fact may be taken by performing the analysis on a molar basis. For this problem Eq. (10.15) written for the vapor reduces to

$$\frac{d}{dz}(N_1) = 0 \tag{10.21}$$

Likewise, Eq. (10.15) written for the gas reduces to

$$\frac{d}{dz}(N_2) = 0 \tag{10.22}$$

The subscript z has been dropped from N_1 and N_2 since it is not needed for this one-dimensional situation. Integration yields $N_2 =$ constant. Since the gas is insoluble in the liquid, $N_2|_{z=0} = 0$, and it follows from Eq. (10.22) that N_2 is identically zero; that is, the gas is stationary.

The molar flux N_1 can be expressed as the sum of a convective and a diffusive component; from Eq. (10.20*b*),

$$N_1 = x_1(N_1 + N_2) - c\mathscr{D}_{12}\frac{dx_1}{dz}$$

But $N_2 = 0$; thus, solving for N_1 yields

$$N_1 = -\frac{c\mathscr{D}_{12}}{1 - x_1}\frac{dx_1}{dz} \tag{10.23}$$

Substitution of this expression in the differential equation Eq. (10.21) gives

$$\frac{d}{dz}\left(-\frac{c\mathscr{D}_{12}}{1-x_1}\frac{dx_1}{dz}\right) = 0 \tag{10.24}$$

The product of the molar concentration c and the binary diffusion coefficient is assumed constant ($\mathscr{D}_{12}$ is also independent of composition). Therefore, the differential equation governing the concentration distribution may be written as

$$\frac{d}{dz}\left(\frac{1}{1-x_1}\frac{dx_1}{dz}\right) = 0 \tag{10.25}$$

which is to be solved subject to the boundary conditions

$$z = 0, \quad x_1 = x_{1,s} \tag{10.26a}$$

$$z = L, \quad x_1 = x_{1,e} \tag{10.26b}$$

Integrating twice and evaluating the constants of integration from the boundary conditions yields the concentration distribution of the vapor,

$$\left(\frac{1-x_1}{1-x_{1,s}}\right) = \left(\frac{1-x_{1,e}}{1-x_{1,s}}\right)^{z/L} \tag{10.27}$$

Alternatively, the concentration distribution of the gas is given by

$$\left(\frac{x_2}{x_{2,s}}\right) = \left(\frac{x_{2,e}}{x_{2,s}}\right)^{z/L} \tag{10.28}$$

The concentration distributions are illustrated in Fig. 10.5. The rate of vapor transport through the pipe is the rate at which the liquid evaporates and may be evaluated as follows:

$$N_1\big|_{z=0} = N_{1,s} = -\frac{c\mathscr{D}_{12}}{1-x_{1,s}}\frac{dx_1}{dz}\bigg|_{z=0}$$

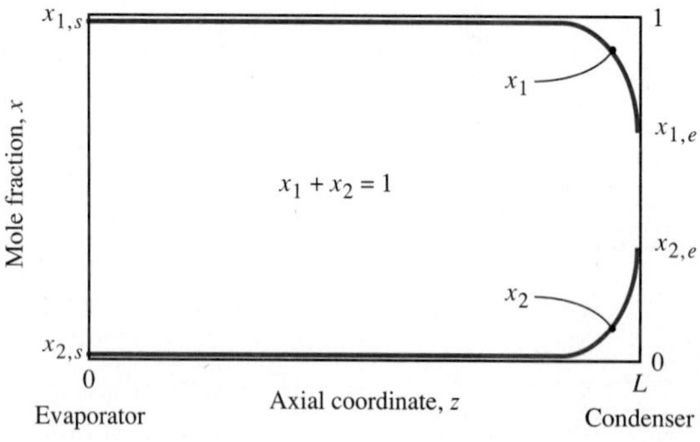

Figure 10.5 Vapor (species 1) and gas (species 2) concentration profiles along a heatpipe.

Differentiation of Eq. (10.27) and evaluation at $z = 0$ gives

$$N_1 \equiv N_{1,s} = \frac{c\mathscr{D}_{12}}{L} \ln \frac{1 - x_{1,e}}{1 - x_{1,s}} \tag{10.29}$$

The rate of heat transfer along the heatpipe is simply $M_1 N_1 h_{fg} (\pi/4) D^2$, where h_{fg} is the enthalpy of vaporization:

$$\frac{\dot{Q}}{(\pi/4)D^2} = \frac{M_1 c\mathscr{D}_{12} h_{fg}}{L} \ln \frac{1 - x_{1,e}}{1 - x_{1,s}} = \frac{M_1 c\mathscr{D}_{12} h_{fg}}{L} \ln \frac{x_{2,e}}{x_{2,s}} \tag{10.30}$$

An Alternative Form for the Vapor Transport Rate

An alternative form of Eq. (10.29) may be obtained by introducing the *logarithmic mean value* of the mole fraction x_2, defined as

$$(x_2)_{lm} = \frac{x_{2,e} - x_{2,s}}{\ln(x_{2,e}/x_{2,s})} \tag{10.31}$$

Algebraic manipulation of Eq. (10.29) and substitution of Eq. (10.31) then yields

$$N_1 = \frac{c\mathscr{D}_{12}}{L(x_2)_{lm}} (x_{1,s} - x_{1,e}) \tag{10.32}$$

For low transfer rates ($N_1 \rightarrow 0$), the value of $(x_2)_{lm}$ approaches its limiting value of unity, and there is a linear Ohm's law type relation between N_1 and $(x_{1,s} - x_{1,e})$. As the concentration difference Δx_1 and transfer rate increase, $(x_2)_{lm}$ decreases, and N_1 is no longer a linear function of Δx_1; this nonlinearity at high mass transfer rates is an important characteristic of high mass transfer rate theory.

Even though we have said that one component is stationary, there is, of course, diffusion of both species. The concentration gradients of components 1 and 2 are equal and opposite, and hence gas 2 diffuses toward the evaporating surface. Although the gas is insoluble, its concentration does not build up at the liquid surface. There is a convective motion in the positive z direction such that at steady state, the net flux N_2, which is the sum of convective component $x_2 N$ and diffusive component $-c\mathscr{D}_{12}(dx_2/dz)$, is exactly zero. A small total pressure gradient is required to overcome wall friction in driving this convective flow along the heatpipe. Convection is, by nature, a bulk movement of the mixture, and hence vapor 1 is also convected in the positive z direction; this was, of course, accounted for in Eq. (10.20*b*) by writing the total flux of vapor 1, N_1, as a sum of convective and diffusive components. This induced convective flow that augments the diffusive flux of vapor 1 is called the *Stefan flow*.

Location of the evaporator and condenser at the ends of the heatpipe allows the use of a simple one-dimensional model of diffusion and convection. The practical consequences of having the condenser at the end of the heatpipe are discussed at the end of Example 10.3. A different concern is whether the one-dimensional analysis is appropriate for the model problem. We have seen that there is a flow along the tube induced by the diffusion process, and we assumed a uniform profile of the velocity $v_z = N_1/c$. In reality, the no-slip condition for a viscous fluid must apply at the walls of the pipe, and the flow pattern in the heatpipe is rather complex. Notwithstanding this complication, our simple one-dimensional model is quite adequate to demonstrate the effect of a noncondensable gas on heatpipe performance.

EXAMPLE 10.3 Effect of Air on Heatpipe Performance

A water heatpipe 70 cm long and of 2 cm inside diameter is taken from a storehouse and its performance checked prior to use. For a heat load of 100 W and an evaporator temperature of 100°C, the condenser temperature is measured to be 95.0°C. An air leak or gas generation is suspected; calculate the amount of air that might have leaked into the heatpipe.

Solution

Given: Test data for a water heatpipe.

Required: Amount of air that has leaked into the heatpipe.

Assumptions: A one-dimensional model is appropriate.

First we determine the expected temperature drop if no air were present. The vapor velocity V is calculated from the heat load, evaluating the steam properties for saturated steam at 1 atm.

$$\rho V\left(\frac{\pi}{4}\right)D^2 h_{fg} = \dot{Q} = 100 \text{ W}$$

$$V = \frac{\dot{Q}}{\rho(\pi/4)D^2 h_{fg}} = \frac{100}{(0.5977)(\pi/4)(0.02)^2(2.257 \times 10^6)} = 0.236 \text{ m/s}$$

The Reynolds number is

$$\text{Re}_D = \frac{VD}{\nu} = \frac{(0.236)(0.02)}{20.1 \times 10^{-6}} = 235$$

The friction factor for fully developed laminar flow is

$$f = \frac{64}{\text{Re}_D} = \frac{64}{235} = 0.272$$

and hence the pressure drop is

$$\Delta P \simeq f\frac{L}{D}\left(\frac{1}{2}\rho V^2\right) = (0.272)\left(\frac{70}{2}\right)(1/2)(0.5977)(0.236)^2 = 0.158 \text{ Pa}$$

Steam tables show that the corresponding temperature drop is quite negligible.

Now consider the situation with air present. We will assume (and check later) that the air concentration at the evaporator end is very small; then, since the evaporator is at 100°C, the total pressure there is 101,330 Pa and varies negligibly along the pipe. From steam tables, the saturation pressure corresponding to 95°C is 84,520 Pa, and the mole fraction of air is

$$x_{2,e} = \frac{P_{2,e}}{P_{\text{total}}} = \frac{101{,}330 - 84{,}520}{101{,}330} = 0.166$$

Equation (10.30) may be rearranged to read

$$\ln \frac{x_{2,e}}{x_{2,s}} = \frac{\dot{Q}L}{M_1 c \mathscr{D}_{12} h_{fg} (\pi/4) D^2}$$

The binary diffusion coefficient for water vapor in air is conveniently given by the formula in Table A.17*a* for $273 < T < 373$ K,

$$\mathscr{D}_{12} = 1.97 \times 10^{-5} \left(\frac{P_0}{P}\right)\left(\frac{T}{T_0}\right)^{1.685} \text{ m}^2/\text{s}; \quad P_0 = 1 \text{ atm}, T_0 = 256 \text{ K}$$

$$\mathscr{D}_{12} = 1.97 \times 10^{-5} \left(\frac{370.5}{256}\right)^{1.685} = 36.7 \times 10^{-6} \text{ m}^2/\text{s}$$

and the total molar concentration c is

$$c = P/\mathscr{R}T = (101{,}330)/(8314)(370.5) = 3.29 \times 10^{-2} \text{ kmol/m}^3$$

where both $\mathscr{D}_{12}$ and c have been evaluated at the average temperature along the heatpipe, 97.5°C $\simeq$ 370.5 K. Let the ratio $x_{2,e}/x_{2,s}$ be denoted by r; then

$$\ln r = \frac{(100)(0.70)}{(18)(3.29 \times 10^{-2})(36.7 \times 10^{-6})(2.257 \times 10^6)(\pi/4)(0.02)^2} = 4.54 \times 10^3$$

which is very large; our assumption of a negligibly small value of $x_{2,s}$ is thus fully justified.

The total moles of air in the heatpipe is the volume integral of the air concentration,

$$W = (\pi/4)D^2 \int_0^L x_2 c \, dz$$

Using Eq. (10.28) we obtain

$$W = (\pi/4)D^2 c x_{2,s} L \int_0^1 r^{z/L} \, d(z/L)$$

$$= (\pi/4)D^2 c x_{2,s} L \frac{r-1}{\ln r} = (\pi/4)D^2 c x_{2,e} \frac{L}{r} \frac{r-1}{\ln r} = \frac{(\pi/4)D^2 c x_{2,e} L}{\ln r}$$

since r is large. Hence,

$$W = \frac{(\pi/4)(0.02)^2(3.29 \times 10^{-2})(0.166)(0.70)}{4.54 \times 10^3} = 2.65 \times 10^{-10} \text{ kmol}$$

The mass of air in the pipe w is equal to WM_2; thus,

$$w = (2.65 \times 10^{-10})(29) = 7.69 \times 10^{-9} \text{ kg} = 7.69 \ \mu\text{g}$$

Comments

1. The fact that such a small amount of air as 7.69 μg would cause a 5°C temperature drop is due to the geometrical arrangement considered. Location of the condenser on

the end of the heatpipe results in direct confrontation between the gas and the vapor attempting to condense. Since a minute amount of gas can completely shut down the system, precise control would be almost impossible. A superior design would be one where the condenser is located along the side wall of the heatpipe; the working fluid can then simply push the gas aside out of the main stream to an unused portion of the condenser. Indeed, this is the principle of the variable-conductance heatpipe described in Section 7.7.3, for which a relatively large amount of gas is present in the heatpipe.

2. Notice the formula used for the diffusion coefficient of water vapor in air. We *cannot* use a Schmidt number $\mathrm{Sc} = \nu/\mathscr{D}_{12} = 0.61$ to obtain $\mathscr{D}_{12}$ as we did in Sections 9.4 and 9.5. For the water vapor–air system, $\mathrm{Sc} = 0.61$ is appropriate only for small concentrations of water vapor (see Exercise 10–7).

10.3.2 Heterogeneous Combustion

In Section 9.3.3 diffusion with a heterogeneous reaction was considered for the special case of a catalytic reaction. The analysis was particularly simple because no net mass was consumed in the reaction, and hence there was no velocity component or associated convective transport normal to the catalyst surface. Similarly, the low-temperature carbon combustion problem of Section 9.5.3 was analyzed on a molar basis to take advantage of the fact that the molar-average velocity normal to the carbon surface was zero: one mole of CO_2 is produced for each mole of O_2 consumed. We now consider a situation where net mass is consumed and there is convective transport normal to the reacting surface, irrespective of whether the analysis is performed on a mass or molar basis. The example problem is that of a small carbon particle entrained and burning in a high-temperature air flow, as shown in Fig. 10.6. At temperatures above about 2300 K, the reaction is

$$2\mathrm{C} + \mathrm{O}_2 \rightarrow 2\,\mathrm{CO}$$

and is in equilibrium (diffusion controlled) with two moles of CO produced for one mole of O_2 consumed: thermodynamic data indicates $m_{\mathrm{O}_2,s} \simeq 0$ and that no carbon dioxide forms in the gas phase. Oxygen is therefore inert in the gas phase. Denoting O_2 as species 1 and assuming quasi-steady, spherically symmetric diffusion,

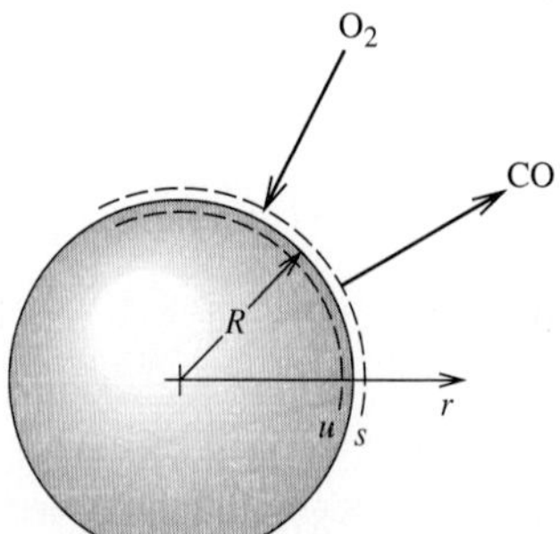

Figure 10.6 Schematic of a carbon particle burning in high-temperature air. The surface reaction is $2\mathrm{C} + \mathrm{O}_2 \rightarrow 2\,\mathrm{CO}$.

Eq. (10.14) reduces to

$$\frac{d}{dr}(r^2 n_1) = 0 \tag{10.33}$$

where the subscript r on n_1 has been dropped for convenience. The mass conservation equation Eq. (10.12b) reduces to

$$\frac{d}{dr}(r^2 \rho v) = \frac{d}{dr}(r^2 n) = 0 \tag{10.34}$$

Integrating Eq. (10.34),

$$r^2 n = \text{Constant} = r^2 n|_{r=R} = R^2 n_s \tag{10.35}$$

where n_s is the mass flux across the s-surface. Equation (10.35) is equivalent to the statement that the mass flow $\dot{m} = 4\pi r^2 n$ is a constant, independent of r. A mass balance on a control volume located between the s- and u-surfaces requires that $n_s = n_u$ since mass is conserved. We denote

$$n_s = n_u = \dot{m}'' \tag{10.36}$$

and term $\dot{m}'' = \dot{m}/A$ the *mass transfer rate* [kg/m^2 s]. This distinctive notation is useful since our major objective is usually to determine the mass transfer rate. Thus, Eq. (10.35) becomes

$$r^2 n = R^2 \dot{m}'' \tag{10.37}$$

Integrating Eq. (10.33),

$$r^2 n_1 = \text{Constant} = R^2 n_{1,s}$$

or, expressing the absolute flux n_1 as a sum of its convective and diffusive components,

$$r^2 \left(m_1 n - \rho \mathscr{D}_{1m} \frac{dm_1}{dr} \right) = R^2 n_{1,s} \tag{10.38}$$

Substitution of $r^2 n = R^2 \dot{m}''$ from Eq. (10.37), rearranging, and integrating with $m_1 = m_{1,s}$ at $r = R$ and $m_1 \to m_{1,e}$ as $r \to \infty$ gives

$$\int_{m_{1,s}}^{m_{1,e}} \frac{dm_1}{m_1 - (n_{1,s}/\dot{m}'')} = \frac{\dot{m}'' R^2}{\rho \mathscr{D}_{1m}} \int_R^\infty \frac{dr}{r^2}$$

where $\rho \mathscr{D}_{1m}$ has been assumed constant. Then,

$$\ln \frac{m_{1,e} - (n_{1,s}/\dot{m}'')}{m_{1,s} - (n_{1,s}/\dot{m}'')} = \frac{\dot{m}'' R}{\rho \mathscr{D}_{1m}}$$

Solving for $\dot{m}''$ and rearranging,

$$\dot{m}'' = \frac{\rho \mathscr{D}_{1m}}{R} \ln \left[1 + \frac{m_{1,e} - m_{1,s}}{m_{1,s} - (n_{1,s}/\dot{m}'')} \right] \tag{10.39}$$

For air, $m_{1,e} = 0.231$, and we have already noted that $m_{1,s} \simeq 0$. The ratio $(n_{1,s}/\dot{m}'')$ is obtained from the stoichiometry of the reaction, as follows:

$$\begin{aligned} &2C + O_2 \rightarrow 2CO \\ &2 \times 12\,\text{kg} + 32\,\text{kg} \rightarrow 2 \times 28\,\text{kg} \end{aligned} \tag{10.40}$$

For each kilogram of carbon consumed, 4/3 kg of oxygen is consumed: thus, for each kilogram of carbon transferred across the u-surface, 4/3 kg of oxygen crosses the s-surface in the negative direction; that is,

$$\frac{n_{1,s}}{n_{C,u}} = -\frac{4}{3}$$

But $n_{C,u} = \dot{m}''$; thus, $(n_{1,s}/\dot{m}'') = -4/3$. Substituting in Eq. (10.39),

$$\dot{m}'' = \frac{\rho\mathscr{D}_{1m}}{R} \ln\left[1 + \frac{0.231 - 0}{0 + 4/3}\right] = \frac{\rho\mathscr{D}_{1m}}{R} \ln[1 + 0.173] = 0.160\,\frac{\rho\mathscr{D}_{1m}}{R} \tag{10.41}$$

which is a particularly simple result for the rate of mass transfer across the particle surface.

Particle Lifetime

The lifetime of the particle can be obtained by performing a mass balance on the particle of volume V and surface area A_s:

$$\frac{d}{dt}(V\rho_{\text{solid}}) = -A_s\dot{m}''$$

where ρ_{solid} is the density of solid carbon. Thus,

$$\frac{d}{dt}\left(\frac{4}{3}\pi R^3 \rho_{\text{solid}}\right) = -(4\pi R^2)\left(0.160\frac{\rho\mathscr{D}_{1m}}{R}\right)$$

$$\frac{dR}{dt} = -0.160\frac{\rho\mathscr{D}_{1m}}{\rho_{\text{solid}}R} \tag{10.42}$$

Integrating with $\rho\mathscr{D}_{1m}$ taken as constant, and $R = R_0$ at $t = 0$, $R = 0$ at $t = \tau$,

$$\int_{R_0}^{0} R\,dR = -\frac{0.160\rho\mathscr{D}_{1m}}{\rho_{\text{solid}}}\int_0^{\tau} dt$$

Hence

$$\tau = \frac{\rho_{\text{solid}}R_0^2}{0.32\rho\mathscr{D}_{1m}} = \frac{\rho_{\text{solid}}D_0^2}{1.28\rho\mathscr{D}_{1m}} \tag{10.43}$$

Property Evaluation

Calculation of τ requires an appropriate reference state for the $\rho\mathscr{D}_{1m}$ product. Since the molecular weight of CO is almost the same as that of air, it is sufficient to use air properties evaluated at the mean film temperature. In order to obtain the mean film temperature, a simultaneous heat and mass transfer problem must be solved to determine the particle temperature, as was done in Example 9.11 for low-temperature oxidation of a carbon particle. Exercise 10–10 requires an approximate heat transfer analysis; a more exact heat transfer analysis is given in Example 10.14.

CO Concentration at the s-Surface

Notice that it is the rate at which O_2 can diffuse to the carbon surface that determines the rate of reaction: hence, the reaction is termed *diffusion-controlled.* We did not use the species conservation equation for the reaction product, CO. If we were to do so, we would obtain an equation identical to Eq. (10.39), with the subscript 2, denoting CO, replacing the subscript 1:

$$\dot{m}'' = \frac{\rho\mathscr{D}_{2m}}{R}\ln\left[1 + \frac{m_{2,e} - m_{2,s}}{m_{2,s} - (n_{2,s}/\dot{m}'')}\right] \qquad \textbf{(10.44)}$$

But since we cannot specify $m_{2,s}$, the concentration of CO at the s-surface, the equation cannot be used to determine $\dot{m}''$. In fact, with $\dot{m}''$ known, this equation allows the determination of $m_{2,s}$. In physical terms, the concentration of the product at the s-surface adjusts itself to the value required for transfer away from the particle at the required rate: diffusion of the product away from the particle does not control the rate of reaction in this situation. Now $m_{2,e} = 0$, since there is no CO far from the particle; also, $n_{2,s}/n_{C,u} = 28/12 = 7/3 = n_{2,s}/\dot{m}''$. Substituting in Eq. (10.44) and using Eq. (10.41) for $\dot{m}''$ gives

$$0.160\mathscr{D}_{1m} = \mathscr{D}_{2m}\ln\left[1 + \frac{0 - m_{2,s}}{m_{2,s} - 7/3}\right] \qquad \textbf{(10.45)}$$

The effective diffusion coefficients for O_2 and CO in the mixture are not too different since their molecular weights are close; assuming $\mathscr{D}_{1m} = \mathscr{D}_{2m}$ and solving gives $m_{2,s} = 0.345$, which is the mass fraction of the reaction product CO at the s-surface. Concentration profiles are shown in Fig. 10.7.

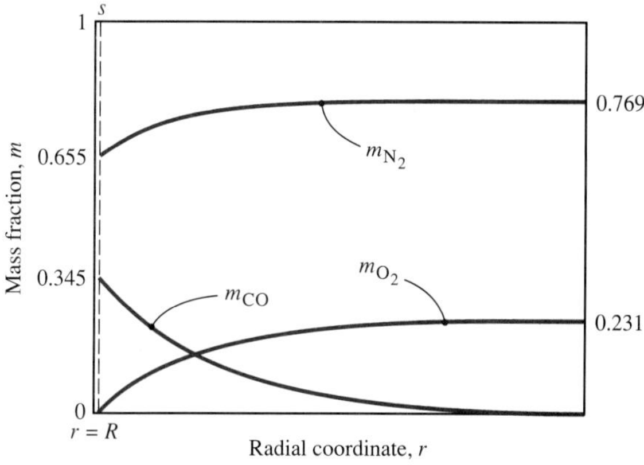

Figure 10.7 Concentration profiles for a carbon particle burning in high-temperature air.

10.3.3 General Chemical Reactions

We now consider the situation where a number of reactions occur. These reactions can be either heterogeneous or homogeneous. As a specific example, we will analyze the combustion of a small carbon particle of radius R entrained in an air stream at moderate temperatures, as shown in Fig. 9.2*d*. If the carbon surface temperature is in the range 1800–2300 K and the ambient air is relatively cool, there is a surface (heterogeneous) reaction in which carbon dioxide is reduced to carbon monoxide,

$$C + CO_2 \rightarrow 2CO$$

and in the gas phase there is a homogeneous reaction in which carbon monoxide is oxidized to carbon dioxide,

$$2CO + O_2 \rightarrow 2CO_2$$

The resulting concentration profiles are shown in Fig. 10.8. The gas-phase reaction actually occurs in a very narrow region called a flame front—the flame one sees in a charcoal-fired barbecue. The gas-phase reaction is exothermic, whereas the surface reaction is endothermic: the resulting temperature profile is also shown in Fig. 10.8. Assuming a quasi-steady state and spherical symmetry, the species conservation equation, Eq. (10.14), written for each species in turn is

$$\frac{1}{r^2}\frac{d}{dr}(r^2 n_{O_2}) = \dot{r}'''_{O_2} \tag{10.46a}$$

$$\frac{1}{r^2}\frac{d}{dr}(r^2 n_{CO_2}) = \dot{r}'''_{CO_2} \tag{10.46b}$$

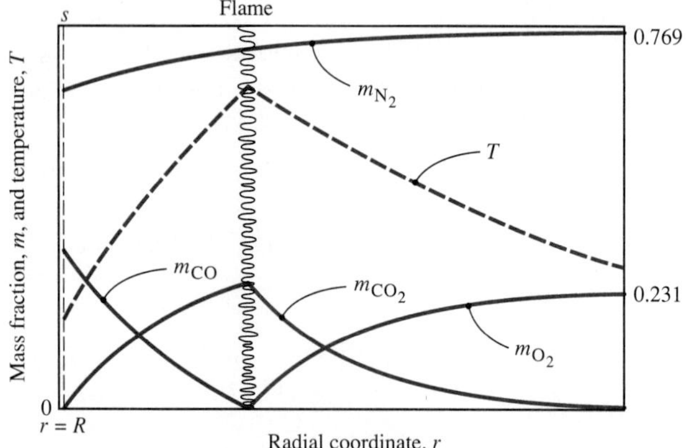

Figure 10.8 Concentration and temperature profiles for a carbon particle burning in air at intermediate temperatures. The surface reaction is $C + CO_2 \rightarrow 2CO$; the reaction in the flame is $2CO + O_2 \rightarrow 2CO_2$.

$$\frac{1}{r^2}\frac{d}{dr}(r^2 n_{CO}) = \dot{r}'''_{CO} \tag{10.46c}$$

$$\frac{1}{r^2}\frac{d}{dr}(r^2 n_{N_2}) = \dot{r}'''_{N_2} = 0, \quad \text{since } N_2 \text{ is inert} \tag{10.46d}$$

where again the subscript r on n_i has been dropped for convenience. The production rates $\dot{r}_i'''$ are given by complex expressions, possibly involving chemical kinetics or equilibrium relations. However, the problem can be solved to obtain the carbon oxidation rate without knowing the $\dot{r}_i'''$ expressions, provided we are prepared to assume that all the species diffusion coefficients are equal. For this chemical system, such an assumption is quite realistic. In order to execute the analysis, three definitions relating to *chemical elements* need to be introduced.

1. The mass fraction of chemical element α in species i is denoted $m_{(\alpha)i}$ and is obtained directly from the chemical formula for the species. For example, $m_{(O)H_2O} = 16/18 = 8/9$, and $m_{(C)CO_2} = 12/44 = 3/11$.
2. The mass fraction of chemical element α at any location in a mixture is given by

$$m_{(\alpha)} = \sum_i m_{(\alpha)i} m_i \tag{10.47}$$

   That is, $m_{(\alpha)}$ is the local mass fraction of element α, irrespective of the chemical species in which it is present.
3. The absolute flux of chemical element α is given by

$$\mathbf{n}_{(\alpha)} = \sum_i m_{(\alpha)i} \mathbf{n}_i \tag{10.48}$$

We now multiply Eq. (10.46*a*) by the mass fraction of element oxygen in species O_2, $m_{(O)O_2}$, Eq. (10.46*b*) by $m_{(O)CO_2}$, Eq. (10.46*c*) by $m_{(O)CO}$, and then add:

$$\frac{1}{r^2}\frac{d}{dr}\left[r^2\left(m_{(O)O_2} n_{O_2} + m_{(O)CO_2} n_{CO_2} + m_{(O)CO} n_{CO}\right)\right]$$
$$= m_{(O)O_2}\dot{r}'''_{O_2} + m_{(O)CO_2}\dot{r}'''_{CO_2} + m_{(O)CO}\dot{r}'''_{CO} \tag{10.49}$$

or

$$\frac{1}{r^2}\frac{d}{dr}\left(r^2 \sum_i m_{(O)i} n_i\right) = \sum_i m_{(O)i}\dot{r}_i''' \tag{10.50}$$

The first term on the right-hand side of Eq. (10.49) is the rate at which element O appears (or disappears) in the form of O_2, as O_2 is produced (or consumed) by the chemical reactions; the second term is the rate at which O appears in the form of CO_2; and so on. Since there cannot be a net creation or destruction of a chemical element, these terms must sum to zero, $\sum_i m_{(O)i}\dot{r}_i''' = 0$. Or, more generally, for any chemical element α,

$$\sum_i m_{(\alpha)i}\dot{r}_i''' = 0 \tag{10.51}$$

Using Eqs. (10.48) and (10.51), Eq. (10.49) becomes

$$\frac{d}{dr}(r^2 n_{(\mathrm{O})}) = 0$$

Integrating once,

$$r^2 n_{(\mathrm{O})} = \text{Constant} = R^2 n_{(\mathrm{O}),s}$$

Substituting back for $n_{(\mathrm{O})}$ from Eq. (10.48) and expressing each absolute species flux as a sum of convective and diffusive components gives

$$r^2 \sum_i m_{(\mathrm{O})i}\left(m_i n - \rho\mathscr{D}_{im}\frac{dm_i}{dr}\right) = R^2 n_{(\mathrm{O}),s}$$

If we now assume all diffusion coefficients are equal, that is $\mathscr{D}_{\mathrm{O}_2m} = \mathscr{D}_{\mathrm{CO}_2m} = \mathscr{D}_{\mathrm{CO},m} = \mathscr{D}$,

$$r^2\left[n\sum_i m_{(\mathrm{O})i}m_i - \rho\mathscr{D}\frac{d}{dr}\sum_i m_{(\mathrm{O})i}m_i\right] = R^2 n_{(\mathrm{O}),s}$$

or

$$r^2\left[n m_{(\mathrm{O})} - \rho\mathscr{D}\frac{dm_{(\mathrm{O})}}{dr}\right] = R^2 n_{(\mathrm{O}),s} \tag{10.52}$$

At this point, the mathematical problem has been reduced to a form identical to that in Section 10.3.2, since Eq. (10.52) is of the same form as Eq. (10.38). We need not repeat the algebra of that analysis: from Eq. (10.39) the result is

$$\dot{m}'' = \frac{\rho\mathscr{D}}{R}\ln\left[1 + \frac{m_{(\mathrm{O}),e} - m_{(\mathrm{O}),s}}{m_{(\mathrm{O}),s} - (n_{(\mathrm{O}),s}/\dot{m}'')}\right] \tag{10.53}$$

In order to use this result to calculate the carbon oxidation rate $\dot{m}''$, we must be able to evaluate $m_{(\mathrm{O}),e}$, $m_{(\mathrm{O}),s}$ and $n_{(\mathrm{O}),s}/\dot{m}''$ from known data. For ambient air,

$$m_{(\mathrm{O}),e} = m_{(\mathrm{O})\mathrm{O}_2}m_{\mathrm{O}_2,e} = m_{\mathrm{O}_2,e} = 0.231$$

At the s-surface,

$$m_{(\mathrm{O}),s} = \sum_i m_{(\mathrm{O})i}m_{i,s} = \frac{1}{1}m_{\mathrm{O}_2,s} + \frac{32}{44}m_{\mathrm{CO}_2,s} + \frac{16}{28}m_{\mathrm{CO},s}$$

In the surface temperature range under consideration, the forward reactions of carbon with oxidizing species are very rapid. Thus, the concentration CO_2 (and O_2) at the

s-surface may be taken to be zero, and then

$$m_{(\mathrm{O}),s} = \frac{16}{28} m_{\mathrm{CO},s}$$

Carbon monoxide is the product of the oxidation reactions: its concentration at the s-surface is unknown. Thus, $m_{(\mathrm{O}),s}$ is unknown, and as a result we cannot use Eq. (10.53) directly to obtain $\dot{m}''$. Instead we proceed as follows. First, Eq. (10.53) is rewritten as

$$m_{(\mathrm{O}),s} - (n_{(\mathrm{O}),s}/\dot{m}'') = [m_{(\mathrm{O}),e} - (n_{(\mathrm{O}),s}/\dot{m}'')] \exp(-\dot{m}''R/\rho\mathscr{D}) \qquad \textbf{(10.54a)}$$

and, recognizing that the analysis could be repeated for the element carbon, we can also write

$$m_{(\mathrm{C}),s} - (n_{(\mathrm{C}),s}/\dot{m}'') = [m_{(\mathrm{C}),e} - (n_{\mathrm{C},s}/\dot{m}'')] \exp(-\dot{m}''R/\rho\mathscr{D}) \qquad \textbf{(10.54b)}$$

Multiplying Eq. (10.54*a*) by ζ and Eq. (10.54*b*) by ξ, adding, and rearranging gives

$$\dot{m}'' = \frac{\rho\mathscr{D}}{R} \ln\left[1 + \frac{(\zeta m_{(\mathrm{O})} + \xi m_{(\mathrm{C})})_e - (\zeta m_{(\mathrm{O})} + \xi m_{(\mathrm{C})})_s}{(\zeta m_{(\mathrm{O})} + \xi m_{(\mathrm{C})})_s - (\zeta n_{(\mathrm{O})} + \xi n_{(\mathrm{C})})_s/\dot{m}''}\right] \qquad \textbf{(10.55)}$$

We are free to choose any values for the multipliers ζ and ξ. Let us see if values can be chosen such that the terms in Eq. (10.55) can be evaluated. The e-state presents no problem. For the s-state,

$$\begin{aligned}(\zeta m_{(\mathrm{O})} + \xi m_{(\mathrm{C})})_s &= \zeta\left(m_{\mathrm{O}_2,s} + \frac{32}{44} m_{\mathrm{CO}_2,s} + \frac{16}{28} m_{\mathrm{CO},s}\right) \\ &\quad + \xi\left(\frac{12}{44} m_{\mathrm{CO}_2,s} + \frac{12}{28} m_{\mathrm{CO},s}\right) \\ &= m_{\mathrm{CO},s}\left(\frac{16}{28}\zeta + \frac{12}{28}\xi\right), \quad \text{since } m_{\mathrm{O}_2,s} = m_{\mathrm{CO}_2,s} = 0\end{aligned}$$

If we choose $\zeta = 12/16$ and $\xi = -1$, then $[(16/28)\zeta + (12/28)\xi] = 0$, and we need not know $m_{\mathrm{CO},s}$ to evaluate the s-state. For this choice of multipliers,

$$\left(\frac{12}{16} m_{(\mathrm{O})} - m_{(\mathrm{C})}\right)_e = \frac{12}{16} m_{\mathrm{O}_2,e} = \frac{3}{4} m_{\mathrm{O}_2,e}$$

To evaluate $n_{(\mathrm{O}),s}$ and $n_{(\mathrm{C}),s}$ we perform element balances on a control volume located between the u- and s-surfaces. Since only the element carbon crosses the u-surface,

$$n_{(\mathrm{O}),s} = n_{(\mathrm{O}),u} = 0; \qquad n_{(\mathrm{C}),s} = n_{(\mathrm{C}),u} = n_s = \dot{m}''$$

Substituting in Eq. (10.55) with $m_{(\mathrm{O}),e} = 0.231$, $m_{(\mathrm{C}),e} = 0$, gives

$$\dot{m}'' = \frac{\rho\mathscr{D}}{R} \ln\left[1 + \frac{(3/4)(0.231) - 0}{0 - (-\dot{m}''/\dot{m}'')}\right] = 0.160 \frac{\rho\mathscr{D}}{R} \qquad \textbf{(10.56)}$$

which is identical to Eq. (10.41), except that Eq. (10.41) requires the diffusion coefficient to be that of O_2 in the mixture, whereas Eq. (10.56) requires $\mathscr{D}$ to be some average of $\mathscr{D}_{O_2,m}$, $\mathscr{D}_{CO_2,m}$, and $\mathscr{D}_{CO,m}$. Apart from this difference, the fact that the detailed reactions and even the final product are not the same has no effect on the oxidation rate.

The general principle underlying the above analysis is that the assumption of equal diffusion coefficients for all species containing the reactants allows the elimination of the species production rate expressions $\dot{r}_i'''$, and leads to an analytical solution in terms of mass fractions of chemical elements. In the combustion literature, this method is often called the *Shvab-Zeldovich transformation*. Although it allows the mass transfer rate to be calculated, the analysis does not yield information as to where the gas-phase reaction actually occurs; that is, in the carbon particle problem, it does not give the radius of the flame front. Exercise 10–11 requires an analysis of this problem that does allow the flame-front radius to be determined.

Property Evaluation

Use of Eq. (10.56) requires evaluation of the $\rho\mathscr{D}$ product. At 1000 K, the binary diffusion coefficients of O_2, CO, and CO_2 in air are $1.52 \times 10^{-4}\,\mathrm{m^2/s}$, $1.54 \times 10^{-4}\,\mathrm{m^2/s}$, and $1.24 \times 10^{-4}\,\mathrm{m^2/s}$, respectively. These values are not too different, and using the value for O_2 will be adequate for most purposes. Evaluation of ρ poses a tricky problem because of the unusual concentration profiles. Since the concentration of CO_2 goes to zero at both the s-surface and at infinity, its presence does not affect evaluation of density at the mean film composition. However, N_2 is the dominant species, and use of the mean film composition should be quite adequate for most purposes. The temperature profile shown in Fig. 10.8 indicates that the choice of an appropriate reference temperature for property evaluation is also difficult. Use of the mean of the ambient temperature and the adiabatic flame temperature for combustion of CO is one possible choice.

EXAMPLE 10.4 Combustion of a Carbon Particle in a Steam-Oxygen Mixture

In a coke gasification process, small carbon particles are entrained in a flow of an oxygen and steam mixture. Determine the effect of the steam content on particle oxidation rate at the inlet of the reactor.

Solution

Given: Carbon particles oxidizing in O_2–H_2O mixtures.

Required: Effect of H_2O mass fraction on particle oxidation rate.

Assumptions: 1. Concentrations of O_2, CO_2, and H_2O at the s-surface are zero.
2. The $\rho\mathscr{D}$ product of the mixture is independent of H_2O content.

Equation (10.55) applies and gives the oxidation rate as

$$\dot{m}'' = \frac{\rho\mathscr{D}}{R}\ln\left[1 + \frac{(\zeta m_{(O)} + \xi m_{(C)})_e - (\zeta m_{(O)} + \xi m_{(C)})_s}{(\zeta m_{(O)} + \xi m_{(C)})_s - (\zeta n_{(O)} + \xi n_{(C)})_s/\dot{m}''}\right]$$

The s-state is evaluated as

$$(\zeta m_{(O)} + \xi m_{(C)})_s = \zeta\left(m_{O_2,s} + \frac{32}{44}m_{CO_2,s} + \frac{16}{28}m_{CO,s} + \frac{16}{18}m_{H_2O,s}\right) + \xi\left(\frac{12}{44}m_{CO_2,s} + \frac{12}{28}m_{(CO),s}\right)$$

If temperatures are sufficiently high, we can assume diffusion-controlled reactions and $m_{O_2,s} = m_{H_2O,s} = m_{CO_2,s} = 0$. Then,

$$(\zeta m_{(O)} + \xi m_{(C)})_s = m_{CO,s}\left(\frac{16}{28}\zeta + \frac{12}{28}\xi\right)$$

Thus, as before, a choice of $\zeta = 12/16$, $\xi = -1$ will eliminate the unknown $m_{CO,s}$. Also as before, $n_{(O),s} = 0$, $n_{(C)},s = \dot{m}''$. Substituting in Eq. (10.55),

$$\dot{m}'' = \frac{\rho\mathscr{D}}{R}\ln\left[1 + \left(\frac{12}{16}m_{(O)} - m_{(C)}\right)_e\right]$$

At the inlet of the reactor, the free stream contains only O_2 and H_2O; hence $m_{CO_2,e} = m_{CO,e} = 0$, and

$$\dot{m}'' = \frac{\rho\mathscr{D}}{R}\ln\left[1 + \frac{12}{16}\left(m_{O_2,e} + \frac{16}{18}m_{H_2O,e}\right)\right]$$

The table shows the effect of steam content $m_{H_2O,e}$ on the dimensionless oxidation rate $\dot{m}''R/\rho\mathscr{D}$.

$m_{H_2O,e}$	$\dot{m}''R/\rho\mathscr{D}$
0	0.560
0.1	0.555
0.2	0.550
0.3	0.545
0.4	0.540
0.5	0.536

Comments

1. The calculated effect of steam content on oxidation rate is small.
2. Since the effect is small, secondary factors related to the effect of steam content on the $\rho\mathscr{D}$ product may play a more important role. The $\rho\mathscr{D}$ product depends on both gas-mixture composition and temperature. As the H_2O and H_2 content of the mixture increases, ρ decreases but the effective value of $\mathscr{D}$ increases. Note, however, we do not have to include $\mathscr{D}_{H_2m}$ when choosing an appropriate value for the assumed equal $\mathscr{D}_{O_2,m}$, $\mathscr{D}_{CO_2,m}$, $\mathscr{D}_{CO,m}$, and $\mathscr{D}_{H_2O,m}$ since H_2 does not contain element (O) or (C).

3. The reactor is a mass exchanger in which the concentrations of the reaction products H_2 and CO increase along the reactor: the above result applies only at the inlet. In a 100% effective exchanger, the outlet mixture will contain H_2 and CO only.

10.3.4 Droplet Evaporation

Low mass transfer rate theory was applied to the evaporation of a water droplet in Example 9.10. We now develop a high mass transfer rate theory for evaporation of a small droplet entrained in a gas flow, as shown in Fig. 10.9*a*. The droplet contains a single component, species 1, and the gas is species 2. The temperature of the droplet surface is T_s (to be determined later), and the corresponding mass fraction of vapor 1 at the s-surface is $m_{1,s} = m_{1,s}(T_s, P)$ from saturation vapor-pressure tables. Assuming a quasi-steady state and spherical symmetry, the governing conservation equation is again Eq. (10.33),

$$\frac{d}{dr}(r^2 n_1) = 0 \tag{10.33}$$

The solution of Eq. (10.33) for $m_1 = m_{1,s}$ at $r = R_1$ and $m_1 = m_{1,e}$ as $r \to \infty$, is Eq. (10.39) with $\mathscr{D}_{1m}$ replaced by $\mathscr{D}_{12}$:

$$\dot{m}'' = \frac{\rho\mathscr{D}_{12}}{R}\ln\left[1 + \frac{m_{1,e} - m_{1,s}}{m_{1,s} - (n_{1,s}/\dot{m}'')}\right] \tag{10.57}$$

If the gas is negligibly soluble in the droplet, then $n_{2,s} = 0$; hence $\dot{m}'' = n_{1,s} + n_{2,s} = n_{1,s}$, and $n_{1,s}/\dot{m}'' = 1$. Equation (10.57) thus becomes

$$\dot{m}'' = \frac{\rho\mathscr{D}_{12}}{R}\ln\left[1 + \frac{m_{1,e} - m_{1,s}}{m_{1,s} - 1}\right] \tag{10.58}$$

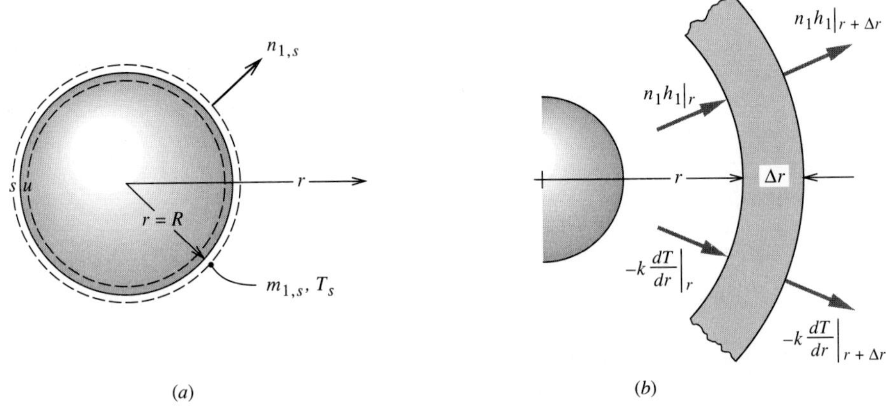

Figure 10.9 (*a*) Schematic of droplet evaporating into a gas. (*b*) Elemental control volume for application of the steady-flow energy equation.

which gives the droplet evaporation rate in terms of the unknown s-surface vapor mass fraction, $m_{1,s}$.

As we learned in Example 9.10, droplet evaporation is a simultaneous heat and mass transfer problem. For quasi-steady evaporation, the enthalpy of vaporization must be supplied by heat transfer from the gas, and the droplet temperature is approximately equal to the psychrometric wet-bulb temperature. To perform a heat transfer analysis, the steady-flow energy equation, Eq. (1.4), is applied to the elemental control volume shown in Fig. 10.9*b* and requires that

$$4\pi r^2\left(n_1h_1 - k\frac{dT}{dr}\right)\bigg|_r - 4\pi r^2\left(n_1h_1 - k\frac{dT}{dr}\right)\bigg|_{r+\Delta r} = 0$$

since $n_2 = 0$: only species 1 transports enthalpy in and out of the control volume. Thus,

$$r^2\left(n_1h_1 - k\frac{dT}{dr}\right) = \text{Constant} = R^2\left(\dot{m}''h_{1,s} - k\frac{dT}{dr}\bigg|_s\right) \tag{10.59}$$

since $n_{1,s} = \dot{m}''$. Also, applying the steady-flow energy equation between the s- and u-surfaces shown in Fig. 10.9*a*,

$$\dot{m}''h_{1,s} - k\frac{dT}{dr}\bigg|_s = \dot{m}''h_{1,u} - k\frac{dT}{dr}\bigg|_u \tag{10.60}$$

Based on Example 9.10, we assume that the initial transient in which the droplet temperature changes from its initial value to a steady value is short, and that the steady value is attained when $-k(dT/dr)|_u \simeq 0$. Then, using Eq. (10.60), Eq. (10.59) becomes

$$r^2\left(n_1h_1 - k\frac{dT}{dr}\right) = R^2\dot{m}''h_{1,u} \tag{10.61}$$

Since there are no chemical reactions, we are free to choose enthalpy datum states as we please. A nice simplification of the analysis is achieved if we take the vapor enthalpy h_1 to be zero at temperature T_s. Also, for simplicity we will assume that the vapor specific heat is independent of temperature. Then, for the vapor,

$$h_1 = \int_{T_s}^{T} c_{p1}\,dT = c_{p1}(T - T_s) \tag{10.62a}$$

and for the liquid at the u-surface,

$$h_{1,u} = -h_{fg} \tag{10.62b}$$

Substituting in Eq. (10.61),

$$r^2\left[n_1c_{p1}(T - T_s) - k\frac{dT}{dr}\right] = -R^2\dot{m}''h_{fg}$$

Rearranging using $n_1 r^2 = \dot{m}'' R^2$ from mass conservation,

$$R^2 \dot{m}''[h_{\text{fg}} + c_{p1}(T - T_s)] = r^2 k \frac{dT}{dr}$$

Separating variables and integrating from R to ∞,

$$\frac{\dot{m}'' R^2}{k} \int_R^\infty \frac{dr}{r^2} = \int_{T_s}^{T_e} \frac{dT}{h_{\text{fg}} + c_{p1}(T - T_s)}$$

where the mixture thermal conductivity has been assumed constant. The result is

$$\dot{m}'' = \frac{k/c_{p1}}{R} \ln\left[1 + \frac{c_{p1}(T_e - T_s)}{h_{\text{fg}}}\right] \quad \textbf{(10.63)}$$

where $c_{p1}(T_e - T_s)/h_{\text{fg}}$ is recognized to be the vapor-phase Jakob number, Ja_v. The Jakob number was introduced in Chapters 3 and 7. Like Eq. (10.58), Eq. (10.63) gives the evaporation rate $\dot{m}''$, this time in terms of the unknown surface temperature T_s. The *simultaneous* heat and mass transfer problem requires the simultaneous solution of Eqs. (10.58) and (10.63) together with the vapor pressure relation,

$$\dot{m}'' = \frac{\rho\mathscr{D}_{12}}{R} \ln\left(1 + \frac{m_{1,e} - m_{1,s}}{m_{1,s} - 1}\right) = \frac{k/c_{p1}}{R} \ln\left(1 + \frac{c_{p1}(T_e - T_s)}{h_{\text{fg}}}\right) \quad \textbf{(10.64}\textbf{\textit{a,b}}\textbf{)}$$

$$m_{1,s} = m_{1,s}(T_s, P) \quad \textbf{(10.64}\textbf{\textit{c}}\textbf{)}$$

which is a set of three equations in the three unknowns $\dot{m}''$, T_s, and $m_{1,s}$. Temperature T_s is the adiabatic vaporization temperature for a droplet evaporating at high mass transfer rates, and will be close to the psychrometric wet-bulb temperature (defined by Eq. (9.57) for the water-air system).

Property Evaluation

To use Eqs. (10.64), the mixture properties ρ, $\mathscr{D}_{12}$, and k must be evaluated at a suitable reference state, since these properties were assumed constant in the analysis (the weak temperature dependence of c_{p1} is of little consequence). Notice that we did not have to assume $c_{p1} = c_{p2} = c_p$ in the analysis. Thus, our result correctly accounts for the difference in specific heats of the two species, and often this difference is large. One possible scheme is to evaluate ρ, $\mathscr{D}_{12}$, and k at the mean film temperature and composition, and c_{p1} at the mean film temperature. However, the validity of this scheme has not been checked against exact numerical solutions or experimental data. An alternative scheme proposed by Hubbard et al. [1] was established by comparing numerical solutions in which all properties were calculated exactly, with a constant-property analysis in which all properties, *including* c_p, were assumed constant at a reference state. A *1/3 rule* result was obtained for which ρ, $\mathscr{D}_{12}$, k, and c_p are evaluated at

$$m_{1,r} = m_{1,s} + (1/3)(m_{1,e} - m_{1,s}) \quad \textbf{(10.65}\textbf{\textit{a}}\textbf{)}$$

$$T_r = T_s + (1/3)(T_e - T_s) \quad \textbf{(10.65}\textbf{\textit{b}}\textbf{)}$$

In using this scheme, c_{p1} in Eq. (10.64a) is set equal to c_p of the mixture evaluated at the reference state. Alternative schemes for property evaluation are available [2,3].

EXAMPLE 10.5 Evaporation of a Water Droplet

A 50 μm–diameter droplet is entrained in a dry air stream at 1 atm, 1650 K. Estimate the droplet lifetime.

Solution

Given: A small water droplet entrained in a dry air stream.

Required: Droplet lifetime.

Assumptions:
1. The initial transient is short.
2. Radiation heat transfer is negligible.
3. Hubbard's 1/3 rule for property evaluation is appropriate.

For quasi-steady evaporation, T_s and $m_{1,s}$ are constant and can be obtained from Eqs. (10.64) by setting $m_{1,e} = 0$ for dry air, and simultaneously solving

$$\ln\left(\frac{1}{1-m_{1,s}}\right) = \frac{k/c_{p1}}{\rho\mathscr{D}_{12}} \ln\left(1 + \frac{c_{p1}(T_e - T_s)}{h_{fg}}\right) \tag{1}$$

$$m_{1,s} = m_{1,s}(T_s, P) \tag{2}$$

We guess $T_s = 350\,\mathrm{K}$, and from PSYCHRO we obtain $m_{1,s} = 0.303$. Using Hubbard's 1/3 rule to evaluate properties, we set $c_{p1} = c_p$ and use the reference state

$$m_{1,r} = 0.303 + (1/3)(0 - 0.303) = 0.202$$

$$T_r = 350 + (1/3)(1650 - 350) = 783\,\mathrm{K}$$

From AIRSTE, mixture properties at the reference state are $\rho = 0.4017\,\mathrm{kg/m^3}$, $c_p = 1296$ J/kg K, $k = 0.0577$ W/m K, $\mathscr{D}_{12} = 1.452 \times 10^{-4}\ \mathrm{m^2/s}$, and from Table A.12*a* at $T = T_s = 350\,\mathrm{K}$, $h_{fg} = 2.316 \times 10^6$ J/kg. Then the left-hand side of Eq. (1) is

$$\ln\left(\frac{1}{1-0.303}\right) = 0.361$$

and the right-hand side of Eq. (1) is

$$\frac{0.0577/1296}{(0.4017)(1.452 \times 10^{-4})} \ln\left(1 + \frac{1296(1650 - 350)}{2.316 \times 10^6}\right) = 0.417$$

Since Eq. (1) is not satisfied, iteration is required and gives the result $T_s = 352.7\,\mathrm{K}$, $m_{1,s} = 0.345$, for which reference state properties include $\rho = 0.3946\ \mathrm{kg/m^3}$, $\mathscr{D}_{12} = 1.459 \times 10^{-4}\ \mathrm{m^2/s}$. From Eq. (10.64*a*), the evaporation rate is

$$\dot{m}'' = \frac{(0.3946)(1.459 \times 10^{-4})}{R} \ln\left(\frac{1}{1-0.345}\right) = \frac{2.44 \times 10^{-5}}{R}\ \mathrm{kg/m^2\,s}$$

A mass balance on the droplet gives

$$\frac{d}{dt}[(4/3)\pi R^3 \rho_l] = -\dot{m}''(4\pi R^2)$$

$$\frac{dR}{dt} = -\frac{(2.44 \times 10^{-5})}{R\rho_l}$$

Integrating with $R = R_0$ at $t = 0$, and $R = 0$ at $t = \tau$ gives the lifetime as

$$\tau = \frac{\rho_l R_0^2}{(2)(2.44 \times 10^{-5})}$$

At 353 K, Table A.8 gives $\rho_l = 971\,\text{kg/m}^3$. Thus,

$$\tau = \frac{(971)(50 \times 10^{-6})^2}{(2)(2.44 \times 10^{-5})} = 0.0497 \text{ s}$$

Comments

1. Rework this example evaluating ρ, $\mathscr{D}_{12}$, and k at the mean film temperature and composition, and c_{p1} as water vapor at the mean film temperature. The result is $T_s = 350.6\,\text{K}$, $m_{1,s} = 0.312$, and $\tau = 0.0466$ s (a 6% shorter lifetime).
2. Vary T_e to see when low mass transfer rate theory becomes invalid.

10.3.5 Droplet Combustion

In many furnaces and combustion chambers liquid fuels are atomized to form a spray of small droplets that burn rapidly and efficiently. Examples include fuel oil in a power plant furnace, kerosene in a gas turbine combustor, and hydrazine in a rocket motor. Although the practical problem involves a spray, it has proven useful to understand the behavior of a single isolated droplet. Analysis of single-droplet combustion assists interpretation of experimental data for single-droplet combustion, and also yields results that can be extended to spray combustion. Of particular importance is the lifetime of the droplet, because the required size of the combustion chamber depends in part on the time required for burning.

Combustion of Liquid Hydrocarbon Fuel Droplets

During combustion of a volatile fuel droplet in air, the fuel vapor and oxygen react in a thin flame, as shown in Fig. 10.10*a*. For hydrocarbon fuels burning in air, the flame diameter is about four to six times the droplet diameter. For a stationary droplet under zero gravity conditions, the flame is spherical; at normal gravity, natural convection produces a plume above the droplet. Heat is transferred from the flame to the droplet and serves to vaporize the fuel. In the flame the vapor reacts with oxygen to form gaseous products, primarily CO_2 and H_2O. The characteristic temperature and concentration profiles are shown in Fig. 10.10*b*. Owing to the very rapid reaction kinetics, there is negligible penetration of the flame by the fuel vapor or oxidant:

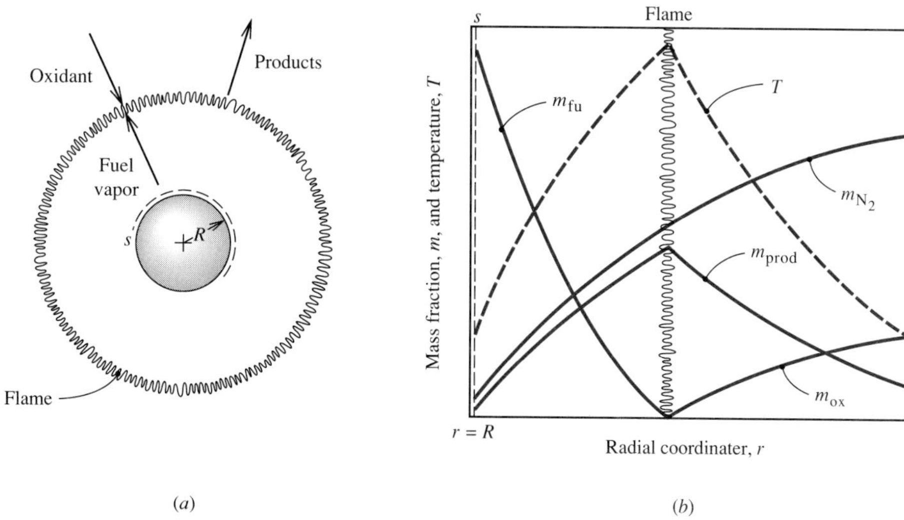

Figure 10.10 A volatile fuel droplet burning in air. (a) Schematic showing the flame surrounding the droplet. (b) Concentration and temperature profiles.

fuel is on one side of the flame, and oxidant is on the other. When a fuel droplet is injected into a combustion chamber, there is a short initial transient during which the droplet heats up, until further conduction into the droplet is negligible and the droplet attains a steady temperature (approximately the wet-bulb temperature). Due to the high temperature of the flame and the volatility of typical liquid hydrocarbon fuels, the wet-bulb temperature is very close to the boiling point of the liquid. The gas-phase reaction at the flame front can be modeled as a single-step reaction,

$$\text{Fuel} + \text{Oxidant} \rightarrow \text{Products} \qquad \textbf{(10.66)}$$

with a constant stoichiometric ratio r and a heat of combustion Δh_c J/kg of fuel. The stoichiometric ratio is typically about 3–4 for hydrocarbon fuels burning in air.[1]

Analysis

This problem involves simultaneous heat and mass transfer. In order to sustain steady combustion, heat must be transferred from the flame to the droplet and the oxidant must diffuse to the flame front. However, if we are prepared to make some simplifying assumptions, we can obtain the combustion rate from a heat transfer analysis only, as will now be shown. We will consider zero-g combustion of a stationary droplet in order to have a spherically symmetric problem: Fig. 10.11 shows the coordinate system and an elemental control volume between radii r and

[1] In this analysis the stoichiometric ratio is defined as the ratio (kg oxidant/kg fuel), and not (kg air/kg fuel) as is the practice in internal-combustion engine analysis.

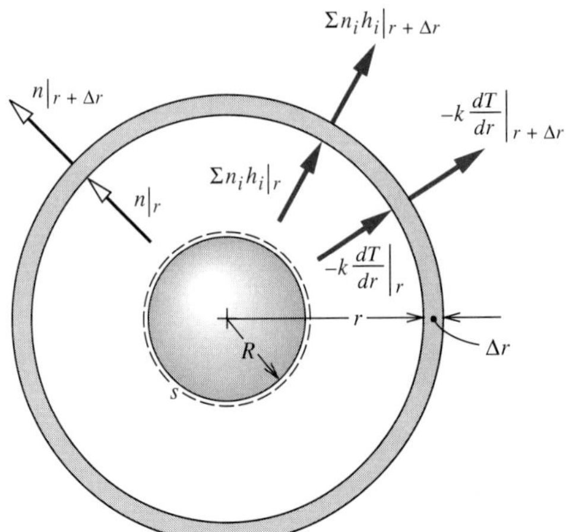

Figure 10.11 A volatile fuel droplet burning in air: elemental control volume for application of the mass conservation equation and the steady-flow energy equation.

$r + \Delta r$. A mass balance on the elemental volume requires that at steady state,

$$4\pi r^2 n|_r - 4\pi r^2 n|_{r+\Delta r} = 0$$

or

$$r^2 n = \text{Constant} = r^2 n|_{r=R} = R^2 \dot{m}'' \tag{10.67}$$

Of course, Eq. (10.67) can also be obtained from the mass conservation equation, Eq. (10.12*b*). Similarly, the steady-flow energy equation applied to the volume requires that[2]

$$4\pi r^2 \left(\sum n_i h_i - k\frac{dT}{dr}\right)_r - 4\pi r^2 \left(\sum n_i h_i - k\frac{dT}{dr}\right)_{r+\Delta r} = 0$$

We have recognized that a convection of enthalpy $n_i h_i$ is associated with the absolute flux of each species, n_i (h_i is the species enthalpy). The conductivity k is the conductivity of the gas mixture. Thermal radiation has been ignored: its effect will be discussed later. Hence,

$$r^2 \left(\sum n_i h_i - k\frac{dT}{dr}\right) = \text{Constant} = R^2 \left(\sum n_{i,s} h_{i,s} - k\frac{dT}{dr}\Big|_s\right) \tag{10.68}$$

Referring to Fig. 10.12, the constant in Eq. (10.68) is conveniently evaluated by noting that the steady-flow energy equation applied between the s- and u-surfaces requires that

$$R^2 \left(\sum n_{i,s} h_{i,s} - k\frac{dT}{dr}\Big|_s\right) = R^2 n_{\text{fu},u} h_u \tag{10.69}$$

[2] The steady-flow energy equation, Eq. (1.4), has been extended to apply to a number of streams of distinct chemical species entering and leaving the control volume.

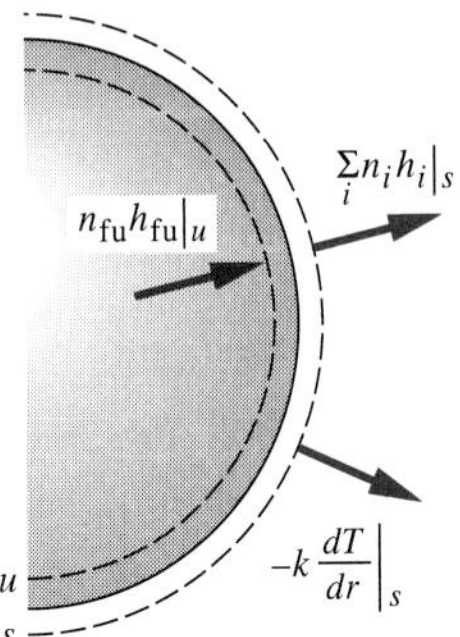

Figure 10.12 Interface energy balance for combustion of a volatile fuel droplet.

because only liquid fuel crosses the u-surface, and $dT/dr|_u = 0$ if the droplet temperature is steady. The enthalpy h_u is the enthalpy of liquid fuel at temperature $T_u = T_s$, and $n_{\text{fu},u} = \dot{m}''$. Equation (10.68) then becomes

$$r^2\left(\sum_i n_i h_i - k\frac{dT}{dr}\right) = R^2\dot{m}''h_u \tag{10.70}$$

The absolute fluxes n_i are now expressed in terms of their convective and diffusive components as

$$n_i = m_i n + j_i = m_i n - \rho\mathscr{D}_{im}\frac{dm_i}{dr}$$

where the $\mathscr{D}_{im}$ are effective binary diffusion coefficients. Substituting in Eq. (10.70),

$$r^2\left(n\sum_i m_i h_i - \sum_i \rho\mathscr{D}_{im}\frac{dm_i}{dr}h_i - k\frac{dT}{dr}\right) = R^2\dot{m}''h_u \tag{10.71}$$

This equation can be simplified as follows. The mixture enthalpy is $h = \sum m_i h_i$; differentiating,

$$\begin{aligned} dh &= \sum m_i dh_i + \sum h_i dm_i \\ &= \sum m_i c_{pi} dT + \sum h_i dm_i \\ &= c_p dT + \sum h_i dm_i \end{aligned}$$

where the mixture specific heat is $c_p = \sum m_i c_{pi}$. Hence,

$$\frac{k}{c_p}\frac{dh}{dr} = k\frac{dT}{dr} + \frac{k}{c_p}\sum_i h_i\frac{dm_i}{dr} \tag{10.72}$$

We now make a key assumption, namely, that the Lewis numbers for all the species are equal to 1. By definition, $\text{Le}_i = \text{Pr}/\text{Sc}_i$. Hence, $\text{Le}_i = 1$ requires that

$$\frac{\text{Pr}}{\text{Sc}_i} = \frac{c_p\mu/k}{\mu/\rho\mathscr{D}_{im}} = \frac{\rho\mathscr{D}_{im}}{k/c_p} = 1$$

That is, $\rho\mathscr{D}_{im} = k/c_p$ for all species i in the mixture. Note that ρ, k, and c_p are *mixture* properties. Then, substituting Eq. (10.72) in Eq. (10.71) and rearranging using Eq. (10.67) gives

$$R^2\dot{m}''h - r^2\frac{k}{c_p}\frac{dh}{dr} = R^2\dot{m}''h_u \tag{10.73}$$

We have succeeded in obtaining a differential equation in terms of a single dependent variable, the mixture enthalpy h. Integrating with $h = h_s$ at $r = R$ and $h = h_e$ as $r \to \infty$,

$$\int_{h_s}^{h_e}\frac{dh}{h - h_u} = \frac{\dot{m}''R^2}{k/c_p}\int_R^{\infty}\frac{dr}{r^2}$$

where we have assumed k/c_p to be constant, independent of r. Then

$$\ln\frac{h_e - h_u}{h_s - h_u} = \frac{\dot{m}''R}{k/c_p}$$

which can be rearranged to give the mass transfer rate,

$$\dot{m}'' = \frac{k/c_p}{R}\ln\left[1 + \frac{h_e - h_s}{h_s - h_u}\right] \tag{10.74}$$

We now assume that, for the purpose of evaluating Eq. (10.74), we can take T_s to equal the boiling point of the fuel, which will be known. (As mentioned earlier, $T_s = T_{\mathrm{WB}}$, but T_{WB} is close to T_{BP} in this situation.) Then $m_{\mathrm{fu},s} \simeq 1$, and Eq. (10.74) can be evaluated to obtain $\dot{m}''$ using standard enthalpy tables to calculate h_e, h_s, and h_u. However, we can obtain a more convenient form of this result as follows.

By definition, the enthalpy of species i is

$$h_i = h_i^0 + \int_{T_0}^{T} c_p\,dT \tag{10.75}$$

where h_i^0 is the *heat of formation* of species i at the datum temperature T_0. Consistent with Eq. (10.66), we will consider a mixture of fuel vapor, oxidant, products, and inerts. We are free to arbitrarily choose the datum temperature T_0, and to assign the heat of combustion (Δh_c J/kg fuel) to the heat of formation of either the fuel, oxidant, or products. If we choose $T_0 = T_s$ and assign the heat of combustion to the oxidant, and if, in addition, we assume that all species specific heats are equal and constant, we obtain

$$h_{\mathrm{fu}} = c_p(T - T_s) \quad \text{(vapor)} \tag{10.76a}$$

$$h_{\mathrm{ox}} = \frac{\Delta h_c}{r} + c_p(T - T_s) \tag{10.76b}$$

$$h_{\mathrm{prod}} = c_p(T - T_s) \tag{10.76c}$$

$$h_{\mathrm{inerts}} = c_p(T - T_s) \tag{10.76d}$$

Then h_e, h_s, and h_u can be evaluated as

$$h_e = \sum m_{i,e} h_{i,e} = m_{\text{ox},e} \frac{\Delta h_c}{r} + m_{\text{ox},e} c_p (T_e - T_s) + m_{\text{inerts},e} c_p (T_e - T_s)$$

since $m_{\text{fu},e} = m_{\text{prod},e} = 0$. But $m_{\text{ox},e} + m_{\text{inerts},e} = 1$; thus,

$$h_e = m_{\text{ox},e} \frac{\Delta h_c}{r} + c_p (T_e - T_s) \tag{10.77a}$$

$$h_s = 0, \quad \text{since there is no oxidant at the } s\text{-surface} \tag{10.77b}$$

$$h_u = h_{\text{fu},u} = -h_{\text{fg}}, \quad \text{the enthalpy of vaporization of the fuel} \tag{10.77c}$$

Substituting into Eq. (10.74) gives the mass transfer rate as

$$\dot{m}'' = \frac{k/c_p}{R} \ln\left[1 + \frac{m_{\text{ox},e} \Delta h_c / r + c_p (T_e - T_s)}{h_{\text{fg}}}\right] \tag{10.78}$$

which is easily evaluated. The mass transfer rate is simply the rate at which the fuel is consumed; a mass balance on the droplet gives the droplet lifetime.

Droplet Lifetime

A mass balance on the droplet requires that

$$\frac{d}{dt}\left(\frac{4}{3}\pi R^3 \rho_l\right) = -4\pi R^2 \dot{m}''$$

Substituting from Eq. (10.78) and rearranging gives the rate of change of droplet diameter D as

$$\frac{dD}{dt} = -\frac{4k/c_p}{\rho_l D} \ln\left[1 + \frac{m_{\text{ox},e} \Delta h_c / r + c_p (T_e - T_s)}{h_{\text{fg}}}\right] \tag{10.79}$$

Integrating with $D = D_0$ at $t = 0$, and $D = 0$ at $t = \tau$ gives the droplet lifetime as

$$\tau = \frac{\rho_l D_0^2}{8\left(\dfrac{k}{c_p}\right) \ln\left[1 + \dfrac{m_{\text{ox},e} \Delta h_c / r + c_p (T_e - T_s)}{h_{\text{fg}}}\right]} \tag{10.80}$$

The droplet lifetime is proportional to droplet diameter squared (as was the case for droplet evaporation and spherical particle sublimation and combustion). Equation (10.80) was first derived by Godsave [4] and by Spalding [5].

Property Evaluation

In deriving Eq. (10.80), we have made a number of assumptions regarding properties:

1. The Lewis numbers of all species are unity, or $\rho \mathscr{D}_{im} = k/c_p$ for all species i (which requires all diffusion coefficients $\mathscr{D}_{im}$ to be equal).

2. All species specific heats are equal.

3. The k/c_p ratio is a constant.

In reality, the diffusion coefficients are not quite equal, the Lewis numbers are not quite unity, and the k/c_p ratio varies, due to both temperature and composition changes, from the droplet surface, across the flame, and into the surrounding air. Thus, Eq. (10.80) will be useful only if some simple and general reference property scheme can be recommended for evaluating k and c_p, one that will give good agreement between theory and experiment. A scheme that works quite well for alkane fuel droplets burning in air was proposed by Law and Williams [6]. The reference temperature is $T_r = (1/2)(T_{\mathrm{BP}} + T_{\mathrm{flame}})$, where T_{flame} is the adiabatic flame temperature. The reference specific heat is $c_{pr} = c_{p\mathrm{fu}}$, and the reference thermal conductivity is $k_r = 0.4k_{\mathrm{fu}} + 0.6k_{\mathrm{air}}$.

The Burning Rate Constant and the Transfer Number

Combustion specialists often quote two parameters that characterize droplet combustion. The *burning rate constant*, β, is defined by the relation

$$\frac{d}{dt}(D^2) = -\beta; \qquad \beta = -\frac{8k}{\rho c_p}\ln\left[1 + \frac{m_{\mathrm{ox},e}\Delta h_c/r + c_p(T_e - T_s)}{h_{\mathrm{fg}}}\right] \tag{10.81}$$

That is, β characterizes the linear decrease of droplet diameter squared for quasi-steady combustion. Figure 10.13 shows typical experimental data plotted as D^2 versus t. After an initial transient there is a linear decrease yielding a line of slope $-\beta$. The goal of experimental studies of single-droplet combustion is usually to determine β: values of β are on the order of 10^{-6} $\mathrm{m^2/s}$ for hydrocarbon fuel droplets burning in air.

The *transfer number* $\mathscr{B}$ is defined as

$$\mathscr{B} = \frac{m_{\mathrm{ox},e}\Delta h_c/r + c_p(T_e - T_s)}{h_{\mathrm{fg}}} \tag{10.82}$$

and Eq. (10.80) can be written more compactly as

$$\tau = \frac{\rho_l D_0^2}{8(k/c_p)\ln(1 + \mathscr{B})} \tag{10.83}$$

In mass transfer theory $\mathscr{B}$ is termed an enthalpy-based *mass transfer driving force*. For hydrocarbon fuel droplets burning in air, $\mathscr{B}$ ranges from 5 to 15.[3] Enthalpy-based mass transfer driving forces are further developed in Sections 10.5.2 and 10.6.2.

Effect of Radiation

Recent research has shown that the effect of radiation from the CO_2 and H_2O products is not negligible for single-droplet combustion [8]. Analysis including the

[3] The parameter $\mathscr{B}$ scales the mass transfer rate; in general use, we can say that the mass transfer rate is low when $\mathscr{B} < 0.1$. Thus, mass transfer rates for combustion of volatile hydrocarbon fuel droplets are very high.

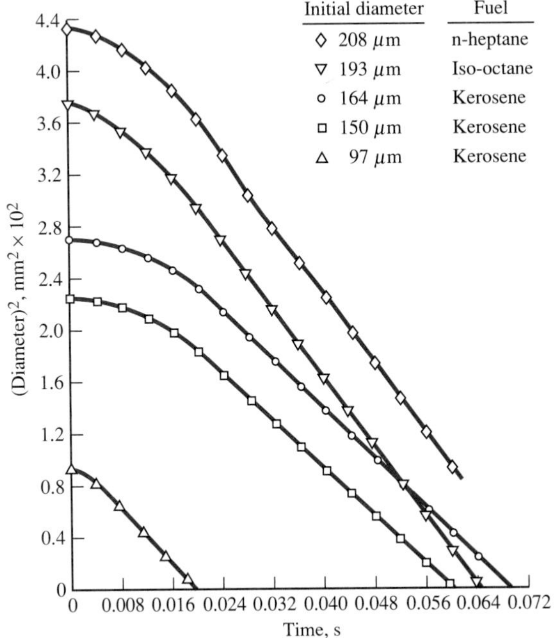

Figure 10.13 Experimental data for hydrocarbon fuel droplets burning in air confirming the "D^2 law" [7].

effect of radiation predicts lower values of the burning rate constant β than when radiation is ignored. Thus, actual droplet lifetimes can be expected to be longer than those predicted by Eq. (10.83). However, when a property evaluation scheme is used to match Eq. (10.83) to experimental data, as was done by Law and Williams [6], the effect of radiation is accounted for. In combustion chambers burning sprays of droplets, radiation heat transfer is large, and our simple single droplet combustion theory does not apply.

EXAMPLE 10.6 Combustion of a n-Hexane Droplet

Estimate the lifetime of a 1 mm–diameter n-hexane droplet burning at nearly zero gravity in air at 1 atm pressure and 300 K. For n-hexane (n-C_6H_{14}), take $\rho_l = 610$ kg/m^3, $h_{fg} = 3.35 \times 10^5$ J/kg, $\Delta h_c = 4.48 \times 10^7$ J/kg, and $T_{BP} = 342$ K. The flame temperature is $T_f = 2320$ K. At the Law and Williams reference temperature of $T_r = (1/2)(T_f + T_{BP})1330$ K, property values for n-hexane vapor include $k = 0.179$ W/m K and $c_p = 4303$ J/kg K.

Solution

Given: A 1 mm–diameter n-hexane droplet burning in air at 1 atm, 300 K.

Required: Droplet lifetime.

Assumptions:
1. Spherical symmetry.
2. Quasi-steady combustion with the droplet at its boiling point.

3. Properties can be evaluated using the reference state scheme of Law and Williams.

Equation (10.83) gives the droplet lifetime τ as

$$\tau = \frac{\rho_l D_0^2}{8(k/c_p)\ln(1+\mathscr{B})}; \qquad \mathscr{B} = \frac{m_{\text{ox},e}\Delta h_c/r + c_p(T_e - T_s)}{h_{\text{fg}}}$$

We first evaluate the transfer number $\mathscr{B}$. The reaction is

$$C_6H_{14} + 9.5\,O_2 \rightarrow 6\,CO_2 + 7\,H_2O$$
$$86.17\,\text{kg} + (9.5)(32)\,\text{kg} \rightarrow$$

Hence, the stoichiometric ratio is $r = (9.5)(32)/86.17 = 3.53$. The mass fraction of oxygen in air $m_{\text{ox},e} = 0.231$, and we take $T_s \simeq T_{\text{BP}} = 342$ K. The Law and Williams reference state scheme requires that c_p be evaluated for pure hexane vapor at $T_r = (1/2)(T_f + T_{\text{BP}}) = 1330$ K, which is given as 4303 J/kg. The transfer number is

$$\mathscr{B} = \frac{(0.231)(4.48\times10^7)/3.53 + 4303(300-342)}{3.35\times10^5} = \frac{(29.3-1.8)\times10^5}{3.35\times10^5} = 8.21$$

The reference state value of the thermal conductivity is

$$k_r = 0.6k_{\text{fu}} + 0.4k_{\text{air}}$$

evaluated at $T_r = 1330$ K. From Table A.7, $k_{\text{air}} = 0.084$ W/m K; hence

$$k_r = (0.4)(0.179) + (0.6)(0.084) = 0.122 \text{ W/m K}$$

The droplet lifetime is then

$$\tau = \frac{(610)(1\times10^{-3})^2}{8(0.122/4303)\ln(1+8.21)} = 1.21 \text{ s}$$

Comments

1. Notice how the sensible heat term $c_p(T_e - T_s)$ is almost negligible compared with the heat of combustion term $m_{\text{ox},e}\Delta h_c/r$. Thus, the air temperature has little effect on droplet lifetime.
2. The Law and Williams reference state scheme is based on experimental data and thus should account for the various shortcomings of our model.
3. Law and Williams [6] also give correction factors for natural or forced convection flow around the droplet.

10.4 HIGH MASS TRANSFER RATE CONVECTION

The analyses in Section 10.3 were examples of *high mass transfer rate theory*, because convection of mass across the interface was properly accounted for. In the heatpipe problem of Example 10.3, the mass transfer rates are *very* high. We can talk in terms of strong blowing at the evaporator end and strong suction at the condenser end. A similar strong suction is characteristic of condensation of steam in power-plant condensers, where the steam always contains a small amount of noncondensable gas. The mass transfer rate for hydrocarbon fuel droplet combustion is also very high; there is a strong blowing away from the droplet surface. Strong blowing is also characteristic of ablation and transpiration cooling, for example, ablation of the heat shield on the Apollo reentry vehicle. Notice that both in the heatpipe evaporator and at the droplet surface, the mass fraction of the evaporating species is close to its limiting value of unity. To simplify the analyses, the only convection considered was that normal to the interface induced by the mass transfer process itself. Forced or natural convection flows along the interface, such as the relative motion between the carbon particle or fuel droplet and the gas flow, were not considered. We now analyze high mass transfer rate forced convection. For this purpose we will use a Couette-flow model and restrict our attention to binary inert mixtures with transfer of a single species. Our purpose is to develop problem-solving procedures for convective mass, momentum, and heat transfer applicable to situations where the low mass transfer theory of Section 9.4 is invalid.

10.4.1 The Couette-Flow Model

In Section 5.2 a Couette-flow model of the high-speed boundary layer was used to give insight into the effect of viscous dissipation on heat transfer. In Section 7.2.4 a Couette-flow model of a vapor boundary layer was used to obtain useful relations for the effects of vapor shear and superheat on film condensation. The Couette-flow model will now be used to examine the essential features of boundary layer flows with mass transfer, in particular, the effects of high mass transfer rates. We will first consider a low-speed Couette flow of an inert binary mixture with only one component (species 1) transferred, and examine mass, momentum, and heat transfer.

Mass Transfer

Figure 10.14 shows the model and coordinate system. For a steady state and no reactions, the species conservation equation, Eq. (10.10), reduces to

$$\frac{\partial n_{1x}}{\partial x} + \frac{\partial n_{1y}}{\partial y} = 0$$

But for a Couette flow, the dependent variables are independent of x; thus, $\partial n_{1x}/\partial x = 0$ and then

$$\frac{\partial n_{1y}}{\partial y} = 0$$

$$n_{1y} = \text{Constant} \tag{10.84}$$

Similarly, the mass conservation equation, Eq. (10.11), reduces to

$$n_y = \text{Constant} \tag{10.85}$$

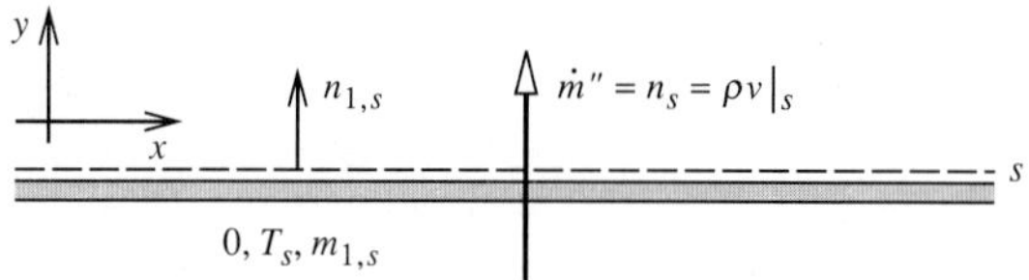

Figure 10.14 Schematic of a Couette-flow model for a boundary layer with mass transfer.

Since the problem is now one-dimensional, the subscript y will be dropped. Then Eq. (10.85) becomes

$$n = \text{Constant} = n|_{y=0} = n_s \equiv \dot{m}'' \qquad \textbf{(10.86)}$$

where $\dot{m}''$ [kg/m^2 s] is the mass transfer rate that was introduced in Section 10.3.2. Substituting in Eq. (10.84),

$$n_1 = m_1 n + j_1 = m_1 \dot{m}'' - \rho \mathscr{D}_{12} \frac{dm_1}{dy} = \text{Constant} = n_{1,s} \qquad \textbf{(10.87)}$$

Rearrangement and integration between the limits of $m_{1,s}$ at $y = 0$ and m_1 at y gives the concentration profile,

$$\int_{m_{1,s}}^{m_1} \frac{dm_1}{m_1 \dot{m}'' - n_{1,s}} = \int_0^y \frac{dy}{\rho \mathscr{D}_{12}}$$

and if $\rho \mathscr{D}_{12}$ is assumed constant,

$$\frac{m_1 - n_{1,s}/\dot{m}''}{m_{1,s} - n_{1,s}/\dot{m}''} = \exp\left(\frac{\dot{m}'' y}{\rho \mathscr{D}_{12}}\right) \qquad \textbf{(10.88)}$$

In many problems of interest we can assume that only one species is transferred, for example, when water evaporates into air. If species 1 is the species transferred, $n_{2,s} = 0$ and $\dot{m}'' = n_s = n_{1,s}$; that is, $n_{1,s}/\dot{m}'' = 1$, and Eq. (10.88) becomes

$$\frac{m_1 - 1}{m_{1,s} - 1} = \exp\left(\frac{\dot{m}'' y}{\rho \mathscr{D}_{12}}\right)$$

which can be rewritten as

$$1 + \frac{m_1 - m_{1,s}}{m_{1,s} - 1} = \exp\left(\frac{\dot{m}'' y}{\rho \mathscr{D}_{12}}\right) \qquad \textbf{(10.89)}$$

This expression for the concentration profile cannot be evaluated immediately since both $m_{1,s}$ and $\dot{m}''$ are required. We proceed as follows: writing Eq. (10.89) for $y = \delta$

where $m_1 = m_{1,e}$ is specified, gives

$$1 + \frac{m_{1,e} - m_{1,s}}{m_{1,s} - 1} = \exp\left(\frac{\dot{m}''\delta}{\rho\mathscr{D}_{12}}\right) \tag{10.90}$$

Case 1: $m_{1,s}$ is specified; for example, the wall at $y = 0$ might be an evaporating liquid at temperature T_s, and $m_{1,s} = m_{1,s}(P, T_s)$ from vapor pressure data. Then Eq. (10.90) gives the evaporation rate $\dot{m}''$, and if it is substituted in Eq. (10.89), the concentration profile can be obtained. In this case, notice that $n_{2,s} = 0$ because the free-stream gas is assumed to be insoluble in the liquid.

Case 2: $\dot{m}''$ is specified; for example, the wall at $y = 0$ might be porous, and species 1 is blown through the wall as is done in transpiration cooling. Then Eq. (10.90) gives $m_{1,s}$, and if it is substituted in Eq. (10.89), the concentration profile can be obtained. In this case, notice that $n_{2,s} = 0$ because the flux of the free-stream component into the wall is zero at steady state.

The Mass Transfer Driving Force and Blowing Factor

Although the Couette-flow problem is solved at this point, it is useful to further rearrange Eq. (10.90). First, recall the definition of the mass transfer conductance $\mathscr{g}_{m1}$: from Eq. (9.36),

$$\mathscr{g}_{m1} = \frac{j_{1,s}}{m_{1,s} - m_{1,e}} \tag{10.91}$$

Notice that $\mathscr{g}_{m1}$ is defined in terms of the *diffusive flux* $j_{1,s}$, rather than the absolute flux $n_{1,s}$ across the surface. By definition,

$$n_{1,s} = m_{1,s}n_s + j_{1,s}$$

and since only species 1 is transferred, $n_{2,s} = 0$, $n_s = n_{1,s} + n_{2,s} = n_{1,s} = \dot{m}''$; hence

$$\dot{m}'' = m_{1,s}\dot{m}'' + j_{1,s} \tag{10.92}$$

Substituting from Eq. (10.91) and rearranging,

$$\dot{m}'' = \mathscr{g}_{m1}\frac{m_{1,e} - m_{1,s}}{m_{1,s} - 1}$$

or

$$\dot{m}'' = \mathscr{g}_{m1}\mathscr{B}_{m1}; \qquad \mathscr{B}_{m1} = \frac{m_{1,e} - m_{1,s}}{m_{1,s} - 1} \tag{10.93}$$

where $\mathscr{B}_{m1}$ is the **mass transfer driving force**. Notice that Eq. (10.93) is not quite in the Ohm's law form, flux equals conductance times driving potential, since the mass transfer driving force $\mathscr{B}_{m1}$ is a nonlinear function of $m_{1,s}$. Also, Eq. (10.93) is quite general: it is based on definitions and applies to any convective situation, not just to a Couette flow. Introducing $\mathscr{B}_{m1}$ into Eq. (10.90) gives

$$1 + \mathscr{B}_{m1} = \exp\left(\frac{\dot{m}''\delta}{\rho\mathscr{D}_{12}}\right)$$

and, solving for $\dot{m}''$,

$$\dot{m}'' = \frac{\rho\mathscr{D}_{12}}{\delta}\ln(1 + \mathscr{B}_{m1}) \qquad \textbf{(10.94)}$$

or

$$\dot{m}'' = \frac{\rho\mathscr{D}_{12}}{\delta}\frac{\ln(1 + \mathscr{B}_{m1})}{\mathscr{B}_{m1}}\mathscr{B}_{m1} \qquad \textbf{(10.95)}$$

Comparing Eqs. (10.93) and (10.95), we can identify

$$\mathscr{g}_{m1} = \frac{\rho\mathscr{D}_{12}}{\delta}\frac{\ln(1 + \mathscr{B}_{m1})}{\mathscr{B}_{m1}} \qquad \textbf{(10.96)}$$

We will find it useful to normalize the mass transfer conductance with its value in the limit of zero mass transfer rate. Equation (10.95) shows that $\dot{m}'' \to 0$ as $\mathscr{B}_{m1} \to 0$, and

$$\lim_{\mathscr{B}_{m1}\to 0}\frac{\ln(1 + \mathscr{B}_{m1})}{\mathscr{B}_{m1}} = \lim_{\mathscr{B}_{m1}\to 0}\frac{\mathscr{B}_{m1} - (1/2)\mathscr{B}_{m1}^2 + \cdots}{\mathscr{B}_{m1}} = 1$$

which, upon substitution in Eq. (10.96), gives

$$\lim_{\dot{m}''\to 0}\mathscr{g}_{m1} = \frac{\rho\mathscr{D}_{12}}{\delta} \equiv \overset{*}{\mathscr{g}}_{m1} \qquad \textbf{(10.97)}$$

The * superscript is used here to denote the limit of zero mass transfer rate, and must not be confused with the * used to denote a molar flux relative to the mass average velocity. Since we very seldom analyze mass convection on a molar basis, there should be little chance of confusion.

Notice the definition of $\mathscr{B}_{m1} = (m_{1,e} - m_{1,s})/(m_{1,s} - 1)$ implies that $\mathscr{B}_{m1} \to 0$ as $m_{1,s} \to m_{1,e}$, that is, as the concentration gradients go to zero, and intuition tells us that this corresponds to the limit $\dot{m}'' \to 0$. It is also instructive to see that as $\dot{m}'' \to 0$, the concentration profile is linear. For $\dot{m}'' \to 0$, Eq. (10.87) becomes

$$-\rho\mathscr{D}_{12}\frac{dm_1}{dy} = \text{Constant}$$

Integrating and applying the boundary conditions gives

$$\frac{m_1 - m_{1,s}}{m_{1,e} - m_{1,s}} = \frac{y}{\delta}$$

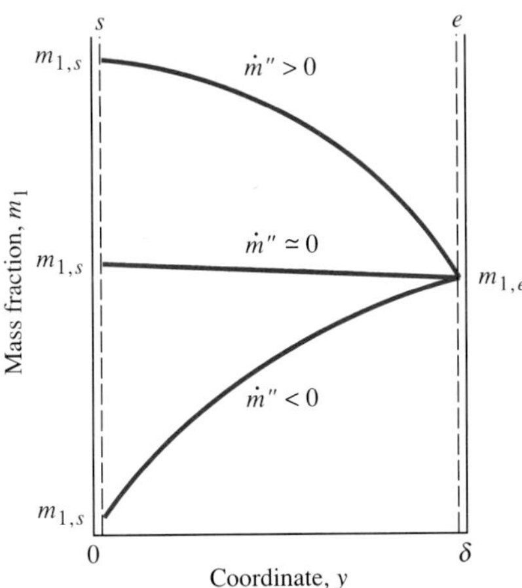

Figure 10.15 Effect of mass transfer on concentration profiles for a Couette flow. For evaporation or transpiration of species 1, $\dot{m}'' > 0$; for condensation of species 1, $\dot{m}'' < 0$.

which is linear, as shown in Fig. 10.15. Then

$$j_{1,s} = -\rho\mathscr{D}_{12}\frac{dm_1}{dy}\bigg|_{y=0} = \frac{\rho\mathscr{D}_{12}}{\delta}(m_{1,s} - m_{1,e}) \quad \textbf{(10.98)}$$

Hence, $g^*_{m1} = j_{1,s}/m_{1,s} - m_{1,e} = \rho\mathscr{D}_{12}/\delta$, as before. With the relation $g^*_{m1} = \rho\mathscr{D}_{12}/\delta$ established, we can rewrite Eq. (10.95) as

$$\dot{m}'' = \underbrace{g^*_{m1}}_{\substack{\text{zero mass}\\ \text{transfer}\\ \text{limit}\\ \text{conductance}}} \times \underbrace{\frac{\ln(1 + \mathscr{B}_{m1})}{\mathscr{B}_{m1}}}_{\substack{\text{blowing}\\ \text{factor}}} \times \underbrace{\mathscr{B}_{m1}}_{\substack{\text{driving}\\ \text{force}}} \quad \textbf{(10.99)}$$

and notice that details of the Couette-flow model do not appear in this result. Recall that our Couette-flow model is intended only as a *model* of a real boundary layer flow, which we have analyzed to examine the effects of mass transfer. We have isolated this effect in Eq. (10.99) as a **blowing factor**, $\ln(1 + \mathscr{B}_{m1})/\mathscr{B}_{m1}$.

In anticipation of the fact that our Couette-flow model blowing factor might be modified using the results of more exact analysis or experimental data, we will write our final result as

$$\dot{m}'' = g_{m1}\mathscr{B}_{m1}; \qquad \mathscr{B}_{m1} = \frac{m_{1,e} - m_{1,s}}{m_{1,s} - 1} \quad \textbf{(10.100)}$$

$$\frac{g_{m1}}{g^*_{m1}} = \frac{\ln(1 + \mathscr{B}_{m1})}{\mathscr{B}_{m1}} \quad \text{(Couette-flow model)} \quad \textbf{(10.101)}$$

where Eq. (10.100) is quite general since it is simply a combination of definitions, and Eq. (10.101) is the Couette-flow result. It sometimes proves useful to work in terms of a parameter that is directly proportional to $\dot{m}''$; we define a **blowing**

parameter

$$B_{m1} = \frac{\dot{m}''}{g^*_{m1}} = \frac{\dot{m}''}{\rho u_e \mathrm{St}^*_m} \tag{10.102}$$

where St_m is the mass transfer Stanton number. Then, for the Couette-flow model,

$$\frac{\mathrm{St}_m}{\mathrm{St}^*_m} = \frac{g_{m1}}{g^*_{m1}} = \frac{B_{m1}}{\exp B_{m1} - 1} \tag{10.103}$$

This form is more convenient for problem solving when $\dot{m}''$ is prescribed, for example, in transpiration cooling.

The preceding Couette-flow analysis is essentially identical to the analysis of diffusion with one component stationary in Section 10.3.1, since the streamwise velocity component u did not enter into the problem. Since species 2 is stationary, Eq. (10.29) with δ replacing L can be rewritten as

$$N_{1,s} = \frac{c\mathscr{D}_{12}}{\delta} \ln\left[1 + \frac{x_{1,e} - x_{1,s}}{x_{1,s} - 1}\right] \tag{10.104}$$

which is the molar equivalent of Eq. (10.94). Equation (10.104) can be viewed as the result of a *stagnant-film* model. Thus, whether we talk of a Couette-flow analysis or a stagnant-film analysis to determine the effects of high mass transfer rates is immaterial: the analyses are essentially identical. Chemical engineers have traditionally preferred to think in terms of a stagnant-film model, while mechanical engineers have preferred a Couette-flow model.[4] Recall also that Eq. (10.104) was derived assuming a constant total molar concentration c, whereas Eq. (10.94) required the generally poorer assumption of a constant mass density ρ.

Momentum Transfer

Considering next momentum transfer, our starting point is the constant-property conservation-of-momentum equation in the form derived in Section 5.4, which remains valid if second-order effects associated with the diffusion velocities are ignored. Equation (5.39) is

$$u\frac{\partial u}{\partial x} + v\frac{\partial u}{\partial y} = \nu\frac{\partial^2 u}{\partial y^2}$$

and, for Couette flow with $\partial u/\partial x = 0$, reduces to

$$v\frac{du}{dy} = \nu\frac{d^2 u}{dy^2}$$

[4] Stagnant-film models in cylindrical and spherical coordinates also give results that can be put in the form of Eq. (10.99).

Multiplying through by ρ and once again noting that $\rho v = n = n_s = \dot{m}''$,

$$\dot{m}''\frac{du}{dy} = \mu\frac{d^2u}{dy^2} \tag{10.105}$$

Integrating with $u = 0$ at $y = 0$ and $u = u_e$ at $y = \delta$ gives the velocity profile as

$$\frac{u}{u_e} = \frac{\exp(\dot{m}''y/\mu) - 1}{\exp(\dot{m}''\delta/\mu) - 1} \tag{10.106}$$

and the shear stress on the wall is

$$\tau_s = \mu\frac{du}{dy}\bigg|_{y=0} = \frac{\dot{m}''u_e}{\exp(\dot{m}''\delta/\mu) - 1} \tag{10.107}$$

Notice that when $\dot{m}''$ is positive and large, $\tau_s \to 0$: we talk about the flow being blown off the wall, giving a zero shear stress. When $\dot{m}''$ is negative and large, $\tau_s \to -\dot{m}''u_e$: in this *strong suction* limit, the shear stress is independent of viscosity and equals the momentum lost by the transferred fluid as it is decelerated from a velocity u_e to zero. The strong suction limit case was used to account for vapor drag on film condensation in Section 7.2.4. Also, the limit of τ_s for $\dot{m}'' \to 0$ is

$$\tau_s^* = \frac{\mu u_e}{\delta} \tag{10.108}$$

which corresponds to a linear velocity profile, as shown in Fig. 10.16. Hence,

$$\frac{\tau_s}{\tau_s^*} = \frac{\dot{m}''\delta/\mu}{\exp(\dot{m}''\delta/\mu) - 1} \tag{10.109}$$

A blowing parameter for momentum transfer is defined by analogy to Eq. (10.102) for mass transfer:

$$B_f = \frac{\dot{m}''u_e}{\tau_s^*} \tag{10.110}$$

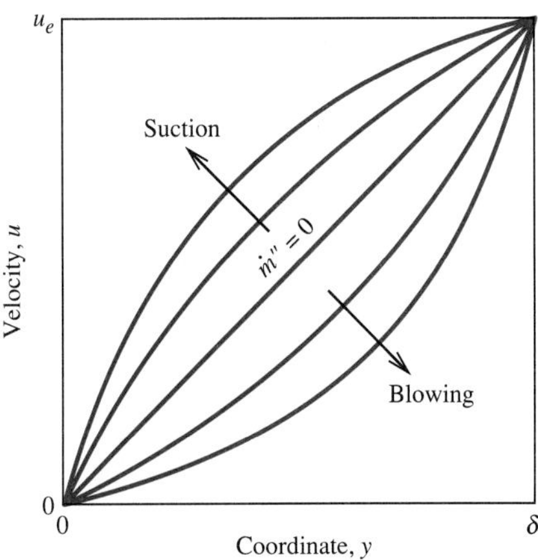

Figure 10.16 Effect of mass transfer on velocity profiles for a Couette flow.

Substituting for τ_s^* from Eq. (10.108) gives $B_f = \dot{m}''\delta/\mu$; then, introducing the skin friction coefficient $C_f = \tau_s/(1/2)\rho u_e^2$, we obtain

$$\frac{C_f}{C_f^*} = \frac{\tau_s}{\tau_s^*} = \frac{B_f}{\exp B_f - 1} \tag{10.111}$$

which is the momentum transfer analogy to the mass transfer result Eq. (10.103).

Heat Transfer

Finally, we consider heat transfer in a low-speed flow, for which we derive the governing conservation equation from first principles. Figure 10.17 shows an elemental control volume of cross-sectional area A and thickness Δy. Energy can flow in and out of the control volume by conduction $-k(dT/dy)$, and species 1 can transport enthalpy $n_1 h_1$. Species 2 is stationary ($n_2 = n_{2,s} = 0$) and does not transport enthalpy. For a Couette flow, there is no x direction temperature gradient, and hence no x direction energy flow. At steady state, the steady-flow energy equation applied to the control volume requires that

$$A\left(n_1 h_1 - k\frac{dT}{dy}\right)_{y+\Delta y} - A\left(n_1 h_1 - k\frac{dT}{dy}\right)_y = 0$$

Dividing by $A\Delta y$ and letting $\Delta y \to 0$,

$$\frac{d}{dy}\left(n_1 h_1 - k\frac{dT}{dy}\right) = 0 \tag{10.112}$$

As before, species conservation requires

$$n_1 = \text{Constant} = n_{1,s} = \dot{m}'' \tag{10.113}$$

For convenience in the analysis that follows, we take the specific heat of species 1 to be independent of temperature, and choose an enthalpy datum state of $h = 0$ at $T = 0$; hence $h_1 = c_{p1}T$. Substituting in Eq. (10.112),

$$\frac{d}{dy}\left(\dot{m}'' c_{p1} T - k\frac{dT}{dy}\right) = 0 \tag{10.114}$$

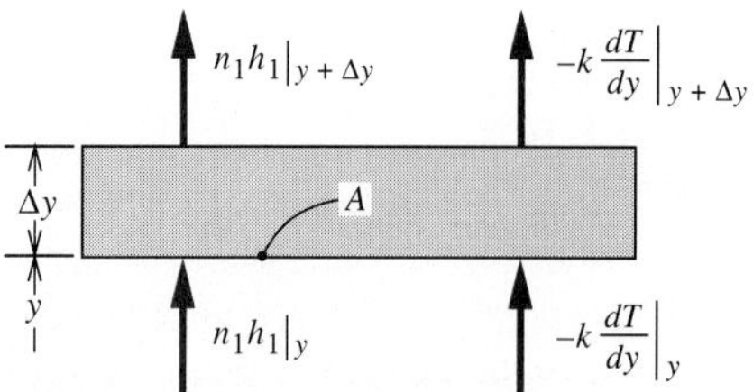

Figure 10.17 Elemental control volume for application of the steady-flow energy equation in a low-speed Couette flow, with transfer of a single chemical species.

Integrating twice with boundary conditions $y = 0$, $T = T_s$; $y = \delta$, $T = T_e$; and k constant, gives the temperature profile as

$$\frac{T - T_s}{T_e - T_s} = \frac{\exp(\dot{m}''c_{p1}/k)y - 1}{\exp(\dot{m}''c_{p1}/k)\delta - 1} \tag{10.115}$$

Differentiation gives the conduction heat flux across the s-surface as

$$-k\frac{dT}{dy}\bigg|_s = \frac{\dot{m}''c_{p1}(T_s - T_e)}{\exp(\dot{m}''c_{p1}/k)\delta - 1} \tag{10.116}$$

Even though there is mass transfer through the wall, we define the heat transfer coefficient by an equation identical to that for an impermeable surface, namely, Eq. (4.3):

$$h_c = \frac{-k(dT/dy)|_s}{T_s - T_e}$$

Substituting Eq. (10.116) then gives

$$h_c = \frac{\dot{m}''c_{p1}}{\exp(\dot{m}''c_{p1}/k)\delta - 1} \tag{10.117}$$

The limit form as $\dot{m}'' \to 0$ is obtained by applying L'Hopital's rule,

$$\lim_{\dot{m}'' \to 0} h_c = h_c^* = \frac{k}{\delta} \tag{10.118}$$

which is simply the value for conduction across a slab δ thick, since in this limit the temperature profile is linear. Combining Eqs. (10.117) and (10.118) gives

$$\frac{h_c}{h_c^*} = \frac{\dot{m}''c_{p1}/h_c^*}{\exp(\dot{m}''c_{p1}/h_c^*) - 1} \tag{10.119}$$

A blowing parameter for heat transfer is now defined as

$$B_h = \frac{\dot{m}''c_{p1}}{h_c^*} \tag{10.120}$$

and Eq. (10.119) becomes

$$\frac{h_c}{h_c^*} = \frac{B_h}{\exp B_h - 1} \tag{10.121}$$

which is the heat transfer analog to Eqs. (10.103) and (10.111). Equation (10.121) allows the heat transfer coefficient h_c for a finite mass transfer rate to be calculated from h_c^*, its value for zero mass transfer, and the mass transfer rate $\dot{m}''$.

To complete the picture for heat transfer, we look at an energy balance on the control volume located between the s- and u-surfaces, as shown in Fig. 10.18. If

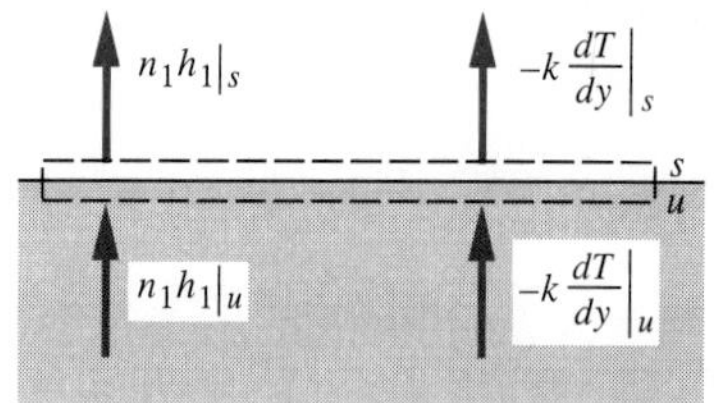

Figure 10.18 Interface energy balance for transfer of a single chemical species.

radiation heat transfer is negligible,

$$-k\frac{dT}{dy}\bigg|_u + n_{1,u}h_{1,u} = -k\frac{dT}{dy}\bigg|_s + n_{1,s}h_{1,s}$$

But $n_{1,s} = n_{1,u} = \dot{m}''$; thus

$$k\frac{dT}{dy}\bigg|_u = k\frac{dT}{dy}\bigg|_s - \dot{m}''(h_{1,s} - h_{1,u}) \tag{10.122}$$

Case 1: If the surface is a liquid evaporating into a gas stream, then $h_{1,s} - h_{1,u} = h_{\text{fg}}$, the enthalpy of vaporization, and with $k(\partial T/\partial y)|_s = h_c(T_e - T_s)$:

$$k\frac{dT}{dy}\bigg|_u = h_c(T_e - T_s) - \dot{m}''h_{\text{fg}} \tag{10.123}$$

Equation (10.123) states that the heat transfer into the liquid (or porous wall filled with liquid) is equal to the convective heat transfer to the surface minus the evaporation rate times vaporization enthalpy. This result is the basis of sweat cooling: the heat transfer into the water is less than that to the surface by the amount $\dot{m}''h_{\text{fg}}$. In addition, h_c is less than h_c^* due to the blowing effect given by Eq. (10.121), and hence the convective heat transfer itself is reduced by the sweat cooling process.

Case 2: If the wall is porous and species 1 is a gas blown through the wall, then $h_{1,s} = h_{1,u}$ and

$$k\frac{dT}{dy}\bigg|_u = h_c(T_e - T_s) \tag{10.124}$$

In this case of *transpiration* cooling, the heat transfer into the wall is reduced by the decrease of h_c due to blowing. If the coolant gas is supplied from a reservoir at temperature T_o, as shown in Fig. 10.19, then an energy balance between the u- and o-surfaces requires

$$-k\frac{dT}{dy}\bigg|_o + n_{1,o}h_{1,o} = -k\frac{dT}{dy}\bigg|_u + n_{1,u}h_{1,u}$$

But $n_{1,o} = n_{1,u} = \dot{m}''$, and if the o-surface is located sufficiently far from the u-surface, the temperature gradient $(dT/dy)|_o$ will be negligible. Then

$$k\frac{dT}{dy}\bigg|_u = \dot{m}''(h_{1,u} - h_{1,o})$$

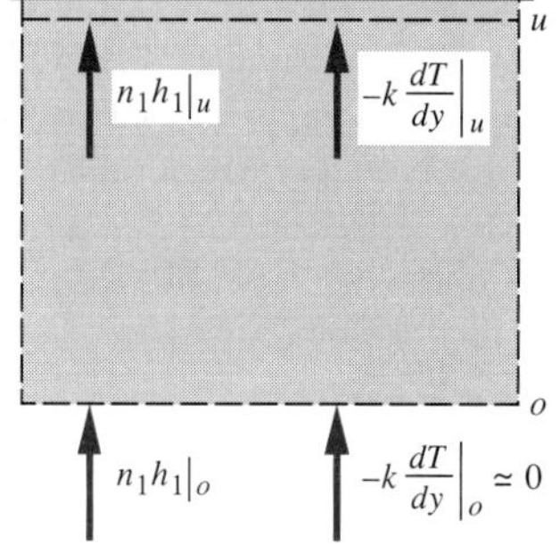

Figure 10.19 Energy balance on a coolant reservoir for transpiration cooling with injection of a single chemical species.

and substituting in Eq. (10.124) gives

$$h_c(T_e - T_s) = \dot{m}''(h_{1,u} - h_{1,o}) \tag{10.125}$$

If the specific heat of the coolant is assumed constant, this result simplifies to

$$h_c(T_e - T_s) = \dot{m}''c_{p1}(T_s - T_o) \tag{10.126}$$

which states that the heat transfer to the wall is equal to the coolant flow rate times its enthalpy change as it flows from the reservoir to the wall surface.

Notice that we did not assume all properties constant in the Couette-flow heat transfer analysis: the specific heats of species 1 and 2 were allowed to be different. Figure 10.20 shows h_c/h_c^* plotted from Eq. (10.119) for various values of c_{p1}, where it is seen that the effect of blowing in reducing heat transfer increases with increasing c_{p1}: one can think of a high-specific-heat gas being more effective in "blocking" the heat from reaching the wall. However, high-specific-heat gases such as helium

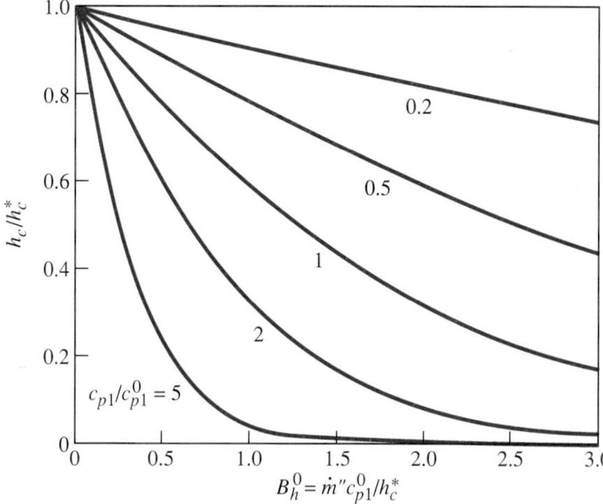

Figure 10.20 The effect of injectant specific heat on heat transfer for injection of single chemical species: a plot of the Couette-flow model result, Eq. (10.119), written as $h_c/h_c^* = (c_{p1}/c_{p1}^0)B_h^0/\exp[(c_{p1}/c_{p1}^0)B_h^0] - 1$; $B_h^0 = \dot{m}''c_{p1}^0/h_c^*$.

also have a low density and high thermal conductivity. Thus, for real air boundary layer flows with helium transpiration, there are additional effects due to density and conductivity (and viscosity) variations across the flow. Also, since $PV = c\mathcal{R}T$, the number of moles of gas stored in a tank of given volume and at a given pressure is independent of species. Thus, a larger tank is required to store a given mass of low-molecular-weight gas, and, in a design trade-off, use of air as the injectant may be advantageous even though the blocking effect per unit mass injection rate is less than that for a low-molecular-weight gas.

Problem Solving

In summary, the Couette-flow model has established the effect of finite mass transfer rates on the mass transfer conductance, the wall shear stress, and the heat transfer coefficient, namely,

$$\frac{g_{m1}}{g_{m1}^*} = \frac{B_{m1}}{\exp B_{m1} - 1}; \qquad B_{m1} = \frac{\dot{m}''}{g_{m1}^*} \tag{10.127a}$$

or

$$\frac{g_{m1}}{g_{m1}^*} = \frac{\ln(1 + \mathcal{B}_{m1})}{\mathcal{B}_{m1}}; \qquad \mathcal{B}_{m1} = \frac{m_{1,e} - m_{1,s}}{m_{1,s} - 1} \tag{10.127b}$$

$$\frac{\tau_s}{\tau_s^*} = \frac{B_f}{\exp B_f - 1}; \qquad B_f = \frac{\dot{m}'' u_e}{\tau_s^*} \tag{10.128}$$

$$\frac{h_c}{h_c^*} = \frac{B_h}{\exp B_h - 1}; \qquad B_h = \frac{\dot{m}'' c_{p1}}{h_c^*} \tag{10.129}$$

The behavior of Eq. (10.127) is shown in Fig. 10.21. Notice that for strong suction $B_{m1} \to -\infty$, while $\mathcal{B}_{m1} \to -1$. Suction occurs in such situations as condensation from a steam-air mixture, and combustion of a solid fuel when the products of combustion are nonvolatile. Of course, if the flow is a binary gas mixture, species 1 cannot be selectively removed through a porous wall; that is, there is no suction problem corresponding to transpiration cooling.

In solving engineering problems, for which the assumption of low mass transfer rate cannot be justified, the procedure is as follows. Values of g_m^*, τ_s^*, and h_c^* are obtained from correlations for an impermeable wall with the same geometry and flow condition, as was done for low mass transfer rate theory in Section 9.4. The correlations in Table 4.10 could have been written as starred quantities (for example, Nu^*, C_f^*) but such a complication was unnecessary in Chapter 4. Equations (10.127)–(10.129) are then used to correct for high mass transfer rates. The blowing factors are based on the Couette-flow model and are only approximately valid for real flows; nevertheless, they are sufficient for many purposes. More accurate blowing factors for specific flows based on more exact analysis can be found in Section 10.4.2. The foregoing procedure works well for forced-flow laminar and turbulent boundary layers and for turbulent duct flows. It is less satisfactory for natural-convection boundary layers and laminar duct flows. Note that the procedure does not apply to the natural-

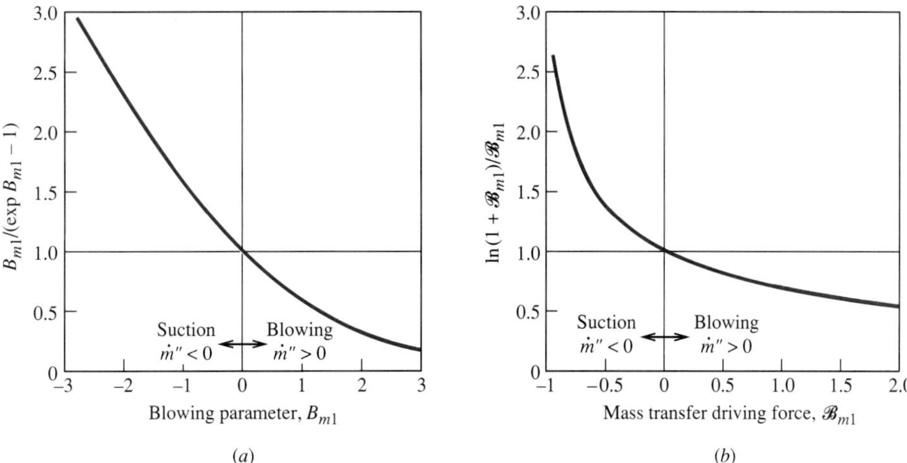

Figure 10.21 Plots of (*a*) Eq. (10.127*a*) and (*b*) Eq. (10.127*b*), showing the effects of suction and blowing on the Couette-flow blowing factor.

convection enclosure flows of items 14 through 17 in Table 4.10, since the heat transfer coefficient is defined for transfer across the cavity, and not for transfer from a surface into fluid.

Variable-Property Effects

Accounting for variable-property effects is not straightforward. A major effect, that of the transferred species specific heat on heat transfer, is accounted for in the blowing factor defined by Eq. (10.129). If we ignore the effect of composition on other properties, namely, density, viscosity, and thermal conductivity, then a simple procedure is to simply evaluate $\mathscr{g}_m^*$, τ_s^*, and h_c^* as if we were using low mass transfer rate theory. For example, in the case of an external flow, the properties will be those of a fluid of free-stream composition evaluated at the mean film temperature. The error in results obtained in this manner depends both on the extent to which variable-property effects are accurately accounted for and on the accuracy of the Couette-flow blowing factors for the particular flow under consideration. These errors may tend to cancel in some situations, and add in others. More accurate procedures for dealing with variable-property effects will be dealt with in Section 10.4.2.

EXAMPLE 10.7 Transpiration Cooling

Air at 1500 K and 1 atm pressure flows at 10 m/s over a porous flat plate that is to be transpiration-cooled. Calculate the rate at which coolant must be injected through the plate at a location 0.2 m from the leading edge if the plate surface is to be maintained at 500 K, and the coolant is supplied from a reservoir beneath the plate at 300 K, if (i) the coolant is air, (ii) the coolant is helium.

Solution

Given: Air at 1500 K flowing over a transpiration cooled flat plate.

Required: Coolant injection rate at $x = 0.2$ m to maintain the plate surface at 500 K.

Assumptions:
1. Couette-flow blowing factors are adequate.
2. The injectant specific heat is constant.
3. Radiation heat transfer is negligible.

The required injection rate is given by Eq. (10.126),

$$h_c(T_e - T_s) = \dot{m}''c_{p1}(T_s - T_o)$$

where the heat transfer coefficient blowing factor is obtained from Eq. (10.119) as

$$\frac{h_c}{h_c^*} = \frac{\dot{m}''c_{p1}/h_c^*}{\exp(\dot{m}''c_{p1}/h_c^*) - 1}$$

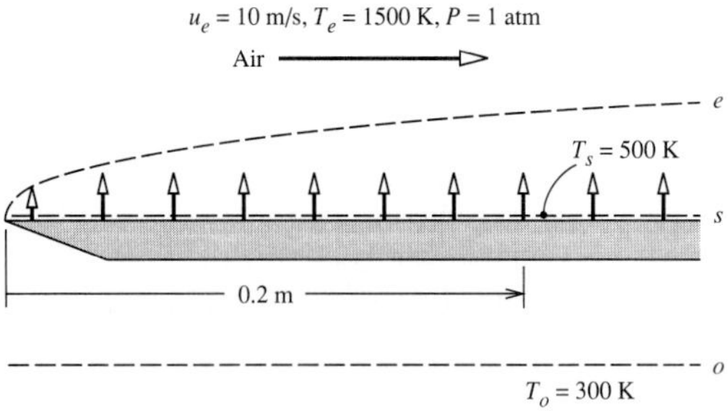

Combining these two equations and solving for $\dot{m}''$ gives

$u_e = 10$ m/s, $T_e = 1500$ K, $P = 1$ atm

Air

e

$T_s = 500$ K

s

0.2 m

o

$T_o = 300$ K

$$\dot{m}'' = \frac{h_c^*}{c_{p1}} \ln\left(1 + \frac{T_e - T_s}{T_s - T_o}\right) \quad \textbf{(1)}$$

In determining h_c^* we evaluate air properties at the mean film temperature, $T_r = 500 + (1/2)(1500 - 500) = 1000$ K. From Table A.7, $\rho = 0.354$ kg/m^3, $\nu = 117.3 \times 10^{-6}$ m^2/s, $c_p = 1130$ J/kg K, Pr $= 0.70$. The Reynolds number is

$$\text{Re}_x = \frac{u_e x}{\nu} = \frac{(10)(0.2)}{(117.3 \times 10^{-6})} = 17{,}050 \quad \text{(laminar flow)}$$

From Eq. (4.56) the local Stanton number and heat transfer coefficient are

$$\text{St}_x^* = 0.332\,\text{Re}_x^{-1/2}\,\text{Pr}^{-2/3}$$
$$= (0.332)(17{,}050)^{-1/2}(0.70)^{-2/3} = 3.23 \times 10^{-3}$$
$$h_c^* = \rho c_p u_e\,\text{St}_x^* = (0.354)(1130)(10)(3.23 \times 10^{-3}) = 12.9\ \text{W/m}^2\,\text{K}$$

(i) *Air injection*. Substituting in Eq. (1) with the assumed constant c_{p1} evaluated at $(1/2)(T_e + T_o) = 900$ K as 1111 J/kg K gives the injection rate as

$$\dot{m}'' = \frac{h_c^*}{c_{p1}} \ln\left(1 + \frac{T_e - T_s}{T_s - T_o}\right)$$
$$= \frac{12.9}{1111} \ln\left(1 + \frac{1500 - 500}{500 - 300}\right) = 0.0208\ \text{kg/m}^2\,\text{s}$$

(ii) *Helium injection*. From Table A.7, $c_{p1} = 5200$ J/kg K, and the required injection rate is

$$\dot{m}'' = \frac{12.9}{5200} \ln\left[1 + \frac{1500 - 500}{500 - 300}\right] = 4.44 \times 10^{-3}\ \text{kg/m}^2\ \text{s}$$

Comments

1. Owing to the high specific heat of helium, the helium injection rate is only 21% of that required for air.
2. Check h_c^* using CONV.
3. To maintain an isothermal plate, Eq. (4.56) shows that $\dot{m}''$ must vary as $x^{-1/2}$.

EXAMPLE 10.8 Sweat Cooling

A wall exposed to intense radiative and convective heating is to be protected by sweat cooling. Water is injected through a porous stainless steel surface at a rate just sufficient to keep the surface wetted. Dry air at 840 K and 1 atm pressure flows past the surface at 100 m/s. Heat transfer experiments with air for the same geometry and flow conditions indicate a local Stanton number of 0.00429. (i) Determine the water supply rate to maintain the surface at 360 K. (ii) If the irradiation is 461 kW/m^2, determine the water supply temperature required to maintain a steady state.

Solution

Given: A wall heated by radiation and convection to be sweat-cooled with water.

Required: Water supply rate and temperature.

Assumptions: 1. The Couette-flow blowing factors are adequate.
2. The flow is a turbulent boundary layer.

Since the wall temperature is specified, the heat and mass transfer problems are uncoupled. We first solve the mass transfer problem to determine the water supply rate, and then solve the heat transfer problem to determine the water supply temperature required to maintain steady conditions.

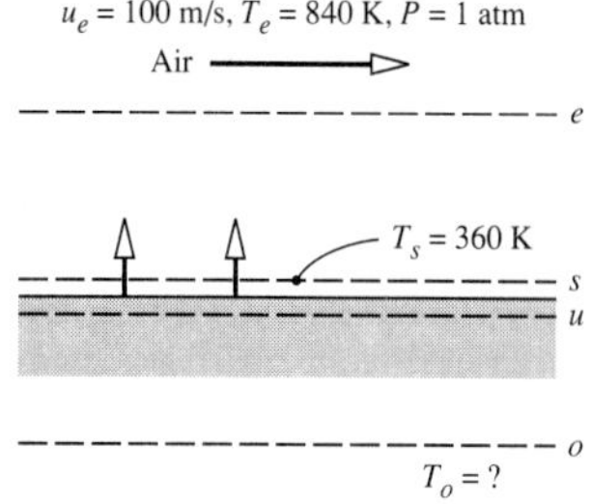

(i) The water supply rate must balance the evaporation rate; from Eq. (10.100) the evaporation rate is

$$\dot{m}'' = g_m \mathcal{B}_{m1}; \qquad \mathcal{B}_{m1} = \frac{m_{1,e} - m_{1,s}}{m_{1,s} - 1}$$

where, from Eq. (10.101), the blowing factor is

$$\frac{g_{m1}}{g_{m1}^*} = \frac{\ln(1 + \mathcal{B}_{m1})}{\mathcal{B}_{m1}}$$

First we calculate the mass transfer driving force $\mathcal{B}_{m1}$. At 360 K, Table A.12*a* gives the saturation vapor pressure of water vapor as 0.6213×10^5 Pa. For 1 atm $= 1.0133 \times 10^5$ Pa total pressure,

$$m_{1,s} = \frac{P_{1,s}}{P_{1,s} + (M_2/M_1)(P - P_{1,s})} = \frac{0.6213}{0.6213 + (29/18)(1.0133 - 0.6213)} = 0.496$$

$$\mathscr{B}_{m1} = \frac{m_{1,e} - m_{1,s}}{m_{1,s} - 1} = \frac{0 - 0.496}{0.496 - 1} = 0.984$$

Next, we must evaluate the mass transfer conductance g^*_{m1}. The data suggest a turbulent boundary layer, for which Eq. (4.64) gives $\text{St}^*_m / \text{St}^* = (\text{Sc}/\text{Pr})^{-0.57}$. Properties are evaluated at the mean film temperature $T_r = (1/2)(360 + 840) = 600$ K. Using AIRSTE to obtain $\text{Pr} = 0.69$ for air and $\text{Sc} = 0.54$ for a dilute H_2O-air mixture gives an estimate of St^*_m as

$$\text{St}^*_m = 0.00429(0.54/0.69)^{-0.57} = 0.00493$$

For air, Table A.7 gives $\rho = 0.589$ kg/m^3, $c_p = 1038$ J/kg K. Thus, the zero-mass-transfer-limit mass transfer conductance is

$$g^*_{m1} = \rho u_e \text{St}^*_m = (0.589)(100)(0.00493) = 0.291 \text{ kg/m}^2\text{ s}$$

and the blowing factor is

$$\frac{g_{m1}}{g^*_{m1}} = \frac{\ln(1 + \mathscr{B}_{m1})}{\mathscr{B}_{m1}} = \frac{\ln(1 + 0.984)}{0.984} = 0.696$$

Thus, the evaporation rate is

$$\dot{m}'' = g_{m1}\mathscr{B}_{m1} = g^*_{m1}\frac{\ln(1 + \mathscr{B}_{m1})}{\mathscr{B}_{m1}}\mathscr{B}_{m1} = (0.291)(0.696)(0.984) = 0.199 \text{ kg/m}^2\text{ s}$$

(ii) The heat transfer problem is conveniently solved by making an energy balance on a control volume located between the s- and o-surfaces: the o-surface is located in the water supply reservoir where conduction is negligible. Since $n_{1,s} = \dot{m}''$,

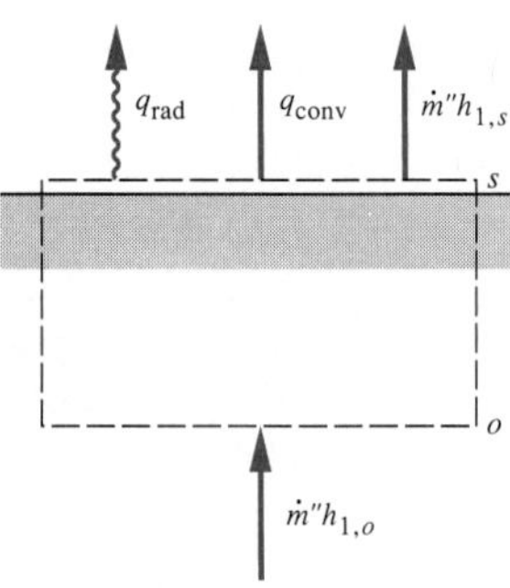

$$q_{rad} + q_{conv} + \dot{m}''h_{1,s} = \dot{m}''h_{1,o}$$

Assuming a constant specific heat for the water, c_{pl}, and writing $q_{conv} = h_c(T_s - T_e)$,

$$\dot{m}''[h_{fg} + c_{pl}(T_s - T_o)] = -q_{rad} + h_c(T_e - T_s) \qquad \textbf{(1)}$$

To evaluate the heat transfer coefficient h_c, we first evaluate h^*_c:

$$h^*_c = \rho u_e c_p \text{St}^* = (0.589)(100)(1038)(0.00429) = 253 \text{ W/m}^2\text{ K}$$

The heat transfer blowing parameter B_h is

$$B_h = \dot{m}''c_{p1}/h^*_c = (0.199)(2003)/(253) = 1.58$$

where c_{p1} is the specific heat of water vapor, and is also evaluated at $T_r = 600$ K. The corrected heat transfer coefficient is then

$$h_c = h^*_c\frac{B_h}{\exp B_h - 1} = (253)\frac{1.58}{\exp(1.58) - 1} = (253)(0.410) = 104 \text{ W/m}^2\text{ K}$$

From Table A.12a, $h_{fg} = 2.291 \times 10^6$ J/kg at 360 K, and from Table A.8, $c_{pl} = 4178$ at a guessed average temperature of ~330 K. Substituting in Eq. (1) gives

$$(0.199)[2.291 \times 10^6 + (4178)(360 - T_o)] = 461 \times 10^3 + 104(840 - 360)$$

Solving, $T_o = 294$ K.

Comments

1. If we assume that the radiation is absorbed and emitted between the s- and u-surfaces (see Section 9.5.1), an energy balance between the u- and o-surfaces gives the conduction heat flux across the u-surface as

$$\begin{aligned} q_{\text{cond},u} &= -k\frac{dT}{dy}\bigg|_u = \dot{m}''c_{pl}(T_o - T_s) \\ &= (0.199)(4178)(294 - 360) \\ &= -54.9 \text{ kW/m}^2 \end{aligned}$$

2. Due to blowing, $\mathscr{g}_{m1}$ is 70% of $\mathscr{g}^*_{m1}$, whereas h_c is only 41% of h^*_c: this behavior is due to the relatively high specific heat of water vapor.

3. An alternative way to pose this problem is to specify the water supply temperature and the irradiation. Then the wall temperature is an unknown, and the heat and mass transfer problems must be solved simultaneously, requiring an iterative process. See, for example, Exercise 10–31.

EXAMPLE 10.9 Effect of an Air Leak on Condenser Performance

Steam at 336 K and 0.115 atm containing 0.62% air by mass flows down at 1 m/s over a horizontal brass condenser tube of 19.1 mm O.D. and 16.1 mm I.D. Coolant water flows through the tube at a bulk velocity of 1.47 m/s. At a location where the bulk coolant temperature is 283 K, estimate the steam condensation rate and compare it with the value for pure steam.

Solution

Given: Steam containing a small amount of air flowing down over a horizontal condenser tube.

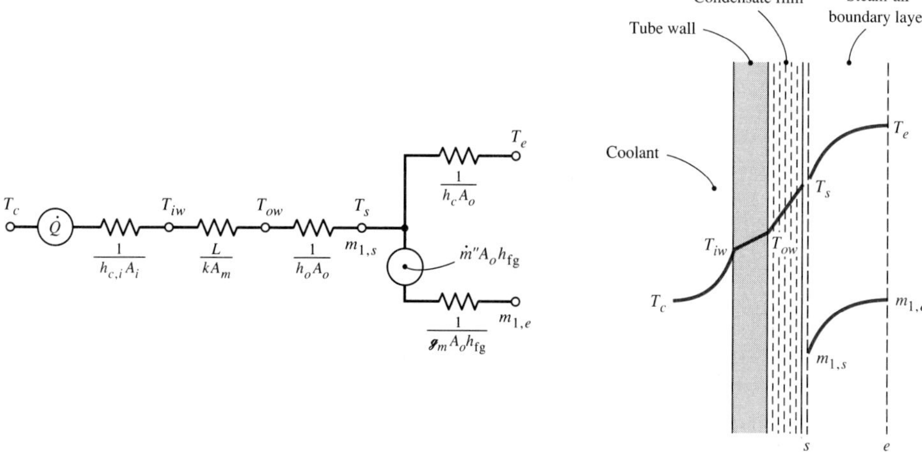

Required: Effect of air on steam condensation rate.

Assumptions:
1. Couette-flow blowing factors are adequate.
2. The presence of the condensate film can be ignored in calculating the steam flow mass transfer conductance.
3. Average values of the transfer coefficients can be used.
4. The isothermal-tube film condensation correlation is adequate.

First we draw an equivalent circuit for this transfer process. It is a hybrid one, involving both heat and mass transfer processes, as shown. The transfer coefficients are all average values. The circuit between T_c and T_s is conventional: $h_{c,i}$ is the inside heat transfer coefficient, L/kA_m is the brass-tube wall resistance, and h_o is the heat transfer coefficient for film condensation. Heat is transferred to the condensate surface both as sensible heat with thermal resistance $1/h_cA_o$ for flow over a cylinder, and as latent heat with the corresponding mass transfer resistance $1/\mathscr{g}_{m1}A_o$. In most situations the sensible heat transfer can be neglected compared with the latent heat transfer. Calculation of the thermal circuit between T_c and T_s is straightforward. For evaluation of fluid properties, reference temperatures of 290 K and 300 K are guessed for $h_{c,i}$ and h_o, respectively. Then, using the correlation Eq. (4.44) for turbulent flow in a tube, $h_{c,i} = 5610$ W/m^2 K, and for laminar film condensation on a horizontal tube, Eq. (7.41) gives $h_o = 17{,}300(T_s - T_{ow})^{-1/4}$ W/m^2 K. Summing the resistances in the circuit gives a total resistance per unit outside area between T_c and T_s of

$$\frac{1}{U} = \frac{1}{h_{c,i}(A_i/A_o)} + \frac{1}{(k/L)[(A_i + A_o)/2A_o]} + \frac{1}{h_o}$$

$$= \frac{1}{(5610)(0.843)} + \frac{1}{(111/1.5 \times 10^{-3})(0.921)} + \frac{1}{(17{,}300)(T_s - T_{ow})^{-1/4}}$$

$$= \frac{1}{4730} + \frac{1}{68{,}200} + \frac{1}{17{,}300(T_s - T_{ow})^{-1/4}}$$

$$U = \frac{10^5}{22.6 + 5.78(T_s - T_{ow})^{1/4}} \tag{1}$$

Next we calculate $\mathscr{g}^*_{m1}$, evaluating properties using the free-stream composition, $m_1 = 0.9938 \simeq 1.0$ (i.e., pure steam) and a guessed reference temperature of 320 K. At 0.115 atm $= 0.1165 \times 10^5$ Pa, the steam is supersaturated, and ρ can be calculated from the ideal gas formula,

$$\rho = PM_1/\mathscr{R}T = (0.1165 \times 10^5)(18)/[(8314)(320)] = 0.0788 \text{ kg/m}^3$$

To calculate ν we first find $\mu = 9.89 \times 10^{-6}$ kg/m s for saturated steam at 320 K from Table A.7; then, since μ is independent of pressure, $\nu = \mu/\rho = 9.89 \times 10^{-6}/0.0788 = 125.5 \times 10^{-6}$ m^2/s for the supersaturated steam. To calculate $\mathscr{D}_{12}$ we use the formula given in Table A.17*a*, $\mathscr{D}_{12} = 1.97 \times 10^{-5}(1/0.115)(320/256)^{1.685} = 249 \times 10^{-6}$ m^2/s. Hence, $\text{Sc} = \nu/\mathscr{D}_{12} = 125.5/249 = 0.504$. The Reynolds number for the steam flow is

$$\text{Re}_D = VD/\nu = (1)(19.1 \times 10^{-3})/(125.5 \times 10^{-6}) = 152.2$$

We will ignore the presence of the condensate film in determining $\mathscr{g}^*_{m1}$: the film is thin compared with the tube diameter, and its surface velocity is small compared with the steam velocity. For forced convection over a cylinder, the mass transfer analog to Eq. (4.71a) is

$$\overline{\mathrm{Sh}}_D^* = 0.3 + \frac{0.62\,\mathrm{Re}^{1/2}\,\mathrm{Sc}^{1/3}}{[1 + (0.4/\mathrm{Sc})^{2/3}]^{1/4}}$$

$$= 0.3 + \frac{(0.62)(152.2)^{1/2}(0.504)^{1/3}}{[1 + (0.4/0.504)^{2/3}]^{1/4}} = 5.51$$

The zero-mass-transfer-limit mass transfer conductance is then

$$\mathscr{g}_{m1}^* = \rho\mathscr{D}_{12}\,\mathrm{Sh}^*/D$$

$$= (0.0788)(249 \times 10^{-6})(5.51)/(19.1 \times 10^{-3}) = 5.66 \times 10^{-3}\ \mathrm{kg/m^2\,s}$$

The rate of latent heat transfer is $\dot{m}''h_{fg}$; from Eqs. (10.100) and (10.101),

$$\dot{m}''h_{fg} = \mathscr{g}_{m1}^*\ln(1 + \mathscr{B}_{m1})h_{fg} = 5.66 \times 10^{-3}\ln\left(1 + \frac{0.9938 - m_{1,s}}{m_{1,s} - 1}\right)h_{fg} \tag{2}$$

If we neglect sensible heat transfer from the steam to the condensate surface, our thermal circuit shows that the current flowing is

$$\dot{Q} = UA_o(T_s - T_c) = -\dot{m}''h_{fg}A_o \tag{3}$$

Substituting Eqs. (1) and (2) in Eq. (3) gives

$$\frac{\dot{Q}}{A_o} = \frac{(T_s - 283)(10^5)}{22.6 + 5.78(T_s - T_{ow})^{1/4}} = -5.66 \times 10^{-3}\ln\left(1 + \frac{0.9938 - m_{1,s}}{m_{1,s} - 1}\right)h_{fg} \tag{4}$$

Equation (4) must be solved by iteration using steam tables for $P_{1,s} = P_{1,\mathrm{sat}}(T_s)$ and $h_{fg}(T_s)$, together with the auxiliary relations

$$m_{1,s} = \frac{P_{1,s}}{P_{1,s} + (29/18)(P - P_{1,s})}; \qquad T_s - T_{ow} = \left(\frac{\dot{Q}/A_o}{17{,}300}\right)^{4/3} \tag{5,6}$$

The result (after some tedious calculations) is $\dot{Q}/A_o = 65{,}500\ \mathrm{W/m^2}$, with $T_s = 303.7$ K and $m_{1,s} = 0.274$. The condensation rate per unit length of tube is

$$(\dot{Q}/A_o)A_o/h_{fg} = (65{,}500)(\pi)(19.1 \times 10^{-3})/(2.43 \times 10^6) = 1.62 \times 10^{-3}\ \mathrm{kg/s\,m}$$

Intermediate temperatures in the circuit can be now calculated and the guessed temperatures for fluid properties revised, if necessary.

We can now check our assumption that the sensible heat transfer is small compared with the latent heat transfer. For pure steam at 320 K, $k = 0.0210$ W/m K, $c_p = 1890$ J/kg K, and Pr $= 0.89$. Using Eq. (4.71a), $\overline{\mathrm{Nu}}_D^* = 6.86$, and hence

$$h_c^* = (k/D)\,\mathrm{Nu}_D^* = (0.0210/1.91 \times 10^{-3})(6.86) = 7.54\ \mathrm{W/m^2\,K}$$

$$\dot{m}'' = -(\dot{Q}/A_o)/h_{fg} = -(65{,}500/2.43 \times 10^6) = -0.0270\ \mathrm{kg/m^2\,s}$$

To correct h_c^* for suction, we use Eq. (10.129):

$$B_h = \dot{m}''c_{p1}/h_c^* = (-0.0270)(1890)/(7.54) = -6.77$$

$$\frac{h_c}{h_c^*} = \frac{B_h}{\exp B_h - 1} = \frac{-6.77}{\exp(-6.77) - 1} = 6.78$$

Notice that the actual heat transfer coefficient is 6.8 times the value for no mass transfer, due to strong suction.

$$h_c = (6.78)(7.54) = 51.1 \text{ W/m}^2 \text{ K}$$

$$h_c(T_e - T_s) = 51.1(336 - 303.7) = 1650 \text{ W/m}^2$$

The sensible heat transfer proves to be only 2.5% of the latent heat transfer, and was justifiably neglected.

For pure condensing steam at 0.115 atm, T_s is the corresponding saturation value; from Table A.12a, $T_s = 322.0$ K. Again an iterative solution is required. Equating the heat flow across the film and across the wall into the coolant,

$$\frac{\dot{Q}}{A_o} = h_o(T_s - T_{ow}) = \left(\frac{1}{h_{c,i}(A_i/A_o)} + \frac{1}{(k/L)[(A_i + A_o)/2A_o]}\right)^{-1}(T_{ow} - T_c) \qquad \textbf{(7)}$$

which reduces to

$$17{,}300(322.0 - T_{ow})^{3/4} = 4423(T_{ow} - 283)$$

The solution is $(322.0 - T_{ow}) = 12.7$ K, and $\dot{Q}/A_o = 17{,}300(322.0 - T_{ow})^{3/4} = 116{,}000$ W/m^2. Thus, the heat transfer rate is reduced to 65,500/116,000 = 56% of the pure steam value by only 0.62% air present in the bulk steam.

Comments

1. Strictly speaking, a new reference temperature for property evaluation should be used in solving Eq. (7).
2. Since power-plant condensers operate at pressures well below atmospheric pressure, air leaks are common. Such condensers are equipped with steam ejectors that continuously pump a steam-air mixture from dead spaces in the condenser to prevent a buildup of air.
3. Notice that the Schmidt number for the bulk steam-air mixture is 0.504, not the value of 0.61 for steam in dilute concentration.
4. The presence of the condensate film was neglected in calculating Re_D, $\overline{\mathrm{Sh}}_D^*$, and $\overline{\mathrm{Nu}}_D^*$.

10.4.2 Improved Blowing Factors

The Couette-flow–based blowing factors developed in Section 10.4.1 are useful for routine engineering problem solving. However, often more accurate results are required, for which these blowing factors may be inadequate. Again we will restrict our attention to binary inert mixtures with transfer of a single species, and our concern is with the effects of the type of flow and variable properties on the blowing factors. Figure 10.22 shows the effects of pressure gradient and variable properties on blowing factors for some laminar boundary layers. Figure 10.23 shows the effect

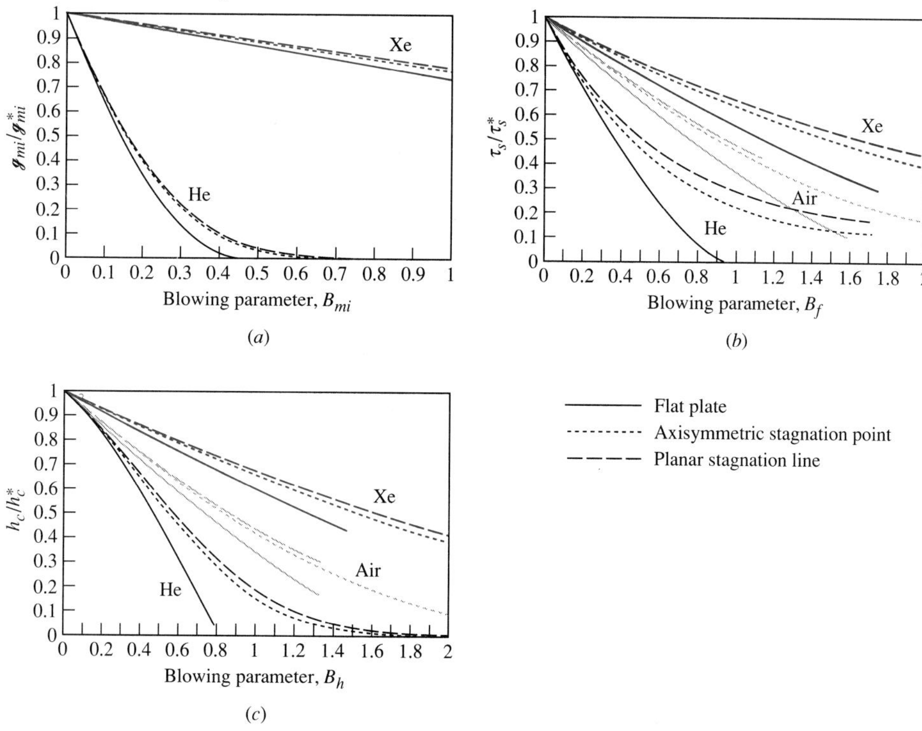

Figure 10.22 Numerical results for the effect of pressure gradient and variable properties on blowing factors for laminar boundary layers: low-speed air flow over a cold wall ($T_s/T_e = 0.1$) with foreign gas injection. (*a*) Mass transfer conductance. (*b*) Wall shear stress. (*c*) Heat transfer coefficient [9,10].

of variable properties for a turbulent boundary layer flow on a flat plate. Clearly, deviations from the Couette-flow blowing factors can be substantial.

A simple procedure for correlating the effects of flow type and variable properties is to use weighting factors in the exponential functions suggested by a constant-property Couette-flow model. Equations (10.127)–(10.129) are generalized as follows. Denoting the injected species as species i, we have

$$\frac{\mathcal{g}_{mi}}{\mathcal{g}_{mi}^*} = \frac{a_{mi}B_{mi}}{\exp(a_{mi}B_{mi}) - 1}; \qquad B_{mi} = \frac{\dot{m}''}{\mathcal{g}_{mi}^*} \tag{10.130a}$$

or

$$\frac{\mathcal{g}_{mi}}{\mathcal{g}_{mi}^*} = \frac{\ln(1 + a_{mi}\mathcal{B}_{mi})}{a_{mi}\mathcal{B}_{mi}}; \qquad \mathcal{B}_{mi} = \frac{\dot{m}''}{\mathcal{g}_{mi}} = \frac{m_{i,e} - m_{i,s}}{m_{i,s} - 1}$$

$$\frac{\tau_s}{\tau_s^*} = \frac{a_{fi}B_f}{\exp(a_{fi}B_f) - 1}; \qquad B_f = \frac{\dot{m}''u_e}{\tau_s^*} \tag{10.130b}$$

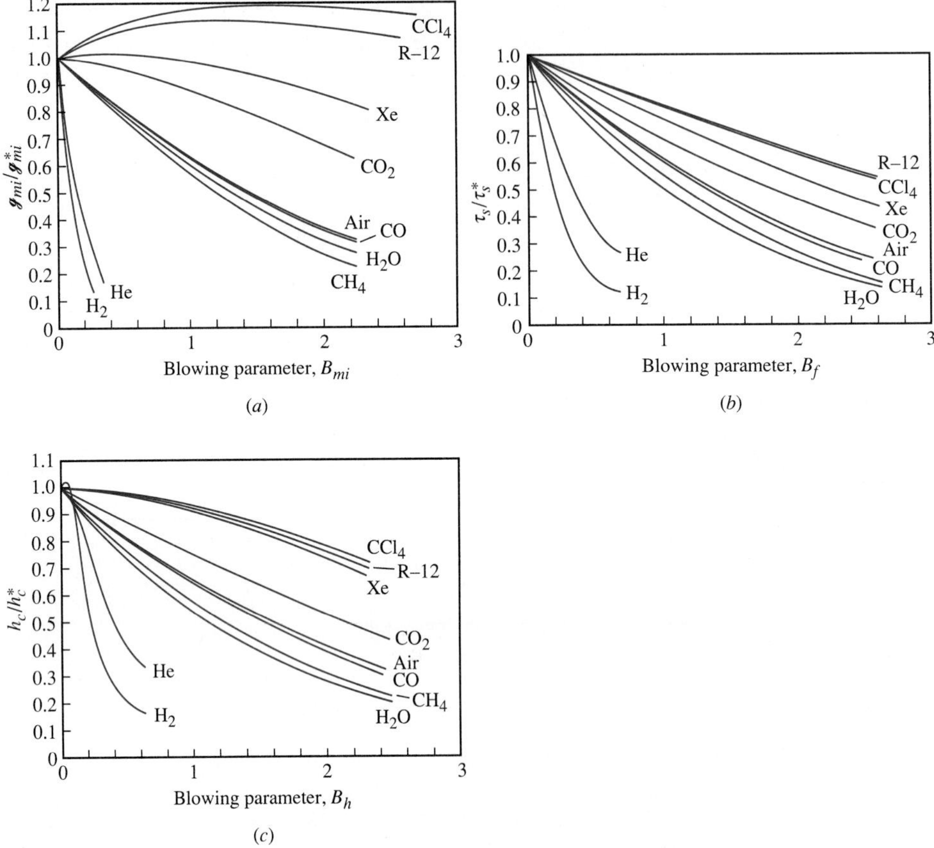

Figure 10.23 Numerical results for the effect of variable properties on blowing factors for a low-speed turbulent air boundary layer on a cold flat plate ($T_s/T_e = 0.2$) with foreign gas injection. (*a*) Mass transfer conductance. (*b*) Wall shear stress. (*c*) Heat transfer coefficient [12].

$$\frac{h_c}{h_c^*} = \frac{a_{hi}B_h}{\exp(a_{hi}B_h) - 1}; \qquad B_h = \frac{\dot{m}''c_{pe}}{h_c^*} \tag{10.130c}$$

Notice that g^*_{mi}, τ^*_s, h^*_c, and c_{pe} are evaluated using properties of the free-stream gas at the mean film temperature, and that Eq. (10.130*c*) is the Couette-flow model result for all properties constant; in particular, the injected and free-stream specific heats are equal. The weighting factor a may be found from exact numerical solutions of boundary layer equations or from experimental data. Some results for laminar and turbulent boundary layers follow.

1. *Laminar boundary layers*. We will restrict our attention to low-speed air flows, for which viscous dissipation and compressibility effects are negligible, and use exact numerical solutions of the self-similar laminar boundary layer equa-

tions [9,10,11]. Least-squares curve fits of the numerical data were obtained using Eqs. (10.130*a*)–(10.130*c*). Then the weighting factors for axisymmetric stagnation-point flow with a cold wall ($T_s/T_e = 0.1$) were correlated as

$$a_{mi} = 1.65(M_{\text{air}}/M_i)^{10/12} \quad \textbf{(10.131a)}$$

$$a_{fi} = 1.38(M_{\text{air}}/M_i)^{5/12} \quad \textbf{(10.131b)}$$

$$a_{hi} = 1.30(M_{\text{air}}/M_i)^{3/12}[c_{pi}/(2.5\mathcal{R}/M_i)] \quad \textbf{(10.131c)}$$

Notice that $c_{pi}/(2.5\mathcal{R}/M_i)$ is unity for monatomic species. For the planar stagnation line and the flat plate, and other values of the temperature ratio T_s/T_e, the values of the species weighting factors are divided by the values given by Eqs. (10.131) to give correction factors G_{mi}, G_{fi}, and G_{hi}, respectively. These correction factors are listed in Table 10.1.

The exponential relation blowing factors cannot accurately represent some of the more anomalous effects of blowing. For example, when a light gas such

Table 10.1 *G* correction factors for foreign gas injection into laminar air boundary layers.

		G_{mi} T_s/T_e			G_{fi} T_s/T_e			G_{hi} T_s/T_e		
Geometry	Species	0.1	0.5	0.9	0.1	0.5	0.9	0.1	0.5	0.9
Flat plate	H	1.14	1.36	1.47	1.30	1.64	1.79	1.15	1.32	—
	H_2	1.03	1.25	1.36	1.19	1.44	1.49	1.56	1.17	1.32
	He	1.051	1.18	1.25	1.34	1.49	1.56	1.18	1.32	—
	Air	—	—	—	1.21	1.27	1.27	1.17	1.21	—
	Xe	1.21	1.13	1.15	1.38	1.35	1.34	1.19	1.18	—
	CCl_4	1.03	0.95	1.00	1.00	1.03	1.03	1.04	1.04	—
Axisymmetric stagnation point	H	1.00	1.04	1.09	1.00	0.62	0.45	1.00	0.94	0.54
	H_2	1.00	1.06	1.06	1.00	0.70	0.62	1.00	1.00	1.01
	He	1.00	1.04	1.03	1.00	0.66	0.56	1.00	1.00	0.95
	C	1.00	1.01	1.00	1.00	0.79	0.69	1.00	0.99	0.87
	CH_4	1.00	1.01	1.00	1.00	0.88	0.84	1.00	1.00	1.00
	O	1.00	0.98	0.97	1.00	0.79	0.70	1.00	0.98	0.95
	H_2O	1.00	1.01	1.00	1.00	0.82	0.73	1.00	1.00	0.99
	Ne	1.00	1.00	0.98	1.00	0.83	0.75	1.00	0.97	0.95
	Air	—	—	—	1.00	0.87	0.82	1.00	0.99	0.97
	A	1.00	0.97	0.94	1.00	0.93	0.91	1.00	0.96	0.95
	CO_2	1.00	0.97	0.95	1.00	0.96	0.94	1.00	0.99	0.97
	Xe	1.00	0.98	0.96	1.00	0.96	1.05	1.00	1.06	0.99
	CCl_4	1.00	0.90	0.83	1.00	1.03	1.07	1.00	0.96	0.93
	I_2	1.00	0.91	0.85	1.00	1.02	1.05	1.00	0.97	0.94
Planar stagnation line	He	0.96	0.98	0.98	0.85	0.53	0.47	0.93	0.91	0.92
	Air	—	—	—	0.94	0.84	0.81	0.94	0.94	—
	Xe	0.92	0.87	0.83	0.90	0.93	0.95	0.93	0.93	—

Based on numerical data of Wortman [11]. Correlations developed by Dr. D. W. Hatfield.

as H_2 is injected, Eq. (10.130*c*) indicates that the effect of blowing is always to reduce heat transfer, due to both the low density and high specific heat of hydrogen. However, at very low injection rates, the heat transfer is actually increased, due to the high thermal conductivity of H_2. For a mixture, $k \simeq \sum x_i k_i$, whereas $c_p = \sum m_i c_{pi}$. At low rates of injection, the mole fraction of H_2 near the wall is much larger than its mass fraction; thus, there is a substantial increase in the mixture conductivity near the wall, but only a small change in the mixture specific heat. An increase in heat transfer results. At higher injection rates, the mass fraction of H_2 is also large, and the effect of high mixture specific heat dominates to cause a decrease in heat transfer.

2. *Turbulent boundary layers*. Here we restrict our attention to air flow along a flat plate for Mach numbers up to 6, and use numerical solutions of boundary layer equations with a mixing length turbulence model [12,13]. Appropriate species weighting factors for $0.2 < T_s/T_e < 2$ are:

$$a_{mi} = 0.79(M_{\text{air}}/M_i)^{1.33} \qquad \textbf{(10.132a)}$$

$$a_{fi} = 0.91(M_{\text{air}}/M_i)^{0.76} \qquad \textbf{(10.132b)}$$

$$a_{hi} = 0.86(M_{\text{air}}/M_i)^{0.73} \qquad \textbf{(10.132c)}$$

In using Eq. (10.130), the limit values for $\dot{m}'' = 0$ are evaluated at the same location along the plate. Whether the injection rate is constant along the plate or varies as $x^{-0.2}$ to give a self-similar boundary layer has little effect on the blowing factors. Thus, Eq. (10.132) has quite general applicability. Notice that the effects of injectant molecular weight are greater for turbulent boundary layers than for laminar ones, and is due to the effect of fluid density on turbulent transport. Also, the injectant specific heat does not appear in a_{hi} as it did for laminar flows. In general, c_{pi} decreases with increasing M_i and is adequately accounted for in the molecular weight ratio.

Reference State Schemes

The reference state approach, in which constant-property solutions are used with properties evaluated at some reference state, is an alternative method for handling variable-property effects. In principle, the reference state is independent of the precise property data used and of the combination of injectant and free-stream species. A reference state approach for the flat plate laminar boundary layer will be given in Section 10.5.

EXAMPLE 10.10 Transpiration Cooling

(i) Rework Example 10.7 for helium injection using the improved blowing factors. (ii) Determine the helium concentration adjacent to the wall.

Solution

Given: Air flow at 1500 K over a transpiration-cooled flat plate.

Required: (i) He injection rate at $x = 0.2$ m to maintain $T_s = 500$ K with $T_o = 300$ K. (ii) The mass fraction $m_{\text{He},s}$.

Assumptions: 1. The boundary layer is laminar.
2. Equation (10.131) and Table 10.1 can be used for the blowing factors.

(i) Combining Eqs. (10.126) and (10.130*c*) and solving for the injection rate gives

$$\dot{m}'' = \frac{h_c^*}{c_{p\,\text{air}} a_{h\text{He}}} \ln\left[1 + \left(\frac{a_{h\text{He}} c_{p\,\text{air}}}{c_{p\text{He}}}\right)\left(\frac{T_e - T_s}{T_s - T_o}\right)\right] \tag{1}$$

From Example 10.7, $h_c^* = 12.9$ W/m^2 K, $c_{p\text{He}} = 5200$ J/kg K, and from Table A.7, $c_{p\,\text{air}} =$ 1130 J/kg K at the mean film temperature of 1000 K. From Eq. (10.131*c*) and Table 10.1 for a flat plate and $T_s/T_e = 0.33$,

$$\begin{aligned} a_{h\text{He}} &= (1.26)(1.30)(M_{\text{air}}/M_{\text{He}})^{3/12}[c_{p\text{He}}/(2.5\mathcal{R}/M_{\text{He}})] \\ &= (1.26)(1.30)(29/4)^{3/12}[5200/(2.5)(8314/4)] \\ &= (1.26)(1.30)(1.64)(1.00) = 2.69 \end{aligned}$$

Substituting in Eq. (1),

$$\begin{aligned} \dot{m}'' &= \frac{12.9}{(1130)(2.69)} \ln\left[1 + \left(\frac{(2.69)(1130)}{5200}\right)\left(\frac{1500 - 500}{500 - 300}\right)\right] \\ &= 5.80 \times 10^{-3} \text{ kg/m}^2\text{ s} \end{aligned}$$

(ii) Equation (10.130*a*) can be rearranged as

$$\dot{m}'' = \left(\frac{g_{m\text{He}}^*}{a_{m\text{He}}}\right) \ln\left(1 + a_{m\text{He}} \frac{m_{\text{He},e} - m_{\text{He},s}}{m_{\text{He},s} - 1}\right) \tag{2}$$

which must be solved for $m_{\text{He},s}$. From Eq. (4.56),

$$\text{St}_{mx}^* = \text{St}_x^*(\text{Sc}/\text{Pr})^{-2/3} \tag{3}$$

For air at the mean film temperature of 1000 K, Tables A.7 and A.17*a* give $\rho = 0.354$ kg/m^3, $\nu = 117.3 \times 10^{-6}$ m^2/s, Pr $= 0.7$, and $\mathcal{D}_{\text{He,air}} = 526 \times 10^{-6}$ m^2/s. Hence,

$$\text{Sc} = \nu/D = 117.3 \times 10^{-6}/526 \times 10^{-6} = 0.223$$

Substituting in Eq. (3) with $\text{St}_x^* = 3.23 \times 10^{-3}$ from Example 10.7,

$$\text{St}_{mx}^* = (3.23 \times 10^{-3})(0.223/0.70)^{-2/3} = 6.92 \times 10^{-3}$$

$$g_{m\text{He}}^* = \rho u_e \text{St}_{mx}^* = (0.354)(10)(6.92 \times 10^{-3}) = 2.45 \times 10^{-2} \text{ kg/m}^2\text{ s}$$

From Eq. (10.131*a*) and Table 10.1 for a flat plate and $T_s/T_e = 0.33$,

$$\begin{aligned} a_{m\text{He}} &= (1.13)(1.65)(M_{\text{air}}/M_{\text{He}})^{10/12} \\ &= (1.13)(1.65)(29/4)^{10/12} = 9.72 \end{aligned}$$

Substituting in Eq. (2) with $m_{\text{He},e} = 0$,

$$5.80 \times 10^{-3} = \left(\frac{2.45 \times 10^{-2}}{9.72}\right) \ln\left(1 + 9.72 \frac{0 - m_{\text{He},s}}{m_{\text{He},s} - 1}\right)$$

Solving, $m_{\text{He},s} = 0.480$.

Comments

1. The injection rate using the improved blowing factor is $(5.80 - 4.44)/4.44 = 31\%$ greater than given by the Couette-flow model blowing factor.
2. Use of the Couette-flow model blowing factor for mass transfer to calculate $m_{\text{He},s}$ will give a very poor result for helium injection (since it does not account for density variations).

10.5 LAMINAR BOUNDARY LAYERS

In Section 5.4 momentum and heat transfer were analyzed for the forced-flow laminar boundary layer on a flat plate. In Section 10.5.1 we extend that analysis to mass transfer. We first examine solutions valid in the limit of zero mass transfer rate, for which the analysis is identical to that of Section 5.4 for heat transfer and yields analogous results. Next we turn to exact self-similar solutions and extend the analysis of Section 5.4.4 to account for blowing or suction at the wall, which allows us to obtain mass transfer solutions valid for high mass transfer rates. In Section 10.5.2 we derive an appropriate form of the energy conservation equation for mixtures and, upon assuming constant properties, obtain two forms of the equation, one in terms of temperature and the other in terms of enthalpy. The advantages of the enthalpy form for simultaneous heat and mass transfer are discussed, and self-similar solutions are obtained for heat transfer that are exactly analogous to those obtained in Section 10.5.1 for mass transfer.

10.5.1 Mass Transfer in Forced Flow along a Flat Plate

Governing Equations

Figure 10.24 shows a schematic of the laminar boundary layer for forced flow along a flat plate, with a constant free-stream velocity u_e and constant mass fraction

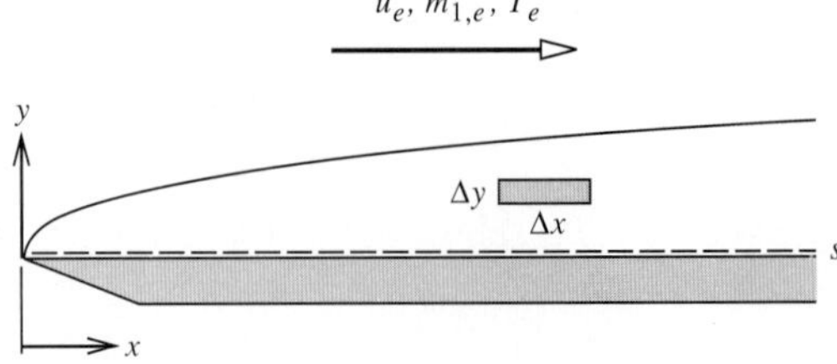

Figure 10.24 Schematic of a laminar boundary layer on a flat plate showing an elemental control volume $\Delta x \Delta y$.

of species 1, $m_{1,e}$. For simplicity we assume steady flow, a binary inert mixture, and constant fluid properties. Appropriate forms of the mass and momentum conservation equations were derived in Section 5.4.1 as

$$\frac{\partial u}{\partial x} + \frac{\partial v}{\partial y} = 0 \tag{5.37}$$

$$u\frac{\partial u}{\partial x} + v\frac{\partial u}{\partial y} = \nu\frac{\partial^2 u}{\partial y^2} \tag{5.38}$$

In using Eq. (5.38), we are ignoring second-order effects associated with the diffusion velocities. The species conservation equation for this flow can be obtained directly from the general species conservation equation derived in Section 10.2.3. However, we will parallel the analysis in Section 5.4.1, and derive the equation from first principles. Figure 10.25 shows an elemental control volume located in the boundary layer. Since there are no chemical reactions, the species conservation principle requires that the net outflow of species 1 by convection and diffusion equal zero:

$$\rho u m_1 \Delta y|_{x+\Delta x} - \rho u m_1 \Delta y|_x + \rho v m_1 \Delta x|_{y+\Delta y} - \rho v m_1 \Delta x|_y - \rho\mathscr{D}_{12}\frac{\partial m_1}{\partial y}\Delta x|_{y+\Delta y} + \rho\mathscr{D}_{12}\frac{\partial m_1}{\partial y}\Delta x|_y = 0$$

Dividing by $\rho\Delta x\Delta y$ and rearranging,

$$\frac{um_1|_{x+\Delta x} - um_1|_x}{\Delta x} + \frac{vm_1|_{y+\Delta y} - vm_1|_y}{\Delta y} = \mathscr{D}_{12}\left[\frac{(\partial m_1/\partial y)_{y+\Delta y} - (\partial m_1/\partial y)_y}{\Delta y}\right]$$

Letting Δx, $\Delta y \to 0$,

$$\frac{\partial}{\partial x}(um_1) + \frac{\partial}{\partial y}(vm_1) = \mathscr{D}_{12}\frac{\partial^2 m_1}{\partial y^2} \tag{10.133}$$

or

$$m_1\frac{\partial u}{\partial x} + u\frac{\partial m_1}{\partial x} + m_1\frac{\partial v}{\partial y} + v\frac{\partial m_1}{\partial y} = \mathscr{D}_{12}\frac{\partial^2 m_1}{\partial y^2} \tag{10.134}$$

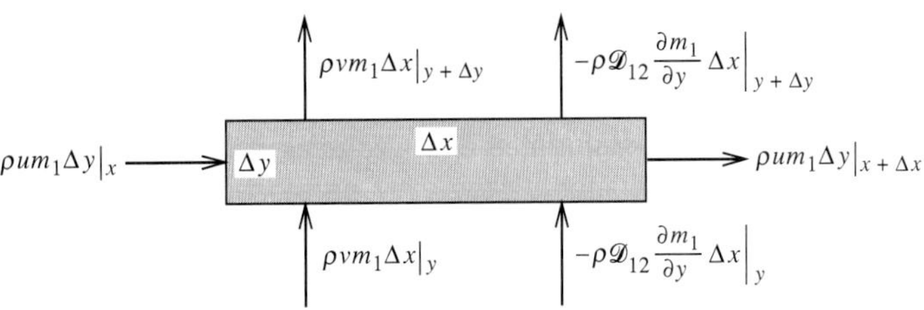

Figure 10.25 Application of the species conservation principle to an elemental control volume in a laminar boundary layer on a flat plate.

Multiplying the continuity equation, Eq. (5.37), by m_1 and subtracting from Eq. (10.134) gives the desired form of the species conservation equation:

$$u\frac{\partial m_1}{\partial x} + v\frac{\partial m_1}{\partial y} = \mathscr{D}_{12}\frac{\partial^2 m_1}{\partial y^2} \tag{10.135}$$

In deriving Eq. (10.135), diffusion in the x direction, $-\rho\mathscr{D}_{12}\partial m_1/\partial x$, has been ignored, since, like conduction $-k\partial T/\partial x$ and the shear stress $-\mu\partial u/\partial x$, it is negligible in a thin boundary layer. Scaling arguments to justify this assumption are given in Section 5.8. Notice that Eq. (10.135) is of analogous form to the energy equation for low-speed flow of a constant-property fluid, Eq. (5.42), with m_1 corresponding to T and $\mathscr{D}_{12}$ to α. Appropriate boundary conditions for Eq. (10.135) are

$$y = 0: \quad m_1 = m_{1,s} \tag{10.136a}$$

$$y \to \infty,\ x = 0: \quad m_1 = m_{1,e} \tag{10.136b}$$

The integral form of the species convection equation can be derived as was done in Section 5.4.3 for the energy equation. But with the analogy between heat and mass transfer clearly established for the differential form of the equation, we will simply write down the result. From Eqs. (5.61) through (5.63),

$$\frac{d\Delta_{m2}}{dx} = \mathrm{St}_{mx} \tag{10.137}$$

where Δ_{m2} is the convection thickness, defined as

$$\Delta_{m2} = \int_0^\infty \left(\frac{u}{u_e}\right)\left(\frac{m_1 - m_{1,e}}{m_{1,s} - m_{1,e}}\right) dy \tag{10.138}$$

and St_{mx} is the local mass transfer Stanton number,

$$\mathrm{St}_{mx} = \frac{g_{mx}}{\rho u_e} = \frac{-\rho\mathscr{D}_{12}(\partial m_1/\partial y)|_0}{\rho u_e(m_{1,s} - m_{1,e})} \tag{10.139}$$

Low Mass Transfer Rate Solutions

We can often obtain the solution of a mass transfer problem from the solution of an analogous heat transfer problem. In the limit of zero mass transfer rate, we have a simple form of the analogy. The heat transfer solutions obtained in Section 5.4 were for an impermeable wall, that is, $v_s = 0$. In general, the flux of species 1 across the s-surface has both convective and diffusive components,

$$n_{1,s} = m_{1,s}\rho v_s + j_{1,s} \tag{10.140}$$

If we set $v_s = 0$, we obtain $n_{1,s} = j_{1,s} = -\rho\mathscr{D}_{12}(\partial m_1/\partial y)|_0$, which is the basis of low mass transfer rate theory. Thus, to obtain low mass transfer rate solutions, we simply replace the Prandtl number by the Schmidt number and the Nusselt number by the Sherwood number in the corresponding heat transfer solution (as we did for correlations in Section 9.4). The plug flow model and unheated starting length solutions obtained in Section 5.4 then become, from Eq. (5.50),

$$\mathrm{Sh}_x = 0.564\,\mathrm{Re}_x^{1/2}\,\mathrm{Sc}^{1/2}(= 0.564\,\mathrm{Pe}_{mx}^{1/2}) \tag{10.141}$$

and from Eq. (5.68),

$$\mathrm{Sh}_x = \frac{0.331\,\mathrm{Re}_x^{1/2}\,\mathrm{Sc}^{1/3}}{[1 - (\xi/x)^{3/4}]^{1/3}} \tag{10.142}$$

The plug flow model is not of much practical use in mass transfer because it is accurate only for very small values of the Schmidt number, and values less than about 0.2 are seldom encountered (see Table 9.2). In the case of heat transfer, liquid metals are characterized by very small Prandtl numbers, and the plug flow model has proven useful. Equation (10.142) applies to discontinuous mass transfer, such as a plate that is bare for $0 < x < \xi$, and coated with naphthalene for $x > \xi$. The plate would also have to be isothermal to have $m_{1,s}$ constant for $x > \xi$. However, discontinuous mass transfer is more often encountered in applications such as transpiration cooling, for which mass transfer (injection) rates tend to be very high and low mass transfer rate theory is invalid.

Self-Similar Solutions

Self-similar solutions of the flat plate laminar boundary layer energy equation were obtained in Section 5.4.4. These solutions were obtained for an impermeable wall, that is, $v_s = 0$; thus, the analogous mass transfer solutions are valid for low mass transfer rates. We will now obtain self-similar solutions for $v_s \neq 0$ that will be appropriate for high mass transfer rates. The forms of the governing equations to be solved are the Blasius equation, Eq. (5.81), and the mass transfer analog to Eq. (5.91), namely,

$$f''' + ff'' = 0 \tag{5.81}$$

$$\phi'' + \mathrm{Sc} f\phi' = 0; \qquad \phi = \frac{m_1 - m_{1,e}}{m_{1,s} - m_{1,e}} \tag{10.143}$$

Primes denote differentiation with respect to the similarity variable $\eta = y(u_e/2\nu x)^{1/2}$, f is the dimensionless stream function, and $f' = u/u_e$ is the dimensionless streamwise velocity. Appropriate boundary conditions are

$$\eta = 0: \quad f' = 0, \phi = 1 \tag{10.144a}$$

$$\eta \to \infty: \quad f' = 1, \phi = 0 \tag{10.144b}$$

Notice that $m_{1,s}$ = constant is required to obtain Eq. (10.143), in the same way T_s = constant was required to obtain Eq. (5.91). If $m_{1,s}$ is not constant, self-similar solutions cannot be obtained.

One further boundary condition is required. Substituting $f'(0) = 0$ in Eq. (5.77) gives

$$v_s = -(u_e \nu/2x)^{1/2} f(0) \tag{10.145}$$

In Section 5.4.4 we set $v_s = 0$ for an impermeable wall, requiring $f(0) = 0$. For finite mass transfer rates we require solutions for $f(0) \neq 0$. The last boundary condition is then

$$\eta = 0: \quad f = \text{Constant} \tag{10.146}$$

where a constant value is required to give f as a function of η only, which is the condition for self-similar solutions. From Eq. (10.145), $v_s \propto x^{-1/2}$ is required for self-similar solutions. However, we shall see later that this condition is not as restrictive as it might first appear. Solution of the Blasius equation for $f(0) \neq 0$ is straightforward, and can be effected by the iteration procedure given in Section 5.4.4. For a given value of $f(0)$, the profile $f(\eta)$ is obtained and used in the integration of Eq. (10.143) to give the profile $\phi(\eta)$ and $\phi'(0)$ for a given value of Sc. The desired result is $\phi'(0)$ as a function of $f(0)$ with Sc as a parameter. The mass transfer Stanton number is obtained from

$$\rho u_e \mathrm{St}_{mx} = g_{mx} = \frac{-\rho \mathscr{D}_{12}(\partial m_1/\partial y)|_0}{m_{1,s} - m_{1,e}} = -\rho \mathscr{D}_{12}\phi'(0)(u_e/2\nu x)^{1/2}$$

Rearranging gives the local Stanton and Sherwood numbers:

$$\mathrm{St}_{mx} = \frac{-\phi'(0)}{2^{1/2}\,\mathrm{Re}_x^{1/2}\,\mathrm{Sc}} \tag{10.147a}$$

$$\mathrm{Sh}_x = \mathrm{Re}_x\,\mathrm{Sc}\,\mathrm{St}_{mx} = \frac{-\phi'(0)\,\mathrm{Re}_x^{1/2}}{2^{1/2}} \tag{10.147b}$$

Of course, Sh_x could have been obtained by analogy from Eq. (5.96).

Figure 10.26 shows dimensionless velocity profiles $f'(\eta)$ for various values of $f(0)$. Negative values of $f(0)$ correspond to blowing (positive mass transfer rates). At $-f(0) = 0.875$, $f''(0) = 0$, indicating a wall shear stress of zero: the boundary layer is then said to have been *blown off* the wall. The boundary layer equations become invalid at boundary layer blow-off. Positive values of $f(0)$ correspond to suction (negative mass transfer rates). As $f(0)$ increases, the boundary layer thins, and $f''(0)$ and the corresponding wall shear stress increase. The boundary layer equations are valid no matter how large $f(0)$ becomes. Figure 10.27 shows concentration profiles $\phi(\eta)$ for various values of $f(0)$, with Sc $= 0.6$ and Sc $= 10$. The effects of blowing and suction on those profiles are similar to those for the velocity profile. The effects of blowing to decrease $\phi'(0)$ and suction to increase $\phi'(0)$ are clearly seen. However, for Sc $= 10$ the effect of blowing in reducing $\phi'(0)$ is more pronounced than for Sc $= 0.6$.

As was done in the Couette-flow analysis of Section 10.4.1, it proves convenient to normalize the mass transfer conductance by its value in the limit of zero mass transfer in order to obtain the mass transfer from a relation of the form $\dot{m}'' = g_m^*(g_m/g_m^*)\mathscr{B}_m$ [cf. Eq. (10.99)].

$$\frac{g_{mx}}{g_{mx}^*} = \frac{\mathrm{St}_{mx}}{\mathrm{St}_{mx}^*} = \frac{\phi'(0)}{\phi'(0)|_{f(0)=0}} \tag{10.148}$$

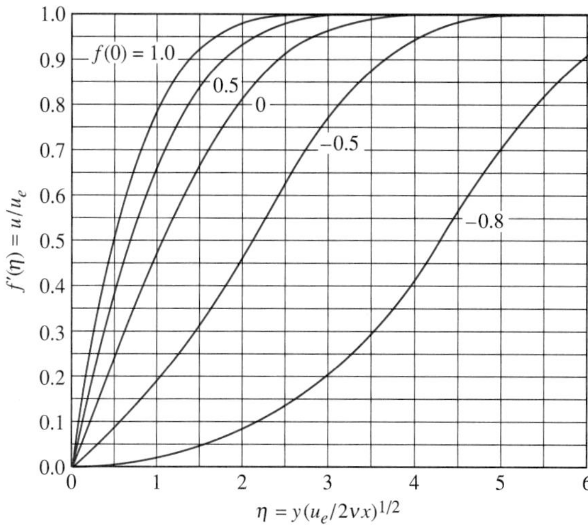

Figure 10.26 Effect of mass transfer on self-similar velocity profiles for a laminar boundary layer on a flat plate: $f(0) = -(\dot{m}''/\rho u_e)(2\,\mathrm{Re}_x)^{1/2}$.

where St^*_{mx} is obtained from the analog of Eq. (5.97) as

$$\mathrm{St}^*_{mx} = 0.332\,\mathrm{Re}_x^{-1/2}\,\mathrm{Sc}^{-2/3}; \qquad \mathrm{Sc} > 0.5 \tag{10.149}$$

The ratio g_{mx}/g^*_{mx} is then a function of Sc and $f(0)$, and from Eq. (10.145) with $\rho v_s = \dot{m}''$,

$$f(0) = -\frac{\dot{m}''}{\rho u_e} 2^{1/2}\,\mathrm{Re}_x^{1/2} \tag{10.150}$$

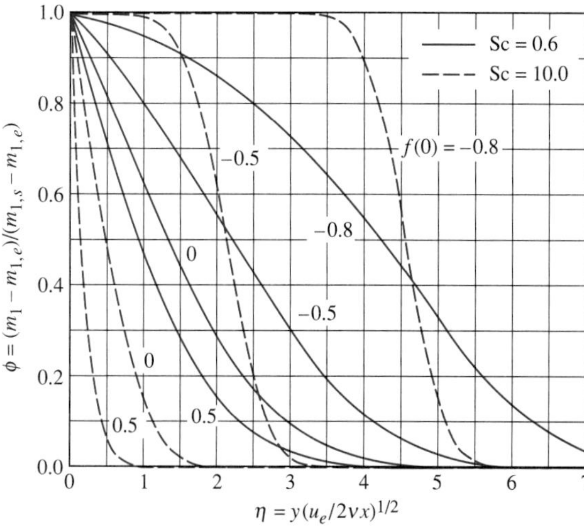

Figure 10.27 Effect of mass transfer on self-similar concentration profiles for a laminar boundary layer on a flat plate for Schmidt numbers equal to 0.6 and 10.0: $f(0) = -(\dot{m}''/\rho u_e)(2\,\mathrm{Re}_x)^{1/2}$.

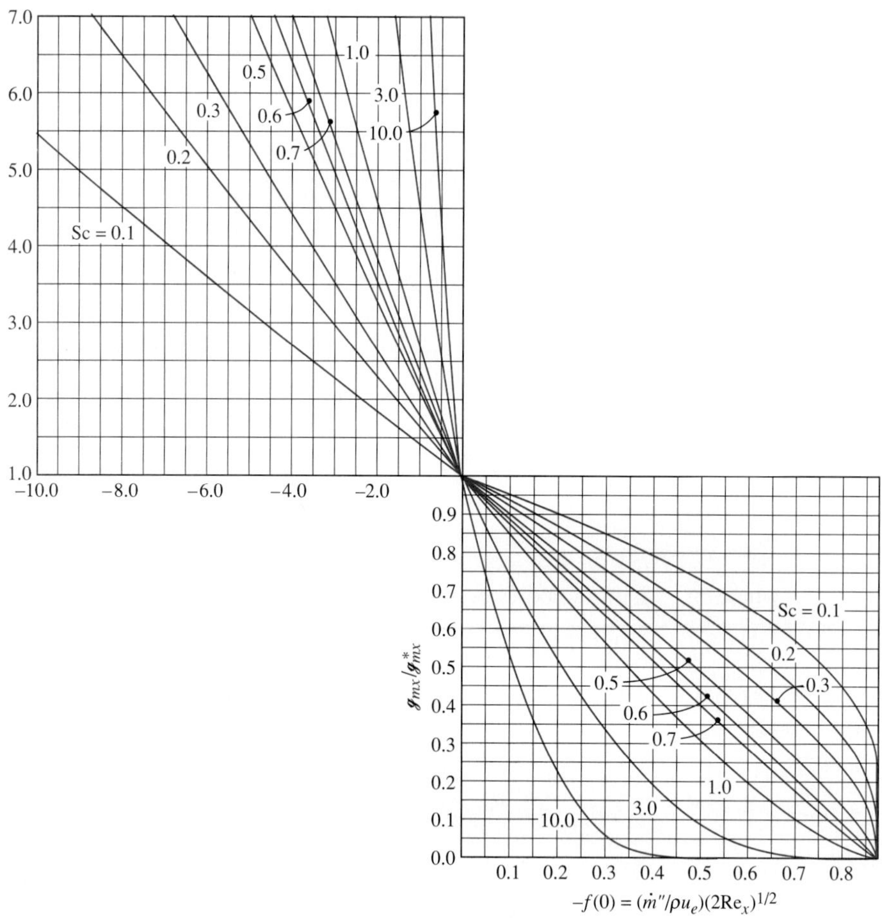

Figure 10.28 Effect of mass transfer on the mass transfer conductance for a laminar boundary layer on a flat plate: g_{mx}/g^*_{mx} versus dimensionless blowing rate $-f(0) = -(\dot{m}''/\rho u_e)(2\,\mathrm{Re}_x)^{1/2}$.

Figure 10.28 shows g_{mx}/g^*_{mx} plotted versus $f(0)$ with Sc as a parameter. The effect of Sc in this plot is large, and for blowing the curve shapes differ for $\mathrm{Sc} < 1$ versus $\mathrm{Sc} > 1$. A plot of the form of Fig. 10.28 is useful when the mass transfer rate is known at the relevant stage of a calculation, for example, in the case of transpiration cooling with a given injection rate. With $\dot{m}''$ known, $f(0)$ can be calculated from Eq. (10.150) and Fig. 10.28 used to obtain g_{mx}/g^*_{mx}. Plots similar to Fig. 10.28 are presented in the pioneering paper of Hartnett and Eckert [14]. An equivalent but different plot is shown in Fig. 10.29, where the abscissa is the blowing parameter $B_m = \dot{m}''/g^*_m$; substituting from Eqs. (10.147) into Eq. (10.150) and rearranging

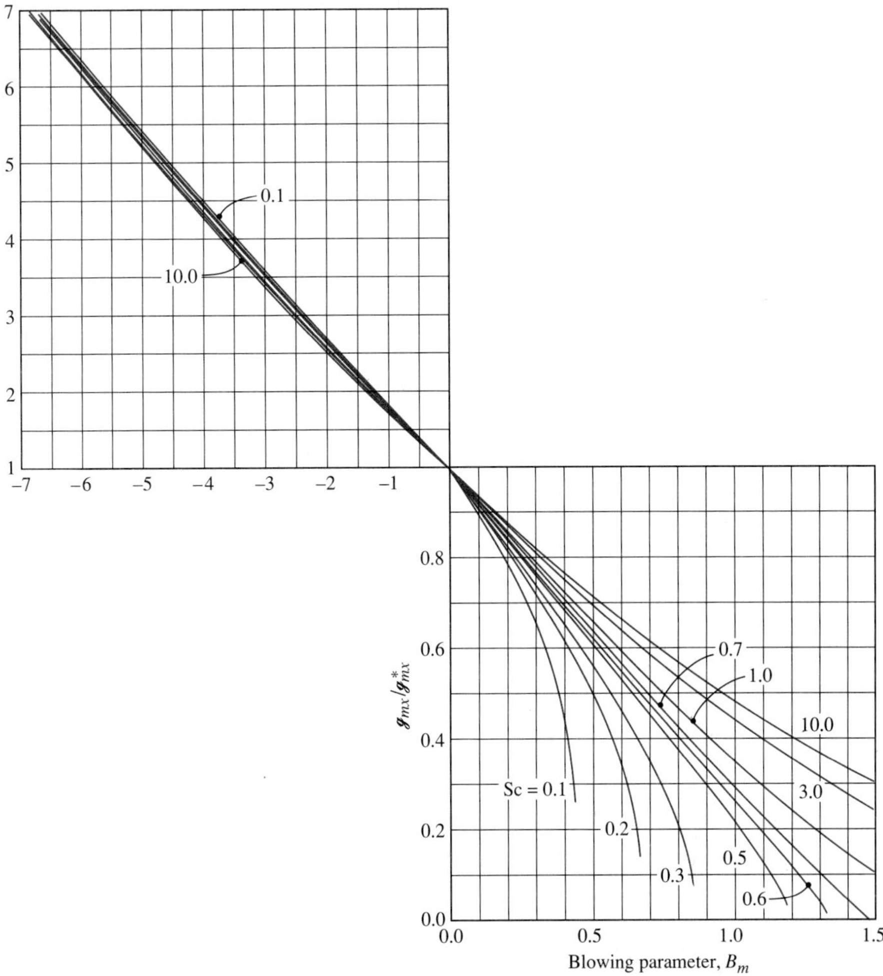

Figure 10.29 Effect of mass transfer on the mass transfer conductance for a laminar boundary layer on a flat plate: g_{mx}/g^*_{mx} versus blowing parameter $B_m = \dot{m}''/g^*_{mx}$.

gives

$$B_m = -\frac{\mathrm{Sc}\, f(0)}{\phi'(0)|_{f(0)=0}} \tag{10.151}$$

The abscissa in Fig. 10.29 is a function of Schmidt number, and the effect of Sc is much less in Fig. 10.29 than in Fig. 10.28, especially for suction.

For problems involving a liquid, species 1, evaporating into a gas, species 2—for example, sweat cooling—the surface temperature T_s might be specified; hence $m_{1,s} = m_{1,s}(T_s, P)$ would be known and the evaporation rate $\dot{m}''$ unknown. Hartnett and Eckert [14] suggest an iterative procedure for such problems making use of a plot like Fig. 10.28. However, such an iterative procedure can be performed once and for all and the results presented in a convenient form. We proceed by making the usual assumption that the gas is insoluble in the liquid; then

$$n_{2,s} = \rho v_s m_{2,s} - \rho \mathscr{D}_{12} \frac{\partial m_2}{\partial y}\bigg|_0 = 0$$

Substituting $m_2 = 1 - m_1$, $\partial m_2/\partial y = -\partial m_1/\partial y$ gives

$$v_s = \frac{-\mathscr{D}_{12}(\partial m_1/\partial y)|_0}{1 - m_{1,s}} \tag{10.152}$$

From Eq. (10.145), $v_s = (u_e \nu/2x)^{1/2} f(0)$; from Eq. (5.73*b*), $\partial/\partial y = (u_e/2\nu x)^{1/2} \partial/\partial\eta$, and from the definition of ϕ, $\partial m_1/\partial y = (m_{1,s} - m_{1,e})\,\partial\phi/\partial y$. Substituting in Eq. (10.152) and rearranging,

$$f(0) = \mathscr{B}_m \frac{\phi'(0)}{\text{Sc}}; \qquad \mathscr{B}_m = \frac{m_{1,e} - m_{1,s}}{m_{1,s} - 1} \tag{10.153}$$

By combining Eqs. (10.153), (10.150), and (10.147*a*), we recover $\dot{m}'' = g_m \mathscr{B}_m$, which was given as Eq. (10.93). The mass transfer driving force $\mathscr{B}_m$ arises naturally in formulating the relation that couples the two governing differential equations, Eqs. (5.81) and (10.143). To every solution $\phi'(0)$ for given $f(0)$ and Sc we can calculate a corresponding value of $\mathscr{B}_m$ from Eq. (10.153) and prepare a plot of g_{mx}/g^*_{mx} versus $\mathscr{B}_m$ with Sc as a parameter, as shown in Fig. 10.30. In problem solving with $m_{1,s}$, and hence $\mathscr{B}_m$, known, Fig. 10.30 gives g_{mx}/g^*_{mx} directly. Then the evaporation rate is obtained from $\dot{m}'' = g^*_{mx}(g_{mx}/g^*_{mx})\mathscr{B}_m$.

Notice that the effect of mass transfer on the wall shear stress can also be obtained from Fig. 10.28. For a Schmidt number of unity, the momentum and species equations, Eqs. (5.81) and (10.143), are of identical form with f' corresponding to ϕ. Thus, the curve for Sc = 1 in Fig. 10.28 also gives $f''(0)/f''(0)|_{f(0)=0} = \tau_s/\tau_s^*$.

Evaluation of Properties

Examples 10.11 and 10.12 illustrate the practical use of the two types of plot for g_{mx}/g^*_{mx}. However, the implications of the constant-property assumption requires careful consideration. Example 10.11 considers a situation where the properties are essentially constant, so that the solution procedure is exact. But such situations are the exception rather than the rule in engineering practice. Indeed, we recognized this fact when introducing improved blowing factors for use with the Couette-flow model in Section 10.4.2. If we wish to use constant-property solutions for engineering problem solving, properties must be evaluated at an appropriate reference state. Simple reference state schemes involve use of the mean film composition and temperature (a “1/2 rule”), or the 1/3 rule of Hubbard et al., given as Eqs. (10.65). Knuth [15,16] developed a reference state scheme based on a Couette-flow model

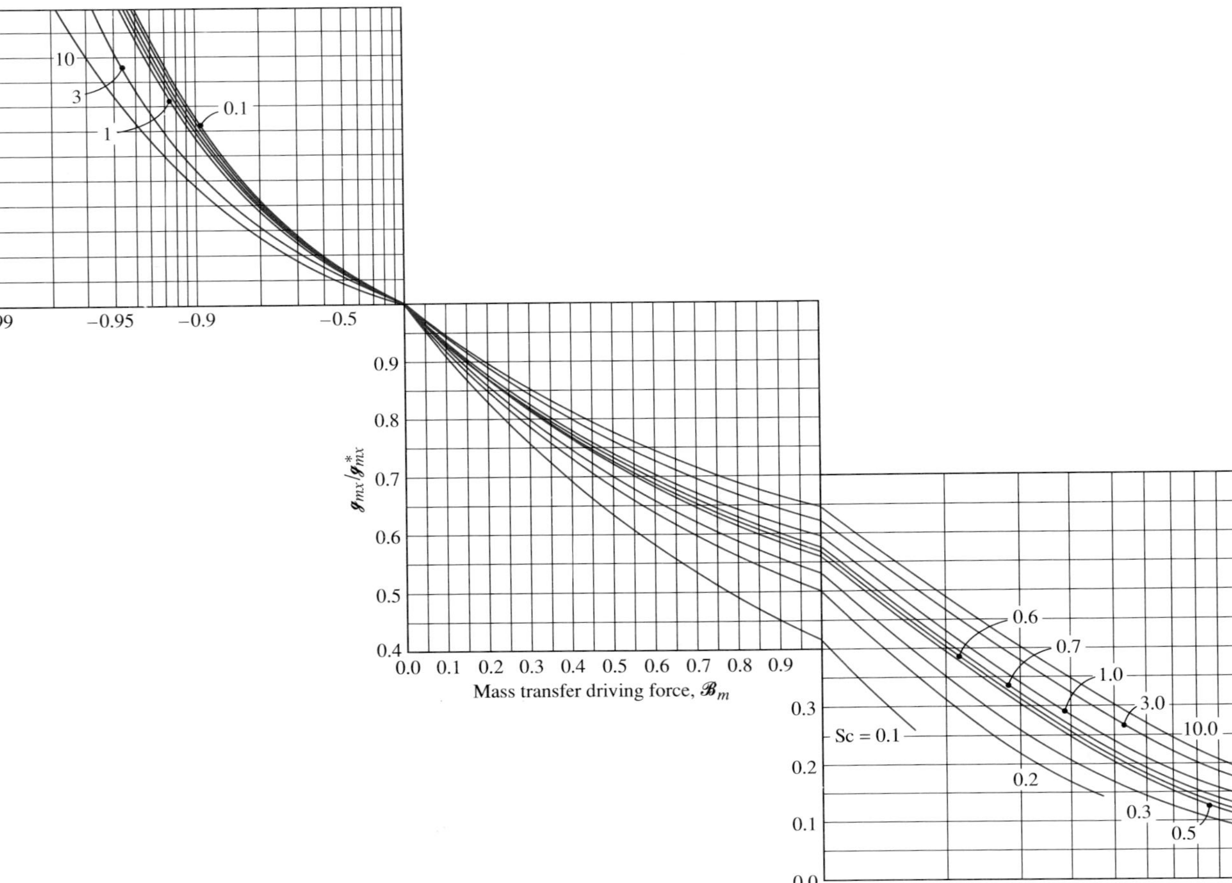

Figure 10.30 Effect of mass transfer on the mass transfer conductance for a laminar boundary layer on a flat plate: g_{mx}/g^*_{mx} versus mass transfer driving force $\mathscr{B}_m = (m_{1,e} - m_{1,s})/(m_{1,s} - 1) = \dot{m}''/g_{mx}$.

of laminar and turbulent flows, in which thermodynamic properties were allowed to vary with composition but transport properties were held constant. For species 1 injected into species 2, the reference mass fraction and temperature are

$$m_{1,r} = 1 - \frac{M_2}{M_2 - M_1} \frac{\ln(M_e/M_s)}{\ln(m_{2,e}M_e/m_{2,s}M_s)} \tag{10.154}$$

$$T_r = 0.5(T_e + T_s) + 0.2r^*(u_e^2/2c_{pr}) + 0.1\left[B_{hr} + (B_{hr} + B_{mr})\frac{c_{p1} - c_{pr}}{c_{pr}}\right](T_s - T_e) \tag{10.155}$$

where r^* is the recovery factor for an impermeable wall [see Section 5.2.2 and cf. Eq. (5.13)]. Use of this scheme requires the calculation of transport properties of gas mixtures. A method based on the kinetic theory of gases is given in Section 9.7. Based on the limited numerical data available at that time, this reference state was shown by Knuth to work well for the laminar boundary layer on a flat plate. Unfortunately, Knuth's scheme is difficult to implement in a hand calculation. Knuth's scheme usually gives values that are between those given by the 1/2 and 1/3 rules. Example 10.12 illustrates use of the 1/2 rule for a situation where property variations are large; Exercises 10–47 and 10–48 require Knuth's scheme to be used. It is also worthwhile to note that some workers have combined the reference state approach for like gas injection, with molecular-weight-dependent empirical factors to account for foreign gas injection. Reasonably good correlations of numerical data for laminar boundary layers have been obtained in this manner [17,18,19].

An issue related to variable-property effects concerns the Schmidt number effect seen in Figs. 10.28 through 10.30. This effect does not have much physical significance. Both Schmidt number and variable-property effects depend on the combination of species under consideration. For gases, variable properties generally have a larger influence on $\mathcal{g}_{mx}/\mathcal{g}_{mx}^*$ than does the Schmidt number, as can be seen by comparing Figs. 10.22*a* and 10.29.

EXAMPLE 10.11 Diffusion of Nitric Oxide in an Air Boundary Layer

Air at 300 K and 1 atm pressure flows at 10 m/s along a porous flat plate, through which air containing 1% NO by mass is injected at a velocity of $3.53 \times 10^{-3}(x)^{-1/2}$ m/s. At 0.2 m from the leading edge, calculate the mass transfer conductance for NO, and the NO concentration at the s-surface. Take $\text{Re}_{tr} = 200{,}000$.

Solution

Given: Air containing a small amount of NO injected into an air boundary layer on a flat plate.

Required: At $x = 0.2$ m, $\mathcal{g}_{mx}$ for NO, and $m_{\text{NO},s}$.

Assumptions: 1. Constant fluid properties.
2. $\text{Re}_{\text{tr}} = 200{,}000$

Since NO is in small concentration and its molecular weight (30) is close to that of air (29), the assumption of constant properties is valid. Air properties at 300 K and 1 atm from Table A.7 are $\rho = 1.177$ kg/m^3 and $\nu = 15.66 \times 10^{-6}$ m^2/s; from Table A.17*a*, $\mathcal{D}_{\text{NO air}} = 0.180 \times 10^{-4}$ m^2/s. The Reynolds and Schmidt numbers are

$$\text{Re}_x = u_e x/\nu = (10)(0.2)/(15.66 \times 10^{-6}) = 127{,}700 < 200{,}000, \text{ laminar}$$

$$\text{Sc} = \nu/\mathcal{D}_{12} = (15.66 \times 10^{-6})/(0.180 \times 10^{-4}) = 0.870$$

The zero-mass-transfer-limit Stanton number is obtained from Eq. (10.149) as

$$\text{St}_{mx}^* = 0.332\,\text{Re}_x^{-1/2}\,\text{Sc}^{-2/3} = (0.332)(127{,}700)^{-1/2}(0.870)^{-2/3} = 1.02 \times 10^{-3}$$

and the zero-mass-transfer-limit conductance is

$$\mathcal{g}_{mx}^* = \rho u_e\,\text{St}_{mx}^* = (1.177)(10)(1.02 \times 10^{-3}) = 1.20 \times 10^{-2} \text{ kg/m}^2\text{ s}$$

In order to obtain the actual Stanton number from Fig. 10.28, we require $f(0)$; from Eq. (10.150),

$$f(0) = -\frac{\dot{m}''}{\rho u_e} 2^{1/2}\,\text{Re}_x^{1/2} = -\left(\frac{v_s}{u_e}\right)(2\,\text{Re}_x)^{1/2}$$

$$v_s = 3.53 \times 10^{-3}(x)^{-1/2} = (3.53 \times 10^{-3})(0.2)^{-1/2} = 7.89 \times 10^{-3} \text{ m/s}$$

$$f(0) = -(7.89 \times 10^{-3}/10)(2 \times 127{,}700)^{1/2} = -0.399$$

From Fig. 10.28, for $\text{Sc} = 0.87$, $f(0) = -0.40$, we obtain $\mathcal{g}_{mx}/\mathcal{g}_{mx}^* = 0.47$. Thus,

$$\mathcal{g}_{mx} = (\mathcal{g}_{mx}/\mathcal{g}_{mx}^*)\mathcal{g}_{mx}^* = (0.47)(1.20 \times 10^{-2}) = 5.64 \times 10^{-3} \text{ kg/m}^2\text{ s}$$

To obtain the mass fraction of NO at the s-surface, we proceed as follows. Denoting NO as species 1, and using Eq. (10.140),

$$n_{1,s} = m_{1,s}\rho v_s + j_{1,s} = m_{1,s}\dot{m}'' + \mathcal{g}_{m1}(m_{1,s} - m_{1,e})$$

$$\dot{m}'' = \rho v_s = (1.177)(7.89 \times 10^{-3}) = 9.29 \times 10^{-3} \text{ kg/m}^2\text{ s}$$

Since the injected air contains 1% NO by mass, the absolute flux of NO across the s-surface is

$$n_{1,s} = n_{1,u} = (0.01)(\dot{m}'') = (0.01)(9.29 \times 10^{-3}) = 9.29 \times 10^{-5} \text{ kg/m}^2\text{ s}$$

Substituting above,

$$9.29 \times 10^{-5} = 9.29 \times 10^{-3} m_{1,s} + (5.64 \times 10^{-3})(m_{1,s} - 0)$$

Solving, $m_{1,s} = 6.22 \times 10^{-3}$ (0.622%).

Comments

1. This example is not an engineering problem: it was constructed to illustrate some important aspects of mass transfer in a laminar boundary layer.
2. Notice that v_s proportional to $x^{-1/2}$ gives a constant value of $f(0)$, as required for a self-similar solution. Also, $m_{1,s}$ is constant along the plate as a result.
3. This is a problem in which *two* species are transferred; nearly all the problems we have considered up to now involved transfer of one species only. We see that, through careful use of Eq. (10.140), handling of more than one transferred species is straightforward. Section 10.6 shows how the mass transfer driving force $\mathcal{B}_m$ defined by Eq. (10.93) can be generalized to situations where more than one species is transferred (see also Exercise 10–28).

EXAMPLE 10.12 Sweat Cooling of a Flat Plate: Water Supply Rate

A flat plate exposed to intense radiative and convective heating is to be sweat-cooled with water. Dry air at 840 K and 1 atm flows at 10 m/s along the plate as a laminar boundary layer. If the plate surface is to be maintained at 360 K, calculate the water supply rate 0.2 m from the leading edge (i) using the improved blowing factors of Section 10.4.2, and (ii) using the constant-property self-similar solution and the 1/2 rule for evaluating properties.

Solution

Given: Air flowing along a radiatively heated, sweat-cooled flat plate.

Required: The water evaporation rate 0.2 m from the leading edge by two methods.

Assumptions: The thermal resistance of the thin film of water on the plate is negligible.

(i) *Using the improved blowing factors*. From Eqs. (10.93) and (10.130*a*),

$$\dot{m}'' = g_{mx}\mathcal{B}_m; \quad \frac{g_m}{g_m^*} = \frac{\ln(1 + a_{m1}\mathcal{B}_m)}{a_{m1}\mathcal{B}_m}$$

We first evaluate $\mathcal{B}_m$. From PSYCHRO, at $T = 360$ K, $P = 1.013\times 10^5$ Pa, $m_{1,s} = m_{1,\text{sat}} = 0.496$. Thus,

$$\mathcal{B}_m = \frac{m_{1,e} - m_{1,s}}{m_{1,s} - 1} = \frac{0 - 0.496}{0.496 - 1} = 0.984$$

The reference temperature for evaluating g_{mx}^* is the mean film temperature, $T_r = 360 + (1/2)(840 - 360) = 600$ K. From Table A.7 for dry air, $\rho = 0.589$ kg/m^3, $\mu = 29.74 \times 10^{-6}$ kg/m s. Also, from Table A.17*a*, $\mathcal{D}_{12} = 92.8 \times 10^{-6}$ m^2/s, and then Sc = 0.54. The Reynolds number is

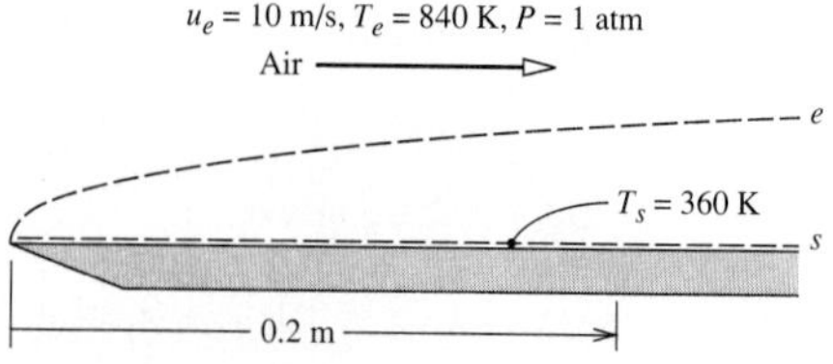

$$\text{Re}_x = u_e\rho x/\mu = (10)(0.589)(0.2)/29.74 \times 10^{-6} = 3.961 \times 10^4$$

From Eq. (10.149), the local zero-mass-transfer-limit Stanton number and conductance are

$$\mathrm{St}_{mx}^{*} = 0.332\,\mathrm{Re}_x^{-1/2}\,\mathrm{Sc}^{-2/3} = (0.332)(3.961 \times 10^4)^{-1/2}(0.54)^{-2/3} = 2.52 \times 10^{-3}$$

$$g_{mx}^{*} = \rho u_e\,\mathrm{St}_{mx}^{*} = (0.589)(10)(2.52 \times 10^{-3}) = 1.48 \times 10^{-2} \text{ kg/m}^2\,\text{s}$$

Using Eq. (10.131*a*),

$$a_{m1} = 1.65(M_{\text{air}}/M_1)^{10/12}G_{m1}$$

Table 10.1 does not contain G factors for H_2O injection on a flat plate. However, comparing the factors for a flat plate and axisymmetric stagnation point, and interpolating between $T_s/T_e = 0.1$ and 0.5, a rough estimate of the G factor is

$$G_{m1} \simeq (1.00)(1.1) = 1.1$$

$$a_{m1} \simeq (1.65)(29/18)^{10/12}(1.1) = 2.7$$

$$\frac{g_{mx}}{g_{mx}^{*}} = \frac{\ln(1 + 2.7 \times 0.984)}{2.7 \times 0.984} = 0.488$$

$$\dot{m}'' = g_{mx}^{*}(g_{mx}/g_{mx}^{*})\mathscr{B}_m = (1.48 \times 10^{-2})(0.488)(0.984) = 7.11 \times 10^{-3} \text{ kg/m}^2\,\text{s}$$

(ii) *Using the self-similar solution and the 1/2 rule*. The reference state for calculating g_{mx}^{*} is $m_{1,r} = (1/2)(0.496 - 0) = 0.248$, $T_r = 360 + (1/2)(840 - 360) = 600$ K. Using AIRSTE, $\rho = 0.512$ kg/m^3, $\mu = 2.71 \times 10^{-5}$ kg/m s, Sc $= 0.564$. The Reynolds number, local Stanton number, and conductance are

$$\mathrm{Re}_x = u_e\rho x/\mu = (10)(0.512)(0.2)/2.71 \times 10^{-5} = 3.78 \times 10^4$$

$$\mathrm{St}_{mx}^{*} = 0.332\,\mathrm{Re}_x^{-1/2}\,\mathrm{Sc}^{-2/3} = (0.332)(3.78 \times 10^4)^{-1/2}(0.564)^{-2/3} = 2.50 \times 10^{-3}$$

$$g_{mx}^{*} = \rho u_e\,\mathrm{St}_{m}^{*} = (0.512)(10)(2.50 \times 10^{-3}) = 1.28 \times 10^{-2} \text{ kg/m}^2\,\text{s}$$

From Fig. 10.30 for Sc $= 0.564$ and $\mathscr{B}_m = 0.984$, $g_{mx}/g_{mx}^{*} = 0.57$. Thus,

$$\dot{m}'' = g_{mx}^{*}(g_{mx}/g_{mx}^{*})\mathscr{B}_m = (1.28 \times 10^{-2})(0.57)(0.984) = 7.18 \times 10^{-3} \text{ kg/m}^2\,\text{s}$$

Comments

1. The difference between the two results for the evaporation rate is 1.0%, which is remarkably small considering we made approximations in each case (G_m was an estimate, and the 1/2 rule is not exact).

2. For a given water reservoir temperature T_o, there is a unique radiative heat flux q_{rad} consistent with a surface temperature of 360 K (see Example 10.13).

10.5.2 Simultaneous Heat and Mass Transfer in Forced Flow along a Flat Plate

We commence by deriving the boundary layer form of the energy conservation equation for a mixture. We assume steady, low-speed, laminar flow and follow the approach used in Section 10.3.5. Referring to Fig. 10.31, the steady-flow energy

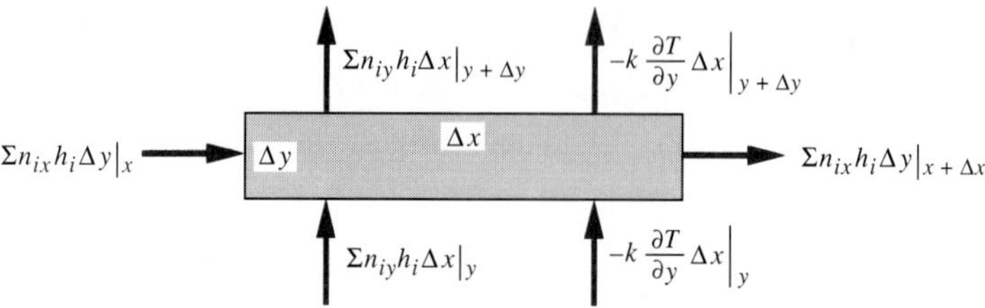

Figure 10.31 Application of the energy conservation principle to an elemental control volume in a flat-plate laminar boundary layer flow of a mixture.

equation, Eq. (1.4), requires that the net outflow of enthalpy equal the heat conducted into the volume:

$$\sum n_{ix}h_i\Delta y|_{x+\Delta x} - \sum n_{ix}h_i\Delta y|_x + \sum n_{iy}h_i\Delta x|_{y+\Delta y} - \sum n_{iy}h_i\Delta x|_y$$
$$= -k\frac{\partial T}{\partial y}\Delta x|_y + k\frac{\partial T}{\partial y}\Delta x|_{y+\Delta y}$$

Dividing by $\Delta x \Delta y$ and rearranging,

$$\frac{\sum n_{ix}h_i|_{x+\Delta x} - \sum n_{ix}h_i|_x}{\Delta x} + \frac{\sum n_{iy}h_i|_{y+\Delta y} - \sum n_{iy}h_i|_y}{\Delta y}$$
$$= \frac{(k\,\partial T/\partial y)_{y+\Delta y} - (k\,\partial T/\partial y)_y}{\Delta y}$$

Letting $\Delta x,\ \Delta y \to 0$,

$$\frac{\partial}{\partial x}\left(\sum n_{ix}h_i\right) + \frac{\partial}{\partial y}\left(\sum n_{iy}h_i\right) = \frac{\partial}{\partial y}\left(k\frac{\partial T}{\partial y}\right) \tag{10.156}$$

In Eq. (10.156), k is the *mixture* conductivity. As in the derivation of Eq. (5.42), conduction in the x direction has been ignored because it is negligible for a thin boundary layer; also, for this low-speed flow along a flat plate, there are no terms that account for viscous dissipation and compression heating.

Equation (10.156) is further rearranged as follows. By definition, $n_{ix} = m_i\rho u + j_{ix}$; hence,

$$\frac{\partial}{\partial x}\left(\sum n_{ix}h_i\right) = \frac{\partial}{\partial x}\sum(m_i\rho u + j_{ix})h_i$$
$$= \frac{\partial}{\partial x}\left(\rho u\sum m_ih_i + \sum j_{ix}h_i\right)$$

$$\frac{\partial}{\partial x}\left(\sum n_{ix}h_i\right) \simeq \frac{\partial}{\partial x}(\rho u h) \tag{10.157a}$$

since $\sum m_i h_i = h$, the mixture enthalpy, and the streamwise diffusion fluxes j_{ix} are negligible in a thin boundary layer. Similarly,

$$\frac{\partial}{\partial y}\left(\sum n_{iy} h_i\right) = \frac{\partial}{\partial y}\left(\rho v h + \sum j_{iy} h_i\right) \tag{10.157b}$$

The second term on the right-hand side of Eq. (10.157b) is not negligible, since it involves diffusion fluxes across the boundary layer. Substituting back in Eq. (10.156) and rearranging,

$$\frac{\partial}{\partial x}(\rho u h) + \frac{\partial}{\partial y}(\rho v h) = \frac{\partial}{\partial y}\left[k\frac{\partial T}{\partial y} - \sum j_{iy} h_i\right] \tag{10.158}$$

Expanding the left-hand side of Eq. (10.158) and using the continuity equation in the usual manner gives

$$\rho u\frac{\partial h}{\partial x} + \rho v\frac{\partial h}{\partial y} = \frac{\partial}{\partial y}\left[k\frac{\partial T}{\partial y} - \sum j_{iy} h_i\right] \tag{10.159}$$

The second term on the right-hand side of Eq. (10.159) is called the *interdiffusion* contribution to energy transport in a mixture, because it accounts for enthalpy transport due to interdiffusion of chemical species in the mixture.[5] For a binary mixture this term becomes

$$-\sum j_{iy} h_i = \rho\mathscr{D}_{12}\frac{\partial m_1}{\partial y}h_1 + \rho\mathscr{D}_{12}\frac{\partial m_2}{\partial y}h_2 = \rho\mathscr{D}_{12}(h_1 - h_2)\frac{\partial m_1}{\partial y} \tag{10.160}$$

since $\partial m_2/\partial y = -\partial m_1/\partial y$. If $h_1 = h_2$, the interdiffusion term is seen to be zero.

Within the limitation of the constant properties assumption, there are two possible approaches to a heat transfer analysis.

Approach 1: Temperature formulation. We assume an inert mixture with all species specific heats equal, $c_{p1} = c_{p2} = \cdots = c_p$. Then, using the same enthalpy datum state for all species gives $\sum j_{iy} h_i = h_i \sum j_{iy} = 0$, $\partial h = c_p \partial T$, and Eq. (10.159) becomes

$$\rho u c_p\frac{\partial T}{\partial x} + \rho v c_p\frac{\partial T}{\partial y} = \frac{\partial}{\partial y}\left(k\frac{\partial T}{\partial y}\right)$$

Finally, the mixture conductivity k is assumed constant to recover Eq. (5.42),

$$u\frac{\partial T}{\partial x} + v\frac{\partial T}{\partial y} = \alpha\frac{\partial^2 T}{\partial y^2} \tag{5.42}$$

[5] Energy transport due to interdiffusion must not be confused with *diffusional conduction*, which is an energy transport similar to heat conduction, but due to concentration gradients rather than a temperature gradient. Diffusional conduction is also called the *diffusion thermo-effect* or *Dufour effect*, and has been ignored in our analysis since it is usually small compared with ordinary conduction.

which is the energy equation for a constant-property pure fluid. The interdiffusion term in the energy equation for a mixture has been eliminated by assuming an inert mixture with equal species specific heats. Boundary conditions for Eq. (5.42) are

$$y = 0: \quad T = T_s \tag{5.88a}$$

$$y \to \infty, x = 0: \quad T = T_e \tag{5.88b}$$

and T_s, T_e are constants for a self-similar solution. As noted before, Eqs. (5.42) and (10.135) are of the same form; that is, the heat and mass transfer problems are analogous. In fact, if $\alpha = \mathscr{D}_{12}$, the solutions are identical, and the normalized temperature profile $(T - T_e)/(T_s - T_e)$ is identical to the normalized mass fraction profile $(m_1 - m_{1,e})/(m_{1,s} - m_{1,e})$. The ratio $\mathscr{D}_{12}/\alpha = \rho c_p \mathscr{D}_{12}/k$ is the Lewis number, which was introduced in Section 9.5.2. When the normalized profiles are identical, it follows from Eqs. (5.96) and (10.147*b*) that the Nusselt number equals the Sherwood number (or, equivalently, the heat transfer and mass transfer Stanton numbers are equal) when the Lewis number is unity. In most gas mixtures the Lewis number is indeed close to unity, but for liquid solutions it can be quite different from unity.

Approach 2: Enthalpy formulation. We assume effective binary diffusion and that the Lewis numbers of all species present are equal to unity, $Le_i = \mathscr{D}_{im}/\alpha = 1$. Notice that in the definition of Lewis number, $\alpha = k/\rho c_p$ is a *mixture property*, and $\mathscr{D}_{im}$ is the species-dependent effective binary diffusion coefficient for species i in the mixture. Implicit in the requirement that $\text{Le}_i = 1$ is that all the $\mathscr{D}_{im}$ are equal. Then, proceeding as we did in Section 10.3.5, we obtain

$$\frac{k}{c_p}\frac{\partial h}{\partial y} = k\frac{\partial T}{\partial y} - \sum j_{iy}h_i, \quad \text{for Le}_i = 1 \tag{10.161}$$

Substituting in Eq. (10.159),

$$\rho u \frac{\partial h}{\partial x} + \rho v \frac{\partial h}{\partial y} = \frac{\partial}{\partial y}\left(\frac{k}{c_p}\frac{\partial h}{\partial y}\right) \tag{10.162}$$

and if we now assume that k/c_p is constant,

$$u\frac{\partial h}{\partial x} + v\frac{\partial h}{\partial y} = \alpha \frac{\partial^2 h}{\partial y^2} \tag{10.163}$$

Equation (10.163) is of the same mathematical form as Eq. (5.42); however, it is of more general validity provided the Lewis number is close to unity, as it is for most gas mixtures. Chemical reactions are allowed, and the interdiffusion term has been retained (albeit approximately). Equation (10.163) is particularly useful for combustion of gaseous or volatile fuels. Thus, we will develop the remainder of our heat transfer analysis using the enthalpy formulation. Appropriate boundary conditions for Eq. (10.163) are

$$y = 0: \quad h = h_s \tag{10.164a}$$

$$y \to \infty, x = 0: \quad h = h_e \tag{10.164b}$$

A relation coupling the momentum and energy equations can be derived that is similar to the relation coupling the momentum and species equations, Eq. (10.153).

We introduce a new concept for this purpose, the *transferred state enthalpy*, h_t, defined by the relation

$$\dot{m}''h_t = \sum n_i h_i|_s - k\left.\frac{\partial T}{\partial y}\right|_s \qquad \textbf{(10.165)}$$

The right-hand side of Eq. (10.165) is the energy flux across the s-surface expressed as a sum of absolute convection and conduction components. The concept of a transferred state enthalpy was introduced by Spalding [20,21] and allows us to conveniently display the analogy between heat and mass transfer. In general, the transferred state is a fictitious state that simply satisfies Eq. (10.165); for the special cases of transpiration or sweat cooling, h_t proves to be equal to the enthalpy in the reservoir. [However, see Eq. (10.194) if there is thermal radiation.] Introducing $n_{i,s} = m_{i,s}\dot{m}'' + j_{i,s}$ into Eq. (10.165) gives

$$\begin{aligned}\dot{m}''h_t &= \dot{m}''h_s + \sum j_i\, h_i|_s - k\left.\frac{\partial T}{\partial y}\right|_s \\ &= \dot{m}''h_s - \frac{k}{c_p}\left.\frac{\partial h}{\partial y}\right|_s, \quad \text{for}\,\mathrm{Le}_i = 1\end{aligned}$$

Introducing a dimensionless enthalpy $\Phi = (h - h_e)/(h_s - h_e)$ and solving for $\dot{m}''$ gives

$$\dot{m}'' = -\frac{h_e - h_s}{h_s - h_t}\frac{k}{c_p}\left.\frac{\partial \Phi}{\partial y}\right|_s \qquad \textbf{(10.166)}$$

or, in terms of self-similar variables using Eqs. (5.77) and (5.73*b*),

$$f(0) = \mathscr{B}_h\frac{\Phi'(0)}{\mathrm{Pr}}; \qquad \mathscr{B}_h = \frac{h_e - h_s}{h_s - h_t} \qquad \textbf{(10.167)}$$

where $\mathscr{B}_h$ is the heat transfer driving force and is analogous to the mass transfer driving force $\mathscr{B}_m$.[6] Equation (10.167) is of the same form as Eq. (10.153).

In order to obtain a complete analogy between heat and mass transfer, the Stanton number must be defined in terms of the sum of the conduction and interdiffusion energy fluxes across the s-surface, and the enthalpy difference across the boundary layer. Using the *heat transfer conductance* g_h introduced in Section 9.4.4,

$$\begin{aligned}\rho u_e \mathrm{St}_x = g_{hx} &= \frac{-k(\partial T/\partial y)|_s + \sum_i j_i h_i|_s}{h_s - h_e} \qquad \textbf{(10.168)} \\ &= \frac{-(k/c_p)(\partial h/\partial y)|_s}{h_s - h_e}, \quad \text{for } \mathrm{Le}_i = 1 \\ &= -\frac{k}{c_p}\Phi'(0)(u_e/2\nu x)^{1/2}\end{aligned}$$

[6] Actually $\mathscr{B}_m$ contains 1 in the denominator, rather than a quantity corresponding directly to h_t. Section 10.6 will define $\mathscr{B}_m$ in terms of a *transferred state mass fraction* $m_{i,t} = n_{i,s}/\dot{m}''$. In the special case of transfer of single species, $m_{i,t} = 1$, giving the result used here.

Recall that assuming $\text{Le}_i = 1$ does not eliminate the interdiffusion term. Rearranging gives the local Stanton number as

$$\text{St}_x = \frac{-\Phi'(0)}{2^{1/2}\,\text{Re}_x^{1/2}\,\text{Pr}} \tag{10.169}$$

which is analogous to Eq. (10.147*a*). Also,

$$\frac{g_{hx}}{g_{hx}^*} = \frac{\text{St}_x}{\text{St}_x^*} = \frac{\Phi'(0)}{\Phi'(0)|_{f(0)=0}} \tag{10.170}$$

and from Eq. (5.97),

$$\text{St}_x^* = 0.332\,\text{Re}_x^{-1/2}\,\text{Pr}^{-2/3}, \qquad \text{Pr} > 0.5 \tag{10.171}$$

As was the case for mass transfer, g_{hx}/g_{hx}^* can be given as a function of $f(0)$ (or, equivalently, $B_h = \dot{m}''/g_{hx}^*$) or of $\mathscr{B}_h$, with Pr as a parameter. The plots are then identical to Figs. (10.28) through (10.30). Notice that Eqs. (10.145), (10.167), and (10.169) combine to give

$$\dot{m}'' = g_{hx}\mathscr{B}_h; \qquad \mathscr{B}_h = \frac{h_e - h_s}{h_s - h_t}; \qquad g_{hx} = \rho u_e\,\text{St}_x \tag{10.172}$$

A troublesome issue concerning this analysis is the use of Prandtl number in Eqs. (10.169) through (10.171). Since we assumed unity Lewis numbers ($\text{Pr} = \text{Sc}_i$) to obtain Eq. (10.169), why not use a Schmidt number in these equations? If all the Le_i were exactly equal to unity, there would be no problem; but in reality $\text{Le}_i \neq 1$, and thus a choice must be made. We choose to use Pr because Eq. (10.169) is the solution of the *energy* equation, which is used to determine *heat transfer*. In particular, we would like our solution to be exact in the limit $\dot{m}'' \to 0$, for which we have a heat transfer problem only: then Eq. (10.171) gives the heat transfer Stanton number exactly.

When using the enthalpy formulation, we see that the analogy between mass and heat transfer results in an analogy between the diffusion flux across the s-surface for mass transfer and the sum of the conduction and interdiffusion fluxes across the s-surface for heat transfer. The driving potential for this sum of conduction and interdiffusion fluxes is the enthalpy difference across the boundary layer $(h_e - h_s)$. For *inert* systems we can derive another form of the analogy between mass and heat transfer, as follows. For an inert mixture, we are free to choose the enthalpy datum state for each species such that the enthalpy of each is zero at the surface temperature T_s. Then Eq. (10.168) becomes

$$\rho u_e\,\text{St}_x = g_{hx} = \frac{-k(\partial T/\partial y)|_s}{-h_e^{T_s}} \tag{10.173}$$

and

$$
\begin{aligned}
h_e^{T_s} &= \sum m_{i,e} h_{i,e}^{T_s} \\
&= \sum m_{i,e}(h_{i,e} - h_{i,s}) \quad \text{independent of datum state} \\
&= \sum m_{i,e} h_{i,e} - \sum m_{i,e} h_{i,s} \\
h_e^{T_s} &= h_e - h_{es}
\end{aligned}
\tag{10.174}
$$

where $h_{es} = \sum m_{i,e} h_{i,s}$ is the enthalpy of a mixture of e-state composition at the s-state temperature. Notice that the right-hand side of Eq. (10.174) is independent of datum states since it involves only enthalpy differences of pure components. Substituting in Eq. (10.173),

$$
\rho u_e \, \mathrm{St}_x = g_{hx} = \frac{-k(\partial T/\partial y)|_s}{h_{es} - h_e}
\tag{10.175}
$$

Thus, we see that the heat transfer Stanton number defined by Eq. (10.168) also gives the conduction heat flux across the s-surface in an inert system, provided that the driving potential is taken to be $(h_{es} - h_e)$, and not $(h_s - h_e)$. This result using the enthalpy formulation is different from the corresponding result obtained using the temperature formulation. In the temperature formulation we assumed all species specific heats equal, $c_{p1} = c_{p2} = c_p$. From Eq. (5.96),

$$
\mathrm{St}_x = \frac{-k(\partial T/\partial y)|_s}{\rho u_e c_p (T_s - T_e)}
\tag{10.176}
$$

whereas from Eq. (10.175),

$$
\mathrm{St}_x = \frac{-k(\partial T/\partial y)|_s}{\rho u_e c_{pe} (T_s - T_e)}
\tag{10.177}
$$

if we take c_p independent of temperature to facilitate the comparison. The temperature formulation requires that we use a specific heat that is some average over the species involved; in contrast, the enthalpy formulation requires that we use the specific heat of the free-stream mixture. The reason for this difference is that the temperature formulation eliminated the interdiffusion term at the outset, whereas the enthalpy formulation retained the interdiffusion term.

The role played by the enthalpy difference $(h_{es} - h_e)$ can be further illuminated if we consider the evaluation of $\mathcal{B}_h$ for transpiration cooling. If an inert gas, species 1, is injected into a flow of inert gas, species 2, then, referring to Eq. (10.165) and Fig. 10.32, an energy balance between the s- and o-surfaces gives

$$
\dot{m}'' h_t = \sum n_i h_i |_s - k \frac{\partial T}{\partial y}\Big|_s = n_1 h_{1,o} = \dot{m}'' h_{1,o}
$$

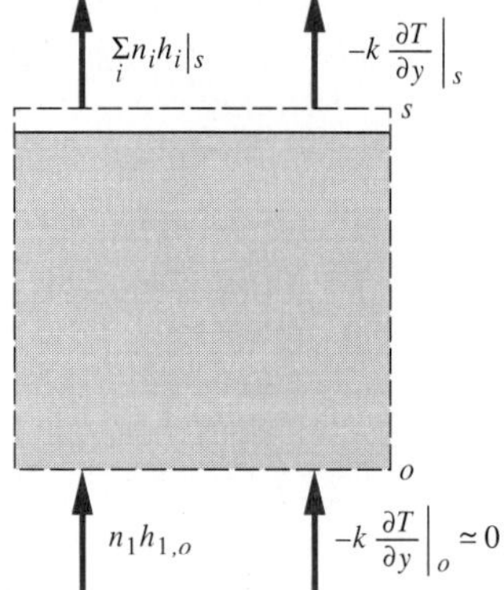

Figure 10.32 Transpiration cooling: the energy conservation principle applied to a control volume located between the s-surface and an o-surface in the coolant reservoir.

Substituting in Eq. (10.172),

$$\mathscr{B}_h = \frac{h_e - h_s}{h_s - h_{1,o}} \tag{10.178}$$

Choosing enthalpy datum states such that $h_1 = h_2 = 0$ at $T = T_s$, and proceeding as before,

$$\mathscr{B}_h = \frac{h_e^{T_s}}{h_{1,o}^{T_s}} = \frac{h_e - h_{es}}{h_{1,s} - h_{1,o}} \tag{10.179}$$

Substituting back in Eq. (10.172),

$$\dot{m}'' = g_h \mathscr{B}_h = \frac{g_h(h_e - h_{es})}{h_{1,s} - h_{1,o}} \tag{10.180}$$

Equation (10.179) gives a driving force for transpiration cooling that is easy to evaluate, since it does not involve the unknown s-surface composition. Further insight can be obtained by rearranging Eq. (10.180) as

$$g_h(h_e - h_{es}) = \dot{m}''(h_{1,s} - h_{1,o})$$

For specific heats independent of temperature and using Eq. (10.177),

$$g_h c_{pe}(T_e - T_s) = -k \frac{\partial T}{\partial y}\Big|_s = \dot{m}'' c_{p1}(T_s - T_o) \tag{10.181}$$

which states that the conduction heat flux across the s-surface equals $\dot{m}''$ times the increase in enthalpy of the coolant. Thus, even in a mixture where there is an interdiffusion contribution to energy transfer across the s-surface, the heat transfer to the wall for transpiration cooling is actually only the conduction component. In the notation of the steady-flow energy equation, Eq. (1.4), the left-hand side of Eq. (10.180) is $\dot{Q}$ and the right-hand side is $\dot{m}\Delta h$ (per unit area).

Examples 10.13 and 10.14 illustrate the use of the enthalpy formulation–based constant-property solutions. As for mass transfer in Section 10.5.1, consideration of real engineering problems requires the use of a suitable reference state for evaluation of properties, and Knuth's scheme, or a 1/2 or 1/3 rule may be used. In problems involving the combustion of volatile hydrocarbon fuels, the scheme of Law and Williams given in Section 10.3.5 may also be useful.

EXAMPLE 10.13 Sweat Cooling of a Flat Plate: Heat Transfer

If, in Example 10.12, the water is supplied from a reservoir at 300 K, determine the radiative heat flux absorbed by the plate. Use the constant-property self-similar solution with the 1/2 rule for evaluating properties.

Solution

Given: Air flowing along a radiatively heated, sweat-cooled flat plate.

Required: The radiative heat flux 0.2 m from the leading edge.

Assumptions:
1. The thermal resistance of the thin film of water on the plate is negligible.
2. A laminar boundary layer.
3. The 1/2 rule is adequate for evaluating properties.
4. Lewis number unity.
5. Species specific heats are independent of temperature.

From Example 10.12, $m_{1,s} = 0.496$, $m_{1,r} = 0.248$, $T_r = 600$ K. Using AIRSTE, $\rho = 0.512$ kg/m^3, $\mu = 2.71 \times 10^{-5}$ kg/m s, Sc $= 0.564$, $c_p = 1275$ J/kg K, $k = 0.0454$ W/m K, Pr $= 0.763$. Also, $\mathrm{Re}_x = 3.78 \times 10^4$. The heat transfer Stanton number and conductance are

$$\mathrm{St}_x^* = 0.332\,\mathrm{Re}_x^{-1/2}\,\mathrm{Pr}^{-2/3} = (0.332)(3.78 \times 10^4)^{-1/2}(0.763)^{-2/3} = 2.05 \times 10^{-3}$$

$$g_{hx}^* = \rho u_e\,\mathrm{St}_x^* = (0.512)(10)(2.05 \times 10^{-3}) = 1.05 \times 10^{-2} \text{ kg/m}^2\text{ s}$$

Since we know $\dot{m}'' = 7.18 \times 10^{-3}$ kg/m^2 s from Example 10.12, Fig. 10.28 will be used to obtain g_h/g_h^*:

$$-f(0) = \frac{\dot{m}''}{\rho u_e} 2^{1/2}\,\mathrm{Re}_x^{1/2} = \frac{7.18 \times 10^{-3}}{(0.512)(10)}(2 \times 3.78 \times 10^4)^{1/2} = 0.386$$

Then, by analogy from Fig. 10.28 for Pr(Sc) $= 0.763$ and $-f(0) = 0.386$, $g_{hx}/g_{hx}^* = 0.52$. Since the system is inert, we can conveniently use Eq. (10.175), which gives

$$k\left.\frac{\partial T}{\partial y}\right|_s = g_{hx}(h_e - h_{es}) \simeq g_{hx}^*(g_{hx}/g_{hx}^*)\bar{c}_{p\,\mathrm{air}}(T_e - T_s)$$

From Table A.7, $\bar{c}_{p\,\mathrm{air}} = (1/2)(1089 + 1007) = 1048$ J/kg K.

$$k\left.\frac{\partial T}{\partial y}\right|_s = (1.05 \times 10^{-2})(0.52)(1048)(840 - 360) = 2747 \text{ W/m}^2$$

Referring to the figure, an energy balance between the s- and o-surfaces with $n_1 = \dot{m}''$ gives

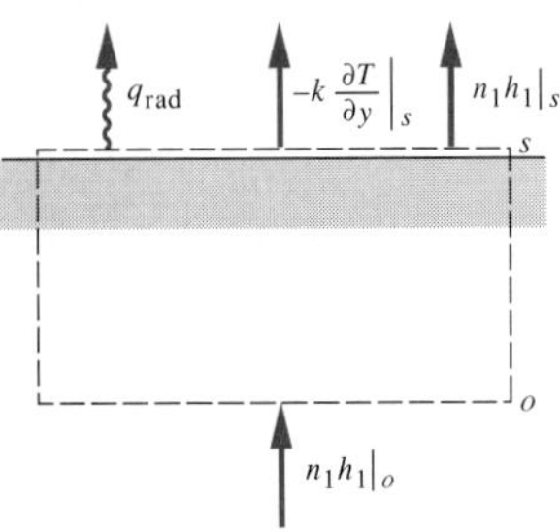

$$\dot{m}''h_1\Big|_s - k\left.\frac{\partial T}{\partial y}\right|_s + q_{\text{rad}} = \dot{m}''h_1\big|_o$$

since only species 1 is transferred. Rearranging,

$$k\left.\frac{\partial T}{\partial y}\right|_s - q_{\text{rad}} = \dot{m}''(h_{1,s} - h_{1,o})$$

$$= \dot{m}''[h_{\text{fg}} + c_{pl}(T_s - T_o)]$$

From Tables A.8 and A.12a, $c_{pl} = 4178$ J/kg K, $h_{\text{fg}} = 2.291 \times 10^6$ J/kg; thus,

$$k\left.\frac{\partial T}{\partial y}\right|_s - q_{\text{rad}} = (7.18 \times 10^{-3})[2.291 \times 10^6 + 4178(360 - 300)]$$

$$2747 - q_{\text{rad}} = 1.825 \times 10^4$$

Solving, $q_{\text{rad}} = 2747 - 1.825 \times 10^4 = -1.55 \times 10^4$ W/m^2 (15.5 kW/m^2)

Comments

1. Solve this problem using the improved blowing factors of Section 10.4.2 and compare results.
2. In this problem T_s was specified and q_{rad} was unknown. If we specify q_{rad} and make T_s the unknown, a complicated iteration process is required. Alternatively, q_{rad} could be varied parametrically to obtain a graph of T_s versus q_{rad}. Either way, sweat-cooling problem solving is not for the fainthearted!
3. Use an air enthalpy table to evaluate $(h_e - h_{es})$ and compare the result with the approximation obtained using $(h_e - h_{es}) \simeq \bar{c}_{p\,\text{air}}(T_e - T_s)$.
4. Although the analytical result used was obtained assuming Le $= 1$ (Pr $=$ Sc), we were careful to use Pr and the heat transfer conductance (rather than Sc and the mass transfer conductance) for these *heat transfer* calculations. We thus ensured that our procedure is exact in the limit of zero mass transfer.
5. Notice that q_{rad} must also vary as $x^{-1/2}$ in order to have a self-similar boundary layer.

EXAMPLE 10.14 Heat Transfer into a Graphite Heat Shield

Graphite is often used as a heat shield material due to its high sublimation temperature (>3500°C). In a particular application, a graphite heat shield is exposed to air at 2000 K and 2 atm flowing at 30 m/s. The flow can be modeled as a laminar boundary layer on a flat plate. Calculate the heat flux into the heat shield at a location 3 cm from the leading edge when the surface temperature has reached 1500 K. The surroundings are black at 1400 K, and graphite has an emittance of 0.88.

Solution

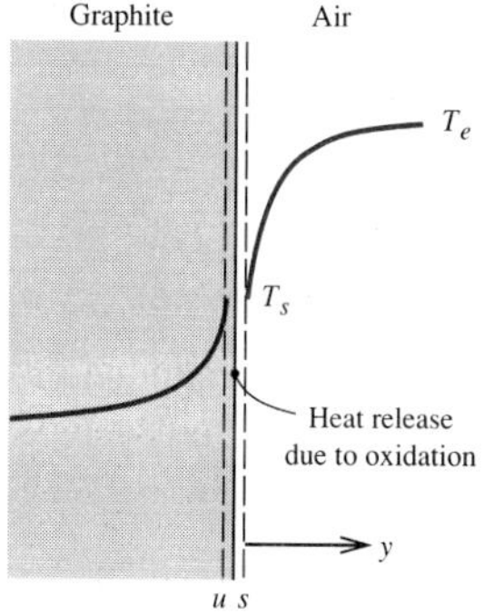

Given: A graphite surface exposed to a hot air flow.

Required: Heat flux at $x = 0.03$ m when the surface temperature is 1500 K.

Assumptions:
1. Diffusion-controlled oxidation to carbon monoxide.
2. Carbon vapor pressure negligible (heterogeneous reaction).
3. Species Lewis numbers unity.
4. Radiation can be modeled as interchange between a small gray body and large black surroundings.

The combustion process is the same as was analyzed in Section 10.3.2: at the specified temperatures, the product of combustion is carbon monoxide, and the reaction is diffusion-controlled with $m_{O_2,s} \simeq 0$.

$$2C + O_2 \rightarrow 2\,CO$$

$$1\,\text{kg} + r\,\text{kg} \rightarrow (1 + r)\,\text{kg}$$

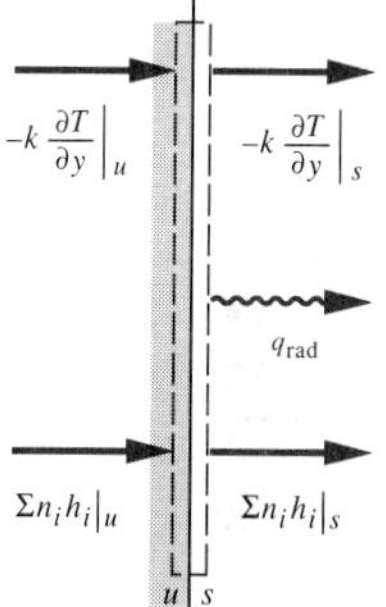

where $r = 32/24 = 1.33$ is the stoichiometric ratio of the reaction. In Section 9.5.3 we made an ad hoc estimate of the heat transfer during combustion of carbon: here a more rigorous approach is used. The surface energy balance is

$$-k\frac{\partial T}{\partial y}\bigg|_u + \sum n_i h_i|_u = -k\frac{\partial T}{\partial y}\bigg|_s + \sum n_i h_{i,s} + q_{\text{rad}}$$

as shown in the figure.

To obtain a better grasp on the physical significance of the terms in this energy balance, it is rearranged as

$$k\frac{\partial T}{\partial y}\bigg|_u = k\frac{\partial T}{\partial y}\bigg|_s + \left(\sum n_i h_i|_u - \sum n_i h_i|_s\right) - q_{\text{rad}} \quad (1)$$

conduction into heat shield; convection from air flow; heat release due to combustion; radiation loss to surroundings

The term accounting for heat release due to combustion is evaluated as follows:

$$\sum n_i h_i|_u = n_{C,u} h_{C,u} = \dot{m}'' h_{C,u}$$

where C denotes solid carbon.

$$\sum n_i h_i|_s = n_{CO,s} h_{CO,s} + n_{O_2,s} h_{O_2,s} = \dot{m}''[(1 + r) h_{CO,s} - r h_{O_2,s}]$$

Thus, the rate of heat release is

$$\left(\sum n_i h_i|_u - \sum n_i h_i|_s\right) = \dot{m}''[h_{C,u} + r h_{O_2,s} - (1 + r) h_{CO,s}] = \dot{m}'' \Delta h_c$$

where Δh_c is the heat of combustion per kilogram of solid carbon at temperature T_s. Also, for a small gray surface in large black surroundings, we can write $q_{\text{rad}} \simeq \sigma\varepsilon(T_s^4 - T_e^4)$. Substituting in Eq. (1),

$$k\frac{\partial T}{\partial y}\bigg|_u = k\frac{\partial T}{\partial y}\bigg|_s + \dot{m}''\Delta h_c - \sigma\varepsilon(T_s^4 - T_e^4) \tag{2}$$

We next calculate the mass transfer rate $\dot{m}''$,

$$\dot{m}'' = \mathscr{g}_{mO_2}\mathscr{B}_{mO_2}$$

where, from Eq. (10.41), $\mathscr{B}_{mO_2} = 0.173$. The 1/2 rule will be used to evaluate properties: $T_r = (1/2)(T_e + T_s) = (1/2)(2000 + 1500) = 1750$ K. Again from Section 10.3.2, $m_{CO,s} = 0.345$. Denoting O_2, CO, and N_2 as species 1, 2, and 3, respectively, $m_{1,r} = (1/2)(0.231 + 0) = 0.116$; $m_{2,r} = (1/2)(0 + 0.345) = 0.173$; $m_{3,r} = (1/2)(0.769 + 0.655) = 0.711$. GASMIX gives the following properties at 2 atm: $\rho = 0.3959$ kg/m^3, $\mu = 5.83 \times 10^{-5}$ kg/m s, $c_p = 1257$ J/kg K, $\text{Sc}_{O_2,m} \simeq 0.73$, Pr = 0.702.

The Reynolds number is

$$\text{Re}_x = u_e\rho x/\mu = (30)(0.3959)(0.03)/(5.83 \times 10^{-5}) = 6112$$

and from Eq. (10.149) the mass transfer Stanton number is

$$\text{St}_{mx}^* = 0.332\ \text{Re}_x^{-1/2}\,\text{Sc}^{-2/3} = (0.332)(6112)^{-1/2}(0.73)^{-2/3} = 5.24 \times 10^{-3}$$

$$\mathscr{g}_{mx}^* = \rho u_e\,\text{St}_m^* = (0.3959)(30)(5.24 \times 10^{-3}) = 6.22 \times 10^{-2}\ \text{kg/m}^2\ \text{s}$$

From Fig. 10.30 with $\mathscr{B}_m = 0.173$, Sc = 0.73, $\mathscr{g}_{mx}/\mathscr{g}_{mx}^* = 0.88$. Thus, $\mathscr{g}_{mx} = (0.88)(6.22 \times 10^{-2}) = 5.47 \times 10^{-2}$ kg/m^2 s. Hence,

$$\dot{m}'' = \mathscr{g}_m\mathscr{B}_m = (5.47 \times 10^{-2})(0.173) = 9.47 \times 10^{-3}\ \text{kg/m}^2\ \text{s}$$

Since the boundary layer is inert, we can use Eq. (10.175) to obtain the s-surface conduction,

$$k\frac{\partial T}{\partial y}\bigg|_s = \mathscr{g}_{hx}(h_e - h_{es}) \simeq \mathscr{g}_{hx}\bar{c}_{p\,\text{air}}(T_e - T_s) \tag{3}$$

where $\bar{c}_{p\,\text{air}}$ is evaluated at $(T_e + T_s)/2$. Using Eqs. (10.149) and (10.171),

$$\mathscr{g}_{hx}^* = \mathscr{g}_{mx}^*(\text{Pr}/\text{Sc})^{-2/3} = (6.22 \times 10^{-2})(0.702/0.730)^{-2/3} = 6.38 \times 10^{-2}\ \text{kg/m}^2\ \text{s}$$

Since we know $\dot{m}''$, we can use Fig. 10.28 to obtain $\mathscr{g}_{hx}/\mathscr{g}_{hx}^*$:

$$-f(0) = (\dot{m}''/\rho u_e)2^{1/2}\,\text{Re}_x^{1/2}$$

$$= [9.47 \times 10^{-3}/(0.3959)(30)](2)^{1/2}(6112)^{1/2} = 8.82 \times 10^{-2}$$

Then, by analogy from Fig. 10.28 for Pr(Sc) = 0.7 and $-f(0) = 8.82 \times 10^{-2}$, $\mathscr{g}_{hx}/\mathscr{g}_{hx}^* = 0.88$. Hence, $\mathscr{g}_{hx} = (0.88)(6.38 \times 10^{-2}) = 5.61 \times 10^{-2}$ kg/m^2 s.

For $T_e = 2000$ K, $T_s = 1500$ K, Table A.7 gives $\bar{c}_{p\,\text{air}} = (1/2)(1244 + 1202) = 1223$ J/kg K. Substituting in Eq. (3),

$$k\frac{\partial T}{\partial y}\bigg|_s = (5.61 \times 10^{-2})(1223)(2000 - 1500) = 3.43 \times 10^4\ \text{W/m}^2$$

An estimate of the heat of combustion Δh_c can be obtained from the data in Table A.25 by subtracting (M_{CO}/M_C) of the value for CO from the value for C (values in Table A.25 are for a CO_2 product per kilogram of fuel):

$$\Delta h_c \simeq 32.78 \times 10^6 - (28/12)(10.11 \times 10^6) \simeq 9.2 \times 10^6 \text{ J/kg}$$

Finally, substituting into Eq. (2) gives the conduction flux into the heat shield:

$$\begin{aligned} k \left.\frac{\partial T}{\partial y}\right|_u &= 3.43 \times 10^4 + (9.47 \times 10^{-3})(9.2 \times 10^6) - (5.67)(0.88)(15^4 - 14^4) \\ &= 3.43 \times 10^4 + 8.71 \times 10^4 - 6.09 \times 10^4 \\ &= 6.05 \times 10^4 \text{ W/m}^2 (61 \text{ kW/m}^2) \end{aligned}$$

Comments

1. The conduction heat flux calculated above would be used as a boundary condition in a finite-difference numerical solution of the heat conduction equation to obtain the transient response of the heat shield.

2. In this situation the surface energy balance is dominated by the radiation heat flux. As T_s increases above 1500 K, q_{rad} increases rapidly and $k\, \partial T/\partial y|_u = 0$ at a value of T_s below 1600 K.

3. Notice that since Pr $\simeq$ Sc, the blowing corrections for mass and heat transfer are identical.

4. The surface recession can cause the heat shield shape to change. Such shape changes are usually tolerated to take advantage of the high sublimation temperature of graphite.

10.6 A GENERAL PROBLEM-SOLVING PROCEDURE FOR CONVECTIVE HEAT AND MASS TRANSFER

In Sections 10.3, 10.4, and 10.5, a number of specific problems involving high mass transfer rates were analyzed. These problems were selected because of their fundamental importance and engineering significance. Section 10.3 introduced diffusion-induced convection (Stefan flow) in the context of some practical evaporation and combustion problems. Section 10.4 used the Couette flow as a model of real convective flows to obtain engineering methods that account for high mass transfer rates. In Section 10.5 convection in a laminar boundary layer on a flat plate was rigorously analyzed, extending the classical momentum and heat transfer analyses to include mass transfer. Each analysis was self-contained, and little was done to generalize the applicability of the results. However, the student should have discerned that the analyses had much in common, and that appropriate generalization should yield results applicable to a wide range of physical situations. For example, the Shvab-Zeldovich transformation used in Section 10.3.3 to analyze carbon particle combustion can also be used to analyze general chemical reactions in the laminar boundary layer of Section 10.5.1. The assumption of unity Lewis number that was used to simplify the energy equation for the laminar boundary layer in Section 10.5.2 can also be used to obtain an alternative result to Eq. (10.63) for droplet evaporation.

In this section we organize the results obtained already to give a general problem-solving procedure for steady convective heat and mass transfer. To develop the procedure, we first carefully discuss the concept of a *transferred state* in Section 10.6.1. The transferred state for enthalpy was introduced in Section 10.5.2, and we now introduce the transferred state for chemical species and elements. The actual problem-solving procedure is presented in Section 10.6.2, and in Section 10.6.3 examples of its application to more complicated problems are given. These problems involve such features as transfer of more than one species, chemical reactions, simultaneous heat and mass transfer, and coupled mass transfer in two adjacent phases.

The development that follows is based mainly on the contributions of D. B. Spalding [20,21,22].

10.6.1 The Transferred State

Mass Transfer

To introduce the concept of a *transferred state* for mass transfer, it is convenient to return to Section 10.4.1, where the mass transfer driving force was defined for a binary inert mixture, with transfer of a single species. We now extend that definition to a multicomponent mixture with transfer of more than one species. The mass transfer conductance for species i is

$$\mathcal{g}_{mi} = \frac{j_{i,s}}{m_{i,s} - m_{i,e}} \tag{10.182}$$

Notice again that $\mathcal{g}_{mi}$ is defined in terms of the *diffusive* flux $j_{i,s}$. Since $\mathcal{g}_{mi}$ is not well behaved if species i reacts in the flow, we will restrict our attention to species that are inert in the flow (but that may be involved in heterogeneous reactions at the wall). The *absolute* flux of species i across the s-surface is, by definition,

$$n_{i,s} = m_{i,s} n_s + j_{i,s} \tag{10.183}$$

where $n_s = \dot{m}''$, the mass transfer rate. We define the transferred-state mass fraction $m_{i,ts}$ as

$$m_{i,ts} = \frac{n_{i,s}}{\dot{m}''} \tag{10.184}$$

for transfer across the s-surface. If we imagine the mass transfer process as a stream of matter flowing through the s-surface, $m_{i,ts}$ is the fraction of the stream that is species i. Substituting Eqs. (10.182) and (10.184) into Eq. (10.183),

$$m_{i,ts}\dot{m}'' = m_{i,s}\dot{m}'' + \mathcal{g}_{mi}(m_{i,s} - m_{i,e})$$

Solving for $\dot{m}''$,

$$\dot{m}'' = \mathcal{g}_{mi} \frac{m_{i,e} - m_{i,s}}{m_{i,s} - m_{i,ts}}$$

or

$$\dot{m}'' = g_{mi}\mathscr{B}_{mi}; \qquad \mathscr{B}_{mi} = \frac{m_{i,e} - m_{i,s}}{m_{i,s} - m_{i,ts}} \tag{10.185}$$

where $\mathscr{B}_{mi}$ is the mass transfer driving force for species i. We can also define a transferred-state mass fraction for transfer across the u-surface as

$$m_{i,tu} = \frac{n_{i,u}}{\dot{m}''} \tag{10.186}$$

In the special case that species i does not react at the wall, that is, between the u- and s-surfaces, $n_{i,u} = n_{i,s}$ and

$$m_{i,tu} = m_{i,ts} \equiv m_{i,t} \tag{10.187}$$

Thus, for problems involving inert species—for example, evaporation and condensation—we need not distinguish between the ts- and tu-states since they are the same, and we simply write $m_{i,t}$ for a transferred-state mass fraction.

We have, of course, implicitly used the transferred-state mass fraction concept in many of the analyses in Sections 10.3 through 10.5. For example, when only a single component is transferred in an inert binary system, as occurs for evaporation of water into air, we can write $n_{1,u} = \dot{m}''$ since only water crosses the u-surface. Thus, $m_{1,tu} = 1$ and, from Eq. (10.187), $m_{1,t} = 1$. Substituting in Eq. (10.185) recovers Eq. (10.93):

$$\dot{m}'' = g_{m1}\mathscr{B}_{m1}; \qquad \mathscr{B}_{m1} = \frac{m_{1,e} - m_{1,s}}{m_{1,s} - 1} \tag{10.93}$$

Figure 10.33 shows heterogeneous oxidation of carbon, as analyzed in Section 10.3.2. At the high temperatures under consideration, the reaction was

$$2C + O_2 \rightarrow 2CO$$

$$2 \times 12\,\text{kg} + 32\,\text{kg} \rightarrow 2 \times 28\,\text{kg}$$

Since only carbon crosses the u-surface, $n_{C,u} = \dot{m}''$, $n_{O_2,u} = 0$, $n_{CO,u} = 0$, and

$$m_{C,tu} = 1; \qquad m_{O_2,tu} = 0; \qquad m_{CO,tu} = 0$$

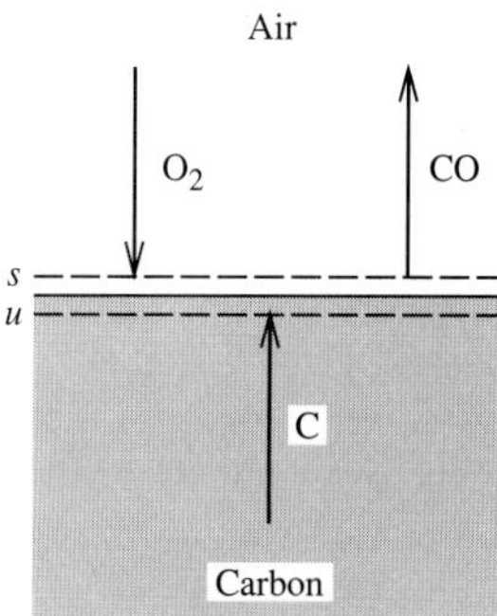

Figure 10.33 Surface species balance for high-temperature combustion of carbon.

To obtain $m_{O_2,ts}$ we note that for each kilogram of carbon crossing the u-surface in the positive direction, 4/3 kg of oxygen crosses the s-surface in the negative direction; thus

$$\frac{n_{O_2,s}}{n_{C,u}} = -\frac{4}{3}$$

But $n_{C,u} = m_{C,tu}\dot{m}'' = (1)\dot{m}'' = \dot{m}''$; hence

$$m_{O_2,ts} = \frac{n_{O_2,s}}{\dot{m}''} = -\frac{4}{3}$$

Substituting in Eq. (10.185) gives

$$\dot{m}'' = g_{mO_2}\frac{m_{O_2,e} - m_{O_2,s}}{m_{O_2,s} + 4/3}$$

which can be compared to Eqs. (10.39) and (10.41). (As before, if the reaction is diffusion-controlled, we can set $m_{O_2,s} \simeq 0$.) We see that use of the transferred-state mass fraction allows us to handle the interface boundary condition in a systematic manner.

We also define a transferred-state mass fraction for chemical element (α). Recall that in Section 10.3.3, we defined the absolute flux of element (α) as $\mathbf{n}_{(\alpha)} = \sum_i m_{(\alpha)i}\mathbf{n}_i$, where $m_{(\alpha)i}$ is the mass fraction of element (α) in species i. Then

$$m_{(\alpha),t} = \frac{n_{(\alpha),s}}{\dot{m}''} = \frac{n_{(\alpha),u}}{\dot{m}''} \tag{10.188}$$

We need not distinguish between ts- and tu-states since $n_{(\alpha),s} = n_{(\alpha),u}$; chemical elements cannot be created or destroyed between the s- and u-surfaces. As an example, consider combustion of carbon in air. Irrespective of the reaction stoichiometry, $n_{(C),u} = \dot{m}''$, $n_{(O),u} = 0$; hence

$$m_{(C),t} = \frac{n_{(C),u}}{\dot{m}''} = 1; \qquad m_{(O),t} = \frac{n_{(O),s}}{\dot{m}''} = \frac{n_{(O),u}}{\dot{m}''} = 0$$

a result that was used in Section 10.3.3. Notice that we can allow chemical reactions to occur both in the flow and at the wall when the analysis is based on chemical elements.

Heat Transfer

The transferred-state enthalpy h_t was introduced in Section 10.5.2 and was defined as

$$\dot{m}''h_t = \sum n_i h_i|_s - k\left.\frac{\partial T}{\partial y}\right|_s \tag{10.165}$$

That is, if energy transfer across the s-surface by convection and conduction is imagined to be due to convection only by a stream of matter crossing the interface at a rate $\dot{m}''$, then h_t is the enthalpy of the stream. In Section 10.5.2, h_t was introduced to obtain a heat transfer result that was of the same form as the mass transfer result of Section 10.5.1. We also saw in Section 10.5.2 that problem solving required evaluating h_t for a specific physical situation by performing an energy balance between the s-surface and the u-surface, or a reservoir o-surface. In transpiration or sweat cooling with negligible surface radiation, h_t proved to be equal to the reservoir enthalpy h_o. For the noncondensable gas problem in a condenser, h_t is evaluated by performing a further energy balance between the u-surface and the bulk coolant.

10.6.2 The Problem-Solving Procedure

Most (but not all) of the results obtained in Sections 10.3 through 10.5 can be cast in the form

$$\dot{m}'' = \mathcal{g}\mathcal{B} \tag{10.189}$$

where $\mathcal{B}$ is a *driving force* based on the mass fraction of a mass species or chemical element, or on enthalpy, and $\mathcal{g}$ is the corresponding conductance. In general, we can write

$$\mathcal{B} = \frac{\mathcal{P}_e - \mathcal{P}_s}{\mathcal{P}_s - \mathcal{P}_{ts}} \tag{10.190}$$

where the **conserved property** $\mathcal{P} = m_i$, $m_{(\alpha)}$, or h. When i is inert in the phase under consideration, or when $\mathcal{P} = m_{(\alpha)}$ or h, $\mathcal{P}_{ts}$ can be replaced by $\mathcal{P}_t$ because the ts- and t-states are identical. Equation (10.189) forms the basis of our problem-solving procedure. In problems involving transfer of more than one species, simultaneous consideration of two adjacent phases, or simultaneous heat and mass transfer, two or more equations of the form of Eq. (10.189) must be solved simultaneously. Also, auxiliary relations such as vapor pressure or solubility data are often required. Two essentially distinct issues are involved in using Eq. (10.189): one is the evaluation of the driving force $\mathcal{B}$, and the other is the evaluation of the conductance $\mathcal{g}$.

Choice of Driving Force

The choice of driving force depends on the type of mass transfer problem under consideration and the information available. The situation may involve evaporation, combustion of a solid fuel, combustion of a volatile fuel, catalysis, and so on. The problem considered most often in Sections 10.3 through 10.5 was mass transfer of a single inert species, either in connection with evaporation, condensation, or transpiration cooling. The driving force was based on mass fraction of the transferred species, $\mathcal{P} = m_1$, with $m_{1,t} = 1$.

If the results of the analysis of carbon combustion in Section 10.3.3 are recast in the form of Eq. (10.189), the driving force is seen to be based on a linear combination of element mass fractions, $\mathcal{P} = \zeta m_{(\mathrm{O})} + \xi m_{(\mathrm{C})}$. In deriving this driving force, the diffusion coefficients of all species containing the elements were assumed equal. Similarly, Section 10.5.2 showed that for an enthalpy-based driving force to be valid, the Lewis numbers, $\mathrm{Le}_i = (k/c_p)/\rho\mathscr{D}_{im}$, of all species in the mixture must equal unity. Since k, c_p, and ρ are all mixture properties, this condition requires all the diffusion coefficients to be equal. In addition, it is easy to show that an enthalpy-based driving force is also valid for a nonreacting mixture of species with all species specific heats equal. The transfer number obtained for combustion of a liquid fuel droplet in Section 10.3.5 was, in fact, an enthalpy-based driving force with the Lewis numbers assumed to be unity, and as a further simplification, the species specific heats were assumed equal and constant.

Evaluation of the Conductance

Evaluation of the conductance $\mathscr{g}$, corresponding to the chosen driving force $\mathscr{B}$, requires identification of the flow configuration, evaluation of a zero mass transfer limit conductance $\mathscr{g}^*$, and its correction using an appropriate blowing factor. If $\mathscr{B}$ is based on the mass fraction of species i, then the corresponding conductance is a mass transfer conductance. For example, for a laminar boundary layer on a flat plate, the local value of $\mathscr{g}^*_{mi}$ is obtained from Eq. (10.149) as

$$\mathscr{g}^*_{mxi} = \rho u_e \mathrm{St}^*_{mxi} = 0.332\,\mathrm{Re}_x^{-1/2}\,\mathrm{Sc}_i^{-2/3}; \qquad \mathrm{Sc}_i > 0.5 \qquad \textbf{(10.191)}$$

where $\mathrm{Sc}_i = \nu/\mathscr{D}_{im}$. If $\mathscr{B}$ is based on chemical element (α), then the local value of $\mathscr{g}^*_{m(\alpha)}$ is obtained from Eq. (10.191) by replacing Sc_i by $\mathrm{Sc}_{(\alpha)} = \nu/\mathscr{D}$, where $\mathscr{D}$ is some average of the $\mathscr{D}_{im}$'s for the species containing element (α). If an enthalpy-based driving force is obtained for an inert mixture by assuming equal species specific heats, then the corresponding zero-mass-transfer-limit conductance is $\mathscr{g}^*_h$. Again, for a laminar boundary layer on a flat plate, the local value of $\mathscr{g}^*_h$ is obtained from Eq. (10.171) as

$$\mathscr{g}^*_{hx} = \rho u_e \mathrm{St}^*_x = 0.332\,\mathrm{Re}_x^{-1/2}\,\mathrm{Pr}^{-2/3}; \qquad \mathrm{Pr} > 0.5 \qquad \textbf{(10.192)}$$

On the other hand, if an enthalpy-based driving force is obtained by assuming unity Lewis numbers, there is the implication $\mathrm{Sc}_i = \mathrm{Pr}$ for all species i and hence $\mathscr{g}^*_{mi} = \mathscr{g}^*_h$. Since usually $\mathrm{Sc}_i \neq \mathrm{Pr}$, evaluation of $\mathscr{g}^*_h$ poses a problem: we usually base it on Pr in order to have an exact result for heat transfer in the limit $\dot{m}'' \to 0$.

Blowing factors to correct $\mathscr{g}^*$ were obtained from various sources in Sections 10.4 and 10.5. In Example 10.8 the blowing factor was obtained from a constant-property Couette-flow model. In Example 10.12, part (i), an improved blowing factor that accounted for variable-property effects and flow configuration was used, and in part (ii) the blowing factor was obtained from an exact self-similar solution for the constant-property laminar boundary layer. It is also possible to use experimental data for blowing factors when available; for example, some such data are available for the combustion of liquid hydrocarbon fuel droplets.

The results for the spherically symmetric flows analyzed in Section 10.3 can also be recast in a form consistent with the foregoing procedure for evaluating the conductance. For example, the analysis of combustion of liquid fuel droplets in Section 10.3.5 gave Eq. (10.74), which can be rearranged as

$$\dot{m}'' = g_h \mathscr{B}_h; \qquad \mathscr{B}_h = \frac{h_e - h_s}{h_s - h_t} \tag{10.193a}$$

$$g_h^* = \frac{k/c_p}{R}; \qquad \frac{g_h}{g_h^*} = \frac{\ln(1 + \mathscr{B}_h)}{\mathscr{B}_h} \tag{10.193b}$$

since $h_u = h_t$ as a result of taking $dT/dr|_u = 0$.

Surface Radiation

It is possible to allow for surface radiation when using the general problem-solving procedure with an enthalpy-based driving force. To do so, we must recognize that the transferred-state enthalpy h_t is always evaluated by performing an energy balance between the s-surface and either the u-surface or a reservoir o-surface. For example, consider simple transpiration cooling with an inert species 1 injected into a flow of species 2, as shown in Fig. 10.34a. An energy balance between the s- and o-surfaces with $n_{1,s} = \dot{m}''$ gives

$$\dot{m}'' h_{1,s} - k \left. \frac{\partial T}{\partial y} \right|_s + q_{\text{rad}} = \dot{m}'' h_{1,o}$$

Using the definition of h_t from Eq. (10.165) and rearranging,

$$h_t = h_{1,o} - (q_{\text{rad}}/\dot{m}'') \tag{10.194}$$

and the enthalpy driving force becomes

$$\mathscr{B}_h = \frac{h_e - h_s}{h_s - h_t} = \frac{h_e - h_s}{(h_s - h_{1,o}) + (q_{\text{rad}}/\dot{m}'')}$$

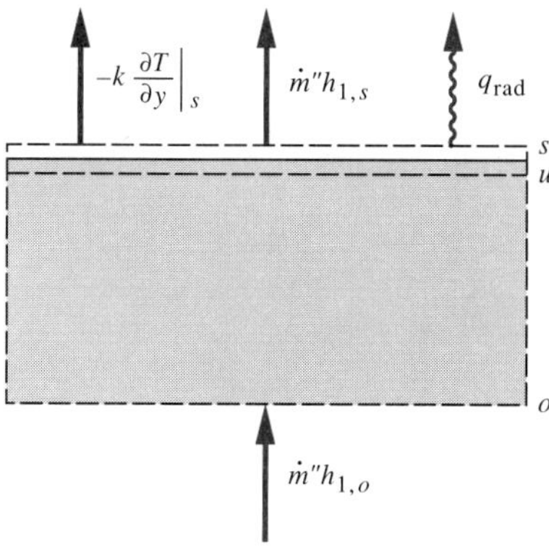

Figure 10.34*a* The surface energy balance with radiation. Transpiration cooling.

As usual, we choose enthalpy datum states such that $h_1 = h_2 = 0$ at $T = T_s$, to obtain

$$\mathcal{B}_h = \frac{h_e^{T_s}}{-h_{1,o}^{T_s} + (q_{\text{rad}}/\dot{m}'')} = \frac{h_e - h_{es}}{(h_{1,s} - h_{1,o}) + (q_{\text{rad}}/\dot{m}'')}$$

Finally, assuming constant species specific heats,

$$\dot{m}'' = g_h \frac{c_{pe}(T_e - T_s)}{c_{p1}(T_s - T_o) + (q_{\text{rad}}/\dot{m}'')} \tag{10.195}$$

We see that radiation heat transfer away from the surface (positive q_{rad}) reduces $\mathcal{B}_h$ and hence reduces the injection rate required to maintain a specified surface temperature.

As a second example, consider drying of a wet textile by combined convective and radiative heating, as shown in Fig. 10.34*b*. An energy balance between the s- and u-surfaces with $n_{1,s} = \dot{m}''$ gives[7]

$$\dot{m}''h_{1,s} - k\left.\frac{\partial T}{\partial y}\right|_s + q_{\text{rad}} = \dot{m}''h_{1,u} + q_{\text{cond}}$$

$$\dot{m}''h_t = \dot{m}''h_{1,u} + q_{\text{cond}} - q_{\text{rad}}$$

and the driving force becomes

$$\mathcal{B}_h = \frac{h_e - h_s}{h_s - h_{1,u} - (q_{\text{cond}}/\dot{m}'') + (q_{\text{rad}}/\dot{m}'')}$$

Again taking $h_1 = h_2 = 0$ at $T = T_s$,

$$\mathcal{B}_h = \frac{h_e - h_{es}}{h_{\text{fg}} - (q_{\text{cond}}/\dot{m}'') + (q_{\text{rad}}/\dot{m}'')} \tag{10.196}$$

In the *constant drying period* the textile is nearly isothermal and $q_{\text{cond}} \simeq 0$. Then, assuming constant species specific heats,

$$\dot{m}'' = g_h \frac{c_{pe}(T_e - T_s)}{h_{\text{fg}} + (q_{\text{rad}}/\dot{m}'')} \tag{10.197}$$

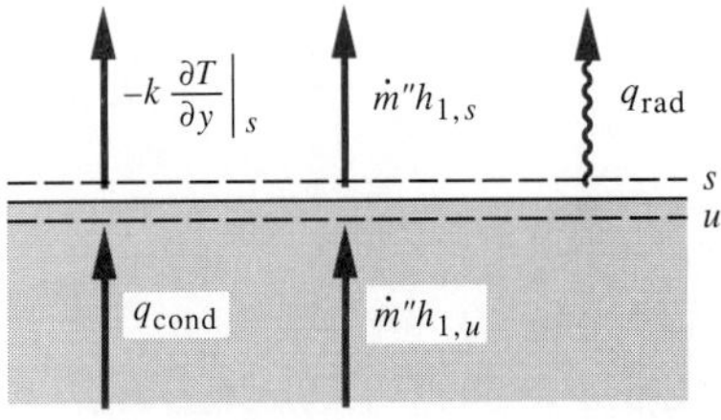

Figure 10.34*b* The surface energy balance with radiation. Textile drying.

[7] Strictly speaking, we should subscript q_{cond} as $q_{\text{cond},u}$ and q_{rad} as $q_{\text{rad},s}$, but to be consistent with Section 9.5.1, we take these subscripts as understood.

In this drying process, q_{rad} is negative, and for a given value of T_s the effect of the radiation flux is to increase the evaporation rate, as expected and desired. Equation (10.197) can be rewritten as

$$\mathcal{g}_h c_{pe}(T_e - T_s) = \dot{m}'' h_{\mathrm{fg}} + q_{\mathrm{rad}}$$

or

$$k \left. \frac{\partial T}{\partial y} \right|_s + (-q_{\mathrm{rad}}) = \dot{m}'' h_{\mathrm{fg}} \tag{10.198}$$

which clearly shows the energy balance implicit in Eq. (10.197).

Variable-Property Effects

A number of schemes are available to account for the effects of variable properties. The schemes recommended in Sections 10.3 through 10.5 are summarized below.

A. Improved blowing factors. For forced-convection laminar or turbulent air boundary layers with injection of a single inert species, the improved blowing factors of Section 10.4.2 should be used. Recall that in this scheme, air properties at the mean film temperature are used to calculate $\mathcal{g}^*$. Also notice that the heat transfer blowing factor gives the conduction component of the s-surface energy flux.

B. General reference states. When constant-property analytical formulas, numerical data, or experimental data are available for the flow configuration of concern, a reference state scheme can be used in which all properties are evaluated at the reference state. Examples include the 1/2 rule (mean film temperature and composition), Hubbard's 1/3 rule (Eqs. 10.65), or Knuth's Couette-flow model–based scheme (Eqs. 10.154 and 10.155).

C. Specific prescriptions. Sometimes specific prescriptions for the problem of concern are available based on experimental data or exact numerical solutions. Examples include the scheme of Law and Williams for liquid hydrocarbon fuel droplet combustion given in Section 10.3.5, and the laminar boundary layer scheme of Simon et al. [19].

D. Couette-flow model. If nothing better is available, the simple Couette-flow model results of Section 10.4.1 should be used. Two possible strategies are (1) calculate all properties at one of the general reference states listed under scheme B above, and (2) evaluate $\mathcal{g}^*$ using the free-stream composition and mean film temperature, and use Eq. (10.119) to account for the effect of unequal specific heats. Since these schemes are rather crude, every effort should be made to obtain appropriate experimental data to validate the results.

Validity of the Procedure

On a first reading, the student may well obtain the impression that this general problem-solving procedure is an ad hoc proposal and does not have a rigorous analytical basis. It is straightforward to provide such a basis, but in so doing one simply repeats elements of the analyses already presented in Sections 10.3 through 10.5.

Spalding [20,21] and Kays and Crawford [22] give a more complete analytical basis for the procedure. In general, the procedure applies to steady, low-speed laminar or turbulent flows of any configuration. The effects of thermal, forced, and pressure diffusion, and diffusional conduction, must be negligible, and multicomponent diffusion must be modeled as effective binary diffusion. Absorption or emission of radiation in the fluid must be negligible. High-speed flows can be treated as a special case if a unity Prandtl number is assumed (see Exercise 10–51). The fluid should be an ideal gas mixture or ideal liquid solution.

Notice that the heat transfer results for the evaporating droplet, Eq. (10.63), and the simple Couette-flow model, Eq. (10.119), cannot be recast in the general form of Eq. (10.189), namely, $\dot{m}'' = g\mathcal{B}$. For these simple flows we were able to account for unequal species specific heats, and took advantage of this fact to obtain useful analytical results. Thus, in using these results, care must be taken to adapt the general problem-solving procedure accordingly.

10.6.3 Applications of the Procedure

There were many examples of problem solving in Sections 10.3 through 10.5; usually the problems were relatively simple. Here, more complicated problems are considered, involving such features as transfer of more than one species, simultaneous heat and mass transfer, complex chemical reactions, and coupled mass transfer in two adjacent phases. The emphasis is on formulating the set of equations to be solved, and evaluating the driving forces.

1. Reaction of carbon with a steam-air mixture. In a number of technological processes, carbon is oxidized by a mixture of air and steam (see Example 10.4). The surface reactions are

$$CO_2 + C \rightarrow 2CO$$

$$H_2O + C \rightarrow CO + H_2$$

At typical process temperatures, these reactions are very rapid and are diffusion-controlled with $m_{CO_2,s} = m_{H_2O,s} = 0$. If we are prepared to assume equal effective binary diffusion coefficients of all species containing the elements (C) and (O), then we can follow the procedure of Section 10.3.3 and choose as conserved property $\mathcal{P}$ the combination of elemental mass fractions $[(12/16)m_{(O)} - m_{(C)}]$, as was done in Example 10.4. Recall that use of this combination avoids the need to know $m_{CO,s}$. Then

$$\mathcal{P}_e = \left(\frac{12}{16}m_{(O)} - m_{(C)}\right)_e = \frac{12}{16}\left(m_{O_2,e} + \frac{16}{18}m_{H_2O,e}\right)$$

$$\mathcal{P}_s = \left(\frac{12}{16}m_{(O)} - m_{(C)}\right)_s = 0$$

$$\mathcal{P}_t = \left(\frac{12}{16}m_{(O)} - m_{(C)}\right)_t = -1$$

$$\dot{m}'' = g_m\mathcal{B}; \qquad \mathcal{B} = \frac{\mathcal{P}_e - \mathcal{P}_s}{\mathcal{P}_s - \mathcal{P}_t} = \frac{12}{16}m_{O_2,e} + \frac{12}{18}m_{H_2O,e} \tag{10.199}$$

The assumption of equal diffusion coefficients is not too bad because we do not have to include $\mathscr{D}_{H_2m}$ in this assumption ($\mathscr{D}_{H_2m}$ is much larger than the other effective binary diffusion coefficients in the mixture).

If there are gas-phase reactions, such as $2CO + O_2 \rightarrow 2CO_2$, Eq. (10.199) remains valid. However, for high bulk gas temperatures there are no gas-phase reactions, and since both O_2 and H_2O are inert in the gas phase, we can then base driving forces on m_{O_2} and m_{H_2O} to write

$$\dot{m}'' = g_{mO_2} \frac{m_{O_2,e} - m_{O_2,s}}{m_{O_2,s} - m_{O_2,ts}} = g_{mH_2O} \frac{m_{H_2O,e} - m_{H_2O,s}}{m_{H_2O,s} - m_{H_2O,ts}} \qquad \textbf{(10.200}\boldsymbol{a,b}\textbf{)}$$

The stoichiometry of the surface reactions are

$$2C + O_2 \rightarrow 2CO \qquad C + H_2O \rightarrow CO + H_2$$

$$1\,\text{kg} + \frac{16}{12}\,\text{kg} \rightarrow \qquad 1\,\text{kg} + \frac{18}{12}\,\text{kg} \rightarrow$$

and thus the mass transfer rate is

$$\dot{m}'' = n_{C,u} = -\left(\frac{12}{16} n_{O_2,s} + \frac{12}{18} n_{H_2O,s}\right)$$

Dividing by $\dot{m}''$,

$$1 = -\frac{3}{4} m_{O_2,ts} - \frac{2}{3} m_{H_2O,ts} \qquad \textbf{(10.200}\boldsymbol{c}\textbf{)}$$

For fast surface reactions, $m_{O_2,s} = m_{H_2O,s} = 0$, and Eqs. (10.200*a*,*b*,*c*) are three equations in the three unknowns $\dot{m}''$, $m_{O_2,ts}$, and $m_{H_2O,ts}$. Appropriate blowing factors must be used to obtain g_m from g_m^*.

At 1600 K, binary diffusion coefficients of O_2, CO, and H_2O in air are 1.52×10^{-4} m^2/s, 1.54×10^{-4} m^2/s, and 2.17×10^{-4} m^2/s, respectively. Thus, assuming equal effective binary diffusion coefficients for these species will give a good result using Eq. (10.199), but in principle, Eqs. (10.200) should give a more accurate result because equal diffusion coefficients for O_2 and H_2O are not assumed. However, accurate blowing factors are not easily obtained because four species are transferred across the *s*-surface, not just one, as considered in Sections 10.4 and 10.5.

2. *Gas absorption*. Many industrial processes employ absorption into a falling liquid film in order to remove, or *scrub*, an unwanted species from a gas stream. Figure 9.1 shows a packed column scrubber, and Exercise 9–30 considered absorption from a pure gas into a laminar falling film. In Chapter 11, gas scrubbers will be analyzed as two-stream mass exchangers using low mass transfer rate theory. Here we consider the application of high mass transfer rate theory to gas absorption to illustrate how transfer between two phases is handled by our general problem-solving procedure. Figure 10.35 shows conditions at a specific location in a packed column where species 1, for example, SO_2, is being absorbed into water. The local bulk concentrations are $m_{1,e}$ and $m_{1,b}$ in the gas and liquid phases, respectively. If the gas stream is saturated with water vapor so that no evaporation takes place, and the

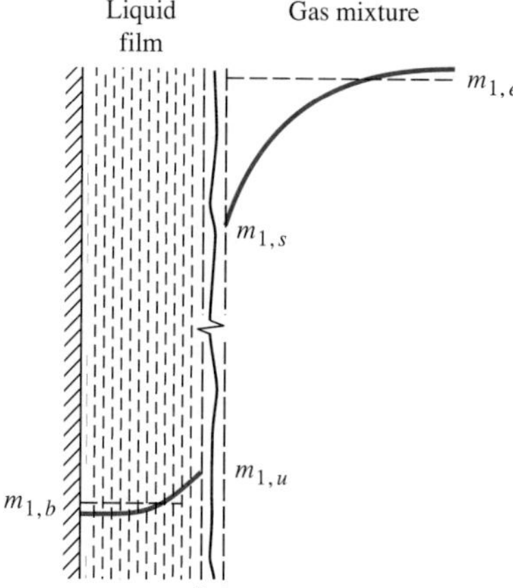

Figure 10.35 Concentration profiles for gas absorption into a falling liquid film in a packed tower.

carrier gas is highly insoluble in water, then $m_{1,t} \simeq 1$ and the gas absorption rate is

$$\dot{m}'' = g_{mG}\frac{m_{1,e} - m_{1,s}}{m_{1,s} - 1} = -g_{mL}\frac{m_{1,b} - m_{1,u}}{m_{1,u} - 1} \qquad \mathbf{(10.201}a,b\mathbf{)}$$

The subscripts G and L on the conductances refer to the gas and liquid phases, respectively. A required auxiliary relation is obtained from solubility data for species 1,

$$m_{1,u} = m_{1,u}(m_{1,s}, P, T) \qquad \mathbf{(10.201}c\mathbf{)}$$

Equations (10.201) are three equations in the three unknowns $\dot{m}''$, $m_{1,s}$, and $m_{1,u}$. Appropriate correlations for g^*_{mG} and g^*_{mL} will be given in Chapter 11.

Gas absorption problems are usually dealt with by chemical engineers, who prefer to work on a molar rather than a mass basis. Thus, in Chapter 11, gas scrubbers are analyzed on a molar basis, rather than the mass basis used in this analysis.

3. *Gas desorption in a direct-contact condenser.* In the open cycle for ocean thermal energy conversion (the *Claude* cycle), steam containing air is condensed onto cold seawater containing dissolved air (see Fig. 7.34). Conditions in the condenser allow air to desorb from the sea water as the steam condenses. Concentration and temperature profiles are shown in Fig. 10.36. We denote the bulk liquid state by subscript b, and gas and liquid side conductances by subscripts G and L, respectively. Species 1 and 2 are H_2O and air. We could formulate this problem as one involving simultaneous transfer of two species and heat. But we can simplify the problem by recognizing that the desorption of air affects the process only indirectly because the air concentration in the liquid phase and the air desorption rate are very small. That is, we can first solve the usual noncondensable gas problem assuming $m_{1,u} = m_{1,t} = 1$, to obtain $\dot{m}''$ and T_s, and then solve the desorption problem. Thus, we write

$$\dot{m}'' = g_{mG}\frac{m_{1,e} - m_{1,s}}{m_{1,s} - 1} = g_{hG}\frac{h_e - h_s}{h_s - h_t} \qquad \mathbf{(10.202}a,b\mathbf{)}$$

$$m_{1,s} = m_{1,s}(T_s, P) \qquad \mathbf{(10.202}c\mathbf{)}$$

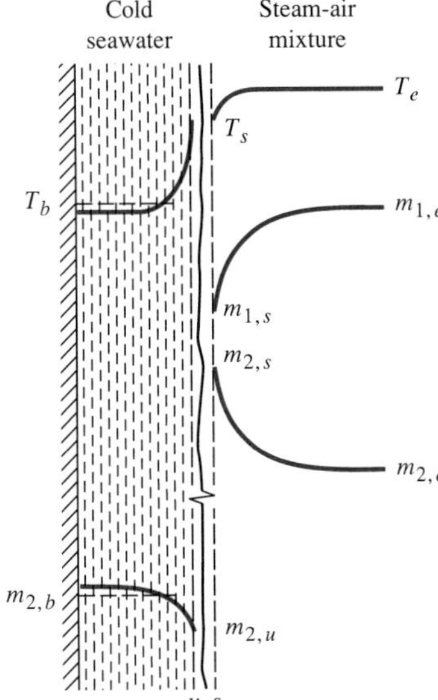

Figure 10.36 Temperature and concentration profiles for direct contact condensation in a condenser for Claude cycle ocean thermal energy conversion.

Evaluation of the heat transfer driving force is simplified by performing an energy balance between the s- and u-surfaces, writing $q_{\text{cond}} = h_{cL}(T_b - T_s)$, and then setting enthalpies equal to zero at T_s, to obtain

$$\mathscr{B}_h = \frac{h_e - h_{es}}{h_{\text{fg}} - (h_{cL}/\dot{m}'')(T_b - T_s)} = \frac{c_{pe}(T_e - T_s)}{h_{\text{fg}} - (h_{cL}/\dot{m}'')(T_b - T_s)}$$

($\dot{m}''$ is negative). The heat transfer coefficient h_{cL} is for heat transfer *into* the bulk liquid: it should not be confused with the heat transfer coefficient of film condensation, which is for heat transfer *across* the liquid film. In a falling-film direct-contact condenser, the film is usually in turbulent flow, and the required h_{cL} can be estimated from the heat transfer analog to the mass transfer correlation given as Eq. (11.45).

Having solved for $\dot{m}''$, m_1, and T_s, we then further simplify the problem by recognizing that air mass-fraction gradients in the liquid phase are very small because the air mass fractions themselves are very small. Hence, if we view the gas desorption process as liquid- and gas-phase resistances in series, the liquid-side resistance dominates and the gas-side resistance can be assumed to be negligible (see the discussion of the mass transfer Biot number in Section 9.4.4). Thus, we need only consider transfer in the liquid phase, and write

$$\dot{m}'' = -\mathscr{g}_{mL}\frac{m_{2,b} - m_{2,u}}{m_{2,u} - m_{2,t}} \qquad \textbf{(10.203}\textbf{\textit{a}}\textbf{)}$$

$$m_{2,u} = m_{2,u}(m_{2,s}, T_s, P), \quad \text{with } m_{2,s} = 1 - m_{1,s} \qquad \textbf{(10.203}\textbf{\textit{b}}\textbf{)}$$

Equation (10.203a) is then rearranged as

$$\dot{m}''m_{2,t} = n_{2,u} = \dot{m}''m_{2,u} - \mathcal{g}_{mL}(m_{2,u} - m_{2,b}) \qquad \textbf{(10.204)}$$

which can be solved for the air desorption rate $n_{2,u}$. An appropriate correlation for $\mathcal{g}_{mL}$ is given by Eq. (11.45).

4. Transpiration cooling with injectant dissociation. A porous wall is to be protected from a high-temperature air flow by transpiration of a dimer A_2. We wish to investigate the effect of dissociation of the dimer, $A_2 \leftrightharpoons 2A$, on heat transfer into the wall, and we will evaluate the driving force for three cases:

(i) Species A_2 does not dissociate at all.

(ii) Species A_2 dissociates wholly within the flow, that is, at temperatures higher than the surface temperature T_s.

(iii) Species A_2 dissociates wholly within the porous wall, that is, at temperatures below T_s.

To make the comparison, it will be sufficient to evaluate the heat transfer driving force $\mathcal{B}_h$ obtained by assuming unity Lewis numbers.

Case (i). Species A_2 is inert: thus, Eq. (10.179) for $\mathcal{B}_h$ can be used. Denoting A_2 as species 1,

$$\mathcal{B}_h = \frac{h_e - h_{es}}{h_{1,s} - h_{1,o}} = \frac{c_{pe}(T_e - T_s)}{c_{p1}(T_s - T_o)} \qquad \textbf{(10.205)}$$

where constant species specific heats have been assumed for convenience, subscript o refers to the reservoir state, and radiation has been ignored.

Case (ii). Let A be species 2 and the air (assumed to be inert) species 3. From Eq. (10.190),

$$\mathcal{B}_h = \frac{h_e - h_s}{h_s - h_t}$$

An energy balance between the s- or u-surface and an o-surface in the reservoir shows that $h_t = h_o$; hence,

$$\mathcal{B}_h = \frac{h_e - h_s}{h_s - h_o} \qquad \textbf{(10.206)}$$

By definition, $h_i = h_i^0 + \int_{T^0}^{T} c_{pi}\, dT$

$$h = \sum m_i h_i, \quad i = 1,\ 2,\ 3$$

If we choose the datum temperature $T^0 = T_s$ and $h_i(T_s) = 0$ for species 1 and 3 (i.e., $h_1^0 = h_3^0 = 0$), then

$$h_1 = \int_{T_s}^{T} c_{p1}\, dT; \qquad h_2 = h_2^{T_s} + \int_{T_s}^{T} c_{p2}\, dT; \qquad h_3 = \int_{T_s}^{T} c_{p3}\, dT$$

and the e-, s-, and o-state enthalpies are

$$h_e = \sum m_{i,e} h_{i,e} = (1)(h_{3,e})$$

$$= \int_{T_s}^{T_e} c_{p3}\, dT \simeq c_{pe}(T_e - T_s) \quad (\text{since } c_{p3} = c_{pe})$$

$h_s = 0$ (since $m_{2,s} = 0$: A_2 dissociation takes place wholly within the flow, i.e., at temperatures above T_s)

$$h_o = \int_{T_s}^{T_o} c_{p1}\, dT = h_{1,o} - h_{1,s} \simeq c_{p1}(T_o - T_s)$$

Substituting in Eq. (10.206),

$$\mathscr{B}_h = \frac{c_{pe}(T_e - T_s)}{c_{p1}(T_s - T_o)}$$

which is the same result as for case (i); hence, the required $\dot{m}''$ and heat transfer to the surface are the same.

Case (iii). Again we choose $T^0 = T_s$, but now we set $h_i(T_s) = 0$ for species 2 and 3:

$$h_1 = h_1^{T_s} + \int_{T_s}^{T} c_{p1}\, dT; \qquad h_2 = \int_{T_s}^{T} c_{p2}\, dT; \qquad h_3 = \int_{T_s}^{T} c_{p3}\, dT$$

$h_e \simeq c_{pe}(T_e - T_s)$, as before

$h_s = 0$ (since $m_{1,s} = 0$: A_2 dissociation takes place wholly within the wall, i.e., at temperatures below T_s)

$$h_o = h_1^{T_s} + \int_{T_s}^{T_o} c_{p1}\, dT \simeq h_1^{T_s} + c_{p1}(T_o - T_s)$$

where $h_1^{T_s}$ is the negative of Δh_d, the heat of dissociation of A_2 at temperature T_s. Substituting in Eq. (10.206),

$$\mathscr{B}_h = \frac{c_{pe}(T_e - T_s)}{c_{p1}(T_s - T_o) + \Delta h_d} \quad (\Delta h_d \text{ is positive}) \tag{10.207}$$

Thus, if the dissociation takes place within the wall, the heat taken up reduces the driving force and injection rate required to balance the surface heating. In contrast, if the dissociation takes place within the boundary layer, there is no effect on heat transfer to the surface. Of course, dissociation in the boundary layer does lower temperatures, and there are second-order effects via temperature-dependent properties. Also, the Lewis number is seldom exactly equal to unity.

5. *Evaporation of a multicomponent droplet.* Liquid hydrocarbon fuels are usually mixtures of a number of components, with different vapor pressures and other thermophysical properties. When such droplets evaporate, the more volatile components

distill off more rapidly than the less volatile components, and the composition of the vapor mixture produced changes with time. Similar phenomena occur in distillation columns of oil-refining operations (and in the production of fine cognac). To illustrate how such problems can be analyzed, we consider the evaporation of an n-component droplet into an inert gas, I. To simplify the problem, we will assume that there is good mixing within the droplet so that the droplet composition and temperature are uniform. When this assumption is made, chemical engineers call the process *batch distillation*; the validity of the assumption will be discussed later. The rate of evaporation is given by

$$\dot{m}'' = g_{mi}\mathcal{B}_{mi} = g_h\mathcal{B}_h; \qquad i = 1, 2, \ldots, n$$

$$\mathcal{B}_{mi} = \frac{m_{i,e} - m_{i,s}}{m_{i,s} - m_{i,t}}; \qquad \mathcal{B}_h = \frac{h_e - h_s}{h_s - h_t} \tag{10.208}$$

$$m_{i,s} = m_{i,s}(m_{i,u}, T_s, P)$$

usually in the form of *Raoult's law*. (Henry's law is a limit form of Raoult's law.) The heat transfer driving force assumes a unity Lewis number, and is simplified by first taking the enthalpy datum state such that species enthalpies are zero at temperature T_s; then

$$\mathcal{B}_h = \frac{h_e - h_{es}}{-h_t^{T_s}} \tag{10.209}$$

An energy balance on a control volume between the u- and s-surfaces allows the transferred-state enthalpy to be evaluated:

$$\dot{m}''h_t^{T_s} = \sum n_i h_{i,u}^{T_s} + q_{\text{cond}}$$

$$h_t^{T_s} = \sum m_{i,t}(-h_{\text{fg}i}) + (q_{\text{cond}}/\dot{m}'')$$

Assuming constant species specific heats and substituting in Eq. (10.209),

$$\mathcal{B}_h \simeq \frac{c_{pe}(T_e - T_s)}{\sum m_{i,t}h_{\text{fg}i} - (q_{\text{cond}}/\dot{m}'')} \tag{10.210}$$

For batch distillation we can use a lumped-capacity model for the droplet of radius R, to write

$$\frac{d}{dt}\left(\frac{4}{3}\pi R^3 \rho_l h\right) = -(4\pi R^2)q_{\text{cond}} \tag{10.211}$$

$$\frac{d}{dt}\left(\frac{4}{3}\pi R^3 \rho_l m_i\right) = -(4\pi R^2)m_{i,t}\dot{m}'', \qquad i = 1, 2, \ldots, n \tag{10.212}$$

with initial conditions

$$t = 0: \quad R = R_0,\ T_s = T_0,\ m_i = m_{i,0} \tag{10.213}$$

Notice that $h = h_u$ and $m_i = m_{i,u}$, since for batch distillation the liquid temperature and composition are assumed to be uniform. Equations (10.211) and (10.212) must be solved numerically, with Eqs. (10.208) solved simultaneously at each time step. However, Eqs. (10.208) are highly nonlinear, and care must be taken to develop a stable and accurate numerical procedure. Gas-phase properties can be evaluated using Hubbard's 1/3 rule extended to multicomponent evaporation.

The batch distillation model is a limiting case corresponding to perfect mixing inside the droplet: the opposite limit is a model that assumes a stationary liquid, and heat and mass transport in the droplet is by conduction and diffusion. Results for these two limiting cases for a binary droplet are compared by Landis and Mills [23]. If there is relative motion between the droplet and the surrounding gas phase, a flow circulation can be induced in the droplet. Flow circulation models, for example, [24,25], give results in between those given by the batch distillation and conduction/diffusion models. However, under some circumstances, droplets may oscillate, with the predominant mode being a shape oscillation. Transport rates in oscillating droplets are much larger than in circulating droplets [26,27,28], and in such situations the batch distillation model may be more appropriate. Notice that our analysis did not require an expression for the heat flux across the u-surface, q_{cond}. The foregoing discussion suggests that it is difficult to specify q_{cond} because the liquid-gas interface does not behave like a solid wall. When a droplet oscillates, the flow inside the droplet is complex and poorly understood: such matters are the subject of current research.

10.7 CLOSURE

In this chapter the theory of simultaneous diffusion and convection was developed, with applications including heatpipes and volatile liquid fuel droplet combustion. High mass transfer rate convection was introduced, initially with emphasis on engineering problem solving. Applications included transpiration and sweat cooling, and the effect of air leaks on condenser performance. Rigorous analysis of high mass transfer rate convection requires the numerical solution of coupled nonlinear governing differential equations. As an example, the solution for the constant-property, forced-convection laminar boundary layer on a flat plate was obtained. The advantage of an enthalpy formulation for simultaneous heat and mass transfer was demonstrated, with special emphasis placed on gaining an understanding of the role played by energy transport due to interdiffusion of species in a mixture. The discussion was rather long and involved, but mastery of these concepts is essential before the student proceeds to the research literature.

The chapter concluded with a general problem-solving procedure for steady convective heat and mass transfer problems. The procedure is based on a generalization of the various analyses presented earlier in the chapter, and it was not necessary to develop the theory underlying the procedure from first principles. Five examples of the solution procedure illustrated its use for more complicated heat and mass transfer problems.

REFERENCES

1. Hubbard, G. L., Denny, V. E., and Mills, A. F., "Droplet evaporation: Effects of transients and variable properties," *Int. J. Heat Mass Transfer*, 18, 1003–1008 (1975).

2. Kent, J. C., "Quasi-steady diffusion controlled droplet evaporation and condensation," *Appl. Sci. Res. Ser. A*, 28, 315–359 (1973).

3. Law, C. K., "Quasi-steady droplet vaporization theory with property variations," *Physics of Fluids*, 18, 1426–1432 (1975).

4. Godsave, G. A. E., "Studies of the combustion of drops in a fuel spray—the burning of single drops of fuel," *Fourth Symposium on Combustion*, Williams & Wilkins, Baltimore, p. 818 (1953).

5. Spalding, D. B., "The combustion of liquid fuels," *Fourth Symposium on Combustion*, Williams & Wilkins, Baltimore, p. 847 (1953).

6. Law, C. K., and Williams, F. A., "Kinetics and convection in the combustion of alkane droplets," *Combustion and Flame*, 19, 393–405 (1972).

7. Nuruzzaman, A. S. M., Martin, G. F., and Hedley, A. B., "Combustion of single droplets and simplified spray systems," *J. Inst. Fuel*, 44, 38–54 (1971).

8. Saitoh, T., Yamazaki, K., and Viskanta, R., "Effect of thermal radiation on transient combustion of a fuel droplet," *J. Thermophysics and Heat Transfer*, 7, 94–100 (1993).

9. Mills, A. F., and Wortman, A., "Two-dimensional stagnation point flows of binary mixtures," *Int. J. Heat Mass Transfer*, 15, 969–987 (1972).

10. Wortman, A., and Mills, A. F., "Accelerating compressible laminar boundary layers of binary gas mixtures," *Archiwum Mechaniki Stosowanej*, 26, 479–497 (1974).

11. Wortman, A., "Mass transfer in self-similar boundary layer flows," Ph.D. dissertation, School of Engineering and Applied Science, University of California, Los Angeles (1969).

12. Landis, R. B., and Mills, A. F., "The calculation of turbulent boundary layers with foreign gas injection," *Int. J. Heat Mass Transfer*, 15, 1905–1932 (1972).

13. Landis, R. B., "Numerical solution of variable property turbulent boundary layers with foreign gas injection," Ph.D. dissertation, School of Engineering and Applied Science, University of California, Los Angeles (1971).

14. Hartnett, J. P., and Eckert, E. R. G., "Mass transfer cooling in a laminar boundary layer with constant fluid properties," *Trans. ASME*, 79, 247–254 (1957).

15. Knuth, E. L., "Use of reference states and constant property solutions in predicting mass-, momentum-, and energy-transfer rates in high speed laminar flows," *Int. J. Heat Mass Transfer*, 6, 1–22 (1963).

16. Knuth, E. L., and Dershin, H., "Use of reference states in predicting transport rates in high-speed turbulent flows with mass transfer," *Int. J. Heat Mass Transfer*, 6, 999–1018 (1963).

17. Baron, J. R., "The binary-mixture boundary layer associated with mass transfer cooling at high speeds," MIT Naval Supersonic Laboratory, Tech. Report 190 (1956).

18. Gross, J. F., Hartnett, J. P., Masson, D. J., and Gazley, C., Jr., "A review of binary laminar boundary layer characteristics," *Int. J. Heat Mass Transfer*, 3, 198–221 (1961).

19. Simon, H. A., Hartnett, J. P., and Liu, C. A., "Transpiration cooling correlations of air and non-air free streams," AIAA Paper no. 68-758, presented at the AIAA 3rd Thermophysics Conference, Los Angeles, June 1968.

20. Spalding, D. B., *Convective Mass Transfer*, McGraw-Hill, New York (1963).

21. Spalding, D. B., "A standard formulation of the steady convective mass-transfer problem," *Int. J. Heat Mass Transfer*, 1, 192–207 (1960).

22. Kays, W. M., and Crawford, M. E., *Convective Heat and Mass Transfer*, 3rd ed., McGraw-Hill, New York (1993).

23. Landis, R. B., and Mills, A. F., "Effect of internal diffusional resistance on the evaporation of binary droplets," Paper B.7.9, 5th International Heat Transfer Conference, Tokyo (1974).

24. Chiang, C. H., Raja, M. S., and Sirignano, W. A., "Numerical analysis of a convecting, vaporizing fuel droplet with variable properties," *Int. J. Heat Mass Transfer*, 35, 1307–1324 (1992).

25. Renksizbulut, M., and Bussman, M., "Multicomponent droplet evaporation at intermediate Reynolds numbers," *Int. J. Heat Mass Transfer*, 36, 2827–2835 (1993).

26. Hijikata, K., Mori, Y., and Kawaguchi, S., "Direct contact condensation of vapor to falling cooled droplets," *Int. J. Heat Mass Transfer*, 27, 1631–1640 (1984).

27. Brunson, R. J., and Welleck, R. M., "Mass transfer within an oscillating liquid droplet," *Can. J. Chem. Eng.*, 48, 267–274 (1970).

28. Mills, A. F., and Hoseyni, M. S., "Diffusive deposition of aerosols in a rising bubble," *Aerosol Science and Technology*, 8, 103–105 (1988).

29. Smirnov, V. A., Verevochkin, G. E., and Brdlick, P. M., "Heat transfer between a jet and a held plate normal to flow," *Int. J. Heat Mass Transfer,* 2, 1–7 (1961).

30. Rose, J. W., "Condensation of a vapor in the presence of a noncondensing gas," *Int. J. Heat Mass Transfer,* 12, 233–237 (1969).

EXERCISES

10–1. Show that in a binary mixture

$$\text{(i) } \nabla x_1 = \frac{M^2}{M_1 M_2}\nabla m_1$$

$$\text{(ii) } \mathbf{j}_1 = \frac{M_1 M_2}{M}\mathbf{J}_1^*$$

10–2. The Chapman-Enskog kinetic theory of gases shows that for ordinary diffusion in a binary mixture, the mass flux relative to the mass average velocity is

$$\mathbf{j}_1 = -\frac{\mathcal{N}^2 m_1 m_2}{\rho}\mathscr{D}_{12}\nabla n_1$$

where m_1 and m_2 are masses of the molecules, and $\mathcal{N}$ and n are number density and number fraction, respectively. Derive from this relation Fick's law in the forms

$$\mathbf{j}_1 = -\rho\mathscr{D}_{12}\nabla m_1; \qquad \mathbf{J}_1^* = -c\mathscr{D}_{12}\nabla x_1$$

10–3. A gas mixture at 1 bar pressure and 300 K contains 20% H_2, 40% O_2, and 40% H_2O by weight. The absolute velocities of each species are 10 m/s, −2 m/s, and 12 m/s, respectively, all in the direction of the z-axis. Calculate $\mathbf{v}$ and $\mathbf{v}^*$ for the mixture and for each species, $\mathbf{n}_i$, $\mathbf{j}_i$, $\mathbf{N}_i$, and $\mathbf{J}_i^*$.

10–4. An end-condenser water heatpipe, 1 m long and with 2 cm inside diameter, is intended to carry a heat load of 100 W when the evaporator temperature is 100°C. If 30 μg of air leak into the heatpipe and the condenser cannot operate below 95°C, determine the new power rating of the heatpipe.

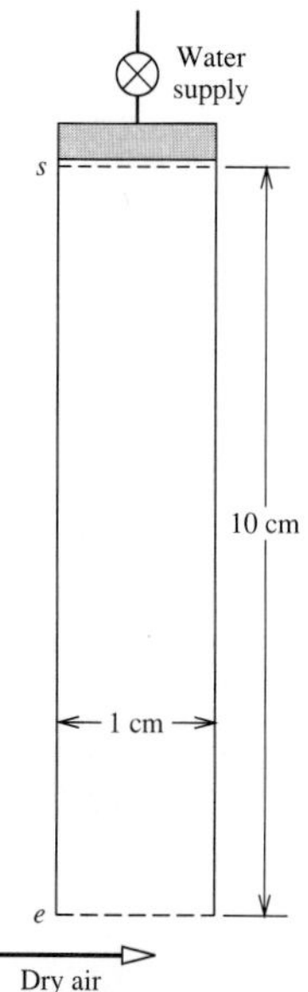

10–5. Water is supplied to a porous plug at the top of a 10 cm–long, 1 cm–diameter tube. The lower end is open, and dry air at 1 atm flows gently over the tube mouth. The water supply rate is controlled to keep the plug surface wet. The complete system is temperature-controlled by a thermostat and can operate in the range 290 K to 370 K. Prepare graphs of the quantities $x_{H_2O,s}$, $n_{H_2O,s}$, $j_{H_2O,s}$, $(x_{air})_{lm}$ as a function of temperature and discuss their behavior.

10–6. Water is contained in the bottom of a 15 cm–long vertical tube maintained at 295 K. Pure helium blows gently over the mouth of the tube at 1 atm pressure. At the s-surface, determine the absolute molar flux and absolute velocity of water vapor. Also determine the water vapor velocity relative to the molar average velocity.

10–7. Calculate the Schmidt number of a saturated mixture of water vapor and air at 300 K containing a trace amount of air.

10–8. In a condenser to recover solvent vapors, 9 kmol of steam (species 1) are condensed for each kmol of solvent (species 2). At the condensate surface $x_{2,s} = 0.85$. The total condensation rate of steam and solvent vapors is 0.002 kmol/m^2 s. Using a one-dimensional model, find the mole fractions x_1 and x_2 as functions of distance from the condensate surface. The total pressure is 3 atm, $T_s = 320$ K, and $\mathscr{D}_{12} = 10 \times 10^{-6}$ m^2/s. (*Hint:* Neither component is stationary.)

10–9. Boron is under consideration as fuel for a solid propellant rocket. Boron particles in air ignite at 1900–2300 K, and, provided the temperature is not too high, the oxidation is heterogeneous to form BO and is diffusion-controlled with $m_{O_2,s} \simeq 0$. In an experimental study, 50 μm–diameter boron particles are ignited in an air stream at 1000 K and 1 atm. Estimate the lifetime of the particles if the particle temperature is measured to be approximately 3000 K. The density of boron at 3000 K is 2370 kg/m^3.

10–10. Dust particles impinging on a graphite heat shield of a hypersonic vehicle can cause carbon particles to be ejected into the high-temperature boundary layer surrounding the vehicle. Such particles may ignite and burn, and in the temperature range of concern (1600–3000 K) diffusion-controlled oxidation to carbon monoxide may be assumed. For an air temperature of 1500 K, estimate particle temperatures and lifetimes for particles with diameters of 1 and 10 μm (see Example 9.11). Take $\varepsilon = 0.9$ and $\Delta h_c = 9.21 \times 10^6$ J/kg carbon.

10–11. The analysis of Section 10.3.3 gives the burning rate of the carbon particle but does not give the location of the flame. If we assume that the flame is thin enough to be represented by a surface at $r = R_f$, the reaction $2CO + O_2 \rightarrow 2CO_2$ can be treated as a surface reaction occurring between the u' and s' surfaces shown in the figure. To determine the location of the flame, first show that the mass transfer rates across the s and s' surfaces are given by

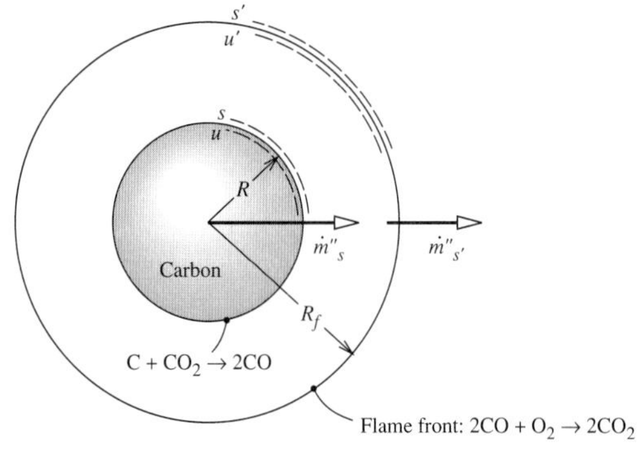

$$\dot{m}_s'' = \frac{R_f}{R_f - R}\frac{\rho\mathscr{D}_{im}}{R}\ln\left[1 + \frac{m_{i,u'} - m_{i,s}}{m_{i,s} - (n_{i,s}/\dot{m}_s'')}\right], \qquad i = CO, CO_2$$

$$\dot{m}_{s'}'' = \frac{\rho\mathscr{D}_{im}}{R_f}\ln\left[1 + \frac{m_{i,e} - m_{i,s'}}{m_{i,s'} - (n_{i,s'}/\dot{m}_{s'}'')}\right], \qquad i = O_2, CO_2$$

Then assume equal effective diffusion coefficients for O_2, CO, and CO_2 and apply mass continuity, equilibrium, and stoichiometric relations at the u-, s-, u'-, and s'-surfaces to relate the unknown species fluxes and concentrations. Hence show that $m_{CO,s} = 0.346$, $m_{CO_2,s'} = 0.293$, $R_f/R = 1.922$, and $\dot{m}_s'' = 0.160\rho\mathscr{D}/R$ (as obtained in Section 10.3.3).

10–12. A spherical droplet of radius R, species 1, evaporates into a stagnant surrounding gas, species 2.

(i) Show that the molar evaporation rate is given by

$$N_{1,s} = \frac{c\mathscr{D}_{12}}{R}\ln\left(1 + \frac{x_{1,e} - x_{1,s}}{x_{1,s} - 1}\right)$$

by solving the governing differential equation in spherical coordinates. The s- and e-surfaces are at the droplet surface and infinity, respectively.

(ii) Show how the preceding result can be obtained from Eq. (10.29) by using the stagnant film model.

(iii) The corresponding result on a mass basis is Eq. (10.58). Discuss why the molar-based result might be preferred for many calculations.

(iv) For evaporation of water into dry air, plot $N_{1,s}R/c\mathscr{D}_{12}$ versus T_s for $0°\text{C} < T_s < 100°\text{C}$.

10–13. Following the analysis of droplet combustion in Section 10.3.5, show that the quasi-steady evaporation rate of a spherical droplet into stagnant surrounding gas can be expressed as

$$\dot{m}'' = \frac{k/c_p}{R}\ln\left(1 + \frac{c_{pe}(T_e - T_s)}{h_{fg}}\right)$$

Carefully list all assumptions made concerning thermophysical properties.

10–14. For a water droplet evaporating into air at 430 K, 1 atm, and 10% RH, compare estimates of the evaporation rate given by

(i) the result of Exercise 10–13.
(ii) Eq. (10.63).
(iii) the result of a constant-property heat transfer analysis.

In all cases use the 1/3 rule for evaluating properties that were assumed constant in the analysis. Since the analyses assume adiabatic vaporization, approximate T_s as the thermodynamic wet-bulb temperature from PSYCHRO.

10–15. A small droplet of water is entrained in a high-temperature air flow at 1 atm, 430 K, and 10% RH. After a short initial transient, heat conduction into the droplet becomes negligible, and the droplet evaporates adiabatically (see Example 9.10). In order to calculate the subsequent evaporation rate, we require the droplet temperature. Compare the following three estimates:

(i) using PSYCHRO.
(ii) using Eq. (9.58).

(iii) using high mass transfer rate theory by combining the results of Exercises 10–12 and Eq. (10.63) with the 1/2 rule reference-state scheme for c and k.

Discuss the basis of each estimate.

10–16. Estimate the lifetime of a 1 mm–diameter n-pentane droplet burning in air at 1 atm pressure and 300 K, at near zero gravity. For n-pentane (n-C_5H_{12}), take $\rho_l = 605$ kg/m^3, $h_{fg} = 3.57 \times 10^5$ J/kg, $\Delta h_c = 4.50 \times 10^7$ J/kg, and $T_{BP} = 309$ K. The flame temperature is $T_f = 2310$ K. At the reference temperature of $(1/2)(T_f + T_{BP}) = 1310$ K, property values for n-pentane vapor include $k = 0.179$ W/m K, $c_p = 4320$ J/kg K.

10–17. Estimate the lifetime of a 1 mm–diameter n-octane droplet burning in air at 1 atm and 300 K, at near zero gravity. For n-octane (n-C_8H_{18}), take $\rho_l = 611$ kg/m^3, $h_{fg} = 3.03 \times 10^5$ J/kg, $\Delta h_c = 4.44 \times 10^7$ J/kg, and $T_{BP} = 399$ K. The flame temperature is $T_f = 2320$ K. At the reference temperature of $(1/2)(T_f + T_{BP}) = 1360$ K, property values of n-octane vapor include $k = 0.113$ W/m K, $c_p = 4280$ J/kg K.

10–18. Forced diffusion plays a key role in the operation of metal-ion lasers, such as the helium-cadmium laser or "blue laser" that is widely used in copying machines and computer printers.

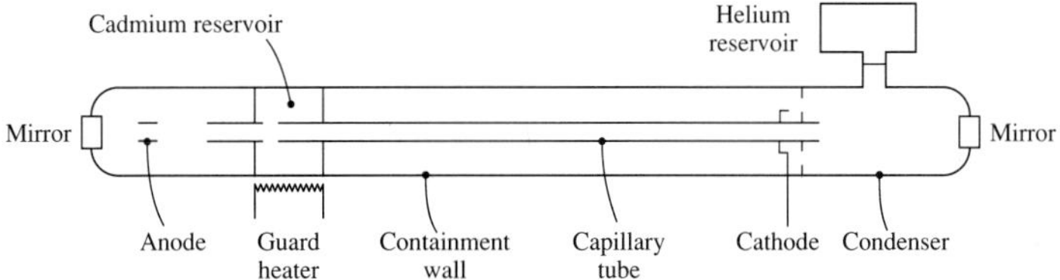

A typical configuration is shown in the figure. The process by which the Cd ions are excited is not well understood, but the relationships between the design parameters required for satisfactory operation are known. The lasing action takes place in the positive column of a He gaseous discharge contained in a capillary tube, where the electric field strength E [V/m] is constant. The Cd vapor pressure is three orders of magnitude less than the He pressure. The electric field causes Cd ions to be transported by forced diffusion from the solid cadmium reservoir to the cathode end of the capillary tube, which is cooled to act as a condenser for the Cd ions. The anode serves to collect electrons at the other end of the tube. With the Cd ions denoted as species 1, the molar flux due to forced diffusion is

$$J_1^{*F} = \mu_1 E c x_1$$

where μ_1 [m^2/V s] is the *mobility* of the ions (and is related to the diffusion coefficient by the Nernst-Einstein equation $\mu_1 = e\mathscr{D}_{1m}/kT$). Notice that $\mu_1 E$ has units of velocity and is called the *cataphoretic velocity*. In this situation it is of the order of 3000 m/s, which is very large indeed. The cadmium vapor is only partially ionized. If we denote the cadmium neutrals as species 2, and

$x = x_1 + x_2$, then we can write $x_1 = \varepsilon x$, where ε is the *ionization fraction* and is of the order of 0.01 in this situation. Cadmium ions are continuously created and destroyed by the collision processes in the plasma of the positive column at a rate $\dot{R}_1''' = -\dot{R}_2'''$, maintaining a constant value of ε.

To analyze steady transport of Cd along the capillary tube, we can assume that radial-direction diffusion phenomena are uncoupled from axial-direction diffusion phenomena. Thus, a one-dimensional analysis is adequate, with concentrations taken as average values over the tube cross section. Proceed as follows.

1. Write down expressions for the absolute fluxes N_1 and N_2.
2. Write down appropriate species conservation equations for species 1 and 2.
3. During steady operation, the helium is stationary. By summing the two species conservation equations and neglecting convection, obtain the equation governing the distribution of total cadmium along the tube as

$$\frac{d}{dz}\left(-c\mathscr{D}\frac{dx}{dz} + \mu_1 E c \varepsilon x\right) = 0$$

   where $\mathscr{D}_{1m} = \mathscr{D}_{2m} = \mathscr{D}$ has been assumed.

4. Let $z^* = z/L$, where L is the length of capillary tube under consideration. Then show that x is the solution of

$$\frac{d}{dz^*}\left(-\frac{dx}{dz^*} + \beta x\right) = 0$$

   where

$$\beta = \frac{\mu_1 E \varepsilon}{\mathscr{D}/L} = \frac{\text{Cataphoretic velocity} \times \text{Ionization fraction}}{\text{Characteristic diffusion velocity}}$$

   In He-Cd lasers, β is of the order of 10^2.

5. Show that the solution for the cathode end, $0 < x < L_1$ is $x = \text{constant} = x_0$, where $x_0 = x_{\text{sat}}(T_0, P)$ for a cadmium reservoir temperature T_0. Also show that $N_c = N_1 + N_2 = \mu_1 E \varepsilon c x_0$ is the cadmium transport to the cathode-end condenser.

6. If the anode region is maintained at a temperature T_a, with $x_a = x_{\text{sat}}(T_a, P)$, show that the solution for the anode end, $0 < z < -L_2$, gives the cadmium flux into the anode region as

$$N_a = \left(\frac{c\mathscr{D}}{L}\right)\left(\frac{\beta}{1 - e^{-\beta}}\right)\left(x_a - x_0 e^{-\beta}\right)$$

The results obtained in steps 5 and 6 above are important design relations for the laser (see Exercise 10–19).

10–19. A helium-cadmium laser of the type analyzed in Exercise 10–18 has a 2 mm–I.D. capillary tube to confine the discharge. Nominal values of relevant parameters include a total pressure of 467 Pa, a cadmium reservoir temperature of 500 K, a positive column-temperature of 700 K, an electric field of 3300 V/m, and an ionization fraction of 0.01.

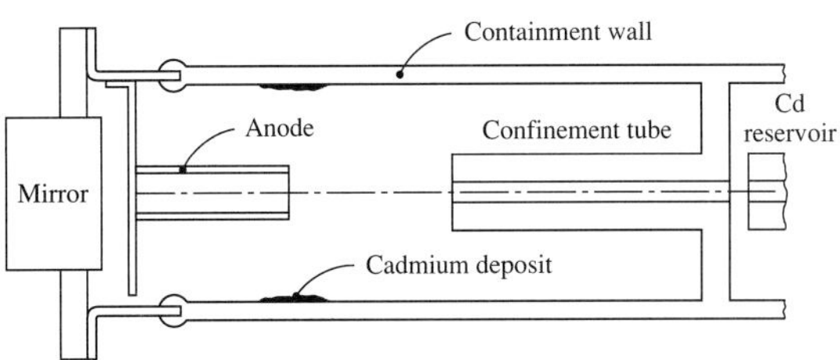

(i) Determine the Cd charge required for a laser lifetime of 10,000 h.

(ii) If the anode-end mirror temperature must not exceed 370 K to avoid damage to the coatings, determine the length of the anode-end confinement capillary tube required for zero Cd flow to the mirror.

(iii) If the laser experiences a "hot shutdown," for example, if the power supply is abruptly interrupted, the field is disestablished while the reservoir is still at 700 K. Make an engineering estimate of the amount of Cd deposited in the anode region for an effective one-dimensional diffusion path of 3.0 cm. Based on experimental data, the time constant for the temperature decay of the reservoir is approximately 3 min.

(iv) When the laser is switched on again, Cd will be pumped out of the anode region. Determine the time required to pump out the amount deposited during the hot shutdown of part (iii) as a function of anode-region temperature T_a in the range 350–400 K. The length of the anode-end confinement capillary tube is 2.5 cm.

Required data for cadmium and the cadmium-helium mixture are:

$$\log_{10} P_{\mathrm{Cd}} = 8.56 - (5720/T), \text{ for } P \text{ in torr, } T \text{ in kelvins}$$

$$\mathscr{D}_{\mathrm{CdHe}} = 5.77 \times 10^{-7} T^{1.75} \text{ m}^2/\text{s at } P = 467 \text{ Pa}$$

$$\mu_{\mathrm{CdHe}} = 6.12 \times 10^{-3} T^{0.75} \text{ m}^2/\text{V s at } P = 467 \text{ Pa}$$

10–20. (i) Show that a stagnant film analysis of high-rate mass transfer from a sphere leads to a result that is identical to the Couette-flow model result, Eq. (10.99).

(ii) Repeat for a long cylinder.

10–21. Rework Example 9.8 for a bath temperature of 360 K.

10–22. A strip of wet textile passes through the central section of a dryer; in this section the temperature of the textile is substantially independent of distance and the drying rate is constant. An air-steam mixture is blown on the top surface of the textile through a perforated-plate distributor. The mixture is at 420 K and 1 atm pressure and contains 8% of water vapor by mass. When the mass velocity of the mixture ρV is 0.8 kg/m^2 s, design charts for the distributor indicate a heat transfer Stanton number of 0.08. Calculate the rate at which water is removed from the textile. (See Exercise 9–46.)

10–23. In a paper-drying unit the wet paper is on the outside of a rotating drum, the inside wall of which is heated by condensing saturated steam at 3 atm pressure. Air at 1 atm pressure, 310 K, and 40% relative humidity is blown onto the paper at a mass velocity ρV of 0.025 kg/m^2 s: design data indicate a mass transfer Stanton number of $\mathrm{St}_m^* = 0.060$. If the overall heat transfer coefficient from the heating steam to the paper surface is estimated to be 75 W/m^2 K, calculate the paper drying rate. (See Exercise 9–47.)

10–24. A wet- and dry-thermocouple installation is to be used to measure the composition of an air-steam mixture under the following conditions: pressure 0.9–1.3 bar, temperature 330–380 K, and relative humidity 10–100%. Write down a set of relations that will allow determination of the mass fraction of water vapor. Discuss any assumptions you make.

10–25. In a turbulent flow over a surface, we can model the turbulent contribution to species diffusive transport using an eddy diffusivity model: we then write Fick's law as

$$j_i = -\rho(\mathscr{D}_{im} + \varepsilon_D)\, dm_i/dy$$

where ε_D is the eddy diffusivity of a chemical species and is a strong function of y, the distance normal to the surface (see Section 5.5). Show that the Couette-flow relation

$$\frac{\mathscr{g}_m}{\mathscr{g}_m^*} = \frac{\ln(1 + \mathscr{B}_m)}{\mathscr{B}_m}; \qquad \mathscr{B}_m = \frac{m_{i,e} - m_{i,s}}{m_{i,s} - n_{i,s}/\dot{m}''}$$

for transfer of an inert species i remains valid if $\int_0^\delta dy/\rho(\mathscr{D}_{im} + \varepsilon_D)$ is independent of $\dot{m}''$.

10–26. Oxyacetylene cutting of steel plate involves heating the plate with an oxygen-acetylene flame to a temperature high enough for the iron to burn: the heat release then sustains the process, and the acetylene supply can be shut off. The cutting is then a process of combustion of iron in almost pure oxygen. The oxides are molten and flow away under the action of gravity and shear forces: there are no volatile oxides to diffuse back into the oxygen flow. Also, at the high temperatures prevailing, there is chemical equilibrium at the s-surface with $m_{O_2,s} \simeq 0$. In a particular cutting process, a jet of 99% oxygen and 1% inerts by mass exits a 4 mm–diameter nozzle held 2 mm from the plate. If the jet velocity is 15 m/s, determine the rate at which iron is consumed. An appropriate correlation for the Sherwood number is [29]

$$\mathrm{Sh}_L = 0.55\,\mathrm{Re}_L^{1/2}\,\mathrm{Sc}^{1/3}$$

where L is the distance of the nozzle from the plate. Assume a plate temperature of 1700 K, a Schmidt number for O_2 and inerts of 0.8, and a Fe_2O_3 oxide.

10–27. A 2.5 cm–diameter cylinder is located in a 2 m/s cross-flow of air at 1 atm and 400 K. The wall in the front stagnation region is porous and is to be maintained at 300 K by air injection at an s-surface velocity of 0.1 m/s. Determine the air reservoir temperature required to maintain a steady state. Neglect radiation heat transfer, and use the improved blowing factors of Section 10.4.2.

10–28. Nitrogen at 1 atm and 800 K flows along a porous flat plate at 100 m/s. A mixture of 50% N_2 and 50% H_2 by mass is transpired through the plate from a reservoir at 300 K. At a location 0.5 m from the leading edge determine

(i) the transpiration rate required to maintain the plate at 400 K.
(ii) the mass fraction of H_2 at the s-surface.

Assume a turbulent boundary layer, and ignore thermal radiation. Use the simple high mass transfer rate theory of Section 10.4.1.

10–29. In a laboratory experiment, dry air at 1350 K and 1 atm flows at 10 m/s over a 3 cm–diameter sweat-cooled cylinder. If the coolant water is supplied from a plenum chamber at 300 K, determine the water supply rate required to balance evaporation at the stagnation line, and the corresponding steady-state surface temperature. Neglect radiation heat transfer.

(i) Use the simple high mass transfer rate theory of Section 10.4.1.
(ii) Use the improved blowing factors given in Section 10.4.2.

10–30. Repeat Exercise 10–29 for the stagnation point of a 3 cm–diameter sphere.

10–31. In a laboratory experiment, dry air at 1350 K and 1 atm flows at 10 m/s along a sweat-cooled flat plate. If the coolant water is supplied from a plenum chamber at 300 K, determine the water supply rate required to balance evaporation at a location 10 cm from the leading edge, and the corresponding surface temperature.

(i) Use the simple high mass transfer rate theory of Section 10.4.1.
(ii) Use the improved blowing factors given in Section 10.4.2.

Assume a laminar boundary layer and negligible radiation heat transfer.

10–32. Repeat Exercise 10–31 for a turbulent boundary layer starting at the leading edge.

10–33. Estimate the lifetime of salt crystals suspended in pure water at 300 K. The crystals can be modeled as spheres of diameter 0.4 mm and with a density of 2170 kg/m^3.

10–34. Air at 1400 K, 1 atm flows across a 2 cm–diameter transpiration-cooled cylinder at 100 m/s. Helium is supplied from a plenum chamber at 300 K. Determine the helium injection rate required to maintain the stagnation line surface temperature at 1000 K. Allow for radiation heat transfer into black surrounds at 1400 K for a surface emittance of 0.7. The simple high mass transfer rate theory of Section 10.4.1 can be used, but comment on its applicability.

10–35. A 1 cm–radius tungsten plug for a missile nose cone is tested in a high-temperature air arc jet. Of concern is the behavior of the plug should the heat shield fail and expose the plug. Test data include a measured surface temperature of 3000 K and a surface recession rate of 0.0481 mm/s. Determine the zero-mass-transfer-limit mass transfer conductance that prevailed in the test. Assume diffusion-controlled oxidation with gaseous W_3O_9 as product and negligible oxygen at the s-surface. Take the density of solid tungsten as 19,000 kg/m^3. Use the simple high mass transfer rate theory of Section 10.4.1, but comment on its applicability.

10–36. A falling-film direct-contact steam condenser has plates spaced 3 cm apart. The steam flow is downward cocurrent with the water films, and at a particular location the following data apply: pressure = 1780 Pa, bulk steam temperature = 295 K, bulk water temperature = 282 K, water flow rate per unit film width = 6.0 kg/m s, steam velocity = 100 m/s, and mass fraction of air in the bulk steam = 0.112%. Determine the local rate of steam condensation. [*Hint:* For the liquid-side heat transfer coefficient, use the heat transfer analog to the correlation given in Exercise 9–32 for gas absorption into a turbulent falling film; and for the gas-side mass transfer conductance, use Eq. (9.47). However, comment on the applicability of these correlations.]

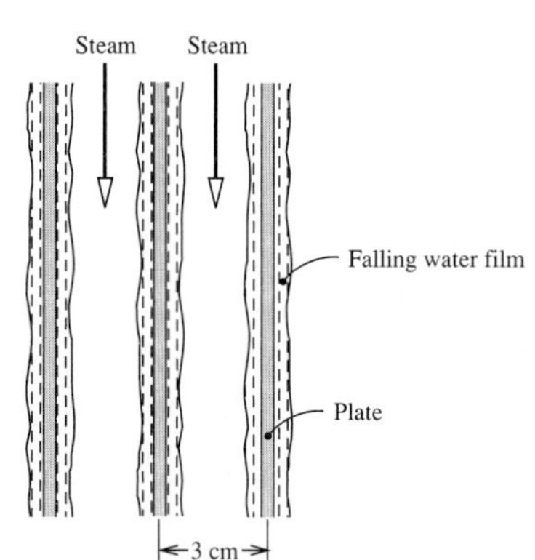

10–37. Steam at 340 K and 21.69 kPa containing 0.85% air by mass flows at 3 m/s down over a horizontal brass condenser tube of 2 cm O.D. and 1 mm wall thickness. Coolant water flows through the tube at a bulk velocity of 2.5 m/s. At a location where the bulk coolant temperature is 290 K, estimate the steam condensation rate and compare it with the value if no air were present.

10–38.
(i) A tungsten plug at 2600 K is exposed to a 40 m/s air stream at 1400 K, 4 atm. Calibration data suggest a mass transfer Stanton number of $\mathrm{St}_m^* = 0.036$. The oxidation product is gaseous W_3O_9, and the concentration of oxygen at the s-surface can be assumed to be very small. Determine the tungsten recession rate in μm/s. For solid tungsten, take $\rho = 19{,}000$ kg/m^3.
(ii) If a porous sintered tungsten plug is tested with nitrogen blown through the surface at 3.0 kg/m^2 s, calculate the new recession rate.
(iii) Compare the N_2 mass fractions at the s-surface for the two situations.

Use the simple high mass transfer rate theory of Section 10.4.1.

10–39. Air at 1 atm and 300 K flows at 10 m/s along a 10 cm–square horizontal slab of dry ice (solid CO_2). If the back side of the slab is well insulated, determine the slab temperature and average sublimation rate. Use a transition Reynolds number of 50,000. The vapor pressure of dry ice is given by $\log_{10} P$ [mm Hg] = 9.9082 − 1367.3/T [K], and the enthalpy of sublimation can be taken as 0.57×10^6 J/kg. Neglect radiation heat transfer and use simple Couette-flow model blowing factors.

10–40. Determine the adiabatic vaporization temperature for small cadmium particles evaporating into pure helium at a total pressure of 3.5 torr in the temperature range T_e = 500–750 K. The vapor pressure of cadmium is given by

$$\log_{10} P \text{ [torr]} = 8.56 - \frac{5720}{T \text{ [K]}}$$

Obtain $\mathscr{D}_{\text{CdHe}}$ using Eq. (9.112) with Lennard-Jones parameters for cadmium of σ = 3.0 Å and ε/k = 3000. Cadmium vapor can be modeled as an ideal monatomic gas for $T < 3000$ K. The enthalpy of sublimation of cadmium can be obtained from the vapor pressure relation using the Clausius-Clapeyron relation. Neglect radiation heat transfer. (*Hint:* Perform the analysis on a molar basis.)

10–41. Modify the result of Section 10.3.5 to account for a relative velocity between the fuel droplet and air stream. Hence obtain the burning-rate constant β as a function of relative velocity V for the combustion of 1 mm–diameter n-hexane droplets burning in air at 1 atm pressure and 300 K. Let $10 < V < 40$ m/s. (*Hint:* Use the same approach as was used for the Couette-flow model of Section 10.4.)

10–42. In a grinding process, iron particles of about 200 μm size are ejected and entrained in an exhaust system. The temperatures attained are sufficient for the particles to ignite and burn on leaving the workpiece. Estimate the burning times of these iron "sparks" if the ambient air is at 300 K and 1 atm. The temperature is not high enough for appreciable evaporation of the metal or the oxide formed, which can be taken to be FeO. It can be assumed that the reaction rate is controlled by diffusion of oxygen through the gas phase. Required data include:

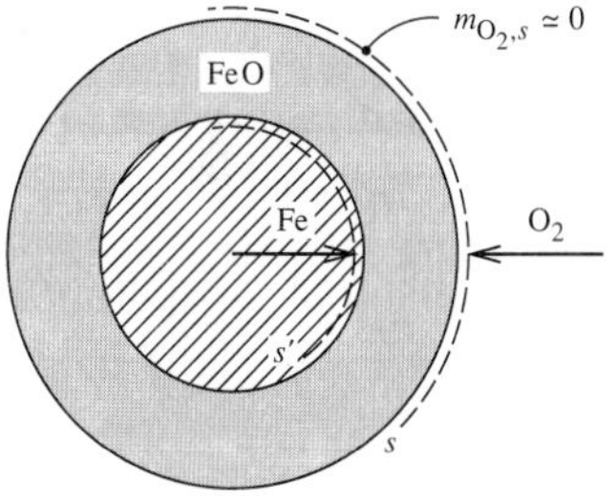

Iron melting temperature and density: 1809 K and 7800 kg/m^3

Oxide melting temperature and density: 1650 K and 3700 kg/m^3

Heat of combustion: 4.58×10^6 J/kg of iron

Emittance of FeO: 0.5

10–43. In a test rig, a steam-air mixture at 4714 Pa and 310 K flows downward at 15 m/s along a 5 cm–square vertical copper plate 3 mm thick. The rear of the plate is cooled by a jet of water at 280 K with a heat transfer coefficient of 4600 W/m^2 K. Estimate the condensation rate as a function of the air content, for $0.001 < m_{\text{air},e} < 0.1$. (*Hint:* To simplify the problem, use average values for the film condensation heat transfer coefficient and the vapor-phase mass transfer conductance.)

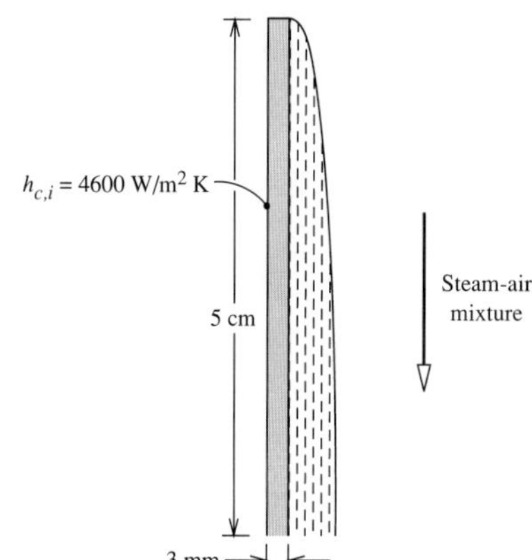

10–44. Air at 300 K, 1 atm flows at 10 m/s along a flat plate also maintained at 300 K. The plate is bare for $0 < x \leq 10$ cm measured from the leading edge, and is coated with a thin layer of naphthalene for $x > 10$ cm. At $x = 20$ cm, calculate Sh_x, St_{mx}, g_{mx}, and $j_{1,s}$. Take $\text{Re}_{tr} = 200{,}000$.

10–45. Following the analysis in Section 5.4.5, derive the integral conservation equations for a laminar natural-convection boundary layer of a gas mixture on a vertical wall in a form suitable for isothermal mass transfer, and valid for high mass transfer rates. Hence obtain the special form of the species equation for transfer of one component only. Also write down an expression for the local Sherwood number valid at low mass transfer rates.

10–46. A porous vertical plate is kept wet with water and maintained at 290 K in dry air at 290 K and 1 atm. Referring to the integral method analysis of natural convection in Section 5.4.5, determine δ, U, Ra_x, Sh_x, g_{mx}, and $j_{1,s}$ at a location 10 cm from the bottom of the plate.

10–47. A flat plate exposed to intense radiative and convective heating is to be sweat cooled with water. Dry air at 840 K and 1 atm flows at 200 m/s along the plate as a laminar boundary layer.

(i) Determine the water supply rate to maintain a surface temperature of 360 K at a location 10 cm from the leading edge.
(ii) If there is a net radiation flux to the plate of $q_{\text{rad}} = -102.3$ kW/m^2, determine the water supply temperature required to maintain a steady state.

Use Knuth's reference-state scheme to account for variable-property effects, in conjunction with the constant-property self-similar solution.

10–48. Air at 1500 K, 1 atm flows at 10 m/s along a porous flat plate that is cooled by injection of helium from a reservoir at 300 K. At a location 20 cm from the leading edge, determine the injection rate required to maintain a surface temperature of 500 K. Also determine the concentration of helium at the plate surface. Use Knuth's reference-state scheme and the constant-property solution for the self-similar laminar boundary layer. Neglect thermal radiation.

10–49. The good performance of the integral method demonstrated in Section 5.4.5 suggests that such methods might yield satisfactory results for the natural-convection noncondensable gas problem. If the molecular weight of the vapor is less than that of the gas, condensation on a vertical wall will produce a downward-flowing vapor boundary layer with motion induced by drag of the liquid film and a downward-directed buoyancy force. Perform an integral method analysis, proceeding as follows.

1. Show that the integral forms of the momentum and species equations for the vapor-phase boundary layer are

$$\frac{d}{dx}\int_0^\infty u^2 dy - v_s u_s = -\nu \left.\frac{\partial u}{\partial y}\right|_s + g\mathcal{M}\int_0^\infty m\, dy$$

$$\frac{d}{dx}\int_0^\infty um\, dy - v_s m_s = -\mathscr{D}_{12}\left.\frac{\partial m}{\partial y}\right|_s$$

where we have written $m = m_2 - m_{2,e}$ and

$$1 - \frac{\rho_e}{\rho} = \mathcal{M}(m_2 - m_{2,e}); \qquad \mathcal{M} = \frac{M_2 - M_1}{M_2 - (M_2 - M_1)m_{2,e}}$$

Subscripts 1 and 2 refer to vapor and gas, respectively. The effect of temperature on mixture density has been neglected.

2. Show that use of a boundary condition requiring the liquid surface to be impermeable to gas relates v_s to $\partial m/\partial y|_s$ as

$$v_s = \frac{\mathscr{D}_{12}}{m_{2,s}}\left.\frac{\partial m}{\partial y}\right|_s$$

3. Following Rose [30], choose the following profiles:

$$u = u_s\left(1 - \frac{y}{\Delta}\right)^2 + U\frac{y}{\Delta}\left(1 - \frac{y}{\Delta}\right)^2$$

$$\frac{m}{m_s} = \left(1 - \frac{y}{\Delta}\right)^2$$

where u_s is the surface velocity of the liquid film obtained from Nusselt's analysis (Eq. 7.4), Δ is the vapor boundary layer thickness, and U is a yet-to-be-determined scaling velocity. Notice that the velocity profile is the sum of two components, the first due to drag by the liquid film and the second due to natural convection. Show that these profiles satisfy appropriate boundary conditions.

4. Substitute the profiles in the governing ordinary differential equations and, proceeding as in Section 5.4.5 with the local condensation rate taken from Nusselt's solution, show that

$$10\,\mathrm{Sc}\frac{\mathrm{Ja}_l}{\mathrm{Pr}_l}\left[\frac{(\rho\mu)_l}{(\rho\mu)_v}\right]\left(\frac{m_{2,e}}{m_{2,s}-m_{2,e}}\right)^2\left(\frac{20}{21}+\frac{m_{2,s}}{m_{2,e}}\mathrm{Sc}\right)$$
$$+\frac{8}{\mathrm{Sc}}\left(\frac{\mathrm{Pr}_l}{\mathrm{Ja}_l}\right)^2\left[\frac{(\rho\mu)_v}{(\rho\mu)_l}\right]\left(\frac{m_{2,s}-m_{2,e}}{m_{2,s}}\right)^2\left[\frac{5}{28}\frac{\mathrm{Ja}_l}{\mathrm{Pr}_l}-\frac{\mathcal{M}(m_{2,s}-m_{2,e})}{3}\right]$$
$$=\frac{100}{21}\frac{m_{2,e}}{m_{2,s}}-2\frac{m_{2,s}-m_{2,e}}{m_{2,s}}+8\,\mathrm{Sc}$$

where $\mathrm{Ja}_l = c_{pl}(T_s - T_w)/h_{\mathrm{fg}}$.

10–50. Referring to Exercise 10–49, obtain the interface temperature and total condensation rate per unit width, for condensation from a steam-air mixture on a 0.05 m–high vertical surface with $P = 1$ atm, $T_e = 375$ K, $m_{2,e} = 0.02$, and $T_w = 360$ K.

10–51. Following the analysis in Section 10.5.2, derive the equation governing total energy conservation $H = h + (1/2)u^2$ for a laminar boundary layer on a flat plate by including transport of kinetic energy, and viscous work. (Refer to Sections 5.2.1 and 5.7.3.) Hence show that for unity Lewis and Prandtl numbers, the equation reduces to a form identical to Eq. (10.162) with static enthalpy h replaced by total enthalpy H.

10–52. A solid-propellant rocket nozzle throat is lined with graphite and has an initial diameter of 5 cm. The elemental composition of the propellant is $m_{(\mathrm{O})} = 0.560$, $m_{(\mathrm{C})} = 0.269$, $m_{(\mathrm{N})} = 0.147$. The combustion chamber pressure is $P_{\mathrm{cc}} = 100$ atm, and the temperature is 3300 K. The characteristic velocity of the propellant is $c^* = 1200$ m/s. After ignition, the surface temperature of the graphite quickly reaches a temperature high enough to give thermodynamic equilibrium at the s-surface, with the concentrations of all oxidizing species equal to zero. Estimate the rate of enlargement of the throat if experimental data for heat transfer to regeneratively cooled liquid-propellant rocket nozzles can be correlated as

$$\mathrm{Nu}_D = 0.020\,\mathrm{Re}_D^{0.8}\,\mathrm{Pr}^{0.33}$$

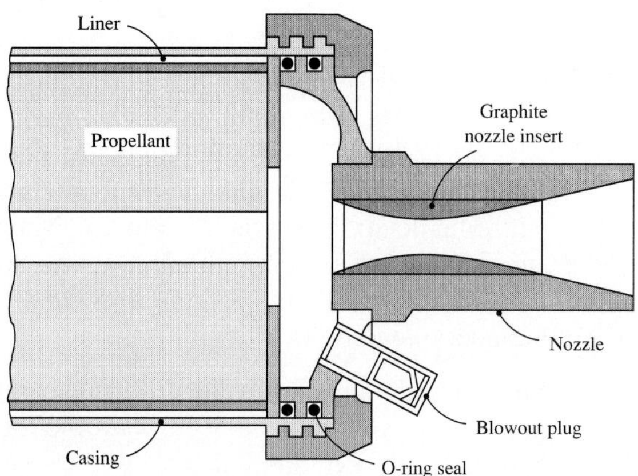

(The mass velocity in the throat of the nozzle is given by $G_{th} = P_{cc}/c^*$.) Take $\mu \simeq 7.5 \times 10^{-5}$ kg/m s and Sc $\simeq 0.7$ to evaluate the conductance.

10–53. In an ammonia scrubber, a mixture of air and ammonia gas flows countercurrent to a falling film of water inside 3 cm–diameter vertical tubes. The exchanger operates at 300 K and 1 atm. The zero-mass-transfer-limit gas-side and liquid-side mass transfer conductances are estimated to be 0.013 kg/m^2 s and 0.13 kg/m^2 s, respectively. At a particular location down the tubes, the bulk mass fractions of ammonia in the air and water streams are 0.62 and 0.09, respectively. Determine the local gas absorption rate. Use constant-property Couette-flow blowing factors.

10–54. In a direct-contact condenser for a pilot open-cycle ocean thermal energy conversion plant, cold seawater films fall down 3 mm–thick parallel plates at a pitch of 4.0 cm. Low-pressure steam flows cocurrently with the seawater. At a particular location down the plates, the following data apply.

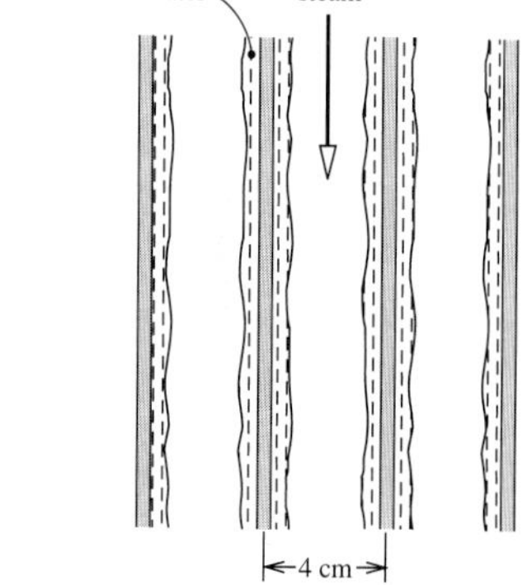

Bulk gas temperature $T_e = 283$ K

Bulk gas air mass fraction $m_{2,e} = 9 \times 10^{-4}$

Bulk liquid temperature $T_s = 278$ K

Bulk liquid air mass fraction $m_{2,b} = 4 \times 10^{-5}$

Total pressure = 1215 Pa

Gas-side mass transfer conductance $g^*_{mG} = 6.0 \times 10^{-3}$ kg/m^2 s

Liquid-side heat transfer conductance $g^*_{hL} = 6.3$ kg/m^2 s

Liquid-side mass transfer conductance $g^*_{mL} = 1.58$ kg/m^2 s

Determine the local condensation rate and air desorption rate. Use constant-property Couette-flow blowing factors.

10–55. Evaporation of a multicomponent droplet is analyzed in Section 10.6.3. If good mixing within the droplet is assumed, the governing equations are Eqs. (10.208), (10.211), and (10.212). Write a computer program to calculate the lifetime of a binary hydrocarbon fuel droplet evaporating into stagnant air. Use your program to obtain the lifetime of a 1 mm–diameter pentane-hexane droplet, initially at 300 K with 50% by mass of each component, in stagnant air at 2000 K and 1 atm. Also plot the bulk composition, bulk temperature, and evaporation rates of each component as a function of time. Property data that can be used include $\rho_l = 610$ kg/m^3; $c_{pl} = 2400$ J/kg K; and for pentane and hexane designated as species 1 and 2, respectively, $M_1 = 72.0$, $M_2 = 86.2$, $h_{fg1} = 3.57 \times 10^5$ J/kg, $h_{fg2} = 3.35 \times 10^5$ J/kg. For the vapor pressures,

$$P = \exp\left[-A\left(\frac{1}{T} - \frac{1}{B}\right)\right] \text{ atm}$$

with $A_1 = 3330$, $A_2 = 3810$, $B_1 = 309$, $B_2 = 314$. Use Raoult's law, namely, $P_{i,s} = x_{i,u}P_{i,s}$ $(x_{i,u} = 1)$. The gas-phase properties can be evaluated using Hubbard's 1/3 rule extended to two evaporating components.

10–56. Iron and ferrous alloys are soluble in molten aluminum. Dissolution rates must be estimated for applications including casting of aluminum, high-temperature containers, and safety analysis of advanced nuclear reactors. For example, in a postulated core meltdown accident in a heavy-water reactor, a pool of molten aluminum may form in the bottom of the reactor vessel. Decay heating of the radioactive materials dissolved in the aluminum could ensure that the pool remains molten ($T_{MP} = 933$ K) for a long time, and thus erosion of the carbon steel vessel may be a serious problem. To obtain a preliminary estimate of the expected erosion rates, consider a pool of aluminum at 1000 K in a vessel with a 5 m–square base, and calculate the erosion rate of the steel base. The solubility of iron in aluminum is approximately

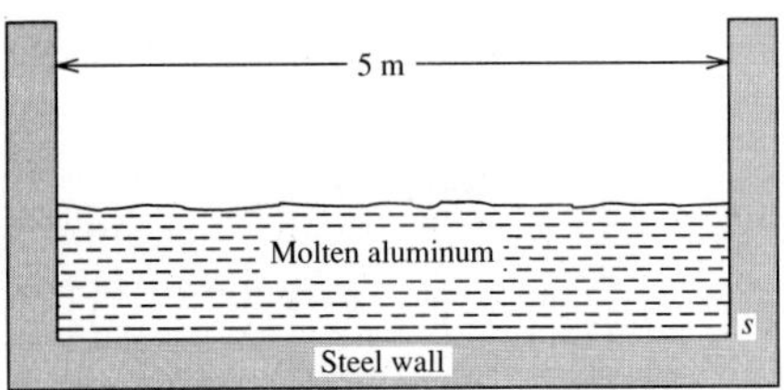

$$m_{1,s} = 0.92 \exp\left[-\left(\frac{T_s K - 1820}{450}\right)^2\right]$$

Also, at 1000 K, take $\rho_s = 2537$ kg/m^3, $\rho_{Al} = 2360$ kg/m^2 s, $\nu_{Al} = 0.47 \times 10^{-6}$ m^2/s, and $Sc_{FeAl} = 260$.

CHAPTER

11

MASS EXCHANGERS

CONTENTS

11.1 INTRODUCTION

Heat exchangers are devices that effect a change in the temperature of one or more flowing streams. Analogously, mass exchangers are devices that effect a change in the composition of one or more flowing streams. In a **catalytic reactor,** the composition of a single stream is changed as it flows over a catalyst that promotes a reaction between components of the stream: the catalytic reactor is a *single-stream* exchanger and is analogous to the single-stream heat exchangers considered in Sections 1.6 and 8.4. A catalytic converter for an automobile is shown in Fig. 11.1. **Filters** and **electrostatic precipitators** are single-stream exchangers in which particles are removed from an air stream. In a **gas scrubber,** an unwanted component of a gas stream is removed by absorption into a liquid stream. In general, the compositions of the two streams vary through a scrubber, so that this is a *two-stream* exchanger, analogous to the two-stream heat exchangers considered in Section 8.5. However, some scrubbers can be analyzed as single-stream exchangers, as will be seen in Section 11.3.2. As was the case for heat exchangers, a large transfer area per unit volume is desirable for mass exchangers. Large surface areas are obtained by spraying liquid into a gas as very small droplets, by bubbling gas through a liquid, or by flowing one or both phases through a packed bed.

Simultaneous heat and mass exchangers are devices that effect changes in both the temperature and the composition of one or more flowing streams. The **adiabatic humidifier** considered in Section 11.5.1 is an example of a single-stream simultaneous heat and mass exchanger; it is also particularly simple since the heat and mass transfer processes can be *uncoupled* and the performance determined by analyzing the mass transfer process only. **Cooling towers** are examples of two-stream simultaneous heat and mass exchangers. In a power plant, the condenser coolant water rejects heat to air in the cooling tower, and is then recycled to the condenser. The water temperature decreases through the tower, and the air temperature and moisture content increase. Counterflow cooling towers are analyzed in Section 11.5.2. Crossflow cooling towers are analyzed in Section 11.5.3. Hand calculations of cooling tower performance can be long and tedious, and are impractical for tower design. The computer program CTOWER performs these calculations efficiently and reliably.

Figure 11.1 A catalytic converter for a diesel engine automobile. (Photograph courtesy Mercedes-Benz AG.)

11.2 SYSTEM BALANCES

In Section 8.3, exchanger energy balances were made on heat exchangers in order to relate inlet and outlet temperatures. Similarly, exchanger, or system, mass balances are made on mass exchangers to relate inlet and outlet compositions. Such balances can be made on a mass or molar basis. In addition, if chemical reactions occur within the system, it is often useful to base such balances on chemical elements, because chemical elements are conserved in chemical reactions. Owing to the greater complexity of mass transfer processes, system balances on mass exchangers play a more important role than energy balances on heat exchangers. System balances on mass exchangers can yield much useful information, and should always be made before considering the design of a mass exchanger to effect the desired mass transfer processes.

11.2.1 Mass, Mole, and Element Balances

At steady state, a *system* balance for a selected chemical species i is made by equating the production rate due to reactions within the system to the rate of net outflow of the species in the various streams entering or leaving the system. For species i we can write

Rate of internal production = Rate of outflow − Rate of inflow

The mass flow rate of the species i in a stream is the total flow in the stream $\dot{m}$ times the mass fraction m_i:

$$\dot{m}_i = m_i \dot{m} \qquad \textbf{(11.1a)}$$

Similarly, the molar flow rate of species i, $\dot{M}_i$, is the total molar flow rate $\dot{M}$ times the mole fraction x_i:

$$\dot{M}_i = x_i \dot{M} \qquad \textbf{(11.1b)}$$

Consider now two or more inlet streams that mix to form a single outlet stream, as shown in Fig. 11.2. The mass balance for a system with k inlet streams is

$$\sum_k (\dot{m})_k = \dot{m}_{\text{out}} \qquad \textbf{(11.2)}$$

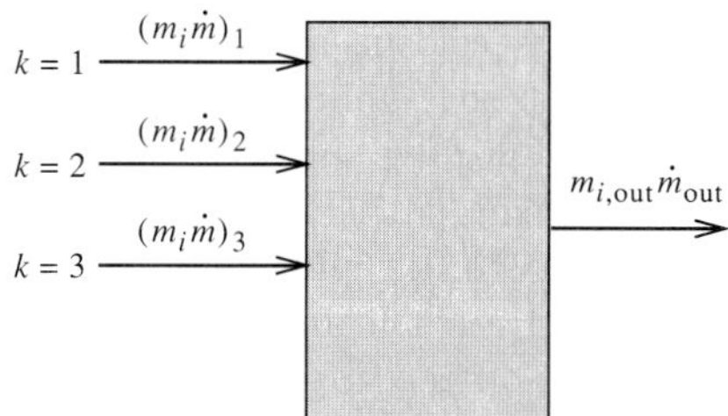

Figure 11.2 A species balance on a system with k inlet streams and one outlet stream.

If the chemical species of interest is inert, there is no internal production, and the species balance for the system takes a simple form,

$$\sum_k (m_i \dot{m})_k = m_{i,\mathrm{out}} \dot{m}_{\mathrm{out}} \tag{11.3a}$$

on a mass basis. On a molar basis,

$$\sum_k (x_i \dot{M})_k = x_{i,\mathrm{out}} \dot{M}_{\mathrm{out}} \tag{11.3b}$$

Often it is useful to perform a system balance on a chemical element, denoted (α), irrespective of which chemical species it is contained in. If nuclear reactions are absent, there cannot be production of chemical elements within the system; thus

$$\text{Rate of inflow of element } (\alpha) = \text{Rate of outflow of element } (\alpha)$$

The mass flow of chemical element (α) in a stream is obtained by multiplying each species flow rate $m_i \dot{m}$ by the mass fraction of chemical element (α) in species i, $m_{(\alpha)i}$, and summing over all species containing (α). Recall that an element mass fraction $m_{(\alpha)i}$ is obtained from the chemical formula of species i and atomic weights. For example, if species i is water, H_2O, and element (α) is hydrogen, (H), $m_{(\alpha)i} = m_{(H)H_2O} = 2/18 = 1/9$. Again, with k inlet streams forming a single outlet stream,

$$\sum_k \left(\left[\sum_i m_{(\alpha)i} m_i \right] \dot{m} \right)_k = \left[\sum_i m_{(\alpha)i} m_{i,\mathrm{out}} \right] \dot{m}_{\mathrm{out}} \tag{11.4}$$

In this text the balances described above will be applied to systems as diverse as power plants, automobile engines, catalytic reactors, and cooling towers. Provided the flows and compositions of all the streams but one are known, one can find the flow and elemental composition of the unknown stream. One need not know what happens in the system. Also, with minimal additional knowledge of what happens in the system—for example, knowledge that no reactions occur or that a particular reaction goes to completion—the molecular composition of the outlet stream can be determined.

11.2.2 A Balance on an Automobile Engine

In most urban areas of the United States, a principal source of nitric oxide in the atmosphere is automobile engine exhaust. Nitric oxide, NO, is itself harmful to humans; additionally, in the presence of sunlight it reacts with oxygen, O_2, to produce nitrogen dioxide, NO_2, which is also harmful. Ozone, O_3, is produced as an intermediate in this reaction, but it reacts very rapidly with any excess NO to produce more NO_2. However, when most of the available NO has been oxidized, which usually occurs some miles downwind of the source, ozone builds up in concentration and produces the characteristic irritating effects of photochemical smog. The precise mechanisms of subsequent reactions are not well understood. One view is that as ozone builds up, it oxidizes unburned hydrocarbons from automobile and industrial emissions. The ozone is replenished by a photochemical reaction involving NO_2,

$$NO_2 + \text{photon} \rightarrow NO + O$$

$$O + O_2 \rightarrow O_3$$

Some of the resulting NO then further reacts with the oxidized hydrocarbons to form particularly dangerous chemical species such as peroxyacetylnitrate (PAN). These species damage plants and are highly irritating (and presumably injurious) to humans. The remaining NO is reoxidized by O_3 to NO_2, and the cycle repeats itself. In strong sunlight, O_3 concentrations in a polluted atmosphere tend to be approximately one-half of the NO_2 concentrations.

Consider a typical automobile that gets 7 km per liter of gasoline (14.3 liters/100 km). An oxygen sensor in the exhaust allows the fuel injection system to maintain an approximately stoichiometric fuel-air mixture. If the gasoline is approximated as octane (C_8H_{18}), one can write

$$C_8H_{18} + 12.5O_2 \rightarrow 8CO_2 + 9H_2O$$

assuming complete combustion. The specific gravity of gasoline can be taken as 0.70, and thus we have the following.

Fuel consumption: (1/7 liter/km)(0.70 kg/liter) $=$ 0.10 kg/km; or, on a molar basis, (0.10/114) $= 8.8 \times 10^{-4}$ kmol/km, since the molecular weight of octane is $(8 \times 12) + 18 = 114$

Oxygen consumption: $(8.8 \times 10^{-4})(12.5) = 1.10 \times 10^{-2}$ kmol/km $=$ 0.35 kg/km

Nitrogen throughput: $(79/21)(1.10 \times 10^{-2}) = 4.14 \times 10^{-2}$ kmol/kg $=$ 1.16 kg/km

Total exhaust flow: $(0.10 + 0.35 + 1.16) = 1.61$ kg/km

Table 11.1 gives the allowable emissions of pollutants according to 1993–94 State of California standards for passenger cars. The unburnt hydrocarbons and carbon monoxide (CO) result from incomplete combustion; the nitrogen oxides (NO_x) result from side reactions between N_2 and O_2 at the high temperatures of the combustion process. The pollutant NO_x is actually emitted largely as NO, with a little NO_2. Because the NO oxidizes after entering the atmosphere, the amount of NO allowed is multiplied by 46/30, the ratio of molecular weights of NO_2 and NO, and added to the amount of NO_2. This sum is then referred to as the allowable emission rate of

Table 11.1 California mobile source emission standards for new passenger cars, 1993–94.

	Allowable Emission	
Pollutant	g/mi	g/km
Hydrocarbons[a]	0.25	0.16
Carbon monoxide	3.4	2.1
NO_x	0.4	0.25
Diesel particulates	0.08	0.05

[a]Excluding methane.

pollutant NO_x. Accordingly, the allowable mass fraction of NO_x in the automobile exhaust is

$$m_{NO_x,out} = (0.25 \times 10^{-3}/1.61) = 1.55 \times 10^{-4}$$

Catalytic converters for automobiles must be sized to reduce the concentration of NO_x in the engine exhaust to this value: the required mass exchanger effectiveness proves to be typically of the order of 99%.

To estimate the total amount of pollutants entering the atmosphere above an urban area such as the Los Angeles basin, we will assume 4 million automobiles in the basin, each traveling, on average, 50 km/day. Then the total exhaust flow into the atmosphere is

$$(1.61 \text{ kg/km})(50 \text{ km/day})(4 \times 10^6 \text{ automobiles}) = 3.22 \times 10^8 \text{ kg/day}$$

and the NO_x flow is

$$(m_{NO_x})(\dot{m}_{exhaust}) = (1.55 \times 10^{-4})(3.22 \times 10^8) = 5.0 \times 10^4 \text{ kg/day}$$

Similarly, the flow rates of hydrocarbons and carbon monoxide can be calculated to be 3.2×10^4 and 4.2×10^5 kg/day, respectively.

11.2.3 A Balance on an Oil-Fired Power Plant

Oil- and coal-fired power plants are the principal sources of the pollutant sulfur dioxide (SO_2) in our atmosphere. The SO_2 primarily results from the oxidation of free sulfur in the fuel. Typically, fuel oils have free sulfur contents ranging from 0.2% to 3% by mass. Consider an 800 MW power plant that burns a fuel oil containing 1% free sulfur by mass. If the fuel oil is approximated as $C_{21}H_{44}$, the main combustion reaction is

$$C_{21}H_{44} + 32O_2 \rightarrow 21CO_2 + 22H_2O$$

The overall efficiency of such a plant is typically 40%, and the heat of combustion of the fuel is approximately 4.2×10^7 J/kg. Thus the fuel flow rate can be estimated as follows.

$$\text{Power output} = (\text{Fuel flow rate})(\text{Heat of combustion})(\text{Efficiency})$$

$$(800 \times 10^6 \text{ J/s}) = (\dot{m}_f \text{ kg/s})(4.2 \times 10^7 \text{ J/kg})(0.40)$$

Solving, $\dot{m}_f = 47.6$ kg/s.

The molecular weight of the fuel is $(21 \times 12) + 44 = 296$; thus, the molar flow rate of fuel is

$$\dot{M}_f = \dot{m}_f/M_f = (47.6 \text{ kg/s})/(296 \text{ kg/kmol}) = 0.161 \text{ kmol/s}$$

From the fuel flow rate we can immediately find the sulfur flow rate:

$$(\dot{m}_S)_f = m_{(S)f}\dot{m}_f = (0.01 \times 47.6 \text{ kg/s}) = 0.476 \text{ kg/s}$$

$$(\dot{M}_S)_f = (\dot{m}_S)_f/M_S = (0.476 \text{ kg/s})/(32 \text{ kg/kmol}) = 0.0149 \text{ kmol/s}$$

The chemical reaction states that for each kmol of fuel, 32 kmol of oxygen are required for stoichiometric combustion:

$$\dot{M}_{O_2} = (32)(0.161 \text{ kmol/s}) = 5.15 \text{ kmol/s}$$

Typically, 10% excess oxygen is used to reduce CO emissions from the stack. The required flow of oxygen is then

$$\dot{M}_{O_2} = (1.1)(5.15) = 5.67 \text{ kmol/s}$$

The nitrogen flow rate is obtained easily since air is approximately 21% O_2 and 79% N_2 by volume: $\dot{M}_{N_2} = (79/21)\dot{M}_{O_2} = (3.76)(5.67) = 21.3$ kmol/s. The stack gas flow rates of the major constituents in the effluent are summarized below.

CO_2	(0.161)(21)	=	3.38 kmol/s	=	149 kg/s
H_2O	(0.161)(22)	=	3.54 kmol/s	=	63.8 kg/s
SO_2		=	0.0149 kmol/s	=	0.954 kg/s
O_2	5.67 − 5.15 − 0.0149	=	0.505 kmol/s	=	16.2 kg/s
N_2		=	21.3 kmol/s	=	596 kg/s
Totals			28.7 kmol/s		826 kg/s

The mean molecular weight of the effluent is $(826/28.7) = 28.8$ kg/kmol. The mass fraction of SO_2 in the effluent is $(0.954/826) = 1.15 \times 10^{-3}$, and the mole fraction is $(0.0149/28.7) = 5.2 \times 10^{-4}$. Small concentrations of pollutants are often expressed as parts per million on a volume (molar) basis: thus, the concentration of SO_2 is 520 ppm. If the effluent is untreated, the flow rate of SO_2 from the plant into the atmosphere is $(0.954)(3600)(24) = 8.2 \times 10^4$ kg/day.

11.2.4 A Balance on an Urban Air Volume

The nature of the air pollution problem in an urban area can be explored by performing a system balance on the air volume above the city. Consider the Los Angeles basin with a surface area of 3200 km^2, for which air pollution is a serious problem owing to the frequent occurrence of an *inversion layer* over the basin. Figure 11.3 shows the temperature distribution in the atmosphere that causes an inversion layer. We will assume that the height of the layer is 500 m and that the average residence time for the air in the layer is one day (that is, a mass of air equal to that contained in the layer flows into, mixes with, and flows out of the layer each day). The volume of system under consideration is $(3200 \times 10^6)(500) = 1.6 \times 10^{12}$ m^3. Air at 1 atm pressure and 300 K has a density of 1.18 kg/m^3, so that the corresponding mass is $(1.6 \times 10^{12})(1.18) = 1.9 \times 10^{12}$ kg. Thus, the mass flow rate $\dot{m}$ through the system is on the order of 1.9×10^{12} kg/day.

The effects of pollutants on plant life and human health are not too well defined as yet. Plant damage has been observed with chronic exposure to as little as 0.03 ppm SO_2. There is evidence that exposure to modest concentrations of such chemical species as CO, NO_x, SO_2, and O_3 is unhealthful. Indeed, it is well established that episodes of high concentrations can kill susceptible segments of a population. For this reason the Los Angeles air quality management authority has established

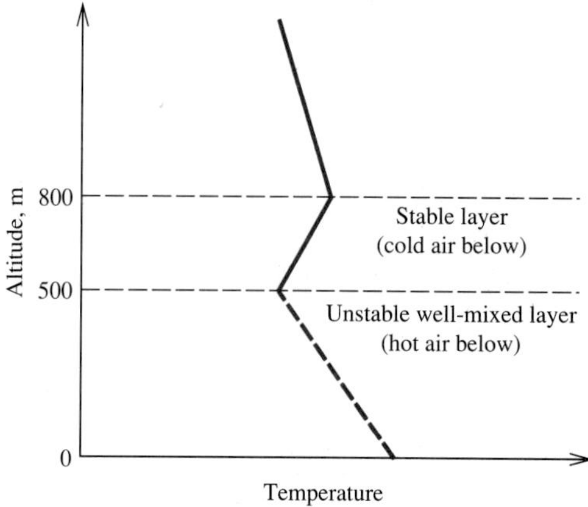

Figure 11.3 Temperature versus altitude variation characteristic of an inversion layer.

what are termed *first-stage alert* pollution levels for CO (40 ppm, 1 h; 20 ppm, 12 h), SO_x (0.5 ppm), and O_3 (0.2 ppm). The *third-stage alert* levels, at which a serious emergency is considered to exist, are 100, 50, 2, and 0.5 ppm, respectively. Species balances on the urban air volume can be used to determine the corresponding allowable flow rates of these pollutants into the volume. For example, to prevent a third-stage alert, the flow rate of SO_x (taken as SO_2) into the air volume must not exceed

$$\dot{m}_{SO_2} = m_{SO_2}\dot{m} = (2 \times 10^{-6})(64/29)(1.9 \times 10^{12}) = 8.4 \times 10^6 \text{ kg/day}$$

Notice that for these dilute mixtures the ppm by volume value is converted to mass fraction by multiplying by $10^{-6}(M_i/M_{\text{air}})$ since $M \simeq M_{\text{air}}$.

To put the preceding result in perspective, the total power-generating capacity in the Los Angeles basin is about 10,000 MW. If all these plants burned fuel with 1% free sulfur, then the results of Section 11.2.3 indicate that the flow of SO_2 into the atmosphere would be $(8.2 \times 10^4)(10{,}000/800) = 1.03 \times 10^6$ kg/day, that is, about 12% of the allowable amount.

11.3 SINGLE-STREAM MASS EXCHANGERS

A single-stream mass exchanger is an exchanger in which there is only one fluid stream, or, if there are two streams, the composition of only one stream varies significantly along the exchanger. Examples of exchangers with only one stream are catalytic reactors (Section 11.3.1), electrostatic precipitators (Section 11.3.3), and filters (Section 11.3.4). Examples of exchangers with two streams, of which the composition of only one varies significantly, include some gas absorption or desorption processes, as described in Section 11.3.2. The analysis of single-stream mass exchangers is particularly simple. In fact, the result is always the familiar effectiveness–number of transfer units relation first derived for single-stream heat exchangers as Eq. (1.59).

11.3.1 Catalytic Reactors

A catalytic reactor is a single-stream exchanger by virtue of the fact that there is no second stream, or adjacent phase, where the reaction products accumulate. Figure 11.4 depicts a model catalytic reactor involving the flow of a single stream with continuous removal of chemical species 1 along its flow path. Since the applications considered in this section all involve a gas stream, the mass flow rate of the stream will be denoted $\dot{m}_G$; m_1 will denote the bulk mass fraction of the transferred species 1 (as for heat exchangers, we do not use a subscript to denote "bulk" in order to keep the notation simple); $j_{1,s}$ is the rate of transfer of species 1 across the s-surface; and $\mathscr{P}\Delta x$ is the area of s-surface between locations x and $x + \Delta x$; that is, $\mathscr{P}$ is the *perimeter* of the transfer surface. By using the perimeter, the result of our analysis will be of general applicability, irrespective of whether the catalyst is in the form of pellets, plates, or a honeycomb matrix. The principle of conservation of species applied to the elemental control volume bounded by the s-surface and located between x and $x + \Delta x$ requires that, at steady state,

$$\text{Rate of inflow of species 1} = \text{Rate of outflow of species 1}$$

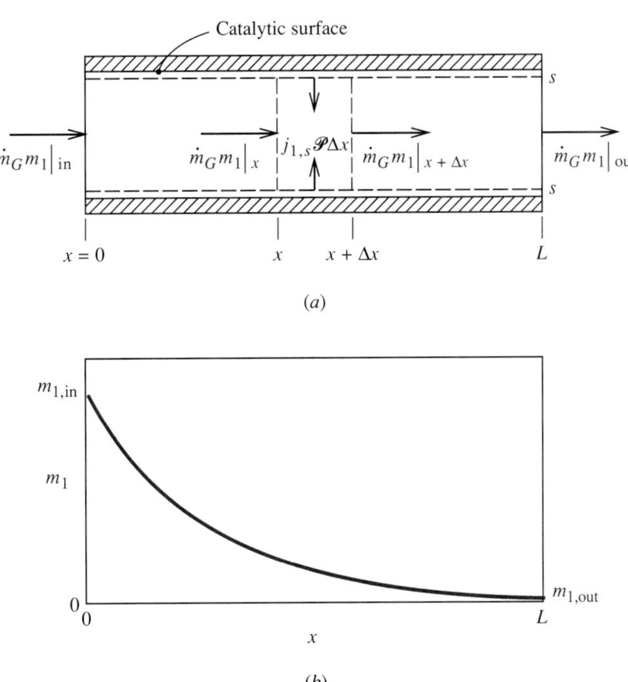

Figure 11.4 A catalytic reactor. (*a*) Schematic showing a species balance on an elemental volume Δx long. (*b*) Reactant concentration variation along the reactor.

since species 1 is inert in the gas phase. It is assumed that there are no *homogeneous* reactions involving species 1. If x-direction diffusion is neglected, then

$$\dot{m}_G m_1|_x + j_{1,s}\mathscr{P}\Delta x = \dot{m}_G m_1|_{x+\Delta x} \tag{11.5}$$

Also, the principle of mass conservation applied to the elemental control volume requires that

$$\dot{m}_G|_x = \dot{m}_G|_{x+\Delta x} = \text{Constant} \tag{11.6}$$

since there is no net mass transfer across the s-surface. The transfer rate of the reactants exactly balances the transfer rate of the products since the catalyst does not enter into the reaction. Then, dividing Eq. (11.5) by $\dot{m}_G\Delta x$ and letting $\Delta x \to 0$,

$$\frac{dm_1}{dx} - \frac{j_{1,s}\mathscr{P}}{\dot{m}_G} = 0 \tag{11.7}$$

Further progress requires an expression for $j_{1,s}$; for a first-order reaction and a catalyst in the form of plates or foil, we can use the result of the analyses in Sections 9.3.3 and 9.4.5. Replacing $m_{1,e}$ in Eq. (9.53) by the bulk value m_1 gives

$$j_{1,s} = -\frac{\dot{m}_1}{A} = \frac{-m_1}{\dfrac{1}{\rho k''} + \dfrac{1}{\mathscr{g}_{m1}}} = -\mathscr{g}_{m1}^{\text{oa}} m_1 \tag{11.8}$$

where $\mathscr{g}_{m1}^{\text{oa}}$ is an *overall mass transfer conductance* for species 1. Substituting in Eq. (11.7) gives

$$\frac{dm_1}{dx} + \frac{\mathscr{g}_{m1}^{\text{oa}}\mathscr{P}}{\dot{m}_G} m_1 = 0 \tag{11.9}$$

which is a first-order ordinary differential equation for $m_1(x)$ The required initial condition is obtained from the composition of the inlet gas mixture as

$$x = 0; \quad m_1 = m_{1,\text{in}}$$

If $\mathscr{g}_{m1}^{\text{oa}}$ can be assumed constant along the exchanger, the solution is

$$\frac{m_1}{m_{1,\text{in}}} = e^{-\mathscr{g}_{m1}^{\text{oa}}\mathscr{P}x/\dot{m}_G}$$

In particular, the outlet concentration of species 1 is obtained by setting $x = L$,

$$\frac{m_{1,\text{out}}}{m_{1,\text{in}}} = e^{-\mathscr{g}_{m1}^{\text{oa}}\mathscr{P}L/\dot{m}_G}$$

and subtracting each side from unity gives

$$\frac{m_{1,\text{in}} - m_{1,\text{out}}}{m_{1,\text{in}}} = 1 - e^{-\mathscr{g}_{m1}^{\text{oa}}\mathscr{P}L/\dot{m}_G} \tag{11.10}$$

The concentration variation is shown in Fig. 11.4. The numerator of the left-hand side of Eq. (11.10) is the actual change in the concentration of species 1, and the denominator ($m_{1,\text{in}} - 0$) is the maximum possible change that could be achieved in an infinitely long exchanger: thus, the left-hand side is recognized to be the *effectiveness* ε of the mass exchanger, analogous to the effectiveness of a heat exchanger. Similarly, the dimensionless exponent on the right-hand side of Eq. (11.10) is recognized to be the number of transfer units N_{tu} of the mass exchanger. Thus, Eq. (11.10) can be written as

$$\varepsilon = 1 - e^{-N_{\text{tu}}} \tag{11.11}$$

which is identical to the result for a single-stream heat exchanger.

Example 11.1 involves the preliminary design of an automobile catalytic converter, in which the catalyst is in the form of oxidized copper foil wrapped in a spiral or packed in wafer form. However, current practice for automobiles is to use either a packed bed of porous catalyst pellets or a ceramic matrix with walls coated with a porous catalyst layer. Mass transfer in porous catalysts was analyzed in Section 9.6. For a pellet in a packed bed, Eq. (9.85) gives the molar flux of species 1 across the s-surface surrounding the pellet as

$$J_{1,s} = -\frac{\dot{M}_1}{S_p} = \frac{-x_{1,e}}{\dfrac{1}{(V_p a_p/S_p)\eta_p k''c} + \dfrac{1}{\mathscr{G}_{m1}}}$$

where V_p and S_p are the pellet volume and s-surface area, a_p is the area of catalyst per unit volume of pellet, and η_p is the pellet effectiveness obtained from Eq. (9.82). Thus, an overall mole transfer conductance can be defined as

$$\frac{1}{\mathscr{G}_{m1}^{\text{oa}}} = \frac{1}{(V_p a_p/S_p)\eta_p k''c} + \frac{1}{\mathscr{G}_{m1}} \tag{11.12}$$

and is used in Eq. (11.11), written on a molar basis, as

$$\varepsilon = \frac{x_{1,\text{in}} - x_{1,\text{out}}}{x_{1,\text{in}}} = 1 - e^{-\mathscr{G}_{m1}^{\text{oa}}\mathscr{P}L/\dot{M}_G} \tag{11.13}$$

Use of Eq. (11.12) is required in Exercise 11–11, for example. Analysis of reactors containing a porous wall matrix is required in Exercises 11–10, 11–12, and 11–13.

As discussed in Section 9.3, CO oxidation on a platinum or cupric oxide catalyst can be approximately modeled as a first-order reaction. Thus, these results can be used for the design of CO oxidation converters. Such converters are used in automobiles in countries where NO emissions are not controlled. Also, oxidation converters are used on two-stroke engines, for which NO emissions are inherently low, and on diesel engines that run on lean fuel-air mixtures. So called "three-way" catalytic converters that promote the reduction of CO and NO in automobile exhaust cannot be properly designed using Eq. (11.13) since the reduction reaction is not first-order. Indeed, the kinetics of this reaction are poorly understood, and current practice is to develop such converters by extensive testing rather than by detailed design.

EXAMPLE 11.1 Removal of Carbon Monoxide from Automobile Exhaust

A 99% effective catalytic converter is to be designed for removing CO from automobile engine exhaust. A suitable catalyst is cupric oxide, and we wish to examine the feasibility of using a packed bed of thin oxidized copper foil wrapped in a spiral or stacked in wafer form, with a passage width of 1 mm. The desired operating temperature is 1200 K, and the rate constant for CO oxidation by CuO is $k'' = 50.6\exp(-9361/T)$ m/s, for T in kelvins. Suggest dimensions of a reactor suitable for an automobile that obtains 8 km/liter at 80 km/h. Assume an air/fuel ratio of 14/1 and exhaust air injection at 10% of the exhaust flow rate.

Solution

Given: Copper foil packed-bed catalytic converter for an automobile that obtains 8 km/liter at 80 km/h.

Required: Dimensions for 99% effective removal of CO from an automobile exhaust.

Assumptions:
1. Air properties can be used for the gas stream.
2. The air/fuel ratio is 14/1, and air is injected into the exhaust at a rate equal to 10% of the exhaust flow rate.
3. A gasoline specific gravity of 0.705.

The first step is to determine the number of transfer units required. Solving Eq. (11.11) for N_{tu} gives

$$N_{tu} = \frac{g_{m1}^{oa}\mathscr{P}L}{\dot{m}_G} = \ln\frac{1}{1-\varepsilon} = \ln\frac{1}{1-0.99} = 4.61$$

Next we evaluate g_{m1}^{oa}; from Eq. (11.8),

$$\frac{1}{g_{m1}^{oa}} = \frac{1}{\rho k''} + \frac{1}{g_{m1}}$$

We approximate ρ by the value for air at 1200 K and 1 atm; $\rho = 0.294$ kg/m^3, and $k'' = 50.6\exp(-9361/1200) = 2.07 \times 10^{-2}$ m/s.

$$g_{m1} = \mathrm{Sh}\rho\mathscr{D}_{1m}/D_h; \qquad D_h = 2b, \quad \text{where } b \text{ is the passage width}$$

From Table 4.5, the Sherwood number for laminar flow between parallel plates with a uniform wall concentration and $L \gg D_h$ is 7.54: there is no Reynolds or Schmidt number dependence. The assumption of laminar flow will be checked later. For $\mathscr{D}_{1m}$ an approximate value is $\mathscr{D}_{\mathrm{CO,air}} = 209 \times 10^{-6}$ m^2/s at 1200 K from Table A.17*a*; then

$$g_{m1} = \frac{7.54\rho\mathscr{D}_{1m}}{2b} = \frac{(7.54)(0.294\ \mathrm{kg/m^3})(209 \times 10^{-6}\ \mathrm{m^2/s})}{(2)(1.0 \times 10^{-3}\ \mathrm{m})} = 0.232\ \mathrm{kg/m^2\,s}$$

$$\frac{1}{g_{m1}^{oa}} = \frac{1}{(0.294)(2.07 \times 10^{-2})} + \frac{1}{0.232} = 164 + 4.3 \quad \text{(rate-limited reaction)}$$

$$g_{m1}^{oa} = 5.94 \times 10^{-3}\ \mathrm{kg/m^2\,s}$$

For a gasoline specific gravity of 0.705, the exhaust flow rate is

$$\begin{aligned}\dot{m}_{\text{exhaust}} &= (80/3600 \text{ km/s})(1/8 \text{ liter/km})(10^{-3} \text{ m}^3/\text{liter})(705 \text{ kg/m}^3) \\ &\quad \times (15 \text{ kg fuel and air/kg fuel}) \\ &= 2.94 \times 10^{-2} \text{ kg/s}\end{aligned}$$

and, with the additional air subsequently added to ensure proper operation of the reactor,

$$\dot{m}_G = (1.1)(2.94 \times 10^{-2}) = 3.23 \times 10^{-2} \text{ kg/s}$$

Then

$$N_{\text{tu}} = 4.61 = \frac{\mathcal{G}_{m1}^{\text{oa}} \mathcal{P} L}{\dot{m}_G} = \frac{(5.94 \times 10^{-3})\mathcal{P} L}{3.23 \times 10^{-2}}$$

Solving, $\mathcal{P} L = 25.1 \text{ m}^2$.

If a cylindrical can of 20 cm diameter is chosen, the perimeter is the number of turns times the average circumference times 2 for both sides of the foil: assuming a 0.1 mm foil thickness, $\mathcal{P} = [0.10/(0.001 + 0.0001)](0.10\pi)(2) = 57.1$ m. Hence, $L = 25.1/57.1 = 0.44$ m (44 cm).

Comments

1. A reactor of this size may not satisfy size limitations: since the reaction is seen to be rate-limited, means for increasing the effective area of catalyst should be explored (e.g., by using a porous catalyst).
2. Check the assumption of laminar flow:

$$\dot{m}_G = \rho V A_c; \qquad V = \frac{\dot{m}_G}{\rho A_c} = \frac{(3.23 \times 10^{-2})}{(0.294)(\pi)(0.1)^2(1.0/1.1)} = 3.85 \text{ m/s}$$

$$\text{Re} = V(2b)/\nu = (3.85)(2)(0.001)/(159 \times 10^{-6}) = 48 < 2800; \text{laminar}$$

3. A spark-ignition engine obtains maximum power on a slightly *rich* mixture, that is, an air/fuel ratio less than the stoichiometric value (~15:1 for gasoline). An automobile fuel-injection system can be designed to take advantage of this characteristic for acceleration. Since there is insufficient O_2 for complete combustion, CO is produced—hence the need for a catalytic converter to reduce CO concentrations in the exhaust gases. The table below shows the relation between the air/fuel ratio and exhaust CO content for typical gasoline.

Percent CO by volume	0.2	0.5	1.0	2.0	3.0	4.0	5.0	6.0
Air/fuel ratio	14.53	14.27	14.10	13.76	13.37	12.99	12.63	12.24

4. The oxidation of CO to CO_2 in the converter is exothermic and releases 10.1×10^6 J/kg CO of heat. An energy balance on the converter will show that the resulting adiabatic temperature rise of the exhaust is approximately 100 K/percent CO by volume. Thus, proper thermal design of the converter is required to ensure its survival. Local burnup of matrix-type converters is a significant problem: when a hot spot develops, the reaction rate increases, and the local temperature can continue to increase.

11.3.2 Gas Absorption

Mass exchangers, in which a chemical species is absorbed into a liquid, are usually two-stream exchangers and will be dealt with in Section 11.4. Under certain circumstances, however, the performance of such exchangers can be described by the simple single-stream mass exchanger equations. Such situations are best illustrated by numerical examples. One such example follows as Example 11.2 and concerns the absorption of a gaseous species into a liquid, in which it subsequently enters into a chemical reaction. The reaction is very fast and maintains a near-zero concentration of the solute species at the s-surface adjacent to the liquid surface. Thus, the liquid-side mass transfer resistance is negligible, and the absorption is *gas side–controlled.* The s-surface concentration can be taken to be zero, and hence does not vary through the exchanger. In an analogous heat exchanger, it is the wall temperature T_s that does not vary through the exchanger, as, for example, in a simple condenser or evaporator.

Exercises 11–16 through 11–19 concern absorption of sparingly soluble gases such as air, CO, and CO_2 into water. Since mole fractions in the liquid phase are very small, usually so too are mole-fraction gradients. Thus, except in unusual circumstances, such gas absorption processes are *liquid side–controlled:* the gas-side mass transfer resistance can be ignored. In addition, the situations in these exercises are such that the u-surface composition varies negligibly along the exchanger so that the single-stream exchanger equations apply. These single-stream mass exchanger problems are, of course, special cases of the more general two-stream mass exchanger, which will be analyzed in Section 11.4. In Sections 11.4.1 and 11.4.2 the concept of gas-side or liquid-side mass transfer control will be carefully developed. However, it is not premature to deal with these special cases here. There are many important applications of gas absorption processes similar to those described above, both in technology and in the environment. Examples include aeration of sewage, carbonation of soft drinks, and aeration of lakes and rivers; the literature on these topics usually assumes liquid-side mass transfer control without discussion, since the validity of such a model is well established.

EXAMPLE 11.2 Scrubbing of Sulfur Dioxide from a Steam Generator Exhaust

The 50,000 kg/h exhaust stream from a steam generator contains 3% sulfur dioxide by volume. To reduce SO_2 air pollution, the gas is passed through a bed of rocks, which can be approximated as 2 cm–diameter spheres, with a volume void fraction of 58%. The bed is 50 m^2 in cross-sectional area, 2 m deep, and at approximately 320 K temperature. Forty percent of the rock surface area is kept wet with an alkaline solution such that the SO_2 concentration adjacent to the wet surface is approximately zero. Estimate the effectiveness of the bed, the mole fraction of SO_2 at the exit, and the kilograms per hour of SO_2 that escape to the atmosphere.

Solution

Given: Packed-bed scrubber for SO_2 absorption.

Required: Bed effectiveness, outlet gas SO_2 mole fraction, and outlet SO_2 flow rate.

Assumptions:
1. 40% of packing wetted.
2. An s-surface concentration of SO_2 equal to zero, giving a gas side–controlled mass transfer process.
3. Gas properties can be approximated by those for air.

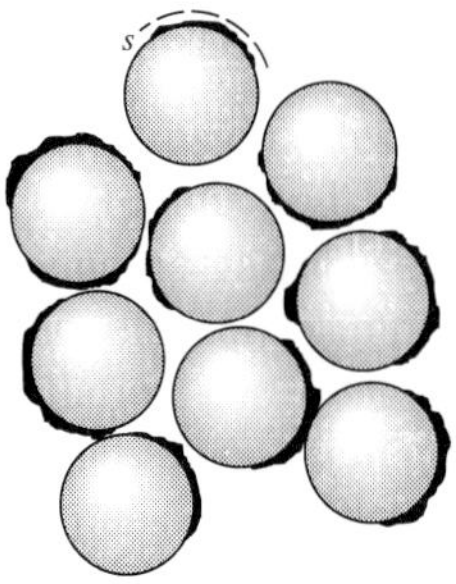

The bed can be modeled as a single-stream mass exchanger because $x_{SO_2 s} \simeq 0$, independent of depth; only the composition change of the exhaust stream is of concern. The number of transfer units for the bed can be calculated from $N_{tu} = \mathscr{G}_m \mathscr{P} H / \dot{M}_G$, where H is the height of the bed. Since the bed is only partially wet, the correlations for a dry packed bed given in Section 4.5.2 will be used to obtain $\mathscr{G}_m$; by analogy, Eq. (4.131) becomes

$$\mathrm{Sh} = (0.5\,\mathrm{Re}^{1/2} + 0.2\,\mathrm{Re}^{2/3})\,\mathrm{Sc}^{1/3}$$

The characteristic length is

$$\mathscr{L} = d_p\left(\frac{\varepsilon_v}{1-\varepsilon_v}\right) = 0.02\left(\frac{0.58}{1-0.58}\right) = 0.0276 \text{ m}$$

and the characteristic velocity is

$$\mathscr{V} = \frac{\dot{m}}{\rho \varepsilon_v A_c} = \frac{(50{,}000/3600)}{(1.106)(0.58)(50)} = 0.433 \text{ m/s}$$

where A_c is the cross-sectional (frontal) area of the bed, and the properties of the exhaust gases have been approximated by those for air at 320 K and 1 atm.

$$\mathrm{Re} = \frac{\mathscr{V}\mathscr{L}\rho}{\mu} = \frac{(0.433)(0.0276)(1.106)}{1.929 \times 10^{-5}} = 685$$

From Table A.17b, Sc = 1.24,

$$\mathrm{Sh} = [0.5(685)^{1/2} + 0.2(685)^{2/3}](1.24)^{1/3} = 30.8$$

$$\mathscr{G}_m = c\mathscr{D}\,\mathrm{Sh}/\mathscr{L} = \mu\,\mathrm{Sh}/M\,\mathrm{Sc}\,\mathscr{L}$$

Thus, $\mathscr{G}_m = (1.929 \times 10^{-5})(30.8)/(29)(1.24)(0.0276) = 5.98 \times 10^{-4}$ kmol/m^2 s.

The molar flow rate of the exhaust gas stream is

$$\dot{M}_G = (50{,}000)/(3600)(29) = 0.479 \text{ kmol/s}$$

Next we must find the perimeter of the transfer area: the specific surface area of the bed is

$$a = \frac{A_p}{V_p}(1-\varepsilon_v) = \frac{6}{d_p}(1-\varepsilon_v) \quad \text{for spheres}$$

$$a = (6/0.02)(1-0.58) = 126 \text{ m}^{-1}$$

$\mathscr{P} = aA_c = (126)(50) = 6300$ m; but only 40% of the rocks are wet, so that the effective perimeter is $(0.4)(6300) = 2520$ m.

$$N_{tu} = \mathscr{G}_m \mathscr{P} H / \dot{M}_G = (5.98 \times 10^{-4})(2520)(2.0)/(0.479) = 6.29$$

Equation (11.11) gives the effectiveness as

$$\varepsilon = 1 - e^{-N_{tu}} = 1 - e^{-6.29} = 1 - 0.00185 = 0.99815$$

The effectiveness of this single-stream exchanger is the actual change in concentration of the gas stream, divided by the maximum possible change. Since the concentration of SO_2 in the outlet stream will be zero in an infinitely long exchanger,

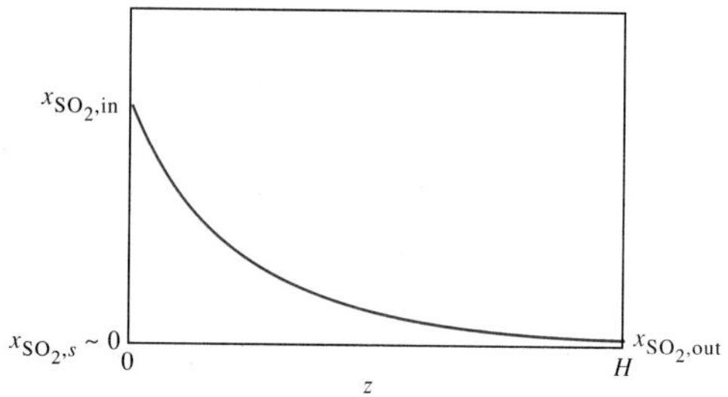

$$\varepsilon = \frac{x_{SO_2,in} - x_{SO_2,out}}{x_{SO_2,in} - 0}$$

Solving for $x_{SO_2,out}$,

$$x_{SO_2,out} = x_{SO_2,in}(1 - \varepsilon) = (0.03)(0.00185) = 5.55 \times 10^{-5}$$

$$\dot{M}_{SO_2,out} = x_{SO_2,out}\dot{M}_G = (5.55 \times 10^{-5})(0.459) = 2.55 \times 10^{-5} \text{ kmol/s}$$

$$\dot{m}_{SO_2,out} = \left(\dot{M}_{SO_2,out}\right)(M_{SO_2}) = (2.55 \times 10^{-5})(64)(3600) = 5.87 \text{ kg/h}$$

Comments

1. Notice that we multiply by the molecular weight of SO_2, not of the exhaust mixture, to obtain the mass flow of SO_2.
2. The effectiveness of this rock bed is very high; it is more convenient to quote its *ineffectiveness*, $1 - \varepsilon = 1 - 0.99815 = 0.00185 = 0.185\%$ (i.e., only 0.185% of the entering SO_2 escapes from the bed).

11.3.3 Electrostatic Precipitators

Electrostatic precipitation is widely used to remove particles from the exhausts of coal-fired power plants and solid waste incinerators. The process is particularly effective in removing particles in the 0.1–10 μm size range. Energy requirements are lower than for other particulate removal systems, such as venturi scrubbers and fabric filters. Although particles in industrial emissions may possess an electric charge, commercial precipitators are based on particle charging in coronas generated upstream or within the precipitator, in order to obtain a maximum charge on the particles. Figure 11.5*a* shows a dry precipitator, and Fig. 11.5*b* shows a weir-type wet-wall precipitator. An electric field set up in the precipitator causes the charged particles to migrate to the collector plates by the phenomenon of *forced diffusion*. The electric field produces a *corona* of gas molecules ionized by high-energy electrons. Excess electrons combine with electronegative gas molecules such as O_2 and SO_2, whereas the negative ions are adsorbed onto particles, which migrate to the grounded plates. Once collected, the particles begin to lose their charge to the plates; this charge transfer completes the electrical circuit to maintain the current flow and voltage drop. In a wet-wall precipitator, the water film serves to prevent dust buildup on the electrodes that would increase resistivity, and also minimizes reentrainment

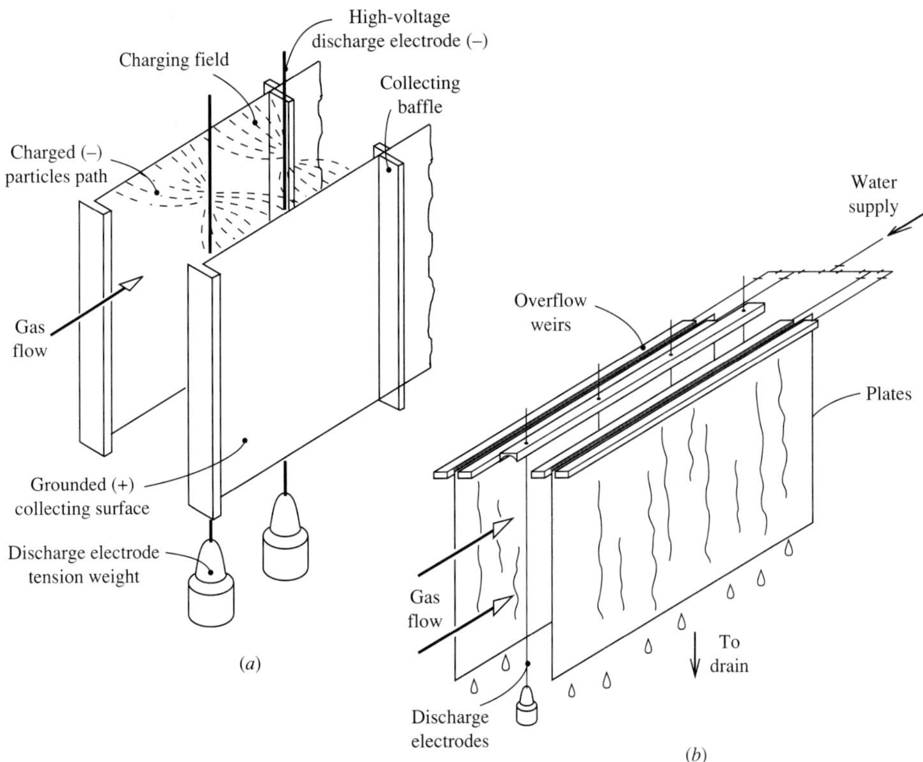

Figure 11.5 Electrostatic precipitators: (*a*) dry wall, (*b*) weir-type wet wall.

of particles into the gas stream. In dry precipitators, dust is removed by periodic *rapping* of the electrodes using mechanical hammers: reentrainment of particles is then a problem, but, on the other hand, dry units do not have the corrosion and scaling problems of wet units. Dry electrostatic precipitators are used for coal-fired power plants and cement kilns; wet units can be found on paint finishing lines, sewage sludge incinerators, and for mist removal downstream of SO_2 scrubbers.

Migration Velocity

The steady migration velocity V^E of a particle due to forced diffusion is obtained from a force balance, in which the electrical force on the particle equals the viscous drag on the particle. For a particle diameter d_p, a particle charge Q [A s], and an electric field E [V/m], the electric force is QE and

$$QE = 3\pi d_p \mu V^E$$

where it has been assumed that the Stokes' drag law, Eq. (4.73), applies. For very small particles it is necessary to introduce a slip correction to Stokes' law to account

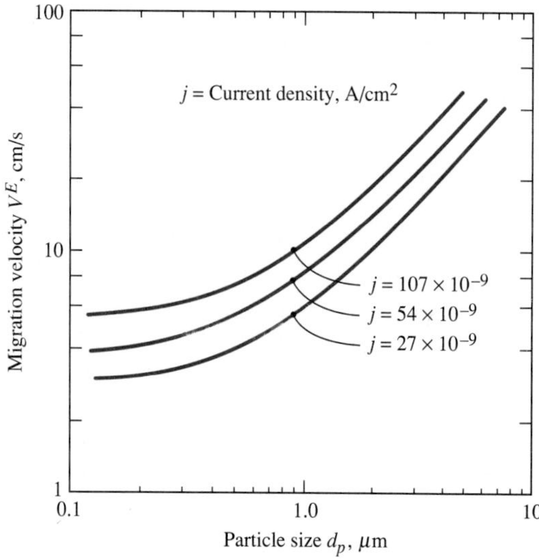

Figure 11.6 Particle migration velocity V^E as a function of particle size and current density at 30 kV [2].

for departure from continuum flow conditions [1]. (See also Section 9.7.3.) Solving for V^E,

$$V^E = \frac{QE}{3\pi d_p \mu} \tag{11.14}$$

The particle charge Q can be calculated from well-known principles of electrostatics, but the theory is complicated [2,3]. Figure 11.6 shows a graph of V^E as a function of particle size for typical precipitator operating conditions.

Precipitator Performance

Figure 11.7 shows an elemental control volume Δx long, used to derive the equation governing particle concentration through a precipitator. The concentration of particles is expressed in terms of a number density $\mathcal{N}$ that has units [m^{-3}] but as a conceptual aid can be assigned units [particles/m^3]. The flow of gas in a commercial precipitator is always turbulent, and the characteristic turbulent mixing velocity is quite large compared with the electrical migration velocity, except very close to the electrodes. Thus, we will assume that the particle concentration $\mathcal{N}$ is

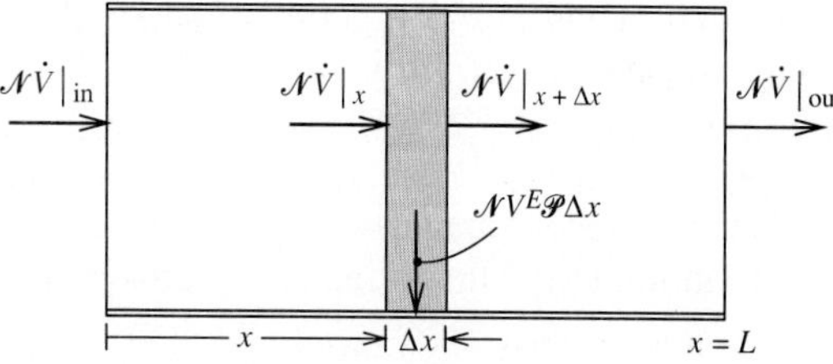

Figure 11.7 Schematic of an electrostatic precipitator showing a particle balance on an elemental volume Δx long.

uniform across the channel. The principle of conservation of particles applied to the elemental control volume requires that, at steady state,

$$\begin{matrix}\text{Rate of inflow} \\ \text{of particles}\end{matrix} = \begin{matrix}\text{Rate of deposition} \\ \text{of particles}\end{matrix} + \begin{matrix}\text{Rate of outflow} \\ \text{of particles}\end{matrix}$$

$$\mathcal{N}\dot{V}|_x = \mathcal{N}V^E\mathcal{P}\Delta x + \mathcal{N}\dot{V}|_{x+\Delta x}$$

where $\dot{V}$ is the volume flow rate of gas, $\mathcal{P}$ is the perimeter of the collecting plates, and V^E is taken as positive toward the plates. Rearranging, dividing by Δx, and letting $\Delta x \to 0$,

$$\frac{d\mathcal{N}}{dx} + \frac{V^E\mathcal{P}}{\dot{V}}\mathcal{N} = 0 \tag{11.15}$$

which is the differential equation governing the particle concentration distribution through the precipitator, $\mathcal{N}(x)$. The required boundary condition is the inlet condition,

$$x = 0: \quad \mathcal{N} = \mathcal{N}_{\text{in}} \tag{11.16}$$

The solution gives $\mathcal{N}(x)$,

$$\mathcal{N} = \mathcal{N}_{\text{in}}e^{-V^E\mathcal{P}x/\dot{V}} \tag{11.17}$$

In particular, at $x = L$, $\mathcal{N} = \mathcal{N}_{\text{out}}$,

$$\mathcal{N}_{\text{out}} = \mathcal{N}_{\text{in}}e^{-V^E\mathcal{P}L/\dot{V}}$$

or

$$\frac{\mathcal{N}_{\text{in}} - \mathcal{N}_{\text{out}}}{\mathcal{N}_{\text{in}}} = 1 - e^{-V^E\mathcal{P}L/\dot{V}} \tag{11.18}$$

In the absence of reentrainment, the particle concentration can be reduced to zero in an infinitely long precipitator. Thus, we recognize that the left-hand side of Eq. (11.18) is the effectiveness[1] of the precipitator as a mass exchanger, and write

$$\varepsilon = 1 - e^{-N_{\text{tu}}}; \qquad N_{\text{tu}} = \frac{V^E\mathcal{P}L}{\dot{V}} \tag{11.19}$$

Notice that the electrical migration velocity can be viewed as a particle transfer conductance; the product $\mathcal{N}V^E$ [particles/m^2 s] gives the local deposition rate.

Equation (11.19) is called the Deutsch-Anderson equation by particle (aerosol) technologists, after E. Anderson, who obtained it experimentally in 1919, and W. Deutsch, who supplied a theoretical derivation in 1922. Equation (11.19) gives the effectiveness for a particular particle size, since the migration velocity is size-dependent. In practice, precipitators collect particles with a wide range of sizes.

[1] The effectiveness is called the *collection efficiency* by aerosol technologists.

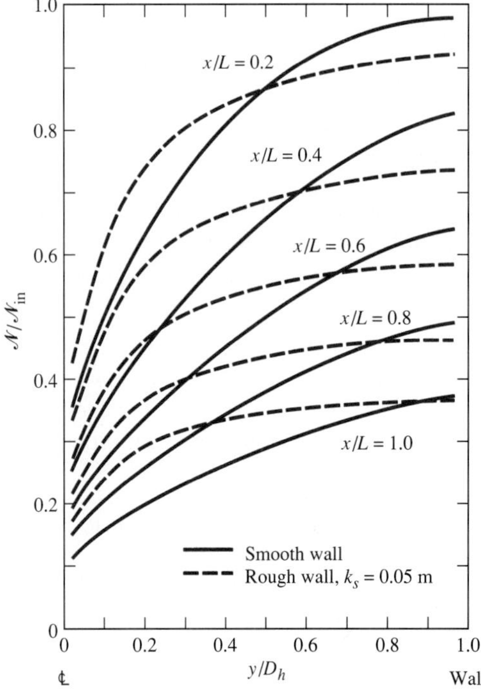

Figure 11.8 Particle concentration profiles in a plate-plate electrostatic precipitator showing the effect of plate roughness [4]. For $V^E \mathscr{P} L/\dot{V} = 10$; coordinate y is measured from the centerplane.

Design calculations can be made by dividing the expected particle size distribution into finite size increments to obtain an overall collection effectiveness.

The assumption of a uniform particle concentration across the channel is, in fact, rather poor. Figure 11.8 shows concentration profiles calculated from more exact theory, where it is seen that the particle concentration increases as the channel wall is approached. Thus, Eq. (11.19) underestimates the effectiveness. Interestingly enough, although the more exact theory is confirmed by laboratory experiments [5], the measured performance of industrial precipitators is always less than that given by Eq. (11.19). The discrepancy is attributed to such problems as particle reentrainment and gas leakage between precipitator sections. Thus, current design practice is to use the Deutsch-Anderson equation with an empirical value of the migration velocity V^E that is less than the theoretical value. Typical design parameters are: plate spacing 0.2–0.28 m, gas velocity 0.8–2.5 m/s, plate height 4–13 m, plate length 0.5 to 2.0 times height, effective migration velocity 3–20 cm/s, and pressure drop 25–125 Pa.

Particle Resistivity

The electrical resistivity of the particles plays a very important role. If the resistance of the dust layer is too low, the electrostatic charge is lost too quickly and collected particles can be reentrained into the gas flow. On the other hand, if the resistance of the layer is too high, the charge is retained at the collecting plates, producing a *back corona* that reduces the degree of ionization and hence lowers the effective migration velocity. Also, since the particles remain strongly attracted to the plate, removal by rapping is difficult. The resistivity of fly ash in the flue gas of coal-fired boilers depends on temperature, and the chemical composition of the ash and gases. Resistivity tends to have a maximum value in the range of about 120–180°C, which coincides with typical flue gas temperatures (below 120°C there is a risk of condensation of sulfuric acid on duct walls; above 180°C the energy loss out of the stack is excessive). Resistivity also decreases with increased sulfur content of the coal. The practical problem is usually one of ensuring that the fly ash resistivity is not too high. For example, if a power plant changes from using a high-sulfur coal to a low-sulfur coal, the performance of the precipitator is adversely affected. One remedy is to chemically condition the flue gases by the addition of a small amount of ammonium salts to reduce the fly ash resistivity. Empirical data for the effect of particle resistivity on effective migration velocity are available [2] and are used by design engineers.

EXAMPLE 11.3 An Electrostatic Precipitator for a Toxic Waste Incinerator

An electrostatic precipitator for a toxic waste incinerator has a plate spacing of 0.2 m and is 20 m long. Gas at 1 atm and 410 K flows at a bulk velocity of 1.5 m/s through the unit. The current density is 27×10^{-9} A/cm^2 at 30 kV. Determine the collection efficiency for particles of sizes 0.5 μm, 0.9 μm, 2.0 μm, and 8.0 μm.

Solution

Given: A 10 m–long electrostatic precipitator operating at 30 kV, giving a current density of 27×10^{-9} A/cm^2.

Required: Collection efficiency for particles of sizes 0.5, 0.9, 2.0, and 8.0 μm.

Assumptions:
1. A uniform particle concentration across the channel (the Deutsch model).
2. The electrical migration velocity is given by the data in Fig. 11.6.

From Eq. (11.19) the effectiveness (collection efficiency) is

$$\varepsilon = 1 - e^{-N_{\text{tu}}}; \qquad N_{\text{tu}} = V^E \mathscr{P} L/\dot{V}$$

For a plate spacing d and precipitator height H, the volume flow rate is $\dot{V} = Hdu_b$, and the perimeter is $\mathscr{P} = 2H$. Hence,

$$N_{\text{tu}} = \frac{V^E \mathscr{P} L}{\dot{V}} = \frac{V^E 2HL}{Hdu_b} = \frac{2V^E L}{du_b}$$

For a particle size of 0.5 μm and current density $j = 27 \times 10^{-9}$ A/cm^2, Fig. 11.6 gives $V^E = 4.0$ cm/s $= 0.040$ m/s. Thus, the number of transfer units is

$$N_{\text{tu}} = \frac{(2)(0.040)(20)}{(0.2)(1.5)} = 5.33$$

giving an effectiveness of

$$\varepsilon = 1 - e^{-5.33} = 1 - 0.00483 = 0.9952$$

The results for all particle sizes are tabulated below.

Particle Size μm	V^E m/s	N_{tu}	$1-\varepsilon$	ε
0.5	0.040	5.33	4.83×10^{-3}	0.99517
0.9	0.055	7.33	6.53×10^{-4}	0.999347
2.0	0.105	14.0	8.32×10^{-7}	0.999999
8.0	0.42	56.0	4.78×10^{-25}	1.000000

Comments

1. The collection efficiency is seen to be very dependent on particle size.
2. For this high-effectiveness exchanger, it is more appropriate to look at $(1-\varepsilon)$, which is the fraction of particles *not* collected.
3. For the larger particles, the very low values of $(1-\varepsilon)$ suggest that, in practice, secondary factors might determine the true collection efficiency (for example, gas leakage or reentrainment while rapping).

11.3.4 Fibrous Filters

A simple and effective method of removing solid particles from a gas stream is to use a filter of some porous material. Commonly used are fabrics of fine fibers woven from cotton or wool, as well as fiberglass and other synthetic materials. Familiar examples include vacuum cleaner bags and air filters on automobiles. Large-scale industrial fabric filters are often called "bag houses" and consist of several large fabric bags. When operating properly, the *collection efficiency* (mass exchanger effectiveness) can exceed 99.99%, with pressure drops ranging from 500 to 1500 Pa. The bags are cleaned by periodic mechanical shaking or using a reverse air flow to dislodge the collected dust layer. Selection of suitable fiber diameters allows particles of a wide range of sizes to be collected.

The process of particle collection by a filter is primarily a combination of diffusion to, interception by, and inertial impaction on the fibers making up the filter. Diffusion is important only for very small particles (see Section 9.7.3), whereas inertial impaction is important only for the larger particles that cannot follow flow streamlines around a fiber. Particles are intercepted by the fiber when their centers come within a particle radius of the fiber, as shown in Fig. 11.9. Analyses of these processes are presented in the aerosol science literature [6,7,8]. The collection efficiency of a single fiber in a bed, γ, is defined as the ratio of particles that are collected to the number that would strike the fiber if the particle path lines were not affected by the presence of the fiber.

The parameters that affect the single-fiber collection efficiency are as follows:

1. The Reynolds number $\mathrm{Re} = \mathscr{V} d_f/\nu$, where $\mathscr{V}$ is a characteristic gas velocity, d_f is the average fiber diameter, and ν is the gas kinematic viscosity. The Reynolds number characterizes the flow around the fiber.

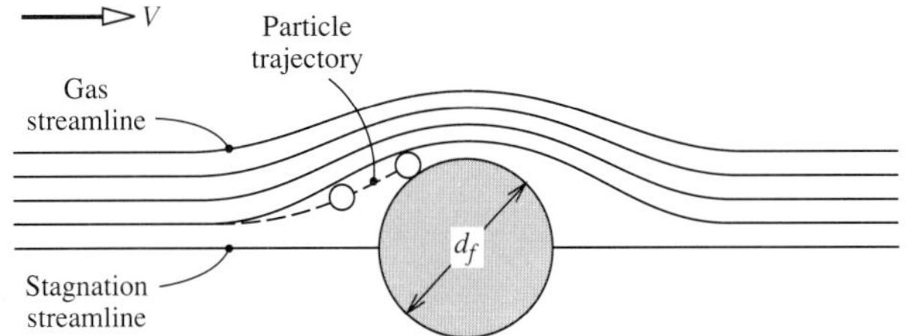

Figure 11.9 Schematic of particle deposition on a filter fiber by inertial impaction.

2. The Peclet number $\text{Pe} = \mathcal{V} d_f/\mathcal{D}$, where $\mathcal{D}$ is the particle diffusion coefficient that was introduced in Section 9.7.3. The Peclet number characterizes the particle concentration boundary layer for diffusional deposition of very small particles.

3. The interception number $R = d_p/d_f$, where d_p is the particle diameter. The interception number characterizes particle deposition by interception.

4. The Stokes number $\text{Stk} = \rho_p d_p^2 \mathcal{V}/18\mu d_f$ where ρ_p is the particle density and μ the gas dynamic viscosity. The Stokes number characterizes inertial effects and hence inertial impaction of large particles that do not follow streamlines.

5. The filter porosity, or volume void fraction, ε_v. The flow field around a fiber depends on the porosity of the filter (and the already defined Reynolds number).

In general, we can write the single-fiber collection efficiency as

$$\gamma = \gamma(\text{Re}, \text{Pe}, R, \text{Stk}, \varepsilon_v)$$

Engineering correlations for γ recognize two Reynolds number regimes: creeping flow for $\text{Re} < 1$, and transitional flow for $1 < \text{Re} < 300$. Also, two regimes of Stokes numbers are appropriate: for $\text{Stk} < 0.5$ impaction is ignored, and for $\text{Stk} > 0.5$ diffusion is ignored. For $\text{Stk} < 0.5$ it is sufficient to consider only the limit $\text{Pe} \gg 1$, which corresponds to a small particle diffusion coefficient and hence a thin particle concentration boundary layer. Section 9.7.3 gives formulas and data for the particle diffusion coefficient that can be used to estimate the Peclet number. Table 11.2 lists recommended formulas for γ.

Figure 11.10 shows a schematic of a filter. The superficial velocity (based on the frontal area, A_{fr}) of the gas flowing through the bed is V, and the free-space velocity (based on flow cross-sectional area) is $\mathcal{V} = V/\varepsilon_v$. The average diameter of the fiber is d_f, and the length of fiber per unit volume of bed is ℓ m/m^3. At steady state, a particle balance on the gas flow in a length of bed Δx long requires that

$$\text{Particle inflow} - \text{Particle outflow} = \text{Rate of particle collection}$$

$$\mathcal{N} VA_{\text{fr}}\big|_x - \mathcal{N} VA_{\text{fr}}\big|_{x+\Delta x} = \gamma(V/\varepsilon_v)\mathcal{N} d_f \ell A_{\text{fr}}\Delta x$$

Table 11.2 Recommended formulas for single-fiber collection efficiency in fibrous filters.

	Viscous Flow, Re < 1	Transitional Flow, 1 < Re < 300
Stk < 0.5 Pe ≫ 1 Impaction negligible	$\gamma = 1.6(\varepsilon_v/H)^{1/3}\,\mathrm{Pe}^{-2/3} + 0.6(\varepsilon_v/H)R^3/(1+R)$, where H is the hydrodynamic factor for Kuwabara flow, $H = -0.5\ln(1-\varepsilon_v) - 0.75 + (1-\varepsilon_v) - 0.25(1-\varepsilon_v)^2$	$\gamma = \gamma_D + \gamma_R + 1.24R^{2/3}/(H\,\mathrm{Pe})^{1/2}$ $\gamma_D = (1 + 0.55\pi\,\mathrm{Pe}^{1/3})/\mathrm{Pe}$ $\gamma_R = [(1+R)^{-1} - (1+R) + 2(1+R)\ln(1+R)]/(2H)$
Stk > 0.5 Diffusion negligible	$\gamma = F_1(R, \mathrm{Stk})F_2(\varepsilon_v)$ $F_1 = R + (0.5 + 0.8R)\mathrm{Stk} - 0.105R\,\mathrm{Stk}^2$ $F_2 = 0.16 + 10.9(1-\varepsilon_v) - 17(1-\varepsilon_v)^2$	$\gamma = 1 - (1-\gamma_R)(1-\gamma_I)$ γ_R as above $\gamma_I = F_3(\mathrm{Re},\mathrm{Stk})F_4(\varepsilon_v)$ $F_3 = \{1 + [1.53 - 0.23\ln\mathrm{Re} + 0.0167(\ln\mathrm{Re})^2]/\mathrm{Stk}\}^{-2}$ $F_4 = F_2/0.16$

where $\mathcal{N}[m^{-3}]$ is the particle concentration. Dividing by $VA_{\mathrm{fr}}\Delta x$ and letting $\Delta x \to 0$,

$$\frac{d\mathcal{N}}{dx} + \frac{d_f \ell \gamma}{\varepsilon_v}\mathcal{N} = 0 \qquad \mathbf{(11.20)}$$

The length of fiber per unit volume ℓ is related to the void fraction as

$$1 - \varepsilon_v = (\pi/4)d_f^2\ell; \qquad \text{hence} \quad \ell = 4(1-\varepsilon_v)/\pi d_f^2$$

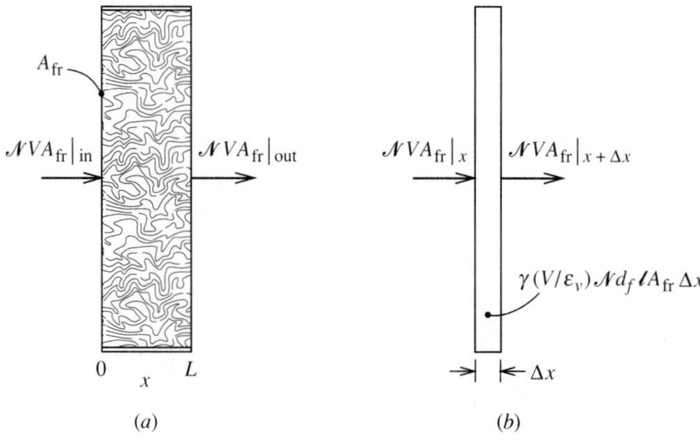

Figure 11.10 (*a*) Schematic of a filter. (*b*) Particle balance on an elemental volume Δx long.

Substituting for ℓ in Eq. (11.20),

$$\frac{d\mathcal{N}}{dx} + \frac{4(1-\varepsilon_v)\gamma}{\varepsilon_v \pi d_f}\mathcal{N} = 0 \tag{11.21}$$

which must be solved subject to the boundary condition

$$x = 0: \quad \mathcal{N} = \mathcal{N}_{\text{in}} \tag{11.22}$$

The solution gives $\mathcal{N}(x)$ as

$$\mathcal{N} = \mathcal{N}_{\text{in}} e^{-4(1-\varepsilon_v)\gamma x/\varepsilon_v \pi d_f} \tag{11.23}$$

Substituting $\mathcal{N} = \mathcal{N}_{\text{out}}$ at $x = L$ and rearranging gives

$$1 - \frac{\mathcal{N}_{\text{out}}}{\mathcal{N}_{\text{in}}} = \frac{\mathcal{N}_{\text{in}} - \mathcal{N}_{\text{out}}}{\mathcal{N}_{\text{in}}} = 1 - e^{-4(1-\varepsilon_v)\gamma L/\varepsilon_v \pi d_f} \tag{11.24}$$

If there is no particle reentrainment, $\mathcal{N}_{\text{out}} \to 0$ in an infinitely thick filter; thus, again we recognize that the left-hand side of Eq. (11.24) is the effectiveness (or collection efficiency) of the filter as a mass exchanger, and we write

$$\varepsilon = 1 - e^{-N_{\text{tu}}}; \qquad N_{\text{tu}} = \frac{4(1-\varepsilon_v)\gamma L}{\varepsilon_v \pi d_f} \tag{11.25}$$

When filters are used to remove radioactive particles, the quantity $\mathcal{N}_{\text{in}}/\mathcal{N}_{\text{out}} = 1/(1-\varepsilon)$ is called the **decontamination factor (DF)**. For example, an effectiveness or collection efficiency of 99.99% corresponds to a decontamination factor of 10^4. When the effectiveness is very high, fewer significant figures are required to specify the decontamination factor to the same accuracy as the effectiveness.

The Computer Program FILTER

FILTER calculates the effectiveness and decontamination factor of fibrous filters according to the theory presented in this section. Spherical particles and air properties are assumed. The formulas for single-fiber collection efficiency are given in Table 11.2.

EXAMPLE 11.4 A Fiberglass Filter for Decontaminating Radioactive Waste Gas

The effluent gas from a process vessel used to treat a radioactive ore is passed through a fiberglass filter of cross-sectional area 1.30 m^2 and thickness 0.5 m. The fibers have a diameter of 30 μm and are packed at a density of 390 kg/m^3. The gas flow is 25 m^3/min at 300 K and 1 atm. Determine the decontamination factor for 0.5 μm particles of average density 5000 kg/m^3.

Solution

Given: Fiberglass filter of thickness 0.5 m.

Required: Decontamination factor for 0.5 μm particles.

Assumptions:
1. No reentrainment of particles.
2. A clean filter.
3. The gas properties can be approximated by those of air.

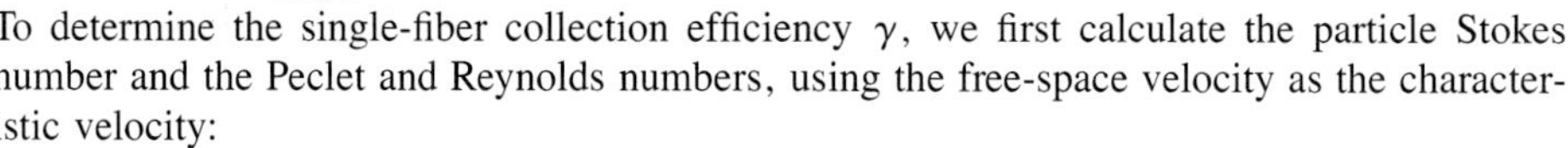

To obtain the decontamination factor, we must first calculate the number of transfer units for the filter; from Eq. (11.25),

$$N_{tu} = \frac{4(1-\varepsilon_v)\gamma L}{\varepsilon_v \pi d_f}$$

With the glass density taken as 2300 kg/m^3, the volume void fraction of the filter is

$$\varepsilon_v = 1 - \frac{390}{2300} = 0.830$$

To determine the single-fiber collection efficiency γ, we first calculate the particle Stokes number and the Peclet and Reynolds numbers, using the free-space velocity as the characteristic velocity:

$$\mathscr{V} = \frac{25}{(60)(1.3)(0.830)} = 0.386 \ \text{m/s}$$

$$\text{Stk} = \frac{\rho_p d_p^2 \mathscr{V}}{18\mu d_f} = \frac{(5000)(0.5\times10^{-6})^2(0.386)}{(18)(18.43\times10^{-6})(30\times10^{-6})} = 0.0485$$

$$\text{Pe} = \frac{\mathscr{V} d_f}{\mathscr{D}} = \frac{(0.386)(30\times10^{-6})}{6.38\times10^{-11}} = 1.815\times10^5$$

where $\mathscr{D}$ has been obtained from Table A.30;

$$\text{Re} = \frac{\mathscr{V} d_f}{\nu} = \frac{(0.386)(30\times10^{-6})}{15.66\times10^{-6}} = 0.739$$

Also, the interception number is

$$R = d_p/d_f = 0.5/30 = 0.0167$$

Since Re < 1.0 and Stk < 0.5, the flow is creeping and impaction is negligible; also Pe $\gg$ 1, and thus from Table 11.2, the collection efficiency of a single fiber is

$$\gamma = 1.6(\varepsilon_v/H)^{1/3}\text{Pe}^{-2/3} + 0.6(\varepsilon_v/H)R^3/(1+R)$$

where

$$H = -0.5\ln(1-\varepsilon_v) - 0.75 + (1-\varepsilon_v) - 0.25(1-\varepsilon_v)^2$$
$$= -0.5\ln(1-0.830) - 0.75 + (1-0.830) - 0.25(1-0.830)^2 = 0.30$$

Hence, we calculate γ as

$$\gamma = 1.6(0.830/0.30)^{1/3}(1.815\times10^5)^{-2/3} + \frac{0.6(0.830/0.30)(0.0168)^3}{(1+0.0168)}$$
$$= 7.01\times10^{-4} + 7.74\times10^{-6} = 7.09\times10^{-4}$$

Then the number of transfer units of the filter is

$$N_{tu} = \frac{4(1-\varepsilon_v)\gamma L}{\varepsilon_v \pi d_f} = \frac{4(1-0.83)(7.09\times10^{-4})(0.5)}{(0.83)(\pi)(30\times10^{-6})} = 3.08$$

giving an effectiveness of

$$\varepsilon = 1 - e^{-N_{tu}} = 1 - e^{-3.08} = 95.4\%$$

or a decontamination factor of

$$\text{DF} = \frac{1}{1-\varepsilon} = 21.8$$

Comments

1. In practice, a number of filter elements are often arranged in series; the first is designed to filter out the largest particles, and succeeding elements filter out smaller particles.
2. Use FILTER to check this result.

11.4 TWO-STREAM MASS EXCHANGERS

Two-stream mass exchangers are widely used in the chemical and process industries and thus are given considerable attention in standard chemical engineering texts. Here we will restrict our attention to countercurrent (counterflow) gas scrubbers, in which a component of a gas stream is removed by absorption into a liquid stream. No chemical reactions will be considered. Thus, the transfer process involves convection from the bulk gas to the gas-liquid interface, dissolution at the interface, and convection into the bulk liquid stream. The solubility of the gas will be described by Henry's law, Eq. (9.13). In the case of systems for which Henry's law is invalid, for example, SO_2-H_2O, an *effective* Henry number will be defined as an average value over the concentration range of interest. Use of Henry's law gives a linear system of equations to be solved, and an analytical result can be obtained that is analogous to Eq. (8.41) for a counterflow heat exchanger. This relatively simple analytical result conveniently displays the essential features of gas scrubber operation.

Gas scrubbers are usually in the form of vertical packed columns (towers), with liquid flowing down over the packing and gas flowing upward (see Fig. 9.1). The design of such columns also requires pressure drop calculations and the establishment of *flooding limits*. A column is said to flood when the gas flow is sufficient to blow the liquid upward against gravity. Correlations for pressure drop and flooding are presented to complete this section.

11.4.1 Overall Mole Transfer Conductances

Figure 11.11 shows a chemical species in a gaseous mixture being absorbed into a nonvolatile liquid, for example, the NH_3-N_2-H_2O system at low temperatures with

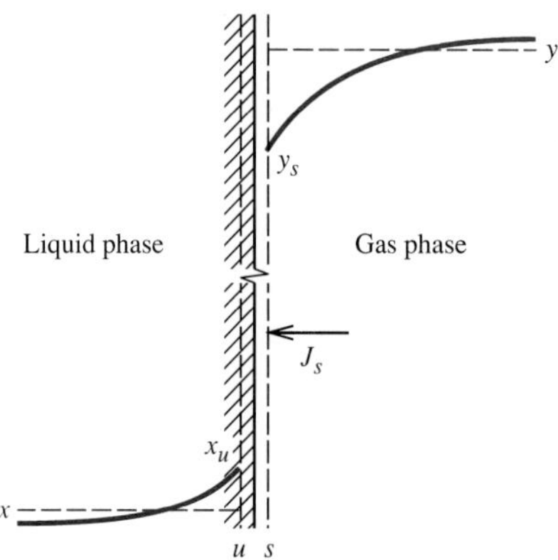

Figure 11.11 Concentration profiles for absorption of a gaseous species into a liquid.

NH_3 the solute and H_2O the solvent. The concentration profiles of the species in question are shown as solid lines, and its bulk concentration in each stream is shown as a dotted line. To avoid excessive subscripting, liquid- and gas-phase mole fractions are denoted x and y, respectively. This notation is commonly used by chemical engineers. Bulk mole fractions are not subscripted, consistent with the conventional notation for two-stream heat exchangers; values at the interface are subscripted s and u as usual. From Henry's law, Eq. (9.13),

$$y_s = \text{He}\, x_u \tag{11.26}$$

The Henry number, He, is obtained from the Henry constant $C_{\text{He}}(T)$ using Eq. (9.14):

$$\text{He} = \frac{C_{\text{He}}(T)}{P} \tag{11.27}$$

where P is total pressure. It is understood that C_{He} and He are properties of the solute-solvent combination.

Actually, the Henry constant is found to vary somewhat with pressure and liquid-phase concentration; however, for sparingly soluble gases the variation is small and can be ignored. Henry constant data for aqueous solutions of sparingly soluble solutes are given in Table A.21. In concentration ranges of practical interest for the NH_3-water and SO_2-water systems, the Henry constant varies appreciably with liquid-phase concentration. Table A.22 gives equilibrium compositions for these systems; often a suitable average value for the Henry constant is sufficient.

Low mass transfer rates will be assumed; thus, from Eq. (9.37), the molar flux across the interface can be written as

$$J_s = \mathcal{G}_G(y - y_s) = \mathcal{G}_L(x_u - x) \tag{11.28}$$

Our sign convention here makes J_s positive for absorption into the liquid (and negative for desorption from the liquid). The subscripts G and L have been introduced to distinguish mole transfer conductances in the gas and liquid phases, and the subscript m has been dropped for convenience. Also, since only one species is transferred, there is no ambiguity in denoting the flux of that species J_s. Substituting

Eq. (11.26) into Eq. (11.28) and rearranging,

$$J_s = \mathcal{G}_G(y - \text{He}\, x_u) = \mathcal{G}_G\left(y - \frac{\text{He}\, J_s}{\mathcal{G}_L} - \text{He}\, x\right)$$

Solving for J_s,

$$J_s = \frac{1}{1/\mathcal{G}_G + \text{He}/\mathcal{G}_L}(y - \text{He}\, x) = \mathcal{G}_G^{\text{oa}}(y - \text{He}\, x) \tag{11.29}$$

where

$$\frac{1}{\mathcal{G}_G^{\text{oa}}} = \frac{1}{\mathcal{G}_G} + \frac{\text{He}}{\mathcal{G}_L} \tag{11.30}$$

and $\mathcal{G}_G^{\text{oa}}$ is called the overall mole transfer conductance for a gas-side driving force $(y - \text{He}\, x)$. The product $(\text{He}\, x)$ may be viewed as a pseudo gas-phase mole fraction: it is the gas-phase mole fraction that would exist at the interface if the liquid phase were at a uniform concentration x.

Equations (11.29) and (11.30) remain valid even when Henry's law is not obeyed: the Henry number is then an *effective* one, as will be illustrated in Example 11.5. Figure 11.12 shows a mass transfer circuit interpretation, which is particularly useful when one has to know y_s in order to obtain the effective Henry number. From the circuit,

$$y_s = y - J_s(1/\mathcal{G}_G) \tag{11.31}$$

or

$$y_s = y - \frac{\mathcal{G}_G^{\text{oa}}}{\mathcal{G}_G}(y - \text{He}\, x) \tag{11.32}$$

and the partial pressure is then $y_s P$.

Equation (11.29) is not a simple analogy to the corresponding heat transfer result: the heat flux from the gas to the liquid is

$$q = \frac{1}{1/h_{cG} + 1/h_{cL}}(T_G - T_L) \tag{11.33}$$

Notice that $q = 0$ for $T_G = T_L$, that is, when the two phases are in thermal equilibrium. For the two phases to be in chemical equilibrium $y = \text{He}\, x$, and, appropriately, Eq. (11.29) gives $J_s = 0$ for $y = \text{He}\, x$. Also, Eq. (11.29) shows that the liquid-side contribution to the overall mass transfer resistance is $\text{He}/\mathcal{G}_L$, not simply $1/\mathcal{G}_L$. For sparingly soluble gases He is very large, and as a result the liquid-side resistance dominates. The reason for this feature is that, since mole fractions are much lower

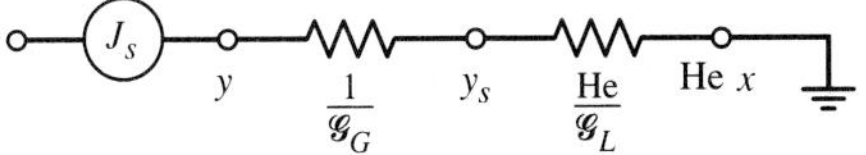

Figure 11.12 Mass transfer circuit for gas absorption (per unit area of interface).

in the liquid phase, so too are mole-fraction gradients and hence diffusion rates (the $c\mathscr{D}$ products and conductances are of the same order in both phases). An analogous situation does not occur for heat transfer because, whereas concentration is discontinuous across the interface, temperature is continuous. Example 11.5 will illustrate the concept of a *controlling* resistance for gas absorption. (See also the discussion of the mass transfer Biot number in Section 9.4.4.)

11.4.2 Mole Transfer Conductances in Packed Columns

Since the number of transfer units of a gas scrubber is proportional to interfacial area, it is common practice to break up the flow of the stream by packing the column with, for example, irregularly shaped solid particles. Typically, these particles range from about 5 to 75 mm in size, are usually made from a ceramic, and are loaded in the column in a random fashion. Some common types are shown in Fig. 11.13. Breaking up the flow also serves to increase the mass transfer conductances. Alternatively, structured packings are used. These packings give lower pressure drops for a given throughput of gas, usually at the expense of a higher installation cost. The column is usually circular in cross section and contains an open space below the packing support to ensure good distribution of the incoming gas stream, and a liquid feed system to spray liquid on the top of the packing (see Fig. 9.1). Typical columns range from 0.3 to 15 m in diameter and from 2 to 50 m in height. From a fluid flow standpoint, problems arise with respect to flooding (gas stream flow rate too high, causing reversal of

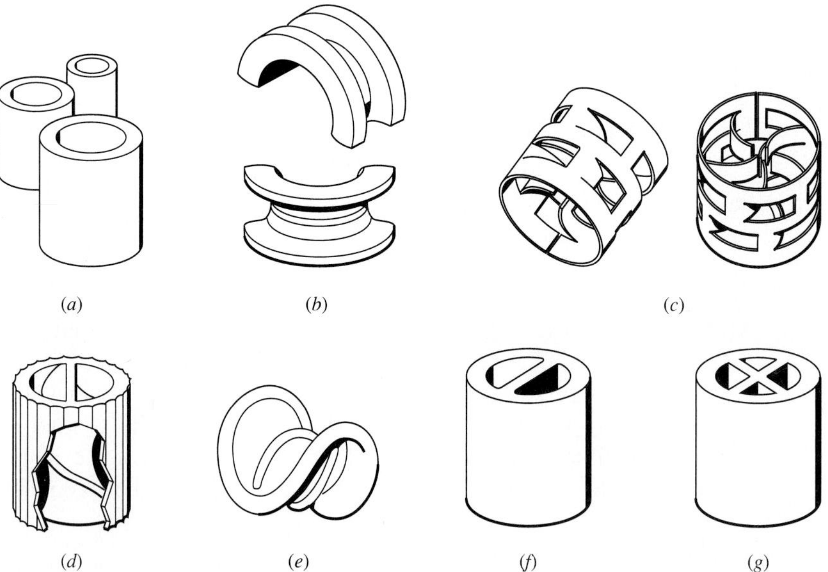

Figure 11.13 Common random packings: (*a*) Raschig rings, (*b*) Intalox saddle, (*c*) Pall ring, (*d*) Cyclohelix spiral ring, (*e*) Berl saddle, (*f*) Lessing ring, (*g*) cross-partition ring.

the liquid downflow), channeling (nonuniform downflow of liquid), entrainment (droplets of liquid carried off at the top as mist), and calculation of pressure drop. Some of these topics will be discussed in Section 11.4.4; here our interest is in calculating mole transfer conductances and the interfacial area.

Random Packings

We first consider random packings such as Raschig rings and Berl saddles. Table 11.3 gives relevant data for some dry packings, including the nominal size D_p, void fraction ε_v, specific surface area a (surface area per unit volume), and drag factor ϕ for pressure drop calculations. As can be imagined, the convective mass transfer processes in a random packing are very complicated. Not only is the geometry of the dry packing, and hence the resulting flow patterns, complex, but also as the liquid flow rate is increased, so the liquid *holdup* in the tower increases, reducing the area of interface for transfer and the cross-sectional area available for gas flow. Holdup depends not only on liquid and gas flow rates, but also on wettability of the packing and hence on surface tensions. One can find many different correlation schemes for random packing conductances in the chemical engineering literature. The correlations given here were developed by Onda and coworkers [9,10,11] and were chosen as a compromise between simplicity, generality, and accuracy. If more accurate correlations are required, the text by Treybal is recommended [12].

Reynolds numbers are defined in terms of superficial mass velocities $\dot{m}/A_{\text{fr}}$, where A_{fr} is the column cross-sectional area, and the reciprocal of the specific surface area, a^{-1}, as characteristic length:

$$\text{Re}_G = \frac{\dot{m}_G}{A_{\text{fr}} a \mu}; \qquad \text{Re}_L = \frac{\dot{m}_L}{A_{\text{fr}} a \mu_l} \qquad \textbf{(11.34}\textbf{\textit{a,b}}\textbf{)}$$

Table 11.3 Characteristics of commercial packings.

	Nominal Size, D_p [m]	Void Fraction, ε_v	Specific Surface Area, a [m^{-1}]	Drag Factor, Φ
Ceramic Raschig rings	0.006	0.73	787	1600
	0.013	0.63	364	580
	0.025	0.73	190	155
	0.038	0.71	125	95
	0.050	0.74	92	65
Metal Raschig rings (1.6 mm wall)	0.013	0.73	387	410
	0.025	0.85	186	137
	0.050	0.92	103	57
Ceramic Berl saddles	0.013	0.63	466	240
	0.025	0.69	249	110
	0.038	0.75	144	65
Interlox saddles (plastic)	0.025	0.91	207	33
	0.050	0.93	108	21

The liquid-side Stanton number is then

$$\mathrm{St}_L = \frac{\mathscr{G}_L}{c_l(\nu_l g)^{1/3}} = C_1 \,\mathrm{Re}_L^{2/3}\,\mathrm{Sc}_l^{-1/2} \tag{11.35}$$

where

$$C_1 = 0.0051(aD_p)^{0.4}$$

The subscript l is used to distinguish liquid-phase properties. The quantity $(\nu_l g)^{1/3}$ has the dimensions of velocity, as is required to form a Stanton number.

The gas-side Sherwood number is

$$\mathrm{Sh}_G = \frac{\mathscr{G}_G}{c\mathscr{D}a} = C_2\,\mathrm{Re}_G^{0.7}\,\mathrm{Sc}^{1/3} \tag{11.36}$$

where

$$C_2 = 2.0(aD_p)^{-2.0} \qquad D_p < 0.013 \text{ m}$$

$$= 5.23(aD_p)^{-2.0} \qquad D_p > 0.013 \text{ m}$$

A correlation is also needed for the transfer area perimeter,

$$\frac{\mathscr{P}/A_{\mathrm{fr}}}{a} = 1 - \exp\left[-1.45\,\mathrm{Re}_G^{0.1}\,\mathrm{We}_L^{0.2}\mathrm{Fr}_L^{-0.05}(\sigma_c/\sigma)^{0.75}\right] \tag{11.37}$$

where the Weber and Froude numbers are defined as

$$\mathrm{We}_L = \frac{(\dot{m}_L/A_{\mathrm{fr}})^2}{\rho_l \sigma a}; \qquad \mathrm{Fr}_L = \frac{(\dot{m}_L/A_{\mathrm{fr}})^2 a}{\rho_l^2 g} \tag{11.38}$$

and σ_c is a critical surface tension for the packing material; values of σ_c are given in Table 11.4. The above correlations are valid for the Reynolds number ranges

$$5 < \mathrm{Re}_G < 1000; \qquad 1 < \mathrm{Re}_L < 40, \quad \text{organic liquids}$$

$$4 < \mathrm{Re}_L < 400, \quad \text{water}$$

and may be used for packings listed in Table 11.3 (but with caution for the Intalox saddles). Most of the experimental data used to develop the correlations fall within

Table 11.4 Critical surface tension for various packing materials.

Packing Material	σ_c [N/m × 10^3]
Carbon	56
Ceramic	61
Glass	73
Polyethylene	33
Polyvinylchloride	40
Steel	75

20% of the values given by the correlation, though some data deviate by as much as 50%.

Structured Packings

Structured packings are used when it is important to have a low gas-flow pressure drop in order to reduce fan power requirements or for vacuum service. Figure 11.14*a* shows a typical commercial structured packing. As a benchmark, it is useful to be able to design a column with a simple vertical-plate packing that would give the lowest possible pressure drop. Figure 11.14*b* shows a packing that has a pitch p, plate thickness t, and falling-film thickness δ. The gas-side Reynolds number is defined in terms of the velocity $\dot{m}_G/\rho A_c$, where A_c is the flow cross-sectional area, and the hydraulic diameter D_h of the flow:

$$\mathrm{Re}_G = \frac{(\dot{m}_G/A_c)D_h}{\mu}$$
$$A_c = A_{\mathrm{fr}}\left(1 - \frac{t + 2\delta}{p}\right); \qquad D_h = 2[p - (t + 2\delta)] \tag{11.39}$$

The gas-side Sherwood number is given by Eq. (9.47):

$$\mathrm{Sh}_G = \frac{\mathcal{G}_G D_h}{c\mathcal{D}} = 0.00814\,\mathrm{Re}_G^{0.83}\,\mathrm{Sc}^{0.44}\,\mathrm{Re}_L^{0.15} \tag{11.40}$$

which is valid for $2000 < \mathrm{Re}_G < 10{,}000; \mathrm{Re}_L < 1200$. The liquid-side Reynolds number is the film Reynolds number introduced for falling films in Chapter 7:

$$\mathrm{Re}_L = \frac{4\Gamma}{\mu_l} = \frac{2(\dot{m}_L/A_{\mathrm{fr}})p}{\mu_l} \tag{11.41}$$

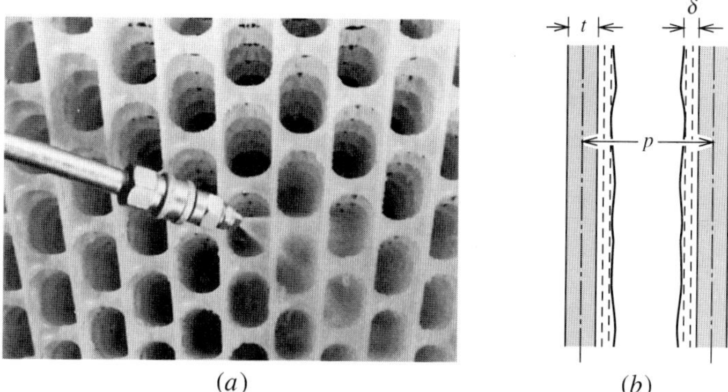

Figure 11.14 Regular packings. (*a*) Munters PN Fill structured packing (showing how it is cleaned). (*b*) Parallel-plate packing.

where Γ is the mass flow rate per unit perimeter. The characteristic velocity $(\nu_l g)^{1/3}$ is used to define the liquid-side Stanton number, and separate correlations are required for the laminar, wavy laminar, and turbulent regimes, as follows.

Laminar regime, $\mathrm{Re}_L < 30$:

$$\mathrm{St}_L = \frac{\mathscr{G}_L}{c_l(\nu_l g)^{1/3}} = 0.233\,\mathrm{Re}_L^{2/3}\,\mathrm{Pe}_m^{-1/2} \tag{11.42}$$

where the mass transfer Peclet number is $\mathrm{Pe}_m = u_\delta H/\mathscr{D}_l$; H is the packing height, and the film surface velocity is

$$u_\delta/(\nu_l g)^{1/3} = 0.413\,\mathrm{Re}_L^{2/3} \tag{11.43}$$

Equation (11.42) can be derived from the results obtained in Exercise 9–30.

Wavy laminar regime [13], $30 < \mathrm{Re}_L < 1100$:

$$\mathrm{St}_L = 0.04\,\mathrm{Re}_L^{0.2}\,\mathrm{Sc}_l^{-0.53} \tag{11.44}$$

Turbulent regime [14], $1100 < \mathrm{Re}_L < 10{,}000$:

$$\mathrm{St}_L = 6.79 \times 10^{-9}\,\mathrm{Re}_L^{n}\,\mathrm{Sc}_l^{-\alpha}\,\mathrm{Ka}_l^{-1/2} \tag{11.45}$$

where

$$n = 3.49\,\mathrm{Ka}_l^{0.068}$$

$$\alpha = 0.36 + 2.43(\sigma\ \mathrm{N/m})$$

and the Kapitza number is $\mathrm{Ka}_l = \nu_l^4\rho_l^3 g/\sigma^3$. Notice the transition Reynolds number is ~ 1100, rather than the value of ~ 1800 used for film condensation. As explained in Section 7.2.2, the lower value corresponds to transition in the outer region of the film, and it is this region that is relevant to gas absorption into the film.

The film thickness is required to calculate the flow cross-sectional area A_c and hydraulic diameter D_h; from reference [15],

$$\frac{\delta}{(\nu_l^2/g)^{1/3}} = 0.909\,\mathrm{Re}_L^{1/3} \qquad \mathrm{Re}_L < 1600 \tag{11.46a}$$

$$= 0.0672\,\mathrm{Re}_L^{2/3} \qquad \mathrm{Re}_L > 1600 \tag{11.46b}$$

Note that these liquid film correlations, Eqs. (11.42) through (11.46), are valid only if the shear stress exerted by the gas flow on the film surface is relatively small. The transfer area perimeter and specific area are obtained directly from the geometry as

$$\frac{\mathscr{P}}{A_{\mathrm{fr}}} = \frac{2}{p} = a \tag{11.47}$$

A great variety of commercial structured packings are used in practice. Manufacturers usually supply performance data in the form of correlations of the conductance times the specific area, rather than separate correlations of the conductance and transfer perimeter.

EXAMPLE 11.5 Controlling Resistances for Gas Absorption

A mixture of air and a trace amount of an inert species flows countercurrent to a falling film of water between parallel plates of 2 mm thickness at a pitch of 5 cm. The air velocity between the plates is 1.8 m/s, and the water flow rate is 0.20 kg/s per unit perimeter. The exchanger operates at 300 K and 1 atm. The water enters saturated with air but contains no species 1. Determine the relative resistances to mass transfer in each phase ($1/\mathcal{G}_G$ and $\text{He}/\mathcal{G}_L$) and the overall mole transfer conductance $\mathcal{G}_G^{\text{oa}}$ when the trace gas is (i) ammonia, (ii) sulfur dioxide, (iii) hydrogen sulfide, and (iv) carbon dioxide.

Solution

Given: Falling water film with a countercurrent flow of air containing an inert species.

Required: Relative resistances to mass transfer in each phase for NH_3, SO_2, H_2S, and CO_2.

Assumptions:
1. Properties of pure air and pure water for the gas and liquid phases, respectively.
2. An *effective* Henry number can be used for the highly soluble SO_2 and NH_3.

The overall mole transfer conductance is given by Eq. (11.30) as

$$\frac{1}{\mathcal{G}_G^{\text{oa}}} = \frac{1}{\mathcal{G}_G} + \frac{\text{He}}{\mathcal{G}_L}$$

First consider the liquid phase. Assuming pure water at 300 K and using Table A.8, $\rho_l = 996 \text{ kg/m}^3$, $\nu_l = 0.87 \times 10^{-6} \text{ m}^2\text{/s}$, $\mu_l = 8.67 \times 10^{-4} \text{ kg/m s}$, $c_l = \rho_l/M = (996 \text{ kg/m}^3)/(18 \text{ kg/kmol}) = 55.3 \text{ kmol/m}^3$. The film Reynolds number is

$$\text{Re}_L = \frac{4\Gamma}{\mu_l} = \frac{(4)(0.20)}{8.67 \times 10^{-4}} = 923$$

Hence, using Eq. (11.44), the liquid-side mole transfer conductance is obtained from

$$\begin{aligned} G_L \text{Sc}_l^{0.53} &= 0.04 c_l (\nu_l g)^{1/3} \text{Re}_L^{0.2} \\ &= (0.04)(55.3)(0.87 \times 10^{-6} \times 9.81)^{1/3}(923)^{0.2} = 0.177 \text{ kmol/m}^2 \text{ s} \end{aligned}$$

The film thickness is calculated from Eq. (11.46*a*):

$$\begin{aligned} \delta &= 0.909 \left(\frac{\nu_l^2}{g} \right)^{1/3} \text{Re}_L^{1/3} = (0.909) \left[\frac{(0.87 \times 10^{-6})^2}{9.81} \right]^{1/3} (923)^{1/3} \\ &= 3.77 \times 10^{-4} \text{ m (0.377 mm)} \end{aligned}$$

Now consider the gas phase and assume its properties to be those of pure air at 300 K and 1 atm; from Table A.7, $\rho = 1.177 \text{ kg/m}^3$, $\mu = 18.43 \times 10^{-6} \text{ kg/m s}$, $c = \rho/M =$

$(1.177\ \text{kg/m}^3)/(29\ \text{kg/kmol}) = 0.0406\ \text{kmol/m}^3$. From Eq. (11.39) the hydraulic diameter is

$$D_h = 2[p - (t + 2\delta)] = 2\{0.05 - [0.002 + (2)(0.000377)]\} = 0.0945\ \text{m}$$

Then the gas-phase Reynolds number is

$$\text{Re}_G = \frac{\rho V D_h}{\mu} = \frac{(1.177)(1.8)(0.0945)}{18.43 \times 10^{-6}} = 10{,}870$$

Using Eq. (11.40), with $\mathscr{D} = \mu/\rho\,\text{Sc}$, the gas-side mole transfer conductance is obtained from

$$\mathscr{G}_G\,\text{Sc}^{0.56} = 0.00814\left(\frac{c\mu}{\rho D_h}\right)\text{Re}_G^{0.83}\,\text{Re}_L^{0.15}$$

$$= 0.00814\left[\frac{(0.0406)(18.43 \times 10^{-6})}{(1.177)(0.0945)}\right](10{,}870)^{0.83}(923)^{0.15}$$

$$= 3.42 \times 10^{-4}\ \text{kmol/m}^2\,\text{s}$$

At 1 atm total pressure, the Henry number equals Henry's constant. From Table A.21, at 300 K the Henry numbers for H_2S and CO_2 are found to be 560 and 1710, respectively. For NH_3 in dilute solution, Henry's law is approximately valid; at 300 K Table A.22*a* gives $P_{NH_3,s}/x_{NH_3,u} \simeq 1.1$. For SO_2, Table A.22*b* shows that the *effective* Henry number at 300 K and $P_{SO_2,s} = 0.003$ atm is $0.003/0.18 \times 10^{-3} = 17$. [The value of $P_{SO_2,s}$ must be initially guessed; subsequently, Eq. (11.32) can be used to find a better value.]

Gas- and liquid-phase Schmidt numbers are obtained from Tables A.17*b* and A.18, respectively. The results are summarized below.

	Absorbed Species			
Derived Quantities	NH_3	SO_2	H_2S	CO_2
Sc (gas)	0.61	1.24	0.94	1.00
$\mathscr{G}_G = 3.42 \times 10^{-4}/\text{Sc}^{0.56}$ $(\times 10^3)$	0.451	0.303	0.354	0.342
Sc_l (liquid)	410	520	430	420
$\mathscr{G}_L = 0.177/\text{Sc}_l^{0.53}$ $(\times 10^3)$	7.30	6.43	7.12	7.21
He	1.1	17.0	560	1710
Gas-side resistance, $1/\mathscr{G}_G$ $(\times 10^{-3})$	2.22	3.30	2.82	2.92
Liquid-side resistance, $\text{He}/\mathscr{G}_L$ $(\times 10^{-3})$	0.15	2.64	78.7	237
Total resistance, $1/\mathscr{G}_G + \text{He}/\mathscr{G}_L$ $(\times 10^{-3})$	2.37	5.94	81.5	240
$\mathscr{G}_G^{oa}$, kmol/m^2 s $(\times 10^3)$	0.422	0.168	0.0123	0.00417

Comments

1. The liquid-side resistance controls the absorption of CO_2 and H_2S, neither resistance controls for SO_2, and the gas-side resistance controls for NH_3.

2. When the liquid-side resistance is very large, common practice is to add a reactant to the solvent (see Example 11.2).

3. The value of $\text{Re}_G(10{,}870)$ is a little higher than the specified upper limit for Eq. (11.40) to be valid (10,000). This modest extrapolation should be satisfactory, since there is no change in flow patterns at these Reynolds numbers.

4. SCRUB (see Section 11.4.3) can be used to check these results.

EXAMPLE 11.6 Overall Mole Transfer Conductance for a Packed Column

Air containing a trace amount of NH_3 flows countercurrent to water in a 1 m^2–cross-sectional-area column packed with 0.038 m–nominal-size ceramic Berl saddles. The flow rates of gas and liquid are 0.35 kg/s and 1.90 kg/s, respectively, and the tower operates at 300 K and 1 atm. Determine the overall mole transfer conductance and transfer perimeter.

Solution

Given: Absorption column packed with Berl saddles.

Required: Overall mole transfer conductance and transfer perimeter for NH_3 absorption into water.

Assumptions: Properties for pure air and pure water can be used for the gas and liquid phases, respectively.

The overall mole transfer conductance is given by Eq. (11.30) as

$$\frac{1}{\mathcal{G}_G^{\text{oa}}} = \frac{1}{\mathcal{G}_G} + \frac{\text{He}}{\mathcal{G}_L}$$

For the liquid phase, assuming pure water at 300 K, $\rho_l = 996$ kg/m^3, $\mu_l = 8.67 \times 10^{-4}$ kg/m s, $\nu_l = 0.87 \times 10^{-6}$ m^2/s, $c_l = 996/18 = 55.3$ kmol/m^3, and $\text{Sc}_l = 410$. From Table 11.3, the specific area of the packing is $a = 144$ m^{-1}. Thus, from Eq. (11.34*b*), the liquid-phase Reynolds number is

$$\text{Re}_L = \frac{\dot{m}_L}{A_{\text{fr}} a \mu_l} = \frac{1.90}{(1)(144)(8.67 \times 10^{-4})} = 15.2$$

From Eq. (11.35), the liquid-side mole transfer conductance is

$$\begin{aligned}\mathcal{G}_L &= 0.0051(aD_p)^{0.4} c_l (\nu_l g)^{1/3}\,\text{Re}_L^{2/3}\,\text{Sc}_l^{-1/2} \\ &= 0.0051(144 \times 0.038)^{0.4}(55.3)(0.87 \times 10^{-6} \times 9.81)^{1/3}(15.2)^{2/3}(410)^{-1/2} \\ &= 3.45 \times 10^{-3}\ \text{kmol/m}^2\ \text{s}\end{aligned}$$

For the gas phase we use properties of pure air at 300 K and 1 atm: $\rho = 1.177$ kg/m^3, $\mu = 18.43 \times 10^{-6}$ kg/m s, $c = 1.177/29 = 0.0406$ kmol/m^3, Sc = 0.61. From Eq. (11.34*a*), the gas-phase Reynolds number is

$$\text{Re}_G = \frac{\dot{m}_G}{A_{\text{fr}} a \mu} = \frac{0.35}{(1)(144)(18.43 \times 10^{-6})} = 132$$

and from Eq. (11.36), the gas-side mole transfer conductance is

$$\begin{aligned}\mathcal{G}_G &= 5.23(aD_p)^{-2} c \left(\frac{\mu}{\rho\,\text{Sc}}\right) a\,\text{Re}_G^{0.7}\,\text{Sc}^{1/3} \\ &= 5.23(144 \times 0.038)^{-2}(0.0406)\frac{18.43 \times 10^{-6}}{1.177 \times 0.61}(144)(132)^{0.7}(0.61)^{1/3} \\ &= 6.78 \times 10^{-4}\ \text{kmol/m}^2\ \text{s}\end{aligned}$$

Then, using Eq. (11.30), we obtain the overall mole transfer conductance as

$$\frac{1}{\mathscr{G}_G^{\text{oa}}} = \frac{1}{\mathscr{G}_G} + \frac{\text{He}}{\mathscr{G}_L}; \qquad \text{He} \simeq 1.1 \text{ from Table A.22}a$$

$$\frac{1}{\mathscr{G}_G^{\text{oa}}} = \frac{1}{6.78 \times 10^{-4}} + \frac{1.1}{3.45 \times 10^{-3}}$$

$$= 1475 + 319 = 1794; \qquad \mathscr{G}_G^{\text{oa}} = 5.57 \times 10^{-4} \text{ kmol/m}^2\text{ s}$$

Equation (11.37) gives the transfer perimeter as

$$\mathscr{P} = aA_{\text{fr}}\left\{1 - \exp[-1.45\,\text{Re}_G^{0.1}\,\text{We}_L^{0.2}\text{Fr}_L^{-0.05}(\sigma_c/\sigma)^{0.75}]\right\}$$

For water at 300 K, Table A.11 gives $\sigma = 71.7 \times 10^{-3}$ N/m, and Table 11.4 gives $\sigma_c = 61 \times 10^{-3}$ N/m. The Weber and Froude numbers are

$$\text{We}_L = \frac{(\dot{m}_L/A_{\text{fr}})^2}{\rho_l \sigma a} = \frac{(1.90/1)^2}{(996)(71.7 \times 10^{-3})(144)} = 3.51 \times 10^{-4}$$

$$\text{Fr}_L = \frac{(\dot{m}_L/A_{\text{fr}})^2 a}{\rho_l^2 g} = \frac{(1.90/1)^2(144)}{(996)^2(9.81)} = 5.34 \times 10^{-5}$$

The perimeter is then

$$\mathscr{P} = (144)(1)\left\{1 - \exp[-1.45(132)^{0.1}(3.51 \times 10^{-4})^{0.2}(5.34 \times 10^{-5})^{-0.05}(61/71.7)^{0.75}]\right\}$$
$$= 72.3\text{m}$$

Comments

1. SCRUB (see Section 11.4.3) can be used to check these results.
2. The gas-side resistance dominates, but the liquid-side resistance is not negligible.

11.4.3 Countercurrent Mass Exchangers

As depicted in Fig. 11.15, a countercurrent[2] mass exchanger has streams flowing in opposite directions. In many situations the liquid stream flows down under the influence of gravity, with countercurrent upflow of the gas stream. For convenience, we separate the gas and liquid into two distinct streams. However, it should be kept in mind that the usual column has multiple paths for the flow of each phase, in order to increase the contact area $\mathscr{P}\Delta z$ and hence the mass transfer rate $J_s\mathscr{P}\Delta z$. The liquid and gas flow rates, $\dot{M}_L$ and $\dot{M}_G$, are both taken as positive. In the usual situation, the species to be absorbed is in small concentration in a carrier gas stream: the amount of gas absorbed is small compared with the stream flow rates, and $\dot{M}_L$ and $\dot{M}_G$ can be taken to be constant.

The exchanger species balance is

$$\dot{M}_G y|_{\text{in}} - \dot{M}_L x|_{\text{out}} = \dot{M}_G y|_{\text{out}} - \dot{M}_L x|_{\text{in}} \tag{11.48}$$

[2] Following chemical engineering practice, we will refer to counterflow mass exchangers as *countercurrent* units.

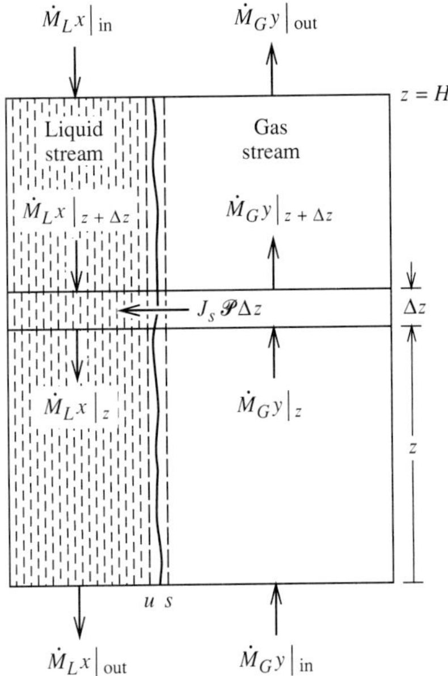

Figure 11.15 Schematic of a countercurrent mass exchanger, showing species balances on an element of packing Δz high.

which is evident upon inspection of Fig. 11.15. A species balance on the gas stream for an element of length Δz between z and $z + \Delta z$ gives

$$\dot{M}_G y|_z = \dot{M}_G y|_{z+\Delta z} + J_s \mathscr{P} \Delta z$$

where J_s is positive for absorption. Taking $\dot{M}_G$ constant, introducing the overall mole transfer conductance from Eq. (11.29), and allowing $\Delta z \to 0$ gives

$$\frac{dy}{dz} = -\frac{\mathscr{G}_G^{oa} \mathscr{P}}{\dot{M}_G}(y - \text{He } x) \tag{11.49}$$

For gas absorption, $y > \text{He } x$, and y is seen to decrease with increasing z. A similar balance for the liquid stream gives

$$\frac{dx}{dz} = -\frac{\mathscr{G}_G^{oa} \mathscr{P}}{\dot{M}_L}(y - \text{He } x) \tag{11.50}$$

Likewise, x decreases with increasing z. This equation is now multiplied by He, and the quantity $\dot{M}_L/\text{He}$ is written $\dot{M}_L^\dagger$. As a result,

$$\frac{d(\text{He } x)}{dz} = -\frac{\mathscr{G}_G^{oa} \mathscr{P}}{\dot{M}_L^\dagger}(y - \text{He } x) \tag{11.51}$$

Subtracting Eq. (11.51) from Eq. (11.49) gives

$$\frac{d}{dz}(y - \text{He } x) = -\mathscr{G}_G^{oa} \mathscr{P} \left(\frac{1}{\dot{M}_G} - \frac{1}{\dot{M}_L^\dagger} \right)(y - \text{He } x) \tag{11.52}$$

Equations (11.49), (11.51), and (11.52) are equivalent forms of the governing differential equation for mass transfer in the column. There is an exact parallelism between these equations and those for a two-stream countercurrent heat exchanger. The boundary conditions required to solve the governing equations are

$$z = 0: \quad y = y_{\text{in}},\ x = x_{\text{out}} \tag{11.53a}$$

$$z = H: \quad y = y_{\text{out}},\ x = x_{\text{in}} \tag{11.53b}$$

Equations (11.48) and (11.52) can be viewed as two equations in essentially eight parameters:

$$\dot{M}_L,\ \dot{M}_G,\ H,\ \mathscr{P},\ x_{\text{in}},\ x_{\text{out}},\ y_{\text{in}},\ y_{\text{out}}$$

(The quantities $\mathscr{G}_G^{\text{oa}}$ and He are not regarded as independent parameters since they are functions of the other parameters as well as temperature and pressure, which are assumed to be known.) Thus, a total of six parameters must be considered as known, leaving two unknowns to be solved for. As was the case for two-stream heat exchangers, the choice of unknowns dictates the method of solution. Recall that in the design of a heat exchanger to meet a specified performance, the inlet and outlet temperatures were specified, and the unknowns were H and $\mathscr{P}$ (the *design* problem). Alternatively, for the determination of the performance of an existing exchanger, H and $\mathscr{P}$ were known and the outlet temperatures were unknown (the *rating* problem). The former analysis led to the concept of log mean temperature difference, and the latter led to the concepts of effectiveness and number of transfer units. However, it was also seen that once an ε–N_{tu} relation had been obtained, it could be inverted to give a N_{tu}–ε relation (either analytically or numerically). This feature, together with the useful physical significance of effectiveness and number of transfer units, suggests that it will be sufficient to solve the rating problem for mass exchangers. Thus, we will assume that the unknowns are x_{out} and y_{out}, and proceed as follows. Equation (11.52) is integrated as

$$\int_{y_{\text{in}}-\text{He}\,x_{\text{out}}}^{y_{\text{out}}-\text{He}\,x_{\text{in}}} d\,\ln(y - \text{He}\,x) = -\left(\frac{1}{\dot{M}_G} - \frac{1}{\dot{M}_L^{\dagger}}\right)(\mathscr{G}_G^{\text{oa}}\mathscr{P}H)$$

$$y_{\text{out}} = \text{He}\,x_{\text{in}} + (y_{\text{in}} - \text{He}\,x_{\text{out}})\exp\left[-\left(\frac{1}{\dot{M}_G} - \frac{1}{\dot{M}_L^{\dagger}}\right)(\mathscr{G}_G^{\text{oa}}\mathscr{P}H)\right] \tag{11.54}$$

To obtain the effectiveness, we must find an expression for $\dot{M}_G(y_{\text{in}} - y_{\text{out}})$, or equivalently, $\dot{M}_L(x_{\text{out}} - x_{\text{in}})$. For convenience, define γ as

$$\gamma = \left(\frac{1}{\dot{M}_G} - \frac{1}{\dot{M}_L^{\dagger}}\right)(\mathscr{G}_G^{\text{oa}}\mathscr{P}H) \tag{11.55}$$

then

$$
\begin{aligned}
y_{\rm in} - y_{\rm out} &= y_{\rm in} - [\text{He } x_{\rm in} + (y_{\rm in} - \text{He } x_{\rm out})e^{-\gamma}] \\
&= (y_{\rm in} - \text{He } x_{\rm in}) - (y_{\rm in} - \text{He } x_{\rm out})e^{-\gamma} \\
&= (y_{\rm in} - \text{He } x_{\rm in}) - \left[y_{\rm in} - \text{He } x_{\rm in} - \frac{\dot{M}_G}{\dot{M}_L^\dagger}(y_{\rm in} - y_{\rm out}) \right] e^{-\gamma} \\
&= (y_{\rm in} - \text{He } x_{\rm in})(1 - e^{-\gamma}) + (y_{\rm in} - y_{\rm out})\frac{\dot{M}_G}{\dot{M}_L^\dagger} e^{-\gamma}
\end{aligned}
$$

where $x_{\rm out}$ has been eliminated by means of the exchanger balance, Eq. (11.48). Rearranging,

$$
\frac{y_{\rm in} - y_{\rm out}}{y_{\rm in} - \text{He } x_{\rm in}} = \frac{1 - e^{-\gamma}}{1 - (\dot{M}_G/\dot{M}_L^\dagger)e^{-\gamma}} \tag{11.56}
$$

Suppose $\dot{M}_G < \dot{M}_L^\dagger$ so that $\gamma > 0$. Then as the column height becomes large, $\gamma \to \infty$, the right-hand side of Eq. (11.56) approaches 1 and $y_{\rm out} \to \text{He } x_{\rm in}$. That is, referring to Fig. 11.16*a*, the streams approach equilibrium at the top of the column. Thus, the left-hand side of Eq. (11.56) is the ratio of the actual absorption from the gas stream to the maximum possible in an infinitely high column, which we call the effectiveness ε for a countercurrent exchanger,

$$
\varepsilon = \frac{\dot{M}_G(y_{\rm in} - y_{\rm out})}{\dot{M}_G(y_{\rm in} - y_{\rm out})_{\rm max}} = \frac{\dot{M}_G(y_{\rm in} - y_{\rm out})}{\dot{M}_G(y_{\rm in} - \text{He } x_{\rm in})} \tag{11.57}
$$

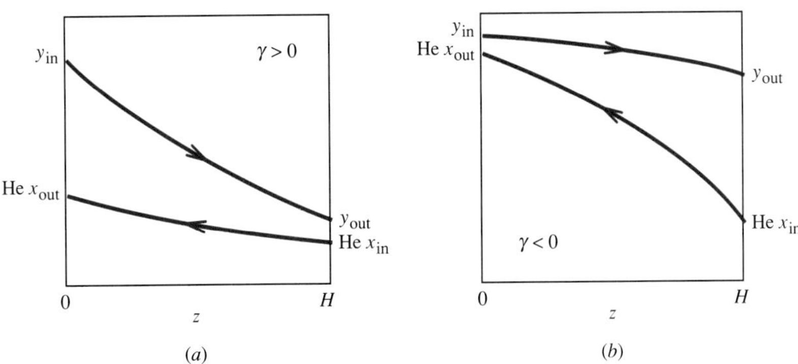

Figure 11.16 Gas-phase and pseudo-gas-phase concentration variations along a countercurrent exchanger: (*a*) $\gamma > 0$ ($\dot{M}_G < \dot{M}_L^\dagger$), (*b*) $\gamma < 0$ ($\dot{M}_G > \dot{M}_L^\dagger$).

and from Eq. (11.56),

$$\varepsilon = \frac{1 - e^{-\gamma}}{1 - (\dot{M}_G/\dot{M}_L^\dagger)e^{-\gamma}} \tag{11.58}$$

For this case with $\dot{M}_G < \dot{M}_L^\dagger$, the performance of the exchanger can be viewed as being limited by the amount of the solute gas carried by the gas stream rather than the capacity of the liquid stream to absorb the gas: y_{out} approaches He x_{in} but He x_{out} does not approach y_{in} as closely, as shown in Fig. 11.16*a*.

Now suppose $\dot{M}_L^\dagger < \dot{M}_G$, so that $\gamma < 0$. Then Eq. (11.54) requires that He $x_{\text{out}} \to y_{\text{in}}$ as $|\gamma| \to \infty$. Referring to Fig. 11.16*b*, the streams approach equilibrium at the bottom of the column. The performance of the exchanger is now limited by the capacity of the liquid stream to absorb gas, which indicates that the effectiveness is more usefully expressed as

$$\varepsilon = \frac{\dot{M}_G(y_{\text{in}} - y_{\text{out}})}{\dot{M}_L(x_{\text{out}} - x_{\text{in}})_{\text{max}}} = \frac{\dot{M}_G(y_{\text{in}} - y_{\text{out}})}{\dot{M}_L^\dagger(y_{\text{in}} - \text{He}\ x_{\text{in}})} = \frac{\dot{M}_L(x_{\text{out}} - x_{\text{in}})}{\dot{M}_L^\dagger(y_{\text{in}} - \text{He}\ x_{\text{in}})} \tag{11.59}$$

Using Eq. (11.56), it is easily shown that

$$\varepsilon = \frac{1 - e^{\gamma}}{1 - (\dot{M}_L^\dagger/\dot{M}_G)e^{\gamma}} \tag{11.60}$$

Equations (11.58) and (11.60) can be written in the more compact forms

$$\varepsilon = \frac{\dot{M}_G}{C_{\text{min}}}\left(\frac{y_{\text{in}} - y_{\text{out}}}{y_{\text{in}} - \text{He}\ x_{\text{in}}}\right) = \frac{\dot{M}_L}{C_{\text{min}}}\left(\frac{x_{\text{out}} - x_{\text{in}}}{y_{\text{in}} - \text{He}\ x_{\text{in}}}\right) \tag{11.61}$$

$$\varepsilon = \frac{1 - e^{-N_{\text{tu}}(1-R_C)}}{1 - R_C e^{-N_{\text{tu}}(1-R_C)}} \tag{11.62}$$

where

$$N_{\text{tu}} = \frac{\mathcal{G}_G^{\text{oa}}\mathcal{P}H}{C_{\text{min}}}; \qquad R_C = \frac{C_{\text{min}}}{C_{\text{max}}} = \frac{\min(\dot{M}_G, \dot{M}_L^\dagger)}{\max(\dot{M}_G, \dot{M}_L^\dagger)}$$

Equation (11.62) is used for determining the performance of a given mass exchanger. In designing an exchanger to obtain a desired effectiveness, the inverse relation is required, which is

$$N_{\text{tu}} = \frac{1}{1 - R_C} \ln \frac{1 - R_C\varepsilon}{1 - \varepsilon} \tag{11.63}$$

Because of their higher effectiveness, countercurrent exchangers are usually preferred over cocurrent exchangers. However, the engineer may well choose a cocurrent exchanger if floor space is limited and a high gas throughput is required: a cocurrent unit has a lower pressure drop and superior liquid distribution, and does not have the flooding limitation of a countercurrent unit. The required ε–N_{tu} and N_{tu}–ε relations are given in Table 8.3. Similarly, equipment configuration constraints may suggest use of a cross-flow unit; again, the required relations are given in Table 8.3.

The Computer Program SCRUB

SCRUB calculates the performance of countercurrent scrubbers for absorption into water. Dilute concentrations and low mass transfer rates are assumed. Packings include Raschig rings, Berl saddles, Interlox saddles, and parallel plates. The "rating" problem is solved; that is, the stream flow rates, inlet compositions, and packing dimensions are specified, and the outlet compositions are calculated. Also, the pressure drop is calculated and the flooding limit is checked, following the recommendations in Section 11.4.4.

EXAMPLE 11.7 Effectiveness of a Packed-Column Ammonia Scrubber

Suppose that ammonia is to be removed from an air stream in the packed column described in Example 11.6. Given that the inlet water is pure and that the inlet air stream contains 3% ammonia by volume, calculate the effectiveness of the column and the outlet air composition if the column is 2 m high.

Solution

Given: A 2 m–high column packed with 0.038 m Berl saddles using 1.90 kg/s of water to scrub NH_3 from a 0.35 kg/s air stream.

Required: Column effectiveness, and y_{out} for $y_{in} = 0.03$.

Assumptions: Properties of pure air and pure water may be used for gas and liquid phases, respectively.

The effectiveness of a countercurrent column is given by Eq. (11.62) as

$$\varepsilon = \frac{1 - e^{-N_{tu}(1-R_C)}}{1 - R_C e^{-N_{tu}(1-R_C)}}; \qquad N_{tu} = \frac{\mathscr{G}_G^{oa}\mathscr{P}H}{C_{min}}; \qquad R_C = \frac{C_{min}}{C_{max}}$$

The molar flow rates and capacity rate ratio are

$$\dot{M}_G = \frac{\dot{m}_G}{M_G} = \frac{0.35 \text{ kg/s}}{29 \text{ kg/kmol}} = 0.0121 \text{ kmol/s}$$

$$\dot{M}_L = \frac{\dot{m}_L}{M_L} = \frac{1.90 \text{ kg/s}}{18 \text{ kg/kmol}} = 0.1056 \text{ kmol/s}$$

$$\dot{M}_L^{\dagger} = \frac{\dot{M}_L}{\text{He}} = \frac{0.1056}{1.1} = 0.0960 \text{ kmol/s}$$

$$C_{min} = \min(\dot{M}_G, \dot{M}_L^{\dagger}) = \min(0.0121, 0.0960) = 0.0121 \text{ kmol/s}$$

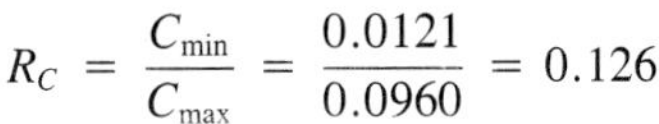

$$R_C = \frac{C_{min}}{C_{max}} = \frac{0.0121}{0.0960} = 0.126$$

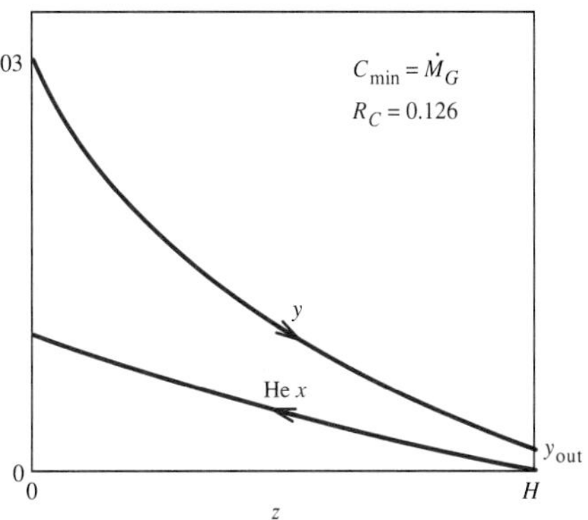

From Example 11.6, the overall mole transfer conductance and transfer perimeter are $\mathcal{G}_G^{oa} = 5.57 \times 10^{-4}$ kmol/m^2 s, $\mathcal{P} = 72.3$ m. Thus, the number of transfer units is

$$N_{tu} = \frac{\mathcal{G}_G^{oa}\mathcal{P}H}{C_{min}} = \frac{(5.57 \times 10^{-4})(72.3)(2)}{0.0121} = 6.66$$

and the effectiveness is

$$\varepsilon = \frac{1 - e^{-N_{tu}(1-R_C)}}{1 - R_C e^{-N_{tu}(1-R_C)}} = \frac{1 - e^{-(6.66)(1-0.126)}}{1 - 0.126e^{-6.68(1-0.126)}} = 99.74\%$$

From Eq. (11.61), noting that $x_{in} = 0$, the mole fraction of ammonia in the outlet air stream is

$$y_{out} = y_{in}(1 - \varepsilon) = 0.03(1 - 0.9974) = 7.78 \times 10^{-5}$$

Comments

1. Use SCRUB to check N_{tu}, ε, and y_{out}.
2. Since the effectiveness is very high, it is more convenient to quote the ineffectiveness as $1 - \varepsilon = 0.259\%$.

11.4.4 Pressure Drop and Flooding in Packed Columns

Random Packings

Figure 11.17 shows typical pressure drop data for random packings. At low gas flow rates ΔP is proportional to velocity squared, even at high liquid flow rates. But as liquid holdup increases, the *loading point* is reached, where ΔP begins to increase more rapidly; finally, in a countercurrent unit, the *flooding limit* is reached, where flow reversal of the liquid occurs. Industrial practice is such that columns usually operate with gas flow rates between the loading point and the flooding limit. Empirical data for pressure drop in this range is usually supplied by the manufacturer of the particular packing. However, most such data are well approximated by the curves in Fig. 11.17, which also shows an approximate flooding limit.

The flooding phenomenon is obviously a complex one. In addition to depending on gas and liquid flow rates, the flooding limit is also found to depend on a packing factor ε_v^3/a, gravity, the inverse cube of surface tension, and other fluid properties. Figure 11.17 shows one widely used correlation. The strong effect of surface tension gives a low flooding limit when water is contaminated by a surfactant.

Structured Packings

Figure 11.18 shows typical pressure drop data for a patented structured packing. For parallel-plate packings, a simple correlation based on work by Bharathan and Wallis [17] is recommended for rough estimates. The idea behind this correlation is that the effect of surface waves on the liquid film of thickness δ is similar to that of

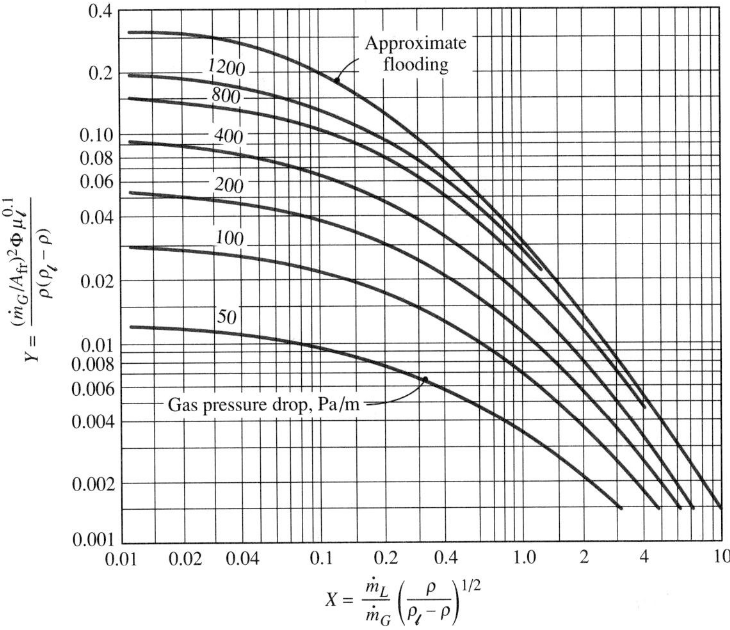

Figure 11.17 Pressure drop and flooding limit in towers with random packings [16]. The drag factor Φ is given in Table 11.3.

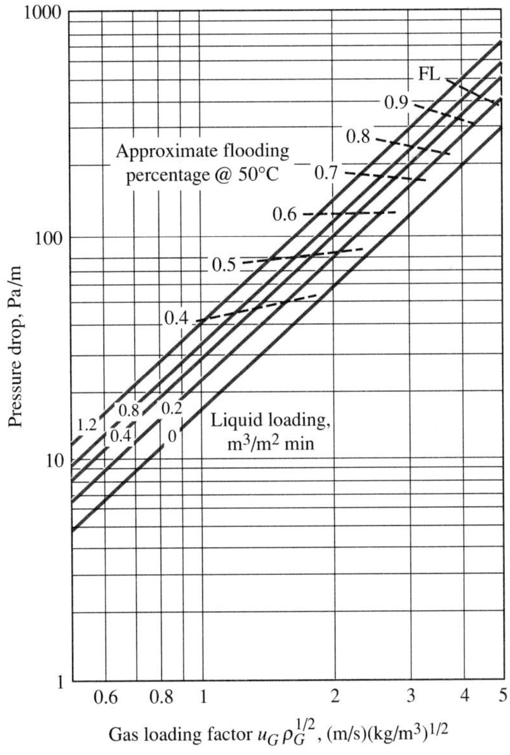

Figure 11.18 Pressure drop and flooding limit for Munters PN Fill structured packing.

wall roughness, and the friction factor is approximately

$$f \simeq 4(0.005 + C\delta^{*n}) \tag{11.64a}$$

$$\log_{10} C = -0.56 + \frac{9.07}{D_h^*}; \qquad n = 1.63 + \frac{4.74}{D_h^*}$$

$$\delta^* = \delta/L_c; \qquad D_h^* = D_h/L_c; \qquad L_c = [\sigma/(\rho_l - \rho)g]^{1/2}$$

Notice that the film thickness and hydraulic diameter are made dimensionless with the characteristic length L_c introduced for nucleate boiling in Section 7.4.3. Also, the friction factor f is defined in terms of a velocity based on the gas-flow cross-sectional area A_c (see Eq. 11.39). For large passages ($D_h^* > 160$), Eq. (11.64a) can be approximated as

$$f = 0.02(1 + 55\delta^{*1.63}) \tag{11.64b}$$

Equation (11.64) is based on data for tubes of circular cross section and may be less reliable for flow between parallel plates. Notice that Fig. 11.18 and Eq. (11.64) apply to downward flow of the liquid phase and upward counterflow of the gas phase.

Flooding in circular tubes has been extensively studied, though there remains considerable controversy as to which of the many available correlations is preferred. Part of the confusion is related to the associated phenomenon of entrainment: at relatively low liquid flow rates, droplets can be stripped from wave crests and be entrained at gas flow rates lower than those required to cause bridging of the gas flow passage by liquid, and consequent flow reversal. Some widely used flooding correlations are, in fact, entrainment correlations. The true flooding limit can be estimated from the correlation developed by Tien and coworkers [18,19]:

$$K_G^{1/2} + 0.8K_L^{1/2} = 2.1 \tanh(0.8D_h^{*1/4}) \tag{11.65}$$

where

$$K_G = \alpha V_G \rho^{1/2}/[g\sigma(\rho_l - \rho)]^{1/4}$$

$$K_L = (1 - \alpha)V_L \rho_l^{1/2}/[g\sigma(\rho_l - \rho)]^{1/4}$$

V_G and V_L are the velocities based on passage cross-sectional area, and α is the fraction of the cross section occupied by the gas flow. In a circular tube, $\alpha \simeq (1 - 2\delta/D)^2$, where δ is the film thickness. The quantities K_G and K_L are dimensionless gas and liquid flow rates: when $(K_G^{1/2} + 0.8K_L^{1/2})$ exceeds the critical value specified by Eq. (11.65), flooding occurs.

EXAMPLE 11.8 Pressure Drop in a Column Packed with Berl Saddles

Determine the pressure drop in a 1 m^2–cross-section, 3 m–high column packed with 0.038 m ceramic Berl saddles, operating at 300 K and 1 atm, when 1.4 kg/s of air flows countercurrent to 4.2 kg/s water.

Solution

Given: Column packed with Berl saddles.

Required: Pressure drop for countercurrent flow if column is 3 m high.

Assumptions: Figure 11.17 is sufficiently accurate.

To use Fig. 11.17, we must calculate the abscissa and ordinate values, X and Y, from the given data. From Tables A.7 and A.8 for 300 K and 1 atm,

$$\rho = 1.177 \text{ kg/m}^3, \qquad \rho_l = 996 \text{ kg/m}^3, \qquad \mu_l = 8.67 \times 10^{-4} \text{ kg/m s}$$

Table 11.3 gives the drag factor for 0.038 m ceramic Berl saddles as $\Phi = 65$. Referring to Fig. 11.17, the abscissa and ordinate values are

$$X = \frac{\dot{m}_L}{\dot{m}_G}\left(\frac{\rho}{\rho_l - \rho}\right)^{1/2} = \frac{4.2}{1.4}\left(\frac{1.177}{996 - 1.177}\right)^{1/2} = 0.103$$

$$Y = \frac{(\dot{m}_G/A_{\text{fr}})^2 \Phi \mu_l^{0.1}}{\rho(\rho_l - \rho)} = \frac{(1.4/1)^2(65)(8.67 \times 10^{-4})^{0.1}}{(1.177)(996 - 1.177)} = 5.38 \times 10^{-2}$$

giving a pressure gradient of $\Delta P/H \simeq 310$ Pa/m. Hence, the column pressure drop is $\Delta P = (310)(3) = 930$ Pa.

Comments

1. Check ΔP using SCRUB. The discrepancy is due to the approximate curve fits of Figure 11.17 used in SCRUB.
2. Figure 11.17 is of very general applicability and thus is not too accurate. If at all possible, the engineer should obtain pressure drop data for the specific packing under consideration. Usually the manufacturer of the packing can supply such data.

11.5 SIMULTANEOUS HEAT AND MASS EXCHANGERS

In general, simultaneous heat and mass exchangers are more difficult to analyze than simple heat or mass exchangers, since coupled differential equations governing conservation of mass species and energy must be solved simultaneously. Direct numerical solution of the differential equations is usually necessary. However, for some exchangers it is possible to simplify the analysis and yet obtain satisfactory results. In the case of a humidifier, the assumption of an adiabatic system reduces the problem to that of a simple single-stream mass exchanger, similar to those of Section 11.3. For counterflow cooling towers, it will be shown that *Merkel's approximation* allows the governing equations to be reduced to a single differential equation in terms of enthalpy, which can be solved easily by numerical integration.

11.5.1 Adiabatic Humidifiers

Figure 11.19 shows examples of adiabatic humidifiers used in air-conditioning systems. The purpose of the humidifier could be to increase the humidity of the air stream; alternatively, the purpose could be to cool the air stream. Since the water is continuously recirculated, with only a small fraction evaporating, the enthalpy of vaporization is supplied by heat transfer from the air stream to the water surface. Various types of packing are used with the objective of obtaining a large water surface area exposed to the air stream without an excessive pressure drop. If the design requires a very low pressure drop for the air stream, a spray chamber may be used in place of the packing. However, such units tend to be very large since the air velocity must be low to prevent water droplets from being blown out of the chamber.

If the amount of makeup water required to replace the evaporated water is relatively small, and if heat losses or gains between the unit and its surroundings are small, the

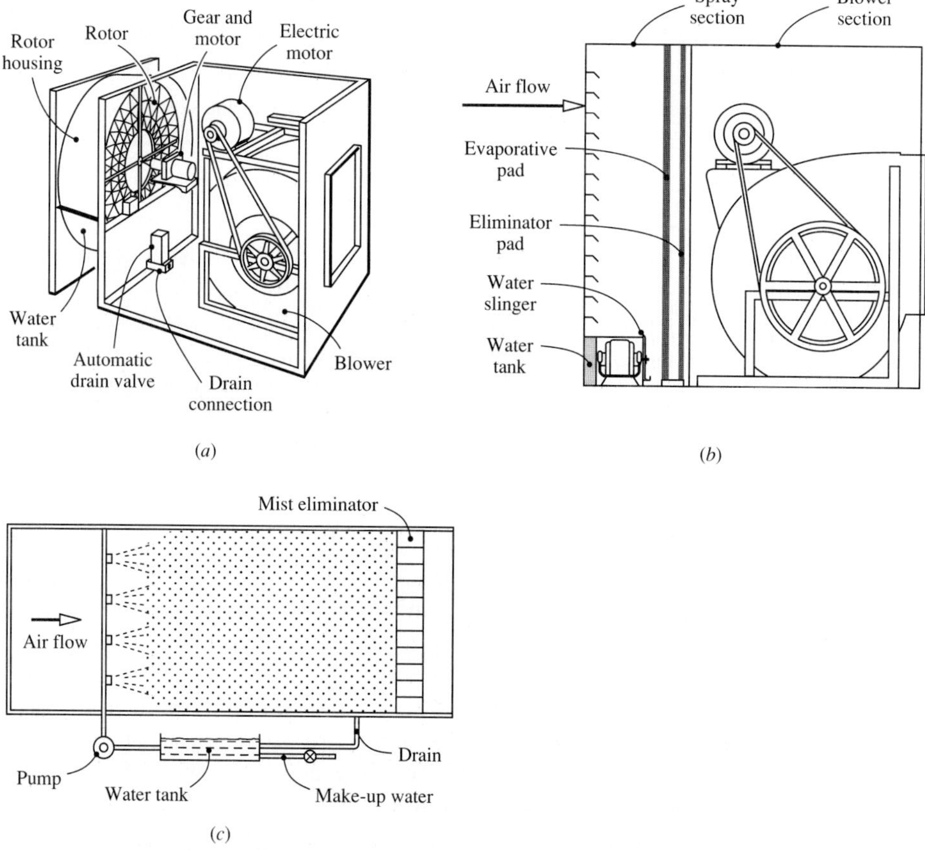

Figure 11.19 Some abiabatic humidifiers used in air conditioning practice. (*a*) Rotary type. (*b*) Slinger type. (*c*) Spray chamber.

process can be assumed to be adiabatic; that is, all the enthalpy of vaporization is supplied from the air itself. At steady state, the recirculated water will approach the wet-bulb temperature of the incoming air, and T_s and $m_{1,s}$ will be unchanging along the exchanger. For engineering purposes it is adequate to assume that the water is at the thermodynamic wet-bulb temperature. Thus, such humidifiers can be classified as single-stream exchangers since only the condition of the air changes along the exchanger. Both heat and mass are transferred in a humidifier, but, in the case of the adiabatic humidifier, the mass transfer and heat transfer can be analyzed separately. In exchangers such as cooling towers, the mass and heat transfer are coupled and must be considered simultaneously; the resulting analysis is more complicated, as will be seen in Section 11.5.2. Low mass transfer rate theory is applicable to the adiabatic humidifiers used in air-conditioning practice. The analysis of heat and mass transfer follows the procedures described in Sections 9.4 and 9.5.

Figure 11.20 depicts an adiabatic humidifier with water, species 1, evaporating into an air stream. The flow rate of the mixture of water vapor and air will be denoted $\dot{m}_G$, m_1 denotes the bulk mass fraction of water vapor, $j_{1,s}$ is the rate of evaporation of water, and $\mathscr{P}$ is the perimeter of transfer surface (s-surface). Application of the principle of conservation of species to the elemental control volume between x and Δx gives an equation identical to Eq. (11.5):

$$\dot{m}_G m_1|_x + j_{1,s}\mathscr{P}\Delta x = \dot{m}_G m_1|_{x+\Delta x}$$

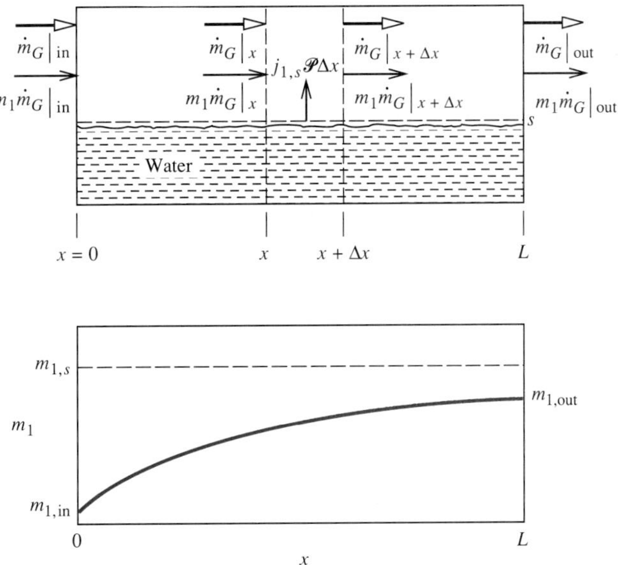

Figure 11.20 Schematic of an adiabatic humidifier showing mass and species balances, and the expected variation of the water vapor mass fraction along the humidifier.

In addition, the principle of mass conservation gives

$$\dot{m}_G|_x + j_{1,s}\mathscr{P}\Delta x = \dot{m}_G|_{x+\Delta x} \tag{11.66}$$

However, for humidifier operation at close to ambient temperatures, the evaporation rate is small compared with the air flow rate, and $\dot{m}_G$ can be *assumed* to be constant: then Eq. (11.66) can be discarded and Eq. (11.7) is obtained as before:

$$\frac{dm_1}{dx} - \frac{j_{1,s}\mathscr{P}}{\dot{m}_G} = 0 \tag{11.67}$$

For the humidifier, the required expression for the evaporation rate $j_{1,s}$ is simple:

$$j_{1,s} = \mathscr{g}_{m1}(m_{1,s} - m_1)$$

where, as stated before, $m_{1,s} = m_{1,\mathrm{sat}}(T_{\mathrm{WB}})$ is constant along the exchanger, and m_1 is the bulk value. Substituting in Eq. (11.67),

$$\frac{dm_1}{dx} + \frac{\mathscr{g}_{m1}\mathscr{P}}{\dot{m}_G}(m_1 - m_{1,s}) = 0 \tag{11.68}$$

Integrating with $m_1 = m_{1,\mathrm{in}}$ at $x = 0$ and $m_1 = m_{1,\mathrm{out}}$ at $x = L$, and rearranging gives

$$\frac{m_{1,\mathrm{out}} - m_{1,\mathrm{in}}}{m_{1,s} - m_{1,\mathrm{in}}} = 1 - e^{-\mathscr{g}_{m1}\mathscr{P}L/\dot{m}_G} \tag{11.69}$$

Again, the left-hand side of this equation is recognized as the exchanger effectiveness since it is the actual increase in moisture content of the air stream divided by the maximum possible increase that could be obtained in an infinitely long exchanger: when m_1 equals $m_{1,s}$, no further evaporation can occur (see Fig. 11.20). The dimensionless exponent on the right-hand side is the number of transfer units of the humidifier. Thus, Eq. (11.11) again applies, with appropriate expressions for ε and N_{tu}.

To obtain the amount of makeup water, $\dot{m}_{\mathrm{add}}$, that must be added, we perform a species balance on the exchanger as a whole:

$$(\dot{m}_G m_1)_{\mathrm{in}} + \dot{m}_{\mathrm{add}} = (\dot{m}_G m_1)_{\mathrm{out}} \tag{11.70}$$

For an approximate result we can take $\dot{m}_G$ constant and write

$$\dot{m}_{\mathrm{add}} \simeq \dot{m}_G(m_{1,\mathrm{out}} - m_{1,\mathrm{in}}) = \dot{m}_G\varepsilon(m_{1,s} - m_{1,\mathrm{in}}) \tag{11.71}$$

Often the purpose of a humidifier is to cool the air rather than to humidify it: it is then called an evaporative or "swamp" cooler. An easy way to determine the outlet air temperature is to recognize that, in an adiabatic humidifier, the wet-bulb temperature of the air does not change as it flows through the exchanger. After calculation of $m_{1,\mathrm{out}}$, the humidity ratio $\omega = m_{1,\mathrm{out}}/(1 - m_{1,\mathrm{out}})$ kg H_2O per kilogram of dry air can be calculated, and a psychrometric chart used to determine the dry-bulb temperature, which is T_{out}. Alternatively, the computer program PSYCHRO can be used with specification of P, T_{WB}, and m_1 in the option that holds P and water vapor concentration constant.

EXAMPLE 11.9 A Laminar-Flow Evaporative Cooler

A solar air-conditioning process under development involves evaporative cooling of the air followed by dehumidification in a silica gel desiccant bed: hot air from a solar air heater is used to regenerate (dry) the silica gel. The only power requirements of the cycle are those of the fans, pumps, and controls. An existing evaporative cooler installation has a cross section 1.1 m high and 0.5 m wide and contains a patented packing. Since fan power and hence pressure drop are key factors in the system design, a laminar-flow device is to be evaluated, in which there is cross-flow of air between vertical plastic plates with falling water films. Find the required length of an 80% effective exchanger if the volumetric air flow rate is 50 m^3/min at 1 atm pressure. The inlet air is at 300 K and 40% RH. Take the plate center-to-center spacing as 9.5 mm, and the combined width of one plate plus two liquid films as 2.5 mm. Also calculate the outlet air temperature.

Solution

Given: Parallel-plate evaporative cooler.

Required: (i) Length for 80% effectiveness when the cooler is required to cool 50 m^3/min of air. (ii) Outlet air temperature.

Assumptions:
1. Negligible heat gain from the surroundings.
2. Laminar flow (check).
3. Properties of pure air may be used.

Geometrical data for the packing are

Number of plates = $(0.5)/(9.5 \times 10^{-3}) = 53$

Perimeter $\mathscr{P} = (2)(53)(1.1) = 116$ m

Flow area $A_c = (1.1)(0.5) - (53)(1.1)(2.5 \times 10^{-3}) = 0.404\ \text{m}^2$

$D_h = 2 \times$ passage width $= (2)(7 \times 10^{-3}) = 1.4 \times 10^{-2}$ m

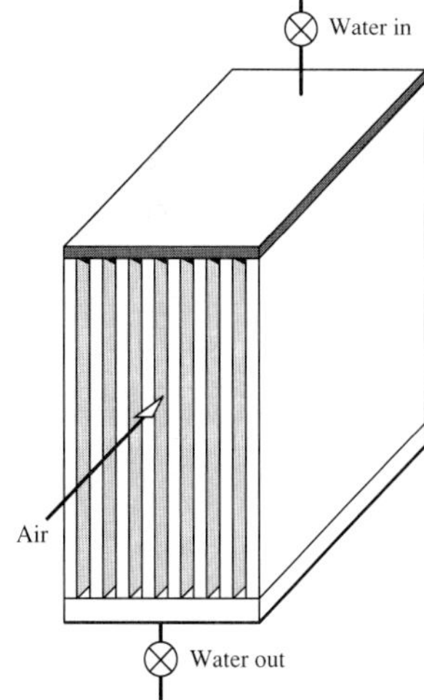

The exchanger is required to be 80% effective; thus, the required number of transfer units is

$$N_{\text{tu}} = \ln \frac{1}{1-\varepsilon} = \ln \frac{1}{1-0.8} = 1.61 = \frac{g_{m1}\mathscr{P}L}{\dot{m}_G}$$

$$L = \frac{1.61\dot{m}_G}{g_{m1}\mathscr{P}}$$

To evaluate g_{m1}, assume laminar flow and no entrance effects; then Sh = 7.54 from Table 4.5. Take properties for pure air at 300 K and 1 atm; then $\rho = 1.177$ kg/m^3, $\nu = 15.66 \times 10^{-6}$ m^2/s. Also, $\text{Sc}_{12} = 0.61$ for the H_2O-air mixture.

$$g_{m1} = 7.54\frac{\rho\mathscr{D}_{12}}{D_h} = 7.54\frac{\rho\nu}{\text{Sc}_{12}D_h} = \frac{(7.54)(1.177)(15.66 \times 10^{-6})}{(0.61)(1.4 \times 10^{-2})}$$

$$= 1.63 \times 10^{-2}\ \text{kg/m}^2\,\text{s}$$

$$L = \frac{1.61\dot{m}_G}{\mathscr{G}_{m1}\mathscr{P}} = \frac{(1.61)(50/60)(1.177)}{(1.63 \times 10^{-2})(116)} = 0.835 \text{ m}$$

To check if the flow is laminar, the Reynolds number is calculated:

$$V = \frac{50/60 \text{ m}^3/\text{s}}{0.404 \text{ m}^2} = 2.06 \text{ m/s}$$

$$\text{Re}_{D_h} = \frac{VD_h}{\nu} = \frac{(2.06)(1.4 \times 10^{-2})}{15.66 \times 10^{-6}} = 1842 < 2800, \quad \text{that is, laminar.}$$

To check the entrance effect we use the mass transfer analog to Eq. (4.51):

$$\overline{\text{Sh}}_{D_h} = 7.54 + \frac{0.03(D_h/L)\,\text{Re}_{D_h}\,\text{Sc}}{1 + 0.016[(D_h/L)\,\text{Re}_{D_h}\,\text{Sc}]^{2/3}} = 7.54 + \frac{(0.03)(18.8)}{1 + 0.016(18.8)^{2/3}} = 8.05$$

New estimates are:

$$\mathscr{G}_{m1} = (8.05/7.54)(1.63 \times 10^{-2}) = 1.74 \times 10^{-2}$$

$$L = (1.63/1.74)(0.835) = 0.782 \text{ m}$$

To obtain T_{out}, we first need to calculate the wet-bulb temperature of the inlet air, as well as $m_{1,\text{in}}$ and $m_{1,s}$. PSYCHRO can be used as follows.

1. Option 1.
   Input: $P = 1.013 \times 10^5$ Pa, $T_{\text{DB}} = 300$ K, RH $= 40\%$
   Output: $T_{\text{WB}} = 290.7$ K, $m_{1,\text{in}} = 8.713 \times 10^{-3}$

2. Option 1.
   Input: $P = 1.013 \times 10^5$ Pa, $T_{\text{DB}} = 290.7$ K, RH $= 100\%$
   Output: $m_{1,s} = 1.239 \times 10^{-2}$

Then, from Eq. (11.69), we obtain the outlet mass fraction as

$$m_{1,\text{out}} = m_{1,\text{in}} + \varepsilon(m_{1,s} - m_{1,\text{in}}) = 0.00871 + 0.8(0.01239 - 0.00871) = 0.01165$$

3. Option 2.
   Input: $P = 1.013 \times 10^5$ Pa, $m_1 = 0.01165$, $T_{\text{WB}} = 290.7$ K
   Output: $T_{\text{DB}} = 292.8 \text{ K} = T_{\text{out}}$

Comments

1. Air-conditioning engineers sometimes use a "temperature effectiveness" to characterize the performance of an evaporative cooler,

$$\varepsilon = \frac{T_{\text{in}} - T_{\text{out}}}{T_{\text{in}} - T_s} = \frac{300 - 292.8}{300 - 290.7} = 0.77 = 77\%$$

   which is a little lower than the mass transfer effectiveness. However, more advanced texts will show that it is more appropriate to use an effectiveness based on enthalpy change to characterize the thermal performance of evaporative coolers.

2. Use Eq. (9.58) to determine T_{out} in step 3 above.

3. Use a psychrometric chart to determine T_{out} and compare the result.
4. Notice that accounting for entrance effects reduces the required length by 6%.
5. Use PSYCHRO to check that the enthalpy of the air is constant through the exchanger, consistent with the adiabatic assumption. Of course, assuming a constant T_{WB} through the exchanger ensures this result.
6. If the effect of ripples on the water film surfaces is ignored, the friction factor can be obtained from Table 4.5 as $f = 96/\text{Re}_{D_h} = 96/1842 = 0.0521$. The pressure drop is then

$$\Delta P = f(L/D_h)(1/2)\rho V^2 = (0.0521)(0.782/1.4 \times 10^{-2})(0.5)(1.177)(2.06)^2$$
$$= 7.3 \text{ Pa}$$

   which is very small. The sum of the inlet and outlet pressure drops is likely to be considerably larger, unless the air ducts have the same cross-section as the exchanger (see Section 8.7.1).
7. See Exercise 11–42 for a performance comparison with a patented packing.

11.5.2 Counterflow Cooling Towers

In a wet cooling tower, water is evaporated into air with the objective of cooling the water stream. Since both the air and the water streams change state along the exchanger, the wet cooling tower is a *two-stream exchanger.* Both natural- and mechanical-draft cooling towers are popular, and examples are shown in Fig. 11.21. Large natural-draft cooling towers are used in power plants for cooling the water supply to the condenser. Smaller mechanical-draft towers are preferred for oil refineries and other process industries, as well as for central air-conditioning systems. Figure 11.21 shows *counterflow* units in which the water flows as thin films down over a suitable packing, and air flows upward. In a natural-draft tower, the air flows upward due to the buoyancy of the warm, moist air leaving the top of the packing. In a mechanical-draft tower, the flow is forced or induced by a fan. Since the air inlet temperature is usually lower than the water inlet temperature, the water is cooled both by evaporation and by sensible heat loss. For usual operating conditions, q_{evap} is considerably larger than q_{conv}.

Exchanger Balances

Figure 11.22 shows the schematic of a counterflow cooling tower packing (or *fill*). The subscript G is used to denote the gas stream, which is a water vapor and air mixture; the subscript L denotes the liquid stream, which is taken to be pure water. Species 1 and 2 are H_2O and air, respectively. The exchanger balances for mass, mass species (H_2O), and energy are

$$\dot{m}_{G,\text{in}} - \dot{m}_{L,\text{out}} = \dot{m}_{G,\text{out}} - \dot{m}_{L,\text{in}} \tag{11.72}$$

$$(\dot{m}_G m_1)_{\text{in}} - \dot{m}_{L,\text{out}} = (\dot{m}_G m_1)_{\text{out}} - \dot{m}_{L,\text{in}} \tag{11.73}$$

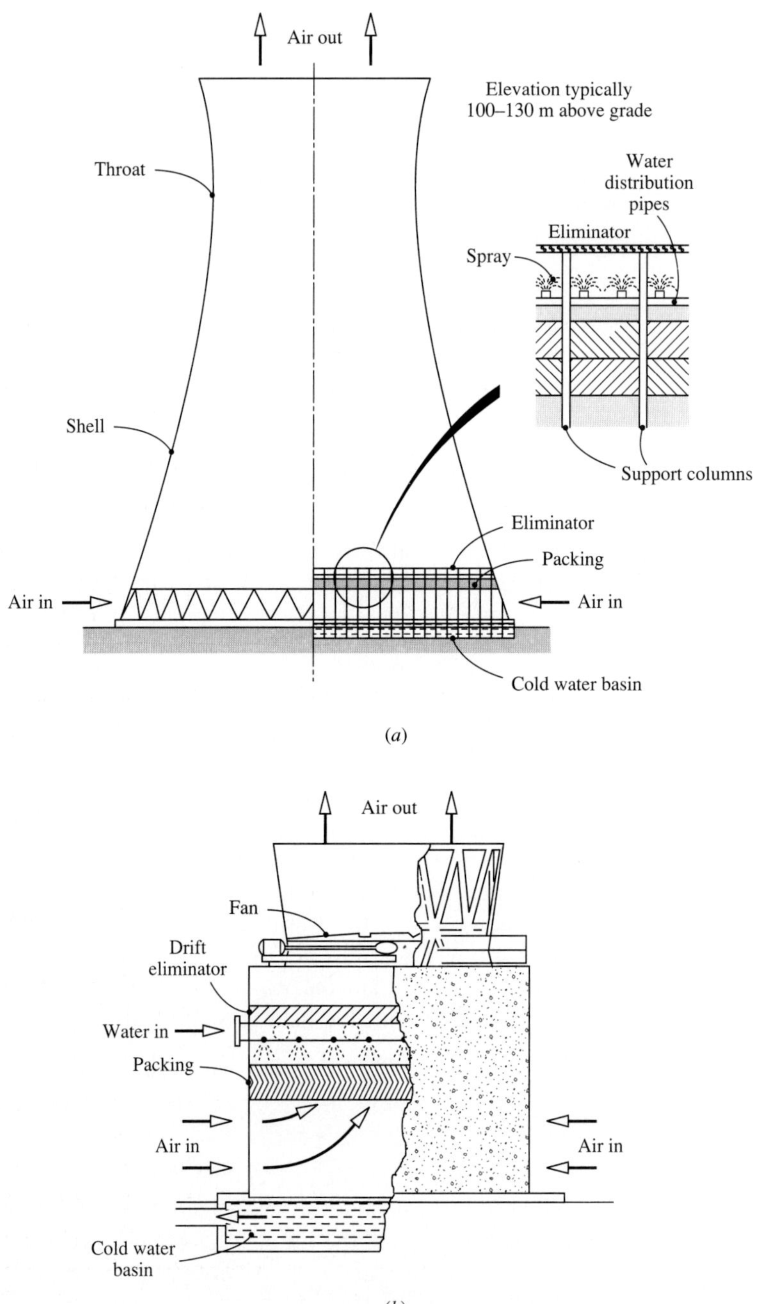

Figure 11.21 Counterflow cooling towers. (*a*) A natural-draft tower for a power plant. (*b*) A mechanical-draft tower for an air-conditioning system.

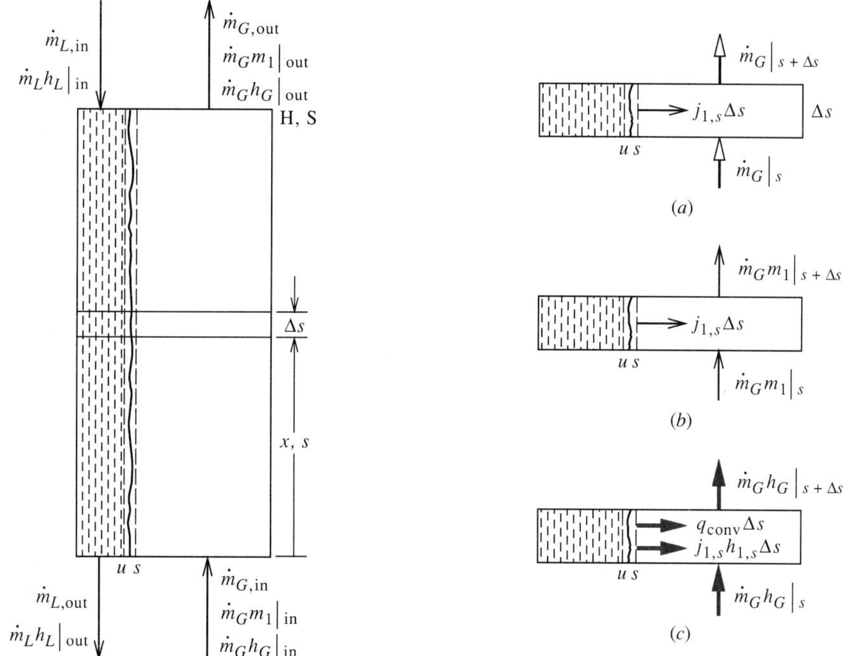

Figure 11.22 Schematic of a counterflow cooling tower packing showing an elemental volume containing a transfer area Δs, and balances on the gas stream: (*a*) mass, (*b*) species, (*c*) energy.

$$(\dot{m}_G h_G)_{\text{in}} - (\dot{m}_L h_L)_{\text{out}} = (\dot{m}_G h_G)_{\text{out}} - (\dot{m}_L h_L)_{\text{in}} \tag{11.74}$$

where the water flow rate, $\dot{m}_L$, is *positive*.[3]

Mixture Enthalpy and Specific Heat

The gas stream is a mixture of water vapor and air. The enthalpy of the mixture is

$$h = \sum_{i=1}^{2} m_i h_i = m_1 h_1 + m_2 h_2 \tag{11.75}$$

Typical enthalpy tables for water vapor and air mixtures are based on a datum temperature of 0°C, at which the enthalpy of dry air and liquid water are set equal to zero. Hence,

$$h_1 = h_{\text{fg}}(0°\text{C}) + \int_0^t c_{p1}\, dt \tag{11.76a}$$

[3] This convention is identical to that used for counterflow heat exchangers in Chapter 8 and for countercurrent mass exchangers in Section 11.4.3. In some texts the convention of a negative water flow rate is used.

$$h_2 = \int_0^t c_{p2}\,dt \tag{11.76b}$$

where t is used to denote temperature in degrees Celsius. Table A.31 and the computer program PSYCHRO are based on these datum states.

The range of temperature encountered in humidifier and cooling tower operation is relatively small, so that the species specific heats can be assumed constant for most purposes. Then

$$h_1 = h_{fg} + c_{p1}t \tag{11.77a}$$

$$h_2 = c_{p2}t \tag{11.77b}$$

Substituting in Eq. (11.75), the mixture enthalpy is

$$h = m_1 h_{fg} + (m_1 c_{p1} + m_2 c_{p2})t$$

or

$$h = m_1 h_{fg} + c_p t \tag{11.78}$$

where $c_p = \sum_{i=1}^{2} m_i c_{pi}$ is the mixture specific heat. For water vapor and air mixtures, the mixture specific heat depends strongly on composition because $c_{p\,H_2O}$ is almost twice $c_{p\,\mathrm{air}}$.

The Governing Differential Equations

For convenience we define the *contact area* in a length of exchanger Δx as $\Delta s = \mathscr{P}\Delta x$, where $\mathscr{P}$ is the transfer area perimeter. Then the total contact area in a packing of height H is

$$S = \int_0^H \mathscr{P}dx = \mathscr{P}H, \quad \text{if } \mathscr{P} \text{ is constant} \tag{11.79}$$

Mass, species, and energy balances on each stream for an element of packing containing contact area Δs gives the differential conservation equations. The mass and species balances are straightforward. For example, the mass balance on an element of the gas stream shown in Fig. 11.22 gives

$$\dot{m}_G|_s + j_{1,s}\Delta s = \dot{m}_G|_{s+\Delta s}$$

Dividing by Δs and letting $\Delta s \to 0$,

$$\frac{d\dot{m}_G}{ds} = j_{1,s} \tag{11.80}$$

Similarly,

$$\frac{d\dot{m}_L}{ds} = j_{1,s} \tag{11.81}$$

$$\frac{d}{ds}(\dot{m}_G m_1) = j_{1,s} \tag{11.82}$$

Low mass transfer rate theory has been used to approximate the rate of evaporation of the water as $j_{1,s}$ [see Eq. (9.38)]. The energy balances require special consideration. The steady-flow energy equation cannot be used in precisely the form of Eq. (1.4) because the stream flow rates $\dot{m}_G$ and $\dot{m}_L$ are not constant. Referring to Fig. 11.22, the energy balance on the gas stream is

$$q_{\text{conv}}\Delta s = \dot{m}_G h_G|_{s+\Delta s} - \dot{m}_G h_G|_s - j_{1,s}h_{1,s}\Delta s$$

where the last term on the right-hand side accounts for the enthalpy added to the control volume by the evaporated water. Dividing by Δs and letting $\Delta s \to 0$,

$$\frac{d}{ds}(\dot{m}_G h_G) = q_{\text{conv}} + j_{1,s}h_{1,s} \tag{11.83}$$

Similarly, for the liquid phase,

$$\frac{d}{ds}(\dot{m}_L h_L) = q_{\text{conv}} + j_{1,s}h_{1,s} \tag{11.84}$$

because energy is conserved between the s- and u-surfaces. It will prove unnecessary in this analysis to write the liquid stream energy balance in terms of u-surface fluxes: Eq. (11.84) will be sufficient. Since $q_{\text{conv}} = h_c(T_s - T_G)$ the energy equation, Eq. (11.83), is in terms of two dependent variables, namely, h and T. Unless we are going to solve the equation directly by numerical means, one variable should be eliminated in favor of the other. Enthalpies are easily written in terms of mass fraction and temperature, but if we go to T as the dependent variable, the mass fraction m_1 will remain as an additional dependent variable. Instead we will eliminate T in favor of h, and after we make some reasonable assumptions, the equation set will reduce to a single equation in terms of only one dependent variable, the enthalpy h.

Simplification of the Governing Equations

Our first step in a rather involved manipulation of Eq. (11.83) is to expand its left-hand side and substitute Eq. (11.80):

$$\frac{d}{ds}(\dot{m}_G h_G) = \dot{m}_G\frac{dh_G}{ds} + h_G\frac{d\dot{m}_G}{ds} = \dot{m}_G\frac{dh_G}{ds} + j_{1,s}h_G \tag{11.85}$$

Thus, Eq. (11.83) becomes

$$\dot{m}_G\frac{dh_G}{ds} = q_{\text{conv}} + j_{1,s}(h_{1,s} - h_G) \tag{11.86}$$

Next the mass and heat transfer conductances are introduced. The evaporation rate is

$$j_{1,s} = g_m(m_{1,s} - m_{1,G}) \tag{11.87}$$

and we will write the convective heat transfer as

$$q_{\text{conv}} = g_h c_{pG}(T_s - T_G) \tag{11.88}$$

where $g_h = h_c/c_{pG}$ is the *heat transfer conductance* introduced in Section 9.4.4. Advanced mass transfer theory (see Section 10.5.2) shows that it is appropriate to use the bulk gas specific heat in the definition of heat transfer conductance, but this point is of no practical significance in the present analysis. Substituting Eqs. (11.87) and (11.88) in Eq. (11.86) gives

$$\dot{m}_G \frac{dh_G}{ds} = g_h c_{pG}(T_s - T_G) + g_m(m_{1,s} - m_{1,G})(h_{1,s} - h_G) \tag{11.89}$$

The Lewis number for dilute water vapor and air mixtures is Le $= \text{Pr/Sc} \simeq 0.69/0.61 = 1.13$, and thus the ratio $g_m/g_h \simeq (1.13)^{2/3} = 1.08$. This result was used in the analysis of the wet- and dry-bulb psychrometer in Section 9.5. Here we will be more bold and take $g_h = g_m$, which is equivalent to assuming a Lewis number of unity.[4] Equation (11.89) then becomes

$$\dot{m}_G \frac{dh_G}{ds} = g_m \left\{ c_{pG}(T_s - T_G) + (m_{1,s} - m_{1,G})(h_{1,s} - h_G) \right\} \tag{11.90}$$

where we choose to use g_m rather than g_h because q_{evap} is usually considerably larger than q_{conv}. We now come to the key step in the analysis, and assert that Eq. (11.90) can be approximated as

$$\dot{m}_G \frac{dh_G}{ds} \simeq g_m(h_s - h_G) \tag{11.91}$$

for typical cooling tower operating conditions, *provided that the usual enthalpy datum states are used*. The validity of the approximation is best demonstrated by a numerical calculation. For example, at the bottom of the tower, possible conditions are $P = 1$ atm, $T_s = 30°\text{C}$, $T_G = 15°\text{C}$, $m_{1,G} = 0$, where dry air has been chosen for convenience. Using PSYCHRO or Tables A.7, A.12*a*, and A.31, $m_{1,s} = 0.0266$ and

$$h_s - h_G = 97.36 - 15.09 = 82.27 \text{ kJ/kg}$$

$$c_{pG}(T_s - T_G) + (m_{1,s} - m_{1,G})(h_{1,s} - h_G)$$
$$= 1.007(30 - 15) + (0.0266 - 0)(2556 - 15) = 82.70 \text{ kJ/kg}$$

The discrepancy is very small. In general, a discrepancy of up to 5% may be expected for usual cooling tower operating conditions. Notice that it is **essential** to use the usual enthalpy datum states, that is, such that enthalpies of dry air and liquid water are zero at 0°C. The magnitude of a term such as $(h_s - h_G)$ depends on the choice of enthalpy datum states since the compositions of the bulk and s-surface mixtures are different. It is possible to arbitrarily choose datum states to obtain $(h_s - h_G) = 0$. On the other hand, a term such as $c_{pG}(T_s - T_G)$ is independent of the choice of datum states.

---

[4] The equality $g_h = g_m$ is often called the *Lewis relation*.

A similar manipulation for the liquid stream gives

$$\dot{m}_L \frac{dh_L}{ds} = \mathscr{g}_m(h_s - h_G) \tag{11.92}$$

Dividing Eq. (11.91) by Eq. (11.92),

$$\frac{dh_G}{dh_L} = \frac{\dot{m}_L}{\dot{m}_G} \tag{11.93}$$

which, if $\dot{m}_L/\dot{m}_G$ is assumed constant, can be integrated from the bottom of the tower where $h_G = h_{G,\text{in}}$ and $h_L = h_{L,\text{out}}$, to give

$$h_G = h_{G,\text{in}} + \frac{\dot{m}_L}{\dot{m}_G}(h_L - h_{L,\text{out}}) \tag{11.94}$$

The definition of the number of transfer units is $N_{\text{tu}} = \mathscr{g}_m S/\dot{m}_L$, where $S = \mathscr{P}H$ is the contact area.[5] Introducing $dN_{\text{tu}} = \mathscr{g}_m ds/\dot{m}_L$ into Eq. (11.91) gives

$$\frac{\dot{m}_G}{\dot{m}_L}\frac{dh_G}{dN_{\text{tu}}} = h_s - h_G \tag{11.95}$$

which can be integrated using Eqs. (11.93) and (11.94):

$$dN_{\text{tu}} = \frac{\dot{m}_G}{\dot{m}_L}\frac{dh_G}{h_s - h_G} = \frac{dh_L}{h_s - h_G}$$

$$N_{\text{tu}} = \int_{h_{L,\text{out}}}^{h_{L,\text{in}}} \frac{dh_L}{h_s - h_G} \tag{11.96}$$

where $h_G = h_{G,\text{in}} + (\dot{m}_L/\dot{m}_G)(h_L - h_{L,\text{out}})$ from Eq. (11.94). Our final assumption (in a long list of assumptions!) is that the liquid-side heat transfer resistance is negligible; then $T_s = T_L$ or

$$h_s(P, T_s) = h_s(P, T_L) \tag{11.97}$$

Equation (11.96) can be integrated numerically using Eq. (11.97), as will be demonstrated in Example 11.10.

This method of calculating the number of transfer units was originally developed by Merkel in 1925 [20]. When the conventional enthalpy datum states are used, the method is accurate up to temperatures of about 60°C. Comparisons with more exact solution procedures seldom show errors greater than 10%. Notice that the method does not give the outlet state of the air; however, in situations encountered in practice, the outlet air can be assumed saturated for the purpose of calculating its density. Before computers became widely used, graphical methods, based on the enthalpy-versus-composition chart for the water–water vapor–air system, were popular for analyzing simultaneous heat and mass exchangers. Spalding [21] gives a comprehensive treatment of such methods. Merkel's method is based on the obser-

[5] This definition is somewhat arbitrary but follows current cooling tower design practice. Some texts define the number of transfer units as $N_{\text{tu}} = \mathscr{g}_m S/\dot{m}_G$.

vation that the mixed-phase isotherms on an enthalpy-composition chart are nearly horizontal below 60°C; assuming these isotherms to be horizontal is equivalent to the approximation made in going from Eq. (11.90) to Eq. (11.91). Merkel's method can be used for higher temperature ranges, or different systems, by defining appropriate enthalpy datum states, that is, states that yield nearly horizontal mixed-phase isotherms in the temperature range under consideration. It is also possible to extend Merkel's method to include a finite liquid-side heat transfer resistance (see Section 11.5.4 and Exercise 11–48), but such refinement is seldom warranted. For a laminar film, the liquid-side resistance is approximately $(\delta/2)/k$, where k is the liquid conductivity and δ the film thickness: for typical operating conditions, the bulk liquid temperature is seldom more than 0.3 K above the interface temperature.

EXAMPLE 11.10 Use of Merkel's Method

A wet cooling tower is required to cool water from 40°C to 26°C when the inlet air is at 10°C, 1 atm, and saturated (a miserable rainy day!). Calculate the required number of transfer units for balanced flow, that is, $\dot{m}_G/\dot{m}_L = 1$.

Solution

Given: Wet countercurrent cooling tower with balanced flow.

Required: Number of transfer units for specified performance.

Assumptions: Merkel's method is appropriate.

Equation (11.96) is to be integrated numerically, with h_G obtained from Eq. (11.94).

$$N_{\text{tu}} = \int_{h_{L,\text{out}}}^{h_{L,\text{in}}} \frac{dh_L}{h_s - h_G}; \qquad h_G = h_{G,\text{in}} + \frac{\dot{m}_L}{\dot{m}_G}(h_L - h_{L,\text{out}})$$

Using Table A.31, $h_{G,\text{in}} = h_{\text{sat}}(10°\text{C}) = 29.15$ kJ/kg, $h_{L,\text{out}} = h_L(26°\text{C}) = 109.07$ kJ/kg. Hence,

$$h_G = 29.15 + (h_L - 109.07)$$

Choosing 2°C intervals for convenient numerical integration, the following table is constructed, with h_L and $h_s = h_s(T_L)$ also obtained from Table A.31.

T_L °C	h_L kJ/kg	h_G kJ/kg	h_s kJ/kg	$h_s - h_G$ kJ/kg	$\dfrac{1}{h_s - h_G}$
26	109.07	29.15	79.12	49.97	0.02001
28	117.43	37.51	87.86	50.35	0.01986
30	125.79	45.87	97.36	51.49	0.01942
32	134.15	54.23	107.67	53.44	0.01871
34	142.50	62.58	118.89	56.31	0.01776
36	150.86	70.94	131.10	60.16	0.01662
38	159.22	79.30	144.39	65.09	0.01536
40	167.58	87.66	158.86	71.20	0.01404

Using the trapezoidal rule,

$$\begin{aligned}\int_{h_{L,\text{out}}}^{h_{L,\text{in}}} \frac{dh_L}{h_s - h_G} &= \frac{8.36}{2}[0.02001 + 2(0.01986 + 0.01942 + 0.01871 \\ &\quad + 0.01776 + 0.01662 + 0.01536) + 0.01404] \\ &= 1.043\end{aligned}$$

From Eq. (11.96), $N_{\text{tu}} = 1.043$. Also, using Table A.31, $T_{G,\text{out}} = 27.9°\text{C}$ for saturated outlet air.

Comments

1. Under usual operating conditions, the air stream becomes saturated a short distance up the packing. However, it continues to take up more water vapor as it flows upward because its temperature (and hence $P_{1,\text{sat}}$) increases.
2. Greater accuracy can be obtained by using smaller integration steps (CTOWER uses 14 steps). However, evaluation of the number of transfer units is not the major source of error in cooling tower design. There is usually a much greater uncertainty associated with the packing mass transfer correlations and the effects of nonuniform water and air flows.

11.5.3 Cross-Flow Cooling Towers

Figure 11.23 shows examples of natural- and mechanical-draft cross-flow cooling towers. Figure 11.24 shows a schematic of a cross-flow tower packing and the coordinate system. To model this cross-flow exchanger, we will assume that both the liquid and gas streams are unidirectional, and that there is no mixing in either stream. The packing has dimensions $X \times W \times H$, and a [m²/m³] is the contact area per unit volume, that is, the area of liquid-gas interface per unit volume. Figure 11.24 also shows an elemental control volume $\Delta x \Delta y \Delta z$. Superficial mass velocities for the gas and liquid streams are defined as

$$G = \frac{\dot{m}_G}{WH}; \qquad L = \frac{\dot{m}_L}{WX} \tag{11.98a}$$

Mass conservation for the gas stream requires that

$$(G|_{x+\Delta x} - G|_x)\Delta y \Delta z = j_{1,s} a \Delta x \Delta y \Delta z$$

Dividing by $\Delta x \Delta y \Delta z$ and letting Δx, Δy, $\Delta z \to 0$, we obtain

$$\frac{\partial G}{\partial x} = a j_{1,s} \tag{11.98b}$$

Similarly, mass conservation for the liquid stream requires that

$$(L|_{y+\Delta y} - L|_y)\Delta x \Delta z = -j_{1,s} a \Delta x \Delta y \Delta z$$

Dividing by $\Delta x \Delta y \Delta z$, and letting Δx, Δy, $\Delta z \to 0$ we obtain

$$\frac{\partial L}{\partial y} = -a j_{1,s} \tag{11.99}$$

(*a*)

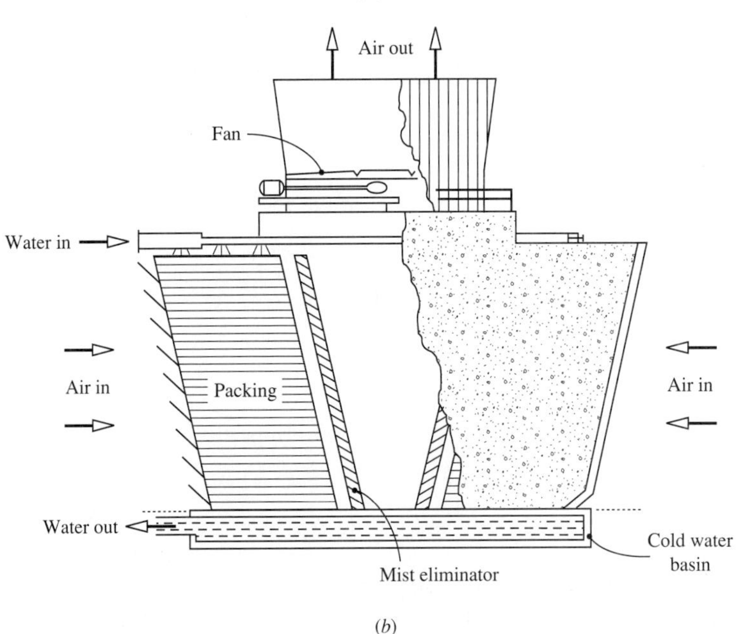

(*b*)

Figure 11.23 Cross-flow towers. (*a*) Natural-draft tower for a power plant. (*b*) A mechanical-draft tower for an oil refinery.

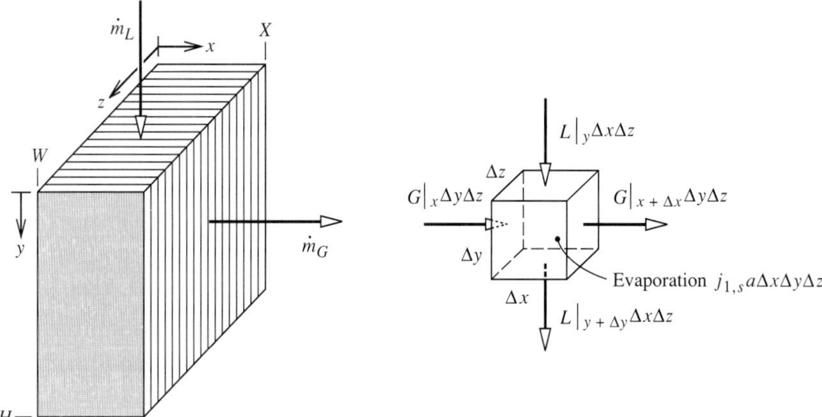

Figure 11.24 Schematic of a cross-flow cooling tower packing showing the coordinate system and mass balances on an elemental volume $\Delta x \Delta y \Delta z$.

The species conservation equation for the gas stream, and the energy conservation equations for the gas and liquid streams are derived in a similar manner:

$$\frac{\partial}{\partial x}(Gm_1) = a\, j_{1,s} \tag{11.100}$$

$$\frac{\partial}{\partial x}(Gh_G) = a[g_h c_{pG}(T_s - T_G) + j_{1,s} h_{1,s}] \tag{11.101}$$

$$\frac{\partial}{\partial y}(Lh_L) = -a[g_h c_{pG}(T_s - T_G) + j_{1,s} h_{1,s}] \tag{11.102}$$

The energy equations can be manipulated as was done for the counterflow case in Section 11.5.2, to obtain

$$\frac{\partial h_G}{\partial x} = \frac{g_m a}{G}(h_s - h_G) \tag{11.103}$$

$$\frac{\partial h_L}{\partial y} = -\frac{g_m a}{L}(h_s - h_G) \tag{11.104}$$

Also, as before, we assume

$$h_s = h_s(h_L) \tag{11.105}$$

The required boundary conditions are the inlet conditions of the streams,

$$x = 0: \quad h_G = h_{G,\text{in}} \tag{11.106a}$$

$$y = 0: \quad h_L = h_{L,\text{in}} \tag{11.106b}$$

Equations (11.103) and (11.104) are a pair of first-order partial differential equations in the two dependent variables h_G and h_L, and can be solved numerically. (Exercise 11–54 requires the formulation of a suitable finite-difference solution procedure.) To determine the tower performance, the average enthalpy of the exit liquid must be determined:

$$\bar{h}_{L,\text{out}} = \frac{1}{X}\int_0^X h_{L,\text{out}}\,dx \tag{11.107}$$

Substituting in an exchanger energy balance on the liquid stream then gives the heat transfer as

$$\dot{Q} = \dot{m}_L(h_{L,\text{in}} - \bar{h}_{L,\text{out}}) \tag{11.108}$$

which is the desired result.

11.5.4 Thermal-Hydraulic Design of Cooling Towers

The thermal-hydraulic design of a mechanical-draft cooling tower is relatively straightforward. The flow rate ratio $\dot{m}_L/\dot{m}_G$ can be specified and varied parametrically to obtain an optimal design, for which the size and cost of the packing is balanced against fan power requirements and operating costs. Data are required for mass transfer conductances and friction for candidate packings. Tables 11.5*a* and 11.5*b*

Table 11.5*a* Packings for counterflow and cross-flow cooling towers: Designations and descriptions.

Counterflow Packings

1. Flat asbestos sheets, pitch 4.45 cm
2. Flat asbestos sheets, pitch 3.81 cm
3. Flat asbestos sheets, pitch 3.18 cm
4. Flat asbestos sheets, pitch 2.54 cm
5. 60° angle corrugated plastic, Munters M12060, pitch 1.17 in
6. 60° angle corrugated plastic, Munters M19060, pitch 1.8 in
7. Vertical corrugated plastic, American Tower Plastics Coolfilm, pitch 1.63 in
8. Horizontal plastic screen, American Tower Plastics Cooldrop, pitch 8 in, 2 in grid
9. Horizontal plastic grid, Ecodyne shape 10, pitch 12 in
10. Angled corrugated plastic, Marley MC67, pitch 1.88 in
11. Dimpled sheets, Toschi Asbestos-Free Cement, pitch 0.72 in
12. Vertical plastic honeycomb, Brentwood Industries Accu-Pack, pitch 1.75 in

Cross-Flow Packings

1. Doron V-bar, 4 in × 8 in spacing
2. Doron V-bar, 8 in × 8 in spacing
3. Ecodyne T-bar, 4 in × 8 in spacing
4. Ecodyne T-bar, 8 in × 8 in spacing
5. Wood lath, parallel to air flow, 4 in × 4 in spacing
6. Wood lath, perpendicular to air flow, 4 in × 4 in spacing
7. Marley α-bar, perpendicular to air flow, 16 in × 4 in spacing
8. Marley ladder, perpendicular to air flow, 8 in × 2 in spacing

give correlations for a selection of packings. In Table 11.5*b*, the mass transfer conductance is correlated as $\mathscr{g}_m a/L$, where a is the transfer area per unit volume and $L = \dot{m}_L/A_{\text{fr}}$ is the superficial mass velocity of the water flow (also called the *water loading* on the packing). Similarly, we define $G = \dot{m}_G/A_{\text{fr}}$. Typical water loadings are 1.8–2.7 kg/m^2 s, and superficial air velocities fall in the range 1.5–4 m/s. No attempt is made to correlate $\mathscr{g}_m$ and a separately, as was done for the mass-exchanger packings considered in Section 11.4.2. The number of transfer units of a packing of

Table 11.5*b* Mass transfer and pressure drop correlations for cooling towers. Data from Lowe and Christie [14] for counterflow packings 1 through 4[a]; all other data from EPRI GS-6730 [15].

Correlations (SI units)

Mass transfer: $$\frac{\mathscr{g}_m a}{L\ [\text{kg/m}^2\ \text{s}]} = C_1 (L^+)^{n_1} (G^+)^{n_2} (T^+_{\text{HW}})^{n_3}$$

where $L^+ = \dfrac{L}{L_0}$, $G^+ = \dfrac{G}{G_0}$, $T^+_{\text{HW}} = \dfrac{1.8 T_{L,\text{in}}\ [°\text{C}] + 32}{110}$

Pressure drop: $$\frac{N}{H \text{ or } X} = C_2 (L^+)^{n_4} (G^+)^{n_5}$$

Packing Number	C_1 m^{-1}	n_1	n_2	n_3	C_2 m^{-1}	n_4	n_5
Counterflow Packings: $L_0 = G_0 = 3.391$ kg/m^2 s							
1	0.289	−0.70	0.70	0.00	2.72	0.35	−0.35
2	0.361	−0.72	0.72	0.00	3.13	0.42	−0.42
3	0.394	−0.76	0.76	0.00	3.38	0.36	−0.36
4	0.459	−0.73	0.73	0.00	3.87	0.52	−0.36
5	2.723	−0.61	0.50	−0.34	19.22	0.34	0.19
6	1.575	−0.50	0.58	−0.40	9.55	0.31	0.05
7	1.378	−0.49	0.56	−0.35	10.10	0.23	−0.04
8	0.558	−0.38	0.48	−0.54	4.33	0.85	−0.60
9	0.525	−0.26	0.58	−0.45	2.36	1.10	−0.64
10	1.312	−0.60	0.62	−0.60	8.33	0.27	−0.14
11	0.755	−0.51	0.93	−0.52	1.51	0.99	0.04
12	1.476	−0.56	0.60	−0.38	6.27	0.31	0.10
Cross-Flow Packings $L_0 = 8.135$ kg/m^2 s, $G_0 = 2.715$ kg/m^2 s							
1	0.161	−0.58	0.52	−0.44	1.44	0.66	−0.73
2	0.171	−0.34	0.32	−0.43	1.97	0.72	−0.82
3	0.184	−0.51	0.28	−0.31	1.38	1.30	0.22
4	0.167	−0.48	0.20	−0.29	1.25	0.89	0.07
5	0.171	−0.58	0.28	−0.29	3.18	0.76	−0.80
6	0.217	−0.51	0.47	−0.34	4.49	0.71	−0.59
7	0.213	−0.41	0.50	−0.42	3.44	0.71	−0.85
8	0.233	−0.45	0.45	−0.48	4.82	0.59	0.16

[a] For packings 1–4, CTOWER calculates pressure drop by interpolation in the original data (which are very sparse). C_2, n_4, and n_5 were obtained from an approximate least-squares curve fit of the data, and are not used in CTOWER.

height H is then

$$N_{\text{tu}} = \frac{g_m S}{\dot{m}_L} = \frac{g_m a H}{L} \tag{11.109}$$

The correlations are in terms of dimensionless mass velocities L^+ and G^+, and a *hot water correction* T^+_{HW}. The hot water correction accounts for a number of factors, such as errors associated with Merkel's method, deviations from low mass transfer rate theory at higher values of T_s, and fluid property dependence on temperature. Frictional resistance to air flow through the packings is correlated as a *loss coefficient* $N = \Delta P/(\rho V^2/2)$ per unit height or depth of packing, as a function of L^+ and G^+. The velocity V is the superficial gas velocity. No hot water correction is required.

In a natural-draft tower, the thermal and hydraulic performance of the tower are coupled, and the flow rate ratio $\dot{m}_L/\dot{m}_G$ $(= L/G)$ cannot be specified a priori. The buoyancy force producing the air flow depends on the state of the air leaving the packing, which in turn depends on L/G and the inlet air and water states. An iterative solution is required to find the operating point of the tower. The buoyancy force available to overcome the shell and packing pressure drops is

$$\Delta P^B = g(\rho_a - \rho_{G,\text{out}})H \tag{11.110}$$

where ρ_a is the ambient air density, and H is usually taken as the distance from the bottom of the packing to the top of the shell. The various pressure drops are conveniently expressed as

$$\Delta P_i = N_i \frac{\rho_{Gi} V_i^2}{2} \tag{11.111}$$

where N_i is the loss coefficient and V_i is the air velocity at the corresponding location. The pressure drops are associated with the shell, the packing, the mist eliminators, supports and pipes, and the water spray below the packing. Some sample correlations are given in Table 11.6.

Water loadings in natural-draft towers typically range from 0.8 to 2.4 kg/m^2 s, and superficial air velocities range from 1 to 2 m/s. The ratio of base diameter to height may be 0.75 to 0.85, and the ratio of throat to base diameter 0.55 to 0.65. The height of the air inlet is usually 0.10 to 0.12 times the base diameter to facilitate air flow into the tower. When a counterflow packing is used, the air flow distribution is not very uniform; for this reason, cross-flow packings are becoming more widely used. However, the assumption of uniform air and water flows in our model of counterflow packing is adequate for most design purposes.

The Computer Program CTOWER

CTOWER calculates the thermal-hydraulic performance of cooling towers. The options are:

1. Counterflow, mechanical draft or natural draft.
2. Cross-flow, mechanical draft.

Table 9.6 Pressure drop correlations for cooling tower shells, sprays, supports, and mist eliminators. N is the loss coefficient defined by Eq. (9.180), with velocity based on cross-sectional area for air flow underneath the packing in items 1–4.

1. Shell (natural-draft counterflow) [14]:

$$N = 0.167\left(\frac{D_B}{b}\right)^2$$

where D_B is the diameter of the shell base and b is the height of the air inlet

2. Spray (natural-draft counterflow) [16]:

$$N = 0.526(Z_p\ [\text{m}] + 1.22)(\dot{m}_L/\dot{m}_G)^{1.32}$$

3. Mist eliminators:

$$N = 2\text{–}4\ (N = 3 \text{ is the nominal value in CTOWER})$$

4. Support columns, pipes, etc. (natural-draft counterflow) [16]:

$$N = 2\text{–}6\ (N = 4 \text{ is the nominal value in CTOWER})$$

5. Fan exit losses for mechanical-draft towers (velocity based on fan exit area)

$$N = 1.0, \text{ forced draft}$$

$$\simeq 0.5, \text{ induced draft, depending on diffuser design}$$

6. Miscellaneous losses for mechanical-draft towers (velocity based on packing cross-sectional area):

$$N \simeq 3$$

($N = 3.5$ is the nominal value in CTOWER to cover items 5 and 6 for mechanical-draft counterflow and cross-flow towers)

The number of transfer units is calculated using Merkel's assumptions and numerical integration. The property data is taken from the ASHRAE Handbook of Fundamentals [25]. Packing performance data are taken from EPRI GS-6370 [23] and Lowe and Christie [22], as given in Table 11.5. Other pressure drops are calculated from the correlations in Table 11.6. For the counterflow towers, the "design" problem is solved; that is, the tower duty is specified, and the required height of packing is calculated. For cross-flow towers, the "rating" problem is solved; that is, for given inlet stream conditions and packing dimensions, the outlet stream conditions are calculated.

CTOWER can be used to perform parametric studies and so help the student understand the response of cooling towers to environmental, duty, and design changes. However, CTOWER should not be used without some thought given to important characteristics of cooling tower behavior. For this purpose, it is useful to consider a graphical representation of Merkel's theory for a counterflow tower. Figure 11.25 shows a chart with moist air enthalpy plotted versus water enthalpy (or, equivalently, water temperature) at 1 atm pressure. The *saturation curve* $h_s(h_u)$, or, equivalently, $h_s(T_s)$, is the enthalpy of saturated air tabulated in the last column of Table A.31. The *operating lines* $h_G(h_L)$ are given by Eq. (11.94) and relate the air enthalpy to

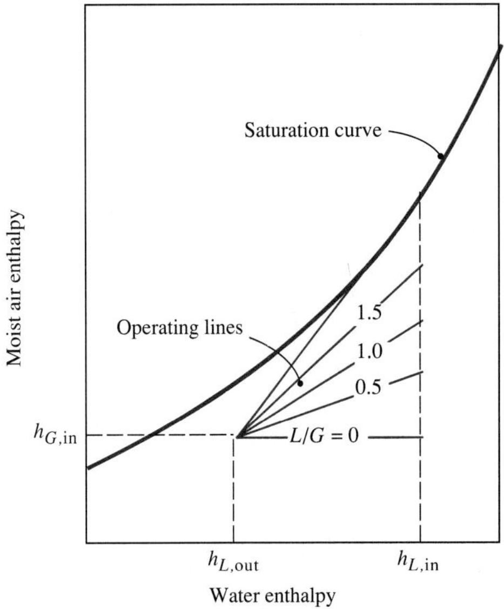

Figure 11.25 Counterflow cooling tower operating lines for various water-to-air flow-rate ratios shown on an enthalpy chart.

the water enthalpy at each location in the packing. The slope of an operating line is L/G. Since the assumption $T_s = T_L$ is made in Merkel's method, vertical lines on the chart connect h_s and h_G at each location in the packing. The driving force for the enthalpy transfer, $(h_s - h_G)$, is the vertical distance between the saturation curve and the operating line. The integral in Eq. (11.96) averages the reciprocal of this distance. Using this chart, a number of observations about cooling tower behavior can be made.

1. Figure 11.25 shows the effect of L/G for fixed water inlet and outlet temperatures, and fixed inlet air temperature and humidity. If we imagine L to be fixed as well, we see that as G decreases, the driving forces decrease, and so a larger NTU is required.
2. The minimum NTU required corresponds to $L/G = 0$, that is, an infinite air flow rate, for which the operating line is horizontal.
3. Due to the curvature of the operating line, it is possible for the operating line to be tangent to the saturation curve. The indicated NTU is then infinite, which tells us that the air flow rate must be increased in order to achieve the desired water cooling range.
4. For a mechanical-draft tower, the optimal value of L/G lies between the two limits described in items 2 and 3 above. If L/G is large, the required height of packing is large, and the capital cost will be excessive. If L/G is small, the required height of packing is small, and the capital cost will be small, but the fan power will be excessive (since fan power is proportional to air volume flow rate times pressure drop). These features are illustrated in Example 11.12.

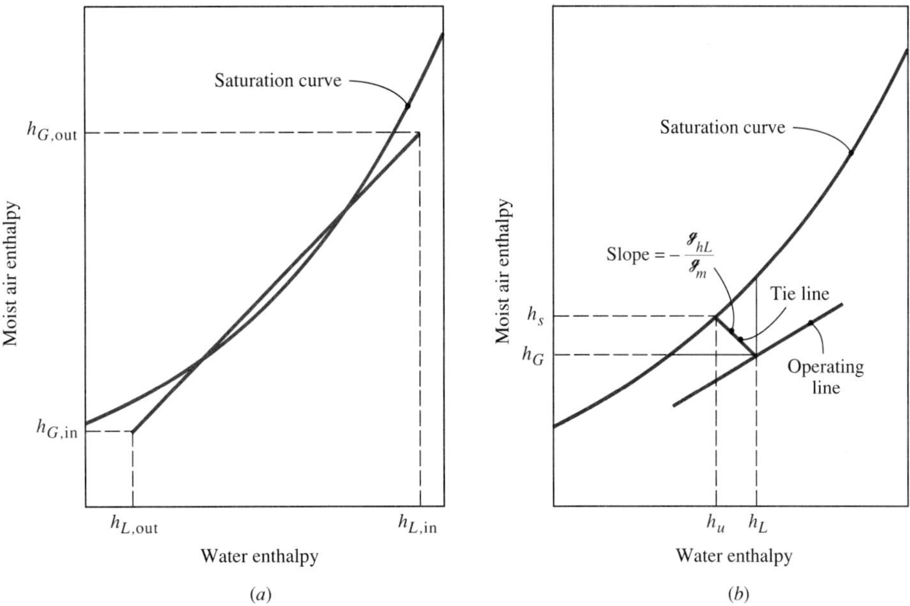

Figure 11.26 (*a*) Thermodynamically admissable end states for counterflow cooling tower operation that are not physically attainable. (*b*) The tie line for counterflow cooling tower operation when the water-side heat transfer resistence is not negligible.

5. Figure 11.26*a* shows an interesting situation, in which water inlet and outlet temperatures, the air inlet condition, and both flow rates are specified. The exchanger energy balance, Eq. (11.74), could show that the outlet air state is thermodynamically possible. However, the operating line must cross the saturation curve to join the inlet and outlet states, so that the mass transfer process required is impossible. Because of the curvature of the saturation line, thermodynamically admissible end states may not be physically attainable.

6. Figure 11.26*b* shows how a nonnegligible water-side heat transfer resistance is represented on the chart. Since the energy transferred to the air stream equals the rate at which heat is transferred from the bulk water to the interface,

$$\mathcal{g}_m(h_s - h_G) \simeq h_{cL}(T_L - T_s) = \mathcal{g}_{hL}(h_L - h_u)$$

if the specific heat of the water c_{pL} is taken to be constant. Thus, the slope of a *tie line* joining corresponding locations on the operating line and saturation curve is

$$\frac{h_s - h_G}{h_u - h_L} = -\frac{\mathcal{g}_{hL}}{\mathcal{g}_m} \tag{11.112}$$

For $\mathcal{g}_{hL}$ very large, the tie line is vertical, and $h_u = h_L$; that is, $T_s = T_L$, as assumed in Merkel's method. If the process is liquid side–controlled, that is,

g_m is very large, then the tie lines are horizontal. It is easy to see how a graphical method could be devised to obtain the NTUs without assuming $T_s = T_L$. It would involve dividing the h_L range into increments and drawing tie lines of the correct slope on the chart to obtain $(h_s - h_G)$ for use in Eq. (11.96). Exercise 11–48 requires the development of a numerical method for this purpose.

Range and Approach

Cooling tower designers and utility engineers have traditionally used two temperature differences to characterize cooling tower operation. The *range* is the difference between the water inlet and outlet temperatures (also called simply the hot and cold water temperatures). The *approach* is the difference between the outlet water temperature and the wet-bulb temperature of the entering (ambient) air. The approach characterizes cooling tower performance; for a given inlet condition, a larger packing will produce a smaller approach to the ambient wet-bulb temperature, and hence a lower water outlet temperature. The approach concept is useful because the ambient dry-bulb temperature has little effect on performance at usual operating conditions (for a specified wet-bulb temperature). Exercise 11–49 is a check of this feature of cooling tower performance.

Cooling Demand Curves

Electric utility engineers have found it convenient to use charts of *cooling demand curves* to evaluate packing specifications. Figure 11.27 is an example of such a chart, on which the required NTU, for a given inlet air wet-bulb temperature and range, is plotted versus L/G with the approach as a parameter. Such a plot is possible

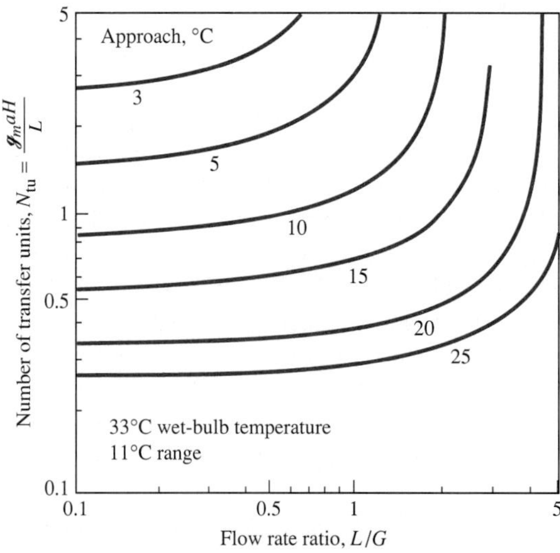

Figure 11.27 Example of cooling demand curves for a specified wet-bulb temperature and range: NTU versus flow rate ratio for a fixed approach.

since the inlet air dry-bulb temperature has only a small effect under usual operating conditions. Now, if it is possible to correlate the mass transfer conductance as

$$\frac{g_m a}{L} = C\left(\frac{L}{G}\right)^{-n} \tag{11.113}$$

the NTU of a packing of height H is

$$\frac{g_m S}{\dot{m}_L} = \frac{g_m a H}{L} = C\left(\frac{L}{G}\right)^{-n} H \tag{11.114}$$

Then Eq. (11.114) can also be plotted on the chart to give the *packing capability line*. For a required approach, the *operating point* of the tower is the intersection of the cooling demand curve and packing capability line. Charts of cooling demand curves have been prepared by the Cooling Tower Institute [26] and by Kelly [27].

Correlations for g_m in the form of Eq. (11.113) do not necessarily fit experimental data well. A dependence $g_m \propto L^{1-n}G^n$ is implied and, in the past, experimental data were often forced to fit such a relation. The correlations for the flat asbestos sheet counterflow packings, items 1–4 in Table 11.5, are examples. An examination of the correlations for the remaining counterflow packings in Table 11.5 show some large deviations: for example, packing 11 gives $n_1 + 1 = -0.51 + 1 = 0.49$, which is much different than $n_2 = 0.93$. (However, $n_2 = 0.93$ is abnormally high since it implies a mass transfer conductance proportional to velocity to the 0.93 power, and this is physically unrealistic.) If the g_m correlation does not have the form of Eq. (11.113), the NTU cannot be plotted as a line on a cooling demand chart.

With the almost universal use of computers and the availability of suitable computer programs, one can expect to see less use of cooling demand charts in the future. Other than CTOWER, computer programs in current use include ESC [28], TEFRI [29], FACTS [30], VERA2D [31], and STAR [32]. These computer programs are based on mathematical models of varying degrees of complexity. Both ESC and TEFRI have one-dimensional models similar to those analyzed in Sections 11.5.2 and 11.5.3, and used by CTOWER. (Even for cross-flow the model can be viewed as one-dimensional since no transverse mixing of flow is allowed.) VERA2D models the water flow as one-dimensional, but allows for two-dimensional air flow: an appropriate form of the momentum conservation equation is solved to determine the flow pattern in the cooling tower. STAR solves complete two-dimensional conservation equations but is a research tool rather than a design tool. Minor differences in the heat and mass transfer models used by the programs are of little consequence. The major sources of error in the predictions made by these programs are related to nonuniform air and water flow, and the correlations of packing mass transfer and pressure drop experimental data. The experimental data are obtained in small-scale test rigs, in which it is impossible to simulate many features of full-size towers—for example, nonuniform flow due to entrance configuration, nonuniform wetting of the packing, and, in the case of counterflow towers, the effect of spray above the packing and rain below the packing. Furthermore, since testing of packings in small-scale test rigs is itself not easy, considerable scatter is seen in such test data. Correlations of the data typically have root mean square errors of 10–20%.

Notwithstanding the limitations of the computer programs described above, it is certain that cooling demand curve charts will soon become obsolete and be replaced by interactive computer software, of which CTOWER is a modest example.

Legionnaires Disease

Legionnaires disease is a form of pneumonia caused by a strain of *legionnella* bacteria (sero group I). Smokers and sick people are particularly vulnerable to the disease. Major outbreaks have occurred at conventions and in hospitals, for which the source of the bacteria has been traced to cooling towers of air-conditioning systems. The bacteria require nutrients such as algae or dead bacteria in sludge, and thrive if iron oxides are present. However, properly designed, installed, and maintained cooling towers have never been implicated in an outbreak of the disease. Key requirements to be met include the following:

1. Mist (drift) eliminators should be effective.
2. The tower should be located so as to minimize the possibility of mist entering a ventilation system.
3. Corrosion in the tower and water lines should be minimized by use of glass fiber, stainless steel, and coated steel.
4. The design should facilitate inspection and cleaning, to allow early detection and remedy of sludge buildup.
5. Water treatment and filtration procedures should meet recommended standards.

EXAMPLE 11.11 A Natural-Draft Cooling Tower for a Power Plant

A hyperbolic-shell, natural-draft counterflow cooling tower has the following specifications:

Tower height = 125 m

Packing diameter = 85 m

Water loading on the packing = 1.75 kg/m^2 s

Height of packing above basin = 11 m

Height of air inlet = 10 m

Water inlet temperature = 45°C

Cooling range = 17°C for ambient air at 5°C, 35% RH, 1000 mbar

Determine the packing height as a function of packing pitch for flat asbestos sheet packing.

Solution

Given: A natural-draft counterflow cooling tower.

Required: Packing height as a function of packing pitch for flat asbestos sheet packing.

Assumptions: 1. Merkel's method for calculating NTU is adequate.
2. The conductance and pressure drop correlations in CTOWER are adequate.

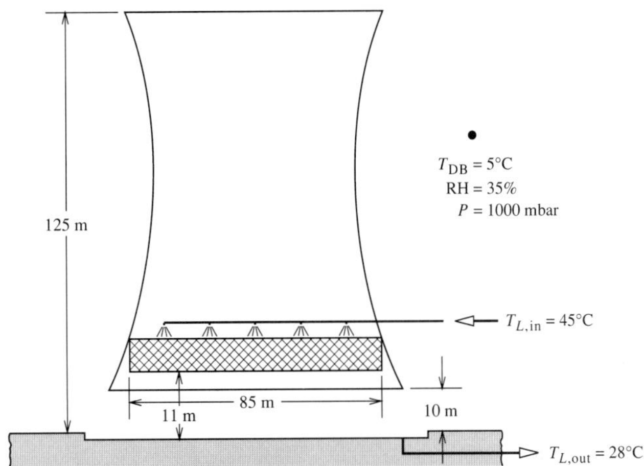

CTOWER will be used to perform the computations. A water loading of 1.75 kg/m^2 s on an 85 m–diameter packing gives a water flow rate of $\dot{m}_L = (1.75)(\pi/4)(85)^2 = 9930$ kg/s. The required inputs are:

Packings 1, 2, 3, and 4

Height of cooling tower = 125 m

Base diameter = 85 m

Height of air inlet = 10 m

Height of spray effect = 11 m

Flow rate of water = 9930 kg/s

Inlet temperature of water = 45°C

Outlet temperature of water = 28°C

Barometric pressure = 1000 mbar

Dry-bulb temperature = 5°C

Relative humidity = 35%

The required results are summarized below.

Packing Number	Pitch, p cm	N_{tu}	Height, H m	H/p
1	4.45	0.683	1.633	36.7
2	3.87	0.682	1.310	33.9
3	3.18	0.683	1.180	37.1
4	2.54	0.682	1.024	40.3

Comments

1. The packing height H decreases as the pitch decreases because the contact area increases linearly with pitch.
2. The cost of the packing is proportional to the area of the plates but will also increase somewhat with decreasing pitch (due to more fastenings, etc.). The area of the plates is proportional to H/p, which is seen to be relatively constant. The irregular behavior is probably due to a lack of precision in the mass transfer conductance correlations.
3. It appears that secondary factors will dictate the choice of packing pitch.

EXAMPLE 11.12 A Mechanical-Draft Cooling Tower for an Oil Refinery

A 5 m–square mechanical-draft counterflow cooling tower is required to cool 45 kg/s of water from 40°C to 20°C when the ambient air is at 10°C, 1 atm, and 35% relative humidity. The packing under consideration is vertical corrugated plastic at 1.63 in pitch, manufactured by American Tower. As a function of air flow rate, determine the required packing height and fan power. Use a fan mechanical efficiency of 75%.

Solution

Given: A mechanical-draft counterflow cooling tower.

Required: Packing height and packing fan power as a function of air flow rate for a corrugated plastic packing.

Assumptions:
1. Merkel's method for calculating the NTU is adequate.
2. The packing correlations in CTOWER based on EPRI GS-6370 are adequate.

CTOWER will be used to perform the computations. The packing is listed as item 7 on the packing menu. The air flow rate is varied from 30 kg/s to 90 kg/s and the results tabulated. The fan power is equal to the air volume flow rate times pressure drop divided by fan efficiency. Taking $\rho_{\text{air}} \simeq 1.25$ kg/m^3 gives power $\simeq 1.067\dot{m}_G\Delta P$.

$\dot{m}_G$ kg/s	N_{tu}	$T_{G,\text{out}}$ °C	H m	ΔP Pa	Power W
30	4.35	37.7	4.06	24.7	791
40	2.27	32.6	1.80	21.2	905
50	1.84	28.9	1.29	25.1	1340
60	1.65	26.1	1.05	30.5	1950
70	1.54	23.8	0.90	36.8	2750
80	1.47	22.0	0.79	44.0	3760
90	1.42	20.5	0.72	51.8	4970

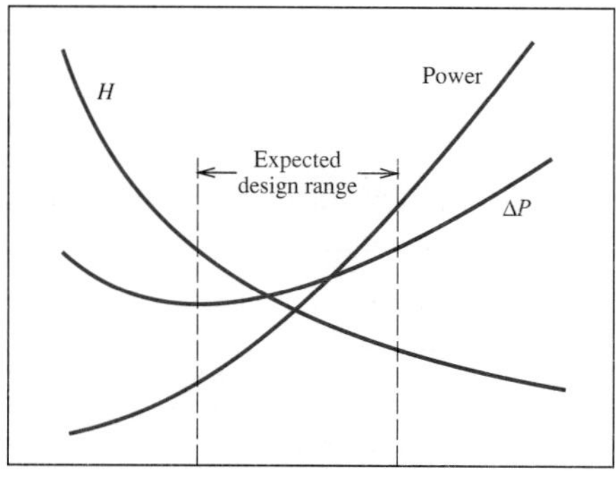

Comments

1. The required packing height steadily decreases as the air flow rate increases.
2. The pressure drop first decreases and then increases with air flow rate, owing to the competing effects of air velocity and packing height.
3. Fan power increases with air flow rate, with the rate of increase becoming rapidly larger at high air flow rates.
4. The design range will reflect a trade-off between packing cost (proportional to height) and operating cost (of which an important component is the fan power).
5. A constant air density of 1.25 kg/m^3 was used to calculate the fan power. In fact, $\rho_{\text{out}} \simeq \rho_{\text{sat}}(T_{G,\text{out}})$ varies with air flow rate, but a constant value is adequate for the purposes of this example.

EXAMPLE 11.13 A Cross-Flow Cooling Tower for an Air-Conditioning System

A large building is to have a new air-conditioning system installed. Since the existing cooling tower is in good working order, it is planned to retain the tower. Performance data for the tower are to be made available to the contractor. The packing is Doron V-bar with a 4 in by 8 in spacing, and is 5 m high, 20 m wide, and 2 m deep. The nominal water and air flow rates are both 100 kg/s. Performance data are required for a water inlet temperature of 38°C and a nominal air inlet condition of 1 atm, 28°C, and 60% RH. Determine the effect of air temperature and humidity on the outlet water temperature.

Solution

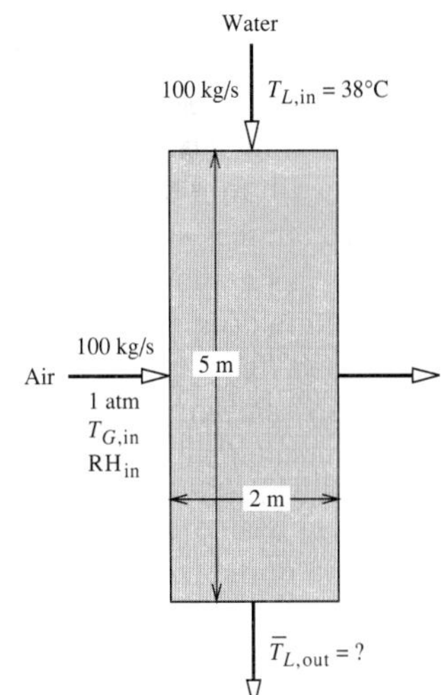

Given: A cross-flow mechanical-draft cooling tower.

Required: Effect of inlet air temperature and humidity on outlet water temperature.

Assumptions:
1. Both streams are unmixed.
2. Merkel's approximation is valid.
3. The conductance correlations in CTOWER (based on EPRI GS-6370) are adequate.

CTOWER can be used for the calculations. The required input is:

Packing 1
Height of packing = 5 m
Width of packing = 20 m
Depth of packing = 2 m
Flow rate of air = 100 kg/s

Flow rate of water = 100 kg/s

Inlet temperature of water = 38.0°C

Barometric pressure = 1013 mbar

Dry-bulb temperature = 28°C (nominal value)

Relative humidity = 60% (nominal value)

The number of transfer units is $N_{tu} = g_m S/\dot{m}_L = 0.396$, and the average outlet water temperatures are tabulated below for a range of inlet air conditions.

	$T_{G,\text{in}}$		
RH_{in}	25°C	28°C	31°C
30%	27.2	27.8	28.5
60%	28.8	29.7	30.8
90%	30.4	31.6	33.0

Comments

1. The cooling tower is but one unit in the air-conditioning system. These data allow the efficiency of the system as a whole to be determined.
2. Study the CTOWER output to see the behavior of the outlet air temperature.
3. The pressure drop for this packing is very small; the fan exit and miscellaneous losses are substantially greater.

11.6 CLOSURE

A great variety of mass exchangers are used in engineering practice. For purposes of analysis, we chose to differentiate single-stream and two-stream steady-flow exchangers. In addition, we differentiated exchangers that transfer mass only and exchangers that transfer both heat and mass. The formula for the effectiveness of a single-stream mass exchanger was found to be identical to that for a single-stream heat exchanger, irrespective of the nature of the physical transfer process—catalytic conversion, scrubbing, precipitating, filtering, or humidifying. In order to analyze two-stream mass exchangers, it was necessary to define an overall mole transfer conductance that properly accounted for a concentration discontinuity at the liquid-gas interface. Special attention was given the role played by the Henry number (or a more general solubility relation) in determining whether the mass transfer processes were gas side– or liquid side–controlled. A special class of two-stream mass exchangers was analyzed, namely, gas scrubbers with gas absorption described by a linear solubility law. An analytical result for the effectiveness of a countercurrent unit was obtained and was of identical form to the corresponding result for a two-stream heat exchanger.

The only example of a two-stream simultaneous heat and mass exchanger considered was the cooling tower, in both counterflow and cross-flow configurations. The analysis was simplified using Merkel's approximation, but, nevertheless, numerical methods were required to determine the final solution. For the counterflow configuration, the "design" problem was solved; that is, the required water inlet temperature and cooling range was specified, and the number of transfer units required to do the task was calculated. For the cross-flow configuration it was more convenient to solve the "rating" problem; that is, the packing and its dimensions were specified, and outlet water and gas stream temperatures were calculated.

Three computer programs accompany Chapter 11: FILTER, SCRUB, and CTOWER. These programs should prove useful for making parametric studies of filter, gas scrubber, and cooling tower performance. These programs are primarily intended as educational tools but may also find use for preliminary design.

The exchangers considered in Chapter 11 could all be analyzed using low mass transfer rate theory, with the analysis of the mass and heat transfer processing following the procedures described in Sections 9.4 and 9.5. An important example of a simultaneous heat and mass exchanger that cannot be analyzed using low mass transfer rate theory is the power plant condenser. As shown in Example 10.9, a small amount of noncondensable gas in steam markedly reduces the condensation rate. In Example 10.9, condensation on a single tube was considered, whereas a typical condenser has many thousands of tubes, with complex steam flow patterns through the tube bank (see Fig. 1.20). Reliable analysis of such condensers requires numerical solution of multidimensional governing equations, and has yet to be successfully accomplished.

REFERENCES

1. Davies, C. N., "Definitive equations for the fluid resistance of spheres," *Proc. Roy. Soc. London,* ser. A, 57, 259–269 (1945).

2. White, H. J., *Industrial Electrostatic Precipitation*, Addison-Wesley, Reading, Mass. (1963).

3. Smith, W. B., and MacDonald, J. R., "Development of a theory for charging of particles by unipolar ions," *J. Aerosol Sci.*, 7, 151–166 (1976).

4. Tsai, R., and Mills, A. F., "Modeling of electrostatic precipitators," *PHOENICS Journal*, 785–810 (1990).

5. Kihm, K., Mitchner, M., and Self, S. A., "Comparison of wire-plate and plate-plate electrostatic precipitators in turbulent flow," *J. Electrostatics*, 19, 21–32 (1987).

6. Fernandez de la Mora, J., "Inertia and interception in the deposition of particles from boundary layers," *Aerosol Sci. Technol.*, 5, 261–266 (1986).

7. Pich, J., "Gas filtration theory," Chap. 1 in *Filtration: Principles and Practice*, eds. M. J. Matteson and C. Orr., Marcel Dekker, New York (1987).

8. Lee, K. W., and Liu, B. Y. H., "Theoretical study of aerosol filtration by fibrous filters," *Aerosol Sci. Technol.*, 1, 147–161 (1982).

9. Onda, K., Takeuchi, H., and Okumoto, Y., "Mass transfer coefficients between gas and liquid phases in packed columns," *J. Chem. Eng. Jpn.*, 1, 56–62 (1968).

10. Perry, R. H., and Chilton, C. H., *Chemical Engineers Handbook*, 5th ed., pp. 18-35–18-39, McGraw-Hill, New York (1973).

11. Foust, A. S., Wenzel, L. A., Clump, C. W., Maus, L., and Anderson, L. B., *Principles of Unit Operations*, 2nd ed., John Wiley & Sons, New York (1980).

12. Treybal, R. E., *Mass-Transfer Operations*, 3rd ed., McGraw-Hill, New York (1980).

13. Emmert, R. E., and Pigford, R. L., "Interface resistance study of gas absorption in falling liquid films," *Chem. Eng. Prog.*, 50, 87–93 (1954).

14. Won, Y. S., and Mills, A. F., "Correlation of the effects of viscosity and surface tension on gas absorption rates into freely falling turbulent liquid films," *Int. J. Heat Mass Transfer*, 25, 223–229 (1982).

15. W. Brotz, "Uber die vorausberechning der absorptions geschwindigkeit von gasen in stramen der fussigkeitsschichten," *Chem. Eng. Tech.*, 26, 470–478 (1954).

16. Eckert, J. S., "How tower packings behave," *Chem. Eng.*, 82, 70–76, April 14 (1975).

17. Bharathan, D., and Wallis, G. B., "Air-water countercurrent annular flow", *Int. J. Multiphase Flow*, 9, 349–366 (1983).

18. Tien, C. L., Chung, K. S., and Liu, C. P., "Flooding in two-phase countercurrent flows: I. Analytical modeling," *PhysicoChemical Hydrodynamics*, 1, 195–207 (1980).

19. Chung, K. S., Liu, C. P., and Tien, C. L., "Flooding in two-phase countercurrent flows: II. Experimental investigations," *PhysicoChemical Hydrodynamics*, 1, 209–220 (1980).

20. Merkel, F., "Verdunstungskühlung," *Forschungsarb. Ing. Wes.*, no. 275 (1925).

21. Spalding, D. B., *Convective Mass Transfer*, McGraw-Hill, New York (1963).

22. Lowe, H. J., and Christie, D. G., "Heat transfer and pressure drop data on cooling tower packings, and model studies of the resistance of natural draft towers to airflow," Paper 113, *International Developments in Heat Transfer*, Proceedings of the International Heat Transfer Conference, Boulder, Colo., August 1961, ASME, New York.

23. Johnson, B. M., ed., *Cooling Tower Performance Prediction and Improvement*, vols. 1 and 2, EPRI GS-6370, Electric Power Research Institute, Palo Alto, Calif. (1990).

24. Singham, J. R., "Natural draft towers," in Hewitt, G. E., coord. ed., *Hemisphere Handbook of Heat Exchanger Design*, Sec. 3.12.3, Hemisphere, New York (1990).

25. *ASHRAE Handbook: 1989 Fundamentals, ASHRAE*, American Society of Heating, Refrigerating, and Air Conditioning Engineers, Atlanta (1989).

26. *Cooling Tower Performance Curves*, Cooling Tower Institute, Houston (1967).

27. Kelly, N. W., *Kelly's Handbook of Cross-flow Cooling Tower Performance*, Neil W. Kelly and Associates, Kansas City, Missouri (1976).

28. Baker, D., *Cooling Tower Performance*, Chemical Publishing Co., New York (1984).

29. Bourillot, C., *TEFRI: Numerical Model for Calculating the Performance of an Evaporative Cooling Tower*, EPRI CS-3212-SR, Electric Power Research Institute, Palo Alto, Calif. (1983).

30. Benton, D. J., *A Numerical Simulation of Heat Transfer in Evaporative Cooling Towers*, WR28-1-900-110, Tennessee Valley Authority (1983).

31. Majumdar, A. K., Singhal, A. K., and Spalding, D. B., *VERA2D-84: A Computer Program for 2-D Analysis of Flow, Heat and Mass Transfer in Evaporative Cooling Towers*, EPRI CS-4073, Electric Power Research Institute, Palo Alto, Calif. (1985).

32. Caytan, Y., "Validation of the two-dimensional numerical model STAR developed for cooling tower design," Proceedings of the 3rd Cooling Tower Workshop, International Association for Hydraulic Research, Budapest, Hungary (1982).

33. Levich, V. G., *Physicochemical Hydrodynamics*, Prentice-Hall, Englewood Cliffs, N.J. (1962).

34. Haberman, W. L., and Morton, R. K., *Experimental Investigation of the Drag and Shape of Air Bubbles Rising in Various Liquids*, U.S. Department of the Navy, Report No. 802, Washington, D.C. (1953).

35. Oldenkamp, R. D., and Katz, B., "Integration of molten carbonate process for control of sulfur oxide emissions into a power plant," ASME Paper 69-WA/APC-6, Winter Annual Meeting, Nov. 16–29 (1969).

EXERCISES

11–1. Propane (C_3H_8) can be used as a fuel in internal combustion engines. Consider an automobile that gets 8 km/liter of liquid propane when cruising at 90 km/h.

(i) Write down the main combustion reaction.

(ii) Calculate the rates at which fuel and oxygen are consumed and at which water vapor and carbon dioxide are produced, and the nitrogen throughput.

(iii) Calculate the rate at which nitric oxide is produced, assuming an emissions concentration of 600 ppm for an engine operating at the stoichiometric air/fuel ratio. Take the specific gravity of propane as 0.5.

11–2. In Section 11.2.4 we obtained an overall balance for an urban air volume under conditions in which pure air flowed into the system at a rate $\dot{m}$, absorbed and mixed with a pollutant species 1 that was generated within the system at a rate $\dot{r}_1$, and then flowed out of the considered volume with a pollutant concentration given by

$$(\dot{m}m_1)_{\text{out}} = (\dot{m}m_1)_{\text{in}} + \dot{r}_1$$

or, since $m_{1,\text{in}}$ was presumed to be zero and $\dot{m}$ was essentially constant, $m_{1,\text{out}} = \dot{r}_1/\dot{m}$. Consider now a balance on an urban air volume over a community that is downwind of a major polluter (this frequently is the case when a sea breeze carries Los Angeles smog inland over Pasadena, California). Suppose a layer of air 600 m thick advances on the community at 3 m/s. When the concentration of CO in the incoming air is 7.5 ppm, calculate the maximum rate (per square kilometer) at which CO may be generated within the community if the CO concentration within its air volume is not to exceed 10.0 ppm. (Take the effective depth of the community as 5000 m.)

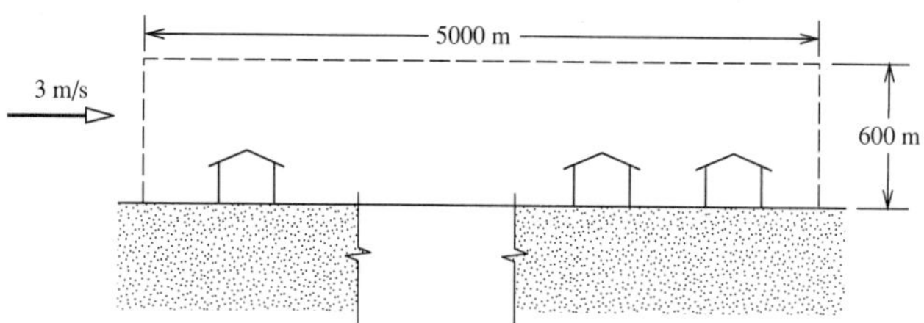

11–3. A demonstration hydrogen-fueled automobile consumes 2.9 kg of fuel per 100 km while traveling at 80 km/h. If the fuel injection system is set to maintain a stoichiometric air/fuel ratio,

(i) write down the combustion reaction.
(ii) calculate the rates at which fuel and oxygen are consumed, the rate of water vapor production, and the nitrogen throughput.
(iii) calculate the allowable mass fraction of NO_x in the exhaust if the automobile is to meet the 1993–94 standards of the U.S. Environmental Protection Agency for California.

11–4. Methanol is a promising alternative fuel for automobiles since fewer smog-causing pollutants are produced. It is currently more expensive than gasoline, but Catalytica Inc. of Mountain View, California, has discovered a promising process to convert inexpensive natural gas to methanol using a mercury-based catalyst (*New York Times*, January 17, 1993). If the process can be successfully commercialized, increased use of methanol as a fuel is likely. Consider an automobile traveling at 80 km/h consuming 21 liters of methanol per 100 km. If the fuel-injection system is controlled to maintain stoichiometric combustion,

(i) write down the combustion reaction.
(ii) calculate the rates at which fuel and oxygen are consumed, the rates of carbon dioxide and water vapor production, and the nitrogen throughput.
(iii) calculate the allowable mass fraction of NO_x in the exhaust if the automobile is to meet the 1993–94 standards of the U.S. Environmental Protection Agency for California.

The specific gravity of methanol can be taken as 0.796.

11–5. A 1000 MW coal-fired power plant burns a coal that has an elemental composition by mass of 3.8% hydrogen, 81.4% carbon, 1.6% nitrogen, and 3.5% oxygen and that also contains 8% ash and 1.7% sulfur. The overall efficiency of the plant is 38%, and the heat of combustion of the coal is 32.3×10^6 J/kg. Ten percent excess air is used to reduce CO emissions. For the effluent from the boiler unit, estimate

(i) the flow rates of the major components.
(ii) the concentration of sulfur oxides, in parts per million by volume, if design guidelines for power plant boilers suggest oxides production of (38 × % sulfur content) lb/ton coal burned.
(iii) the flow rate of particulates if design guidelines for pulverized coal boiler units suggest particulates production of (16 × % ash content) lb/ton coal burned.

11–6. In an open-cycle ocean thermal-energy conversion pilot plant, warm surface seawater at 300 K enters an upcomer 25 m below sea level and is delivered to the evaporator at a little less than the barometric height ($\simeq$10 m). As the water flows up the upcomer, air comes out of solution: bubbles originally in the water grow, and new bubbles nucleate on suitable nucleation sites. At a height of $\simeq$9 m, a primary holding tank is situated and is maintained at a pressure of 10,000 Pa by a vacuum exhaust system. Estimate the composition and volume (m^3/m^3 water) of the vapor-gas mixture vented if the maximum possible amount of air is removed. The warm seawater can be taken to be saturated with air at 1 bar, and dissolved air in seawater can be modeled as a single species with an effective Henry constant of 85,000 bar at 300 K.

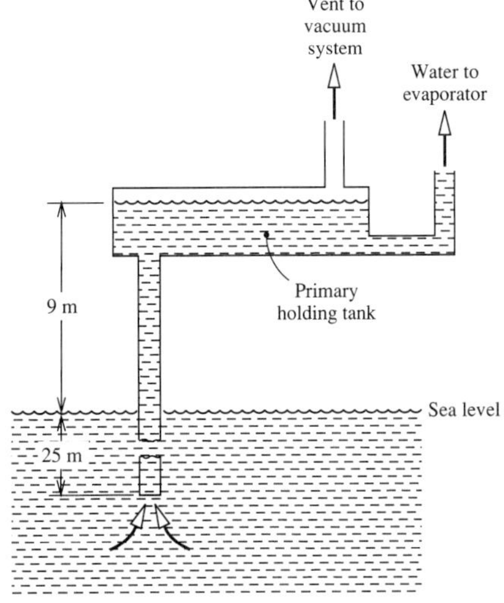

11–7. A clean room for an electronics manufacturing process is of length 20 m, width 10 m, and height 3 m. The air-conditioning system supplies 100 m^3/min of fresh, clean air to the room. At time $t = 0$, a faulty unit commences releasing a contaminant at a rate of 600 μg/min. Assuming the air in the room to be well mixed, determine the concentration of contaminants in the air as a function of time, and its final steady value.

11–8. Determine the surface areas required for the catalytic reactor described in Example 11.1 if

(i) the passage width is 0.7 mm.
(ii) the required effectiveness is 0.999.
(iii) the vehicle gets only 5 km/liter.

Compare your results with the result obtained in Example 11.1, and sketch qualitatively the effects of the three parameters cited above.

11–9. Exhaust gases from an automobile flow through the first stage of a catalytic converter at a rate of 0.04 kg/s. The inlet gases contain 1000 ppm NO and an estimated 5–10% CO content. Catalytic reduction of the NO takes place in 0.4 cm–diameter spherical porous catalyst pellets with an effective rate constant based on s-surface area of $k''_{\text{eff}} = 0.2$ m/s (see Exercise 9–55) packed with a volume void fraction of 0.3. Space considerations suggest a 15 cm outside diameter for the converter. Assuming an operating temperature and pressure of 850 K, 1 atm, and a mean molecular weight for the exhaust gases of 29, determine

(i) the NTU required to reduce 99% of the NO.
(ii) the overall mass transfer coefficient.
(iii) the required s-surface transfer area.
(iv) the required length of converter.
(v) the inlet and outlet NO mass fractions.
(vi) Why are the catalyst pellets made porous? What information would you require in order to calculate the effective rate constant?

11–10. A catalytic converter for an automobile contains a matrix formed of porous ceramic coated with a metallic catalyst. The flow passages are approximately 1.5 mm square, and the walls are 0.2 mm thick. At 80 km/h the gasoline consumption is 14.3 liters/100 km, and the exhaust gases are at 800 K and 1.2 bar pressure and contain 0.13% NO by weight. Suggest suitable dimensions for a converter that will reduce the NO mass fraction to 0.008%. The effective rate constant k''_{eff} for the reaction at 800 K is 0.15 m/s, which accounts for the additional catalyst area in the porous ceramic. Take an air/fuel ratio of 14:1.

11–11. Determine the length of a 10 cm–diameter catalytic converter necessary to obtain a 20-fold reduction in CO emission from an automobile engine as it develops 34.0 kW at a specific fuel consumption of 0.2 kg/kW h and an air/fuel ratio of 14:1. The unit is to operate at 850 K, the concentration of CO in the exhaust is 1% by volume, and the exhaust back pressure is 120 kPa. Air is injected into the converter at 10% of the exhaust flow rate. The catalyst is in the form of 0.3 cm–diameter spherical pellets of copper oxide on alumina with 100 m^2/cm^3

catalytic surface area. The pellets are packed to give a volume void fraction of 0.25. Assume a pellet effectiveness of 2.0%.

11–12. A catalytic converter for an automobile has passages of equilateral triangular cross section with 1 mm sides. The porous passage walls are 0.2 mm thick and have a void fraction of 0.75, a tortuosity factor of 5.0, a mean pore radius of 1 μm, and a surface area per unit volume of 6×10^5 cm^2/cm^3. The converter is 10 cm in diameter and 10 cm long. The converter is tested with an exhaust flow rate of 0.060 kg/s at 1 atm and 900 K. The platinum-based catalyst reduces nitric oxide in a reaction that may be approximately modeled as first-order with an activation energy of 17 kcal/mol and a preexponential factor of 70 m/s. Determine the decontamination factor for the converter [DF = $1/(1 - \varepsilon)$].

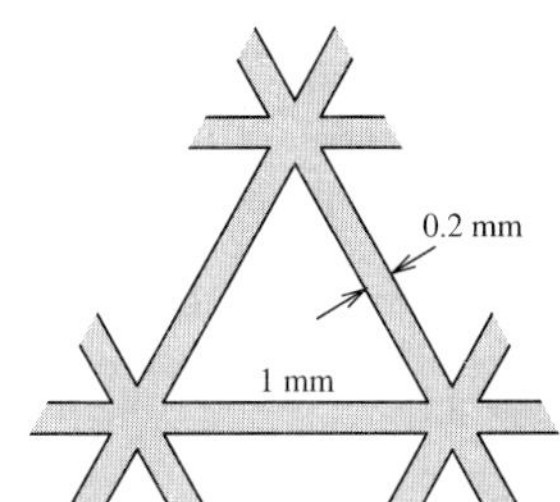

11–13. A catalytic converter for a luxury automobile has a matrix formed of porous ceramic coated with a metallic catalyst. The flow passages are 1.7 mm square, and the walls are 0.25 mm thick. At 90 km/h the gasoline consumption is 20.9 liters/100 km, and the exhaust gas is at 850 K, 1.15 bar, and contains 3.5% CO by weight. Suggest suitable dimensions for a converter that will reduce the CO concentration to 0.1%. The effective rate constant for the reaction at 850 K is $k''_{\text{eff}} = 0.106$ m/s, which accounts for the additional catalyst area in the porous walls. The air/fuel ratio is 13.2:1, with air injected into the reactor at 10% of the exhaust flow rate. Also calculate the pressure drop. Take a fuel density of 705 kg/m^3.

11–14. A catalytic converter for removing CO from automobile engine exhaust is to be fabricated from oxidized copper-clad metal foil wound in spiral form. The design condition is 80 km/h, 20 liters/100 km, and 10% excess air. The diameter of the unit must not exceed 15 cm, and the flow in the converter should be laminar. The converter is to be 98% effective when operating at 1150 K and 1 atm. Determine the required length of the converter, and the pressure drop, as a function of foil pitch. Take the rate constant for CO oxidation as $500 \exp(-9400/T)$ m/s for T in kelvins, and an air/fuel ratio of 15:1.

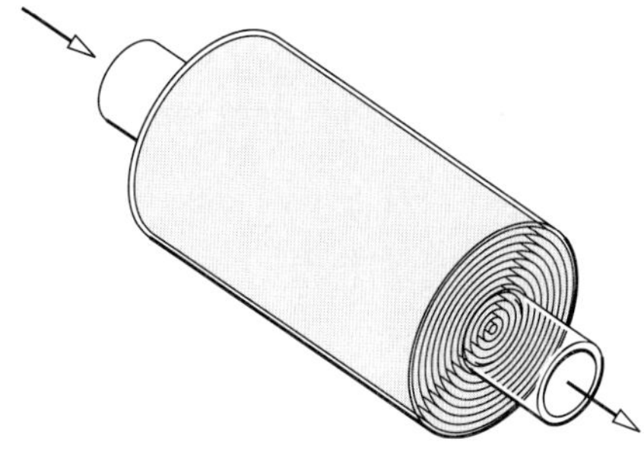

11–15. A 99.5% effective converter is required for an automobile engine when it develops 36.4 kW at a specific fuel consumption of 0.18 kg/kW h and an air/fuel ratio of 13.4:1. The converter is to operate at 820 K, the concentration of CO in the exhaust is 3% by volume, and the exhaust back pressure is 115 kPa. Air is injected into the converter at 10% of the exhaust flow rate. The catalyst is in the form of 0.25 cm–diameter spherical pellets of copper oxide on alumina with 100 m^2/cm^3 catalytic surface area. The pellets are packaged in a 12 cm–diameter can to give a volume void fraction of 0.27. Determine the length of the packing. Assume a pellet effectiveness of 2.5%.

11–16. In an experimental rig, gas is absorbed from a CO_2 stream into a turbulent water film falling inside a vertical tube. The tube has a 2.05 cm inner diameter and a height of 1.98 m. For a particular test, the system was maintained at 25°C, and the water flow rate was 5.3×10^{-5} m^3/s. The bulk inlet and outlet CO_2 concentrations in the water were measured as 0.0153 $kmol/m^3$ and 0.0242 $kmol/m^3$, respectively. The CO_2 gas was at 1 atm pressure and can be assumed to be saturated with water vapor. Calculate the average mole transfer conductance $\overline{\mathscr{G}}_m$ for mass transfer into the liquid film. Equation 11.46 can be used to calculate the average film thickness.

11–17. In a pilot plant for open-cycle ocean thermal-energy conversion (Claude cycle), warm water enters an upcomer 30 m below sea level and flows at 3 m/s to the evaporator located at the barometric height of approximately 10 m above sea level. The inlet seawater can be assumed to be saturated with air at 300 K and 1 atm pressure and contains air bubbles of radius $R_0 = 20$–150 μm. As the bubbles travel through the upcomer, air is absorbed into the water below sea level and subsequently desorbed above sea level. In addition, the bubbles grow due to the decrease in hydrostatic pressure. By performing an analysis on a molar basis, obtain an expression for bubble radius $R(z)$ as a function of height above the upcomer inlet. Hence prepare graphs of $R(z)$ for $20 < R_0 < 150$ μm. Since seawater is relatively unclean, small bubbles tend to behave as solid spheres; thus the Levich expression for the liquid-side mole transfer conductance is appropriate [33],

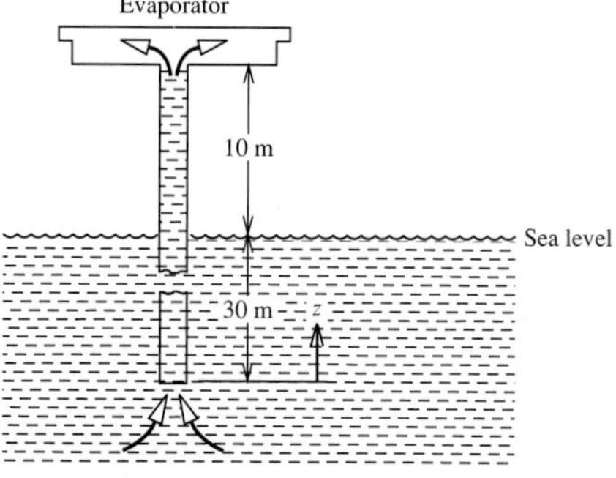

$$\mathrm{Sh} = (2\mathrm{Pe}/\pi)^{1/2}; \qquad \mathrm{Pe} \gg 1$$

where the Sherwood and Peclet numbers have diameter as characteristic length. For very small bubbles Stokes' law, Eq. (4.73), can be used to give the velocity of bubbles relative to the seawater. However, for larger bubbles ($R =$ 2–5 mm) Stokes' law does not apply, and a constant value of 0.23 m/s is appropriate [34].

11–18. Water at 330 K flows down the inside of a 4 cm–I.D., 2 m–high tube at 223 kg/h. Air at 1 atm pressure flows up the tube at 0.1 m/s and contains 5% carbon monoxide by volume at the inlet. If the inlet water contains no CO, determine the bulk concentration of CO in the outlet water. See Section 11.4.2 for appropriate liquid-side mole transfer conductance correlations.

11–19. A laboratory desorption column consists of a bundle of 15, 2 cm–O.D., thin-wall stainless steel tubes, 0.7 m high, in a glass shell, with the liquid able to flow down both the insides and outsides of the tubes. Water at 300 K containing 0.0238 kmol/m^3 dissolved CO_2 is fed to the tower at a rate of 400 kg/h. If a vacuum pump maintains the pressure inside the tower at 50 mm Hg absolute, calculate the column effectiveness and the concentration of CO_2 in the outlet water. See Section 11.4.2 for appropriate liquid-side mole transfer conductance correlations.

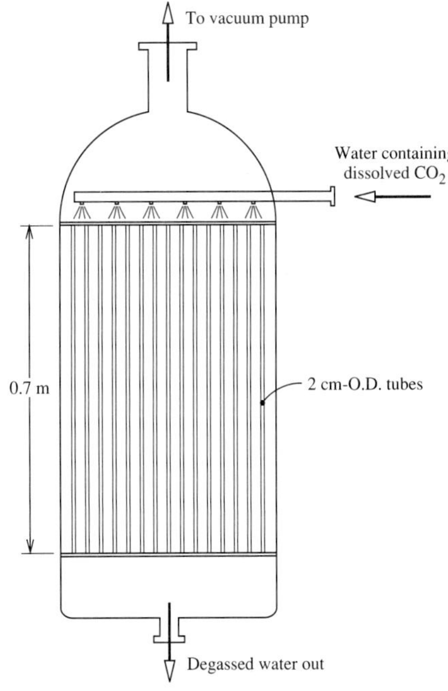

11–20. An electrostatic precipitator has a plate spacing of 0.25 m and is 25 m long. Gas at 1 atm and 300 K flows at a bulk velocity of 2 m/s through the precipitator. Calculate the effectiveness of the unit for removing particles of size

(i) 0.3 μm.
(ii) 1 μm.
(iii) 3 μm.

Use the migration velocity from Fig. 11.6, with $j = 54 \times 10^{-9}$ A/cm^2.

11–21. An electrostatic precipitator was designed to be 98.5% effective for an effective migration velocity of 6 cm/s and a gas velocity of 1.5 m/s. Determine the ineffectiveness of the unit when the following changes are made to the operating conditions:

(i) The migration velocity is decreased to 4 cm/s.
(ii) The migration velocity is increased to 8 cm/s.
(iii) The gas velocity is decreased to 1 m/s.
(iv) The gas velocity is increased to 2 m/s.

11–22. An electrostatic precipitator is to treat a gas stream flowing at 6000 m^3/min. If it is to be 98% effective in removing 0.5 μm size particles, calculate the required plate area, and the number of plates, if plates are available in 3 m lengths and are 10 m high. The operating current density is 60×10^{-9} A/cm^2

at 30 kV. The gas velocity should not exceed 2.0 m/s, and a plate spacing of 0.25 m is appropriate.

11–23. A filter has a cross-sectional area of 1 m^2, has a void fraction of 0.8, and is 0.5 m thick. Air at 290 K and 1 atm containing particles flows through the filter at 20 m^3/min. Prepare a graph of decontamination factor versus fiber diameter ($30 < d_f < 100$ μm) for particle diameters of 0.2, 0.5, and 1.5 μm. Assume a particle density of 4800 kg/m^3 and a fiber density of 1800 kg/m^3.

11–24. A filter containing 25 μm fibers, packed at a density of 380 kg/m^3, is required to remove particulate matter from an air flow at 320 K, 1 atm. The flow rate is 22 m^3/min, and the filter must be accommodated in a 1.3 m–diameter duct. Prepare a graph of decontamination factor versus filter thickness (2 cm $< L <$ 30 cm) for particle diameters of 0.1, 0.5, 1, and 2 μm. Assume a fiber density of 2000 kg/m^3 and a particle density of 5100 kg/m^3.

11–25. A fiberglass filter is required to process 40 m^3/min of air at 310 K and 1.03 bar, and yield a decontamination factor of 15 for 0.5 μm particles of 5500 kg/m^3 density. The packing available has 40 μm–diameter fibers packed at a density of 450 kg/m^3. The air flows in a 1.2 m–square duct, and the filter should fill the duct to minimize pressure drop. Determine the required thickness of filter.

11–26. Rederive Eq. (11.29) by eliminating x_u in Eq. (11.28). Also show that Eq. (11.29) is valid even if the Henry number is not constant, that is, if the relation is nonlinear in form, say $y_s = f(x_u)x_u$.

11–27. Paralleling the derivations in Section 11.4.3, develop the ε–N_{tu} relations for a cocurrent mass exchanger.

11–28. Ammonia gas is to be scrubbed from an air stream by a countercurrent flow of water in a packed column. The column is 1 m in diameter and 3 m high, and is packed with 0.038 m–nominal size ceramic Raschig rings. The gas and liquid flow rates are 2500 and 5000 kg/h, respectively, and the column operates at 310 K and 1 atm. The entering air stream contains 5% ammonia by volume, whereas the entering water is pure.

(i) Calculate the column effectiveness and the outlet ammonia concentration.
(ii) How much is the outlet ammonia concentration reduced if the column height is doubled?
(iii) How much is the outlet ammonia concentration reduced if the outlet gas stream is passed through a second 3 m–high column, supplied with pure water?

11–29. Experiments with a new type of packing in a packed column were performed with dilute NH_3 in air absorbed into water at 300 K and 1 atm. At the test flow rates, $\dot{M}_G/\dot{M}_L = 2$ and the countercurrent effectiveness is measured to be 0.81. Estimate the number of transfer units of the column for the following two situations.

(i) Absorption of acetone from a dilute mixture with air into water at 300 K. Use an effective Henry number of 1.2.

(ii) Absorption of carbon monoxide from a dilute mixture with air into water at 330 K.

11–30. A falling-film tower has a shell diameter of 2.5 m; contains 900, 4.75 cm–I.D. tubes; and is to be used for scrubbing SO_2 from an air stream. Each tube is to process 50 kg/h of gas in countercurrent flow with 600 kg/h of water. The tower operates at 1 atm pressure, and cooling water circulated in the shell maintains the film temperature at approximately 290 K. If the inlet water is pure and the inlet air contains 3% SO_2 by volume, determine the tower height required to reduce the outlet SO_2 mole fraction to 0.006. Since Henry's law is invalid over this concentration range, use an appropriate numerical integration procedure.

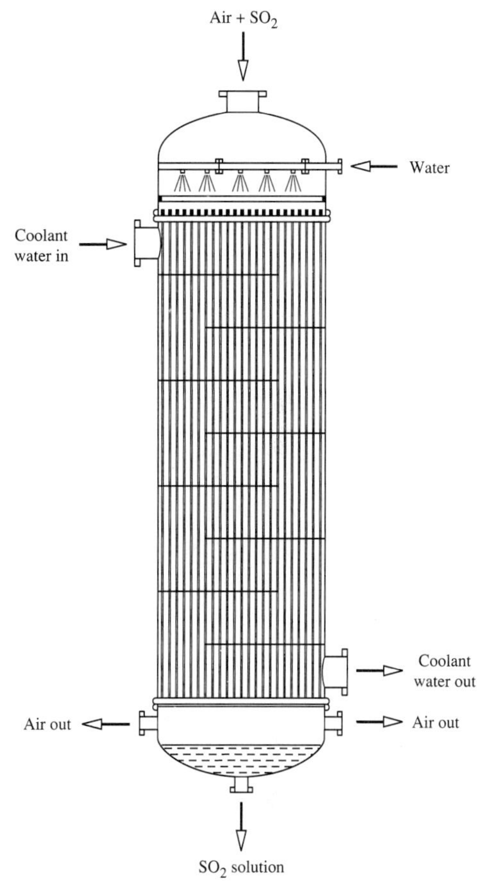

11–31. An existing countercurrent column for recovery of SO_2 from industrial exhaust gas is a packed tower 20 m high. The gas contains 6% SO_2 by volume, and is scrubbed at atmospheric pressure by water at 295 K flowing over the packing at a rate 50% greater than that which would be required by an infinite countercurrent unit to recover 85% of the SO_2. Experimental data indicate that the existing unit recovers 85% of the SO_2 in the entering gas stream. A study is being made of ways to reduce the air pollution still further. One proposal is to build a second, duplicate tower alongside the first and have gas from the top of the first tower enter the top of the second tower and flow cocurrent with the falling water in the new unit. It is proposed to continue the same gas and water flow rates, but the water will enter the top of the new tower, be collected at its base, and be pumped to the top of the existing one. The entering gas will continue to flow into the base of the existing unit and will exit from the base of the new one. Find the outlet SO_2 mole fraction and the percent recovery that would be obtained if the proposal were adopted.

11–32. Reevaluate the performance of the ammonia scrubber in Example 11.7 for cocurrent flow.

11–33. An absorption column is to be designed to recover 99% of the ammonia in an air-ammonia stream by absorption into water. The entering stream contains 3.0% NH_3 by volume. The column is maintained at 290 K by cooling coils, and the pressure is 1 atm.

(i) What is the minimum water flow rate in kmol/kmol of entering gas stream?
(ii) For a water flow rate 50% greater than the minimum, how many transfer units are required for a countercurrent unit?

11–34. Air containing 10% NH_3 by volume is to be scrubbed by water in a countercurrent column 2 m high and with a 1.2 m^2 cross-sectional area. The flow rates of the gas and liquid are 1.0 kg/s and 3.0 kg/s, respectively, and the column operates at 310 K, 1 atm. Calculate the column effectiveness, outlet NH_3 concentration, and pressure drop, for 0.025 m–nominal size

(i) ceramic Raschig rings.
(ii) steel Raschig rings.
(iii) ceramic Berl saddles.
(iv) polyvinylchloride Interlox saddles.

Repeat for a gas flow rate of 2.0 kg/s.

11–35. A mixture of air and 5% inert species by volume is to be scrubbed by water in a 1 m–diameter, 3 m–high countercurrent column packed with 0.025 m–nominal size ceramic Raschig rings. The column operates at 295 K, 1 atm, and the flow rates of gas and liquid are 0.8 kg/s and 2.1 kg/s, respectively. Determine the Henry number, C_{min}, relative resistances to mass transfer in each phase ($1/\mathcal{G}_G$ and $\text{He}/\mathcal{G}_L$), the overall mole transfer conductance $\mathcal{G}_G^{\text{oa}}$, the number of transfer units N_{tu}, and the column effectiveness ε, when the inert species is

(i) carbon monoxide.
(ii) carbon dioxide.
(iii) hydrogen sulfide.
(iv) sulfur dioxide.
(v) ammonia.

Tabulate your results and comment on significant trends.

11–36. Water at 290 K containing 1.11×10^{-4} kmol/m^3 O_2 is to be oxygenated by a pure oxygen stream at 1.02 atm. An available packed column has a 0.5 m^2 cross-sectional area, is 2.4 m high, and is packed with 0.013 m–nominal size ceramic Raschig rings. Prepare performance maps in which the column effectiveness and the gas-side pressure drop are tabulated as a function of water and gas flow rates. Hence recommend operating parameters that will likely be economical and feasible. (The density and viscosity of the oxygen stream can be approximated by air values in order to use SCRUB.)

11–37. Oldenkamp [35] has described a "dry" scrubbing process to remove SO_2 from flue gas, in which the SO_2 is absorbed into a molten mixture of carbonates of lithium (32%), sodium (33%), and potassium (35%) at 700 K. The absorption step is followed by a complex sequence of operations to recover the carbonates in essentially pure form: here we consider only the design of the SO_2 absorber. The table below gives equilibrium data for the air-SO_2-M_2CO_3 system, where M_2CO_3 designates the mixture of carbonates.

x_u	$He \times 10^4$	$y_s \times 10^4$
0	3.12	0
0.1	3.47	0.347
0.2	3.90	0.784
0.3	4.46	1.34
0.4	5.20	2.08
0.5	6.24	3.12
0.6	7.80	4.68

Notice that the Henry number is much less than unity, suggesting a gas side–controlled mass transfer process; also, He is not constant.

Design a SO_2 absorber to reduce the SO_2 in the flue gas of the oil-fired power plant described in Section 11.2.3, from a mole fraction of 5.2×10^{-4} to 1.5×10^{-4}. The packing should be in the form of closely packed concentric cylinders with the M_2CO_3 liquid films flowing downward, and countercurrent flow of flue gas up through the annular spaces. In order to fit into the stack of the power plant, the exchanger cannot exceed 15 m in diameter. Laminar flow of the flue gas would be desirable to have a low pressure drop. It will be sufficient to use a constant value of the Henry number. The absorber is to operate at 700 K, and the flue gas is at 1 atm pressure.

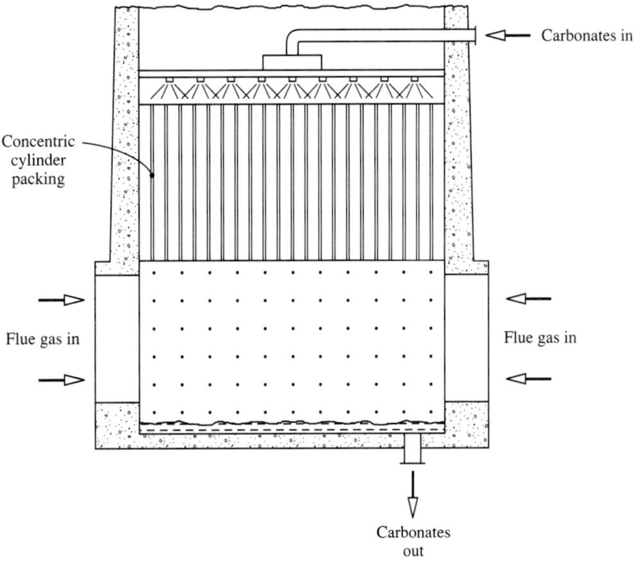

11–38. Redesign the absorber of Exercise 11–37 for cocurrent operation.

11–39. A low-pressure-drop humidifier has a packing of vertical plastic plates with falling water films. The plate center-to-center spacing is 8 mm, and the combined width of one plate plus two liquid films can be taken as 2 mm. A 70 m^3/min flow of air at 1 atm pressure, 300 K, and 30% relative humidity is to be processed. Suggest dimensions for the unit if it is to be 85% effective, have a height equal to twice its width, and have laminar air flow.

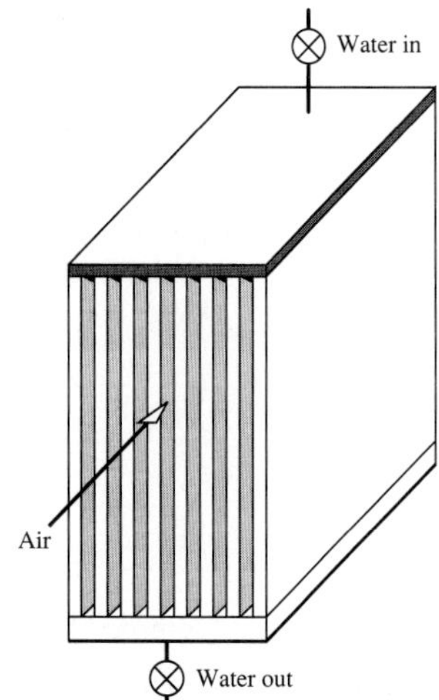

11–40. In an experimental rig, a water film flows down the inside of a 4 cm–I.D. vertical tube, 3 m high. Air flows up the tube at 3 m/s and enters at 318 K, 10% RH, and 1 atm. The water inlet temperature is adjusted to equal the wet-bulb temperature of the inlet air, which is 294 K. The outside of the tube is well insulated. If the mass fraction of water vapor in the outlet air is measured to be 0.0142, determine

(i) the effectiveness of the exchanger.
(ii) the number of transfer units.
(iii) the constant C in the correlation $\overline{\mathrm{Sh}}_D = C\,\mathrm{Re}_D^{0.8}\mathrm{Sc}^{0.4}$ for mass transfer between the water surface and the air stream.
(iv) an estimate of the makeup water required for steady operation.

11–41. A cross-flow humidifier has a packing of 1 m–high vertical plates with falling water films. A 100 m^3/min flow of air at 310 K, 1 atm, and 15% RH is to be processed. If an 85% effective unit is required, determine possible dimensions and pressure drop as a function of plate pitch (5 mm $< p <$ 10 mm). The combined width of one plate and two films can be taken as 2 mm.

11–42. An existing evaporative cooler for a solar air-conditioning process has a cross section of 1.0 m × 0.50 m (chosen to match an associated heat exchanger) and has 18 in–thick Munters Humi-Kool packing made out of CELdek material (a

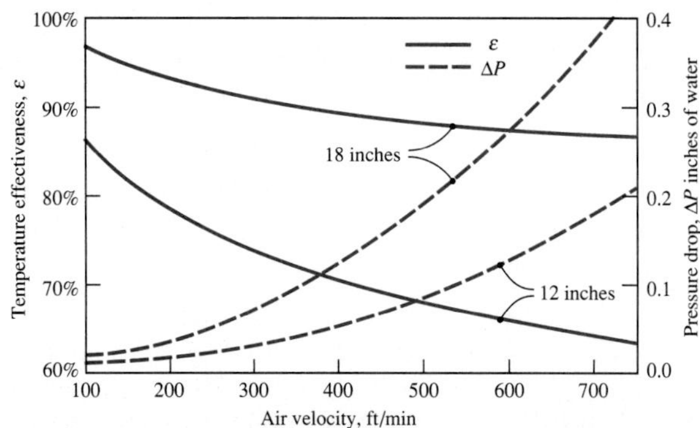

cellulose paper impregnated with insoluble anti-rot salts, rigidifying saturants, and wetting agents). The packing has a cross-fluted configuration, which induces good mixing between the water and air. The manufacturer's data for temperature effectiveness, $\varepsilon = (T_{\text{in}} - T_{\text{out}})/(T_{\text{in}} - T_s)$, and air flow pressure drop are given in the figure. Since pressure drop is a key factor in the system design, it has been suggested that a laminar-flow humidifier, with cross-flow of air through a parallel-plate packing that supports vertical falling water films, may prove to be a superior design. Investigate this possibility for a volumetric air flow rate of 56 m^3/min at 1 atm, and inlet air at 37°C and 40% RH.

11–43. Show that the liquid-side heat transfer coefficient for a laminar falling film on a cooling tower packing is approximately $k/(\delta/2)$ where δ is the film thickness. Hence, determine the difference between T_L and T_s for the following conditions: $\text{Re}_L = 300$, $T_L = 310$ K, $g_{m1} = 0.01$ kg/m^2 s, $T_G = 295$ K, $P = 1$ atm. (*Hint:* Use the solution of an appropriate transient heat conduction problem.)

11–44. Air, at an average temperature of 310 K and 1 atm, flows countercurrent to water at an average temperature of 320 K in a cooling tower packed with parallel plates 2 mm thick, at a pitch of 15 mm. If the superficial air velocity is 5 m/s and the water flow rate is 20 kg/s per square meter of tower cross-sectional area, determine the pressure drop of the air for a 2 m–high packing. Also check to see that the flooding limit is not exceeded. Use the correlations in Section 11.4.4.

11–45. Perform a hand calculation using Merkel's method to determine the number of transfer units for Example 11.12 when $\dot{m}_G/\dot{m}_L = 1$.

11–46. An operating hyperbolic-shell, natural-draft cooling tower is 110 m high, with an air inlet 11 m high. The packing has diameter 72 m and is 15 m above the catch basin. The packing is 2 m–high flat asbestos sheets at 4.45 cm (1 3/4 inches) pitch. The water has an inlet temperature of 44°C and is loaded onto the packing at 2.1 kg/s per square meter of cross-sectional area. Determine the outlet water temperature for an ambient air condition of $P = 990$ mbar, $T = 16$°C, and RH = 40%.

11–47. A 5 m–square mechanical-draft counterflow cooling tower is required to cool 45 kg/s of water from 40°C to 20°C, when the ambient air is at 1 atm, 10°C, and 35% relative humidity. As a function of air flow rate, determine the required packing height and fan power. The packing is in the form of asbestos sheets at a pitch of 3.81 cm (1.5 in). Use a fan efficiency of 80%.

11–48. In Section 11.5.2 a method was developed for determining the number of transfer units of a counterflow cooling tower. A key assumption was that the liquid-side heat transfer resistance was negligible. Modify the method to account for a finite liquid-side resistance.

11–49. A 5 m–square mechanical-draft counterflow cooling tower is required to cool 45 kg/s of water from 40°C to 20°C with an air flow of 50 kg/s. The packing is vertical corrugated plastic at 1.63 in pitch, manufactured by American Tower. Prepare a map of required packing height versus inlet dry- and wet-bulb temperatures in the range 0–20°C at 1000 mbar pressure. Comment on the significance of the results.

11–50. Use CTOWER to generate a cooling demand curve for the following conditions: $P = 1000$ mbar, $T_{\text{WB,in}} = 10°\text{C}$, range $= 15°\text{C}$, approach $= 5°\text{C}$, and an air superficial mass velocity of 1 kg/m² s. Show that $T_{\text{DB,in}}$ has little effect on the curve. Also show the packing capability line for 6.0 m–high asbestos sheets of pitch 3.18 cm (item 3 of Table 11.5), and hence determine the corresponding operating point.

11–51. A cross-flow cooling tower for an engineering school air-conditioning system is to be retrofitted with a new packing. The space available for the packing in one module of a four-module system is 4 m high, 5 m wide, and 2 m deep. The nominal water and air flow rates are 25 kg/s and 30 kg/s, respectively. For a water inlet temperature of 40°C and an air inlet condition of 1000 mbar, $T_{\text{DB}} = 25°\text{C}$, $T_{\text{WB}} = 23°\text{C}$, investigate the performance of the eight cross-flow packings listed in Table 11.5. Tabulate the cooling range achieved and the air flow pressure drop. Comment on the results.

11–52. A new type of packing has been tested in a small-scale test rig containing a 28 cm × 28 cm square cross-section, 0.9 m–high packing element. The following data were measured: $\dot{m}_{G,\text{in}} = 0.178$ kg/s; $\dot{m}_{L,\text{in}} = 0.550$ kg/s; $T_{\text{DB,in}} = 24.3°\text{C}$, $T_{\text{WB,in}} = 14.5°\text{C}$; $T_{\text{WB,out}} = 25.2°\text{C}$; $T_{L,\text{in}} = 28.90°\text{C}$, $T_{L,\text{out}} = 26.03°\text{C}$. Also, $P = 1$ atm.

(i) Perform an energy balance to assess the accuracy of the data.
(ii) Determine the NTU obtained in the test.
(iii) The manufacturer has proposed the following correlation for the mass transfer conductance times specific area product:

$$\mathcal{g}_m a = 1.46 L^{0.39} G^{0.5} \text{ kg/m}^3 \text{ s}$$

Compare the measured $\mathcal{g}_m a$ with the manufacturer's value.

11–53. A closed-form analytical solution can be obtained for the performance of a counterflow cooling tower if the saturation curve is linearized over the interval of concern. Then we can write $h_s = \xi + \eta h_u$, where ξ and η can be determined from the chord joining $h_{s,\text{in}}$ and $h_{s,\text{out}}$, or as a tangent to the h_s curve. Derive an expression for the number of transfer units that is similar in form to Eq. (11.63) for a countercurrent mass exchanger.

11–54. Develop a finite-difference numerical scheme to solve Eqs. (11.103) and (11.104) for a cross-flow cooling tower. Supply all necessary auxiliary data so that a computer programmer could write a program similar to the cross-flow option of CTOWER, based on the information you provide.

11–55. Data for the performance of cooling tower packings are often not of high precision. The test rigs are, of necessity, rather complex, and some of the measurements are difficult to make. Tabulated below are sample counterflow test data for a commercial packing 0.915 m high, with a frontal area of 0.093 m^2. Also given are estimated errors in each measurement.

	$\dot{m}_{L,\text{in}}$ kg/s	$\dot{m}_{G,\text{in}}$ kg/s	$T_{L,\text{in}}$ °C	$T_{L,\text{out}}$ °C	$T_{\text{DB,in}}$ °C	$T_{\text{WB,in}}$ °C	P mbar
1	0.232	0.161	31.1	23.8	22.9	16.2	990
2	0.232	0.088	31.1	26.8	23.2	16.3	990
3	0.579	0.136	31.0	28.1	23.6	16.6	990
Error	±2%	±3%	±0.2 K	±0.5 K	±0.2 K	±0.2 K	±5 mbar

The error in $T_{L,\text{out}}$ is large because air- and water-flow maldistribution causes the outlet water temperature to be nonuniform, and an appropriate average value must be obtained. Using CTOWER, propagate the measurement errors to obtain the expected errors in $\mathcal{g}_m a/L$. Discuss the trends that can be observed.

11–56. Using the result of Exercise 11–48, rework Example 11.10 accounting for the liquid-side heat transfer resistance. Consider a parallel-plate packing at a pitch of 3.18 cm, with $L = G = 2.5$ kg/m^2 s. Use Eqs. (11.40) and (11.44) for the gas-side and liquid-side resistances, respectively. Evaluate all properties at 300 K and 1 atm, and take the width of the plates plus two films as 6 mm.

APPENDIX

A

PROPERTY DATA

LIST OF TABLES

Table A.1*a* Solid metals: Melting point and thermal properties at 300 K

Metal (% composition)	T_{MP} K	ρ kg/m^3	c J/kg K	k W/m K	α^a m^2/s$\times 10^6$
Aluminum					
Pure	933	2702	903	237	97.1
Duralumin (4.4 Cu, 1.0 Mg, 0.75 Mn, 0.4 Si)	775	2770	875	174	71.8
Alloy 195, cast (4.5 Cu)		2790	883	168	68.1
Beryllium	1550	1850	1825	200	59.2
Bismuth	545	9780	122	7.9	6.59
Cadmium	594	8650	231	97	48.4
Copper					
Pure	1358	8933	385	401	117
Electrolytic tough pitch (Cu + Ag 99.90 minimum)		8950	385	386	112
Commercial bronze (10A1)	1293	8800	420	52	14.1
Brass (30 Zn)	1188	8530	380	111	34.2
German silver (15 Ni, 22 Zn)		8618	410	116	32.8
Constantan (40 Ni)		8920	420	22.7	6.06
Constantan (45 Ni)		8860		23	
Gold	1336	19300	129	317	127
Iron					
Pure	1810	7870	447	80.2	22.8
Armco (99.75 pure)		7870	447	72.7	20.7
Cast (4 C)		7272	420	51	16.7
Carbon steels					
AISI 1010 (0.1 C, 0.4 Mn)		7830	434	64	18.8
AISI 1042, annealed (0.42 C, 0.64 Mn, 0.063 Ni, 0.13 Cu)		7840	460	50	13.9
AISI 4130, hardened and tempered (0.3 C, 0.5 Mn, 0.3 Si, 0.95 Cr, 0.5 Mo)		7840	460	43	11.9
Stainless steels					
AISI 302 (18–8) (0.15 C, 2 Mn, 1 Si, 16–18 Cr, 6–8 Ni)		8055	480	15	3.88
AISI 304 (0.08 C, 2 Mn, 1 Si, 18–20 Cr, 8–10 Ni)	1670	7900	477	15	3.98

(Continued)

Table A.1*a* (*Concluded*)

Metal (% composition)	T_{MP} K	ρ kg/m^3	c J/kg K	k W/m K	α m^2/s$\times 10^6$
Stainless steels *(continued)*					
AISI 316 (0.08 C, 2 Mn, 1 Si, 16–18 Cr, 10–14 Ni, 2–3 Mo)		8238	468	13	3.37
AISI 410 (0.15 C, 1 Mn, 1 Si, 11.5–13 Cr)		7770	460	25	7.00
Lead	601	11340	129	35.3	24.1
Magnesium					
Pure	929	1740	1024	156	87.6
Alloy A8 (8 Al, 0.5 Zn)					
Molybdenum	2894	10240	251	138	53.6
Nickel					
Pure	1728	8900	444	91	23.0
Inconel-X-750 (15.5 Cr, 1 Nb, 2.5 Ti, 0.7 Al, 7 Fe)	1665	8510	439	11.7	3.13
Nichrome (20 Cr)	1672	8314	460	13	3.40
Nimonic 75 (20 Cr, 0.4 Ti)		8370	461	11.7	3.03
Hasteloy B (38 Mo, 5 Fe)		9240	381	12.2	3.47
Cupro-Nickel (50 Cu)		8800	421	19.5	5.26
Chromel-P (10 Cr)		8730		17	
Alumel (2 Mn, 2 Al)		8600		48	
Palladium	1827	12020	244	71.8	24.5
Platinum					
Pure	2045	21450	133	71.6	25.1
60 Pt–40 Rh (40 Rh)	1800	16630	162	47	174
Silicon	1685	2330	712	148	89.2
Silver	1235	10500	235	429	174
Tantalum	3269	16600	140	57.5	24.7
Tin	505	7310	227	66.6	40.1
Titanium					
Pure	1993	4500	522	21.9	9.32
Ti-6Al-4V		4420	610	5.8	2.15
Ti-2Al-2 Mn		4510	466	8.4	4.0
Tungsten	3660	19300	132	174	68.3
Zinc	693	7140	389	116	41.8
Zirconium					
Pure	2125	6570	278	22.7	12.4
Zircaloy-4 (1.2–1.75 Sn, 0.18–0.24 Fe, 0.07–0.13 Cr)		6560	285	14.2	7.60

[a] This table and subsequent ones are to read as $\alpha \times 10^6 = 97.1$, that is, $\alpha = 97.1 \times 10^{-6}$ m^2/s.

Table A.1*b* Solid metals: Temperature dependence of thermal conductivity k [W/m K] (see Table A.1*a* for metal compositions)

Metal	Temperature, K								
	200	300	400	500	600	800	1000	1200	1500
Aluminum									
Pure	237	237	240	236	231	218			
Duralumin	138	174	187	188					
Alloy 195, cast		168	174	180	185				
Copper									
Pure	413	401	393	386	379	366	352	339	
Commercial bronze	42	52	52	55					
Brass	74	111	134	143	146	150			
German silver		116	135	145	147				
Gold	323	317	311	304	298	284	270	255	
Iron									
Armco	81	73	66	59	53	42	32	29	31
Cast		51	44	39	36	27	23		
Carbon steels									
AISI 1010		64	59	54	49	39	31		
AISI 1042		52	50	48	45	37	29	26	30
AISI 4130		43	42	41	40	37	31	27	31
Stainless steels									
AISI 302		15	17	19	20	23	25		
AISI 304	13	15	17	18	20	23	25		
AISI 316		13	15	17	18	21	24		
AISI 410	25	25	26	27	27	29			
Lead	37	35	34	33	31				
Magnesium									
Pure	199	156	153	151	149	146			
Alloy A8			84						
Nickel									
Pure	105	91	80	72	66	68	72	76	83
Inconel-X-750	10.3	11.7	13.5	15.1	17.0	20.5	24.0	27.6	30.0
Nichrome		13	14	16	17	21			
Platinum	73	72	72	72	73	76	79	83	90
Silver	420	429	425	419	412	396	379	361	
Tantalum	58	58	58	59	59	59	60	61	62
Tin	73	67	62	60					
Titanium									
Pure	25	22	20	20	19	19	21	22	25
Ti-6Al-4V		5.8							
Tungsten	185	174	159	146	137	125	118	112	106
Zirconium									
Pure	25	23	22	21	21	21	23	26	29
Zircaloy-4	13.3	14.2	15.2	16.2	17.2	19.2	21.2	23.2	

Table A.1*c* Solid metals: Temperature dependence of specific heat capacity c [J/kg K] (see Table A.1*a* for metal compositions)

Metal	Temperature, K								
	200	300	400	500	600	800	1000	1200	1500
Aluminum									
Pure	798	903	949	996	1033	1146			
Duralumin		875							
Alloy 195, cast		883							
Copper									
Pure	356	385	397	412	417	433	451	480	
Commercial bronze	785	420	460	500					
Brass	360	380	395	410	425				
German silver		410							
Gold	124	129	131	133	135	140	145	155	
Iron									
Armco	384	447	490	530	574	680	975	609	634
Cast		420							
Carbon steels									
AISI 1010		434	487	520	559	685	1168		
AISI 1042			500	530	570	700	1430		
AISI 4130			500	530	570	690	840		
Stainless steels									
AISI 302		480	512	531	559	585	606		
AISI 304	402	477	515	539	557	582	611	640	682
AISI 316		468	504	528	550	576	602		
AISI 410		460							
Lead	125	129	132	136	142				
Magnesium									
Pure	934	1024	1074	1170	1170	1267			
Alloy A8			1000						
Nickel									
Pure	383	444	485	500	512	530	562	594	616
Inconel-X-750	372	439	473	490	510	546	626		
Nichrome			480	500	525	545			
Platinum	125	133	136	139	141	146	152	157	165
Silver	225	232	239	244	250	262	277	292	
Tantalum	133	140	144	145	146	149	152	155	160
Tin	215	227	243						
Titanium									
Pure	405	522	551	572	591	633	675	680	686
Ti-6Al-4V		610							
Tungsten	122	132	137	140	142	145	148	152	157
Zirconium									
Pure	264	278	300	312	322	342	362	344	344
Zircaloy-4			300	314	327	348	369		

Table A.2 Solid dielectrics: Thermal properties

Dielectric	T K	ρ kg/m^3	c J/kg K	k W/m K	α m^2/s$\times 10^6$
Aluminum oxide, Al_2O_3					
Sapphire	300	3970	765	46	15.2
Alumina	300	3970	765	36	11.9
	400		940	27	7.2
	600		1110	16	3.6
	1000		1225	7.6	1.6
	1500			5.4	
Carbon					
Diamond (type IIb)	300	3300	510	1300	772
ATJ-S graphite	300	1810	1300	98	42
	1000		1926	55	16
	2000		2139	38	9.8
	3000		2180	33	8.4
Pyrolytic graphite	300	2210	709	1950	1240
k parallel to layers	600		1406	892	287
	1000		1793	534	135
	2000		2043	262	58
k perpendicular to layers	300	2210	709	5.7	4.1
	600		1406	2.68	2.1
	1000		1793	1.60	1.67
	2000		2043	0.81	1.51
Graphite fiber epoxy	200	1400	640	8.7	9.7
(25% volume) composite	300		935	11.1	8.5
k parallel to fibers	400		1220	13.0	7.6
k perpendicular to fibers	200	1400	640	0.68	0.76
	300		935	0.87	0.66
	400		1220	1.1	0.64
Carbon-carbon weave	300	1860	810	110	73
	1000		1800	56	17
	2000		2140	36.5	9.2
	3000		2220	34.5	8.4
	4000		2260	34	8.1
	5000		2270	34	8.1
Ice	273	910	1930	2.22	1.26
Plastics					
Cellulose acetate	300	1300	1510	0.24	0.12
Neoprene rubber	300	1250	1930	0.19	0.079
Phenolic, filled	300	1760	1260	0.50	0.23
Polyamide (nylon)	300	1140	1670	0.24	0.13
Polyethylene (high density)	300	960	2090	0.33	0.16
Polypropylene	300	1170	1930	0.17	0.075
Polyvinylchloride	300	1714	1050	0.092	0.051
Teflon	300	2200	1050	0.35	0.15
	400			0.45	0.19

(Continued)

Table A.2 ***(Concluded)***

Dielectric	T K	ρ kg/m^3	c J/kg K	k W/m K	α m^2/s$\times 10^6$
Silicon dioxide, SiO_2	200	2650		16.4	
Crystalline (quartz)	300		745	10.4	5.3
k parallel to c-axis	400		885	7.6	3.2
	600		1075	5.0	1.7
k perpendicular to c-axis	200	2650		9.5	
	300		745	6.2	3.1
	400		885	4.7	2.0
	600		1075	3.4	1.2
Polycrystalline (fused silica glass)	300	2220	745	1.38	0.83
	400		905	1.51	0.75
	600		1040	1.75	0.75
	800		1105	2.17	0.88
	1000		1155	2.87	1.11
	1200		1195	4.0	1.51
Titanium dioxide, TiO_2 (rutile)	300	4157	710	8.4	2.8
	600		880	5.0	1.4
	1200		945	3.3	0.84
Uranium oxide, UO_2	300	10,890	240	7.9	3.0
	500		265	6.0	2.1
	1000		305	3.9	1.2
	1500		325	2.6	0.79
	2000		355	2.3	0.59
	2500		405	2.5	0.57

Table A.3 Insulators and building materials: Thermal properties

	T K	ρ kg/m^3	c J/kg K	k W/m K	α m^2/s$\times 10^6$
Asbestos paper, laminated and corrugated	300	190		0.078	
4 ply	320			0.085	
	340			0.091	
	360			0.097	
	380			0.101	
8 ply	300	300		0.068	
	320			0.073	
	340			0.077	
	360			0.080	
	380			0.083	
Brick					
B&W K-28 insulating	600			0.03	
	1300			0.04	
Chrome	400	3010	835	2.3	0.92
	800			2.5	
	1200			2.0	
Fireclay	400	2645	960	0.9	0.35
	800			1.4	
	1200			1.7	
	1600			1.8	
Common	300	1920	835	0.72	0.45
Face	300	2083		1.3	
Concrete					
Stone 1–2–4 mix	300	2100	880	1.4	0.75
Cork	300	160	1680	0.043	0.16
Cotton	300	80	1300	0.06	0.58
Glass					
Fused silica	300	2220	745	1.38	0.83
Borosilicate (Pyrex)	300	2640	800	1.09	0.51
Soda-lime (25 Na_2O, 10 CaO)	300	2400	840	0.88	0.44
Cellular glass	240			0.048	
	260			0.051	
	280			0.054	
	300	145		0.058	
	320			0.063	
	340			0.067	
Fiberglass, paper-faced batt	300	16	835	0.046	3.4
	300	40		0.035	
	260			0.029	
	280			0.033	
	300	28		0.038	
	320			0.043	
	340			0.048	
	360			0.054	
	380			0.060	
	400			0.066	

(Continued)

Table A.3 ***(Concluded)***

	T K	ρ kg/m^3	c J/kg K	k W/m K	α m^2/s$\times 10^6$
Loose fill					
Cellulose, wood or paper pulp	290			0.038	
	300	45		0.039	
	310			0.042	
Vermiculite, expanded	240			0.058	
	260			0.061	
	280			0.064	
	300	122		0.069	
	320			0.074	
	240			0.052	
	260			0.056	
	280			0.059	
	300	80		0.063	
	320			0.068	
Magnesia	300	270		0.062	
(85 %)	350			0.068	
	400			0.073	
	450			0.078	
	500			0.082	
Paper	300	930	2500	0.13	0.056
Polystyrene, rigid	240			0.023	
	260			0.024	
	280			0.026	
	300	30–60	1210	0.028	0.4–0.8
	320			0.030	
Polyurethane, rigid foam	300	70		0.026	
Rubber					
Hard	270	1200	2010	0.15	0.062
Neoprene	300	1250	1930	0.19	0.079
Rigid foamed	260			0.028	
	280			0.030	
	300	70		0.032	
	320			0.034	
Snow	273	110		0.049	
		500		0.190	
Soil					
Dry	300	1500	1900	1.0	0.35
Wet	300	1900	2200	2.0	0.5
Woods					
Oak, parallel to grain	300	820	2400	0.35	0.18
perpendicular to grain	300	820	2400	0.21	0.11
White Pine, parallel to grain	300	500	2800	0.24	0.17
perpendicular to grain	300	500	2800	0.10	0.071
Wool, sheep	300	145		0.05	

Table A.4 Thermal conductivity of selected materials at cryogenic temperatures

Material	Temperature, K				
	5	10	30	100	200
Metals					
Aluminum (2024–T4)	3.5	7.7	21	50	72
Brass				71	94
Copper (OFHC)			950	430	400
Carbon steel (C1020)	4.3	12	34	64	70
Stainless steel (303)	0.29	0.71	3.5	9.0	12
(304)	0.16	0.82	3.3	9.5	13
Nonmetals					
Glass, Pyrex				0.56	0.89
Glass, Phoenix		0.11	0.17	0.55	
Nylon-66	0.018	0.033	0.21		
Polyethylene	0.040	0.15	0.75		
Silicone rubber				0.18	0.21
Teflon			0.18	0.62	1.0
Vacuum grease (Dow-Corning silicone)	0.021				
Varnish (G.E. #7031, thermosetting)	0.065	0.075	0.15	0.25	0.36

Table A.5*a* Total hemispherical emittance at $T_s \simeq 300$ K, and solar absorptance[a]

Material and Surface Condition	Total Hemispherical Emittance	Solar Absorptance
Aluminum		
Foil, as received	0.05	
Foil, bright dipped	0.03	0.10
Vacuum-deposited on duPont Mylar	0.03	0.10
Alloy 6061, as received	0.04	0.37
Alloy 7075–T6, sandblasted with 60 mesh silicon carbide grit	0.30	0.55
Weathered alloy 75S–T6	0.20	0.54
Aluminized silicone resin paint, Dow-Corning XP–310	0.20	0.27
Hard-anodized	0.80	0.23
Soft-anodized	0.76	0.55
Roofing	0.24	
Asbestos		
Board	0.93	
Cloth	0.87	
Slate	0.94	
Asphalt	0.88	
Brass		
Oxidized	0.60	
Polished	0.04	
Brick	0.90	0.63
Carbon		
Graphite, crushed on sodium silicate	0.88	0.96
Lampblack	0.92	
Chromium		
Bright plate	0.16	
Heated 50 hr at 870 K	0.18	0.78
Coal	0.78	
Concrete, rough	0.91	0.60
Copper		
Electroplated	0.03	0.47
Black oxidized in Ebanol C	0.16	0.91
Oxidized plate	0.76	
Earthenware		
Glazed	0.90	
Matte	0.93	
Frost, rime	0.99	
Glass		
Polished	0.87–0.92	
Pyrex	0.80	
Smooth	0.91	
Second-surface mirror	0.81	0.13
Gold		
On stainless steel	0.09	
On 3M tape Y8194	0.025	

(Continued)

Table A.5*a* (*Continued*)

Material and Surface Condition	Total Hemispherical Emittance	Solar Absorptance
Granite	0.44	
Gravel	0.30	
Ice		
Crystal	0.96	
Smooth	0.97	
Inconel X		
Bright	0.21	0.90
Oxidized 4 hr at 1270 K	0.72	
Oxidized 10 hr at 980 K	0.79	
Iron		
ARMCO, bright	0.12	
ARMCO, oxidized	0.30	
Cast, oxidized	0.57	
Rusted	0.83	
Wrought, polished	0.29	
Wrought, dull	0.91	
Limestone	0.92	
Magnesium oxide	0.72	
Marble		
Polished	0.89	
Smooth	0.56	
White	0.92	
Mortar, lime	0.90	
Mylar, duPont film, aluminized on second surface		
6 μm thick	0.37	
25 μm thick	0.63	
75 μm thick	0.81	
Nickel		
Electroplated	0.03	0.22
Tabor solar absorber, electro-oxidized on copper		
110–30	0.05	0.85
125–30	0.11	0.85
Oak, planed	0.88	
Paints		
Black		
Parson's optical	0.92	0.97
Silicone high heat	0.90	0.94
Epoxy	0.87	0.95
Gloss	0.90	
Enamel, heated 1000 hr at 650 K	0.80	
Silver Chromatone	0.24	0.20
White		
Acrylic resin	0.90	0.26
Gloss	0.85	
Epoxy	0.85	0.25

(Continued)

Table A.5*a* (Concluded)

Material and Surface Condition	Total Hemispherical Emittance	Solar Absorptance
Paper		
Roofing	0.88	
White	0.86	
Plaster, rough	0.89	
Platinum-coated stainless steel	0.12	
Refractory		
Black	0.94	
White	0.90	
Rubber	0.88	
Sand	0.75	
Sandstone, red	0.59	
Silica		
Sintered, powdered, fused	0.82	0.08
Second-surface mirror		
Aluminized	0.81	0.14
Silvered	0.81	0.07
Silver		
Polished	0.02	
Plated on nickel on stainless steel	0.08	
Heated 300 hr at 650 K	0.15	
Slate	0.85	
Snow, fresh	0.82	0.13
Soil	0.94	
Spruce, sanded	0.80	
Stainless steel		
AISI 312, heated 300 hr at 530 K	0.26	
AISI 301, with armco black oxide	0.75	0.89
AISI 410, heated to 980 K	0.15	0.76
AISI 303, sandblasted heavily with 80 mesh aluminum oxide grit	0.42	0.85
Teflon	0.85	0.12
Titanium		
75 A	0.12	
75 A, oxidized 300 hr at 730 K in air	0.21	0.80
C–110 M, oxidized 100 hr at 700 K in air	0.06	0.52
C–110 M, oxidized 300 hr at 730 K in air	0.20	0.77
Tungsten, polished	0.03	
Water	0.90	0.98
White potassium zirconium silicate spacecraft coating	0.87	0.13
Zinc, blackened by Tabor solar collector electrochemical treatment, 120–20	0.14	0.89

[a] Since the solar spectrum outside the earth's atmosphere is different from that at ground level, appropriate values of solar absorptance are a little different. The values in this table are for extraterrestrial conditions, except those for brick, concrete, snow, and water.

Table A.5*b* Temperature variation of total hemispherical emittance for selected surfaces

Material and Surface Condition	Temperature, K						
	200	400	600	800	1000	1200	1400
Aluminum, polished foil	0.03	0.05	0.06	0.07			
Aluminum alloy 245T, polished	0.03	0.04	0.05	0.06	0.08		
Aluminum oxide, Al_2O_3		0.78	0.69	0.61	0.54	0.49	0.42
Cadmium sulphide, CdS	0.56	0.27	0.16				
Carbon		0.83	0.82	0.81	0.81	0.80	0.80
Copper, polished	0.02	0.03	0.03	0.04	0.04	0.05	0.05
Copper, oxidized at 1000 K			0.50	0.58	0.80		
Gold, polished foil	0.02	0.03	0.05	0.06	0.07	0.08	
Iron, polished	0.05	0.09	0.14	0.20	0.25	0.30	0.35
Iron oxides					0.82	0.82	0.80
Magnesium oxide, MgO	0.72	0.73	0.62	0.52	0.47	0.41	0.36
Molybdenum, polished	0.05	0.07	0.08	0.10	0.12	0.14	0.17
Monel metal, polished	0.14	0.15	0.17	0.19	0.22	0.26	0.34
Sodium chloride, NaCl	0.44	0.24					
Nickel, polished	0.08	0.10	0.11	0.12	0.15	0.17	0.21
Platinum, polished	0.05	0.07	0.09	0.11	0.13	0.15	0.17
Silver, polished	0.02	0.02	0.03	0.03	0.03	0.04	0.04
Silicon carbide, SiC	0.84	0.84	0.83	0.83	0.83	0.83	0.83
Silicon oxide, SiO_2, fused	0.72	0.79					
Tantalum, polished	0.01	0.03	0.05	0.08	0.11	0.13	0.16
Titanium alloy A110-AT, polished	0.14	0.17	0.21	0.23	0.26	0.28	0.29
Tungsten, polished		0.03	0.05	0.07	0.09	0.12	0.16
Zirconium oxide, ZrO_2	0.84	0.73	0.63	0.51	0.44	0.42	0.40
Zinc sulphide	0.56	0.30					

Table A.6*a* Spectral and total absorptances of metals for normal incidence

Spectral absorptance, normal incidence, $1 < \lambda < 25\mu m$

$$\alpha(\lambda, T_s) \simeq A\left[\frac{(1 + \lambda^2/\lambda_{12}^2)^{1/2} - 1}{\lambda^2/2\lambda_{12}^2}\right]^{1/2} + \frac{B}{C + \lambda^2}$$

Total absorptance, normal incidence, $330 \text{ K} < T_e < 2200 \text{ K}$

$$\alpha(T_s, T_e) \simeq A\left[\frac{(1 + 0.078C_2^2/\lambda_{12}^2T_e^2)^{1/2} - 1}{0.039C_2^2/\lambda_{12}^2T_e^2}\right]^{1/2} + \frac{18.6(BT_e^2/C_2^2)}{1 + 18.6(CT_e^2/C_2^2)}$$

where $C_2 = hc_o/k = 149,387\ \mu\text{m K}$

Metal	Parameter ($T_s \simeq 300$ K)			
	A	B	C	λ_{12}
Aluminum foil	0.0165	0.23	8.9	14
Cadmium, 99.99%, rolled plate	0.054	2.15	3.2	9
Chromium, polished electroplate	0.076	1.58	3.9	3
Columbium, 99.99%, rolled plate	0.15	0.29	≃0	1
Copper, 99.99%, polished	0.018	0.077	3.2	45
Gold, 99.99%, polished	0.020	0.056	1.4	45
Indium, 99.99%, scraped	0.060	0.24	1.3	6
Inconel X, rolled plate	0.44	0.036	≃0	1
Lead, 99.99%, scraped	0.16	0.39	1.1	4
Manganese, 99.99%, polished	0.19	4.8	11	8
Molybdenum, 99.99%	0.033	0.36	≃0	7
Nickel, 99.99%, polished	0.029	0.83	2.4	5
Platinum, 99.99%, cold rolled	0.038	0.42	≃0	4
Rhodium, polished electroplate	0.06	1.27	10	6
Silver, polished electroplate	0.011	0.16	11	70
Stainless steel, 303, lapped	≃0.71	≃0	≃0	≃0.125
Tin, 99.99%, rolled plate	0.052	0.56	0.8	7
Titanium, polished electroplate	0.13	2.9	8.3	12
Titanium, 99%, lapped	0.09	6.5	15	12
Tungsten, 99.99%, lapped	0.05	0.49	0.3	3
Vanadium, 99.99%, rolled plate	0.17	0.66	0.93	1
Zinc, 99.9%	0.036	0.26	≃0	8
Zirconium, 99.99%, rolled plate	0.64	4	35	1

From *Advances in Thermophysical Properties at Extreme Temperatures and Pressures,* Am. Soc. Mech. Engrs., New York: 1965, pp. 189–199.

Table A.6*b* Spectral absorptances at room temperature and an angle of incidence of 25° from the normal [for nonconductors $\alpha(25°) \simeq \alpha(\text{hemispherical})$]

	Wavelength, λ [μm]										
	0.3	0.35	0.4	0.45	0.5	0.6	0.7	0.8	1.0	1.5	2.0
Bright metals:											
Aluminum	0.05	0.05	0.07	0.07	0.08	0.11	0.12	0.14	0.08	0.04	0.035
Chromium	0.52	0.48	0.43	0.40	0.39	0.37	0.37	0.37	0.40	0.34	0.26
Copper					0.53	0.23	0.14	0.10	0.06	0.032	0.029
Gold	0.80	0.78	0.75	0.74	0.60	0.17	0.11	0.08	0.043	0.034	0.027
Stainless steel		0.61	0.57	0.54	0.53	0.49	0.46	0.44	0.34	0.28	0.25
Titanium	0.71	0.65	0.59	0.56	0.53	0.48	0.47	0.43	0.44	0.40	0.34
Paints and coatings:											
3M black velvet	0.97	→	→	→	→	→	→	→	→	→	0.96
Hard-anodized aluminum	0.95	0.94	0.93	→	→	→	→	→	0.92	0.90	0.85
Anodized titanium					0.53	0.48	0.48	0.48	0.50	0.50	0.52
White epoxy paint			0.60	0.12	0.10	0.15	0.21	0.09	0.08	0.30	0.43
Flame sprayed alumina	0.60	0.48	0.34	0.29	0.27	0.24	0.23	0.23	0.25	0.32	0.51
Aluminum paint			0.25	0.25	0.25	0.26	0.29	0.31	0.28	0.25	0.23

	Wavelength, λ [μm]										
	3	4	5	6	8	10	12	15	20	30	40
Bright metals:											
Aluminum	0.029	0.026	0.023	0.021	0.019	0.018	0.017	0.015	0.014	0.013	0.012
Chromium	0.19	0.145	0.110	0.088	0.078	0.065	0.059	0.05	0.047	0.036	0.030
Copper	0.029	0.022	0.021	0.020	0.020	0.018	0.018	0.018	0.018		
Gold	0.025	0.023	0.023	0.022	0.022	0.020	0.020	0.020	0.020		
Stainless steel	0.20	0.17	0.15	0.14	0.12	0.11	0.10	0.09	0.077	0.062	0.053
Titanium	0.29	0.24	0.22	0.20	0.17	0.15	0.14	0.13	0.11	0.09	0.08
Paints and coatings:											
3M black velvet	0.96	0.96	0.95	0.96	0.91	0.95	0.95	0.94	0.94	0.97	0.97
Hard-anodized aluminum	0.92	0.74	0.70	0.83	0.96	0.98	0.84	0.83	0.80	0.80	
Anodized titanium	0.89	0.76	0.76	0.82	0.83	0.90	0.91	0.88	0.85		
White epoxy paint	0.93	0.90	0.90	0.91	0.93	0.91	0.93	0.90	0.84	0.81	0.80
Flame sprayed alumina	0.73	0.53	0.62	0.88	0.98	0.98	0.74	0.79	0.75		
Aluminum paint	0.23	0.23	0.22	0.22	0.26	0.22	0.21	0.20	0.19		

Table A.7 Gases[a]: Thermal properties

Gas	T K	k W/m K	ρ kg/m³	c_p J/kg K	$\mu \times 10^{6\,b}$ kg/m s	$\nu \times 10^{6\,b}$ m²/s	Pr
Air	150	0.0158	2.355	1017	10.64	4.52	0.69
(82 K BP)	200	0.0197	1.767	1009	13.59	7.69	0.69
	250	0.0235	1.413	1009	16.14	11.42	0.69
	260	0.0242	1.360	1009	16.63	12.23	0.69
	270	0.0249	1.311	1009	17.12	13.06	0.69
	280	0.0255	1.265	1008	17.60	13.91	0.69
	290	0.0261	1.220	1007	18.02	14.77	0.69
	300	0.0267	1.177	1005	18.43	15.66	0.69
	310	0.0274	1.141	1005	18.87	16.54	0.69
	320	0.0281	1.106	1006	19.29	17.44	0.69
	330	0.0287	1.073	1006	19.71	18.37	0.69
	340	0.0294	1.042	1007	20.13	19.32	0.69
	350	0.0300	1.012	1007	20.54	20.30	0.69
	360	0.0306	0.983	1007	20.94	21.30	0.69
	370	0.0313	0.956	1008	21.34	22.32	0.69
	380	0.0319	0.931	1008	21.75	23.36	0.69
	390	0.0325	0.906	1009	22.12	24.42	0.69
	400	0.0331	0.883	1009	22.52	25.50	0.69
	500	0.0389	0.706	1017	26.33	37.30	0.69
	600	0.0447	0.589	1038	29.74	50.50	0.69
	700	0.0503	0.507	1065	33.03	65.15	0.70
	800	0.0559	0.442	1089	35.89	81.20	0.70
	900	0.0616	0.392	1111	38.65	98.60	0.70
	1000	0.0672	0.354	1130	41.52	117.3	0.70
	1500	0.0926	0.235	1202	53.82	229.0	0.70
	2000	0.1149	0.176	1244	64.77	368.0	0.70
Ammonia	250	0.0198	0.842	2200	8.20	9.70	0.91
(239.7 K BP)	300	0.0246	0.703	2200	10.1	14.30	0.90
	400	0.0364	0.520	2270	13.8	26.60	0.86
	500	0.0511	0.413	2420	17.6	42.50	0.83
Argon	150	0.0096	3.28	527	12.5	3.80	0.68
(77.4 K BP)	200	0.0125	2.45	525	16.3	6.65	0.68
	250	0.0151	1.95	523	19.7	10.11	0.68
	300	0.0176	1.622	521	22.9	14.1	0.68
	400	0.0223	1.217	520	28.6	23.5	0.67
	500	0.0265	0.973	520	33.7	34.6	0.66
	600	0.0302	0.811	520	38.4	47.3	0.66
	800	0.0369	0.608	520	46.6	76.6	0.66
	1000	0.0427	0.487	520	54.2	111.2	0.66
	1500	0.0551	0.324	520	70.6	218.0	0.67

(Continued)

Table A.7 ***(Continued)***

Gas	T K	k W/m K	ρ kg/m^3	c_p J/kg K	$\mu \times 10^{6\,b}$ kg/m s	$\nu \times 10^{6\,b}$ m^2/s	Pr
Carbon dioxide (195 K subl.)	250	0.01435	2.15	782	12.8	5.97	0.70
	300	0.01810	1.788	844	15.2	8.50	0.71
	400	0.0259	1.341	937	19.6	14.6	0.71
	500	0.0333	1.073	1011	23.5	21.9	0.71
	600	0.0407	0.894	1074	27.1	30.3	0.71
	800	0.0544	0.671	1168	33.4	49.8	0.72
	1000	0.0665	0.537	1232	38.8	72.3	0.72
	1500	0.0945	0.358	1329	51.5	143.8	0.72
	2000	0.1176	0.268	1371	61.9	231.0	0.72
Refrigerant-12 (243 K BP)	260	0.00769	5.668	560	10.98	1.938	0.80
	280	0.00868	5.263	582	11.80	2.242	0.79
	300	0.00970	4.912	602	12.60	2.566	0.78
	320	0.0107	4.605	621	13.39	2.908	0.77
	340	0.0118	4.334	639	14.16	3.268	0.77
	360	0.0129	4.093	655	14.92	3.644	0.76
	380	0.0140	3.878	671	15.66	4.038	0.75
	400	0.0151	3.684	685	16.39	4.449	0.74
	420	0.0162	3.509	698	17.10	4.875	0.74
	440	0.0173	3.349	710	17.80	5.314	0.73
	460	0.0184	3.203	721	18.48	5.770	0.72
Refrigerant-113 (320.7 K BP)	320.7	0.00866	7.120	651	10.77	1.513	0.81
	340	0.00958	6.716	668	11.24	1.674	0.78
	360	0.01056	6.343	688	11.69	1.843	0.76
	380	0.01154	6.009	706	12.11	2.015	0.74
	400	0.01254	5.709	724	12.53	2.195	0.72
	420	0.01359	5.437	742	12.93	2.378	0.71
	440	0.01483	5.190	758	13.35	2.572	0.68
	460	0.01623	4.964	774	13.78	2.776	0.66
Helium (4.3 K BP)	50	0.046	0.974	5200	6.46	6.63	0.73
	100	0.072	0.487	5200	9.94	20.4	0.72
	150	0.096	0.325	5200	13.0	40.0	0.70
	200	0.116	0.244	5200	15.6	64.0	0.70
	250	0.133	0.195	5200	17.9	92.0	0.70
	300	0.149	0.1624	5200	20.1	124.0	0.70
	400	0.178	0.1218	5200	24.4	200.0	0.71
	500	0.205	0.0974	5200	28.2	290.0	0.72
	600	0.229	0.0812	5200	31.7	390.0	0.72
	800	0.273	0.0609	5200	37.8	620.0	0.72
	1000	0.313	0.0487	5200	43.3	890.0	0.72

(Continued)

Table A.7 *(Continued)*

Gas	T K	k W/m K	ρ kg/m^3	c_p J/kg K	$\mu \times 10^6$ [b] kg/m s	$\nu \times 10^6$ [b] m^2/s	Pr
Hydrogen (20.3 K BP)	20	0.0158	1.219	10400	1.08	0.893	0.72
	40	0.0302	0.6094	10300	2.06	3.38	0.70
	60	0.0451	0.4062	10660	2.87	7.06	0.68
	80	0.0621	0.3047	11790	3.57	11.7	0.68
	100	0.0805	0.2437	13320	4.21	17.3	0.70
	150	0.125	0.1625	16170	5.60	34.4	0.73
	200	0.158	0.1219	15910	6.81	55.8	0.68
	250	0.181	0.0975	15250	7.91	81.1	0.67
	300	0.198	0.0812	14780	8.93	109.9	0.67
	400	0.227	0.0609	14400	10.8	177.6	0.69
	500	0.259	0.0487	14350	12.6	258.1	0.70
	600	0.299	0.0406	14400	14.3	350.9	0.69
	800	0.385	0.0305	14530	17.4	572.5	0.66
	1000	0.423	0.0244	14760	20.5	841.2	0.72
	1500	0.587	0.0164	16000	25.6	1560	0.70
	2000	0.751	0.0123	17050	30.9	2510	0.70
Mercury (630 K BP)	650	0.0100	3.761	104	64.08	17.04	0.67
	700	0.0108	3.493	104	69.25	19.83	0.67
	800	0.0124	3.056	104	79.45	26.00	0.67
	900	0.0139	2.716	104	89.30	32.87	0.67
	1000	0.0154	2.445	104	98.67	40.36	0.67
	1200	0.0181	2.037	104	115.9	56.93	0.67
	1400	0.0206	1.746	104	132.1	75.68	0.67
	1600	0.0231	1.528	104	148.3	97.11	0.67
	1800	0.0258	1.358	104	165.1	121.5	0.67
	2000	0.0282	1.222	104	180.9	148.0	0.67
Nitrogen (77.4 K BP)	150	0.0157	2.276	1050	10.3	4.53	0.69
	200	0.0197	1.707	1045	13.1	7.65	0.69
	250	0.0234	1.366	1044	15.5	11.3	0.69
	300	0.0267	1.138	1043	17.7	15.5	0.69
	400	0.0326	0.854	1047	21.5	25.2	0.69
	500	0.0383	0.683	1057	25.1	36.7	0.69
	600	0.044	0.569	1075	28.3	49.7	0.69
	800	0.055	0.427	1123	34.2	80.0	0.70
	1000	0.066	0.341	1167	39.4	115.6	0.70
	1500	0.091	0.228	1244	51.5	226.0	0.70
	2000	0.114	0.171	1287	61.9	362.0	0.70

(Continued)

Table A.7 *(Concluded)*

Gas	T K	k W/m K	ρ kg/m^3	c_p J/kg K	$\mu \times 10^6$ [b] kg/m s	$\nu \times 10^6$ [b] m^2/s	Pr
Oxygen	150	0.0148	2.60	890	11.4	4.39	0.69
(90.2 K BP)	200	0.0192	1.949	900	14.7	7.55	0.69
	250	0.0234	1.559	910	17.8	11.4	0.69
	300	0.0274	1.299	920	20.6	15.8	0.69
	400	0.0348	0.975	945	25.4	26.1	0.69
	500	0.042	0.780	970	29.9	38.3	0.69
	600	0.049	0.650	1000	33.9	52.5	0.69
	800	0.062	0.487	1050	41.1	84.5	0.70
	1000	0.074	0.390	1085	47.6	122.0	0.70
	1500	0.101	0.260	1140	62.1	239	0.70
	2000	0.126	0.195	1180	74.9	384	0.70
Saturated steam	273.15	0.0182	0.0048	1850	7.94	1655	0.81
(not at 1 atm)	280	0.0186	0.0076	1850	8.29	1091	0.83
	290	0.0192	0.0142	1860	8.69	612	0.84
	300	0.0198	0.0255	1870	9.09	356.5	0.86
	310	0.0204	0.0436	1890	9.49	217.7	0.88
	320	0.0210	0.0715	1890	9.89	138.3	0.89
	330	0.0217	0.1135	1910	10.3	90.7	0.91
	340	0.0223	0.1741	1930	10.7	61.4	0.92
	350	0.0230	0.2600	1950	11.1	42.6	0.94
	360	0.0237	0.3783	1980	11.5	30.4	0.96
	370	0.0246	0.5375	2020	11.9	22.1	0.98
	373.15	0.0248	0.5977	2020	12.0	20.1	0.98
	380	0.0254	0.7479	2057	12.3	16.4	1.00
Superheated	400	0.0277	0.555	1900	14.0	25.2	0.96
steam	500	0.0365	0.441	1947	17.7	40.1	0.94
(373.2 K BP)	600	0.046	0.366	2003	21.4	58.5	0.93
	800	0.066	0.275	2130	28.1	102.3	0.91
	1000	0.088	0.220	2267	34.3	155.8	0.88
	1500	0.148	0.146	2594	49.1	336.0	0.86
	2000	0.206	0.109	2832	62.7	575.0	0.86

[a] At 1 atm pressure unless otherwise noted.

[b] This table and subsequent ones are to be read as $\nu \times 10^6 = 4.52$, that is, $\nu = 4.52 \times 10^{-6}$ m^2/s.

Table A.8 Dielectric liquids: Thermal properties

Saturated Liquid (Melting point) (Boiling point) (Latent heat at BP)	T K	k W/m K	ρ kg/m^3	c_p J/kg K	$\mu \times 10^4$ kg/m s	$\nu \times 10^6$ m^2/s	Pr
Ammonia	220	0.547	705	4480	3.35	0.475	2.75
(195 K MP)	230	0.547	696	4480	2.82	0.405	2.31
(240 K BP)	240	0.547	683	4480	2.42	0.355	1.99
(1.37×10^6 J/kg)	250	0.547	670	4500	2.14	0.320	1.76
	260	0.544	657	4550	1.93	0.293	1.61
	270	0.540	642	4620	1.74	0.271	1.49
	280	0.533	631	4710	1.60	0.253	1.41
	290	0.522	616	4800	1.44	0.234	1.33
	300	0.510	602	4900	1.31	0.217	1.26
	310	0.496	587	4990	1.19	0.202	1.19
	320	0.481	572	5080	1.08	0.188	1.14
Carbon dioxide	220	0.080	1170	1850	1.39	0.119	3.22
(195 K subl.)	230	0.096	1130	1900	1.33	0.118	2.64
(0.57×10^6 J/kg)	240	0.1095	1090	1950	1.28	0.117	2.27
	250	0.1145	1045	2000	1.21	0.1155	2.11
	260	0.113	1000	2100	1.14	0.1135	2.11
	270	0.1075	945	2400	1.04	0.1105	2.33
	280	0.100	885	2850	0.925	0.1045	2.64
	290	0.090	805	4500	0.657	0.094	3.78
	300	0.076	670	11000	0.549	0.082	7.95
Engine oil, unused	280	0.147	895	1810	21900	2450	27000
(SAE 50)	290	0.146	889	1850	10900	1230	13900
	300	0.1445	883	1900	5030	570	6600
	310	0.1435	877	1950	2500	285	3400
	320	0.1425	871	1990	1370	157	1910
	330	0.1415	865	2030	796	92	1140
	340	0.1405	859	2070	515	60	760
	350	0.139	854	2120	350	41	530
	360	0.138	848	2160	255	30.1	400
	370	0.137	842	2200	189	22.5	300
	380	0.136	837	2250	147	17.6	245
	390	0.135	832	2290	112	13.5	191
	400	0.134	826	2330	88.4	10.7	154
	410	0.133	820	2380	71.3	8.7	128
	420	0.132	815	2420	57.9	7.1	106

(Continued)

Table A.8 ***(Continued)***

Saturated Liquid (Melting point) (Boiling point) (Latent heat at BP)	T K	k W/m K	ρ kg/m^3	c_p J/kg K	$\mu \times 10^4$ kg/m s	$\nu \times 10^6$ m^2/s	Pr
Refrigerant-12 (CCl_2F_2)	220	0.0675	1552	880	4.94	0.318	6.4
(115 K MP)	230	0.0680	1528	885	4.39	0.287	5.7
(243 K BP)	240	0.0695	1502	890	3.94	0.262	5.0
(0.165×10^6 J/kg)	250	0.0705	1473	905	3.55	0.241	4.6
	260	0.0715	1442	915	3.23	0.224	4.1
	270	0.0725	1407	930	3.04	0.216	3.9
	280	0.073	1370	945	2.85	0.208	3.7
	290	0.0725	1332	960	2.68	0.201	3.6
	300	0.071	1298	980	2.54	0.196	3.5
	310	0.0695	1263	995	2.44	0.193	3.5
	320	0.0675	1222	1015	2.32	0.190	3.5
Refrigerant-113	260	0.0830	1648.5	895	12.32	0.747	13.3
($Cl_2CFCClF_2$)	280	0.0787	1603.4	933	8.85	0.552	10.5
(236 K MP)	300	0.0747	1557.1	958	6.64	0.426	8.52
(320.7 K BP)	320	0.0707	1509.1	983	5.20	0.345	7.23
(0.144×10^6 J/kg)	320.7	0.0705	1507.3	984	5.16	0.342	7.20
	340	0.0664	1459.0	1000	4.19	0.287	6.31
	360	0.0624	1406.0	1029	3.44	0.245	5.67
	380	0.0583	1349.5	1059	2.89	0.214	5.25
	400	0.0543	1287.5	1109	2.46	0.191	5.02
	420	0.0498	1217.8	1176	2.10	0.172	4.96
	440	0.0448	1135.7	1268	1.75	0.154	4.95
	460	0.0386	1029.4	1381	1.33	0.129	4.76
Nitrogen	70	0.151	841	2025	2.17	0.258	2.91
(63.3 K MP)	77.4	0.137	809	2060	1.62	0.200	2.43
(77.4 K BP)	80	0.132	796	2070	1.48	0.186	2.32
(0.200×10^6 J/kg)	90	0.114	746	2130	1.10	0.147	2.05
	100	0.097	689	2310	0.87	0.126	2.07
	110	0.080	620	2710	0.71	0.115	2.42
	120	0.063	525	4350	0.48	0.091	3.30
Oxygen	60	0.19	1280	1660	5.89	0.46	5.1
(55 K MP)	70	0.17	1220	1666	3.78	0.31	3.7
(90 K BP)	80	0.16	1190	1679	2.50	0.21	2.6
(0.213×10^6 J/kg)	90	0.15	1140	1694	1.60	0.14	1.8
	100	0.14	1110	1717	1.22	0.11	1.50

(Continued)

Table A.8 ***(Concluded)***

Saturated Liquid (Melting point) (Boiling point) (Latent heat at BP)	T K	k W/m K	ρ kg/m^3	c_p J/kg K	$\mu \times 10^4$ kg/m s	$\nu \times 10^6$ m^2/s	Pr
Therminol 60[a]	230	0.132	1040	1380	6210	597	6490
(205 K MP)	250	0.131	1030	1460	686	66.6	765
(561 K 10% BP)	300	0.129	995	1640	63.8	6.41	81.1
	350	0.125	960	1820	21.5	2.24	31.3
	400	0.120	924	1990	10.8	1.17	17.9
	450	0.115	888	2160	6.62	0.745	12.4
	500	0.108	849	2320	4.59	0.541	9.86
	550	0.100	808	2470	3.47	0.429	8.57
Water	275	0.556	1000	4217	17.00	1.70	12.9
(273 K MP)	280	0.568	1000	4203	14.50	1.45	10.7
(373 K BP)	285	0.580	1000	4192	12.50	1.25	9.0
(2.26×10^6 J/kg)	290	0.591	999	4186	11.00	1.10	7.8
	295	0.602	998	4181	9.68	0.97	6.7
	300	0.611	996	4178	8.67	0.87	5.9
	310	0.628	993	4174	6.95	0.70	4.6
	320	0.641	989	4174	5.84	0.59	3.8
	330	0.652	985	4178	4.92	0.50	3.2
	340	0.661	980	4184	4.31	0.44	2.7
	350	0.669	973	4190	3.79	0.39	2.4
	360	0.676	967	4200	3.29	0.34	2.0
	370	0.680	960	4209	2.95	0.31	1.81
	373.15	0.681	958	4212	2.85	0.30	1.76
	380	0.683	953	4220	2.67	0.28	1.65
	390	0.684	945	4234	2.44	0.26	1.51
	400	0.685	937	4250	2.25	0.24	1.40
	420	0.684	919	4290	1.93	0.21	1.21
	440	0.679	899	4340	1.71	0.19	1.09
	460	0.670	879	4400	1.49	0.17	0.98
	480	0.657	857	4490	1.37	0.16	0.94
	500	0.638	837	4600	1.26	0.15	0.91
	520	0.607	820	4770	1.15	0.14	0.90
	540	0.577	806	5010	1.05	0.13	0.91
	560	0.547	796	5310	0.955	0.12	0.93
	580	0.516	787	5590	0.866	0.11	0.94

[a] Registered trademark of Monsanto Chemical Company, St. Louis; also sold under the brand name "Santotherm."

Table A.9 Liquid metals: Thermal properties

Liquid metal (Melting point) (Boiling point) (Latent heat at BP)	T K	k W/m K	ρ kg/m^3	c_p J/kg K	$\mu \times 10^4$ kg/m s	$\nu \times 10^6$ m^2/s	Pr
Lead	650	16.7	10530	158	23.9	0.227	0.023
(600 K MP)	700	17.5	10470	156	21.1	0.202	0.019
(2020 K BP)	800	19.0	10350	155	17.2	0.166	0.014
(0.850×10^6 J/kg)	900	20.4	10230	155	14.9	0.146	0.011
Lithium	500	43.7	514	4340	5.31	1.033	0.053
(453 K MP)	600	46.1	503	4230	4.26	0.847	0.039
(1613 K BP)	700	48.4	493	4190	3.58	0.726	0.031
(19.5×10^6 J/kg)	800	50.7	483	4170	3.10	0.642	0.025
	900	55.9	473	4160	2.47	0.522	0.018
Mercury	300	8.4	13530	140	14.9	0.110	0.025
(234 K MP)	400	9.8	13280	140	11.3	0.085	0.016
(630 K BP)	500	11.0	13040	140	9.78	0.075	0.012
(0.292×10^6 J/kg)	600	12.1	12780	140	8.31	0.065	0.010
Potassium	400	45.5	814	800	4.9	0.60	0.0086
(337 K MP)	500	43.6	790	790	2.8	0.35	0.0050
(1049 K BP)	600	41.6	765	780	2.1	0.28	0.0040
(2.02×10^6 J/kg)	700	39.5	741	770	1.9	0.25	0.0036
	800	36.8	717	750	1.6	0.23	0.0034
	900	34.4	692	740	1.5	0.21	0.0031
Sodium	500	79.2	900	1335	4.2	0.47	0.0071
(371 K MP)	600	74.7	868	1310	3.1	0.36	0.0055
(1156 K BP)	700	70.1	840	1280	2.5	0.30	0.0046
(3.86×10^6 J/kg)	800	65.7	813	1260	2.2	0.27	0.0042
	900	62.1	792	1255	2.0	0.25	0.0040
	1000	59.3	772	1255	1.8	0.23	0.0038
	1100	56.7	753	1255	1.6	0.21	0.0035

Table A.10*a* Volume expansion coefficients for liquids

Liquid	T K	$\beta \times 10^3$ 1/K	Liquid	T K	$\beta \times 10^3$ 1/K
Ammonia	293	2.45	Hydrogen	20.3	15.1
Engine oil	273	0.70	Mercury	273	0.18
(SAE 50)	430	0.70		550	0.18
Ethylene glycol	273	0.65	Nitrogen	70	4.9
$C_2H_4(OH)_2$	373	0.65		77.4	5.7
Refrigerant-12	240	1.85		80	5.9
	260	2.10		90	7.2
	280	2.35		100	9.0
	300	2.75		110	12
	320	3.5		120	24
Refrigerant-113	260	1.3	Oxygen	89	2.0
	280	1.4	Sodium	366	0.27
	300	1.5	Therminol 60	230	0.79
	320	1.7		250	0.75
	340	1.8		300	0.70
	360	2.0		350	0.70
	380	2.2		400	0.76
	400	2.5		450	0.84
	420	3.1		500	0.96
	440	4.0		550	1.1
	460	6.2			
Glycerin	280	0.47			
$C_3H_5(OH)_3$	300	0.48			
	320	0.50			

Table A.10*b* Density and volume expansion coefficient of water

T K	ρ kg/m^3	$\beta \times 10^6$ 1/K
273.15	999.8679	−68.05
274.00	999.9190	−51.30
275.00	999.9628	−32.74
276.00	999.9896	−15.30
277.00	999.9999	1.16
278.00	999.9941	16.78
279.00	999.9727	31.69
280.00	999.9362	46.04
285.00	999.5417	114.1
290.00	998.8281	174.0
295.00	997.8332	227.5
300.00	996.5833	276.1
310.00	993.4103	361.9
320.00	989.12	436.7
330.00	984.25	504.0
340.00	979.43	566.0
350.00	973.71	624.4
360.00	967.12	697.9
370.00	960.61	728.7
373.15	957.85	750.1
380.00	953.29	788
390.00	945.17	841
400.00	937.21	896
450.00	890.47	1129
500.00	831.26	1432

Table A.11 Surface tension

Fluid	T K	$\sigma \times 10^3$ N/m	Fluid	T K	$\sigma \times 10^3$ N/m
Water	275	75.3	Oxygen *(continued)*	90	13.5
	280	74.8		100	11.1
	290	73.7	Potassium	400	110
	300	71.7		500	105
	310	70.0		600	97
	320	68.3		700	90
	330	66.6		800	83
	340	64.9		900	76
	350	63.2	Sodium	500	175
	360	61.4		700	160
	370	59.5		900	140
	373.15	58.9		1100	120
	380	57.6	Nitrogen	68	11.00
	390	55.6		70	10.53
	400	53.6		72	10.07
	420	49.4		74	9.62
	440	45.1		76	9.16
	460	40.7		77.4	8.85
	480	36.2		78	8.72
	500	31.6		80	8.27
	550	19.7		82	7.84
	600	8.4		84	7.42
	647.30	0.0		86	6.99
Ammonia	220	39		88	6.57
	240	34		90	6.16
	260	30	Refrigerant-12	180	25.6
	280	25		190	24.0
	300	20		200	22.4
	320	16		210	20.9
Carbon dioxide	248	9.1		220	19.4
	293	1.2		230	17.9
Hydrogen	15	2.8		240	16.5
	20	2.0		250	15.0
	25	1.1		260	13.6
Lead	600	470		270	12.3
	700	452		280	10.9
	800	437		290	9.66
	900	421		300	8.39
Lithium	500	390		310	7.15
	700	360		320	5.97
	900	335	Refrigerant-113	270	20.4
Mercury	300	470		280	19.2
	400	450		290	18.1
	500	430		300	17.0
	600	400		310	15.9
	700	380		320	14.7
Oxygen	60	20.7		330	13.7
	70	18.3		340	12.6
	80	16.0			

Table A.12*a* Thermodynamic properties of saturated steam

T K	$P \times 10^{-5}$ Pa	v m^3/kg	ρ kg/m^3	$h_{fg} \times 10^{-6}$ J/kg
273.15	0.00610	206.4	0.00484	2.501
274.00	0.00649	194.6	0.00514	2.499
275.00	0.00698	181.8	0.00550	2.496
276.00	0.00750	169.8	0.00589	2.494
277.00	0.00805	158.8	0.00630	2.492
278.00	0.00863	148.6	0.00673	2.490
279.00	0.00925	139.1	0.00719	2.488
280.00	0.00991	130.4	0.00767	2.486
281.00	0.01061	122.2	0.00818	2.484
282.00	0.01136	114.6	0.00873	2.482
283.00	0.01215	107.6	0.00929	2.479
284.00	0.01299	101.0	0.00990	2.476
285.00	0.01388	94.75	0.01055	2.473
286.00	0.01482	89.06	0.01123	2.471
287.00	0.01582	83.73	0.01194	2.468
288.00	0.01688	78.75	0.01270	2.466
289.00	0.01800	74.09	0.01350	2.463
290.00	0.01918	69.74	0.01434	2.461
291.00	0.02043	65.68	0.01523	2.459
292.00	0.02176	61.89	0.01616	2.456
293.00	0.02315	58.35	0.01714	2.454
294.00	0.02463	55.05	0.01817	2.451
295.00	0.02619	51.96	0.01925	2.449
296.00	0.02783	49.07	0.02038	2.447
297.00	0.02957	46.37	0.02157	2.444
298.00	0.03139	43.82	0.02282	2.442
299.00	0.03331	41.42	0.02414	2.439
300.00	0.03533	39.15	0.02554	2.437
301.00	0.03746	37.05	0.02700	2.434
302.00	0.03971	35.07	0.02851	2.432
303.00	0.04206	33.21	0.03011	2.430
304.00	0.04454	31.46	0.03179	2.427
305.00	0.04714	29.81	0.03355	2.425
306.00	0.04987	28.26	0.03539	2.423
307.00	0.05274	26.81	0.03730	2.421
308.00	0.05576	25.44	0.03931	2.418
309.00	0.05892	24.16	0.04139	2.416
310.00	0.06224	22.95	0.04357	2.414
311.00	0.06572	21.81	0.04585	2.412
312.00	0.06936	20.73	0.04824	2.409

(Continued)

Table A.12*a* (Continued)

T K	$P \times 10^{-5}$ Pa	v m^3/kg	ρ kg/m^3	$h_{fg} \times 10^{-6}$ J/kg
313.00	0.07318	19.72	0.05071	2.407
314.00	0.07717	18.75	0.05333	2.404
315.00	0.08135	17.83	0.05609	2.401
316.00	0.08573	16.97	0.05893	2.399
317.00	0.09031	16.16	0.06188	2.396
318.00	0.09511	15.39	0.06498	2.394
319.00	0.10012	14.66	0.06821	2.391
320.00	0.10535	13.98	0.07153	2.389
321.00	0.11082	13.33	0.07502	2.387
322.00	0.11652	12.72	0.07862	2.384
323.00	0.12247	12.14	0.08237	2.382
324.00	0.12868	11.59	0.08628	2.379
325.00	0.13514	11.06	0.09042	2.377
326.00	0.14191	10.56	0.09470	2.375
327.00	0.14896	10.09	0.09911	2.372
328.00	0.15630	9.644	0.1037	2.370
329.00	0.16395	9.219	0.1085	2.367
330.00	0.17192	8.817	0.1134	2.365
331.00	0.18021	8.434	0.1186	2.363
332.00	0.18885	8.072	0.1239	2.360
333.00	0.19783	7.727	0.1294	2.358
334.00	0.20718	7.400	0.1351	2.355
335.00	0.2169	7.090	0.1410	2.353
336.00	0.2270	6.794	0.1472	2.351
337.00	0.2375	6.512	0.1536	2.348
338.00	0.2484	6.244	0.1602	2.346
339.00	0.2597	5.987	0.1670	2.343
340.00	0.2715	5.741	0.1742	2.341
341.00	0.2837	5.509	0.1815	2.339
342.00	0.2964	5.288	0.1891	2.336
343.00	0.3096	5.077	0.1970	2.334
344.00	0.3233	4.876	0.2051	2.332
345.00	0.3375	4.684	0.2135	2.329
346.00	0.3521	4.500	0.2222	2.326
347.00	0.3673	4.325	0.2312	2.324
348.00	0.3831	4.158	0.2405	2.321
349.00	0.3994	3.999	0.2501	2.319
350.00	0.4164	3.847	0.2599	2.316
351.00	0.4339	3.701	0.2702	2.313
352.00	0.4520	3.562	0.2807	2.311

(Continued)

Table A.12a (Concluded)

T K	$P \times 10^{-5}$ Pa	v m^3/kg	ρ kg/m^3	$h_{fg} \times 10^{-6}$ J/kg
353.00	0.4708	3.429	0.2916	2.308
354.00	0.4902	3.301	0.3029	2.306
355.00	0.5103	3.179	0.3146	2.303
356.00	0.5310	3.062	0.3266	2.301
357.00	0.5525	2.951	0.3389	2.299
358.00	0.5747	2.844	0.3516	2.296
359.00	0.5976	2.742	0.3647	2.294
360.00	0.6213	2.644	0.3782	2.291
361.00	0.6457	2.550	0.3922	2.288
362.00	0.6710	2.460	0.4065	2.285
363.00	0.6970	2.373	0.4214	2.283
364.00	0.7240	2.291	0.4365	2.280
365.00	0.7518	2.212	0.4521	2.277
366.00	0.7804	2.136	0.4682	2.274
367.00	0.8100	2.063	0.4847	2.272
368.00	0.8405	1.993	0.5018	2.269
369.00	0.8719	1.925	0.5195	2.267
370.00	0.9044	1.861	0.5373	2.265
371.00	0.9377	1.798	0.5562	2.263
372.00	0.9722	1.738	0.5754	2.260
373.00	1.0076	1.681	0.5949	2.257
373.15	1.0133	1.673	0.5977	2.257
380.00	1.2875	1.337	0.7479	2.238
390.00	1.7952	0.9800	1.020	2.211
400.00	2.4563	0.7308	1.368	2.183
410.00	3.303	0.5535	1.807	2.154
420.00	4.371	0.4254	2.351	2.124
430.00	5.701	0.3311	3.020	2.093
440.00	7.335	0.2609	3.833	2.059
450.00	9.322	0.2082	4.803	2.025
460.00	11.708	0.1671	5.984	1.990
470.00	14.551	0.1353	7.391	1.953
480.00	17.908	0.1109	9.017	1.914
490.00	21.839	0.09172	10.90	1.872
500.00	26.401	0.07573	13.20	1.827
510.00	31.676	0.06374	15.69	1.779
520.00	37.726	0.05427	18.43	1.729
530.00	44.618	0.04639	21.56	1.676
540.00	52.420	0.03919	25.52	1.621
550.00	61.200	0.03175	31.50	1.563

Table A.12*b* Thermodynamic properties of saturated ammonia

T K	P kPa	v m^3/kg	$h_{fg} \times 10^{-6}$ J/kg
224	42.98	2.5055	1.414
226	48.27	2.2479	1.409
228	54.09	2.0212	1.404
230	60.48	1.8213	1.398
232	67.46	1.6445	1.392
234	75.10	1.4878	1.387
236	83.42	1.3487	1.381
238	92.50	1.2248	1.375
240	102.29	1.1143	1.369
242	112.96	1.0155	1.363
244	124.52	0.9271	1.358
246	137.05	0.8478	1.351
248	150.58	0.7765	1.345
250	165.07	0.7123	1.339
252	180.68	0.6544	1.333
254	197.50	0.6021	1.327
256	215.47	0.5548	1.320
258	234.80	0.5118	1.314
260	255.41	0.4728	1.307
262	277.46	0.4374	1.301
264	300.98	0.4051	1.294
266	326.04	0.3757	1.287
268	352.80	0.3487	1.280
270	381.12	0.3241	1.273
272	411.23	0.3016	1.267
274	443.24	0.2809	1.259
276	477.14	0.2619	1.252
278	512.97	0.2445	1.245
280	550.86	0.2285	1.238
282	590.87	0.2136	1.230
284	633.16	0.2000	1.223
286	677.70	0.1873	1.215
288	724.66	0.1756	1.207
290	774.15	0.1648	1.199
292	826.05	0.1548	1.191
294	880.67	0.1455	1.183
296	938.00	0.1369	1.175
298	998.03	0.1289	1.167
300	1061.35	0.1214	1.159
302	1127.39	0.1144	1.150

(Continued)

Table A.12*b* (Concluded)

T K	P kPa	v m^3/kg	$h_{fg} \times 10^{-6}$ J/kg
304	1195.88	0.1079	1.141
306	1268.41	0.1019	1.133
308	1344.02	0.0962	1.124
310	1423.12	0.0909	1.115
312	1505.21	0.0860	1.106
314	1590.85	0.0814	1.097
316	1680.86	0.0770	1.087
318	1773.70	0.0729	1.078
320	1871.02	0.0690	1.068

Table A.12*c* Thermodynamic properties of saturated nitrogen

T K	P kPa	v m^3/kg	$h_{fg} \times 10^{-6}$ J/kg
77.4	101.3	0.2209	0.1995
80	137.8	0.1656	0.1962
85	228.0	0.1028	0.1892
90	358.7	0.06681	0.1813
95	538.1	0.04486	0.1726
100	777.2	0.03123	0.1623
105	1085	0.02227	0.1509
110	1473	0.01613	0.1373
115	1954	0.01156	0.1201
120	2537	0.008101	0.0962
125	3236	0.004889	0.0495
126	3392	0.003216	0

Table A.12*d* Thermodynamic properties of saturated mercury

T K	P MPa	v m^3/kg	$h_{fg} \times 10^{-6}$ J/kg
385	0.00007	229.1	0.2971
390	0.00009	182.0	0.2971
395	0.00011	146.0	0.2970
400	0.00014	117.6	0.2969
405	0.00018	94.76	0.2968
410	0.00022	77.00	0.2967
415	0.00027	63.15	0.2966
420	0.00033	51.84	0.2965
425	0.00041	42.72	0.2964
430	0.00049	35.46	0.2963
435	0.00060	29.51	0.2962
440	0.00073	24.68	0.2961
445	0.00088	20.74	0.2960
450	0.00105	17.53	0.2959
460	0.00150	12.68	0.2957
470	0.00208	9.223	0.2955
480	0.00287	6.863	0.2954
490	0.00390	5.135	0.2952
500	0.00524	3.902	0.2950
520	0.00913	2.331	0.2946
540	0.01527	1.459	0.2942
560	0.02460	0.9370	0.2938
580	0.03815	0.6210	0.2935
600	0.05749	0.4276	0.2931
620	0.08450	0.3010	0.2927
640	0.12117	0.2180	0.2923
660	0.16969	0.1602	0.2920
680	0.23281	0.1201	0.2916
700	0.31357	0.09179	0.2912
720	0.41544	0.07133	0.2908
740	0.54145	0.05629	0.2905
760	0.69616	0.04501	0.2901
780	0.88308	0.03645	0.2897
800	1.10570	0.02989	0.2893
820	1.36881	0.02477	0.2890
840	1.67803	0.02072	0.2886
860	2.03709	0.01750	0.2882
880	2.44991	0.01491	0.2878
900	2.92017	0.01282	0.2874
920	3.45295	0.01109	0.2871
940	4.05586	0.00967	0.2867
960	4.72617	0.00849	0.2863
980	5.47829	0.00749	0.2859
1000	6.30516	0.00665	0.2856

Table A.12*e* Thermodynamic properties of saturated refrigerant-12 (dichlorodifluoromethane)

T K	P MPa	v m^3/kg	$h_{fg} \times 10^{-6}$ J/kg
185	0.0032	3.8378	0.1889
190	0.0049	2.6628	0.1869
195	0.0071	1.8891	0.1849
200	0.0100	1.3682	0.1829
205	0.0138	1.0099	0.1809
210	0.0188	0.7582	0.1789
215	0.0252	0.5782	0.1769
220	0.0331	0.4473	0.1748
225	0.0430	0.3507	0.1729
230	0.0552	0.2782	0.1707
235	0.0700	0.2233	0.1687
240	0.0876	0.1810	0.1666
245	0.1086	0.1482	0.1644
250	0.1333	0.1222	0.1622
255	0.1621	0.1017	0.1599
260	0.1955	0.08534	0.1577
265	0.2340	0.07204	0.1553
270	0.2779	0.06120	0.1529
275	0.3278	0.05229	0.1505
280	0.3842	0.04491	0.1479
285	0.4476	0.03876	0.1452
290	0.5185	0.03360	0.1425
295	0.5975	0.02925	0.1397
300	0.6851	0.02555	0.1368
305	0.7818	0.02239	0.1338
310	0.8883	0.01968	0.1306
315	1.0051	0.01734	0.1273
320	1.1329	0.01531	0.1237
325	1.2723	0.01354	0.1200
330	1.4239	0.01199	0.1161
335	1.5883	0.01062	0.1118
340	1.7664	0.009417	0.1073
345	1.9586	0.008342	0.1024
350	2.1659	0.007378	0.0970
355	2.3890	0.006510	0.0911
360	2.6286	0.005721	0.0845
365	2.8857	0.004994	0.0770
370	3.1609	0.004311	0.0680
375	3.4552	0.003582	0.0529

Table A.12*f* Thermodynamic properties of saturated refrigerant-113 (trichlorotrifluoroethane)

T K	P MPa	v m^3/kg	$h_{fg} \times 10^{-6}$ J/kg
250	0.004350	2.536	0.1683
260	0.007676	1.492	0.1613
270	0.012896	0.9189	0.1588
280	0.020759	0.5893	0.1561
290	0.032165	0.3917	0.1533
300	0.048173	0.2686	0.1503
310	0.069991	0.1895	0.1473
320	0.098962	0.1369	0.1440
320.71	0.10133	0.1339	0.1438
330	0.13657	0.1011	0.1407
340	0.18437	0.07618	0.1372
350	0.24405	0.05835	0.1335
360	0.31737	0.04536	0.1297
380	0.51233	0.02844	0.1216
400	0.78490	0.01852	0.1125
420	1.1524	0.01237	0.1021
440	1.6345	0.008328	0.08963
460	2.2559	0.005185	0.07306

Table A.13*a* Aqueous ethylene glycol solutions: Thermal properties

	Percent Glycol, by Mass (Freezing Point)				
T K	20 (263.7 K)	30 (257.6 K)	40 (248.7 K)	50 (237.6 K)	60 (213.2 K)
Thermal Conductivity, k [W/m K]					
250	—	—	0.456	0.425	0.400
260	—	0.488	0.456	0.423	0.397
270	0.513	0.488	0.456	0.422	0.394
280	0.519	0.491	0.456	0.419	0.393
290	0.522	0.493	0.456	0.418	0.389
300	0.525	0.494	0.456	0.418	0.388
310	0.531	0.495	0.456	0.417	0.385
320	0.538	0.497	0.456	0.416	0.381
Density, ρ [kg/m^3]					
250	—	—	1069.0	1084.0	1098.5
260	—	1051.6	1066.0	1080.5	1094.0
270	1034.1	1048.1	1061.5	1076.0	1089.0
280	1031.1	1044.1	1057.0	1071.0	1085.0
290	1026.6	1039.1	1052.1	1066.0	1079.0
300	1022.6	1035.6	1047.1	1060.5	1073.0
310	1018.6	1031.1	1043.1	1054.1	1066.5
320	1014.1	1025.6	1037.1	1048.1	1060.0
Specific Heat, c_p [J/kg K]					
250	—	—	—	3480	3260
260	—	—	3710	3490	3300
270	4050	3930	3720	3510	3330
280	4030	3920	3740	3550	3360
290	4020	3930	3750	3570	3400
300	4010	3940	3760	3600	3430
310	4010	3950	3780	3620	3460
320	4020	3960	3800	3650	3500
Dynamic Viscosity, μ [kg/m s $\times 10^3$]					
250	—	—	—	29.0	45.0
260	—	—	11.2	16.2	20.0
270	3.6	5.0	7.0	9.5	14.1
280	2.4	3.4	4.9	6.2	8.8
290	1.8	2.3	3.1	4.1	5.6
300	1.4	1.8	2.3	3.0	4.0
310	1.1	1.4	1.8	2.3	3.0
320	0.9	1.1	1.5	1.8	2.4

Table A.13b Aqueous sodium chloride solutions: Thermal properties

T	Percent NaCl, by Mass				
K	5	10	15	20	25
Freezing Point [K]					
	270.2	266.6	262.2	256.7	264.3
Thermal Conductivity, k [W/m K]					
260	—	—	—	0.434	—
270	—	0.504	0.478	0.449	0.420
280	0.542	0.516	0.490	0.464	0.438
290	0.560	0.534	0.508	0.481	0.454
300	0.575	0.550	0.525	0.498	0.470
Density, ρ [kg/m^3]					
260	—	—	—	1160	—
270	—	1076	1115	1156	1189
280	1035	1073	1111	1153	1184
290	1033	1070	1108	1150	1180
300	1032	1069	1107	1149	1178
Specific Heat, c_p [J/kg K]					
260	—	—	—	3365	—
270	—	3680	3515	3380	3280
280	3920	3705	3535	3395	3290
290	3930	3720	3555	3415	3300
300	3940	3730	3575	3425	3310
Dynamic Viscosity, μ [kg/m s$\times 10^3$]					
260	—	—	—	4.70	—
270	—	2.30	2.60	3.05	3.83
280	1.50	1.65	1.95	2.25	2.60
290	1.17	1.30	1.50	1.75	2.03
300	0.90	1.00	1.18	1.38	1.65

Table A.14*a* Dimensions of commercial pipes [mm] (ASA standard)

Nominal Pipe Size (≃ I.D., in)		Schedule 5	10	40	80	160	XX Strong
$\frac{1}{4}$	O.D.	13.716	13.716	13.716	13.716		
	Wall	1.245	1.651	2.235	3.023		
	I.D.	11.227	10.414	9.246	7.671		
$\frac{3}{8}$	O.D.	17.145	17.145	17.145	17.145		
	Wall	1.245	1.651	2.311	3.200		
	I.D.	14.656	13.843	12.522	10.744		
$\frac{1}{2}$	O.D.	21.336	21.336	21.336	21.336	21.336	21.336
	Wall	1.651	2.108	2.769	3.734	4.750	7.468
	I.D.	18.034	17.120	15.799	13.868	11.836	6.401
$\frac{3}{4}$	O.D.	26.670	26.670	26.670	26.670	26.670	26.670
	Wall	1.651	2.108	2.870	3.912	5.534	7.823
	I.D.	23.368	22.454	20.930	18.847	15.596	11.024
1	O.D.	33.401	33.401	33.401	33.401	33.401	33.401
	Wall	1.651	2.769	3.378	4.547	6.350	9.093
	I.D.	30.099	27.864	26.645	24.308	20.701	15.215
$1\frac{1}{2}$	O.D.	48.260	48.260	48.260	48.260	48.260	48.260
	Wall	1.651	2.769	3.683	5.080	7.137	10.160
	I.D.	44.958	42.723	40.894	38.100	33.985	27.940
2	O.D.	60.325	60.325	60.325	60.325	60.325	60.325
	Wall	1.651	2.769	3.912	5.537	8.712	11.074
	I.D.	57.023	54.788	52.502	49.251	42.901	38.176
3	O.D.			88.900	88.900	88.900	88.900
	Wall			5.486	7.62	11.125	15.240
	I.D.			77.927	73.660	66.650	58.420
4	O.D.			114.300	114.300	114.300	114.300
	Wall			6.020	8.560	13.487	17.120
	I.D.			102.260	97.180	87.325	80.061
5	O.D.			141.300	141.300	141.300	141.300
	Wall			6.553	9.525	15.875	19.050
	I.D.			128.194	122.250	109.550	103.200
6	O.D.			168.275	168.275	168.275	168.275
	Wall			7.150	10.973	18.237	21.946
	I.D.			153.975	146.329	131.801	124.384
8	O.D.			219.075	219.075		
	Wall			8.179	12.700		
	I.D.			202.717	193.675		
10	O.D.			273.050	273.050		
	Wall			9.271	15.062		
	I.D.			254.508	242.926		
12	O.D.			323.850	323.850		
	Wall			10.312	17.450		
	I.D.			303.255	288.950		

Table A.14*b* Dimensions of commercial tubes [mm] (ASTM standard)

Nominal Size (≃O.D., in)	O.D.	Gage (BWG)	Wall	I.D.
$\frac{3}{16}$	4.775	20	0.889	2.997
$\frac{1}{4}$	6.350	22	0.711	4.928
		20	0.889	4.572
		18	1.245	3.861
$\frac{5}{16}$	7.950	20	0.889	6.172
		18	1.245	5.461
$\frac{3}{8}$	9.525	20	0.889	7.747
		18	1.245	7.036
		16	1.651	6.223
$\frac{1}{2}$	12.700	20	0.889	10.922
		18	1.245	10.211
		16	1.651	9.398
		14	2.108	8.484
$\frac{5}{8}$	15.875	18	1.245	13.386
		16	1.651	12.573
		14	2.108	11.659
$\frac{3}{4}$	19.050	20	0.889	17.272
		18	1.245	16.561
		16	1.651	15.748
		14	2.108	14.834
$\frac{7}{8}$	22.225	16	1.651	18.923
1	25.400	20	0.889	23.622
		18	1.245	22.911
		16	1.651	22.098
		14	2.108	21.184
		12	2.769	19.863
		10	3.404	18.593
$1\frac{1}{4}$	31.750	18	1.245	29.261
		16	1.651	28.448
		14	2.108	27.534
		12	2.769	26.213
		10	3.404	24.943
$1\frac{1}{2}$	38.100	18	1.245	35.611
		16	1.651	34.798
		14	2.108	33.884
		12	2.769	32.563
		10	3.404	31.293
		8	4.191	29.718
2	50.800	18	1.245	48.311
		16	1.651	47.498
		14	2.108	46.584
		12	2.769	45.263
		10	3.404	43.993
		8	4.191	42.418

Table A.14*c* Dimensions of seamless steel tubes for tubular heat exchangers [mm] (DIN 28 180)

Outside diameter	Wall thickness 1.2	1.6	2.0	2.6	3.2
	Unalloyed and alloy steel tubes				
16	x	x	x		
20		x	x	x	
25		x	x	x	x
30		x	x	x	x
38			x	x	x
	Austenitic stainless steel tubes				
16	x	x	x		
20	x	x	x	x	
25		x	x	x	x
30		x	x	x	x
38		x	x	x	x

Table A.14*d* Dimensions of wrought copper and copper alloy tubes for condensers and heat exchangers [mm] (DIN 1785-83)

Outside diameter	Wall thickness 0.75	1.0	1.25	1.5	2.0
8	x	x	x		
10	x	x	x		
11	x	x	x		
12	x	x	x		
14	x	x	x		
15	x	x	x		
16	x	x	x	x	
18		x	x	x	
19		x	x	x	x
20		x	x	x	x
22		x	x	x	x
23		x	x	x	x
24		x	x	x	x
25		x	x	x	x
28		x	x	x	x
30		x	x	x	x
32		x	x	x	x
35		x	x	x	x
					x

Table A.14*e* Dimensions of seamless cold drawn stainless steel tubes [mm] (LN 9398)[a]

	Wall thickness								
Outside diameter	0.5	0.6	0.8	1.0	1.2	1.6	2.0	2.5	3.2
5	x		x						
6	x		x	x					
8	x		x	x	x				
10	x		x		x	x			
12			x		x		x		
14				x					
16			x	x	x		x	x	
18		x				x			
20			x	x	x	x		x	x
22			x						
25				x	x	x		x	
28				x					
32			x		x		x		
36									
40			x			x			
45			x			x	x		
50						x	x		

[a] Sizes in conformance with ISO 2964 R20.

Table A.14*f* Dimensions of seamless drawn wrought aluminum alloy tubes [mm] (LN 9223)[a]

	Wall thickness					
Outside diameter	0.8	1.0	1.2	1.6	2.0	2.5
12	x					
14						
16	x					
18						
20	x					
22						
25	x					
28						
32	x	x	x			
36						
40	x					
45		x	x			
50		x	x	x		
56						
63		x	x		x	
70						
80				x		x

[a] Sizes in conformance with ISO 2964, R20.

Table A.15 U.S. standard atmosphere

Altitude m	Temperature K	Pressure Pa	Density Ratio ρ/ρ_0	Gravity m/s^2	Mean Free Path m	Molecular Weight
0	288.150	1.01325+5	1.0000+0	9.8066	6.6328−8	28.964
200	286.850	9.89454+4	9.8094−1	9.8060	6.7617−8	28.964
400	285.550	9.66114+4	9.6216−1	9.8054	6.8936−8	28.964
600	284.250	9.43223+4	9.4366−1	9.8048	7.0288−8	28.964
800	282.951	9.20775+4	9.2543−1	9.8042	7.1672−8	28.964
1000	281.651	8.98762+4	9.0748−1	9.8036	7.3090−8	28.964
2000	275.154	7.95014+4	8.2168−1	9.8005	8.0723−8	28.964
3000	268.659	7.01211+4	7.4225−1	9.7974	8.9361−8	28.964
4000	262.166	6.16604+4	6.6885−1	9.7943	9.9166−8	28.964
5000	255.676	5.40482+4	6.0117−1	9.7912	1.1033−7	28.964
6000	249.187	4.72176+4	5.3887−1	9.7882	1.2309−7	28.964
7000	242.700	4.11052+4	4.8165−1	9.7851	1.3771−7	28.964
8000	236.215	3.56516+4	4.2921−1	9.7820	1.5453−7	28.964
9000	229.733	3.08007+4	3.8128−1	9.7789	1.7396−7	28.964
10,000	223.252	2.64999+4	3.3756−1	9.7759	1.9649−7	28.964
20,000	216.650	5.52930+3	7.2579−2	9.7452	9.1387−7	28.964
30,000	226.509	1.19703+3	1.5029−2	9.7147	4.4134−6	28.964
40,000	250.350	2.87143+2	3.2618−3	9.6844	2.0335−5	28.964
47,400	270.650	1.10220+2	1.1581−3	9.6620	5.7272−5	28.964
50,000	270.650	7.97790+1	8.3827−4	9.6542	7.9125−5	28.964
52,000	270.650	6.22283+1	6.5386−4	9.6481	1.0144−4	28.964
60,000	255.722	2.24606+1	2.4973−4	9.6241	2.6560−4	28.964
70,000	219.700	5.52047+0	7.1457−5	9.5941	9.2821−4	28.964
80,000	180.65	1.0366+0	1.632−5	9.564	4.065−3	28.964
90,000	180.65	1.6438−1	2.588−6	9.535	2.563−2	28.96
100,000	210.02	3.0075−2	4.060−7	9.505	1.629−1	28.88
105,000	233.90	1.4318−2	1.728−7	9.490	3.810−1	28.75
110,000	257.00	7.3544−3	8.024−8	9.476	8.150−1	28.56
115,000	303.78	4.1224−3	3.774−8	9.461	1.719+0	28.32
120,000	349.49	2.5217−3	1.988−8	9.447	3.233+0	28.07
125,000	442.35	1.6863−3	1.041−8	9.432	6.118+0	27.81
130,000	533.80	1.2214−3	6.195−9	9.417	1.019+1	27.58
135,000	624.30	9.3330−4	4.017−9	9.403	1.560+1	27.37
140,000	714.22	7.4104−4	2.770−9	9.388	2.248+1	27.20
145,000	803.74	6.0560−4	2.001−9	9.374	3.095+1	27.05
150,000	892.79	5.0617−4	1.498−9	9.360	4.114+1	26.92
155,000	957.94	4.2992−4	1.181−9	9.345	5.197+1	26.79
160,000	1022.23	3.6943−4	1.159−9	9.331	6.454+1	26.66

Table A.16 Selected physical constants

Universal gas constant

$$\mathscr{R} = 8314.3 \text{ J/kmol K}$$

$$= 1.987 \text{ kcal/kmol K (thermochemical calorie)}$$

Standard gravitational acceleration

$$g = 9.80665 \text{ m/s}^2$$

Atomic mass unit

$$u = 1.66057 \times 10^{-27} \text{ kg}$$

Avogadro's number

$$\mathscr{A} = 6.02252 \times 10^{26} \text{ molecules/kmol}$$

Boltzmann constant

$$k = 1.38054 \times 10^{-23} \text{ J/K molecule}$$

Planck's constant

$$h = 6.6256 \times 10^{-34} \text{ J s}$$

Speed of light *in vacuo*

$$c_0 = 2.997925 \times 10^8 \text{ m/s}$$

Stefan-Boltzmann constant

$$\sigma = 2\pi^5 k^4/15 h^3 c_0^2 = 5.670 \times 10^{-8} \text{ W/m}^2 \text{ K}^4$$

Elementary electric charge

$$e = 1.60 \times 10^{-19} \text{ A s}$$

Table A.17*a* Diffusion coefficients in air at 1 atm (1.013×10^5 Pa)[a]

	Binary Diffusion Coefficient [$m^2/s \times 10^4$]							
T [K]	O_2	CO_2	CO	C_7H_{16}	H_2	NO	SO_2	He
200	0.095	0.074	0.098	0.036	0.375	0.088	0.058	0.363
300	0.188	0.157	0.202	0.075	0.777	0.180	0.126	0.713
400	0.325	0.263	0.332	0.128	1.25	0.303	0.214	1.14
500	0.475	0.385	0.485	0.194	1.71	0.443	0.326	1.66
600	0.646	0.537	0.659	0.270	2.44	0.603	0.440	2.26
700	0.838	0.684	0.854	0.354	3.17	0.782	0.576	2.91
800	1.05	0.857	1.06	0.442	3.93	0.978	0.724	3.64
900	1.26	1.05	1.28	0.538	4.77	1.18	0.887	4.42
1000	1.52	1.24	1.54	0.641	5.69	1.41	1.06	5.26
1200	2.06	1.69	2.09	0.881	7.77	1.92	1.44	7.12
1400	2.66	2.17	2.70	1.13	9.90	2.45	1.87	9.20
1600	3.32	2.75	3.37	1.41	12.5	3.04	2.34	11.5
1800	4.03	3.28	4.10	1.72	15.2	3.70	2.85	13.9
2000	4.80	3.94	4.87	2.06	18.0	4.48	3.36	16.6

[a] Owing to the practical importance of water vapor–air mixtures, engineers have used convenient empirical formulas for $\mathscr{D}_{H_2O\ air}$. A formula that has been widely used for many years, and is used in this text is

$$\mathscr{D}_{H_2O\ air} = 1.97 \times 10^{-5}\left(\frac{P_0}{P}\right)\left(\frac{T}{T_0}\right)^{1.685}\ m^2/s; \qquad 273\ K < T < 373\ K$$

where $P_0 = 1$ atm; $T_0 = 256$ K. More recently the following formula has found increasing use [Marrero, T. R., and Mason, E. A., "Gaseous diffusion coefficients," *J. Phys. Chem. Ref. Data,* 1, 3–118 (1972)]:

$$\mathscr{D}_{H_2O\ air} = 1.87 \times 10^{-10}\frac{T^{2.072}}{P}; \qquad 280\ K < T < 450\ K$$

$$= 2.75 \times 10^{-9}\frac{T^{1.632}}{P}; \qquad 450\ K < T < 1070\ K$$

for P in atmospheres and T in kelvins. Over the temperature range 290–330 K, the discrepancy between the two formulas is less than 2.5%. For small concentrations of water vapor in air, the older formula gives a constant value of $Sc_{H_2O\ air} = 0.61$ over the temperature range 273–373 K. On the other hand, the Marrero and Mason formula gives values of $Sc_{H_2O\ air}$ that vary from 0.63 at 280 K to 0.57 at 373 K.

Table A.17*b* Schmidt number for vapors in dilute mixture in air at normal temperature, enthalpy of vaporization, and boiling point at 1 atm[a]

Vapor	Chemical Formula	Sc[b]	h_{fg} J/kg × 10^{-6}	T_{BP} K
Acetone	CH_3COCH_3	1.42	0.527	329
Ammonia	NH_3	0.61	1.370	240
Benzene	C_6H_6	1.79	0.395	354
Carbon dioxide	CO_2	1.00	0.398	194
Carbon monoxide	CO	0.77	0.217	81
Chlorine	Cl_2	1.42	0.288	238
Ethanol	CH_3CH_2OH	1.32	0.854	352
Helium	He	0.22		4.3
Heptane	C_7H_{16}	2.0	0.340	372
Hydrogen	H_2	0.20	0.454	20.3
Hydrogen sulfide	H_2S	0.94	0.548	213
Methanol	CH_3OH	0.98	1.100	338
Naphthalene[c]	$C_{10}H_8$	2.35		491
Nitric oxide	NO	0.87	0.465	121
Octane	C_8H_{18}	2.66	0.303	399
Oxygen	O_2	0.83	0.214	90.6
Pentane	C_5H_{12}	1.49	0.357	309
Sulfur dioxide	SO_2	1.24	0.398	263
Water vapor	H_2O	0.61	2.257	373

[a]With the Clausius-Clapeyron relation, one may estimate vapor pressure as

$$P_{sat} \simeq \exp\left\{-\frac{M h_{fg}}{\mathcal{R}}\left(\frac{1}{T} - \frac{1}{T_{BP}}\right)\right\} \text{ atm}, \quad \text{for } T \sim T_{BP}$$

[b]The Schmidt number is defined as $Sc = \mu/\rho\mathcal{D} = \nu/\mathcal{D}$. Since the vapors are in small concentrations, values for μ, ρ, and ν can be taken as pure air values.

[c]From a recent study by Cho, C., Irvine, T. F., Jr., and Karni, J., "Measurement of the diffusion coefficient of naphthalene into air," *Int. J. Heat Mass Transfer,* 35, 957–966 (1992). Also, $h_{sg} = 0.567 \times 10^6$ J/kg at 300 K.

Table A.18 Schmidt numbers for dilute solution in water at 300 K[a]

Solute	Sc	M
Helium	120	4.003
Hydrogen	190	2.016
Nitrogen	280	28.02
Water	340	18.016
Nitric oxide	350	30.01
Carbon monoxide	360	28.01
Oxygen	400	32.00
Ammonia	410	17.03
Carbon dioxide	420	44.01
Hydrogen sulfide	430	34.08
Ethylene	450	28.05
Methane	490	16.04
Nitrous oxide	490	44.02
Sulfur dioxide	520	64.06
Sodium chloride	540	58.45
Sodium hydroxide	490	40.00
Acetic acid	620	60.05
Acetone	630	58.08
Methanol	640	32.04
Ethanol	640	46.07
Chlorine	670	70.90
Benzene	720	78.11
Ethylene glycol	720	62.07
n-Propanol	730	60.09
i-Propanol	730	60.09
Propane	750	44.09
Aniline	800	93.13
Benzoic acid	830	122.12
Glycerol	1040	92.09
Sucrose	1670	342.3

[a]For other temperatures use $Sc/Sc_{300\text{ K}} \simeq (\mu^2/\rho T)/(\mu^2/\rho T)_{300\text{ K}}$, where μ and ρ are for water, and T is absolute temperature. For chemically similar solutes of different molecular weights use $Sc_2/Sc_1 \simeq (M_2/M_1)^{0.4}$. A table of $(\mu^2/\rho T)/(\mu^2/\rho T)_{300\text{ K}}$ for water follows.

T [K]	$(\mu^2/\rho T)/(\mu^2/\rho T)_{300\text{ K}}$
290	1.66
300	1.00
310	0.623
320	0.429
330	0.296
340	0.221
350	0.167
360	0.123
370	0.097

From Spalding, D. B., *Convective Mass Transfer*, McGraw-Hill, New York (1963).

Table A.19 Diffusion coefficients in solids, $\mathscr{D} = \mathscr{D}_0 \exp(-E_a/\mathscr{R}T)$

System	$\mathscr{D}_0$ m^2/s	E_a kJ/kmol
Oxygen–Pyrex glass	6.19×10^{-8}	4.69×10^4
Oxygen–fused silica glass	2.61×10^{-9}	3.77×10^4
Oxygen–titanium	5.0×10^{-3}	2.13×10^5
Oxygen–titanium alloy (Ti-6Al-4V)	5.82×10^{-2}	2.59×10^5
Oxygen–zirconium	4.68×10^{-5}	7.06×10^5
Hydrogen–iron	7.60×10^{-8}	5.60×10^3
Hydrogen–α-titanium	1.80×10^{-6}	5.18×10^4
Hydrogen–β-titanium	1.95×10^{-7}	2.78×10^4
Hydrogen–zirconium	1.09×10^{-7}	4.81×10^4
Hydrogen–Zircaloy-4	1.27×10^{-5}	6.05×10^5
Deuterium–Pyrex glass	6.19×10^{-8}	4.69×10^4
Deuterium–fused silica glass	2.61×10^{-9}	3.77×10^4
Helium–Pyrex glass	4.76×10^{-8}	2.72×10^4
Helium–fused silica glass	5.29×10^{-8}	2.55×10^4
Helium–borosilicate glass	1.94×10^{-9}	2.34×10^4
Neon–borosilicate glass	1.02×10^{-10}	3.77×10^4
Carbon–FCC iron	2.3×10^{-5}	1.378×10^5
Carbon–BCC iron	1.1×10^{-6}	8.75×10^4

Various sources.

Table A.20 Selected atomic weights[a]

Aluminum	Al	26.98	Molybdenum	Mo	95.94
Antimony	Sb	121.76	Neodymium	Nd	144.24
Argon	Ar	39.95	Neon	Ne	20.18
Arsenic	As	74.92	Nickel	Ni	58.71
Barium	Ba	137.34	Niobium	Nb	92.91
Beryllium	Be	9.012	Nitrogen	N	14.007
Bismuth	Bi	208.98	Oxygen	O	15.999
Boron	B	10.81	Palladium	Pd	106.4
Bromine	Br	79.90	Phosphorus	P	30.97
Cadmium	Cd	112.40	Platinum	Pt	195.09
Calcium	Ca	40.08	Plutonium	Pu	242
Carbon	C	12.01	Potassium	K	39.10
Cesium	Cs	132.91	Radium	Ra	226
Chlorine	Cl	35.45	Radon	Rn	222
Chromium	Cr	51.996	Rhenium	Re	186.2
Cobalt	Co	58.93	Rhodium	Rh	102.90
Copper	Cu	63.55	Rubidium	Rb	85.47
Fluorine	F	18.998	Selenium	Se	78.96
Gadolinium	Gd	157.25	Silicon	Si	28.09
Gallium	Ga	69.72	Silver	Ag	107.87
Germanium	Ge	72.59	Sodium	Na	22.99
Gold	Au	196.97	Strontium	Sr	87.62
Hafnium	Hf	178.5	Sulfur	S	32.06
Helium	He	4.003	Tantalum	Ta	180.95
Hydrogen	H	1.008	Tellurium	Te	127.60
Indium	In	114.82	Thallium	Tl	204.37
Iodine	I	126.90	Thorium	Th	232.04
Iron	Fe	55.85	Tin	Sn	118.69
Krypton	Kr	83.8	Titanium	Ti	47.90
Lead	Pb	207.19	Tungsten	W	183.85
Lithium	Li	6.939	Uranium	U	238.03
Magnesium	Mg	24.31	Xenon	Xe	131.30
Manganese	Mn	54.94	Zinc	Zn	65.37
Mercury	Hg	200.59	Zirconium	Zr	91.22

[a] Based on the isotope $C^{12} = 12$.

Selected from the *Handbook of Chemistry and Physics*, 50th ed., Chemical Rubber Co., Cleveland (1969).

Table A.21 Henry constants C_{He} for dilute aqueous solutions at moderate pressures ($P_{i,s}/x_{i,u}$ in atm, or in bar = 10^5 Pa, within the accuracy of the data)

Solute	290 K	300 K	310 K	320 K	330 K	340 K
H_2S	440	560	700	830	980	1,140
CO_2	1,280	1,710	2,170	2,720	3,220	—
O_2	38,000	45,000	52,000	57,000	61,000	65,000
H_2	67,000	72,000	75,000	76,000	77,000	76,000
CO	51,000	60,000	67,000	74,000	80,000	84,000
Air	62,000	74,000	84,000	92,000	99,000	104,000
N_2	76,000	89,000	101,000	110,000	118,000	124,000

Table A.22*a* Equilibrium compositions for the NH_3-water system

$P_{i,s}$ atm	$x_{i,u}$ 290 K	300 K	310 K	320 K	330 K
0.02	0.030	0.019	0.012	0.008	0.006
0.04	0.056	0.036	0.024	0.016	0.012
0.06	0.078	0.052	0.035	0.024	0.017
0.08	0.096	0.064	0.046	0.032	0.023
0.1	0.11	0.079	0.056	0.040	0.029
0.2	0.18	0.14	0.099	0.057	0.052
0.4	0.26	0.21	0.16	0.12	0.092
0.6	0.31	0.26	0.20	0.16	0.13
0.8	0.35	0.29	0.23	0.19	0.15
1.0	—	0.32	0.27	0.22	0.17

Table A.22*b* Equilibrium compositions for the SO_2-water system[a]

$P_{i,s}$ atm	$x_{i,u} \times 10^3$ 290 K	300 K	310 K	320 K
0.001	0.12	0.084	0.059	0.042
0.003	0.25	0.18	0.13	0.093
0.01	0.62	0.42	0.31	0.22
0.03	1.4	1.1	0.73	0.51
0.1	4.1	2.9	2.0	1.4
0.3	11.0	7.9	5.6	3.9
1.0	33.0	24.0	18.0	12.0

[a] Notice that Henry's law is invalid for the water-SO_2 system, even at very dilute concentrations.

Table A.23 Solubility and permeability of gases in solids

Gas	Solid	Temperature K	$\mathscr{S}$ [m³(STP)/m³ atm] or $\mathscr{S}'^{a}$	Permeability[b] m³(STP)/m² s (atm/m)
H_2	Vulcanized rubber	300	$\mathscr{S} = 0.040$	0.34×10^{-10}
	Vulcanized neoprene	290	$\mathscr{S} = 0.051$	0.053×10^{-10}
	Silicone rubber	300		4.2×10^{-10}
	Natural rubber	300		0.37×10^{-10}
	Polyethylene	300		0.065×10^{-10}
	Polycarbonate	300		0.091×10^{-10}
	Fused silica	400	$\mathscr{S}' \simeq 0.035$	
		800	$\mathscr{S}' \simeq 0.030$	
	Nickel	360	$\mathscr{S}' = 0.202$	
		440	$\mathscr{S}' = 0.192$	
He	Silicone rubber	300		2.3×10^{-10}
	Natural rubber	300		0.24×10^{-10}
	Polycarbonate	300		0.11×10^{-10}
	Nylon 66	300		0.0076×10^{-10}
	Teflon	300		0.047×10^{-10}
	Fused silica	300	$\mathscr{S}' \simeq 0.018$	
		800	$\mathscr{S}' \simeq 0.026$	
	Pyrex glass	300	$\mathscr{S}' \simeq 0.006$	
		800	$\mathscr{S}' \simeq 0.024$	
	7740 glass	470	$\mathscr{S} = 0.0084$	4.6×10^{-13}
	(94% $SiO_2 + B_2O_3 + P_2O_5$	580	$\mathscr{S} = 0.0038$	1.6×10^{-12}
	5% $Na_2O + Li_2 + K_2O$	720	$\mathscr{S} = 0.0046$	6.4×10^{-12}
	1% other oxides)			
	7056 glass	390	$\mathscr{S} = 0.0039$	1.2×10^{-14}
	(90% $SiO_2 + B_2O_3 + P_2O_5$	680	$\mathscr{S} = 0.0059$	1.0×10^{-12}
	8% $Na_2O + Li_2 + K_2O$			
	1% PbO, 5% other oxides)			
O_2	Vulcanized rubber	300	$\mathscr{S} = 0.070$	0.15×10^{-10}
	Silicone rubber	300		3.8×10^{-10}
	Natural rubber	300		0.18×10^{-10}
	Polyethylene	300		4.2×10^{-12}
	Polycarbonate	300		0.011×10^{-10}
	Silicone-polycarbonate copolymer (57% silicone)	300		1.2×10^{-10}
	Ethyl cellulose	300		0.09×10^{-10}
N_2	Vulcanized rubber	300	$\mathscr{S} = 0.035$	0.054×10^{-10}
	Silicone rubber	300		1.9×10^{-10}
	Natural rubber	300		0.062×10^{-10}
	Silicone-polycarbonate copolymer (57% silicone)	300		0.53×10^{-10}
	Teflon	300		0.019×10^{-10}

(Continued)

Table A.23 ***(Concluded)***

Gas	Solid	Temperature K	$\mathscr{S}$ [m^3(STP)/m^3 atm] or $\mathscr{S}'$[a]	Permeability[b] m^3(STP)/m^2 s (atm/m)
CO_2	Vulcanized rubber	300	$\mathscr{S} = 0.90$	1.0×10^{-10}
	Silicone rubber	300		21×10^{-10}
	Natural rubber	300		1.0×10^{-10}
	Silicone-polycarbonate copolymer (57% silicone)	300		7.4×10^{-10}
	Nylon 66	300		0.0013×10^{-10}
H_2O	Cellophane	310		$0.91 - 1.8 \times 10^{-10}$
Ne	Fused silica	300–1200	$\mathscr{S}' \simeq 0.002$	
Ar	Fused silica	900–1200	$\mathscr{S}' \simeq 0.01$	

[a] Solubility $\mathscr{S}$ = Volume of solute gas (0°C, 1 atm) dissolved in unit volume of solid when the gas is at 1 atm partial pressure. Solubility coefficient $\mathscr{S}' = c_{1,u}/c_{1,s}$.

[b] Permeability $\mathscr{P}_{12} = \mathscr{D}_{12}\mathscr{S}$.

From various sources, including Geankoplis, C. J., *Transport Processes and Unit Operations,* 3rd ed., Prentice-Hall, Englewood Cliffs, N.J. (1993); Doremus, R. H., *Glass Science,* Wiley, New York (1973); Altemose, V. O., "Helium diffusion through glass," *J. Appl. Phys.*, 32, 1309–1316 (1961).

Table A.24 Solubility of inorganic compounds in water[a]

Solute	Formula	Solid Phase	T [K] 273.15	280	290	300	310	320	330	340	350	360	370	373.15
Aluminum sulfate	$Al_2(SO_4)_3$	$18H_2O$	31.2	32.8	35.5	39.1	44.3	50.3	57.0	63.9	70.8	78.3	84.6	89.0
Calcium bicarbonate	$Ca(HCO_3)_2$	—	16.15	16.30	16.53	16.75	16.98	17.20	17.43	17.65	17.88	18.10	18.33	18.40
Calcium chloride	$CaCl_2$	$6H_2O$	59.5	63.3	71.5	93.3	137.2	—	—	—	—	—	—	—
	$CaCl_2$	$2H_2O$	—	—	—	—	—	—	134.6	140.2	145.3	150.9	157.0	159.0
Calcium hydroxide	$Ca(OH)_2$	—	0.185	0.179	0.168	0.157	0.145	0.132	0.120	0.109	0.098	0.088	0.080	0.077
Potassium chloride	KCl	—	27.6	29.9	33.1	36.1	39.1	41.8	44.6	47.4	50.2	53.1	55.8	56.7
Potassium nitrate	KNO_3	—	13.3	18.5	28.2	41.3	58.2	78.7	102.3	129.2	159.2	191.6	232.1	246.0
Potassium sulfate	K_2SO_4	—	7.35	8.63	10.51	12.38	14.20	15.95	17.64	19.25	20.88	22.36	23.69	24.1
Sodium bicarbonate	$NaHCO_3$	—	6.9	7.76	9.14	10.63	12.20	13.90	15.79	—	—	—	—	—
Sodium carbonate	Na_2O_3	$10H_2O$	7	10.8	18.7	33.4	—	—	—	—	—	—	—	—
	Na_2CO_3	$1H_2O$	—	—	—	—	49.1	47.8	46.7	46.2	45.9	45.7	45.55	45.5
Sodium chloride	NaCl	—	35.7	35.8	35.9	36.2	36.5	36.9	37.2	37.6	38.2	38.8	39.5	39.8
Sodium nitrate	$NaNO_2$	—	73	78	85	93	101	111	121	132	144	159	175	180
Sodium sulfate	Na_2SO_4	$10H_2O$	5.0	7.7	16.1	34.1	—	—	—	—	—	—	—	—
	Na_2SO_4	$7H_2O$	19.5	26.7	39.6	—	—	—	—	—	—	—	—	—
	Na_2SO_4	—	—	—	—	—	49.6	47.4	45.7	43.7	44.0	43.3	42.7	42.5

[a] Solubility expressed in kilograms of anhydrous substance that is soluble in 100 kg water.
Adapted from *Handbook of Chemistry,* 10th ed., McGraw-Hill, New York (1961).

Table A.25 Combustion data

Fuel		Heats of Combustion		Stoichiometric Combustion (no excess air)						
				Required Air kg/kg_{fuel}			Major Flue Gases kg/kg_{fuel}			
Substance	M	Higher Heat (liquid H_2O) $J/kg \times 10^{-6}$	Lower Heat (vapor H_2O) $J/kg \times 10^{-6}$	O_2	N_2	Air	CO_2	H_2O	N_2	SO_2
C	12.01	32.78	32.78	2.66	8.86	11.53	3.66	—	8.86	
H_2	2.016	142.1	120.1	7.94	26.41	34.34	—	8.94	26.41	
CO	28.01	10.11	10.11	0.57	1.90	2.47	1.57	—	1.90	
CH_4	16.041	55.54	50.06	3.99	13.28	17.27	2.74	2.25	13.28	
C_2H_6	30.067	51.92	47.52	3.73	12.39	16.12	2.93	1.80	12.39	
C_3H_8	44.092	50.38	46.39	3.63	12.07	15.70	2.99	1.63	12.07	
C_4H_{10}	58.118	49.56	45.78	3.58	11.91	15.49	3.03	1.55	11.91	
C_5H_{12}	72.144	49.06	45.40	3.55	11.81	15.35	3.05	1.50	11.81	
C_2H_4	28.051	50.34	47.21	3.42	11.39	14.81	3.14	1.29	11.39	
C_3H_6	42.077	48.94	45.80	3.42	11.39	14.81	3.14	1.29	11.39	
C_4H_8	56.102	48.47	45.35	3.42	11.39	14.81	3.14	1.29	11.39	
C_5H_{10}	70.128	48.18	45.04	3.42	11.39	14.81	3.14	1.29	11.39	
C_6H_6	78.107	42.36	40.66	3.07	10.22	13.30	3.38	0.69	10.22	
C_7H_8	92.132	42.89	40.98	3.13	10.40	13.53	3.34	0.78	10.40	
C_8H_{10}	106.158	43.38	41.31	3.17	10.53	13.70	3.32	0.85	10.53	
C_2H_2	26.036	50.01	48.32	3.07	10.22	13.30	3.38	0.69	10.22	
$C_{10}H_8$	128.162	40.24	38.86	3.00	9.97	12.96	3.43	0.56	9.97	
CH_3OH	32.041	23.86	21.12	1.50	4.98	6.48	1.37	1.13	4.98	
C_2H_5OH	46.067	30.61	27.75	2.08	6.93	9.02	1.92	1.17	6.93	
NH_3	17.031	22.49	18.61	1.41	4.69	6.10	—	1.59	5.51	
S	32.06	9.264	9.264	1.00	3.29	4.29	—	—	3.29	2.00
H_2S	34.076	16.51	15.22	1.41	4.69	6.20	—	0.53	4.69	1.88

Adapted from *Fuel Flue Gases*, American Gas Association, New York (1940).

Table A.26 Force constants for the Lennard-Jones potential model

Species	σ Å	ε/k K	Species	σ Å	ε/k K	Species	σ Å	ε/k K
Al	2.655	2750	CH_3CCH	4.761	252	Li_2O	3.561	1827
AlO	3.204	542	C_3H_8	5.118	237	Mg	2.926	1614
Al_2	2.940	2750	n-C_3H_7OH	4.549	577	N	3.298	71
Air	3.711	79	n-C_4H_{10}	4.687	531	NH_3	2.900	558
Ar	3.542	93	iso-C_4H_{10}	5.278	330	NO	3.492	117
C	3.385	32	n-C_5H_{12}	5.784	341	N_2	3.798	71
CCl_2	4.692	213	C_6H_{12}	6.182	297	N_2O	3.828	232
CCl_2F_2	5.25	253	n-C_6H_{14}	5.949	399	Na	3.567	1375
CCl_4	5.947	323	Cl	3.613	131	NaCl	4.186	1989
CH	3.370	69	Cl_2	4.217	316	NaOH	3.804	1962
$CHCl_3$	5.389	340	H	2.708	37	Na_2	4.156	1375
CH_3OH	3.626	482	HCN	3.630	569	Ne	2.820	33
CH_4	3.758	149	HCl	3.339	345	O	3.050	107
CN	3.856	75	H_2	2.827	60	OH	3.147	80
CO	3.690	92	H_2O^a	2.641	809.1	O_2	3.467	107
CO_2	3.941	195	H_2O^b	3.737	32	S	3.839	847
CS_2	4.483	467	H_2O_2	4.196	289	SO	3.993	301
C_2	3.913	79	H_2S	3.623	301	SO_2	4.112	335
C_2H_2	4.033	232	He	2.551	10	Si	2.910	3036
C_2H_4	4.163	225	Hg	2.969	750	SiO	3.374	569
C_2H_6	4.443	216	I_2	5.160	474	SiO_2	3.706	2954
C_2H_5OH	4.530	363	Li	2.850	1899	UF_6	5.967	237
C_2N_2	4.361	349	LiO	3.334	450	Xe	4.047	231
$C_2H_2CHCH_3$	4.678	299	Li_2	3.200	1899	Zn	2.284	1393

[a] For μ and k.

[b] For $\mathscr{D}$.

Taken largely from Svehla, R. A., *Estimated Viscosities and Thermal Conductivities of Gases at High Temperatures*, NASA TR R-132 (1962).

Table A.27 Molar volume at the normal boiling point

Species	$\tilde{V}_b$ m³/kmol × 10³	T_b K	Species	$\tilde{V}_b$ m³/kmol × 10³	T_b K
Hydrogen	14.3	21	Carbon tetrachloride	102	350
Oxygen	25.6	90	Methyl chloride	50.6	249
Nitrogen	31.2	77	Ethyl acetate	106	350
Air	29.9	79	Acetic acid	64.1	337
Carbon monoxide	30.7	82	Acetone	77.5	329
Carbon dioxide	34.0	195	Ethylpropyl ether	129	335
Sulfur dioxide	44.8	263	Dimethyl ether	63.8	250
Nitric oxide	23.6	121	*n*-Propyl alcohol	81.8	370
Nitrous oxide	36.4	185	Methanol	42.5	338
Ammonia	25.8	240	Methane	37.7	112
Water	18.9	373	Propane	74.5	229
Hydrogen sulfide	32.9	212	Heptane	162	372
Chlorine	48.4	239	Ethylene	49.4	169
Bromine	53.2	332	Acetylene	42.0	190
Iodine	715	458	Benzene	96.5	353
Hydrochloric acid	30.6	188	Fluorobenzene	102	358
Methyl formate	62.8	305	Chlorobenzene	115	405
Bromobenzene	120	429	Iodobenzene	130	462
Dichlorodifluoromethane	80.7	245			

Table A.28 Collision integrals for the Lennard-Jones potential model

kT/ε	$\Omega_\mu = \Omega_k$	$\Omega_{\mathcal{D}}$	kT/ε	$\Omega_\mu = \Omega_k$	$\Omega_{\mathcal{D}}$	kT/ε	$\Omega_\mu = \Omega_k$	$\Omega_{\mathcal{D}}$
0.30	2.785	2.662	1.60	1.279	1.167	3.80	0.9811	0.8942
0.35	2.628	2.476	1.65	1.264	1.153	3.90	0.9755	0.8888
0.40	2.492	2.318	1.70	1.248	1.140	4.00	0.9700	0.8836
0.45	2.368	2.184	1.75	1.234	1.128	4.10	0.9649	0.8788
0.50	2.257	2.066	1.80	1.221	1.116	4.20	0.9600	0.8740
0.55	2.156	1.966	1.85	1.209	1.105	4.30	0.9553	0.8694
0.60	2.065	1.877	1.90	1.197	1.094	4.40	0.9507	0.8652
0.65	1.982	1.798	1.95	1.186	1.084	4.50	0.9464	0.8610
0.70	1.908	1.729	2.00	1.175	1.075	4.60	0.9422	0.8568
0.75	1.841	1.667	2.10	1.156	1.057	4.70	0.9382	0.8530
0.80	1.780	1.612	2.20	1.138	1.041	4.80	0.9343	0.8492
0.85	1.725	1.562	2.30	1.122	1.026	4.90	0.9305	0.8456
0.90	1.675	1.517	2.40	1.107	1.012	5.0	0.9269	0.8422
0.95	1.629	1.476	2.50	1.093	0.9996	6.0	0.8963	0.8124
1.00	1.587	1.439	2.60	1.081	0.9878	7.0	0.8727	0.7896
1.05	1.549	1.406	2.70	1.069	0.9770	8.0	0.8538	0.7712
1.10	1.514	1.375	2.80	1.058	0.9672	9.0	0.8379	0.7556
1.15	1.482	1.346	2.90	1.048	0.9576	10.0	0.8242	0.7424
1.20	1.452	1.320	3.00	1.039	0.9490	20.0	0.7432	0.6640
1.25	1.424	1.296	3.10	1.030	0.9406	30.0	0.7005	0.6232
1.30	1.399	1.273	3.20	1.022	0.9328	40.0	0.6718	0.5960
1.35	1.375	1.253	3.30	1.014	0.9256	50.0	0.6504	0.5756
1.40	1.353	1.233	3.40	1.007	0.9186	60.0	0.6335	0.5596
1.45	1.333	1.215	3.50	0.9999	0.9120	70.0	0.6194	0.5464
1.50	1.314	1.198	3.60	0.9932	0.9058	80.0	0.6076	0.5352
1.55	1.296	1.182	3.70	0.9870	0.8998	90.0	0.5973	0.5256
						100.0	0.5882	0.5170

Taken from Hirschfelder, J. O., Curtiss, C. F., and Bird, R. B., *Molecular Theory of Gases and Liquids*, John Wiley & Sons, New York (1954).

Table A.29 Constant-pressure specific heat capacity c_p [J/kg K]: Database in GASMIX

		Temperature [K]								
Species	Formula	200	300	400	600	800	1000	1500	2000	2500
Air		1009	1005	1009	1038	1089	1130	1202	1244	1285
Aluminum	Al	820	793	783	776	773	772	771	771	771
Aluminum	Al_2	646	714	764	795	785	770	750	748	756
Aluminum oxide	AlO	686	719	756	807	838	866	961	1051	1095
Argon	Ar	520	520	520	520	520	520	520	520	520
Carbon	C	1741	1735	1733	1732	1731	1731	1733	1745	1769
Refrigerant-12	CCl_2F_2	485	601	681	774	820	844	871	881	885
Carbon tetrachloride	CCl_4	455	543	596	628	670	681	693	697	699
Methane	CH_4	2087	2226	2525	3256	3923	4475	5396	5884	6157
Ethanol	C_2H_5OH	1810	1972	2126	2408	2652	2857	3262	3540	
n-Pentane	n-C_5H_{12}	1983	2216	2438	2843	3188	3467	4011	4380	
Cyclohexane	C_6H_{12}		1272	1781	2676	3319	3769	4405		
n-Hexane	n-C_6H_{14}	1983	2215	2433	2833	3170	3444	3981	4341	
Carbon monoxide	CO	1039	1040	1048	1087	1139	1183	1257	1294	1315
Carbon dioxide	CO_2	735	844	937	1074	1168	1232	1329	1371	1397
Chlorine	Cl_2	447	479	498	515	523	528	535	542	516
Hydrogen	H_2	15910	14780	14400	14400	14530	14760	16000	17050	17779
Water	H_2O	1851	1870	1900	2003	2130	2267	2594	2832	2992
Hydrogen peroxide	H_2O_2	1084	1271	1424	1637	1759	1848	2009		
Hydrogen sulfide	H_2S	980	1004	1044	1143	1248	1344	1511	1604	1660
Hydrogen chloride	HCl	799	799	800	811	836	868	934	976	1003
Helium	He	5200	5200	5200	5200	5200	5200	5200	5200	5200
Mercury	Hg	104	104	104	104	104	104	104	104	104
Iodine	I_2	142	145	147	148	149	150	156	168	178
Nitrogen	N_2	1045	1043	1047	1075	1123	1167	1244	1287	1307
Nitrous oxide	N_2O	764	877	970	1099	1187	1247	1327	1363	1381
Ammonia	NH_3	1982	2096	2273	2659	3008	3317	3908	4277	4499
Neon	Ne	1030	1030	1030	1030	1030	1030	1030	1030	1030
Nitric oxide	NO	1014	995	998	1041	1092	1133	1192	1221	1238
Oxygen	O	1421	1390	1343	1320	1312	1370	1303	1302	1303
Hydroxyl	OH	1810	1763	1743	1736	1759	1804	1936	2039	2110
Oxygen	O_2	900	920	945	1000	1050	1085	1140	1180	1214
Sulfur dioxide	SO_2	568	624	679	766	819	851	890	909	920
Uranium hexafluoride	UF_6		369	399	422	432	439			
Xenon	Xe	158	158	158	158	158	158	158	158	158

Various sources, primarily Chase, M. W., Jr., et al., *JANAF Thermochemical Tables*, 3rd ed., American Chemical Society (1985), and Rossini, F. D., et al., *Selected Values of Properties of Hydrocarbons*, National Bureau of Standards Circular C461 (1947).

Table A.30 Slip factor C, Brownian diffusion coefficient $\mathcal{D}_p$, and particle Schmidt number Sc_p for spherical particles in air at 1 atm and 300 K

d_p μm	C	$\mathcal{D}_p$ m²/s	Sc_p
0.001	222.3	5.30×10^{-6}	2.96
0.002	111.4	1.33×10^{-6}	1.18×10^{1}
0.005	44.9	2.14×10^{-7}	7.31×10^{1}
0.01	22.7	5.42×10^{-8}	2.89×10^{2}
0.02	11.7	1.39×10^{-8}	1.12×10^{3}
0.05	5.07	2.42×10^{-9}	6.48×10^{3}
0.1	2.92	6.95×10^{-10}	2.25×10^{4}
0.2	1.89	2.26×10^{-10}	6.94×10^{4}
0.5	1.34	6.38×10^{-11}	2.45×10^{5}
1	1.17	2.78×10^{-11}	5.62×10^{5}
2	1.08	1.29×10^{-11}	1.21×10^{6}
5	1.03	4.93×10^{-12}	3.18×10^{6}
10	1.02	2.43×10^{-12}	6.46×10^{6}

Table A.31 Thermodynamic properties of water vapor–air mixtures at 1 atm

Temp °C	Saturation Mass Fraction	Specific Volume m³/kg: Dry Air	Specific Volume m³/kg: Saturated Air	Enthalpy[a,b] kJ/kg: Liquid Water	Enthalpy[a,b] kJ/kg: Dry Air	Enthalpy[a,b] kJ/kg: Saturated Air
10	0.007608	0.8018	0.8054	42.13	10.059	29.145
11	0.008136	0.8046	0.8086	46.32	11.065	31.481
12	0.008696	0.8075	0.8117	50.52	12.071	33.898
13	0.009289	0.8103	0.8148	54.71	13.077	36.401
14	0.009918	0.8131	0.8180	58.90	14.083	38.995
15	0.01058	0.8160	0.8212	63.08	15.089	41.684
16	0.01129	0.8188	0.8244	67.27	16.095	44.473
17	0.01204	0.8217	0.8276	71.45	17.101	47.367
18	0.01283	0.8245	0.8309	75.64	18.107	50.372
19	0.01366	0.8273	0.8341	79.82	19.113	53.493
20	0.01455	0.8302	0.8374	83.99	20.120	56.736
21	0.01548	0.8330	0.8408	88.17	21.128	60.107
22	0.01647	0.8359	0.8441	92.35	22.134	63.612
23	0.01751	0.8387	0.8475	96.53	23.140	67.259
24	0.01861	0.8415	0.8510	100.71	24.147	71.054
25	0.01978	0.8444	0.8544	104.89	25.153	75.004
26	0.02100	0.8472	0.8579	109.07	26.159	79.116
27	0.02229	0.8500	0.8615	113.25	27.166	83.400
28	0.02366	0.8529	0.8650	117.43	28.172	87.862
29	0.02509	0.8557	0.8686	121.61	29.178	92.511
30	0.02660	0.8586	0.8723	125.79	30.185	97.357
31	0.02820	0.8614	0.8760	129.97	31.191	102.408
32	0.02987	0.8642	0.8798	134.15	32.198	107.674
33	0.03164	0.8671	0.8836	138.32	33.204	113.166
34	0.03350	0.8699	0.8874	142.50	34.211	118.893
35	0.03545	0.8728	0.8914	146.68	35.218	124.868
36	0.03751	0.8756	0.8953	150.86	36.224	131.100
37	0.03967	0.8784	0.8994	155.04	37.231	137.604
38	0.04194	0.8813	0.9035	159.22	38.238	144.389
39	0.04432	0.8841	0.9077	163.40	39.245	151.471
40	0.04683	0.8870	0.9119	167.58	40.252	158.862
41	0.04946	0.8898	0.9162	171.76	41.259	166.577
42	0.05222	0.8926	0.9206	175.94	42.266	174.630
43	0.05512	0.8955	0.9251	180.12	43.273	183.037
44	0.05817	0.8983	0.9297	184.29	44.280	191.815
45	0.06137	0.9012	0.9343	188.47	45.287	200.980
46	0.06472	0.9040	0.9391	192.65	46.294	210.550
47	0.06842	0.9068	0.9439	196.83	47.301	220.543
48	0.07193	0.9097	0.9489	201.01	48.308	230.980
49	0.07580	0.9125	0.9539	205.19	49.316	241.881

[a] The enthalpies of dry air and liquid water are set equal to zero at a datum temperature of 0°C.

[b] The enthalpy of an unsaturated water vapor–air mixture can be calculated as $h = h_{\text{dry air}} + (m_1/m_{1,\text{sat}})(h_{\text{sat}} - h_{\text{dry air}})$.

APPENDIX

B

UNITS, CONVERSION FACTORS, AND MATHEMATICS

LIST OF TABLES

Table B.1*a* Base and supplementary SI units

Quantity	Name	Symbol
Length	meter	m
Mass	kilogram	kg
Time	second	s
Electric current	ampere	A
Thermodynamic temperature	kelvin	K
Amount of substance[a]	mole	mol
Luminous intensity	candela	cd
Plane angle	radian	rd
Solid angle	steradian	sr

[a] The mole is the gram mole of the cgs units system.

Table B.1*b* Derived SI units

Quantity	Name	Symbol	Definition
Frequency	hertz	Hz	s^{-1}
Force	newton	N	kg m/s^2
Pressure, stress	pascal	Pa	N/m^2
Energy, work	joule	J	N m
Power	watt	W	J/s
Electric charge, quantity of electricity	coulomb	C	A s
Electric potential, electromotive force	volt	V	J/C
Electrical capacitance	farad	F	C/V
Electrical resistance	ohm	Ω	V/A
Electrical conductance	siemens	S	Ω^{-1}
Magnetic flux	weber	Wb	V s
Magnetic flux density	tesla	T	Wb/m^2
Inductance	henry	H	Wb/A
Luminous flux	lumen	lm	cd sr
Illuminance	lux	lx	lm/m^2
Celsius temperature	degree Celsius	°C	1°C = 1 K[a]

[a] Celsius temperature $T - T_0$, where T is in kelvins and $T_0 = 273.15$ K. The unit *degree Celsius* is equal to the unit *kelvin*.

Table B.1*c* Recognized non-SI units

Quantity	Name	Symbol	Definition
Time	minute	min	60 s
	hour	h	60 m
	day	d	24 h
Plane angle	degree	°	$(\pi/180)$ rad
	minute	′	$(1/60)°$
	second	″	$(1/60)'$
Volume	liter	l	10^{-3} m^3
Mass	ton (metric)	t	10^3 kg
Energy	electron volt	eV	1.60219×10^{-19} J
Mass of atoms	atomic mass unit	u	1.66057×10^{-27} kg
Length	astronomical unit	AU	149597.870×10^6 m
	parsec	pc	206,265 AU
Pressure	bar	bar	10^5 Pa

Adapted from NZS 6501:1092 "Units of Measurement," Standards Association of New Zealand, Wellington.

Table B.1*d* Multiples of SI units

Factor	Prefix	Symbol
10^{18}	exa	E
10^{15}	peta	P
10^{12}	tera	T
10^9	giga	G
10^6	mega	M
10^3	kilo	k
10^2	hecto	h
10	deca	da
10^{-1}	deci	d
10^{-2}	centi	c
10^{-3}	milli	m
10^{-6}	micro	μ
10^{-9}	nano	n
10^{-12}	pico	p
10^{-15}	femto	f
10^{-18}	atto	a

Table B.2 Conversion factors

Temperature	0.555 K/°R T[°R] = T[°F] + 459.67	T[K] = T[°C] + 273.15
Length	0.3048 m/ft 2.54 cm/in	1609 m/mi
Velocity	0.3048 (m/s)/(ft/s) 0.4470 (m/s)/(mph)	0.2778 (m/s)/(km/h) 1.6093 (km/h)/(mph)
Volume	2.832×10^{-2} m^3/ft^3 3.785×10^{-3} m^3/gal 42 gal/bbl (oil)	10^{-3} m^3/liter 4.545×10^{-3} m^3/Imperial gal.
Mass	0.4536 kg/lb	14.59 kg/slug
Force	4.448 N/lb$_f$	10^{-5} N/dyne
Stress	47.88 (N/m^2)/(lb$_f$/ft^2) 6895 (N/m^2)/psi	10^{-1} (N/m^2)/(dyne/cm^2)
Pressure	6895 Pa/psi 1.0133×10^5 Pa/atm 760 torr/atm	10^5 Pa/bar 133.3 Pa/torr
Energy, work	1055 J/Btu 4187 J/kcal[a] 1.6021×10^{-19} J/ev	10^{-7} J/erg 1.356 J/ft lb$_f$
Power	0.2931 W/(Btu/hr)	0.7457 kW/hp
Heat flux	3.155 (W/m^2)/(Btu/ft^2 hr)	4.187×10^4 (W/m^2)/(cal/cm^2 s)
Heat transfer coefficient	5.678 (W/m^2 K)/(Btu/ft^2 hr °F)	4.187×10^4 (W/m^2 K)/(cal/cm^2 s °C)
Mass flux	1.3563×10^{-3} (kg/m^2 s)/(lb/ft^2 hr)	10^{-1} (kg/m^{2}s)/(g/cm^2 s)
Mole flux	1.3563×10^{-3} (kmol/m^2 s)/(lb mole/ft^2 hr)	10^{-1} (kmol/m^2 s)/(g mole/cm^2 s)
Density	16.018 (kg/m^3)/(lb/ft^3) 515.3 (kg/m^3)/(slug/ft^3)	10^3 (kg/m^3)/(g/cm^3)
Enthalpy	2326 (J/kg)/(Btu/lb)	4187 (J/kg)/(cal/g)
Specific heat	4187 (J/kg K)/(Btu/lb °F)	4187 (J/kg K)/(cal/g °C)
Dynamic viscosity	47.88 (kg/m s)/(lb$_f$s/ft^2) 10^{-3} (kg/m s)/cp	10^{-1} (kg/m s)/poise 1 (kg/m s)/(Ns/m^2)
Diffusivity	2.581×10^{-5} (m^2/s)/(ft^2/hr)	10^{-4} (m^2/s)/(cm^2/s)
Thermal conductivity	1.731 (W/m K)/(Btu/hr ft °F)	418.7 (W/m K)/(cal/s cm °C)

[a] I.T. calorie (International Steam Table calorie). Also in use is the thermochemical calorie, for which there are 4184 J/kcal.

Table B.3 Bessel functions

The differential equation

$$\frac{d^2y}{dx^2} + \frac{1}{x}\frac{dy}{dx} + \left(a^2 - \frac{n^2}{x^2}\right)y = 0$$

has the solution

$$y = AJ_n(ax) + BY_n(ax) \qquad \text{for } n = 0, \text{ or an integer}$$

where J_n and Y_n are order n *Bessel functions* of the first and second kinds, respectively.

The differential equation

$$\frac{d^2y}{dx^2} + \frac{1}{x}\frac{dy}{dx} + \left(-a^2 - \frac{n^2}{x^2}\right)y = 0$$

has the solution

$$y = CI_n(ax) + DK_n(ax)$$

where I_n and K_n are order n *modified Bessel functions* of the first and second kinds, respectively.

Rules for differentiation are as follows:

$$\frac{d}{dx}[J_0(ax)] = -aJ_1(ax)$$

$$\frac{d}{dx}[Y_0(ax)] = -aY_1(ax)$$

$$\frac{d}{dx}[I_0(ax)] = aI_1(ax)$$

$$\frac{d}{dx}[K_0(ax)] = -aK_1(ax)$$

Table B.3*a* Bessel functions of the first and second kinds, orders 0 and 1

x	$J_0(x)$	$J_1(x)$	$Y_0(x)$	$Y_1(x)$
0.0	1.00000	0.00000	$-\infty$	$-\infty$
0.2	+0.99002	+0.09950	−1.0811	−3.3238
0.4	+0.96039	+0.19603	−0.60602	−1.7809
0.6	+0.91200	+0.28670	−0.30851	−1.2604
0.8	+0.84629	+0.36884	−0.08680	−0.97814
1.0	+0.76520	+0.44005	+0.08825	−0.78121
1.2	+0.67113	+0.49830	+0.22808	−0.62113
1.4	+0.56686	+0.54195	+0.33790	−0.47915
1.6	+0.45540	+0.56990	+0.42043	−0.34758
1.8	+0.33999	+0.58152	+0.47743	−0.22366
2.0	+0.22389	+0.57672	+0.51038	−0.10703
2.2	+0.11036	+0.55596	+0.52078	+0.00149
2.4	+0.00251	+0.52019	+0.51042	+0.10049
2.6	−0.09680	+0.47082	+0.48133	+0.18836
2.8	−0.18503	+0.40971	+0.43591	+0.26355
3.0	−0.26005	+0.33906	+0.37685	+0.32467
3.2	−0.32019	+0.26134	+0.30705	+0.37071
3.4	−0.36430	+0.17923	+0.22962	+0.40101
3.6	−0.39177	+0.09547	+0.14771	+0.41539
3.8	−0.40256	+0.01282	+0.06540	+0.41411
4.0	−0.39715	−0.06604	−0.01694	+0.39792
4.2	−0.37656	−0.13864	−0.09375	+0.36801
4.4	−0.34226	−0.20278	−0.16333	+0.32597
4.6	−0.29614	−0.25655	−0.22345	+0.27374
4.8	−0.24042	−0.29850	−0.27230	+0.21357
5.0	−0.17760	−0.32760	−0.30851	+0.14786
5.2	−0.11029	−0.34322	−0.33125	+0.07919
5.4	−0.04121	−0.34534	−0.34017	+0.01013
5.6	+0.02697	−0.33433	−0.33544	−0.05681
5.8	+0.09170	−0.31103	−0.31775	−0.11923
6.0	+0.15065	−0.27668	−0.28819	−0.17501
6.2	+0.20175	−0.23292	−0.24830	−0.22228
6.4	+0.24331	−0.18164	−0.19995	−0.25955
6.6	+0.27404	−0.12498	−0.14523	−0.28575
6.8	+0.29310	−0.06252	−0.08643	−0.30019
7.0	+0.30007	−0.00468	−0.02595	−0.30267
7.2	+0.29507	+0.05432	+0.03385	−0.29342
7.4	+0.27859	+0.10963	+0.09068	−0.27315
7.6	+0.25160	+0.15921	+0.14243	−0.24280
7.8	+0.25541	+0.20136	+0.18722	−0.20389

(Continued)

Table B.3a ***(Concluded)***

x	$J_0(x)$	$J_1(x)$	$Y_0(x)$	$Y_1(x)$
8.0	+0.17165	+0.23464	+0.22352	−0.15806
8.2	+0.12222	+0.25800	+0.25011	−0.10724
8.4	+0.06916	+0.27079	+0.26622	−0.05348
8.6	+0.01462	+0.27275	+0.27146	−0.00108
8.8	−0.03923	+0.26407	+0.26587	+0.05436
9.0	−0.09033	+0.24531	+0.24994	+0.10431
9.2	−0.13675	+0.21471	+0.22449	+0.14911
9.4	−0.17677	+0.18163	+0.19074	+0.18714
9.6	−0.20898	+0.13952	+0.15018	+0.21706
9.8	−0.23227	+0.09284	+0.10453	+0.23789
10.0	−0.24594	+0.04347	+0.05567	+0.24902

Table B.3*b* Modified Bessel functions of the first and second kinds, orders 0 and 1

x	$e^{-x}I_0(x)$	$e^{-x}I_1(x)$	$e^{x}K_0(x)$	$e^{x}K_1(x)$
0.0	1.00000	0.00000		
0.2	0.82693	0.08228	2.14075	5.83338
0.4	0.69740	0.13676	1.66268	3.25867
0.6	0.59932	0.17216	1.41673	2.37392
0.8	0.52414	0.19449	1.25820	1.91793
1.0	0.46575	0.20791	1.14446	1.63615
1.2	0.41978	0.21525	1.05748	1.44289
1.4	0.38306	0.21850	0.98806	1.30105
1.6	0.35331	0.21901	0.93094	1.19186
1.8	0.32887	0.21772	0.88283	1.10480
2.0	0.30850	0.21526	0.84156	1.03347
2.2	0.29131	0.21208	0.80565	0.97377
2.4	0.27662	0.20848	0.77401	0.92291
2.6	0.26391	0.20465	0.74586	0.87896
2.8	0.25280	0.20073	0.72060	0.84053
3.0	0.24300	0.19682	0.69776	0.80656
3.2	0.23426	0.19297	0.67697	0.77628
3.4	0.22643	0.18922	0.65795	0.74907
3.6	0.21934	0.18560	0.64045	0.72446
3.8	0.21290	0.18210	0.62429	0.70206
4.0	0.20700	0.17875	0.60929	0.68157
4.2	0.20157	0.17553	0.59533	0.66274
4.4	0.19656	0.17245	0.58230	0.64535
4.6	0.19191	0.16949	0.57008	0.62924
4.8	0.18758	0.16667	0.55861	0.61425
5.0	0.18354	0.16397	0.54780	0.60027
5.2	0.17974	0.16138	0.53760	0.58718
5.4	0.17618	0.15890	0.52795	0.57490
5.6	0.17282	0.15652	0.51881	0.56335
5.8	0.16965	0.15424	0.51012	0.55246
6.0	0.16665	0.15205	0.50186	0.54217
6.2	0.16381	0.14994	0.49399	0.53243
6.4	0.16110	0.14792	0.48647	0.52318
6.6	0.15853	0.14597	0.47929	0.51440
6.8	0.15608	0.14409	0.47242	0.50604
7.0	0.15373	0.14228	0.46584	0.49807
7.2	0.15149	0.14054	0.45953	0.49046
7.4	0.14935	0.13886	0.45346	0.48318
7.6	0.14729	0.13723	0.44763	0.47622
7.8	0.14532	0.13566	0.44202	0.46955

(Continued)

Table B.3*b* (*Concluded*)

x	$e^{-x}I_0(x)$	$e^{-x}I_1(x)$	$e^{x}K_0(x)$	$e^{x}K_1(x)$
8.0	0.14343	0.13414	0.43662	0.46314
8.2	0.14160	0.13267	0.43141	0.45699
8.4	0.13985	0.13124	0.42638	0.45108
8.6	0.13816	0.12986	0.42152	0.44539
8.8	0.13653	0.12852	0.41683	0.43991
9.0	0.13495	0.12722	0.41229	0.43462
9.2	0.13343	0.12596	0.40790	0.42952
9.4	0.13196	0.12473	0.40364	0.42459
9.6	0.13054	0.12354	0.39951	0.41983
9.8	0.12916	0.12238	0.39551	0.41522
10.0	0.12783	0.12126	0.39163	0.41076

Table B.4 The complementary error function

$$\operatorname{erfc} \eta = 1 - \frac{2}{\pi^{1/2}} \int_0^{\eta} e^{-u^2}\, du$$

η	erfc η	η	erfc η	η	erfc η
0.00	1.0000	0.76	0.2825	1.52	0.03159
0.02	0.9774	0.78	0.2700	1.54	0.02941
0.04	0.9549	0.80	0.2579	1.56	0.02737
0.06	0.9324	0.82	0.2462	1.58	0.02545
0.08	0.9099	0.84	0.2349	1.60	0.02365
0.10	0.8875	0.86	0.2239	1.62	0.02196
0.12	0.8652	0.88	0.2133	1.64	0.02038
0.14	0.8431	0.90	0.2031	1.66	0.01890
0.16	0.8210	0.92	0.1932	1.68	0.01751
0.18	0.7991	0.94	0.1837	1.70	0.01621
0.20	0.7773	0.96	0.1746	1.72	0.01500
0.22	0.7557	0.98	0.1658	1.74	0.01387
0.24	0.7343	1.00	0.1573	1.76	0.01281
0.26	0.7131	1.02	0.1492	1.78	0.01183
0.28	0.6921	1.04	0.1413	1.80	0.01091
0.30	0.6714	1.06	0.1339	1.82	0.01006
0.32	0.6509	1.08	0.1267	1.84	0.00926
0.34	0.6306	1.10	0.1198	1.86	0.00853
0.36	0.6107	1.12	0.1132	1.88	0.00784
0.38	0.5910	1.14	0.1069	1.90	0.00721
0.40	0.5716	1.16	0.10090	1.92	0.00662
0.42	0.5525	1.18	0.09516	1.94	0.00608
0.44	0.5338	1.20	0.08969	1.96	0.00557
0.46	0.5153	1.22	0.08447	1.98	0.00511
0.48	0.4973	1.24	0.07950	2.00	0.00468
0.50	0.4795	1.26	0.07476	2.10	0.002980
0.52	0.4621	1.28	0.07027	2.20	0.001863
0.54	0.4451	1.30	0.06599	2.30	0.001143
0.56	0.4284	1.32	0.06194	2.40	0.000689
0.58	0.4121	1.34	0.05809	2.50	0.000407
0.60	0.3961	1.36	0.05444	2.60	0.000236
0.62	0.3806	1.38	0.05098	2.70	0.000134
0.64	0.3654	1.40	0.04772	2.80	0.000075
0.66	0.3506	1.42	0.04462	2.90	0.000041
0.68	0.3362	1.44	0.04170	3.00	0.000022
0.70	0.3222	1.46	0.03895	3.20	0.000006
0.72	0.3086	1.48	0.03635	3.40	0.000002
0.74	0.2953	1.50	0.03390	3.60	0.000000

For large η, $\operatorname{erfc} \eta = \frac{1}{\pi^{1/2}} e^{-\eta^2} \left[\frac{1}{\eta} - \frac{1}{2\eta^3} + \frac{1 \times 3}{2^2 \eta^5} - \frac{1 \times 3 \times 5}{2^3 \eta^7} + \cdots \right]$

APPENDIX

C

CHARTS

LIST OF FIGURES

Heat Conduction

Thermal Radiation

Heat Exchangers

C.4*a* LMTD correction factor for a heat exchanger with one shell pass and 2, 4, 6, . . . tube passes.

C.4*b* LMTD correction factor for a cross-flow heat exchanger with both fluids unmixed.

C.4*c* LMTD correction factor for a cross-flow heat exchanger with both fluids mixed.

C.4*d* LMTD correction factor for a cross-flow heat exchanger with two tube passes (unmixed) and one shell pass (mixed).

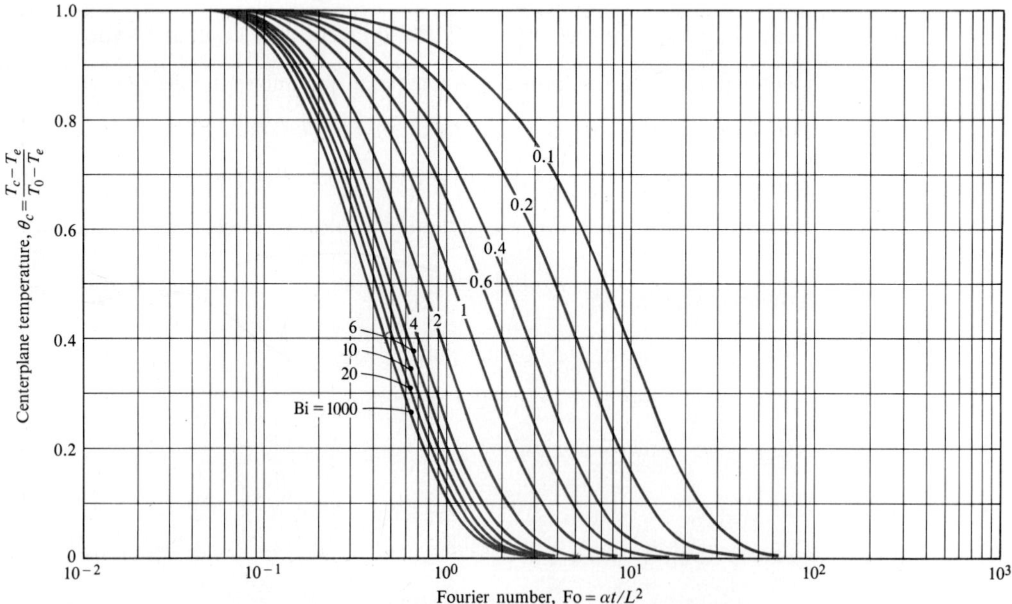

Figure C.1*a* Centerplane temperature response for a convectively cooled slab; Bi $= h_c L/k$, where L is the slab half-width.

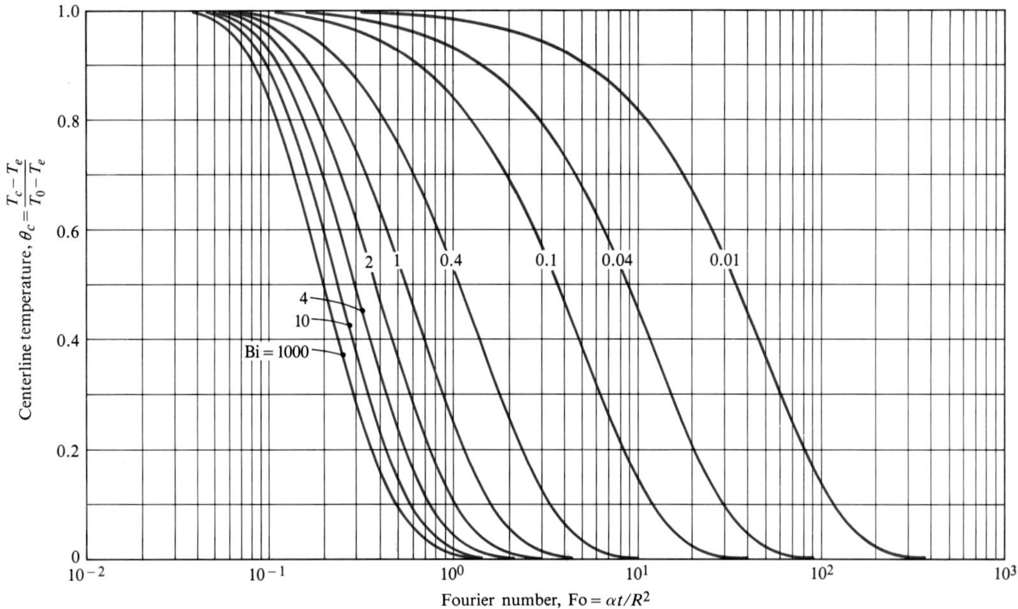

Figure C.1*b* Centerline temperature response for a convectively cooled cylinder; Bi $= h_c R/k$.

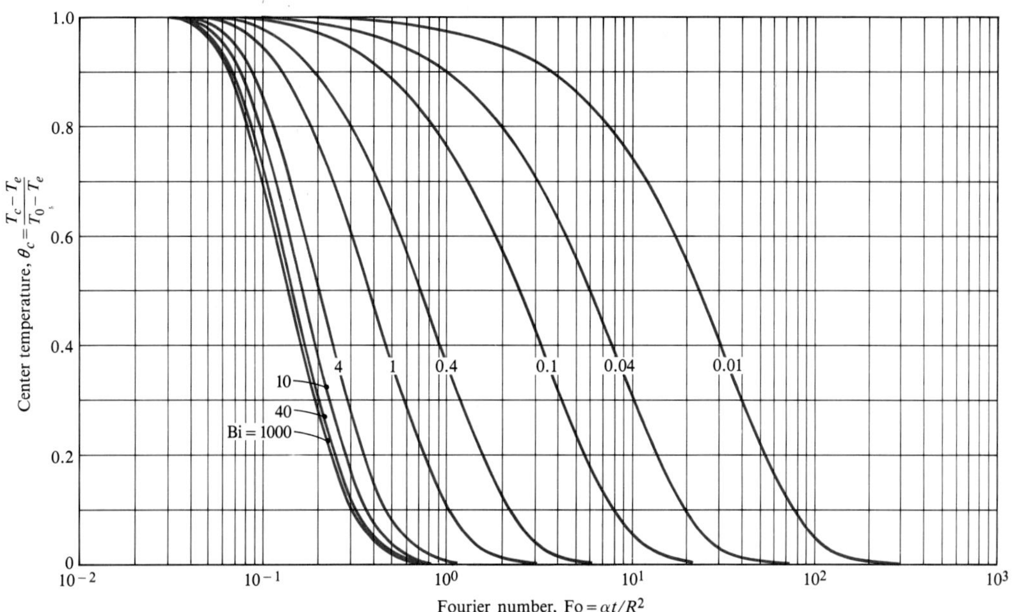

Figure C.1*c* Center temperature response for a convectively cooled sphere; Bi $= h_c R/k$.

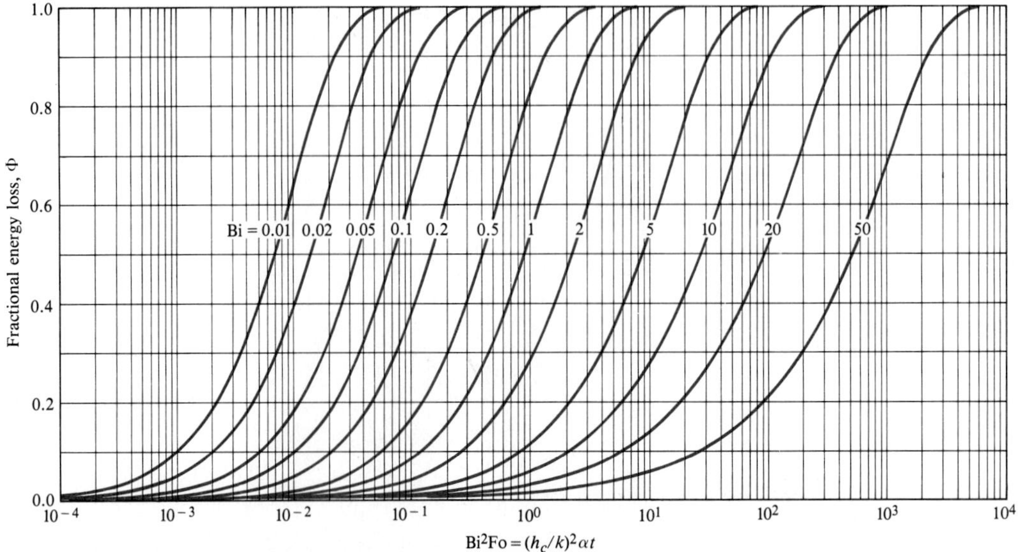

Figure C.2*a* Fractional energy loss for a convectively cooled slab; Bi = h_cL/k, where L is the slab half-width.

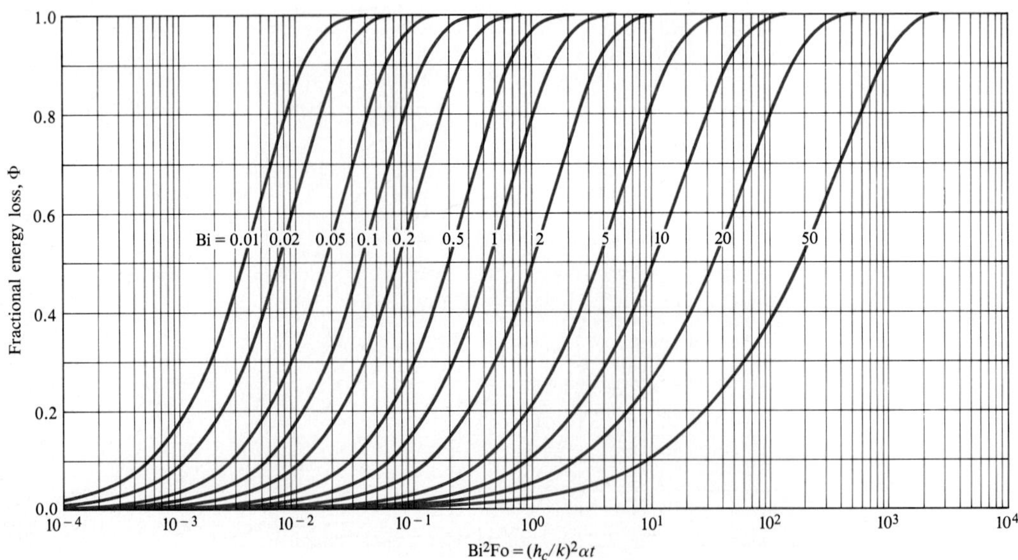

Figure C.2*b* Fractional energy loss for a convectively cooled cylinder; Bi = h_cR/k.

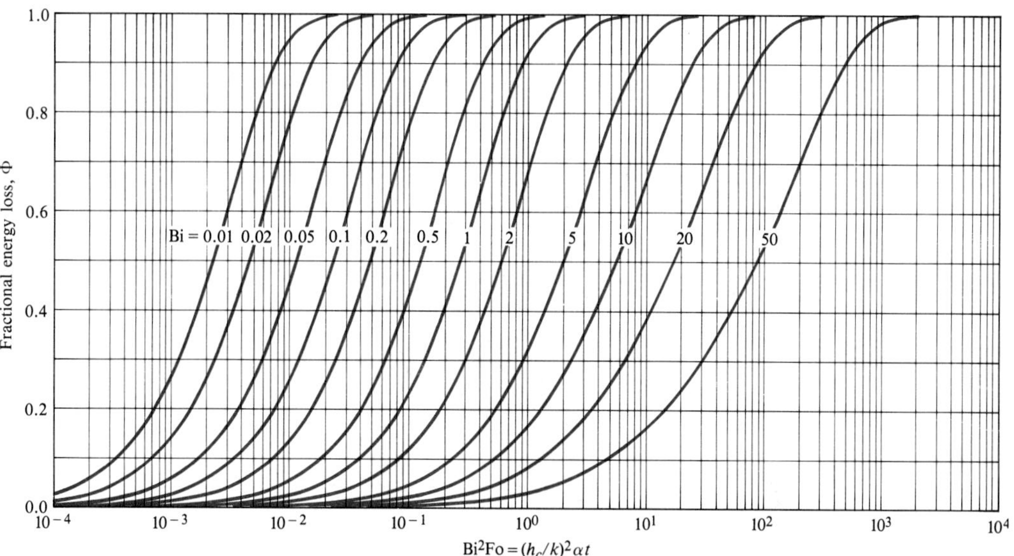

Figure C.2*c* Fractional energy loss for a convectively cooled sphere; Bi $= h_cR/k$.

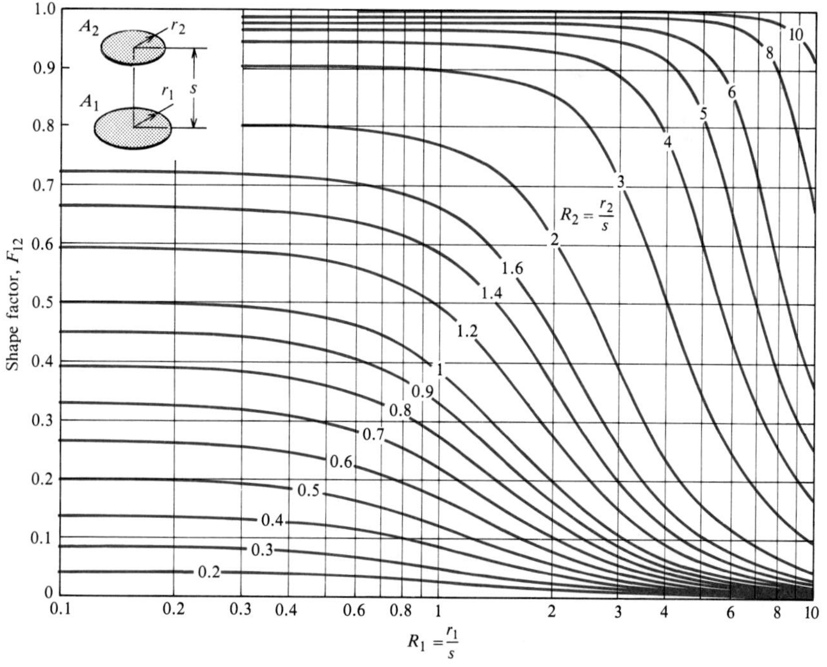

Figure C.3*a* Shape (view) factor for coaxial parallel disks.

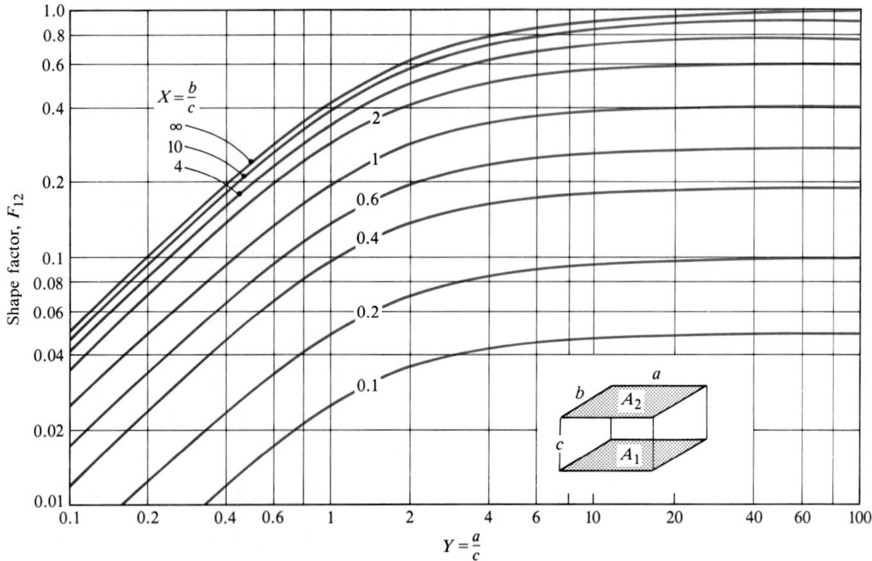

Figure C.3*b* Shape (view) factor for opposite rectangles.

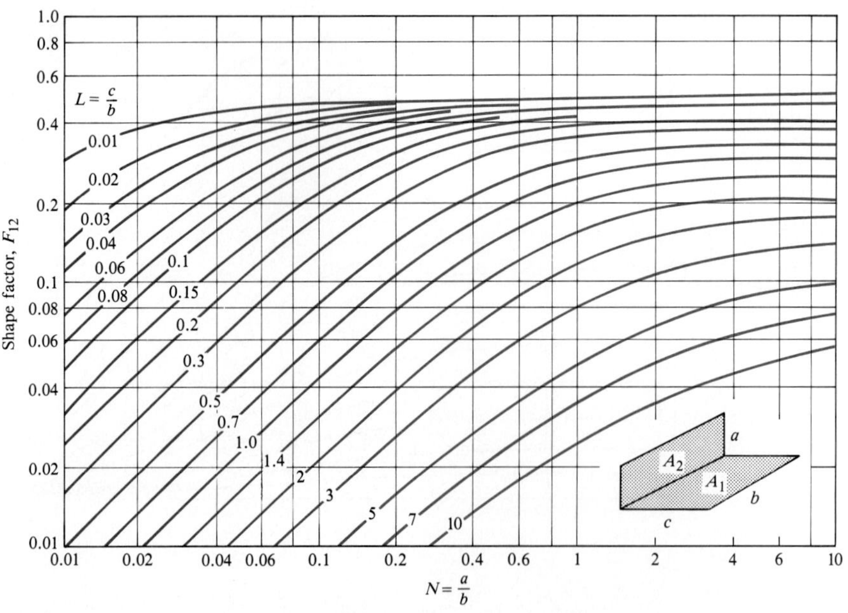

Figure C.3*c* Shape (view) factor for adjacent rectangles.

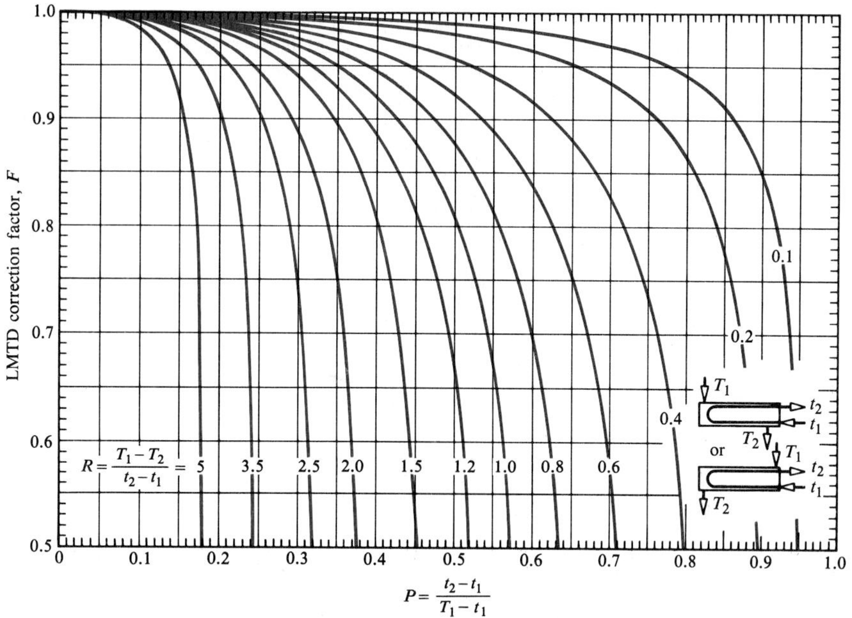

Figure C.4*a* LMTD correction factor for a heat exchanger with one shell pass and 2, 4, 6, . . . tube passes.

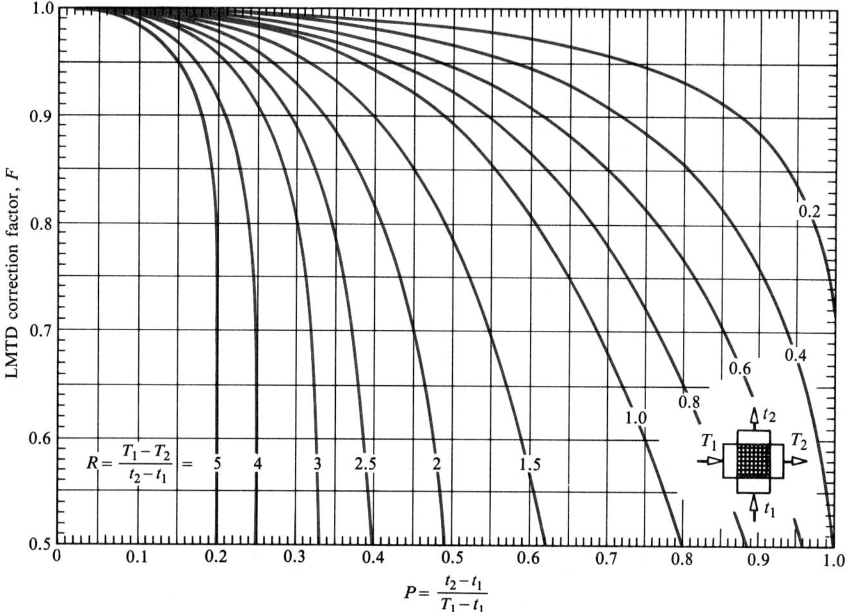

Figure C.4*b* LMTD correction factor for a cross-flow heat exchanger with both fluids unmixed.

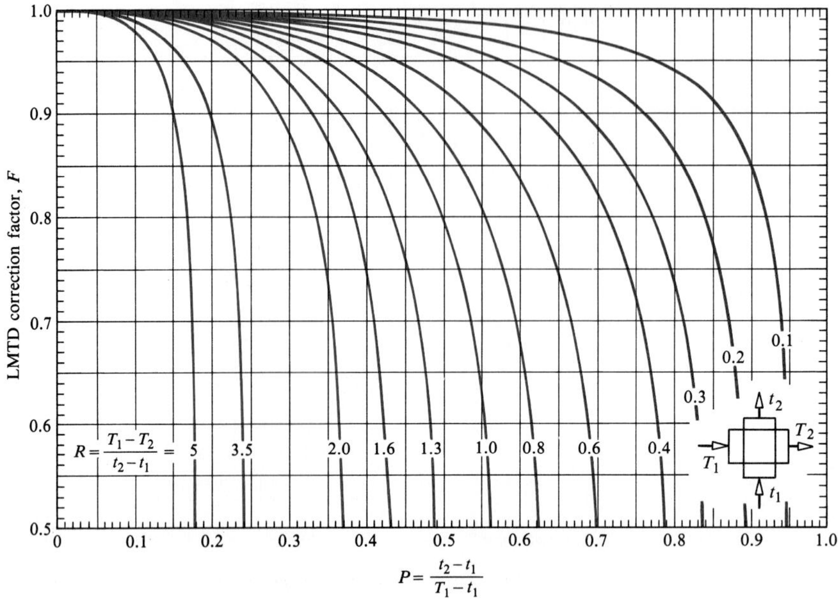

Figure C.4*c* LMTD correction factor for a cross-flow heat exchanger with both fluids mixed.

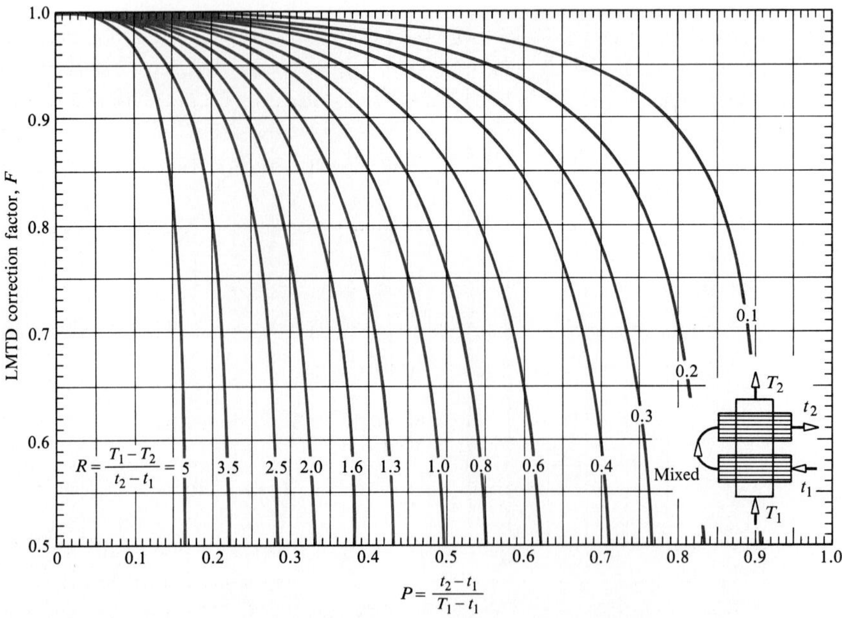

Figure C.4*d* LMTD correction factor for a cross-flow heat exchanger with two tube passes (unmixed) and one shell pass (mixed).

BIBLIOGRAPHY

GENERAL

There are many general heat transfer texts that can be consulted by the student seeking a different perspective on the material presented in *Heat and Mass Transfer*. Following is a selection of books that the student might find useful.

Chapman, A. J., *Fundamentals of Heat Transfer,* Macmillan, New York (1987).

Holman, J. P., *Heat Transfer,* 7th ed., McGraw-Hill, New York (1990).

Incropera, F. C., and DeWitt, D. P., *Fundamentals of Heat and Mass Transfer,* 3rd ed., John Wiley & Sons, New York (1990).

Kreith, F., and Bohn, M. S., *Principles of Heat Transfer,* 4th ed., Harper & Row, New York (1986).

Lienhard, J. H., *A Heat Transfer Textbook,* 2nd ed., Prentice-Hall, Englewood Cliffs, N.J. (1987).

Özisik, M. N., *Heat Transfer,* McGraw-Hill, New York (1985).

White, F. M., *Heat and Mass Transfer,* Addison-Wesley, Reading, Mass. (1988).

Somewhat more advanced texts include:

Bird, R. B., Stewart, W. E., and Lightfoot, E. N., *Transport Phenomena,* John Wiley & Sons, New York (1960).

Eckert, E. R. G., and Drake, R. M., Jr., *Analysis of Heat and Mass Transfer,* McGraw-Hill, New York (1972).

Gebhart, B., *Heat Transfer,* 2nd ed., McGraw-Hill, New York (1971).

Kutateladze, S. S., *Fundamentals of Heat Transfer,* Edward Arnold (Publishers) Ltd., London (1963). This book is translated from the original Russian language second edition.

Rohsenow, W. M., and Choi, H. Y., *Heat, Mass, and Momentum Transfer,* Prentice-Hall, Englewood Cliffs, N.J. (1961).

Useful reference works include:

Rohsenow, W. M., Hartnett, J. P., and Ganic, E. N., eds., *Handbook of Heat Transfer Fundamentals,* 2nd ed., McGraw-Hill, New York (1985).
Jakob, M., *Heat Transfer,* 2 vols., John Wiley & Sons, New York (1949).
McAdams, W. H., *Heat Transmission,* 3rd ed., McGraw-Hill, New York (1954).

An introduction to temperature measurement is given by:

Holman, J. P., *Experimental Methods for Engineers,* 5th ed., McGraw-Hill, New York (1989).

CHAPTER 2

Conduction in nuclear fuel elements is dealt with in:

El-Wakil, M. M., *Nuclear Heat Transport,* 3rd printing, American Nuclear Society, LaGrange Park, Ill. (1981), Chap. 13.

More detailed treatments of fins are given in:

Schneider, P. J., *Conduction Heat Transfer,* Addison-Wesley, Reading, Mass. (1955).
Kern, D. Q., and Kraus, A. D., *Extended Surface Heat Transfer,* McGraw-Hill, New York (1972).

CHAPTER 3

A treatise on the mathematical analysis of heat conduction is:

Carslaw, H. S., and Jaeger, J. C., *Conduction of Heat in Solids,* 2nd ed., Oxford University Press, New York (1959).

It contains a large number of solutions for steady and transient heat conduction problems.

Transform methods are emphasized in:

Özisik, M. N., *Heat Conduction,* John Wiley & Sons, New York (1980).

More elementary heat conduction texts are:

Grigull, U., and Sander, H., *Heat Conduction,* Springer-Verlag and Hemisphere, Washington, D.C. (1984).
Myer, G. E., *Analytical Methods in Conduction Heat Transfer,* Genium, Schenectady, N.Y. (1987).

A comprehensive compilation of conduction shape factors is given by:

Hahne, E., and Grigull, U., "Formfactor und Formwiderstand der Stationären Mehrdimensionalen Wärmeleitung," *Int. J. Heat Mass Transfer,* 18, 751–767 (1975).

More advanced treatments of numerical methods for heat conduction may be found in:

Patankar, S. V., *Numerical Heat Transfer and Fluid Flow,* Hemisphere, Washington, D.C., and McGraw-Hill, New York (1980).

Minkowycz, W. J., Sparrow, E. M., Schneider, G. E., and Pletcher, R. H., *Handbook of Numerical Heat Transfer,* John Wiley & Sons, New York (1988).

CHAPTER 4

Useful compilations of correlation formulas for convective heat transfer may be found in the following reference works.

Hewitt, G. F., coord. ed., *Hemisphere Handbook of Heat Exchanger Design, 2: Fluid Mechanics and Heat Transfer,* Hemisphere, New York (1990).

Kakac, S., Shah, R. H., and Aung, W., *Handbook of Single Phase Convective Heat Transfer,* John Wiley & Sons, New York (1987).

Norris, R. H., et al., eds., *Heat Transfer and Fluid Flow Data Books,* General Electric Co., Schenectady, N.Y. (1943, with supplements to current date).

Rohsenow, W. M., Hartnett, J. P., and Ganic, E. N., eds., *Handbook of Heat Transfer Fundamentals,* 2nd ed., McGraw-Hill, New York (1985).

Žukauskas, A., and Ulinskas, R., *Heat Transfer in Tube Banks in Crossflow,* Hemisphere, New York (1988).

CHAPTER 5

Advanced texts on convective heat transfer are:

Bejan, A., *Convective Heat Transfer,* John Wiley & Sons, New York (1984).

Burmeister, L. C., *Convection Heat Transfer,* 2nd ed., John Wiley & Sons, New York (1993).

Gebhart, B., Jaluria, Y., Mahajan, R. L., and Sammakia, B., *Buoyancy-Induced Flows and Transport,* Hemisphere, New York (1988).

Kays, W. M., and Crawford, M. E., *Convective Heat and Mass Transfer,* 3rd ed., McGraw-Hill, New York (1993).

Useful supporting fluid mechanics texts are:

Schlichting, H., *Boundary Layer Theory,* 7th ed. (trans. J. Kestin), McGraw-Hill, New York (1979).

White, F. M., *Viscous Fluid Flow,* 2nd ed., McGraw-Hill, New York (1991).

CHAPTER 6

Recommended advanced texts on radiation heat transfer are:

Brewster, M. Q., *Thermal Radiative Transfer and Properties,* John Wiley & Sons, New York (1992).

Edwards, D. K., *Radiation Heat Transfer Notes,* Hemisphere, Washington, D.C. (1981).

Hottel, H. C., and Sarofim, A. F., *Radiative Transfer,* McGraw-Hill, New York (1967).

Modest, M. F., *Radiative Heat Transfer,* McGraw-Hill, New York (1993).

Siegel, R., and Howell, J. R., *Thermal Radiation Heat Transfer,* 3rd ed., Hemisphere, Washington, D.C. (1992).

Sparrow, E. M., and Cess, R. D., *Radiation Heat Transfer,* augm. ed., Hemisphere, Washington, D.C. (1978).

Solar radiation is treated in:

Beckman, W. A., Klein, S. A., and Duffie, J. A., *Solar Heating Design by the F-Chart Method,* John Wiley & Sons, New York (1977).

Duffie, J. A., and Beckman, W. A., *Solar Engineering of Thermal Processes,* John Wiley & Sons, New York (1980).

Kreith, F., and Kreider, J. F., *Principles of Solar Engineering,* Hemisphere/McGraw-Hill, Washington, D.C. (1978).

Design methods for solar collectors are given by:

Edwards, D. K., *Solar Collector Design,* Franklin Institute Press, Philadelphia (1978).

CHAPTER 7

Reference works for condensation, boiling, and two-phase flow include:

Carey, V. P., *Liquid-Vapor Phase-Change Phenomena,* Hemisphere, Washington, D.C. (1992).

Chisholm, D., *Two-phase Flow in Pipelines and Heat Exchangers,* George Godwin, London (1983).

Collier, J. G., *Convective Boiling and Condensation,* 2nd ed., McGraw-Hill, New York (1981).

Dwyer, D. E., *Boiling Liquid-Metal Heat Transfer,* American Nuclear Society, Hinsdale, Ill. (1976).

Hahne, E., and Grigull, U., eds., *Heat Transfer in Boiling,* Academic Press, New York (1977).

Hsu, Y. Y., and Graham, R. W., *Transport Processes in Boiling and Two-Phase Systems,* Hemisphere/McGraw-Hill, Washington, D.C. (1976).

van Stalen, S., and Cole, R., *Boiling Phenomena,* 2 vols., Hemisphere, Washington, D.C.

Methods for calculating pressure drop in two-phase flows are summarized in:

Griffith, P., "Two-Phase Flow," Chap. 11 in Rohsenow, W. M., Hartnett, J. P., and Ganic, E. W., eds., *Handbook of Heat Transfer Fundamentals,* 2nd ed., McGraw-Hill, New York (1985).

Reference works for heatpipes include:

Chi, S. W., *Heat Pipe Theory and Practice,* Hemisphere, Washington, D.C. (1976).

Dunn, P., and Reay, D. A., *Heat Pipes,* 4th ed., Pergamon Press, Oxford (1994).

Ivanovskii, M. N., Sorokin, V. P., and Yagodkin, I. V., *The Physical Principles of Heat Pipes,* trans. ed. G. Rice, Oxford University Press, Oxford (1982).

Terpstra, M., and van Veen, J. G., *Heat Pipes Construction and Application,* Elsevier Applied Science, London (1987).

CHAPTER 8

Useful reference books are:

Afgan, N., and Schlunder, E. U., *Heat Exchanger Design and Theory Source-Book,* Scripta/McGraw-Hill, New York (1974).

Barron, R. F., *Cryogenic Systems,* 2nd ed., Oxford University Press, New York (1986).

Chereminisoff, N. P., ed., *Handbook of Heat and Mass Transfer, Vol. 1: Heat Transfer Operations,* Gulf, Houston (1986).

Fraas, A. P., *Heat Exchanger Design,* 2nd ed., John Wiley & Sons, New York (1989).

Green, D. W., and Maloney, J. O., eds., *Perry's Chemical Engineers Handbook,* 6th ed., McGraw-Hill, New York (1984).

Hausen, H., *Heat Transfer in Counterflow, Parallel Flow and Cross Flow,* McGraw-Hill, New York (1983).

Hewitt, G. F., coord. ed., *Hemisphere Handbook of Heat Exchanger Design,* Hemisphere, New York (1990).

Kays, W. M., and London, A. L., *Compact Heat Exchangers,* 3rd ed., McGraw-Hill, New York (1984).

Kutz, M., ed., *Mechanical Engineers Handbook,* John Wiley & Sons, New York (1986).

Martin, H., *Heat Exchangers,* Hemisphere, Washington, D.C. (1992).

Shah, R. K., and Mueller, A. C., "Heat Exchangers," Chap. 4 in Rohsenow, W. M., Hartnett, J. P., and Ganic, E. N., eds., *Handbook of Heat Transfer Applications,* McGraw-Hill, New York (1985).

Standards of Tubular Exchangers Manufacturers Association, 6th ed. (1978).

CHAPTER 9

Most modern treatments of diffusive mass transfer derive from:

Bird, R. B., "Theory of Diffusion," in *Advances in Chemical Engineering,* vol. 1, Academic Press, New York (1956) (see pp. 170 et seq.).

Essentially the same material, together with worked examples, is given by:

Bird, R. B., Stewart, W. E., and Lightfoot, E. N., *Transport Phenomena,* John Wiley & Sons, New York (1960).

A text that has a treatment of mass transfer similar to that in this text is:

Lienhard, J. H., *A Heat Transfer Text Book,* 2nd ed., Prentice-Hall, Englewood Cliffs, N.J. (1987).

Texts written for chemical engineers include:

Cussler, E. L., *Diffusion: Mass Transfer in Fluid Systems,* Cambridge University Press, London (1984).

Geankoplis, C. J., *Transport Processes and Unit Operations,* 3rd ed., Prentice-Hall, Englewood Cliffs, N.J. (1993).

Hines, A. L., and Maddox, R. N., *Mass Transfer,* Prentice-Hall, Englewood Cliffs, N.J. (1985).

Sherwood, T. K., Pigford, R. L., and Wilke, C. R., *Mass Transfer,* McGraw-Hill, New York (1975).

Skelland, A. P., *Diffusional Mass Transfer,* R. E. Krieger, Melbourne, Fla. (1985).

The text by Geankoplis contains a nice introduction to the drying of porous solids.

More advanced texts containing analytical solutions for many steady and transient mass diffusion problems include:

Crank, J., *The Mathematics of Diffusion,* 2nd ed., Clarendon Press, Oxford (1975).

Gebhart, B. J., *Heat Conduction and Mass Diffusion,* McGraw-Hill, New York (1993).

CHAPTER 10

The fundamentals of diffusion in a moving medium are nicely presented in:

Bird, R. B., Stewart, W. E., and Lightfoot, E. N., *Transport Phenomena,* John Wiley & Sons, New York (1960).

The Spalding formulation of steady convective heat and mass transfer is presented in detail in:

Kays, W. M., and Crawford, M. E., *Convective Heat and Mass Transfer,* 3rd ed., McGraw-Hill, New York (1993).

A wide range of engineering problems involving convective mass transfer are treated using the simple Reynolds flux model in:

Spalding, D. B., *Convective Mass Transfer,* McGraw-Hill, New York (1963).

A number of interesting advanced mass transfer problems are analyzed in:

Rosner, D. E., *Transport Processes in Chemical Reacting Flow Systems,* Butterworth-Heinemann, Stoneham, Mass. (1986).

CHAPTER 11

Texts written for chemical engineers dealing with mass exchangers include:

Geankoplis, C. J., *Transport Processes and Unit Operations,* 3rd ed., Prentice-Hall, Englewood Cliffs, N.J. (1993).

Treybal, R. E., *Mass Transfer Operations,* 3rd ed., McGraw-Hill, New York (1980).

Walas, S. M., *Chemical Process Equipment: Selection and Design,* Butterworth-Heinemann, Stoneham, Mass. (1988).

A great variety of pollution control equipment is nicely described in:

Liptak, B. G., *Municipal Waste Disposal in the 1990's,* Chilton, Radnor, Pa. (1991).

Additional references for electrostatic precipitators include:

Robinson, M., "Electrostatic Precipitation," in Strauss, W., ed., *Air Pollution Control: Part I,* Wiley-Interscience, New York (1971).

Strauss, W., *Industrial Gas Cleaning,* Pergamon Press, New York (1975).

Design methods for electrostatic precipitators are described by:

Turner, J. H., Lawless, P. A., Yamamoto, T., Coy, D. W., Greiner, G. P., McKenna, J. D., and Vatavuk, W. M., "Sizing and Costing of Electrostatic Precipitators. Part I: Sizing Considerations," *J. Air Pollut. Control Assoc.,* 38, 458–471 (1988); "Part II: Costing Considerations," *J. Air Pollut. Control Assoc.,* 38, 715–726 (1988).

Design methods for gas absorption are given by:

Zenz, F. A., "Design of Gas Absorption Towers," Sec. 3.2 in Schweitzer, P. A., ed., *Handbook of Separation Techniques for Chemical Engineers,* McGraw-Hill, New York (1979).

A complete reference on filters is:

Matteson, M. J., and Orr, C., eds., *Filtration: Principles and Practice,* Marcel Dekker, New York (1987).

Reference works for humidifiers and cooling towers include:

American Society of Heating, Refrigerating and Air Conditioning Engineers, *ASHRAE 1992 Systems and Equipment Handbook (SI Version)*, ASHRAE, Atlanta (1992).

Baker, D., *Cooling Tower Performance,* Chemical Publishing Co., New York (1984).

Cheremisinoff, N. P., and Cheremisinoff, P. N., *Cooling Towers: Selection, Design and Practice,* Ann Arbor Science Publishers, Ann Arbor, Mich. (1981).

Hill, G. B., Pring, E. S., and Osborn, P. D., *Cooling Towers: Principles and Practice,* 3rd ed., Butterworth-Heinemann, London (1990).

Johnson, B. M., ed., *Cooling Tower Performance Prediction and Improvement,* vols. 1 and 2, EPRI GS-6370, Electric Power Research Institute, Palo Alto, Calif. (1990).

Kelly, N. W., *Kelly's Handbook of Cross-Flow Cooling Tower Performance,* Neil W. Kelly and Associates, Kansas City, Mo. (1976).

Singham, J. R., "Natural Draft Towers" and "Mechanical Draft Towers," Secs. 3.12.2 and 3.12.4 in Hewitt, G. F., coord. ed., *Hemisphere Handbook of Heat Exchange Design,* Hemisphere, New York (1990).

Drying theory and practice are described in:

Keey, R. B., *Drying Principles and Practice,* Pergamon Press, Elmsford, N.Y. (1972).

Keey, R. B., *Drying of Loose and Particulate Materials,* Hemisphere, New York (1992).

Mujamdar, A. S., ed., *Handbook of Industrial Drying,* Marcel Dekker, New York (1987).

Strumillo, C., and Kudra, T., *Drying: Principles, Applications and Design,* Gordon & Breach, New York (1986).

APPENDIX A

The most comprehensive compilation of thermophysical property data available is:

Touloukian, Y. S., and Ho, C. Y., eds., *Thermophysical Properties of Matter,* 13 vols., IFI/Plenum Press, New York (1970–1977; addenda continue to be issued).

Other useful sources include:

American Society of Heating, Refrigerating and Air Conditioning Engineers, *ASHRAE Handbook of Fundamentals,* ASHRAE, New York (1981).

American Society of Metals, *Metals Handbook, Vol. 1: Properties and Selection of Metals,* 8th ed., ASM, Metals Park, Ohio (1961).

Eckert, E. R. G., and Drake, R. M., Jr., *Analysis of Heat and Mass Transfer,* McGraw-Hill, New York (1972).

Hewitt, G. F., coord. ed., *Hemisphere Handbook of Heat Exchanger Design,* Hemisphere, New York (1990).

Irvine, T. F., and Hartnett, J. P., eds., *Steam and Air Tables in SI Units,* Hemisphere/McGraw-Hill, Washington, D.C. (1976).

McAdams, W. H., *Heat Transmission,* 3rd ed., McGraw-Hill, New York (1954).

Norris, R. H., et al., eds., *Heat Transfer and Fluid Flow Data Books,* General Electric Co., Schenectady, N.Y. (1943, with supplements to current date).

Vargaftik, N. B., *Handbook of Physical Properties of Liquids and Gases,* 2nd ed., Hemisphere, Washington, D.C. (1983).

HEAT AND MASS TRANSFER JOURNALS

The two most widely read heat and mass transfer journals are:

International Journal of Heat and Mass Transfer
Journal of Heat Transfer (Transactions of the American Society of Mechanical Engineers)

Other relevant journals include:

AIAA Journal of Thermophysics and Heat Transfer
AIChE Journal
ASHRAE Journal
Aerosol Science and Technology
Chemical Engineering Progress
Combustion Science and Technology
Experimental Heat Transfer
Experimental Thermal and Fluid Science
Heat and Fluid Flow
Heat Transfer Engineering
Heat Transfer: Japanese Research
Heat Transfer: Recent Contents (titles)
Heat Transfer: Soviet Research
International Communications in Heat and Mass Transfer
Journal of Aerosol Science
Journal of Enhanced Heat Transfer
Journal of Solar Energy Engineering
Letters in Heat and Mass Transfer
Nuclear Engineering and Design
Numerical Heat Transfer
PHOENICS Journal
Previews of Heat and Mass Transfer (Abstracts)
Wärme und Stoffübertragung

NOMENCLATURE

A	area, m^2; amplitude, m
A_c	cross-sectional area; area for flow, m^2
A_{eff}	effective area, m^2
A_f	fin surface area, m^2
A_{fr}	frontal area, m^2
A_p	prime (unfinned) area, profile area of a straight fin, particle surface area, m^2
a	surface area per unit volume, m^{-1}; speed of sound, m/s
$\mathscr{A}$	Avogadro's number, molecules/kmol
B	blowing parameter
BP	boiling point
Bi	Biot number, Eqs. (1.40) and (9.50)
Bo	Boussinesq number, Eq. (4.31)
Br	Brinkman number, Eq. (4.24)
$\mathscr{B}$	mass transfer driving force
b	height of air inlet for a natural-draft cooling tower, m
C	flow thermal capacity (flow rate times specific heat), W/K; cost, \$; thermal capacity, J/K; electrical capacitance, F; slip correction factor, Eq. (9.128)
C_D	drag coefficient, Eq. (4.68)
C_{He}	Henry constant, bar or atm
C_R	thermal capacity for a regenerator, W/K
C_f	skin friction coefficient, Eq. (4.14)
C_{fb}	constant in film boiling correlations
$C_{\max}$	constant in boiling peak heat flux correlations
$C_{\min}$	constant in boiling minimum heat flux correlations
C_{nb}	constant in nucleate boiling correlation
$\mathscr{C}$	annual cost, \$/yr
c	specific heat, J/kg K; average molecular speed, m/s; unit cost (e.g., per unit transfer area), \$/m^2; molar concentration, kmol/m^3
c_p	specific heat at constant pressure, J/kg K; pumping power unit cost, \$/W h
c_v	specific heat at constant volume, J/kg K
$\mathscr{c}$	speed of light, m/s; unit annual cost (e.g., per unit transfer area), \$/m^2 yr
$\mathscr{c}_0$	speed of light in a vacuum, m/s
$\mathscr{c}_{pL}$	annual cost of pumping power per unit length, \$/m yr

D	diameter, m
D_B	shell base diameter for a natural-draft cooling tower, m
D_p	nominal size of a random packing, m
D_h	hydraulic diameter, m
DF	decontamination factor
$\mathscr{D}_{12}$	binary diffusion coefficient, m^2/s
$\mathscr{D}_{im}$	effective binary diffusion coefficient, m^2/s
$\mathscr{D}_p$	particle Brownian diffusion coefficient, m^2/s
d	molecule diameter, Å
d_f	fiber diameter, μm
d_p	particle diameter, m or μm
E	energy, J; emissive power, W/m^2; voltage, V; electric field, V/m
E_a	activation energy, kcal/mol or kJ/kmol
$\mathscr{E}$	dimensionless thermoeconomic parameter, Eq. (8.98)
e	elementary electric charge, A s
Ec	Eckert number, Eq. (4.23)
Eu	Euler number, Eq. (4.17)
F	force, N; function
F_{ij}	shape (view) factor
Fo	Fourier number, Eq. (3.35)
Fr	Froude number, Eq. (11.38)
$\mathscr{F}$	LMTD correction factor
$\mathscr{F}_{ij}$	transfer factor
f	dimensionless stream function; friction factor, Eq. (4.15) or (4.119); sound frequency, Hz; friction coefficient, kg/s
G	irradiation, W/m^2; mass velocity, kg/m^2 s; gas stream superficial mass velocity, kg/m^2 s; correction factor
G_0	solar constant, kW/m^2
$\mathscr{G}$	area × shape factor product, m^2
$\mathscr{G}_m$	mole transfer conductance, $kmol/m^2$ s
Gr	Grashof number, Eq. (4.27)
g	gravitational acceleration, m/s^2
$\mathbf{g}$	gravity vector, m/s^2
$\mathscr{g}_m$	mass transfer conductance, kg/m^2 s
$\mathscr{g}_h$	heat transfer conductance, kg/m^2 s
H	elevation or height, m; total enthalpy, J/kg; hydrodynamic factor for Kuwabara flow, Table 11.2
He	Henry number, Eq. (9.13)
ΔH_c	heat of combustion per kmol, J/kmol
h	heat transfer coefficient, W/m^2 K; enthalpy, J/kg; characteristic roughness height, m; metric coefficient
h_b	boiling heat transfer coefficient, W/m^2 K
h_c	convective heat transfer coefficient, W/m^2 K
h_i	interfacial conductance, W/m^2 K; interfacial heat transfer coefficient, W/m^2 K
h_r	radiative heat transfer coefficient, W/m^2 K
h_{fg}	enthalpy of vaporization, J/kg
h'_{fg}	enthalpy of vaporization plus subcooling correction, J/kg
h_{fs}	enthalpy of solidification, J/kg
h_{sg}	enthalpy of sublimation, J/kg
Δh_c	heat of combustion per unit mass, J/kg
Δh_d	heat of dissociation per unit mass, J/kg
$\mathscr{h}$	Planck's constant, J s
I	intensity, W/m^2 sr; modified Bessel function of first kind; electrical current, A
i	annual interest rate
$\mathbf{i}$	unit vector in the x direction
J	radiosity, W/m^2; Bessel function of first kind; diffusion molar flux, $kmol/m^2$ s
$\mathscr{J}$	molecular flux, molecules/m^2 s; particle flux, particles/m^2 s
Ja	Jakob number, Eq. (3.90)
j	superficial velocity, m/s; diffusion mass flux, kg/m^2 s
$\mathbf{j}$	unit vector in the y direction
K	modified Bessel function of second kind; parameter in forced-convection condensation correlation, Eq. (7.101); dimensionless flow rate in flooding correlation, Eq. (11.65)

K_e, K_c expansion and contraction coefficients, Eqs. (8.72) and (8.73)
Ka Kapitza number, Eq. (11.45)
Kn Knudsen number, Eq. (9.129)
k thermal conductivity, W/m K; absorptive index
k_s equivalent sand grain roughness, m
k'' rate constant for a heterogeneous reaction, m/s if first-order
$\mathscr{k}$ Boltzmann constant, J/K
$\mathbf{k}$ unit vector in the z direction
L length, m; liquid stream superficial mass velocity, kg/m^2 s
L_c characteristic length defined by Eq. (7.63), $[\sigma/\rho_l - \rho_v)g]^{1/2}$, m
L_{eff} effective heatpipe length, m
Le Lewis number, Eq. (9.57)
$\mathscr{L}$ characteristic length; effective beam length, m
$\mathscr{L}_m$ mean beam length, m
$\mathscr{L}_m^o$ geometric mean beam length, m
ℓ Prandtl mixing length, m; collision mean free path, m
ℓ_t transport mean free path, m
M molecular weight, kg/kmol
$\dot{M}$ molar flow rate, kmol/s
$\dot{M}_L^\dagger$ molar flow rate of liquid stream divided by Henry number, kmol/s
$\mathscr{M}$ figure of merit, Eq. (7.134)
m mass fraction
$\dot{m}$ mass flow rate, kg/s
$\dot{m}''$ mass flow rate per unit area across a phase interface (mass transfer rate), kg/m^2 s
$\mathscr{m}$ mass of a molecule, kg
MP melting point
N number of plates; number of tube rows transverse to flow; absolute molar flux, kmol/m^2 s; number of velocity heads
N_{tu} number of transfer units, Eqs. (8.33), (11.62), and (11.95)
N_{tu}^1 one-side number of transfer units, Eq. (8.81)
Nu Nusselt number, Eqs. (4.19) and (7.20)
$\mathscr{N}$ molecule number density, molecules/m^3; particle number density, particles/m^3
n absolute mass flux, kg/m^2 s; refractive index; loan period, yr
$\mathscr{n}$ number fraction
$\mathbf{N}$ absolute molar flux vector in a mixture, kmol/m^2 s
$\mathbf{n}$ absolute mass flux vector in a mixture, kg/m^2 s
P pressure, Pa
P_E equivalent broadening pressure ratio, Eq. (6.98)
P_L S_L/D
P_T S_T/D
Pe Peclet number, Eq. (4.22), Section 11.3.4
Pr Prandtl number, Eq. (4.18)
p pitch, m; momentum, kg m/s
$\mathscr{P}$ perimeter, m; permeability, m^3 (STP)/m^2 s (atm/m); conserved property
Q thermal energy, J; particle charge, A s
$\dot{Q}$ rate of heat transfer into a system, rate of heat flow, W
$\dot{Q}_v$ internal heat source, W
$\dot{Q}_v'''$ internal heat source per unit volume, W/m^3
$\mathbf{q}$ heat flux vector, W/m^2
q heat flux, W/m^2
R radius, m; gas constant J/kg K; thermal resistance, K/W; electrical resistance, Ω; interception number, Section 11.3.4
R_C capacity ratio, Eq. (8.34)
R_c critical bubble radius, m
R_f fouling factor, [W/m^2 K]$^{-1}$
R_R ratio of regenerator matrix thermal capacity to flow thermal capacity, Eq. (8.62)
$\dot{R}'''$ molar rate of species production in a homogeneous reaction, kmol/m^3 s
Ra Rayleigh number, Eq. (4.29)
Re Reynolds number, Eqs. (4.13) and (7.18)

$\mathscr{R}$ universal gas constant, J/kmol K
r radial coordinate, m; recovery factor; annual charge (annuity), yr^{-1}
r_e effective pore radius of a catalyst, μm
r_p pore radius of a wick, m
$\dot{r}'''$ mass rate of species production in a homogeneous reaction, kg/m^3 s
S conduction shape factor, m; surface area, m^2; contact area per unit length, m^2
S_p pellet surface area, m^2
S' surface area per unit width, m
S_L longitudinal pitch, m
S_T transverse pitch, m
Sc Schmidt number, Section 9.2.3
Sh Sherwood number, Eq. (9.41)
St Stanton number, Eqs. (4.21) and (9.41)
Stk Stokes number, Section 11.3.4
$\mathscr{S}$ solubility, m^3(STP)/m^3 atm
$\mathscr{S}'$ solubility coefficient
T temperature, K or °C
T^+_{HW} hot water correction factor for cooling tower packings
t time, s; thickness, m; temperature, °C
t_c time constant, s
U overall heat transfer coefficient, W/m^2 K; internal energy, J
u specific internal energy, J/kg; velocity component in x direction, m/s; orthogonal curvilinear coordinate
u_b bulk velocity, m/s
V velocity, m/s; volume, m^3
V^E electrical migration velocity, m/s
v specific volume, m^3/kg; velocity component in y direction, m/s
v_t characteristic turbulence speed, m/s
v^* friction velocity, m/s
$\bar{v}$ average molecular speed, m/s
$\mathscr{V}$ characteristic velocity, m/s; annual value, \$/yr
υ value of heat energy, \$/W h
$\mathbf{V}$ velocity vector, m/s
$\mathbf{v}$ mass-average velocity vector in a mixture
$\mathbf{v}^*$ mole-average velocity vector in a mixture
W width of a surface, m; work done on a system, J; mass, kg; molar content of a system, kmol
$\dot{W}$ rate of doing work, W
We Weber number, Eqs. (7.136) and (11.38)
w mass content of a system, kg; velocity component in z direction, m/s
x rectangular coordinate, m; vapor mass quality; mole fraction; liquid-phase mole fraction
Y Bessel function of second kind
y rectangular coordinate, m; gas-phase mole fraction
Z_p height of packing above basin in a natural-draft cooling tower
z rectangular coordinate, m; elevation, m

GREEK SYMBOLS

α thermal diffusivity, m^2/s; absorptance; fraction of cross section occupied by gas flow
β thermal coefficient of volume expansion, K^{-1}; fin parameter; sound intensity, dB; geometric mesh factor; burning rate constant, m^2/s
β' heat transfer area per unit volume (plate-fin exchanger), m^{-1}
Γ flow rate per unit width, kg/m s
γ single-fiber collection efficiency
Δ finite increment; thermal boundary layer thickness, m
Δ_2 energy thickness, m, Eq. (5.62)
Δ_{m2} convection thickness for mass transfer, Eq. (10.138)
δ film thickness, m; hydrodynamic boundary layer thickness, m

δ_1 displacement thickness, m
δ_2 momentum thickness, m
δ_f equivalent stagnant film thickness, m
ε emittance; heat exchanger effectiveness; mass exchanger effectiveness; maximum energy of attraction for molecules, J/molecule; ionization fraction
$\Delta\varepsilon_c$ effectiveness correction for axial conduction
ε_M eddy diffusivity of momentum (eddy viscosity), m^2/s
ε_H eddy diffusivity of heat, m^2/s
ε_v void fraction
ζ dimensionless time; dimensionless velocity, $u/(2RT)^{1/2}$
η dimensionless spatial coordinate; similarity variable
η_f fin efficiency, Eq. (2.42)
η_p pump efficiency; catalyst pellet effectiveness
η_t total surface efficiency, Eq. (2.46)
Θ angle of exposure for a wheel-type regenerator, rad
θ angle, rad; contact angle, °; dimensionless temperature
κ absorption coefficient, m^{-1}; Darcy permeability, m^2; von Karman's constant
Λ Thiele modulus, Eq. (9.81)
λ eigenvalue; wavelength, μm; axial conduction parameter, Eq. (8.54) or (8.56)
μ dynamic viscosity, kg/m s; ion mobility, m^2/V s
ν kinematic viscosity, m^2/s; wavenumber, cm^{-1}
ν_f frequency, $Hz(s^{-1})$
ξ unheated starting length, m; dimensionless spatial coordinate; general spatial variable, m
ρ density, kg/m^3; reflectance
σ surface tension, N/m; Stefan-Boltzmann constant, $W/m^2\ K^4$; core-to-frontal-area ratio; electrical conductivity, $\Omega^{-1}\ m^{-1}$; collision diameter, Å
σ_c condensation coefficient; critical surface tension, N/m; collision cross section, m^2
σ_e evaporation coefficient
σ_t transport cross section, m^2
τ shear stress, N/m^2; time period, s; transmittance; annual operating time, h; tortuosity factor
Φ fractional loss of energy, Eq. (3.73); arrangement factors for tube banks, Eqs. (4.116) and (4.117); forced-convection boiling parameter, Eq. (7.93); dimensionless enthalpy; drag factor for random packing
Φ_{ij} factor in Wilke's mixture rule, Eq. (9.118)
ϕ angle, rad; dimensionless mass fraction; association parameter for a solvent
χ fin parameter, Eq. (2.42); tube bank pressure drop correction factor, Eq. (4.119); Martinelli parameter, Eq. (7.99); dynamic shape factor, Eq. (9.132)
Ψ Prandtl number function for natural convection, Eq. (4.84)
ψ stream function, m^2/s; tube bank pitch factor, Eq. (4.115)
Ω angular velocity, rad/s
Ω_μ collision integral for viscosity and thermal conductivity
$\Omega_{\mathscr{D}}$ collision integral for diffusion
ω solid angle, sr; angular velocity, rad/s; humidity ratio

SUBSCRIPTS

a radiation-emitting gaseous chemical species; annual
aw adiabatic wall
b bulk or mixed mean value for a stream; blackbody; boiling point

C	cold side or stream; capillary
c	convection; condenser; centerline; critical; coolant
conv	convection
e	external; free-stream; evaporator
es	free-stream composition at the surface temperature
F	friction
FC	forced convection
f	fin; fixed; forced; friction
fr	frontal
fu	fuel
G	gravity; gas stream
g	gas
H	hot side or stream; Hemholtz instability
h	based on enthalpy
i	inside; internal; interfacial; initial; species i
j	species j
L	liquid stream
l	liquid
lm	logarithmic mean value
M	momentum; matrix
MP	melting point
m	mean; m-surface; mass transfer
N	normal
n	natural
o	outside; reservoir
opt	optimal
ox	oxidant
p	pumping; catalyst pellet
prod	products of a reaction
r	radiation; reservoir; reference
rad	radiation
S	scaling value
s	solar; salvage; s-surface (in a fluid, adjacent to an interface or wall)
sat	saturated
T	thermal; Taylor instability
TP	two-phase
t	turbulent; taxes; transferred state
tr	transition
u	u-surface (in a condensed phase adjacent to an interface)
v	vapor phase
w	at a solid wall; wall material
0	initial
12	species 1 and 2 in a binary mixture
∞	far away; far upstream
(α)	chemical element
λ	spectral value (hemispherically averaged for surface radiation properties)

SUPERSCRIPTS

B	buoyancy
b	black
C	cold stream
E	forced diffusion in an electric field
H	hot stream
i	internal
oa	overall
SP	single phase
T	reference temperature
0	reference state; geometric value
$*$	reduced value; dimensionless; relative to mole-average velocity; limit of zero mass transfer
$+$	directed outward from a surface; dimensionless (turbulence)
$-$	directed toward a surface
$'$	fluctuating component; differentiation with respect to η; per unit length
$''$	per unit area
$'''$	per unit volume

OVERSCORES

$\bar{\ }$	average
$\dot{\ }$	per unit time
$\tilde{\ }$	per kmol
$\hat{\ }$	relative

INDEX

G

H

S

T

U

V

W

X

Z